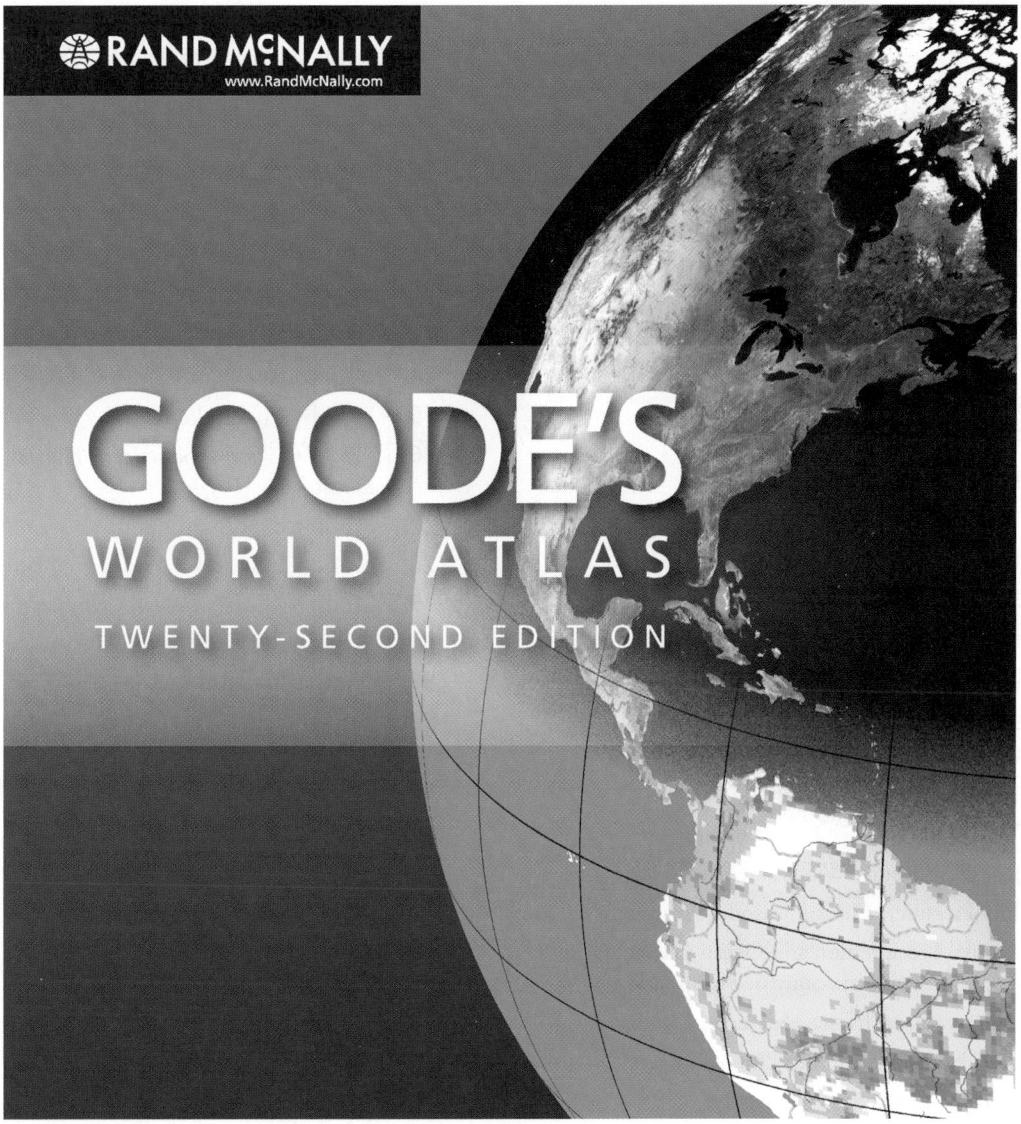
RAND McNALLY
www.RandMcNally.com
GOODE'S
WORLD ATLAS
TWENTY-SECOND EDITION

Editor
Howard Veregin, Ph.D.

www.RandMcNally.com

Introduction

5 Foreword
6 Goode's Atlas Through Time
7 Preface
8 Basic Earth Properties
9 Cartography and Geospatial Technology
10 Map Scale
11 Map Projections
12 Map Projections Used in Goode's World Atlas
14 Thematic Map Types in Goode's World Atlas
16 Reference Map Legend
17 Solar System / The Seasons

The World

18 World • Political
20 World • Physical
22 World • Plate Tectonics / Volcanoes / Earthquakes
24 World • Landforms / Gravity Anomaly
26 Indian and Pacific Oceans Floor
28 Atlantic Ocean Floor
29 Arctic Ocean Floor
30 World • Climate Regions
32 World • January, July Temperature
33 World • Temperature Range / Polar Region Temperature
34 World • January Pressure, Winds / Seasonal Precipitation
35 World • July Pressure, Winds / Seasonal Precipitation
36 World • Annual Precipitation and Ocean Currents / Precipitation Change / Variability of Precipitation
38 World • Temperature Change / Sea Level Change
39 World • CO_2 Emissions / Kyoto Protocol
40 World • Oceanic Environments / Sea Ice
42 World • Natural Vegetation
44 World • Soil Taxonomy
46 World • Terrestrial Biomes
48 World • Population Density
50 World • Birth Rate / Death Rate
51 World • Natural Increase / Urbanized Population
52 World • Gross Domestic Product / Literacy
53 World • Languages / Religions
54 World • HIV Infection / Tuberculosis
55 World • Malaria / Physicians
56 World • Life Expectancy / Undernourishment
57 World • Food Aid / Drinking Water
58 World • Agricultural Production / Agricultural Yield
59 World • Wheat / Tea, Rye
60 World • Maize (Corn) / Coffee, Oats
61 World • Barley, Cocoa Beans / Rice, Millet and Grain Sorghum
62 World • Potatoes, Cassava / Sugar, Spices
63 World • Fruits / Tobacco, Fisheries
64 World • Vegetable Oils
65 World • Natural Fibers, Rubber / Beer and Wine
66 World • Cattle / Pigs
67 World • Sheep / Poultry
68 World • Copper / Tin, Bauxite
69 World • Lead, Lithium / Zinc, Coltan
70 World • Iron Ore and Ferroalloys
71 World • Steel / Precious Metals
72 World • Energy Balance / Electrical Energy Production
73 World • Hydroelectricity / Alternative Energy
74 World • Petroleum / Natural Gas
75 World • Coal / Uranium
76 World • Wood Production / Humid Tropical Forest Loss
77 World • Exports / Imports
78 World • Drug Use / Prison Population
79 World • Military Power / Women's Rights
80 World • Alliances
81 World • Refugees / Major Conflicts
82 World • Telecommunications / Internet / Shipping Lanes
84 World • Time Zones / Hours of Daylight

North America

85 North America • Political
86 North America • Physical
87 North America • Land Cover
88 North America • Satellite Image / Precipitation / Landforms / Vegetation
89 North America • Population Density / Natural Hazards / Energy / Minerals, Agriculture
90 United States and Canada • Temperature / Precipitation / Evapotranspiration
91 United States and Canada • Solar Radiation / Wind Speed / El Niño / Sea Level Rise / Tornadoes / Storm Frequency
92 United States and Canada • Natural Vegetation
94 United States and Canada • Agriculture
96 United States and Canada • Geology
98 United States and Canada • Federal Lands and Interstate Highways
100 United States and Canada • Transportation / Territorial Expansion
101 United States and Canada • Minerals
102 United States and Canada • Population Density
103 United States • Demographics
106 United States • Environmental Issues
107 United States • Labor Structure / Manufacturing

108 United States • Political
110 United States • Physical
112 Northwestern United States
114 North Central United States
116 Northeastern United States
118 Southwestern United States / San Diego
120 Central United States
122 South Central United States / Houston
124 Southeastern United States
126 Alaska
127 Hawaii • Precipitation / Vegetation / Population Density / Environments
128 Canada • Political
130 Canada • Physical
132 Southwestern Canada / Vancouver
134 South Central Canada
136 Southeastern Canada
138 Eastern Canada / Québec
140 Middle America / Aruba-Netherlands Antilles / Bermuda
142 The West Indies / Lesser Antilles / Puerto Rico
144 Mexico
146 Central Mexico
148 Yucatan Peninsula, Guatemala, Belize and El Salvador
149 Honduras, Nicaragua and Costa Rica
150 Panama / Panama Canal
151 United States Cities • Las Vegas / Los Angeles / Portland / Sacramento / San Francisco / Seattle
152 United States Cities • Atlanta / Boston / Cleveland / New York-Philadelphia / Pittsburgh / Washington-Baltimore
154 United States Cities • Chicago-Milwaukee / Detroit / Kansas City / Minneapolis-St. Paul / St. Louis
155 United States Cities • Dallas-Ft. Worth / Denver / Miami / Orlando / Phoenix / Tampa-St. Petersburg
156 Canada Cities • Calgary / Edmonton / Montréal / Ottawa / Toronto / Winnipeg
157 Middle America Cities and Regional • Guadalajara / Guatemala City / Havana / Mexico City / Monterrey / St. Croix / San Juan / Virgin Islands

South America

158 South America • Political
159 South America • Physical
160 South America • Land Cover
161 South America • Satellite Image / Precipitation / Landforms / Vegetation
162 South America • Population Density / Natural Hazards / Energy / Minerals, Agriculture
163 South America Cities • Bogotá-Medellín / Caracas / Lima / Recife / Santiago
164 Northern South America
166 Amazon Basin
168 Central South America / Buenos Aires
170 Peru and Ecuador / Galapagos Islands
171 Southern South America
172 Southeastern Brazil / Rio de Janeiro / São Paulo
173 Río de la Plata Region

Europe

174 Europe • Political
176 Europe • Physical
178 Europe • Land Cover
180 Europe • Satellite Image / Precipitation / Landforms / Vegetation
181 Europe • Population Density / Natural Hazards / Energy / Minerals, Agriculture
182 Europe • Geology
184 Western Europe
186 Eastern Europe
188 Southern Europe and the Mediterranean
190 The British Isles and Iceland / Faroe Islands / London / Shetland Islands-Orkney Islands
192 Southern Scandinavia and the Baltic States
194 Central Europe
196 France / Marseille / Paris / The Riviera
198 Spain and Portugal / Azores / Canary Islands / Lisbon / Madrid
200 Italy, Greece and the Balkans / Crete / Malta
202 Western Russia and the Dnieper Basin
204 Europe Cities • Athens / Berlin / Birmingham-Nottingham / Copenhagen / İstanbul / Manchester-Liverpool / Naples / Rome / Vienna
205 Europe Cities • Amsterdam-Rotterdam / Antwerp-Brussels / Essen-Düsseldorf / Moscow / St. Petersburg

Asia

206 Asia • Political
208 Asia • Physical
210 Asia • Land Cover
212 Asia • Satellite Image / Landforms
213 Asia • Precipitation / Vegetation
214 Asia • Population Density / Energy
215 Asia • Natural Hazards / Minerals, Agriculture
216 Northern Eurasia • Population Density / Ethnicity
217 Middle East • Population Density / Ethnicity
218 Russia
220 Southwestern Asia
222 Eastern Asia
224 Southeastern Asia / Taiwan
226 Central Asia

227 The Caucasus
228 The Middle East
230 The Persian Gulf
232 Iran, Pakistan and Afghanistan
234 Northern India
236 Southern India and Sri Lanka
237 India, Pakistan, Bangladesh and Sri Lanka • Population Density / Minerals, Agriculture
238 Southern China
240 Northeastern China
242 China and Taiwan • Population Density / Minerals, Agriculture
243 Korea
244 Japan / Okinawa
245 Southern Japan / Tōkyō / Ōsaka
246 Southeast Asia / Singapore
248 Malaysia and Central Indonesia
250 The Philippines / Batan Islands-Babuyan Islands
251 Asia Cities • Baghdād / Delhi / Karāchi / Kolkata / Kuala Lumpur / Manila / Mumbai / Seoul / T'aipei / Tehrān
252 Asia Cities • Bangkok / Beijing / Dhaka / Hong Kong / Jakarta / Jerusalem-Tel Aviv / Shanghai

Africa

253 Africa • Political
254 Africa • Physical
255 Africa • Land Cover
256 Africa • Satellite Image / Precipitation / Landforms / Vegetation
257 Africa • Population Density / Natural Hazards / Energy / Minerals, Agriculture
258 Northwestern Africa
260 Western Africa / Southern Nigeria
262 Central Africa
264 Southern Africa / Mauritius-Reunion
266 Northeastern Africa
267 The Rift Valley
268 Africa Cities and Regional • Cairo / Lagos / Nile Valley / Suez Canal
269 Africa Cities • Addis Ababa / Algiers / Casablanca / Johannesburg

Australia, New Zealand and Oceania

270 Australia and New Zealand • Political
272 Australia and New Zealand • Physical
274 Australia and New Zealand • Land Cover / Natural Hazards / Energy
275 Australia • Satellite Image / Precipitation / Landforms / Vegetation / Population Density / Minerals, Agriculture
276 Southeastern Australia
277 Northeastern Australia and Papua New Guinea
278 New Zealand / Auckland / Melbourne / Sydney / Wellington
279 Pacific Islands • Fiji / Guam / New Caledonia-Vanuatu / Sakishimi Shotō / Samoa-American Samoa / Solomon Islands / Tahiti
280 Oceania

The Oceans and Polar Regions

282 Pacific Ocean
284 Atlantic Ocean
286 Indian Ocean
287 Southern Ocean and Antarctica
288 Arctic Ocean

Tables and Indexes

289 World Political Information Table
294 World Comparisons
295 Principal Cities of the World
296 Glossary of Foreign Geographic Terms
297 Abbreviations of Geographic Names and Terms / Pronunciation of Geographic Names
298 Sources
300 Subject Index
301 Pronouncing Index

Goode's World Atlas

Formerly *Goode's School Atlas*

Made in U.S.A.

Library of Congress Catalog Card Number 99-38535

08 **WA** 15 14

Cover Image

Top half. Blue Marble Next Generation monthly composite image for August. Blue Marble Next Generation images are derived from MODIS data at a spatial resolution of 500 meters. MODIS (Moderate Resolution Imaging Spectroradiometer) sensors on board the Terra and Aqua satellites provide global coverage every one to two days in 36 spectral bands. Source: NASA Visible Earth program (http://visibleearth.nasa.gov/).

Bottom half. Map of humid tropical forest loss for the period 2000-2005, derived from MODIS and Landsat imagery. Humid tropical forest loss is estimated to be over 27 million hectares for this period. Source: Hansen, M.C., Stehman, S.V., Potapov, P.V., Loveland, T.R., Townshend, J.R.G., DeFries, R.S., Pittman, K.W., Stolle, F., Steininger, M.K., Carroll, M., and Dimiceli, C. (2008). Humid tropical forest clearing from 2000 to 2005 quantified using multi-temporal and multi-resolution remotely sensed data. PNAS,105(27), 9439-9444. (http://globalmonitoring.sdstate.edu/projects/gfm/humidtropics/data.html).

The cover image illustrates how remote sensing data, coupled with geographic information systems for analysis and display, are increasingly being used to map and monitor changes in the global environment.

Interest in geography has increased dramatically in the last few decades.

Perhaps it is because of efforts by those who teach geography or study it. Perhaps, because of instant global communications and the Internet, we're all more aware of global events. Maybe it's globalization, or recent wars, or global terrorism. Perhaps, because of environmental concerns, we feel a responsibility to better understand and manage Earth and its resources.

Whatever the reasons, this renewed interest in geography is serious. Billions of dollars are being spent every year collecting geographic data. Globally, tens of thousands of organizations of all kinds — government, business, academic, non-profit — have recognized that many of the problems they face must be understood geographically.

In emergencies, government agencies at every level need geographic information about the hurricanes, wildfires, earthquakes, tsunamis, storm surges, and floods they must respond to. They also need to know about the geography of political trouble spots, famines, droughts, terrorism, the narcotics trade, energy resources, shipping, war fighting, and a long list of other topics.

But spending for geographic information goes far beyond governments. Geographic information is also essential to understanding commerce, business, history, military campaigns, migrations, exploration, evangelization, cultural diffusion, origins of civilizations, distribution of organisms and their ecology, agriculture, climate change, natural resources, transportation patterns, productivity, epidemiology, conservation, election results, and many, many other topics. Organizations throughout the world recognize this, pay for geographic information, and hire people to manage and analyze it for them.

Geography matters! The community of people who rely on it is growing every day.

Geography today is a multidisciplinary science. Our ability to collect geographic data scientifically is exploding and we have begun to acquire the methods needed to effectively manage this data explosion for the entire planet. This includes new tools like satellite images of Earth, geographic information system (GIS) software to process and display enormous volumes of data, and global positioning system (GPS) devices that can accurately determine locations anywhere on Earth.

Goode's World Atlas is part of this revolutionary growth. The atlas has long been a staple of the college classroom, educating students about important geographic issues of the day. The current 22nd edition, which you hold in your hands, makes extensive use of digital geographic information of the kind I refer to above. It focuses on important contemporary issues like globalization, global climate change, food security, and environmental degradation. It uses GIS to integrate information and render it for cartographic display.

Goode's World Atlas helps us understand our world and our place in it. It helps us interpret stories in the news, understand international conflicts, evaluate foreign competition for jobs, make informed decisions about free trade or immigration, respond to changing oil prices or possible climate change, and think through complex domestic and foreign policy issues.

Goode's World Atlas is an essential guidebook to the new geography, helping us sort through and decipher patterns in a flood of geographic data. It helps us make sense of these data, and it provides authoritative cartographic interpretations of complex geographic issues.

I have studied and worked with geographic information for more than forty years. My experiences have given me insights into what our world was and is, and what we can make of it in the future. I think that geography can give us new eyes with which to see the world, so that — as the poet remarked — after all our traveling we return home and see it for the first time. I believe that Goode's World Atlas is an invaluable component of this learning process. The Atlas continues to evolve and adapt, but remains rooted in its original function — helping us develop geographic understanding and knowledge as a way to make sense of our world.

Jack Dangermond

President, ESRI

Goode's Atlas is named for John Paul Goode, who created the atlas and served as its editor for many editions. Goode was one of the first U.S. academic cartographers. He was born in rural Minnesota in 1862, received his bachelor's degree from the University of Minnesota in 1889, and earned his doctorate in economic geography from the University of Pennsylvania in 1903. He spent much of his professional career at the University of Chicago. Among his many accomplishments he is perhaps best known for the development of the Interrupted Homolosine projection, which he first presented at the Association of American Geographers meeting in 1923, and which has been used extensively in Goode's Atlas and in many other geographic publications to the present day.

John Paul Goode

The Homolosine is a composite of two projections, the Mollweide (Homolographic) and the Sinusoidal. Goode interrupted the Homolosine over the oceans to minimize distortion of shapes over continental land masses. Lines of latitude on the Homolosine are straight lines, to facilitate analysis of comparative latitudes. Also, the projection is equal area. Goode was a strong proponent of equal area projections and an equally strong opponent of the Mercator projection, widely used in the early part of the 20th century for world maps. As Goode stated in the introduction to the 1st edition of the atlas (1923, p. x), the distortion of area on the Mercator projection is so extreme that "it becomes pedagogically a crime to use Mercator's map" for studies of areal distributions such as population density, rainfall, or sizes of countries.

Under Goode's editorship the atlas doubled in size. The 1st edition of Goode's School Atlas contained 96 pages of maps. The 4th edition (1932), the last edition that Goode would edit before his death, contained 174 pages of maps. Goode introduced many of the thematic map topics that are still found in the atlas today, including world economic maps of agricultural commodities, minerals, energy, and international trade. These topics reflect Goode's interest and training in economic geography.

Goode remained the only name on Goode's School Atlas until the 8th edition (1949), on which Edward B. Espenshade, Jr., was credited with numerous updates and revisions. Espenshade was then named editor for the 9th edition (1953). Espenshade was one of Goode's students and spent his academic career at Northwestern University in Evanston, Illinois. The 9th edition was significant in many respects. It boasted a new title, Goode's World Atlas, and contained many of the features of the modern atlas.

In particular, Espenshade made extensive use of maps compiled by experts in specific subdisciplines of geography. Examples include natural vegetation by A. W. Küchler, physiography by Erwin Raisz, climate regions by Glenn Trewartha, and agricultural regions by Derwent Whittlesey. By relying on the research of these and other scholars, Espenshade was able to incorporate the latest advances in the study of geographical phenomena. Espenshade also oversaw the creation of a new reference map series, which included hand-drawn shaded relief for the first time in the atlas. These reference maps were introduced in the 11th edition (1960).

Joel L. Morrison, then at the University of Wisconsin, joined Espenshade as associate editor on the 14th edition (1974). Morrison, who had a distinguished career in academia and the federal government, was affiliated with the atlas through the 19th edition (1995). In the 1970s and 1980s the atlas saw numerous innovations, including the introduction of ocean floor shaded relief maps, reference maps of major world cities, a continent environments map series, and the first use of cartograms.

The 19th edition was Espenshade's last as editor. On that edition, John C. Hudson assumed the role of associate editor. Hudson, a distinguished academic geographer at Northwestern University, then took on the role of editor for the 20th edition. Hudson introduced many new thematic maps, including world ecoregions, origins of plants, refugees, conflicts, and oceanic environments.

Howard Veregin was named editor for the 21st edition. Veregin was on the geography faculty at the University of Minnesota, then moved to Rand McNally where he currently serves as director of geographic information services. Veregin created the Goode's Editorial Board to help reorient the atlas in relation to modern geographic scholarship and pedagogy. With the 22nd edition the atlas became all-digital for the first time, with most maps produced using geographic information systems (GIS) technology. Major innovations for the 22nd edition include a new digital reference map series (the first new series since the 11th edition), many new thematic maps, and an updated design.

Throughout its history Goode's Atlas has adapted to changes in cartographic technology, map design, and geographic curricula. However, it has always maintained the pedagogical foundation that John Paul Goode established in the 1st edition in 1923. It should be seen first and foremost as a work of scholarship, incorporating the latest insights into geographical research and knowledge. It is also a fascinating portrait of almost nine decades of evolution in geography and cartography.

Robert B. McMaster, Ph.D.
Susanna A. McMaster, Ph.D.
University of Minnesota

The 22nd edition of Goode's World Atlas blends dramatic new maps and exceptional cartography with the strong traditions that have made Goode's Atlas a standard for over 85 years.

The 22nd edition features new thematic maps that focus on topics important to modern geography, including global climate change, sea level rise, CO_2 emissions, polar ice fluctuations, forest loss, extreme weather events, infectious diseases, water resources, and energy production. These maps have been produced with the latest digital sources integrated within geographic information systems (GIS) technology to deliver a contemporary portrait of the planet. We have also retained and updated the new maps introduced in the 21st edition, including HIV infection, military power, women's rights, and food aid. Other thematic maps and graphs have been updated using the same standards and quality requirements that have always been a defining feature of Goode's World Atlas.

The 22nd edition also delivers over 160 pages of new, digitally produced reference maps, providing detailed coverage of all continents. We have paid particular attention to expanding our coverage of Africa, Asia, and Central and South America. The new reference maps were produced using state-of-the-art GIS technology to integrate digital data sources and render them for cartographic display. Underlying these maps is Rand McNally's proprietary digital world database, the same trusted source used in many of our other world atlases.

At Rand McNally we take pride in the quality of our cartography and the rigorous standards we set for the research underlying each map. For the 22nd edition we worked closely with our Editorial Advisory Board to select new map topics, assess cartographic approaches, and identify new atlas features. Longtime Goode's users will see numerous changes to the atlas, including the use of more contemporary color palettes and graphic treatments to improve clarity, readability, and aesthetics. In addition we have implemented changes that bring more consistency to each section of the atlas.

Needless to say this atlas would not have been possible without the efforts of a very talented cartographic development team, who conducted basic research, developed new thematic maps, created the new reference maps, designed the maps and page layouts, performed quality assurance, and helped work through countless editorial decisions. I include their names here in alphabetical order.

Robert Argersinger, Greg Babiak, Genna Davis, Marzee Eckhoff, Brett Gover, Justin Griffin, Rob Harris, Michael Healy, Susan Hudson, Valbona Kokoshi, Marc Kugel, Brian Lash, Felix A. Lopez, Andy Lotter, Nina Lusterman, Donna McGrath, Rob Merrill, Joerg Metzner, Angela Mrotek, Darren Raffel, Amy J. Ruggles, Damon Sather, Dave Simmons, Andy Skinner, Jeff Thomas, Raymond T. Tobiaski, Tom Vitacco, Steve Wiertz, Yanyan Zhang

A brief acknowledgment such as this cannot really do justice to the thousands of hours of effort expended by these and other Rand McNally subject matter experts. Nor does this list include Rand McNally employees who worked on previous editions of the atlas, and to whom the current atlas owes much.

I would also like to acknowledge the work of the Editorial Advisory Board, who participated in discussions of the new directions we were planning to take, and made significant contributions in terms of content and design. We are indebted to the Board for helping us refocus Goode's Atlas in relation to modern geographic scholarship and teaching. The Goode's Editorial Advisory Board members are listed on the title page of this atlas.

While Goode's Atlas continues to change with the times, it remains the same accurate and reliable educational resource that J. Paul Goode originally intended it to be. We at Rand McNally remain committed to providing you with the most trusted tools to help you and your students open your classrooms to the world.

Howard Veregin

Howard Veregin, Ph.D., Editor

Skokie, Illinois

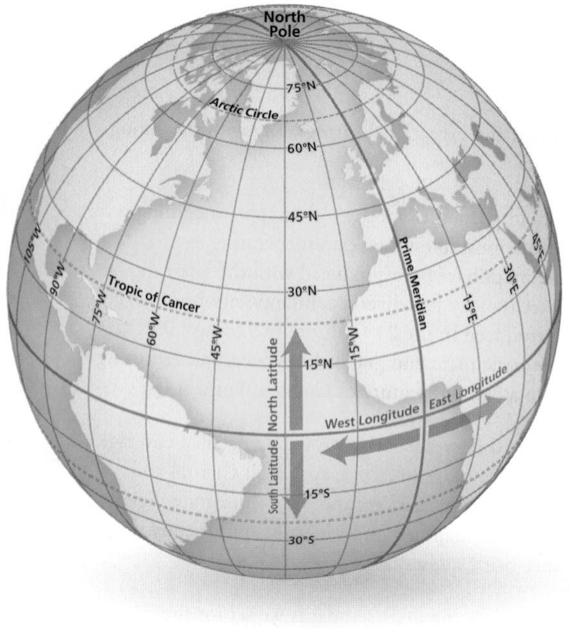
North Pole
75°N
Arctic Circle
60°N
45°N
Prime Meridian
30°N
Tropic of Cancer
15°N
North Latitude
West Longitude
East Longitude
South Latitude
15°S
30°S
105°W
90°W
75°W
60°W
45°W
15°W
15°E
30°E
45°E

View of Earth centered on 30° N, 30° W

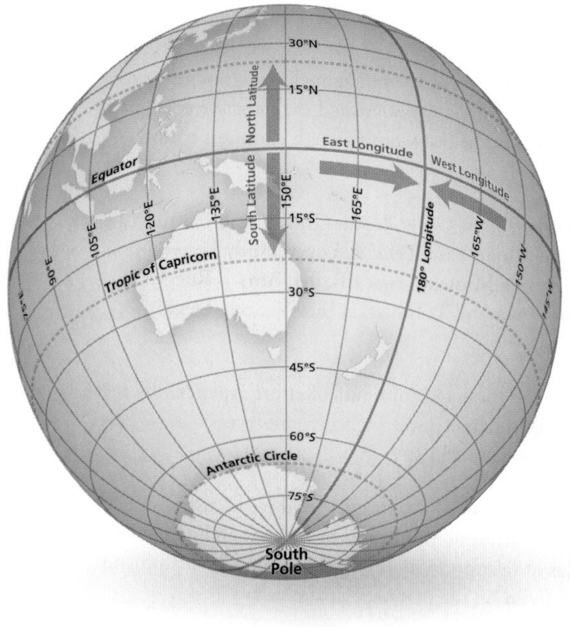
30°N
15°N
North Latitude
East Longitude
West Longitude
Equator
South Latitude
15°S
Tropic of Capricorn
30°S
45°S
60°S
Antarctic Circle
75°S
South Pole
90°E
105°E
120°E
135°E
150°E
165°E
180° Longitude
165°W
150°W
135°W

View of Earth centered on 30° S, 150° E

Basic Earth Properties

Earth is essentially spherical in shape. The North and South Poles are aligned with Earth's axis of rotation. The Equator is equidistant between the Poles and divides Earth into northern and southern hemispheres.

Latitude and longitude identify the locations of features on Earth's surface. Latitude is the angle north or south of the Equator. Longitude is the angle east or west of the Prime (Greenwich) Meridian. A meridian is a line of longitude extending from the North Pole to the South Pole. The Prime Meridian is the meridian passing through the Royal Observatory in Greenwich, England. This location for the Prime Meridian was adopted at the International Meridian Conference in Washington, D.C., in 1884.

Latitude and longitude are usually given in degrees, minutes and seconds. There are 60 minutes in a degree and 60 seconds in a minute. The symbols °, ' and " represent degrees, minutes and seconds, respectively. For latitude the symbols N and S indicate degrees north or south of the Equator. The latitude of the Equator is 0° and the latitudes of the North and South Poles are 90° N and 90° S. For longitude the symbols E and W indicate degrees east or west of the Prime Meridian. Longitude ranges from 0° at the Prime Meridian to 180° E or W. The meridian at 180° E is the same as the meridian at 180° W, and this meridian is the approximate location of the International Date Line. This meridian and the Prime Meridian divide Earth into eastern and western hemispheres.

A latitude-longitude coordinate pair defines the location of a feature on Earth. As an example, the Rand McNally building in Skokie, Illinois, has the coordinates 42° 3' 37" N, 87° 45' 39" W. Often the number of seconds are omitted from the coordinates if a high level of precision is not required.

Lines of latitude are also known as parallels. Two parallels of special importance are the Tropic of Cancer and the Tropic of Capricorn, at approximately 23° 30' N and S respectively. This angle coincides with the inclination of Earth's axis relative to its orbital plane around the Sun. The Tropics are the lines of latitude where the noon sun is directly overhead on the solstices. Two other important parallels are the Arctic Circle and the Antarctic Circle, at approximately 66° 30' N and S respectively. These lines mark the most northerly and southerly points at which the Sun can be seen on the solstices.

The Geographic Grid

The geographic grid is the grid of latitude-longitude lines on Earth. The following are some important characteristics of the grid.

Lines of longitude (meridians) are equal in length and meet at the Poles.

Lines of latitude (parallels) are parallel to each other and equally spaced along meridians.

The length of parallels decreases as one gets closer to the Poles. For example, the length of the parallel at 60° latitude is one-half the length of the Equator.

Meridians get closer together with increasing distance from the Equator, and finally converge at the Poles.

Parallels and meridians meet at right angles.

Cartography and Geospatial Technology

Geography's subject matter includes the people, landforms, climate, and other physical and human phenomena that make up Earth's environments and give unique character to different places. Geographers construct maps to visualize how these phenomena vary over geographic space. Maps help geographers understand and explain phenomena and their interactions.

The art and science of mapmaking is known as cartography. Although maps were once drawn by hand, they are now usually created using digital technology. This technology includes GIS (geographic information systems), as well as GPS (global positioning system) and remote sensing. Collectively these are known as geospatial technologies.

GIS is a specialized type of software that enables the integration, processing, analysis, and display of digital geographic information. It combines an underlying database with spatial analysis tools and cartographic rendering capabilities. First developed in the 1960s, GIS has evolved rapidly in the last decade with advances in computer processing power and the increased availability of digital geographic information. Applications of GIS have also diversified rapidly as more users have recognized its utility for solving geographic problems.

In cartography, GIS has redefined how maps are made. Since the underlying data for a map is stored in a database, a map is just one of many possible data representations. Many different maps can be produced from one database, based on different permutations of attributes, for different map scales, geographic areas, or time periods, and using different map treatments. GIS also enhances the efficiency of map production. For example, map symbology can be driven off stored attributes, selection and generalization can be conducted using defined rules, map text can be placed automatically, and index creation can be automated. All of the maps in this edition of Goode's World Atlas are digital, and the vast majority have been created using GIS software.

GIS also greatly enhances the ability to integrate data from a variety of sources and process these data for specific mapping purposes. In this sense GIS is closely related to other geospatial technologies such as GPS and remote sensing. GPS is a satellite-based system for capturing precise information about locations on Earth's surface. Originally developed by and for the military, GPS is now the underlying technology behind personal navigation devices and location-based services. GPS has revolutionized the field of surveying and is used in a wide variety of fields where accurate coordinates are needed.

Remote sensing refers to the collection of data about Earth from satellites and aircraft. Advances in remote sensing have greatly magnified the volume and types of geographic data available for mapping. Many of the maps in this atlas are derived from remote sensing imagery, including the maps of land cover, gravity, sea level change, sea ice, and forest loss. Modern remote sensing systems are designed to focus on specific portions of the electromagnetic spectrum, some of which cannot be seen by the human eye. This capability allows very specific geographic phenomena to be imaged and analyzed.

The diagram below illustrates how geospatial technology can be used to map and monitor changes in the global environment. This edition of Goode's World Atlas reflects this growing awareness and capability by incorporating the latest digital data sources wherever possible.

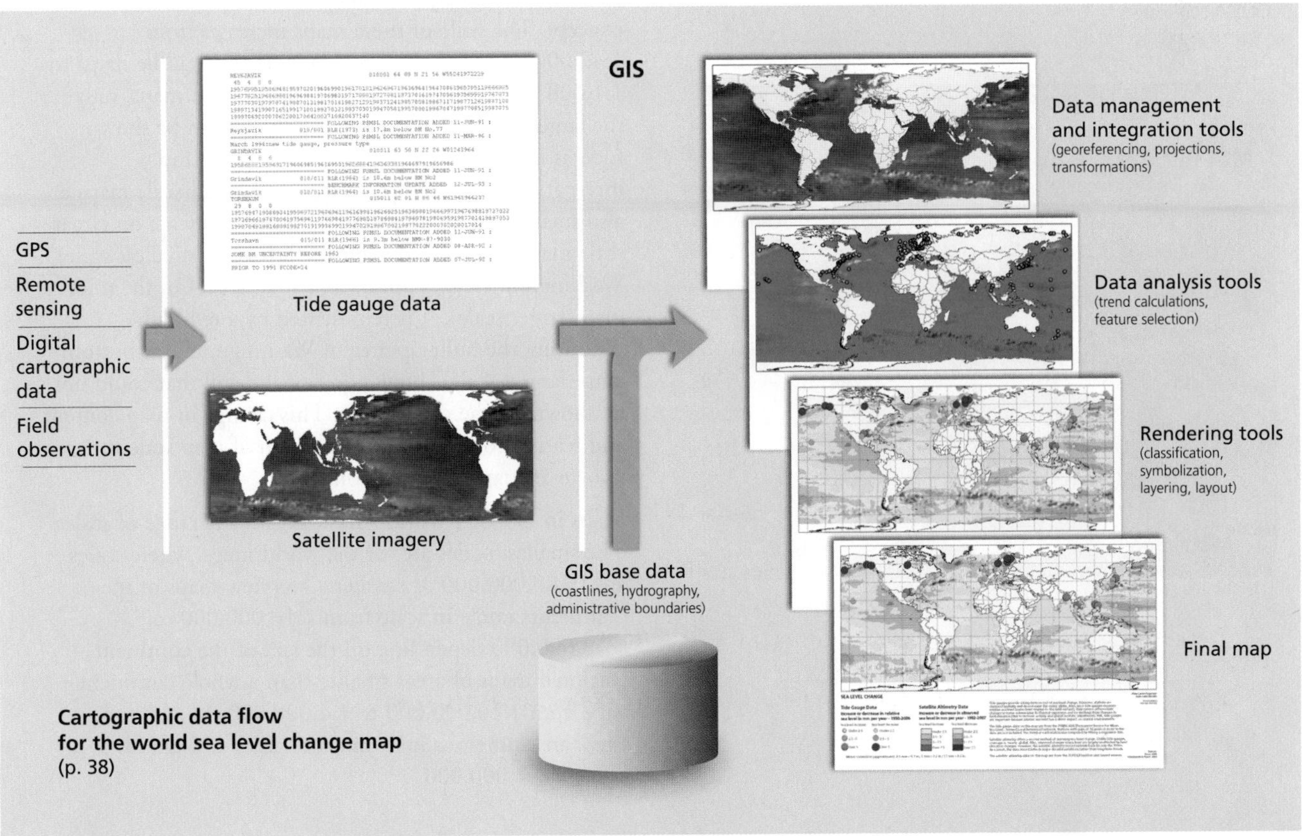
GPS
Remote sensing
Digital cartographic data
Field observations
Tide gauge data
Satellite imagery
GIS
GIS base data
(coastlines, hydrography, administrative boundaries)
Data management and integration tools
(georeferencing, projections, transformations)
Data analysis tools
(trend calculations, feature selection)
Rendering tools
(classification, symbolization, layering, layout)
Final map

Cartographic data flow for the world sea level change map (p. 38)

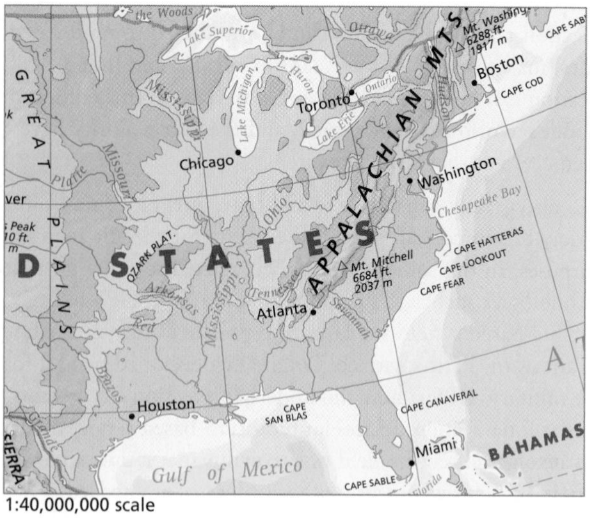
Lake Superior
Ottawa
Mt. Washington 6288 ft. 1917 m
Boston
CAPE COD
Toronto
Ontario
Lake Michigan
Huron
Lake Erie
Chicago
Missouri
Mississippi
Platte
APPALACHIAN MTS.
Washington
Chesapeake Bay
Ohio
GREAT PLAINS
STATES
OZARK PLAT.
Mt. Mitchell 6684 ft. 2037 m
CAPE HATTERAS
CAPE LOOKOUT
CAPE FEAR
Tennessee
Arkansas
Red
Atlanta
Savannah
Brazos
Houston
CAPE SAN BLAS
CAPE CANAVERAL
Miami
BAHAMAS
Gulf of Mexico
CAPE SABLE
Florida
1:40,000,000 scale

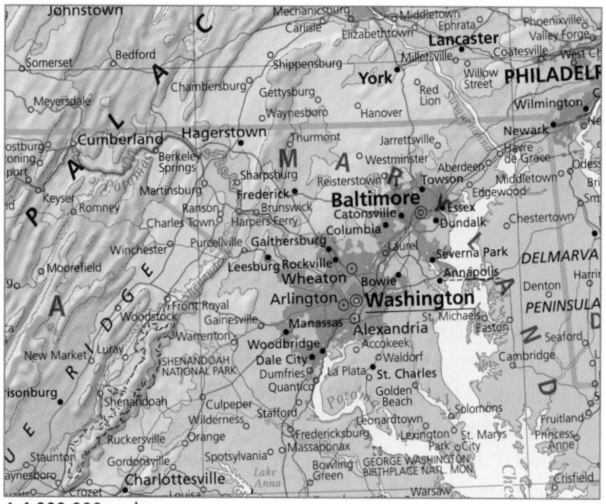
Johnstown
Mechanicsburg
Carlisle
Middletown
Elizabethtown
Ephrata
Lancaster
Phoenixville
Valley Forge
Bedford
Somerset
Shippensburg
Millersville
Coatesville
York
Willow Street
PHILADELPHIA
Chambersburg
Gettysburg
Waynesboro
Hanover
Red Lion
Wilmington
Meyersdale
Cumberland
Hagerstown
Thurmont
Jarrettsville
Newark
Havre de Grace
Berkeley Springs
Sharpsburg
Westminster
Aberdeen
Reisterstown
Towson
Middletown
Martinsburg
Frederick
Edgewood
Keyser
Romney
Ranson
Brunswick
Baltimore
Essex
Charles Town
Harpers Ferry
Catonsville
Columbia
Dundalk
Chestertown
Purcellville
Gaithersburg
Winchester
Laurel
Moorefield
Leesburg
Rockville
Severna Park
DELMARVA
Wheaton
Bowie
Annapolis
Arlington
Washington
Denton
Front Royal
PENINSULA
Gainesville
Manassas
St. Michaels
Woodstock
Alexandria
Easton
Warrenton
Woodbridge
Accokeek
Seaford
New Market
Luray
Dale City
Waldorf
Cambridge
SHENANDOAH NATIONAL PARK
Dumfries
La Plata
St. Charles
Quantico
Golden Beach
Shenandoah
Culpeper
Stafford
Solomons
Wilderness
Leonardtown
Fruitland
Orange
Fredericksburg
Lexington Park
St. Marys City
Princess Anne
Ruckersville
Massaponax
Staunton
Gordonsville
Spotsylvania
Bowling Green
GEORGE WASHINGTON BIRTHPLACE NAT'L MON.
Charlottesville
Lake Anna
Crisfield
Crozet
Warsaw
1:4,000,000 scale

River
Cooksville
BALTIMORE
Essex
FORT MCHENRY NAT'L MON.
Damascus Mtn.
Catonsville
MARYLAND
Ellicott City
Arbutus
Dundalk
Lansdowne
Clarksburg
Triadelphia Reservoir
Edgemere
Columbia
Germantown
Highland
Glen Burnie
Riviera Beach
Olney
Savage
Severn
Burtonsville
FORT GEORGE G. MEADE
Lake Shore
Rockville
Aspen Hill
Laurel
Severna Park
North Potomac
Wheaton
Beltsville
Odenton
Patuxent
Arnold
Potomac
Silver Spring
Greenbelt
Crofton
BROAD NECK
Bethesda
College Park
Bowie
Lanham
Annapolis
McLean
DIST. OF COLUMBIA
Hyattsville
Falls Church
Mitchellville
Londontown
Largo
THOMAS POINT
Arlington
WASHINGTON
Mayo
Suitland
Hillcrest Heights
KENT POINT
Alexandria
Camp Springs
Upper Marlboro
ANDREWS AIR FORCE BASE
Deale
Franconia
Groveton
1:1,000,000 scale

Map Scale

Map scale is the ratio of distance on a map to distance on Earth's surface. For example, if two towns on a map are separated by a distance of 1 inch, and these towns are actually 1 mile apart, then the scale of the map is 1 inch to 1 mile.

The statement "1 inch to 1 mile" is a verbal scale. Verbal scales are simple and intuitive, but it can be difficult to compare verbal scales for different maps on which different linear units are used, such as kilometers instead of miles. A more flexible way to express map scale is the representative fraction. To construct a representative fraction, the numerator and denominator are first converted to the same units. For example, since there are 63,360 inches in a mile, the verbal scale "1 inch to 1 mile" can be expressed as "1 inch to 63,360 inches". Next the unit names are dropped and the scale is expressed as a ratio, in this case 1:63,360. This means that 1 linear unit on the map represents 63,360 linear units on Earth, whether those units are inches, miles, kilometers, or some other unit of measurement.

Map scale can also be represented in graphical form. Many maps contain a graphic scale (or bar scale) showing real-world units such as miles or kilometers. The bar scale is usually subdivided to allow easy calculation of distance on the map. However, using a bar scale to measure distance can result in significant errors, especially on small-scale maps covering large areas. This is due to the distortion of distances on the map, as discussed in the map projection section below.

Map scale determines the amount of detail that can be portrayed on a map. The maps on this page illustrate this concept. The scale of these maps increases from 1:40,000,000 (top map) to 1:4,000,000 (middle map) to 1:1,000,000 (bottom map). On small-scale maps, only the largest and most important features can be shown, such as large cities, major rivers and lakes, and international boundaries. Features on small-scale maps are also smaller and more generalized than they are on larger-scale maps. For example, on the top map (smallest scale), Washington, D.C., appears as a small dot. On the middle map (larger scale), it is represented by a red blob indicating the built-up area of Washington. The bottom map (largest scale) shows additional detail that could not be shown on the other maps. This change in map content and feature complexity as a function of map scale is known as map generalization.

Maps in Goode's World Atlas have a wide range of scales. The smallest scales are for the world maps, where scales are 1:100,000,000 or smaller. Overview maps of the continents range in scale from 1:16,000,000 to 1:40,000,000 depending on the size of the continent. Regional maps of areas smaller than a whole continent vary from 1:16,000,000 to 1:4,000,000. In addition there are numerous inset maps of cities and islands at a scale of 1:1,000,000.

Map Projections

A map projection is a geometric representation of Earth's surface on a flat surface. Since Earth is roughly spherical, a map projection is needed to produce any flat map, whether a page in this atlas or a computer-generated map of driving directions on www.RandMcNally.com. Hundreds of projections have been developed since the dawn of cartography. A limitation of all of these projections is that they introduce geometric distortion. Some projections distort shape, others distort area, and all distort distance to some degree.

In order to choose an appropriate projection for a particular map, cartographers must pay careful attention to the properties that are distorted and the properties that are preserved by the projection. If shape is preserved, the projection is "conformal." On conformal projections the shapes of geographic features agree with their shapes on Earth. However, a limitation of conformal projections is that they necessarily distort area. This means that the sizes of the geographic features on the map will not be directly comparable. Some will be too large and others too small.

If areas are correctly represented, the projection is "equal area." On equal area projections the sizes of features on the map are directly comparable and in correct proportion to their sizes on Earth. However, in order to achieve this effect, equal area projections distort shape. No projection can preserve both shape and area simultaneously. Some projections preserve neither shape nor area, but instead balance shape and area distortion, creating a compromise projection.

The term "equidistant" is often used for projections that preserve distance. However this can be misleading since distance can only be preserved selectively, such as along specific meridians or parallels. No projection correctly preserves distance in all directions at all locations. Since distance is closely related to scale, one implication is that map scale is often only approximate and may not apply to the entire coverage area of a map. This problem is especially acute for small-scale maps covering large areas.

The projection selected for a particular map depends on the relative importance of different types of distortion, which in turn depends on the purpose of the map. For example, world maps showing phenomena that vary with area, such as population density, often use an equal area projection to give an accurate depiction of the importance of each region.

Map projections are created using mathematical procedures. To illustrate the general principles of projections without using mathematics, we can view a projection as the geometric transfer of information from a globe to a flat projection surface, such as a sheet of paper. If we allow the paper to be rolled in different ways, we can derive three basic types of map projections called cylindrical, conic, and azimuthal.

For cylindrical projections, the sheet of paper is rolled into a tube and wrapped around the globe so that it is tangent (touching) along a circle such as the Equator (see figure below). Information from the globe is transferred to the tube, and the tube is then unrolled to produce the final flat map.

Conic projections use a cone rather than a cylinder. The figure shows the cone tangent to the globe along a line of latitude with the apex of the cone over the North Pole. The line of tangency is called the standard parallel of the projection. Azimuthal projections use a flat projection surface that is tangent to the globe at a single point, such as the North Pole (see figure below).

In general, map distortion increases with distance away from the point or line of tangency. This is why maps of equatorial, mid-latitude, and polar regions often use cylindrical, conic and azimuthal projections, respectively.

The projection surface model is useful for illustrating how projections are developed. However, each of the three projection surfaces actually represents scores of individual projections. There are, for example, many projections with the term "cylindrical" in the name, each of which has the same basic rectangular shape, but different spacings of parallels and meridians.

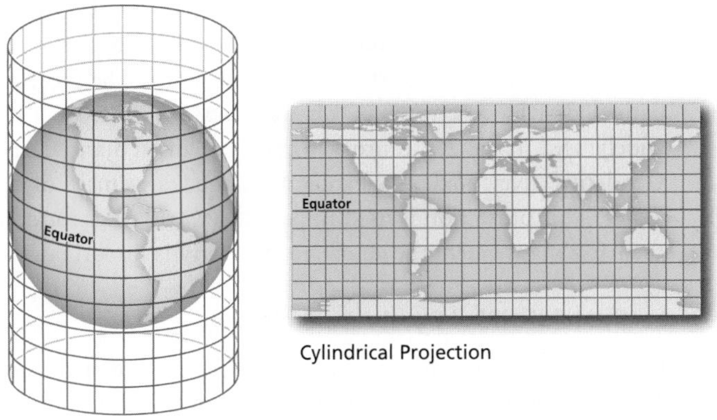
Equator
Equator

Cylindrical Projection

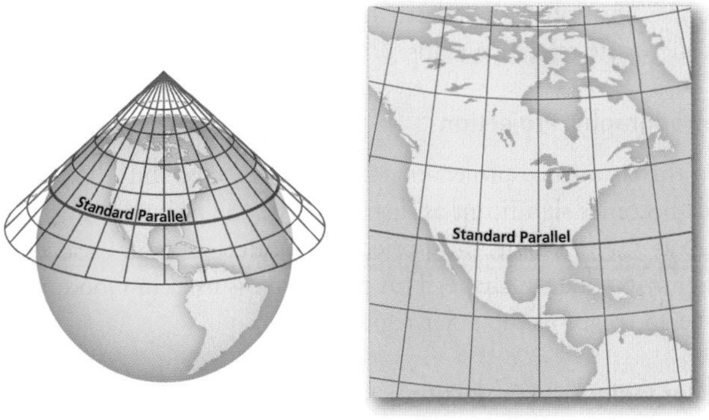
Standard Parallel
Standard Parallel

Conic Projection

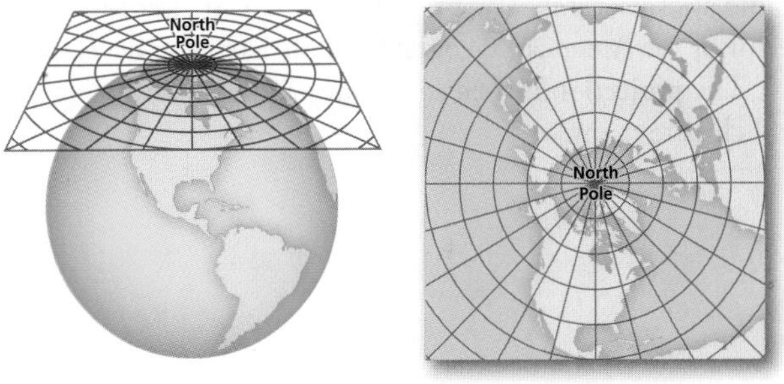
North Pole
North Pole

Azimuthal Projection

Map Projections Used in Goode's World Atlas

Of the hundreds of projections that have been developed, only a fraction are in everyday use. The main projections used in Goode's World Atlas are described below.

Lambert Conformal Conic Projection

On this conic projection, spacing between parallels increases with distance away from the standard parallel, which allows the geometric property of shape (but not area) to be preserved. The projection is named after Johann Lambert, an 18th century mathematician who developed some of the most important projections in use today. It became widely used in the United States in the 20th century following its adoption for many state mapping programs. This projection is used extensively in Goode's World Atlas for larger-scale reference maps.

Albers Equal Area Conic Projection

On this conic projection, spacing between parallels decreases with distance away from the standard parallel, which allows the geometric property of area (but not shape) to be preserved. The projection is named after Heinrich Albers, who developed it in 1805. It became widely used in the 20th century, when the United States Coast and Geodetic Survey made it a standard for equal area maps of the United States. This projection is used in Goode's World Atlas for continent thematic maps where the equal area property is important.

Lambert Azimuthal Equal Area Projection

On this azimuthal projection, area is preserved, but at the expense of significant shape distortion as distance from the point of tangency increases. This projection is most appropriate for areas of roughly circular shape. This projection, like the Lambert Conformal Conic, is named after Johann Lambert. It is used in Goode's World Atlas for smaller-scale reference maps.

Stereographic Projection

On this azimuthal projection, shape is preserved, but distortion of area becomes significant as distance from the point of tangency increases. As a result, this projection is often used for areas that are roughly circular in shape. This projection is used in Goode's World Atlas for maps of the polar regions.

Miller Cylindrical Projection

This cylindrical projection is neither conformal nor equal area. However, it is a useful compromise projection to show Earth in a simple, rectangular form. One problem is that polar areas exhibit significant exaggeration of area, a problem common to many cylindrical projections. The projection is named after Osborn Miller, director of the American Geographical Society, who developed it in 1942. The projection is used in Goode's World Atlas for many of the world climate maps.

Lambert Conformal Conic Projection

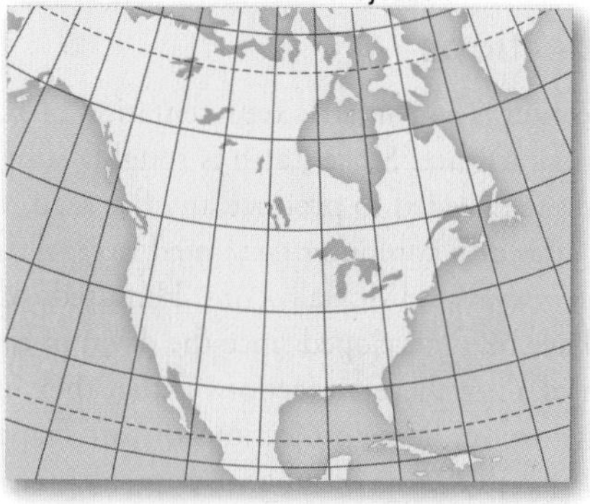

Albers Equal Area Conic Projection

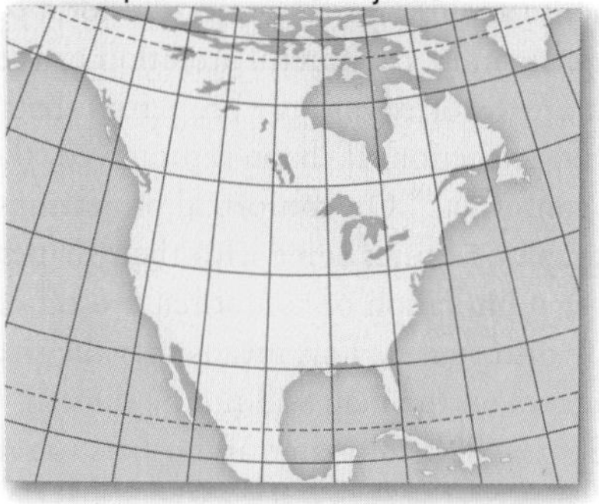

Lambert Azimuthal Equal Area Projection

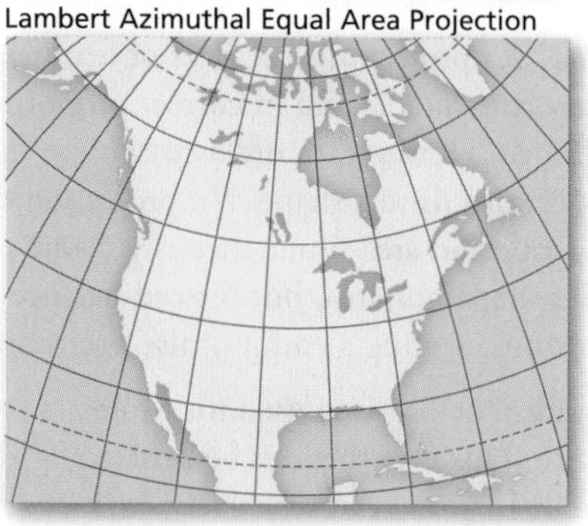

Stereographic Projection

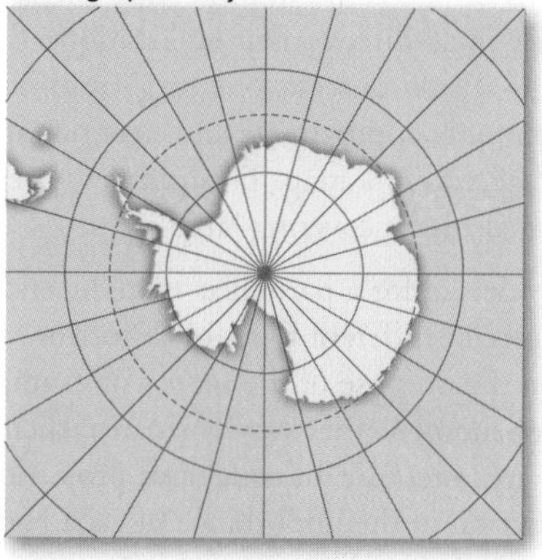

Miller Cylindrical Projection

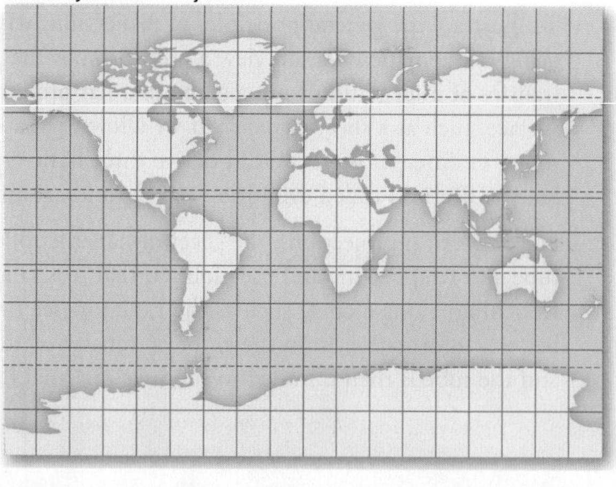

Plate Carrée Projection

This cylindrical projection is neither conformal nor equal area. Its main utility lies in the fact that it shows lines of latitude as evenly spaced lines on the map. This allows for effective thematic map display for phenomena that are measured at regular intervals of latitude. The projection is used in Goode's World Atlas for world climate change maps.

Plate Carrée Projection

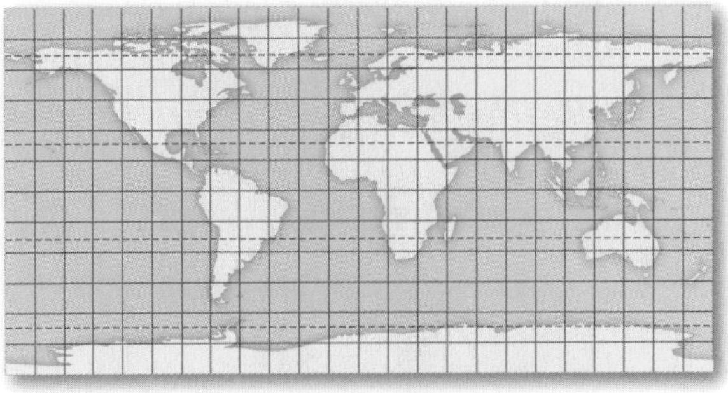

Sinusoidal Projection

The straight, evenly-spaced parallels on this pseudocylindrical projection resemble the parallels on cylindrical projections. Unlike cylindrical projections, however, meridians are curved and converge at the poles. This causes significant shape distortion in polar regions. The projection is therefore not conformal, although it is equal area. The Sinusoidal is the oldest-known pseudocylindrical projection, dating to the 16th century. It is not used extensively in Goode's World Atlas. However, along with the Mollweide projection, it is the basis for the Goode's Interrupted Homolosine projection described below.

Sinusoidal Projection

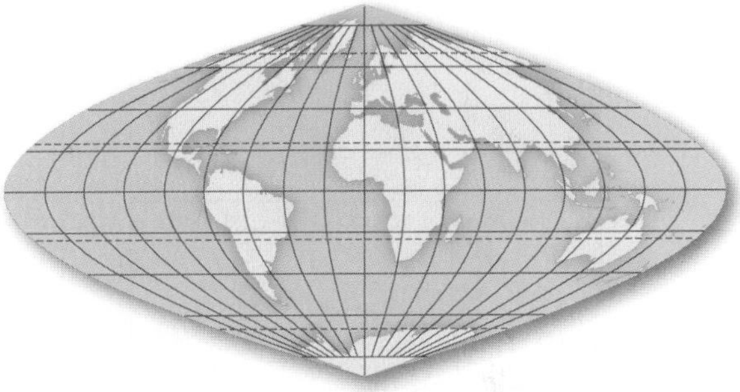

Mollweide Projection

The Mollweide (or Homolographic) projection resembles the Sinusoidal but has less shape distortion in polar areas due to its elliptical form. Like the Sinusoidal projection, it is equal area but not conformal. It is one of several pseudocylindrical projections developed in the 19th century, and is named after Karl Mollweide, an astronomer and mathematician, who developed it in 1805. It is not used extensively in Goode's World Atlas. However, along with the Sinusoidal projection, it is the basis for the Goode's Interrupted Homolosine projection described below.

Mollweide Projection

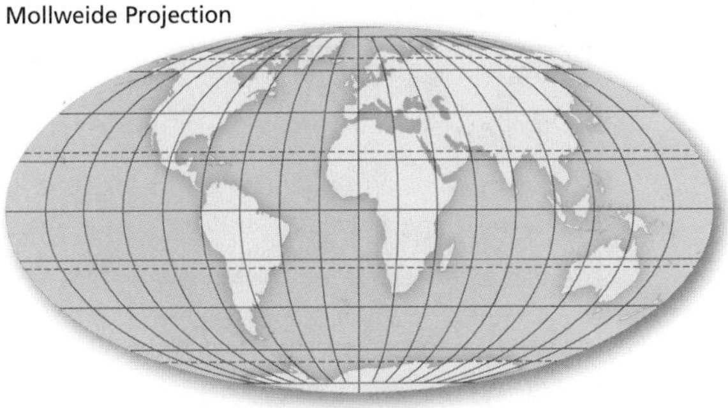

Goode's Interrupted Homolosine Equal Area Projection

This projection is a fusion of the Sinusoidal projection between 40° 44' N and S, and the Mollweide projection between these parallels and the Poles. The projection is equal area but not conformal. The unique appearance of the projection is due to the introduction of discontinuities in oceanic regions, the goal of which is to reduce distortion for continental land masses. A condensed version of the projection also exists in which the Atlantic Ocean is compressed to help maximize the scale of the map on the page. The Goode's Interrupted Homolosine projection is named after J. Paul Goode of the University of Chicago, who developed it in 1923. Goode was an advocate of interrupted projections and, as editor of Goode's School Atlas, promoted their use in education. This projection is used extensively in Goode's World Atlas for world thematic maps.

Goode's Interrupted Homolosine Equal Area Projection

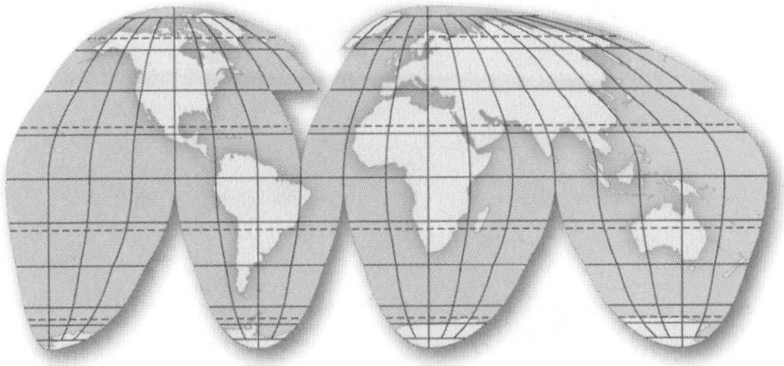

Robinson Projection

This pseudocylindrical projection resembles the Mollweide projection except that polar regions are flattened and stretched out. While neither conformal nor equal area, the Robinson projection manages to balance shape and area distortion in an effective way. The projection was developed in 1963 by Arthur Robinson of the University of Wisconsin, at the request of Rand McNally. The Robinson projection is widely used in Goode's World Atlas for world thematic maps where the interrupted nature of the Goode's Homolosine projection would be inappropriate.

Robinson Projection

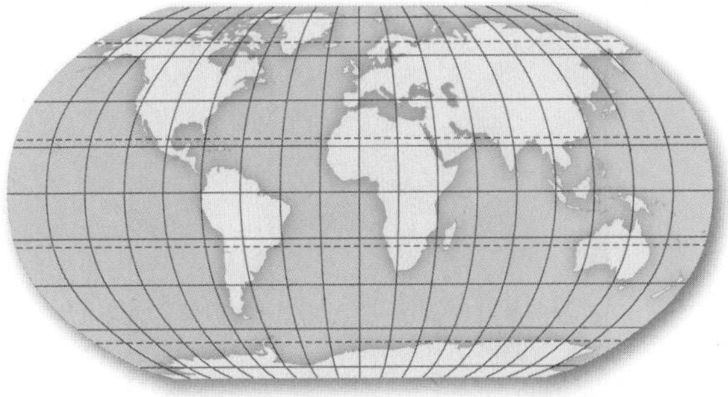

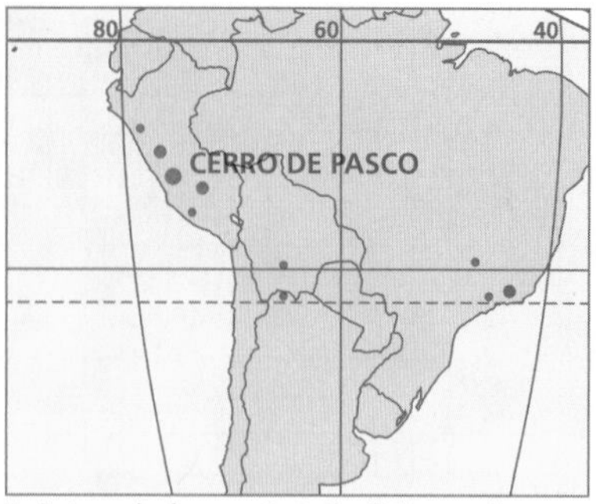
80
60
40
CERRO DE PASCO

Point symbol map: Detail of Zinc and Coltan (p. 69)

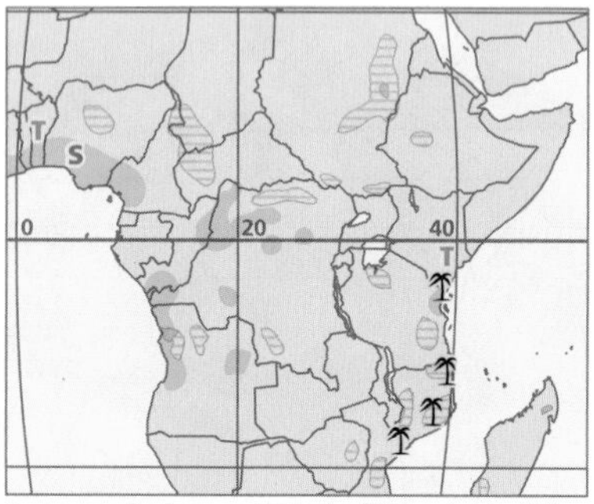
S
0
20
40
T

Area symbol map: Detail of Vegetable Oils (p. 64)

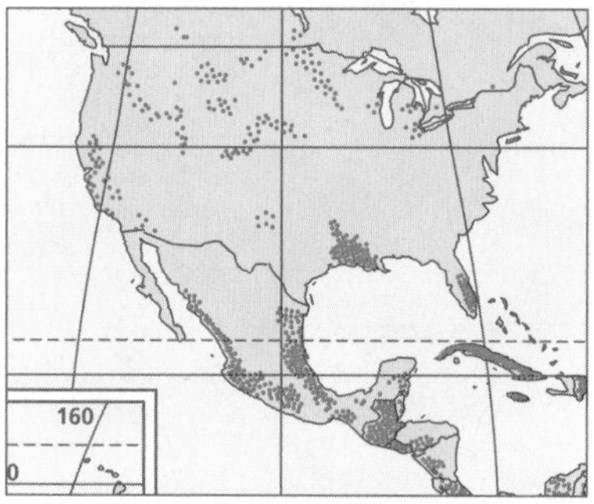
160

Dot map: Detail of Sugar, Spices (p. 62)

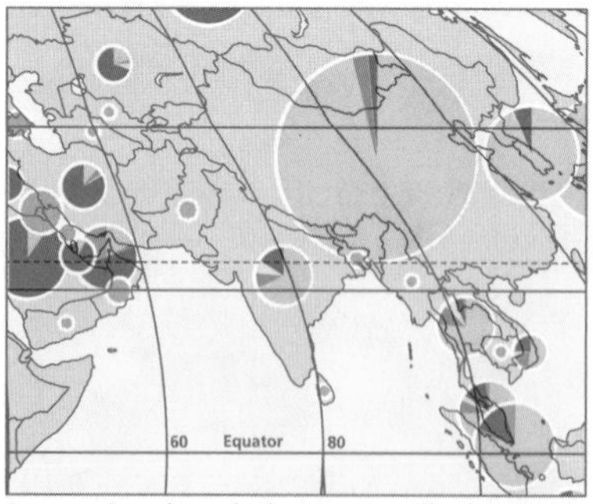
60
Equator
80

Proportional symbol map: Detail of Exports (p. 77)

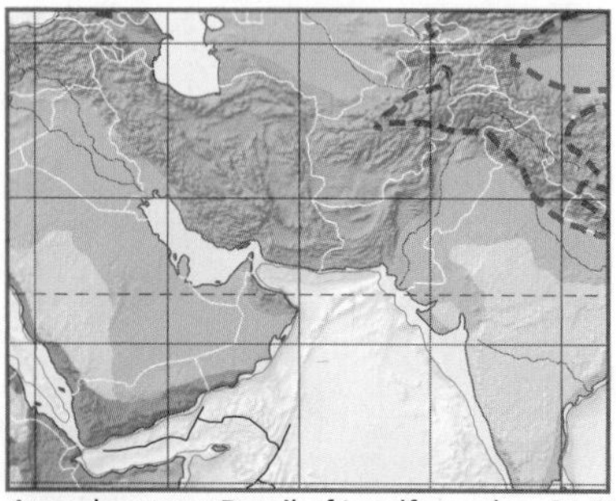

Area class map: Detail of Landforms (pp. 24-25)

Thematic Map Types in Goode's World Atlas

Thematic maps depict a single theme such as population density, agricultural productivity or annual precipitation. The selected theme is presented on a base of locational information, such as coastlines, country boundaries, and major drainage features.

Goode's World Atlas contains many different types of thematic maps. The characteristics of each are summarized below.

Point Symbol Maps

Point symbol maps are perhaps the simplest type of thematic map. They show features that occur at discrete locations. Examples include earthquakes, nuclear power plants, and mineral-producing areas. The Zinc and Coltan map (p. 69) is an example of a point symbol map. Different colors represent the two different materials, and various symbol sizes show relative importance.

Area Symbol Maps

Area symbol maps are useful for delineating regions of interest. For example, the Vegetable Oils maps (p. 64) show major oil-producing regions in different colors. Some point symbols also appear on this map, for less extensive oil crops.

Dot Maps

Dot maps show a distribution using a pattern of dots, where each dot represents a certain quantity or amount. For example, on the Sugar and Spice map (p. 62), each dot represents 20,000 metric tons of sugar produced. The different dot colors represent different sources of sugar (cane vs. beet). Dot maps are an effective way of representing the variable density of geographic phenomena.

Proportional Symbol Maps

Proportional symbol maps portray numeric quantities, such as total toxic chemical releases per state, or the total value of agricultural goods produced by country. The symbols on these maps — usually circles — are drawn such that the size the symbol is proportional to the value at that location. For example, the Exports map (p. 77) shows the value of goods exported by each country in the world, in billions of U.S. dollars. Proportional symbols are frequently subdivided based on the percentage of individual components making up the total. The Exports map uses wedges of different color to show the percentages of various types of exports, such as manufactured articles and raw materials.

Area Class Maps

Area class maps divide Earth into zones based on categories of a particular geographic phenomenon. For example, the Landforms map (pp. 24-25) divides Earth into seven unique structural regions based on landform type and origin. Other examples of area class maps in Goode's World Atlas include soil taxonomy (pp. 44-45), terrestrial biomes (pp. 46-47), and natural vegetation (pp. 42-43).

Flow Line Maps

Flow line maps show flows between locations. Usually the thickness of the flow lines is proportional to the volume of the flow. Flows may be physical commodities like petroleum or less tangible quantities like information. The flow lines on the Communication Network Infrastructure map (pp. 82-83) represent bandwidth usage in gigabits per second. Note that the locations of flow lines may not accurately represent the actual physical route.

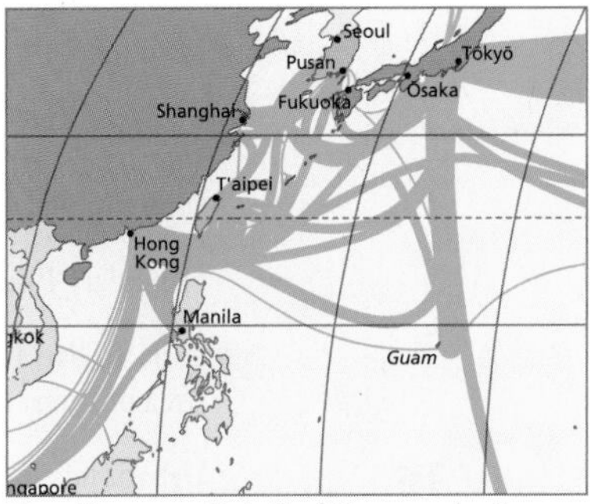
Seoul
Tōkyō
Pusan
Shanghai
Fukuoka
Ōsaka
T'aipei
Hong Kong
Manila
Guam

Flow line map: Detail of Communication Network Infrastructure (pp. 82-83)

Choropleth Maps

Choropleth maps apply distinctive colors to predefined areas, such as counties or states, to represent different quantities in each area. The quantities shown are usually rates, percentages, or densities. For example, the Birth Rate map (p. 50) shows the annual number of births per one thousand people for each country.

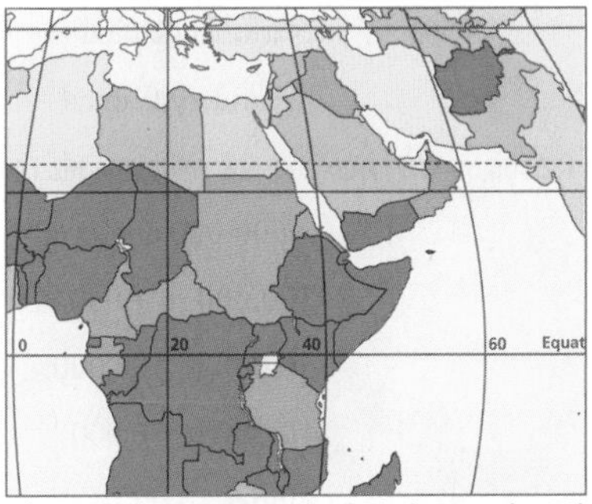
0
20
40
60
Equat

Choropleth map: Detail of Birth Rate (p. 50)

Isoline Maps

Isoline maps are used to portray quantities that vary continuously over space. These maps are frequently used for climate variables such as precipitation and temperature. For example, the January Temperature map for the North Polar region (p. 33) contains isolines at intervals of 5° C. Colors are also used to assist map interpretation.

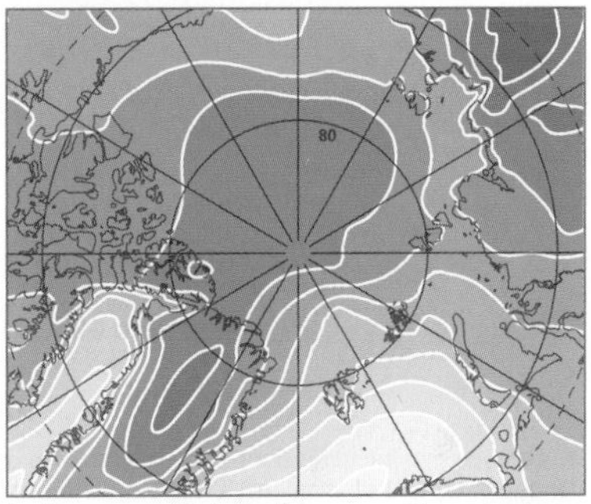
80

Isoline map: Detail of January Temperature (p. 33)

Grid-Based Maps

Grid-based maps rely on data points occurring at regular intervals in a two-dimensional grid. Some grid-based maps are actually digital images, analogous to the pictures captured by digital cameras. These maps are created from a very fine grid of cells called pixels, each of which is assigned a color that corresponds to a specific value or range of values. The population density maps in this atlas (pp. 48-49, for example) are examples of this type. Other grid-based maps are based on data integrated over a coarser grid, such as the map showing temperature change for 5-degree grid cells (p. 38) and the tornado map showing the frequency of tornadoes within 1-degree grid cells (p. 91). Grid-based mapping is increasingly being used to map environmental phenomena observable from remote sensing systems.

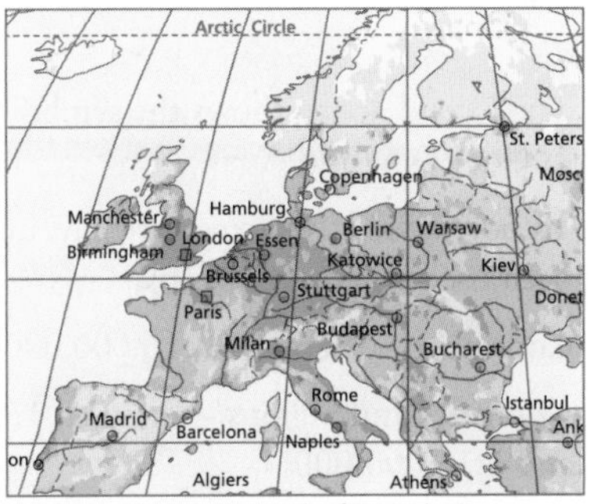
Arctic Circle
St. Peters
Mosc
Copenhagen
Manchester
Hamburg
Birmingham
London
Essen
Berlin
Warsaw
Brussels
Katowice
Kiev
Paris
Stuttgart
Donet
Milan
Budapest
Bucharest
Rome
Istanbul
Madrid
Barcelona
Naples
Ank
Algiers
Athens

Grid-based map: Detail of Population Density (pp. 48-49)

Cartograms

Cartograms are maps on which shapes and areas have been deliberately distorted. The cartograms in this atlas draw each country as a rectangle whose size is proportional to the population of the country. This means that the countries with the largest areas are those with the largest populations, regardless of actual country area. Cartograms make explicit the relationship between the mapped variable and the size of the affected population. As an example, consider the HIV cartogram (p. 54). Both Chad and Nigeria have relatively high rates of HIV infection, but Nigeria is much larger than Chad on the cartogram, since Nigeria's population is much larger. This informs the cartogram reader that the population affected by HIV is much larger in Nigeria.

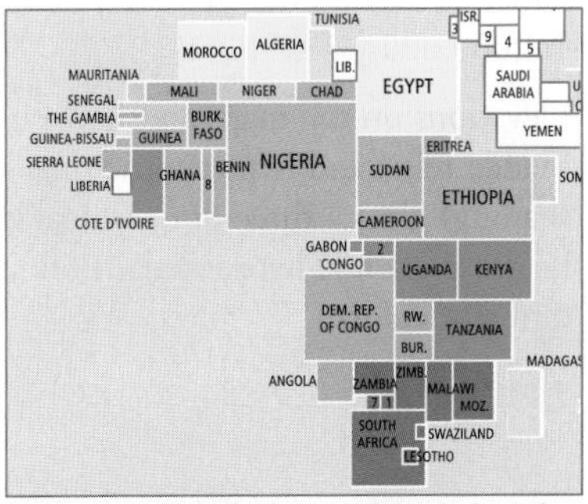
TUNISIA
ISR.
MOROCCO
ALGERIA
LIB.
MAURITANIA
EGYPT
SENEGAL
MALI
NIGER
CHAD
SAUDI ARABIA
THE GAMBIA
BURK. FASO
YEMEN
GUINEA-BISSAU
GUINEA
ERITREA
SIERRA LEONE
NIGERIA
LIBERIA
GHANA
BENIN
SUDAN
ETHIOPIA
COTE D'IVOIRE
CAMEROON
GABON
CONGO
UGANDA
KENYA
DEM. REP. OF CONGO
RW.
TANZANIA
BUR.
ANGOLA
ZAMBIA
ZIMB.
MALAWI
MOZ.
SOUTH AFRICA
SWAZILAND
LESOTHO

Cartogram: Detail of HIV Infection (p. 54)

Political Boundaries

- International
- Disputed or Unrecognized
- Secondary (State, Provincial, etc.)
- International Boundary over Water
- Secondary Boundary over Water
- Park, Indian Reservation, Area of Interest
- Urbanized Area

Populated Places

- TŌKYŌ — National Capital
- Boise — Secondary Capital

1:2,000,000, 1:1,000,000 and 1:500,000 Inset Maps

- 1,000,000 and over
- 100,000 to 1,000,000
- 50,000 to 100,000
- 10,000 to 50,000
- Under 10,000
- Neighborhood, Section of City

Other Reference Maps

- 1,000,000 and over
- 250,000 to 1,000,000
- 100,000 to 250,000
- 25,000 to 100,000
- Under 25,000

Note: Type size indicates the relative importance of the city. On the continent physical maps, city populations and relative importance are not differentiated.

Cultural Features

- Dam
- Point of Interest
- Ruins
- PALESTINE — Cultural or Historic Region

Transportation

- Major Road
- Minor Road
- Railroad
- Airport

Land Features

- Peak, Spot Height
- Pass
- Sand
- Contours

Elevation

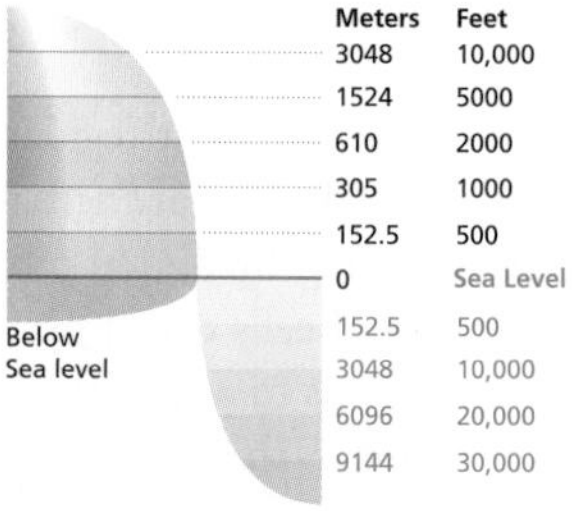
Meters Feet
3048 10,000
1524 5000
610 2000
305 1000
152.5 500
0 Sea Level
Below Sea level
152.5 500
3048 10,000
6096 20,000
9144 30,000

Note: The 500 foot contour is not shown on the small-scale oceans and polar regions maps.

Lakes and Reservoirs

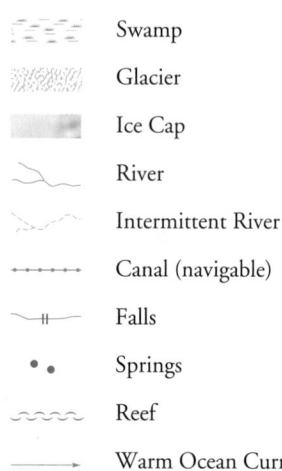

- Fresh Water
- Fresh Water: Intermittent
- Dry Lake
- Salt Water
- Salt Water: Intermittent

Other Water features

- Swamp
- Glacier
- Ice Cap
- River
- Intermittent River
- Canal (navigable)
- Falls
- Springs
- Reef
- Warm Ocean Current
- Cold Ocean Current

The legend above shows the symbols used for reference maps in Goode's World Atlas.

To portray relative areas correctly, uniform map scales have been used wherever possible:

Continents – Between 1:16,000,000 and 1:40,000,000

Countries and regions – Between 1:4,000,000 and 1:16,000,000

World, polar areas and oceans – 1:40,000,000 and smaller

City and island inset maps – 1:500,000, 1:1,000,000 and 1:2,000,000

Elevations on the maps are shown using a combination of shaded relief and hypsometric tints. Shaded relief (or hill-shading) gives a three-dimensional impression of the landscape, while hypsometric tints show elevation ranges in different colors.

The choice of names for mapped features is complicated by the fact that a variety of languages and alphabets are used throughout the world. A local-names policy is used in Goode's World Atlas for populated places and local physical features. For some major features, an English form of the name is used with the local name, e.g., Vienna (Wien) and Naples (Napoli). In countries where more than one official language is used, names are given in the dominant local language. For large physical features spanning international borders, the conventional English form of the name is used. In cases where a non-Roman alphabet is used, names have been transliterated according to accepted practice.

Selected features are also listed in the Index, which includes a pronunciation guide. A list of foreign geographic terms is provided in the Glossary.

THE SOLAR SYSTEM

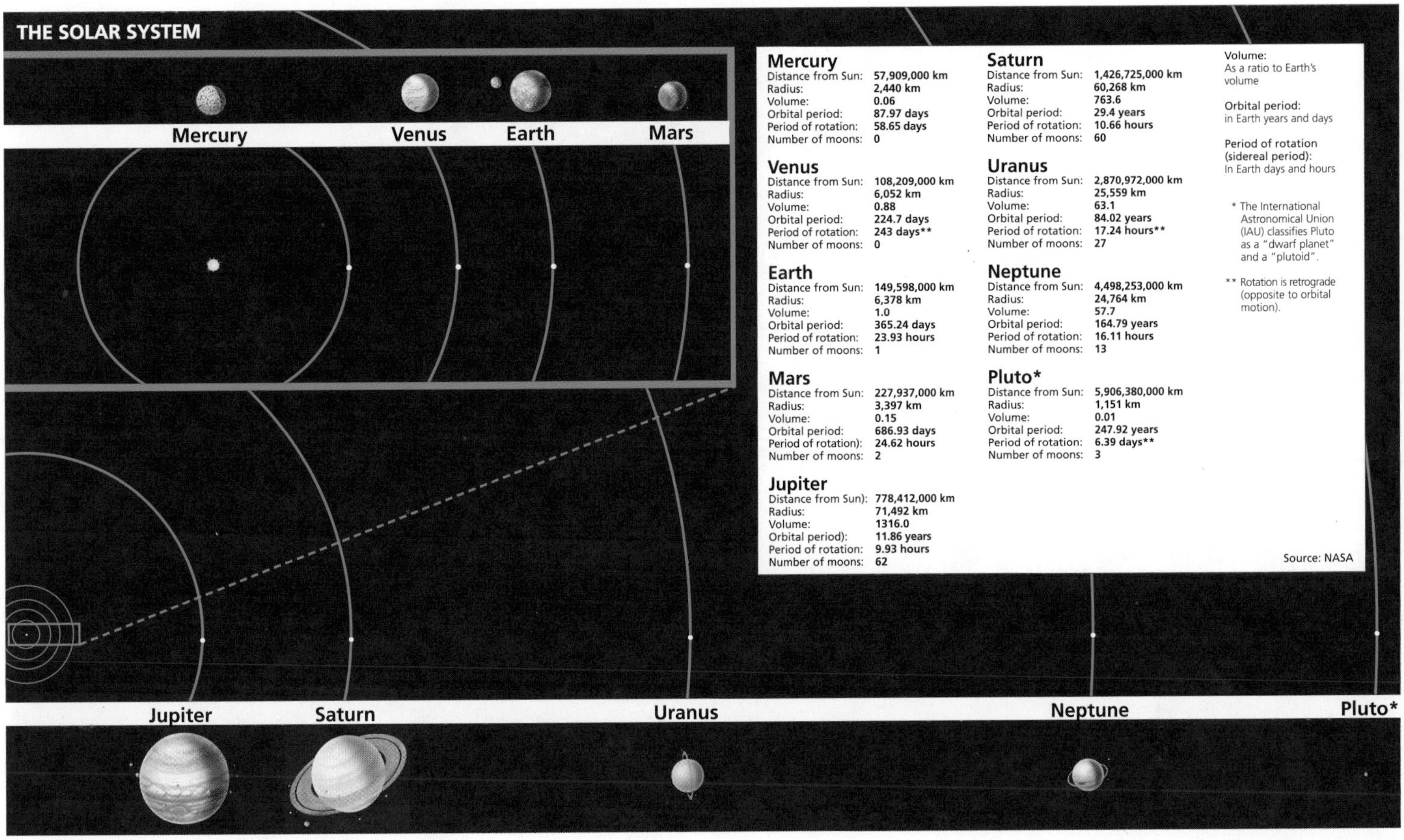
Mercury
Venus
Earth
Mars
Jupiter
Saturn
Uranus
Neptune
Pluto*

Planet	Distance from Sun	Radius	Volume	Orbital period	Period of rotation	Number of moons
Mercury	57,909,000 km	2,440 km	0.06	87.97 days	58.65 days	0
Venus	108,209,000 km	6,052 km	0.88	224.7 days	243 days**	0
Earth	149,598,000 km	6,378 km	1.0	365.24 days	23.93 hours	1
Mars	227,937,000 km	3,397 km	0.15	686.93 days	24.62 hours	2
Jupiter	778,412,000 km	71,492 km	1316.0	11.86 years	9.93 hours	62
Saturn	1,426,725,000 km	60,268 km	763.6	29.4 years	10.66 hours	60
Uranus	2,870,972,000 km	25,559 km	63.1	84.02 years	17.24 hours**	27
Neptune	4,498,253,000 km	24,764 km	57.7	164.79 years	16.11 hours	13
Pluto*	5,906,380,000 km	1,151 km	0.01	247.92 years	6.39 days**	3

Volume: As a ratio to Earth's volume

Orbital period: in Earth years and days

Period of rotation (sidereal period): In Earth days and hours

* The International Astronomical Union (IAU) classifies Pluto as a "dwarf planet" and a "plutoid".

** Rotation is retrograde (opposite to orbital motion).

Source: NASA

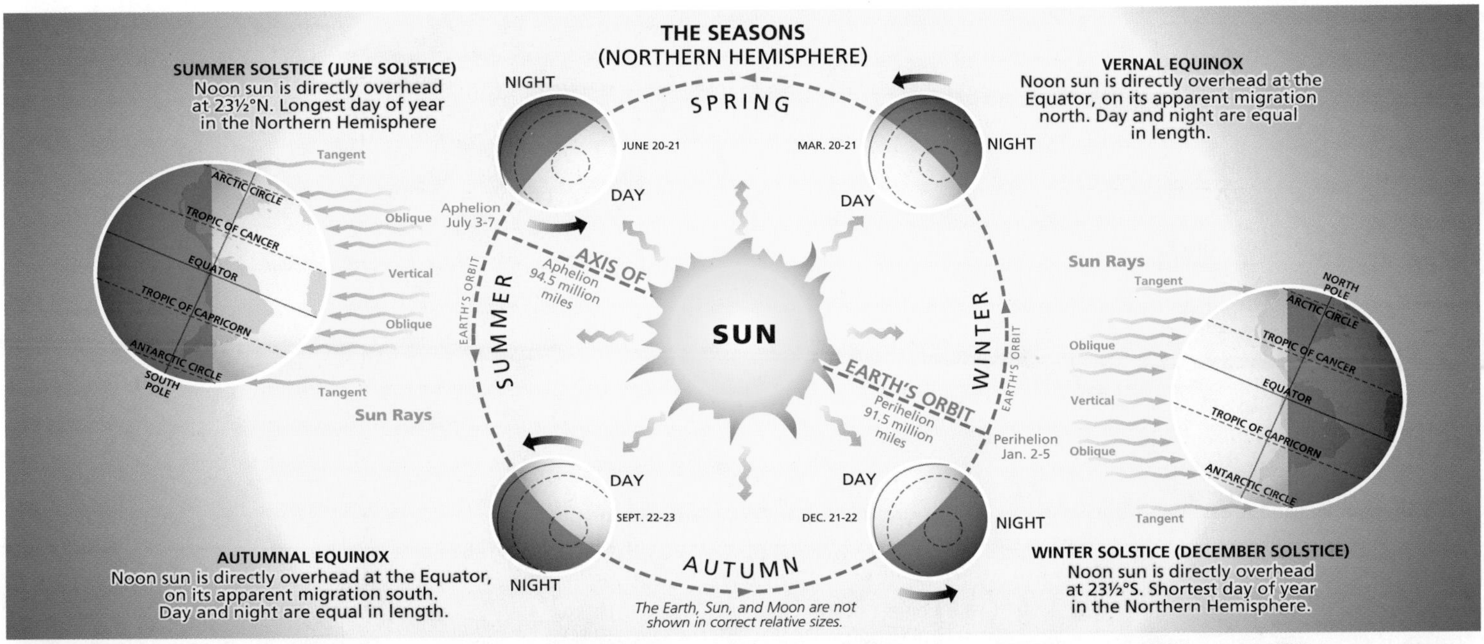
THE SEASONS
(NORTHERN HEMISPHERE)
SUMMER SOLSTICE (JUNE SOLSTICE)
Noon sun is directly overhead at 23½°N. Longest day of year in the Northern Hemisphere
VERNAL EQUINOX
Noon sun is directly overhead at the Equator, on its apparent migration north. Day and night are equal in length.
AUTUMNAL EQUINOX
Noon sun is directly overhead at the Equator, on its apparent migration south. Day and night are equal in length.
WINTER SOLSTICE (DECEMBER SOLSTICE)
Noon sun is directly overhead at 23½°S. Shortest day of year in the Northern Hemisphere.
SPRING
SUMMER
AUTUMN
WINTER
SUN
NIGHT
DAY
JUNE 20-21
MAR. 20-21
SEPT. 22-23
DEC. 21-22
Aphelion July 3-7
Aphelion 94.5 million miles
AXIS OF
EARTH'S ORBIT
Perihelion 91.5 million miles
Perihelion Jan. 2-5
ARCTIC CIRCLE
TROPIC OF CANCER
EQUATOR
TROPIC OF CAPRICORN
ANTARCTIC CIRCLE
SOUTH POLE
NORTH POLE
Tangent
Oblique
Vertical
Sun Rays
The Earth, Sun, and Moon are not shown in correct relative sizes.

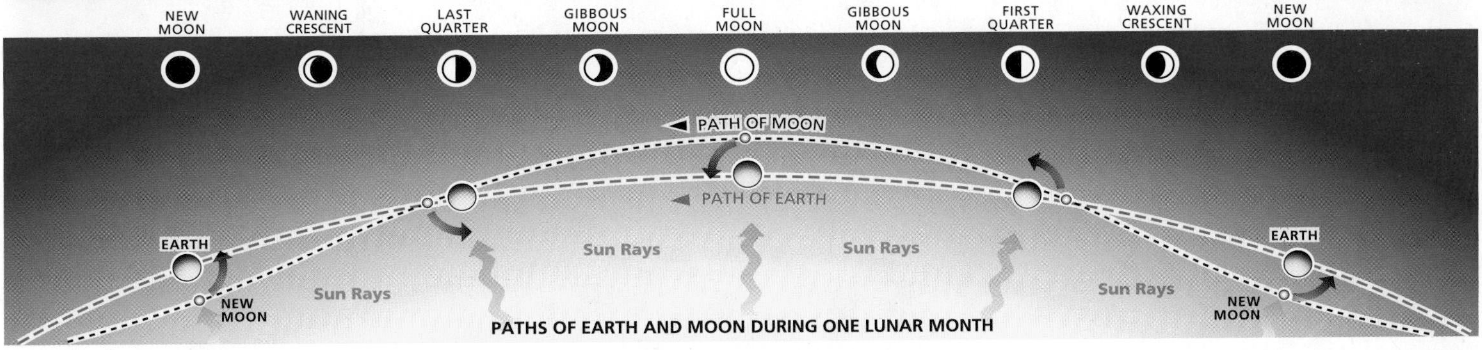
NEW MOON
WANING CRESCENT
LAST QUARTER
GIBBOUS MOON
FULL MOON
GIBBOUS MOON
FIRST QUARTER
WAXING CRESCENT
NEW MOON
PATH OF MOON
PATH OF EARTH
EARTH
Sun Rays
PATHS OF EARTH AND MOON DURING ONE LUNAR MONTH

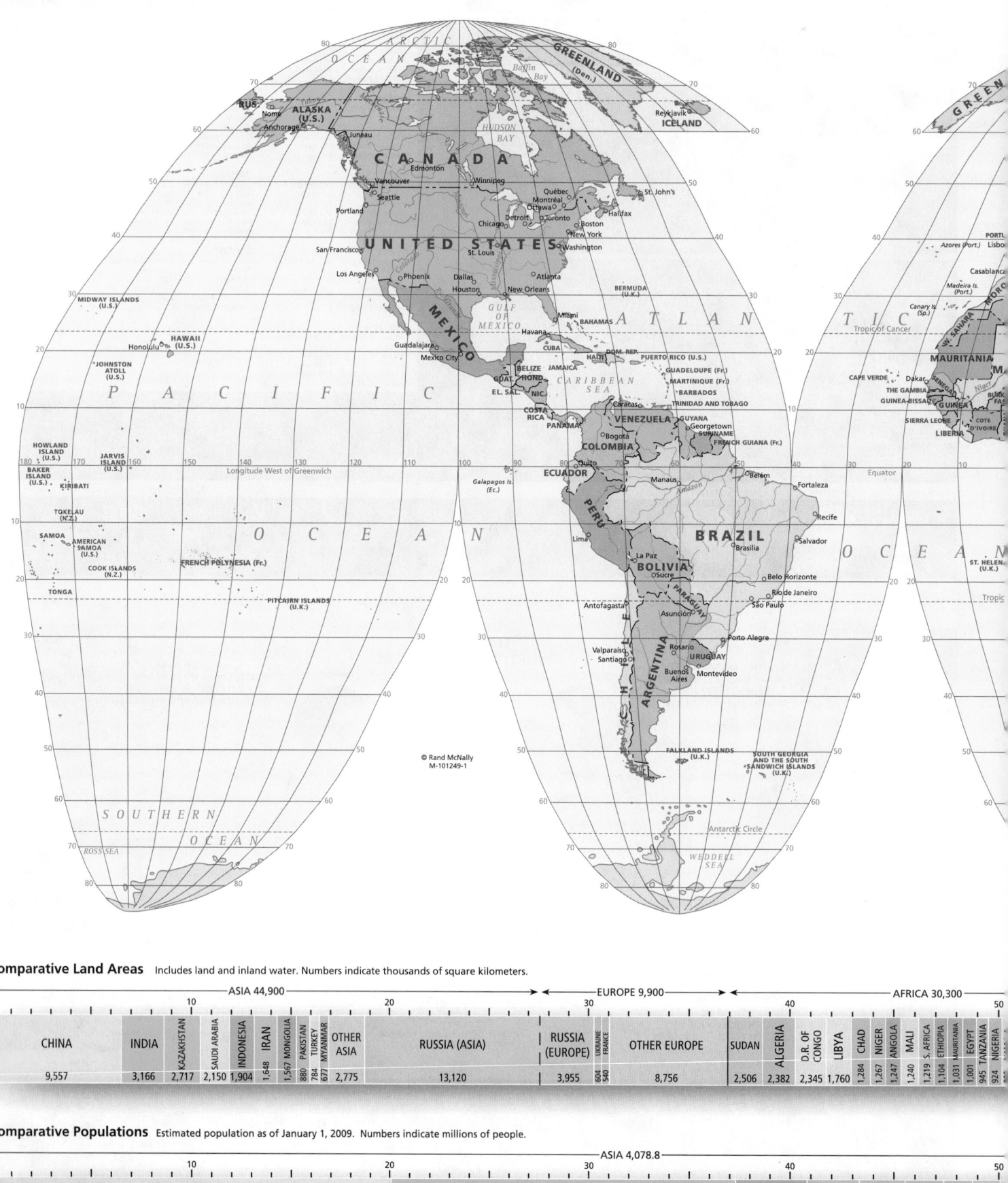

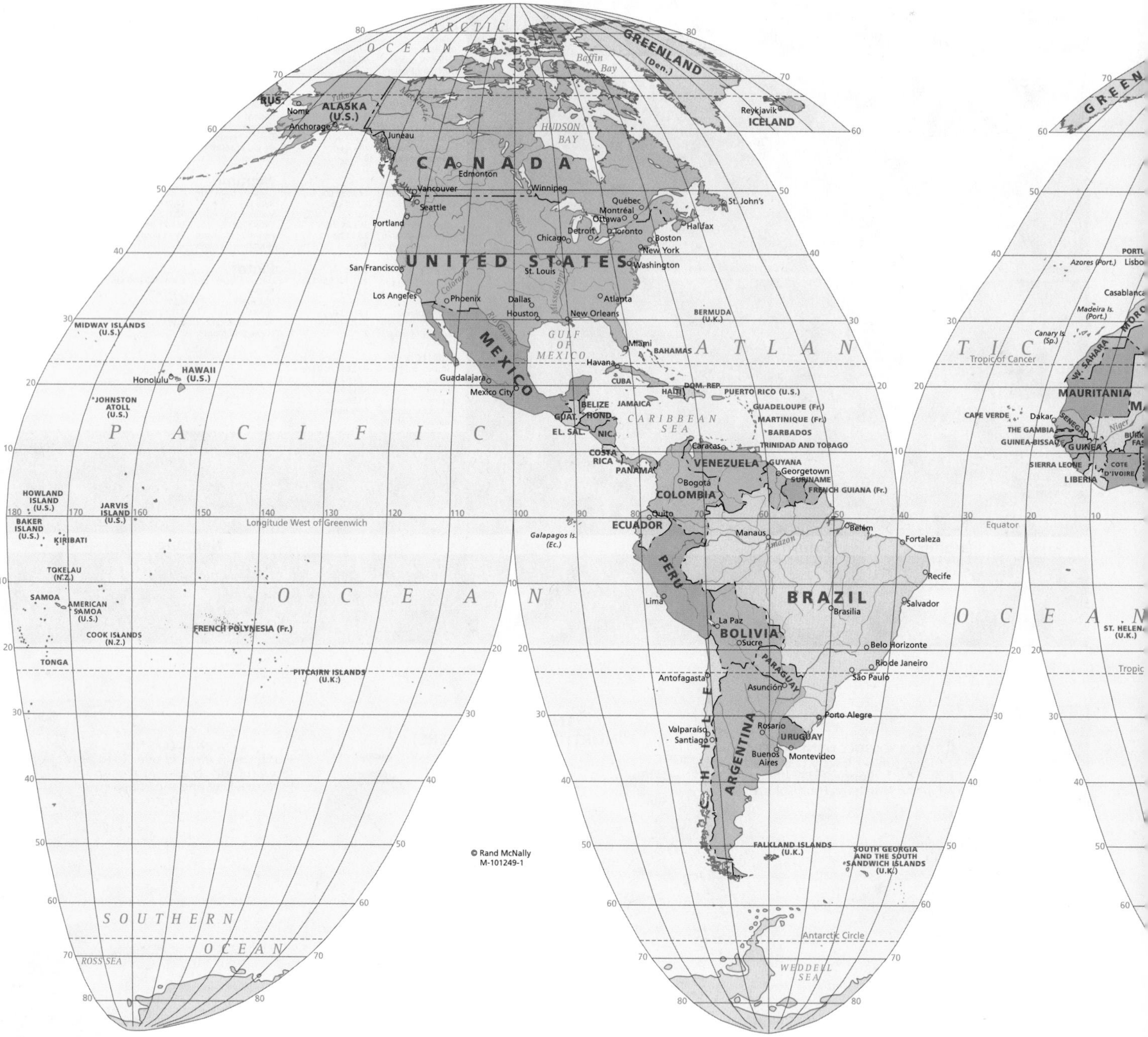
ARCTIC OCEAN
GREENLAND (Den.)
Baffin Bay
RUS.
Nome
ALASKA (U.S.)
Anchorage
Juneau
HUDSON BAY
Reykjavík
ICELAND
CANADA
Edmonton
Vancouver
Winnipeg
Seattle
Québec
Montréal
Ottawa
St. John's
Halifax
Portland
Detroit
Toronto
Chicago
Boston
New York
UNITED STATES
San Francisco
St. Louis
Washington
Los Angeles
Phoenix
Dallas
Atlanta
Houston
New Orleans
BERMUDA (U.K.)
MIDWAY ISLANDS (U.S.)
GULF OF MEXICO
Miami
BAHAMAS
MEXICO
Havana
CUBA
ATLANTIC
Tropic of Cancer
HAWAII (U.S.)
Honolulu
Guadalajara
Mexico City
DOM. REP.
HAITI
PUERTO RICO (U.S.)
GUADELOUPE (Fr.)
MARTINIQUE (Fr.)
BARBADOS
TRINIDAD AND TOBAGO
JOHNSTON ATOLL (U.S.)
BELIZE
JAMAICA
HOND.
GUAT.
EL SAL.
NIC.
CARIBBEAN SEA
PACIFIC
Caracas
COSTA RICA
PANAMA
VENEZUELA
GUYANA
Georgetown
SURINAME
FRENCH GUIANA (Fr.)
Bogotá
COLOMBIA
HOWLAND ISLAND (U.S.)
BAKER ISLAND (U.S.)
JARVIS ISLAND (U.S.)
KIRIBATI
Longitude West of Greenwich
Quito
ECUADOR
Galapagos Is. (Ec.)
Belém
Fortaleza
Equator
Manaus
Amazon
TOKELAU (N.Z.)
SAMOA
AMERICAN SAMOA (U.S.)
COOK ISLANDS (N.Z.)
FRENCH POLYNESIA (Fr.)
TONGA
OCEAN
PERU
Lima
BRAZIL
Brasília
Recife
Salvador
La Paz
BOLIVIA
Sucre
Belo Horizonte
Rio de Janeiro
São Paulo
PARAGUAY
Asunción
Antofagasta
PITCAIRN ISLANDS (U.K.)
Porto Alegre
Rosario
URUGUAY
Valparaíso
Santiago
Buenos Aires
Montevideo
ARGENTINA
CHILE
FALKLAND ISLANDS (U.K.)
SOUTH GEORGIA AND THE SOUTH SANDWICH ISLANDS (U.K.)
© Rand McNally
M-101249-1
SOUTHERN OCEAN
ROSS SEA
Antarctic Circle
WEDDELL SEA
Azores (Port.)
Madeira Is. (Port.)
Canary Is. (Sp.)
Casablanca
W. SAHARA
MAURITANIA
CAPE VERDE
Dakar
THE GAMBIA
GUINEA-BISSAU
GUINEA
SIERRA LEONE
LIBERIA
COTE D'IVOIRE
Niger
ST. HELENA (U.K.)
Tropic

Comparative Land Areas Includes land and inland water. Numbers indicate thousands of square kilometers.

ASIA 44,900 — EUROPE 9,900 — AFRICA 30,300

CHINA	9,557
INDIA	3,166
KAZAKHSTAN	2,717
SAUDI ARABIA	2,150
INDONESIA	1,904
IRAN	1,648
MONGOLIA	1,567
PAKISTAN	880
TURKEY	784
MYANMAR	677
OTHER ASIA	2,775
RUSSIA (ASIA)	13,120
RUSSIA (EUROPE)	3,955
UKRAINE	604
FRANCE	540
OTHER EUROPE	8,756
SUDAN	2,506
ALGERIA	2,382
D.R. OF CONGO	2,345
LIBYA	1,760
CHAD	1,284
NIGER	1,267
ANGOLA	1,247
MALI	1,240
S. AFRICA	1,219
ETHIOPIA	1,104
MAURITANIA	1,031
EGYPT	1,001
TANZANIA	945
NIGERIA	924

Comparative Populations Estimated population as of January 1, 2009. Numbers indicate millions of people.

ASIA 4,078.8

CHINA	1,341.8
INDIA	1,157.1
INDONESIA	238.9
PAKISTAN	174.5
BANGLADESH	155.0
JAPAN	127.2
PHILIPPINES	97.0
VIETNAM	86.5

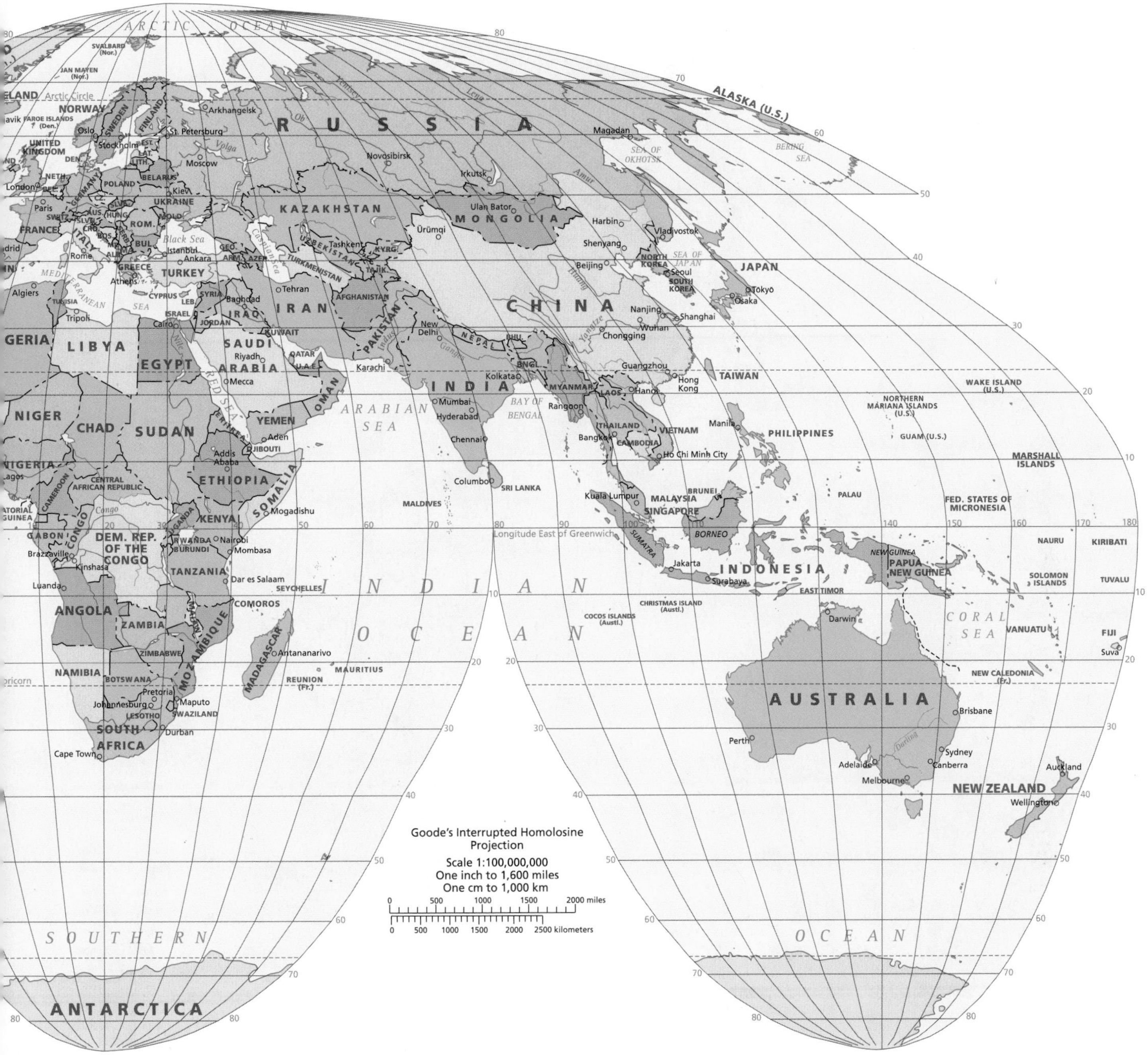
ARCTIC OCEAN
SVALBARD (Nor.)
JAN MAYEN (Nor.)
Arctic Circle
NORWAY
FAROE ISLANDS (Den.)
UNITED KINGDOM
DEN.
SWEDEN
FINLAND
Oslo
Stockholm
St. Petersburg
Arkhangelsk
London
Paris
FRANCE
POLAND
BELARUS
UKRAINE
Kiev
Moscow
ROM.
BUL.
ITALY
Rome
GREECE
Athens
Black Sea
Istanbul
Ankara
TURKEY
CYPRUS
SYRIA
LEB.
ISRAEL
JORDAN
Cairo
Baghdad
IRAQ
KUWAIT
Tehran
IRAN
MEDITERRANEAN SEA
Algiers
TUNISIA
Tripoli
LIBYA
EGYPT
SAUDI ARABIA
Riyadh
Mecca
QATAR
U.A.E.
OMAN
YEMEN
Aden
DJIBOUTI
RED SEA
ERITREA
SUDAN
CHAD
NIGER
NIGERIA
Lagos
CAMEROON
CENTRAL AFRICAN REPUBLIC
ETHIOPIA
Addis Ababa
SOMALIA
Mogadishu
KENYA
Nairobi
Mombasa
UGANDA
RWANDA
BURUNDI
GABON
CONGO
Brazzaville
Kinshasa
DEM. REP. OF THE CONGO
TANZANIA
Dar es Salaam
SEYCHELLES
COMOROS
Luanda
ANGOLA
ZAMBIA
ZIMBABWE
MOZAMBIQUE
MADAGASCAR
Antananarivo
MAURITIUS
REUNION (Fr.)
NAMIBIA
BOTSWANA
Pretoria
Johannesburg
Maputo
SWAZILAND
LESOTHO
Durban
SOUTH AFRICA
Cape Town
RUSSIA
Novosibirsk
Irkutsk
Magadan
SEA OF OKHOTSK
BERING SEA
ALASKA (U.S.)
KAZAKHSTAN
UZBEKISTAN
TURKMENISTAN
KYRG.
TAJIK.
Tashkent
AFGHANISTAN
PAKISTAN
Karachi
Caspian Sea
Ürümqi
Ulan Bator
MONGOLIA
CHINA
Harbin
Vladivostok
Shenyang
Beijing
NORTH KOREA
SOUTH KOREA
Seoul
SEA OF JAPAN
JAPAN
Tokyo
Osaka
Nanjing
Shanghai
Wuhan
Chongqing
Guangzhou
Hong Kong
TAIWAN
New Delhi
NEPAL
BHU.
BNGL.
INDIA
Kolkata
Mumbai
Hyderabad
Chennai
ARABIAN SEA
BAY OF BENGAL
MYANMAR
Rangoon
LAOS
Hanoi
THAILAND
Bangkok
CAMBODIA
VIETNAM
Ho Chi Minh City
Manila
PHILIPPINES
SRI LANKA
Colombo
MALDIVES
Kuala Lumpur
MALAYSIA
SINGAPORE
BRUNEI
SUMATRA
BORNEO
Jakarta
Surabaya
INDONESIA
EAST TIMOR
NEW GUINEA
PAPUA NEW GUINEA
NORTHERN MARIANA ISLANDS (U.S.)
GUAM (U.S.)
WAKE ISLAND (U.S.)
MARSHALL ISLANDS
PALAU
FED. STATES OF MICRONESIA
NAURU
KIRIBATI
SOLOMON ISLANDS
TUVALU
CHRISTMAS ISLAND (Austl.)
COCOS ISLANDS (Austl.)
INDIAN OCEAN
Longitude East of Greenwich
Darwin
CORAL SEA
VANUATU
FIJI
Suva
NEW CALEDONIA (Fr.)
AUSTRALIA
Brisbane
Perth
Sydney
Canberra
Adelaide
Melbourne
Auckland
NEW ZEALAND
Wellington
Goode's Interrupted Homolosine Projection
Scale 1:100,000,000
One inch to 1,600 miles
One cm to 1,000 km
0 500 1000 1500 2000 miles
0 500 1000 1500 2000 2500 kilometers
SOUTHERN OCEAN
ANTARCTICA

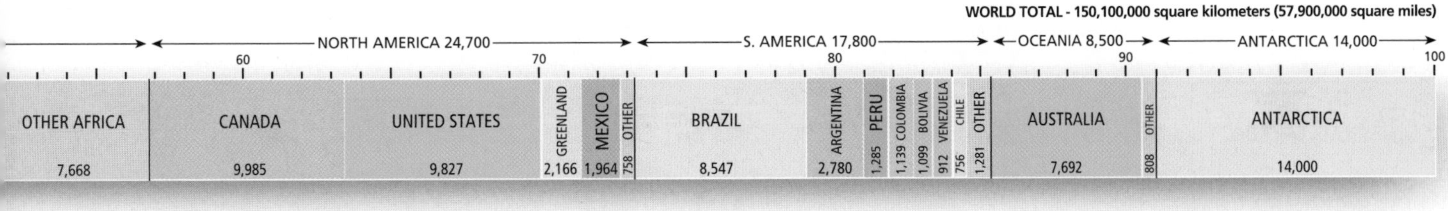
WORLD TOTAL - 150,100,000 square kilometers (57,900,000 square miles)
NORTH AMERICA 24,700
S. AMERICA 17,800
OCEANIA 8,500
ANTARCTICA 14,000
60
70
80
90
100
OTHER AFRICA 7,668
CANADA 9,985
UNITED STATES 9,827
GREENLAND 2,166
MEXICO 1,964
OTHER 758
BRAZIL 8,547
ARGENTINA 2,780
PERU 1,285
COLOMBIA 1,139
BOLIVIA 1,099
VENEZUELA 912
CHILE 756
OTHER 1,281
AUSTRALIA 7,692
OTHER 808
ANTARCTICA 14,000

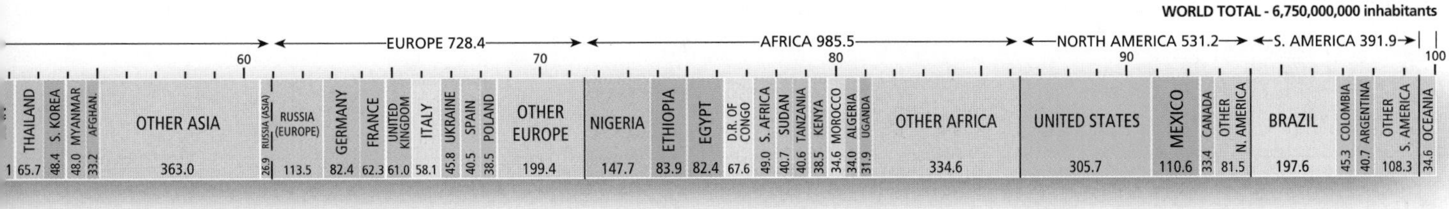
WORLD TOTAL - 6,750,000,000 inhabitants
EUROPE 728.4
AFRICA 985.5
NORTH AMERICA 531.2
S. AMERICA 391.9
60
70
80
90
100
THAILAND 65.7
S. KOREA 48.4
MYANMAR 48.0
AFGHAN. 33.2
OTHER ASIA 363.0
RUSSIA (ASIA) 26.9
RUSSIA (EUROPE) 113.5
GERMANY 82.4
FRANCE 62.3
UNITED KINGDOM 61.0
ITALY 58.1
UKRAINE 45.8
SPAIN 40.5
POLAND 38.5
OTHER EUROPE 199.4
NIGERIA 147.7
ETHIOPIA 83.9
EGYPT 82.4
D.R. OF CONGO 67.6
S. AFRICA 49.0
SUDAN 40.7
TANZANIA 40.6
KENYA 38.5
MOROCCO 34.6
ALGERIA 34.0
UGANDA 31.9
OTHER AFRICA 334.6
UNITED STATES 305.7
MEXICO 110.6
CANADA 33.4
OTHER N. AMERICA 81.5
BRAZIL 197.6
COLOMBIA 45.3
ARGENTINA 40.7
OTHER S. AMERICA 108.3
OCEANIA 34.6

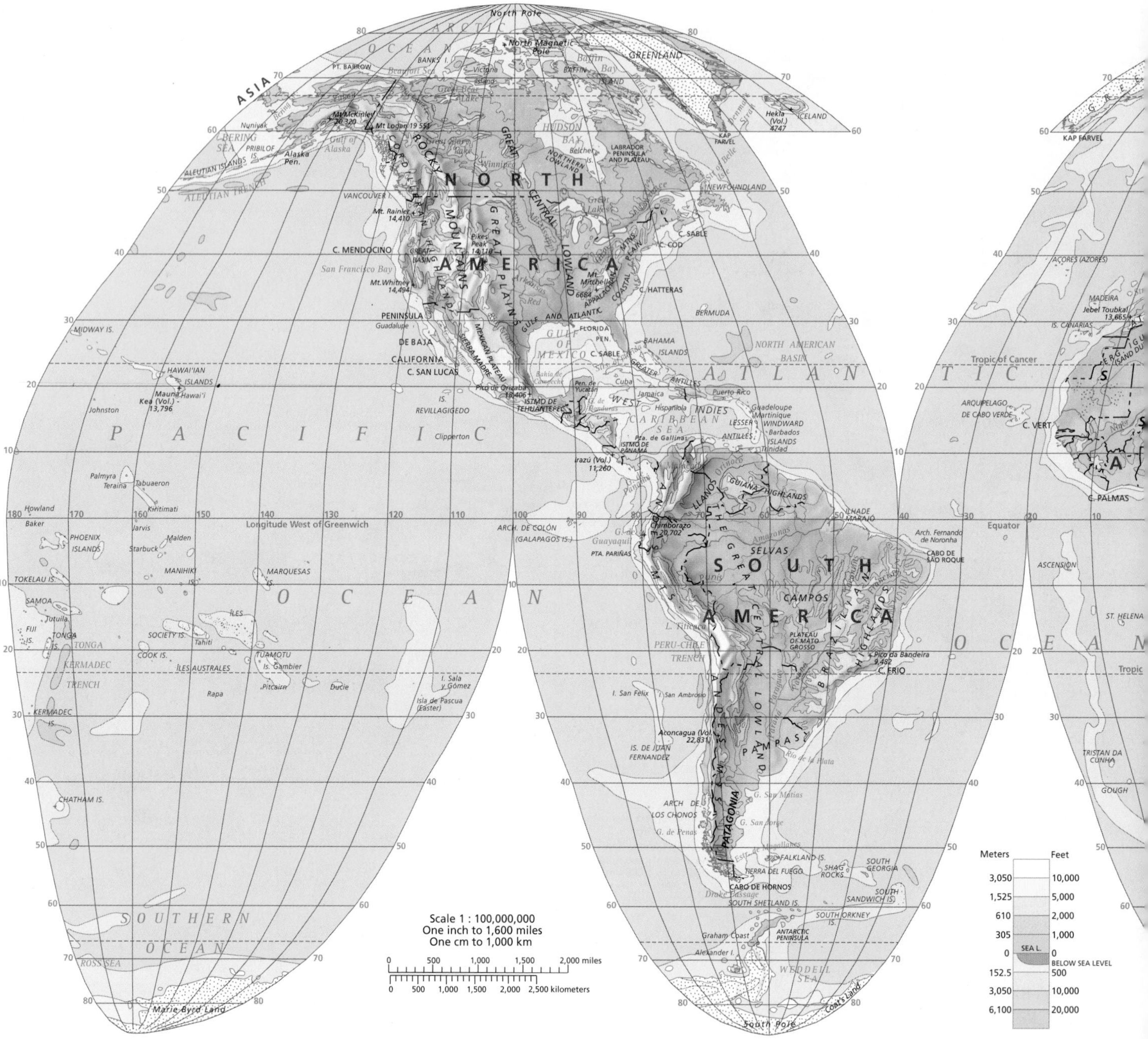
North Pole
ARCTIC OCEAN
North Magnetic Pole
GREENLAND
ASIA
NORTH AMERICA
SOUTH AMERICA
PACIFIC OCEAN
ATLANTIC OCEAN
SOUTHERN OCEAN
Mt McKinley 20,320
Mt Logan 19,551
Mt. Rainier 14,410
Pikes Peak 14,110
Mt. Whitney 14,494
Mt. Mitchell 6684
Pico de Orizaba 18,406
Irazú (Vol.) 11,260
Chimborazo 20,702
Aconcagua (Vol.) 22,831
Pico da Bandeira 9,492
Mauna Kea (Vol.) 13,796
Hekla (Vol.) 4747
Jebel Toubkal 13,665
Tropic of Cancer
Equator
Longitude West of Greenwich
South Pole
Scale 1 : 100,000,000
One inch to 1,600 miles
One cm to 1,000 km
0 500 1,000 1,500 2,000 miles
0 500 1,000 1,500 2,000 2,500 kilometers
Meters
Feet
3,050 10,000
1,525 5,000
610 2,000
305 1,000
0 SEA L. 0
BELOW SEA LEVEL
152.5 500
3,050 10,000
6,100 20,000

Land Elevations in Profile

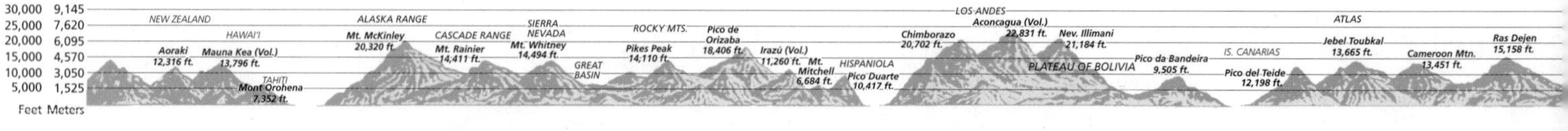
OCEANIA
NORTH AMERICA
SOUTH AMERICA
AFRICA
30,000 9,145
25,000 7,620
20,000 6,095
15,000 4,570
10,000 3,050
5,000 1,525
Feet Meters
NEW ZEALAND
Aoraki 12,316 ft.
HAWAI'I
Mauna Kea (Vol.) 13,796 ft.
TAHITI
Mont Orohena 7,352 ft.
ALASKA RANGE
Mt. McKinley 20,320 ft.
CASCADE RANGE
Mt. Rainier 14,411 ft.
SIERRA NEVADA
Mt. Whitney 14,494 ft.
GREAT BASIN
ROCKY MTS.
Pikes Peak 14,110 ft.
Pico de Orizaba 18,406 ft.
Irazú (Vol.) 11,260 ft.
Mt. Mitchell 6,684 ft.
HISPANIOLA
Pico Duarte 10,417 ft.
LOS ANDES
Chimborazo 20,702 ft.
Aconcagua (Vol.) 22,831 ft.
Nev. Illimani 21,184 ft.
PLATEAU OF BOLIVIA
Pico da Bandeira 9,505 ft.
IS. CANARIAS
Pico del Teide 12,198 ft.
ATLAS
Jebel Toubkal 13,665 ft.
Cameroon Mtn. 13,451 ft.
Ras Dejen 15,158 ft.

Ocean Depths in Profile

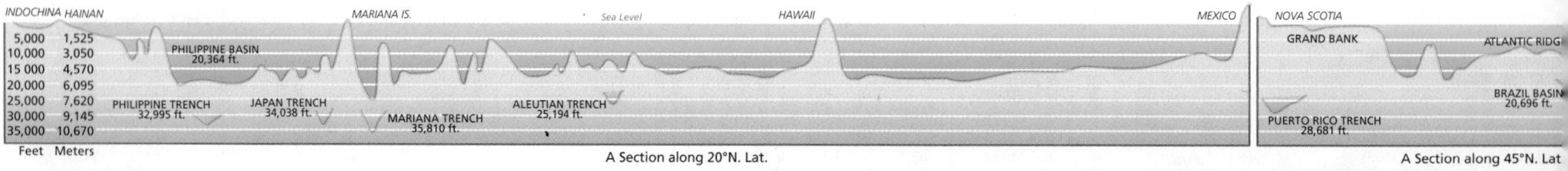
PACIFIC OCEAN
ATLANTIC O
INDOCHINA HAINAN
MARIANA IS.
Sea Level
HAWAII
MEXICO
NOVA SCOTIA
5,000 1,525
10,000 3,050
15 000 4,570
20,000 6,095
25,000 7,620
30,000 9,145
35,000 10,670
Feet Meters
PHILIPPINE BASIN 20,364 ft.
PHILIPPINE TRENCH 32,995 ft.
JAPAN TRENCH 34,038 ft.
MARIANA TRENCH 35,810 ft.
ALEUTIAN TRENCH 25,194 ft.
GRAND BANK
ATLANTIC RIDGE
BRAZIL BASIN 20,696 ft.
PUERTO RICO TRENCH 28,681 ft.

A Section along 20°N. Lat.

A Section along 45°N. Lat

North Pole
ARCTIC OCEAN
SVALBARD
FRANTSA-IOSIFA
NOVAYA ZEMLYA
BARENTS SEA
NORD KAPP
Jan Mayen
Arctic Circle
ICELAND
FAROE IS.
BRITISH ISLES
EUROPE
ASIA
N. AMERICA
Ostrov Vrangelya
St. Lawrence
POLUOSTROV KAMCHATKA
Klyuchevskaya (Vol.) 15,584
SEA OF OKHOTSK
BERING SEA
ALEUTIAN IS.
SAKHALIN
KURIL ISLANDS
HOKKAIDO
JAPAN TRENCH
SEA OF JAPAN
KOREAN PEN.
HONSHU
Fuji-san (Vol.) 12,388
KYUSHU
EAST CHINA SEA
NANSEI SHOTO
TAIWAN
BONIN IS.
Hsinkao Shan 13,113
HAINAN
SOUTH CHINA SEA
LUZON
PHILIPPINES
PHILIPPINE TRENCH
MINDANAO
MARIANA ISLANDS
Guam
MARIANA TRENCH
WAKE
MARSHALL ISLANDS
CAROLINE ISLANDS
GILBERT ISLANDS
Nauru
TUVALU
FIJI IS.
Viti Levu
WEST SIBERIAN PLAIN
URALS
STANOVOY KHREBET
GREATER KHINGAN RANGE
MANCHURIAN PLAIN
PLATEAU OF MONGOLIA
GOBI DESERT
ALTAI MTS.
TARIM BASIN
PAMIRS
HINDU KUSH
PLATEAU OF TIBET
KUNLUN SHAN
Mt. Everest 29,028
HIMALAYAS
NORTH CHINA PLAIN
Yellow Sea
Aral Sea
CASPIAN DEPRESSION
Black Sea
Gora El'brus 18,510
CAUCASUS
Mt. Ararat 16,854
PLATEAU OF ASIA MINOR
Qolleh-ye Damavand 18,386
Cyprus
Crete
SYRIAN DESERT
PLATEAU OF IRAN
AN NAFUD
PLATEAU AND PENINSULA OF ARABIA
RA'S AL HADD
PENINSULA OF INDIA
DECCAN PLATEAU
BAY OF BENGAL
ARABIAN SEA
LAKSHADWEEP
ANDAMAN ISLANDS
NICOBAR IS.
ISTHMUS OF KRA
INDOCHINA PENINSULA
MALAY PENINSULA
SRI LANKA
C. COMORIN
MALDIVE ISLANDS
Gulf of Thailand
Kinabalu 13,455
BORNEO
Celebes Sea
Halmahera
EAST INDIES
Moluccas
Banda Sea
Java Sea
JAVA
SUMATRA
SUNDA ISLANDS
JAVA TRENCH
Flores
Timor Sea
Arafura Sea
Puncak Jaya 16,503
NEW GUINEA
New Ireland
New Britain
SOLOMON ISLANDS
C. YORK
Gulf of Carpentaria
CORAL SEA
GREAT BARRIER REEF
NEW HEBRIDES
NEW CALEDONIA
NORTH WEST CAPE
GT. SANDY DESERT
WESTERN PLATEAU
GT. VICTORIA DESERT
AUSTRALIA
THE GREAT PLAINS
GREAT DIVIDING RANGE
Shark Bay
C. LEEUWIN
Great Australian Bight
Spencer Gulf
Bass Str.
Mt. Kosciuszko 7,313
C. HOWE
TASMANIA
SOUTH EAST CAPE
TASMAN SEA
NORTH CAPE
NORTH ISLAND
NEW ZEALAND
Aoraki 12,316
SOUTH ISLAND
Stewart
BOUNTY IS.
ANTIPODES
AUCKLAND IS.
Campbell
MACQUARIE IS.
BALLENY IS.
Wilkes Land
Victoria Land
Ross Sea
DAVIS SEA
South Pole
Mont Blanc 15,771
ALPS
PYRENEES
Corsica
Sardinia
Mt. Etna (Vol.) 10,902
BALKAN PEN.
MEDITERRANEAN SEA
BARKA PLATEAU
LIBYAN DESERT
NUBIAN DESERT
TASSILI N'ADJER (PLAT.)
OASES OF FEZZAN
SAHARA
SUDAN
Lake Chad
GREAT
AFRICA
Ras Dashen Terara 15,158
ETHIOPIAN PLATEAU
Gulf of Aden
Socotra
GEES GWARDAFUY
ADAMAWA HIGHLANDS
Cameroon Mtn. 13,451
BIOKO
CENTRAL PLATEAU
Lake Victoria
Kilimanjaro 19,340
Zanzibar
Lake Tanganyika
ALDABRA IS.
COMORO IS.
C. d' Ambre
MADAGASCAR
Mozambique Channel
C. FRIO
KALAHARI DESERT
Mont aux Sources 10,822
GREAT KARROO
C. OF GOOD HOPE
C. AGULHAS
C. STE. MARIE
Réunion
Mauritius
MASCARENE IS.
Rodrigues
AMIRANTE IS.
CHAGOS ARCH.
DIEGO GARCIA
COCOS IS.
CHRISTMAS
INDIAN OCEAN
Longitude East of Greenwich
ÎLE AMSTERDAM
ÎLE ST. PAUL
ÎLES CROZET
ÎLES KERGUELEN
Heard
PRINCE EDWARD IS.
BOUVETØYA
SOUTHERN OCEAN
Antarctic Circle
Enderby Land
ANTARCTICA

Goode's Interrupted Homolosine Projection

For Glossary of Foreign Geographical Terms see page 296

M-100962-1

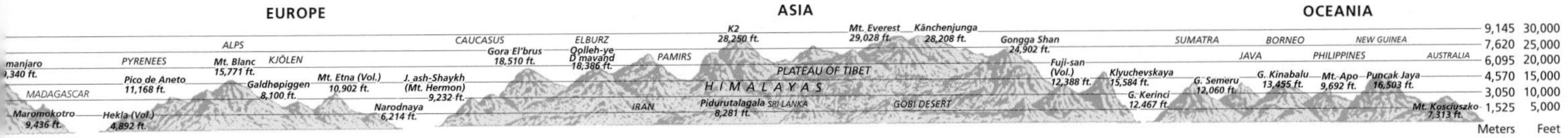
EUROPE
ASIA
OCEANIA
ALPS
CAUCASUS
ELBURZ
K2 28,250 ft.
Mt. Everest 29,028 ft.
Känchenjunga 28,208 ft.
Gongga Shan 24,902 ft.
SUMATRA
BORNEO
NEW GUINEA
JAVA
PHILIPPINES
AUSTRALIA
Gora El'brus 18,510 ft.
Qolleh-ye Damavand 18,386 ft.
PAMIRS
PYRENEES
Mt. Blanc 15,771 ft.
KJÖLEN
Pico de Aneto 11,168 ft.
Galdhøpiggen 8,100 ft.
Mt. Etna (Vol.) 10,902 ft.
J. ash-Shaykh (Mt. Hermon) 9,232 ft.
Narodnaya 6,214 ft.
MADAGASCAR
Maromokotro 9,436 ft.
Hekla (Vol.) 4,892 ft.
PLATEAU OF TIBET
HIMALAYAS
IRAN
Pidurutalagala 8,281 ft.
SRI LANKA
GOBI DESERT
Fuji-san (Vol.) 12,388 ft.
Klyuchevskaya 15,584 ft.
G. Kerinci 12,467 ft.
G. Semeru 12,060 ft.
G. Kinabalu 13,455 ft.
Mt. Apo 9,692 ft.
Puncak Jaya 16,503 ft.
Mt. Kosciuszko 7,313 ft.
9,145 30,000
7,620 25,000
6,095 20,000
4,570 15,000
3,050 10,000
1,525 5,000
Meters Feet

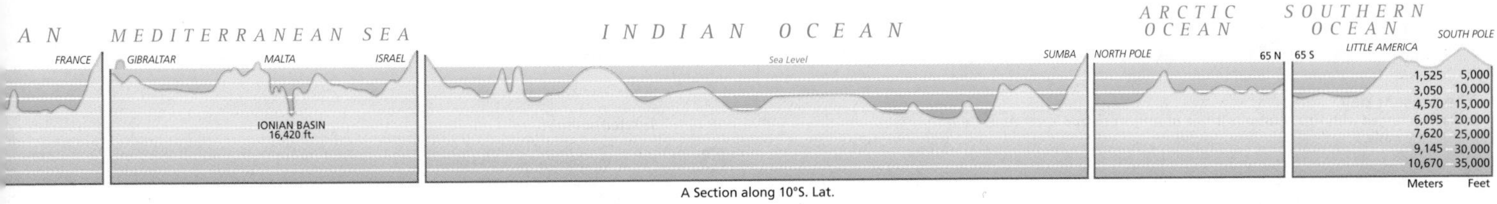
MEDITERRANEAN SEA
INDIAN OCEAN
ARCTIC OCEAN
SOUTHERN OCEAN
FRANCE
GIBRALTAR
MALTA
ISRAEL
IONIAN BASIN 16,420 ft.
Sea Level
SUMBA
NORTH POLE
65 N
65 S
LITTLE AMERICA
SOUTH POLE
1,525 5,000
3,050 10,000
4,570 15,000
6,095 20,000
7,620 25,000
9,145 30,000
10,670 35,000
Meters Feet

A Section along 10°S. Lat.

EVOLUTION OF THE CONTINENTS

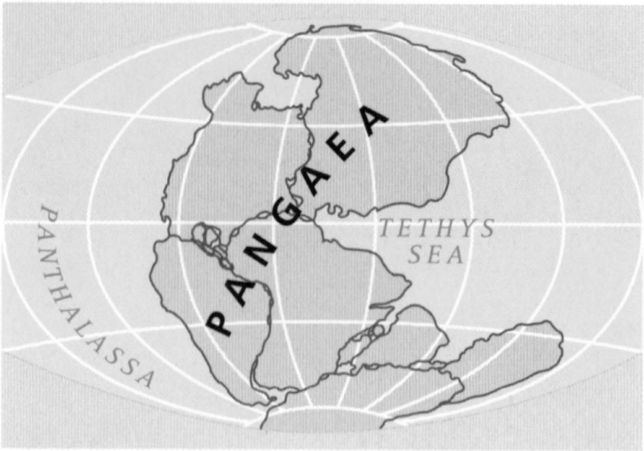
PANGAEA
PANTHALASSA
TETHYS SEA

225 million years ago
The supercontinent of Pangaea exists and Panthalassa forms the ancestral ocean. Tethys Sea separates Eurasia and Africa.

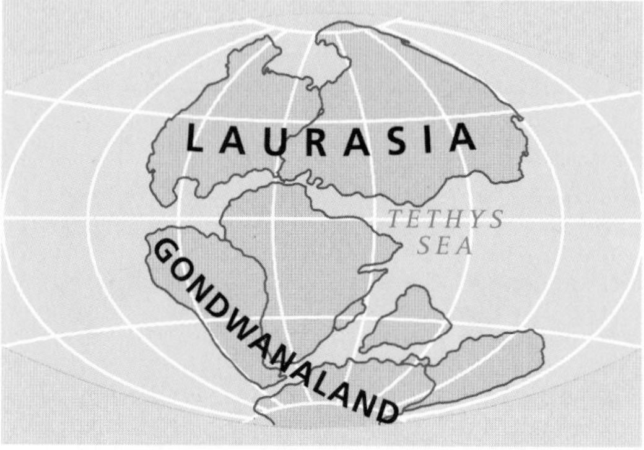
LAURASIA
TETHYS SEA
GONDWANALAND

180 million years ago
Pangaea splits, Laurasia drifts north. Gondwanaland breaks into South America/Africa, India, and Australia/Antarctica.

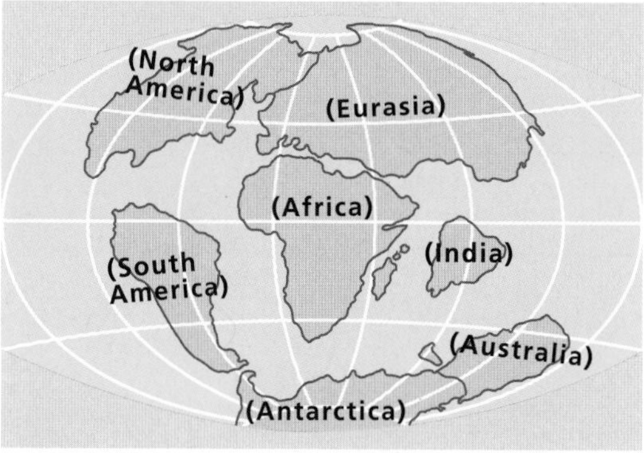
(North America)
(Eurasia)
(Africa)
(South America)
(India)
(Australia)
(Antarctica)

65 million years ago
Ocean basins take shape as South America and India move from Africa and the Tethys Sea closes to form the Mediterranean Sea.

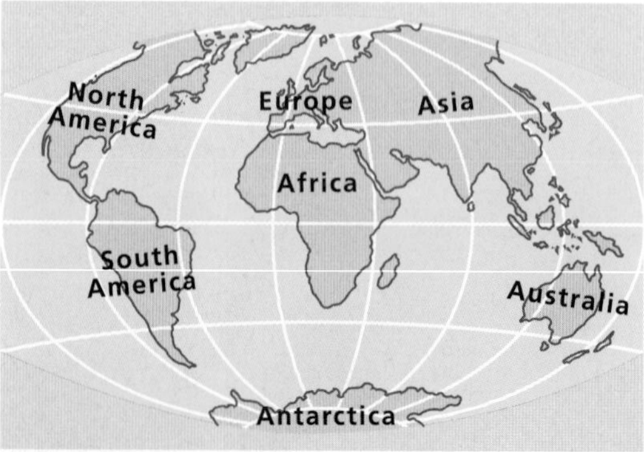
North America
Europe
Asia
Africa
South America
Australia
Antarctica

The present day
India has merged with Asia, Australia is free of Antarctica, and North America is free of Eurasia.

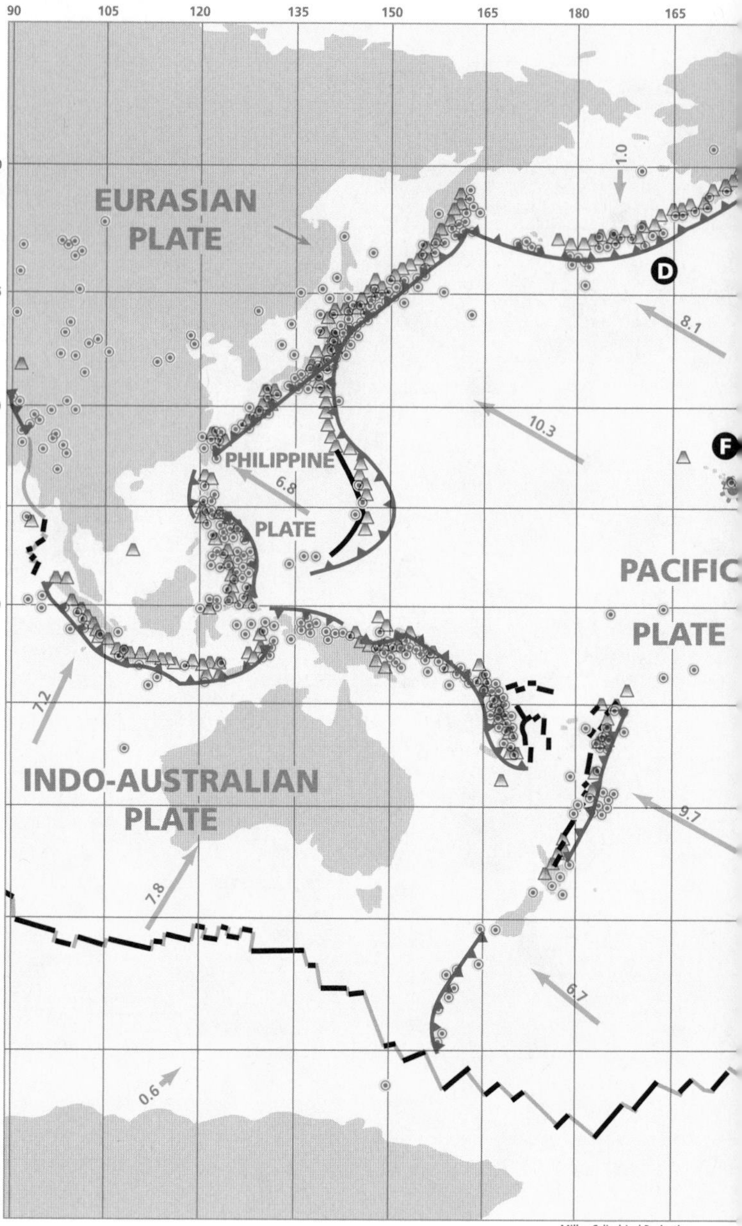
90
105
120
135
150
165
180
165
60
45
30
15
0
15
30
45
60
EURASIAN PLATE
PHILIPPINE PLATE
PACIFIC PLATE
INDO-AUSTRALIAN PLATE
1.0
8.1
10.3
6.8
7.2
9.7
7.8
6.7
0.6
D
F

Miller Cylindrical Projection
Scale 1:128,000,000
One inch to 2,020 miles
One cm to 1,280 km

PLATE TECTONICS

Types of plate boundaries
See text at right for explanation

- Divergent
- Convergent
- Transform

Other map symbols

- Direction of plate movement
- 6.7 — Length of arrow is proportional to the amount of plate movement (number indicates centimeters of movement per year)
- Earthquake of magnitude 7.5 and above (from 10 A.D. to the present)
- Volcano (eruption since 1900)
- Selected hot spots
- A — Key to text descriptions and diagrams

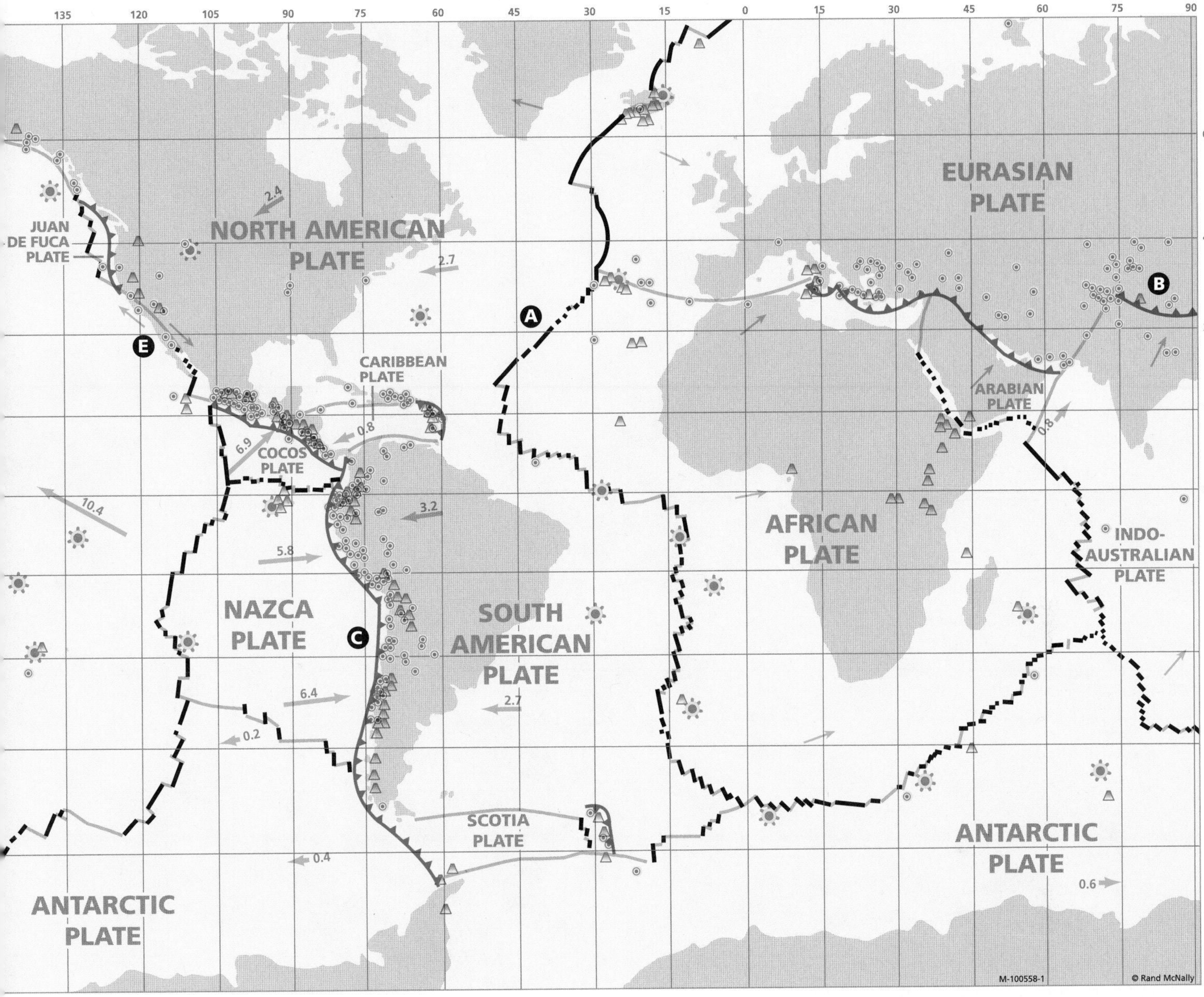
135
120
105
90
75
60
45
30
15
0
15
30
45
60
75
90
60
45
30
15
0
15
30
45
60
EURASIAN PLATE
NORTH AMERICAN PLATE
JUAN DE FUCA PLATE
2.4
2.7
A
B
E
CARIBBEAN PLATE
ARABIAN PLATE
0.8
0.8
COCOS PLATE
6.9
10.4
3.2
AFRICAN PLATE
INDO-AUSTRALIAN PLATE
5.8
NAZCA PLATE
C
SOUTH AMERICAN PLATE
6.4
2.7
0.2
SCOTIA PLATE
0.4
ANTARCTIC PLATE
ANTARCTIC PLATE
0.6
M-100558-1
© Rand McNally

Plate tectonic theory describes the motions of the lithosphere, the outer surface of which forms the Earth's crust. The theory originated with scientist Alfred Wegener's work on continental drift in the early part of the 20th century. According to plate tectonic theory, the lithosphere is composed of distinct plates that move relative to each other as a result of convection currents deep within the Earth's mantle. The largest of these plates and their movements are shown on the map above.

There are three main types of plate boundaries.

Divergent plate boundaries occur where two adjacent plates move away from each other. As the plates separate, upwelling magma from the mantle solidifies, and new crust is formed. (See diagram to the right.) These boundaries frequently make up oceanic ridge zones, such as the Mid-Atlantic Ridge (symbol A on map above). This spreading explains why North and South America have separated from Eurasia and Africa over time, as shown on the map series to the left. The Mid-Atlantic Ridge is actually part of a much larger subaqueous divergent boundary system that encircles the Earth.

Convergent plate boundaries occur where two adjacent plates collide with one another. When two continental plates collide, the resulting compression of lithospheric material causes large mountain ranges to form. The Himalayas, for example, were formed by the collision of the Eurasian and Indo-Australian Plates (symbol B on map above).

In other cases one plate is forced (subducted) under the other and the lithospheric material from the descending plate is recycled within the mantle. These areas are called **subduction zones**.

Subduction zones occur when a continental plate collides with an oceanic plate. An example occurs along the west coast of South America where the Nazca Plate is being subducted under the South American Plate, creating the long, deep Peru-Chile trench and the Andes mountain chain (symbol C on map above). This area is part of a much larger ring of convergent plate boundaries circling the Pacific and known as the Ring of Fire. Volcanoes and earthquakes are common features in this region.

Subduction zones can also occur when two oceanic plates collide. Intense volcanic activity in these areas eventually results in the formation of long, volcanic island chains. (See diagram to the right.) The Aleutian Islands of Alaska are one example (symbol D on map above).

Transform boundaries occur when two plates slide laterally past each other with no divergence or convergence. Commonly they offset the active spreading ridges of divergent boundaries on the ocean floor. The San Andreas fault zone of California is an example of a terrestrial transform boundary (symbol E on map above).

Volcanoes and earthquakes do not occur only at plate boundaries. At certain isolated **hot spots**, upwelling magma rises to the surface to create tall volcanoes. Over time, as the plate moves, long islands chains are formed. The Hawai'ian Islands are one such example (symbol F on map above).

The rate of movement of tectonic plates is very slow, on the order of several centimeters per year. Over geological time, these small movements accumulate and cause fragmentation and reformation of continental land masses, as shown in the map series to the left. The process is still underway, which implies that the arrangement of the continents millions of years from now will be quite different from what it is today.

Convergent plate boundary
Island arc subduction zone
(symbol D on map)

Divergent plate boundary
Oceanic ridge
(symbol A on map)

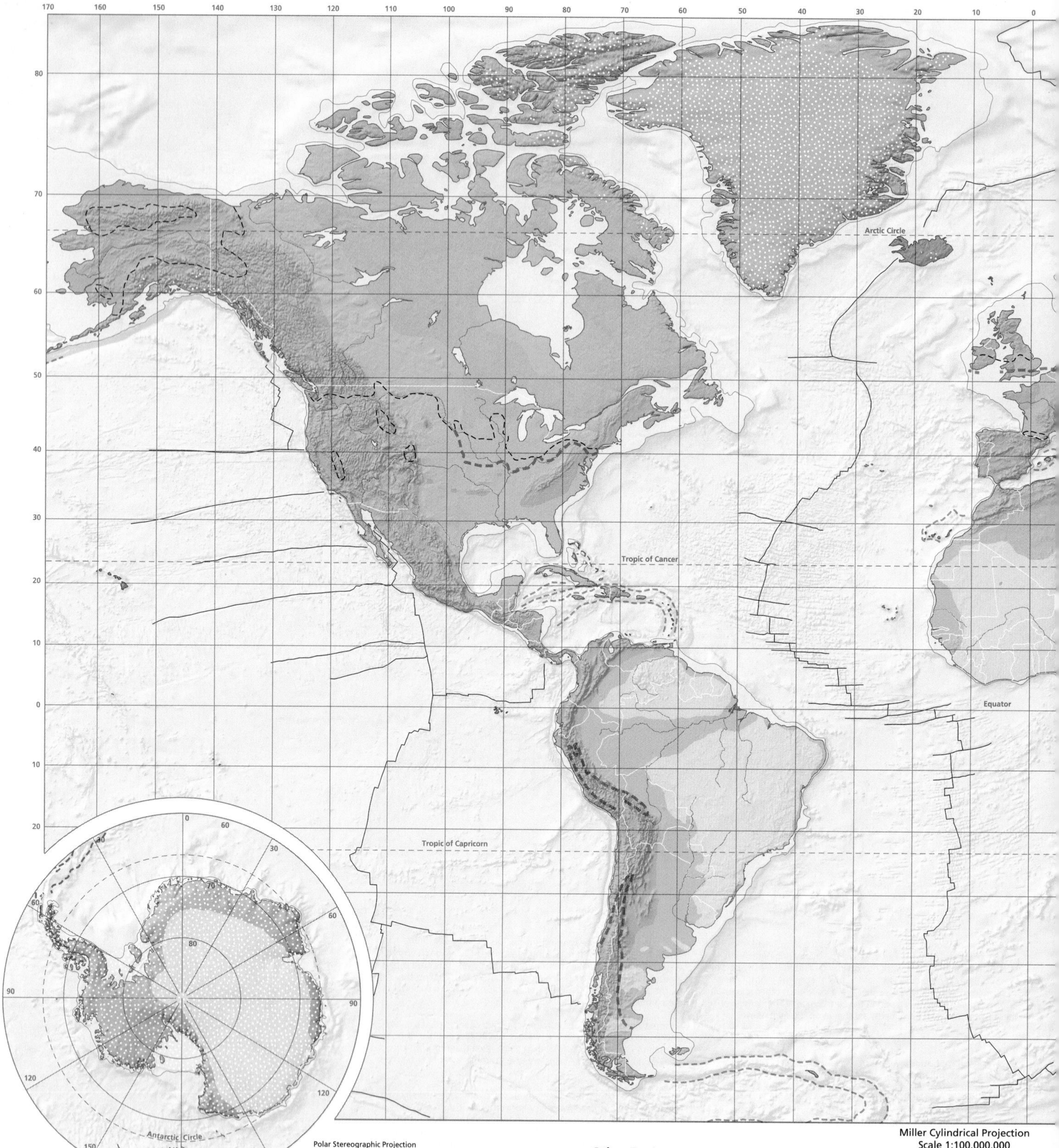
Arctic Circle
Tropic of Cancer
Equator
Tropic of Capricorn
Antarctic Circle
West
East

Polar Stereographic Projection
Scale 1:100,000,000

Miller Cylindrical Projection
Scale 1:100,000,000
One inch to 1,580 miles
One cm to 1,000 km

0 500 1000 1500 Miles
0 500 1000 1500 2000 Kilometers

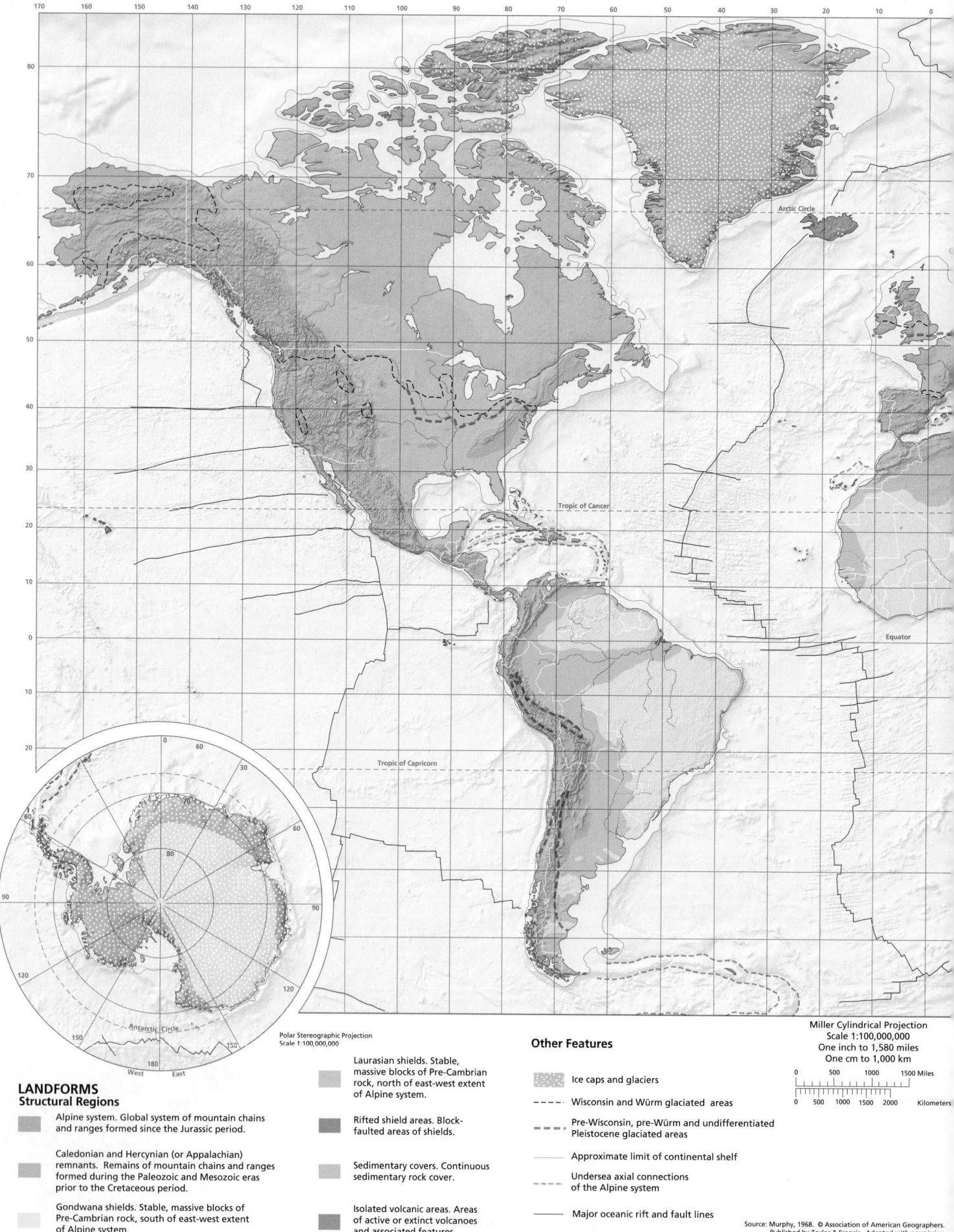

LANDFORMS
Structural Regions

- Alpine system. Global system of mountain chains and ranges formed since the Jurassic period.
- Caledonian and Hercynian (or Appalachian) remnants. Remains of mountain chains and ranges formed during the Paleozoic and Mesozoic eras prior to the Cretaceous period.
- Gondwana shields. Stable, massive blocks of Pre-Cambrian rock, south of east-west extent of Alpine system.
- Laurasian shields. Stable, massive blocks of Pre-Cambrian rock, north of east-west extent of Alpine system.
- Rifted shield areas. Block-faulted areas of shields.
- Sedimentary covers. Continuous sedimentary rock cover.
- Isolated volcanic areas. Areas of active or extinct volcanoes and associated features.

Other Features

- Ice caps and glaciers
- Wisconsin and Würm glaciated areas
- Pre-Wisconsin, pre-Würm and undifferentiated Pleistocene glaciated areas
- Approximate limit of continental shelf
- Undersea axial connections of the Alpine system
- Major oceanic rift and fault lines

Source: Murphy, 1968. © Association of American Geographers. Published by Taylor & Francis. Adapted with permission of the Association of American Geographers.

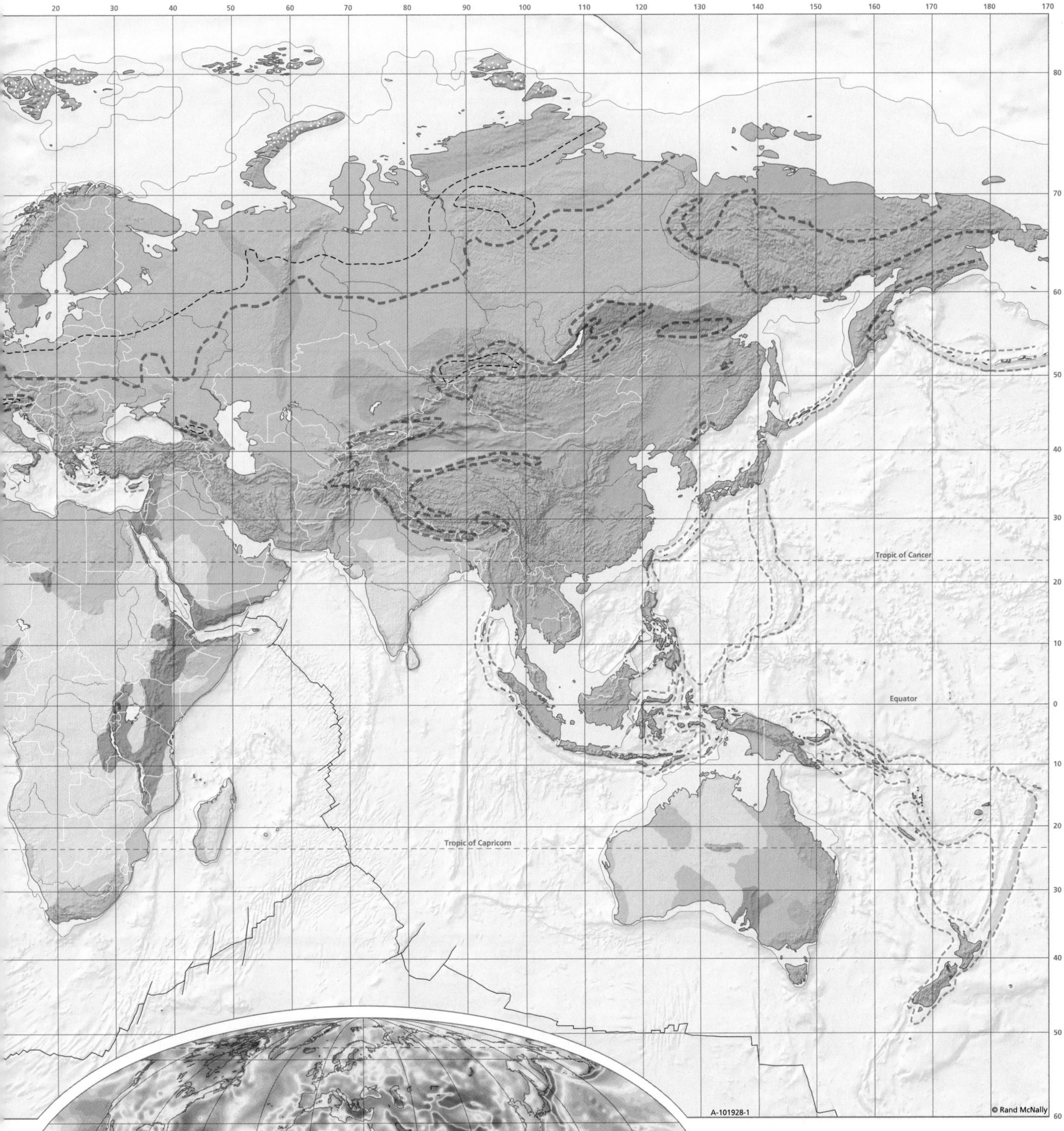
20 30 40 50 60 70 80 90 100 110 120 130 140 150 160 170 180 170
80
70
60
50
40
30
20
10
0
10
20
30
40
50
60
Tropic of Cancer
Equator
Tropic of Capricorn
A-101928-1
© Rand McNally

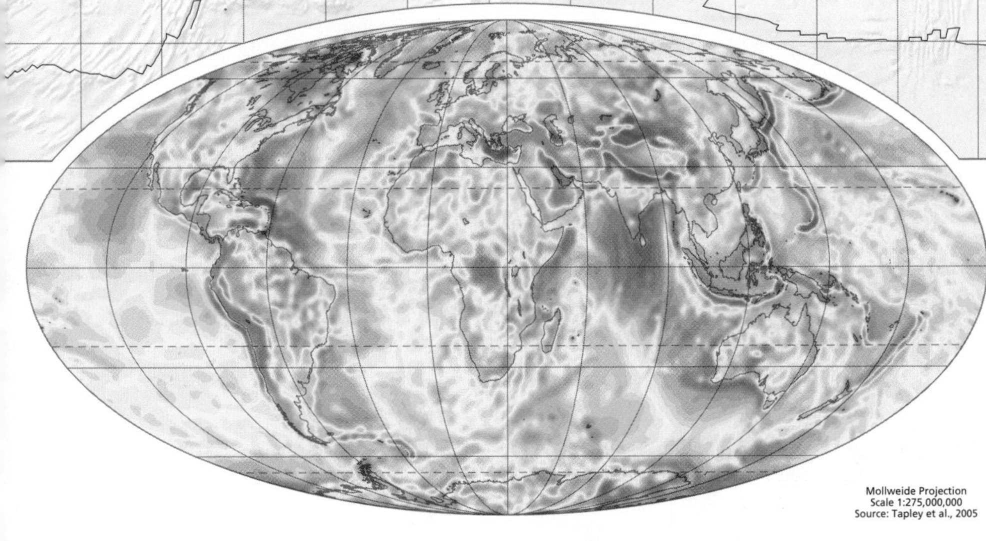

Mollweide Projection
Scale 1:275,000,000
Source: Tapley et al., 2005

GRAVITY ANOMALY

mGal 75 50 25 0 -25 -50 -75

This map is based on the GGM02C gravity anomaly model. This model was derived from over a year of GRACE (Gravity Recovery And Climate Experiment) satellite data coupled with terrestrially-based gravity observations. The gravity anomaly is the difference between observed gravity and the standard gravity on a reference ellipsoid. It is measured in milligals (mGal), defined as 10^{-5} m/s^2 or approximately one millionth of the standard acceleration on the Earth's surface. Values above zero have higher than standard gravity, and vice versa.

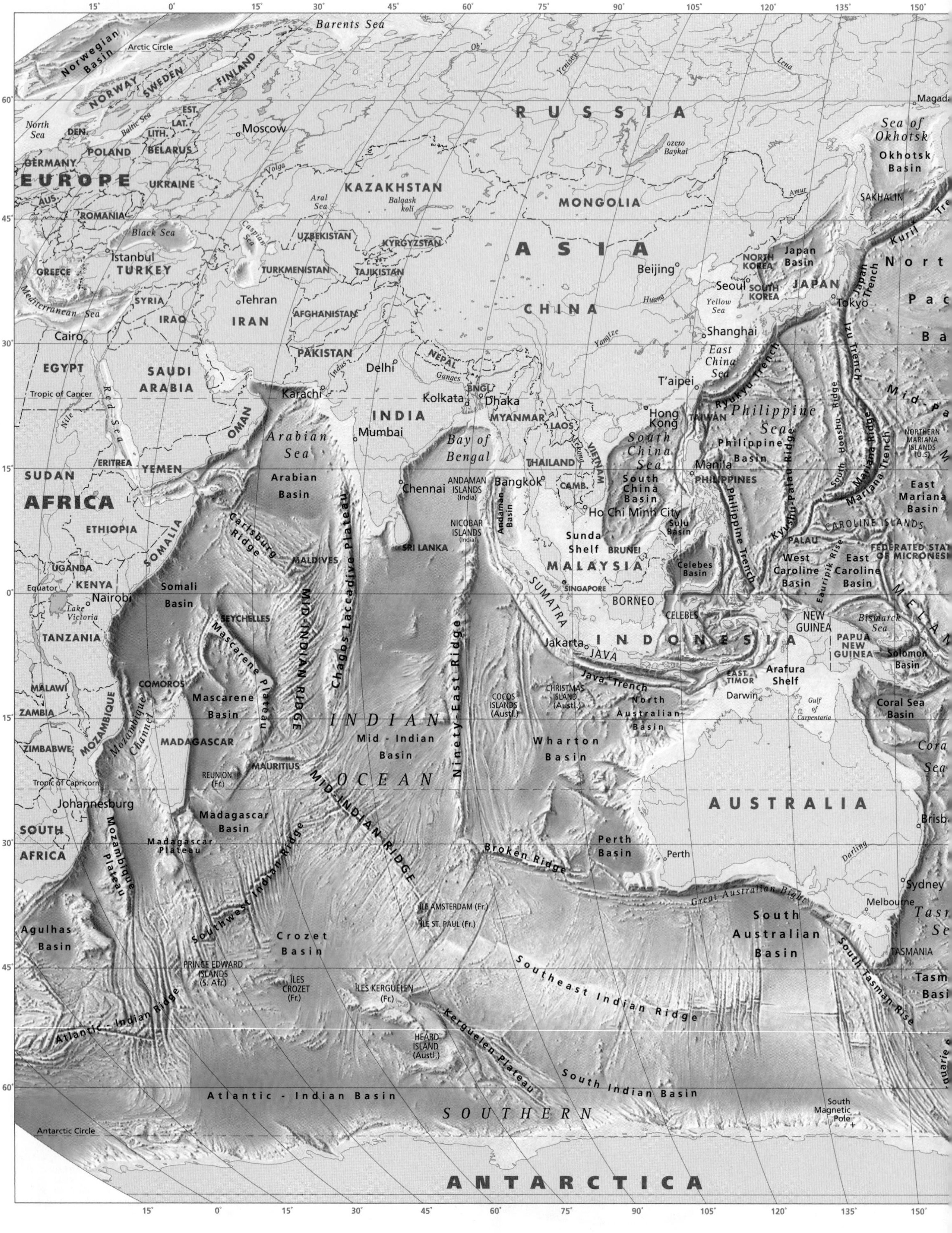

Barents Sea
Norwegian Basin
Arctic Circle
NORWAY
SWEDEN
FINLAND
North Sea
DEN.
Baltic Sea
EST.
LAT.
LITH.
Moscow
POLAND
BELARUS
GERMANY
EUROPE
UKRAINE
AUS.
ROMANIA
Black Sea
Istanbul
TURKEY
GREECE
Mediterranean Sea
SYRIA
IRAQ
Tehran
IRAN
Cairo
EGYPT
SAUDI ARABIA
Tropic of Cancer
Red Sea
Nile
OMAN
ERITREA
YEMEN
SUDAN
AFRICA
ETHIOPIA
SOMALIA
UGANDA
KENYA
Equator
Nairobi
Lake Victoria
TANZANIA
MALAWI
ZAMBIA
MOZAMBIQUE
ZIMBABWE
Tropic of Capricorn
Johannesburg
SOUTH AFRICA
Ob'
Yenisey
Lena
RUSSIA
Volga
Aral Sea
Caspian Sea
KAZAKHSTAN
Balqash köli
ozero Baykal
MONGOLIA
Amur
UZBEKISTAN
KYRGYZSTAN
TURKMENISTAN
TAJIKISTAN
AFGHANISTAN
ASIA
CHINA
Huang
Yangtze
Beijing
NORTH KOREA
SOUTH KOREA
Seoul
Japan Basin
JAPAN
Tokyo
Yellow Sea
Shanghai
East China Sea
PAKISTAN
Indus
Delhi
NEPAL
Ganges
BNGL.
Karachi
Kolkata
Dhaka
INDIA
MYANMAR
LAOS
Mekong
VIETNAM
T'aipei
Hong Kong
TAIWAN
Mumbai
Arabian Sea
Bay of Bengal
South China Sea
THAILAND
Bangkok
CAMB.
Ho Chi Minh City
Manila
PHILIPPINES
Chennai
ANDAMAN ISLANDS (India)
NICOBAR ISLANDS (India)
Andaman Basin
SRI LANKA
Sunda Shelf
BRUNEI
MALAYSIA
SINGAPORE
SUMATRA
BORNEO
CELEBES
Celebes Basin
Sulu Basin
South China Basin
Jakarta
JAVA
INDONESIA
Java Trench
EAST TIMOR
Darwin
Arafura Shelf
Gulf of Carpentaria
Magadan
Sea of Okhotsk
Okhotsk Basin
SAKHALIN
Kuril Trench
Japan Trench
Izu Trench
North Pacific Basin
Ryukyu Trench
Philippine Sea
Philippine Basin
Philippine Trench
Kyushu-Palau Ridge
South Honshu Ridge
Mariana Ridge
Mariana Trench
NORTHERN MARIANA ISLANDS (U.S.)
East Mariana Basin
Mid-Pacific Mountains
CAROLINE ISLANDS
PALAU
FEDERATED STATES OF MICRONESIA
West Caroline Basin
Eauripik Rise
East Caroline Basin
MELANESIA
NEW GUINEA
Bismarck Sea
PAPUA NEW GUINEA
Solomon Basin
Coral Sea Basin
Coral Sea
AUSTRALIA
Brisbane
Darling
Sydney
Melbourne
Perth
Great Australian Bight
South Australian Basin
Tasman Sea
TASMANIA
Tasman Basin
South Tasman Rise
Macquarie Ridge
Arabian Basin
Carlsberg Ridge
MALDIVES
Chagos-Laccadive Plateau
MID-INDIAN RIDGE
Somali Basin
SEYCHELLES
Mascarene Plateau
COMOROS
Mascarene Basin
Mozambique Channel
MADAGASCAR
REUNION (Fr.)
MAURITIUS
INDIAN OCEAN
Mid-Indian Basin
Ninety-East Ridge
COCOS ISLANDS (Austl.)
CHRISTMAS ISLAND (Austl.)
North Australian Basin
Wharton Basin
Madagascar Basin
Madagascar Plateau
Mozambique Plateau
Southwest Indian Ridge
Broken Ridge
Perth Basin
ÎLE AMSTERDAM (Fr.)
ÎLE ST. PAUL (Fr.)
Agulhas Basin
Crozet Basin
PRINCE EDWARD ISLANDS (S. Afr.)
ÎLES CROZET (Fr.)
ÎLES KERGUELEN (Fr.)
Southeast Indian Ridge
Atlantic-Indian Ridge
Kerguelen Plateau
HEARD ISLAND (Austl.)
South Indian Basin
Atlantic-Indian Basin
SOUTHERN
South Magnetic Pole
Antarctic Circle
ANTARCTICA
15° 0° 15° 30° 45° 60° 75° 90° 105° 120° 135° 150°
60° 45° 30° 15° 0° 15° 30° 45° 60°

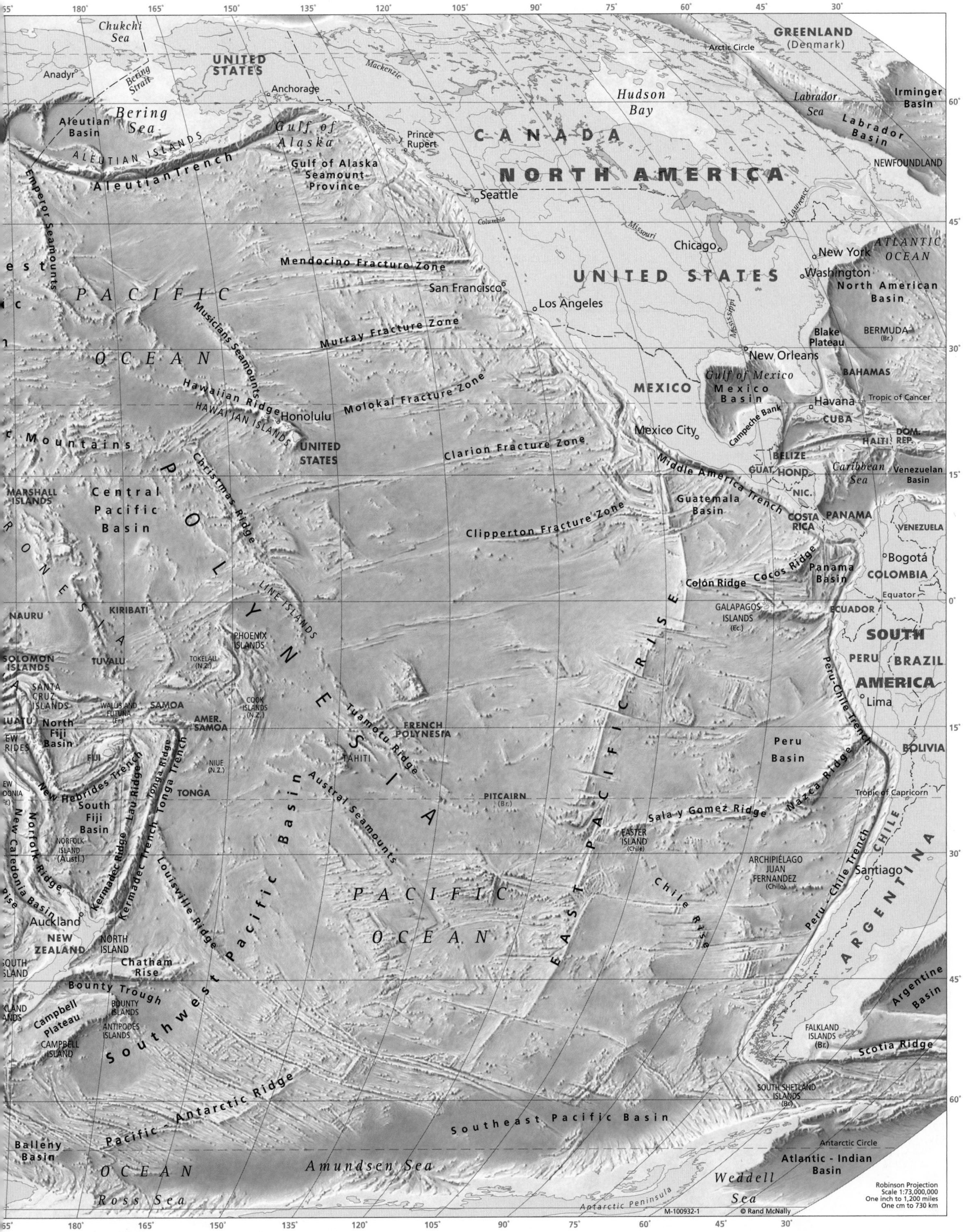
Chukchi Sea
Anadyr
Bering Strait
UNITED STATES
Anchorage
Mackenzie
Hudson Bay
GREENLAND (Denmark)
Arctic Circle
Irminger Basin
Labrador Sea
Labrador Basin
Aleutian Basin
Bering Sea
ALEUTIAN ISLANDS
Gulf of Alaska
Prince Rupert
CANADA
NORTH AMERICA
NEWFOUNDLAND
Aleutian Trench
Emperor Seamounts
Gulf of Alaska Seamount Province
Seattle
Columbia
Missouri
St. Lawrence
Chicago
New York
ATLANTIC OCEAN
Washington
UNITED STATES
Mendocino Fracture Zone
San Francisco
Los Angeles
North American Basin
PACIFIC
Musicians Seamounts
Murray Fracture Zone
Mississippi
Blake Plateau
BERMUDA (Br.)
OCEAN
New Orleans
Hawaiian Ridge
HAWAIIAN ISLANDS
Molokai Fracture Zone
Gulf of Mexico
Mexico Basin
MEXICO
BAHAMAS
Tropic of Cancer
Honolulu
Havana
CUBA
Campeche Bank
Mexico City
UNITED STATES
Clarion Fracture Zone
Mountains
Christmas Ridge
Middle America Trench
BELIZE
HAITI
DOM. REP.
GUAT.
HOND.
Caribbean Sea
Venezuelan Basin
MARSHALL ISLANDS
Central Pacific Basin
POLYNESIA
NIC.
Guatemala Basin
COSTA RICA
PANAMA
VENEZUELA
Clipperton Fracture Zone
LINE ISLANDS
Panama Basin
Bogotá
COLOMBIA
Cocos Ridge
Colón Ridge
Equator
NAURU
KIRIBATI
GALAPAGOS ISLANDS (Ec.)
ECUADOR
PHOENIX ISLANDS
SOUTH AMERICA
SOLOMON ISLANDS
TUVALU
TOKELAU (N.Z.)
PERU
BRAZIL
EAST PACIFIC RISE
Peru-Chile Trench
SANTA CRUZ ISLANDS
WALLIS AND FUTUNA (Fr.)
SAMOA
COOK ISLANDS (N.Z.)
Lima
North Fiji Basin
AMER. SAMOA
Tuamotu Ridge
FRENCH POLYNESIA
FIJI
TAHITI
NIUE (N.Z.)
Peru Basin
BOLIVIA
New Hebrides Trench
Lau Ridge
Tonga Ridge
Tonga Trench
TONGA
South Fiji Basin
Austral Seamounts
PITCAIRN (Br.)
Nazca Ridge
Tropic of Capricorn
Sala y Gomez Ridge
New Caledonia Basin
Norfolk Ridge
NORFOLK ISLAND (Austl.)
Kermadec Ridge
Kermadec Trench
Louisville Ridge
EASTER ISLAND (Chile)
CHILE
ARCHIPIÉLAGO JUAN FERNANDEZ (Chile)
Santiago
Chile Rise
PACIFIC OCEAN
Southwest Pacific Basin
ARGENTINA
Peru-Chile Trench
Auckland
NEW ZEALAND
NORTH ISLAND
SOUTH ISLAND
Chatham Rise
Bounty Trough
Argentine Basin
Campbell Plateau
BOUNTY ISLANDS
ANTIPODES ISLANDS
CAMPBELL ISLAND
FALKLAND ISLANDS (Br.)
Scotia Ridge
SOUTH SHETLAND ISLANDS (Br.)
Pacific - Antarctic Ridge
Southeast Pacific Basin
Balleny Basin
Antarctic Circle
Atlantic - Indian Basin
OCEAN
Amundsen Sea
Weddell Sea
Ross Sea
Antarctic Peninsula
M-100932-1
© Rand McNally
Robinson Projection
Scale 1:73,000,000
One inch to 1,200 miles
One cm to 730 km
180°
165°
150°
135°
120°
105°
90°
75°
60°
45°
30°
15°
0°

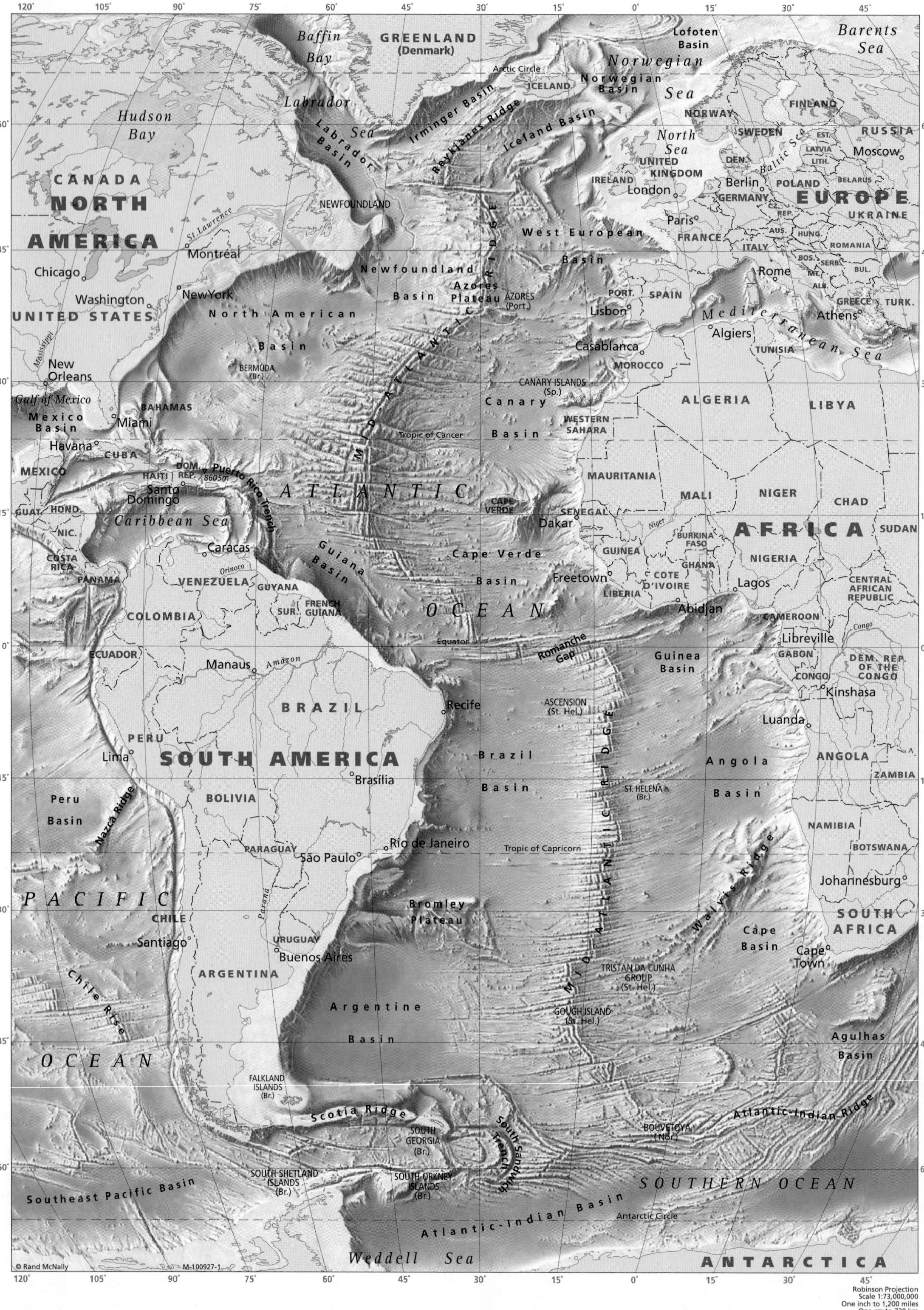

120° 105° 90° 75° 60° 45° 30° 15° 0° 15° 30° 45°
60° 45° 30° 15° 0° 15° 30° 45° 60°
Baffin Bay
GREENLAND (Denmark)
Lofoten Basin
Barents Sea
Arctic Circle
ICELAND
Norwegian Sea
Norwegian Basin
Labrador Sea
Labrador Basin
Irminger Basin
Reykjanes Ridge
Iceland Basin
NORWAY
FINLAND
RUSSIA
Hudson Bay
North Sea
SWEDEN
Baltic Sea
EST.
LATVIA
LITH.
DEN.
Moscow
UNITED KINGDOM
IRELAND
London
Berlin
POLAND
BELARUS
CANADA
NORTH AMERICA
NEWFOUNDLAND
GERMANY
CZ. REP.
EUROPE
UKRAINE
St. Lawrence
Paris
AUS.
HUNG.
ROMANIA
West European Basin
FRANCE
ITALY
Montreal
BOS.
SERB.
BUL.
Chicago
Newfoundland Basin
Azores Plateau
AZORES (Port.)
Rome
MT.
ALB.
Washington
New York
PORT.
SPAIN
GREECE
TURK.
UNITED STATES
North American Basin
Lisbon
Mediterranean Sea
Athens
Algiers
Mississippi
MID-ATLANTIC RIDGE
TUNISIA
New Orleans
BERMUDA (Br.)
Casablanca
MOROCCO
CANARY ISLANDS (Sp.)
Gulf of Mexico
BAHAMAS
Canary Basin
ALGERIA
LIBYA
Mexico Basin
Miami
WESTERN SAHARA
Havana
CUBA
Tropic of Cancer
MEXICO
DOM. REP.
HAITI
Puerto Rico Trench
8605m
MAURITANIA
Santo Domingo
ATLANTIC
MALI
NIGER
CHAD
GUAT.
HOND.
CAPE VERDE
SENEGAL
Caribbean Sea
Dakar
AFRICA
SUDAN
NIC.
Niger
BURKINA FASO
COSTA RICA
Caracas
Guiana Basin
Cape Verde Basin
GUINEA
GHANA
NIGERIA
PANAMA
Orinoco
VENEZUELA
GUYANA
Freetown
COTE D'IVOIRE
Lagos
CENTRAL AFRICAN REPUBLIC
SUR.
FRENCH GUIANA
LIBERIA
Abidjan
COLOMBIA
OCEAN
CAMEROON
Congo
Equator
Libreville
Romanche Gap
GABON
ECUADOR
Guinea Basin
DEM. REP. OF THE CONGO
Manaus
Amazon
CONGO
Kinshasa
ASCENSION (St. Hel.)
BRAZIL
Recife
Luanda
PERU
Lima
SOUTH AMERICA
Brazil Basin
Angola Basin
ANGOLA
ZAMBIA
Brasília
ST. HELENA (Br.)
Peru Basin
BOLIVIA
Nazca Ridge
NAMIBIA
Walvis Ridge
PARAGUAY
Rio de Janeiro
Tropic of Capricorn
BOTSWANA
São Paulo
Johannesburg
PACIFIC
Paraná
Bromley Plateau
SOUTH AFRICA
CHILE
Cape Basin
Santiago
URUGUAY
Cape Town
Buenos Aires
TRISTAN DA CUNHA GROUP (St. Hel.)
Chile Rise
ARGENTINA
Argentine Basin
GOUGH ISLAND (St. Hel.)
Agulhas Basin
OCEAN
FALKLAND ISLANDS (Br.)
Scotia Ridge
South Sandwich Trench
Atlantic-Indian Ridge
SOUTH GEORGIA (Br.)
BOUVETOYA (Nor.)
SOUTH SHETLAND ISLANDS (Br.)
SOUTH ORKNEY ISLANDS (Br.)
SOUTHERN OCEAN
Southeast Pacific Basin
Atlantic-Indian Basin
Antarctic Circle
Weddell Sea
ANTARCTICA
© Rand McNally
M-100927-1

Robinson Projection
Scale 1:73,000,000
One inch to 1,200 miles
One cm to 730 km

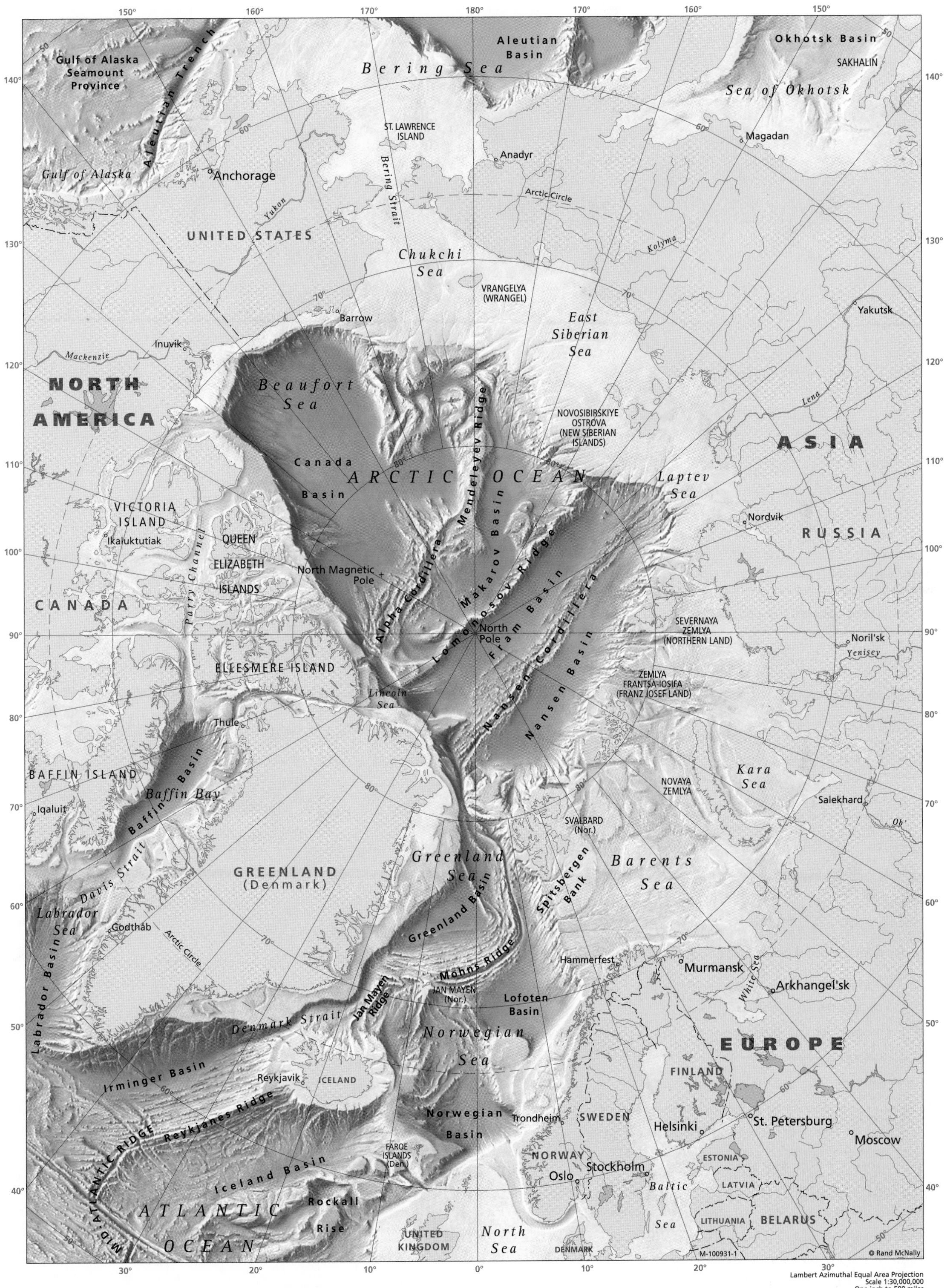
Gulf of Alaska Seamount Province
Aleutian Trench
Gulf of Alaska
Anchorage
Bering Sea
Aleutian Basin
Okhotsk Basin
SAKHALIN
Sea of Okhotsk
Magadan
ST. LAWRENCE ISLAND
Anadyr
Bering Strait
Arctic Circle
Yukon
UNITED STATES
Chukchi Sea
Kolyma
VRANGELYA (WRANGEL)
Barrow
East Siberian Sea
Yakutsk
Inuvik
Mackenzie
NORTH AMERICA
Beaufort Sea
Mendeleyev Ridge
NOVOSIBIRSKIYE OSTROVA (NEW SIBERIAN ISLANDS)
Lena
ASIA
Canada Basin
ARCTIC OCEAN
Laptev Sea
VICTORIA ISLAND
Ikaluktutiak
QUEEN ELIZABETH ISLANDS
Parry Channel
Makarov Basin
Nordvik
RUSSIA
North Magnetic Pole
Alpha Cordillera
Lomonosov Ridge
Fram Basin
Gakkel Cordillera
CANADA
North Pole
SEVERNAYA ZEMLYA (NORTHERN LAND)
Noril'sk
Yenisey
ELLESMERE ISLAND
Lincoln Sea
Nansen Cordillera
Nansen Basin
ZEMLYA FRANTSA-IOSIFA (FRANZ JOSEF LAND)
Thule
BAFFIN ISLAND
Baffin Basin
Baffin Bay
Kara Sea
NOVAYA ZEMLYA
Salekhard
Iqaluit
SVALBARD (Nor.)
Ob'
Davis Strait
GREENLAND (Denmark)
Greenland Sea
Barents Sea
Spitsbergen Bank
Labrador Sea
Godthåb
Greenland Basin
Labrador Basin
Mohns Ridge
Hammerfest
Murmansk
White Sea
Arkhangel'sk
Jan Mayen Ridge
JAN MAYEN (Nor.)
Lofoten Basin
Denmark Strait
Norwegian Sea
EUROPE
Irminger Basin
Reykjavik
ICELAND
FINLAND
Reykjanes Ridge
Norwegian Basin
Trondheim
SWEDEN
Helsinki
St. Petersburg
Moscow
MID-ATLANTIC RIDGE
FAROE ISLANDS (Den.)
NORWAY
ESTONIA
Stockholm
Oslo
Iceland Basin
Baltic Sea
LATVIA
ATLANTIC OCEAN
Rockall Rise
LITHUANIA
BELARUS
UNITED KINGDOM
North Sea
DENMARK
M-100931-1
© Rand McNally

Lambert Azimuthal Equal Area Projection
Scale 1:30,000,000
One inch to 500 miles
One cm to 300 cm

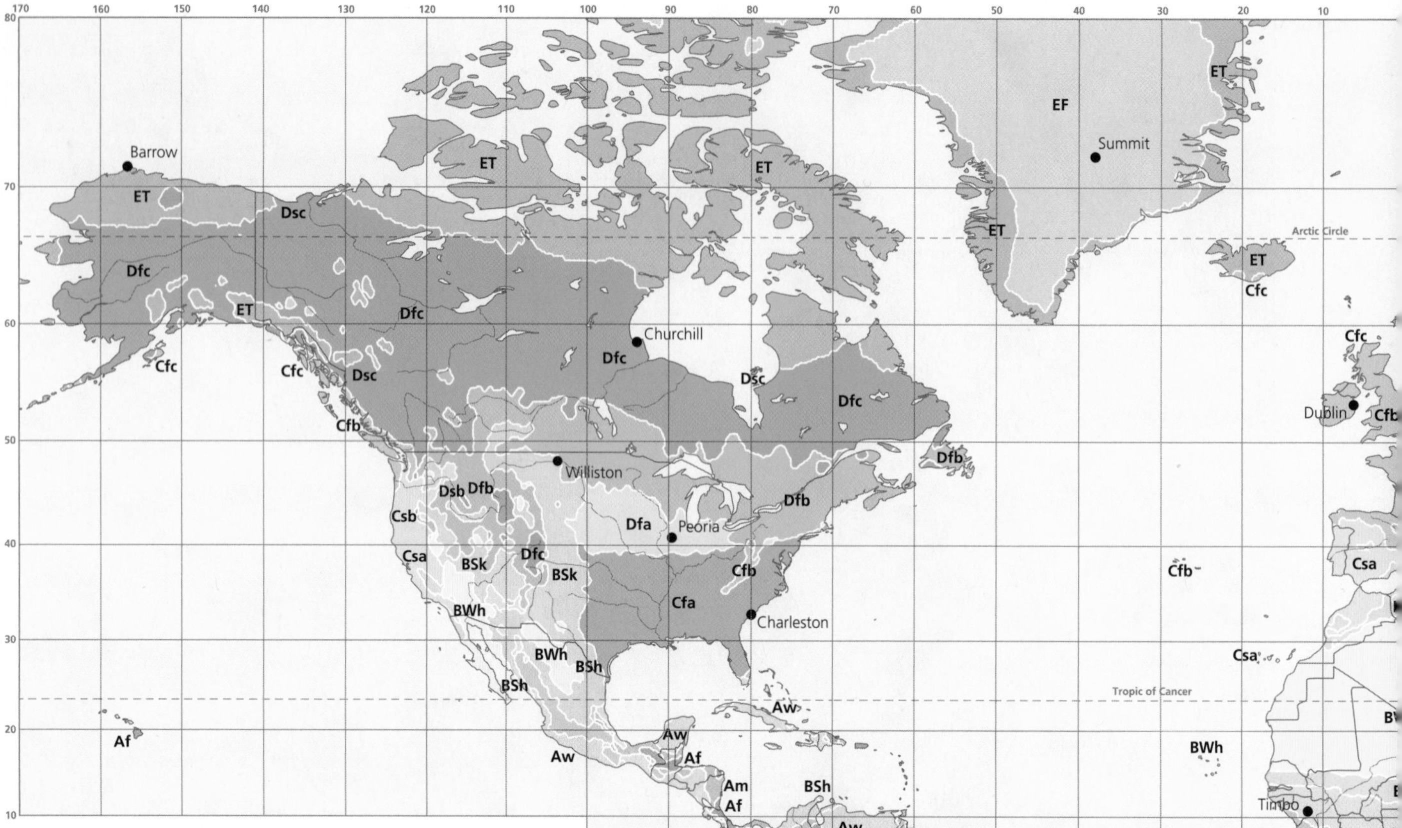

170
160
150
140
130
120
110
100
90
80
70
60
50
40
30
20
10
Barrow
Summit
Churchill
Williston
Peoria
Charleston
Dublin
Timbo
Arctic Circle
Tropic of Cancer
Equator
Tropic of Capricorn
EF
ET
Dsc
Dfc
Cfc
Cfb
Dfb
Dsb
Csb
Csa
BSk
BWh
BSh
Dfa
Cfa
Aw
Af
Am
Cwa
BWk

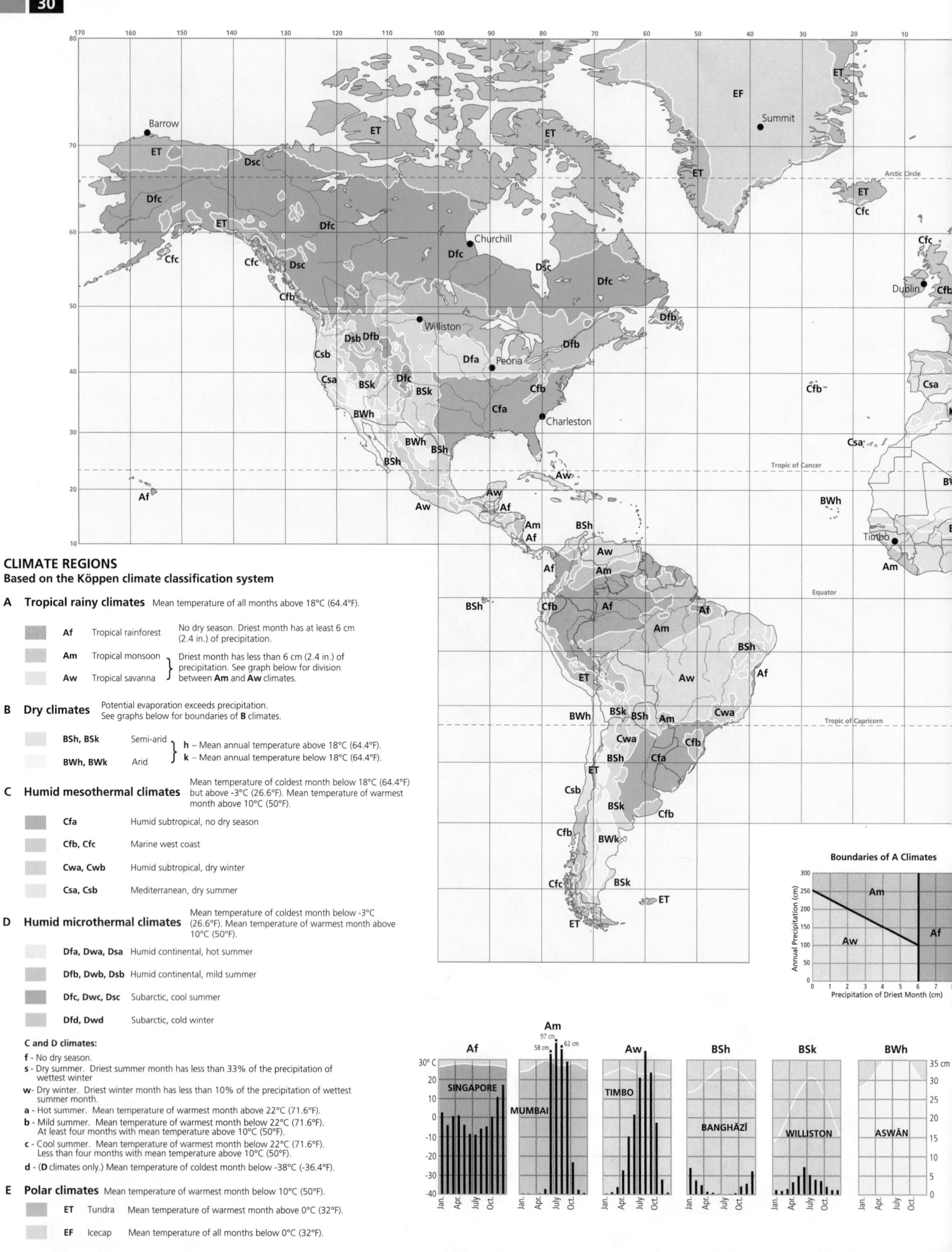

CLIMATE REGIONS

Based on the Köppen climate classification system

A Tropical rainy climates Mean temperature of all months above 18°C (64.4°F).

- **Af** Tropical rainforest — No dry season. Driest month has at least 6 cm (2.4 in.) of precipitation.
- **Am** Tropical monsoon
- **Aw** Tropical savanna

(Am, Aw:) Driest month has less than 6 cm (2.4 in.) of precipitation. See graph below for division between **Am** and **Aw** climates.

B Dry climates Potential evaporation exceeds precipitation. See graphs below for boundaries of **B** climates.

- **BSh, BSk** Semi-arid
- **BWh, BWk** Arid

h – Mean annual temperature above 18°C (64.4°F).
k – Mean annual temperature below 18°C (64.4°F).

C Humid mesothermal climates Mean temperature of coldest month below 18°C (64.4°F) but above -3°C (26.6°F). Mean temperature of warmest month above 10°C (50°F).

- **Cfa** Humid subtropical, no dry season
- **Cfb, Cfc** Marine west coast
- **Cwa, Cwb** Humid subtropical, dry winter
- **Csa, Csb** Mediterranean, dry summer

D Humid microthermal climates Mean temperature of coldest month below -3°C (26.6°F). Mean temperature of warmest month above 10°C (50°F).

- **Dfa, Dwa, Dsa** Humid continental, hot summer
- **Dfb, Dwb, Dsb** Humid continental, mild summer
- **Dfc, Dwc, Dsc** Subarctic, cool summer
- **Dfd, Dwd** Subarctic, cold winter

C and D climates:

- **f** - No dry season.
- **s** - Dry summer. Driest summer month has less than 33% of the precipitation of wettest winter
- **w**- Dry winter. Driest winter month has less than 10% of the precipitation of wettest summer month.
- **a** - Hot summer. Mean temperature of warmest month above 22°C (71.6°F).
- **b** - Mild summer. Mean temperature of warmest month below 22°C (71.6°F). At least four months with mean temperature above 10°C (50°F).
- **c** - Cool summer. Mean temperature of warmest month below 22°C (71.6°F). Less than four months with mean temperature above 10°C (50°F).
- **d** - (**D** climates only.) Mean temperature of coldest month below -38°C (-36.4°F).

E Polar climates Mean temperature of warmest month below 10°C (50°F).

- **ET** Tundra — Mean temperature of warmest month above 0°C (32°F).
- **EF** Icecap — Mean temperature of all months below 0°C (32°F).

Boundaries of A Climates

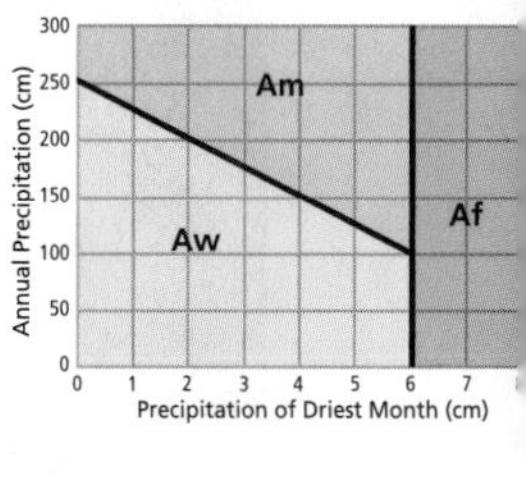

Annual Precipitation (cm)
300
250
200
150
100
50
0
0 1 2 3 4 5 6 7
Precipitation of Driest Month (cm)
Am
Aw
Af

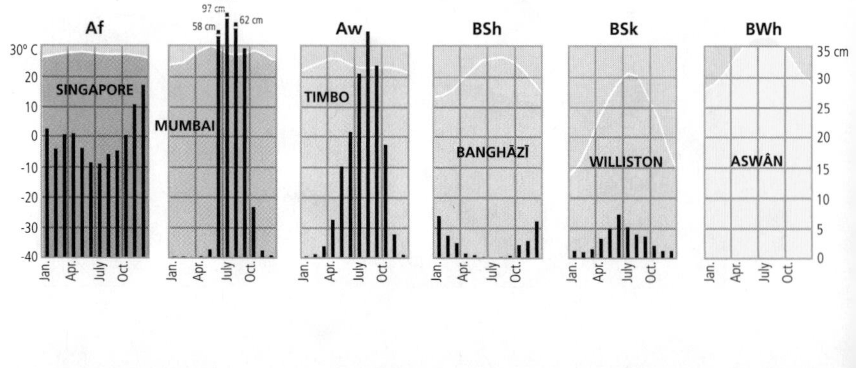

Af
Am
Aw
BSh
BSk
BWh
97 cm
58 cm
62 cm
30° C
20
10
0
-10
-20
-30
-40
35 cm
30
25
20
15
10
5
0
SINGAPORE
MUMBAI
TIMBO
BANGHĀZĪ
WILLISTON
ASWÂN
Jan.
Apr.
July
Oct.

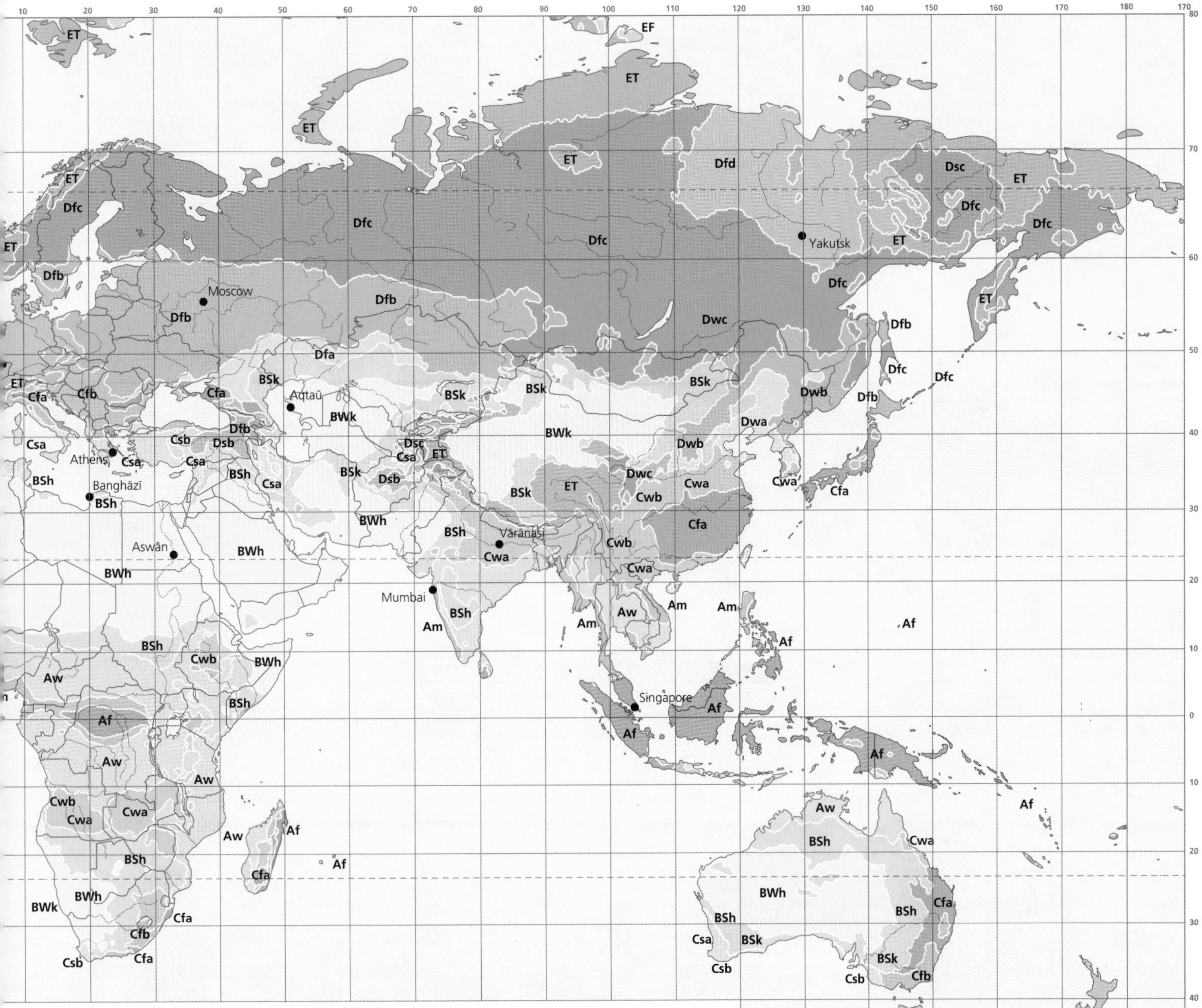
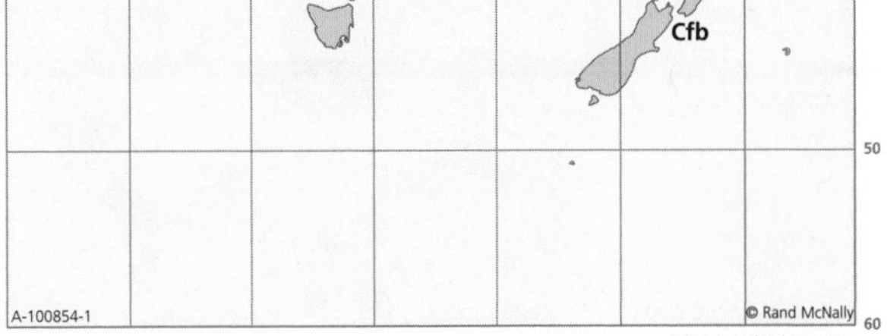
EF
ET
Dfd
Dsc
Dfc
Dfb
Dwc
Dwb
Dwa
Dfa
Dsb
Dsc
BSk
BWk
BSh
BWh
Csa
Csb
Cfa
Cfb
Cwa
Cwb
Am
Aw
Af
Moscow
Yakutsk
Aqtaū
Athens
Banghāzī
Aswân
Vārānasi
Mumbai
Singapore
A-100854-1
© Rand McNally

Miller Cylindrical Projection
Scale 1:100,000,000
One inch to 1,580 miles
One cm to 1,000 km
Derived from New et al., 2000
based on Trewartha, 1968

Boundaries of B Climates
Precipitation evenly distributed

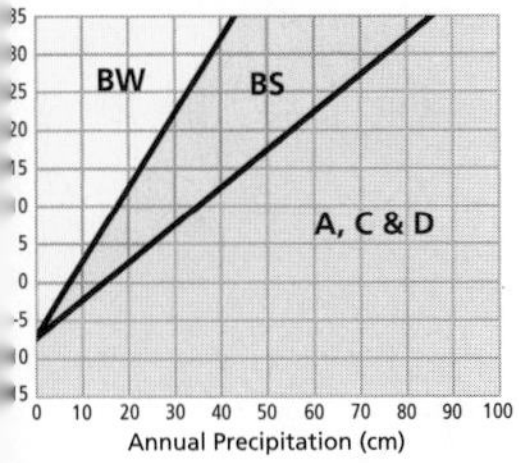
BW
BS
A, C & D
Annual Precipitation (cm)

Boundaries of B Climates
Summer concentration of precipitation

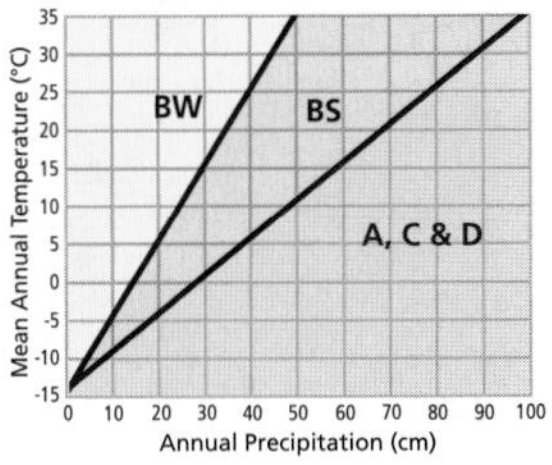
BW
BS
A, C & D
Mean Annual Temperature (°C)
Annual Precipitation (cm)

Boundaries of B Climates
Winter concentration of precipitation

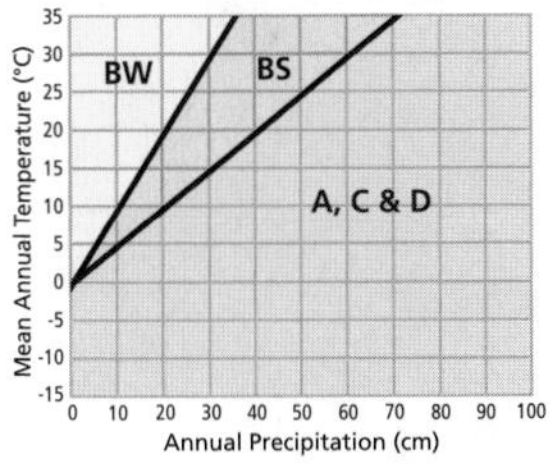
BW
BS
A, C & D
Mean Annual Temperature (°C)
Annual Precipitation (cm)

Curves show temperature in degrees Celsius (°C)
Vertical bars show rainfall In centimeters (cm)

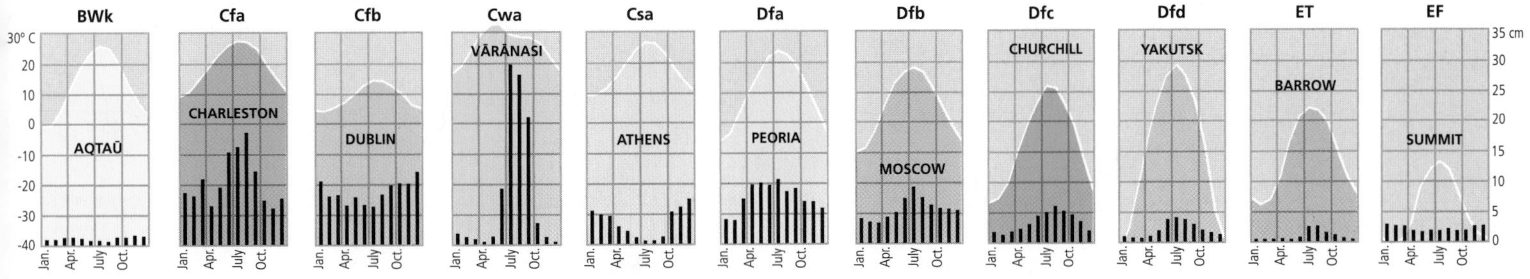
BWk
Cfa
Cfb
Cwa
Csa
Dfa
Dfb
Dfc
Dfd
ET
EF
30° C
35 cm
AQTAŪ
CHARLESTON
DUBLIN
VĀRĀNASI
ATHENS
PEORIA
MOSCOW
CHURCHILL
YAKUTSK
BARROW
SUMMIT
Jan.
Apr.
July
Oct.

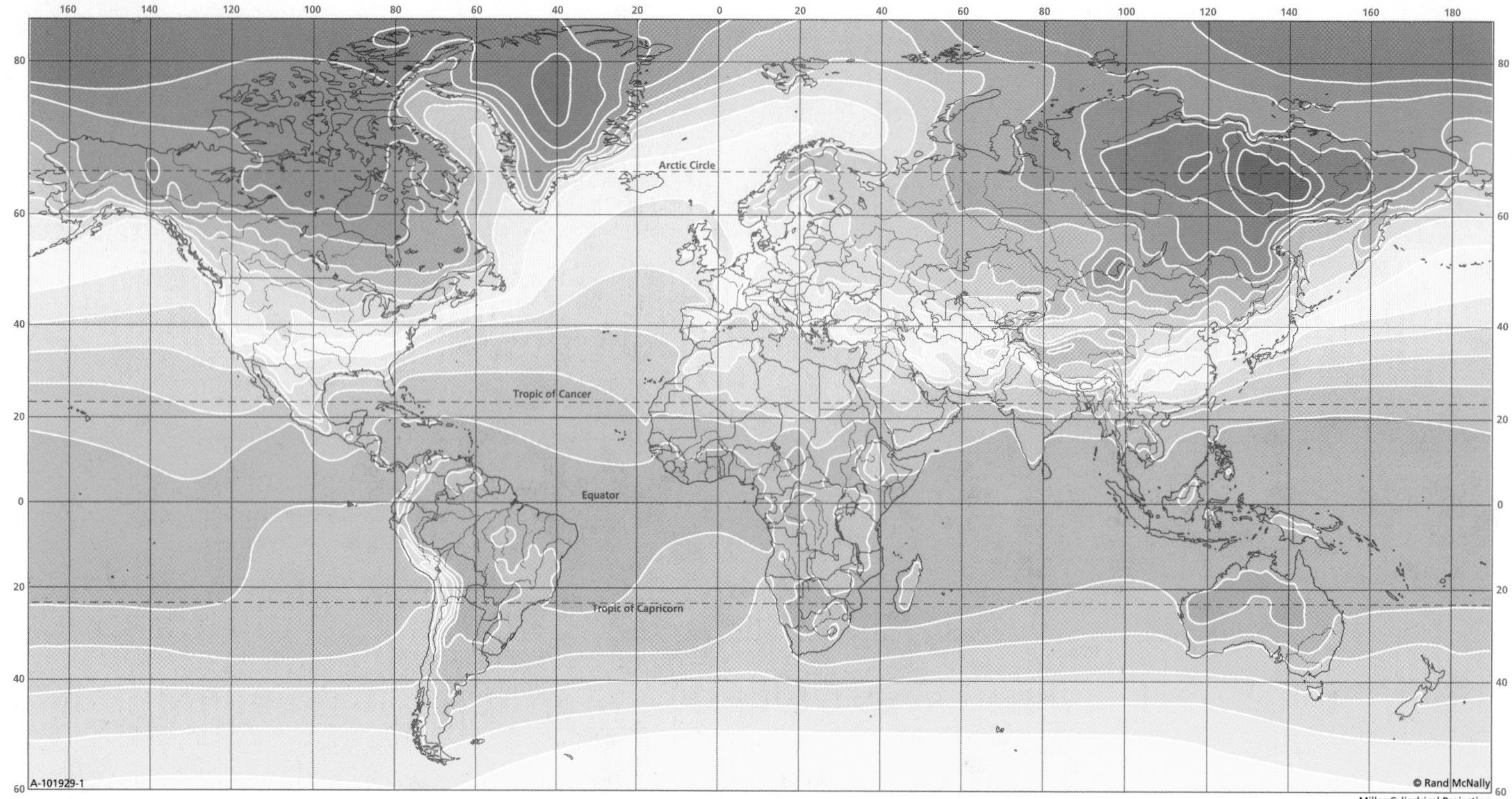

Arctic Circle
Tropic of Cancer
Equator
Tropic of Capricorn
A-101929-1
© Rand McNally

AVERAGE JANUARY TEMPERATURE

°C -45 -40 -35 -30 -25 -20 -15 -10 -5 0 5 10 15 20 25 30

°F -49 -40 -31 -22 -13 -4 5 14 23 32 41 50 59 68 77 86

Miller Cylindrical Projection
Scale 1:200,000,000
Sources: New et al., 2000; NOAA

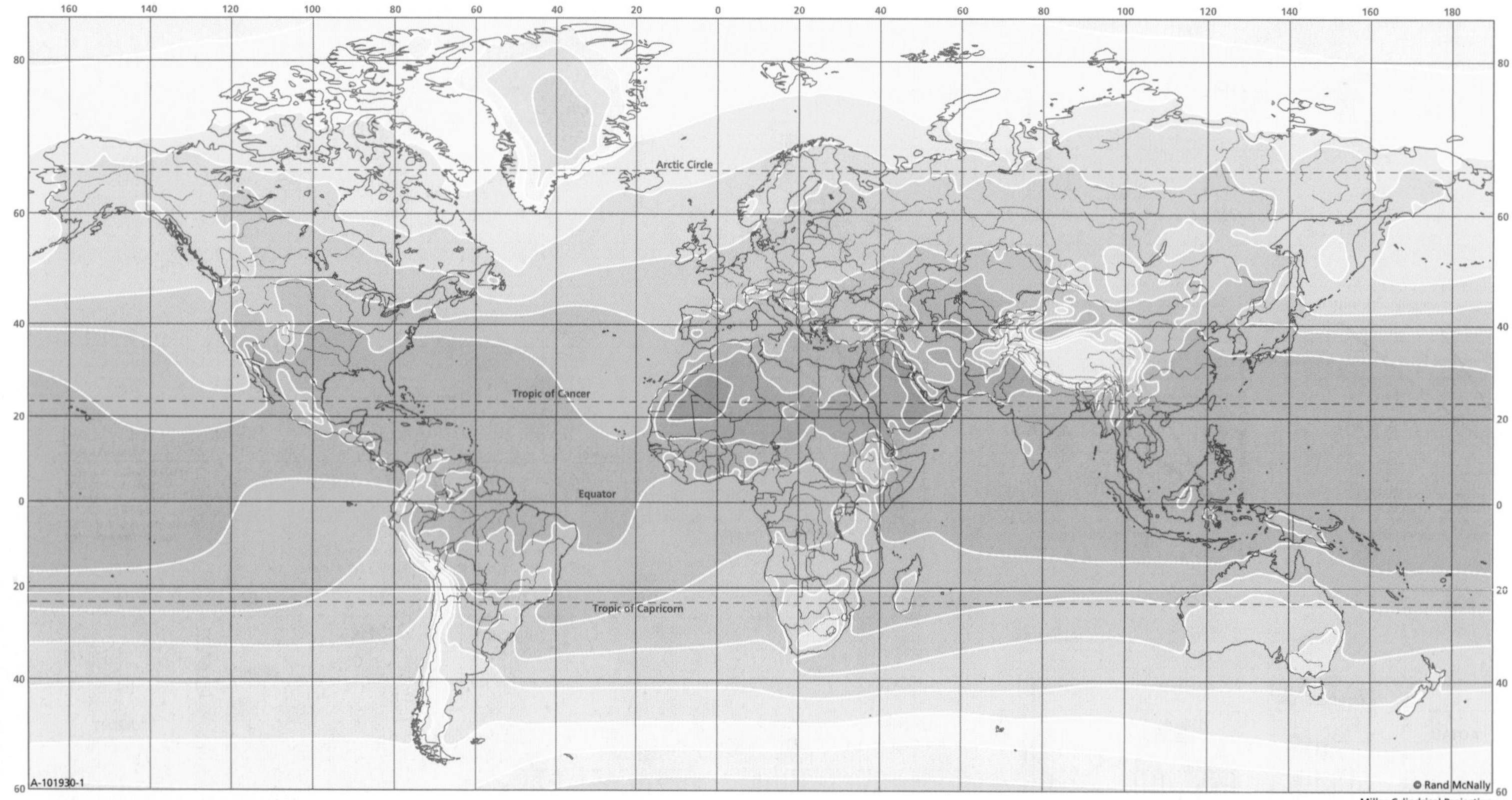

Arctic Circle
Tropic of Cancer
Equator
Tropic of Capricorn
A-101930-1
© Rand McNally

AVERAGE JULY TEMPERATURE

°C -10 -5 0 5 10 15 20 25 30 35

°F 14 23 32 41 50 59 68 77 86 95

Miller Cylindrical Projection
Scale 1:200,000,000
Sources: New et al., 2000; NOAA

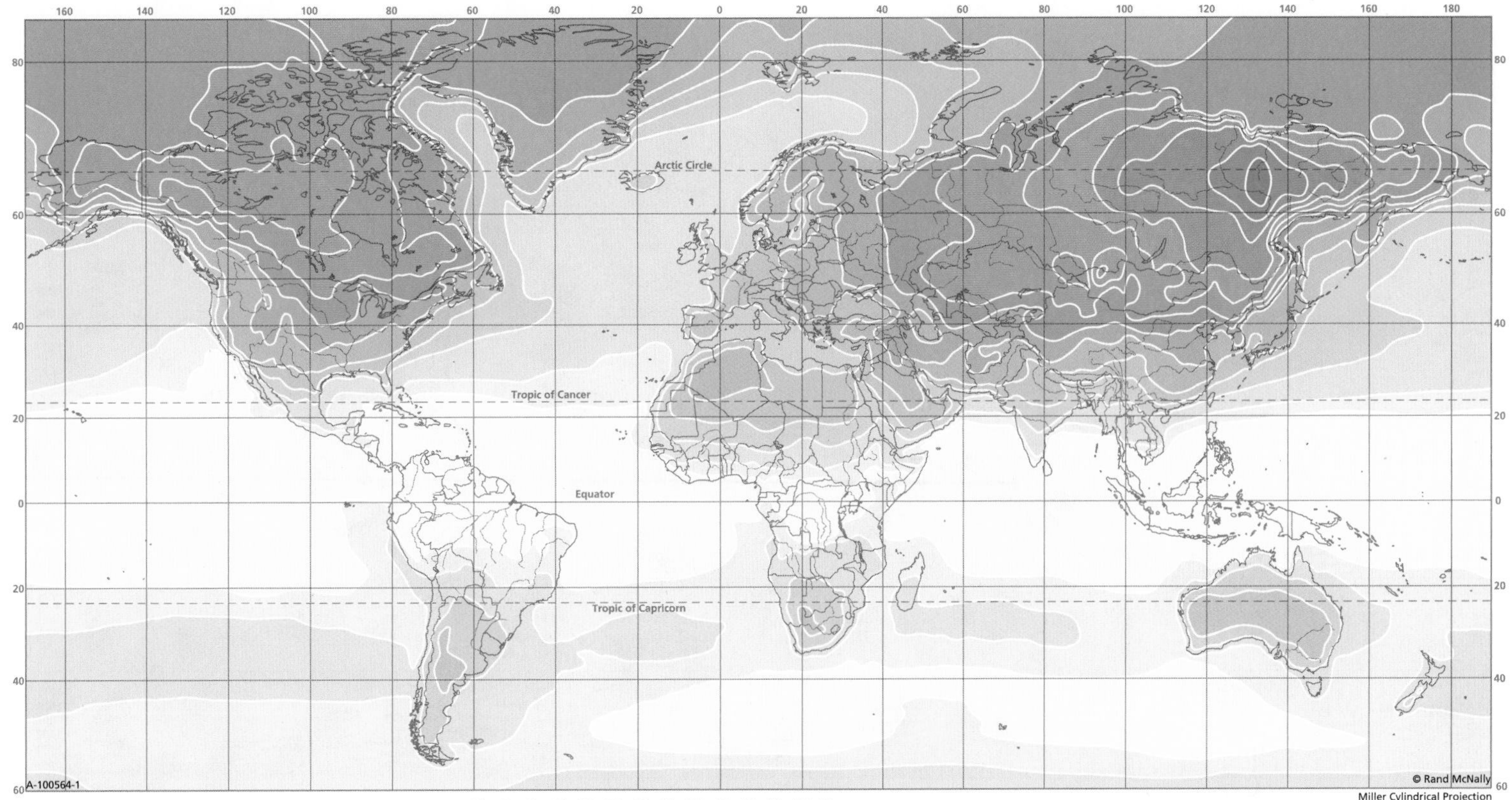
Arctic Circle
Tropic of Cancer
Equator
Tropic of Capricorn
A-100564-1
© Rand McNally

Miller Cylindrical Projection
Scale 1:200,000,000
Sources: New et al., 2000; NOAA

AVERAGE ANNUAL TEMPERATURE RANGE

C° 3 5 10 15 20 25 30 35 40 45 50 55 60

F° 5 9 18 27 36 45 54 63 72 81 90 99 108

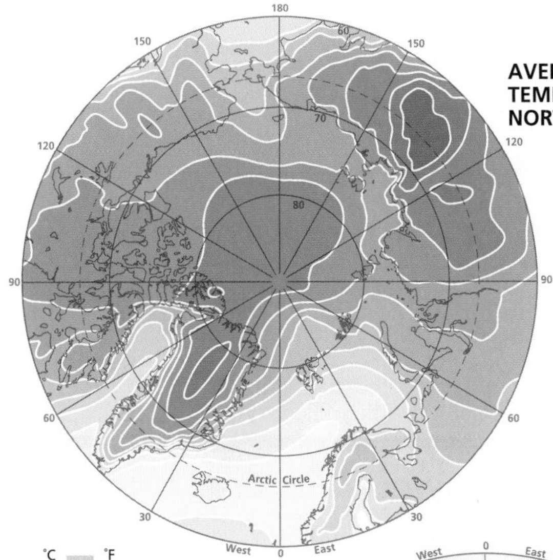
Arctic Circle
West
East

AVERAGE JANUARY TEMPERATURE, NORTH POLAR REGION

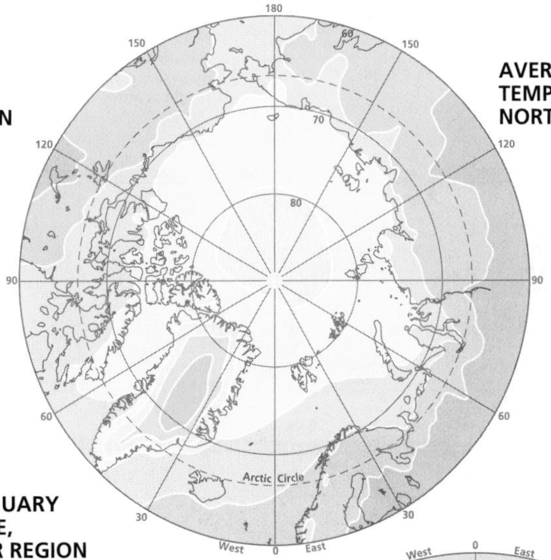
Arctic Circle
West
East

AVERAGE JULY TEMPERATURE, NORTH POLAR REGION

°C	°F
15	59
10	50
5	41
0	32
-5	23
-10	14
-15	5
-20	-4
-25	-13
-30	-22
-35	-31
-40	-40
-45	-49
-50	-58
-55	-67
-60	-76
-65	-85

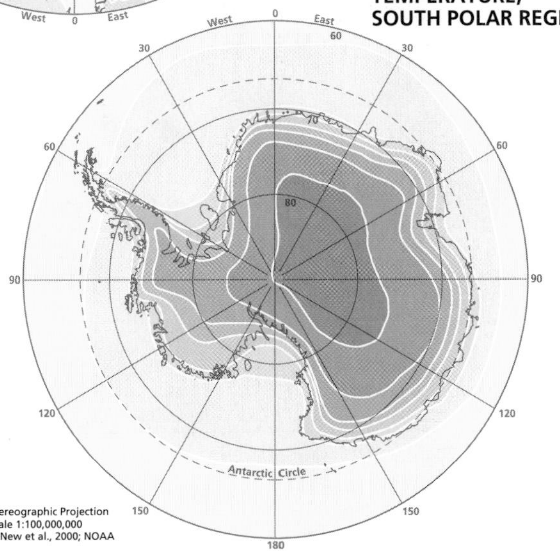
West
East
Antarctic Circle

AVERAGE JANUARY TEMPERATURE, SOUTH POLAR REGION

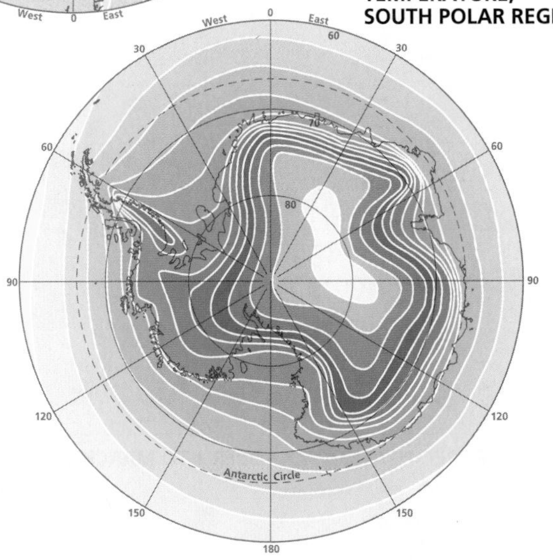
West
East
Antarctic Circle

AVERAGE JULY TEMPERATURE, SOUTH POLAR REGION

Polar Stereographic Projection
Scale 1:100,000,000
Sources: New et al., 2000; NOAA

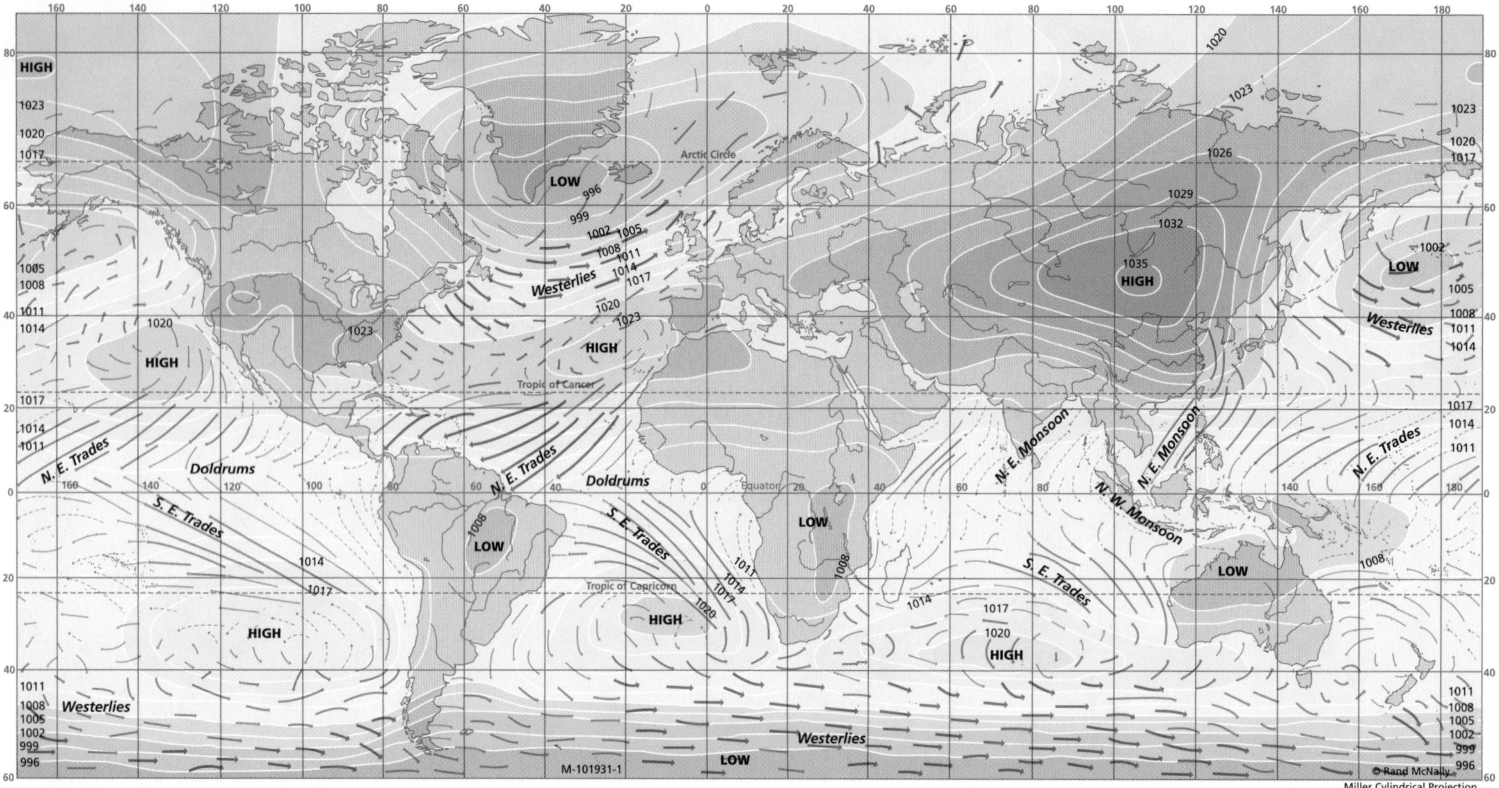

HIGH
LOW
Arctic Circle
Tropic of Cancer
Equator
Tropic of Capricorn
Westerlies
N. E. Trades
S. E. Trades
Doldrums
N. E. Monsoon
N. W. Monsoon
1035
1032
1029
1026
1023
1020
1017
1014
1011
1008
1005
1002
999
996
M-101931-1
© Rand McNally

Miller Cylindrical Projection
Scale 1:200,000,000

JANUARY PRESSURE AND PREDOMINANT WINDS

Atmospheric Pressure

in millibars (mb)

Normal sea-level pressure (1013.25 mb)

1032 1026 1020 1014 1008 1002 996

Isobars on map at intervals of 3 millibars

Wind Speed

Kilometers per hour (kph)	Miles per hour (mph)
0-16	0-10
19-24	10-15
24-40	15-25
Over 40	Over 25

Direction of arrow indicates dominant wind direction.
Length of arrow indicates steadiness of wind.

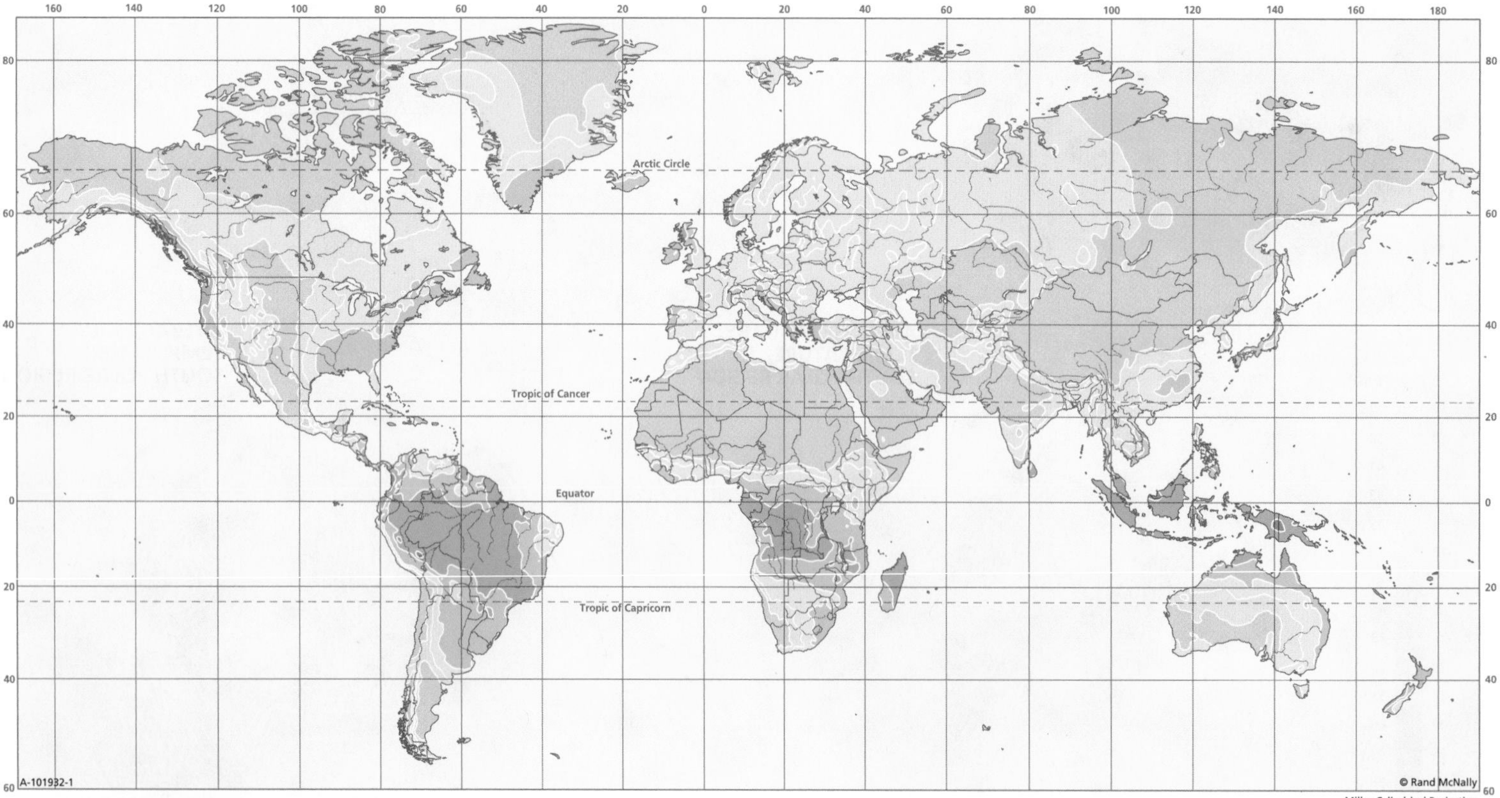

Arctic Circle
Tropic of Cancer
Equator
Tropic of Capricorn
A-101932-1
© Rand McNally

Miller Cylindrical Projection
Scale 1:200,000,000
Source: New et al., 2000

AVERAGE PRECIPITATION - OCTOBER 1 TO MARCH 31

Centimeters	12.5	25	50	100	200
Inches	5	10	20	40	80

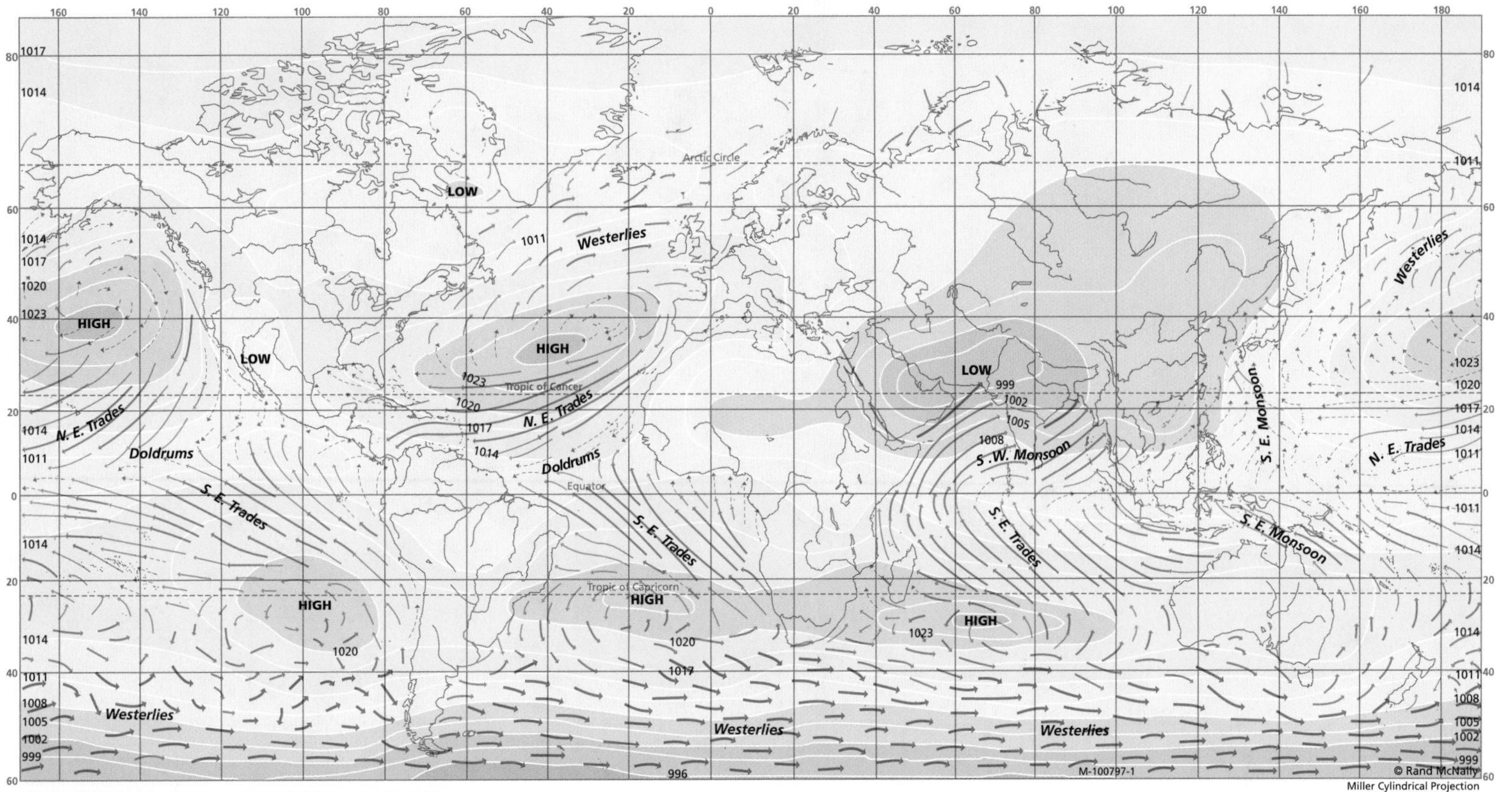

Arctic Circle
Tropic of Cancer
Equator
Tropic of Capricorn
LOW
HIGH
Westerlies
N. E. Trades
S. E. Trades
Doldrums
S. W. Monsoon
S. E. Monsoon
1017
1014
1011
1020
1023
1008
1005
1002
999
996
M-100797-1
© Rand McNally

Miller Cylindrical Projection
Scale 1:200,000,000

JULY PRESSURE AND PREDOMINANT WINDS

Atmospheric Pressure
in millibars (mb)

Normal sea-level pressure (1013.25 mb)

1026 1020 1014 1008 1002 996

Isobars on map at intervals of 3 millibars

Wind Speed

Kilometers per hour (kph)	Miles per hour (mph)
0-16	0-10
19-24	10-15
24-40	15-25
Over 40	Over 25

Direction of arrow indicates dominant wind direction.
Length of arrow indicates steadiness of wind.

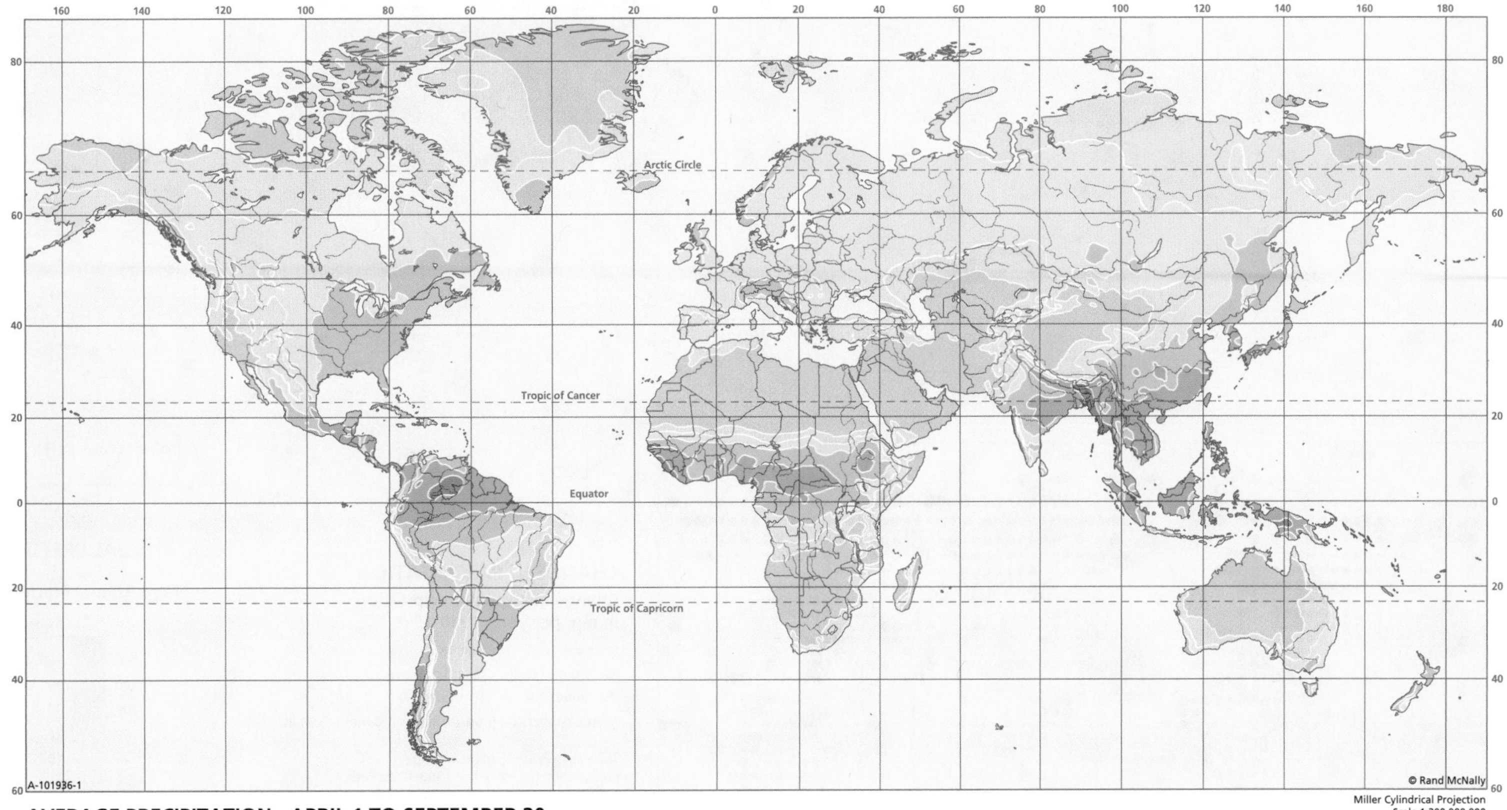

Arctic Circle
Tropic of Cancer
Equator
Tropic of Capricorn
A-101936-1
© Rand McNally

Miller Cylindrical Projection
Scale 1:200,000,000
Source: New et al., 2000

AVERAGE PRECIPITATION - APRIL 1 TO SEPTEMBER 30

12.5 25 50 100 200 Centimeters

5 10 20 40 80 Inches

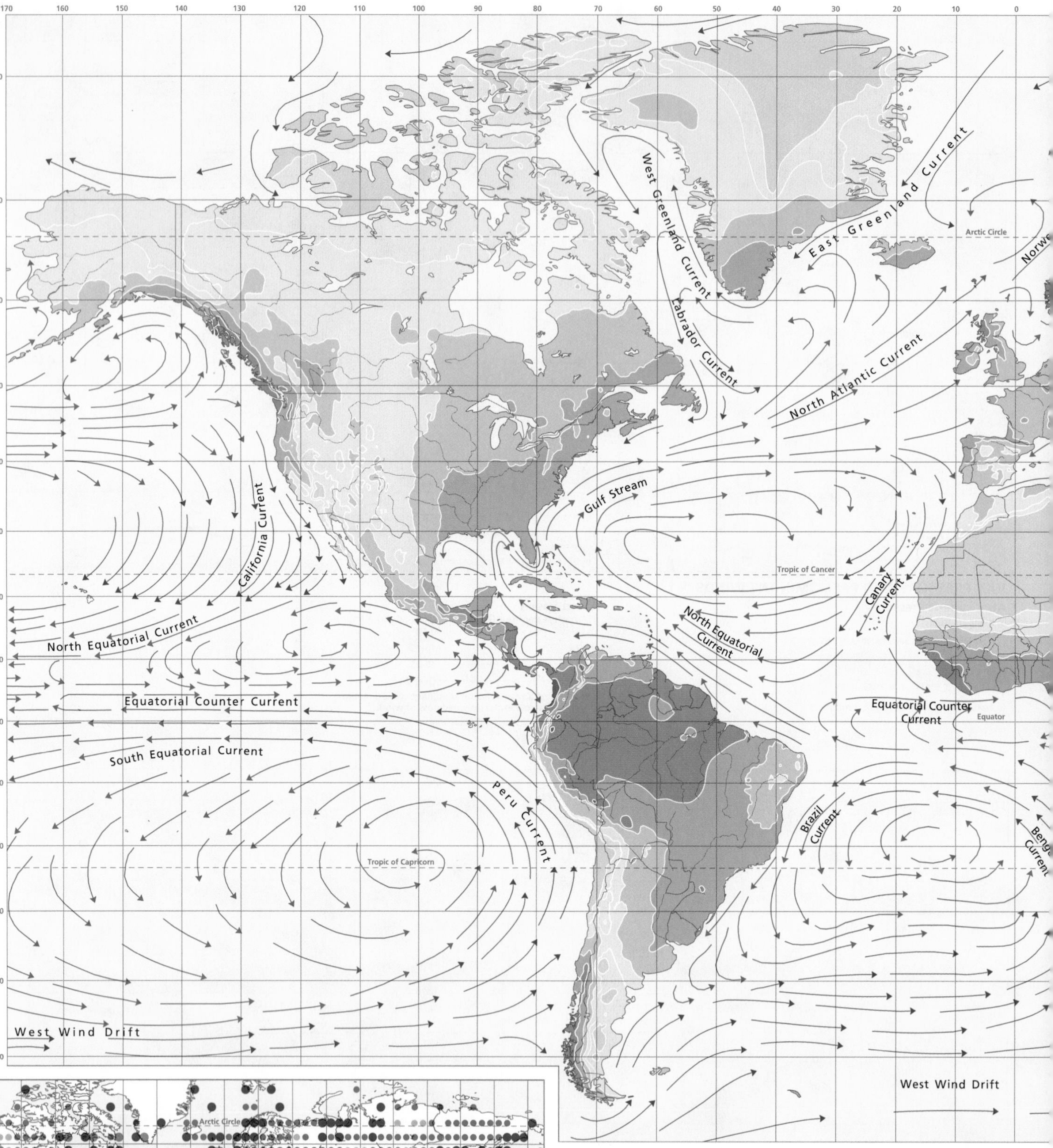
West Greenland Current
East Greenland Current
Arctic Circle
Labrador Current
North Atlantic Current
Gulf Stream
California Current
Tropic of Cancer
Canary Current
North Equatorial Current
North Equatorial Current
Equatorial Counter Current
Equatorial Counter Current
Equator
South Equatorial Current
Peru Current
Brazil Current
Tropic of Capricorn
West Wind Drift
West Wind Drift

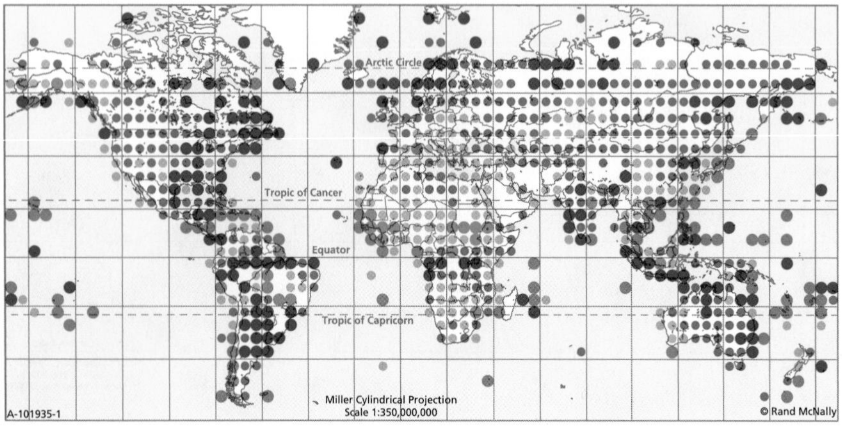
Arctic Circle
Tropic of Cancer
Equator
Tropic of Capricorn
Miller Cylindrical Projection
Scale 1:350,000,000
A-101935-1
© Rand McNally

CHANGE IN PRECIPITATION

Change in annual precipitation in mm per year, 1900-1998

Increase	Decrease
● 1.0 or more	● 1.0 or more
• Under 1.0	• Under 1.0

Metric conversion (approximate): 1.0 mm = 0.04 in.

This map shows the trend in annual precipitation for the period 1900-1998. Each symbol on the map is a 5-degree by 5-degree grid cell. The trend for each cell was computed by fitting a regression line to the data.

Derived from Hulme, 1998

ANNUAL PRECIPITATI

Average Annual Precipitatior

Cm	Inches
400	160
200	80
100	40
50	20
25	10
12.5	5

Source: New et al., 2000

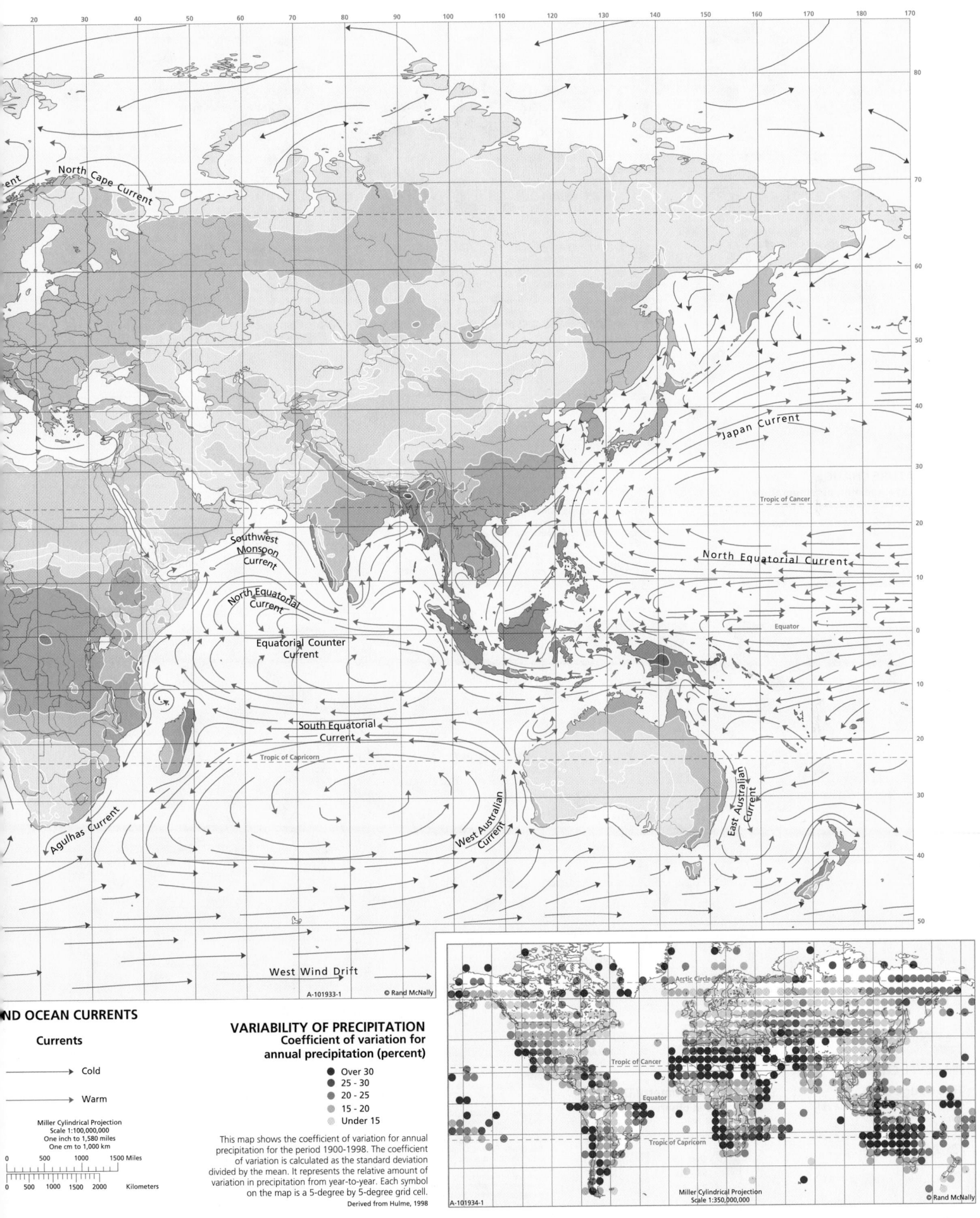

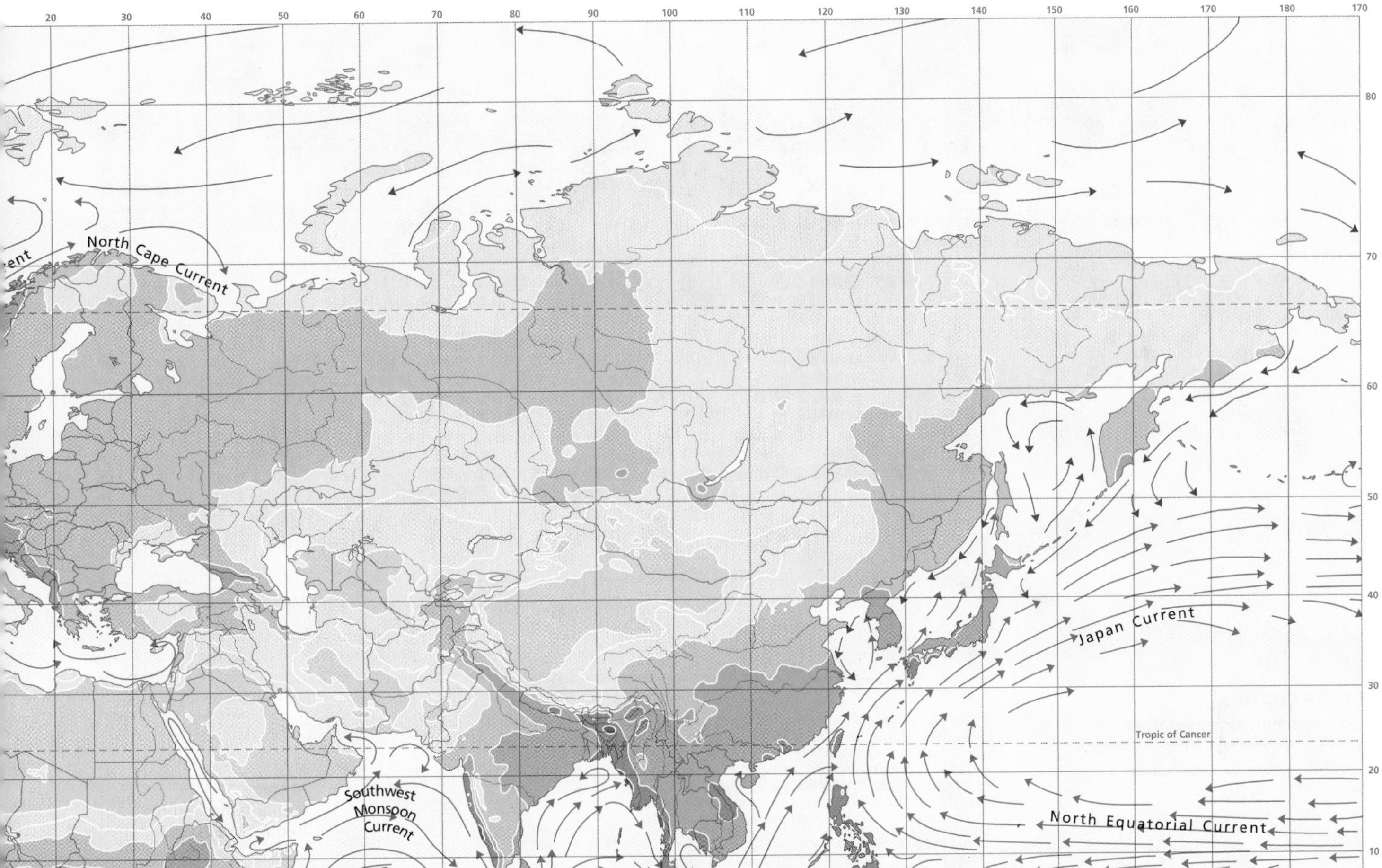
North Cape Current
Southwest Monsoon Current
North Equatorial Current
Equatorial Counter Current
South Equatorial Current
Agulhas Current
West Australian Current
East Australian Current
Japan Current
North Equatorial Current
West Wind Drift
Tropic of Cancer
Equator
Tropic of Capricorn
A-101933-1
© Rand McNally

ND OCEAN CURRENTS

Currents

Cold

Warm

Miller Cylindrical Projection
Scale 1:100,000,000
One inch to 1,580 miles
One cm to 1,000 km

0 500 1000 1500 Miles

0 500 1000 1500 2000 Kilometers

VARIABILITY OF PRECIPITATION

Coefficient of variation for annual precipitation (percent)

- Over 30
- 25 - 30
- 20 - 25
- 15 - 20
- Under 15

This map shows the coefficient of variation for annual precipitation for the period 1900-1998. The coefficient of variation is calculated as the standard deviation divided by the mean. It represents the relative amount of variation in precipitation from year-to-year. Each symbol on the map is a 5-degree by 5-degree grid cell.

Derived from Hulme, 1998

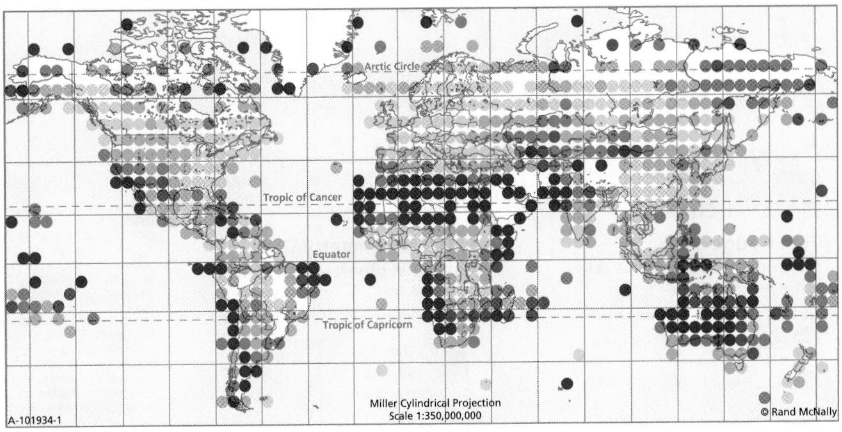
Arctic Circle
Tropic of Cancer
Equator
Tropic of Capricorn
Miller Cylindrical Projection
Scale 1:350,000,000
A-101934-1
© Rand McNally

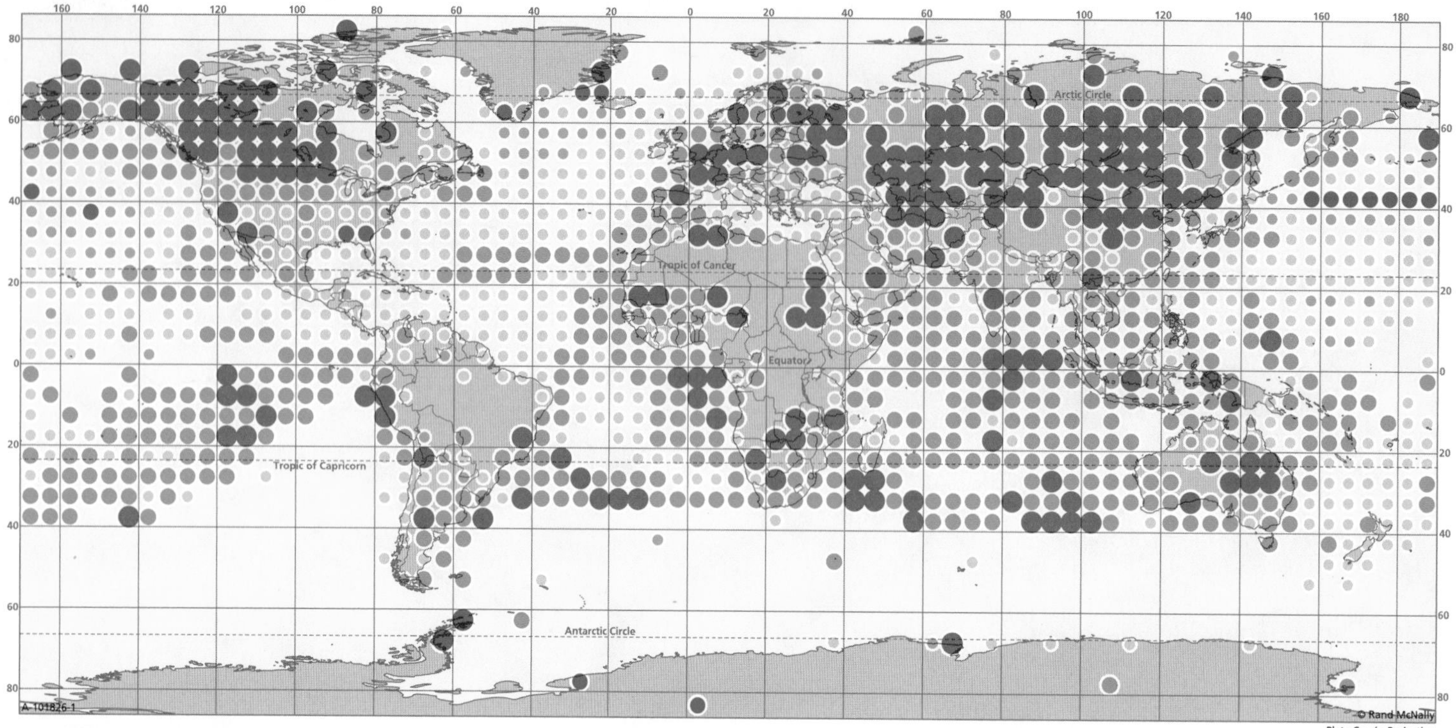

Arctic Circle
Tropic of Cancer
Equator
Tropic of Capricorn
Antarctic Circle
A-101826-1
© Rand McNally

TEMPERATURE CHANGE

Change in average annual temperature

in Celsius degrees (C°) per decade, 1950-2006

Temperature increase
- Over 0.2
- 0.1 - 0.2
- Under 0.1

Temperature decrease
- Over 0.1
- Under 0.1

Temperature conversion (approximate): 0.1 C° = 0.18 F°; 0.2 C° = 0.36 F°

This map is derived from the HadCRUT3 temperature anomaly dataset. The anomaly for a given year is the difference in temperature from the baseline period of 1961-1990. Each symbol on the map is a 5-degree by 5-degree grid cell. Cells with a gap of 10 years or more in the record are not included. The trend for each cell was computed by fitting a regression line to the data.

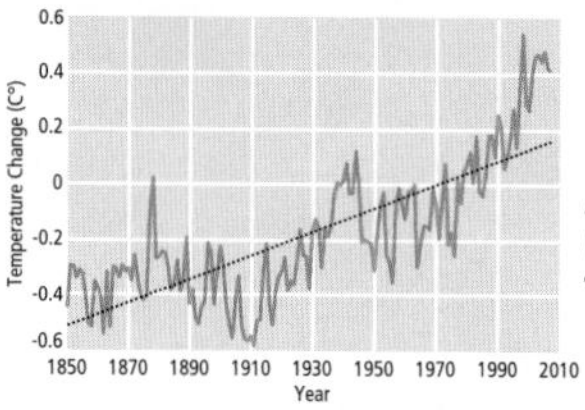

Temperature Change (C°)
0.6
0.4
0.2
0
-0.2
-0.4
-0.6
1850 1870 1890 1910 1930 1950 1970 1990 2010
Year

Average Annual Global Temperature Trend, 1850-2007

Plate Carrée Projection
Scale 1:200,000,000
Derived from Brohan et al., 2006

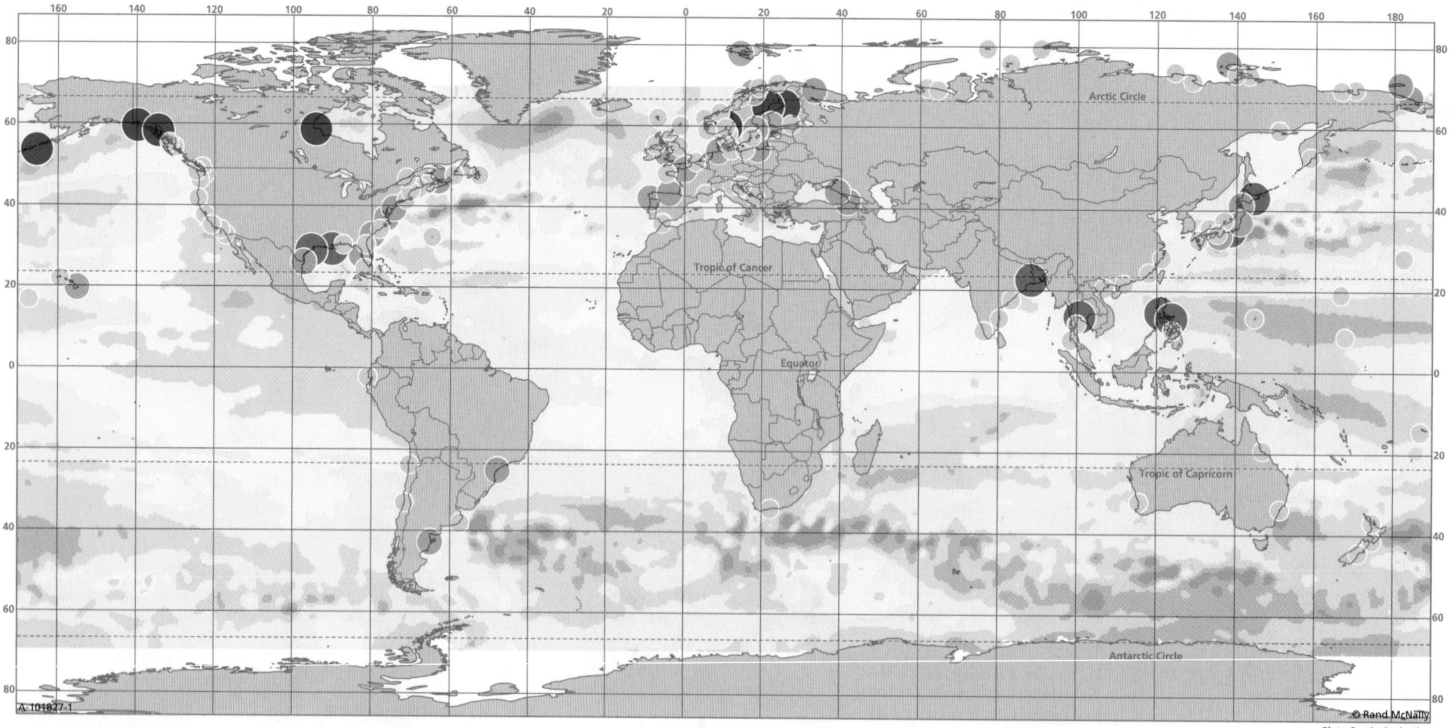

Arctic Circle
Tropic of Cancer
Equator
Tropic of Capricorn
Antarctic Circle
A-101827-1
© Rand McNally

SEA LEVEL CHANGE

Tide Gauge Data

Change in relative sea level

in mm per year, 1950-2006

Sea level increase
- Over 5.0
- 2.5 - 5.0
- Under 2.5

Sea level decrease
- Over 5.0
- 2.5 - 5.0
- Under 2.5

Satellite Altimetry Data

Change in observed sea level

in mm per year, 1992-2007

Sea level increase
- Over 7.5
- 5.0 - 7.5
- 2.5 - 5.0
- Under 2.5

Sea level decrease
- Over 7.5
- 5.0 - 7.5
- 2.5 - 5.0
- Under 2.5

Metric conversion (approximate): 2.5 mm = 0.1 in.; 5.0 mm = 0.2 in.; 7.5 mm = 0.3 in.

Tide gauges provide a long-term record of sea level change. The record extends for 200 years in some cases. However, stations are clustered spatially and do not cover the entire globe. Also, since tide gauges measure relative sea level (water level relative to the land surface), they cannot differentiate changes in water volume (due to thermal expansion and ice melting) from changes in land elevation (due to tectonic activity and glacial isostatic adjustment). Still, tide gauges are important because relative sea level has a direct impact on coastal environments.

The tide gauge data on this map are from the PSMSL-RLR (Permanent Service for Mean Sea Level - Revised Local Reference) network. Stations with gaps of 10 years or more in the data are not included. The trend at each station was computed by fitting a regression line.

Satellite altimetry offers a second method of assessing sea level change. Unlike tide gauges, coverage is nearly global. Also, observed changes in sea level are largely unaffected by land elevation changes. However, the satellite altimetry record extends back to only the 1990s. As a result, the data record reflects major decadal variations rather than long-term trends.

The satellite altimetry data on this map are from the TOPEX/Poseidon and Jason1 sensors.

Plate Carrée Projection
Scale 1:200,000,000
Sources: NOAA Laboratory for Satellite Altimetry; Woodworth and Player, 2003

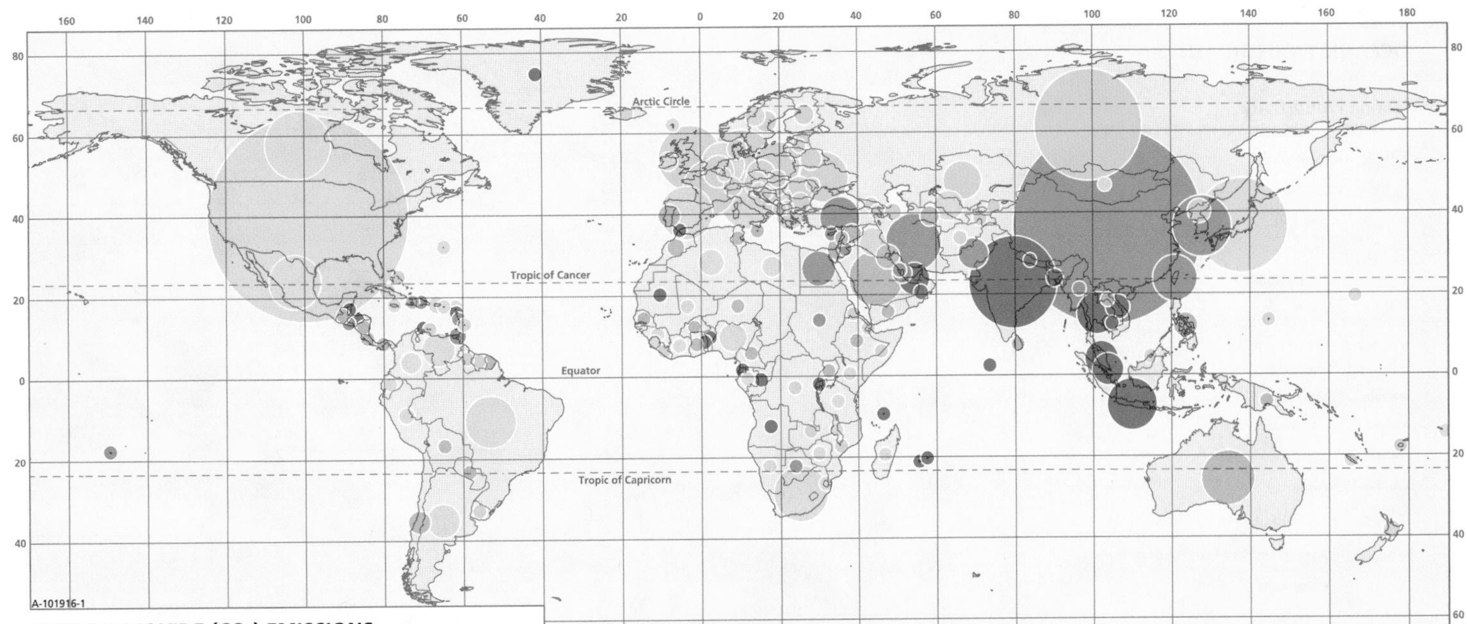
Arctic Circle
Tropic of Cancer
Equator
Tropic of Capricorn
Antarctic Circle
A-101916-1
© Rand McNally

Plate Carrée Projection
Scale 1:200,000,000
Sources: Energy Information Administration; Keeling et al., 2001

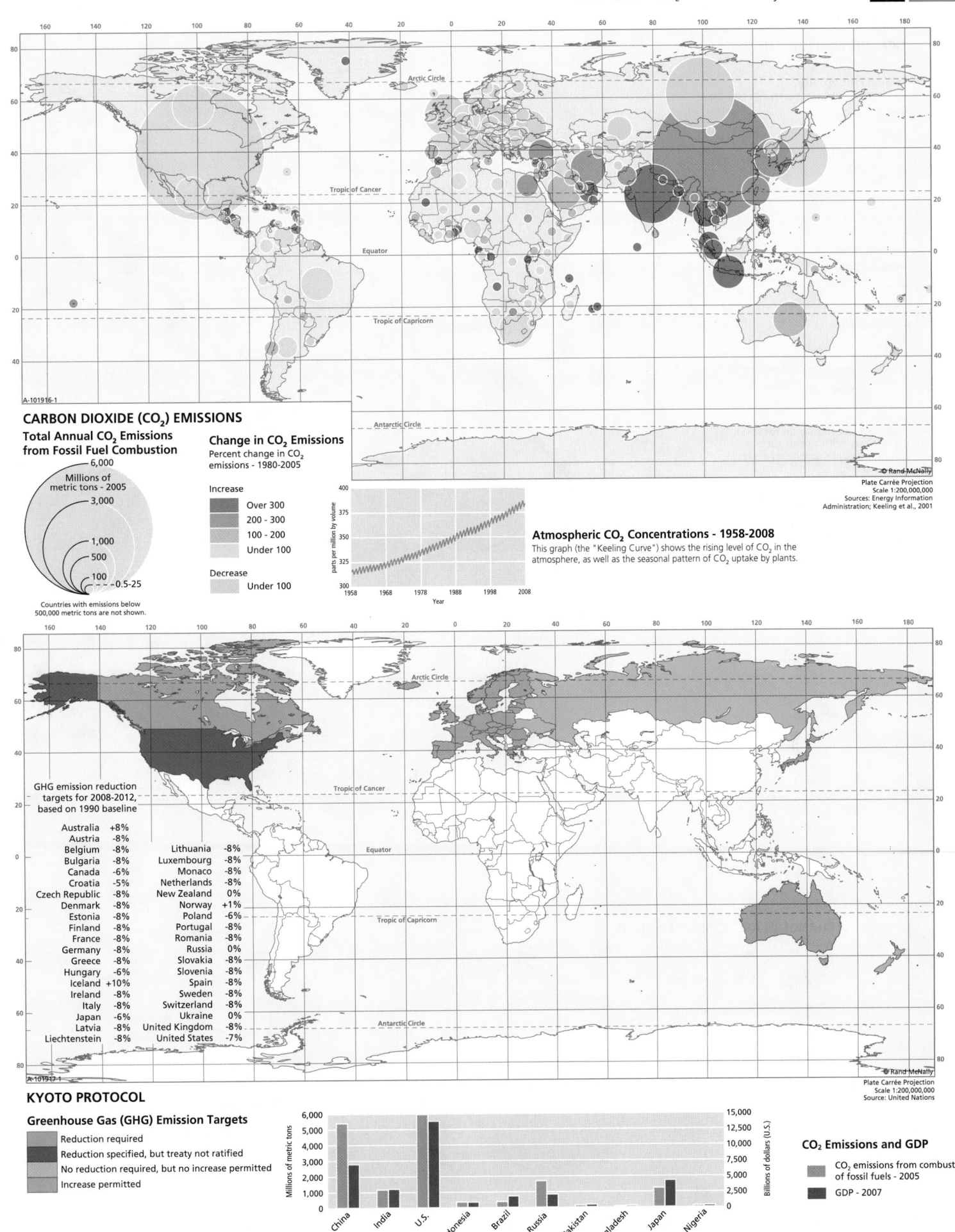

CARBON DIOXIDE (CO_2) EMISSIONS

Total Annual CO_2 Emissions from Fossil Fuel Combustion

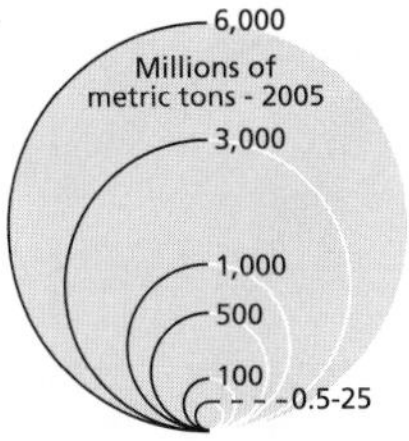
6,000
Millions of metric tons - 2005
3,000
1,000
500
100
0.5-25

Countries with emissions below 500,000 metric tons are not shown.

Change in CO_2 Emissions

Percent change in CO_2 emissions - 1980-2005

Increase
- Over 300
- 200 - 300
- 100 - 200
- Under 100

Decrease
- Under 100

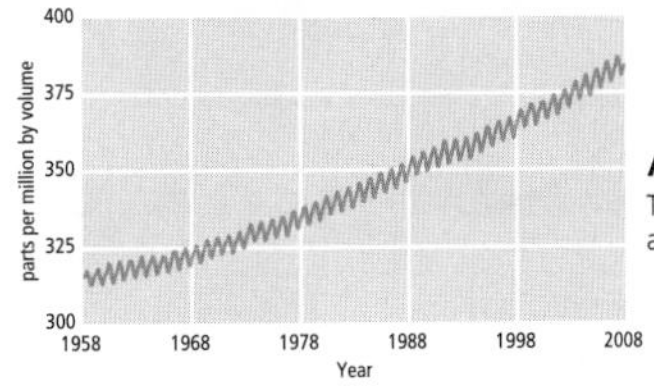
400
375
350
325
300
parts per million by volume
1958
1968
1978
1988
1998
2008
Year

Atmospheric CO_2 Concentrations - 1958-2008

This graph (the "Keeling Curve") shows the rising level of CO_2 in the atmosphere, as well as the seasonal pattern of CO_2 uptake by plants.

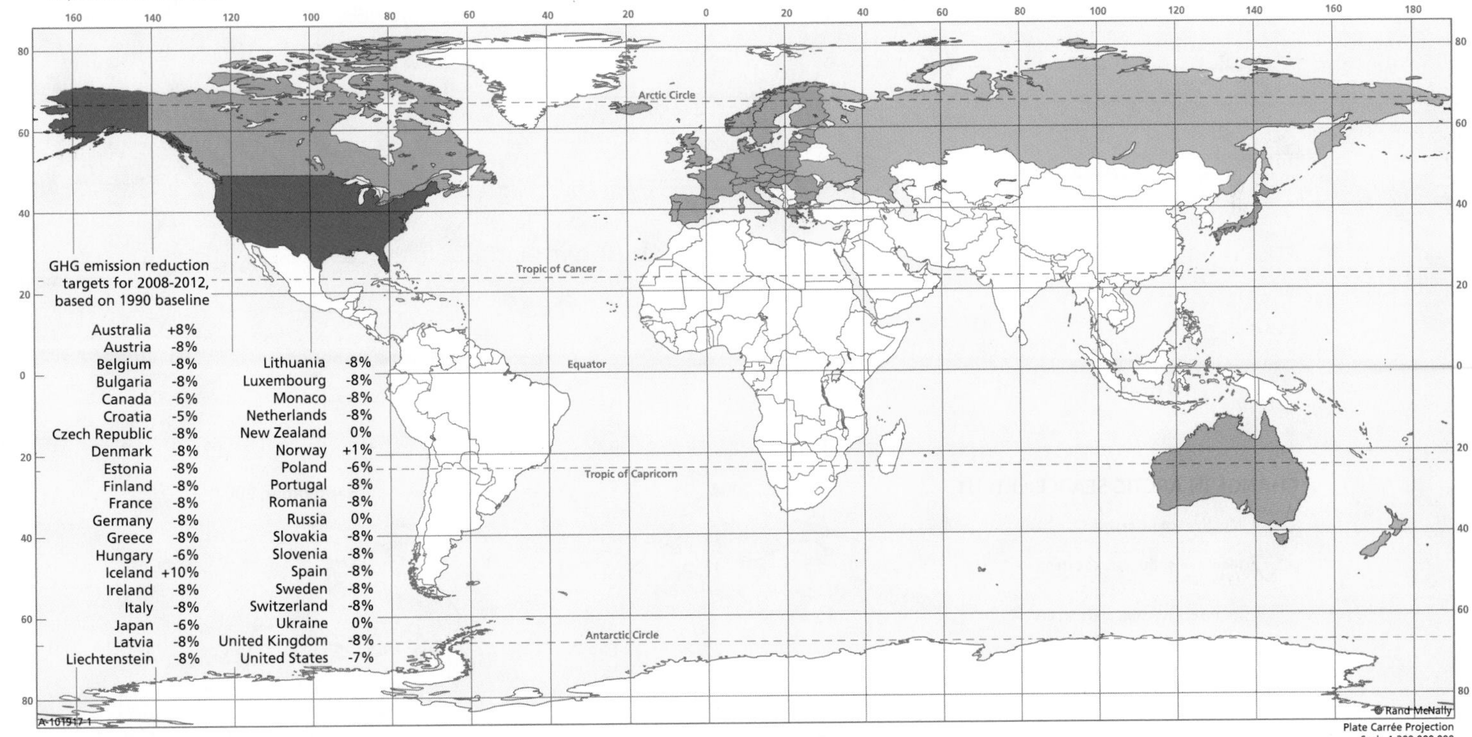
Arctic Circle
Tropic of Cancer
Equator
Tropic of Capricorn
Antarctic Circle
A-101917-1
© Rand McNally

GHG emission reduction targets for 2008-2012, based on 1990 baseline

Australia	+8%	Lithuania	-8%
Austria	-8%	Luxembourg	-8%
Belgium	-8%	Monaco	-8%
Bulgaria	-8%	Netherlands	-8%
Canada	-6%	New Zealand	0%
Croatia	-5%	Norway	+1%
Czech Republic	-8%	Poland	-6%
Denmark	-8%	Portugal	-8%
Estonia	-8%	Romania	-8%
Finland	-8%	Russia	0%
France	-8%	Slovakia	-8%
Germany	-8%	Slovenia	-8%
Greece	-8%	Spain	-8%
Hungary	-6%	Sweden	-8%
Iceland	+10%	Switzerland	-8%
Ireland	-8%	Ukraine	0%
Italy	-8%	United Kingdom	-8%
Japan	-6%	United States	-7%
Latvia	-8%		
Liechtenstein	-8%		

Plate Carrée Projection
Scale 1:200,000,000
Source: United Nations

KYOTO PROTOCOL

Greenhouse Gas (GHG) Emission Targets

- Reduction required
- Reduction specified, but treaty not ratified
- No reduction required, but no increase permitted
- Increase permitted

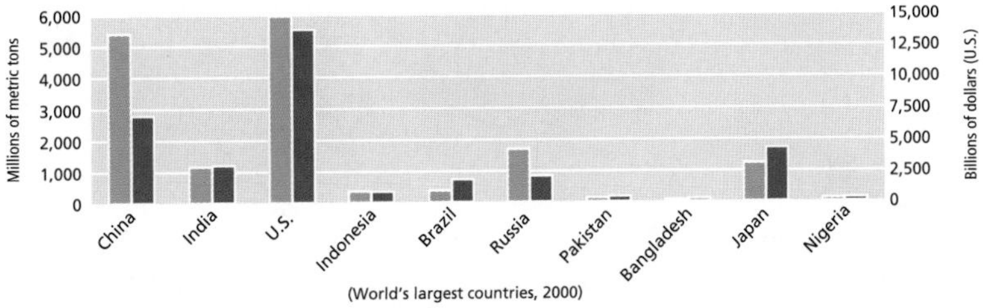
6,000
5,000
4,000
3,000
2,000
1,000
0
Millions of metric tons
15,000
12,500
10,000
7,500
5,000
2,500
0
Billions of dollars (U.S.)
China
India
U.S.
Indonesia
Brazil
Russia
Pakistan
Bangladesh
Japan
Nigeria

(World's largest countries, 2000)

CO_2 Emissions and GDP

- CO_2 emissions from combustion of fossil fuels - 2005
- GDP - 2007

OCEANIC ENVIRONMENTS

Marine Productivity
Milligrams of carbon per square meter per day

- Over 500
- 250-500
- 150-250
- 100-150
- Under 100

Velocity of current
Nautical miles per day

- Over 36
- 24 - 36
- 12 - 24
- Under 12
- Areas of upwelling cold water
- Average limits of sea ice or drift ice
- Coral reefs

Atmospheric heat gain (or loss) by contact with ocean surface
Calories per square centimeter per year

- + 80,000
- + 60,000
- + 40,000
- 0
- - 40,000
- - 60,000

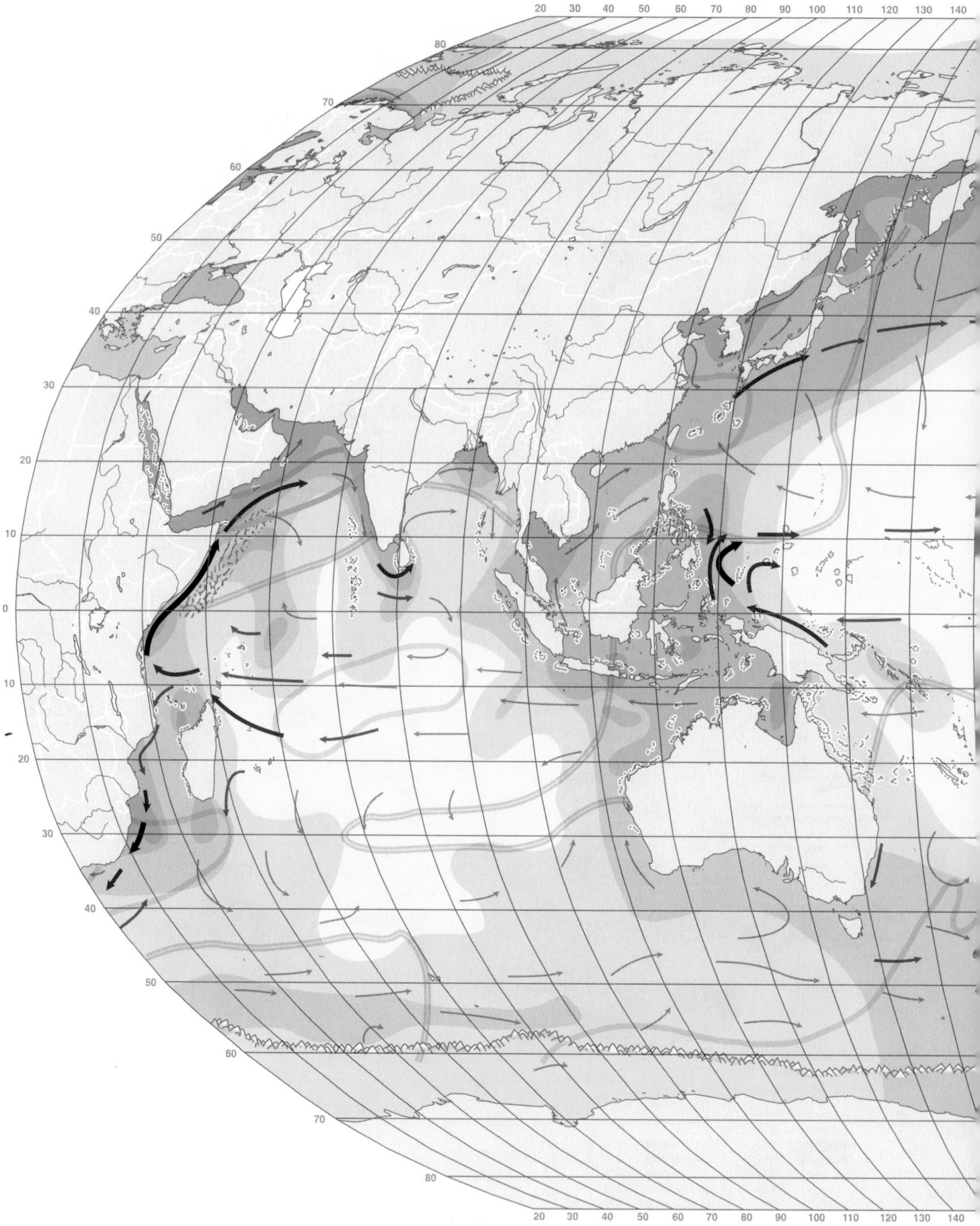

Robinson Projection
Scale 1:110,000,000
One inch to 1,750 miles
One cm to 1,100 km

0 500 1000 1500 2000 Miles
0 1000 2000 3000 Kilometers

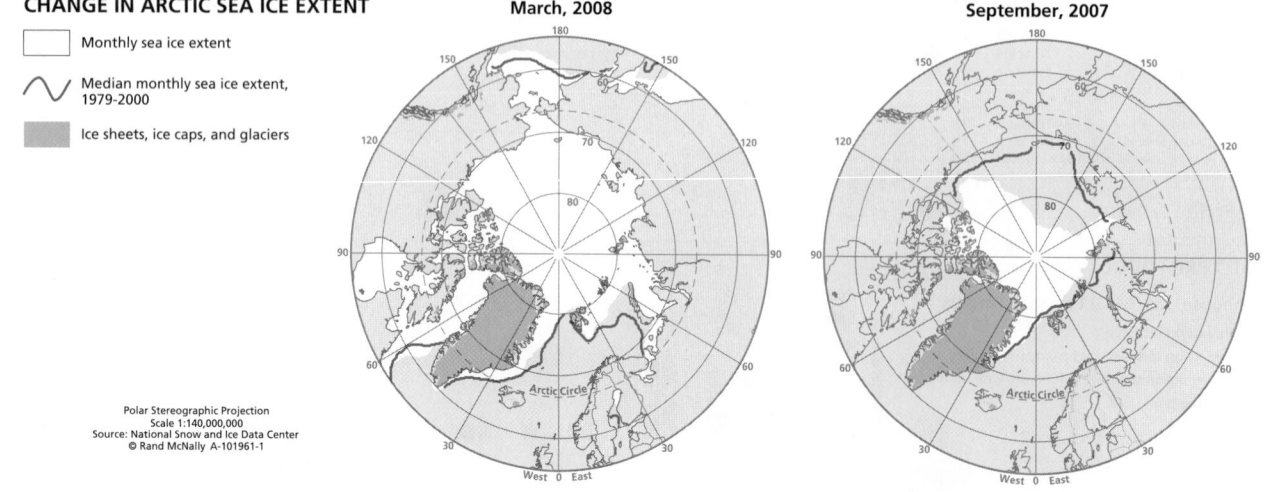

CHANGE IN ARCTIC SEA ICE EXTENT

- Monthly sea ice extent
- Median monthly sea ice extent, 1979-2000
- Ice sheets, ice caps, and glaciers

March, 2008

180
150
120
90
60
30
70
80
Arctic Circle
West 0 East

September, 2007

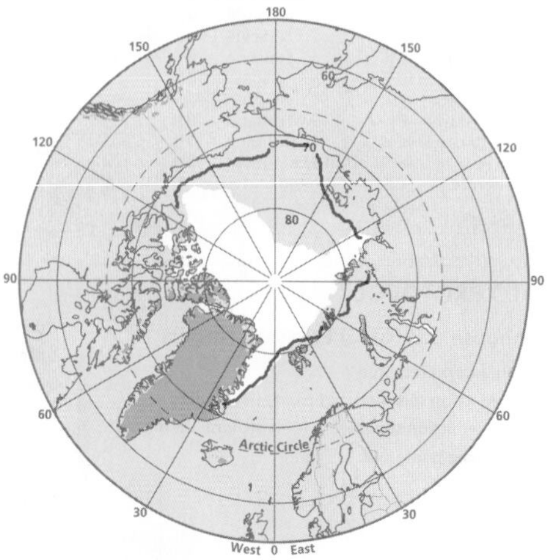
180
150
120
90
60
30
70
80
Arctic Circle
West 0 East

Polar Stereographic Projection
Scale 1:140,000,000
Source: National Snow and Ice Data Center

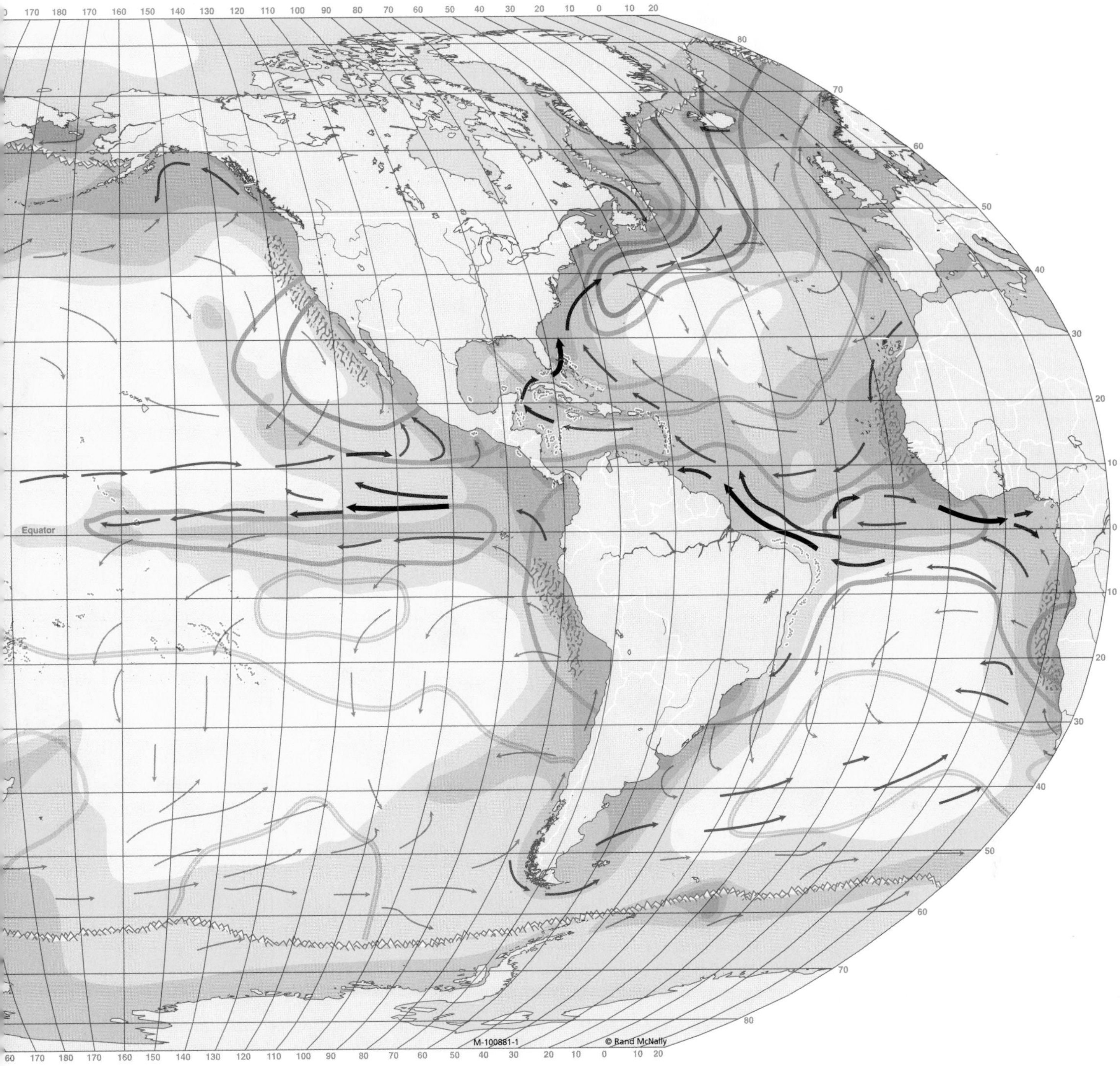
Equator
M-100881-1
© Rand McNally

CHANGE IN ANTARCTIC SEA ICE EXTENT

- Monthly sea ice extent
- Median monthly sea ice extent, 1979-2000
- Ice sheets, ice caps, and glaciers

February, 2008

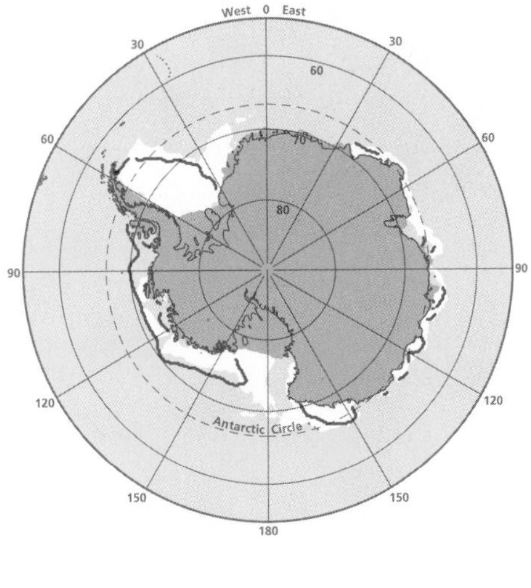
West 0 East
Antarctic Circle

September, 2007

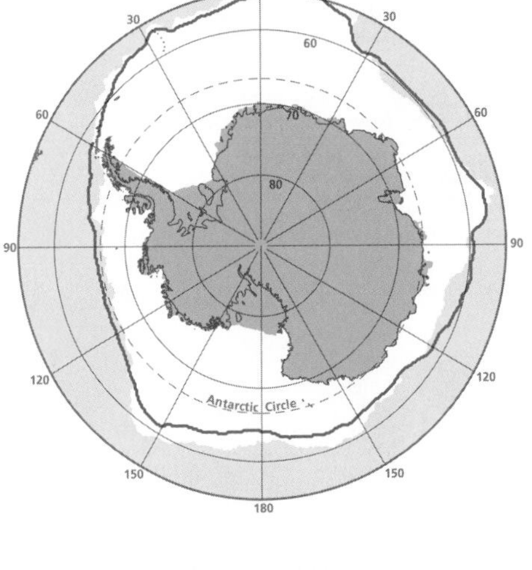
West 0 East
Antarctic Circle

Polar Stereographic Projection
Scale 1:140,000,000
Source: National Snow and Ice Data Center

NATURAL VEGETATION

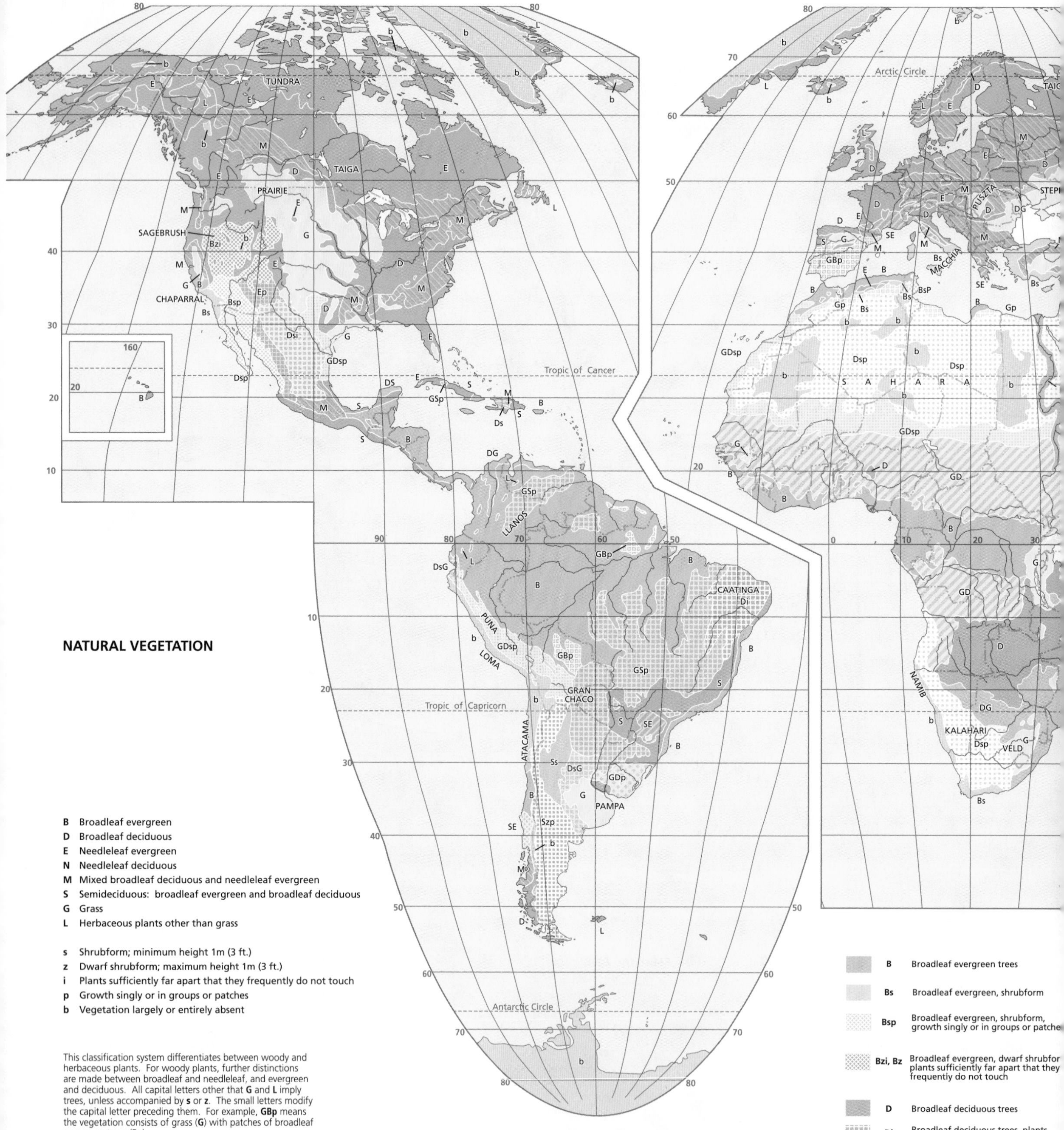
TUNDRA
TAIGA
PRAIRIE
SAGEBRUSH
CHAPARRAL
Tropic of Cancer
LLANOS
PUNA
LOMA
GRAN CHACO
CAATINGA
ATACAMA
PAMPA
Tropic of Capricorn
Antarctic Circle
Arctic Circle
PUSZTA
MACCHIA
S A H A R A
NAMIB
KALAHARI
VELD

- **B** Broadleaf evergreen
- **D** Broadleaf deciduous
- **E** Needleleaf evergreen
- **N** Needleleaf deciduous
- **M** Mixed broadleaf deciduous and needleleaf evergreen
- **S** Semideciduous: broadleaf evergreen and broadleaf deciduous
- **G** Grass
- **L** Herbaceous plants other than grass

- **s** Shrubform; minimum height 1m (3 ft.)
- **z** Dwarf shrubform; maximum height 1m (3 ft.)
- **i** Plants sufficiently far apart that they frequently do not touch
- **p** Growth singly or in groups or patches
- **b** Vegetation largely or entirely absent

This classification system differentiates between woody and herbaceous plants. For woody plants, further distinctions are made between broadleaf and needleleaf, and evergreen and deciduous. All capital letters other that **G** and **L** imply trees, unless accompanied by **s** or **z**. The small letters modify the capital letter preceding them. For example, **GBp** means the vegetation consists of grass (**G**) with patches of broadleaf evergreen trees (**Bp**).

B Broadleaf evergreen trees

Bs Broadleaf evergreen, shrubform

Bsp Broadleaf evergreen, shrubform, growth singly or in groups or patche

Bzi, Bz Broadleaf evergreen, dwarf shrubfor plants sufficiently far apart that they frequently do not touch

D Broadleaf deciduous trees

Di Broadleaf deciduous trees, plants sufficiently far apart that they frequently do not touch

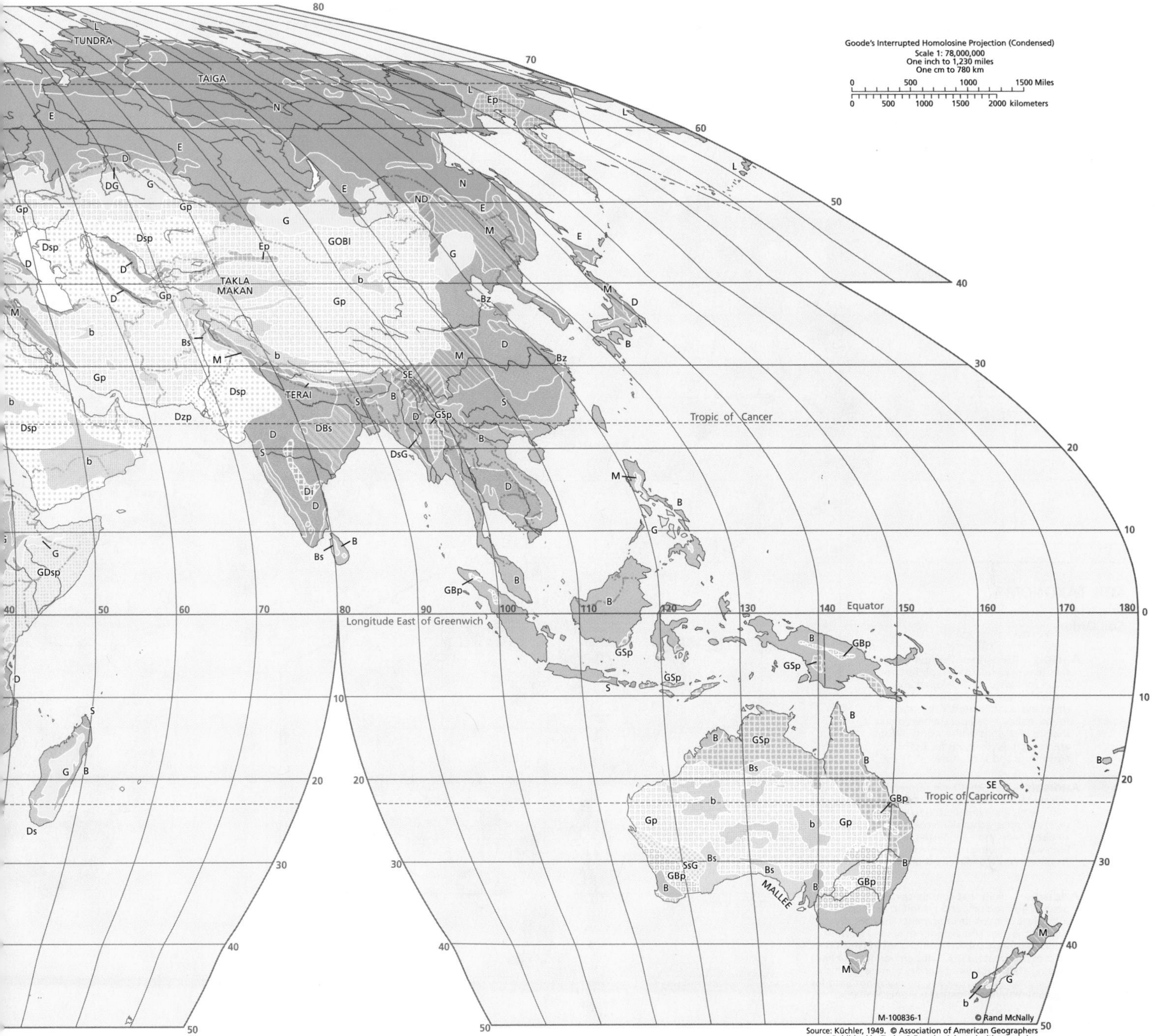
Goode's Interrupted Homolosine Projection (Condensed)
Scale 1: 78,000,000
One inch to 1,230 miles
One cm to 780 km
0 500 1000 1500 Miles
0 500 1000 1500 2000 kilometers
TUNDRA
TAIGA
GOBI
TAKLA MAKAN
TERAI
MALLEE
Tropic of Cancer
Equator
Tropic of Capricorn
Longitude East of Greenwich
M-100836-1
© Rand McNally

Source: Küchler, 1949. © Association of American Geographers
Published by Taylor & Francis. Adapted with permission of the Association of American Geographers.

- **Ds** Broadleaf deciduous, shrubform
- **Dsi** Broadleaf deciduous, shrubform, plants sufficiently far apart that they frequently do not touch
- **Dsp** Broadleaf deciduous, shrubform, growth singly or in groups or patches
- **Dzp** Broadleaf deciduous, dwarf shrubform, growth singly or in groups or patches
- **DsG** Broadleaf deciduous, shrubform Grass and other herbaceous plants
- **DG** Broadleaf deciduous trees Grass and other herbaceous plants
- **DBs** Broadleaf deciduous trees Broadleaf evergreen, shrubform
- **E** Needleleaf evergreen trees
- **Ep** Needleleaf evergreen trees, growth singly or in groups or patches
- **G** Grass and other herbaceous plants
- **Gp** Grass and other herbaceous plants, growth singly or in groups or patches
- **GBp** Grass and other herbaceous plants Broadleaf evergreen trees, growth singly or in groups or patches
- **GD** Grass and other herbaceous plants Broadleaf deciduous trees
- **GDp** Grass and other herbaceous plants Broadleaf deciduous trees, growth singly or in groups or patches
- **GDsp** Grass and other herbaceous plants Broadleaf deciduous, shrubform, growth singly or in groups or patches
- **GSp** Grass and other herbaceous plants Semideciduous: broadleaf evergreen and broadleaf deciduous trees, growth singly or in groups or patches
- **L** Herbaceous plants other than grass
- **M** Mixed broadleaf deciduous and needleleaf evergreen trees
- **N** Needleleaf deciduous trees
- **ND** Needleleaf deciduous trees Broadleaf deciduous trees
- **S** Semideciduous: broadleaf evergreen and broadleaf deciduous trees
- **Ss** Semideciduous: broadleaf evergreen and broadleaf deciduous, shrubform
- **SsG** Semideciduous: broadleaf evergreen and broadleaf deciduous, shrubform Grass and other herbaceous plants
- **Szp** Semideciduous: broadleaf evergeen and broadleaf deciduous, dwarf shrubform, growth singly or in groups or patches
- **SE** Semideciduous: broadleaf evergreen and broadleaf deciduous trees Needleleaf evergreen trees
- **b** Vegetation largely or entirely absent

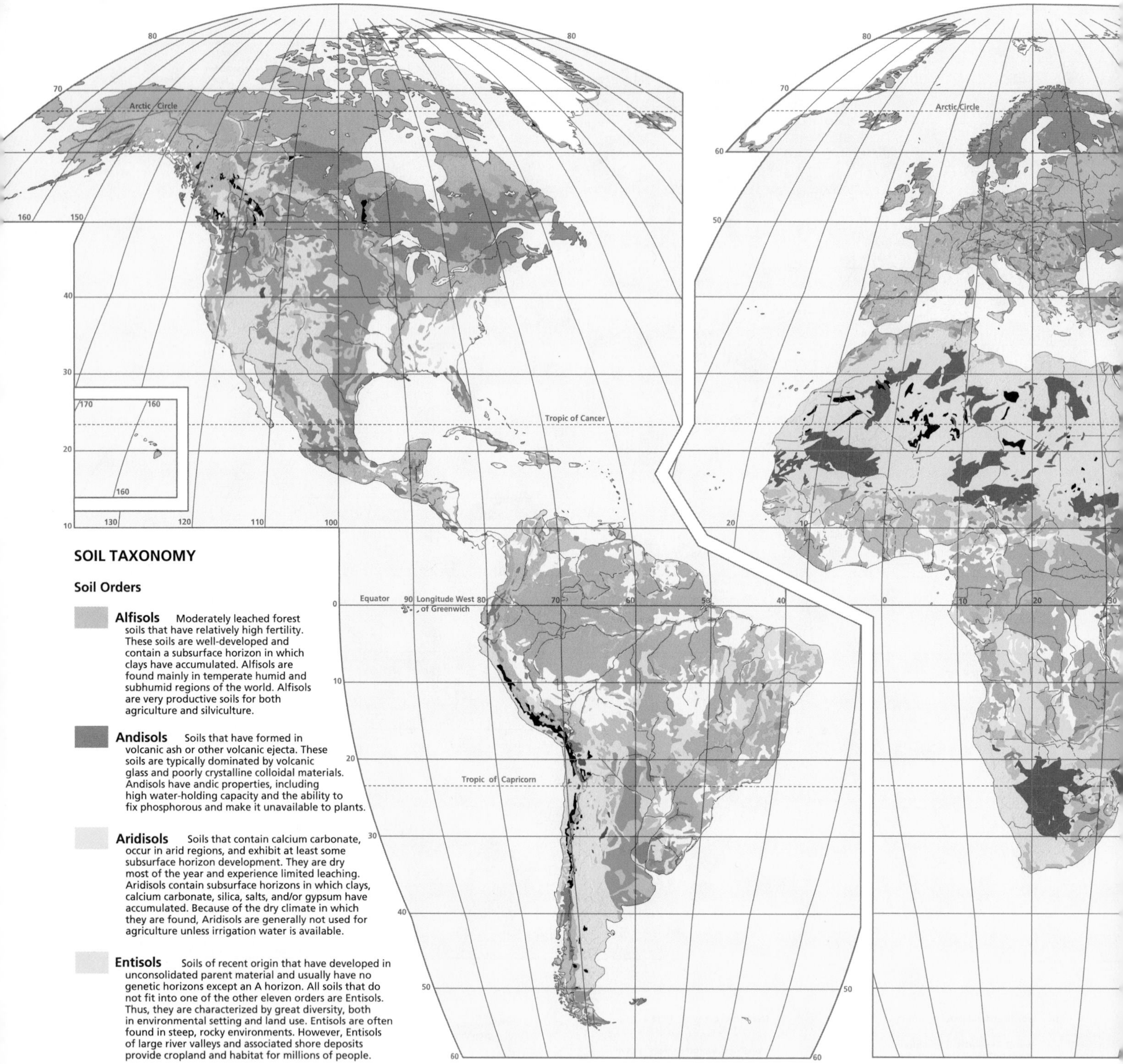
Arctic Circle
Tropic of Cancer
Equator
Longitude West of Greenwich
Tropic of Capricorn

SOIL TAXONOMY

Soil Orders

Alfisols Moderately leached forest soils that have relatively high fertility. These soils are well-developed and contain a subsurface horizon in which clays have accumulated. Alfisols are found mainly in temperate humid and subhumid regions of the world. Alfisols are very productive soils for both agriculture and silviculture.

Andisols Soils that have formed in volcanic ash or other volcanic ejecta. These soils are typically dominated by volcanic glass and poorly crystalline colloidal materials. Andisols have andic properties, including high water-holding capacity and the ability to fix phosphorous and make it unavailable to plants.

Aridisols Soils that contain calcium carbonate, occur in arid regions, and exhibit at least some subsurface horizon development. They are dry most of the year and experience limited leaching. Aridisols contain subsurface horizons in which clays, calcium carbonate, silica, salts, and/or gypsum have accumulated. Because of the dry climate in which they are found, Aridisols are generally not used for agriculture unless irrigation water is available.

Entisols Soils of recent origin that have developed in unconsolidated parent material and usually have no genetic horizons except an A horizon. All soils that do not fit into one of the other eleven orders are Entisols. Thus, they are characterized by great diversity, both in environmental setting and land use. Entisols are often found in steep, rocky environments. However, Entisols of large river valleys and associated shore deposits provide cropland and habitat for millions of people.

Gelisols Soils of very cold climates that contain permafrost within 2 m (6.5 ft.) of the surface. These soils are limited to the high-latitude polar regions and high mountain elevations. Gelisols show relatively little morphological development. Low soil temperatures cause soil-forming processes such as decomposition of organic materials to proceed slowly. As a result, Gelisols store large quantities of organic carbon. Because of the extreme environment in which they are found, Gelisols support only a small fraction of the world's population. The frozen condition of Gelisol landscapes makes them sensitive to human activities.

Histosols Soils that are composed mainly of organic materials. They contain at least 20 to 30 percent organic matter by weight and are more than 40 cm (15.75 in.) thick. Most Histosols form in settings such as wetlands where restricted drainage inhibits the decomposition of plant and animal remains, allowing these organic materials to accumulate over time. As a result, Histosols are ecologically important because of the large quantities of carbon they contain. Histosols are often referred to as peats and mucks and are mined for fuel and horticultural products.

Inceptisols Soils that exhibit minimal horizon development. They are more developed than Entisols, but still lack the features that are characteristic of other soil orders. Inceptisols are widely distributed and occur under a wide range of ecological settings. They are often found on fairly steep slopes, young geomorphic surfaces, and on resistant parent materials. Land use varies considerably with Inceptisols.

Mollisols Soils of grassland ecosystems. These soils are characterized by a thick, dark surface horizon that results from the long-term addition of organic materials derived from plant roots. Mollisols primarily occur in the mid-latitudes and are extensive in prairie regions. Mollisols are among some of the most important and productive agricultural soils in the world.

Oxisols Highly-weathered soils that are found primarily in the intertropical regions of the world. These soils contain few weatherable minerals and are often rich in iron and aluminum oxide minerals. Most Oxisols have extremely low native fertility, resulting from very low nutrient reserves, high phosphorus retention by oxide minerals, and low cation exchange capacity. Oxisols can be quite productive with inputs of lime and fertilizers.

Spodosols Acid soils characterized by a subsurface accumulation of humus that is complexed with aluminum and iron. Spodosols often occur under coniferous forest in cool, moist climates. Because they are naturally infertile, Spodosols require additions of lime in order to be productive agriculturally.

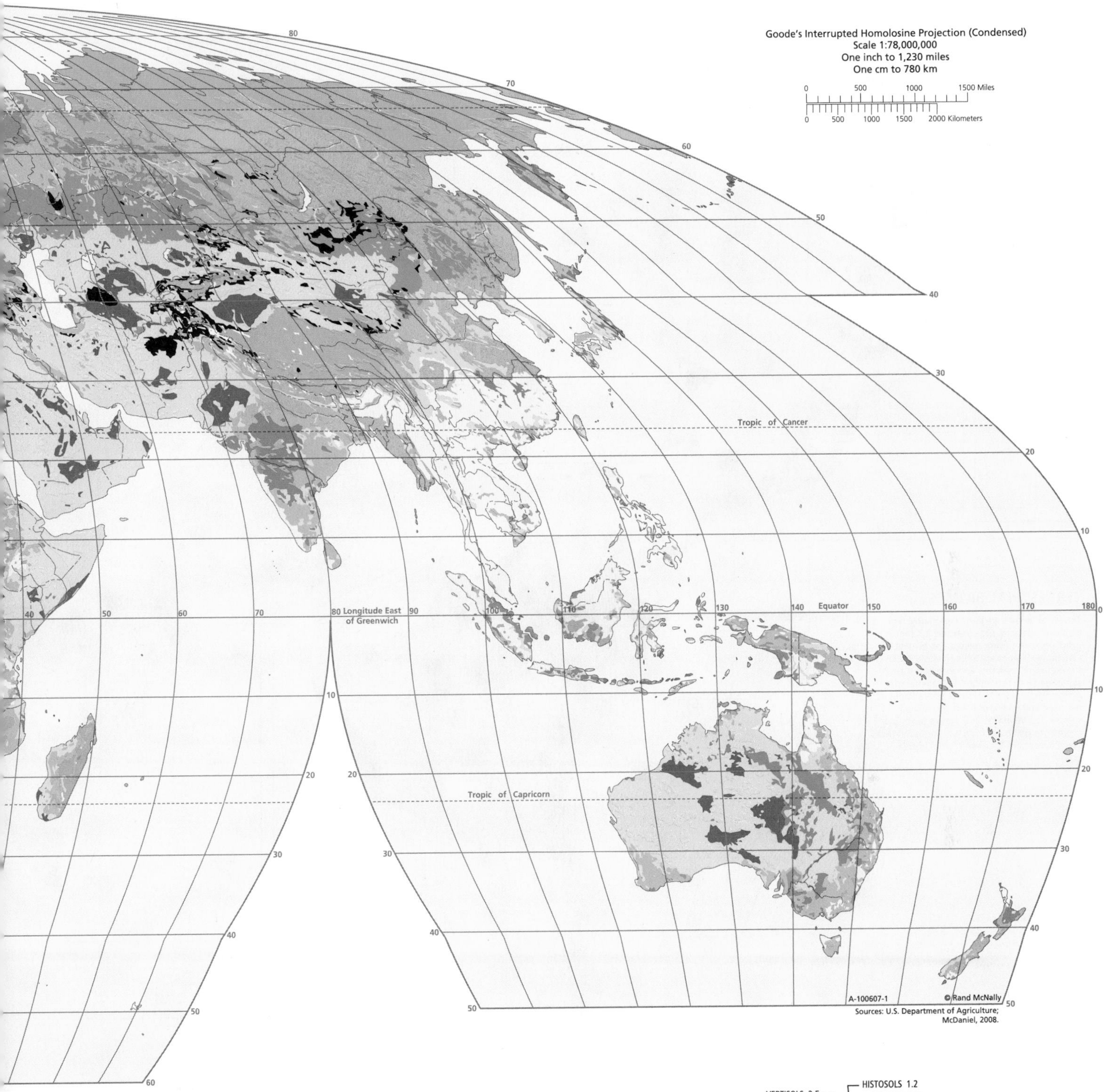
Goode's Interrupted Homolosine Projection (Condensed)
Scale 1:78,000,000
One inch to 1,230 miles
One cm to 780 km
0 500 1000 1500 Miles
0 500 1000 1500 2000 Kilometers
80
70
60
50
40
30
20
10
0
Tropic of Cancer
Tropic of Capricorn
Equator
80 Longitude East of Greenwich
90
100
110
120
130
140
150
160
170
180
A-100607-1
© Rand McNally
Sources: U.S. Department of Agriculture; McDaniel, 2008.

Ultisols Strongly leached, acid forest soils with relatively low native fertility. They are found primarily in humid temperate and tropical areas of the world, typically on older, stable landscapes where intense weathering of primary minerals has occurred. Because of the favorable climate regimes in which they are typically found, Ultisols often support productive forests. However, high acidity and limited availability of nutrients makes them poorly suited to agriculture without the use of fertilizer and lime.

Vertisols Clay-rich soils that shrink and swell with changes in moisture content. This shrink-swell action creates serious engineering problems and generally prevents formation of distinct, well-developed horizons.

Ice/Glacier

Rocky land

Salt

Shifting sands

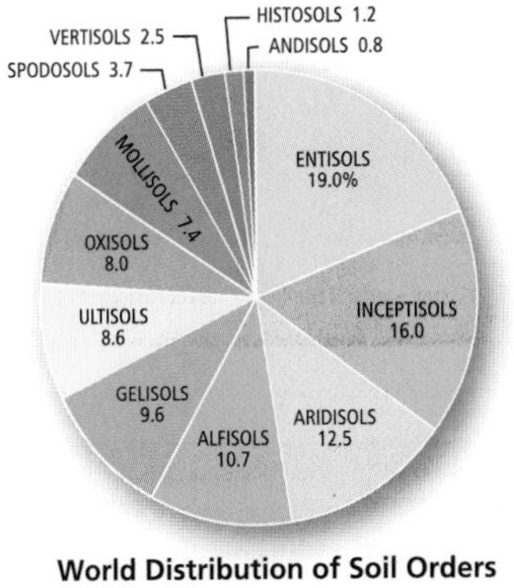
HISTOSOLS 1.2
VERTISOLS 2.5
ANDISOLS 0.8
SPODOSOLS 3.7
MOLLISOLS 7.4
ENTISOLS 19.0%
OXISOLS 8.0
INCEPTISOLS 16.0
ULTISOLS 8.6
GELISOLS 9.6
ALFISOLS 10.7
ARIDISOLS 12.5

World Distribution of Soil Orders

World Total - 123,826,000 sq. km

TERRESTRIAL BIOMES

Terrestrial biomes are large geographic regions within which living organisms exhibit similar adaptations to environmental and climatic conditions. Biomes are a broad classification of the earth's ecosystems, and may be further subdivided into ecoregions. Ecoregions are geographic areas with distinct assemblages of natural communities of plant and animal species. The World Wildlife Fund's terrestrial ecoregions database, from which this map was derived, contains fourteen biomes and 867 ecoregions.

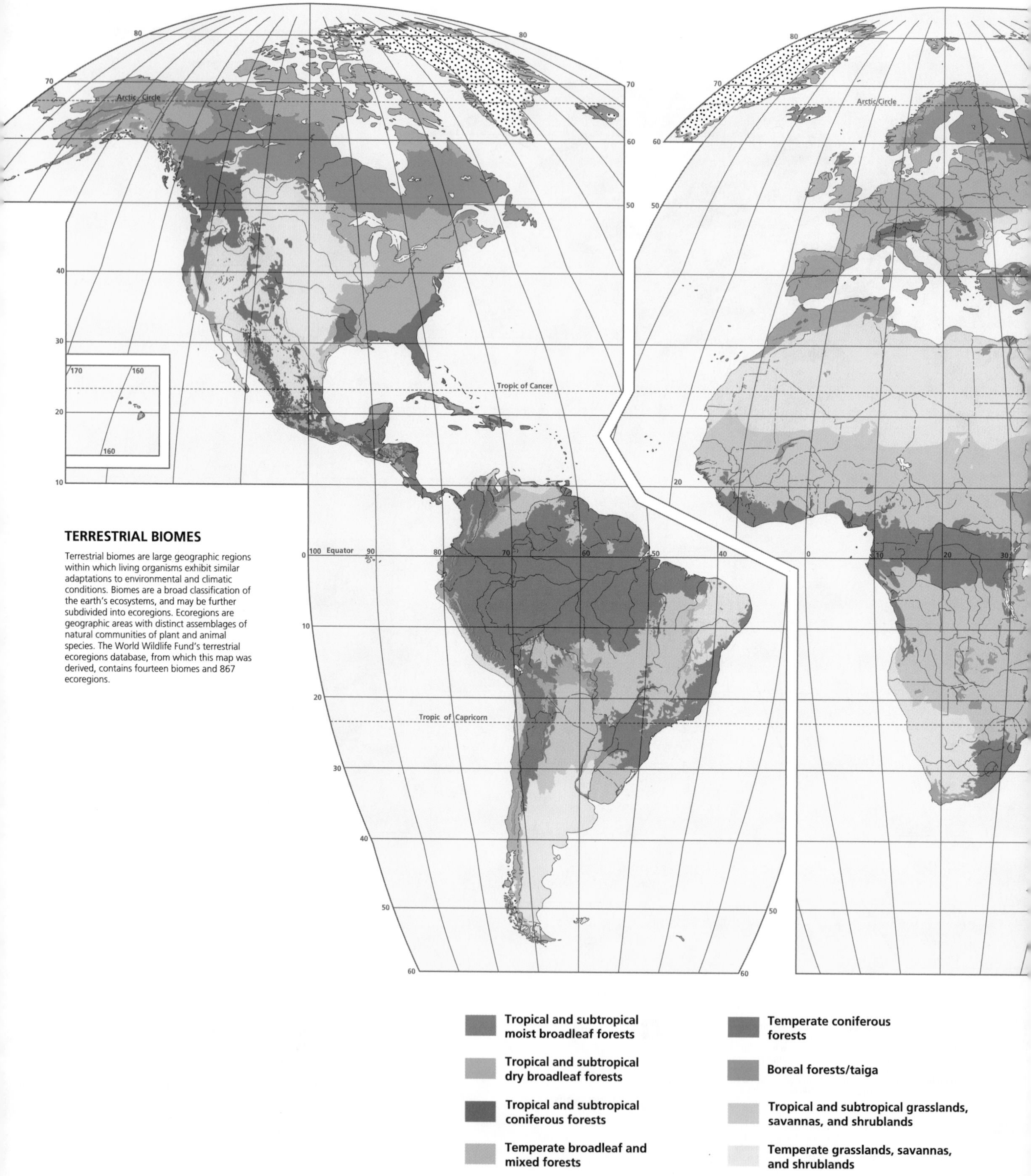

Arctic Circle
Tropic of Cancer
Equator
Tropic of Capricorn
Tropical and subtropical moist broadleaf forests
Tropical and subtropical dry broadleaf forests
Tropical and subtropical coniferous forests
Temperate broadleaf and mixed forests
Temperate coniferous forests
Boreal forests/taiga
Tropical and subtropical grasslands, savannas, and shrublands
Temperate grasslands, savannas, and shrublands

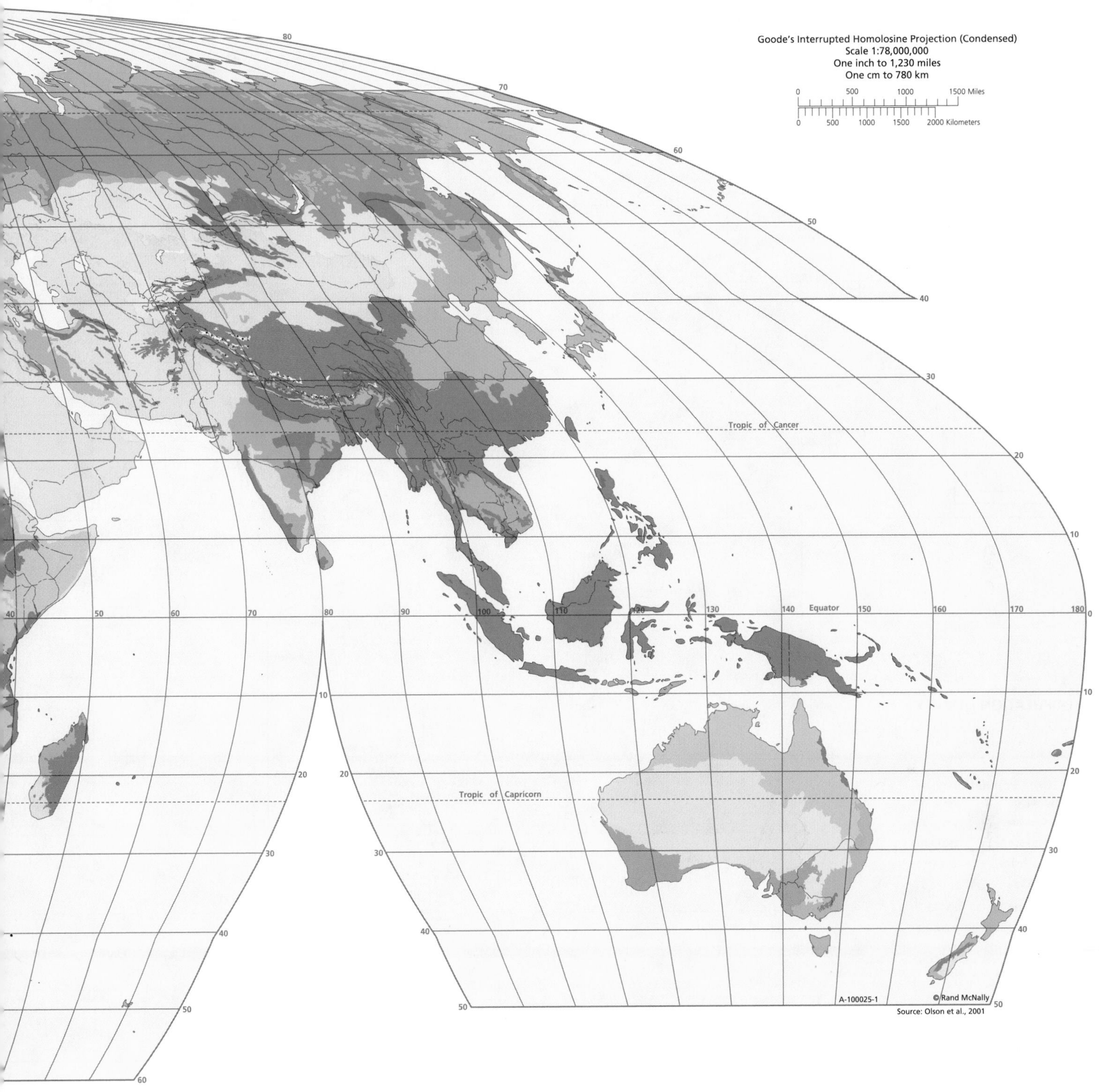

Goode's Interrupted Homolosine Projection (Condensed)
Scale 1:78,000,000
One inch to 1,230 miles
One cm to 780 km
0 500 1000 1500 Miles
0 500 1000 1500 2000 Kilometers
Tropic of Cancer
Equator
Tropic of Capricorn
80
70
60
50
40
30
20
10
0
40 50 60 70 80 90 100 110 120 130 140 150 160 170 180
A-100025-1
© Rand McNally
Source: Olson et al., 2001

Flooded grasslands and savannas

Montane grasslands and shrublands

Tundra

Mediterranean forests, woodlands, and scrub

Deserts and xeric shrublands

Mangroves

Rocks and ice

POPULATION DENSITY

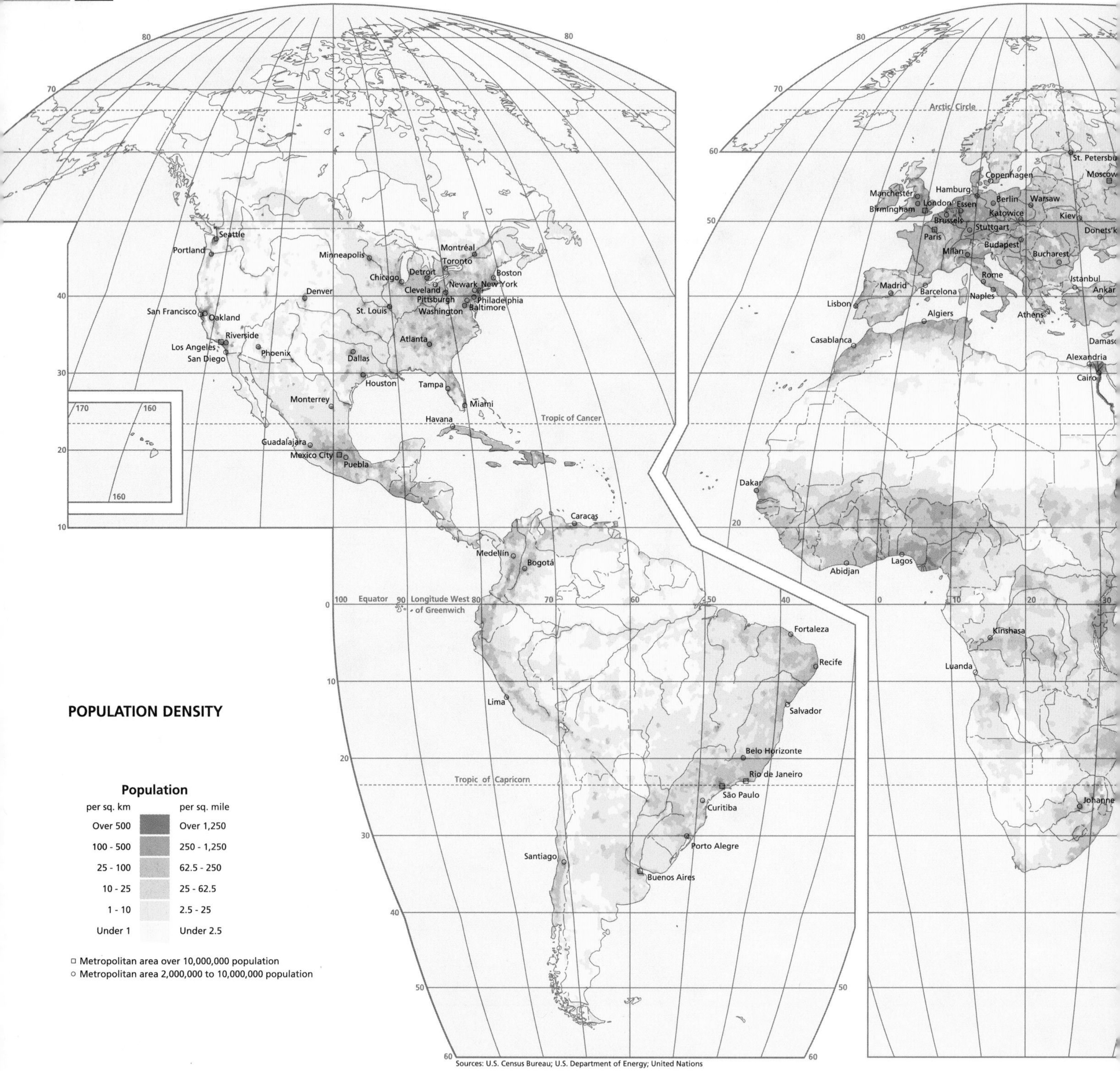
Population
per sq. km
per sq. mile
Over 500
Over 1,250
100 - 500
250 - 1,250
25 - 100
62.5 - 250
10 - 25
25 - 62.5
1 - 10
2.5 - 25
Under 1
Under 2.5
□ Metropolitan area over 10,000,000 population
○ Metropolitan area 2,000,000 to 10,000,000 population
Arctic Circle
Tropic of Cancer
Tropic of Capricorn
Equator
Longitude West of Greenwich
Seattle
Portland
Minneapolis
Montréal
Toronto
Detroit
Chicago
Boston
Cleveland
Newark
New York
Pittsburgh
Philadelphia
Baltimore
Washington
Denver
St. Louis
San Francisco
Oakland
Riverside
Los Angeles
San Diego
Phoenix
Atlanta
Dallas
Houston
Tampa
Miami
Monterrey
Havana
Guadalajara
Mexico City
Puebla
Caracas
Medellín
Bogotá
Fortaleza
Recife
Lima
Salvador
Belo Horizonte
Rio de Janeiro
São Paulo
Curitiba
Porto Alegre
Santiago
Buenos Aires
St. Petersbu
Moscow
Copenhagen
Manchester
Birmingham
Hamburg
London
Essen
Berlin
Warsaw
Katowice
Kiev
Brussels
Paris
Stuttgart
Donets'k
Budapest
Milan
Bucharest
Madrid
Barcelona
Rome
Istanbul
Lisbon
Naples
Ankar
Algiers
Athens
Casablanca
Damasc
Alexandria
Cairo
Dakar
Lagos
Abidjan
Kinshasa
Luanda
Johanne
Sources: U.S. Census Bureau; U.S. Department of Energy; United Nations

Largest Countries of the World 1950, 2000, 2050

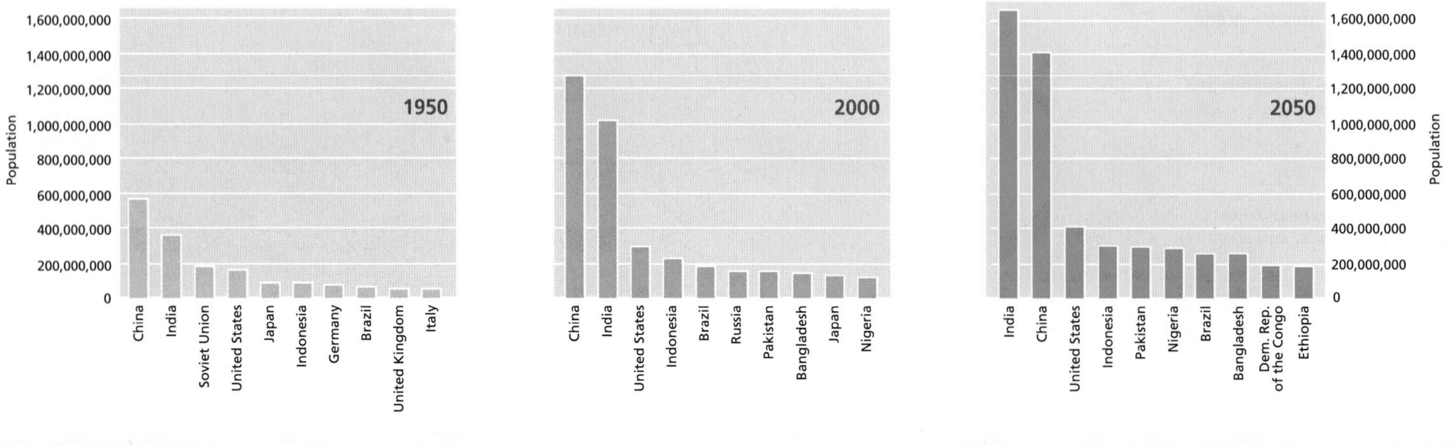
Population
1,600,000,000
1,400,000,000
1,200,000,000
1,000,000,000
800,000,000
600,000,000
400,000,000
200,000,000
0
1950
China
India
Soviet Union
United States
Japan
Indonesia
Germany
Brazil
United Kingdom
Italy
2000
China
India
United States
Indonesia
Brazil
Russia
Pakistan
Bangladesh
Japan
Nigeria
2050
India
China
United States
Indonesia
Pakistan
Nigeria
Brazil
Bangladesh
Dem. Rep. of the Congo
Ethiopia

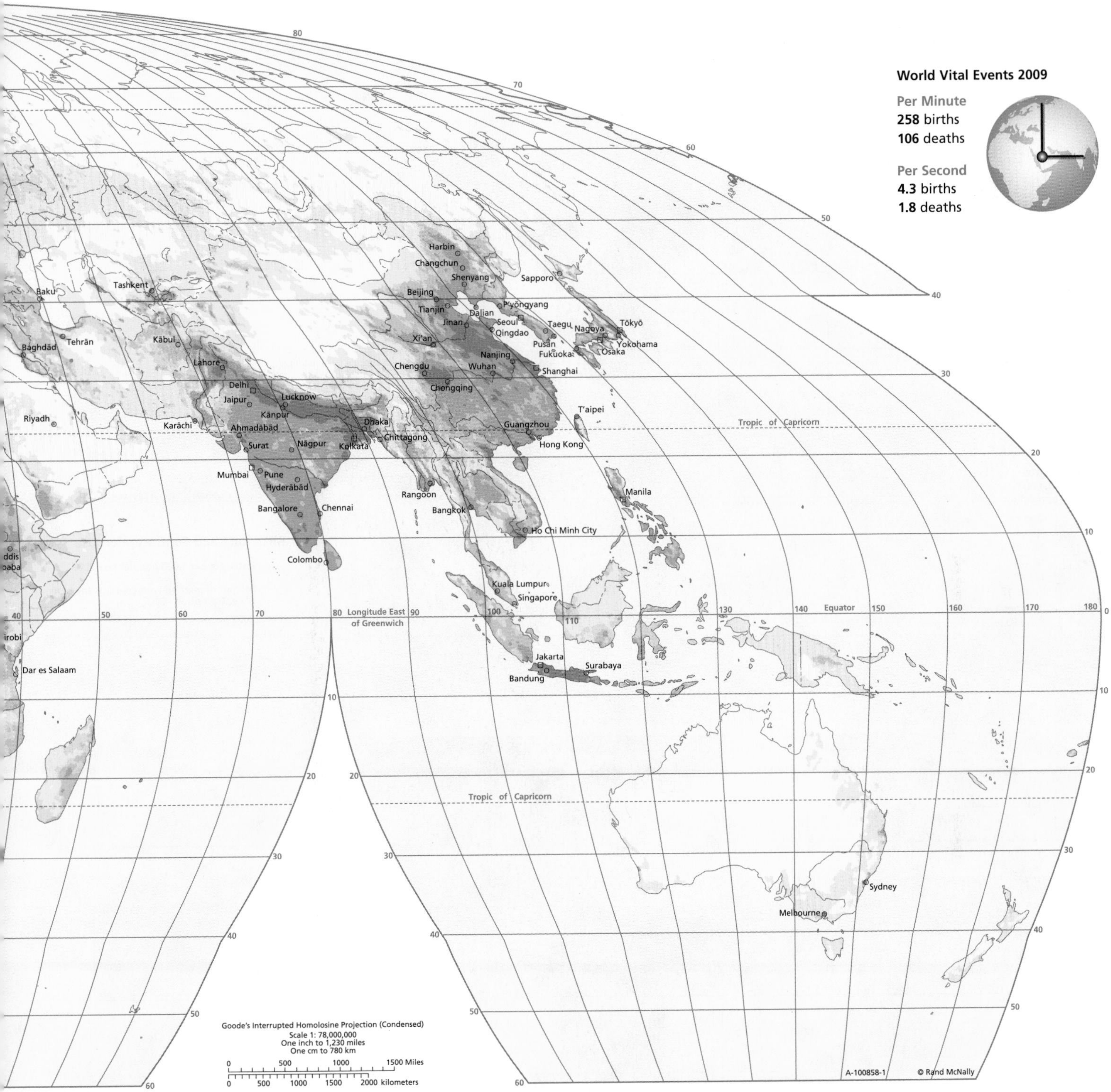

Baku
Tashkent
Tehrān
Baghdād
Kābul
Lahore
Delhi
Jaipur
Lucknow
Kānpur
Riyadh
Karāchi
Ahmadābād
Surat
Nāgpur
Kolkata
Dhaka
Chittagong
Mumbai
Pune
Hyderābād
Bangalore
Chennai
Colombo
Harbin
Changchun
Shenyang
Beijing
Tianjin
Dalian
P'yŏngyang
Jinan
Seoul
Qingdao
Taegu
Nagoya
Tōkyō
Yokohama
Ōsaka
Pusan
Fukuoka
Sapporo
Xi'an
Chengdu
Nanjing
Wuhan
Shanghai
Chongqing
T'aipei
Guangzhou
Hong Kong
Rangoon
Bangkok
Manila
Ho Chi Minh City
Kuala Lumpur
Singapore
Jakarta
Bandung
Surabaya
Sydney
Melbourne
Dar es Salaam
Tropic of Capricorn
Equator
Longitude East of Greenwich
Goode's Interrupted Homolosine Projection (Condensed)
Scale 1: 78,000,000
One inch to 1,230 miles
One cm to 780 km
0 500 1000 1500 Miles
0 500 1000 1500 2000 kilometers
A-100858-1
© Rand McNally

World Vital Events 2009

Per Minute
258 births
106 deaths

Per Second
4.3 births
1.8 deaths

Age and Sex Composition 2000

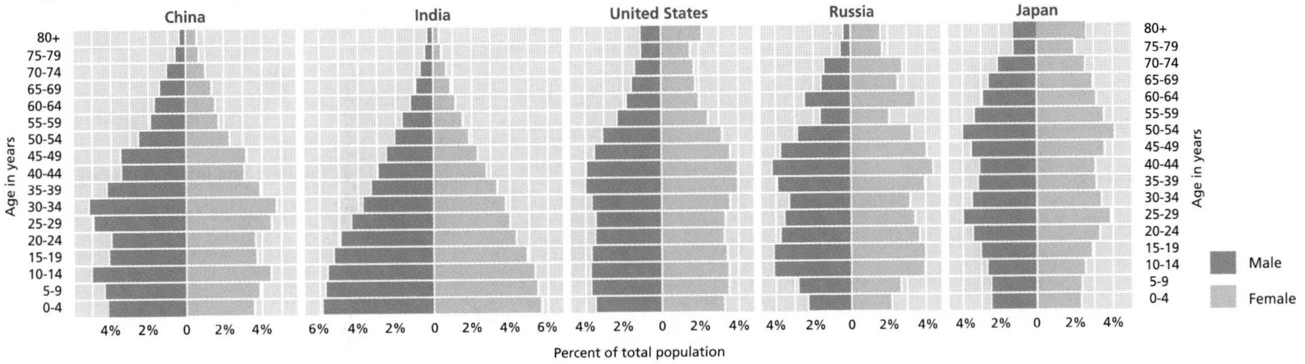

China
India
United States
Russia
Japan
80+
75-79
70-74
65-69
60-64
55-59
50-54
45-49
40-44
35-39
30-34
25-29
20-24
15-19
10-14
5-9
0-4
Age in years
Male
Female
Percent of total population

BIRTH RATE

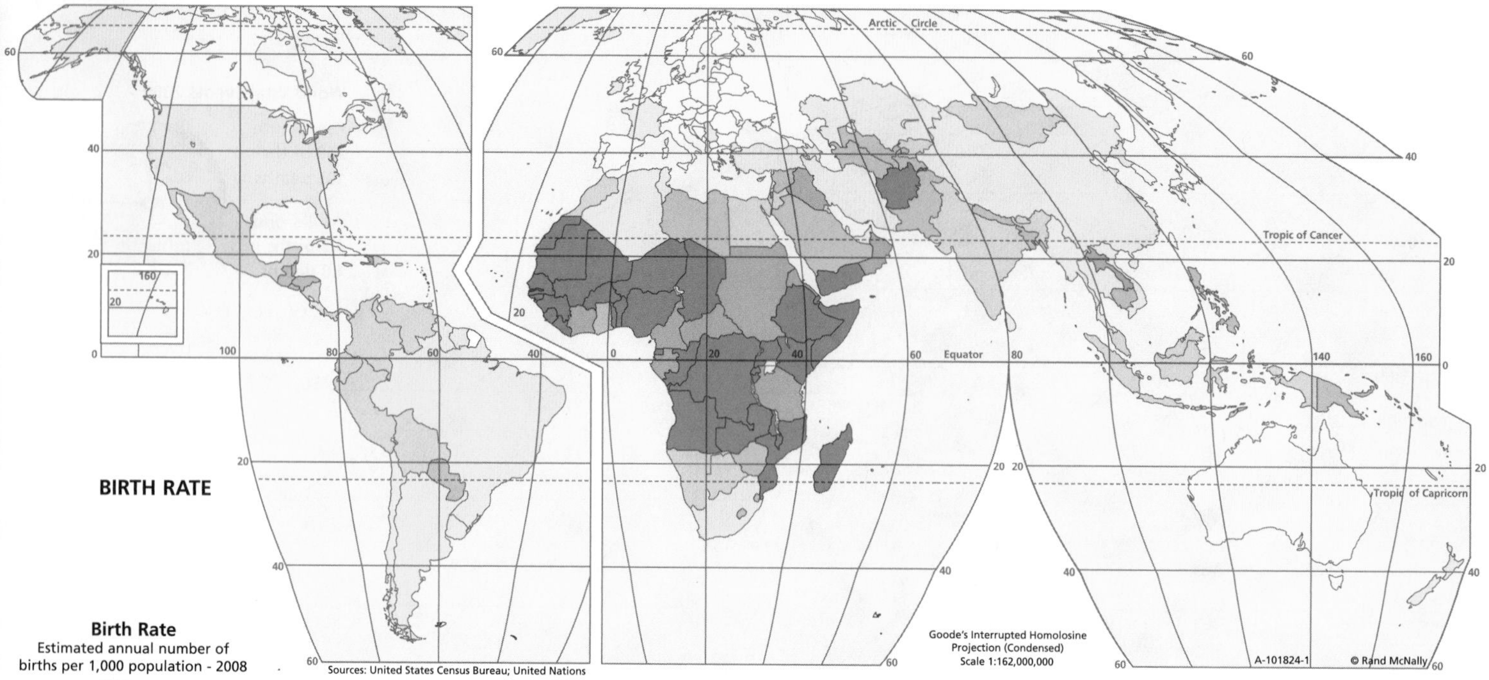
Arctic Circle
Tropic of Cancer
Equator
Tropic of Capricorn
Goode's Interrupted Homolosine Projection (Condensed)
Scale 1:162,000,000
A-101824-1
© Rand McNally
Sources: United States Census Bureau; United Nations

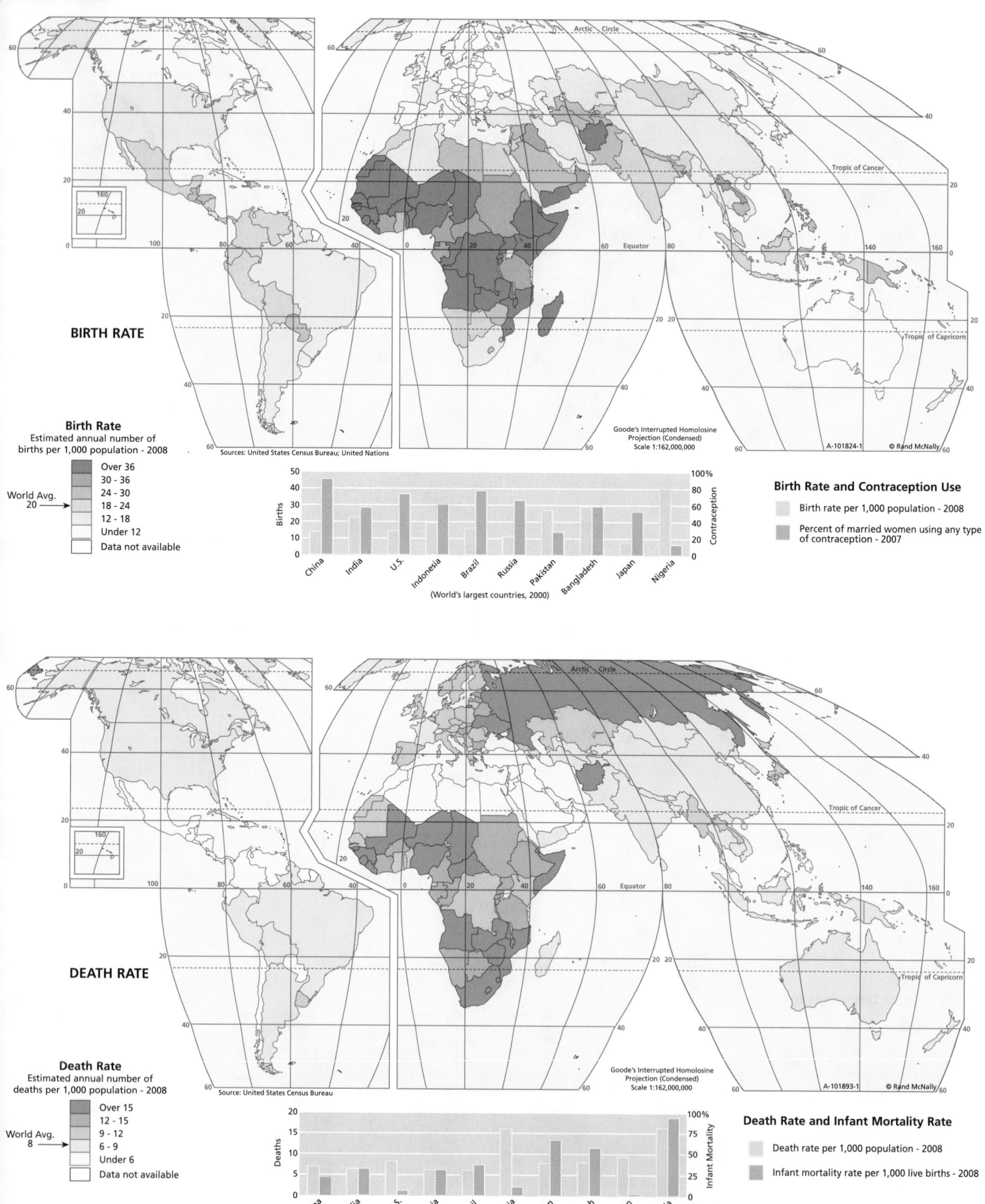

Birth Rate
Estimated annual number of births per 1,000 population - 2008

- Over 36
- 30 - 36
- 24 - 30
- 18 - 24 (World Avg. 20)
- 12 - 18
- Under 12
- Data not available

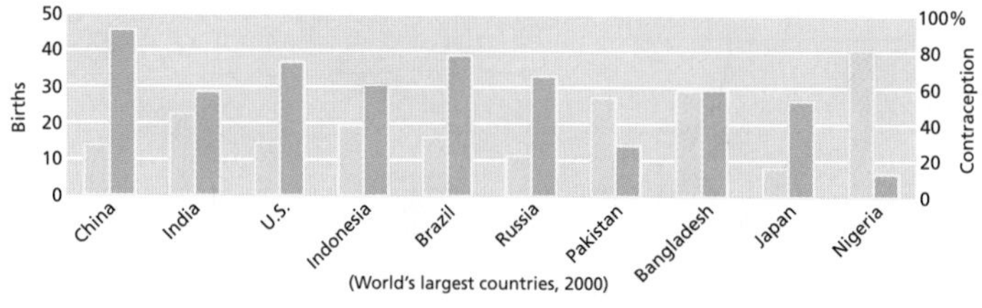
Births
Contraception
China
India
U.S.
Indonesia
Brazil
Russia
Pakistan
Bangladesh
Japan
Nigeria
(World's largest countries, 2000)

Birth Rate and Contraception Use

- Birth rate per 1,000 population - 2008
- Percent of married women using any type of contraception - 2007

DEATH RATE

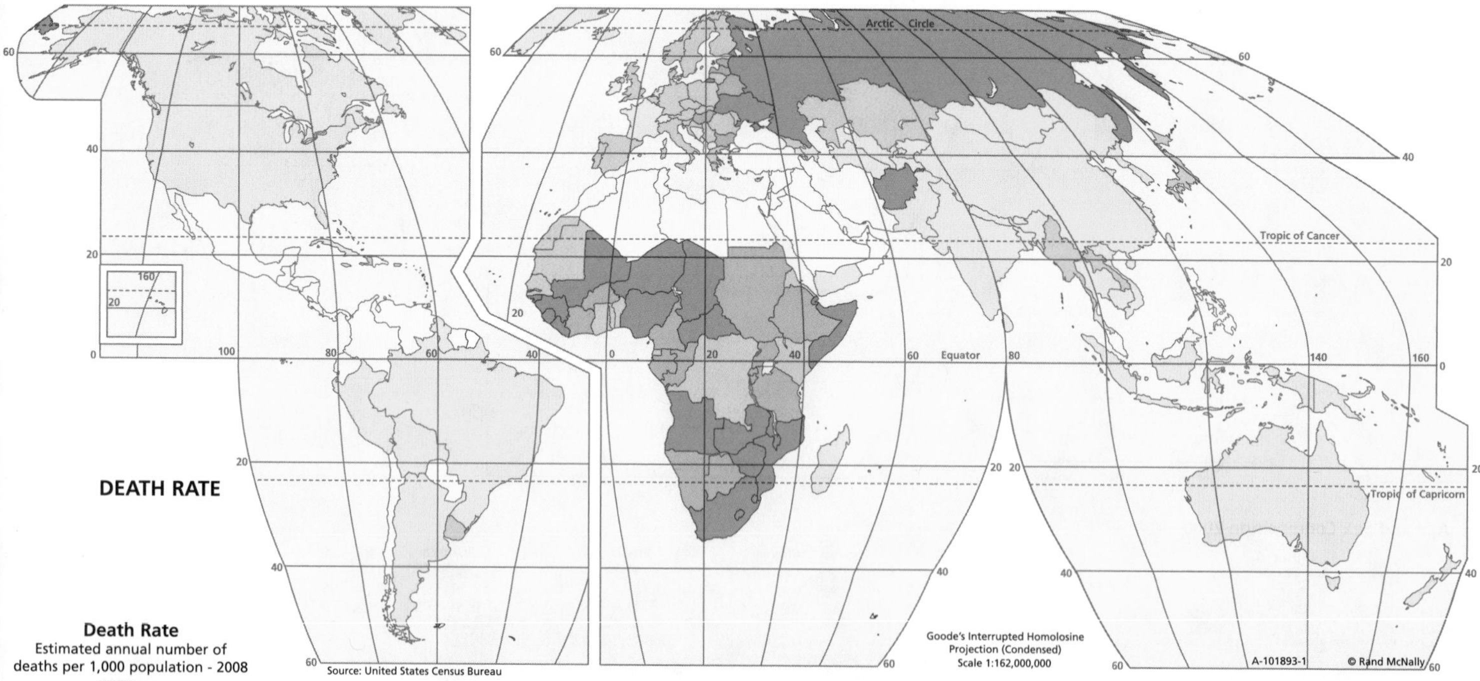
Arctic Circle
Tropic of Cancer
Equator
Tropic of Capricorn
Goode's Interrupted Homolosine Projection (Condensed)
Scale 1:162,000,000
A-101893-1
© Rand McNally
Source: United States Census Bureau

Death Rate
Estimated annual number of deaths per 1,000 population - 2008

- Over 15
- 12 - 15
- 9 - 12
- 6 - 9 (World Avg. 8)
- Under 6
- Data not available

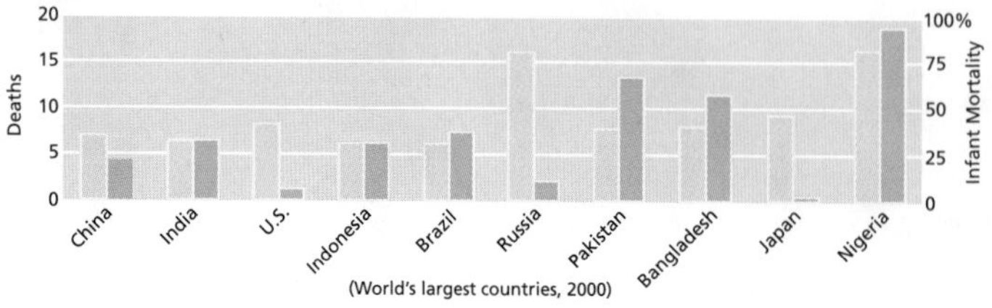
Deaths
Infant Mortality
China
India
U.S.
Indonesia
Brazil
Russia
Pakistan
Bangladesh
Japan
Nigeria
(World's largest countries, 2000)

Death Rate and Infant Mortality Rate

- Death rate per 1,000 population - 2008
- Infant mortality rate per 1,000 live births - 2008

NATURAL INCREASE

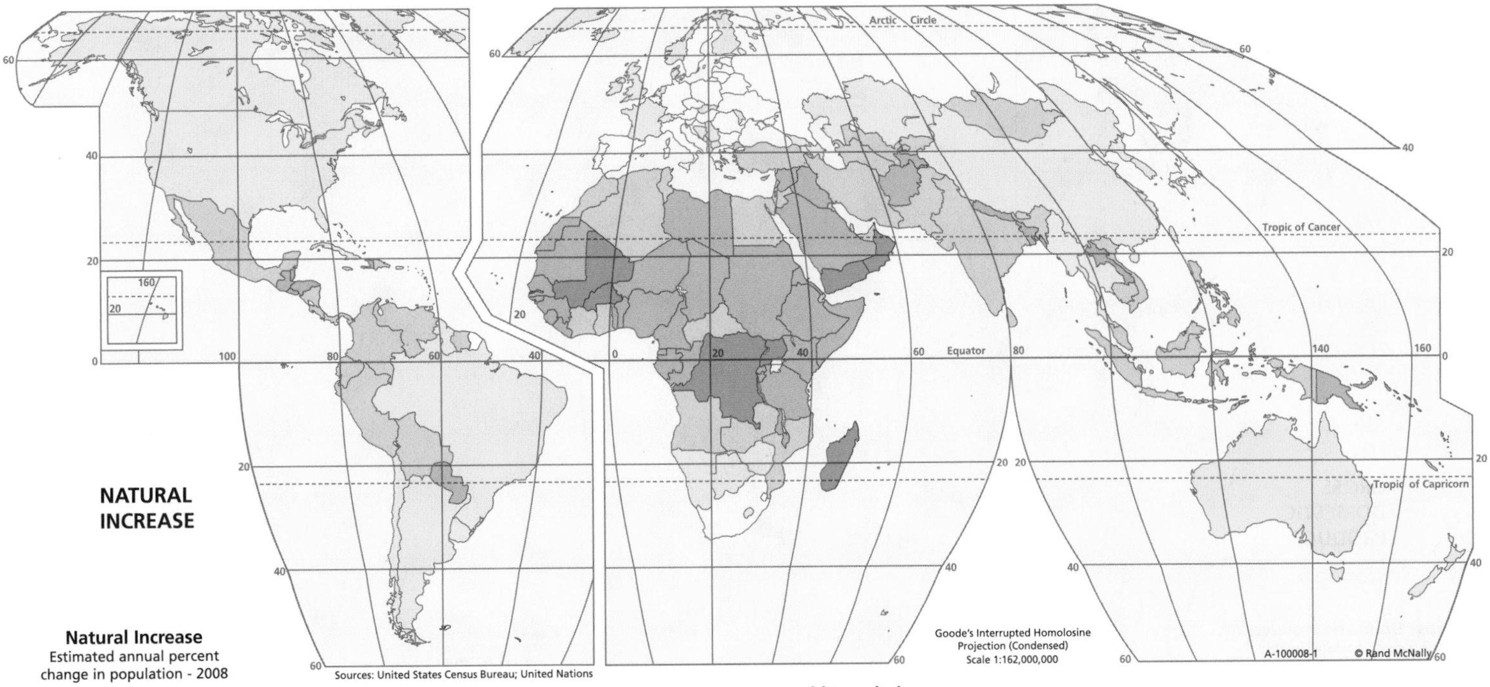
Arctic Circle
Tropic of Cancer
Equator
Tropic of Capricorn
Goode's Interrupted Homolosine Projection (Condensed)
Scale 1:162,000,000
A-100008-1
© Rand McNally

Sources: United States Census Bureau; United Nations

Natural Increase

Estimated annual percent change in population - 2008

- Over 3%
- 2 - 3%
- 1 - 2% (World Avg. 1.2%)
- 0 - 1%
- 0 - 1% decrease
- Data not available

World Population

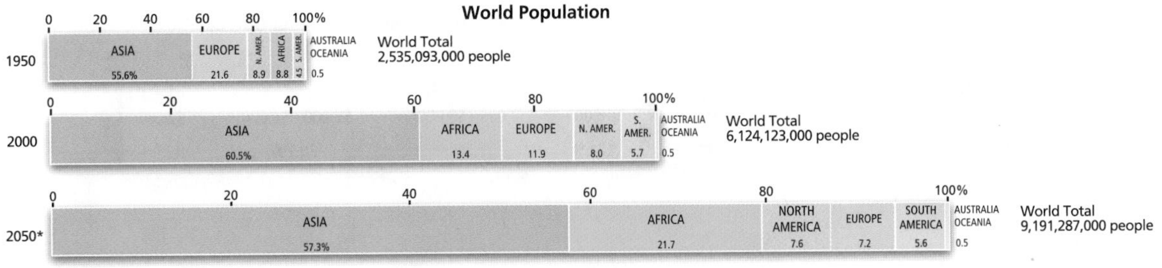
1950
ASIA 55.6%
EUROPE 21.6
N. AMER. 8.9
AFRICA 8.8
S. AMER. 4.5
AUSTRALIA OCEANIA 0.5
World Total 2,535,093,000 people
2000
ASIA 60.5%
AFRICA 13.4
EUROPE 11.9
N. AMER. 8.0
S. AMER. 5.7
AUSTRALIA OCEANIA 0.5
World Total 6,124,123,000 people
2050*
ASIA 57.3%
AFRICA 21.7
NORTH AMERICA 7.6
EUROPE 7.2
SOUTH AMERICA 5.6
AUSTRALIA OCEANIA 0.5
World Total 9,191,287,000 people

* Projected

URBANIZED POPULATION

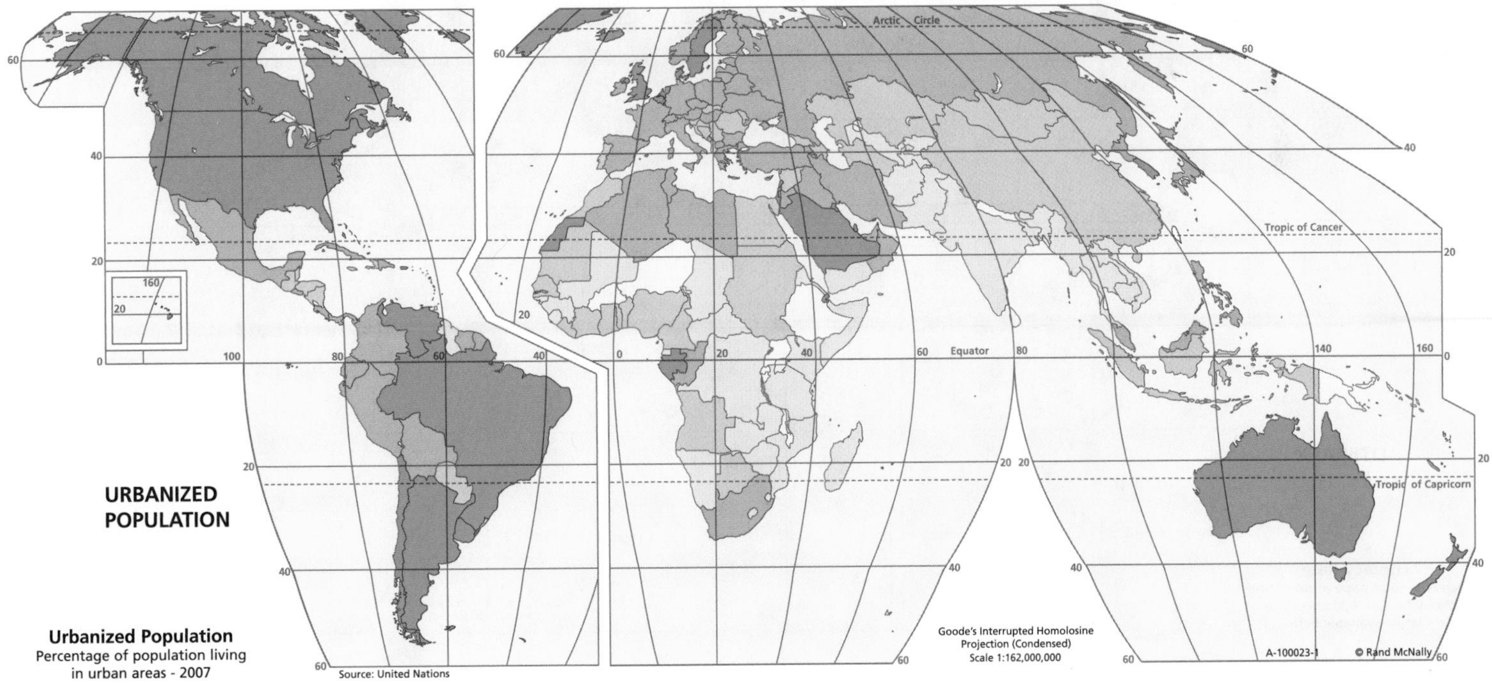
Arctic Circle
Tropic of Cancer
Equator
Tropic of Capricorn
Goode's Interrupted Homolosine Projection (Condensed)
Scale 1:162,000,000
A-100023-1
© Rand McNally

Source: United Nations

Urbanized Population

Percentage of population living in urban areas - 2007

- Over 80%
- 60% - 80%
- 40% - 60% (World Avg. 49.4%)
- 20% - 40%
- Under 20%
- Data not available

Percentage of Population in each Urbanized Area Category - 2007

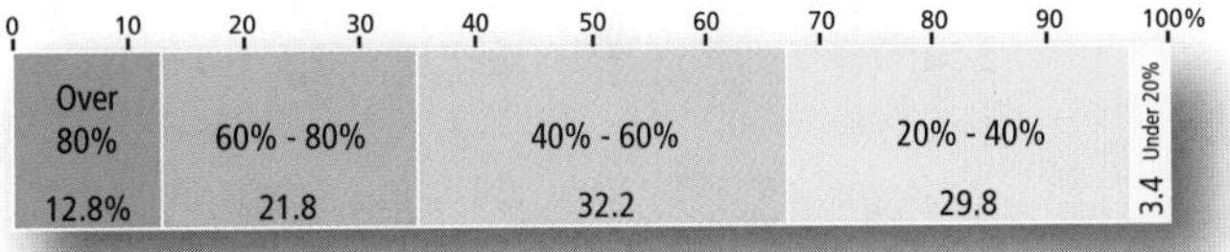
Over 80% 12.8%
60% - 80% 21.8
40% - 60% 32.2
20% - 40% 29.8
Under 20% 3.4

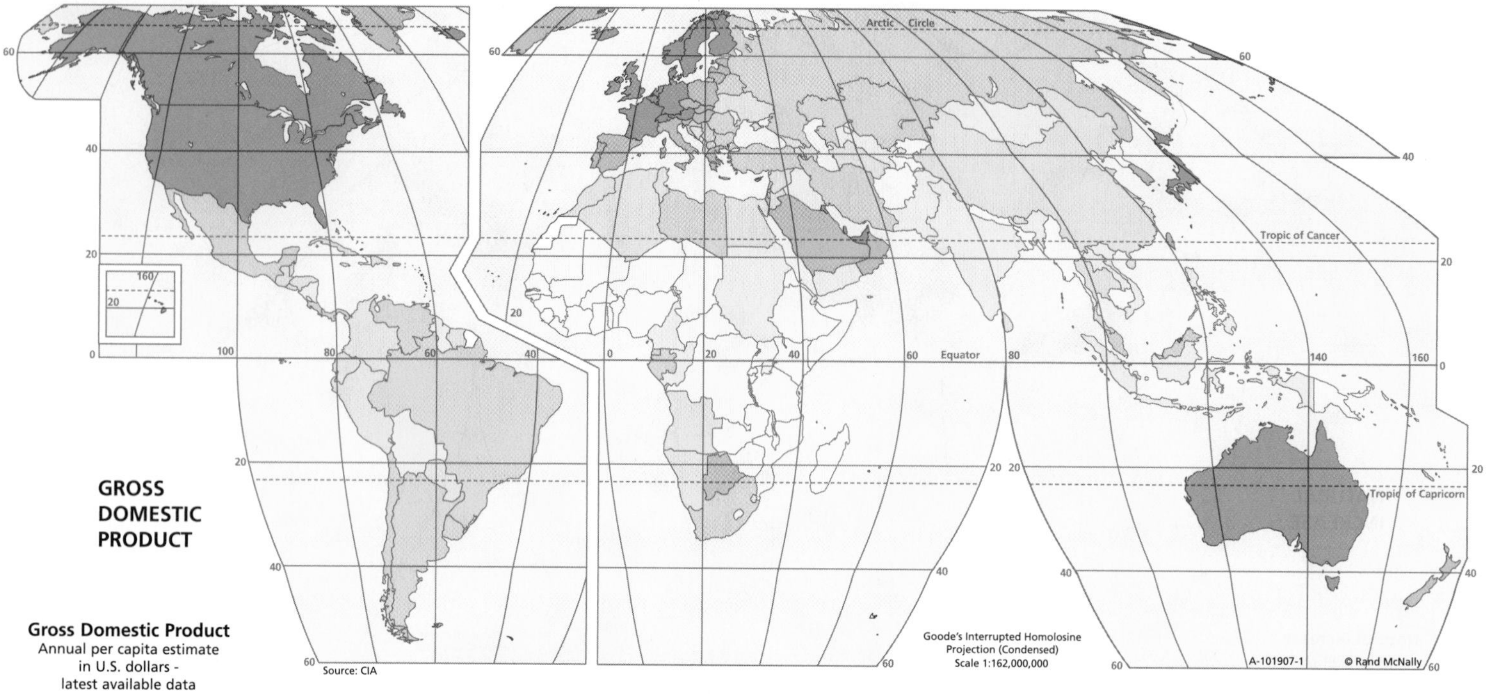
Arctic Circle
Tropic of Cancer
Equator
Tropic of Capricorn
Source: CIA
Goode's Interrupted Homolosine
Projection (Condensed)
Scale 1:162,000,000
A-101907-1
© Rand McNally

GROSS DOMESTIC PRODUCT

Gross Domestic Product
Annual per capita estimate in U.S. dollars - latest available data

- Over $32,000
- $16,000 - $32,000
- $8,000 - $16,000 (World Avg. $10,000)
- $4,000 - $8,000
- $2,000 - $4,000
- Under $2,000
- Data not available

Percentage of World Population in each Per Capita GDP Category

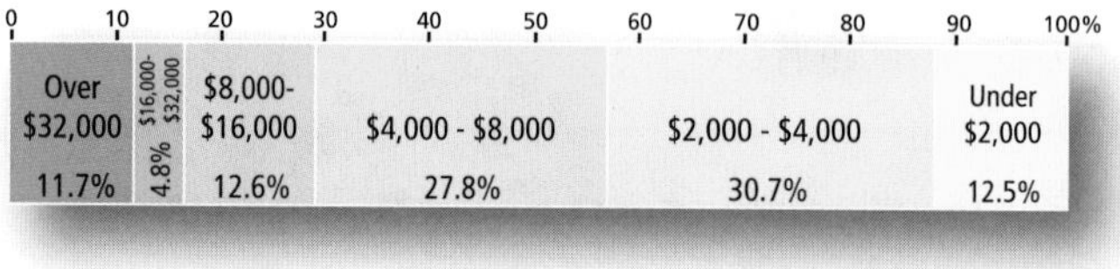
0
10
20
30
40
50
60
70
80
90
100%
Over $32,000
11.7%
$16,000-$32,000
4.8%
$8,000-$16,000
12.6%
$4,000 - $8,000
27.8%
$2,000 - $4,000
30.7%
Under $2,000
12.5%

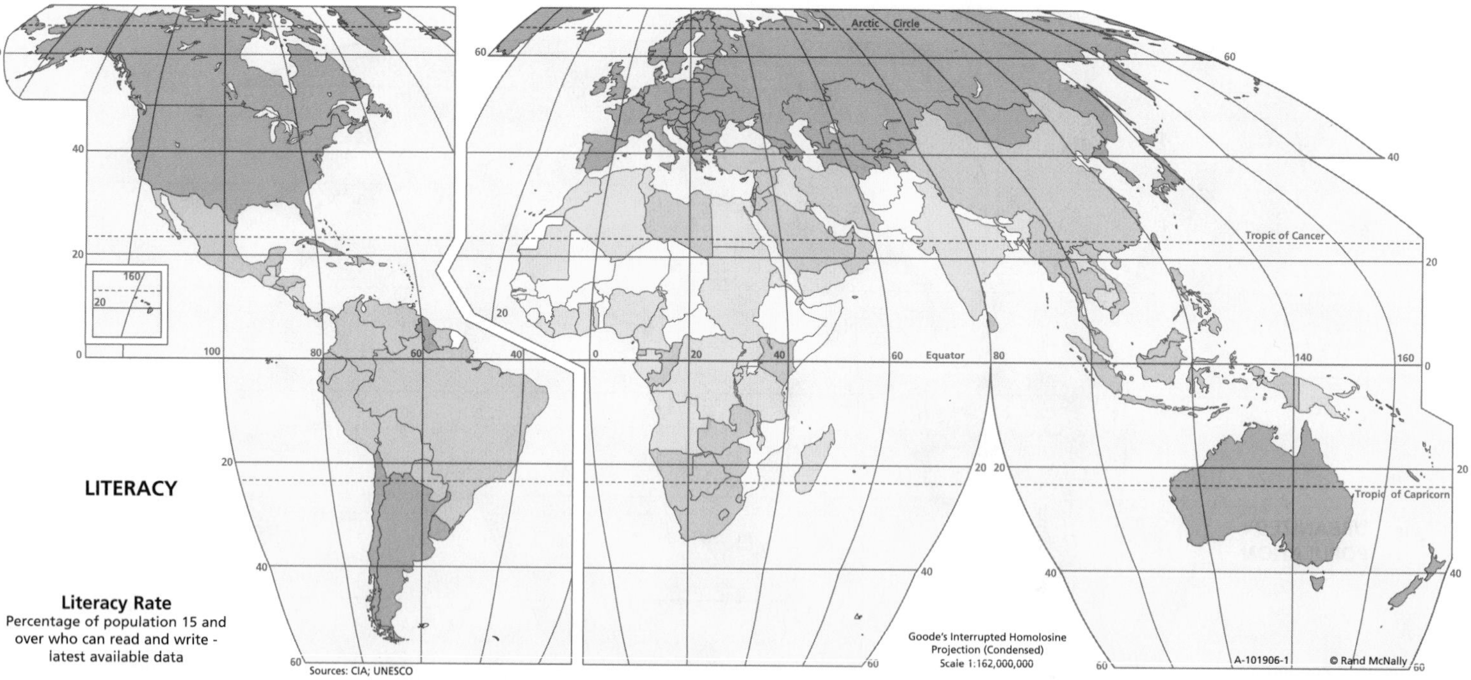
Arctic Circle
Tropic of Cancer
Equator
Tropic of Capricorn
Sources: CIA; UNESCO
Goode's Interrupted Homolosine
Projection (Condensed)
Scale 1:162,000,000
A-101906-1
© Rand McNally

LITERACY

Literacy Rate
Percentage of population 15 and over who can read and write - latest available data

- Over 95%
- 75 - 95% (World Avg. 82%)
- 50 - 75%
- Under 50%
- Data not available

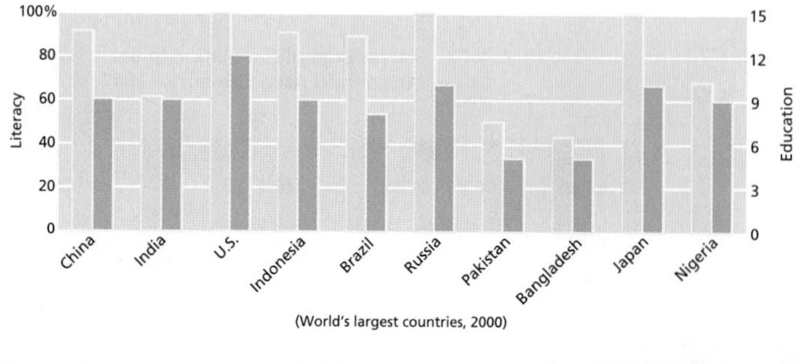
100%
80
60
40
20
0
Literacy
15
12
9
6
3
0
Education
China
India
U.S.
Indonesia
Brazil
Russia
Pakistan
Bangladesh
Japan
Nigeria
(World's largest countries, 2000)

Literacy and Compulsory Education

- Literacy rate
- Years of compulsory education

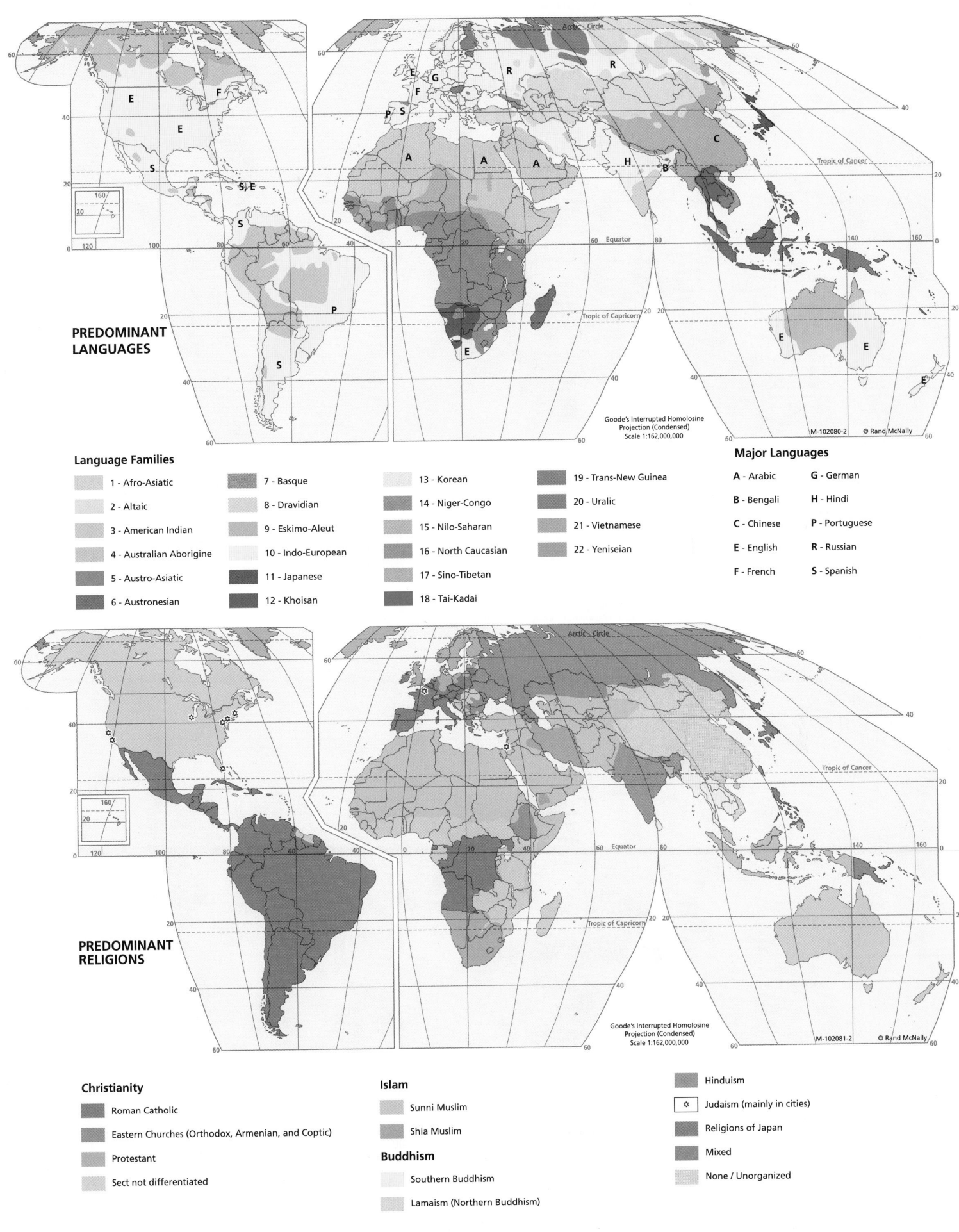

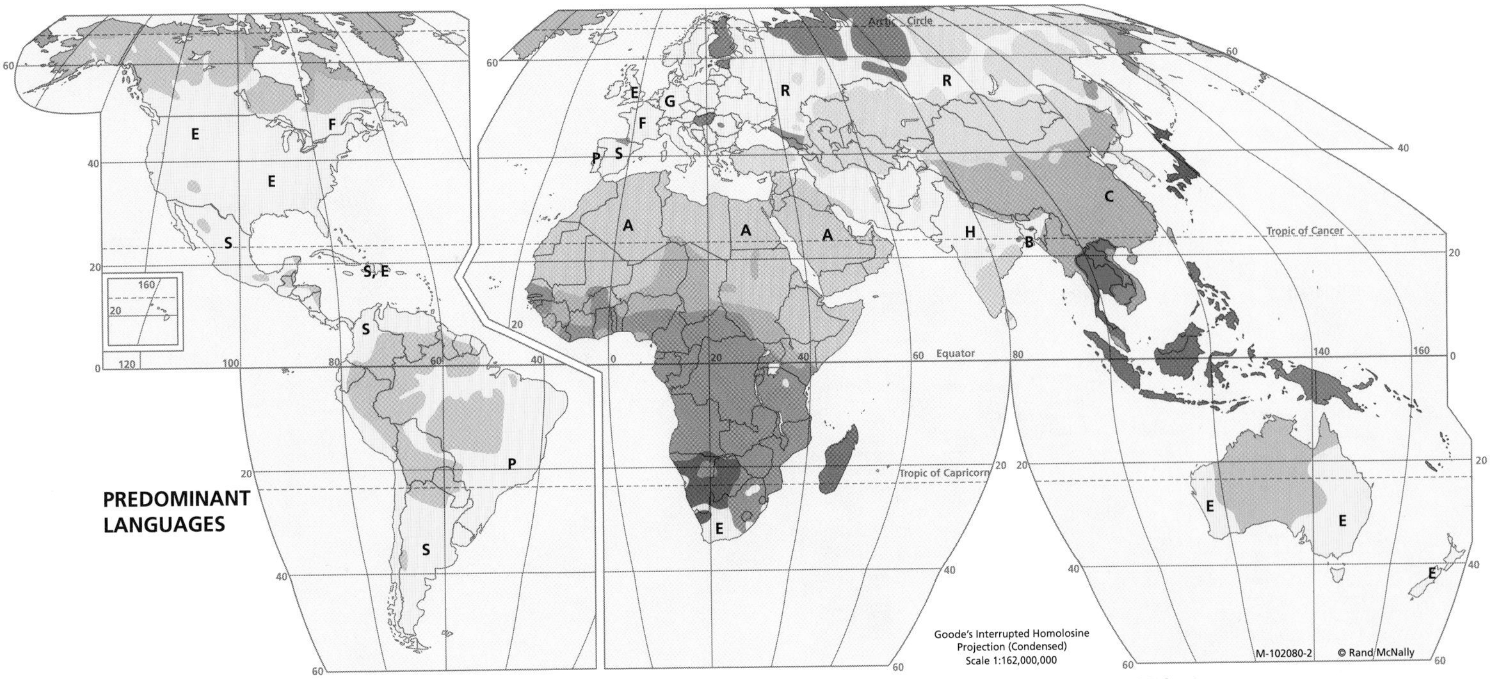
PREDOMINANT LANGUAGES
Arctic Circle
Tropic of Cancer
Equator
Tropic of Capricorn
E
F
E
S
S, E
S
P
S
E
G
F
P
S
R
R
A
A
A
H
B
C
E
E
E
E
Goode's Interrupted Homolosine Projection (Condensed) Scale 1:162,000,000
M-102080-2
© Rand McNally

Language Families

- 1 - Afro-Asiatic
- 2 - Altaic
- 3 - American Indian
- 4 - Australian Aborigine
- 5 - Austro-Asiatic
- 6 - Austronesian
- 7 - Basque
- 8 - Dravidian
- 9 - Eskimo-Aleut
- 10 - Indo-European
- 11 - Japanese
- 12 - Khoisan
- 13 - Korean
- 14 - Niger-Congo
- 15 - Nilo-Saharan
- 16 - North Caucasian
- 17 - Sino-Tibetan
- 18 - Tai-Kadai
- 19 - Trans-New Guinea
- 20 - Uralic
- 21 - Vietnamese
- 22 - Yeniseian

Major Languages

- A - Arabic
- B - Bengali
- C - Chinese
- E - English
- F - French
- G - German
- H - Hindi
- P - Portuguese
- R - Russian
- S - Spanish

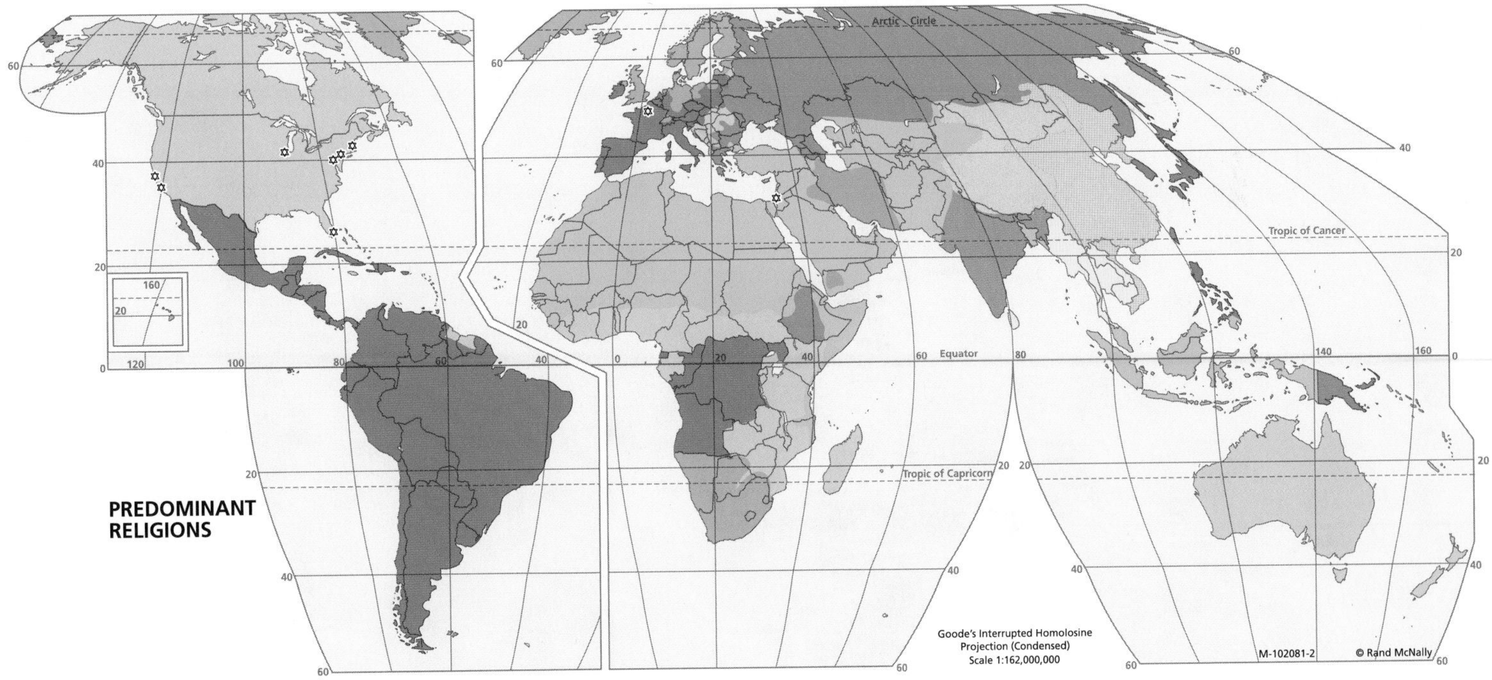
PREDOMINANT RELIGIONS
Arctic Circle
Tropic of Cancer
Equator
Tropic of Capricorn
Goode's Interrupted Homolosine Projection (Condensed) Scale 1:162,000,000
M-102081-2
© Rand McNally

Christianity

- Roman Catholic
- Eastern Churches (Orthodox, Armenian, and Coptic)
- Protestant
- Sect not differentiated

Islam

- Sunni Muslim
- Shia Muslim

Buddhism

- Southern Buddhism
- Lamaism (Northern Buddhism)

- Hinduism
- ✡ Judaism (mainly in cities)
- Religions of Japan
- Mixed
- None / Unorganized

HIV INFECTION

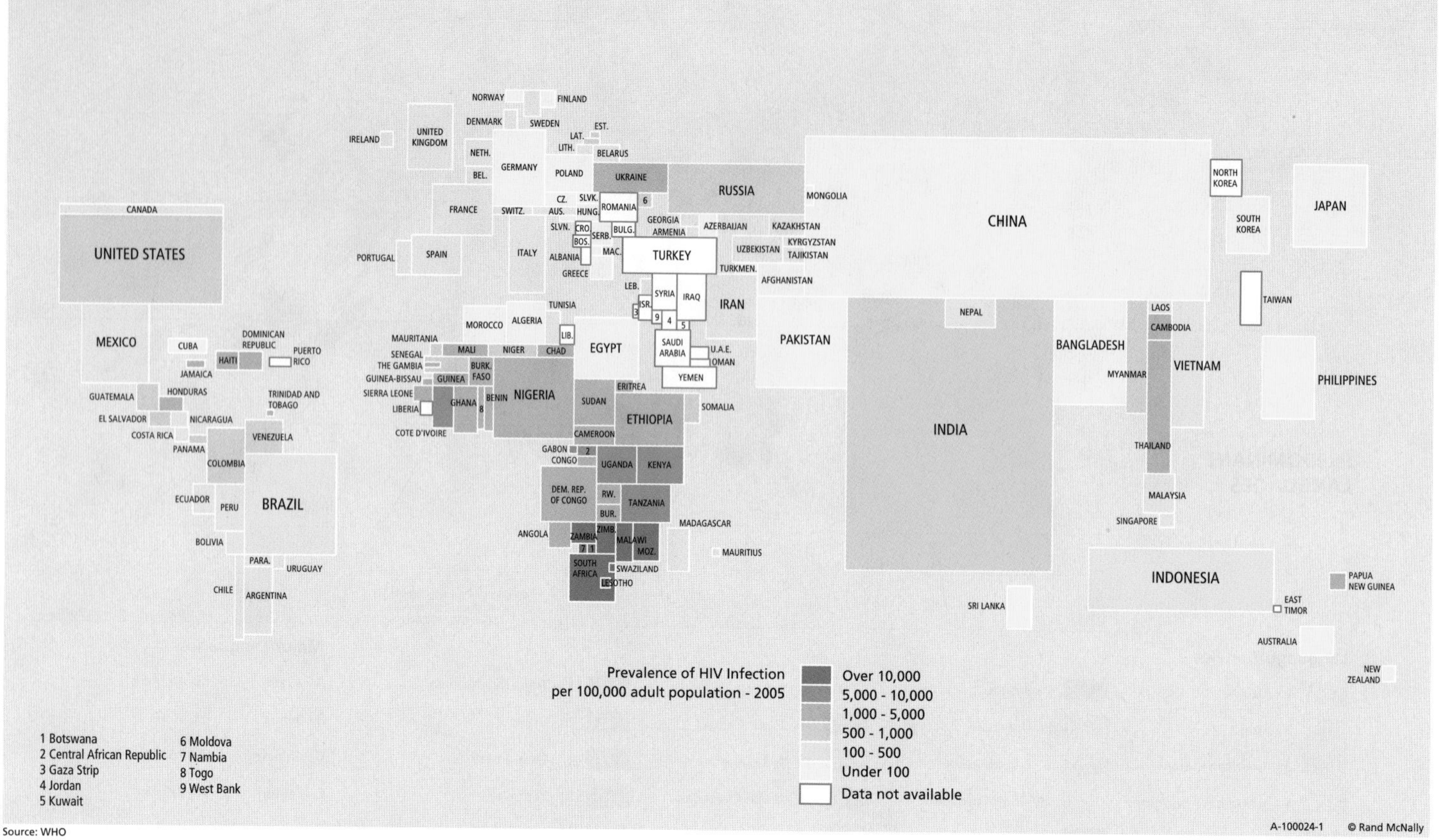
NORWAY
FINLAND
DENMARK
SWEDEN
EST.
LAT.
LITH.
BELARUS
IRELAND
UNITED KINGDOM
NETH.
GERMANY
POLAND
BEL.
UKRAINE
RUSSIA
MONGOLIA
CHINA
NORTH KOREA
JAPAN
SOUTH KOREA
CANADA
UNITED STATES
FRANCE
SWITZ.
CZ.
SLVK.
AUS.
HUNG.
ROMANIA
6
SLVN.
CRO.
SERB.
BOS.
BULG.
GEORGIA
ARMENIA
AZERBAIJAN
KAZAKHSTAN
UZBEKISTAN
KYRGYZSTAN
TAJIKISTAN
PORTUGAL
SPAIN
ITALY
ALBANIA
MAC.
TURKEY
TURKMEN.
AFGHANISTAN
GREECE
LEB.
SYRIA
IRAQ
IRAN
ISR.
3
9
4
5
TAIWAN
TUNISIA
MOROCCO
ALGERIA
LIB.
EGYPT
SAUDI ARABIA
U.A.E.
OMAN
YEMEN
PAKISTAN
NEPAL
BANGLADESH
LAOS
CAMBODIA
VIETNAM
PHILIPPINES
MYANMAR
INDIA
THAILAND
MALAYSIA
SINGAPORE
MEXICO
CUBA
DOMINICAN REPUBLIC
PUERTO RICO
HAITI
JAMAICA
GUATEMALA
HONDURAS
TRINIDAD AND TOBAGO
EL SALVADOR
NICARAGUA
COSTA RICA
PANAMA
VENEZUELA
COLOMBIA
ECUADOR
PERU
BRAZIL
BOLIVIA
PARA.
URUGUAY
CHILE
ARGENTINA
MAURITANIA
SENEGAL
THE GAMBIA
GUINEA-BISSAU
SIERRA LEONE
LIBERIA
COTE D'IVOIRE
MALI
NIGER
CHAD
BURK. FASO
GUINEA
GHANA
8
BENIN
NIGERIA
SUDAN
ERITREA
ETHIOPIA
SOMALIA
CAMEROON
GABON
2
CONGO
UGANDA
KENYA
DEM. REP. OF CONGO
RW.
BUR.
TANZANIA
ANGOLA
ZAMBIA
ZIMB.
MALAWI
MOZ.
7
1
SOUTH AFRICA
SWAZILAND
LESOTHO
MADAGASCAR
MAURITIUS
SRI LANKA
INDONESIA
PAPUA NEW GUINEA
EAST TIMOR
AUSTRALIA
NEW ZEALAND
Prevalence of HIV Infection per 100,000 adult population - 2005
Over 10,000
5,000 - 10,000
1,000 - 5,000
500 - 1,000
100 - 500
Under 100
Data not available
1 Botswana
2 Central African Republic
3 Gaza Strip
4 Jordan
5 Kuwait
6 Moldova
7 Nambia
8 Togo
9 West Bank
A-100024-1
© Rand McNally

Source: WHO

Size of each country is proportional to its population

Countries with populations under 1,000,000 are not shown.

☐ = 25,000,000 people

TUBERCULOSIS

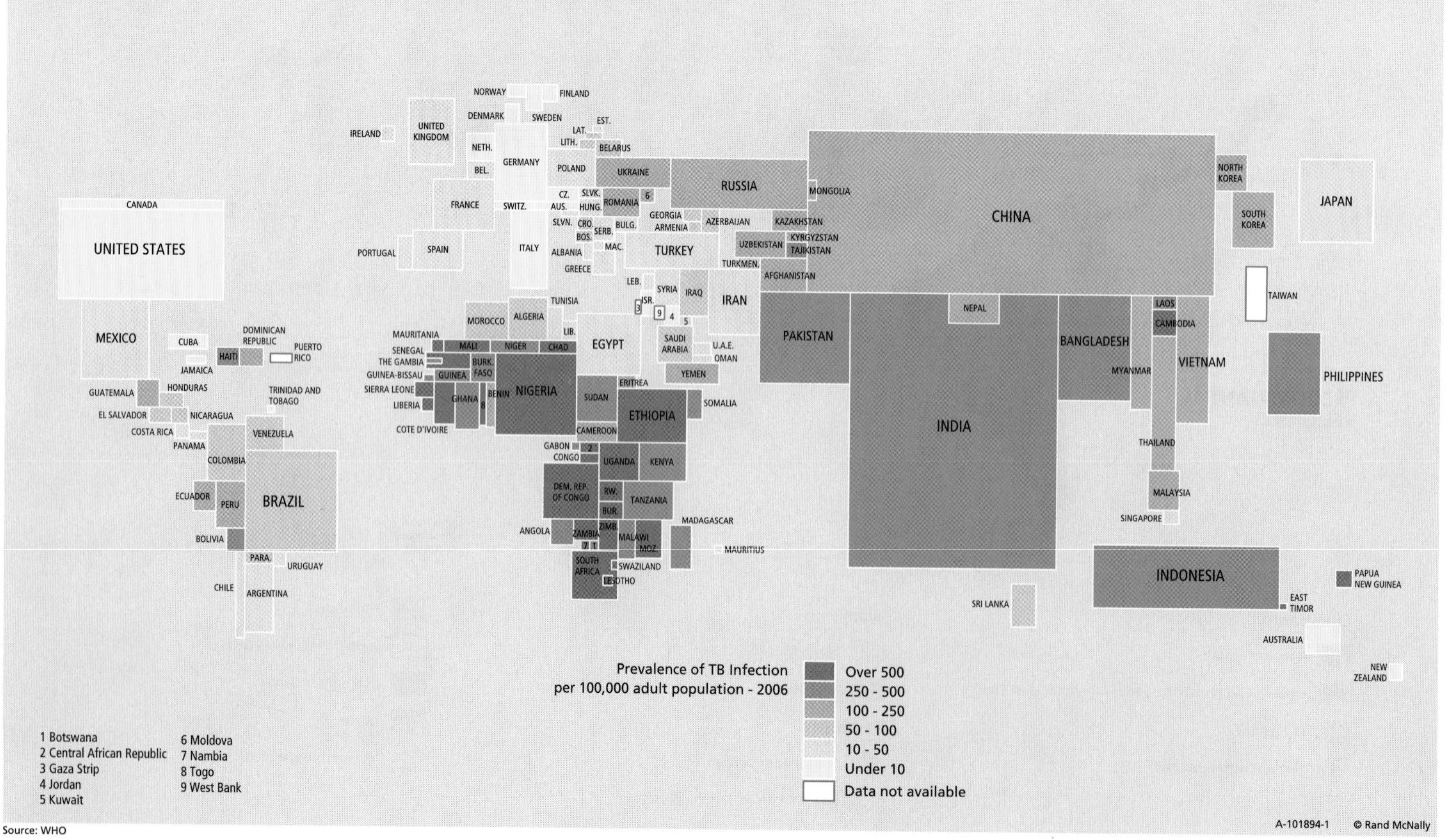
NORWAY
FINLAND
DENMARK
SWEDEN
EST.
LAT.
LITH.
BELARUS
IRELAND
UNITED KINGDOM
NETH.
GERMANY
POLAND
BEL.
UKRAINE
RUSSIA
MONGOLIA
CHINA
NORTH KOREA
JAPAN
SOUTH KOREA
CANADA
UNITED STATES
FRANCE
SWITZ.
CZ.
SLVK.
AUS.
HUNG.
ROMANIA
6
SLVN.
CRO.
SERB.
BOS.
BULG.
GEORGIA
ARMENIA
AZERBAIJAN
KAZAKHSTAN
UZBEKISTAN
KYRGYZSTAN
TAJIKISTAN
PORTUGAL
SPAIN
ITALY
ALBANIA
MAC.
TURKEY
TURKMEN.
AFGHANISTAN
GREECE
LEB.
SYRIA
IRAQ
IRAN
ISR.
3
9
4
5
TAIWAN
TUNISIA
MOROCCO
ALGERIA
LIB.
EGYPT
SAUDI ARABIA
U.A.E.
OMAN
YEMEN
PAKISTAN
NEPAL
BANGLADESH
LAOS
CAMBODIA
VIETNAM
PHILIPPINES
MYANMAR
INDIA
THAILAND
MALAYSIA
SINGAPORE
MEXICO
CUBA
DOMINICAN REPUBLIC
PUERTO RICO
HAITI
JAMAICA
GUATEMALA
HONDURAS
TRINIDAD AND TOBAGO
EL SALVADOR
NICARAGUA
COSTA RICA
PANAMA
VENEZUELA
COLOMBIA
ECUADOR
PERU
BRAZIL
BOLIVIA
PARA.
URUGUAY
CHILE
ARGENTINA
MAURITANIA
SENEGAL
THE GAMBIA
GUINEA-BISSAU
SIERRA LEONE
LIBERIA
COTE D'IVOIRE
MALI
NIGER
CHAD
BURK. FASO
GUINEA
GHANA
8
BENIN
NIGERIA
SUDAN
ERITREA
ETHIOPIA
SOMALIA
CAMEROON
GABON
2
CONGO
UGANDA
KENYA
DEM. REP. OF CONGO
RW.
BUR.
TANZANIA
ANGOLA
ZAMBIA
ZIMB.
MALAWI
MOZ.
7
1
SOUTH AFRICA
SWAZILAND
LESOTHO
MADAGASCAR
MAURITIUS
SRI LANKA
INDONESIA
PAPUA NEW GUINEA
EAST TIMOR
AUSTRALIA
NEW ZEALAND
Prevalence of TB Infection per 100,000 adult population - 2006
Over 500
250 - 500
100 - 250
50 - 100
10 - 50
Under 10
Data not available
1 Botswana
2 Central African Republic
3 Gaza Strip
4 Jordan
5 Kuwait
6 Moldova
7 Nambia
8 Togo
9 West Bank
A-101894-1
© Rand McNally

Source: WHO

MALARIA

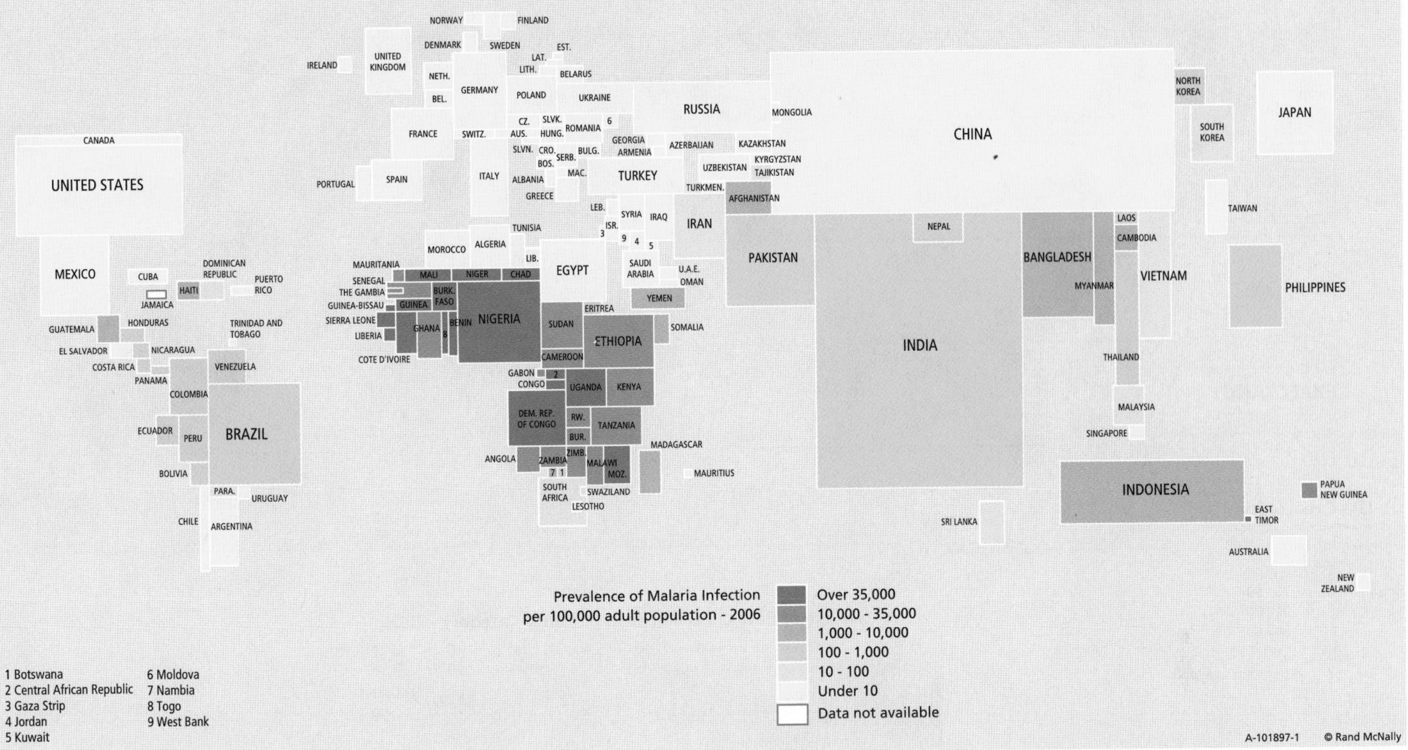
NORWAY
FINLAND
DENMARK
SWEDEN
EST.
LAT.
LITH.
IRELAND
UNITED KINGDOM
NETH.
BELARUS
BEL.
GERMANY
POLAND
UKRAINE
RUSSIA
MONGOLIA
CHINA
NORTH KOREA
SOUTH KOREA
JAPAN
CANADA
UNITED STATES
FRANCE
SWITZ.
AUS.
CZ.
SLVK.
HUNG.
ROMANIA
6
SLVN.
CRO.
SERB.
BOS.
BULG.
MAC.
GEORGIA
ARMENIA
AZERBAIJAN
KAZAKHSTAN
KYRGYZSTAN
UZBEKISTAN
TAJIKISTAN
PORTUGAL
SPAIN
ITALY
ALBANIA
GREECE
TURKEY
TURKMEN.
AFGHANISTAN
TAIWAN
LEB.
SYRIA
ISR.
IRAQ
IRAN
NEPAL
LAOS
CAMBODIA
TUNISIA
MOROCCO
ALGERIA
LIB.
EGYPT
SAUDI ARABIA
U.A.E.
OMAN
YEMEN
PAKISTAN
BANGLADESH
MYANMAR
VIETNAM
PHILIPPINES
MEXICO
CUBA
DOMINICAN REPUBLIC
PUERTO RICO
HAITI
JAMAICA
MAURITANIA
SENEGAL
THE GAMBIA
GUINEA-BISSAU
SIERRA LEONE
LIBERIA
COTE D'IVOIRE
MALI
NIGER
CHAD
GUINEA
BURK. FASO
GHANA
BENIN
NIGERIA
8
ERITREA
SUDAN
ETHIOPIA
SOMALIA
INDIA
GUATEMALA
HONDURAS
EL SALVADOR
NICARAGUA
COSTA RICA
PANAMA
TRINIDAD AND TOBAGO
VENEZUELA
COLOMBIA
CAMEROON
GABON
CONGO
2
UGANDA
KENYA
THAILAND
MALAYSIA
SINGAPORE
ECUADOR
PERU
BRAZIL
BOLIVIA
DEM. REP. OF CONGO
RW.
BUR.
TANZANIA
ANGOLA
ZAMBIA
ZIMB.
MALAWI
MOZ.
MADAGASCAR
MAURITIUS
SOUTH AFRICA
SWAZILAND
LESOTHO
PARA.
URUGUAY
CHILE
ARGENTINA
INDONESIA
PAPUA NEW GUINEA
EAST TIMOR
SRI LANKA
AUSTRALIA
NEW ZEALAND
Prevalence of Malaria Infection per 100,000 adult population - 2006
Over 35,000
10,000 - 35,000
1,000 - 10,000
100 - 1,000
10 - 100
Under 10
Data not available
1 Botswana
2 Central African Republic
3 Gaza Strip
4 Jordan
5 Kuwait
6 Moldova
7 Nambia
8 Togo
9 West Bank
A-101897-1
© Rand McNally

Source: WHO

The maps on these two pages are called **cartograms**. On these cartograms, the size of each country is proportional to its total population. This means that the countries with the largest areas are those with the largest populations. The shapes of countries must be distorted in order to achieve this proportional representation. Here, each country is shown as a rectangle in order to facilitate size comparisons.

One advantage of these cartograms is that they reveal the relationship between the mapped variable and the affected population. Consider the example of Chad and Nigeria. Both have relatively high rates of HIV infection (between 1,000 and 5,000 cases per 100,000 population). But Nigeria is much larger than Chad on the cartogram, which informs the reader that the population affected by HIV is much larger in Nigeria.

PHYSICIANS

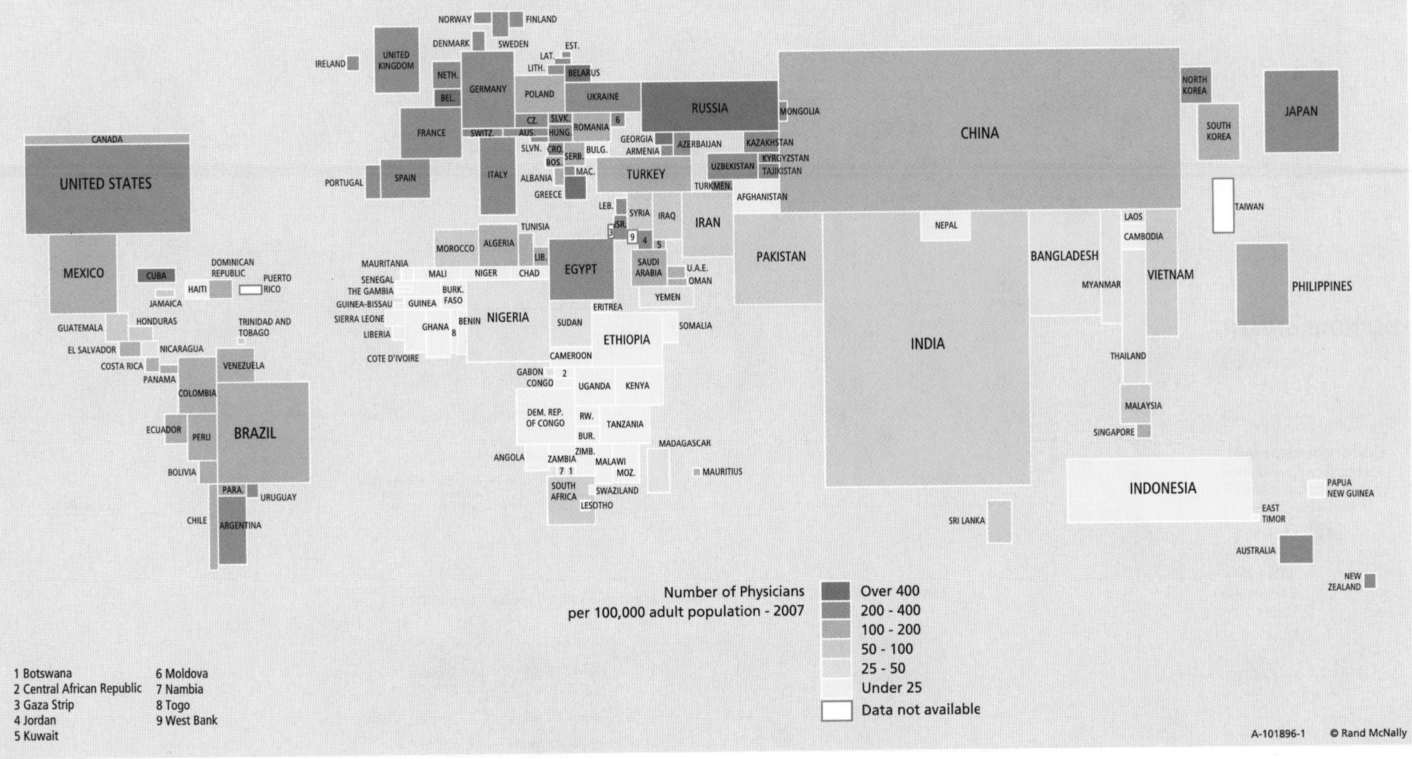
NORWAY
FINLAND
DENMARK
SWEDEN
EST.
LAT.
LITH.
IRELAND
UNITED KINGDOM
NETH.
BELARUS
BEL.
GERMANY
POLAND
UKRAINE
RUSSIA
MONGOLIA
CHINA
NORTH KOREA
SOUTH KOREA
JAPAN
CANADA
UNITED STATES
FRANCE
SWITZ.
AUS.
CZ.
SLVK.
HUNG.
ROMANIA
6
SLVN.
CRO.
SERB.
BOS.
BULG.
MAC.
GEORGIA
ARMENIA
AZERBAIJAN
KAZAKHSTAN
KYRGYZSTAN
UZBEKISTAN
TAJIKISTAN
PORTUGAL
SPAIN
ITALY
ALBANIA
GREECE
TURKEY
TURKMEN.
AFGHANISTAN
TAIWAN
LEB.
SYRIA
ISR.
IRAQ
IRAN
NEPAL
LAOS
CAMBODIA
TUNISIA
MOROCCO
ALGERIA
LIB.
EGYPT
SAUDI ARABIA
U.A.E.
OMAN
YEMEN
PAKISTAN
BANGLADESH
MYANMAR
VIETNAM
PHILIPPINES
MEXICO
CUBA
DOMINICAN REPUBLIC
PUERTO RICO
HAITI
JAMAICA
MAURITANIA
SENEGAL
THE GAMBIA
GUINEA-BISSAU
SIERRA LEONE
LIBERIA
COTE D'IVOIRE
MALI
NIGER
CHAD
GUINEA
BURK. FASO
GHANA
BENIN
NIGERIA
8
ERITREA
SUDAN
ETHIOPIA
SOMALIA
INDIA
GUATEMALA
HONDURAS
EL SALVADOR
NICARAGUA
COSTA RICA
PANAMA
TRINIDAD AND TOBAGO
VENEZUELA
COLOMBIA
CAMEROON
GABON
CONGO
2
UGANDA
KENYA
THAILAND
MALAYSIA
SINGAPORE
ECUADOR
PERU
BRAZIL
BOLIVIA
DEM. REP. OF CONGO
RW.
BUR.
TANZANIA
ANGOLA
ZAMBIA
ZIMB.
MALAWI
MOZ.
MADAGASCAR
MAURITIUS
SOUTH AFRICA
SWAZILAND
LESOTHO
PARA.
URUGUAY
CHILE
ARGENTINA
INDONESIA
PAPUA NEW GUINEA
EAST TIMOR
SRI LANKA
AUSTRALIA
NEW ZEALAND
Number of Physicians per 100,000 adult population - 2007
Over 400
200 - 400
100 - 200
50 - 100
25 - 50
Under 25
Data not available
1 Botswana
2 Central African Republic
3 Gaza Strip
4 Jordan
5 Kuwait
6 Moldova
7 Nambia
8 Togo
9 West Bank
A-101896-1
© Rand McNally

Source: WHO

LIFE EXPECTANCY

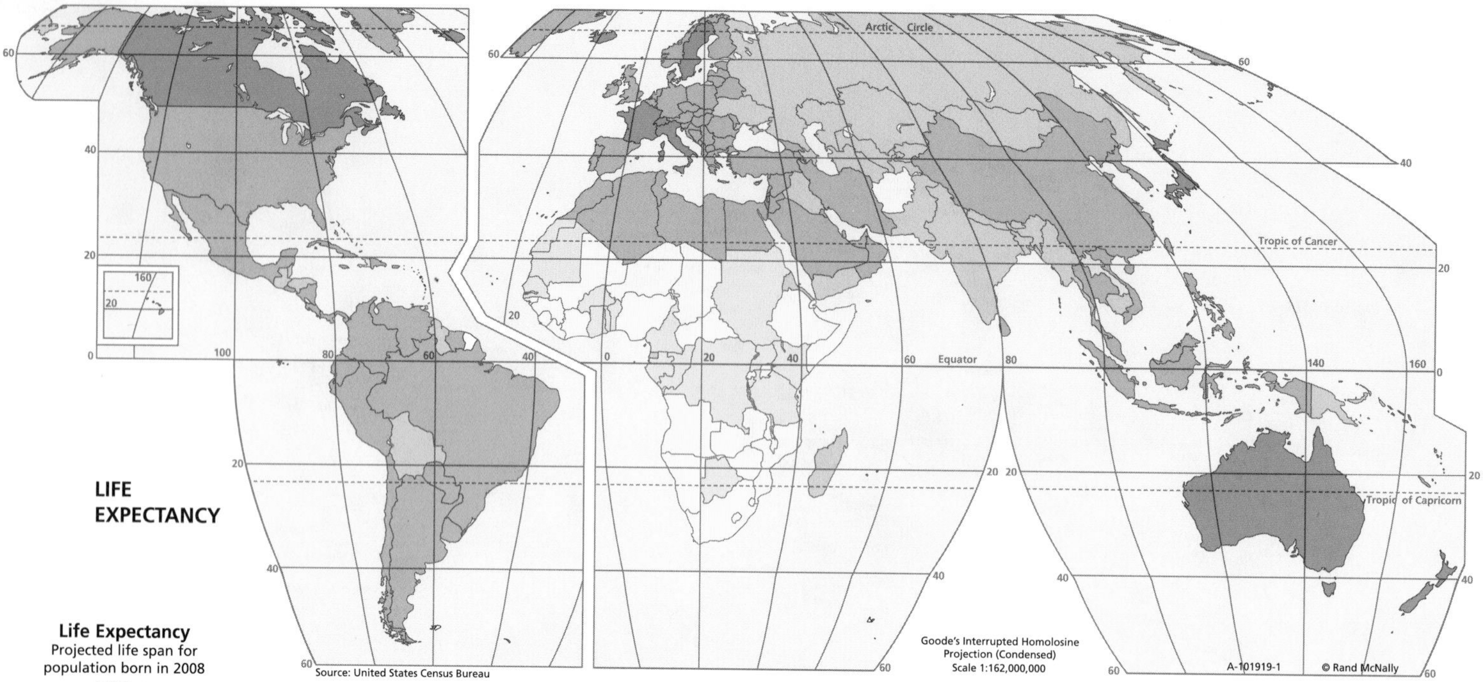

Arctic Circle
Tropic of Cancer
Equator
Tropic of Capricorn
Source: United States Census Bureau
Goode's Interrupted Homolosine Projection (Condensed)
Scale 1:162,000,000
A-101919-1
© Rand McNally

Life Expectancy

Projected life span for population born in 2008

- Over 80
- 70 - 80
- 60 - 70 (World Avg. 66)
- 50 - 60
- Under 50
- Data not available

Percentage of Births in each Life Expectancy Category - 2008

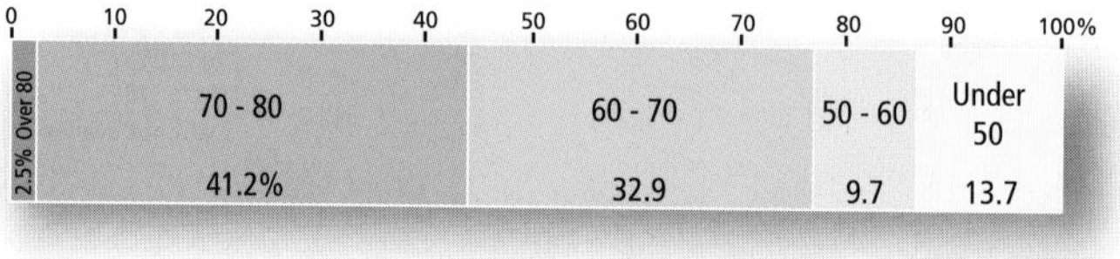

0 10 20 30 40 50 60 70 80 90 100%
2.5% Over 80
70 - 80
41.2%
60 - 70
32.9
50 - 60
9.7
Under 50
13.7

UNDERNOURISHMENT

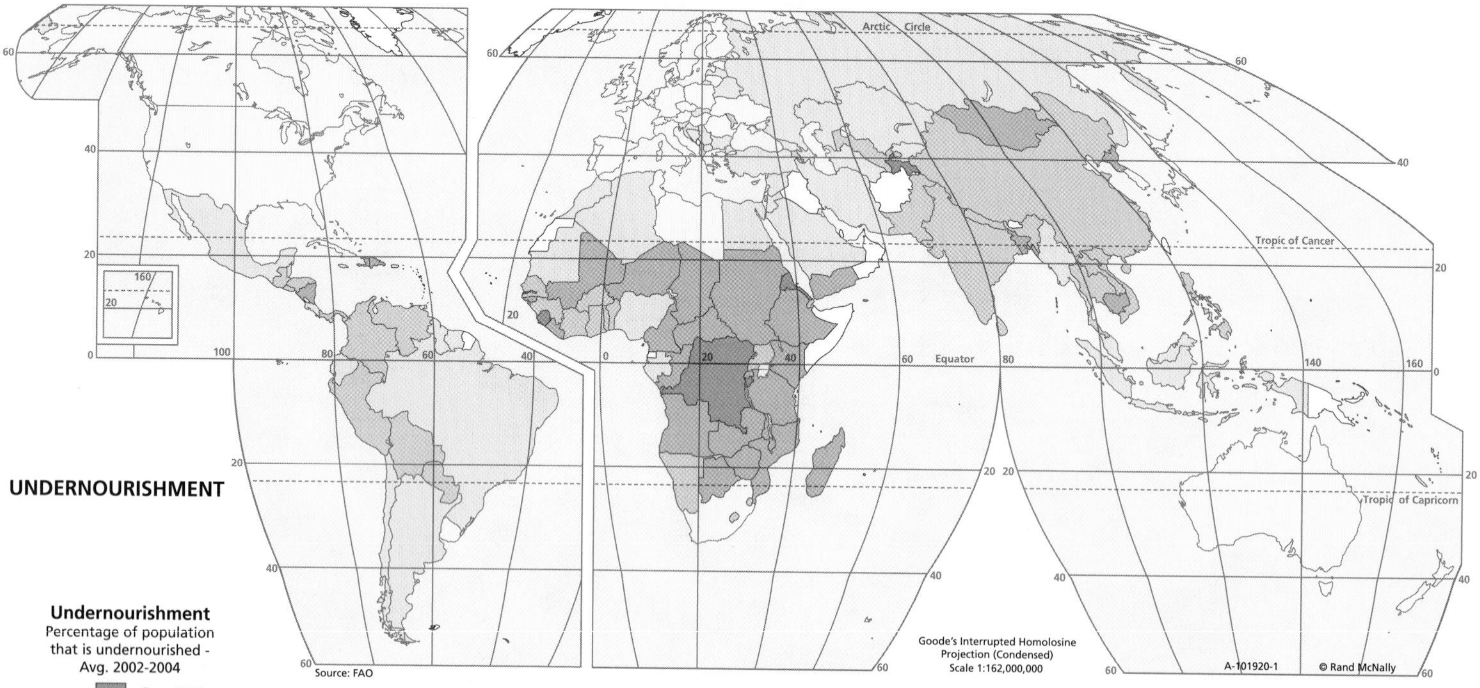

Arctic Circle
Tropic of Cancer
Equator
Tropic of Capricorn
Source: FAO
Goode's Interrupted Homolosine Projection (Condensed)
Scale 1:162,000,000
A-101920-1
© Rand McNally

Undernourishment

Percentage of population that is undernourished - Avg. 2002-2004

- Over 50%
- 25% - 50%
- 10% - 25%
- 2.5% - 10%
- Under 2.5%
- Data not available

Undernourished People World Total* - 825,900,000 people - Avg. 2002-2004

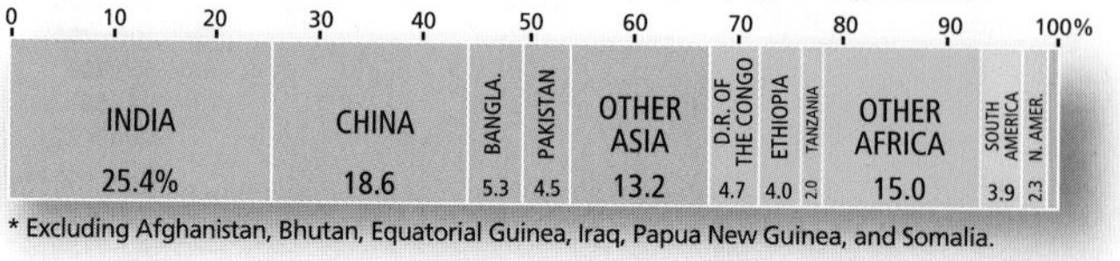

0 10 20 30 40 50 60 70 80 90 100%
INDIA
25.4%
CHINA
18.6
BANGLA.
5.3
PAKISTAN
4.5
OTHER ASIA
13.2
D.R. OF THE CONGO
4.7
ETHIOPIA
4.0
TANZANIA
2.0
OTHER AFRICA
15.0
SOUTH AMERICA
3.9
N. AMER.
2.3

* Excluding Afghanistan, Bhutan, Equatorial Guinea, Iraq, Papua New Guinea, and Somalia.

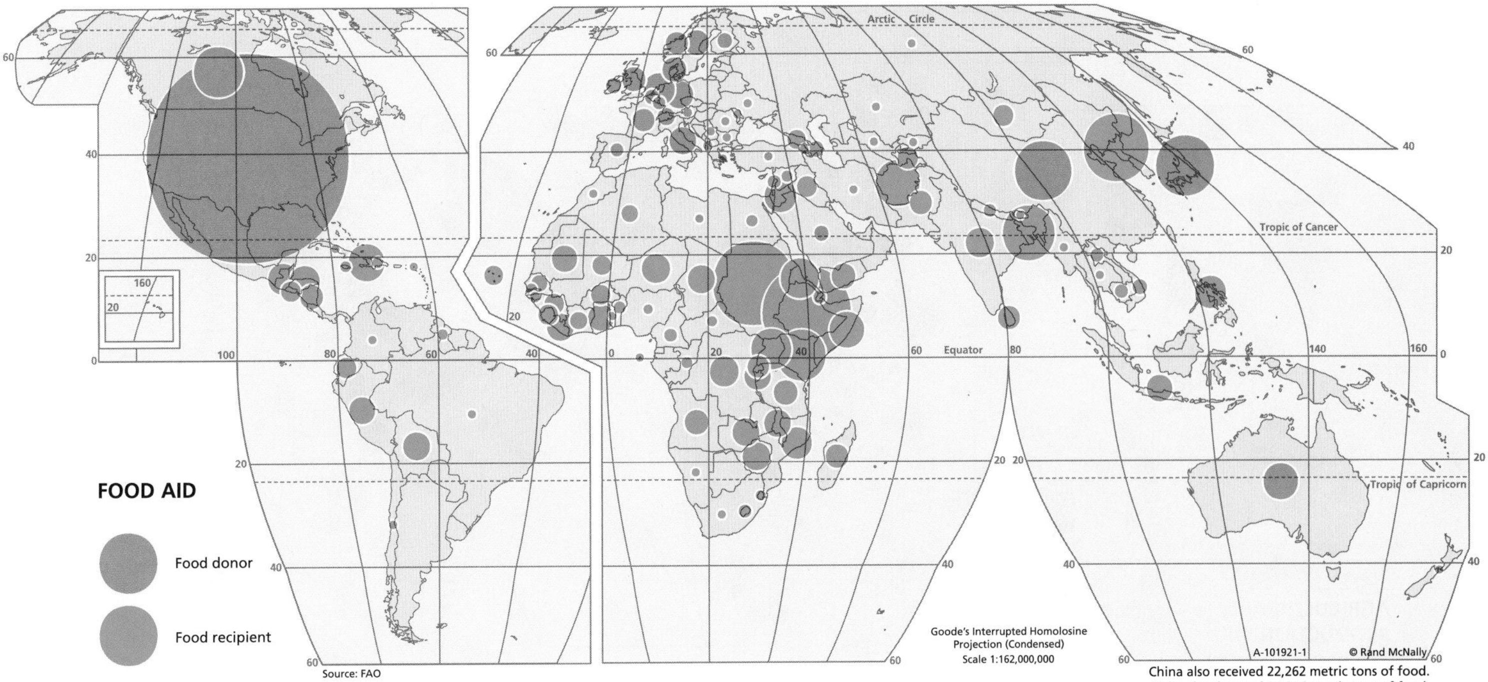
FOOD AID
Food donor
Food recipient
Arctic Circle
Tropic of Cancer
Equator
Tropic of Capricorn
Source: FAO
Goode's Interrupted Homolosine Projection (Condensed)
Scale 1:162,000,000
A-101921-1
© Rand McNally

China also received 22,262 metric tons of food.
India also donated 24,286 metric tons of food.

Amount of Food Donated or Received
in metric tons - Avg. 2004-2006

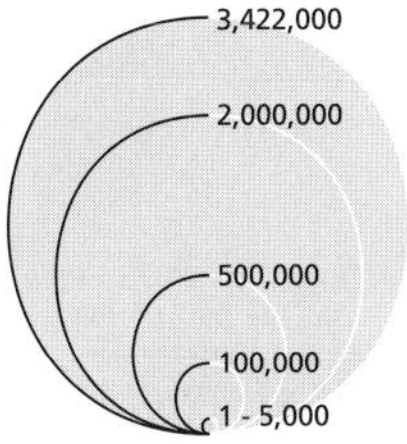
3,422,000
2,000,000
500,000
100,000
1 - 5,000

Aid by Type of Food

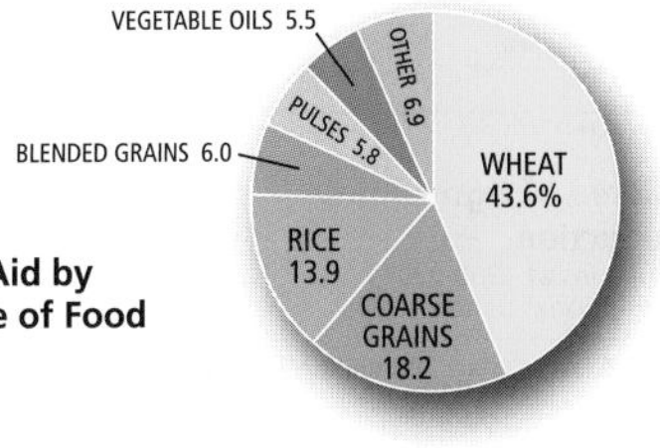
WHEAT 43.6%
COARSE GRAINS 18.2
RICE 13.9
BLENDED GRAINS 6.0
PULSES 5.8
VEGETABLE OILS 5.5
OTHER 6.9

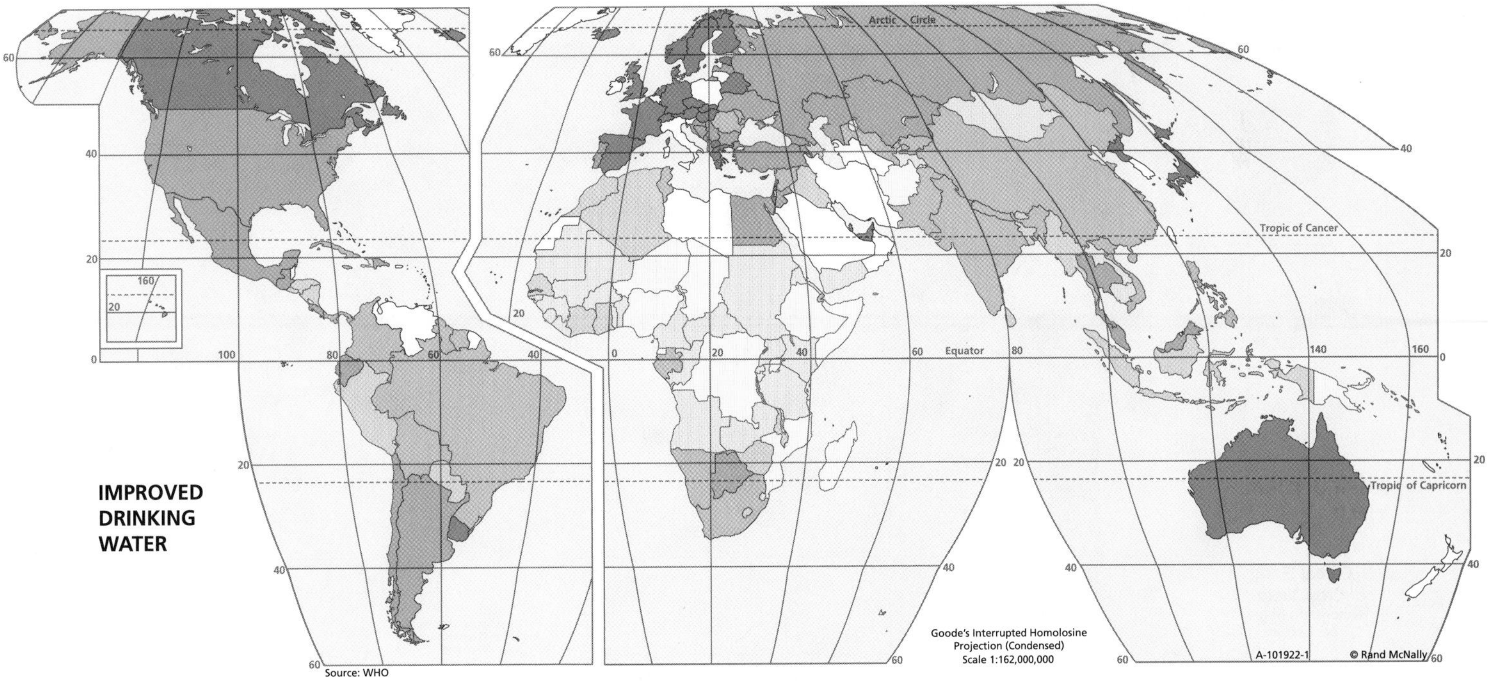
IMPROVED DRINKING WATER
Arctic Circle
Tropic of Cancer
Equator
Tropic of Capricorn
Source: WHO
Goode's Interrupted Homolosine Projection (Condensed)
Scale 1:162,000,000
A-101922-1
© Rand McNally

Drinking Water
Percentage of population with sustainable access to improved drinking water - 2006

- 100%
- 95 - 99%
- 85 - 95%
- 75 - 85%
- 50 - 75%
- Under 50%
- Data not available

Percentage of World Population in Each Drinking Water Access Category - 2006

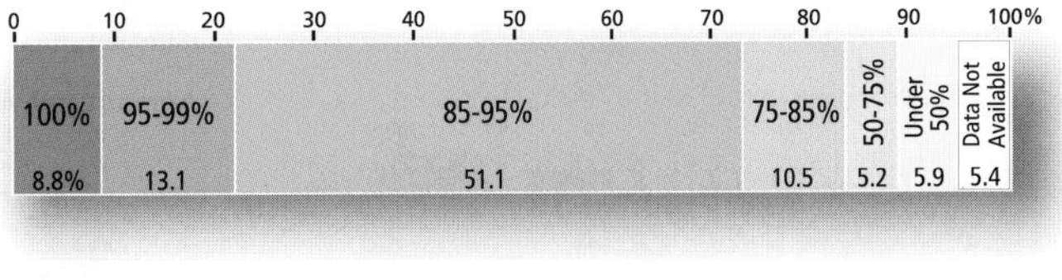
0 10 20 30 40 50 60 70 80 90 100%
100% 8.8%
95-99% 13.1
85-95% 51.1
75-85% 10.5
50-75% 5.2
Under 50% 5.9
Data Not Available 5.4

AGRICULTURAL PRODUCTION

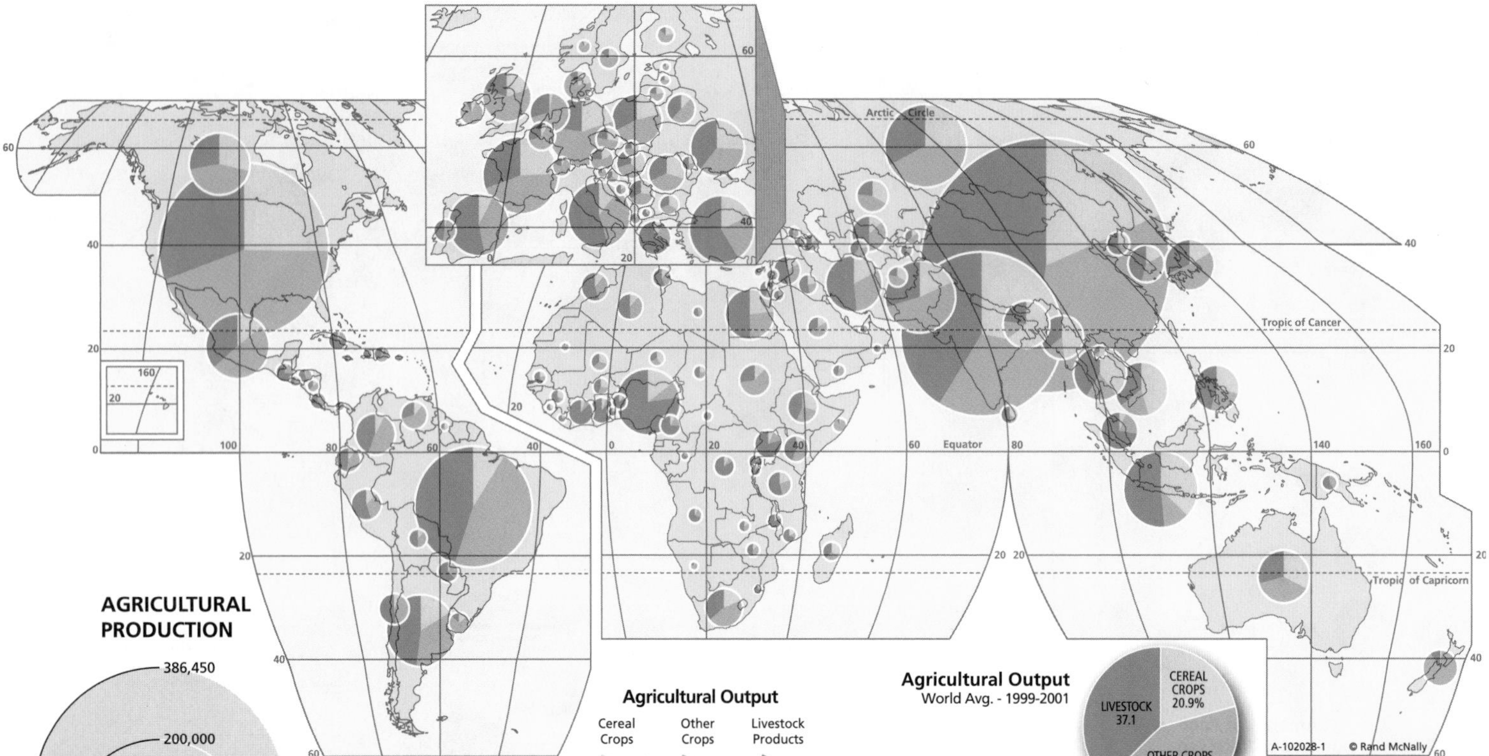
Arctic Circle
Tropic of Cancer
Equator
Tropic of Capricorn
Source: FAO
A-102028-1 © Rand McNally
Goode's Interrupted Homolosine Projection (Condensed)
Scale 1:162,000,000

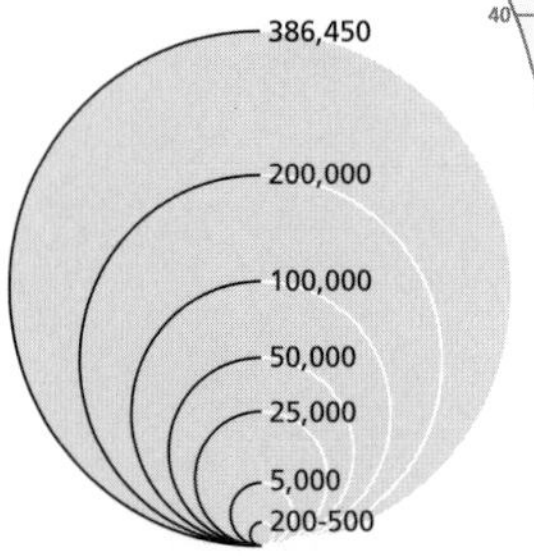
386,450
200,000
100,000
50,000
25,000
5,000
200-500

Volume of Agricultural Production

in millions of 1999-2001 FAO international dollars - 2005

If value is less than 200 million international dollars, no symbol is shown.

Taiwan figures are included with China.

Agricultural Output

Cereal Crops
Other Crops
Livestock Products

Agricultural Output

World Avg. - 1999-2001

CEREAL CROPS 20.9%
OTHER CROPS 42.0
LIVESTOCK 37.1

Agricultural Production

World Total - 1,665,529,219,000 1999-2001 FAO international dollars - 2005

0 10 20 30 40 50 60 70 80 90 100%
CHINA 23.2%
INDIA 9.6
INDON. 2.1
OTHER ASIA 14.2
UNITED STATES 10.9
OTHER N. AMER. 3.7
BRAZIL 5.2
OTHER S. AMER. 4.2
RUSSIA 2.5
FRANCE 2.2
OTHER EUROPE 12.8
AFRICA 7.7

AGRICULTURAL YIELD

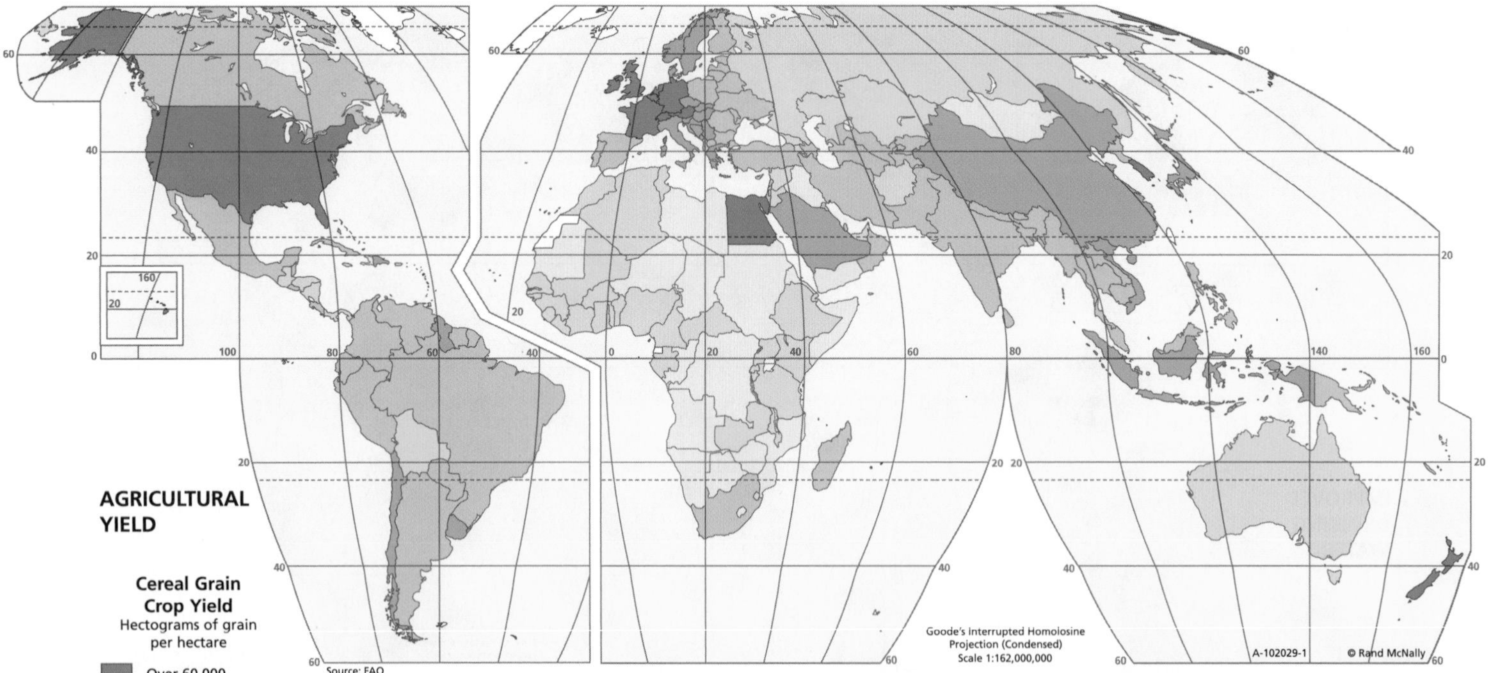
Source: FAO
Goode's Interrupted Homolosine Projection (Condensed)
Scale 1:162,000,000
A-102029-1 © Rand McNally

Cereal Grain Crop Yield

Hectograms of grain per hectare

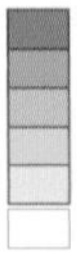
Over 60,000
40,000 - 60,000
20,000 - 40,000
10,000 - 20,000
Under 10,000
Data not available

Tractors

World Total* - 28,494,000 tractors - 2006

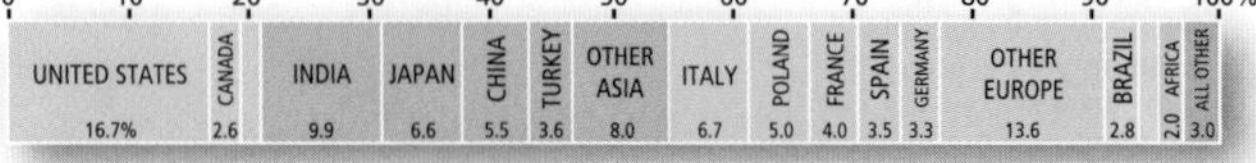
0 10 20 30 40 50 60 70 80 90 100%
UNITED STATES 16.7%
CANADA 2.6
INDIA 9.9
JAPAN 6.6
CHINA 5.5
TURKEY 3.6
OTHER ASIA 8.0
ITALY 6.7
POLAND 5.0
FRANCE 4.0
SPAIN 3.5
GERMANY 3.3
OTHER EUROPE 13.6
BRAZIL 2.8
AFRICA 2.0
ALL OTHER 3.0

* Excluding data for Jordan, Paraguay, and Suriname.

Arable Land

World Total - 1,421,200,000 hectares - 2005

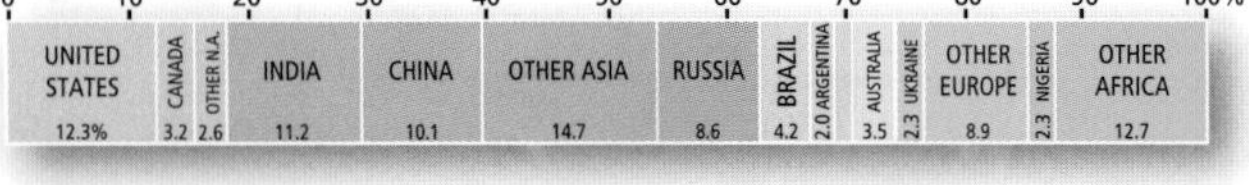
0 10 20 30 40 50 60 70 80 90 100%
UNITED STATES 12.3%
CANADA 3.2
OTHER N.A. 2.6
INDIA 11.2
CHINA 10.1
OTHER ASIA 14.7
RUSSIA 8.6
BRAZIL 4.2
ARGENTINA 2.0
AUSTRALIA 3.5
UKRAINE 2.3
OTHER EUROPE 8.9
NIGERIA 2.3
OTHER AFRICA 12.7

Agricultural Water Withdrawals

World Total** - 2,692,100,000,000 cubic meters per year - latest available data

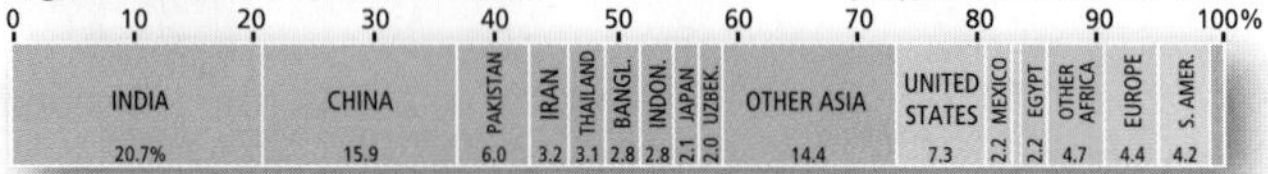
0 10 20 30 40 50 60 70 80 90 100%
INDIA 20.7%
CHINA 15.9
PAKISTAN 6.0
IRAN 3.2
THAILAND 3.1
BANGL. 2.8
INDON. 2.8
JAPAN 2.0
UZBEK.
OTHER ASIA 14.4
UNITED STATES 7.3
MEXICO 2.2
EGYPT 2.2
OTHER AFRICA 4.7
EUROPE 4.4
S. AMER. 4.2

** Excluding data for Slovakia and the former Yugoslav republics.

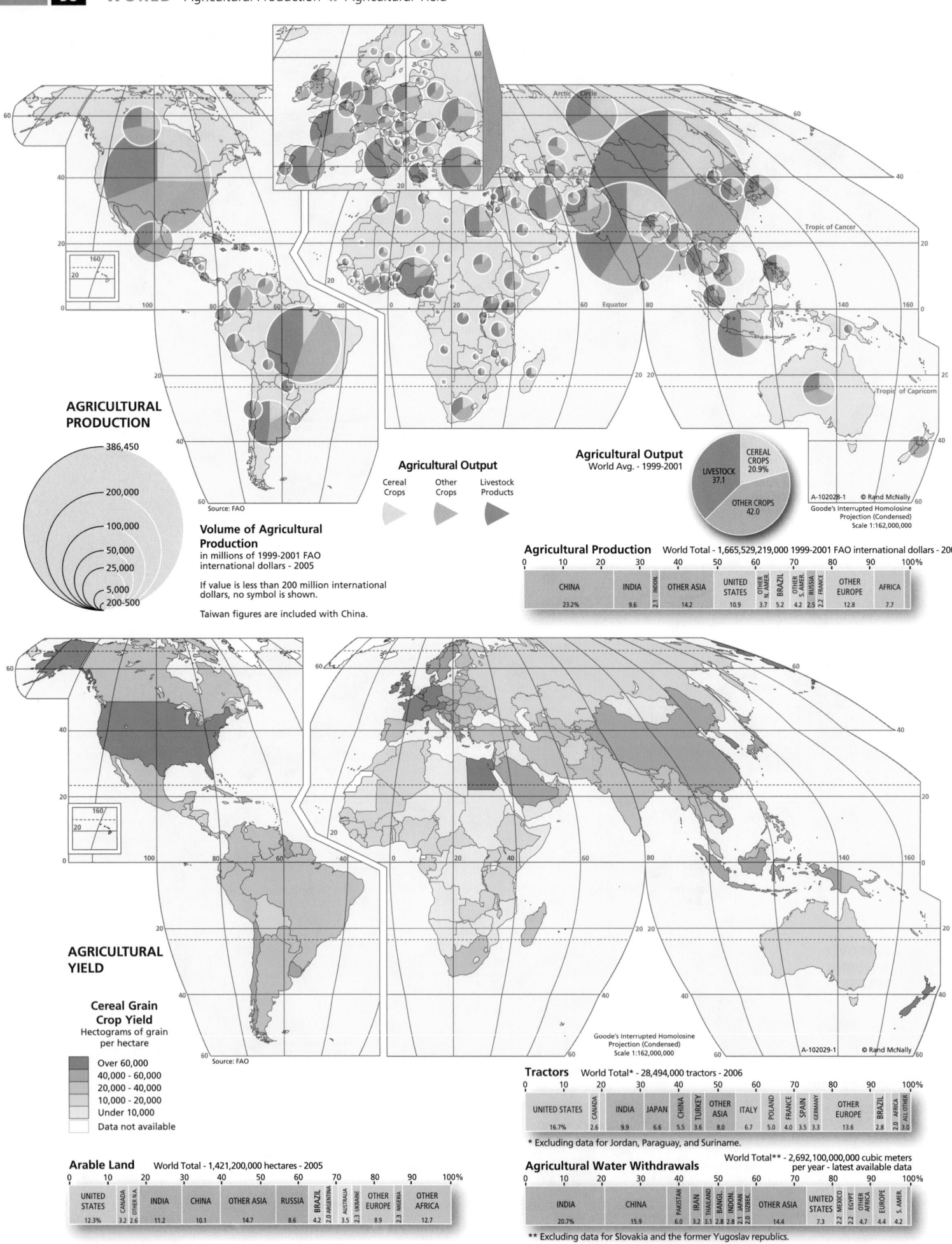

WHEAT

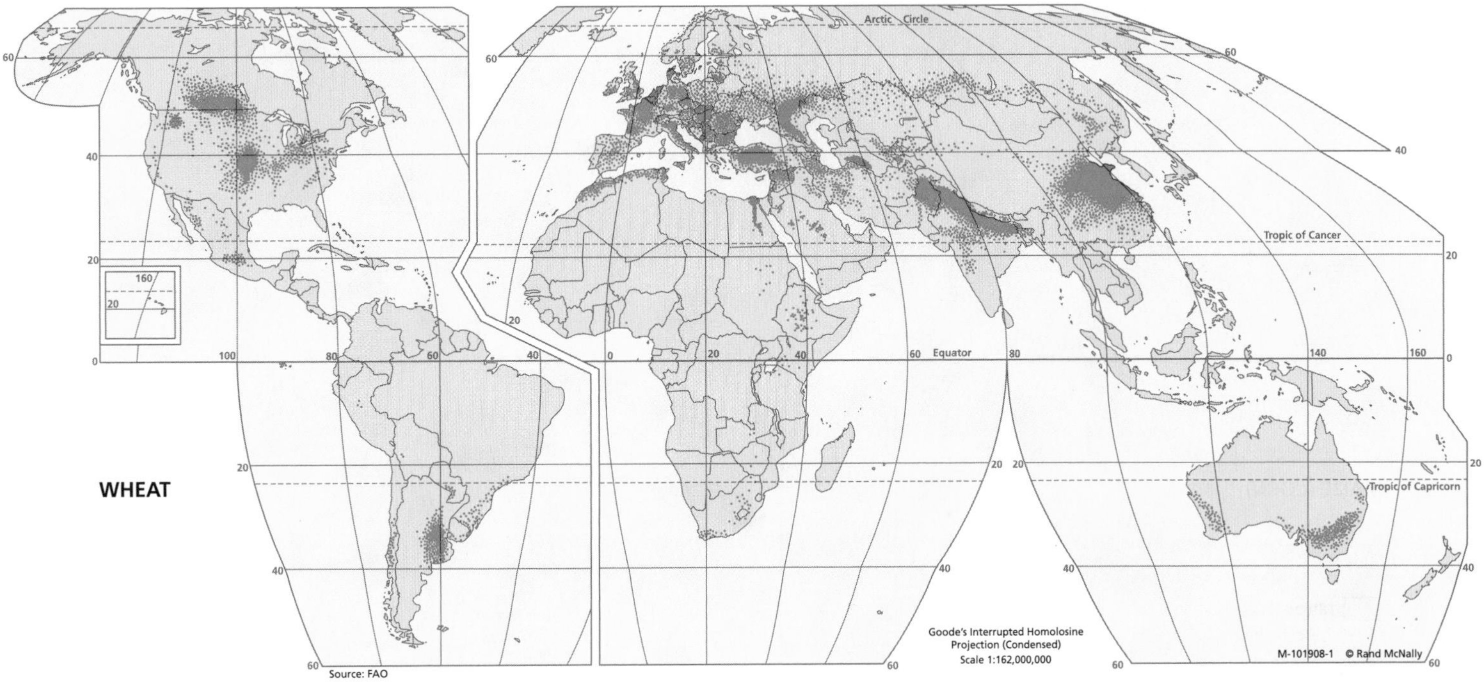
Arctic Circle
Tropic of Cancer
Equator
Tropic of Capricorn
Goode's Interrupted Homolosine
Projection (Condensed)
Scale 1:162,000,000
M-101908-1 © Rand McNally

Source: FAO

Wheat
Each dot represents 2,000,000 bushels (54,400 metric tons) per year

Wheat Imports World Imports - 119,058,000 metric tons - Avg. 2003-2005

ITALY	SPAIN	BELGIUM	NETH.	OTHER EUROPE	BRAZIL	OTHER S. AMERICA	JAPAN	CHINA	INDONESIA	S. KOREA	PHILIPPINES	IRAQ	OTHER ASIA	ALGERIA	EGYPT	NIGERIA	MOROCCO	OTHER AFRICA	MEXICO	OTHER N. AMERICA	ALL OTHER
5.7%	4.4	2.8	2.7	8.8	4.6	4.2	4.5	4.2	3.5	3.0	2.0	2.0	14.7	4.5	4.4	2.4	2.1	8.4	3.0	3.5	4.6

Wheat World Production - 622,582,000 metric tons - Avg. 2004-2006

CHINA	INDIA	TURKEY	PAKISTAN	IRAN	OTHER ASIA	UNITED STATES	CANADA	RUSSIA	FRANCE	GERMANY	UKRAINE	U.K.	OTHER EUROPE	AUSTRALIA	ARGENTINA	AFRICA
15.7%	11.3	3.3	3.3	2.3	6.5	9.4	4.3	7.4	6.0	3.8	2.7	2.4	10.9	3.0	2.3	3.7

Wheat Exports World Exports - 117,329,000 metric tons - Avg. 2003-2005

UNITED STATES	CANADA	FRANCE	GERMANY	UKRAINE	U.K.	OTHER EUROPE	AUSTRALIA	ARGEN-TINA	RUSSIA	KAZAKHSTAN	INDIA	OTHER ASIA
24.3%	11.6	13.4	3.7	2.7	2.5	7.2	11.9	7.5	6.4	2.5	2.0	2.6

TEA, RYE

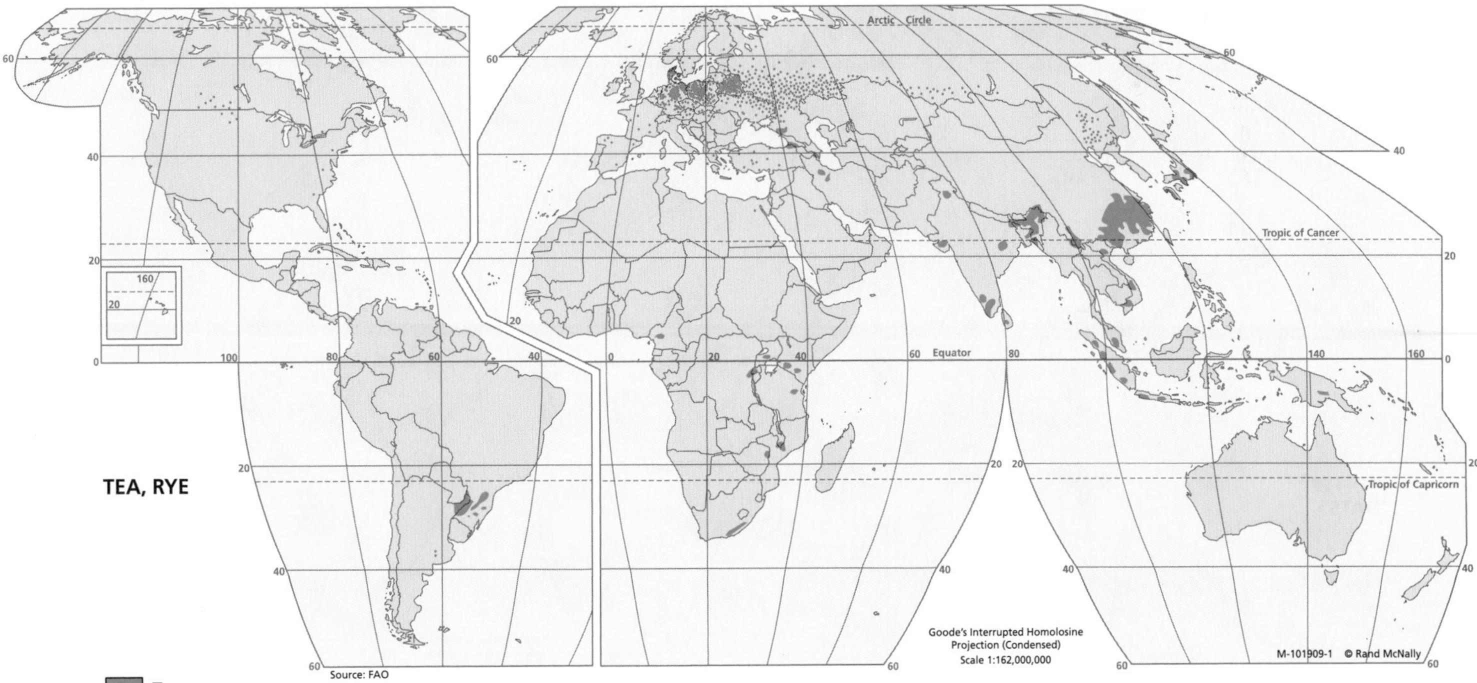
Arctic Circle
Tropic of Cancer
Equator
Tropic of Capricorn
Goode's Interrupted Homolosine
Projection (Condensed)
Scale 1:162,000,000
M-101909-1 © Rand McNally

Source: FAO

Tea
Major producing areas

Rye
Each dot represents 1,000,000 bushels (25,400 metric tons) per year

Tea World Production - 3,529,000 metric tons - Avg. 2004-2006

CHINA	INDIA	SRI LANKA	TURKEY	INDONESIA	VIETNAM	JAPAN	OTHER ASIA	KENYA	OTHER AFRICA	S. AMERICA
26.5%	24.4	8.8	5.9	4.8	3.7	2.8	5.7	9.1	5.3	2.7

Rye World Production - 15,383,000 metric tons - Avg. 2004-2006

RUSSIA	POLAND	GERMANY	BELARUS	UKRAINE	OTHER EUROPE	CHINA	OTHER ASIA	CANADA
20.5%	22.3	20.1	7.9	7.7	10.7	4.2	2.2	2.3

Tea Imports World Imports - 1,426,000 metric tons - Avg. 2003-2005

RUSSIA	UNITED KINGDOM	GERMANY	POLAND	NETH.	OTHER EUROPE	PAKISTAN	U.A.E.	JAPAN	CHINA	AFGHAN.	OTHER ASIA	UNITED STATES	OTHER	MOROCCO	OTHER AFRICA	ALL OTHER
12.2%	10.9	3.1	2.2	2.1	6.5	8.4	3.9	3.6	2.5	2.0	16.8	6.9	1.5	3.3	10.8	3.4

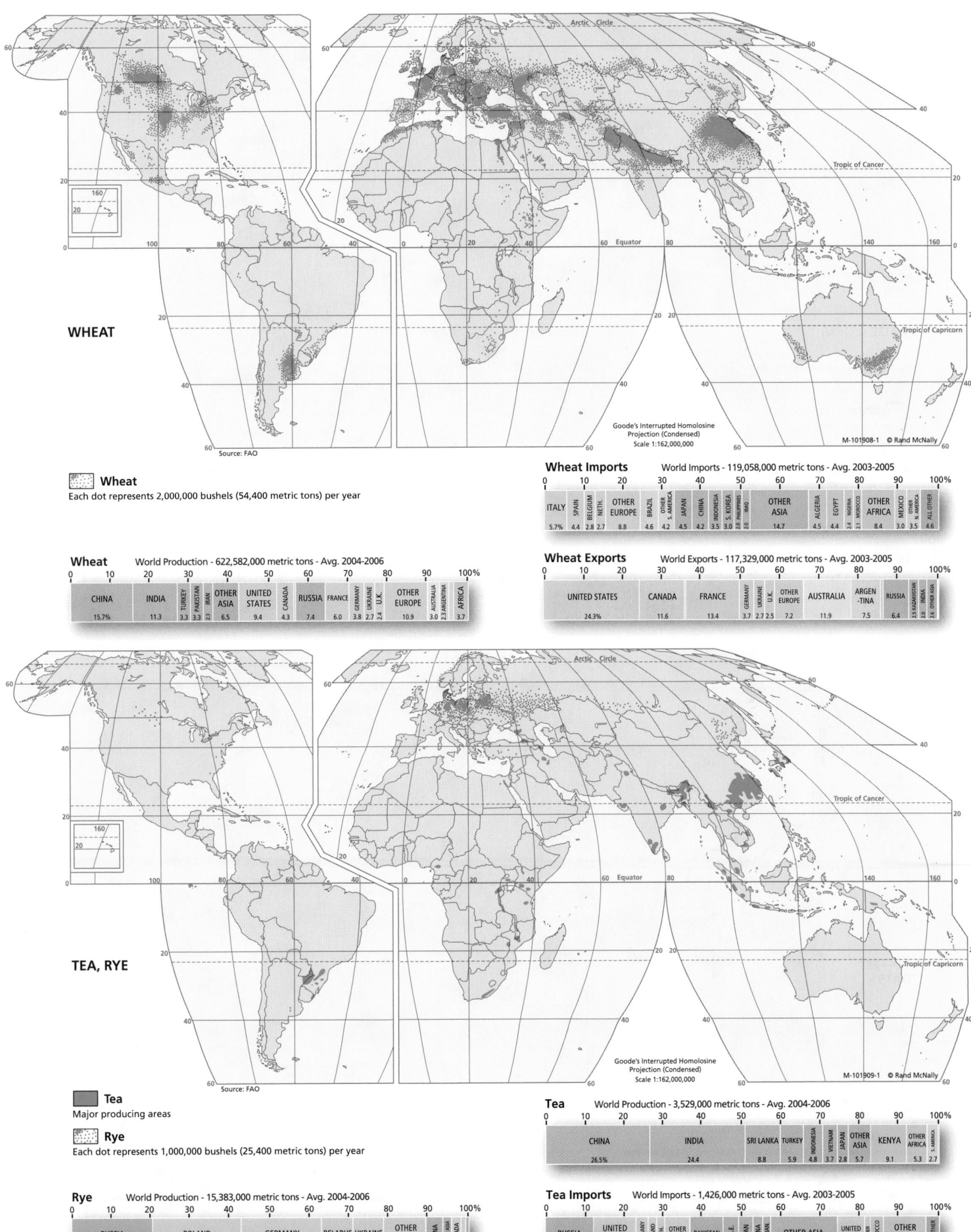

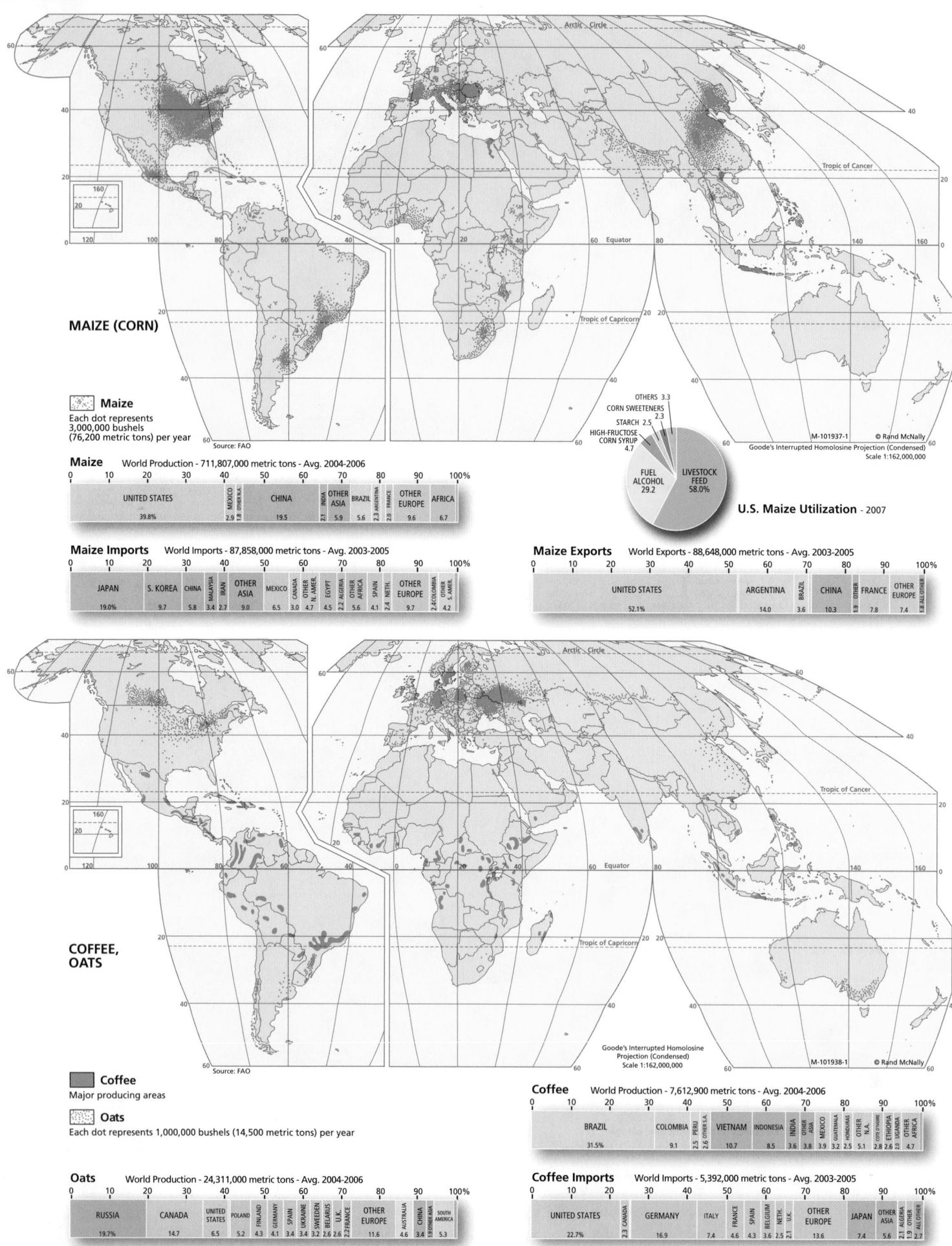

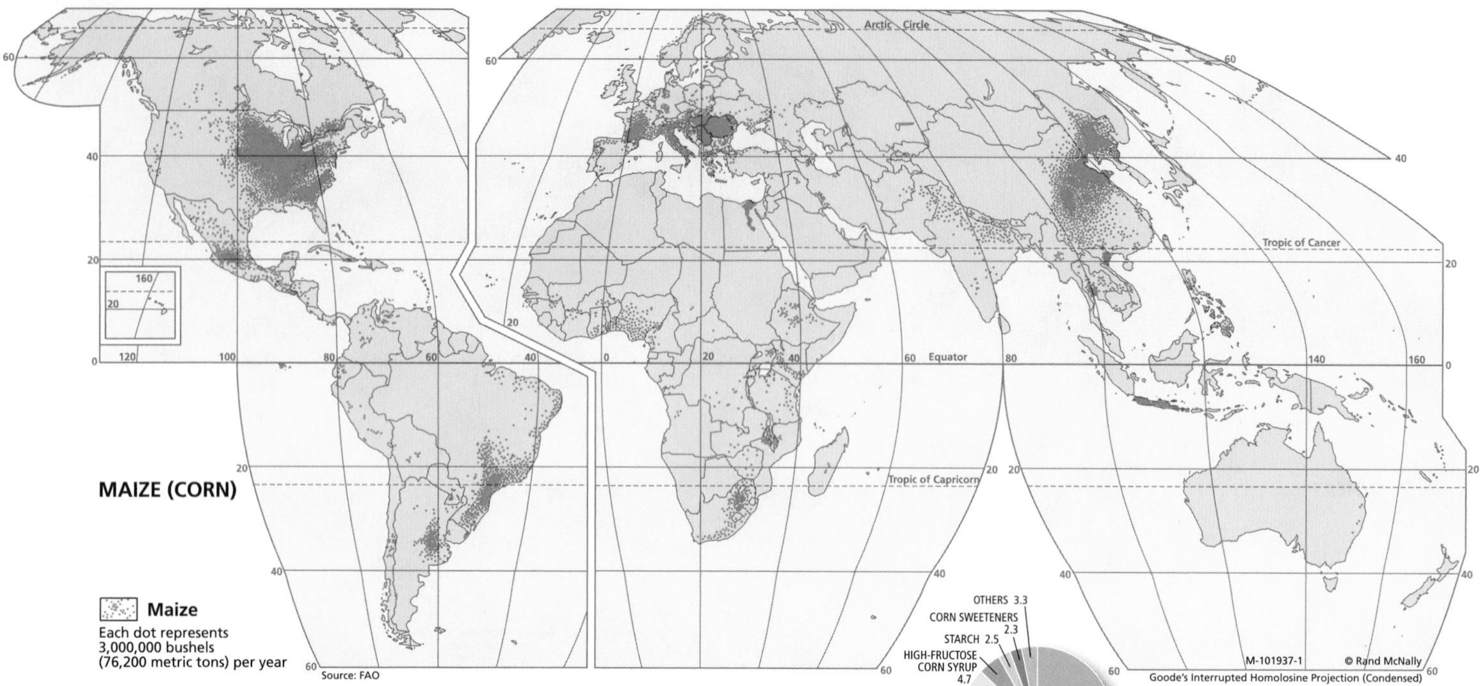
MAIZE (CORN)
Arctic Circle
Tropic of Cancer
Equator
Tropic of Capricorn
Maize
Each dot represents 3,000,000 bushels (76,200 metric tons) per year
Source: FAO
M-101937-1
© Rand McNally
Goode's Interrupted Homolosine Projection (Condensed)
Scale 1:162,000,000

Maize World Production - 711,807,000 metric tons - Avg. 2004-2006

UNITED STATES	MEXICO	OTHER N.A.	CHINA	INDIA	OTHER ASIA	BRAZIL	ARGENTINA	FRANCE	OTHER EUROPE	AFRICA
39.8%	2.9	1.8	19.5	2.1	5.9	5.6	2.3	2.0	9.6	6.7

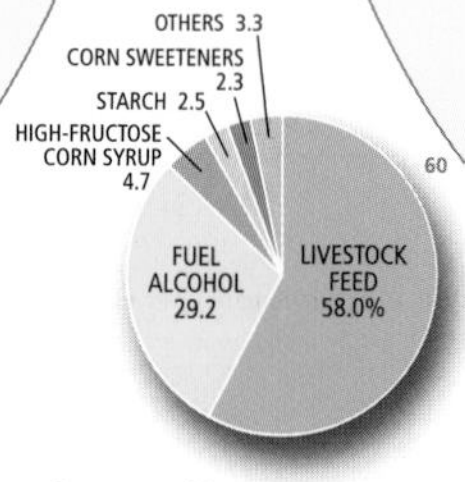
OTHERS 3.3
CORN SWEETENERS 2.3
STARCH 2.5
HIGH-FRUCTOSE CORN SYRUP 4.7
FUEL ALCOHOL 29.2
LIVESTOCK FEED 58.0%

U.S. Maize Utilization - 2007

Maize Imports World Imports - 87,858,000 metric tons - Avg. 2003-2005

JAPAN	S. KOREA	CHINA	MALAYSIA	IRAN	OTHER ASIA	MEXICO	CANADA	OTHER N. AMER.	EGYPT	ALGERIA	OTHER AFRICA	SPAIN	NETH.	OTHER EUROPE	COLOMBIA	OTHER S. AMER.
19.0%	9.7	5.8	3.4	2.7	9.0	6.5	3.0	4.7	4.5	2.2	5.6	4.1	2.4	9.7	2.4	4.2

Maize Exports World Exports - 88,648,000 metric tons - Avg. 2003-2005

UNITED STATES	ARGENTINA	BRAZIL	CHINA	OTHER	FRANCE	OTHER EUROPE	ALL OTHER
52.1%	14.0	3.6	10.3	1.9	7.8	7.4	1.8

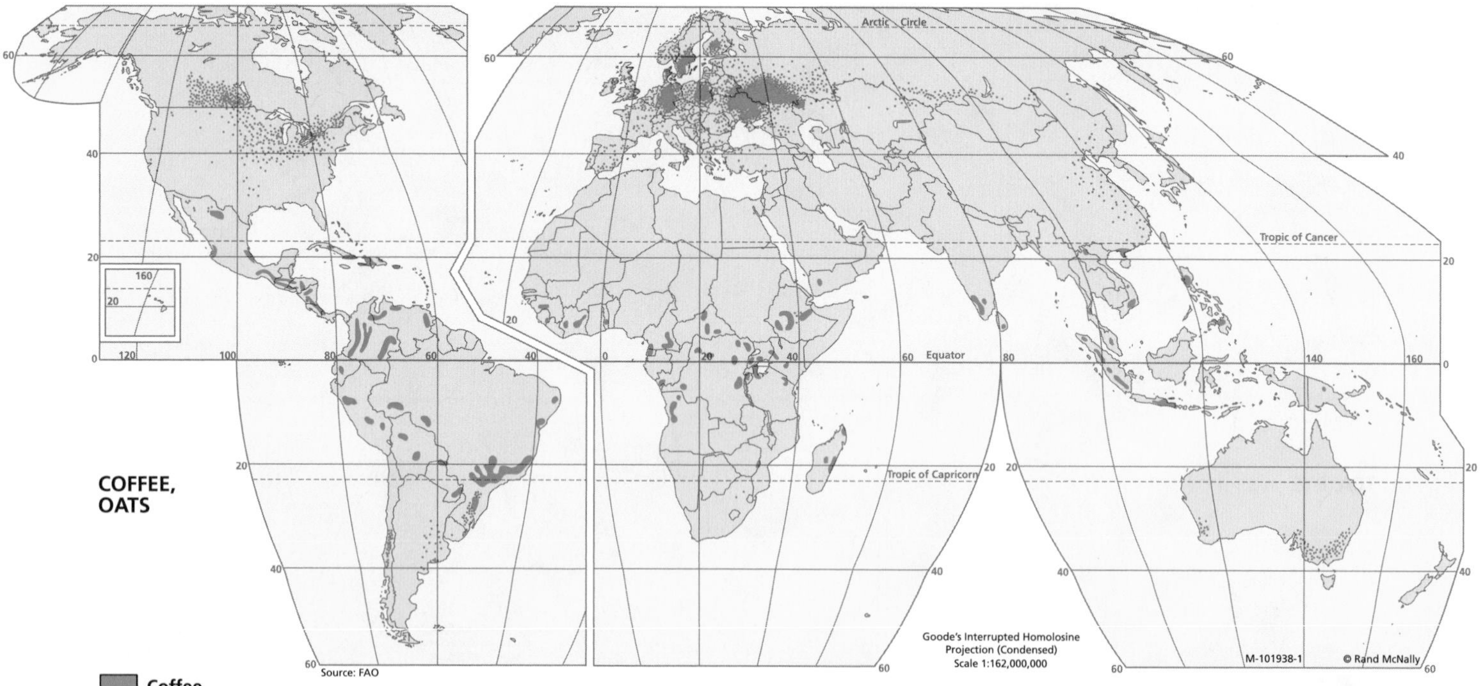
COFFEE, OATS
Arctic Circle
Tropic of Cancer
Equator
Tropic of Capricorn
Source: FAO
Goode's Interrupted Homolosine Projection (Condensed)
Scale 1:162,000,000
M-101938-1
© Rand McNally

Coffee
Major producing areas

Oats
Each dot represents 1,000,000 bushels (14,500 metric tons) per year

Coffee World Production - 7,612,900 metric tons - Avg. 2004-2006

BRAZIL	COLOMBIA	PERU	OTHER S.A.	VIETNAM	INDONESIA	INDIA	OTHER ASIA	MEXICO	GUATEMALA	HONDURAS	OTHER N.A.	COTE D'IVOIRE	ETHIOPIA	UGANDA	OTHER AFRICA
31.5%	9.1	2.5	2.6	10.7	8.5	3.6	3.8	3.9	3.2	2.5	5.1	2.8	2.6	2.0	4.7

Oats World Production - 24,311,000 metric tons - Avg. 2004-2006

RUSSIA	CANADA	UNITED STATES	POLAND	FINLAND	GERMANY	SPAIN	UKRAINE	SWEEDEN	BELARUS	U.K.	FRANCE	OTHER EUROPE	AUSTRALIA	CHINA	OTHER ASIA	SOUTH AMERICA
19.7%	14.7	6.5	5.2	4.3	4.1	3.4	3.4	3.2	2.6	2.6	2.2	11.6	4.6	3.4	1.9	5.3

Coffee Imports World Imports - 5,392,000 metric tons - Avg. 2003-2005

UNITED STATES	CANADA	GERMANY	ITALY	FRANCE	SPAIN	BELGIUM	NETH.	U.K.	OTHER EUROPE	JAPAN	OTHER ASIA	ALGERIA	OTHER	ALL OTHER
22.7%	2.3	16.9	7.4	4.6	4.3	3.6	2.5	2.1	13.6	7.4	5.6	2.1	1.9	2.7

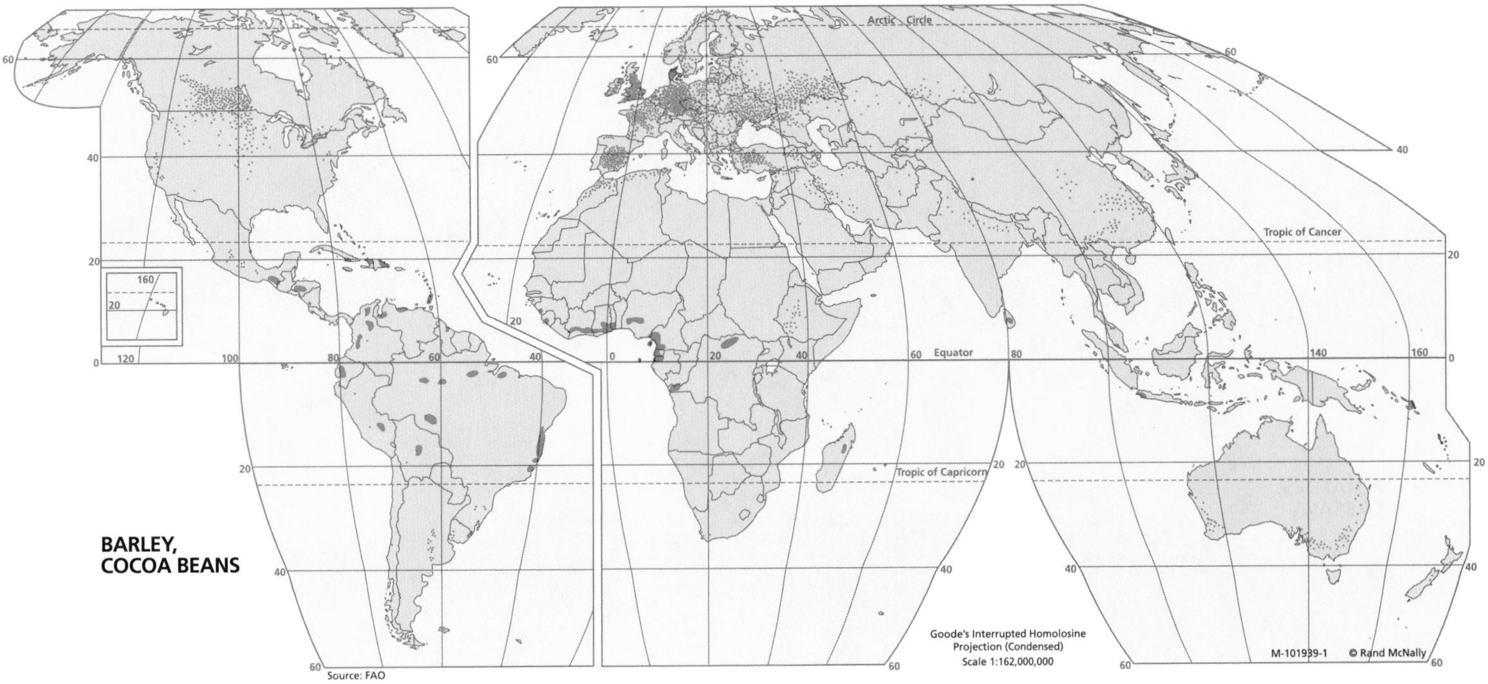
Arctic Circle
Tropic of Cancer
Equator
Tropic of Capricorn
Goode's Interrupted Homolosine
Projection (Condensed)
Scale 1:162,000,000
M-101939-1
© Rand McNally
Source: FAO

BARLEY, COCOA BEANS

Barley
Each dot represents 3,000,000 bushels (65,400 metric tons) per year

Cocoa Beans
Major producing areas

Barley World Production - 144,809,000 metric tons - Avg. 2004-2006

RUSSIA	GERMANY	FRANCE	UKRAINE	SPAIN	U.K.	DENMARK	POLAND	OTHER EUROPE	CANADA	UNITED STATES	TURKEY	CHINA	IRAN	OTHER ASIA	AUSTRALIA	AFRICA	ALL OTHER
11.8%	8.4	7.3	7.2	5.4	3.8	2.5	2.4	13.0	8.2	3.7	6.5	2.3	2.0	4.4	4.9	3.9	1.4

Cocoa Beans World Production - 4,017,000 metric tons - Avg. 2004-2006

COTE D'IVOIRE	GHANA	NIGERIA	CAMEROON	OTHER AF.	INDONESIA	OTHEER ASIA	BRAZIL	ECUADOR	OTHER S.A.	N. AMERICA	ALL OTHER
34.6%	18.3	11.1	4.2	2.7	14.9	1.3	5.0	2.3	2.1	2.3	1.2

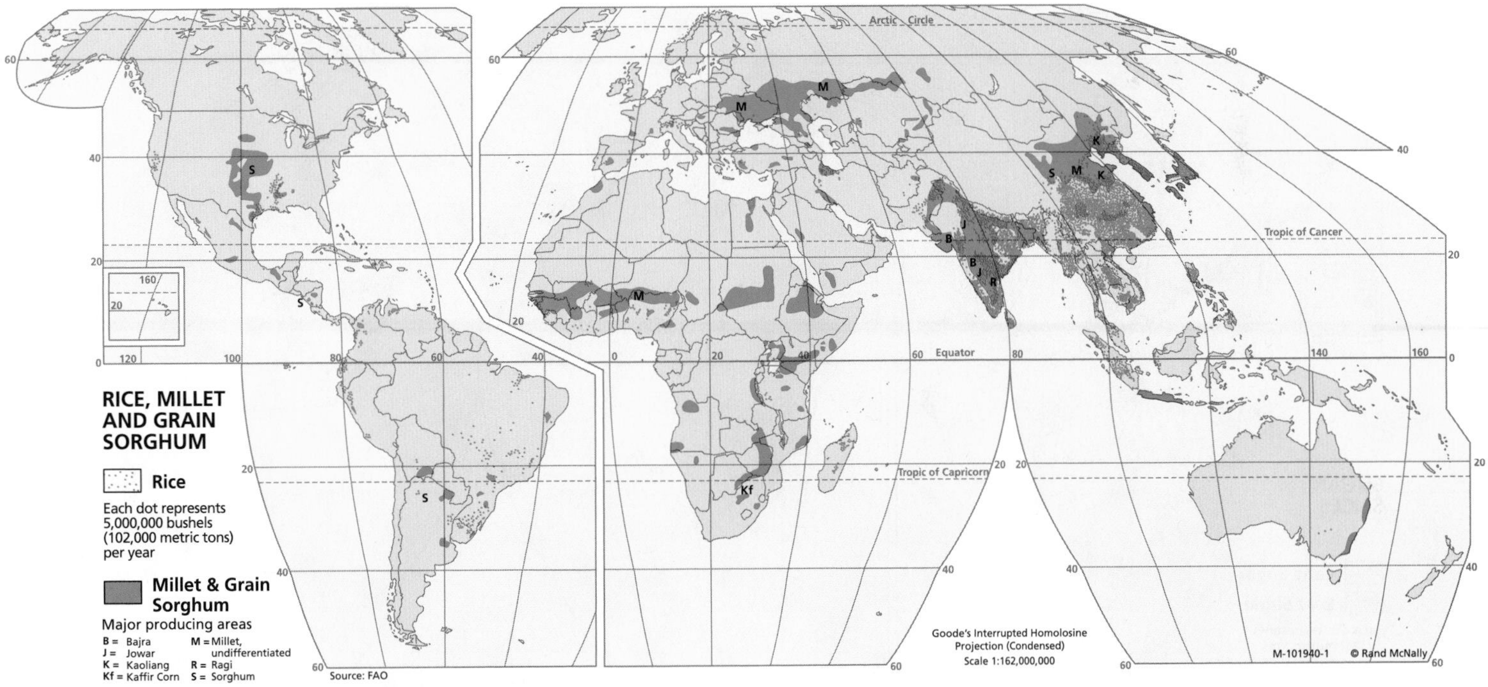
Arctic Circle
Tropic of Cancer
Equator
Tropic of Capricorn
Goode's Interrupted Homolosine
Projection (Condensed)
Scale 1:162,000,000
M-101940-1
© Rand McNally
Source: FAO

RICE, MILLET AND GRAIN SORGHUM

Rice
Each dot represents 5,000,000 bushels (102,000 metric tons) per year

Millet & Grain Sorghum
Major producing areas

B = Bajra
J = Jowar
K = Kaoliang
Kf = Kaffir Corn
M = Millet, undifferentiated
R = Ragi
S = Sorghum

Rice World Production - 624,479,000 metric tons - Avg. 2004-2006

CHINA	INDIA	INDONESIA	BANGL.	VIETNAM	THAILAND	MYANMAR	PHILIPPINES	OTHER ASIA	BRAZIL	OTHER S.A.	AFRICA	ALL OTHER
28.9%	21.3	8.7	6.4	5.8	4.7	4.0	2.4	8.4	2.0	1.7	3.2	2.5

Rice Imports World Imports - 26,906,000 metric tons - Avg. 2003-2005

NIGERIA	SENEGAL	S. AFRICA	COTE D'IVOIRE	OTHER AFRICA	PHILIPPINES	IRAN	BNGL.	S. ARABIA	N. KOREA	INDONESIA	JAPAN	CHINA	IRAQ	OTHER ASIA	BRAZIL		U.K.	OTHER EUROPE	CUBA	OTHER N. AMER.	ALL OTHER
5.2%	3.2	2.8	2.8	15.4	4.1	3.8	3.5	3.5	2.8	2.7	2.7	2.4	2.4	14.5	3.0		2.1	9.0	2.0	7.9	3.1

Millet & Grain Sorghum World Production - 88,629,000 metric tons - Avg. 2004-2006

INDIA	CHINA	OTHER	NIGERIA	SUDAN	NIGER	BURKINA F.	ETHIOPIA	MALI	OTHER AFRICA	UNITED STATES	MEXICO	ARGENTINA	BRAZIL		ALL OTHER
19.8%	4.8	2.1	18.5	5.3	3.8	2.9	2.8	2.0	9.8	11.1	6.8	2.8	2.0		3.8

Rice Exports* World Exports - 28,749,000 metric tons - Avg. 2003-2005

THAILAND	VIETNAM	INDIA	PAKISTAN	CHINA	OTHER ASIA	UNITED STATES	EGYPT	ITALY	OTHER	URUGUAY	OTHER
30.1%	15.2	14.2	7.6	4.8	2.1	12.4	2.9	2.3	3.1	2.3	2.3

* including reexports

POTATOES, CASSAVA

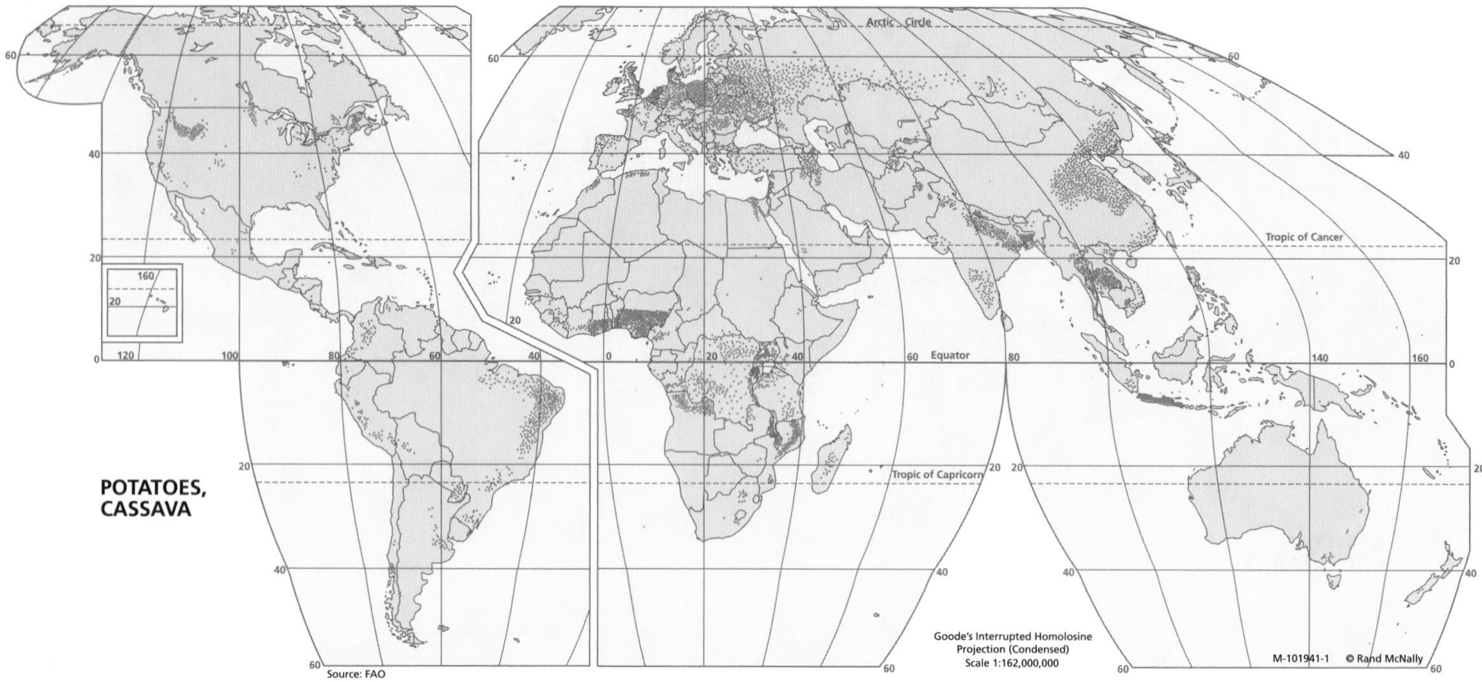
Arctic Circle
Tropic of Cancer
Equator
Tropic of Capricorn
Goode's Interrupted Homolosine Projection (Condensed)
Scale 1:162,000,000
M-101941-1 © Rand McNally
Source: FAO

Potatoes
Each dot represents 100,000 metric tons average annual production

Cassava
Each dot represents 100,000 metric tons average annual production

Potatoes World Production - 323,418,000 metric tons - Avg. 2004-2006

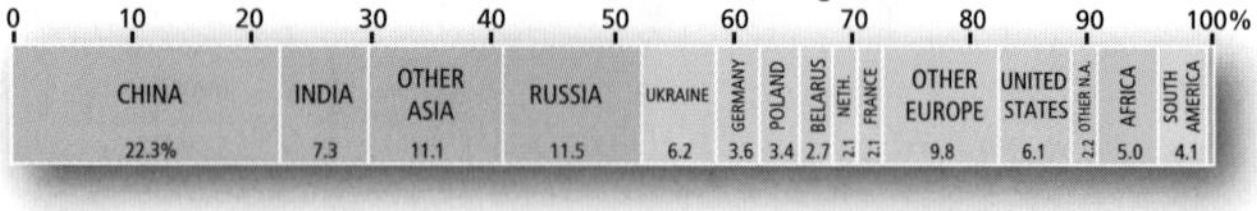
CHINA 22.3%
INDIA 7.3
OTHER ASIA 11.1
RUSSIA 11.5
UKRAINE 6.2
GERMANY 3.6
POLAND 3.4
BELARUS 2.7
NETH. 2.1
FRANCE 2.1
OTHER EUROPE 9.8
UNITED STATES 6.1
OTHER N.A. 2.2
AFRICA 5.0
SOUTH AMERICA 4.1

Cassava World Production - 214,400,000 metric tons - Avg. 2004-2006

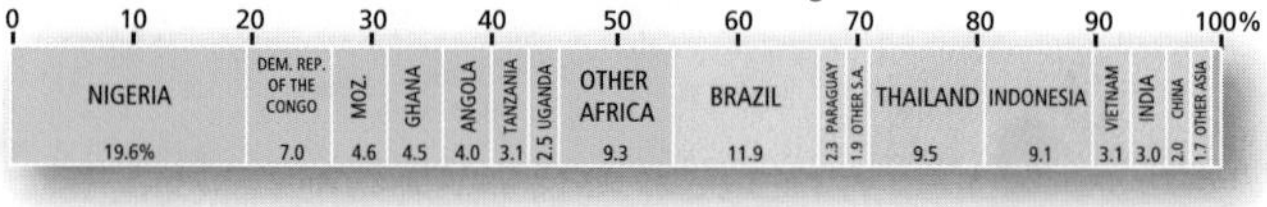
NIGERIA 19.6%
DEM. REP. OF THE CONGO 7.0
MOZ. 4.6
GHANA 4.5
ANGOLA 4.0
TANZANIA 3.1
2.5 UGANDA
OTHER AFRICA 9.3
BRAZIL 11.9
2.3 PARAGUAY
1.9 OTHER S.A.
THAILAND 9.5
INDONESIA 9.1
VIETNAM 3.1
INDIA 3.0
CHINA 2.0
1.7 OTHER ASIA

SUGAR, SPICES

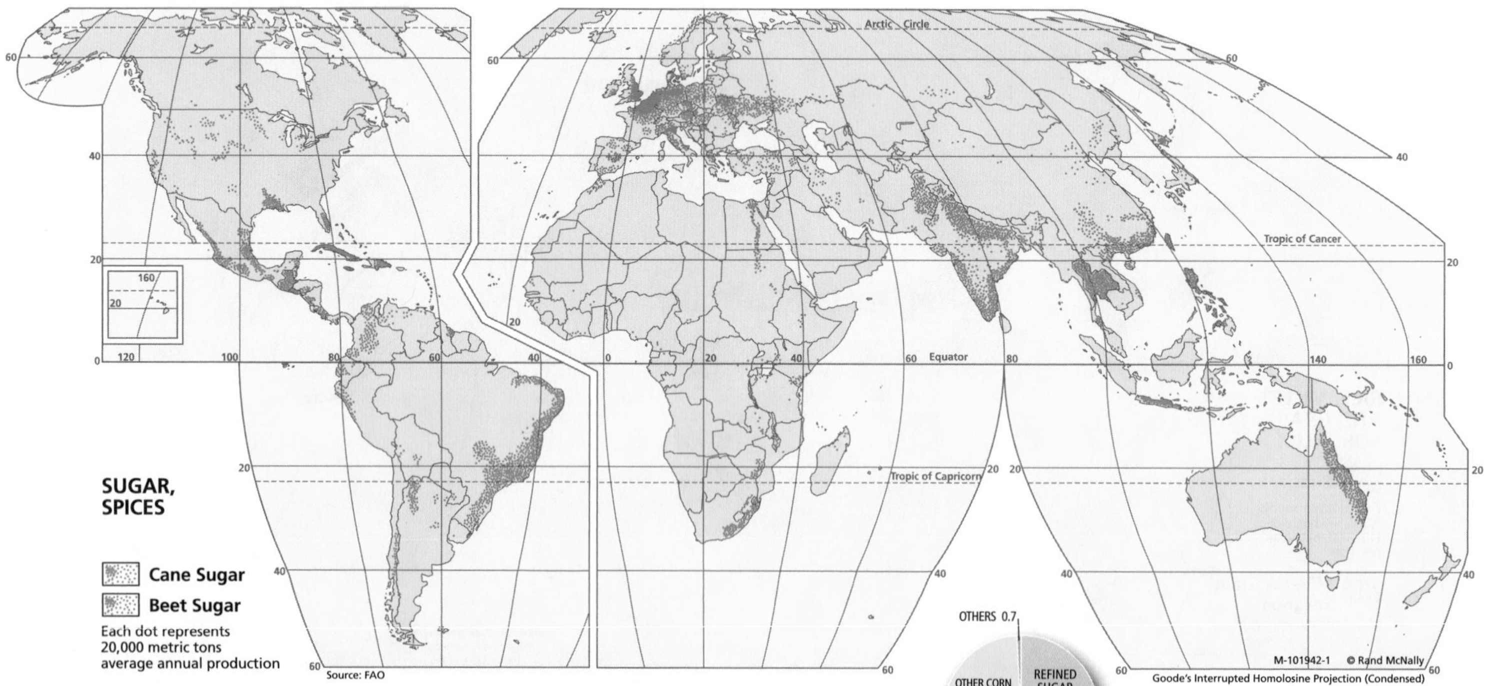
Arctic Circle
Tropic of Cancer
Equator
Tropic of Capricorn
M-101942-1 © Rand McNally
Goode's Interrupted Homolosine Projection (Condensed)
Scale 1:162,000,000
Source: FAO

Cane Sugar
Beet Sugar
Each dot represents 20,000 metric tons average annual production

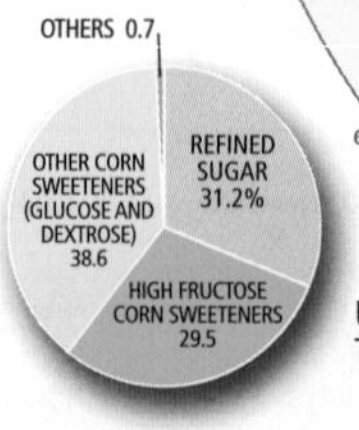
OTHERS 0.7
REFINED SUGAR 31.2%
OTHER CORN SWEETENERS (GLUCOSE AND DEXTROSE) 38.6
HIGH FRUCTOSE CORN SWEETENERS 29.5

U.S. Sweetener Consumption Per Person
Total - 90.9 kilograms - Avg. 2004-2006

Cane Sugar World Production - 112,294,000 metric tons - Avg. 2004-2006

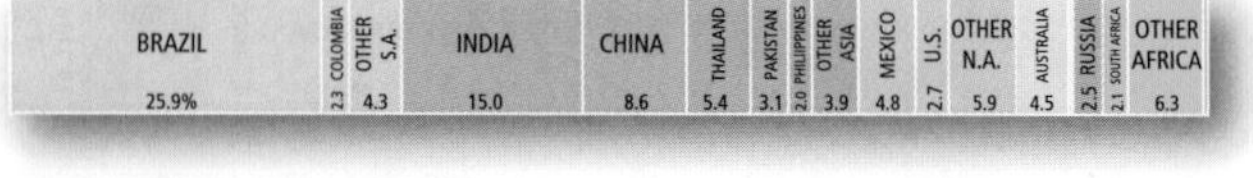
BRAZIL 25.9%
2.3 COLOMBIA
OTHER S.A. 4.3
INDIA 15.0
CHINA 8.6
THAILAND 5.4
PAKISTAN 3.1
2.0 PHILIPPINES
OTHER ASIA 3.9
MEXICO 4.8
2.7 U.S.
OTHER N.A. 5.9
AUSTRALIA 4.5
2.5 RUSSIA
2.1 SOUTH AFRICA
OTHER AFRICA 6.3

Beet Sugar World Production - 36,870,000 metric tons - Avg. 2004-2006

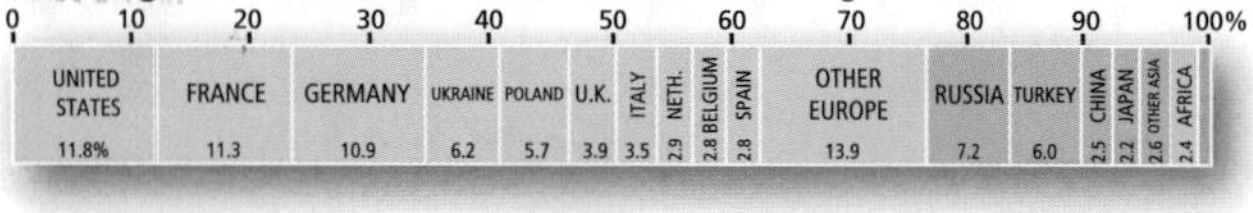
UNITED STATES 11.8%
FRANCE 11.3
GERMANY 10.9
UKRAINE 6.2
POLAND 5.7
U.K. 3.9
ITALY 3.5
2.9 NETH.
2.8 BELGIUM
2.8 SPAIN
OTHER EUROPE 13.9
RUSSIA 7.2
TURKEY 6.0
2.5 CHINA
2.2 JAPAN
2.6 OTHER ASIA
2.4 AFRICA

Spices World Total - 7,306,000 metric tons - Avg. 2004-2006

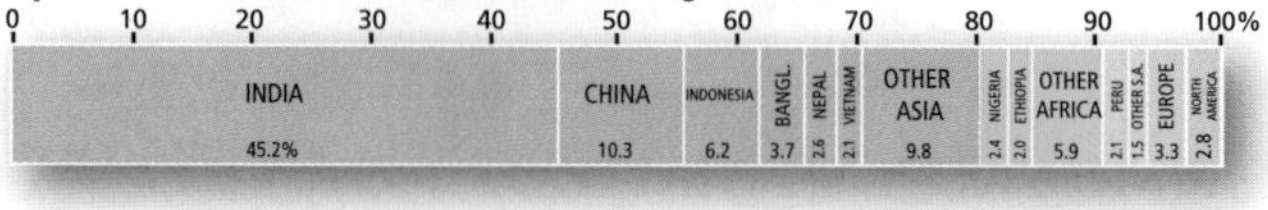
INDIA 45.2%
CHINA 10.3
INDONESIA 6.2
BANGL. 3.7
2.6 NEPAL
2.1 VIETNAM
OTHER ASIA 9.8
2.4 NIGERIA
2.0 ETHIOPIA
OTHER AFRICA 5.9
2.1 PERU
1.5 OTHER S.A.
EUROPE 3.3
2.8 NORTH AMERICA

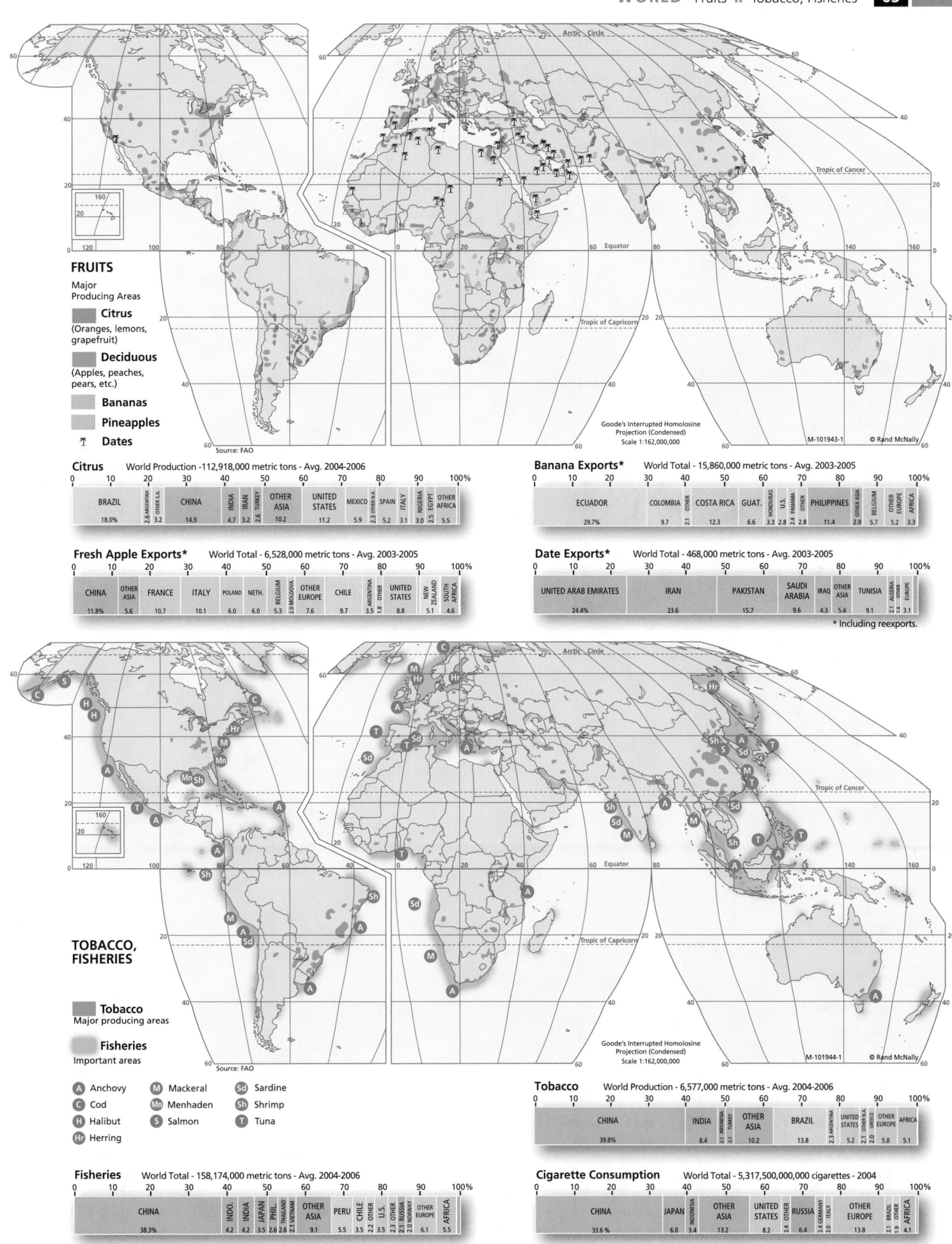

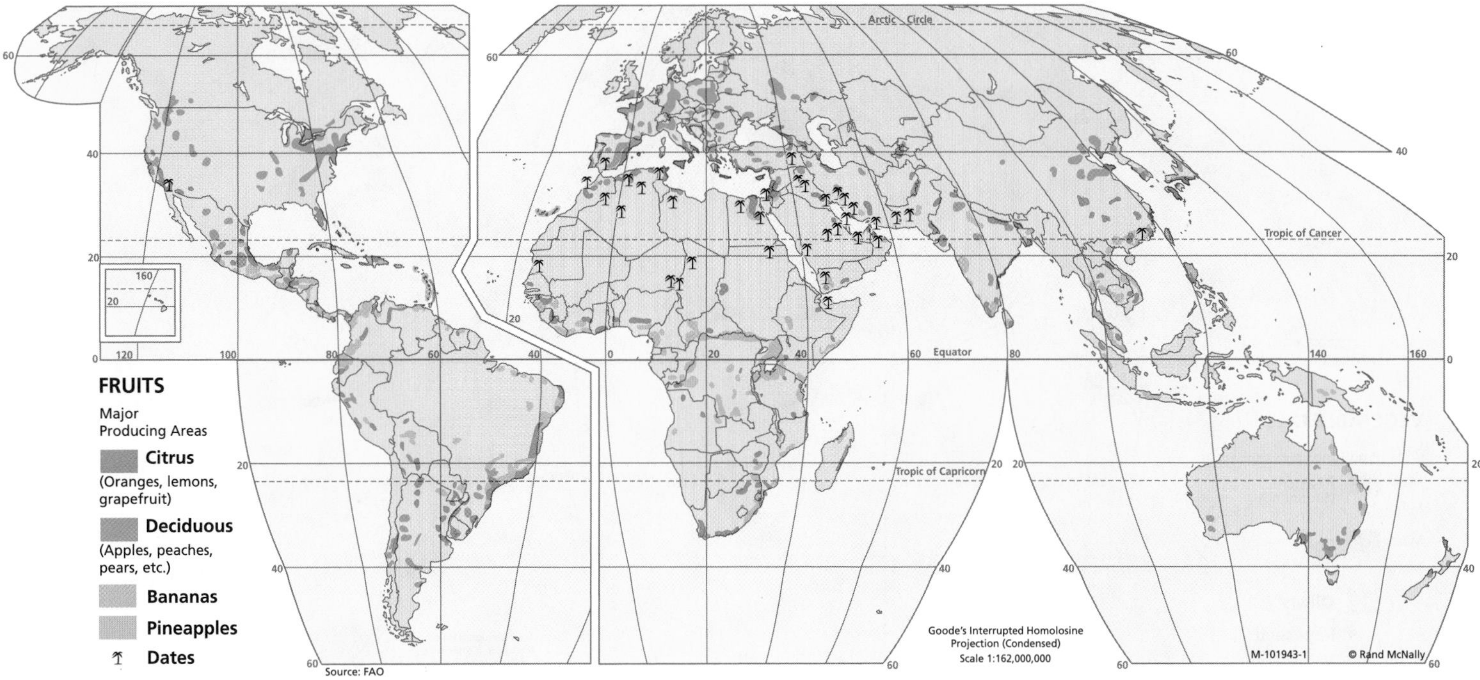
Arctic Circle
Tropic of Cancer
Equator
Tropic of Capricorn
FRUITS
Major Producing Areas
Citrus
(Oranges, lemons, grapefruit)
Deciduous
(Apples, peaches, pears, etc.)
Bananas
Pineapples
Dates
Source: FAO
Goode's Interrupted Homolosine Projection (Condensed)
Scale 1:162,000,000
M-101943-1
© Rand McNally

Citrus World Production -112,918,000 metric tons - Avg. 2004-2006

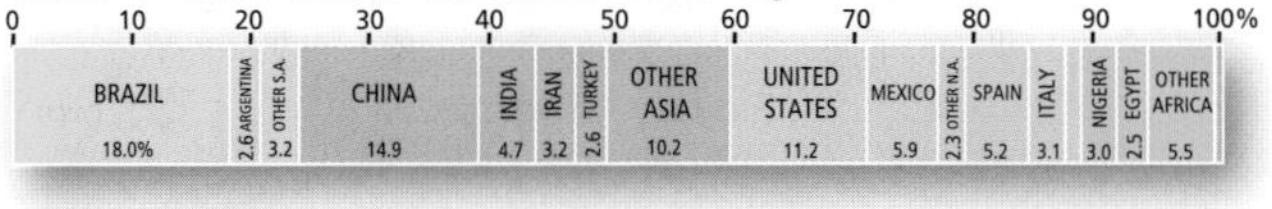
BRAZIL 18.0%
ARGENTINA 2.6
OTHER S.A. 3.2
CHINA 14.9
INDIA 4.7
IRAN 3.2
TURKEY 2.6
OTHER ASIA 10.2
UNITED STATES 11.2
MEXICO 5.9
OTHER N.A. 2.3
SPAIN 5.2
ITALY 3.1
NIGERIA 3.0
EGYPT 2.5
OTHER AFRICA 5.5

Banana Exports* World Total - 15,860,000 metric tons - Avg. 2003-2005

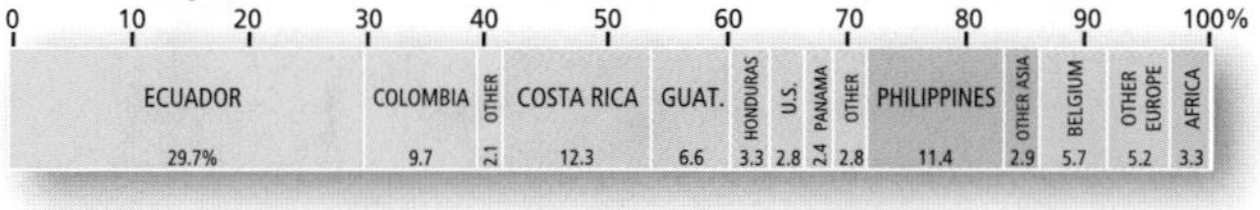
ECUADOR 29.7%
COLOMBIA 9.7
OTHER 2.1
COSTA RICA 12.3
GUAT. 6.6
HONDURAS 3.3
U.S. 2.8
PANAMA 2.4
OTHER 2.8
PHILIPPINES 11.4
OTHER ASIA 2.9
BELGIUM 5.7
OTHER EUROPE 5.2
AFRICA 3.3

Fresh Apple Exports* World Total - 6,528,000 metric tons - Avg. 2003-2005

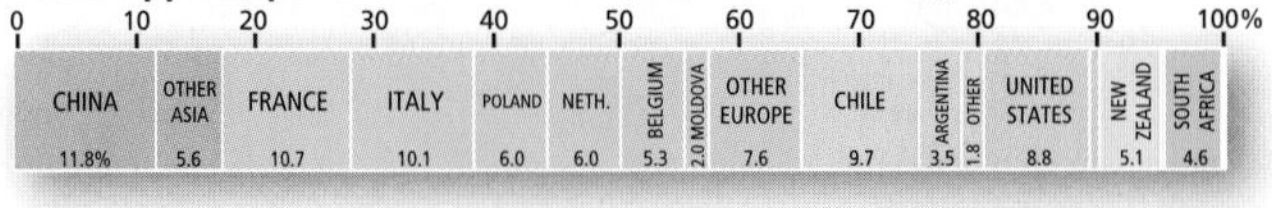
CHINA 11.8%
OTHER ASIA 5.6
FRANCE 10.7
ITALY 10.1
POLAND 6.0
NETH. 6.0
BELGIUM 5.3
MOLDOVA 2.0
OTHER EUROPE 7.6
CHILE 9.7
ARGENTINA 3.5
OTHER 1.8
UNITED STATES 8.8
NEW ZEALAND 5.1
SOUTH AFRICA 4.6

Date Exports* World Total - 468,000 metric tons - Avg. 2003-2005

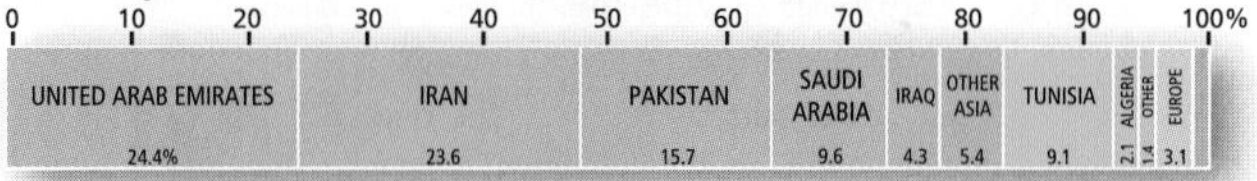
UNITED ARAB EMIRATES 24.4%
IRAN 23.6
PAKISTAN 15.7
SAUDI ARABIA 9.6
IRAQ 4.3
OTHER ASIA 5.4
TUNISIA 9.1
ALGERIA 2.1
OTHER 1.4
EUROPE 3.1

* Including reexports.

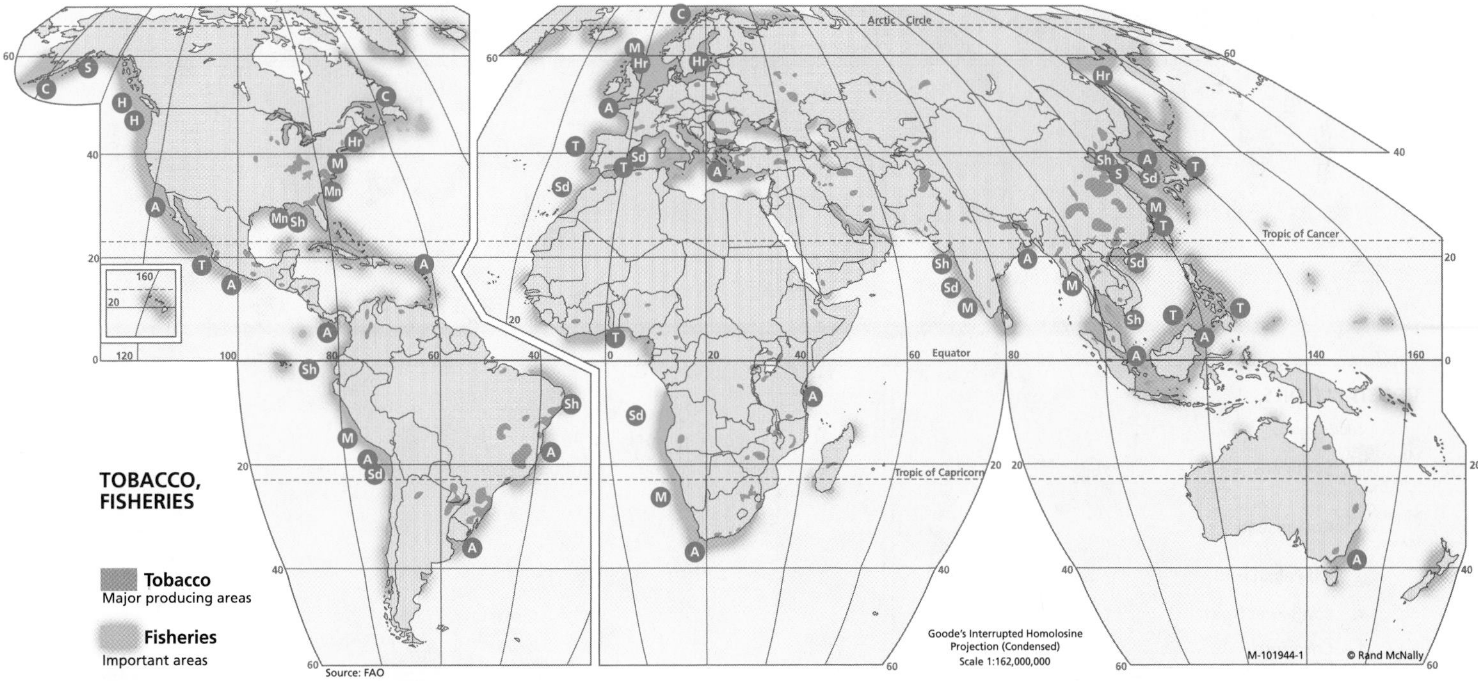
Arctic Circle
Tropic of Cancer
Equator
Tropic of Capricorn
TOBACCO, FISHERIES
Tobacco
Major producing areas
Fisheries
Important areas
Source: FAO
Goode's Interrupted Homolosine Projection (Condensed)
Scale 1:162,000,000
M-101944-1
© Rand McNally

A Anchovy
C Cod
H Halibut
Hr Herring
M Mackeral
Mn Menhaden
S Salmon
Sd Sardine
Sh Shrimp
T Tuna

Tobacco World Production - 6,577,000 metric tons - Avg. 2004-2006

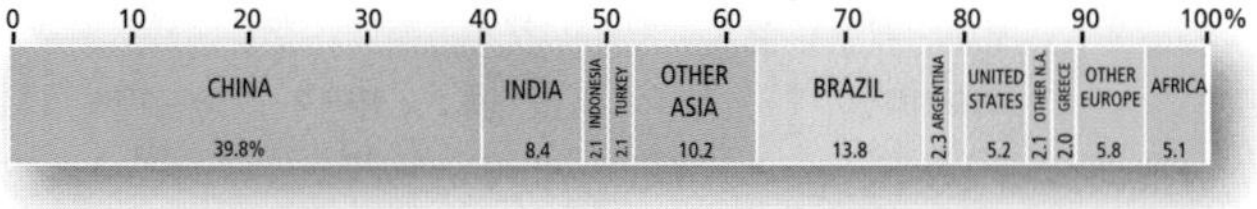
CHINA 39.8%
INDIA 8.4
INDONESIA 2.1
TURKEY 2.1
OTHER ASIA 10.2
BRAZIL 13.8
ARGENTINA 2.3
UNITED STATES 5.2
OTHER N.A. 2.1
GREECE 2.0
OTHER EUROPE 5.8
AFRICA 5.1

Fisheries World Total - 158,174,000 metric tons - Avg. 2004-2006

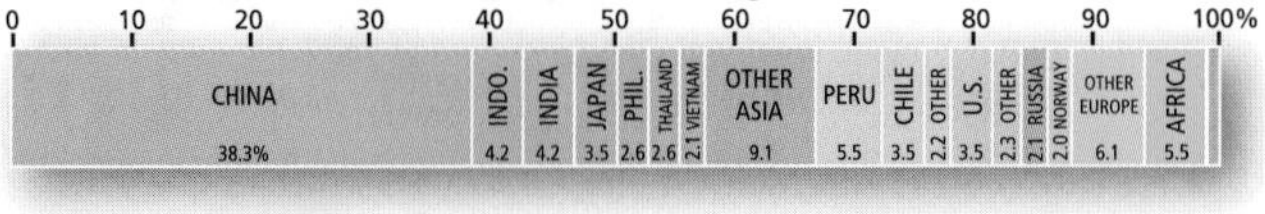
CHINA 38.3%
INDO. 4.2
INDIA 4.2
JAPAN 3.5
PHIL. 2.6
THAILAND 2.6
VIETNAM 2.1
OTHER ASIA 9.1
PERU 5.5
CHILE 3.5
OTHER 2.2
U.S. 3.5
OTHER 2.3
RUSSIA 2.1
NORWAY 2.0
OTHER EUROPE 6.1
AFRICA 5.5

Cigarette Consumption World Total - 5,317,500,000,000 cigarettes - 2004

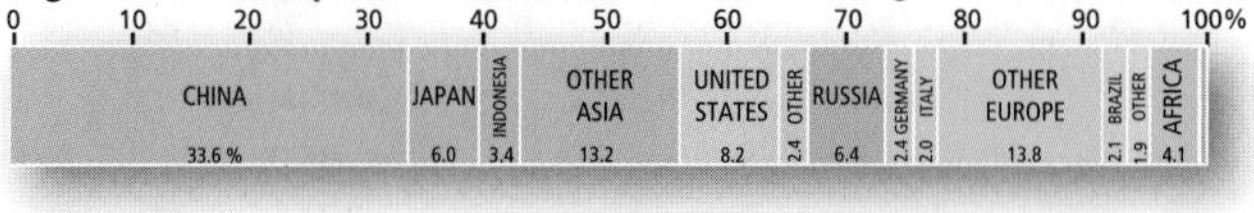
CHINA 33.6 %
JAPAN 6.0
INDONESIA 3.4
OTHER ASIA 13.2
UNITED STATES 8.2
OTHER 2.4
RUSSIA 6.4
GERMANY 2.4
ITALY 2.0
OTHER EUROPE 13.8
BRAZIL 2.1
OTHER 1.9
AFRICA 4.1

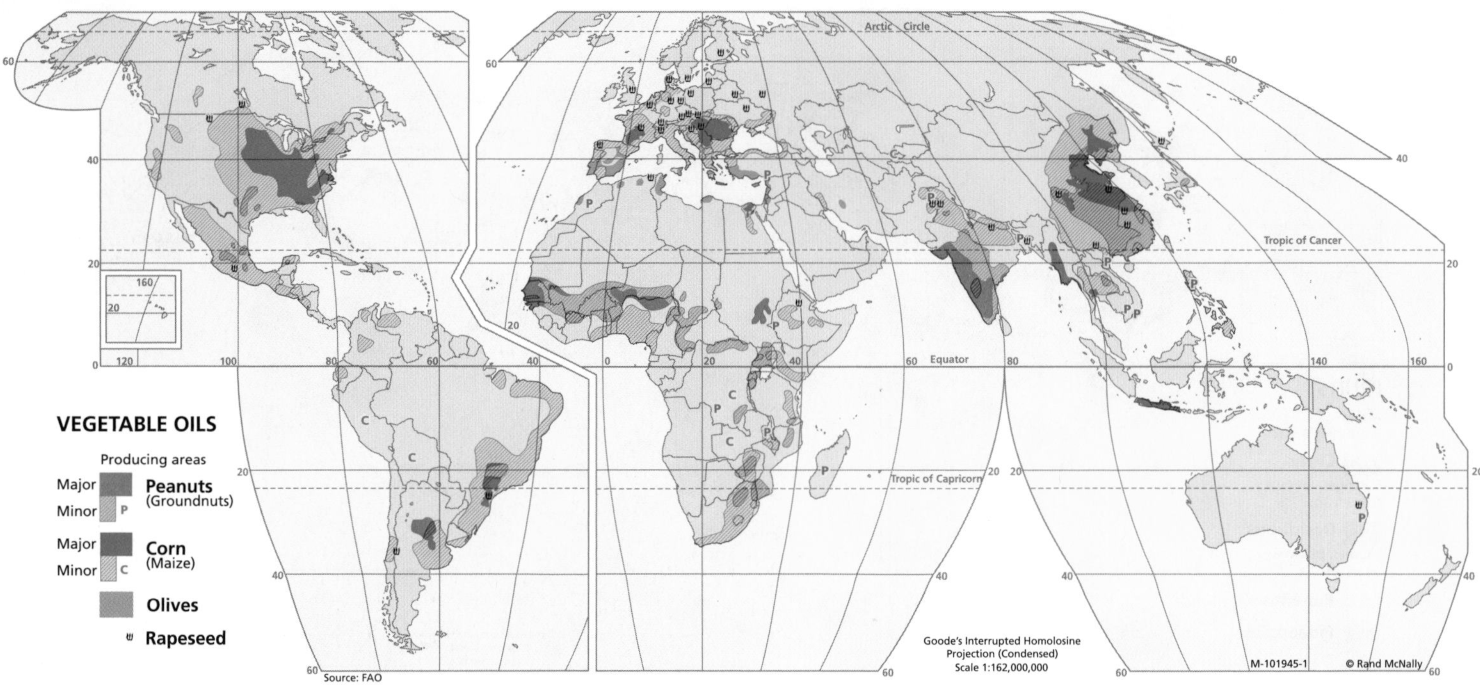
VEGETABLE OILS
Producing areas
Major
Minor
Peanuts (Groundnuts)
Major
Minor
Corn (Maize)
Olives
Rapeseed
Arctic Circle
Tropic of Cancer
Equator
Tropic of Capricorn
Source: FAO
Goode's Interrupted Homolosine Projection (Condensed)
Scale 1:162,000,000
M-101945-1
© Rand McNally

Peanut Oil World Production - 5,190,000 metric tons - Avg. 2004-2006

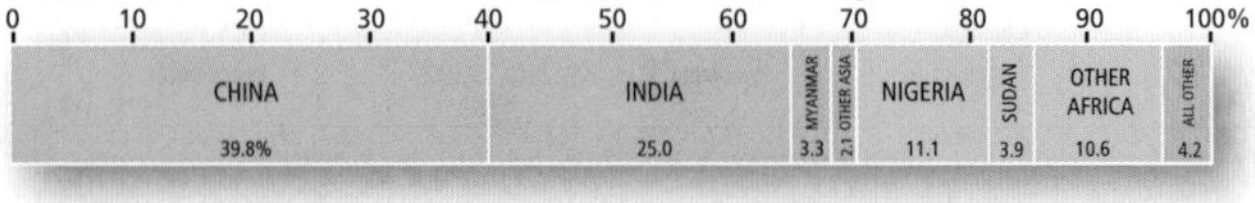
CHINA 39.8%
INDIA 25.0
MYANMAR 3.3
OTHER ASIA 2.1
NIGERIA 11.1
SUDAN 3.9
OTHER AFRICA 10.6
ALL OTHER 4.2

Canola (Rapeseed) Oil World Production - 16,019,000 metric tons - Avg. 2004-2006

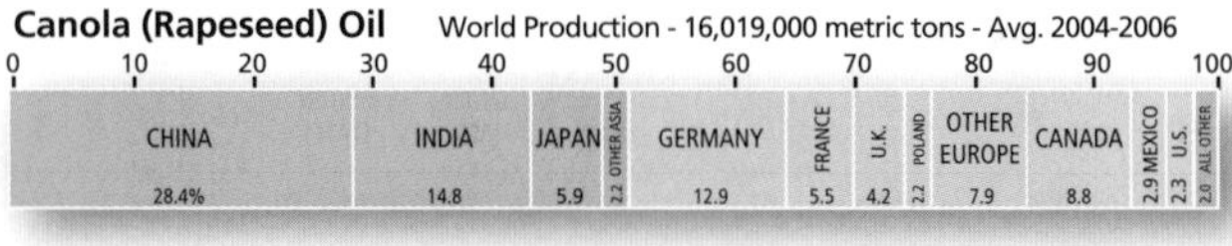
CHINA 28.4%
INDIA 14.8
JAPAN 5.9
OTHER ASIA 2.2
GERMANY 12.9
FRANCE 5.5
U.K. 4.2
POLAND 2.2
OTHER EUROPE 7.9
CANADA 8.8
MEXICO 2.9
U.S. 2.3
ALL OTHER 2.0

Corn Oil World Production - 2,078,000 metric tons - Avg. 2004-2006

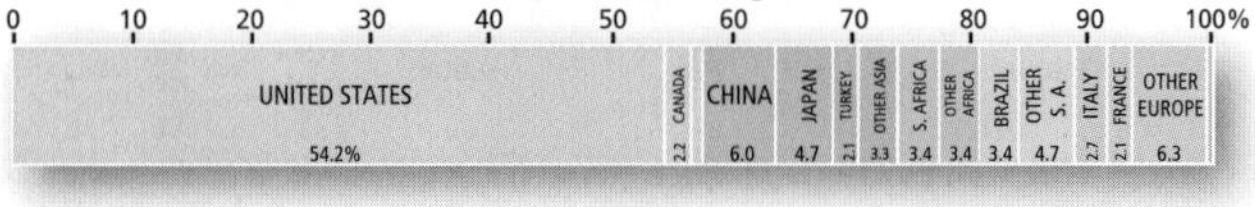
UNITED STATES 54.2%
CANADA 2.2
CHINA 6.0
JAPAN 4.7
TURKEY 2.1
OTHER ASIA 3.3
S. AFRICA 3.4
OTHER AFRICA 3.4
BRAZIL 3.4
OTHER S.A. 4.7
ITALY 2.7
FRANCE 2.1
OTHER EUROPE 6.3

Olive Oil World Production - 2,686,000 metric tons - Avg. 2004-2006

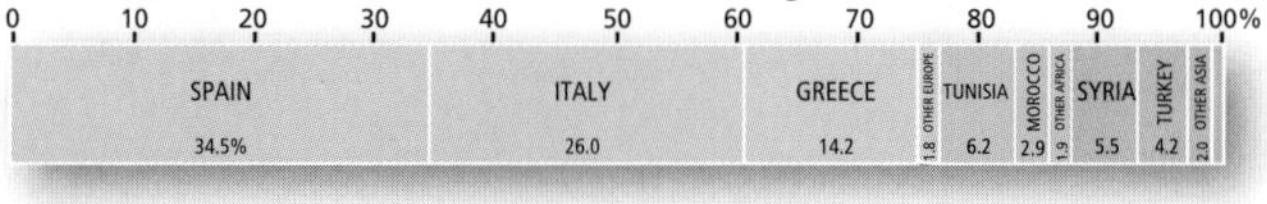
SPAIN 34.5%
ITALY 26.0
GREECE 14.2
OTHER EUROPE 1.8
TUNISIA 6.2
MOROCCO 2.9
OTHER AFRICA 1.9
SYRIA 5.5
TURKEY 4.2
OTHER ASIA 2.8

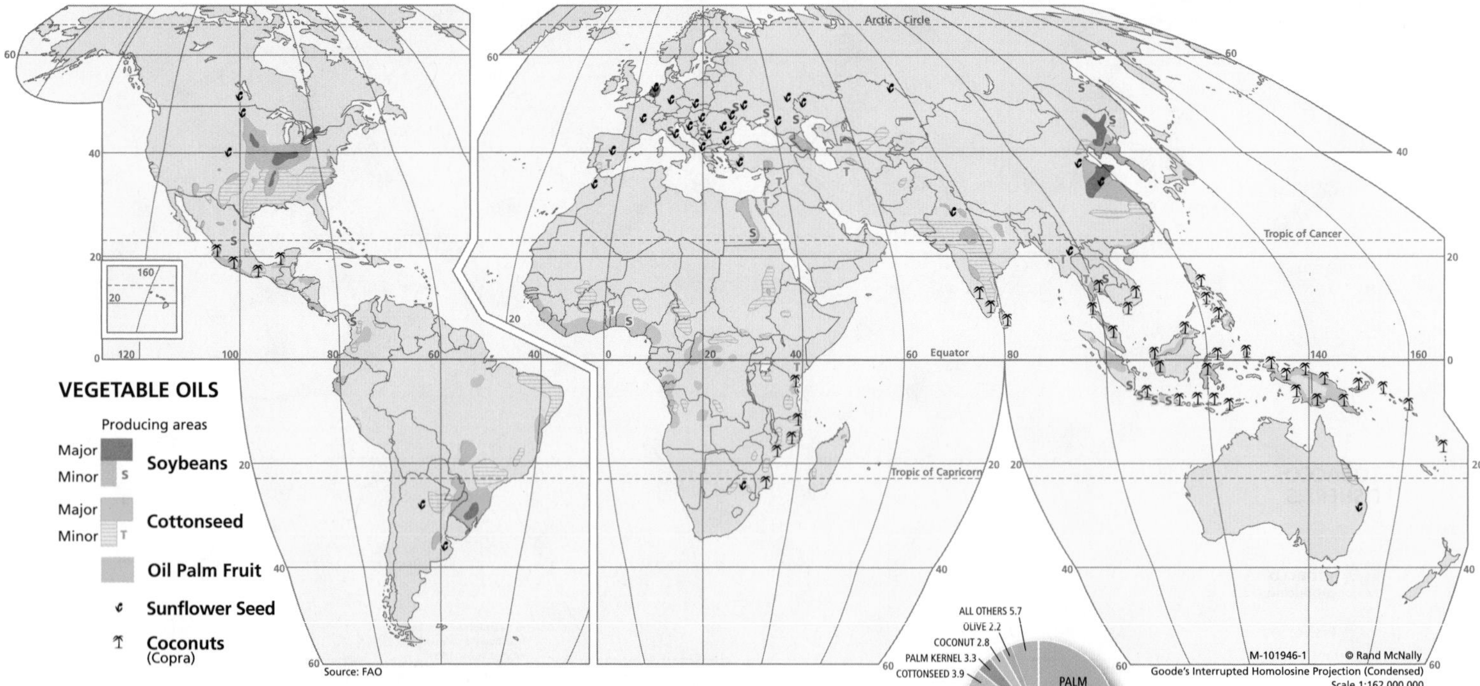
VEGETABLE OILS
Producing areas
Major
Minor
Soybeans
Major
Minor
Cottonseed
Oil Palm Fruit
Sunflower Seed
Coconuts (Copra)
Arctic Circle
Tropic of Cancer
Equator
Tropic of Capricorn
Source: FAO
M-101946-1
© Rand McNally
Goode's Interrupted Homolosine Projection (Condensed)
Scale 1:162,000,000

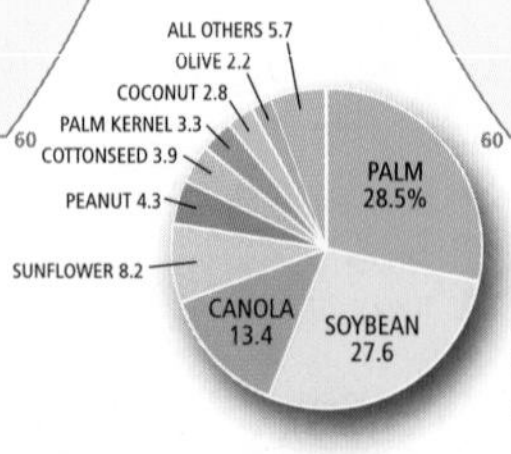
ALL OTHERS 5.7
OLIVE 2.2
COCONUT 2.8
PALM KERNEL 3.3
COTTONSEED 3.9
PEANUT 4.3
SUNFLOWER 8.2
CANOLA 13.4
SOYBEAN 27.6
PALM 28.5%

Vegetable Oils
World Production - 119,860,000 metric tons - Avg. 2004-2006

Soybean Oil World Production - 33,054,000 metric tons - Avg. 2004-2006

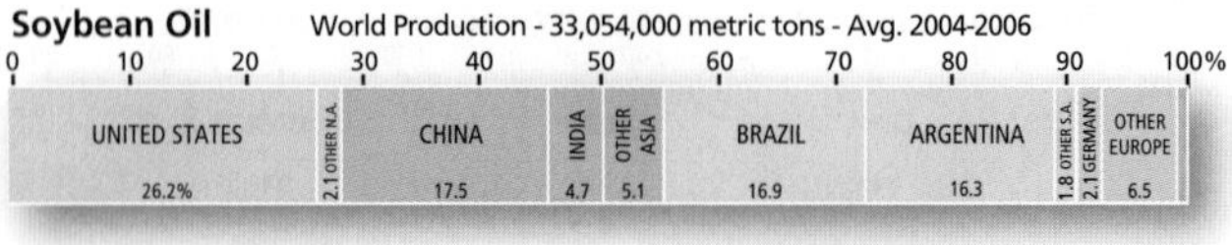
UNITED STATES 26.2%
OTHER N.A. 2.1
CHINA 17.5
INDIA 4.7
OTHER ASIA 5.1
BRAZIL 16.9
ARGENTINA 16.3
OTHER S.A. 1.8
GERMANY 2.1
OTHER EUROPE 6.5

Palm Oil World Production - 34,219,000 metric tons - Avg. 2004-2006

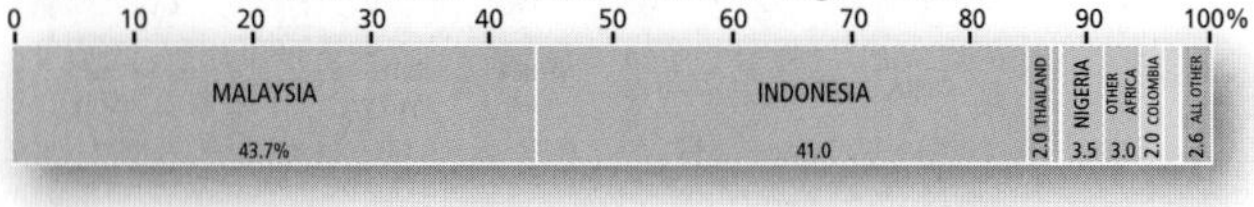
MALAYSIA 43.7%
INDONESIA 41.0
THAILAND 2.0
NIGERIA 3.5
OTHER AFRICA 3.0
COLOMBIA 2.0
ALL OTHER 2.6

Sunflower Oil World Production - 9,925,000 metric tons - Avg. 2004-2006

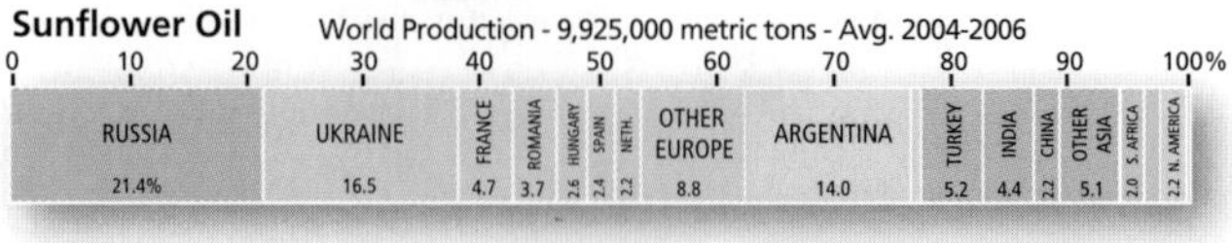
RUSSIA 21.4%
UKRAINE 16.5
FRANCE 4.7
ROMANIA 3.7
HUNGARY 2.6
SPAIN 2.4
NETH. 2.2
OTHER EUROPE 8.8
ARGENTINA 14.0
TURKEY 5.2
INDIA 4.4
CHINA 2.2
OTHER ASIA 5.1
S. AFRICA 2.0
N. AMERICA 2.2

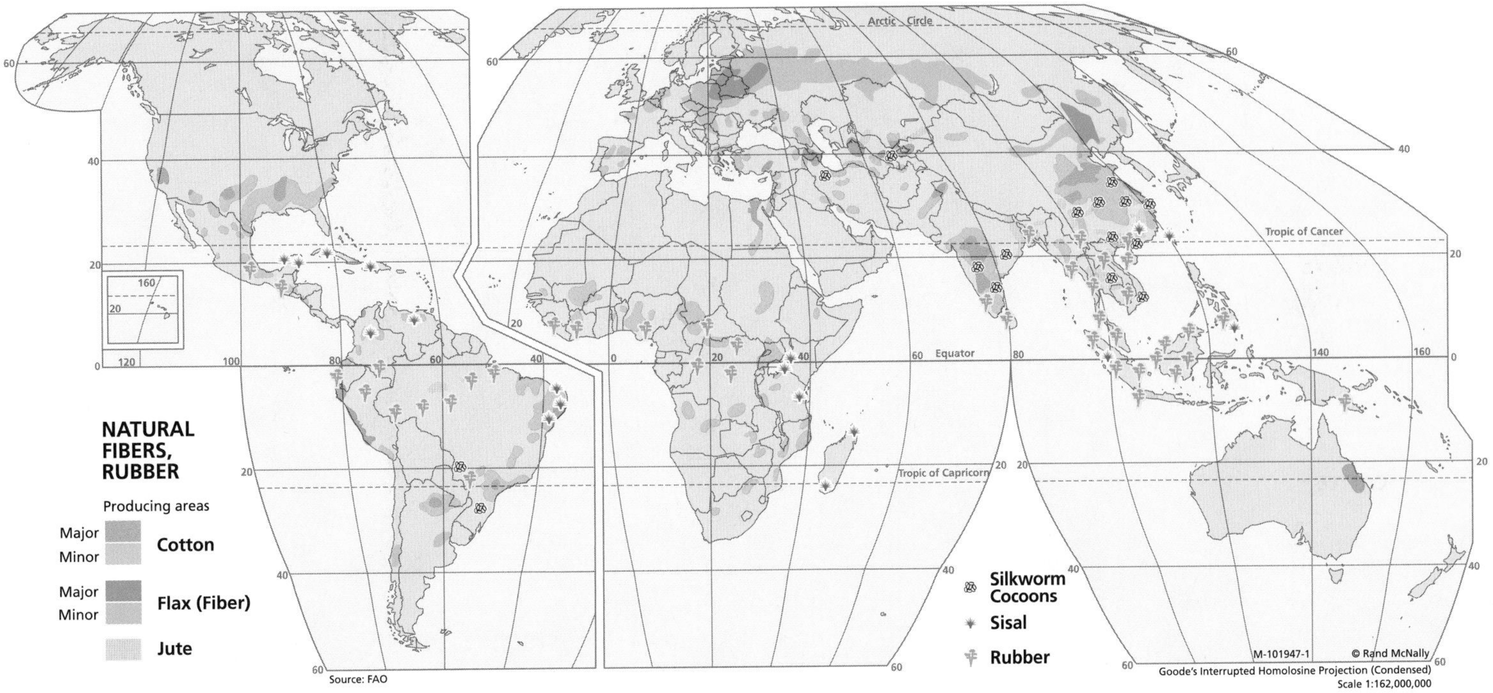
NATURAL FIBERS, RUBBER
Producing areas
Major
Minor
Cotton
Major
Minor
Flax (Fiber)
Jute
Silkworm Cocoons
Sisal
Rubber
Arctic Circle
Tropic of Cancer
Equator
Tropic of Capricorn
Source: FAO
M-101947-1
© Rand McNally
Goode's Interrupted Homolosine Projection (Condensed)
Scale 1:162,000,000

Cotton (Lint) World Production - 24,714,000 metric tons - Avg. 2004-2006

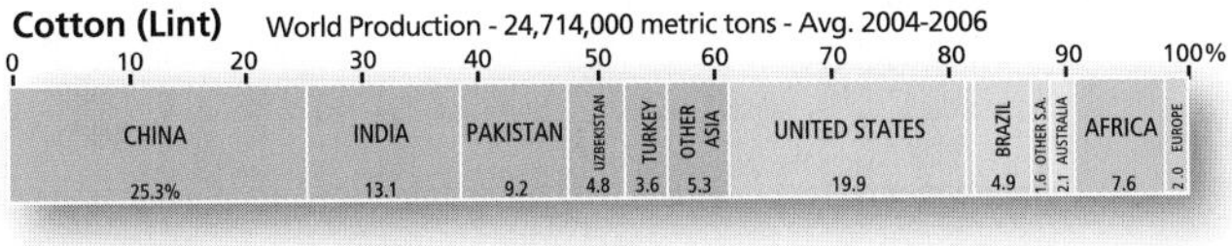
CHINA 25.3%
INDIA 13.1
PAKISTAN 9.2
UZBEKISTAN 4.8
TURKEY 3.6
OTHER ASIA 5.3
UNITED STATES 19.9
BRAZIL 4.9
OTHER S.A. 1.6
AUSTRALIA 2.1
AFRICA 7.6
EUROPE 2.0

Jute (and Substitutes) World Production - 3,015,000 metric tons - Avg. 2004-2006

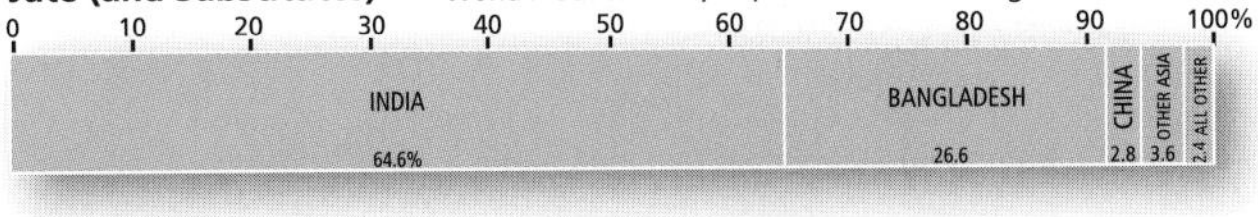
INDIA 64.6%
BANGLADESH 26.6
CHINA 2.8
OTHER ASIA 3.6
ALL OTHER 2.4

Silkworm Cocoons World Production - 547,000 metric tons - Avg. 2004-2007

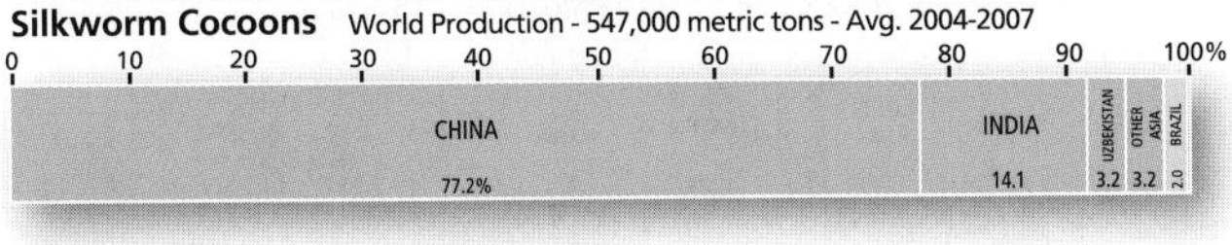
CHINA 77.2%
INDIA 14.1
UZBEKISTAN 3.2
OTHER ASIA 3.2
BRAZIL 2.0

Flax (Fiber and Tow) World Production - 995,000 metric tons - Avg. 2004-2006

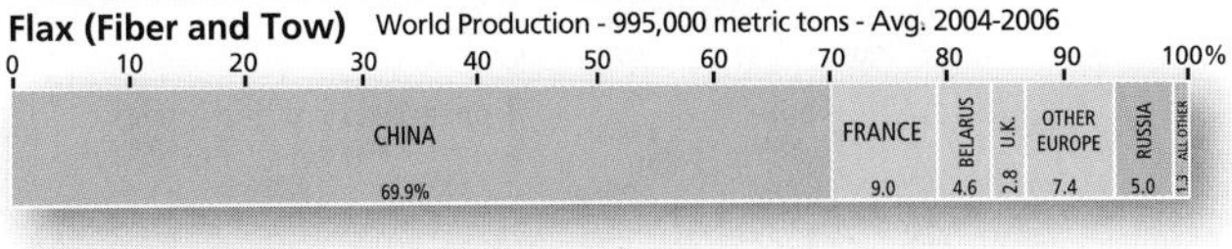
CHINA 69.9%
FRANCE 9.0
BELARUS 4.6
U.K. 2.8
OTHER EUROPE 7.4
RUSSIA 5.0
ALL OTHER 1.3

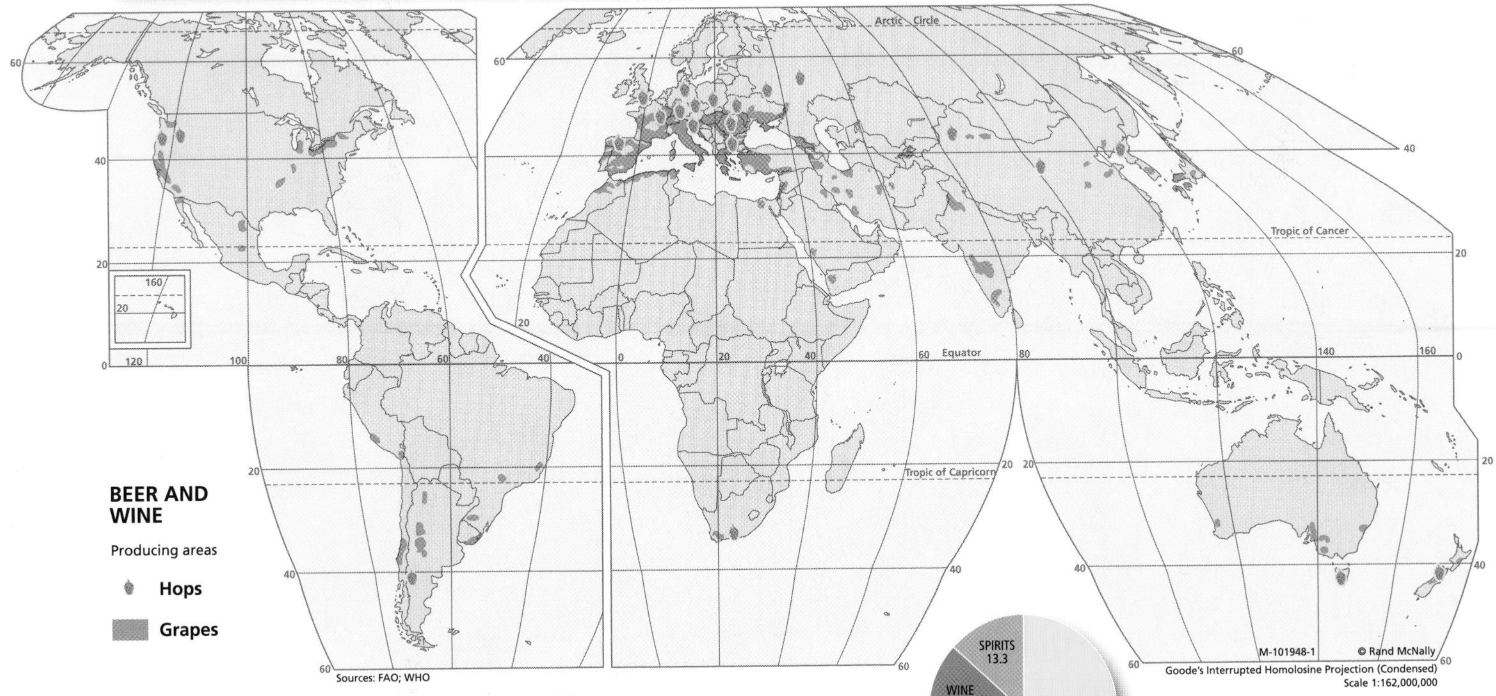
BEER AND WINE
Producing areas
Hops
Grapes
Arctic Circle
Tropic of Cancer
Equator
Tropic of Capricorn
Sources: FAO; WHO
M-101948-1
© Rand McNally
Goode's Interrupted Homolosine Projection (Condensed)
Scale 1:162,000,000

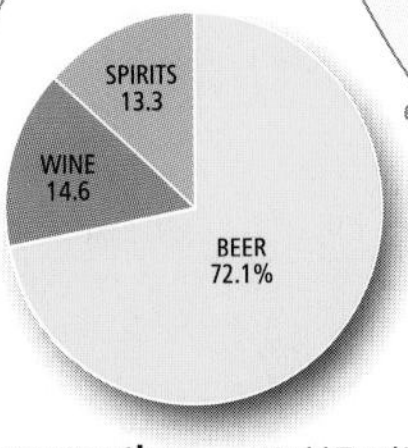
SPIRITS 13.3
WINE 14.6
BEER 72.1%

World Consumption of Alcoholic Beverages

Beer World Production - 151,251,000 metric tons - Avg. 2004-2006

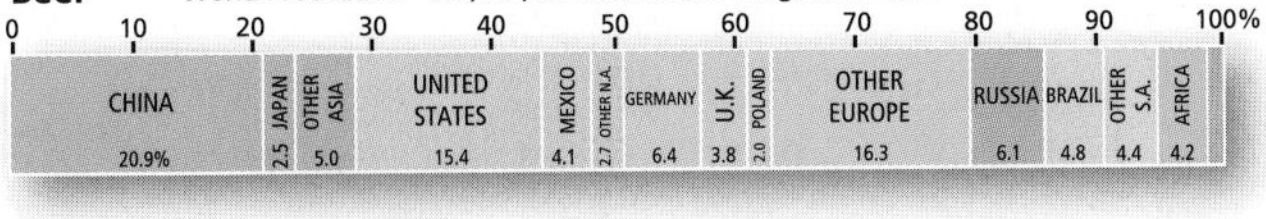
CHINA 20.9%
JAPAN 2.5
OTHER ASIA 5.0
UNITED STATES 15.4
MEXICO 4.1
OTHER N.A. 2.7
GERMANY 6.4
U.K. 3.8
POLAND 2.0
OTHER EUROPE 16.3
RUSSIA 6.1
BRAZIL 4.8
OTHER S.A. 4.4
AFRICA 4.2

Wine World Production - 29,119,000 metric tons - Avg. 2004-2006

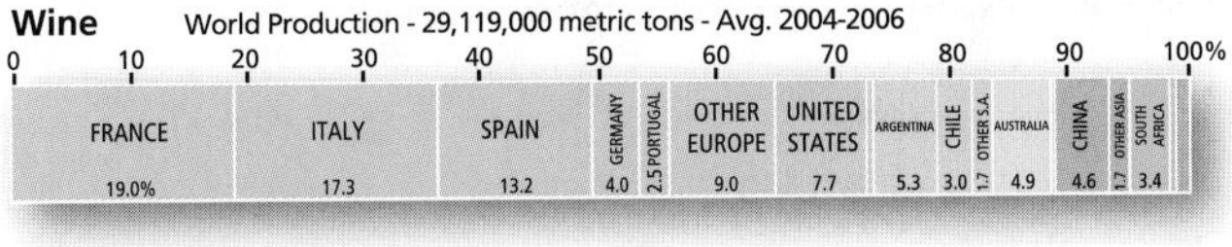
FRANCE 19.0%
ITALY 17.3
SPAIN 13.2
GERMANY 4.0
PORTUGAL 2.5
OTHER EUROPE 9.0
UNITED STATES 7.7
ARGENTINA 5.3
CHILE 3.0
OTHER S.A. 1.7
AUSTRALIA 4.9
CHINA 4.6
OTHER ASIA 1.7
SOUTH AFRICA 3.4

Alcohol Consumption World Total* - 19,932,522,000 liters - 2004

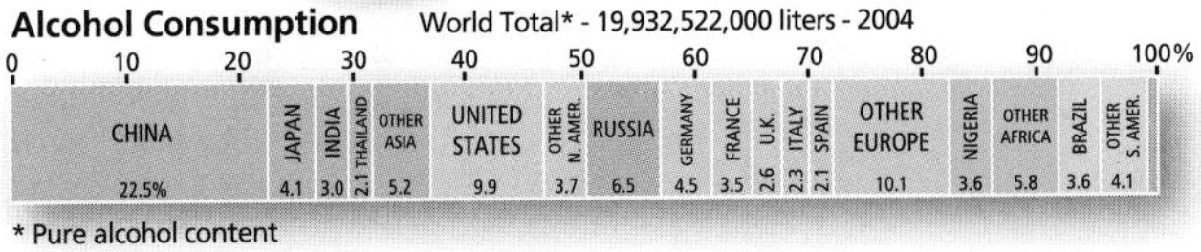
CHINA 22.5%
JAPAN 4.1
INDIA 3.0
THAILAND 2.1
OTHER ASIA 5.2
UNITED STATES 9.9
OTHER N. AMER. 3.7
RUSSIA 6.5
GERMANY 4.5
FRANCE 3.5
U.K. 2.6
ITALY 2.3
SPAIN 2.1
OTHER EUROPE 10.1
NIGERIA 3.6
OTHER AFRICA 5.8
BRAZIL 3.6
OTHER S. AMER. 4.1

* Pure alcohol content

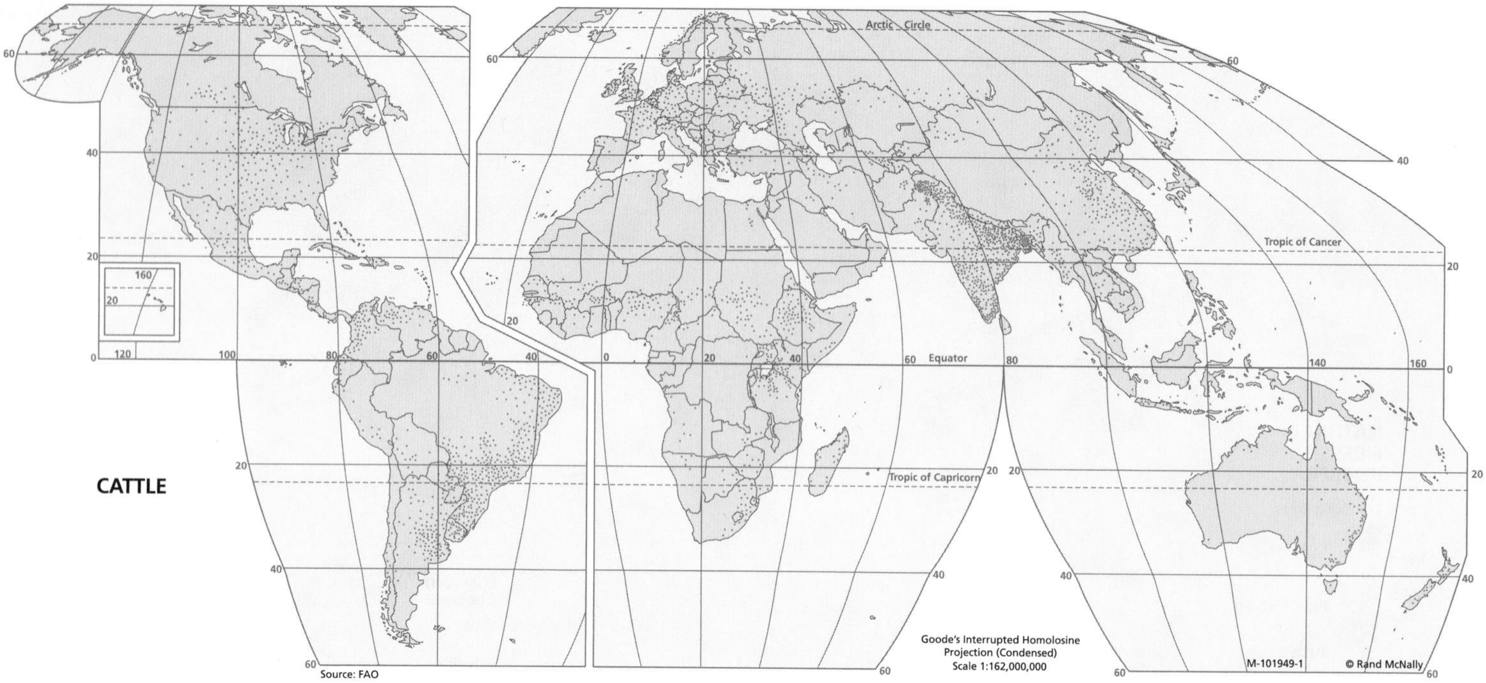
Arctic Circle
Tropic of Cancer
Equator
Tropic of Capricorn
Source: FAO
Goode's Interrupted Homolosine
Projection (Condensed)
Scale 1:162,000,000
M-101949-1
© Rand McNally

CATTLE

Cattle
Each dot represents 500,000 head

Cattle World Total - 1,373,367,000 head - Avg. 2004-2006

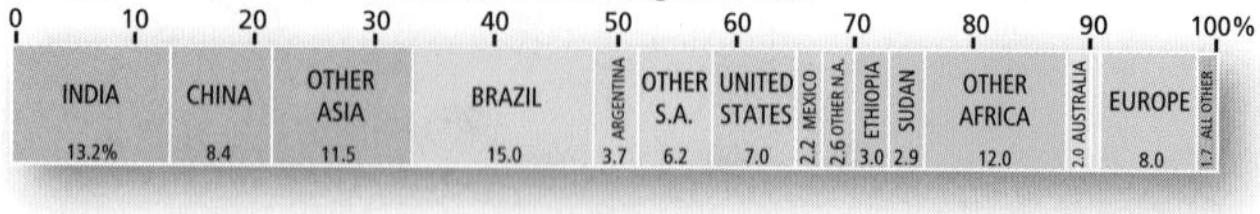
INDIA 13.2%
CHINA 8.4
OTHER ASIA 11.5
BRAZIL 15.0
ARGENTINA 3.7
OTHER S.A. 6.2
UNITED STATES 7.0
2.2 MEXICO
2.6 OTHER N.A.
ETHIOPIA 3.0
SUDAN 2.9
OTHER AFRICA 12.0
2.0 AUSTRALIA
EUROPE 8.0
1.7 ALL OTHER

Beef and Veal Imports World Total* - 7,837,000 metric tons - Avg. 2003-2005

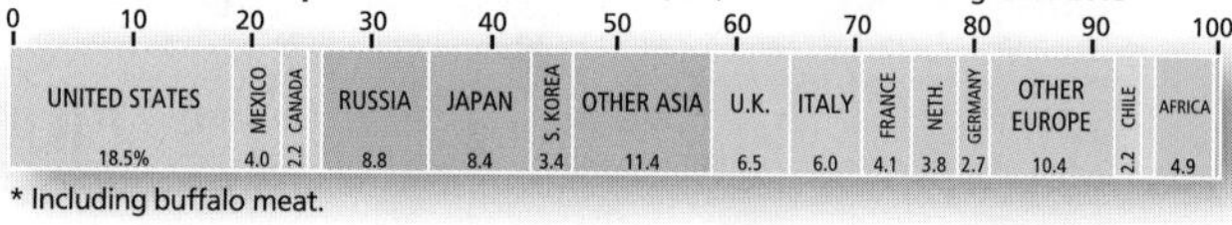
UNITED STATES 18.5%
MEXICO 4.0
2.2 CANADA
RUSSIA 8.8
JAPAN 8.4
S. KOREA 3.4
OTHER ASIA 11.4
U.K. 6.5
ITALY 6.0
FRANCE 4.1
NETH. 3.8
GERMANY 2.7
OTHER EUROPE 10.4
CHILE 2.2
AFRICA 4.9

* Including buffalo meat.

Beef and Veal Exports World Total*† - 8,330,000 metric tons - Avg. 2003-2005

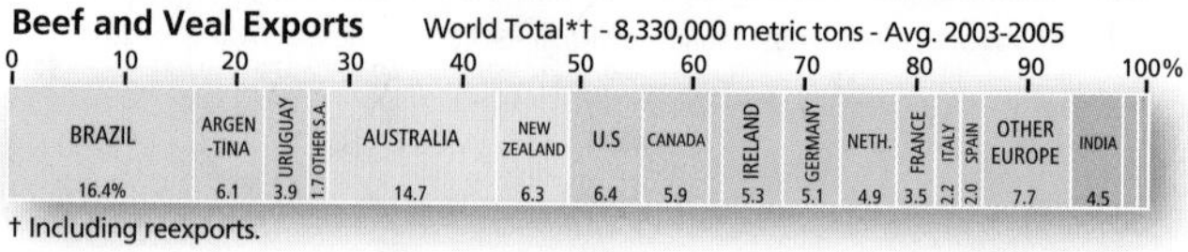
BRAZIL 16.4%
ARGEN-TINA 6.1
URUGUAY 3.9
1.7 OTHER S.A.
AUSTRALIA 14.7
NEW ZEALAND 6.3
U.S 6.4
CANADA 5.9
IRELAND 5.3
GERMANY 5.1
NETH. 4.9
FRANCE 3.5
ITALY 2.2
SPAIN 2.0
OTHER EUROPE 7.7
INDIA 4.5

† Including reexports.

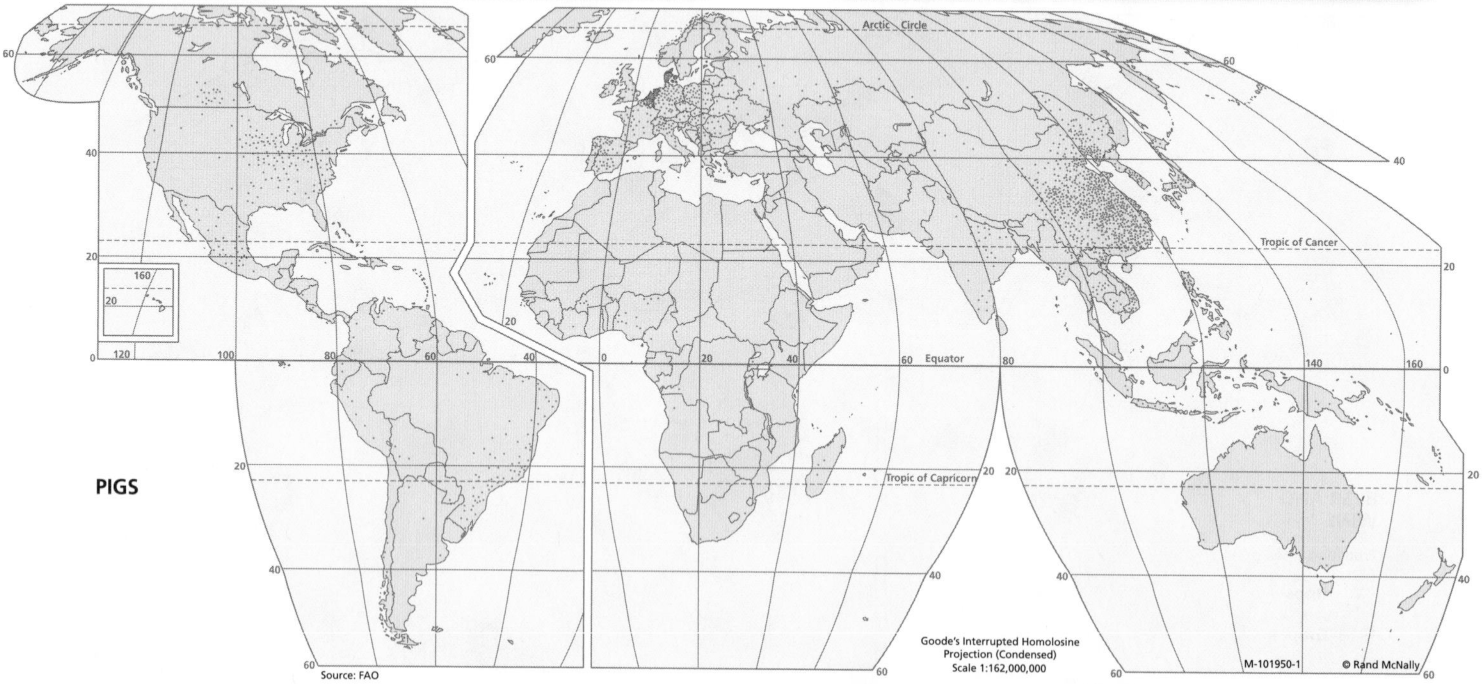
Arctic Circle
Tropic of Cancer
Equator
Tropic of Capricorn
Source: FAO
Goode's Interrupted Homolosine
Projection (Condensed)
Scale 1:162,000,000
M-101950-1
© Rand McNally

PIGS

Pigs
Each dot represents 500,000 head

Pigs World Total - 969,741,000 head - Avg. 2004-2006

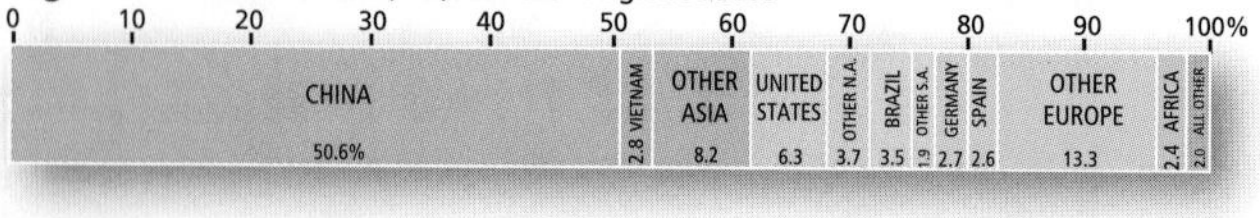
CHINA 50.6%
2.8 VIETNAM
OTHER ASIA 8.2
UNITED STATES 6.3
OTHER N.A. 3.7
BRAZIL 3.5
1.9 OTHER S.A.
GERMANY 2.7
SPAIN 2.6
OTHER EUROPE 13.3
2.4 AFRICA
2.0 ALL OTHER

Pork Imports World Total - 6,653,000 metric tons - Avg. 2003-2005

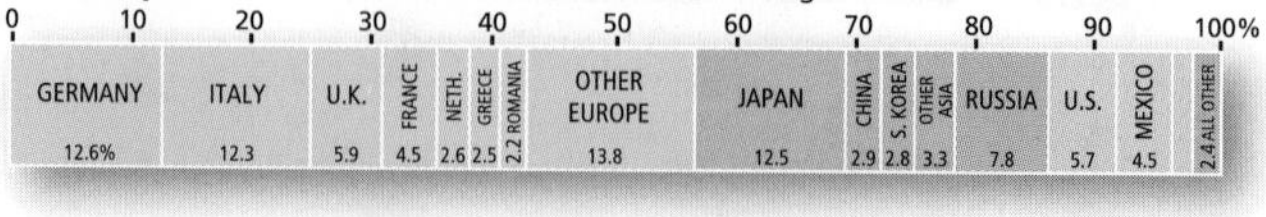
GERMANY 12.6%
ITALY 12.3
U.K. 5.9
FRANCE 4.5
NETH. 2.6
GREECE 2.5
2.2 ROMANIA
OTHER EUROPE 13.8
JAPAN 12.5
CHINA 2.9
S. KOREA 2.8
OTHER ASIA 3.3
RUSSIA 7.8
U.S. 5.7
MEXICO 4.5
2.4 ALL OTHER

Pork Exports World Total** - 7,002,000 metric tons - Avg. 2003-2005

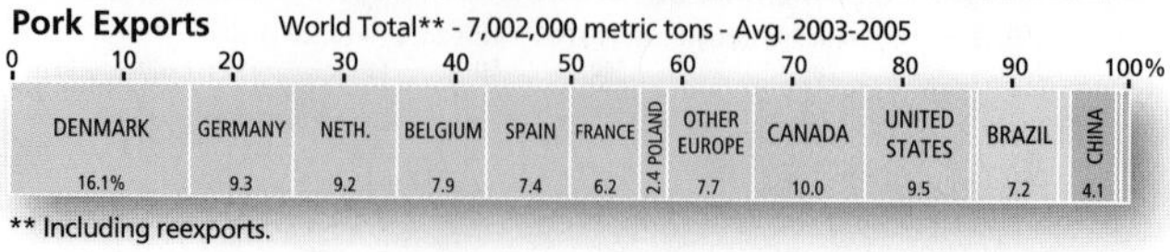
DENMARK 16.1%
GERMANY 9.3
NETH. 9.2
BELGIUM 7.9
SPAIN 7.4
FRANCE 6.2
2.4 POLAND
OTHER EUROPE 7.7
CANADA 10.0
UNITED STATES 9.5
BRAZIL 7.2
CHINA 4.1

** Including reexports.

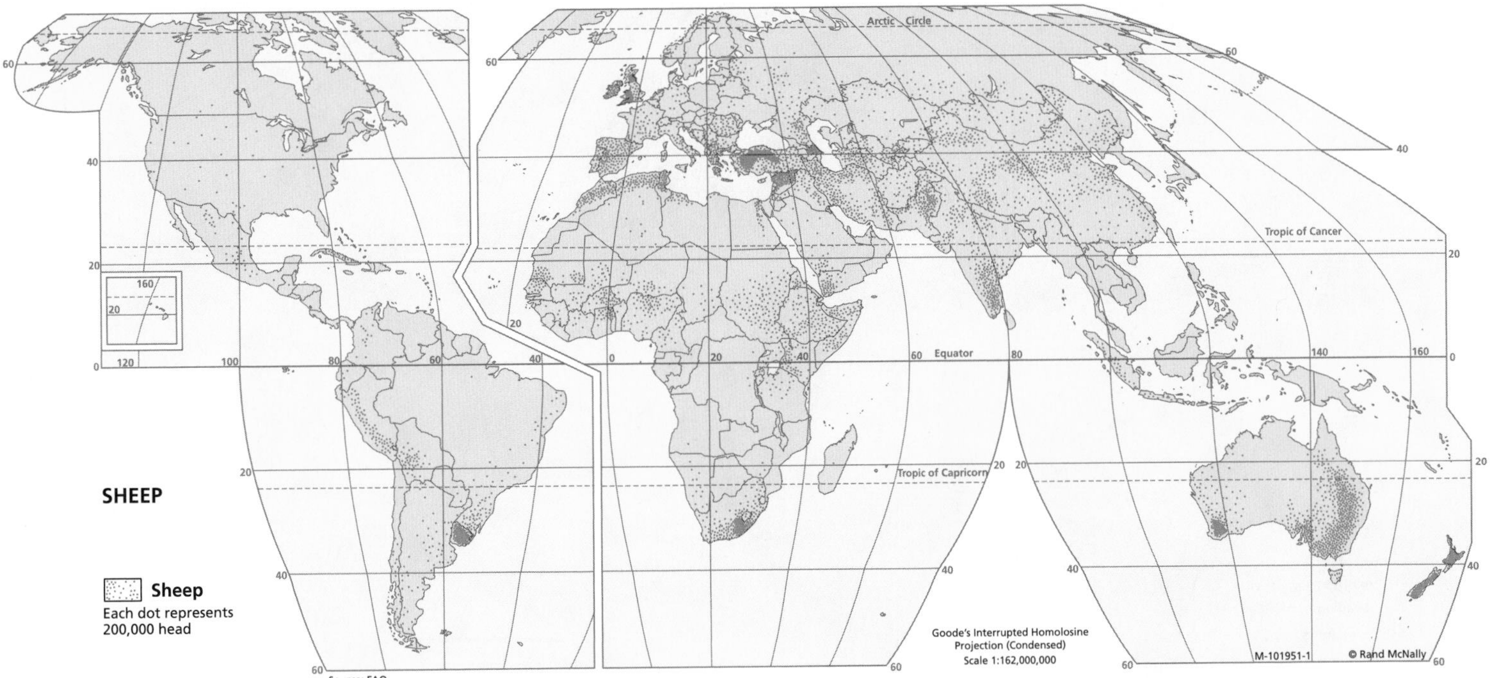
Arctic Circle
Tropic of Cancer
Equator
Tropic of Capricorn
SHEEP
Sheep
Each dot represents
200,000 head
Goode's Interrupted Homolosine
Projection (Condensed)
Scale 1:162,000,000
M-101951-1
© Rand McNally

Source: FAO

Sheep World Total - 1,087,072,000 head - Avg. 2004-2006

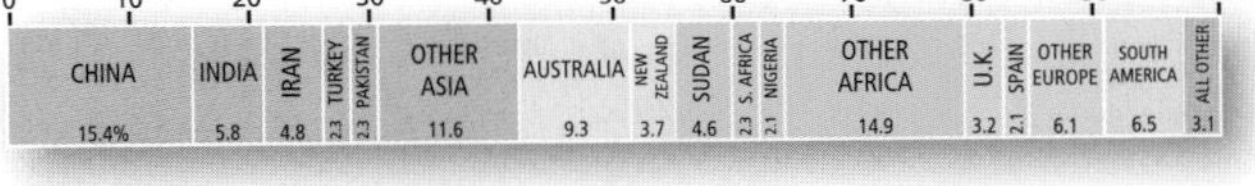
0 10 20 30 40 50 60 70 80 90 100%
CHINA 15.4%
INDIA 5.8
IRAN 4.8
TURKEY 2.3
PAKISTAN 2.3
OTHER ASIA 11.6
AUSTRALIA 9.3
NEW ZEALAND 3.7
SUDAN 4.6
S. AFRICA 2.3
NIGERIA 2.1
OTHER AFRICA 14.9
U.K. 3.2
SPAIN 2.1
OTHER EUROPE 6.1
SOUTH AMERICA 6.5
ALL OTHER 3.1

Wool (Raw) World Production - 2,186,000 metric tons - Avg. 2004-2006

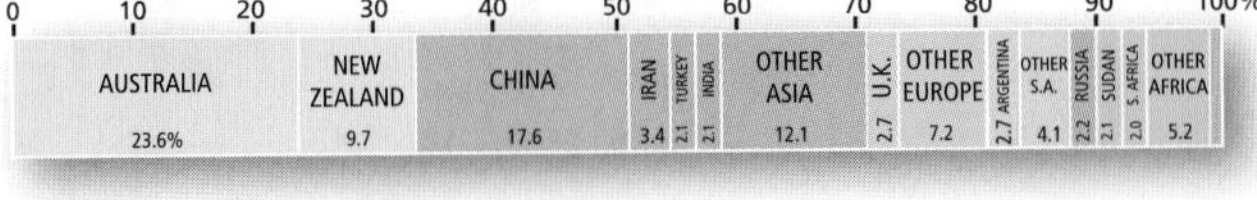
0 10 20 30 40 50 60 70 80 90 100%
AUSTRALIA 23.6%
NEW ZEALAND 9.7
CHINA 17.6
IRAN 3.4
TURKEY 2.1
INDIA 2.1
OTHER ASIA 12.1
U.K. 2.7
OTHER EUROPE 7.2
ARGENTINA 2.7
OTHER S.A. 4.1
RUSSIA 2.2
SUDAN 2.1
S. AFRICA 2.0
OTHER AFRICA 5.2

Wool Imports (Raw) World Total - 508,000 metric tons - Avg. 2003-2005

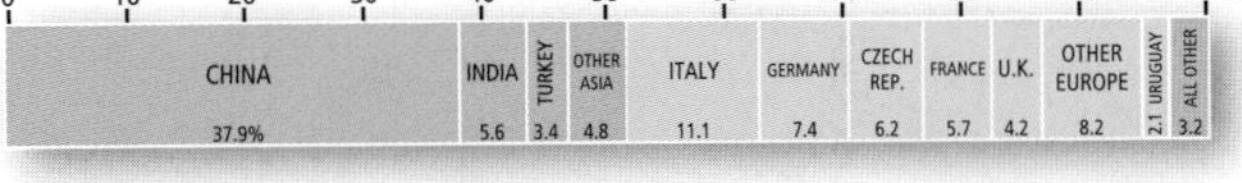
0 10 20 30 40 50 60 70 80 90 100%
CHINA 37.9%
INDIA 5.6
TURKEY 3.4
OTHER ASIA 4.8
ITALY 11.1
GERMANY 7.4
CZECH REP. 6.2
FRANCE 5.7
U.K. 4.2
OTHER EUROPE 8.2
URUGUAY 2.1
ALL OTHER 3.2

Wool Exports (Raw) World Total* - 560,000 metric tons - Avg. 2003-2005

0 10 20 30 40 50 60 70 80 90 100%
AUSTRALIA 59.5%
NEW ZEALAND 7.8
S. AFRICA 3.6
GERMANY 3.1
OTHER EUROPE 14.0
ARG. 2.3
OTHER S.A. 2.4
ASIA 4.9
ALL OTHER 1.7

* Including reexports.

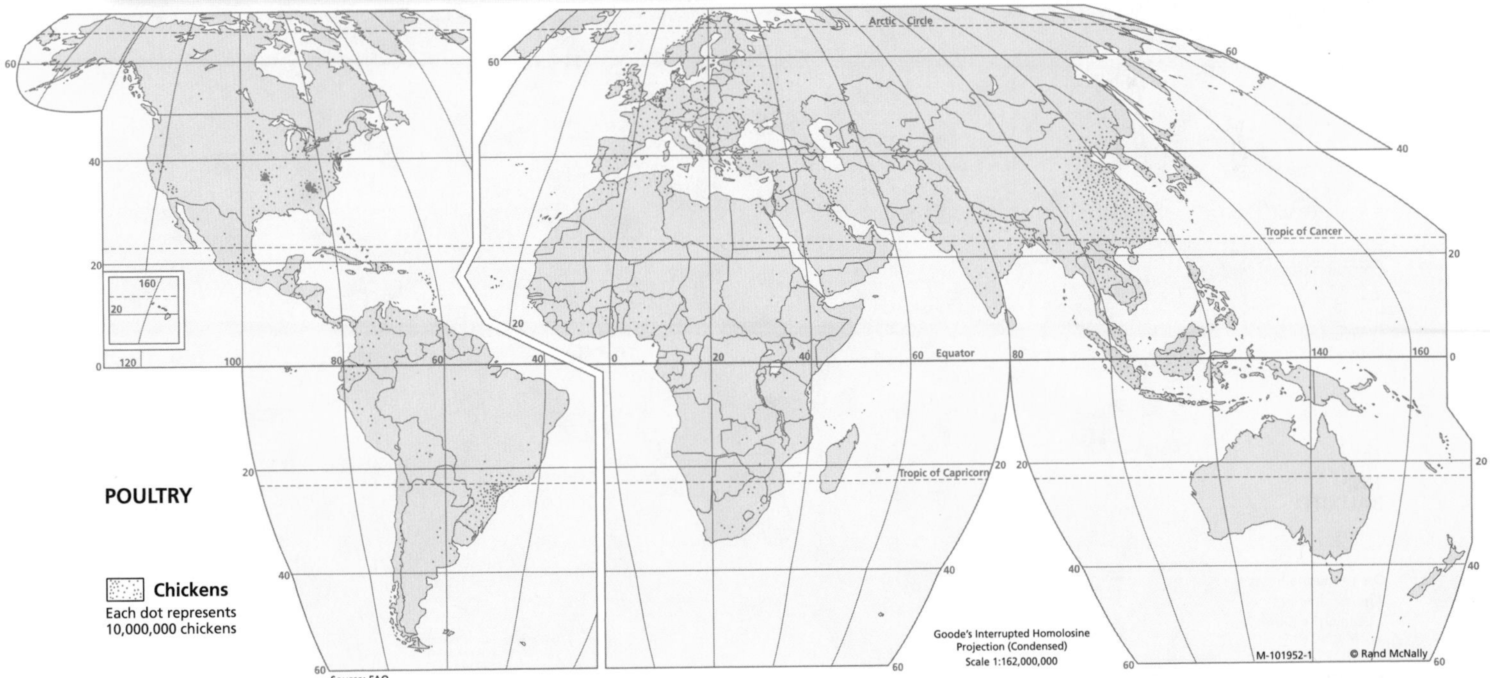
Arctic Circle
Tropic of Cancer
Equator
Tropic of Capricorn
POULTRY
Chickens
Each dot represents
10,000,000 chickens
Goode's Interrupted Homolosine
Projection (Condensed)
Scale 1:162,000,000
M-101952-1
© Rand McNally

Source: FAO

Chickens World Total - 16,617,237,000 head - Avg. 2004-2006

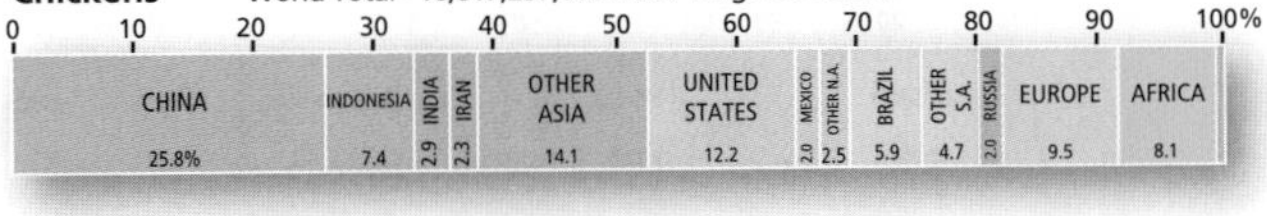
0 10 20 30 40 50 60 70 80 90 100%
CHINA 25.8%
INDONESIA 7.4
INDIA 2.9
IRAN 2.3
OTHER ASIA 14.1
UNITED STATES 12.2
MEXICO 2.0
OTHER N.A. 2.5
BRAZIL 5.9
OTHER S.A. 4.7
RUSSIA 2.0
EUROPE 9.5
AFRICA 8.1

Ducks World Total - 1,043,493,000 head - Avg. 2004-2006

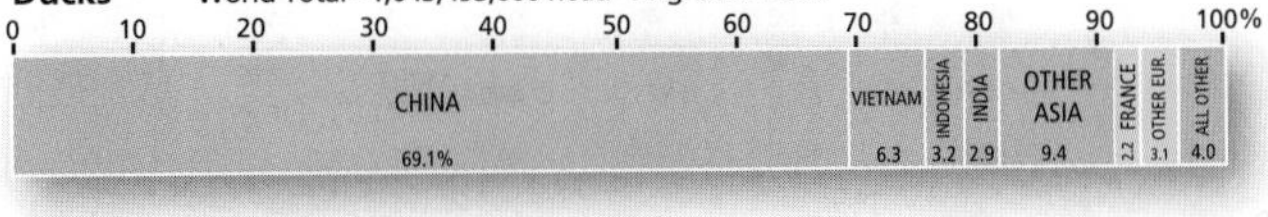
0 10 20 30 40 50 60 70 80 90 100%
CHINA 69.1%
VIETNAM 6.3
INDONESIA 3.2
INDIA 2.9
OTHER ASIA 9.4
FRANCE 2.2
OTHER EUR. 3.1
ALL OTHER 4.0

Hen Eggs World Production - 59,789,000 eggs - Avg. 2004-2006

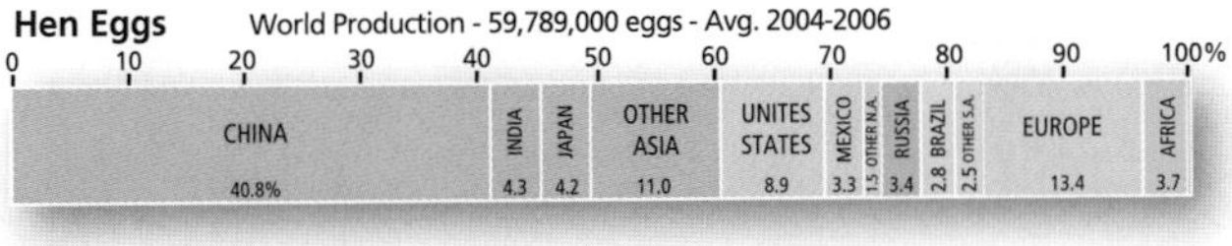
0 10 20 30 40 50 60 70 80 90 100%
CHINA 40.8%
INDIA 4.3
JAPAN 4.2
OTHER ASIA 11.0
UNITES STATES 8.9
MEXICO 3.3
OTHER N.A. 1.5
RUSSIA 3.4
BRAZIL 2.8
OTHER S.A. 2.5
EUROPE 13.4
AFRICA 3.7

Turkeys World Total - 465,310,000 head - Avg. 2004-2006

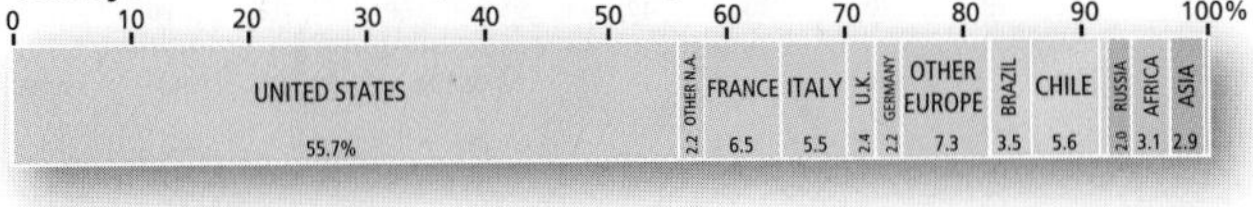
0 10 20 30 40 50 60 70 80 90 100%
UNITED STATES 55.7%
OTHER N.A. 2.2
FRANCE 6.5
ITALY 5.5
U.K. 2.4
GERMANY 2.2
OTHER EUROPE 7.3
BRAZIL 3.5
CHILE 5.6
RUSSIA 2.0
AFRICA 3.1
ASIA 2.9

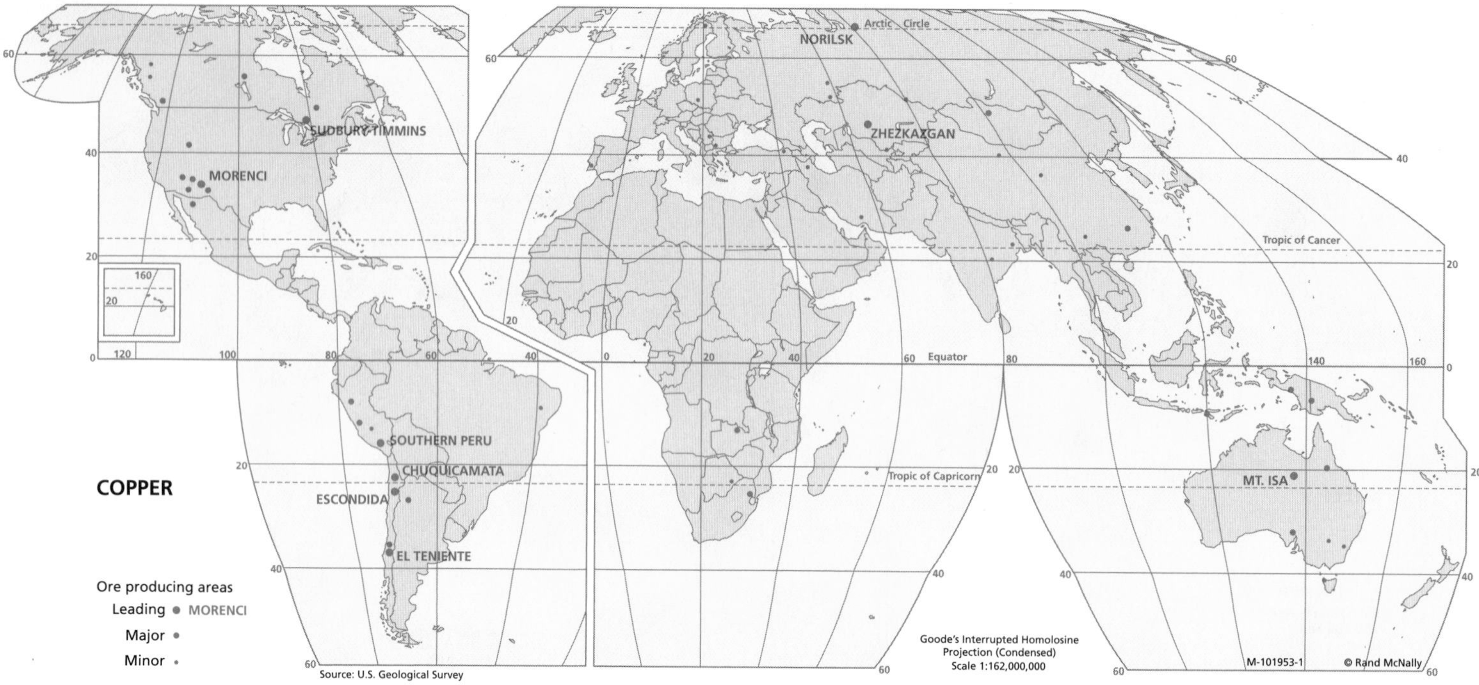
COPPER
Ore producing areas
Leading • MORENCI
Major •
Minor •
NORILSK
ZHEZKAZGAN
SUDBURY-TIMMINS
MORENCI
SOUTHERN PERU
CHUQUICAMATA
ESCONDIDA
EL TENIENTE
MT. ISA
Arctic Circle
Tropic of Cancer
Equator
Tropic of Capricorn
Goode's Interrupted Homolosine Projection (Condensed) Scale 1:162,000,000
M-101953-1
© Rand McNally
Source: U.S. Geological Survey

Copper Reserves World Total - 941,300,000 metric tons - 2006

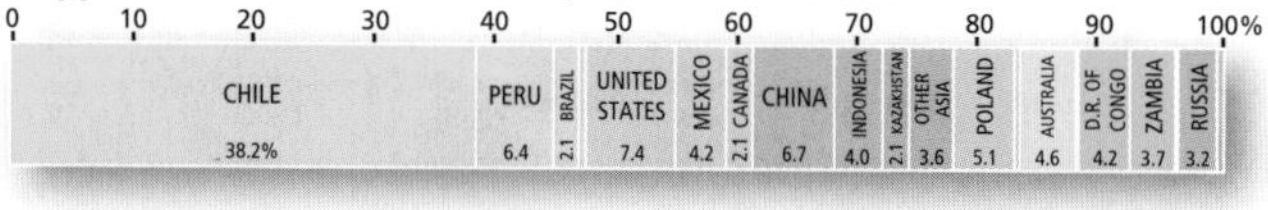
0 10 20 30 40 50 60 70 80 90 100%
CHILE 38.2%
PERU 6.4
BRAZIL 2.1
UNITED STATES 7.4
MEXICO 4.2
CANADA 2.1
CHINA 6.7
INDONESIA 4.0
KAZAKHSTAN 2.1
OTHER ASIA 3.6
POLAND 5.1
AUSTRALIA 4.6
D.R. OF CONGO 4.2
ZAMBIA 3.7
RUSSIA 3.2

Copper World Mine Production - 14,961,000 metric tons (metal content) - Avg. 2004-2006

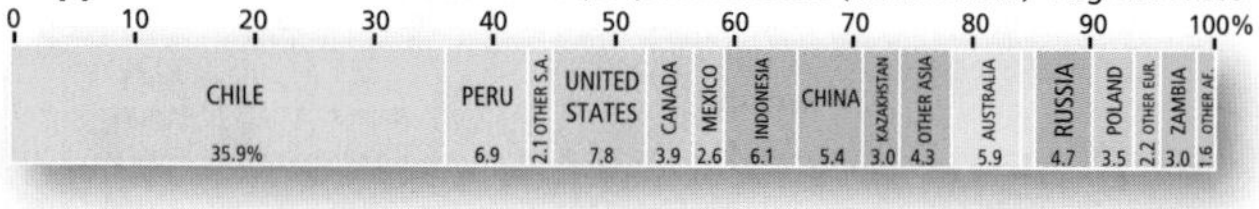
0 10 20 30 40 50 60 70 80 90 100%
CHILE 35.9%
PERU 6.9
OTHER S.A. 2.1
UNITED STATES 7.8
CANADA 3.9
MEXICO 2.6
INDONESIA 6.1
CHINA 5.4
KAZAKHSTAN 3.0
OTHER ASIA 4.3
AUSTRALIA 5.9
RUSSIA 4.7
POLAND 3.5
OTHER EUR. 2.2
ZAMBIA 3.0
OTHER AF. 1.6

Refined Copper World Total - 16,662,000 metric tons - Avg. 2004-2006

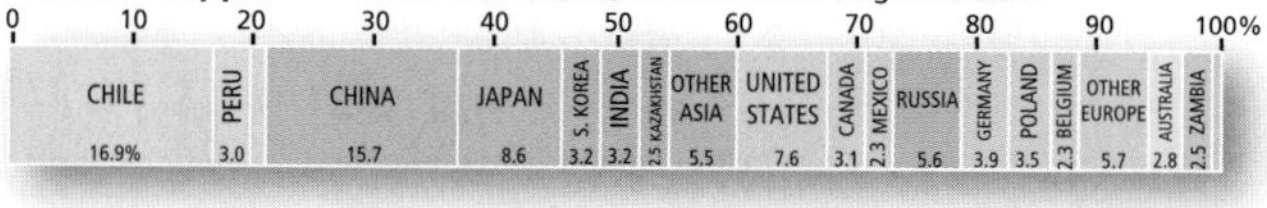
0 10 20 30 40 50 60 70 80 90 100%
CHILE 16.9%
PERU 3.0
CHINA 15.7
JAPAN 8.6
S. KOREA 3.2
INDIA 3.2
KAZAKHSTAN 2.5
OTHER ASIA 5.5
UNITED STATES 7.6
CANADA 3.1
MEXICO 2.3
RUSSIA 5.6
GERMANY 3.9
POLAND 3.5
BELGIUM 2.3
OTHER EUROPE 5.7
AUSTRALIA 2.8
ZAMBIA 2.5

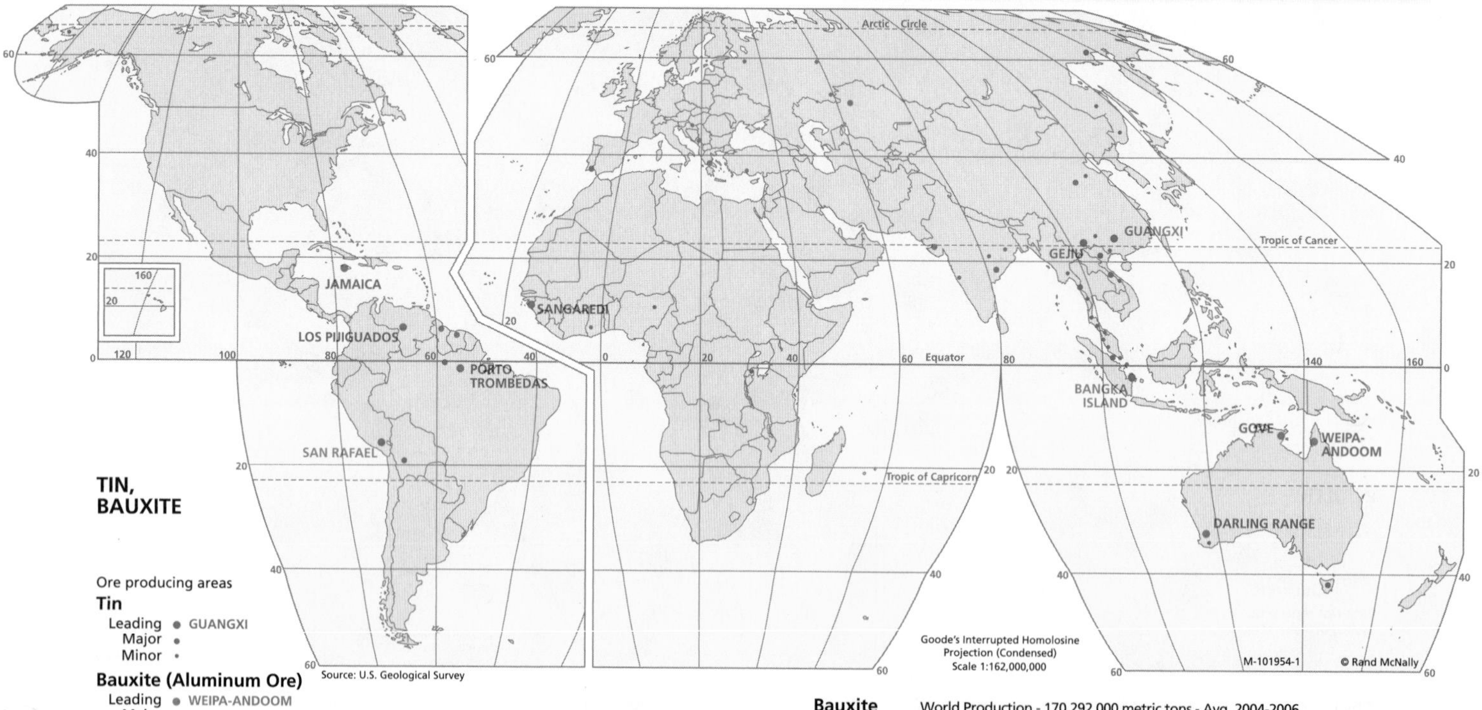
TIN, BAUXITE
Ore producing areas
Tin
Leading • GUANGXI
Major •
Minor •
Bauxite (Aluminum Ore)
Leading • WEIPA-ANDOOM
Major •
Minor •
GUANGXI
GEJIU
JAMAICA
SANGAREDI
LOS PIJIGUADOS
PORTO TROMBEDAS
SAN RAFAEL
BANGKA ISLAND
GOVE
WEIPA-ANDOOM
DARLING RANGE
Arctic Circle
Tropic of Cancer
Equator
Tropic of Capricorn
Goode's Interrupted Homolosine Projection (Condensed) Scale 1:162,000,000
M-101954-1
© Rand McNally
Source: U.S. Geological Survey

Bauxite World Production - 170,292,000 metric tons - Avg. 2004-2006

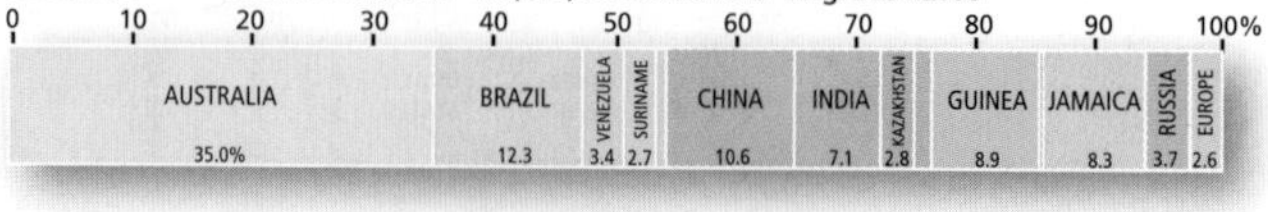
0 10 20 30 40 50 60 70 80 90 100%
AUSTRALIA 35.0%
BRAZIL 12.3
VENEZUELA 3.4
SURINAME 2.7
CHINA 10.6
INDIA 7.1
KAZAKHSTAN 2.8
GUINEA 8.9
JAMAICA 8.3
RUSSIA 3.7
EUROPE 2.6

Tin World Mine Production - 301,300 metric tons (metal content) - Avg. 2004-2006

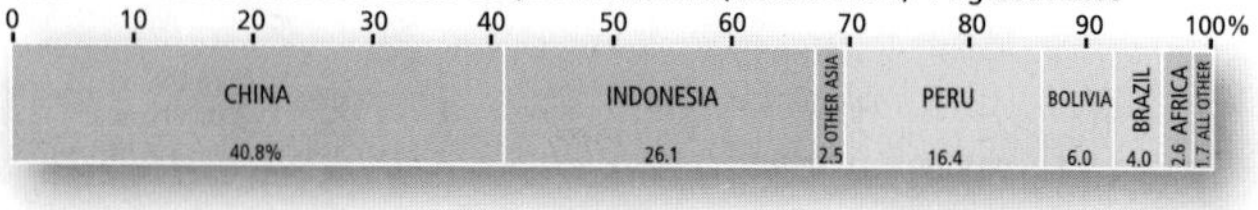
0 10 20 30 40 50 60 70 80 90 100%
CHINA 40.8%
INDONESIA 26.1
OTHER ASIA 2.5
PERU 16.4
BOLIVIA 6.0
BRAZIL 4.0
AFRICA 2.6
ALL OTHER 1.7

Aluminum World Production - 31,820,000 metric tons - Avg. 2004-2006

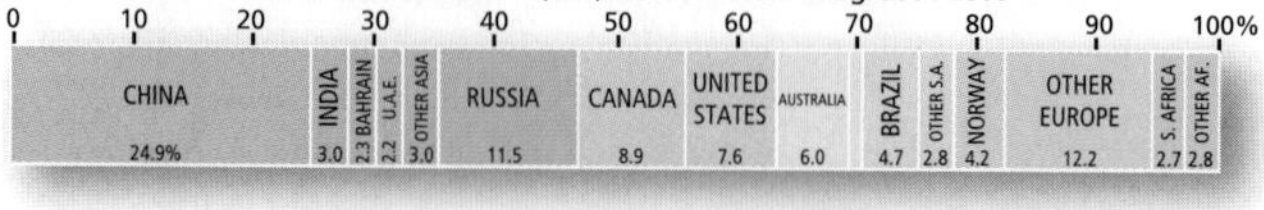
0 10 20 30 40 50 60 70 80 90 100%
CHINA 24.9%
INDIA 3.0
BAHRAIN 2.3
U.A.E. 2.2
OTHER ASIA 3.0
RUSSIA 11.5
CANADA 8.9
UNITED STATES 7.6
AUSTRALIA 6.0
BRAZIL 4.7
OTHER S.A. 2.8
NORWAY 4.2
OTHER EUROPE 12.2
S. AFRICA 2.7
OTHER AF. 2.8

LEAD, LITHIUM

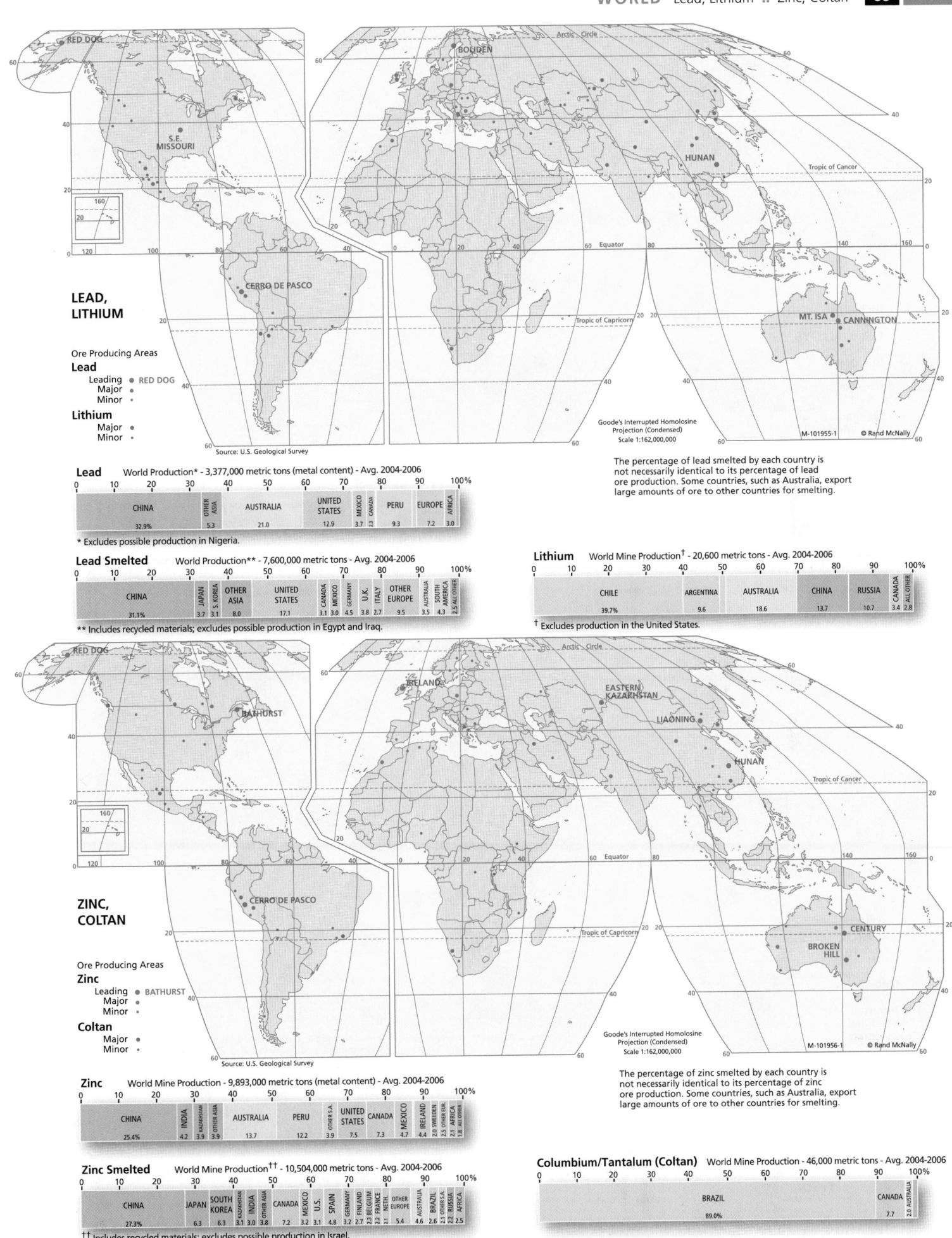

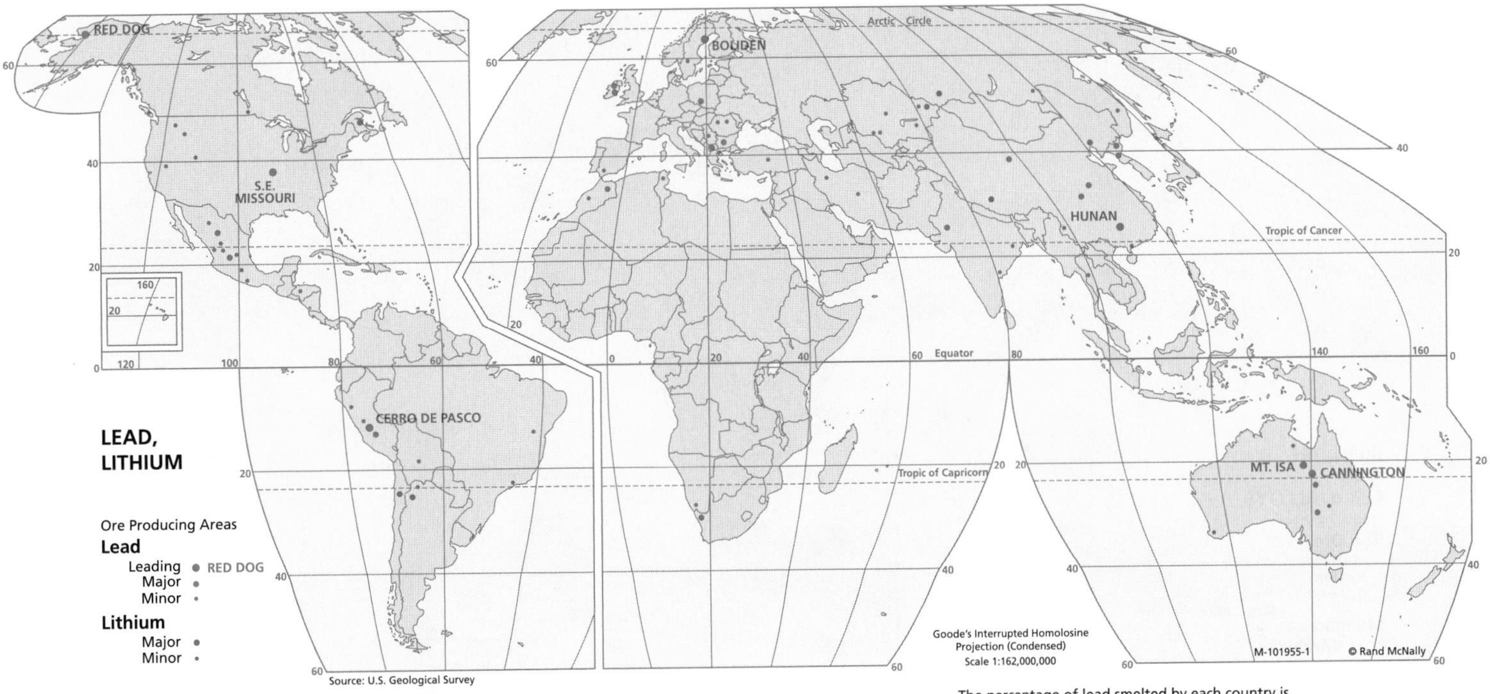
RED DOG
BOLIDEN
Arctic Circle
S.E. MISSOURI
HUNAN
Tropic of Cancer
Equator
CERRO DE PASCO
Tropic of Capricorn
MT. ISA
CANNINGTON
Ore Producing Areas
Lead
Leading ● RED DOG
Major
Minor
Lithium
Major
Minor
Source: U.S. Geological Survey
Goode's Interrupted Homolosine Projection (Condensed)
Scale 1:162,000,000
M-101955-1
© Rand McNally

The percentage of lead smelted by each country is not necessarily identical to its percentage of lead ore production. Some countries, such as Australia, export large amounts of ore to other countries for smelting.

Lead World Production* - 3,377,000 metric tons (metal content) - Avg. 2004-2006

China	Other Asia	Australia	United States	Mexico	Canada	Peru	Europe	Africa
32.9%	5.3	21.0	12.9	3.7	2.3	9.3	7.2	3.0

* Excludes possible production in Nigeria.

Lead Smelted World Production** - 7,600,000 metric tons - Avg. 2004-2006

China	Japan	S. Korea	Other Asia	United States	Canada	Mexico	Germany	U.K.	Italy	Other Europe	Australia	South America	All Other
31.1%	3.7	3.1	8.0	17.1	3.1	3.0	4.5	3.8	2.7	9.5	3.5	4.3	2.5

** Includes recycled materials; excludes possible production in Egypt and Iraq.

Lithium World Mine Production† - 20,600 metric tons - Avg. 2004-2006

Chile	Argentina	Australia	China	Russia	Canada	All Other
39.7%	9.6	18.6	13.7	10.7	3.4	2.8

† Excludes production in the United States.

ZINC, COLTAN

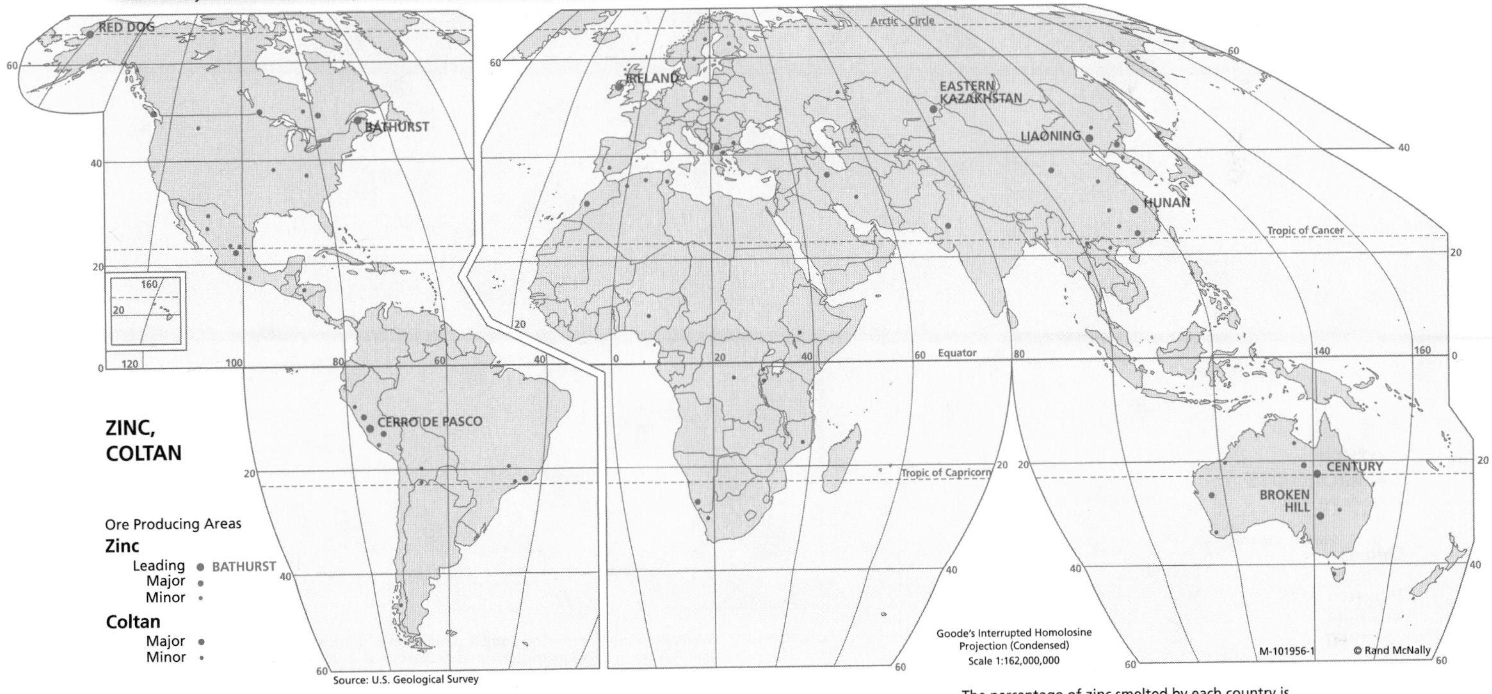
RED DOG
Arctic Circle
IRELAND
EASTERN KAZAKHSTAN
LIAONING
BATHURST
HUNAN
Tropic of Cancer
Equator
CERRO DE PASCO
Tropic of Capricorn
CENTURY
BROKEN HILL
Ore Producing Areas
Zinc
Leading ● BATHURST
Major
Minor
Coltan
Major
Minor
Source: U.S. Geological Survey
Goode's Interrupted Homolosine Projection (Condensed)
Scale 1:162,000,000
M-101956-1
© Rand McNally

The percentage of zinc smelted by each country is not necessarily identical to its percentage of zinc ore production. Some countries, such as Australia, export large amounts of ore to other countries for smelting.

Zinc World Mine Production - 9,893,000 metric tons (metal content) - Avg. 2004-2006

China	India	Kazakhstan	Other Asia	Australia	Peru	Other S.A.	United States	Canada	Mexico	Ireland	Sweden	Other Eur.	Africa	All Other
25.4%	4.2	3.9	3.9	13.7	12.2	3.9	7.5	7.3	4.7	4.4	2.0	2.5	2.1	1.8

Zinc Smelted World Mine Production†† - 10,504,000 metric tons - Avg. 2004-2006

China	Japan	South Korea	Kazakhstan	India	Other Asia	Canada	Mexico	U.S.	Spain	Germany	Finland	Belgium	France	Neth.	Other Europe	Australia	Brazil	Other S.A.	Russia	Africa
27.3%	6.3	6.3	3.1	3.0	3.8	7.2	3.2	3.1	4.8	3.2	2.7	2.3	2.2	2.1	5.4	4.6	2.6	2.1	2.2	2.5

†† Includes recycled materials; excludes possible production in Israel.

Columbium/Tantalum (Coltan) World Mine Production - 46,000 metric tons - Avg. 2004-2006

Brazil	Canada	Australia
89.0%	7.7	2.0

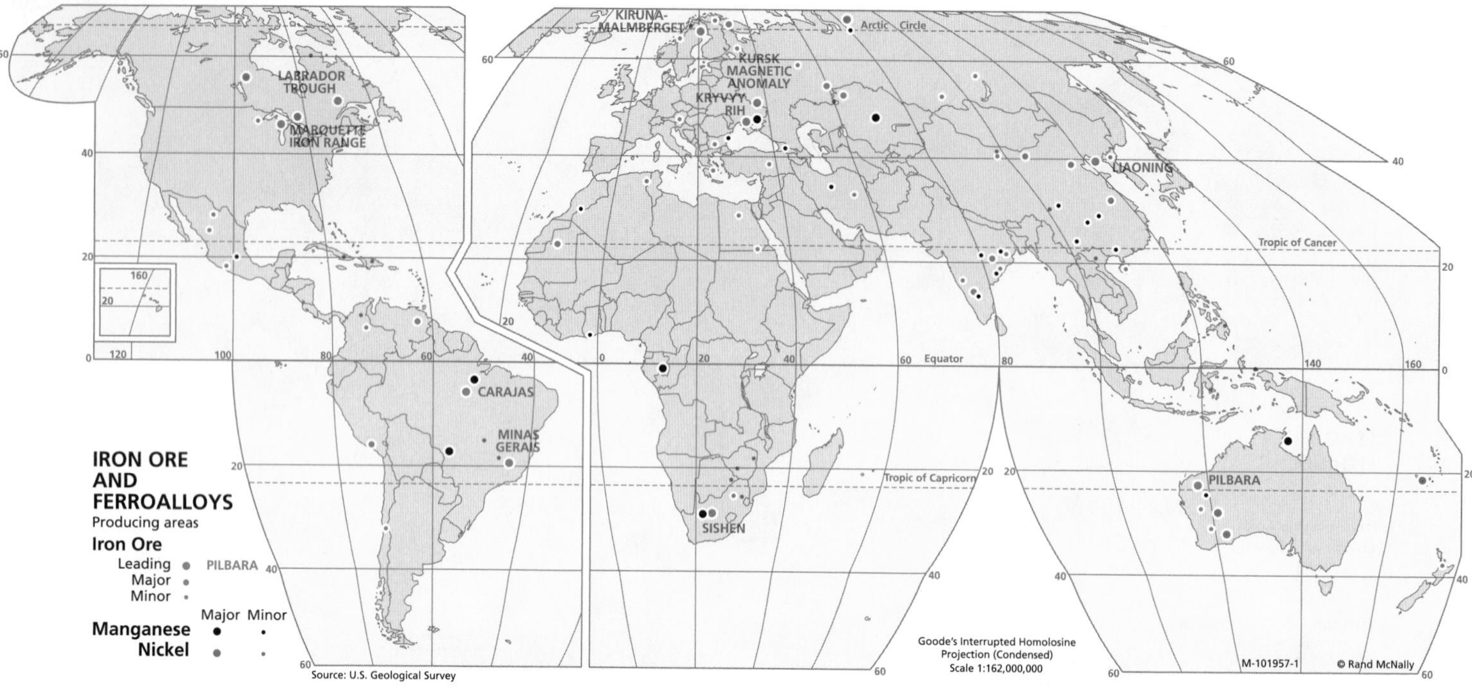
IRON ORE AND FERROALLOYS
Producing areas
Iron Ore
Leading PILBARA
Major
Minor
Major Minor
Manganese
Nickel
KIRUNA-MALMBERGET
KURSK MAGNETIC ANOMALY
KRYVYY RIH
LIAONING
LABRADOR TROUGH
MARQUETTE IRON RANGE
CARAJAS
MINAS GERAIS
SISHEN
PILBARA
Arctic Circle
Tropic of Cancer
Equator
Tropic of Capricorn
Goode's Interrupted Homolosine Projection (Condensed)
Scale 1:162,000,000
M-101957-1
© Rand McNally
Source: U.S. Geological Survey

Manganese World Production* - 10,944,000 metric tons (metal content) - Avg. 2004-2006

SOUTH AFRICA	GABON	GHANA	AUSTRALIA	CHINA	INDIA	KAZAKH.	BRAZIL	UKRAINE	ALL OTHER
19.2%	11.4	5.4	16.0	12.8	7.2	5.1	12.4	7.3	3.2

* Excluding possible production in Cuba, Panama, and Sudan.

Nickel World Production - 1,501,000 metric tons (metal content) - Avg. 2004-2006

RUSSIA	CANADA	CUBA	DOM. REP.	AUSTRALIA	NEW CALEDONIA	INDONESIA	CHINA	PHILIPPINES	COLOMBIA	BRAZIL	S. AFRICA	BOTSWANA	EUROPE
21.1%	13.8	4.9	3.2	12.5	7.4	9.1	5.1	2.3	5.7	4.6	2.8	2.5	3.0

Iron Ore World Production** - 842,891,000 metric tons (metal content) - Avg. 2004-2006

BRAZIL	OTHER S.A.	AUSTRALIA	CHINA	INDIA	OTHER ASIA	RUSSIA	UKRAINE	OTHER EUR.	U.S.	CANADA	S. AFRICA
22.6%	2.8	19.1	17.3	10.6	2.9	6.8	4.5	2.1	4.0	2.3	3.0

** Excluding possible production in Cuba.

Iron Ore Reserves World Total - 183,200,000,000 metric tons (metal content) - 2006

BRAZIL	VENEZUELA	RUSSIA	AUSTRALIA	UKRAINE	SWEDEN	CHINA	KAZAKH.	INDIA	U.S.	OTHER N.A.	ALL OTHER
22.4%	2.0	16.9	13.6	10.9	2.7	8.2	4.0	3.4	2.5	1.9	10.6

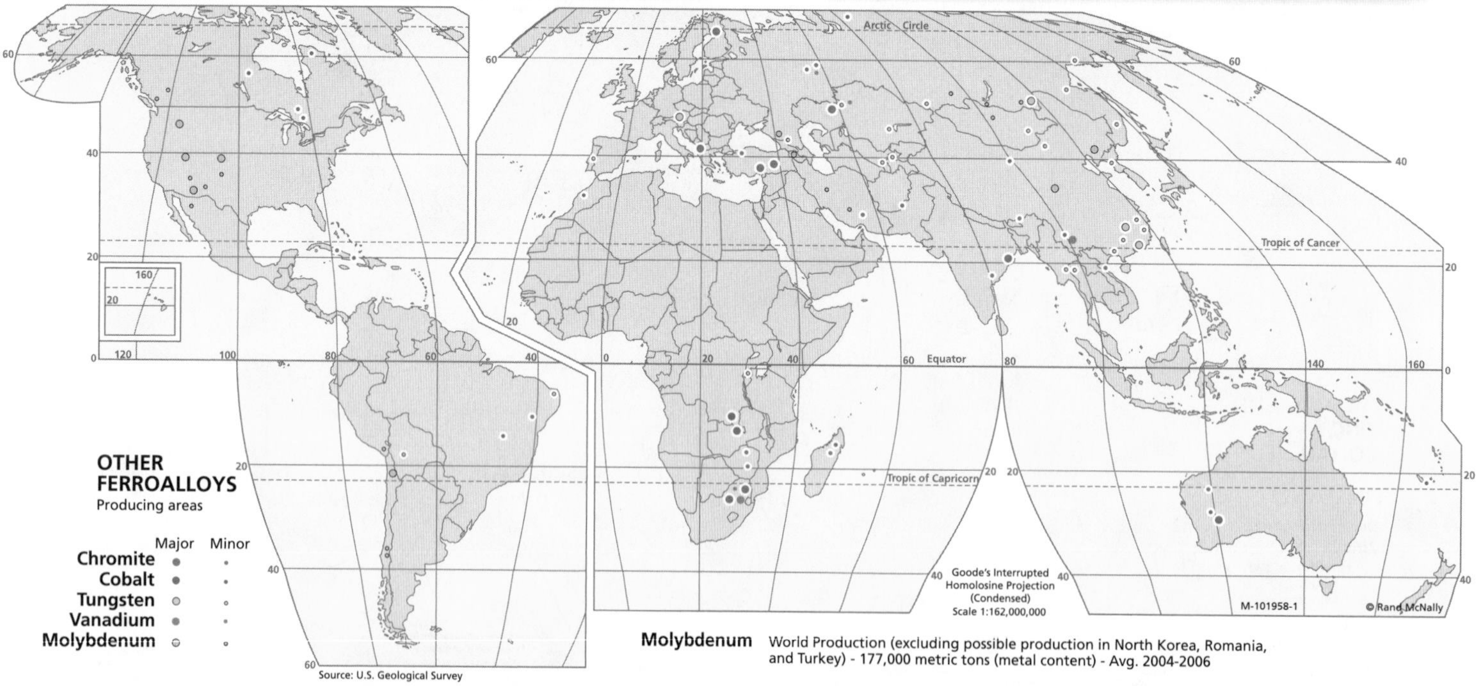
OTHER FERROALLOYS
Producing areas
Major Minor
Chromite
Cobalt
Tungsten
Vanadium
Molybdenum
Arctic Circle
Tropic of Cancer
Equator
Tropic of Capricorn
Goode's Interrupted Homolosine Projection (Condensed)
Scale 1:162,000,000
M-101958-1
© Rand McNally
Source: U.S. Geological Survey

Molybdenum World Production (excluding possible production in North Korea, Romania, and Turkey) - 177,000 metric tons (metal content) - Avg. 2004-2006

Chromite World Production - 18,792, 000 metric tons - Avg. 2004-2006

SOUTH AFRICA	ZIMBABWE	KAZAKHSTAN	INDIA	TURKEY	OTHER ASIA	RUSSIA	BRAZIL	FINLAND	ALL OTHER
40.0%	3.3	18.5	17.4	4.3	3.9	3.7	3.2	3.0	1.5

Cobalt World Production† - 63,000 metric tons - Avg. 2004-2006

DEM. REP. OF THE CONGO	ZAMBIA	MOROCCO	AUSTRALIA	NEW CAL.	CANADA	CUBA	RUSSIA	CHINA	ALL OTHER
38.4%	14.4	2.5	11.4	3.4	9.4	5.9	7.8	3.0	1.9

† Excluding possible production in Kyrgyzstan, Nigeria, Peru, and Turkey.

Tungsten World Production†† - 94,000 metric tons - Avg. 2004-2006

CHINA	RUSSIA	EUROPE	ALL OTHER
88.6	4.9	2.2	3.2

†† Excluding possible production in Kyrgyzstan, Nigeria, Peru, and Turkey.

Vanadium World Mine Production‡ - 55,000 metric tons (metal content) - Avg. 2004-2006

SOUTH AFRICA	CHINA	OTHER ASIA	RUSSIA
41.3%	30.7	2.8	25.0

‡ Excluding possible production in Germany and several other European countries.

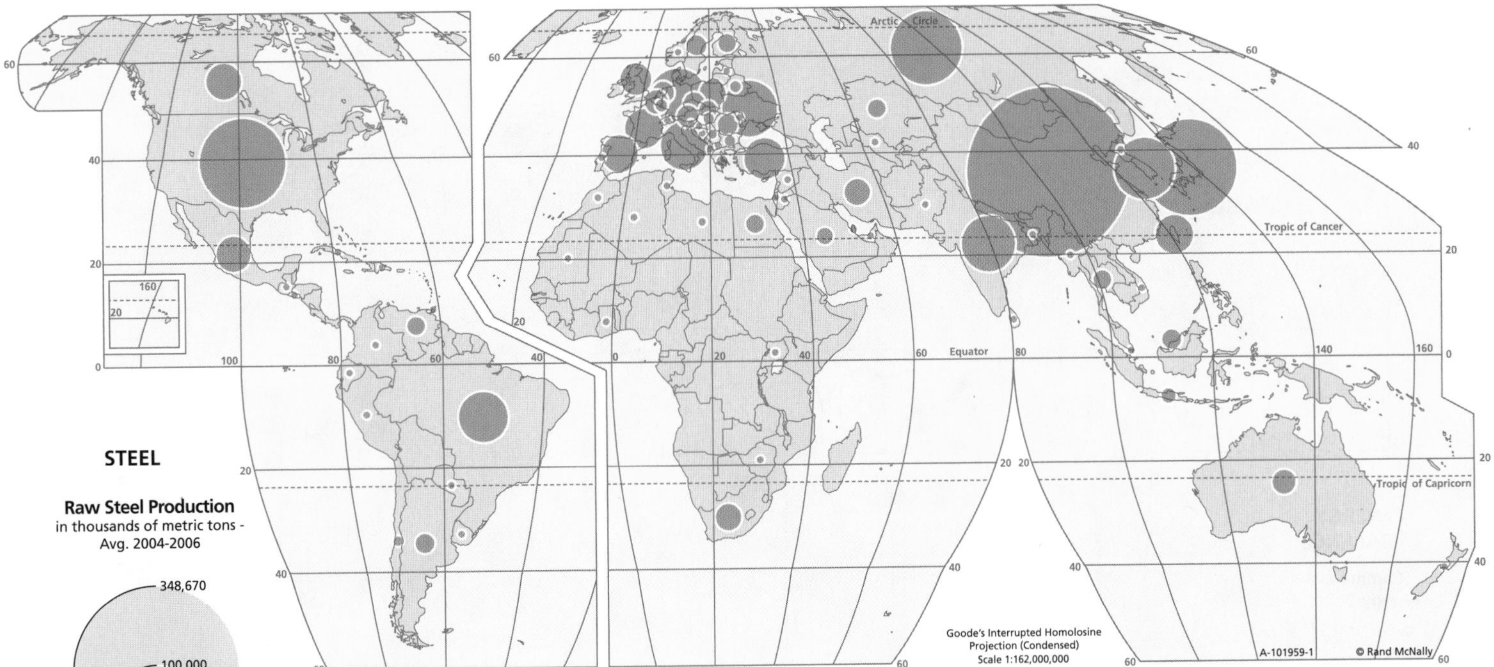

Arctic Circle
Tropic of Cancer
Equator
Tropic of Capricorn
Goode's Interrupted Homolosine Projection (Condensed) Scale 1:162,000,000
A-101959-1
© Rand McNally

Sources: United Nations; U.S. Geological Survey

STEEL

Raw Steel Production
in thousands of metric tons - Avg. 2004-2006

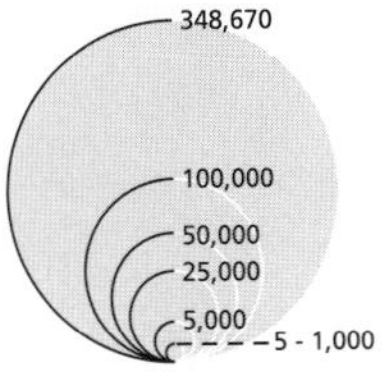

348,670
100,000
50,000
25,000
5,000
5 - 1,000

Iron and Steel Exports World Total* - $357,262,000,000 (U.S. dollars) - 2006

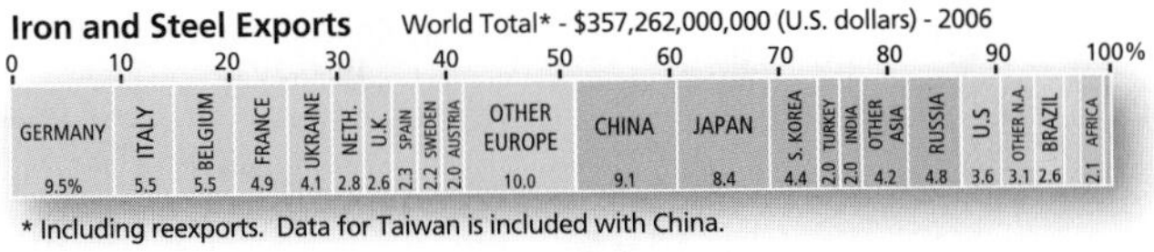

0 10 20 30 40 50 60 70 80 90 100%
GERMANY 9.5%
ITALY 5.5
BELGIUM 5.5
FRANCE 4.9
UKRAINE 4.1
NETH. 2.8
U.K. 2.6
SPAIN 2.3
SWEDEN 2.2
AUSTRIA 2.0
OTHER EUROPE 10.0
CHINA 9.1
JAPAN 8.4
S. KOREA 4.4
TURKEY 2.0
INDIA 2.0
OTHER ASIA 4.2
RUSSIA 4.8
U.S 3.6
OTHER N.A. 3.1
BRAZIL 2.6
AFRICA 2.1

* Including reexports. Data for Taiwan is included with China.

Steel Production Trends 1945-2000

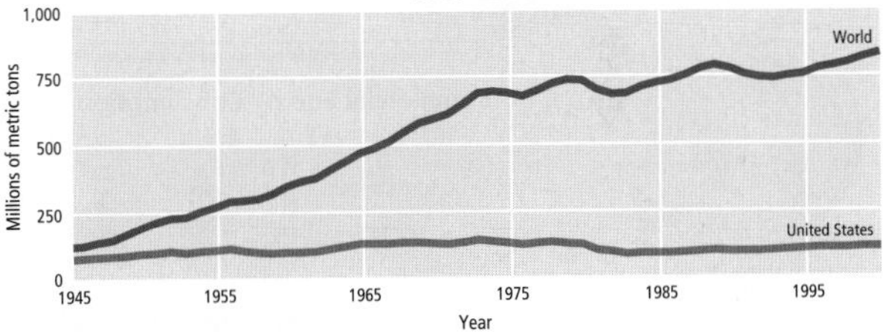

1,000
750
500
250
0
Millions of metric tons
World
United States
1945
1955
1965
1975
1985
1995
Year

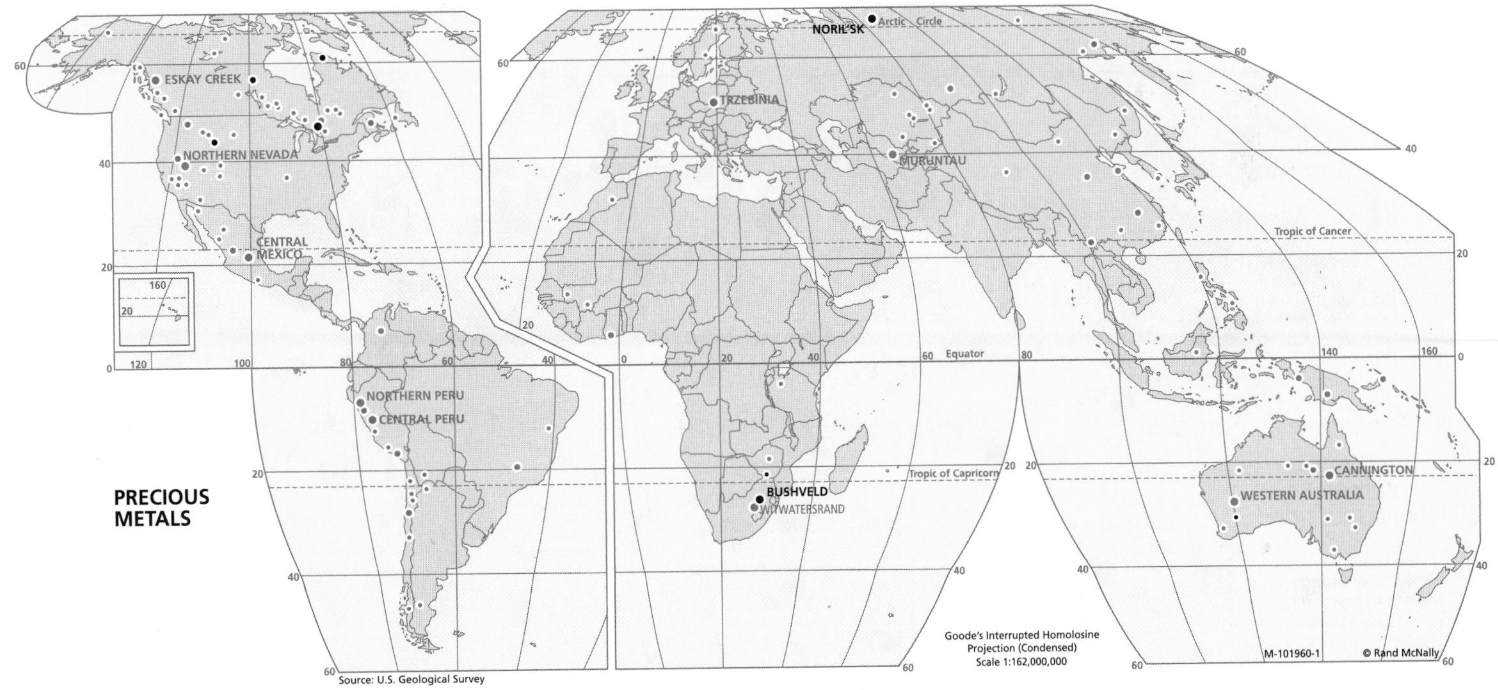

NORIL'SK
Arctic Circle
ESKAY CREEK
TRZEBINIA
NORTHERN NEVADA
MURUNTAU
CENTRAL MEXICO
Tropic of Cancer
Equator
NORTHERN PERU
CENTRAL PERU
Tropic of Capricorn
BUSHVELD
WITWATERSRAND
CANNINGTON
WESTERN AUSTRALIA
Goode's Interrupted Homolosine Projection (Condensed) Scale 1:162,000,000
M-101960-1
© Rand McNally

PRECIOUS METALS

Source: U.S. Geological Survey

Gold Producing Areas
Leading ● MURUNTAU
Major ●
Minor •

Silver Producing Areas
Leading ● CANNINGTON
Major ●
Minor •

Platinum Producing Areas
Leading ● NORIL'SK
Major ●
Minor •

Silver World Production - 20,200 metric tons - Avg. 2004-2006

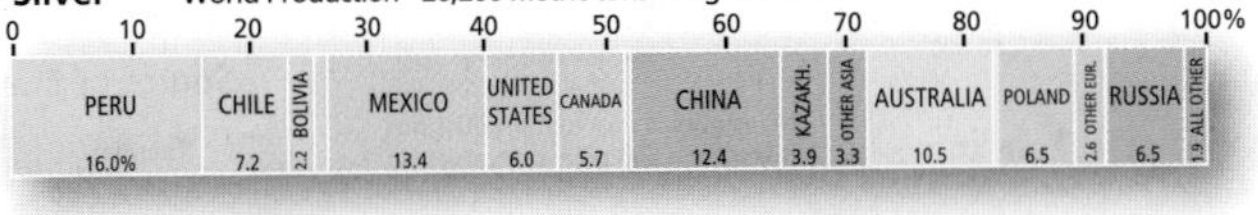

0 10 20 30 40 50 60 70 80 90 100%
PERU 16.0%
CHILE 7.2
BOLIVIA 2.2
MEXICO 13.4
UNITED STATES 6.0
CANADA 5.7
CHINA 12.4
KAZAKH. 3.9
OTHER ASIA 3.3
AUSTRALIA 10.5
POLAND 6.5
OTHER EUR. 2.6
RUSSIA 6.5
ALL OTHER 1.9

Gold World Production - 2,500 metric tons - Avg. 2004-2006

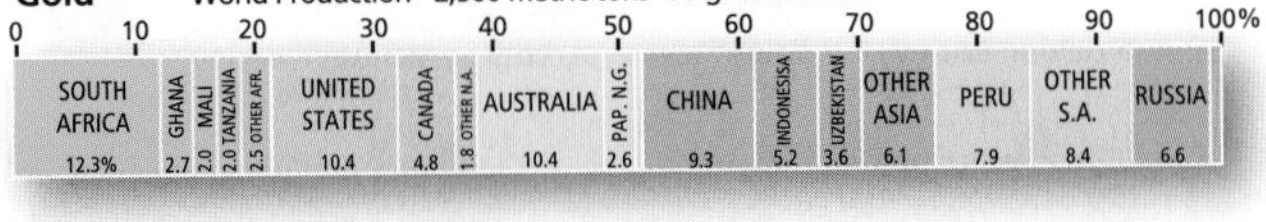

0 10 20 30 40 50 60 70 80 90 100%
SOUTH AFRICA 12.3%
GHANA 2.7
MALI 2.0
TANZANIA 2.0
OTHER AFR. 2.5
UNITED STATES 10.4
CANADA 4.8
OTHER N.A. 1.8
AUSTRALIA 10.4
PAP. N.G. 2.6
CHINA 9.3
INDONESIA 5.2
UZBEKISTAN 3.6
OTHER ASIA 6.1
PERU 7.9
OTHER S.A. 8.4
RUSSIA 6.6

Platinum-Group Metals World Production* - 500 metric tons - Avg. 2004-2006

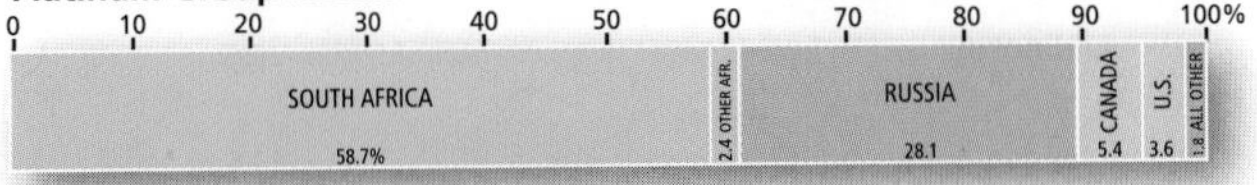

0 10 20 30 40 50 60 70 80 90 100%
SOUTH AFRICA 58.7%
OTHER AFR. 2.4
RUSSIA 28.1
CANADA 5.4
U.S. 3.6
ALL OTHER 1.8

* Excluding possible production in China, Indonesia, and Philippines.

Taiwan figures are included with China.

Botswana, Lesotho, Namibia and Swaziland figures are included with South Africa.

Montenegro figures are included with Serbia.

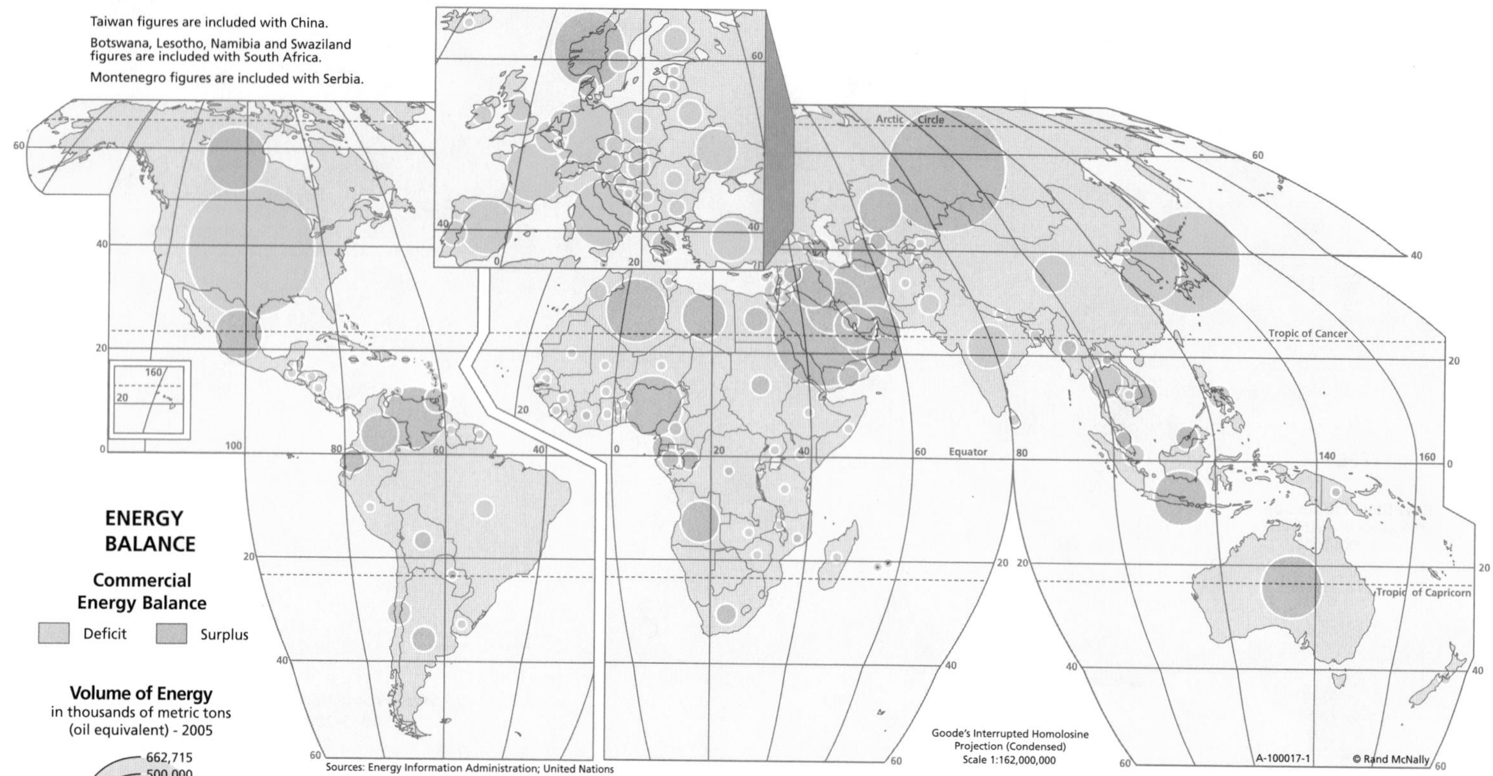
Arctic Circle
Tropic of Cancer
Equator
Tropic of Capricorn
Goode's Interrupted Homolosine Projection (Condensed)
Scale 1:162,000,000
A-100017-1
© Rand McNally
Sources: Energy Information Administration; United Nations

ENERGY BALANCE

Commercial Energy Balance

Deficit   Surplus

Volume of Energy
in thousands of metric tons (oil equivalent) - 2005

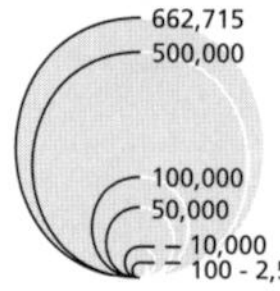
662,715
500,000
100,000
50,000
10,000
100 - 2,500

No symbol is shown if value is less than 100,000 metric tons.

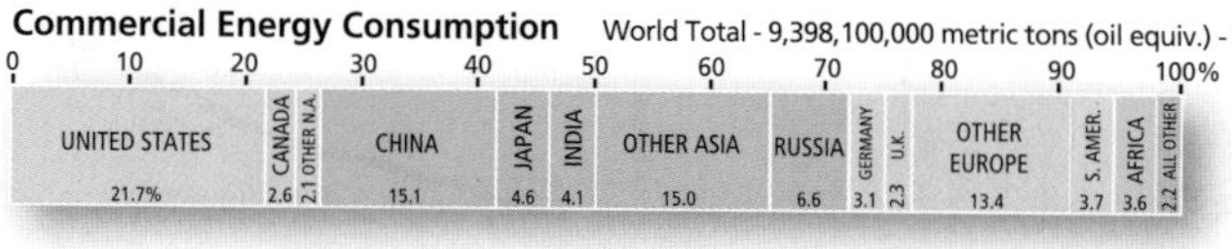
Commercial Energy Consumption
World Total - 9,398,100,000 metric tons (oil equiv.) - 2005
0 10 20 30 40 50 60 70 80 90 100%
UNITED STATES 21.7%
CANADA 2.6
OTHER N.A. 2.1
CHINA 15.1
JAPAN 4.6
INDIA 4.1
OTHER ASIA 15.0
RUSSIA 6.6
GERMANY 3.1
U.K. 2.3
OTHER EUROPE 13.4
S. AMER. 3.7
AFRICA 3.6
ALL OTHER 2.2

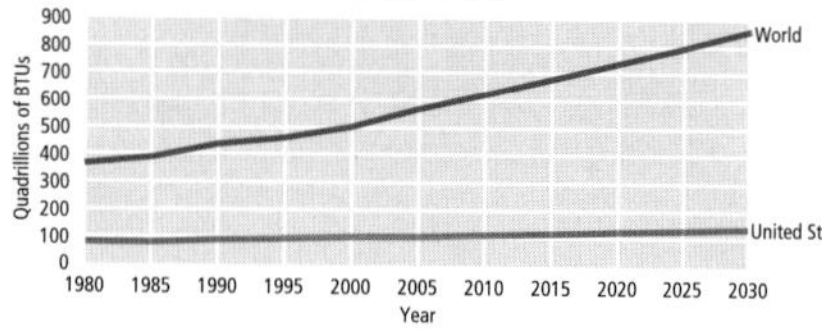
Energy Consumption Trends
1980-2030
Quadrillions of BTUs
900
800
700
600
500
400
300
200
100
0
World
United States
1980 1985 1990 1995 2000 2005 2010 2015 2020 2025 2030
Year

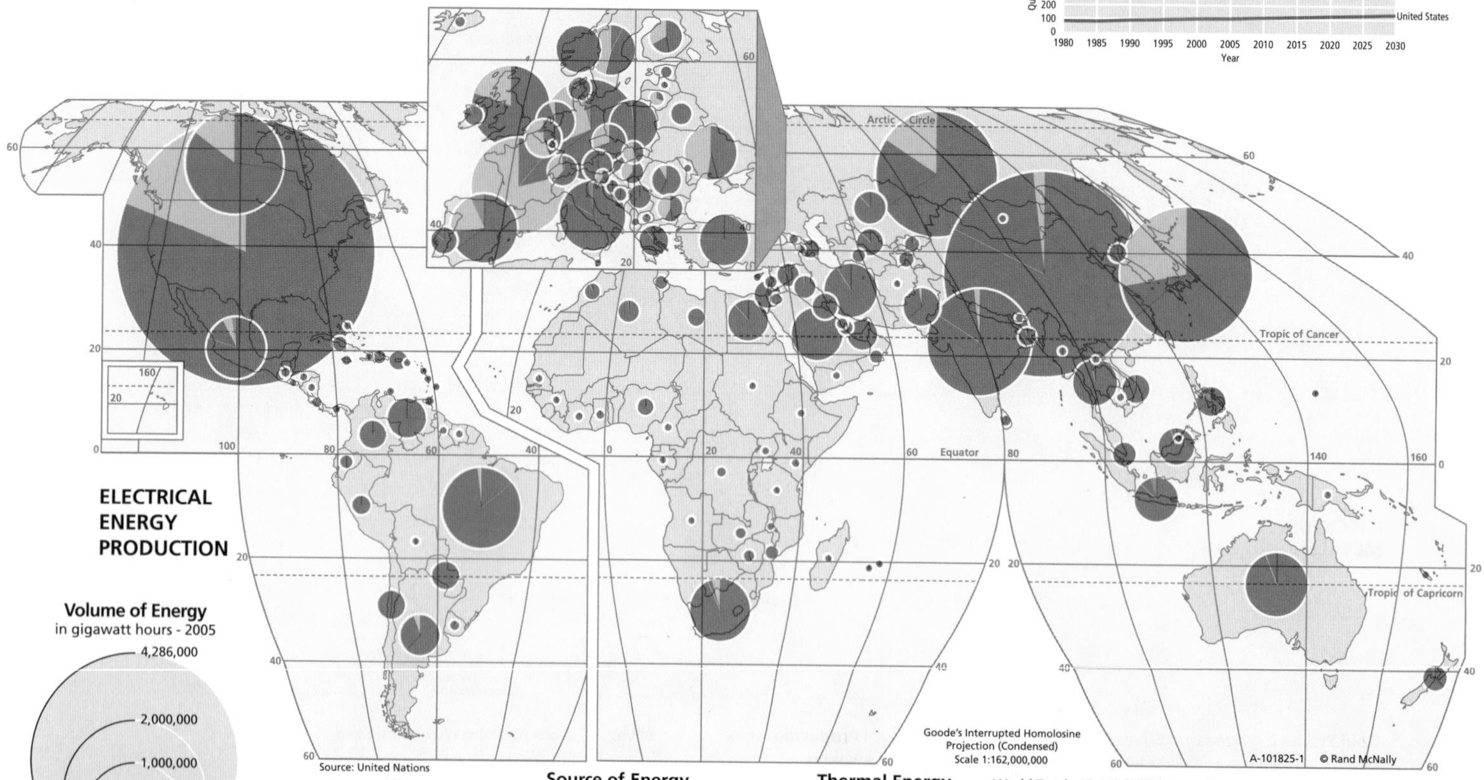
Arctic Circle
Tropic of Cancer
Equator
Tropic of Capricorn
Goode's Interrupted Homolosine Projection (Condensed)
Scale 1:162,000,000
A-101825-1
© Rand McNally
Source: United Nations

ELECTRICAL ENERGY PRODUCTION

Volume of Energy
in gigawatt hours - 2005

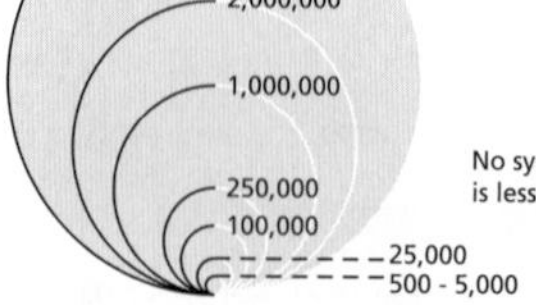
4,286,000
2,000,000
1,000,000
250,000
100,000
25,000
500 - 5,000

No symbol is shown if production is less than 500 gigawatt hours.

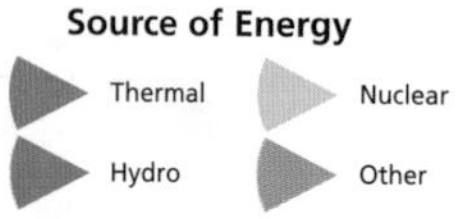
Source of Energy
Thermal
Nuclear
Hydro
Other

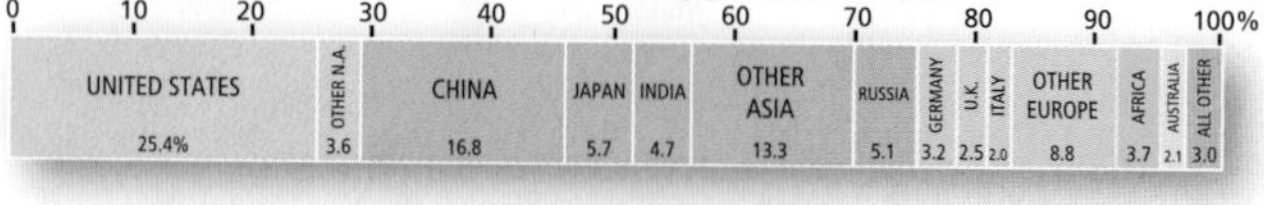
Thermal Energy
World Total - 12,413,000 gigawatt hours - 2005
0 10 20 30 40 50 60 70 80 90 100%
UNITED STATES 25.4%
OTHER N.A. 3.6
CHINA 16.8
JAPAN 5.7
INDIA 4.7
OTHER ASIA 13.3
RUSSIA 5.1
GERMANY 3.2
U.K. 2.5
ITALY 2.0
OTHER EUROPE 8.8
AFRICA 3.7
AUSTRALIA 2.1
ALL OTHER 3.0

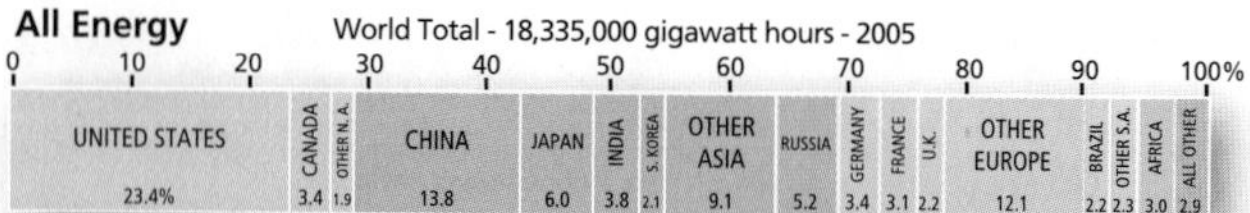
All Energy
World Total - 18,335,000 gigawatt hours - 2005
0 10 20 30 40 50 60 70 80 90 100%
UNITED STATES 23.4%
CANADA 3.4
OTHER N.A. 1.9
CHINA 13.8
JAPAN 6.0
INDIA 3.8
S. KOREA 2.1
OTHER ASIA 9.1
RUSSIA 5.2
GERMANY 3.4
FRANCE 3.1
U.K. 2.2
OTHER EUROPE 12.1
BRAZIL 2.2
OTHER S.A. 2.3
AFRICA 3.0
ALL OTHER 2.9

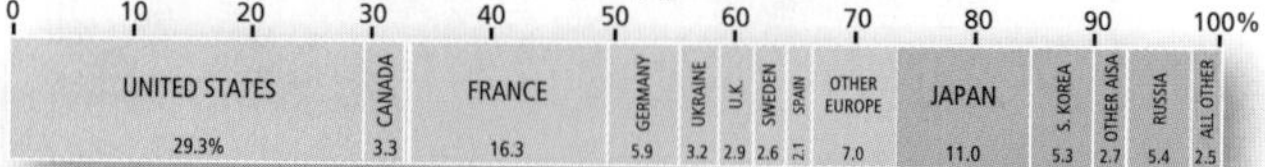
Nuclear Energy
World Total - 2,768,000 gigawatt hours - 2005
0 10 20 30 40 50 60 70 80 90 100%
UNITED STATES 29.3%
CANADA 3.3
FRANCE 16.3
GERMANY 5.9
UKRAINE 3.2
U.K. 2.9
SWEDEN 2.6
SPAIN 2.1
OTHER EUROPE 7.0
JAPAN 11.0
S. KOREA 5.3
OTHER ASIA 2.7
RUSSIA 5.4
ALL OTHER 2.5

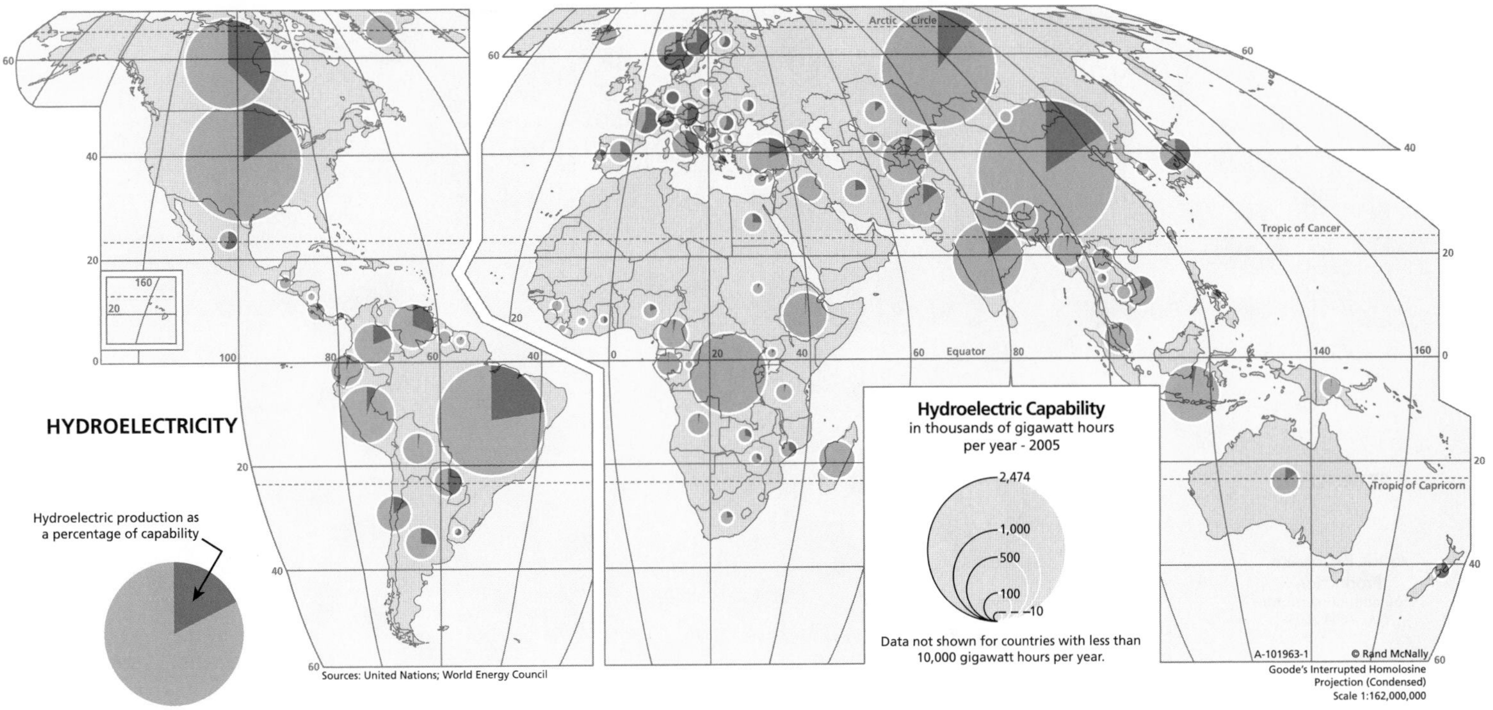
HYDROELECTRICITY
Hydroelectric production as a percentage of capability
Hydroelectric Capability
in thousands of gigawatt hours
per year - 2005
2,474
1,000
500
100
10
Data not shown for countries with less than 10,000 gigawatt hours per year.
Arctic Circle
Tropic of Cancer
Equator
Tropic of Capricorn
Sources: United Nations; World Energy Council
A-101963-1
© Rand McNally
Goode's Interrupted Homolosine
Projection (Condensed)
Scale 1:162,000,000

Hydroelectric Capability

World Total* - 16,494,000 gigawatt hours/year

CHINA	INDIA	INDONESIA	OTHER ASIA	UNITED STATES	CANADA	RUSSIA	BRAZIL	PERU	OTHER S. AMER.	D.R. OF THE CONGO	OTHER AFRICA	EUROPE
15.0%	4.0	2.4	13.1	10.6	5.9	10.1	9.0	2.4	7.0	4.7	6.5	6.3

* Technically exploitable capability.

Hydroelectricity

World Production - 2,996,000 gigawatt hours - 2005

CANADA	UNITED STATES	CHINA	INDIA	JAPAN	OTHER ASIA	BRAZIL	VENEZ.	OTHER S. AMER.	RUSSIA	NORWAY	SWEDEN	OTHER EUROPE	AFRICA
12.1%	9.7	13.3	3.3	2.9	7.8	11.3	2.5	6.3	5.8	4.6	2.4	11.7	3.0

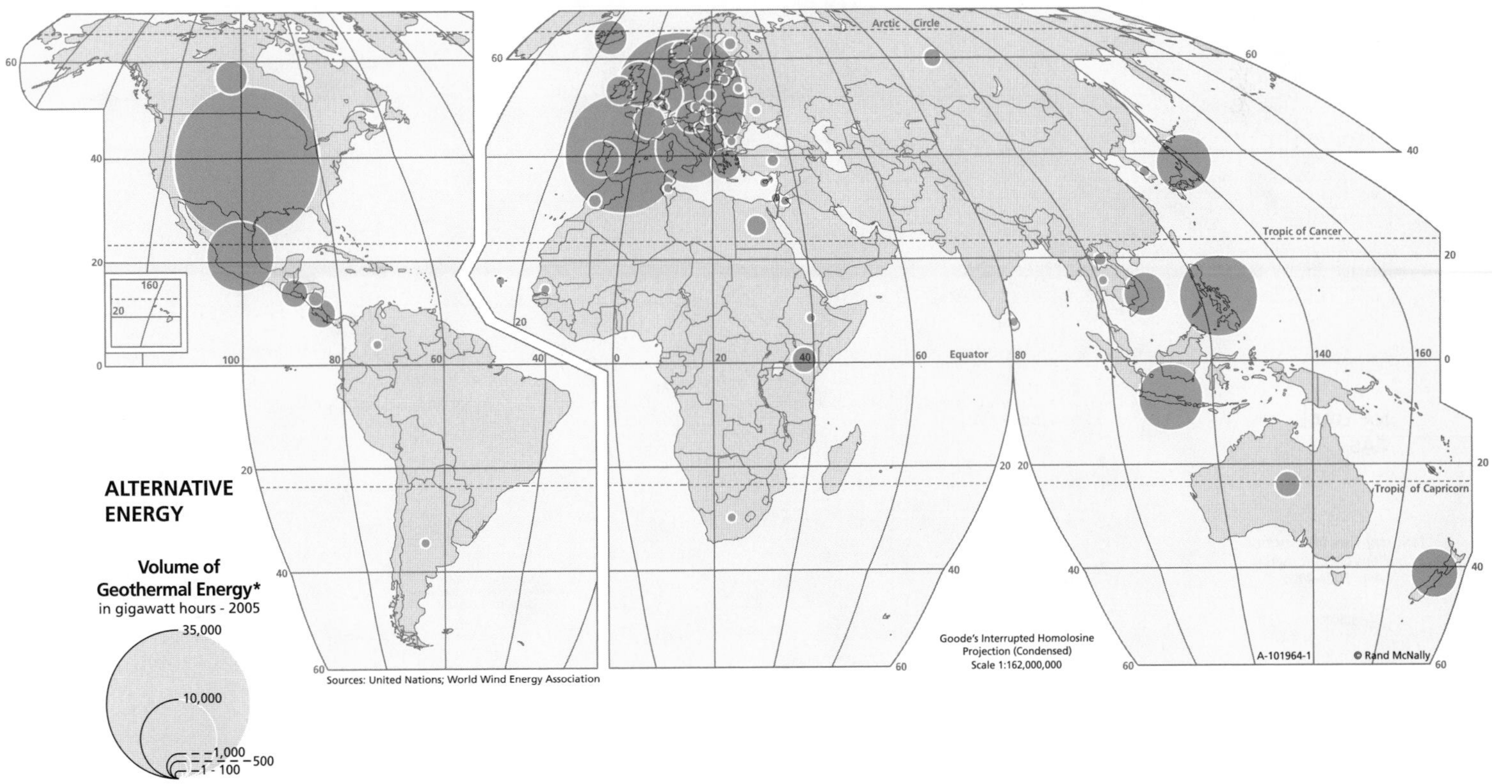
ALTERNATIVE ENERGY
Volume of Geothermal Energy*
in gigawatt hours - 2005
35,000
10,000
1,000
500
1 - 100
Arctic Circle
Tropic of Cancer
Equator
Tropic of Capricorn
Sources: United Nations; World Wind Energy Association
Goode's Interrupted Homolosine
Projection (Condensed)
Scale 1:162,000,000
A-101964-1
© Rand McNally

Geothermal Energy

World Production* - 158,000 gigawatt hours - 2005

UNITED STATES	MEXICO	OTHER N.A.	GERMANY	SPAIN	ITALY	DENMARK	OTHER EUROPE	PHIL.	INDON.	JAPAN	OTHER ASIA	N.Z.
22.3%	4.7	2.5	18.0	13.5	4.9	4.2	10.2	6.3	4.2	3.0	2.1	2.2

* May include other sources of energy, such as solar or wind energy.

Wind Energy

World Installed Capacity - 59 gigawatts - 2005

GERMANY	SPAIN	DENMARK	ITALY	U.K.	NETH.	OTHER EUROPE	UNITED STATES	INDIA	CHINA	OTHER ASIA
31.2%	17.0	5.3	2.9	2.3	2.1	8.5	15.5	7.5	2.1	2.3

PETROLEUM

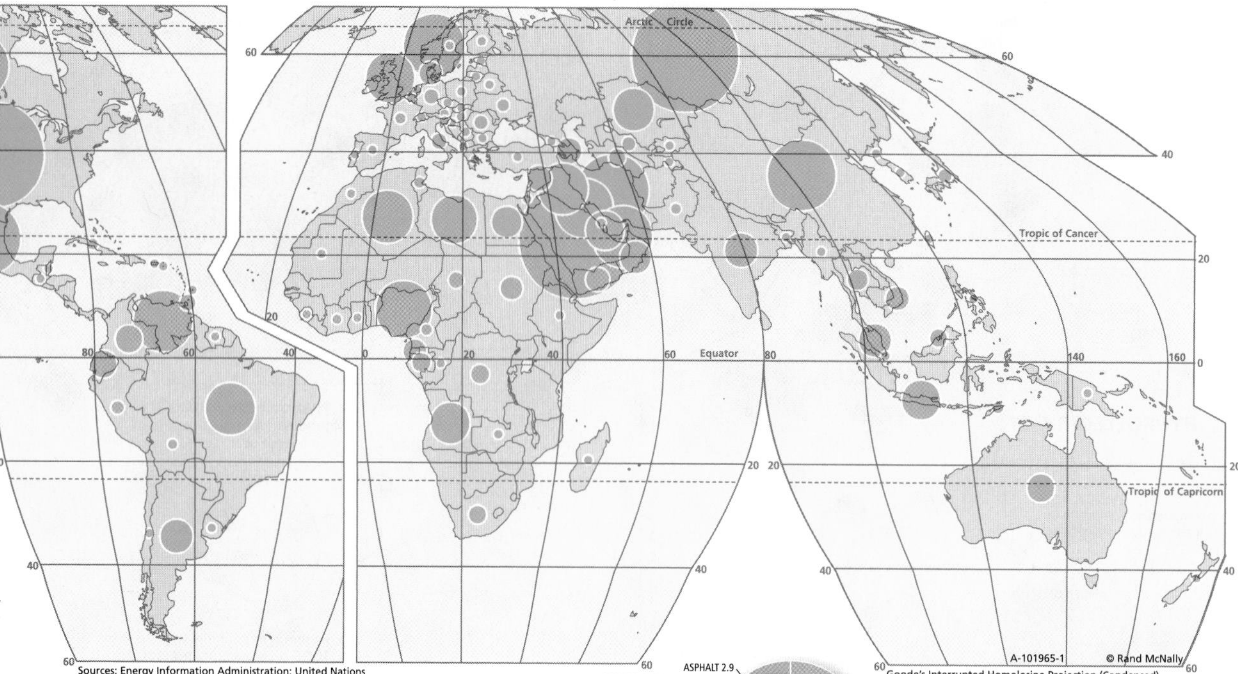
Arctic Circle
Tropic of Cancer
Equator
Tropic of Capricorn
A-101965-1
© Rand McNally

Sources: Energy Information Administration; United Nations

Goode's Interrupted Homolosine Projection (Condensed)
Scale 1:162,000,000

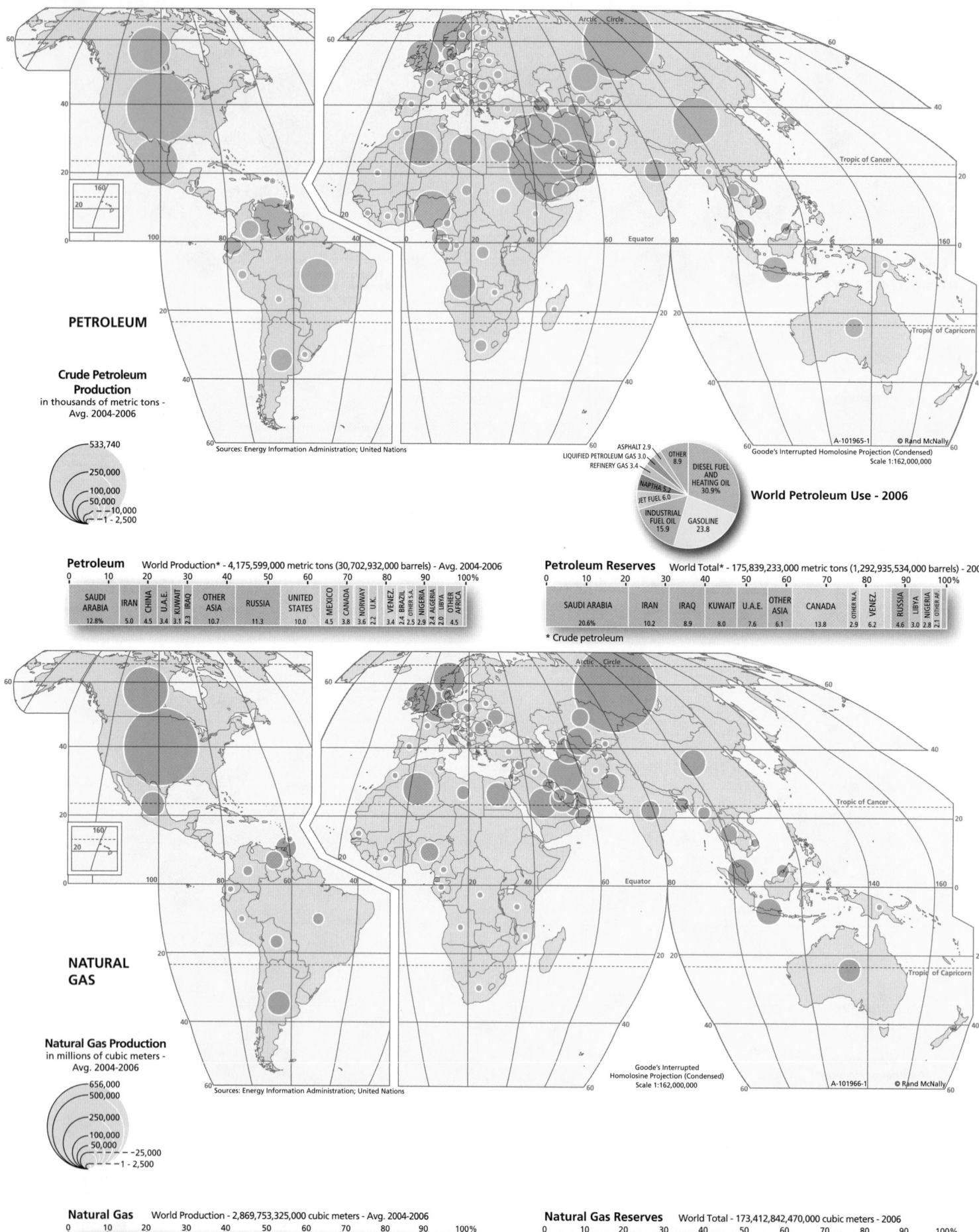

Crude Petroleum Production
in thousands of metric tons - Avg. 2004-2006

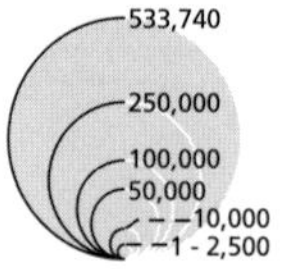
533,740
250,000
100,000
50,000
10,000
1 - 2,500

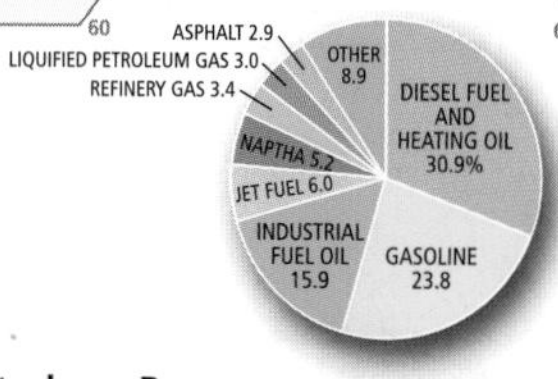
ASPHALT 2.9
LIQUIFIED PETROLEUM GAS 3.0
REFINERY GAS 3.4
NAPTHA 5.2
JET FUEL 6.0
INDUSTRIAL FUEL OIL 15.9
GASOLINE 23.8
DIESEL FUEL AND HEATING OIL 30.9%
OTHER 8.9

World Petroleum Use - 2006

Petroleum World Production* - 4,175,599,000 metric tons (30,702,932,000 barrels) - Avg. 2004-2006

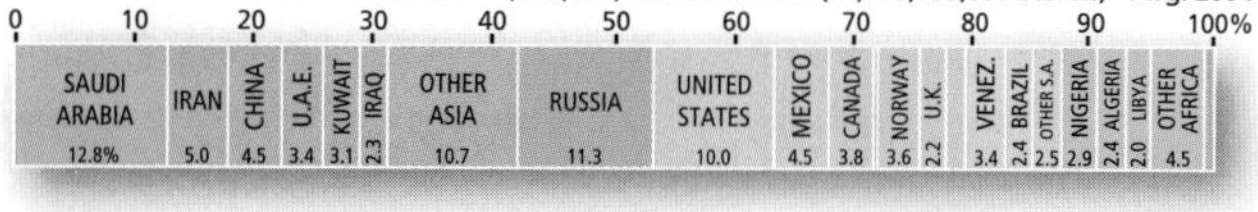
0 10 20 30 40 50 60 70 80 90 100%
SAUDI ARABIA 12.8%
IRAN 5.0
CHINA 4.5
U.A.E. 3.4
KUWAIT 3.1
IRAQ 2.3
OTHER ASIA 10.7
RUSSIA 11.3
UNITED STATES 10.0
MEXICO 4.5
CANADA 3.8
NORWAY 3.6
U.K. 2.2
VENEZ. 3.4
BRAZIL 2.4
OTHER S.A. 2.5
NIGERIA 2.9
ALGERIA 2.4
LIBYA 2.0
OTHER AFRICA 4.5

Petroleum Reserves World Total* - 175,839,233,000 metric tons (1,292,935,534,000 barrels) - 2006

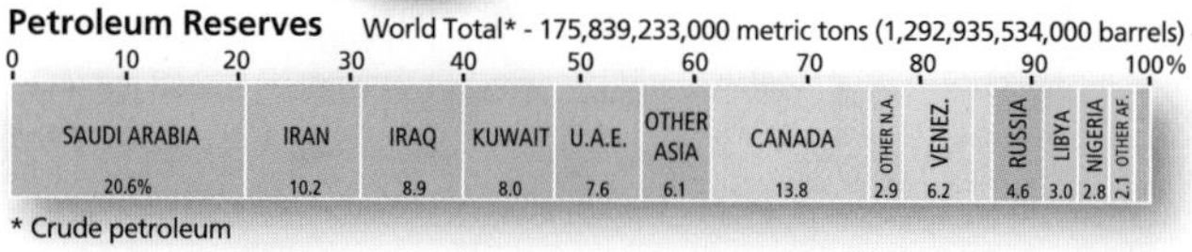
0 10 20 30 40 50 60 70 80 90 100%
SAUDI ARABIA 20.6%
IRAN 10.2
IRAQ 8.9
KUWAIT 8.0
U.A.E. 7.6
OTHER ASIA 6.1
CANADA 13.8
OTHER N.A. 2.9
VENEZ. 6.2
RUSSIA 4.6
LIBYA 3.0
NIGERIA 2.8
OTHER AF. 2.1

* Crude petroleum

NATURAL GAS

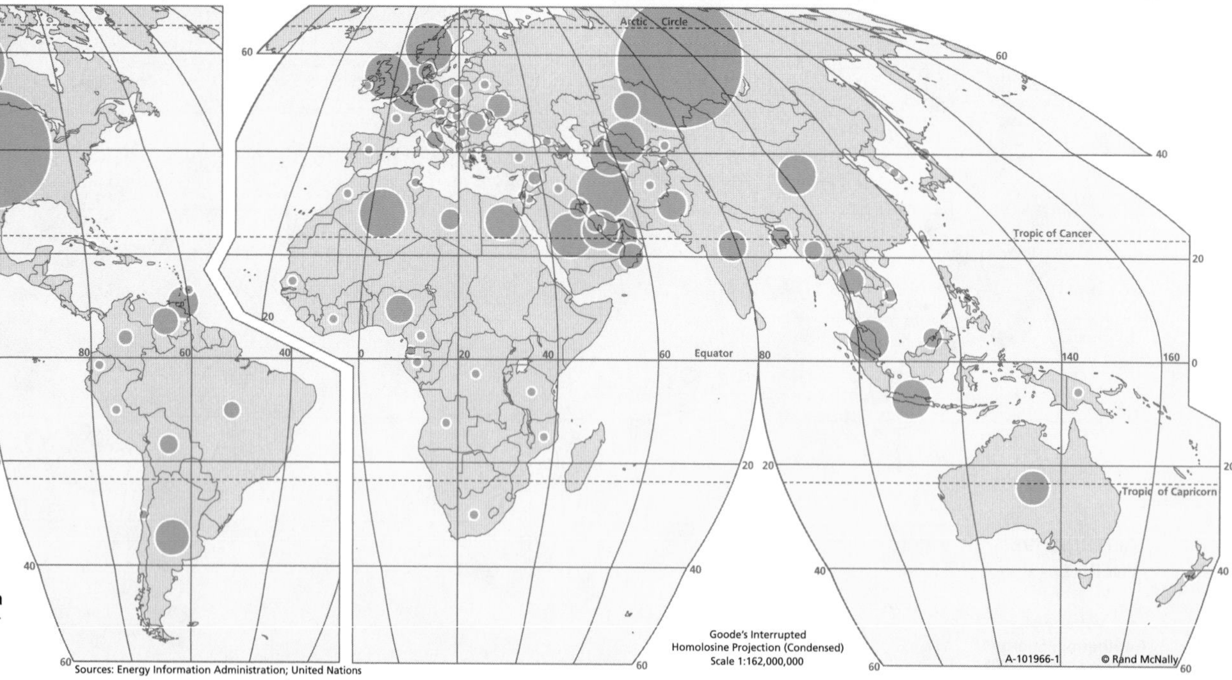
Arctic Circle
Tropic of Cancer
Equator
Tropic of Capricorn
A-101966-1
© Rand McNally

Goode's Interrupted Homolosine Projection (Condensed)
Scale 1:162,000,000

Sources: Energy Information Administration; United Nations

Natural Gas Production
in millions of cubic meters - Avg. 2004-2006

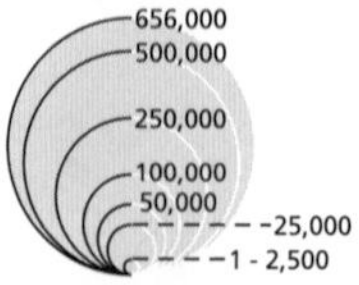
656,000
500,000
250,000
100,000
50,000
25,000
1 - 2,500

Natural Gas World Production - 2,869,753,325,000 cubic meters - Avg. 2004-2006

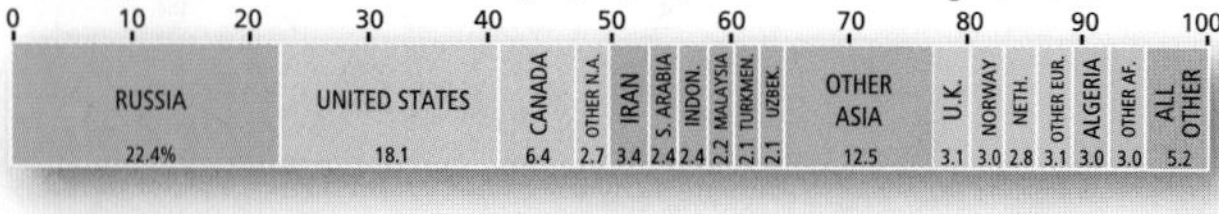
0 10 20 30 40 50 60 70 80 90 100%
RUSSIA 22.4%
UNITED STATES 18.1
CANADA 6.4
OTHER N.A. 2.7
IRAN 3.4
S. ARABIA 2.4
INDON. 2.4
MALAYSIA 2.2
TURKMEN. 2.1
UZBEK. 2.1
OTHER ASIA 12.5
U.K. 3.1
NORWAY 3.0
NETH. 2.8
OTHER EUR. 3.1
ALGERIA 3.0
OTHER AF. 3.0
ALL OTHER 5.2

Natural Gas Reserves World Total - 173,412,842,470,000 cubic meters - 2006

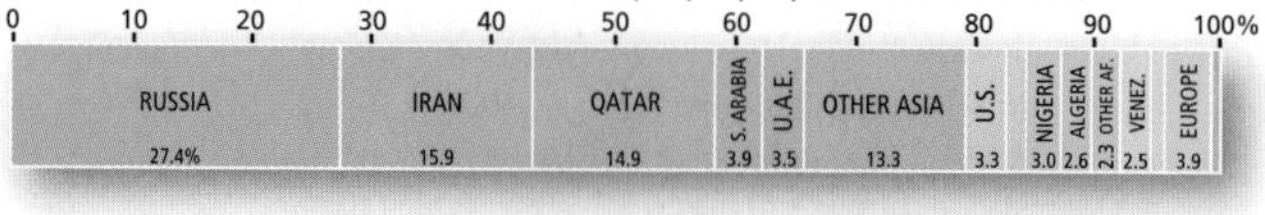
0 10 20 30 40 50 60 70 80 90 100%
RUSSIA 27.4%
IRAN 15.9
QATAR 14.9
S. ARABIA 3.9
U.A.E. 3.5
OTHER ASIA 13.3
U.S. 3.3
NIGERIA 3.0
ALGERIA 2.6
OTHER AF. 2.3
VENEZ. 2.5
EUROPE 3.9

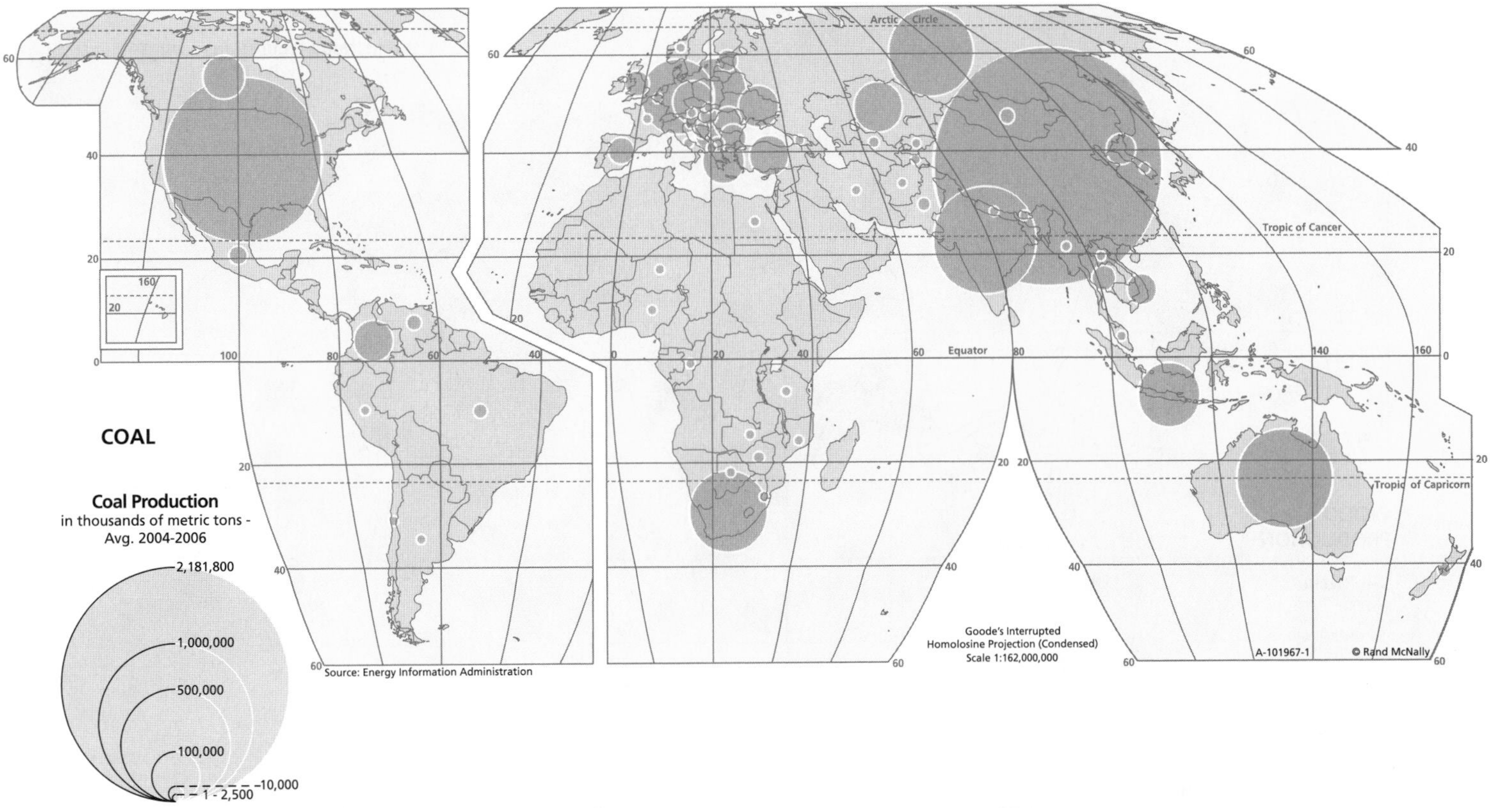
Arctic Circle
Tropic of Cancer
Equator
Tropic of Capricorn
COAL
Coal Production
in thousands of metric tons - Avg. 2004-2006
2,181,800
1,000,000
500,000
100,000
10,000
1 - 2,500
Source: Energy Information Administration
Goode's Interrupted Homolosine Projection (Condensed)
Scale 1:162,000,000
A-101967-1
© Rand McNally

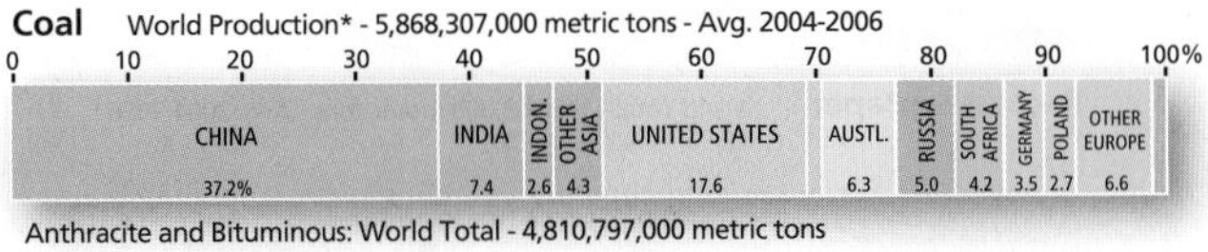
Coal World Production* - 5,868,307,000 metric tons - Avg. 2004-2006
0 10 20 30 40 50 60 70 80 90 100%
CHINA 37.2%
INDIA 7.4
INDON. 2.6
OTHER ASIA 4.3
UNITED STATES 17.6
AUSTL. 6.3
RUSSIA 5.0
SOUTH AFRICA 4.2
GERMANY 3.5
POLAND 2.7
OTHER EUROPE 6.6
Anthracite and Bituminous: World Total - 4,810,797,000 metric tons

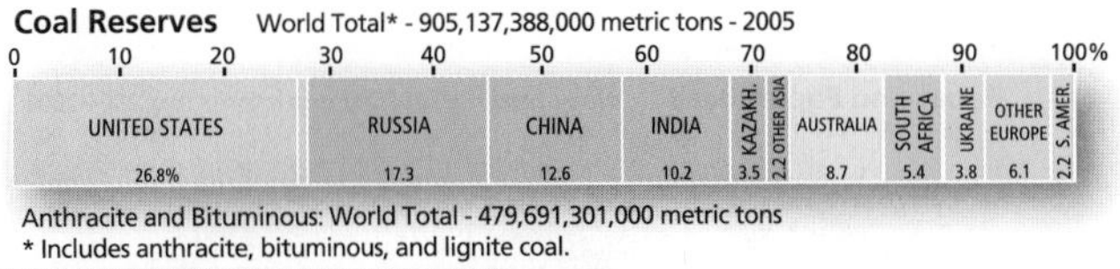
Coal Reserves World Total* - 905,137,388,000 metric tons - 2005
0 10 20 30 40 50 60 70 80 90 100%
UNITED STATES 26.8%
RUSSIA 17.3
CHINA 12.6
INDIA 10.2
KAZAKH. 3.5
OTHER ASIA 2.2
AUSTRALIA 8.7
SOUTH AFRICA 5.4
UKRAINE 3.8
OTHER EUROPE 6.1
S. AMER. 2.2
Anthracite and Bituminous: World Total - 479,691,301,000 metric tons
* Includes anthracite, bituminous, and lignite coal.

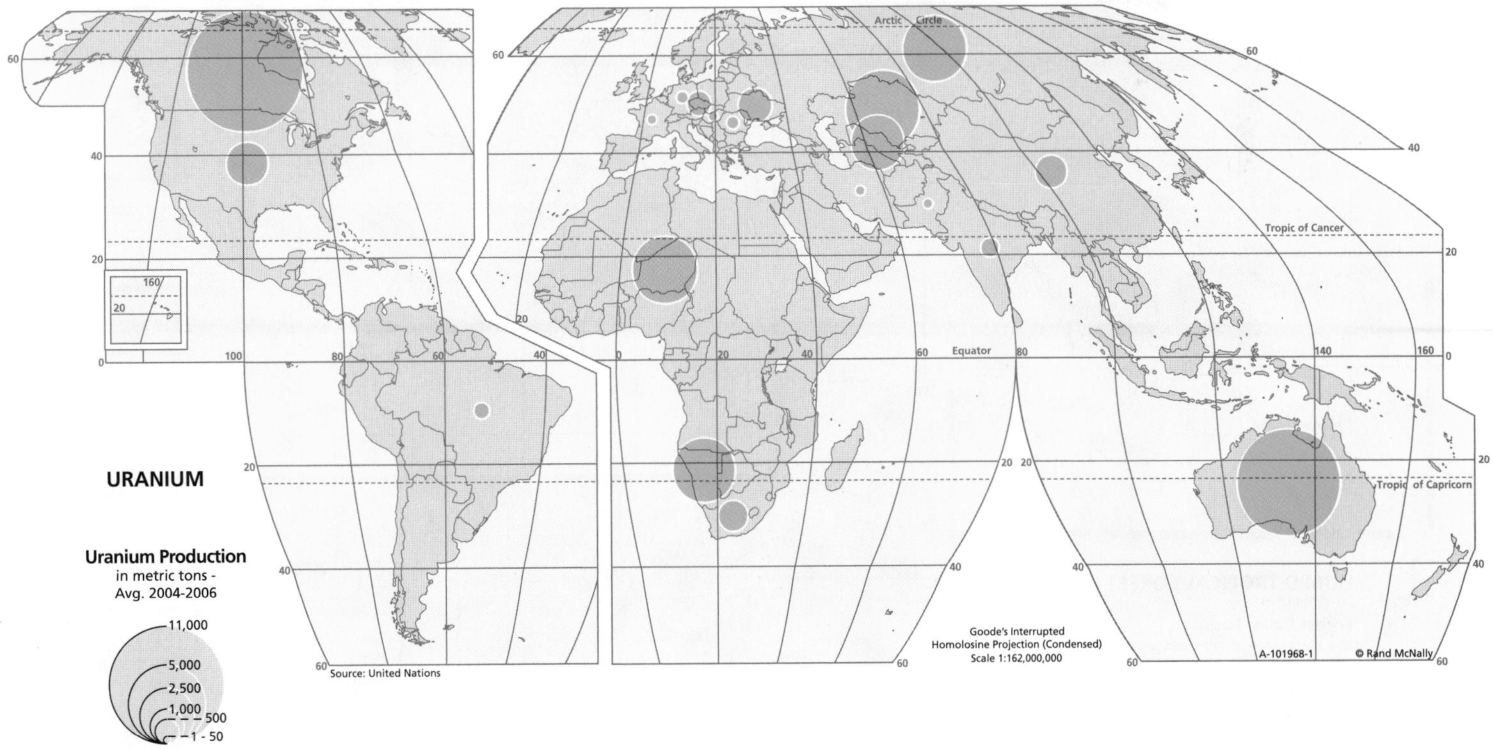
Arctic Circle
Tropic of Cancer
Equator
Tropic of Capricorn
URANIUM
Uranium Production
in metric tons - Avg. 2004-2006
11,000
5,000
2,500
1,000
500
1 - 50
Source: United Nations
Goode's Interrupted Homolosine Projection (Condensed)
Scale 1:162,000,000
A-101968-1
© Rand McNally

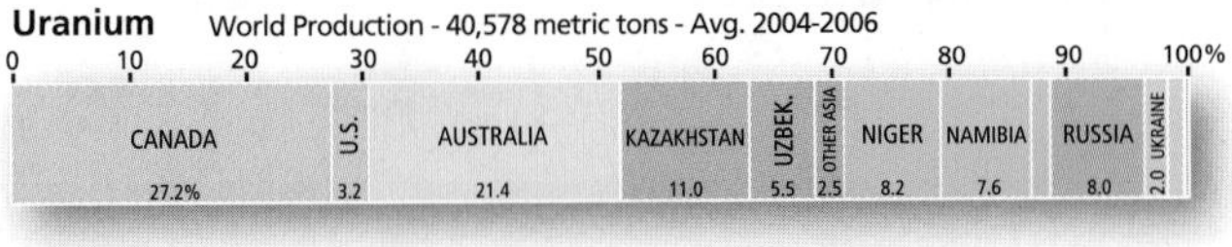
Uranium World Production - 40,578 metric tons - Avg. 2004-2006
0 10 20 30 40 50 60 70 80 90 100%
CANADA 27.2%
U.S. 3.2
AUSTRALIA 21.4
KAZAKHSTAN 11.0
UZBEK. 5.5
OTHER ASIA 2.5
NIGER 8.2
NAMIBIA 7.6
RUSSIA 8.0
UKRAINE 2.0

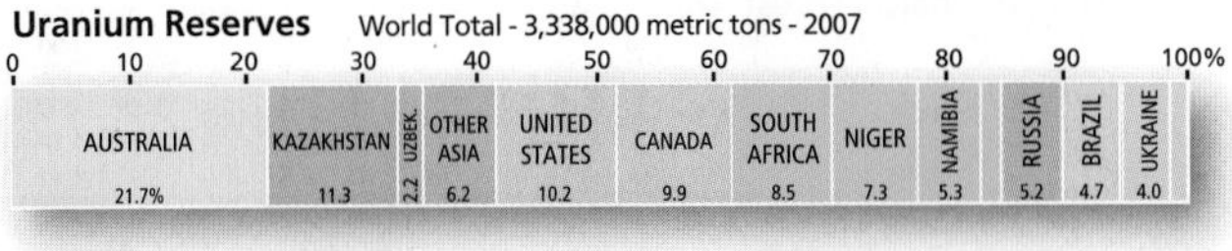
Uranium Reserves World Total - 3,338,000 metric tons - 2007
0 10 20 30 40 50 60 70 80 90 100%
AUSTRALIA 21.7%
KAZAKHSTAN 11.3
UZBEK. 2.2
OTHER ASIA 6.2
UNITED STATES 10.2
CANADA 9.9
SOUTH AFRICA 8.5
NIGER 7.3
NAMIBIA 5.3
RUSSIA 5.2
BRAZIL 4.7
UKRAINE 4.0

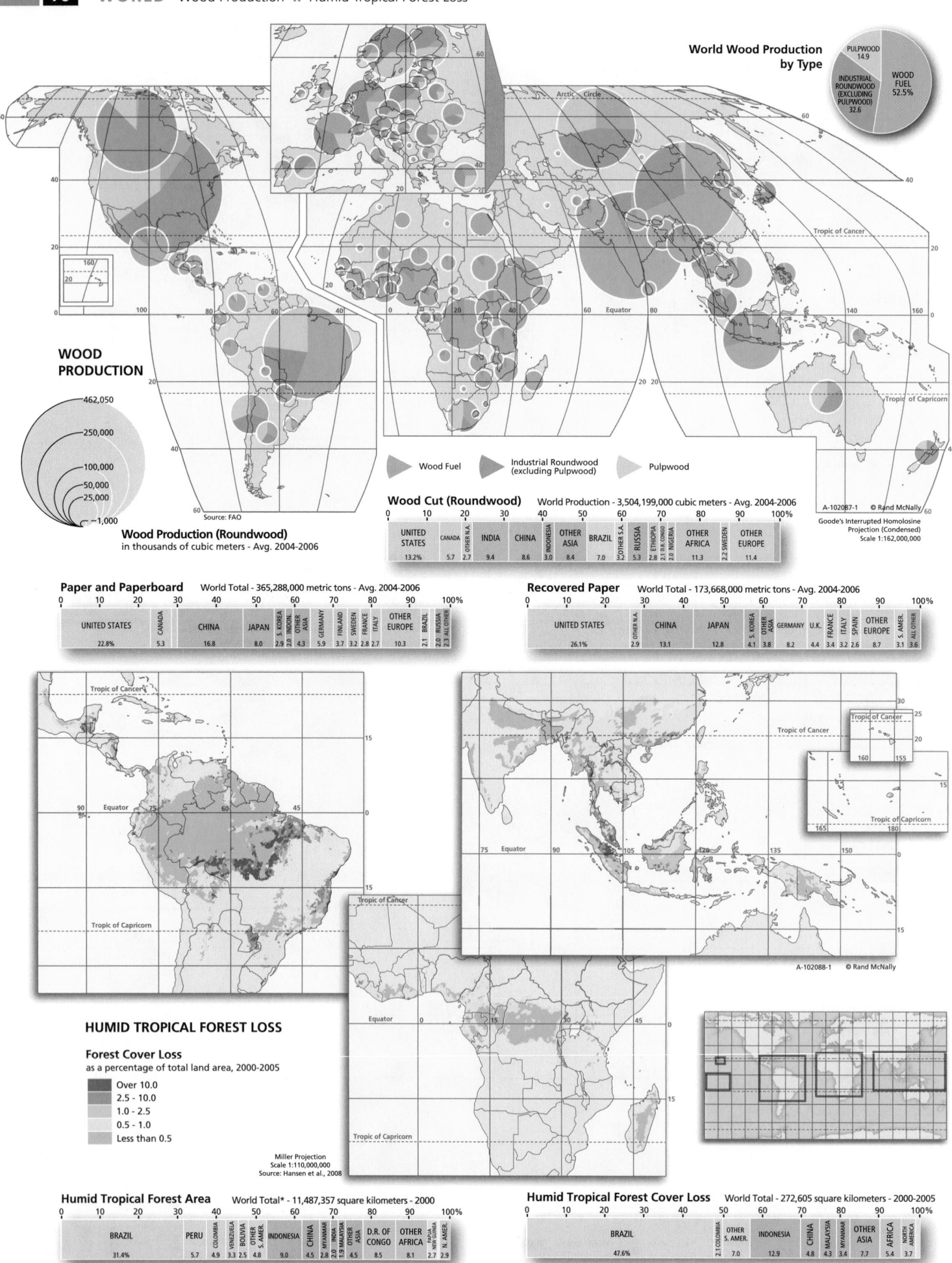

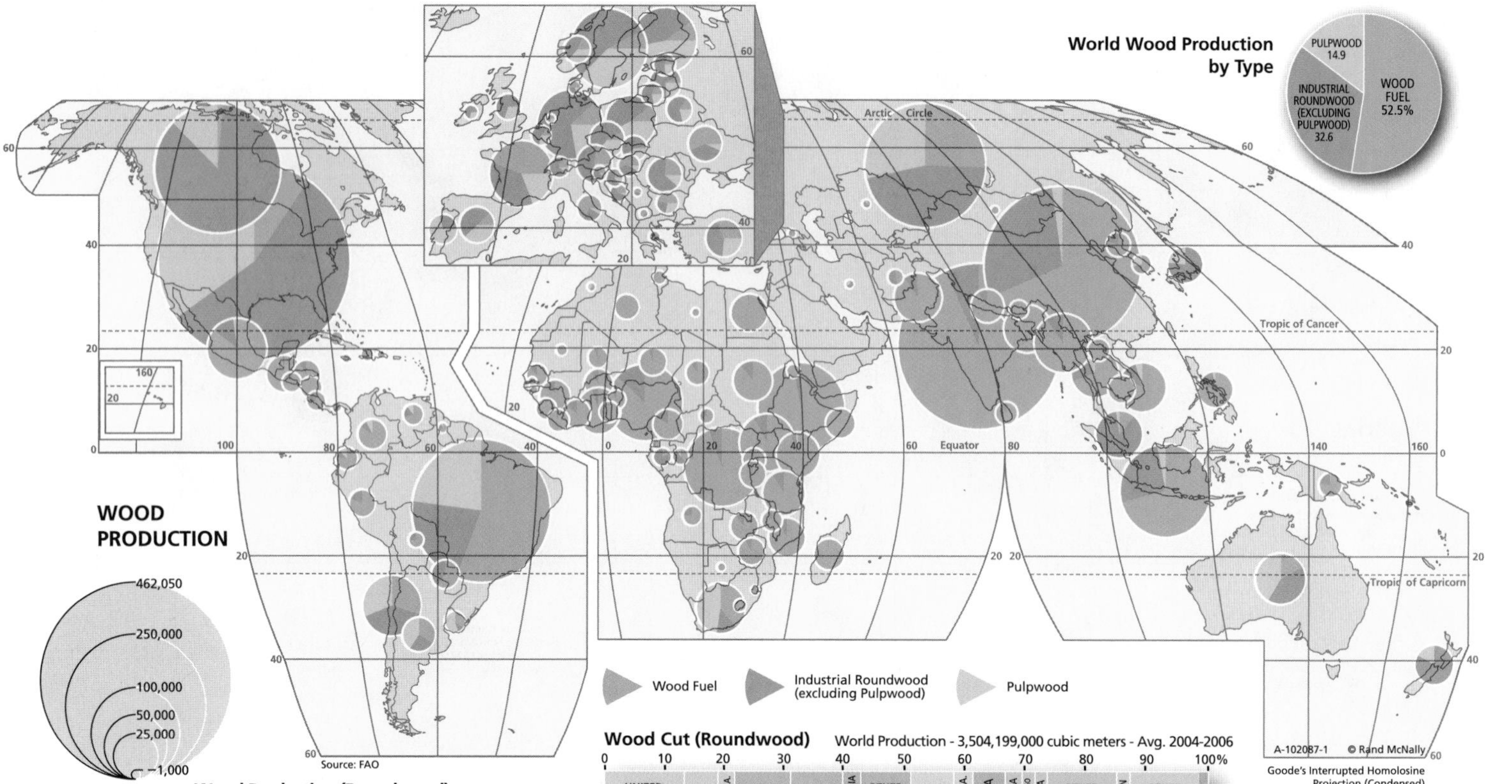
World Wood Production by Type
PULPWOOD 14.9
INDUSTRIAL ROUNDWOOD (EXCLUDING PULPWOOD) 32.6
WOOD FUEL 52.5%
Arctic Circle
Tropic of Cancer
Equator
Tropic of Capricorn
WOOD PRODUCTION
462,050
250,000
100,000
50,000
25,000
1,000
Wood Fuel
Industrial Roundwood (excluding Pulpwood)
Pulpwood
Source: FAO
A-102087-1 © Rand McNally
Goode's Interrupted Homolosine Projection (Condensed) Scale 1:162,000,000

Wood Production (Roundwood)
in thousands of cubic meters - Avg. 2004-2006

Wood Cut (Roundwood) World Production - 3,504,199,000 cubic meters - Avg. 2004-2006

UNITED STATES	CANADA	OTHER N.A.	INDIA	CHINA	INDONESIA	OTHER ASIA	BRAZIL	OTHER S.A.	RUSSIA	ETHIOPIA	D.R. CONGO	NIGERIA	OTHER AFRICA	SWEDEN	OTHER EUROPE
13.2%	5.7	2.7	9.4	8.6	3.0	8.4	7.0	3.2	5.3	2.8	2.1	2.0	11.3	2.2	11.4

Paper and Paperboard World Total - 365,288,000 metric tons - Avg. 2004-2006

UNITED STATES	CANADA	CHINA	JAPAN	S. KOREA	INDON.	OTHER ASIA	GERMANY	FINLAND	SWEDEN	FRANCE	ITALY	OTHER EUROPE	BRAZIL	RUSSIA	ALL OTHER
22.8%	5.3	16.8	8.0	2.9	2.0	4.3	5.9	3.7	3.2	2.8	2.7	10.3	2.1	2.0	2.3

Recovered Paper World Total - 173,668,000 metric tons - Avg. 2004-2006

UNITED STATES	OTHER N.A.	CHINA	JAPAN	S. KOREA	OTHER ASIA	GERMANY	U.K.	FRANCE	ITALY	SPAIN	OTHER EUROPE	S. AMER.	ALL OTHER
26.1%	2.9	13.1	12.8	4.1	3.8	8.2	4.4	3.4	3.2	2.6	8.7	3.1	3.6

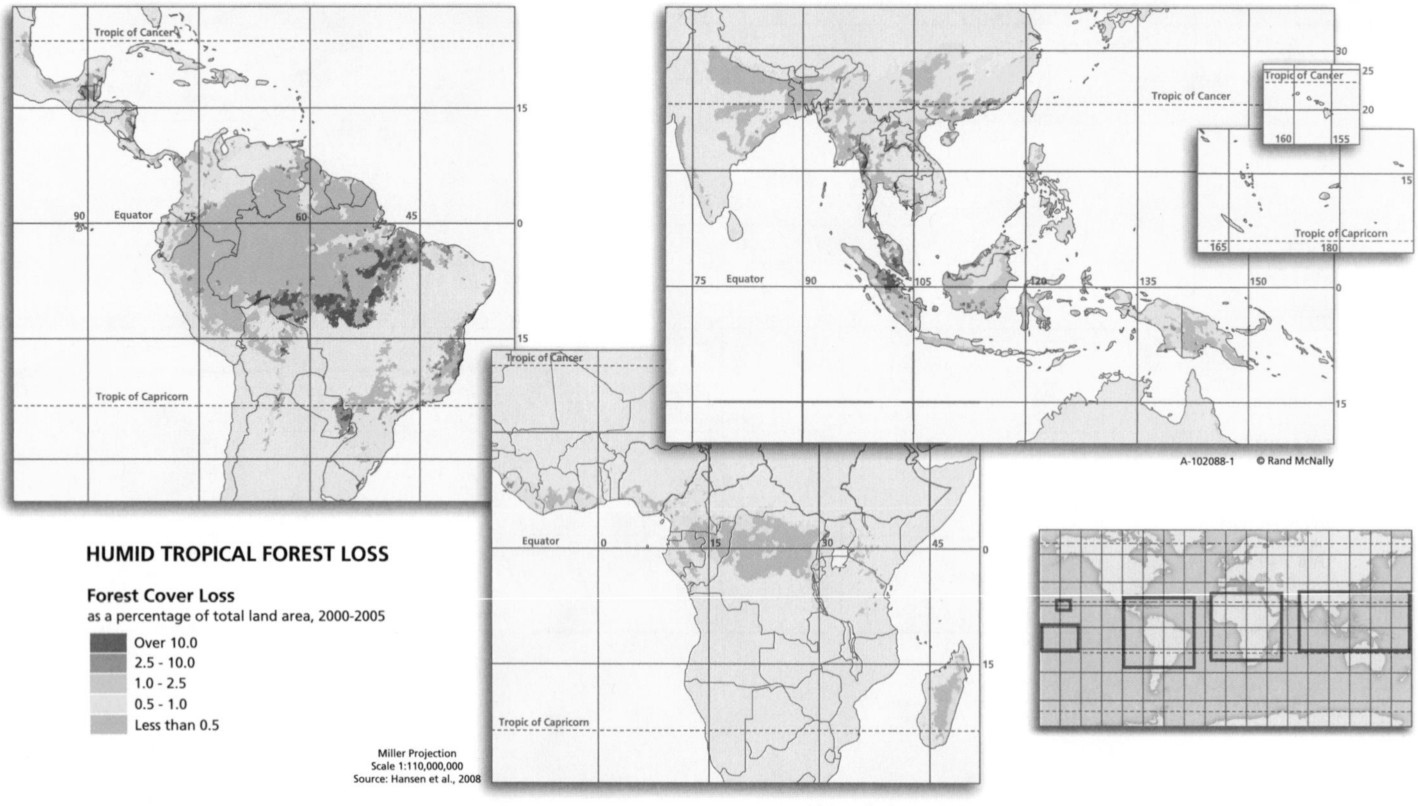
Tropic of Cancer
Equator
Tropic of Capricorn
A-102088-1 © Rand McNally

HUMID TROPICAL FOREST LOSS

Forest Cover Loss
as a percentage of total land area, 2000-2005

- Over 10.0
- 2.5 - 10.0
- 1.0 - 2.5
- 0.5 - 1.0
- Less than 0.5

Miller Projection
Scale 1:110,000,000
Source: Hansen et al., 2008

Humid Tropical Forest Area World Total* - 11,487,357 square kilometers - 2000

BRAZIL	PERU	COLOMBIA	VENEZUELA	BOLIVIA	OTHER S. AMER.	INDONESIA	CHINA	MYANMAR	INDIA	MALAYSIA	OTHER ASIA	D.R. OF CONGO	OTHER AFRICA	PAPUA NEW GUINEA	N. AMER.
31.4%	5.7	4.9	3.3	2.5	4.8	9.0	4.5	2.8	2.0	1.9	4.5	8.5	8.1	2.7	2.9

* Defined as areas with tree canopy cover of 25% or more.

Humid Tropical Forest Cover Loss World Total - 272,605 square kilometers - 2000-2005

BRAZIL	COLOMBIA	OTHER S. AMER.	INDONESIA	CHINA	MALAYSIA	MYANMAR	OTHER ASIA	AFRICA	NORTH AMERICA
47.6%	2.1	7.0	12.9	4.8	4.3	3.4	7.7	5.4	3.7

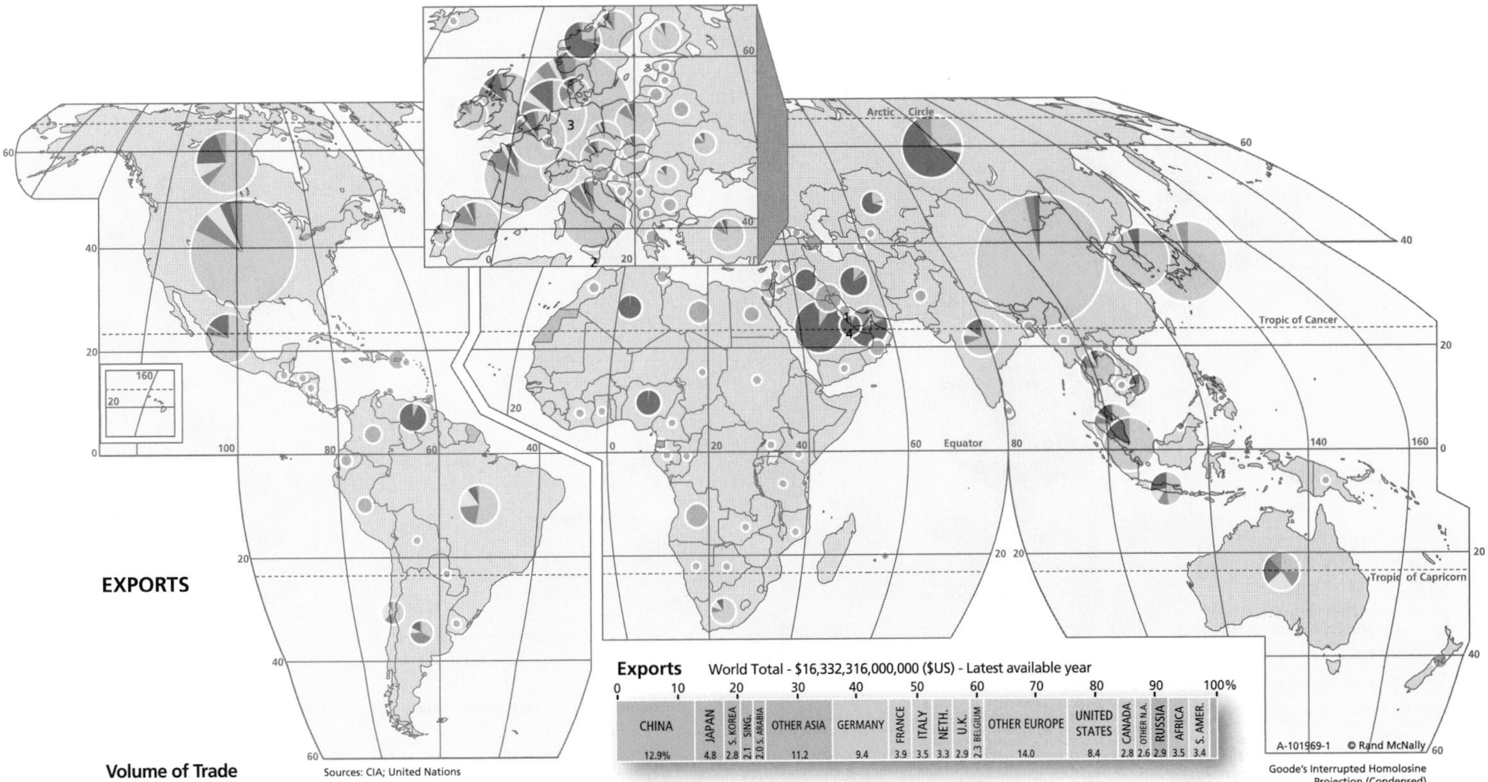
EXPORTS
Arctic Circle
Tropic of Cancer
Equator
Tropic of Capricorn
Exports World Total - $16,332,316,000,000 ($US) - Latest available year
0 10 20 30 40 50 60 70 80 90 100%
CHINA 12.9%
JAPAN 4.8
S. KOREA 2.8
SING. 2.1
S. ARABIA 2.0
OTHER ASIA 11.2
GERMANY 9.4
FRANCE 3.9
ITALY 3.5
NETH. 3.3
U.K. 2.9
BELGIUM 2.3
OTHER EUROPE 14.0
UNITED STATES 8.4
CANADA 2.8
OTHER N.A. 2.6
RUSSIA 2.9
AFRICA 3.5
S. AMER. 3.4
Sources: CIA; United Nations
A-101969-1 © Rand McNally

Goode's Interrupted Homolosine Projection (Condensed)
Scale 1:162,000,000

Volume of Trade
in billions of U.S. dollars - latest available year

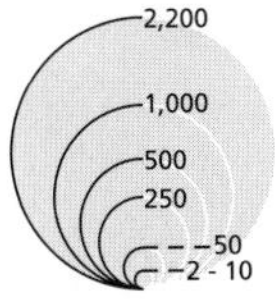
2,200
1,000
500
250
50
2 - 10

If volume of trade is less than 50 billion dollars, composition of trade is undifferentiated.
If volume of trade is less than 2 billion dollars, no symbol is shown.

Composition of Trade

Manufactured Articles | Food, Beverage, & Tobacco | Raw Materials | Fuel & Related Products | All Other | Undifferentiated

Data not available

China figures include Taiwan, Hong Kong, and Macau.
France figures include overseas departments.

Country Key
1 Bahrain
2 Malta
3 Netherlands
4 Qatar

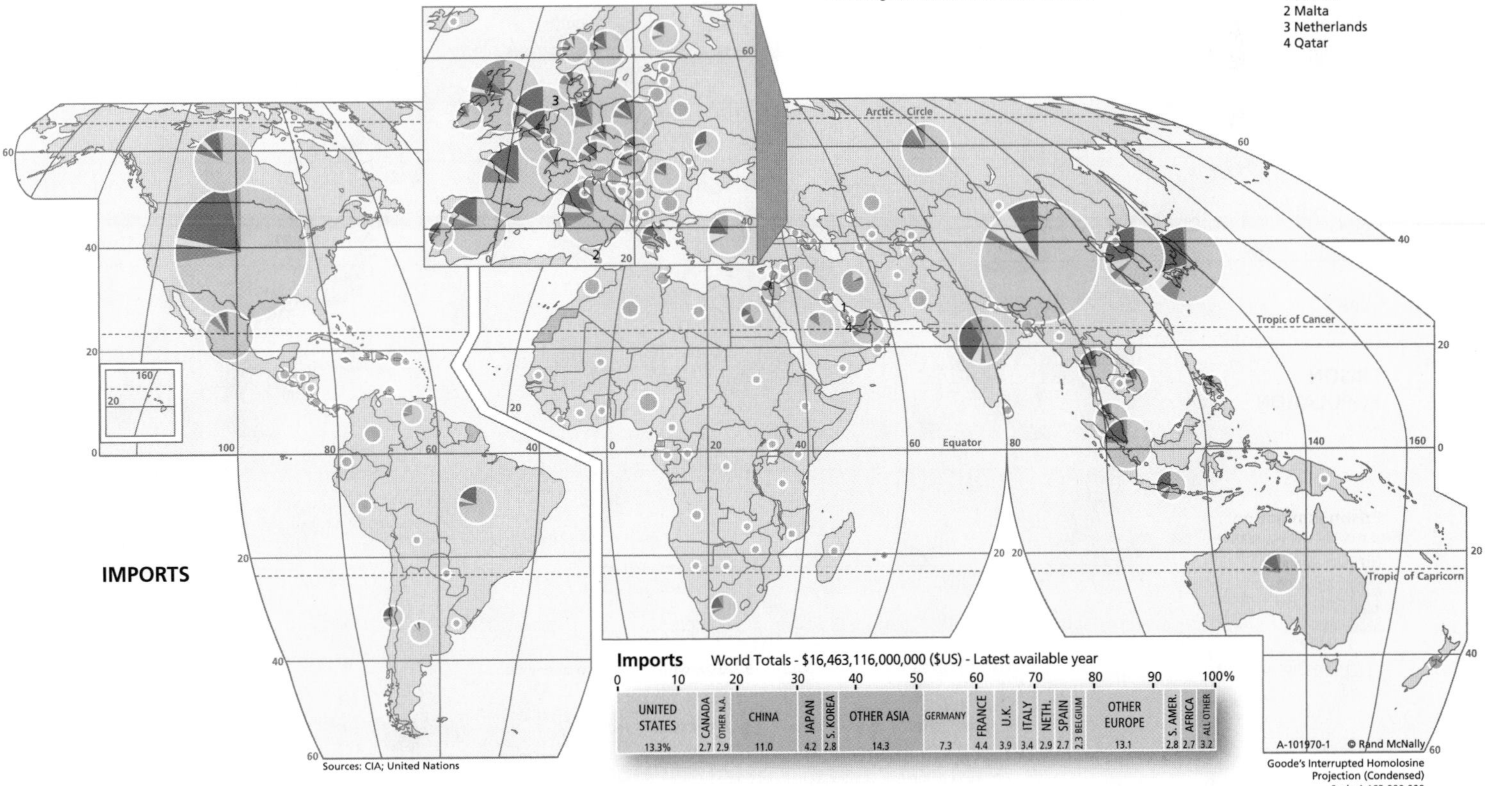
IMPORTS
Arctic Circle
Tropic of Cancer
Equator
Tropic of Capricorn
Imports World Totals - $16,463,116,000,000 ($US) - Latest available year
0 10 20 30 40 50 60 70 80 90 100%
UNITED STATES 13.3%
CANADA 2.7
OTHER N.A. 2.9
CHINA 11.0
JAPAN 4.2
S. KOREA 2.8
OTHER ASIA 14.3
GERMANY 7.3
FRANCE 4.4
U.K. 3.9
ITALY 3.4
NETH. 2.9
SPAIN 2.7
BELGIUM 2.3
OTHER EUROPE 13.1
S. AMER. 2.8
AFRICA 2.7
ALL OTHER 3.2
Sources: CIA; United Nations
A-101970-1 © Rand McNally

Goode's Interrupted Homolosine Projection (Condensed)
Scale 1:162,000,000

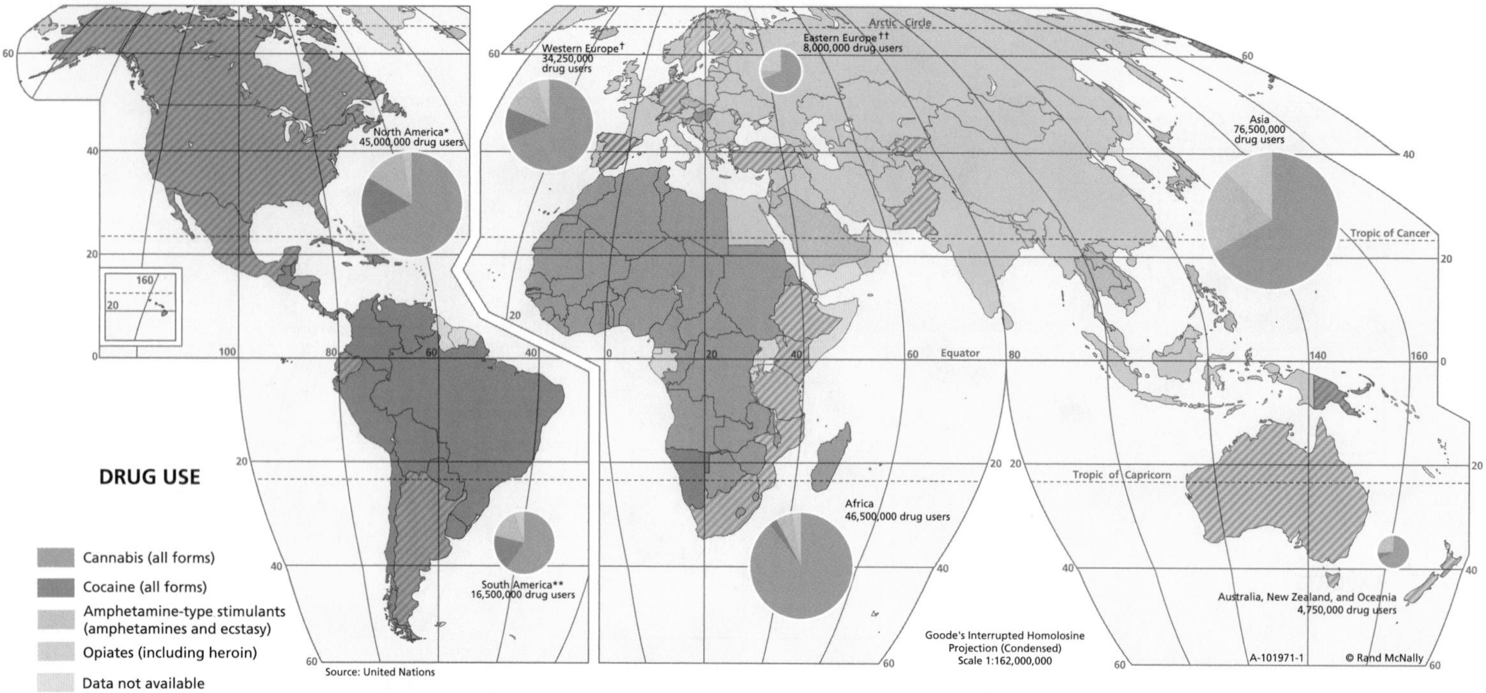
Arctic Circle
Western Europe†
34,250,000 drug users
Eastern Europe††
8,000,000 drug users
North America*
45,000,000 drug users
Asia
76,500,000 drug users
Tropic of Cancer
Equator
Tropic of Capricorn
Africa
46,500,000 drug users
South America**
16,500,000 drug users
Australia, New Zealand, and Oceania
4,750,000 drug users
Goode's Interrupted Homolosine Projection (Condensed)
Scale 1:162,000,000
A-101971-1
© Rand McNally
Source: United Nations

DRUG USE

- Cannabis (all forms)
- Cocaine (all forms)
- Amphetamine-type stimulants (amphetamines and ecstasy)
- Opiates (including heroin)
- Data not available

Circles show annual prevalence of drug use - 2006. Circle size is proportional to number of drug users per continent.

Country tints show primary drug or drugs of abuse, based on drug treatment statistics - most current available year.

Other drugs, such as hallucinogens, depressants, and inhalants, are not included.

* Includes Canada, the United States, and Mexico.
** Includes Central America and the Caribbean.
† Includes Cyprus and Turkey; excludes Russia, Belarus, Moldova, and Ukraine.
†† Includes Russia, Belarus, Moldova, and Ukraine.

Cocaine Potential World Production - 984 metric tons - 2006

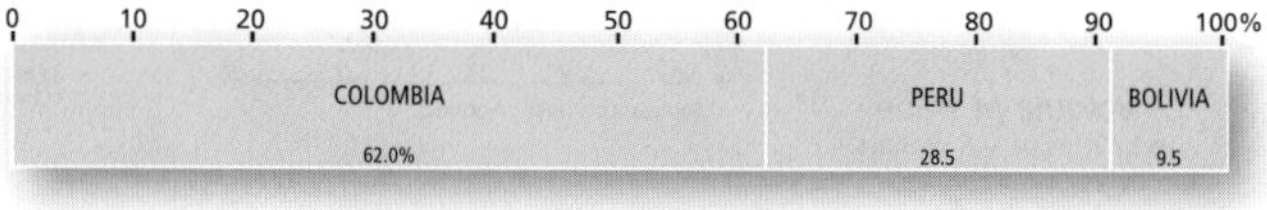
0 10 20 30 40 50 60 70 80 90 100%
COLOMBIA 62.0%
PERU 28.5
BOLIVIA 9.5

Cannabis World Production - 41,400 metric tons - 2006

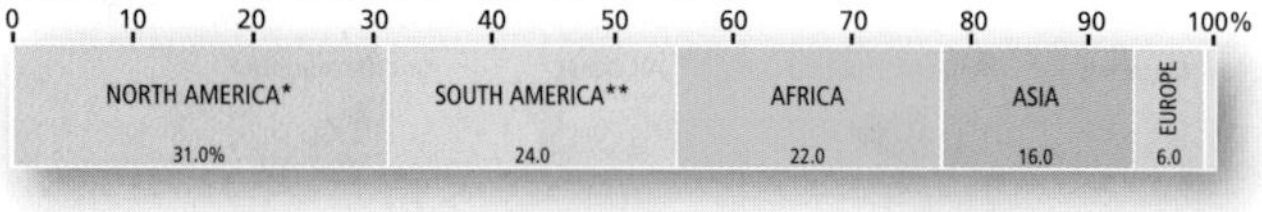
0 10 20 30 40 50 60 70 80 90 100%
NORTH AMERICA* 31.0%
SOUTH AMERICA** 24.0
AFRICA 22.0
ASIA 16.0
EUROPE 6.0

Opium Potential World Production (Dry Opium) - 6,611 metric tons - 2006

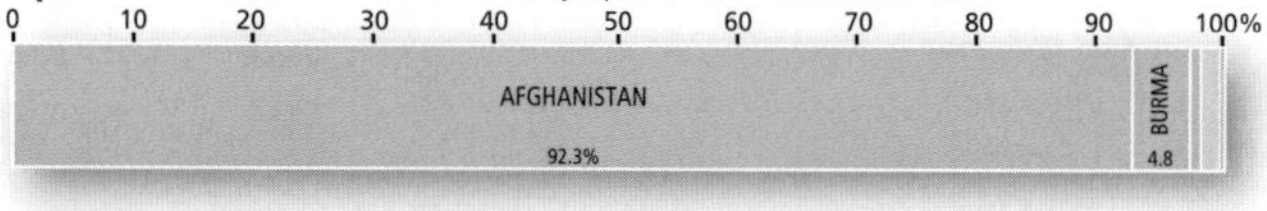
0 10 20 30 40 50 60 70 80 90 100%
AFGHANISTAN 92.3%
BURMA 4.8

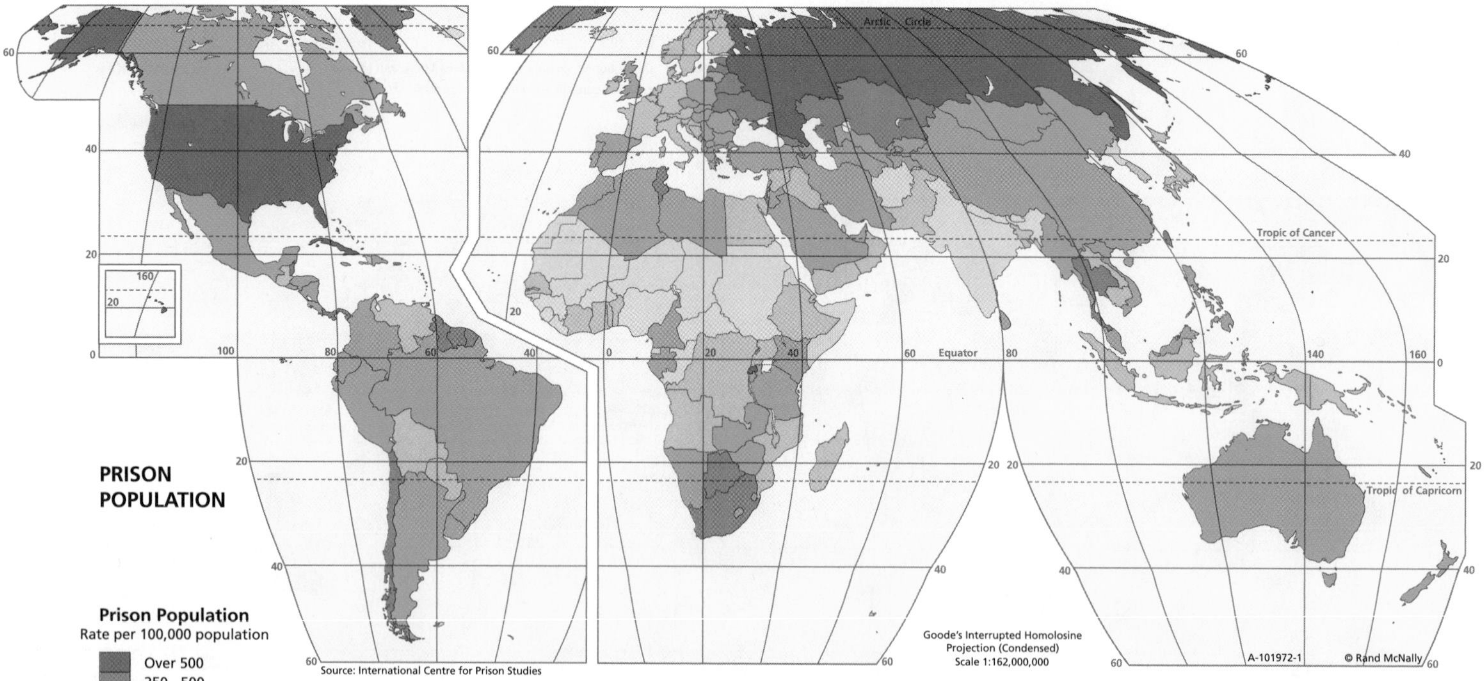
Arctic Circle
Tropic of Cancer
Equator
Tropic of Capricorn
Goode's Interrupted Homolosine Projection (Condensed)
Scale 1:162,000,000
A-101972-1
© Rand McNally
Source: International Centre for Prison Studies

PRISON POPULATION

Prison Population
Rate per 100,000 population

- Over 500
- 250 - 500
- 100 - 250
- 50 - 100
- 0 - 50
- Data not available

Prison Population World Total‡ - 9,720,000 prisoners - latest available data

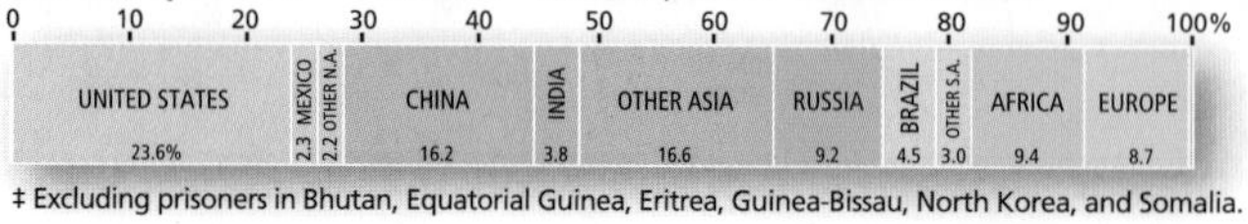
0 10 20 30 40 50 60 70 80 90 100%
UNITED STATES 23.6%
MEXICO 2.3
OTHER N.A. 2.2
CHINA 16.2
INDIA 3.8
OTHER ASIA 16.6
RUSSIA 9.2
BRAZIL 4.5
OTHER S.A. 3.0
AFRICA 9.4
EUROPE 8.7

‡ Excluding prisoners in Bhutan, Equatorial Guinea, Eritrea, Guinea-Bissau, North Korea, and Somalia.

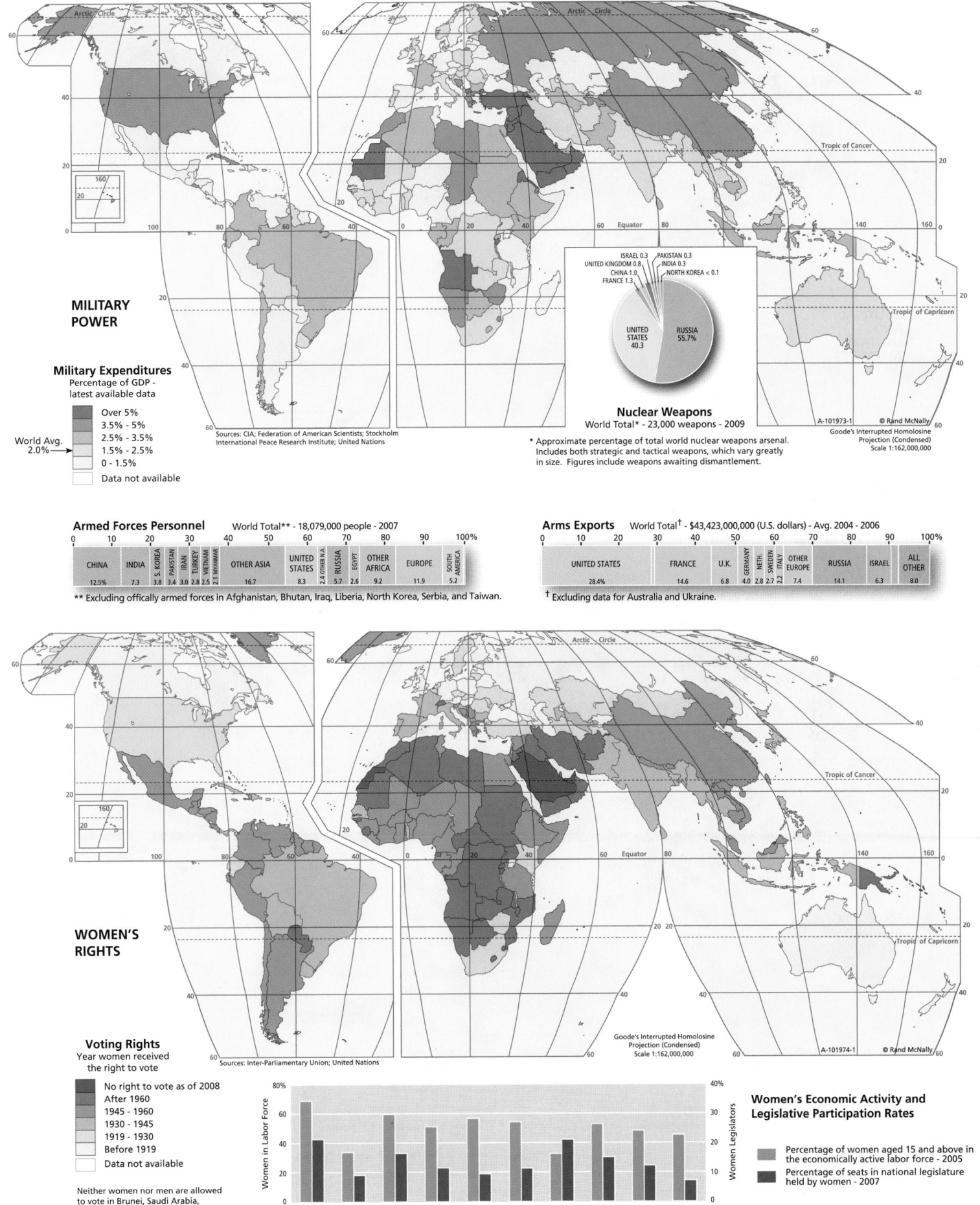

MILITARY POWER

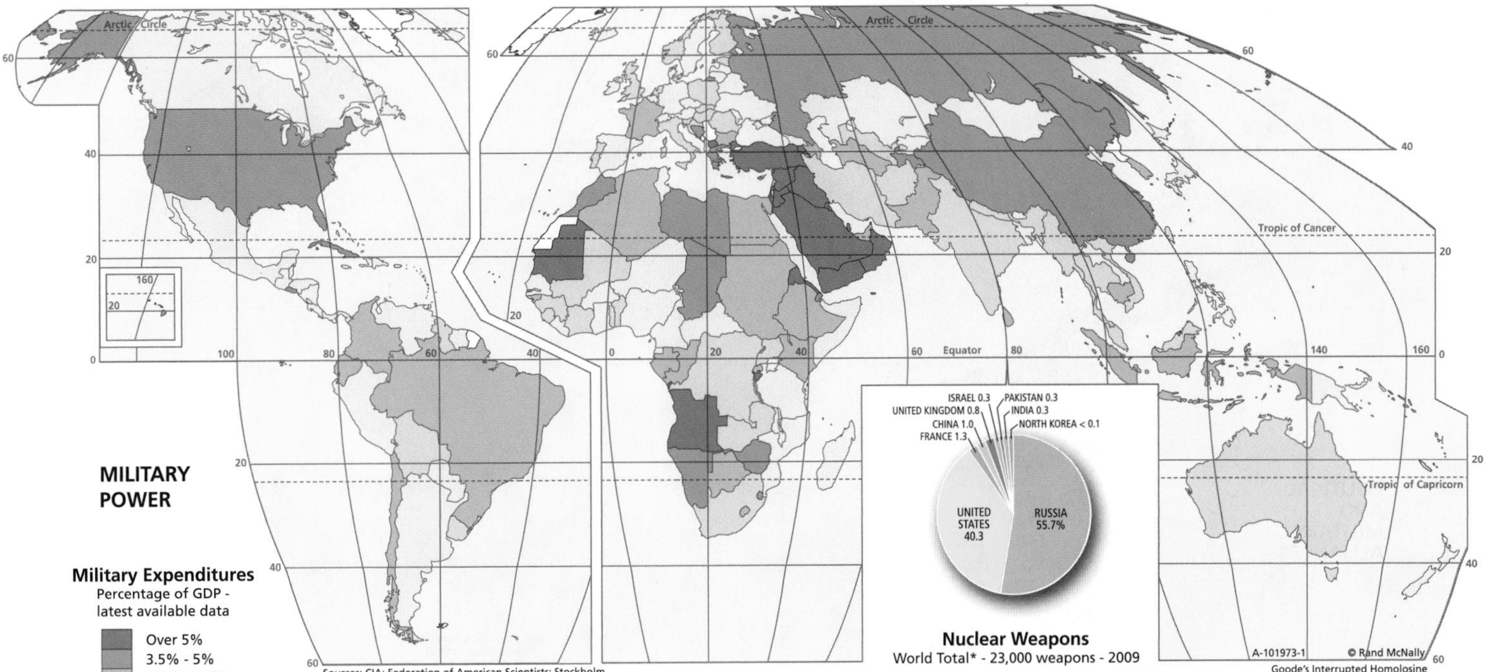

Arctic Circle
Tropic of Cancer
Equator
Tropic of Capricorn
A-101973-1
© Rand McNally
Goode's Interrupted Homolosine Projection (Condensed) Scale 1:162,000,000
ISRAEL 0.3
PAKISTAN 0.3
UNITED KINGDOM 0.8
INDIA 0.3
CHINA 1.0
NORTH KOREA < 0.1
FRANCE 1.3
UNITED STATES 40.3
RUSSIA 55.7%

Sources: CIA; Federation of American Scientists; Stockholm International Peace Research Institute; United Nations

Nuclear Weapons

World Total* - 23,000 weapons - 2009

* Approximate percentage of total world nuclear weapons arsenal. Includes both strategic and tactical weapons, which vary greatly in size. Figures include weapons awaiting dismantlement.

Military Expenditures

Percentage of GDP - latest available data

- Over 5%
- 3.5% - 5%
- 2.5% - 3.5%
- 1.5% - 2.5% (World Avg. 2.0%)
- 0 - 1.5%
- Data not available

Armed Forces Personnel

World Total** - 18,079,000 people - 2007

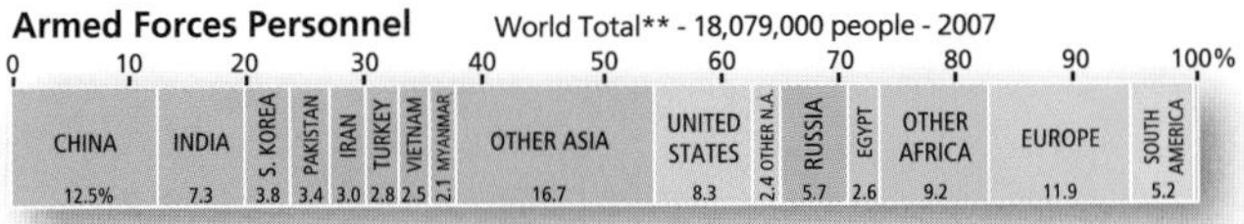

0 10 20 30 40 50 60 70 80 90 100%
CHINA 12.5%
INDIA 7.3
S. KOREA 3.8
PAKISTAN 3.4
IRAN 3.0
TURKEY 2.8
VIETNAM 2.5
MYANMAR 2.1
OTHER ASIA 16.7
UNITED STATES 8.3
OTHER N.A. 2.4
RUSSIA 5.7
EGYPT 2.6
OTHER AFRICA 9.2
EUROPE 11.9
SOUTH AMERICA 5.2

** Excluding offically armed forces in Afghanistan, Bhutan, Iraq, Liberia, North Korea, Serbia, and Taiwan.

Arms Exports

World Total† - $43,423,000,000 (U.S. dollars) - Avg. 2004 - 2006

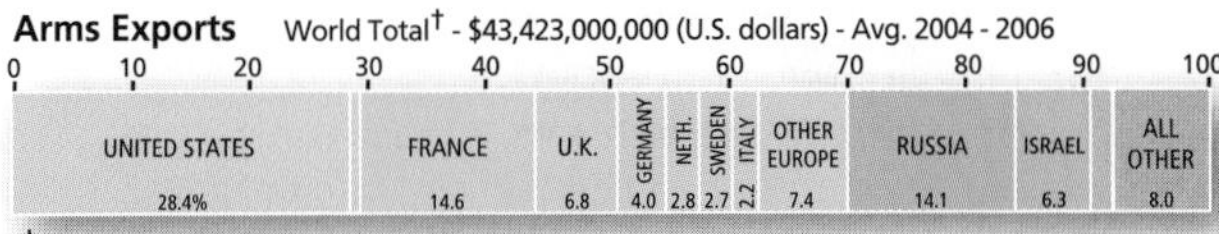

0 10 20 30 40 50 60 70 80 90 100%
UNITED STATES 28.4%
FRANCE 14.6
U.K. 6.8
GERMANY 4.0
NETH. 2.8
SWEDEN 2.7
ITALY 2.2
OTHER EUROPE 7.4
RUSSIA 14.1
ISRAEL 6.3
ALL OTHER 8.0

† Excluding data for Australia and Ukraine.

WOMEN'S RIGHTS

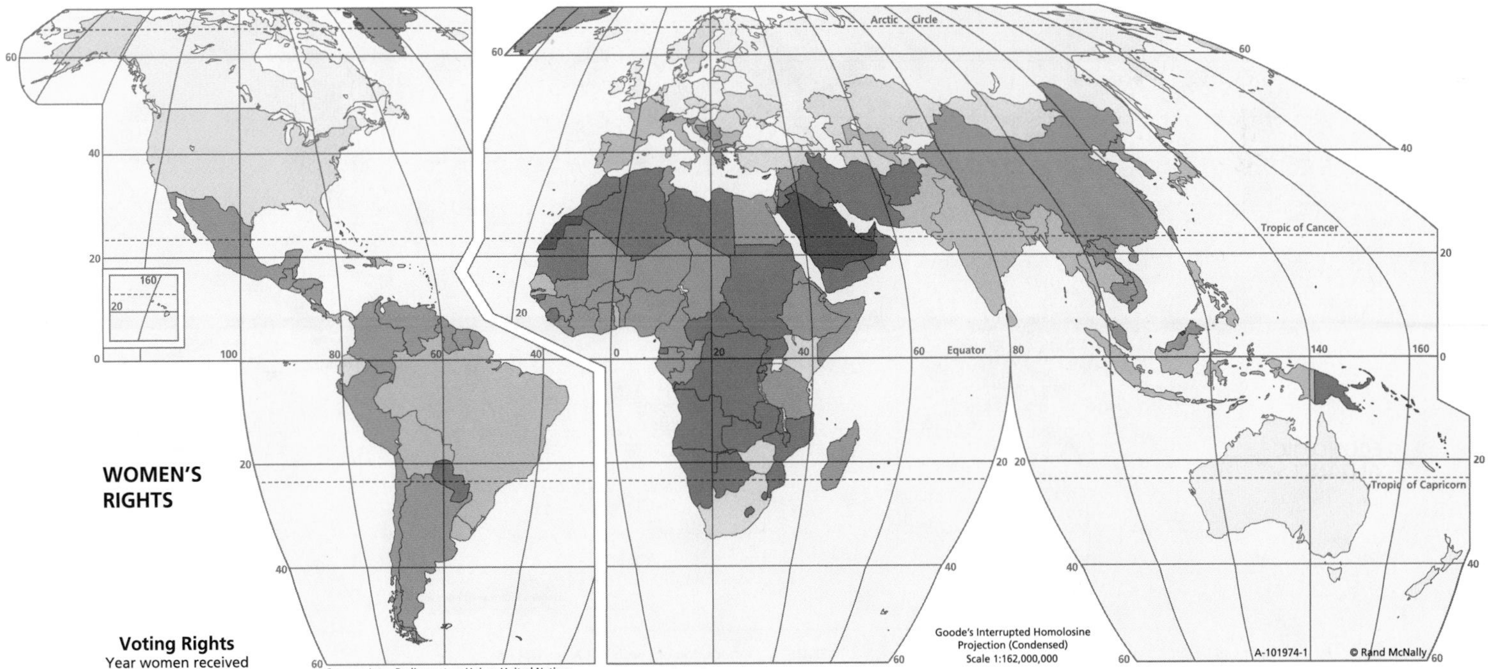

Arctic Circle
Tropic of Cancer
Equator
Tropic of Capricorn
Goode's Interrupted Homolosine Projection (Condensed) Scale 1:162,000,000
A-101974-1
© Rand McNally

Sources: Inter-Parliamentary Union; United Nations

Voting Rights

Year women received the right to vote

- No right to vote as of 2008
- After 1960
- 1945 - 1960
- 1930 - 1945
- 1919 - 1930
- Before 1919
- Data not available

Neither women nor men are allowed to vote in Brunei, Saudi Arabia, United Arab Emirates, or Western Sahara.

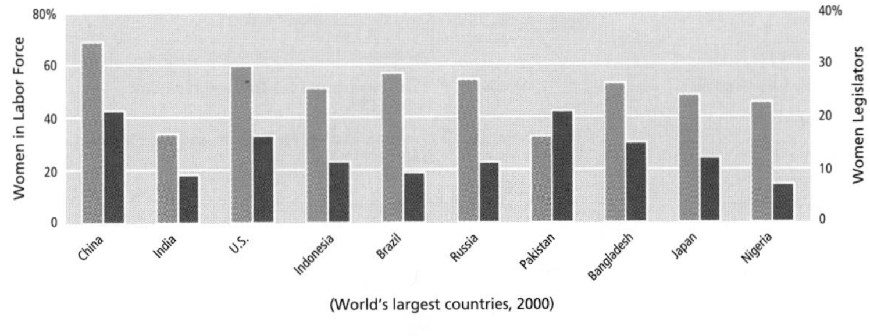

80%
60
40
20
0
Women in Labor Force
40%
30
20
10
0
Women Legislators
China
India
U.S.
Indonesia
Brazil
Russia
Pakistan
Bangladesh
Japan
Nigeria
(World's largest countries, 2000)

Women's Economic Activity and Legislative Participation Rates

- Percentage of women aged 15 and above in the economically active labor force - 2005
- Percentage of seats in national legislature held by women - 2007

POLITICAL AND MILITARY ALLIANCES

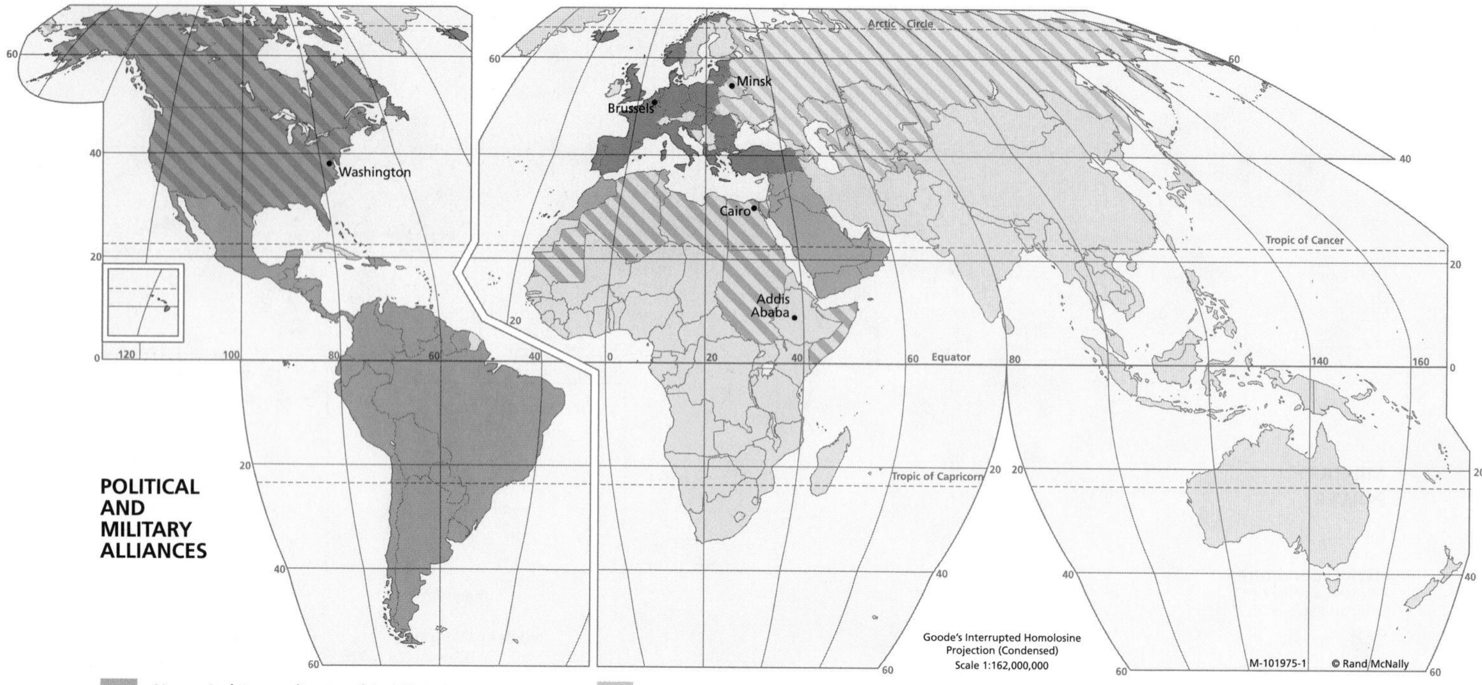
Arctic Circle
Minsk
Brussels
Washington
Cairo
Addis Ababa
Tropic of Cancer
Equator
Tropic of Capricorn
Goode's Interrupted Homolosine Projection (Condensed)
Scale 1:162,000,000
M-101975-1
© Rand McNally

AL	**Arab League** (League of Arab States) Founded 1945. Headquarters in Cairo, Egypt.
OAS	**Organization of American States** Founded 1948. Headquarters in Washington, D.C., United States.
NATO	**North Atlantic Treaty Organization** Founded 1949. Headquarters in Brussels, Belgium.
PFP	**Partnership for Peace Program**
CIS	**Commonwealth of Independent States** Founded 1991. Headquarters in Minsk, Belarus.
AU	**African Union** Founded 2000. Headquarters in Addis Ababa, Ethiopia.
	Not affiliated with above organizations.

ECONOMIC ALLIANCES

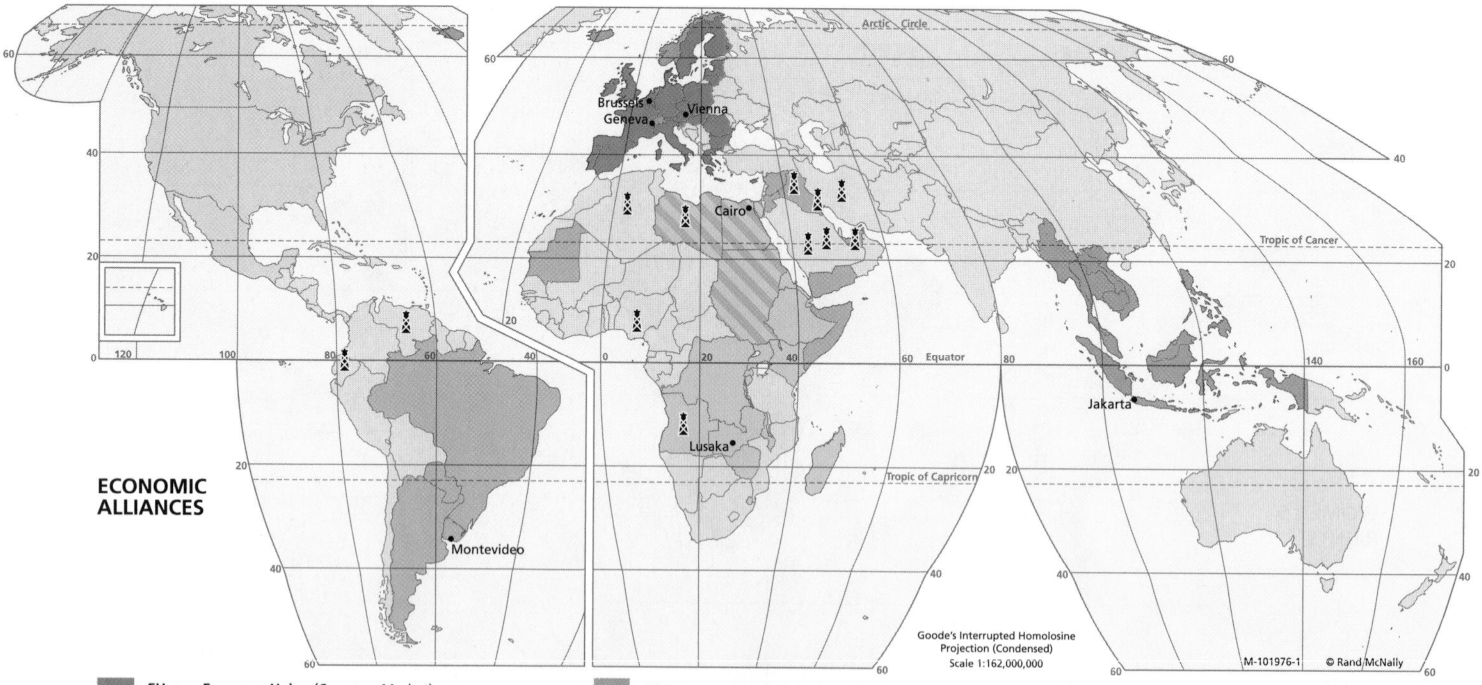
Arctic Circle
Brussels
Geneva
Vienna
Cairo
Tropic of Cancer
Equator
Jakarta
Lusaka
Tropic of Capricorn
Montevideo
Goode's Interrupted Homolosine Projection (Condensed)
Scale 1:162,000,000
M-101976-1
© Rand McNally

EU	**European Union** (Common Market) Founded 1957. Headquarters in Brussels, Belgium.
EFTA	**European Free Trade Association** Founded 1960. Headquarters in Geneva, Switzerland.
OPEC	**Organization of Petroleum Exporting Countries** Founded 1960. Headquarters in Vienna, Austria.
CAEU	**Council of Arab Economic Unity** Founded 1964. Headquarters in Cairo, Egypt. Includes Arab Common Market countries.
ASEAN	**Association of Southeast Asian Nations** Founded 1967. Headquarters in Jakarta, Indonesia.
MERCOSUR	**Southern Common Market** Founded 1991. Headquarters in Montevideo, Uruguay.
NAFTA	**North American Free Trade Agreement** Signed 1992.
COMESA	**Common Market for Eastern and Southern Africa** Founded 1994. Headquarters in Lusaka, Zambia.
	Not affiliated with above organizations.

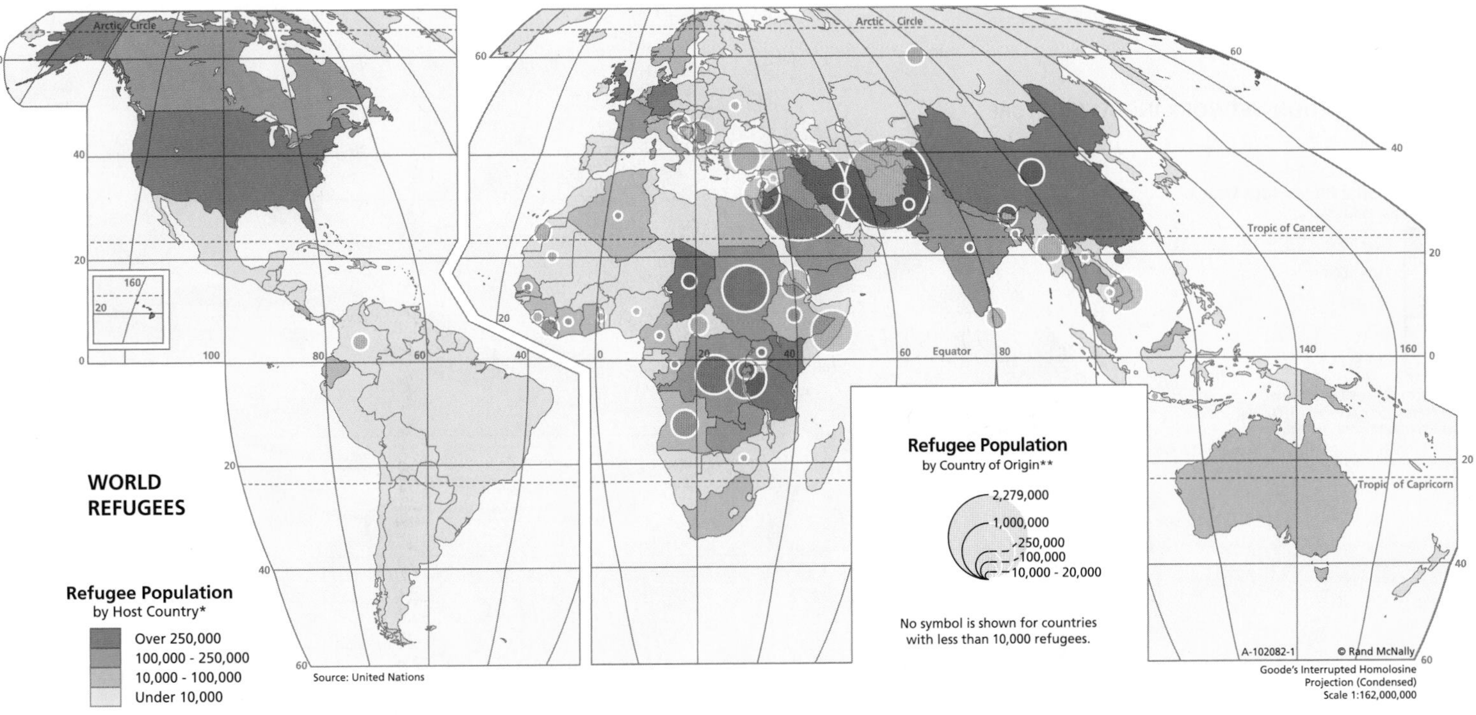
WORLD REFUGEES
Refugee Population
by Host Country*
Over 250,000
100,000 - 250,000
10,000 - 100,000
Under 10,000
Source: United Nations
Refugee Population
by Country of Origin**
2,279,000
1,000,000
250,000
100,000
10,000 - 20,000
No symbol is shown for countries with less than 10,000 refugees.
Arctic Circle
Tropic of Cancer
Equator
Tropic of Capricorn
A-102082-1
© Rand McNally
Goode's Interrupted Homolosine Projection (Condensed)
Scale 1:162,000,000

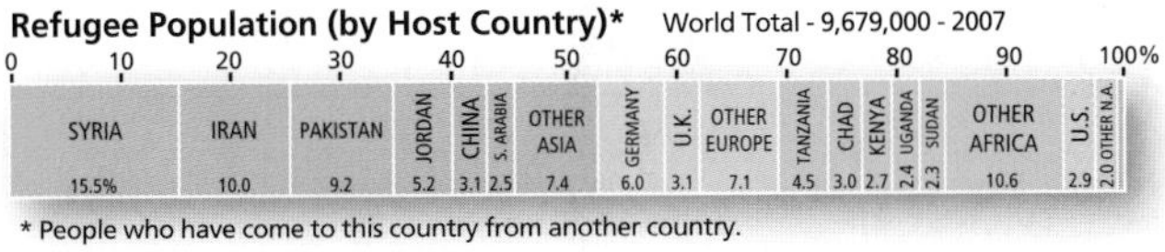
Refugee Population (by Host Country)*
World Total - 9,679,000 - 2007
0 10 20 30 40 50 60 70 80 90 100%
SYRIA 15.5%
IRAN 10.0
PAKISTAN 9.2
JORDAN 5.2
CHINA 3.1
S. ARABIA 2.5
OTHER ASIA 7.4
GERMANY 6.0
U.K. 3.1
OTHER EUROPE 7.1
TANZANIA 4.5
CHAD 3.0
KENYA 2.7
UGANDA 2.4
SUDAN 2.3
OTHER AFRICA 10.6
U.S. 2.9
OTHER N.A. 2.0
* People who have come to this country from another country.

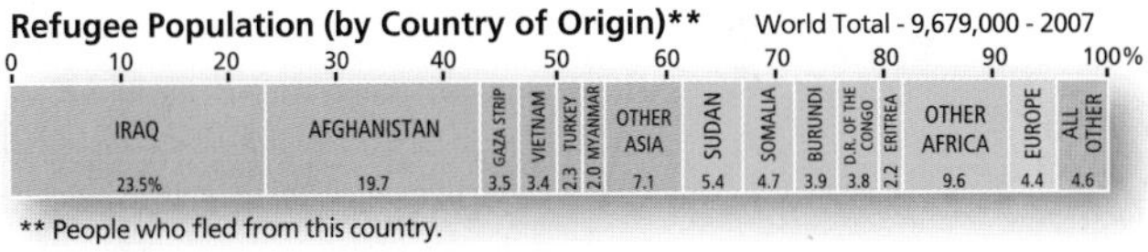
Refugee Population (by Country of Origin)**
World Total - 9,679,000 - 2007
0 10 20 30 40 50 60 70 80 90 100%
IRAQ 23.5%
AFGHANISTAN 19.7
GAZA STRIP 3.5
VIETNAM 3.4
TURKEY 2.3
MYANMAR 2.0
OTHER ASIA 7.1
SUDAN 5.4
SOMALIA 4.7
BURUNDI 3.9
D.R. OF THE CONGO 3.8
ERITREA 2.2
OTHER AFRICA 9.6
EUROPE 4.4
ALL OTHER 4.6
** People who fled from this country.

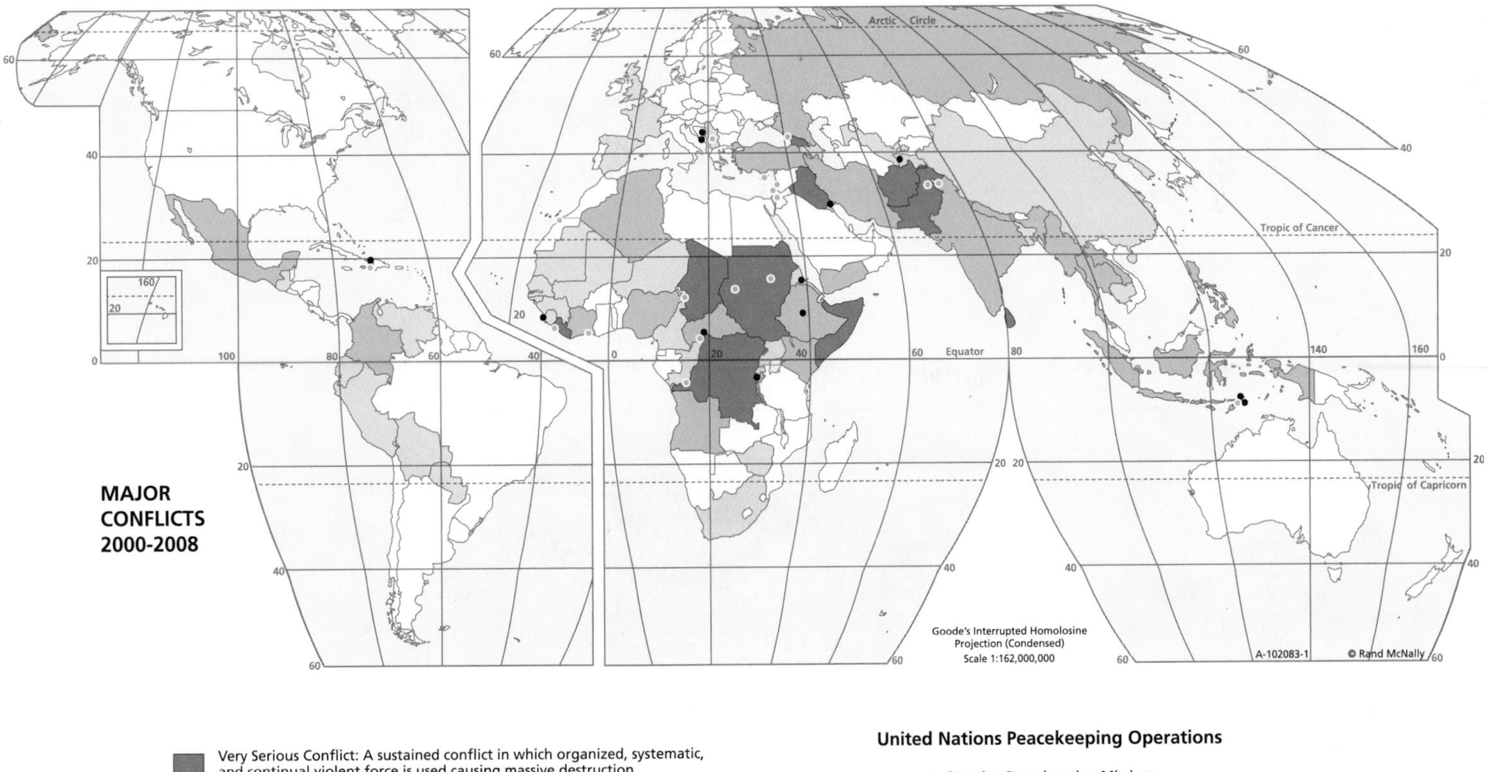
MAJOR CONFLICTS 2000-2008
Arctic Circle
Tropic of Cancer
Equator
Tropic of Capricorn
Goode's Interrupted Homolosine Projection (Condensed)
Scale 1:162,000,000
A-102083-1
© Rand McNally

Very Serious Conflict: A sustained conflict in which organized, systematic, and continual violent force is used causing massive destruction.

Serious Conflict: Severe crisis where organized violence is used regularly.

Hot Spot: A tense situation in which at least one of the parties uses violence in sporadic incidents.

United Nations Peacekeeping Operations

- Ongoing Peacekeeping Missions
- Completed Peacekeeping Missions

COMMUNICATION NETWORK INFRASTRUCTURE

International Bandwidth Usage

Gigabits per second (Gbps) - 2007

- Over 1000
- 250 - 1000
- 50 - 250
- 1 - 50
- Less than 1

Capacity deployed by carriers, internet service providers (ISPs), and enterprises to carry internet, voice, and private network traffic across international borders.

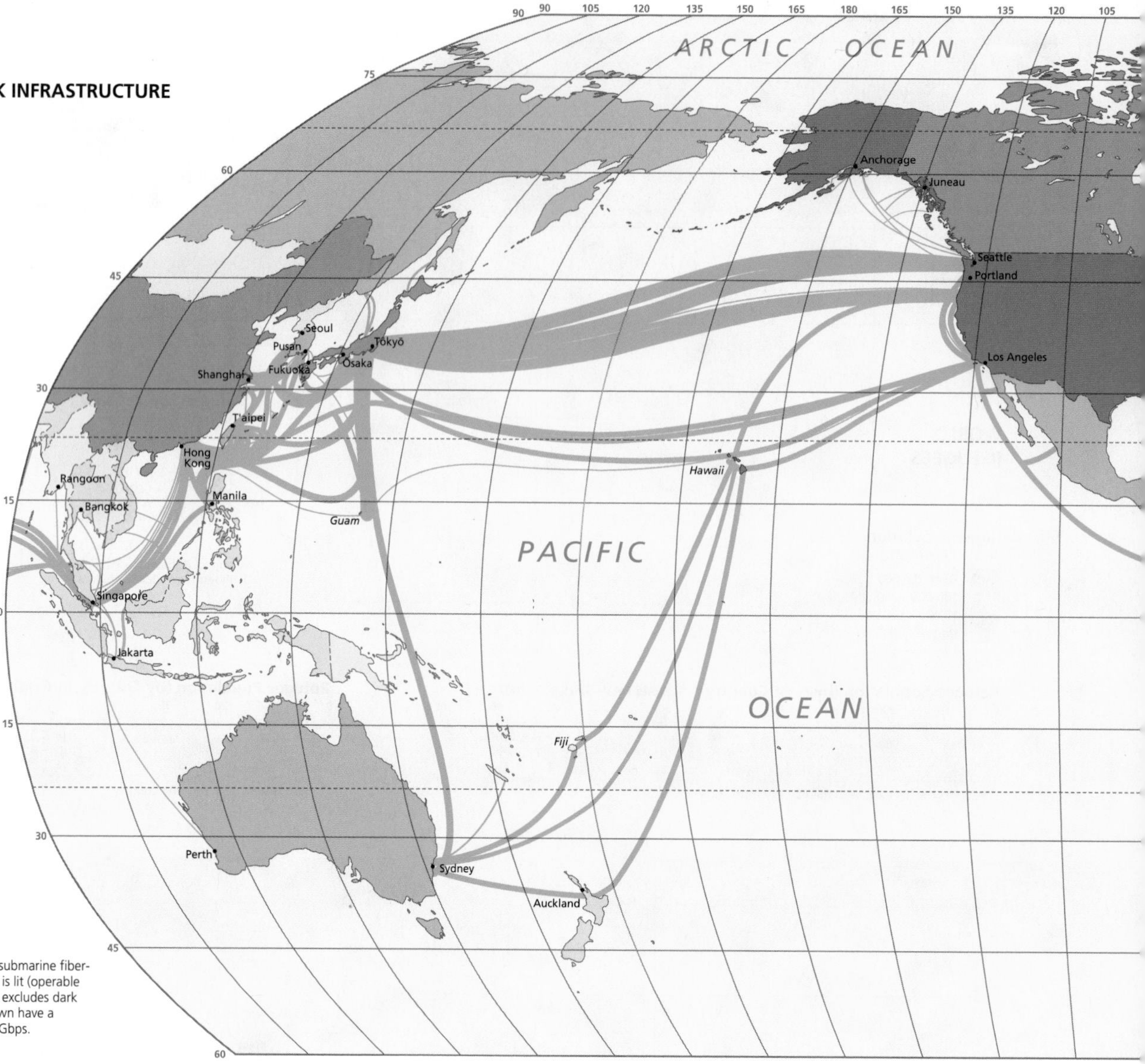

ARCTIC OCEAN
PACIFIC
OCEAN
Anchorage
Juneau
Seattle
Portland
Los Angeles
Hawaii
Seoul
Pusan
Tokyo
Fukuoka
Osaka
Shanghai
Taipei
Hong Kong
Rangoon
Bangkok
Manila
Guam
Singapore
Jakarta
Fiji
Perth
Sydney
Auckland

Submarine Cable Capacity

Lit capacity of submarine cables, in Gigabits per second (Gbps) - 2008

- Over 500
- 50 - 500
- 10 - 50

Line thickness is proportional to lit capacity of submarine fiber-optic cables. Lit capacity includes all cable that is lit (operable and capable of transmitting a light signal), but excludes dark fiber (inactive or inoperable cable). Cables shown have a maximum upgradeable capacity of at least 10 Gbps.

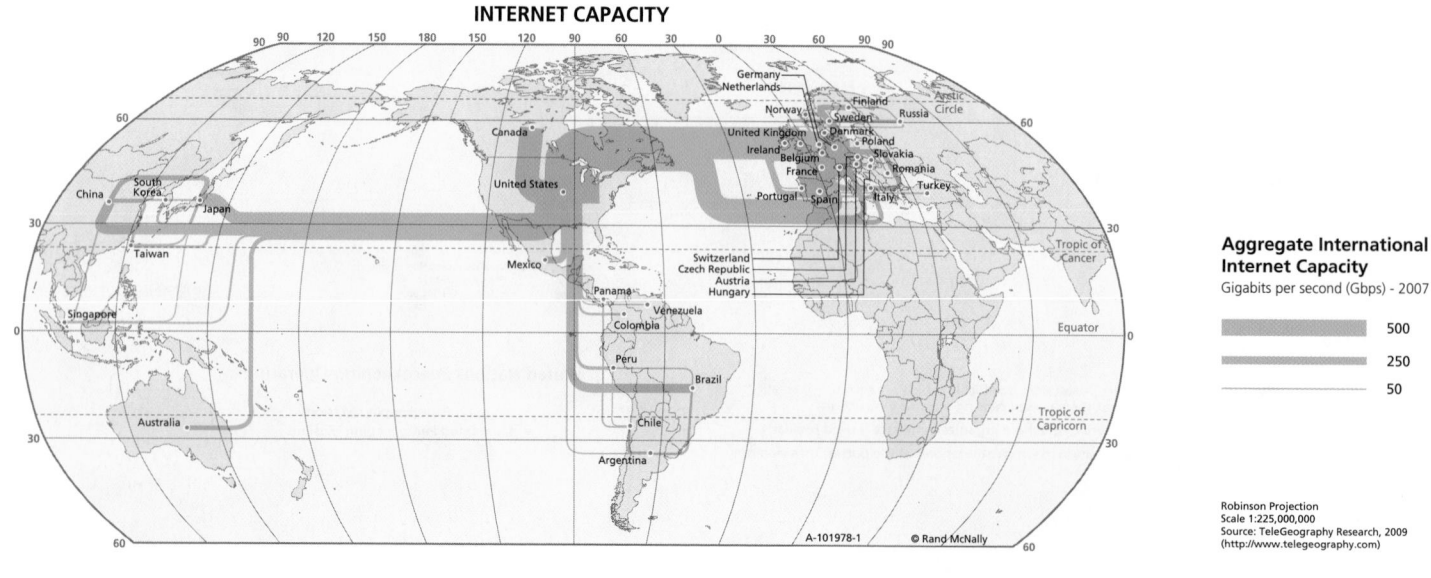

INTERNET CAPACITY

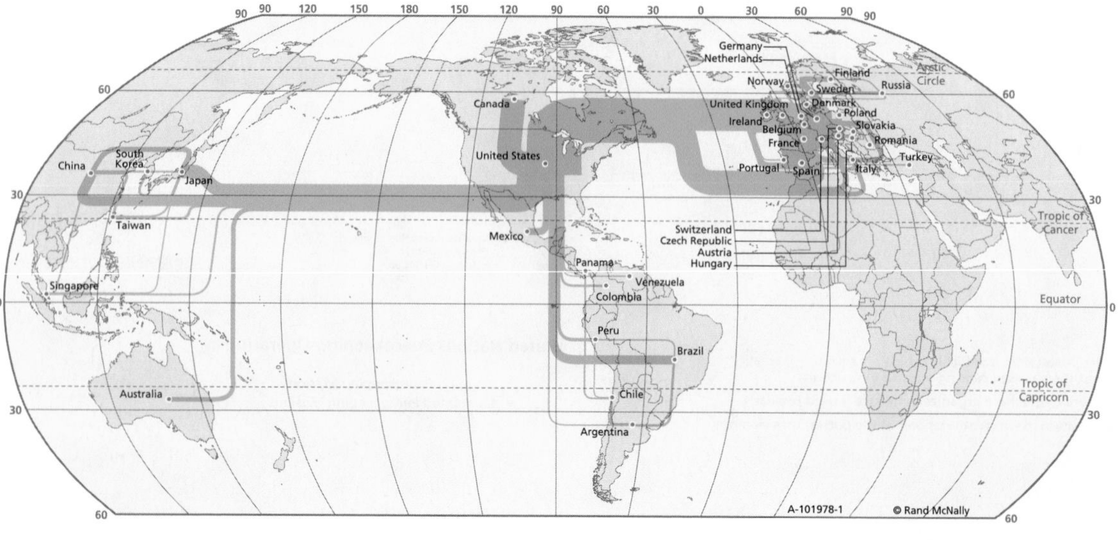

Germany
Netherlands
Finland
Norway
Sweden
Russia
Arctic Circle
Denmark
United Kingdom
Ireland
Poland
Belgium
Slovakia
France
Romania
Turkey
Portugal
Spain
Italy
Canada
United States
China
South Korea
Japan
Taiwan
Mexico
Switzerland
Czech Republic
Austria
Hungary
Tropic of Cancer
Panama
Venezuela
Colombia
Singapore
Equator
Peru
Brazil
Tropic of Capricorn
Australia
Chile
Argentina
A-101978-1
© Rand McNally

Aggregate International Internet Capacity

Gigabits per second (Gbps) - 2007

- 500
- 250
- 50

Robinson Projection
Scale 1:225,000,000
Source: TeleGeography Research, 2009
(http://www.telegeography.com)

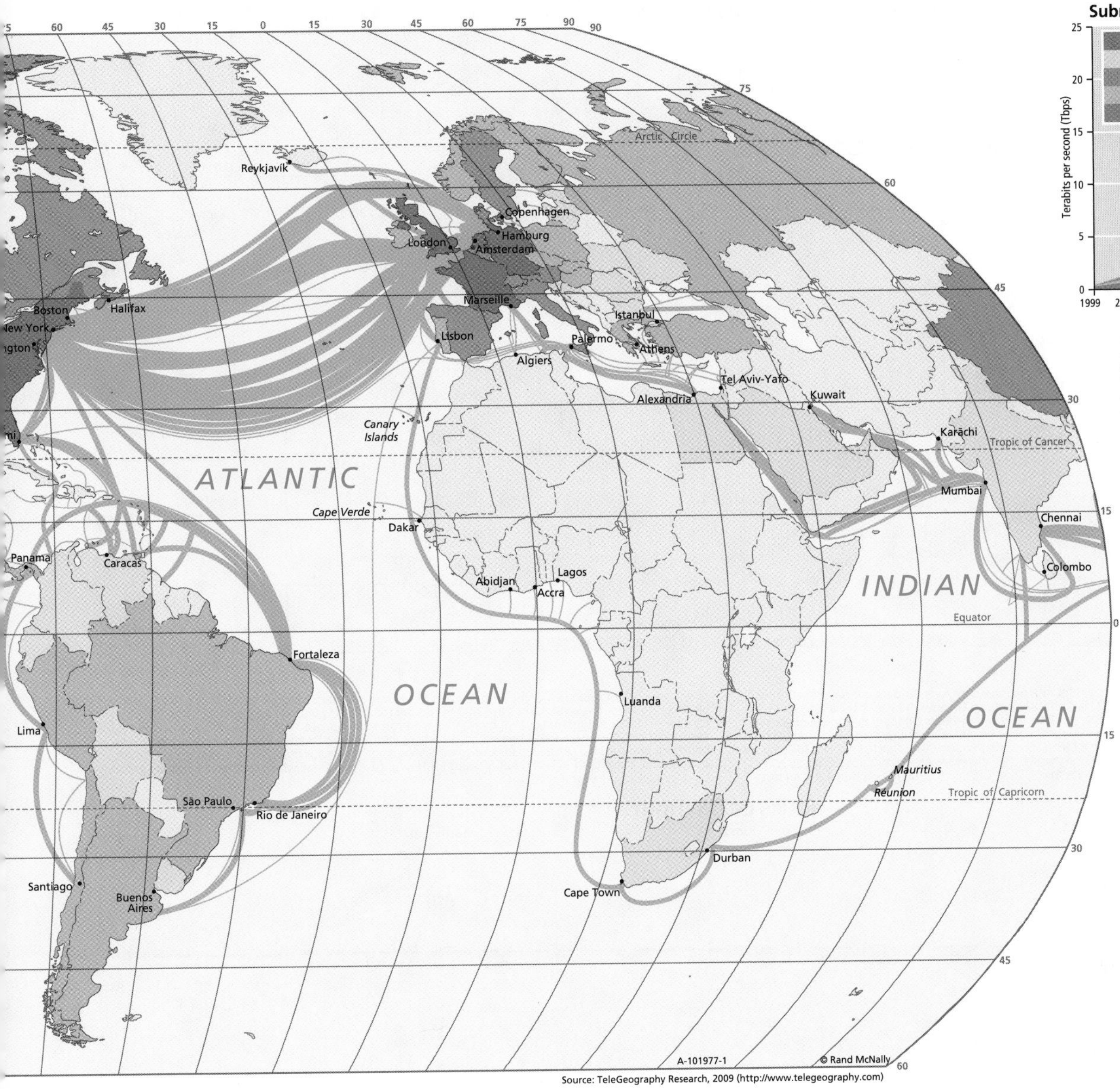
ATLANTIC
OCEAN
INDIAN
OCEAN
Arctic Circle
Tropic of Cancer
Equator
Tropic of Capricorn
Reykjavík
Copenhagen
Hamburg
London
Amsterdam
Marseille
Lisbon
Istanbul
Palermo
Athens
Algiers
Tel Aviv-Yafo
Alexandria
Kuwait
Karāchi
Mumbai
Chennai
Colombo
Canary Islands
Cape Verde
Dakar
Abidjan
Lagos
Accra
Luanda
Mauritius
Réunion
Durban
Cape Town
Boston
Halifax
New York
ngton
Panama
Caracas
Fortaleza
Lima
São Paulo
Rio de Janeiro
Santiago
Buenos Aires
A-101977-1
© Rand McNally

Source: TeleGeography Research, 2009 (http://www.telegeography.com)

Submarine Cable Capacity by Route

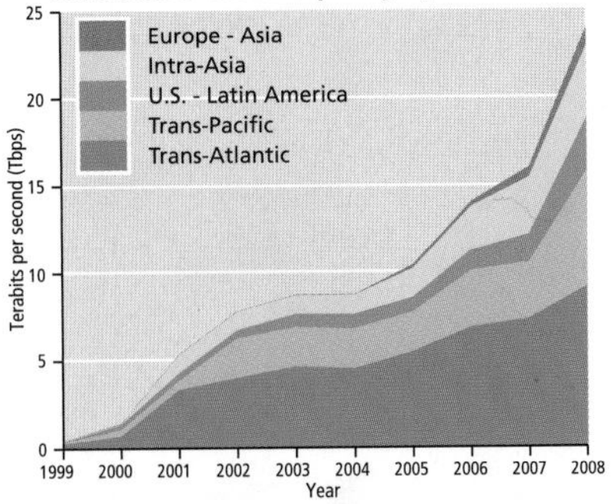
Europe - Asia
Intra-Asia
U.S. - Latin America
Trans-Pacific
Trans-Atlantic
Terabits per second (Tbps)
0
5
10
15
20
25
1999
2000
2001
2002
2003
2004
2005
2006
2007
2008
Year

Figures denote lit capacity of submarine fiber-optic cables. Trans-Pacific capacity excludes cables linking the United States to Australia and New Zealand. Trans-Atlantic capacity excludes cables linking Europe to South America. Intra-Asia capacity includes cables with landings in both Hong Kong and Japan. Europe-Asia capacity reflects available capacity between the Middle East and Europe, and excludes Europe-Asia capacity routed via Russia or the United States.

Robinson Projection
Scale 1:100,000,000
One inch to 1,580 miles
One cm to 1,000 km

0 500 1000 1500 2000 Miles
0 500 1000 1500 2000 2500 Kilometers

SHIPPING LANES

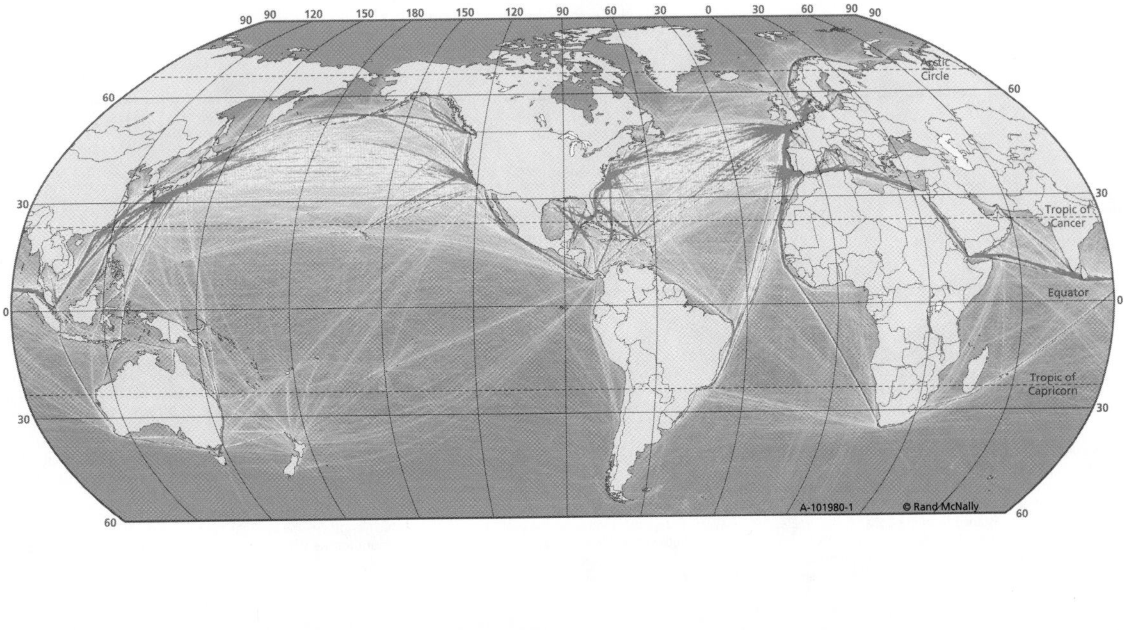
Arctic Circle
Tropic of Cancer
Equator
Tropic of Capricorn
A-101980-1
© Rand McNally

Relative Frequency of Ship Traffic

Highest
Lowest

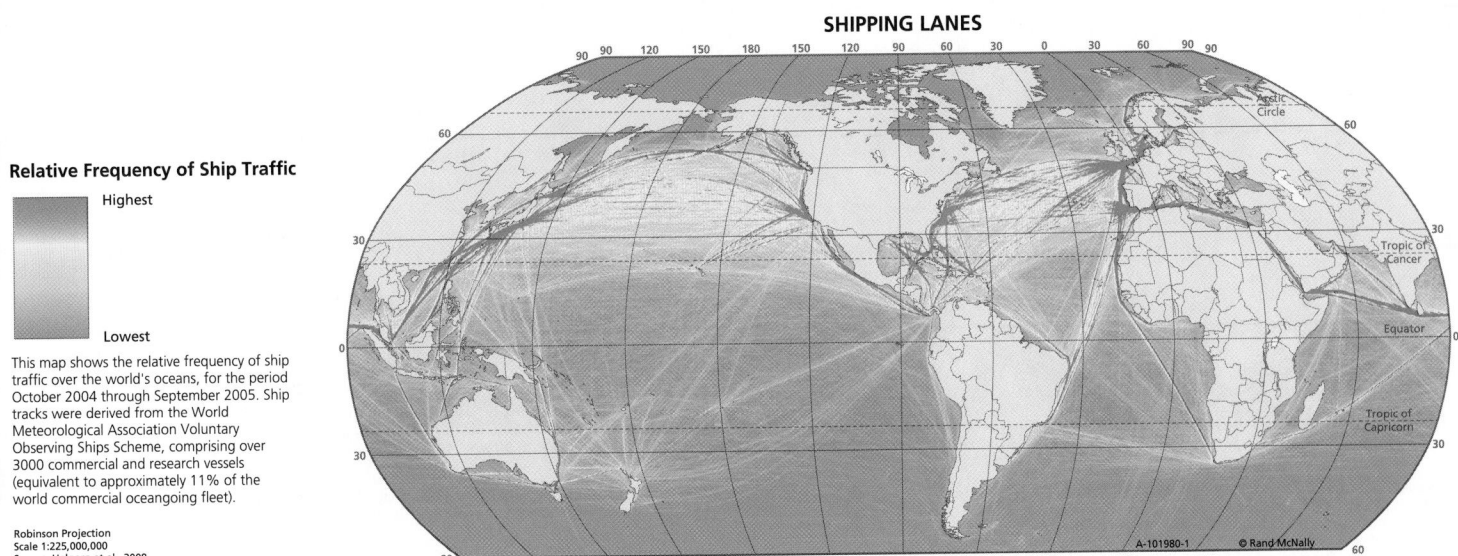

This map shows the relative frequency of ship traffic over the world's oceans, for the period October 2004 through September 2005. Ship tracks were derived from the World Meteorological Association Voluntary Observing Ships Scheme, comprising over 3000 commercial and research vessels (equivalent to approximately 11% of the world commercial oceangoing fleet).

Robinson Projection
Scale 1:225,000,000
Source: Halpern et al., 2008

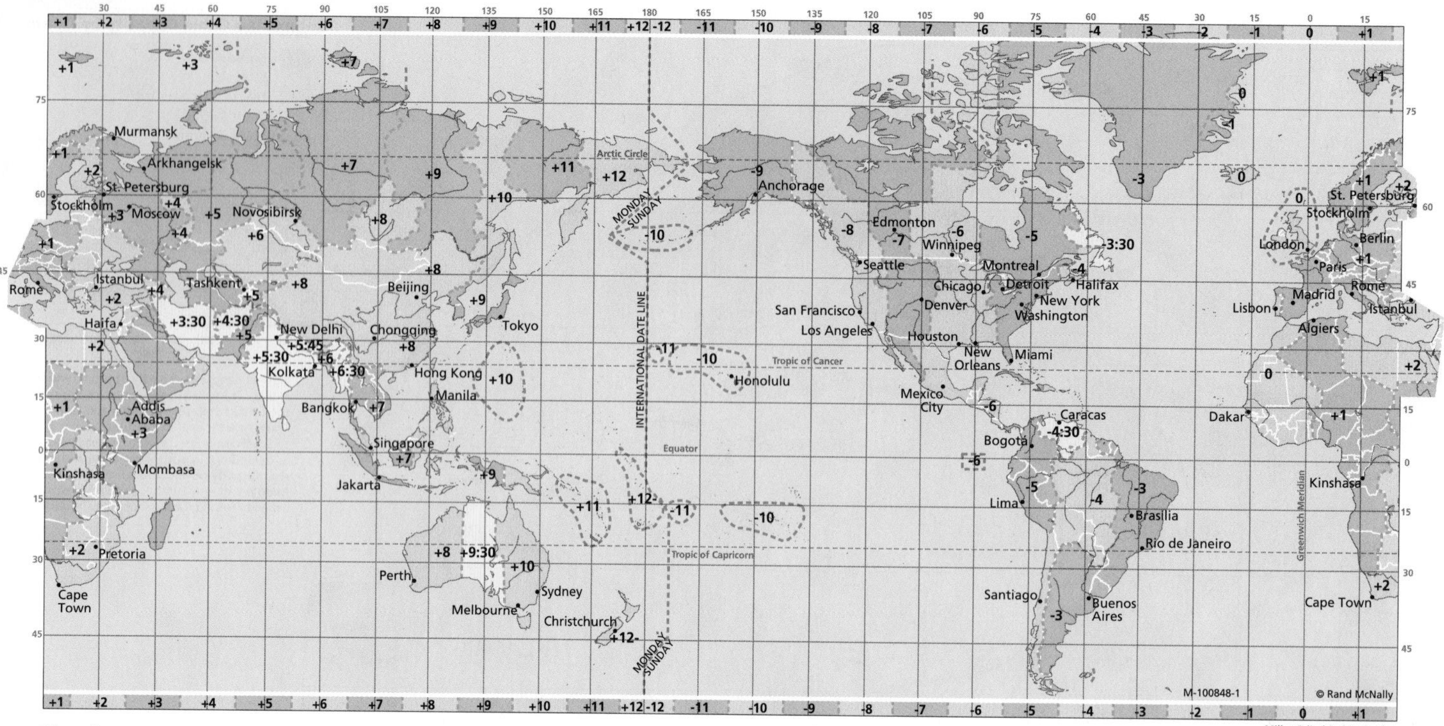
Murmansk
Arkhangelsk
St. Petersburg
Stockholm
Moscow
Novosibirsk
Istanbul
Rome
Haifa
Tashkent
New Delhi
Kolkata
Beijing
Chongqing
Tokyo
Hong Kong
Manila
Bangkok
Singapore
Jakarta
Addis Ababa
Mombasa
Kinshasa
Pretoria
Cape Town
Perth
Sydney
Melbourne
Christchurch
Arctic Circle
MONDAY SUNDAY
INTERNATIONAL DATE LINE
Tropic of Cancer
Equator
Tropic of Capricorn
Anchorage
Honolulu
Edmonton
Winnipeg
Seattle
San Francisco
Los Angeles
Denver
Chicago
Detroit
Montreal
Halifax
New York
Washington
Houston
New Orleans
Miami
Mexico City
Caracas
Bogota
Lima
Brasília
Rio de Janeiro
Santiago
Buenos Aires
Dakar
London
Paris
Berlin
Madrid
Lisbon
Algiers
Greenwich Meridian
M-100848-1
© Rand McNally

Miller Cylindrical Projection
Scale 1:225,000,000

Time Zones

Coordinated Universal Time (UTC) is the standard for international time zones and the official reference for standard time across the world. Although UTC has officially replaced Greenwich Mean Time (GMT), both terms are widely employed and, in casual usage, are essentially synonymous. On the time zone map above, the numbers along the top and bottom edges indicate the time difference, in hours, from UTC. The first time zone, with a value of 0, is centered on the Prime Meridian running through Greenwich, England. To compute standard time at any location, add the value on the map to UTC at Greenwich. For example, Chicago is in time zone UTC -6, which means it is 6 hours earlier than UTC at Greenwich. This means that if it is noon at Greenwich then it is 6 a.m. in Chicago.

To ensure synchronization with the Sun's location, time zone boundaries should follow lines of longitude very precisely. However this is rarely the case, and most time zones boundaries are very irregular. They are often constrained to follow international or internal administrative boundaries, and may be shifted east or west for various reasons. Discontinuities sometimes exist where time changes by more than one hour across a zone boundary, and the UTC difference for some time zones is less than a full hour. To make matters even more complicated, these time zones do not account for Daylight Savings Time, which is observed in some jurisdictions for part of the year.

HOURS OF DAYLIGHT

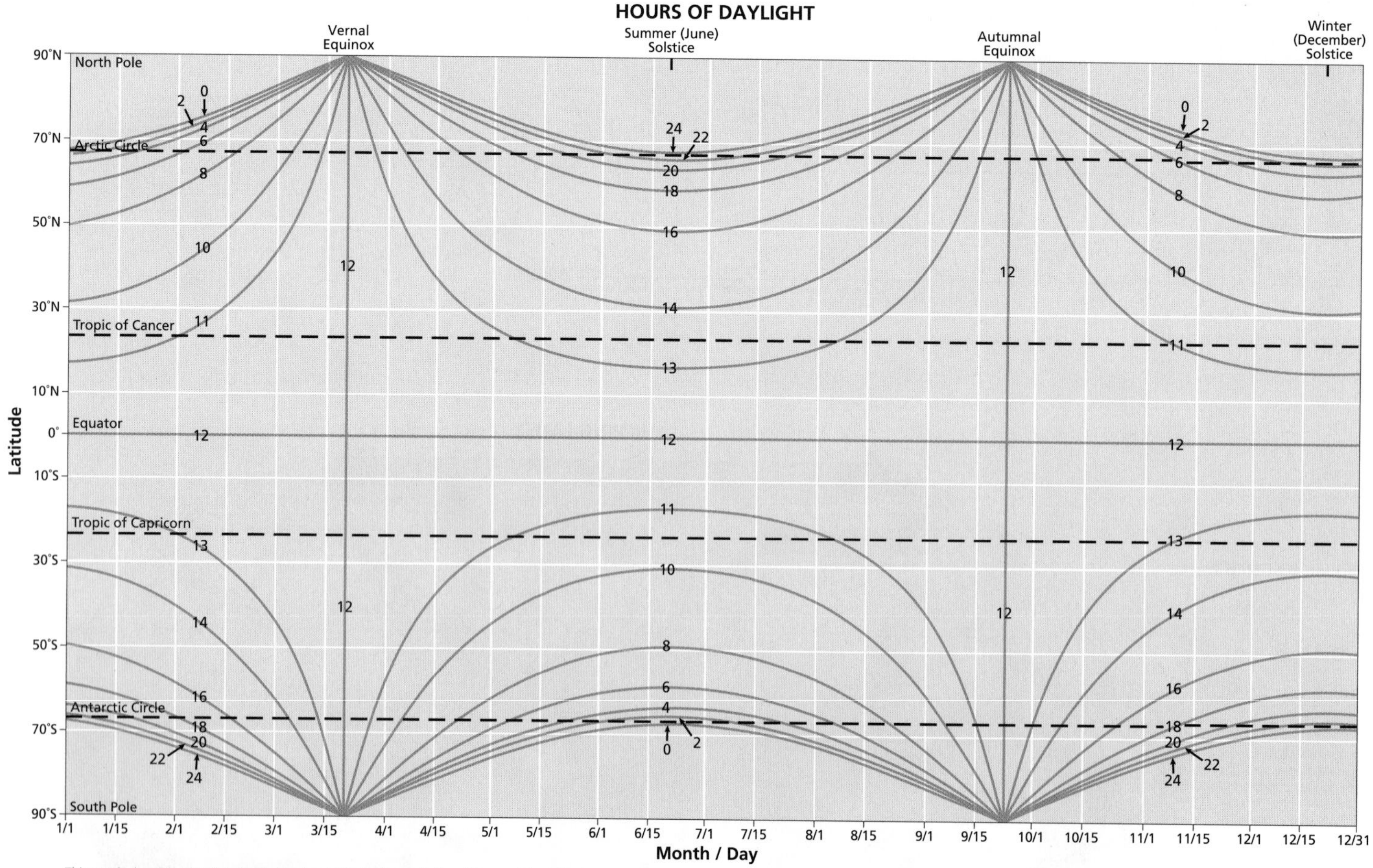
Vernal Equinox
Summer (June) Solstice
Autumnal Equinox
Winter (December) Solstice
North Pole
Arctic Circle
Tropic of Cancer
Equator
Tropic of Capricorn
Antarctic Circle
South Pole
Latitude
Month / Day

This graph shows hours of daylight at various latitudes for each day of the year.The following are some important patterns evident on the graph.

- The Equator experiences 12 hours of daylight every day of the year.
- Every point on the Earth experiences 12 hours of daylight at the vernal and autumnal equinoxes.
- The greater the distance from the Equator, the greater the variability in daylight length over the year.
- In the northern hemisphere, daylight length is greater than 12 hours between the vernal and autumnal equinoxes (the northern hemisphere summer) and less than 12 hours between the autumnal and vernal equinoxes (the northern hemisphere winter). The opposite pattern occurs in the southern hemisphere.
- Areas north of the Arctic Circle and south of the Antarctic Circle experience an annual pattern with periods of total darkness and periods of continuous daylight.

The data used to create this graph do not account for refraction of the Sun's rays by the Earth's atmosphere, which lengthens the daylight period slightly. The calculations are based on the center of the Sun, and do not account for the size of the solar disk, which also extends the daylight period by several minutes.

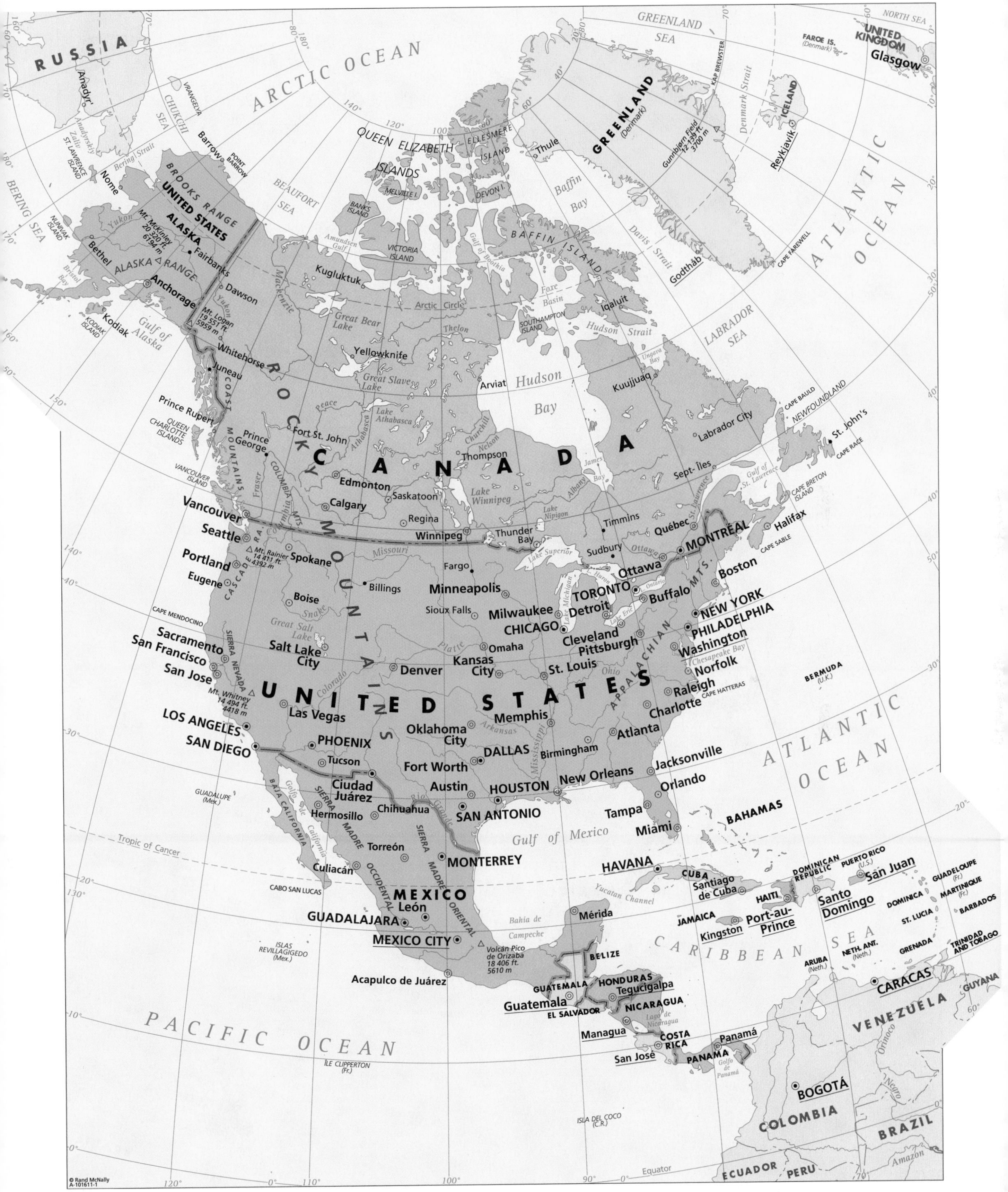

ARCTIC OCEAN
RUSSIA
Anadyr'
BERING SEA
Bering Strait
ST. LAWRENCE ISLAND
NUNIVAK ISLAND
CHUKCHI SEA
BEAUFORT SEA
Barrow
POINT BARROW
Nome
Bethel
BROOKS RANGE
UNITED STATES
ALASKA
Mt. McKinley 20 320 ft. 6194 m
Fairbanks
ALASKA RANGE
Anchorage
Bristol Bay
Kodiak
KODIAK ISLAND
Gulf of Alaska
Dawson
Mt. Logan 19 551 ft. 5959 m
Whitehorse
Juneau
COAST MOUNTAINS
Prince Rupert
QUEEN CHARLOTTE ISLANDS
Prince George
VANCOUVER ISLAND
Vancouver
Seattle
Mt. Rainier 14 411 ft. 4392 m
Spokane
Portland
Eugene
CAPE MENDOCINO
Sacramento
San Francisco
San Jose
SIERRA NEVADA
Mt. Whitney 14 494 ft. 4418 m
LOS ANGELES
SAN DIEGO
GUADALUPE (Mex.)
BAJA CALIFORNIA
Golfo de California
QUEEN ELIZABETH ISLANDS
ELLESMERE ISLAND
MELVILLE I.
DEVON I.
BANKS ISLAND
VICTORIA ISLAND
Amundsen Gulf
Kugluktuk
Great Bear Lake
Arctic Circle
Yellowknife
Great Slave Lake
Lake Athabasca
Fort St. John
ROCKY MOUNTAINS
CANADA
Edmonton
Calgary
Saskatoon
Regina
Winnipeg
Thompson
Lake Winnipeg
Arviat
Hudson Bay
James Bay
Thunder Bay
Lake Nipigon
Lake Superior
Timmins
Sudbury
Québec
MONTRÉAL
Ottawa
TORONTO
Halifax
CAPE SABLE
CAPE BRETON ISLAND
Gulf of St. Lawrence
Sept- Îles
Labrador City
Kuujjuaq
Ungava Bay
Hudson Strait
Iqaluit
Foxe Basin
SOUTHAMPTON ISLAND
BAFFIN ISLAND
Gulf of Boothia
Baffin Bay
Davis Strait
Thule
GREENLAND (Denmark)
Gunnbjørn Field 12 139 ft. 3700 m
KAP BREWSTER
Godthåb
CAPE FAREWELL
GREENLAND SEA
Denmark Strait
ICELAND
Reykjavík
FAROE IS. (Denmark)
UNITED KINGDOM
Glasgow
NORTH SEA
LABRADOR SEA
CAPE BAULD
NEWFOUNDLAND
St. John's
CAPE RACE
ATLANTIC OCEAN
Fargo
Billings
Boise
Great Salt Lake
Salt Lake City
Minneapolis
Sioux Falls
Milwaukee
Detroit
CHICAGO
Cleveland
Pittsburgh
Buffalo
Boston
NEW YORK
PHILADELPHIA
Washington
Chesapeake Bay
Norfolk
Raleigh
CAPE HATTERAS
Charlotte
Omaha
Kansas City
Denver
St. Louis
APPALACHIAN MTS.
UNITED STATES
Las Vegas
Memphis
Atlanta
Birmingham
Oklahoma City
PHOENIX
Tucson
DALLAS
Fort Worth
Austin
HOUSTON
SAN ANTONIO
New Orleans
Jacksonville
Orlando
Tampa
Miami
BERMUDA (U.K.)
BAHAMAS
Ciudad Juárez
Chihuahua
Hermosillo
SIERRA MADRE OCCIDENTAL
SIERRA MADRE ORIENTAL
Torreón
MONTERREY
Culiacán
CABO SAN LUCAS
Tropic of Cancer
Gulf of Mexico
MEXICO
León
GUADALAJARA
MEXICO CITY
Volcán Pico de Orizaba 18 406 ft. 5610 m
Acapulco de Juárez
ISLAS REVILLAGIGEDO (Mex.)
Bahía de Campeche
Mérida
Yucatan Channel
HAVANA
CUBA
Santiago de Cuba
JAMAICA
Kingston
HAITI
Port-au-Prince
DOMINICAN REPUBLIC
Santo Domingo
PUERTO RICO (U.S.)
San Juan
GUADELOUPE (Fr.)
DOMINICA
MARTINIQUE (Fr.)
ST. LUCIA
BARBADOS
GRENADA
TRINIDAD AND TOBAGO
ARUBA (Neth.)
NETH. ANT. (Neth.)
CARIBBEAN SEA
BELIZE
GUATEMALA
Guatemala
EL SALVADOR
HONDURAS
Tegucigalpa
NICARAGUA
Lago de Nicaragua
Managua
COSTA RICA
San José
PANAMA
Panamá
Golfo de Panamá
CARACAS
VENEZUELA
GUYANA
Orinoco
BOGOTÁ
COLOMBIA
BRAZIL
Amazon
ECUADOR
PERU
Equator
PACIFIC OCEAN
ÎLE CLIPPERTON (Fr.)
ISLA DEL COCO (C.R.)
© Rand McNally A-101611-1

0 200 400 600 800 1000 1200 Miles

0 200 400 600 800 1000 1200 1400 1600 1800 2000 Kilometers

Lambert Azimuthal Equal Area Projection
Scale 1:40,000,000
One inch to 640 miles
One cm to 400 km

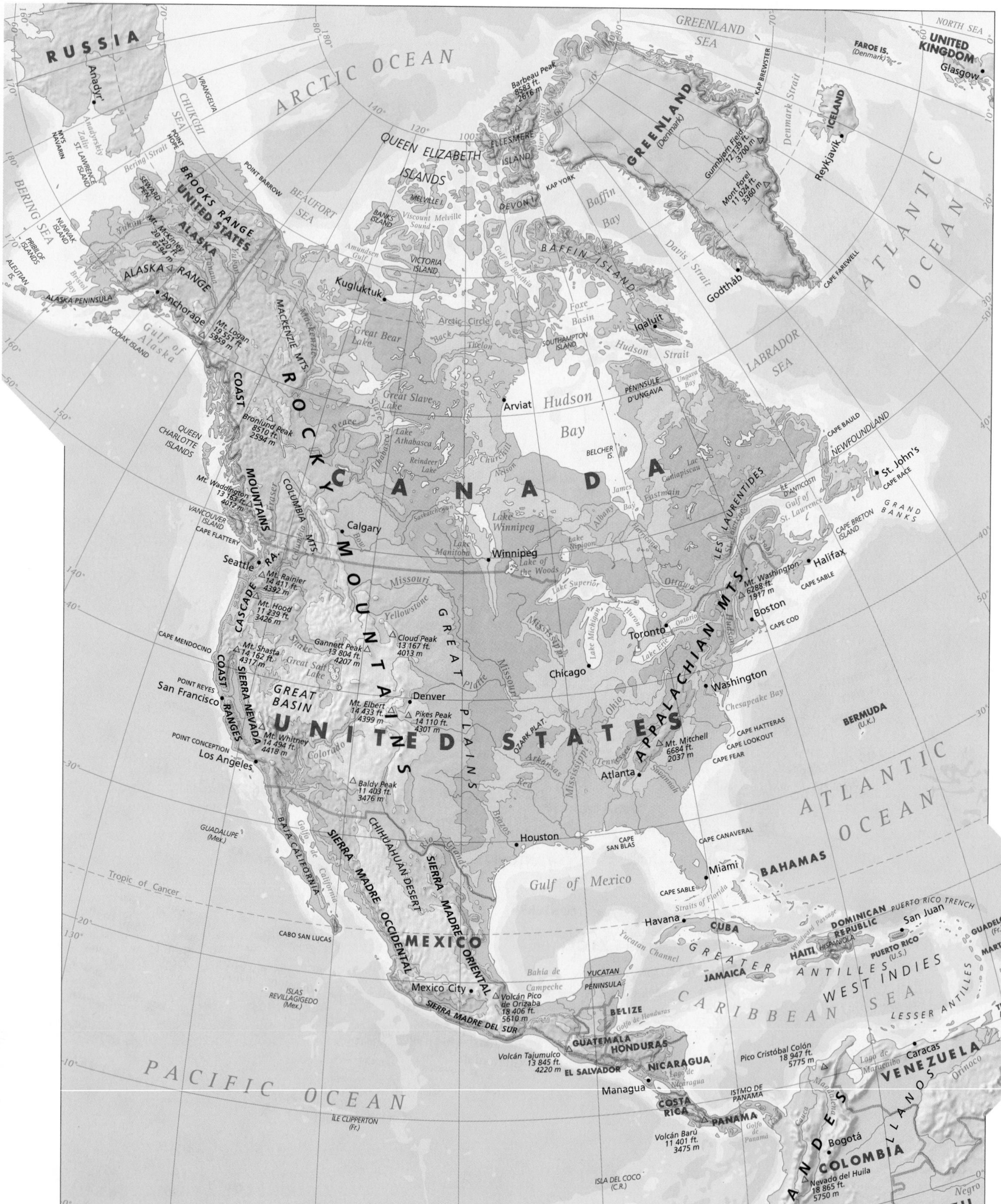

RUSSIA
Anadyr'
ARCTIC OCEAN
GREENLAND SEA
NORTH SEA
UNITED KINGDOM
Glasgow
FAROE IS. (Denmark)
WRANGELYA
CHUKCHI SEA
MYS NAVARIN
ST. LAWRENCE ISLAND
Bering Strait
POINT HOPE
POINT BARROW
SEWARD PEN.
BERING SEA
NUNIVAK ISLAND
PRIBILOF ISLANDS
ALEUTIAN IS.
BROOKS RANGE
UNITED STATES
ALASKA
Mt. McKinley 20 320 ft. 6194 m
Yukon
ALASKA RANGE
ALASKA PENINSULA
Anchorage
KODIAK ISLAND
Gulf of Alaska
Mt. Logan 19 551 ft. 5959 m
BEAUFORT SEA
QUEEN ELIZABETH ISLANDS
ELLESMERE ISLAND
Barbeau Peak 8583 ft. 2616 m
GREENLAND (Denmark)
KAP BREWSTER
Gunnbjørn Fjeld 12 139 ft. 3700 m
Mont Forel 11 024 ft. 3360 m
Denmark Strait
ICELAND
Reykjavík
ATLANTIC OCEAN
KAP YORK
DEVON I.
MELVILLE I.
BANKS ISLAND
Viscount Melville Sound
Amundsen Gulf
VICTORIA ISLAND
Gulf of Boothia
Baffin Bay
BAFFIN ISLAND
Davis Strait
Godthåb
CAPE FAREWELL
Kugluktuk
Arctic Circle
Great Bear Lake
Back
Thelon
Foxe Basin
SOUTHAMPTON ISLAND
Iqaluit
Hudson Strait
LABRADOR SEA
MACKENZIE MTS.
Mackenzie
COAST
ROCKY MOUNTAINS
Great Slave Lake
Peace
Arviat
Hudson Bay
PENINSULE D'UNGAVA
Ungava Bay
Bronlund Peak 8510 ft. 2594 m
QUEEN CHARLOTTE ISLANDS
Lake Athabasca
Reindeer Lake
Churchill
Nelson
BELCHER IS.
CANADA
Lac Caniapiscau
CAPE BAULD
NEWFOUNDLAND
St. John's
CAPE RACE
Mt. Waddington 13 163 ft. 4012 m
MOUNTAINS
COLUMBIA MTS.
Fraser
VANCOUVER ISLAND
CAPE FLATTERY
Calgary
Saskatchewan
Lake Winnipeg
James Bay
Eastmain
Albany
LES LAURENTIDES
ÎLE D'ANTICOSTI
Gulf of St. Lawrence
GRAND BANKS
CAPE BRETON ISLAND
Lake Manitoba
Winnipeg
Lake of the Woods
Lake Nipigon
Seattle
Mt. Rainier 14 411 ft. 4392 m
Missouri
Lake Superior
Ottawa
Mt. Washington 6288 ft. 1917 m
Halifax
CAPE SABLE
CASCADE
Mt. Hood 11 239 ft. 3426 m
Yellowstone
GREAT PLAINS
Lake Michigan
Huron
Ontario
Toronto
Boston
CAPE COD
CAPE MENDOCINO
Mt. Shasta 14 162 ft. 4317 m
Snake
Gannett Peak 13 804 ft. 4207 m
Cloud Peak 13 167 ft. 4013 m
Great Salt Lake
Lake Erie
APPALACHIAN MTS.
Chicago
Washington
Chesapeake Bay
POINT REYES
San Francisco
COAST RANGES
SIERRA NEVADA
GREAT BASIN
Denver
Platte
Mt. Elbert 14 433 ft. 4399 m
Pikes Peak 14 110 ft. 4301 m
UNITED STATES
OZARK PLAT.
Ohio
Mt. Mitchell 6684 ft. 2037 m
CAPE HATTERAS
CAPE LOOKOUT
CAPE FEAR
BERMUDA (U.K.)
POINT CONCEPTION
Mt. Whitney 14 494 ft. 4418 m
Los Angeles
Colorado
Arkansas
Tennessee
Mississippi
Atlanta
Savannah
Baldy Peak 11 403 ft. 3476 m
Red
Brazos
Houston
CAPE SAN BLAS
CAPE CANAVERAL
GUADALUPE (Mex.)
BAJA CALIFORNIA
Golfo de California
SIERRA MADRE OCCIDENTAL
CHIHUAHUAN DESERT
SIERRA MADRE ORIENTAL
Rio Grande
Miami
BAHAMAS
Tropic of Cancer
Gulf of Mexico
CAPE SABLE
Straits of Florida
PUERTO RICO TRENCH
Havana
CUBA
DOMINICAN REPUBLIC
San Juan
HAITI
HISPANIOLA
PUERTO RICO (U.S.)
GUADELOUPE (Fr.)
MARTINIQUE (Fr.)
BARBADOS
CABO SAN LUCAS
MEXICO
Yucatan Channel
GREATER ANTILLES
JAMAICA
WEST INDIES
Bahía de Campeche
YUCATAN PENINSULA
Mexico City
Volcán Pico de Orizaba 18 406 ft. 5610 m
ISLAS REVILLAGIGEDO (Mex.)
SIERRA MADRE DEL SUR
BELIZE
Golfo de Honduras
CARIBBEAN SEA
LESSER ANTILLES
TRINIDAD AND TOBAGO
GUATEMALA
HONDURAS
Volcán Tajumulco 13 845 ft. 4220 m
EL SALVADOR
NICARAGUA
Lago de Nicaragua
Pico Cristóbal Colón 18 947 ft. 5775 m
Lago de Maracaibo
Caracas
VENEZUELA
GUYANA
Orinoco
PACIFIC OCEAN
Managua
ISTMO DE PANAMA
COSTA RICA
PANAMA
Golfo de Panamá
Magdalena
LLANOS
ÎLE CLIPPERTON (Fr.)
Volcán Barú 11 401 ft. 3475 m
ANDES
Bogotá
COLOMBIA
Nevado del Huila 18 865 ft. 5750 m
Cauca
ISLA DEL COCO (C.R.)
Negro
BRAZIL
Equator
ECUADOR
PERU
Amazon
© Rand McNally A-101612-1

0 200 400 600 800 1000 1200 Miles

0 200 400 600 800 1000 1200 1400 1600 1800 2000 Kilometers

Lambert Azimuthal Equal Area Projection
Scale 1:40,000,000
One inch to 640 miles
One cm to 400 km

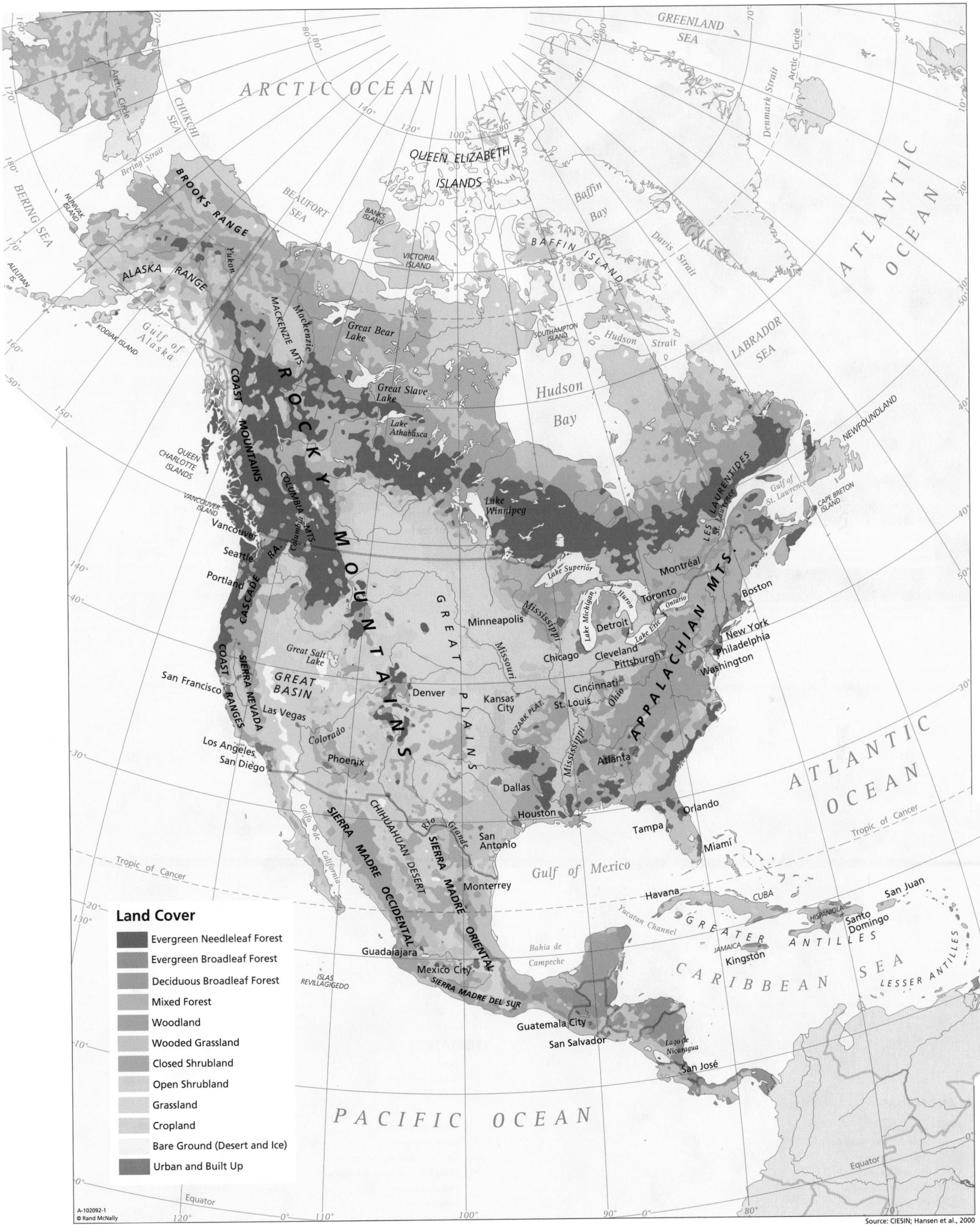
ARCTIC OCEAN
GREENLAND SEA
CHUKCHI SEA
BERING SEA
Bering Strait
NUNIVAK ISLAND
ALEUTIAN IS.
BEAUFORT SEA
QUEEN ELIZABETH ISLANDS
BANKS ISLAND
VICTORIA ISLAND
BAFFIN ISLAND
Baffin Bay
Davis Strait
Denmark Strait
Arctic Circle
ATLANTIC OCEAN
BROOKS RANGE
Yukon
ALASKA RANGE
KODIAK ISLAND
Gulf of Alaska
MACKENZIE MTS.
Mackenzie
Great Bear Lake
SOUTHAMPTON ISLAND
Hudson Strait
LABRADOR SEA
Great Slave Lake
Lake Athabasca
Hudson Bay
COAST MOUNTAINS
ROCKY MOUNTAINS
QUEEN CHARLOTTE ISLANDS
COLUMBIA MTS.
Columbia
VANCOUVER ISLAND
Vancouver
Seattle
Portland
CASCADE RA.
Lake Winnipeg
NEWFOUNDLAND
LES LAURENTIDES
St. Lawrence
Gulf of St. Lawrence
CAPE BRETON ISLAND
Lake Superior
Montréal
Boston
Toronto
Lake Ontario
L. Huron
Lake Michigan
Detroit
Lake Erie
APPALACHIAN MTS.
New York
Philadelphia
Washington
Pittsburgh
Cleveland
Chicago
Minneapolis
Mississippi
Missouri
GREAT PLAINS
Great Salt Lake
COAST RANGES
SIERRA NEVADA
GREAT BASIN
San Francisco
Las Vegas
Denver
Kansas City
Cincinnati
St. Louis
Ohio
OZARK PLAT.
Colorado
Los Angeles
San Diego
Phoenix
Dallas
Atlanta
Houston
Orlando
Tampa
Miami
San Antonio
Rio Grande
Golfo de California
SIERRA MADRE OCCIDENTAL
CHIHUAHUAN DESERT
SIERRA MADRE ORIENTAL
Monterrey
Gulf of Mexico
Tropic of Cancer
Havana
CUBA
Yucatan Channel
GREATER ANTILLES
HISPANIOLA
Santo Domingo
San Juan
JAMAICA
Kingston
LESSER ANTILLES
CARIBBEAN SEA
Bahia de Campeche
Guadalajara
Mexico City
ISLAS REVILLAGIGEDO
SIERRA MADRE DEL SUR
Guatemala City
San Salvador
Lago de Nicaragua
San José
PACIFIC OCEAN
Equator
Land Cover
Evergreen Needleleaf Forest
Evergreen Broadleaf Forest
Deciduous Broadleaf Forest
Mixed Forest
Woodland
Wooded Grassland
Closed Shrubland
Open Shrubland
Grassland
Cropland
Bare Ground (Desert and Ice)
Urban and Built Up
A-102092-1
© Rand McNally
Source: CIESIN; Hansen et al., 2000

0 200 400 600 800 1000 1200 Miles

0 200 400 600 800 1000 1200 1400 1600 1800 2000 Kilometers

Lambert Azimuthal Equal Area Projection
Scale 1:40,000,000
One inch to 640 miles
One cm to 400 km

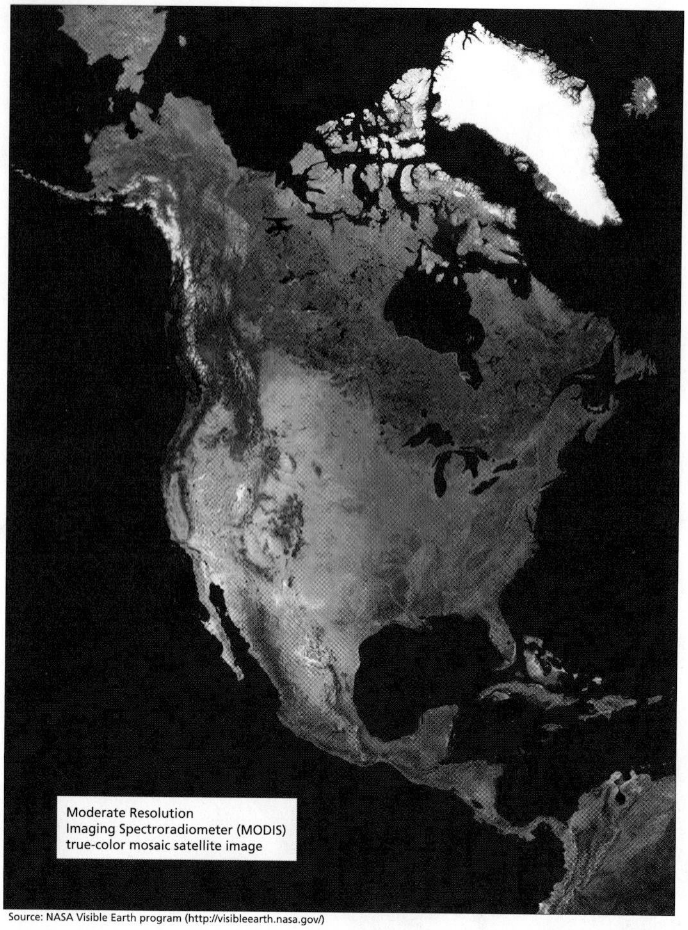
Moderate Resolution
Imaging Spectroradiometer (MODIS)
true-color mosaic satellite image

Source: NASA Visible Earth program (http://visibleearth.nasa.gov/)

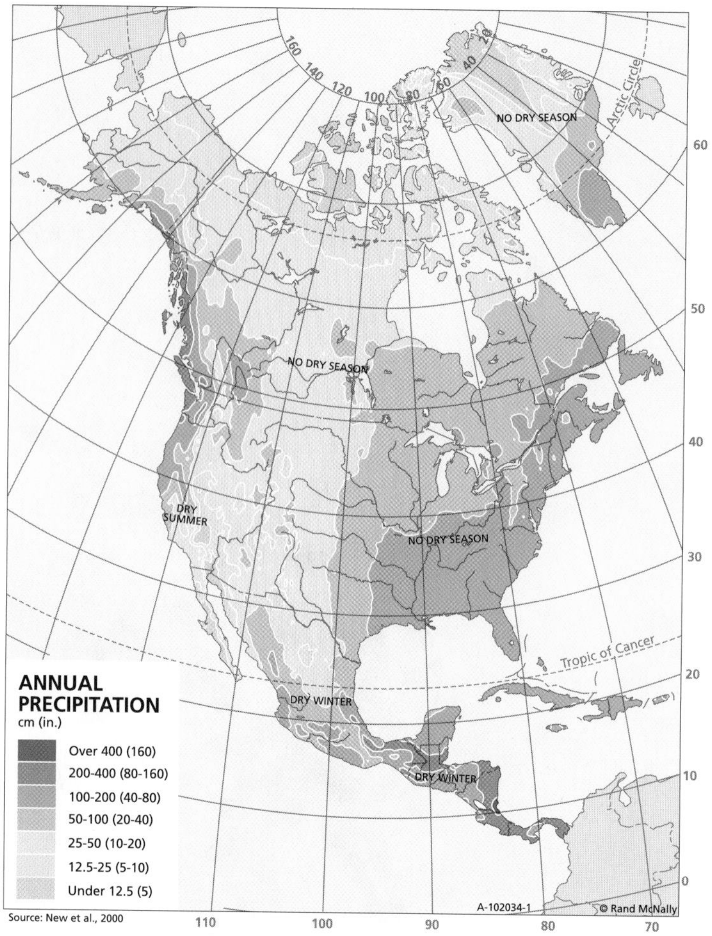
ANNUAL PRECIPITATION
cm (in.)
Over 400 (160)
200-400 (80-160)
100-200 (40-80)
50-100 (20-40)
25-50 (10-20)
12.5-25 (5-10)
Under 12.5 (5)
NO DRY SEASON
NO DRY SEASON
DRY SUMMER
NO DRY SEASON
DRY WINTER
DRY WINTER
Arctic Circle
Tropic of Cancer
A-102034-1
© Rand McNally

Source: New et al., 2000

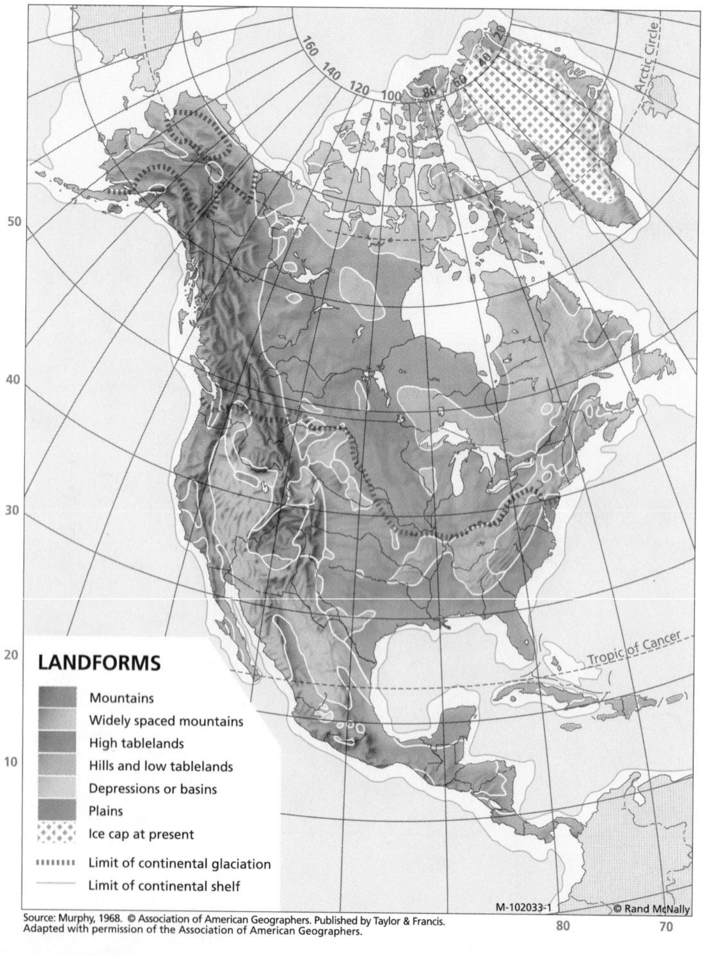
LANDFORMS
Mountains
Widely spaced mountains
High tablelands
Hills and low tablelands
Depressions or basins
Plains
Ice cap at present
Limit of continental glaciation
Limit of continental shelf
Arctic Circle
Tropic of Cancer
M-102033-1
© Rand McNally

Source: Murphy, 1968. © Association of American Geographers. Published by Taylor & Francis. Adapted with permission of the Association of American Geographers.

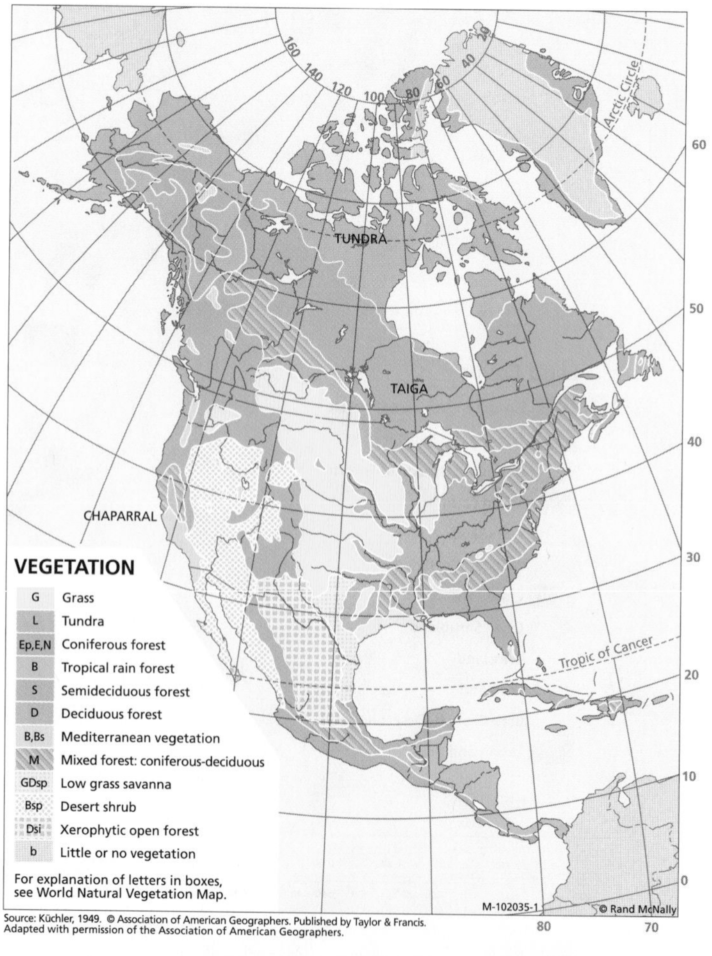
VEGETATION
G Grass
L Tundra
Ep,E,N Coniferous forest
B Tropical rain forest
S Semideciduous forest
D Deciduous forest
B,Bs Mediterranean vegetation
M Mixed forest: coniferous-deciduous
GDsp Low grass savanna
Bsp Desert shrub
Dsi Xerophytic open forest
b Little or no vegetation
For explanation of letters in boxes, see World Natural Vegetation Map.
TUNDRA
TAIGA
CHAPARRAL
Arctic Circle
Tropic of Cancer
M-102035-1
© Rand McNally

Source: Küchler, 1949. © Association of American Geographers. Published by Taylor & Francis. Adapted with permission of the Association of American Geographers.

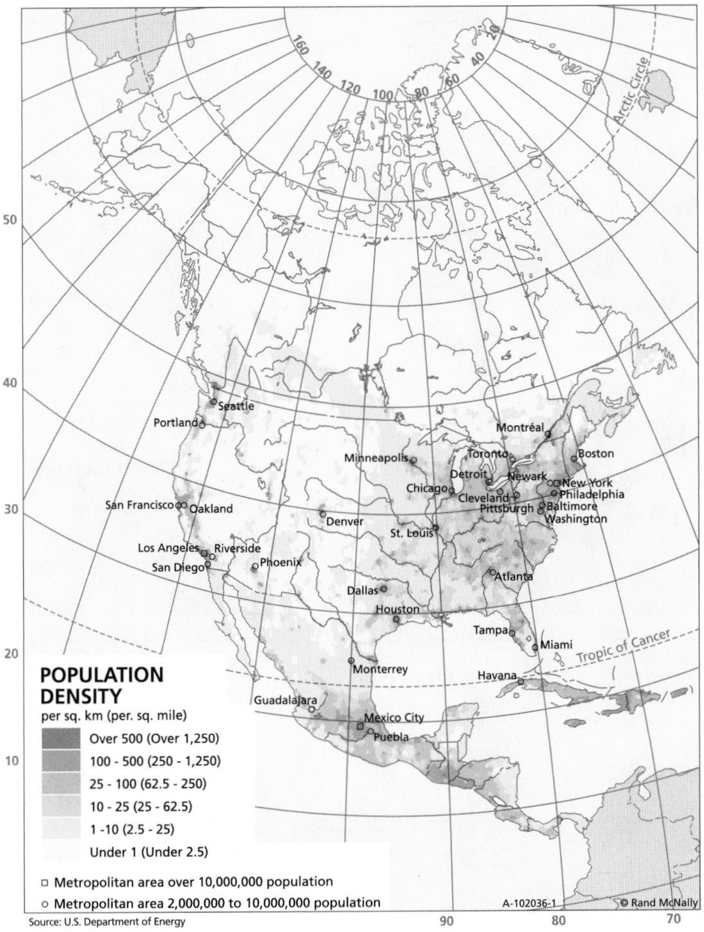

POPULATION DENSITY
per sq. km (per. sq. mile)
Over 500 (Over 1,250)
100 - 500 (250 - 1,250)
25 - 100 (62.5 - 250)
10 - 25 (25 - 62.5)
1 -10 (2.5 - 25)
Under 1 (Under 2.5)
Metropolitan area over 10,000,000 population
Metropolitan area 2,000,000 to 10,000,000 population
Seattle
Portland
San Francisco
Oakland
Los Angeles
Riverside
San Diego
Phoenix
Denver
Minneapolis
Chicago
Detroit
Toronto
Montréal
Boston
Newark
New York
Philadelphia
Baltimore
Washington
Cleveland
Pittsburgh
St. Louis
Atlanta
Dallas
Houston
Tampa
Miami
Monterrey
Havana
Guadalajara
Mexico City
Puebla
Arctic Circle
Tropic of Cancer
A-102036-1
© Rand McNally

Source: U.S. Department of Energy

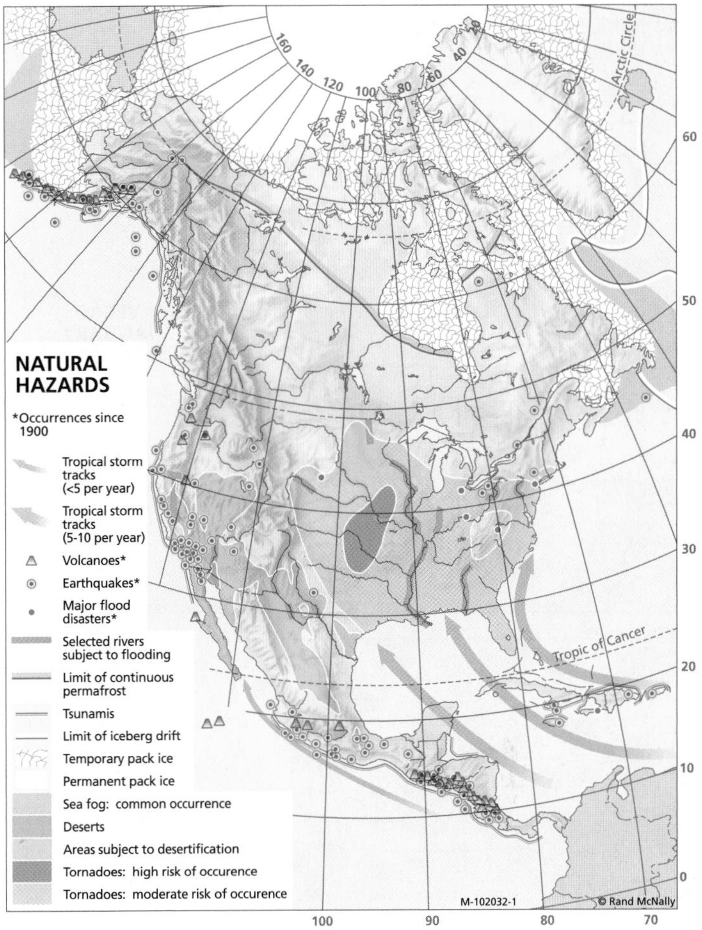

NATURAL HAZARDS
*Occurrences since 1900
Tropical storm tracks (<5 per year)
Tropical storm tracks (5-10 per year)
Volcanoes*
Earthquakes*
Major flood disasters*
Selected rivers subject to flooding
Limit of continuous permafrost
Tsunamis
Limit of iceberg drift
Temporary pack ice
Permanent pack ice
Sea fog: common occurrence
Deserts
Areas subject to desertification
Tornadoes: high risk of occurence
Tornadoes: moderate risk of occurence
Arctic Circle
Tropic of Cancer
M-102032-1
© Rand McNally

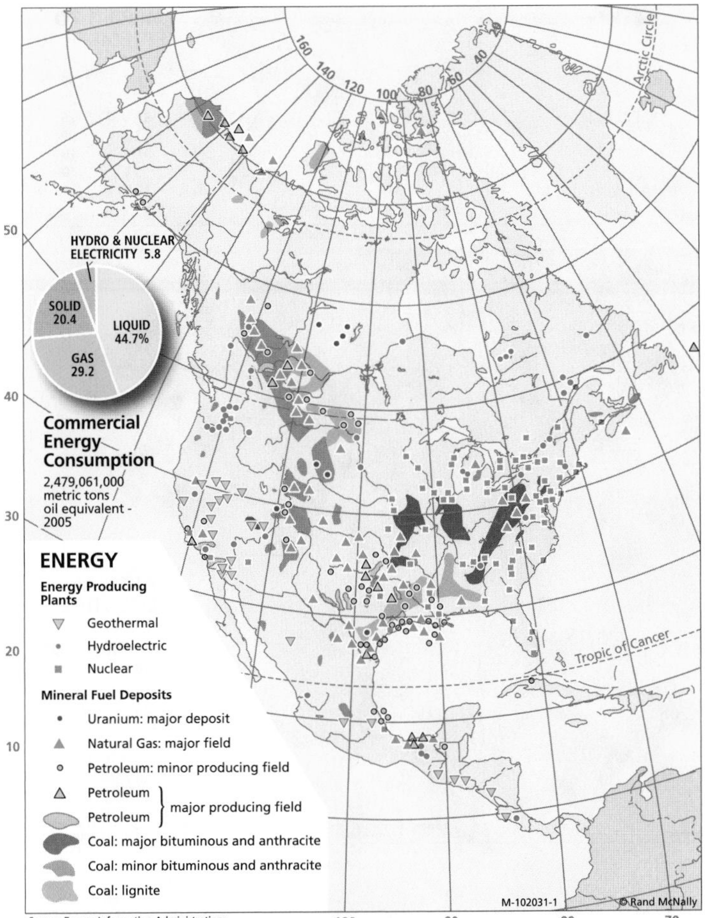

HYDRO & NUCLEAR ELECTRICITY 5.8
SOLID 20.4
LIQUID 44.7%
GAS 29.2
Commercial Energy Consumption
2,479,061,000 metric tons oil equivalent - 2005
ENERGY
Energy Producing Plants
Geothermal
Hydroelectric
Nuclear
Mineral Fuel Deposits
Uranium: major deposit
Natural Gas: major field
Petroleum: minor producing field
Petroleum
Petroleum
major producing field
Coal: major bituminous and anthracite
Coal: minor bituminous and anthracite
Coal: lignite
Arctic Circle
Tropic of Cancer
M-102031-1
© Rand McNally

Source: Energy Information Administration; United Nations

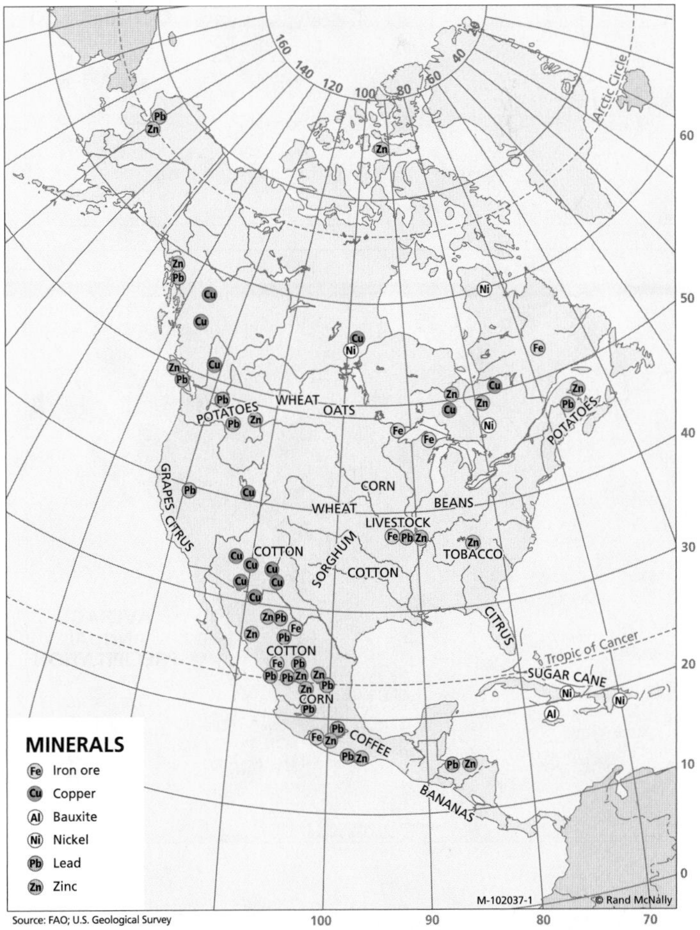

MINERALS
Iron ore
Copper
Bauxite
Nickel
Lead
Zinc
WHEAT
OATS
POTATOES
POTATOES
GRAPES
CITRUS
CORN
WHEAT
BEANS
LIVESTOCK
TOBACCO
COTTON
SORGHUM
COTTON
CITRUS
COTTON
CORN
SUGAR CANE
COFFEE
BANANAS
Arctic Circle
Tropic of Cancer
M-102037-1
© Rand McNally

Source: FAO; U.S. Geological Survey

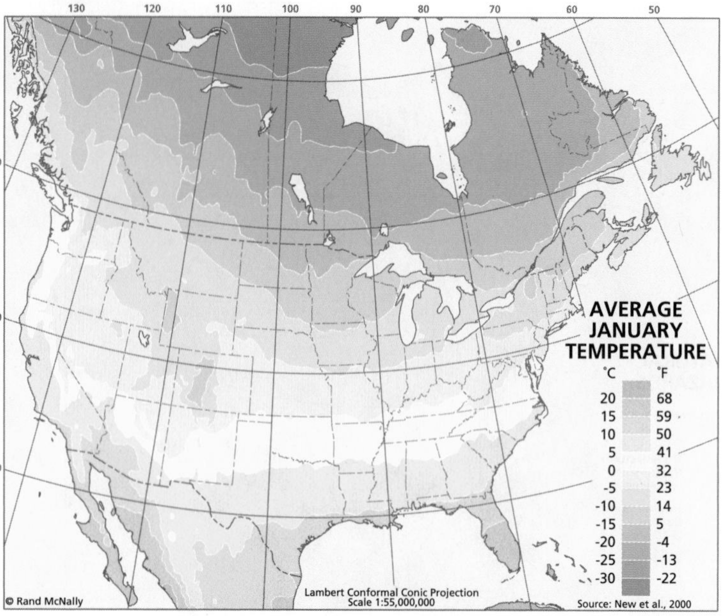
AVERAGE JANUARY TEMPERATURE
°C °F
20 68
15 59
10 50
5 41
0 32
-5 23
-10 14
-15 5
-20 -4
-25 -13
-30 -22
© Rand McNally
Lambert Conformal Conic Projection
Scale 1:55,000,000
Source: New et al., 2000

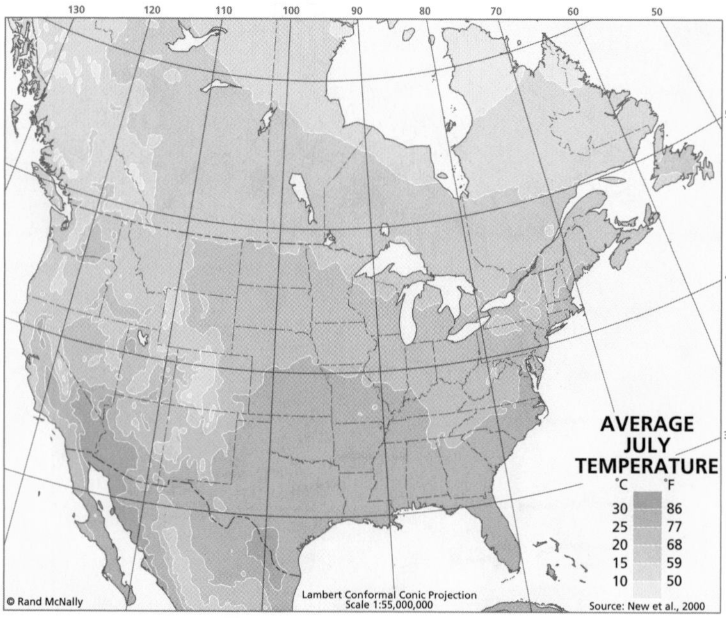
AVERAGE JULY TEMPERATURE
°C °F
30 86
25 77
20 68
15 59
10 50
© Rand McNally
Lambert Conformal Conic Projection
Scale 1:55,000,000
Source: New et al., 2000

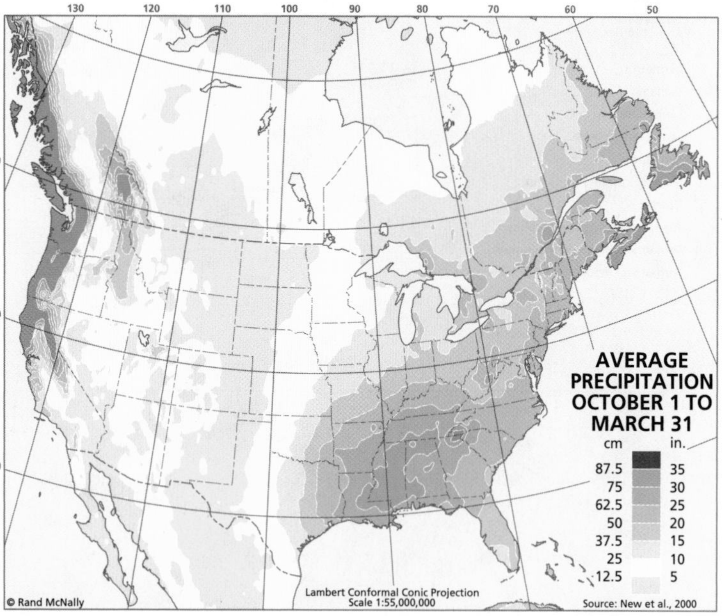
AVERAGE PRECIPITATION OCTOBER 1 TO MARCH 31
cm in.
87.5 35
75 30
62.5 25
50 20
37.5 15
25 10
12.5 5
© Rand McNally
Lambert Conformal Conic Projection
Scale 1:55,000,000
Source: New et al., 2000

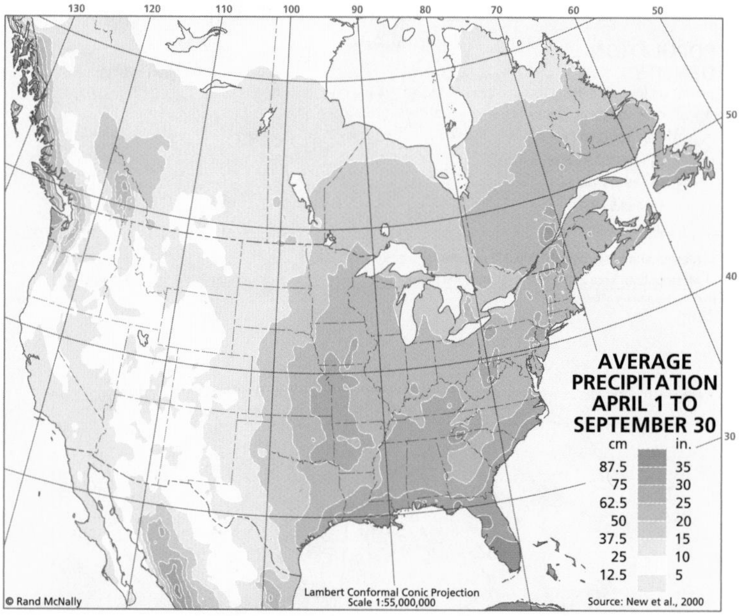
AVERAGE PRECIPITATION APRIL 1 TO SEPTEMBER 30
cm in.
87.5 35
75 30
62.5 25
50 20
37.5 15
25 10
12.5 5
© Rand McNally
Lambert Conformal Conic Projection
Scale 1:55,000,000
Source: New et al., 2000

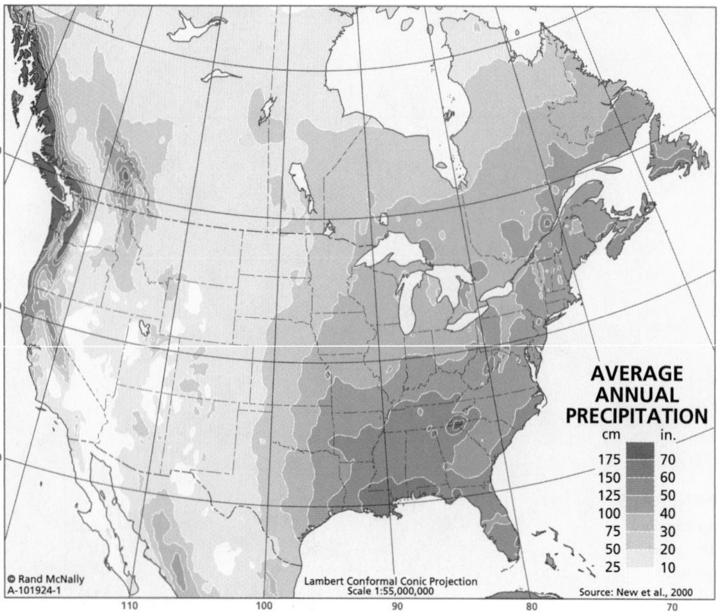
AVERAGE ANNUAL PRECIPITATION
cm in.
175 70
150 60
125 50
100 40
75 30
50 20
25 10
© Rand McNally
A-101924-1
Lambert Conformal Conic Projection
Scale 1:55,000,000
Source: New et al., 2000

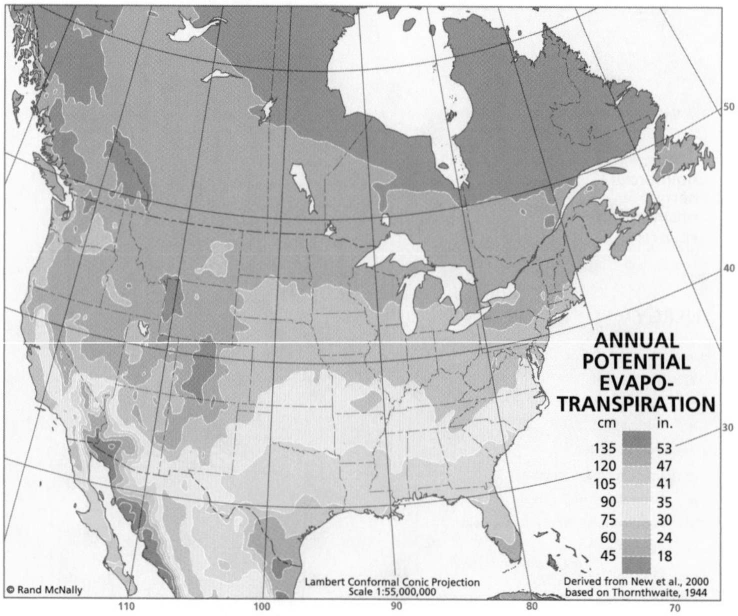
ANNUAL POTENTIAL EVAPO-TRANSPIRATION
cm in.
135 53
120 47
105 41
90 35
75 30
60 24
45 18
© Rand McNally
Lambert Conformal Conic Projection
Scale 1:55,000,000
Derived from New et al., 2000 based on Thornthwaite, 1944

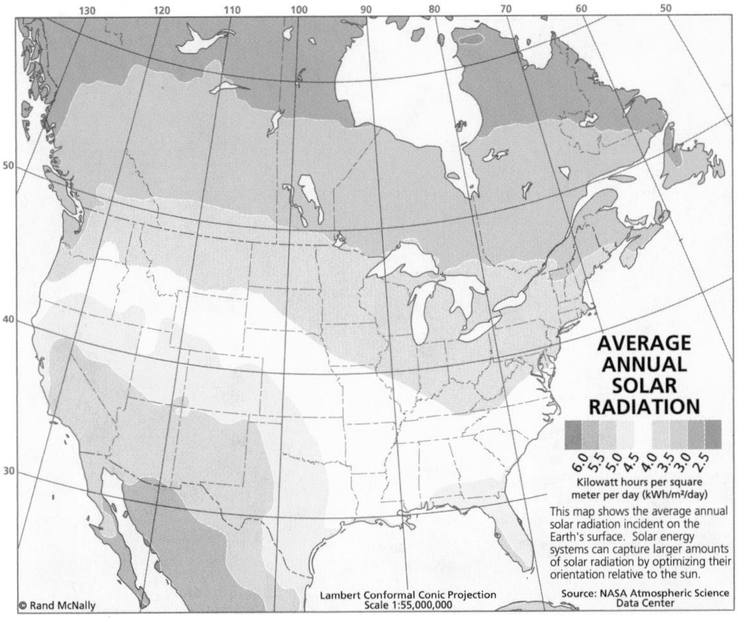
AVERAGE ANNUAL SOLAR RADIATION
6.0 5.5 5.0 4.5 4.0 3.5 3.0 2.5
Kilowatt hours per square meter per day (kWh/m²/day)
This map shows the average annual solar radiation incident on the Earth's surface. Solar energy systems can capture larger amounts of solar radiation by optimizing their orientation relative to the sun.
Lambert Conformal Conic Projection Scale 1:55,000,000
Source: NASA Atmospheric Science Data Center
© Rand McNally

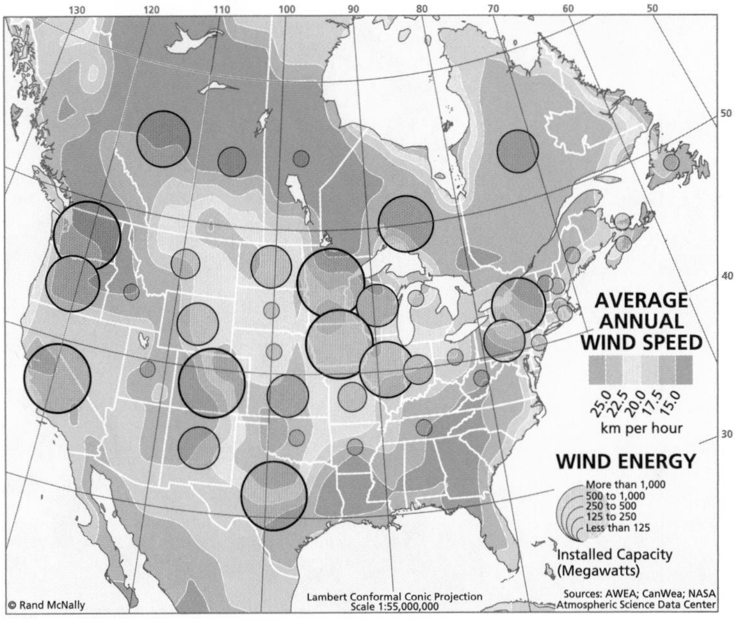
AVERAGE ANNUAL WIND SPEED
25.0 22.5 20.0 17.5 15.0
km per hour
WIND ENERGY
More than 1,000
500 to 1,000
250 to 500
125 to 250
Less than 125
Installed Capacity (Megawatts)
Lambert Conformal Conic Projection Scale 1:55,000,000
Sources: AWEA; CanWea; NASA Atmospheric Science Data Center
© Rand McNally

EL NIÑO CLIMATE ANOMALIES

These two maps show temperature and precipitation anomalies associated with the 1982-83 El Niño-Southern Oscillation (ENSO) event, one of strongest such events on record. The maps compare temperature and precipitation values for the 1982-83 winter season (October 1, 1982 through March 31, 1983) to winter averages for the 1961-90 baseline period. A positive anomaly indicates a higher than average temperature or precipitation value for 1982-83, while a negative anomaly indicates a lower than average value.

Derived from Brohan et al., 2006, and Hulme, 1998

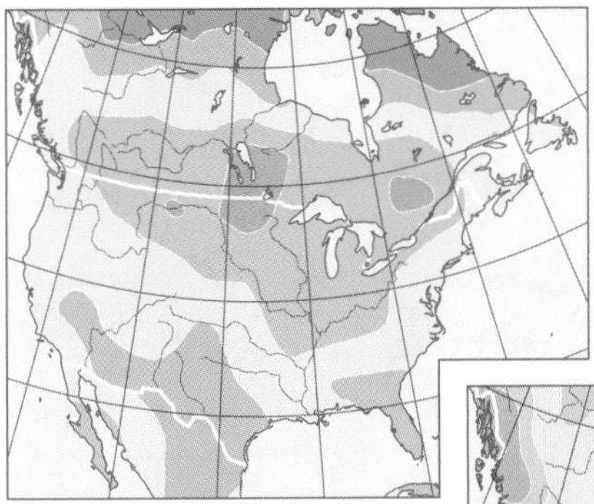

Difference in average monthly winter temperature 1982-83 versus 1961-90

C°	F°
2.0	3.6
1.0	1.8
0.0	0.0
-1.0	-1.8

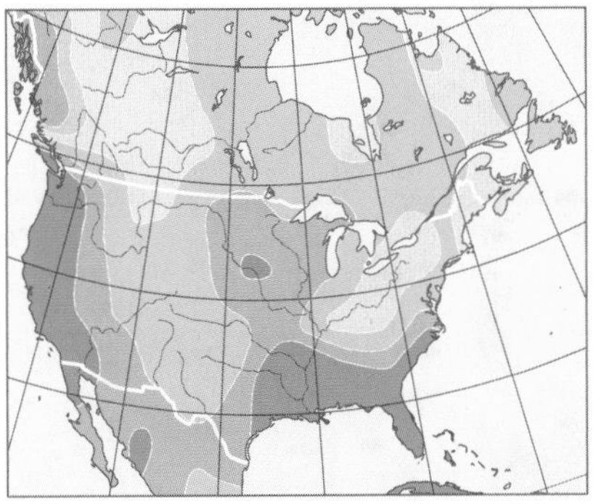

Difference in average monthly winter precipitation 1982-83 versus 1961-90

cm	in.
2.0	0.8
1.0	0.4
0.0	0.0
-1.0	-0.4
-2.0	-0.8

© Rand McNally
Lambert Conformal Conic Projection
Scale 1:100,000,000

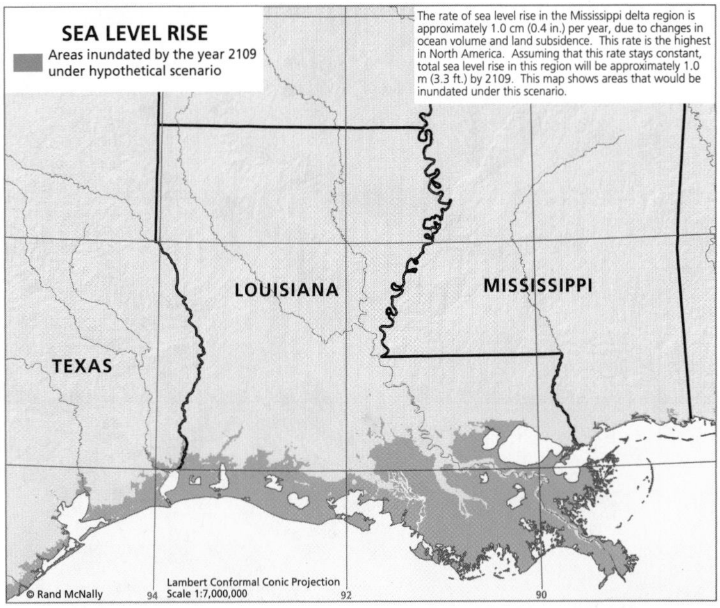
SEA LEVEL RISE
Areas inundated by the year 2109 under hypothetical scenario
The rate of sea level rise in the Mississippi delta region is approximately 1.0 cm (0.4 in.) per year, due to changes in ocean volume and land subsidence. This rate is the highest in North America. Assuming that this rate stays constant, total sea level rise in this region will be approximately 1.0 m (3.3 ft.) by 2109. This map shows areas that would be inundated under this scenario.
LOUISIANA
MISSISSIPPI
TEXAS
Lambert Conformal Conic Projection Scale 1:7,000,000
© Rand McNally

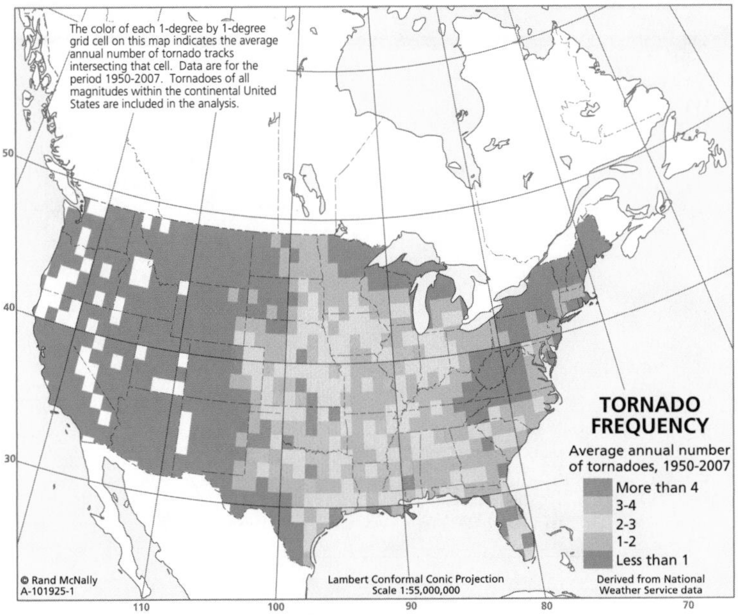
The color of each 1-degree by 1-degree grid cell on this map indicates the average annual number of tornado tracks intersecting that cell. Data are for the period 1950-2007. Tornadoes of all magnitudes within the continental United States are included in the analysis.
TORNADO FREQUENCY
Average annual number of tornadoes, 1950-2007
More than 4
3-4
2-3
1-2
Less than 1
Lambert Conformal Conic Projection Scale 1:55,000,000
Derived from National Weather Service data
© Rand McNally
A-101925-1

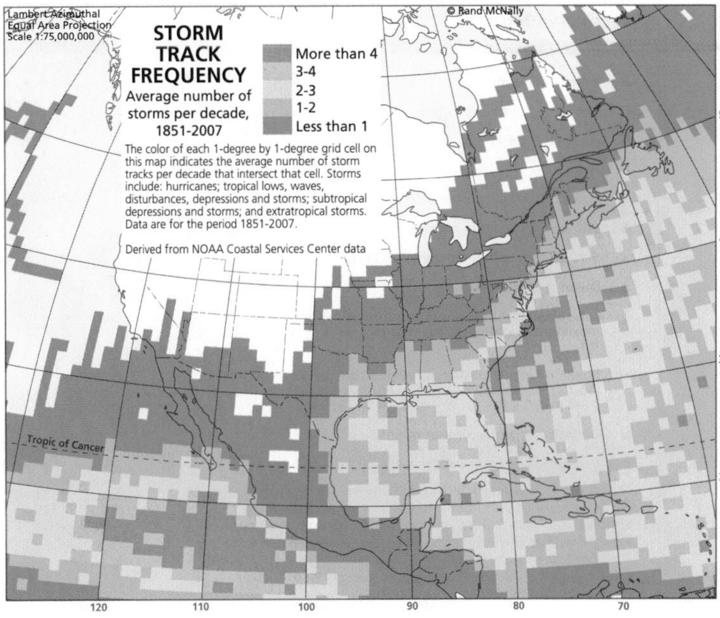
Lambert Azimuthal Equal Area Projection Scale 1:75,000,000
© Rand McNally
STORM TRACK FREQUENCY
Average number of storms per decade, 1851-2007
More than 4
3-4
2-3
1-2
Less than 1
The color of each 1-degree by 1-degree grid cell on this map indicates the average number of storm tracks per decade that intersect that cell. Storms include: hurricanes; tropical lows, waves, disturbances, depressions and storms; subtropical depressions and storms; and extratropical storms. Data are for the period 1851-2007.
Derived from NOAA Coastal Services Center data
Tropic of Cancer

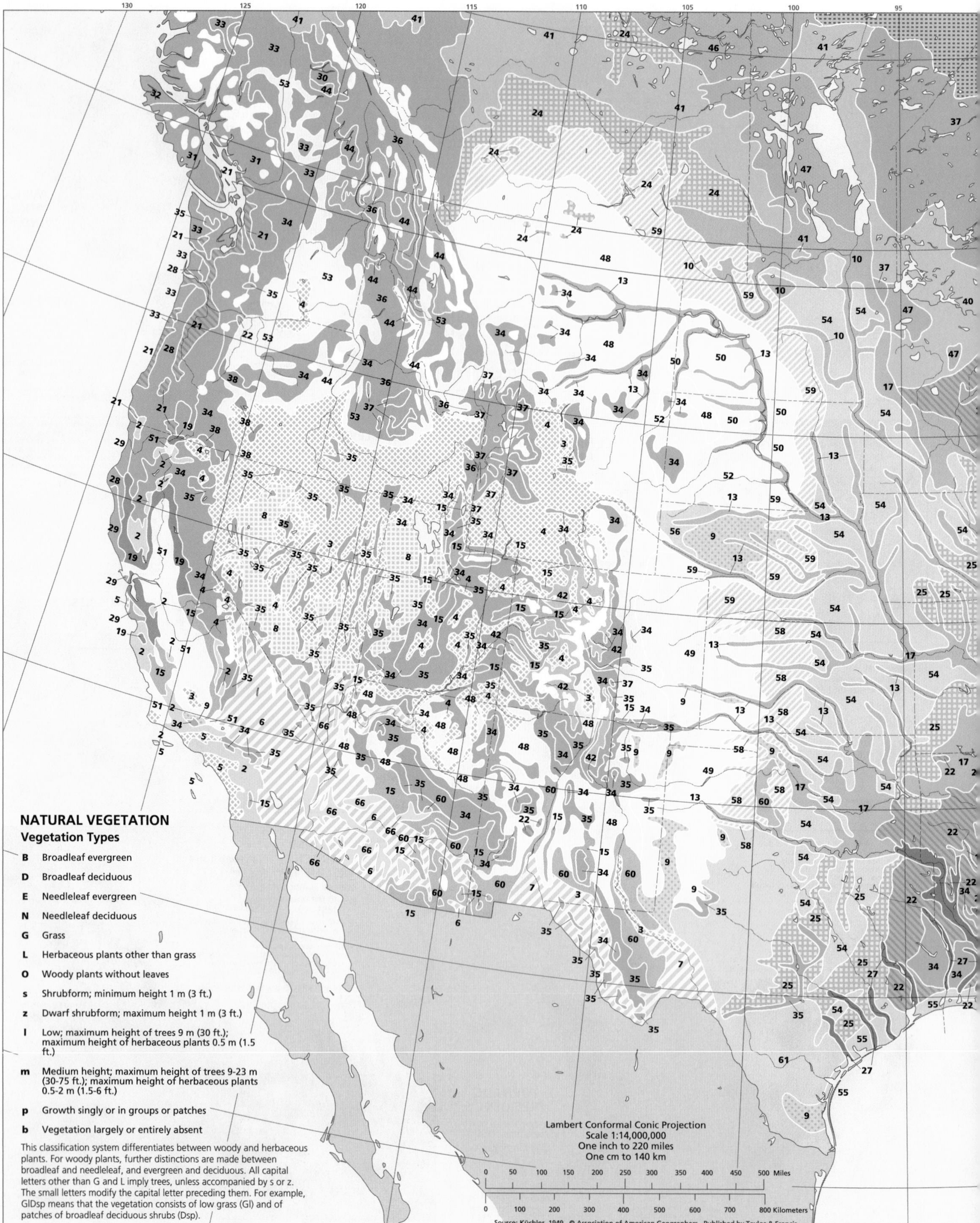

NATURAL VEGETATION
Vegetation Types
B Broadleaf evergreen
D Broadleaf deciduous
E Needleleaf evergreen
N Needleleaf deciduous
G Grass
L Herbaceous plants other than grass
O Woody plants without leaves
s Shrubform; minimum height 1 m (3 ft.)
z Dwarf shrubform; maximum height 1 m (3 ft.)
l Low; maximum height of trees 9 m (30 ft.); maximum height of herbaceous plants 0.5 m (1.5 ft.)
m Medium height; maximum height of trees 9-23 m (30-75 ft.); maximum height of herbaceous plants 0.5-2 m (1.5-6 ft.)
p Growth singly or in groups or patches
b Vegetation largely or entirely absent
This classification system differentiates between woody and herbaceous plants. For woody plants, further distinctions are made between broadleaf and needleleaf, and evergreen and deciduous. All capital letters other than G and L imply trees, unless accompanied by s or z. The small letters modify the capital letter preceding them. For example, GlDsp means that the vegetation consists of low grass (Gl) and of patches of broadleaf deciduous shrubs (Dsp).
Lambert Conformal Conic Projection
Scale 1:14,000,000
One inch to 220 miles
One cm to 140 km
0 50 100 150 200 250 300 350 400 450 500 Miles
0 100 200 300 400 500 600 700 800 Kilometers
Source: Küchler, 1949. © Association of American Geographers. Published by Taylor & Francis. Adapted with permission of the Association of American Geographers.

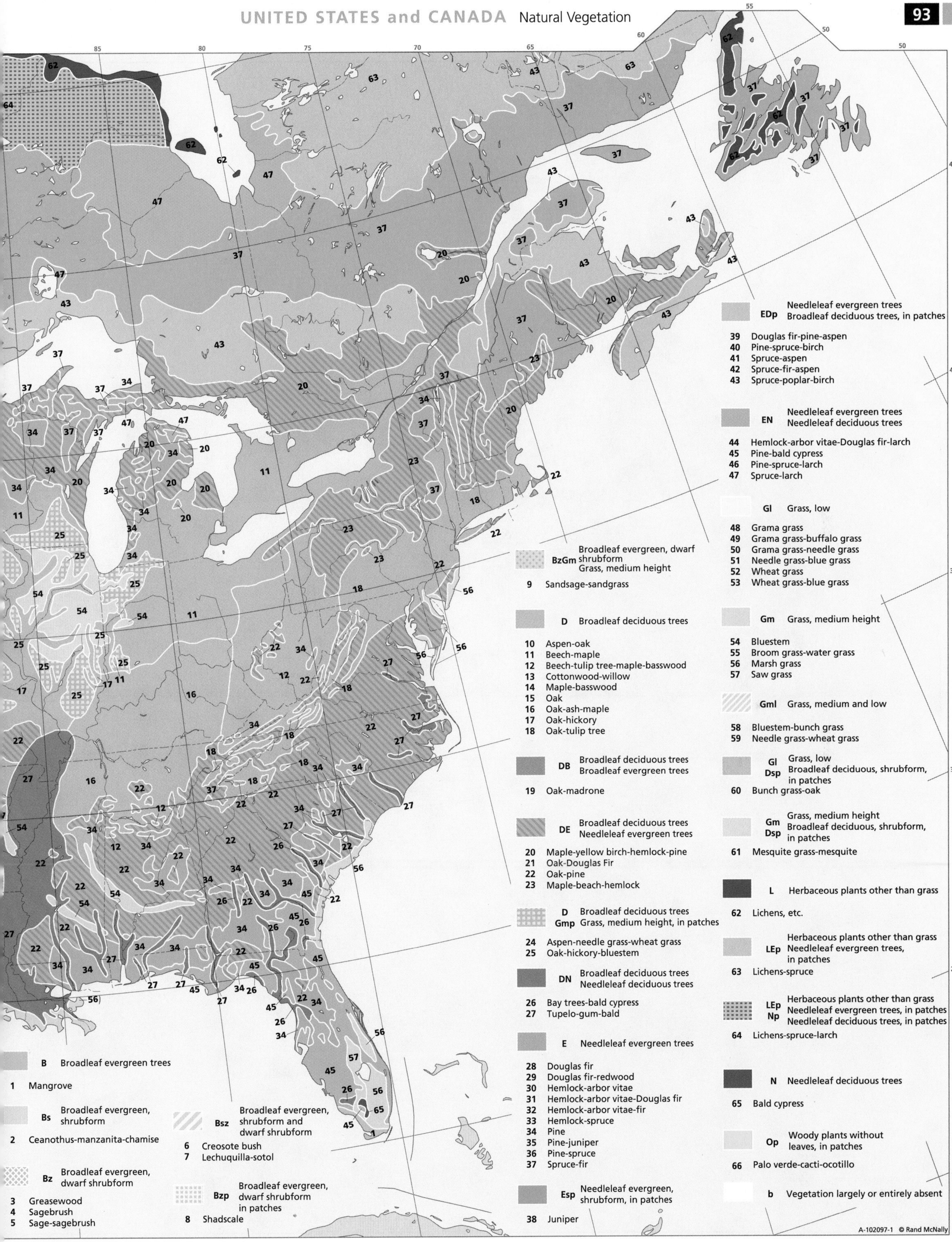
B Broadleaf evergreen trees
1 Mangrove
Bs Broadleaf evergreen, shrubform
2 Ceanothus-manzanita-chamise
Bz Broadleaf evergreen, dwarf shrubform
3 Greasewood
4 Sagebrush
5 Sage-sagebrush
Bsz Broadleaf evergreen, shrubform and dwarf shrubform
6 Creosote bush
7 Lechuquilla-sotol
Bzp Broadleaf evergreen, dwarf shrubform in patches
8 Shadscale
BzGm Broadleaf evergreen, dwarf shrubform Grass, medium height
9 Sandsage-sandgrass
D Broadleaf deciduous trees
10 Aspen-oak
11 Beech-maple
12 Beech-tulip tree-maple-basswood
13 Cottonwood-willow
14 Maple-basswood
15 Oak
16 Oak-ash-maple
17 Oak-hickory
18 Oak-tulip tree
DB Broadleaf deciduous trees Broadleaf evergreen trees
19 Oak-madrone
DE Broadleaf deciduous trees Needleleaf evergreen trees
20 Maple-yellow birch-hemlock-pine
21 Oak-Douglas Fir
22 Oak-pine
23 Maple-beach-hemlock
D Gmp Broadleaf deciduous trees Grass, medium height, in patches
24 Aspen-needle grass-wheat grass
25 Oak-hickory-bluestem
DN Broadleaf deciduous trees Needleleaf deciduous trees
26 Bay trees-bald cypress
27 Tupelo-gum-bald
E Needleleaf evergreen trees
28 Douglas fir
29 Douglas fir-redwood
30 Hemlock-arbor vitae
31 Hemlock-arbor vitae-Douglas fir
32 Hemlock-arbor vitae-fir
33 Hemlock-spruce
34 Pine
35 Pine-juniper
36 Pine-spruce
37 Spruce-fir
Esp Needleleaf evergreen, shrubform, in patches
38 Juniper
EDp Needleleaf evergreen trees Broadleaf deciduous trees, in patches
39 Douglas fir-pine-aspen
40 Pine-spruce-birch
41 Spruce-aspen
42 Spruce-fir-aspen
43 Spruce-poplar-birch
EN Needleleaf evergreen trees Needleleaf deciduous trees
44 Hemlock-arbor vitae-Douglas fir-larch
45 Pine-bald cypress
46 Pine-spruce-larch
47 Spruce-larch
Gl Grass, low
48 Grama grass
49 Grama grass-buffalo grass
50 Grama grass-needle grass
51 Needle grass-blue grass
52 Wheat grass
53 Wheat grass-blue grass
Gm Grass, medium height
54 Bluestem
55 Broom grass-water grass
56 Marsh grass
57 Saw grass
Gml Grass, medium and low
58 Bluestem-bunch grass
59 Needle grass-wheat grass
Gl Dsp Grass, low Broadleaf deciduous, shrubform, in patches
60 Bunch grass-oak
Gm Dsp Grass, medium height Broadleaf deciduous, shrubform, in patches
61 Mesquite grass-mesquite
L Herbaceous plants other than grass
62 Lichens, etc.
LEp Herbaceous plants other than grass Needleleaf evergreen trees, in patches
63 Lichens-spruce
LEp Np Herbaceous plants other than grass Needleleaf evergreen trees, in patches Needleleaf deciduous trees, in patches
64 Lichens-spruce-larch
N Needleleaf deciduous trees
65 Bald cypress
Op Woody plants without leaves, in patches
66 Palo verde-cacti-ocotillo
b Vegetation largely or entirely absent
A-102097-1 © Rand McNally

130
120
110
100
50
40
30
Lambert Conformal Conic Projection
Scale 1:15,000,000
One inch to 237 miles
One cm to 150 km
0 100 200 300 Miles
0 100 200 300 400 Kilometers
© Rand McNally
M-100990-1

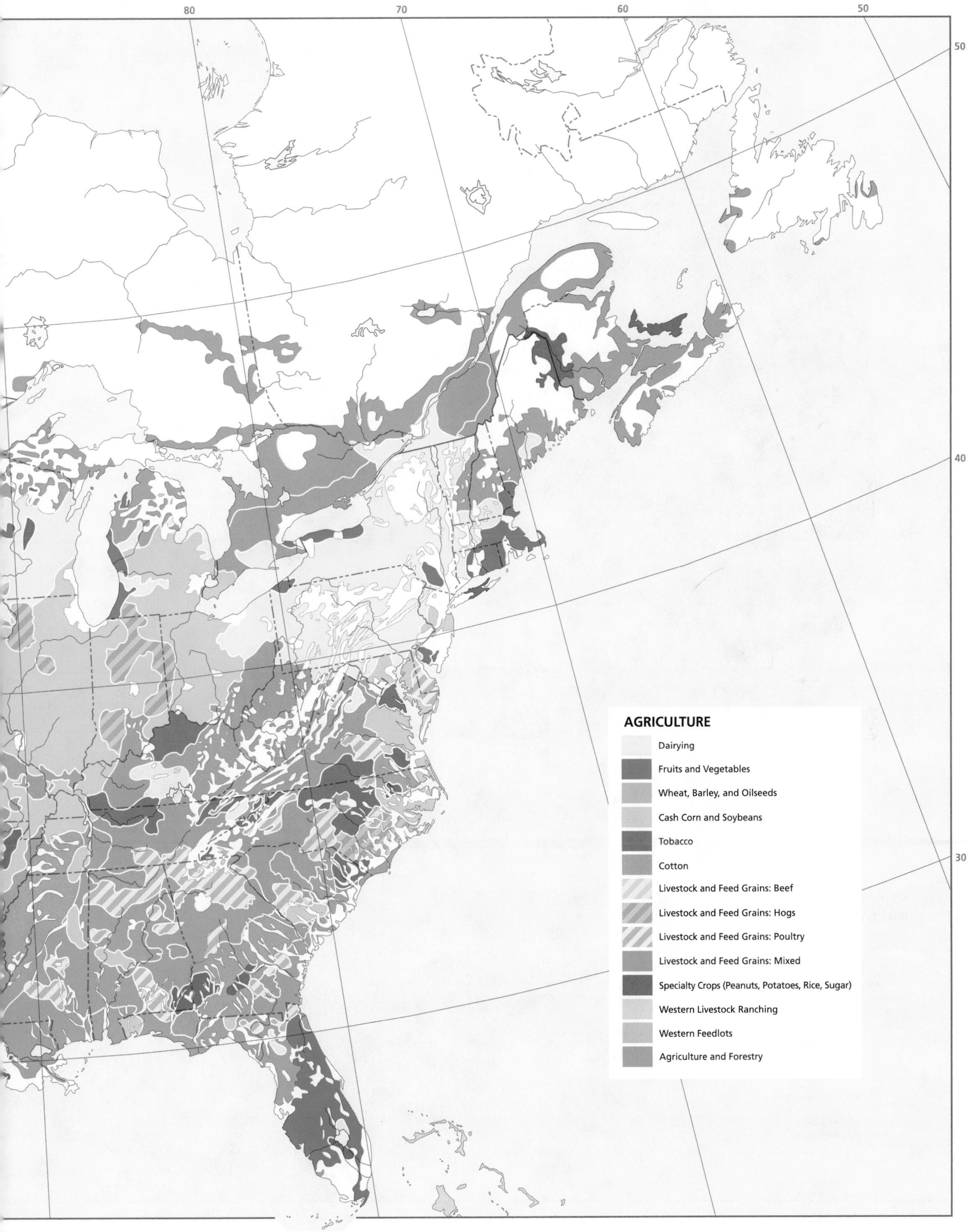
80
70
60
50
50
40
30
AGRICULTURE
Dairying
Fruits and Vegetables
Wheat, Barley, and Oilseeds
Cash Corn and Soybeans
Tobacco
Cotton
Livestock and Feed Grains: Beef
Livestock and Feed Grains: Hogs
Livestock and Feed Grains: Poultry
Livestock and Feed Grains: Mixed
Specialty Crops (Peanuts, Potatoes, Rice, Sugar)
Western Livestock Ranching
Western Feedlots
Agriculture and Forestry

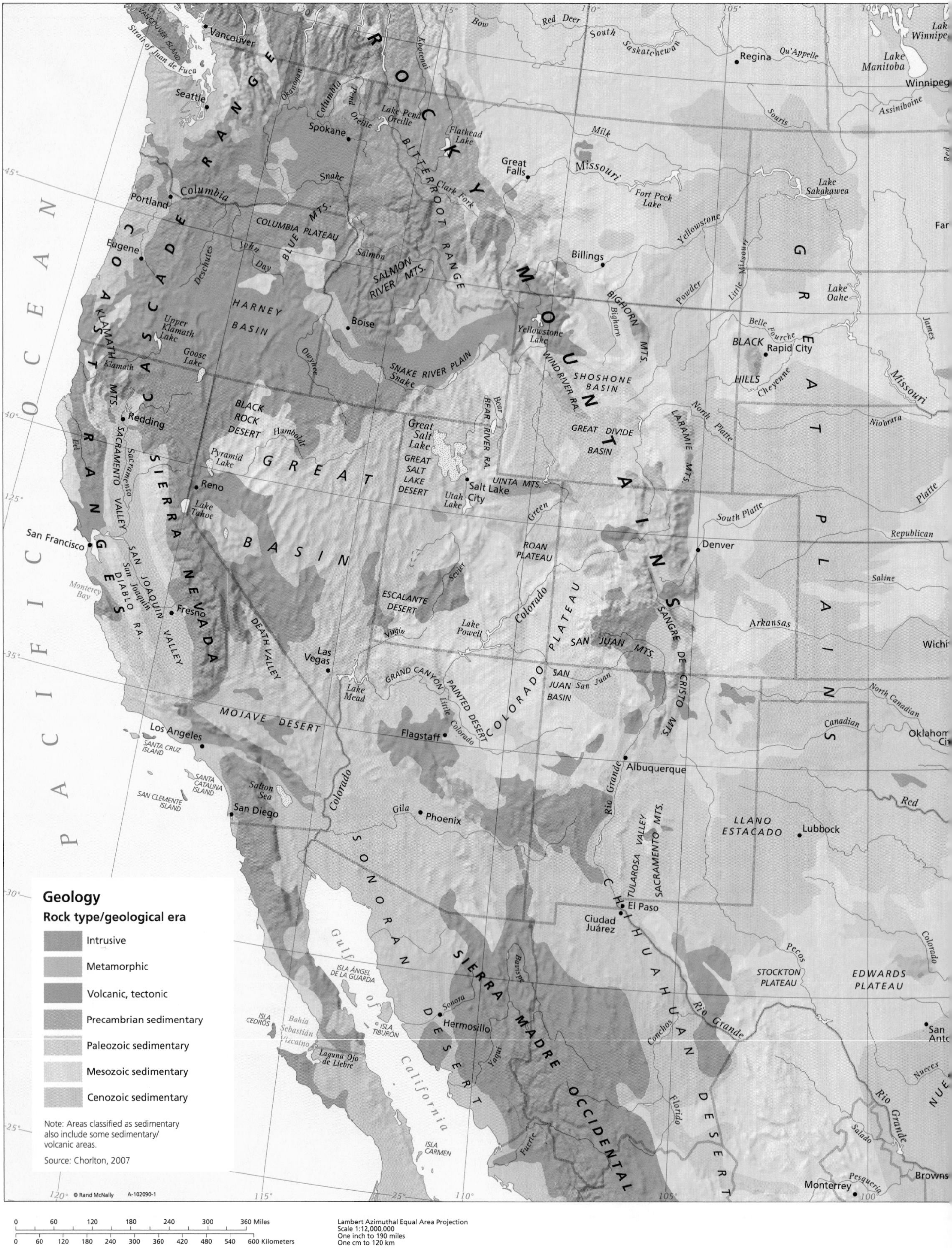
Geology
Rock type/geological era
Intrusive
Metamorphic
Volcanic, tectonic
Precambrian sedimentary
Paleozoic sedimentary
Mesozoic sedimentary
Cenozoic sedimentary
Note: Areas classified as sedimentary also include some sedimentary/volcanic areas.
Source: Chorlton, 2007
© Rand McNally A-102090-1
PACIFIC OCEAN
ROCKY MOUNTAINS
CASCADE RANGE
COAST RANGES
SIERRA NEVADA
GREAT BASIN
GREAT PLAINS
COLORADO PLATEAU
COLUMBIA PLATEAU
SNAKE RIVER PLAIN
HARNEY BASIN
BLACK ROCK DESERT
GREAT SALT LAKE DESERT
BITTERROOT RANGE
SALMON RIVER MTS.
BLUE MTS.
KLAMATH MTS.
SACRAMENTO VALLEY
SAN JOAQUIN VALLEY
DIABLO RA.
DEATH VALLEY
MOJAVE DESERT
SONORAN DESERT
CHIHUAHUAN DESERT
SIERRA MADRE OCCIDENTAL
BIGHORN MTS.
WIND RIVER RA.
SHOSHONE BASIN
GREAT DIVIDE BASIN
LARAMIE MTS.
UINTA MTS.
BEAR RIVER RA.
ROAN PLATEAU
ESCALANTE DESERT
GRAND CANYON
PAINTED DESERT
SAN JUAN MTS.
SAN JUAN BASIN
SANGRE DE CRISTO MTS.
TULAROSA VALLEY
SACRAMENTO MTS.
LLANO ESTACADO
STOCKTON PLATEAU
EDWARDS PLATEAU
BLACK HILLS
VANCOUVER ISLAND
Strait of Juan de Fuca
SANTA CRUZ ISLAND
SANTA CATALINA ISLAND
SAN CLEMENTE ISLAND
ISLA CEDROS
ISLA ÁNGEL DE LA GUARDA
ISLA TIBURÓN
ISLA CARMEN
Bahía Sebastián Vizcaíno
Laguna Ojo de Liebre
Gulf of California
Monterey Bay
Vancouver
Seattle
Spokane
Portland
Eugene
Boise
Redding
Reno
San Francisco
Fresno
Las Vegas
Los Angeles
San Diego
Phoenix
Flagstaff
Salt Lake City
Denver
Albuquerque
El Paso
Ciudad Juárez
Hermosillo
Lubbock
Great Falls
Billings
Rapid City
Regina
Winnipeg
Monterrey
Columbia
Snake
Missouri
Yellowstone
Colorado
Rio Grande
Green
Gila
Pecos
Arkansas
Platte
South Platte
North Platte
Republican
Saline
Canadian
North Canadian
Red
Niobrara
Cheyenne
Belle Fourche
Powder
Little Missouri
Bighorn
Milk
South Saskatchewan
Red Deer
Bow
Qu'Appelle
Souris
Assiniboine
James
Salmon
John Day
Deschutes
Owyhee
Humboldt
Sevier
Virgin
Sacramento
San Joaquin
Eel
Klamath
Kootenai
Okanogan
Pend Oreille
Clark Fork
Bear
Little Colorado
San Juan
Conchos
Yaqui
Sonora
Fuerte
Florido
Bavispe
Nueces
Salado
Pesquería
Lake Pend Oreille
Flathead Lake
Fort Peck Lake
Lake Sakakawea
Lake Oahe
Lake Manitoba
Lake Winnipeg
Yellowstone Lake
Upper Klamath Lake
Goose Lake
Pyramid Lake
Lake Tahoe
Great Salt Lake
Utah Lake
Lake Powell
Lake Mead
Salton Sea
125° 120° 115° 110° 105° 100°
45° 40° 35° 30° 25°
0 60 120 180 240 300 360 Miles
0 60 120 180 240 300 360 420 480 540 600 Kilometers
Lambert Azimuthal Equal Area Projection
Scale 1:12,000,000
One inch to 190 miles
One cm to 120 km

Lake St. Joseph
Lac Seul
Winnipeg
Lake Nipigon
Lake of the Woods
Rainy
Upper Red Lake
Lower Lake
Leech Lake
Duluth
Mille Lacs Lake
Mississippi
St. Croix
Minneapolis
Minnesota
Sioux Falls
ISLE ROYALE
MICHIPICOTEN ISLAND
LAKE SUPERIOR
KEWEENAW PEN.
Sault Ste. Marie
Missinaibi
Mattagami
Abitibi
Lake Abitibi
Harricana
Lac Saint-Jean
Saguenay
Réservoir Gouin
Saint-Maurice
LES LAURENTIDES
Québec
Réservoir Cabonga
Gatineau
Ottawa
Montréal
Richelieu
Lake Nipissing
Saint John
Grand Lake
Penobscot
Bay of Fundy
Portland
Gulf of Maine
Lake Champlain
ADIRONDACK MTS.
St. Lawrence
Lake Simcoe
Georgian Bay
MANITOULIN ISLAND
LAKE HURON
BEAVER ISLAND
Green Bay
DOOR PEN.
LAKE MICHIGAN
Saginaw Bay
LAKE ONTARIO
Buffalo
Connecticut
Boston
CAPE COD
NANTUCKET ISLAND
MARTHA'S VINEYARD
CATSKILL MTS.
Hartford
Hudson
Lake Winnebago
Wisconsin
Milwaukee
Muskegon
Grand
Lake St. Clair
Detroit
LAKE ERIE
Allegheny
LONG ISLAND
New York
Delaware
Cleveland
Chicago
Rock
Fox
Kankakee
Maumee
Pittsburgh
Philadelphia
Susquehanna
Delaware Bay
DELMARVA PENINSULA
Omaha
Des Moines
Des Moines
Mississippi
Illinois
Wabash
Columbus
Scioto
Ohio
Potomac
Washington
Chesapeake Bay
Indianapolis
Cincinnati
Kaskaskia
Kanawha
James
Kansas
Topeka
Kansas City
Missouri
St. Louis
Kentucky
New
Norfolk
Lake of the Ozarks
OZARK PLATEAU
Ohio
Roanoke
HATTERAS ISLAND
Raleigh
BLUE RIDGE
Holston
Yadkin
Neuse
Nashville
Cumberland
CUMBERLAND PLATEAU
ALLEGHENY PLATEAU
APPALACHIAN MOUNTAINS
PIEDMONT
Charlotte
Cape Fear
Great Pee Dee
White
Arkansas
BOSTON MTS.
Memphis
Tennessee
Santee
Lake Marion
Atlanta
Savannah
J. Strom Thurmond Res.
Charleston
OUACHITA MTS.
Little Rock
Arkansas
Ouachita
Mississippi
Yazoo
Birmingham
Coosa
Tombigbee
Flint
Oconee
Ocmulgee
Altamaha
HILTON HEAD ISLAND
Dallas
Sabine
Red
Trinity
Pearl
Alabama
Chattahoochee
St. Marys
Jacksonville
Toledo Bend Reservoir
Neches
Brazos
Mobile
Mobile
Apalachicola
Suwannee
St. Johns
Lake Pontchartrain
New Orleans
MISSISSIPPI DELTA
Houston
PLAINS
GALVESTON ISLAND
Tampa
Tampa Bay
Lake Okeechobee
Miami
ABACO
GRAND BAHAMA
BAHAMA ISLANDS
ELEUTHERA
NEW PROVIDENCE
CAT ISLAND
SAN SALVADOR
GULF OF MEXICO
ATLANTIC OCEAN
CAPE SABLE
KEY LARGO
KEY WEST
FLORIDA KEYS
Straits of Florida
ANDROS
MANGROVE CAY
GREAT EXUMA
LONG ISLAND
CROOKED ISLAND
Tropic of Cancer
95°
90°
85°
80°
75°
70°
65°
45°
40°
35°
30°
25°
The geological information on this map is highly generalized, and is intended for use at scales of 1:10,000,000 or smaller.

125°
120°
115°
110°
105°
100°
50°
45°
40°
35°
30°
Saskatoon
Calgary
Vancouver
Victoria
PACIFIC RIM NATL. PARK
GLACIER N.P.
KOOTENAY N.P.
BANFF N.P.
BLACKFOOT
C.F.B. Suffield
RIDING MOUNTAIN N.P.
Regina
Winnipeg
MAKAH
QUINAULT
OLYMPIC N.P.
Olympic
Mt. Baker
Okanogan
NORTH CASCADES N.P.
Colville
COLVILLE
PIEGAN
WATERTON LAKES N.P.
BLOOD
GRASSLANDS N.P.
Seattle
Snoqualmie
Wenatchee
SPOKANE
Kaniksu
Kootenai
GLACIER N.P.
BLACKFEET
Camp P. Shilo
Spokane
Coeur d'Alene
MT. RAINIER N.P.
Gifford Pinchot
YAKAMA
Yakima Firing Center
COEUR D'ALENE
St. Joe
Lolo
FLATHEAD
Flathead
Lewis and Clark
ROCKY BOYS
FORT BELKNAP
FORT PECK
TURTLE MOUNTAIN
Portland
Siuslaw
Mt. Hood
Clearwater
UMATILLA
Umatilla
NEZ PERCE
HELLS CANYON N.M.
Wallowa
Charles M. Russell
FT. BERTHOLD
SPIRIT LAKE
WARM SPRINGS
Nez Perce
Bitterroot
Helena
Butte
Lewis and Clark
Little Missouri
THEODORE ROOSEVELT NATL. PARK
Bismarck
Willamette
Payette
Whitman
Ochoco
Sheyenne
Umpqua
Deschutes
Malheur
Salmon-Challis
Beaverhead Deerlodge
Gallatin
BIGHORN CANYON N.R.A.
Custer
NORTHERN CHEYENNE
CRATER LAKE N.P.
Winema
CROW
Custer
STANDING ROCK
Siskiyou
Rogue River
Boise
Boise
SAWTOOTH N.R.A.
Targhee
YELLOWSTONE N.P.
Grand River
CHEYENNE RIVER
Fremont
Malheur
Sawtooth
GRAND TETON N.P.
Teton
Shoshone
Bighorn
Black Hills
REDWOOD N.P.
Klamath
Hart Mountain
HAGERMAN FOSSIL BEDS N.M.
CRATERS OF THE MOON N.M.
FORT HALL
DEVILS TOWER N.M.
Rapid City
BADLANDS N.P.
Ft. Pierre
Buffalo Gap
CROW CREEK
PIPESTONE N.M.
Modoc
DUCK VALLEY
WIND RIVER
Thunder Basin
LOWER BRULE
Sioux Falls
Shasta
Sheldon
Caribou
Six Rivers
Trinity
Humboldt/ Toiyabe
Bridger
PINE RIDGE
ROSEBUD
YANKTON
Lassen
LASSEN VOLCANIC N.P.
Cache
Medicine Bow
Oglala
Nebraska
Mendocino
PYRAMID LAKE
Plumas
AGATE FOSSIL BEDS N.M.
SANTEE
Humboldt/ Toiyabe
Utah Test & Training Range
Salt Lake City
FOSSIL BUTTE N.M.
FLAMING GORGE N.R.A.
Medicine Bow
WINNEBAGO
Tahoe
Reno
Wasatch
Cheyenne
SCOTTS BLUFF N.M.
Nebraska
OMAHA
POINT REYES N.S.
Sacramento
Eldorado
WALKER RIVER
GOSHUTE
Uinta
Ashley
UINTAH AND OURAY
Routt
Roosevelt
Pawnee
Omaha
Dugway Proving Grounds
DINOSAUR N.M.
ROCKY MOUNTAIN N.P.
San Francisco
San Jose
Stanislaus
Humboldt/ Toiyabe
GREAT BASIN N.P.
Manti-La-Sal
Grand Mesa
White River
Arapaho
Denver
YOSEMITE N.P.
Inyo
Humboldt/ Toiyabe
Fishlake
CANYONLANDS N.P.
Sierra
Nellis Air Bombing Range
CAPITOL REEF N.P.
Gunnison
Pike
KICKAPOO
POTTAWATOMIE
Ft. Riley
KINGS CANYON N.P.
SEQUOIA N.P.
Nevada Test Site
Dixie
Dixie
San Isabel
Ft. Carson
Los Padres
Hunter Liggett Mil. Res.
TULE RIVER
DEATH VALLEY N.P.
GRAND STAIRCASE-ESCALANTE N.M.
GLEN CANYON N.R.A.
Uncompahgre
CANYONS OF THE ANCIENTS N.M.
Rio Grande
Sequoia
Desert
ZION N.P.
UTE MTN.
San Juan
SOUTHERN UTE
CARRIZO PLAIN N.M.
China Lake N.W.C.
Las Vegas
GRAND CANYON-PARASHANT N.M.
Kaibab
NAVAJO N.M.
Wichita
Comanche
Vandenberg A.F.B.
Edwards A.F.B.
GRAND CANYON N.P.
NAVAJO
JICARILLA APACHE
Cimarron
Ft. Irwin
MOJAVE N. PRES.
LAKE MEAD N.R.A.
HUALAPAI
HOPI
Carson
CAPULIN VOLCANO N.M.
Los Padres
CANYON DE CHELLY N.M.
CHANNEL ISLANDS NATL. PARK
Angeles
San Bernardino
Twentynine Palms M.C.B.
Santa Fe
Kiowa
Rita Blanca
OSAGE
Los Angeles
Cibola
Tulsa
Camp Pendleton
CHEMEHUEVI VALLEY
Prescott
Coconino
ZUNI
FT. UNION N.M.
Black Kettle
JOSHUA TREE N.P.
Sitgreaves
EL MALPAIS N.M.
Albuquerque
Oklahoma City
Cleveland
COLORADO RIVER
Chocolate Mtn. Gunnery Range
ACOMA
San Diego
CABRILLO N.M.
Phoenix
Tonto
FORT APACHE
Apache
Yuma Proving Ground
Kofa
GILA RIVER
Cibola
Ft. Sill
SAN CARLOS
Lincoln
CHICKASAW N.R.A.
Goldwater A.F. Range
SONORAN DESERT N.M.
Gila
White Sands Missile Range
Lubbock
Cabeza Prieta
MESCALERO
ORGAN PIPE CACTUS N.M.
TOHONO O'ODHAM
Tucson
Lincoln
Lyndon B. Johnson
Coronado
WHITE SANDS N.M.
CARLSBAD CAVERNS N.P.
Dallas
Fort Bliss
El Paso
GUADALUPE MOUNTAINS N.P.
Ft. Hood
BIG BEND NATL. PARK
AMISTAD N.R.A.
San Antonio
Corpus Christi
PADRE ISLAND N.S.
70°
170°
160°
150°
140°
130°
60°
NOATAK N. PRES.
GATES OF THE ARCTIC N.P. & PRES.
IVVAVIK N.P.
TUKTUT NOGAIT N.P.
CAPE KRUSENSTERN N.M.
Arctic
KOBUK VALLEY N.P.
VUNTUT N.P.
Arctic Circle
Selawik
BERING LAND BRIDGE N. PRES.
Kanuti
Yukon Flats
YUKON-CHARLEY RIVERS N. PRES.
Koyukuk
WHITE MTS. N.R.A.
Innoko
Nowitna
Fairbanks
Fort Wainwright
Fort Greeley
NAHANNI N.P.
DENALI N.P. & PRES.
Tetlin
Yukon Delta
WRANGELL-ST. ELIAS N.P. & PRES.
Whitehorse
KLUANE N.P.
Anchorage
LAKE CLARK N.P. & PRES.
Kenai
Chugach
Juneau
Togiak
KENAI FORDS N.P.
Tongass
GLACIER BAY N.P. & PRES.
Becharof
KATMAI N.P. & PRES.
ADMIRALTY N.M.
Tongass
Alaska Peninsula
MISTY FJORDS N.M.
ANIAKCHAK N.M. & PRES.
Kodiak
Izembek
Alaska Peninsula
ANNETTE ISLAND
160°
150°
140°
Kilauea Point
22°
Honolulu
HALEAKALĀ N.P.
20°
Hakalau Forest
HAWAI'I VOLCANOES N.P.
160°
158°
156°
105°

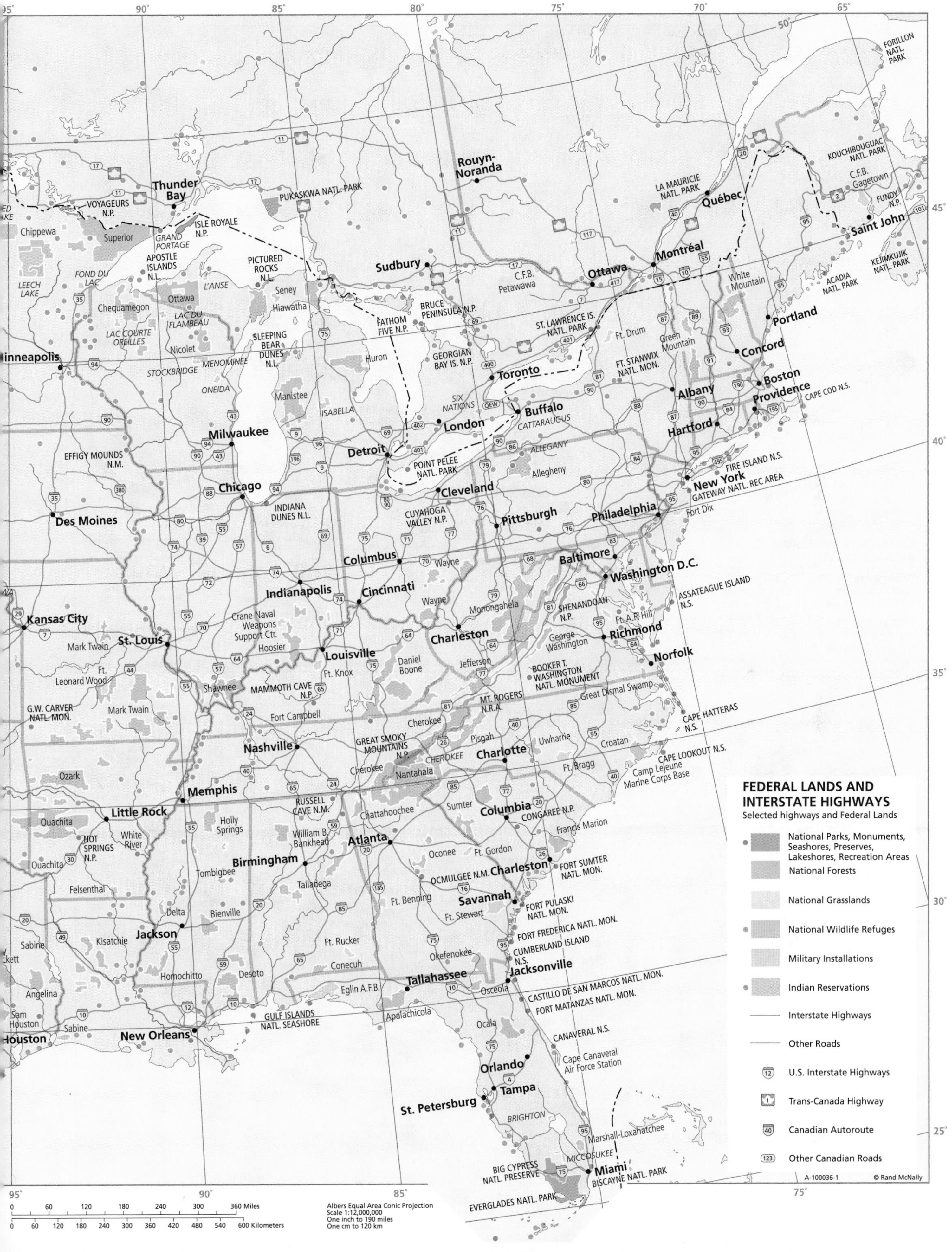

FEDERAL LANDS AND INTERSTATE HIGHWAYS
Selected highways and Federal Lands
National Parks, Monuments, Seashores, Preserves, Lakeshores, Recreation Areas
National Forests
National Grasslands
National Wildlife Refuges
Military Installations
Indian Reservations
Interstate Highways
Other Roads
U.S. Interstate Highways
Trans-Canada Highway
Canadian Autoroute
Other Canadian Roads
A-100036-1
© Rand McNally
Albers Equal Area Conic Projection
Scale 1:12,000,000
One inch to 190 miles
One cm to 120 km
360 Miles
600 Kilometers
Thunder Bay
Rouyn-Noranda
Sudbury
Ottawa
Montréal
Québec
Saint John
Portland
Concord
Boston
Providence
Albany
Hartford
New York
Philadelphia
Baltimore
Washington D.C.
Richmond
Norfolk
Toronto
Buffalo
London
Detroit
Cleveland
Pittsburgh
Columbus
Cincinnati
Indianapolis
Chicago
Milwaukee
Minneapolis
Des Moines
Kansas City
St. Louis
Louisville
Charleston
Nashville
Memphis
Little Rock
Charlotte
Columbia
Atlanta
Birmingham
Charleston
Savannah
Jackson
Jacksonville
Tallahassee
New Orleans
Houston
Orlando
Tampa
St. Petersburg
Miami
VOYAGEURS N.P.
ISLE ROYALE N.P.
GRAND PORTAGE
APOSTLE ISLANDS N.L.
PICTURED ROCKS N.L.
PUKASKWA NATL. PARK
FATHOM FIVE N.P.
BRUCE PENINSULA N.P.
GEORGIAN BAY IS. N.P.
ST. LAWRENCE IS. NATL. PARK
LA MAURICIE NATL. PARK
FORILLON NATL. PARK
KOUCHIBOUGUAC NATL. PARK
FUNDY N.P.
KEJIMKUJIK NATL. PARK
ACADIA NATL. PARK
C.F.B. Gagetown
C.F.B. Petawawa
SLEEPING BEAR DUNES N.L.
INDIANA DUNES N.L.
POINT PELEE NATL. PARK
CUYAHOGA VALLEY N.P.
EFFIGY MOUNDS N.M.
FT. STANWIX NATL. MON.
FIRE ISLAND N.S.
GATEWAY NATL. REC AREA
CAPE COD N.S.
ASSATEAGUE ISLAND N.S.
SHENANDOAH N.P.
BOOKER T. WASHINGTON NATL. MONUMENT
MT. ROGERS N.R.A.
MAMMOTH CAVE N.P.
GREAT SMOKY MOUNTAINS N.P.
CAPE HATTERAS N.S.
CAPE LOOKOUT N.S.
RUSSELL CAVE N.M.
G.W. CARVER NATL. MON.
HOT SPRINGS N.P.
CONGAREE N.P.
FORT SUMTER NATL. MON.
OCMULGEE N.M.
FORT PULASKI NATL. MON.
FORT FREDERICA NATL. MON.
CUMBERLAND ISLAND N.S.
CASTILLO DE SAN MARCOS NATL. MON.
FORT MATANZAS NATL. MON.
CANAVERAL N.S.
GULF ISLANDS NATL. SEASHORE
BIG CYPRESS NATL. PRESERVE
EVERGLADES NATL. PARK
BISCAYNE NATL. PARK
Superior
Chippewa
Chequamegon
Ottawa
Nicolet
Hiawatha
Seney
Huron
Manistee
Allegheny
Wayne
Monongahela
Hoosier
Shawnee
Daniel Boone
Jefferson
Cherokee
Pisgah
Nantahala
Chattahoochee
Sumter
Francis Marion
Uwharrie
Croatan
Oconee
Talladega
William B. Bankhead
Tombigbee
Holly Springs
Bienville
Delta
Homochitto
Desoto
Conecuh
Apalachicola
Ocala
Osceola
Ozark
Ouachita
Mark Twain
Kisatchie
Sabine
Angelina
Sam Houston
Felsenthal
Marshall-Loxahatchee
Okefenokee
White Mountain
Green Mountain
Ft. Drum
Fort Dix
Ft. A.P. Hill
Ft. Bragg
Camp Lejeune Marine Corps Base
Ft. Knox
Fort Campbell
Crane Naval Weapons Support Ctr.
Ft. Leonard Wood
Ft. Benning
Ft. Gordon
Ft. Stewart
Ft. Rucker
Eglin A.F.B.
Cape Canaveral Air Force Station
George Washington
Great Dismal Swamp
LEECH LAKE
FOND DU LAC
LAC COURTE OREILLES
LAC DU FLAMBEAU
L'ANSE
MENOMINEE
STOCKBRIDGE
ONEIDA
ISABELLA
SIX NATIONS
CATTARAUGUS
ALLEGANY
CHEROKEE
MICCOSUKEE
BRIGHTON

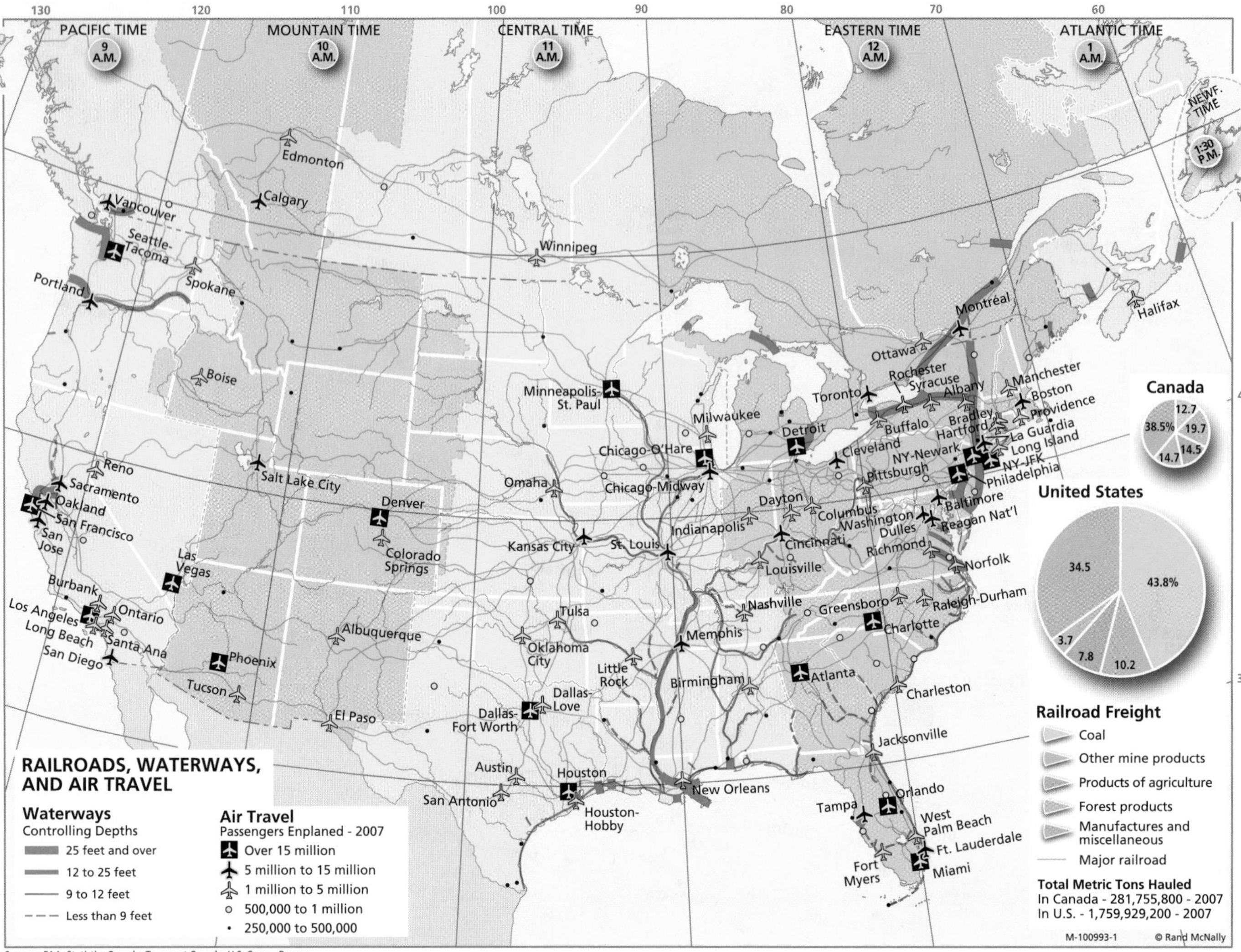
PACIFIC TIME 9 A.M.
MOUNTAIN TIME 10 A.M.
CENTRAL TIME 11 A.M.
EASTERN TIME 12 A.M.
ATLANTIC TIME 1 A.M.
NEWF. TIME 1:30 P.M.
Edmonton
Calgary
Vancouver
Seattle-Tacoma
Portland
Spokane
Winnipeg
Boise
Reno
Sacramento
Oakland
San Francisco
San Jose
Salt Lake City
Las Vegas
Burbank
Ontario
Los Angeles
Long Beach
Santa Ana
San Diego
Phoenix
Tucson
Albuquerque
El Paso
Denver
Colorado Springs
Minneapolis-St. Paul
Omaha
Kansas City
St. Louis
Tulsa
Oklahoma City
Little Rock
Dallas-Love
Dallas-Fort Worth
Austin
San Antonio
Houston
Houston-Hobby
New Orleans
Memphis
Birmingham
Nashville
Chicago-O'Hare
Chicago-Midway
Milwaukee
Indianapolis
Detroit
Cleveland
Dayton
Columbus
Cincinnati
Louisville
Pittsburgh
Toronto
Buffalo
Rochester
Syracuse
Ottawa
Montréal
Albany
Manchester
Boston
Providence
Bradley
Hartford
La Guardia
Long Island
NY-Newark
NY-JFK
Philadelphia
Baltimore
Washington Dulles
Reagan Nat'l
Richmond
Norfolk
Greensboro
Raleigh-Durham
Charlotte
Atlanta
Charleston
Jacksonville
Orlando
Tampa
West Palm Beach
Ft. Lauderdale
Fort Myers
Miami
Halifax
Canada
38.5%
12.7
19.7
14.5
14.7
United States
34.5
43.8%
3.7
7.8
10.2
Railroad Freight
Coal
Other mine products
Products of agriculture
Forest products
Manufactures and miscellaneous
Major railroad
Total Metric Tons Hauled
In Canada - 281,755,800 - 2007
In U.S. - 1,759,929,200 - 2007
RAILROADS, WATERWAYS, AND AIR TRAVEL
Waterways
Controlling Depths
25 feet and over
12 to 25 feet
9 to 12 feet
Less than 9 feet
Air Travel
Passengers Enplaned - 2007
Over 15 million
5 million to 15 million
1 million to 5 million
500,000 to 1 million
250,000 to 500,000
M-100993-1
© Rand McNally

Sources: FAA; Statistics Canada; Transport Canada; U.S. Census Bureau

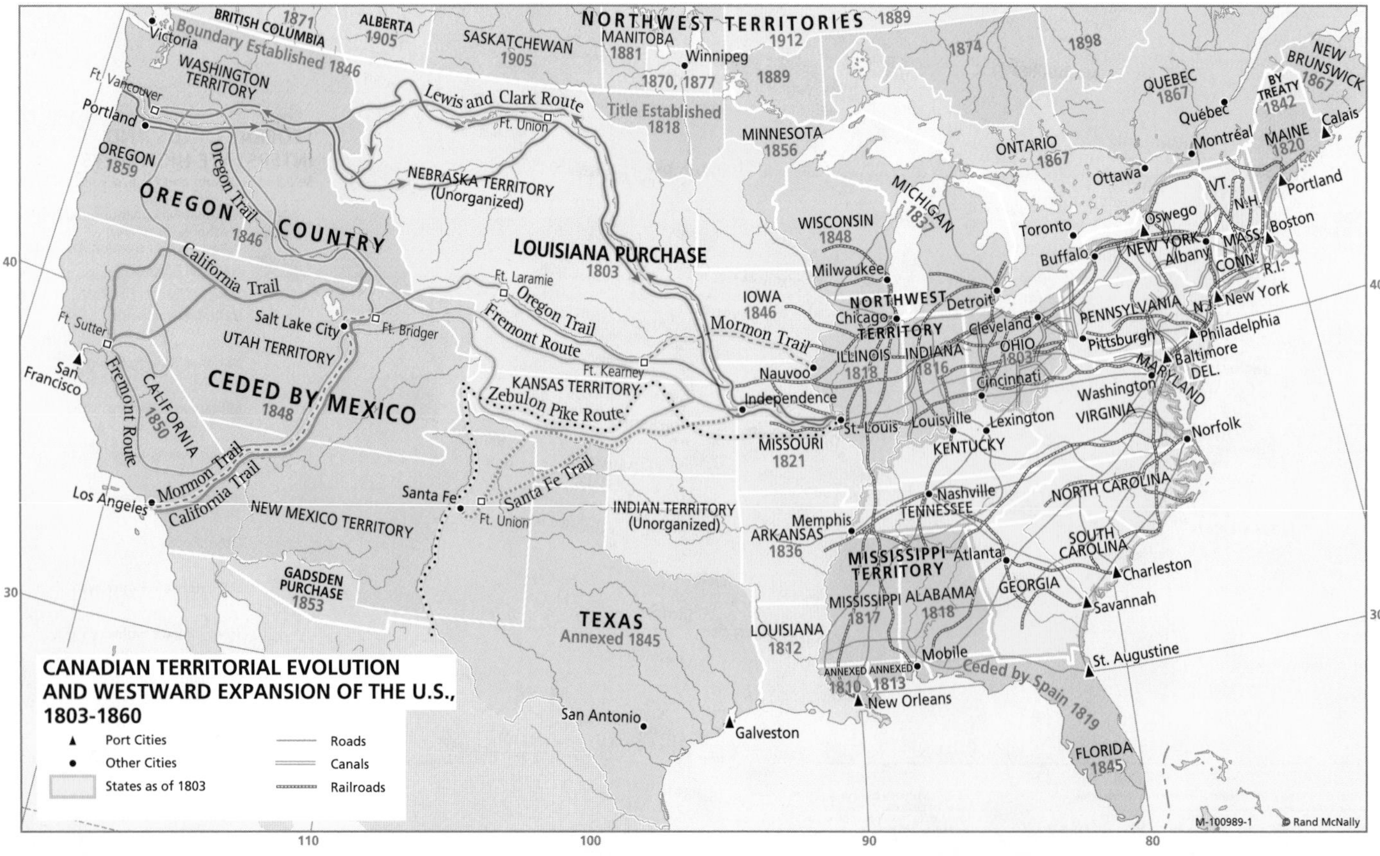
BRITISH COLUMBIA 1871
ALBERTA 1905
SASKATCHEWAN 1905
MANITOBA 1881
NORTHWEST TERRITORIES 1889
1912
1874
1898
Victoria
Boundary Established 1846
WASHINGTON TERRITORY
Ft. Vancouver
Portland
OREGON 1859
OREGON COUNTRY 1846
Oregon Trail
Lewis and Clark Route
Ft. Union
Winnipeg
1870, 1877
1889
Title Established 1818
MINNESOTA 1856
NEBRASKA TERRITORY (Unorganized)
LOUISIANA PURCHASE 1803
California Trail
Salt Lake City
Ft. Bridger
UTAH TERRITORY
Ft. Sutter
San Francisco
Fremont Route
CALIFORNIA 1850
CEDED BY MEXICO 1848
Mormon Trail
California Trail
Los Angeles
Ft. Laramie
Oregon Trail
Fremont Route
Ft. Kearney
KANSAS TERRITORY
Zebulon Pike Route
Santa Fe
Ft. Union
Santa Fe Trail
NEW MEXICO TERRITORY
GADSDEN PURCHASE 1853
INDIAN TERRITORY (Unorganized)
TEXAS Annexed 1845
San Antonio
Galveston
IOWA 1846
Nauvoo
Independence
Mormon Trail
MISSOURI 1821
St. Louis
ARKANSAS 1836
Memphis
LOUISIANA 1812
ANNEXED 1810
ANNEXED 1813
New Orleans
Mobile
Ceded by Spain 1819
FLORIDA 1845
MISSISSIPPI TERRITORY
MISSISSIPPI 1817
ALABAMA 1818
Atlanta
GEORGIA
Savannah
St. Augustine
Charleston
SOUTH CAROLINA
NORTH CAROLINA
TENNESSEE
Nashville
KENTUCKY
Louisville
Lexington
VIRGINIA
Norfolk
Washington
MARYLAND
DEL.
Baltimore
Philadelphia
PENNSYLVANIA
Pittsburgh
OHIO 1803
Cincinnati
Cleveland
INDIANA 1816
ILLINOIS 1818
NORTHWEST TERRITORY
Chicago
Detroit
Milwaukee
WISCONSIN 1848
MICHIGAN 1837
Toronto
Buffalo
ONTARIO 1867
Ottawa
QUEBEC 1867
Québec
Montréal
BY TREATY 1842
MAINE 1820
NEW BRUNSWICK 1867
Calais
Portland
VT.
N.H.
MASS.
Boston
R.I.
CONN.
NEW YORK
Albany
Oswego
New York
N.J.
CANADIAN TERRITORIAL EVOLUTION AND WESTWARD EXPANSION OF THE U.S., 1803-1860
Port Cities
Other Cities
States as of 1803
Roads
Canals
Railroads
M-100989-1
© Rand McNally

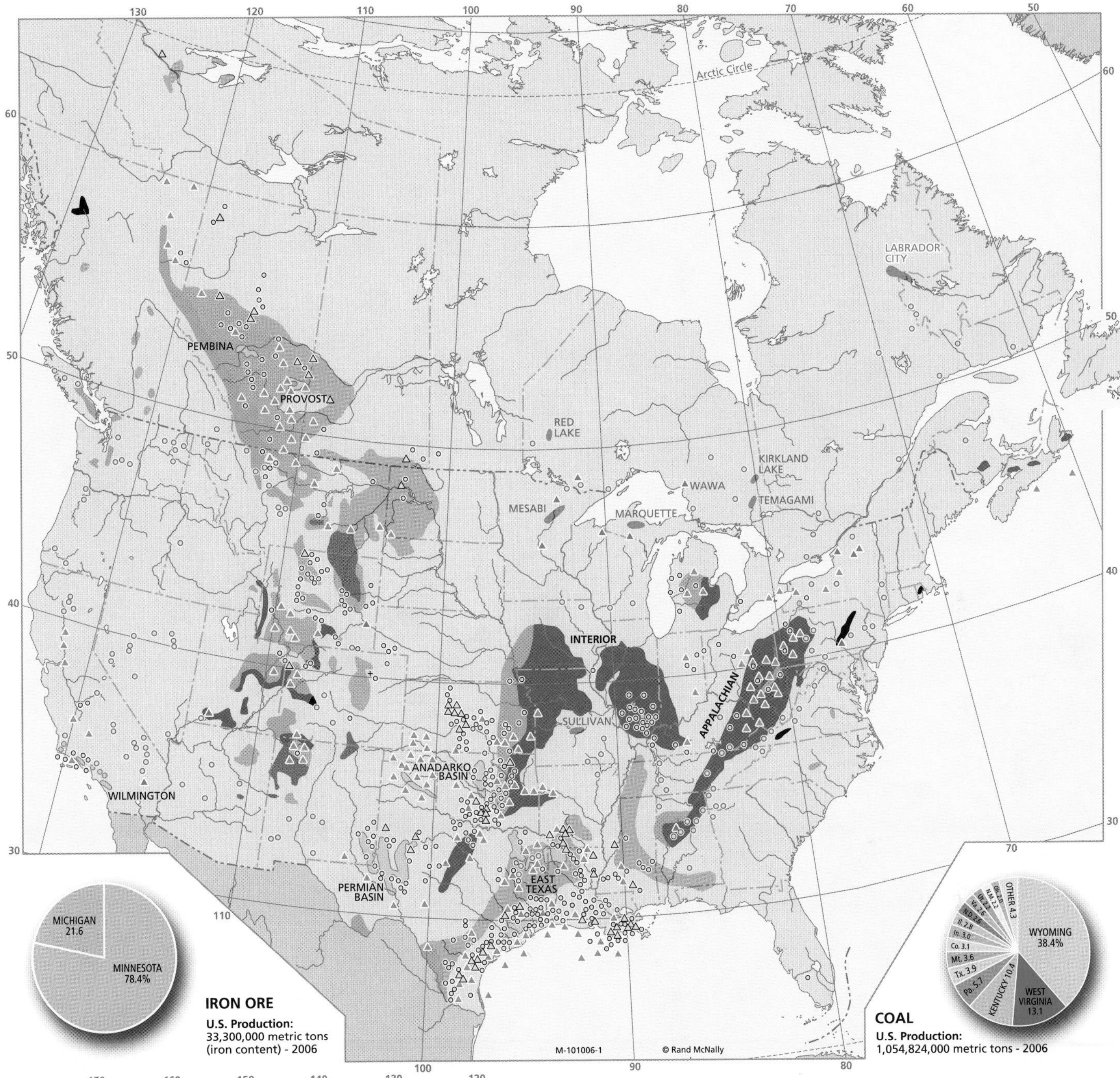

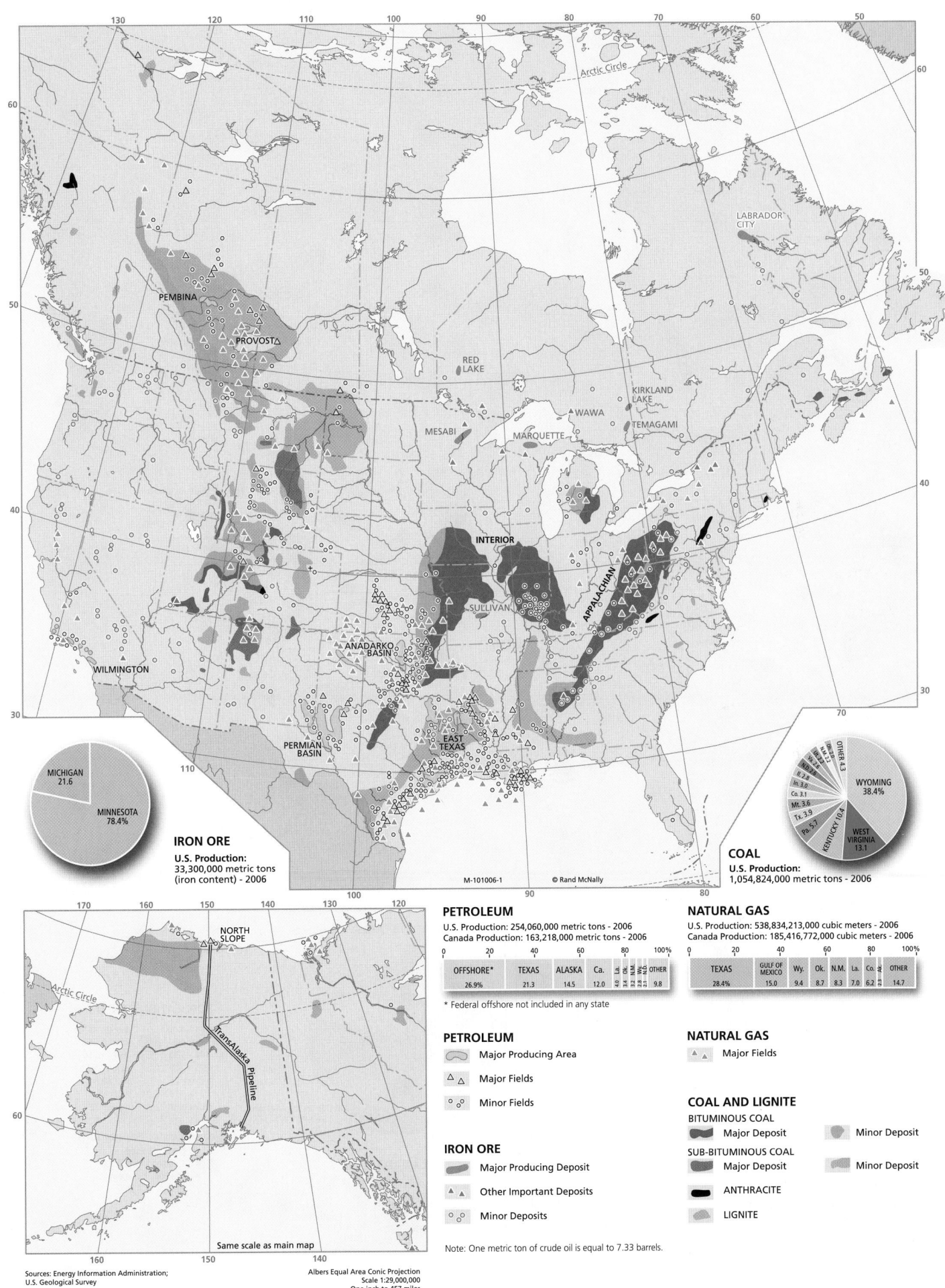

Arctic Circle
PEMBINA
PROVOST
RED LAKE
KIRKLAND LAKE
WAWA
TEMAGAMI
MESABI
MARQUETTE
LABRADOR CITY
INTERIOR
SULLIVAN
APPALACHIAN
ANADARKO BASIN
EAST TEXAS
PERMIAN BASIN
WILMINGTON
MICHIGAN 21.6
MINNESOTA 78.4%
WYOMING 38.4%
WEST VIRGINIA 13.1
KENTUCKY 10.4
Pa. 5.7
Tx. 3.9
Mt. 3.6
Co. 3.1
In. 3.0
Il. 2.8
N.D. 2.6
Va. 2.6
Ut. 2.2
N.M. 2.2
Oh. 2.0
OTHER 4.3
M-101006-1
© Rand McNally

IRON ORE

U.S. Production:
33,300,000 metric tons (iron content) - 2006

COAL

U.S. Production:
1,054,824,000 metric tons - 2006

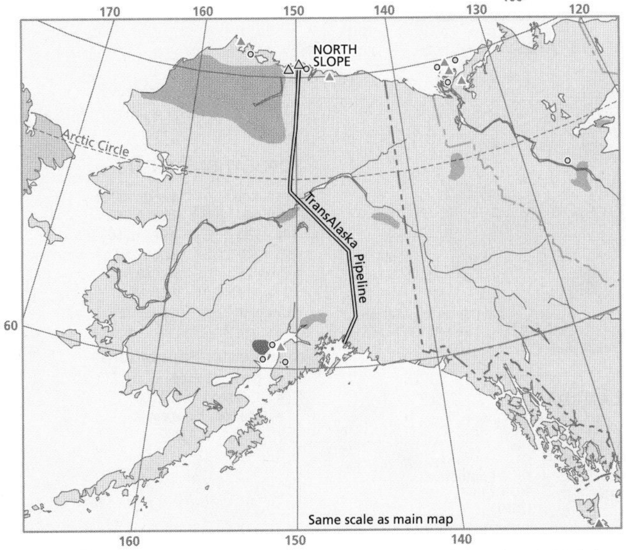

NORTH SLOPE
Arctic Circle
TransAlaska Pipeline
Same scale as main map

Sources: Energy Information Administration; U.S. Geological Survey

Albers Equal Area Conic Projection
Scale 1:29,000,000
One inch to 457 miles
One cm to 290 km

PETROLEUM

U.S. Production: 254,060,000 metric tons - 2006
Canada Production: 163,218,000 metric tons - 2006

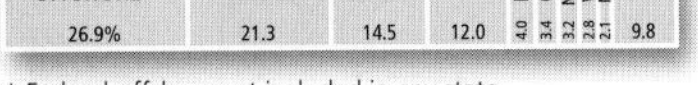

0
20
40
60
80
100%
OFFSHORE* 26.9%
TEXAS 21.3
ALASKA 14.5
Ca. 12.0
La. 4.0
Ok. 3.4
N.M. 3.2
Wy. 2.8
N.D. 2.1
OTHER 9.8

* Federal offshore not included in any state

NATURAL GAS

U.S. Production: 538,834,213,000 cubic meters - 2006
Canada Production: 185,416,772,000 cubic meters - 2006

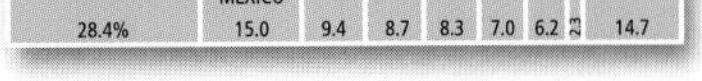

0
20
40
60
80
100%
TEXAS 28.4%
GULF OF MEXICO 15.0
Wy. 9.4
Ok. 8.7
N.M. 8.3
La. 7.0
Co. 6.2
Ak. 2.3
OTHER 14.7

PETROLEUM

- Major Producing Area
- Major Fields
- Minor Fields

IRON ORE

- Major Producing Deposit
- Other Important Deposits
- Minor Deposits

NATURAL GAS

- Major Fields

COAL AND LIGNITE

BITUMINOUS COAL

- Major Deposit
- Minor Deposit

SUB-BITUMINOUS COAL

- Major Deposit
- Minor Deposit

- ANTHRACITE
- LIGNITE

Note: One metric ton of crude oil is equal to 7.33 barrels.

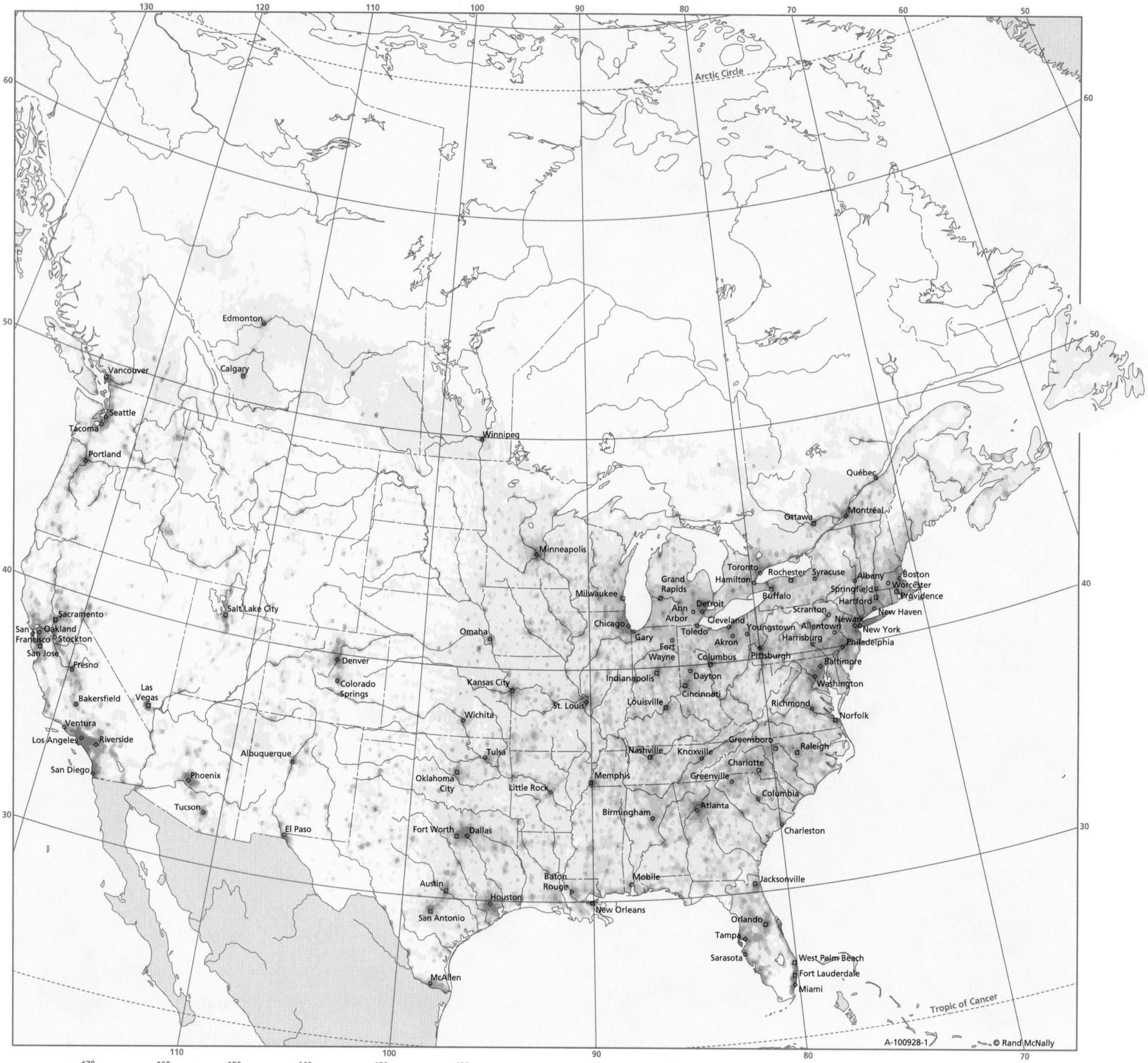
Arctic Circle
Tropic of Cancer
Edmonton
Calgary
Vancouver
Seattle
Tacoma
Portland
Winnipeg
Québec
Montréal
Ottawa
Minneapolis
Toronto
Hamilton
Rochester
Syracuse
Albany
Boston
Worcester
Springfield
Providence
Hartford
New Haven
Buffalo
Grand Rapids
Milwaukee
Detroit
Ann Arbor
Toledo
Cleveland
Scranton
Youngstown
Allentown
Newark
New York
Harrisburg
Philadelphia
Chicago
Gary
Fort Wayne
Akron
Columbus
Pittsburgh
Baltimore
Washington
Indianapolis
Dayton
Cincinnati
Louisville
Richmond
Norfolk
Salt Lake City
Sacramento
San Francisco
Oakland
Stockton
San Jose
Fresno
Bakersfield
Las Vegas
Ventura
Los Angeles
Riverside
San Diego
Omaha
Denver
Colorado Springs
Kansas City
St. Louis
Wichita
Albuquerque
Phoenix
Tucson
Tulsa
Oklahoma City
Little Rock
Memphis
Nashville
Knoxville
Greensboro
Raleigh
Charlotte
Greenville
Columbia
Atlanta
Birmingham
Charleston
El Paso
Fort Worth
Dallas
Austin
San Antonio
Houston
Baton Rouge
New Orleans
Mobile
Jacksonville
Orlando
Tampa
Sarasota
West Palm Beach
Fort Lauderdale
Miami
McAllen
A-100928-1
© Rand McNally

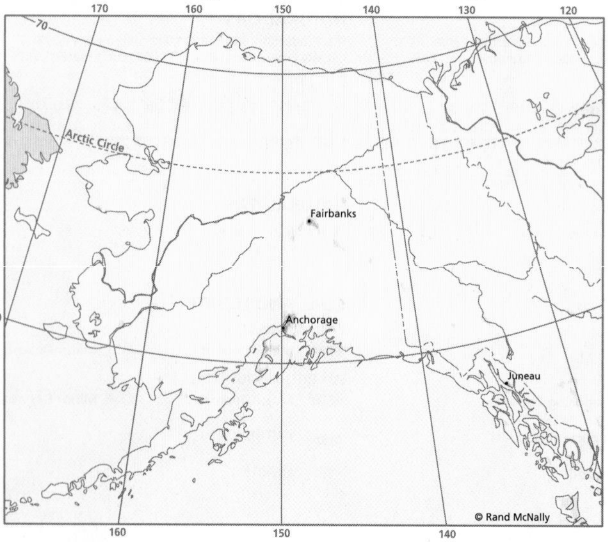
Arctic Circle
Fairbanks
Anchorage
Juneau
© Rand McNally

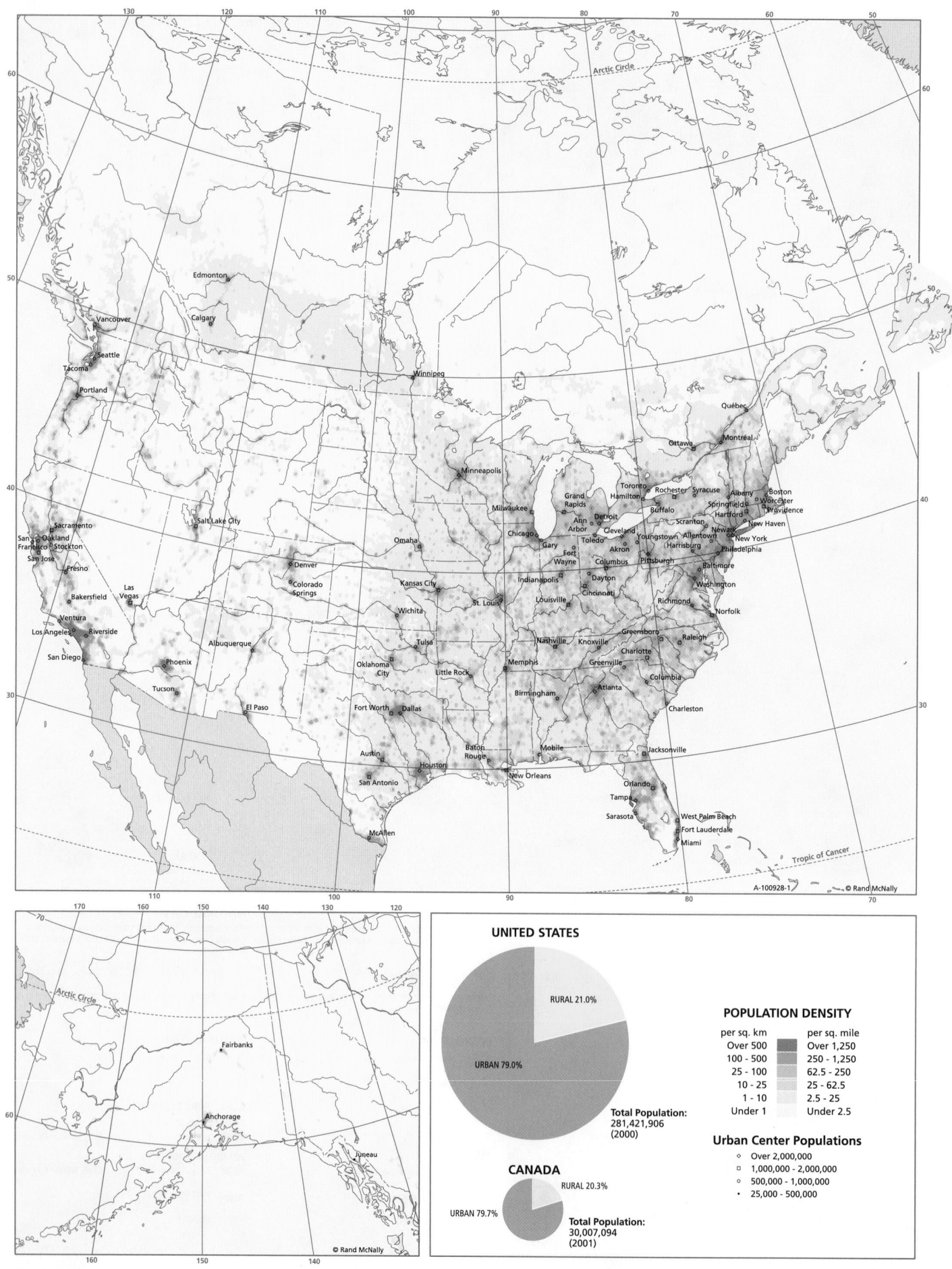

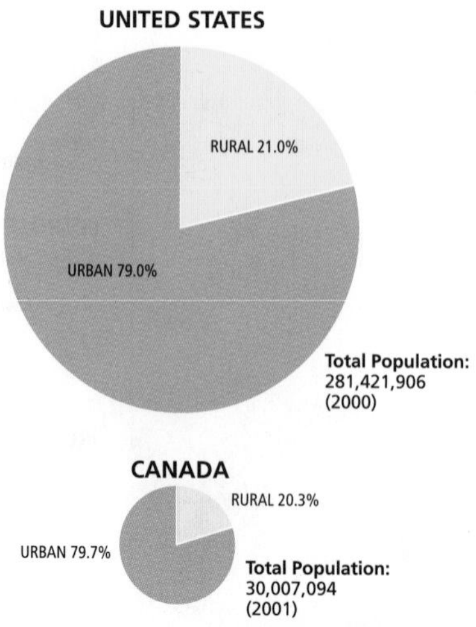
UNITED STATES
RURAL 21.0%
URBAN 79.0%
Total Population: 281,421,906 (2000)
CANADA
RURAL 20.3%
URBAN 79.7%
Total Population: 30,007,094 (2001)

POPULATION DENSITY

per sq. km	per sq. mile
Over 500	Over 1,250
100 - 500	250 - 1,250
25 - 100	62.5 - 250
10 - 25	25 - 62.5
1 - 10	2.5 - 25
Under 1	Under 2.5

Urban Center Populations

- Over 2,000,000
- 1,000,000 - 2,000,000
- 500,000 - 1,000,000
- 25,000 - 500,000

Sources: Census of Canada; U.S. Census Bureau; U.S. Department of Energy

Albers Equal Area Conic Projection
Scale 1:29,000,000
One inch to 457 miles
One cm to 290 km

WHITE POPULATION

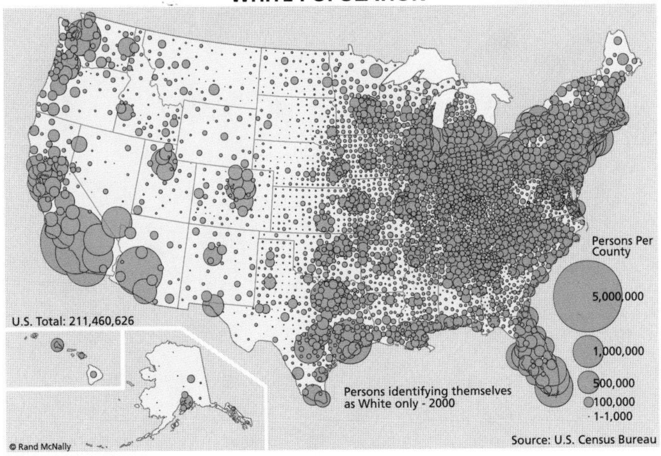
U.S. Total: 211,460,626
Persons Per County
5,000,000
1,000,000
500,000
100,000
1-1,000
Persons identifying themselves as White only - 2000
Source: U.S. Census Bureau
© Rand McNally

AFRICAN AMERICAN POPULATION

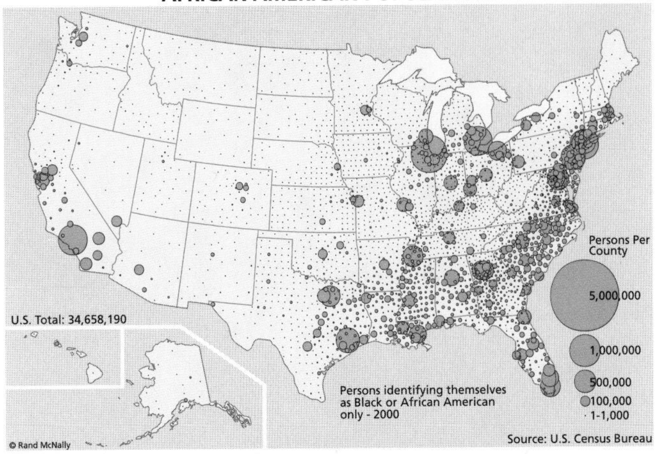
U.S. Total: 34,658,190
Persons Per County
5,000,000
1,000,000
500,000
100,000
1-1,000
Persons identifying themselves as Black or African American only - 2000
Source: U.S. Census Bureau
© Rand McNally

ASIAN POPULATION

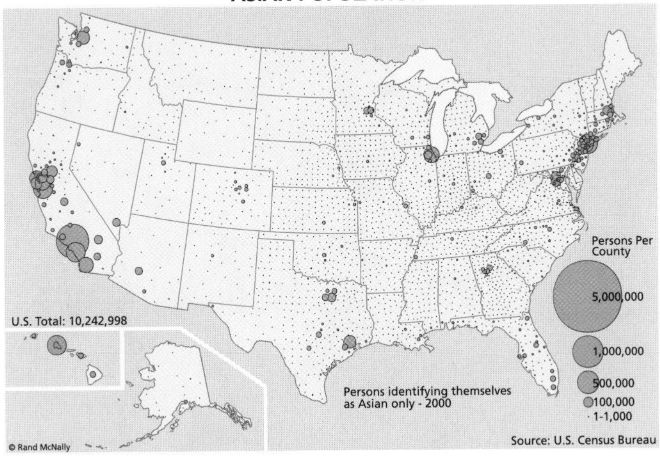
U.S. Total: 10,242,998
Persons Per County
5,000,000
1,000,000
500,000
100,000
1-1,000
Persons identifying themselves as Asian only - 2000
Source: U.S. Census Bureau
© Rand McNally

AMERICAN INDIAN AND ALASKA NATIVE POPULATION

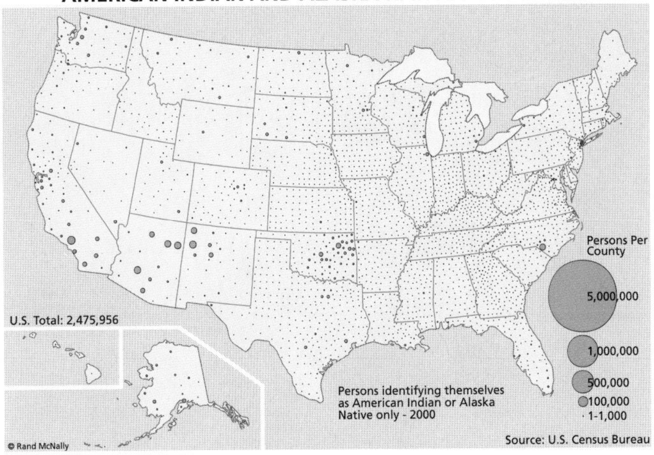
U.S. Total: 2,475,956
Persons Per County
5,000,000
1,000,000
500,000
100,000
1-1,000
Persons identifying themselves as American Indian or Alaska Native only - 2000
Source: U.S. Census Bureau
© Rand McNally

NATIVE HAWAIIAN AND PACIFIC ISLANDER POPULATION

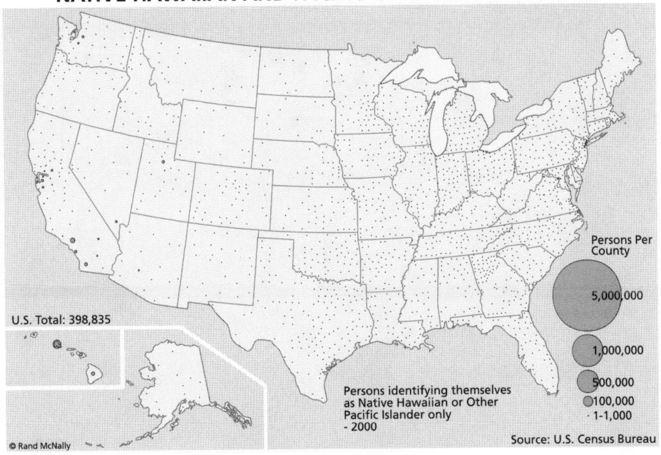
U.S. Total: 398,835
Persons Per County
5,000,000
1,000,000
500,000
100,000
1-1,000
Persons identifying themselves as Native Hawaiian or Other Pacific Islander only - 2000
Source: U.S. Census Bureau
© Rand McNally

SOME OTHER RACE

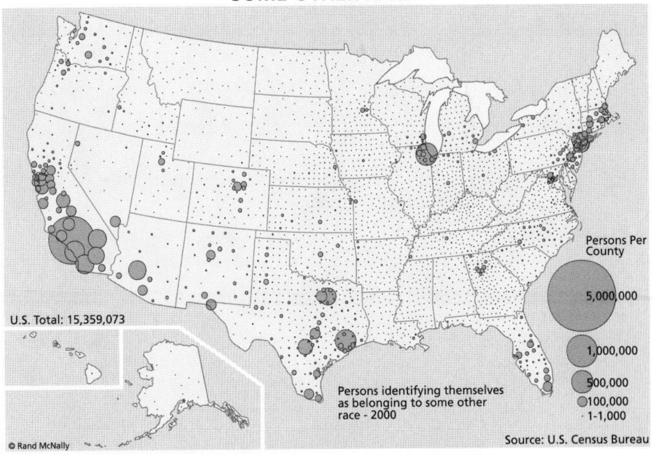
U.S. Total: 15,359,073
Persons Per County
5,000,000
1,000,000
500,000
100,000
1-1,000
Persons identifying themselves as belonging to some other race - 2000
Source: U.S. Census Bureau
© Rand McNally

TWO OR MORE RACES

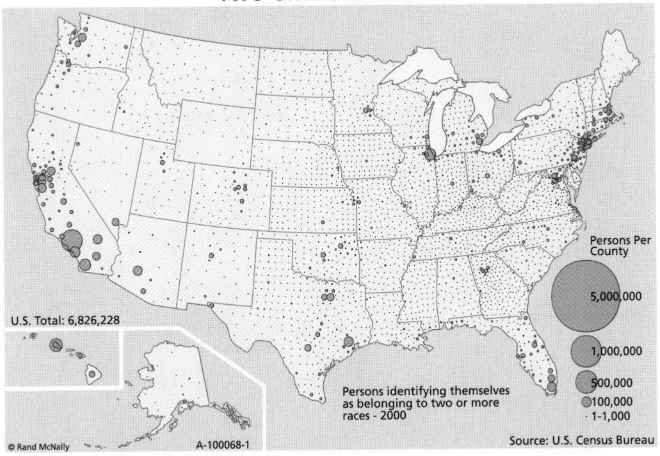
U.S. Total: 6,826,228
Persons Per County
5,000,000
1,000,000
500,000
100,000
1-1,000
Persons identifying themselves as belonging to two or more races - 2000
Source: U.S. Census Bureau
© Rand McNally
A-100068-1

HISPANIC POPULATION (ANY RACE)

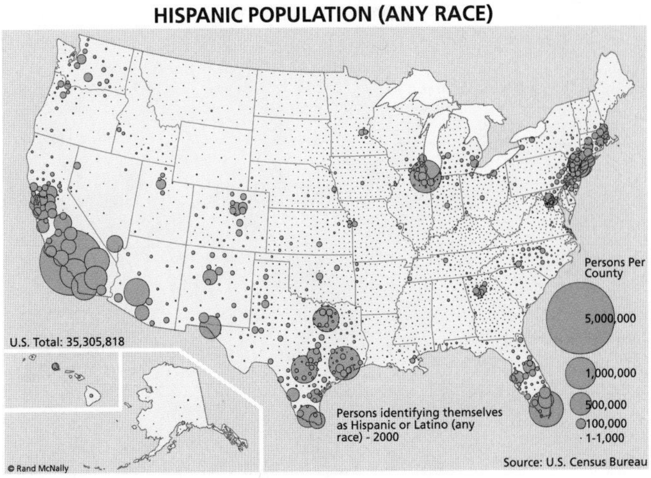
U.S. Total: 35,305,818
Persons Per County
5,000,000
1,000,000
500,000
100,000
1-1,000
Persons identifying themselves as Hispanic or Latino (any race) - 2000
Source: U.S. Census Bureau
© Rand McNally

POPULATION CHANGE

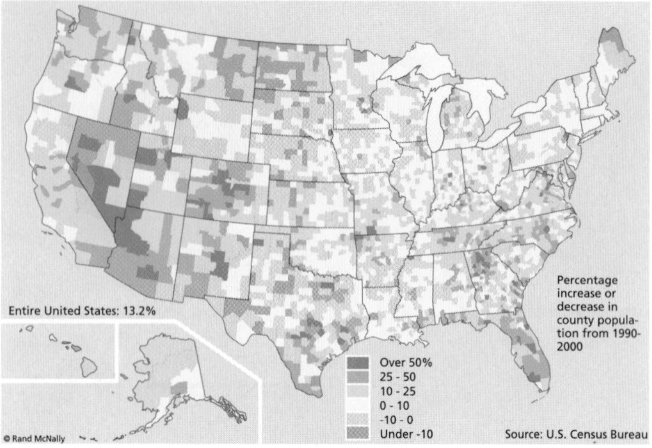
Entire United States: 13.2%
Percentage increase or decrease in county population from 1990-2000
Over 50%
25 - 50
10 - 25
0 - 10
-10 - 0
Under -10
© Rand McNally
Source: U.S. Census Bureau

INTER-STATE POPULATION SHIFTS

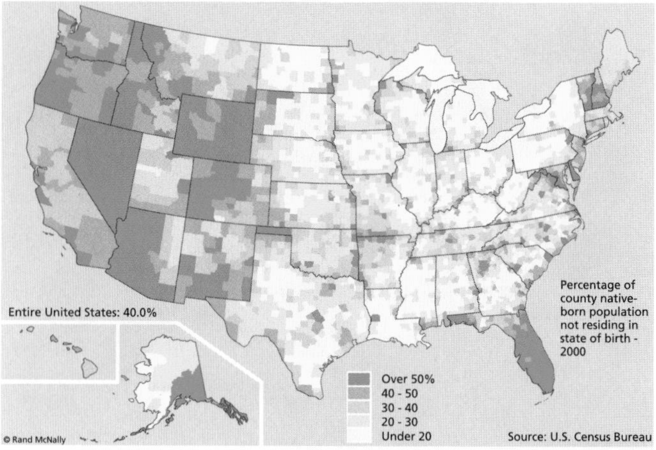
Entire United States: 40.0%
Percentage of county native-born population not residing in state of birth - 2000
Over 50%
40 - 50
30 - 40
20 - 30
Under 20
© Rand McNally
Source: U.S. Census Bureau

POPULATION UNDER 18

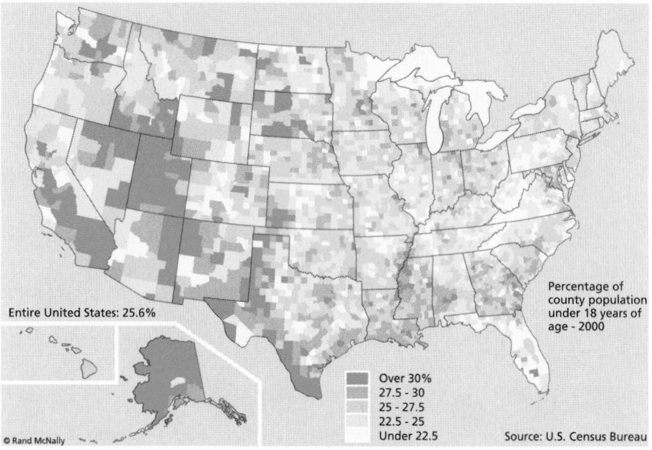
Entire United States: 25.6%
Percentage of county population under 18 years of age - 2000
Over 30%
27.5 - 30
25 - 27.5
22.5 - 25
Under 22.5
© Rand McNally
Source: U.S. Census Bureau

POPULATION 65 AND OVER

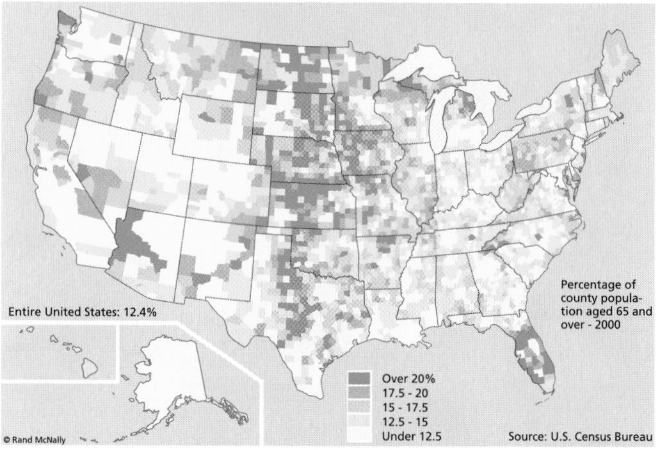
Entire United States: 12.4%
Percentage of county population aged 65 and over - 2000
Over 20%
17.5 - 20
15 - 17.5
12.5 - 15
Under 12.5
© Rand McNally
Source: U.S. Census Bureau

EDUCATIONAL ATTAINMENT RATE

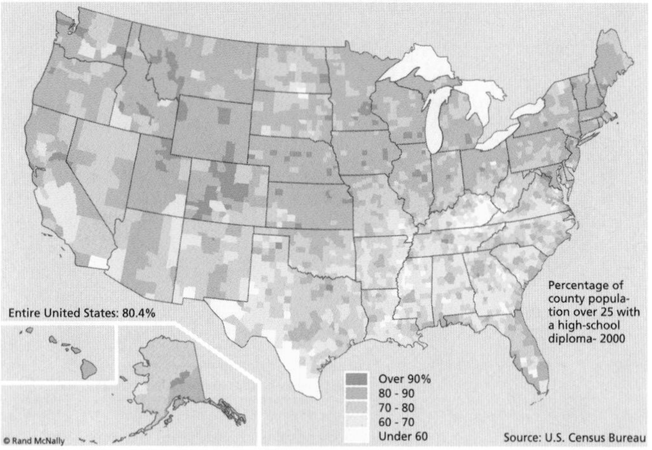
Entire United States: 80.4%
Percentage of county population over 25 with a high-school diploma- 2000
Over 90%
80 - 90
70 - 80
60 - 70
Under 60
© Rand McNally
Source: U.S. Census Bureau

COLLEGE ENROLLMENT RATE

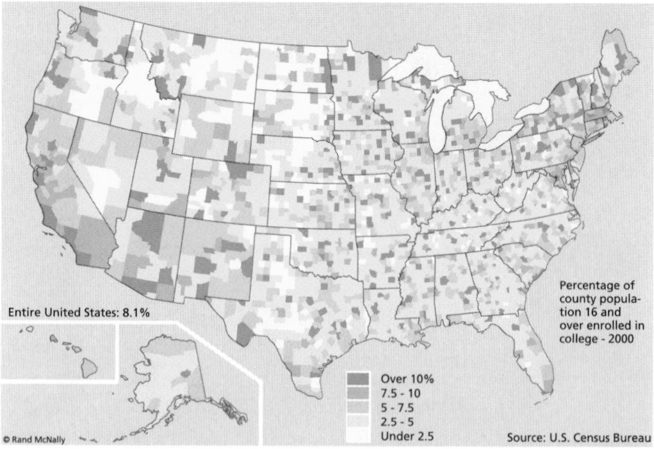
Entire United States: 8.1%
Percentage of county population 16 and over enrolled in college - 2000
Over 10%
7.5 - 10
5 - 7.5
2.5 - 5
Under 2.5
© Rand McNally
Source: U.S. Census Bureau

COMMUTING TIME

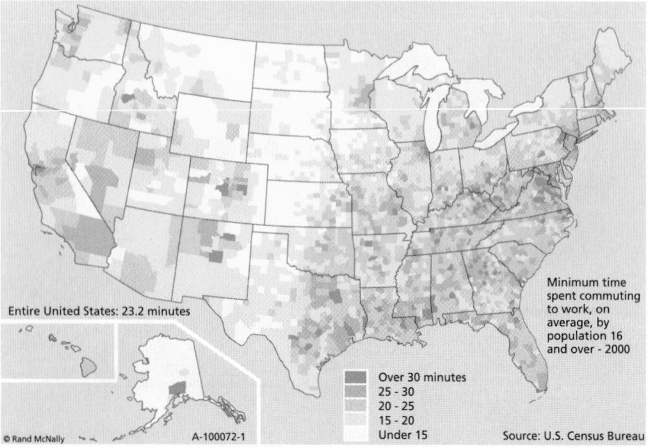
Entire United States: 23.2 minutes
Minimum time spent commuting to work, on average, by population 16 and over - 2000
Over 30 minutes
25 - 30
20 - 25
15 - 20
Under 15
© Rand McNally
A-100072-1
Source: U.S. Census Bureau

MEDIAN DECADE OF HOUSE CONSTRUCTION

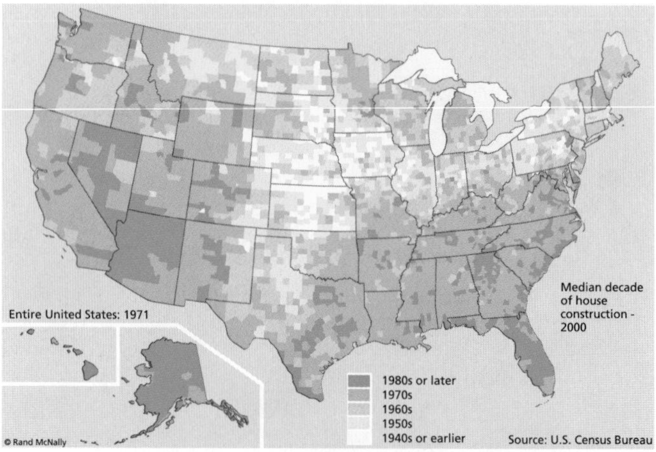
Entire United States: 1971
Median decade of house construction - 2000
1980s or later
1970s
1960s
1950s
1940s or earlier
© Rand McNally
Source: U.S. Census Bureau

WOMEN'S MEDIAN EARNINGS

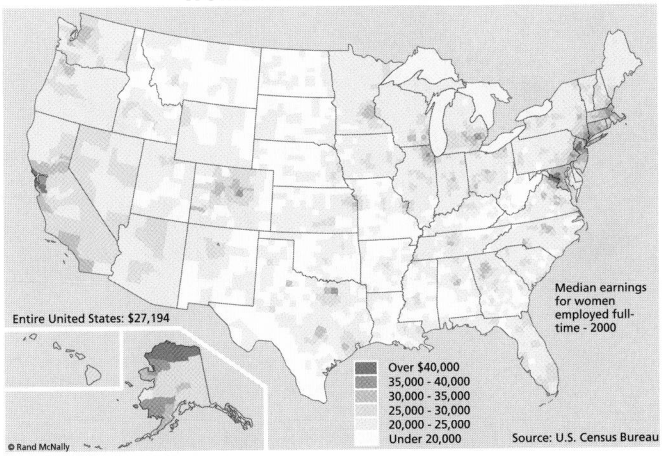
Entire United States: $27,194
Median earnings for women employed full-time - 2000
Over $40,000
35,000 - 40,000
30,000 - 35,000
25,000 - 30,000
20,000 - 25,000
Under 20,000
Source: U.S. Census Bureau
© Rand McNally

MEN'S MEDIAN EARNINGS

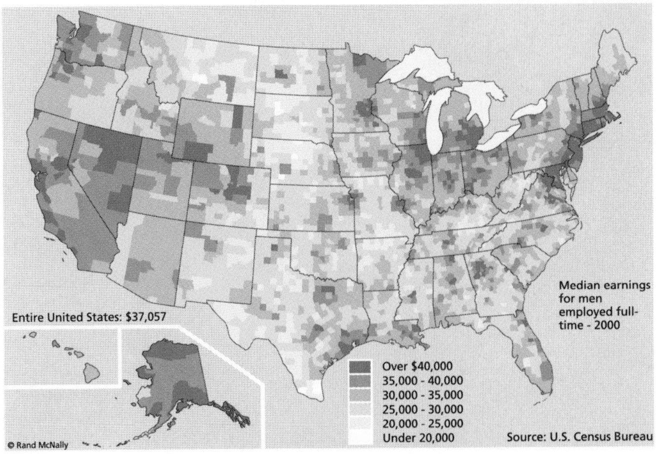
Entire United States: $37,057
Median earnings for men employed full-time - 2000
Over $40,000
35,000 - 40,000
30,000 - 35,000
25,000 - 30,000
20,000 - 25,000
Under 20,000
Source: U.S. Census Bureau
© Rand McNally

RATIO OF WOMEN'S TO MEN'S EARNINGS

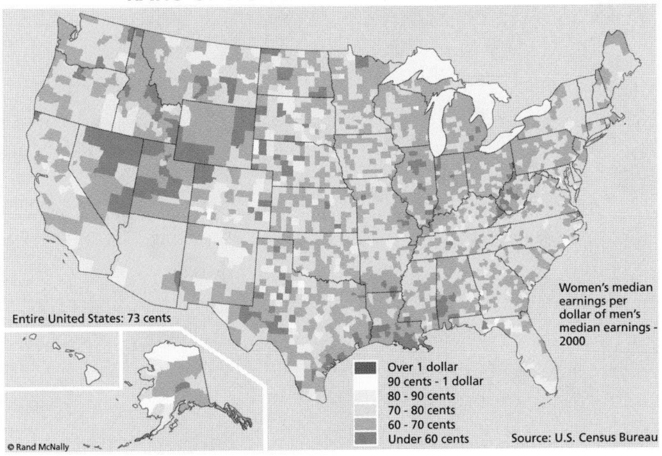
Entire United States: 73 cents
Women's median earnings per dollar of men's median earnings - 2000
Over 1 dollar
90 cents - 1 dollar
80 - 90 cents
70 - 80 cents
60 - 70 cents
Under 60 cents
Source: U.S. Census Bureau
© Rand McNally

MEDIAN HOUSEHOLD INCOME

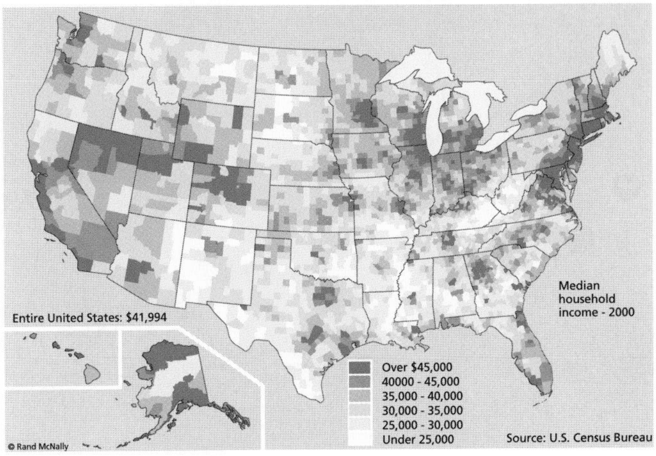
Entire United States: $41,994
Median household income - 2000
Over $45,000
40000 - 45,000
35,000 - 40,000
30,000 - 35,000
25,000 - 30,000
Under 25,000
Source: U.S. Census Bureau
© Rand McNally

HOUSEHOLDS HEADED BY WOMEN

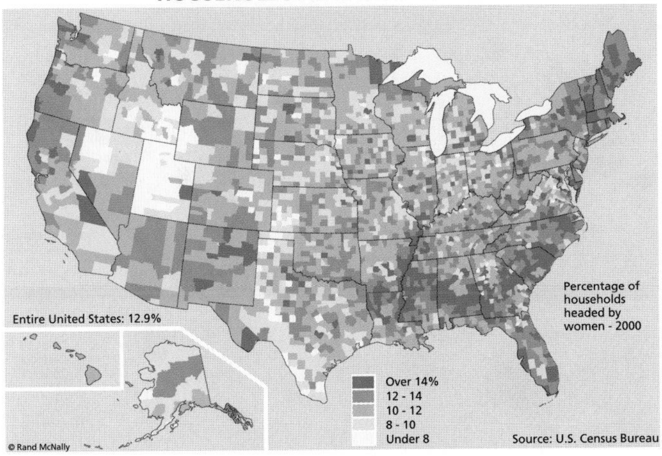
Entire United States: 12.9%
Percentage of households headed by women - 2000
Over 14%
12 - 14
10 - 12
8 - 10
Under 8
Source: U.S. Census Bureau
© Rand McNally

CHILDREN LIVING IN POVERTY

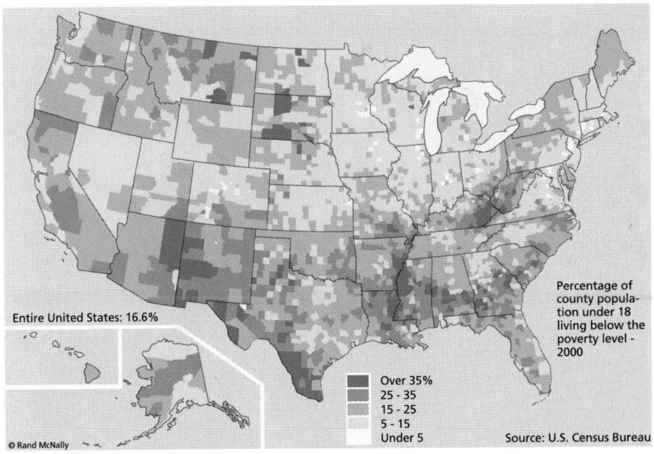
Entire United States: 16.6%
Percentage of county population under 18 living below the poverty level - 2000
Over 35%
25 - 35
15 - 25
5 - 15
Under 5
Source: U.S. Census Bureau
© Rand McNally

UNEMPLOYMENT RATE

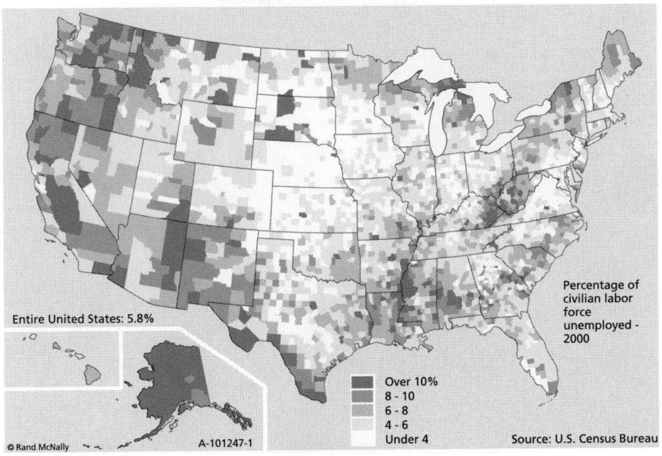
Entire United States: 5.8%
Percentage of civilian labor force unemployed - 2000
Over 10%
8 - 10
6 - 8
4 - 6
Under 4
Source: U.S. Census Bureau
© Rand McNally
A-101247-1

NON-ENGLISH SPEAKING HOUSEHOLDS

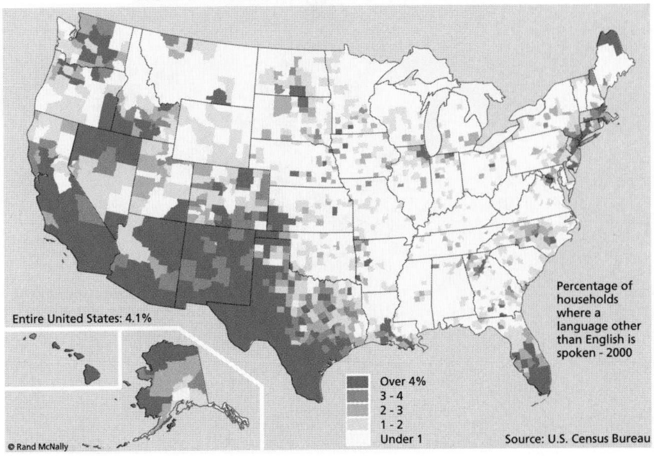
Entire United States: 4.1%
Percentage of households where a language other than English is spoken - 2000
Over 4%
3 - 4
2 - 3
1 - 2
Under 1
Source: U.S. Census Bureau
© Rand McNally

AQUIFERS

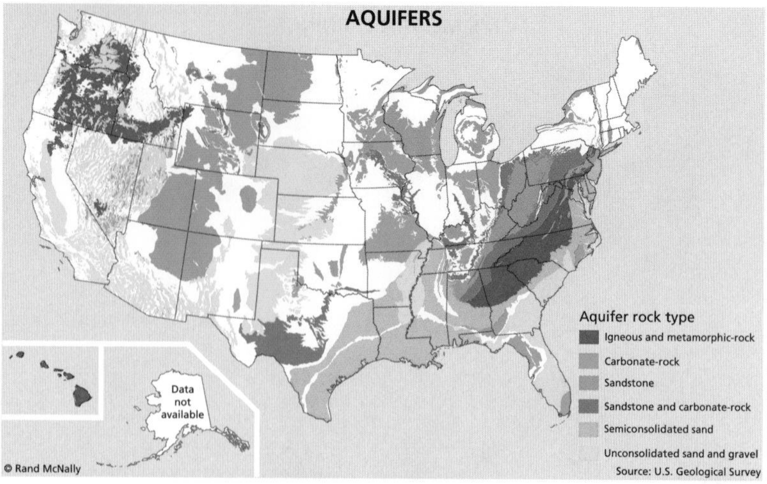
Data not available
Aquifer rock type
Igneous and metamorphic-rock
Carbonate-rock
Sandstone
Sandstone and carbonate-rock
Semiconsolidated sand
Unconsolidated sand and gravel
© Rand McNally
Source: U.S. Geological Survey

WATER USE

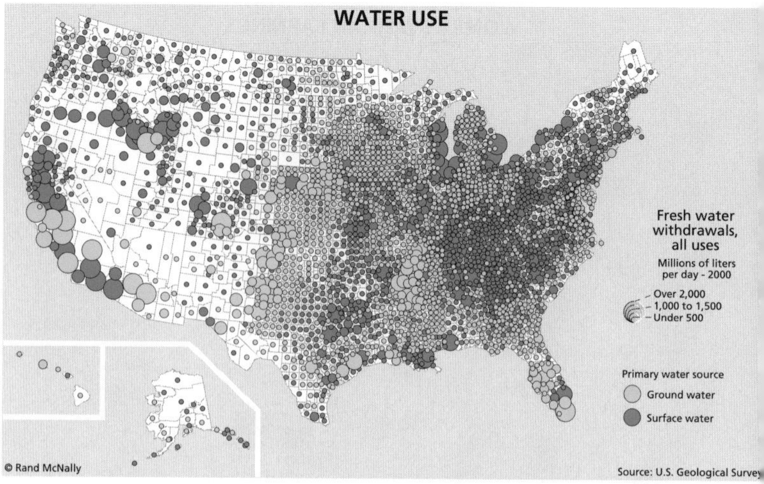
Fresh water withdrawals, all uses
Millions of liters per day - 2000
Over 2,000
1,000 to 1,500
Under 500
Primary water source
Ground water
Surface water
© Rand McNally
Source: U.S. Geological Survey

RADIOACTIVE WASTE

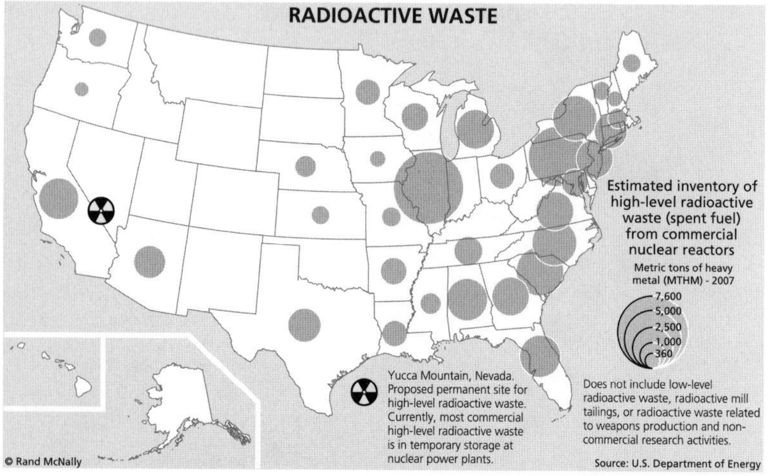
Estimated inventory of high-level radioactive waste (spent fuel) from commercial nuclear reactors
Metric tons of heavy metal (MTHM) - 2007
7,600
5,000
2,500
1,000
360
Yucca Mountain, Nevada. Proposed permanent site for high-level radioactive waste. Currently, most commercial high-level radioactive waste is in temporary storage at nuclear power plants.
Does not include low-level radioactive waste, radioactive mill tailings, or radioactive waste related to weapons production and non-commercial research activities.
© Rand McNally
Source: U.S. Department of Energy

CO_2 EMISSIONS

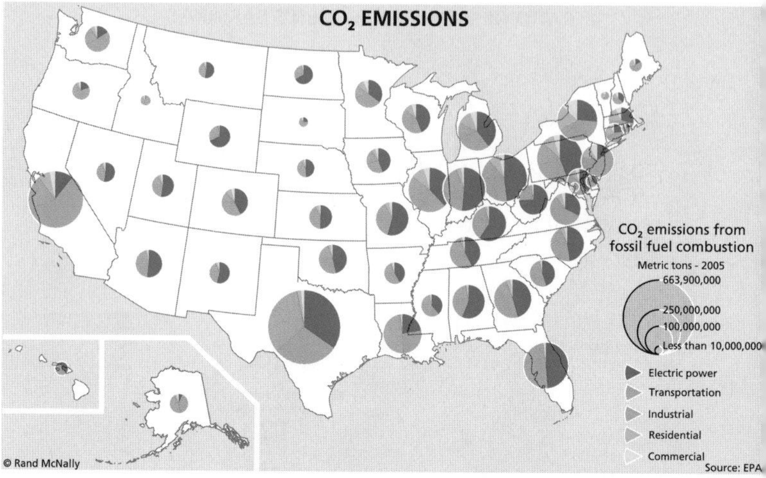
CO_2 emissions from fossil fuel combustion
Metric tons - 2005
663,900,000
250,000,000
100,000,000
Less than 10,000,000
Electric power
Transportation
Industrial
Residential
Commercial
© Rand McNally
Source: EPA

AIR QUALITY

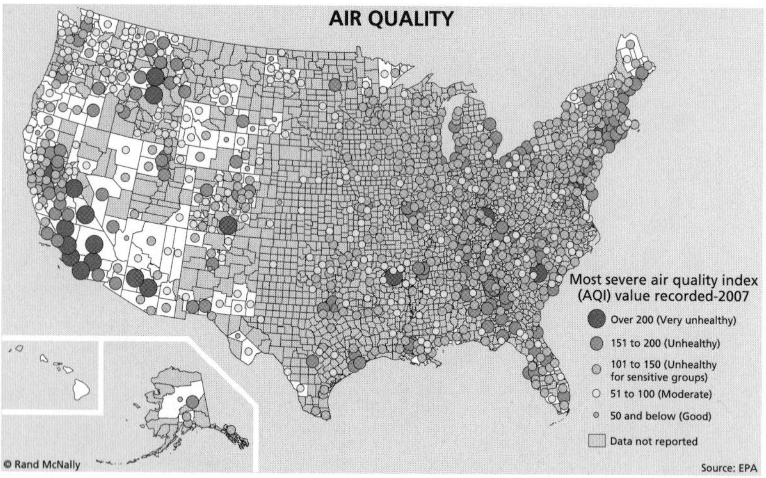
Most severe air quality index (AQI) value recorded-2007
Over 200 (Very unhealthy)
151 to 200 (Unhealthy)
101 to 150 (Unhealthy for sensitive groups)
51 to 100 (Moderate)
50 and below (Good)
Data not reported
© Rand McNally
Source: EPA

TOXIC CHEMICAL RELEASES

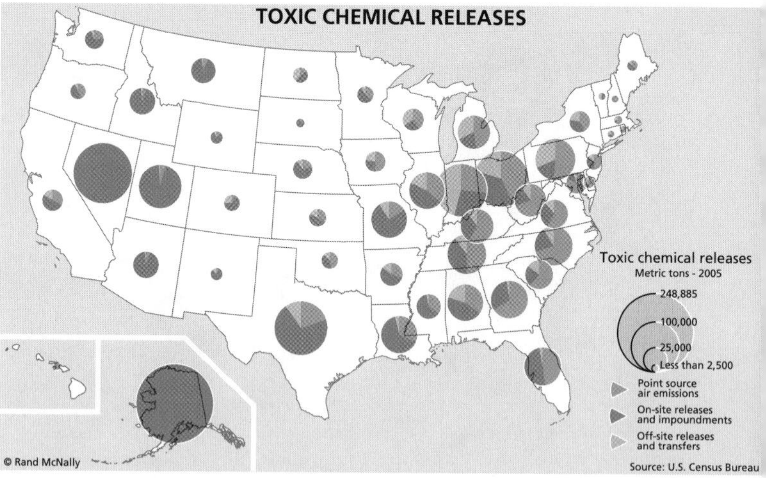
Toxic chemical releases
Metric tons - 2005
248,885
100,000
25,000
Less than 2,500
Point source air emissions
On-site releases and impoundments
Off-site releases and transfers
© Rand McNally
Source: U.S. Census Bureau

SUPERFUND SITES

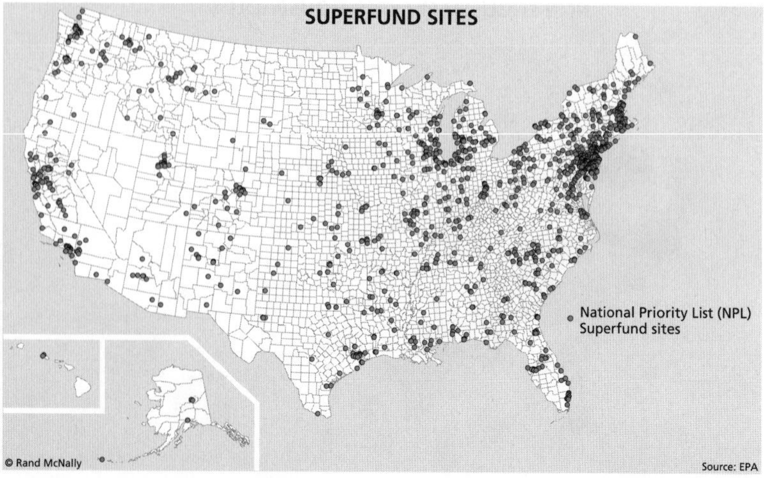
National Priority List (NPL) Superfund sites
© Rand McNally
Source: EPA

HAZARDOUS WASTE

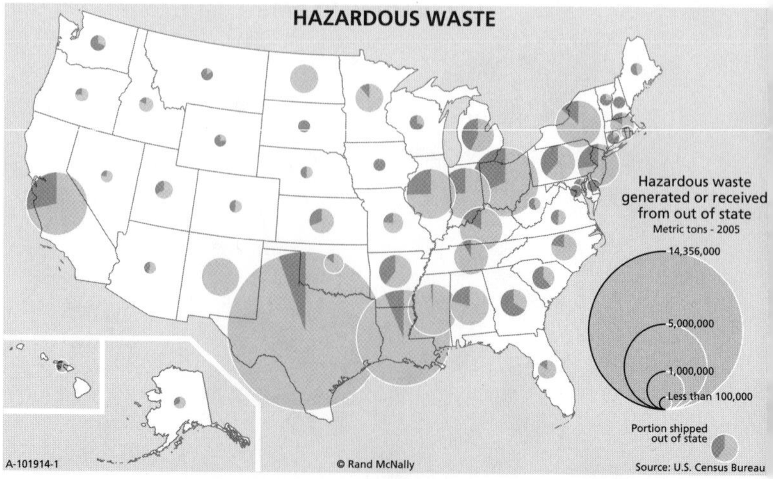
Hazardous waste generated or received from out of state
Metric tons - 2005
14,356,000
5,000,000
1,000,000
Less than 100,000
Portion shipped out of state
A-101914-1
© Rand McNally
Source: U.S. Census Bureau

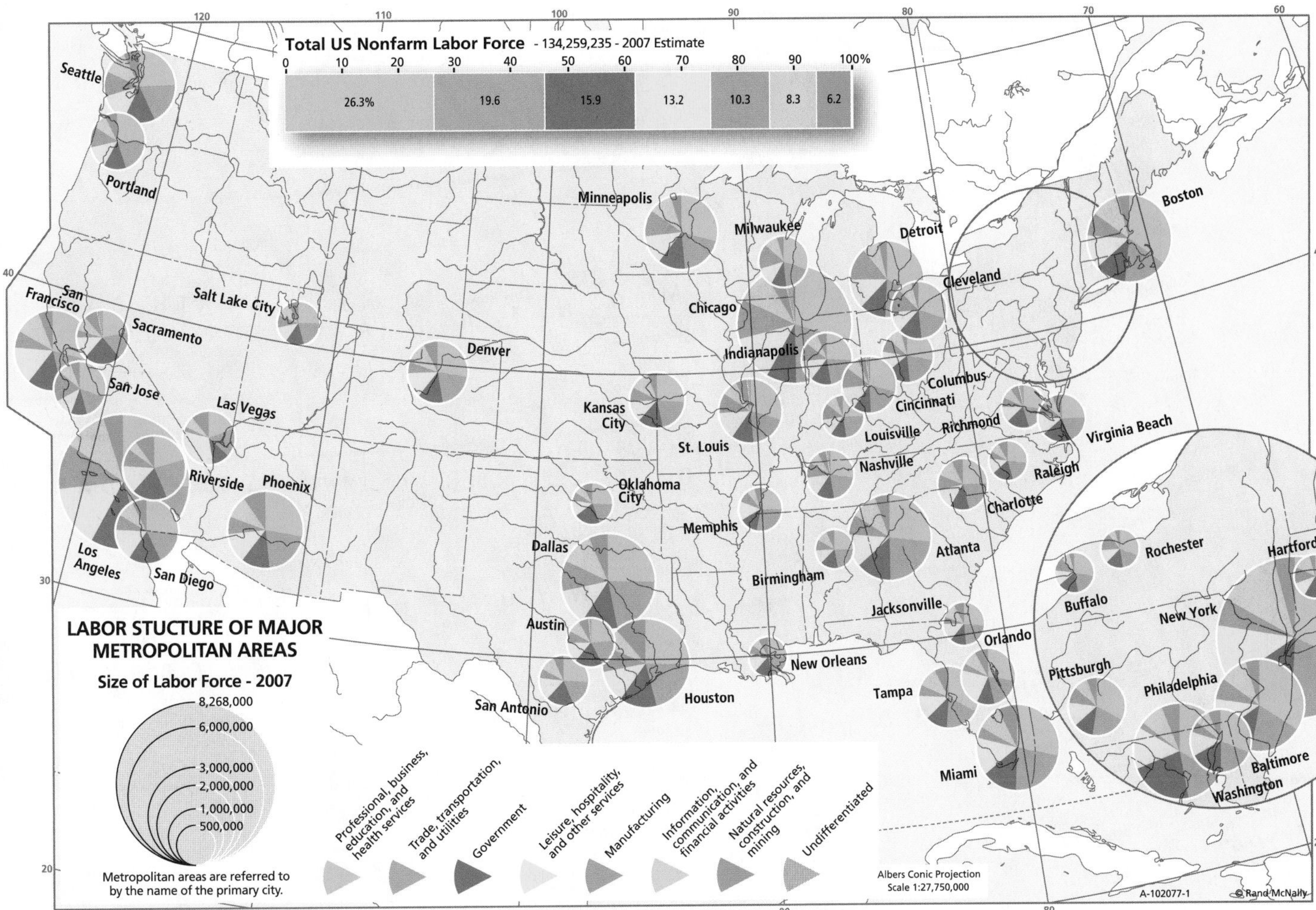

Total US Nonfarm Labor Force - 134,259,235 - 2007 Estimate
0 10 20 30 40 50 60 70 80 90 100%
26.3% 19.6 15.9 13.2 10.3 8.3 6.2
LABOR STUCTURE OF MAJOR METROPOLITAN AREAS
Size of Labor Force - 2007
8,268,000
6,000,000
3,000,000
2,000,000
1,000,000
500,000
Metropolitan areas are referred to by the name of the primary city.
Professional, business, education, and health services
Trade, transportation, and utilities
Government
Leisure, hospitality, and other services
Manufacturing
Information, communication, and financial activities
Natural resources, construction, and mining
Undifferentiated
Seattle
Portland
San Francisco
Sacramento
San Jose
Salt Lake City
Las Vegas
Riverside
Phoenix
Los Angeles
San Diego
Denver
Minneapolis
Milwaukee
Chicago
Detroit
Cleveland
Boston
Indianapolis
Columbus
Cincinnati
Louisville
Richmond
Virginia Beach
Kansas City
St. Louis
Nashville
Raleigh
Oklahoma City
Memphis
Charlotte
Dallas
Atlanta
Birmingham
Austin
Jacksonville
New Orleans
Orlando
Houston
San Antonio
Tampa
Miami
Rochester
Hartford
Buffalo
New York
Providence
Pittsburgh
Philadelphia
Baltimore
Washington
Albers Conic Projection
Scale 1:27,750,000
A-102077-1
© Rand McNally

Source: Bureau of Labor Statistics

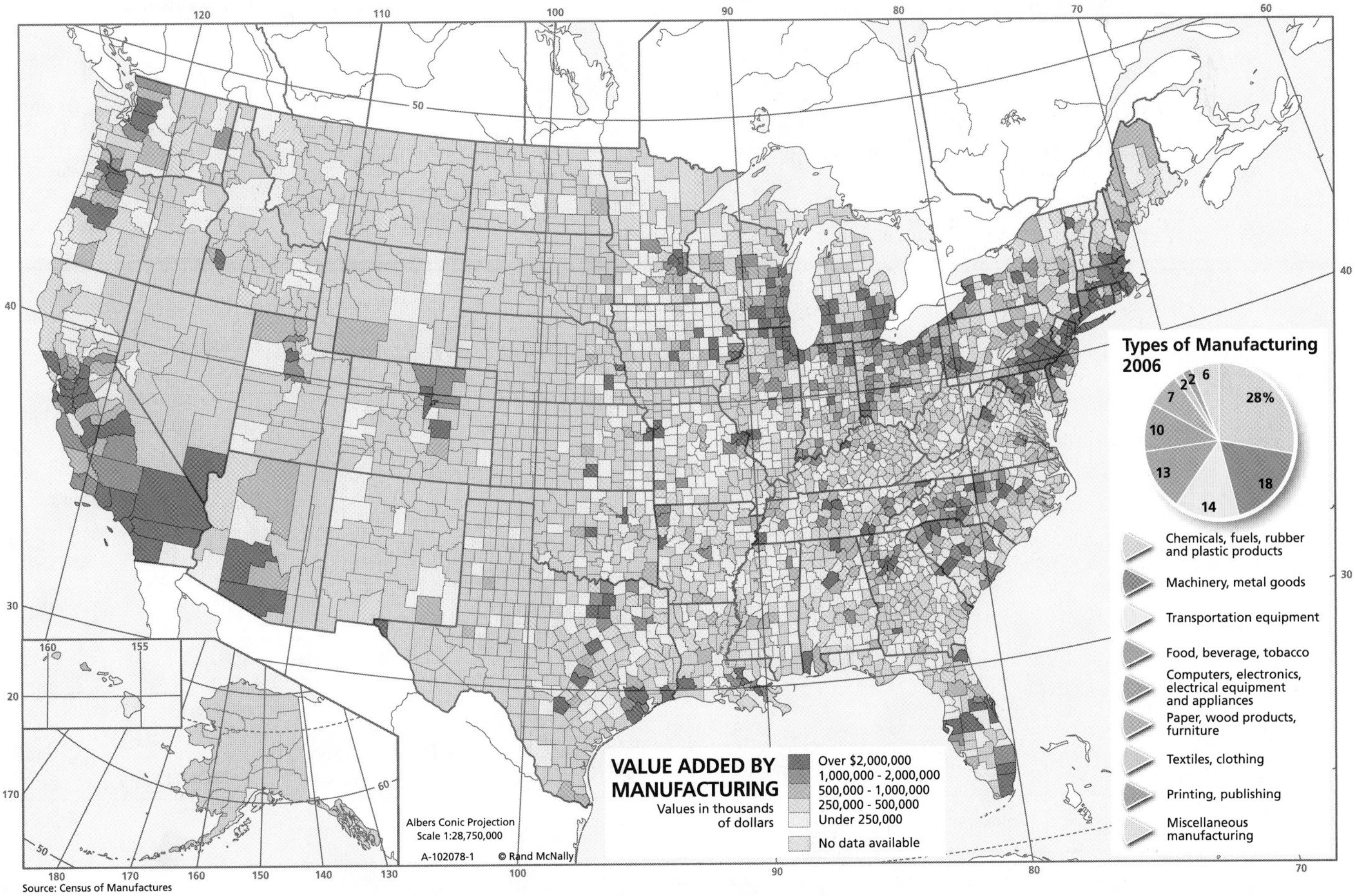

VALUE ADDED BY MANUFACTURING
Values in thousands of dollars
Over $2,000,000
1,000,000 - 2,000,000
500,000 - 1,000,000
250,000 - 500,000
Under 250,000
No data available
Types of Manufacturing 2006
28%
18
14
13
10
7
2
2
6
Chemicals, fuels, rubber and plastic products
Machinery, metal goods
Transportation equipment
Food, beverage, tobacco
Computers, electronics, electrical equipment and appliances
Paper, wood products, furniture
Textiles, clothing
Printing, publishing
Miscellaneous manufacturing
Albers Conic Projection
Scale 1:28,750,000
A-102078-1
© Rand McNally

Source: Census of Manufactures

BRITISH COLUMBIA
ALBERTA
SASKATCHEWAN
MANITOBA
CANADA
Calgary
Vancouver
Victoria
Lethbridge
Regina
Winnipeg
Lake Winnipeg
Lake Manitoba
Assiniboine
CAPE FLATTERY
Port Angeles
OLYMPIC NATL. PARK
Bellingham
Everett
NORTH CASCADES NATL. PARK
Bremerton
Seattle
Olympia
Tacoma
WASHINGTON
Wenatchee
Spokane
Mt. Rainier 14 411 ft. 4392 m
MT. RAINIER NATL. PARK
Yakima
Astoria
Longview
Vancouver
Portland
Kennewick
Walla Walla
Pullman
Coeur d'Alene
Lewiston
Columbia
CASCADE RANGE
ROCKY MOUNTAINS
GLACIER NATL. PARK
Shelby
Kalispell
Havre
Glasgow
Missouri
Missoula
Great Falls
Lewistown
Fort Peck Lake
MONTANA
Helena
Anaconda
Butte
Bozeman
Livingston
Billings
Glendive
Miles City
Yellowstone
BITTERROOT RANGE
Granite Peak 12 799 ft. 3901 m
YELLOWSTONE NATIONAL PARK
Williston
Minot
Devils Lake
Grafton
Grand Forks
Lake Sakakawea
Souris
NORTH DAKOTA
Dickinson
Jamestown
Fargo
Bismarck
Salem
The Dalles
Mt. Hood 11 239 ft. 3426 m
Pendleton
Corvallis
BLUE MTS.
Baker City
Eugene
John Day
Bend
OREGON
Coos Bay
CRATER LAKE NATIONAL PARK
Burns
Medford
Klamath Falls
Salmon
SALMON RIVER MOUNTAINS
IDAHO
Caldwell
Boise
Nampa
Mountain Home
Hyndman Peak 12 009 ft. 3660 m
Rexburg
Idaho Falls
Snake
Owyhee
Twin Falls
Pocatello
Sheridan
Cody
Cloud Peak 13 167 ft. 4013 m
GRAND TETON NATL. PARK
Jackson
Gannett Peak 13 804 ft. 4207 m
Gillette
Powder
Spearfish
Mobridge
Aberdeen
Lake Oahe
Watertown
SOUTH DAKOTA
Pierre
Huron
Brookings
BLACK HILLS
Harney Peak 7242 ft. 2207 m
Rapid City
BADLANDS NATIONAL PARK
Hot Springs
Mitchell
Riverton
Casper
WYOMING
Rawlins
Green
Rock Springs
Laramie
Cheyenne
Valentine
Yankton
Niobrara
Norfolk
Scottsbluff
North Platte
NEBRASKA
Columbus
Grand Island
Kearney
Hastings
Platte
McCook
Republican
REDWOOD NATIONAL PARK
Eureka
CAPE MENDOCINO
Klamath
Mt. Shasta 14 162 ft. 4317 m
Redding
LASSEN VOLCANIC NATIONAL PARK
Eel
Fort Bragg
Chico
Ukiah
Yuba City
Sacramento
SIERRA NEVADA
Santa Rosa
Oakland
San Francisco
Stockton
San Jose
Santa Cruz
Modesto
YOSEMITE NATL. PARK
Monterey
Salinas
Fresno
Bishop
KINGS CANYON NATL. PARK
Mt. Whitney 14 494 ft. 4418 m
SEQUOIA NATL. PARK
Visalia
Porterville
COAST RANGES
CALIFORNIA
San Luis Obispo
Bakersfield
Lompoc
Lancaster
Santa Barbara
Oxnard
CHANNEL ISLANDS NATIONAL PARK
LOS ANGELES
Long Beach
Anaheim
Riverside
Palm Springs
JOSHUA TREE NATIONAL PARK
Salton Sea
Oceanside
SAN DIEGO
El Centro
TIJUANA
Mexicali
Ensenada
Winnemucca
Humboldt
Elko
NEVADA
Reno
Lake Tahoe
Carson City
Hawthorne
Ely
GREAT BASIN NATL. PARK
Tonopah
DEATH VALLEY NATIONAL PARK
Las Vegas
Lake Mead
Henderson
Barstow
Great Salt Lake
Logan
Ogden
Salt Lake City
Tooele
Provo
UTAH
Price
ARCHES NATL. PARK
CANYONLANDS NATL. PARK
Moab
CAPITOL REEF NATL. PARK
Cedar City
BRYCE CANYON NATL. PARK
ZION NATL. PARK
St. George
Lake Powell
Virgin
Steamboat Springs
ROCKY MOUNTAIN NATL. PARK
Fort Collins
Greeley
Sterling
Longs Peak 14 255 ft. 4345 m
Boulder
South Platte
Grand Junction
Mt. Elbert 14 433 ft. 4399 m
Denver
Pikes Peak 14 110 ft. 4301 m
Colorado Springs
COLORADO
Montrose
Canon City
Pueblo
Arkansas
Blanca Peak 14 345 ft. 4372 m
MESA VERDE NATL. PARK
Cortez
Durango
Trinidad
Colby
Saline
Salina
Hays
Garden City
KANSAS
Hutchinson
Dodge City
Wichita
Liberal
GRAND CANYON NATIONAL PARK
Kingman
Humphreys Peak 12 633 ft. 3851 m
Lake Havasu City
Flagstaff
Winslow
PETRIFIED FOREST NATIONAL PARK
Prescott
ARIZONA
Colorado
Mesa
PHOENIX
Baldy Peak 11 403 ft. 3476 m
Globe
Gila
Yuma
Casa Grande
SAGUARO NATIONAL PARK
Tucson
Nogales
Nogales
San Juan
Farmington
Los Alamos
Gallup
Santa Fe
Las Vegas
Tucumcari
Albuquerque
NEW MEXICO
Clovis
Socorro
Roswell
Silver City
Alamogordo
Hobbs
Pecos
Deming
Las Cruces
Carlsbad
El Paso
CIUDAD JUÁREZ
GUADALUPE MTS. NATIONAL PARK
Woodward
Enid
Canadian
Pampa
Oklahoma City
Amarillo
OKLAHOMA
Norman
Altus
Lawton
Plainview
Wichita Falls
Lubbock
Brazos
Abilene
Fort Worth
Big Spring
Midland
Odessa
San Angelo
Colorado
Killeen
Fort Stockton
TEXAS
Austin
Kerrville
BIG BEND NATIONAL PARK
Del Rio
SAN ANTONIO
Eagle Pass
Nueces
Laredo
Kingsville
Nuevo Laredo
Rio Grande
McAllen
Reynosa
Brownsville
NUEVO LEÓN
MONTERREY
TAMPICO
COAHUILA
Monclova
Torreón
DURANGO
CHIHUAHUA
Chihuahua
Conchos
SIERRA MADRE OCCIDENTAL
MEXICO
SONORA
Sonora
Hermosillo
Yaqui
Ciudad Obregón
Fuerte
Los Mochis
SINALOA
Gulf of California
BAJA CALIFORNIA
BAJA CALIF. SUR
GUADALUPE
PACIFIC OCEAN
125°
120°
115°
110°
105°
100°
50°
45°
40°
35°
30°
25°
© Rand McNally
A-101613-1

0 60 120 180 240 300 360 Miles

0 60 120 180 240 300 360 420 480 540 600 Kilometers

Lambert Azimuthal Equal Area Projection
Scale 1:12,000,000
One inch to 190 miles
One cm to 120 km

CANADA
ONTARIO
QUÉBEC
Lake Seul
Lake Nipigon
Lake of the Woods
Rainy Lake
Winnipeg
Réservoir Gouin
Saguenay
Québec
Trois-Rivières
Saint Maurice
St. Lawrence
MONTRÉAL
Sherbrooke
Presque Isle
NEW BRUNSWICK
Fredericton
Saint John
Calais
Bay of Fundy
N.S.
P.E.I.
Thunder Bay
ISLE ROYALE NATIONAL PARK
International Falls
VOYAGEURS NATL. PARK
LAKE SUPERIOR
Missinaibi
Abitibi
Mattagami
Harricana
Sudbury
Lake Nipissing
Ottawa
Gatineau
Sault Ste. Marie
Georgian Bay
Ottawa
Ogdensburg
Bemidji
Hibbing
Houghton
Marquette
Duluth
Superior
Ashland
Fergus Falls
Brainerd
MINNESOTA
St. Cloud
Willmar
Minneapolis
St. Paul
Minnesota
Mankato
Rochester
Austin
Sioux Falls
MICHIGAN
LAKE MICHIGAN
LAKE HURON
Petoskey
Alpena
Traverse City
Saginaw Bay
Rhinelander
Escanaba
Wausau
Eau Claire
WISCONSIN
Appleton
Green Bay
Winona
Oshkosh
Sheboygan
La Crosse
Mount Pleasant
Saginaw
Grand Rapids
Flint
Port Huron
Milwaukee
Lansing
Detroit
Windsor
Lake St. Clair
Madison
Racine
Kenosha
Janesville
Kalamazoo
Battle Creek
Ann Arbor
Wisconsin
Mason City
Des Moines
Waterloo
Dubuque
Sioux City
Fort Dodge
Cedar Rapids
Rockford
Elgin
CHICAGO
IOWA
Ames
Iowa City
Aurora
Joliet
Gary
South Bend
Toledo
Missouri
Davenport
Moline
Kankakee
Fort Wayne
Wabash
Omaha
Council Bluffs
Des Moines
Galesburg
Kokomo
Lafayette
Muncie
Burlington
Peoria
Champaign
INDIANA
Illinois
Bloomington
ILLINOIS
Lincoln
Kirksville
Quincy
Decatur
Indianapolis
Beatrice
St. Joseph
Springfield
Terre Haute
Bloomington
Kansas City
Manhattan
Columbia
St. Charles
Mississippi
Kaskaskia
Vincennes
Louisville
Topeka
Lawrence
Jefferson City
St. Louis
Belleville
Evansville
Emporia
MISSOURI
Rolla
Carbondale
Owensboro
Pittsburg
Joplin
Cape Girardeau
Springfield
Sikeston
Paducah
Winfield
Ponca City
Tulsa
Stillwater
Muskogee
OKLAHOMA
McAlester
Ardmore
TORONTO
LAKE ONTARIO
Watertown
ADIRONDACK MTS.
Plattsburgh
Burlington
Montpelier
VT.
Mt. Washington 6288 ft. 1917 m
N.H.
Concord
MAINE
Bangor
Augusta
Lewiston
Portland
Portsmouth
Gulf of Maine
Connecticut
Manchester
Lowell
Boston
MASS.
Worcester
Providence
R.I.
New Bedford
Hamilton
Niagara Falls
Rochester
Syracuse
Utica
Albany
Buffalo
NEW YORK
Binghamton
Springfield
Hartford
CONN.
New Haven
London
Elmira
Bridgeport
NEW YORK
LAKE ERIE
Erie
Jamestown
Williamsport
Scranton
Newark
N.J.
Trenton
Cleveland
Youngstown
PENN.
Allentown
Reading
PHILADELPHIA
Akron
Pittsburgh
Harrisburg
Lancaster
Atlantic City
Mansfield
Canton
Lima
OHIO
Hagerstown
Dover
MD.
DEL.
Delaware Bay
Columbus
Wheeling
Baltimore
Annapolis
Salisbury
Dayton
Springfield
Morgantown
WEST VIRGINIA
Washington
Parkersburg
SHENANDOAH NATL. PARK
Harrisonburg
Fredericksburg
Charlottesville
Cincinnati
Ohio
Charleston
Ashland
Huntington
Richmond
Chesapeake Bay
James
Frankfort
Lexington
Richmond
Beckley
Lynchburg
Petersburg
Virginia Beach
Norfolk
Newport News
Roanoke
Blacksburg
VIRGINIA
Elizabeth City
KENTUCKY
Somerset
Danville
MAMMOTH CAVE NATL. PARK
Bowling Green
Cumberland
Johnson City
Winston-Salem
Greensboro
Rocky Mount
Greenville
CAPE HATTERAS
Hopkinsville
Clarksville
Kingsport
Durham
Raleigh
Mt. Mitchell 6684 ft. 2037 m
Oak Ridge
Knoxville
Asheville
NORTH CAROLINA
Fayetteville
Jacksonville
CAPE LOOKOUT
Nashville
TENNESSEE
Murfreesboro
GREAT SMOKY MTS. NATL. PARK
APPALACHIAN MTS.
Charlotte
Cape Fear
Pee Dee
Wilmington
CAPE FEAR
Springdale
Fayetteville
Jonesboro
White
Jackson
Columbia
Spartanburg
Rock Hill
Florence
Memphis
Chattanooga
Greenville
SOUTH CAROLINA
Sumter
Myrtle Beach
Fort Smith
Arkansas
Conway
ARKANSAS
Huntsville
Dalton
Gainesville
Rome
Columbia
Florence
Decatur
Athens
Aiken
Santee
Little Rock
Pine Bluff
Tupelo
Gadsden
Coosa
Atlanta
Augusta
Savannah
Charleston
Hot Springs
Clarksdale
Birmingham
Anniston
Oconee
Lagrange
Macon
Hope
Greenville
Columbus
Tuscaloosa
GEORGIA
Paris
Ouachita
MISSISSIPPI
ALABAMA
Auburn
Savannah
Sherman
Texarkana
El Dorado
Yazoo
Selma
Columbus
Altamaha
Denton
DALLAS
Shreveport
Monroe
Vicksburg
Meridian
Jackson
Montgomery
Ocmulgee
Brunswick
Longview
Red
Tombigbee
Flint
Albany
Corsicana
Tyler
Sabine
Natchitoches
Alabama
Chattahoochee
Waycross
Trinity
Toledo Bend Res.
Natchez
Pearl
Dothan
Valdosta
Jacksonville
Waco
Hattiesburg
St. Augustine
Alexandria
Tallahassee
Lufkin
Mobile
FLORIDA
Suwannee
LOUISIANA
Pensacola
Apalachicola
Gainesville
Daytona Beach
Huntsville
Neches
Baton Rouge
Gulfport
Biloxi
Panama City
Ocala
Bryan
TEXAS
Lafayette
CAPE SAN BLAS
Titusville
Beaumont
Lake Charles
New Orleans
CAPE CANAVERAL
Orlando
Melbourne
HOUSTON
Port Arthur
New Iberia
Houma
Lakeland
Tampa
Fort Pierce
Galveston
Lake Jackson
St. Petersburg
West Palm Beach
ABACO
GRAND BAHAMA
Victoria
Sarasota
Port Charlotte
Lake Okeechobee
Fort Lauderdale
Fort Myers
ELEUTHERA
Miami
Nassau
NEW PROVIDENCE
SAN SALVADOR
Corpus Christi
GULF OF MEXICO
EVERGLADES NATIONAL PARK
BISCAYNE NATL. PARK
ANDROS
CAT ISLAND
RUM CAY
BAHAMAS
GREAT EXUMA
LONG ISLAND
CROOKED ISLAND
Key West
Tropic of Cancer
ATLANTIC OCEAN
95°
90°
85°
80°
75°
70°
65°
45°
40°
35°
30°
25°

BRITISH COLUMBIA
Vancouver
VANCOUVER ISLAND
Strait of Juan de Fuca
CAPE FLATTERY
Mt. Baker 10 785 ft. 3287 m
Mt. Olympus 7969 ft. 2429 m
Seattle
WASHINGTON
CAPE DISAPPOINTMENT
Mt. St. Helens 8364 ft. 2549 m
Mt. Rainier 14 411 ft. 4392 m
Mt. Adams 12 276 ft. 3742 m
Portland
Columbia
Mt. Hood 11 239 ft. 3426 m
Eugene
OREGON
CASCADE RANGE
CAPE BLANCO
Upper Klamath Lake
Goose Lake
KLAMATH MTS.
Mt. Shasta 14 162 ft. 4317 m
CAPE MENDOCINO
Eagle Peak 9892 ft. 3015 m
Redding
Lassen Peak 10 457 ft. 3187 m
POINT ARENA
SACRAMENTO VALLEY
SIERRA NEVADA
Reno
Lake Tahoe
POINT REYES
San Francisco
COAST RANGES
SAN JOAQUIN VALLEY
DIABLO RA.
Monterey Bay
Fresno
Mt. Ritter 13 140 ft. 4005 m
Mt. Whitney 14 494 ft. 4418 m
CALIFORNIA
DEATH VALLEY
POINT CONCEPTION
Los Angeles
SANTA CRUZ ISLAND
SANTA CATALINA ISLAND
SAN CLEMENTE ISLAND
San Diego
MOJAVE DESERT
San Gorgonio Mountain 11 499 ft. 3505 m
Salton Sea
PACIFIC OCEAN
Spokane
COLUMBIA PLATEAU
BLUE MTS.
Rock Creek Butte 9106 ft. 2776 m
HARNEY BASIN
Drake Peak 8407 ft. 2562 m
BLACK ROCK DESERT
GREAT NEVADA BASIN
Boundary Peak 13 143 ft. 4006 m
Wheeler Peak 13 063 ft. 3982 m
Las Vegas
Lake Mead
ALBERTA
Red Deer
Bow
South Saskatchewan
SASKATCHEWAN
Regina
Qu'Appelle
Lake Winnipeg
Lake Manitoba
Winnipeg
MANITOBA
Assiniboine
Souris
CANADA
ROCKY MOUNTAINS
BITTERROOT RANGE
Flathead Lake
Lake Pend Oreille
Trapper Peak 10 157 ft. 3096 m
Mt. Haggin 10 607 ft. 3233 m
Great Falls
Missouri
Milk
Fort Peck Lake
MONTANA
Yellowstone
Billings
Koch Peak 11 286 ft. 3440 m
Granite Peak 12 799 ft. 3901 m
SALMON RIVER MTS.
Borah Peak 12 662 ft. 3859 m
Boise
IDAHO
Hyndman Peak 12 009 ft. 3660 m
SNAKE RIVER PLAIN
Grand Teton 13 770 ft. 4197 m
Yellowstone Lake
Francs Peak 13 153 ft. 4009 m
BIGHORN MTS.
Cloud Peak 13 167 ft. 4013 m
Gannett Peak 13 804 ft. 4207 m
WIND RIVER RA.
SHOSHONE BASIN
WYOMING
GREAT DIVIDE BASIN
Medicine Bow Peak 12 013 ft. 3662 m
LARAMIE MTS.
Matterhorn 10 839 ft. 3304 m
BEAR RIVER RA.
Great Salt Lake
GREAT SALT LAKE DESERT
Salt Lake City
Utah Lake
UINTA MTS.
Kings Peak 13 528 ft. 4123 m
UTAH
ROAN PLATEAU
Delano Peak 12 173 ft. 3710 m
ESCALANTE DESERT
Lake Powell
Colorado
COLORADO PLATEAU
GRAND CANYON
PAINTED DESERT
Humphreys Peak 12 633 ft. 3851 m
Flagstaff
ARIZONA
Baldy Peak 11 403 ft. 3476 m
Phoenix
Gila
Chiricahua Peak 9798 ft. 2986 m
Longs Peak 14 255 ft. 4345 m
Grays Peak 14 270 ft. 4349 m
Denver
South Platte
COLORADO
Mt. Elbert 14 433 ft. 4399 m
Pikes Peak 14 110 ft. 4301 m
Uncompahgre Peak 14 309 ft. 4361 m
SAN JUAN MTS.
Blanca Peak 14 345 ft. 4372 m
SANGRE DE CRISTO MTS.
SAN JUAN BASIN
Wheeler Peak 13 161 ft. 4011 m
Santa Fe Baldy 12 622 ft. 3847 m
NEW MEXICO
Albuquerque
Rio Grande
TULAROSA VALLEY
SACRAMENTO MTS.
El Paso
Ciudad Juárez
Guadalupe Peak 8749 ft. 2667 m
LLANO ESTACADO
Lubbock
NORTH DAKOTA
Lake Sakakawea
SOUTH DAKOTA
Lake Oahe
BLACK HILLS
Harney Peak 7242 ft. 2207 m
Rapid City
Belle Fourche
Cheyenne
NEBRASKA
North Platte
Niobrara
Platte
Republican
KANSAS
Saline
Arkansas
Wichita
GREAT PLAINS
North Canadian
Canadian
Oklahoma City
OKLA
Red
TEXAS
Pecos
STOCKTON PLATEAU
EDWARDS PLATEAU
San Antonio
Nueces
Brownsville
SONORAN DESERT
Gulf of California
ISLA ÁNGEL DE LA GUARDA
ISLA TIBURÓN
Hermosillo
Sonora
SIERRA MADRE OCCIDENTAL
CHIHUAHUA
CHIHUAHUAN DESERT
MEXICO
COAHUILA
NUEVO LEÓN
Monterrey
DURANGO
SINALOA
BAJA CALIFORNIA
BAJA CALIF. SUR
GUADALUPE
ISLA CEDROS
Bahía Sebastián Vizcaíno
Laguna Ojo de Liebre
ISLA CARMEN
A-101614-1
© Rand McNally

0 60 120 180 240 300 360 Miles
0 60 120 180 240 300 360 420 480 540 600 Kilometers

Lambert Azimuthal Equal Area Projection
Scale 1:12,000,000
One inch to 190 miles
One cm to 120 km

ONTARIO
QUÉBEC
LES LAURENTIDES
Québec
Lake St. Joseph
Lake Nipigon
Lac Seul
Winnipeg
Lake of the Woods
Rainy
Upper Red Lake
Lake Abitibi
Abitibi
Mattagami
Missinaibi
Harricana
Réservoir Gouin
Réservoir Cabonga
Lac Saint-Jean
Saguenay
Saint-Maurice
Gatineau
Richelieu
Ottawa
Montréal
Lake Nipissing
Georgian Bay
Lake Simcoe
ISLE ROYALE
MICHIPICOTEN ISLAND
KEWEENAW POINT
KEWEENAW PEN.
LAKE SUPERIOR
Sault Ste. Marie
MANITOULIN ISLAND
BEAVER ISLAND
LAKE HURON
Saginaw Bay
LAKE MICHIGAN
Green Bay
DOOR PEN.
Lake St. Clair
LAKE ONTARIO
LAKE ERIE
NEW BRUNSWICK
Grand Lake
Saint John
Bay of Fundy
N.S.
P.E.I.
Mt. Katahdin 5268 ft. 1606 m
MAINE
Penobscot
Mt. Washington 6288 ft. 1917 m
Portland
Gulf of Maine
VT.
N.H.
Lake Champlain
ADIRONDACK MTS.
MASS.
Boston
CAPE COD
NANTUCKET ISLAND
MARTHA'S VINEYARD
CONN.
R.I.
Hartford
CATSKILL MTS.
NEW YORK
Buffalo
LONG ISLAND
New York
NEW JERSEY
Philadelphia
CAPE MAY
Delaware Bay
DEL.
DELMARVA PENINSULA
MARYLAND
Washington
Chesapeake Bay
CAPE CHARLES
Potomac
Susquehanna
PENNSYLVANIA
Pittsburgh
Allegheny
ALLEGHENY PLATEAU
APPALACHIAN MOUNTAINS
MINNESOTA
Duluth
Leech Lake
Mille Lacs Lake
Mississippi
Minneapolis
Minnesota
St. Croix
WISCONSIN
Lake Winnebago
Wisconsin
Milwaukee
MICHIGAN
Muskegon
Grand
Detroit
Sioux Falls
IOWA
Des Moines
Omaha
Chicago
Rock
Fox
Kankakee
Maumee
Cleveland
OHIO
Columbus
Scioto
Wabash
INDIANA
Indianapolis
Cincinnati
ILLINOIS
Illinois
Kaskaskia
WEST VIRGINIA
Spruce Knob 4863 ft. 1482 m
Keeney Knob 3927 ft. 1197 m
Kanawha
James
VIRGINIA
Norfolk
Roanoke
HATTERAS ISLAND
CAPE HATTERAS
Raleigh
NORTH CAROLINA
Neuse
CAPE LOOKOUT
Charlotte
Cape Fear
CAPE FEAR
Mt. Rogers 5729 ft. 1746 m
Mt. Mitchell 6684 ft. 2037 m
BLUE RIDGE
Yadkin
Kansas
Topeka
Kansas City
Missouri
St. Louis
MISSOURI
Lake of the Ozarks
OZARK PLATEAU
White
BOSTON MTS.
Arkansas
KENTUCKY
Kentucky
Ohio
CUMBERLAND PLATEAU
Cumberland
Nashville
Holston
TENNESSEE
Clingmans Dome 6643 ft. 2025 m
Memphis
Tennessee
ARKANSAS
OUACHITA MTS.
Little Rock
Ouachita
MISSISSIPPI
Yazoo
Pearl
ALABAMA
Tombigbee
Birmingham
Coosa
Alabama
Mobile
Atlanta
Flint
Chattahoochee
Apalachicola
GEORGIA
Oconee
Ocmulgee
Altamaha
J. Strom Thurmond Res.
Savannah
SOUTH CAROLINA
Great Pee Dee
Lake Marion
Santee
CAPE ROMAIN
Charleston
HILTON HEAD ISLAND
St. Marys
Jacksonville
Suwannee
St. Johns
FLORIDA
CAPE CANAVERAL
CAPE SAN BLAS
Tampa
Tampa Bay
Lake Okeechobee
Miami
CAPE ROMANO
CAPE SABLE
KEY LARGO
KEY WEST
FLORIDA KEYS
Straits of Florida
Dallas
Sabine
Trinity
Red
Toledo Bend Reservoir
Neches
Brazos
LOUISIANA
Lake Pontchartrain
New Orleans
MISSISSIPPI DELTA
Houston
PLAINS
GALVESTON ISLAND
GULF OF MEXICO
ATLANTIC OCEAN
ABACO
GRAND BAHAMA
ELEUTHERA
NEW PROVIDENCE
CAT ISLAND
SAN SALVADOR
ANDROS
BAHAMAS
GREAT EXUMA
MANGROVE CAY
LONG ISLAND
CROOKED ISLAND
Tropic of Cancer
95°
90°
85°
80°
75°
70°
65°
50°
45°
40°
35°
30°
25°

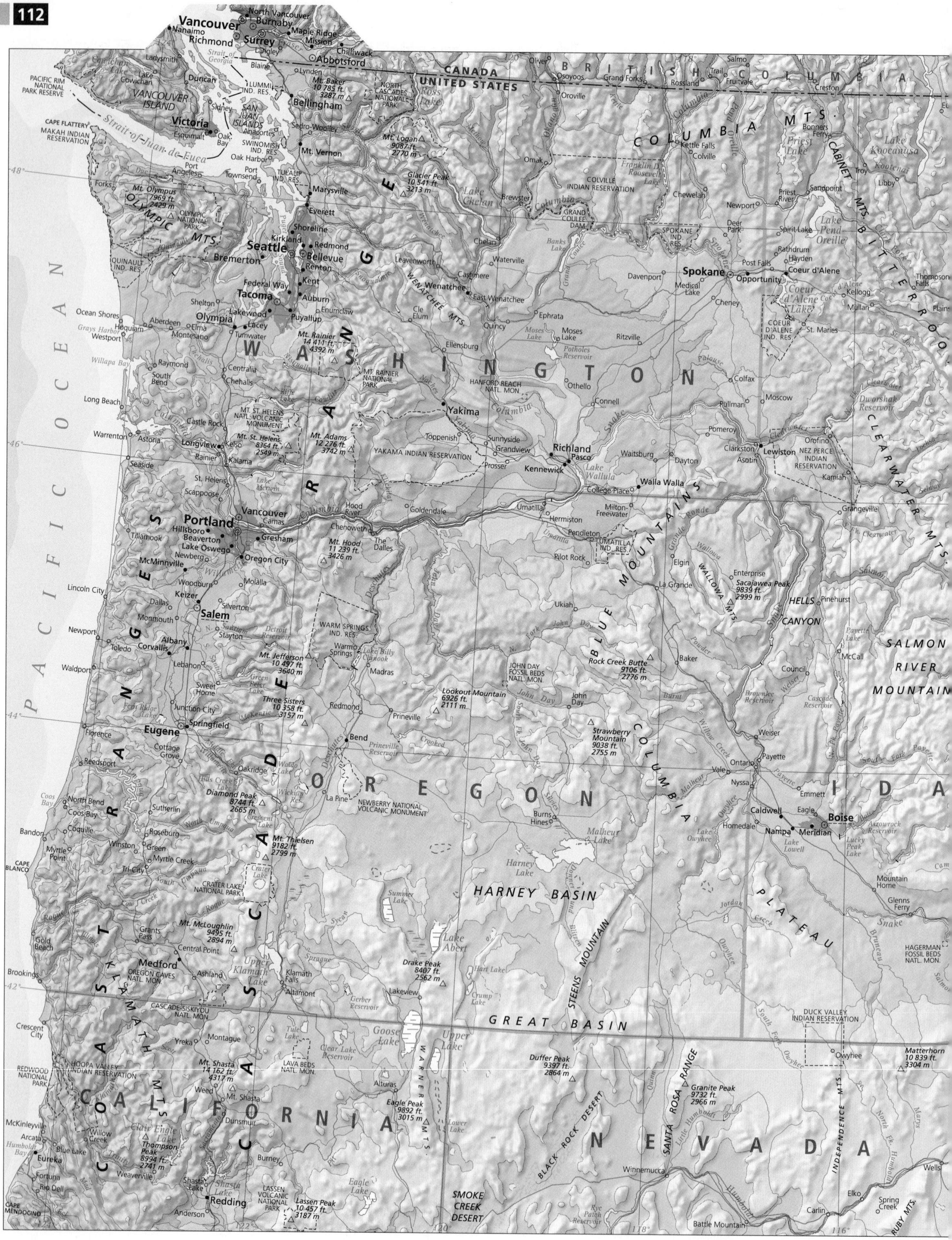

CANADA
UNITED STATES
BRITISH COLUMBIA
Vancouver
North Vancouver
Burnaby
Nanaimo
Richmond
Surrey
Maple Ridge
Mission
Chilliwack
Langley
Abbotsford
Strait of Georgia
Blaine
Lynden
Ladysmith
Lake Cowichan
Duncan
PACIFIC RIM NATIONAL PARK RESERVE
VANCOUVER ISLAND
LUMMI IND. RES.
Mt. Baker 10 785 ft. 3287 m
NORTH CASCADES NATIONAL PARK
Ross Lake
Bellingham
Sidney
SAN JUAN ISLANDS
Victoria
Esquimalt
Oak Bay
Anacortes
CAPE FLATTERY
MAKAH INDIAN RESERVATION
Strait of Juan de Fuca
Sedro-Woolley
SWINOMISH IND. RES.
Mt. Logan 9087 ft. 2770 m
Oak Harbor
Mt. Vernon
Port Angeles
Port Townsend
TULALIP IND. RES.
Glacier Peak 10 541 ft. 3213 m
Forks
Mt. Olympus 7969 ft. 2429 m
OLYMPIC NATIONAL PARK
OLYMPIC MTS.
Marysville
Everett
Shoreline
Kirkland
Redmond
Seattle
Bellevue
Bremerton
Renton
Kent
Federal Way
Tacoma
Auburn
QUINAULT IND. RES.
Shelton
Lakewood
Puyallup
Enumclaw
Olympia
Lacey
Tumwater
Ocean Shores
Hoquiam
Aberdeen
Elma
Montesano
Westport
Grays Harbor
Mt. Rainier 14 411 ft. 4392 m
Willapa Bay
Raymond
South Bend
Centralia
Chehalis
MT. RAINIER NATIONAL PARK
Long Beach
MT. ST. HELENS NATL. VOLCANIC MONUMENT
Castle Rock
Mt. St. Helens 8364 ft. 2549 m
Mt. Adams 12 276 ft. 3742 m
Warrenton
Astoria
Longview
Kelso
Rainier
Kalama
Seaside
St. Helens
Scappoose
Lake Merwin
Oliver
Osoyoos
Grand Forks
Rossland
Trail
Salmo
Fruitvale
Creston
Oroville
COLUMBIA MTS.
Kettle Falls
Colville
Bonners Ferry
Priest Lake
Lake Koocanusa
CABINET MTS.
Troy
Libby
Omak
Franklin D. Roosevelt Lake
COLVILLE INDIAN RESERVATION
Chewelah
Priest River
Sandpoint
Lake Chelan
Brewster
GRAND COULEE DAM
Newport
Deer Park
Spirit Lake
Lake Pend Oreille
SPOKANE IND. RES.
Banks Lake
Chelan
Rathdrum
Hayden
Waterville
Leavenworth
Post Falls
Coeur d'Alene
Cashmere
Davenport
Spokane
Opportunity
Wenatchee
East Wenatchee
WENATCHEE MTS.
Medical Lake
Cheney
Kellogg
Thompson Falls
Coeur d'Alene Lake
Mullan
Cle Elum
Ephrata
Quincy
Moses Lake
St. Maries
COEUR D'ALENE IND. RES.
Ritzville
Ellensburg
Potholes Reservoir
WASHINGTON
HANFORD REACH NATL. MON.
Othello
Colfax
Connell
Moscow
Yakima
Pullman
Dworshak Reservoir
Toppenish
Sunnyside
Grandview
Pomeroy
YAKAMA INDIAN RESERVATION
Prosser
Richland
Pasco
Kennewick
Waitsburg
Dayton
Clarkston
Lewiston
Asotin
Orofino
NEZ PERCE INDIAN RESERVATION
Kamiah
Lake Wallula
Walla Walla
College Place
CLEARWATER MTS.
BITTERROOT
Hood River
Vancouver
Camas
Goldendale
Umatilla
Milton-Freewater
Hermiston
Grangeville
Portland
Gresham
Hillsboro
Beaverton
Lake Oswego
Tillamook
Newberg
Oregon City
Chenoweth
The Dalles
Mt. Hood 11 239 ft. 3426 m
Pendleton
UMATILLA IND. RES.
Pilot Rock
McMinnville
Woodburn
Molalla
Elgin
Enterprise
Sacajawea Peak 9839 ft. 2999 m
La Grande
WALLOWA MTS.
BLUE MOUNTAINS
Lincoln City
Keizer
Silverton
Dallas
Salem
Monmouth
WARM SPRINGS IND. RES.
Ukiah
HELLS CANYON
Pinehurst
Newport
Stayton
Detroit Reservoir
Toledo
Albany
Corvallis
Warm Springs
Lake Billy Chinook
Mt. Jefferson 10 497 ft. 3640 m
Rock Creek Butte 9106 ft. 2776 m
Baker
McCall
SALMON RIVER MOUNTAINS
Waldport
Lebanon
Madras
JOHN DAY FOSSIL BEDS NATL. MON.
Council
Sweet Home
Green Peter Lake
Three Sisters 10 358 ft. 3157 m
Lookout Mountain 6926 ft. 2111 m
John Day
Brownlee Reservoir
Cascade Reservoir
Junction City
Redmond
Prineville
Florence
Eugene
Springfield
Bend
Strawberry Mountain 9038 ft. 2755 m
Weiser
Cottage Grove
Prineville Reservoir
Payette
Reedsport
Oakridge
Ontario
Vale
OREGON
Nyssa
Emmett
IDAHO
Coos Bay
North Bend
Diamond Peak 8744 ft. 2665 m
La Pine
NEWBERRY NATIONAL VOLCANIC MONUMENT
Caldwell
Eagle
Boise
Sutherlin
Burns
Hines
Homedale
Nampa
Meridian
Bandon
Coquille
Roseburg
Mt. Thielsen 9182 ft. 2799 m
Malheur Lake
Lake Owyhee
Arrowrock Reservoir
Lucky Peak Lake
Lake Lowell
CAPE BLANCO
Myrtle Point
Winston
Green
Myrtle Creek
Tri-City
Harney Lake
Crater Lake
CRATER LAKE NATIONAL PARK
Mountain Home
HARNEY BASIN
PLATEAU
Summer Lake
Glenns Ferry
Gold Beach
Grants Pass
Mt. McLoughlin 9495 ft. 2894 m
Lake Abert
HAGERMAN FOSSIL BEDS NATL. MON.
Central Point
Medford
Upper Klamath Lake
Drake Peak 8407 ft. 2562 m
STEENS MOUNTAIN
Brookings
OREGON CAVES NATL. MON.
Ashland
Klamath Falls
Hart Lake
Altamont
Lakeview
Crump Lake
Gerber Reservoir
CASCADE-SISKIYOU NATL. MON.
GREAT BASIN
DUCK VALLEY INDIAN RESERVATION
Crescent City
Tule Lake
Goose Lake
Upper Lake
Yreka
Montague
Clear Lake Reservoir
Owyhee
Matterhorn 10 839 ft. 3304 m
REDWOOD NATIONAL PARK
HOOPA VALLEY INDIAN RESERVATION
Mt. Shasta 14 162 ft. 4317 m
LAVA BEDS NATL. MON.
Duffer Peak 9397 ft. 2864 m
SANTA ROSA RANGE
Granite Peak 9732 ft. 2966 m
Weed
Mt. Shasta
Alturas
WARNER MTS.
BLACK ROCK DESERT
INDEPENDENCE MTS.
CALIFORNIA
Eagle Peak 9892 ft. 3015 m
McKinleyville
Dunsmuir
Clair Engle Lake
Arcata
Willow Creek
Thompson Peak 8994 ft. 2741 m
Humboldt Bay
Blue Lake
Eureka
Lower Lake
NEVADA
Burney
Fortuna
Weaverville
Winnemucca
Wells
Rio Dell
Shasta Lake
Eagle Lake
SMOKE CREEK DESERT
Elko
Spring Creek
CAPE MENDOCINO
Redding
LASSEN VOLCANIC NATIONAL PARK
Lassen Peak 10 457 ft. 3187 m
Rye Patch Reservoir
Carlin
Battle Mountain
RUBY MTS.
Anderson
KLAMATH MTS.
COAST RANGES
CASCADE RANGE
PACIFIC OCEAN
48°
46°
44°
42°
122°
120°
118°
116°

0 20 40 60 80 100 120 Miles

0 20 40 60 80 100 120 140 160 180 200 Kilometers

Lambert Conformal Conic Projection
Scale 1:4,000,000
One inch to 64 miles
One cm to 40 km

Wood Mountain 3350 ft. 1021 m
Rockglen
GRASSLANDS NATL. PARK
SASKATCHEWAN
ALBERTA
CANADA
UNITED STATES
WATERTON LAKES NATL. PARK
Cardston
BLOOD INDIAN RESERVE
Milk River
Milk
Mt. Cleveland 10 466 ft. 3190 m
GLACIER NATIONAL PARK
LEWIS RA.
Sunburst
BLACKFEET INDIAN RESERVATION
Cut Bank
Browning
Shelby
Chester
Fresno Reservoir
Battle Creek
Chinook
Harlem
Havre
ROCKY BOY'S IND. RES.
FORT BELKNAP INDIAN RESERVATION
Nelson Reservoir
Frenchman
Malta
Lake Bowdoin
Scobey
Plentywood
Medicine Lake
FORT PECK INDIAN RESERVATION
Culbertson
Poplar
Porcupine Cr.
Glasgow
Wolf Point
Missouri
Sidney
Lake Elwell
Marias
Big Sandy Cr.
Columbia Falls
Hungry Horse Reservoir
SWAN RA.
Conrad
Teton
Fort Peck Lake
Beaver Cr.
Flathead Lake
LEWIS AND CLARK RA.
Choteau
Fort Benton
Missouri
Polson
FLATHEAD INDIAN RESERVATION
Ronan
Sun
Great Falls
Ulm
Jordan
Redwater
Musselshell
Box Elder Creek
Glendive
Cascade
Arrow Creek
Judith
Red Mountain 9411 ft. 2868 m
Lewistown
MONTANA
Terry
Yellowstone
O'Fallon Creek
Missouri
LITTLE BELT MTS.
Belt Creek
Smith
Flatwillow Creek
Missoula
East Missoula
Lolo
BIG BELT MTS.
Clark Fork
Miles City
Baker
East Helena
Canyon Ferry Lake
Helena
White Sulphur Springs
Harlowton
Roundup
Lolo Pass 5235 ft. 1596 m
Stevensville
Mt. Edith 9501 ft. 2896 m
Musselshell
Forsyth
Rosebud Creek
Pumpkin Cr.
Deer Lodge
Philipsburg
Hamilton
Townsend
Bitterroot
Rock Cr.
Mt. Haggin 10 607 ft. 3233 m
Anaconda
Boulder
CRAZY MTS.
Sweet Grass Cr.
Worden
Colstrip
Powder
Butte
Billings
Hardin
Lame Deer
ANACONDA RA.
Whitehall
Three Forks
Manhattan
Belgrade
Big Timber
Yellowstone
Lockwood
Crow Agency
NORTHERN CHEYENNE IND. RES.
Boxelder Creek
Trapper Peak 10 157 ft. 3096 m
Big Hole
Bozeman
Livingston
Laurel
Columbus
LITTLE BIGHORN BATTLEFIELD NATL. MON.
Jefferson
Tweedy Mountain 11 154 ft. 3400 m
PIONEER MTS.
Ennis Lake
Stillwater
Mt. Wood 12 320 ft. 3755 m
CROW INDIAN RESERVATION
Tongue
Homer Youngs Peak 10 621 ft. 3237 m
Beaverhead
Ruby
Madison
Big Sky
Gallatin
Red Lodge
Granite Peak 12 799 ft. 3901 m
Bighorn Lake
Little Bighorn
Little Missouri
Mt. McGuire 10 082 ft. 3073 m
Dillon
Salmon
Gardiner
Ranchester
Little Powder
Clark Canyon Reservoir
BEAVERHEAD MTS.
Koch Peak 11 286 ft. 3440 m
Hebgen Lake
Mt. Holmes 10 336 ft. 3150 m
Clarks Fork Yellowstone
Lovell
Sheridan
Clear Creek
DEVILS TOWER NATL. MON.
Keyhole Reservoir
LEMHI RANGE
Lemhi
Red Rock
Lima Reservoir
West Yellowstone
ABSAROKA RANGE
Powell
Trout Peak 12 244 ft. 3732 m
Story
Shoshone
BIGHORN MTS.
BIGHORN
Greybull
Cody
Buffalo Bill Reservoir
Buffalo
Gillette
Moorcroft
YELLOWSTONE NATIONAL PARK
Cloud Peak 13 167 ft. 4013 m
Challis
Basin
Nowood
LOST RIVER RANGE
Shoshone Lake
Yellowstone Lake
BASIN
Borah Peak 12 662 ft. 3859 m
Camas Creek
Francs Peak 13 153 ft. 4009 m
Gooseberry Creek
Worland
Castle Peak 11 815 ft. 3601 m
Ashton
St. Anthony
GRAND TETON NATIONAL PARK
Jackson Lake
Henrys Fork
Kaycee
Powder
Belle Fourche
IDAHO
Rexburg
Driggs
Grand Teton 13 770 ft. 4197 m
Owl Creek
Thermopolis
Antelope Creek
Sun Valley
Hyndman Peak 12 009 ft. 3660 m
Rigby
Snake
Gros Ventre
Dubois
WIND RIVER INDIAN RESERVATION
Dry Fork Cheyenne
Arco
Big Lost
Idaho Falls
Ketchum
Hailey
Ammon
Jackson
Boysen Reservoir
Badwater Cr.
Bellevue
Shelley
Palisades Reservoir
Downs Mountain 13 349 ft. 4069 m
WIND RIVER RANGE
Wind
SHOSHONE
South Fork Powder
Little Wood
CRATERS OF THE MOON NATL. MON.
Doubletop Peak 11 682 ft. 3561 m
Gannett Peak 13 804 ft. 4207 m
Fort Washakie
Riverton
Casper
North Platte
Douglas
Blackfoot
Mills
Evansville
Glenrock
SNAKE RIVER PLAIN
American Falls Reservoir
FORT HALL IND. RES.
Blackfoot Reservoir
Fremont Lake
BASIN
Wood
Shoshone
Aberdeen
Chubbuck
Pinedale
Lander
Glendo Reservoir
Gooding
American Falls
Pocatello
WYOMING
Wendell
Jerome
Lake Walcott
Portneuf
Afton
Sweetwater
Alcova Reservoir
McCammon
Soda Springs
Wind River Peak 13 192 ft. 4021 m
Twin Falls
Heyburn
Rupert
Meade Peak 9957 ft. 3035 m
Pathfinder Reservoir
Kimberly
Burley
BEAR RIVER RANGE
South Pass 7549 ft. 2301 m
Wheatland
Seminoe Reservoir
Goose Creek
Raft
Montpelier
Medicine Bow
Laramie
Wheatland Reservoir
Malad City
Bear
Hams Fork
Fontenelle Reservoir
Big Sandy
GREAT DIVIDE BASIN
Hanna
Preston
Bear Lake
FOSSIL BUTTE NATL. MON.
Kemmerer
Rawlins
Smithfield
Elk Mountain 11 156 ft. 3400 m
Tremonton
Logan
Diamondville
Green
Bitter Cr.
MEDICINE BOW MTS.
Wellsville
Rock Springs
Saratoga
Medicine Bow Peak 12 013 ft. 3662 m
Laramie
Brigham City
Green River
North Platte
Great Salt Lake
North Ogden
Fort Bridger
Lyman
Bridger Peak 11 004 ft. 3354 m
Evanston
Flaming Gorge Reservoir
Roy
Ogden
South Ogden
PARK RA.
COLORADO
Pilot Peak 10 716 ft. 3266 m
GREAT SALT LAKE DESERT
Clearfield
Layton
Kaysville
Blacks Fork
Cache la Poudre
Weber
UTAH
Bountiful
Kings Peak 13 528 ft. 4123 m
Little Snake
Hagues Peak 13 560 ft. 4133 m
Salt Lake City
UINTA MTS.
Craig
ROCKY MOUNTAIN NATIONAL PARK
West Wendover
Wendover
West Valley
Murray
UINTAH AND OURAY IND. RES.
DINOSAUR NATL. MON.
Yampa
Steamboat Springs
Longs Peak 14 255 ft. 4345 m
Tooele
West Jordan
Sandy
Heber City
Vernal
Rand McNally A-101621-1

CANADA
UNITED STATES
SASKATCHEWAN
MANITOBA
NORTH DAKOTA
SOUTH DAKOTA
NEBRASKA
MONTANA
WYOMING
COLORADO
MIN
GREAT PLAINS
BADLANDS
SAND HILLS
BLACK HILLS
Turtle Mountain 2516 ft. 767 m
Harney Peak 7242 ft. 2207 m
Estevan
Oxbow
Melita
Boissevain
Killarney
Morden
Winkler
Morris
Altona
Pembina
Walhalla
Scobey
Plentywood
Crosby
Bowbells
Mohall
Bottineau
Rolla
Belcourt
Kenmare
Lake Darling
Langdon
Cavalier
Hallock
Warroad
Roseau
FORT PECK INDIAN RESERVATION
TURTLE MOUNTAIN INDIAN RESERVATION
Medicine Lake
Wolf Point
Poplar
Culbertson
Tioga
Stanley
Williston
Burlington
Minot
Rugby
Towner
Cando
Leeds
Park River
Grafton
Stephen
Thief Lake
Agassiz Pool
Devils Lake
Lakota
Warren
Thief River Falls
RED LAKE IND. RES.
Sidney
Watford City
New Town
Lake Sakakawea
FORT BERTHOLD IND. RES.
THEODORE ROOSEVELT NATIONAL PARK (NORTH UNIT)
Velva
SPIRIT LAKE INDIAN RESERVATION
Grand Forks
East Grand Forks
Red Lake Falls
Larimore
Harvey
Fessenden
New Rockford
Thompson
Crookston
Garrison
Northwood
Underwood
Washburn
Carrington
Cooperstown
Mayville
Hillsboro
Fosston
Glendive
Beulah
Hazen
THEODORE ROOSEVELT NATIONAL PARK (SOUTH UNIT)
Wilton
Horsehead Lake
Lake Ashtabula
Jamestown Reservoir
Ada
Mahnomen
WHITE EARTH INDIAN RESERVATION
Terry
Beach
Belfield
Dickinson
Hebron
Glen Ullin
Mandan
Bismarck
Jamestown
Valley City
Casselton
West Fargo
Fargo
Moorhead
Hawley
Detroit Lakes
Park Rapids
Frazee
Long Lake
Baker
Marmarth
Bowman
Hettinger
Lemmon
Linton
Wishek
Edgeley
La Moure
Enderlin
Lisbon
Barnesville
Pelican Rapids
Perham
Wade
Fort Yates
Ashley
Ellendale
Oakes
Wahpeton
Breckenridge
Fergus Falls
Hankinson
Elbow Lake
Alexandria
McLaughlin
STANDING ROCK INDIAN RESERVATION
Eureka
Britton
LAKE TRAVERSE INDIAN RESERVATION
Sisseton
Wheaton
Lake Traverse
Morris
Glenwood
Mobridge
Selby
Ipswich
Aberdeen
Groton
Columbia Road Reservoir
Waubay Lake
Big Stone Lake
Lake Oahe
CHEYENNE RIVER INDIAN RESERVATION
Webster
Ortonville
Benson
Appleton
Faith
Gettysburg
Milbank
DEVILS TOWER NATL. MON.
Belle Fourche Reservoir
Newell
Belle Fourche
Redfield
Clark
Watertown
Madison
Dawson
Montevideo
Keyhole Reservoir
Sundance
Spearfish
Sturgis
Onida
Clear Lake
Granite Falls
Renville
Moorcroft
Lead
Deadwood
Highmore
Miller
Lake Poinsett
Canby
Minneota
Pierre
Fort Pierre
Lake Sharpe
CROW CREEK INDIAN RESERVATION
Huron
De Smet
Arlington
Brookings
Marshall
Box Elder
Rapid City
Newcastle
Wall
Philip
LOWER BRULE INDIAN RESERVATION
Wessington Springs
Lake Preston
Volga
Elkton
Tyler
Tracy
JEWEL CAVE NATL. MON.
Custer
BADLANDS NATIONAL PARK
Presho
Flandreau
PIPESTONE NATL. MON.
Pipestone
Slayton
WIND CAVE NATIONAL PARK
Murdo
Chamberlain
Kimball
Howard
Madison
Hot Springs
White River
Mitchell
Salem
Dell Rapids
Hartford
Luverne
Lakefield
PINE RIDGE INDIAN RESERVATION
Edgemont
Lake Francis Case
Alexandria
Sioux Falls
Brandon
Adrian
Worthing
Winner
Parkston
Parker
Rock Rapids
Sibley
Mission
Martin
Rosebud
ROSEBUD INDIAN RESERVATION
Wounded Knee
Pine Ridge
Platte
Armour
Gregory
Freeman
Lennox
Canton
Rock Valley
Sheldon
Lusk
Chadron
Rushville
Crawford
Glendo Reservoir
Valentine
Wagner
YANKTON IND. RES.
Tyndall
Centerville
Beresford
Sioux Center
Orange City
Yankton
Vermillion
Elk Point
Le Mars
Cherokee
Lewis and Clark Lake
Fort Laramie
AGATE FOSSIL BEDS NATL. MON.
Hemingford
Bassett
Ainsworth
Atkinson
O'Neill
SANTEE INDIAN RESERVATION
Bloomfield
Hartington
Ponca
South Sioux City
Sioux City
Dakota City
Torrington
Alliance
Plainview
Randolph
Laurel
WINNEBAGO IND. RES.
Wakefield
Wayne
Pender
OMAHA IND. RES.
Ida Grove
Morrill
Mitchell
Scottsbluff
SCOTTS BLUFF NATL. MON.
Gering
Bayard
Pierce
Neligh
Tilden
Norfolk
Battle Creek
Wisner
Onawa
Denison
Charter Oak
Bridgeport
Burwell
West Point
Lyons
Oakland
Dunlap
Madison
Tekamah
Woodbine
Oshkosh
Sargent
Albion
Ord
Greeley
Harlan
Pine Bluffs
Kimball
Lake McConaughy
Cheyenne
Sidney
Chappell
Ogallala
Sutherland
North Platte
Broken Bow
Ansley
Fullerton
Columbus
Schuyler
Fremont
Blair
Missouri Valley
Omaha
Council Bluffs
Crook
Julesburg
St. Paul
Central City
David City
Wahoo
La Vista
Gretna
Papillion
Bellevue
Ashland
Glenwood
Stromsburg
Ravenna
Grant
Gothenburg
Cozad
Lexington
Elm Creek
Kearney
Gibbon
Grand Island
Aurora
York
Waverly
Plattsmouth
Sterling
Haxtun
Holyoke
Greeley
Doniphan
Seward
Milford
Lincoln
Shenandoah
Missouri
Yellowstone
Little Missouri
Cheyenne
White
Niobrara
North Platte
South Platte
Platte
Loup
James
Sheyenne
Red
Big Sioux
A-101620-1 © Rand McNally

0 20 40 60 80 100 120 Miles

0 20 40 60 80 100 120 140 160 180 200 Kilometers

Lambert Conformal Conic Projection
Scale 1:4,000,000
One inch to 64 miles
One cm to 40 km

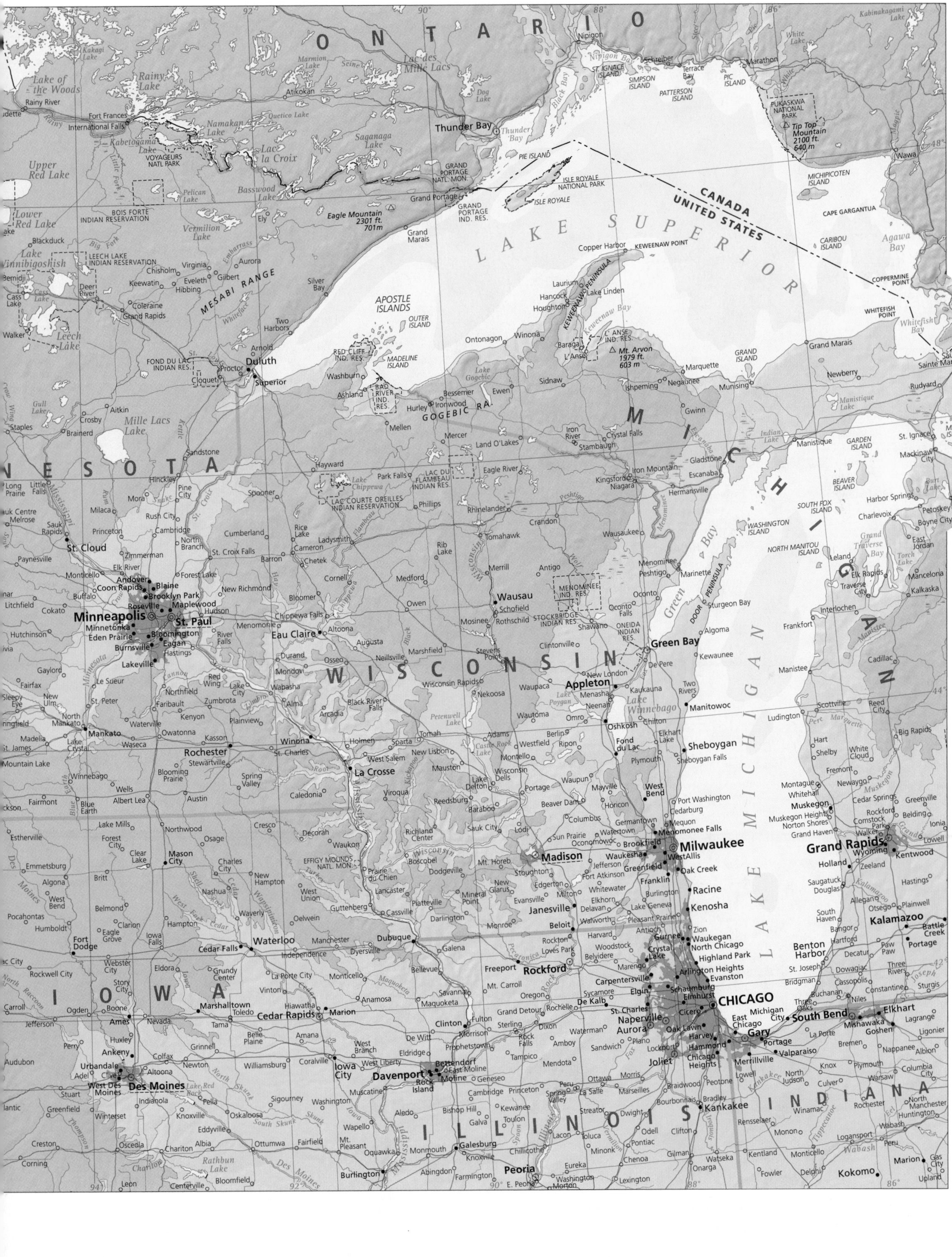
ONTARIO
LAKE SUPERIOR
CANADA
UNITED STATES
MINNESOTA
WISCONSIN
MICHIGAN
IOWA
ILLINOIS
INDIANA
LAKE MICHIGAN
LAKE HURON
Lake of the Woods
Rainy Lake
Rainy River
International Falls
Fort Frances
Kakagi Lake
Marmion Lake
Lac des Mille Lacs
Atikokan
Dog Lake
Nipigon
Nipigon Bay
Schreiber
Terrace Bay
Marathon
White Lake
Kabinakagami Lake
ST. IGNACE ISLAND
SIMPSON ISLAND
PATTERSON ISLAND
PIC ISLAND
PUKASKWA NATIONAL PARK
Tip Top Mountain 2100 ft. 640 m
Wawa
Namakan Lake
Quetico Lake
Kabetogama Lake
VOYAGEURS NATL PARK
Saganaga Lake
Lac la Croix
Thunder Bay
PIE ISLAND
GRAND PORTAGE NATL. MON.
Grand Portage
GRAND PORTAGE IND. RES.
ISLE ROYALE NATIONAL PARK
ISLE ROYALE
MICHIPICOTEN ISLAND
CAPE GARGANTUA
Agawa Bay
CARIBOU ISLAND
COPPERMINE POINT
WHITEFISH POINT
Whitefish Bay
Sault Sainte Marie
Upper Red Lake
Lower Red Lake
Pelican Lake
Basswood Lake
BOIS FORTE INDIAN RESERVATION
Ely
Eagle Mountain 2301 ft. 701m
Grand Marais
Vermilion Lake
Blackduck
Lake Winnibigoshish
LEECH LAKE INDIAN RESERVATION
Bemidji
Cass Lake
Deer River
Chisholm
Virginia
Aurora
Gilbert
Eveleth
Keewatin
Hibbing
MESABI RANGE
Silver Bay
Coleraine
Grand Rapids
Two Harbors
Walker
Leech Lake
FOND DU LAC INDIAN RES.
Cloquet
Proctor
Duluth
Superior
Arnold
APOSTLE ISLANDS
OUTER ISLAND
RED CLIFF IND. RES.
MADELINE ISLAND
Washburn
Ashland
BAD RIVER IND. RES.
Copper Harbor
KEWEENAW POINT
KEWEENAW PENINSULA
Keweenaw Bay
Laurium
Lake Linden
Hancock
Houghton
Ontonagon
Winona
Baraga
L'Anse
L'ANSE IND. RES.
Mt. Arvon 1979 ft. 603 m
Grand Marais
Marquette
Negaunee
Ishpeming
Munising
Newberry
Sainte Marie
Rudyard
Sidnaw
Lake Gogebic
Bessemer
Ironwood
Hurley
Ewen
GOGEBIC RA.
Gwinn
Manistique Lake
Aitkin
Gull Lake
Crosby
Staples
Brainerd
Mille Lacs Lake
Mellen
Mercer
Land O'Lakes
Iron River
Crystal Falls
Stambaugh
Indian Lake
Manistique
GARDEN ISLAND
St. Ignace
MACKINAC ISLAND
Mackinaw City
Cheboygan
Sandstone
Hayward
Park Falls
LAC DU FLAMBEAU INDIAN RES.
Eagle River
Iron Mountain
Kingsford
Niagara
Gladstone
Escanaba
Hermansville
BEAVER ISLAND
Harbor Springs
Indian River
Long Prairie
Little Falls
Hinckley
Pine City
Spooner
LAC COURTE OREILLES INDIAN RESERVATION
Phillips
Rhinelander
Crandon
SOUTH FOX ISLAND
Charlevoix
Petoskey
Boyne City
Sauk Centre
Melrose
Milaca
Mora
Rush City
Sauk Rapids
Princeton
Cambridge
Cumberland
Rice Lake
Ladysmith
Tomahawk
Wausaukee
WASHINGTON ISLAND
NORTH MANITOU ISLAND
East Jordan
St. Cloud
Paynesville
Zimmerman
North Branch
St. Croix Falls
Barron
Cameron
Chetek
Rib Lake
Merrill
Antigo
Menominee
Peshtigo
Marinette
Leland
Elk Rapids
Mancelona
Traverse City
Kalkaska
Monticello
Elk River
Andover
Blaine
Coon Rapids
Brooklyn Park
Forest Lake
New Richmond
Cornell
Medford
Wausau
Schofield
MENOMINEE IND. RES.
Oconto
DOOR PENINSULA
Sturgeon Bay
Interlochen
Litchfield
Buffalo
Cokato
Minneapolis
Roseville
Maplewood
Hudson
St. Paul
Minnetonka
Eden Prairie
Bloomington
Burnsville
Eagan
Hastings
Lakeville
Bloomer
Chippewa Falls
Owen
Mosinee
Rothschild
STOCKBRIDGE INDIAN RES.
Oconto Falls
Shawano
ONEIDA INDIAN RES.
Algoma
Frankfort
Hutchinson
Menomonie
Eau Claire
Altoona
River Falls
Augusta
Marshfield
Stevens Point
Clintonville
Green Bay
De Pere
Kewaunee
Manistee
Cadillac
Gaylord
Le Sueur
Durand
Mondovi
Osseo
Neillsville
New London
Appleton
Kaukauna
Two Rivers
Manitowoc
Fairfax
New Ulm
St. Peter
Red Wing
Lake City
Northfield
Faribault
Zumbrota
Wabasha
Alma
Wisconsin Rapids
Nekoosa
Black River Falls
Waupaca
Menasha
Neenah
Lake Poygan
Lake Winnebago
Ludington
Scottville
Reed City
Mankato
North Mankato
Waterville
Kenyon
Plainview
Arcadia
Petenwell Lake
Wautoma
Omro
Oshkosh
Chilton
Elkhart Lake
Big Rapids
Madelia
St. James
Lake Crystal
Waseca
Owatonna
Kasson
Rochester
Stewartville
Winona
St. Charles
Holmen
Sparta
Tomah
Adams
Castle Rock Lake
Westfield
Berlin
Ripon
Fond du Lac
Sheboygan
Sheboygan Falls
Hart
Shelby
White Cloud
Mountain Lake
Winnebago
Wells
Albert Lea
Blooming Prairie
Spring Valley
Austin
La Crosse
West Salem
Mauston
New Lisbon
Wisconsin Dells
Lake Delton
Montello
Portage
Waupun
Mayville
Plymouth
West Bend
Fremont
Newaygo
Fairmont
Blue Earth
Caledonia
Viroqua
Reedsburg
Baraboo
Beaver Dam
Horicon
Columbus
Port Washington
Cedarburg
Montague
Whitehall
Muskegon
Muskegon Heights
Norton Shores
Grand Haven
Cedar Springs
Greenville
Rockford
Comstock Park
Walker
Belding
Ionia
Lake Mills
Estherville
Forest City
Northwood
Osage
Cresco
Decorah
Waukon
Richland Center
Sauk City
Lodi
Sun Prairie
Oconomowoc
Watertown
Germantown
Mequon
Menomonee Falls
Brookfield
Milwaukee
West Allis
Waukesha
Grand Rapids
Wyoming
Kentwood
Lowell
Emmetsburg
Clear Lake
Mason City
Charles City
New Hampton
EFFIGY MOUNDS NATL. MON.
Boscobel
Prairie du Chien
Dodgeville
Mt. Horeb
Madison
Stoughton
Jefferson
Fort Atkinson
Greenfield
Oak Creek
Franklin
Holland
Zeeland
Hastings
Algona
Britt
Nashua
West Union
Lancaster
New Glarus
Edgerton
Milton
Whitewater
Burlington
Racine
Saugatuck
Douglas
Allegan
Otsego
Plainwell
West Bend
Belmond
Waverly
Oelwein
Guttenberg
Cassville
Platteville
Mineral Point
Evansville
Janesville
Elkhorn
Delavan
Lake Geneva
Kenosha
South Haven
Kalamazoo
Pocahontas
Humboldt
Clarion
Eagle Grove
Hampton
Darlington
Monroe
Beloit
Walworth
Pleasant Prairie
Zion
Bangor
Battle Creek
Fort Dodge
Iowa Falls
Waterloo
Cedar Falls
Manchester
Dubuque
Dyersville
Independence
Galena
Rockton
Loves Park
Harvard
Woodstock
Antioch
Gurnee
Waukegan
North Chicago
Highland Park
Benton Harbor
Hartford
Paw Paw
Portage
Decatur
Dowagiac
Three Rivers
St. Joseph
Webster City
Rockwell City
Eldora
Grundy Center
La Porte City
Monticello
Bellevue
Freeport
Rockford
Belvidere
Crystal Lake
Marengo
Arlington Heights
Evanston
Carpentersville
Elgin
Schaumburg
Elmhurst
Bridgman
Cassopolis
Constantine
Sturgis
Carroll
Ogden
Boone
Story City
Vinton
Anamosa
Savanna
Mt. Carroll
Oregon
Sycamore
De Kalb
St. Charles
CHICAGO
Buchanan
Niles
Jefferson
Ames
Nevada
Marshalltown
Toledo
Tama
Hiawatha
Cedar Rapids
Marion
Maquoketa
Clinton
Fulton
Grand Detour
Rochelle
Naperville
Aurora
Cicero
Oak Lawn
Harvey
Hammond
Gary
East Chicago
Michigan City
South Bend
Mishawaka
Elkhart
Goshen
Lagrange
Ligonier
Huxley
Ankeny
Perry
Belle Plaine
Amana
West Branch
De Witt
Morrison
Sterling
Rock Falls
Dixon
Amboy
Waterman
Sandwich
Plano
Lockport
Joliet
Chicago Heights
Merrillville
Valparaiso
La Porte
Bremen
Nappanee
Albion
Audubon
Urbandale
Clive
Altoona
Colfax
Newton
Grinnell
Williamsburg
Coralville
Iowa City
West Liberty
Eldridge
Bettendorf
East Moline
Davenport
Moline
Rock Island
Prophetstown
Tampico
Mendota
Ottawa
Morris
Peotone
Lowell
North Judson
Knox
Plymouth
Warsaw
Columbia City
Adel
West Des Moines
Des Moines
Indianola
Pella
Knoxville
Sigourney
Washington
Muscatine
Geneseo
Cambridge
Princeton
Kewanee
Spring Valley
Peru
La Salle
Marseilles
Streator
Dwight
Braidwood
Bourbonnais
Bradley
Kankakee
Culver
Winamac
Rochester
North Manchester
Huntington
Wabash
Stuart
Greenfield
Winterset
Oskaloosa
Eddyville
Aledo
Wapello
Bishop Hill
Galva
Toulon
Lacon
Toluca
Odell
Clifton
Rensselaer
Logansport
Peru
Creston
Corning
Osceola
Chariton
Albia
Ottumwa
Fairfield
Mt. Pleasant
Oquawka
Monmouth
Galesburg
Knoxville
Chillicothe
Minonk
Pontiac
Chenoa
Gilman
Watseka
Onarga
Kentland
Monon
Monticello
Fowler
Delphi
Marion
Gas City
Upland
Kokomo
Rathbun Lake
Leon
Centerville
Bloomfield
Burlington
Abingdon
Farmington
Peoria
E. Peoria
Washington
Morton
Eureka
Lexington
Lake Red Rock
Lake Michigan
GREEN BAY
Lake Superior

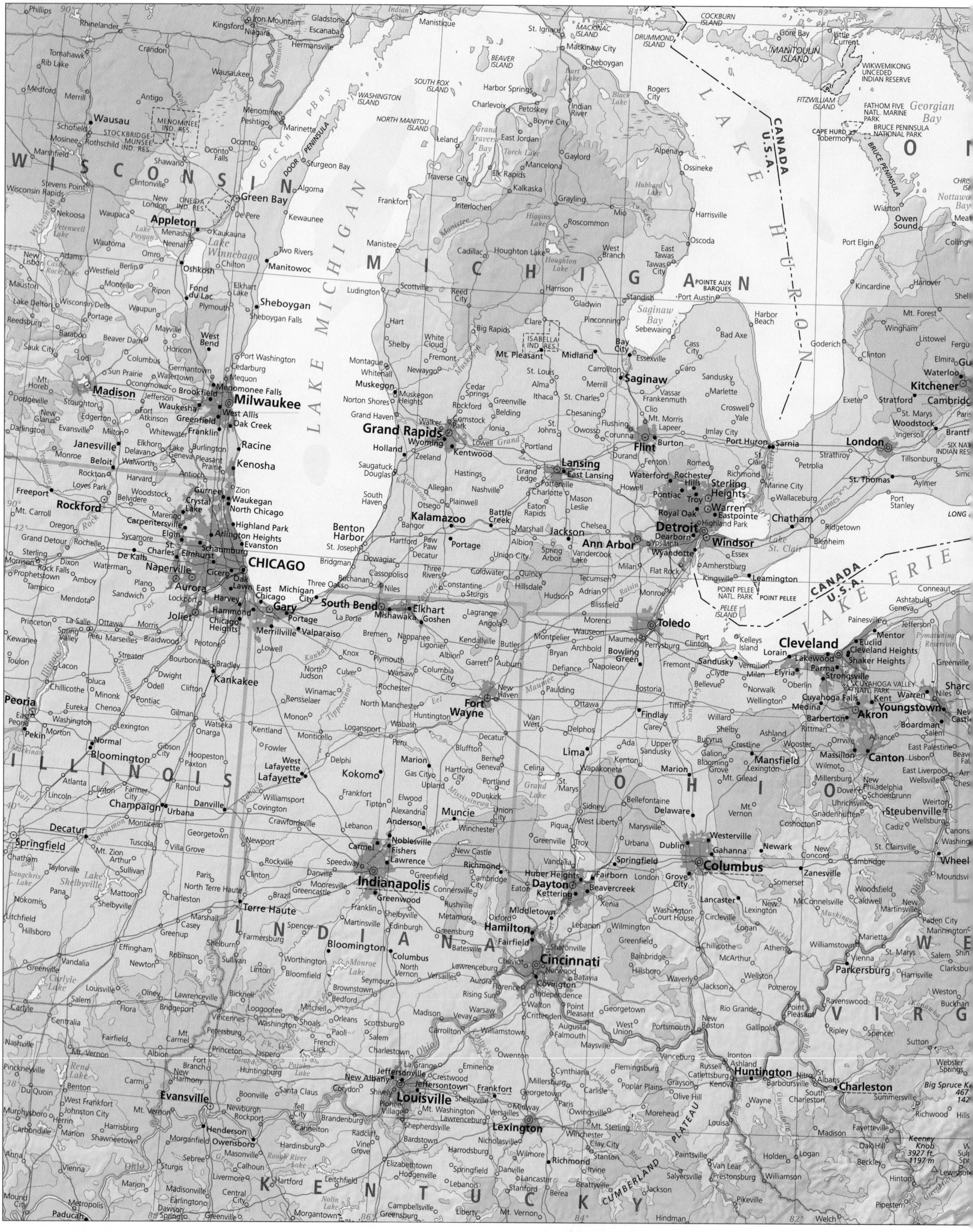
WISCONSIN
MICHIGAN
ILLINOIS
INDIANA
OHIO
KENTUCKY
LAKE MICHIGAN
LAKE HURON
LAKE ERIE
CANADA
U.S.A.
Green Bay
Georgian Bay
Saginaw Bay
Lake Winnebago
Lake St. Clair
DOOR PENINSULA
BRUCE PENINSULA
MANITOULIN ISLAND
DRUMMOND ISLAND
COCKBURN ISLAND
MACKINAC ISLAND
BEAVER ISLAND
SOUTH FOX ISLAND
NORTH MANITOU ISLAND
WASHINGTON ISLAND
FITZWILLIAM ISLAND
PELEE ISLAND
POINT PELEE NATL. PARK
FATHOM FIVE NATL. MARINE PARK
BRUCE PENINSULA NATIONAL PARK
WIKWEMIKONG UNCEDED INDIAN RESERVE
MENOMINEE IND. RES.
STOCKBRIDGE-MUNSEE IND. RES.
ONEIDA IND. RES.
ISABELLA IND. RES.
CUYAHOGA VALLEY NATL. PARK
CUMBERLAND PLATEAU
Big Spruce Knob 4671 ft. 1424 m
Keeney Knob 3927 ft. 1197 m
Milwaukee
Madison
Green Bay
Appleton
Oshkosh
Sheboygan
Manitowoc
Wausau
Racine
Kenosha
Janesville
Beloit
Rockford
CHICAGO
Naperville
Aurora
Joliet
Gary
Hammond
South Bend
Elkhart
Kankakee
Peoria
Bloomington
Normal
Champaign
Urbana
Decatur
Springfield
Danville
Grand Rapids
Kalamazoo
Lansing
East Lansing
Flint
Saginaw
Ann Arbor
Detroit
Windsor
Dearborn
Warren
Sterling Heights
Pontiac
Port Huron
Sarnia
London
Kitchener
Chatham
Toledo
Cleveland
Akron
Canton
Youngstown
Mansfield
Columbus
Dayton
Springfield
Cincinnati
Hamilton
Lima
Findlay
Fort Wayne
Muncie
Anderson
Kokomo
Lafayette
West Lafayette
Indianapolis
Bloomington
Terre Haute
Evansville
Henderson
Owensboro
Louisville
Lexington
Frankfort
Huntington
Charleston
Parkersburg
Steubenville
Wheeling
Lambert Conformal Conic Projection

0 20 40 60 80 100 120 Miles

0 20 40 60 80 100 120 140 160 180 200 Kilometers

Lambert Conformal Conic Projection
Scale 1:4,000,000
One inch to 64 miles
One cm to 40 km

QUÉBEC
ONTARIO
NEW YORK
VERMONT
NEW HAMPSHIRE
MAINE
MASS.
CONN.
R.I.
PENNSYLVANIA
NEW JERSEY
MARYLAND
DEL.
VIRGINIA
CANADA
U.S.A.
LAKE ONTARIO
ATLANTIC OCEAN
ADIRONDACK MTS.
CATSKILL MTS.
ALLEGHENY PLAT.
APPALACHIAN MOUNTAINS
BLUE RIDGE
GREEN MTS.
WHITE MTS.
TACONIC RA.
DELMARVA PENINSULA
Long Island Sound
Chesapeake Bay
Delaware Bay
Cape Cod Bay
Nantucket Sound
Rhode Island Sound
Massachusetts Bay
Bigelow Bight
Lake Champlain
Lake George
Oneida Lake
Lake Simcoe
Great Sacandaga Lake
MONTRÉAL
Ottawa
Gatineau
Laval
Longueuil
Sherbrooke
TORONTO
Mississauga
Hamilton
St. Catharines
Oshawa
Kingston
Peterborough
Barrie
Buffalo
Rochester
Syracuse
Utica
Albany
Schenectady
Binghamton
Burlington
Portland
Manchester
Lowell
Lawrence
Boston
Worcester
Springfield
Hartford
Providence
New Haven
Bridgeport
Stamford
Waterbury
New London
NEW YORK
Newark
Jersey City
Paterson
Yonkers
Trenton
Allentown
Scranton
Wilkes-Barre
Reading
Harrisburg
Lancaster
York
PHILADELPHIA
Wilmington
Camden
Atlantic City
Vineland
Dover
Baltimore
Washington
Alexandria
Arlington
Annapolis
Richmond
Norfolk
Virginia Beach
Newport News
Hampton
Portsmouth
Pittsburgh
Altoona
Johnstown
State College
Williamsport
Erie
Jamestown
Charlottesville
Harrisonburg
Lynchburg
Petersburg
Mt. Marcy 5344 ft. 1629 m
Mt. Mansfield 4393 ft. 1339 m
Mt. Washington 6288 ft. 1917 m
Dorset Peak 3770 ft. 1149 m
Coburn Mountain 3718 ft. 1133 m
Spruce Knob 4863 ft. 1482 m
CAPE ANN
CAPE COD
MONTAUK POINT
CAPE MAY
CAPE HENLOPEN
CAPE CHARLES
LONG ISLAND
BLOCK ISLAND
MARTHA'S VINEYARD
NANTUCKET ISLAND
ASSATEAGUE ISLAND
PARRAMORE ISLAND
AMHERST ISLAND
ST. REGIS IND. RES.
CATTARAUGUS INDIAN RESERVATION
ALLEGANY IND. RES.
ST. LAWRENCE ISLANDS NATIONAL PARK
SHENANDOAH NATIONAL PARK
GEORGE WASHINGTON BIRTHPLACE NAT'L. MON.
© Rand McNally A-101618-1
a
QUEBEC
CANADA
UNITED STATES
NEW BRUNSWICK
MAINE
N.H.
Gulf of Maine
Bangor
Augusta
Lewiston
Portland
Fredericton
Moosehead Lake
Mt. Katahdin 5268 ft. 1606 m
Mt. Carleton 2680 ft. 817 m
Mt. Washington 6288 ft. 1917 m
Coburn Mountain 3718 ft. 1133 m
ACADIA NATIONAL PARK
MT. DESERT ISLAND
GRAND MANAN ISLAND
VINALHAVEN ISLAND
© Rand McNally A-101619-1

Inset map a
Lambert Conformal Conic Projection
Scale 1:6,000,000
One inch to 96 miles
One cm to 60 km

NEVADA
CALIFORNIA
PACIFIC OCEAN
SIERRA NEVADA
COAST RANGES
SACRAMENTO VALLEY
SAN JOAQUIN VALLEY
GREAT BASIN
MOJAVE DESERT
DIABLO RANGE
SANTA LUCIA RA.
CHANNEL ISLANDS
INYO MTS.
SHOSHONE MTS.
TOIYABE RA.
BIG SMOKY VALLEY
RUBY MTS.
STILLWATER RANGE
DEATH VALLEY
PANAMINT RA.
AMARGOSA DESERT
SPRING MTS.
SAN BERNARDINO MTS.
IMPERIAL VALLEY
BAJA CALIFORNIA
USA
MEXICO
Sacramento
San Francisco
Oakland
San Jose
Stockton
Modesto
Fresno
Bakersfield
Reno
Carson City
Las Vegas
LOS ANGELES
SAN DIEGO
TIJUANA
Mexicali
Santa Barbara
Ventura (San Buenaventura)
Long Beach
Anaheim
Riverside
Santa Ana
San Bernardino
Pasadena
Glendale
Burbank
Oxnard
Santa Rosa
Salinas
Visalia
Merced
Chico
Redding
Eureka
Mt. Whitney 14 494 ft. 4418 m
Death Valley
North Las Vegas
Paradise
Henderson
Salton Sea
Mono Lake
Lake Tahoe
Pyramid Lake
Walker Lake
Honey Lake
Lake Almanor
Clear Lake
Monterey Bay
Santa Barbara Channel
Gulf of Santa Catalina
Point Arena
Point Reyes
Point Conception
Point Arguello
Point Buchon
YOSEMITE NATIONAL PARK
KINGS CANYON NATIONAL PARK
SEQUOIA NATIONAL PARK
DEATH VALLEY NATIONAL PARK
JOSHUA TREE NATIONAL PARK
CHANNEL ISLANDS NATIONAL PARK
PINNACLES NATL. MON.
Anthony Peak 6954 ft. 2120 m
Sutter Buttes 2116 ft. 645 m
Donner Pass 7087 ft. 2160 m
Wheeler Peak 11 664 ft. 3555 m
Excelsior Mountain 12 446 ft. 3794 m
Mt. Ritter 13 140 ft. 4005 m
Boundary Peak 13 143 ft. 4006 m
White Mountain Peak 14 246 ft. 4342 m
North Palisade 14 242 ft. 4341 m
Telescope Peak 11 049 ft. 3368 m
Charleston Peak 11 918 ft. 3633 m
Mt. Pinos 8831 ft. 2692 m
Mt. Wilson 5710 ft. 1740 m
Junipero Serra Peak 5862 ft. 1787 m
Summit Mountain 10 461 ft. 3189 m
Currant Mountain 11 513 ft. 3509 m
Arc Dome 11 773 ft. 3588 m
Mt. Jefferson 11 942 ft. 3640 m
Troy Peak 11 298 ft. 3444 m
Kawich Peak 9404 ft. 2866 m
Yucca Mountain 6584 ft. 2007 m
Toro Peak 8717 ft. 2357 m
Palomar Mountain 6140 ft. 1871 m
© Rand McNally A-101622-1
a
Oceanside
Escondido
Carlsbad
Encinitas
Vista
San Marcos
Poway
Santee
El Cajon
La Mesa
National City
Chula Vista
Coronado
Imperial Beach
La Jolla
PACIFIC OCEAN
UNITED STATES
MEXICO
BAJA CALIFORNIA
Black Mountain 1552 ft. 473 m
San Miguel Mountain 2566 ft. 782 m
© Rand McNally A-101623-1

Inset map a
Lambert Conformal Conic Projection
Scale 1:1,000,000
One inch to 16 miles
One cm to 10 km

0 20 40 60 80 100 120 Miles
0 20 40 60 80 100 120 140 160 180 200 Kilometers

Lambert Conformal Conic Projection
Scale 1:4,000,000
One inch to 64 miles
One cm to 40 km

UTAH
COLORADO
ARIZONA
NEW MEXICO
SONORA
CHIHUAHUA
TEXAS
USA
MEXICO
GREAT SALT LAKE DESERT
SEVIER DESERT
ESCALANTE DESERT
SNAKE RANGE
WASATCH PLATEAU
ROAN PLATEAU
COLORADO PLATEAU
UNCOMPAHGRE PLAT.
ROCKY MTS.
SAWATCH RA.
SANGRE DE CRISTO MTS.
SAN JUAN MTS.
SAN LUIS VALLEY
GLEN CANYON
MARBLE CANYON
KAIBAB PLATEAU
KANAB PLATEAU
SHIVWITS PLATEAU
GRAND CANYON
COCONINO PLATEAU
PAINTED DESERT
NAVAJO INDIAN RESERVATION
CHUSKA MTS.
SAN JUAN BASIN
HOPI INDIAN RESERVATION
MOGOLLON RIM
HUALAPAI MTS.
SONORAN DESERT
DESIERTO DE ALTAR
BLACK RANGE
SAN ANDRES MTS.
TULAROSA VALLEY
PELONCILLO MTS.
TOHONO O'ODHAM INDIAN RESERVATION
Provo
Orem
Millcreek
Tooele
American Fork
Spanish Fork
Santaquin
Nephi
Mt. Nebo 11 928 ft. 3636 m
Mt. Pleasant
Price
Castle Dale
Huntington
Ephraim
Gunnison
Salina
Ferron
Green River
Delta
Fillmore
Richfield
Monroe
Milford
Beaver
Circleville
Panguitch
Parowan
Cedar City
Escalante
St. George
Hurricane
Kanab
Page
Mesquite
Moab
Monticello
Bluff
Mexican Hat
Roosevelt
Duchesne
Dinosaur
Meeker
Kremmling
Granby
Hot Sulphur Springs
Boulder
Idaho Springs
Evergreen
Vail
Leadville
Glenwood Springs
Aspen
Buena Vista
Salida
Gunnison
Montrose
Delta
Fruita
Palisade
Grand Junction
Ouray
Dove Creek
Cortez
Durango
Pagosa Springs
Monte Vista
Alamosa
La Jara
Tierra Amarilla
Aztec
Farmington
Shiprock
Teec Nos Pos
Kayenta
Chinle
Moenkopi
Hotevilla
Grand Canyon
Gallup
Thoreau
Grants
Rio Rancho
Albuquerque
Moriarty
Los Lunas
Belen
Jarales
Socorro
Magdalena
Santa Fe
Espanola
Los Alamos
Sanders
Holbrook
Winslow
Flagstaff
Williams
Ash Fork
Kingman
Laughlin
Bullhead City
Needles
Lake Havasu City
Sedona
Cottonwood
Prescott
St. Johns
Show Low
Springerville
Alpine
Payson
Wickenburg
Salome
Quartzsite
Surprise
Sun City
Glendale
Scottsdale
PHOENIX
Tempe
Mesa
Chandler
Apache Junction
Superior
Globe
San Carlos
Florence
Casa Grande
Eloy
Gila Bend
Ajo
Yuma
Sonoyta
Oracle
Tucson
Green Valley
Benson
Willcox
Tombstone
Sierra Vista
Bisbee
Nogales
Naco
Douglas
Agua Prieta
Safford
Clifton
Glenwood
Silver City
Bayard
Lordsburg
Deming
Columbus
Las Palomas
Las Cruces
Hatch
Truth or Consequences (Hot Springs)
Anthony
El Paso
CIUDAD JUÁREZ
Animas
Monument Peak 9196 ft. 2803 m
Mt. Powell 13 448 ft. 4099 m
Grays Peak 14 270 ft. 4349 m
Mt. Lincoln 14 286 ft. 4354 m
Mt. Massive 14 421 ft. 4396 m
Snowmass Mountain 14 092 ft. 4295 m
Mt. Elbert 14 433 ft. 4399 m
La Plata Peak 14 337 ft. 4370 m
Castle Peak 14 265 ft. 4348 m
Mt. Yale 14 196 ft. 4327 m
Monarch Pass 11 312 ft. 3448 m
Uncompahgre Peak 14 309 ft. 4361 m
Mt. Wilson 14 246 ft. 4342 m
Wolf Creek Pass 10 856 ft. 3309 m
Summit Peak 13 300 ft. 4054 m
Mt. Peale 12 721 ft. 3877 m
Abajo Peak 11 360 ft. 3463 m
Mt. Marvine 11 610 ft. 3539 m
Mt. Ellen 11 522 ft. 3512 m
Delano Peak 12 173 ft. 3710 m
Indian Peak 9790 ft. 2984 m
Wheeler Peak 13 063 ft. 3982 m
Ship Rock 7178 ft. 2188 m
Humphreys Peak 12 633 ft. 4263 m
Mt. Taylor 11 301 ft. 3036 m
Baldy Peak 11 403 ft. 3476 m
Mazatzal Peak 7903 ft. 2409 m
Whitewater Baldy 10 895 ft. 3321 m
Mt. Graham 10 720 ft. 3267 m
Chiricahua Peak 9798 ft. 2986 m
GREAT BASIN NATL. PARK
TIMPANOGOS CAVE NATL. MON.
UINTAH AND OURAY INDIAN RESERVATION
ARCHES NATIONAL PARK
COLORADO NATL. MON.
BLACK CANYON OF THE GUNNISON NATL. PARK
CANYONLANDS NATL. PARK
CAPITOL REEF NATIONAL PARK
NATURAL BRIDGES NATL. MON.
CANYONS OF THE ANCIENTS NATL. MON.
HOVENWEEP NATL. MON.
MESA VERDE NATL. PARK
UTE MOUNTAIN IND. RES.
SOUTHERN UTE IND. RES.
AZTEC RUINS NATL. MON.
GREAT SAND DUNES NATL. PARK
JICARILLA APACHE IND. RES.
CEDAR BREAKS NATL. MON.
ZION NATIONAL PARK
BRYCE CANYON NATIONAL PARK
GRAND STAIRCASE-ESCALANTE NATL. MON.
RAINBOW BRIDGE NATL. MON.
KAIBAB IND. RES.
PIPE SPRING NATL. MON.
NAVAJO NATL. MON.
GRAND CANYON-PARASHANT NATL. MON.
GRAND CANYON NATIONAL PARK
HAVASUPAI IND. RES.
HUALAPAI IND. RES.
CANYON DE CHELLY NATL. MON.
WUPATKI NATL. MON.
SUNSET CRATER NATL. MON.
WALNUT CANYON NATL. MON.
PETRIFIED FOREST NATIONAL PARK
ZUNI PUEBLO
RAMAH IND. RES.
EL MORRO NATL. MON.
EL MALPAIS NATL. MON.
ACOMA PUEBLO
LAGUNA PUEBLO
CANONCITO IND. RES.
PETROGLYPH NATL. MON.
ISLETA PUEBLO
ALAMO IND. RES.
SALINAS PUEBLO MISSIONS NATL. MON.
BANDELIER NATL. MON.
FORT MOJAVE IND. RES.
TUZIGOOT NATL. MON.
MONTEZUMA CASTLE NATL. MON.
AGUA FRIA NATL. MON.
FORT APACHE INDIAN RESERVATION
COLORADO RIVER INDIAN RESERVATION
FORT MCDOWELL IND. RES.
SALT RIVER IND. RES.
TONTO NATL. MON.
SAN CARLOS INDIAN RESERVATION
GILA RIVER IND. RES.
CASA GRANDE RUINS NATL. MON.
IRONWOOD FOREST NATL. MON.
SAGUARO NATL. PARK
SAGUARO NATIONAL PARK
ORGAN PIPE CACTUS NATL. MON.
SAN XAVIER IND. RES.
FORT YUMA IND. RES.
GILA CLIFF DWELLINGS NATL. MON.
CHIRICAHUA NATL. MON.
WHITE SANDS NATL. MON.
Utah Lake
Strawberry Reservoir
Sevier Lake
Sevier Bridge Reservoir
Little Salt Lake
Lake Powell
Lake Mead
Lake Mohave
Lake Havasu
Blue Mesa Reservoir
Navajo Reservoir
Abiquiu Reservoir
Theodore Roosevelt Lake
San Carlos Reservoir
Painted Rock Reservoir
Elephant Butte Reservoir
Willcox Playa
Lake Granby
Colorado
Green
San Juan
Little Colorado
Rio Grande
Gila
Salt
Puerco
Zuni
Gunnison
Dolores
San Rafael
Dirty Devil
Fremont
Virgin
Bill Williams
Santa Cruz
San Pedro
San Simon
San Francisco
Chama
Conejos
Duchesne
White
Price
Hayden Rhodes Aqueduct
Tucson Aqueduct
114°
112°
110°
108°
38°
34°
32°

Steamboat Springs
Yampa
Hagues Peak 13 560 ft. 4133 m
Fort Collins
Wellington
ROCKY MOUNTAIN NATIONAL PARK
Loveland
Windsor
Greeley
Crow Creek
Crook
Sterling
Haxtun
Holyoke
Grant
Imperial
Frenchman Creek
Red Willow Creek
Gothenburg
Cozad
Lexington
Kearney
Gibbon
Grand Island
Doniphan
Ravenna
Platte
Hastings
Holdrege
Minden
Hot Sulphur Springs
Kremmling
Granby
Longs Peak 14 255 ft. 4345 m
Berthoud
Longmont
South Platte
Fort Lupton
Fort Morgan
Brush
Akron
Yuma
Wray
Benkelman
Republican
Cambridge
Arapahoe
McCook
Beaver Cr.
Alma
Harlan County Lake
Red Cloud
Superior
Colorado
Mt. Powell 13 448 ft. 4099 m
Grays Peak 14 270 ft. 4349 m
Boulder
Lafayette
Westminster
Brighton
Thornton
Arvada
Golden
Idaho Springs
Denver
Aurora
Lakewood
Englewood
Evergreen
Littleton
Centennial
Parker
Castle Rock
Eagle
Vail
Mt. Lincoln 14 286 ft. 4354 m
Snowmass Mtn. 14 092 ft. 4295 m
Mt. Massive 14 421 ft. 4396 m
Leadville
Aspen
Mt. Elbert 14 433 ft. 4399 m
La Plata Peak 14 337 ft. 4370 m
Castle Peak 14 265 ft. 4348 m
SAWATCH RA.
Mt. Yale 14 196 ft. 4327 m
Buena Vista
FLORISSANT FOSSIL BEDS NATL. MON.
Pikes Peak 14 110 ft. 4301 m
Manitou Springs
Colorado Springs
Monarch Pass 11 312 ft. 3448 m
Gunnison
Salida
Cripple Creek
Arikaree
South Fork Republican
West Bijou Creek
Limon
Burlington
Goodland
Colby
Oberlin
Norton
Phillipsburg
Mankato
Prairie Dog Creek
North Fork Solomon
Waconda Lake
Beloit
Stockton
South Fork Solomon
Solomon
Oakley
WaKeeney
Saline
Lincoln
Big Sandy Creek
Cheyenne Wells
Sharon Springs
Big Creek
Smoky Hill
Hays
Wilson Lake
Russell
Cedar Bluff Reservoir
Ellsworth
ROCKY
COLORADO
Canon City
Arkansas
Fountain Cr.
Horse Cr.
Eads
Scott City
Ness City
KANSAS
Hoisington
Pueblo
Huerfano
Rocky Ford
John Martin Res.
La Junta
Lamar
Syracuse
Arkansas
Lakin
Garden City
Great Bend
Ellinwood
Lyons
Larned
Pawnee
Hutchinson
South Hutchinson
Kinsley
Cimarron
Dodge City
Cheney Reservoir
Pratt
Kingman
Saguache Cr.
SAN JUAN MTS.
SANGRE DE CRISTO MTS.
GREAT SAND DUNES NATL. PARK
SAN LUIS VALLEY
Monte Vista
Wolf Creek Pass 10 856 ft. 3309 m
Alamosa
Blanca Peak 14 345 ft. 4372 m
Walsenburg
Apishapa
Purgatoire
Two Butte Creek
Summit Peak 13 300 ft. 4054 m
La Jara
Culebra Peak 14 047 ft. 4282 m
Springfield
Ulysses
Sublette
Minneola
North Fork Cimarron
Medicine Lodge
Chikaskia
Trinidad
JICARILLA APACHE INDIAN RESERVATION
Raton Pass 7835 ft. 2388 m
Raton
CAPULIN VOLCANO NATL. MON.
Hugoton
Meade
Liberal
Crooked Cr.
Tierra Amarilla
Wheeler Peak 13 161 ft. 4011 m
Cimarron
Boise City
Hooker
Beaver
Buffalo
Salt Fork Arkansas
Alva
Medford
Chama
TAOS PUEBLO
Taos
Abiquiu Reservoir
Cimarron
Ranchos de Taos
Angel Fire
Springer
Goodwell
Guymon
Cherokee
Grande
Clayton
Coldwater Creek
Perryton
Woodward
Waynoka
SANTA CLARA PUEBLO
SAN JUAN PUEBLO
Espanola
Truchas Peak 13 102 ft. 3993 m
Canadian
Carrizo
Ute Creek
Stratford
Palo Duro Creek
Wolf Creek
Shattuck
Enid
Los Alamos
SAN ILDEFONSO PUEBLO
BANDELIER NATL. MON.
NAMBE PUEBLO
POJOAQUE PUEBLO
FORT UNION NATL. MON.
Mora
Rita Blanca Cr.
Dalhart
Cactus
Spearman
Fairview
Canton Lake
Hennessey
JEMEZ PUEBLO
ZIA PUEBLO
COCHITI PUEBLO
Santa Fe
Seiling
North Canadian
SANTO DOMINGO
Mora
Pecos
Las Vegas
Dumas
Stinnett
Canadian
Watonga
Kingfisher
LAGUNA PUEBLO
SANTA ANA PUEBLO
SAN FELIPE PUEBLO
Rio Rancho
SANDIA PUEBLO
PETROGLYPH NATL. MON.
CANONCITO IND. RES.
Albuquerque
Lake Meredith
Borger
Fritch
ALIBATES FLINT QUARRIES NATL. MON.
Panhandle
Pampa
Thomas
Okarche
Geary
Canadian
Logan
Tucumcari
Wheeler
Clinton
El Reno
Elk City
Weatherford
ISLETA PUEBLO
Moriarty
Vega
Amarillo
Sayre
Cordell
Washita
LAGUNA PUEBLO
Los Lunas
Belen
Jarales
Santa Rosa
Claude
Shamrock
Erick
North Fk. Red
Hobart
Carnegie
Anadarko
Chickasha
NEW MEXICO
Canyon
Prairie Dog Town Fork Red
Clarendon
Salt Fork
Wellington
Mangum
Red
WICHITA MTS.
OKLAHOMA
Hereford
Vaughn
Lake Sumner
Fort Sumner
Friona
Dimmitt
Tulia
Memphis
Hollis
Altus
Snyder
Cache
Marlow
Lawton
Duncan
SALINAS PUEBLO MISSIONS NATL. MON.
Clovis
Texico
Farwell
White
Childress
Red
Frederick
Portales
Muleshoe
Plainview
Quanah
Walters
Comanche
Grandfield
Waurika Lake
Waurika
LLANO ESTACADO
Hale Center
Pease
Vernon
Burkburnett
Carrizozo
Sierra Blanca Peak 11 973 ft. 3649 m
SACRAMENTO MTS.
Capitan Peak 10 083 ft. 3073 m
Capitan
Littlefield
Abernathy
Floydada
Paducah
Electra
Wichita Falls
Iowa Park
Tongue
Wichita
Lake Kemp
Henrietta
Ruidoso Downs
Ruidoso
Rio Hondo
Roswell
Shallowater
Lubbock
Ralls
Levelland
Seymour
Lake Arrowhead
MESCALERO APACHE IND. RES.
TULAROSA VALLEY
Dexter
Tatum
Wolfforth
Slaton
TEXAS
Bowie
Munday
Olney
WHITE SANDS NATL. MON.
Alamogordo
Rio Felix
Hagerman
Brownfield
Post
Double Mountain Fork Brazos
Salt Fork Brazos
North Fork Brazos
Lake Bridgeport
Artesia
Lovington
Denver City
Sulphur Springs Draw
Tahoka
Haskell
Lake Stamford
Clear Fk. Brazos
Jacksboro
Graham
Bridgeport
Brantley Tank
Stamford
Possum Kingdom Lake
Hobbs
Seminole
Lamesa
Hamlin
Anson
Hubbard Creek Res.
Mineral Wells
Snyder
Albany
Breckenridge
Weatherford
GUADALUPE MTS.
Carlsbad
Eunice
Monument Draw
Colorado
Brazos
Loving
CARLSBAD CAVERNS NATIONAL PARK
Roscoe
Merkel
Abilene
Ranger
Sweetwater
Baird
Cisco
Andrews
Big Spring
Colorado City
Clyde
Eastland
Granbury
Jal
106°
104°
102°
100°
98°
40°
38°
36°
34°
32°

0 20 40 60 80 100 120 Miles

0 20 40 60 80 100 120 140 160 180 200 Kilometers

Lambert Conformal Conic Projection
Scale 1:4,000,000
One inch to 64 miles
One cm to 40 km

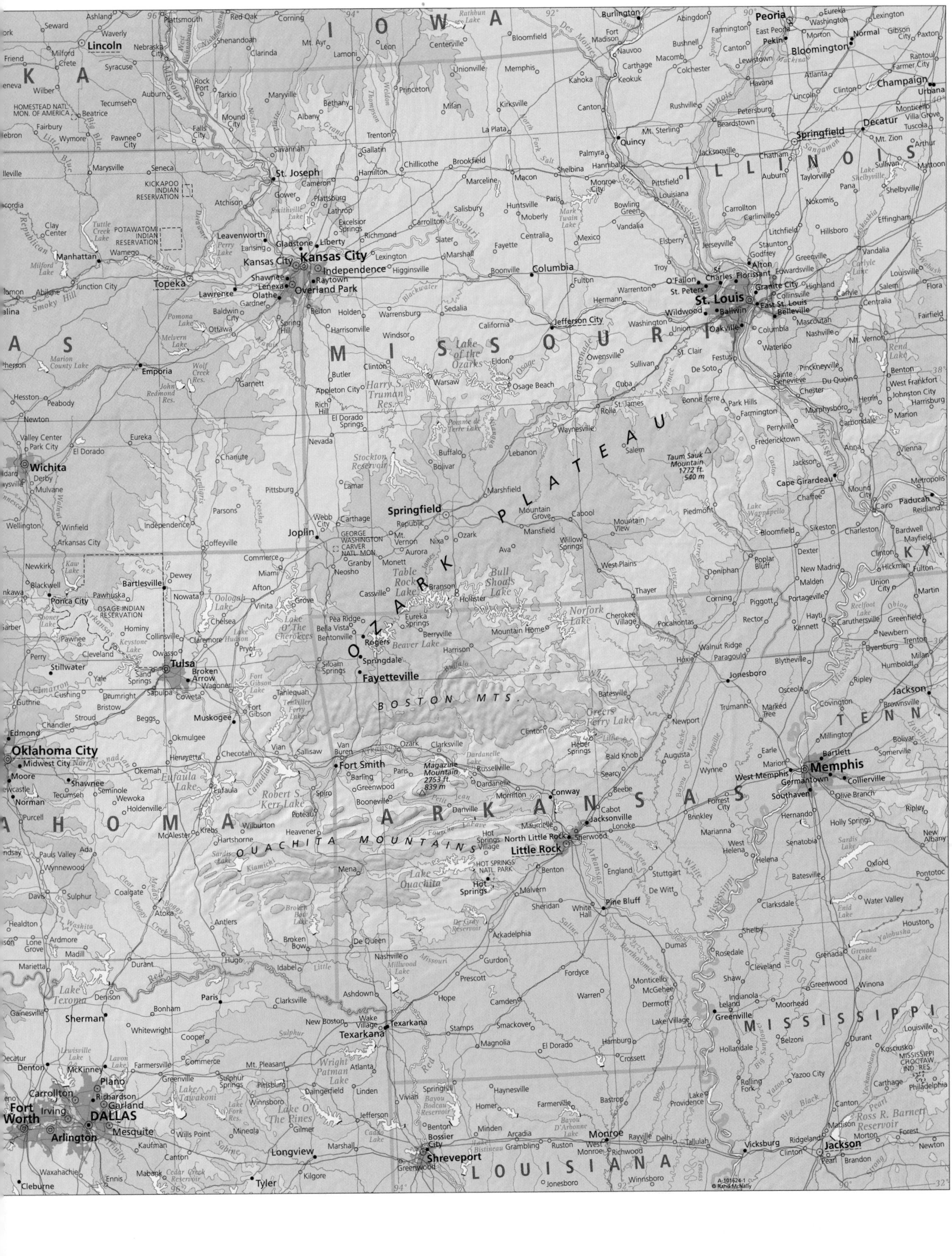
IOWA
MISSOURI
ILLINOIS
KANSAS
OKLAHOMA
ARKANSAS
TENN
KY
MISSISSIPPI
LOUISIANA
OZARK PLATEAU
BOSTON MTS.
OUACHITA MOUNTAINS
Taum Sauk Mountain 1772 ft. 540 m
Magazine Mountain 2753 ft. 839 m
Lake of the Ozarks
Harry S. Truman Res.
Table Rock Lake
Bull Shoals Lake
Norfork Lake
Beaver Lake
Lake O' The Cherokees
Greers Ferry Lake
Lake Ouachita
Lake Texoma
Robert S. Kerr Lake
Eufaula Lake
HOMESTEAD NATL. MON. OF AMERICA
KICKAPOO INDIAN RESERVATION
POTAWATOMI INDIAN RESERVATION
OSAGE INDIAN RESERVATION
GEORGE WASHINGTON CARVER NATL. MON.
HOT SPRINGS NATL. PARK
MISSISSIPPI CHOCTAW IND. RES.
Lincoln
Topeka
Kansas City
Independence
Overland Park
St. Joseph
Wichita
Columbia
Jefferson City
St. Louis
Springfield
Joplin
Peoria
Bloomington
Champaign
Decatur
Quincy
Tulsa
Oklahoma City
Fort Smith
Fayetteville
Little Rock
North Little Rock
Pine Bluff
Memphis
Jonesboro
Texarkana
Shreveport
Monroe
Jackson
DALLAS
Fort Worth
Arlington
Tyler
Longview
Cape Girardeau
Paducah
A-101624-1
© Rand McNally

WHITE SANDS NATL. MON.
MESCALERO APACHE INDIAN RESERVATION
Hagerman
Brownfield
Post
Haskell
Jacksboro
Alamogordo
SACRAMENTO MTS.
TULAROSA VALLEY
NEW MEXICO
Artesia
Lovington
Denver City
Tahoka
Lake Stamford
Graham
Possum Kingdom Lake
Mineral Wells
Brantley Tank
Hobbs
Seminole
Lamesa
Stamford
Hamlin
Hubbard Creek Reservoir
Snyder
Anson
Albany
Breckenridge
Weatherford
GUADALUPE MTS.
Carlsbad
CARLSBAD CAVERNS NATIONAL PARK
Loving
Eunice
Monument Draw
Andrews
Roscoe
Merkel
Abilene
Sweetwater
Ranger
Colorado City
Clyde
Baird
Cisco
Eastland
Big Spring
Anthony
El Paso
CIUDAD JUÁREZ
GUADALUPE MOUNTAINS NATIONAL PARK
Guadalupe Peak 8749 ft. 2667 m
Red Bluff Reservoir
Jal
Midland
Stanton
Stephenville
West Odessa
Odessa
Kermit
Winters
De Leon
Proctor Lake
Dublin
Sterling City
Comanche
Coleman
Santa Anna
Brownwood
Hamilton
Ballinger
Guadalupe Bravos
Sierra Blanca
Pecos
Monahans
Crane
San Angelo
Twin Buttes Reservoir
Goldthwaite
Big Lake
Eden
Van Horn
McCamey
Brady
San Saba
Copperas Cove
Fort Stockton
Lampasas
DAVIS MTS.
Mt. Livermore 8379 ft 2554 m
UNITED STATES
MEXICO
Menard
Ozona
TEXAS
Mason
Lake Buchanan
Llano
EDWARDS PLATEAU
Alpine
Marfa
STOCKTON PLATEAU
Sonora
Junction
Marble Falls
Lake Lyndon B. Johnson
Cathedral Mountain 6868 ft. 2093 m
Sanderson
Fredericksburg
Johnson City
Rocksprings
Kerrville
San Marcos
DESIERTO DE CHIHUAHUA
Rio Grande
Bravo del Norte
Presidio
Ojinaga
Coyame
Laguna del Cuervo
Conchos
Emory Peak 7825 ft. 2385 m
BIG BEND NATIONAL PARK
Amistad Reservoir
Del Rio
Ciudad Acuña
Brackettville
Medina Lake
Boerne
New Braunfels
Leon Valley
Universal City
Schertz
Converse
Kirby
SAN ANTONIO
Hondo
Uvalde
Aldama
Presa Luis L. León
CHIHUAHUA
SIERRA MADRE ORIENTAL
Jiménez
La Pryor
Poteet
Floresville
Pleasanton
Karnes City
Pearsall
Piedras Negras
Eagle Pass
Dilley
Crystal City
Choke Canyon Reservoir
Meoqui
Rosales
Delicias
Zaragoza
Nava
Morelos
Allende
Carrizo Springs
Cotulla
George West
Saucillo
Naica
Laguna del Guaje
Encinal
Ciudad Camargo
Melchor Múzquiz
Nueva Rosita
Sabinas
Presa Venustiano Carranza
Freer
San Diego
Alice
Presa de la Boquilla
COAHUILA
BOLSÓN DE MAPIMÍ
Nuevo Laredo
Laredo
Kingsville
Hidalgo del Parral
San Francisco del Oro
Santa Bárbara
Ciudad Anáhuac
Hebbronville
Premont
Falfurrias
Cuatrociénegas
Nadadores
San Buenaventura
Lampazos de Naranjo
Escalón
Laguna del Rey
Frontera
Monclova
Castaños
Candela
Zapata
SIERRA MADRE OCCIDENTAL
Villa Ocampo
Falcon Reservoir
Sabinas Hidalgo
Villaldama
Nueva Ciudad Guerrero
Mier
Rio Grande City
Ciudad Miguel Alemán
Ciudad Camargo
Edinburg
Mission
McAllen
Pharr
Weslaco
Tlahualillo de Zaragoza
Santa María del Oro
Bermejillo
Francisco I. Madero
San Pedro de las Colonias
Cerralvo
Presa Marte R. Gómez
Reynosa
Hidalgo
Presa Lázaro Cárdenas
Mapimí
Villa de García
NUEVO LEÓN
Río Bravo
Gómez Palacio
Torreón
Santa Catarina
San Nicolás de los Garza
Lerdo
Matamoros
DURANGO
MONTERREY
Guadalupe
China
Valle Hermoso
Ramos Arizpe
Arteaga
PARQUE NACIONAL CUMBRES DE MONTERREY
Cadereyta Jiménez
Nazas
Parras de la Fuente
Santiago Papasquiaro
Rodeo
Presa Francisco Zarco
Cerro Las Nopaleras 10 236 ft 3120 m
Viesca
General Cepeda
Saltillo
Allende
TAMAULIPAS
Nuevo Ideal
Laguna Santiaguillo
Velardeña
Cuencamé de Ceniceros
San Juan del Río
General Terán
Montemorelos
Peñón Blanco
Cerro La Bandera 10 597 m 3230 ft.
Canatlán
Francisco I. Madero
ZACATECAS
Galeana
Linares
San Fernando
Pecos
Colorado
Concho
Llano
Guadalupe
Medina
Nueces
Salado
Sabinas
San Juan
Conchos
Florido
Nazas
Aguanaval
104°
102°
100°
98°
106°
32°
30°
28°
26°

0 20 40 60 80 100 120 Miles

0 20 40 60 80 100 120 140 160 180 200 Kilometers

Lambert Conformal Conic Projection
Scale 1:4,000,000
One inch to 64 miles
One cm to 40 km

ARK.
MISSISSIPPI
LOUISIANA
GULF OF MEXICO
Decatur
Denton
McKinney
Lewisville Lake
Plano
Carrollton
Richardson
Garland
Fort Worth
Irving
DALLAS
Arlington
Mesquite
Farmersville
Commerce
Greenville
Mt. Pleasant
Wright Patman Lake
Atlanta
Sulphur Springs
Pittsburg
Daingerfield
Lake O' The Pines
Linden
Winnsboro
Caddo Lake
Jefferson
Gilmer
Mineola
Wills Point
Kaufman
Canton
Longview
Marshall
Shreveport
Bossier City
Greenwood
Kilgore
Tyler
Waxahachie
Ennis
Cedar Creek Res.
Mabank
Athens
Lake Palestine
Carthage
Henderson
Cleburne
Itasca
Hillsboro
Italy
Corsicana
Jacksonville
Tenaha
Center
Toledo Bend Reservoir
Meridian
Hubbard
Mexia
Fairfield
Palestine
Rusk
Nacogdoches
San Augustine
Waco
Hewitt
Mart
Groesbeck
McGregor
Grapeland
Buffalo
Crockett
Lufkin
Diboll
Moody
Marlin
Killeen
Temple
Belton
Franklin
Hearne
Madisonville
Trinity
Lake Livingston
Jasper
Newton
Cameron
Bartlett
Bryan
College Station
Huntsville
Woodville
Livingston
Kirbyville
Georgetown
Taylor
Round Rock
Pflugerville
Rockdale
Caldwell
Somerville Lake
Somerville
Navasota
Lake Conroe
Willis
Conroe
Cleveland
Kountze
Silsbee
Lumberton
Austin
Elgin
Bastrop
Giddings
Brenham
The Woodlands
Kingwood
Humble
Hempstead
Dayton
Liberty
Beaumont
Wallisville Lake
Nederland
Port Neches
Port Arthur
Orange
Vinton
Sabine Lake
Smithville
Lockhart
Luling
Sealy
Columbus
Eagle Lake
HOUSTON
Missouri City
Sugar Land
Rosenberg
Pearland
Baytown
League City
Texas City
Galveston Bay
BOLIVAR PENINSULA
Galveston
GALVESTON ISLAND
Gonzales
Hallettsville
Wharton
West Columbia
Angleton
Lake Jackson
Freeport
West Bay
Yoakum
Cuero
El Campo
Edna
Lake Texana
Bay City
Victoria
Palacios
Coleto Creek Reservoir
Goliad
Port Lavaca
Matagorda Bay
MATAGORDA ISLAND
Beeville
Refugio
San Antonio Bay
Copano Bay
Rockport
Sinton
SAN JOSE ISLAND
Aransas Pass
Portland
Ingleside
Corpus Christi
Corpus Christi Bay
MUSTANG ISLAND
Robstown
PADRE ISLAND
Laguna Madre
Raymondville
Harlingen
San Benito
Brownsville
Matamoros
BARRA DE LOS AMERICANOS
El Dorado
Crossett
Springhill
Haynesville
Bayou Bodcau Reservoir
Homer
Bayou D'Arbonne Lake
Farmerville
Bastrop
Benton
Minden
Arcadia
Grambling
Ruston
Monroe
West Monroe
Rayville
Delhi
Richwood
Tallulah
Vicksburg
Lake Providence
Rolling Fork
Yazoo City
MISSISSIPPI CHOCTAW IND. RES.
Carthage
Canton
Madison
Ross R. Barnett Reservoir
Ridgeland
Jackson
Clinton
Pearl
Brandon
Morton
Forest
Lake Bistineau
Jonesboro
Winnsboro
Crystal Springs
Coushatta
Mansfield
Winnfield
Olla
Port Gibson
Hazlehurst
Mendenhall
Magee
Natchitoches
Many
Lake Iatt
Jena
Catahoula Lake
Jonesville
Ferriday
Vidalia
Natchez
Fayette
Brookhaven
Collins
Hattiesburg
Colfax
Ball
Pineville
Alexandria
Marksville
McComb
Magnolia
Columbia
Purvis
Tylertown
Lumberton
Leesville
New Llano
Lecompte
Glenmora
Bunkie
Woodville
Kentwood
Franklinton
Poplarville
Bogalusa
De Ridder
Oakdale
Jackson
Zachary
Baker
Amite
Merryville
Mamou
Ville Platte
New Roads
Baton Rouge
Hammond
Covington
Picayune
Gulfport
Long Beach
Bay St. Louis
Opelousas
Eunice
Kinder
De Quincy
Ponchatoula
Mandeville
Slidell
Lake Maurepas
Lake Pontchartrain
Plaquemine
Gonzales
Westlake
Sulphur
Lake Charles
Jennings
Crowley
Rayne
Lafayette
St. Martinville
Donaldsonville
La Place
Kenner
Metairie
Marrero
Gretna
New Orleans
Chalmette
Lake Borgne
Chandeleur Sound
Calcasieu Lake
Grand Lake
White Lake
Lake Arthur
Kaplan
Abbeville
New Iberia
Jeanerette
Franklin
Morgan City
Thibodaux
Lake Salvador
Raceland
Breton Sound
Patterson
Houma
Larose
Cut Off
Port Sulphur
Vermilion Bay
MARSH ISLAND
Atchafalaya Bay
POINT AU FER ISLAND
Caillou Bay
Terrebonne Bay
Timbalier Bay
Barataria Bay
Grand Isle
Buras
© Rand McNally A-101625-1
a
Moss Hill
The Woodlands
New Caney
Spring
Tomball
Porter
Kingwood
Westfield
Huffman
Dayton
Liberty
Hockley
Cypress
Humble
Lake Houston
Crosby
Aldine
Jersey Village
Sheldon
Sheldon Reservoir
Mont Belvieu
Wallisville Lake
Wallisville
Addicks Reservoir
Highlands
Katy
Hunters Creek Village
Jacinto City
Channelview
Barker Reservoir
HOUSTON
Galena Park
Baytown
West University Place
Bellaire
Pasadena
Deer Park
Mission Bend
Sugar Land
South Houston
La Porte
Trinity Bay
Stafford
Missouri City
Richmond
Rosenberg
Fresno
Seabrook
Kemah
Pearland
Friendswood
SMITH POINT
Smith Point
League City
Galveston Bay
Dickinson
Smithers Lake
Manvel
BOLIVAR PENINSULA
Needville
Alvin
Algoa
Santa Fe
Texas City
Port Bolivar
La Marque
Guy
Hitchcock
Galveston
Liverpool
Gulf of Mexico
Eagle Nest Lake
Danbury
West Bay
GALVESTON ISLAND
© Rand McNally A-101626-1

Inset map a
Lambert Conformal Conic Projection
Scale 1:1,000,000
One inch to 16 miles
One cm to 10 km

MISSOURI
KENTUCKY
ARKANSAS
TENNESSEE
MISSISSIPPI
ALABAMA
GEORGIA
LOUISIANA
FLORIDA
CUMBERLAND PLATEAU
APPALACHIAN
GULF OF MEXICO
OZARK PLAT.
Memphis
Nashville
Knoxville
Chattanooga
Huntsville
Birmingham
Atlanta
Montgomery
Jackson
Baton Rouge
New Orleans
Mobile
Pensacola
Tallahassee
Lafayette
Macon
Columbus
Tuscaloosa
Hattiesburg
Biloxi
Gulfport
Meridian
Vicksburg
Tupelo
Florence
Decatur
Gadsden
Anniston
Auburn
Dothan
Albany
Paducah
Clarksville
Bowling Green
Hopkinsville
Murfreesboro
Jonesboro
Greenville
Houma
Metairie
Panama City
Fort Walton Beach
Lake Pontchartrain
Kentucky Lake
Lake Barkley
Mississippi
Tennessee
Ohio
Alabama
Chandeleur Sound
CHANDELEUR ISLAND
Breton Sound
Atchafalaya Bay
Mississippi Sound
CAPE SAN BLAS
CAPE ST. GEORGE
ST. GEORGE ISLAND
Apalachee Bay
GREAT SMOKY MTS. NAT'L PARK
Clingmans Dome 6643 ft. 2025 m
MAMMOTH CAVE NAT'L PARK
RUSSELL CAVE NAT'L MON.
MISSISSIPPI CHOCTAW IND. RES.
OCMULGEE NAT'L MON.
Ross R. Barnett Reservoir
W. F. George Reservoir
Lake Seminole
Martin Lake
Lewis Smith Lake
Wheeler Lake
Wilson Lake
Guntersville Lake
Sardis Lake
Enid Lake
Grenada Lake
Lake Sidney Lanier
Lake Sinclair
Dale Hollow Lake
Norris Lake
Lake Cumberland
Cherokee Lake
Chickamauga Lake
Reelfoot Lake
Wappapello Lake
Norfork Lake
Greers Ferry Lake
HORN ISLAND
DAUPHIN ISLAND
SANTA ROSA ISLAND
POINT AU FER ISLAND
MARSH ISLAND
92°
90°
88°
86°
84°
36°
32°
© Rand McNally
A-101627-1

0 20 40 60 80 100 120 Miles
0 20 40 60 80 100 120 140 160 180 200 Kilometers

Lambert Conformal Conic Projection
Scale 1:4,000,000
One inch to 64 miles
One cm to 40 km

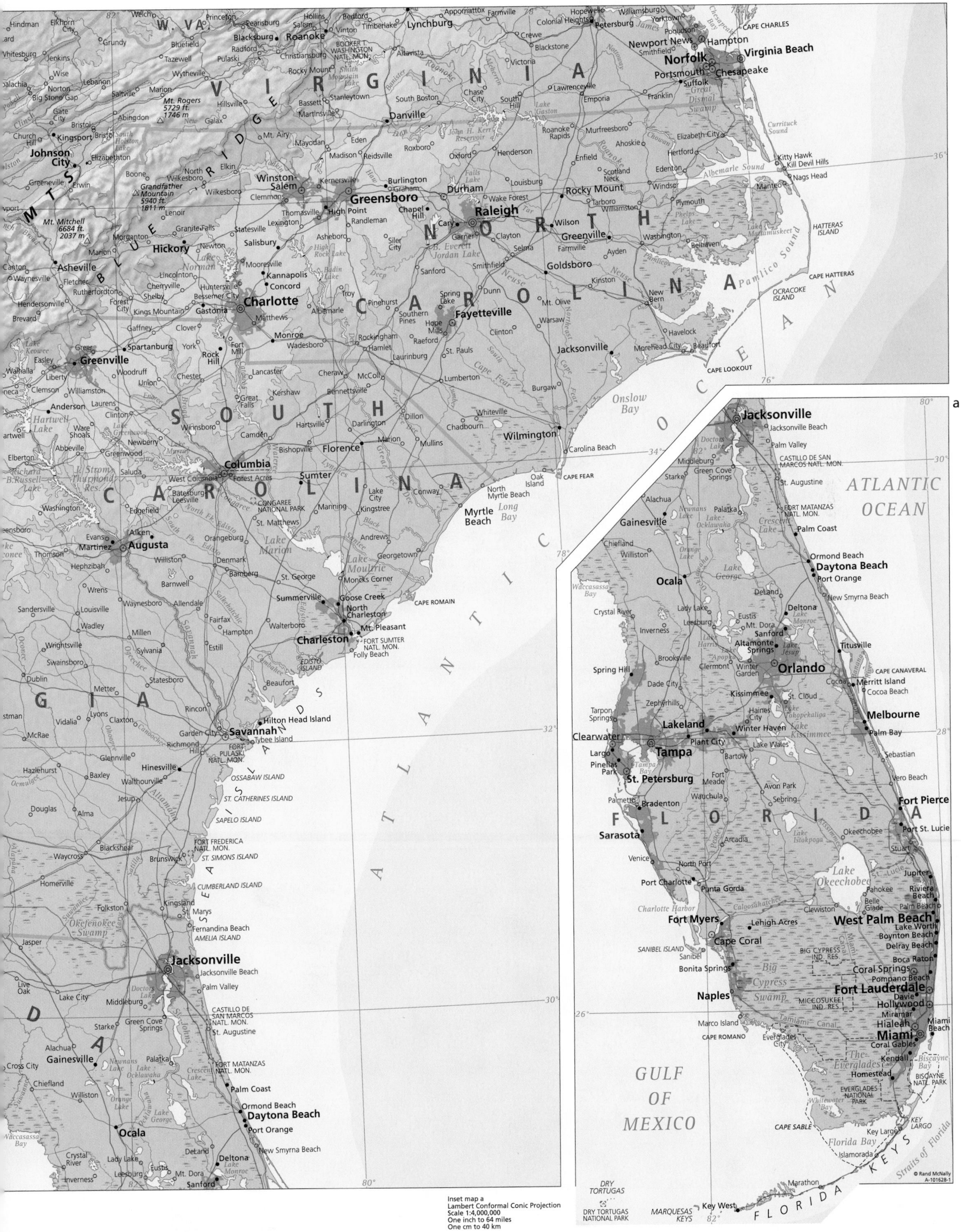
VIRGINIA
W. VA.
NORTH CAROLINA
SOUTH CAROLINA
GEORGIA
FLORIDA
BLUE RIDGE MTS.
ATLANTIC OCEAN
SEA ISLANDS
GULF OF MEXICO
FLORIDA KEYS
Straits of Florida
Lynchburg
Roanoke
Blacksburg
Danville
Norfolk
Virginia Beach
Chesapeake
Portsmouth
Newport News
Hampton
Suffolk
Petersburg
Appomattox
Farmville
Williamsburg
Yorktown
Colonial Heights
Hopewell
Emporia
Franklin
South Boston
Martinsville
Bassett
Stanleytown
Galax
Marion
Wytheville
Pulaski
Radford
Christiansburg
Salem
Vinton
Bedford
Rocky Mount
Altavista
Crewe
Blackstone
Victoria
Lawrenceville
Chase City
South Hill
Abingdon
Bristol
Kingsport
Johnson City
Elizabethton
Big Stone Gap
Norton
Wise
Lebanon
Saltville
Tazewell
Bluefield
Princeton
Welch
Mt. Rogers 5729 ft. 1746 m
Grandfather Mountain 5940 ft. 1811 m
Mt. Mitchell 6684 ft. 2037 m
BOOKER T. WASHINGTON NATL. MON.
Smith Mountain Lake
Lake Gaston
John H. Kerr Reservoir
Great Dismal Swamp
Currituck Sound
Albemarle Sound
Pamlico Sound
Kitty Hawk
Kill Devil Hills
Nags Head
Manteo
Elizabeth City
Hertford
Edenton
Windsor
Plymouth
Washington
Belhaven
Lake Mattamuskeet
HATTERAS ISLAND
CAPE HATTERAS
OCRACOKE ISLAND
CAPE LOOKOUT
CAPE CHARLES
Winston-Salem
Greensboro
High Point
Burlington
Graham
Durham
Chapel Hill
Raleigh
Cary
Garner
Clayton
Wake Forest
Louisburg
Henderson
Oxford
Roxboro
Reidsville
Eden
Madison
Mayodan
Mt. Airy
Elkin
North Wilkesboro
Wilkesboro
Boone
Lenoir
Morganton
Hickory
Newton
Statesville
Mooresville
Kannapolis
Concord
Salisbury
Lexington
Thomasville
Asheboro
Charlotte
Gastonia
Matthews
Monroe
Wadesboro
Rockingham
Hamlet
Laurinburg
Southern Pines
Pinehurst
Sanford
Fayetteville
Hope Mills
Raeford
Lumberton
St. Pauls
Dunn
Smithfield
Selma
Goldsboro
Wilson
Rocky Mount
Tarboro
Williamston
Greenville
Farmville
Ayden
Kinston
New Bern
Havelock
Morehead City
Beaufort
Jacksonville
Burgaw
Whiteville
Chadbourn
Wilmington
Carolina Beach
CAPE FEAR
Oak Island
Onslow Bay
Long Bay
Asheville
Hendersonville
Brevard
Canton
Waynesville
Marion
Shelby
Kings Mountain
Forest City
Rutherfordton
Spartanburg
Greenville
Greer
Easley
Clemson
Anderson
Walhalla
Laurens
Union
Gaffney
Rock Hill
York
Chester
Lancaster
Camden
Columbia
West Columbia
Forest Acres
Sumter
Florence
Darlington
Hartsville
Bennettsville
Cheraw
Dillon
Marion
Mullins
Conway
Myrtle Beach
North Myrtle Beach
Georgetown
Kingstree
Manning
Lake Marion
Lake Moultrie
Moncks Corner
Goose Creek
Summerville
North Charleston
Charleston
Mt. Pleasant
Folly Beach
FORT SUMTER NATL. MON.
CAPE ROMAIN
CONGAREE NATIONAL PARK
Orangeburg
St. Matthews
Bamberg
Denmark
Barnwell
Aiken
North Augusta
Augusta
Martinez
Evans
Hephzibah
Waynesboro
Allendale
Hampton
Walterboro
Beaufort
Hilton Head Island
Savannah
Tybee Island
FORT PULASKI NATL. MON.
Richmond Hill
Hinesville
Statesboro
Vidalia
Lyons
Claxton
Metter
Swainsboro
Dublin
Jesup
Brunswick
St. Simons Island
FORT FREDERICA NATL. MON.
OSSABAW ISLAND
ST. CATHERINES ISLAND
SAPELO ISLAND
CUMBERLAND ISLAND
Kingsland
St. Marys
Fernandina Beach
AMELIA ISLAND
Waycross
Blackshear
Okefenokee Swamp
Folkston
Homerville
Douglas
Alma
Hazlehurst
Baxley
Jacksonville
Jacksonville Beach
Palm Valley
CASTILLO DE SAN MARCOS NATL. MON.
St. Augustine
FORT MATANZAS NATL. MON.
Palm Coast
Ormond Beach
Daytona Beach
Port Orange
New Smyrna Beach
Deltona
Sanford
Gainesville
Ocala
Palatka
Starke
Middleburg
Lake City
Orlando
Kissimmee
St. Cloud
Titusville
CAPE CANAVERAL
Merritt Island
Cocoa
Cocoa Beach
Melbourne
Palm Bay
Sebastian
Vero Beach
Fort Pierce
Port St. Lucie
Stuart
Jupiter
Lake Okeechobee
Okeechobee
West Palm Beach
Palm Beach
Lake Worth
Boynton Beach
Delray Beach
Boca Raton
Pompano Beach
Coral Springs
Fort Lauderdale
Hollywood
Miramar
Hialeah
Miami
Miami Beach
Coral Gables
Kendall
Homestead
BISCAYNE NATL. PARK
EVERGLADES NATIONAL PARK
The Everglades
Florida Bay
Key Largo
Islamorada
Marathon
Key West
DRY TORTUGAS NATIONAL PARK
MARQUESAS KEYS
CAPE SABLE
CAPE ROMANO
Everglades City
Marco Island
Naples
Bonita Springs
Big Cypress Swamp
BIG CYPRESS IND. RES.
MICCOSUKEE IND. RES.
Tamiami Canal
Cape Coral
Fort Myers
Lehigh Acres
SANIBEL ISLAND
Charlotte Harbor
Punta Gorda
Port Charlotte
North Port
Venice
Sarasota
Bradenton
Palmetto
St. Petersburg
Tampa
Tampa Bay
Clearwater
Largo
Pinellas Park
Tarpon Springs
Spring Hill
Lakeland
Plant City
Winter Haven
Bartow
Lake Wales
Avon Park
Sebring
Arcadia
Wauchula
Crystal River
Inverness
Leesburg
Lady Lake
Eustis
Mt. Dora
Altamonte Springs
Winter Garden
Clermont
Dade City
Zephyrhills
Waccasassa Bay
Chiefland
Williston
DeLand
Lake George
© Rand McNally A-101628-1

Inset map a
Lambert Conformal Conic Projection
Scale 1:4,000,000
One inch to 64 miles
One cm to 40 km

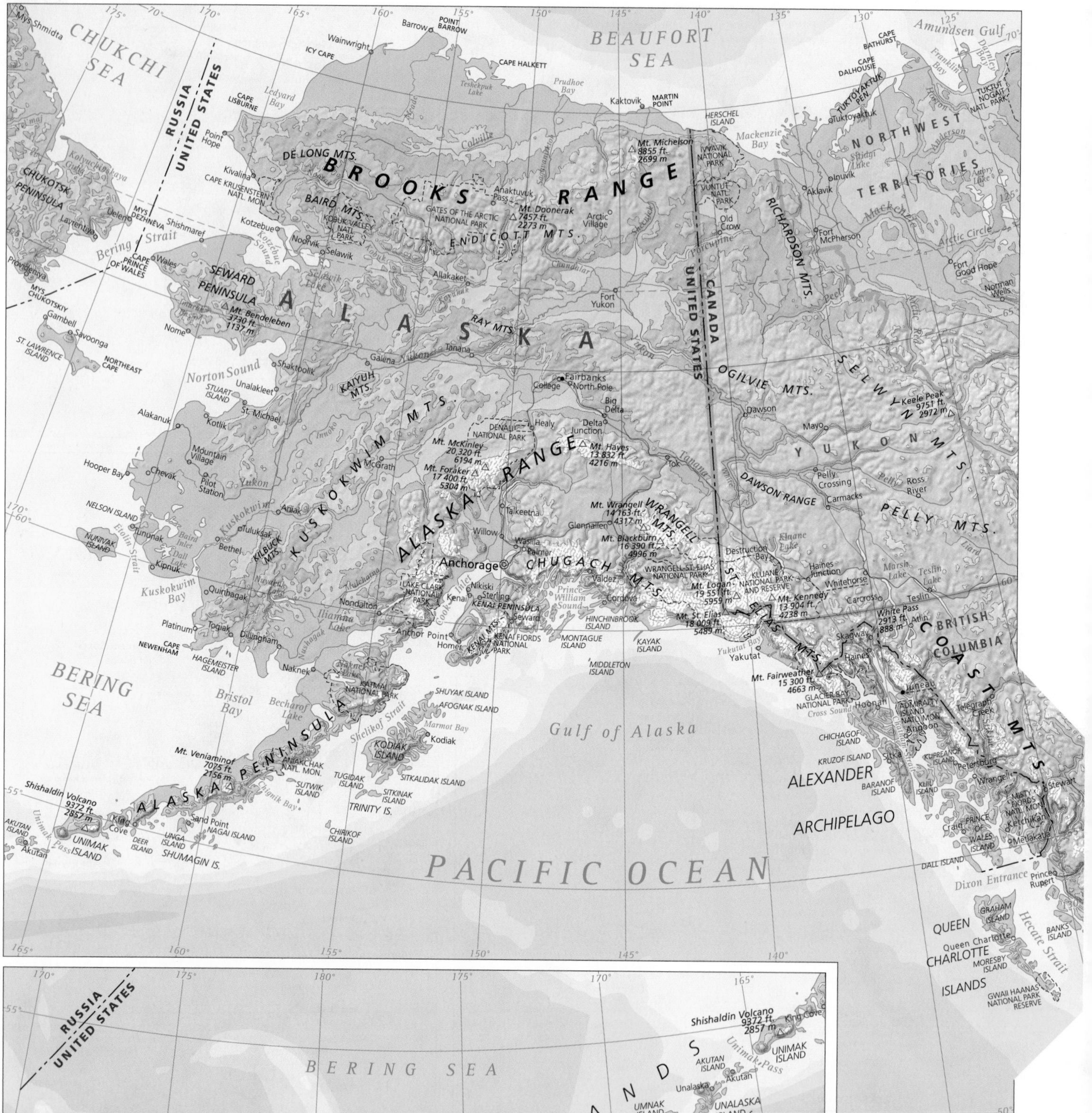
BEAUFORT SEA
CHUKCHI SEA
RUSSIA
UNITED STATES
CANADA
BROOKS RANGE
ALASKA
DE LONG MTS.
BAIRD MTS.
ENDICOTT MTS.
GATES OF THE ARCTIC NATIONAL PARK
KOBUK VALLEY NATL. PARK
CAPE KRUSENSTERN NATL. MON.
Mt. Michelson 8855 ft. 2699 m
Mt. Doonerak 7457 ft. 2273 m
IVVAVIK NATIONAL PARK
VUNTUT NATL. PARK
RICHARDSON MTS.
NORTHWEST TERRITORIES
Amundsen Gulf
CAPE BATHURST
CAPE DALHOUSIE
TUKTOYAKTUK PEN.
Tuktoyaktuk
TUKTUT NOGAIT NATL. PARK
Mackenzie Bay
HERSCHEL ISLAND
Inuvik
Aklavik
Fort McPherson
Arctic Circle
Fort Good Hope
Norman Wells
Old Crow
Kaktovik
MARTIN POINT
Prudhoe Bay
Teshekpuk Lake
CAPE HALKETT
Barrow
POINT BARROW
Wainwright
ICY CAPE
Ledyard Bay
CAPE LISBURNE
Point Hope
Kivalina
Noatak
Kotzebue
Kotzebue Sound
Noorvik
Selawik
Selawik Lake
Anaktuvuk Pass
Arctic Village
Allakaket
Fort Yukon
Chandalar
Porcupine
Colville
Mys Shmidta
CHUKOTSK PENINSULA
Kolyuchinskaya Guba
Lavrentiya
Uelen
MYS DEZHNEVA
Bering Strait
Shishmaref
CAPE PRINCE OF WALES
Wales
Provideniya
MYS CHUKOTSKIY
SEWARD PENINSULA
Mt. Bendeleben 3730 ft. 1137 m
Nome
Gambell
Savoonga
ST. LAWRENCE ISLAND
NORTHEAST CAPE
Norton Sound
STUART ISLAND
Unalakleet
Shaktoolik
St. Michael
Kotlik
Alakanuk
KAIYUH MTS.
Galena
Tanana
RAY MTS.
Fairbanks
College
North Pole
Big Delta
Delta Junction
Healy
DENALI NATIONAL PARK
Mt. McKinley 20 320 ft. 6194 m
Mt. Foraker 17 400 ft. 5304 m
Mt. Hayes 13 832 ft. 4216 m
ALASKA RANGE
KUSKOKWIM MTS.
McGrath
Tok
Tanana
Yukon
OGILVIE MTS.
Dawson
SELWYN MTS.
Keele Peak 9751 ft. 2972 m
Mayo
YUKON MTS.
Pelly Crossing
Ross River
DAWSON RANGE
Carmacks
PELLY MTS.
Mountain Village
Hooper Bay
Chevak
Pilot Station
NELSON ISLAND
NUNIVAK ISLAND
Etolin Strait
Tununak
Tuluksak
Aniak
Bethel
KILBUCK MTS.
Kipnuk
Kuskokwim Bay
Quinhagak
Platinum
Togiak
Dillingham
CAPE NEWENHAM
HAGEMEISTER ISLAND
Talkeetna
Willow
Wasilla
Palmer
Anchorage
Glennallen
Mt. Wrangell 14 163 ft. 4317 m
WRANGELL MTS.
Mt. Blackburn 16 390 ft. 4996 m
WRANGELL-ST. ELIAS NATIONAL PARK
CHUGACH MTS.
Valdez
Cordova
Prince William Sound
HINCHINBROOK ISLAND
MONTAGUE ISLAND
KAYAK ISLAND
MIDDLETON ISLAND
Mt. Logan 19 551 ft. 5959 m
KLUANE NATIONAL PARK AND RESERVE
Mt. Kennedy 13 904 ft. 4238 m
Mt. St. Elias 18 009 ft. 5489 m
ST. ELIAS MTS.
Destruction Bay
Kluane Lake
Haines Junction
Whitehorse
Carcross
Teslin
Teslin Lake
Marsh Lake
White Pass 2913 ft. 888 m
Atlin
BRITISH COLUMBIA
COAST MTS.
Skagway
Haines
Juneau
Yakutat
Yakutat Bay
Mt. Fairweather 15 300 ft. 4663 m
GLACIER BAY NATIONAL PARK
Cross Sound
Hoonah
ADMIRALTY ISLAND NATL. MON.
Angoon
Telegraph Creek
CHICHAGOF ISLAND
KRUZOF ISLAND
Sitka
BARANOF ISLAND
KUPREANOF ISLAND
Petersburg
Wrangell
KUIU ISLAND
Stewart
MISTY FJORDS NATL. MON.
Ketchikan
Metlakatla
Craig
PRINCE OF WALES ISLAND
ALEXANDER ARCHIPELAGO
DALL ISLAND
Dixon Entrance
Prince Rupert
GRAHAM ISLAND
Queen Charlotte
QUEEN CHARLOTTE ISLANDS
MORESBY ISLAND
BANKS ISLAND
Hecate Strait
GWAII HAANAS NATIONAL PARK RESERVE
LAKE CLARK NATIONAL PARK
Nondalton
Iliamna Lake
Nikiski
Kenai
Sterling
KENAI PENINSULA
Seward
KENAI MTS.
KENAI FJORDS NATIONAL PARK
Cook Inlet
Anchor Point
Homer
Nushagak Bay
Naknek
KATMAI NATIONAL PARK
Becharof Lake
Bristol Bay
BERING SEA
SHUYAK ISLAND
AFOGNAK ISLAND
Marmot Bay
Shelikof Strait
Kodiak
KODIAK ISLAND
SITKALIDAK ISLAND
TUGIDAK ISLAND
SITKINAK ISLAND
TRINITY IS.
ANIAKCHAK NATL. MON.
SUTWIK ISLAND
CHIRIKOF ISLAND
ALASKA PENINSULA
Mt. Veniaminof 7075 ft. 2156 m
Chignik Bay
Shishaldin Volcano 9372 ft. 2857 m
King Cove
Sand Point
NAGAI ISLAND
UNGA ISLAND
DEER ISLAND
SHUMAGIN IS.
AKUTAN ISLAND
Akutan
Unimak Pass
UNIMAK ISLAND
Gulf of Alaska
PACIFIC OCEAN
A-101615-1
© Rand McNally

a

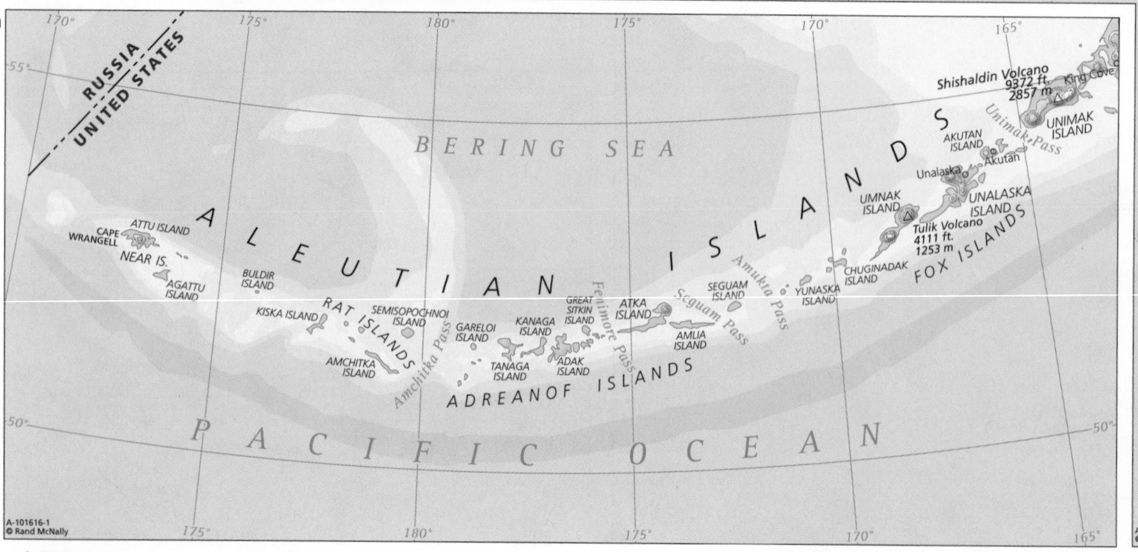
RUSSIA
UNITED STATES
BERING SEA
ALEUTIAN ISLANDS
Shishaldin Volcano 9372 ft. 2857 m
King Cove
UNIMAK ISLAND
Unimak Pass
AKUTAN ISLAND
Akutan
Unalaska
UNALASKA ISLAND
UMNAK ISLAND
Tulik Volcano 4111 ft. 1253 m
FOX ISLANDS
CHUGINADAK ISLAND
YUNASKA ISLAND
Amukta Pass
SEGUAM ISLAND
Seguam Pass
AMLIA ISLAND
ATKA ISLAND
GREAT SITKIN ISLAND
Fenimore Pass
ADAK ISLAND
KANAGA ISLAND
TANAGA ISLAND
GARELOI ISLAND
SEMISOPOCHNOI ISLAND
Amchitka Pass
ADREANOF ISLANDS
RAT ISLANDS
KISKA ISLAND
AMCHITKA ISLAND
BULDIR ISLAND
AGATTU ISLAND
ATTU ISLAND
CAPE WRANGELL
NEAR IS.
PACIFIC OCEAN
A-101616-1
© Rand McNally

Inset map a
Lambert Azimuthal Equal Area Projection
Scale 1:12,000,000
One inch to 190 miles
One cm to 120 km

0 60 120 180 240 300 360 Miles
0 60 120 180 240 300 360 420 480 540 600 Kilometers

Lambert Azimuthal Equal Area Projection
Scale 1:12,000,000
One inch to 190 miles
One cm to 120 km

a

b

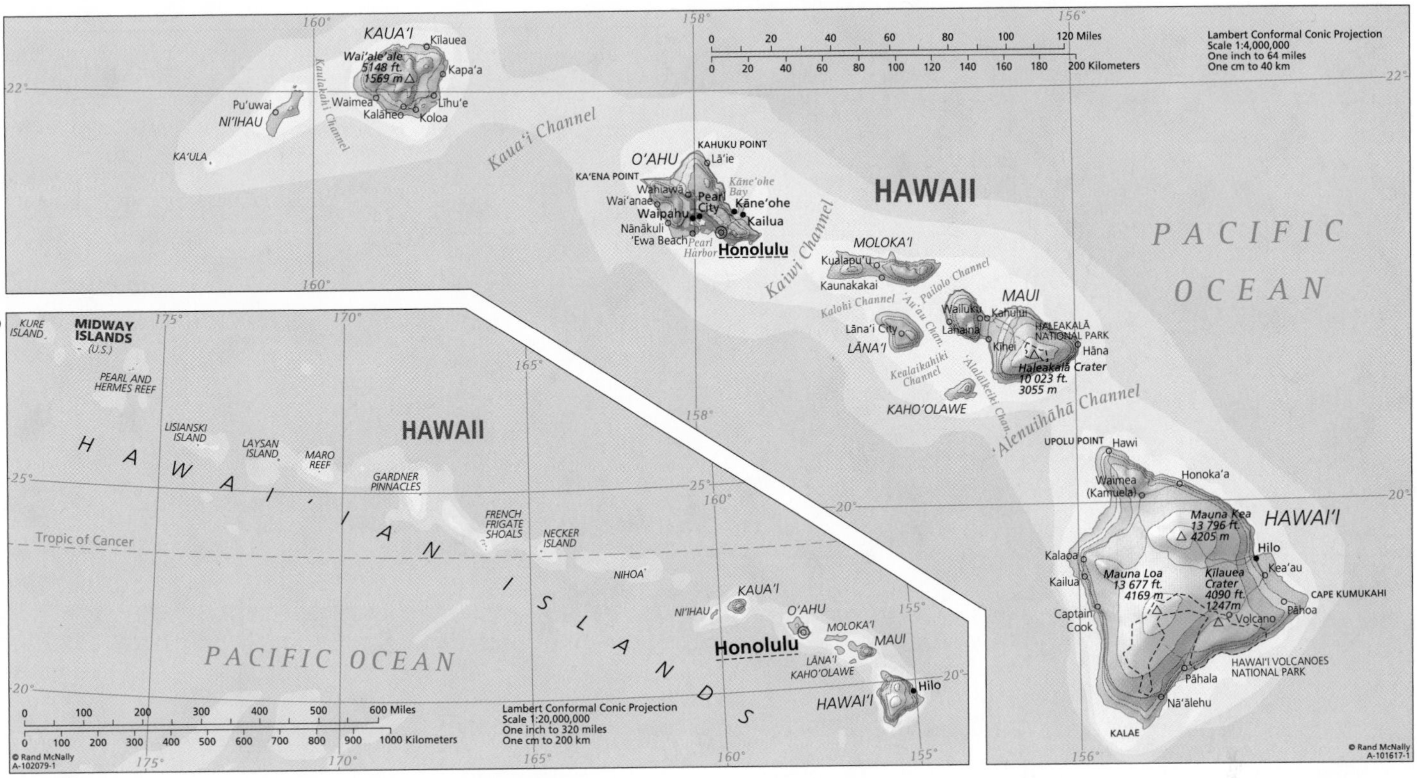

KAUA'I
Kilauea
Wai'ale'ale 5148 ft. 1569 m
Kapa'a
Waimea
Līhu'e
Kalāheo
Koloa
Pu'uwai
NI'IHAU
KA'ULA
Kaulakahi Channel
Kaua'i Channel
O'AHU
KAHUKU POINT
Lā'ie
KA'ENA POINT
Wahiawā
Wai'anae
Pearl City
Waipahu
Kāne'ohe Bay
Kāne'ohe
Kailua
Nānākuli
'Ewa Beach
Pearl Harbor
Honolulu
Kaiwi Channel
HAWAII
MOLOKA'I
Kualapu'u
Kaunakakai
Kalohi Channel
Pailolo Channel
'Au'au Chan.
MAUI
Wailuku
Kahului
Lahaina
Lāna'i City
LĀNA'I
Kīhei
HALEAKALĀ NATIONAL PARK
Hāna
Haleakalā Crater 10 023 ft. 3055 m
Kealaikahiki Channel
KAHO'OLAWE
'Alalākeiki Chan.
'Alenuihāhā Channel
UPOLU POINT
Hawi
Honoka'a
Waimea (Kamuela)
Mauna Kea 13 796 ft. 4205 m
HAWAI'I
Hilo
Kea'au
Kalaoa
Kailua
Mauna Loa 13 677 ft. 4169 m
Kilauea Crater 4090 ft. 1247m
Volcano
Pāhoa
CAPE KUMUKAHI
Captain Cook
Pāhala
HAWAI'I VOLCANOES NATIONAL PARK
Nā'ālehu
KALAE
PACIFIC OCEAN
0 20 40 60 80 100 120 Miles
0 20 40 60 80 100 120 140 160 180 200 Kilometers
Lambert Conformal Conic Projection
Scale 1:4,000,000
One inch to 64 miles
One cm to 40 km
© Rand McNally
A-101617-1
KURE ISLAND
MIDWAY ISLANDS (U.S.)
PEARL AND HERMES REEF
LISIANSKI ISLAND
LAYSAN ISLAND
MARO REEF
GARDNER PINNACLES
FRENCH FRIGATE SHOALS
NECKER ISLAND
NIHOA
HAWAII
HAWAIIAN ISLANDS
Tropic of Cancer
KAUA'I
NI'IHAU
O'AHU
MOLOKA'I
MAUI
LĀNA'I
KAHO'OLAWE
HAWAI'I
Honolulu
Hilo
PACIFIC OCEAN
0 100 200 300 400 500 600 Miles
0 100 200 300 400 500 600 700 800 900 1000 Kilometers
Lambert Conformal Conic Projection
Scale 1:20,000,000
One inch to 320 miles
One cm to 200 km
© Rand McNally
A-102079-1

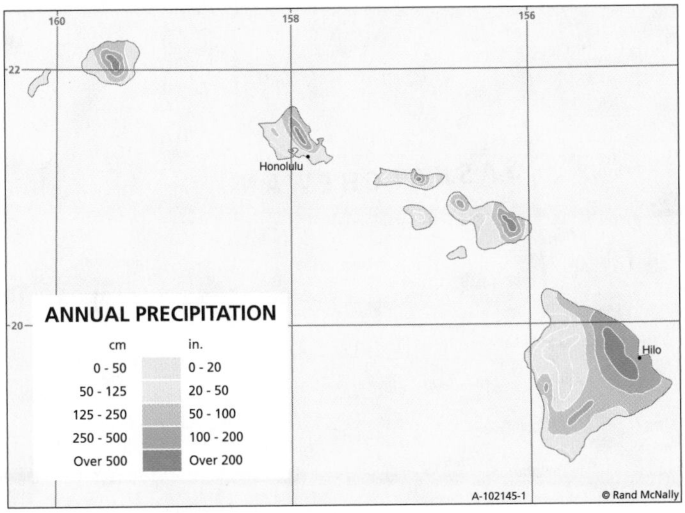

Honolulu
Hilo
ANNUAL PRECIPITATION
cm in.
0 - 50 0 - 20
50 - 125 20 - 50
125 - 250 50 - 100
250 - 500 100 - 200
Over 500 Over 200
A-102145-1
© Rand McNally

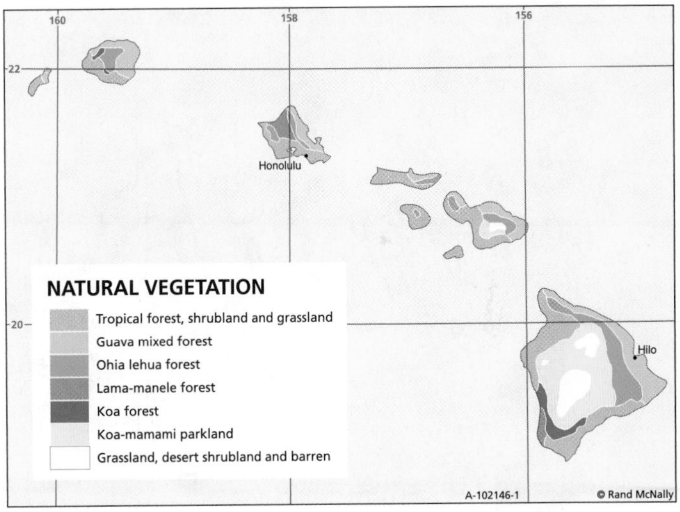

Honolulu
Hilo
NATURAL VEGETATION
Tropical forest, shrubland and grassland
Guava mixed forest
Ohia lehua forest
Lama-manele forest
Koa forest
Koa-mamami parkland
Grassland, desert shrubland and barren
A-102146-1
© Rand McNally

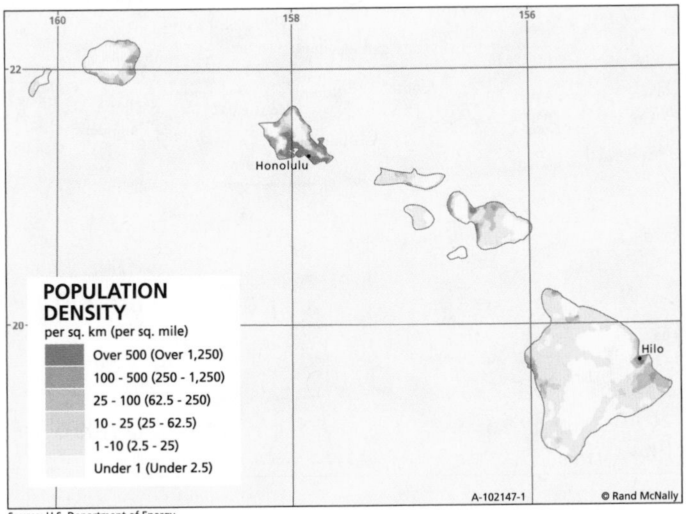

Honolulu
Hilo
POPULATION DENSITY
per sq. km (per sq. mile)
Over 500 (Over 1,250)
100 - 500 (250 - 1,250)
25 - 100 (62.5 - 250)
10 - 25 (25 - 62.5)
1 -10 (2.5 - 25)
Under 1 (Under 2.5)
A-102147-1
© Rand McNally

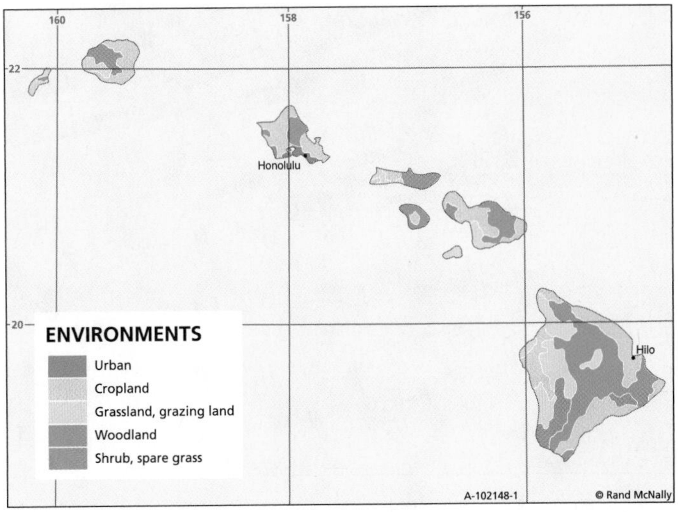

Honolulu
Hilo
ENVIRONMENTS
Urban
Cropland
Grassland, grazing land
Woodland
Shrub, spare grass
A-102148-1
© Rand McNally

Source: U.S. Department of Energy

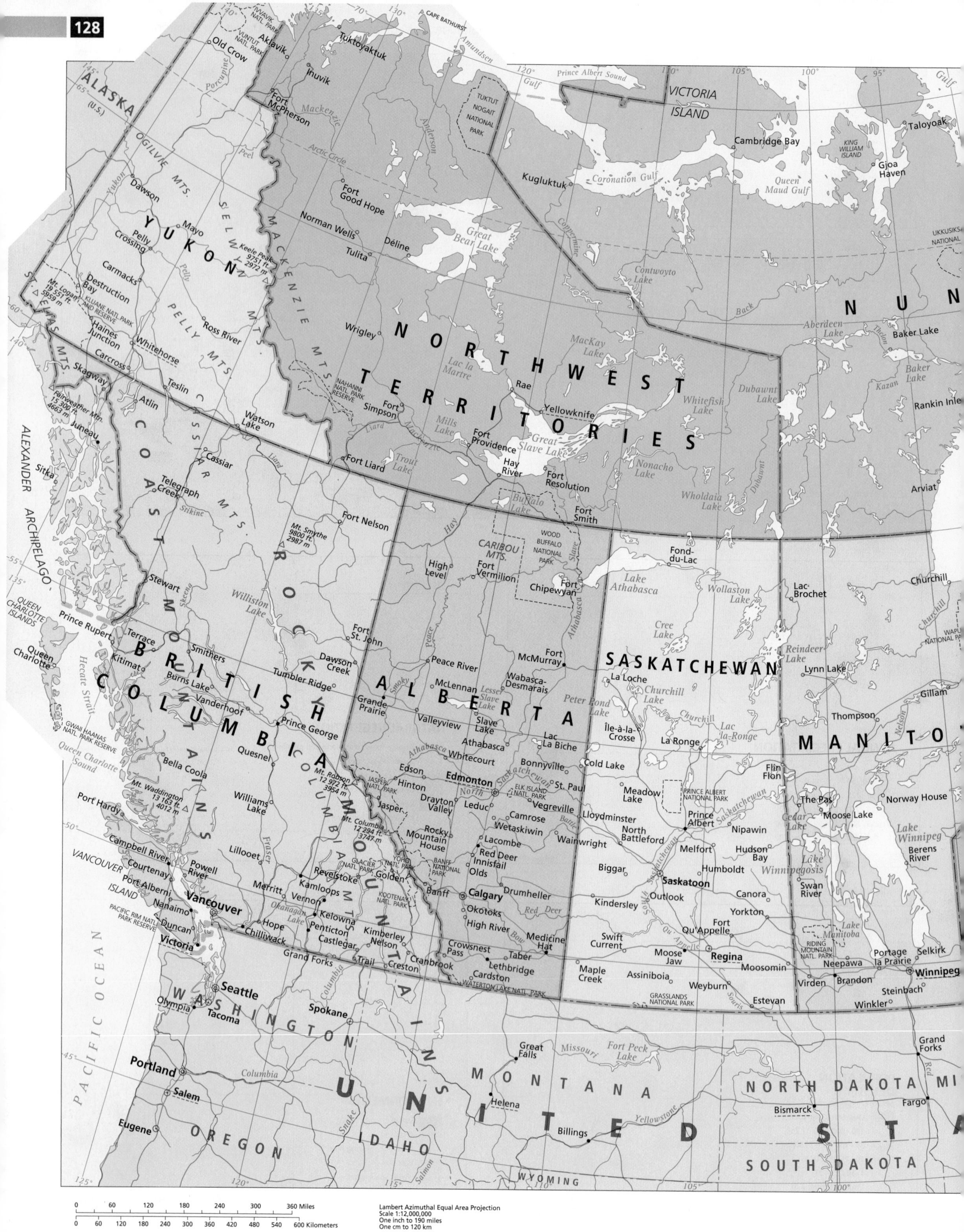
ALASKA (U.S.)
YUKON
NORTHWEST TERRITORIES
NUN
BRITISH COLUMBIA
ALBERTA
SASKATCHEWAN
MANITO
UNITED STA
WASHINGTON
OREGON
IDAHO
MONTANA
WYOMING
NORTH DAKOTA
SOUTH DAKOTA
MI
VICTORIA ISLAND
KING WILLIAM ISLAND
VANCOUVER ISLAND
QUEEN CHARLOTTE ISLANDS
ALEXANDER ARCHIPELAGO
PACIFIC OCEAN
OGILVIE MTS.
SELWYN MTS.
MACKENZIE MTS.
PELLY MTS.
CASSIAR MTS.
COAST MOUNTAINS
ROCKY MOUNTAINS
COLUMBIA MTS.
ST. ELIAS MTS.
CARIBOU MTS.
CAPE BATHURST
IVVAVIK NATL. PARK
VUNTUT NATL. PARK
TUKTUT NOGAIT NATIONAL PARK
UKKUSIKSALIK NATIONAL
KLUANE NATL PARK AND RESERVE
NAHANNI NATL. PARK RESERVE
WOOD BUFFALO NATIONAL PARK
GWAII HAANAS NATL. PARK RESERVE
PACIFIC RIM NATL. PARK RESERVE
JASPER NATL. PARK
BANFF NATIONAL PARK
YOHO NATL. PARK
GLACIER NATL. PARK
KOOTENAY NATL. PARK
ELK ISLAND NATL. PARK
WATERTON LAKE NATL. PARK
PRINCE ALBERT NATIONAL PARK
GRASSLANDS NATIONAL PARK
RIDING MOUNTAIN NATL. PARK
WAPUSK NATIONAL PA
Keele Peak 9751 ft. 2972 m
Mt. Logan 19 551 ft. 5959 m
Fairweather Mtn. 15 300 ft. 4663 m
Mt. Smythe 9800 ft. 2987 m
Mt. Robson 12 972 ft. 3954 m
Mt. Columbia 12 294 ft. 3747 m
Mt. Waddington 13 163 ft. 4012 m
Amundsen Gulf
Prince Albert Sound
Coronation Gulf
Queen Maud Gulf
Arctic Circle
Great Bear Lake
Great Slave Lake
Contwoyto Lake
Aberdeen Lake
Baker Lake
MacKay Lake
Lac la Martre
Whitefish Lake
Dubawnt Lake
Mills Lake
Trout Lake
Nonacho Lake
Wholdaia Lake
Buffalo Lake
Lake Athabasca
Wollaston Lake
Cree Lake
Reindeer Lake
Peter Pond Lake
Lac la Ronge
Lesser Slave Lake
Williston Lake
Cedar Lake
Lake Winnipeg
Lake Winnipegosis
Lake Manitoba
Fort Peck Lake
Okanagan Lake
Hecate Strait
Queen Charlotte Sound
Mackenzie
Anderson
Coppermine
Back
Thelon
Kazan
Dubawnt
Porcupine
Peel
Yukon
Pelly
Liard
Stikine
Skeena
Hay
Slave
Peace
Smoky
Athabasca
North Saskatchewan
South Saskatchewan
Battle
Red Deer
Bow
Churchill
Nelson
Qu'Appelle
Souris
Fraser
Columbia
Snake
Salmon
Missouri
Yellowstone
Red
Aklavik
Old Crow
Tuktoyaktuk
Inuvik
Fort McPherson
Dawson
Mayo
Pelly Crossing
Carmacks
Destruction Bay
Haines Junction
Whitehorse
Carcross
Ross River
Skagway
Teslin
Atlin
Watson Lake
Juneau
Sitka
Cassiar
Telegraph Creek
Fort Good Hope
Norman Wells
Tulita
Déline
Kugluktuk
Cambridge Bay
Taloyoak
Gjoa Haven
Wrigley
Fort Simpson
Rae
Yellowknife
Fort Providence
Hay River
Fort Resolution
Fort Liard
Fort Smith
Baker Lake
Rankin Inlet
Arviat
Fort Nelson
High Level
Fort Vermilion
Fort Chipewyan
Fond-du-Lac
Lac Brochet
Churchill
Stewart
Prince Rupert
Terrace
Kitimat
Smithers
Burns Lake
Vanderhoof
Queen Charlotte
Fort St. John
Dawson Creek
Tumbler Ridge
Peace River
McLennan
Grande Prairie
Valleyview
Slave Lake
Fort McMurray
Wabasca-Desmarais
La Loche
Lynn Lake
Gillam
Thompson
Prince George
Quesnel
Bella Coola
Williams Lake
Port Hardy
Athabasca
Lac La Biche
Whitecourt
Edson
Hinton
Jasper
Edmonton
Drayton Valley
Leduc
Bonnyville
Cold Lake
St. Paul
Vegreville
Camrose
Wetaskiwin
Île-à-la-Crosse
La Ronge
Flin Flon
The Pas
Norway House
Moose Lake
Meadow Lake
Lloydminster
North Battleford
Prince Albert
Nipawin
Melfort
Hudson Bay
Berens River
Rocky Mountain House
Lacombe
Red Deer
Innisfail
Olds
Wainwright
Biggar
Humboldt
Saskatoon
Swan River
Campbell River
Powell River
Courtenay
Port Alberni
Nanaimo
Vancouver
Duncan
Victoria
Lillooet
Revelstoke
Golden
Merritt
Kamloops
Vernon
Kelowna
Penticton
Hope
Chilliwack
Banff
Calgary
Okotoks
High River
Drumheller
Kindersley
Outlook
Canora
Yorkton
Castlegar
Nelson
Kimberley
Grand Forks
Trail
Creston
Cranbrook
Crowsnest Pass
Lethbridge
Cardston
Taber
Medicine Hat
Swift Current
Moose Jaw
Fort Qu'Appelle
Regina
Maple Creek
Assiniboia
Weyburn
Estevan
Moosomin
Virden
Brandon
Neepawa
Portage la Prairie
Selkirk
Winnipeg
Steinbach
Winkler
Seattle
Olympia
Tacoma
Spokane
Great Falls
Helena
Billings
Portland
Salem
Eugene
Bismarck
Fargo
Grand Forks
0 60 120 180 240 300 360 Miles
0 60 120 180 240 300 360 420 480 540 600 Kilometers

Lambert Azimuthal Equal Area Projection
Scale 1:12,000,000
One inch to 190 miles
One cm to 120 km

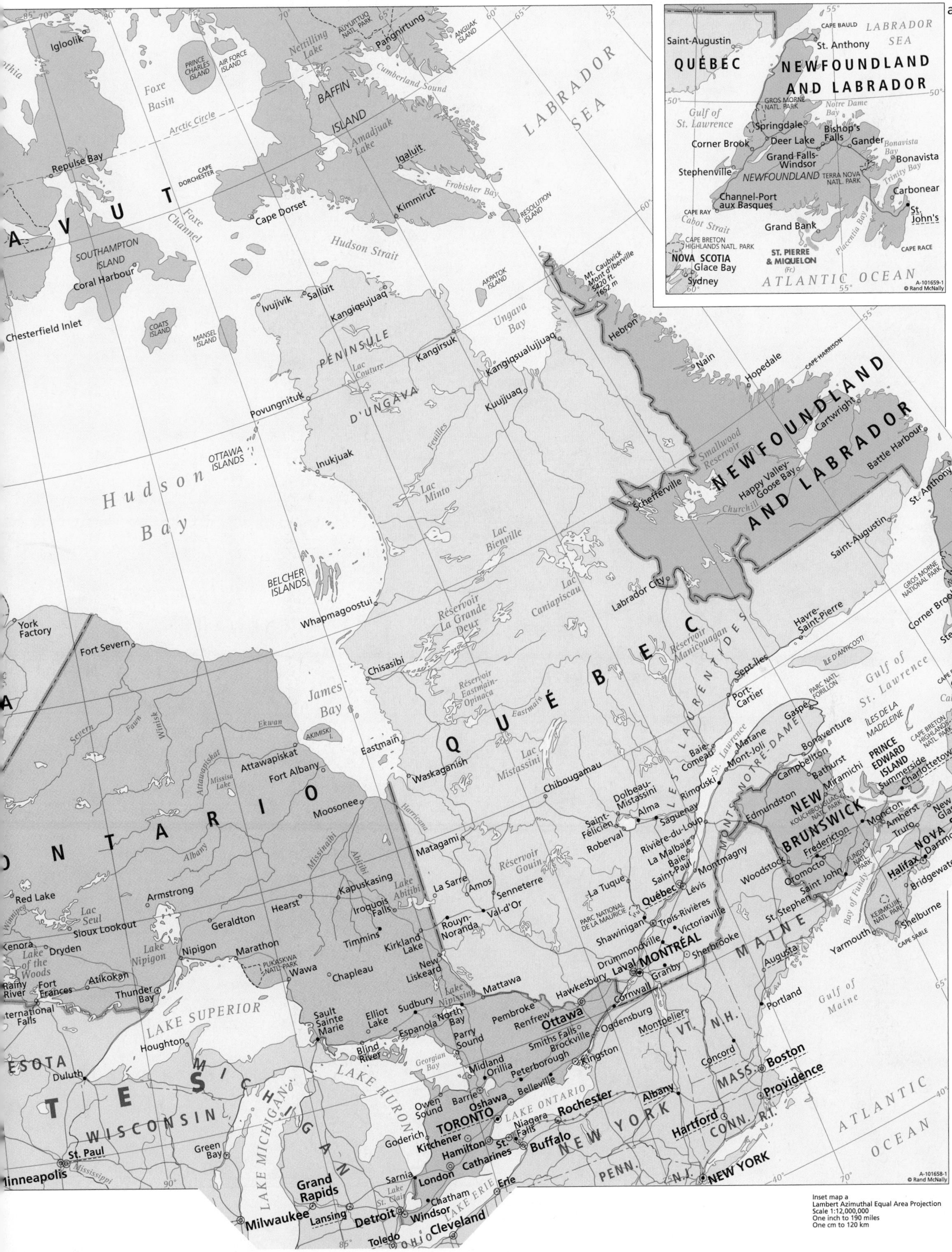
a
Saint-Augustin
QUÉBEC
NEWFOUNDLAND AND LABRADOR
LABRADOR SEA
CAPE BAULD
St. Anthony
GROS MORNE NATL. PARK
Notre Dame Bay
Gulf of St. Lawrence
Springdale
Bishop's Falls
Corner Brook
Deer Lake
Gander
Bonavista Bay
Bonavista
Grand Falls-Windsor
Stephenville
NEWFOUNDLAND
TERRA NOVA NATL. PARK
Trinity Bay
Carbonear
Channel-Port aux Basques
CAPE RAY
St. John's
Cabot Strait
Grand Bank
Placentia Bay
CAPE BRETON HIGHLANDS NATL. PARK
NOVA SCOTIA
Glace Bay
Sydney
ST. PIERRE & MIQUELON (Fr.)
CAPE RACE
ATLANTIC OCEAN
A-101659-1 © Rand McNally
Igloolik
Foxe Basin
PRINCE CHARLES ISLAND
AIR FORCE ISLAND
Nettilling Lake
AUYUITTUQ NATL. PARK
Pangnirtung
ANGIJAK ISLAND
Cumberland Sound
BAFFIN ISLAND
Arctic Circle
Amadjuak Lake
LABRADOR SEA
Repulse Bay
CAPE DORCHESTER
Iqaluit
Frobisher Bay
Kimmirut
RESOLUTION ISLAND
Foxe Channel
Cape Dorset
Hudson Strait
SOUTHAMPTON ISLAND
Coral Harbour
AKPATOK ISLAND
Mt. Caubvick Mont d'Iberville 5420 ft. 1652 m
Ivujivik
Salluit
Kangiqsujuaq
Chesterfield Inlet
COATS ISLAND
MANSEL ISLAND
Ungava Bay
Hebron
PÉNINSULE D'UNGAVA
Lac Couture
Kangirsuk
Kangiqsualujjuaq
Nain
Hopedale
CAPE HARRISON
Povungnituk
Kuujjuaq
Cartwright
NEWFOUNDLAND AND LABRADOR
Smallwood Reservoir
OTTAWA ISLANDS
Inukjuak
Hudson Bay
Lac Minto
Schefferville
Happy Valley-Goose Bay
Battle Harbour
Churchill
St. Anthony
Saint-Augustin
Lac Bienville
Springdale
BELCHER ISLANDS
Lac Caniapiscau
Labrador City
GROS MORNE NATIONAL PARK
Deer Lake
Whapmagoostui
Réservoir La Grande Deux
Corner Brook
Stephenville
York Factory
Havre-Saint-Pierre
Réservoir Manicouagan
Fort Severn
Chisasibi
ÎLE D'ANTICOSTI
Sept-Îles
Gulf of St. Lawrence
Réservoir Eastmain-Opinaca
Port-Cartier
Channel-Port aux Basques
Cabot Strait
James Bay
QUÉBEC
LES LAURENTIDES
Severn
Fawn
Winisk
Ekwan
AKIMISKI
Eastmain
PARC NATL. FORILLON
Gaspé
ÎLES DE LA MADELEINE
CAPE BRETON HIGHLANDS NATL. PARK
Glace Bay
Sydney
Attawapiskat
Fort Albany
Missisa Lake
Waskaganish
Lac Mistassini
Matane
Mont-Joli
Bonaventure
Baie-Comeau
Campbellton
PRINCE EDWARD ISLAND
Charlottetown
Summerside
Chibougamau
Rimouski
Bathurst
Miramichi
ONTARIO
Moosonee
Harricana
Dolbeau-Mistassini
Alma
Saguenay
Edmundston
KOUCHIBOUGUAC NATL. PARK
NEW BRUNSWICK
Moncton
Amherst
New Glasgow
NOVA SCOTIA
Saint-Félicien
Roberval
Rivière-du-Loup
La Malbaie
Baie-Saint-Paul
Albany
Missinaibi
Abitibi
Matagami
Réservoir Gouin
MONTS NOTRE-DAME
Fredericton
Truro
Halifax
Dartmouth
FUNDY NATL. PARK
Montmagny
Woodstock
Oromocto
Saint John
Bridgewater
Red Lake
Armstrong
Kapuskasing
Lac Seul
Hearst
Iroquois Falls
Lake Abitibi
La Sarre
Amos
Senneterre
La Tuque
Québec
Lévis
Bay of Fundy
KEJIMKUJIK NATL. PARK
Shelburne
Sioux Lookout
Geraldton
Rouyn-Noranda
Val-d'Or
PARC NATIONAL DE LA MAURICIE
Trois-Rivières
Victoriaville
St. Stephen
Kenora
Dryden
Lake Nipigon
Nipigon
Marathon
Timmins
Kirkland Lake
Shawinigan
Drummondville
MONTRÉAL
Sherbrooke
MAINE
Yarmouth
CAPE SABLE
Lake of the Woods
PUKASKWA NATL. PARK
Wawa
Chapleau
New Liskeard
Laval
Granby
Augusta
Rainy River
Fort Frances
Atikokan
Thunder Bay
Lake Nipissing
Mattawa
Hawkesbury
Cornwall
Portland
Gulf of Maine
International Falls
LAKE SUPERIOR
Sault Sainte Marie
Elliot Lake
Sudbury
North Bay
Pembroke
Renfrew
Ottawa
Ogdensburg
Montpelier
VT.
N.H.
Espanola
Parry Sound
Smiths Falls
Brockville
Houghton
Blind River
Georgian Bay
Kingston
Concord
Boston
MINNESOTA
Duluth
MICHIGAN
Midland
Orillia
Peterborough
MASS.
Providence
LAKE HURON
Belleville
Owen Sound
Barrie
Oshawa
LAKE ONTARIO
Rochester
Albany
ATLANTIC OCEAN
WISCONSIN
TORONTO
NEW YORK
Hartford
CONN.
R.I.
Kitchener
Niagara Falls
Goderich
Hamilton
St. Catharines
Buffalo
St. Paul
Green Bay
LAKE MICHIGAN
Minneapolis
Mississippi
PENN.
N.J.
NEW YORK
A-101658-1 © Rand McNally
Sarnia
London
Erie
Grand Rapids
Lake St. Clair
Chatham
LAKE ERIE
Milwaukee
Lansing
Detroit
Windsor
Cleveland
Toledo
OHIO

Inset map a
Lambert Azimuthal Equal Area Projection
Scale 1:12,000,000
One inch to 190 miles
One cm to 120 km

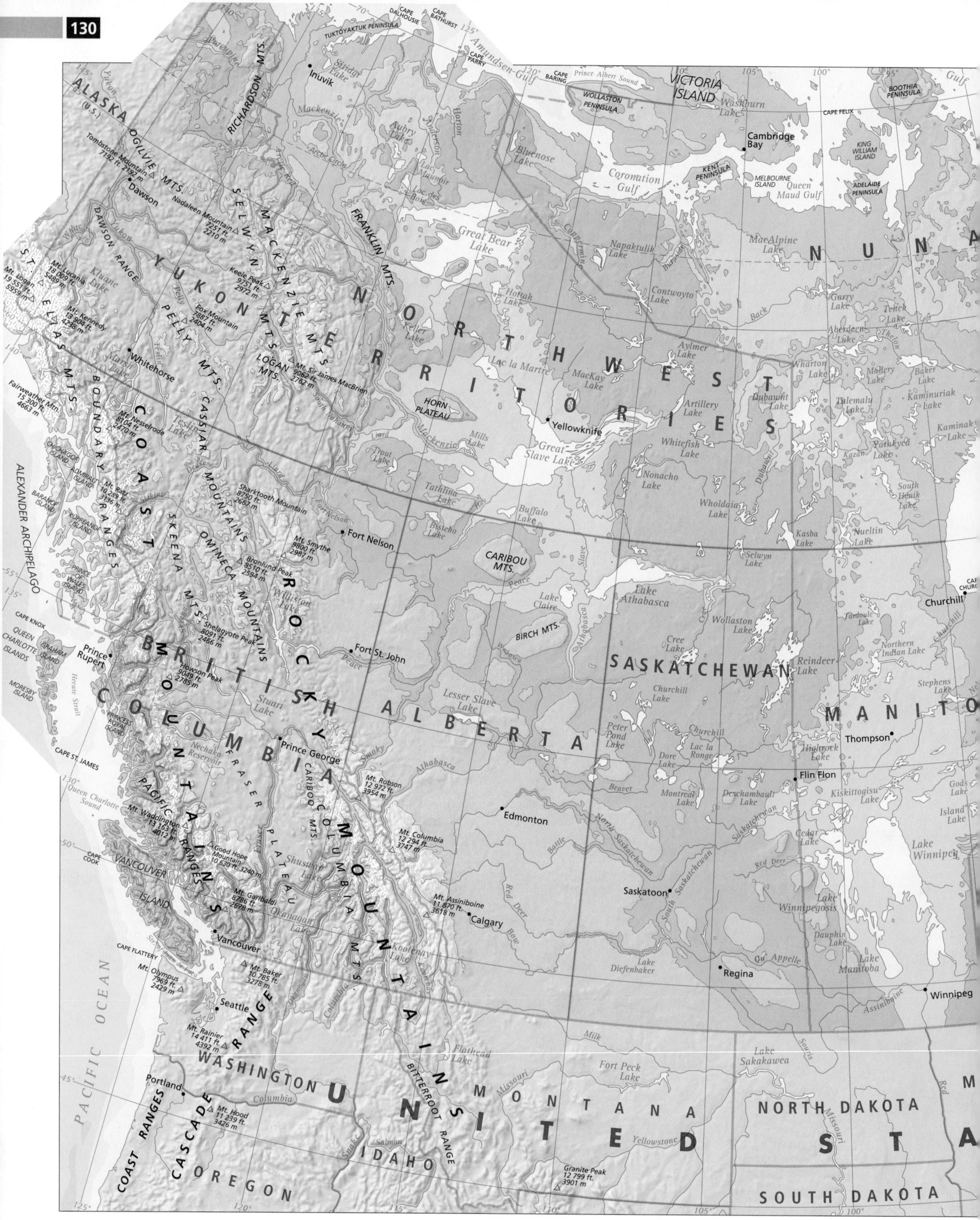
NORTHWEST TERRITORIES
NUNAVUT
YUKON TERRITORY
BRITISH COLUMBIA
ALBERTA
SASKATCHEWAN
MANITOBA
UNITED STATES
WASHINGTON
OREGON
IDAHO
MONTANA
NORTH DAKOTA
SOUTH DAKOTA
ALASKA (U.S.)
PACIFIC OCEAN
VICTORIA ISLAND
Inuvik
Dawson
Whitehorse
Yellowknife
Cambridge Bay
Fort Nelson
Fort St. John
Prince Rupert
Prince George
Edmonton
Calgary
Saskatoon
Regina
Winnipeg
Thompson
Flin Flon
Churchill
Vancouver
Seattle
Portland
ROCKY MOUNTAINS
COAST MOUNTAINS
CASSIAR MOUNTAINS
MACKENZIE MTS.
OGILVIE MTS.
RICHARDSON MTS.
FRANKLIN MTS.
SELWYN MTS.
PELLY MTS.
LOGAN MTS.
ST. ELIAS MTS.
SKEENA MTS.
OMINECA MOUNTAINS
CARIBOO MTS.
COLUMBIA MOUNTAINS
PACIFIC RANGES
CASCADE RANGE
COAST RANGES
BITTERROOT RANGE
FRASER PLATEAU
HORN PLATEAU
CARIBOU MTS.
BIRCH MTS.
Great Bear Lake
Great Slave Lake
Lake Athabasca
Reindeer Lake
Lake Winnipeg
Lake Winnipegosis
Lake Manitoba
Lesser Slave Lake
Coronation Gulf
Queen Maud Gulf
Amundsen Gulf
Mt. Logan 19 551 ft. 5959 m
Mt. Robson 12 972 ft. 3954 m
Mt. Rainier 14 411 ft. 4392 m
Mt. Hood 11 239 ft. 3426 m
Granite Peak 12 799 ft. 3901 m

0 60 120 180 240 300 360 Miles

0 60 120 180 240 300 360 420 480 540 600 Kilometers

Lambert Azimuthal Equal Area Projection
Scale 1:12,000,000
One inch to 190 miles
One cm to 120 km

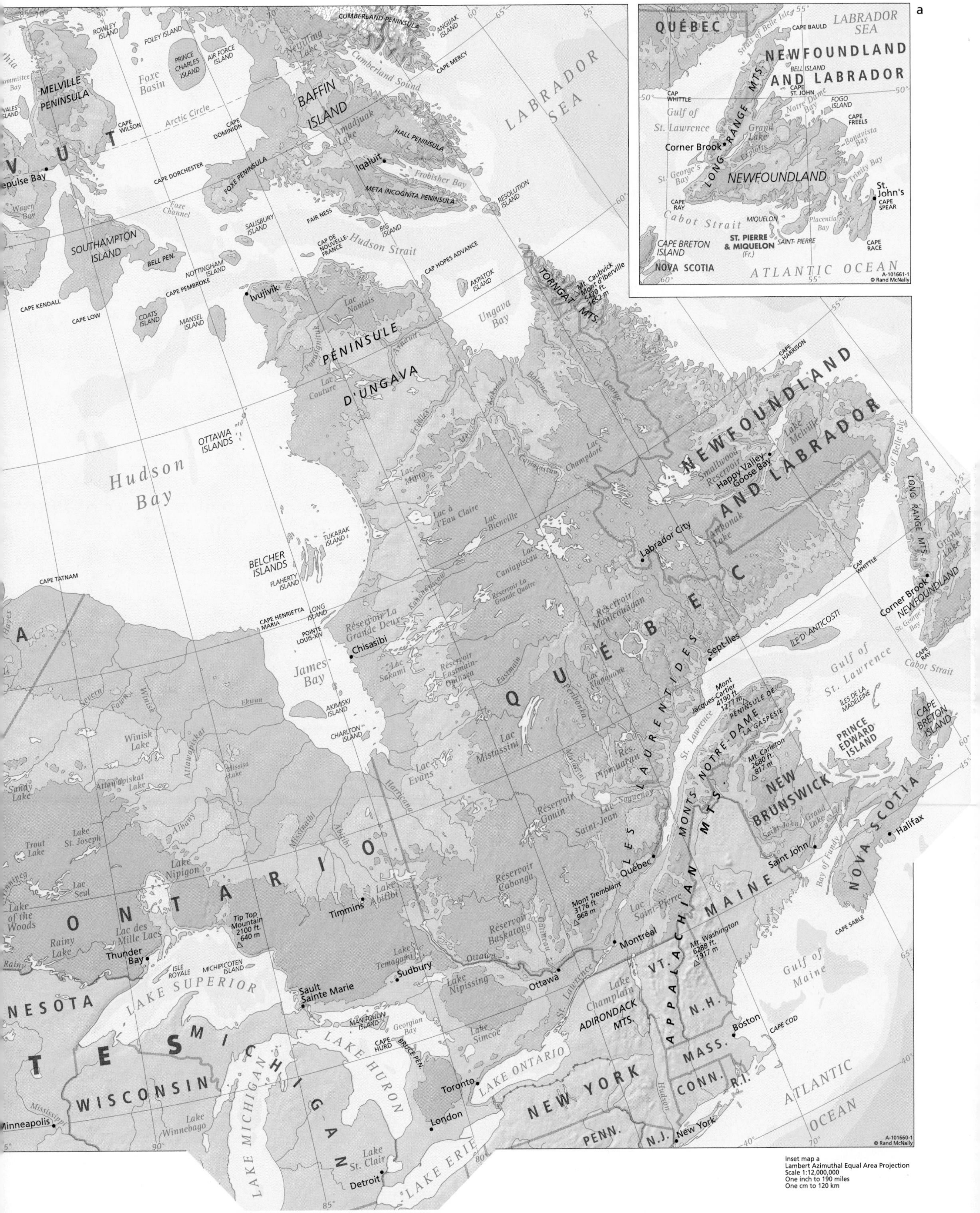
a
QUÉBEC
LABRADOR SEA
CAPE BAULD
Strait of Belle Isle
NEWFOUNDLAND AND LABRADOR
BELL ISLAND
CAPE ST. JOHN
CAPE WHITTLE
Gulf of St. Lawrence
LONG RANGE MTS.
Notre Dame Bay
FOGO ISLAND
CAPE FREELS
Grand Lake
Corner Brook
Bonavista Bay
Exploits
NEWFOUNDLAND
Trinity Bay
St. George's Bay
St. John's
CAPE SPEAR
CAPE RAY
Cabot Strait
MIQUELON
Placentia Bay
ST. PIERRE & MIQUELON (Fr.)
SAINT-PIERRE
CAPE BRETON ISLAND
CAPE RACE
NOVA SCOTIA
ATLANTIC OCEAN
A-101661-1
© Rand McNally
CUMBERLAND PENINSULA
ROWLEY ISLAND
FOLEY ISLAND
PRINCE CHARLES ISLAND
AIR FORCE ISLAND
Nettilling Lake
ANGIJAK ISLAND
Cumberland Sound
CAPE MERCY
Foxe Basin
MELVILLE PENINSULA
BAFFIN ISLAND
LABRADOR SEA
Arctic Circle
CAPE WILSON
CAPE DOMINION
Amadjuak Lake
HALL PENINSULA
Repulse Bay
CAPE DORCHESTER
FOXE PENINSULA
Iqaluit
Frobisher Bay
META INCOGNITA PENINSULA
Wager Bay
Foxe Channel
RESOLUTION ISLAND
SALISBURY ISLAND
FAIR NESS
BIG ISLAND
SOUTHAMPTON ISLAND
CAPE DE NOUVELLE-FRANCE
Hudson Strait
BELL PEN.
NOTTINGHAM ISLAND
CAPE HOPES ADVANCE
CAPE PEMBROKE
AKPATOK ISLAND
Ivujivik
TORNGAT MTS.
Mt. Caubvick Mont d'Iberville 5420 ft. 1652 m
CAPE KENDALL
CAPE LOW
COATS ISLAND
MANSEL ISLAND
Lac Nantais
Ungava Bay
PÉNINSULE D'UNGAVA
Lac Couture
CAPE HARRISON
NEWFOUNDLAND AND LABRADOR
George
Lake Melville
OTTAWA ISLANDS
Lac Champdoré
Smallwood Reservoir
Happy Valley-Goose Bay
Caniapiscau
Hudson Bay
Lac Minto
Lac à l'Eau Claire
Lac Bienville
Atikonak Lake
Strait of Belle Isle
LONG RANGE MTS.
Grand Lake
TUKARAK ISLAND
Labrador City
BELCHER ISLANDS
FLAHERTY ISLAND
Lac Caniapiscau
Réservoir La Grande Quatre
CAPE WHITTLE
Corner Brook
NEWFOUNDLAND
St. George's Bay
CAPE TATNAM
CAPE HENRIETTA MARIA
LONG ISLAND
Réservoir La Grande Deux
Réservoir Manicouagan
QUÉBEC
POINTE LOUIS-XIV
Chisasibi
Sept-Îles
ÎLE D'ANTICOSTI
CAPE RAY
Cabot Strait
Gulf of St. Lawrence
James Bay
Lac Sakami
Réservoir Eastmain-Opinaca
Eastmain
Lac Manouane
Mont Jacques-Cartier 4190 ft. 1277 m
PÉNINSULE DE LA GASPÉSIE
ÎLES DE LA MADELEINE
CAPE BRETON ISLAND
Severn
Fawn
Winisk
Ekwan
AKIMISKI ISLAND
Péribonka
LES LAURENTIDES
MONTS NOTRE-DAME
CHARLTON ISLAND
Winisk Lake
Attawapiskat
Lac Mistassini
Mt. Carleton 2680 ft. 817 m
PRINCE EDWARD ISLAND
Missisa Lake
Attawapiskat Lake
Lac Evans
Rés. Pipmuacan
NEW BRUNSWICK
Sandy Lake
Harricana
Saguenay
Réservoir Gouin
Grand Lake
NOVA SCOTIA
Albany
Lac Saint-Jean
Saint John
Halifax
Lake St. Joseph
Trout Lake
Missinaibi
Abitibi
Lake Nipigon
ONTARIO
Lac Seul
Réservoir Cabonga
Québec
Bay of Fundy
Lake Abitibi
Timmins
Mont Tremblant 3176 ft. 968 m
APPALACHIAN MTS.
MAINE
Lake of the Woods
Tip Top Mountain 2100 ft. 640 m
Lac des Mille Lacs
Réservoir Baskatong
Lac Saint-Pierre
CAPE SABLE
Rainy Lake
Thunder Bay
Lake Temagami
Gatineau
Montréal
Mt. Washington 6288 ft. 1917 m
Gulf of Maine
Rainy
ISLE ROYALE
MICHIPICOTEN ISLAND
Sudbury
Ottawa
VT.
LAKE SUPERIOR
Lake Nipissing
Ottawa
St. Lawrence
Lake Champlain
Sault Sainte Marie
N.H.
MINNESOTA
ADIRONDACK MTS.
MANITOULIN ISLAND
Georgian Bay
Lake Simcoe
Boston
CAPE COD
MICHIGAN
CAPE HURD
BRUCE PEN.
LAKE HURON
MASS.
LAKE ONTARIO
Toronto
WISCONSIN
NEW YORK
CONN.
R.I.
ATLANTIC OCEAN
Mississippi
Lake Winnebago
London
Hudson
Minneapolis
LAKE MICHIGAN
PENN.
N.J.
New York
A-101660-1
© Rand McNally
Lake St. Clair
LAKE ERIE
Detroit

Inset map a
Lambert Azimuthal Equal Area Projection
Scale 1:12,000,000
One inch to 190 miles
One cm to 120 km

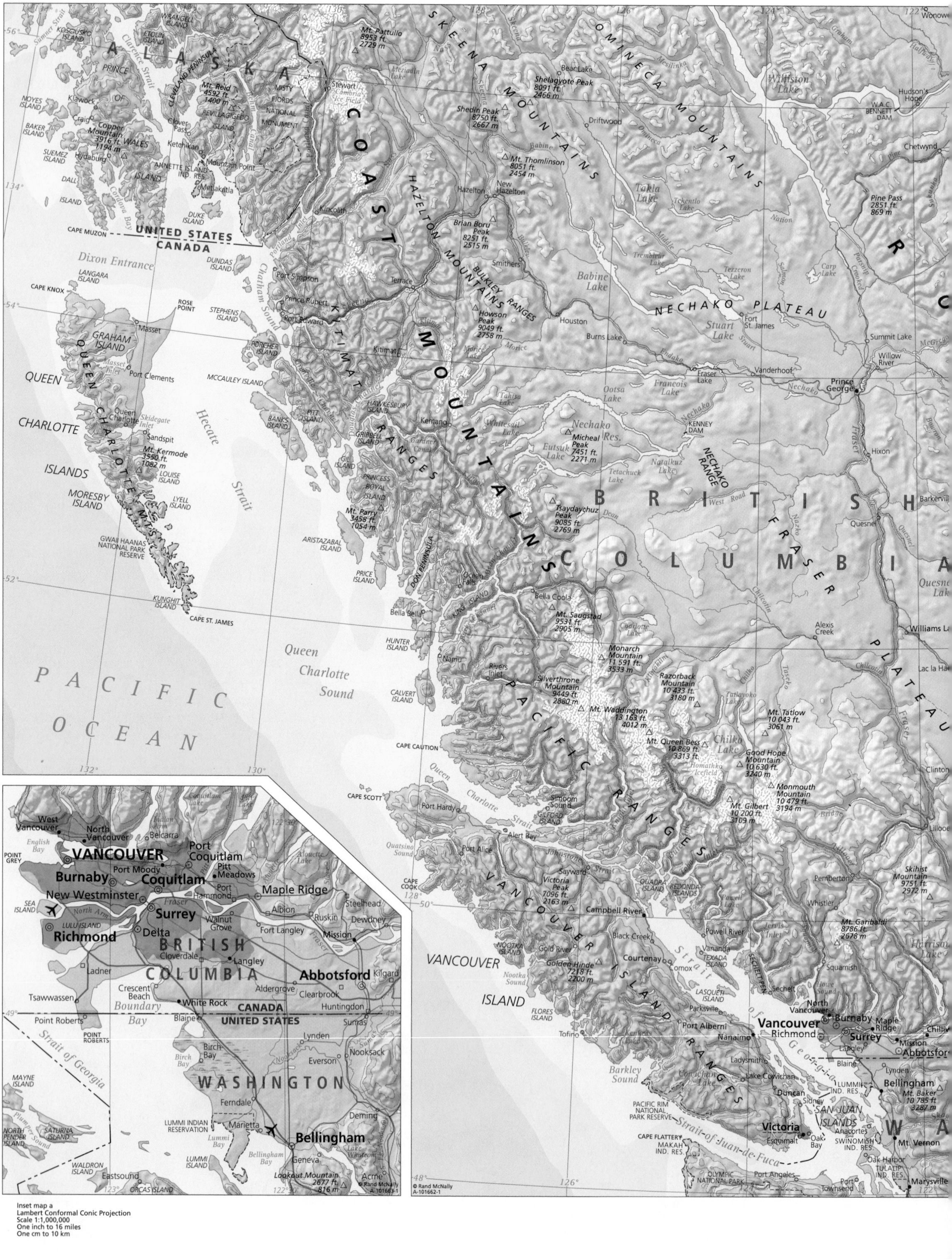
ALASKA
UNITED STATES
CANADA
BRITISH COLUMBIA
COAST MOUNTAINS
SKEENA MOUNTAINS
OMINECA MOUNTAINS
HAZELTON MOUNTAINS
BULKLEY RANGES
KITIMAT RANGES
PACIFIC RANGES
VANCOUVER ISLAND RANGES
QUEEN CHARLOTTE MTS.
NECHAKO PLATEAU
NECHAKO RANGE
FRASER PLATEAU
QUEEN CHARLOTTE ISLANDS
PACIFIC OCEAN
Dixon Entrance
Hecate Strait
Queen Charlotte Sound
Queen Charlotte Strait
Chatham Sound
Strait of Georgia
Strait of Juan de Fuca
Barkley Sound
Nootka Sound
Quatsino Sound
VANCOUVER ISLAND
GRAHAM ISLAND
MORESBY ISLAND
PRINCE OF WALES ISLAND
REVILLAGIGEDO ISLAND
MISTY FIORDS NATIONAL MONUMENT
GWAII HAANAS NATIONAL PARK RESERVE
PACIFIC RIM NATIONAL PARK RESERVE
OLYMPIC NATIONAL PARK
CAPE MUZON
CAPE KNOX
ROSE POINT
CAPE ST. JAMES
CAPE CAUTION
CAPE SCOTT
CAPE COOK
CAPE FLATTERY
Mt. Pattullo 8953 ft. 2729 m
Mt. Reid 4592 ft. 1400 m
Copper Mountain 3916 ft. 1194 m
Shelagyote Peak 8091 ft. 2466 m
Shedin Peak 8750 ft. 2667 m
Mt. Thomlinson 8051 ft. 2454 m
Brian Boru Peak 8251 ft. 2515 m
Howson Peak 9049 ft. 2758 m
Pine Pass 2851 ft. 869 m
Mt. Kermode 3550 ft. 1082 m
Micheal Peak 7451 ft. 2271 m
Tsaydaychuz Peak 9085 ft. 2769 m
Mt. Parry 3458 ft. 1054 m
Mt. Saugstad 9531 ft. 2905 m
Monarch Mountain 11 591 ft. 3533 m
Silverthrone Mountain 9449 ft. 2880 m
Razorback Mountain 10 433 ft. 3180 m
Mt. Waddington 13 163 ft. 4012 m
Mt. Tatlow 10 043 ft. 3061 m
Mt. Queen Bess 10 869 ft. 3313 m
Good Hope Mountain 10 630 ft. 3240 m
Monmouth Mountain 10 479 ft. 3194 m
Mt. Gilbert 10 200 ft. 3109 m
Skihist Mountain 9751 ft. 2972 m
Victoria Peak 7096 ft. 2163 m
Golden Hinde 7218 ft. 2200 m
Mt. Garibaldi 8786 ft. 2678 m
Mt. Baker 10 785 ft. 3287 m
Ketchikan
Metlakatla
Stewart
Prince Rupert
Port Simpson
Port Edward
Kitimat
Terrace
Hazelton
Smithers
Houston
Burns Lake
Fort St. James
Vanderhoof
Prince George
Quesnel
Bella Coola
Bella Bella
Ocean Falls
Masset
Port Clements
Sandspit
Queen Charlotte
Port Hardy
Alert Bay
Port Alice
Campbell River
Courtenay
Powell River
Gold River
Port Alberni
Tofino
Nanaimo
Duncan
Victoria
Vancouver
Richmond
Burnaby
Surrey
Abbotsford
Bellingham
Port Angeles
Williston Lake
Babine Lake
Takla Lake
Stuart Lake
Fraser Lake
François Lake
Ootsa Lake
Nechako Res.
Chilko Lake
W.A.C. BENNETT DAM
KENNEY DAM

Inset map a

VANCOUVER
North Vancouver
West Vancouver
Burnaby
Coquitlam
Port Coquitlam
Port Moody
Pitt Meadows
Maple Ridge
New Westminster
Richmond
Delta
Surrey
Langley
Abbotsford
Mission
Cloverdale
White Rock
Tsawwassen
Point Roberts
Boundary Bay
BRITISH COLUMBIA
CANADA
UNITED STATES
WASHINGTON
Blaine
Lynden
Everson
Nooksack
Ferndale
Bellingham
Strait of Georgia
LUMMI INDIAN RESERVATION
Lookout Mountain 2677 ft. 816 m

Inset map a
Lambert Conformal Conic Projection
Scale 1:1,000,000
One inch to 16 miles
One cm to 10 km

ALBERTA
SASK.
ROCKY MOUNTAINS
COLUMBIA MOUNTAINS
CARIBOO MOUNTAINS
SELKIRK MOUNTAINS
MONASHEE MTS.
PURCELL MTS.
CASCADE RANGE
WASHINGTON
MONTANA
IDAHO
CANADA
UNITED STATES
LEWIS RA.
SWAN RA.
CABINET MTS.
Edmonton
Calgary
Red Deer
Lethbridge
Medicine Hat
Grande Prairie
Kamloops
Kelowna
Penticton
Vernon
Great Falls
Dikonville
Peace River
Grimshaw
Nampa
Fairview
Fort St. John
Taylor
Dawson Creek
Rycroft
Falher
McLennan
Grouard Mission
Lesser Slave Lake
High Prairie
Slave Lake
Wabasca-Desmarais
Utikuma Lake
Kimiwan Lake
Cardinal Lake
Calling Lake
Lac La Biche
Athabasca
Boyle
Cold Lake
Primrose Lake
Peter Pond Lake
Buffalo Narrows
Kazan Lake
Canoe Lake
Winefred Lake
Bonnyville
Muriel Lake
Beaver
Frog Lake
St. Paul
Brightsand Lake
St. Walburg
Smoky Lake
Tumbler Ridge
Beaverlodge
Sturgeon Lake
Valleyview
Fox Creek (Iosegun Lake)
Whitecourt
Barrhead
Westlock
Mayerthorpe
Lac Sainte Anne
St. Albert
Fort Saskatchewan
Lamont
ELK ISLAND NATL. PARK
Vegreville
Vermilion
Lloydminster
Maidstone
Lashburn
Manito Lake
Sherwood Park
Beaverhill Lake
Spruce Grove
Wabamun Lake
Entwistle
Chip Lake
Edson
Grande Cache
Hinton
Drayton Valley
Leduc
Camrose
Viking
Wainwright
Pigeon Lake
Wetaskiwin
Ponoka
SAMSON INDIAN RESERVE
Buffalo Lake
Macklin
Provost
Rimbey
Gull Lake
Lacombe
Stettler
Castor
Sullivan Lake
Rocky Mountain House
Nordegg
Sylvan Lake
Gough Lake
Innisfail
Bowden
Olds
Three Hills
Hanna
Alsask
Didsbury
Drumheller
Crossfield
Beiseker
Airdrie
STONEY IND. RES.
Cochrane
Strathmore
Bassano
Brooks
TSUU T'INA NATION INDIAN RESERVE
SIKSIKA INDIAN RESERVE
Okotoks
High River
Lake Newell
Redcliff
Nanton
Stavely
Claresholm
Bow Island
Taber
Coaldale
Pakowki Lake
PEIGAN IND. RES.
Fort Macleod
BLOOD IND. RES.
Stirling
Crowsnest Pass
Pincher Creek
Fernie
WATERTON LAKES NATL. PARK
St. Mary Reservoir
Cardston
Milk River
Sunburst
Cut Bank
Shelby
Chester
Lake Elwell
BLACKFEET INDIAN RESERVATION
Browning
Conrad
Choteau
Ulm
GLACIER NATIONAL PARK
Mt. Cleveland 10 466 ft. 3190 m
Eureka
Whitefish
Columbia Falls
Kalispell
Evergreen
Hungry Horse Res.
Flathead Lake
Polson
FLATHEAD INDIAN RESERVATION
Mt. Chown 10 928 ft. 3331 m
Mt. Robson 12 972 ft. 3954 m
JASPER NATIONAL PARK
Yellowhead Pass 3717 ft. 1133 m
Jasper
McBride
Valemount
Mt. Sir Wilfrid Laurier 11 549 ft. 3520 m
Mitchell Lake
Mt. Brazeau 11 385 ft. 3470 m
Mt. Alberta 11 873 ft. 3619 m
Mt. Columbia 12 294 ft. 3747 m
Murtle Lake
Blue River
Mt. Forbes 11 850 ft. 3612 m
Howse Peak 10 794 ft. 3290 m
Kinbasket Lake
Mt. Chapman 10 089 ft. 3075 m
Mt. Freshfield 10 958 ft. 3340 m
BANFF NATIONAL PARK
Mt. St. Bride 10 866 ft. 3312 m
Mt. Sir Sandford 11 591 ft. 3533 m
YOHO NATL. PARK
Kicking Horse Pass 5338 ft. 1627 m
Lake Louise
Canim Lake
Mahood Lake
Clearwater
One Hundred Mile House
Green Lake
Bonaparte Lake
Barrière
Adams Lake
Pukeashun Mountain 7556 ft. 2303 m
MT. REVELSTOKE NATIONAL PARK
GLACIER NATIONAL PARK
Golden
Banff
Canmore
KOOTENAY NATL. PARK
Mt. Assiniboine 11 870 ft. 3618 m
Mt. Sir Douglas 11 175 ft. 3406 m
Mt. King George 11 227 ft. 3422 m
Mt. Joffre 11 250 ft. 3429 m
Shuswap Lake
Revelstoke
Sicamous
Chase
Cache Creek
Salmon Arm
Radium Hot Springs
Invermere
Mt. Odin 9751 ft. 2972 m
Mt. Farnham 11 342 ft. 3457 m
Fairmont Hot Springs
Enderby
Armstrong
Upper Arrow Lake
Nakusp
Columbia Lake
Okanagan Lake
Merritt
Kootenay Lake
Kimberley
Crowsnest Pass 4449 ft. 1356 m
Lower Arrow Lake
Peachland
Cranbrook
Summerland
Nelson
Princeton
Castlegar
Salmo
Oliver
Trail
Fruitvale
Creston
Grand Forks
Rossland
Osoyoos
Oroville
NORTH CASCADES NATL. PARK
Ross Lake
Mt. Logan 9087 ft. 2770 m
Bonners Ferry
Lake Koocanusa
Kettle Falls
Colville
Troy
Libby
Priest Lake
Sandpoint
Omak
COLVILLE INDIAN RESERVATION
Franklin D. Roosevelt Lake
Chewelah
Priest River
Newport
Lake Pend Oreille
Spirit Lake
Brewster
Lake Chelan
GRAND COULEE DAM
SPOKANE IND. RES.
Deer Park
Glacier Peak 10 541 ft. 3213 m
Peace
Smoky
Wapiti
Little Smoky
Simonette
Kakwa
Berland
Athabasca
McLeod
Pembina
North Saskatchewan
Brazeau
Battle
Ribstone Creek
Eyehill Cr.
Red Deer
Sounding Creek
Berry Creek
Clearwater
Ram
Rosebud
Bow
Little Bow
Highwood
Oldman
South Saskatchewan
Milk
Belly
Waterhen
Dillon
Christina
House
Muskwa
Wabasca
Lesser Slave
Kiskatinaw
Murray
Pine
Beatton
Fraser
Thompson
South Thompson
North Thompson
Columbia
Kootenay
Kootenai
Kettle
Similkameen
Okanogan
Methow
Pend Oreille
Clark Fork
Flathead
South Fork Flathead
Marias
Teton
Sun
Missouri
Elk
Kettle
120°
118°
116°
114°
112°
110°
56°
54°
52°
50°
48°

0 20 40 60 80 100 120 Miles
0 20 40 60 80 100 120 140 160 180 200 Kilometers

Lambert Conformal Conic Projection
Scale 1:4,000,000
One inch to 64 miles
One cm to 40 km

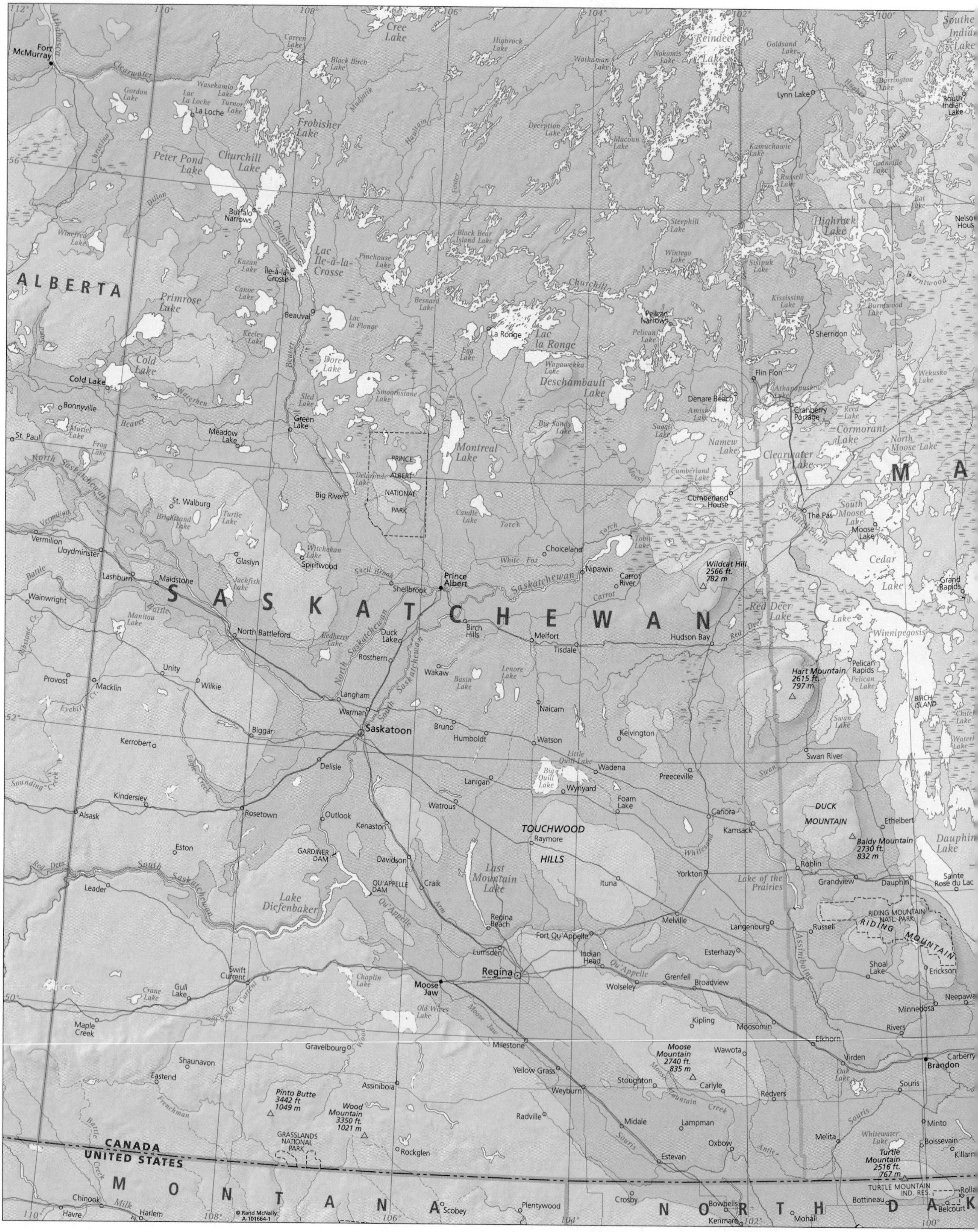
ALBERTA
SASKATCHEWAN
MA
CANADA
UNITED STATES
MONTANA
NORTH DAK
Fort McMurray
Athabasca
Clearwater
Gordon Lake
Wasekamio Lake
Lac La Loche
La Loche
Turnor Lake
Careen Lake
Cree Lake
Black Birch Lake
Mudjatik
Frobisher Lake
Churchill Lake
Peter Pond Lake
Christina
Buffalo Narrows
Dillon
Winefred Lake
Kazan Lake
Île-à-la-Crosse
Lac Île-à-la-Crosse
Churchill
Pinehouse Lake
Primrose Lake
Canoe Lake
Beauval
Lac la Plonge
Keeley Lake
Cold Lake
Dore Lake
Smoothstone Lake
Sled Lake
Waterhen
Bonnyville
Beaver
Green Lake
Meadow Lake
Muriel Lake
Frog Lake
St. Paul
North Saskatchewan
St. Walburg
Brightsand Lake
Turtle Lake
Vermilion
Lloydminster
Glaslyn
Witchekan Lake
Spiritwood
Big River
Delaronde Lake
PRINCE ALBERT NATIONAL PARK
Montreal Lake
Candle Lake
Torch
White Fox
Choiceland
Nipawin
Carrot River
Carrot
Tobin Lake
Wildcat Hill 2566 ft. 782 m
Lashburn
Maidstone
Jackfish Lake
Shell Brook
Shellbrook
Prince Albert
Saskatchewan
Battle
Wainwright
Manitou Lake
North Battleford
Redberry Lake
Duck Lake
Birch Hills
Melfort
Tisdale
Hudson Bay
Red Deer
Red Deer Lake
Rosthern
Wakaw
Lenore Lake
Basin Lake
Unity
Provost
Macklin
Wilkie
Eyehill
Langham
Warman
South Saskatchewan
Saskatoon
Biggar
Bruno
Humboldt
Naicam
Watson
Kelvington
Kerrobert
Delisle
Little Quill Lake
Big Quill Lake
Wadena
Lanigan
Wynyard
Preeceville
Swan River
Swan
Sounding Creek
Kindersley
Eagle Creek
Foam Lake
Watrous
Alsask
Rosetown
Outlook
Kenaston
TOUCHWOOD HILLS
Raymore
Canora
Kamsack
DUCK MOUNTAIN
Ethelbert
Baldy Mountain 2730 ft. 832 m
Dauphin Lake
Eston
GARDINER DAM
Davidson
Whitesand
Yorkton
Roblin
Red Deer
South Saskatchewan
Leader
QU'APPELLE DAM
Craik
Last Mountain Lake
Ituna
Lake of the Prairies
Grandview
Dauphin
Sainte Rose du Lac
Lake Diefenbaker
Qu'Appelle
Arm
Regina Beach
Melville
Langenburg
Russell
RIDING MOUNTAIN NATL. PARK
RIDING MOUNTAIN
Fort Qu'Appelle
Lumsden
Indian Head
Esterhazy
Assiniboine
Swift Current
Chaplin Lake
Regina
Moose Jaw
Grenfell
Broadview
Shoal Lake
Erickson
Crane Lake
Gull Lake
Wolseley
Neepawa
Old Wives Lake
Minnedosa
Maple Creek
Kipling
Moosomin
Rivers
Gravelbourg
Wood
Milestone
Elkhorn
Moose Mountain 2740 ft. 835 m
Wawota
Virden
Shaunavon
Eastend
Yellow Grass
Stoughton
Brandon
Carberry
Pinto Butte 3442 ft 1049 m
Assiniboia
Weyburn
Carlyle
Redvers
Souris
Frenchman
Wood Mountain 3350 ft. 1021 m
Radville
Midale
Lampman
Minto
Battle Creek
GRASSLANDS NATIONAL PARK
Oxbow
Melita
Whitewater Lake
Boissevain
Rockglen
Estevan
Antler
Turtle Mountain 2516 ft. 767 m
Killarney
TURTLE MOUNTAIN IND. RES.
Chinook
Milk
Havre
Harlem
Scobey
Plentywood
Crosby
Bowbells
Kenmare
Mohall
Bottineau
Belcourt
Rolla
Highrock Lake
Wathaman Lake
Reindeer Lake
Nokomis Lake
Goldsand Lake
Lynn Lake
Barrington Lake
South Indian Lake
Deception Lake
Macoun Lake
Kamuchawie Lake
Russell Lake
Granville Lake
Rat Lake
Steephill Lake
Black Bear Island Lake
Wintego Lake
Sisipuk Lake
Highrock Lake
Burntwood
Kississing Lake
Besnard Lake
Pelican Narrows
Pelican Lake
La Ronge
Lac la Ronge
Egg Lake
Sherridon
Wapawekka Lake
Deschambault Lake
Flin Flon
Athapapuskow Lake
Wekusko Lake
Denare Beach
Amisk Lake
Cranberry Portage
Reed Lake
Big Sandy Lake
Suggi Lake
Namew Lake
Cormorant Lake
North Moose Lake
Clearwater Lake
Cumberland Lake
Cumberland House
The Pas
South Moose Lake
Moose Lake
Cedar Lake
Grand Rapids
Winnipegosis
Hart Mountain 2615 ft. 797 m
Pelican Rapids
Pelican Lake
BIRCH ISLAND
Swan Lake
© Rand McNally A-101664-1

0 20 40 60 80 100 120 Miles
0 20 40 60 80 100 120 140 160 180 200 Kilometers

Lambert Conformal Conic Projection
Scale 1:4,000,000
One inch to 64 miles
One cm to 40 km

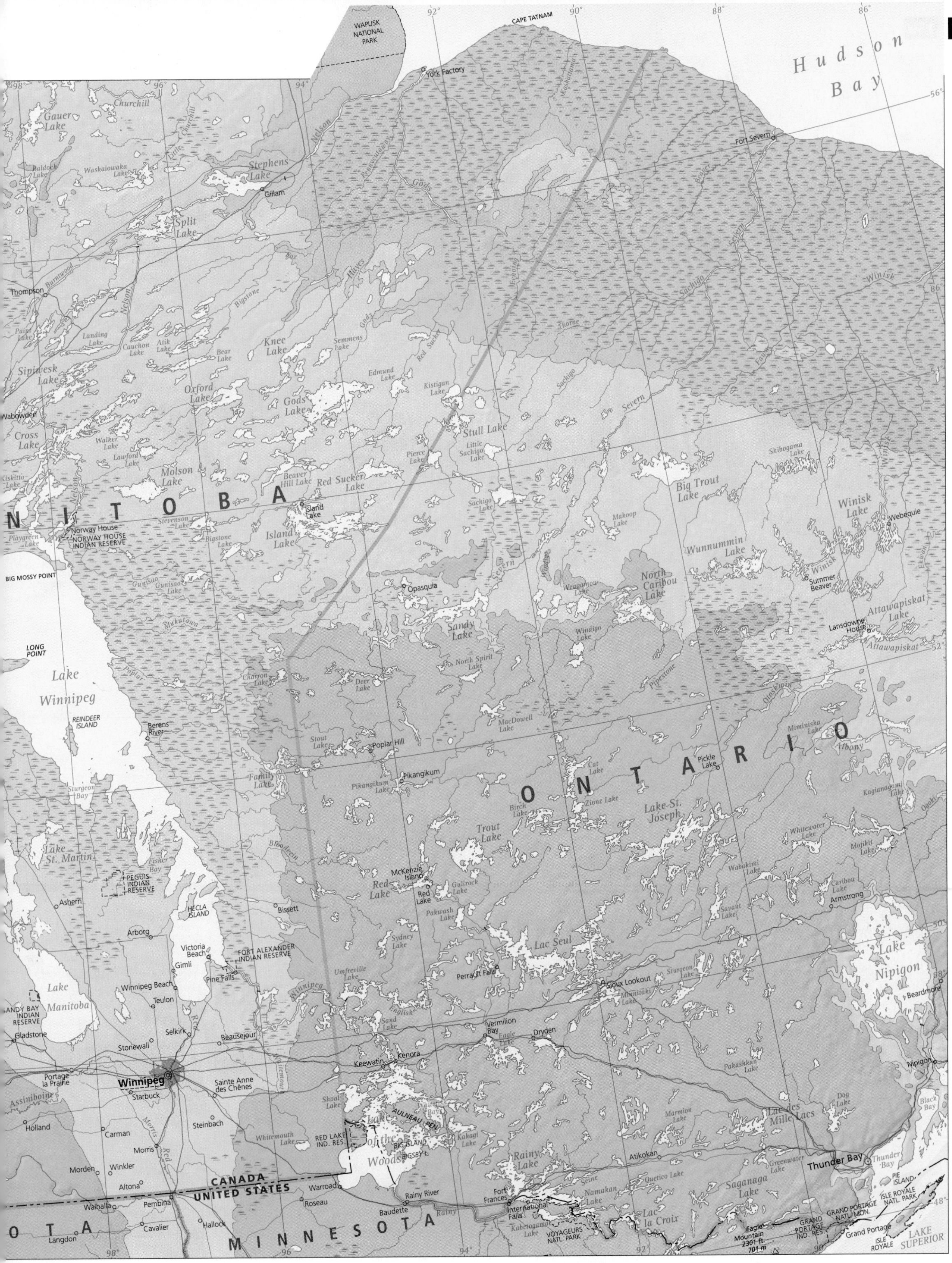

Hudson Bay
CAPE TATNAM
WAPUSK NATIONAL PARK
York Factory
Fort Severn
Churchill
Gauer Lake
Baldock Lake
Waskaiowaka Lake
Stephens Lake
Gillam
Split Lake
Nelson
Little Churchill
Pennycutaway
Gods
Kaskattama
Beaver
Severn
Winisk
Sachigo
Thorne
Fawn
Thompson
Burntwood
Bigstone
Hayes
Paint Lake
Landing Lake
Cauchon Lake
Atik Lake
Bear Lake
Knee Lake
Semmens Lake
Red Sucker
Edmund Lake
Kistigan Lake
Sipiwesk Lake
Oxford Lake
Gods Lake
Stull Lake
Wabowden
Cross Lake
Walker Lake
Lawford Lake
Molson Lake
Pierce Lake
Little Sachigo Lake
Shibogama Lake
Kiskitto Lake
Beaver Hill Lake
Red Sucker Lake
Big Trout Lake
MANITOBA
Norway House
NORWAY HOUSE INDIAN RESERVE
Island Lake
Stevenson Lake
Sachigo Lake
Makoop Lake
Winisk Lake
Webequie
Playgreen Lake
Bigstone Lake
Wunnummin Lake
BIG MOSSY POINT
Gunisao Lake
Opasquia
Windigo
Weagamow Lake
North Caribou Lake
Summer Beaver
Attawapiskat Lake
Mukutawa
Sandy Lake
Windigo Lake
Lansdowne House
Attawapiskat
LONG POINT
Lake Winnipeg
North Spirit Lake
Charron Lake
Deer Lake
Pipestone
Otoskwin
REINDEER ISLAND
Berens River
MacDowell Lake
Miminiska Lake
Albany
Stout Lake
Poplar Hill
Pickle Lake
ONTARIO
Family Lake
Pikangikum
Pikangikum Lake
Cat Lake
Sturgeon Bay
Birch Lake
Zionz Lake
Lake St. Joseph
Kagianagami Lake
Ogoki
Trout Lake
Whitewater Lake
Mojikit Lake
Lake St. Martin
Bloodvein
Fisher Bay
PEGUIS INDIAN RESERVE
McKenzie Island
Red Lake
Gullrock Lake
Wabakimi Lake
Caribou Lake
Armstrong
Ashern
HECLA ISLAND
Bissett
Pakwash Lake
Savant Lake
Arborg
Sydney Lake
Lac Seul
Lake Nipigon
Victoria Beach
FORT ALEXANDER INDIAN RESERVE
Gimli
Pine Falls
Umfreville Lake
Perrault Falls
Sturgeon Lake
Lake Manitoba
Winnipeg Beach
Sioux Lookout
Minnitaki Lake
Beardmore
SANDY BAY INDIAN RESERVE
Teulon
Winnipeg
English
Sand Lake
Gladstone
Selkirk
Beausejour
Vermilion Bay
Eagle Lake
Dryden
Stonewall
Keewatin
Kenora
Pakashkan Lake
Nipigon
Portage la Prairie
Sainte Anne des Chênes
Whitemouth
Assiniboine
Starbuck
Shoal Lake
Dog Lake
Black Bay
AULNEAU PEN.
Lac des Mille Lacs
Marmion Lake
Holland
Steinbach
Lake of the Woods
Kakagi Lake
Carman
Whitemouth Lake
RED LAKE IND. RES.
BIG ISLAND
BIGSBY I.
Morris
Rainy Lake
Atikokan
Thunder Bay
Morden
Winkler
Greenwater Lake
Quetico Lake
Altona
CANADA
UNITED STATES
Warroad
Rainy River
Seine
Namakan Lake
Saganaga Lake
PIE ISLAND
ISLE ROYALE NATL. PARK
Walhalla
Pembina
Roseau
Baudette
Fort Frances
International Falls
GRAND PORTAGE NATL. MON.
DAKOTA
MINNESOTA
Cavalier
Hallock
Kabetogama Lake
VOYAGEURS NATL. PARK
Lac la Croix
Eagle Mountain 2301 ft. 701 m
GRAND PORTAGE IND. RES.
Grand Portage
ISLE ROYALE
LAKE SUPERIOR
Langdon
Red
Winnipeg
98°
96°
94°
92°
90°
88°
86°
56°
54°
52°
50°
48°

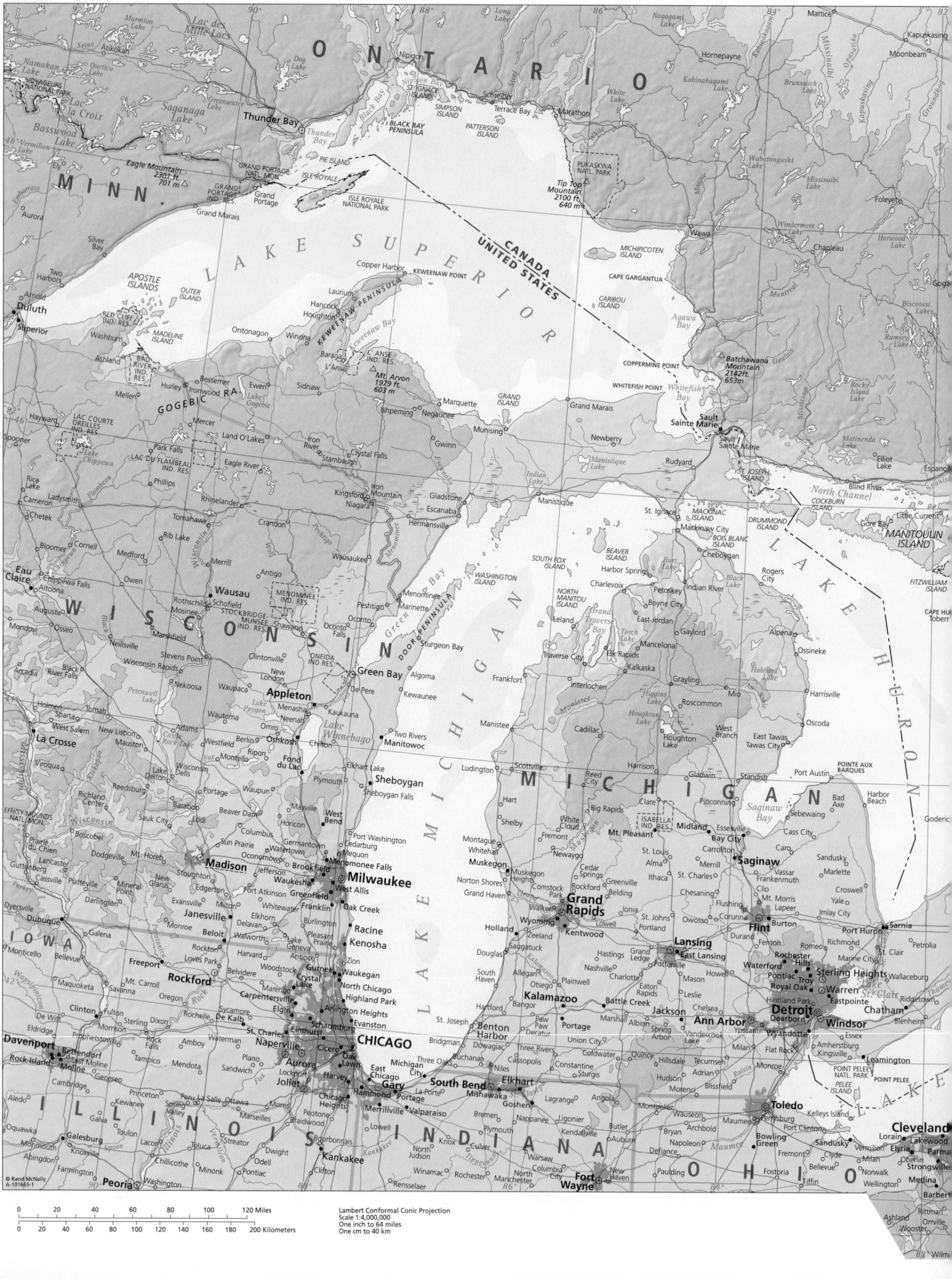
ONTARIO
MINN.
WISCONSIN
MICHIGAN
IOWA
ILLINOIS
INDIANA
OHIO
LAKE SUPERIOR
LAKE MICHIGAN
LAKE HURON
CANADA
UNITED STATES
Thunder Bay
Nipigon
Schreiber
Terrace Bay
Marathon
Hornepayne
Mattice
Kapuskasing
Moonbeam
Wawa
Chapleau
Foleyet
Sault Sainte Marie
Sault Sainte Marie
Elliot Lake
Blind River
Little Current
MANITOULIN ISLAND
Atikokan
Lac des Mille Lacs
Lac la Croix
Saganaga Lake
Basswood Lake
VOYAGEURS NATIONAL PARK
Eagle Mountain 2301 ft. 701 m
GRAND PORTAGE NATL. MON.
GRAND PORTAGE IND. RES.
Grand Portage
Grand Marais
ISLE ROYALE
ISLE ROYALE NATIONAL PARK
PIE ISLAND
Black Bay
BLACK BAY PENINSULA
ST. IGNACE ISLAND
SIMPSON ISLAND
PATTERSON ISLAND
PUKASKWA NATL. PARK
Tip Top Mountain 2100 ft. 640 m
MICHIPICOTEN ISLAND
CAPE GARGANTUA
CARIBOU ISLAND
Agawa Bay
Batchawana Mountain 2142 ft. 653 m
COPPERMINE POINT
WHITEFISH POINT
Whitefish Bay
Ely
Aurora
Silver Bay
Two Harbors
Duluth
Superior
APOSTLE ISLANDS
OUTER ISLAND
MADELINE ISLAND
RED CLIFF IND. RES.
BAD RIVER IND. RES.
Washburn
Ashland
Mellen
Hurley
Ironwood
Bessemer
GOGEBIC RA.
Ontonagon
Copper Harbor
KEWEENAW POINT
KEWEENAW PENINSULA
Laurium
Hancock
Houghton
Keweenaw Bay
Baraga
L'Anse
L'ANSE IND. RES.
Mt. Arvon 1979 ft. 603 m
Ewen
Sidnaw
Marquette
Ishpeming
Negaunee
GRAND ISLAND
Munising
Grand Marais
Newberry
Gwinn
Iron River
Crystal Falls
Stambaugh
Iron Mountain
Kingsford
Niagara
Gladstone
Escanaba
Hermansville
Manistique
Rudyard
St. Ignace
Mackinaw City
MACKINAC ISLAND
BOIS BLANC ISLAND
DRUMMOND ISLAND
COCKBURN ISLAND
North Channel
ST. JOSEPH ISLAND
Cheboygan
Rogers City
BEAVER ISLAND
Harbor Springs
Petoskey
Charlevoix
Boyne City
East Jordan
Gaylord
Alpena
Ossineke
SOUTH FOX ISLAND
NORTH MANITOU ISLAND
Leland
Grand Traverse Bay
Traverse City
Elk Rapids
Mancelona
Kalkaska
Grayling
Mio
Harrisville
Frankfort
Interlochen
Roscommon
Oscoda
Manistee
Cadillac
Houghton Lake
West Branch
East Tawas
Tawas City
Ludington
Scottville
Reed City
Harrison
Gladwin
Standish
Port Austin
POINTE AUX BARQUES
Hart
Big Rapids
Clare
Pinconning
Saginaw Bay
Bad Axe
Harbor Beach
Shelby
White Cloud
Mt. Pleasant
ISABELLA IND. RES.
Midland
Essexville
Bay City
Sebewaing
Cass City
Montague
Whitehall
Fremont
Newaygo
St. Louis
Alma
Carrollton
Saginaw
Muskegon
Muskegon Heights
Norton Shores
Grand Haven
Cedar Springs
Rockford
Greenville
Belding
Ithaca
St. Charles
Merrill
Vassar
Frankenmuth
Sandusky
Marlette
Comstock Park
Walker
Grand Rapids
Wyoming
Kentwood
Lowell
Ionia
St. Johns
Portland
Chesaning
Owosso
Flushing
Corunna
Clio
Mt. Morris
Lapeer
Flint
Burton
Imlay City
Port Huron
Sarnia
Holland
Zeeland
Saugatuck
Douglas
Lansing
East Lansing
Grand Ledge
Hastings
Nashville
Charlotte
Potterville
Mason
Howell
Durand
Fenton
Romeo
Richmond
Marine City
St. Clair
Rochester Hills
Waterford
Pontiac
Troy
Royal Oak
Sterling Heights
Warren
Wallaceburg
Eastpointe
Lake St. Clair
Highland Park
Detroit
Dearborn
Windsor
Chatham
South Haven
Allegan
Plainwell
Otsego
Kalamazoo
Eaton Rapids
Leslie
Battle Creek
Jackson
Chelsea
Ann Arbor
Ypsilanti
Wyandotte
Hartford
Bangor
Paw Paw
Portage
Marshall
Albion
Spring Arbor
Vandercook Lake
Essex
Amherstburg
Kingsville
Leamington
St. Joseph
Benton Harbor
Decatur
Three Rivers
Union City
Coldwater
Quincy
Hillsdale
Tecumseh
Milan
Flat Rock
Bridgman
Dowagiac
Cassopolis
Constantine
Sturgis
Adrian
Hudson
Morenci
Blissfield
POINT PELEE NATL. PARK
POINT PELEE
PELEE ISLAND
Michigan City
Niles
Buchanan
Three Oaks
South Bend
Elkhart
Mishawaka
Goshen
Lagrange
Angola
Toledo
Maumee
Perrysburg
Kelleys Island
Montpelier
Wauseon
Archbold
Bryan
Napoleon
Defiance
Bowling Green
Port Clinton
Sandusky
Fremont
Vermilion
Lorain
Elyria
Cleveland
Lakewood
Parma
Strongsville
Medina
Barberton
Norwalk
Wellington
Bellevue
Clyde
Milan
Oberlin
Fostoria
Tiffin
Paulding
Rittman
Orrville
Ashland
Wooster
Hayward
LAC COURTE OREILLES IND. RES.
Spooner
Rice Lake
Cameron
Ladysmith
Chetek
Bloomer
Cornell
Eau Claire
Chippewa Falls
Altoona
Augusta
Osseo
Mondovi
Park Falls
LAC DU FLAMBEAU IND. RES.
Mercer
Land O'Lakes
Eagle River
Phillips
Rhinelander
Tomahawk
Crandon
Rib Lake
Medford
Merrill
Antigo
Wausau
Schofield
Rothschild
Mosinee
STOCKBRIDGE-MUNSEE IND. RES.
MENOMINEE IND. RES.
Wausaukee
Menominee
Marinette
Peshtigo
Oconto
Oconto Falls
Shawano
Green Bay
DOOR PENINSULA
Sturgeon Bay
WASHINGTON ISLAND
Marshfield
Neillsville
Stevens Point
Wisconsin Rapids
Nekoosa
Clintonville
New London
ONEIDA IND. RES.
Green Bay
De Pere
Algoma
Kewaunee
Black River Falls
Arcadia
Waupaca
Appleton
Menasha
Neenah
Kaukauna
Lake Winnebago
Wautoma
Oshkosh
Omro
Two Rivers
Manitowoc
Holmen
Sparta
Tomah
West Salem
La Crosse
New Lisbon
Mauston
Adams
Westfield
Berlin
Ripon
Chilton
Fond du Lac
Elkhart Lake
Sheboygan
Sheboygan Falls
Plymouth
Viroqua
Lake Delton
Wisconsin Dells
Montello
Portage
Reedsburg
Richland Center
Baraboo
Waupun
Beaver Dam
Mayville
West Bend
Horicon
EFFIGY MOUNDS NATL. MON.
Sauk City
Lodi
Columbus
Port Washington
Cedarburg
Mequon
Boscobel
Prairie du Chien
Sun Prairie
Watertown
Oconomowoc
Germantown
Menomonee Falls
Brookfield
Madison
Waukesha
Milwaukee
West Allis
Greenfield
Oak Creek
Franklin
Lancaster
Guttenberg
Cassville
Platteville
Dodgeville
Mt. Horeb
Mineral Point
New Glarus
Stoughton
Jefferson
Edgerton
Fort Atkinson
Whitewater
Burlington
Racine
Kenosha
Darlington
Evansville
Milton
Janesville
Elkhorn
Delavan
Dyersville
Dubuque
Monroe
Beloit
Walworth
Lake Geneva
Pleasant Prairie
Galena
Monticello
Bellevue
Freeport
Rockton
Loves Park
Rockford
Harvard
Antioch
Zion
Gurnee
Waukegan
Woodstock
Crystal Lake
Belvidere
Maquoketa
Savanna
Mt. Carroll
Marengo
Carpentersville
Elgin
North Chicago
Highland Park
Arlington Heights
Evanston
Oregon
Sycamore
De Kalb
Schaumburg
Elmhurst
Clinton
Fulton
Sterling
Dixon
Rochelle
St. Charles
Naperville
Cicero
CHICAGO
Oak Lawn
De Witt
Morrison
Prophetstown
Eldridge
Davenport
Bettendorf
Rock Island
Moline
East Moline
Geneseo
Rock Falls
Amboy
Waterman
Tampico
Mendota
Plano
Sandwich
Aurora
Lockport
Joliet
Harvey
East Chicago
Gary
Hammond
Portage
La Porte
Chicago Heights
Merrillville
Valparaiso
Cambridge
Aledo
Princeton
Peru
La Salle
Ottawa
Spring Valley
Morris
Peotone
Braidwood
Marseilles
Galva
Toulon
Quawka
Galesburg
Monmouth
Knoxville
Abingdon
Lacon
Toluca
Streator
Dwight
Odell
Pontiac
Chillicothe
Minonk
Farmington
Peoria
Washington
Bourbonnais
Kankakee
Clifton
Lowell
Bremen
Nappanee
Plymouth
Ligonier
Kendallville
Butler
Auburn
Knox
Culver
North Judson
Winamac
Rochester
Warsaw
Columbia City
North Manchester
Fort Wayne
New Haven
Rensselaer
© Rand McNally A-101665-1
0 20 40 60 80 100 120 Miles
0 20 40 60 80 100 120 140 160 180 200 Kilometers
Lambert Conformal Conic Projection
Scale 1:4,000,000
One inch to 64 miles
One cm to 40 km

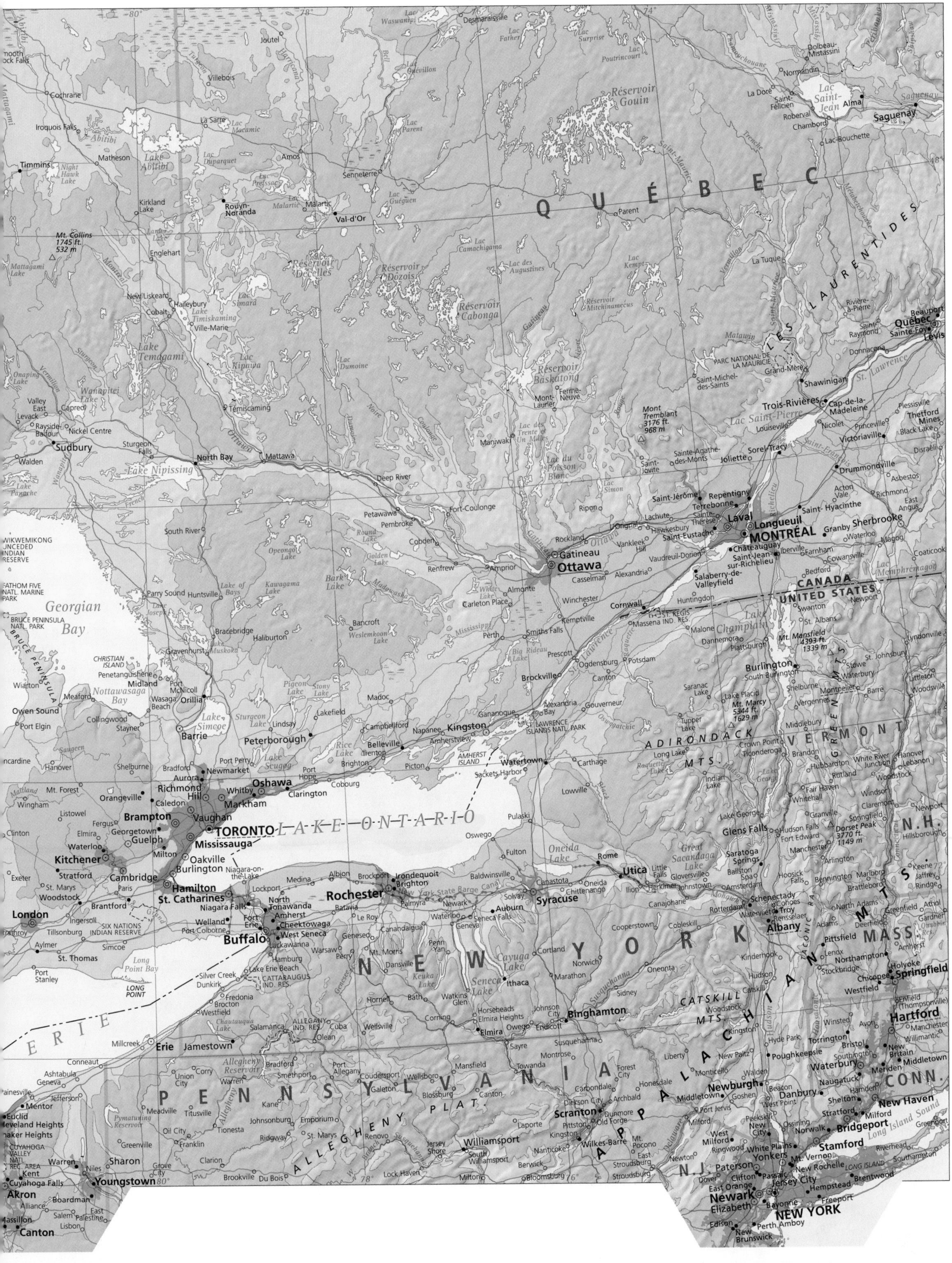
QUÉBEC
LES LAURENTIDES
Réservoir Gouin
Lac Saint-Jean
Saguenay
Alma
Dolbeau-Mistassini
Normandin
Roberval
Chambord
La Doré
Saint-Félicien
Lac-Bouchette
La Tuque
Parent
Val-d'Or
Malartic
Amos
Senneterre
Rouyn-Noranda
Lac Abitibi
Lake Abitibi
Timmins
Matheson
Iroquois Falls
Cochrane
La Sarre
Villebois
Joutel
Desmaraisville
Lac Waswanipi
Lac Quévillon
Réservoir Decelles
Réservoir Dozois
Réservoir Cabonga
Réservoir Baskatong
Lac Kempt
Réservoir Mitchinamecus
Mt. Collins 1745 ft. 532 m
Kirkland Lake
Englehart
New Liskeard
Haileybury
Cobalt
Ville-Marie
Lake Timiskaming
Lake Temagami
Lac Kipawa
Témiscaming
Sudbury
Sturgeon Falls
North Bay
Mattawa
Lake Nipissing
Valley East
Capreol
Levack
Rayside-Balfour
Nickel Centre
Walden
Deep River
Petawawa
Pembroke
Cobden
Renfrew
Arnprior
Fort-Coulonge
Mont-Laurier
Ferme-Neuve
Maniwaki
Mont Tremblant 3176 ft. 968 m
Saint-Michel-des-Saints
PARC NATIONAL DE LA MAURICIE
Grand-Mère
Shawinigan
Trois-Rivières
Cap-de-la-Madeleine
Lac Saint-Pierre
Louiseville
Nicolet
Victoriaville
Plessisville
Thetford Mines
Black Lake
Disraeli
Drummondville
Asbestos
Acton Vale
Richmond
East Angus
Sorel-Tracy
Joliette
Sainte-Agathe-des-Monts
Saint-Jovite
Saint-Jérôme
Repentigny
Terrebonne
Sainte-Thérèse
Saint-Eustache
Laval
Longueuil
MONTRÉAL
Saint-Hyacinthe
Granby
Sherbrooke
Waterloo
Magog
Coaticook
Lac Memphrémagog
Cowansville
Farnham
Bedford
Châteauguay
Saint-Jean-sur-Richelieu
Iberville
Vaudreuil-Dorion
Salaberry-de-Valleyfield
Huntingdon
Québec
Sainte-Foy
Beauport
Lévis
Saint-Raymond
Donnacona
Rivière-à-Pierre
St. Lawrence
Gatineau
Ottawa
Rockland
Lachute
Hawkesbury
Vankleek Hill
Alexandria
Casselman
Winchester
Cornwall
Kemptville
Smiths Falls
Perth
Carleton Place
Almonte
Prescott
Ogdensburg
Brockville
Gananoque
Kingston
Napanee
Amherstview
Belleville
Trenton
Picton
Brighton
Campbellford
Peterborough
Lindsay
Lakefield
Madoc
Bancroft
Haliburton
Bracebridge
Huntsville
Parry Sound
Gravenhurst
Orillia
Barrie
Lake Simcoe
Midland
Penetanguishene
Collingwood
Stayner
Wasaga Beach
Owen Sound
Wiarton
Georgian Bay
BRUCE PENINSULA
Nottawasaga Bay
Port Elgin
Hanover
Shelburne
Orangeville
Mt. Forest
Wingham
Listowel
Kitchener
Waterloo
Guelph
Cambridge
Stratford
Woodstock
Brantford
London
St. Thomas
Aylmer
Tillsonburg
Ingersoll
SIX NATIONS INDIAN RESERVE
Simcoe
Long Point Bay
LONG POINT
Port Stanley
Brampton
Vaughan
Richmond Hill
Markham
Aurora
Newmarket
Bradford
Whitby
Oshawa
Clarington
Port Hope
Cobourg
TORONTO
Mississauga
Oakville
Burlington
Milton
Hamilton
St. Catharines
Niagara Falls
Niagara-on-the-Lake
Welland
Fort Erie
Port Colborne
LAKE ONTARIO
CANADA
UNITED STATES
Massena
Malone
Plattsburgh
Lake Champlain
Burlington
Mt. Mansfield 4393 ft. 1339 m
Mt. Marcy 5344 ft. 1629 m
Lake Placid
Saranac Lake
Tupper Lake
ADIRONDACK MTS.
Watertown
Sackets Harbor
Carthage
Lowville
Pulaski
Oswego
Fulton
Oneida Lake
Rome
Utica
Syracuse
Auburn
Rochester
Irondequoit
Brighton
Brockport
Lockport
North Tonawanda
Amherst
Cheektowaga
Buffalo
West Seneca
Lackawanna
Hamburg
Dunkirk
Fredonia
Jamestown
Salamanca
Olean
Hornell
Corning
Elmira
Ithaca
Cortland
Binghamton
Norwich
Oneonta
NEW YORK
Albany
Schenectady
Troy
Saratoga Springs
Glens Falls
Lake George
CATSKILL MTS.
Kingston
Poughkeepsie
Newburgh
Middletown
White Plains
Yonkers
Mt. Vernon
New Rochelle
NEW YORK
Newark
Jersey City
Paterson
Elizabeth
Bayonne
Hempstead
Stamford
Bridgeport
New Haven
Waterbury
Hartford
Springfield
Pittsfield
VERMONT
GREEN MTS.
Rutland
N.H.
MASS.
CONN.
Long Island Sound
APPALACHIAN MTS.
PENNSYLVANIA
ALLEGHENY PLATEAU
Erie
Scranton
Wilkes-Barre
Williamsport
Youngstown
Akron
Canton
Sharon
LAKE ERIE

QUÉBEC
LES LAURENTIDES
Lac Mistassini
Mistassini
Lac Waconichi
Chibougamau
Lac Chibougamau
Lac Manouane
Lac Péribonka
Réservoir Pipmuacan
Réservoir Outardes Quatre
Réservoir Manicouagan
Réservoir Manic Trois
Réservoir Manic Deux
Lac Sainte-Anne
Grand lac Germain
Lac Fournier
Lac Magpie
Sept-Îles
Port-Cartier
Havre-Saint-Pierre
RÉSERVE DE PARC NATIONAL DE L'ARCHIPEL-DE-MINGAN
Détroit de Jacques-Cartier
POINTE DE L' OUEST
Rivière-Pentecôte
Baie-Trinité
POINTE DES MONTS
Baie-Comeau
RÉSERVE INDIENNE DE BETSIAMITES
Forestville
Normandin
Dolbeau-Mistassini
La Doré
Saint-Félicien
Lac Saint-Jean
Roberval
Chambord
Alma
Saguenay
Lac-Bouchette
Lac Pouthincourt
Tadoussac
St. Lawrence
Détroit d'Honguedo
Mont Jacques-Cartier 4190 ft. 1277 m
Cap-Chat
Matane
Sainte-Félicité
Mont-Joli
Rimouski
Le Bic
PARC NATIONAL FORILLON
Gaspé
CAP GASPÉ
Baie de Gaspé
PÉNINSULE DE LA GASPÉSIE
MONTS NOTRE DAME MOUNTAINS
Lac Matapédia
Saint-Gabriel
Amqui
Causapscal
Percé
Chandler
New Richmond
Bonaventure
Nouvelle
Dalhousie
Matapédia
Campbellton
Chaleur Bay
New Carlisle
MISCOU ISLAND
Trois-Pistoles
Saint-Siméon
Rivière-du-Loup
Clermont
La Malbaie
Baie-Saint-Paul
La Tuque
Cabano
Dégelis
Lac Témiscouata
Pohénégamook
Saint-Pacôme
La Pocatière
ÎLE AUX COUDRES
Beaupré
ÎLE D'ORLÉANS
Québec
Beauport
Lévis
Sainte-Foy
Saint-Raymond
Rivière-à-Pierre
PARC NATIONAL DE LA MAURICIE
Grand-Mère
Shawinigan
Donnacona
Montmagny
Saint-Pamphile
Trois-Rivières
Cap-de-la-Madeleine
Louiseville
Nicolet
Lac Saint-Pierre
Sorel-Tracy
Saint-Anselme
Saint-Joseph-de-Beauce
Plessisville
Princeville
Victoriaville
Thetford Mines
Black Lake
Saint-Georges
Drummondville
Disraeli
Asbestos
Lac Saint-François
Acton Vale
Richmond
Saint-Hyacinthe
Granby
Sherbrooke
Waterloo
Cowansville
Magog
Lac-Mégantic
Scotstown
East Angus
Coaticook
Lac Memphrémagog
CANADA
U.S.A.
Newport
APPALACHIAN MOUNTAINS
Coburn Mountain 3718 ft. 1133 m
Jacquet River
Kedgwick
Beresford
Caraquet
ÎLE LAMÈQUE
Bathurst
Tracadie-Sheila
Mt. Carleton 2680 ft. 817 m
Madawaska
Edmundston
Saint-Basile
Fort Kent
Van Buren
St. Léonard
Grand Falls
Eagle Lake
Square Lake
Limestone
Caribou
Plaster Rock
Washburn
Fort Fairfield
Ashland
Presque Isle
Perth-Andover
Miramichi Bay
Miramichi
KOUCHIBOUGUAC NATL. PARK
NEW BRUNSWICK
Bath
Mars Hill
Monticello
Doaktown
Rogersville
Hartland
Houlton
Woodstock
Bouctouche
U.S.A.
CANADA
Patten
Island Falls
Chamberlain Lake
Chesuncook Lake
Mt. Katahdin 5268 ft. 1606 m
Moosehead Lake
Rockwood
Millinocket
Fredericton
Grand Lake
Chipman
Salmon
Moncton
Dieppe
Shediac
Oromocto
Spednic Lake
Medway
Danforth
Greenville
Brownville Junction
McAdam
Fredericton Junction
Oromocto Lake
Sussex
Petitcodiac
Sackville
Amherst
Milo
Howland
Lincoln
Bingham
Dexter
Dover-Foxcroft
Hampton
FUNDY NATIONAL PARK
Alma
Chignecto Bay
Springhill
Oxford
COBEQUID MTS.
Parrsboro
Rangeley
Errol
Calais
St. Stephen
St. George
St. John
St. Andrews
Passamaquoddy Bay
Eastport
Lubec
Grand Manan
GRAND MANAN ISLAND
Bay of Fundy
Minas Channel
Minas Basin
Cobequid Bay
Truro
Brookfield
Canning
Wolfville
Kentville
Hantsport
Windsor
Middleton
Annapolis Royal
Bridgetown
Elmsdale
Musquodoboit Harbour
Dartmouth
Halifax
NOVA
Digby
Western Shore
New Germany
KEJIMKUJIK NATIONAL PARK
Lake Rossignol
Bridgewater
Lunenburg
Mahone Bay
Liverpool
Yarmouth
Wedgeport
Shelburne
Lockeport
Clark's Harbour
CAPE SABLE
LONG ISLAND
St. Marys Bay
PRINCE EDWARD ISLAND
Tignish
Alberton
O'Leary
Malpeque Bay
Egmont Bay
PRINCE EDWARD ISLAND NATL. PARK
Summerside
Port Borden
Mt. Stewart
Cape Tormentine
Charlottetown
Port Elgin
Northumberland Strait
Westville
Stellarton
Pictou
VERMONT
MAINE
Guildhall
Groveton
St. Johnsbury
Lyndonville
Montpelier
Barre
Littleton
Lancaster
Berlin
Mexico
Rumford
Farmington
Madison
Skowhegan
Pittsfield
Bangor
Orono
Brewer
Graham Lake
Mooselookmeguntic Lake
WHITE MTS.
Mt. Washington 6288 ft. 1917 m
Woodsville
North Woodstock
Conway
Norway
South Paris
Mechanic Falls
Bridgton
Auburn
Lewiston
Winthrop
Augusta
Gardiner
Waterville
Bucksport
Searsport
Belfast
Camden
Ellsworth
Bar Harbor
Winter Harbor
ACADIA NATL. PARK
MT. DESERT ISLAND
Machias
Jonesport
Rockland
Penobscot Bay
Waldoboro
Wiscasset
Bath
Brunswick
Lisbon Falls
VINALHAVEN ISLAND
ISLE AU HAUT
Boothbay Harbor
NEW HAMPSHIRE
Plymouth
Hanover
Lebanon
Lake Winnipesaukee
Ossipee
Claremont
Newport
Franklin
Laconia
Westbrook
Portland
Sanford
Saco
Old Orchard Beach
Biddeford
Canterbury
Concord
Rochester
Kennebunk
Somersworth
Dover
Hillsborough
Manchester
Raymond
Exeter
Kittery
Portsmouth
Bigelow Bight
Gulf of Maine
Jaffrey
Rindge
Nashua
Derry
Amesbury
Newburyport
Haverhill
Gardner
Fitchburg
Leominster
Lowell
Lawrence
CAPE ANN
Gloucester
Salem
Lexington
Lynn
MASS.
Worcester
Newton
Boston
Massachusetts Bay
Framingham
Cambridge
Auburn
Quincy
Woonsocket
Norwood
Brockton
R.I.
Providence
Attleboro
Cranston
Taunton
Plymouth
70°
68°
66°
64°
72°
74°
50°
48°
44°

0 20 40 60 80 100 120 Miles
0 20 40 60 80 100 120 140 160 180 200 Kilometers

Lambert Conformal Conic Projection
Scale 1:4,000,000
One inch to 64 miles
One cm to 40 km

LABRADOR SEA
Gulf of St. Lawrence
ATLANTIC OCEAN
Cabot Strait
Strait of Belle Isle
NEWFOUNDLAND AND LABRADOR
NEWFOUNDLAND
LONG RANGE MOUNTAINS
ÎLE D' ANTICOSTI
POINTE DE L' EST
Red Bay
CAPE BAULD
L'ANSE-AUX-MEADOWS
Pistolet Bay
St. Anthony
Hare Bay
GROAIS ISLAND
GREY ISLANDS
Roddickton
BELL ISLAND
HORSE ISLANDS
ST. JOHN ISLAND
Port Saunders
Blue Mountain 2129 ft. 649 m
Saint-Augustin
Mutton Bay
CAP DU GROS MÉCATINA
ÎLE DU PETIT MÉCATINA
Lac Robertson
Natashquan
La Romaine
CAP WHITTLE
White Bay
CAPE ST. JOHN
La Scie
Notre Dame Bay
NEW WORLD ISLAND
FOGO ISLAND
Fogo
Durrell
Hamilton Sound
GROS MORNE NATIONAL PARK
Gros Morne 2644 ft. 806 m
Rocky Harbour
Springdale
Robert's Arm
Sandy Lake
Carmanville
CAPE FREELS
Lewisporte
New-Wes-Valley
Bay of Islands
Deer Lake
Botwood
Glenwood
Gander
Lark Harbour
Pasadena
Badger
Bishop's Falls
Gander Lake
Hare Bay
Bonavista Bay
Corner Brook
Grand Lake
Buchans
Grand Falls-Windsor
Glovertown
Bonavista
LONG POINT
Port au Port Bay
Stephenville
Red Indian Lake
TERRA NOVA NATIONAL PARK
Catalina
CAPE ST. GEORGE
St. George's Bay
Victoria Lake
Jeddore Lake
RANDOM ISLAND
GRATES POINT
Shoal Harbour
Bay de Verde
Trinity Bay
Conception Bay
Meelpaeg Lake
CAPE ANGUILLE
Pouch Cove
Torbay
Carbonear
Harbour Grace
Wabana
St. John's
CAPE SPEAR
Bay Roberts
Brigus
CAPE RAY
Channel-Port aux Basques
Isle aux Morts
Burgeo
Hermitage Bay
Harbour Breton
Belle Bay
AVALON PENINSULA
Witless Bay
Fox Harbour
BRUNETTE ISLAND
Fortune Bay
BURIN PENINSULA
Placentia Bay
Grand Bank
Marystown
MIQUELON
Fortune
Burin
St. Lawrence
Branch
St. Mary's Bay
CAPE ST. MARY'S
LANGLADE
Saint-Pierre
SAINT-PIERRE
SAINT PIERRE & MIQUELON (Fr.)
St. Shotts
CAPE RACE
LA GROSSE ÎLE
ÎLE DE L' EST
ÎLE DU CAP AUX MEULES
ÎLES DE LA MADELEINE (Que.)
Cap-aux-Meules
ÎLE DU HAVRE AUBERT
CAPE NORTH
Aspy Bay
Dingwall
CAPE BRETON HIGHLANDS NATL. PARK
St. Ann's Bay
Sydney Mines
North Sydney
New Waterford
Glace Bay
Sydney
SCATARIE ISLAND
Souris
Georgetown
Port Hood
Murray Harbour
St. Georges Bay
Bras d'Or Lake
Louisbourg
Antigonish
St. Peters
Port Hawkesbury
ISLE MADAME
CAPE BRETON ISLAND
Chedabucto Bay
Canso
SABLE ISLAND (N.S.)
62°
60°
58°
56°
54°
52°
50°
48°
46°
44°
© Rand McNally A-101666-1
a
LES LAURENTIDES
Saint-Joachim
Sainte-Anne-de-Beaupré
Beaupré
Château-Richer
ÎLE D'ORLÉANS
Lac-Beauport
L'Ange-Gardien
Boischatel
Saint-Raymond
Shannon
Charlesbourg
Beauport
Sainte-Catherine-de-la-Jacques-Cartier
Val-Bélair
Sainte-Pétronille
Chute-Panet
Loretteville
Vanier
QUÉBEC
L'Ancienne-Lorette
Lévis
Beaumont
Sainte-Foy
Sillery
Pont-Rouge
Cap-Rouge
Saint-Romuald
Saint-Basile
Neuville
Charny
Saint-Jean-Chrysostome
Portneuf
Donnacona
Saint-Nicolas
Saint-Rédempteur
Saint-Henri
Cap-Santé
St. Lawrence (Saint-Laurent)
Saint-Antoine-de-Tilly
Saint-Apollinaire
Saint-Anselme
Rivière-Bois-Clair
Sainte-Croix
Saint-Agapit
Laurier-Station
Saint-Flavien
Saint-Gilles
Dosquet
Saint-Bernard
47°
46°30'
© Rand McNally A-101667-1

Inset map a
Lambert Conformal Conic Projection
Scale 1:1,000,000
One inch to 16 miles
One cm to 10 km

Oxnard
Long Beach
LOS ANGELES
Anaheim
CALIFORNIA
SAN DIEGO
Oceanside
TIJUANA
Mexicali
Ensenada
Humphreys Peak 12 633 ft. 3851 m
Baldy Peak 11 403 ft. 3476 m
PHOENIX
ARIZONA
Tucson
NEW MEXICO
Albuquerque
El Paso
CIUDAD JUÁREZ
Nogales
UNITED STATES
KANSAS
MISSOURI
Owensboro
ILL.
Springfield
Tulsa
Nashville
Amarillo
Oklahoma City
OZARK PLATEAU
ARKANSAS
Fort Smith
OKLAHOMA
Memphis
TENN.
Little Rock
Huntsville
Lubbock
Wichita Falls
Lake Texoma
Pine Bluff
Birmingham
Abilene
Fort Worth
DALLAS
Midland
TEXAS
Tyler
Shreveport
Monroe
ALA.
Jackson
Waco
EDWARDS PLATEAU
MISSISSIPPI
Austin
SAN ANTONIO
Beaumont
Baton Rouge
LOUISIANA
Mobile
HOUSTON
Lake Charles
New Orleans
Galveston
Corpus Christi
Laredo
Nuevo Laredo
Piedras Negras
Monclova
Brownsville
Reynosa
MONTERREY
Matamoros
Gulf of Mexico
ISLA GUADALUPE
ISLA ÁNGEL DE LA GUARDA
ISLA CEDROS
PUNTA EUGENIA
ISLA TIBURÓN
SONORAN DESERT
Golfo de California
BAJA CALIFORNIA
Hermosillo
Santa Rosalía
Chihuahua
Ciudad Obregón
SIERRA MADRE OCCIDENTAL
CHIHUAHUAN DESERT
Hidalgo del Parral
Los Mochis
Gómez Palacio
Tropic of Cancer
ISLA SANTA MARGARITA
Culiacán
La Paz
ISLA CERRALVO
Torreón
MEXICO
CABO SAN LUCAS
Mazatlán
Durango
SIERRA MADRE ORIENTAL
Ciudad Victoria
Matehuala
Zacatecas
ISLAS TRES MARÍAS
Ciudad Mante
Aguascalientes
San Luis Potosí
Tepic
Tampico
Ciudad Valles
GUADALAJARA
LEÓN
Puerto Vallarta
Querétaro
Tuxpan de Rodríguez Cano
ISLAS REVILLAGIGEDO (Mex.)
Zamora de Hidalgo
Celaya
MEXICO CITY
Guzmán
Morelia
Volcán Pico de Orizaba 18 406 ft. 5610 m
Xalapa
Uruapan del Progreso
Toluca de Lerdo
Veracruz
Tecomán
Puebla de Zaragoza
Córdoba
Bahía de Campeche
Lázaro Cárdenas
Iguala
San Andrés Tuxtla
Ciudad del Carmen
Zihuatanejo
SIERRA MADRE DEL SUR
Coatzacoalcos
Mérida
Cancún
YUCATAN PENINSULA
CABO CATOCHE
Chetumal
Villahermosa
Acapulco de Juárez
Oaxaca de Juárez
Tuxtla Gutiérrez
San Cristóbal de las Casas
Belize City
Belmopan
BELIZE
Salina Cruz
Golfo de Tehuantepec
Comitán de Domínguez
Golfo de Honduras
La Ceiba
MIDDLE AMERICA TRENCH
Volcán Tajumulco 13 845 ft. 4220 m
GUATEMALA
Quetzaltenango
San Pedro Sula
Tapachula
Guatemala
Tegucigalpa
Santa Ana
San Miguel
San Salvador
EL SALVADOR
León
Managua
PACIFIC OCEAN
ÎLE CLIPPERTON (Fr.)
ISLA DEL COCO (C.R.)
© Rand McNally A-101668-1

a

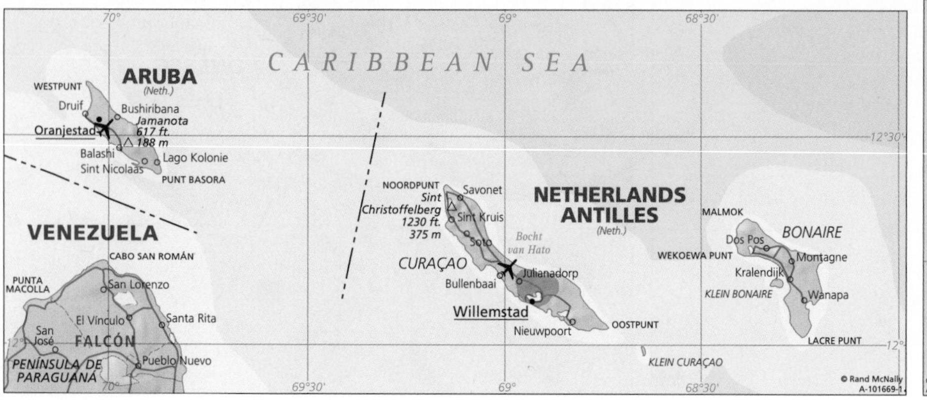
CARIBBEAN SEA
ARUBA (Neth.)
WESTPUNT
Druif
Bushiribana
Oranjestad
Jamanota 617 ft. 188 m
Balashi
Lago Kolonie
Sint Nicolaas
PUNT BASORA
VENEZUELA
CABO SAN ROMÁN
PUNTA MACOLLA
San Lorenzo
El Vínculo
Santa Rita
San José
FALCÓN
PENÍNSULA DE PARAGUANÁ
Pueblo Nuevo
NOORDPUNT
Savonet
Sint Christoffelberg 1230 ft. 375 m
Sint Kruis
Soto
NETHERLANDS ANTILLES (Neth.)
Bocht van Hato
CURAÇAO
Julianadorp
Bullenbaai
Willemstad
Nieuwpoort
OOSTPUNT
KLEIN CURAÇAO
MALMOK
Dos Pos
BONAIRE
WEKOEWA PUNT
Montagne
Kralendijk
KLEIN BONAIRE
Wanapa
LACRE PUNT
© Rand McNally A-101669-1

Inset map a
Lambert Conformal Conic Projection
Scale 1:2,000,000
One inch to 32 miles
One cm to 20 km

b
BERMUDA
(U.K.)
ST. GEORGE'S ISLAND
St. George
ST. DAVID'S ISLAND
Harrington Sound
Castle Harbour
Flatts
Town Hill
259 ft.
79 m
Hamilton
Great Sound
SOMERSET ISLAND
HIGH POINT
ATLANTIC OCEAN
© Rand McNally
A-101670-1
W.VA.
Roanoke
Richmond
Norfolk
Chesapeake Bay
Virginia Beach
VIRGINIA
Knoxville
Mt. Mitchell
6684 ft.
2037 m
Raleigh
Charlotte
NORTH CAROLINA
Chattanooga
APPALACHIAN MTS.
CAPE HATTERAS
Fayetteville
CAPE LOOKOUT
Wilmington
CAPE FEAR
SOUTH CAROLINA
Atlanta
Columbia
Columbus
GEORGIA
Charleston
Montgomery
Savannah
Jacksonville
Tallahassee
FLORIDA
CAPE CANAVERAL
Orlando
St. Petersburg
Tampa
Lake Okeechobee
West Palm Beach
Fort Lauderdale
Miami
CAPE SABLE
Key West
Straits of Florida
BERMUDA
(U.K.)
Hamilton
ATLANTIC OCEAN
Tropic of Cancer
ABACO
GRAND BAHAMA
Nassau
ELEUTHERA
NEW PROVIDENCE
ANDROS
CAT ISLAND
BAHAMAS
SAN SALVADOR
MANGROVE CAY
GREAT EXUMA
LONG ISLAND
CROOKED ISLAND
ACKLINS
MAYAGUANA
TURKS AND CAICOS ISLANDS
(U.K.)
Grand Turk
GREAT INAGUA
HAVANA
Matanzas
Pinar del Río
Santa Clara
CAYO COCO
CAYO ROMANO
CAYO GUAJABA
Golfo de Batabanó
Cienfuegos
CUBA
Camagüey
Holguín
ISLA DE LA JUVENTUD
Bayamo
Guantánamo
Windward Passage
Manzanillo
CABO CRUZ
Pico Turquino
6470 ft.
1972 m
Santiago de Cuba
CAYMAN ISLANDS
(U.K.)
George Town
GRAND CAYMAN
GREATER ANTILLES
Cap-Haïtien
Santiago de los Caballeros
Pico Duarte
10 417 ft.
3175 m
Gonaïves
HAITI
HISPANIOLA
DOMINICAN REPUBLIC
Port-au-Prince
Santo Domingo
San Pedro de Macorís
San Juan
Ponce
PUERTO RICO
(U.S.)
PUERTO RICO TRENCH
Montego Bay
JAMAICA
Kingston
Spanish Town
VIRGIN ISLANDS
(U.S.)
BRITISH VIRGIN ISLANDS
(U.K.)
ANGUILLA
(U.K.)
Charlotte Amalie
ST. CROIX
(U.S.)
LEEWARD ISLANDS
ANTIGUA AND BARBUDA
ST. KITTS AND NEVIS
MONTSERRAT
(U.K.)
GRANDE-TERRE
GUADELOUPE
(Fr.)
Basse-Terre
MARIE-GALANTE
DOMINICA
Roseau
MARTINIQUE
(Fr.)
Fort-de-France
WEST INDIES
Castries
ST. LUCIA
BARBADOS
Kingstown
Bridgetown
ST. VINCENT AND THE GRENADINES
GRENADA
LESSER ANTILLES
WINDWARD ISLANDS
CARIBBEAN SEA
HONDURAS
CABO GRACIAS A DIOS
Tegucigalpa
ISLA DE PROVIDENCIA
(Col.)
NICARAGUA
Lago de Nicaragua
Bluefields
ISLA DE SAN ANDRÉS
(Col.)
COSTA RICA
Volcán Irazú
11 260 ft.
3432 m
Puntarenas
San José
Puerto Limón
Cerro Chirripó
12 530 ft.
3819 m
Volcán Barú
11 401 ft.
3475 m
David
Colón
Panama Canal
ISTMO DE PANAMÁ
Panama
Santiago
Golfo de Chiriquí
PEN. DE AZUERO
Golfo de Panamá
ISLA DE COIBA
PUNTA MARIATO
ARUBA
(Neth.)
NETHERLANDS ANTILLES
(Neth.)
BONAIRE
Willemstad
CURAÇAO
ISLA LA ORCHILA
ISLA BLANQUILLA
ISLA DE MARGARITA
ISLA LA TORTUGA
TOBAGO
TRINIDAD AND TOBAGO
Port of Spain
TRINIDAD
PUNTA GALLINAS
PEN. DE LA GUAJIRA
Golfo de Venezuela
Punto Fijo
Santa Marta
Barranquilla
Soledad
Pico Cristóbal Colón
18 947 ft.
5775 m
MARACAIBO
Cabimas
Lago de Maracaibo
Puerto Cabello
CARACAS
Barquisimeto
Carúpano
Cumaná
Barcelona
Golfo de Paria
Cartagena
Valle de la Pascua
Acarigua
El Tigre
Sincelejo
Magangué
Valera
Calabozo
Ciudad Bolívar
Ciudad Guayana
Montería
Mérida
Pico Bolívar
16 427 ft.
5007 m
San Fernando de Apure
Cerro Mato
6312 ft.
1863 m
Embalse de Guri
Ocaña
Cúcuta
San Cristóbal
VENEZUELA
Georgetown
Barrancabermeja
Bucaramanga
Angel Falls
Mt. Roraima
9432 ft.
2875 m
GUYANA
MEDELLÍN
LLANOS
Puerto Ayacucho
GUIANA HIGHLANDS
PAKARAIMA MTS.
SURINAME
Sogamoso
La Dorada
Manizales
Nevado del Tolima
17 110 ft.
5215 m
Tunja
CABO CORRIENTES
Cartago
BOGOTÁ
San Fernando de Atabapo
Cerro Marahuaca
8461 ft.
2579 m
BRAZIL
Buenaventura
Ibagué
Villavicencio
Boa Vista
ISLA DE MALPELO
(Col.)
Palmira
COLOMBIA
CALI
Nevado del Huila
18 865 ft.
5750 m

0 80 160 240 320 400 480 Miles
0 80 160 240 320 400 480 560 640 720 800 Kilometers

Lambert Azimuthal Equal Area Projection
Scale 1:16,000,000
One inch to 256 miles
One cm to 160 km

Inset map b
Lambert Conformal Conic Projection
Scale 1:1,000,000
One inch to 16 miles
One cm to 10 km

a

ATLANTIC OCEAN
BAHAMAS
GULF OF MEXICO
UNITED STATES
FLORIDA
Lake Okeechobee
Big Cypress Swamp
The Everglades
Tamiami Canal
Biscayne Bay
Florida Bay
FLORIDA KEYS
Straits of Florida
Venice
Port Charlotte
Punta Gorda
Fort Myers
Cape Coral
Bonita Springs
Naples
Marco Island
CAPE ROMANO
CAPE SABLE
Clewiston
Belle Glade
Stuart
Jupiter
Riviera Beach
West Palm Beach
Boynton Beach
Coral Springs
Boca Raton
Pompano Beach
Fort Lauderdale
Hollywood
Hialeah
Miami Beach
Miami
Homestead
KEY LARGO
Key Largo
Marathon
Key West
MARQUESAS KEYS
DRY TORTUGAS
GREAT SALE CAY
LITTLE ABACO
Cooper's Town
West End
Freeport
GRAND BAHAMA
McLeans Town
Marsh Harbour
MOORE'S ISLAND
ABACO
Sandy Point
SOUTHWEST POINT
Northwest Providence Channel
Northeast Providence Channel
BIMINI ISLANDS
Alice Town
BERRY ISLANDS
Nicholl's Town
Nassau
ROSE ISLAND
NEW PROVIDENCE
Andros Town
ANDROS
Dunmore Town
ELEUTHERA
Governor's Harbour
Rock Sound
EAST END POINT
Arthur's Town
CAT ISLAND
New Bight
HAWKS NEST POINT
Port Howe
Exuma Sound
GREAT GUANA CAY
EXUMA CAYS
Kemps Bay
CISTERN POINT
TONGUE OF THE OCEAN
GREAT BAHAMA BANK
CAPE SANTA MARIA
RUM CAY
GREAT EXUMA
LITTLE EXUMA
LONG ISLAND
Deadman's Cay
Clarence Town
JUMENTOS CAYS
RAGGED ISLAND RANGE
CAPE VERDE
RAGGED ISLAND
Santaren Channel
CAY SAL
CAY SAL BANK
ANGUILLA CAYS
Nicholas Channel
Old Bahama Channel
Tropic of Cancer
GREATER ANTILLES
ARCH. DE SABANA
ARCH. DE CAMAGÜEY
CAYO FRAGOSO
CAYO COCO
CAYO ROMANO
CAYO GUAJABA
CAYO SABINAL
CUBA
HAVANA (LA HABANA)
CIUDAD DE LA HABANA
LA HABANA
PINAR DEL RÍO
La Esperanza
Minas de Matahambre
Artemisa
Candelaria
Güira de Melena
Güines
Los Palacios
Pinar del Río
Guane
PUNTA GORDA
Golfo de Batabanó
CABO DE SAN ANTONIO
CABO CORRIENTES
CAYOS DE SAN FELIPE
Nueva Gerona
La Fe
Ensenada de la Siguanea
ISLA DE LA JUVENTUD
ARCHIPIÉLAGO DE LOS CANARREOS
CAYO LARGO
Matanzas
Cárdenas
Jovellanos
Colón
MATANZAS
Ensenada de la Broa
Jagüey Grande
PENÍNSULA DE ZAPATA
Quemado de Güines
Sagua la Grande
VILLA CLARA
Caibarién
Santa Clara
Aguada de Pasajeros
CIENFUEGOS
Palmira
Cienfuegos
Placetas
Yaguajay
Bahía de Cochinos (Bay of Pigs)
Pico San Juan 3740 ft. 1140 m
Trinidad
SANCTI SPÍRITUS
Sancti Spíritus
CIEGO DE ÁVILA
Morón
Ciego de Ávila
Esmeralda
Tunas de Zaza
Júcaro
Golfo de Ana María
CAMAGÜEY
Florida
Minas
Nuevitas
Camagüey
Vertientes
ARCHIPIÉLAGO DE LOS JARDINES DE LA REINA
Santa Cruz del Sur
Martí
LAS TUNAS
Las Tunas
Puerto Padre
Gibara
Holguín
HOLGUÍN
PUNTA DE MULAS
Banes
Bahía de Nipe
Cueto
Mayarí
Sagua de Tánamo
Golfo de Guacanayabo
Manzanillo
Bayamo
Jiguaní
Campechuela
GRANMA
SANTIAGO DE CUBA
GUANTÁNAMO
Guantánamo
Palma Soriano
Niquero
SIERRA MAESTRA
CABO CRUZ
Pico Turquino 6470 ft. 1972 m
Santiago de Cuba
Caimanera
Bahía de Guantánamo
LITTLE CAYMAN
CAYMAN BRAC
GRAND CAYMAN
CAYMAN ISLANDS (U.K.)
George Town
CAYMAN TRENCH
JAMAICA
Montego Bay
Falmouth
Ocho Rios
SOUTH NEGRIL POINT
Savanna-la-Mar
Mt. Denham 3235 ft. 986 m
Spanish Town
Mandeville
May Pen
Port Antonio
Blue Mountain Peak 7402 ft. 2256 m
Kingston
MORANT POINT
PORTLAND POINT
Portland Bight
Jamaica Channel
NAVASSA I. (U.S.; claimed by Haiti)
MORANT CAYS (Jam.)
PEDRO CAYS (Jam.)
CARIBBEAN SEA
PUERTO RICO (U.S.)
ATLANTIC OCEAN
CARIBBEAN SEA
PUNTA AGUJEREADA
Isabela
Hatillo
Arecibo
Vega Baja
San Juan
Aguadilla
PUNTA HIGÜERO
Manatí
Vega Alta
Bayamón
Guaynabo
Carolina
Río Grande
CABEZAS DE SAN JUAN
Fajardo
ISLA DE CULEBRA
Florida
Lares
San Sebastián
Utuado
Trujillo Alto
El Toro 3524 ft. 1074 m
Ceiba
PUNTA PUERCA
Caguas
Juncos
Mayagüez
Hormigueros
Cerro de Punta 4390 ft. 1338 m
CORDILLERA CENTRAL
Cidra
San Lorenzo
Punta Santiago
Humacao
SIERRA DE CAYEY
Cayey
Coamo
San Germán
Yauco
Cabo Rojo
Guayanilla
Juana Díaz
Guayama
Yabucoa
Guánica
Ponce
Salinas
Santa Isabel
PUNTA PETRONA
PUNTA BREA
CABO ROJO
ISLA CAJA DE MUERTOS
Canal de la Mona
Sonda de Vieques
Pasaje de Vieques
Virgin Passage
ISLA DE VIEQUES
PUNTA ESTE
© Rand McNally
A-101690-1

Inset map a
Lambert Conformal Conic Projection
Scale 1:2,000,000
One inch to 32 miles
One cm to 20 km

b
PUERTO RICO TRENCH
ATLANTIC OCEAN
PUERTO RICO (U.S.)
San Juan
Aguadilla
Arecibo
PUNTA HIGÜERO
Bayamón
Cerro de Punta 4390 ft. 1338 m
Mayagüez
Caguas
Fajardo
Humacao
CORD. CENTRAL
ISLA DE MONA
Canal de la Mona
CABO ROJO
Ponce
Guayama
ISLA DE CULEBRA
ISLA DE VIEQUES
Charlotte Amalie
ST. THOMAS
ST. JOHN
TORTOLA
Road Town
VIRGIN GORDA
ANEGADA
VIRGIN ISLANDS (U.S.)
BRITISH VIRGIN ISLANDS (U.K.)
Frederiksted
Christiansted
ST. CROIX
Anegada Passage
SOMBRERO
ANGUILLA (U.K.)
DOG I.
SCRUB I.
The Valley
ST.-MARTIN (France)
SINT MAARTEN
Philipsburg
ST.-BARTHÉLEMY
NETHERLANDS ANTILLES (Neth.)
SABA
SINT EUSTATIUS
ST. KITTS AND NEVIS
ST. CHRISTOPHER
Basseterre
Charlestown
NEVIS
Nevis Peak 3232 ft. 985 m
REDONDA
BARBUDA
ANTIGUA AND BARBUDA
St. John's
ANTIGUA
Boggy Peak 1319 ft. 402 m
MONTSERRAT (U.K.)
Soufrière Hills
Guadeloupe Passage
LEEWARD ISLANDS
LESSER ANTILLES
VENEZUELAN BASIN
CARIBBEAN SEA
POINTE DE LA GRANDE VIGIE
GRANDE-TERRE
BASSE-TERRE
LA DÉSIRADE
Pointe-à-Pitre
Les Abymes
Soufrière 4813 ft. 1467 m
GUADELOUPE (France)
Basse-Terre
Trois-Rivières
MARIE-GALANTE
Grand-Bourg
LES SAINTES
ISLA LAS AVES (Venezuela)
Dominica Passage
Marigot
Morne Diablotins 4747 ft. 1447 m
DOMINICA
Roseau
Martinique Passage
MARTINIQUE (France)
Montagne Pelée 4583 ft. 1397 m
La Trinité
Fort-de-France
Le Lamentin
POINTE DES SALINES
St. Lucia Channel
POINTE DU CAP
Castries
Mt. Gimie 3117 ft. 950 m
ST. LUCIA
Vieux Fort
St. Vincent Passage
Soufrière 4049 ft. 1234 m
Georgetown
ST. VINCENT
Kingstown
ST. VINCENT AND THE GRENADINES
BEQUIA
MUSTIQUE
CANOUAN
UNION I.
CARRIACOU
GRENADINES
RONDE I.
Mt. St. Catherine 2756 ft. 840 m
Grenville
St. George's
GRENADA
WINDWARD ISLANDS
BARBADOS
Speightstown
Mt. Hillaby 1115 ft. 340 m
Bridgetown
TOBAGO
Charlotteville
Scarborough
TRINIDAD AND TOBAGO
El Cerro del Aripo 3084 ft. 940 m
GALERA POINT
PENÍNSULA DE PARIA
Dragons Mouths
Morvant
Port of Spain
Arima
Sangre Grande
Güiria
Gulf of Paria
TRINIDAD
Rio Claro
San Fernando
Point Fortin
GALEOTA POINT
Pedernales
Serpents Mouth
DELTA DEL ORINOCO
Caño Macareo
Caño Mariusa
ISLA TOBEJUBA
VENEZUELA
Tucupita
DELTA AMACURO
SAN SALVADOR (WATLING ISLAND)
SAMANA CAY
CROOKED ISLAND
NORTH EAST POINT
LONG CAY
Bight of Acklins
ACKLINS
SALINA POINT
Mayaguana Passage
MAYAGUANA
Caicos Passage
TURKS AND CAICOS ISLANDS (U.K.)
Kew
NORTH CAICOS
PROVIDENCIALES
MIDDLE CAICOS
EAST CAICOS
WEST CAICOS
CAICOS ISLANDS
LITTLE INAGUA
Grand Turk
TURKS ISLANDS
Turks Island Passage
NORTH EAST POINT
PALACCA POINT
Mouchoir Passage
SEAL CAYS
MOUCHOIR BANK
Matthew Town
Lake Rosa
GREAT INAGUA
Silver Bank Passage
SILVER BANK
Baracoa
PUNTA DE QUEMADO
Windward Passage
ÎLE DE LA TORTUE
NAVIDAD BANK
Port-de-Paix
Monte Cristi
CABO ISABELA
CAP DU MÔLE
Cap-Haïtien
Puerto Plata
CABO MACORÍS
CAP À FOUX
Limbé
Fort-Liberté
CABO FRANCÉS VIEJO
Yaque del Norte
Dajabón
Mao
Tamboril
Bahía Escocesa
Gonaïves
Santiago de los Caballeros
Moca
San Francisco de Macorís
Nagua
CABO SAMANÁ
Morne Bonhomme 5866 ft. 1788 m
HISPANIOLA
Sánchez
Samaná
Golfe de la Gonâve
Saint-Marc
Hinche
Pico Duarte 10 417 ft. 3175 m
La Vega
Yuna
Bahía de Samaná
Canal de Saint-Marc
Verrettes
DOMINICAN REPUBLIC
ÎLE DE LA GONÂVE
Comendador
Bonao
Artibonite
HAITI
Alto Bandera 8629 ft. 2630 m
San Juan de la Maguana
El Seibo
CABO ENGAÑO
GRANDE CAYEMITE
Canal du Sud
Jérémie
Sabana Yegua
Villa Altagracia
Santo Domingo
Soco
Higüey
Consuelo
Port-au-Prince
Anse-d'Hainault
Léogâne
Lago Enriquillo
Bajos de Haina
La Romana
POINTE FANCHON
Pic Macaya 7700 ft. 2347 m
Pétion-Ville
Neiba
Azua
San Pedro de Macorís
Petit-Goâve
Aquin
Baní
Bahía de Ocoa
San Cristóbal
Bahía de Yuma
Jacmel
Morne La Selle 8773 ft. 2674 m
Barahona
PUNTA PALENQUE
Coteaux
Les Cayes
ÎLE À VACHE
POINTE ABACOU
ISLA SAONA
Canal de la Mona
Pedernales
Enriquillo
CABO FALSO
CABO BEATA
ISLA BEATA
CARIBBEAN SEA
© Rand McNally A-101689-1
© Rand McNally A-101691-1

0 25 50 75 100 125 150 Miles
0 25 50 75 100 125 150 175 200 225 250 Kilometers

Lambert Conformal Conic Projection
Scale 1:5,000,000
One inch to 80 miles
One cm to 50 km

Inset map b
Lambert Conformal Conic Projection
Scale 1:5,000,000
One inch to 80 miles
One cm to 50 km

SAN DIEGO
CALIFORNIA
TIJUANA
Tecate
Mexicali
Yuma
San Luis Río Colorado
Ensenada
Gila Bend
Casa Grande
ARIZONA
Tucson
NEW MEXICO
Silver City
Alamogordo
SACRAMENTO MTS.
Artesia
Hobbs
Lamesa
Lordsburg
Deming
Las Cruces
Carlsbad
Big Spring
UNITED
El Paso
CIUDAD JUÁREZ
Guadalupe Pk. 8749 ft. 2667 m
Kermit
Midland
Odessa
Pecos
Big Lake
DESIERTO DE ALTAR
Sonoyta
ISLA MONTAGUE
Pico del Diablo 10 171 ft. 3100 m
PUNTA COLNETT
San Felipe
Puerto Peñasco
Chiricahua Peak 9798 ft. 2986 m
Sierra Vista
Nogales
Douglas
Agua Prieta
Colonia Vicente Guerrero
Lázaro Cárdenas
CABO SAN QUINTÍN
Caborca
Santa Ana
Magdalena de Kino
Cananea
Nacozari de García
Janos
Ascensión
Van Horn
Fort Stockton
Ozo
STOCKTON PLATEAU
Guadalupe Bravos
Nuevo Casas Grandes
Ahumada
Alpine
Marfa
Sanderson
Golfo de California
Benjamín Hill
Loma La Atravesada 2264 ft. 690 m
PUNTA SAN ANTONIO
Puerto Libertad
Carbó
Moctezuma
Nácori Chico
Buenaventura
Presidio
Ojinaga
BAJA CALIFORNIA
SONORA
SIERRA MADRE OCCIDENTAL
ISLA ÁNGEL DE LA GUARDA
ISLA TIBURÓN
Bahía Kino
Hermosillo
Madera
CHIHUAHUA
Chihuahua
Aldama
Guerrero
Cuauhtémoc
ISLA CEDROS
Bahía Sebastián Vizcaíno
Presa Alvaro Obregón
Cerro El Nopal 10 039 ft. 3060 m
Ocampo
Meoqui
Delicias
PUNTA EUGENIA
Guerrero Negro
Volcán de las Tres Vírgenes 6299 ft. 1920 m
Guaymas
Empalme
San Juanito
Saucillo
Ciudad Camargo
Melchor Múzquiz
DESIERTO DE VIZCAÍNO
Santa Rosalía
Vícam
Esperanza
San Francisco de Borja
Nonoava
COAHUILA
PUNTA SAN ROQUE
San Ignacio
PUNTA CHIVATO
Ciudad Obregón
Villa Juárez
Navojoa
Urique
Jiménez
Cuatrociénegas
PUNTA ABREOJOS
Mulegé
Huatabampo
Álamos
Guachochi
San Francisco del Oro
Hidalgo del Parral
Santa Bárbara
Monclova
Choix
El Fuerte
La Purísima
Loreto
ISLA CARMEN
San Blas
Los Mochis
Santa María del Oro
Tlahualilo de Zaragoza
San Pedro de las Colonias
BAJA CALIFORNIA SUR
Villa Insurgentes
ISLA SANTA MAGDALENA
Ciudad Constitución
Topolobampo
Guasave
Mocorito
Guamúchil
Gómez Palacio
Matamoros
Torreón
Parras de la Fuente
DURANGO
Santiago Papasquiaro
Nazas
Cuencamé de Ceniceros
ISLA SAN JOSÉ
CABO SAN LÁZARO
Bahía Magdalena
ISLA SANTA MARGARITA
ISLA DEL ESPÍRITU SANTO
Culiacán
Cerro La Bandera 10 597 ft. 3230 m
Canatlán
Costa Rica
Guadalupe Victoria
Concepción del Oro
ISLA CERRALVO
La Paz
SINALOA
Tayoltita
Durango
Juan Aldama
ZACATECAS
Río Grande
La Cruz
Tropic of Cancer
PUNTA ARENA
El Salto
Sombrerete
Todos Santos
Santiago
Mazatlán
Villa Unión
Fresnillo
San José del Cabo
San Lucas
CABO SAN LUCAS
Laguna del Caimanero
Escuinapa de Hidalgo
MEX
Valparaíso
Zacatecas
Jerez de García Salinas
Acaponeta
Loreto
AGUASCALIENTES
NAYARIT
Tuxpan
Santiago Ixcuintla
Tlaltenango de Sánchez Román
Aguascalientes
ISLA MARÍA MADRE
ISLA MARÍA MAGDALENA
Tepic
Lagos de Moreno
Nochistlán
PACIFIC OCEAN
ISLAS TRES MARÍAS
ISLA MARÍA CLEOFAS
Compostela
LEÓN
Tepatitlán de Morelos
San Francisco del Rincón
PUNTA DE MITA
Puerto Vallarta
Bahía de Banderas
Tequila
GUADALAJARA
CABO CORRIENTES
Ameca
Ocotlán
JALISCO
La Barca
Laguna de Chapala
Tomatlán
Autlán de Navarro
Guzmán
Zamora de Hidalgo
Uruapan del Progreso
Nevado de Colima 13 911 ft. 4240 m
Pico de Tancítaro 12 664 ft. 3860 m
ISLAS REVILLAGIGEDO
ISLA SAN BENEDICTO
ISLA ROCA PARTIDA
Cihuatlán
Colima
MICH
ISLA SOCORRO
Manzanillo
COLIMA
Apatzingán de la Constitución
Nueva Italia
Tecomán
ISLA CLARIÓN
Coahuayana
PUNTA TEJUPAN
Arteaga
Lázaro Cárdenas
Ixtapa
Zihuatanejo
118° 116° 114° 112° 110° 108° 106° 104° 102°
32° 30° 28° 26° 24° 22° 20° 18° 16°

0 40 80 120 160 200 240 Miles

0 40 80 120 160 200 240 280 320 360 400 Kilometers

Lambert Azimuthal Equal Area Projection
Scale 1:8,000,000
One inch to 128 miles
One cm to 80 km

GULF OF MEXICO
Bahía de Campeche
CARIBBEAN SEA
Yucatan Channel
Tropic of Cancer
MIDDLE AMERICA TRENCH
Golfo de Tehuantepec
STATES
TEXAS
EDWARDS PLATEAU
LOUISIANA
MISSISSIPPI
ALABAMA
GEORGIA
FLORIDA
NUEVO LEÓN
TAMAULIPAS
SAN LUIS POTOSÍ
GUANAJUATO
QUERÉTARO
HIDALGO
VERACRUZ
TLAXCALA
MÉXICO
MORELOS
PUEBLA
GUERRERO
OAXACA
TABASCO
CHIAPAS
CAMPECHE
YUCATÁN
QUINTANA ROO
YUCATAN PENINSULA
BELIZE
GUATEMALA
HONDURAS
EL SAL-
SIERRA MADRE ORIENTAL
SIERRA MADRE DEL SUR
SIERRA MADRE
ISTMO DE TEHUANTEPEC
DALLAS
Fort Worth
Stamford
Abilene
Cleburne
Tyler
Longview
Shreveport
Ruston
Monroe
Tallulah
Vicksburg
Jackson
Meridian
Troy
Albany
Ozark
Andalusia
Dothan
Bainbridge
Corsicana
Jacksonville
Natchitoches
Brookhaven
Natchez
Laurel
Hattiesburg
Brownwood
Waco
Nacogdoches
Toledo Bend Res.
Alexandria
McComb
Mobile
Tallahassee
San Angelo
Lufkin
Sam Rayburn Res.
De Ridder
Bogalusa
Pensacola
Fort Walton Beach
Killeen
Brady
Temple
Lake Livingston
Opelousas
Baton Rouge
Hammond
Biloxi
Pascagoula
Panama City
Bryan
Huntsville
Conroe
Lake Charles
Lafayette
New Orleans
Gulfport
Lake Pontchartrain
Sonora
Austin
Fredericksburg
Kerrville
Beaumont
Port Arthur
New Iberia
Thibodaux
Houma
Morgan City
CAPE SAN BLAS
ST. GEORGE ISLAND
HOUSTON
Baytown
San Marcos
Sugar Land
Texas City
Galveston
MARSH ISLAND
POINT AU FER ISLAND
SAN ANTONIO
Del Rio
Uvalde
Lake Jackson
GALVESTON ISLAND
Freeport
Eagle Pass
Victoria
Cotulla
Beeville
Allende
MATAGORDA ISLAND
SAN JOSE ISLAND
Corpus Christi
Kingsville
Nuevo Laredo
Laredo
Ciudad Anáhuac
PADRE ISLAND
Laguna Madre
Zapata
Falcon Reservoir
Sabinas Hidalgo
Raymondville
McAllen
Harlingen
Brownsville
Reynosa
Cerralvo
Matamoros
Guadalupe
MONTERREY
Valle Hermoso
Saltillo
Montemorelos
Linares
San Fernando
BARRA DE LOS AMERICANOS
Hidalgo
Abasolo
Ciudad Victoria
Soto la Marina
Aldama
Tula
Ciudad Mante
Cuauhtémoc
Cerritos
Ciudad Valles
Tampico
San Luis Potosí
Rioverde
Pánuco
Laguna de Tamiahua
CABO ROJO
Tantoyuca
Tamazunchale
Tamiahua
San Luis de la Paz
Tuxpan de Rodríguez Cano
Poza Rica de Hidalgo
Papantla de Olarte
Salamanca
Ixmiquilpan
Querétaro
EL TAJÍN
Celaya
Pachuca de Soto
Martínez de la Torre
Acámbaro
Tulancingo
Teziutlán
Morelia
MEXICO CITY
Ciudad Netzahualcóyotl
Xalapa
Toluca de Lerdo
Tlaxcala de Xicohténcatl
Volcán Pico de Orizaba 18 406 ft. 5610 m
Veracruz
Nevado de Toluca 15 387 ft. 4690 m
PUEBLA DE ZARAGOZA
Córdoba
Alvarado
PUNTA ROCA PARTIDA
Cuernavaca
Orizaba
Tehuacán
Tierra Blanca
San Andrés Tuxtla
Taxco de Alarcón
Coatzacoalcos
Iguala
Tuxtepec
Minatitlán
Chilpancingo de los Bravo
Huajuapan de León
Asunción Nochixtlán
Tlapa de Comonfort
Oaxaca de Juárez
Tecpan de Galeana
Putla de Guerrero
San Marcos
Acapulco de Juárez
Ometepec
Miahuatlán de Porfirio Díaz
Santiago Pinotepa Nacional
Santiago Jamiltepec
Puerto Escondido
PUNTA CORNETA
Puerto Ángel
Salina Cruz
Juchitán de Zaragoza
Matías Romero
Ixtepec
Tuxtla Gutiérrez
Cintalapa
Chahuites
Tonalá
Pijijiapan
Mapastepec
Huixtla
Tapachula
Mazatenango
Escuintla
Guatemala
Quetzaltenango
Huehuetenango
Volcán Tajumulco 13 845 ft. 4220 m
Salamá
Cobán
El Estor
Chiquimula
Zacapa
COPÁN
Cerro El Pital 8957 ft. 2730 m
Santa Rosa de Copán
Cerro Las Minas 9347 ft. 2849 m
Siguatepeque
Comayagua
Tegucigalpa
Villahermosa
Paraíso
Frontera
Jonuta
Ciudad del Carmen
Laguna de Términos
Sabancuy
Escárcega
Champotón
Seybaplaya
PUNTA MORRO
Campeche
Hopelchén
Dzitbalché
Maxcanú
Hunucmá
Progreso
Mérida
Rio Lagartos
CABO CATOCHE
Panabá
Temax
Tizimín
Cancún
Playa del Carmen
Cozumel
ISLA COZUMEL
Valladolid
CHICHÉN ITZÁ
Ticul
UXMAL
Tekax
Peto
Tulum
TULUM
Felipe Carrillo Puerto
Bahía Chetumal
Chetumal
Corozal
AMBERGRIS CAY
Belize City
TURNEFFE ISLANDS
Belmopan
Dangriga
Emiliano Zapata
Tenosique
Teapa
PALENQUE
TIKAL
Lago Petén Itzá
La Libertad
San Benito
San Cristóbal de las Casas
Comitán de Domínguez
Presa de la Angostura
Sán Luis
Punta Gorda
PUNTA NEGRA
Golfo de Honduras
Puerto Barrios
Livingston
Puerto Cortés
La Ceiba
Tela
San Pedro Sula
Santa Rita
Yoro
Guaimaca
ISLAS DE LA BAHÍA (Hond.)
© Rand McNally A-101672-1

Costa Rica
Ensenada del Pabellón
Quila
Coacoyole
Cosalá
San Miguel de Cruces
Cerro La Bandera 10 597 ft. 3230 m
Francisco I. Madero
Canatlán
El Dorado
Higuera de Abuya
DURANGO
Guadalupe Victoria
CHIHUAHUAN
Concepción del Oro
COAHUILA
Conitaca
Otinapa
Juan Aldama
Miguel Auza
DESERT
Culiacán
Ajoya
Tayoltita
Durango
Francisco Murguía
Santa Ana
La Cruz
SINALOA
El Salto
Presa Guadalupe Victoria
Nombre de Dios
Vicente Guerrero
Río Grande
Vanegas
NUEV
LEÓN
Dimas
La Ciudad
Sombrerete
Catorce
La Paz
Matehuala
Cerro Grande 10 433 ft. 3180 m
El Quelite
Súchil
Saín Alto
Cañitas de Felipe Pescador
Tropic of Cancer
Mezquital
SIERRA MADRE OCCIDENTAL
SIERRA MADRE
Mier y Noriega
Agua Caliente
Cebollas
ZACATECAS
Santo Domingo
Mazatlán
Concordia
Jiménez del Téul
Villa de Cos
Illescas
Villa Unión
Fresnillo
Charcas
San Ignacio
Rosario
Trujillo
Laguna del Caimanero
Escuinapa de Hidalgo
Mineral de Cucharas
Coronado
El Huisache
Ramos
Valparaíso
Jerez de García Salinas
Zacatecas
Troncoso
Teacapan
Huejuquilla El Alto
Guadalupe
Salinas de Hidalgo
SAN LUIS POTOSÍ
Villa de Arista
Acaponeta
Huazamota
Cerro Lechuguilla 8136 ft. 2480 m
Ciudad Cuauhtémoc
Ojocaliente
Cerritos
Tecuala
Villanueva
Pinos
Monte Escobedo
Rincón de Romos
Loreto
Cerro Los Huacales 9777 ft. 2980 m
San Luis Potosí
Soledad Díez Gutiérrez
Rosamorada
Laguna Agua Brava
Colotlán
NAYARIT
ISLA SAN JUANITO
ISLA ISABELA
Tuxpan
Ruiz
Villa Guerrero
AGUASCALIENTES
Aguascalientes
Rioverde
Santiago Ixcuintla
Tlaltenango de Sánchez Román
Villa de Arriaga
Santa María del Río
ISLA MARÍA MADRE
Calvillo
ISLAS TRES MARÍAS
ISLA MARÍA MAGDALENA
Cerro Grande 9186 ft. 2800 m
Matanzas
San Blas
Tepic
Jalpa
Villa Hidalgo
Teocaltiche
Lagos de Moreno
San Felipe
San Diego de la Unión
Cerro el Vigía 8990 ft. 2740 m
Téul de González Ortega
Juchipila
Nochistlán
San Luis de la Paz
ISLA MARÍA CLEOFAS
Compostela
Volcán Ceboruco 7480 ft. 2280 m
Moyahua
LEÓN
Dolores Hidalgo
Las Varas
Ixtlán del Río
Yahualica
Jalostotitlán
San Francisco del Rincón
Guanajuato
Ahuacatlán
San Juan de Abajo
Cerro Peña Gorda 8399 ft. 2560 m
Tequila
San Miguel El Alto
Silao
Presa Begonias
PUNTA DE MITA
Magdalena
Tesistán
Zapopan
Tepatitlán de Morelos
Ciudad Manuel Doblado
GUANAJUATO
QUERÉ
Bahía de Banderas
Puerto Vallarta
Tala
GUADALAJARA
Arandas
Irapuato
Mascota
Tlaquepaque
Querétaro
CABO CORRIENTES
Ameca
Lago de Atotonilco
Acatlán de Juárez
Tototlán
Salamanca
Celaya
Cerro la Tetilla 8793 ft. 2680 m
Atenguillo
Cocula
Apaseo El Grande
Talpa de Allende
Chapala
Ocotlán
Valle de Santiago
Cortazar
San Juan del Río
JALISCO
La Barca
La Piedad de Cabadas
Laguna de Yuriria
Salvatierra
Zacoalco de Torres
Peñamiller de Nado 10 909 ft. 3325 m
Ayutla
Laguna de Sayula
Laguna de Chapala
Uriangato
Jerécuaro
Juchitlán
Sahuayo de José María Morelos
Penjamillo de Degollado
Moroleón
Unión de Tula
Acámbaro
Tomatlán
Zamora de Hidalgo
Sayula
Autlán de Navarro
Cotija de la Paz
Zacapu
Lago de Cuitzeo
Cerro de Enmedio 11 775 ft. 3589 m
Venustiano Carranza
Ciudad Hidalgo
Purificación
Tamazula de Gordiano
Paracho de Verduzco
Quiroga
Guzmán
Nevado de Colima 13 911 ft. 4240 m
Lago de Pátzcuaro
Morelia
Paricutín 9186 ft. 2800 m
Cuautitlán
Tecalitlán
Pico de Tancítaro 12 664 ft. 3860 m
Pátzcuaro
Uruapan del Progreso
Heroica Zitácuaro
Cihuatlán
Comala
Ario de Rosales
Tacámbaro de Codallos
MÉ
Colima
Tepalcatepec
Manzanillo
COLIMA
MICHOACÁN
Apatzingán de la Constitución
La Huacana
Laguna de Cuyutlán
Tecomán
Nueva Italia de Ruiz
Tejupilco de Hidalgo
Armería
Coalcomán de Matamoros
Presa del Infiernillo
Nocupétaro
Coahuayana
Cerro El Tejocote 8268 ft. 2520 m
Aguililla
Churumuco de Morelos
Huetamo de Núñez
Aquila
Balsas
Arteaga
PUNTA TEJUPAN
Coyuca de Catalán
Ciudad Altamirano
Arcelia
Colmeneros
Playa Azul
La Unión
Lázaro Cárdenas
Bahía Petacalco
Cerro La Cruz 7251 ft. 2210 m
Tlacotepe
Ixtapa
Zihuatanejo
Petatlán
Cerro Baúl 11 201 ft. 3414 m
PACIFIC OCEAN
Técpan de Galeana
GUE
Atoyac de Álva
Coyuca de Benítez
Laguna Mitla
EAST PACIFIC RISE
MIDDLE AMERICA TRENCH
106°
104°
102°
24°
22°
20°
18°
16°
© Rand McNally
A-101672-1

0 20 40 60 80 100 120 Miles

0 20 40 60 80 100 120 140 160 180 200 Kilometers

Lambert Conformal Conic Projection
Scale 1:4,000,000
One inch to 64 miles
One cm to 40 km

GULF OF MEXICO
MEXICO BASIN
Tropic of Cancer
CAMPECHE BANK
Bahía de Campeche
Golfo de Tehuantepec
TAMAULIPAS
HIDALGO
TLAXCALA
MEXICO CITY
MORELOS
PUEBLA
VERACRUZ
OAXACA
TABASCO
CHIAPAS
SIERRA MADRE DE CHIAPAS
MADRE DEL SUR
ISTMO DE TEHUANTEPEC
BARRA DE LOS AMERICANOS
Laguna Madre
BARRA SOTO LA MARINA
Hidalgo
Santander Jiménez
La Luz
Abasolo
Las Guayabas
Ciudad Victoria
Presa Vicente Guerrero
Soto la Marina
La Pesca
Jaumave
Llera de Canales
Tula
Xicoténcatl
Aldama
Laguna de San Andrés
Ciudad Mante
González
Manuel
Antiguo Morelos
Cuauhtémoc
Ciudad del Maíz
Ciudad Madero
Tampico
Ebano
Ciudad Valles
Cárdenas
Pánuco
Laguna de Tamiahua
CAYO NUEVO
Ozuluama
Tamapatz
Tempoal de Sánchez
Magozal
CABO ROJO
ISLA DE LOBOS
Tantoyuca
Tamiahua
Tamazunchale
Jalpan de Serra
Cerro Azul
Huejutla de Reyes
Tuxpan de Rodríguez Cano
Santiago de la Peña
Jacala
Chicontepec de Tejeda
Zimapán
Tihuatlán
Zacualtipan
Poza Rica de Hidalgo
Papantla de Olarte
Gutiérrez Zamora
Ixmiquilpan
EL TAJÍN
Huichapan
Actopan
Nautla
Huauchinango
Pachuca de Soto
Tulancingo
Cuetzalan del Progreso
Martínez de la Torre
Tepeji de Ocampo
Zacatlán
Misantla
Teziutlán
Chignahuapan
TEOTIHUACÁN
Santiago Teyahualco
San Juan Teotihuacán
Tlalnepantla
Ciudad Netzahualcóyotl
Perote
Xalapa
Apizaco
Libres
Cerro Cofre de Perote 13 944 ft. 4250 m
San Martín Texmelucan
Tlaxcala de Xicohténcatl
Veracruz
Toluca de Lerdo
Vol. Iztaccíhuatl 17 159 ft. 5230 m
Cholula de Rivadabia
PUEBLA DE ZARAGOZA
PUNTA ANTÓN LIZARDO
Nev. de Toluca 15 387 ft. 4690 m
Amecameca de Juárez
San Andrés Cholula
Tepeaca
Volcán Pico de Orizaba 18 406 ft. 5610 m
Córdoba
Cuernavaca
Yautepec
Volcán Popocatépetl 17 930 ft. 5465 m
Tecamachalco
Orizaba
Alvarado
PUNTA ROCA PARTIDA
Nuevo Progreso
Emiliano Zapata
Cuautla
Tlacotalpan
Laguna Machona
Laguna Mecoacán
Frontera
San Andrés Tuxtla
Izúcar de Matamoros
Tierra Blanca
Paraíso
Comalcalco
Tehuacán
Cosamaloapan de Carpio
Cerro Santa Marta 5577 ft. 1700 m
Iguala
Chiautla de Tapia
San Gabriel Chilac
Acatlán de Osorio
Coatzacoalcos
Villahermosa
Huitzuco de los Figueroa
Presa Miguel Alemán
Huautla
Tuxtepec
Minatitlán
Cárdenas
Balsas Sur
Mezcala
Petlalcingo
Isla
Acayucan
Las Choapas
Macuspana
San Juan Evangelista
Olinalá
Huajuapan de León
Cuicatlán
Playa Vicente
Laguna Rosario
Chilapa de Álvarez
Tamazulapan del Progreso
Teapa
Tlapa de Comonfort
Chilpancingo de los Bravo
Pichucalco
Silacayoapan
Asunción Nochixtlán
Jesús Carranza
Tlalixtaquilla
Santiago Choapan
Ixtlán de Juárez
Petalcingo
Chilón
Santa María Asunción Tlaxiaco
Huitzo
Presa Netzahualcóyotl
Simojovel
Metlatonoc
Cerro Pedro de Olla 10 991 ft. 3350 m
Copainalá
Malinaltepec
Oaxaca de Juárez
Putla de Guerrero
Matías Romero
San Cristóbal de las Casas
Ayutla de los Libres
Ocotlán de Morelos
Cerro Las Flores 7054 ft. 2150 m
Ocozocuautla
Chiapa de Corzo
Cintalapa
Tuxtla Gutiérrez
San Marcos
Xochistlahuaca
Cerro Los Bolones 9154 ft. 2790 m
Cruz Grande
Ometepec
Ejutla de Crespo
Nejapa de Madero
Ixtepec
Juchitán de Zaragoza
Santo Domingo Zanatepec
Venustiano Carranza
Laguna Chautengo
Presa Benito Juárez
Las Rosas
Cuajinicuilapa
Santiago Pinotepa Nacional
Miahuatlán de Porfirio Díaz
Santo Domingo Tehuantepec
Chahuites
Arriaga
Villa Flores
Santiago Jamiltepec
Salina Cruz
Laguna Superior
Laguna Inferior
Mar Muerto
Tonalá
Cerro Tres Picos 8366 ft. 2550 m
Revolución Mexicana
Presa de la 16 Angostura
Río Grande
Puerto Arista
Lagunas de Chacagua
Santa María Colotepec
Puerto Escondido
San Pedro Pochutla
Pijijiapan
Cerro el Triunfo 8038 ft. 2450 m
PUNTA CORNETA
Puerto Ángel
Mapastepec
98°
96°
94°
92°
24°
22°
20°
18°

GULF OF MEXICO
CAMPECHE BANK
CAYOS ARCAS
Bahía de Campeche
CARIBBEAN SEA
PACIFIC OCEAN
MIDDLE AMERICA TRENCH
MEXICO
YUCATÁN
YUCATAN PENINSULA
CAMPECHE
QUINTANA ROO
TABASCO
CHIAPAS
GUATEMALA
BELIZE
HONDURAS
EL SALVADOR
NICARAGUA
SIERRA MADRE
MAYA MOUNTAINS
CORDILLERA OPALACA
MONTAÑAS DE COMAYAGUA
Mérida
Progreso
Campeche
Cancún
Cozumel
Chetumal
Ciudad del Carmen
Villahermosa
Tuxtla Gutiérrez
San Cristóbal de las Casas
Tapachula
Quetzaltenango
Guatemala
Belize City
Belmopan
San Pedro Sula
Tegucigalpa
San Salvador
Santa Ana
San Miguel
Soyapango
La Ceiba
Puerto Cortés
Puerto Barrios
Cobán
Flores
Escárcega
Valladolid
Tizimín
Chichén Itzá
Tikal
Palenque
Copán
Cerro el Triunfo 8038 ft. 2450 m
Volcán Tacaná 13,428 ft. 4093 m
Volcán Tajumulco 13,845 ft. 4220 m
Volcán Santa María 12,375 ft. 3772 m
Volcán de Fuego 12,346 ft. 3763 m
Victoria Peak 3675 ft. 1120 m
Cerro El Pital 8957 ft. 2730 m
Volcán de San Vicente 7156 ft. 2181 m
Pico Bonito 7989 ft. 2435 m
Pico Pijol 7487 ft. 2282 m
Montaña El Chile 7402 ft. 2256 m
Volcán Cosigüina 2818 ft. 859 m
Lago Petén Itzá
Lago de Izabal
Lago de Atitlán
Golfo de Fonseca
Gulf of Honduras
Bahía de Amatique
Laguna de Términos
Bahía Chetumal
BANCO CHINCHORRO
TURNEFFE ISLANDS
ISLAS DE LA BAHÍA
ISLA COZUMEL
ISLA MUJERES
ISLA CONTOY
CABO CATOCHE
92°
90°
88°
20°
18°
16°
14°
© Rand McNally A-101673-1

0 20 40 60 80 100 120 Miles
0 20 40 60 80 100 120 140 160 180 200 Kilometers

Lambert Conformal Conic Projection
Scale 1:4,000,000
One inch to 64 miles
One cm to 40 km

BELIZE
GUATEMALA
HONDURAS
EL SALVADOR
NICARAGUA
COSTA RICA
PANAMA
CARIBBEAN SEA
PACIFIC OCEAN
MIDDLE AMERICA TRENCH
COCOS RIDGE
Gulf of Honduras
ISLAS DE LA BAHÍA
ISLA DE ROATÁN
ISLA DE GUANAJA
ISLA DE UTILA
Golfo de Fonseca
Lago de Managua
Lago de Nicaragua
ISLA DE OMETEPE
ISLA ZAPATERA
Golfo de Nicoya
Golfo de Papagayo
Golfo de Santa Elena
PENÍNSULA DE NICOYA
PENÍNSULA DE OSA
Golfo Dulce
Bahía de Coronado
ISLAS DEL MAÍZ
CAYOS MISKITOS
COSTA DE MOSQUITOS
CORDILLERA ISABELIA
CORDILLERA DARIENSE
CORDILLERA CHONTALEÑA
CORDILLERA DE GUANACASTE
CORDILLERA DE TALAMANCA
CORDILLERA OPALACA
MONTAÑAS DE COMAYAGUA
CABO GRACIAS A DIOS
CABO CAMARÓN
CABO DE HONDURAS
PUNTA PATUCA
CABO SANTA ELENA
CABO VELAS
CABO BLANCO
CABO MATAPALO
PUNTA BURICA
PUNTA COSIGÜINA
Laguna de Caratasca
Laguna de Perlas
Tegucigalpa
San Pedro Sula
La Ceiba
Puerto Cortés
Tela
Trujillo
Choloma
El Progreso
Comayagua
Siguatepeque
Santa Rosa de Copán
Juticalpa
Catacamas
Danlí
Choluteca
Puerto Lempira
Managua
León
Chinandega
Granada
Masaya
Matagalpa
Jinotega
Estelí
Bluefields
Puerto Cabezas
San Salvador
Soyapango
San Miguel
Usulután
San José
Heredia
Alajuela
Cartago
Puntarenas
Liberia
Puerto Limón
Pico Bonito 7989 ft. 2435 m
Cerro Las Minas 9347 ft. 2849 m
Cerro El Pital 8957 ft. 2730 m
Volcán Cosigüina 2818 ft. 859 m
Volcán San Cristóbal 5725 ft. 1745 m
Volcán Momotombo 4199 ft. 1280 m
Vol. Mombacho 4413 ft. 1345 m
Volcán Concepción 5282 ft. 1610 m
Volcán Miravalles 6654 ft. 2028 m
Volcán Arenal 5358 ft. 1633 m
Volcán Poás 8871 ft. 2704 m
Vol. Irazú 11 260 ft. 3432 m
Cerro Chirripó 12 530 ft. 3819 m
Cerro Kámuk 11 660 ft. 3554 m
Volcán Barú 11 401 ft. 3475 m
© Rand McNally A-101675-1

0 20 40 60 80 100 120 Miles
0 20 40 60 80 100 120 140 160 180 200 Kilometers

Lambert Conformal Conic Projection
Scale 1:4,000,000
One inch to 64 miles
One cm to 40 km

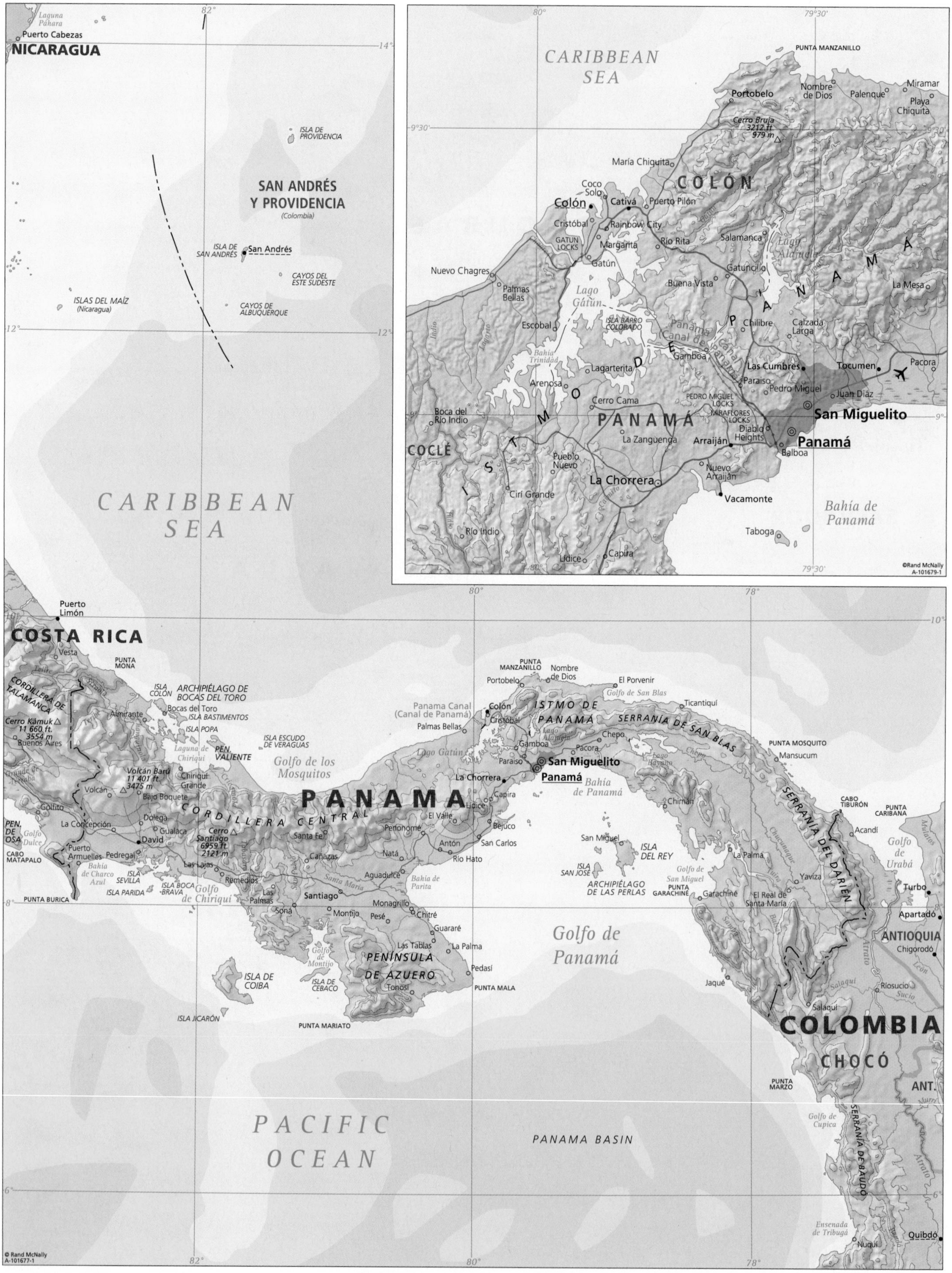
NICARAGUA
Puerto Cabezas
SAN ANDRÉS Y PROVIDENCIA
(Colombia)
ISLA DE PROVIDENCIA
ISLA DE SAN ANDRÉS
San Andrés
CAYOS DEL ESTE SUDESTE
ISLAS DEL MAÍZ
(Nicaragua)
CAYOS DE ALBUQUERQUE
CARIBBEAN SEA
Puerto Limón
COSTA RICA
PANAMA
Golfo de los Mosquitos
ARCHIPIÉLAGO DE BOCAS DEL TORO
CORDILLERA CENTRAL
Volcán Barú 11 401 ft. 3475 m
David
Golfo de Chiriquí
ISLA DE COIBA
PENÍNSULA DE AZUERO
Santiago
Chitré
Las Tablas
PUNTA MALA
PUNTA MARIATO
Colón
Panama Canal (Canal de Panamá)
San Miguelito
Panamá
Bahía de Panamá
Golfo de Panamá
ISLA DEL REY
ARCHIPIÉLAGO DE LAS PERLAS
SERRANÍA DE SAN BLAS
Golfo de San Blas
SERRANÍA DEL DARIÉN
COLOMBIA
CHOCÓ
ANTIOQUIA
Golfo de Urabá
Turbo
Quibdó
PACIFIC OCEAN
PANAMA BASIN
© Rand McNally A-101677-1
a
CARIBBEAN SEA
PUNTA MANZANILLO
Portobelo
Nombre de Dios
COLÓN
Colón
Cristóbal
GATUN LOCKS
Lago Gatún
ISTMO DE PANAMÁ
PANAMÁ
Gamboa
PEDRO MIGUEL LOCKS
MIRAFLORES LOCKS
San Miguelito
Panamá
Balboa
La Chorrera
Arraiján
Vacamonte
Taboga
Bahía de Panamá
COCLÉ
©Rand McNally A-101679-1

0 20 40 60 80 100 120 Miles
0 20 40 60 80 100 120 140 160 180 200 Kilometers

Lambert Conformal Conic Projection
Scale 1:4,000,000
One inch to 64 miles
One cm to 40 km

Inset map a
Lambert Conformal Conic Projection
Scale 1:1,000,000
One inch to 16 miles
One cm to 10 km

a

Camano
Port Susan
Warm Beach
Arlington
CAMANO ISLAND
North Marysville
TULALIP INDIAN RESERVATION
Granite Falls
Marysville
Freeland
WHIDBEY ISLAND
EVERETT NAVAL STATION
Lake Stevens
Everett
Port Ludlow
Mukilteo
Snohomish
Port Gamble
Puget Sound
Mill Creek
Monroe
PORT GAMBLE INDIAN RESERVATION
Lynnwood
Mountlake Terrace
BANGOR NAVAL SUBMARINE BASE
Edmonds
Shoreline
Bothell
Poulsbo
Kenmore
Woodinville
PORT MADISON INDIAN RESERVATION
Lake Forest Park
Lake Washington
Silverdale
Kirkland
Redmond
Bainbridge Island
Sammamish
BAINBRIDGE ISLAND
SEATTLE
Bellevue
Elliott Bay
Lake Sammamish
Bremerton
Mercer Island
PUGET SOUND NAVAL SHIPYARD
Port Orchard
White Center
Issaquah
Sunnyslope
Renton
Burien
Tukwila
Vashon
SeaTac
VASHON ISLAND
Des Moines
Maple Valley
Kent
Gig Harbor
MAURY ISLAND
Covington
Federal Way
Browns Point
Auburn
Black Diamond
Algona
MUCKLESHOOT IND. RES.
Tacoma
PUYALLUP IND. RES.
University Place
Sumner
Enumclaw
MCNEIL ISLAND
Lakewood
Lake Tapps
Puyallup
MCCHORD AIR FORCE BASE
South Hill
ANDERSON ISLAND
Buckley
Parkland
Dupont
Spanaway
Orting
Elk Plain
FORT LEWIS
NISQUALLY IND. RES.
© Rand McNally A-101646-1

b

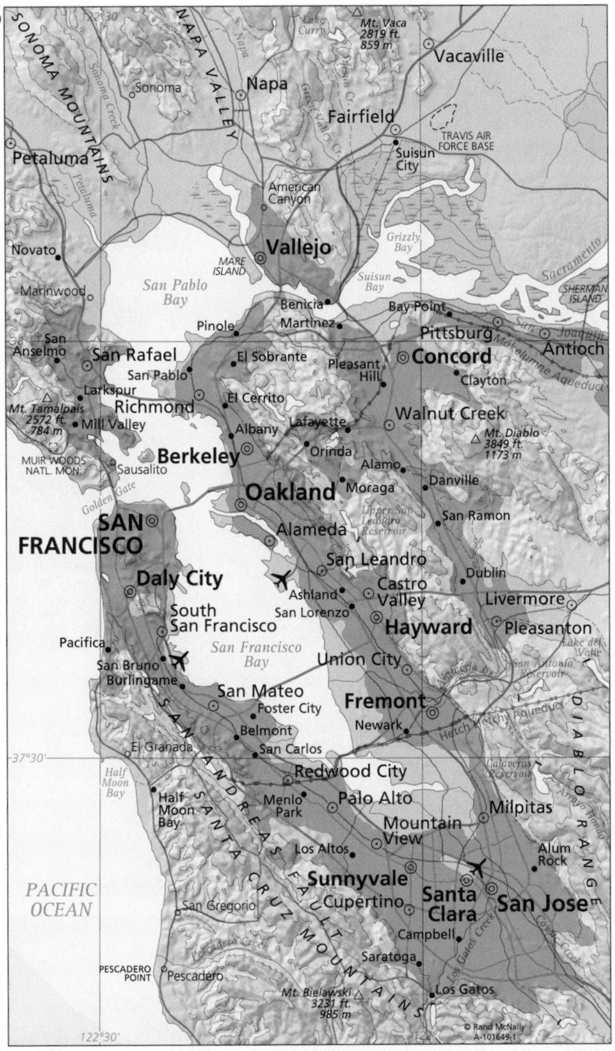
SONOMA MOUNTAINS
NAPA VALLEY
Mt. Vaca 2819 ft. 859 m
Vacaville
Sonoma
Napa
Fairfield
TRAVIS AIR FORCE BASE
Petaluma
Suisun City
American Canyon
Novato
Vallejo
Grizzly Bay
MARE ISLAND
San Pablo Bay
Suisun Bay
Sacramento
Marinwood
Benicia
Martinez
Bay Point
SHERMAN ISLAND
Pinole
Pittsburg
Antioch
San Anselmo
San Rafael
El Sobrante
Concord
Pleasant Hill
San Pablo
Larkspur
Clayton
Richmond
El Cerrito
Mt. Tamalpais 2572 ft. 784 m
Mill Valley
Albany
Lafayette
Walnut Creek
Mt. Diablo 3849 ft. 1173 m
Orinda
Berkeley
MUIR WOODS NATL. MON.
Sausalito
Alamo
Golden Gate
Danville
Oakland
Moraga
SAN FRANCISCO
Alameda
San Ramon
San Leandro
Dublin
Daly City
Castro Valley
Ashland
Livermore
South San Francisco
San Lorenzo
Pleasanton
Hayward
Pacifica
San Francisco Bay
Union City
San Bruno
Burlingame
San Mateo
Foster City
Fremont
Newark
Belmont
San Carlos
El Granada
Redwood City
Half Moon Bay
Menlo Park
Palo Alto
Milpitas
Mountain View
Los Altos
Alum Rock
PACIFIC OCEAN
Sunnyvale
Cupertino
Santa Clara
San Jose
San Gregorio
Campbell
Saratoga
PESCADERO POINT
Pescadero
Mt. Bielawski 3231 ft. 985 m
Los Gatos
SAN ANDREAS FAULT
SANTA CRUZ MOUNTAINS
DIABLO RANGE
© Rand McNally A-101649-1

c

North Auburn
Lincoln
Newcastle
Auburn
Rocklin
Roseville
Citrus Heights
Folsom Lake
Rio Linda
North Highlands
Orangevale
El Dorado Hills
Fair Oaks
Folsom
Arcade
Carmichael
West Sacramento
Arden
Rancho Cordova
SACRAMENTO
Rosemont
Parkway
Florin
Elk Grove
Hood
© Rand McNally A-101651-1

d

Ridgefield
Battle Ground
Scappoose
WASHINGTON
SAUVIE ISLAND
Vancouver Lake
Orchards South
Minnehaha
North Plains
Vancouver
Camas
Washougal
Cedar Mill
PORTLAND
Troutdale
Hillsboro
West Slope
Wood Village
Aloha
Gresham
Beaverton
Milwaukie
Tigard
Lake Oswego
Clackamas
Oak Grove
Tualatin
West Linn
Gladstone
Sherwood
Oregon City
Wilsonville
OREGON
Newberg
Canby
Estacada
© Rand McNally A-101650-1

e

SANTA SUSANA MOUNTAINS
SAN BERNARDINO MOUNTAINS
Moorpark
San Fernando
Mt. San Antonio 10 064 ft. 3068 m
Cajon Mountain 5310 ft. 1618 m
Simi Valley
Somis
Mt. Wilson 5710 ft. 1740 m
Cucamonga Peak 8859 ft. 2700 m
Lake Arrowhead
Crestline
SIMI HILLS
SAN FERNANDO VALLEY
La Canada Flintridge
SAN GABRIEL MOUNTAINS
Big Bear Lake
Running Springs
Keller Peak 7882 ft. 2402 m
Thousand Oaks
Van Nuys
Altadena
Burbank
Pasadena
Monrovia
Azusa
Glendora
Highland
Calabasas
Agoura Hills
Glendale
Arcadia
Fontana
Rialto
SANTA MONICA MOUNTAINS
Hollywood
Alhambra
Baldwin Park
Covina
Claremont
Upland
San Bernardino
Topanga
West Hollywood
Beverly Hills
Ontario
Rancho Cucamonga
Bloomington
Malibu
LOS ANGELES
East Los Angeles
El Monte
West Covina
Pomona
Colton
Loma Linda
Redlands
Yucaipa
POINT DUME
Santa Monica
Culver City
Montebello
Diamond Bar
Chino
Mira Loma
Rubidoux
South Gate
Pico Rivera
Whittier
MORONGO INDIAN RESERVATION
Inglewood
Chino Hills
Pedley
Santa Monica Bay
La Habra
Riverside
Moreno Valley
CHINO HILLS
Hawthorne
Downey
Brea
Norco
Beaumont
Banning
El Segundo
Yorba Linda
Compton
Norwalk
La Mirada
Woodcrest
Manhattan Beach
PACIFIC OCEAN
Bellflower
Buena Park
Fullerton
Corona
Redondo Beach
Colorado River Aqueduct
SOBOBA IND. RES.
Torrance
Carson
Lakewood
Anaheim
Lake Mathews
Perris
San Jacinto
PALOS VERDES POINT
Lomita
Garden Grove
Orange
Rancho Palos Verdes
Long Beach
Westminster
Tustin
Homeland
Hemet
Valle Vista
Santa Ana
Santiago Peak 5687 ft. 1733 m
Sun City
San Pedro
San Pedro Bay
POINT FERMIN
Fountain Valley
Irvine
Lake Elsinore
Huntington Beach
Costa Mesa
Lake Forest
Rancho Santa Margarita
Wildomar
Newport Beach
Mission Viejo
Laguna Hills
Murrieta
Skinner Reservoir
Laguna Beach
Laguna Niguel
Temecula
Dana Point
San Juan Capistrano
DANA POINT
San Clemente
CAMP PENDLETON MARINE CORPS BASE
PECHANGA INDIAN RESERVATION
SANTA ANA MOUNTAINS
SAN JACINTO VALLEY
PERRIS VALLEY
© Rand McNally A-101647-1
f
NELLIS AIR FORCE BASE
North Las Vegas
Sunrise Manor
LAS VEGAS
Paradise
Winchester
Spring Valley
Lake Mead
Henderson
HOOVER DAM
Boulder City
© Rand McNally A-101648-1

0 5 10 15 20 25 30 Miles

0 5 10 15 20 25 30 35 40 45 50 Kilometers

Lambert Conformal Conic Projection
Scale 1:1,000,000
One inch to 16 miles
One cm to 10 km

a

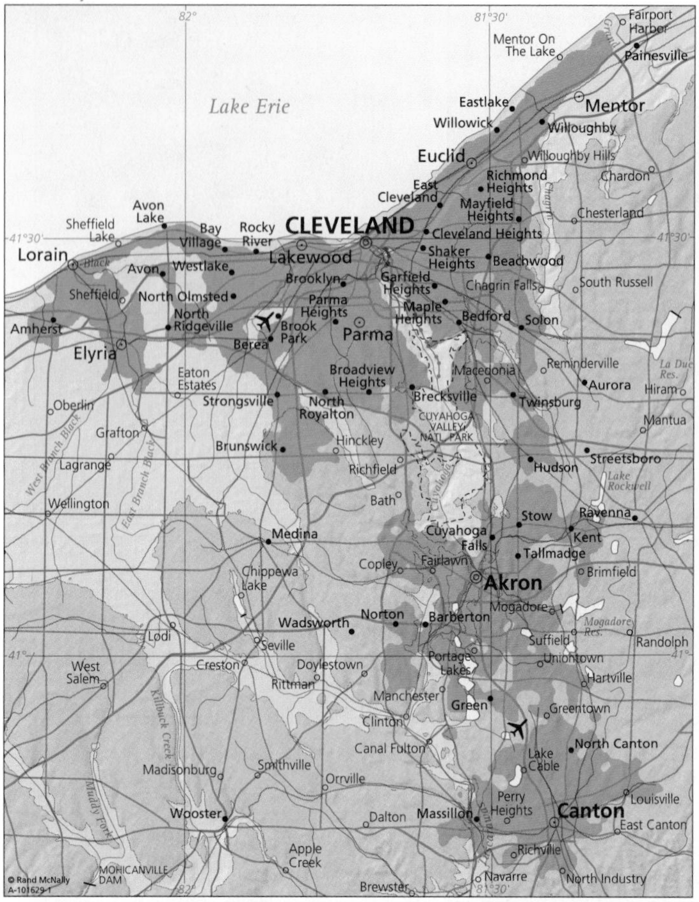
Lake Erie
CLEVELAND
Fairport Harbor
Painesville
Mentor On The Lake
Mentor
Eastlake
Willowick
Willoughby
Willoughby Hills
Euclid
Richmond Heights
Chardon
East Cleveland
Mayfield Heights
Chesterland
Avon Lake
Sheffield Lake
Bay Village
Rocky River
Lakewood
Cleveland Heights
Lorain
Westlake
Avon
Shaker Heights
Beachwood
Brooklyn
Garfield Heights
Chagrin Falls
South Russell
Sheffield
North Olmsted
Parma Heights
Maple Heights
Bedford
Solon
Amherst
North Ridgeville
Brook Park
Parma
Berea
Elyria
Eaton Estates
Broadview Heights
Macedonia
Reminderville
Aurora
Hiram
La Due Res.
Oberlin
Strongsville
North Royalton
Brecksville
Twinsburg
CUYAHOGA VALLEY NATL. PARK
Mantua
Grafton
Brunswick
Hinckley
Streetsboro
Lagrange
Richfield
Hudson
Lake Rockwell
Wellington
Bath
Stow
Ravenna
Medina
Cuyahoga Falls
Kent
Tallmadge
Chippewa Lake
Copley
Fairlawn
Akron
Brimfield
Norton
Mogadore
Wadsworth
Barberton
Mogadore Res.
Lodi
Seville
Suffield
Randolph
West Salem
Creston
Doylestown
Portage Lakes
Uniontown
Rittman
Hartville
Manchester
Green
Greentown
Clinton
Canal Fulton
North Canton
Lake Cable
Madisonburg
Smithville
Orrville
Perry Heights
Louisville
Wooster
Dalton
Massillon
Canton
East Canton
Apple Creek
Richville
MOHICANVILLE DAM
Navarre
North Industry
Brewster
West Branch Black
East Branch Black
Killbuck Creek
Muddy Fork
82°
81°30'
41°30'
41°
© Rand McNally A-101629-1

b

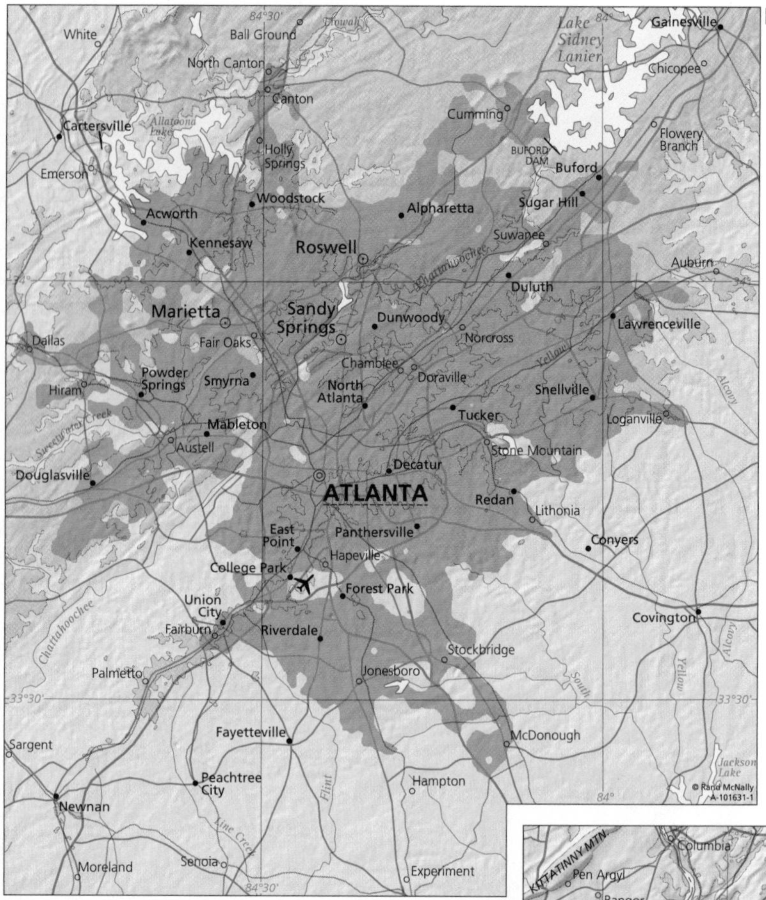
ATLANTA
White
Ball Ground
Lake Sidney Lanier
Gainesville
North Canton
Canton
Chicopee
Cartersville
Allatoona Lake
Cumming
Flowery Branch
Holly Springs
BUFORD DAM
Buford
Emerson
Woodstock
Sugar Hill
Acworth
Alpharetta
Suwanee
Kennesaw
Roswell
Auburn
Duluth
Marietta
Sandy Springs
Dunwoody
Lawrenceville
Dallas
Fair Oaks
Norcross
Chamblee
Powder Springs
Smyrna
Doraville
Snellville
Hiram
North Atlanta
Tucker
Loganville
Mableton
Austell
Stone Mountain
Douglasville
Decatur
Redan
Lithonia
East Point
Panthersville
Hapeville
Conyers
College Park
Forest Park
Union City
Fairburn
Riverdale
Covington
Palmetto
Stockbridge
Jonesboro
Chattahoochee
Sweetwater Creek
Yellow
Alcovy
South
Flint
Line Creek
Fayetteville
McDonough
Sargent
Peachtree City
Hampton
Jackson Lake
Newnan
Moreland
Senoia
Experiment
84°30'
84°
33°30'
© Rand McNally A-101631-1

c

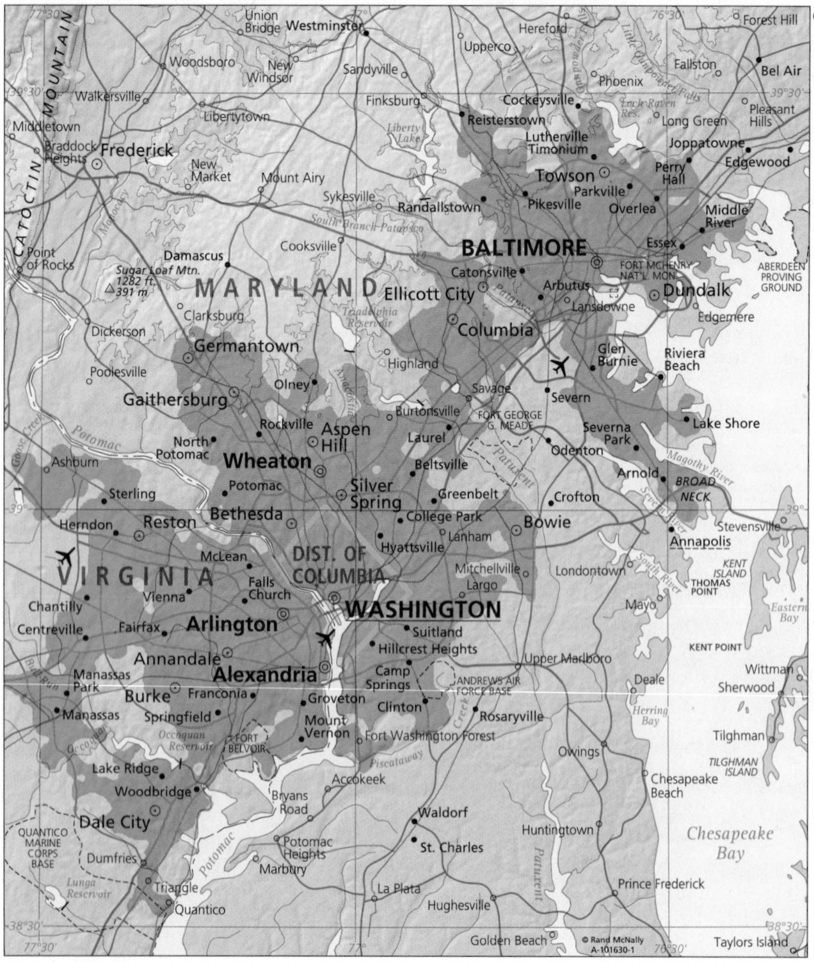
BALTIMORE
WASHINGTON
MARYLAND
VIRGINIA
DIST. OF COLUMBIA
CATOCTIN MOUNTAIN
Union Bridge
Westminster
Hereford
Forest Hill
Woodsboro
New Windsor
Upperco
Sandyville
Fallston
Bel Air
Phoenix
Walkersville
Finksburg
Cockeysville
Pleasant Hills
Loch Raven Res.
Long Green
Middletown
Libertytown
Liberty Lake
Reisterstown
Lutherville Timonium
Joppatowne
Braddock Heights
Frederick
Edgewood
New Market
Mount Airy
Towson
Perry Hall
Sykesville
Parkville
Randallstown
Pikesville
Overlea
Middle River
South Branch Patapsco
Point of Rocks
Damascus
Cooksville
Essex
Sugar Loaf Mtn. 1282 ft. 391 m
Catonsville
FORT MCHENRY NAT'L MON.
ABERDEEN PROVING GROUND
Arbutus
Ellicott City
Dundalk
Lansdowne
Clarksburg
Triadelphia Reservoir
Edgemere
Dickerson
Columbia
Germantown
Glen Burnie
Riviera Beach
Highland
Poolesville
Olney
Savage
Severn
Gaithersburg
Burtonsville
FORT GEORGE G. MEADE
Lake Shore
Rockville
Aspen Hill
Laurel
Severna Park
North Potomac
Odenton
Ashburn
Wheaton
Beltsville
Arnold
Magothy River
Potomac
Silver Spring
Crofton
BROAD NECK
Sterling
Greenbelt
Bethesda
Herndon
Reston
College Park
Bowie
Stevensville
Lanham
Annapolis
Hyattsville
McLean
Mitchellville
KENT ISLAND
Londontown
THOMAS POINT
Largo
Vienna
Falls Church
Chantilly
Arlington
Suitland
Mayo
Eastern Bay
Centreville
Fairfax
Hillcrest Heights
Annandale
Upper Marlboro
KENT POINT
Alexandria
Camp Springs
Wittman
Manassas Park
ANDREWS AIR FORCE BASE
Deale
Burke
Franconia
Sherwood
Groveton
Clinton
Rosaryville
Manassas
Springfield
Mount Vernon
Herring Bay
Fort Washington Forest
Occoquan Reservoir
FORT BELVOIR
Tilghman
Owings
TILGHMAN ISLAND
Lake Ridge
Accokeek
Chesapeake Beach
Woodbridge
Bryans Road
Waldorf
Dale City
Huntingtown
Chesapeake Bay
QUANTICO MARINE CORPS BASE
Potomac Heights
Dumfries
St. Charles
Marbury
Lunga Reservoir
Triangle
La Plata
Prince Frederick
Quantico
Hughesville
Golden Beach
Taylors Island
Patuxent
Piscataway
77°30'
77°
76°30'
39°30'
39°
38°30'
© Rand McNally A-101630-1

d

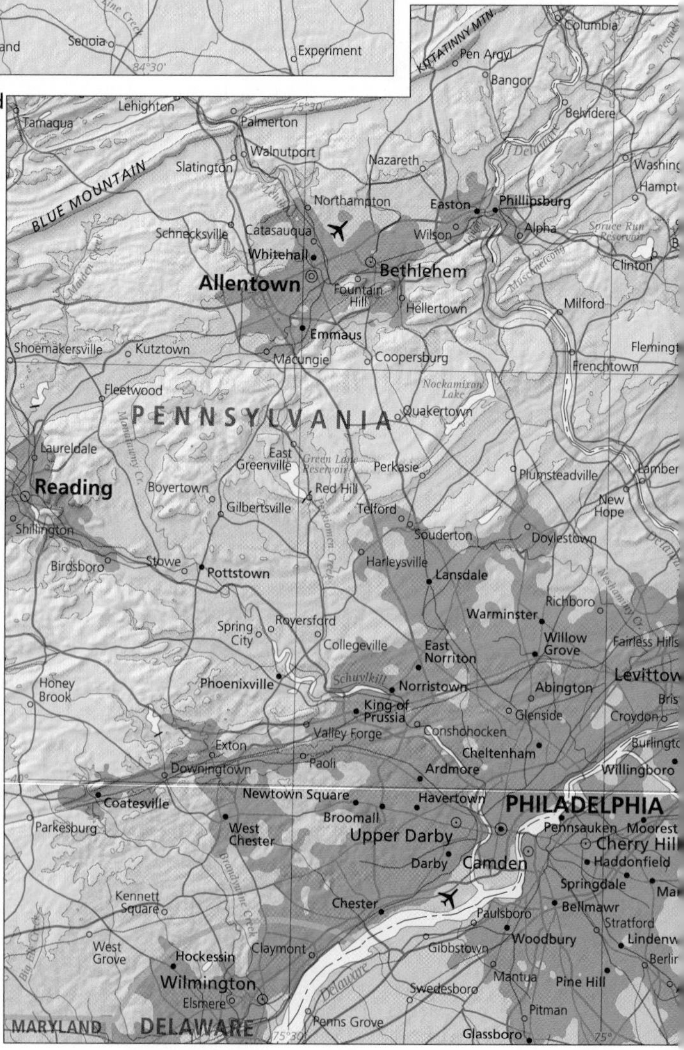
PHILADELPHIA
PENNSYLVANIA
KITTATINNY MTN.
BLUE MOUNTAIN
Columbia
Pen Argyl
Bangor
Belvidere
Tamaqua
Lehighton
Palmerton
Walnutport
Nazareth
Slatington
Northampton
Easton
Phillipsburg
Schnecksville
Catasauqua
Wilson
Alpha
Whitehall
Bethlehem
Allentown
Fountain Hill
Hellertown
Milford
Emmaus
Shoemakersville
Kutztown
Macungie
Coopersburg
Frenchtown
Fleetwood
Nockamixon Lake
Quakertown
Laureldale
East Greenville
Green Lane Reservoir
Perkasie
Plumsteadville
Reading
Boyertown
Red Hill
New Hope
Shillington
Gilbertsville
Telford
Souderton
Doylestown
Birdsboro
Stowe
Pottstown
Harleysville
Lansdale
Richboro
Warminster
Royersford
Spring City
Collegeville
East Norriton
Willow Grove
Honey Brook
Phoenixville
Norristown
Abington
Schuylkill
King of Prussia
Glenside
Valley Forge
Conshohocken
Exton
Paoli
Cheltenham
Downingtown
Ardmore
Coatesville
Newtown Square
Havertown
Parkesburg
West Chester
Broomall
Upper Darby
Pennsauken
Darby
Camden
Haddonfield
Springdale
Kennett Square
Chester
Bellmawr
Paulsboro
Stratford
West Grove
Hockessin
Claymont
Gibbstown
Woodbury
Wilmington
Mantua
Pine Hill
Elsmere
Swedesboro
Pitman
Penns Grove
Glassboro
MARYLAND
DELAWARE
Delaware
Brandywine Creek
75°30'
75°

0 5 10 15 20 25 30 Miles
0 5 10 15 20 25 30 35 40 45 50 Kilometers

Lambert Conformal Conic Projection
Scale 1:1,000,000
One inch to 16 miles
One cm to 10 km

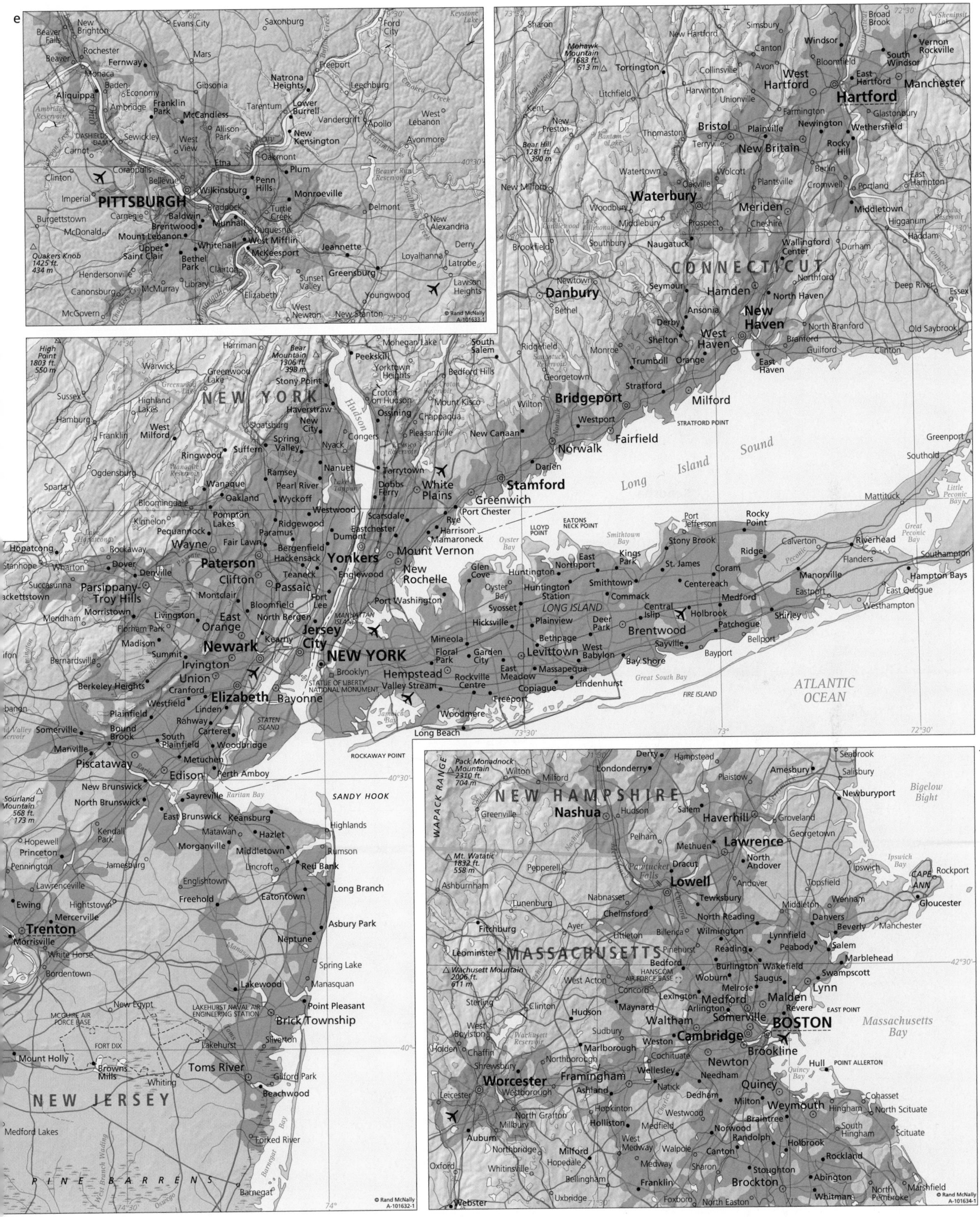
e
PITTSBURGH
Beaver Falls
New Brighton
Evans City
Saxonburg
Ford City
Rochester
Fernway
Mars
Freeport
Beaver
Monaca
Natrona Heights
Leechburg
Aliquippa
Baden
Economy
Gibsonia
Ambridge
Franklin Park
McCandless
Tarentum
Lower Burrell
Vandergrift
Apollo
West Lebanon
Allison Park
Sewickley
New Kensington
Carnot
West View
Avonmore
Oakmont
Etna
Plum
Coraopolis
Bellevue
Clinton
Penn Hills
Imperial
Wilkinsburg
Monroeville
Burgettstown
Carnegie
Baldwin
Turtle Creek
Delmont
McDonald
Brentwood
Munhall
New Alexandria
Mount Lebanon
Duquesne
West Mifflin
Quakers Knob 1425 ft. 434 m
Upper Saint Clair
Whitehall
McKeesport
Jeannette
Derry
Bethel Park
Clairton
Loyalhanna
Latrobe
Hendersonville
Library
Greensburg
Sunset Valley
Canonsburg
McMurray
Elizabeth
Youngwood
Lawson Heights
McGovern
West Newton
New Stanton
© Rand McNally A-101633-1
CONNECTICUT
Sharon
Simsbury
Broad Brook
New Hartford
Windsor
Vernon Rockville
Mohawk Mountain 1683 ft. 513 m
Torrington
Canton
Avon
South Windsor
Collinsville
Bloomfield
West Hartford
East Hartford
Manchester
Litchfield
Harwinton
Unionville
Hartford
Kent
New Preston
Farmington
Glastonbury
Bristol
Plainville
Newington
Wethersfield
Thomaston
Terryville
New Britain
Rocky Hill
Bear Hill 1281 ft. 390 m
Berlin
Watertown
Wolcott
Plantsville
Cromwell
Portland
East Hampton
New Milford
Oakville
Waterbury
Meriden
Middletown
Woodbury
Middlebury
Prospect
Cheshire
Higganum
Haddam
Southbury
Naugatuck
Wallingford Center
Durham
Brookfield
Newtown
Seymour
Northford
Deep River
Danbury
Hamden
North Haven
Essex
Bethel
Ansonia
New Haven
North Branford
Old Saybrook
Derby
Shelton
West Haven
Branford
Guilford
Clinton
Ridgefield
Monroe
Trumbull
Orange
East Haven
Georgetown
Stratford
Milford
Wilton
Bridgeport
Westport
STRATFORD POINT
Fairfield
Norwalk
New Canaan
Darien
Stamford
Greenwich
Long Island Sound
Greenport
Southold
Mattituck
Little Peconic Bay
NEW YORK
High Point 1803 ft. 550 m
Harriman
Bear Mountain 1306 ft. 398 m
Mohegan Lake
South Salem
Peekskill
Warwick
Greenwood Lake
Yorktown Heights
Bedford Hills
Stony Point
Croton on Hudson
Mount Kisco
Sussex
Highland Lakes
Haverstraw
Hudson
Ossining
Chappaqua
Hamburg
New City
Pleasantville
West Milford
Congers
Franklin
Spring Valley
Nyack
Ringwood
Suffern
Nanuet
Tarrytown
Ogdensburg
Ramsey
White Plains
Sparta
Wanaque
Pearl River
Dobbs Ferry
Oakland
Wyckoff
Bloomingdale
Westwood
Scarsdale
Rye
Port Chester
Kinnelon
Pompton Lakes
Ridgewood
Eastchester
Harrison
Pequannock
Paramus
Dumont
Mamaroneck
LLOYD POINT
EATONS NECK POINT
Port Jefferson
Rocky Point
Hopatcong
Rockaway
Wayne
Fair Lawn
Bergenfield
Yonkers
Mount Vernon
Oyster Bay
Smithtown Bay
Kings Park
Stony Brook
Calverton
Riverhead
Great Peconic Bay
Stanhope
Wharton
Dover
Denville
Paterson
Hackensack
Teaneck
Englewood
New Rochelle
Glen Cove
Huntington
East Northport
St. James
Coram
Ridge
Southampton
Flanders
Succasunna
Parsippany-Troy Hills
Clifton
Passaic
Fort Lee
Oyster Bay
Huntington Station
Smithtown
Centereach
Manorville
Hampton Bays
Montclair
Port Washington
Commack
Medford
Eastport
East Quogue
Morristown
Bloomfield
Syosset
LONG ISLAND
Central Islip
Holbrook
Westhampton
Mendham
Livingston
East Orange
North Bergen
MANHATTAN ISLAND
Hicksville
Plainview
Deer Park
Patchogue
Shirley
Florham Park
Kearny
Jersey City
Bethpage
Brentwood
Bellport
Madison
Newark
NEW YORK
Mineola
Sayville
Bernardsville
Summit
Irvington
Floral Park
Garden City
Levittown
West Babylon
Bayport
Union
Brooklyn
Hempstead
East Meadow
Massapequa
Bay Shore
Berkeley Heights
Cranford
Elizabeth
STATUE OF LIBERTY NATIONAL MONUMENT
Valley Stream
Rockville Centre
Copiague
Lindenhurst
Great South Bay
Westfield
Bayonne
Freeport
FIRE ISLAND
ATLANTIC OCEAN
Plainfield
Linden
Rahway
STATEN ISLAND
Jamaica Bay
Woodmere
Somerville
Bound Brook
Carteret
Long Beach
Manville
South Plainfield
Woodbridge
ROCKAWAY POINT
Piscataway
Metuchen
Edison
Perth Amboy
Sourland Mountain 568 ft. 173 m
New Brunswick
Sayreville
Raritan Bay
SANDY HOOK
North Brunswick
East Brunswick
Keansburg
Highlands
Kendall Park
Matawan
Hazlet
Hopewell
Morganville
Rumson
Princeton
Middletown
Pennington
Jamesburg
Lincroft
Red Bank
Englishtown
Lawrenceville
Long Branch
Freehold
Eatontown
Ewing
Hightstown
Mercerville
Asbury Park
Trenton
Morrisville
Neptune
White Horse
Spring Lake
Bordentown
Lakewood
Manasquan
New Egypt
McGUIRE AIR FORCE BASE
LAKEHURST NAVAL AIR ENGINEERING STATION
Point Pleasant
Brick Township
FORT DIX
Lakehurst
Silverton
Mount Holly
Browns Mills
Toms River
Gilford Park
Whiting
NEW JERSEY
Beachwood
Medford Lakes
Forked River
PINE BARRENS
Barnegat Bay
Barnegat
© Rand McNally A-101632-1
f
Pack Monadnock Mountain 2310 ft. 704 m
Derry
Hampstead
Seabrook
Wilton
Milford
Londonderry
Amesbury
Salisbury
WAPACK RANGE
Plaistow
NEW HAMPSHIRE
Newburyport
Bigelow Bight
Nashua
Hudson
Salem
Greenville
Haverhill
Groveland
Pelham
Methuen
Lawrence
Georgetown
Mt. Watatic 1832 ft. 558 m
Dracut
North Andover
Ipswich Bay
Ipswich
Rockport
Pepperell
Pawtucket Falls
Lowell
CAPE ANN
Andover
Topsfield
Ashburnham
Tewksbury
Wenham
Gloucester
Lunenburg
Nabnasset
Middleton
Chelmsford
North Reading
Danvers
Fitchburg
Ayer
Littleton
Billerica
Wilmington
Lynnfield
Beverly
Manchester
Peabody
Salem
Leominster
MASSACHUSETTS
Pinehurst
Reading
Marblehead
Wachusett Mountain 2006 ft. 611 m
Bedford
Burlington
Wakefield
Swampscott
HANSCOM AIR FORCE BASE
Saugus
West Acton
Woburn
Melrose
Lynn
Concord
Lexington
Medford
Malden
Sterling
Clinton
Maynard
Arlington
Revere
Hudson
EAST POINT
Waltham
Somerville
BOSTON
Massachusetts Bay
West Boylston
Sudbury
Wachusett Reservoir
Cambridge
Holden
Chaffin
Marlborough
Weston
Brookline
Shrewsbury
Northborough
Cochituate
Newton
Hull
POINT ALLERTON
Wellesley
Quincy Bay
Worcester
Framingham
Needham
Quincy
Westborough
Ashland
Natick
Dedham
Cohasset
Leicester
Milton
Weymouth
Hingham
North Scituate
Hopkinton
Westwood
North Grafton
Braintree
Millbury
Holliston
Medfield
Norwood
South Hingham
Scituate
Auburn
West Medway
Randolph
Holbrook
Northbridge
Milford
Walpole
Canton
Rockland
Oxford
Hopedale
Medway
Sharon
Stoughton
Whitinsville
Bellingham
Franklin
Brockton
Abington
Marshfield
North Pembroke
Uxbridge
Whitman
Foxboro
North Easton
Webster
© Rand McNally A-101634-1

0 5 10 15 20 25 30 Miles

0 5 10 15 20 25 30 35 40 45 50 Kilometers

Lambert Conformal Conic Projection
Scale 1:1,000,000
One inch to 16 miles
One cm to 10 km

a

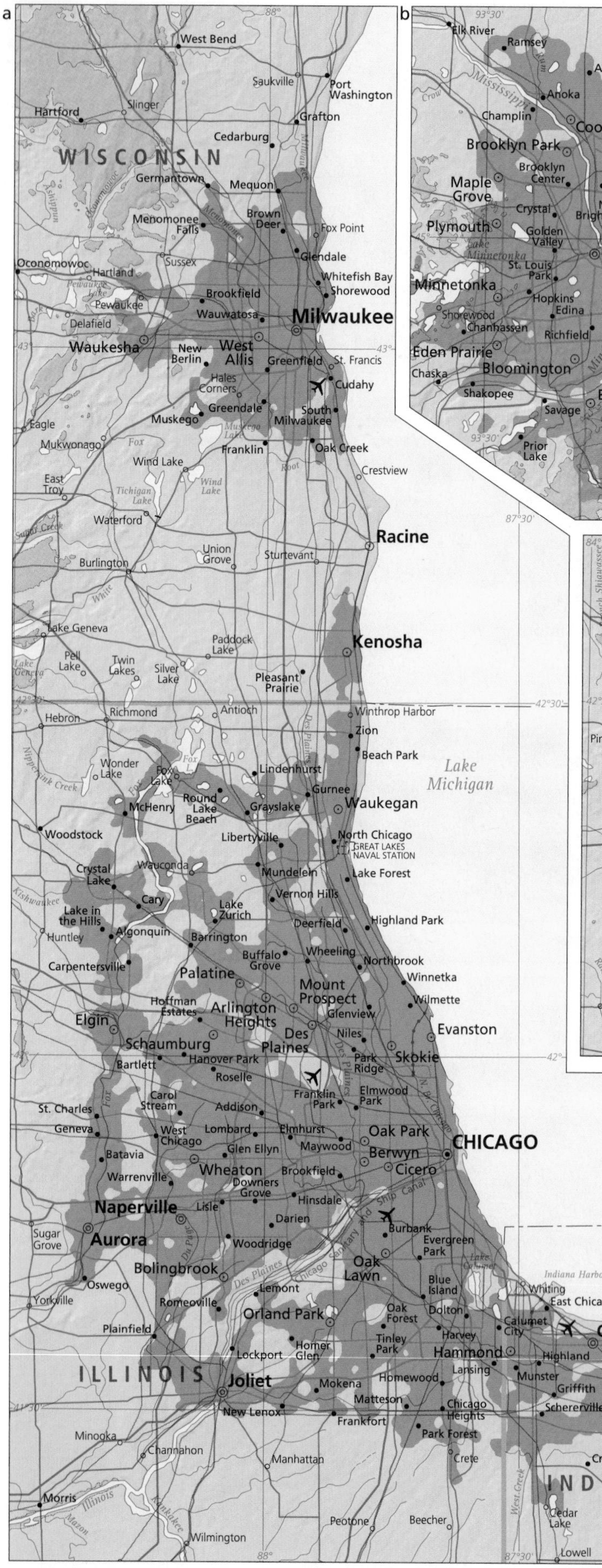
West Bend
Saukville
Port Washington
Hartford
Slinger
Grafton
Cedarburg
WISCONSIN
Germantown
Mequon
Menomonee Falls
Brown Deer
Fox Point
Glendale
Oconomowoc
Hartland
Sussex
Whitefish Bay
Shorewood
Pewaukee
Brookfield
Delafield
Wauwatosa
Milwaukee
Waukesha
New Berlin
West Allis
Greenfield
St. Francis
Cudahy
Hales Corners
Greendale
South Milwaukee
Muskego
Eagle
Mukwonago
Franklin
Oak Creek
Wind Lake
Crestview
East Troy
Waterford
Racine
Union Grove
Sturtevant
Burlington
Lake Geneva
Kenosha
Paddock Lake
Pell Lake
Twin Lakes
Silver Lake
Pleasant Prairie
Hebron
Richmond
Antioch
Winthrop Harbor
Zion
Beach Park
Lake Michigan
Wonder Lake
Fox Lake
Lindenhurst
Gurnee
Round Lake Beach
McHenry
Grayslake
Waukegan
Woodstock
Libertyville
North Chicago
GREAT LAKES NAVAL STATION
Crystal Lake
Wauconda
Mundelein
Lake Forest
Cary
Lake Zurich
Vernon Hills
Lake in the Hills
Huntley
Algonquin
Deerfield
Highland Park
Barrington
Carpentersville
Buffalo Grove
Wheeling
Northbrook
Palatine
Winnetka
Mount Prospect
Wilmette
Hoffman Estates
Arlington Heights
Glenview
Elgin
Des Plaines
Niles
Evanston
Schaumburg
Hanover Park
Park Ridge
Skokie
Bartlett
Roselle
Franklin Park
Elmwood Park
St. Charles
Carol Stream
Addison
Geneva
West Chicago
Lombard
Elmhurst
Oak Park
CHICAGO
Glen Ellyn
Maywood
Berwyn
Batavia
Wheaton
Brookfield
Cicero
Warrenville
Downers Grove
Hinsdale
Naperville
Lisle
Darien
Burbank
Sugar Grove
Aurora
Woodridge
Evergreen Park
Bolingbrook
Oak Lawn
Indiana Harbor
Oswego
Lemont
Blue Island
Whiting
East Chicago
Yorkville
Romeoville
Orland Park
Oak Forest
Dolton
Calumet City
Gary
Plainfield
Tinley Park
Harvey
Lockport
Homer Glen
Hammond
Highland
Lake Station
Portage
ILLINOIS
Joliet
Mokena
Lansing
Homewood
Munster
Hobart
Griffith
Matteson
Chicago Heights
Schererville
New Lenox
Frankfort
Merrillville
Minooka
Park Forest
Channahon
Manhattan
Crete
Crown Point
INDIANA
Morris
Peotone
Beecher
Cedar Lake
Wilmington
Hebron
Lowell
© Rand McNally A-101641-1

b

Elk River
Ramsey
Forest Lake
Andover
Anoka
Blaine
Champlin
Coon Rapids
Lino Lakes
Brooklyn Park
Brooklyn Center
Fridley
Shoreview
Maple Grove
Crystal
New Brighton
Vadnais Heights
White Bear Lake
Plymouth
Golden Valley
Roseville
Maplewood
MINNEAPOLIS
Oakdale
St. Louis Park
Minnetonka
St. Paul
Hopkins
Edina
Woodbury
Shorewood
Chanhassen
Richfield
West St. Paul
South St. Paul
Eden Prairie
Eagan
Cottage Grove
Chaska
Bloomington
Inver Grove Heights
Shakopee
Burnsville
Savage
Apple Valley
Rosemount
Prior Lake
Lakeville
Farmington
© Rand McNally A-101642-1

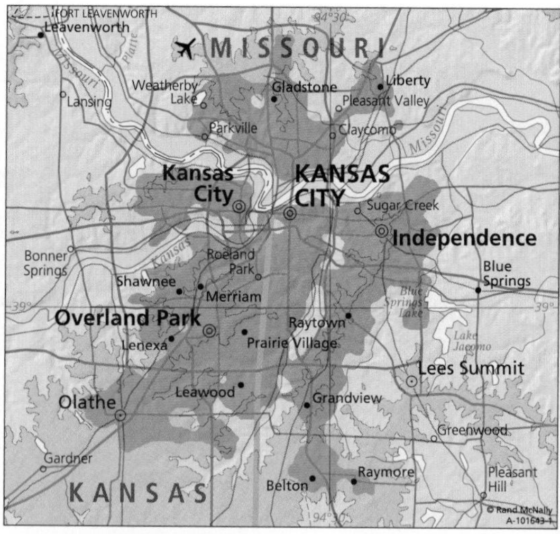
FORT LEAVENWORTH
Leavenworth
MISSOURI
Liberty
Weatherby Lake
Gladstone
Pleasant Valley
Lansing
Parkville
Claycomo
Kansas City
KANSAS CITY
Sugar Creek
Independence
Bonner Springs
Roeland Park
Shawnee
Merriam
Blue Springs
Overland Park
Raytown
Lenexa
Prairie Village
Lees Summit
Olathe
Leawood
Grandview
Greenwood
Gardner
Raymore
Pleasant Hill
KANSAS
Belton
© Rand McNally A-101643-1

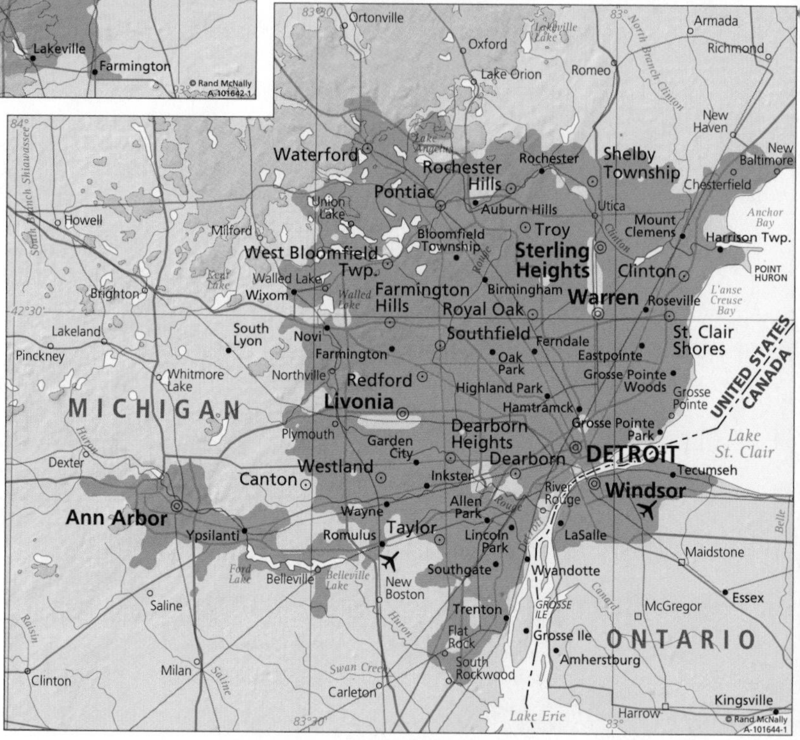
Ortonville
Armada
Oxford
Richmond
Lake Orion
Romeo
New Haven
Waterford
Rochester Hills
Rochester
Shelby Township
New Baltimore
Pontiac
Chesterfield
Union Lake
Auburn Hills
Utica
Howell
Milford
Troy
Mount Clemens
Harrison Twp.
Bloomfield Township
West Bloomfield Twp.
Sterling Heights
Clinton
Walled Lake
Farmington Hills
Birmingham
Wixom
Brighton
Royal Oak
Warren
Roseville
Lakeland
South Lyon
Novi
Southfield
Ferndale
St. Clair Shores
Pinckney
Farmington
Oak Park
Eastpointe
Northville
Redford
Grosse Pointe Woods
Whitmore Lake
Highland Park
Grosse Pointe
MICHIGAN
Livonia
Hamtramck
UNITED STATES
CANADA
Plymouth
Dearborn Heights
Grosse Pointe Park
Lake St. Clair
Dexter
Garden City
DETROIT
Tecumseh
Westland
Dearborn
Canton
Inkster
Windsor
Ann Arbor
Wayne
River Rouge
Allen Park
Ypsilanti
Romulus
Taylor
Lincoln Park
LaSalle
Maidstone
Belleville
Southgate
Wyandotte
New Boston
Essex
Saline
McGregor
Trenton
Grosse Ile
ONTARIO
Flat Rock
Amherstburg
Milan
South Rockwood
Clinton
Carleton
Kingsville
Lake Erie
Harrow
© Rand McNally A-101644-1

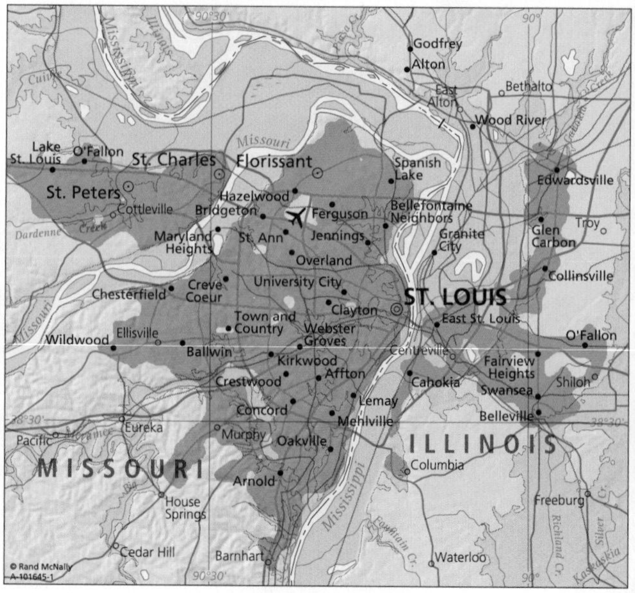
Godfrey
Alton
East Alton
Bethalto
Wood River
Lake St. Louis
O'Fallon
St. Charles
Florissant
Spanish Lake
St. Peters
Hazelwood
Edwardsville
Cottleville
Bridgeton
Ferguson
Bellefontaine Neighbors
Maryland Heights
St. Ann
Jennings
Granite City
Glen Carbon
Troy
Overland
Collinsville
Chesterfield
Creve Coeur
University City
Clayton
ST. LOUIS
Town and Country
East St. Louis
Wildwood
Ellisville
Webster Groves
O'Fallon
Ballwin
Centreville
Kirkwood
Fairview Heights
Crestwood
Affton
Cahokia
Shiloh
Swansea
Concord
Lemay
Belleville
Eureka
Mehlville
Pacific
Murphy
Oakville
ILLINOIS
MISSOURI
Columbia
Arnold
House Springs
Freeburg
Cedar Hill
Barnhart
Waterloo
© Rand McNally A-101645-1

0 5 10 15 20 25 30 Miles
0 5 10 15 20 25 30 35 40 45 50 Kilometers

Lambert Conformal Conic Projection
Scale 1:1,000,000
One inch to 16 miles
One cm to 10 km

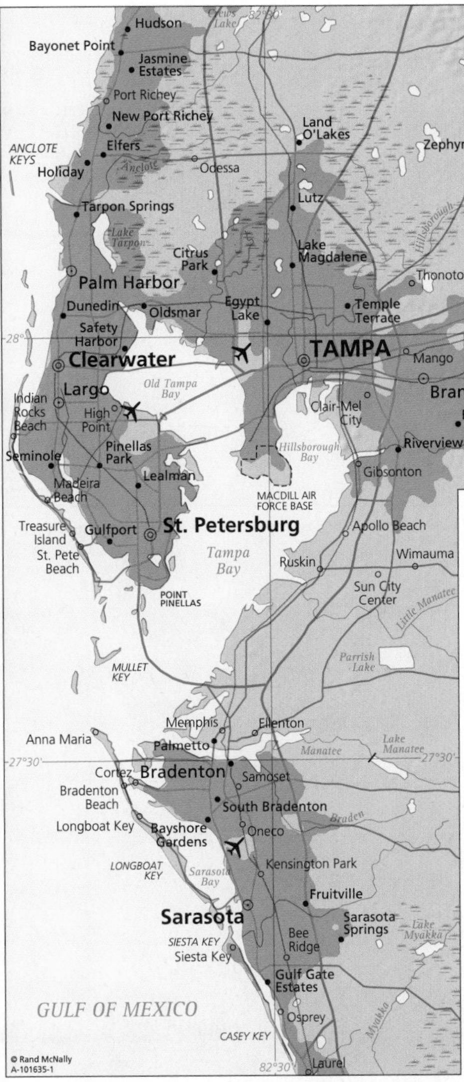
Hudson
Bayonet Point
Jasmine Estates
Port Richey
New Port Richey
ANCLOTE KEYS
Elfers
Holiday
Land O'Lakes
Zephyrhills
Odessa
Crystal Springs
Tarpon Springs
Lutz
Gibsonia
Kathleen
Lake Tarpon
Citrus Park
Lake Magdalene
Thonotosassa
Lakeland
Palm Harbor
Dunedin
Oldsmar
Egypt Lake
Temple Terrace
Winston
Eaton Park
Safety Harbor
Dover
Plant City
Clearwater
TAMPA
Mango
Medulla
Highland City
Largo
Old Tampa Bay
Brandon
Indian Rocks Beach
High Point
Clair-Mel City
Bloomingdale
Mulberry
Bartow
Seminole
Pinellas Park
Hillsborough Bay
Riverview
Madeira Beach
Lealman
Gibsonton
MACDILL AIR FORCE BASE
Treasure Island
Gulfport
St. Petersburg
Apollo Beach
St. Pete Beach
Tampa Bay
Ruskin
Wimauma
POINT PINELLAS
Sun City Center
Little Manatee
MULLET KEY
Parrish Lake
Memphis
Ellenton
Anna Maria
Palmetto
Lake Manatee
Manatee
Cortez
Bradenton
Samoset
Bradenton Beach
South Bradenton
Braden
Longboat Key
Bayshore Gardens
Oneco
LONGBOAT KEY
Kensington Park
Sarasota Bay
Fruitville
Sarasota
Sarasota Springs
Lake Myakka
SIESTA KEY
Siesta Key
Bee Ridge
Gulf Gate Estates
GULF OF MEXICO
Osprey
CASEY KEY
Myakka
Laurel
© Rand McNally A-101635-1

b

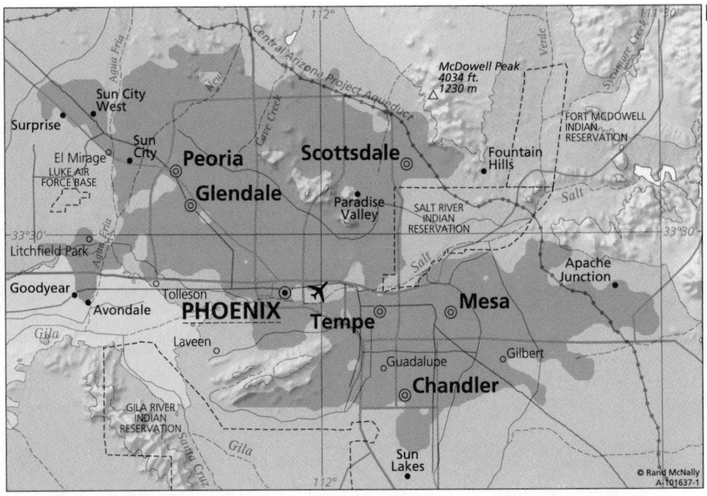
Central Arizona Project Aqueduct
McDowell Peak 4034 ft. 1230 m
Sun City West
Surprise
Sun City
El Mirage
LUKE AIR FORCE BASE
Peoria
Scottsdale
Fountain Hills
FORT MCDOWELL INDIAN RESERVATION
Glendale
Paradise Valley
SALT RIVER INDIAN RESERVATION
Salt
Litchfield Park
Apache Junction
Goodyear
Tolleson
Avondale
PHOENIX
Tempe
Mesa
Gila
Laveen
Guadalupe
Gilbert
Chandler
GILA RIVER INDIAN RESERVATION
Sun Lakes
© Rand McNally A-101637-1

c

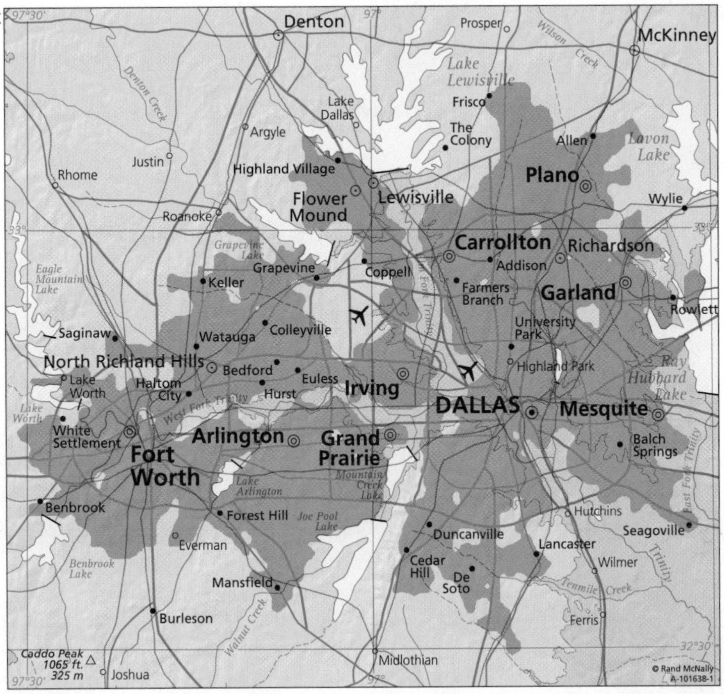
Denton
Prosper
McKinney
Lake Lewisville
Frisco
Lake Dallas
The Colony
Argyle
Allen
Lavon Lake
Justin
Highland Village
Plano
Rhome
Flower Mound
Lewisville
Wylie
Roanoke
Grapevine Lake
Carrollton
Richardson
Eagle Mountain Lake
Grapevine
Addison
Keller
Coppell
Farmers Branch
Garland
Rowlett
University Park
Saginaw
Watauga
Colleyville
North Richland Hills
Bedford
Euless
Highland Park
Ray Hubbard Lake
Lake Worth
Haltom City
Hurst
Irving
DALLAS
Mesquite
White Settlement
Arlington
Grand Prairie
Balch Springs
Fort Worth
Mountain Creek Lake
Lake Arlington
Benbrook
Forest Hill
Joe Pool Lake
Hutchins
Seagoville
Duncanville
Everman
Lancaster
Benbrook Lake
Cedar Hill
De Soto
Wilmer
Mansfield
Ferris
Burleson
Caddo Peak 1065 ft. 325 m
Joshua
Midlothian
© Rand McNally A-101638-1

d

Jupiter
Juno Beach
Palm Beach Gardens
North Palm Beach
Riviera Beach
Lake Park
West Palm Beach
Royal Palm Beach
Palm Beach
West Palm Beach Canal
Wellington
Palm Springs
Greenacres
Lake Worth
Lantana
Boynton Beach
The Everglades
Delray Beach
Villages of Oriole
Highland Beach
Hillsboro Canal
Boca Raton
Deerfield Beach
Coral Springs
Parkland
Lighthouse Point
Coconut Creek
Margate
Pompano Beach
Tamarac
North Lauderdale
Lauderdale Lakes
Oakland Park
Sunrise
Fort Lauderdale
Lauderhill
Plantation
Weston
Davie
Dania
South New River Canal
Cooper City
Pembroke Pines
Hollywood
Miramar
Hallandale
North Miami Beach
Aventura
Miami Gardens
Opa Locka
Miami Lakes
North Miami
Miami Shores
Hialeah Gardens
Hialeah
Miami Springs
Brownsville
Miami Beach
Tamiami Canal
Sweetwater
MIAMI
Westchester
ATLANTIC OCEAN
Coral Gables
Tamiami
EVERGLADES NAT'L PARK
Glenvar Heights
South Miami
Key Biscayne
CAPE FLORIDA
Kendall
Kendale Lakes
Pinecrest
Richmond Heights
Perrine
Cutler Ridge
Biscayne Bay
South Miami Heights
Princeton
BISCAYNE NATIONAL PARK
Leisure City
Homestead
Florida City
ELLIOTT KEY
© Rand McNally A-101640-1

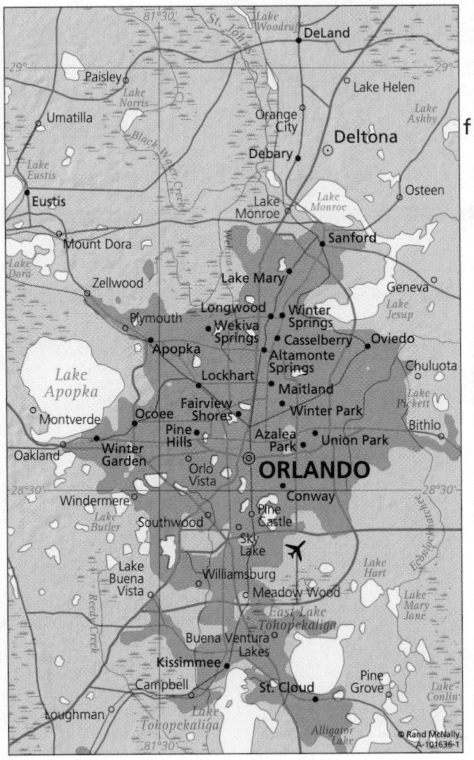
Lake Woodruff
DeLand
Paisley
Lake Norris
Lake Helen
Umatilla
Orange City
Lake Ashby
Deltona
Debary
Lake Eustis
Eustis
Lake Monroe
Osteen
Mount Dora
Sanford
Lake Dora
Zellwood
Lake Mary
Geneva
Longwood
Winter Springs
Lake Jesup
Plymouth
Wekiva Springs
Casselberry
Oviedo
Apopka
Altamonte Springs
Chuluota
Lake Apopka
Lockhart
Maitland
Fairview Shores
Winter Park
Montverde
Ocoee
Pine Hills
Bithlo
Oakland
Winter Garden
Azalea Park
Union Park
Orlo Vista
ORLANDO
Conway
Windermere
Lake Butler
Southwood
Pine Castle
Sky Lake
Lake Hart
Williamsburg
Lake Buena Vista
Meadow Wood
Lake Mary Jane
East Lake Tohopekaliga
Buena Ventura Lakes
Kissimmee
Campbell
St. Cloud
Pine Grove
Lake Conlin
Loughman
Lake Tohopekaliga
Alligator Lake
© Rand McNally A-101636-1

f

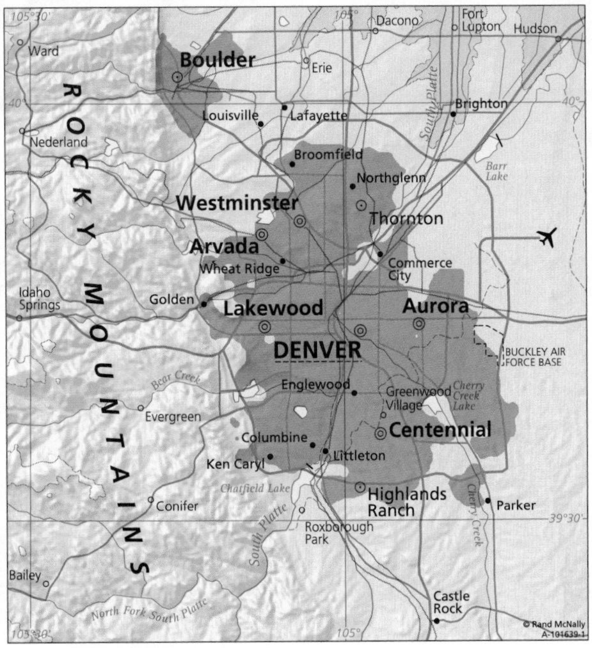
Ward
Dacono
Fort Lupton
Hudson
Boulder
Erie
Louisville
Lafayette
Brighton
Nederland
Broomfield
Barr Lake
Northglenn
Westminster
Thornton
ROCKY MOUNTAINS
Arvada
Wheat Ridge
Commerce City
Idaho Springs
Golden
Lakewood
Aurora
DENVER
BUCKLEY AIR FORCE BASE
Englewood
Cherry Creek Lake
Greenwood Village
Evergreen
Centennial
Columbine
Littleton
Ken Caryl
Chatfield Lake
Highlands Ranch
Conifer
Parker
Roxborough Park
Bailey
North Fork South Platte
Castle Rock
© Rand McNally A-101639-1

0 5 10 15 20 25 30 Miles
0 5 10 15 20 25 30 35 40 45 50 Kilometers

Lambert Conformal Conic Projection
Scale 1:1,000,000
One inch to 16 miles
One cm to 10 km

a
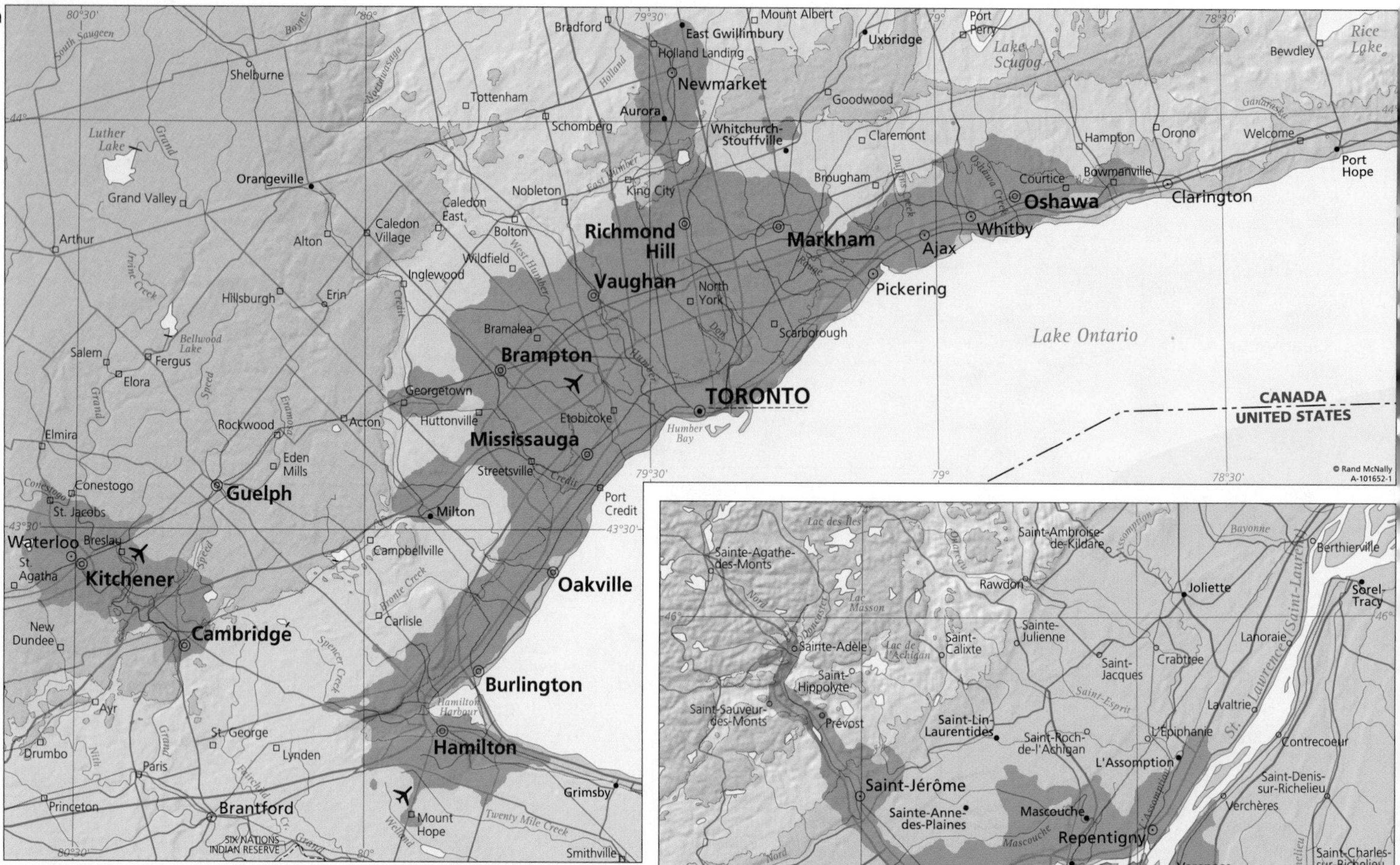
East Gwillimbury
Mount Albert
Uxbridge
Port Perry
Lake Scugog
Rice Lake
Bewdley
Bradford
Holland Landing
Newmarket
Shelburne
Tottenham
Goodwood
Aurora
Schomberg
Whitchurch-Stouffville
Claremont
Hampton
Orono
Welcome
Port Hope
Luther Lake
Orangeville
Grand Valley
Brougham
Courtice
Bowmanville
Oshawa
Clarington
Caledon East
Nobleton
King City
Whitby
Arthur
Alton
Caledon Village
Bolton
Richmond Hill
Markham
Ajax
Wildfield
Inglewood
Vaughan
North York
Pickering
Hillsburgh
Erin
Bramalea
Scarborough
Lake Ontario
Salem
Fergus
Elora
Brampton
Georgetown
TORONTO
Rockwood
Acton
Huttonville
Etobicoke
Humber Bay
CANADA
UNITED STATES
Elmira
Eden Mills
Mississauga
Streetsville
Conestogo
Guelph
Port Credit
St. Jacobs
Milton
Waterloo
Breslau
Campbellville
St. Agatha
Kitchener
Oakville
New Dundee
Carlisle
Cambridge
Burlington
Ayr
Hamilton Harbour
St. George
Hamilton
Drumbo
Lynden
Paris
Grimsby
Mount Hope
Princeton
Brantford
SIX NATIONS INDIAN RESERVE
Smithville
Twenty Mile Creek
© Rand McNally A-101652-1

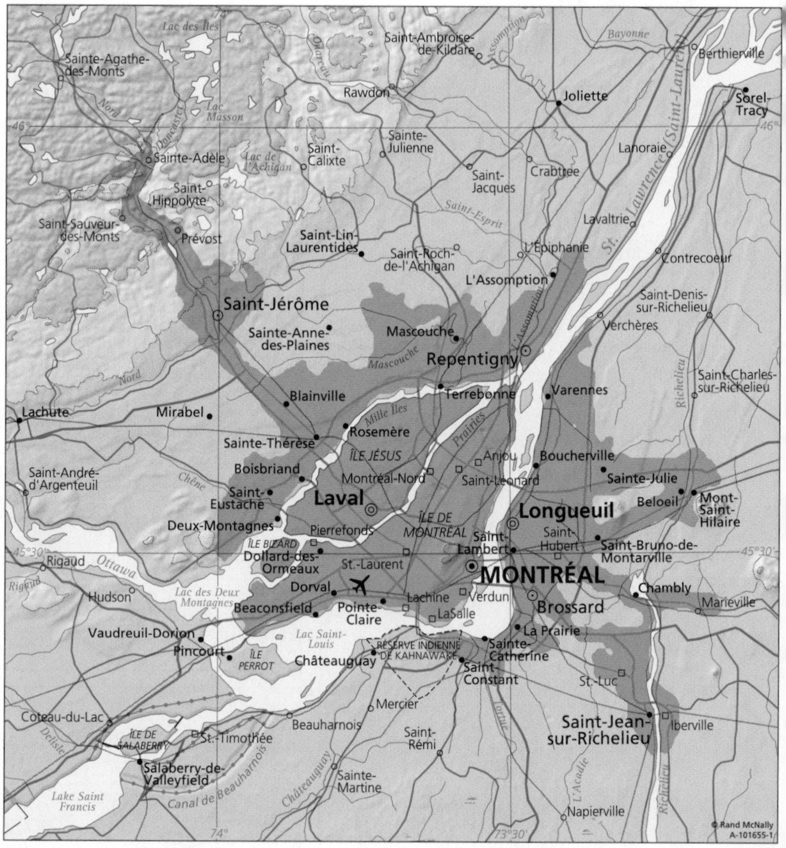
Lac des Îles
Saint-Ambroise-de-Kildare
Bayonne
Berthierville
Sainte-Agathe-des-Monts
Rawdon
Joliette
Sorel-Tracy
Lac Masson
Sainte-Adèle
Saint-Calixte
Sainte-Julienne
Lanoraie
Saint-Jacques
Crabtree
Saint-Hippolyte
Saint-Sauveur-des-Monts
Prévost
Saint-Lin-Laurentides
Saint-Roch-de-l'Achigan
Lavaltrie
L'Épiphanie
Contrecoeur
L'Assomption
Saint-Jérôme
Saint-Denis-sur-Richelieu
Verchères
Sainte-Anne-des-Plaines
Mascouche
Repentigny
Saint-Charles-sur-Richelieu
Lachute
Mirabel
Blainville
Terrebonne
Varennes
Rosemère
Sainte-Thérèse
ÎLE JÉSUS
Saint-André-d'Argenteuil
Boisbriand
Anjou
Boucherville
Sainte-Julie
Montréal-Nord
Saint-Léonard
Saint-Eustache
Laval
Beloeil
Mont-Saint-Hilaire
Longueuil
Deux-Montagnes
Pierrefonds
ÎLE DE MONTRÉAL
Saint-Lambert
Saint-Hubert
Saint-Bruno-de-Montarville
ÎLE BIZARD
Dollard-des-Ormeaux
St.-Laurent
Rigaud
MONTRÉAL
Dorval
Verdun
Hudson
Chambly
Lac des Deux Montagnes
Beaconsfield
Lachine
Brossard
Marieville
Pointe-Claire
LaSalle
Lac Saint-Louis
La Prairie
Vaudreuil-Dorion
Pincourt
RÉSERVE INDIENNE DE KAHNAWAKE
Sainte-Catherine
ÎLE PERROT
Châteauguay
Saint-Constant
St.-Luc
Coteau-du-Lac
Mercier
Saint-Jean-sur-Richelieu
Iberville
Beauharnois
ÎLE DE SALABERRY
St.-Timothée
Saint-Rémi
Salaberry-de-Valleyfield
Sainte-Martine
Lake Saint Francis
Canal de Beauharnois
Napierville
© Rand McNally A-101655-1

c

Thurso
QUÉBEC
Buckingham
Masson-Angers
Rockland
Quyon
Chelsea
Mont King 1129 ft. 344 m
Gatineau
Cumberland
Clarence Creek
Orléans
Sarsfield
Arnprior
Galetta
Hull
Rockcliffe Park
Vanier
Navan
Gloucester
Aylmer
OTTAWA
Carlsbad Springs
Vars
Pakenham
Carp
Nepean
Blossom Park
Lac Deschênes
Kanata
Limoges
ONTARIO
Embrun
Stittsville
Metcalfe
Russell
Almonte
Manotick
Clayton
Clayton Lake
Richmond
Ashton
Vernon
North Gower
Osgoode
Carleton Place
© Rand McNally A-101653-1

d

Airdrie
Irricana
Keoma
Cochrane
Kathyrn
Balzac
Bow
Dalroy
Elbow
CALGARY
TSUU T'INA NATION INDIAN RESERVE
Shepard
Langdon
Priddis
Midnapore
Sheep
Highwood
Okotoks
© Rand McNally A-101654-1

e
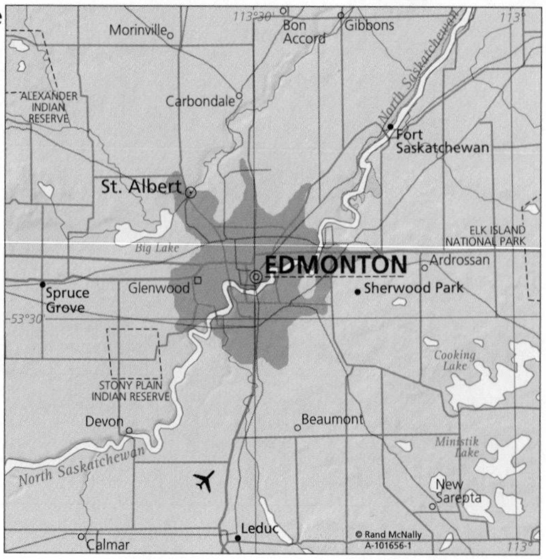
Bon Accord
Gibbons
Morinville
ALEXANDER INDIAN RESERVE
Carbondale
Fort Saskatchewan
North Saskatchewan
St. Albert
ELK ISLAND NATIONAL PARK
Big Lake
Ardrossan
EDMONTON
Spruce Grove
Glenwood
Sherwood Park
Cooking Lake
STONY PLAIN INDIAN RESERVE
Beaumont
Devon
Miquelon Lake
New Sarepta
Leduc
Calmar
© Rand McNally A-101656-1

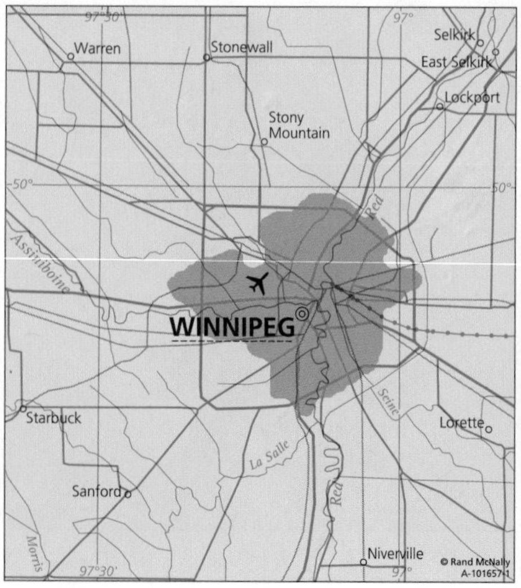
Selkirk
Warren
Stonewall
East Selkirk
Lockport
Stony Mountain
Assiniboine
Red
WINNIPEG
Starbuck
Seine
Lorette
La Salle
Sanford
Morris
Niverville
© Rand McNally A-101657-1

0 5 10 15 20 25 30 Miles
0 5 10 15 20 25 30 35 40 45 50 Kilometers

Lambert Conformal Conic Projection
Scale 1:1,000,000
One inch to 16 miles
One cm to 10 km

a

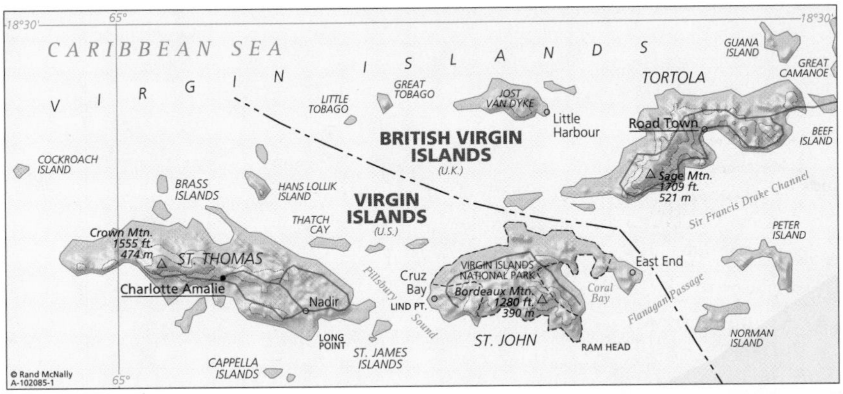
CARIBBEAN SEA
VIRGIN ISLANDS
TORTOLA
GUANA ISLAND
GREAT CAMANOE
GREAT TOBAGO
LITTLE TOBAGO
JOST VAN DYKE
Little Harbour
Road Town
BEEF ISLAND
BRITISH VIRGIN ISLANDS
(U.K.)
COCKROACH ISLAND
BRASS ISLANDS
HANS LOLLIK ISLAND
THATCH CAY
VIRGIN ISLANDS
(U.S.)
Sage Mtn. 1709 ft. 521 m
Sir Francis Drake Channel
PETER ISLAND
Crown Mtn. 1555 ft. 474 m
ST. THOMAS
Charlotte Amalie
Nadir
Cruz Bay
LIND PT.
VIRGIN ISLANDS NATIONAL PARK
Bordeaux Mtn. 1280 ft. 390 m
Coral Bay
East End
Flanagan Passage
Pillsbury Sound
LONG POINT
ST. JAMES ISLANDS
ST. JOHN
RAM HEAD
NORMAN ISLAND
CAPPELLA ISLANDS
© Rand McNally A-102085-1

b

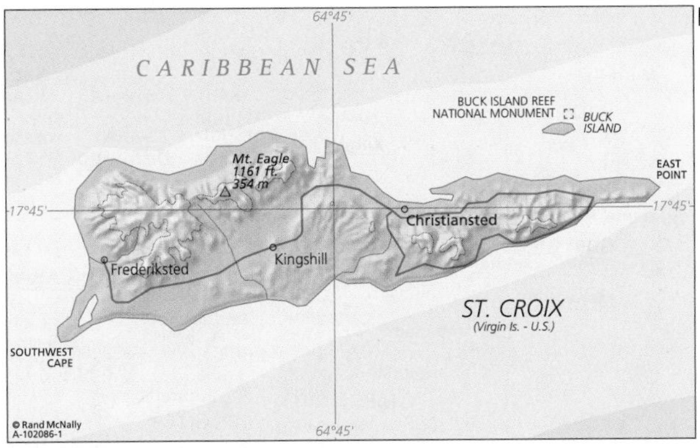
CARIBBEAN SEA
BUCK ISLAND REEF NATIONAL MONUMENT
BUCK ISLAND
EAST POINT
Mt. Eagle 1161 ft. 354 m
Christiansted
Frederiksted
Kingshill
ST. CROIX
(Virgin Is. - U.S.)
SOUTHWEST CAPE
© Rand McNally A-102086-1

c

GULF OF MEXICO
HAVANA
(LA HABANA)
Bahía de La Habana
Guanabacoa
Campo Florido
CIUDAD DE LA HABANA
Marianao
Playa Baracoa
Santa María del Rosario
Jaruco
Tapaste
Santiago de las Vegas
Bauta
Caimito
Guanajay
San José de las Lajas
San Antonio de los Baños
Bejucal
LA HABANA
Ceiba del Agua
La Salud
San Antonio de las Vegas
Güines
Quivicán
San Felipe
Alquízar
Güira de Melena
Melena del Sur
Mayabeque
© Rand McNally A-101682-1

d

CHIMALTENANGO
San Martín Jilotepeque
San Raimundo
San Pedro Ayampuc
San Juan Sacatepéquez
GUATEMALA
Chinautla
Chimaltenango
Mixco
Sumpango
Guatemala
SACATEPÉQUEZ
KAMINALJUYÚ
Canalitos
Antigua Guatemala
Santa Catarina Pinula
San José Pinula
Villa Nueva
Ciudad Vieja
Petapa
Fraijanes
Alotenango
Amatitlán
Villa Canales
Volcán de Agua 12 336 ft. 3760 m
Lago de Amatitlán
San Vicente Pacaya
Palín
Santa Elena Barillas
Volcán de Pacaya 8346 ft. 2544 m
Los Dolores
Santa Cruz Naranjo
El Jocotillo
Escuintla
Barberena
ESCUINTLA
SANTA ROSA
© Rand McNally A-102084-1

e

ATLANTIC OCEAN
Bahía de San Juan
San Juan
Vega Baja
Dorado
Toa Baja
PUNTA VACÍA TALEGA
Vega Alta
Toa Alta
Cataño
Loíza Aldea
Laguna San José
Bayamón
Río Piedras
Carolina
Río Grande
Corozal
Guaynabo
Canóvanas
Trujillo Alto
El Yunque 3494 ft. 1065 m
Naranjito
Aguas Buenas
Gurabo
SIERRA DE LUQUILLO
El Toro 3524 ft. 1074 m
Juncos
Orocovis
La Torrecilla 3094 ft. 943 m
Comerío
Caguas
Las Piedras
Barranquitas
Cidra
Lago de Cidra
San Lorenzo
Cerro La Santa 2963 ft. 903 m
Aibonito
Humacao
CORDILLERA CENTRAL
Cayey
SIERRA DE CAYEY
Coamo
Lago Carite
© Rand McNally A-101686-1

f

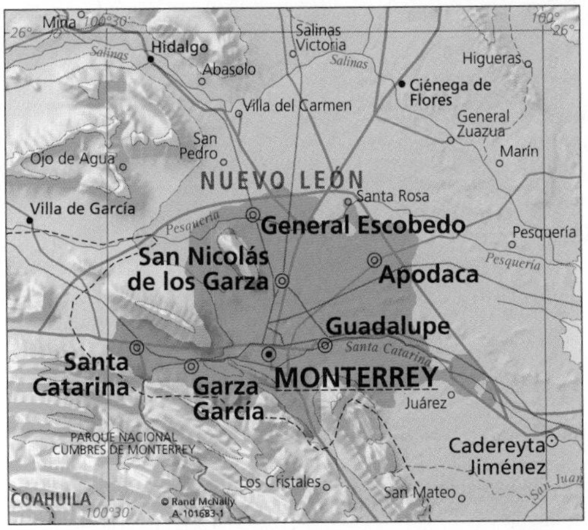
Mina
Salinas Victoria
Hidalgo
Abasolo
Higueras
Ciénega de Flores
Villa del Carmen
General Zuazua
San Pedro
Marín
Ojo de Agua
NUEVO LEÓN
Santa Rosa
Villa de García
General Escobedo
Pesquería
San Nicolás de los Garza
Apodaca
Guadalupe
Santa Catarina
Garza García
MONTERREY
Juárez
PARQUE NACIONAL CUMBRES DE MONTERREY
Cadereyta Jiménez
Los Cristales
San Mateo
COAHUILA
© Rand McNally A-101683-1

g

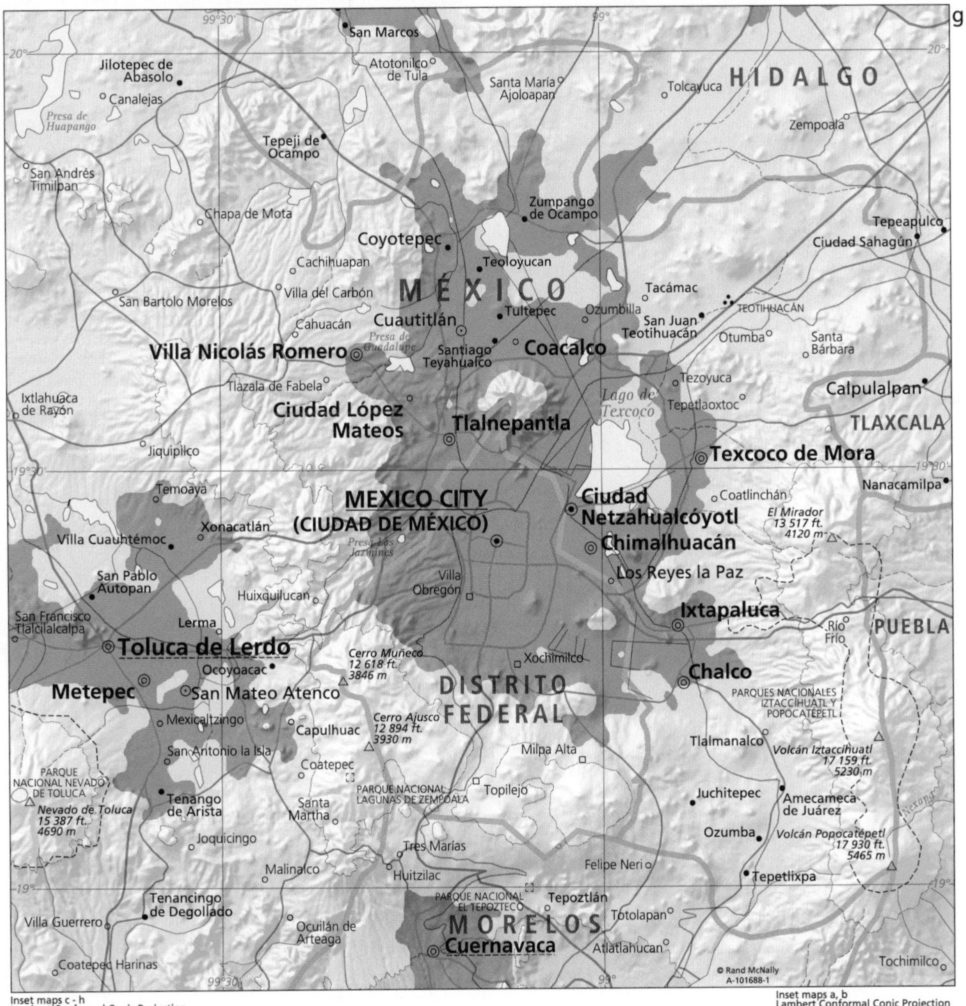
San Marcos
Jilotepec de Abasolo
Atotonilco de Tula
Santa María Ajoloapan
Tolcayuca
HIDALGO
Canalejas
Presa de Huapango
Zempoala
Tepeji de Ocampo
San Andrés Timilpan
Zumpango de Ocampo
Chapa de Mota
Tepeapulco
Ciudad Sahagún
Coyotepec
Cachihuapan
Teoloyucan
Villa del Carbón
MÉXICO
Tacámac
San Bartolo Morelos
Cuautitlán
Tultepec
Ozumbilla
TEOTIHUACÁN
San Juan Teotihuacán
Cahuacán
Santiago Teyahualco
Coacalco
Otumba
Santa Bárbara
Villa Nicolás Romero
Tezoyuca
Tlazala de Fabela
Tepetlaoxtoc
Calpulalpan
Ixtlahuaca de Rayón
Ciudad López Mateos
Tlalnepantla
Lago de Texcoco
TLAXCALA
Jiquipilco
Texcoco de Mora
Temoaya
Coatlinchán
Nanacamilpa
MEXICO CITY
(CIUDAD DE MÉXICO)
Ciudad Netzahualcóyotl
El Mirador 13 517 ft. 4120 m
Xonacatlán
Villa Cuauhtémoc
Chimalhuacán
Villa Obregón
Los Reyes la Paz
San Pablo Autopan
Huixquilucan
Ixtapaluca
San Francisco Tlalcilalcalpa
Lerma
Río Frío
PUEBLA
Toluca de Lerdo
Ocoyoacac
Cerro Muñeco 12 618 ft. 3846 m
Xochimilco
Chalco
Metepec
San Mateo Atenco
DISTRITO FEDERAL
PARQUES NACIONALES IZTACCÍHUATL Y POPOCATÉPETL
Mexicaltzingo
Cerro Ajusco 12 894 ft. 3930 m
Capulhuac
San Antonio la Isla
Milpa Alta
Tlalmanalco
Volcán Iztaccíhuatl 17 159 ft. 5230 m
PARQUE NACIONAL NEVADO DE TOLUCA
Coatepec
Nevado de Toluca 15 387 ft. 4690 m
Tenango de Arista
Santa Martha
PARQUE NACIONAL LAGUNAS DE ZEMPOALA
Topilejo
Juchitepec
Amecameca de Juárez
Joquicingo
Ozumba
Volcán Popocatépetl 17 930 ft. 5465 m
Tres Marías
Malinalco
Huitzilac
Felipe Neri
Tepetlixpa
Tenancingo de Degollado
PARQUE NACIONAL EL TEPOZTECO
Tepoztlán
Villa Guerrero
Ocuilán de Arteaga
MORELOS
Totolapan
Cuernavaca
Coatepec Harinas
Atlatlahucan
Tochimilco
© Rand McNally A-101688-1

h

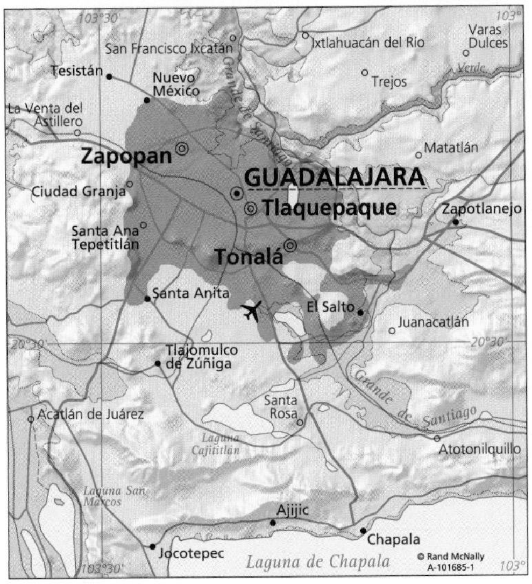
Varas Dulces
San Francisco Ixcatán
Ixtlahuacán del Río
Tesistán
Nuevo México
Trejos
La Venta del Astillero
Matatlán
Zapopan
GUADALAJARA
Ciudad Granja
Tlaquepaque
Zapotlanejo
Santa Ana Tepetitlán
Tonalá
Santa Anita
El Salto
Juanacatlán
Tlajomulco de Zúñiga
Grande de Santiago
Santa Rosa
Acatlán de Juárez
Laguna Cajititlán
Atotonilquillo
Laguna San Marcos
Ajijic
Chapala
Jocotepec
Laguna de Chapala
© Rand McNally A-101685-1

0 5 10 15 20 25 30 Miles
0 5 10 15 20 25 30 35 40 45 50 Kilometers

Inset maps c - h
Lambert Conformal Conic Projection
Scale 1:1,000,000
One inch to 16 miles
One cm to 10 km

Inset maps a, b
Lambert Conformal Conic Projection
Scale 1:500,000
One inch to 8 miles
One cm to 5 km

Gulf of Mexico
HAVANA
CUBA
Mérida
MEXICO
BELIZE
GUATEMALA
HONDURAS
Guatemala
EL SALVADOR
NICARAGUA
Managua
COSTA RICA
Panamá
PANAMA
JAMAICA
Kingston
HAITI
Port-au-Prince
DOMINICAN REPUBLIC
Santo Domingo
San Juan
PUERTO RICO (U.S.)
GUADELOUPE (Fr.)
MARTINIQUE (Fr.)
ST. LUCIA
BARBADOS
GRENADA
ARUBA (Neth.)
TRINIDAD AND TOBAGO
CARIBBEAN SEA
ATLANTIC OCEAN
Tropic of Cancer
Barranquilla
Cartagena
MARACAIBO
CARACAS
Barcelona
Barquisimeto
Ciudad Bolívar
Ciudad Guayana
Cúcuta
San Cristóbal
VENEZUELA
Orinoco
Georgetown
GUYANA
Paramaribo
SURINAME
Cayenne
FRENCH GUIANA (Fr.)
CABO ORANGE
MEDELLÍN
Bucaramanga
Manizales
BOGOTÁ
LLANOS
ISLA DEL COCO (C.R.)
Buenaventura
CALI
COLOMBIA
Boa Vista
Pico da Neblina 9888 ft. 3014 m
Macapá
Esmeraldas
QUITO
ECUADOR
Chimborazo 20 702 ft. 6310 m
Japurá
Negro
Amazon
Equator
Putumayo
MANAUS
Santarém
Belém
São Luís
Parnaíba
Fortaleza
ARCHIPIÉLAGO DE COLÓN (GALAPAGOS ISLANDS) (Ec.)
GUAYAQUIL
Cuenca
Iquitos
SELVAS
Loja
Marañón
Tapajós
Xingu
Marabá
Teresina
Parnaíba
CABO DE SÃO ROQUE
Natal
Campina Grande
João Pessoa
Caruaru
RECIFE
Represa de Sobradinho
Juazeiro
Tocantins
Juruá
Purus
Madeira
Ucayali
Chiclayo
Cajamarca
Trujillo
Pucallpa
Rio Branco
Porto Velho
BRAZIL
Nevado Huascarán 22 133 ft. 6746 m
PERU
Huancayo
Lima
Guaporé
Maceió
Aracaju
Cusco
BOLIVIA
Trinidad
PUNTA CARRETA
Ica
ANDES
Puno
La Paz
Cochabamba
SANTA CRUZ DE LA SIERRA
Araguaia
São Francisco
SALVADOR
Itabuna
Cuiabá
BRASÍLIA
GOIÂNIA
Arequipa
Oruro
Montes Claros
Arica
Potosí
Sucre
Corumbá
BELO HORIZONTE
Iquique
Tarija
PARAGUAY
Pico da Bandeira 9505 ft. 2897 m
Vitória
Campos
ILHAS MARTIN VAZ (Braz.)
GRAN CHACO
SÃO PAULO
CABO FRIO
RIO DE JANEIRO
Tropic of Capricorn
Antofagasta
Salta
Santos
Asunción
CURITIBA
ISLA SAN FÉLIX (Chile)
San Miguel de Tucumán
Salado
Corrientes
Posadas
Santiago del Estero
Paraná
Uruguay
Florianópolis
PACIFIC OCEAN
Santa Maria
PORTO ALEGRE
Coquimbo
CÓRDOBA
Santa Fe
Pelotas
Salto
Paysandú
Cerro Aconcagua 22 831 ft. 6959 m
Mendoza
Rio Grande
Valparaíso
Rosario
URUGUAY
ARCHIPIÉLAGO JUAN FERNÁNDEZ (Chile)
SANTIAGO
BUENOS AIRES
La Plata
MONTEVIDEO
Concepción
PAMPA
CABO SAN ANTONIO
Bahía Blanca
Mar del Plata
Neuquén
Negro
Valdivia
CHILE
ARGENTINA
Osorno
Golfo San Matías
Puerto Montt
ATLANTIC OCEAN
ARCHIPIÉLAGO DE LOS CHONOS
Comodoro Rivadavia
Golfo San Jorge
CABO TRES PUNTAS
Monte San Valentín 13 314 ft. 4058 m
PATAGONIA
FALKLAND ISLANDS (U.K.)
Río Gallegos
Stanley
Punta Arenas
SCOTIA SEA
CAPE HORN
Drake Passage
SOUTH GEORGIA AND THE SOUTH SANDWICH ISLANDS (U.K.)
SOUTH SHETLAND IS. (U.K.)
SOUTH ORKNEY IS. (U.K.)
SOUTHERN OCEAN
ANTARCTIC PENINSULA
Antarctic Circle
© Rand McNally
A-101692-1

0 200 400 600 800 1000 1200 Miles

0 200 400 600 800 1000 1200 1400 1600 1800 2000 Kilometers

Lambert Azimuthal Equal Area Projection
Scale 1:40,000,000
One inch to 640 miles
One cm to 400 km

Tropic of Cancer
Gulf of Mexico
Havana
CUBA
Yucatan Channel
YUCATAN PENINSULA
MEXICO
BELIZE
GUATEMALA
HONDURAS
EL SALVADOR
NICARAGUA
Lago de Nicaragua
COSTA RICA
PANAMA
ISTMO DE PANAMÁ
Golfo de Panamá
JAMAICA
HAITI
DOMINICAN REPUBLIC
HISPANIOLA
PUERTO RICO TRENCH
PUERTO RICO (U.S.)
GREATER ANTILLES
WEST INDIES
CARIBBEAN SEA
LESSER ANTILLES
GUADELOUPE (Fr.)
MARTINIQUE (Fr.)
BARBADOS
TRINIDAD AND TOBAGO
ATLANTIC OCEAN
Pico Cristóbal Colón 18 947 ft. 5775 m
Maracaibo
Lago de Maracaibo
Pico Bolívar 16 427 ft. 5007 m
Caracas
Orinoco
VENEZUELA
GUYANA
SURINAME
FRENCH GUIANA (Fr.)
Paramaribo
CABO ORANGE
Mt. Roraima 9432 ft. 2875 m
LLANOS
Bogotá
COLOMBIA
Nevado del Tolima 17 110 ft. 5215 m
Nevado del Huila 18 865 ft. 5750 m
Pico da Neblina 9888 ft. 3014 m
ISLA DEL COCO (C.R.)
ISLA DE MALPELO (Col.)
PUNTA GALERA
Cayambe 18 996 ft. 5790 m
Quito
ECUADOR
ANDES
Equator
Represa Balbina
Amazon
Negro
Japurá
Putumayo
Manaus
Belém
Fortaleza
ILHA FERNANDO DE NORONHA
CABO DE SÃO ROQUE
PONTA DO SEIXAS
Recife
ARCHIPIÉLAGO DE COLÓN (GALAPAGOS ISLANDS) (Ec.)
Golfo de Guayaquil
Chimborazo 20 702 ft. 6310 m
Iquitos
PUNTA PARIÑAS
Marañón
Juruá
Purus
Madeira
Tapajós
Xingu
SELVAS
BRAZIL
Represa de Sobradinho
Tocantins
Araguaia
São Francisco
Nevado Huascarán 22 133 ft. 6746 m
Porto Velho
PERU
Ucayali
Madre de Dios
Beni
Guaporé
Lima
Salvador
Pico das Almas 6024 ft. 1836 m
PLANALTO DO MATO GROSSO
Brasília
PUNTA CARRETA
Nevado Coropuna 20 686 ft. 6305 m
Lago Titicaca
Nevado Illampu 21 066 ft. 6421 m
La Paz
Grande
Nevado Sajama 21 463 ft. 6542 m
ALTIPLANO
BOLIVIA
Paraguay
SERRA DO ESPINHAÇO
Belo Horizonte
PERU-CHILE TRENCH
DESIERTO DE ATACAMA
Volcán San Pedro 20 161 ft. 6145 m
CHACO
PARAGUAY
Paraná
Pico da Bandeira 9505 ft. 2897 m
ILHAS MARTIN VAZ (Braz.)
São Paulo
Rio de Janeiro
CABO FRIO
Tropic of Capricorn
Volcán Llullaillaco 22 110 ft. 6739 m
GRAN CHACO
Asunción
Iguassu Falls
ISLA SAN FÉLIX (Chile)
Nevado Ojos del Salado 22 615 ft. 6893 m
San Miguel de Tucumán
Salado
Uruguay
Porto Alegre
CABO BASCUÑÁN
Cerro General Manuel Belgrano 20 505 ft. 6250 m
Córdoba
PACIFIC OCEAN
Lagoa dos Patos
Cerro Aconcagua 22 831 ft. 6959 m
URUGUAY
ARCHIPIÉLAGO JUAN FERNÁNDEZ (Chile)
Santiago
Buenos Aires
Montevideo
Río de la Plata
PAMPA
CABO SAN ANTONIO
PUNTA LAVAPIÉ
Colorado
Bahía Blanca
Negro
CHILE
ARGENTINA
Monte Tronador 11 453 ft. 3491 m
Golfo San Matías
ISLA GRANDE DE CHILOÉ
Chubut
CABO DOS BAHÍAS
ARCHIPIÉLAGO DE LOS CHONOS
Golfo San Jorge
CABO TRES PUNTAS
Monte San Valentín 13 314 ft. 4058 m
PATAGONIA
FALKLAND ISLANDS (U.K.)
EAST FALKLAND
WEST FALKLAND
ISLA WELLINGTON
ISLA DESOLACIÓN
Punta Arenas
TIERRA DEL FUEGO
ISLA SANTA INÉS
ISLA HOSTE
CAPE HORN (CABO DE HORNOS)
CAPE DISAPPOINTMENT
SOUTH GEORGIA
SOUTH GEORGIA AND THE SOUTH SANDWICH ISLANDS (U.K.)
SOUTH SANDWICH TRENCH
SOUTH SANDWICH IS.
SCOTIA SEA
Drake Passage
SOUTH SHETLAND IS. (U.K.)
SOUTH ORKNEY IS. (U.K.)
ANTARCTIC PENINSULA
SOUTHERN OCEAN
Antarctic Circle
© Rand McNally A-101693-1

0 200 400 600 800 1000 1200 Miles

0 200 400 600 800 1000 1200 1400 1600 1800 2000 Kilometers

Lambert Azimuthal Equal Area Projection
Scale 1:40,000,000
One inch to 640 miles
One cm to 400 km

Tropic of Cancer
CARIBBEAN SEA
ATLANTIC OCEAN
Barranquilla
Caracas
Orinoco
Medellín
Bogotá
LLANOS
ANDES
GALAPAGOS ISLANDS
Quito
Equator
Guayaquil
Manaus
Amazon
SELVAS
Fortaleza
Recife
Ucayali
CORDILLERA ORIENTAL
Lima
PLANALTO DO MATO GROSSO
Salvador
Lago Titicaca
La Paz
ALTIPLANO
Goiânia
Brasília
São Francisco
DESIERTO DE ATACAMA
Belo Horizonte
Paraná
GRAN CHACO
Asunción
Rio de Janeiro
São Paulo
Tropic of Capricorn
Curitiba
PACIFIC OCEAN
Córdoba
Porto Alegre
Santiago
Buenos Aires
Montevideo
PAMPA
ATLANTIC OCEAN
PATAGONIA
FALKLAND ISLANDS
TIERRA DEL FUEGO
SOUTH GEORGIA
SCOTIA SEA
Drake Passage
SOUTH SHETLAND IS.
SOUTH SANDWICH IS.
SOUTH ORKNEY IS.
SOUTHERN OCEAN
Antarctic Circle
Land Cover
Evergreen Needleleaf Forest
Evergreen Broadleaf Forest
Deciduous Broadleaf Forest
Mixed Forest
Woodland
Wooded Grassland
Closed Shrubland
Open Shrubland
Grassland
Cropland
Bare Ground (Desert and Ice)
Urban and Built Up
A-102093-1
© Rand McNally

Source: CIESIN; Hansen et al., 2000

0 200 400 600 800 1000 1200 Miles
0 200 400 600 800 1000 1200 1400 1600 1800 2000 Kilometers

Lambert Azimuthal Equal Area Projection
Scale 1:40,000,000
One inch to 640 miles
One cm to 400 km

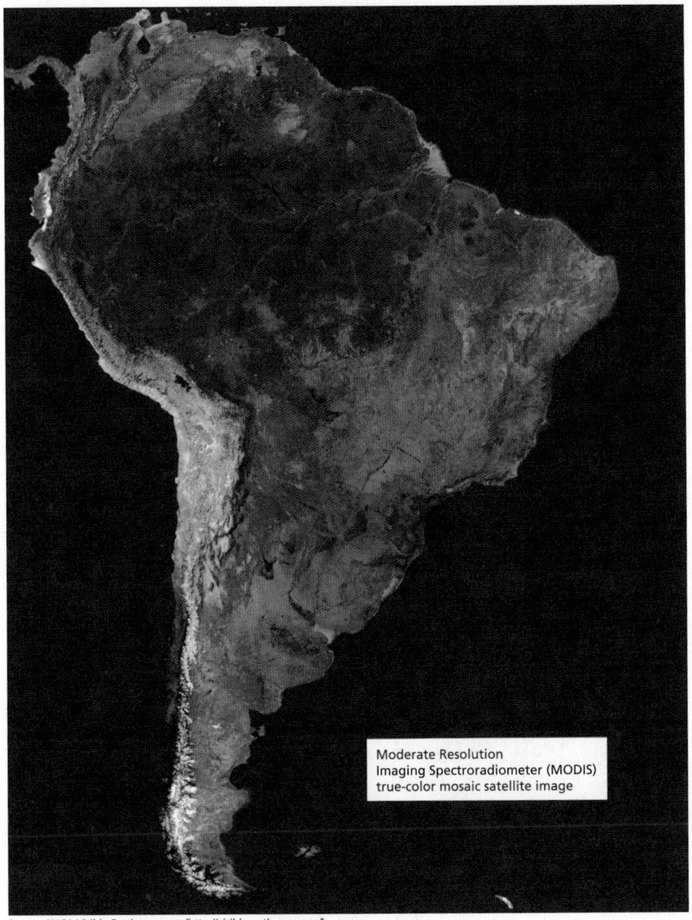
Moderate Resolution
Imaging Spectroradiometer (MODIS)
true-color mosaic satellite image

Source: NASA Visible Earth program (http://visibleearth.nasa.gov/)

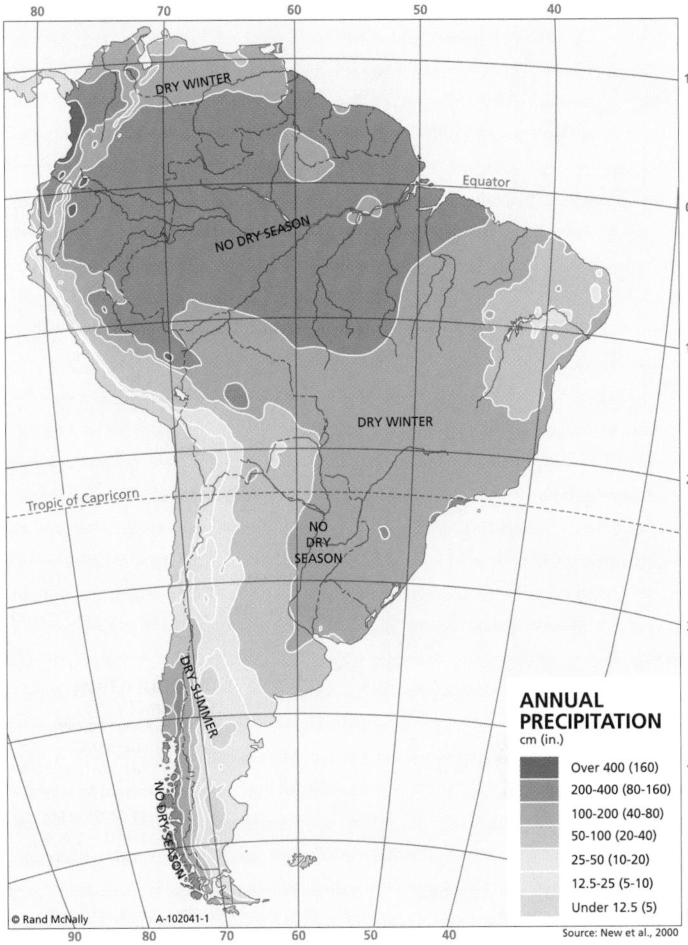
DRY WINTER
Equator
NO DRY SEASON
DRY WINTER
Tropic of Capricorn
NO DRY SEASON
DRY SUMMER
NO DRY SEASON
ANNUAL PRECIPITATION
cm (in.)
Over 400 (160)
200-400 (80-160)
100-200 (40-80)
50-100 (20-40)
25-50 (10-20)
12.5-25 (5-10)
Under 12.5 (5)
© Rand McNally
A-102041-1

Source: New et al., 2000

Equator
Tropic of Capricorn
LANDFORMS
Mountains
Widely spaced mountains
High tablelands
Hills and low tablelands
Depressions or basins
Plains
Limit of continental shelf
© Rand McNally
M-102038-1

Source: Murphy, 1968. © Association of American Geographers. Published by Taylor & Francis. Adapted with permission of the Association of American Geographers.

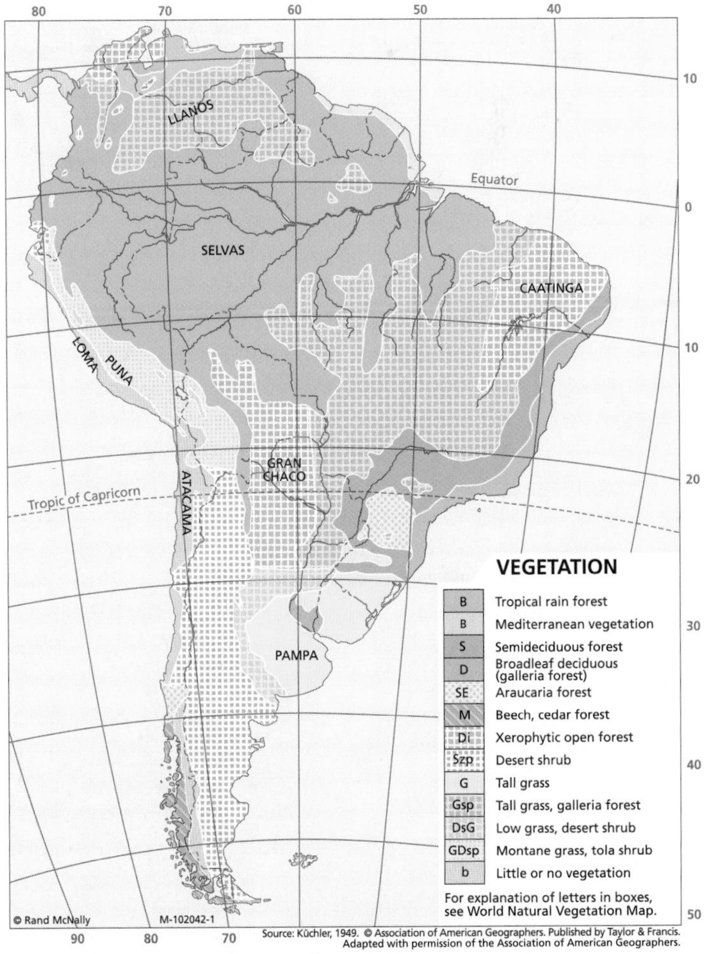
LLANOS
Equator
SELVAS
CAATINGA
LOMA
PUNA
GRAN CHACO
ATACAMA
Tropic of Capricorn
PAMPA
VEGETATION
B Tropical rain forest
B Mediterranean vegetation
S Semideciduous forest
D Broadleaf deciduous (galleria forest)
SE Araucaria forest
M Beech, cedar forest
Di Xerophytic open forest
Szp Desert shrub
G Tall grass
Gsp Tall grass, galleria forest
DsG Low grass, desert shrub
GDsp Montane grass, tola shrub
b Little or no vegetation
For explanation of letters in boxes, see World Natural Vegetation Map.
© Rand McNally
M-102042-1

Source: Küchler, 1949. © Association of American Geographers. Published by Taylor & Francis. Adapted with permission of the Association of American Geographers.

POPULATION DENSITY
per sq. km (per sq. mile)
Over 500 (Over 1,250)
100 - 500 (250 - 1,250)
25 - 100 (62.5 - 250)
10 - 25 (25 - 62.5)
1 -10 (2.5 - 25)
Under 1 (Under 2.5)
Metropolitan area over 10,000,000 population
Metropolitan area 2,000,000 to 10,000,000 population
Caracas
Medellín
Bogotá
Equator
Fortaleza
Recife
Lima
Salvador
Belo Horizonte
Tropic of Capricorn
São Paulo
Rio de Janeiro
Curitiba
Porto Alegre
Santiago
Buenos Aires
© Rand McNally
A-102043-1
Source: U.S. Department of Energy

NATURAL HAZARDS
Volcanoes*
Earthquakes*
Major flood disasters*
Tsunamis
Limit of iceberg drift
Deserts
Areas subject to desertification
*Occurrences since 1900
Equator
Tropic of Capricorn
© Rand McNally
M-102039-1

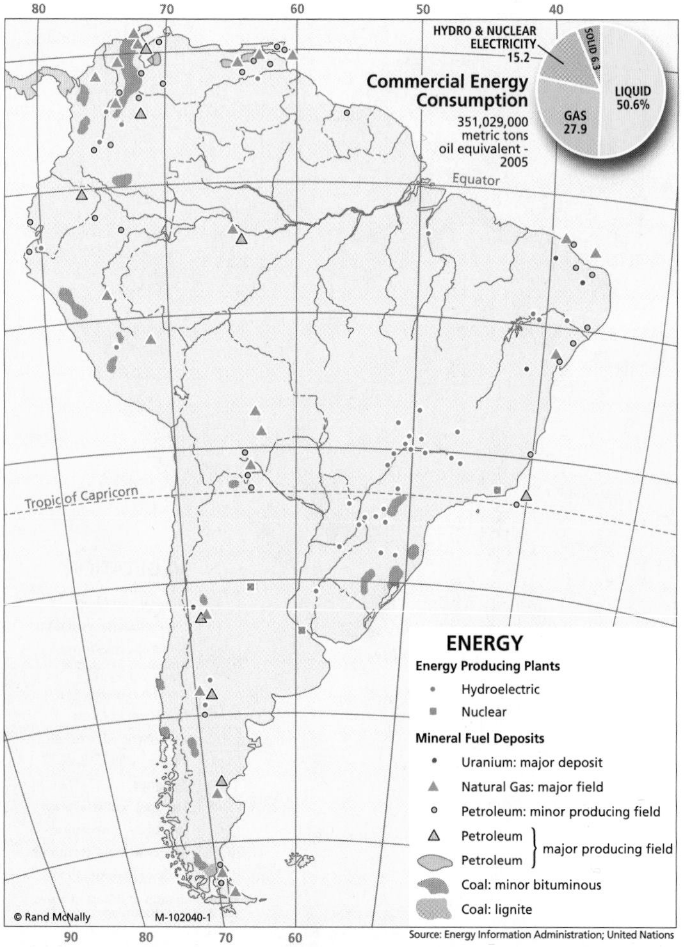

Commercial Energy Consumption
351,029,000 metric tons oil equivalent - 2005
HYDRO & NUCLEAR ELECTRICITY 15.2
SOLID 6.3
LIQUID 50.6%
GAS 27.9
ENERGY
Energy Producing Plants
Hydroelectric
Nuclear
Mineral Fuel Deposits
Uranium: major deposit
Natural Gas: major field
Petroleum: minor producing field
Petroleum
Petroleum
major producing field
Coal: minor bituminous
Coal: lignite
Equator
Tropic of Capricorn
© Rand McNally
M-102040-1
Source: Energy Information Administration; United Nations

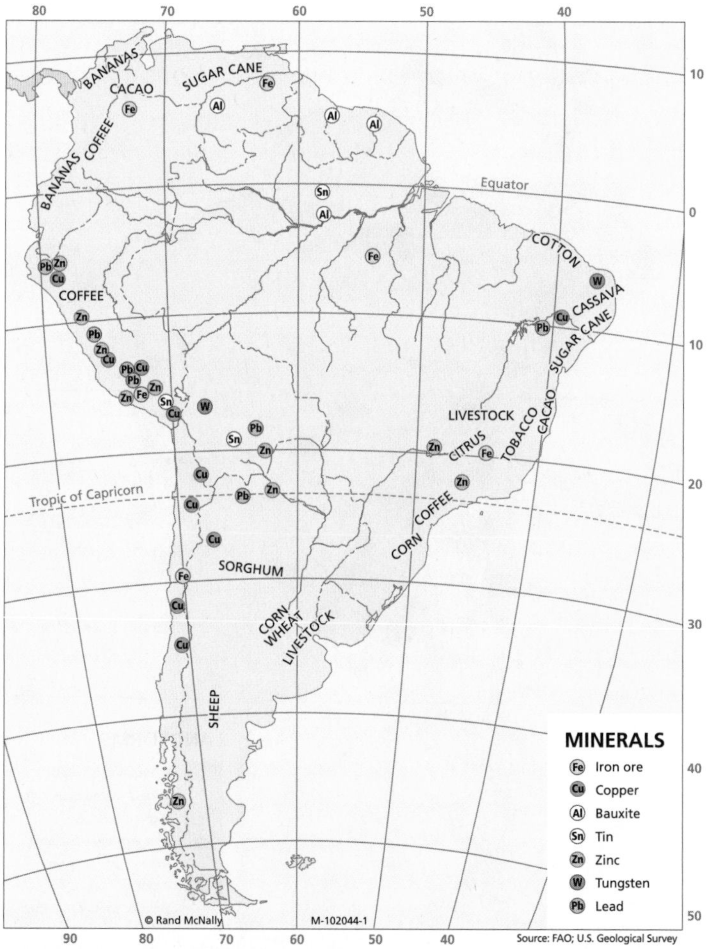

MINERALS
Iron ore
Copper
Bauxite
Tin
Zinc
Tungsten
Lead
BANANAS
CACAO
SUGAR CANE
COFFEE
BANANAS
COTTON
CASSAVA
SUGAR CANE
COFFEE
LIVESTOCK
CITRUS
TOBACCO
CACAO
COFFEE
CORN
SORGHUM
CORN
WHEAT
LIVESTOCK
SHEEP
Equator
Tropic of Capricorn
© Rand McNally
M-102044-1
Source: FAO; U.S. Geological Survey

a

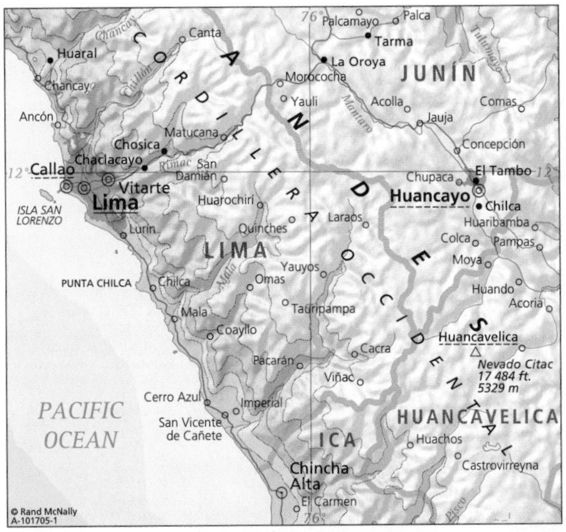
Canta
Palcamayo
Palca
Tarma
Huaral
CORDILLERA OCCIDENTAL
ANDES
La Oroya
Chancay
Morococha
JUNÍN
Yauli
Acolla
Comas
Ancón
Jauja
Matucana
Concepción
Chosica
Chaclacayo
San Damián
Chupaca
El Tambo
Callao
Vitarte
Lima
Huancayo
Chilca
ISLA SAN LORENZO
Huarochirí
Laraos
Huaribamba
Lurín
Quinches
Colca
Pampas
LIMA
Yauyos
Moya
PUNTA CHILCA
Chilca
Omas
Huando
Tauripampa
Acoria
Mala
Coayllo
Huancavelica
Cacra
Nevado Citac 17 484 ft. 5329 m
Pacarán
Viñac
Cerro Azul
Imperial
PACIFIC OCEAN
San Vicente de Cañete
HUANCAVELICA
ICA
Huachos
Chincha Alta
Castrovirreyna
El Carmen
© Rand McNally A-101705-1

b

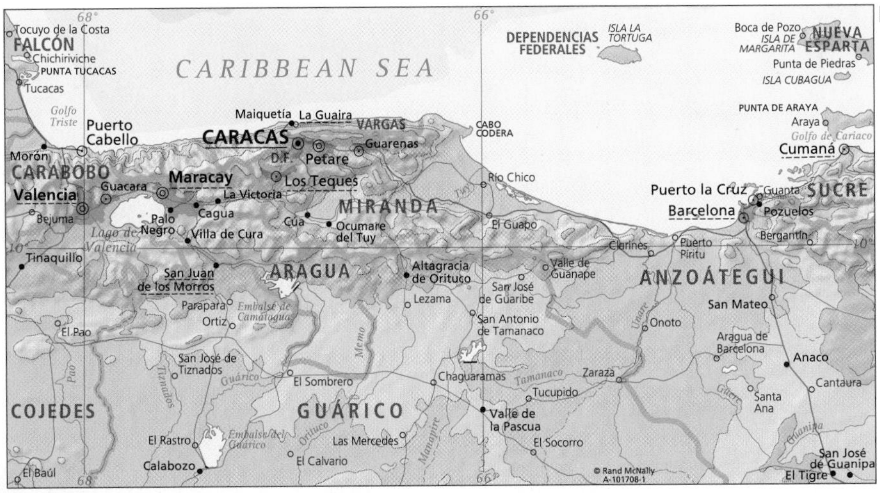
Tocuyo de la Costa
FALCÓN
Chichiriviche
PUNTA TUCACAS
Tucacas
CARIBBEAN SEA
DEPENDENCIAS FEDERALES
ISLA LA TORTUGA
Boca de Pozo
NUEVA ESPARTA
ISLA DE MARGARITA
Punta de Piedras
ISLA CUBAGUA
PUNTA DE ARAYA
Araya
Golfo de Cariaco
Golfo Triste
Puerto Cabello
Maiquetía
La Guaira
VARGAS
CARACAS
Guarenas
CABO CODERA
Cumaná
Morón
CARABOBO
D.F.
Petare
Valencia
Guacara
Maracay
Los Teques
Río Chico
Puerto la Cruz
Guanta
SUCRE
La Victoria
MIRANDA
Barcelona
Pozuelos
Bejuma
Palo Negro
Cagua
Cúa
Ocumare del Tuy
El Guapo
Lago de Valencia
Villa de Cura
Bergantín
Clarines
Puerto Píritu
Tinaquillo
ARAGUA
San Juan de los Morros
Altagracia de Orituco
Valle de Guanape
ANZOÁTEGUI
San José de Guaribe
Parapara
Embalse de Camatagua
Lezama
San Mateo
Ortiz
San Antonio de Tamanaco
Onoto
El Pao
Aragua de Barcelona
San José de Tiznados
Anaco
El Sombrero
Chaguaramas
Zaraza
Cantaura
Tucupido
Santa Ana
COJEDES
GUÁRICO
Valle de la Pascua
El Rastro
Embalse del Guárico
Las Mercedes
El Socorro
Calabozo
El Calvario
San José de Guanipa
El Baúl
El Tigre
© Rand McNally A-101708-1

c

Golfo de Cupica
Páramo Frontino 13 386 ft. 4080 m
Antioquia
Cisneros
Yolombó
Maceo
SERRANÍA DE BAUDÓ
Don Matías
Puerto Berrío
Bello
Barbosa
Urrao
MEDELLÍN
Copacabana
Itagüí
Envigado
Embalse Playas
Embalse del Peñol
SANTANDER
Caldas
Rionegro
ANTIOQUIA
Puerto Boyacá
Vélez
Salgar
Fredonia
Santa Barbara
Ensenada de Tribugá
Bolívar
Jericó
Abejorral
Sonsón
Nuquí
Quibdó
Andes
BOYACÁ
Chiquinquirá
CABO CORRIENTES
Aguadas
Pensilvania
Puerto Salgar
Riosucio
Salamina
La Dorada
La Palma
Ubaté
CHOCÓ
RISARALDA
CALDAS
CUNDINAMARCA
Istmina
Anserma
Neira
Mariquita
Honda
Apía
Manizales
Pacho
Condoto
Cerro Tamaná 13 780 ft. 4200 m
Chinchiná
Nevado del Ruiz 17 717 ft. 5400 m
Fresno
Villeta
Zipaquirá
PACIFIC OCEAN
Pizarro
Armero
Líbano
Chía
La Virginia
Dos Quebradas
Facatativá
Embalse de Tominé
Pereira
Madrid
Cartago
Venadillo
Funza
Quimbaya
Nevado del Tolima 17 110 ft. 5215 m
Soacha
Montenegro
BOGOTÁ
Armenia
Calarcá
Ibagué
Pichimá
Roldanillo
Zarzal
QUINDÍO
El Calvario
Fusagasugá
DIST. CAP.
Bolívar
Caicedonia
Girardot
San Juan
Sevilla
Rovira
Flandes
Melgar
Trujillo
Bugalagrande
TOLIMA
Espinal
Carmen de Apicalá
Tuluá
Guamo
Acacías
Buenaventura
Darién
Buga
San Antonio
Ortega
Villarrica
Cerro Nevado 14 961 ft. 4560 m
Guamal
PUNTA MAGDALENA
Restrepo
Guacarí
Purificación
Golfo de Tortugas
Chaparral
Embalse de Hidroprado
VALLE DEL CAUCA
El Cerrito
Natagaima
Yumbo
Ataco
Palmira
Granada
CALI
Pradera
Dolores
Candelaria
Florida
Colombia
META
Jamundí
Puerto Tejada
Nevado del Huila 18 865 ft. 5750 m
Aipe
Uribe
Baraya
CAUCA
Buenos Aires
Santander
HUILA
Tello
CORDILLERA OCCIDENTAL
CORDILLERA CENTRAL
CORDILLERA ORIENTAL
© Rand McNally A-101706-1

d

Areia Branca
Grossos
São Bento do Norte
Mossoró
Macau
Touros
RIO GRANDE DO NORTE
Pendências
Afonso Bezerra
Pedro Avelino
João Câmara
CABO DE SÃO ROQUE
Maxaranguape
Ipanguaçu
Upanema
Assu
Angicos
Taipu
Ceará-Mirim
Lajes
Natal
Augusto Severo
São Rafael
Macaíba
Santana do Matos
São Paulo do Potengi
Cerro Corá
São Tomé
Monte Alegre
São José do Mipibu
Jucurutu
Florânia
Currais Novos
Arês
Santa Cruz
São José do Campestre
Goianinha
Jardim de Piranhas
Caicó
Cruzeta
Acari
Santo Antônio
Pedro Velho
Canguaretama
Japi
Jardim do Seridó
Picuí
Cuité
Araruna
Tacima
Caiçara
São João do Sabugi
Parelhas
Solânea
Bananeiras
Pirpirituba
Rio Tinto
São Mamede
Santa Luzia
Cubati
Guarabira
Mamanguape
Cabedelo
Areia
Patos
Esperança
Alagoa Grande
Santa Rita
João Pessoa
Pocinhos
Alagoa Nova
Sapé
Bayeux
Teixeira
Taperoá
Campina Grande
Ingá
Desterro
Pedras de Fogo
Itapetim
Serra Branca
PARAÍBA
Itabaiana
Alhandra
Aroeiras
Timbaúba
Aliança
São José do Egito
Sumé
Goiana
São Vicente Ferrer
Nazaré da Mata
Abreu e Lima
Limoeiro
Carpina
Monteiro
Santa Cruz do Capibaribe
Olinda
Camaragibe
Sertânia
Jaboatão
RECIFE
Gravatá
Vitória de Santo Antão
Muribeca dos Guararapes
Poção
Caruaru
Bezerros
Cabo
Pesqueira
Belo Jardim
Amaraji
Escada
Arcoverde
Pedra
Cachoeirinha
Altinho
Bonito
São Bento do Una
Ribeirão
Lajedo
Panelas
Buíque
Palmares
PERNAMBUCO
Maraial
Barreiros
Quipapá
Garanhuns
Canhotinho
Palmeirina
Colônia Leopoldina
Maragoji
São José da Laje
Águas Belas
Correntes
Porto Calvo
Bom Conselho
União dos Palmares
Porto de Pedras
Santana do Ipanema
Murici
São Luís do Quitunde
Quebrângulo
Olho d'Água das Flores
Palmeira dos Índios
Viçosa
Rio Largo
Barra de Santo Antônio
Atalaia
Major Isidoro
ALAGOAS
Pilar
Maceió
Batalha
Arapiraca
Anádia
Boca da Mata
Marechal Deodoro
São Miguel dos Campos
Traipu
Junqueiro
São Brás
Igreja Nova
Itabi
Porto Real do Colégio
Coruripe
Propriá
Penedo
Aquidabã
Neópolis
Piaçabuçu
SERGIPE
Muribeca
Brejo Grande
Ilha das Flores
Nossa Senhora das Dores
Capela
Japaratuba
Maruim
Laranjeiras
ATLANTIC OCEAN
Aracaju
Barra dos Coqueiros
São Cristóvão
Itaporanga d'Ajuda
© Rand McNally A-101709-1

e

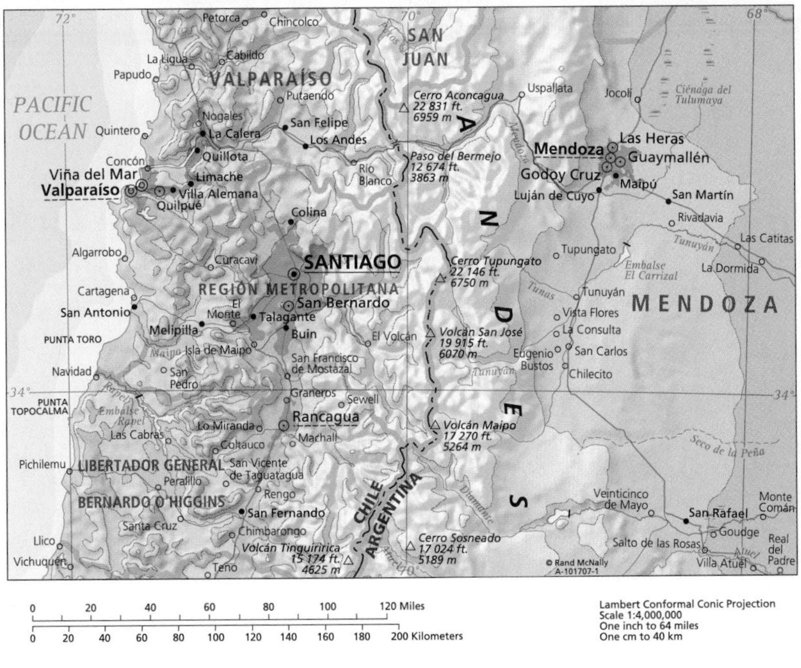
Petorca
Chincolco
SAN JUAN
La Ligua
Cabildo
Papudo
VALPARAÍSO
Putaendo
Uspallata
Jocolí
Ciénaga del Tulumaya
PACIFIC OCEAN
Cerro Aconcagua 22 831 ft. 6959 m
Nogales
Quintero
La Calera
San Felipe
Los Andes
Concón
Quillota
Paso del Bermejo 12 674 ft. 3863 m
Mendoza
Las Heras
Guaymallén
Viña del Mar
Limache
Río Blanco
Godoy Cruz
Valparaíso
Villa Alemana
Maipú
Quilpué
Luján de Cuyo
San Martín
Colina
Rivadavia
ANDES
Tupungato
Las Catitas
Algarrobo
Curacaví
SANTIAGO
Cerro Tupungato 22 146 ft. 6750 m
Embalse El Carrizal
La Dormida
Cartagena
REGIÓN METROPOLITANA
San Bernardo
Tunuyán
MENDOZA
San Antonio
El Monte
Talagante
Vista Flores
Melipilla
Buin
El Volcán
Volcán San José 19 915 ft. 6070 m
La Consulta
PUNTA TORO
Isla de Maipo
Eugenio Bustos
San Carlos
San Francisco de Mostazal
Navidad
San Pedro
Chilecito
PUNTA TOPOCALMA
Graneros
Sewell
Rancagua
Las Cabras
Lo Miranda
Volcán Maipo 17 270 ft. 5264 m
Machalí
Coltauco
Pichilemu
LIBERTADOR GENERAL
San Vicente de Tagua Tagua
Peralillo
Rengo
BERNARDO O'HIGGINS
CHILE
ARGENTINA
Veinticinco de Mayo
Monte Comán
San Fernando
San Rafael
Santa Cruz
Chimbarongo
Goudge
Real del Padre
Lolol
Cerro Sosneado 17 024 ft. 5189 m
Salto de las Rosas
Vichuquén
Volcán Tinguiririca 15 174 ft. 4625 m
Teno
Villa Atuel
© Rand McNally A-101707-1

0 20 40 60 80 100 120 Miles
0 20 40 60 80 100 120 140 160 180 200 Kilometers

Lambert Conformal Conic Projection
Scale 1:4,000,000
One inch to 64 miles
One cm to 40 km

COLOMBIAN BASIN
ISLA DE PROVIDENCIA (Col.)
ISLA DE SAN ANDRÉS (Col.)
CAYOS DEL ESTE SUDESTE
ISLAS DEL MAÍZ (Nic.)
CARIBBEAN SEA
PUNTA GALLINAS
PENÍNSULA DE LA GUAJIRA
PENÍNSULA DE PARAGUANÁ
Oranjestad
ARUBA (Neth.)
NETHERLANDS ANTILLES (Neth.)
CURAÇAO
BONAIRE
Willemstad
LESSER
ISLA BLANQUILLA
ISLA LA ORCHILA
ISLAS LOS ROQUES
ISLA LA TORTUGA
Riohacha
Punto Fijo
Golfo de Venezuela
CABO DE LA AGUJA
Barranquilla
Santa Marta
Ciénaga
Soledad
Maicao
Puerto Cumarebo
Coro
San Juan de los Cayos
Capatárida
San Luis
MARACAIBO
Altagracia
Cabimas
Ciudad Ojeda
Pico Cristóbal Colón 18 947 ft. 5775 m
Valledupar
Villa del Rosario
Sabanalarga
Cartagena
Fundación
Arjona
Siquisique
Puerto Cabello
CARACAS
Guarenas
Puerto la Cruz
San Felipe
Los Teques
Valencia
Maracay
Barcelona
Puerto Limón
COSTA RICA
Portobelo
ISTMO DE PANAMÁ
San Jacinto
San Onofre
Plato
Machiques
Lago de Maracaibo
Carora
Barquisimeto
Yaritagua
Lago de Valencia
San Juan de los Morros
Altagracia de Orituco
Aragua de Barcelona
Colón
Panama Canal
Almirante
Volcán Barú 11 401 ft. 3475 m
Golfo de los Mosquitos
El Carmen de Bolívar
Sincelejo
Magangué
Mompós
Mene Grande
El Tocuyo
Tinaco
San Carlos
Acarigua
Zaraza
Lórica
Sincé
Trujillo
Valera
Embalse del Guárico
Las Mercedes
Valle de la Pascua
La Chorrera
Panamá
Lago Bayano
SERRANÍA DEL DARIÉN
Cereté
Sahagún
El Banco
Encontrados
San Carlos del Zulia
Guanare
Portuguesa
El Baúl
PANAMA
David
Penonomé
Acandí
Montería
Golfo de Urabá
San Marcos
Mérida
Pico Bolívar 16 427 ft. 5007 m
Barinas
Guanarito
Calabozo
Santa María de Ipire
Santiago
Aguadulce
La Palma
ISLA DEL REY
Ayapel
Aguachica
Ocaña
La Fría
CORDILLERA MÉRIDA
Ciudad Bolivia
Arismendi
Ciudad de Nutrias
San Fernando de Apure
Las Bonitas
PUNTA BURICA
Golfo de Chiriquí
Soná
Chitré
Las Tablas
PEN. DE AZUERO
Yaviza
Turbo
Caucasia
Nechí
Simití
Cúcuta
Rubio
San Cristóbal
Apurito
Achaguas
Mantecal
Apure
Arauca
Mapire
ISLA DE COIBA
Golfo de Panamá
Jaqué
PUNTA MALA
PUNTA MARIATO
Chigorodó
Riosucio
Atrato
Puerto Wilches
Pamplona
Bucaramanga
Guasdualito
Cabruta
Cerro Mato 6112 ft. 1863 m
Ituango
Barrancabermeja
Floridablanca
Arauca
Elorza
Capanaparo
La Urbana
Orinoco
PUNTA MARZO
Frontino
Yarumal
Antioquia
Amalfi
Puerto Berrío
Socorro
San Gil
Málaga
Tame
Puerto Rondón
Cravo Norte
Casanare
VENEZUELA
Bello
MEDELLÍN
Urrao
Itagüí
Santa Bárbara
Bolívar
Sonsón
Aguadas
Chiquinquirá
Duitama
Tunja
Sogamoso
Paz de Ariporo
Trinidad
Puerto Carreño
Puerto Ayacucho
Cerro Yaví 8009 ft. 2441 m
Nuquí
Quibdó
CABO CORRIENTES
Salamina
Manizales
La Dorada
Honda
Yopal
Meta
Samariapo
Tomo
Istmina
Pereira
Nev. del Tolima 17 110 ft. 5215 m
Zipaquirá
Orocué
PACIFIC OCEAN
CORDILLERA OCCIDENTAL
CORDILLERA CENTRAL
CORDILLERA ORIENTAL
Cartago
Armenia
Ibagué
Facatativá
BOGOTÁ
Cháviva
Vichada
Sucuaro
Fusagasugá
Girardot
Espinal
Sevilla
Tuluá
Buga
Palmira
Villavicencio
Puerto López
San Fernando de Atabapo
Cerro Marahuaca 8461 ft. 2579 m
ISLA DE MALPELO (Col.)
Buenaventura
Yumbo
CALI
Pradera
Puerto Tejada
Cerro Nevado 14 961 ft. 4560 m
San Martín
Puerto Limón
Guaviare
Uvá
LLANOS
ISLA GORGONA
Nevado del Huila 18 865 ft. 5750 m
Neiva
Uribe
SERRANÍA DE LA MACARENA
COLOMBIA
La Esmeralda
Mosquera
Guapi
Popayán
Timbío
Garzón
San Vicente del Caguán
San José del Guaviare
Inírida
Victorino
Maroa
Casiquiare
Tumaco
Patía
Pitalito
Bolívar
Calamar
Negro
Tigre
San Carlos de Río Negro
Siapa
CABO MANGLARES
Barbacoas
La Unión
Puerto Rico
Florencia
Pico Tamacuarí 7677 ft. 2340 m
El Diviso
Túquerres
Pasto
Miraflores
Içana
Cuiarí
Esmeraldas
PUNTA GALERA
San Lorenzo
Nevado Cumbal 15 630 ft. 4764 m
Tulcán
Ipiales
Vaupés
Mitú
Pico da Neblina 9888 ft. 3014 m
Cucuí
Puerto Asís
Tres Esquinas
Santa Rita
Caquetá
Caguán
Apaporis
Iauaretê
Uaupés
Içana
Atuntaqui
Otavalo
Ibarra
Cayambe
Aguarico
Macujer
Lérida
QUITO
Santo Domingo de los Colorados
CABO PASADO
Bahía de Caráquez
Chone
Cayambe 18 996 ft. 5790 m
Machachi
Puerto Leguízamo
Taraquá
Tapuruquara
Manta
Calceta
CABO SAN LORENZO
Portoviejo
Latacunga
Cotopaxi 19 347 ft. 5897 m
Tena
Puerto Francisco de Orellana
Napo
Nuevo Rocafuerte
Putumayo
Curicuriari
Quevedo
Ambato
Chimborazo 20 702 ft. 6310 m
Baños
Puyo
La Chorrera
La Pedrera
Jipijapa
Paján
Vinces
Guaranda
ECUADOR
Tarqui
Daule
Babahoyo
Riobamba
Sangay 17 159 ft. 5230 m
El Encanto
Japurá
Uneiuxi
Maraã
GUAYAQUIL
Milagro
Curaray
Arica
Japurá
PUNTA SANTA ELENA
Salinas
Santa Elena
Alausí
Macas
Pastaza
Putumayo
Lago Amanã
Golfo de Guayaquil
ISLA PUNÁ
Naranjal
Santa Clotilde
Napo
Corrientes
Tigre
Tonantins
Fonte Boa
Cuenca
Azogues
Tarapacá
Santo Antônio do Içá
Amazon (Solimões)
Içá
Machala
Sígsig
Andoas
Zarumilla
Tumbes
Pasaje
Gualaquiza
Santa Rosa
Pebas
Alvarães
Lago Tefé
Tefé
Zorritos
Saraguro
Intuto
São Paulo de Olivença
Amazon (Amazonas)
Morona
Pastaza
Máncora
Catamayo
Loja
Iquitos
Caballococha
Jutaí
Juruá
Catacocha
Zamora
Leticia
AMAZONAS
Macará
Nauta
San Joaquín de Omaguas
PERU
Talara
PUNTA PARIÑAS
Negritos
Sullana
Cerro Viejo 12 907 ft. 3934 m
Marañón
Yavari
Javari
Benjamin Constant
Concórdia
Carauari
Jutaí
Chulucanas
Paita
Piura
Castilla
Huancabamba
Requena
Bahía de Sechura
Sechura
Cerro Bravo 12 359 ft. 3767 m
ANDES
CORDILLERA CENTRAL
Lagunas
Itui
PERU-CHILE TRENCH
DESIERTO DE SECHURA
Olmos
Jaén
PUNTA NEGRA
Moyobamba
Yurimaguas
Ucayali
SELVAS
Salas
Cutervo
Chachapoyas
Lamas
Tarapoto
Bolognesi
Eirunepé
Juruá
Tapauá
Canutama
Ferreñafe
Lambayeque
Chiclayo
Huallaga
Chepén
Cajamarca
Juanjuí
Bellavista
Orellana
Juruá
Envira
Lábrea
Pacasmayo
Puerto Chicama
Cruzeiro do Sul
Tarauacá
Fortaleza do Ituxi
Pauini
Purus
CHAN CHAN
Trujillo
Salaverry
Tayabamba
ACRE
Feijó
Ituxi
© Rand McNally
A-101694-1

0 50 100 150 200 250 300 Miles

0 50 100 150 200 250 300 350 400 450 500 Kilometers

Lambert Azimuthal Equal Area Projection
Scale 1:10,000,000
One inch to 160 miles
One cm to 100 km

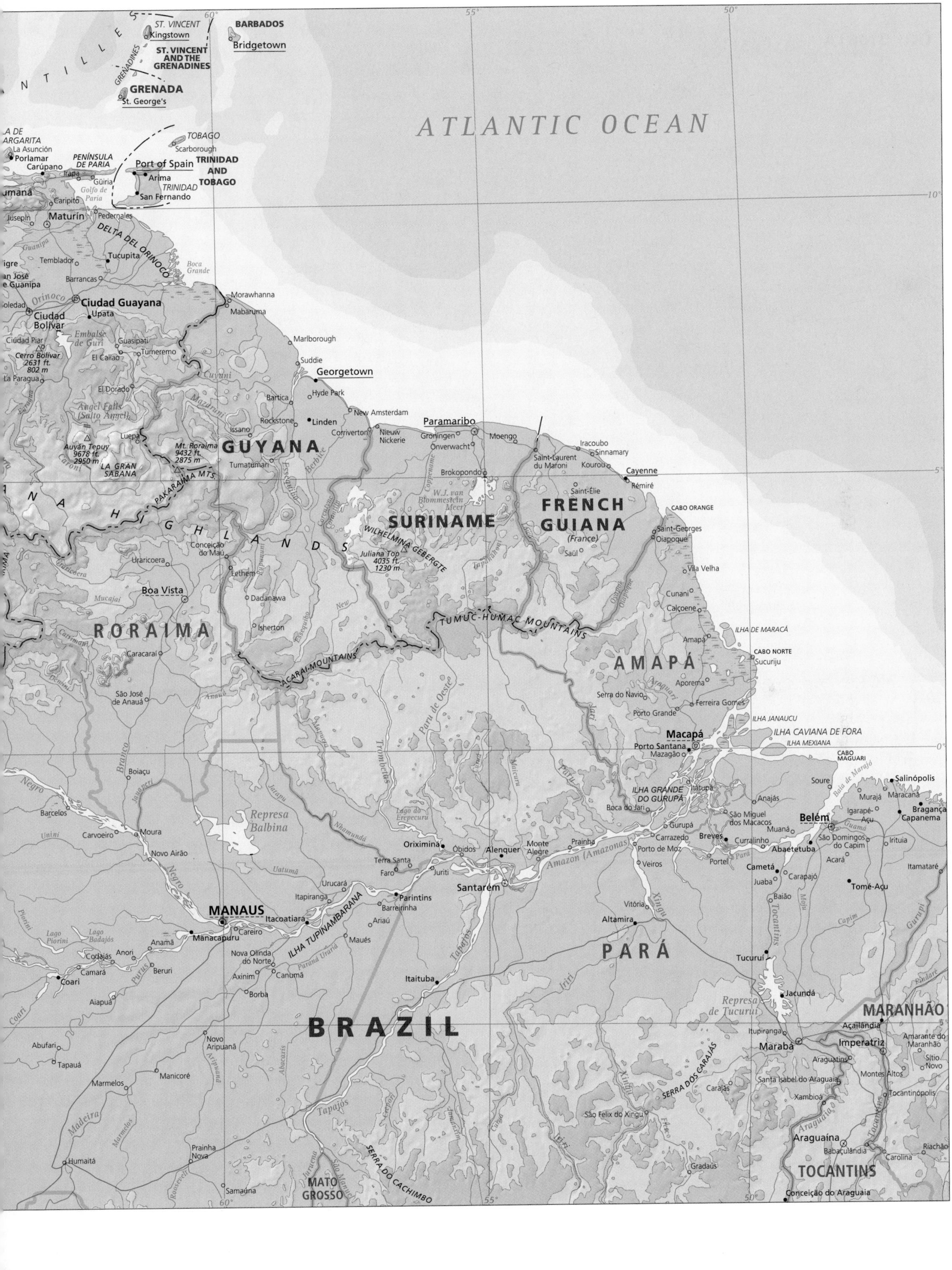
ATLANTIC OCEAN
ST. VINCENT
Kingstown
ST. VINCENT AND THE GRENADINES
BARBADOS
Bridgetown
GRENADINES
GRENADA
St. George's
TOBAGO
Scarborough
Port of Spain
TRINIDAD AND TOBAGO
Arima
TRINIDAD
San Fernando
La Asunción
Porlamar
Carúpano
PENÍNSULA DE PARIA
Irapa
Güiria
Golfo de Paria
Caripito
Maturín
Pedernales
DELTA DEL ORINOCO
Boca Grande
Tucupita
Temblador
Barrancas
Orinoco
Ciudad Guayana
Ciudad Bolívar
Upata
Embalse de Guri
Ciudad Piar
Guasipati
El Callao
Tumeremo
Cerro Bolívar 2631 ft. 802 m
La Paragua
El Dorado
Angel Falls (Salto Angel)
Auyán Tepuy 9678 ft. 2950 m
LA GRAN SABANA
Caroni
Luepa
Mt. Roraima 9432 ft. 2875 m
PAKARAIMA MTS.
Morawhanna
Mabaruma
Marlborough
Suddie
Georgetown
Cuyuni
Mazaruni
Bartica
Hyde Park
Rockstone
Linden
Issano
New Amsterdam
Corriverton
GUYANA
Tumatumari
Essequibo
Berbice
Paramaribo
Nieuw Nickerie
Groningen
Onverwacht
Moengo
Brokopondo
Coppename
W.J. van Blommestein Meer
SURINAME
WILHELMINA GEBERGTE
Juliana Top 4035 ft. 1230 m
Tapanahoni
Iracoubo
Sinnamary
Saint-Laurent du Maroni
Kourou
Cayenne
Rémire
Saint-Élie
FRENCH GUIANA
(France)
Saül
CABO ORANGE
Saint-Georges
Oiapoque
Vila Velha
Cunani
Calçoene
TUMUC-HUMAC MOUNTAINS
ILHA DE MARACÁ
Amapá
CABO NORTE
Sucuriju
AMAPÁ
Aporema
Serra do Navio
Ferreira Gomes
Porto Grande
Jari
ILHA JANAUCU
Macapá
ILHA CAVIANA DE FORA
ILHA MEXIANA
Porto Santana
Mazagão
CABO MAGUARI
Soure
Baía de Marajó
Salinópolis
Maracanã
Bragança
Capanema
Murajá
Igarapé-Açu
ILHA GRANDE DO GURUPÁ
Boca do Jari
Anajás
São Miguel dos Macacos
Belém
Muaná
Gurupá
Carrazedo
Breves
Curralinho
Abaetetuba
São Domingos do Capim
Irituia
Porto de Moz
Veiros
Portel
Cametá
Acará
Itamataré
Juaba
Carapajó
Tomé-Açu
Baião
Moju
Tocantins
Capim
Gurupi
HIGHLANDS
Conceição do Maú
Uraricoera
Lethem
Dadanawa
Boa Vista
Mucajaí
Isherton
RORAIMA
ACARAI MOUNTAINS
Caracaraí
São José de Anauá
Branco
Boiaçu
Jauaperi
Negro
Barcelos
Carvoeiro
Moura
Unini
Novo Airão
Represa Balbina
Uatumã
Jatapu
Nhamundá
Lago do Erepecuru
Trombetas
Paru de Oeste
Oriximiná
Óbidos
Alenquer
Monte Alegre
Prainha
Amazon (Amazonas)
Terra Santa
Faro
Juriti
Santarém
Urucará
Itapiranga
Parintins
Barreirinha
ILHA TUPINAMBARANA
MANAUS
Itacoatiara
Manacapuru
Careiro
Anamã
Anori
Codajás
Camará
Coari
Beruri
Purus
Nova Olinda do Norte
Maués
Ariaú
Axinim
Canumã
Borba
Aiapuã
Itaituba
Tapajós
Vitória
Xingu
Altamira
PARÁ
Tucuruí
Represa de Tucuruí
Jacundá
Iriri
MARANHÃO
Açailândia
Imperatriz
Itupiranga
Marabá
Amarante do Maranhão
Araguatins
Sítio Novo
Montes Altos
Santa Isabel do Araguaia
Tocantinópolis
Xambioá
SERRA DOS CARAJÁS
Carajás
São Felix do Xingu
Fresco
Araguaína
Babaçulândia
Carolina
Riachão
Gradaús
TOCANTINS
Conceição do Araguaia
BRAZIL
Novo Aripuanã
Aripuanã
Abacaxis
Abufari
Tapauá
Marmelos
Manicoré
Madeira
Humaitá
Prainha Nova
Roosevelt
Samaúna
Juruena
MATO GROSSO
SERRA DO CACHIMBO
Jamanxim
60°
55°
50°
10°
5°
0°

COLOMBIA
VENEZUELA
GUYANA
SURINAME
FRENCH GUIANA (Fr.)
GUIANA HIGHLANDS
TUMUC-HUMAC MOUNTAINS
ACARAI MOUNTAINS
RORAIMA
AMAZONAS
SELVAS
ACRE
RONDÔNIA
PERU
BOLIVIA
CHILE
BRA
MATO GROSSO
MATO GROSSO DO SUL
PLANALTO DO MATO GROSSO
ANDES
CORD. OCCIDENTAL
CORD. REAL
ALTIPLANO
DESIERTO DE ATACAMA
PACIFIC OCEAN
GRAN CHACO
SERRA DO CACHIMBO
SERRA DOS APIACÁS
SERRA FORMOSA
SERRA DO NORTE
SERRA DO TOMBADOR
SERRA DOS PACAÁS NOVOS
SERRANÍA DE HUANCHACA
ILHA TUPINAMBARANA
Represa Balbina
Pico Tamacuarí 7677 ft. 2340 m
Pico da Neblina 9888 ft. 3014 m
Nevado Auzangate 20 945 ft. 6384 m
Nevado Culijón 19 199 ft. 5852 m
Nevado Cololo 19 406 ft. 5915 m
Nevado Illampu 21 066 ft. 6421 m
Volcán Misti 19 101 ft. 5822 m
Volcán Tutupaca 19 078 ft. 5815 m
Nevado de Illimani 21 184 ft. 6457 m
Cerro Tunari 16 519 ft. 5035 m
Nevado Sajama 21 463 ft. 6542 m
Cerro Chochís 4232 ft. 1290 m
Lago Titicaca
Lago Poopó
Pantanal de São Lourenço
Pantanal do Rio Negro
Bañados del Izozog
Equator
MANAUS
Porto Velho
Rio Branco
La Paz
El Alto
SANTA CRUZ DE LA SIERRA
Cochabamba
Sucre
Oruro
Arequipa
Juliaca
Puno
Tacna
Arica
Moquegua
Ilo
Cuiabá
Várzea Grande
Rondonópolis
Santarém
Parintins
Itacoatiara
Manacapuru
Tefé
Coari
Leticia
Tabatinga
Benjamin Constant
São Gabriel da Cachoeira
Barcelos
Moura
Novo Airão
Oriximiná
Óbidos
Alenquer
Monte Alegre
Prainha
Itaituba
Humaitá
Lábrea
Boca do Acre
Cruzeiro do Sul
Eirunepé
Feijó
Sena Madureira
Manuel Urbano
Xapuri
Brasiléia
Cobija
Puerto Rico
Abunã
Ariquemes
Ji-Paraná
Vilhena
Guajará-Mirim
Guayaramerín
Riberalta
Trinidad
Santa Ana
Magdalena
San Ignacio de Moxo
Rurrenabaque
Reyes
San Borja
Puerto Maldonado
Puerto Heath
Iñapari
Montero
Cáceres
Corumbá
Puerto Suárez
Roboré
San José de Chiquitos
San Ignacio de Velasco
Concepción
San Matías
Sinop
Alta Floresta
Cachimbo
Diamantino
Nortelândia
Alto Paraguai
Nobres
Rosário Oeste
Barra do Bugres
Primavera do Leste
Poxoréu
Jaciara
Barão de Melgaço
Poconé
Alto Garças
Alto Araguaia
Pedro Gomes
Coxim
Rio Verde de Mato Grosso
Lambert Azimuthal Equal Area Projection
Scale 1:10,000,000
One inch to 160 miles
One cm to 100 km
0 50 100 150 200 250 300 Miles
0 50 100 150 200 250 300 350 400 450 500 Kilometers

ATLANTIC OCEAN
Equator
AMAPÁ
PARÁ
MARANHÃO
CEARÁ
RIO GRANDE DO NORTE
PARAÍBA
PERNAMBUCO
ALAGOAS
SERGIPE
PIAUÍ
TOCANTINS
BAHIA
GOIÁS
DISTRITO FEDERAL
MINAS GERAIS
ESPÍRITO SANTO
ZIL
Macapá
Belém
São Luís
Fortaleza
Teresina
Natal
João Pessoa
RECIFE
Maceió
Aracaju
SALVADOR
Palmas
BRASÍLIA
GOIÂNIA
ILHA DE MARACÁ
CABO NORTE
ILHA JANAUCU
ILHA CAVIANA DE FORA
ILHA MEXIANA
CABO MAGUARI
ILHA GRANDE DO GURUPÁ
Represa de Tucuruí
SERRA DOS CARAJÁS
SERRA DO ESTRONDO
CHAPADA DAS MANGABEIRAS
SERRA GRANDE
CHAPADA DO ARARIPE
Represa de Sobradinho
SERRA DO RONCADOR
SERRA DOURADA
SERRA DO ESPINHAÇO
Pico das Almas 6024 ft. 1836 m
ILHA DE ITAPARICA
PONTA DO SEIXAS
CABO DE SÃO ROQUE
PONTA DA BALEIA
Represa de Três Marias
Represa da Emborcação
Represa de São Simão
Calçoene
Amapá
Sucuriju
Aporema
Ferreira Gomes
Serra do Navio
Porto Grande
Porto Santana
Mazagão
Soure
Salinópolis
Bragança
Capanema
Abaetetuba
Breves
Cametá
Tomé-Açu
Tucuruí
Jacundá
Marabá
Imperatriz
Açailândia
Araguaína
Carolina
Balsas
Caxias
Codó
Bacabal
Pedreiras
Presidente Dutra
Barra do Corda
Grajaú
Floriano
Picos
Oeiras
Parnaíba
Sobral
Crateús
Quixadá
Maracanaú
Juazeiro do Norte
Crato
Mossoró
Caicó
Patos
Campina Grande
Caruaru
Olinda
Garanhuns
Petrolina
Juazeiro
Paulo Afonso
Arapiraca
Feira de Santana
Alagoinhas
Camaçari
Jequié
Vitória da Conquista
Itabuna
Ilhéus
Porto Seguro
Teófilo Otoni
Governador Valadares
Montes Claros
Barreiras
Bom Jesus da Lapa
Guanambi
Brumado
Xique-Xique
Irecê
Jacobina
Porto Nacional
Gurupi
Paracatu
Unaí
Luziânia
Anápolis
Formosa
Uberlândia
Itumbiara
Catalão
Patos de Minas
Rio Verde
Jataí
Barra do Garças
Diamantina
Pirapora
Curvelo
Nanuque
São Mateus
Itapetinga
Canavieiras
Valença
Santo Amaro
Santo Antônio de Jesus
Ribeira do Pombal
Senhor do Bonfim
Euclides da Cunha
Lagarto
Estância
Penedo
Propriá
Palmares
Salgueiro
Ouricuri
Serra Talhada
Arcoverde
Sousa
Cajazeiras
Iguatu
Tianguá
Camocim
Piripiri
Campo Maior
Timon
Chapadinha
Coroatá
Pinheiro
Viana
Colinas
Porangatu
Goianésia
Ceres
Inhumas
Iporá
Mineiros
Morrinhos
Araguari
Patrocínio
João Pinheiro
Januária
Janaúba
Almenara
Itamaraju
Caravelas
50°
45°
40°
35°
0°
5°
10°
15°

BOLIVIA
PERU
CHILE
ARGENTINA
PARAGUAY
URUGUAY
PACIFIC OCEAN
MATO GROSSO
A N D E S
CORDILLERA OCCIDENTAL
CORDILLERA REAL
ALTIPLANO
DESIERTO DE ATACAMA
PUNA DE ATACAMA
GRAN CHACO
CORDILLERA DOMEYKO
PERU-CHILE TRENCH
Tropic of Capricorn
Lago Titicaca
Lago Poopó
SALAR DE UYUNI
SALAR DE ATACAMA
SALINAS GRANDES
SALINAS DE AMBARGASTA
Bañados del Izozog
Pantanal de São Lourenço
Pantanal do Rio Negro
SERRA DA BODOQUENA
SERRA DE MARACAJU
COXILHA DE SANTANA
Embalse Salto Grande
Río de la Plata
Bahía Samborombón
Bañados del Atuel
Laguna Llancanelo
SIERRA DE SAN LUIS
P A M P A
JUJUY
SALTA
FORMOSA
CHACO
TUCUMÁN
SANTIAGO DEL ESTERO
CATAMARCA
LA RIOJA
SAN JUAN
SANTA FE
CORRIENTES
ENTRE RÍOS
CÓRDOBA
SAN LUIS
MENDOZA
BUENOS AIRES
LA PAMPA
Nevado Cololo 19 406 ft. 5915 m
Nevado Coropuna 20 686 ft. 6305 m
Nevado Illampu 21 066 ft. 6421 m
Volcán Misti 19 101 ft. 5822 m
Volcán Tutupaca 19 078 ft. 5815 m
Nevado de Illimani 21 184 ft. 6457 m
Cerro Tunari 16 519 ft. 5035 m
Nevado Sajama 21 463 ft. 6542 m
Cerro Chochís 4232 ft. 1290 m
Cerro Chorolque 18 419 ft. 5614 m
Volcán San Pedro 20 161 ft. 6145 m
Cerro Lípez 19 452 ft. 5929 m
Cerro Coyaguaima 18 596 ft. 5668 m
Volcán Licancábur 19 409 ft. 5916 m
Cerro Púlar 20 449 ft. 6233 m
Volcán Llullaillaco 22 110 ft. 6739 m
Volcán Antofalla 21 027 ft. 6409 m
Cerro Galán 19 396 ft. 5912 m
Nevado Ojos del Salado 22 615 ft. 6893 m
Cerro Bonete Grande 22 546 ft. 6872 m
Cerro General Manuel Belgrano 20 505 ft. 6250 m
Cerro Las Tórtolas 20 735 ft. 6320 m
Cerro Mercedario 22 208 ft. 6769 m
Cerro Aconcagua 22 831 ft. 6959 m
Cerro Tupungato 22 146 ft. 6750 m
Volcán Maipo 17 270 ft. 5264 m
Cerro Nevado 12 575 ft. 3833 m
Volcán Domuyo 15 427 ft. 4702 m
La Paz
El Alto
Cochabamba
SANTA CRUZ DE LA SIERRA
Sucre
Potosí
Oruro
Tarija
Arequipa
Juliaca
Puno
Tacna
Arica
Iquique
Antofagasta
Calama
Asunción
SANTIAGO
Valparaíso
Viña del Mar
Mendoza
San Juan
Córdoba
CÓRDOBA
Rosario
Santa Fe
Paraná
Corrientes
Resistencia
Formosa
Posadas
Salta
San Salvador de Jujuy
San Miguel de Tucumán
Santiago del Estero
San Fernando del Valle de Catamarca
La Rioja
San Luis
Río Cuarto
BUENOS AIRES
La Plata
MONTEVIDEO
Corumbá
Cuiabá
Várzea Grande
Coquimbo
La Serena
Copiapó
Rancagua
Talca
Chillán
Curicó
Concordia
Paysandú
Salto
Uruguaiana
PUNTA ANGAMOS
PUNTA SAN PEDRO
PUNTA MEDIO
CABO BASCUÑÁN
PUNTA LENGUA DE VACA
PUNTA TOPOCALMA
CABO CARRANZA
CABO SAN ANTONIO
15°
20°
25°
30°
35°
60°
65°
70°
75°

0 50 100 150 200 250 300 Miles

0 50 100 150 200 250 300 350 400 450 500 Kilometers

Lambert Azimuthal Equal Area Projection
Scale 1:10,000,000
One inch to 160 miles
One cm to 100 km

BRASÍLIA
DISTRITO FEDERAL
Goiânia
GOIÂNIA
GOIÁS
Anápolis
Luziânia
Formosa
Unaí
Paracatu
Montes Claros
Pirapora
Januária
São Francisco
Janaúba
Rio Verde
Jataí
Mineiros
Itumbiara
Catalão
Uberlândia
Araguari
Patos de Minas
Uberaba
Araxá
BRAZIL
MATO GROSSO DO SUL
Campo Grande
MINAS GERAIS
SERRA DO ESPINHAÇO
BAHIA
Porto Seguro
Itapetinga
Canavieiras
Almenara
Teófilo Otoni
Nanuque
Governador Valadares
ESPÍRITO SANTO
São Mateus
Linhares
Colatina
Vitória
Vila Velha
Guarapari
Cachoeiro de Itapemirim
Ipatinga
Sete Lagoas
BELO HORIZONTE
Divinópolis
Franca
Ribeirão Preto
Barretos
São José do Rio Preto
Catanduva
Araçatuba
Andradina
Presidente Prudente
SÃO PAULO
Marília
Bauru
Araraquara
São Carlos
Rio Claro
Piracicaba
Limeira
Campinas
Jundiaí
Sorocaba
SÃO PAULO
Santo André
Santos
São José dos Campos
Taubaté
Varginha
Poços de Caldas
Juiz de Fora
RIO DE JANEIRO
Campos
Volta Redonda
Barra Mansa
Nova Iguaçu
Niterói
Petrópolis
Teresópolis
Nova Friburgo
Macaé
Pico da Bandeira 9505 ft. 2897 m
Pico das Agulhas Negras 9144 ft. 2787 m
SERRA DA MANTIQUEIRA
CABO DE SÃO TOMÉ
CABO FRIO
ILHA GRANDE
ILHA DE SÃO SEBASTIÃO
Tropic of Capricorn
PARANÁ
Londrina
Maringá
Arapongas
Apucarana
Umuarama
Campo Mourão
Ponta Grossa
Cascavel
Foz do Iguaçu
Guarapuava
CURITIBA
Paranaguá
SERRA DO PARANAPIACABA
SERRA DO MAR
SERRA GERAL
Joinville
Blumenau
Itajaí
SANTA CATARINA
Chapecó
Lages
Florianópolis
ILHA DE SANTA CATARINA
Criciúma
Tubarão
Laguna
RIO GRANDE DO SUL
Passo Fundo
Caxias do Sul
Novo Hamburgo
São Leopoldo
Canoas
PORTO ALEGRE
Santa Maria
Lagoa dos Patos
Pelotas
Rio Grande
Lagoa Mirim
Santa Vitória do Palmar
Cerro Catedral 1686 ft. 514 m
CABO POLONIO
Rocha
Maldonado
BRAZIL BASIN
ATLANTIC OCEAN
© Rand McNally A-101696-1

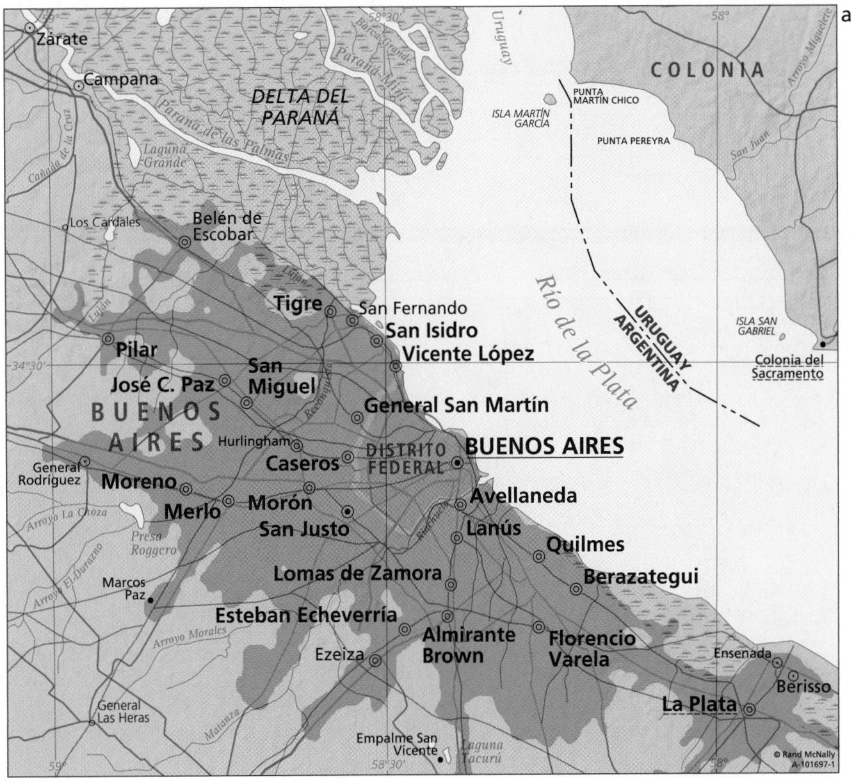
a
Zárate
Campana
DELTA DEL PARANÁ
COLONIA
PUNTA MARTÍN CHICO
ISLA MARTÍN GARCÍA
PUNTA PEREYRA
Belén de Escobar
Los Cardales
Tigre
San Fernando
San Isidro
Vicente López
Río de la Plata
URUGUAY ARGENTINA
ISLA SAN GABRIEL
Colonia del Sacramento
Pilar
José C. Paz
San Miguel
General San Martín
BUENOS AIRES
Hurlingham
Caseros
DISTRITO FEDERAL
BUENOS AIRES
General Rodríguez
Moreno
Merlo
Morón
San Justo
Avellaneda
Lanús
Quilmes
Berazategui
Lomas de Zamora
Marcos Paz
Esteban Echeverría
Almirante Brown
Florencio Varela
Ezeiza
Ensenada
Berisso
La Plata
General Las Heras
Empalme San Vicente
© Rand McNally A-101697-1

Inset map a
Lambert Conformal Conic Projection
Scale 1:1,000,000
One inch to 16 miles
One cm to 10 km

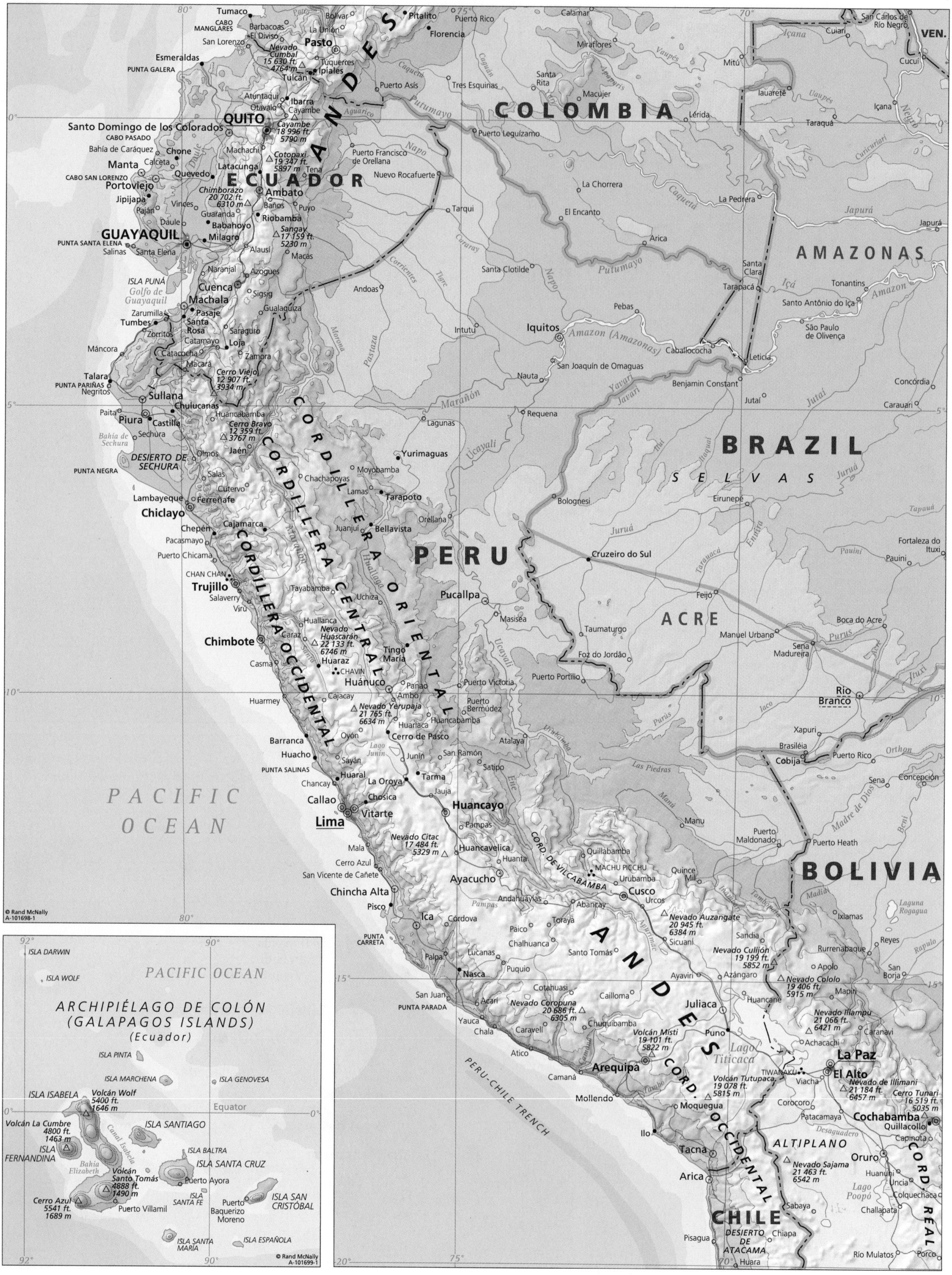
COLOMBIA
ECUADOR
PERU
BRAZIL
BOLIVIA
CHILE
VEN.
AMAZONAS
SELVAS
ACRE
ANDES
CORDILLERA ORIENTAL
CORDILLERA CENTRAL
CORDILLERA OCCIDENTAL
CORD. DE VILCABAMBA
CORD. OCCIDENTAL
CORD. REAL
ALTIPLANO
PACIFIC OCEAN
PERU-CHILE TRENCH
DESIERTO DE SECHURA
DESIERTO DE ATACAMA
Tumaco
CABO MANGLARES
Barbacoas
El Diviso
San Lorenzo
Esmeraldas
PUNTA GALERA
Pasto
Nevado Cumbal 15 630 ft. 4764 m
Túquerres
Ipiales
Tulcán
Atuntaqui
Ibarra
Otavalo
Cayambe
QUITO
Cayambe 18 996 ft. 5790 m
Santo Domingo de los Colorados
CABO PASADO
Bahía de Caráquez
Chone
Manta
Calceta
Machachi
Cotopaxi 19 347 ft. 5897 m
Tena
Latacunga
Quevedo
CABO SAN LORENZO
Portoviejo
Chimborazo 20 702 ft. 6310 m
Ambato
Jipijapa
Vinces
Baños
Puyo
Paján
Guaranda
Riobamba
Daule
Babahoyo
GUAYAQUIL
Milagro
Sangay 17 159 ft. 5230 m
PUNTA SANTA ELENA
Salinas
Santa Elena
Alausí
Macas
Naranjal
Azogues
ISLA PUNÁ
Golfo de Guayaquil
Cuenca
Sígsig
Machala
Pasaje
Zarumilla
Tumbes
Santa Rosa
Gualaquiza
Zorritos
Saraguro
Catamayo
Loja
Máncora
Catacocha
Zamora
Macará
Cerro Viejo 12 907 ft. 3934 m
Talara
PUNTA PARIÑAS
Negritos
Sullana
Chulucanas
Paita
Piura
Castilla
Huancabamba
Cerro Bravo 12 359 ft. 3767 m
Bahía de Sechura
Sechura
Olmos
Jaén
PUNTA NEGRA
Salas
Cutervo
Lambayeque
Ferreñafe
Chiclayo
Chepén
Cajamarca
Pacasmayo
Puerto Chicama
CHAN CHAN
Trujillo
Salaverry
Virú
Chimbote
Casma
Huarmey
Barranca
Huacho
PUNTA SALINAS
Chancay
Huaral
Callao
Lima
Vitarte
Chosica
La Oroya
Tarma
Jauja
Huancayo
Pampas
Mala
Cerro Azul
San Vicente de Cañete
Chincha Alta
Pisco
Ica
PUNTA CARRETA
Córdova
Palpa
Nasca
Lucanas
Puquio
San Juan
PUNTA PARADA
Acarí
Yauca
Chala
Caravelí
Atico
Camaná
Arequipa
Mollendo
Moquegua
Ilo
Tacna
Arica
Pisagua
Chiapa
Huara
Pitalito
Puerto Rico
Florencia
Puerto Asís
Tres Esquinas
Santa Rita
Macujer
Calamar
Miraflores
Mitú
Iauareté
Içana
Cuiari
San Carlos de Río Negro
Cucuí
Lérida
Taraquá
Puerto Leguízamo
Puerto Francisco de Orellana
Nuevo Rocafuerte
Tarqui
La Chorrera
El Encanto
La Pedrera
Japurá
Arica
Santa Clara
Tarapacá
Tonantins
Santo Antônio do Içá
São Paulo de Olivença
Santa Clotilde
Andoas
Pebas
Iquitos
Intutu
Caballococha
Leticia
San Joaquín de Omaguas
Nauta
Benjamin Constant
Requena
Jutaí
Concórdia
Carauari
Lagunas
Yurimaguas
Moyobamba
Chachapoyas
Lamas
Tarapoto
Eirunepé
Bolognesi
Juanjuí
Bellavista
Orellana
Cruzeiro do Sul
Fortaleza do Ituxi
Pauini
Tayabamba
Uchiza
Pucallpa
Masisea
Feijó
Huallanca
Nevado Huascarán 22 133 ft. 6746 m
Caraz
Tingo María
Huaraz
CHAVÍN
Taumaturgo
Manuel Urbano
Sena Madureira
Boca do Acre
Foz do Jordão
Puerto Portillo
Huánuco
Panao
Ambo
Cajacay
Puerto Victoria
Puerto Bermúdez
Nevado Yerupajá 21 765 ft. 6634 m
Huariaca
Huancabamba
Oyón
Cerro de Pasco
Atalaya
Lago Junín
Sayán
Junín
San Ramón
Satipo
Rio Branco
Xapuri
Brasiléia
Cobija
Puerto Rico
Sena
Concepción
Nevado Citac 17 484 ft. 5329 m
Huancavelica
Huanta
Quillabamba
MACHU PICCHU
Urubamba
Ayacucho
Manu
Puerto Maldonado
Puerto Heath
Quince Mil
Andahuaylas
Abancay
Cusco
Urcos
Nevado Auzangate 20 945 ft. 6384 m
Sicuani
Ixiamas
Laguna Rogagua
Toraya
Paico
Chalhuanca
Santo Tomás
Sandia
Nevado Cullijón 19 199 ft. 5852 m
Rurrenabaque
Reyes
Apolo
San Borja
Ayaviri
Azángaro
Nevado Cololo 19 406 ft. 5915 m
Mapiri
Cotahuasi
Cailloma
Nevado Coropuna 20 686 ft. 6305 m
Chuquibamba
Juliaca
Huancané
Nevado Illampu 21 066 ft. 6421 m
Caranavi
Volcán Misti 19 101 ft. 5822 m
Puno
Lago Titicaca
Achacachi
La Paz
TIWANAKU
El Alto
Viacha
Nevado de Illimani 21 184 ft. 6457 m
Volcán Tutupaca 19 078 ft. 5815 m
Corocoro
Cerro Tunari 16 519 ft. 5035 m
Patacamaya
Cochabamba
Quillacollo
Capinota
Oruro
Nevado Sajama 21 463 ft. 6542 m
Huanuni
Uncía
Lago Poopó
Colquechaca
Challapata
Sabaya
Río Mulatos
Porco
Napo
Putumayo
Caquetá
Marañón
Ucayali
Amazon (Amazonas)
Yavari
Juruá
Purus
Madre de Dios
Huallaga
Pastaza
Tigre
Curaray
Corrientes
Morona
Mantaro
Apurímac
Urubamba
Inambari
Tambopata
Madidi
Beni
Tambo
Desaguadero
Içá
Japurá
Jutaí
Tarauacá
Envira
Tapauá
Pauini
Iaco
Acre
Ituxi
Orthon
Las Piedras
Manu
Ene
Pampas
Aguarico
Vaupés
Uaupés
Apaporis
Içana
Negro
Curicuriari
0°
5°
10°
15°
20°
70°
75°
80°
© Rand McNally A-101698-1
a
PACIFIC OCEAN
ARCHIPIÉLAGO DE COLÓN (GALAPAGOS ISLANDS) (Ecuador)
ISLA DARWIN
ISLA WOLF
ISLA PINTA
ISLA MARCHENA
ISLA GENOVESA
ISLA ISABELA
Volcán Wolf 5400 ft. 1646 m
Equator
Volcán La Cumbre 4800 ft. 1463 m
ISLA SANTIAGO
ISLA FERNANDINA
Canal Isabela
ISLA BALTRA
ISLA SANTA CRUZ
Bahía Elizabeth
Volcán Santo Tomás 4888 ft. 1490 m
Puerto Ayora
Cerro Azul 5541 ft. 1689 m
ISLA SANTA FE
Puerto Baquerizo Moreno
ISLA SAN CRISTÓBAL
Puerto Villamil
ISLA SANTA MARÍA
ISLA ESPAÑOLA
92°
90°
0°
© Rand McNally A-101699-1

Inset map a
Lambert Conformal Conic Projection
Scale 1:6,000,000
One inch to 96 miles
One cm to 60 km

0 50 100 150 200 250 300 Miles
0 50 100 150 200 250 300 350 400 450 500 Kilometers

Lambert Azimuthal Equal Area Projection
Scale 1:10,000,000
One inch to 160 miles
One cm to 100 km

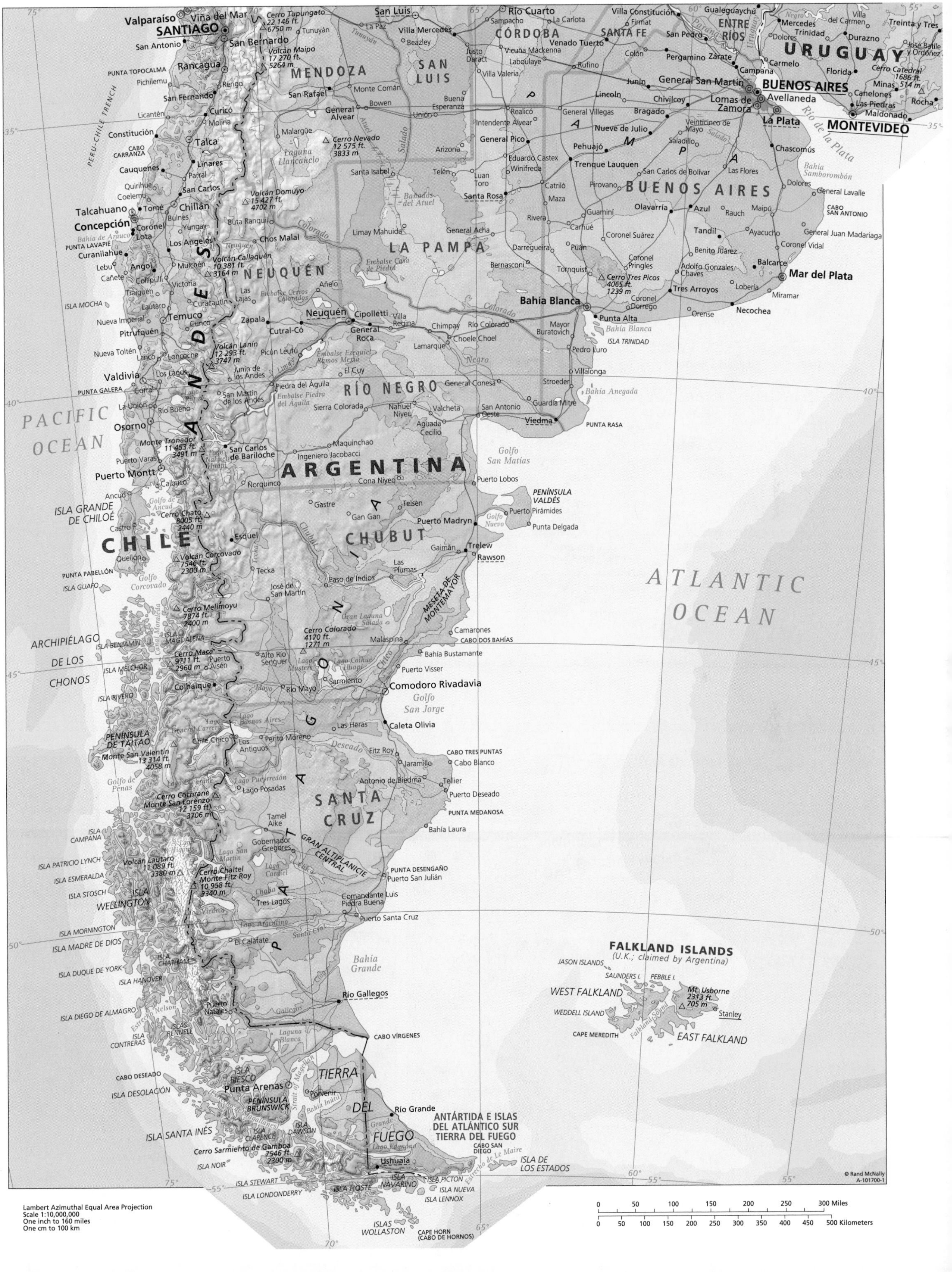
PACIFIC OCEAN
ATLANTIC OCEAN
ARGENTINA
CHILE
URUGUAY
A N D E S
P A M P A
P A T A G O N I A
TIERRA DEL FUEGO
MENDOZA
SAN LUIS
CÓRDOBA
SANTA FE
ENTRE RÍOS
LA PAMPA
NEUQUÉN
RÍO NEGRO
CHUBUT
SANTA CRUZ
BUENOS AIRES
ANTÁRTIDA E ISLAS DEL ATLÁNTICO SUR TIERRA DEL FUEGO
PERU-CHILE TRENCH
SANTIAGO
Valparaíso
Viña del Mar
San Antonio
Rancagua
San Bernardo
Cerro Tupungato 22 146 ft. 6750 m
Volcán Maipo 17 270 ft. 5264 m
Tunuyán
PUNTA TOPOCALMA
Pichilemu
Rengo
San Fernando
Curicó
Licantén
Molina
Constitución
Talca
CABO CARRANZA
Linares
Cauquenes
Parral
Quirihue
Coelemu
San Carlos
Talcahuano
Tomé
Chillán
Concepción
Coronel
Lota
Bulnes
Yungay
Bahía de Arauco
PUNTA LAVAPIÉ
Curanilahue
Los Angeles
Lebu
Angol
Cañete
Collipulli
Mulchén
Victoria
Traiguén
Curacautín
ISLA MOCHA
Lautaro
Temuco
Nueva Imperial
Cunco
Pitrufquén
Nueva Toltén
Lanco
Loncoche
Valdivia
Los Lagos
PUNTA GALERA
Corral
La Unión
Río Bueno
Osorno
Monte Tronador 11 453 ft. 3491 m
Puerto Varas
Puerto Montt
Calbuco
Ancud
Golfo de Ancud
ISLA GRANDE DE CHILOÉ
Cerro Chato 8005 ft. 2440 m
Castro
Quellón
Volcán Corcovado 7546 ft. 2300 m
PUNTA PABELLÓN
ISLA GUAFO
Golfo Corcovado
Cerro Melimoyu 7874 ft. 2400 m
ARCHIPIÉLAGO DE LOS CHONOS
ISLA BENJAMÍN
ISLA MAGDALENA
Cerro Maca 9711 ft. 2960 m
Puerto Aisén
ISLA MELCHOR
Coihaique
ISLA RIVERO
PENÍNSULA DE TAITAO
Chile Chico
Monte San Valentín 13 314 ft. 4058 m
Golfo de Penas
Lago Cochrane
Cerro Cochrane Monte San Lorenzo 12 159 ft. 3706 m
ISLA CAMPANA
ISLA PATRICIO LYNCH
ISLA ESMERALDA
ISLA STOSCH
ISLA WELLINGTON
Volcán Lautaro 11 089 ft. 3380 m
Cerro Chaltel Monte Fitz Roy 10 958 ft. 3340 m
ISLA MORNINGTON
ISLA MADRE DE DIOS
ISLA CHATHAM
ISLA DUQUE DE YORK
ISLA HANOVER
ISLA DIEGO DE ALMAGRO
Puerto Natales
ISLAS RENNELL
Estrecho Nelson
ISLA CONTRERAS
CABO DESEADO
ISLA DESOLACIÓN
ISLA RIESCO
Punta Arenas
PENÍNSULA BRUNSWICK
Strait of Magellan
ISLA SANTA INÉS
ISLA CLARENCE
ISLA DAWSON
Cerro Sarmiento de Gamboa 7546 ft. 2300 m
ISLA NOIR
ISLA STEWART
ISLA LONDONDERRY
ISLA HOSTE
ISLA NAVARINO
ISLA PICTON
ISLA NUEVA
ISLA LENNOX
ISLAS WOLLASTON
CAPE HORN (CABO DE HORNOS)
Ushuaia
Porvenir
Bahía Inútil
Río Grande
CABO SAN DIEGO
Estrecho de Le Maire
ISLA DE LOS ESTADOS
San Luis
La Paz
Villa Mercedes
Beazley
Mendoza
San Rafael
Monte Comán
Bowen
General Alvear
Malargüe
Cerro Nevado 12 575 ft. 3833 m
Laguna Llancanelo
Buena Esperanza
Unión
Arizona
Santa Isabel
Telén
Luan Toro
Bañados del Atuel
Santa Rosa
Volcán Domuyo 15 427 ft. 4702 m
Buta Ranquil
Chos Malal
Limay Mahuida
General Acha
Embalse Casa de Piedra
Volcán Callaquén 10 381 ft. 3164 m
Las Lajas
Embalse Cerros Colorados
Añelo
Neuquén
Cipolletti
Villa Regina
Zapala
Cutral-Có
General Roca
Chimpay
Río Colorado
Choele Choel
Lamarque
Volcán Lanín 12 293 ft. 3747 m
Junín de los Andes
Picún Leufú
Embalse Ezequiel Ramos Mexía
El Cuy
Piedra del Águila
Embalse Piedra del Águila
San Martín de los Andes
Sierra Colorada
Nahuel Niyeu
Valcheta
Aguada Cecilio
General Conesa
San Antonio Oeste
Viedma
Guardia Mitre
Stroeder
San Carlos de Bariloche
Lago Nahuel Huapí
Maquinchao
Ingeniero Jacobacci
Norquinco
Cona Niyeu
Golfo San Matías
Puerto Lobos
PENÍNSULA VALDÉS
Puerto Pirámides
Punta Delgada
Golfo Nuevo
Puerto Madryn
Gastre
Telsen
Gan Gan
Esquel
Trelew
Rawson
Gaimán
Tecka
Las Plumas
Paso de Indios
José de San Martín
MESETA DE MONTEMAYOR
Gran Laguna Salada
Cerro Colorado 4170 ft. 1271 m
Camarones
CABO DOS BAHÍAS
Malaspina
Bahía Bustamante
Alto Río Senguer
Lago Musters
Lago Colhué Huapí
Puerto Visser
Sarmiento
Río Mayo
Comodoro Rivadavia
Golfo San Jorge
Lago Buenos Aires
Las Heras
Caleta Olivia
Los Antiguos
Perito Moreno
Fitz Roy
CABO TRES PUNTAS
Cabo Blanco
Jaramillo
Lago Pueyrredón
Lago Posadas
Antonio de Biedma
Tellier
Puerto Deseado
PUNTA MEDANOSA
Tamel Aike
Bahía Laura
Gobernador Gregores
GRAN ALTIPLANICIE CENTRAL
Lago San Martín
Lago Cardiel
PUNTA DESENGAÑO
Puerto San Julián
Lago Viedma
Tres Lagos
Comandante Luis Piedra Buena
Puerto Santa Cruz
Lago Argentino
El Calafate
Bahía Grande
Río Gallegos
Laguna Blanca
CABO VÍRGENES
Río Cuarto
Sampacho
La Carlota
Villa Constitución
Firmat
Gualeguaychú
Venado Tuerto
Vicuña Mackenna
Justo Daract
Laboulaye
Villa Valeria
Rufino
San Pedro
Colón
Pergamino
Zárate
Campana
Junín
General San Martín
BUENOS AIRES
Avellaneda
Lomas de Zamora
La Plata
Lincoln
Chivilcoy
Bragado
Realicó
General Villegas
Intendente Alvear
Nueve de Julio
Veinticinco de Mayo
General Pico
Pehuajó
Saladillo
Eduardo Castex
Winifreda
Trenque Lauquen
San Carlos de Bolívar
Las Flores
Chascomús
Catriló
Pirovano
Dolores
Bahía Samborombón
General Lavalle
CABO SAN ANTONIO
Maza
Olavarría
Azul
Rauch
Maipú
Guaminí
Rivera
Carhué
Coronel Suárez
Tandil
Ayacucho
General Juan Madariaga
Darregueira
Puán
Benito Juárez
Coronel Vidal
Bernasconi
Tornquist
Cerro Tres Picos 4065 ft. 1239 m
Coronel Pringles
Adolfo Gonzales Chaves
Balcarce
Mar del Plata
Tres Arroyos
Lobería
Miramar
Bahía Blanca
Coronel Dorrego
Orense
Necochea
Punta Alta
ISLA TRINIDAD
Mayor Buratovich
Pedro Luro
Villalonga
Bahía Anegada
PUNTA RASA
Mercedes
Trinidad
Dolores
Carmelo
Durazno
Villa del Carmen
Treinta y Tres
José Batlle y Ordóñez
Cerro Catedral 1686 ft. 514 m
Florida
Minas
Canelones
Las Piedras
Rocha
Maldonado
MONTEVIDEO
Río de la Plata
Río Uruguay
Paraná
FALKLAND ISLANDS (U.K.; claimed by Argentina)
JASON ISLANDS
SAUNDERS I.
PEBBLE I.
WEST FALKLAND
Mt. Usborne 2313 ft. 705 m
WEDDELL ISLAND
Stanley
CAPE MEREDITH
Falkland Sound
EAST FALKLAND
© Rand McNally A-101700-1
0 50 100 150 200 250 300 Miles
0 50 100 150 200 250 300 350 400 450 500 Kilometers

Lambert Azimuthal Equal Area Projection
Scale 1:10,000,000
One inch to 160 miles
One cm to 100 km

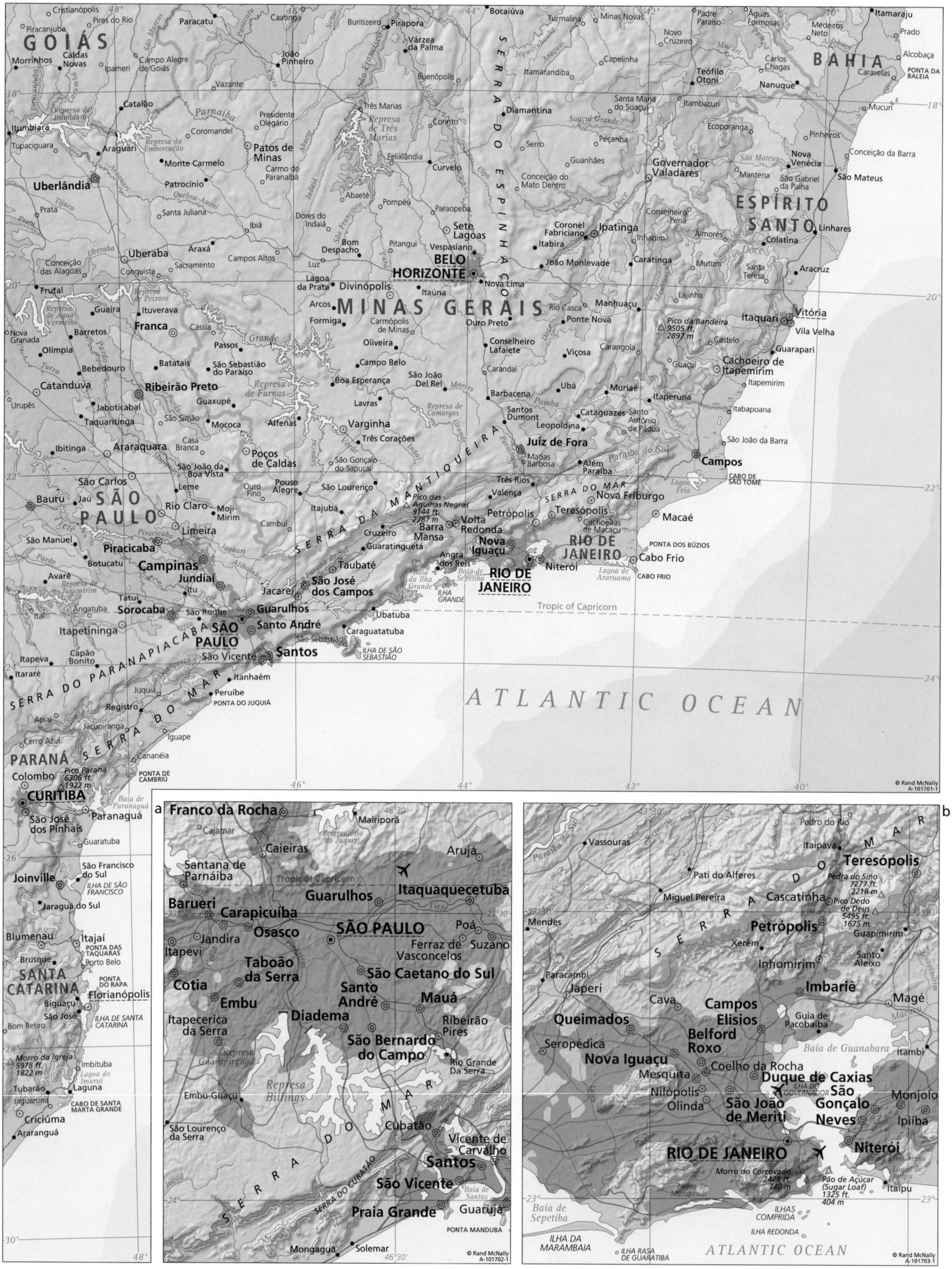
GOIÁS
MINAS GERAIS
SÃO PAULO
RIO DE JANEIRO
ESPÍRITO SANTO
BAHIA
PARANÁ
SANTA CATARINA
ATLANTIC OCEAN
SERRA DO ESPINHAÇO
SERRA DA MANTIQUEIRA
SERRA DO MAR
SERRA DO PARANAPIACABA
Tropic of Capricorn
BELO HORIZONTE
RIO DE JANEIRO
SÃO PAULO
CURITIBA
Vitória
Florianópolis
Cristianópolis
Piracanjuba
Pires do Rio
Morrinhos
Caldas Novas
Ipameri
Campo Alegre de Goiás
Paracatu
Caatinga
Buritizeiro
Pirapora
Várzea da Palma
João Pinheiro
Bocaiúva
Turmalina
Minas Novas
Padre Paraíso
Águas Formosas
Itamaraju
Medeiros Neto
Prado
Alcobaça
Caravelas
PONTA DA BALEIA
Novo Cruzeiro
Carlos Chagas
Teófilo Otoni
Nanuque
Capelinha
Itamarandiba
Buenópolis
Vazante
Catalão
Represa de Itumbiara
Itumbiara
Tupaciguara
Araguari
Coromandel
Presidente Olegário
Três Marias
Represa de Três Marias
Corinto
Diamantina
Santa Maria do Suaçuí
Itambacuri
Mucuri
Ecoporanga
Pinheiros
Conceição da Barra
Represa da Emborcação
Patos de Minas
Monte Carmelo
Carmo do Paranaíba
Felixlândia
Curvelo
Serro
Peçanha
Guanhães
Governador Valadares
Nova Venécia
São Mateus
Mantena
São Gabriel da Palha
Uberlândia
Patrocínio
Prata
Santa Juliana
Ibiá
Abaeté
Dores do Indaiá
Pompéu
Paraopeba
Sete Lagoas
Conceição do Mato Dentro
Conselheiro Pena
Linhares
Coronel Fabriciano
Ipatinga
Inhapim
Aimorés
Colatina
Bom Despacho
Pitangui
Vespasiano
Itabira
Araxá
Uberaba
Campos Altos
Conceição das Alagoas
Conquista
Sacramento
Luz
João Monlevade
Caratinga
Mutum
Santa Teresa
Aracruz
Lagoa da Prata
Divinópolis
Itaúna
Nova Lima
Lajinha
Frutal
Represa de Peixoto
Represa de Água Vermelha
Guaíra
Ituverava
Arcos
Rio Casca
Manhuaçu
Nova Granada
Barretos
Franca
Cássia
Formiga
Carmópolis de Minas
Ouro Preto
Ponte Nova
Pico da Bandeira 9505 ft. 2897 m
Itaquari
Vila Velha
Olímpia
Passos
Oliveira
Conselheiro Lafaiete
Viçosa
Carangola
Castelo
Guarapari
Bebedouro
Batatais
São Sebastião do Paraíso
Campo Belo
Carandaí
Guaçuí
Cachoeiro de Itapemirim
Itapemirim
Catanduva
Ribeirão Preto
Boa Esperança
São João Del Rei
Represa de Furnas
Barbacena
Ubá
Muriaé
Itaperuna
Urupês
Jaboticabal
Guaxupé
Lavras
Represa de Camargos
Taquaritinga
São Simão
Mococa
Alfenas
Varginha
Santos Dumont
Cataguazes
Leopoldina
Santo Antônio de Pádua
Itabapoana
Três Corações
São João da Barra
Ibitinga
Araraquara
Casa Branca
Poços de Caldas
São Gonçalo do Sapucaí
Juiz de Fora
Matias Barbosa
Além Paraíba
Campos
São Carlos
São João da Boa Vista
CABO DE SÃO TOMÉ
Leme
Ouro Fino
Pouso Alegre
São Lourenço
Três Rios
Valença
Bauru
Jaú
Rio Claro
Moji-Mirim
Itajubá
Pico das Agulhas Negras 9144 ft. 2787 m
Petrópolis
Teresópolis
Nova Friburgo
Macaé
Cambuí
Volta Redonda
Barra Mansa
Cachoeiras de Macacu
São Manuel
Piracicaba
Limeira
Cruzeiro
Guaratinguetá
Nova Iguaçu
PONTA DOS BÚZIOS
Botucatu
Campinas
Taubaté
Angra dos Reis
Baía de Sepetiba
Baía da Ilha Grande
Niterói
Cabo Frio
Avaré
Jundiaí
Itu
Jacareí
São José dos Campos
ILHA GRANDE
Lagoa de Araruama
CABO FRIO
Tatuí
Sorocaba
São Roque
Guarulhos
Angatuba
Itaí
Itapetininga
Santo André
Ubatuba
Caraguatatuba
São Sebastião
ILHA DE SÃO SEBASTIÃO
Capão Bonito
Itapeva
São Vicente
Santos
Itararé
Itanhaém
Juquiá
Peruíbe
PONTA DO JUQUIÁ
Registro
Apiaí
Jacupiranga
Cerro Azul
Iguape
Cananéia
PONTA DE CAMBRIU
Colombo
Pico Paraná 6306 ft. 1922 m
Baía de Paranaguá
São José dos Pinhais
Paranaguá
Guaratuba
Joinville
São Francisco do Sul
ILHA DE SÃO FRANCISCO
Jaraguá do Sul
Blumenau
Itajaí
PONTA DAS TAQUARAS
Brusque
Porto Belo
PONTA DO RAPA
Biguaçu
São José
ILHA DE SANTA CATARINA
Bom Retiro
Morro da Igreja 5978 ft. 1822 m
Imbituba
Lagoa do Imaruí
Tubarão
Laguna
Jaguaruna
CABO DE SANTA MARTA GRANDE
Criciúma
Araranguá
© Rand McNally A-101701-1
a
Franco da Rocha
Mairiporã
Cajamar
Reservatório do Juqueri
Caieiras
Arujá
Santana de Parnaíba
Guarulhos
Itaquaquecetuba
Barueri
Carapicuíba
Osasco
SÃO PAULO
Poá
Jandira
Itapevi
Ferraz de Vasconcelos
Suzano
Taboão da Serra
São Caetano do Sul
Cotia
Embu
Santo André
Mauá
Diadema
Itapecerica da Serra
Ribeirão Pires
Represa Guarapiranga
São Bernardo do Campo
Rio Grande Da Serra
Represa Billings
Embu-Guaçu
São Lourenço da Serra
Cubatão
Vicente de Carvalho
Santos
São Vicente
Baía de Santos
SERRA DO CUBATÃO
Praia Grande
Guarujá
PONTA MANDUBA
Mongaguá
Solemar
© Rand McNally A-101702-1
b
Vassouras
Pedro do Rio
Itaipava
Teresópolis
Pati do Alferes
Pedra do Sino 7277 ft. 2218 m
Miguel Pereira
Cascatinha
Pico Dedo de Deus 5495 ft. 1675 m
Mendes
Petrópolis
Guapimirim
Xerém
Santo Aleixo
Inhomirim
Paracambi
Japeri
Imbariê
Magé
Cava
Campos Elísios
Queimados
Guia de Pacobaíba
Belford Roxo
Seropédica
Baía de Guanabara
Itambi
Nova Iguaçu
Coelho da Rocha
Mesquita
Duque de Caxias
Nilópolis
ILHA DO GOVERNADOR
São Gonçalo
Monjolo
Olinda
São João de Meriti
Neves
Ipiíba
Niterói
RIO DE JANEIRO
Morro do Corcovado 2428 ft. 740 m
Pão de Açúcar (Sugar Loaf) 1325 ft. 404 m
Itaipu
Baía de Sepetiba
ILHAS COMPRIDA
ILHA REDONDA
ILHA DA MARAMBAIA
ILHA RASA DE GUARATIBA
ATLANTIC OCEAN
© Rand McNally A-101703-1

0 30 60 90 120 150 180 Miles
0 30 60 90 120 150 180 210 240 270 300 Kilometers

Lambert Conformal Conic Projection
Scale 1:6,000,000
One inch to 96 miles
One cm to 60 km

Inset maps a, b
Lambert Conformal Conic Projection
Scale 1:1,000,000
One inch to 16 miles
One cm to 10 km

SANTIAGO DEL ESTERO
CHACO
GRAN CHACO
PARAGUAY
MISIONES
CORRIENTES
MESOPOTAMIA
BRAZIL
RIO GRANDE DO SUL
SANTA FE
ENTRE RÍOS
URUGUAY
CÓRDOBA
PAMPA
ARGENTINA
BUENOS AIRES
LA PAMPA
RÍO NEGRO
ATLANTIC OCEAN
COXILHA DE SANTANA
CUCHILLA GRANDE
Río de la Plata
BUENOS AIRES
MONTEVIDEO
CÓRDOBA
Rosario
Santa Fe
Paraná
Corrientes
Resistencia
Posadas
Santa Rosa
La Plata
Mar del Plata
Bahía Blanca
Pelotas
Rio Grande
Uruguaiana
Concordia
Salto
Paysandú
Tacuarembó
Rivera
Santana do Livramento
Encarnación
Santa Maria
Alegrete
Bagé
Melo
Jaguarão
Río Cuarto
Villa María
San Nicolás de los Arroyos
Pergamino
Junín
Tandil
Azul
Olavarría
Necochea
Tres Arroyos
Campana
Zárate
Morón
General San Martín
Avellaneda
Lomas de Zamora
Colonia del Sacramento
Punta del Este
Maldonado
Minas
Florida
Durazno
Mercedes
Fray Bentos
Gualeguaychú
Concepción del Uruguay
Goya
Reconquista
Esquina
Rafaela
Cerro Catedral 1686 ft. 514 m
Cerro Tres Picos 4065 ft. 1239 m
Laguna Mar Chiquita
Laguna Merín
Lagoa dos Patos
Lagoa Mirim
Esteros del Iberá
Embalse Salto Grande
Bahía Samborombón
PUNTA PIEDRAS
PUNTA RASA
CABO SAN ANTONIO
CABO CORRIENTES
CABO POLONIO
CABO SANTA MARÍA
ISLA TRINIDAD
© Rand McNally A-101704-1

0 30 60 90 120 150 180 Miles
0 30 60 90 120 150 180 210 240 270 300 Kilometers

Lambert Conformal Conic Projection
Scale 1:6,000,000
One inch to 96 miles
One cm to 60 km

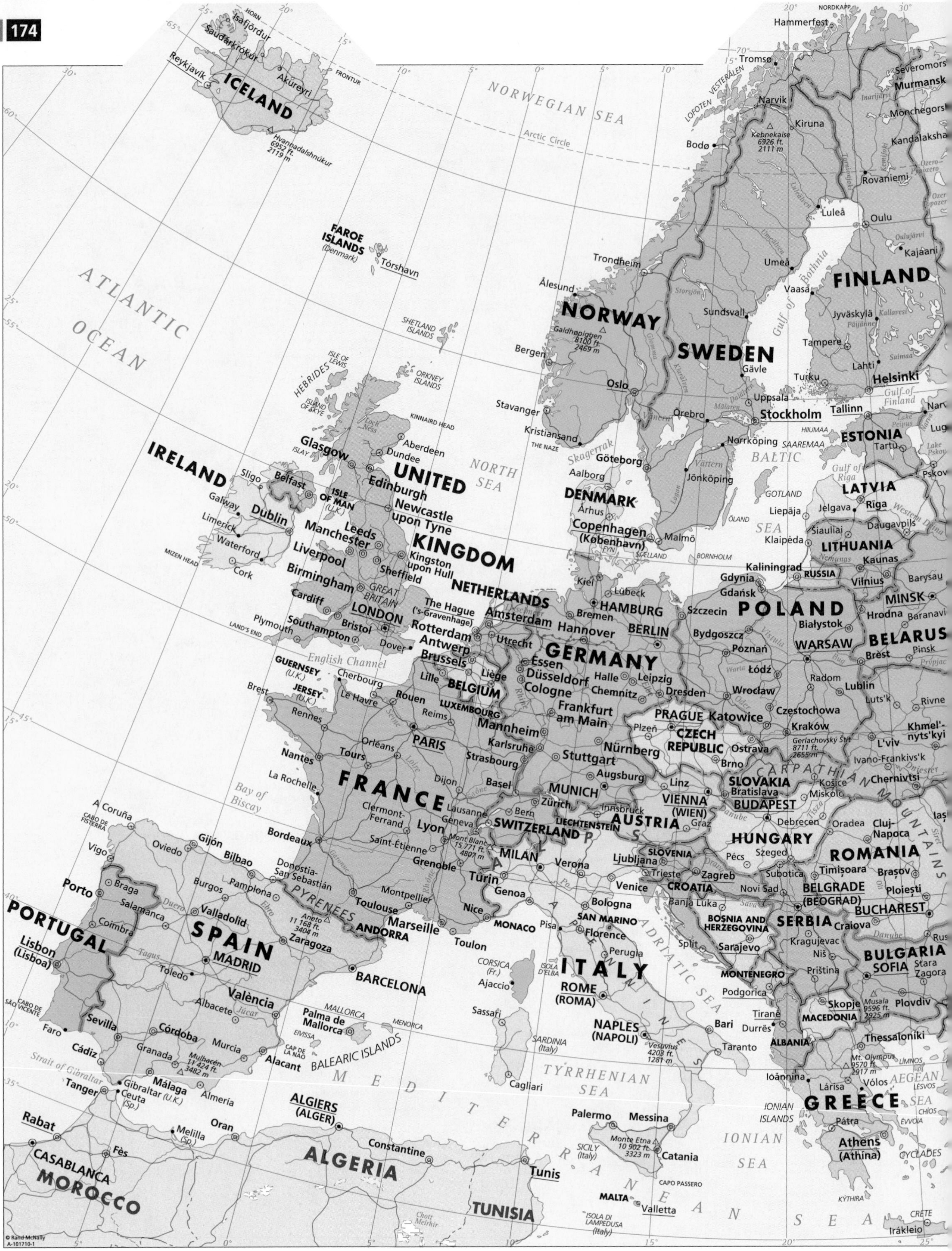
ICELAND
Reykjavík
Ísafjörður
Sauðárkrókur
Akureyri
HORN
FRONTUR
Hvannadalshnúkur
6952 ft.
2119 m
NORWEGIAN SEA
Arctic Circle
ATLANTIC
OCEAN
FAROE ISLANDS
(Denmark)
Tórshavn
SHETLAND ISLANDS
ORKNEY ISLANDS
HEBRIDES
ISLE OF LEWIS
ISLAND OF SKYE
KINNAIRD HEAD
NORTH SEA
IRELAND
Sligo
Belfast
Galway
Dublin
Limerick
Waterford
Cork
MIZEN HEAD
UNITED
KINGDOM
Glasgow
Aberdeen
Dundee
Edinburgh
Newcastle upon Tyne
ISLE OF MAN
(U.K.)
Leeds
Manchester
Liverpool
Kingston upon Hull
Sheffield
Birmingham
GREAT BRITAIN
Cardiff
Bristol
LONDON
Plymouth
Southampton
Dover
LAND'S END
English Channel
GUERNSEY
(U.K.)
JERSEY
(U.K.)
NORWAY
Ålesund
Trondheim
Bergen
Galdhøpiggen
8100 ft.
2469 m
Stavanger
Oslo
Kristiansand
THE NAZE
Skagerrak
Bodø
Tromsø
Hammerfest
NORDKAPP
LOFOTEN
VESTERÅLEN
Narvik
Kiruna
Kebnekaise
6926 ft.
2111 m
SWEDEN
Sundsvall
Gävle
Uppsala
Stockholm
Örebro
Norrköping
Jönköping
Göteborg
Malmö
Luleå
Umeå
Gulf of Bothnia
FINLAND
Oulu
Kajaani
Rovaniemi
Vaasa
Jyväskylä
Tampere
Lahti
Turku
Helsinki
Gulf of Finland
Severomorsk
Murmansk
Monchegorsk
Kandalaksha
Tallinn
ESTONIA
Tartu
HIIUMAA
SAAREMAA
BALTIC
SEA
GOTLAND
ÖLAND
BORNHOLM
Gulf of Riga
LATVIA
Rīga
Liepāja
Jelgava
Daugavpils
Šiauliai
LITHUANIA
Klaipėda
Kaunas
Vilnius
Kaliningrad
RUSSIA
DENMARK
Aalborg
Århus
Copenhagen
(København)
FYN
SJÆLLAND
Kiel
Lübeck
HAMBURG
Bremen
Hannover
BERLIN
NETHERLANDS
Amsterdam
The Hague
('s-Gravenhage)
Rotterdam
Utrecht
Antwerp
Brussels
BELGIUM
Liège
Lille
LUXEMBOURG
GERMANY
Essen
Düsseldorf
Cologne
Halle
Leipzig
Chemnitz
Dresden
Frankfurt am Main
Mannheim
Karlsruhe
Stuttgart
Nürnberg
Augsburg
MUNICH
POLAND
Szczecin
Gdynia
Gdańsk
Bydgoszcz
Poznań
Białystok
WARSAW
Łódź
Wrocław
Radom
Lublin
Częstochowa
Katowice
Kraków
MINSK
Hrodna
Baranavichy
BELARUS
Brest
Pinsk
Barysau
Rivne
Luts'k
Khmel'nyts'kyi
L'viv
Ivano-Frankivs'k
Chernivtsi
Gerlachovský Štít
8711 ft.
2655 m
PRAGUE
Plzeň
CZECH REPUBLIC
Ostrava
Brno
SLOVAKIA
Bratislava
Košice
Miskolc
Linz
VIENNA
(WIEN)
AUSTRIA
Innsbruck
Graz
LIECHTENSTEIN
BUDAPEST
HUNGARY
Debrecen
Pécs
Szeged
CARPATHIAN MOUNTAINS
ROMANIA
Oradea
Cluj-Napoca
Iași
Timișoara
Brașov
Ploiești
BUCHAREST
Craiova
FRANCE
Brest
Cherbourg
Le Havre
Rouen
Reims
Rennes
PARIS
Orléans
Nantes
Tours
Loire
Seine
Dijon
Strasbourg
Basel
La Rochelle
Bay of Biscay
Clermont-Ferrand
Lyon
Saint-Étienne
Grenoble
Bordeaux
Toulouse
Montpellier
Marseille
Toulon
Nice
MONACO
ANDORRA
PYRENEES
Aneto
11 168 ft.
3404 m
Mont Blanc
15 771 ft.
4807 m
SWITZERLAND
Bern
Zürich
Lausanne
Geneva
MILAN
Turin
Genoa
Verona
Venice
Bologna
SAN MARINO
Pisa
Florence
Perugia
ITALY
ROME
(ROMA)
NAPLES
(NAPOLI)
Vesuvius
4203 ft.
1281 m
Bari
Taranto
CORSICA
(Fr.)
Ajaccio
ISOLA D'ELBA
SARDINIA
(Italy)
Sassari
Cagliari
TYRRHENIAN SEA
ADRIATIC SEA
APENNINES
Palermo
Messina
Catania
Monte Etna
10 902 ft.
3323 m
SICILY
(Italy)
CAPO PASSERO
MALTA
Valletta
ISOLA DI LAMPEDUSA
(Italy)
SLOVENIA
Ljubljana
Trieste
CROATIA
Zagreb
Split
Banja Luka
BOSNIA AND HERZEGOVINA
Sarajevo
SERBIA
Novi Sad
Subotica
BELGRADE
(BEOGRAD)
Kragujevac
Niš
Priština
MONTENEGRO
Podgorica
Tiranë
Durrës
ALBANIA
Skopje
MACEDONIA
Musala
9596 ft.
2925 m
BULGARIA
SOFIA
Stara Zagora
Plovdiv
Thessaloniki
Mt. Olympus
9570 ft.
2917 m
Ioánnina
Lárisa
Vólos
GREECE
Pátra
Athens
(Athína)
IONIAN ISLANDS
IONIAN SEA
AEGEAN SEA
LÉSVOS
CHÍOS
ÉVVOIA
CYCLADES
KÝTHIRA
CRETE
Irákleio
PORTUGAL
Porto
Braga
Coimbra
Lisbon
(Lisboa)
CABO DE SÃO VICENTE
Faro
SPAIN
A Coruña
CABO DE FISTERRA
Vigo
Oviedo
Gijón
Bilbao
Donostia-San Sebastián
Pamplona
Burgos
Valladolid
Salamanca
Zaragoza
MADRID
Toledo
Albacete
València
Sevilla
Córdoba
Murcia
Granada
Mulhacén
11 424 ft.
3482 m
Málaga
Cádiz
Almería
Alacant
CAP DE LA NAO
BARCELONA
MALLORCA
Palma de Mallorca
MENORCA
EIVISSA
BALEARIC ISLANDS
Strait of Gibraltar
Gibraltar (U.K.)
Tanger
Ceuta
(Sp.)
Melilla
(Sp.)
MEDITERRANEAN SEA
Rabat
CASABLANCA
Fès
MOROCCO
Oran
ALGIERS
(ALGER)
Constantine
ALGERIA
Tunis
TUNISIA
Chott Melrhir
© Rand McNally
A-101710-1

0 80 160 240 320 400 480 Miles
0 80 160 240 320 400 480 560 640 720 800 Kilometers

Lambert Conformal Conic Projection
Scale 1:16,000,000
One inch to 256 miles
One cm to 160 km

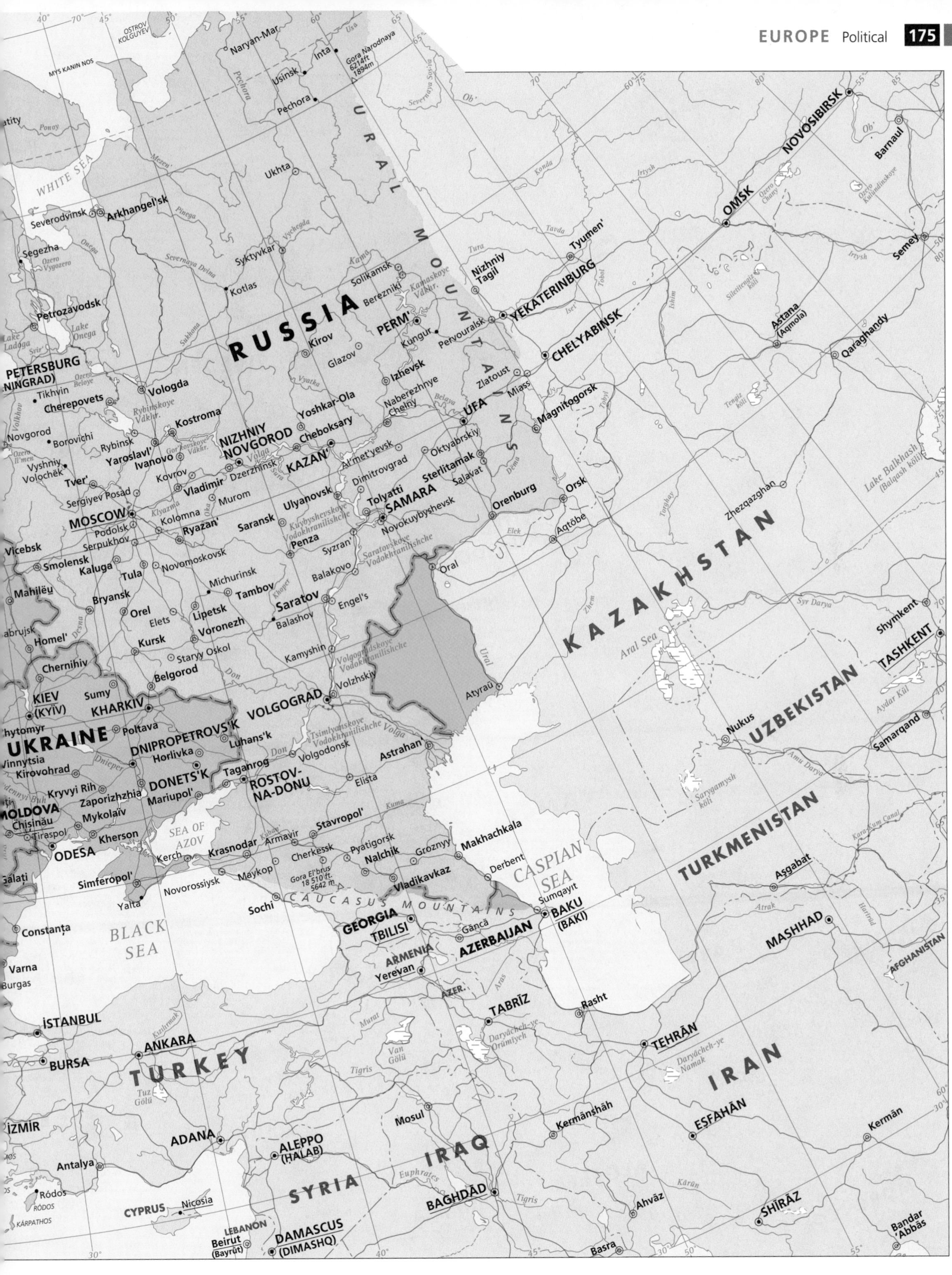
RUSSIA
KAZAKHSTAN
UKRAINE
MOLDOVA
UZBEKISTAN
TURKMENISTAN
GEORGIA
ARMENIA
AZERBAIJAN
AZER.
TURKEY
SYRIA
IRAQ
IRAN
CYPRUS
LEBANON
AFGHANISTAN
U R A L M O U N T A I N S
C A U C A S U S M O U N T A I N S
WHITE SEA
BLACK SEA
SEA OF AZOV
CASPIAN SEA
Aral Sea
Lake Balkhash (Balqash köli)
OSTROV KOLGUYEV
MYS KANIN NOS
Naryan-Mar
Inta
Usinsk
Pechora
Gora Narodnaya 6214ft 1894m
Ukhta
Severodvinsk
Arkhangel'sk
Segezha
Ozero Vygozero
Petrozavodsk
Lake Onega
Lake Ladoga
Syktyvkar
Kotlas
Solikamsk
Berezniki
Kamskoye Vdkhr.
PERM'
Kungur
Kirov
Glazov
Izhevsk
Pervouralsk
YEKATERINBURG
Nizhniy Tagil
Tyumen'
CHELYABINSK
Zlatoust
Miass
Magnitogorsk
UFA
Naberezhnye Chelny
Oktyabrskiy
Sterlitamak
Salavat
Al'met'yevsk
Dimitrovgrad
Tolyatti
SAMARA
Novokuybyshevsk
Orenburg
Orsk
Aqtöbe
Oral
Atyraū
PETERSBURG
NINGRAD)
Tikhvin
Cherepovets
Vologda
Rybinskoye Vdkhr.
Kostroma
Novgorod
Borovichi
Rybinsk
Yaroslavl'
Ivanovo
NIZHNIY NOVGOROD
Cheboksary
Yoshkar-Ola
KAZAN'
Vyshniy Volochëk
Tver'
Kovrov
Vladimir
Dzerzhinsk
Sergiyev Posad
MOSCOW
Kolomna
Murom
Ryazan'
Saransk
Ulyanovsk
Podolsk
Serpukhov
Penza
Syzran'
Vicebsk
Smolensk
Kaluga
Tula
Novomoskovsk
Mahilëu
Bryansk
Michurinsk
Tambov
Orel
Elets
Lipetsk
Voronezh
Saratov
Engel's
Balakovo
Balashov
Homel'
Kursk
Staryy Oskol
Kamyshin
Chernihiv
Belgorod
Volgogradskoye Vodokhranilishche
Volzhskiy
KIEV (KYÏV)
Sumy
KHARKIV
VOLGOGRAD
Zhytomyr
Poltava
DNIPROPETROVS'K
Luhans'k
Vinnytsia
Kirovohrad
Horlivka
DONETS'K
Taganrog
ROSTOV-NA-DONU
Volgodonsk
Astrahan'
Elista
Kryvyi Rih
Zaporizhzhia
Mariupol'
Chişinău
Mykolaiv
Tiraspol
Kherson
ODESA
Stavropol'
Krasnodar
Armavir
Cherkessk
Pyatigorsk
Nalchik
Groznyy
Makhachkala
Kerch
Galați
Simferopol'
Yalta
Novorossiysk
Maykop
Gora El'brus 18 510 ft 5642 m
Vladikavkaz
Derbent
Sochi
Sumqayıt
BAKU (BAKI)
Constanța
GEORGIA TBILISI
Gäncä
Yerevan
Varna
Burgas
İSTANBUL
TABRĪZ
Rasht
BURSA
ANKARA
Van Gölü
Daryācheh-ye Orūmīyeh
TEHRĀN
Daryācheh-ye Namak
Tuz Gölü
İZMİR
Mosul
Kermānshāh
ESFAHĀN
Kermān
ADANA
ALEPPO (HALAB)
Antalya
Ródos
RÓDOS
KÁRPATHOS
Nicosia
BAGHDĀD
Ahvāz
SHĪRĀZ
Bandar 'Abbās
Beirut (Bayrūt)
DAMASCUS (DIMASHQ)
Basra
NOVOSIBIRSK
Barnaul
OMSK
Ozero Chany
Ozero Kulundinskoye
Semey
Astana (Aqmola)
Qaraghandy
Tengiz köli
Siletitengiz köli
Zhezqazghan
Shymkent
TASHKENT
Aydar Kŭl
Samarqand
Nukus
Sarygamysh köli
Aşgabat
MASHHAD
Amu Darya
Syr Darya
Kara-Kum Canal
Atrak
Harirud
Volga
Don
Dnieper
Ural
Kama
Vyatka
Belaya
Sura
Oka
Desna
Khoper
Kuma
Terek
Aras
Murat
Kızılırmak
Tigris
Euphrates
Kārūn
Pechora
Mezen'
Pinega
Vychegda
Severnaya Dvina
Sukhona
Onega
Volkhov
Ob'
Severnaya Sos'va
Konda
Irtysh
Tavda
Tura
Tobol
Iset'
Ishim
Torghay
Emba
Zhem
Elek
Dema
Ponoy
Svir'
Saratovskoye Vodokhranilishche
Kuybyshevskoye Vodokhranilishche
Tsimlyanskoye Vodokhranilishche
Gor'kovskoye Vdkhr.
Ozero Il'men
Ozero Beloye
Apatity
Kabarda

ICELAND
Reykjavík
Hekla 4892 ft. 1491 m
Hvannadalshnúkur 6952 ft. 2119 m
HORN
FRONTUR
NORWEGIAN SEA
Arctic Circle
NORWEGIAN BASIN
FAROE ISLANDS (Denmark)
ATLANTIC OCEAN
SHETLAND ISLANDS
MAINLAND
RONA
CAPE WRATH
ISLE OF LEWIS
HEBRIDES
ORKNEY ISLANDS
Moray Firth
ISLAND OF SKYE
KINNAIRD HEAD
ISLAND OF MULL
JURA
ISLAY
Ben Nevis 4406 ft. 1343 m
Glasgow
IRELAND
ISLE OF MAN (U.K.)
UNITED KINGDOM
GREAT BRITAIN
IRISH SEA
Dublin
Carrauntoohil 3406 ft. 1038 m
MIZEN HEAD
Shannon
Snowdon 3560 ft. 1085 m
Manchester
St. George's Channel
NORTH SEA
London
LAND'S END
WEST EUROPEAN BASIN
English Channel
Strait of Dover
GUERNSEY (U.K.)
JERSEY (U.K.)
NORWAY
FRØYA
HITRA
SMØLA
Trondheim
Snøhetta 7500 ft. 2286 m
Helagsfjället 5892 ft. 1796 m
Storsjön
Galdhøpiggen 8100 ft. 2469 m
Glittertinden 8087 ft. 2465 m
Bergen
Gaustatoppen 6178 ft. 1883 m
Oslo
THE NAZE
Skagerrak
Kattegat
Vänern
Vättern
Mälaren
Göteborg
Stockholm
SWEDEN
VEGA
LOFOTEN
VESTERÅLEN
Vestfjorden
Sulitjelma 6280 ft. 1914 m
Kebnekaise 6926 ft. 2111 m
Torneträsk
Hornavan
L A P L A N D
Hammerfest
SØRØYA
NORDKAPP
RINGVASSØYA
SENJA
Haltiatunturi 4357 ft. 1328 m
Murmansk
Inarijärvi
Oulu
Oulujärvi
Gulf of Bothnia
FINLAND
Kallavesi
Pihlajavesi
Päijänne
Saimaa
Helsinki
St. Petersburg (Leningrad)
Gulf of Finland
ÅLAND
HIIUMAA
SAAREMAA
ESTONIA
Lake Peipus
Gulf of Riga
BALTIC SEA
GOTLAND
LATVIA
Rīga
ÖLAND
LITHUANIA
Nemunas
Western Dvina
DENMARK
Copenhagen
FYN
SJÆLLAND
LOLLAND
BORNHOLM
RÜGEN
Gdańsk
RUSSIA
Minsk
BELARUS
Hamburg
Elbe
NETHERLANDS
IJsselmeer
Amsterdam
Waal
Mittellandkanal
Berlin
POLAND
Vistula
Warta
Oder
Bug
Warsaw
Prypjac'
Brussels
BELGIUM
GERMANY
Rhine
Frankfurt am Main
LUXEMBOURG
Paris
Seine
Marne
Prague
CZECH REPUBLIC
Kraków
Gerlachovský Štít 8711 ft. 2655 m
Dniester
Loire
FRANCE
Bay of Biscay
Danube
Munich
Inn
Vienna (Wien)
SLOVAKIA
CARPATHIAN MOUNTAINS
Zürich
Lake Constance
LIECH.
AUSTRIA
Budapest
SWITZERLAND
Lake Geneva
Saône
Puy de Sancy 6184 ft. 1885 m
Lyon
Mont Blanc 15,771 ft. 4807 m
Matterhorn 14,692 ft. 4478 m
Großglockner 12,457 ft. 3797 m
Triglav 9396 ft. 2864 m
Piz Bernina 13,287 ft. 4050 m
A L P S
HUNGARY
Balaton
ROMANIA
Vârful Moldoveanu 8346 ft. 2544 m
Mureș
TRANSYLVANIAN ALPS
CABO ORTEGAL
CABO DE FISTERRA
Bordeaux
Dordogne
Garonne
MASSIF CENTRAL
Rhône
Milan
Po
SLOVENIA
Zagreb
CROATIA
Drava
Sava
Bilbao
CORDILLERA CANTABRICA
I B E R I A N
P E N I N S U L A
Duero
Ebro
PYRENEES
Aneto 11 168 ft. 3404 m
ANDORRA
Golfe du Lion
MONACO
Marseille
LIGURIAN SEA
SAN MARINO
ADRIATIC SEA
BOSNIA AND HERZEGOVINA
Belgrade (Beograd)
SERBIA
Bucharest
Danube
PORTUGAL
SPAIN
Lisbon (Lisboa)
Tagus
Madrid
Barcelona
CAP CORSE
CORSICA (France)
Monte Cinto 8878 ft. 2706 m
ISOLA D'ELBA
ITALY
A P E N N I N E S
Corno Grande 9560 ft. 2914 m
Rome (Roma)
Naples (Napoli)
Vesuvius 4203 ft. 1281 m
MONTENEGRO
BALKAN
BULGARIA
Sofia
Musala 9596 ft. 2925 m
CABO DE SÃO VICENTE
Guadiana
SIERRA MORENA
Júcar
València
Guadalquivir
MALLORCA
MENORCA
EIVISSA
CAP DE LA NAO
BALEARIC ISLANDS
SARDINIA (Italy)
ALBANIA
MACEDONIA
PENINSULA
THÁSOS
LÍMNOS
Mt. Olympus 9570 ft. 2917 m
PINDOS OROS
Strait of Otranto
Golfo di Taranto
Mulhacén 11 424 ft. 3482 m
Strait of Gibraltar
Gibraltar (U.K.)
TYRRHENIAN SEA
CAPO SPARTIVENTO
IONIAN ISLANDS
GREECE
AEGEAN SEA
LÉSVOS
CHÍOS
ÉVVOIA
M E D I T E R R A N E A N S E A
Algiers (Alger)
Palermo
SICILY
Monte Etna 10 902 ft. 3323 m
IONIAN SEA
Parnassós 8061 ft. 2457 m
Athens (Athína)
CYCLADES
Casablanca
ALGERIA
CAP BON
Tunis
CAPO PASSERO
MOROCCO
Chott ech Chergui
MALTA
ÁKRA TAÍNARO
ÁKRA MALÉAS
KÝTHIRA
SEA OF CRETE
CRETE
ATLAS MOUNTAINS
Irhil M'Goun 13 356 ft. 4071 m
TUNISIA
Chott Melrhir
ISOLA DI LAMPEDUSA (Italy)
© Rand McNally
A-101731-1
0 80 160 240 320 400 480 Miles
0 80 160 240 320 400 480 560 640 720 800 Kilometers

Lambert Conformal Conic Projection
Scale 1:16,000,000
One inch to 256 miles
One cm to 160 km

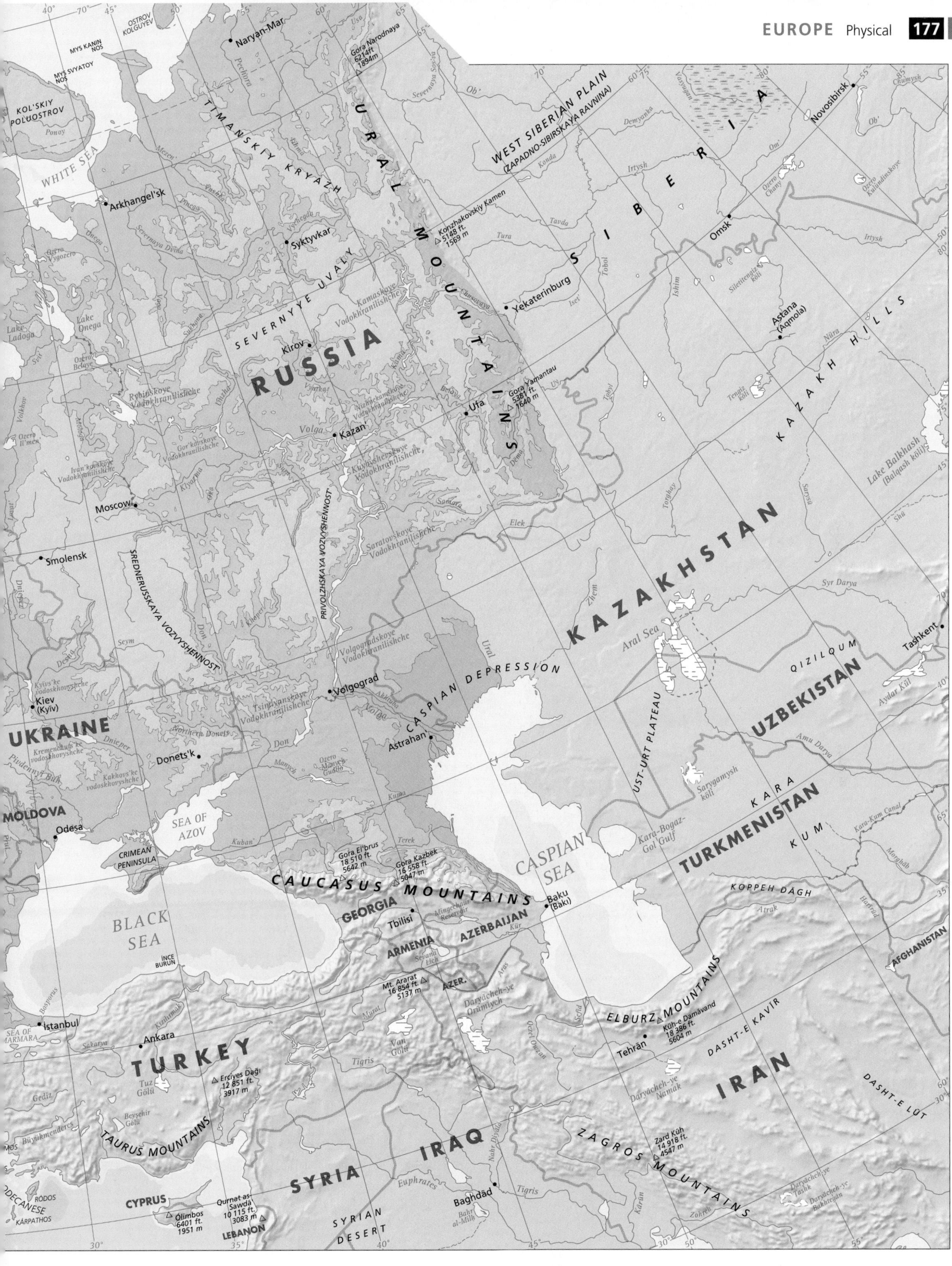
OSTROV KOLGUYEV
Naryan-Mar
Gora Narodnaya 6214ft 1894m
MYS KANIN NOS
MYS SVYATOY NOS
KOL'SKIY POLUOSTROV
Ponoy
WHITE SEA
Pechora
Usa
Severnaya Sos'va
Ob'
TIMANSKIY KRYAZH
URAL MOUNTAINS
WEST SIBERIAN PLAIN (ZAPADNO-SIBIRSKAYA RAVNINA)
SIBERIA
Novosibirsk
Chumysh
Demyanka
Vasyugan
Konda
Irtysh
Om'
Ozero Chany
Ozero Kulundinskoye
Mezen'
Arkhangel'sk
Pinega
Onega
Vychegda
Severnaya Dvina
Syktyvkar
Konzhakovskiy Kamen 5148 ft. 1569 m
Tavda
Tura
Omsk
Ozero Vygozero
Tobol
Ishim
Siletitengiz Köli
Kamaskoye Vodokhranilishche
Chusovaya
Yekaterinburg
Iset'
Astana (Aqmola)
Lake Onega
Lake Ladoga
SEVERNYYE UVALY
Kirov
RUSSIA
Vyatka
Nizhnekamskoye Vodokhranilishche
Belaya
Gora Yamantau 5381 ft. 1640 m
Ufa
Ural
Nura
KAZAKH HILLS
Tengiz Köli
Rybinskoye Vodokhranilishche
Volkhov
Ozero Il'men'
Gor'kovskoye Vodokhranilishche
Volga
Kazan'
Kuybyshevskoye Vodokhranilishche
Ivan'kovskoye Vodokhranilishche
Moscow
Oka
Samara
Lake Balkhash (Balqash köli)
Torghay
Sarysu
Shū
Elek
Smolensk
KAZAKHSTAN
SREDNERUSSKAYA VOZVYSHENNOST'
PRIVOLZHSKAYA VOZVYSHENNOST'
Saratovskoye Vodokhranilishche
Zhem
Syr Darya
Dnieper
Seym
Desna
Don
Khoper
Volgogradskoye Vodokhranilishche
Aral Sea
Tashkent
Kyivs'ke vodoskhovyshche
Kiev (Kyiv)
Volgograd
CASPIAN DEPRESSION
QIZILQUM
UZBEKISTAN
Aydar Kŭl
UKRAINE
Tsimlyanskoye Vodokhranilishche
Northern Donets
Akhtuba
UST-URT PLATEAU
Amu Darya
Kremenchuts'ke vodoskhovyshche
Donets'k
Astrahan
Ozero Manych Gudilo
Manych
Pivdennyy Buh
Kakhovs'ke vodoskhovyshche
Sarygamysh köli
KARA KUM
TURKMENISTAN
MOLDOVA
Odesa
SEA OF AZOV
Kuma
Kara-Kum Canal
Prut
Kuban'
Terek
Kara-Bogaz-Gol Gulf
Murghab
CRIMEAN PENINSULA
Gora El'brus 18 510 ft. 5642 m
Gora Kazbek 16 558 ft. 5047 m
CASPIAN SEA
CAUCASUS MOUNTAINS
Baku (Bakı)
KOPPEH DAGH
BLACK SEA
GEORGIA
Tbilisi
Mingechaur Reservoir
Kür
AZERBAIJAN
Atrak
Harīrūd
ARMENIA
Sevana Lich
AFGHANISTAN
İNCE BURUN
Mt. Ararat 16 854 ft. 5137 m
AZER.
Aras
ELBURZ MOUNTAINS
Bosporus
Daryācheh-ye Orūmīyeh
Kūh-e Damāvand 18 386 ft. 5604 m
DASHT-E KAVĪR
İstanbul
SEA OF MARMARA
Kızılırmak
Sakarya
Ankara
Murat
Van Gölü
Sefīd
Qezel Owzan
Tehrān
TURKEY
Tuz Gölü
Erciyes Daği 12 851 ft. 3917 m
Tigris
Daryācheh-ye Namak
IRAN
DASHT-E LŪT
Gediz
Beyşehir Gölü
Büyükmenderes
TAURUS MOUNTAINS
IRAQ
ZAGROS MOUNTAINS
Zard Kūh 14 918 ft. 4547 m
Nahr Diyālā
SYRIA
Euphrates
Baghdād
RÓDOS
DODECANESE
KÁRPATHOS
CYPRUS
Ólimbos 6401 ft. 1951 m
Qurnat as Sawdā' 10 115 ft. 3083 m
LEBANON
SYRIAN DESERT
Bahr al-Milh
Kārūn
Zohreh
Daryācheh-ye Tashk
Daryācheh-ye Bakhtegān

NORWEGIAN SEA
Arctic Circle
ATLANTIC OCEAN
FAROE ISLANDS
SHETLAND ISLANDS
ORKNEY ISLANDS
HEBRIDES
Trondheim
Gulf of Bothnia
Helsinki
Gulf of Finland
St. Petersbu
Oslo
Stockholm
SAAREMAA
BALTIC SEA
GOTLAND
ÖLAND
Göteborg
NORTH SEA
Glasgow
Dublin
IRISH SEA
Leeds
Manchester
Birmingham
London
Copenhagen
BORNHOLM
Riga
Western Dvina
Minsk
Hamburg
Amsterdam
Elbe
Berlin
Warsaw
Bug
Łódź
Oder
Brussels
English Channel
Essen
Rhine
Frankfurt am Main
Prague
Katowice
Seine
Paris
Mannheim
Loire
Stuttgart
Danube
Munich
CARPATHIAN MOUNTAINS
Dniester
Vienna
Bay of Biscay
Budapest
Bordeaux
Lyon
MASSIF CENTRAL
Rhône
ALPS
Turin
Milan
CORDILLERA CANTÁBRICA
PYRENEES
Ebro
Porto
TRANSYLVANIAN ALPS
Belgrade
Bucharest
Marseille
LIGURIAN SEA
APENNINES
ADRIATIC SEA
Madrid
Tagus
Lisbon
Barcelona
CORSICA
Rome
Sofia
SIERRA MORENA
Sevilla
València
SARDINIA
Naples
BALEARIC ISLANDS
MEDITERRANEAN SEA
TYRRHENIAN SEA
PINDOS OROS
AEGEAN SEA
Strait of Gibraltar
IONIAN ISLANDS
IONIAN SEA
SICILY
Athens
CYCLADES
MALTA
SEA OF CRETE
CRETE
A-102094-1
© Rand McNally

0 80 160 240 320 400 480 Miles
0 80 160 240 320 400 480 560 640 720 800 Kilometers

Lambert Conformal Conic Projection
Scale 1:16,000,000
One inch to 256 miles
One cm to 160 km

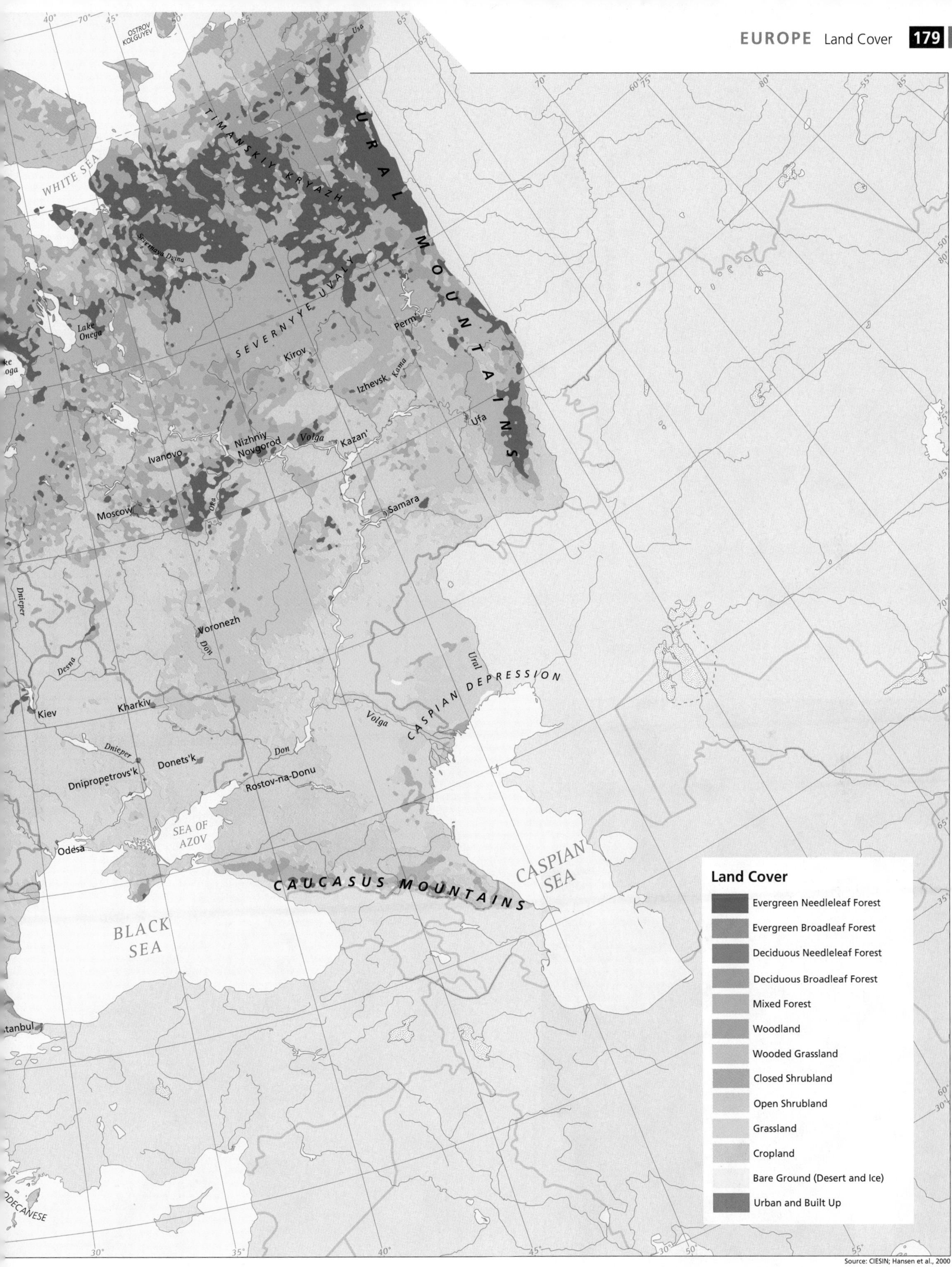
OSTROV KOLGUYEV
WHITE SEA
TIMANSKIY KRYAZH
URAL MOUNTAINS
Usa
Severnaya Dvina
SEVERNYYE UVALY
Lake Onega
Perm'
Kirov
Izhevsk
Kama
Ufa
Nizhniy Novgorod
Volga
Kazan'
Ivanovo
Moscow
Oka
Samara
Dnieper
Voronezh
Don
Desna
Ural
CASPIAN DEPRESSION
Kiev
Kharkiv
Volga
Dnieper
Don
Donets'k
Dnipropetrovs'k
Rostov-na-Donu
SEA OF AZOV
Odesa
CASPIAN SEA
CAUCASUS MOUNTAINS
BLACK SEA
Istanbul
DODECANESE
Land Cover
Evergreen Needleleaf Forest
Evergreen Broadleaf Forest
Deciduous Needleleaf Forest
Deciduous Broadleaf Forest
Mixed Forest
Woodland
Wooded Grassland
Closed Shrubland
Open Shrubland
Grassland
Cropland
Bare Ground (Desert and Ice)
Urban and Built Up

Source: CIESIN; Hansen et al., 2000

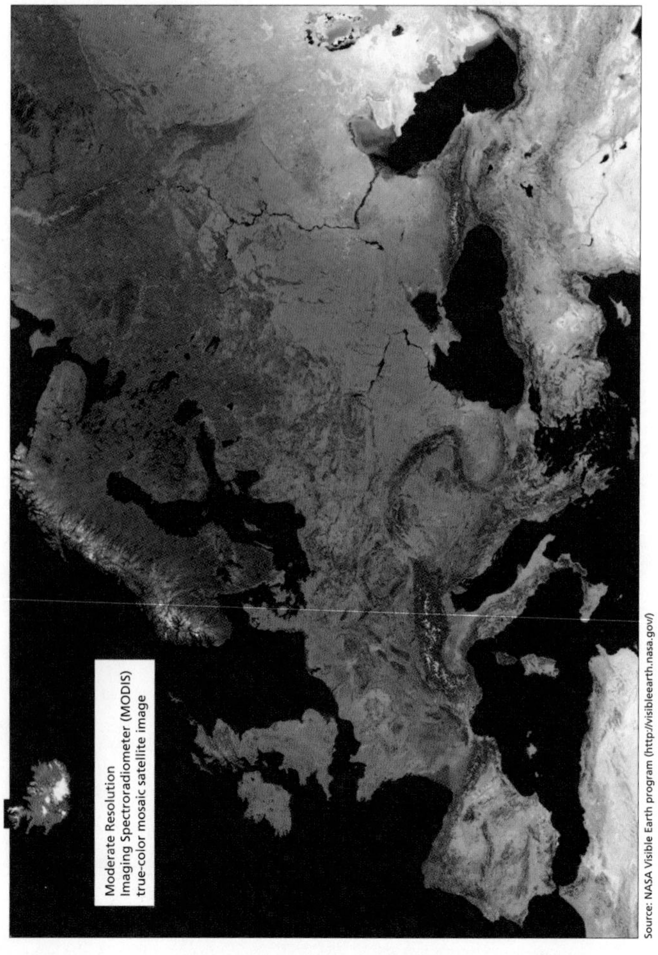
Moderate Resolution
Imaging Spectroradiometer (MODIS)
true-color mosaic satellite image
Source: NASA Visible Earth program (http://visibleearth.nasa.gov/)

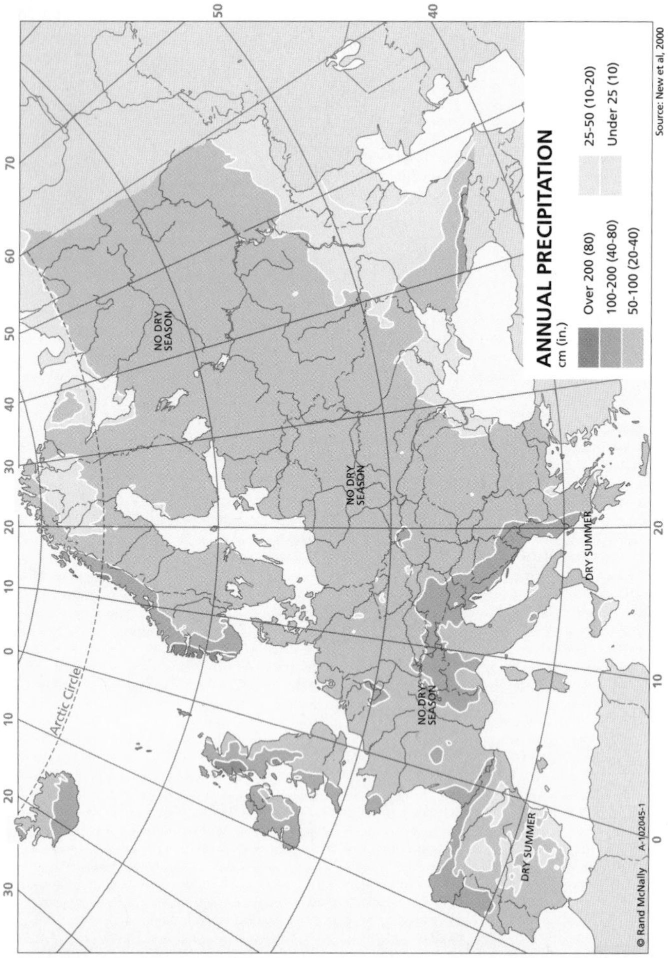
ANNUAL PRECIPITATION
cm (in.)
Over 200 (80)
100-200 (40-80)
50-100 (20-40)
25-50 (10-20)
Under 25 (10)
NO DRY SEASON
DRY SUMMER
Arctic Circle
© Rand McNally A-102045-1
Source: New et al, 2000

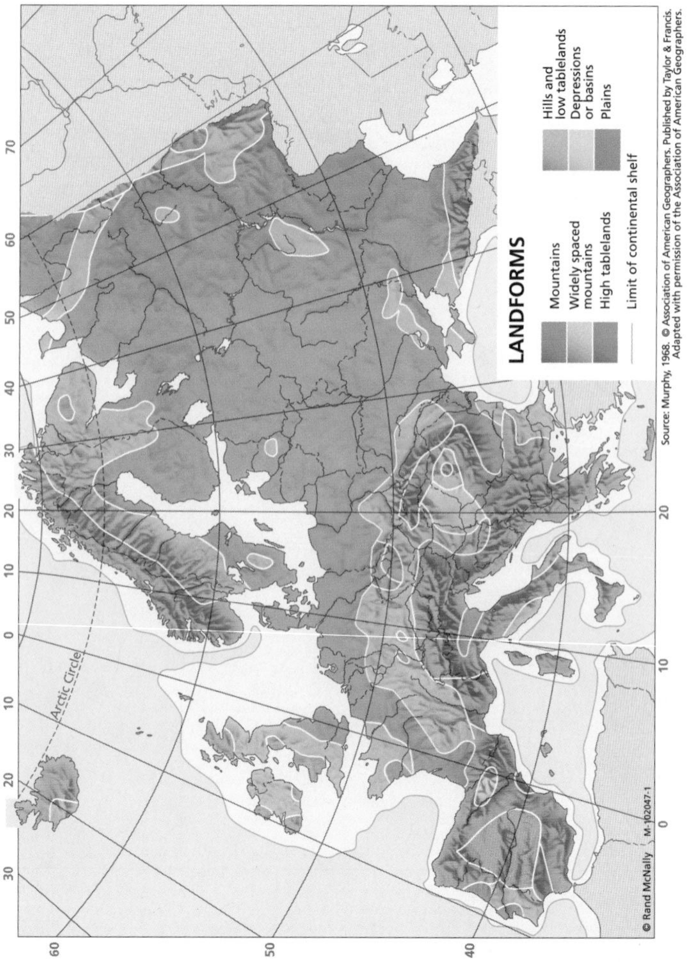
LANDFORMS
Mountains
Widely spaced mountains
High tablelands
Hills and low tablelands
Depressions or basins
Plains
Limit of continental shelf
Arctic Circle
© Rand McNally M-102047-1
Source: Murphy, 1968. © Association of American Geographers. Published by Taylor & Francis. Adapted with permission of the Association of American Geographers.

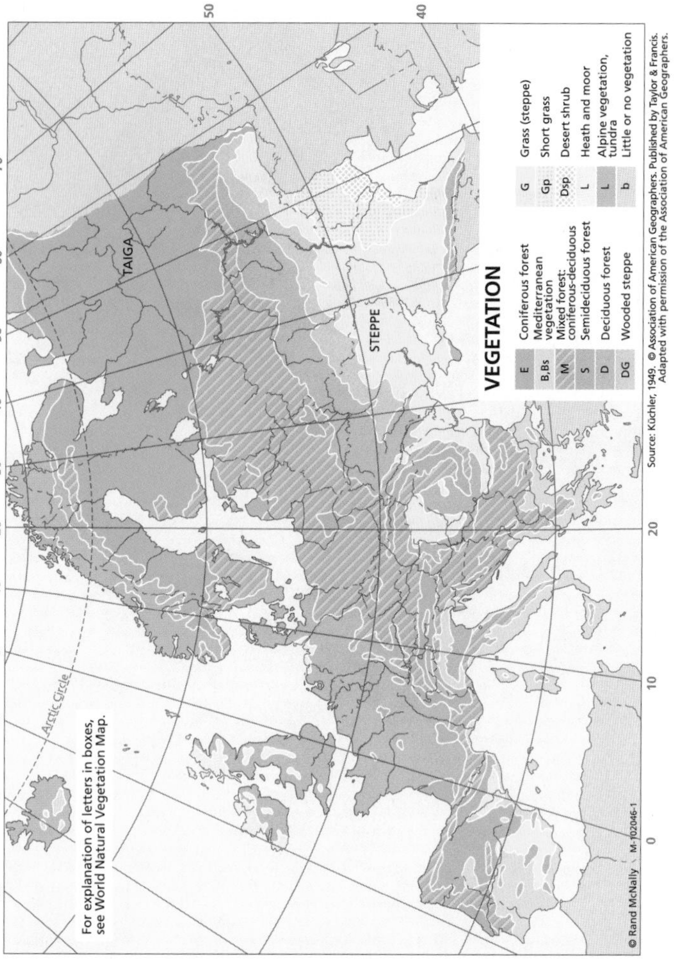
VEGETATION
E Coniferous forest
B,Bs Mediterranean vegetation
M Mixed forest: coniferous-deciduous
S Semideciduous forest
D Deciduous forest
DG Wooded steppe
G Grass (steppe)
Gp Short grass
Dsp Desert shrub
L Heath and moor
L Alpine vegetation, tundra
b Little or no vegetation
TAIGA
STEPPE
For explanation of letters in boxes, see World Natural Vegetation Map.
Arctic Circle
© Rand McNally M-102046-1
Source: Küchler, 1949. © Association of American Geographers. Published by Taylor & Francis. Adapted with permission of the Association of American Geographers.

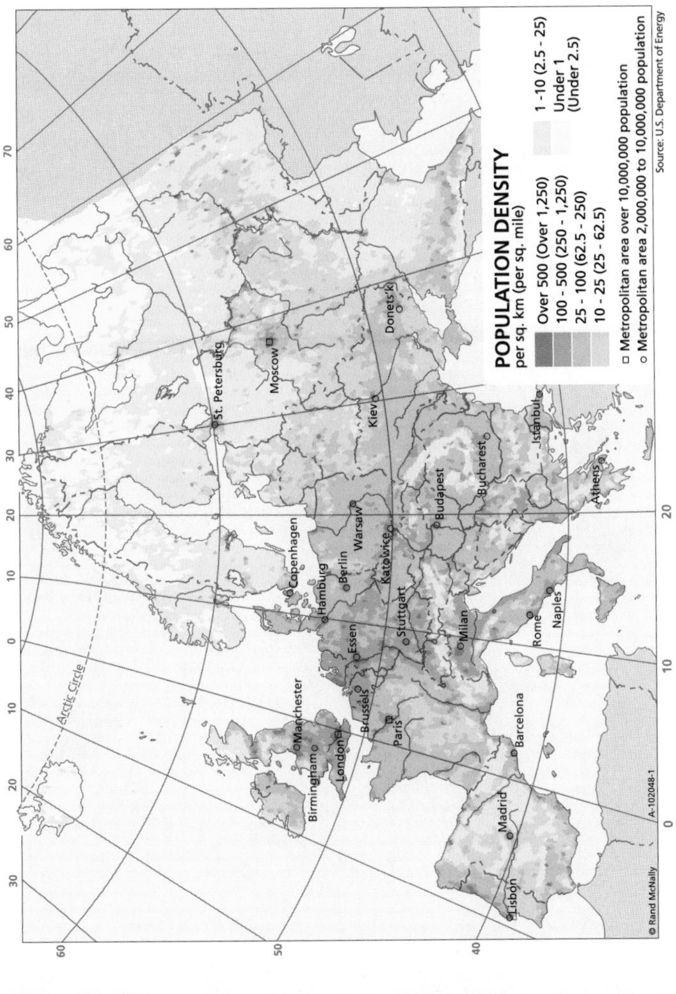
POPULATION DENSITY
per sq. km (per sq. mile)
Over 500 (Over 1,250)
100 - 500 (250 - 1,250)
25 - 100 (62.5 - 250)
10 - 25 (25 - 62.5)
1 -10 (2.5 - 25)
Under 1 (Under 2.5)
□ Metropolitan area over 10,000,000 population
○ Metropolitan area 2,000,000 to 10,000,000 population
Source: U.S. Department of Energy
Arctic Circle
St. Petersburg
Moscow
Kiev
Donetsk
Copenhagen
Hamburg
Berlin
Warsaw
Katowice
Essen
Stuttgart
Milan
Budapest
Bucharest
Istanbul
Athens
Rome
Naples
Manchester
Birmingham
London
Brussels
Paris
Barcelona
Madrid
Lisbon
© Rand McNally
A-102048-1

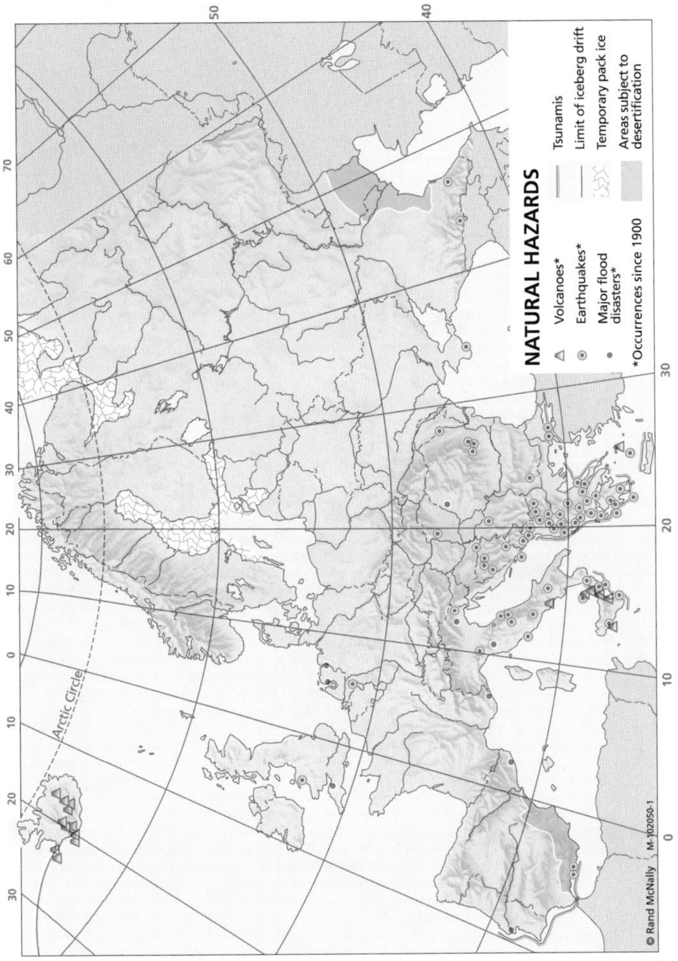
NATURAL HAZARDS
Volcanoes*
Earthquakes*
Major flood disasters*
*Occurrences since 1900
Tsunamis
Limit of iceberg drift
Temporary pack ice
Areas subject to desertification
Arctic Circle
© Rand McNally
M-102050-1

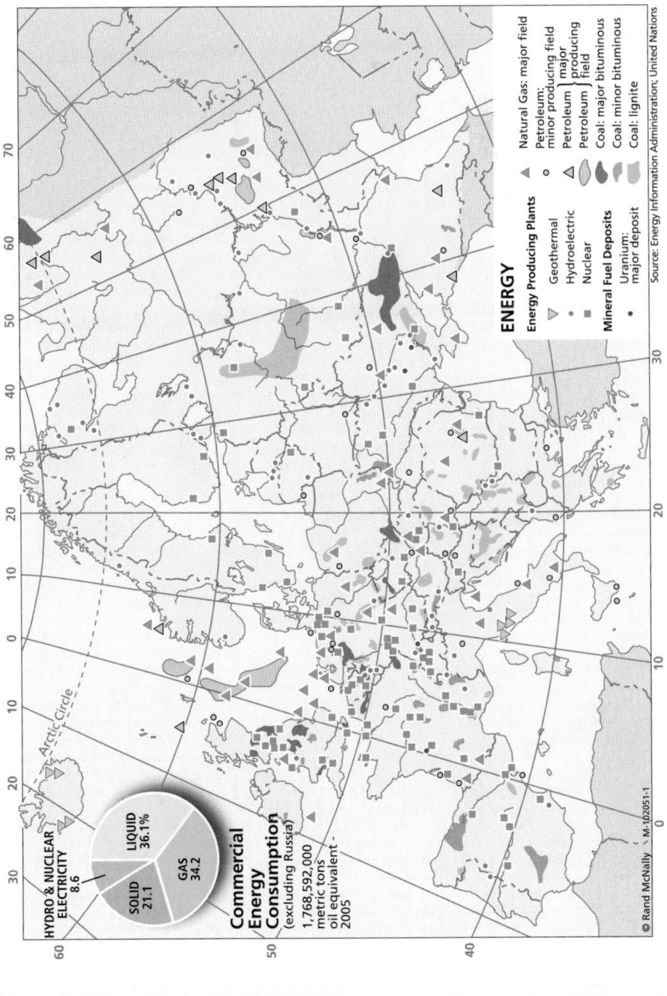
ENERGY
Energy Producing Plants
Geothermal
Hydroelectric
Nuclear
Mineral Fuel Deposits
Uranium: major deposit
Natural Gas: major field
Petroleum: minor producing field
Petroleum
Petroleum
major producing field
Coal: major bituminous
Coal: minor bituminous
Coal: lignite
Source: Energy Information Administration; United Nations
Commercial Energy Consumption
(excluding Russia)
1,768,592,000 metric tons oil equivalent - 2005
HYDRO & NUCLEAR ELECTRICITY 8.6
LIQUID 36.1%
GAS 34.2
SOLID 21.1
Arctic Circle
© Rand McNally
M-102051-1

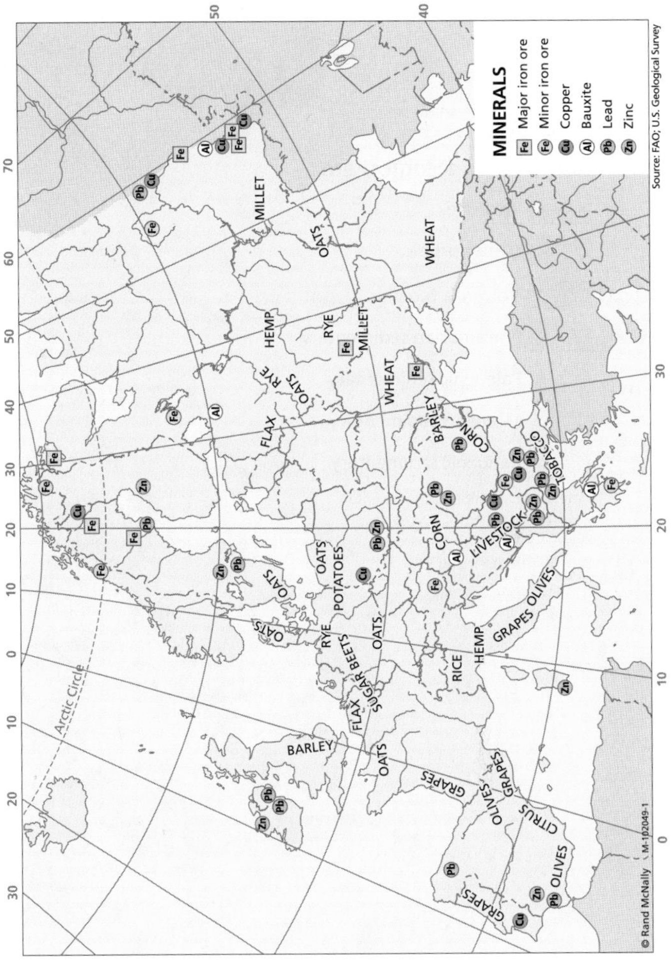
MINERALS
Fe Major iron ore
Fe Minor iron ore
Cu Copper
Al Bauxite
Pb Lead
Zn Zinc
Source: FAO; U.S. Geological Survey
MILLET
OATS
WHEAT
HEMP
RYE
FLAX
BARLEY
CORN
TOBACCO
LIVESTOCK
POTATOES
SUGAR BEETS
RICE
GRAPES
OLIVES
CITRUS
Arctic Circle
© Rand McNally
M-102049-1

Geology
Rock type/geological era
Intrusive
Metamorphic
Volcanic, tectonic
Precambrian sedimentary
Paleozoic sedimentary
Mesozoic sedimentary
Cenozoic sedimentary
Note: Areas classified as sedimentary also include some sedimentary/volcanic areas.
Source: Chorlton, 2007
The geological information on this map is highly generalized, and is intended for use at scales of 1:10,000,000 or smaller.
Arctic Circle
NORWEGIAN SEA
NORTH SEA
ATLANTIC OCEAN
Bay of Biscay
English Channel
IRISH SEA
BALTIC SEA
Gulf of Bothnia
Gulf of Finland
MEDITERRANEAN SEA
LIGURIAN SEA
TYRRHENIAN SEA
ADRIATIC SEA
IONIAN SEA
AEGEAN SEA
SEA OF CRET
Strait of Gibraltar
FAROE ISLANDS
SHETLAND ISLANDS
ORKNEY ISLANDS
HEBRIDES
SAAREMAA
GOTLAND
ÖLAND
BORNHOLM
CORSICA
SARDINIA
SICILY
MALTA
BALEARIC ISLANDS
IONIAN ISLANDS
CYCLADES
CRETE
ALPS
APENNINES
PYRENEES
CORDILLERA CANTÁBRICA
SIERRA MORENA
MASSIF CENTRAL
CARPATHIAN MOUNTAINS
TRANSYLVANIAN ALPS
PINDOS ÓROS
ATLAS MOUNTAINS
Reykjavík
Trondheim
Oslo
Stockholm
Helsinki
St. Petersbu
Göteborg
Copenhagen
Riga
Minsk
Glasgow
Dublin
Manchester
Leeds
Birmingham
London
Amsterdam
Hamburg
Berlin
Brussels
Essen
Frankfurt am Main
Mannheim
Stuttgart
Munich
Paris
Prague
Warsaw
Łódź
Katowice
Vienna
Budapest
Belgrade
Bucharest
Sofia
Athens
Bordeaux
Lyon
Marseille
Turin
Milan
Rome
Naples
Porto
Lisbon
Madrid
Barcelona
València
Sevilla
Algiers
Casablanca
Lake Peipu
Western Dvina
Vistula
Elbe
Oder
Bug
Rhine
Seine
Loire
Dordogne
Rhône
Danube
Dniester
Drava
Sava
Po
Duero
Tagus
Ebro

0 80 160 240 320 400 480 Miles
0 80 160 240 320 400 480 560 640 720 800 Kilometers

Lambert Conformal Conic Projection
Scale 1:16,000,000
One inch to 256 miles
One cm to 160 km

OSTROV KOLGUYEV
Pechora
Usa
URAL MOUNTAINS
TIMANSKIY KRYAZH
WHITE SEA
Ob'
WEST SIBERIAN PLAIN
SIBERIA
Novosibirsk
Irtysh
Ozero Kulundinskoye
Omsk
Severnaya Dvina
SEVERNYYE UVALY
Tobol
Yekaterinburg
Perm'
Lake Onega
Kirov
Kama
Izhevsk
KAZAKH HILLS
Astana
Tobyl
Ufa
Nizhniy Novgorod
Volga
Kazan'
Ivanovo
Lake Balkhash
Samara
Oka
Moscow
Shū
Dnieper
Don
Voronezh
Zhem
Syr Darya
Ural
Aral Sea
Tashkent
Desna
CASPIAN DEPRESSION
QIZILQUM
Kiev
Kharkiv
Northern Donets
Volga
UST-URT PLATEAU
Dnieper
Don
Donets'k
Dnipropetrovs'k
Rostov-na-Donu
Amu Darya
Kuma
KARA KUM
SEA OF AZOV
Prut
Odesa
Kuban'
Kara-Bogaz-Gol Gulf
CASPIAN SEA
Morghāb
CAUCASUS MOUNTAINS
KOPPEH DAGH
Harīrūd
BLACK SEA
Tbilisi
Aras
ELBURZ MOUNTAINS
İstanbul
Kızılırmak
DASHT-E KAVĪR
Sakarya
Ankara
Tehrān
Tigris
DASHT-E LŪT
TAURUS MOUNTAINS
ZAGROS MOUNTAINS
Euphrates
Baghdād
Tigris
Kārūn
DODECANESE
CYPRUS
SYRIAN DESERT
© Rand McNally
A-102091-1

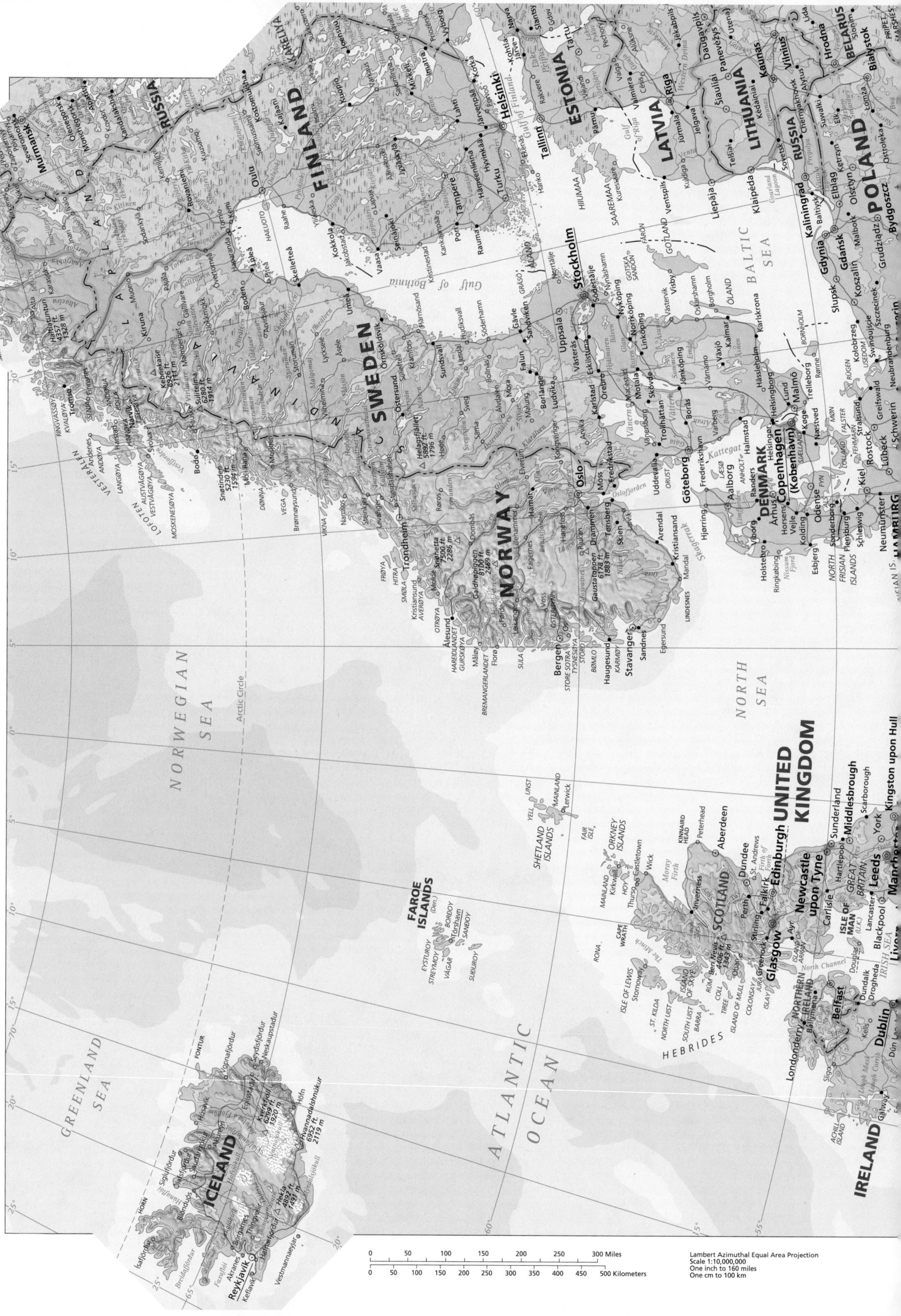

BARENTS SEA
NORWEGIAN SEA
GREENLAND SEA
ATLANTIC OCEAN
NORTH SEA
BALTIC SEA
Gulf of Bothnia
Gulf of Finland
Gulf of Riga
Arctic Circle
ICELAND
Reykjavík
FAROE ISLANDS (Den.)
Tórshavn
SHETLAND ISLANDS
ORKNEY ISLANDS
HEBRIDES
UNITED KINGDOM
SCOTLAND
NORTHERN IRELAND
IRELAND
Dublin
Belfast
Glasgow
Edinburgh
Aberdeen
Newcastle upon Tyne
Leeds
Kingston upon Hull
NORWAY
Oslo
Bergen
Stavanger
Trondheim
SWEDEN
Stockholm
Göteborg
Malmö
Uppsala
FINLAND
Helsinki
Tampere
Turku
Oulu
RUSSIA
Murmansk
KARELIYA
ESTONIA
Tallinn
LATVIA
Riga
LITHUANIA
Vilnius
Kaunas
BELARUS
POLAND
Gdańsk
Kaliningrad
DENMARK
Copenhagen (København)
Århus
Aalborg
LAPLAND

0 50 100 150 200 250 300 Miles

0 50 100 150 200 250 300 350 400 450 500 Kilometers

Lambert Azimuthal Equal Area Projection
Scale 1:10,000,000
One inch to 160 miles
One cm to 100 km

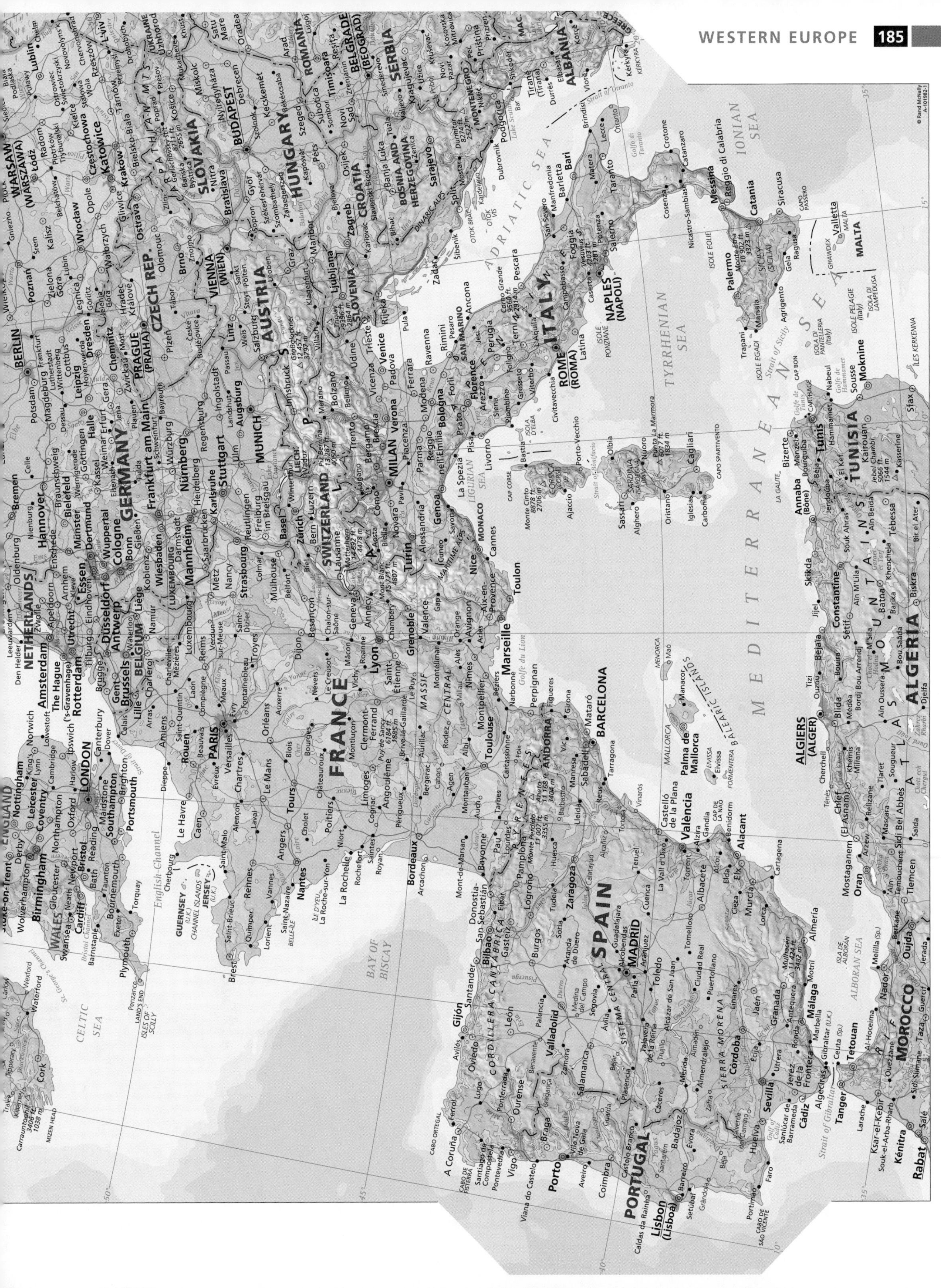
SPAIN
PORTUGAL
FRANCE
ITALY
GERMANY
SWITZERLAND
AUSTRIA
CZECH REP.
SLOVAKIA
HUNGARY
SLOVENIA
CROATIA
BOSNIA AND HERZEGOVINA
SERBIA
MONTENEGRO
ALBANIA
ROMANIA
NETHERLANDS
BELGIUM
LUXEMBOURG
ANDORRA
MONACO
SAN MARINO
LIECH.
MALTA
MOROCCO
ALGERIA
TUNISIA
ENGLAND
WALES
MEDITERRANEAN SEA
TYRRHENIAN SEA
ADRIATIC SEA
IONIAN SEA
LIGURIAN SEA
ALBORAN SEA
BAY OF BISCAY
CELTIC SEA
English Channel
Strait of Gibraltar
Strait of Sicily
Strait of Otranto
BALEARIC ISLANDS
PYRENEES
ALPS
CARPATHIAN MTS.
ATLAS MOUNTAINS
SIERRA MORENA
MASSIF CENTRAL
DINARIC ALPS
MADRID
Lisbon (Lisboa)
PARIS
LONDON
BERLIN
ROME (ROMA)
VIENNA (WIEN)
PRAGUE (PRAHA)
BUDAPEST
BRUSSELS
Amsterdam
BARCELONA
MILAN
MUNICH
NAPLES (NAPOLI)
Marseille
Lyon
Bordeaux
Toulouse
Sevilla
València
Zaragoza
Porto
Bilbao
Málaga
ALGIERS (ALGER)
Tunis
Tanger
Rabat
Sarajevo
Zagreb
Ljubljana
Tiranë (Tirana)
BELGRADE (BEOGRAD)
WARSAW (WARSZAWA)
Venice
Florence
Genoa
Turin
Palermo
Catania
Cagliari
Valletta
© Rand McNally A-101982-1

BARENTS SEA
WHITE SEA
PECHORSKOYE MORE
NORWEGIAN SEA
BALTIC SEA
Gulf of Bothnia
Gulf of Finland
Gulf of Riga
U R A L M T S.
TIMANSKIY KRYAZH
BOL'SHEZEMEL'SKAYA TUNDRA
SEVERNYYE UVALY
KOL'SKIY POLUOSTROV
POLUOSTROV KANIN
OSTROV KOLGUYEV
NORWAY
SWEDEN
FINLAND
ESTONIA
LATVIA
LITHUANIA
BELARUS
POLAND
RUSSIA
KOMI
KARELIYA
UDMURTIYA
TATARSTAN
CHUVASHIYA
MARIY-EL
MORDOVIYA
BASHKORTOSTAN
MOSCOW
ST. PETERSBURG (LENINGRAD)
Helsinki
Stockholm
Tallinn
Riga
Vilnius
MINSK
WARSAW (WARSZAWA)
NIZHNIY NOVGOROD (GORKIY)
KAZAN'
PERM'
UFA
SAMARA
Arkhangel'sk (Archangel)
Murmansk

0 50 100 150 200 250 300 Miles
0 50 100 150 200 250 300 350 400 450 500 Kilometers

Lambert Azimuthal Equal Area Projection
Scale 1:10,000,000
One inch to 160 miles
One cm to 100 km

KAZAKHSTAN
UKRAINE
TURKEY
ROMANIA
BULGARIA
GREECE
GEORGIA
ARMENIA
AZERBAIJAN
IRAN
IRAQ
SYRIA
CYPRUS
MOLDOVA
KALMYKIYA
DAGESTAN
CHECHNYA
BLACK SEA
CASPIAN SEA
SEA OF AZOV
MEDITERRANEAN SEA
SEA OF MARMARA
AEGEAN SEA
SEA OF CRETE
SYRIAN DESERT
CASPIAN DEPRESSION
CARPATHIAN MTS
TRANSYLVANIAN ALPS
BALKAN MTS
RHODOPE MTS
CAUCASUS MOUNTAINS
LESSER CAUCASUS
TAURUS MOUNTAINS
ELBURZ MOUNTAINS
MESOPOTAMIA
CRIMEAN PENINSULA (KRYMS'KYY PIVOSTRIV)
DODECANESE
CYCLADES
KIEV (KYYIV)
BUCHAREST (BUCUREŞTI)
ANKARA
TBILISI
YEREVAN
BAKU (BAKI)
TEHRĀN
İSTANBUL
İZMIR
BURSA
ADANA
ODESA
KHARKIV
DNIPROPETROVS'K
DONETS'K
ZAPORIZHZHIA
ROSTOV-NA-DONU
VOLGOGRAD
Saratov
Astrakhan'
Makhachkala
Sevastopol'
Simferopol'
Chişinău
Constanţa
Varna
Burgas
Sochi
Trabzon
Samsun
Erzurum
Van
Tabrīz
Mosul (Al-Mawşil)
Kirkuk
ALEPPO (HALAB)
Nicosia
A ADYGEA
B KARACHAYEVO-CHERKESIYA
C KARBARDINO-BALKARIYA
D SEVERNAYA OSETIYA-ALANIYA (NORTH OSSETIA)
E INGUSHETIYA
Gora El'brus 18,510 ft. 5642 m
Mt. Ararat 16,854 ft. 5137 m
Kolleh-ye Damāvand 18,386 ft. 5604 m
North Cyprus declared itself an independent Turkish Republic in 1983.
© Rand McNally A-101981-1

ATLANTIC OCEAN
BAY OF BISCAY
FRANCE
SPAIN
PORTUGAL
ANDORRA
SWITZERLAND
GERMANY
AUSTRIA
ITALY
SLOVENIA
MONACO
SAN MARINO
LIECH.
BELGIUM
LUXEMBOURG
CZECH
MOROCCO
ALGERIA
TUNISIA
MALTA
PARIS
MADRID
Lisbon (Lisboa)
ROME (ROMA)
ALGIERS (ALGER)
Tunis
Tripoli (Ṭarābulus)
Rabat
Bern
PRAGUE (PRAHA)
NAPLES (NAPOLI)
BARCELONA
Valletta
Marseille
Lyon
Toulouse
Bordeaux
Nantes
Nice
Toulon
Montpellier
Strasbourg
Frankfurt am Main
MUNICH
Stuttgart
Nürnberg
Mannheim
Zürich
Geneva
MILAN
Turin
Genoa
Florence
Bologna
Venice
Verona
Palermo
Catania
Cagliari
Sassari
Porto
Coimbra
Bilbao
Valladolid
Zaragoza
València
Palma de Mallorca
Alacant
Murcia
Córdoba
Sevilla
Granada
Málaga
Cádiz
Gibraltar (U.K.)
Ceuta (Sp.)
Melilla (Sp.)
Tanger
Tetouan
Fès
Meknès
CASABLANCA
Oujda
Oran
Constantine
Annaba (Bône)
Sfax
Sousse
Biskra
Touggourt
Ouargla
Ghardaïa
Béchar
El Golea
Timimoun
Ghadāmis
PYRENEES
MASSIF CENTRAL
CORDILLERA CANTÁBRICA
SISTEMA CENTRAL
SIERRA MORENA
ATLAS MOUNTAINS
MOYEN ATLAS
HAUT ATLAS
RIF
ALPS
APENNINES
MARITIME ALPS
Golfe du Lion
LIGURIAN SEA
TYRRHENIAN SEA
ADRIATIC
MEDITERRANEAN
ALBORAN SEA
Strait of Gibraltar
Strait of Bonifacio
Strait of Sicily
Gulf of Cádiz
Golfe de Tunis
Golfe de Hammamet
Golfe de Gabès
BALEARIC ISLANDS
MENORCA
MALLORCA
EIVISSA
FORMENTERA
CORSICA (CORSE) (Fr.)
SARDINIA (SARDEGNA) (Italy)
SICILY (SICILIA)
GRAND ERG OCCIDENTAL
GRAND ERG ORIENTAL
'ERG IGUÎDI
'ERG CHECH
SAHARA
TRIPOLITANIA
AL-HAMADAH AL-HAMRA'
HAMADA DE TINRHERT
Chott Melrhir
Chott Jerid
Mont Blanc 15 771 ft. 4807 m
Matterhorn 14 692 ft. 4478 m
Aneto 11 168 ft. 3404 m
Mulhacén 11 424 ft. 3482 m
Monte Etna 10 902 ft. 3323 m
Puy de Sancy 6184 ft. 1885 m
Monte Perdido 11 007 ft. 3355 m
Monte Cinto 8878 ft. 2706 m
Punta La Marmora 6017 ft. 1834 m
Jebel Chambi 5066 ft. 1544 m
Djebel Aïssa 7333 ft. 2235 m
Corno Grande 9560 ft. 2914 m
Vesuvio 4203 ft. 1281 m
© Rand McNally A-101877-1
0 50 100 150 200 250 300 Miles
0 50 100 150 200 250 300 350 400 450 500 Kilometers

Lambert Azimuthal Equal Area Projection
Scale 1:10,000,000
One inch to 160 miles
One cm to 100 km

POLAND
Opole
Częstochowa
Katowice
Kraków
Tarnów
Rzeszów
Gliwice
Ostrava
Bielsko-Biała
Przemyśl
Stalowa Wola
Olomouc
Zlín
Znojmo
VIENNA (WIEN)
Bratislava
SLOVAKIA
Nitra
Banská Bystrica
Gerlachovský štít 8711 ft. 2655 m
Poprad
Prešov
Košice
Uzhhorod
Mukacheve
Khust
Miskolc
Nyíregyháza
Debrecen
Győr
BUDAPEST
Székesfehérvár
Szombathely
Zalaegerszeg
Kecskemét
Szolnok
HUNGARY
Békéscsaba
Kaposvár
Pécs
Szeged
Maribor
Bjelovar
Zagreb
CROATIA
Osijek
Slavonski Brod
Sombor
Subotica
Novi Sad
Zrenjanin
BELGRADE (BEOGRAD)
Bihać
Banja Luka
Tuzla
BOSNIA AND HERZEGOVINA
Zenica
Sarajevo
Šibenik
Split
Mostar
Durmitor 8274 ft. 2522 m
MONTENEGRO
Nikšić
Dubrovnik
Podgorica
Lake Scutari
Bar
Shkodër
Smederevo
Valjevo
Kragujevac
Užice
SERBIA
Priboj
Kruševac
Niš
Novi Pazar
Kosovska Mitrovica
Leskovac
Peć
Priština
Prizren
Kumanovo
Skopje
MACEDONIA
Prilep
Bitola
Tirane
Durrës
Elbasan
ALBANIA
Korçë
Vlorë
Lviv
Ternopil'
Stryi
Drohobych
Chortkiv
Ivano-Frankivs'k
Kolomyia
Chernivtsi
Brody
Dubno
Shepetivka
Berdychiv
Bila Tserkva
Starokostiantyniv
Khmel'nyts'kyi
Vinnytsia
Zhmerynka
Kam'ianets'-Podil's'kyi
Mohyliv-Podil's'kyi
Uman'
Cherkasy
Smila
Kremenchuk
Kirovohrad
Oleksandriia
Dniprodzerzhyns'k
DNIPROPETROVS'K
Kramators'k
Luhans'k
Krasnyi Luch
Horlivka
Makiivka
DONETS'K
Shakhty
Volgodonsk
Novocherkassk
ROSTOV-NA-DONU
Taganrog
Azov
Zernograd
Salsk
Zaporizhzhia
Marhanets'
Nikopol'
Kryvyi Rih
Inhulets'
UKRAINE
Pervomais'k
Kotovs'k
Voznesens'k
Tokmak
Melitopol'
Mariupol'
Berdians'k
RUSSIA
Tikhoretsk
Kanevskaya
Timashevsk
Kropotkin
Armavir
Krasnodar
Slavyansk-na-Kubani
Labinsk
Krymsk
Belorechensk
Maykop
ADYGEA
KARACHAYEVO-CHERKESIYA
Gora El'brus 18 510 ft. 5642 m
CAUCASUS MOUNTAINS
Tuapse
Sochi
Gagra
Sokhumi
GEORGIA
Zugdidi
Poti
Batumi
Nova Kakhovka
Kherson
Mykolaïv
ODESA
Chișinău
Tiraspol
MOLDOVA
Bălți
Rîbnița
Soroca
Botoșani
Suceava
Iași
Roman
Vaslui
Bacău
Piatra-Neamț
Baia Mare
Satu Mare
Vârful Pietrosu 7556 ft. 2303 m
Oradea
Cluj-Napoca
Turda
Târgu Mureș
ROMANIA
Mediaș
Vârful Moldoveanu 8346 ft. 2544 m
Sibiu
Brașov
Focșani
Galați
Brăila
Buzău
Tulcea
Izmail
Arad
Timișoara
Lugoj
Hunedoara
Petroșani
Reșița
TRANSYLVANIAN ALPS
Râmnicu Vâlcea
Ploiești
Târgu Jiu
Pitești
Târgoviște
Slobozia
Călărași
BUCHAREST (BUCUREȘTI)
Constanța
Mangalia
Drobeta-Turnu Severin
Craiova
Vidin
Giurgiu
Ruse
Razgrad
Shumen
Dobrich
Varna
Pleven
BULGARIA
Vratsa
Gabrovo
Veliko Tarnovo
Sliven
Yambol
Burgas
BALKAN MOUNTAINS
SOFIA (SOFIYA)
Pernik
Musala 9596 ft. 2925 m
Plovdiv
Stara Zagora
Dimitrovgrad
Blagoevgrad
Kŭrdzhali
RHODOPE MTS.
Kirklareli
Edirne
Lüleburgaz
Çorlu
Tekirdağ
Keşan
Komotini
Drama
Serres
Kavála
Alexandroúpoli
THÁSOS
SAMOTHRÁKI
Thessaloníki
Kateríni
Véroia
Kozáni
Mt. Olympus 9570 ft. 2917 m
Lárisa
Vólos
Tríkala
Ioánnina
Kérkyra
KÉRKYRA
IONIAN ISLANDS
Préveza
Árta
LEFKÁDA
Agrínio
Lamía
Parnassós 8061 ft. 2457 m
GREECE
Chalkída
Athens (Athína)
Peiraias
Pátra
KEFALLONIÁ
ZÁKYNTHOS
PELOPONNESUS
Pýrgos
Tripoli
Árgos
Kalamáta
Spárti
ÁKRA TAÍNARO
KÝTHIRA
ÁKRA MALÉAS
MYRTÓÖN PÉLAGOS
CYCLADES
NÁXOS
PÁROS
MÍLOS
THÍRA
ÍOS
KÉA
ÁNDROS
TÍNOS
SKÝROS
ÉVVOIA
VÓREIOI SPORADES
LÍMNOS
GÖKÇEADA
LÉSVOS
Mytilíni
CHÍOS
SÁMOS
IKARÍA
KÓS
DODECANESE
Ródos
RÓDOS
KÁRPATHOS
KÁSOS
SEA OF CRETE
Chaniá
Ídi Óros 8051 ft. 2454 m
Irákleio
CRETE (KRÍTI)
GÁVDOS
AEGEAN SEA
Thermaikós Kólpos
Gulf of Corinth
Manfredonia
Foggia
Barletta
Bari
Matera
Brindisi
Potenza
Taranto
Lecce
Otranto
Strait of Otranto
Golfo di Taranto
Cosenza
Crotone
Catanzaro
Messina
Reggio di Calabria
IONIAN SEA
MEDITERRANEAN SEA
BLACK SEA
SEA OF AZOV
Dzhankoi
Krasnoperekops'k
CRIMEAN PENINSULA (KRYMS'KYI PIVOSTRIV)
Kerch
Feodosiia
Yevpatoriia
Simferopol'
Sevastopol'
Yalta
Anapa
Novorossiysk
Dnistrovs'kyi lyman
Bilhorod-Dnistrovs'kyi
İNCE BURUN
Sinop
Bafra
Samsun
Trabzon
Rize
Kaçkar Daği 12 900 ft. 3932 m
Giresun
Ordu
Gümüşhane
Bayburt
Şebinkarahisar
Zonguldak
Ereğli
Bartın
Karabük
Kastamonu
Merzifon
Amasya
Çorum
Turhal
Zile
Tokat
Erzincan
Sivas
Çankırı
Düzce
Bolu
İSTANBUL
Bosporus
SEA OF MARMARA
İzmit
Sakarya
Beypazarı
Kırıkkale
Yozgat
Yerköy
Kırşehir
ANKARA
Polatlı
Gelibolu
Çanakkale
Bandırma
BURSA
İnegöl
Mustafakemalpaşa
Balıkesir
Eskişehir
Kütahya
Tavşanlı
TURKEY
Tuz Gölü
Aksaray
Nevşehir
Kayseri
Erciyes Daği 12 851 ft. 3917 m
Niğde
Malatya
Elbistan
Adıyaman
Edremit
Bergama
Akhisar
Manisa
İZMİR
Salihli
Uşak
Afyon
Akşehir
Konya
Ödemiş
Nazilli
Aydın
Denizli
Isparta
Burdur
Muğla
Milas
Bodrum
Fethiye
Antalya
Manavgat
Alanya
Karaman
Ereğli
TAURUS MOUNTAINS
Kahramanmaraş
Gaziantep
Osmaniye
Kozan
ADANA
Ceyhan
Kilis
Tarsus
İçel
İskenderun
Hatay
Silifke
Anamur
Antalya Körfezi
MEGÍSTI
ALEPPO (HALAB)
Idlib
SYRIA
Latakia (Al-Lādhiqīyah)
Jablah
Ḥamāh
Ḥimṣ
Ṭarṭūs
Qurnat as-Sawdā' 10 115 ft. 3083 m
Tripoli (Ṭarābulus)
Beirut (Bayrūt)
LEBANON
DAMASCUS (DIMASHQ)
Mt. Hermon 9232 ft. 2814 m
Ṣaydā
Tyre (Ṣūr)
As-Suwaydā'
'Akko
Haifa
Irbid
ISRAEL
Nābulus
Az-Zarqā'
Netanya
'Ammān
Tel Aviv-Yafo
Reḥovot
Jerusalem (Yerushalayim)
Dead Sea
Gaza
Be'er Sheva'
JORDAN
Ma'ān
Nicosia
Gazimağusa
Ólimbos 6401 ft. 1951 m
Larnaka
Lemesós
Néa Páfos
CYPRUS
North Cyprus declared itself an independent Turkish Republic in 1983.
Ⓐ Gaza Strip is administered by the Palestinian Authority following unilateral withdrawal by Israel in 2005.
Ⓑ West Bank is controlled by Israel and parts are administered by the Palestinian Authority.
Ⓒ Golan Heights has been unilaterally annexed by Israel.
Al-Baydā'
Darnah
CYRENE
Al-Marj
Tūkrah
Tubruq (Tobruk)
At-Tamīmī
Banghāzī
Qaminis
Sulūq
CYRENAICA
Khalīj Surt
Surt
Ajdābiyā
An-Nawfalīyah
Al-'Uqaylah
Marsá al-Burayqah
LIBYA
LIBYAN DESERT (AṢ-ṢAḤRĀ' AL-LĪBĪYAH)
Waddān
Marādah
Awjilah
Salûm
Sîdi Barrâni
Mersa Matruh
El-Alamein
ALEXANDRIA (EL-ISKANDARÎYA)
Rashîd
Damietta
Port Said
El-Manṣûra
El-Mahalla el-Kubra
Ismailia
Damanhûr
Tanta
Zagazig
Suez
CAIRO (EL-QÂHIRA)
EL-GÎZA
QATTARA DEPRESSION
EGYPT
El-Fayoum
Beni Suef
El-Fashn
Maghâgha
Beni Mazâr
Samâlût
El-Minya
Siwa
Bawîti
WESTERN DESERT
Nile
SINAI PENINSULA
Mt. Sinai 7497 ft. 2285 m
Gulf of Suez
Gulf of Aqaba
Eilat
Al-'Aqabah
Jabal al-Lawz 7884 ft. 2403 m
SAUDI ARABIA
El-Ṭûr
Sharm el-Sheikh
RÂS MOHAMMED
RED SEA
El-'Arîsh

a

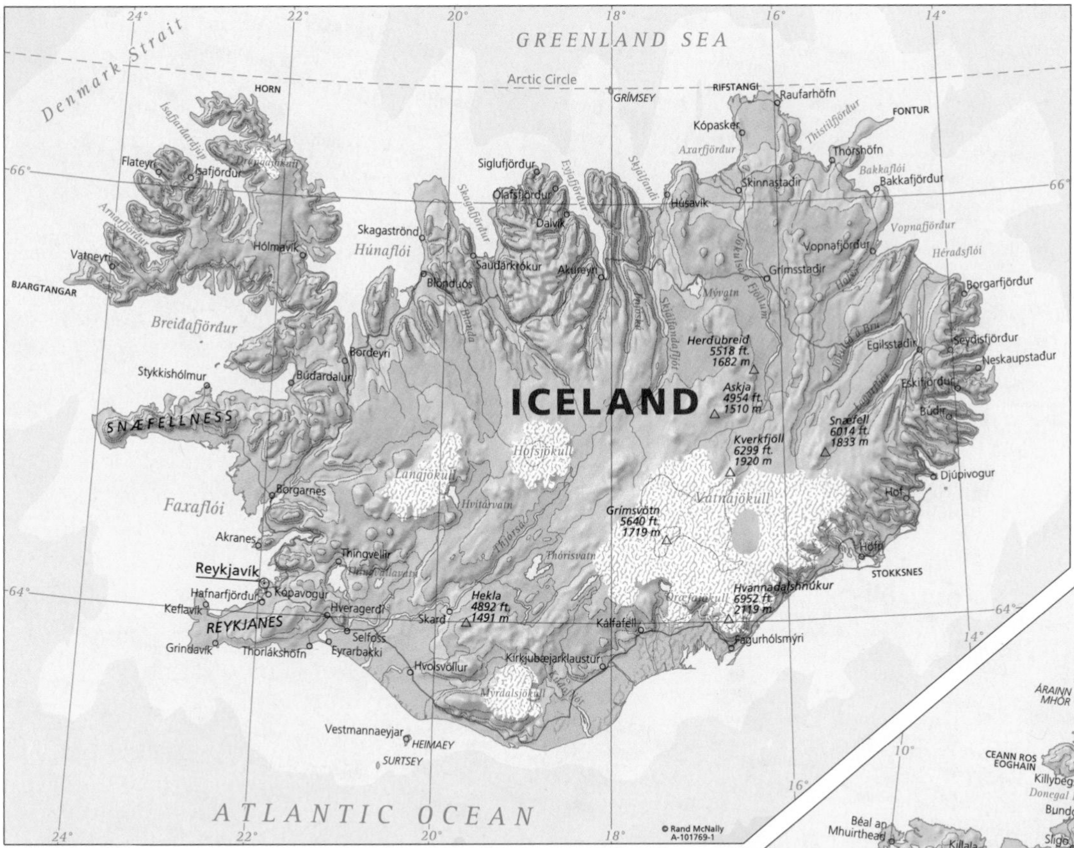
GREENLAND SEA
Denmark Strait
Arctic Circle
GRÍMSEY
RIFSTANGI
Raufarhöfn
HORN
FONTUR
Kópasker
Axarfjörður
Thistilfjörður
Thorshöfn
Bakkaflói
Bakkafjörður
Flateyri
Ísafjörður
Siglufjörður
Ólafsfjörður
Dalvík
Skinnastaður
Húsavík
Skagaströnd
Húnaflói
Vopnafjörður
Héraðsflói
Vatneyri
Hólmavík
Blönduós
Sauðárkrókur
Akureyri
Grímsstaðir
Borgarfjörður
BJARGTANGAR
Mývatn
Breiðafjörður
Herðubreið
5518 ft.
1682 m
Seyðisfjörður
Egilsstaðir
Neskaupstaður
Borðeyri
Stykkishólmur
Búðardalur
Askja
4954 ft.
1510 m
Eskifjörður
ICELAND
SNÆFELLNESS
Búðir
Kverkfjöll
6299 ft.
1920 m
Snæfell
6014 ft.
1833 m
Hofsjökull
Langjökull
Djúpivogur
Borgarnes
Hvítárvatn
Vatnajökull
Hof
Faxaflói
Grímsvötn
5640 ft.
1719 m
Akranes
Höfn
Thingvellir
Reykjavík
STOKKSNES
Hafnarfjörður
Kópavogur
Keflavík
Hekla
4892 ft.
1491 m
Hvannadalshnúkur
6952 ft.
2119 m
REYKJANES
Hveragerði
Skarð
Kálfafell
Grindavík
Thorlákshöfn
Selfoss
Eyrarbakki
Fagurhólsmýri
Hvolsvöllur
Kirkjubæjarklaustur
Mýrdalsjökull
Vestmannaeyjar
HEIMAEY
SURTSEY
ATLANTIC OCEAN
© Rand McNally
A-101769-1

b

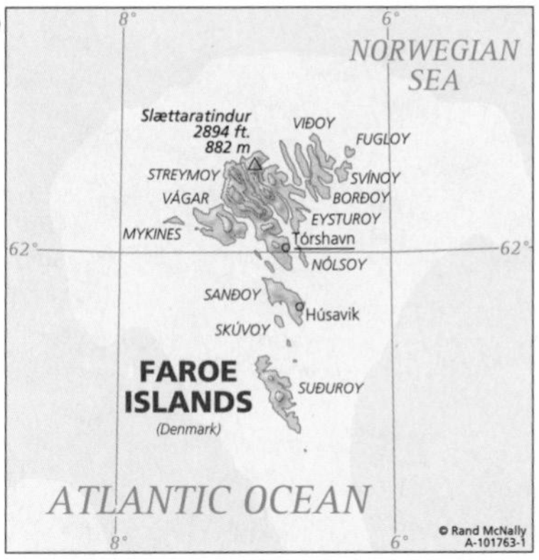
NORWEGIAN
SEA
Slættaratindur
2894 ft.
882 m
VIÐOY
FUGLOY
STREYMOY
SVÍNOY
VÁGAR
BORÐOY
EYSTUROY
MYKINES
Tórshavn
NÓLSOY
SANDOY
Húsavík
SKÚVOY
FAROE
ISLANDS
(Denmark)
SUÐUROY
ATLANTIC OCEAN
© Rand McNally
A-101763-1

c

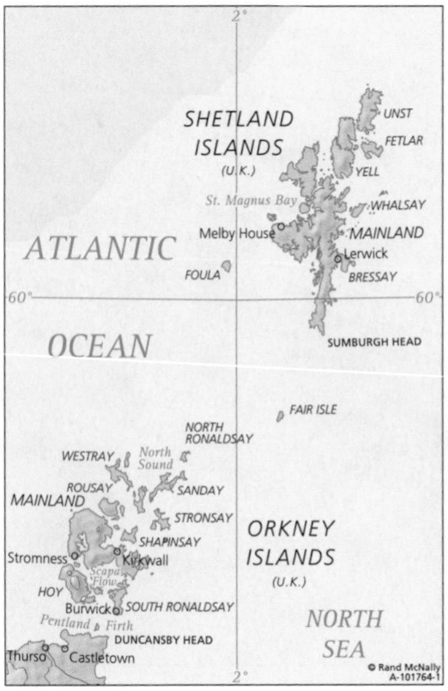
SHETLAND
ISLANDS
(U.K.)
UNST
FETLAR
YELL
St. Magnus Bay
WHALSAY
ATLANTIC
Melby House
MAINLAND
Lerwick
BRESSAY
FOULA
OCEAN
SUMBURGH HEAD
FAIR ISLE
NORTH
RONALDSAY
WESTRAY
North
Sound
ROUSAY
SANDAY
MAINLAND
STRONSAY
ORKNEY
ISLANDS
(U.K.)
SHAPINSAY
Stromness
Kirkwall
Scapa
Flow
HOY
Burwick
SOUTH RONALDSAY
Pentland Firth
DUNCANSBY HEAD
NORTH
SEA
Thurso
Castletown
© Rand McNally
A-101764-1

Carloway
ISLE OF
LEWIS
Stornoway
The Minch
SCARP
TARANSAY
Tarbert
ST. KILDA
HEBRIDES
PABBAY
ENSAY
The Little Minch
NORTH
UIST
Uig
BENBECULA
ISLAND
OF SKYE
Torridon
RAASAY
Plockton
SOUTH
UIST
Kyle of
Lochalsh
Lochboisdale
CANNA
ERISKAY
BARRA
RUM
Mallaig
SANDRAY
EIGG
MINGULAY
SEA OF
THE HEBRIDES
COLL
Tobermory
TIREE
ISLAND
OF MULL
IONA
Firth of Lorn
SCARBA
COLONSAY
Lochgilphead
JURA
Port Askaig
Sound of Jura
ISLAY
Port Ellen
ARRAN
MALIN
HEAD
Campbeltown
RATHLIN
ISLAND
Ballycastle
Gaoth
Dobhair
Buncrana
Moville
ÁRAINN
MHÓR
An Earagail
2467 ft.
752 m
Coleraine
North Channel
Letterkenny
Londonderry
Ballymena
Lifford
Strabane
Larne
CEANN ROS
EOGHAIN
Donegal
NORTHERN
IRELAND
Antrim
Newtownabbey
Killybegs
Omagh
Belfast
Bangor
Donegal Bay
Lough Neagh
Bundoran
Dungannon
Lisburn
Béal an
Mhuirthead
Enniskillen
Lurgan
Strangford
Lough
Sligo
Armagh
Banbridge
Killala
Monaghan
Downpatrick
ACHILL
ISLAND
Ballina
Newry
MOURNE
MTS.
Newcastle
Carrick-on-
Shannon
Castleblayney
CLARE ISLAND
Castlebar
Boyle
Cavan
INISHTURK
Swinford
Dundalk
Dundalk Bay
INISHBOFIN
Westport
Claremorris
Castlerea
Bailieborough
Ardee
Longford
Ballinrobe
Roscommon
Kells
Drogheda
Clifden
CONNEMARA
Tuam
Navan
Balbriggan
Mullingar
Athlone
Lough
Ree
Swords
GARUMNA
Galway
Ballinasloe
Maynooth
Dublin
(Baile Átha Cliath)
Edenderry
Grand Canal
Royal Canal
Galway Bay
Tullamore
Dún Laoghaire
INIS MÓR
INIS
MEÁIN
Loughrea
Droichead Nua
Naas
Bray
Birr
Portlaoise
IRELAND
Roscrea
Lugnaquillia
Mountain
3031 ft.
924 m
Ennis
Athy
WICKLOW MTS.
Wicklow
Nenagh
Arklow
Carlow
Kilrush
Shannon
LOOP
HEAD
Thurles
Tullow
Foynes
Limerick
Kilkenny
Gorey
Listowel
Tipperary
Cashel
Newcastle
West
Tralee
Ráth Luirc
Enniscorthy
An Daingean
Castleisland
Clonmel
Carrick-
on-Suir
Wexford
AN BLASCAOD MÓR
Newmarket
New
Ross
Rosslare
Mitchelstown
Waterford
Carrauntoohil
3406 ft.
1038 m
Mallow
Dingle Bay
Fermoy
Killarney
CARNSORE
POINT
VALENCIA ISLAND
Tramore
Cahersiveen
Kenmare
Dungarvan
Cork
Macroom
Cobh
Youghal
Passage
West
Fishguard
Bantry
ST. DAVID'S
HEAD
St. David's
DURSEY
ISLAND
Bantry Bay
Clonakilty
Kinsale
Cork
Harbour
St. George's Channel
Haverfordwest
Milford Haven
Pembroke
MIZEN
HEAD
CLEAR
ISLAND
Skibbereen
OLD HEAD
OF KINSALE
ATLANTIC
OCEAN
CELTIC
SEA
Truro
Camborne
Penzance
Falmouth
LAND'S END
ISLES OF
SCILLY
LIZARD
POINT
© Rand McNally
A-101759-1

0 20 40 60 80 100 120 Miles

0 20 40 60 80 100 120 140 160 180 200 Kilometers

Lambert Conformal Conic Projection
Scale 1:4,000,000
One inch to 64 miles
One cm to 40 km

d
Luton
Stevenage
Hitchin
Leighton Buzzard
Dunstable
Aylesbury
Hemel Hempstead
Harpenden
Welwyn Garden City
Hertford
Hatfield
St. Albans
Cheshunt
Harlow
Bishop's Stortford
Sawbridgeworth
Chelmsford
Braintree
Halstead
Witham
Maldon
Watford
Rickmansworth
High Wycombe
Beaconsfield
Slough
Maidenhead
Windsor
Bracknell
Staines
Wokingham
Woking
Farnborough
Aldershot
Guildford
Leith Hill 965 ft. 294 m
Dorking
Reigate
Crawley
Horsham
East Grinstead
Royal Tunbridge Wells
Tonbridge
Sevenoaks
Maidstone
Chatham
Gillingham
Rochester
Gravesend
Dartford
Basildon
Southend-on-Sea
Canvey Island
Brentwood
Billericay
Grays
LONDON
Harrow
Barnet
Enfield
Brent
Ealing
Hounslow
Richmond
Wandsworth
Lambeth
Lewisham
Greenwich
Croydon
Bromley
Epsom
Ewell
Ashtead
Leatherhead
Caterham
Banstead
CHILTERN HILLS
THE WEALD
VALE OF KENT
Thames
© Rand McNally
Scotland
Moray Firth
Durness
Thurso
Wick
Inverness
Aberdeen
Dundee
Perth
Stirling
Glasgow
Edinburgh
Kirkcaldy
GREAT BRITAIN
Newcastle upon Tyne
Sunderland
Middlesbrough
Carlisle
ISLE OF MAN (U.K.)
Douglas
IRISH SEA
Blackpool
Preston
Leeds
Bradford
York
Kingston upon Hull
Liverpool
Manchester
Sheffield
UNITED KINGDOM
NORTH SEA
Nottingham
Stoke-on-Trent
Derby
Leicester
Coventry
Birmingham
Wolverhampton
WALES
Norwich
Great Yarmouth
Peterborough
Cambridge
Ipswich
Northampton
ENGLAND
Oxford
Gloucester
Hereford
Cardiff
Swansea
Bristol
Bath
Reading
LONDON
Southampton
Portsmouth
Brighton
Bournemouth
Exeter
Plymouth
English Channel
Bristol Channel
FRANCE
Dunkerque
Calais
Lille
Amiens
Dieppe
BELGIUM
Brussels (Bruxelles)
Antwerp (Antwerpen)
Gent
Liège
NETHERLANDS
Amsterdam
The Hague ('s-Gravenhage)
Rotterdam
Utrecht
Groningen
Eindhoven
GERMANY
Düsseldorf
Duisburg
Aachen
LUXEMBOURG
Luxembourg
ARDENNES
FRISIAN ISLANDS

Inset map d
Lambert Conformal Conic Projection
Scale 1:1,000,000
One inch to 16 miles
One cm to 10 km

NORWAY
SWEDEN
DENMARK
GERMANY
POLAND
NORTH SEA
BALTIC
Gulf
Skagerrak
Kattegat
Oslo
Stockholm
Copenhagen (København)
Göteborg
Bergen
Stavanger
Kristiansand
Aalborg
Århus
Odense
Malmö
Helsingborg
Uppsala
Västerås
Örebro
Norrköping
Linköping
Jönköping
Kalmar
Karlskrona
Karlstad
Gävle
Falun
Borlänge
Sundsvall
Hamar
Lillehammer
Drammen
Bærum
Asker
Lillestrøm
Fredrikstad
Sarpsborg
Moss
Skien
Porsgrunn
Tønsberg
Sandefjord
Arendal
Haugesund
Sandnes
Egersund
Trollhättan
Uddevalla
Borås
Halmstad
Kristianstad
Lund
Trelleborg
Esbjerg
Kolding
Vejle
Horsens
Herning
Silkeborg
Viborg
Randers
Holstebro
Frederikshavn
Hjørring
Helsingør
Hillerød
Roskilde
Køge
Slagelse
Næstved
Svendborg
Nykøbing
Flensburg
Kiel
Neumünster
Lübeck
Rostock
Stralsund
Greifswald
Wismar
Schwerin
HAMBURG
Cuxhaven
Bremerhaven
Wilhelmshaven
Emden
Koszalin
Słupsk
Gdynia
Gdańsk (Danzig)
Glittertinden 8087 ft. 2465 m
Galdhøpiggen 8100 ft. 2469 m
Gaustatoppen 6178 ft. 1883 m
Sånfjället 4190 ft. 3736 m
JOTUNHEIMEN
DOVRE-FJELL
Vänern
Vättern
Mälaren
Hjälmaren
Mjøsa
GOTLAND
ÖLAND
BORNHOLM (Denmark)
JUTLAND
SJÆLLAND
FYN
LOLLAND
FALSTER
RÜGEN
USEDOM
Aalborg Bugt
Jammerbugten
Deutsche Bucht
Mecklenburger Bucht
Pomeranian Bay
Öresund
Stettin Lagoon
NORTH FRISIAN ISLANDS
HELGOLAND
NETH.
MECKLENBURG
POMERANIA

0 20 40 60 80 100 120 Miles
0 20 40 60 80 100 120 140 160 180 200 Kilometers

Lambert Conformal Conic Projection
Scale 1:4,000,000
One inch to 64 miles
One cm to 40 km

FINLAND
ESTONIA
LATVIA
LITHUANIA
BELARUS
RUSSIA
RUSSIA
KALININGRADSKAYA OBLAST'
LENINGRADSKAYA OBLAST'
PSKOVSKAYA OBLAST'
NOVGORODSKAYA OBLAST'
TVERSKAYA OBLAST'
SMOLENSKAYA OBLAST'
VALDAYSKAYA VOZVYSHENNOST'
KARELIYA
KARELIA
VICEBSK
MINSK
HRODNA
MAHILËU
HOMEL'
MAZURY
of Bothnia
Gulf of Finland
Gulf of Riga
SEA
ALAND SEA
Lake Ladoga (Ladozhskoye Ozero)
Lake Peipus
Lake Pskov
Courland Lagoon
Vistula Lagoon
Gulf of Gdańsk
Narva laht
Vyborgskiy Zaliv
Saimaa
Haukivesi
Pihlajavesi
Päijänne
Ozero Il'men'
HIIUMAA
SAAREMAA
ÅLAND
OSTROV GOGLAND
OSTROV MOSHCHNYY
OSTROV VALAAM
OSTROV MANTSINSARI
NAISSAAR
OSMUSSAAR
VORMSI
MUHU
KIHNU
RUHNU SAAR
ABRUKA SAAR
GOTSKA SANDÖN
FÅRÖN
KOLKASRAGS
AKMEŅRAGS
Helsinki
Tallinn
Riga
Vilnius
Minsk
ST. PETERSBURG (LENINGRAD)
Tampere
Turku
Espoo
Vantaa
Lahti
Kouvola
Kotka
Hamina
Pori
Rauma
Mikkeli
Jyväskylä
Savonlinna
Lappeenranta
Imatra
Vyborg
Hämeenlinna
Hyvinkää
Järvenpää
Porvoo
Lohja
Salo
Hanko
Mariehamn
Narva
Kohtla-Järve
Sillamäe
Rakvere
Paldiski
Haapsalu
Kärdla
Kuressaare
Pärnu
Viljandi
Tartu
Valga
Võru
Põlva
Pskov
Novgorod
Luga
Kingisepp
Ivangorod
Slantsy
Gdov
Kolpino
Pushkin
Gatchina
Kronshtadt
Lomonosov
Petrodvorets
Sestroretsk
Zelenogorsk
Vsevolozhsk
Kirishi
Tosno
Volkhov
Tikhvin
Staraya Russa
Ostrov
Porkhov
Velikiye Luki
Nevel'
Ventspils
Liepāja
Jūrmala
Jelgava
Valmiera
Cēsis
Daugavpils
Rēzekne
Jēkabpils
Madona
Gulbene
Alūksne
Talsi
Tukums
Bauska
Klaipėda
Šiauliai
Panevėžys
Kaunas
Marijampolė
Alytus
Utena
Kėdainiai
Ukmergė
Tauragė
Telšiai
Mažeikiai
Kaliningrad
Sovetsk
Chernyakhovsk
Gusev
Baltiysk
Elbląg
Olsztyn
Suwałki
Augustów
Hrodna
Lida
Maladzečna
Barysau
Navapolack
Polack
Vicebsk
Orša
Mahilëu
Ašmjany
Smarhon'
Vilejka
Salihorsk
© Rand McNally A-101765-1

DENMARK
NETHERLANDS
BELGIUM
LUXEMBOURG
GERMANY
FRANCE
SWITZERLAND
LIECHTENSTEIN
AUSTRIA
CZECH REPUBLIC
ITALY
SLOVENIA
NORTH SEA
BALTIC
FRISIAN ISLANDS
Deutsche Bucht
HAMBURG
BERLIN
Bremen
Hannover
Kiel
Lübeck
Rostock
Schwerin
Magdeburg
Braunschweig
Wolfsburg
Bielefeld
Münster
Osnabrück
Essen
Dortmund
Duisburg
Bochum
Düsseldorf
Wuppertal
Cologne (Köln)
Bonn
Aachen
Halle
Leipzig
Dresden
Chemnitz
Erfurt
Kassel
Göttingen
Frankfurt am Main
Wiesbaden
Mainz
Mannheim
Heidelberg
Karlsruhe
Stuttgart
Würzburg
Nürnberg
Regensburg
Augsburg
MUNICH (MÜNCHEN)
Ingolstadt
Freiburg im Breisgau
Saarbrücken
Amsterdam
Rotterdam
Utrecht
Groningen
Liège
Luxembourg
Strasbourg
Metz
Nancy
Basel
Zürich
Bern
Geneva (Genève)
Lausanne
Innsbruck
Salzburg
Linz
Graz
Szczecin
PRAGUE (PRAHA)
Plzeň
Ljubljana
Zugspitze 9718 ft. 2962 m
Hoher Dachstein 9826 ft. 2995 m
Großglockner 12 457 ft. 3797 m
Jungfrau 13 642 ft. 4158 m
Matterhorn 14 692 ft. 4478 m
Mont Blanc 15 771 ft. 4807 m
Sněžka 5256 ft. 1602 m
Klínovec 4081 ft. 1244 m
Botrange 2277 ft. 694 m

0 20 40 60 80 100 120 Miles
0 20 40 60 80 100 120 140 160 180 200 Kilometers

Lambert Conformal Conic Projection
Scale 1:4,000,000
One inch to 64 miles
One cm to 40 km

RUSSIA
LITHUANIA
BELARUS
POLAND
UKRAINE
SLOVAKIA
HUNGARY
ROMANIA
SERBIA
CROATIA
MOLD.
SEA
Gulf of Gdansk
Kaliningrad
Baltiysk
Svetlyy
Gvardeysk
Chernyakhovsk
Gusev
Zheleznodorozhnyy
Ozërsk
Marijampolė
Alytus
Lazdijai
Varėna
Eišiškės
Šalčininkai
Druskininkai
Maladzečna
Minsk
MINSK
Lida
Hrodna
HRODNA
Navahrudak
Stoubcy
Baranavičy
Slonim
Vaukavysk
Masty
Sluck
Salihorsk
Pinsk
Kobryn
Brest
BREST
Luninec
Mikaševičy
HOMEL'
Mozyr
Łeba
Puck
Hel
Wejherowo
Gdynia
Sopot
Gdańsk (Danzig)
Słupsk
Lębork
Kościerzyna
Tczew
Malbork
Elbląg
Braniewo
Starogard Gdański
Kwidzyn
Ostróda
Olsztyn
Iława
Szczytno
Kętrzyn
Giżycko
Ełk
Suwałki
Augustów
Grajewo
Mońki
Sokółka
Białystok
Grudziądz
Świecie
Chojnice
Szczecinek
Chełmno
Brodnica
Działdowo
Nidzica
Mława
Ostrołęka
Łomża
Łapy
Zambrów
Bielsk Podlaski
Hajnówka
Bydgoszcz
Toruń
Piła
Inowrocław
Włocławek
Płock
Ciechanów
Pułtusk
Wyszków
Ostrów Mazowiecka
Treblinka
Siemiatycze
Gniezno
Poznań
Swarzędz
Września
Konin
Koło
Kutno
Sochaczew
Łowicz
WARSAW (WARSZAWA)
Nowy Dwór Mazowiecki
Legionowo
Wołomin
Otwock
Pruszków
Grodzisk Mazowiecki
Żyrardów
Mińsk Mazowiecki
Siedlce
Biała Podlaska
Łuków
Skierniewice
Łódź
Zgierz
Pabianice
Kalisz
Ostrów Wielkopolski
Sieradz
Piotrków Trybunalski
Tomaszów Mazowiecki
Radom
Puławy
Lublin
Świdnik
Chełm
Zamość
Kielce
Ostrowiec Świętokrzyski
Skarżysko-Kamienna
Starachowice
Kraśnik
Stalowa Wola
Sandomierz
Tarnobrzeg
Biłgoraj
Mielec
Leżajsk
Jarosław
Przemyśl
Rzeszów
Dębica
Tarnów
Bochnia
Kraków
Wieliczka
Oświęcim (Auschwitz)
Katowice
Bytom
Sosnowiec
Zabrze
Gliwice
Tychy
Rybnik
Racibórz
Bielsko-Biała
Nowy Sącz
Nowy Targ
Zakopane
Krosno
Sanok
Częstochowa
Opole
Wrocław
Oleśnica
Oława
Brzeg
Wałbrzych
Świdnica
Kluczbork
Lubliniec
Radomsko
Wieluń
Bełchatów
Leszno
Kościan
Jarocin
Krotoszyn
Ostrava
Opava
Havířov
Karviná
Frýdek-Místek
Olomouc
Přerov
Prostějov
Kroměříž
Zlín
Brno
Uherské Hradiště
Břeclav
Hodonín
Žilina
Martin
Ružomberok
Liptovský Mikuláš
Poprad
Prešov
Košice
Michalovce
Humenné
Bardejov
Banská Bystrica
Zvolen
Trenčín
Prievidza
Nitra
Trnava
Bratislava
Lučenec
Rimavská Sobota
Gerlachovský štít 8711 ft. 2655 m
Dumbier 6703 ft. 2043 m
NÍZKE TATRY
CARPATHIAN MOUNTAINS
BESKIDS
MAZURY
PRIPET MARSHES
VOLYN'
RIVNE
ZHYTOMYR
L'VIV
TERNOPIL'
KHMEL'NYTS'KYI
IVANO-FRANKIVS'K
ZAKARPATTIA
CHERNIVTSI
Kovel'
Luts'k
Rivne
Dubno
Kostopil'
Novohrad-Volyns'kyi
Volodymyr-Volyns'kyi
Novovolyns'k
Chervonohrad
L'viv
Drohobych
Stryi
Truskavets'
Boryslav
Sambir
Kalush
Ivano-Frankivs'k
Kolomyia
Ternopil'
Khmel'nyts'kyi
Kam'ianets'-Podil's'kyi
Chernivtsi
Uzhhorod
Mukacheve
Berehove
Khust
hora Hoverla 6762 ft. 2061 m
Miskolc
Nyíregyháza
Debrecen
Eger
Salgótarján
Gyöngyös
Budapest
Vác
Esztergom
Komárno
Győr
Tatabánya
Székesfehérvár
Veszprém
Szombathely
Zalaegerszeg
Dunaújváros
Kecskemét
Szolnok
Cegléd
Nagykőrös
Kiskunfélegyháza
Szeged
Hódmezővásárhely
Békéscsaba
Gyula
Oradea
Arad
Kaposvár
Pécs
Kekes 3327 ft. 1014 m
Balaton
Subotica
Sombor
Osijek
Satu Mare
Baia Mare
Sighetu Marmației
Zalău
Cluj-Napoca
Bistrița
Târgu Mureș
Turda
Câmpia Turzii
Alba Iulia
Deva
Sibiu
Brașov
Suceava
Botoșani
Piatra-Neamț
Vârful Pietrosu 7556 ft. 2303 m
Vârful Bihor 6066 ft. 1849 m
© Rand McNally A-101781-1

UNITED KINGDOM
CELTIC SEA
HARTLAND POINT
Bude
Taunton
Yeovil
Salisbury
Winchester
Crawley
Horsham
Royal Tunbridge Wells
Ashford
Folkestone
Dover
Southampton
Portsmouth
Chichester
Brighton
Hastings
Bexhill
Eastbourne
BEACHY HEAD
Worthing
Bournemouth
Poole
Dorchester
Weymouth
Cowes
Ryde
Newport
ISLE OF WIGHT
Lyme Bay
Exeter
Honiton
Lyme Regis
Exmouth
DARTMOOR
Torquay
Paignton
Plymouth
Bodmin
Camborne
Truro
Penzance
LAND'S END
Falmouth
LIZARD POINT
Salcombe
START POINT
Strait of Dover
English Channel
Nieuwpoort
Dunkerque
Calais
Gent
Lokeren
Sint-Niklaas
Lier
Mechelen
Leuven
Brussels (Bruxelles)
Tienen
Waterloo
Halle
BELGIUM
Namur
Nivelles
La Louvière
Mons
Charleroi
Dinant
Roeselare
Izegem
Aalst
Ieper
Kortrijk
Tourcoing
Roubaix
Tournai
Lille
Grayelines
Marquise
Saint-Omer
Armentières
Boulogne-sur-Mer
Lumbres
Étaples
Lillers
Béthune
Bruay-en-Artois
Lens
Liévin
Hesdin
Berck
Douai
Valenciennes
Maubeuge
Saint-Pol-sur-Ternoise
Arras
Cambrai
Caudry
Hautmont
Bohain-en-Vermandois
Fourmies
Fumay
Hirson
Revin
Abbeville
Doullens
Albert
Péronne
Eu
Dieppe
Amiens
Saint-Quentin
Charleville-Mézières
Sedan
Neufchâtel-en-Bray
Aumale
Fécamp
Montdidier
Chauny
Noyon
Laon
Rethel
Bolbec
Yvetot
Lillebonne
Montivilliers
Le Havre
Gournay-en-Bray
Beauvais
Rouen
Saint-Étienne-du-Rouvray
Clermont
Compiègne
Soissons
Reims
Honfleur
Trouville-sur-Mer
Pont-Audemer
Elbeuf
Louviers
Gisors
Méru
Creil
Chantilly
Crépy-en-Valois
Villers-Cotterêts
Château-Thierry
Épernay
Châlons-en-Champagne
CHAMPAGNE
Baie de la Seine
ALDERNEY
CAP DE LA HAGUE
POINTE DE BARFLEUR
Cherbourg
Valognes
GUERNSEY (U.K.)
St. Peter Port
SARK
CHANNEL ISLANDS
Passage de la Déroute
Barneville-Carteret
St. Helier
JERSEY (U.K.)
Bayeux
Ouistreham
Caen
Lisieux
Saint-Lô
Coutances
NORMANDY
Bernay
Évreux
Vernon
Gaillon
Pontoise
Saint-Denis
Mantes-la-Jolie
Argenteuil
PARIS
Boulogne-Billancourt
Versailles
Palaiseau
Saint-Maur-des-Fossés
Meaux
Coulommiers
Vitry-le-François
Saint-Dizier
Falaise
Vire
Flers
Argentan
Granville
Bréhal
Avranches
L'Aigle
Dreux
Rambouillet
Évry
Melun
Nangis
Provins
Romilly-sur-Seine
Maintenon
COLLINES DU PERCHE
Golfe de Saint-Malo
Perros-Guirec
Paimpol
Saint-Pol-de-Léon
Lannion
Morlaix
Guingamp
Saint-Brieuc
Lamballe
Étables
Saint-Malo
Cancale
Dinard
Dinan
ÎLE DE MOLÈNE
Saint-Renan
Landerneau
Brest
Carhaix-Plouguer
Iroise
Douarnenez
POINTE DU RAZ
Quimper
Concarneau
Quimperlé
Guilvinec
Lorient
Hennebont
Pontivy
Loudéac
Ploërmel
BRITTANY
Vannes
Auray
ÎLE DE GROIX
Redon
Fougères
Rennes
Vitré
Mayenne
Évron
Laval
Alençon
Nogent-le-Rotrou
La Ferté-Bernard
Chartres
Étampes
Fontainebleau
Montereau-Faut-Yonne
Nemours
Troyes
Sens
Bar-sur-Seine
Saint-Florentin
Châtillon-sur-Seine
Pithiviers
Montargis
Joigny
Auxerre
Châteaudun
Le Mans
Orléans
Châteauneuf-sur-Loire
Sablé-sur-Sarthe
Château-Gontier
Châteaubriant
Segré
La Flèche
Vendôme
Beaugency
Blois
Gien
Montbard
Avallon
Clamecy
Cosne-sur-Loire
Romorantin-Lanthenay
Amboise
Tours
Bléré
Angers
Trélazé
Longué
Saumur
Bourgueil
Chinon
Loudun
Loches
Vierzon
Mehun-sur-Yèvre
Bourges
Nevers
Saulieu
Beaune
Autun
Decize
Pontchâteau
Guérande
BELLE-ÎLE
Ancenis
Saint-Nazaire
Nantes
Carquefou
Rezé
Vertou
La Baule-Escoublac
Pornic
Clisson
Cholet
Thouars
Lac de Grand Lieu
ÎLE DE NOIRMOUTIER
Les Herbiers
Mauléon
Saint-Jean-de-Monts
ÎLE D'YEU
Challans
La Roche-sur-Yon
Saint-Gilles-Croix-de-Vie
Bressuire
Parthenay
Châtellerault
Issoudun
Saint-Florent-sur-Cher
Châteauroux
Saint-Amand-Mont-Rond
Montceau-les-Mines
Le Creusot
Chalon-sur-Saône
Les Sables-d'Olonne
Luçon
Fontenay-le-Comte
Poitiers
Argenton-sur-Creuse
La Châtre
FRANCE
Moulins
Paray-le-Monial
Montmorillon
Montluçon
Commentry
Mâcon
Niort
Pertuis Breton
ÎLE DE RÉ
La Rochelle
Rochefort
ÎLE D'OLÉRON
Ruffec
Bellac
Guéret
Vichy
Roanne
Villefranche-sur-Saône
BAY OF BISCAY
Saint-Jean-d'Angély
Saint-Junien
Aubusson
Riom
Tarare
Villeurbanne
Lyon
La Tremblade
Saintes
Cognac
Rochechouart
Limoges
Clermont-Ferrand
Thiers
Feurs
Vénissieux
Royan
Pierre-Buffière
Montbrison
Ambert
Saint-Chamond
Saint-Étienne
Firminy
Vienne
Angoulême
Nontron
Ussel
Issoire
Puy de Sancy 6184 ft. 1885 m
Gironde
Saint-Vivien-de-Médoc
Mirambeau
Chalais
Thiviers
Uzerche
Égletons
Tulle
Brioude
Annonay
Lac de Carcans
Saint-Laurent-et-Benon
Saint-Astier
Périgueux
Brive-la-Gaillarde
Mauriac
Plomb du Cantal 6096 ft. 1858 m
MASSIF CENTRAL
Le Puy
Valence
Montignac
Mussidan
Mérignac
Libourne
Bergerac
Saint-Cyprien
Souillac
Sarlat-la-Canéda
Aurillac
Saint-Flour
Mont Mézenc 5751 ft. 1753 m
Pessac
Bordeaux
Talence
Mur-de-Barrez
Bassin d'Arcachon
Arcachon
La Teste-de-Buch
Gourdon
Aumont-Aubrac
Langogne
Privas
Montélimar
Figeac
Lauzun
Decazeville
Belin
Langon
Marmande
Villeneuve-sur-Lot
Cahors
Espalion
Mende
CÉVENNES
Bourg-Saint-Andéol
Bollène
Bagnols-sur-Cèze
Orange
Biscarrosse
Bazas
Tonneins
Villefranche-de-Rouergue
Rodez
LANDES
Labouheyre
Mimizan-les-Bains
Agen
Moissac
La Grand'Combe
Alès
Millau
Carmaux
Le Vigan
Avignon
Morcenx
Roquefort
Castelsarrasin
Montauban
Gaillac
Albi
Saint-Affrique
Nîmes
Beaucaire
Mont-de-Marsan
GASCONY
Fleurance
Lunel
Arles
Adour
Dax
Aire-sur-l'Adour
Auch
Graulhet
Lodève
Montpellier
Castres
Colomiers
Toulouse
Mazamet
Pézenas
Le Grau-du-Roi
Étang de Vaccarès
Santander
Santoña
Llanes
San Vicente de la Barquera
Laredo
Castro-Urdiales
Bermeo
Donostia-San Sebastián
Biarritz
Bayonne
Anglet
Orthez
Torrelavega
Portugalete
Barakaldo
Bilbao
Arizgoiti
Saint-Jean-de-Luz
Irun
Renteria
Gave de Pau
Muret
Canal du Midi
Béziers
Sète
Bassin de Thau
Agde
Peña Prieta 8320 ft. 2536 m
CORDILLERA CANTÁBRICA
Durango
Eibar
Bergara
Reinosa
Embalse del Ebro
Tolosa
Oloron-Sainte-Marie
Pau
Tarbes
Lourdes
Bagnères-de-Bigorre
Saint-Gaudens
Pamiers
Castelnaudary
Carcassonne
Limoux
Narbonne
Golfe du Lion
Guardo
Beasain
Alsasua
Pamplona
Saint-Girons
Foix
Lavelanet
Carrión
Pisuerga
SPAIN
Gasteiz
Miranda de Ebro
Estella
PYRENEES
Aneto 11,168 ft. 3404 m
Rivesaltes
Étang de Leucate
OLD CASTILLE
Haro
Embalse de Yesa
Monte Perdido 11,007 ft. 3355 m
Ax-les-Thermes
Perpignan
Argelès-sur-Mer
Port-Vendres
Burgos
Logroño
Tafalla
Jaca
Sabiñánigo
ANDORRA
Andorra la Vella
Puigcerdà
Prades
Llançà
CAP DE CREUS
La Seu d'Urgell
Figueres
Golfe de Roses
Ripoll
Olot
Banyoles
Berga
Ter
0 20 40 60 80 100 120 Miles
0 20 40 60 80 100 120 140 160 180 200 Kilometers
Lambert Conformal Conic Projection
Scale 1:4,000,000
One inch to 64 miles
One cm to 40 km

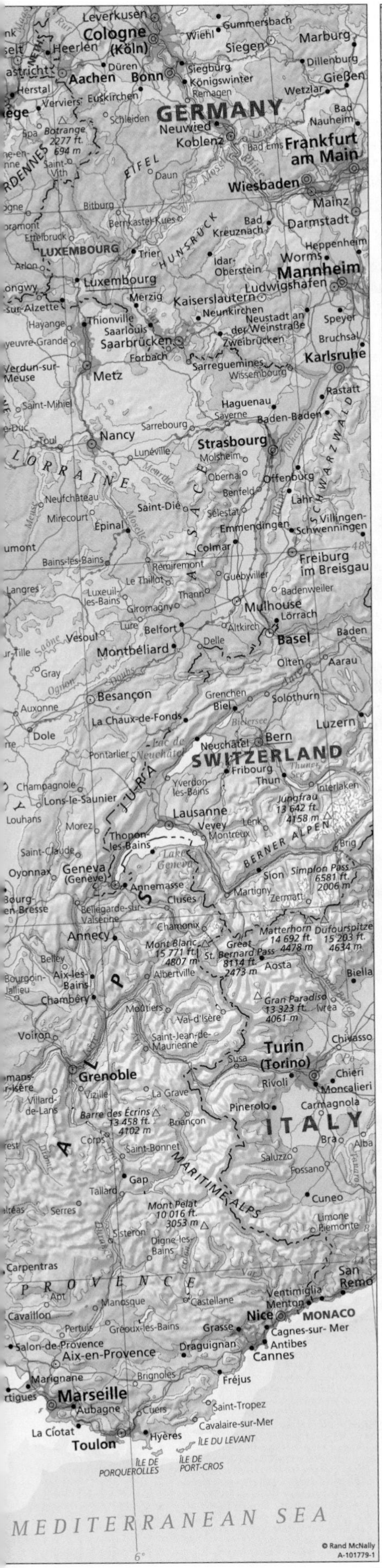
Leverkusen
Cologne (Köln)
Gummersbach
Marburg
Wiehl
Siegen
Heerlen
Düren
Siegburg
Dillenburg
Aachen
Bonn
Königswinter
Gießen
Herstal
Euskirchen
Remagen
Wetzlar
Verviers
Schleiden
GERMANY
Bad Nauheim
Neuwied
Botrange 2277 ft. 694 m
Koblenz
Bad Ems
Frankfurt am Main
Saint-Vith
EIFEL
Daun
Wiesbaden
ARDENNES
Mainz
Bitburg
Bernkastel-Kues
Bad Kreuznach
Darmstadt
Ettelbruck
HUNSRÜCK
LUXEMBOURG
Trier
Heppenheim
Arlon
Idar-Oberstein
Worms
Mannheim
Luxembourg
Ludwigshafen
Longwy
Merzig
Kaiserslautern
Esch-sur-Alzette
Neunkirchen
Speyer
Hayange
Thionville
Saarlouis
Neustadt an der Weinstraße
Saarbrücken
Bruchsal
Forbach
Zweibrücken
Verdun-sur-Meuse
Sarreguemines
Karlsruhe
Metz
Wissembourg
Saint-Mihiel
Haguenau
Rastatt
Saverne
Baden-Baden
Sarrebourg
Nancy
Toul
Strasbourg
Lunéville
Molsheim
LORRAINE
Obernai
Offenburg
SCHWARZWALD
Benfeld
Neufchâteau
Saint-Dié
Lahr
Sélestat
Mirecourt
Villingen-Schwenningen
Épinal
Emmendingen
Colmar
Bains-les-Bains
Freiburg im Breisgau
Remiremont
Guebwiller
ALSACE
Le Thillot
Badenweiler
Langres
Luxeuil-les-Bains
Thann
Giromagny
Mulhouse
Lörrach
Lure
Belfort
Altkirch
Vesoul
Delle
Basel
Montbéliard
Baden
Olten
Aarau
Gray
Besançon
Grenchen
Solothurn
Biel
Auxonne
La Chaux-de-Fonds
Luzern
Dole
Neuchâtel
Bern
Pontarlier
SWITZERLAND
Fribourg
Champagnole
Yverdon-les-Bains
Thun
Interlaken
Lons-le-Saunier
JURA
Jungfrau 13 642 ft. 4158 m
Lausanne
Louhans
Vevey
Montreux
Morez
Thonon-les-Bains
BERNER ALPEN
Saint-Claude
Lake Geneva
Oyonnax
Geneva (Genève)
Sion
Simplon Pass 6581 ft. 2006 m
Annemasse
Martigny
Bourg-en-Bresse
Bellegarde-sur-Valserine
Cluses
Zermatt
Chamonix
Matterhorn 14 692 ft. 4478 m
Dufourspitze 15 203 ft. 4634 m
Annecy
Mont Blanc 15 771 ft. 4807 m
Great St. Bernard Pass 8114 ft. 2473 m
Aosta
Belley
Aix-les-Bains
Albertville
Biella
Chambéry
Moûtiers
Gran Paradiso 13 323 ft. 4061 m
Ivrea
Val-d'Isère
Voiron
Saint-Jean-de-Maurienne
Turin (Torino)
Chivasso
Grenoble
Susa
Chieri
Rivoli
Moncalieri
Vizille
La Grave
Villard-de-Lans
Barre des Écrins 13 458 ft. 4102 m
Pinerolo
Carmagnola
Briançon
ITALY
Bra
Alba
Corps
Saint-Bonnet
Saluzzo
Gap
Fossano
Tallard
MARITIME ALPS
Cuneo
Serres
Mont Pelat 10 016 ft. 3053 m
Limone Piemonte
Sisteron
Digne-les-Bains
Carpentras
PROVENCE
Apt
Manosque
Castellane
Ventimiglia
San Remo
Menton
Cavaillon
Pertuis
Gréoux-les-Bains
Grasse
Nice
MONACO
Salon-de-Provence
Draguignan
Cagnes-sur-Mer
Aix-en-Provence
Antibes
Cannes
Marignane
Brignoles
Fréjus
Marseille
Aubagne
Cuers
Saint-Tropez
La Ciotat
Cavalaire-sur-Mer
Toulon
Hyères
ÎLE DU LEVANT
ÎLE DE PORQUEROLLES
ÎLE DE PORT-CROS
MEDITERRANEAN SEA
© Rand McNally A-101779-1

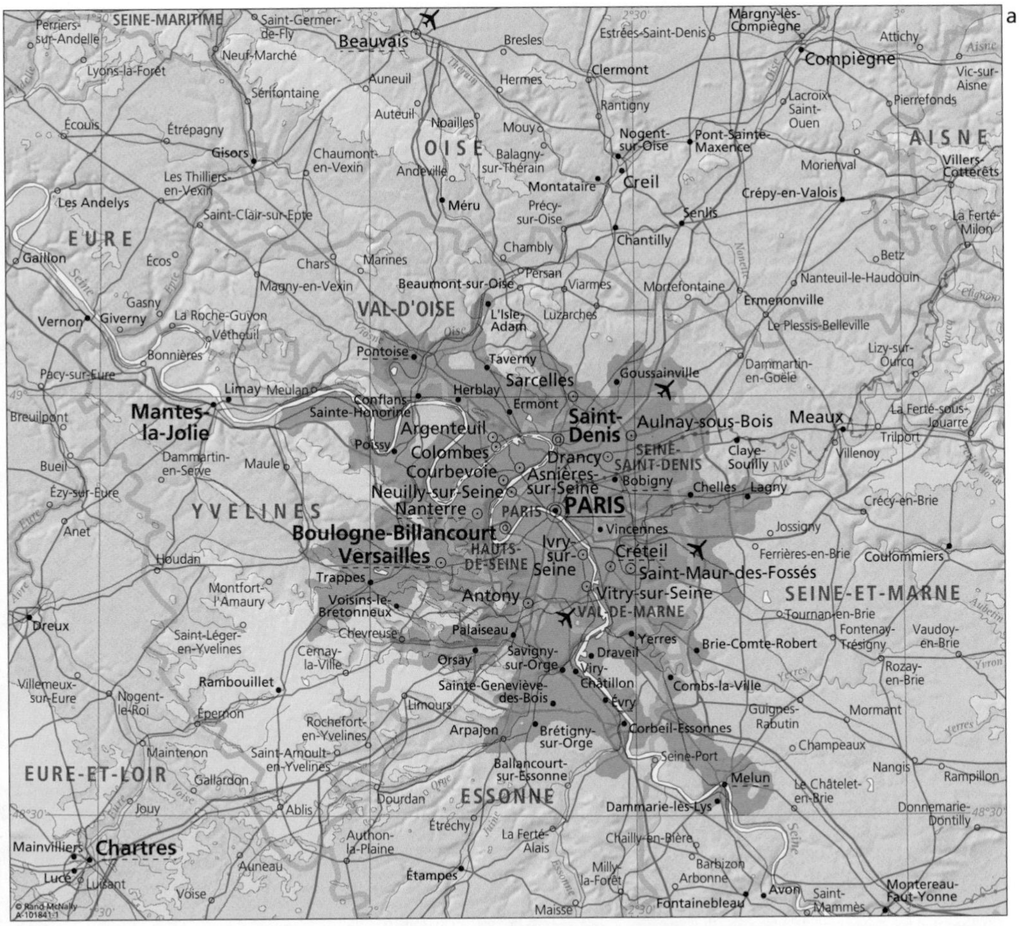
a
Perriers-sur-Andelle
SEINE-MARITIME
Saint-Germer-de-Fly
Beauvais
Margny-lès-Compiègne
Estrées-Saint-Denis
Attichy
Neuf-Marché
Bresles
Compiègne
Lyons-la-Forêt
Clermont
Vic-sur-Aisne
Auneuil
Hermes
Lacroix-Saint-Ouen
Pierrefonds
Sérifontaine
Rantigny
Écouis
Auteuil
Noailles
Étrépagny
Mouy
Nogent-sur-Oise
Pont-Sainte-Maxence
AISNE
Gisors
Chaumont-en-Vexin
OISE
Balagny-sur-Thérain
Villers-Cotterêts
Andeville
Morienval
Les Thilliers-en-Vexin
Montataire
Creil
Les Andelys
Méru
Précy-sur-Oise
Crépy-en-Valois
Saint-Clair-sur-Epte
Senlis
La Ferté-Milon
EURE
Chantilly
Chambly
Gaillon
Écos
Chars
Marines
Persan
Betz
Nanteuil-le-Haudouin
Gasny
Magny-en-Vexin
Beaumont-sur-Oise
Viarmes
Mortefontaine
Ermenonville
Vernon
Giverny
La Roche-Guyon
VAL-D'OISE
L'Isle-Adam
Luzarches
Le Plessis-Belleville
Vétheuil
Lizy-sur-Ourcq
Bonnières
Pontoise
Taverny
Goussainville
Dammartin-en-Goële
Pacy-sur-Eure
Sarcelles
Limay
Meulan
Herblay
Ermont
Conflans-Sainte-Honorine
Mantes-la-Jolie
Saint-Denis
Aulnay-sous-Bois
Meaux
La Ferté-sous-Jouarre
Breuilpont
Argenteuil
Poissy
Colombes
Drancy
SEINE-SAINT-DENIS
Claye-Souilly
Trilport
Dammartin-en-Serve
Courbevoie
Asnières-sur-Seine
Bobigny
Villenoy
Bueil
Maule
Neuilly-sur-Seine
Chelles
Lagny
Ézy-sur-Eure
Nanterre
PARIS
Crécy-en-Brie
YVELINES
Boulogne-Billancourt
Vincennes
Jossigny
Anet
HAUTS-DE-SEINE
Ivry-sur-Seine
Créteil
Ferrières-en-Brie
Coulommiers
Houdan
Versailles
Saint-Maur-des-Fossés
Trappes
Montfort-l'Amaury
Voisins-le-Bretonneux
Antony
Vitry-sur-Seine
SEINE-ET-MARNE
VAL-DE-MARNE
Tournan-en-Brie
Dreux
Saint-Léger-en-Yvelines
Chevreuse
Palaiseau
Yerres
Fontenay-Trésigny
Vaudoy-en-Brie
Cernay-la-Ville
Orsay
Savigny-sur-Orge
Draveil
Brie-Comte-Robert
Rozay-en-Brie
Villemeux-sur-Eure
Rambouillet
Nogent-le-Roi
Viry-Châtillon
Sainte-Geneviève-des-Bois
Combs-la-Ville
Limours
Évry
Mormant
Épernon
Guignes-Rabutin
Rochefort-en-Yvelines
Arpajon
Corbeil-Essonnes
Brétigny-sur-Orge
Champeaux
Maintenon
Saint-Arnoult-en-Yvelines
Nangis
Rampillon
Seine-Port
Gallardon
Ballancourt-sur-Essonne
EURE-ET-LOIR
Melun
Le Châtelet-en-Brie
Dourdan
ESSONNE
Jouy
Ablis
Donnemarie-Dontilly
Dammarie-lès-Lys
Étréchy
Authon-la-Plaine
La Ferté-Alais
Mainvilliers
Chartres
Chailly-en-Bière
Lucé
Auneau
Milly-la-Forêt
Barbizon
Étampes
Arbonne
Lucrant
Voise
Avon
Saint-Mammès
Montereau-Fault-Yonne
Maisse
Fontainebleau
© Rand McNally A-101841-1

b
Mézel
Saint-André-les-Alpes
Annot
Puget-Théniers
Touët-sur-Var
Clans
Lantosque
Col de Brouis 2884 ft. 879 m
La Brillanne
La Bégude-Blanche
ALPES-DE-HAUTE-PROVENCE
Barrême
Les Scaffarels
Entrevaux
GORGES DU CIANS
La Tour
Peira-Cava
Oraison
Sénez
Villars-sur-Var
Utelle
Col de Braus 3287 ft. 1002 m
Sospel
Vergons
Ascros
Puimoisson
Sommet de la Bernarde 6368 ft. 1941 m
Sallagriffon
Saint-Auban
Harpille 5531 ft. 1686 m
Roquestéron
Lucéram
Valensole
Moustiers-Sainte-Marie
Castellane
ALPES-MARITIMES
Levens
ITALY
FRANCE
Riez
La Palud
Thorenc
Cime du Cheiron 5830 ft. 1777 m
Contes
L'Escarène
Gréolières
Tourrette-Levens
Menton
Gréoux-les-Bains
Les Salles-sur-Verdon
Caille
Coursegoules
Carros
Beausoleil
Gattières
Monaco
MONACO
Aiguines
La Martre
Monte Carlo
Quinson
Comps-sur-Artuby
Saint-Vallier-de-Thiey
Gourdon
Vence
Nice
ALPES DE PROVENCE
PROVENCE
Cagnes-sur-Mer
Saint-Laurent-du-Var
Mons
Saint-Cézaire-sur-Siagne
Grasse
Seillans
Biot
Baie des Anges
Aups
Fayence
Mouans-Sartoux
Tavernes
Bargemon
Ampus
Montauroux
Mougins
Varages
Barjols
Callas
Antibes
Le Cannet
Vallauris
Salernes
Villecroze
Draguignan
CAP D'ANTIBES
Brue-Auriac
Mandelieu-la-Napoule
Cannes
Cotignac
VAR
Trans-en-Provence
Bagnols-en-Forêt
ÎLES DE LÉRINS
Carcès
Lorgues
Les Arcs
Le Muy
Puget-sur-Argens
ESTEREL
Saint-Maximin-la-Sainte-Baume
Anthéor
Vidauban
CÔTE D'AZUR
Le Thoronet
Fréjus
Montagne de la Loube 2753 ft. 839 m
Roquebrune-sur-Argens
Saint-Raphaël
Agay
CAP DU DRAMONT
SEA
Tourves
Brignoles
Le Luc
Saint-Aygulf
La Roquebrussanne
Gonfaron
La Nartelle
Signes
Carnoules
MAURES
La Garde-Freinet
Sainte-Maxime
Méounes-lès-Montrieux
Pignans
Puget-Ville
Grimaud
Saint-Tropez
Cuers
Cogolin
La Foux
Pierrefeu-du-Var
Collobrières
Solliès-Pont
Cavalaire-sur-Mer
Ramatuelle
CAP CAMARAT
La Valette-du-Var
Bormes-les-Mimosas
Toulon
La Crau
MEDITERRANEAN
CAP LARDIER
La Garde
Hyères
La Londe
Le Lavandou
Le Rayol-Canadel-sur-Mer
La Seyne-sur-Mer
Hyères-Plage
CAP BÉNAT
Carqueiranne
ÎLE DU LEVANT
CAP SICIÉ
Giens
Porquerolles
Port-Cros
ÎLES D'HYÈRES
Golfe du Lion
ÎLE DE PORQUEROLLES
ÎLE DE PORT-CROS
© Rand McNally A-101842-1
c
Aix-en-Provence
Montagne Sainte-Victoire 3317 ft. 1011 m
Berre-l'Étang
Rognac
BOUCHES-DU-RHÔNE
Étang de Berre
Fuveau
Gardanne
Marignane
Simiane
Châteauneuf
CHAÎNE DE L'ÉTOILE
Auriol
L'Estaque
Roquevaire
Allauch
Carry-le-Rouet
Gémenos
Marseille
Aubagne
La Penne-sur-Huveaune
Mazargues
Cassis
CAP CROISETTE
Golfe du Lion
ÎLE DE RIOU
La Ciotat
© Rand McNally A-101843-1

Inset maps a - c
Lambert Conformal Conic Projection
Scale 1:1,000,000
One inch to 16 miles
One cm to 10 km

BAY OF BISCAY
ATLANTIC
OCEAN
ALBORAN SEA
Gulf of Cadiz
Strait of Gibraltar
PORTUGAL
SPAIN
MOROCCO
CORDILLERA CANTÁBRICA
SISTEMA IBÉRICO
SISTEMA CENTRAL
SIERRA MORENA
CORDILLERA PENIBÉTICA
OLD CASTILE
NEW CASTILE
LEÓN
ANDALUCIA
LA MANCHA
COSTA DEL SOL
MONTES DE TOLEDO
MONTES DE LEÓN
SIERRA DE GATA
SIERRA DE GUADALUPE
SIERRA DE GREDOS
SA. DE GUADARRAMA
SIERRA NEVADA
SIERRA DE SEGURA
MADRID
Lisbon (Lisboa)
Porto
Braga
Coimbra
Setúbal
Faro
Évora
Vigo
A Coruña
Ferrol
Santiago de Compostela
Lugo
Ourense
Pontevedra
Gijón
Oviedo
Avilés
Santander
Bilbao
Donostia-San Sebastián
Gasteiz
Pamplona
Logroño
Burgos
León
Palencia
Valladolid
Zamora
Salamanca
Segovia
Ávila
Toledo
Cuenca
Zaragoza
Soria
Cáceres
Badajoz
Mérida
Ciudad Real
Albacete
Córdoba
Jaén
Sevilla
Huelva
Cádiz
Jerez de la Frontera
Málaga
Granada
Almería
Murcia
Algeciras
Gibraltar (U.K.)
Ceuta (Sp.)
Melilla (Sp.)
Tanger
Tetouan
Nador
Al-Hoceima
Peña Prieta 8320 ft. 2536 m
Pico de Peñalara 7972 ft. 2430 m
Estrela 6539 ft. 1993 m
Javalambre 6627 ft. 2020 m
Sagra 7815 ft. 2382 m
Mulhacén 11 424 ft. 3482 m
CABO ORTEGAL
CABO DE PEÑAS
CABO DE FISTERRA
CABO MONDEGO
CABO CARVOEIRO
CABO DA ROCA
CABO ESPICHEL
CABO DE SÃO VICENTE
CABO DE SANTA MARIA
CABO TRAFALGAR
CABO DE GATA
PUNTA DE ALMINA
CAP SPARTEL
CAP DES TROIS FOURCHES
ISLA DE ALBORÁN (Sp.)
ISLAS CHAFARINAS (Sp.)
© Rand McNally
A-101778-1
0 20 40 60 80 100 120 Miles
0 20 40 60 80 100 120 140 160 180 200 Kilometers

FRANCE
Toulouse
Montpellier
Carcassonne
Narbonne
Perpignan
Golfe du Lion
Béziers
Sète
Agde
Albi
Castres
Arles
Salon-de-Provence
Martigues
Lunel
Auch
Tarbes
Pau
Lourdes
Foix
Pamiers
Limoux
Castelnaudary
Mazamet
Lodève
Gaillac
Graulhet
Colomiers
Muret
Fleurance
Mont-de-Marsan
Aire-sur-l'Adour
Bagnères-de-Bigorre
Saint-Gaudens
Saint-Girons
Lavelanet
Rivesaltes
Ax-les-Thermes
Prades
Argelès-sur-Mer
Port-Vendres
PYRENEES
ANDORRA
Andorra la Vella
Monte Perdido 11 007 ft. 3355 m
Aneto 11 168 ft. 3404 m
Puigcerdà
La Seu d'Urgell
Llança
CAP DE CREUS
Golf de Roses
Figueres
Olot
Banyoles
Ripoll
Berga
Manlleu
Vic
Girona
CAP DE BEGUR
Palamós
Sant Feliu de Guíxols
Blanes
Calella
Mataró
Badalona
BARCELONA
L'Hospitalet de Llobregat
Granollers
Terrassa
Sabadell
Manresa
Igualada
Molins de Rei
Gavà
Vilafranca del Penedès
Vilanova i la Geltrú
El Vendrell
Valls
Reus
Tarragona
Cambrils
CATALUNYA
Huesca
Barbastro
Monzón
Balaguer
Tàrrega
Lleida
Fraga
Embalse de Mequinenza
Caspe
Alcañiz
Golf de Sant Jordi
Tortosa
CAP DE TORTOSA
Amposta
Sant Carles de la Ràpita
Alcanar
Vinaròs
Benicarló
Morella
Torreblanca
Castelló de la Plana
Almassora
Borriana
Onda
La Vall d'Uixó
Llíria
Sagunt
Burjassot
Golf de València
València
L'Albufera
Sueca
Cullera
Tavernes de la Valldigna
Xàtiva
Gandia
Oliva
Dénia
Ontinyent
CAP DE LA NAO
Serra d'Aitana 5112 ft. 1558 m
Alcoi
Benidorm
La Vila Joiosa
Elda
Novelda
Alacant
Elx
Torrevella
Mar Menor
CABO DE PALOS
La Unión
BALEARIC ISLANDS (ILLES BALEARS)
Ciutadella de Menorca
Alaior
Maó
MENORCA (MINORCA)
CAP DE FORMENTOR
Puig Major 4741 ft. 1445 m
Alcúdia
Badia d'Alcúdia
Inca
Sa Pobla
Artà
MALLORCA (MAJORCA)
Palma de Mallorca
Calvià
Llucmajor
Manacor
Felanitx
Badia de Palma
CAP DE SES SALINES
ILLA DE CABRERA
EIVISSA (IBIZA)
Sant Joan de Labritja
Sant Antoni de Portmany
Santa Eulària del Riu
Eivissa
FORMENTERA
MEDITERRANEAN SEA
ALGIERS (ALGER)
Dellys
Tizi Ouzou
'Aïn Benian
Boudouaou
Bordj Menaïel
Bou Ismaïl
Boufarik
Cherchell
Lakhdaria
Ténès
El Affroun
Blida
Bouira
Médéa
Sour el Ghozlane
Le Guelta
Miliana
Khemis Miliana
'Aïn Defla
Chlef (El Asnam)
Oued Fodda
DAHRA
TELL
Sidi Ali
Bou Kadir
Oued Rhiou
Ksar el- Boukhari
Sidi Aïssa
Mostaganem
Aïn Tedelès
Idjioula
Theniet el Had
CAP FERRAT
Arzew
Golfe d'Arzew
Mers el Kebir
Gdyel
Relizane
Zemmora
ALGERIA
'Aïn el Hadjel
Oran (Ouahran)
Tissemsilt
Mohammadia
Tiaret
Mehdia
'Aïn Oussera
Sig
Tighenif
Mascara
Sebkha d'Oran
Sfizef
Sidi Bel Abbès
Ksar Chellala
Sougueur
ATLAS MOUNTAINS
Ben Badis

Lambert Conformal Conic Projection
Scale 1:4,000,000
One inch to 64 miles
One cm to 40 km

a

Guadarrama
Colmenar Viejo
Torrejón del Rey
Villalba La Estación
Embalse de El Pardo
San Lorenzo de El Escorial
Torrelodones
Tres Cantos
San Sebastián de los Reyes
Alalpardo
Valdeavero
Guadalajara
CASTILLA-LA MANCHA
El Escorial
Galapagar
Las Matas
Alcobendas
Meco
Azuqueca de Henares
Embalse de Valmayor
El Pardo
Ajalvir
Camarma de Esteruelas
Paracuellos
Las Rozas de Madrid
Alcalá de Henares
Villanueva de la Cañada
Majadahonda
Santorcaz
Pozuelo de Alarcón
Coslada
Torrejón de Ardoz
Quijorna
San Fernando de Henares
Corpa
Villaviciosa de Odón
MADRID
Mejorada del Campo
Olmeda de las Fuentes
Villanueva de Perales
Alcorcón
Leganés
Rivas-Vaciamadrid
Campo Real
Navalcarnero
Móstoles
Getafe
Arganda del Rey
Orusco de Tajuña
Fuenlabrada
MADRID
Tielmes
El Álamo
Pinto
San Martín de la Vega
Batres
Griñón
Parla
Morata de Tajuña
Valmojado
Cubas de la Sagra
Valdemoro
Villarejo de Salvanés
Las Ventas de Retamosa
Ugena
Ciempozuelos
Valdelaguna
Chinchón
El Viso de San Juan
Illescas
Esquivias
Seseña Nuevo
Villaconejos
Belmonte de Tajo
Colmenar de Oreja
Villamanrique de Tajo
CASTILLA-LA MANCHA
Numancia de la Sagra
Aranjuez
Cabañas de la Sagra
Tagus (Tajo)
© Rand McNally A-101844-1

b

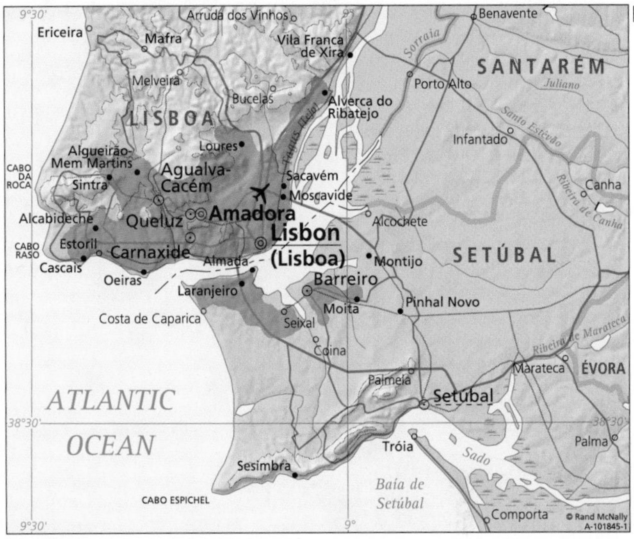
Arruda dos Vinhos
Benavente
Ericeira
Mafra
Vila Franca de Xira
SANTARÉM
Porto Alto
Melveira
Bucelas
Alverca do Ribatejo
LISBOA
Infantado
Loures
CABO DA ROCA
Algueirão-Mem Martins
Agualva-Cacém
Sintra
Sacavém
Moscavide
Canha
Alcabideche
Queluz
Amadora
Lisbon (Lisboa)
Alcochete
CABO RASO
Estoril
Carnaxide
Cascais
Almada
Montijo
SETÚBAL
Oeiras
Laranjeiro
Barreiro
Moita
Pinhal Novo
Costa de Caparica
Seixal
Coina
Marateca
ÉVORA
Palmela
ATLANTIC OCEAN
Setúbal
Tróia
Palma
Sado
Sesimbra
Baía de Setúbal
CABO ESPICHEL
Comporta
© Rand McNally A-101845-1

c

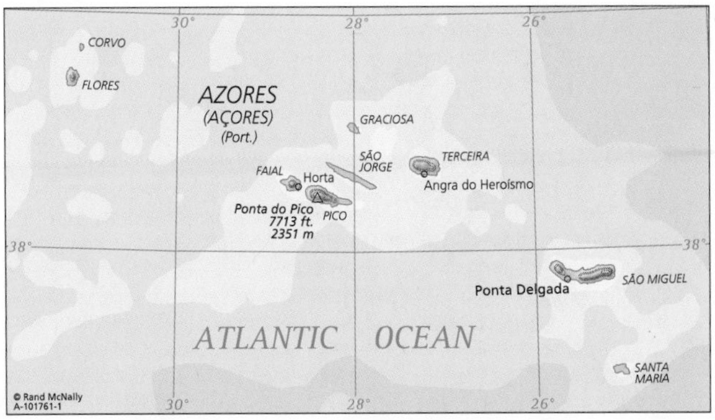
CORVO
FLORES
AZORES (AÇORES) (Port.)
GRACIOSA
SÃO JORGE
TERCEIRA
FAIAL
Horta
Angra do Heroísmo
Ponta do Pico 7713 ft. 2351 m
PICO
SÃO MIGUEL
Ponta Delgada
ATLANTIC OCEAN
SANTA MARIA
© Rand McNally A-101761-1

d

ATLANTIC OCEAN
ISLA ALEGRANZA
LANZAROTE
Arrecife
Santa Cruz de la Palma
LA PALMA
Santa Cruz de Tenerife
Pico del Teide 12 198 ft. 3718 m
Puerto de la Cruz
Las Palmas de Gran Canaria
FUERTEVENTURA
LA GOMERA
TENERIFE
GRAN CANARIA
Tarfaya
CAP JUBY
MOR.
EL HIERRO
CANARY ISLANDS (ISLAS CANARIAS) (Spain)
WESTERN SAHARA
Laayoune (El Aaiún)
© Rand McNally A-101760-1

Inset maps a, b
Lambert Conformal Conic Projection
Scale 1:1,000,000
One inch to 16 miles
One cm to 10 km

Inset maps c, d
Lambert Conformal Conic Projection
Scale 1:8,000,000
One inch to 128 miles
One cm to 80 km

SWITZERLAND
AUSTRIA
SLOVENIA
CROATIA
BOSNIA AND HERZEGOVINA
FRANCE
MONACO
ITALY
SAN MARINO
VATICAN CITY
MALTA
ALPS
APPENNINO LIGURE
DOLOMITI
CARNIC ALPS
JULIAN ALPS
DINARIC ALPS
VELEBIT
HOHE TAUERN
RHAETIAN ALPS
ALPI OROBIE
MARITIME ALPS
APENNINES
LIGURIAN SEA
TYRRHENIAN SEA
ADRIATIC SEA
SEA OF CRETE
MEDITERRANEAN SEA
Golfo di Genova
Golfo di Venezia
Gulf of Trieste
Golfo di Gaeta
Golfo di Napoli
Golfo di Salerno
Golfo di Policastro
Golfo di Taranto
Golfo di Manfredonia
Golfo di Squillace
Golfo di Sant'Eufemia
Golfo di Catania
Golfo di Noto
Golfo di Gela
Strait of Sicily
Strait of Bonifacio
Malta Channel
MILAN (MILANO)
ROME (ROMA)
NAPLES (NAPOLI)
Turin (Torino)
Genoa (Genova)
Florence (Firenze)
Venice (Venezia)
Bologna
Verona
Trieste
Ljubljana
Zagreb
Sarajevo
Palermo
Catania
Messina
Bari
Cagliari
Valletta
CORSICA (CORSE) (France)
SARDINIA (SARDEGNA) (Italy)
SICILY (SICILIA)
CRETE (KRÍTI) (Greece)
Irákleio
Chaniá
Réthymno
Monte Cinto 8878 ft. 2706 m
Monte Incudine 7001 ft. 2134 m
Punta La Marmora 6017 ft. 1834 m
Corno Grande 9560 ft. 2914 m
Monte Amaro 9170 ft. 2795 m
Vesuvius 4203 ft. 1281 m
Monte Etna 10 902 ft. 3323 m
Mont Blanc 15 771 ft. 4807 m
Gran Paradiso 13 323 ft. 4061 m
Triglav 9396 ft. 2864 m
Ⓐ Kosovo unilaterally declared its independence from Serbia in 2008.
0 20 40 60 80 100 120 Miles
0 20 40 60 80 100 120 140 160 180 200 Kilometers
Lambert Conformal Conic Projection
Scale 1:4,500,000
One inch to 71 miles
One cm to 45 km
Inset map b
Lambert Conformal Conic Projection
Scale 1:2,000,000
One inch to 32 miles
One cm to 20 km
© Rand McNally A-101846-1
© Rand McNally A-101762-1

ROMANIA
SERBIA
BULGARIA
MACEDONIA
ALBANIA
GREECE
TURKEY
MONTENEGRO
MOLDOVA
UKRAINE
KOSOVO-METOHIJA
VOJVODINA
BALKAN PENINSULA
TRANSYLVANIA
TRANSYLVANIAN ALPS
CARPAŢII MERIDIONALI
WALLACHIA
STARA PLANINA
SREDNA GORA
RHODOPE MOUNTAINS
YILDIZ DAĞLARI
THRACE
PINDOS ÓROS
THESSALY
PELOPONNESUS
BLACK SEA
AEGEAN SEA
IONIAN SEA
SEA OF MARMARA (MARMARA DENIZI)
SEA OF CRETE
MYRTÓON PÉLAGOS
THRAKIKÓ PÉLAGOS
IONIAN ISLANDS (IÓNIOI NÍSOI)
CYCLADES (KIKLÁDES)
DODECANESE (DODEKANISOY)
VÓREIOI SPORADES
Strait of Otranto
Thermaïkós Kólpos
Strymonikós Kólpos
Saronikós Kólpos
Argolikós Kólpos
Lakonikós Kólpos
Messiniakós Kólpos
Kyparissiakós Kólpos
Gulf of Corinth (Korinthiakós Kólpos)
Patraïkós Kólpos
Pagasitikós Kólpos
Amvrakikós Kólpos
Edremit Körfezi
Gemlik Körfezi
Kuşadası Körfezi
Gökova Körfezi
Saros Körfezi
İzmir Körfezi
BELGRADE (BEOGRAD)
BUCHAREST (BUCUREŞTI)
SOFIA (SOFIYA)
Skopje
Tiranë
Athens (Athína)
İSTANBUL
İZMIR
BURSA
Thessaloníki
Plovdiv
Varna
Constanţa
Braşov
Ploieşti
Craiova
Timişoara
Novi Sad
Niš
Priština
Podgorica
Edirne
Denizli
Pátra
Lárisa
Vârful Moldoveanu 8346 ft. 2544 m
Musala 9596 ft. 2925 m
Mt. Olympus 9570 ft. 2917 m
Mt. Athos 6660 ft. 2030 m
Parnassos 8061 ft. 2457 m
Uludağ 8343 ft. 2543 m
Korab 9035 ft. 2754 m
Durmitor 8274 ft. 2522 m
Botev 7795 ft. 2376 m
Boz Dağ 7936 ft. 2419 m
© Rand McNally A-101796-1

RUSSIA
ESTONIA
LATVIA
LITHUANIA
BELARUS
MOSCOW (MOSKVA)
MINSK
VOLOGODSKAYA OBLAST'
YAROSLAVSKAYA OBLAST'
IVANOVSKAYA OBLAST'
VLADIMIRSKAYA OBLAST'
RYAZANSKAYA OBLAST'
LIPETSKAYA OBLAST'
MOSKOVSKAYA OBLAST'
TUL'SKAYA OBLAST'
KALUZHSKAYA OBLAST'
ORLOVSKAYA OBLAST'
KURSKAYA OBLAST'
BRYANSKAYA OBLAST'
SMOLENSKAYA OBLAST'
TVERSKAYA OBLAST'
NOVGORODSKAYA OBLAST'
PSKOVSKAYA OBLAST'
LENINGRADSKAYA OBLAST'
VALDAYSKAYA VOZVYSHENNOST'
SREDNERUSSKAYA VOZVYSHENNOST'
VICEBSK
MAHILËU
HOMEL'
MINSK
HRODNA
BREST
PRIPET MARSHES
Cherepovets
Rybinsk
Yaroslavl'
Vladimir
Ryazan'
Lipetsk
Voronezh
Tula
Kaluga
Orël
Kursk
Bryansk
Smolensk
Tver'
Novgorod
Pskov
Tartu
Daugavpils
Vilnius
Vicebsk
Mahilëu
Homel'
Babrujsk
Baranavičy
Pinsk
Lake Peipus
Ozero Il'men'
Rybinskoye Vodokhranilishche
Volga
Dnieper
Western Dvina

0 20 40 60 80 100 120 Miles

0 20 40 60 80 100 120 140 160 180 200 Kilometers

Lambert Conformal Conic Projection
Scale 1:4,000,000
One inch to 64 miles
One cm to 40 km

UKRAINE
MOLDOVA
ROMANIA
BLACK SEA
SEA OF AZOV
KIEV (KYIV)
KHARKIV
ODESA
DNIPROPETROVS'K
DONETS'K
ZAPORIZHZHIA
ROSTOV-NA-DONU
CHIŞINĂU
KRASNODARSKIY KRAY
ROSTOVSKAYA OBLAST'
BELGORODSKAYA OBLAST'
VORONEZHSKAYA OBLAST'
RESPUBLIKA KRYM
CRIMEAN PENINSULA (KRYMS'KYI PIVOSTRIV)
CARPATHIAN MOUNTAINS
CAUCASUS MOUNTAINS
KERCHENS'KYI PIVOSTRIV
Sevastopol' MISTO SEVASTOPOL'
Simferopol
Kerch
Mariupol'
Luhans'k
Kherson
Mykolaiv
Kryvyi Rih
Poltava
Sumy
Chernihiv
Cherkasy
Kirovohrad
Vinnytsia
Zhytomyr
Khmel'nyts'kyi
Rivne
Constanța
Galați
Iași
Bacău
Krasnodar
Novorossiysk
Taganrog
Belgorod
The Pridnestrovian Moldavian Republic has unilaterally declared its independence from Moldova.
© Rand McNally A-101788-1

a

Blackpool
Preston
Blackburn
Burnley
Bradford
Leeds
Halifax
Dewsbury
Huddersfield
Rochdale
Bolton
Wigan
Oldham
Liverpool
St. Helens
Manchester
Salford
Stockport
Warrington
Birkenhead
Widnes
Runcorn
Ellesmere Port
Chester
Macclesfield
Southport
IRISH SEA
Crosby
Bootle
Wallasey
Huyton-with-Roby
Sale
LANCASHIRE PLAIN
CHESHIRE PLAIN
ENGLAND
WALES
Kinder Scout 2087 ft. 636 m
HIGH PEAK
Pendle Hill 1827 ft. 557 m
Boulsworth Hill 1699 ft. 518 m
© Rand McNally A-101847-1

b

Newcastle-under-Lyme
Stoke-on-Trent
Nottingham
Derby
Stapleford
Stafford
Burton upon Trent
Rugeley
CANNOCK CHASE
VALE OF TRENT
Cannock
Lichfield
Tamworth
Leicester
Walsall
Sutton Coldfield
Wolverhampton
West Bromwich
Dudley
Nuneaton
Birmingham
Coventry
Solihull
Kidderminster
Rugby
Redditch
Loughborough
© Rand McNally A-101848-1

c

LAZIO
VATICAN CITY
ROME (ROMA)
CAMPAGNA DI ROMA
Guidonia-Montecelio
Tivoli
Fiumicino
Lido di Ostia
Frascati
Ciampino
COLLI ALBANI
Pomezia
Aprilia
Velletri
TYRRHENIAN SEA
Latina
Anzio
Nettuno
Monte Pellecchia 4488 ft. 1368 m
© Rand McNally A-101849-1

d

BRANDENBURG
Oranienburg
Bernau bei Berlin
BERLIN
Potsdam
Spandau
Neuenhagen
Kleinmachnow
Teltow
Ludwigsfelde
Königs Wusterhausen
Zossen
Strausberg
Eberswalde-Finow
© Rand McNally A-101850-1

e

Helsingør
Helsingborg
SKÅNE
FREDERIKSBORG
Hillerød
Landskrona
Øresund
SWEDEN
DENMARK
Kongens Lyngby
Gentofte
KØBENHAVNS KOMMUNE
Frederiksberg
Copenhagen (København)
KØBENHAVN
Malmö
ROSKILDE
Roskilde
AMAGER
Køge Bugt
© Rand McNally A-101851-1

f

STEREA ELLADA
ATTIKÍ
Párnitha 4636 ft. 1413 m
Acharnés
Elefsína
Amaroúsion
Peiraiás
Athens (Athína)
Salamína
SALAMÍNA
Saronikós Kólpos
Glyfáda
© Rand McNally A-101854-1

g
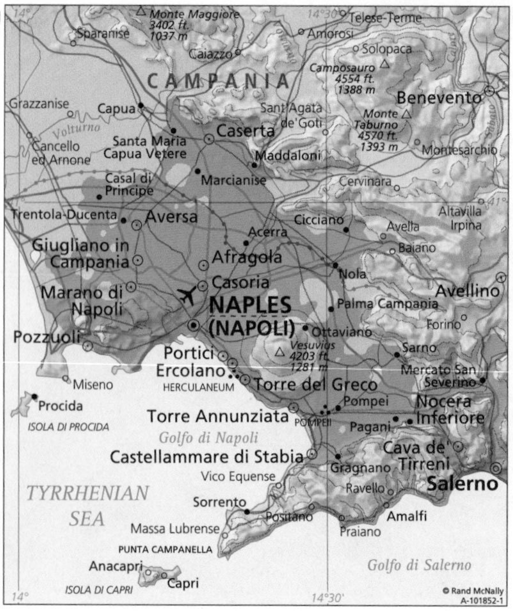
CAMPANIA
Monte Maggiore 3402 ft. 1037 m
Camposauro 4554 ft. 1388 m
Monte Taburno 4570 ft. 1393 m
Benevento
Caserta
Aversa
Giugliano in Campania
Afragola
Casoria
NAPLES (NAPOLI)
Marano di Napoli
Pozzuoli
Nola
Avellino
Vesuvius 4203 ft. 1281 m
Portici
Ercolano
Torre del Greco
Torre Annunziata
Nocera Inferiore
Golfo di Napoli
Castellammare di Stabia
Cava de' Tirreni
Salerno
TYRRHENIAN SEA
Sorrento
Capri
Golfo di Salerno
© Rand McNally A-101852-1

h
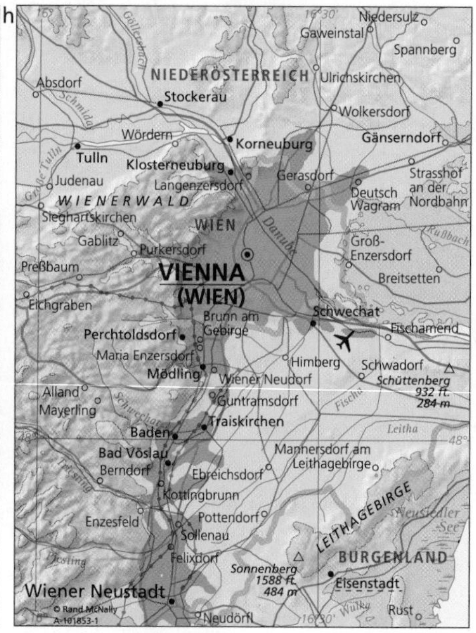
NIEDERÖSTERREICH
Stockerau
Korneuburg
Klosterneuburg
WIENERWALD
WIEN
VIENNA (WIEN)
Schwechat
Mödling
Baden
LEITHAGEBIRGE
BURGENLAND
Wiener Neustadt
Eisenstadt
© Rand McNally A-101853-1

j

BLACK SEA
İSTANBUL
Sarıyer
Beykoz
Üsküdar
Bakırköy
Yeşilköy
Sultanbeyli
SEA OF MARMARA (MARMARA DENIZI)
KIZIL ADALAR
Pendik
KOCAELI
© Rand McNally A-101855-1

0 5 10 15 20 25 30 Miles
0 5 10 15 20 25 30 35 40 45 50 Kilometers

Lambert Conformal Conic Projection
Scale 1:1,000,000
One inch to 16 miles
One cm to 10 km

a

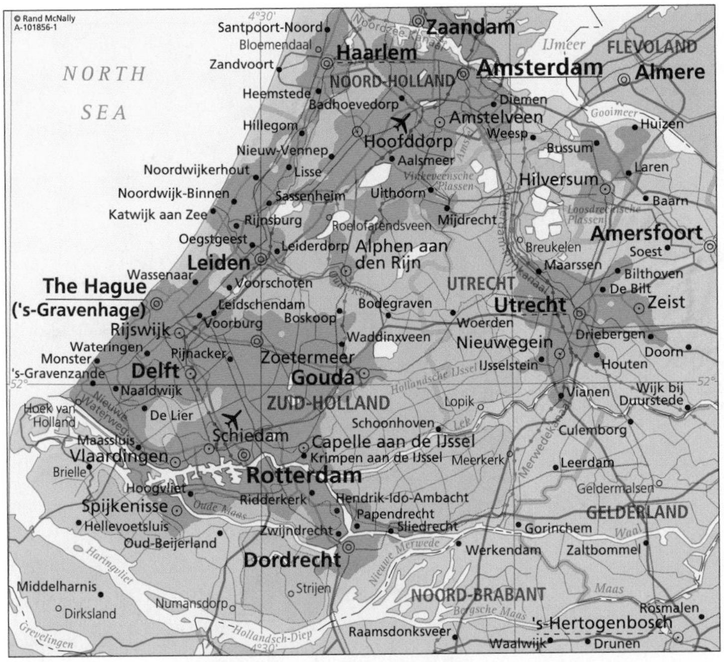
© Rand McNally A-101856-1
NORTH SEA
Santpoort-Noord
Bloemendaal
Zaandam
Haarlem
IJmeer
FLEVOLAND
Zandvoort
Amsterdam
Almere
Heemstede
NOORD-HOLLAND
Badhoevedorp
Diemen
Amstelveen
Gooimeer
Hillegom
Hoofddorp
Weesp
Huizen
Nieuw-Vennep
Bussum
Noordwijkerhout
Lisse
Aalsmeer
Laren
Noordwijk-Binnen
Sassenheim
Uithoorn
Hilversum
Baarn
Katwijk aan Zee
Rijnsburg
Roelofarendsveen
Mijdrecht
Oegstgeest
Leiderdorp
Alphen aan den Rijn
Breukelen
Amersfoort
Soest
Wassenaar
Leiden
Maarssen
Bilthoven
The Hague ('s-Gravenhage)
Voorschoten
UTRECHT
De Bilt
Leidschendam
Bodegraven
Utrecht
Zeist
Voorburg
Boskoop
Woerden
Rijswijk
Waddinxveen
Nieuwegein
Driebergen
Wateringen
Pijnacker
Zoetermeer
Doorn
Monster
's-Gravenzande
Delft
Gouda
IJsselstein
Houten
Naaldwijk
ZUID-HOLLAND
Lopik
Wijk bij Duurstede
Vianen
Hoek van Holland
De Lier
Schoonhoven
Culemborg
Schiedam
Maassluis
Capelle aan de IJssel
Vlaardingen
Krimpen aan de IJssel
Meerkerk
Leerdam
Brielle
Rotterdam
Hoogvliet
Geldermalsen
Spijkenisse
Ridderkerk
Hendrik-Ido-Ambacht
GELDERLAND
Papendrecht
Hellevoetsluis
Zwijndrecht
Sliedrecht
Gorinchem
Oud-Beijerland
Dordrecht
Werkendam
Zaltbommel
Haringvliet
Middelharnis
Strijen
Dirksland
Numansdorp
NOORD-BRABANT
Maas
Rosmalen
Hollandsch Diep
Raamsdonksveer
's-Hertogenbosch
Grevelingen
Waalwijk
Drunen
4°30'
52°

b

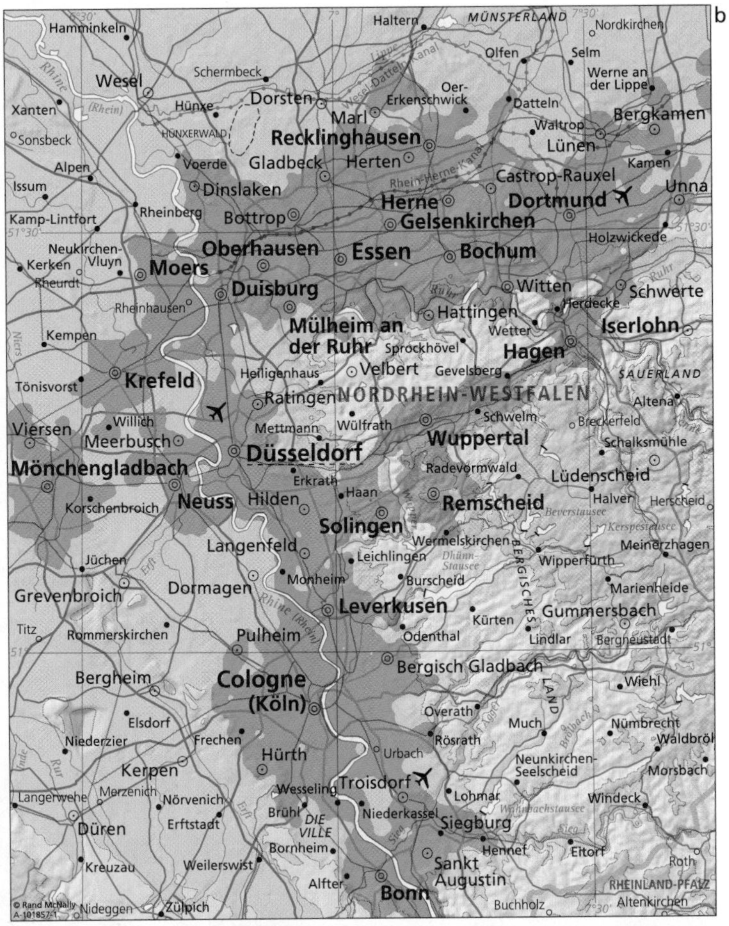
Hamminkeln
Haltern
MÜNSTERLAND
Nordkirchen
Olfen
Selm
Schermbeck
Werne an der Lippe
Wesel
Dorsten
Oer-Erkenschwick
Datteln
Xanten
Hünxe
Marl
Waltrop
Bergkamen
Sonsbeck
Recklinghausen
Lünen
Alpen
Voerde
Gladbeck
Herten
Kamen
Issum
Dinslaken
Castrop-Rauxel
Unna
Kamp-Lintfort
Rheinberg
Bottrop
Herne
Dortmund
Gelsenkirchen
Neukirchen-Vluyn
Holzwickede
Kerken
Oberhausen
Essen
Bochum
Moers
Witten
Rheurdt
Duisburg
Schwerte
Herdecke
Rheinhausen
Hattingen
Mülheim an der Ruhr
Iserlohn
Kempen
Wetter
Sprockhövel
Hagen
Velbert
Heiligenhaus
Gevelsberg
SAUERLAND
Tönisvorst
Krefeld
Ratingen
NORDRHEIN-WESTFALEN
Altena
Schwelm
Breckerfeld
Willich
Mettmann
Wülfrath
Viersen
Meerbusch
Wuppertal
Schalksmühle
Düsseldorf
Mönchengladbach
Radevormwald
Erkrath
Lüdenscheid
Haan
Halver
Herscheid
Korschenbroich
Neuss
Hilden
Remscheid
Solingen
Meinerzhagen
Langenfeld
Wermelskirchen
Jüchen
Leichlingen
Wipperfürth
Monheim
Burscheid
Grevenbroich
Dormagen
Marienheide
Leverkusen
Gummersbach
Kürten
Titz
Rommerskirchen
Pulheim
Odenthal
Lindlar
Bergneustadt
Bergisch Gladbach
Bergheim
Cologne (Köln)
Wiehl
Overath
Elsdorf
Much
Nümbrecht
Niederzier
Frechen
Rösrath
Waldbröl
Hürth
Urbach
Neunkirchen-Seelscheid
Kerpen
Morsbach
Troisdorf
Langerwehe
Merzenich
Nörvenich
Wesseling
Lohmar
Windeck
Brühl
Düren
Erftstadt
Niederkassel
Siegburg
Bornheim
Hennef
Eitorf
Kreuzau
Weilerswist
Sankt Augustin
Roth
Alfter
Bonn
RHEINLAND-PFALZ
Nideggen
Zülpich
Buchholz
Altenkirchen
© Rand McNally A-101857-1

c

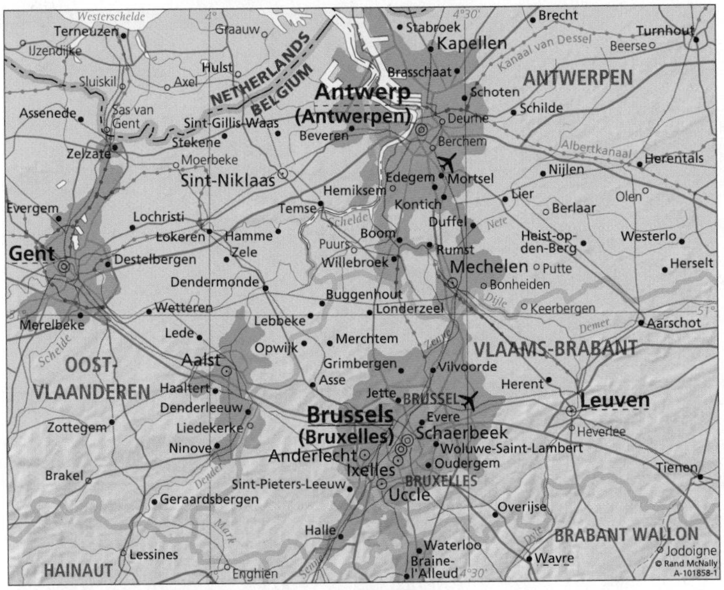
Westerschelde
Brecht
Terneuzen
Graauw
Stabroek
Turnhout
Uzendijke
NETHERLANDS
BELGIUM
Kapellen
Beerse
Hulst
Brasschaat
ANTWERPEN
Sluiskil
Axel
Antwerp (Antwerpen)
Schoten
Assenede
Sas van Gent
Schilde
Sint-Gillis-Waas
Deurne
Zelzate
Stekene
Beveren
Berchem
Moerbeke
Herentals
Sint-Niklaas
Edegem
Mortsel
Nijlen
Hemiksem
Lier
Evergem
Temse
Kontich
Berlaar
Olen
Lochristi
Duffel
Lokeren
Hamme
Puurs
Heist-op-den-Berg
Westerlo
Gent
Zele
Boom
Destelbergen
Rumst
Willebroek
Mechelen
Putte
Herselt
Dendermonde
Bonheiden
Buggenhout
Wetteren
Londerzeel
Keerbergen
Merelbeke
Lebbeke
Aarschot
Lede
Merchtem
Opwijk
VLAAMS-BRABANT
Aalst
Grimbergen
OOST-VLAANDEREN
Vilvoorde
Asse
Haaltert
Herent
Jette
Brussel
Leuven
Denderleeuw
Brussels (Bruxelles)
Evere
Liedekerke
Schaerbeek
Heverlee
Zottegem
Ninove
Anderlecht
Woluwe-Saint-Lambert
Oudergem
Ixelles
Brakel
Tienen
Sint-Pieters-Leeuw
Geraardsbergen
Uccle
Overijse
BRUXELLES
Halle
BRABANT WALLON
Lessines
Waterloo
Jodoigne
Wavre
HAINAUT
Enghien
Braine-l'Alleud
© Rand McNally A-101858-1

d

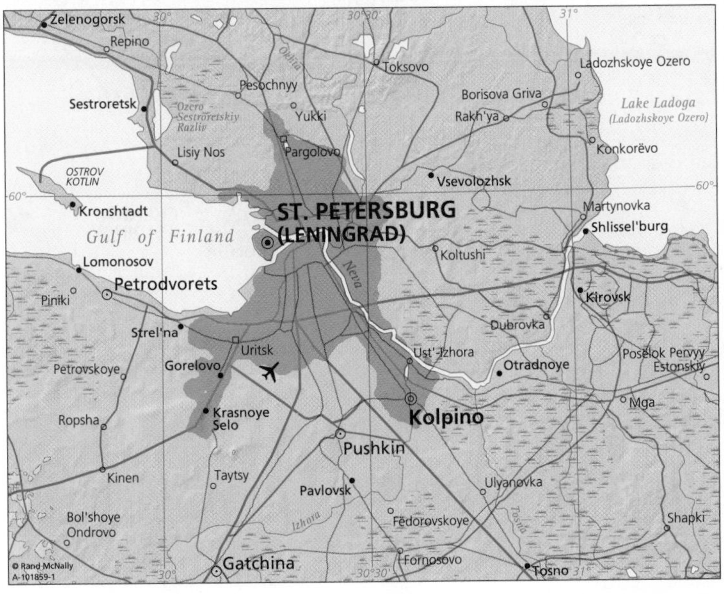
Zelenogorsk
Repino
Toksovo
Ladozhskoye Ozero
Pesochnyy
Sestroretsk
Borisova Griva
Lake Ladoga (Ladozhskoye Ozero)
Ozero Sestroretskiy Razliv
Yukki
Rakh'ya
Pargolovo
Konkorëvo
Lisiy Nos
OSTROV KOTLIN
Vsevolozhsk
Martynovka
Kronshtadt
ST. PETERSBURG (LENINGRAD)
Shlissel'burg
Gulf of Finland
Koltushi
Lomonosov
Petrodvorets
Piniki
Kirovsk
Strel'na
Dubrovka
Uritsk
Ust'-Izhora
Poselok Pervyy Estonskiy
Petrovskoye
Gorelovo
Otradnoye
Neva
Mga
Krasnoye Selo
Ropsha
Kolpino
Pushkin
Kinen
Taytsy
Ulyanovka
Pavlovsk
Bol'shoye Ondrovo
Fëdorovskoye
Shapki
Gatchina
Fornosovo
Tosno
© Rand McNally A-101859-1

e

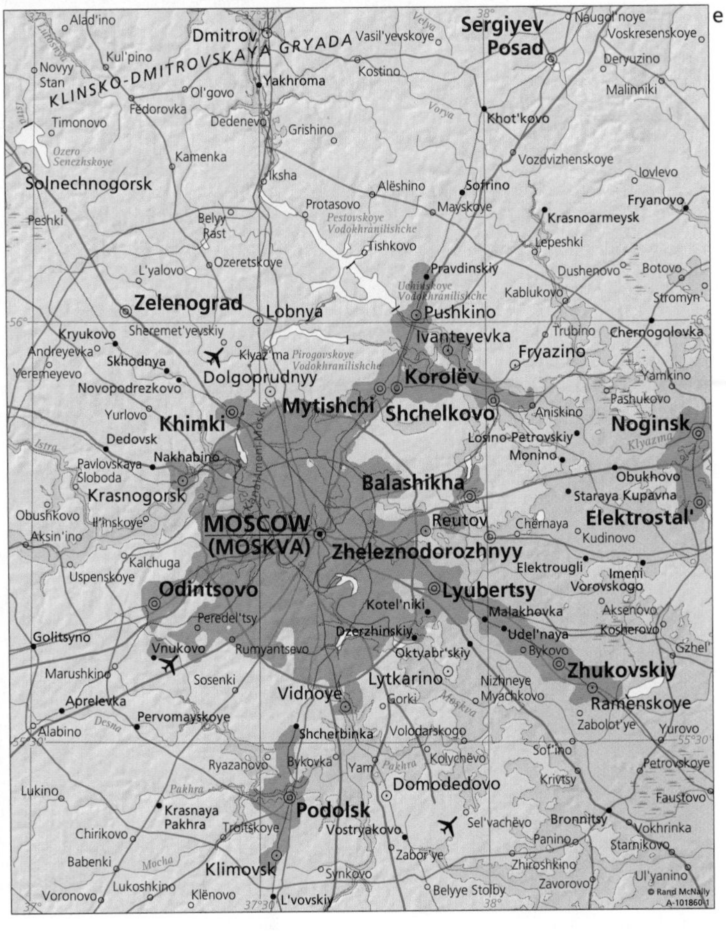
Alad'ino
Dmitrov
Vasil'yevskoye
Sergiyev Posad
Naugol'noye
Voskresenskoye
Kul'pino
KLINSKO-DMITROVSKAYA GRYADA
Novyy Stan
Yakhroma
Kostino
Deryuzino
Fedorovka
Ol'govo
Malinniki
Dedenevo
Khot'kovo
Timonovo
Grishino
Ozero Senezhskoye
Kamenka
Vozdvizhenskoye
Iksha
Iovlevo
Solnechnogorsk
Alëshino
Sofrino
Fryanovo
Protasovo
Mayskoye
Pestovskoye Vodokhranilishche
Peshki
Krasnoarmeysk
Belyy Rast
Tishkovo
Lepeshki
Ozeretskoye
L'yalovo
Pravdinskiy
Dushenovo
Botovo
Zelenograd
Lobnya
Kablukovo
Pushkino
Stromyn'
Kryukovo
Sheremet'yevskiy
Ivanteyevka
Trubino
Chernogolovka
Andreyevka
Klyaz'ma
Fryazino
Skhodnya
Yeremeyevo
Dolgoprudnyy
Korolëv
Yamkino
Novopodrezkovo
Pashukovo
Mytishchi
Yurlovo
Khimki
Shchelkovo
Aniskino
Noginsk
Dedovsk
Losino-Petrovskiy
Nakhabino
Monino
Pavlovskaya Sloboda
Balashikha
Obukhovo
Krasnogorsk
Staraya Kupavna
Obushkovo
Reutov
Chernaya
Elektrostal'
Aksin'ino
Il'inskoye
MOSCOW (MOSKVA)
Kudinovo
Zheleznodorozhnyy
Kalchuga
Elektrougli
Imeni Vorovskogo
Uspenskoye
Odintsovo
Lyubertsy
Kotel'niki
Aksenovo
Malakhovka
Peredel'tsy
Kosherovo
Golitsyno
Dzerzhinskiy
Udel'naya
Vnukovo
Rumyantsevo
Bykovo
Gzhel'
Oktyabr'skiy
Marushkino
Lytkarino
Zhukovskiy
Sosenki
Nizhneye Myachkovo
Vidnoye
Aprelevka
Gorki
Ramenskoye
Pervomayskoye
Alabino
Shcherbinka
Volodarskogo
Zabolot'ye
Yurovo
Ryazanovo
Bykovka
Kolychëvo
Yam
Sof'ino
Petrovskoye
Lukino
Krasnaya Pakhra
Domodedovo
Krivtsy
Faustovo
Podolsk
Chirikovo
Troitskoye
Vostryakovo
Sel'vachëvo
Bronnitsy
Vokhrinka
Panino
Babenki
Klimovsk
Zabor'ye
Starnikovo
Zhiroshkino
Synkovo
Lukoshkino
Zavorovo
Voronovo
Klënovo
L'vovskiy
Belyye Stolby
Ul'yanino
© Rand McNally A-101860-1

0 5 10 15 20 25 30 Miles

0 5 10 15 20 25 30 35 40 45 50 Kilometers

Lambert Conformal Conic Projection
Scale 1:1,000,000
One inch to 16 miles
One cm to 10 km

ARCTIC OCEAN
ATLANTIC OCEAN
NORWEGIAN SEA
BARENTS SEA
KARA SEA
SEVERNAYA ZEMLYA
FRANZ JOSEF LAND
SVALBARD (Norway)
NOVAYA ZEMLYA
OSTROV KOLGUYEV
FAROE ISLANDS (Den.)
AZORES (Port.)
Arctic Circle
Khatanga
Dudinka
Noril'sk
Salekhard
Gora Narodnaya 6214 ft. 1894 m
Khanty-Mansiysk
R U S
IRELAND
UNITED KINGDOM
LONDON
NORTH SEA
NORWAY
SWEDEN
FINLAND
Stockholm
DENMARK
NETH.
BELG.
PARIS
FRANCE
BERLIN
GERMANY
BALTIC SEA
EST.
LATVIA
LITH.
RUS.
Arkhangel'sk
Lake Onega
Pechora
URAL MTS.
ST. PETERSBURG (LENINGRAD)
PORTUGAL
SPAIN
MADRID
Rhine
SWITZ.
CZECH REP.
POLAND
BELARUS
AUSTRIA
SLVK.
HUNGARY
MOSCOW
Volga
Oka
Nizhniy Tagil
YEKATERINBURG
Tyumen'
Irtysh
Ob'
Yenisey
Podkamennaya
Tomsk
Krasnoyarsk
CORSICA
CROATIA
BUDAPEST
KIEV
UKRAINE
MOLD.
CASABLANCA
MOROCCO
ALGIERS
ITALY
ROME
SARDINIA
MEDITERRANEAN SEA
BOS.
SERBIA
ROMANIA
BUCHAREST
Danube
Don
Magnitogorsk
Orsk
CHELYABINSK
OMSK
NOVOSIBIRSK
Barnaul
Novokuzneck
Kyzyl
Tobol
Ishim
SICILY
ALB.
MACE.
BULGARIA
Astana
ALGERIA
TUNISIA
BLACK SEA
CAUCASUS MTS.
Gora El'brus 18 510 ft. 5642 m
Ural
Qaraghandy
Semey
KAZAKHSTAN
Aral Sea
Zhezqazghan
Caspian Sea
İSTANBUL
GREECE
ANKARA
İZMİR
TURKEY
GEORGIA
ARM.
TBILISI
Yerevan
AZER.
BAKU
Lake Balkhash
Balqash
ALTAY MTS.
CRETE
Nukus
UZBEKISTAN
Syr Darya
ALMATY
TIEN SHAN
ÜRÜMQI
Bishkek
KYRGYZSTAN
TASHKENT
Tarim
LIBYA
CYPRUS
Nicosia
ALEPPO
Beirut
LEBANON
SYRIA
TABRĪZ
TURKMENISTAN
Aşgabat
MASHHAD
Samarqand
Dushanbe
TAJIKISTAN
Kashi
ISRAEL
Jerusalem
DAMASCUS
Euphrates
Tigris
TEHRĀN
Kūh-e Damvand 18 386 ft. 5604 m
Amu Darya
CAIRO
'Ammān
JORDAN
IRAQ
BAGHDĀD
K2 28 250 ft. 8611 m
KUNLUN SHAN
Tropic of Cancer
EGYPT
ES̄FAHĀN
Kabul
AFGHANISTAN
Basra
Kuwait
IRAN
Indus
Islāmābād
NIGER
KUWAIT
LAHORE
SHĪRĀZ
Quetta
SAUDI ARABIA
BAHRAIN
RIYADH
Persian Gulf
DELHI
PAKISTAN
Lhasa
Mt. Everest 29 028 ft. 8848 m
Brahmaputra
NEPAL
BHUTAN
Lake Nasser
Nile
Medina
QATAR
Doha
New Delhi
HIMALAYA
CHAD
JIDDAH
Mecca
RED SEA
Abu Dhabi
U.A.E.
HYDERĀBĀD
Kathmandu
Thimphu
Yamuna
KĀNPUR
Ganges
Muscat
KARĀCHI
PATNA
DHAKA
Lake Chad
RA'S AL-HADD
Khartoum
OMAN
AHMADĀBĀD
INDIA
KOLKATA (CALCUTTA)
BNGL.
Chari
SUDAN
Blue Nile
White Nile
ERITREA
Narmada
NĀGPUR
YEMEN
SANAA
MUMBAI (BOMBAY)
Godāvari
HYDERĀBĀD
Aden
ARABIAN SEA
Krishna
CENTRAL AFRICAN REPUBLIC
DJIBOUTI
Gulf of Aden
SOCOTRA
BAY OF BENGAL
ADDIS ABABA
ETHIOPIA
Mountain Nile
Oubangui
Congo
CONGO
BENGALŪRU
CHENNAI (MADRAS)
Cochin
SOMALIA
Lake Rudolf
SRI LANKA
CAPE COMORIN
Colombo
DONDRA HEAD
DEMOCRATIC REPUBLIC OF THE CONGO
UGANDA
KENYA
Male'
MALDIVES
RWANDA
Equator
BURUNDI
Lake Victoria
NAIROBI
INDIAN OCEAN
Kasai
Lake Tanganyika
TANZANIA
DAR ES SALAAM
ANGOLA
SEYCHELLES
ZAMBIA
CHAGOS ARCHIPELAGO (B.I.O.T.)
0 200 400 600 800 1000 1200 Miles
0 200 400 600 800 1000 1200 1400 1600 1800 2000 Kilometers

Lambert Azimuthal Equal Area Projection
Scale 1:40,000,000
One inch to 640 miles
One cm to 400 km

UNITED STATES
BERING SEA
ST. LAWRENCE ISLAND
MYS DEZHNEVA
MYS NAVARIN
ALEUTIAN ISLANDS
OSTROV VRANGELYA
EAST SIBERIAN SEA
NEW SIBERIAN ISLANDS
LAPTEV SEA
Cherskiy
Anadyr'
CAPE OLYUTORSKI
Arctic Circle
Indigirka
Kolyma
Tiksi
Gora Pobeda 10 325 ft. 3147 m
Zhigansk
Magadan
Klyuchevskaya Sopka Vulkan 15 584 ft. 4750 m
Petropavlovsk-Kamchatskiy
MYS LOPATKA
Lena
Vilyuy
Yakutsk
Aldan
Lensk
Tunguska
SEA OF OKHOTSK
SAKHALIN
KURIL ISLANDS
Nikolayevsk-na-Amure
Komsomol'sk-na-Amure
Yuzhno-Sakhalinsk
Lake Baikal
Chita
Irkutsk
Ulan-Ude
Argun'
Amur
Ussuri
Khabarovsk
PACIFIC OCEAN
MIDWAY ISLANDS (U.S.)
SAPPORO
HOKKAIDŌ
QIQIHAR
HARBIN
Vladivostok
Ulaanbaatar
Kerulen
MONGOLIA
CHANGCHUN
SEA OF JAPAN
Sendai
HONSHŪ
SHENYANG
NORTH KOREA
P'YŎNGYANG
TŌKYŌ
Fuji-san 12 388 ft. 3776 m
BEIJING
GOBI DESERT
Huang
SEOUL
SOUTH KOREA
NAGOYA
ŌSAKA
JAPAN
TIANJIN
QINGDAO
PUSAN
SHIKOKU
Yinchuan
TAIYUAN
JINAN
LANZHOU
ZHENGZHOU
KYŪSHŪ
Kagoshima
CHINA
XI'AN
NANJING
SHANGHAI
BONIN ISLANDS (Japan)
VOLCANO ISLANDS (Japan)
WAKE ISLAND (U.S.)
Tropic of Cancer
Yalong
CHENGDU
WUHAN
Yangtze
EAST CHINA SEA
RYUKYU ISLANDS
Naha
OKINAWA-JIMA
NANCHANG
CHONGQING
CHANGSHA
GUIYANG
FUZHOU
T'AIPEI
TAIWAN
MICRONESIA
KUNMING
PHILIPPINE SEA
MARIANA IS.
NORTHERN MARIANA ISLANDS (U.S.)
MARSHALL ISLANDS
GUANGZHOU
HONG KONG
SAIPAN
TINIAN
GUAM (U.S.)
Mandalay
Ha Noi
Haikou
ESCARPADA POINT
MYANMAR (BURMA)
LAOS
HAINAN DAO
LUZON
PHILIPPINES
CAROLINE ISLANDS
POHNPEI
YANGON
Vientiane
Mekong
Da Nang
THAILAND
VIETNAM
MANILA
MINDORO
SAMAR
KOSRAE
KIRIBATI
BANGKOK
SOUTH CHINA SEA
Cebu
CAMBODIA
PALAU
FEDERATED STATES OF MICRONESIA
ANDAMAN ISLANDS (India)
ANDAMAN SEA
NEGROS
MINDANAO
Phnom Penh
HO CHI MINH CITY (SAIGON)
Davao
SULU SEA
NAURU
NICOBAR ISLANDS (India)
MUI CA MAU
Gunong Kinabalu 13 455 ft. 4101 m
CELEBES SEA
Equator
Bandar Seri Begawan
BRUNEI
MALAYSIA
MELANESIA
MEDAN
KUALA LUMPUR
Manado
NEW IRELAND
TANJUNG D'URVILLE
Jayapura
BISMARCK SEA
BOUGAINVILLE
CHOISEUL
SINGAPORE
SUMATRA
Balikpapan
BORNEO
MOLUCCAS
Puncak Jaya 16 503 ft. 5030 m
NEW GUINEA
NEW BRITAIN
SANTA ISABEL
Padang
CELEBES
MALAITA
Gunung Kerinci 12 467 ft. 3800 m
PALEMBANG
Banjarmasin
BANDA SEA
PAPUA NEW GUINEA
SOLOMON SEA
SANTA CRUZ IS.
SAN CRISTOBAL
SOLOMON ISLANDS
JAVA SEA
MAKASSAR
INDONESIA
TANJUNG VALS
Port Moresby
JAKARTA
SURABAYA
FLORES SEA
Dili
EAST TIMOR
ARAFURA SEA
JAVA
BALI
FLORES
SUMBAWA
SUMBA
TIMOR
MELVILLE ISLAND
CHRISTMAS ISLAND (Austl.)
TIMOR SEA
AUSTRALIA
© Rand McNally A-101711-1

ARCTIC OCEAN
ATLANTIC OCEAN
NORWEGIAN SEA
BARENTS SEA
KARA SEA
NORTH SEA
BALTIC SEA
Bay of Biscay
MEDITERRANEAN SEA
BLACK SEA
CASPIAN SEA
Aral Sea
RED SEA
Persian Gulf
Gulf of Oman
Gulf of Aden
ARABIAN SEA
BAY OF BENGAL
INDIAN OCEAN
AZORES (Port.)
FAROE ISLANDS (Den.)
SVALBARD (Norway)
FRANZ JOSEF LAND
SEVERNAYA ZEMLYA
NOVAYA ZEMLYA
TAYMYR PENINSULA
YAMAL PENINSULA
OSTROV KOLGUYEV
MYS KANIN NOS
NORDKAPP
Arctic Circle
NORWEGIAN TRENCH
Galdhøpiggen 8100 ft. 2469 m
IRELAND
MIZEN HEAD
UNITED KINGDOM
London
NORWAY
SWEDEN
FINLAND
DENMARK
NETH.
BELG.
GERMANY
FRANCE
Paris
Mont Blanc 15 771 ft. 4807 m
SWITZ.
ALPS
AUSTRIA
CZECH REP.
SLVK.
POLAND
HUNGARY
CARPATHIAN MTS.
CROATIA
BOS.
SERBIA
ROMANIA
MOLD.
UKRAINE
BELARUS
LITH.
RUS.
LATVIA
EST.
St. Petersburg (Leningrad)
Lake Ladoga
Lake Onega
Moscow
Volga
Oka
Dnieper
Don
Danube
Elbe
Vistula
Rhine
Loire
Sukhona
Pechora
Kama
PORTUGAL
CABO DE FISTERRA
CABO DE SÃO VICENTE
SPAIN
Madrid
Mulhacén 11 424 ft. 3482 m
PYRENEES
CORSICA
SARDINIA
ITALY
Rome
Vesuvius 4203 ft. 1281 m
Monte Etna 10 902 ft. 3323 m
ALB.
MACE.
BULGARIA
GREECE
Mt. Olympus 9570 ft. 2917 m
CRETE
İstanbul
TURKEY
Erciyes Daği 12 851 ft. 3917 m
CYPRUS
LEBANON
SYRIA
Damascus
ISRAEL
JORDAN
Dead Sea
IRAQ
Baghdād
Euphrates
Tigris
GEORGIA
ARM.
AZER.
Baku
Mt. Ararat 16 854 ft. 5137 m
CAUCASUS MTS.
Gora El'brus 18 510 ft. 5642 m
Sea of Azov
CASPIAN DEPRESSION
URAL MTS.
Gora Narodnaya 6214 ft. 1894 m
Yekaterinburg
Ural
Tobol
Ishim
Irtysh
RUSSIA
WEST SIBERIAN PLAIN
Novosibirsk
Noril'sk
Yenisey
KAZAKHSTAN
Qaraghandy
KAZAKH HILLS
Lake Balkhash
Zhaysang köli
ALTAY MTS.
SAYAN
Mt. Belukha 14 783 ft. 4506 m
Üvs nuur
UZBEKISTAN
UST-URT PLAT.
QIZILQUM
Syr Darya
Amu Darya
Tashkent
Almaty
Ürümqi
KYRGYZSTAN
TIEN SHAN
Jengish Chokusu 24 406 ft. 7439 m
Tarim
TARIM PENDI
KARA KUM
TURKMENISTAN
Pik Imeni Ismail Samani 24 590 ft. 7495 m
PAMIRS
TAJIKISTAN
HINDU KUSH
K2 28 250 ft. 8611 m
ALTUN SHAN
KUNLUN SHAN
PLATEAU OF TIBET
Tehran
Kūh-e Damāvand 18 386 ft. 5604 m
DASHT-E KAVĪR
DASHT-E LŪT
ZAGROS MOUNTAINS
IRAN
AFGHANISTAN
Kabul
PAKISTAN
Lahore
Delhi
GREAT INDIAN DESERT
Karāchi
HIMALAYAS
NEPAL
Annapūrna 26 545 ft. 8091 m
Mt. Everest 29 028 ft. 8848 m
Lhasa
Brahmaputra
BHUTAN
Ganges
Yamuna
Chambal
Narmada
Godāvari
Krishna
INDIA
Dhaka
BNGL.
Kolkata (Calcutta)
DECCAN
Mumbai (Bombay)
WESTERN GHATS
EASTERN GHATS
Bengalūru
LAKSHADWEEP
Anai Mudi 8842 ft. 2695 m
CAPE COMORIN
SRI LANKA
Colombo
Pidurutalagala 8281 ft. 2524 m
DONDRA HEAD
MALDIVES
CHAGOS ARCHIPELAGO (B.I.O.T.)
SEYCHELLES
SOCOTRA (Yemen)
GEES GWARDAFUY
RA'S FARTAK
RA'S AL-HADD
KUWAIT
AN-NAFŪD
SAUDI ARABIA
ARABIAN PENINSULA
Riyadh
Jiddah
AL-HIJAZ
'ASĪR
QATAR
U.A.E.
Jabal ash-Shām 9957 ft. 3035 m
OMAN
RUB' AL-KHALI
YEMEN
Sanaa
MOROCCO
ATLAS MOUNTAINS
ALGERIA
TUNISIA
Chott Melrhir
Chott Jerid
Khalij Surt
LIBYA
EGYPT
Cairo
Tropic of Cancer
Lake Nasser
Nile
NIGER
Emi Koussi 11 204 ft. 3415 m
CHAD
Lake Chad
Chari
SUDAN
Jabal Marrah 10 072 ft. 3070 m
White Nile
Blue Nile
Atbarah
ERITREA
Ras Dejen 15 158 ft. 4620 m
Lake Tana
DJIBOUTI
Addis Ababa
ETHIOPIA
SOMALIA
CENTRAL AFRICAN REPUBLIC
Mbomou
Uele
Congo
CONGO
Mountain Nile
UGANDA
Lake Albert
Lake Rudolf
Margherita Peak 16 763 ft. 5109 m
Lake Edward
DEMOCRATIC REPUBLIC OF THE CONGO
KENYA
Mt. Kenya 17 058 ft. 5199 m
Equator
RWANDA
BURUNDI
Lake Victoria
Lake Tanganyika
Kilimanjaro 19 340 ft. 5895 m
TANZANIA
Dar es Salaam
Sankuru
Kasai
ANGOLA
Lake Mweru
ZAMBIA

0 200 400 600 800 1000 1200 Miles
0 200 400 600 800 1000 1200 1400 1600 1800 2000 Kilometers

Lambert Azimuthal Equal Area Projection
Scale 1:40,000,000
One inch to 640 miles
One cm to 400 km

UNITED STATES
CHUKOTSK PEN.
MYS DEZNEVA
OSTROV VRANGELYA
Gulf of Anadyr
MYS NAVARIN
NEW SIBERIAN ISLANDS
EAST SIBERIAN SEA
BERING SEA
ALEUTIAN ISLANDS
Arctic Circle
Anadyr
Gora Ledyanaya 8406 ft. 2562 m
MYS OLYUTORSKIY
LAPTEV SEA
Indigirka
Kolyma
Gora Pobeda 10 325 ft. 3147 m
CHERSKIY MTS.
KAMCHATKA PENINSULA
KOMANDORSKI ISLANDS
ALEUTIAN TRENCH
CAPE WRANGELL
VERKHOYANSK MTS.
Lena
Magadan
Klyuchevskaya Sopka 15 584 ft. 4750 m
Vulkan
Petropavlovsk-Kamchatskiy
MYS LOPATKA
A S I A
Vilyuy
Yakutsk
Aldan
DZHUGDZHUR RANGE
MYS YELIZAVETY
SEA OF OKHOTSK
KURIL TRENCH
Tunguska
S I B E R I A
STANOVOY RANGE
STANOVOY MTS.
SAKHALIN
MYS TERPENIYA
KURIL ISLANDS
P A C I F I C O C E A N
MIDWAY ISLANDS (U.S.)
HAWAI'IAN ISLANDS
NECKER RIDGE
Lake Baikal
Komsomol'sk-na-Amure
SIKHOTE-ALIN
Tatar Strait
Amur
Argun
Irkutsk
GREATER KHINGAN RANGE
Harbin
Sapporo
HOKKAIDŌ
JAPAN TRENCH
Ulaanbaatar
Kerulen
MONGOLIA
SEA OF JAPAN
HONSHŪ
NORTH KOREA
Tōkyō
JAPAN
Tropic of Cancer
G O B I D E S E R T
Beijing
Seoul
SOUTH KOREA
CHENGSHAN JIAO
Fuji-san 12 388 ft. 3776 m
YELLOW SEA
SHIKOKU
IZU-SHOTO
IZU TRENCH
Qinghai Hu
Huang
CHEJU-DO
KYŪSHŪ
WAKE ISLAND (U.S.)
CHINA
Xi'an
QIN LING
EAST CHINA SEA
BONIN IS. (Japan)
Shanghai
Three Gorges Reservoir
Yangtze
OKINAWA-JIMA
VOLCANO ISLANDS (Japan)
MARIANA TRENCH
Chongqing
RYUKYU ISLANDS
RYUKYU TRENCH
T'aipei
TAIWAN
Yu Shan 13 114 ft. 3997 m
Taiwan Strait
PHILIPPINE SEA
NORTHERN MARIANA ISLANDS (U.S.)
MARIANA ISLANDS
M I C R O N E S I A
BIKINI
ENEWETAK
KWAJALEIN
MARSHALL ISLANDS
MAJURO
Hong Kong
Luzon Strait
SAIPAN
TINIAN
MYANMAR (BURMA)
Red
Gulf of Tonkin
ESCARPADA POINT
LUZON
GUAM (U.S.)
LAOS
HAINAN DAO
PHILIPPINES
CHALLENGER DEEP
C A R O L I N E I S L A N D S
SOUTH CHINA SEA
Manila
CHUUK
POHNPEI
KOSRAE
KIRIBATI
Yangon
THAILAND
SAMAR
MINDORO
YAP
YAP TRENCH
TARAWA
Bangkok
VIETNAM
PHILIPPINE TRENCH
CAMBODIA
PANAY
FEDERATED STATES OF MICRONESIA
ANDAMAN ISLANDS (India)
ANDAMAN SEA
Tonle Sap
PALAWAN
PALAU
Ho Chi Minh City (Saigon)
NEGROS
Gulf of Thailand
SULU SEA
Mt. Apo 9692 ft. 2954 m
MINDANAO
NAURU
NICOBAR ISLANDS (India)
MUI CA MAU
Gunong Kinabalu 13 455 ft. 4101 m
TINACA POINT
MALAY PENINSULA
CELEBES SEA
Equator
BRUNEI
Strait of Malacca
HALMAHERA
MANUS ISLAND
NEW HANOVER
NEW IRELAND
M E L A N E S I A
MALAYSIA
IRAN MTS.
PULAU WAIGEO
BIAK
TANJUNG D'URVILLE
Jayapura
BISMARCK SEA
PULAU SIMEULUE
SINGAPORE
Singapore
Kapuas
BORNEO
Makassar Strait
MOLUCCA SEA
PULAU YAPEN
Sepik
NEW BRITAIN
BOUGAINVILLE
CHOISEUL
PULAU NIAS
SUMATRA
MOLUCCAS
CERAM
SANTA ISABEL
SOLOMON ISLANDS
Bukit Raya 7474 ft. 2278 m
CELEBES
Puncak Jaya 16 503 ft. 5030 m
NEW GUINEA
Mt. Wilhelm 14 793 ft. 4509 m
NEW BRITAIN TRENCH
NEW GEORGIA
MALAITA
PULAU SIBERUT
G R E A T E R S U N D A I S L A N D S
BURU
BANDA SEA
KEPULAUAN ARU
PAPUA NEW GUINEA
GUADALCANAL
SAN CRISTOBAL
Gunung Kerinci 12 467 ft. 3800 m
TANJUNG PUTING
TANJUNG SELATAN
Makassar
PULAU BUTON
Fly
SOLOMON SEA
SANTA CRUZ IS.
SAN CRISTOBAL TRENCH
RENNELL
I N D O N E S I A
PULAU WETAR
PULAU YAMDENA
TANJUNG VALS
JAVA SEA
FLORES SEA
Jakarta
JAVA
BALI
LESSER SUNDA ISLANDS
ARAFURA SEA
CAPE YORK
EAST TIMOR
Gunung Semeru 12 060 ft. 3676 m
LOMBOK
SUMBAWA
FLORES
TIMOR
MELVILLE ISLAND
AUSTRALIA
JAVA TRENCH
CHRISTMAS ISLAND (Aust.)
SUMBA
TIMOR SEA
CAPE ARNHEM
© Rand McNally A-101712-1

ARCTIC OCEAN
SEVERNAYA ZEMLYA
ATLANTIC OCEAN
Arctic Circle
BARENTS SEA
KARA SEA
URAL MOUNTAINS
Ob'
Yenisey
WEST SIBERIAN PLAIN
Yekaterinburg
Chelyabinsk
Ural
Irtysh
SAYAN
ALTAY MTS.
MEDITERRANEAN SEA
Istanbul
BLACK SEA
CAUCASUS MTS.
CASPIAN SEA
Ankara
Aral Sea
UST-URT PLAT.
KAZAKH HILLS
Lake Balkhash
Syr Darya
QIZILQUM
Almaty
Tashkent
TIEN SHAN
KARA KUM
Dushanbe
Amu Darya
Euphrates
Tigris
Tabrīz
Tel Aviv-Yafo
Damascus
Tehrān
Mashhad
'Ammān
Baghdād
DASHT-E KAVIR
ZAGROS MTS.
Eşfahān
DASHT-E LŪT
PAMIRS
TARIM PENDI
ALTUN SHAN
HINDU KUSH
KUNLUN SHAN
HIMALAYA
PLATEAU OF TIBET
Islāmābād
Lahore
Brahmaputra
Tropic of Cancer
AN-NAFŪD
AL-HIJAZ
RED SEA
Riyadh
Persian Gulf
Gulf of Oman
Jiddah
Mecca
'ASĪR
RUB' AL-KHALI
Indus
GREAT INDIAN DESERT
Delhi
Kathmandu
Hyderābād
Karāchi
Kānpur
Ganges
Dhaka
Ahmadābād
Indore
Kolkata
DECCAN
Sanaa
Mumbai
WESTERN GHATS
Hyderābād
EASTERN GHATS
ARABIAN SEA
BAY OF BENGAL
Gulf of Aden
Bengalūru
Chennai
Cochin
Colombo
INDIAN OCEAN
Equator
A-102095-1
© Rand McNally

0 200 400 600 800 1000 1200 Miles
0 200 400 600 800 1000 1200 1400 1600 1800 2000 Kilometers

Lambert Azimuthal Equal Area Projection
Scale 1:40,000,000
One inch to 640 miles
One cm to 400 km

Land Cover
Evergreen Needleleaf Forest
Evergreen Broadleaf Forest
Deciduous Needleleaf Forest
Deciduous Broadleaf Forest
Mixed Forest
Woodland
Wooded Grassland
Closed Shrubland
Open Shrubland
Grassland
Cropland
Bare Ground (Desert and Ice)
Urban and Built Up
LAPTEV SEA
NEW SIBERIAN ISLANDS
EAST SIBERIAN SEA
OSTROV VRANGELYA
BERING SEA
Arctic Circle
Kolyma
CHERSKIY MTS.
VERKHOYANSK MTS.
Lena
SREDINNY KHREBET
SEA OF OKHOTSK
SAKHALIN
KURIL ISLANDS
DZHUGDZHUR RANGE
STANOVOY RANGE
STANOVOY MTS.
SIBERIA
Lake Baikal
Amur
SIKHOTE-ALIN'
GREATER KHINGAN RANGE
Harbin
Sapporo
HOKKAIDŌ
Changchun
Shenyang
SEA OF JAPAN
HONSHŪ
Tokyō
PACIFIC OCEAN
GOBI DESERT
Beijing
Baotou
Tianjin
Seoul
Ōsaka
Nagoya
Qingdao
Jinan
Taiyuan
Huang
YELLOW SEA
Fukuoka
SHIKOKU
KYŪSHŪ
Xi'an
QIN LING
Nanjing
Shanghai
EAST CHINA SEA
RYUKYU ISLANDS
Chengdu
Yangtze
Chongqing
Nanchang
Tropic of Cancer
T'aipei
Guiyang
Fuzhou
Taiwan Strait
PHILIPPINE SEA
MARIANA ISLANDS
MICRONESIA
Guangzhou
Xi
Hong Kong
Ha Noi
HAINAN DAO
LUZON
Salween
Mekong
SOUTH CHINA SEA
Manila
CAROLINE ISLANDS
MINDORO
SAMAR
Bangkok
ANDAMAN ISLANDS
ANDAMAN SEA
PALAWAN
NEGROS
Gulf of Thailand
Ho Chi Minh City
SULU SEA
MINDANAO
NICOBAR ISLANDS
Equator
CELEBES SEA
Medan
Kuala Lumpur
IRAN MTS.
SUMATRA
Singapore
BORNEO
MOLUCCAS
PEGUNUNGAN MAOKE
NEW GUINEA
Palu
SUNDA
CELEBES
GREATER
Palembang
ISLANDS
BANDA SEA
Makassar
JAVA SEA
Bandar Lampung
FLORES SEA
Jakarta
Semarang
Surabaya
LESSER SUNDA ISLANDS
ARAFURA SEA
JAVA
Denpasar
TIMOR SEA
Source: CIESIN; Hansen et al., 2000

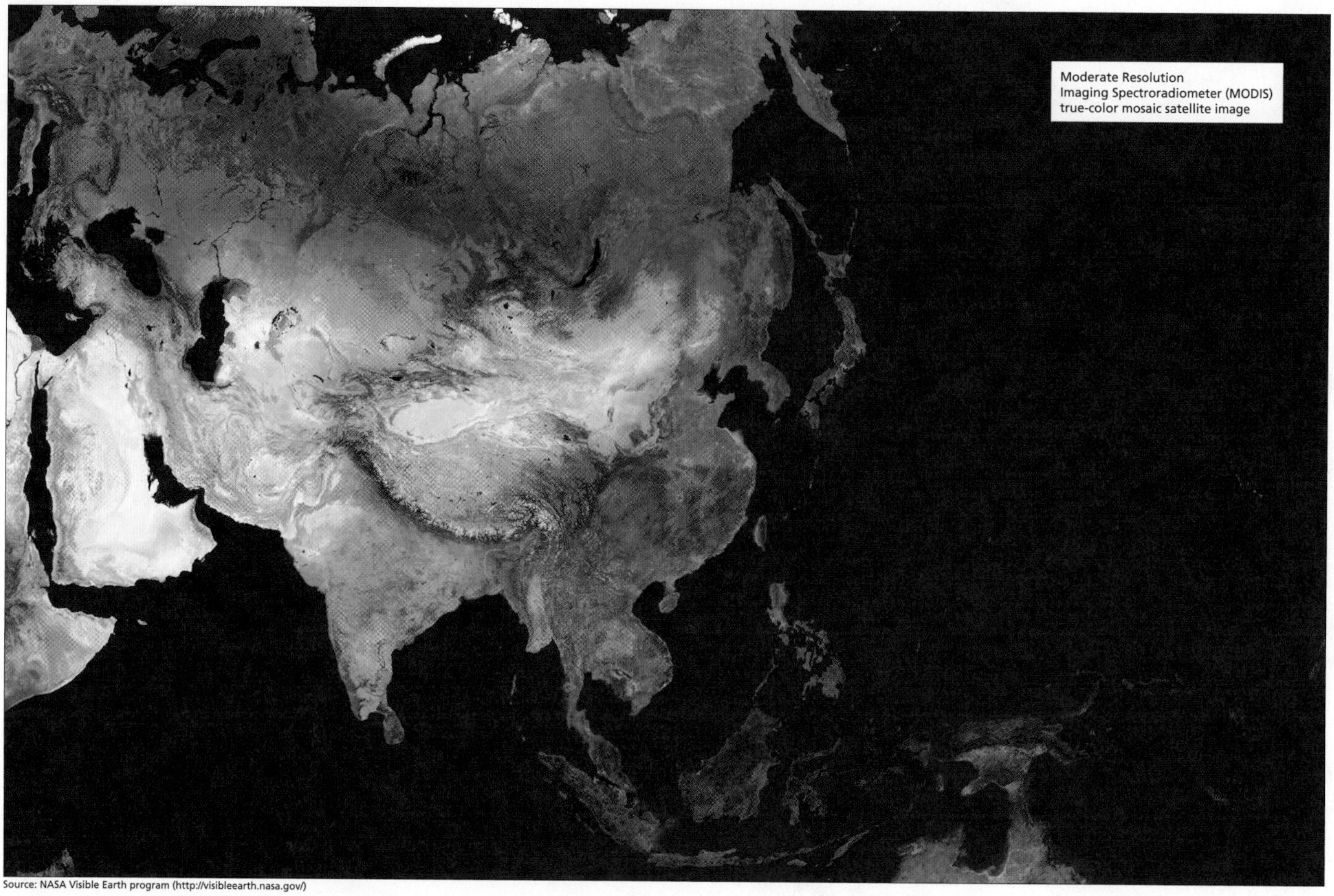
Moderate Resolution
Imaging Spectroradiometer (MODIS)
true-color mosaic satellite image

Source: NASA Visible Earth program (http://visibleearth.nasa.gov/)

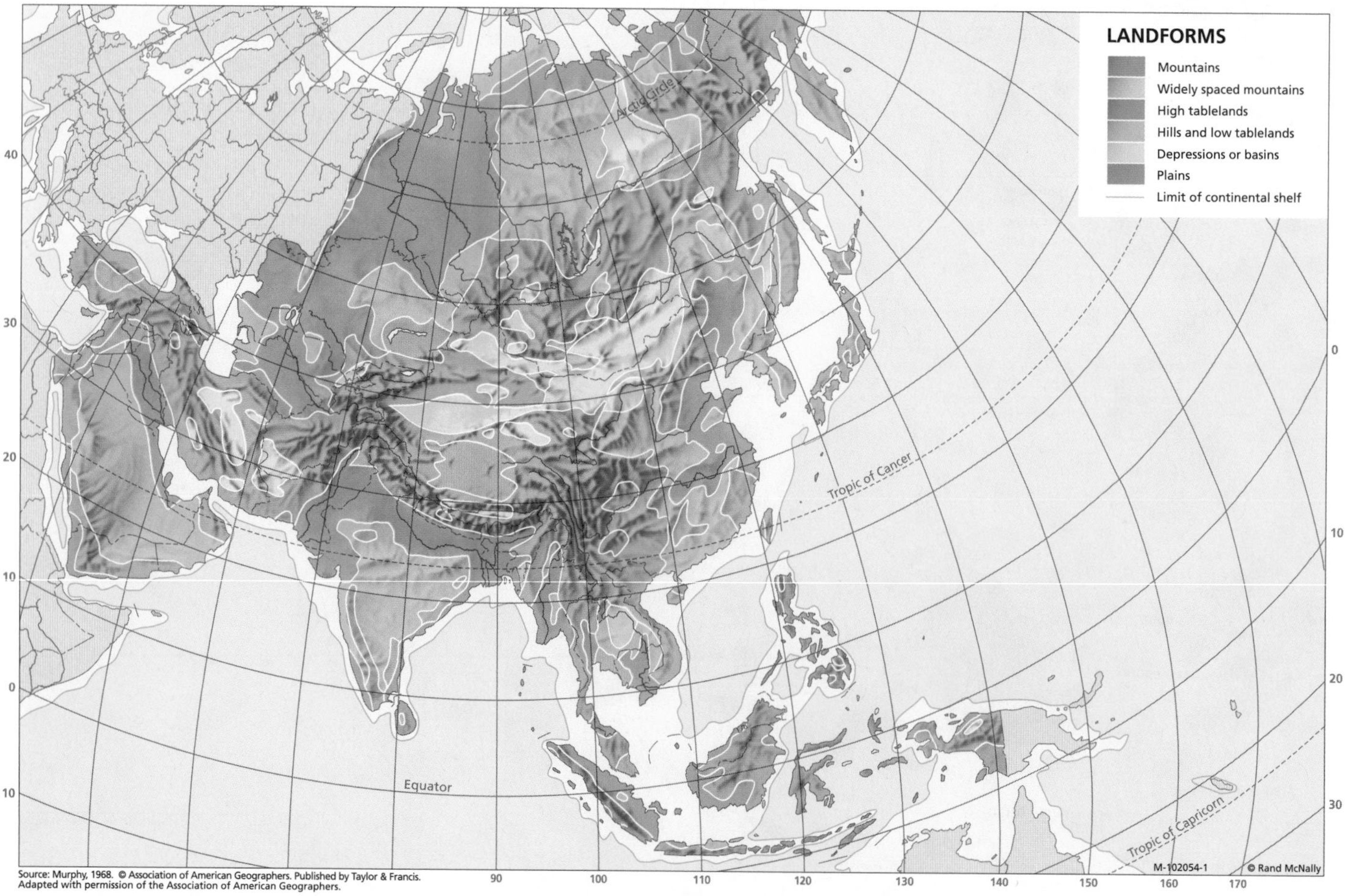
LANDFORMS
Mountains
Widely spaced mountains
High tablelands
Hills and low tablelands
Depressions or basins
Plains
Limit of continental shelf
Arctic Circle
Tropic of Cancer
Equator
Tropic of Capricorn
40
30
20
10
0
10
0
10
20
30
90
100
110
120
130
140
150
160
170
M-102054-1
© Rand McNally

Source: Murphy, 1968. © Association of American Geographers. Published by Taylor & Francis. Adapted with permission of the Association of American Geographers.

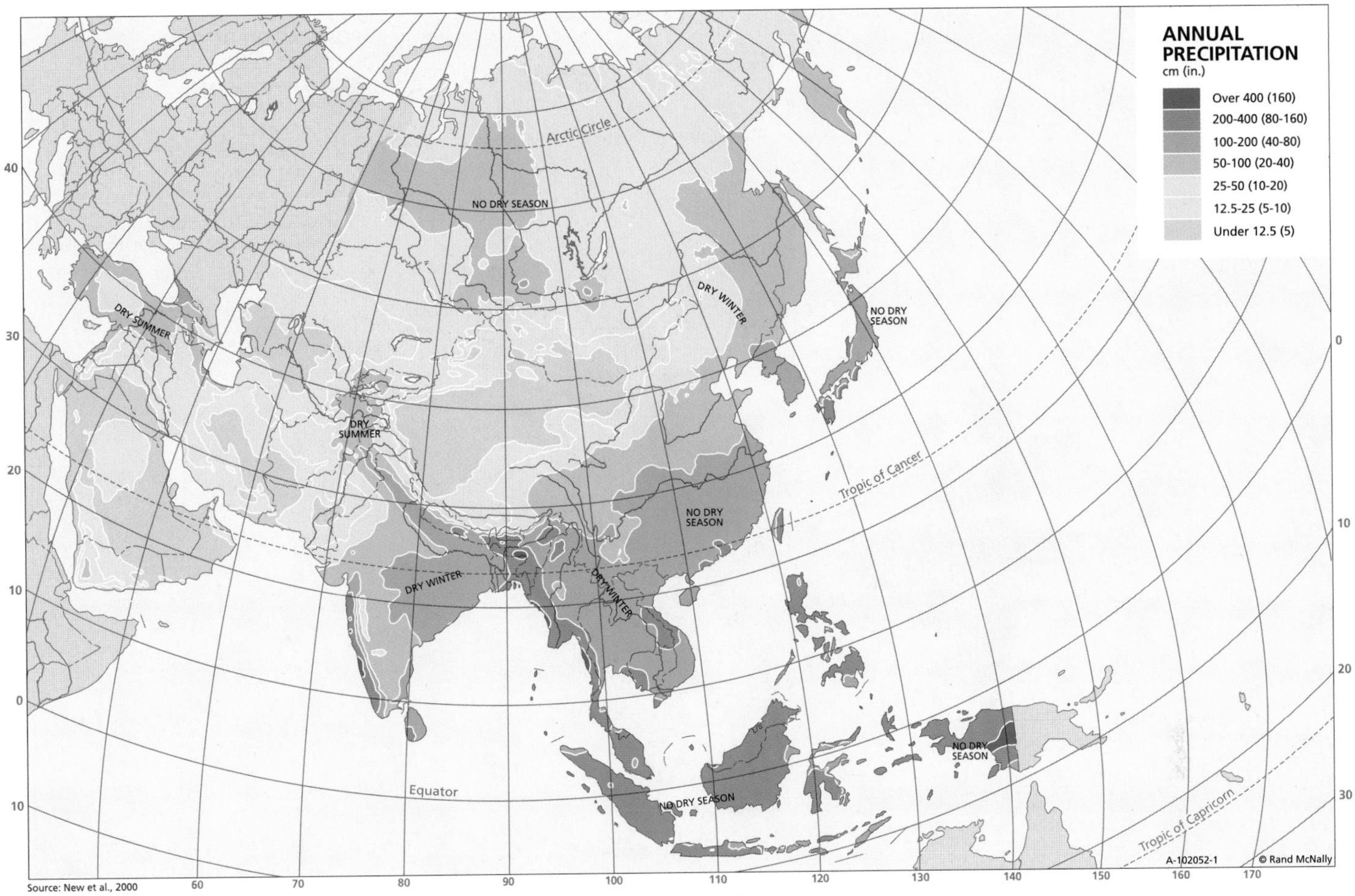
ANNUAL PRECIPITATION
cm (in.)
Over 400 (160)
200-400 (80-160)
100-200 (40-80)
50-100 (20-40)
25-50 (10-20)
12.5-25 (5-10)
Under 12.5 (5)
Arctic Circle
NO DRY SEASON
DRY SUMMER
DRY WINTER
NO DRY SEASON
DRY SUMMER
Tropic of Cancer
NO DRY SEASON
DRY WINTER
DRY WINTER
NO DRY SEASON
Equator
NO DRY SEASON
Tropic of Capricorn
A-102052-1
© Rand McNally
40
30
20
10
0
10
0
10
20
30

Source: New et al., 2000
60 70 80 90 100 110 120 130 140 150 160 170

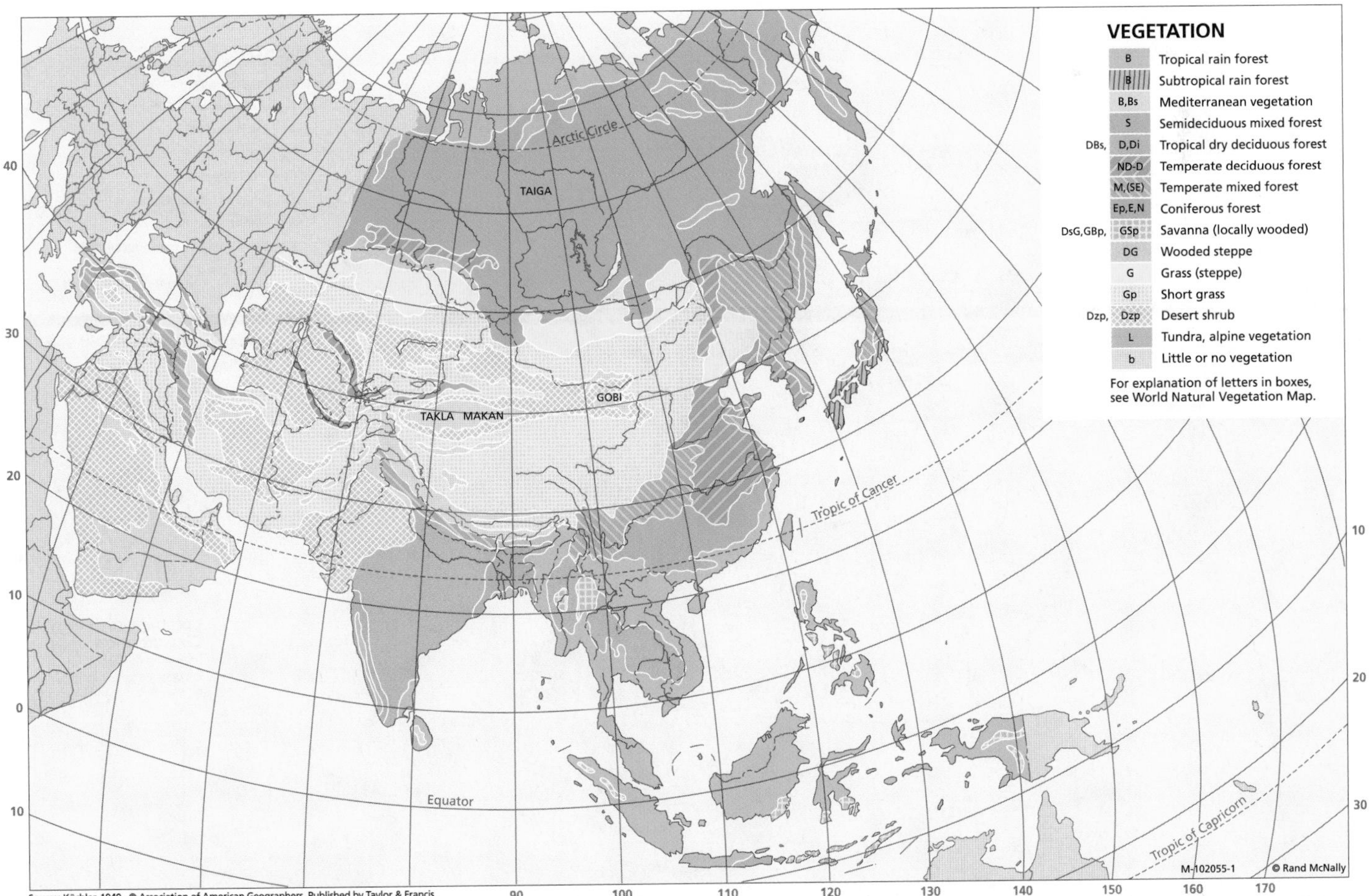
VEGETATION
B Tropical rain forest
B Subtropical rain forest
B,Bs Mediterranean vegetation
S Semideciduous mixed forest
DBs, D,Di Tropical dry deciduous forest
ND-D Temperate deciduous forest
M,(SE) Temperate mixed forest
Ep,E,N Coniferous forest
DsG,GBp, GSp Savanna (locally wooded)
DG Wooded steppe
G Grass (steppe)
Gp Short grass
Dzp, Dzp Desert shrub
L Tundra, alpine vegetation
b Little or no vegetation
For explanation of letters in boxes, see World Natural Vegetation Map.
Arctic Circle
TAIGA
GOBI
TAKLA MAKAN
Tropic of Cancer
Equator
Tropic of Capricorn
M-102055-1
© Rand McNally
40
30
20
10
0
10
10
20
30

Source: Küchler, 1949. © Association of American Geographers. Published by Taylor & Francis. Adapted with permission of the Association of American Geographers.
90 100 110 120 130 140 150 160 170

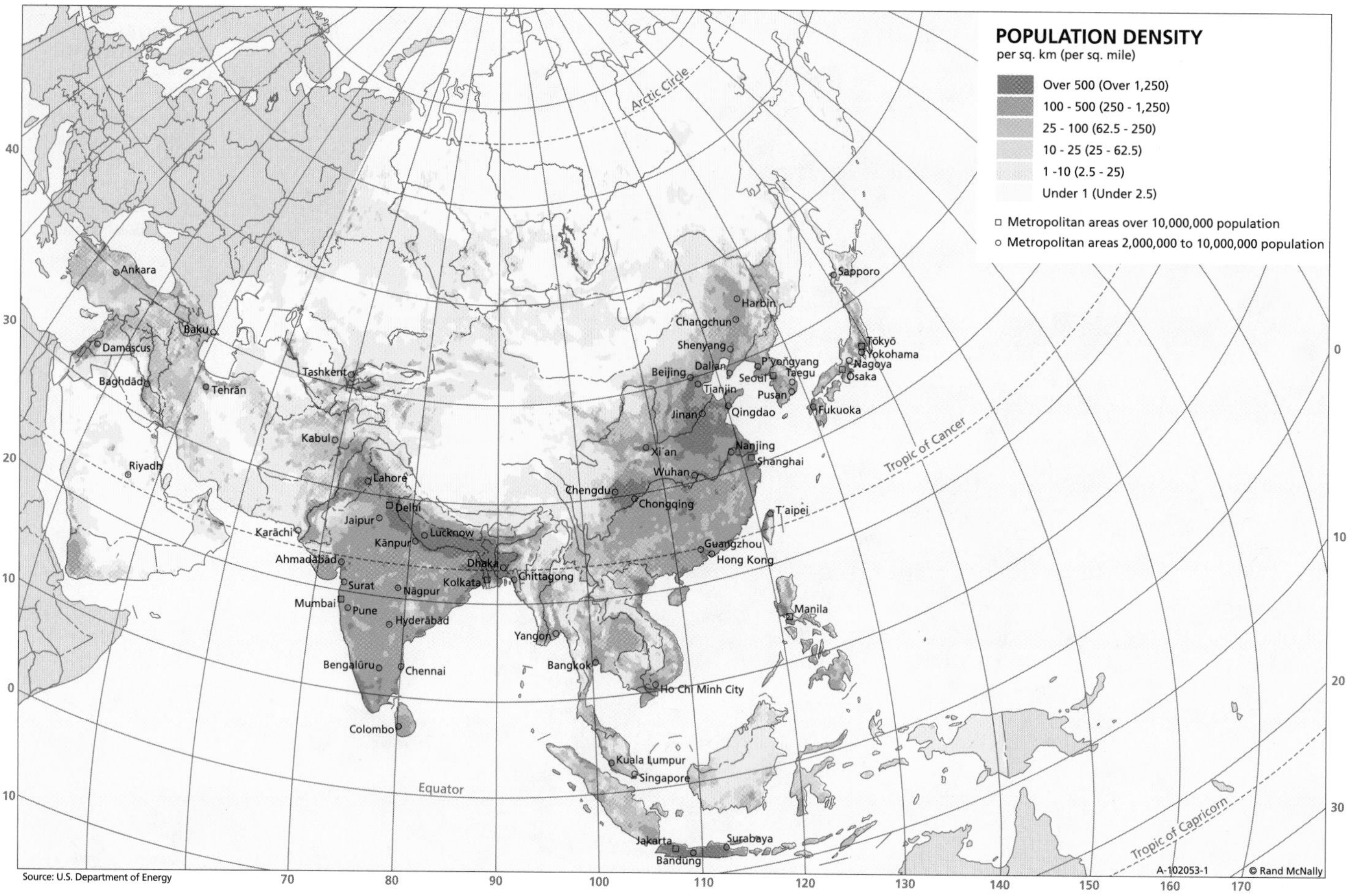
POPULATION DENSITY
per sq. km (per sq. mile)
Over 500 (Over 1,250)
100 - 500 (250 - 1,250)
25 - 100 (62.5 - 250)
10 - 25 (25 - 62.5)
1 -10 (2.5 - 25)
Under 1 (Under 2.5)
Metropolitan areas over 10,000,000 population
Metropolitan areas 2,000,000 to 10,000,000 population
Arctic Circle
Tropic of Cancer
Equator
Tropic of Capricorn
Ankara
Baku
Damascus
Baghdad
Tehrān
Tashkent
Kabul
Riyadh
Lahore
Delhi
Karāchi
Jaipur
Lucknow
Kānpur
Ahmadābād
Dhaka
Kolkata
Chittagong
Surat
Nāgpur
Mumbai
Pune
Hyderābād
Bengalūru
Chennai
Colombo
Yangon
Bangkok
Ho Chi Minh City
Kuala Lumpur
Singapore
Jakarta
Bandung
Surabaya
Manila
Chengdu
Chongqing
Xi'an
Wuhan
Nanjing
Shanghai
T'aipei
Guangzhou
Hong Kong
Harbin
Changchun
Shenyang
Beijing
Dalian
Tianjin
Jinan
Qingdao
P'yŏngyang
Seoul
Taegu
Pusan
Sapporo
Tōkyō
Yokohama
Nagoya
Ōsaka
Fukuoka
Source: U.S. Department of Energy
A-102053-1
© Rand McNally

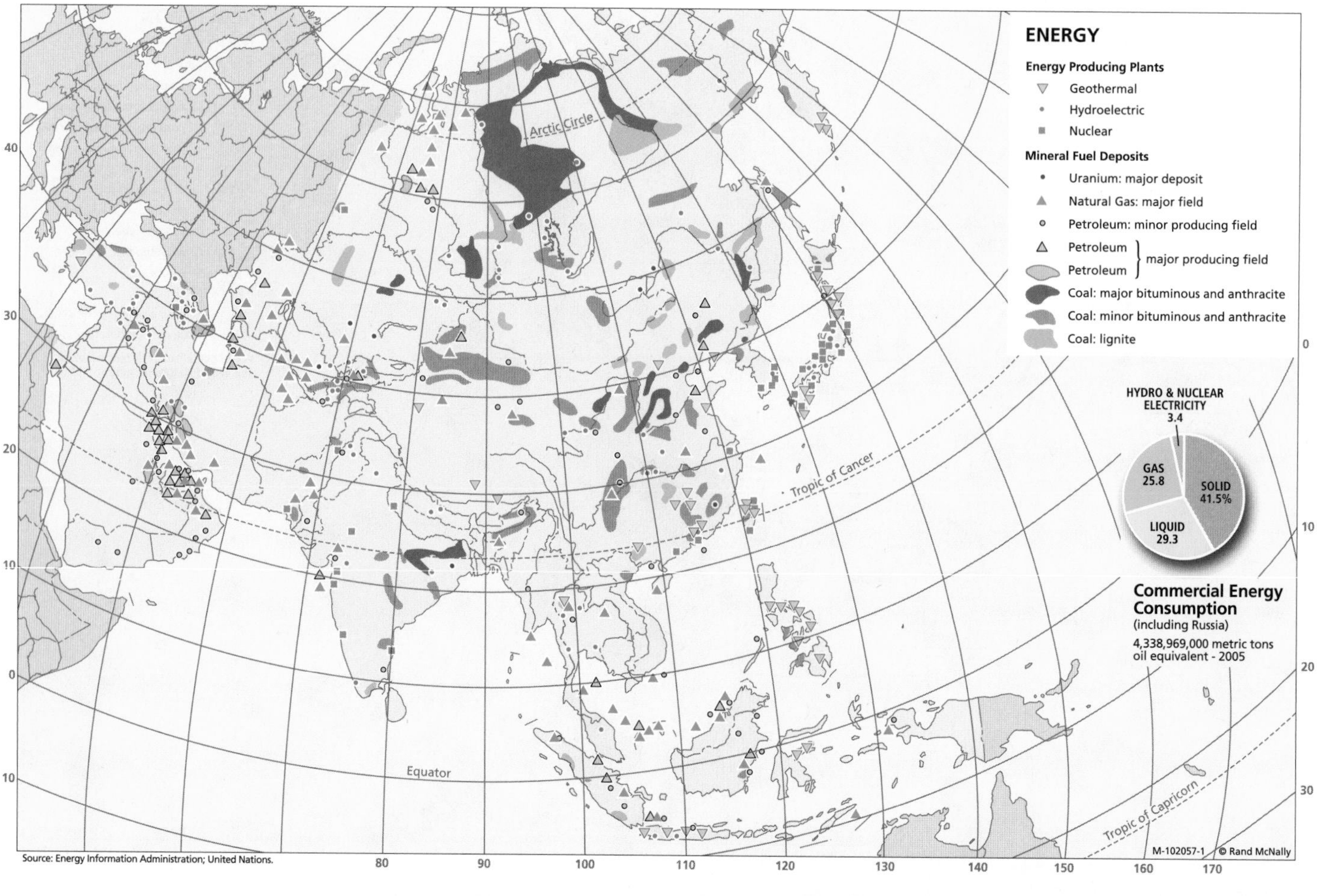
ENERGY
Energy Producing Plants
Geothermal
Hydroelectric
Nuclear
Mineral Fuel Deposits
Uranium: major deposit
Natural Gas: major field
Petroleum: minor producing field
Petroleum
Petroleum
major producing field
Coal: major bituminous and anthracite
Coal: minor bituminous and anthracite
Coal: lignite
Arctic Circle
Tropic of Cancer
Equator
Tropic of Capricorn
HYDRO & NUCLEAR ELECTRICITY 3.4
GAS 25.8
SOLID 41.5%
LIQUID 29.3
Commercial Energy Consumption
(including Russia)
4,338,969,000 metric tons oil equivalent - 2005
Source: Energy Information Administration; United Nations.
M-102057-1
© Rand McNally

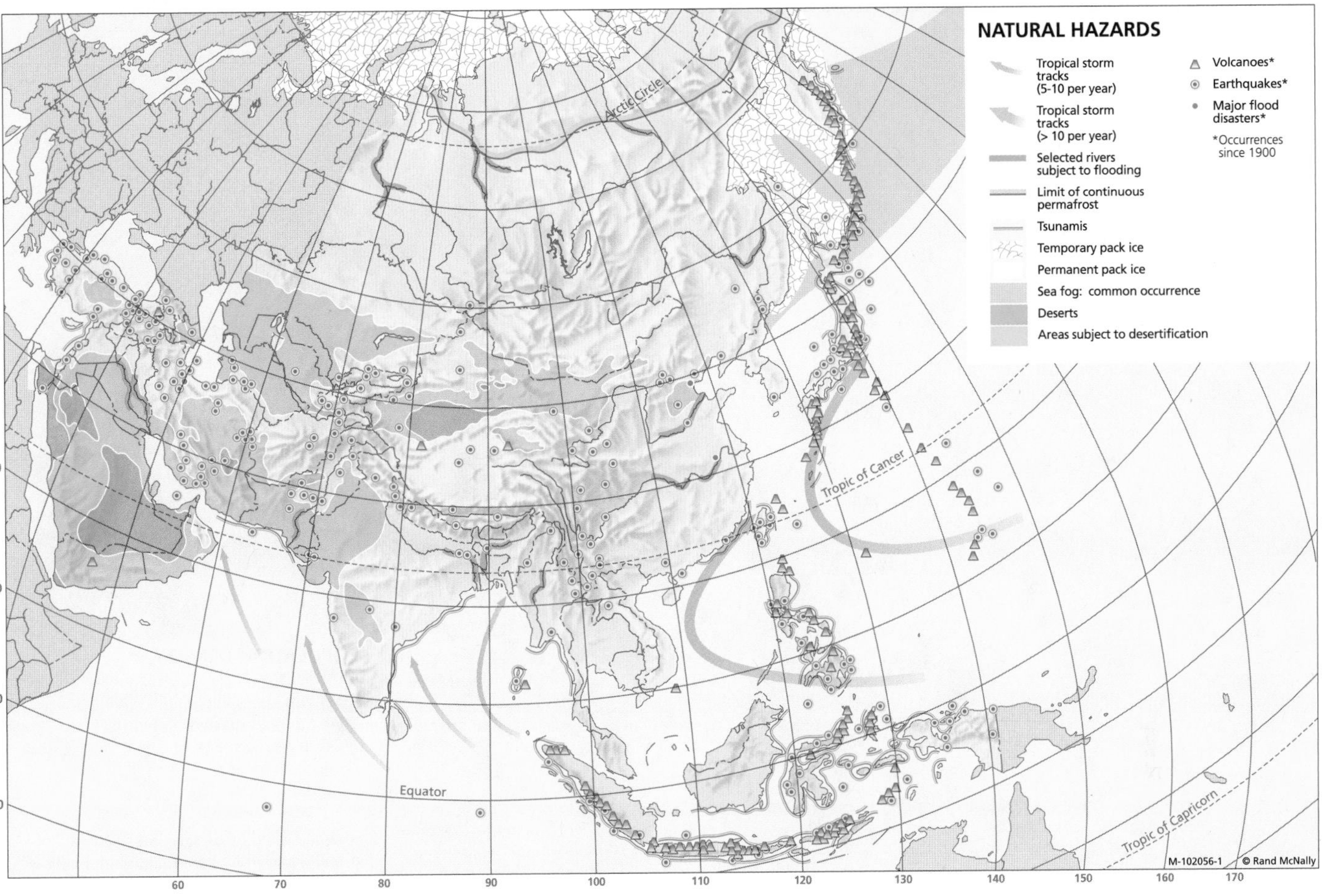
NATURAL HAZARDS
Tropical storm tracks (5-10 per year)
Tropical storm tracks (> 10 per year)
Selected rivers subject to flooding
Limit of continuous permafrost
Tsunamis
Temporary pack ice
Permanent pack ice
Sea fog: common occurrence
Deserts
Areas subject to desertification
Volcanoes*
Earthquakes*
Major flood disasters*
*Occurrences since 1900
Arctic Circle
Tropic of Cancer
Equator
Tropic of Capricorn
M-102056-1
© Rand McNally

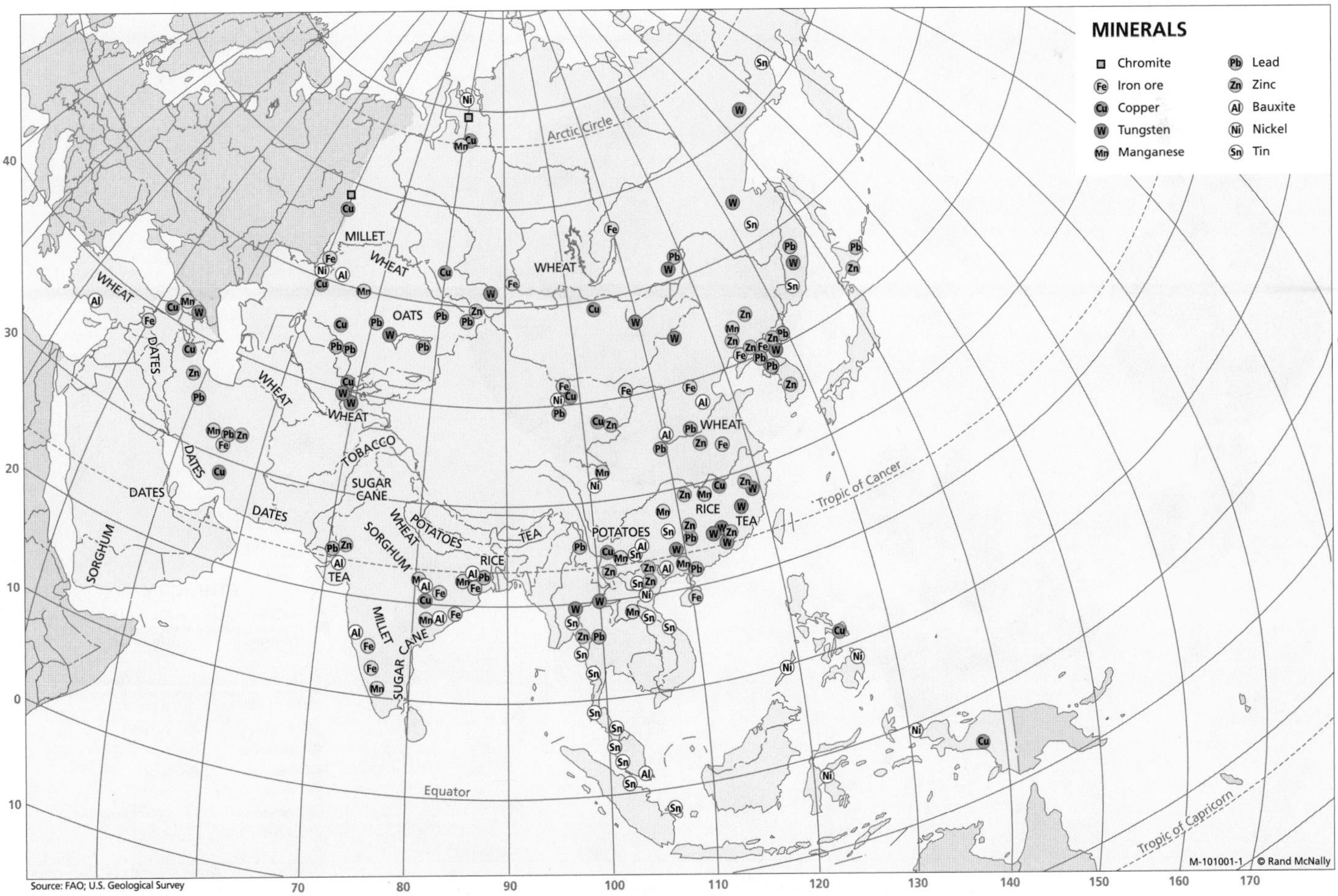
MINERALS
Chromite
Fe Iron ore
Cu Copper
W Tungsten
Mn Manganese
Pb Lead
Zn Zinc
Al Bauxite
Ni Nickel
Sn Tin
MILLET
WHEAT
OATS
DATES
SORGHUM
TOBACCO
SUGAR CANE
POTATOES
RICE
TEA
Arctic Circle
Tropic of Cancer
Equator
Tropic of Capricorn
M-101001-1
© Rand McNally

Source: FAO; U.S. Geological Survey

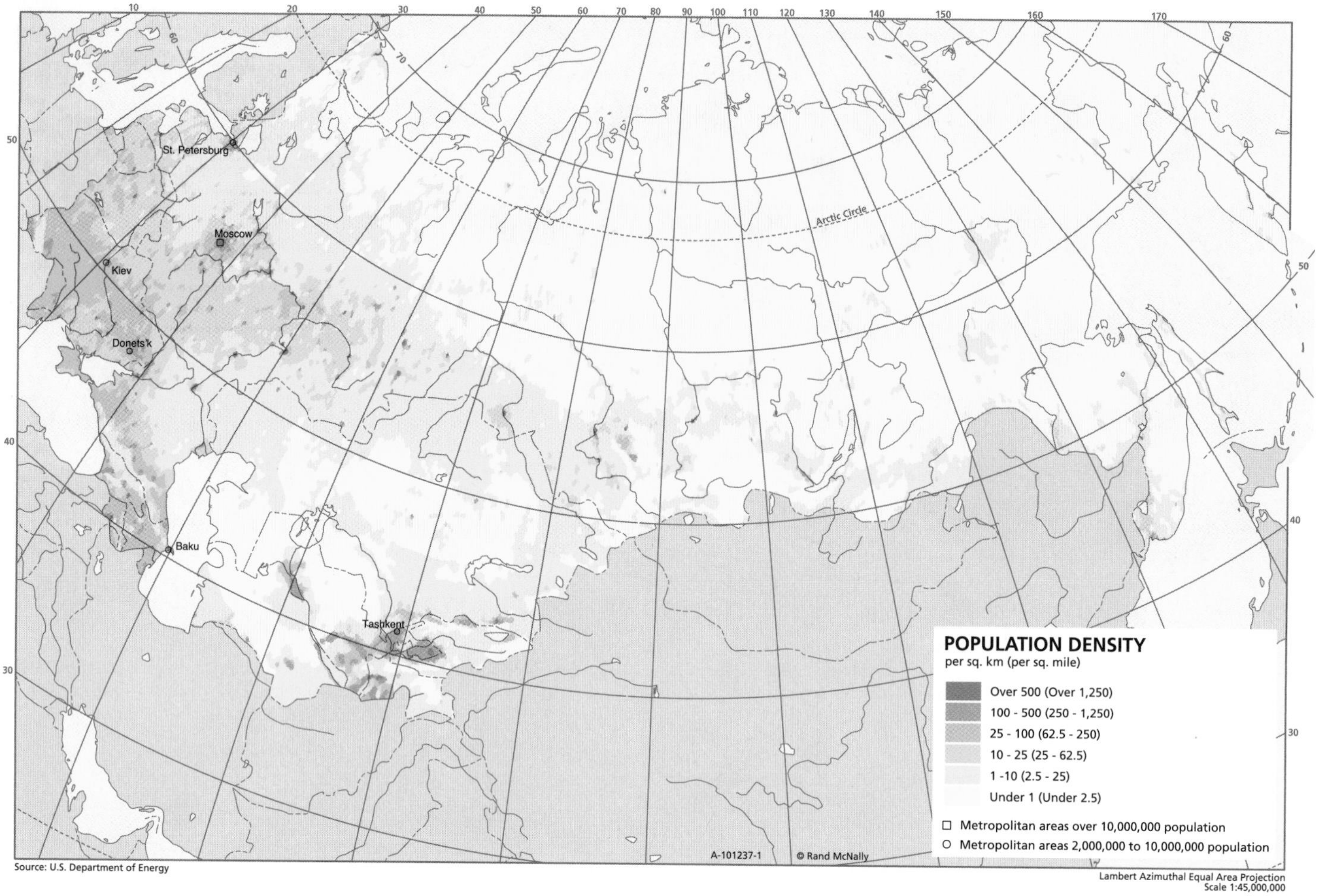
St. Petersburg
Moscow
Kiev
Donets'k
Baku
Tashkent
Arctic Circle
POPULATION DENSITY
per sq. km (per sq. mile)
Over 500 (Over 1,250)
100 - 500 (250 - 1,250)
25 - 100 (62.5 - 250)
10 - 25 (25 - 62.5)
1 -10 (2.5 - 25)
Under 1 (Under 2.5)
Metropolitan areas over 10,000,000 population
Metropolitan areas 2,000,000 to 10,000,000 population
A-101237-1
© Rand McNally
Lambert Azimuthal Equal Area Projection
Scale 1:45,000,000

Source: U.S. Department of Energy

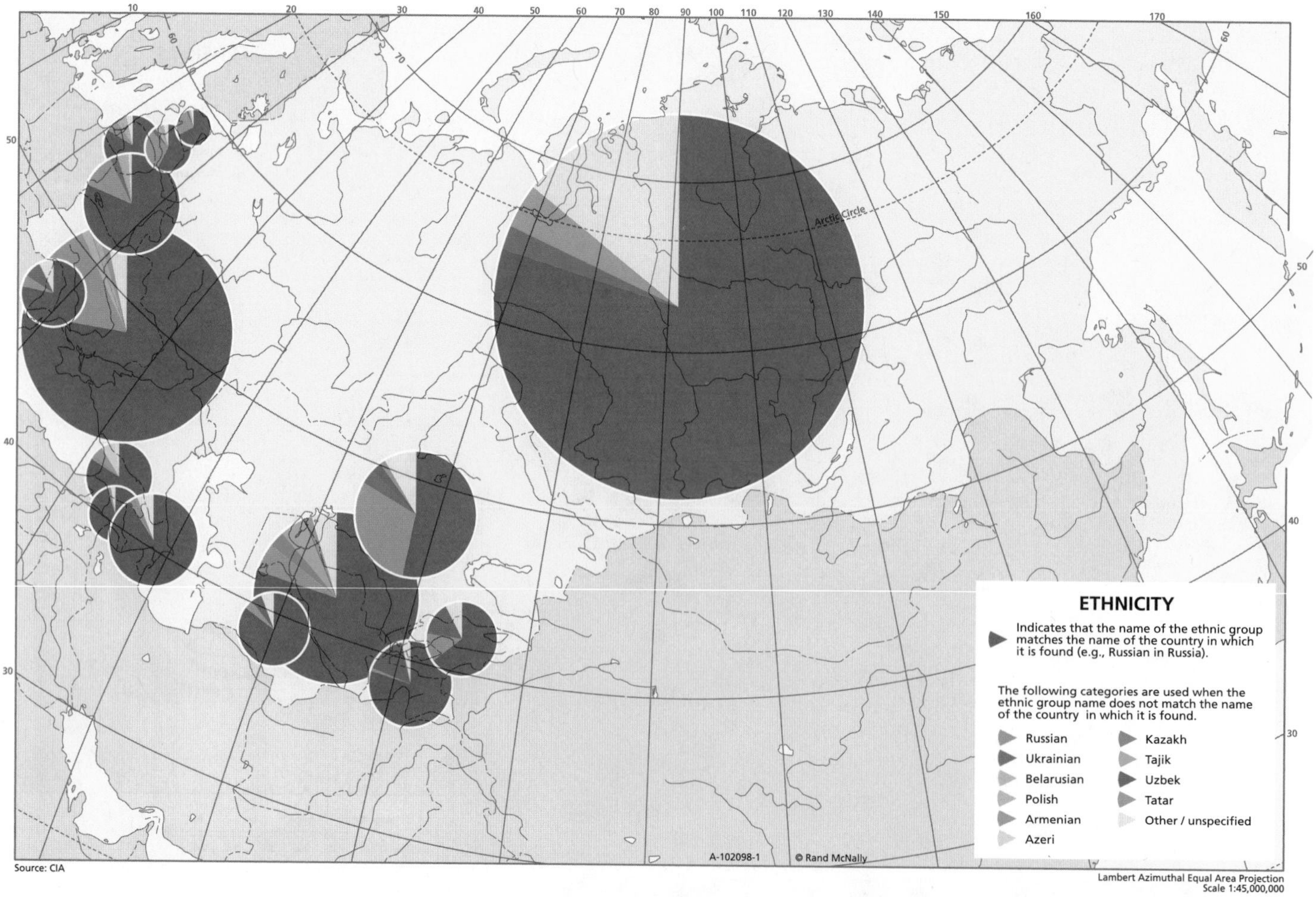
Arctic Circle
ETHNICITY
Indicates that the name of the ethnic group matches the name of the country in which it is found (e.g., Russian in Russia).
The following categories are used when the ethnic group name does not match the name of the country in which it is found.
Russian
Ukrainian
Belarusian
Polish
Armenian
Azeri
Kazakh
Tajik
Uzbek
Tatar
Other / unspecified
A-102098-1
© Rand McNally
Lambert Azimuthal Equal Area Projection
Scale 1:45,000,000

Source: CIA

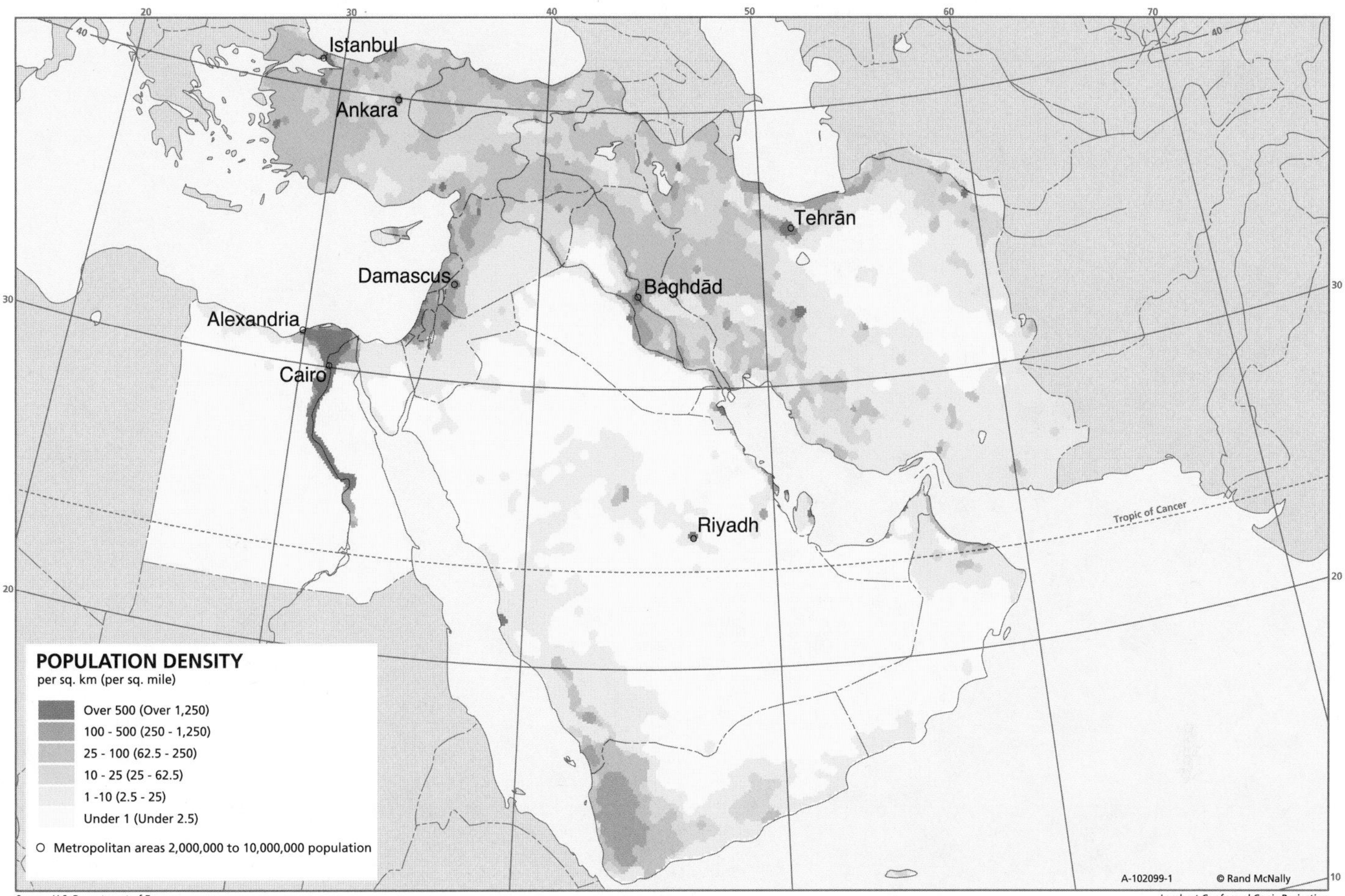
Istanbul
Ankara
Tehrān
Damascus
Baghdād
Alexandria
Cairo
Riyadh
Tropic of Cancer
POPULATION DENSITY
per sq. km (per sq. mile)
Over 500 (Over 1,250)
100 - 500 (250 - 1,250)
25 - 100 (62.5 - 250)
10 - 25 (25 - 62.5)
1 -10 (2.5 - 25)
Under 1 (Under 2.5)
Metropolitan areas 2,000,000 to 10,000,000 population
A-102099-1
© Rand McNally

Source: U.S. Department of Energy

Lambert Conformal Conic Projection
Scale 1:29,000,000

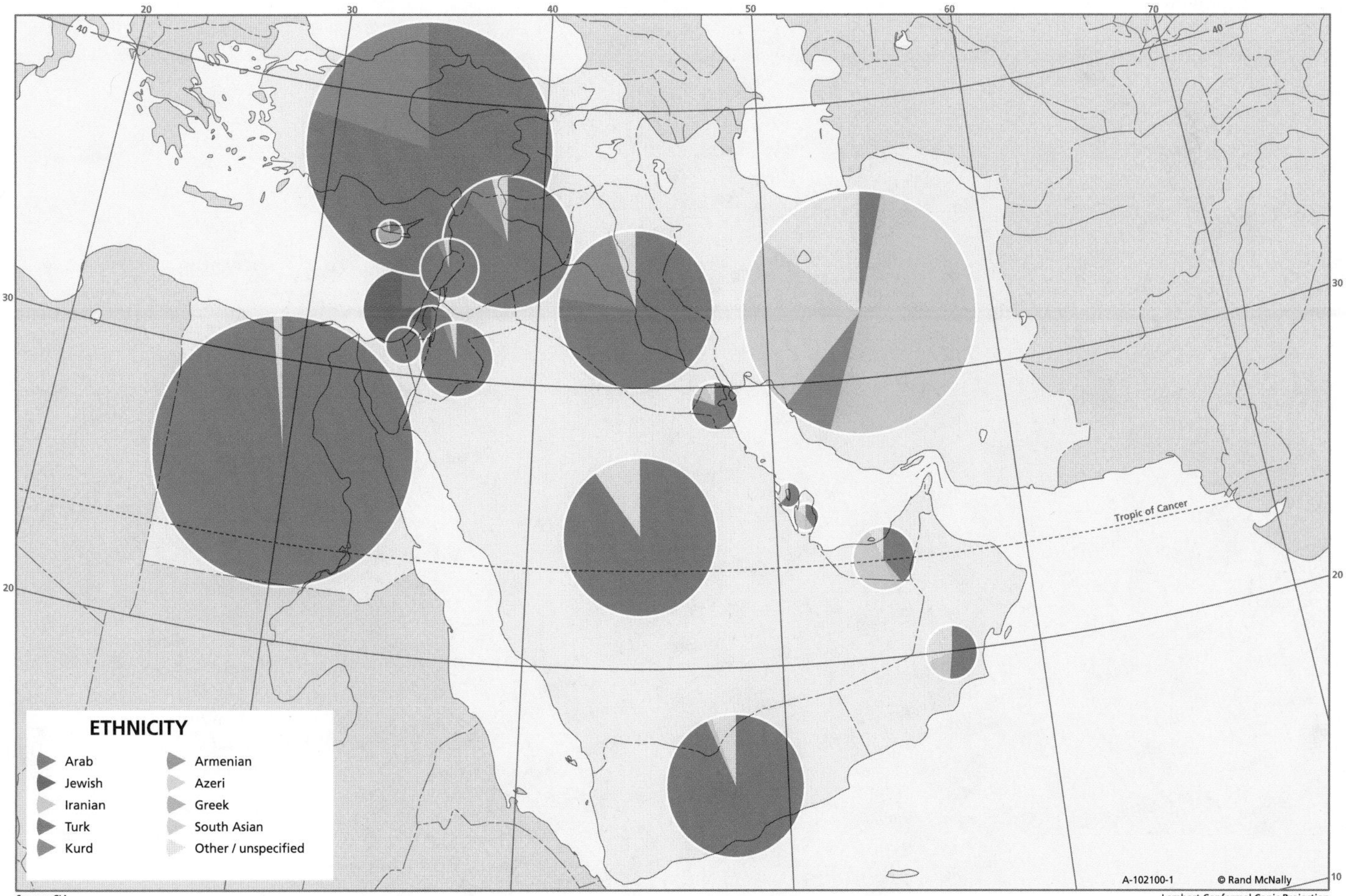
Tropic of Cancer
ETHNICITY
Arab
Jewish
Iranian
Turk
Kurd
Armenian
Azeri
Greek
South Asian
Other / unspecified
A-102100-1
© Rand McNally

Source: CIA

Lambert Conformal Conic Projection
Scale 1:29,000,000

NORWEGIAN BASIN
NORWEGIAN SEA
ARCTIC
SPITSBERGEN
NORDAUSTLANDET
SVALBARD (Nor.)
EDGEØYA
BJØRNØYA (Nor.)
FRANZ JOSEF LAND (ZEMLYA FRANTSA-IOSIFA)
FAROE ISLANDS (Den.)
Tórshavn
SHETLAND IS. (U.K.)
ORKNEY ISLANDS
NORTH SEA
LOFOTEN
VESTERÅLEN
NORWAY
SWEDEN
FINLAND
Stavanger
Kristiansand
Bergen
Oslo
Trondheim
Tromsø
Narvik
Kiruna
Luleå
Skagerrak
Göteborg
Stockholm
Uppsala
Norrköping
Gävle
Gulf of Bothnia
Vaasa
Oulu
Kajaani
Tampere
Turku
Helsinki
Lahti
Rovaniemi
NORDKAPP
BARENTS TROUGH
BARENTS SEA
Murmansk
Monchegorsk
Apatity
Kandalaksha
KOL'SKIY POLUOSTROV
WHITE SEA
Severodvinsk
Arkhangel'sk
OSTROV KOLGUYEV
MYS KANIN NOS
NOVAYA ZEMLYA
KARA SEA
OSTROV SHMIDTA
OSTROV PIONER
OSTROV VIZE
OSTROV BELYY
OSTROV SIBIRYAKOVA
OSTROV VAYGACH
Dikson
SEVERNA
Pechorskoye More
POLUOSTROV YAMAL
BALTIC SEA
ÖLAND
GOTLAND
SAAREMAA
HIIUMAA
Tallinn
Gulf of Finland
ESTONIA
Narva
Tartu
Gulf of Riga
Riga
LATVIA
Liepāja
Klaipėda
LITHUANIA
Kaliningrad
RUS.
POL.
Kaunas
Vilnius
Daugavpils
Pskov
Luga
ST. PETERSBURG (LENINGRAD)
Tikhvin
Petrozavodsk
Cherepovets
Novgorod
Borovichi
Vologda
Hrodna
Babruĭsk
MINSK
BELARUS
Pinsk
Homel'
Vicebsk
Mahilëu
Smolensk
MOSCOW (MOSKVA)
Tver'
Yaroslavl'
Rybinsk
Ivanovo
Kostroma
Vladimir
Kaluga
Bryansk
Tula
Kolomna
Ryazan'
Dzerzhinsk
NIZHNIY NOVGOROD
Chernihiv
Kursk
Orel
Elets
Sumy
KIEV (Kyïv)
Poltava
UKRAINE
KHARKIV
DNIPROPETROVS'K
Zaporizhzhia
Mariupol'
DONETS'K
Luhans'k
Belgorod
Staryy Oskol
Voronezh
Lipetsk
Tambov
Penza
Saransk
Balashov
Kamyshin
Saratov
Balakovo
SEA OF AZOV
Kerch
Krasnodar
ROSTOV-NA-DONU
VOLGOGRAD
Maykop
Elista
Sochi
Stavropol'
Gora El'brus 18 510 ft. 5642 m
Cherkessk
Nalchik
Sokhumi
Kutaisi
Batumi
BLACK SEA
CAUCASUS MTS.
GEORGIA
TBILISI
Groznyy
Makhachkala
Derbent
Astrahan'
CASPIAN DEPRESSION
Atyraū
Gyumri
Erzurum
Yerevan
ARMENIA
Mt. Ararat 16 854 ft. 5137 m
Gäncä
Mingäçevir
AZERBAIJAN
Xankändi
AZER.
TURKEY
Van
TABRIZ
Orūmiyeh
Marāgheh
Ardabil
Lankaran
Rasht
Zanjān
Qazvin
Sanandaj
IRAQ
IRAN
Gorgān
Sumqayıt
BAKU (BAKI)
CASPIAN SEA
Aqtaū
Türkmenbaşy
Balkanabat
Gyzylarbat
TURKMENISTAN
KARA KUM
KOPET DAG
Aşgabat
Gonbad-e Kāvūs
Qūchān
Sabzevār
MASHHAD
Tejen
Mary
Kara-Kum Canal
Türkmenabat
Buxoro
Qarshi
Samarqand
Dushanbe
TAJIKISTAN
AFGHAN.
PAMIRS
Pik Imeni Ismail Samani 24 590 ft. 7495 m
Khujand
Farg'ona
Namangan
Andijon
Osh
KYRGYZSTAN
Naryn
Bishkek
Balykchy
Karakol
Jengish Chokusu 24 406 ft. 7439 m
TIEN SHAN
TARIM PENDI
Aksu
Kashi
Shache
Korla
Bosten Hu
ÜRÜMQI
Changji
Qitai
Yining
ALMATY
Taraz
Shymkent
TASHKENT
Türkistan
QIZILQUM
UZBEKISTAN
Nukus
Daşoguz
Urganch
Qŭnghirot
UST-URT PLATEAU
Aral Sea
Qyzylorda
Bayqongyr (Baikonur)
Syr Darya
Amu Darya
KAZAKHSTAN
Zhezqazghan
KAZAKH HILLS
Qaraghandy
Temirtaū
Astana (Aqmola)
Ekibastuz
Pavlodar
Arqalyq
Tengiz köli
Lake Balkhash (Balqash köli)
Balqash
Taldyqorghan
Qapshaghay bögeni
Ayakōz
Zhaysang köli
Alakol köli
Tacheng
Karamay
Ebinur Hu
Manas Hu
Usu
Semey
Öskemen
Zyryanovsk
Ridder
Mt. Belukha 14 783 ft. 4506 m
ALTAY MTS.
MONGOL
Ölgiy
Uliungur Hu
Rubtsovsk
Gorno-Altaysk
Bijsk
Barnaul
Sayanogorsk
Novokuznetsk
Abakan
Prokopyevsk
Belovo
Leninsk-Kuznetskiy
Kemerovo
Yurga
NOVOSIBIRSK
Kuybyshev
OMSK
Petropavlovsk
Kōkshetaū
Shchūchīnsk
Qostanay
Rūdnyy
Zhetiqara
Orsk
Aqtōbe
Oral
Orenburg
Magnitogorsk
Kurgan
CHELYABINSK
Zlatoust
Salavat
Sterlitamak
UFA
SAMARA
Tolyatti
Syzran'
Dimitrovgrad
Ulyanovsk
Al'met'yevsk
Naberezhnye Chelny
KAZAN'
Cheboksary
Yoshkar-Ola
Izhevsk
Glazov
Kirov
PERM'
Kungur
Pervouralsk
YEKATERINBURG
Nizhniy Tagil
Tyumen'
Tobol'sk
Ishim
Tara
Serov
URAL MTS.
Berezniki
SEVERNYYE UVALY
Syktyvkar
Kotlas
Ukhta
Pechora
TIMANSKIY KRYAZH
Gora Narodnaya 6214 ft. 1894 m
Vorkuta
Naryan-Mar
Arctic Circle
Salekhard
Yar-Sale
Nadym
WEST SIBERIAN PLAIN
(ZAPADNO-SIBIRSKAYA RAVNINA)
Tarko-Sale
Khanty-Mansiysk
Surgut
Nizhnevartovsk
Kargasok
Kolpashevo
Tomsk
Asino
Achinsk
Anzhero-Sudzhensk
Vasyugan'ye
Turukhansk
Igarka
Tazovskiy
Dudinka
Noril'sk
Ozero Pyasino
RUSSIA

OCEAN
LAPTEV SEA
EAST SIBERIAN SEA
CHUKCHI SEA
Bering Strait
BERING SEA
SEA OF OKHOTSK
SEA OF JAPAN
PACIFIC OCEAN
YELLOW SEA
Korea Bay
Bo Hai
Gulf of Anadyr
OSTROV VRANGELYA
NEW SIBERIAN ISLANDS (NOVOSIBIRSKIYE OSTROVA)
OSTROV KOTEL'NYY
OSTROV BOL'SHOY LYAKHOVSKIY
OSTROV NOVAYA SIBIR'
OSTROV FADDEYEVSKIY
OSTROV BENNETTA
OSTROV OKTYABR'SKOY REVOLYUTSII
OSTROV BOL'SHEVIK
OSTROV MALYY TAYMYR
MYS CHELYUSKIN
POLUOSTROV TAYMYR
OSTROV BOL'SHOY BEGICHEV
SEVERO-SIBIRSKAYA NIZMENNOST'
ASIA
RUSSIA
CHUKOTSK PENINSULA
ST. LAWRENCE ISLAND (U.S.)
ST. MATTHEW ISLAND (U.S.)
MYS NAVARIN
ANADYRSKOYE PLOSKOGOR'YE
KORYAKSKOYE NAGOR'YE
Gora Ledyanaya 8402 ft. 2562 m
MYS OLYUTORSKIY
ALEUTIAN BASIN
OSTROV KARAGINSKIY
YUKAGIRSKOYE PLOSKOGOR'YE
KHREBET CHERSKOGO
Gora Pobeda 10,325 ft. 3147 m
VERKHOYANSKIY KHREBET
KAMCHATKA PENINSULA
SREDINNYY KHREBET
Vulkan Klyuchevskaya Sopka 15,584 ft. 4750 m
KOMANDORSKI ISLANDS
Petropavlovsk-Kamchatskiy
KHREBET DZHUGDZHUR
SHANTARSKIYE OSTROVA
SAKHALIN
Yuzhno-Sakhalinsk
OKHOTSK BASIN
KURIL ISLANDS
ALDANSKOYE NAGOR'YE
STANOVOY KHREBET
STANOVOYE NAGOR'YE
Gora Skalistyy Golets 8094 ft. 2467 m
YABLONOVYY KHREBET
Lake Baikal
SAYAN MTS.
GREATER KHINGAN RANGE (DA HINGGAN LING)
SIKHOTE-ALIN'
Yakutsk
Magadan
Mirnyy
Bratsk
Krasnoyarsk
Irkutsk
Angarsk
Ulan-Ude
Chita
Komsomol'sk-na-Amure
Khabarovsk
Blagoveshchensk
Vladivostok
Nakhodka
Ulaanbaatar
Darhan
MONGOLIA
HANGAYN NURUU
GOBI DESERT
CHINA
QIQIHAR
HARBIN
CHANGCHUN
JILIN
SHENYANG
FUSHUN
ANSHAN
DALIAN
TANGSHAN
TIANJIN
BEIJING
BAOTOU
Hohhot
Datong
Zhangjiakou
Chifeng
Hailar
NORTH KOREA
P'YONGYANG
SOUTH KOREA
SEOUL (SOUL)
PUSAN
Taegu
Kwangju
JAPAN
HOKKAIDO
SAPPORO
Hakodate
HONSHU
TOKYO
YOKOHAMA
NAGOYA
KYOTO
KOBE
OSAKA
HIROSHIMA
FUKUOKA
Kitakyushu
SHIKOKU
KYUSHU
Fuji-san 12,388 ft. 3776 m
Hami
© Rand McNally
A-101732-1

0 100 200 300 400 500 600 Miles
0 100 200 300 400 500 600 700 800 900 1000 Kilometers

Lambert Azimuthal Equal Area Projection
Scale 1:20,000,000
One inch to 320 miles
One cm to 200 km

TURKEY
Antalya
Isparta
Konya
Aksaray
Kayseri
Kahramanmaraş
Karaman
ADANA
Gaziantep
Alanya
TAURUS MTS.
İçel
Malatya
Elazığ
Diyarbakır
Muş
Bitlis
Van
Siverek
Siirt
Şanlıurfa
Mardin
Ağri
Mt. Ararat 16 854 ft. 5137 m
ARMENIA
Yerevan
AZER.
AZERBAIJAN
Yevlax
Xankändi
Sumqayıt
BAKU (BAKI)
Salyan
Länkäran
Khvoy
Orūmīyeh
TABRĪZ
Ardabīl
Marāgheh
Mahābād
Saqqez
Rasht
Zanjān
Qazvīn
Karaj
TEHRĀN
CASPIAN SEA
Kara-Bogaz-Gol Gulf
Türkmenbaşy
Balkanabat
TURKMENISTAN
Gumdag
Gyzylarbat
Bäherden
Aşgabat
KARA KUM
Türkmenabat
UZBEK
Navo
Buxoro
Urganch
Amu Darya
KOPETDAGH
Mary
Tejen
Bayram-Alï
Atamyr
Kara-Kum Canal
Gonbad-e Kāvūs
Gorgān
Quchān
Āmol
Sārī
ELBURZ MOUNTAINS
Shāhrūd
Kūh-e Damāvand 18 386 ft. 5604 m
Sabzevār
MASHHAD
Neyshābūr
Torbat-e Heydarīyeh
Meymaneh
Gușgy
SELSELEH-YE S
Herāt
Harirud
Farāh
AFGH
Hatay
İskenderun
Latakia (Al-Lādhiqīyah)
Nicosia
CYPRUS
Lemesós
ALEPPO (HALAB)
Al-Hasakah
Tartūs
Tripoli (Tarābulus)
LEBANON
Beirut
Hamāh
Hims
Dayr az-Zawr
SYRIA
Tudmur
MEDITERRANEAN SEA
Şaydā
Haifa
ISRAEL
Tel Aviv-Yafo
Port Said
Jerusalem
Gaza
Tanta
Ismailia
Suez
CAIRO (EL-QÂHIRA)
DAMASCUS (DIMASHQ)
Irbid
Az-Zarqā'
'Ammān
Dead Sea
JORDAN
Ma'ān
Elat
Al-'Aqabah
SINAI PENINSULA
Gulf of Suez
Gulf of Aqaba
EGYPT
Mosul
Arbīl
Kirkuk
As-Sulaymānīyah
Sanandaj
Tikrīt
Kermānshāh
Hamadān
Ar-Ramādī
Ba'qūbah
BAGHDĀD
IRAQ
MESOPOTAMIA
SYRIAN DESERT
Karbalā'
Al-Hillah
Al-Kūt
An-Najaf
Ad-Dīwānīyah
Al-'Amārah
As-Samāwah
An-Nāsirīyah
Ar'ar
Sakākah
Al-Jawf
Rafhā'
Basra
Ābādān
Arāk
Qom
Daryācheh-ye Namak
Kāshān
Borūjerd
Khorramābād
Dezfūl
Masjed-e Soleymān
Najafābād
ESFAHĀN
Zard Kūh 14 918 ft. 4547 m
ZAGROS MOUNTAINS
Ahvāz
Behbahān
Qomsheh
DASHT-E KAVĪR
IRAN
Tabas
Qā'en
Yazd
DASHT-E LŪT
Kermān
Rafsanjān
Zābol
Hāmūn-e Sāberī
Daryācheh-ye Hāmūn
Zāhedān
Bam
Hāmūn-i-Lora
Hāmūn-i-Mashkel
Hāmūn-e Jaz Mūrīān
BALU
Īrānshahr
Jabal al-Lawz 7884 ft. 2403 m
Tabūk
AN-NAFŪD
Hurghada
Qena
Quseir
El-Uqsor
Edfu
Kom Ombo
Aswân
Lake Nasser
KUWAIT
BŪBIYĀN
Kuwait (Al-Kuwayt)
Kāzerūn
SHĪRĀZ
Būshehr
Daryācheh-ye Bakhtegān
Jahrom
Persian Gulf
Lār
Bandar 'Abbās
Ḥā'il
Wādī ar-Rummah
Buraydah
Ar-Rass
'Unayzah
Az-Zilfī
Al-Majma'ah
Umm Lajj
AD-DAHNĀ'
Ad-Dammām
Al-Jubayl
Al-Khubar
BAHRAIN
Al-Manāmah
JAZĪREH-YE LĀVĀN
JAZĪREH-YE QESHM
JAZĪREH-YE KĪSH
Bandar-e Lengeh
Strait of Hormuz
Al-Hufūf
QATAR
Doha
OMAN
DUBAI
Ash-Shāriqah
Abu Dhabi
Al-'Ayn
Bandar Beheshtī
Turbat
Gwādar
Gwātar Bay
Gulf of Oman
Ṣuḥār
Al-Khābūrah
Muscat
Maṭraḥ
Tropic of Cancer
Yanbu'al-Baḥr
Medina (Al-Madīnah)
RIYADH (AR-RIYĀD)
Ad-Dawādimī
'Afīf
NAJD
Ad-Dilam
JABAL TUWAYQ
Rābigh
Halā'ib
NUBIAN DESERT
JIDDAH
Mecca (Makkah)
At-Tā'if
AL-HIJAZ
SAUDI ARABIA
ARABIAN PENINSULA
UNITED ARAB EMIRATES
Jabal ash-Shām 9957 ft. 3035 m
Adam
Ṣūr
RA'S AL-ḤADD
RED SEA
Būr Sūdān
SUDAN
Sawākin
Sinkāt
Tawkar
Qal'at Bīshah
Wādī ad-Dawāsir
RUB' AL-KHALI
OMAN
MAṢĪRAH
Khalīj Maṣīrah
'ASĪR
Khamīs Mushayt
Abhā
ERITREA
Aroma
Kassala
Keren
Akordat
DAHLAK ARCHIPELAGO
Jīzān
Abū 'Arīsh
Şa'dah
Teseney
Barentu
Massawa
Asmara
Khashm al-Qirbah
'Umrān
Ḥajjah
Jabal an-Nabī Shu'ayb 12 008 ft. 3660 m
SANAA (ṢAN'Ā')
YEMEN
Saywūn
Shibām
ḤAḌRAMAWT
Ṣalālah
JUZUR AL-ḤALLĀNĪYAH
Ghubbat al-Qamar
Al-Ghaydah
RA'S FARTAK
ARABIAN
Al-Qaḍārif
Aksum
Ādīgrat
Mek'elē
Ras Dejen 15 158 ft. 4620 m
Al-Ḥudaydah
Dhamār
Rida'
Zabīd
Ibb
Al-Mukallā
Ash-Shiḥr
Al-Mukhā
Ta'izz
Al-Ḥawrah
DANAKIL
Aseb
Bab el-Mandeb
Gonder
Tana Hāyk'
Bahir Dar
Aden ('Adan)
Gulf of Aden
DJIBOUTI
Djibouti
Dikhil
Saylac
Mot'a
Desē
Talo 14 478 ft. 4413 m
Debre Mark'os
'ABD AL-KŪRĪ (Yemen)
Hadībū
SOCOTRA (Yemen)
Caluula
GEES GWARDAFUY
Boosaaso
Qandala
ARABIAN
ADDIS ABABA (ĀDĪS ĀBEBA)
Nek'emtē
Dirē Dawa
Boorama
Berbera
Shimbiris 7897 ft. 2407 m
Ceerigaabo
RAAS XAAFUUN
Giyon
Hārer
Jījiga
Hargeysa
Burco
Nazrēt
Āsela
Jīma
Hosa'ina
ETHIOPIA
SOMALIA
Degeh Bur
RIFT VALLEY
Sodo
Āwasa
Goba
OGADEN
Laascaanood
Arba Minch
Dīla
Kibre Mengist
Nēgelē
Wabē Gestro
Wabē Shebelē
Werdēr
K'ebrī Dehar
Eyl
Gaalkacyo
Ch'ew Bahir (Lake Stefanie)
Genalē
CARLSBERG RIDGE
SOMALI BASIN
30°
35°
40°
45°
50°
55°
60°
20°
15°
10°

0 80 160 240 320 400 480 Miles
0 80 160 240 320 400 480 560 640 720 800 Kilometers

Lambert Azimuthal Equal Area Projection
Scale 1:16,000,000
One inch to 256 miles
One cm to 160 km

TASHKENT
Namangan
KYRGYZSTAN
Andijon
Osh
Farg'ona
Jizzax
Khujand
Istaravshan
Samarqand
Qarshi
TAJIKISTAN
Pik Imeni Ismail Samani 24 590 ft. 7495 m
Dushanbe
Denov
Qŭrghonteppa
Kŭlob
Khorugh
PAMIRS
Termiz
Feyzābād
Kondoz
Kholm
Khānābād
Mazār-e Sharīf
HINDU KUSH
KARAKORAM RANGE
K2 (Qogir Feng) 28 250 ft. 8611 m
Chitrāl
Gilgit
Koh-e Folādī 16 847 ft. 5135 m
Chārīkār
Kabul
Khyber Pass 3501 ft. 1067 m
Mingāora
Mardān
Jalālābād
Peshāwar
Islāmābād
Srīnagar
JAMMU AND KASHMIR
Ghaznī
Gardeyz
RĀWALPINDI
Jammu
Siālkot
Gujrānwāla
HIMĀCHAL PRADESH
Sargodha
LAHORE
Amritsar
FAISALĀBĀD
Jhang Sadar
Jalandhar
LUDHIĀNA
Shimla
Chandīgarh
UTTARAKHAND
Dehra Dūn
PUNJAB
Patiāla
Nanda Devi 25 645 ft. 7817 m
Chaman
Quetta
Dera Ghāzi Khān
MULTĀN
Sahāranpur
HARYĀNA
Hisār
Muzaffarnagar
MEERUT
Sibi
Bahāwalpur
DELHI
New Delhi
Rāmpur
Bareilly
PAKISTAN
Khānpur
Jacobābād
Shikārpur
Khuzdār
Sukkur
Lārkāna
KIRTHAR RANGE
Dādu
Nawābshāh
GREAT INDIAN DESERT
Bīkāner
Sīkar
Alwar
Alīgarh
Mathura
Shāhjahānpur
LUCKNOW
Fīrozābād
Etāwah
ĀGRA
JAIPUR
Jodhpur
Ajmer
Gwalior
KĀNPUR
UTTAR PRADESH
RĀJASTHĀN
Bārmer
Pāli
Jhānsi
Allahābād
HYDERĀBĀD
Mīrpur Khās
KARĀCHI
Bhīlwāra
Kota
Udaipur
INDIA
Sāgar
VINDHYA RANGE
Satna
Mirzāpur
Pālanpur
RANN OF KUTCH
Pātan
Gāndhinagar
Bhuj
GUJARĀT
AHMADĀBĀD
Nadiād
Ujjain
BHOPĀL
Jabalpur
MADHYA PRADESH
INDORE
Gulf of Kachchh
Jāmnagar
Rājkot
Vadodara
Bhāvnagar
KĀTHIĀWĀR PENINSULA
Jūnāgadh
Porbandar
Verāval
SŪRAT
Gulf of Khambhāt
Daman
Dhule
Jālgaon
Akola
Amrāvati
NĀGPUR
Bilāspur
CHHATTISGARH
Raipur
Bhilai
Mālegaon
Aurangābād
Chandrapur
Nāshik
Thāne
MAHĀRĀSHTRA
Parbhani
Nānded
DECCAN
Ahmadnagar
Nizāmābād
Karīmnagar
Warangal
MUMBAI (BOMBAY)
Pune
Satāra
Solāpur
HYDERĀBĀD
Rājahmundry
Kākināda
Vishākhapatnam
Vizianagaram
Gulbarga
Sāngli
Kolhāpur
Belgaum
Panaji
GOA
Hubli-Dhārwār
Rāichūr
Hospet
Bellary
ANDHRA PRADESH
Kurnool
Guntūr
Vijayawāda
Machilīpatnam
KARNĀTAKA
Anantapur
Nellore
Dāvangere
Shimoga
Cuddapah
BENGALŪRU (BANGALORE)
CHENNAI (MADRAS)
Tumkūr
Vellore
Mangalore
Mysore
Pondicherry
COROMANDEL COAST
Salem
Coimbatore
Tiruchchirāppalli
Thanjāvūr
Kozhikode
KERALA
Anai Mudi 8842 ft. 2695 m
Madurai
Thrissur
Cochin
TAMIL NADU
Kollam
Tuticorin
Tirunelveli
Thiruvananthapuram
Nāgercoil
CAPE COMORIN
Gulf of Mannar
Palk Strait
Jaffna
Jaffna Lagoon
Trincomalee
Anuradhapura
Batticaloa
Kandy
Pidurutalagala 8281 ft. 2524 m
SRI LANKA
Negombo
Colombo
Sri Jayewardenepura Kotte
Galle
DONDRA HEAD
KILTTĀN ISLAND
ĀNDROTT ISLAND
KAVARATTI ISLAND
KALPENI ISLAND
LAKSHADWEEP
LAKSHADWEEP (India)
MINICOY ISLAND
Eight Degree Channel
MALDIVES
LAKSHADWEEP SEA
SEA
BASIN
TARIM PENDI
TAKLA MAKAN DESERT
Kashi
Shache
Pishan
Hotan
Yutian
Minfeng
Qiemo
XINJIANG
Ruoqiang
ALTUN SHAN
Lenghu
Mangya
TSAIDAM BASIN
Golmud
Dulan
Tianjun
Xining
Qinghai Hu (Koko Nor)
QINGHAI
Ngoring Hu
BAYAN HAR SHAN
Darlag
Huang (Yellow)
KUNLUN SHAN
Muztag 25 338 ft. 7723 m
CHINA
Qumarlêb
Garqu Yan
Geladandong 21 722 ft. 6621 m
Dogai Coring
PLATEAU OF TIBET
Yushu
TANGGULA SHAN
Dêgê
SICHUAN
Changmar
Duomula
Amdo
Nagqu
Qamdo
Mekong
XIZANG (TIBET)
Zhaxigang
Ngangla Ringco
Taguke
Zhari Namco
Siling Co
Nam Co
Gyaring Co
Tangra Yumco
Namjagbarwa Feng 25 443 ft. 7755 m
NYAINQÊNTANGLHA SHAN
GANGDISÊ SHAN
Zhongba
Saga
Sa'gya
Lhasa
Yamzho Yumco
Puma Yumco
Kula Kangri 24 784 ft. 7554 m
Brahmaputra
ARUNACHAL PRADESH
HIMALAYAS
Annapurna 26 545 ft. 8091 m
Mt. Everest 29 028 ft. 8848 m
SIKKIM
Thimphu
BHUTAN
Dibrugarh
Itānagar
Sibsāgar
Jorhāt
Mahendranagar
NEPAL
Pokhara
Kathmandu
Gangtok
Shiliguri
ASSAM
Nagaon
Dimāpur
NAGALAND
Myitkyinā
Wirgañj
Wirātnagar
Jalpāiguri
Guwāhāti
Dispur
Kohima
Gorakhpur
Muzaffarpur
Darbhanga
Dinājpur
MEGHĀLAYA
Shillong
Silchar
Imphāl
Bhamo
VĀRĀNASI (BENARES)
PATNA
Rangpur
Sylhet
MANIPUR
Katha
BIHĀR
Bhāgalpur
Mymensingh
Agartala
Aizawl
MYANMAR (BURMA)
Gaya
Rājshāhi
BANGLADESH
DHAKA
TRIPURA
MIZORAM
Shwebo
JHARKHAND
Dhanbād
Āsansol
Khulna
Barisāl
CHITTAGONG
Monywa
Maymyo
Mandalay
Rānchi
WEST BENGAL
Bhātpāra
Mt. Victoria 10 016 ft. 3053 m
Myingyan
Paletwa
Jamshedpur
KOLKATA (CALCUTTA)
Meiktila
Taunggyi
Raurkela
Kharagpur
Yenangyaung
Pyinmana
Sambalpur
Bāleshwar
Bhadrak
Sittwe
Nay Pyi Taw
Cuttack
Toungoo
Bhubaneshwar
ORISSA
RAMREE ISLAND
ARAKAN YOMA
Prome
Puri
CHEDUBA ISLAND
Brahmapur
Tharrawaddy
Jagdalpur
Henzada
Pathein
YANGON (RANGOON)
BAY OF BENGAL
PREPARIS ISLAND
COCO ISLANDS
Coco Channel
NORTH ANDAMAN
MIDDLE ANDAMAN
SOUTH ANDAMAN
ANDAMAN ISLANDS
ANDAMAN SEA
Port Blair
LITTLE ANDAMAN
ANDAMAN AND NICOBAR ISLANDS (India)
Ten Degree Channel
CAR NICOBAR ISLAND
NICOBAR ISLANDS
KATCHALL ISLAND
LITTLE NICOBAR
GREAT NICOBAR
© Rand McNally
A-101741-1

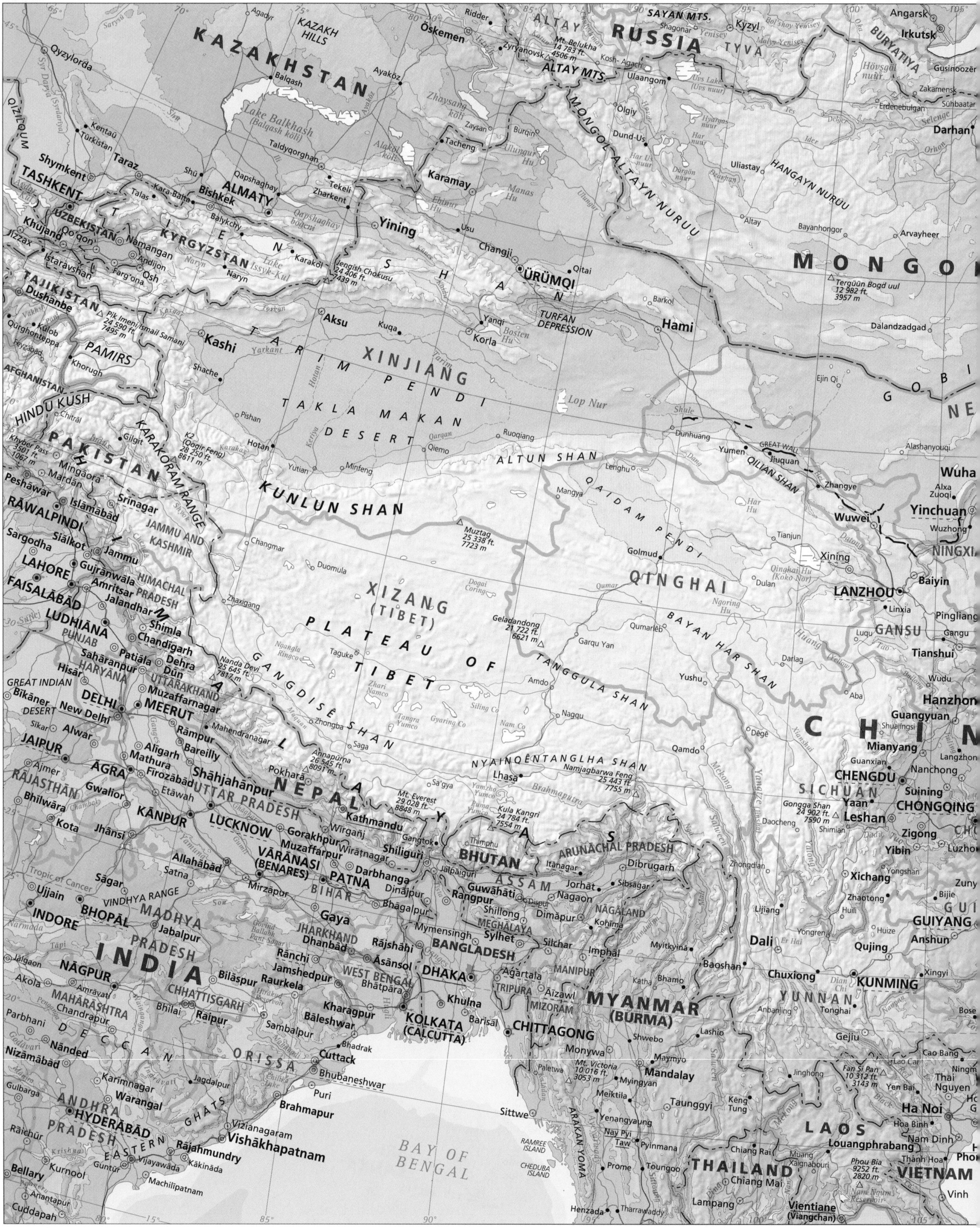
KAZAKHSTAN
KAZAKH HILLS
RUSSIA
SAYAN MTS.
ALTAY
ALTAY MTS.
TYVA
BURYATIYA
MONGOLIA
MONGOL ALTAYN NURUU
HANGAYN NURUU
GOBI
Lake Balkhash (Balqash köli)
Zhaysang köli
Alakol köli
Lop Nur
Bosten Hu
Ebinur Hu
Issyk-Kul
Qapshaghay bögeni
Uvs Lake (Uvs nuur)
Hövsgöl nuur
Qyzylorda
Kentau
Türkistan
Shymkent
Taraz
TASHKENT
Talas
Kara-Balta
Bishkek
ALMATY
Qapshaghay
Tekeli
Zharkent
Taldyqorghan
Balqash
Ayaköz
Agadyr
Öskemen
Ridder
Zyryanovsk
Mt. Belukha 14 783 ft. 4506 m
Kosh-Agach
Shagonar
Kyzyl
Ulaangom
Ölgiy
Dund-Us
Uliastay
Altay
Bayanhongor
Arvayheer
Angarsk
Irkutsk
Gusinoozërsk
Zakamensk
Erdenebulgan
Sühbaatar
Darhan
Zaysan
Burqin
Tacheng
Karamay
Shü
Balykchy
UZBEKISTAN
Qo'qon
Khujand
Jizzax
Namangan
KYRGYZSTAN
Andijon
Farg'ona
Osh
Istaravshan
Naryn
Karakol
Jengish Chokusu 24 406 ft. 7439 m
TAJIKISTAN
Dushanbe
Kulob
Qurghonteppa
Pik Imeni Ismail Samani 24 590 ft. 7495 m
PAMIRS
Khorugh
AFGHANISTAN
HINDU KUSH
Chitral
Gilgit
Khyber Pass 3501 ft. 1067 m
PAKISTAN
Mingaora
Mardan
Peshawar
Islāmābād
RAWALPINDI
Sargodha
Siālkot
Jammu
LAHORE
Gujrānwāla
FAISALĀBĀD
Amritsar
Jalandhar
LUDHIĀNA
PUNJAB
Shimla
Chandīgarh
Patiāla
Dehra Dūn
HARYANA
UTTARAKHAND
Hisār
GREAT INDIAN DESERT
Bīkāner
DELHI
New Delhi
MEERUT
Muzaffarnagar
Sīkar
Alwar
JAIPUR
Ajmer
RAJASTHAN
Bhīlwāra
Kota
Aligarh
Mathura
ĀGRA
Firozābād
Etāwah
Gwalior
Jhānsi
KĀNPUR
LUCKNOW
Shāhjahānpur
Bareilly
Rāmpur
UTTAR PRADESH
Allahābād
Sāgar
Tropic of Cancer
Ujjain
INDORE
BHOPĀL
VINDHYA RANGE
MADHYA PRADESH
Jabalpur
INDIA
NĀGPUR
Jalgaon
Akola
Amrāvati
MAHĀRĀSHTRA
Chandrapur
Parbhani
Nānded
Nizāmābād
DECCAN
Karīmnagar
Warangal
Gulbarga
ANDHRA PRADESH
HYDERĀBĀD
Rāichūr
Kurnool
Bellary
Anantapur
Cuddapah
EASTERN GHATS
Guntūr
Vijayawāda
Rājahmundry
Kākināda
Machilīpatnam
Vizianagaram
Vishākhapatnam
Brahmapur
Puri
Bhubaneshwar
Cuttack
Bhadrak
ORISSA
Jagdalpur
Sambalpur
Bāleshwar
Kharagpur
CHHATTISGARH
Raipur
Bhilai
Bilāspur
Raurkela
JHARKHAND
Rānchi
Jamshedpur
Dhanbād
Āsansol
WEST BENGAL
Bhātpāra
KOLKATA (CALCUTTA)
Gaya
BIHAR
PATNA
Darbhanga
Muzaffarpur
VĀRĀNASI (BENARES)
Gorakhpur
Mirzāpur
Satna
Bhāgalpur
Dinājpur
Rājshāhi
BANGLADESH
DHAKA
Khulna
Barisal
Mymensingh
Sylhet
Rangpur
Jalpāiguri
Shiliguri
Gangtok
Thimphu
BHUTAN
ASSAM
Guwāhāti
Dispur
Shillong
MEGHALAYA
Nagaon
Jorhāt
Sibsagar
Dibrugarh
Itānagar
ARUNĀCHAL PRADESH
Dīmāpur
NĀGĀLAND
Kohīma
Silchar
Imphāl
MANIPUR
Āgartala
TRIPURA
Āīzawl
MIZORAM
CHITTAGONG
MYANMAR (BURMA)
Myitkyina
Bhamo
Katha
Shwebo
Lashio
Monywa
Maymyo
Mandalay
Mt. Victoria 10 016 ft. 3053 m
Paletwa
Myingyan
Meiktila
Taunggyi
Yenangyaung
Nay Pyi Taw
Pyinmana
Sittwe
RAMREE ISLAND
CHEDUBA ISLAND
ARAKAN YOMA
Prome
Toungoo
Henzada
Tharrawaddy
BAY OF BENGAL
NEPAL
Kathmandu
Pokharā
Annapūrna 26 545 ft. 8091 m
Mahendranagar
Wirāṭnagar
Mt. Everest 29 028 ft. 8848 m
Kula Kangri 24 784 ft. 7554 m
Nanda Devi 25 645 ft. 7817 m
HIMALAYAS
HIMACHAL PRADESH
JAMMU AND KASHMIR
Srinagar
KARAKORAM RANGE
K2 (Qogir Feng) 28 250 ft. 8611 m
TIEN SHAN
Yining
Usu
Changji
ÜRÜMQI
Qitai
Barkol
Hami
TURFAN DEPRESSION
Yanqi
Korla
Kuqa
Aksu
Kashi
Yarkant
Shache
TARIM PENDI
XINJIANG
TAKLA MAKAN DESERT
Pishan
Hotan
Yutian
Minfeng
Qiemo
Ruoqiang
ALTUN SHAN
KUNLUN SHAN
Muztag 25 338 ft. 7723 m
Changmar
Duomula
XIZANG (TIBET)
PLATEAU OF TIBET
Zhaxigang
Taguke
GANGDISÊ SHAN
Zhongba
Saga
Sa'gya
Lhasa
NYAINQÊNTANGHLA SHAN
Namjagbarwa Feng 25 443 ft. 7755 m
Brahmaputra
Nagqu
Nam Co
Siling Co
Amdo
TANGGULA SHAN
Geladandong 21 722 ft. 6621 m
Dogai Coring
Qamdo
Dêgê
Yushu
Garqu Yan
Qumarlêb
BAYAN HAR SHAN
QINGHAI
QAIDAM PENDI
Golmud
Mangya
Lenghu
Dunhuang
Yumen
Jiuquan
QILIAN SHAN
GREAT WALL
Zhangye
Wuwei
Xining
Qinghai Hu (Koko Nor)
Dulan
Tianjun
LANZHOU
Linxia
GANSU
Tianshui
Pingliang
Baiyin
NINGXIA
Yinchuan
Wuzhong
Alxa Zuoqi
Alashanyouqi
Ejin Qi
Dalandzadgad
Tergüün Bogd uul 12 982 ft. 3957 m
Darlag
Aba
Guangyuan
Mianyang
CHENGDU
Nanchong
SICHUAN
Suining
CHONGQING
Ya'an
Leshan
Gongga Shan 24 902 ft. 7590 m
Daocheng
Zigong
Yibin
Luzhou
Xichang
Zhaotong
Bijie
GUIYANG
Anshun
Qujing
KUNMING
Chuxiong
Dali
Lijiang
Zhongdian
YUNNAN
Baoshan
Tonghai
Gejiu
Jinghong
CHINA
Fan Si Pan 10 312 ft. 3143 m
Lao Cai
Yen Bai
Ha Noi
Hoa Binh
Nam Dinh
Thanh Hoa
Vinh
VIETNAM
LAOS
Louangphrabang
Xaignabouri
Vientiane (Viangchan)
Phou Bia 9252 ft. 2820 m
THAILAND
Chiang Rai
Chiang Mai
Lampang
Keng Tung
Mekong
Yangtze
Salween

0 80 160 240 320 400 480 Miles
0 80 160 240 320 400 480 560 640 720 800 Kilometers

Lambert Azimuthal Equal Area Projection
Scale 1:16,000,000
One inch to 256 miles
One cm to 160 km

RUSSIA
MANCHURIA
GREATER KHINGAN RANGE (DA HINGGAN LING)
LESSER KHINGAN RANGE
SIKHOTE-ALIN'
YABLONOVYY KHREBET
Lake Baikal (Ozero Baykal)
Ulan-Ude
Chita
Nerchinsk
Shilka
Baley
Sherlovaya Gora
Borzya
Petrovsk-Zabaykal'skiy
Skovorodino
Mohe
Magdagachi
Gulian
Shimanovsk
Svobodnyy
Belogorsk
Blagoveshchensk
Heihe
Raychikhinsk
Birobidzhan
Khabarovsk
Litovko
Sovetskaya Gavan'
Yitulihe
Nunjiang
Nahe
Bei'an
Yichun
Bugt
Hailar
Manzhouli
Hulun Nur
Choybalsan
Ulaanbaatar
Kerulen
Zalantun
QIQIHAR
HEILONGJIANG
Yi'an
Hailun
Hegang
Jiamusi
Fujin
Tieli
Suihua
Shuangyashan
Anda
Yilan
Boli
Mishan
Jixi
Dalnerechensk
Lesozavodsk
Spassk-Dal'niy
Dal'negorsk
Kavalerovo
Arxan
Ulanhot
Tailai
HARBIN
Acheng
Shuangcheng
Hulan
Baicheng
JILIN
Fuyu
Mudanjiang
Tongyu
Dehui
Jiutai
Nong'an
Arsen'yev
Ussuriysk
Artëm
Vladivostok
Nakhodka
Buyant-Uhaa
Ulaan-Uul
Erenhot
Xilinhot
Linxi
Qagan Nur
Bairin Zuoqi
Tongliao
CHANGCHUN
Gongzhuling
Siping
Liaoyuan
Huadian
Dunhua
Yanji
Hunjiang
Paektu-san 9 003 ft. 2744 m
Ch'ŏngjin
Sonid Youqi
Duolun
Chifeng
Fuxin
Tieling
FUSHUN
Beipiao
LIAONING
SHENYANG
Benxi
Tonghua
Hyesan
Kanggye
Kimch'aek
Chaoyang
Jinzhou
ANSHAN
NORTH KOREA
Hamhŭng
Hŭngdŏki-dong
Yingkou
Liaodong Wan
Dandong
Sinŭiju
Wŏnsan
Wafangdian
P'YONGYANG
Korea Bay
Namp'o
Both North Korea and South Korea claim to be the sole legitimate government of Korea.
ULLŬNG-DO (S.Kor.)
Bayan Obo
MONGOLIA
Wuyuan
BAOTOU
Hohhot
Zhangjiakou
Chengde
Jining
Xuanhua
BEIJING
Qinhuangdao
Changli
Huang (Yellow)
Jungar Qi
Datong
Zhuozhou
TANGSHAN
TIANJIN
Lüshun
DALIAN
Kaesŏng
SEOUL (SŎUL)
INCH'ŎN
SOUTH KOREA
Ch'ŏngju
Taejŏn
Chŏnju
Kwangju
Mokp'o
Masan
Taegu
P'ohang
PUSAN
Cheju
CHEJU-DO
Korea Strait
Bo Hai
Cangzhou
Baoding
HEBEI
Ningwu
Wutai
Wushenqi
GREAT WALL
Yulin
Yangquan
SHIJIAZHUANG
TAIYUAN
Yuci
Dezhou
Suide
SHAANXI
Fenyang
Pingyao
SHANXI
Xingtai
Handan
Changzhi
TAIHANG SHAN
JINAN
Zibo
Weifang
Yantai
Weihai
CHENGSHAN JIAO
Laiyang
SHANDONG BANDAO
QINGDAO
Tai'an
SHANDONG
Yishui
YELLOW SEA
Yan'an
Qingyang
Linfen
Huangling
Anyang
Xinxiang
Jiaozuo
Jining
Yanzhou
Tongchuan
Yuncheng
Kaifeng
Lianyungang
Weinan
Luoyang
ZHENGZHOU
Xuzhou
JIANGSU
Baoji
XI'AN
Pingdingshan
Xuchang
Qingjiang
Yancheng
QIN LING
Ningshan
HENAN
Suzhou
Fuyang
Hongze Hu
Bengbu
Xinghua
Ankang
Nanyang
Huainan
NANJING
Yangzhou
Zhenjiang
HEFEI
Changzhou
Wuxi
CHANGXING DAO
Xinyang
DABIE SHAN
Xiangfan
Three Gorges Reservoir
Lu'an
Wuhu
ANHUI
Suzhou
SHANGHAI
Daxian
HUBEI
Anlu
Xiaogan
Jiaxing
HANGZHOU
ZHOUSHAN DAO
WUHAN
Anqing
Shaoxing
NINGBO
Wanxian
Yichang
Enshi
Shashi
Huangshi
Jingdezhen
Jinhua
Dongyang
Taizhou
Fuling
Jinshi
Yueyang
Poyang Hu
Quzhou
ZHEJIANG
Changde
NANCHANG
Shangrao
Lishui
HUNAN
Yiyang
JIANGXI
Fuzhou
Wenzhou
Yuanling
CHANGSHA
Yichun
Longquan
Xiangtan
Pingxiang
Shaowu
Jian'ou
Tongren
Shaoyang
Ji'an
Nanping
Hongjiang
Hengyang
Jinggangshan
WUYI SHAN
Yong'an
FUZHOU
MATSU TAO
Duyun
Yongzhou
Ganzhou
FUJIAN
T'AIPEI
Quanzhou
Chenzhou
NAN LING
Guilin
Nanxiong
Hsinchu
Zhangzhou
T'aichung
QUEMOY
Xiamen
Yü Shan 13 114 ft. 3997 m
Mangchang
Hechi
Shaoguan
Liucheng
Qingtang
Meizhou
Hexian
Liuzhou
Chao'an
Chiai
Taiwan Strait
GUANGXI
Wuzhou
GUANGDONG
GUANGZHOU
T'ainan
TAIWAN
Guiping
Dongguan
Shantou
China and many other countries do not recognize the existence of Taiwan as a separate country.
Litang
Foshan
Lufeng
KAOHSIUNG
Shenzhen
Nanning
Yulin
Jiangmen
HONG KONG (XIANGGANG)
OLUAN PI
Macau (Aomen)
Bashi Channel
Beihai
Zhanjiang
LEIZHOU BANDAO
Haikou
PRATAS ISLAND (Occupied by Taiwan; claimed by China)
Gulf of Tonkin
HAINAN DAO
HAINAN
Dongfang
Wuzhi Shan 6125 ft. 1867 m
Baoting
Sanya
SOUTH CHINA SEA
SOUTH CHINA BASIN
BATAN ISLANDS
Basco
Luzon Strait
Balintang Channel
BABUYAN ISLANDS
ESCARPADA POINT
CAPE BOJEADOR
Laoag
PHILIPPINES
Vigan
LUZON
Tuguegarao City
EAST CHINA SEA
SEA OF JAPAN
Tatar Strait
SAKHALIN
Makarov
Dolinsk
Yuzhno-Sakhalinsk
Korsakov
Kholmsk
Nevel'sk
SEA OF OKHOTSK
OSTROV ITURUP (ETOROFU-TŌ)
OSTROV URUP
KURIL ISLANDS
OSTROV KUNASHIR (KUNASHIRI-TŌ)
OSTROV SHIKOTAN (SHIKOTAN-TŌ)
MALAYA KURIL'SKAYA GRYADA (HABOMAI-SHOTŌ)
La Perouse Strait
Wakkanai
RISHIRI-TŌ
Asahi-dake 7513 ft. 2290 m
Asahikawa
HOKKAIDŌ
Kushiro
Obihiro
The islands known in Japan as the Northern Territories and in Russia as the Southern Kuril Islands are occupied by Russia and claimed by Japan.
SAPPORO
Tomakomai
Hakodate
Aomori
Hachinohe
Morioka
Akita
Sakata
Sendai
Yamagata
Fukushima
SADO
Kōriyama
Niigata
Iwaki
Nagaoka
Utsunomiya
Nagano
Mito
Toyama
Maebashi
TŌKYŌ
Kanazawa
HONSHU
YOKOHAMA
Fuji-san 12 388 ft. 3776 m
OKI-SHOTŌ
Tottori
KYŌTO
NAGOYA
Hamamatsu
Matsue
KŌBE
ŌSAKA
Okayama
Wakayama
MIYAKE-JIMA
IZU-SHOTŌ
HACHIJŌ-JIMA
AOGA-SHIMA
HIROSHIMA
Takamatsu
Matsuyama
JAPAN
Kitakyūshū
Kōchi
SHIKOKU
FUKUOKA
KYŪSHŪ
Sasebo
Kumamoto
Nagasaki
Miyazaki
Kagoshima
TANEGA-SHIMA
YAKU-SHIMA
NAKANO-SHIMA
SUWANOSE-JIMA
AMAMI-Ō-SHIMA
Naze
TOKUNO-SHIMA
OKINO-ERABU-SHIMA
RYUKYU ISLANDS (NANSEI-SHOTŌ)
OKINAWA-JIMA
Naha
KUME-JIMA
KITA-DAITŌ-JIMA
MINAMI-DAITŌ-JIMA (Japan)
OKINO-DAITŌ-JIMA (Japan)
Tropic of Cancer
MIYAKO-JIMA
Hirara
ISHIGAKI-SHIMA
IRIOMOTE-JIMA
SAKISHIMA-SHOTŌ
PHILIPPINE SEA
PHILIPPINE BASIN
OKINO-TORI-SHIMA (Japan)
© Rand McNally A-101749-1

CHINA
YUNNAN
GUANGXI
GUANGDONG
Tonghai
Gejiu
Tropic of Cancer
Bose
Nanning
Cao Bang
Lao Cai
Fan Si Pan 10 312 ft. 3143 m
Yen Bai
Thai Nguyen
Ningming
Beihai
Jiangmen
Shenzhen
Macau (Aomen)
HONG KONG (XIANGGANG)
KAOHSIUNG
TAIWAN
OLUAN PI
Bashi Channel
BATAN ISLANDS
Basco
Luzon Strait
Balintang Channel
BABUYAN ISLANDS
PRATAS ISLAND (Occupied by Taiwan; claimed by China)
Zhanjiang
LEIZHOU BANDAO
Haikou
Ha Noi
Hon Gai
Hai Phong
Hoa Binh
Nam Dinh
Thanh Hoa
Gulf of Tonkin
HAINAN DAO
HAINAN
Wuzhi Shan 6125 ft. 1867 m
Dongfang
Baoting
Sanya
Vinh
Jinghong
Mt. Victoria 10 016 ft. 3053 m
Paletwa
Maymyo
Mandalay
Myingyan
Meiktila
Taunggyi
Keng Tung
Yenangyaung
Sittwe
MYANMAR (BURMA)
ARAKAN YOMA
Nay Pyi Taw
Pyinmana
RAMREE ISLAND
CHEDUBA ISLAND
Chiang Rai
Chiang Mai
Louangphrabang
LAOS
Muang Xaignabouri
Phou Bia 9252 ft. 2820 m
Vientiane (Viangchan)
Toungoo
Prome
Lampang
Henzada
Tharrawaddy
Bago
Pathein
YANGON (RANGOON)
Gulf of Martaban
Pha-an
Mawlamyine
Phitsanulok
Khon Kaen
Udon Thani
Savannakhet
Dong Hoi
CHAÎNE ANNAMITIQUE
Hue
Da Nang
PARACEL ISLANDS (Occupied by China; claimed by Taiwan and Vietnam)
THAILAND
Nakhon Sawan
Nakhon Ratchasima
Ubon Ratchathani
Pakxé
Ye
PREPARIS ISLAND
COCO IS.
Coco Channel
NORTH ANDAMAN
MIDDLE ANDAMAN
SOUTH ANDAMAN
ANDAMAN ISLANDS
Port Blair
LITTLE ANDAMAN
INDOCHINA
Phra Nakhon Si Ayutthaya
Dawei
Nakhon Pathom
BANGKOK (KRUNG THEP)
Chon Buri
Siĕmréab
Stœng Trêng
Tonle Sap
Bătdâmbâng
Kâmpóng Thum
Play Ku
VIETNAM
Quy Nhon
KADAN KYUN
Mergui
Chanthaburi
CAMBODIA
Krâchéh
MUI KE GA
Nha Trang
ANDAMAN SEA
MERGUI ARCHIPELAGO
KO CHANG
KO KUT
Phnom Penh
Da Lat
HO CHI MINH CITY (SAIGON)
Bien Hoa
Phan Thiet
Chumphon
Kâmpóng Saôm
Kâmpôt
Mekong
Vung Tau
Ten Degree Channel
ANDAMAN AND NICOBAR ISLANDS (India)
ISTHMUS OF KRA
Kawthaung
DAO PHU QUOC
Rach Gia
Can Tho
Bac Lieu
Gulf of Thailand
KO SAMUI
Surat Thani
CAR NICOBAR ISLAND
NICOBAR ISLANDS
Nakhon Si Thammarat
MUI CA MAU
CON SON
KATCHALL ISLAND
KO PHUKET
LITTLE NICOBAR
GREAT NICOBAR
Great Channel
Thale Sap Songkhla
Trang
Songkhla
Hat Yai
Yala
Kota Bharu
MALAY PENINSULA
Banda Aceh
PULAU LANGKAWI
Alor Setar
George Town
PULAU PINANG
Kuala Terengganu
Lhokseumawe
Taiping
Gunung Tahan 7175 ft. 2187 m
Ipoh
MALAYSIA
Langsa
Binjai
MEDAN
Tebingtinggi
Pematangsiantar
PULAU SIMEULUE
Strait of Malacca
Klang
KUALA LUMPUR
Melaka
Batu Pahat
Johor Bahru
SINGAPORE
Dumai
KEPULAUAN BANYAK
Sibolga
PULAU NIAS
Pekanbaru
Equator
KEPULAUAN BATU
Bukittinggi
Padang
PULAU SIBERUT
PULAU SIPURA
PULAU PAGAI UTARA
PULAU PAGAI SELATAN
KEPULAUAN MENTAWAI
Selat Mentawai
PEGUNUNGAN BARISAN
Gunung Kerinci 12 467 ft. 3800 m
Jambi
Tembilahan
Indragiri
Kampar
Rokan
SUMATRA (SUMATERA)
Prabumulih
PALEMBANG
Bengkulu
Gunung Dempo 10 364 ft. 3159 m
Kotabumi
Bandar Lampung
PULAU ENGGANO
TANJUNG CINA
Selat Sunda
PULAU TIOMAN
KEPULAUAN LINGGA
PULAU LINGGA
PULAU SINGKEP
PULAU BANGKA
Tanjungpandan
BELITUNG
Selat Gelasa
SUNDA SHELF
PULAU LAUT
NATUNA BESAR
KEPULAUAN ANAMBAS
SOUTH CHINA BASIN
SOUTH CHINA SEA
NAMYIT ISLAND
NANSHAN ISLAND
SPRATLY ISLANDS (Claimed by China, Malaysia, the Philippines, Taiwan and Vietnam)
PALAWAN
Puerto Princesa
BALABAC ISLAND
Balabac Strait
PULAU BANGGI
CAGAYAN SULU ISLAND
SULU SEA
Kudat
Kota Kinabalu
Gunung Kinabalu 13 455 ft. 4101 m
Sandakan
Bandar Seri Begawan
BRUNEI
Miri
Lahad Datu
Tawau
Gunung Murud 7946 ft. 2422 m
Bintulu
Tarakan
TANJUNG DATU
Sibu
MALAYSIA
Kuching
UPPER KAPUAS MTS.
IRAN MTS.
Singkawang
Pontianak
Sintang
PEG. MULLER
BORNEO (KALIMANTAN)
PEG. SCHWANER
Bukit Raya 7474 ft. 2278 m
Samarinda
Balikpapan
Bontang
TANJUNG MANGKALIHAT
KEPULAUAN KARIMATA
Ketapang
Selat Karimata
Palangkaraya
Banjarbaru
Banjarmasin
TANJUNG PUTING
TANJUNG SELATAN
PULAU LAUT
INDONESIA
Makassar Strait (Selat Makasar)
GREATER SUNDA ISLANDS
JAVA SEA (LAUT JAWA)
KEPULAUAN LAUT KECIL
PULAU MASALEMBU BESAR
KEPULAUAN KARIMUNJAWA
PULAU BAWEAN
KEPULAUAN KANGEAN
JAKARTA
Bogor
Sukabumi
BANDUNG
Cirebon
SEMARANG
Gunung Slamet 11 247 ft. 3428 m
Surakarta
SURABAYA
MADURA
Sumenep
Pasuruan
Kediri
Malang
Yogyakarta
JAVA (JAWA)
Gunung Semeru 12 060 ft. 3676 m
Jember
BALI
BALI SEA
Singaraja
Denpasar
Gunung Rinjani 12 224 ft. 3726 m
LOMBOK
Mataram
SUMBAWA
Sumbawa Besar
Raba
FLORES
Ende
Larantuka
LESSER SUNDA ISLANDS
SAVU SEA
SUMBA
Waikabubak
Kupang
PULAU SAWU
PULAU ROTI
FLORES SEA
PULAU TANAHJAMPEA
PULAU KALAO
PULAU KALAOTOA
PULAU SELAYAR
PULAU KABAENA
PULAU MUNA
PULAU BUTON
MAKASSAR (UJUNGPANDANG)
Watampone
Teluk Bone
Kendari
Palopo
Majene
Parepare
Bulu Rantekombola 11 335 ft. 3455 m
CELEBES (SULAWESI)
Teluk Tolo
PULAU PELENG
Luwuk
Palu
KEPULAUAN TOGIAN
Teluk Tomini
Tolitoli
CELEBES SEA
CELEBES BASIN
TAWITAWI ISLAND
JOLO ISLAND
Jolo
BASILAN ISLAND
Zamboanga
Pagadian
Dumaguete
NEGROS
Panay Gulf
Bacolod
Iloilo
PANAY
Roxas
SIBUYAN SEA
CUYO ISLAND
DUMARAN ISLAND
CULION ISLAND
BUSUANGA ISLAND
MINDORO
TABLAS ISLAND
Calapan
Naga
Lucena
Lamon Bay
POLILLO ISLANDS
MANILA
San Fernando
Angeles
Quezon City
Cabanatuan
LUZON
SIERRA MADRE
Dagupan
Baguio
Mt. Pulog 9626 ft. 2934 m
Vigan
Laoag
CAPE BOJEADOR
ESCARPADA POINT
Tuguegarao City
INDIAN OCEAN
JAVA TRENCH
CHRISTMAS ISLAND (Austl.)
COCOS ISLANDS (Austl.)
ASHMORE AND CARTIER ISLANDS (Austl.)
95°
100°
105°
110°
115°
120°
15°
10°
5°
0°
5°
10°
© Rand McNally A-101748-1

0 80 160 240 320 400 480 Miles
0 80 160 240 320 400 480 560 640 720 800 Kilometers

Lambert Azimuthal Equal Area Projection
Scale 1:16,000,000
One inch to 256 miles
One cm to 160 km

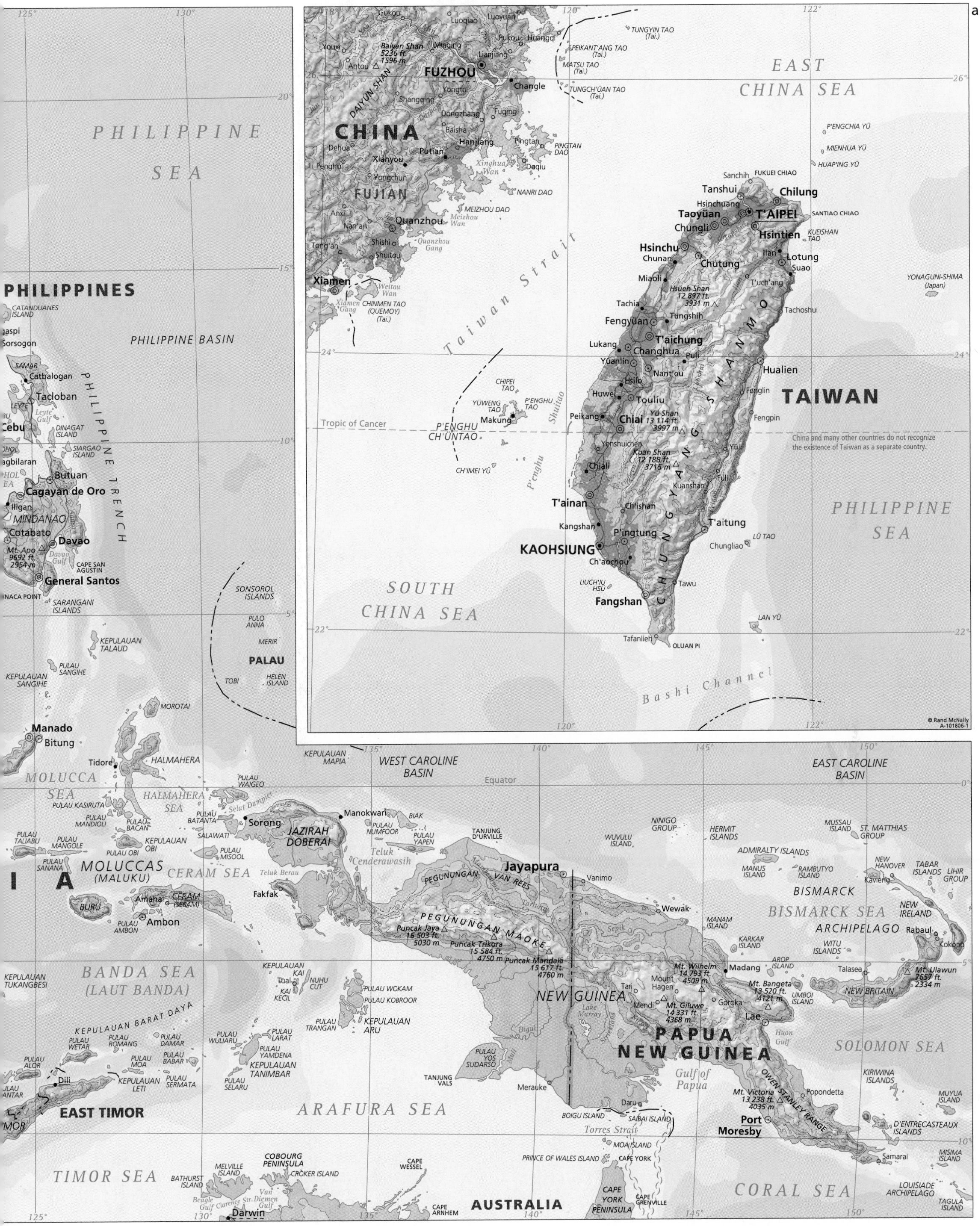
a
CHINA
FUJIAN
FUZHOU
Gukou
Luoqiao
Luoyuan
Pukou
Huangqi
Youxi
Baiyan Shan 5236 ft. 1596 m
Minqing
Lianjiang
Antou
DAIYUN SHAN
Changle
Yongtai
Shanggang
Fuqing
Dongzhang
Baisha
Hanjiang
Pingtan
PINGTAN DAO
Dehua
Putian
Xianyou
Penghu
Yongchun
Xinghua Wan
Daqiu
NANRI DAO
MEIZHOU DAO
Meizhou Wan
Anxi
Nan'an
Quanzhou
Shishi
Quanzhou Gang
Tong'an
Shuitou
Xiamen
Weitou Wan
Xiamen Gang
CHINMEN TAO (QUEMOY) (Tai.)
TUNGYIN TAO (Tai.)
PEIKANT'ANG TAO (Tai.)
MATSU TAO (Tai.)
TUNGCH'ÜAN TAO (Tai.)
EAST CHINA SEA
P'ENGCHIA YÜ
MIENHUA YÜ
HUAP'ING YÜ
Sanchih
FUKUEI CHIAO
Tanshui
Chilung
Hsinchuang
Taoyüan
T'AIPEI
SANTIAO CHIAO
Chungli
Hsintien
KUEISHAN TAO
Hsinchu
Chunan
Ilan
Lotung
Suao
Chutung
Miaoli
Hsüeh Shan 12 897 ft. 3931 m
T'uch'ang
YONAGUNI-SHIMA (Japan)
Tachia
Tachoshui
Fengyüan
Tungshih
T'aichung
Lukang
Changhua
Puli
Yüanlin
Nant'ou
Hualien
Hsilo
Huwei
Touliu
Fenglin
TAIWAN
Taiwan Strait
CHIPEI TAO
YÜWENG TAO
P'ENGHU TAO
Makung
P'ENGHU CH'ÜNTAO
CH'IMEI YÜ
Shuitao
P'enghu
Tropic of Cancer
Peikang
Chiai
Yü Shan 13 114 ft. 3997 m
Fengpin
China and many other countries do not recognize the existence of Taiwan as a separate country.
Yenshuichen
Kuan Shan 12 188 ft. 3715 m
Yüli
Chiali
Fuli
Kuanshan
T'ainan
Ch'ishan
CHUNGYANG SHANMO
PHILIPPINE SEA
Kangshan
P'ingtung
T'aitung
LÜ TAO
Chungliao
KAOHSIUNG
Ch'aochou
SOUTH CHINA SEA
LIUCH'IU HSÜ
Tawu
Fangshan
LAN YÜ
Tafanlieh
OLUAN PI
Bashi Channel
© Rand McNally A-101806-1
PHILIPPINE SEA
PHILIPPINES
CATANDUANES ISLAND
Sorsogon
PHILIPPINE BASIN
SAMAR
Catbalogan
Tacloban
LEYTE
Leyte Gulf
Cebu
DINAGAT ISLAND
SIARGAO ISLAND
PHILIPPINE TRENCH
Butuan
Cagayan de Oro
Iligan
MINDANAO
Cotabato
Davao
Mt. Apo 9692 ft. 2954 m
Davao Gulf
CAPE SAN AGUSTIN
General Santos
SARANGANI ISLANDS
SONSOROL ISLANDS
PULO ANNA
MERIR
PALAU
TOBI
HELEN ISLAND
KEPULAUAN TALAUD
PULAU SANGIHE
KEPULAUAN SANGIHE
MOROTAI
Manado
Bitung
Tidore
HALMAHERA
MOLUCCA SEA
HALMAHERA SEA
PULAU KASIRUTA
PULAU MANDIOLI
PULAU BACAN
PULAU WAIGEO
PULAU BATANTA
Selat Dampier
Sorong
SALAWATI
JAZIRAH DOBERAI
Manokwari
BIAK
PULAU NUMFOOR
PULAU YAPEN
KEPULAUAN MAPIA
WEST CAROLINE BASIN
Equator
EAST CAROLINE BASIN
PULAU TALIABU
PULAU MANGOLE
PULAU OBI
KEPULAUAN OBI
PULAU MISOOL
PULAU SANANA
MOLUCCAS (MALUKU)
CERAM SEA
Teluk Berau
Teluk Cenderawasih
TANJUNG D'URVILLE
Jayapura
Vanimo
PEGUNUNGAN VAN REES
NINIGO GROUP
WUVULU ISLAND
HERMIT ISLANDS
MUSSAU ISLAND
ST. MATTHIAS GROUP
ADMIRALTY ISLANDS
MANUS ISLAND
RAMBUTYO ISLAND
NEW HANOVER
TABAR ISLANDS
LIHIR GROUP
Kavieng
BISMARCK SEA
BISMARCK ARCHIPELAGO
NEW IRELAND
Rabaul
Kokopo
Fakfak
BURU
Amahai
CERAM (SERAM)
PULAU AMBON
Ambon
Wewak
MANAM ISLAND
KARKAR ISLAND
WITU ISLANDS
PEGUNUNGAN MAOKE
Puncak Jaya 16 503 ft. 5030 m
Puncak Trikora 15 584 ft. 4750 m
Puncak Mandala 15 617 ft. 4760 m
Sepik
Mt. Wilhelm 14 793 ft. 4509 m
Madang
AROP ISLAND
Talasea
Mt. Ulawun 7657 ft. 2334 m
NEW BRITAIN
KEPULAUAN TUKANGBESI
BANDA SEA (LAUT BANDA)
KEPULAUAN KAI
Tual
KAI KECIL
NUHU CUT
PULAU WOKAM
PULAU KOBROOR
NEW GUINEA
Tari
Mount Hagen
Mendi
Mt. Giluwe 14 331 ft. 4368 m
Goroka
Mt. Bangeta 13 520 ft. 4121 m
UMBOI ISLAND
Lae
Lake Murray
KEPULAUAN BARAT DAYA
PULAU WETAR
PULAU ROMANG
PULAU DAMAR
PULAU WULIARU
PULAU LARAT
PULAU TRANGAN
KEPULAUAN ARU
Digul
PAPUA NEW GUINEA
Huon Gulf
SOLOMON SEA
PULAU ALOR
PULAU MOA
PULAU BABAR
PULAU YAMDENA
KEPULAUAN TANIMBAR
PULAU YOS SUDARSO
Fly
Gulf of Papua
KIRIWINA ISLANDS
Dili
KEPULAUAN LETI
PULAU SERMATA
PULAU SELARU
TANJUNG VALS
Merauke
Mt. Victoria 13 238 ft. 4035 m
Popondetta
OWEN STANLEY RANGE
MUYUA ISLAND
EAST TIMOR
ARAFURA SEA
Daru
BOIGU ISLAND
SAIBAI ISLAND
Port Moresby
D'ENTRECASTEAUX ISLANDS
Torres Strait
MOA ISLAND
PRINCE OF WALES ISLAND
CAPE YORK
Samarai
MISIMA ISLAND
TIMOR SEA
BATHURST ISLAND
MELVILLE ISLAND
COBOURG PENINSULA
CROKER ISLAND
CAPE WESSEL
CAPE YORK PENINSULA
CAPE GRENVILLE
CORAL SEA
LOUISIADE ARCHIPELAGO
TAGULA ISLAND
Beagle Gulf
Clarence Str.
Van Diemen Gulf
Darwin
CAPE ARNHEM
AUSTRALIA

Inset map a
Lambert Conformal Conic Projection
Scale 1:4,000,000
One inch to 64 miles
One cm to 40 km

Lambert Azimuthal Equal Area Projection
Scale 1:12,000,000
One inch to 190 miles
One cm to 120 km

0 60 120 180 240 300 360 Miles
0 60 120 180 240 300 360 420 480 540 600 Kilometers

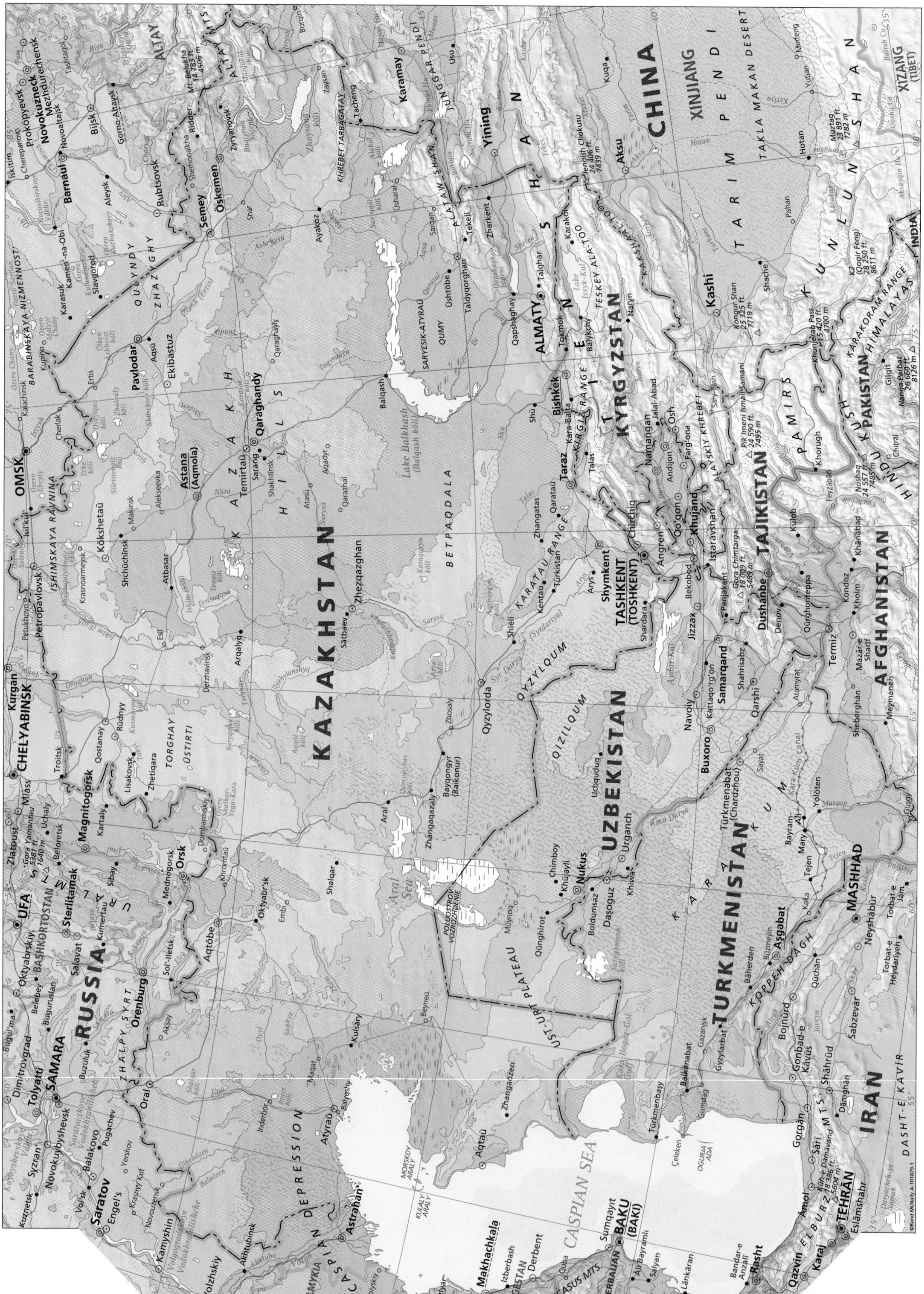
KAZAKHSTAN
RUSSIA
CHINA
UZBEKISTAN
TURKMENISTAN
KYRGYZSTAN
TAJIKISTAN
AFGHANISTAN
PAKISTAN
IRAN
INDIA
XINJIANG
XIZANG (TIBET)
CASPIAN SEA
Aral Sea
Lake Balkhash (Balqash köli)
KAZAKH HILLS
TIAN SHAN
TARIM PENDI
TAKLA MAKAN DESERT
KUNLUN SHAN
PAMIRS
HINDU KUSH
KARAKORAM RANGE
HIMALAYAS
QYZYLQUM
QIZILQUM
KARA KUM
BETPAQDALA
UST-URT PLATEAU
CASPIAN DEPRESSION
ALTAY
ALMATY
ASTANA (Aqmola)
TASHKENT (TOSHKENT)
Bishkek
Dushanbe
Asgabat
BAKU (BAKI)
TEHRAN
OMSK
CHELYABINSK
UFA
SAMARA
Orenburg
Qaraghandy
Semey
Öskemen
Pavlodar
Aqtöbe
Atyrau
Aqtau
Qyzylorda
Shymkent
Taraz
Samarqand
Buxoro
Nukus
Urganch
MASHHAD
Kashi
Aksu
Yining
Hotan
Karamay

©Rand McNally A-101879-1

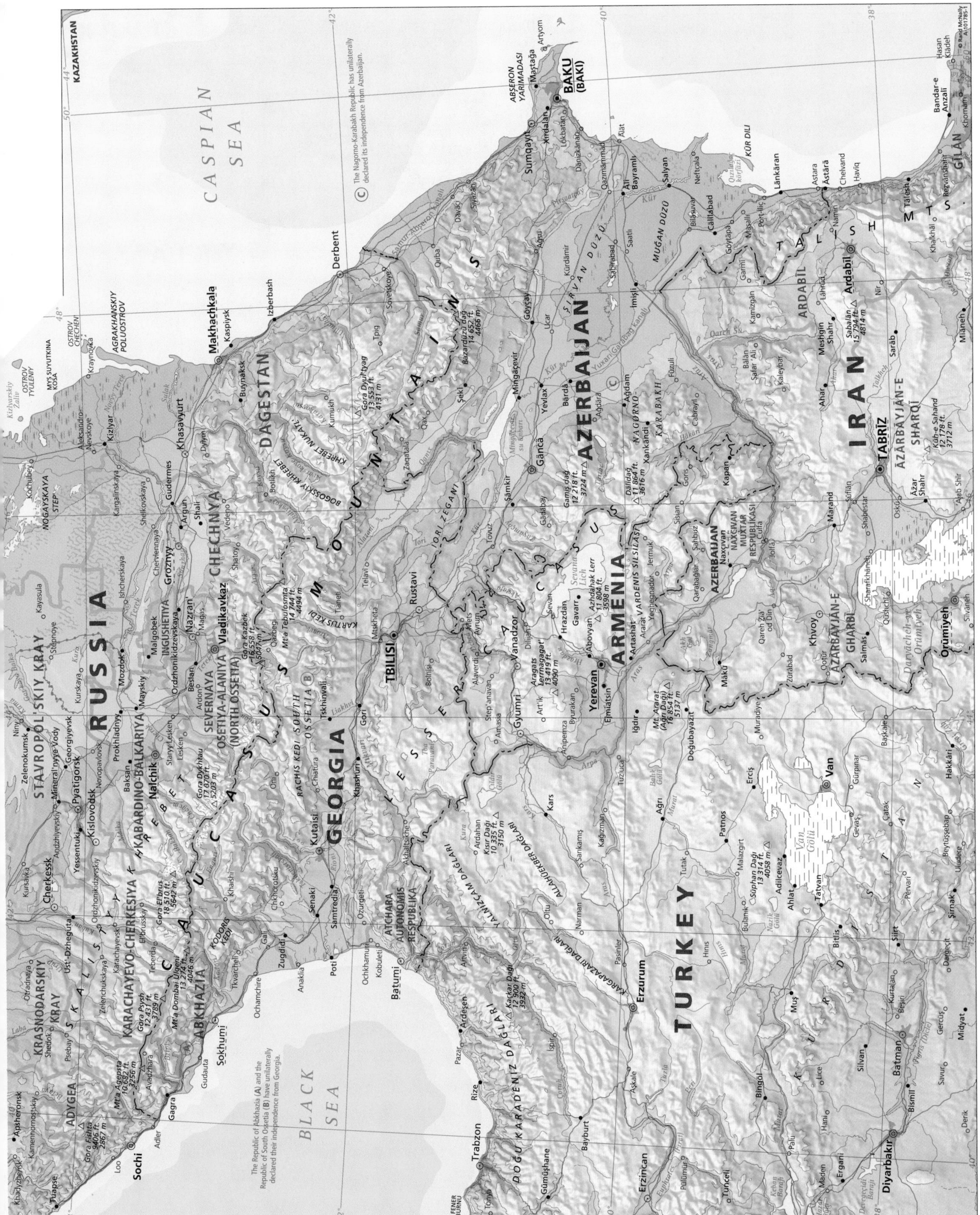
RUSSIA
GEORGIA
ARMENIA
AZERBAIJAN
TURKEY
IRAN
KAZAKHSTAN
CASPIAN SEA
BLACK SEA
CAUCASUS MOUNTAINS
LESSER CAUCASUS
KURDISTAN
TALISH MTS.
DAGESTAN
CHECHNYA
INGUSHETIYA
SEVERNAYA OSETIYA-ALANIYA (NORTH OSSETIA)
KABARDINO-BALKARIYA
KARACHAYEVO-CHERKESIYA
STAVROPOL'SKIY KRAY
KRASNODARSKIY KRAY
ADYGEA
ABKHAZIA
SOUTH OSSETIA
ATCHARA AUTONOMIS RESPUBLIKA
NAXÇIVAN MUXTAR RESPUBLIKASI
NAGORNO-KARABAKH
ARDABĪL
ĀZARBĀYJĀN-E SHARQĪ
ĀZARBĀYJĀN-E GHARBĪ
GĪLĀN
TBILISI
YEREVAN
BAKU (BAKI)
TABRĪZ
Makhachkala
Derbent
Groznyy
Vladikavkaz
Nal'chik
Sochi
Sokhumi
Batumi
Kutaisi
Gyumri
Gäncä
Sumqayıt
Länkäran
Ardabīl
Orūmīyeh
Van
Erzurum
Trabzon
Kars
Diyarbakır
Van Gölü
Gora El'brus 18,510 ft. 5642 m
Gora Kazbek 16,558 ft. 5047 m
Mt. Ararat (Ağrı Dağı) 16,854 ft. 5137 m
The Republic of Abkhazia (A) and the Republic of South Ossetia (B) have unilaterally declared their independence from Georgia.
The Nagorno-Karabakh Republic has unilaterally declared its independence from Azerbaijan.
© Rand McNally

0 20 40 60 80 100 120 Miles

0 20 40 60 80 100 120 140 160 180 200 Kilometers

Lambert Conformal Conic Projection
Scale 1:4,000,000
One inch to 64 miles
One cm to 40 km

TURKEY
SYRIA
LEBANON
ISRAEL
JORDAN
EGYPT
SAUDI
CYPRUS
WEST BANK
GOLAN HEIGHTS
GAZA STRIP
MEDITERRANEAN SEA
TAURUS MOUNTAINS
SINAI PENINSULA
GEBEL EL-TÎH
NEGEV
SYRIAN DESERT
AL-HAMĀD
WĀDĪ AS-SIRHĀN
BEKAA VALLEY
MIDYAN
AL-BUSAYTĀ'
AT-TAWĪL
AL-'URAYQ
AL-HUFRAH
HOLOT HALUZA
North Cyprus declared itself an independent Turkish Republic in 1983.
(A) Gaza Strip is administered by the Palestinian Authority following unilateral withdrawal by Israel in 2005.
(B) West Bank is controlled by Israel and parts are administered by the Palestinian Authority.
(C) Golan Heights has been unilaterally annexed by Israel.
Antalya Körfezi
Alanya
Ermenek
Mut
Kuzucubelen
Tarsus
ADANA
İçel
Erdemli
Silifke
Gazipaşa
Gülnar
İNCEKUM BURNU
Anamur
ANAMUR BURNU
İslâhiye
Gaziantep
Nizip
Birecik
Sürüç
Şanlıurfa
Viranşehir
Derik
Oğuzeli
Barak
Kilis
Jarābulus
Ceylanpınar
Akçakale
Ra's al-'Ayn
Dörtyol
İskenderun
İskenderun Körfezi
Kırıkhan
Sennyol
Reyhanlı
AKINCI BURNU
Hatay
Samandağı
A'zāz
'Afrīn
Al-Bāb
Manbij
Dār Ta'izzah
ALEPPO (HALAB)
As-Safīrah
Maskanah
Buhayrat al-Asad
Ar-Raqqah
Euphrates
Dulq Maghār
Salūq 'Atīq
Tall Tamir
Idlib
Jisr ash-Shughūr
RA'S AL-BASĪT
Arīhā
Al-Haffah
Ma'arrat an-Nu'mān
Latakia (Al-Lādhiqīyah)
Jablah
As-Saqlabīyah
Sūrān
Bāniyās
Hamāh
Salamīyah
Suwaydah
Dayr az-Zawr
Al-Kawm
At-Tayyibah
Busayrah
Al-Mayādīn
As-Sukhnah
Tartūs
Tall Kalakh
Buhayrat Qattīnah
Hims
Shinshār
Furqlus
Tripoli (Tarābulus)
Qurnat as-Sawdā' 10 115 ft. 3083 m
Al-Qusayr
Al-Hirmil
Tudmur
PALMYRA
Al-Batrūn
Jubayl
BYBLOS
Bsharrī
Al-Qaryatayn
Beirut (Bayrūt)
Jūnīyah
Ba'labakk
BAALBEK
An-Nabk
B'abdā
Zahlah
Az-Zabadānī
Al-Qutayfah
Sab 'Ābār
Saydā
DAMASCUS (DIMASHQ)
Dūmā
Jaramānah
Dārayyā
Qatanā
Khān Abū Shāmāt
Mt. Hermon 9232 ft. 2814 m
Al-Kiswah
Tyre (Sūr)
Jwayyā
Qiryat Shemona
At-Tanf
'Akāshāt
Har Meron 3963 ft. 1208 m
Nahariyya
'Akko
Zefat
Al-Qunaytirah
Al-Mismīyah
Lāhithah
Haifa (Hefa)
Tirat Karmel
Teverya
Sea of Galilee
Fīq
Nazareth (Nazerat)
Dor
CAESAREA
'Afula
Irbid
Dar'ā
As-Suwaydā'
Jabal ad-Durūz 5909 ft. 1801 m
Ar-Ramthā
Salkhad
Ar-Rutbah
Hadera
Jenin
Bet She'an
Netanya
Tūlkarm
Tūbās
'Ajlūn
Al-Mafraq
Herzliyya
Qalqīliya
Jarash
Mahattat al-Hafīf
Tel Aviv-Yafo
Nābulus
Petah Tiqwa
Rishon LeZiyyon
As-Salt
Az-Zarqā'
Jabal 'Unayzah 3084 ft. 940 m
Rehovot
Rām Allāh
Arīhā (Jericho)
Ar-Rusayfah
Ammān
Ashdod
Ashqelon
Bethlehem (Bayt Lahm)
Jerusalem (Yerushalayim)
Mādabā
Azraq ash-Shīshān
Al-Jīzah
Port Said (Bûr Sa'îd)
Bûr Fu'ad
Gaza (Ghazzah)
Qiryat Gat
Al-Khalīl
Yattah
Dead Sea
Dhībān
Turayf
Khān Yūnis
Rafah
Be'er Sheva'
MASADA
'Arad
Al-Mazra'ah
Al-Hadīthah
Kāf
Khalîg el-Tîna
Rommâni
El-Arish
Sabkhet el-Bardawîl
Al-Karak
Al-Qatrānah
Al-Jalāmīd
El-Qantara el-Sharqîya
Dimona
Ismailia
Nizzana
Yeroham
El-Quseima
At-Tafīlah
Wādī Bā'ir
Wādī al-Hasah
Sabkhat Hazawzā
Mizpé Ramon
Gebel Yi'allaq 3589 ft. 1094 m
Wadi el-Arish
Suez Canal
Great Bitter Lake
Ash-Shawbak
Al-'Īsāwīyah
PETRA
Suez (El-Suweis)
Bur Taufiq
Al-Jafr
QĀ' AL-JAFR
Ma'ān
Wādī Hadraj
El-Kuntilla
Ra's an-Naqb
Nakhl
Wadi el-Aqaba
Wādī al-Ghinah
Sakākah
Makhfar al-Quwayrah
El-Thamad
Qārah
Al-Jawf
Elat
Al-'Aqabah
Jabal Ramm 5755 ft. 1754 m
Abu Zenîma
GEBEL EL-'IGMA
Haql
Al-Mudawwarah
Gulf of Suez (Khalîg el-Suweis)
Râs el-Gineina 5335 ft. 1626 m
Nuweiba
Gulf of Aqaba
Al-Bi'r
Mt. Sinai (Gebel Mûsa) 7497 ft. 2285 m
Jabal al-Lawz 7884 ft. 2403 m
Wādī Fajr
Dahab
Al-Bad'
Gebel Abu Khashaba 4797 ft. 1462 m
El-Tûr
Maqnā
Tabūk
Al-Qalībah
Nabq
Strait of Tiran
'Aynūnah
Ash-Sharmah
Sharm el-Sheikh
Madîq Gûbâl
TĪRĀN
SANĀFĪR
RED SEA
Gemsa
RÂS MOHAMMED
Yermasoyia
Kólpos Khrisokhoús
KORUÇAM BURNU
Dipkarpaz
ZAFER BURNU
Ziyamet
Güzelyurt Körfezi
Girne
Güzelyurt
İskele
Nicosia
Strovolos
Gazimağusa Körfezi
Gazimağusa
Polis
Ólimbos 6401 ft. 1951 m
Larnaka
Kólpos Lárnakos
AKROTÍRION PIDÁLION
Néa Páfos
Lemesós
Akrotíri
AKROTÍRION GÁTAS
0 20 40 60 80 100 120 Miles
0 20 40 60 80 100 120 140 160 180 200 Kilometers
Lambert Conformal Conic Projection
Scale 1:4,000,000
One inch to 64 miles
One cm to 40 km

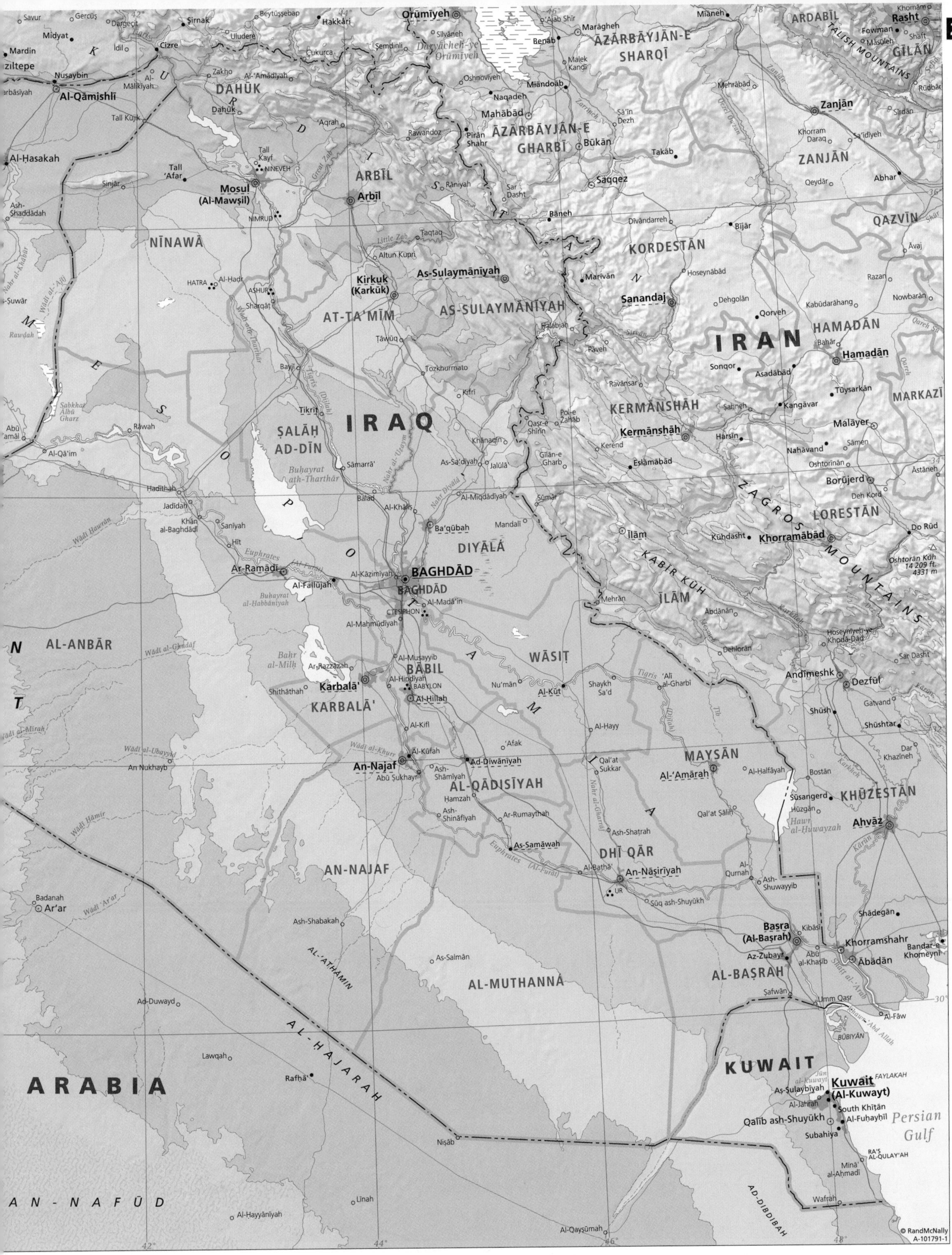
IRAQ
IRAN
KUWAIT
ARABIA
AN-NAFŪD
Persian Gulf
Baghdād
BAGHDĀD
Mosul (Al-Mawṣil)
Arbīl
Kirkuk (Karkūk)
As-Sulaymānīyah
Sanandaj
Hamadān
Kermānshāh
Khorramābād
Zanjān
Rasht
Orūmīyeh
Daryācheh-ye Orūmīyeh
Mahābād
Saqqez
Ar-Ramādī
Al-Fallūjah
Karbalā'
An-Najaf
Al-Ḥillah
Ad-Dīwānīyah
Al-Kūt
Al-'Amārah
An-Nāṣirīyah
As-Samāwah
Baṣra (Al-Baṣrah)
Khorramshahr
Ābādān
Ahvāz
Dezfūl
Kuwait (Al-Kuwayt)
Al-Qāmishlī
Al-Ḥasakah
Tikrīt
Sāmarrā'
Ba'qūbah
Ḥadīthah
Rafḥā'
Ar'ar
Nişāb
DAHŪK
ARBĪL
NĪNAWĀ
AT-TA'MĪM
AS-SULAYMĀNĪYAH
ṢALĀḤ AD-DĪN
DIYĀLĀ
AL-ANBĀR
BĀBIL
KARBALĀ'
WĀSIṬ
AL-QĀDISĪYAH
MAYSĀN
DHĪ QĀR
AN-NAJAF
AL-MUTHANNÁ
AL-BAṢRAH
ĀZARBĀYJĀN-E SHARQĪ
ĀZARBĀYJĀN-E GHARBĪ
ZANJĀN
QAZVĪN
GĪLĀN
ARDABĪL
KORDESTĀN
HAMADĀN
KERMĀNSHĀH
ĪLĀM
LORESTĀN
MARKAZĪ
KHŪZESTĀN
ZAGROS MOUNTAINS
KABIR KUH
TALISH MOUNTAINS
AL-ḤAJARAH
AL-'ATHĀMIN
AD-DIBDIBAH
KURDISTAN
MESOPOTAMIA
Oshtorān Kūh 14 209 ft. 4331 m
Euphrates
Tigris
Buḥayrat ath-Tharthār
Baḥr al-Milḥ
Hawr al-Ḥuwayzah
© RandMcNally A-101791-1

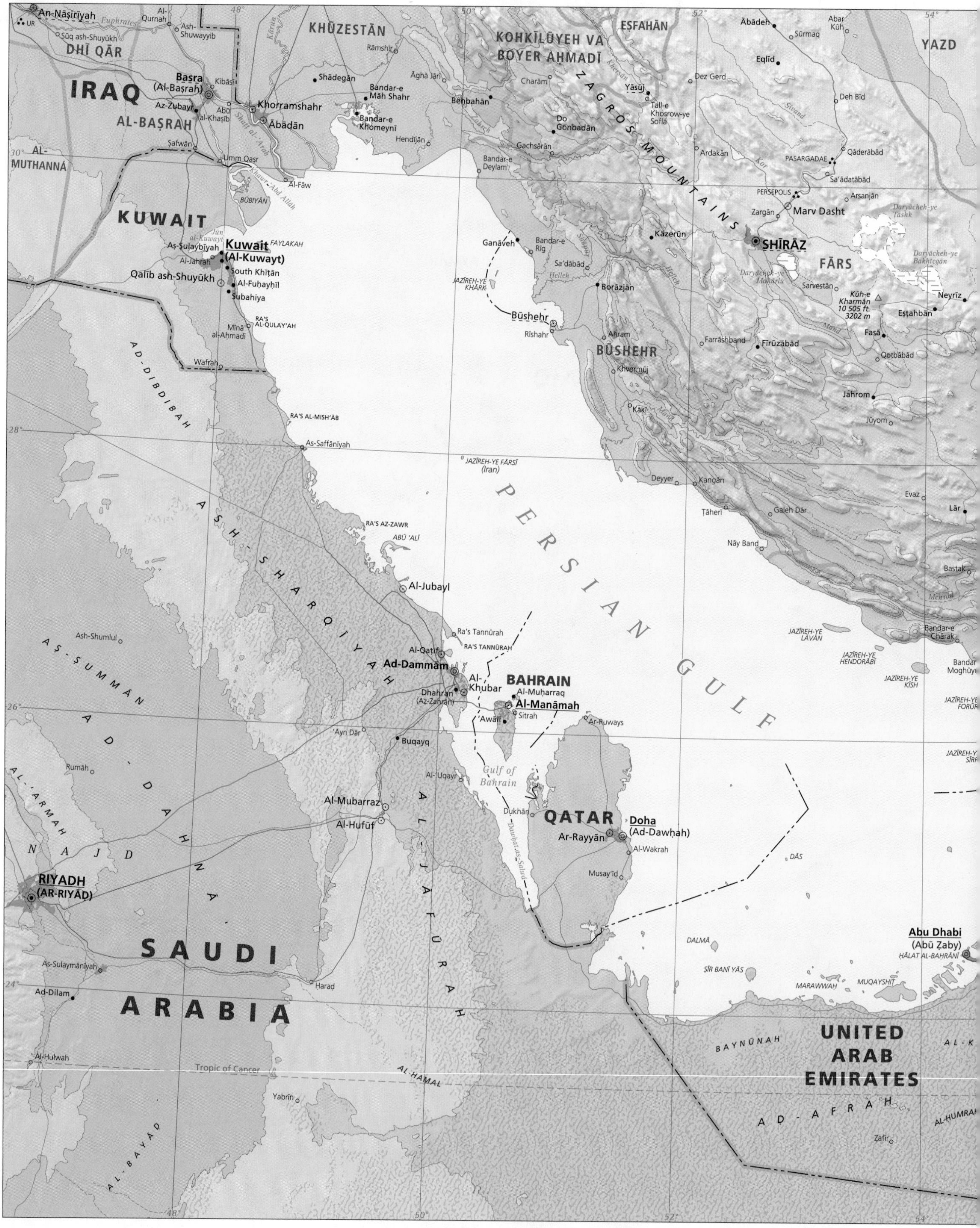
IRAQ
KUWAIT
SAUDI ARABIA
BAHRAIN
QATAR
UNITED ARAB EMIRATES
PERSIAN GULF
ZAGROS MOUNTAINS
KHŪZESTĀN
KOHKĪLŪYEH VA BOYER AHMADĪ
ESFAHĀN
YAZD
FĀRS
BŪSHEHR
DHĪ QĀR
AL-BAŞRAH
AL-MUTHANNĀ
An-Nāşirīyah
UR
Sūq ash-Shuyūkh
Al-Qurnah
Ash-Shuwayyib
Basra (Al-Başrah)
Kibāsī
Az-Zubayr
Abū al-Khaşīb
Khorramshahr
Ābādān
Şafwān
Umm Qaşr
Al-Fāw
BŪBIYĀN
Shādegān
Rāmshīr
Āghā Jārī
Bandar-e Māh Shahr
Bandar-e Khomeynī
Hendījān
Behbahān
Charām
Yāsūj
Tall-e Khosrow-ye Soflā
Do Gonbadān
Gachsārān
Bandar-e Deylam
Dez Gerd
Ābādeh
Sūrmaq
Abar Kūh
Eqlīd
Deh Bīd
Ardakān
Qāderābād
PASARGADAE
Sa'ādatābād
PERSEPOLIS
Arsanjān
Zargān
Marv Dasht
Daryācheh-ye Tashk
Daryācheh-ye Bakhtegān
Daryācheh-ye Mahārlū
SHĪRĀZ
Kāzerūn
Sarvestān
Kūh-e Kharmān 10 505 ft. 3202 m
Neyrīz
Eşţahbān
Fasā
Fīrūzābād
Farrāshband
Qotbābād
Jahrom
Jūyom
Ganāveh
Bandar-e Rīg
Sa'dābād
Borāzjān
JAZĪREH-YE KHĀRK
Būshehr
Rīshahr
Ahram
Khvormūj
Kākī
Deyyer
Kangān
Ţāherī
Galeh Dār
Evaz
Lār
Nāy Band
Bastak
Bandar-e Chārak
Bandar Moghūyeh
JAZĪREH-YE LĀVĀN
JAZĪREH-YE HENDORĀBĪ
JAZĪREH-YE KĪSH
JAZĪREH-YE FORŪR
JAZĪREH-YE SĪRRĪ
JAZĪREH-YE FĀRSĪ (Iran)
Jūn al-Kuwayt
FAYLAKAH
Aş-Şulaybīyah
Kuwait (Al-Kuwayt)
Al-Jahrah
South Khīţān
Qalib ash-Shuyūkh
Al-Fuḩayḩīl
Subahiya
Mīnā' al-Aḩmadī
RA'S AL-QULAY'AH
Wafrah
AD-DIBDIBAH
RA'S AL-MISH'ĀB
As-Saffānīyah
ASH-SHARQĪYAH
RA'S AZ-ZAWR
ABŪ 'ALĪ
Al-Jubayl
Ra's Tannūrah
RA'S TANNŪRAH
Al-Qaţīf
Ad-Dammām
Al-Khubar
Dhahran (Az-Zahrān)
Al-Muḩarraq
Al-Manāmah
'Awālī
Sitrah
Ar-Ruways
'Ayn Dār
Buqayq
Al-'Uqayr
Gulf of Bahrain
Dukhān
Doha (Ad-Dawḩah)
Ar-Rayyān
Al-Wakrah
Musay'īd
Dawḩat as-Salwá
DĀS
Al-Mubarraz
Al-Hufūf
AL-JĀFŪRAH
Ash-Shumlul
AŞ-ŞUMMĀN
AD-DAHNĀ'
Rumāh
AL-'ARMAH
NAJD
RIYADH (AR-RIYĀḌ)
As-Sulaymānīyah
Haraḍ
Ad-Dilam
Al-Hulwah
Tropic of Cancer
AL-HAMAL
Yabrīn
AL-BAYAḌ
DALMĀ
ŞĪR BANĪ YĀS
MARAWWAH
MUQAYSHIŢ
Abu Dhabi (Abū Zaby)
HĀLAT AL-BAHRĀNĪ
BAYNŪNAH
AL-K
AḌ-ḌAFRAH
AL-HUMRAH
Zafir
30°
28°
26°
24°
48°
50°
52°
54°

0 20 40 60 80 100 120 Miles
0 20 40 60 80 100 120 140 160 180 200 Kilometers

Lambert Conformal Conic Projection
Scale 1:4,000,000
One inch to 64 miles
One cm to 40 km

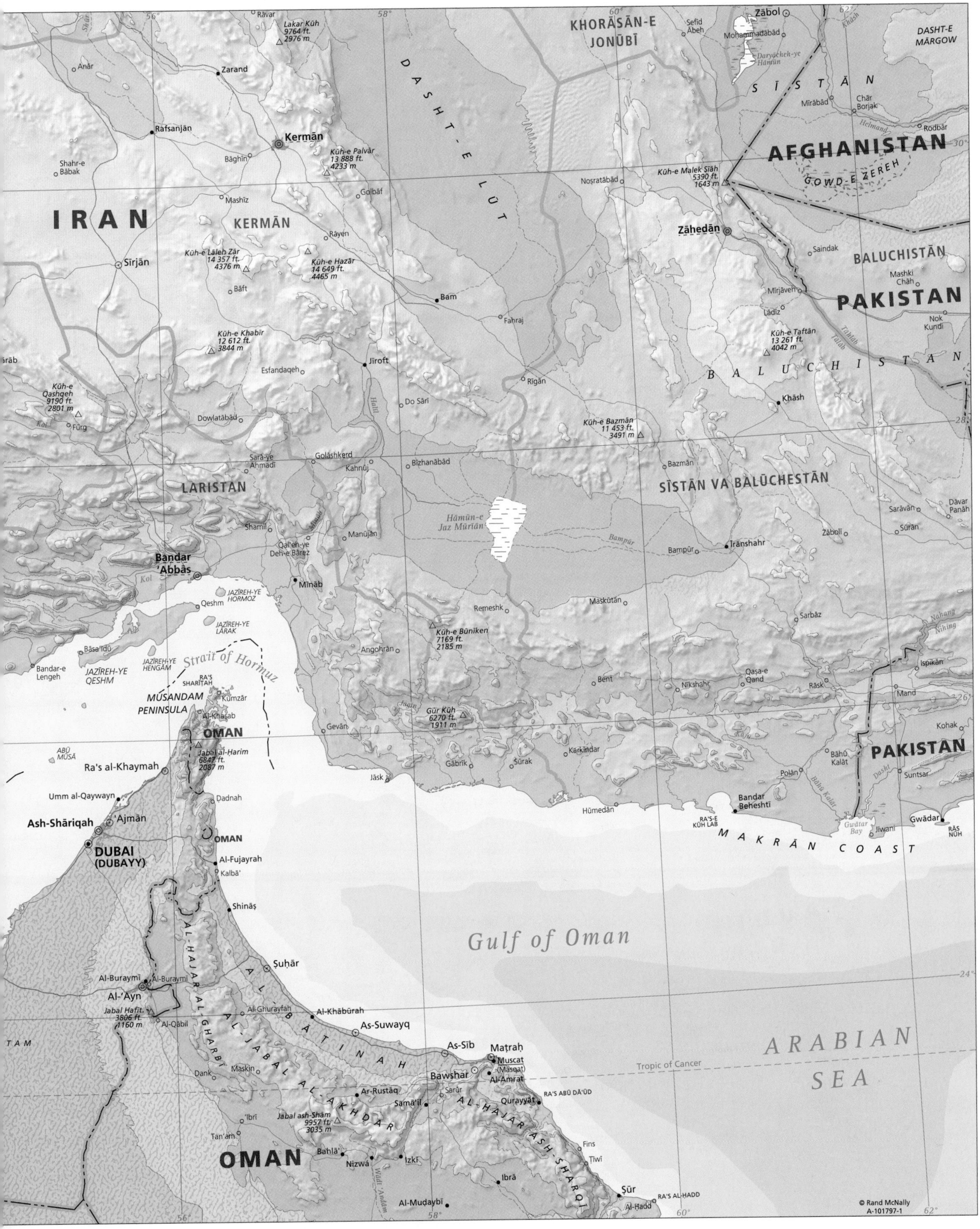
IRAN
KERMĀN
Kermān
Zarand
Rafsanjān
Anār
Shahr-e Bābak
Bāghīn
Golbāf
Mashīz
Rāyen
Sīrjān
Bāft
Bam
Fahraj
Jīroft
Esfandaqeh
Do Sārī
Rīgān
Dowlatābād
Fūrg
Lakar Kūh 9764 ft. 2976 m
Kūh-e Palvār 13 888 ft. 4233 m
Kūh-e Laleh Zār 14 357 ft. 4376 m
Kūh-e Hazār 14 649 ft. 4465 m
Kūh-e Khabīr 12 612 ft. 3844 m
Kūh-e Qashqeh 9190 ft. 2801 m
DASHT-E LŪT
KHORĀSĀN-E JONŪBĪ
Sefīd Ābeh
Zābol
Mohammadābād
Daryācheh-ye Hāmūn
SĪSTĀN
Mīrābād
Chār Borjak
Helmand
Rodbār
DASHT-E MĀRGOW
AFGHANISTAN
GOWD-E ZEREH
Kūh-e Malek Sīāh 5390 ft. 1643 m
Nosratābād
Zāhedān
Saindak
BALUCHISTĀN
Mashki Chāh
Mīrjāveh
Lādīz
PAKISTAN
Nok Kundi
Kūh-e Taftān 13 261 ft. 4042 m
BALUCHISTAN
Khāsh
Kūh-e Bazmān 11 453 ft. 3491 m
Bazmān
SĪSTĀN VA BALŪCHESTĀN
Sarāvān
Dāvar Panāh
Zāboli
Sūrān
Īrānshahr
Bampūr
Hāmūn-e Jaz Mūrīān
Bizhanābād
Golāshkerd
Kahnūj
Sarā-ye Ahmadī
LARISTAN
Shamīl
Manūjān
Qal'eh-ye Deh-e Bārez
Bandar 'Abbās
Mīnāb
Jazīreh-ye Hormoz
Qeshm
Jazīreh-ye Lārak
Bāsa'īdū
Bandar-e Lengeh
Jazīreh-ye Qeshm
Jazīreh-ye Hengām
Strait of Hormuz
Ra's Sharītah
Kumzār
Musandam Peninsula
Al-Khasab
OMAN
Jabal al-Harim 6847 ft. 2087 m
Abū Mūsā
Ra's al-Khaymah
Umm al-Qaywayn
Ash-Shāriqah
'Ajmān
DUBAI (DUBAYY)
Dadnah
Al-Fujayrah
Kalbā'
Shināṣ
Ṣuhār
Al-Buraymī
Al-'Ayn
Jabal Hafit 3806 ft. 1160 m
Al-Qābil
Al-Ghurayfah
Al-Khābūrah
As-Suwayq
AL-BĀTINAH
AL-HAJAR AL-GHARBĪ
AL-JABAL AL-AKHDAR
AL-HAJAR ASH-SHARQĪ
As-Sīb
Maṭraḥ
Muscat (Masqaṭ)
Bawshar
Al-Amrat
Ra's Abū Dā'ūd
Qurayyāt
Dank
Maskin
Ar-Rustāq
Samā'il
Sarūr
'Ibrī
Tan'am
Jabal ash-Sham 9957 ft. 3035 m
OMAN
Bahlā'
Nizwá
Izkī
Ibrā
Fins
Tīwī
Ṣūr
Ra's al-Hadd
Al-Hadd
Al-Mudaybī
Wādī 'Andam
Remeshk
Kūh-e Būniken 7169 ft. 2185 m
Angohrān
Maskūtān
Sarbāz
Bent
Nīkshahr
Qasa-e Qand
Rāsk
Ispīkān
Mand
Kohak
Gevān
Gūr Kūh 6270 ft. 1911 m
Kārkindar
Gābrīk
Sūrak
Jāsk
Hūmedān
Ra's-e Kūh Lab
Bandar Beheshtī
Polān
Bāhū Kalāt
PAKISTAN
Suntsar
Gwādar
Jiwani
Gwātar Bay
Ra's Nūh
MAKRĀN COAST
Gulf of Oman
ARABIAN SEA
Tropic of Cancer
© Rand McNally A-101797-1

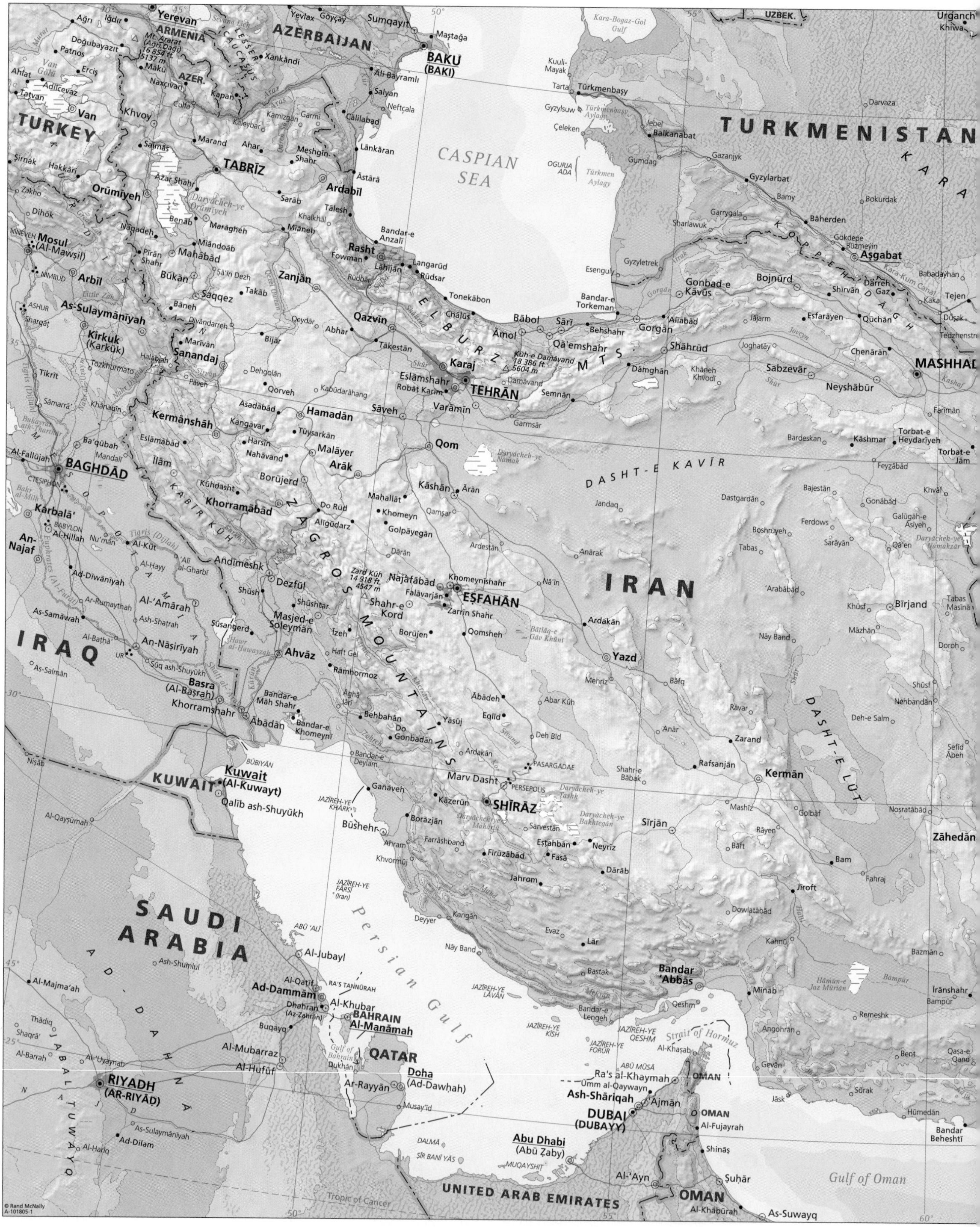
Yerevan
ARMENIA
Mt. Ararat (Ağrı Dağı) 16 854 ft. 5137 m
AZERBAIJAN
BAKU (BAKI)
Sumqayıt
Maştağa
Yevlax
Göyçay
Xankändi
Ağrı
Iğdır
Doğubayazıt
Patnos
Erciş
Ahlat
Adilcevaz
Tatvan
Van
Mākū
AZER.
Naxçıvan
Culfa
Kapan
Āli-Bayramlı
Salyan
Neftçala
Cälilabad
Länkäran
Kaleybar
Kamizgān
Garmi
TURKEY
Khvoy
Salmās
Marand
Ahar
Meshgin Shahr
TABRĪZ
Şırnak
Hakkâri
Āzar Shahr
Orūmīyeh
Daryācheh-ye Orūmīyeh
Sarāb
Ardabīl
Āstārā
Zakho
Dihōk
Benāb
Marāgheh
Khalkhāl
Mīāneh
Naqadeh
MINEVEH
Mosul (Al-Mawşil)
Mīāndoāb
Mahābād
Pīrān Shahr
NIMRUD
Arbīl
Būkān
Sā'īn Dezh
Saqqez
Takāb
Zanjān
Rasht
Fowman
Bandar-e Anzalī
Lāhījān
Langarūd
Rūdsar
Rūdbār
ASHUR
Shargāţ
As-Sulaymānīyah
Bāneh
Dīvāndarreh
Kirkuk (Karkūk)
Marīvān
Sanandaj
Halabjah
Tozkhurmato
Tikrit
Sāmarrā'
Qeydār
Abhar
Qazvīn
Bījār
Tākestān
Karaj
ELBURZ MTS
Tonekābon
Chālūs
Āmol
Bābol
Sārī
Behshahr
Qā'emshahr
Kūh-e Damāvand 18 386 ft. 5604 m
Damāvand
Dehgolān
Pāveh
Qorveh
Kabūdarāhang
Eslāmshahr
Robāţ Karīm
TEHRĀN
Varāmīn
Semnān
Garmsār
Dāmghān
Shāhrūd
Khāneh Khvodī
Asadābād
Hamadān
Sāveh
Kermānshāh
Khānaqīn
Kangāvar
Tūysarkān
Ba'qūbah
Eslāmābād
Mandalī
Harsīn
Nahāvand
Malāyer
Arāk
Qom
Daryācheh-ye Namak
Īlām
Al-Fallūjah
BAGHDĀD
CTESIPHON
Borūjerd
Kūhdasht
Khorramābād
Kāshān
Ārān
Maḩallāt
Khomeyn
Qamşar
Golpāyegān
Do Rūd
Alīgūdarz
Karbalā'
BABYLON
Al-Ḩillah
An-Najaf
Nu'mān
Al-Kūt
Dārān
Ardestān
Anārak
Andīmeshk
Dezfūl
Shūsh
Shūshtar
Zard Kūh 14 918 ft. 4547 m
Najafābād
Khomeynīshahr
Falāvarjān
EŞFAHĀN
Nā'īn
Shahr-e Kord
Zarrīn Shahr
Ad-Dīwānīyah
Ar-Rumaythah
Al-'Amārah
Ash-Shaţrah
As-Samāwah
An-Nāşirīyah
UR
Sūsangerd
Masjed-e Soleymān
Īzeh
Borūjen
Qomsheh
Ardakān
Yazd
Al-Baţḩā'
IRAQ
As-Salmān
Sūq ash-Shuyūkh
Hawr al-Huwayzah
Ahvāz
Haft Gel
Rāmhormoz
Basra (Al-Başrah)
Khorramshahr
Ābādān
Bandar-e Māh Shahr
Bandar-e Khomeynī
Āghā Jārī
Behbahān
Do Gonbadān
Yāsūj
Ābādeh
Eqlīd
Abar Kūh
Mehrīz
Bāfq
Deh Bīd
Ardakān
PASARGADAE
Marv Dasht
PERSEPOLIS
Bandar-e Deylam
BŪBIYĀN
Kuwait (Al-Kuwayt)
KUWAIT
Qalib ash-Shuyūkh
Nişāb
Al-Qayşūmah
Ganāveh
JAZĪREH-YE KHĀRK
Būshehr
Kāzerūn
SHĪRĀZ
Borāzjān
Daryācheh-ye Tashk
Daryācheh-ye Bakhtegān
Sarvestān
Eşţahbān
Neyrīz
Ahram
Farrāshband
Fīrūzābād
Fasā
Khvormūj
Dārāb
Jahrom
Sīrjān
JAZĪREH-YE FĀRSĪ (Iran)
Persian Gulf
SAUDI ARABIA
AD DAHNĀ
Deyyer
Kangān
Evaz
Lār
Nāy Band
ABŪ 'ALĪ
Al-Jubayl
Ash-Shumlūl
Al-Majma'ah
Al-Qaţīf
RA'S TANNŪRAH
Ad-Dammām
Al-Khubar
Dhahran (Az-Zahran)
BAHRAIN
Al-Manāmah
Buqayq
Gulf of Bahrain
QATAR
Al-Mubarraz
Al-Hufūf
Dukhān
Doha (Ad-Dawḩah)
Ar-Rayyān
Musay'īd
Thādiq
Shaqrā'
Al-Barrah
Al-'Uyaynah
RIYADH (AR-RIYĀD)
JABAL TUWAYQ
As-Sulaymānīyah
Ad-Dilam
Al-Hariq
Tropic of Cancer
DALMĀ
ŞĪR BANĪ YĀS
MUQAYSHIT
Abu Dhabi (Abū Z̧aby)
UNITED ARAB EMIRATES
Bastak
JAZĪREH-YE LĀVĀN
Bandar-e Lengeh
JAZĪREH-YE KĪSH
JAZĪREH-YE FORŪR
ABŪ MŪSĀ
JAZĪREH-YE QESHM
Qeshm
Bandar 'Abbās
Mīnāb
Strait of Hormuz
Al-Khaşab
Ra's al-Khaymah
OMAN
Umm al-Qaywayn
Ash-Shāriqah
'Ajmān
DUBAI (DUBAYY)
Al-Fujayrah
Shināş
Al-'Ayn
Şuḩār
OMAN
Al-Khābūrah
As-Suwayq
Gulf of Oman
CASPIAN SEA
Kara-Bogaz-Gol Gulf
Kuuli-Mayak
Tarta
Türkmenbaşy
Gyzylsuw
Türkmenbaşy Aylagy
Çeleken
OGURJA ADA
Türkmen Aylagy
Jebel
Balkanabat
Gumdag
TURKMENISTAN
KARA KUM
UZBEK.
Urganch
Khiwa
Darvaza
Gazanjyk
Gyzylarbat
Bamy
Bokurdak
Garrygala
Sharlawuk
Bäherden
KOPPEH DAGH
Gökdepe
Büzmeýin
Aşgabat
Babadayhan
Gyzyletrek
Esenguly
Bandar-e Torkeman
Gonbad-e Kāvūs
Bojnūrd
Shīrvān
Darreh Gaz
Kara-Kum Canal
Kaka
Tejen
Gorgān
Ālīābād
Jājarm
Esfarāyen
Qūchān
Duşak
Tedzhenstroy
Joghatāy
Chenārān
MASHHAD
Sabzevār
Neyshābūr
Kashaf
Farīmān
Torbat-e Ḩeydarīyeh
Kāshmar
Bardeskan
Torbat-e Jām
Feyẕābād
DASHT-E KAVĪR
Jandaq
Dastgardān
Bajestān
Gonābād
Khvāf
Galūgāh-e Āsīyeh
Ferdows
Boshrūyeh
Sarāyān
Qā'en
Daryācheh-ye Namakzār
Tabas
IRAN
'Arabābād
Khūsf
Bīrjand
Tabas Masīnā
Nāy Band
Māzhān
Doroh
Shūsf
Nehbandān
Rāvar
DASHT-E LŪT
Deh-e Salm
Anār
Zarand
Rafsanjān
Kermān
Shahr-e Bābak
Sefīd Ābeh
Mashīz
Golbāf
Noşratābād
Zāhedān
Rāyen
Bāft
Bam
Fahraj
Jīroft
Dowlatābād
Kahnūj
Bazmān
Hāmūn-e Jaz Mūrīān
Bampūr
Īrānshahr
Remeshk
Angohrān
Bent
Qaşa-e Qand
Gevān
Sūrak
Jāsk
Hūmedān
Bandar Beheshtī
ZAGROS MOUNTAINS
KABIR KUH
MESOPOTAMIA
Tigris (Dijlah)
Euphrates (Al-Furāt)
© Rand McNally A-101805-1

0 40 80 120 160 200 240 Miles

0 40 80 120 160 200 240 280 320 360 400 Kilometers

Lambert Azimuthal Equal Area Projection
Scale 1:8,000,000
One inch to 128 miles
One cm to 80 km

UZBEKISTAN
KAZAKHSTAN
KYRGYZSTAN
TAJIKISTAN
AFGHANISTAN
PAKISTAN
CHINA
INDIA
TASHKENT (TOSHKENT)
Chirchiq
Namangan
Andijon
Osh
Khujand
Buxoro
Samarqand
Navoiy
Jizzax
Qarshi
Türkmenabat (Chardzhou)
Mary
Dushanbe
Kŭlob
Qŭrghonteppa
Khorugh
Termiz
Kashi
Shache
Yecheng
XINJIANG
TARIM PENDI
KUNLUN SHAN
PAMIRS
HINDU KUSH
KARAKORAM RANGE
HIMALAYAS
NORTHERN AREAS
NORTH-WEST FRONTIER
JAMMU AND KASHMIR
AZAD KASHMIR
FEDERALLY ADMINISTERED TRIBAL AREAS
Lenin Peak 23 406 ft. 7134 m
Pik Imeni Ismail Samani 24 590 ft. 7495 m
Gora Chimtarga 18 009 ft. 5489 m
Kongur Shan 25 325 ft. 7719 m
Khunjerab Pass 15 420 ft. 4700 m
K2 (Qogir Feng) 28 250 ft. 8611 m
Rakaposhi 25 551 ft. 7788 m
Nanga Parbat 26 660 ft. 8126 m
Tirich Mir 25 230 ft. 7690 m
Koh-e Foladi 16 847 ft. 5135 m
Khyber Pass 3501 ft. 1067 m
Mazār-e Sharīf
Kondoz
Baghlān
Feyzābād
Kabul (Kābol)
Jalalabad
Herāt
Ghazni
Kandahār
Farāh
Chaman
Quetta
Peshāwar
Islāmābād
RĀWALPINDI
Srīnagar
Leh
Jammu
Sialkot
Gujrāt
Gujrānwāla
LAHORE
Amritsar
LUDHIĀNA
Jalandhar
Shimla
Chandīgarh
FAISALĀBĀD
Sargodha
Jhang Sadar
MULTĀN
Bahāwalpur
Dera Ghāzi Khān
Dera Ismāīl Khān
PUNJAB
HIMĀCHAL PRADESH
HARYĀNA
DELHI
New Delhi
SULAIMAN RANGE
TOBA KĀKAR RANGE
BALUCHISTĀN
SIND
RĀJASTHĀN
GREAT INDIAN DESERT
THAR DESERT
Bīkāner
Jodhpur
JAIPUR
Ajmer
Udaipur
Kota
Sukkur
Jacobābād
Lārkāna
Nawābshāh
HYDERĀBĀD
KARĀCHI
Gwādar
MAKRĀN COAST
CENTRAL MAKRĀN RANGE
KIRTHAR RANGE
ARAVALLI RANGE
RANN OF KUTCH
GUJARĀT
AHMADĀBĀD
Vadodara
MADHYA PRADESH
INDORE
Ujjain
Tropic of Cancer
ARABIAN SEA
DASHT-E MĀRGOW
RĪGESTĀN
SELSELEH-YE SAFĪD KOH

TAJIKISTAN
Taxkorgan Tajik Zizhixian
Khunjerab Pass 15 420 ft. 4700 m
TURK.
AFGHANISTAN
PAKISTAN
HINDU KUSH
KARAKORAM RANGE
NORTHERN AREAS
NORTH-WEST FRONTIER
JAMMU AND KASHMIR
AZAD KASHMIR
PIR PANJAL RANGE
FEDERALLY ADMINISTERED TRIBAL AREAS
TOBA KĀKAR RANGE
SULAIMAN RANGE
RĪGESTĀN
BALUCHISTĀN
SIND
KĪRTHAR RANGE
PUNJAB
HIMĀCHAL PRADESH
UTTARANCHAL
HARYĀNA
UTTAR PRADESH
RĀJASTHĀN
THAR DESERT
GREAT INDIAN DESERT
ARAVALLI RANGE
RANN OF KUTCH
GUJARĀT
KĀTHIĀWĀR PENINSULA
MADHYA PRADESH
VINDHYA RANGE
MAHĀRĀSHTRA
DAMĀN AND DIU
DĀDRA AND NAGAR HAVELI
Tropic of Cancer
ARABIAN SEA
Gulf of Kachchh
Gulf of Khambhāt
Sonmiāni Bay
Tirich Mīr 25 230 ft. 7690 m
Rakaposhi 25 551 ft. 7788 m
K2 (Qogir Feng) 28 250 ft. 8611 m
Nanga Parbat 26 660 ft. 8126 m
Muztag 23 891 ft. 7282 m
Koh-e Folādī 16 847 ft. 5135 m
Khyber Pass 3501 ft. 1067 m
Kangrinboqê Feng 22 028 ft. 6714 m
Kāmet 25 446 ft. 7756 m
Nanda Devi 25 645 ft. 7817 m
Kalsūbai 5400 ft. 1646 m
Kabul (Kābol)
Kandahār
Ghazni
Herāt
Meymaneh
Baghlān
Jalālābād
Quetta
Peshāwar
Islāmābād
RĀWALPINDI
Srīnagar
Gilgit
Jammu
Siālkot
Gujrānwāla
Sargodha
FAISALĀBĀD
LAHORE
Amritsar
LUDHIĀNA
Chandīgarh
Shimla
MULTĀN
Bahāwalpur
Dehra Dūn
MEERUT
DELHI
New Delhi
Sukkur
Lārkāna
Hyderābād
HYDERĀBĀD
KARĀCHI
Bīkāner
Jodhpur
JAIPUR
Ajmer
ĀGRA
Gwalior
KĀNPUR
LUCKNOW
Bareilly
Alīgarh
Udaipur
Kota
AHMADĀBĀD
Vadodara
Rājkot
Jamnagar
Bhāvnagar
SŪRAT
Veraval
Ujjain
INDORE
BHOPĀL
Jabalpur
Sāgar
Jhānsi
Nāshik
Mālegaon
NĀGPUR
Chandrapur
Akola
Amrāvati
Jālna
Aurangābād
© Rand McNally A-101821-1
65°
70°
75°
80°
20°
25°
30°
35°

0 40 80 120 160 200 240 Miles

0 40 80 120 160 200 240 280 320 360 400 Kilometers

Lambert Azimuthal Equal Area Projection
Scale 1:8,000,000
One inch to 128 miles
One cm to 80 km

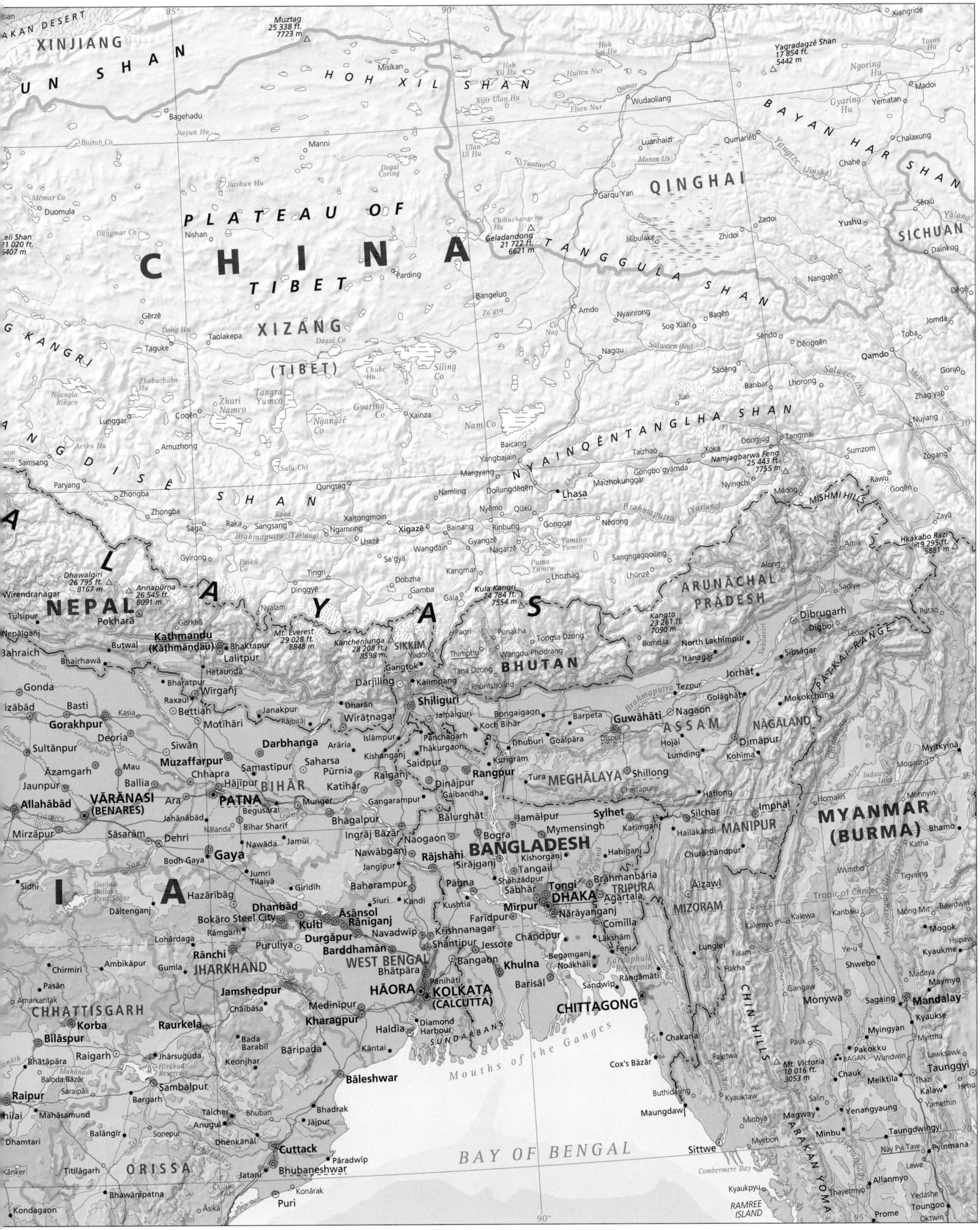
XINJIANG
HOH XIL SHAN
PLATEAU OF TIBET
CHINA
XIZANG (TIBET)
QINGHAI
SICHUAN
BAYAN HAR SHAN
TANGGULA SHAN
NYAINQÊNTANGLHA SHAN
GANGDISÊ SHAN
HIMALAYAS
NEPAL
Kathmandu (Kathmandau)
Mt. Everest 29 028 ft. 8848 m
Kanchenjunga 28 208 ft. 8598 m
Dhawalgiri 26 795 ft. 8167 m
Annapūrna 26 545 ft. 8091 m
Muztag 25 338 ft. 7723 m
Geladandong 21 722 ft. 6621 m
Kula Kangri 24 784 ft. 7554 m
Namjagbarwa Feng 25 443 ft. 7755 m
Kangto 23 261 ft. 7090 m
Hkakabo Razi 19 295 ft. 5881 m
Mt. Victoria 10 016 ft. 3053 m
Lhasa
Xigazê
SIKKIM
Gangtok
BHUTAN
Thimphu
ARUNĀCHAL PRADESH
ASSAM
NĀGĀLAND
MANIPUR
MIZORAM
TRIPURA
MEGHĀLAYA
BANGLADESH
DHAKA
Chittagong
Rājshāhi
Khulna
Sylhet
BIHĀR
PATNA
JHARKHAND
Rānchi
WEST BENGAL
KOLKATA (CALCUTTA)
HĀORA
CHHATTISGARH
ORISSA
Bhubaneshwar
Cuttack
VĀRĀNASI (BENARES)
Allahābād
Gorakhpur
Gaya
Dhanbād
Āsānsol
Jamshedpur
Shiliguri
Darjiling
Guwāhāti
Shillong
Imphāl
Dibrugarh
MYANMAR (BURMA)
Mandalay
Sittwe
CHIN HILLS
ARAKAN YOMA
PĀTKAI RANGE
MISHMI HILLS
Mouths of the Ganges
SUNDARBANS
BAY OF BENGAL
RAMREE ISLAND
Tropic of Cancer

GUJARAT
SŪRAT
Gulf of Khambhāt
Navsāri
Bilimora
Bulsār
Damān
DAMĀN AND DIU
Silvassa
DĀDRA AND NAGAR HAVELI
Nāshik
Mālegaon
Dhule
Jālgaon
Bhusāwal
Burhānpur
Nepa Nagar
Shāhāda
Nandurbār
Shirpur
Amalner
Chopda
Achalpur
Warud
Pāndhurna
Gondia
NĀGPUR
Kāmthi
Tumsar
Bhandāra
Umred
Wardha
Hinganghāt
Amrāvati
Akola
Khāmgaon
Buldāna
Chikhli
Shegaon
Malkāpur
Anjangaon
Akot
Kātol
Arvi
Pulgaon
Kāranja
Wāshīm
Digras
Yavatmāl
Wani
Chandrapur
Ballālpur
Durg
Raipur
Bhilai
Rāj Nāndgaon
Dhamtari
Dalli Rājhara
Kānker
Mahāsamund
Bālāngīr
Sambalpur
Bargarh
Saraipāli
Baloda Bāzār
Tālcher
Bhuban
Anugul
Dhenkānāl
Cuttack
Bhubaneshwar
Jatani
Puri
ORISSA
Titilāgarh
Sonepur
Bhawānipatna
Asika
Brahmapur
Chatrapur
Mahendra Giri 4925 ft. 1501 m
Ichchhāpuram
Sompeta
Rāyagada
Korāput
Jeypore
Pārvatīpuram
Parlākimidi
Sālūr
Bobbili
Narasannapeta
Srīkākulam
Vizianagaram
Vishākhapatnam
Anakāpalle
Narsīpatnam
Tuni
Elamanchili
Kondagaon
Nowrangapur
Jagdalpur
CHHATTISGARH
Kothāpet
Manuguru
Bhadrāchalam
Yellandu
Kottagūdem
Rājahmundry
Pithapuram
Kākināda
Yanam
PONDICHERRY
Amalāpuram
Narasapur
Tādepallegūdem
Elūru
Kolleru Lake
Vijayawāda
Gudivāda
Machilīpatnam
Tenāli
Ponnūru Nidubrolu
Bāpatla
Chīrāla
FALSE DIVI POINT
Guntūr
Ongole
Kandukūr
Kāvali
Nellore
Gūdūr
Sri Kālahasti
Pulicat Lake
Puttūr
Tiruvottiyūr
CHENNAI (MADRAS)
Tāmbaram
Kānchipuram
Madurāntakam
Tiruvannāmalai
Pondicherry
Cuddalore
Neyveli
Chidambaram
Kāraikāl
Nāgappattinam
Vedāranniyam
POINT CALIMERE
Palk Strait
Palk Bay
Jaffna
Jaffna Lagoon
Kilinochchi
Rāmeswaram
PĀMBAN ISLAND
MANNAR ISLAND
Mannar
Vavuniya
Trincomalee
Anurādhapura
SRI LANKA
Polonnaruwa
Batticaloa
Puttalam
Chilaw
Kurunegala
Matale
Kandy
Negombo
Colombo
Pidurutalagala 8281 ft. 2524 m
Badulla
Amparai
Dehiwala-Mount Lavinia
Sri Jayewardenepura Kotte
Moratuwa
Kalutara
Ratnapura
Ambalangoda
Galle
Tangalla
Hambantota
Matara
DONDRA HEAD
Gulf of Mannar
Tuticorin
Tirunelveli
Tiruchchendūr
Kovilpatti
Rāmanāthapuram
Sankarankovil
Tenkāsi
Rājapālaiyam
Aruppukkottai
Virudunagar
Madurai
Devakottai
Kāraikkudi
Pudukkottai
Dindigul
Tiruchchirāppalli
Thanjāvūr
Pattukkottai
Kumbakonam
Māyūram
Nāmakkal
Karūr
Salem
Āttūr
Villupuram
Dharmapuri
Kollegāl
Mettūr
Idappādi
Erode
Tiruppur
Coimbatore
Pollāchi
Palani
Vālpārai
Udagamandalam
Coonoor
TAMIL NĀDU
Anai Mudi 8842 ft. 2695 m
Kambam
Cochin
Shertallai
Kottayam
Alappuzha
Kāyankulam
Kollam
Varkkallai
Thiruvananthapuram
Neyyāttinkara
Nāgercoil
CAPE COMORIN
KERALA
Thrissur
Chāvakkād
Ponnāni
Tirūr
Beypore
Kozhikode
Manjeri
Pālghāt
Badagara
Mahe
PONDICHERRY
Cannanore
Taliparamba
Gundlupet
Nanjangūd
Mysore
Malavalli
Madikeri
Mandya
Hole Narsipur
Rāmanagaram
Kanakapura
Hosūr
Krishnagiri
Vāniyambādi
Āmbūr
Vellore
Arcot
Chittoor
Kolār
BENGALŪRU (BANGALORE)
Hassan
Mangalore
Kāsaragod
Mūlki
Chikmagalūr
Tiptūr
Tumkūr
Chik Ballāpur
Madanapalle
Tirupati
Rāyachoti
Kadiri
Gauribidanūr
Kadūr
Udupi
Coondapoor
Bhadra Reservoir
Tarikere
Bhatkal
Shimoga
Bhadrāvati
Hindupur
Sira
Hiriyūr
Chitradurga
Challakere
Sāgar
Linganamakki Reservoir
Kumta
Kārwār
Sirsi
Hāveri
Harihar
Dāvangere
Rāyadrug
KARNĀTAKA
Bellary
Anantapur
Dharmavaram
Cuddapah
Rājampet
Proddatūr
Jammalamadugu
Tādpatri
Gooty
Guntakal
Nandyāl
ANDHRA PRADESH
Nandikotkūr
Mārkāpur
Chilakalūrupet
Kurnool
Gadwāl
Wanparti
Emmiganūru
Ādoni
Hospet
HAMPI
Tungabhadra Res.
Gangāwati
Gajendragarh
Ilkal
Guledagudda
Bāgalkot
Gokāk
Bail Hongal
Belgaum
Hubli-Dhārwār
Gadag
Lakshmeshwar
Dandeli
Panaji
GOA
Marmagao
Madgaon
Sāvantvādi
Mālvan
Nipāni
Kolhāpur
Ichalkaranji
Jamkhandi
Sāngli
Ratnāgiri
Karād
Sāngola
Chiplūn
Sātāra
Wai
Mahād
Phaltan
Pandharpur
Solāpur
Akalkot
Indi
Bijāpur
Tālikota
Rāichūr
Gulbarga
Shāhābād
Yādgīr
Nārāyanpet
Mahbūbnagar
Nāgārjuna Sāgar
Mācherla
INDIA
Nalgonda
Sūrāpet
Jaggayyapeta
Tāndūr
HYDERĀBĀD
Khammam
Bhongīr
Jangaon
Zahīrābād
Aland
Homnābād
Basavakalyān
Bīdar
Medak
Siddipet
Warangal
DECCAN
Kāmāreddi
Nizāmābād
Karīmnagar
Jagtiāl
Mancherāl
Belampalli
Nirmal
Koratla
Bodhan
Dīglūr
Udgīr
Lātūr
Osmānābād
Bārsi
Ambājogāi
Parli
Bīr
Nānded
Dharmābād
Bhaisa
Basmat
Parbhani
Ahmadnagar
MAHĀRĀSHTRA
Jālna
Hingoli
Pusad
Ādilābād
Aurangābād
Kannad
Chālisgaon
Manmād
Satāna
Deolāli
Igatpuri
Kopargaon
Srīrāmpur
Sangamner
Kalsūbai 5400 ft. 1646 m
Vasai
Thāne
Kalyān
Ulhāsnagar
Panvel
MUMBAI (BOMBAY)
Lonāvale
Pune
Daund
Bārāmati
WESTERN GHĀTS
EASTERN GHĀTS
MALABAR COAST
COROMANDEL COAST
ARABIAN SEA
BAY OF BENGAL
LAKSHADWEEP SEA
INDIAN OCEAN
BITRA ISLAND
CHETTLATT ISLAND
KILTĀN ISLAND
KADAMATT ISLAND
AGATTI ISLAND
PITTI ISLAND
ĀNDROTT ISLAND
KAVARATTI ISLAND
Kavaratti
KALPENI ISLAND
LAKSHADWEEP (LACCADIVE ISLANDS)
Nine Degree Channel
LAKSHADWEEP
MINICOY ISLAND
Eight Degree Channel
KANDUFURI
MULADU
NAGURI
HANIMADU
MAKUNUDU
NUMARA
KANDUTE
FARUKOLU
WADU
KENDIKOLU
WATARU
DIFURI
MALDIVES
MALE ATOLL
Male
20°
15°
10°
5°
75°
80°
85°
© Rand McNally A-101815-1

0 40 80 120 160 200 240 Miles

0 40 80 120 160 200 240 280 320 360 400 Kilometers

Lambert Azimuthal Equal Area Projection
Scale 1:8,000,000
One inch to 128 miles
One cm to 80 km

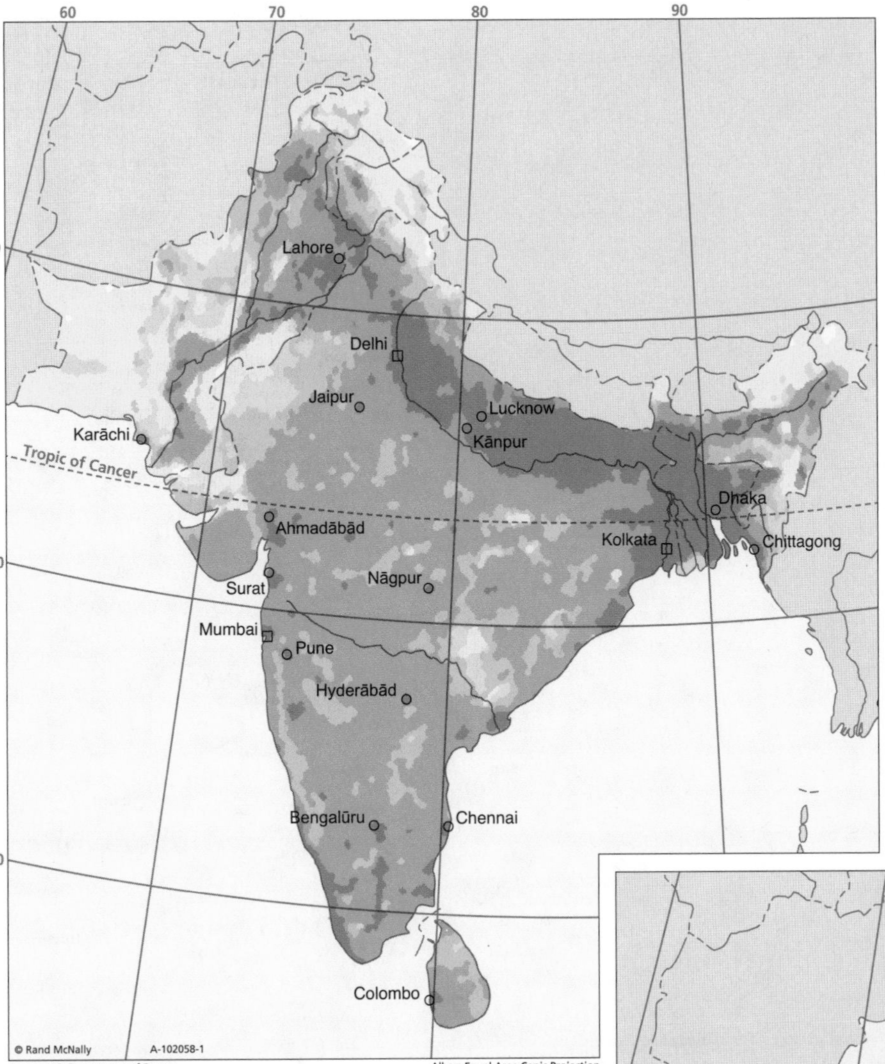
60
70
80
90
30
20
10
Lahore
Delhi
Jaipur
Lucknow
Kānpur
Karāchi
Tropic of Cancer
Dhaka
Ahmadābād
Kolkata
Chittagong
Surat
Nāgpur
Mumbai
Pune
Hyderābād
Bengalūru
Chennai
Colombo
© Rand McNally
A-102058-1

Source: U.S. Department of Energy

Albers Equal Area Conic Projection
Scale 1:28,700,000

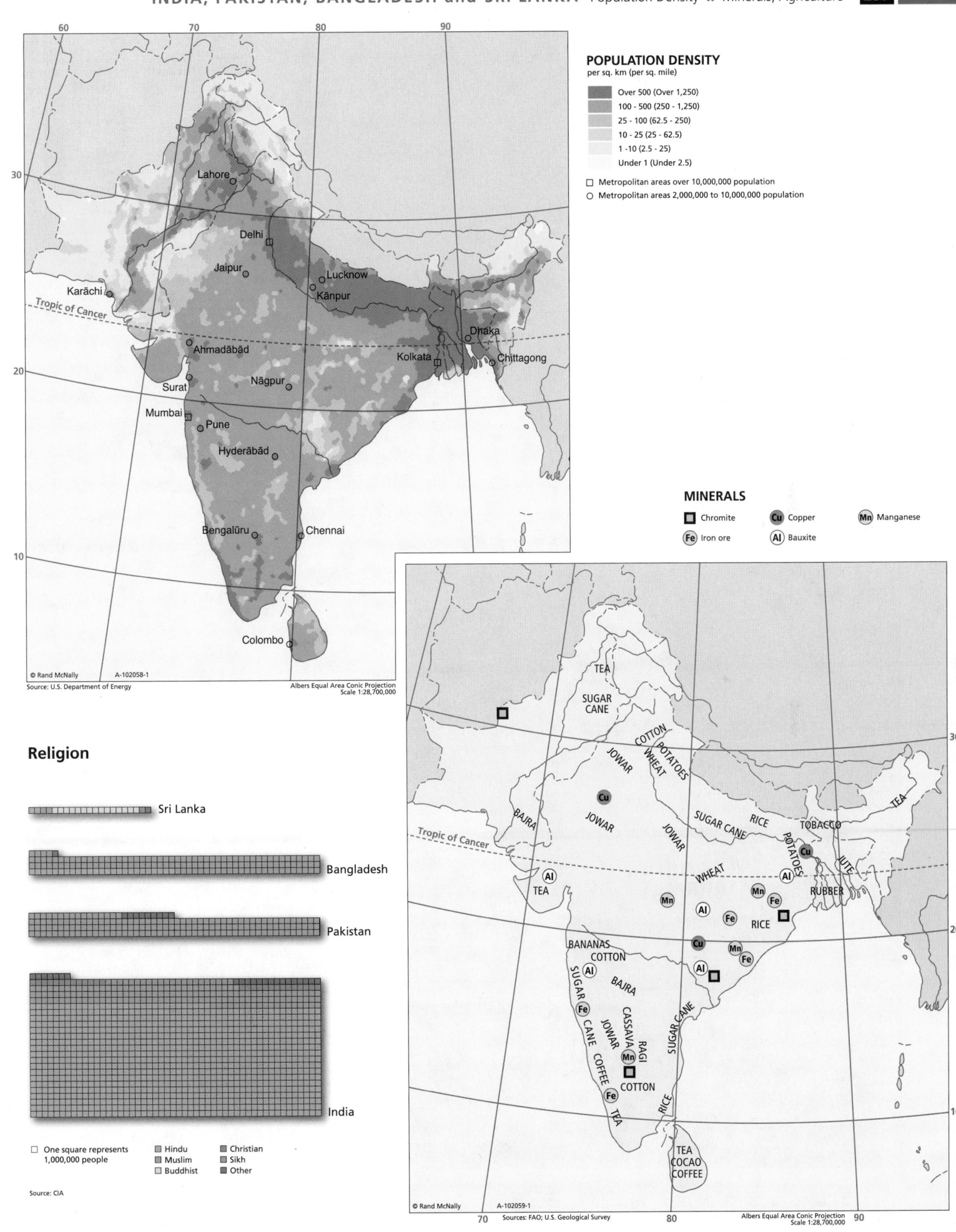

POPULATION DENSITY

per sq. km (per sq. mile)

- Over 500 (Over 1,250)
- 100 - 500 (250 - 1,250)
- 25 - 100 (62.5 - 250)
- 10 - 25 (25 - 62.5)
- 1 -10 (2.5 - 25)
- Under 1 (Under 2.5)

□ Metropolitan areas over 10,000,000 population
○ Metropolitan areas 2,000,000 to 10,000,000 population

MINERALS

Chromite
Cu Copper
Mn Manganese
Fe Iron ore
Al Bauxite

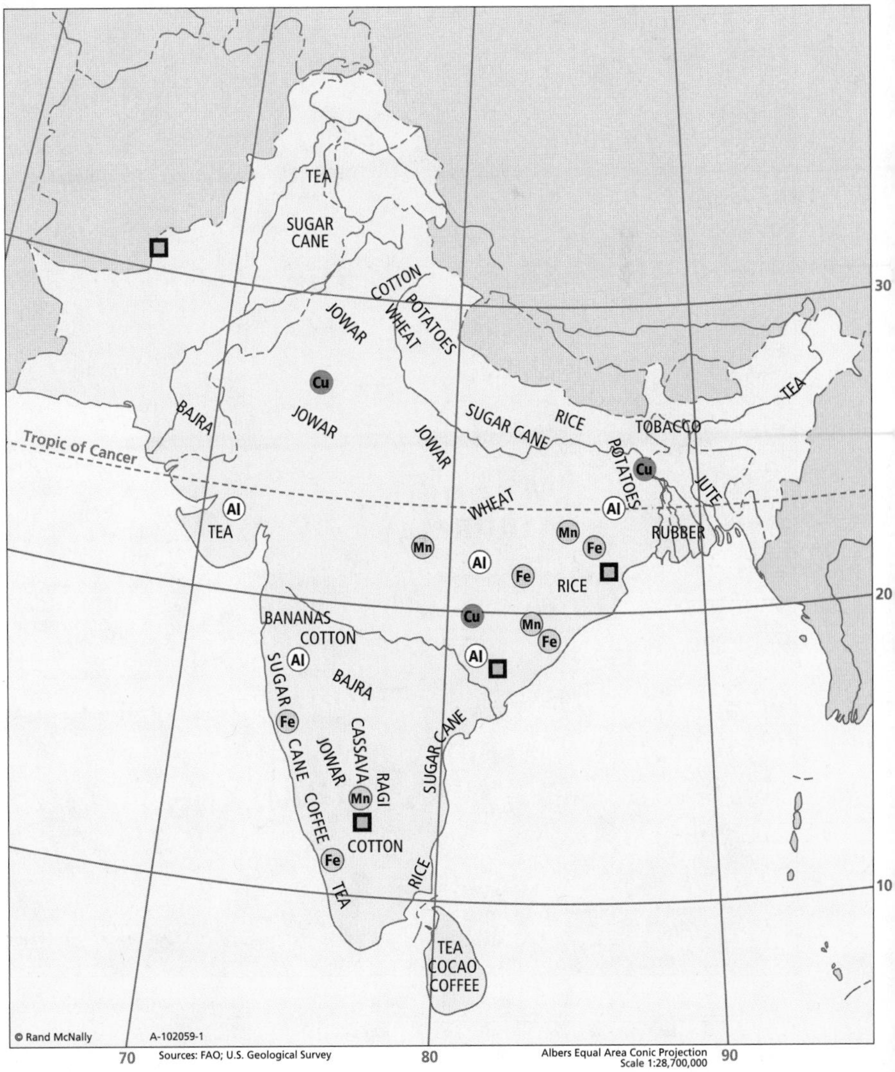
TEA
SUGAR CANE
COTTON
POTATOES
WHEAT
JOWAR
Cu
BAJRA
JOWAR
JOWAR
SUGAR CANE
RICE
TOBACCO
TEA
Tropic of Cancer
POTATOES
Cu
JUTE
WHEAT
Al
Al
TEA
Mn
Mn
Fe
RUBBER
Al
Fe
RICE
Cu
Mn
Fe
BANANAS
COTTON
Al
Al
BAJRA
SUGAR CANE
Fe
CASSAVA
JOWAR
RAGI
SUGAR CANE
Mn
COFFEE
COTTON
Fe
RICE
TEA
TEA
COCAO
COFFEE
30
20
10
70
80
90
© Rand McNally
A-102059-1

Sources: FAO; U.S. Geological Survey

Albers Equal Area Conic Projection
Scale 1:28,700,000

Religion

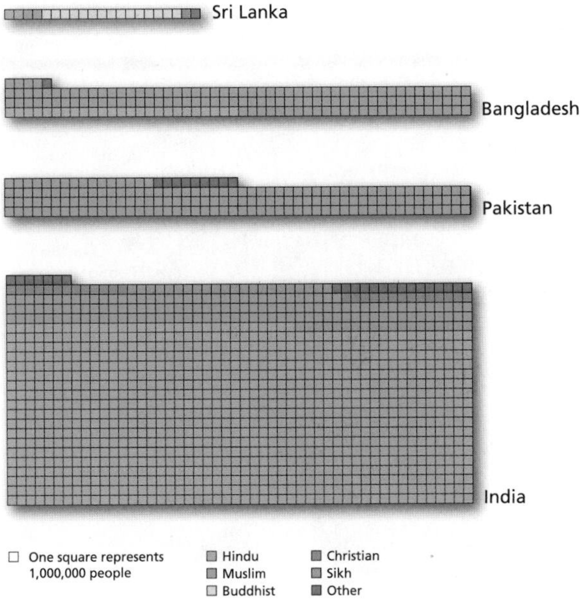
Sri Lanka
Bangladesh
Pakistan
India

□ One square represents 1,000,000 people

Hindu
Muslim
Buddhist
Christian
Sikh
Other

Source: CIA

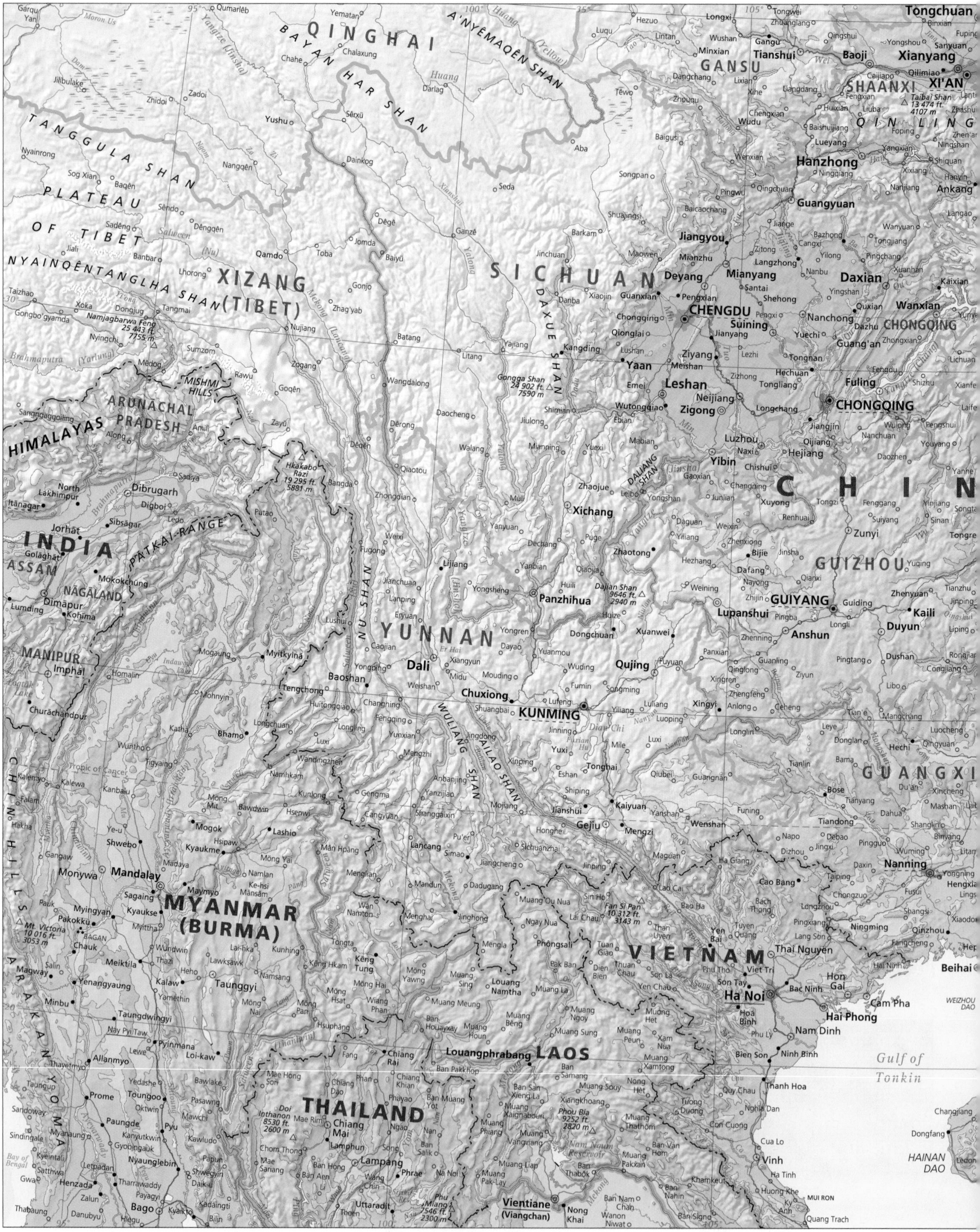
QINGHAI
BAYAN HAR SHAN
A'NYÊMAQÊN SHAN
GANSU
SHAANXI
XI'AN
QIN LING
TANGGULA SHAN
PLATEAU OF TIBET
NYAINQÊNTANGLHA SHAN
XIZANG (TIBET)
SICHUAN
DAXUE SHAN
CHENGDU
CHONGQING
DALIANG SHAN
ARUNACHAL PRADESH
HIMALAYAS
MISHMI HILLS
INDIA
ASSAM
NAGALAND
MANIPUR
PATKAI RANGE
CHIN HILLS
ARAKAN YOMA
YUNNAN
NU SHAN
WULIANG SHAN
AILAO SHAN
KUNMING
GUIZHOU
GUIYANG
GUANGXI
CHINA
MYANMAR (BURMA)
THAILAND
LAOS
VIETNAM
Ha Noi
Gulf of Tonkin
HAINAN DAO
Tongchuan
Xianyang
Baoji
Tianshui
Hanzhong
Guangyuan
Ankang
Jiangyou
Mianyang
Deyang
Daxian
Wanxian
Nanchong
Suining
Leshan
Zigong
Neijiang
Luzhou
Yibin
Fuling
Xichang
Panzhihua
Zhaotong
Lijiang
Dali
Baoshan
Chuxiong
Qujing
Anshun
Zunyi
Kaili
Duyun
Nanning
Mandalay
Dibrugarh
Imphal
Myitkyina
Bhamo
Lashio
Taunggyi
Chiang Mai
Chiang Rai
Louangphrabang
Vientiane (Viangchan)
Hai Phong
Nam Dinh
Thai Nguyen
Beihai
Gongga Shan 24 902 ft. 7590 m
Namjagbarwa Feng 25 443 ft. 7755 m
Hkakabo Razi 19 295 ft. 5881 m
Taibai Shan 13 474 ft. 4107 m
Fan Si Pan 10 312 ft. 3143 m
Mt. Victoria 10 016 ft. 3053 m
Doi Inthanon 8530 ft. 2600 m
Phou Bia 9252 ft. 2820 m
Dajian Shan 9646 ft. 2940 m
Tropic of Cancer

0 40 80 120 160 200 240 Miles
0 40 80 120 160 200 240 280 320 360 400 Kilometers

Lambert Azimuthal Equal Area Projection
Scale 1:8,000,000
One inch to 128 miles
One cm to 80 km

HENAN
ZHENGZHOU
Kaifeng
Luoyang
Pingdingshan
Xuchang
Luohe
Nanyang
Zaozhuang
Lianyungang
Xuzhou
Bozhou
Suzhou
Qingjiang
Yancheng
JIANGSU
Bengbu
Huainan
Fuyang
ANHUI
HEFEI
Ma'anshan
NANJING
Yangzhou
Zhenjiang
Nantong
Changzhou
Wuxi
Suzhou
Changshu
SHANGHAI
CHONGMING DAO
Wuhu
Tongling
Anqing
Huzhou
Jiaxing
HANGZHOU
Shaoxing
NINGBO
ZHEJIANG
ZHOUSHAN QUNDAO
EAST CHINA SEA
Jinhua
Quzhou
Wenzhou
Lishui
Taizhou
Xiangfan
Shiyan
Laohekou
Three Gorges Reservoir
HUBEI
Yichang
Jingmen
Xiaogan
WUHAN
Shashi
Huangshi
Xianning
DABIE SHAN
Jiujiang
Jingdezhen
NANCHANG
Shangrao
Yueyang
Changde
HUNAN
CHANGSHA
Xiangtan
Zhuzhou
Yichun
Pingxiang
Hengyang
Shaoyang
Ji'an
JIANGXI
Fuzhou
WUYI SHAN
FUJIAN
Nanping
FUZHOU
Sanming
Putian
Quanzhou
Zhangzhou
Xiamen
Longyan
Ganzhou
Chenzhou
Shaoguan
NAN LING
GUANGDONG
Meizhou
Chao'an
Shantou
Huizhou
GUANGZHOU (CANTON)
Dongguan
Foshan
Zhaoqing
Jiangmen
Zhongshan
Macau (Aomen)
XINJIULONG
HONGKONG (XIANGGANG)
Maoming
Zhanjiang
Wuzhou
Guilin
Yulin
LEIZHOU BANDAO
Haikou
HAINAN
Sanya
TAIWAN
T'AIPEI
Chilung
Taoyüan
Hsinchu
T'aichung
Changhua
Chiai
T'ainan
KAOHSIUNG
Hualien
T'aitung
Taiwan Strait
Tropic of Cancer
China and many other countries do not recognize the existence of Taiwan as a separate country.
PRATAS ISLAND (Occupied by Taiwan; claimed by China)
SOUTH CHINA SEA
PHILIPPINE SEA
Luzon Strait
Bashi Channel
BATAN ISLANDS
BABUYAN ISLANDS
LUZON
PHILIPPINES
Aparri
Laoag
Vigan
CHEJU-DO (South Korea)
Cheju
SAKISHIMA-SHOTO (Japan)
© Rand McNally
A-101803-1

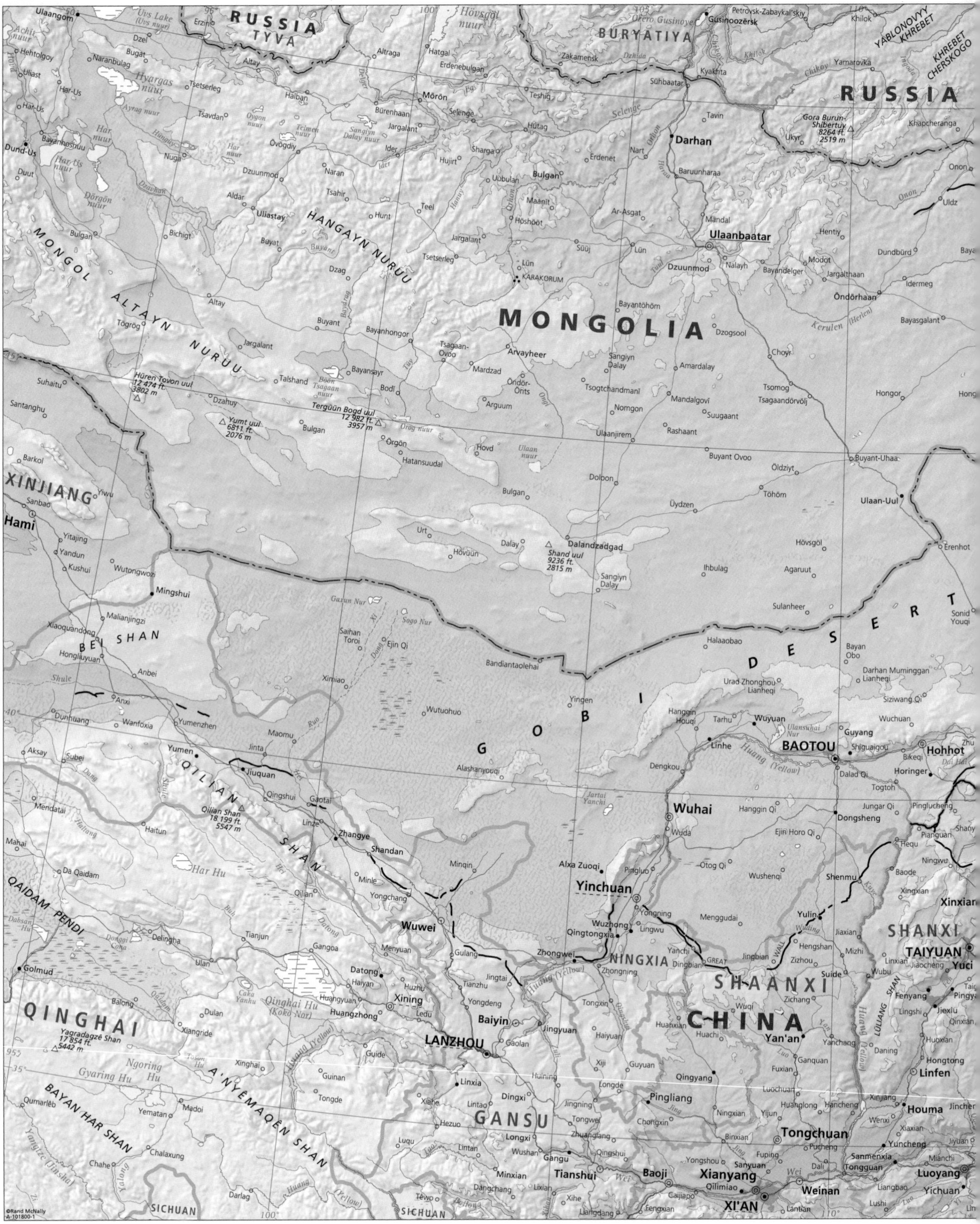
RUSSIA
TYVA
BURYATIYA
RUSSIA
YABLONOVYY KHREBET
KHREBET CHERSKOGO
MONGOLIA
HANGAYN NURUU
MONGOL ALTAYN NURUU
XINJIANG
GOBI DESERT
BEI SHAN
QILIAN SHAN
QAIDAM PENDI
QINGHAI
A'NYÊMAQÊN SHAN
BAYAN HAR SHAN
GANSU
NINGXIA
SHAANXI
SHANXI
CHINA
SICHUAN
LÜLIANG SHAN
GREAT WALL
Ulaangom
Uvs Lake (Uvs nuur)
Erzin
Hövsgöl nuur
Ozero Gusinoye
Gusinoozërsk
Petrovsk-Zabaykal'skiy
Khilok
Yamarovka
Achit nuur
Hehtolgoy
Naranbulag
Bugat
Dzel
Altraga
Hatgal
Erdenebulgan
Zakamensk
Dzhida
Chikoy
Kyakhta
Sühbaatar
Üliast
Har-Us
Hyargas nuur
Tsetserleg
Altay
Halban
Mörön
Teshig
Selenge
Tavin
Darhan
Gora Burun-Shibertuy 8264 ft. 2519 m
Ukyr
Khapcheranga
Ayrag nuur
Tsavdan
Oygon nuur
Telmen nuur
Sangiyn Dalay nuur
Bürenhaan
Jargalant
Hutag
Bayannuur
Har nuur
Nuga
Övögdiy
Ider
Hujirt
Sharga
Erdenet
Nart
Orhon
Dund-Us
Har Us nuur
Duut
Dzuunmod
Naran
Bulgan
Baruunharaa
Onon
Dörgön nuur
Dzavhan
Tsahir
Uubulan
Aldar
Uliastay
Hunt
Teel
Maanit
Ar-Asgat
Uldz
Höshööt
Mandal
Bichigt
Buyat
Buyant
Jargalant
Süüj
Lün
Ulaanbaatar
Hentiy
Dundbürd
Bulgan
Tsetserleg
Dzag
Lün
Karakorum
Dzuunmod
Nalayh
Bayandelger
Modot
Jargalthaan
Idermeg
Altay
Bayantöhöm
Öndörhaan
Tögrög
Baydrag
Buyant
Bayanhongor
Kerulen (Herlen)
Bayasgalant
Dzogsool
Jargalant
Tsagaan-Ovoo
Arvayheer
Choyr
Mardzad
Sangiyn Dalay
Amardalay
Bayansayr
Böön Tsagaan nuur
Talshand
Bodĭ
Öndör-Önts
Tsogtchandmanĭ
Tsomog
Tsagaandörvölj
Hongor
Suhaitu
Hüren Tovon uul 12 474 ft. 3802 m
Dzahuy
Tergüün Bogd uul 12 982 ft. 3957 m
Arguum
Nomgon
Mandalgovĭ
Suugaant
Santanghu
Yumt uul 6811 ft. 2076 m
Bulgan
Orog nuur
Ulaanjirem
Rashaant
Örgön
Hovd
Ulaan nuur
Hatansuudal
Buyant Ovoo
Buyant-Uhaa
Barkol
Öldziyt
Dolbon
Yiwu
Sanbao
Hami
Bulgan
Töhöm
Ulaan-Uul
Üydzen
Yitajing
Yandun
Kushui
Urt
Hövüün
Dalay
Dalandzadgad
Shand uul 9236 ft. 2815 m
Hövsgöl
Erenhot
Ihbulag
Agaruut
Wutongwozi
Mingshui
Sangiyn Dalay
Gaxun Nur
Sogo Nur
Sulanheer
Sonid Youqi
Malianjingzi
Xiaoquandong
Saihan Toroi
Ejin Qi
Halaobao
Bayan Obo
Hongliuyuan
Bandiantaolehai
Darhan Muminggan Lianheqi
Shule
Anbei
Ximiao
Urad Zhonghou Lianheqi
Anxi
Yingen
Siziwang Qi
Dunhuang
Wanfoxia
Yumenzhen
Wutuohuo
Hanggin Houqi
Tarhu
Wuyuan
Ulansuhai Nur
Wuchuan
Guyang
Maomu
Aksay
Subei
Yumen
Jinta
Linhe
BAOTOU
Shiguaigou
Hohhot
Bikeqi
Jiuquan
Alashanyouqi
Dengkou
Huang (Yellow)
Dalad Qi
Horinger
Qingshui
Gaotai
Jartai Yanchi
Togtoh
Mendatai
Qilian Shan 18 199 ft. 5547 m
Wuhai
Hanggin Qi
Jungar Qi
Pinglucheng
Linze
Haltang
Haitun
Wuda
Dongsheng
Zhangye
Shandan
Ejin Horo Qi
Pianguan
Mahai
Hequ
Minqin
Alxa Zuoqi
Pingluo
Otog Qi
Wushenqi
Ningwu
Da Qaidam
Har Hu
Minle
Yinchuan
Shenmu
Baode
Qilian
Yongchang
Xingxian
Xinxian
Yongning
Menggudai
Yulin
Dabsan Hu
Wuwei
Wuzhong
Lingwu
Jiaxian
Delingha
Tianjun
Qingtongxia
Hengshan
Mizhi
Gangoa
Menyuan
Gulang
Zhongwei
Yanchi
Dingbian
Jingbian
Zizhou
Linxian
TAIYUAN
Golmud
Ulan
Datong
Zhongning
Suide
Jiaocheng
Yuci
Haiyan
Huzhu
Jingtai
Tianzhu
Wubu
Fenyang
Pingyao
Caka Yanhu
Qinghai Hu (Koko Nor)
Huangyuan
Xining
Yongdeng
Tongxin
Zichang
Lingshi
Jiexiu
Balong
Dulan
Huangzhong
Ledu
Baiyin
Jingyuan
Haiyuan
Wuqi
Huachi
Yan'an
Qinxian
Xiangride
LANZHOU
Gaolan
Huanxian
Yanchang
Huoxian
Yagradagzê Shan 17 854 ft. 5442 m
Xinghai
Guide
Xiji
Guyuan
Ganquan
Daning
Ngoring Hu
Gyaring Hu
Guinan
Linxia
Huining
Qingyang
Fuxian
Hongtong
Linfen
Tongde
Longde
Luochuan
Xiahe
Lintao
Dingxi
Pingliang
Xinjiang
Qumarlêb
Madoi
Yematan
Jingning
Huanglong
Hancheng
Houma
Jincheng
Hezuo
Tongwei
Ningxian
Yijun
Wenxi
Zhuanglang
Chongxin
Binxian
Tongchuan
Xiaxian
Luqu
Longxi
Lintan
Wushan
Gangu
Qingshui
Fuping
Pucheng
Yuncheng
Chalaxung
Yongshou
Sanyuan
Sanmenxia
Mianchi
Chahe
Minxian
Tianshui
Baoji
Xianyang
Dali
Tongguan
Luoyang
Yangtze (Jinsha)
Yalong
Lixian
Qilimiao
Weinan
Liangbao
Darlag
Tewo
Dangchang
Xihe
Caijiapo
XI'AN
Lushi
Yichuan
Lantian
Liangdang
Fengxian
SICHUAN
©Rand McNally A-101800-1

0 40 80 120 160 200 240 Miles

0 40 80 120 160 200 240 280 320 360 400 Kilometers

Lambert Azimuthal Equal Area Projection
Scale 1:8,000,000
One inch to 128 miles
One cm to 80 km

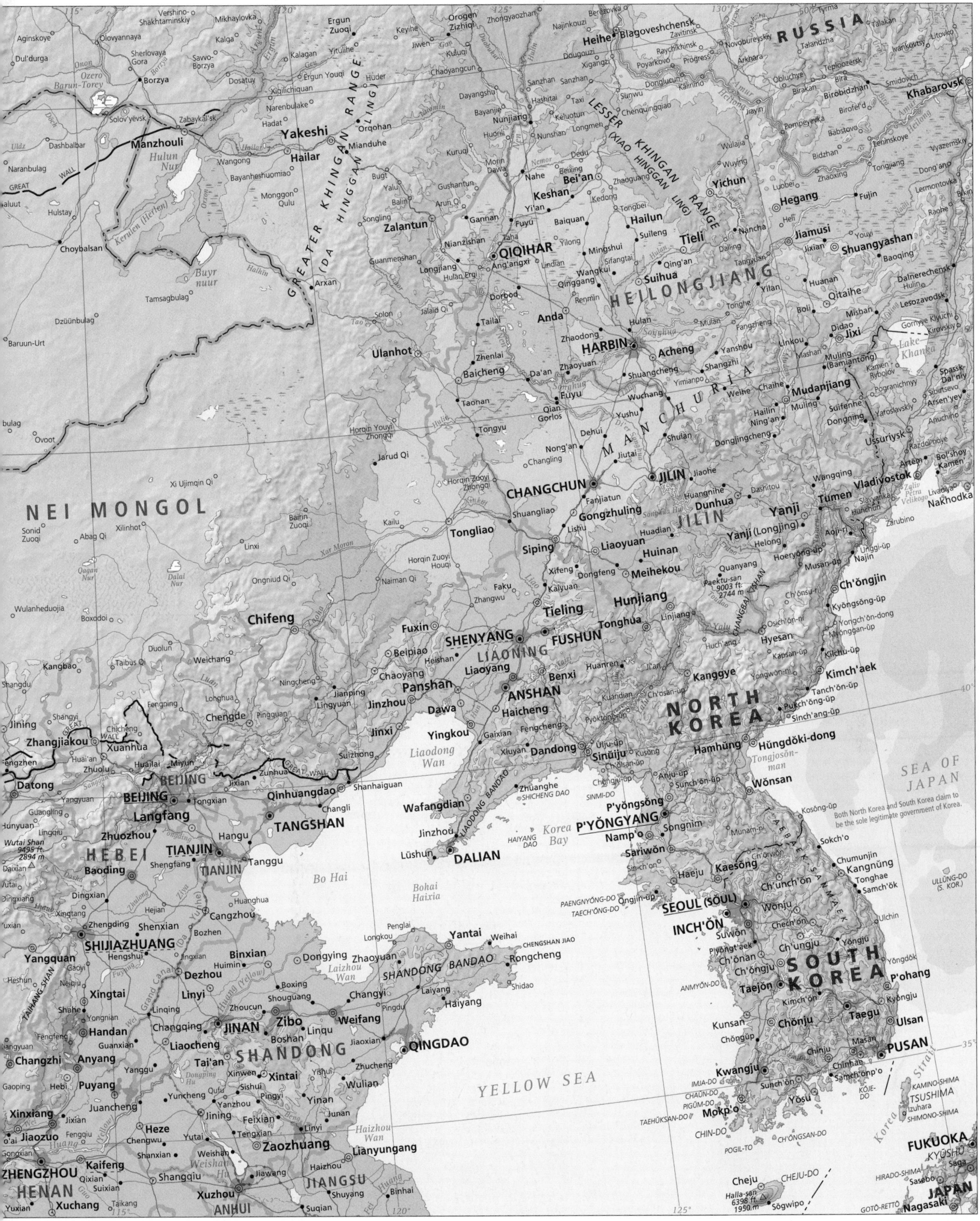
RUSSIA
Blagoveshchensk
Heihe
Khabarovsk
Birobidzhan
Dalnerechensk
Lesozavodsk
Spassk-Dal'niy
Ussuriysk
Vladivostok
Nakhodka
Lake Khanka
Borzya
Zabaykal'sk
Manzhouli
Hulun Nur
Hailar
Yakeshi
Choybalsan
Buyr nuur
GREATER KHINGAN RANGE
(DA HINGGAN LING)
LESSER KHINGAN RANGE
(XIAO HINGGAN LING)
Heihe
Nunjiang
Bei'an
Keshan
Yichun
Hegang
Jiamusi
Shuangyashan
Jixi
Qitaihe
Tieli
Hailun
Suihua
Zalantun
QIQIHAR
Anda
HARBIN
Acheng
HEILONGJIANG
Ulanhot
Baicheng
Mudanjiang
MANCHURIA
CHANGCHUN
JILIN
Gongzhuling
Siping
Liaoyuan
Tonghua
Yanji
Tumen
Tongliao
NEI MONGOL
Chifeng
Fuxin
SHENYANG
FUSHUN
Tieling
Benxi
ANSHAN
LIAONING
Jinzhou
Panshan
Yingkou
Dandong
Liaodong Wan
LIAODONG BANDAO
DALIAN
Lüshun
Wafangdian
Korea Bay
NORTH KOREA
Sinŭiju
Hamhŭng
Ch'ŏngjin
Wŏnsan
P'YŎNGYANG
Nampo
Sariwŏn
Haeju
Kaesŏng
SEOUL (SŎUL)
INCH'ŎN
Suwŏn
Ch'ŏnan
Taejŏn
Ch'ŏngju
SOUTH KOREA
Taegu
Ulsan
P'ohang
PUSAN
Kwangju
Mokp'o
Cheju
CHEJU-DO
Chŏnju
Kunsan
SEA OF JAPAN
Both North Korea and South Korea claim to be the sole legitimate government of Korea.
Korea Strait
TSUSHIMA
FUKUOKA
KYŪSHŪ
Nagasaki
JAPAN
YELLOW SEA
Bo Hai
Bohai Haixia
Laizhou Wan
Haizhou Wan
SHANDONG BANDAO
BEIJING
TIANJIN
Tanggu
TANGSHAN
Qinhuangdao
Langfang
HEBEI
Baoding
SHIJIAZHUANG
Zhangjiakou
Datong
Chengde
Cangzhou
Dezhou
JINAN
Zibo
Weifang
Yantai
Weihai
QINGDAO
SHANDONG
Linyi
Zaozhuang
Lianyungang
Xuzhou
JIANGSU
ANHUI
HENAN
ZHENGZHOU
Kaifeng
Xuchang
Anyang
Handan
Xingtai
Changzhi
Xinxiang
Jiaozuo
Heze
Jining
Tai'an
Liaocheng
Yangquan
GREAT WALL
Paektu-san 9003 ft. 2744 m
Halla-san 6398 ft. 1950 m

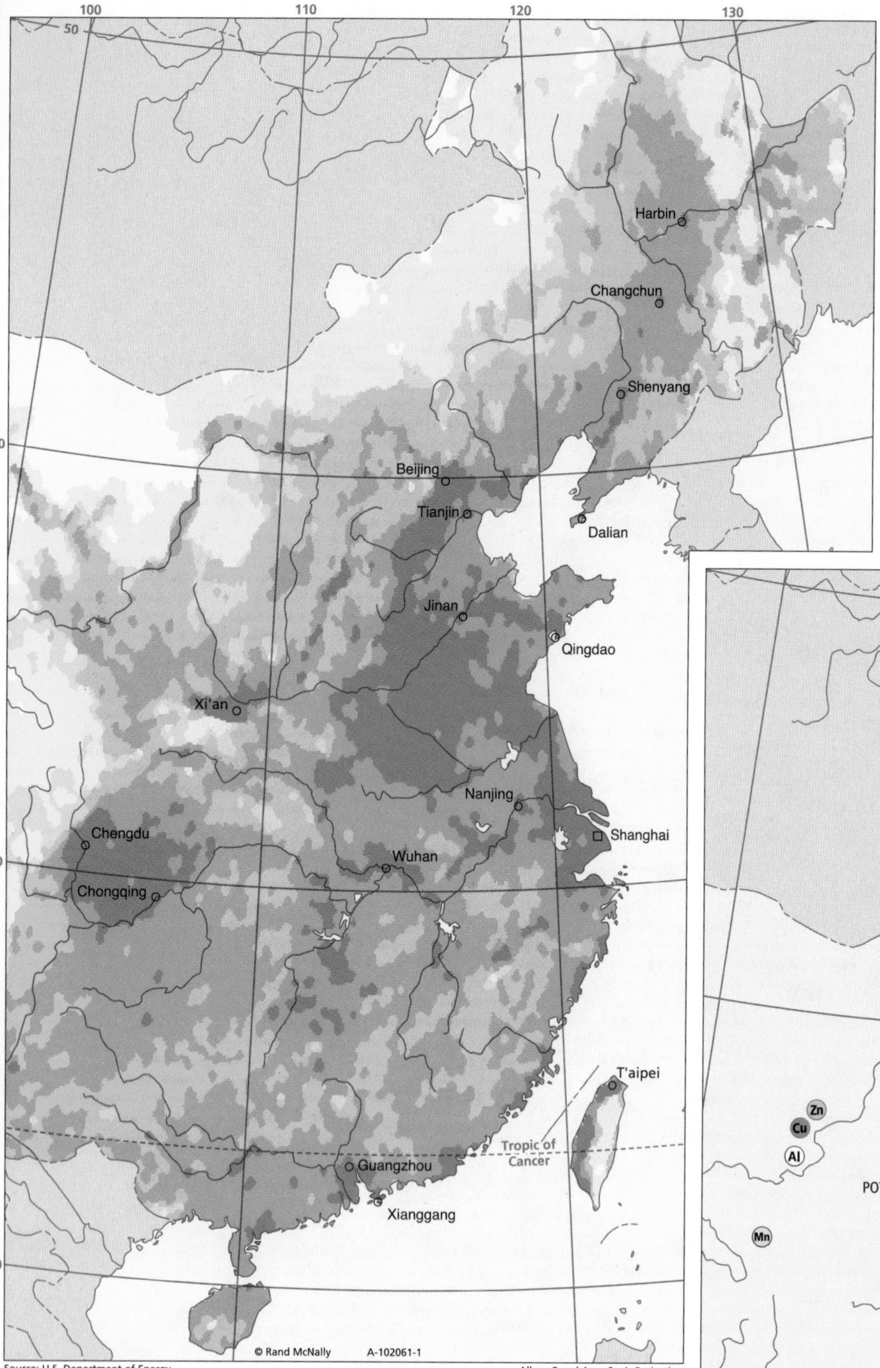

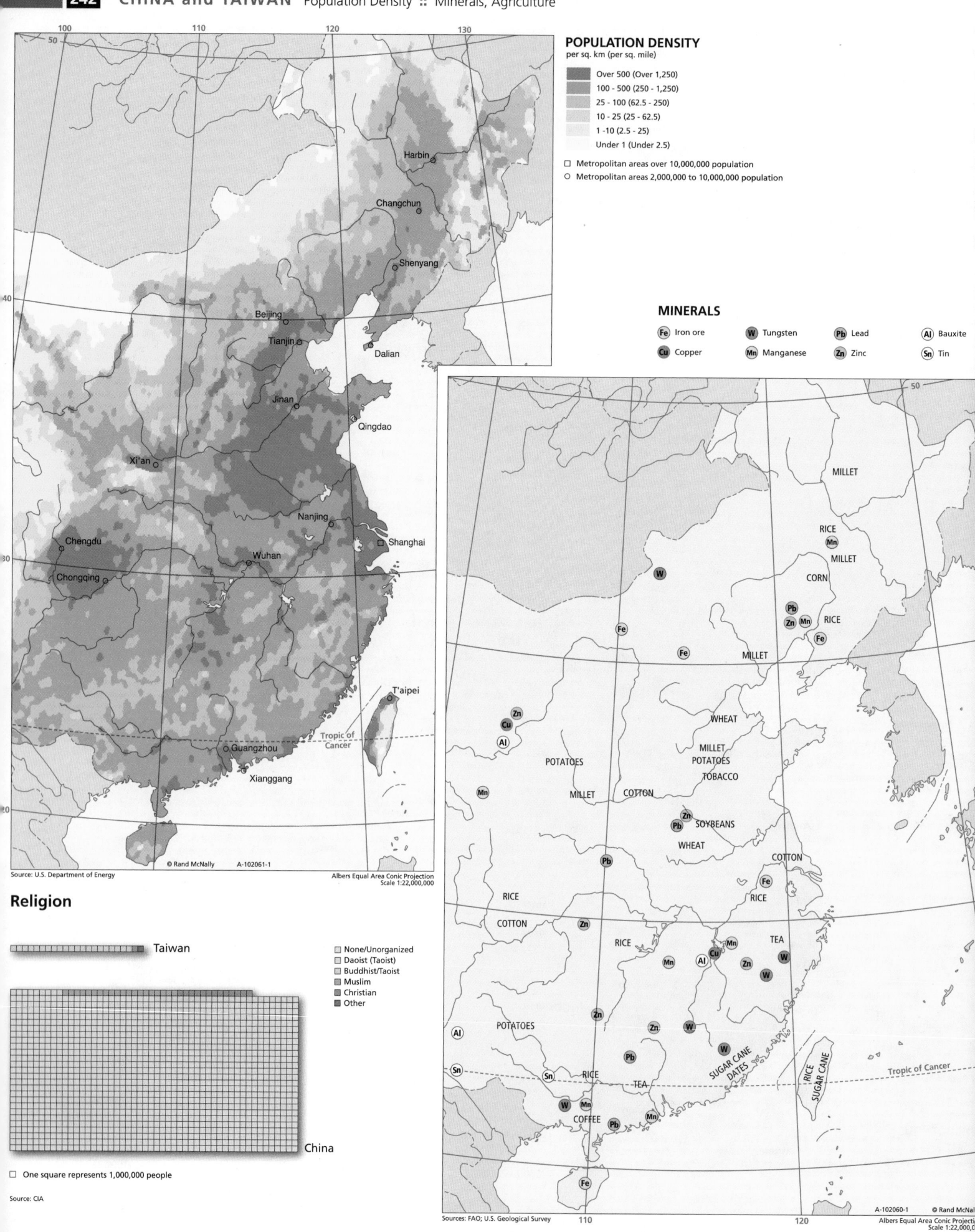

POPULATION DENSITY
per sq. km (per sq. mile)
Over 500 (Over 1,250)
100 - 500 (250 - 1,250)
25 - 100 (62.5 - 250)
10 - 25 (25 - 62.5)
1 -10 (2.5 - 25)
Under 1 (Under 2.5)
Metropolitan areas over 10,000,000 population
Metropolitan areas 2,000,000 to 10,000,000 population
100
110
120
130
50
40
30
20
Harbin
Changchun
Shenyang
Beijing
Tianjin
Dalian
Jinan
Qingdao
Xi'an
Nanjing
Chengdu
Shanghai
Wuhan
Chongqing
T'aipei
Tropic of Cancer
Guangzhou
Xianggang
© Rand McNally
A-102061-1

Source: U.S. Department of Energy

Albers Equal Area Conic Projection
Scale 1:22,000,000

Religion

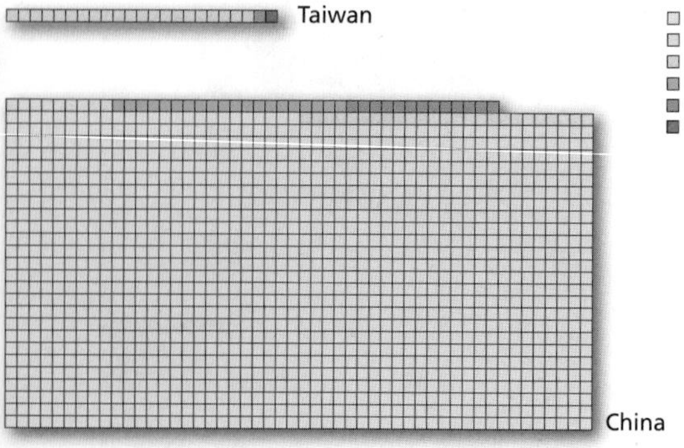

Taiwan
None/Unorganized
Daoist (Taoist)
Buddhist/Taoist
Muslim
Christian
Other
China

One square represents 1,000,000 people

Source: CIA

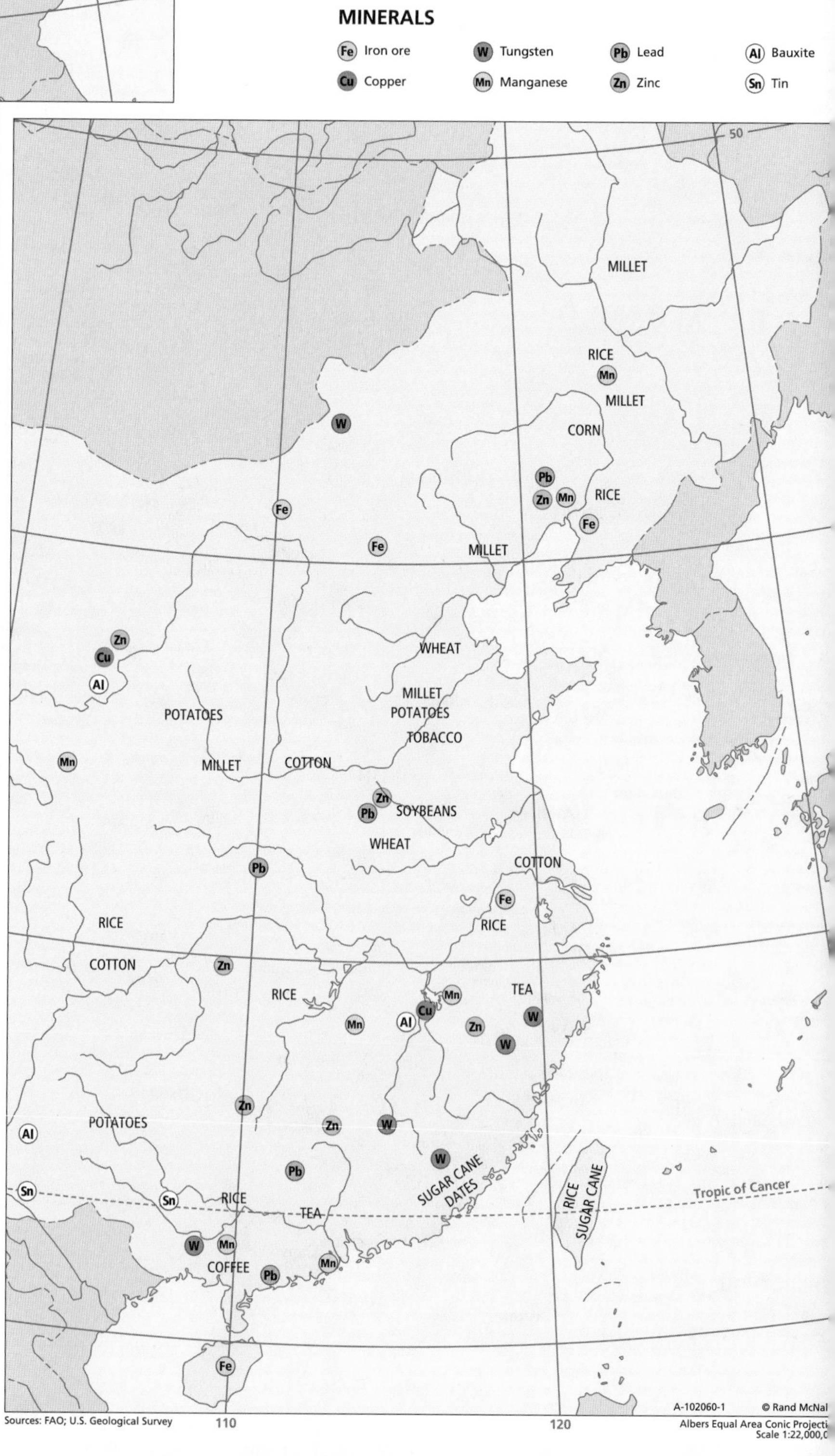

MINERALS
Fe Iron ore
Cu Copper
W Tungsten
Mn Manganese
Pb Lead
Zn Zinc
Al Bauxite
Sn Tin
50
MILLET
RICE
MILLET
CORN
RICE
MILLET
WHEAT
MILLET
POTATOES
TOBACCO
POTATOES
MILLET
COTTON
SOYBEANS
WHEAT
COTTON
RICE
RICE
COTTON
RICE
TEA
POTATOES
SUGAR CANE
DATES
RICE
SUGAR CANE
Tropic of Cancer
RICE
TEA
COFFEE
A-102060-1
© Rand McNally
110
120

Sources: FAO; U.S. Geological Survey

Albers Equal Area Conic Projection
Scale 1:22,000,000

CHINA
LIAONING
JILIN
RUSSIA
NORTH KOREA
SOUTH KOREA
JAPAN
SEA OF JAPAN
YELLOW SEA
Korea Bay
Sŏjosŏn-man
Tongjosŏn-man
Kyŏnggi-man
Korea Strait
Both North Korea and South Korea claim to be the sole legitimate government of Korea.
Faku
Kaiyuan
Kaiyuancheng
Dongfeng
Meihekou
Huinan
Tieling
Daguijia
Qingyuan
Liuhe
Jingyu
Fusong
Quanyang
Erdaobaihe
Helong
Yanji
Tumen
Shixian
Mijiang
Hunchun
Kraskino
Pos'yet
Yanji (Longjing)
Aoji-ri
Hoeryŏng-ŭp
Unggi-ŭp
Najin
NAJIN SŎNBONG-SI
Musan-ŭp
HAMGYŎNG-BUKTO
Pugŏ-ri
Xinmin
Hushitai
Dashiqiao
SHENYANG
FUSHUN
Nanzamu
Sujiatun
Chenxiangtun
Xinbin
Tonghua
Hunjiang
Yayuan
Linjiang
Paektu-san 9003 ft. 2744 m
CHANGBAI SHAN
Kwanmo-bong 8337 ft. 2541 m
Ch'ŏnsu-ri
Nanam
Ch'ŏngjin
Kyŏngsŏng-ŭp
Chuŭl-li
Liudaogou
Osich'ŏn-ni
Benxi
Erhulai
Hengdu
Huch'ang
Sinp'a-ŭp
Samsu-ŭp
Hyesan
Sŏman
Yongch'ŏn-dong
Myŏnggan-ŭp
Liaoyang
ANSHAN
Tianshifu
Balidianzi
Huanren
Manp'o
Ji'an
YANGGANG-DO
Kanggup'o
Kapsan-ŭp
Nanfen
Nangnim-ŭp
Kanggye
Kilchu-ŭp
MUSU-DAN
Ximucheng
Qingchengzi
Kuandian
Ch'osan-ŭp
CHAGANG-DO
Puksubaek-san 8274 ft. 2522 m
P'ungsan-ŭp
Yongwŏn-ni
Kimch'aek
HAMGYŎNG SANJULGI
Tanch'ŏn-ŭp
Fengcheng
Suzigou
Xiuyan
Pyŏktong-ŭp
Sakchu-ŭp
KANGNAM-SANMAEK
Toan
Sŏyang-ni
HAMGYŎNG-NAMDO
Pukch'ŏng-ŭp
Sinhŭng-ŭp
Sinp'o
Sinch'ang-ŭp
Dandong
Ŭiju-ŭp
Sinŭiju
P'YŎNGAN-BUKTO
Hŭich'ŏn
Myohyang-san 6263 ft. 1909 m
NANGNIM SANJULGI
Oro-ŭp
Hongwŏn-ŭp
Samho-dong
Hamhŭng
Hŭngdŏki-dong
Gushan
Kusŏng
Kujang-ŭp
Tŏkch'ŏn-ŭp
Zhuanghe
Ch'ŏlsan-ŭp
Sŏnch'ŏn-ŭp
Chŏngju-ŭp
Kaech'ŏn-ŭp
SHICHENG DAO
KA-DO
SINMI-DO
Sinanju-hodonglagu
Anju-ŭp
P'YŎNGAN-NAMDO
Pukch'ang-ŭp
Kŭmya-ŭp
Kowŏn-ŭp
CHANGSHAN QUNDAO
Sunch'ŏn-ŭp
Unsan-ŭp
P'yŏngsŏng
P'yŏch'ang-ni
Munch'ŏn-ŭp
Wŏnsan
HAIYANG DAO
P'YŎNGYANG
P'YŎNGYANG-SI
Kangdong-ŭp
Anbyŏn-ŭp
Hamjong-ni
Kangsŏ
NAMP'O-SI
Namp'o
Songnim
P'ŏptong-ŭp
KANGWŎN-DO
Koksan-ŭp
Hoeyang-ŭp
Hwangju-ŭp
Munam-ni
Kosŏng-ŭp
CH'O-DO
Sariwŏn
HWANGHAE-BUKTO
Sep'o-ŭp
Chaeryŏng-ŭp
Ich'ŏn-ŭp
Sin-ch'ŏn
T'AEBAEK-SANMAEK
Kansŏng
Changyŏn-ŭp
P'yŏngsan-ŭp
Ch'ŏrwŏn
Kimhwa
Sokch'o
CHANGSAN-GOT
HWANGHAE-NAMDO
Kŭmch'ŏn-ŭp
Hwach'ŏn
Inje
Haeju
KAESŎNG-SI
PAENGNYŎNG-DO
Ongjin-ŭp
Kaesŏng
P'anmunjŏm-ni
Chumunjin
TAECH'ŎNG-DO
Yŏnan-ŭp
Munsan
Ch'unch'ŏn
Kangnŭng
Ŭijŏngbu
KANGWŎN-DO
Kyebang-san 5174 ft. 1577 m
SUNWI-DO
KANGHWA-DO
SEOUL (SŎUL)
Tonghae
Samch'ŏk
ULLŬNG-DO (South Korea)
INCH'ŎN
INCH'ŎN-GWANGYŎKSI
Sŏngnam
Anyang
KYŎNGGI-DO
Wŏnju
Suwŏn
Osan
Yŏngwŏl
TŎKCHŎK-KUNDO
TAEBU-DO
Chech'ŏn
Ulchin
P'yŏngt'aek
Ch'ungju
CH'UNGCH'ŎNG-BUKTO
Ch'ŏnan
T'aean
Sŏsan
Ŏnyang
Yŏngju
Yesan
Ch'ŏngju
Yech'ŏn
Yŏngyang
Hongsŏng
Choch'iwŏn
CH'UNGCH'ŎNG-NAMDO
Andong
ANMYŎN-DO
Kongju
Ŭisŏng
Yŏngdŏk
Taejŏn
Poryŏng
Puyŏ
TAEJŎN-GWANGYŎKSI
Sangju
KYŎNGSANG-BUKTO
OEYŎN-DO
Kanggyŏng
Yŏngdong
CHANGGI-AP
SOBAEK-SANMAEK
Kimch'ŏn
P'ohang
Waegwan
Sŏch'ŏn
Yŏngch'ŏn
Kunsan
Iksan
Taegu
TAEGU-GWANGYŎKSI
Kyŏngju
Kyŏngsan
Kimje
Chŏnju
CHŎLLA-BUKTO
Kŏch'ang
WI-DO
Chŏngŭp
Ulsan
ULSAN-GWANGYŎKSI
Miryang
Namwŏn
Chiri-san 6283 ft. 1915 m
KYŎNGSANG-NAMDO
Samnangjin
Yangsan
ANMA-DO
Ch'angwŏn
Kimhae
Tongnae
KWANGJU-GWANGYŎKSI
Koksŏng
Chinju
Masan
BUSAN-GWANGYŎKSI
IMJA-DO
Songjŏng
Kwangju
Hadong
Chinhae
PUSAN
Naju
CHŎLLA-NAMDO
Sunch'ŏn
Kwangyang
Samch'ŏnp'o
CHAŬN-DO
Mokp'o
Pŏlgyo
Ch'ungmu
PIGŬM-DO
Yŏngam
Yŏsu
NAMHAE-DO
KŎJE-DO
HŬKSAN-CHEDO
MI-SHIMA
Changhŭng
TOLSAN-DO
YOKCHI-DO
TAEHŬKSAN-DO
Kohŭng
Haenam
KŬMO-DO
TSUSHIMA (Japan)
KAMINO-SHIMA
Chindo
CHIN-DO
KOGŬM-DO
OERARO-DO
HAJO-DO
NOHWA-DO
WAN-DO
SAENGIL-TO
CH'O-DO
Izuhara
SHIMONO-SHIMA
POGIL-TO
SOAN-DO
CH'ŎNGSAN-DO
KŎMUN-DO
HONSHŪ
YAMAGUCHI
Shimonoseki
Ube
Kitakyūshū
124°
126°
128°
130°
34°
36°
38°
40°
42°
© Rand McNally A-101809-1

0 20 40 60 80 100 120 Miles

0 20 40 60 80 100 120 140 160 180 200 Kilometers

Lambert Conformal Conic Projection
Scale 1:4,000,000
One inch to 64 miles
One cm to 40 km

RUSSIA
CHINA
MANCHURIA
HEILONGJIANG
JILIN
NORTH KOREA
SOUTH KOREA
JAPAN
SEA OF JAPAN
SEA OF OKHOTSK
PACIFIC OCEAN
PHILIPPINE SEA
EAST CHINA SEA
SIKHOTE-ALIN'
Tatar Strait
La Perouse Strait
Korea Strait
Tsugaru-kaikyō
SAKHALIN
Zaliv Aniva
Lake Khanka
Zaliv Petra Velikogo
Jiamusi
Shuangyashan
Jixi
Qitaihe
Mudanjiang
Yanji
Tumen
Vladivostok
Ussuriysk
Nakhodka
Ch'ŏngjin
Kimch'aek
Dunhua
Dalnerechensk
Yuzhno-Sakhalinsk
Korsakov
Kholmsk
Wakkanai
HOKKAIDŌ
SAPPORO
Asahikawa
Kitami
Obihiro
Kushiro
Nemuro
Otaru
Ebetsu
Tomakomai
Muroran
Hakodate
Asahi-dake 7513 ft. 2290 m
The islands known in Japan as the Northern Territories and in Russia as the Southern Kuril Islands are occupied by Russia and claimed by Japan.
OSTROV KUNASHIR (KUNASHIRI-TŌ) (Rus.)
Aomori
Hachinohe
Hirosaki
Mutsu
Akita
Morioka
Miyako
Sakata
Tsuruoka
Yamagata
Sendai
Ishinomaki
Niigata
Fukushima
Kōriyama
Iwaki
SADO
NOTO-HANTŌ
Kanazawa
Toyama
Nagano
Utsunomiya
Maebashi
Hitachinaka
Mito
Tsukuba
SAITAMA
TŌKYŌ
KAWASAKI
YOKOHAMA
Chiba
Yokosuka
Chōshi
HONSHŪ
Fuji-san 12,388 ft. 3776 m
NAGOYA
Gifu
Shizuoka
Hamamatsu
KYŌTO
ŌSAKA
KŌBE
Himeji
Okayama
Kurashiki
Fukuyama
HIROSHIMA
Matsue
Tottori
Wakayama
Tokushima
Takamatsu
Matsuyama
Kōchi
SHIKOKU
Kitakyūshū
FUKUOKA
Kumamoto
Ōita
Nagasaki
KYŪSHŪ
Miyazaki
Kagoshima
Miyakonojō
Ulsan
PUSAN
P'ohang
TSUSHIMA
OKI-SHOTŌ
IZU-SHOTŌ
ŌSUMI-SHOTŌ
RYUKYU ISLANDS (NANSEI-SHOTŌ)
Kumano-nada
Enshū-nada
AMAMI-ŌSHIMA
AMAMI-SHOTŌ
OKINAWA-SHOTŌ
OKINAWA-JIMA
Naha
Okinawa
Ginowan
Nago
© Rand McNally A-101814-1

0 40 80 120 160 200 240 Miles
0 40 80 120 160 200 240 280 320 360 400 Kilometers

Lambert Conformal Conic Projection
Scale 1:8,000,000
One inch to 128 miles
One cm to 80 km

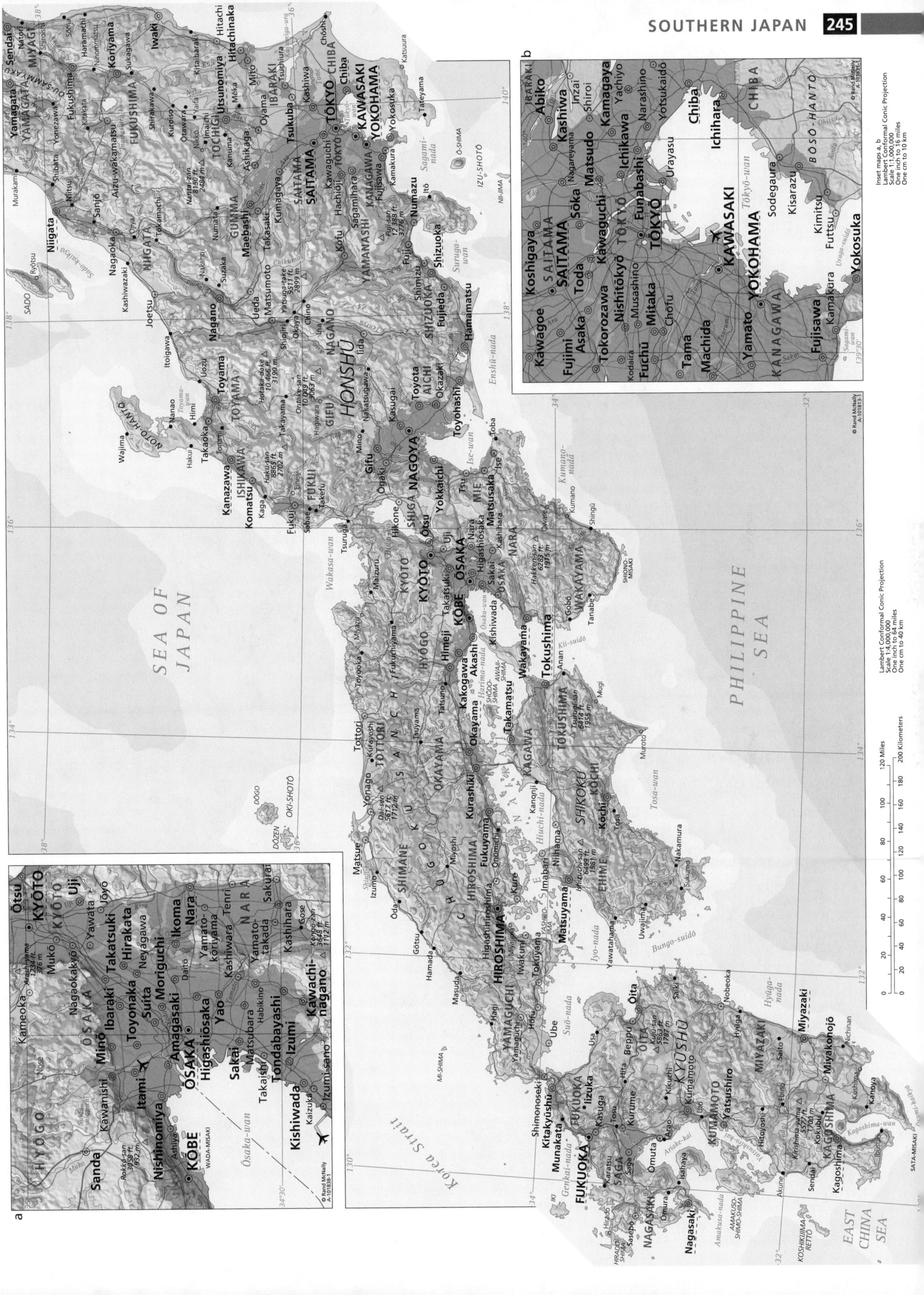
SEA OF JAPAN
PHILIPPINE SEA
EAST CHINA SEA
Korea Strait
HONSHŪ
SHIKOKU
KYŪSHŪ
TŌKYŌ
YOKOHAMA
NAGOYA
ŌSAKA
KYŌTO
KŌBE
HIROSHIMA
FUKUOKA
Lambert Conformal Conic Projection
Scale 1:4,000,000
One inch to 64 miles
One cm to 40 km
Inset maps a, b
Lambert Conformal Conic Projection
Scale 1:1,000,000
One inch to 16 miles
One cm to 10 km
© Rand McNally

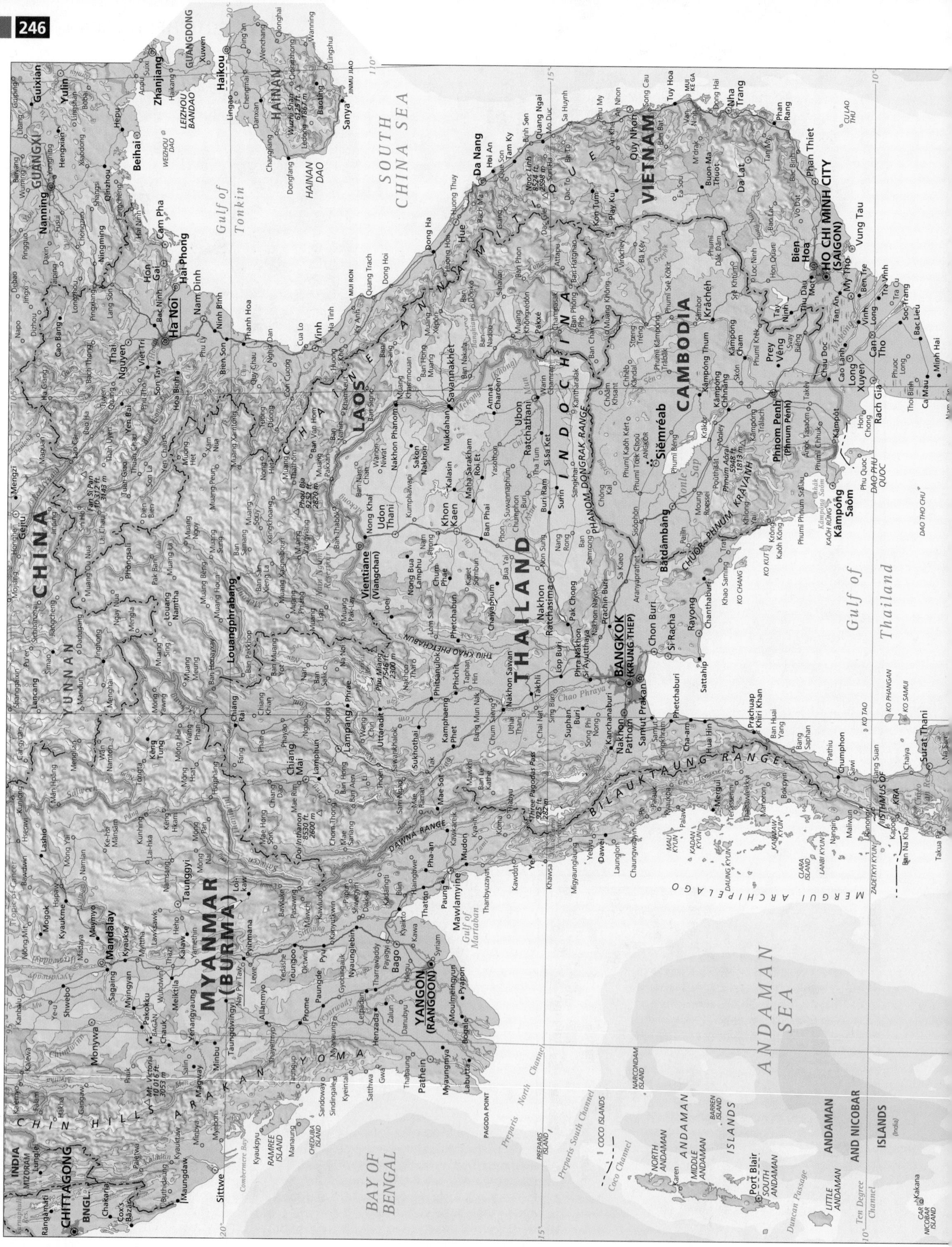
CHINA
GUANGXI
YUNNAN
GUANGDONG
HAINAN
MYANMAR (BURMA)
THAILAND
LAOS
VIETNAM
CAMBODIA
INDIA
BNGL.
I N D O C H I N A
SOUTH CHINA SEA
Gulf of Tonkin
Gulf of Thailand
ANDAMAN SEA
BAY OF BENGAL
Gulf of Martaban
ANDAMAN AND NICOBAR ISLANDS (India)
MERGUI ARCHIPELAGO
BILAUKTAUNG RANGE
DAWNA RANGE
PHANOM DONGRAK RANGE
CHUŎR PHNUM KRÂVANH
THIU KHAO PHETCHABUN
ARAKAN YOMA
CHIN HILLS
ISTHMUS OF KRA
Nanning
Beihai
Haikou
Zhanjiang
Sanya
Kunming
Ha Noi
Hai Phong
Da Nang
Hue
Vientiane (Viangchan)
Louangphrabang
BANGKOK (KRUNG THEP)
Chiang Mai
Nakhon Ratchasima
Phnom Penh (Phnum Pénh)
Siĕmréab
HO CHI MINH CITY (SAIGON)
Nha Trang
Quy Nhon
Mandalay
Nay Pyi Taw
YANGON (RANGOON)
Mawlamyine
Sittwe
CHITTAGONG
Port Blair
Three Pagodas Pass

0 40 80 120 160 200 240 Miles
0 40 80 120 160 200 240 280 320 360 400 Kilometers

Lambert Azimuthal Equal Area Projection
Scale 1:8,000,000
One inch to 128 miles
One cm to 80 km

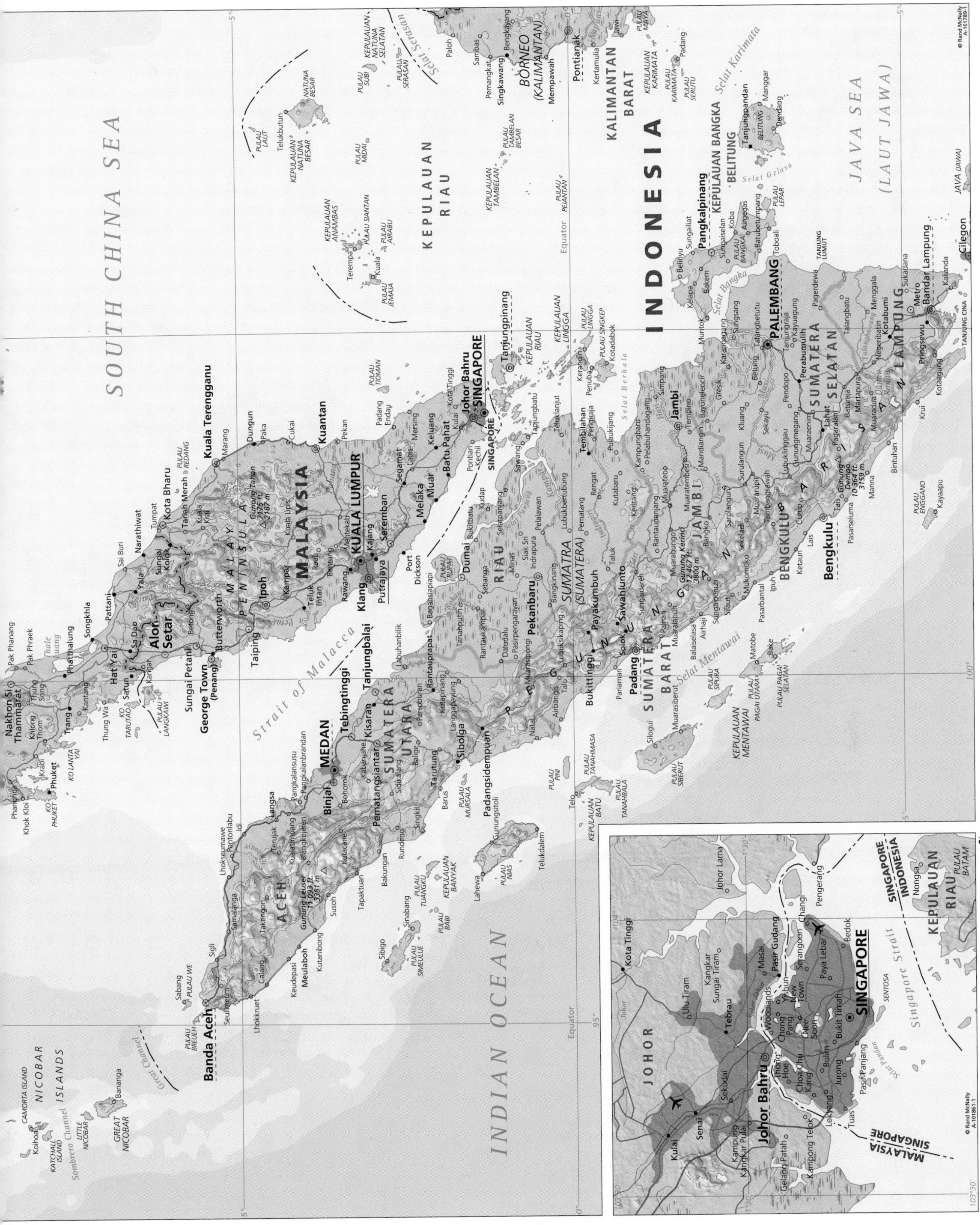
SOUTH CHINA SEA
INDIAN OCEAN
JAVA SEA
(LAUT JAWA)
Strait of Malacca
INDONESIA
MALAYSIA
KUALA LUMPUR
SINGAPORE
MEDAN
PALEMBANG
Padang
Pekanbaru
Jambi
Bengkulu
Bandar Lampung
Pangkalpinang
Pontianak
Banda Aceh
Kuala Terengganu
Kuantan
Ipoh
Klang
Putrajaya
Johor Bahru
Tanjungpinang
George Town (Penang)
Alor Setar
Kota Bharu
Melaka
Seremban
SUMATERA UTARA
SUMATERA BARAT
SUMATERA SELATAN
RIAU
JAMBI
BENGKULU
LAMPUNG
ACEH
KEPULAUAN RIAU
KEPULAUAN BANGKA BELITUNG
KALIMANTAN BARAT
BORNEO (KALIMANTAN)
SUMATRA (SUMATERA)
NICOBAR ISLANDS
KEPULAUAN MENTAWAI
Equator
JOHOR
SINGAPORE
MALAYSIA
Singapore Strait
© Rand McNally A-101789-1
© Rand McNally A-101861-1
a

Inset map a
Lambert Conformal Conic Projection
Scale 1:1,000,000
One inch to 16 miles
One cm to 10 km

SOUTH CHINA SEA
THAILAND
Songkhla
Hat Yai
Pattani
Sai Buri
Satun
Sa Dao
Yala
Narathiwat
Kangar
Sungai Kolok
Tumpat
Kota Bharu
PULAU LANGKAWI
Alor Setar
Sungai Petani
PULAU REDANG
Tanah Merah
George Town (Penang)
Butterworth
Kuala Krai
Kuala Terengganu
Marang
MALAY PENINSULA
Taiping
Dungun
Ipoh
Gunong Tahan 7175 ft. 2187 m
Paka
Kampar
Kuala Lipis
Cukai
MALAYSIA
Teluk Intan
Raub
Bentung
Kuantan
Rawang
Pekan
Mentekab
Klang
KUALA LUMPUR
Kajang
Putrajaya
Seremban
Tanjungbalai
Labuhanbilik
Strait of Malacca
Port Dickson
Segamat
Labis
Padang Endau
PULAU TIOMAN
Mersing
Melaka
Muar
Keluang
Batu Pahat
Kota Tinggi
Pontian Kechil
Kulai
Johor Bahru
SINGAPORE
Rantauprapat
Kotapinang
Bagansiapiapi
PULAU RUPAT
Dumai
Bukitbatu
SUMATERA UTARA
Rantaukampar
Sebanga
Kudap
Dalududu
RIAU
Selatpanjang
Pasirpengarayan
Minas
Siak Sri Indrapura
Muarasipongi
Pekanbaru
Pelalawan
Bangkinang
Lubuksikaping
Talu
SUMATRA (SUMATERA)
Lubukbertubung
Payakumbuh
Bukittinggi
Pariaman
Padang
Sawahlunto
Solok
Pematang
Tembilahan
Taluk
Rengat
Pengiraja
Kutabaru
Kentang
Pulaukijang
Sungaidareh
SUMATERA BARAT
Kampungbaru
Pelabuhandagang
Rantaupanjang
Painan
Muarabungo
Muaratebo
Muaralabuh
Gunung Kerinci 12 467 ft. 3800 m
Muaratembesi
Jambi
JAMBI
Balaiselasa
Airhaji
Sungaipenuh
Mandiangin
Tempino
PULAU SIPURA
Bangko
Bayunglencir
Sarolangun
PULAU PAGAI UTARA
Silaut
Sekeladi
Gresik
Mukomuko
Surulangun
Kluang
Matobe
KEPULAUAN MENTAWAI
Pasarbantal
Muararupit
Sekayu
Betung
Ipuh
Tambangsawah
Bake
BENGKULU
Lubuklinggau
Pendopo
PALEMBANG
PULAU PAGAI SELATAN
Ketaun
Curup
Gunungmegang
Tanjungraja
Kayuagung
Lais
Muaraenim
Perabumulih
Bengkulu
SUMATERA SELATAN
Pagerdewa
Tanjung Lumut
Tais
Lahat
Pagaralam
Gunung Dempo 10 364 ft. 3159 m
Baturaja
Martapura
Pasarseluma
Muaradua
Talangbatu
Manna
Negeribatin
Menggala
Kotabumi
Bintuhan
LAMPUNG
Danau Ranau
Sukadana
Metro
PULAU ENGGANO
Krui
Pringsewu
Bandar Lampung
Kayaapu
Kotaagung
Kalianda
TANJUNG CINA
Cilegon
KRAKATOA
Selat Sunda
Serang
JAKARTA
Karawang
Indramayu
Labuhan
DEPOK
Cikampek
PULAU PANAITAN
BANTEN
Bogor
Cianjur
JAWA BARAT
Klangenang
Cirebon
TANJUNG CANGKUANG
Sukabumi
Majalaya
Tegal
Pekalongan
Pelabuhanratu
BANDUNG
Garut
Jampang-kulon
Tasikmalaya
Gunung Slamet 11 247 ft. 3428 m
Sindangbarang
Karangnunggal
Cijulang
Cilacap
INDIAN OCEAN
PULAU LAUT
KEPULAUAN NATUNA
NATUNA BESAR
Telukbutun
KEPULAUAN ANAMBAS
Terempa
PULAU SIANTAN
PULAU JEMAJA
Kuala
PULAU AIRABU
PULAU MIDAI
PULAU SUBI
KEPULAUAN NATUNA SELATAN
KEPULAUAN RIAU
PULAU SERASAN
Selat Serasan
TANJONG DATU
SINGAPORE
Tanjungpinang
Sawang
Tanjungbatu
KEPULAUAN RIAU
KEPULAUAN LINGGA
Teluklanjut
KEPULAUAN TAMBELAN
PULAU TAMBELAN BESAR
PULAU PEJANTAN
Equator
Kerandin
Penuba
PULAU LINGGA
PULAU SINGKEP
Kotadabok
Selat Berhala
Simpang
Belinyu
Sungailiat
Muntok
Kelapa
Pangkalpinang
Bakem
Selat Bangka
KEPULAUAN BANGKA BELITUNG
Sungaiselan
Koba
Karangagung
Sungsang
PULAU BANGKA
Airgegas
Talangbetutu
Batubetumpang
Tanjungpandan
Toboali
PULAU LEPAR
Selat Gelasa
BELITUNG
Manggar
Dendang
Selat Karimata
Lutong
Miri
Niah
Bintulu
Tubau
Mukah
PULAU BRUIT
Igan
Balingian
Belaga
SARAWAK
Paloh
Sibu
MALAYSIA
Kanowit
Kapit
Sarikei
Rajang
Sematan
Paloh
Kuching
Saratok
UPPER KAPUAS MTS.
Sambas
Bau
Simunjan
Betong
Siluas
Pemangkat
Sri Aman
Singkawang
Gunung Niut 5581 ft. 1701 m
Nangabadau
Nangaobat
Bengkayang
Seriari
Putussibau
Mualang
Ngabang
Kapuas
Nangaraun
Balaisepuah
Semitau
PEGUNUNGAN MULLER
Sosok
Mempawah
Sanggau
Pontianak
Sintang
Longgunt
Kalasin
KALIMANTAN BARAT
Meliau
Nangapinoh
Tanjungpusu
Kertamulia
Jenu
PEGUNUNGAN SCHWANER
Rangantemiang
Jawi
Bukit Raya 7474 ft. 2278 m
Teratak
Kotabaharu
Sungaipinang
PULAU MAYA
Telukbatang
Tewah
KEPULAUAN KARIMATA
Sukadana
Nangalangki
Kualakurun
Sandai
PULAU KARIMATA
Teluk Sukadana
Nangatayap
Bakumpai
PULAU SERUTU
Padang
Kualamanjual
KALIMANTAN TENGAH
Ketapang
Serengka
Panahan
Memala
Batutinggi
Kualapesaguan
Panopah
Mabau
Palangkaraya
Kendawangan
Sukaraja
Kotawaringin
Sampit
PULAU BAWAL
Pangkalanbuun
Telegapulang
Matua
Mendawai
TANJUNG PUTING
Kumai
JAVA SEA (LAUT JAWA)
GREATER SUNDA
KEPULAUAN KARIMUNJAWA
Tambak
PULAU BAWEAN
INDON
Jepara
Pati
Rembang
Kudus
Tuban
Bangkalan
Sumenep
JAWA TENGAH
SEMARANG
Cepu
Gresik
Blega
MADURA
Pamekasan
Boja
Purwokerto
Salatiga
JAWA TIMUR
SURABAYA
Magelang
Surakarta
Jombang
Bangil
Selat Madura
Purworejo
Madiun
Kediri
Pasuruan
Probolinggo
Ponorogo
Yogyakarta
Malang
Situbondo
YOGYAKARTA
Tulungagung
Lumajang
Jember
Pacitan
Blitar
Gunung Semeru 12 060 ft. 3676 m
Genteng
JAVA (JAWA)
CHRISTMAS ISLAND (Austl.)
Settlement
105°
110°
100°
5°
10°
0 40 80 120 160 200 240 Miles
0 40 80 120 160 200 240 280 320 360 400 Kilometers
Lambert Azimuthal Equal Area Projection
Scale 1:8,000,000
One inch to 128 miles
One cm to 80 km

SULU SEA
PHILIPPINES
PHILIPPINE SEA
Moro Gulf
Davao Gulf
MINDANAO
Cotabato
Kidapawan
Mt. Apo 9692 ft. 2954 m
Davao
Mati
Digos
Tacurong
Koronadal
Lebak
Polomolok
General Santos
Kiamba
CAPE SAN AGUSTIN
TINACA POINT
SARANGANI ISLANDS
PULAU MIANGAS
ZAMBOANGA PENINSULA
Zamboanga
OLUTANGA ISLAND
Isabela
Lamitan
Maluso
BASILAN ISLAND
PANGUTARAN ISLAND
Pangutaran
SULU ARCHIPELAGO
Jolo
Parang
JOLO ISLAND
TONGQUIL ISLAND
Siasi
SIASI ISLAND
TAWITAWI ISLAND
Balimbing
SIBUTU ISLAND
Sitangkai
Tungku
TANJONG LABIAN
Cagayan de Tawi-Tawi
CAGAYAN SULU ISLAND
PULAU BALAMBANGAN
PULAU BANGGI
PULAU JAMBONGAN
Kudat
Kota Belud
Tenghilan
Gunong Kinabalu 13 455 ft. 4101 m
Kota Kinabalu
Ranau
Papar
Sandakan
Beluran
Tambunan
SABAH
Kampung Litang
Beaufort
Labuan
Bandar Seri Begawan
Melalap
Keningau
Tenom
Kuamut
Lahad Datu
Susul
Tomani
Lawas
Pensiangan
BRUNEI
Tawau
Semporna
PULAU SEBATIK
Gunong Murud 7946 ft. 2422 m
Kabinu
Sesibu
Bareo
Tanahmerah
PULAU BUNYU
Long Akah
Malinau
Sesayap
Lama
Punan
PULAU TARAKAN
Tarakan
Bengara
Lemutan
Tanjungselor
Napaku
Muarapangean
Datumakuta
Tanjungbatu
IRAN MOUNTAINS
Longsegah
Tanjungredep
Kong Kemul 6736 ft. 2053 m
Muaralasan
Longnawan
Kubumesaai
KALIMANTAN TIMUR
Dumaring
Batuputih
Punan
BORNEO (KALIMANTAN)
Muarawahau
Kelolokan
Lembak
Talok
TANJUNG MANGKALIHAT
Sepasu
Longbangun
Merah
Muaraancalung
Mamehaktebo
Danompari
Longiram
Bontang
Muarakaman
Santan
Intu
Tenggarong
Samarinda
Muaratuhup
Sungaitiram
Muarabenangin
Muaratewe
Samboja
Balikpapan
Muarapayang
Muaratunan
Buntok
Dayu
Tanahgrogot
TANJUNG ARU
Jangeru
Amuntai
Barabai
Sampanahan
Kandangan
Mehakit
KALIMANTAN SELATAN
Banjarmasin
Kotabaru
Batulicin
Banjarbaru
PULAU SEBUKU
Pagatan
Pelaihari
Satui
Kintap
Karambu
PULAU LAUT
Batakan
TANJUNG SELATAN
KEPULAUAN LAUT KECIL
PULAU MATASIRI
PULAU KERAMIAN
PULAU KALAMBAU
ISLANDS
PULAU MASALEMBU BESAR
ESIA
KEPULAUAN KANGEAN
PULAU KANGEAN
Arjasa
PULAU SEPANJANG
BALI SEA (LAUT BALI)
BALI
Singaraja
Banyuwangi
Negara
Penebel
Klungkung
Denpasar
Karangasem
Bayan
Gunung Rinjani 12 224 ft. 3726 m
Mataram
Praya
Taliwang
NUSA PENIDA
LOMBOK
Selat Lombok
PULAU MOYO
Gunung Tambora 9350 ft. 2850 m
Sumbawa Besar
Teluk Saleh
Dompu
Raba
Bima
SUMBAWA
Plampang
Lunyuk
Bangkulua
NUSA TENGGARA BARAT
LESSER SUNDA ISLANDS
PULAU SANGEANG
PULAU KOMODO
Labuhanbajo
Ruteng
Reo
FLORES
Mborong
Bajawa
Ende
Maumere
Nangahale
Larantuka
Selat Sumba
PULAU PALOE
PULAU ADONARA
Balaurin
Kalabahi
PULAU ALOR
Lewoleba
PULAU SOLOR
PULAU LOMBLEN
PULAU PANTAR
NUSA TENGGARA TIMUR
Waikelo
Memboro
Waikabubak
Waingapu
Payeti
Tosi
Lewapaku
SUMBA
Melolo
Lalindi
Baing
SAVU SEA (LAUT SAWU)
PULAU RAIJUA
Seba
PULAU SAWU
PULAU ROTI
Baa
Batuidu
Tudameda
Kupang
PULAU SEMAU
Baun
Bokong
Kolbano
Pariti
Lelogama
TIMOR
Soe
Besikama
Naikliu
Kefamenanu
Atambua
Tilomar
Pante Makasar
Atapupu
Selat Ombai
Maubara
Aileu
Dili
Fatuberliu
Manatuto
Uatolari
Betano
Lautem
EAST TIMOR
Taramana
PULAU ATAURO
Selat Wetar
TIMOR SEA
Tutuala
KEPULAUAN LETI
PULAU SERMATA
PULAU WETAR
Lioppa
Ilwaki
Arwala
PULAU KISAR
Kaiwatu
PULAU MOA
Hila
PULAU ROMANG
KEPULAUAN BARAT DAYA
PULAU DAMAR
Wulur
FLORES SEA (LAUT FLORES)
PULAU TANAHJAMPEA
Ujung
PULAU KALAO
PULAU BONERATE
PULAU KALAOTOA
PULAU KAKABIA
PULAU BATUATA
Benteng
PULAU SELAYAR
Bulukumba
Jeneponto
Takalar
Bantaeng
Sungguminasa
MAKASSAR (UJUNGPANDANG)
Maros
Pangkajene
Sinjai
Uloanlamuru
Watampone
Towari
SULAWESI SELATAN
Singkang
Pampanua
Parepare
Pinrang
Rapang
Enrekang
Majene
Polewali
Bulu Rantekombola 11 335 ft. 3455 m
Rantepao
Palopo
Tolala
Mamasa
Malunda
Mamuju
CELEBES (SULAWESI)
Masamba
Bonebone
Wotu
Malili
Sampaga
Karosa
Lariang
Pangian
Donggala
Palu
Toboli
Parigi
Sausu
Kasimbar
Toribulu
Mapaga
Teluk Tomini
PULAU UNAUNA
KEPULAUAN TOGIAN
Bongka
Uebonti
Bunta
Lubu
Luwuk
Malik
Teku
Poso
Tojo
Namo
Lempe
Lemoro
SULAWESI TENGAH
Batui
Danau Poso
Tentena
Tokala
Kabali
Tamanusi
Pendolo
Kolonodale
Bahu-mbelu
Wosu
Danau Matana
Saroako
Danau Towuti
Siumbatu
Lasolo
Malamala
Sanggona
Teluk Bone
Tamboli
Kolaka
SULAWESI TENGGARA
Torobuku
Kendari
PULAU WOWONI
Langara
Kolono
Lawowa
Buapinang
Laloa
Raha
PULAU MUNA
PULAU BUTON
Bonelipu
Bone
Lawele
Sikeli
PULAU KABAENA
Baubau
Sampolawa
PULAU WANGIWANGI
KEPULAUAN TUKANGBESI
PULAU KALEDUPA
PULAU TOMEA
Papalia
PULAU BINONGKO
PULAU MANUI
Teluk Tolo
Dongi
Selat Peleng
PULAU PELENG
Mantok
KEPULAUAN BANGGAI
Banggai
Kembani
PULAU BANGGAI
Taduno
PULAU BANGKULU
Toliitoli
Buol
Paleleh
Bukit Malino 8015 ft. 2443 m
Sikutu
Bangkir
Tomini
Moutong
Tilamuta
GORONTALO
Bumbulan
Sumalata
Boroko
Isimu
Gorontalo
Bilato
Taludaa
Kompot
Kotamobagu
Bintauna
Lombagin
Amurang
Belang
Munte
Sabang
Bajo
Sidoan
Manado
Tondano
Bitung
Kema
TANJUNG TORAWITAN
SULAWESI UTARA
KEPULAUAN SANGIHE
Tahuna
PULAU SANGIHE
Ngalipaeng
PULAU SIAU
PULAU TAHULANDANG
PULAU KARAKALONG
KEPULAUAN NANUSA
PULAU KARAKELONG
Beo
KEPULAUAN TALAUD
PULAU SALEBABU
PULAU KABURUANG
CELEBES SEA (LAUT SULAWESI)
MOLUCCA SEA (LAUT MALUKU)
Equator
MALUKU UTARA
MOLUCCAS
MOROTAI
Berebere
Wayabula
Galela
Tobelo
Akelamo
Kau
Ternate
Tidore
HALMAHERA
Mesa
Mafa
Cenga
PULAU KASIRUTA
Labuha
Gani
PULAU MANDIOLI
PULAU BACAN
Selat Obi
KEPULAUAN OBI
Laiwui
PULAU OBI
Wai
KEPULAUAN SULA
Todeli
PULAU TALIABU
PULAU MANGOLE
Lekitobi
Auponhia
Talaga
PULAU SANANA
CERAM SEA (LAUT SERAM)
PULAU BOANO
Piru
CERAM (SERAM)
Amahai
Waeputih
PULAU KELANG
Namlea
Selat Manipa
BURU
PULAU MANIPA
Wamsasi
Ambon
Namrole
PULAU AMBON
PULAU AMBELAU
MOLUCCAS (MALUKU)
BANDA SEA (LAUT BANDA)
KEPULAUAN PENYU
KEPULAUAN LUCIPARA
115°
120°
125°
5°
0°
10°
© Rand McNally A-101777-1

PHILIPPINES
SOUTH CHINA SEA
PHILIPPINE SEA
SULU SEA
CELEBES SEA
SIBUYAN SEA
VISAYAN SEA
BOHOL SEA
Samar Sea
Camotes Sea
Moro Gulf
Davao Gulf
Illana Bay
Iligan Bay
Macajalar Bay
Ragay Gulf
Lagonoy Gulf
Tayabas Bay
Lamon Bay
Manila Bay
Lingayen Gulf
Baler Bay
Dingalan Bay
Panay Gulf
Asid Gulf
Honda Bay
Sibuguey Bay
Lianga Bay
Mindoro Strait
Tablas Strait
Polillo Strait
Linapacan Strait
Jintotolo Channel
San Bernardino Strait
Surigao Strait
Tañon Strait
Cebu Strait
Guimaras Strait
Canigao Channel
Basilan Strait
Balabac Strait
North Balabac Strait
Palawan Passage
Cuyo West Pass
Cuyo East Pass
Sibutu Passage
Babuyan Channel
Lake Taal
Lake Sultan Alonto
LUZON
MINDORO
PANAY
NEGROS
CEBU
BOHOL
LEYTE
SAMAR
MASBATE
MARINDUQUE
PALAWAN
MINDANAO
BORNEO
MALAYSIA
CORD. CENTRAL
SIERRA MADRE
BONDOC PENINSULA
ZAMBOANGA PENINSULA
SULU ARCHIPELAGO
BATAAN PENINSULA
CUYO ISLANDS
CAGAYAN ISLANDS
CALAMIAN GROUP
POLILLO ISLANDS
LUBANG ISLANDS
CATANDUANES ISLAND
SIBUYAN ISLAND
TABLAS ISLAND
JOLO GROUP
PANGUTARAN GROUP
TAPUL GROUP
SAMALES GROUP
TAWITAWI ISLAND
BASILAN ISLAND
DINAGAT ISLAND
SIARGAO ISLAND
CAMIGUIN ISLAND
SIQUIJOR ISLAND
GUIMARAS ISLAND
SCARBOROUGH REEF
MANILA
Quezon City
Pasig
Cavite
Meycouayan
Malolos
San Fernando
Angeles
Olongapo
Mariveles
Baguio
Dagupan
Lingayen
San Fernando
Vigan
Laoag
Aparri
Tuguegarao City
Cabanatuan
Tarlac
Batangas
Lipa
Lucena
Naga
Legaspi
Sorsogon
Masbate
Catbalogan
Tacloban
Roxas
Iloilo
Bacolod
Dumaguete
Cebu
Mandaue
Lapu-Lapu
Tagbilaran
Puerto Princesa
Surigao
Butuan
Cagayan de Oro
Iligan
Ozamis
Dipolog
Pagadian
Marawi
Malaybalay
Cotabato
Davao
Digos
General Santos
Zamboanga
Isabela
Jolo
Sandakan
Lahad Datu
Mt. Pulog 9626 ft. 2934 m
Mt. Pinatubo 5840 ft. 1780 m
Mt. Sicapoo 7329 ft. 2234 m
Mt. Palanan 3976 ft. 1212 m
Mt. Halcon 8481 ft. 2585 m
Mt. Baco 8159 ft. 2487 m
Mayon Volcano 8077 ft. 2462 m
Victoria Peaks 5607 ft. 1709 m
Mt. Mantalingajan 6841 ft. 2085 m
Mt. Kaatoan 9501 ft. 2896 m
Mt. Apo 9692 ft. 2954 m
Mt. Busa 6834 ft. 2083 m
BATAN ISLANDS
BABUYAN ISLANDS
Luzon Strait
Bashi Channel
Balintang Channel
Basco
Calayan
© Rand McNally A-101808-1
© Rand McNally A-101807-1
0 30 60 90 120 150 180 Miles
0 30 60 90 120 150 180 210 240 270 300 Kilometers

Lambert Conformal Conic Projection
Scale 1:6,000,000
One inch to 96 miles
One cm to 60 km

a

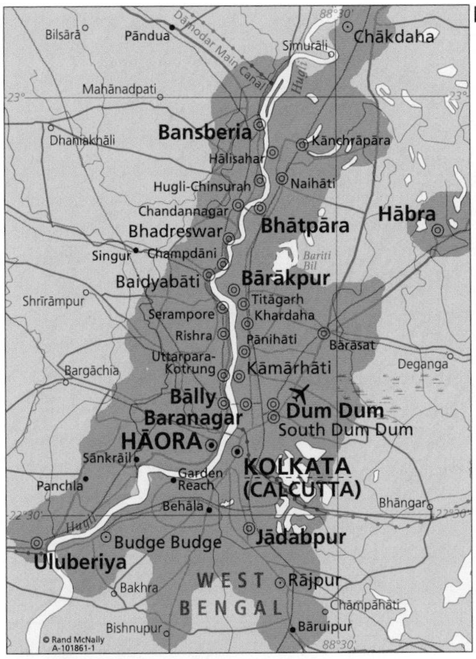

Bilsārā
Pāndua
Damodar Main Canal
Chākdaha
Simurāli
Hugli
Mahānadpati
Bansberia
Kānchrāpāra
Dhaniakhāli
Hālisahar
Naihāti
Hugli-Chinsurah
Chandannagar
Hābra
Bhadreswar
Bhātpāra
Singur
Champdāni
Bariti Bil
Baidyabāti
Bārākpur
Shrīrāmpur
Titāgarh
Serampore
Khardaha
Rishra
Pānihāti
Bārāsat
Uttarpara-Kotrung
Bargāchia
Kāmārhāti
Deganga
Bally
Baranagar
Dum Dum
South Dum Dum
HĀORA
Sānkrāil
KOLKATA (CALCUTTA)
Panchla
Garden Reach
Behāla
Bhāngar
Budge Budge
Jādabpur
Uluberiya
WEST BENGAL
Bakhra
Rājpur
Chāmpāhāti
Bishnupur
Bāruipur
© Rand McNally A-101861-1

b

Kepong
Batu Caves
Sungai Buluh
Setapak
Kuala Ampang
KUALA LUMPUR
Meru
Ampang
Shah Alam
KUALA LUMPUR
Petaling Jaya
Hulu Langat
Klang
Puchong
Serdang
Kajang
Putrajaya
Semenyih
Teluk Panglima Garang
Kampung Pulau Ibul
Beroga
Bangi
Jenjarum
Kampung Dengkil
Telok Datok
Chondoi
Langat
Mantin
Banting
Salak
NEGERI SEMBILAN
Morib
SELANGOR
Tiroi
Kampung Batu Laut
Sepang
Seremban
Kampung Janginl
Sungai Pelik
Kampong Mambau
Tanah Merah
Strait of Malacca
Kampung Cuah
Siliau
Port Dickson
© Rand McNally A-101836-1

c

Kaesŏng
P'anmunjŏm-ni
Paeya
Changdan
KAESŎNG-SI
NORTH KOREA
SOUTH KOREA
Tongduch'ŏn
Sinŭp
Munsan
P'och'ŏn
Ch'ŏnggot
KYŎNGGI-DO
Songu-ri
Yangju
Ŭijŏngbu
Han-gang
Koyang
Mikŭm
Kimp'o
SEOUL (SŎUL)
Kuri
INCH'ŎN-GWANGYŎKSI
Hanam
YONGJONG-DO
Bucheon
SŎUL-T'ŬKPYŎLSI
INCH'ŎN
Kwangmyŏng
Shihŭng
Kwach'ŏn
Sŏngnam
Kanghwa-man
Anyang
Kunp'o
Ŭiwang
Ansan
TAEBU-DO
Suwŏn
Yongin
© Rand McNally A-101835-1

d

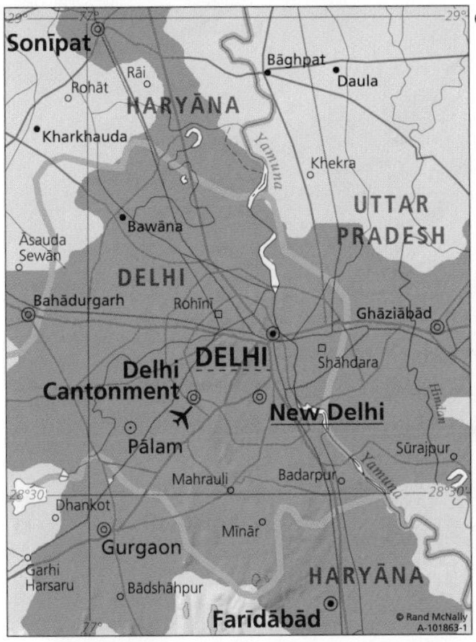

Sonīpat
Bāghpat
Daula
Rohāt
Rāi
HARYĀNA
Kharkhauda
Yamuna
Khekra
UTTAR PRADESH
Bawāna
Āsauda Sewan
DELHI
Bahādurgarh
Rohini
Ghāziābād
Delhi Cantonment
DELHI
Shāhdara
New Delhi
Pālam
Sūrajpur
Mahrauli
Badarpur
Dhankot
Gurgaon
Mīnār
Garhi Harsaru
Bādshāhpur
HARYĀNA
Farīdābād
© Rand McNally A-101863-1

e

FUKUEI CHIAO
Shihmen
Kangtzup'ing
Sanchih
EAST CHINA SEA
Chinshan
Tanshui
Wanli
Ch'ihsing Shan 3547 ft. 1081 m
Paotouts'o
Chilung
Link'ou
Tayüan
Sangchungshih
Hsichih
Hsinchuang
T'AIPEI
Taoyüan
Huoshaoliao
T'aipeihsien
Shihting
Chungli
Hsintien
Pingchen
Sanhsia
P'inglin
Tahsi
Wulai
Chiamu Shan 5190 ft. 1582 m
Lungt'an
Chiaopah
Chiaohsi
Kuanhsi
Ilan
T'aman Shan 5190 ft. 1582 m
Chuangwei
Lotung
Kunghsi
Sanhsing
© Rand McNally A-101830-1

f

San Fernando
San Rafael
BULACAN
Guagua
Lubao
Baliuag
Angat
Norzagaray
Macabebe
PAMPANGA
Calumpit
Malolos
Guiguinto
Orani
Samal
Balagtas
Bocaue
Meycauayan
Novaliches Watershed
San Mateo
Balanga
Malabon
Quezon City
Caloocan
Orion
Manila Bay
MANILA
Makati
Antipolo
Limay
Pasig
Taytay
BATAAN
RIZAL
Angono
Cabcaben
Cavite
Bacoor Bay
Parañaque
Binangonan
Noveleta
Bacoor
CORREGIDOR ISLAND
Rosario
Imus
South Channel
Tanza
Laguna de Bay
CAVITE
Muntinglupa
Naic
Carmona
LAGUNA
Ternate
Dasmariñas
© Rand McNally A-101834-1

g

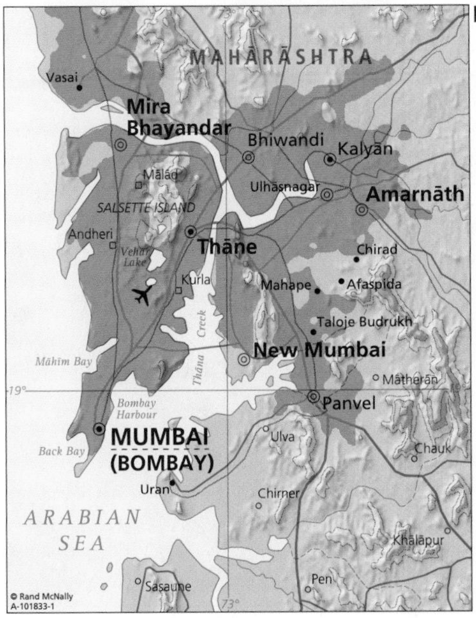

MAHĀRĀSHTRA
Vasai
Mira Bhayandar
Bhiwandi
Kalyān
Mālād
Ulhāsnagar
Amarnāth
SALSETTE ISLAND
Andheri
Vihār Lake
Thāne
Chirad
Kurla
Mahape
Afaspida
Taloje Budrukh
Thāna Creek
New Mumbai
Māhim Bay
Mātherān
Bombay Harbour
Panvel
Back Bay
MUMBAI (BOMBAY)
Ulva
Chauk
Uran
Chirner
ARABIAN SEA
Khālāpur
Pen
Sasaune
© Rand McNally A-101833-1

h

Bassām
Sālih Hasan
Arab Yahūdah
Muhammad Bāqir
Hayy ath Thawrah
Khalaf Laftah
BAGHDĀD
Al-Kāzimīyah
AL-ANBĀR
BAGHDĀD
Hayy ad Dūrah
Abū Ghurayb
Midhat
Tigris (Dijlah)
Khamīs ash Shāhin
'Uwayrij
Majid Sha'lān
DIYĀLĀ
Al Hājj Ja'far
Sālih Muhammad
Abū Muhammad
Qal'at Hamīd al Habash
BĀBIL
Zukaytūn
Al-Madā'in
© Rand McNally A-101864-1

j

BALUCHISTĀN
Hab Chauki
North Karāchi
Gadap
Malir Cantonment
North Nazimabad
Orangi
Malir Cantonment
Pinjrapur
Baldia
KARĀCHI
SIND
Mauripur
Lyari
Jamshed
Kiamari
Landhi
Korangi
Bin Qasim
Clifton
ARABIAN SEA
© Rand McNally A-101865-1

k

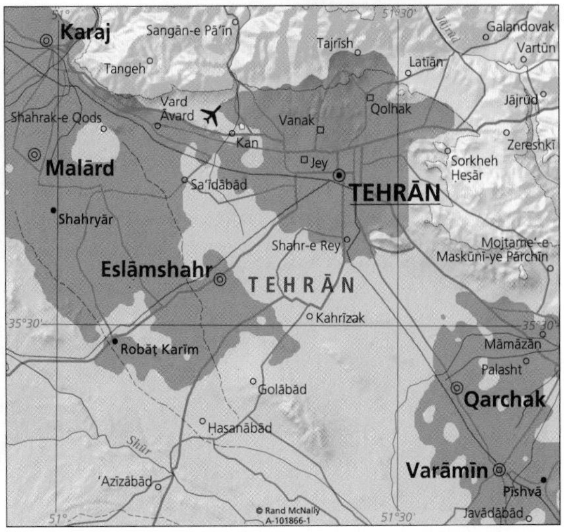

Karaj
Sangān-e Pā'īn
Galandovak
Vartūn
Tajrīsh
Tangeh
Latiān
Vard Āvard
Qolhak
Jājrūd
Shahrak-e Qods
Vanak
Kan
Zereshk
Malārd
Jey
Sorkheh Heşār
Sa'īdābād
TEHRĀN
Shahryār
Shahr-e Rey
Mojtame'-e Maskūnī-ye Pārchīn
Eslāmshahr
TEHRĀN
Kahrīzak
Robāţ Karīm
Māmāzān
Palasht
Qarchak
Golābād
Hasanābād
Shūr
Varāmīn
'Azīzābād
Pīshvā
Javādābād
© Rand McNally A-101866-1

0 5 10 15 20 25 30 Miles

0 5 10 15 20 25 30 35 40 45 50 Kilometers

Lambert Conformal Conic Projection
Scale 1:1,000,000
One inch to 16 miles
One cm to 10 km

a

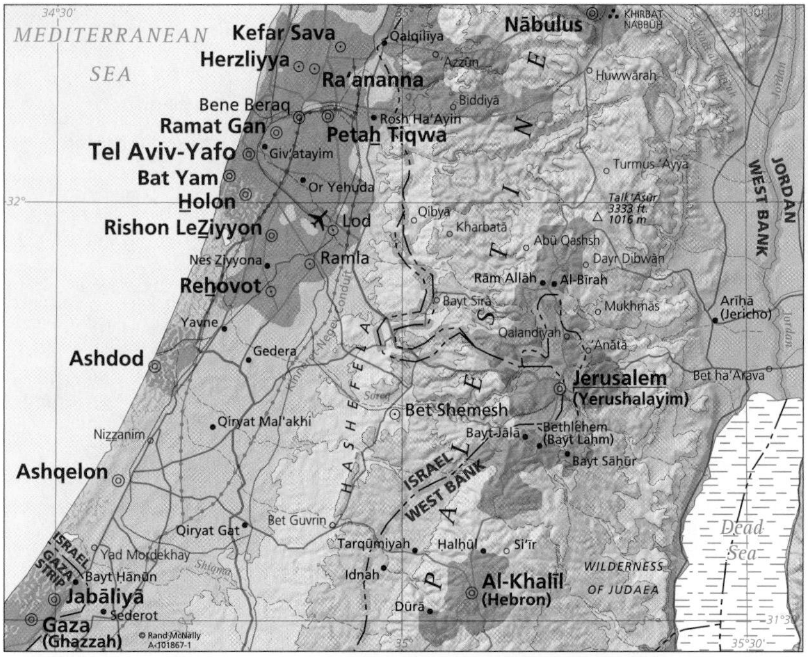
MEDITERRANEAN
SEA
Kefar Sava
Herzliyya
Ra'ananna
Qalqiliya
Nābulus
KHIRBAT NABBUR
'Azzūn
Huwwārah
Biddiyā
Bene Beraq
Rosh Ha'Ayin
Ramat Gan
Petah Tiqwa
Tel Aviv-Yafo
Giv'atayim
Turmus 'Ayya
Bat Yam
Or Yehuda
Holon
Tall 'Asur 3333 ft. 1016 m
Rishon LeZiyyon
Lod
Qibyā
Kharbatā
Abū Qashsh
Nes Ziyyona
Ramla
Dayr Dibwān
Rehovot
Rām Allāh
Al-Bīrah
Bayt Sīrā
Mukhmās
Yavne
Qalandiyah
'Anāta
Arīhā (Jericho)
Ashdod
Gedera
Jerusalem (Yerushalayim)
Bet ha'Arava
Bet Shemesh
Qiryat Mal'akhi
Nizzanim
Bethlehem (Bayt Lahm)
Bayt Jālā
Bayt Sāhūr
Ashqelon
ISRAEL
WEST BANK
HASHEFELA
PALESTINE
Bet Guvrin
Qiryat Gat
Dead Sea
ISRAEL
GAZA STRIP
Yad Mordekhay
Tarqūmiyah
Halhūl
Si'īr
Bayt Hānūn
Idnah
Al-Khalīl (Hebron)
WILDERNESS OF JUDAEA
Jabāliyā
Sederot
Dūrā
Gaza (Ghazzah)
JORDAN
WEST BANK
32°
31°30'
34°30'
35°
35°30'
© Rand McNally A-101867-1

b

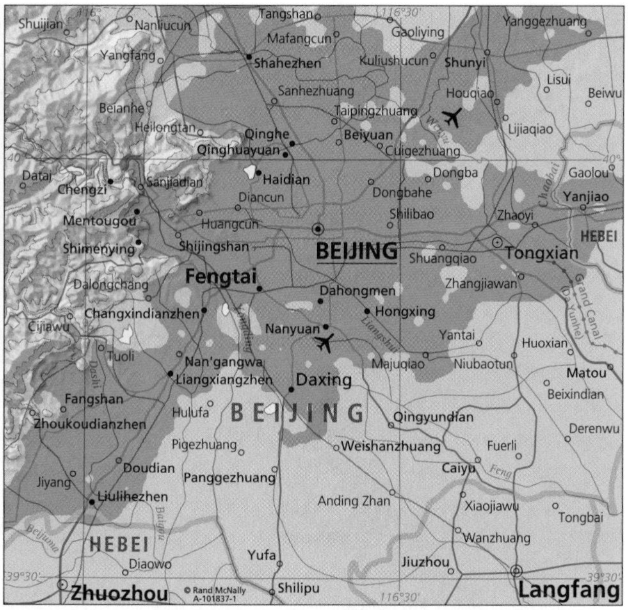
Shuijian
Nanliucun
Tangshan
Gaoliying
Yanggezhuang
Mafangcun
Yangfang
Shahezhen
Kuliushucun
Shunyi
Sanhezhuang
Lisui
Beiwu
Houqiao
Beianhe
Taipingzhuang
Heilongtan
Qinghe
Beiyuan
Lijiaqiao
Qinghuayuan
Cuigezhuang
Datai
Haidian
Dongba
Gaolou
Chengzi
Sanjiadian
Diancun
Dongbahe
Yanjiao
Mentougou
Huangcun
Shilibao
Zhaoyi
HEBEI
Shimenying
Shijingshan
BEIJING
Shuangqiao
Tongxian
Dalongchang
Fengtai
Zhangjiawan
Changxindianzhen
Dahongmen
Hongxing
Cijiawu
Nanyuan
Yantai
Huoxian
Tuoli
Majuqiao
Niubaotun
Matou
Nan'gangwa
Liangxiangzhen
Daxing
Beixindian
Fangshan
Zhoukoudianzhen
Hulufa
BEIJING
Qingyundian
Derenwu
Pigezhuang
Weishanzhuang
Fuerli
Doudian
Caiyu
Jiyang
Panggezhuang
Liulihezhen
Anding Zhan
Xiaojiawu
Tongbai
Wanzhuang
HEBEI
Diaowo
Yufa
Jiuzhou
Zhuozhou
Shilipu
Langfang
40°
39°30'
116°
116°30'
© Rand McNally A-101837-1

c

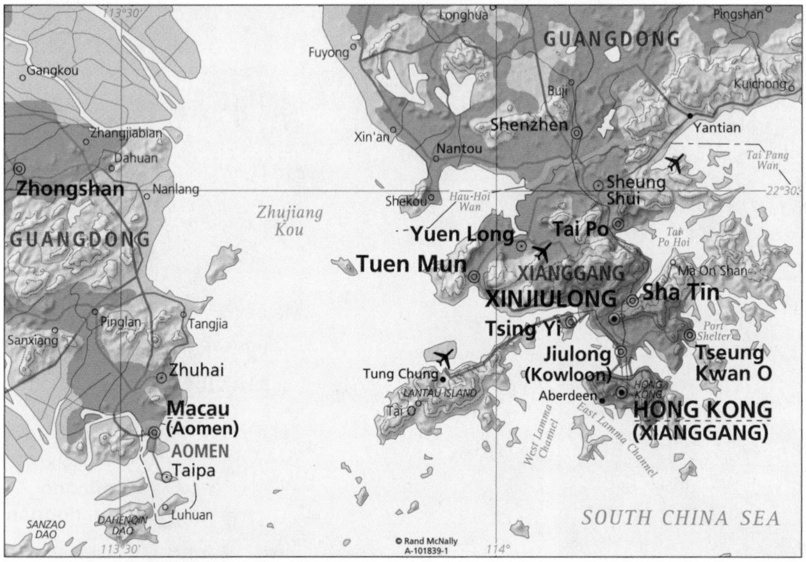
Longhua
Pingshan
GUANGDONG
Fuyong
Gangkou
Buji
Kuichong
Shenzhen
Yantian
Xin'an
Zhangjiabian
Nantou
Dahuan
Tai Pang Wan
Zhongshan
Nanlang
Sheung Shui
Shekou
Hau Hoi Wan
Zhujiang Kou
GUANGDONG
Yuen Long
Tai Po
Tai Po Hoi
Tuen Mun
XIANGGANG
Ma On Shan
XINJIULONG
Sha Tin
Tsing Yi
Port Shelter
Pinglan
Tangjia
Sanxiang
Jiulong (Kowloon)
Tseung Kwan O
Zhuhai
Tung Chung
LANTAU ISLAND
Aberdeen
HONG KONG
Tai O
West Lamma Channel
East Lamma Channel
HONG KONG (XIANGGANG)
Macau (Aomen)
AOMEN
Taipa
SANZAO DAO
DAHENGQIN DAO
Luhuan
SOUTH CHINA SEA
22°30'
113°30'
114°
© Rand McNally A-101839-1

d

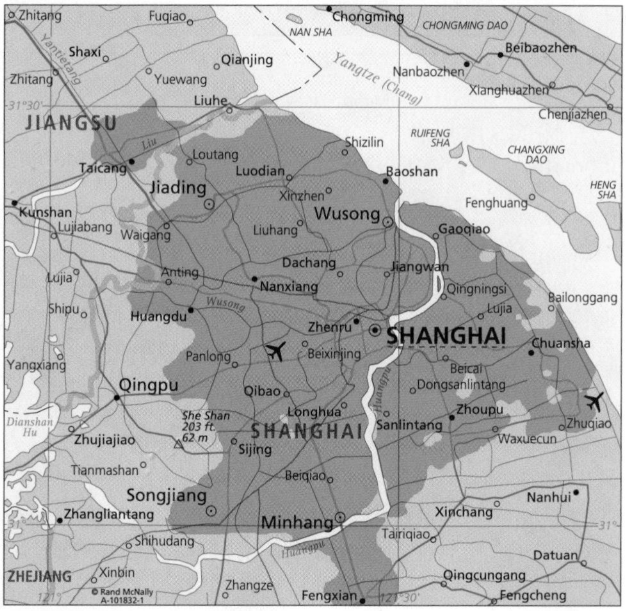
Zhitang
Fuqiao
Chongming
NAN SHA
CHONGMING DAO
Beibaozhen
Shaxi
Qianjing
Zhitang
Yuewang
Yangtze (Chang)
Nanbaozhen
Xianghuazhen
Liuhe
Chenjiazhen
JIANGSU
RUIFENG SHA
CHANGXING DAO
Shizilin
Loutang
Taicang
Luodian
Baoshan
HENG SHA
Jiading
Xinzhen
Fenghuang
Kunshan
Lujiabang
Wusong
Waigang
Liuhang
Gaoqiao
Dachang
Jiangwan
Lujia
Anting
Nanxiang
Qingningsi
Bailonggang
Shipu
Lujia
Huangdu
Zhenru
SHANGHAI
Chuansha
Yangxiang
Panlong
Beixinjing
Beicai
Dongsanlintang
Qingpu
Qibao
Longhua
Zhoupu
Zhuqiao
She Shan 203 ft. 62 m
Sanlintang
Dianshan Hu
Zhujiajiao
SHANGHAI
Waxuecun
Sijing
Tianmashan
Beiqiao
Songjiang
Nanhui
Zhangliantang
Minhang
Xinchang
Shihudang
Tairiqiao
Datuan
ZHEJIANG
Xinbin
Qingcungang
Zhangze
Fengxian
Fengcheng
31°30'
31°
121°
121°30'
© Rand McNally A-101832-1

e

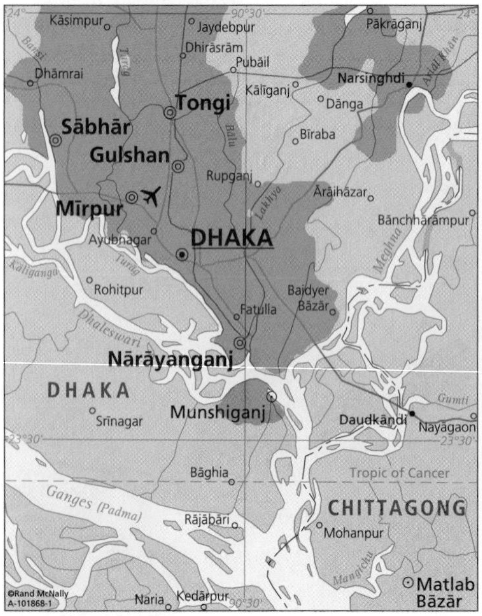
Kāsimpur
Jaydebpur
Pākrāganj
Dhīrāsrām
Pubāil
Dhāmrai
Narsinghdi
Kāliganj
Dānga
Tongi
Sābhār
Bīraba
Gulshan
Rupganj
Mīrpur
Ārāihāzar
Bānchhārāmpur
Āyubnagar
DHAKA
Rohitpur
Baidyer Bāzār
Fatulla
Nārāyanganj
DHAKA
Gumti
Srīnagar
Munshiganj
Daudkāndi
Nayāgaon
Tropic of Cancer
Bāghia
Ganges (Padma)
Rājābāri
CHITTAGONG
Mohanpur
Matlab Bāzār
Naria
Kedārpur
24°
23°30'
90°30'
©Rand McNally A-101868-1

f

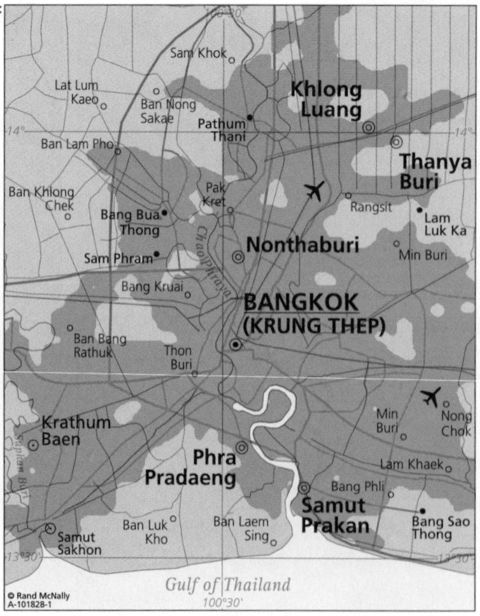
Sam Khok
Lat Lum Kaeo
Ban Nong Sakae
Khlong Luang
Pathum Thani
Ban Lam Pho
Thanya Buri
Pak Kret
Ban Khlong Chek
Bang Bua Thong
Rangsit
Lam Luk Ka
Nonthaburi
Sam Phram
Min Buri
Bang Kruai
BANGKOK (KRUNG THEP)
Ban Bang Rathuk
Thon Buri
Min Buri
Nong Chok
Krathum Baen
Phra Pradaeng
Lam Khaek
Bang Phli
Samut Prakan
Bang Sao Thong
Ban Luk Kho
Ban Laem Sing
Samut Sakhon
Gulf of Thailand
14°
13°30'
100°30'
© Rand McNally A-101828-1

g

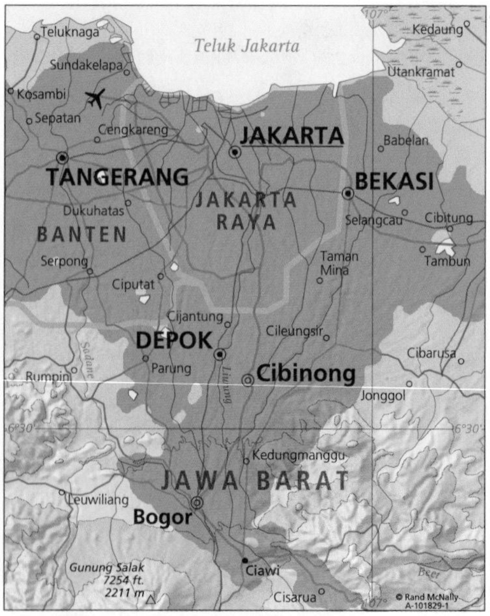
Teluknaga
Kedaung
Sundakelapa
Teluk Jakarta
Utankramat
Kosambi
Sepatan
Cengkareng
JAKARTA
Babelan
TANGERANG
BEKASI
JAKARTA RAYA
Dukuhatas
Selangcau
Cibitung
BANTEN
Serpong
Taman Mina
Tambun
Ciputat
Cijantung
Cileungsir
DEPOK
Cibarusa
Parung
Rumpin
Cibinong
Jonggol
Kedungmanggu
JAWA BARAT
Leuwiliang
Bogor
Gunung Salak 7254 ft. 2211 m
Ciawi
Cisarua
6°30'
107°
© Rand McNally A-101829-1

0 5 10 15 20 25 30 Miles
0 5 10 15 20 25 30 35 40 45 50 Kilometers

Lambert Conformal Conic Projection
Scale 1:1,000,000
One inch to 16 miles
One cm to 10 km

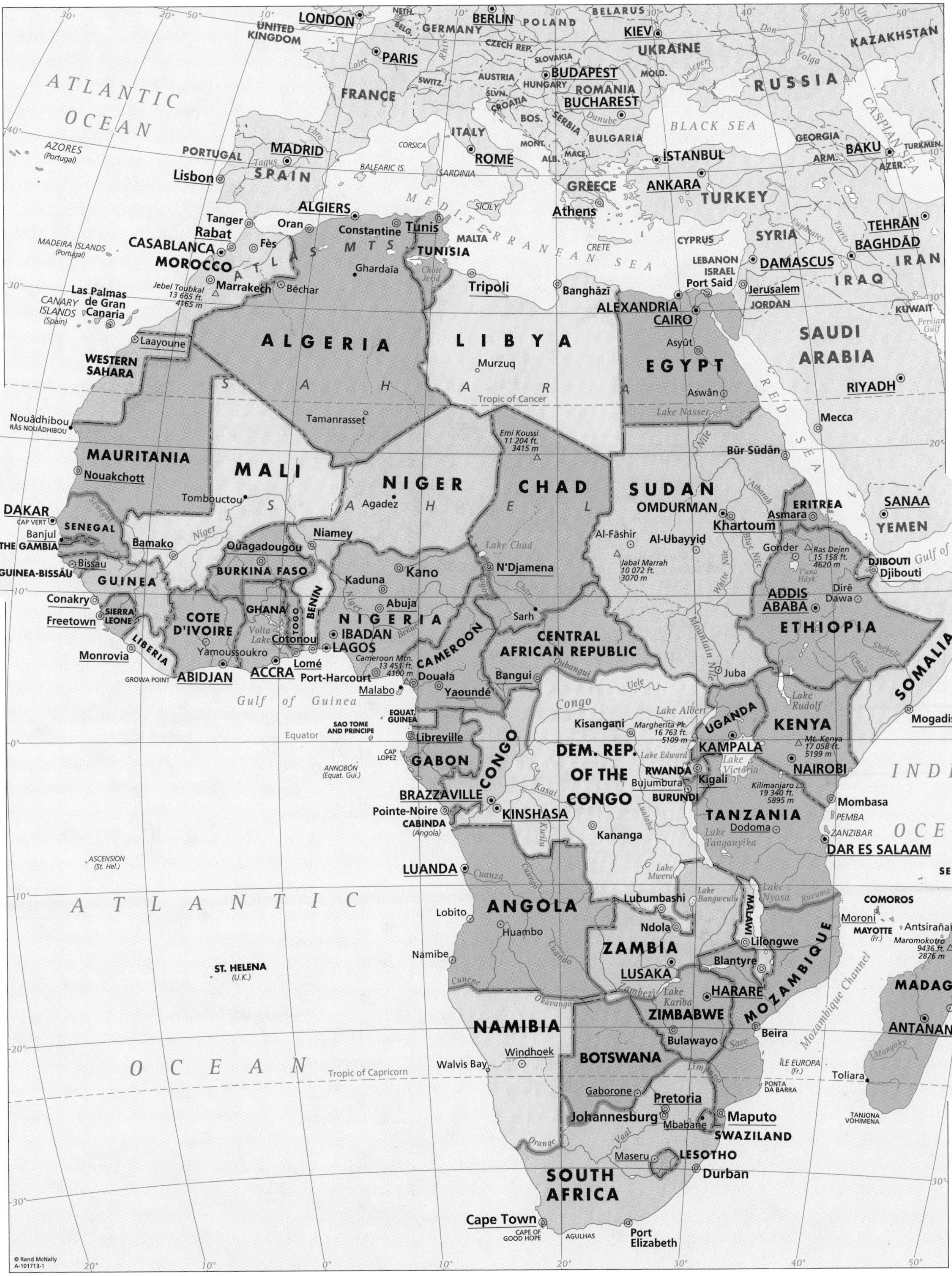
ATLANTIC OCEAN
AZORES (Portugal)
PORTUGAL
Lisbon
MADRID
SPAIN
LONDON
UNITED KINGDOM
NETH.
BELG.
PARIS
FRANCE
SWITZ.
GERMANY
BERLIN
POLAND
CZECH REP.
SLOVAKIA
AUSTRIA
HUNGARY
BUDAPEST
SLVN.
CROATIA
BOS.
SERBIA
ROMANIA
BUCHAREST
MOLD.
BELARUS
KIEV
UKRAINE
RUSSIA
KAZAKHSTAN
BLACK SEA
CASPIAN SEA
GEORGIA
ARM.
BAKU
AZER.
TURKMEN.
BULGARIA
MONT.
ALB.
MACE.
ITALY
ROME
CORSICA
SARDINIA
BALEARIC IS.
SICILY
MALTA
GREECE
Athens
ISTANBUL
ANKARA
TURKEY
CRETE
CYPRUS
SYRIA
LEBANON
ISRAEL
DAMASCUS
Jerusalem
JORDAN
IRAQ
BAGHDĀD
TEHRĀN
IRAN
KUWAIT
SAUDI ARABIA
RIYADH
Mecca
YEMEN
SANAA
MEDITERRANEAN SEA
Tanger
Rabat
CASABLANCA
Fès
MOROCCO
Marrakech
Béchar
Oran
ALGIERS
Constantine
Tunis
TUNISIA
Ghardaïa
Tripoli
Banghāzī
ALEXANDRIA
CAIRO
Port Said
MADEIRA ISLANDS (Portugal)
CANARY ISLANDS (Spain)
Las Palmas de Gran Canaria
Jebel Toubkal 13 665 ft. 4165 m
ATLAS MTS.
Laayoune
WESTERN SAHARA
ALGERIA
LIBYA
Murzuq
EGYPT
Asyût
Aswân
Lake Nasser
Tropic of Cancer
SAHARA
Tamanrasset
Nouâdhibou
RÂS NOUÂDHIBOU
MAURITANIA
Nouakchott
MALI
Tombouctou
NIGER
Agadez
CHAD
Emi Koussi 11 204 ft. 3415 m
SUDAN
OMDURMAN
Khartoum
Bûr Sûdân
RED SEA
ERITREA
Asmara
CAPE VERDE
Praia
DAKAR
CAP VERT
SENEGAL
Banjul
THE GAMBIA
Bissau
GUINEA-BISSAU
GUINEA
Conakry
Freetown
SIERRA LEONE
LIBERIA
Monrovia
Bamako
Niamey
Ouagadougou
BURKINA FASO
COTE D'IVOIRE
Yamoussoukro
ABIDJAN
GROWA POINT
GHANA
Volta Lake
ACCRA
TOGO
Lomé
BENIN
Cotonou
NIGERIA
IBADAN
LAGOS
Abuja
Kaduna
Kano
Lake Chad
N'Djamena
Al-Fāshir
Al-Ubayyid
Jabal Marrah 10 072 ft. 3070 m
Gonder
Ras Dejen 15 158 ft. 4620 m
DJIBOUTI
Djibouti
Gulf of Aden
GEES GWARDAFUY
Dire Dawa
ADDIS ABABA
ETHIOPIA
SOMALIA
Mogadishu
Port-Harcourt
Cameroon Mtn. 13 451 ft. 4100 m
CAMEROON
Douala
Yaoundé
Malabo
Gulf of Guinea
Sarh
CENTRAL AFRICAN REPUBLIC
Bangui
Juba
EQUAT. GUINEA
SAO TOME AND PRINCIPE
Equator
Libreville
GABON
CAP LOPEZ
ANNOBÓN (Equat. Gui.)
CONGO
BRAZZAVILLE
KINSHASA
Pointe-Noire
CABINDA (Angola)
Kisangani
Margherita Pk. 16 763 ft. 5109 m
Lake Albert
UGANDA
KAMPALA
Lake Edward
DEM. REP. OF THE CONGO
RWANDA
Kigali
Bujumbura
BURUNDI
Lake Rudolf
KENYA
Mt. Kenya 17 058 ft. 5199 m
NAIROBI
Lake Victoria
Kilimanjaro 19 340 ft. 5895 m
Mombasa
PEMBA
ZANZIBAR
DAR ES SALAAM
TANZANIA
Dodoma
Lake Tanganyika
Kananga
INDIAN OCEAN
ASCENSION (St. Hel.)
LUANDA
Lobito
ANGOLA
Huambo
Namibe
Lubumbashi
Lake Mweru
Lake Bangweulu
Lake Nyasa
MALAWI
Lilongwe
Blantyre
Ndola
ZAMBIA
LUSAKA
Lake Kariba
HARARE
ZIMBABWE
Bulawayo
MOZAMBIQUE
Mozambique Channel
Beira
SEYCHELLES
COMOROS
Moroni
MAYOTTE (Fr.)
TANJONA BOBAOMBY
Antsirañana
Maromokotro 9436 ft. 2876 m
MADAGASCAR
Toamasina
ANTANANARIVO
Toliara
TANJONA VOHIMENA
ÎLE EUROPA (Fr.)
PONTA DA BARRA
ST. HELENA (U.K.)
NAMIBIA
Windhoek
Walvis Bay
Tropic of Capricorn
BOTSWANA
Gaborone
Pretoria
Johannesburg
Mbabane
Maputo
SWAZILAND
Maseru
LESOTHO
Durban
SOUTH AFRICA
Cape Town
CAPE OF GOOD HOPE
AGULHAS
Port Elizabeth
© Rand McNally
A-101713-1

0 200 400 600 800 1000 1200 Miles

0 200 400 600 800 1000 1200 1400 1600 1800 2000 Kilometers

Lambert Azimuthal Equal Area Projection
Scale 1:40,000,000
One inch to 640 miles
One cm to 400 km

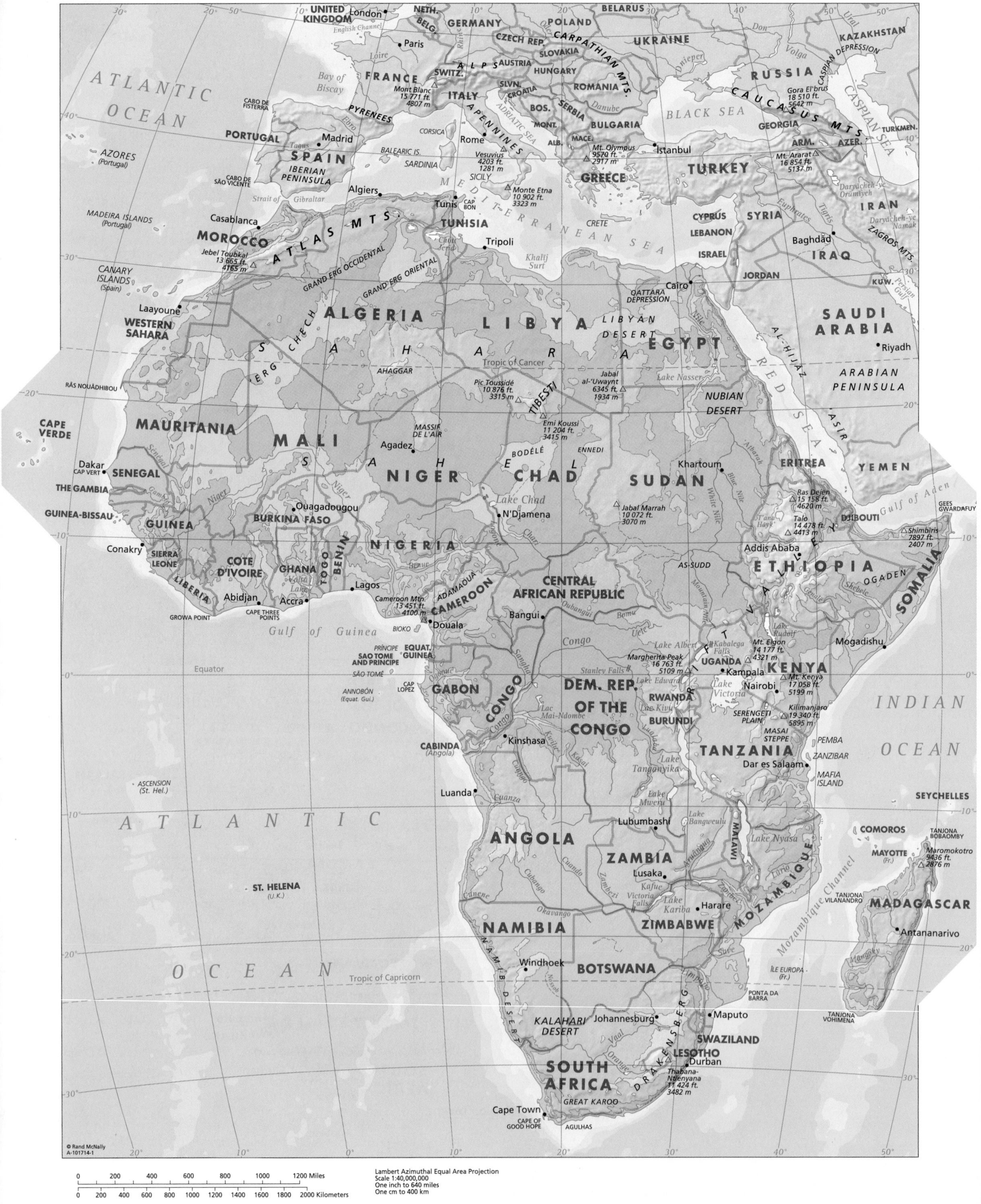
ATLANTIC OCEAN
UNITED KINGDOM
London
English Channel
NETH.
BELG.
GERMANY
POLAND
BELARUS
CZECH REP.
CARPATHIAN MTS.
UKRAINE
KAZAKHSTAN
CASPIAN DEPRESSION
Paris
FRANCE
Loire
Bay of Biscay
SWITZ.
ALPS
AUSTRIA
SLOVAKIA
HUNGARY
ROMANIA
RUSSIA
Mont Blanc 15 771 ft. 4807 m
ITALY
SLVN.
CROATIA
BOS.
SERBIA
Danube
CAUCASUS MTS.
Gora El'brus 18 510 ft. 5642 m
CASPIAN SEA
CABO DE FISTERRA
PYRENEES
APENNINES
BLACK SEA
GEORGIA
TURKMEN.
PORTUGAL
Madrid
CORSICA
Rome
ADRIATIC SEA
MONT.
BULGARIA
ARM.
AZER.
AZORES (Portugal)
SPAIN
IBERIAN PENINSULA
BALEARIC IS.
SARDINIA
Vesuvius 4203 ft. 1281 m
ALB.
MACE.
Mt. Olympus 9570 ft. 2917 m
Istanbul
TURKEY
Mt. Ararat 16 854 ft. 5137 m
CABO DE SÃO VICENTE
SICILY
GREECE
Strait of Gibraltar
Algiers
Monte Etna 10 902 ft. 3323 m
MEDITERRANEAN SEA
Tunis
CAP BON
TUNISIA
CRETE
CYPRUS
SYRIA
LEBANON
IRAN
Daryācheh-ye Orūmīyeh
MADEIRA ISLANDS (Portugal)
Casablanca
MOROCCO
ATLAS MTS.
Tripoli
Khalīj Surt
ISRAEL
Baghdād
IRAQ
ZAGROS MTS.
Jebel Toubkal 13 665 ft. 4165 m
GRAND ERG OCCIDENTAL
GRAND ERG ORIENTAL
CANARY ISLANDS (Spain)
Cairo
JORDAN
KUW.
Persian Gulf
QATTARA DEPRESSION
Laayoune
WESTERN SAHARA
ALGERIA
LIBYA
LIBYAN DESERT
EGYPT
SAUDI ARABIA
Riyadh
SAHARA
Tropic of Cancer
AL-HIJAZ
AHAGGAR
Lake Nasser
RED SEA
ARABIAN PENINSULA
RĀS NOUĀDHIBOU
Pic Toussidé 10 876 ft. 3315 m
TIBESTI
Jabal al-'Uwaynt 6345 ft. 1934 m
NUBIAN DESERT
CAPE VERDE
MAURITANIA
MALI
MASSIF DE L'AÏR
Emi Koussi 11 204 ft. 3415 m
ENNEDI
Agadez
BODÉLÉ
SAHEL
Khartoum
ERITREA
YEMEN
Dakar
CAP VERT
SENEGAL
NIGER
CHAD
SUDAN
THE GAMBIA
Lake Chad
Ras Dejen 15 158 ft. 4620 m
Gulf of Aden
GEES GWARDAFUY
Ouagadougou
N'Djamena
Jabal Marrah 10 072 ft. 3070 m
Talo 14 478 ft. 4413 m
DJIBOUTI
GUINEA-BISSAU
GUINEA
BURKINA FASO
Shimbiris 7897 ft. 2407 m
Conakry
SIERRA LEONE
NIGERIA
Addis Ababa
COTE D'IVOIRE
GHANA
TOGO
BENIN
AS-SUDD
ETHIOPIA
OGADEN
LIBERIA
Lagos
CENTRAL AFRICAN REPUBLIC
SOMALIA
Abidjan
Accra
Cameroon Mtn. 13 451 ft. 4100 m
ADAMAOUA
CAMEROON
Bangui
GROWA POINT
CAPE THREE POINTS
Gulf of Guinea
BIOKO
Douala
RIFT VALLEY
Mogadishu
PRÍNCIPE
EQUAT. GUINEA
SAO TOME AND PRINCIPE
Congo
Lake Albert
Kabalega Falls
Mt. Elgon 14 177 ft. 4321 m
Margherita Peak 16 763 ft. 5109 m
UGANDA
KENYA
SÃO TOMÉ
Equator
Stanley Falls
Kampala
Lake Edward
Mt. Kenya 17 058 ft. 5199 m
ANNOBÓN (Equat. Gui.)
CAP LOPEZ
GABON
CONGO
DEM. REP. OF THE CONGO
RWANDA
Lake Victoria
Nairobi
INDIAN OCEAN
Lac Mai-Ndombe
BURUNDI
Kilimanjaro 19 340 ft. 5895 m
SERENGETI PLAIN
MASAI STEPPE
PEMBA
Kinshasa
CABINDA (Angola)
TANZANIA
ZANZIBAR
Dar es Salaam
Lake Tanganyika
MAFIA ISLAND
ASCENSION (St. Hel.)
Luanda
Cuanza
SEYCHELLES
Lake Mweru
Lubumbashi
Lake Bangweulu
ATLANTIC OCEAN
ANGOLA
MALAWI
Lake Nyasa
COMOROS
TANJONA BOBAOMBY
ZAMBIA
MAYOTTE (Fr.)
Maromokotro 9436 ft. 2876 m
Lusaka
ST. HELENA (U.K.)
Victoria Falls
Lake Kariba
Harare
MOZAMBIQUE
Mozambique Channel
TANJONA VILANANDRO
MADAGASCAR
Okavango
NAMIBIA
ZIMBABWE
Antananarivo
NAMIB DESERT
Windhoek
BOTSWANA
Tropic of Capricorn
ÎLE EUROPA (Fr.)
PONTA DA BARRA
TANJONA VOHIMENA
KALAHARI DESERT
Johannesburg
Maputo
DRAKENSBERG
SWAZILAND
LESOTHO
Durban
SOUTH AFRICA
Thabana-Ntlenyana 11 424 ft. 3482 m
GREAT KAROO
Cape Town
CAPE OF GOOD HOPE
AGULHAS
© Rand McNally A-101714-1

0 200 400 600 800 1000 1200 Miles

0 200 400 600 800 1000 1200 1400 1600 1800 2000 Kilometers

Lambert Azimuthal Equal Area Projection
Scale 1:40,000,000
One inch to 640 miles
One cm to 400 km

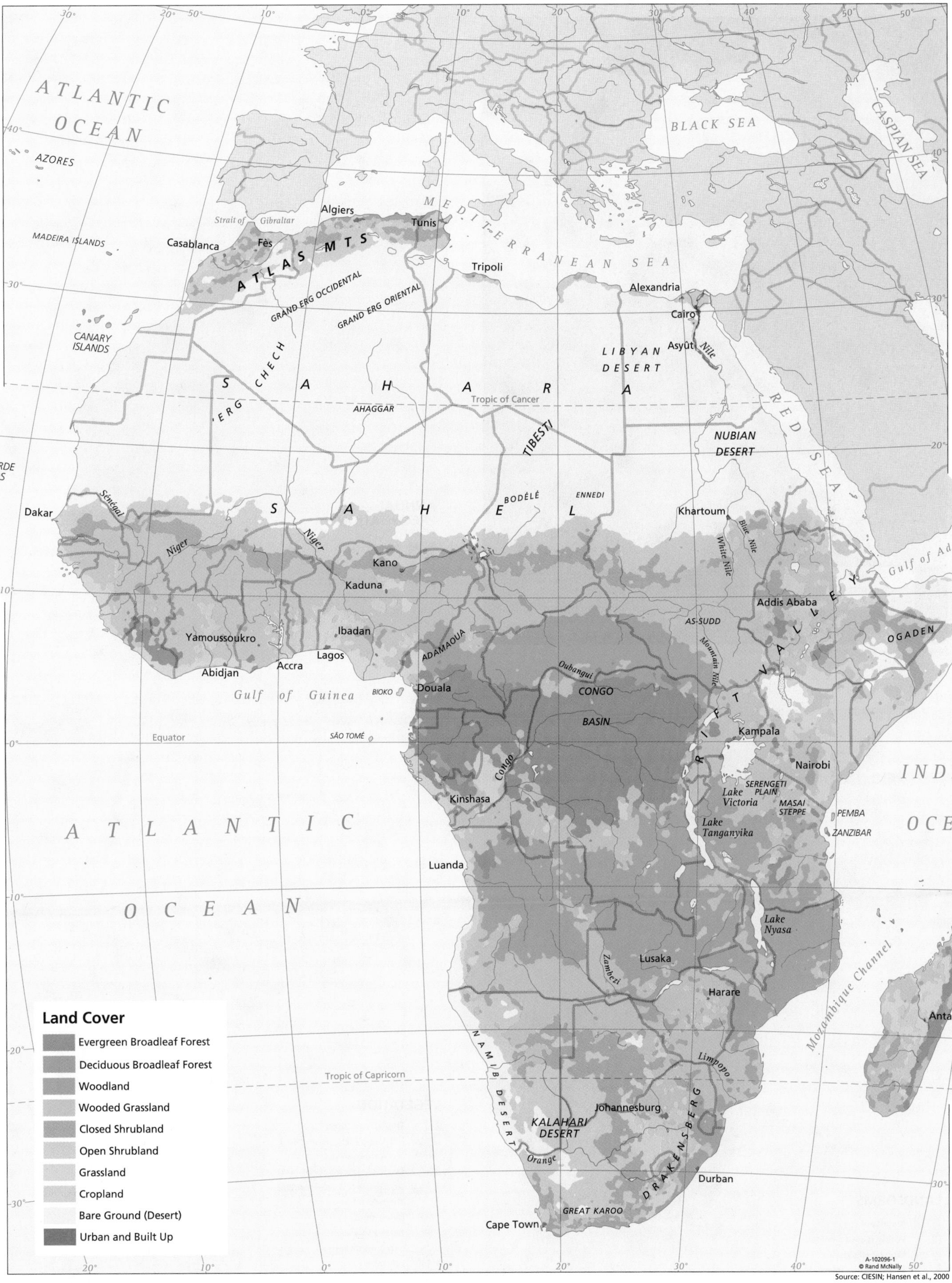
ATLANTIC OCEAN
AZORES
MADEIRA ISLANDS
CANARY ISLANDS
CAPE VERDE ISLANDS
BLACK SEA
CASPIAN SEA
MEDITERRANEAN SEA
Strait of Gibraltar
Algiers
Tunis
Casablanca
Fès
ATLAS MTS
Tripoli
GRAND ERG OCCIDENTAL
GRAND ERG ORIENTAL
ERG CHECH
SAHARA
AHAGGAR
Tropic of Cancer
TIBESTI
LIBYAN DESERT
Alexandria
Cairo
Asyût
Nile
RED SEA
NUBIAN DESERT
BODÉLÉ
ENNEDI
SAHEL
Dakar
Sénégal
Niger
Kano
Kaduna
Khartoum
Blue Nile
White Nile
Gulf of Aden
Addis Ababa
AS-SUDD
Mountain Nile
RIFT VALLEY
OGADEN
Ibadan
Yamoussoukro
Lagos
Accra
Abidjan
ADAMAOUA
Gulf of Guinea
BIOKO
Douala
Oubangui
CONGO BASIN
SÃO TOMÉ
Equator
Kampala
Congo
Nairobi
Kinshasa
SERENGETI PLAIN
Lake Victoria
MASAI STEPPE
INDIAN OCEAN
PEMBA
ZANZIBAR
Lake Tanganyika
ATLANTIC OCEAN
Luanda
Lake Nyasa
Lusaka
Zambezi
Mozambique Channel
Harare
Antananarivo
NAMIB DESERT
Limpopo
Tropic of Capricorn
Johannesburg
KALAHARI DESERT
DRAKENSBERG
Orange
Durban
GREAT KAROO
Cape Town
Land Cover
Evergreen Broadleaf Forest
Deciduous Broadleaf Forest
Woodland
Wooded Grassland
Closed Shrubland
Open Shrubland
Grassland
Cropland
Bare Ground (Desert)
Urban and Built Up
A-102096-1
© Rand McNally

Source: CIESIN; Hansen et al., 2000

0 200 400 600 800 1000 1200 Miles

0 200 400 600 800 1000 1200 1400 1600 1800 2000 Kilometers

Lambert Azimuthal Equal Area Projection
Scale 1:40,000,000
One inch to 640 miles
One cm to 400 km

Moderate Resolution Imaging Spectroradiometer (MODIS) true-color mosaic satellite image

Source: NASA Visible Earth program (http://visibleearth.nasa.gov/)

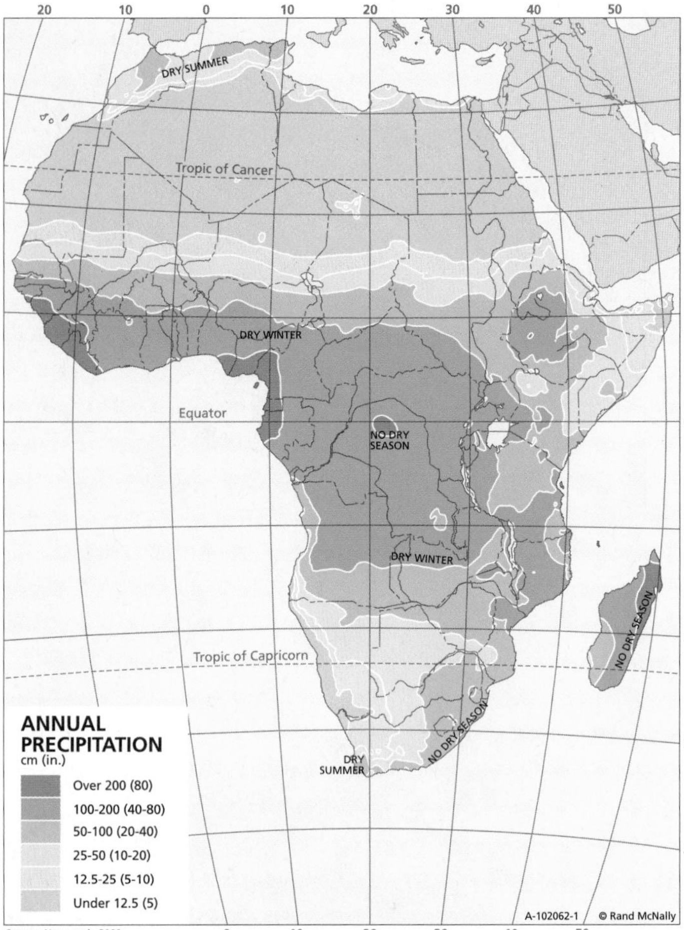

DRY SUMMER
Tropic of Cancer
DRY WINTER
Equator
NO DRY SEASON
DRY WINTER
NO DRY SEASON
Tropic of Capricorn
NO DRY SEASON
DRY SUMMER
ANNUAL PRECIPITATION
cm (in.)
Over 200 (80)
100-200 (40-80)
50-100 (20-40)
25-50 (10-20)
12.5-25 (5-10)
Under 12.5 (5)
A-102062-1
© Rand McNally

Source: New et al., 2000

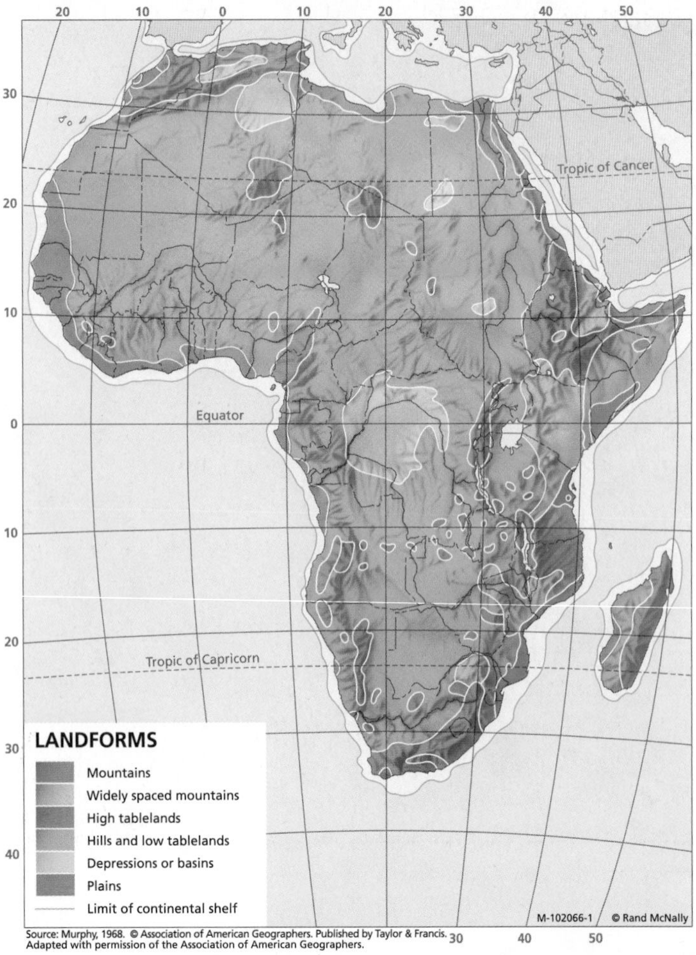

Tropic of Cancer
Equator
Tropic of Capricorn
LANDFORMS
Mountains
Widely spaced mountains
High tablelands
Hills and low tablelands
Depressions or basins
Plains
Limit of continental shelf
M-102066-1
© Rand McNally

Source: Murphy, 1968. © Association of American Geographers. Published by Taylor & Francis. Adapted with permission of the Association of American Geographers.

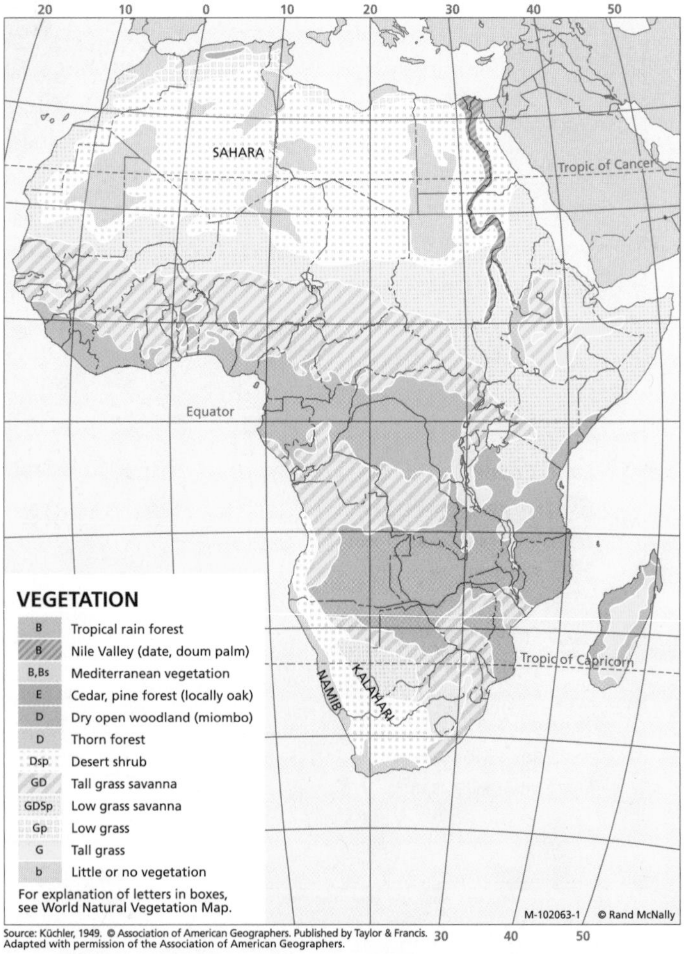

SAHARA
Tropic of Cancer
Equator
NAMIB
KALAHARI
Tropic of Capricorn
VEGETATION
B Tropical rain forest
B Nile Valley (date, doum palm)
B,Bs Mediterranean vegetation
E Cedar, pine forest (locally oak)
D Dry open woodland (miombo)
D Thorn forest
Dsp Desert shrub
GD Tall grass savanna
GDSp Low grass savanna
Gp Low grass
G Tall grass
b Little or no vegetation
For explanation of letters in boxes, see World Natural Vegetation Map.
M-102063-1
© Rand McNally

Source: Küchler, 1949. © Association of American Geographers. Published by Taylor & Francis. Adapted with permission of the Association of American Geographers.

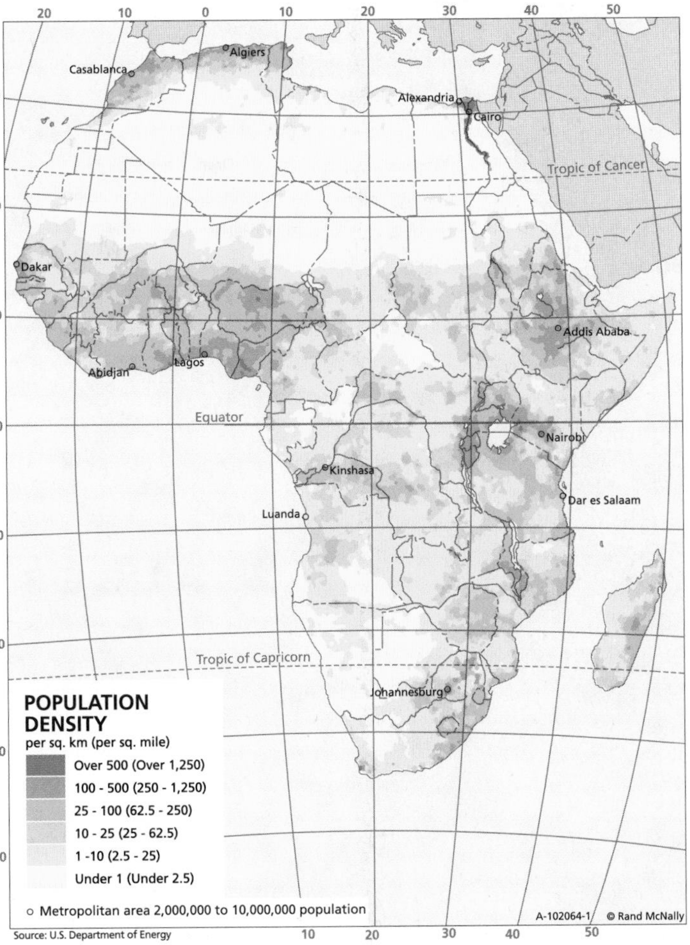
POPULATION DENSITY
per sq. km (per sq. mile)
Over 500 (Over 1,250)
100 - 500 (250 - 1,250)
25 - 100 (62.5 - 250)
10 - 25 (25 - 62.5)
1 -10 (2.5 - 25)
Under 1 (Under 2.5)
o Metropolitan area 2,000,000 to 10,000,000 population
Casablanca
Algiers
Alexandria
Cairo
Dakar
Abidjan
Lagos
Addis Ababa
Nairobi
Kinshasa
Luanda
Dar es Salaam
Johannesburg
Equator
Tropic of Cancer
Tropic of Capricorn
A-102064-1
© Rand McNally

Source: U.S. Department of Energy

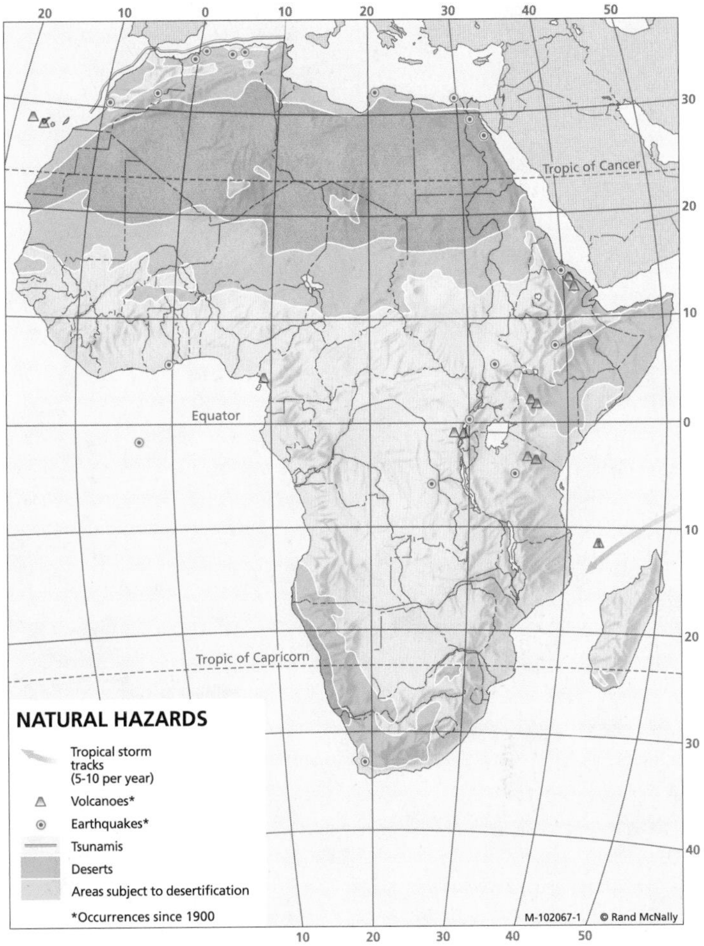
NATURAL HAZARDS
Tropical storm tracks (5-10 per year)
Volcanoes*
Earthquakes*
Tsunamis
Deserts
Areas subject to desertification
*Occurrences since 1900
Equator
Tropic of Cancer
Tropic of Capricorn
M-102067-1
© Rand McNally

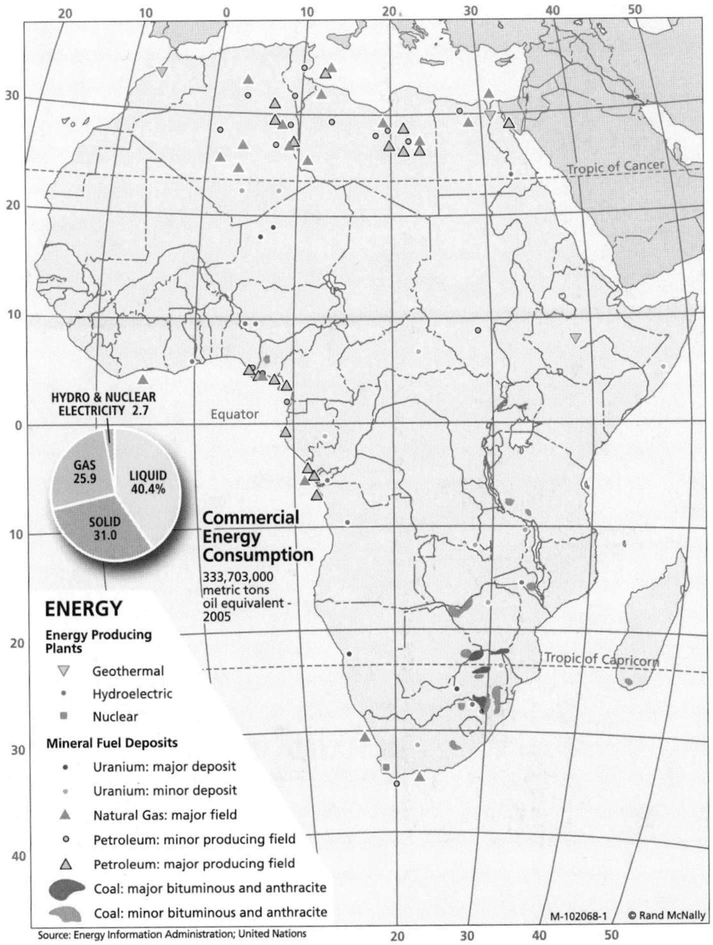
ENERGY
Energy Producing Plants
Geothermal
Hydroelectric
Nuclear
Mineral Fuel Deposits
Uranium: major deposit
Uranium: minor deposit
Natural Gas: major field
Petroleum: minor producing field
Petroleum: major producing field
Coal: major bituminous and anthracite
Coal: minor bituminous and anthracite
HYDRO & NUCLEAR ELECTRICITY 2.7
GAS 25.9
LIQUID 40.4%
SOLID 31.0
Commercial Energy Consumption
333,703,000 metric tons oil equivalent - 2005
Equator
Tropic of Cancer
Tropic of Capricorn
M-102068-1
© Rand McNally

Source: Energy Information Administration; United Nations

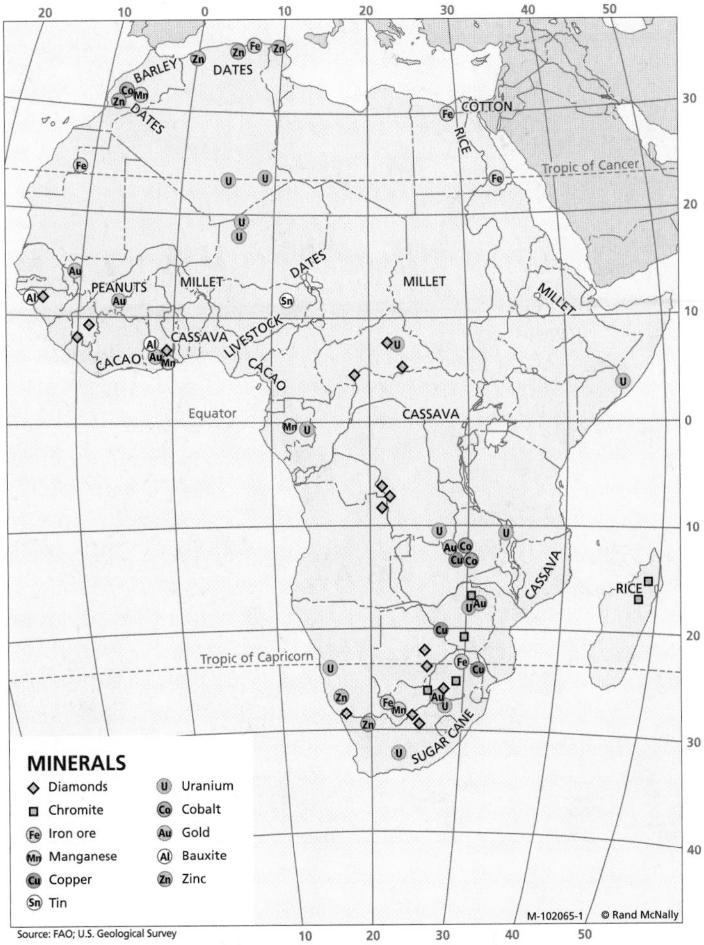
MINERALS
Diamonds
Chromite
Iron ore
Manganese
Copper
Tin
Uranium
Cobalt
Gold
Bauxite
Zinc
BARLEY
DATES
COTTON
RICE
PEANUTS
MILLET
CASSAVA
LIVESTOCK
CACAO
SUGAR CANE
Equator
Tropic of Cancer
Tropic of Capricorn
M-102065-1
© Rand McNally

Source: FAO; U.S. Geological Survey

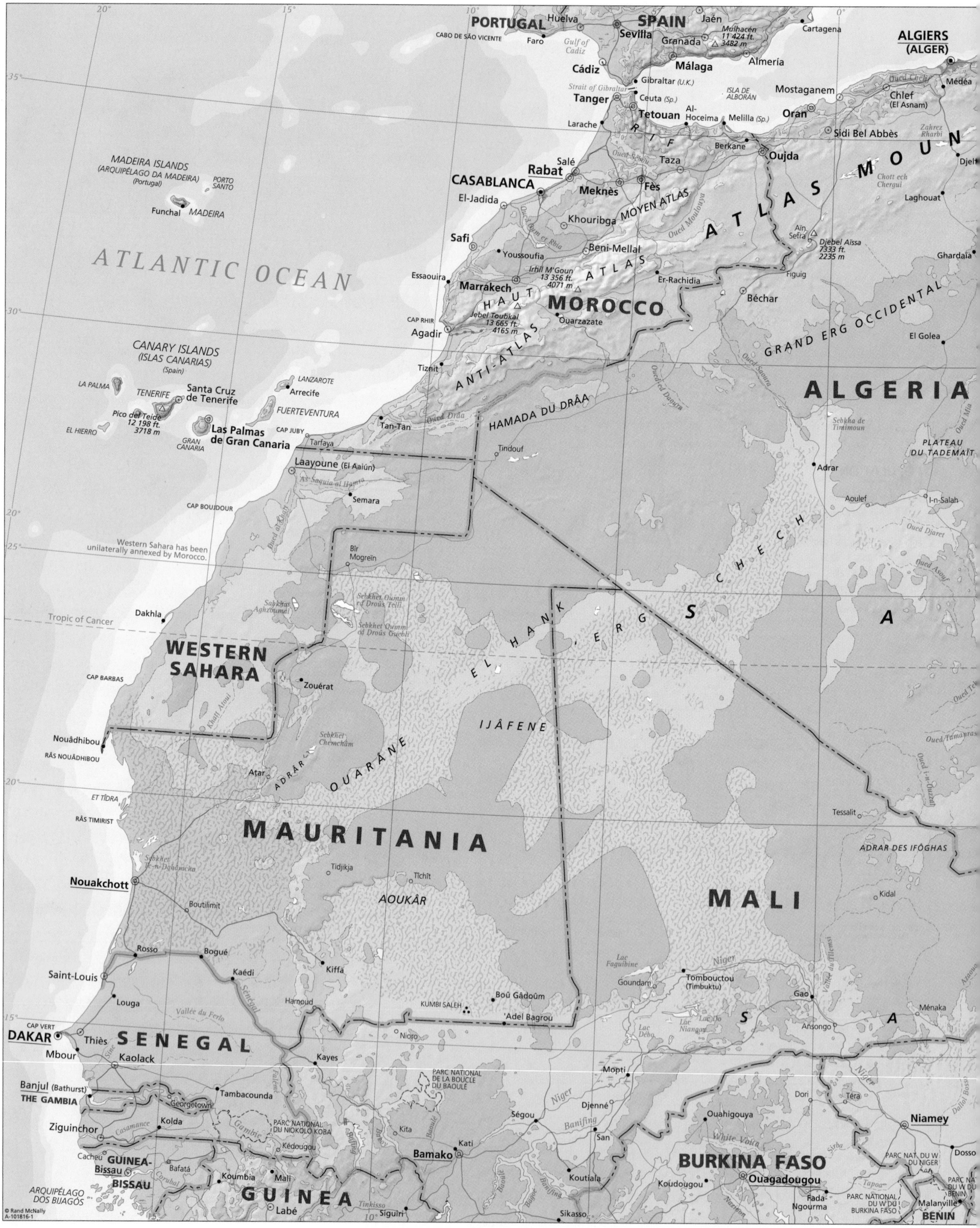
PORTUGAL
Huelva
SPAIN
Jaén
Sevilla
Granada
Mulhacén 11 424 ft. 3482 m
Cartagena
CABO DE SÃO VICENTE
Faro
Gulf of Cadiz
ALGIERS (ALGER)
Cádiz
Málaga
Almería
Gibraltar (U.K.)
Strait of Gibraltar
Médéa
Tanger
Ceuta (Sp.)
ISLA DE ALBORÁN
Mostaganem
Chlef (El Asnam)
Tetouan
Al-Hoceima
Melilla (Sp.)
Oran
Larache
RIF
Sidi Bel Abbès
Berkane
Oujda
MADEIRA ISLANDS (ARQUIPÉLAGO DA MADEIRA) (Portugal)
PORTO SANTO
Funchal
MADEIRA
Salé
Taza
Rabat
CASABLANCA
Meknès
Fès
Chott ech Chergui
El-Jadida
Khouribga
MOYEN ATLAS
Laghouat
ATLAS MOUN
Safi
Beni-Mellal
Aïn Sefra
Djebel Aïssa 7333 ft. 2235 m
Youssoufia
Ghardaïa
ATLANTIC OCEAN
Irhil M'Goun 13 356 ft. 4071 m
Essaouira
Marrakech
HAUT ATLAS
Er-Rachidia
Figuig
Béchar
MOROCCO
Jebel Toubkal 13 665 ft. 4165 m
Ouarzazate
CAP RHIR
Agadir
GRAND ERG OCCIDENTAL
El Golea
CANARY ISLANDS (ISLAS CANARIAS) (Spain)
Tiznit
ANTI-ATLAS
LANZAROTE
Arrecife
LA PALMA
TENERIFE
Santa Cruz de Tenerife
ALGERIA
FUERTEVENTURA
Pico del Teide 12 198 ft. 3718 m
Las Palmas de Gran Canaria
EL HIERRO
GRAN CANARIA
CAP JUBY
Tan-Tan
HAMADA DU DRÂA
Oued Drâa
Tarfaya
Sebkha de Timimoun
PLATEAU DU TADEMAÏT
Tindouf
Laayoune (El Aaiún)
Adrar
Semara
Aoulef
In-Salah
CAP BOUJDOUR
Oued Djaret
Western Sahara has been unilaterally annexed by Morocco.
Bir Mogreïn
ERG CHECH
Sebkhet Oumm ed Droûs Telli
Sebkhet Oumm ed Droûs Guebli
Dakhla
Tropic of Cancer
EL HANK
S
A
WESTERN SAHARA
CAP BARBAS
Zouérat
IJÂFENE
Sebkhet Chemchâm
Nouâdhibou
RÂS NOUÂDHIBOU
Atar
ADRAR
OUARÂNE
ET TÎDRA
RÂS TIMIRIST
Tessalit
MAURITANIA
ADRAR DES IFÔGHAS
Tidjikja
Tîchît
Nouakchott
AOUKÂR
MALI
Kidal
Boutilimit
Rosso
Bogué
Lac Faguibine
Saint-Louis
Kaédi
Kiffa
Goundam
Tombouctou (Timbuktu)
Louga
Hamoud
KUMBI SALEH
Boû Gâdoûm
Gao
Ménaka
Vallée du Ferlo
'Adel Bagrou
CAP VERT
DAKAR
Thiès
SENEGAL
Nioro
Ansongo
Mbour
Kaolack
Kayes
Mopti
PARC NATIONAL DE LA BOUCLE DE LA BAOULÉ
Banjul (Bathurst)
THE GAMBIA
Tambacounda
Georgetown
Djenné
Dori
Téra
Ziguinchor
Kolda
PARC NATIONAL DU NIOKOLO KOBA
Ségou
Ouahigouya
Niamey
Kédougou
Kita
San
Cacheu
GUINEA-BISSAU
Bissau
Bafatá
Bamako
Kati
BURKINA FASO
PARC NAT. DU W DU NIGER
Dosso
Koumbia
Mali
Koutiala
Koudougou
Ouagadougou
ARQUIPÉLAGO DOS BIJAGÓS
GUINEA
Labé
Siguiri
Sikasso
Fada-Ngourma
PARC NATIONAL DU W DU BURKINA FASO
Malanville
BENIN
© Rand McNally A-101816-1

0 60 120 180 240 300 360 Miles
0 60 120 180 240 300 360 420 480 540 600 Kilometers

Lambert Azimuthal Equal Area Projection
Scale 1:12,000,000
One inch to 190 miles
One cm to 120 km

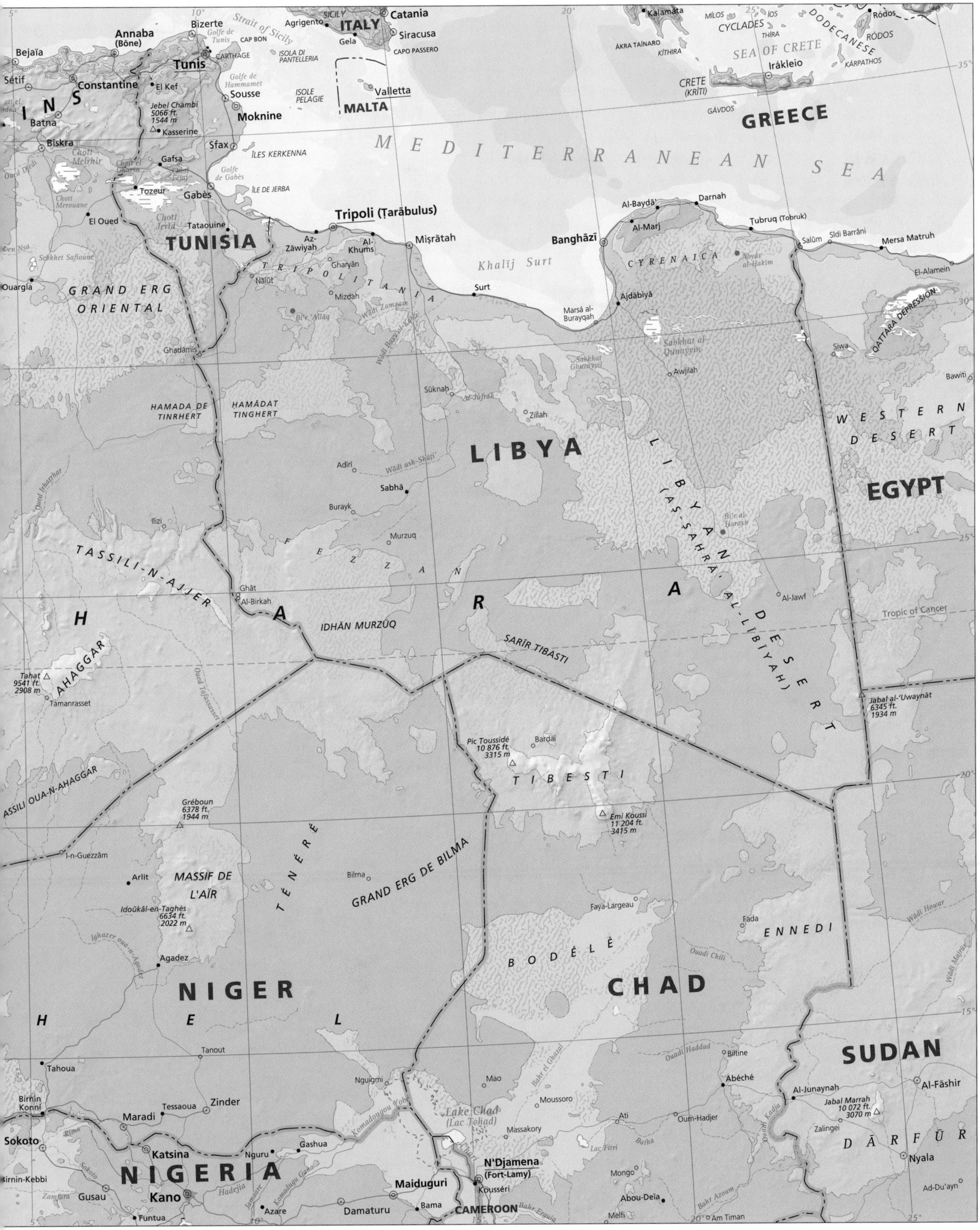

ITALY
SICILY
Catania
Siracusa
Agrigento
Gela
CAPO PASSERO
Strait of Sicily
MALTA
Valletta
ISOLE PELAGIE
ISOLA DI PANTELLERIA
CAP BON
CARTHAGE
Bizerte
Tunis
Golfe de Tunis
Annaba (Bône)
Bejaïa
Sétif
Constantine
El Kef
Golfe de Hammamet
Sousse
Moknine
Jebel Chambi 5066 ft. 1544 m
Kasserine
Batna
Biskra
Sfax
ÎLES KERKENNA
Gafsa
Chott Melrhir
Chott el Gharsa
Chott el Fejaj
Golfe de Gabès
Tozeur
Gabès
ÎLE DE JERBA
Chott Merouane
Chott el Jerid
El Oued
Tataouine
TUNISIA
Sebkhet Safioune
Ouargla
GRAND ERG ORIENTAL
Nālūt
Ghadāmis
HAMADA DE TINRHERT
HAMĀDAT TINGHERT
Illizi
TASSILI-N-AJJER
Ghāt
Al-Birkah
AHAGGAR
Tahat 9541 ft. 2908 m
Tamanrasset
Oued Tafassasset
TASSILI OUA-N-AHAGGAR
Gréboun 6378 ft. 1944 m
I-n-Guezzâm
Arlit
MASSIF DE L'AÏR
Idoûkâl-en-Taghès 6634 ft. 2022 m
Agadez
NIGER
TÉNÉRÉ
Tanout
Tahoua
Birnin Konni
Maradi
Tessaoua
Zinder
Sokoto
Katsina
Nguru
Gashua
NIGERIA
Birnin-Kebbi
Gusau
Kano
Funtua
Azare
Damaturu
Maiduguri
Bama
Nguigmi
Lake Chad (Lac Tchad)
N'Djamena (Fort-Lamy)
Kousséri
CAMEROON
MEDITERRANEAN SEA
Tripoli (Ṭarābulus)
Az-Zāwiyah
Al-Khums
Gharyān
Mizdah
TRIPOLITANIA
Bi'r 'Allaq
Miṣrātah
Khalīj Surt
Surt
Sūknah
Al-Jufrah
Zillah
LIBYA
Adīrī
Wādī ash-Shāṭi'
Sabhā
Burayk
Murzuq
FEZZAN
IDHĀN MURZŪQ
SAHARA
SARĪR TIBASTI
Pic Toussidé 10 876 ft. 3315 m
Bardaï
TIBESTI
Emi Koussi 11 204 ft. 3415 m
GRAND ERG DE BILMA
Bilma
Faya-Largeau
BODÉLÉ
CHAD
Fada
ENNEDI
Ouadi Chili
Ouadi Haddad
Biltine
Abéché
Mao
Moussoro
Bahr el Ghazal
Massakory
Ati
Oum-Hadjer
Lac Fitri
Batha
Mongo
Abou-Deïa
Melfi
Am Timan
Bahr Azoum
SAHEL
Banghāzī
Al-Bayḍā'
Al-Marj
Darnah
Tubruq (Tobruk)
CYRENAICA
Abyār al-Ḥakīm
Marsá al-Burayqah
Ajdābiyā
Sabkhat al-Qunayyin
Sabkhat Ghuzayyil
Awjilah
LIBYAN DESERT (AṢ-ṢAḤRĀ' AL-LĪBIYAH)
Bi'r al-Harash
Al-Jawf
Jabal al-'Uwaynāt 6345 ft. 1934 m
Salûm
Sîdi Barrâni
Mersa Matruh
El-Alamein
QATTARA DEPRESSION
Siwa
Bawiti
WESTERN DESERT
EGYPT
Tropic of Cancer
SUDAN
Wādī Howar
Wādī Magrur
Al-Junaynah
Al-Fāshir
Jabal Marrah 10 072 ft. 3070 m
Zalingei
DĀRFŪR
Nyala
Ad-Du'ayn
GREECE
Kalamata
ÁKRA TAÍNARO
KÍTHIRA
MÍLOS
CYCLADES
IOS
THÍRA
DODECANESE
Ródos
RÓDOS
KÁRPATHOS
SEA OF CRETE
Irákleio
CRETE (KRÍTI)
GÁVDOS

MAURITANIA
SENEGAL
MALI
GUINEA
CAPE VERDE
THE GAMBIA
GUINEA-BISSAU
SIERRA LEONE
LIBERIA
COTE D'IVOIR (IVORY COAST)
Nouakchott
Saint-Louis
Thiès
CAP VERT
DAKAR
Mbour
Kaolack
Banjul (Bathurst)
Ziguinchor
Bissau
Conakry
Freetown
Monrovia
Bamako
ABIDJAN
SANTO ANTÃO
Mindelo
SÃO VICENTE
SÃO NICOLAU
SAL
BOA VISTA
SANTIAGO
MAIO
BRAVA
FOGO
Praia
ET TÎDRA
RÂS TIMIRIST
AOUKÂR
Tidjikja
Tichît
Boutilimit
Rosso
Bogué
Kiffa
Kaédi
Hamoud
KUMBI SALEH
Boû Gâdoûm
'Adel Bagrou
Louga
Vallée du Ferlo
Nioro
PARC NATIONAL DE LA BOUCLE DU BAOULÉ
Kayes
Tambacounda
Georgetown
PARC NATIONAL DU NIOKOLO KOBA
Kolda
Kédougou
Cacheu
Bafatá
ARQUIPÉLAGO DOS BIJAGÓS
Mali
Koumbia
Labé
Fria
Kindia
Kabala
Bintimani 6381 ft. 1945 m
Kissidougou
Makeni
Lunsar
Koidu-Sefadu
Bo
Kenema
SHERBRO ISLAND
Voinjama
Nzérékoré
Kakata
Buchanan
Zwedru
Gbanga
Greenville
Harper
GROWA POINT
Kita
Kati
Ségou
Djenné
Mopti
San
Koutiala
Siguiri
Kankan
Sikasso
Bobo-Dioulasso
Banfora
Ferkéssédougou
Korhogo
Pic de Tibé 4934 ft. 1504 m
Mont Nimba 5748 ft. 1752 m
Man
Daloa
Bouaké
Lac de Kossou
Yamoussoukro
Abengourou
Gagnoa
Divo
Anyama
Grand-Bassam
San-Pédro
PARC NATIONAL DE LA COMOÉ
Lac Faguibine
Gounda
Lac Débo
Niger
Sénégal
© Rand McNally A-101876-1

a

NIGERIA
BENIN
CAMEROON
NIGER DELTA
Bight of Benin
Gulf of Guinea
Bight of Biafra
OBAN HILLS
Ilorin
Oke Odde
Igbaja
Ajasse
Offa
Egbe
Omu-Aran
Ogbomosho
Otu
Iseyin
Igana
Oyo
Oyan
Ila
Osi
Ikirun
Ilobu
Oshogbo
Ijero-Ekiti
Ikole
Inisa
Ede
Fiditi
Iwo
Oke-Mesi
Igede-Ekiti
Ado-Ekiti
Kabba
Ilesha
Eruwa
Igbo-Ora
Lalupon
Gbongan
Ife
Ilawe-Ekiti
Ikerre
Ikare
Oka
Okene
Ajaokuta
Ketou
Igbasa-Odo
IBADAN
Ikire
Emure-Ekiti
Owo
Igarra
Zangnanado
Abeokuta
Akure
Oke-Igbo
Idanre
Ondo
Idah
Pobè
Ilaro
Ishara
Ijebu-Igbo
Auchi
Shagamu
Ijebu-Ode
Ifon
Sakété
Adjohoun
Otta
Ikorodu
Ore
Ekpoma
Uromi
Ubiaja
Nsukka
Aku
Eha-Amufu
Porto-Novo
Ado
Ikeja
Epe
Lac Nokoué
Lagos Lagoon
Mushin
Okitipupa
Siluko
Enugu
Udi
Amagunze
Cotonou
Badagri
LAGOS
Benin City
Asaba
Ogwashi-Uku
Awka
Abakaliki
Onitsha
Nnewi
Mberubu
Awgu
Obubra
Ikom
Ozubulu
Sapele
Abraka
Obiaruku
Ihiala
Okigwi
Afikpo
Ajasso
Kwale
Orlu
Okwe
Ugep
Oguta
Bende
Ozoro
Warri
Ughelli
Ivorogbo
Owerri
Umuahia
Amaigbo
Arochukwu
Omoko
Forcados
Burutu
Oban
Ikot-Ekpene
Ahoada
Elele
Aba
Uyo
Calabar
Abak
Yenagoa
Port-Harcourt
Ikang
Degema
Okrika
Oron
Abonnema
Buguma
Bori
Eket
Opobo
Nembe
Opobo Town
Mbonge
Brass
Bonny
Ouémé
Ogun
Niger
Anambra
Benue
Cross
Imo
Osse
© Rand McNally A-101799-1

Inset map a
Lambert Conformal Conic Projection
Scale 1:4,000,000
One inch to 64 miles
One cm to 40 km

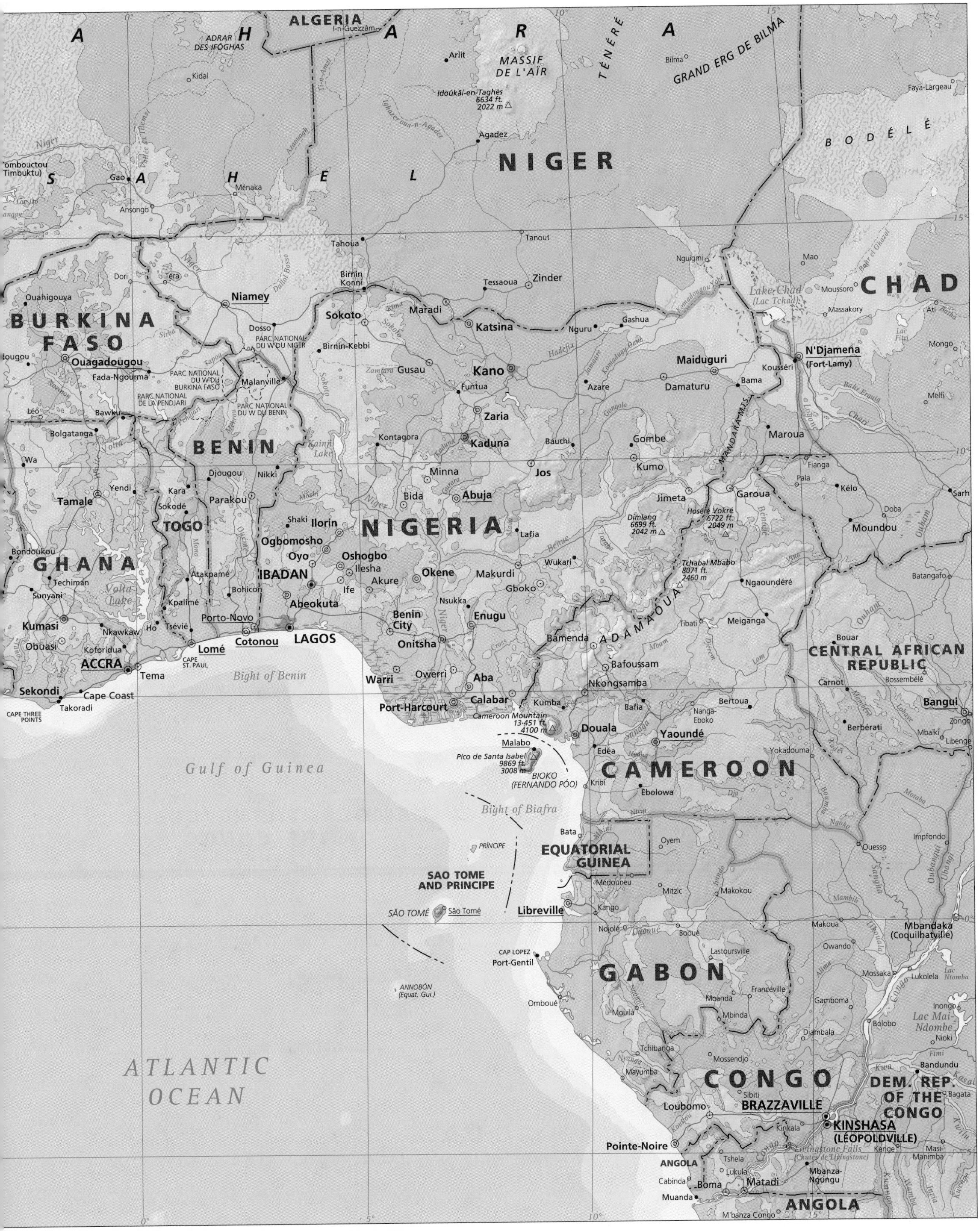
ALGERIA
I-n-Guezzâm
ADRAR DES IFÔGHAS
Kidal
Arlit
MASSIF DE L'AÏR
Idoûkâl-en-Taghès 6634 ft. 2022 m
TÉNÉRÉ
Bilma
GRAND ERG DE BILMA
Faya-Largeau
Agadez
NIGER
BODÉLÉ
Tombouctou (Timbuktu)
Gao
Ménaka
Ansongo
S A H E L
Tahoua
Tanout
Nguigmi
Mao
Dori
Téra
Birnin Konni
Tessaoua
Zinder
Lake Chad (Lac Tchad)
Moussoro
CHAD
Ouahigouya
Niamey
Sokoto
Maradi
Massakory
Ati
BURKINA FASO
Dosso
PARC NATIONAL DU W DU NIGER
Katsina
Nguru
Gashua
Birnin-Kebbi
Ouagadougou
Fada-Ngourma
PARC NATIONAL DU W DU BURKINA FASO
PARC NATIONAL DE LA PENDJARI
Malanville
Gusau
Kano
Maiduguri
N'Djamena (Fort-Lamy)
Kousséri
Mongo
Funtua
Azare
Damaturu
Bama
Léo
Bawku
PARC NATIONAL DU W DU BENIN
Zaria
Melfi
Bolgatanga
BENIN
Kainji Lake
Kontagora
Kaduna
Bauchi
Gombe
Maroua
Wa
Djougou
Nikki
Minna
Jos
Kumo
MANDARA MTS.
Fianga
Pala
Kélo
Sarh
Tamale
Yendi
Kara
Parakou
Bida
Abuja
Jimeta
Garoua
Sokodé
TOGO
Shaki
Ilorin
NIGERIA
Lafia
Dimlang 6699 ft. 2042 m
Hosséré Vokré 6722 ft. 2049 m
Doba
Moundou
Bondoukou
GHANA
Ogbomosho
Oyo
Oshogbo
Ilesha
Atakpamé
IBADAN
Ife
Akure
Okene
Makurdi
Wukari
Tchabal Mbabo 8071 ft. 2460 m
Techiman
Volta Lake
Sunyani
Bohicon
Abeokuta
Gboko
Ngaoundéré
Batangafo
Kpalimé
Nsukka
ADAMAOUA
Kumasi
Nkawkaw
Ho
Tsévié
Porto-Novo
Benin City
Enugu
Tibati
Meiganga
Obuasi
Koforidua
Lomé
Cotonou
LAGOS
Onitsha
Bamenda
Bouar
ACCRA
Tema
CAPE ST. PAUL
Bight of Benin
CENTRAL AFRICAN REPUBLIC
Bafoussam
Carnot
Bossembélé
Warri
Owerri
Aba
Nkongsamba
Sekondi
Cape Coast
Takoradi
CAPE THREE POINTS
Port-Harcourt
Calabar
Kumba
Bafia
Bertoua
Bangui
Cameroon Mountain 13,451 ft. 4100 m
Nanga-Eboko
Douala
Yaoundé
Berbérati
Zongo
Mbaïki
Libenge
Malabo
Edéa
Pico de Santa Isabel 9869 ft. 3008 m
Yokadouma
Gulf of Guinea
BIOKO (FERNANDO PÓO)
Kribi
CAMEROON
Ebolowa
Bight of Biafra
Bata
Oyem
Impfondo
PRÍNCIPE
EQUATORIAL GUINEA
Ouesso
SAO TOME AND PRINCIPE
Médouneu
Mitzic
Makokou
SÃO TOMÉ
São Tomé
Libreville
Kango
Ndjolé
Booué
Makoua
Mbandaka (Coquilhatville)
Owando
Lastoursville
CAP LOPEZ
Port-Gentil
GABON
Mossaka
Lukolela
ANNOBÓN (Equat. Gui.)
Franceville
Omboué
Moanda
Gamboma
Mouila
Mbinda
Inongo
Lac Mai-Ndombe
Bolobo
Djambala
Nioki
Tchibanga
Mossendjo
Bandundu
ATLANTIC OCEAN
Mayumba
CONGO
DEM. REP. OF THE CONGO
Bagata
Sibiti
Loubomo
BRAZZAVILLE
Kinkala
KINSHASA (LÉOPOLDVILLE)
Pointe-Noire
Livingstone Falls (Chutes de Livingstone)
Kenge
Masi-Manimba
ANGOLA
Tshela
Lukula
Mbanza-Ngungu
Cabinda
Boma
Matadi
Muanda
M'banza Congo
ANGOLA

0 60 120 180 240 300 360 Miles
0 60 120 180 240 300 360 420 480 540 600 Kilometers

Lambert Azimuthal Equal Area Projection
Scale 1:12,000,000
One inch to 190 miles
One cm to 120 km

NIGER
CHAD
NIGERIA
CENTRAL AFRICAN REPUBLIC
CAMEROON
EQUATORIAL GUINEA
SAO TOME AND PRINCIPE
GABON
CONGO
DEMOCRATIC REPUBLIC OF THE CONGO (ZAIRE)
ANGOLA
D Ā R F Ū R
S A H E L
A D A M A O U A
MASSIF DES BONGO
MANDARA MTS.
K A T A N G A
M O N T S   M I T U M B A
ATLANTIC OCEAN
Gulf of Guinea
Bight of Biafra
Lake Chad (Lac Tchad)
Lac Mai-Ndombe
Lac Ntomba
Lake Mweru
PARC NATIONAL DU BAMINGUI-BANGORAN
N'Djamena (Fort-Lamy)
Abuja
Yaoundé
Bangui
Malabo
Libreville
São Tomé
BRAZZAVILLE
KINSHASA (LÉOPOLDVILLE)
LUANDA
Kano
Kaduna
Zaria
Katsina
Maiduguri
Maroua
Garoua
Ngaoundéré
Douala
Port-Harcourt
Enugu
Calabar
Pointe-Noire
Port-Gentil
Mbandaka (Coquilhatville)
Kisangani
Kananga
Mbuji-Mayi (Bakwanga)
Kolwezi
Likasi (Jadotville)
Malanje
Bukavu
Cameroon Mountain 13 451 ft. 4100 m
Pico de Santa Isabel 9869 ft. 3008 m
Jabal Marrah 10 072 ft. 3070 m
Ugoma 9780 ft. 2981 m
Tchabal Mbabo 8071 ft. 2460 m
Dimlang 6699 ft. 2042 m
Hosséré Vokré 6722 ft. 2049 m
Congo
Oubangui
Kasai
Chari
Logone
Benue
Sangha
Lualaba
Cuanza
© Rand McNally A-101812-1

0 60 120 180 240 300 360 Miles
0 60 120 180 240 300 360 420 480 540 600 Kilometers

Lambert Azimuthal Equal Area Projection
Scale 1:12,000,000
One inch to 190 miles
One cm to 120 km

ERITREA
RED SEA
YEMEN
HADRAMAWT
DJIBOUTI
Djibouti
Gulf of Aden
The Republic of Somaliland unilaterally declared its independence from Somalia in 1991.
SUDAN
ETHIOPIA
ADDIS ABABA (ĀDĪS ĀBEBA)
RIFT VALLEY
DANAKIL
AHMAR MTS.
OGADEN
SOMALIA
Mogadishu (Muqdisho)
KENYA
NAIROBI
UGANDA
KAMPALA
RWANDA
Kigali
BURUNDI
Bujumbura
TANZANIA
DAR ES SALAAM
Dodoma
ZAMBIA
MALAWI
MOZAMBIQUE
INDIAN OCEAN
SEYCHELLES
Lake Victoria
Lake Rudolf
Lake Tanganyika
Lake Nyasa
Lake Albert
Lake Kyoga
Lake Rukwa
Lake Mweru Wantipa
Lake Bangweulu
CHALBI DESERT
SERENGETI PLAIN
MASAI STEPPE
Mt. Kenya (Kirinyaga) 17 058 ft. 5199 m
Kilimanjaro 19 340 ft. 5895 m
Mt. Meru 14 977 ft. 4565 m
Mt. Elgon 14 177 ft. 4321 m
Margherita Peak 16 763 ft. 5109 m
Ras Dejen 15 158 ft. 4620 m
Guna Terara 13 881 ft. 4231 m
Talo 14 478 ft. 4413 m
Batu 14 131 ft. 4307 m
Guge 13 780 ft. 4200 m
Gara Muleta 11 171 ft. 3405 m
Tulu Welel 10 833 ft. 3302 m
Gurage 12 208 ft. 3721 m
Shimbiris 7897 ft. 2407 m
Gamud 8156 ft. 2486 m
Kinyeti 10 456 ft. 3187 m
Moroto 10 118 ft. 3084 m
Ng'iro 9029 ft. 2752 m
Hanang 11 211 ft. 3417 m
Al-Hudaydah
Aden ('Adan)
Ta'izz
Al-Mukalla
Dire Dawa
Hargeysa
Berbera
Kismaayo
Mombasa
Malindi
Zanzibar
Kisumu
Nakuru
Eldoret
Meru
Thika
Machakos
Arusha
Moshi
Tanga
Mwanza
Tabora
Kigoma
Morogoro
Mbeya
Iringa
Juba
Gulu
Lira
Jinja
Entebbe
Al-Ubayyid
Wad Madani
Gonder
Bahir Dar
Mek'ele
Aksum
Harer
Jijiga
Nazret
Jima
Awasa
Arba Minch'
Moyale
Mandera
Marka
Baydhabo
Jawhar
Garissa
Lamu
Wajir
Marsabit
Lodwar
Kitale
Mbale
Masindi
Bukoba
Musoma
Shinyanga
Singida
Sumbawanga
Songea
Mtwara
Lindi
PEMBA
ZANZIBAR
MAFIA ISLAND
GROUPE D'ALDABRA
ATOLL DE COSMOLEDO
ASSOMPTION
ASTOVE
ST. PIERRE
ATOLL DE PROVIDENCE
CABO DELGADO
KIPENGERE RANGE
TSAVO EAST NATL. PARK
TSAVO WEST NATL. PARK
SERENGETI NATIONAL PARK
RUAHA NATL. PARK
MURCHISON FALLS NATIONAL PARK

LUANDA
PONTA DAS PALMEIRINHAS
N'dalatando
Malanje
Saurimo
Kamina
Bukama
Lake Mweru
DEM. REP. OF THE CONGO
KATANGA
Kawambwa
Kasenga
Lubudi
Dilolo
Kolwezi
Likasi (Jadotville)
Lubumbashi (Élisabethville)
Mansa
Lake Bangweulu
Gabela
Sumbe
Luena
Camacupa
Kuito
PLANALTO DO BIÉ
Morro de Môco 8596 ft. 2620 m
Lobito
Benguela
Caála
Huambo (Nova Lisboa)
Cubal
Ganda
CABO DE SANTA MARIA
ANGOLA
Solwezi
Chingola
Mufulira
Kitwe
Ndola
Luanshya
Busanga Swamp
Lukanga Swamp
ZAMBIA
Kabwe
Menongue
Lubango
Namibe
Tombua
Kaoma
Mumbwa
Mongu
KAFUE NATIONAL PARK
LUSAKA
LOWER ZAMBEZI NATIONAL PARK
Mazabuka
Monze
MANA POOLS NATIONAL PARK
Kariba
Lake Kariba
Choma
Cubia
PONTA DA MARCA
Katima Mulilo
Victoria Falls
Livingstone
CAPRIVI STRIP
VICTORIA FALLS NATIONAL PARK
Hwange
HWANGE NATIONAL PARK
Oshakati
Ondangwa
OVAMBOLAND
Rundu
Opuwo
CAPE FRIA
KAOKO VELD
ETOSHA NATIONAL PARK
Tsumeb
Grootfontein
KAUKAU VELD
Gumare
OKAVANGO DELTA
Maun
ZIMBA
Gweru
Bulawayo
Khorixas
Otjiwarongo
Lake Ngami
MAKGADIKGADI
Brandberg 8461 ft. 2579 m
NAMIBIA
NAMIB
Omaruru
Erongo 7562 ft. 2305 m
DAMARALAND
Okahandja
Francistown
Ghanzi
CENTRAL KALAHARI GAME RESERVE
Selebi-Phikwe
Serowe
Palapye
Gobabis
Swakopmund
Windhoek
Walvis Bay
BOTSWANA
Mahalapye
Rehoboth
Tropic of Capricorn
Polokwane (Pietersburg)
KGALAGADI TRANSFRONTIER PARK
NAMIB NAUKLUFT PARK
DESERT
Molepolole
Mochudi
Potgietersrus
Thabazimbi
Mariental
Khakhea
Gaborone
HOLLANDSBIRD ISLAND
Kanye
GEMSBOK NATIONAL PARK
KALAHARI GEMSBOK NATIONAL PARK
Pretoria (Tshwane)
Rustenburg
GREAT NAMAQUALAND
Tshabong
Mafikeng
Johannesburg
Soweto
Witbank
Carletonville
Vereeniging
Keetmanshoop
KALAHARI
Lüderitz
Vryburg
Klerksdorp
Standerton
ATLANTIC OCEAN
Schroffenstein 7224 ft. 2202 m
GHAAPPLATO
Kuruman
Kroonstad
Newcastl
DESERT
Welkom
Karasburg
Warrenton
Bethlehem
Harrismith
Upington
Ladysmith
Oranjemund
Kimberley
Ficksburg
AUGRABIES FALLS NATIONAL PARK
Kakamas
Bloemfontein
Maseru
LITTLE NAMAQUALAND
Kenhardt
LESOTHO
Thabana Ntlenyana
Springbok
Prieska
Hopetown
SOUTH AFRICA
De Aar
Colesberg
Kokstad
Carnarvon
Aliwal North
Victoria West
Middelburg
Umtata
Calvinia
Kompasberg 8209 ft. 2502 m
Queenstown
Lambert's Bay
Clanwilliam
NUWEVELDBERGE
Graaff-Reinet
Cradock
Beaufort West
King William's Town
GREAT KAROO
Somerset East
Saldanha
Piketberg
Willowmore
ADDO ELEPHANT NATIONAL PARK
East London
Grahamstown
Atlantis
Worcester
Oudtshoorn
LITTLE KAROO
Uitenhage
Port Alfred
Cape Town
Paarl
Swellendam
George
Port Elizabeth
Simon's Town
Mosselbaai
CAPE OF GOOD HOPE
AGULHAS
CAPE ST. FRANCIS
0 60 120 180 240 300 360 Miles
0 60 120 180 240 300 360 420 480 540 600 Kilometers

Lambert Azimuthal Equal Area Projection
Scale 1:12,000,000
One inch to 190 miles
One cm to 120 km

TANZANIA
MALAWI
MOZAMBIQUE
MADAGASCAR
SEYCHELLES
COMOROS
MAYOTTE
(France; claimed by Comoros)
Mbeya
Tukuyu
Karonga
Kasama
Chinsali
Nganda
8550 ft.
2606 m
Songea
Lake Nyasa
Mzuzu
Bangweulu Swamps
Mpika
MUCHINGA MOUNTAINS
KIPENGERE RANGE
LIKOMA ISLAND
SOUTH LUANGWA NATIONAL PARK
Kasungu
Lichinga
Chipata
Lilongwe
Mangochi
Lake Malombe
Ulóngué
Cuamba
Albufeira Cahora Bassa
Luangwa
Songo
Zomba
Lake Chilwa
Gurué
Blantyre
Sapitwa
9849 ft.
3002 m
Tete
Nsanje
Mocuba
Chinhoyi
HARARE
(SALISBURY)
Chitungwiza
Marondera
Cantandica
Inyangani
8504 ft.
2592 m
Rusape
Marromeu
Mutare
Chimoio
Monte Binga
7995 ft.
2437 m
Dondo
Beira
Masvingo
GREAT ZIMBABWE
Chiredzi
GONAREZHOU NATIONAL PARK
GREAT LIMPOPO TRANSFRONTIER PARK
PARQUE NACIONAL DO LIMPOPO
Tzaneen
KRUGER NATIONAL PARK
Chókwe
Chibuto
Xai-Xai
Nelspruit
Matola
Maputo
(Lourenço Marques)
ILHA DA INHACA
Mbabane
Lobamba
SWAZILAND
Lake Sibayi
Vryheid
ZULULAND
Lake St. Lucia
Richards Bay
Pietermaritzburg
Durban
Port Shepstone
Lindi
Mtwara
CABO DELGADO
Mocímboa da Praia
Mueda
ILHA QUIRIMBA
Pemba
Montepuez
Namapa
Nacala
Nampula
Ilha de Moçambique
Angoche
ILHA ANGOCHE
Moma
Quelimane
Vilankulo
ILHA DO BAZARUTO
PONTA SÃO SEBASTIÃO
PONTA DA BARRA FALSA
Massinga
Maxixe
PONTA DA BARRA
Mozambique Channel
BASSAS DA INDIA
(France; claimed by Madagascar)
ÎLE EUROPA
(France; claimed by Madagascar)
ÎLE JUAN DE NOVA
(France; claimed by Madagascar)
GROUPE D'ALDABRA
ASSOMPTION
ATOLL DE COSMOLEDO
ASTOVE
ATOLL DE PROVIDENCE
ST. PIERRE
ATOLL DE FARQUHAR
ÎLES GLORIEUSES
(France; claimed by Madagascar)
NJAZIDJA
Moroni
NZWANI
Mutsamudu
MWALI
ARCHIPEL DES COMORES
Mamoudzou
TANJONA BOBAOMBY
TANJONA ANORONTANY
Antsiranana
(Diégo-Suarez)
NOSY MITSIO
NOSY BE
Ambanja
Maromokotro
9436 ft.
2876 m
Sambava
NOSY LAVA
Antsohihy
Antalaha
TANJONA ANGONTSY
SAIKANOSY MASOALA
Mahajanga
Marovoay
TANJONA VILANANDRO
NOSY CHESTERFIELD
Besalampy
Mananara Avaratra
Tsaratanana
NOSY SAINTE MARIE
Soanierana Ivongo
Amparafaravola
Ambatondrazaka
Maintirano
Toamasina
NOSY BARREN
Tsiroanomandidy
ANTANANARIVO
Soavinandriana
Tsiafajavona
8668 ft.
2642 m
Antanifotsy
Belo Tsiribihina
Antsirabe
Mahanoro
Morondava
Ambositra
Nosy-Varika
Mananjary
Fianarantsoa
Morombe
TANJONA ANKABOA
Manakara
Ankazoabo
Ihosy
Boby
8720 ft.
2658 m
Farafangana
Toliara
Vangaindrano
Tropic of Capricorn
Betioky
Tolañaro
(Faradofay)
Tsiombe
Ambovombe
TANJONA VOHIMENA
INDIAN OCEAN
© Rand McNally
A-101819-1
a
INDIAN OCEAN
POINTE L'HORTAL
Triolet
Rivière du Rempart
Port Louis
Rose Hill
Curepipe
Mahébourg
MAURITIUS
REUNION
(France)
Saint-Denis
Le Port
Saint-André
Saint-Paul
Saint-Benoît
Piton des Neiges
10 072 ft.
3070 m
Piton de la Fournaise
8635 ft.
2632 m
Saint-Pierre
Saint-Louis
Saint-Joseph
MASCARENE ISLANDS
© Rand McNally
A-101782-1

Inset map a
Lambert Conformal Conic Projection
Scale 1:4,000,000
One inch to 64 miles
One cm to 40 km

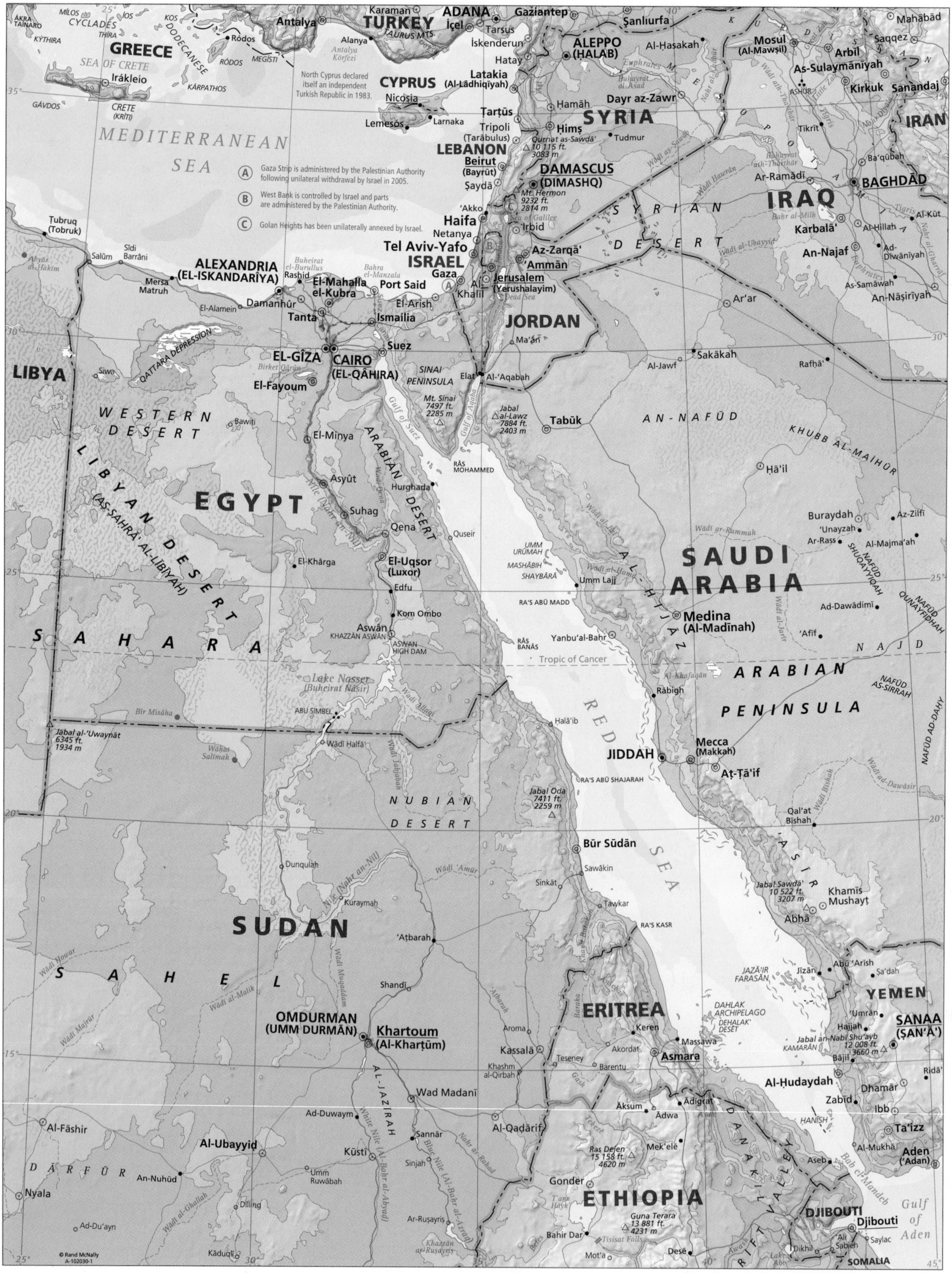

TURKEY
GREECE
CYPRUS
SYRIA
LEBANON
ISRAEL
JORDAN
IRAQ
IRAN
LIBYA
EGYPT
SAUDI ARABIA
SUDAN
ERITREA
ETHIOPIA
YEMEN
DJIBOUTI
SOMALIA
MEDITERRANEAN SEA
SEA OF CRETE
RED SEA
Gulf of Aden
Gulf of Suez
Gulf of Aqaba
Tropic of Cancer
Antalya
Alanya
Karaman
ADANA
İçel
Tarsus
Gaziantep
Şanlıurfa
İskenderun
Hatay
Rodos
Irákleio
CRETE (KRÍTI)
CYCLADES
DODECANESE
North Cyprus declared itself an independent Turkish Republic in 1983.
Nicosia
Lemesós
Larnaka
ALEPPO (HALAB)
Latakia (Al-Lādhiqīyah)
Ţarţūs
Tripoli (Ţarābulus)
Hamāh
Ḩimş
Dayr az-Zawr
Al-Ḩasakah
Tudmur
Beirut (Bayrūt)
Şaydā
DAMASCUS (DIMASHQ)
Qurnat as-Sawdā' 10 115 ft. 3083 m
Mt. Hermon 9232 ft. 2814 m
Mosul (Al-Mawşil)
Arbīl
As-Sulaymānīyah
Kirkuk
Sanandaj
Saqqez
Mahābād
Tikrīt
Ba'qūbah
BAGHDĀD
Ar-Ramādī
Karbalā'
Al-Ḩillah
Al-Kūt
An-Najaf
Ad-Dīwānīyah
As-Samāwah
An-Nāşirīyah
SYRIAN DESERT
Ⓐ Gaza Strip is administered by the Palestinian Authority following unilateral withdrawal by Israel in 2005.
Ⓑ West Bank is controlled by Israel and parts are administered by the Palestinian Authority.
Ⓒ Golan Heights has been unilaterally annexed by Israel.
'Akko
Haifa
Netanya
Tel Aviv-Yafo
Irbid
Az-Zarqā'
'Ammān
Jerusalem (Yerushalayim)
Gaza
Dead Sea
Ma'ān
Ar'ar
Sakākah
Al-Jawf
Rafḩā'
Ţubruq (Tobruk)
Salûm
Sîdi Barrâni
Mersa Matruh
ALEXANDRIA (EL-ISKANDARÎYA)
Rashîd
El-Mahalla el-Kubra
Port Said
El-'Arîsh
Damanhûr
El-Alamein
Tanta
Ismailia
Suez
EL-GÎZA
CAIRO (EL-QÂHIRA)
El-Fayoum
QATTARA DEPRESSION
Siwa
SINAI PENINSULA
Elat
Al-'Aqabah
Mt. Sinai 7497 ft. 2285 m
Jabal al-Lawz 7884 ft. 2403 m
Tabūk
AN-NAFŪD
KHUBB AL-MAIHŪR
Ḩā'il
WESTERN DESERT
LIBYAN DESERT (AS-ŞAḨRĀ' AL-LĪBĪYAH)
ARABIAN DESERT
Bawiti
El-Mînya
Asyût
Hurghada
RÂS MOHAMMED
Suhag
Qena
Quseir
El-Khârga
El-Uqsor (Luxor)
Edfu
Kom Ombo
Aswân
ASWAN HIGH DAM
Lake Nasser (Buheirat Nâsir)
ABU SIMBEL
Wâdi Halfā
Bîr Misâha
Jabal al-'Uwaynāt 6345 ft. 1934 m
SAHARA
Buraydah
'Unayzah
Ar-Rass
Az-Zilfī
Al-Majma'ah
Umm Lajj
Yanbu'al-Bahr
Medina (Al-Madīnah)
AL-ḨIJĀZ
Ad-Dawādimī
'Afīf
NAJD
ARABIAN PENINSULA
Rābigh
Halā'ib
JIDDAH
Mecca (Makkah)
Aţ-Ţā'if
Jabal Oda 7411 ft. 2259 m
NUBIAN DESERT
Qal'at Bīshah
Būr Sūdān
Sawākin
Sinkāt
Ţawkar
'ASĪR
Jabal Sawdā' 10 522 ft. 3207 m
Khamis Mushayţ
Abhā
Jīzān
Abū 'Arīsh
Şa'dah
Dunqulah
Kuraymah
'Aţbarah
Shandī
OMDURMAN (UMM DURMĀN)
Khartoum (Al-Kharţūm)
SAHEL
Aroma
Kassalā
Keren
Massawa
Asmara
DAHLAK ARCHIPELAGO
Al-Ḩudaydah
SANAA (ŞAN'Ā')
Jabal an-Nabī Shu'ayb 12 008 ft. 3660 m
Dhamār
Ibb
Ta'izz
Aden ('Adan)
Al-Fāshir
Al-Ubayyid
DĀRFŪR
An-Nuhūd
Nyala
Ad-Du'ayn
Wad Madanī
Ad-Duwaym
Al-Qaḑārif
Sannār
Kūstī
Sinjah
Umm Ruwābah
Dilling
Ar-Ruşayriş
Kāduqlī
AL-JAZĪRAH
Aksum
Adwa
Adigrat
Mek'elē
Ras Dejen 15 158 ft. 4620 m
Gonder
Bahir Dar
Guna Terara 13 881 ft. 4231 m
Assab
Djibouti
Dikhil
Ali Sabieh
Saylac
Bab el-Mandeb
Desē
Mot'a
RIFT VALLEY
© Rand McNally A-102030-1

0 60 120 180 240 300 360 Miles
0 60 120 180 240 300 360 420 480 540 600 Kilometers

Lambert Azimuthal Equal Area Projection
Scale 1:12,000,000
One inch to 190 miles
One cm to 120 km

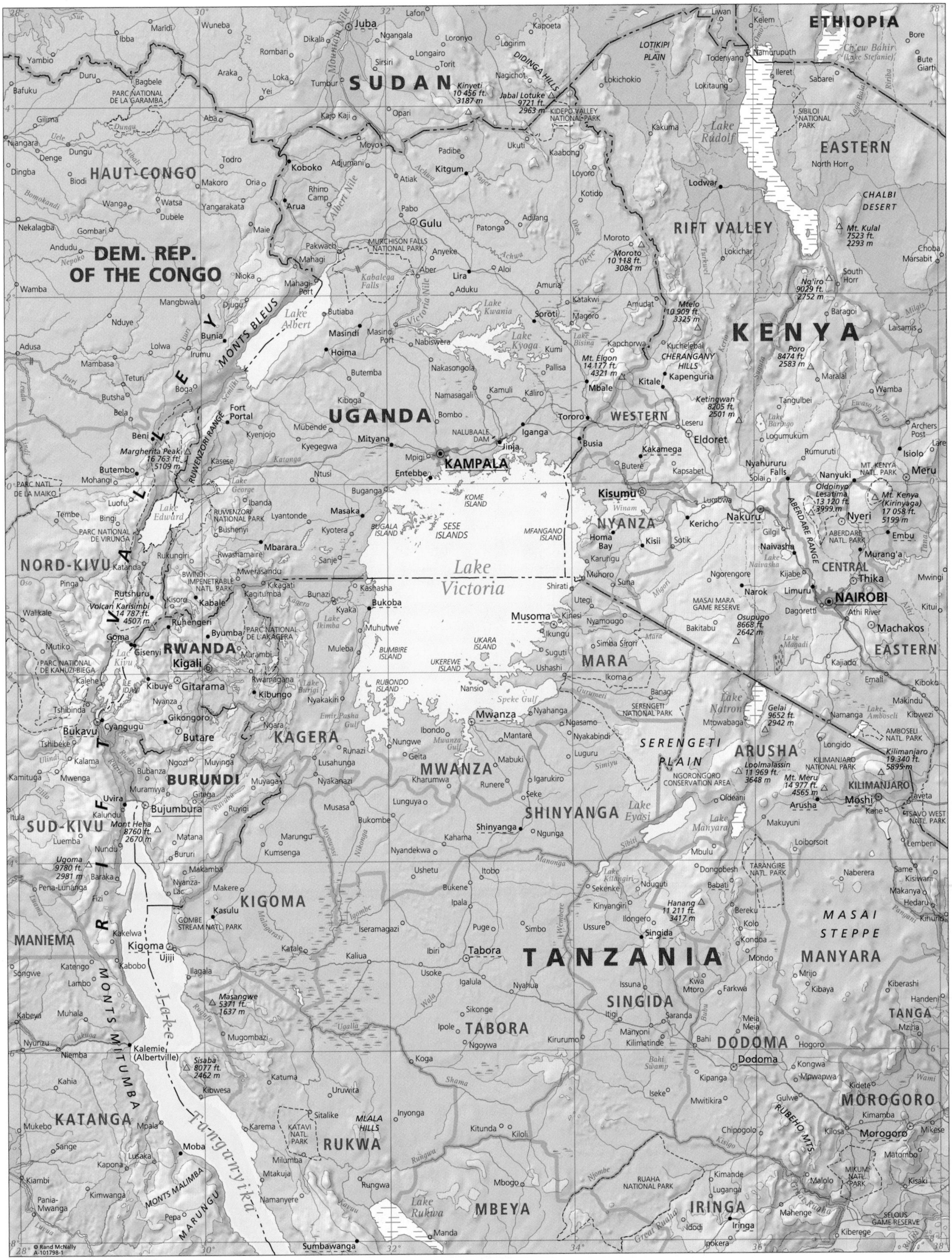
ETHIOPIA
SUDAN
UGANDA
KENYA
TANZANIA
RWANDA
BURUNDI
DEM. REP. OF THE CONGO
HAUT-CONGO
NORD-KIVU
SUD-KIVU
MANIEMA
KATANGA
RIFT VALLEY
EASTERN
WESTERN
NYANZA
CENTRAL
MARA
KAGERA
MWANZA
SHINYANGA
ARUSHA
KILIMANJARO
KIGOMA
SINGIDA
MANYARA
TANGA
DODOMA
MOROGORO
TABORA
RUKWA
MBEYA
IRINGA
Lake Victoria
Lake Rudolf
Lake Albert
Lake Kyoga
Lake Edward
Lake George
Lac Kivu
Lake Tanganyika
Lake Rukwa
Lake Natron
Lake Eyasi
Lake Manyara
Lake Magadi
Lake Naivasha
Lake Baringo
Lake Kwania
Lake Bisina
Lake Amboseli
Lake Kitangiri
Ch'ew Bahir (Lake Stefanie)
Juba
Kampala
Nairobi
Kigali
Bujumbura
Dodoma
Mwanza
Kisumu
Nakuru
Eldoret
Entebbe
Jinja
Gulu
Lira
Soroti
Mbale
Tororo
Kitale
Lodwar
Moroto
Arusha
Moshi
Tabora
Kigoma
Ujiji
Kalemie (Albertville)
Goma
Bukavu
Uvira
Butare
Gitarama
Ruhengeri
Musoma
Bukoba
Shinyanga
Singida
Iringa
Morogoro
Sumbawanga
Mpanda
Masaka
Mbarara
Fort Portal
Hoima
Masindi
Arua
Koboko
Bunia
Beni
Butembo
Thika
Nyeri
Embu
Machakos
Meru
Isiolo
Marsabit
Naivasha
Narok
Kericho
Kisii
Homa Bay
Kakamega
Busia
Kinyeti 10 456 ft. 3187 m
Jabal Lotuke 9721 ft. 2963 m
Moroto 10 118 ft. 3084 m
Mtelo 10 909 ft. 3325 m
Mt. Elgon 14 177 ft. 4321 m
Mt. Kulal 7523 ft. 2293 m
Ng'iro 9029 ft. 2752 m
Poro 8474 ft. 2583 m
Ketingwan 8205 ft. 2501 m
Mt. Kenya (Kirinyaga) 17 058 ft. 5199 m
Oldoinyo Lesatima 13 120 ft. 3999 m
Osupuko 8668 ft. 2642 m
Gelai 9652 ft. 2942 m
Kilimanjaro 19 340 ft. 5895 m
Mt. Meru 14 977 ft. 4565 m
Loolmalassin 11 969 ft. 3648 m
Hanang 11 211 ft. 3417 m
Margherita Peak 16 763 ft. 5109 m
Volcan Karisimbi 14 787 ft. 4507 m
Mont Heha 8760 ft. 2670 m
Ugoma 9780 ft. 2981 m
Masangwe 5371 ft. 1637 m
Sisaba 8077 ft. 2462 m
RUWENZORI RANGE
MONTS BLEUS
MONTS MITUMBA
MONTS MALIMBA
MARUNGU
ABERDARE RANGE
RUBEHO MTS.
CHERANGANY HILLS
DIDINGA HILLS
MLALA HILLS
CHALBI DESERT
SERENGETI PLAIN
MASAI STEPPE
LOTIKIPI PLAIN
KIDEPO VALLEY NATIONAL PARK
MURCHISON FALLS NATIONAL PARK
RUWENZORI NATIONAL PARK
BWINDI IMPENETRABLE NATL. PARK
PARC NATIONAL DE L'AKAGERA
PARC NATIONAL DE VIRUNGA
PARC NATIONAL DE KAHUZI-BIEGA
PARC NATIONAL DE LA GARAMBA
PARC NATL DE LA MAIKO
SIBILOI NATIONAL PARK
MT. KENYA NATL. PARK
ABERDARE NATL. PARK
MASAI MARA GAME RESERVE
SERENGETI NATIONAL PARK
NGORONGORO CONSERVATION AREA
AMBOSELI NATL. PARK
KILIMANJARO NATIONAL PARK
TSAVO WEST NATL. PARK
TARANGIRE NATL. PARK
GOMBE STREAM NATL. PARK
KATAVI NATL. PARK
RUAHA NATIONAL PARK
MIKUMI NATL. PARK
SELOUS GAME RESERVE
Albert Nile
Victoria Nile
Kabalega Falls
Nyahururu Falls
Spawn Gulf
Speke Gulf
Emin Pasha Gulf
Mwanza Gulf
SESE ISLANDS
UKEREWE ISLAND
NALUBAALE DAM
© Rand McNally A-101798-1

0 30 60 90 120 150 180 Miles
0 30 60 90 120 150 180 210 240 270 300 Kilometers

Lambert Conformal Conic Projection
Scale 1:6,000,000
One inch to 96 miles
One cm to 60 km

a

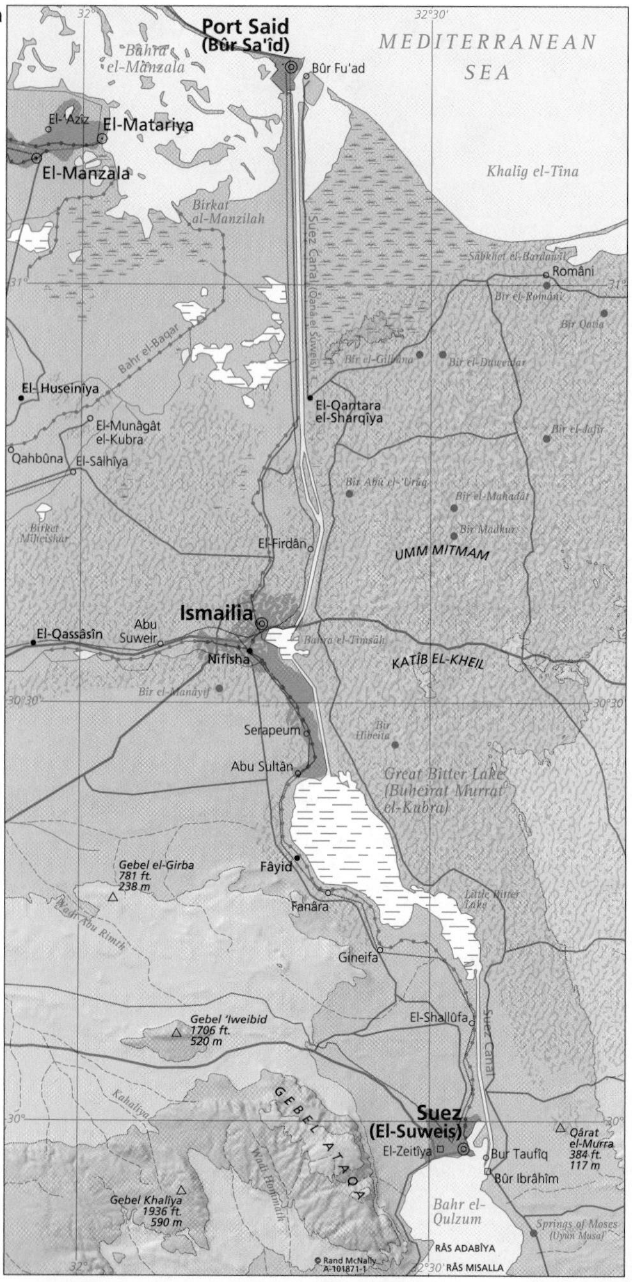
Port Said
(Bûr Sa'îd)
Bûr Fu'ad
MEDITERRANEAN SEA
Bahra el-Manzala
El-'Aziz
El-Matariya
El-Manzala
Birkat al-Manzilah
Khalîg el-Tîna
Sabkhet el-Bardawil
Români
Bîr el-Români
Bîr Qatia
Bahr el-Baqar
Bîr el-Gifgâfa
Bîr el-Duweidar
El-Huseiniya
El-Qantara el-Sharqiya
El-Munâgât el-Kubra
Qahbûna
El-Sâlhîya
Bîr el-Jafir
Bîr Abu el-'Uruq
Bîr el-Mahadât
Bîr Madkur
Birkat Mihetshar
El-Firdân
UMM MITMAM
Ismailia
El-Qassâsîn
Abu Suweir
Nifîsha
Bahra el-Timsâh
KATÎB EL-KHEIL
Bîr el-Manâyif
Serapeum
Bîr Hibeita
Abu Sultân
Great Bitter Lake (Buheirat Murrat el-Kubra)
Gebel el-Girba 781 ft. 238 m
Fâyid
Fanâra
Little Bitter Lake
Wadi Abu Rimth
Gineifa
Gebel 'Iweibid 1706 ft. 520 m
El-Shallûfa
Suez Canal
GEBEL ATAQA
Suez (El-Suweis)
El-Zeitîya
Bur Taufîq
Bûr Ibrâhîm
Qârat el-Murra 384 ft. 117 m
Gebel Khalîya 1936 ft. 590 m
Bahr el-Qulzum
Springs of Moses ('Uyun Musa)
RÂS ADABÎYA
RÂS MISALLA
© Rand McNally A-101871-1

b

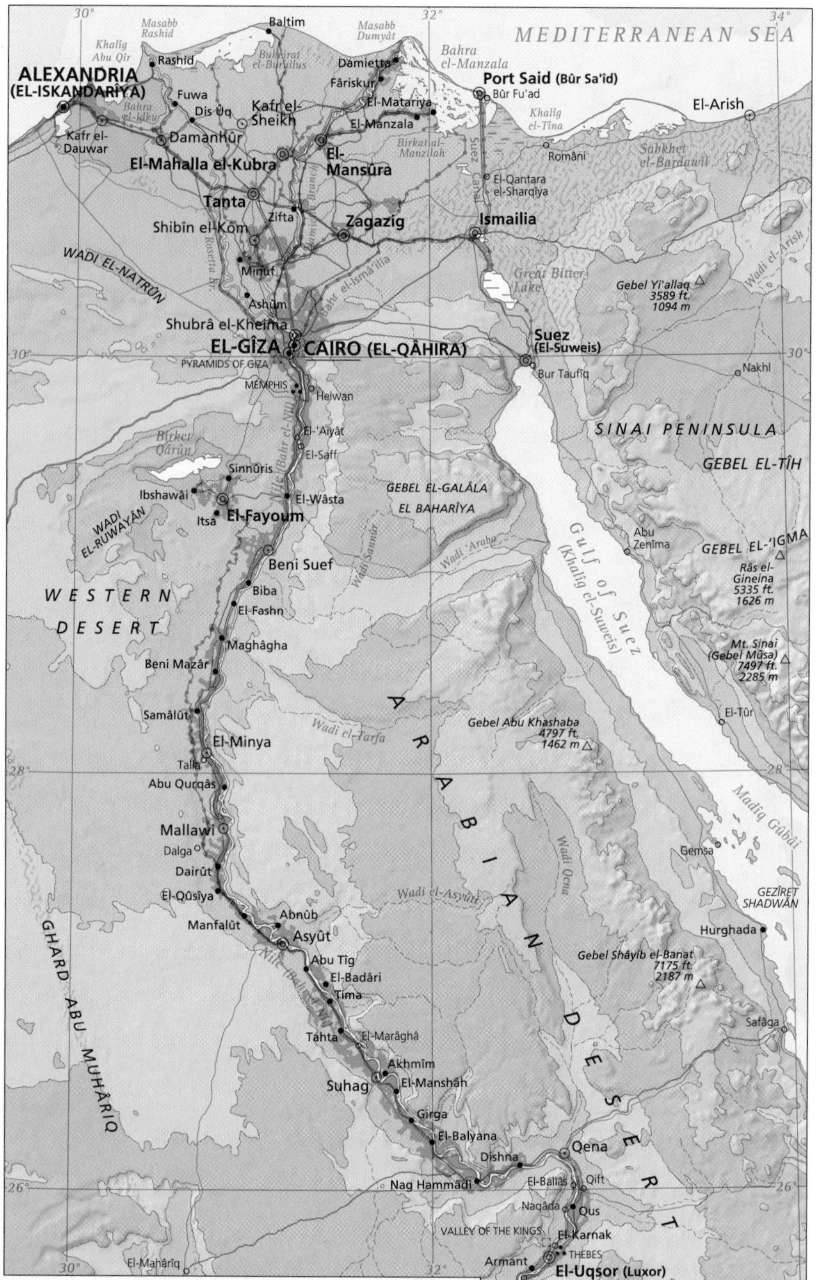
MEDITERRANEAN SEA
Baltim
Masabb Rashid
Masabb Dumyât
Khalîg Abu Qîr
Rashîd
Buheirat el-Burullus
Damietta
Fâriskur
Bahra el-Manzala
ALEXANDRIA
(EL-ISKANDARÎYA)
Fuwa
Dis Ûq
Kafr el-Sheikh
El-Matariya
El-Manzala
Port Said (Bûr Sa'îd)
Bûr Fu'ad
El-Arish
Kafr el-Dauwar
Damanhûr
Khalîg el-Tîna
Români
Sabkhet el-Bardawil
El-Mahalla el-Kubra
El-Mansûra
El-Qantara el-Sharqiya
Tanta
Zifta
Zagazig
Ismailia
Shibîn el-Kôm
WADI EL-NATRÛN
Minûf
Great Bitter Lake
Gebel Yi'allaq 3589 ft. 1094 m
Ashmûn
Shubrâ el-Kheima
EL-GÎZA
CAIRO (EL-QÂHIRA)
Suez (El-Suweis)
Bur Taufîq
PYRAMIDS OF GIZA
MEMPHIS
Helwan
Nakhl
SINAI PENINSULA
GEBEL EL-TÎH
El-'Aiyât
El-Saff
Birket Qârûn
Sinnûris
Ibshawâi
El-Wâsta
GEBEL EL-GALÂLA EL BAHARÎYA
Itsa
El-Fayoum
WADI EL-RUWAYÂN
Gulf of Suez (Khalîg el-Suweis)
Abu Zenîma
GEBEL EL-'IGMA
Beni Suef
Râs el-Gineina 5335 ft. 1626 m
WESTERN DESERT
Biba
El-Fashn
Maghâgha
Mt. Sinai (Gebel Mûsa) 7497 ft. 2285 m
Beni Mazâr
ARABIAN DESERT
Samâlût
El-Minya
Wadi el-Tarfa
Gebel Abu Khashaba 4797 ft. 1462 m
El-Tûr
Tala
Abu Qurqâs
Madîq Gûbâl
Mallawi
Dalga
Gemsa
Dairût
El-Qûsîya
GEZÎRET SHADWÂN
Manfalût
Abnûb
Asyût
Hurghada
GHARD ABU MUHÂRIQ
Abu Tig
El-Badâri
Gebel Shâyib el-Banat 7175 ft. 2187 m
Tima
Tahta
El-Marâgha
Safâga
Akhmîm
Suhag
El-Manshâh
Girga
El-Balyana
Dishna
Qena
Nag Hammâdi
El-Ballâs
Qift
Naqâda
Qus
VALLEY OF THE KINGS
Karnak
THEBES
Armant
El-Uqsor (Luxor)
El-Mahâriq
El-'Idîsât
Asfûn el-Matâ'na
El-Deir
Isna
Edfu
El-Ridîsîya Bahri
Silwa Bahari
El-Nasser
Bimbân
Kom Ombo
Darâw
Aswân
KHAZZÂN ASWÂN
ASWAN HIGH DAM
Tropic of Cancer
Lake Nasser (Buheirat Nâsir)
© Rand McNally A-101801-1

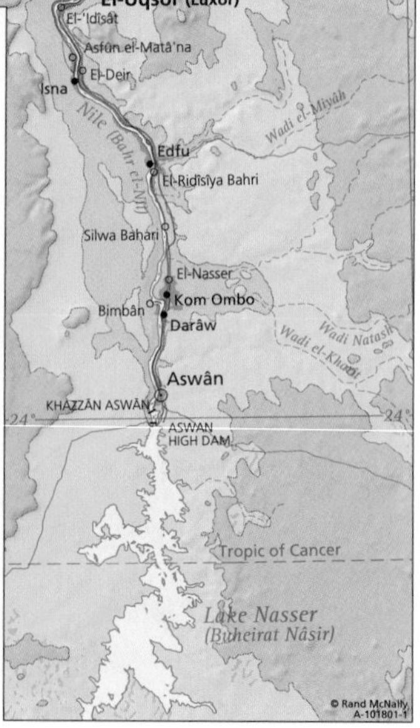

c

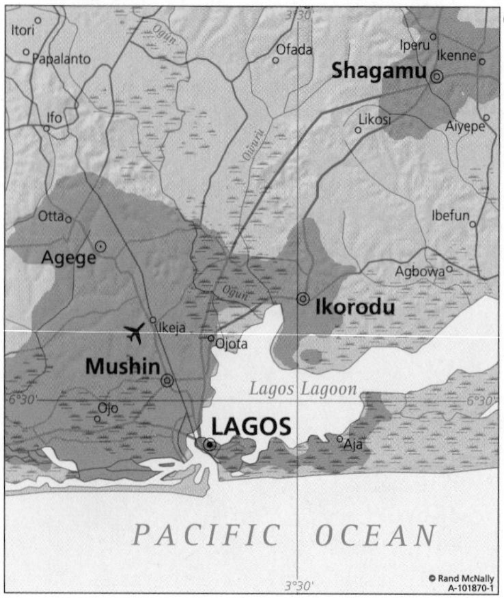
Itori
Papalanto
Ofada
Iperu
Ikenne
Shagamu
Ifo
Likosi
Aiyepe
Otta
Ibefun
Agege
Agbowa
Ikorodu
Ikeja
Ojota
Mushin
Lagos Lagoon
Ojo
LAGOS
Aja
PACIFIC OCEAN
© Rand McNally A-101870-1

d

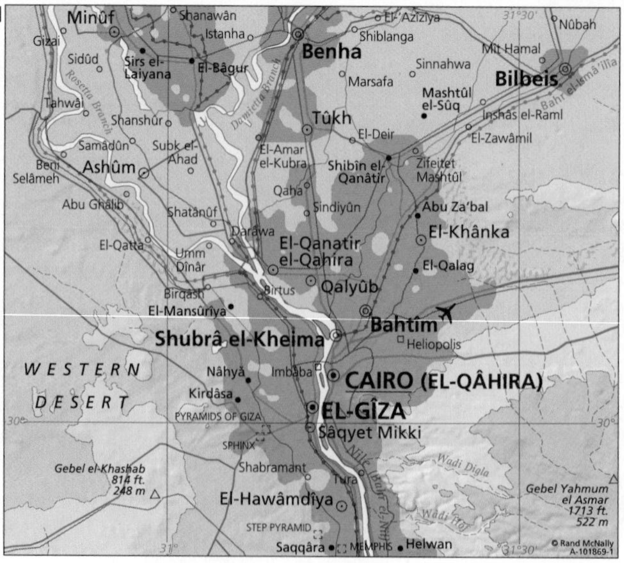
Minûf
Shanawân
Istanha
'Aziziya
Nûbah
Gizai
Shiblanga
Benha
Sidûd
Sirs el-Laiyana
El-Bâgur
Mit Hamal
Sinnahwa
Marsafa
Bilbeis
Mashtûl el-Sûq
Tahwâi
Tûkh
Inshâs el-Raml
Shanshûr
El-Deir
El-Zawâmil
Samadûn
Subk el-Ahad
El-Amar el-Kubra
Shibîn el-Qanâtir
Zifeitei Mashtûl
Beni Selâmeh
Ashmûn
Qaha
Abu Ghâlib
Shatânûf
Sindiyûn
Abu Za'bal
El-Khânka
El-Qatta
Umm Dînâr
El-Qanatir el-Qahira
El-Qalag
Birqâsh
Birtus
Qalyûb
El-Mansûrîya
Bahtîm
Shubrâ el-Kheima
Heliopolis
WESTERN DESERT
Nâhyâ
Imbâba
CAIRO (EL-QÂHIRA)
Kirdâsa
PYRAMIDS OF GIZA
EL-GÎZA
Sâqyet Mikki
SPHINX
Wadi Digla
Gebel el-Khashab 814 ft. 248 m
Shabramant
Tura
El-Hawâmdîya
Gebel Yahmum el Asmar 1713 ft. 522 m
STEP PYRAMID
Saqqâra
MEMPHIS
Helwan
© Rand McNally A-101869-1

0 5 10 15 20 25 30 Miles
0 5 10 15 20 25 30 35 40 45 50 Kilometers

Inset maps a,c,d
Lambert Conformal Conic Projection
Scale 1:1,000,000
One inch to 16 miles
One cm to 10 km

Inset map b
Lambert Conformal Conic Projection
Scale 1:4,000,000
One inch to 64 miles
One cm to 40 km

a

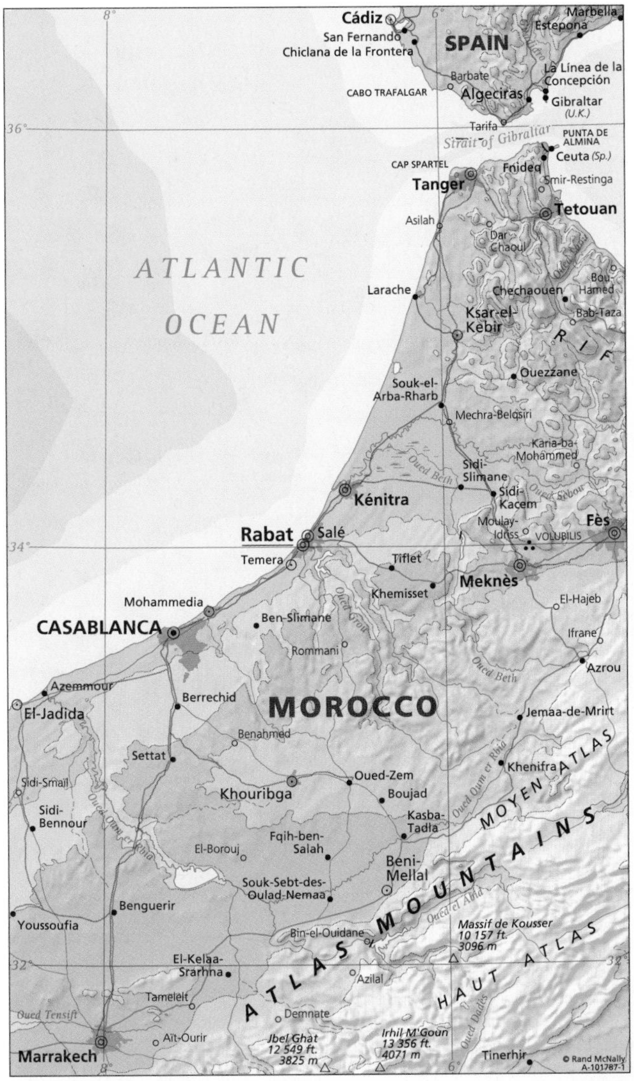
Cádiz
San Fernando
Chiclana de la Frontera
SPAIN
Marbella
Estepona
CABO TRAFALGAR
Barbate
La Línea de la Concepción
Algeciras
Gibraltar (U.K.)
Tarifa
Strait of Gibraltar
PUNTA DE ALMINA
Ceuta (Sp.)
CAP SPARTEL
Fnideq
Tanger
Smir-Restinga
Tetouan
Asilah
ATLANTIC
OCEAN
Dar Chaoui
Larache
Bou-Hamed
Chechaouen
Bab-Taza
Ksar-el-Kebir
RIF
Ouezzane
Souk-el-Arba-Rharb
Mechra-Belqsiri
Karia-ba-Mohammed
Oued Beth
Sidi-Slimane
Oued Sebou
Kénitra
Sidi-Kacem
Moulay-Idriss
VOLUBILIS
Fès
Rabat
Salé
Temera
Tiflet
Khemisset
Meknès
Mohammedia
El-Hajeb
Ben-Slimane
CASABLANCA
Ifrane
Rommani
Azrou
Azemmour
Oued Beth
El-Jadida
Berrechid
MOROCCO
Jemaa-de-Mrirt
Benahmed
Settat
Khenifra
Sidi-Smail
Oued-Zem
MOYEN ATLAS
Khouribga
Boujad
Sidi-Bennour
Kasba-Tadla
Fqih-ben-Salah
El-Borouj
Beni-Mellal
ATLAS MOUNTAINS
Souk-Sebt-des-Oulad-Nemaa
Benguerir
Youssoufia
Bin-el-Ouidane
Massif de Kousser 10 157 ft. 3096 m
El-Kelaa-Srarhna
Azilal
HAUT ATLAS
Tameleit
Oued Tensift
Demnate
Marrakech
Aït-Ourir
Jbel Ghat 12 549 ft. 3825 m
Irhil M'Goun 13 356 ft. 4071 m
Tinerhir
© Rand McNally A-101787-1

b

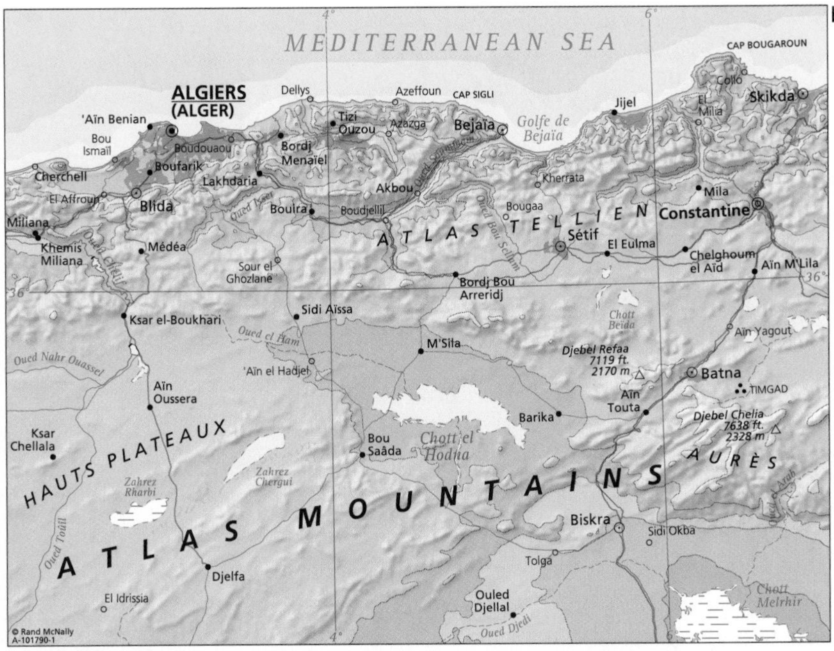
MEDITERRANEAN SEA
CAP BOUGAROUN
ALGIERS (ALGER)
Dellys
Azeffoun
CAP SIGLI
Jijel
Collo
El Milia
Skikda
'Aïn Benian
Bou Ismaïl
Tizi Ouzou
Azazga
Béjaïa
Golfe de Bejaïa
Boudouaou
Bordj Menaïel
Boufarik
Cherchell
Lakhdaria
Akbou
Kherrata
Mila
El Affroun
Blida
Bouira
Boudjellil
Bougaa
Constantine
Miliana
Khemis Miliana
Médéa
ATLAS TELLIEN
Sétif
El Eulma
Chelghoum el Aïd
Aïn M'Lila
Sour el Ghozlane
Bordj Bou Arreridj
Ksar el-Boukhari
Sidi Aïssa
Chott Beida
Aïn Yagout
Oued Nahr Ouassel
M'Sila
Djebel Refaa 7119 ft. 2170 m
'Aïn el Hadjel
Batna
TIMGAD
Aïn Oussera
Aïn Touta
Barika
Djebel Chelia 7638 ft. 2328 m
Ksar Chellala
HAUTS PLATEAUX
Bou Saâda
Chott el Hodna
AURÈS
Zahrez Chergui
Zahrez Rharbi
ATLAS MOUNTAINS
Biskra
Sidi Okba
Djelfa
Tolga
El Idrissia
Ouled Djellal
Chott Melrhir
© Rand McNally A-101790-1

c

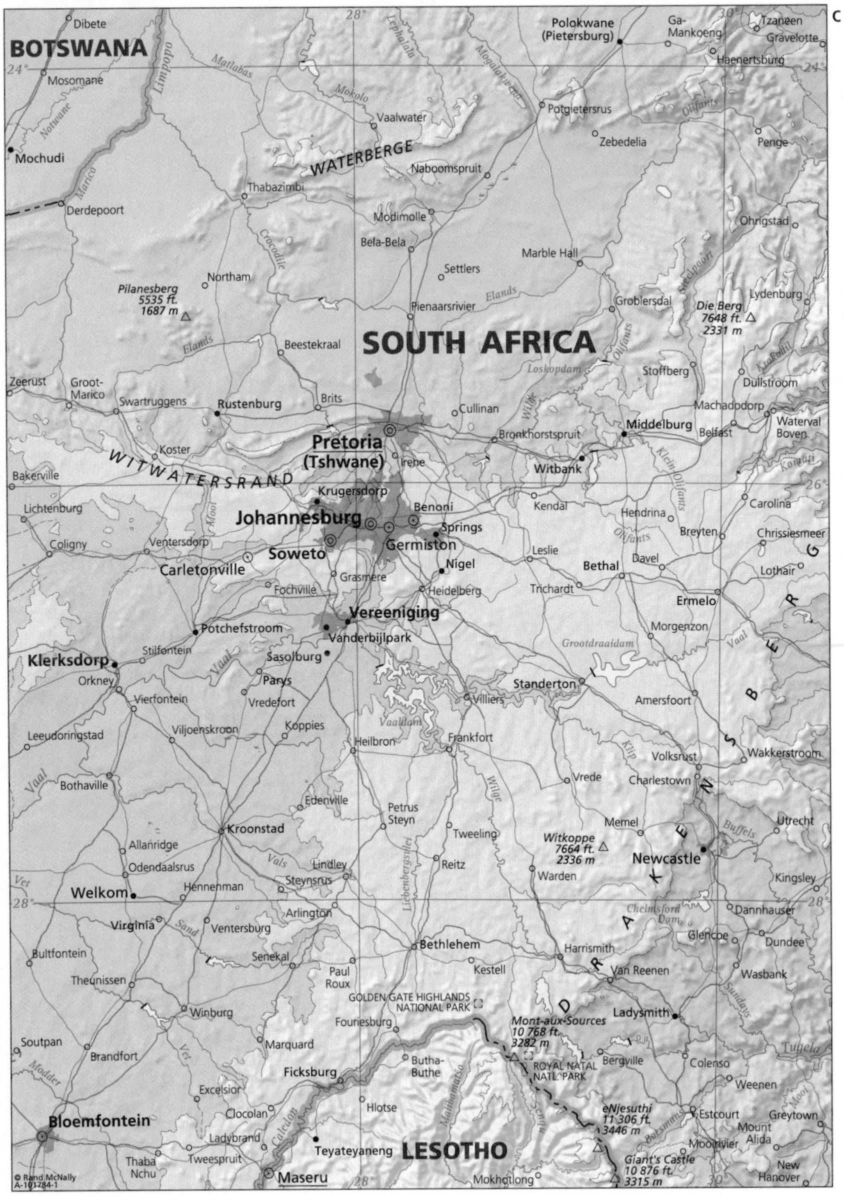
Dibete
Polokwane (Pietersburg)
Ga-Mankoeng
Tzaneen
Gravelotte
BOTSWANA
Haenertsburg
Mosomane
Limpopo
Vaalwater
Potgietersrus
Zebedelia
Mochudi
WATERBERGE
Naboomspruit
Penge
Thabazimbi
Derdepoort
Modimolle
Ohrigstad
Bela-Bela
Marble Hall
Settlers
Northam
Pilanesberg 5535 ft. 1687 m
Pienaarsrivier
Groblersdal
Lydenburg
Die Berg 7648 ft. 2331 m
Beestekraal
SOUTH AFRICA
Stoffberg
Zeerust
Groot-Marico
Dullstroom
Swartruggens
Rustenburg
Brits
Cullinan
Machadodorp
Pretoria (Tshwane)
Bronkhorstspruit
Middelburg
Belfast
Waterval Boven
Koster
Irene
WITWATERSRAND
Witbank
Bakerville
Krugersdorp
Lichtenburg
Benoni
Kendal
Hendrina
Carolina
Johannesburg
Springs
Breyten
Chrissiesmeer
Ventersdorp
Coligny
Soweto
Germiston
Leslie
Lothair
Carletonville
Nigel
Bethal
Davel
Fochville
Grasmere
Heidelberg
Trichardt
Ermelo
Vereeniging
Potchefstroom
Vanderbijlpark
Morgenzon
Klerksdorp
Stilfontein
Sasolburg
Grootdraaidam
Orkney
Parys
Standerton
Vierfontein
Vredefort
Villiers
Amersfoort
Viljoenskroon
Koppies
Leeudoringstad
Heilbron
Frankfort
Volksrust
Wakkerstroom
Bothaville
Vrede
Charlestown
Edenville
Petrus Steyn
Memel
Utrecht
Kroonstad
Tweeling
Witkoppe 7664 ft. 2336 m
Newcastle
Allanridge
Reitz
Odendaalsrus
Lindley
Warden
Kingsley
Hennenman
Steynsrus
Welkom
Arlington
Chelmsford Dam
Dannhauser
Virginia
Ventersburg
Glencoe
Dundee
Senekal
Bethlehem
Harrismith
Bultfontein
Paul Roux
Kestell
Van Reenen
Wasbank
Theunissen
GOLDEN GATE HIGHLANDS NATIONAL PARK
DRAKENSBERG
Winburg
Fouriesburg
Ladysmith
Mont-aux-Sources 10 768 ft. 3282 m
Soutpan
Marquard
Bergville
Colenso
Brandfort
Butha-Buthe
ROYAL NATAL NATL. PARK
Ficksburg
Excelsior
Weenen
Hlotse
Clocolan
eNjesuthi 11 306 ft. 3446 m
Estcourt
Greytown
Bloemfontein
Ladybrand
Mount Alida
Mooi River
Teyateyaneng
LESOTHO
Thaba Nchu
Tweespruit
Giant's Castle 10 876 ft. 3315 m
New Hanover
Maseru
Mokhotlong
© Rand McNally A-101784-1

d

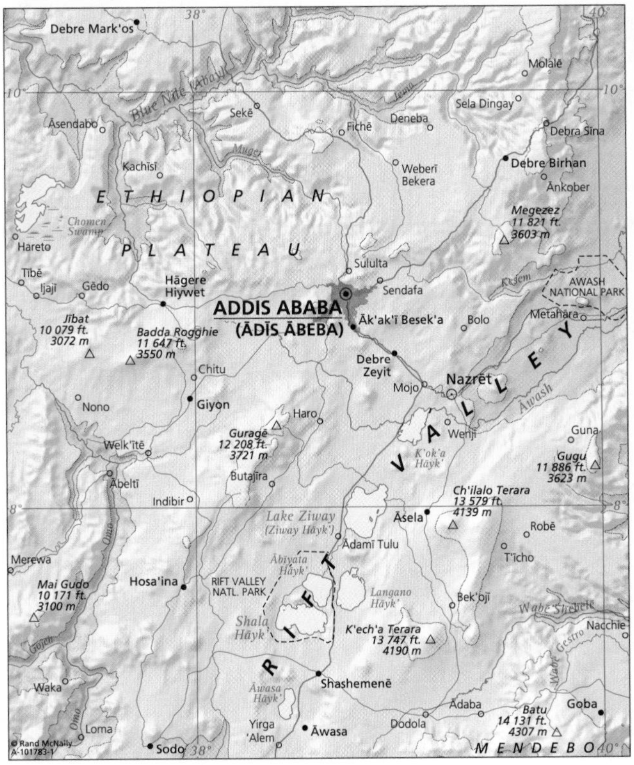
Debre Mark'os
Molalē
Blue Nile (Abay)
Sela Dingay
Āsendabo
Sekē
Fichē
Deneba
Debra Sina
Kachīsī
Weberī Bekera
Debre Birhan
Ankober
ETHIOPIAN
PLATEAU
Chomen Swamp
Megezez 11 821 ft. 3603 m
Hareto
Sululta
Tībē
Gēdo
Hāgere Hiywet
Sendafa
AWASH NATIONAL PARK
ADDIS ABABA (ĀDĪS ĀBEBA)
Āk'ak'ī Besek'a
Bolo
Metahara
Jībat 10 079 ft. 3072 m
Badda Rogghie 11 647 ft. 3550 m
Debre Zeyit
Chitu
Mojo
Nazrēt
Nono
Giyon
Haro
Awash
Gurage 12 208 ft. 3721 m
Wenjī
Guna
Welk'ītē
Gugu 11 886 ft. 3623 m
Butajīra
K'ok'a Hāyk'
Ābeltī
Indibir
Ch'ilalo Terara 13 579 ft. 4139 m
Lake Ziway (Ziway Hāyk')
Āsela
Robē
Ādamī Tulu
Merewa
T'icho
Abiyata Hāyk'
RIFT VALLEY
Mai Gudo 10 171 ft. 3100 m
Hosa'ina
RIFT VALLEY NATL. PARK
Langano Hāyk'
Bek'bjī
Wabē Shebelē
Shala Hāyk'
K'ech'a Terara 13 747 ft. 4190 m
Nacchie
Waka
Shashemenē
Āwasa Hāyk'
Ādaba
Batu 14 131 ft. 4307 m
Goba
Loma
Yirga 'Alem
Āwasa
Dodola
Sodo
MENDEBO
© Rand McNally A-101783-1

0 20 40 60 80 100 120 Miles

0 20 40 60 80 100 120 140 160 180 200 Kilometer

Lambert Azimuthal Equal Area Projection
Scale 1:4,000,000
One inch to 64 miles
One cm to 40 km

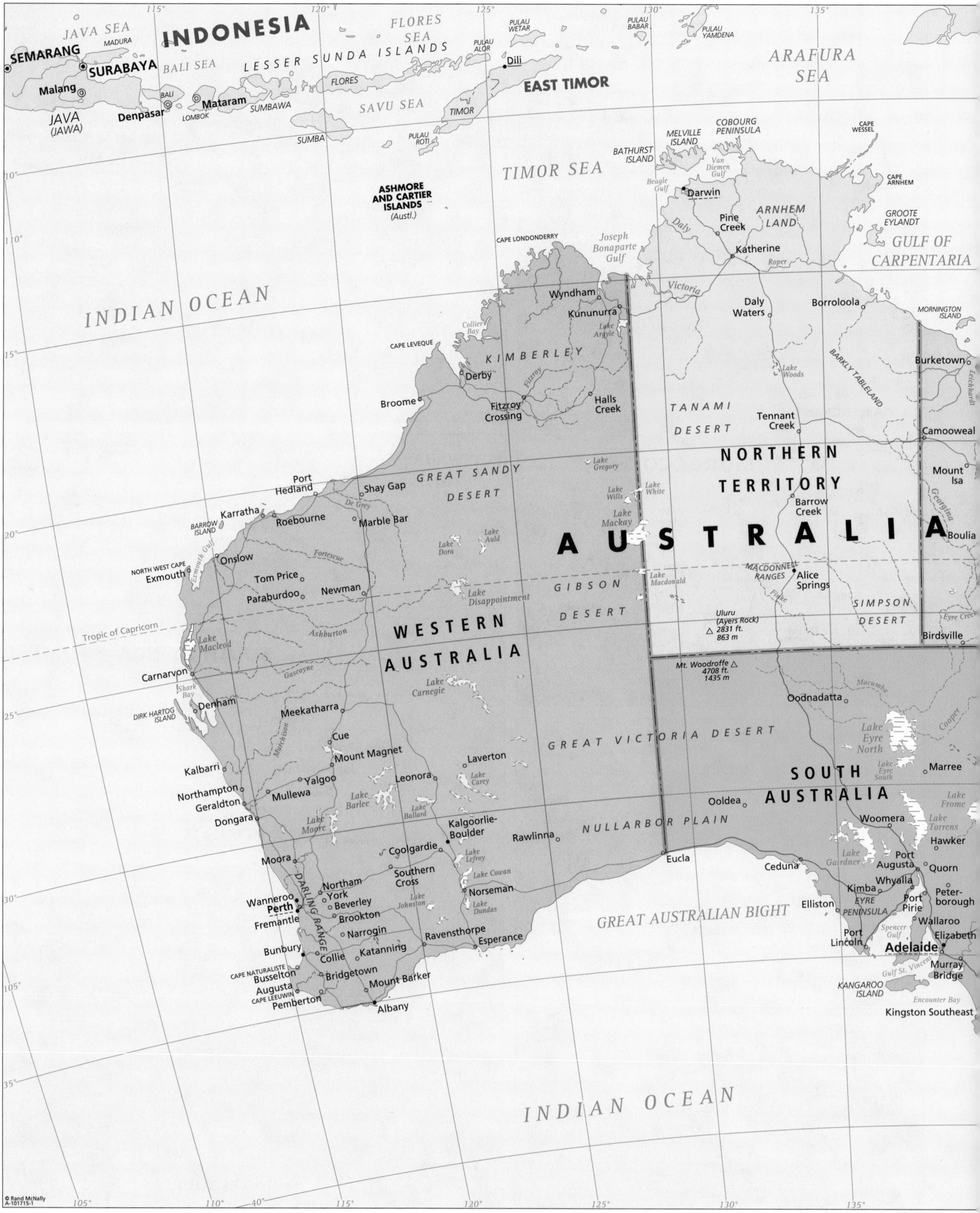

INDONESIA
JAVA SEA
MADURA
SEMARANG
SURABAYA
Malang
JAVA
(JAWA)
BALI SEA
BALI
Denpasar
LOMBOK
Mataram
SUMBAWA
LESSER SUNDA ISLANDS
FLORES SEA
FLORES
SAVU SEA
SUMBA
PULAU ROTI
TIMOR
PULAU ALOR
PULAU WETAR
Dili
EAST TIMOR
PULAU BABAR
PULAU YAMDENA
ARAFURA SEA
TIMOR SEA
ASHMORE AND CARTIER ISLANDS
(Austl.)
MELVILLE ISLAND
BATHURST ISLAND
COBOURG PENINSULA
CAPE WESSEL
Van Diemen Gulf
Beagle Gulf
Darwin
CAPE ARNHEM
ARNHEM LAND
Pine Creek
GROOTE EYLANDT
Daly
Katherine
Roper
GULF OF CARPENTARIA
CAPE LONDONDERRY
Joseph Bonaparte Gulf
Victoria
Wyndham
Kununurra
Lake Argyle
Daly Waters
Borroloola
MORNINGTON ISLAND
INDIAN OCEAN
Collier Bay
CAPE LEVEQUE
KIMBERLEY
Derby
Fitzroy
Broome
Fitzroy Crossing
Halls Creek
Lake Woods
BARKLY TABLELAND
Burketown
Leichhardt
TANAMI DESERT
Tennant Creek
Camooweal
NORTHERN TERRITORY
Mount Isa
Lake Gregory
GREAT SANDY DESERT
Port Hedland
Shay Gap
De Grey
Lake Wills
Lake White
Barrow Creek
Georgina
Karratha
BARROW ISLAND
Roebourne
Marble Bar
Lake Mackay
AUSTRALIA
Boulia
Lake Dora
Lake Auld
Exmouth Gulf
Onslow
Fortescue
NORTH WEST CAPE
Exmouth
Tom Price
Paraburdoo
Newman
Lake Disappointment
GIBSON DESERT
Lake Macdonald
MACDONNELL RANGES
Alice Springs
Finke
SIMPSON DESERT
Eyre Creek
Tropic of Capricorn
Lake Macleod
Ashburton
WESTERN AUSTRALIA
Uluru (Ayers Rock)
2831 ft.
863 m
Birdsville
Carnarvon
Gascoyne
Shark Bay
Lake Carnegie
Mt. Woodroffe
4708 ft.
1435 m
Macumba
DIRK HARTOG ISLAND
Denham
Meekatharra
Oodnadatta
Cooper
Murchison
Cue
Mount Magnet
GREAT VICTORIA DESERT
Lake Eyre North
Lake Eyre South
Laverton
Kalbarri
Yalgoo
Leonora
Lake Carey
SOUTH AUSTRALIA
Marree
Northampton
Geraldton
Mullewa
Lake Barlee
Lake Ballard
Ooldea
Lake Frome
Dongara
Lake Moore
Kalgoorlie-Boulder
Rawlinna
NULLARBOR PLAIN
Woomera
Lake Torrens
Coolgardie
Lake Lefroy
Hawker
Moora
Eucla
Lake Gairdner
Port Augusta
Quorn
Ceduna
DARLING RANGE
Northam
Southern Cross
Lake Cowan
Whyalla
Kimba
Peterborough
Wanneroo
York
Norseman
EYRE PENINSULA
Port Pirie
Perth
Beverley
Lake Johnston
Lake Dundas
Elliston
Fremantle
Brookton
GREAT AUSTRALIAN BIGHT
Wallaroo
Narrogin
Spencer Gulf
Elizabeth
Bunbury
Collie
Katanning
Ravensthorpe
Esperance
Port Lincoln
Adelaide
CAPE NATURALISTE
Busselton
Bridgetown
Gulf St. Vincent
Murray Bridge
Augusta
CAPE LEEUWIN
Mount Barker
KANGAROO ISLAND
Pemberton
Albany
Encounter Bay
Kingston Southeast
INDIAN OCEAN
115°
120°
125°
130°
135°
105°
110°
10°
15°
20°
25°
30°
35°
40°

0 80 160 240 320 400 480 Miles

0 80 160 240 320 400 480 560 640 720 800 Kilometers

Lambert Azimuthal Equal Area Projection
Scale 1:16,000,000
One inch to 256 miles
One cm to 160 km

NEW GUINEA
PAPUA NEW GUINEA
Fly
Gulf of Papua
Torres Strait
Port Moresby
SOLOMON SEA
BOUGAINVILLE
CHOISEUL
VELLA LAVELLA
SANTA ISABEL
NEW GEORGIA
RENDOVA ISLAND
SOLOMON ISLANDS
MALAITA
Honiara
GUADALCANAL
SAN CRISTOBAL
RENNELL
SANTA CRUZ ISLANDS
D'ENTRECASTEAUX ISLANDS
MUYUA ISLAND
LOUISIADE ARCHIPELAGO
TAGULA ISLAND
PRINCE OF WALES ISLAND
CAPE YORK
CAPE GRENVILLE
Weipa
CAPE YORK PENINSULA
CAPE MELVILLE
Cooktown
Laura
Mitchell
Mareeba
Cairns
Atherton
Innisfail
Karumba
Normanton
GREAT BARRIER REEF
HINCHINBROOK ISLAND
Ingham
Gilbert
Townsville
Burdekin
Bowen
Charters Towers
Cloncurry
Flinders
Richmond
Collinsville
Hughenden
Mackay
Sutton
Sarina
GREAT DIVIDING RANGE
Winton
Aramac
Clermont
Longreach
Ilfracombe
Emerald
Rockhampton
Fitzroy
Diamantina
Thomson
Barcaldine
Mount Morgan
Gladstone
GREAT ARTESIAN BASIN
Blackall
Barcoo
Bundaberg
SANDY CAPE
Hervey Bay
QUEENSLAND
FRASER ISLAND
Maryborough
Lake Yamma Yamma
Dawson
Quilpie
Charleville
Gympie
Nambour
GREY RANGE
Roma
Maranoa
Dalby
Innamincka
Brisbane
Saint George
Toowoomba
Ipswich
Southport
Cunnamulla
Warrego
Culgoa
Macintyre
Warwick
Lismore
Moree
Tenterfield
Grafton
Inverell
Glen Innes
Bourke
Barwon
Narrabri
Namoi
Coffs Harbour
Armidale
Darling
Macquarie
Wilcannia
Cobar
Tamworth
Kempsey
Nyngan
Broken Hill
NEW SOUTH WALES
Dubbo
Port Macquarie
Maitland
Cessnock
Newcastle
Orange
Lachlan
Gosford
Wentworth
Parramatta
Mildura
Hay
Griffith
Campbelltown
Sydney
Loxton
Wagga Wagga
Goulburn
Wollongong
Murrumbidgee
Pinnaroo
Murray
Swan Hill
Deniliquin
AUSTRALIAN CAPITAL TERRITORY
JERVIS BAY TERRITORY
Canberra
Bordertown
Albury
Shepparton
Wodonga
Cooma
Bendigo
Horsham
Mt. Kosciuszko 7313 ft. 2229 m
Stawell
Maryborough
Ararat
Bega
Ballarat
VICTORIA
Hamilton
Geelong
Melbourne
Bairnsdale
CAPE HOWE
Portland
Warrnambool
CAPE OTWAY
KING ISLAND
Bass Strait
FLINDERS ISLAND
FURNEAUX GROUP
CAPE BARREN ISLAND
Smithton
Burnie
Devonport
TASMANIA
Launceston
Queenstown
Mt. Ossa 5305 ft. 1617 m
New Norfolk
Hobart
Port Arthur
SOUTH EAST CAPE
CORAL SEA
CORAL SEA ISLANDS TERRITORY (Austl.)
MELLISH REEF
ÎLES CHESTERFIELD
ÎLE DE SABLE
WRECK REEF
CAYE DE L'OBSERVATOIRE
CATO ISLAND
TORRES ISLANDS
BANKS ISLANDS
ESPIRITU SANTO
MALAKULA
VANUATU
NEW HEBRIDES
AMBRYM
ÉPI
ÉFATÉ
Port Vila
ERROMANGO
TANNA
ANATOM
NEW CALEDONIA
LOYALTY ISLANDS
LIFOU
MARE
NEW CALEDONIA (France)
Nouméa
ÎLE DES PINS
Tropic of Capricorn
PACIFIC OCEAN
LORD HOWE ISLAND (Austl.)
TASMAN SEA

a

NORTH CAPE
Whangarei
Auckland
Manukau
Hamilton
Tauranga
EAST CAPE
NORTH ISLAND
Rotorua
Lake Taupo
NEW ZEALAND
New Plymouth
Gisborne
CAPE EGMONT
Napier
Hastings
Wanganui
CAPE FAREWELL
Palmerston North
TASMAN SEA
Nelson
Lower Hutt
Westport
Blenheim
Wellington
Greymouth
SOUTH ISLAND
SOUTHERN ALPS
Rakaia
Christchurch
Aoraki 12 316 ft. 3754 m
Timaru
PACIFIC OCEAN
Alexandra
Clutha
WEST CAPE
Dunedin
BOUNTY TROUGH
Invercargill
STEWART ISLAND
© Rand McNally A-101717-1

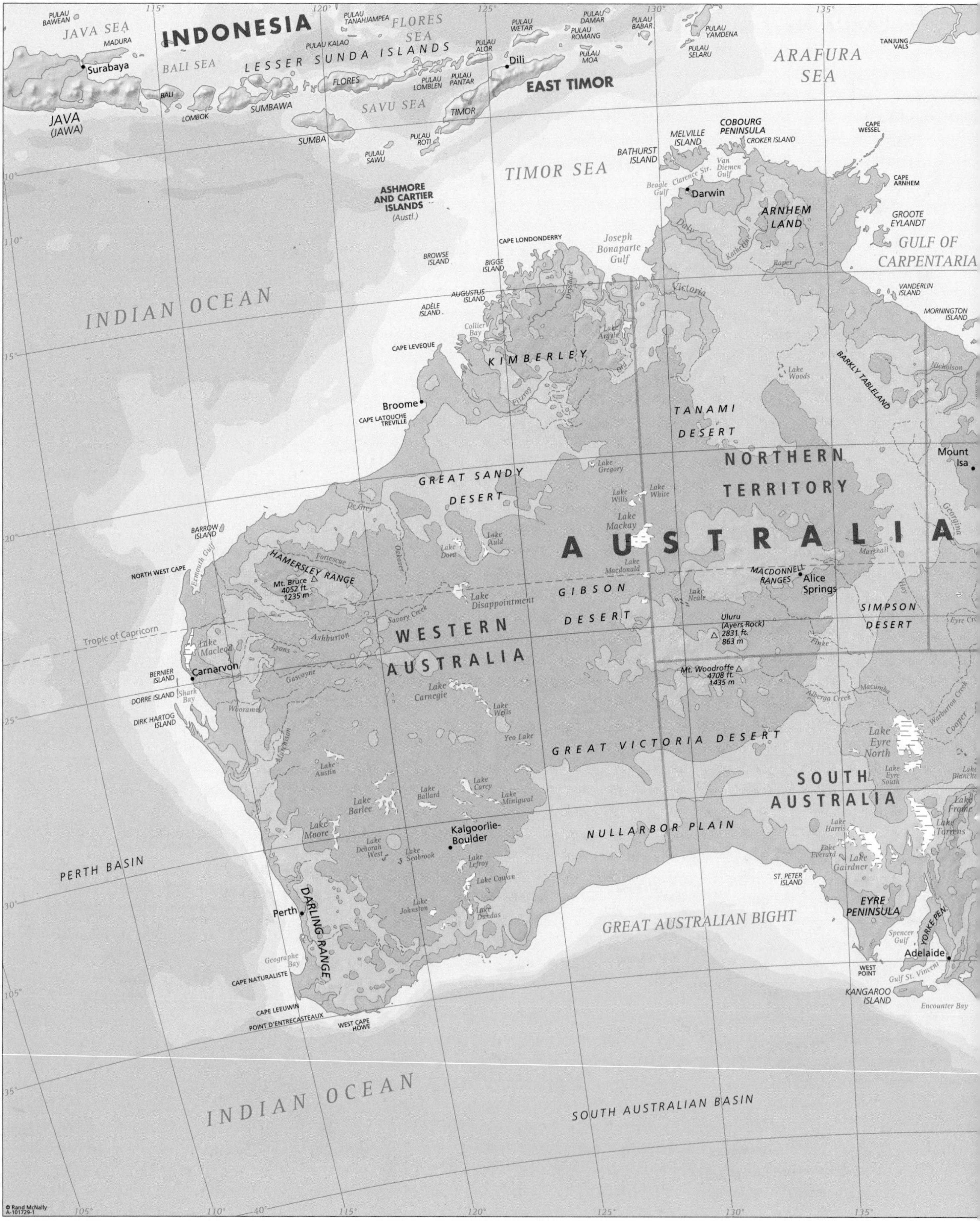

INDONESIA
JAVA SEA
PULAU BAWEAN
MADURA
Surabaya
JAVA (JAWA)
BALI SEA
BALI
LOMBOK
SUMBAWA
LESSER SUNDA ISLANDS
PULAU KALAO
PULAU TANAHJAMPEA
FLORES SEA
FLORES
SUMBA
SAVU SEA
PULAU SAWU
PULAU ROTI
TIMOR
PULAU LOMBLEN
PULAU PANTAR
PULAU ALOR
Dili
EAST TIMOR
PULAU WETAR
PULAU DAMAR
PULAU ROMANG
PULAU MOA
PULAU BABAR
PULAU YAMDENA
PULAU SELARU
ARAFURA SEA
TANJUNG VALS
TIMOR SEA
ASHMORE AND CARTIER ISLANDS (Austl.)
BATHURST ISLAND
MELVILLE ISLAND
COBOURG PENINSULA
CROKER ISLAND
Clarence Str.
Van Diemen Gulf
Beagle Gulf
Darwin
ARNHEM LAND
CAPE WESSEL
CAPE ARNHEM
GROOTE EYLANDT
GULF OF CARPENTARIA
Daly
Katherine
Roper
Joseph Bonaparte Gulf
CAPE LONDONDERRY
BROWSE ISLAND
BIGGE ISLAND
Drysdale
Victoria
VANDERLIN ISLAND
MORNINGTON ISLAND
AUGUSTUS ISLAND
ADÈLE ISLAND
Collier Bay
Lake Argyle
CAPE LEVEQUE
KIMBERLEY
Ord
Fitzroy
INDIAN OCEAN
Broome
CAPE LATOUCHE TREVILLE
BARKLY TABLELAND
Nicholson
Lake Woods
TANAMI DESERT
NORTHERN TERRITORY
Mount Isa
Lake Gregory
GREAT SANDY DESERT
De Grey
Lake Wills
Lake White
Lake Mackay
Georgina
BARROW ISLAND
Exmouth Gulf
NORTH WEST CAPE
HAMERSLEY RANGE
Fortescue
Mt. Bruce 4052 ft. 1235 m
Oakover
Lake Dora
Lake Auld
AUSTRALIA
Lake Macdonald
MACDONNELL RANGES
Alice Springs
Marshall
Hay
Lake Disappointment
GIBSON DESERT
Lake Neale
SIMPSON DESERT
Savory Creek
Tropic of Capricorn
WESTERN AUSTRALIA
Ashburton
Lyons
Lake Macleod
Uluru (Ayers Rock) 2831 ft. 863 m
Finke
Eyre Creek
BERNIER ISLAND
Carnarvon
Gascoyne
DORRE ISLAND
Shark Bay
DIRK HARTOG ISLAND
Wooramel
Mt. Woodroffe 4708 ft. 1435 m
Alberga Creek
Macumba
Warburton Creek
Cooper
Lake Carnegie
Lake Wells
Yeo Lake
Murchison
Lake Eyre North
Lake Eyre South
GREAT VICTORIA DESERT
Lake Austin
Lake Blanche
SOUTH AUSTRALIA
Lake Barlee
Lake Ballard
Lake Carey
Lake Minigwal
Lake Frome
Lake Torrens
Lake Moore
Lake Deborah West
Lake Seabrook
Kalgoorlie-Boulder
NULLARBOR PLAIN
Lake Harris
Lake Everard
Lake Gairdner
Lake Lefroy
Lake Cowan
ST. PETER ISLAND
PERTH BASIN
Lake Johnston
Lake Dundas
EYRE PENINSULA
Perth
DARLING RANGE
GREAT AUSTRALIAN BIGHT
Spencer Gulf
YORKE PEN.
Adelaide
Geographe Bay
WEST POINT
Gulf St. Vincent
CAPE NATURALISTE
KANGAROO ISLAND
Encounter Bay
CAPE LEEUWIN
POINT D'ENTRECASTEAUX
WEST CAPE HOWE
INDIAN OCEAN
SOUTH AUSTRALIAN BASIN
© Rand McNally A-101729-1

0 80 160 240 320 400 480 Miles

0 80 160 240 320 400 480 560 640 720 800 Kilometers

Lambert Azimuthal Equal Area Projection
Scale 1:16,000,000
One inch to 256 miles
One cm to 160 km

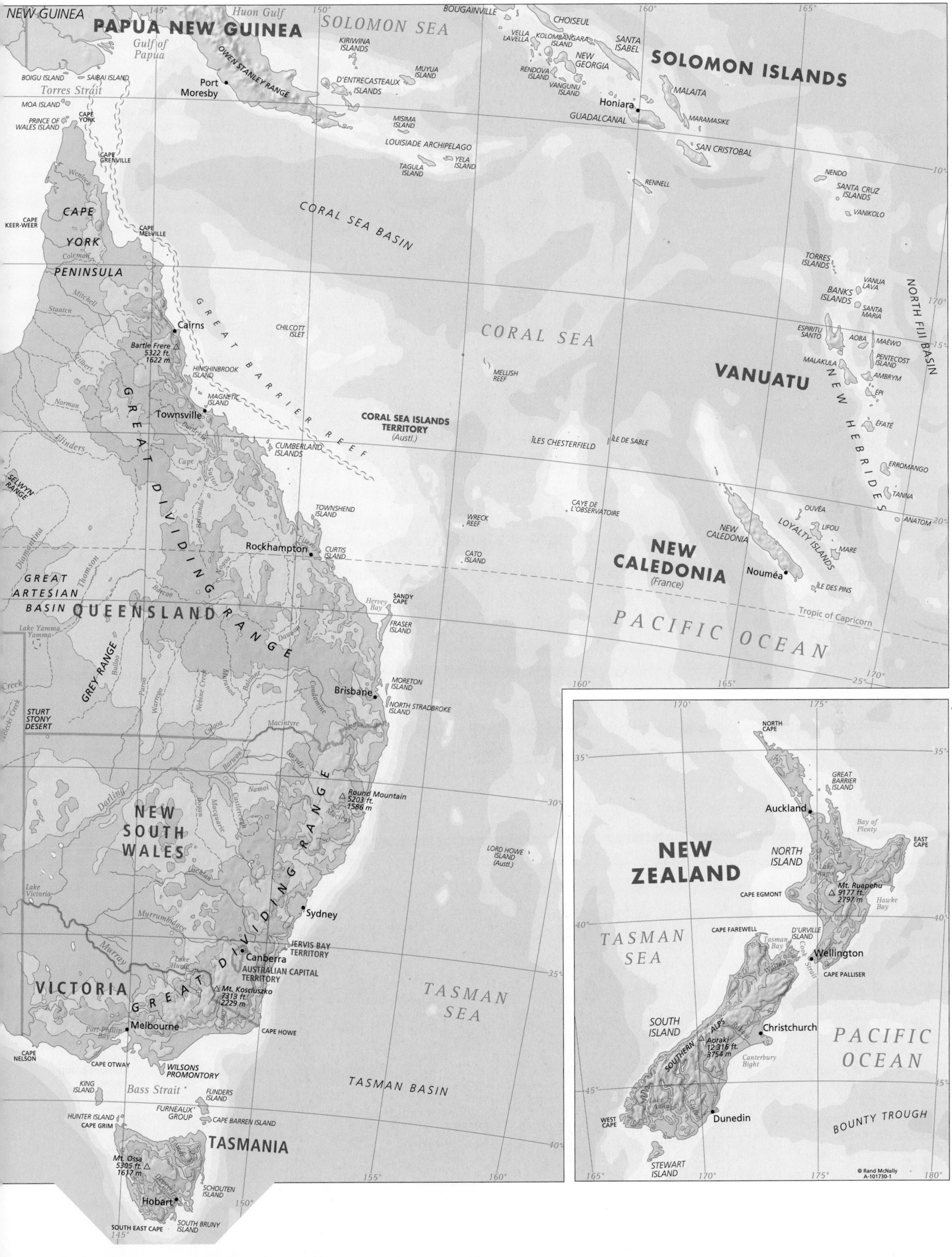
NEW GUINEA
PAPUA NEW GUINEA
Huon Gulf
SOLOMON SEA
BOUGAINVILLE
CHOISEUL
Gulf of Papua
OWEN STANLEY RANGE
KIRIWINA ISLANDS
VELLA LAVELLA
KOLOMBANGARA ISLAND
SANTA ISABEL
NEW GEORGIA
SOLOMON ISLANDS
BOIGU ISLAND
SAIBAI ISLAND
Torres Strait
Port Moresby
D'ENTRECASTEAUX ISLANDS
MUYUA ISLAND
RENDOVA ISLAND
VANGUNU ISLAND
MALAITA
Honiara
MOA ISLAND
GUADALCANAL
PRINCE OF WALES ISLAND
CAPE YORK
MISIMA ISLAND
MARAMASIKE
LOUISIADE ARCHIPELAGO
SAN CRISTOBAL
CAPE GRENVILLE
YELA ISLAND
TAGULA ISLAND
NENDO
RENNELL
SANTA CRUZ ISLANDS
VANIKOLO
Wenlock
CAPE
CORAL SEA BASIN
CAPE KEER-WEER
CAPE MELVILLE
YORK
Coleman
PENINSULA
TORRES ISLANDS
VANUA LAVA
Mitchell
BANKS ISLANDS
SANTA MARIA
NORTH FIJI BASIN
Staaten
GREAT BARRIER REEF
Cairns
CHILCOTT ISLET
CORAL SEA
ESPIRITU SANTO
AOBA
MAEWO
Bartle Frere 5322 ft. 1622 m
PENTECOST ISLAND
MALAKULA
Gilbert
HINCHINBROOK ISLAND
MELLISH REEF
AMBRYM
VANUATU
NEW HEBRIDES
ÉPI
MAGNETIC ISLAND
Norman
Townsville
CORAL SEA ISLANDS TERRITORY (Austl.)
ÉFATÉ
Burdekin
Flinders
CUMBERLAND ISLANDS
ÎLES CHESTERFIELD
ÎLE DE SABLE
Cape
ERROMANGO
Suttor
SELWYN RANGE
TANNA
Belyando
CAYE DE L'OBSERVATOIRE
OUVÉA
TOWNSHEND ISLAND
WRECK REEF
ANATOM
LIFOU
LOYALTY ISLANDS
NEW CALEDONIA
Diamantina
MARE
Rockhampton
CURTIS ISLAND
CATO ISLAND
NEW CALEDONIA (France)
Thomson
Nouméa
GREAT ARTESIAN BASIN
Nogoa
ÎLE DES PINS
Barcoo
SANDY CAPE
Hervey Bay
Tropic of Capricorn
QUEENSLAND
FRASER ISLAND
PACIFIC OCEAN
Lake Yamma Yamma
Dawson
GREY RANGE
Bulloo
MORETON ISLAND
Creek
Paroo
Warrego
Nebine Creek
Maranoa
Balonne
Condamine
Brisbane
NORTH STRADBROKE ISLAND
Strzelecki Creek
STURT STONY DESERT
Culgoa
Macintyre
Barwon
Gwydir
Namoi
Round Mountain 5203 ft. 1586 m
Darling
NEW SOUTH WALES
Bogan
Macquarie
Castlereagh
Macleay
LORD HOWE ISLAND (Austl.)
Lachlan
Lake Victoria
Sydney
Murrumbidgee
Murray
JERVIS BAY TERRITORY
Canberra
AUSTRALIAN CAPITAL TERRITORY
Lake Hume
VICTORIA
Mt. Kosciuszko 7313 ft. 2229 m
TASMAN SEA
Snowy
Melbourne
Port Phillip Bay
CAPE HOWE
CAPE NELSON
CAPE OTWAY
WILSONS PROMONTORY
TASMAN BASIN
KING ISLAND
Bass Strait
FLINDERS ISLAND
HUNTER ISLAND
FURNEAUX GROUP
CAPE BARREN ISLAND
CAPE GRIM
TASMANIA
Mt. Ossa 5305 ft. 1617 m
Derwent
SCHOUTEN ISLAND
Hobart
SOUTH BRUNY ISLAND
SOUTH EAST CAPE
145°
150°
155°
160°
165°
170°
10°
15°
20°
25°
30°
35°
40°
175°
NORTH CAPE
35°
GREAT BARRIER ISLAND
Auckland
Bay of Plenty
EAST CAPE
NEW ZEALAND
NORTH ISLAND
Waikato
Lake Taupo
Mt. Ruapehu 9177 ft. 2797 m
CAPE EGMONT
Hawke Bay
40°
TASMAN SEA
CAPE FAREWELL
D'URVILLE ISLAND
Tasman Bay
Cook Strait
Wellington
Wairau
CAPE PALLISER
SOUTH ISLAND
SOUTHERN ALPS
Christchurch
Aoraki 12 316 ft. 3754 m
PACIFIC OCEAN
Canterbury Bight
45°
Lake Anau
Clutha
WEST CAPE
Dunedin
BOUNTY TROUGH
STEWART ISLAND
165°
180°
© Rand McNally A-101730-1

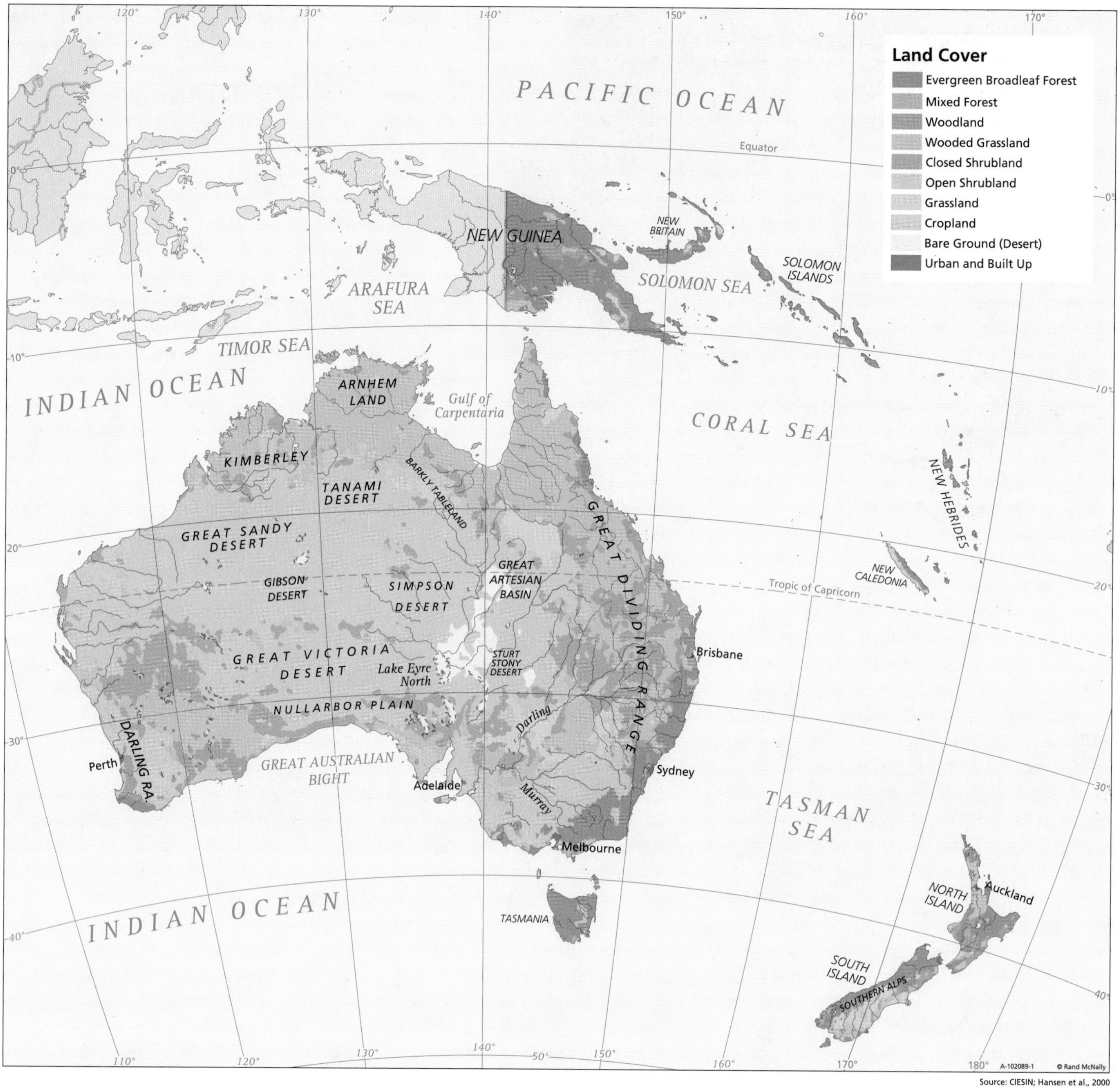
Land Cover
Evergreen Broadleaf Forest
Mixed Forest
Woodland
Wooded Grassland
Closed Shrubland
Open Shrubland
Grassland
Cropland
Bare Ground (Desert)
Urban and Built Up
PACIFIC OCEAN
Equator
NEW GUINEA
NEW BRITAIN
SOLOMON ISLANDS
SOLOMON SEA
ARAFURA SEA
TIMOR SEA
INDIAN OCEAN
ARNHEM LAND
Gulf of Carpentaria
CORAL SEA
KIMBERLEY
TANAMI DESERT
BARKLY TABLELAND
GREAT DIVIDING RANGE
NEW HEBRIDES
GREAT SANDY DESERT
GREAT ARTESIAN BASIN
NEW CALEDONIA
GIBSON DESERT
SIMPSON DESERT
Tropic of Capricorn
GREAT VICTORIA DESERT
Lake Eyre North
STURT STONY DESERT
Brisbane
NULLARBOR PLAIN
Darling
DARLING RA.
Perth
GREAT AUSTRALIAN BIGHT
Adelaide
Sydney
Murray
TASMAN SEA
Melbourne
NORTH ISLAND
Auckland
TASMANIA
INDIAN OCEAN
SOUTH ISLAND
SOUTHERN ALPS
A-102089-1
© Rand McNally

Source: CIESIN; Hansen et al., 2000

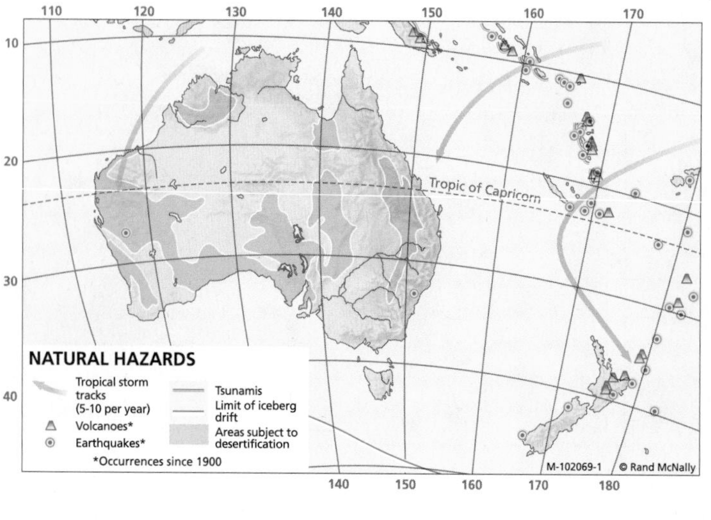
Tropic of Capricorn
NATURAL HAZARDS
Tropical storm tracks (5-10 per year)
Volcanoes*
Earthquakes*
Tsunamis
Limit of iceberg drift
Areas subject to desertification
*Occurrences since 1900
M-102069-1
© Rand McNally

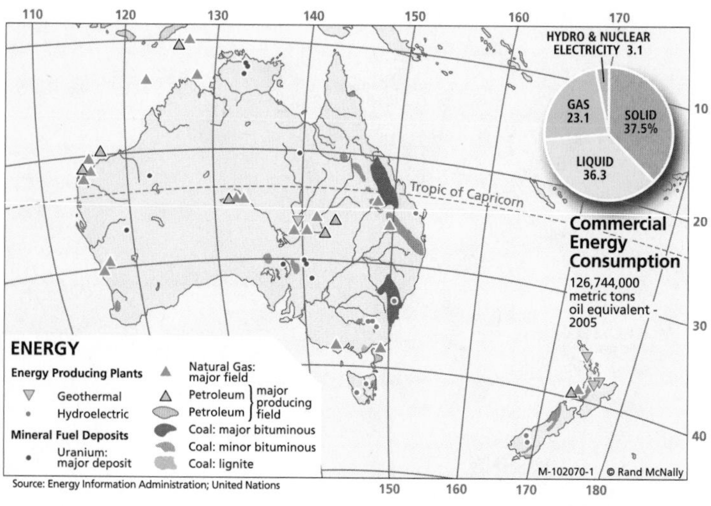
HYDRO & NUCLEAR ELECTRICITY 3.1
GAS 23.1
SOLID 37.5%
LIQUID 36.3
Commercial Energy Consumption
126,744,000 metric tons oil equivalent - 2005
Tropic of Capricorn
ENERGY
Energy Producing Plants
Geothermal
Hydroelectric
Mineral Fuel Deposits
Uranium: major deposit
Natural Gas: major field
Petroleum
Petroleum
major producing field
Coal: major bituminous
Coal: minor bituminous
Coal: lignite
M-102070-1
© Rand McNally

Source: Energy Information Administration; United Nations

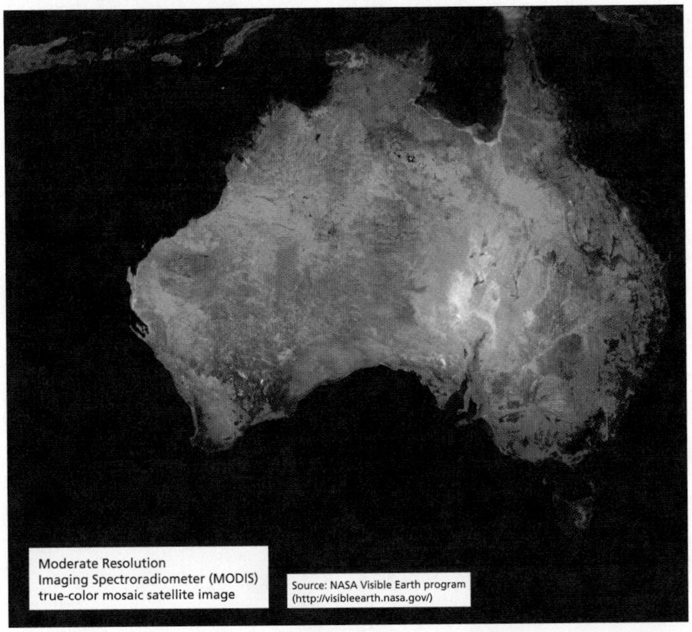
Moderate Resolution
Imaging Spectroradiometer (MODIS)
true-color mosaic satellite image
Source: NASA Visible Earth program
(http://visibleearth.nasa.gov/)

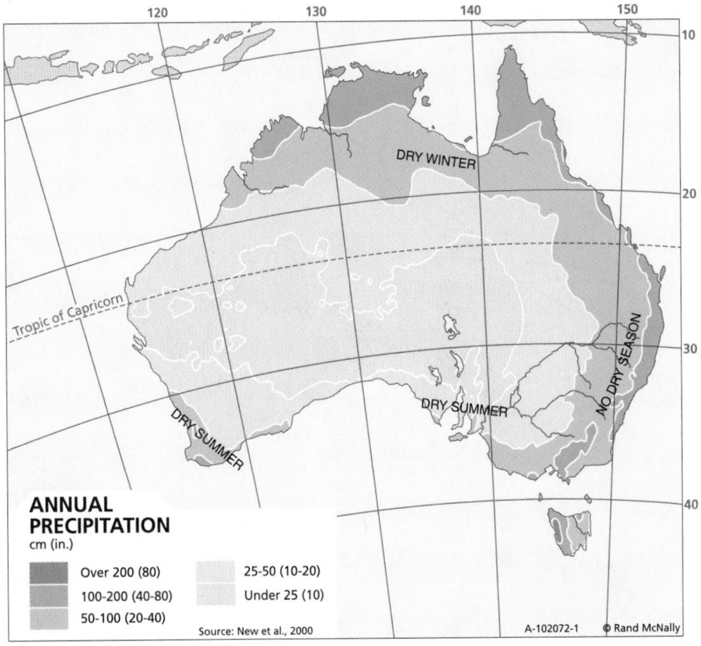
120
130
140
150
10
20
30
40
DRY WINTER
Tropic of Capricorn
NO DRY SEASON
DRY SUMMER
DRY SUMMER
ANNUAL
PRECIPITATION
cm (in.)
Over 200 (80)
100-200 (40-80)
50-100 (20-40)
25-50 (10-20)
Under 25 (10)
Source: New et al., 2000
A-102072-1
© Rand McNally

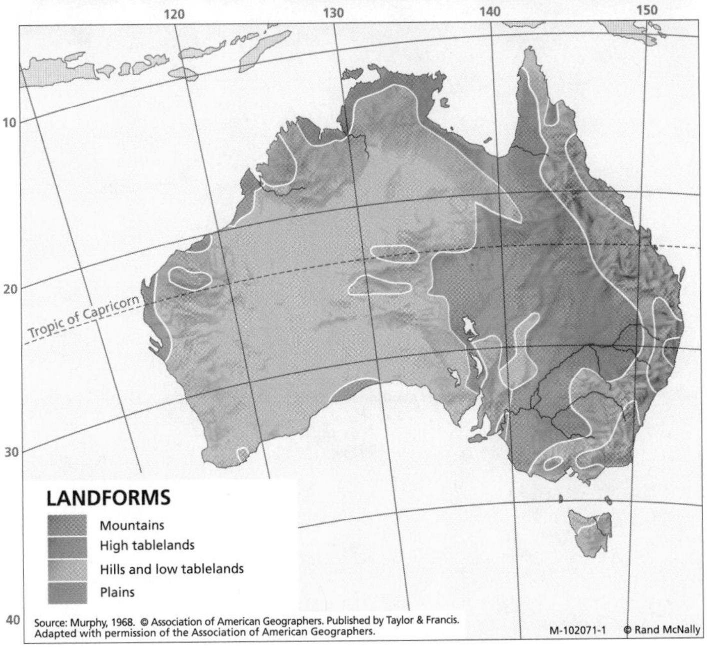
120
130
140
150
10
20
30
40
Tropic of Capricorn
LANDFORMS
Mountains
High tablelands
Hills and low tablelands
Plains
Source: Murphy, 1968. © Association of American Geographers. Published by Taylor & Francis.
Adapted with permission of the Association of American Geographers.
M-102071-1
© Rand McNally

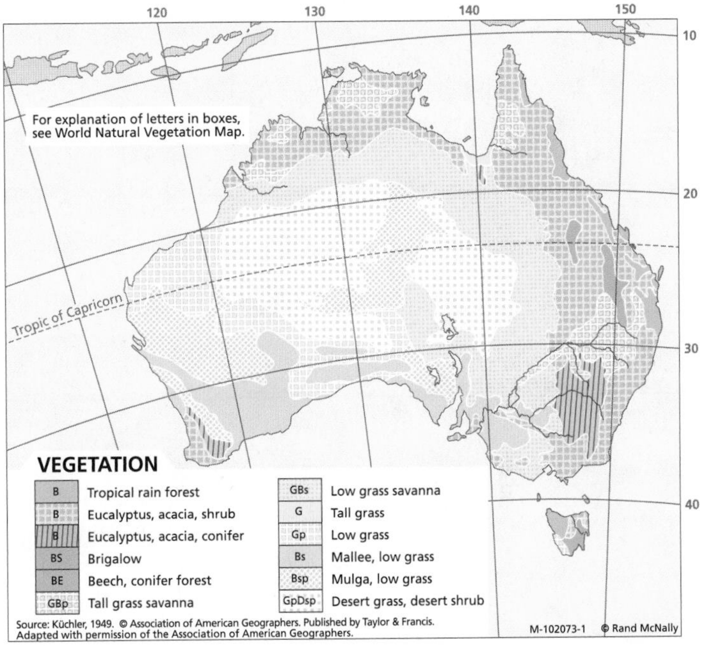
120
130
140
150
10
20
30
40
For explanation of letters in boxes,
see World Natural Vegetation Map.
Tropic of Capricorn
VEGETATION
B Tropical rain forest
B Eucalyptus, acacia, shrub
B Eucalyptus, acacia, conifer
BS Brigalow
BE Beech, conifer forest
GBp Tall grass savanna
GBs Low grass savanna
G Tall grass
Gp Low grass
Bs Mallee, low grass
Bsp Mulga, low grass
GpDsp Desert grass, desert shrub
Source: Küchler, 1949. © Association of American Geographers. Published by Taylor & Francis.
Adapted with permission of the Association of American Geographers.
M-102073-1
© Rand McNally

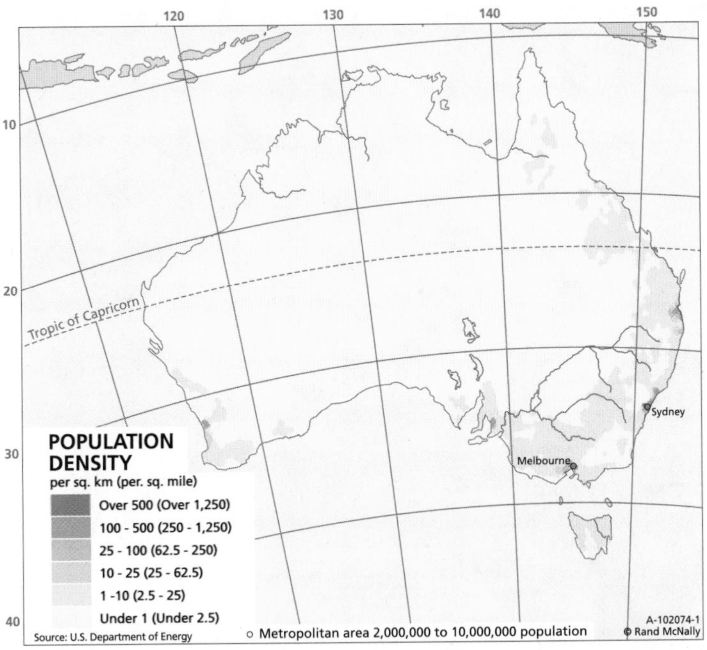
120
130
140
150
10
20
30
40
Tropic of Capricorn
Sydney
Melbourne
POPULATION
DENSITY
per sq. km (per. sq. mile)
Over 500 (Over 1,250)
100 - 500 (250 - 1,250)
25 - 100 (62.5 - 250)
10 - 25 (25 - 62.5)
1 -10 (2.5 - 25)
Under 1 (Under 2.5)
Source: U.S. Department of Energy
○ Metropolitan area 2,000,000 to 10,000,000 population
A-102074-1
© Rand McNally

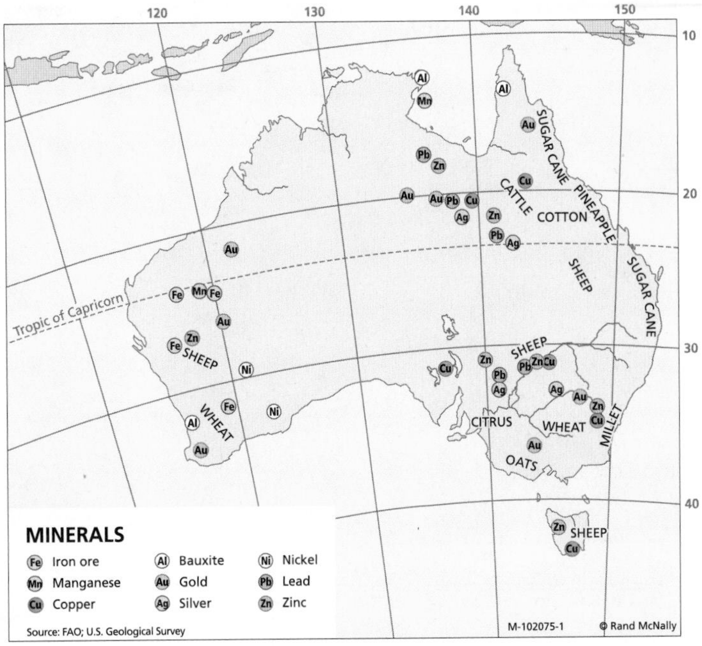
120
130
140
150
10
20
30
40
SUGAR CANE
CATTLE
COTTON
PINEAPPLE
SHEEP
SUGAR CANE
Tropic of Capricorn
SHEEP
SHEEP
WHEAT
CITRUS
WHEAT
MILLET
OATS
SHEEP
MINERALS
Fe Iron ore
Mn Manganese
Cu Copper
Al Bauxite
Au Gold
Ag Silver
Ni Nickel
Pb Lead
Zn Zinc
Source: FAO; U.S. Geological Survey
M-102075-1
© Rand McNally

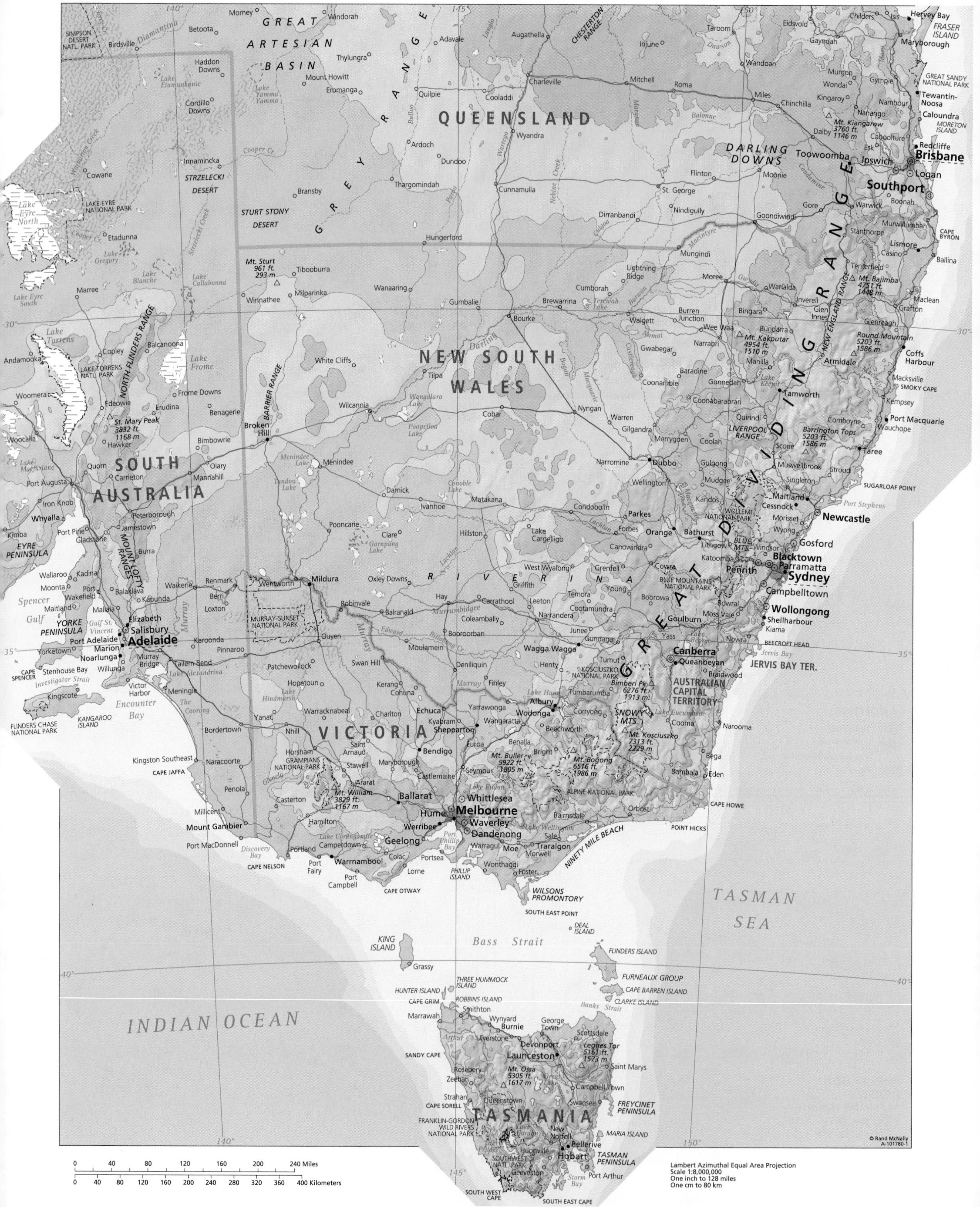

GREAT ARTESIAN BASIN
GREY RANGE
QUEENSLAND
NEW SOUTH WALES
SOUTH AUSTRALIA
VICTORIA
TASMANIA
RIVERINA
GREAT DIVIDING RANGE
DARLING DOWNS
STRZELECKI DESERT
STURT STONY DESERT
SIMPSON DESERT NATL. PARK
LAKE EYRE NATIONAL PARK
Lake Eyre North
Lake Eyre South
Lake Torrens
LAKE TORRENS NATL. PARK
NORTH FLINDERS RANGE
Lake Frome
Lake Blanche
Lake Callabonna
Lake Gregory
BARRIER RANGE
MOUNT LOFTY RANGES
St. Mary Peak 3832 ft. 1168 m
Mt. Sturt 961 ft. 293 m
EYRE PENINSULA
YORKE PENINSULA
Spencer Gulf
Gulf St. Vincent
Investigator Strait
KANGAROO ISLAND
FLINDERS CHASE NATIONAL PARK
CAPE SPENCER
Encounter Bay
Lake Alexandrina
The Coorong
CAPE JAFFA
MURRAY-SUNSET NATIONAL PARK
GRAMPIANS NATIONAL PARK
Mt. William 3829 ft. 1167 m
Discovery Bay
CAPE NELSON
CAPE OTWAY
PHILLIP ISLAND
Port Phillip Bay
WILSONS PROMONTORY
SOUTH EAST POINT
NINETY MILE BEACH
POINT HICKS
CAPE HOWE
Mt. Buller 5922 ft. 1805 m
Mt. Bogong 6516 ft. 1986 m
ALPINE NATIONAL PARK
Mt. Kosciuszko 7313 ft. 2229 m
KOSCIUSZKO NATIONAL PARK
SNOWY MTS.
Bimberi Pk. 6276 ft. 1913 m
AUSTRALIAN CAPITAL TERRITORY
JERVIS BAY TER.
BEECROFT HEAD
BLUE MOUNTAINS NATIONAL PARK
BLUE MTS.
WOLLEMI NATIONAL PARK
LIVERPOOL RANGE
Barrington Tops 5203 ft. 1586 m
SUGARLOAF POINT
Port Stephens
SMOKY CAPE
Round Mountain 5203 ft. 1586 m
Mt. Kaputar 4954 ft. 1510 m
NEW ENGLAND RANGE
Mt. Bajimba 4751 ft. 1448 m
CAPE BYRON
Mt. Kiangarow 3760 ft. 1146 m
MORETON ISLAND
GREAT SANDY NATIONAL PARK
FRASER ISLAND
CHESTERTON RANGE
TASMAN SEA
INDIAN OCEAN
Bass Strait
KING ISLAND
DEAL ISLAND
FLINDERS ISLAND
FURNEAUX GROUP
CAPE BARREN ISLAND
CLARKE ISLAND
Banks Strait
THREE HUMMOCK ISLAND
HUNTER ISLAND
ROBBINS ISLAND
CAPE GRIM
SANDY CAPE
CAPE SORELL
Legges Tor 5161 ft. 1573 m
Mt. Ossa 5305 ft. 1617 m
FREYCINET PENINSULA
MARIA ISLAND
TASMAN PENINSULA
FRANKLIN-GORDON WILD RIVERS NATIONAL PARK
SOUTHWEST NATL. PARK
Storm Bay
SOUTH WEST CAPE
SOUTH EAST CAPE
Brisbane
Sydney
Melbourne
Adelaide
Canberra
Hobart
Newcastle
Wollongong
Geelong
Ballarat
Bendigo
Launceston
Toowoomba
Southport
Ipswich
Broken Hill
Mildura
Wagga Wagga
Albury
Dubbo
Tamworth
Port Macquarie
Coffs Harbour
Whyalla
Port Augusta
Port Pirie
Mount Gambier
Parramatta
Blacktown
Penrith
Campbelltown
Gosford
Devonport
Burnie
140°
145°
150°
30°
35°
40°
© Rand McNally A-101780-1
0 40 80 120 160 200 240 Miles
0 40 80 120 160 200 240 280 320 360 400 Kilometers
Lambert Azimuthal Equal Area Projection
Scale 1:8,000,000
One inch to 128 miles
One cm to 80 km

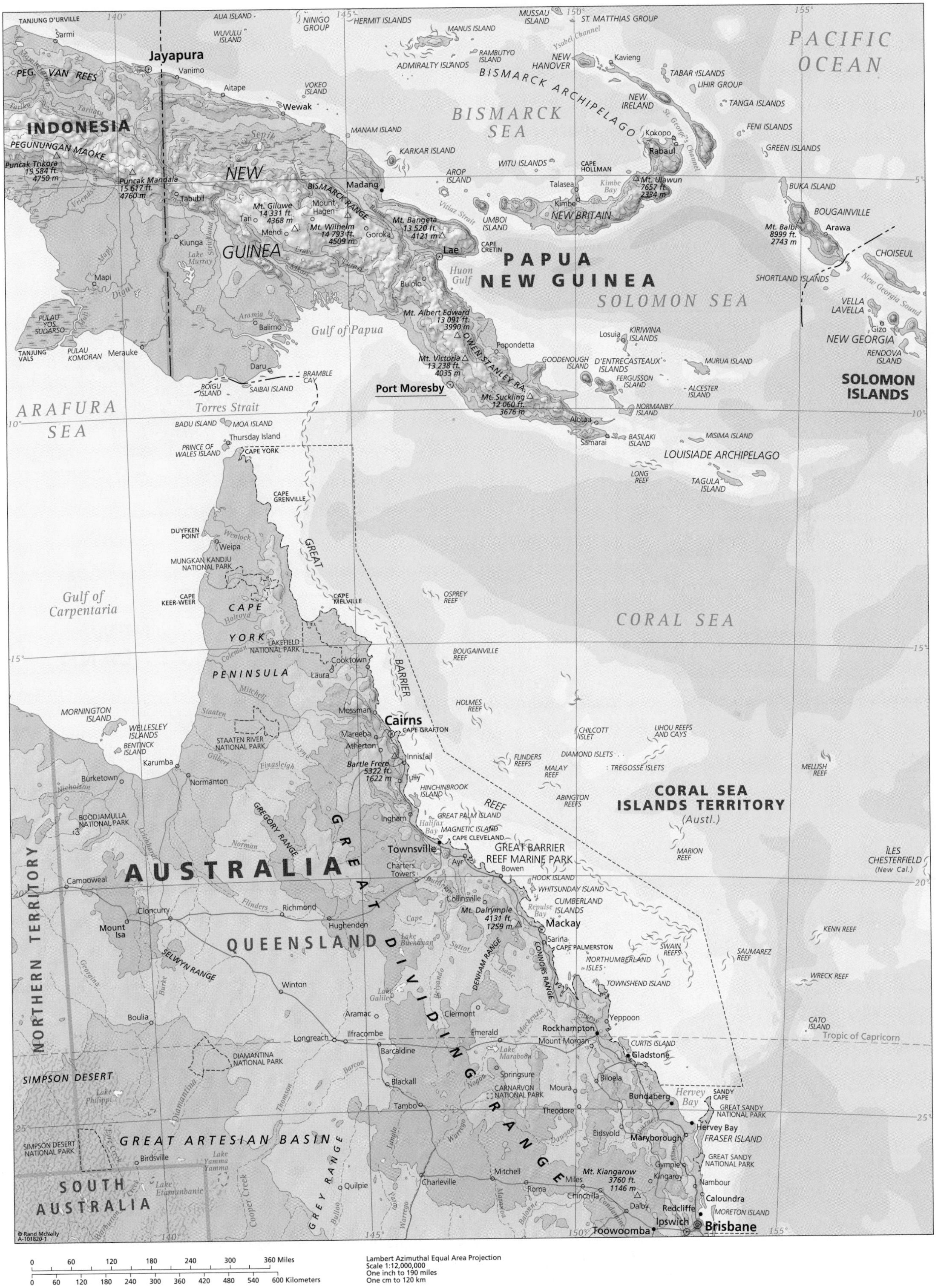
PACIFIC OCEAN
BISMARCK SEA
BISMARCK ARCHIPELAGO
INDONESIA
PAPUA NEW GUINEA
NEW GUINEA
SOLOMON SEA
SOLOMON ISLANDS
ARAFURA SEA
Gulf of Papua
Torres Strait
Gulf of Carpentaria
CORAL SEA
CORAL SEA ISLANDS TERRITORY
(Austl.)
AUSTRALIA
QUEENSLAND
NORTHERN TERRITORY
SOUTH AUSTRALIA
CAPE YORK PENINSULA
GREAT DIVIDING RANGE
GREAT BARRIER REEF
GREAT BARRIER REEF MARINE PARK
GREAT ARTESIAN BASIN
SIMPSON DESERT
Tropic of Capricorn
Jayapura
Vanimo
Aitape
Wewak
Sarmi
TANJUNG D'URVILLE
PEG. VAN REES
PEGUNUNGAN MAOKE
Puncak Trikora 15 584 ft. 4750 m
Puncak Mandala 15 617 ft. 4760 m
Tabubil
Kiunga
Mapi
Merauke
PULAU YOS SUDARSO
PULAU KOMORAN
TANJUNG VALS
Madang
Mount Hagen
Mt. Giluwe 14 331 ft. 4368 m
Mt. Wilhelm 14 793 ft. 4509 m
Mt. Bangeta 13 520 ft. 4121 m
Tari
Mendi
Goroka
Lae
Bulolo
Huon Gulf
Balimo
Daru
BISMARCK RANGE
Mt. Albert Edward 13 091 ft. 3990 m
Mt. Victoria 13 238 ft. 4035 m
Mt. Suckling 12 060 ft. 3676 m
OWEN STANLEY RA.
Popondetta
Port Moresby
Alotau
Samarai
AUA ISLAND
WUVULU ISLAND
NINIGO GROUP
HERMIT ISLANDS
MANUS ISLAND
ADMIRALTY ISLANDS
RAMBUTYO ISLAND
MUSSAU ISLAND
ST. MATTHIAS GROUP
NEW HANOVER
Kavieng
NEW IRELAND
TABAR ISLANDS
LIHIR GROUP
TANGA ISLANDS
FENI ISLANDS
GREEN ISLANDS
Kokopo
Rabaul
VOKEO ISLAND
MANAM ISLAND
KARKAR ISLAND
AROP ISLAND
WITU ISLANDS
CAPE HOLLMAN
Talasea
Kimbe
Mt. Ulawun 7657 ft. 2334 m
NEW BRITAIN
Vitiaz Strait
UMBOI ISLAND
CAPE CRETIN
BUKA ISLAND
BOUGAINVILLE
Arawa
Mt. Balbi 8999 ft. 2743 m
CHOISEUL
SHORTLAND ISLANDS
VELLA LAVELLA
Gizo
NEW GEORGIA
RENDOVA ISLAND
Losuia
KIRIWINA ISLANDS
GOODENOUGH ISLAND
D'ENTRECASTEAUX ISLANDS
FERGUSSON ISLAND
MURUA ISLAND
ALCESTER ISLAND
NORMANBY ISLAND
BASILAKI ISLAND
MISIMA ISLAND
LOUISIADE ARCHIPELAGO
LONG REEF
TAGULA ISLAND
BOIGU ISLAND
SAIBAI ISLAND
BRAMBLE CAY
BADU ISLAND
MOA ISLAND
Thursday Island
PRINCE OF WALES ISLAND
CAPE YORK
CAPE GRENVILLE
DUYFKEN POINT
Weipa
MUNGKAN KANDJU NATIONAL PARK
CAPE KEER-WEER
CAPE MELVILLE
LAKEFIELD NATIONAL PARK
Cooktown
Laura
OSPREY REEF
BOUGAINVILLE REEF
HOLMES REEF
MORNINGTON ISLAND
WELLESLEY ISLANDS
BENTINCK ISLAND
Karumba
Burketown
Normanton
STAATEN RIVER NATIONAL PARK
Mossman
Cairns
CAPE GRAFTON
Mareeba
Atherton
Innisfail
Bartle Frere 5322 ft. 1622 m
Tully
HINCHINBROOK ISLAND
Ingham
GREAT PALM ISLAND
MAGNETIC ISLAND
CAPE CLEVELAND
Townsville
Charters Towers
Ayr
Bowen
CHILCOTT ISLET
LIHOU REEFS AND CAYS
DIAMOND ISLETS
FLINDERS REEFS
MALAY REEF
TREGOSSE ISLETS
ABINGTON REEFS
MELLISH REEF
MARION REEF
ÎLES CHESTERFIELD (New Cal.)
BOODJAMULLA NATIONAL PARK
GREGORY RANGE
Camooweal
Mount Isa
Cloncury
Richmond
Hughenden
Collinsville
Mt. Dalrymple 4131 ft. 1259 m
HOOK ISLAND
WHITSUNDAY ISLAND
CUMBERLAND ISLANDS
Mackay
Sarina
CAPE PALMERSTON
NORTHUMBERLAND ISLES
SWAIN REEFS
TOWNSHEND ISLAND
KENN REEF
SAUMAREZ REEF
WRECK REEF
CATO ISLAND
SELWYN RANGE
DENHAM RANGE
CONNORS RANGE
Winton
Boulia
Aramac
Clermont
Yeppoon
Rockhampton
Mount Morgan
Emerald
Longreach
Ilfracombe
Barcaldine
CURTIS ISLAND
Gladstone
DIAMANTINA NATIONAL PARK
Springsure
Blackall
CARNARVON NATIONAL PARK
Moura
Biloela
Bundaberg
Hervey Bay
SANDY CAPE
GREAT SANDY NATIONAL PARK
Tambo
Theodore
Hervey Bay
FRASER ISLAND
Eidsvold
Maryborough
SIMPSON DESERT NATIONAL PARK
Birdsville
GREY RANGE
Quilpie
Charleville
Mitchell
Roma
Chinchilla
Mt. Kiangarow 3760 ft. 1146 m
Miles
Gympie
Kingaroy
Nambour
Caloundra
Dalby
Redcliffe
MORETON ISLAND
Ipswich
Brisbane
Toowoomba
© Rand McNally A-101820-1
0 60 120 180 240 300 360 Miles
0 60 120 180 240 300 360 420 480 540 600 Kilometers

Lambert Azimuthal Equal Area Projection
Scale 1:12,000,000
One inch to 190 miles
One cm to 120 km

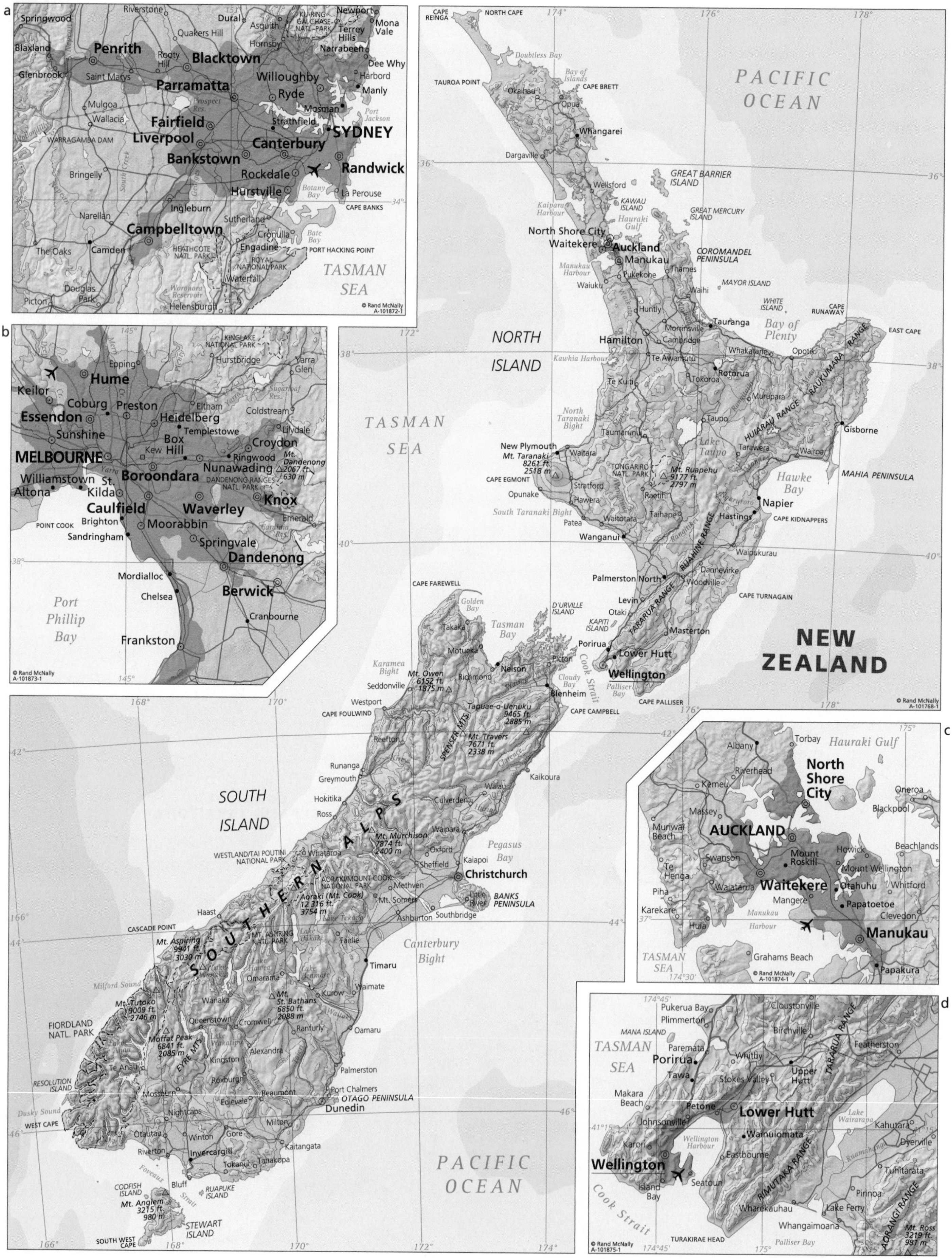
a
Riverstone
Dural
KU-RING-GAI CHASE NATL. PARK
Newport
Mona Vale
Springwood
Quakers Hill
Terrey Hills
Hornsby
Narrabeen
Blaxland
Penrith
Rooty Hill
Blacktown
Dee Why
Glenbrook
Saint Marys
Willoughby
Harbord
Parramatta
Ryde
Manly
Prospect Res.
Mulgoa
Mosman
Port Jackson
Wallacia
Fairfield
Strathfield
SYDNEY
WARRAGAMBA DAM
Liverpool
Canterbury
Bankstown
Rockdale
Randwick
Bringelly
Hurstville
Botany Bay
La Perouse
CAPE BANKS
Ingleburn
Narellan
Sutherland
Campbelltown
Cronulla
Bate Bay
The Oaks
Camden
HEATHCOTE NATL. PARK
Engadine
PORT HACKING POINT
ROYAL NATIONAL PARK
Woronora Reservoir
Waterfall
TASMAN SEA
Douglas Park
Picton
Helensburgh
© Rand McNally A-101872-1
b
KINGLAKE NATIONAL PARK
Epping
Hurstbridge
Yarra Glen
Hume
Sugarloaf Res.
Keilor
Coburg
Preston
Eltham
Essendon
Heidelberg
Coldstream
Sunshine
Templestowe
Lilydale
Box Hill
Croydon
MELBOURNE
Kew
Ringwood
Mt. Dandenong 2067 ft. 630 m
Williamstown
Nunawading
Altona
St. Kilda
Boroondara
DANDENONG RANGES NATL. PARK
Caulfield
Waverley
Knox
POINT COOK
Brighton
Moorabbin
Emerald
Sandringham
Cardinia Res.
Springvale
Dandenong
Mordialloc
Chelsea
Berwick
Port Phillip Bay
Cranbourne
Frankston
© Rand McNally A-101873-1
CAPE REINGA
NORTH CAPE
Doubtless Bay
Bay of Islands
CAPE BRETT
TAUROA POINT
Okaihau
Opua
PACIFIC OCEAN
Whangarei
Dargaville
GREAT BARRIER ISLAND
Wellsford
KAWAU ISLAND
Kaipara Harbour
Hauraki Gulf
GREAT MERCURY ISLAND
North Shore City
Waitekere
Auckland
Manukau
COROMANDEL PENINSULA
Manukau Harbour
Pukekohe
Thames
Waiuku
Waihi
MAYOR ISLAND
Huntly
WHITE ISLAND
CAPE RUNAWAY
Tauranga
Morrinsville
Bay of Plenty
EAST CAPE
NORTH ISLAND
Hamilton
Cambridge
Whakatane
Opotiki
Kawhia Harbour
Te Awamutu
RAUKUMARA RANGE
Te Kuiti
Rotorua
Tokoroa
Murupara
TASMAN SEA
North Taranaki Bight
Taupo
HUIARAU RANGE
Gisborne
Taumarunui
Lake Taupo
New Plymouth
Waitara
Mt. Taranaki 8261 ft. 2518 m
TONGARIRO NATL. PARK
Mt. Ruapehu 9177 ft. 2797 m
Wairoa
CAPE EGMONT
MAHIA PENINSULA
Hawke Bay
Opunake
Stratford
Hawera
Raetihi
Napier
South Taranaki Bight
Taihape
Hastings
CAPE KIDNAPPERS
Patea
Waitotara
Wanganui
Waipukurau
RUAHINE RANGE
Palmerston North
Dannevirke
CAPE FAREWELL
Woodville
Golden Bay
D'URVILLE ISLAND
Levin
CAPE TURNAGAIN
Tasman Bay
KAPITI ISLAND
Otaki
Takaka
TARARUA RANGE
Masterton
Motueka
NEW ZEALAND
Porirua
Lower Hutt
Karamea Bight
Nelson
Picton
Wellington
Mt. Owen 6152 ft. 1875 m
Richmond
Cook Strait
Seddonville
Cloudy Bay
Palliser Bay
Blenheim
Westport
CAPE PALLISER
CAPE FOULWIND
CAPE CAMPBELL
Tapuae-o-Uenuku 9465 ft. 2885 m
Reefton
SPENSER MTS.
Mt. Travers 7671 ft. 2338 m
Runanga
Greymouth
Kaikoura
SOUTH ISLAND
Hokitika
Culverden
Ross
Waiau
Mt. Murchison 7874 ft. 2400 m
Waipara
WESTLAND/TAI POUTINI NATIONAL PARK
Whataroa
Oxford
Pegasus Bay
Sheffield
Kaiapoi
AORAKI/MOUNT COOK NATIONAL PARK
Christchurch
Methven
Aoraki (Mt. Cook) 12,316 ft. 3754 m
Mt. Somers
Little River
BANKS PENINSULA
Haast
Ashburton
Southbridge
CASCADE POINT
Lake Tekapo
Mt. Aspiring 9941 ft. 3030 m
MT. ASPIRING NATL. PARK
Fairlie
SOUTHERN ALPS
Canterbury Bight
Timaru
Milford Sound
Omarama
Mt. Tutoko 9009 ft. 2746 m
Mt. St. Bathans 6850 ft. 2088 m
Waimate
Kurow
Wanaka
FIORDLAND NATL. PARK
Queenstown
Cromwell
Ranfurly
Oamaru
Moffat Peak 6841 ft. 2085 m
EYRE MTS.
Alexandra
Kingston
Te Anau
Palmerston
RESOLUTION ISLAND
Roxburgh
Port Chalmers
OTAGO PENINSULA
Mossburn
Beaumont
Edievale
Dunedin
Dusky Sound
Nightcaps
Milton
WEST CAPE
Otautau
Winton
Gore
Kaitangata
Riverton
Invercargill
Tokanui
Tahakopa
PACIFIC OCEAN
Foveaux Strait
Bluff
CODFISH ISLAND
RUAPUKE ISLAND
Mt. Anglem 3215 ft. 980 m
STEWART ISLAND
SOUTH WEST CAPE
© Rand McNally A-101768-1
c
Albany
Torbay
Hauraki Gulf
Riverhead
North Shore City
Kumeu
Oneroa
Massey
Blackpool
Muriwai Beach
AUCKLAND
Beachlands
Mount Roskill
Howick
Swanson
Te Henga
Mount Wellington
Waitakere
Otahuhu
Whitford
Piha
Waiatarua
Mangere
Papatoetoe
Karekare
Manukau Harbour
Clevedon
Huia
Manukau
TASMAN SEA
Grahams Beach
Papakura
© Rand McNally A-101874-1
d
Pukerua Bay
Cloustonville
Plimmerton
MANA ISLAND
Birchville
TASMAN SEA
Paremata
Featherston
Porirua
Whitby
TARARUA RANGE
Tawa
Upper Hutt
Stokes Valley
Makara Beach
Petone
Lower Hutt
Lake Wairarapa
Johnsonville
Kahutara
Wainuiomata
Wellington Harbour
Karori
Pyerville
Eastbourne
Wellington
Tuhitarata
Seatoun
RIMUTAKA RANGE
Island Bay
Pirinoa
Cook Strait
Lake Ferry
Wharekauhau
AORANGI RANGE
Whangaimoana
Mt. Ross 3219 ft. 981 m
TURAKIRAE HEAD
Palliser Bay
© Rand McNally A-101875-1

0 30 60 90 120 150 180 Miles
0 30 60 90 120 150 180 210 240 270 300 Kilometers

Lambert Azimuthal Equal Area Projection
Scale 1:6,000,000
One inch to 96 miles
One cm to 60 km

Inset maps a - d
Lambert Conformal Conic Projection
Scale 1:1,000,000
One inch to 16 miles
One cm to 10 km

a

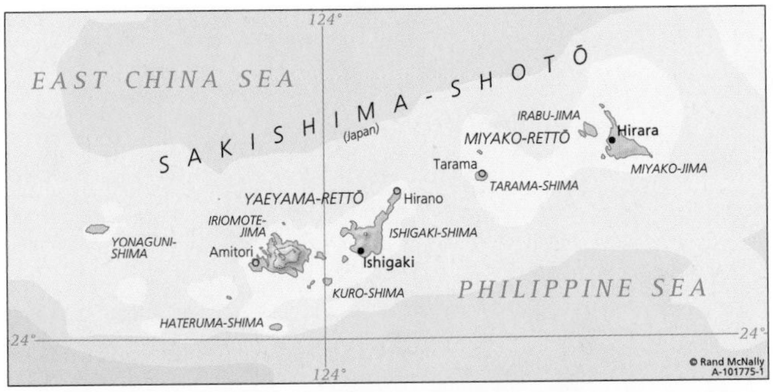
EAST CHINA SEA
SAKISHIMA-SHOTŌ
(Japan)
IRABU-JIMA
MIYAKO-RETTŌ
Hirara
MIYAKO-JIMA
Tarama
TARAMA-SHIMA
YAEYAMA-RETTŌ
Hirano
IRIOMOTE-JIMA
ISHIGAKI-SHIMA
YONAGUNI-SHIMA
Amitori
Ishigaki
KURO-SHIMA
PHILIPPINE SEA
HATERUMA-SHIMA
124°
24°
© Rand McNally A-101775-1

b

PACIFIC OCEAN
172°
170°
CAPE MULINU'U
SAVAI'I
Fagamalo
A'opo
Pu'apu'a
Falelima
Mauga Silisili 6096 ft. 1858 m
SAMOA
CAPE ASUISUI
Taga
Apolima Strait
Apia
UPOLU
AMERICAN SAMOA
(U.S.)
Falelatai
Ti'avea
Lotofaga
Salani
14°
SAMOA ISLANDS
Pago Pago
TUTUILA
Leone
Nu'uuli
Ofu
OLOSEGA
OFU
Tau
TAU
MANUA ISLANDS
© Rand McNally A-101773-1

c

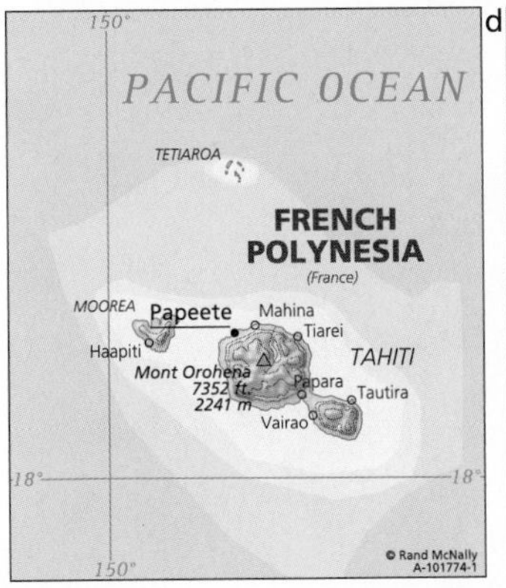
PACIFIC OCEAN
150°
TETIAROA
FRENCH POLYNESIA
(France)
MOOREA
Papeete
Mahina
Tiarei
Haapiti
Mont Orohena 7352 ft. 2241 m
TAHITI
Papara
Tautira
Vairao
18°
© Rand McNally A-101774-1

d

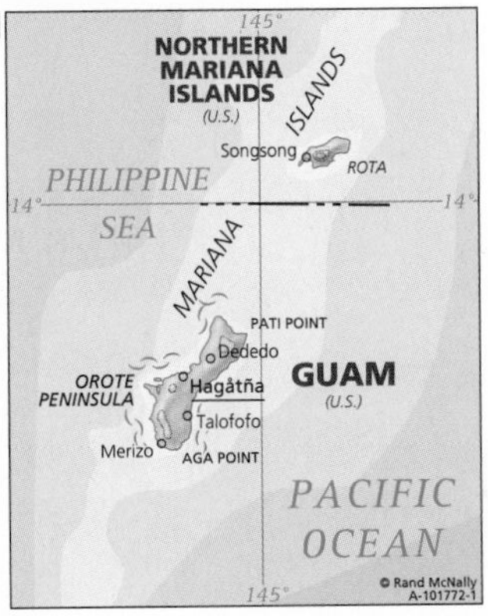
145°
NORTHERN MARIANA ISLANDS
(U.S.)
ISLANDS
Songsong
ROTA
PHILIPPINE SEA
14°
MARIANA
PATI POINT
Dededo
GUAM
(U.S.)
OROTE PENINSULA
Hagåtña
Talofofo
Merizo
AGA POINT
PACIFIC OCEAN
© Rand McNally A-101772-1

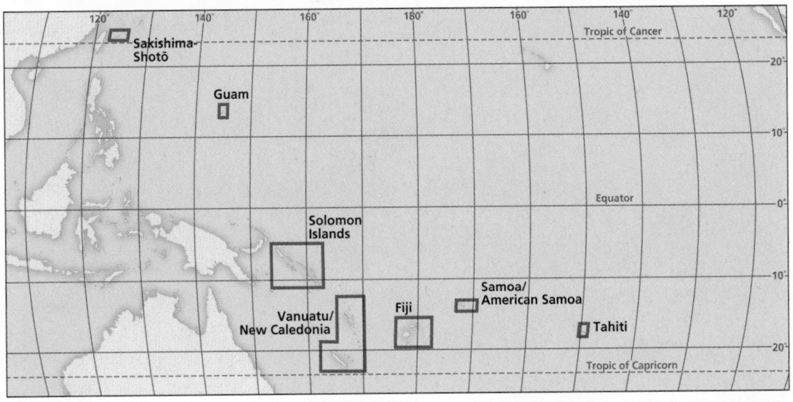
120°
140°
160°
180°
160°
140°
120°
Tropic of Cancer
Sakishima-Shotō
Guam
20°
10°
Equator
0°
Solomon Islands
10°
Samoa/ American Samoa
Fiji
Vanuatu/ New Caledonia
Tahiti
20°
Tropic of Capricorn

e

154°
156°
158°
160°
162°
Lemankoa
BUKA ISLAND
Sohano
Dios
PAPUA NEW GUINEA
Mt. Balbi 8999 ft. 2743 m
Wakunai
Amun
ONTONG JAVA
6°
Torokina
Vito
Mt. Takuan 7251 ft. 2210 m
RONCADOR REEF
CAPE TOROKINA
Arawa
Empress Augusta Bay
Taki
PACIFIC OCEAN
Jaba
Buin
BOUGAINVILLE
Taro
CHOISEUL
BRADLEY REEFS
FAURO I.
Sasamungga
SOLOMON ISLANDS
SHORTLAND I.
SHORTLAND ISLANDS
Luti
VAGHENA ISLAND
MONO ISLAND
Kia
VELLA LAVELLA
Maravari
KOLOMBANGARA ISLAND
New Georgia Sound
SANTA ISABEL
DAI
8°
RANONGGA ISLAND
Gizo
NEW GEORGIA
Buala
SIMBO ISLAND
VANGUNU ISLAND
Susubona
Blanche Channel
SAN JORGE ISLAND
Sepi
Dala
RENDOVA ISLAND
FLORIDA IS.
Auki
MALAITA
TETEPARE ISLAND
NGGATOKAE I.
PAVUVU I.
Tulagi
NGGELA SULE
NGGELA PILE
Bina
Oteotea
MBOROKUA
SAVO I.
Yandina
Visale
CAPE ESPERANCE
Indispensable Strait
RUSSELL IS.
MARAMASIKE
Tangarare
Honiara
Aola
SOLOMON SEA
Mt. Makarakomburu 8028 ft. 2447 m
Mbola
ULAWA ISLAND
Avu Avu
Inakona
10°
GUADALCANAL
UKI NI MASI ISLAND
Kirakira
Makira Harbour
SAN CRISTOBAL
© Rand McNally A-101766-1

g

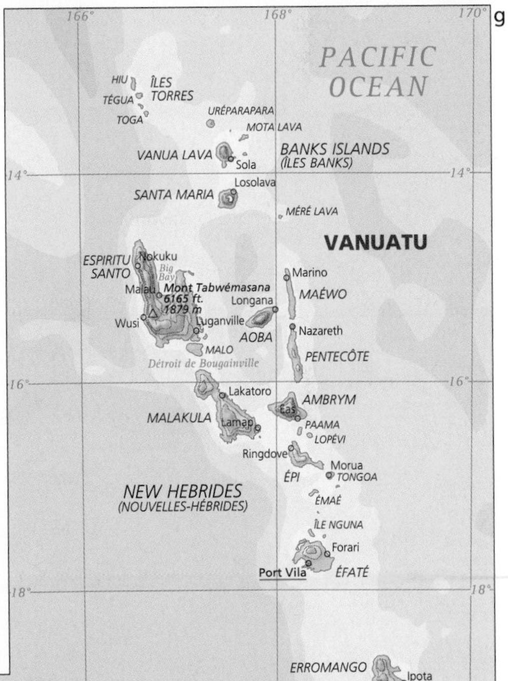
166°
168°
170°
PACIFIC OCEAN
HIU
ÎLES TORRES
TÉGUA
TOGA
URÉPARAPARA
MOTA LAVA
VANUA LAVA
Sola
BANKS ISLANDS (ÎLES BANKS)
14°
Losolava
SANTA MARIA
MÉRÉ LAVA
ESPIRITU SANTO
Nokuku
Big Bay
VANUATU
Malau
Mont Tabwémasana 6165 ft. 1879 m
Marino
MAÉWO
Longana
Wusi
Luganville
AOBA
Nazareth
MALO
PENTECÔTE
Détroit de Bougainville
16°
Lakatoro
AMBRYM
MALAKULA
Lamap
Eas
PAAMA
LOPÉVI
Ringdove
Morua
ÉPI
TONGOA
NEW HEBRIDES (NOUVELLES-HÉBRIDES)
ÉMAÉ
ÎLE NGUNA
Forari
Port Vila
ÉFATÉ
18°
ERROMANGO
Ipota

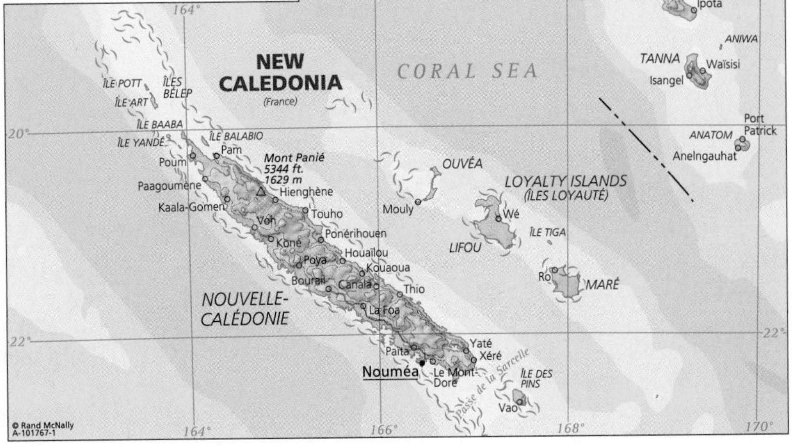
164°
166°
168°
170°
ANIWA
TANNA
Waïsisi
Isangel
NEW CALEDONIA
(France)
CORAL SEA
ÎLE POTT
ÎLES BELEP
ÎLE ART
ÎLE BAABA
Port Patrick
ANATOM
20°
ÎLE BALABIO
ÎLE YANDÉ
Pam
Anelngauhat
Mont Panié 5344 ft. 1629 m
Poum
OUVÉA
Hienghène
Paagoumène
LOYALTY ISLANDS (ÎLES LOYAUTÉ)
Kaala-Gomen
Touho
Voh
Mouly
Wé
Koné
Ponérihouen
ÎLE TIGA
Poya
Houaïlou
LIFOU
Kouaoua
Bourail
Canala
Ro
MARÉ
Thio
NOUVELLE-CALÉDONIE
La Foa
22°
Païta
Yaté
Xéré
Nouméa
Le Mont-Dore
Passe de la Sarcelle
ÎLE DES PINS
Vao
© Rand McNally A-101767-1

f

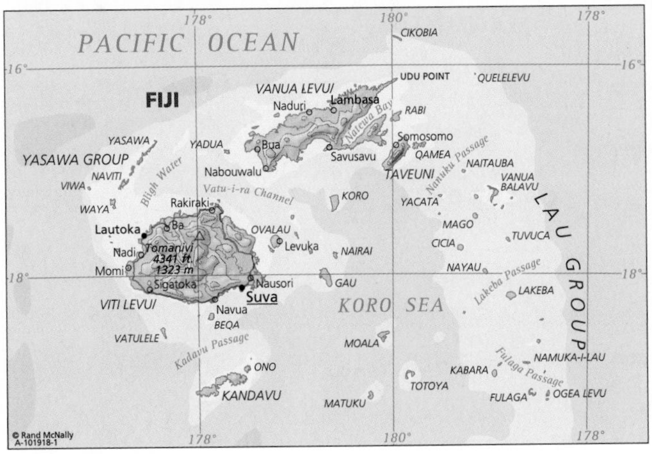
178°
180°
178°
CIKOBIA
PACIFIC OCEAN
16°
UDU POINT
QUELELEVU
VANUA LEVU
FIJI
Naduri
Lambasa
Natewa Bay
RABI
YASAWA
YADUA
Bua
Somosomo
YASAWA GROUP
Savusavu
QAMEA
NAITAUBA
NAVITI
Bligh Water
Nabouwalu
TAVEUNI
Nanuku Passage
VANUA BALAVU
VIWA
Vatu-i-ra Channel
WAYA
Rakiraki
KORO
YACATA
Lautoka
Ba
OVALAU
MAGO
Nadi
Tomanivi 4341 ft. 1323 m
Levuka
NAIRAI
CICIA
TUVUCA
Momi
NAYAU
Lakeba Passage
LAU GROUP
18°
Sigatoka
Nausori
GAU
LAKEBA
VITI LEVU
Suva
KORO SEA
Navua
BEQA
VATULELE
Kadavu Passage
MOALA
NAMUKA-I-LAU
ONO
KABARA
Fulaga Passage
KANDAVU
TOTOYA
MATUKU
FULAGA
OGEA LEVU
© Rand McNally A-101918-1

Inset maps a - d
Lambert Azimuthal Equal Area Projection
Scale 1:4,000,000
One inch to 64 miles
One cm to 40 km

Inset maps e - g
Lambert Azimuthal Equal Area Projection
Scale 1:8,000,000
One inch to 128 miles
One cm to 80 km

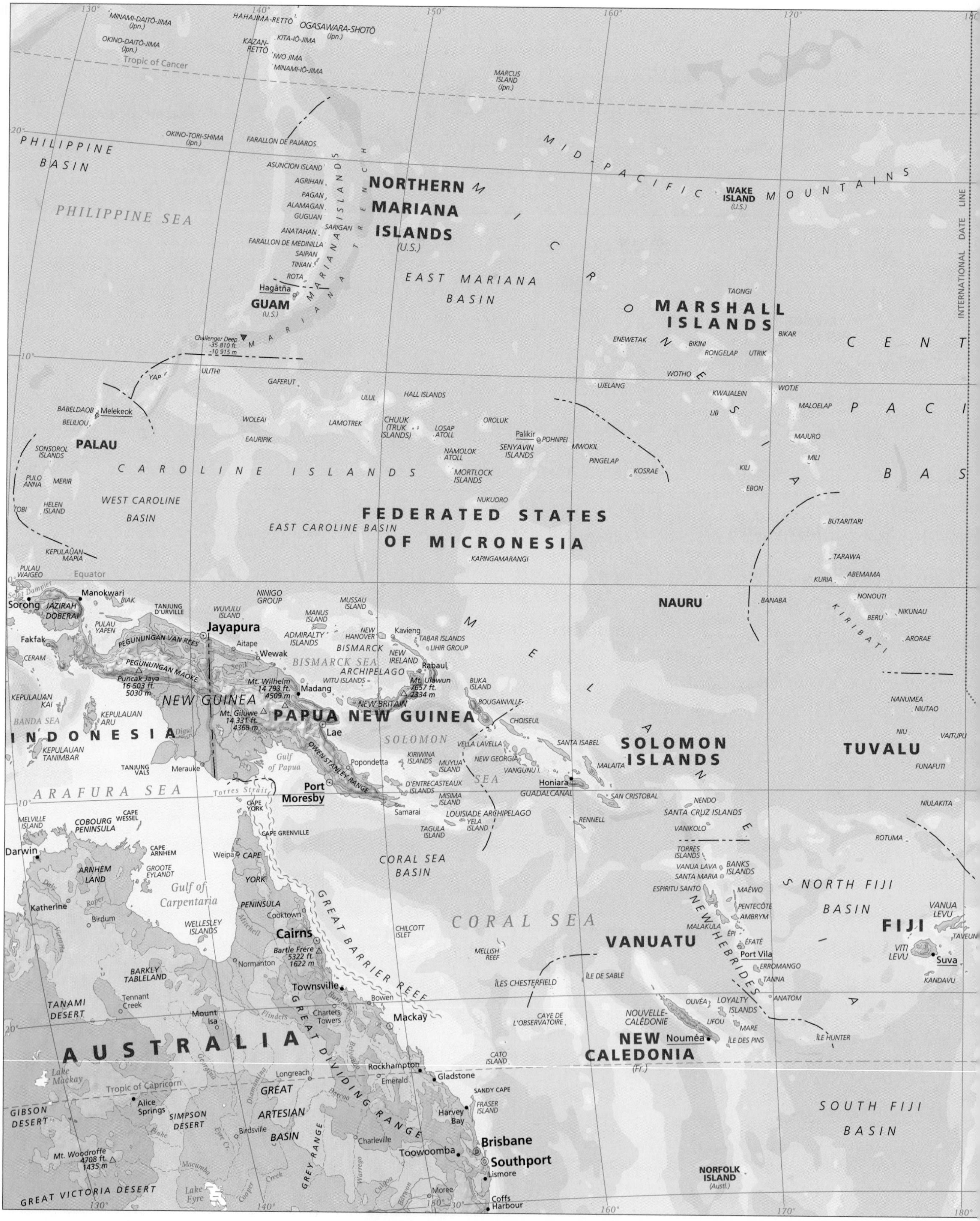
MINAMI-DAITŌ-JIMA (Jpn.)
OKINO-DAITŌ-JIMA (Jpn.)
Tropic of Cancer
HAHAJIMA-RETTŌ
OGASAWARA-SHOTŌ (Jpn.)
KITA-IŌ-JIMA
KAZAN-RETTŌ
IWO JIMA
MINAMI-IŌ-JIMA
MARCUS ISLAND (Jpn.)
OKINO-TORI-SHIMA (Jpn.)
FARALLON DE PAJAROS
PHILIPPINE BASIN
MID-PACIFIC MOUNTAINS
ASUNCION ISLAND
AGRIHAN
PAGAN
ALAMAGAN
GUGUAN
SARIGAN
ANATAHAN
FARALLON DE MEDINILLA
SAIPAN
TINIAN
ROTA
NORTHERN MARIANA ISLANDS (U.S.)
MARIANA ISLANDS
MARIANA TRENCH
PHILIPPINE SEA
WAKE ISLAND (U.S.)
INTERNATIONAL DATE LINE
EAST MARIANA BASIN
Hagåtña
GUAM (U.S.)
MICRONESIA
TAONGI
MARSHALL ISLANDS
Challenger Deep -35 810 ft. -10 915 m
ENEWETAK
BIKINI
RONGELAP
UTRIK
BIKAR
CENTRAL PACIFIC BASIN
YAP
ULITHI
GAFERUT
WOTHO
UJELANG
KWAJALEIN
WOTJE
ULUL
HALL ISLANDS
LIB
MALOELAP
BABELDAOB
Melekeok
BELILIOU
WOLEAI
LAMOTREK
CHUUK (TRUK ISLANDS)
LOSAP ATOLL
OROLUK
Palikir
POHNPEI
MWOKIL
MAJURO
EAURIPIK
SONSOROL ISLANDS
PALAU
NAMOLOK ATOLL
SENYAVIN ISLANDS
PINGELAP
MILI
CAROLINE ISLANDS
MORTLOCK ISLANDS
KOSRAE
KILI
PULO ANNA
MERIR
EBON
TOBI
HELEN ISLAND
WEST CAROLINE BASIN
NUKUORO
FEDERATED STATES OF MICRONESIA
BUTARITARI
EAST CAROLINE BASIN
KEPULAUAN MAPIA
KAPINGAMARANGI
TARAWA
PULAU WAIGEO
Equator
KURIA
ABEMAMA
Selat Dampier
Manokwari
BIAK
NINIGO GROUP
MUSSAU ISLAND
NAURU
BANABA
NONOUTI
Sorong
JAZIRAH DOBERAI
TANJUNG D'URVILLE
WUVULU ISLAND
MANUS ISLAND
KIRIBATI
BERU
NIKUNAU
PULAU YAPEN
Jayapura
Fakfak
PEGUNUNGAN VAN REES
Aitape
ADMIRALTY ISLANDS
NEW HANOVER
Kavieng
TABAR ISLANDS
LIHIR GROUP
ARORAE
CERAM
Wewak
BISMARCK ARCHIPELAGO
NEW IRELAND
Sepik
BISMARCK SEA
Rabaul
PEGUNUNGAN MAOKE
Puncak Jaya 16 503 ft. 5030 m
Mt. Wilhelm 14 793 ft. 4509 m
WITU ISLANDS
Mt. Ulawun 7657 ft. 2334 m
BUKA ISLAND
MELANESIA
Madang
KEPULAUAN KAI
NEW GUINEA
NEW BRITAIN
BOUGAINVILLE
NANUMEA
BANDA SEA
KEPULAUAN ARU
Mt. Giluwe 14 331 ft. 4368 m
PAPUA NEW GUINEA
NIUTAO
INDONESIA
Digul
Lae
CHOISEUL
SOLOMON SEA
NIU
VAITUPU
KEPULAUAN TANIMBAR
Gulf of Papua
OWEN STANLEY RANGE
VELLA LAVELLA
SANTA ISABEL
SOLOMON ISLANDS
TUVALU
Fly
KIRIWINA ISLANDS
NEW GEORGIA
TANJUNG VALS
Merauke
Popondetta
MUYUA ISLAND
VANGUNU I.
MALAITA
FUNAFUTI
ARAFURA SEA
Torres Strait
Port Moresby
D'ENTRECASTEAUX ISLANDS
Honiara
GUADALCANAL
CAPE YORK
MISIMA ISLAND
SAN CRISTOBAL
NENDO
NIULAKITA
MELVILLE ISLAND
CAPE WESSEL
COBOURG PENINSULA
Samarai
LOUISIADE ARCHIPELAGO
SANTA CRUZ ISLANDS
YELA ISLAND
RENNELL
VANIKOLO
TAGULA ISLAND
CAPE GRENVILLE
ROTUMA
Darwin
CAPE ARNHEM
Weipa
CAPE YORK
CORAL SEA BASIN
TORRES ISLANDS
ARNHEM LAND
GROOTE EYLANDT
VANUA LAVA
BANKS ISLANDS
YORK
SANTA MARIA
Daly
Gulf of Carpentaria
ESPIRITU SANTO
MAÉWO
NORTH FIJI BASIN
Katherine
Roper
PENINSULA
PENTECÔTE
VANUA LEVU
Cooktown
AMBRYM
Birdum
WELLESLEY ISLANDS
Mitchell
GREAT BARRIER REEF
CORAL SEA
NEW HEBRIDES
FIJI
Victoria
Cairns
CHILCOTT ISLET
MALAKULA
ÉPI
TAVEUNI
VANUATU
ÉFATÉ
Bartle Frere 5322 ft. 1622 m
MELLISH REEF
Port Vila
VITI LEVU
Suva
Normanton
ERROMANGO
BARKLY TABLELAND
TANNA
KANDAVU
ÎLE DE SABLE
Townsville
ÎLES CHESTERFIELD
ANATOM
TANAMI DESERT
Tennant Creek
Burdekin
Bowen
OUVÉA
LOYALTY ISLANDS
Flinders
Charters Towers
Mackay
NOUVELLE-CALÉDONIE
LIFOU
Mount Isa
CAYE DE L'OBSERVATOIRE
GREAT DIVIDING RANGE
MARE
NEW CALEDONIA (Fr.)
Nouméa
ÎLE DES PINS
ÎLE HUNTER
AUSTRALIA
Belyando
CATO ISLAND
Lake Mackay
Georgina
Rockhampton
Diamantina
Longreach
Emerald
Gladstone
Tropic of Capricorn
GREAT
ARTESIAN
BASIN
Barcoo
SANDY CAPE
FRASER ISLAND
Alice Springs
SOUTH FIJI BASIN
GIBSON DESERT
SIMPSON DESERT
Harvey Bay
Eyre Cr.
Finke
Birdsville
GREY RANGE
Charleville
Brisbane
Mt. Woodroffe 4708 ft. 1435 m
Toowoomba
Southport
Macumba
Warrego
Lismore
NORFOLK ISLAND (Austl.)
Lake Eyre
Cooper Creek
Culgoa
Barwon
Moree
GREAT VICTORIA DESERT
Coffs Harbour
130°
140°
150°
160°
170°
180°
20°
10°
30°

0 150 300 450 600 750 Miles
0 150 300 450 600 750 900 1,050 Kilometers

Lambert Azimuthal Equal Area Projection
Scale 1:27,000,000
One inch to 426 miles
One cm to 270 km

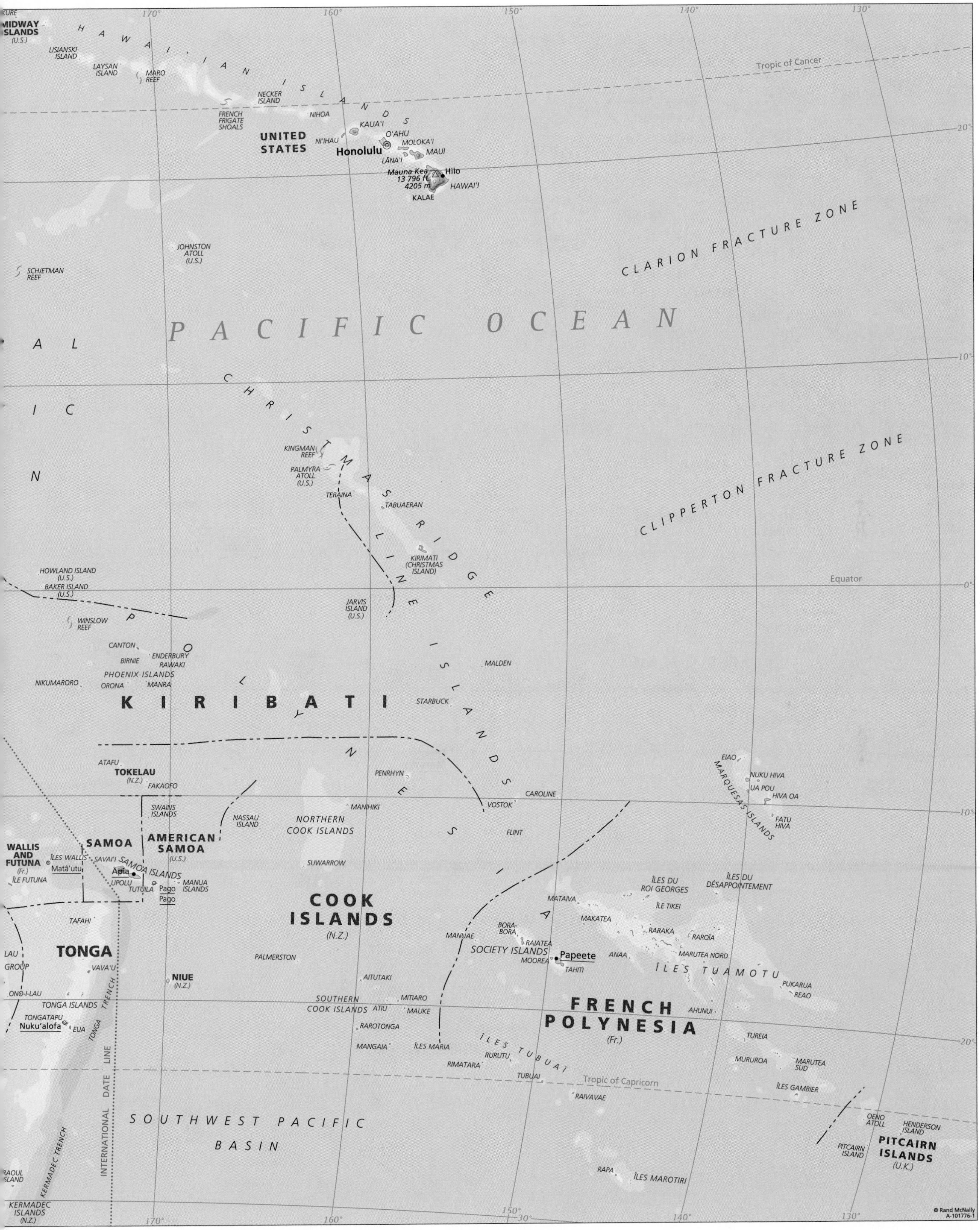
MIDWAY ISLANDS (U.S.)
HAWAIIAN ISLANDS
LISIANSKI ISLAND
LAYSAN ISLAND
MARO REEF
NECKER ISLAND
FRENCH FRIGATE SHOALS
NIHOA
KAUA'I
O'AHU
NI'IHAU
MOLOKA'I
MAUI
LĀNA'I
UNITED STATES
Honolulu
Mauna Kea 13 796 ft 4205 m
Hilo
HAWAI'I
KALAE
Tropic of Cancer
JOHNSTON ATOLL (U.S.)
SCHJETMAN REEF
CLARION FRACTURE ZONE
PACIFIC OCEAN
CHRISTMAS RIDGE
KINGMAN REEF
PALMYRA ATOLL (U.S.)
TERAINA
TABUAERAN
KIRIMATI (CHRISTMAS ISLAND)
CLIPPERTON FRACTURE ZONE
HOWLAND ISLAND (U.S.)
BAKER ISLAND (U.S.)
Equator
JARVIS ISLAND (U.S.)
WINSLOW REEF
CANTON
ENDERBURY
BIRNIE
RAWAKI
PHOENIX ISLANDS
NIKUMARORO
ORONA
MANRA
POLYNESIA
MALDEN
KIRIBATI
STARBUCK
LINE ISLANDS
ATAFU
TOKELAU (N.Z.)
FAKAOFO
PENRHYN
EIAO
MARQUESAS ISLANDS
NUKU HIVA
UA POU
HIVA OA
FATU HIVA
CAROLINE
VOSTOK
SWAINS ISLANDS
MANIHIKI
NASSAU ISLAND
NORTHERN COOK ISLANDS
FLINT
WALLIS AND FUTUNA (Fr.)
ÎLES WALLIS
Matâ'utu
ÎLE FUTUNA
SAMOA
SAVAI'I
SAMOA ISLANDS
Apia
UPOLU
AMERICAN SAMOA (U.S.)
TUTUILA
Pago Pago
MANUA ISLANDS
SUWARROW
ÎLES DU ROI GEORGES
ÎLES DU DÉSAPPOINTEMENT
ÎLE TIKEI
MATAIVA
COOK ISLANDS (N.Z.)
MAKATEA
TAFAHI
BORA-BORA
RAIATEA
RARAKA
RAROÏA
MANUAE
SOCIETY ISLANDS
MOOREA
Papeete
TAHITI
ANAA
MARUTEA NORD
ÎLES TUAMOTU
LAU GROUP
TONGA
VAVA'U
PALMERSTON
NIUE (N.Z.)
AITUTAKI
PUKARUA
REAO
ONO-I-LAU
TONGA ISLANDS
TONGA TRENCH
SOUTHERN COOK ISLANDS
ATIU
MITIARO
MAUKE
FRENCH POLYNESIA (Fr.)
AHUNUI
TONGATAPU
Nuku'alofa
EUA
RAROTONGA
MANGAIA
ÎLES MARIA
ÎLES TUBUAI
RURUTU
RIMATARA
TUBUAI
TUREIA
MURUROA
MARUTEA SUD
ÎLES GAMBIER
Tropic of Capricorn
RAIVAVAE
INTERNATIONAL DATE LINE
SOUTHWEST PACIFIC BASIN
OENO ATOLL
HENDERSON ISLAND
PITCAIRN ISLAND
PITCAIRN ISLANDS (U.K.)
KERMADEC TRENCH
RAOUL ISLAND
KERMADEC ISLANDS (N.Z.)
RAPA
ÎLES MAROTIRI
170°
160°
150°
140°
130°
20°
10°
0°
10°
20°
30°
© Rand McNally A-101776-1

RUSSIA
CHELYABINSK
YEKATERINBURG
URAL MTS.
OMSK
NOVOSIBIRSK
Lake Baikal
SEA OF OKHOTSK
KAMCHATKA PEN.
BERING SEA
ST. LAWRENCE ISLAND
KOMANDORSKI ISLANDS
Petropavlovsk-Kamchatskiy
ALEUTIAN ISLANDS
Arctic Circle
Astana
KAZAKHSTAN
Ulaanbaatar
MONGOLIA
SAKHALIN
KURIL ISLANDS
Khabarovsk
HARBIN
Vladivostok
ALMATY
KYRG.
TASHKENT
ÜRÜMQI
SHENYANG
SAPPORO
HOKKAIDO
TAJIK.
BEIJING
NORTH KOREA
SEA OF JAPAN
HONSHŪ
PAK.
Huang
SOUTH KOREA
SEOUL
PUSAN
JAPAN
TŌKYŌ
JAPAN CURRENT
CHINA
ŌSAKA
SHIKOKU
HIMALAYAS
NANJING
KYŪSHŪ
WUHAN
SHANGHAI
Mt. Everest 29 028 ft. 8848 m
NEPAL
CHONGQING
EAST CHINA SEA
RYUKYU ISLANDS
BONIN IS.
Kathmandu
BHU.
INDIA
DHAKA
T'AIPEI
Tropic of Cancer
BNGL.
HONG KONG
TAIWAN
KOLKATA
MYANMAR
Ha Noi
INTERNATIONAL DATE LINE
LAOS
SOUTH CHINA SEA
PHILIPPINE SEA
MICRONESIA
YANGON
HAINAN DAO
LUZON
MARIANA IS.
NORTHERN MARIANA ISLANDS (U.S.)
Bay of Bengal
THAILAND
BANGKOK
MANILA
ANDAMAN ISLANDS (India)
ANDAMAN SEA
CAMBODIA
VIETNAM
GUAM (U.S.)
NORTH EQUATORIAL CURRENT
HO CHI MINH CITY
PHILIPPINES
Gulf of Thailand
MINDANAO
MARSHALL ISLANDS
NICOBAR ISLANDS (India)
SULU SEA
CAROLINE ISLANDS
MALAYSIA
BRUNEI
PALAU
FEDERATED STATES OF MICRONESIA
Kuala Lumpur
MEDAN
CELEBES SEA
SINGAPORE
HOWLAND ISLAND (U.S.)
SUMATRA
Equator
BORNEO
BAKER ISLAND (U.S.)
POLYNESIA
INDONESIA
NAURU
BANABA
CELEBES
MOLUCCAS
Makassar
BISMARCK SEA
NEW IRELAND
GILBERT ISLANDS
KIRIBATI
PHOENIX ISLANDS
JAVA SEA
BANDA SEA
NEW GUINEA
MELANESIA
JAKARTA
SURABAYA
Lae
NEW BRITAIN
JAVA
BALI
FLORES SEA
FLORES
PAPUA NEW GUINEA
SOLOMON SEA
SOLOMON ISLANDS
ARAFURA SEA
SUMBA
SAVU SEA
EAST TIMOR
TIMOR
TIMOR SEA
TUVALU
TOKELAU (N.Z.)
Port Moresby
CHRISTMAS ISLAND (Austl.)
COCOS ISLANDS (Austl.)
Darwin
Gulf of Carpentaria
WALLIS AND FUTUNA (Fr.)
SAMOA
AMERICAN SAMOA (U.S.)
CORAL SEA
VANUATU
NEW HEBRIDES
Cairns
FIJI
TONGA
NIUE (N.Z.)
NEW CALEDONIA (Fr.)
LOYALTY IS.
Tropic of Capricorn
INDIAN OCEAN
AUSTRALIA
GREAT DIVIDING RANGE
EAST AUSTRALIAN CURRENT
WEST AUSTRALIAN CURRENT
Brisbane
NORFOLK ISLAND (Austl.)
Perth
Great Australian Bight
Sydney
Canberra
NORTH ISLAND
Adelaide
Murray
Auckland
ÎLE AMSTERDAM (Fr.)
Mt. Kosciuszko 7313 ft. 2229 m
Melbourne
TASMAN SEA
NEW ZEALAND
ÎLE ST.-PAUL (Fr.)
Bass Strait
TASMANIA
Hobart
Wellington
SOUTH ISLAND
Christchurch
CHATHAM ISLANDS (N.Z.)
Dunedin
ÎLES KERGUÉLEN (Fr.)
STEWART ISLAND
BOUNTY IS.
SNARES ISLANDS
ANTIPODES IS.
HEARD AND MCDONALD ISLANDS (Austl.)
AUCKLAND ISLANDS
CAMPBELL ISLAND
© Rand McNally A-102151-1

0 300 600 900 1200 1500 Miles

0 300 600 900 1200 1500 1800 2100 Kilometers

Mollweide Projection
Scale 1:55,000,000
One inch to 868 miles
One cm to 550 km

Mt. McKinley
20 320 ft.
6194 m
ALASKA
Anchorage
Gulf of Alaska
KODIAK ISLAND
COAST MTS.
Prince Rupert
QUEEN CHARLOTTE ISLANDS
VANCOUVER ISLAND
Vancouver
Seattle
Portland
CASCADE RA.
Edmonton
Calgary
CANADA
ROCKY MTS.
GREAT PLAINS
Hudson Bay
Nelson
Albany
Winnipeg
Lake Superior
Lake Huron
Lake Michigan
L. Ontario
Lake Erie
Missouri
Mississippi
Snake
Great Salt Lake
Colorado
Ohio
Ottawa
MONTRÉAL
TORONTO
CHICAGO
NEW YORK
PHILADELPHIA
Washington
APPALACHIAN MTS.
LABRADOR CURRENT
NEWFOUNDLAND
CAPE BRETON ISLAND
NORTH ATLANTIC CURRENT
UNITED KINGDOM
IRELAND
AZORES (PORT.)
GULF STREAM
BERMUDA (U.K.)
San Francisco
SIERRA NEVADA
UNITED STATES
LOS ANGELES
CHANNEL IS.
SAN DIEGO
CALIFORNIA CURRENT
DALLAS
Rio Grande
HOUSTON
SIERRA MADRE OCCIDENTAL
GULF OF MEXICO
Miami
BAHAMAS
ATLANTIC OCEAN
Tropic of Cancer
MONTERREY
HAVANA
MEXICO
GUADALAJARA
MEXICO CITY
CUBA
JAMAICA
HAITI
DOM. REP.
PUERTO RICO (U.S.)
SANTO DOMINGO
NORTH EQUATORIAL CURRENT
BELIZE
GUAT.
HONDURAS
Guatemala
EL SAL.
NICARAGUA
Managua
CARIBBEAN SEA
HAWAI'IAN ISLANDS (U.S.)
ISLAS REVILLAGIGEDO (Mex.)
NORTH EQUATORIAL CURRENT
MARACAIBO
TRINIDAD AND TOBAGO
CARACAS
COSTA RICA
Panamá
PANAMA
VENEZUELA
GUY.
SUR.
FR. GUI. (Fr.)
MEDELLÍN
CALI
BOGOTÁ
COLOMBIA
CHRISTMAS ISLAND
EQUATORIAL COUNTER CURRENT
QUITO
Equator
GALAPAGOS ISLANDS (Ec.)
ECUADOR
GUAYAQUIL
Amazon
MANAUS
LINE ISLANDS
SOUTH EQUATORIAL CURRENT
PERU
BRAZIL
SELVAS
ANDES
VOSTOK
MARQUESAS ISLANDS
FLINT
Lima
PERU CURRENT
ÎLES DU ROI GEORGES
ÎLES DU DÉSAPPOINTEMENT
SOCIETY IS.
TAHITI
ÎLES TUAMOTU
La Paz
BOLIVIA
BRASÍLIA
ÎLES MARIA
FRENCH POLYNESIA (Fr.)
ÎLES TUBUAI
CHACO
Tropic of Capricorn
Antofagasta
PARAGUAY
PITCAIRN ISLANDS (U.K.)
HENDERSON ISLAND
Asunción
SÃO PAULO
EASTER ISLAND (ISLA DE PASCUA) (Chile)
ISLA SALA Y GÓMEZ (Chile)
ISLA SAN FÉLIX (Chile)
ISLA SAN AMBROSIO (Chile)
CHILE
GRAN
ARGENTINA
CÓRDOBA
PORTO ALEGRE
SANTIAGO
PAMPA
URUG.
MONTEVIDEO
ARCHIPIÉLAGO JUAN FERNÁNDEZ (Chile)
Cerro Aconcagua
22 831 ft.
6959 m
Concepción
BUENOS AIRES
Bahía Blanca
ISLA GRANDE DE CHILOÉ
ARCHIPIÉLAGO DE LOS CHONOS
PATAGONIA
Golfo San Jorge
ATLANTIC OCEAN
WEST WIND DRIFT
FALKLAND ISLANDS (U.K.)
SOUTH GEORGIA AND THE SOUTH SANDWICH ISLANDS (U.K.)
WEST WIND DRIFT
Punta Arenas
TIERRA DEL FUEGO
CABO DE HORNOS

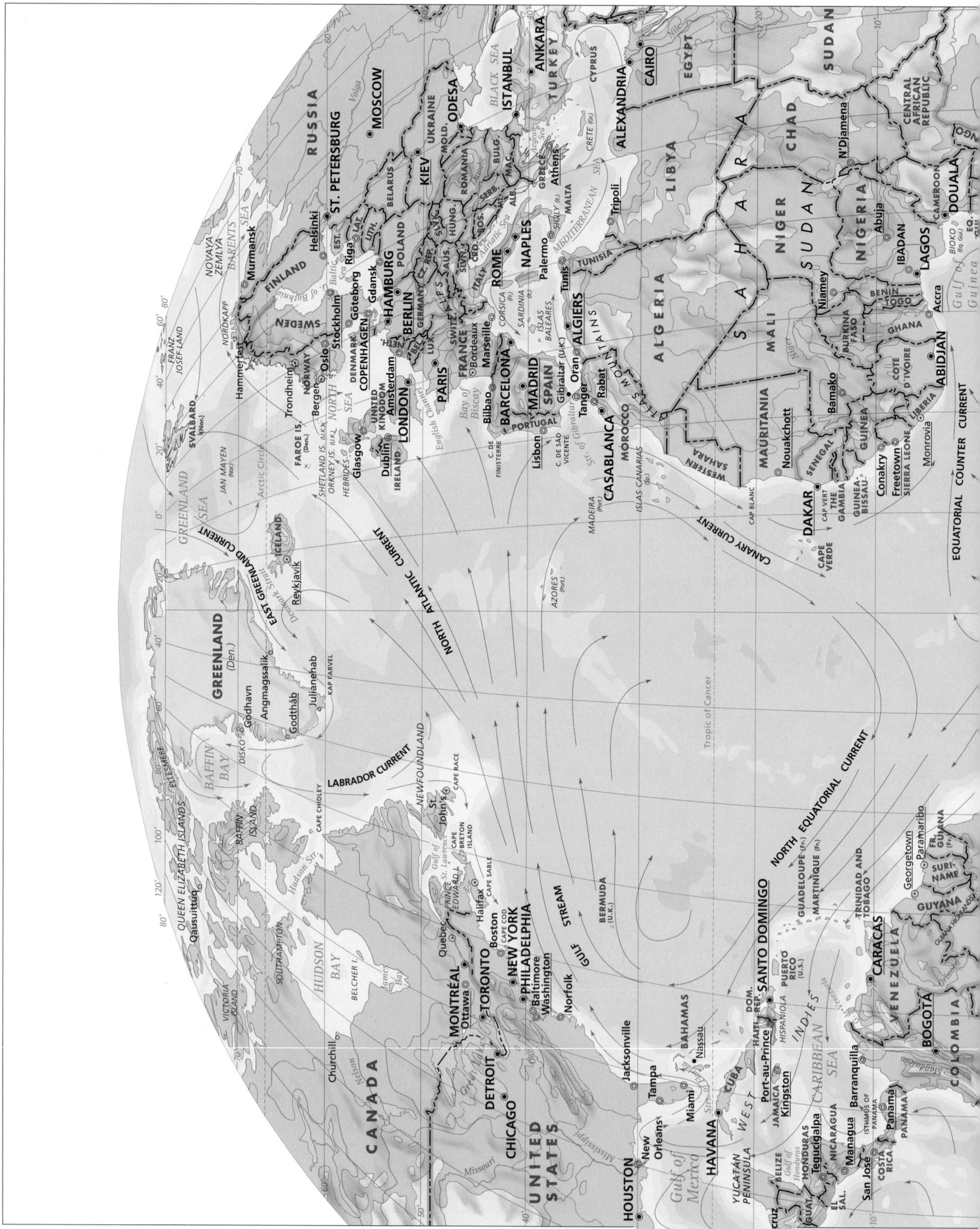
RUSSIA
MOSCOW
ST. PETERSBURG
Murmansk
Helsinki
FINLAND
SWEDEN
NORWAY
Stockholm
Oslo
Bergen
Trondheim
Hammerfest
Göteborg
COPENHAGEN
DENMARK
Gdansk
Riga
EST.
LAT.
LITH.
BELARUS
UKRAINE
KIEV
MOLD.
ODESA
ROMANIA
BULG.
MAC.
GREECE
Athens
ALB.
SERB.
BOS.
CRO.
SLVN.
AUS.
HUNG.
SLVK.
CZ. REP.
POLAND
HAMBURG
BERLIN
GERMANY
SWITZ.
LUX.
BEL.
NETH.
Amsterdam
UNITED KINGDOM
LONDON
Dublin
IRELAND
Glasgow
FRANCE
PARIS
Bordeaux
Marseille
ITALY
ROME
NAPLES
Palermo
CORSICA (Fr.)
SARDINIA (It.)
SICILY (It.)
MALTA
ISLAS BALEARES
SPAIN
MADRID
BARCELONA
Bilbao
PORTUGAL
Lisbon
Gibraltar (U.K.)
Tanger
Rabat
CASABLANCA
MOROCCO
Oran
ALGIERS
Tunis
TUNISIA
ALGERIA
LIBYA
Tripoli
EGYPT
CAIRO
ALEXANDRIA
TURKEY
ANKARA
ISTANBUL
CYPRUS
CRETE (Gr.)
BLACK SEA
MEDITERRANEAN SEA
SUDAN
CHAD
N'Djamena
CENTRAL AFRICAN REPUBLIC
CAMEROON
DOUALA
NIGERIA
Abuja
IBADAN
LAGOS
BENIN
TOGO
GHANA
Accra
NIGER
Niamey
BURKINA FASO
CÔTE D'IVOIRE
ABIDJAN
LIBERIA
Monrovia
SIERRA LEONE
Freetown
GUINEA
Conakry
GUINEA-BISSAU
THE GAMBIA
SENEGAL
DAKAR
CAPE VERDE
MALI
Bamako
MAURITANIA
Nouakchott
WESTERN SAHARA
SAHARA
ATLAS MOUNTAINS
ISLAS CANARIAS (Sp.)
MADEIRA (Port.)
AZORES (Port.)
CANARY CURRENT
EQUATORIAL COUNTER CURRENT
NORTH EQUATORIAL CURRENT
NORTH ATLANTIC CURRENT
GULF STREAM
LABRADOR CURRENT
EAST GREENLAND CURRENT
Gulf of Guinea
FAROE IS. (Den.)
SHETLAND IS. (U.K.)
ORKNEY IS. (U.K.)
HEBRIDES
NORTH SEA
Bay of Biscay
English Channel
C. DE FINISTERRE
C. DE SÃO VICENTE
CAP BLANC
CAP VERT
SVALBARD (Nor.)
FRANZ JOSEF LAND
NOVAYA ZEMLYA
BARENTS SEA
NORDKAPP
JAN MAYEN (Nor.)
GREENLAND SEA
Arctic Circle
ICELAND
Reykjavik
Denmark Strait
GREENLAND (Den.)
Angmagssalik
Godhavn
Godthåb
Julianehab
KAP FARVEL
DISKO
BAFFIN BAY
BAFFIN ISLAND
ELLESMERE
QUEEN ELIZABETH ISLANDS
Qausuittuq
VICTORIA ISLAND
SOUTHAMPTON
HUDSON BAY
Hudson Str.
CAPE CHIDLEY
BELCHER I.
James Bay
Churchill
CANADA
NEWFOUNDLAND
St. John's
CAPE RACE
CAPE BRETON ISLAND
Gulf of St. Lawrence
PRINCE EDWARD I.
Halifax
CAPE SABLE
Quebec
MONTRÉAL
Ottawa
TORONTO
DETROIT
CHICAGO
Boston
CAPE COD
NEW YORK
PHILADELPHIA
Baltimore
Washington
Norfolk
UNITED STATES
Missouri
Mississippi
Jacksonville
Tampa
Miami
New Orleans
HOUSTON
BERMUDA (U.K.)
Tropic of Cancer
BAHAMAS
Nassau
CUBA
HAVANA
HAITI
Port-au-Prince
DOM. REP.
SANTO DOMINGO
HISPANIOLA
PUERTO RICO (U.S.)
JAMAICA
Kingston
WEST INDIES
CARIBBEAN SEA
GUADELOUPE (Fr.)
MARTINIQUE (Fr.)
TRINIDAD AND TOBAGO
Gulf of Mexico
YUCATAN PENINSULA
MEXICO
Veracruz
BELIZE
GUAT.
EL SAL.
HONDURAS
Tegucigalpa
NICARAGUA
Managua
COSTA RICA
San José
PANAMA
Panama
ISTHMUS OF PANAMA
Barranquilla
COLOMBIA
BOGOTÁ
VENEZUELA
CARACAS
GUYANA
Georgetown
SURINAME
Paramaribo
FR. GUIANA (Fr.)

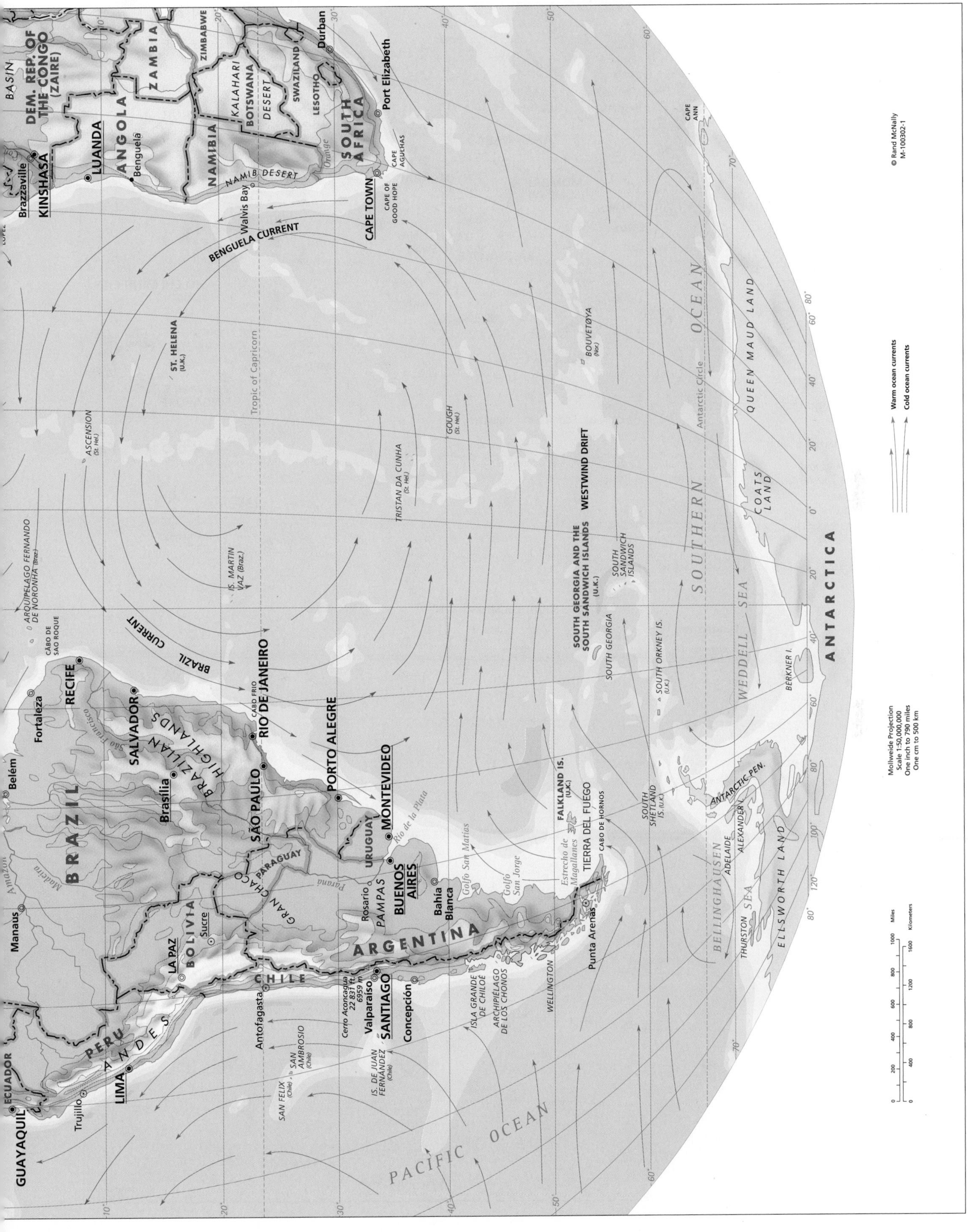
DEM. REP. OF THE CONGO (ZAIRE)
ANGOLA
ZAMBIA
ZIMBABWE
NAMIBIA
BOTSWANA
KALAHARI DESERT
SWAZILAND
LESOTHO
SOUTH AFRICA
Durban
Port Elizabeth
KINSHASA
Brazzaville
LUANDA
Benguela
NAMIB DESERT
Walvis Bay
Orange
CAPE TOWN
CAPE OF GOOD HOPE
CAPE AGULHAS
BENGUELA CURRENT
CAPE ANN
ST. HELENA (U.K.)
BOUVETØYA (Nor.)
QUEEN MAUD LAND
SOUTHERN OCEAN
Antarctic Circle
Tropic of Capricorn
ASCENSION (St. Hel.)
GOUGH (St. Hel.)
TRISTAN DA CUNHA (St. Hel.)
WESTWIND DRIFT
COATS LAND
ARQUIPÉLAGO FERNANDO DE NORONHA (Braz.)
CABO DE SAO ROQUE
IS. MARTIN VAZ (Braz.)
SOUTH GEORGIA AND THE SOUTH SANDWICH ISLANDS (U.K.)
SOUTH SANDWICH ISLANDS
SOUTH GEORGIA
SOUTH ORKNEY IS. (U.K.)
ANTARCTICA
WEDDELL SEA
BERKNER I.
BRAZIL CURRENT
RECIFE
Fortaleza
Belém
SALVADOR
BRAZILIAN HIGHLANDS
São Francisco
CABO FRIO
RIO DE JANEIRO
PORTO ALEGRE
SÃO PAULO
Brasília
BRAZIL
Amazon
Madeira
Manaus
MONTEVIDEO
Río de la Plata
URUGUAY
PARAGUAY
GRAN CHACO
Paraná
Rosario
PAMPAS
BUENOS AIRES
Bahía Blanca
Golfo San Matías
Golfo San Jorge
FALKLAND IS. (U.K.)
Estrecho de Magallanes
TIERRA DEL FUEGO
CABO DE HORNOS
SOUTH SHETLAND IS. (U.K.)
ANTARCTIC PEN.
ALEXANDER I.
ADELAIDE
BELLINGHAUSEN SEA
THURSTON
ELLSWORTH LAND
BOLIVIA
Sucre
LA PAZ
ARGENTINA
CHILE
Antofagasta
Cerro Aconcagua 22 831 ft. 6959 m
Valparaíso
SANTIAGO
Concepción
ISLA GRANDE DE CHILOÉ
ARCHIPIÉLAGO DE LOS CHONOS
WELLINGTON
Punta Arenas
PERU
ANDES
LIMA
Trujillo
ECUADOR
GUAYAQUIL
SAN FÉLIX (Chile)
SAN AMBROSIO (Chile)
IS. DE JUAN FERNÁNDEZ (Chile)
PACIFIC OCEAN
Warm ocean currents
Cold ocean currents
Mollweide Projection
Scale 1:50,000,000
One inch to 790 miles
One cm to 500 km
Miles
Kilometers
© Rand McNally
M-100302-1

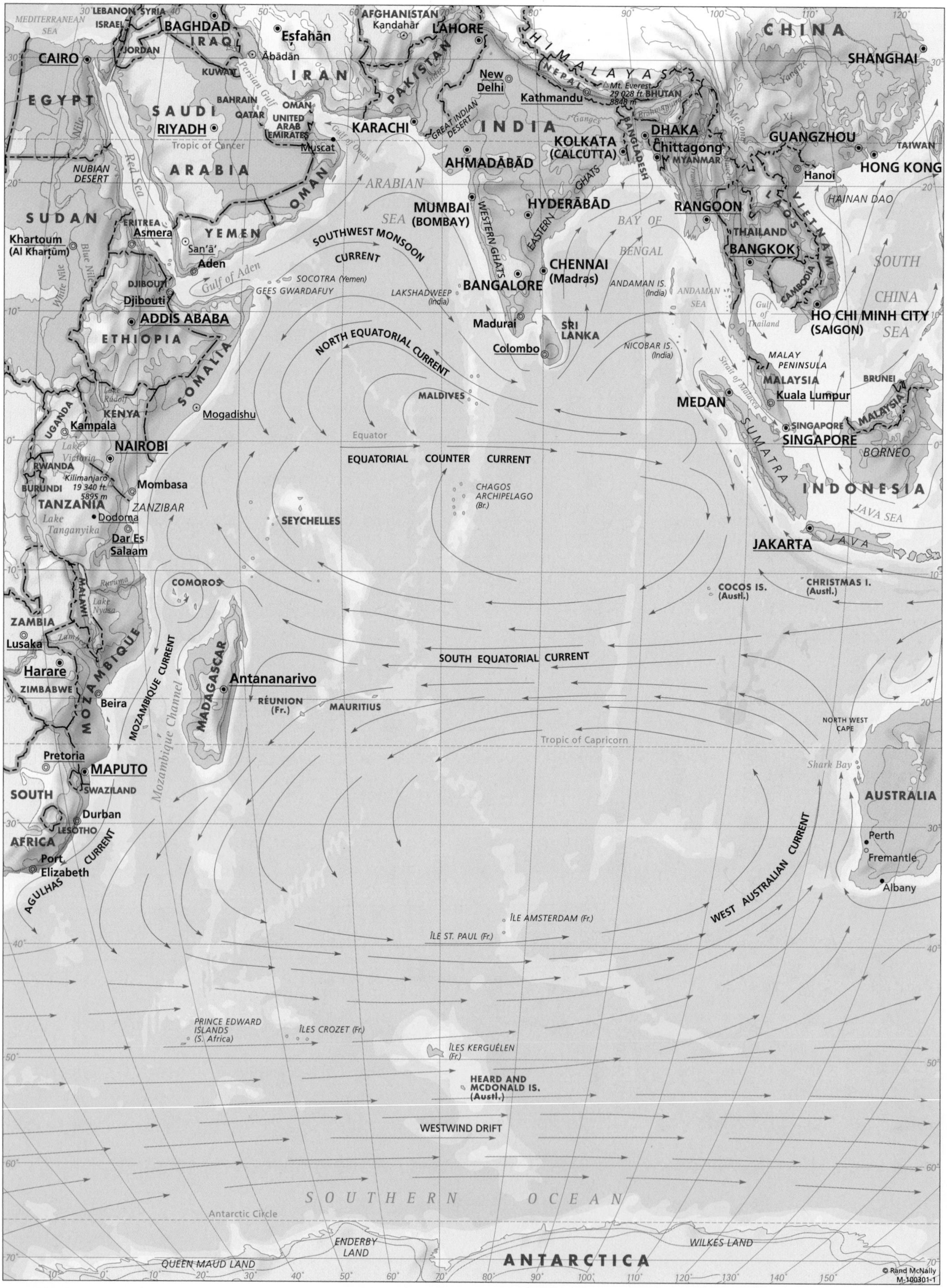
MEDITERRANEAN SEA
LEBANON
SYRIA
ISRAEL
JORDAN
BAGHDAD
IRAQ
KUWAIT
Esfahān
Ābādān
IRAN
AFGHANISTAN
Kandahār
LAHORE
PAKISTAN
CHINA
SHANGHAI
HIMALAYAS
NEPAL
Mt. Everest 29 028 ft. 8848 m
BHUTAN
New Delhi
Kathmandu
CAIRO
EGYPT
SAUDI
RIYADH
ARABIA
BAHRAIN
QATAR
UNITED ARAB EMIRATES
OMAN
Persian Gulf
Gulf of Oman
Muscat
KARACHI
GREAT INDIAN DESERT
INDIA
Ganges
Brahmaputra
BANGLADESH
DHAKA
Chittagong
MYANMAR
KOLKATA (CALCUTTA)
AHMADĀBĀD
GUANGZHOU
TAIWAN
HONG KONG
Hanoi
HAINAN DAO
Tropic of Cancer
NUBIAN DESERT
Nile
Red Sea
ARABIAN SEA
SUDAN
Khartoum (Al Kharṭūm)
Blue Nile
White Nile
ERITREA
Asmera
YEMEN
San'ā'
Aden
Gulf of Aden
MUMBAI (BOMBAY)
WESTERN GHATS
EASTERN GHATS
HYDERĀBĀD
BAY OF BENGAL
RANGOON
THAILAND
LAOS
VIETNAM
BANGKOK
CAMBODIA
SOUTH CHINA SEA
SOUTHWEST MONSOON CURRENT
SOCOTRA (Yemen)
GEES GWARDAFUY
DJIBOUTI
Djibouti
LAKSHADWEEP (India)
BANGALORE
CHENNAI (Madras)
ANDAMAN IS. (India)
ANDAMAN SEA
Gulf of Thailand
HO CHI MINH CITY (SAIGON)
ADDIS ABABA
ETHIOPIA
Madurai
SRI LANKA
Colombo
NORTH EQUATORIAL CURRENT
NICOBAR IS. (India)
MALAY PENINSULA
MALAYSIA
BRUNEI
Kuala Lumpur
Strait of Malacca
SOMALIA
Lake Rudolf
KENYA
MALDIVES
MEDAN
Mogadishu
UGANDA
Kampala
Lake Victoria
NAIROBI
Equator
EQUATORIAL COUNTER CURRENT
SUMATRA
SINGAPORE
BORNEO
RWANDA
BURUNDI
Kilimanjaro 19 340 ft. 5895 m
Mombasa
CHAGOS ARCHIPELAGO (Br.)
INDONESIA
TANZANIA
ZANZIBAR
Dodoma
Lake Tanganyika
Dar Es Salaam
SEYCHELLES
JAVA SEA
JAVA
JAKARTA
Rovuma
MALAWI
Lake Nyasa
COMOROS
COCOS IS. (Austl.)
CHRISTMAS I. (Austl.)
ZAMBIA
Lusaka
Zambezi
Harare
ZIMBABWE
MOZAMBIQUE
MOZAMBIQUE CURRENT
Mozambique Channel
MADAGASCAR
Antananarivo
SOUTH EQUATORIAL CURRENT
Beira
RÉUNION (Fr.)
MAURITIUS
NORTH WEST CAPE
Tropic of Capricorn
Pretoria
MAPUTO
SWAZILAND
Shark Bay
SOUTH AFRICA
AUSTRALIA
Durban
LESOTHO
Perth
Fremantle
Albany
Port Elizabeth
AGULHAS CURRENT
WEST AUSTRALIAN CURRENT
ÎLE AMSTERDAM (Fr.)
ÎLE ST. PAUL (Fr.)
PRINCE EDWARD ISLANDS (S. Africa)
ÎLES CROZET (Fr.)
ÎLES KERGUÉLEN (Fr.)
HEARD AND MCDONALD IS. (Austl.)
WESTWIND DRIFT
SOUTHERN OCEAN
Antarctic Circle
ENDERBY LAND
WILKES LAND
QUEEN MAUD LAND
ANTARCTICA
© Rand McNally M-100301-1

0 200 400 600 800 1000 Miles
0 400 800 1200 1600 Kilometers

Mollweide Projection
Scale 1:50,000,000
One inch to 790 miles
One cm to 500 km

Warm ocean currents
Cold ocean currents

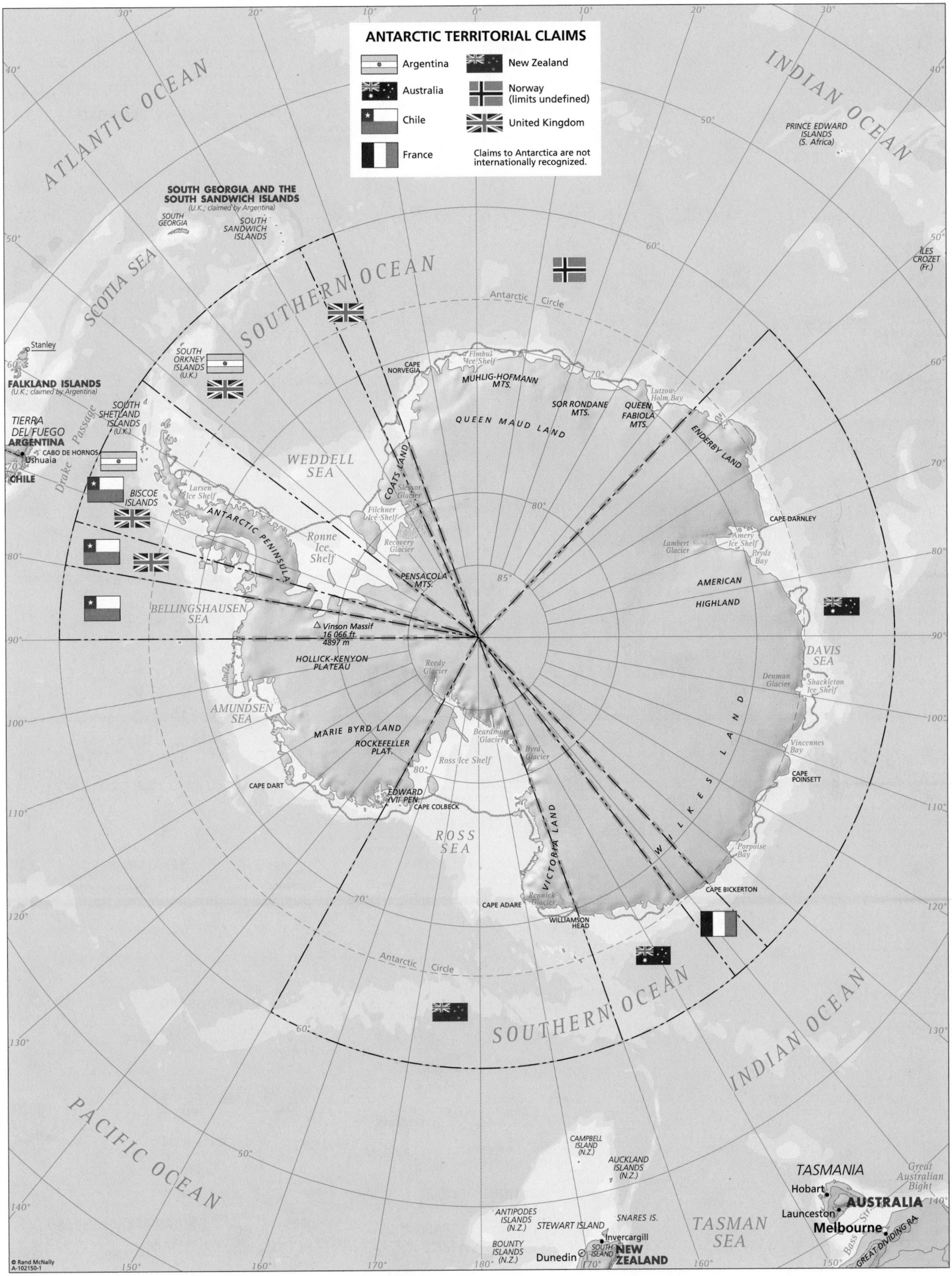
ANTARCTIC TERRITORIAL CLAIMS
Argentina
Australia
Chile
France
New Zealand
Norway (limits undefined)
United Kingdom
Claims to Antarctica are not internationally recognized.
ATLANTIC OCEAN
INDIAN OCEAN
PACIFIC OCEAN
SOUTHERN OCEAN
SCOTIA SEA
WEDDELL SEA
BELLINGSHAUSEN SEA
AMUNDSEN SEA
ROSS SEA
DAVIS SEA
TASMAN SEA
SOUTH GEORGIA AND THE SOUTH SANDWICH ISLANDS
(U.K.; claimed by Argentina)
SOUTH GEORGIA
SOUTH SANDWICH ISLANDS
PRINCE EDWARD ISLANDS (S. Africa)
ÎLES CROZET (Fr.)
Stanley
FALKLAND ISLANDS
(U.K.; claimed by Argentina)
SOUTH ORKNEY ISLANDS (U.K.)
SOUTH SHETLAND ISLANDS (U.K.)
TIERRA DEL FUEGO
ARGENTINA
CABO DE HORNOS
Ushuaia
CHILE
Drake Passage
BISCOE ISLANDS
Antarctic Circle
CAPE NORVEGIA
Fimbul Ice Shelf
MUHLIG-HOFMANN MTS.
SOR RONDANE MTS.
QUEEN FABIOLA MTS.
Lutzow-Holm Bay
QUEEN MAUD LAND
ENDERBY LAND
COATS LAND
Slessor Glacier
Filchner Ice Shelf
Recovery Glacier
Ronne Ice Shelf
Larsen Ice Shelf
ANTARCTIC PENINSULA
PENSACOLA MTS.
CAPE DARNLEY
Lambert Glacier
Amery Ice Shelf
Prydz Bay
AMERICAN HIGHLAND
Vinson Massif 16 066 ft. 4897 m
HOLLICK-KENYON PLATEAU
Reedy Glacier
Denman Glacier
Shackleton Ice Shelf
MARIE BYRD LAND
ROCKEFELLER PLAT.
Beardmore Glacier
Byrd Glacier
Ross Ice Shelf
WILKES LAND
Vincennes Bay
CAPE POINSETT
CAPE DART
EDWARD VII PEN.
CAPE COLBECK
VICTORIA LAND
Porpoise Bay
CAPE BICKERTON
Rennick Glacier
CAPE ADARE
WILLIAMSON HEAD
CAMPBELL ISLAND (N.Z.)
AUCKLAND ISLANDS (N.Z.)
ANTIPODES ISLANDS (N.Z.)
STEWART ISLAND
SNARES IS.
BOUNTY ISLANDS (N.Z.)
Invercargill
Dunedin
SOUTH ISLAND
NEW ZEALAND
TASMANIA
Hobart
Launceston
AUSTRALIA
Melbourne
Bass Str.
GREAT DIVIDING RA.
Great Australian Bight
© Rand McNally
A-102150-1

0 200 400 600 800 1000 1200 Miles

0 200 400 600 800 1000 1200 1400 1600 1800 2000 Kilometers

Lambert Azimuthal Equal Area Projection
Scale 1:40,000,000
One inch to 640 miles
One cm to 400 km

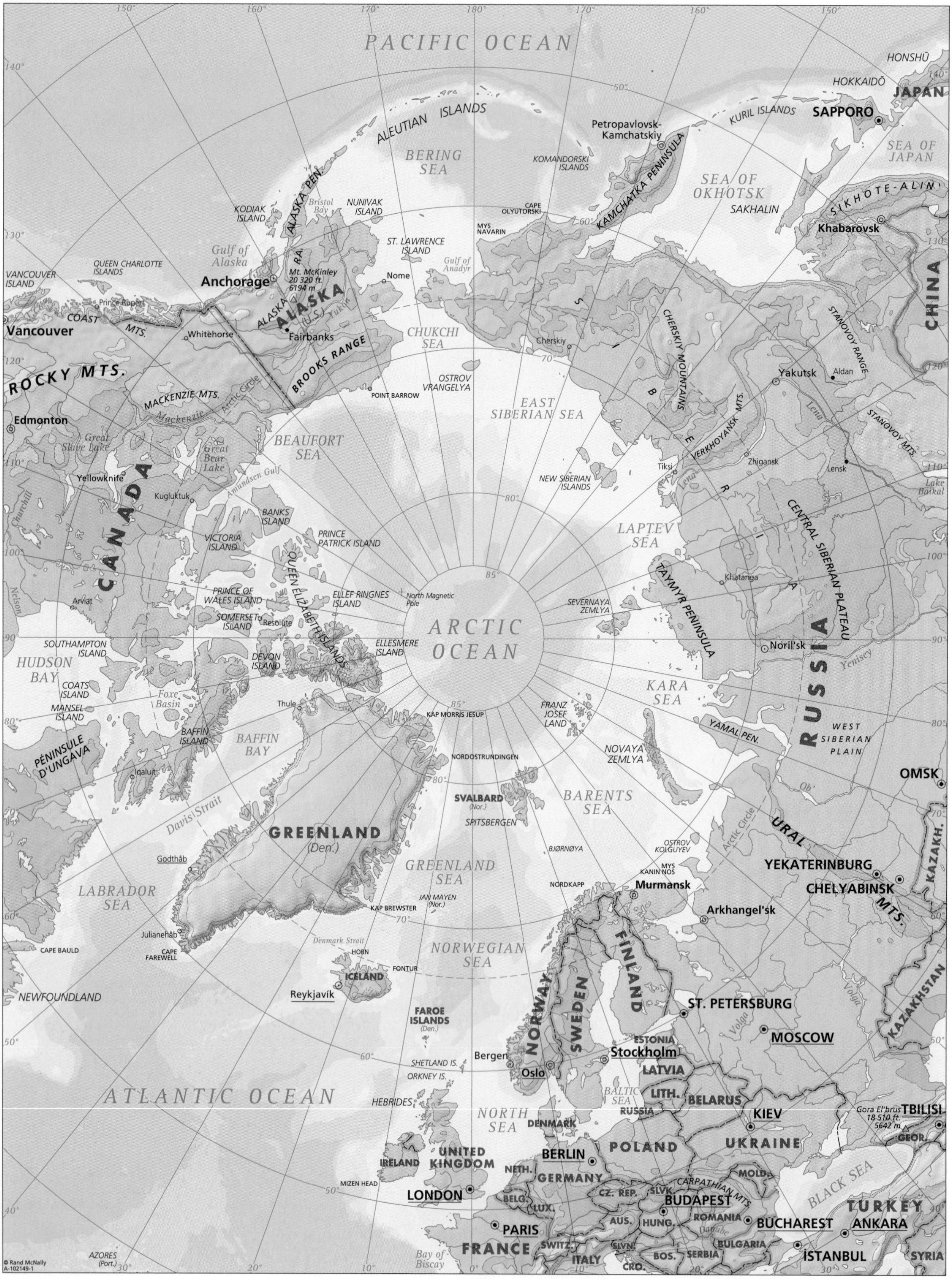
PACIFIC OCEAN
ARCTIC OCEAN
ATLANTIC OCEAN
BERING SEA
ALEUTIAN ISLANDS
SEA OF OKHOTSK
SEA OF JAPAN
HONSHŪ
HOKKAIDŌ
JAPAN
SAPPORO
KURIL ISLANDS
Petropavlovsk-Kamchatskiy
KAMCHATKA PENINSULA
KOMANDORSKI ISLANDS
CAPE OLYUTORSKI
MYS NAVARIN
SAKHALIN
SIKHOTE-ALIN
Khabarovsk
CHINA
KODIAK ISLAND
ALASKA PEN.
Bristol Bay
NUNIVAK ISLAND
ST. LAWRENCE ISLAND
Gulf of Anadyr
Gulf of Alaska
QUEEN CHARLOTTE ISLANDS
VANCOUVER ISLAND
Anchorage
Mt. McKinley 20 320 ft. 6194 m
ALASKA RA.
ALASKA (U.S.)
Yukon
Nome
Prince Rupert
COAST MTS.
Vancouver
Whitehorse
Fairbanks
BROOKS RANGE
CHUKCHI SEA
ROCKY MTS.
POINT BARROW
OSTROV VRANGELYA
MACKENZIE MTS.
Mackenzie
Arctic Circle
Edmonton
Great Slave Lake
Great Bear Lake
BEAUFORT SEA
EAST SIBERIAN SEA
Cherskiy
SIBERIA
CHERSKIY MOUNTAINS
STANOVOY RANGE
Yakutsk
Aldan
STANOVOY MTS.
VERKHOYANSK MTS.
Lena
Zhigansk
Lensk
Tiksi
NEW SIBERIAN ISLANDS
Lake Baikal
Yellowknife
CANADA
Kugluktuk
Amundsen Gulf
Churchill
BANKS ISLAND
VICTORIA ISLAND
PRINCE PATRICK ISLAND
QUEEN ELIZABETH ISLANDS
LAPTEV SEA
CENTRAL SIBERIAN PLATEAU
Nelson
Arviat
PRINCE OF WALES ISLAND
ELLEF RINGNES ISLAND
North Magnetic Pole
SOMERSET ISLAND
Resolute
TAYMYR PENINSULA
Khatanga
SEVERNAYA ZEMLYA
SOUTHAMPTON ISLAND
HUDSON BAY
DEVON ISLAND
ELLESMERE ISLAND
Noril'sk
Yenisey
RUSSIA
COATS ISLAND
Foxe Basin
MANSEL ISLAND
Thule
KAP MORRIS JESUP
FRANZ JOSEF LAND
KARA SEA
YAMAL PEN.
WEST SIBERIAN PLAIN
PÉNINSULE D'UNGAVA
BAFFIN ISLAND
BAFFIN BAY
NORDOSTRUNDINGEN
NOVAYA ZEMLYA
OMSK
Iqaluit
Ob'
Davis Strait
SVALBARD (Nor.)
SPITSBERGEN
BARENTS SEA
GREENLAND (Den.)
URAL MTS.
KAZAKH.
Godthåb
BJØRNØYA
OSTROV KOLGUYEV
MYS KANIN NOS
YEKATERINBURG
CHELYABINSK
GREENLAND SEA
LABRADOR SEA
JAN MAYEN (Nor.)
NORDKAPP
Murmansk
KAP BREWSTER
Arkhangel'sk
Julianehåb
Denmark Strait
CAPE BAULD
CAPE FAREWELL
HORN
NORWEGIAN SEA
FONTUR
ICELAND
Reykjavík
NEWFOUNDLAND
FINLAND
SWEDEN
NORWAY
KAZAKHSTAN
FAROE ISLANDS (Den.)
ST. PETERSBURG
Volga
MOSCOW
ESTONIA
Bergen
Oslo
Stockholm
SHETLAND IS.
ORKNEY IS.
LATVIA
BALTIC SEA
LITH.
BELARUS
HEBRIDES
RUSSIA
NORTH SEA
DENMARK
KIEV
Gora El'brus 18 510 ft. 5642 m
TBILISI
GEOR.
UNITED KINGDOM
IRELAND
BERLIN
POLAND
UKRAINE
NETH.
GERMANY
MOLD.
MIZEN HEAD
LONDON
CZ. REP.
SLVK.
CARPATHIAN MTS.
BLACK SEA
BELG.
LUX.
BUDAPEST
TURKEY
AUS.
HUNG.
ROMANIA
BUCHAREST
ANKARA
PARIS
Danube
BULGARIA
SWITZ.
SLVN.
FRANCE
ITALY
BOS.
SERBIA
İSTANBUL
SYRIA
CRO.
Bay of Biscay
AZORES (Port.)
© Rand McNally A-102149-1

0 200 400 600 800 1000 1200 Miles

0 200 400 600 800 1000 1200 1400 1600 1800 2000 Kilometers

Lambert Azimuthal Equal Area Projection
Scale 1:40,000,000
One inch to 640 miles
One cm to 400 km

This table gives the area, population, population density, political status, capital, and predominant languages for every country in the world. The political units listed are categorized by political status in the 'Form of Government and Ruling Power' column of the table, as follows:

A independent countries;

B internally independent political entities which are under the protection of another country in matters of defense and foreign affairs;

C colonies and other dependent political units;

D the major administrative subdivisions of Australia, Canada, China, the United Kingdom, and the United States.

For comparison, the table also includes the continents and the world. All footnotes appear at the end of the table.

The populations are estimates for January 1, 2009, made by Rand McNally on the basis of official data, United States Census Bureau estimates, and other available information. Area figures include inland water.

Region or political division	Est. Pop. 1/1/09	Area sq. km.	Area sq. mi.	Pop. per sq. km.	Pop. per sq. mi.	Form of Government and Ruling Power	Political Status	Capital	Predominant Languages
Afars and Issas *see Djibouti*									
† **Afghanistan**	33,170,000	652,090	251,773	51	132	Transitional	A	Kabul (Kābol)	Dari, Pashto, Uzbek, Turkmen
Africa	985,490,000	30,300,000	11,700,000	33	84				
Alabama	4,685,000	135,765	52,419	35	89	State (U.S.)	D	Montgomery	English
Alaska	690,000	1,717,854	663,267	0.4	1.0	State (U.S.)	D	Juneau	English, indigenous
† **Albania**	3,630,000	28,748	11,100	126	327	Republic	A	Tiranë	Albanian, Greek
Alberta	3,570,000	661,848	255,541	5.4	14	Province (Canada)	D	Edmonton	English
† **Algeria**	33,975,000	2,381,741	919,595	14	37	Republic	A	Algiers (Alger)	Arabic, Berber dialects, French
American Samoa	65,000	199	77	327	844	Unincorporated territory (U.S.)	C	Pago Pago	Samoan, English
† **Andorra**	83,000	468	181	177	459	Parliamentary co-principality (Spanish and French)	B	Andorra la Vella	Catalan, Spanish (Castilian), French
† **Angola**	12,665,000	1,246,700	481,354	10	26	Republic	A	Luanda	Portuguese, indigenous
Anguilla	14,000	96	37	146	378	Overseas territory (U.K. protection)	B	The Valley	English
Anhui	61,865,000	139,000	53,668	445	1,153	Province (China)	D	Hefei	Chinese (Mandarin)
Antarctica	*(1)*	14,000,000	5,400,000						
† **Antigua and Barbuda**	85,000	442	171	192	497	Parliamentary state	A	St. John's	English, local dialects
Aomen *see Macau*									
† **Argentina**	40,700,000	2,780,400	1,073,519	15	38	Republic	A	Buenos Aires	Spanish, English, Italian, German, French
Arizona	6,535,000	295,254	113,998	22	57	State (U.S.)	D	Phoenix	English
Arkansas	2,870,000	137,732	53,179	21	54	State (U.S.)	D	Little Rock	English
† **Armenia**	2,965,000	29,800	11,506	99	258	Republic	A	Yerevan	Armenian, Russian
Aruba	100,000	193	75	518	1,333	Self-governing territory (Netherlands protection)	B	Oranjestad	Dutch, Papiamento, English, Spanish
Ascension	1,000	88	34	11	29	Dependency (St. Helena)	C	Georgetown	English
Asia	4,078,790,000	44,900,000	17,300,000	91	236				
† **Australia**	21,135,000	7,692,030	2,969,910	2.7	7.1	Federal parliamentary state	A	Canberra	English, indigenous
Australian Capital Territory	340,000	2,360	911	144	373	Territory (Australia)	D	Canberra	English
† **Austria**	8,210,000	83,858	32,378	98	254	Federal republic	A	Vienna (Wien)	German
† **Azerbaijan**	8,205,000	86,600	33,437	95	245	Republic	A	Baku (Bakı)	Azeri, Russian, Armenian
† **Bahamas**	310,000	13,939	5,382	22	58	Parliamentary State	A	Nassau	English, Creole
† **Bahrain**	725,000	691	267	1,049	2,715	Monarchy	A	Manama (Al-Manāmah)	Arabic, English, Farsi, Urdu
† **Bangladesh**	155,045,000	143,998	55,598	1,077	2,789	Republic	A	Dhaka	Bangla, English
† **Barbados**	285,000	430	166	663	1,717	Parliamentary state	A	Bridgetown	English
Beijing (Peking)	16,010,000	16,800	6,487	953	2,468	Autonomous city (China)	D	Beijing	Chinese (Mandarin)
† **Belarus**	9,665,000	207,600	80,155	47	121	Republic	A	Minsk	Belarussian, Russian
Belau *see Palau*									
† **Belgium**	10,410,000	30,528	11,787	341	883	Constitutional monarchy	A	Brussels (Bruxelles)	Dutch (Flemish), French, German
† **Belize**	305,000	22,966	8,867	13	34	Parliamentary state	A	Belmopan	English, Spanish, Mayan, Garifuna
† **Benin**	8,660,000	112,622	43,484	77	199	Republic	A	Porto-Novo and Cotonou	French, Fon, Yoruba, indigenous
Bermuda	68,000	54	21	1,259	3,238	Overseas territory (U.K.)	C	Hamilton	English
† **Bhutan**	685,000	46,500	17,954	15	38	Monarchy (Indian protection)	B	Thimphu	Dzongkha, Tibetan and Nepalese dialects
† **Bolivia**	9,690,000	1,098,581	424,165	8.8	23	Republic	A	La Paz and Sucre	Aymara, Quechua, Spanish
† **Bosnia and Herzegovina**	4,605,000	51,197	19,767	90	233	Republic	A	Sarajevo	Bosnian, Croatian, Serbian
† **Botswana**	1,970,000	581,730	224,607	3.4	8.8	Republic	A	Gaborone	English, Tswana
† **Brazil**	197,550,000	8,547,404	3,300,172	23	60	Federal republic	A	Brasília	Portuguese, Spanish, English, French
British Columbia	4,380,000	944,735	364,764	4.6	12	Province (Canada)	D	Victoria	English
British Indian Ocean Territory	*(1)*	60	23			Overseas territory (U.K.)	C		English
British Virgin Islands	24,000	151	58	159	414	Overseas territory (U.K.)	C	Road Town	English
† **Brunei**	385,000	5,765	2,226	67	173	Monarchy	A	Bandar Seri Begawan	Malay, English, Chinese
† **Bulgaria**	7,235,000	110,994	42,855	65	169	Republic	A	Sofia (Sofiya)	Bulgarian, Turkish
† **Burkina Faso**	15,500,000	274,200	105,869	57	146	Republic	A	Ouagadougou	French, indigenous
Burma *see Myanmar*									
† **Burundi**	8,840,000	27,830	10,745	318	823	Republic	A	Bujumbura	French, Kirundi, Swahili
California	36,955,000	423,970	163,696	87	226	State (U.S.)	D	Sacramento	English
† **Cambodia**	14,365,000	181,035	69,898	79	206	Constitutional monarchy	A	Phnom Penh (Phnum Pénh)	Khmer, French
† **Cameroon**	18,670,000	475,440	183,568	39	102	Republic	A	Yaoundé	English, French, indigenous
† **Canada**	33,350,000	9,984,670	3,855,103	3.3	8.7	Federal parliamentary state	A	Ottawa	English, French
† **Cape Verde**	430,000	4,033	1,557	107	276	Republic	A	Praia	Portuguese, Crioulo
Cayman Islands	48,000	264	102	182	471	Overseas territory (U.K.)	C	George Town	English
† **Central African Republic**	4,480,000	622,984	240,536	7.2	19	Republic	A	Bangui	French, Sango, Arabic, indigenous
Ceylon *see Sri Lanka*									
† **Chad**	10,220,000	1,284,000	495,755	8.0	21	Republic	A	N'Djamena (Fort-Lamy)	Arabic, French, indigenous
Channel Islands	155,000	194	75	799	2,067	Two crown dependencies (U.K. protection)			English, French
† **Chile**	16,530,000	756,096	291,930	22	57	Republic	A	Santiago	Spanish
† **China (incl. Hong Kong and Macau)** *(2)*	1,341,820,000	9,557,172	3,690,045	140	364	Socialist republic	A	Beijing	Chinese dialects
Chongqing	28,430,000	82,400	31,815	345	894	Autonomous city (China)	D	Chongqing	Chinese (Mandarin)
Christmas Island	1,500	135	52	11	29	External territory (Australia)	C	Settlement	English, Chinese, Malay
Cocos (Keeling) Islands	600	14	5.4	43	111	External territory (Australia)	C	West Island	English, Cocos-Malay, Malay
† **Colombia**	45,330,000	1,138,914	439,737	40	103	Republic	A	Bogotá	Spanish
Colorado	4,965,000	269,601	104,094	18	48	State (U.S.)	D	Denver	English
† **Comoros (excl. Mayotte)**	740,000	2,235	863	331	857	Federal Islamic republic	A	Moroni	Arabic, French, Comoran
† **Congo**	3,960,000	342,000	132,047	12	30	Republic	A	Brazzaville	French, Lingala, Kikongo, indigenous
† **Congo, Democratic Republic of the (Zaire)**	67,590,000	2,345,095	905,446	29	75	Republic	A	Kinshasa (Léopoldville)	French, Kikongo, Lingala, Swahili, Tshiluba, Kingwana

Region or political division	Est. Pop. 1/1/09	Area sq. km.	Area sq. mi.	Pop. per sq. km.	Pop. per sq. mi.	Form of Government and Ruling Power	Political Status	Capital	Predominant Languages
Connecticut	3,520,000	14,357	5,543	245	635	State (U.S.)	D	Hartford	English
Cook Islands	12,000	236	91	51	132	Self-governing territory (New Zealand protection)	B	Avarua	English, Maori
† Costa Rica	4,225,000	51,100	19,730	83	214	Republic	A	San José	Spanish
† Cote d'Ivoire (Ivory Coast)	20,395,000	322,463	124,504	63	164	Republic	A	Abidjan and Yamoussoukro	French, Dioula and other indigenous
† Croatia	4,490,000	56,538	21,829	79	206	Republic	A	Zagreb	Croatian
† Cuba	11,440,000	110,861	42,804	103	267	Socialist republic	A	Havana (La Habana)	Spanish
† Cyprus	795,000	9,251	3,572	86	223	Republic	A	Nicosia	Greek, Turkish, English
† Czech Republic	10,215,000	78,866	30,450	130	335	Republic	A	Prague (Praha)	Czech, Slovak
Delaware	880,000	6,447	2,489	136	354	State (U.S.)	D	Dover	English
† Denmark	5,495,000	43,096	16,640	128	330	Constitutional monarchy	A	Copenhagen (København)	Danish
District of Columbia	595,000	177	68	3,362	8,750	Federal district (U.S.)	D	Washington	English
† Djibouti	510,000	23,200	8,958	22	57	Republic	A	Djibouti	French, Arabic, Somali, Afar
† Dominica	73,000	751	290	97	252	Republic	A	Roseau	English, French
† Dominican Republic	9,580,000	48,511	18,730	197	511	Republic	A	Santo Domingo	Spanish
† East Timor (Timor-Leste)	1,120,000	14,874	5,743	75	195	Republic	A	Dili	Portuguese, Tetum, Bahasa Indonesia (Malay)
† Ecuador	14,465,000	283,561	109,484	51	132	Republic	A	Quito	Spanish, Quechua, indigenous
† Egypt	82,400,000	1,001,449	386,662	82	213	Socialist republic	A	Cairo (El Qâhira)	Arabic
Ellice Islands *see Tuvalu*									
† El Salvador	7,125,000	21,041	8,124	339	877	Republic	A	San Salvador	Spanish, Nahua
England	51,135,000	130,422	50,356	392	1,015	Administrative division (U.K.)	D	London	English
† Equatorial Guinea	625,000	28,051	10,831	22	58	Republic	A	Malabo	Spanish, indigenous, English
† Eritrea	5,575,000	117,600	45,406	47	123	Republic	A	Asmera	Tigre, Kunama, Cushitic dialects, Nora Bana, Arabic
† Estonia	1,305,000	45,227	17,462	29	75	Republic	A	Tallinn	Estonian, Latvian, Lithuanian, Russian
† Ethiopia	83,870,000	1,104,300	426,373	76	197	Federal republic	A	Addis Ababa (Ādīs Ābeba)	Amharic, Tigrinya, Orominga, Guaraginga, Somali, Arabic
Europe	728,420,000	9,900,000	3,800,000	74	192				
Falkland Islands *(3)*	3,000	12,173	4,700	0.2	0.6	Overseas territory (U.K.)	C	Stanley	English
Faroe Islands	49,000	1,399	540	35	91	Self-governing territory (Danish protection)	B	Tórshavn	Danish, Faroese
† Fiji	940,000	18,274	7,056	51	133	Republic	A	Suva	English, Fijian, Hindustani
† Finland	5,250,000	338,145	130,559	16	40	Republic	A	Helsinki	Finnish, Swedish, Lapp, Russian
Florida	18,430,000	170,304	65,755	108	280	State (U.S.)	D	Tallahassee	English
† France (excl. Overseas Departments)	62,260,000	539,965	208,482	115	299	Republic	A	Paris	French
French Guiana	210,000	83,534	32,253	2.5	6.5	Overseas department (France)	C	Cayenne	French
French Polynesia	285,000	4,000	1,544	71	185	Overseas territory (France)	C	Papeete	French, Tahitian
Fujian	36,025,000	120,000	46,332	300	778	Province (China)	D	Fuzhou	Chinese dialects
† Gabon	1,500,000	267,668	103,347	5.6	15	Republic	A	Libreville	French, Fang, indigenous
† Gambia, The	1,760,000	10,689	4,127	165	426	Republic	A	Banjul (Bathurst)	English, Malinke, Wolof, Fula, indigenous
Gansu	26,385,000	450,000	173,746	59	152	Province (China)	D	Lanzhou	Chinese (Mandarin), Mongolian, Tibetan dialects
Gaza Strip	1,525,000	360	139	4,236	10,971	Israeli territory with limited self-government			Arabic
Georgia	9,740,000	153,910	59,425	63	164	State (U.S.)	D	Atlanta	English
† Georgia	4,625,000	69,700	26,911	66	172	Republic	A	Tbilisi	Georgian, Russian, Armenian, Azeri
† Germany	82,350,000	357,022	137,847	231	597	Federal republic	A	Berlin	German
† Ghana	23,610,000	238,533	92,098	99	256	Republic	A	Accra	English, Akan and other indigenous
Gibraltar	28,000	6.0	2.3	4,667	12,174	Overseas territory (U.K.)	C	Gibraltar	English, Spanish, Italian, Portuguese
Gilbert Islands *see Kiribati*									
Golan Heights	41,000	1,176	454	35	90	Occupied by Israel			Arabic, Hebrew
Great Britain *see United Kingdom*									
† Greece	10,730,000	131,957	50,949	81	211	Republic	A	Athens (Athína)	Greek, English, French
Greenland	58,000	2,166,086	836,331	0.03	0.07	Self-governing territory (Danish protection)	B	Godthåb	Danish, Greenlandic, Inuit dialects
† Grenada	91,000	344	133	265	684	Parliamentary state	A	St. George's	English, French
Guadeloupe (incl. St. Barthelemy and St. Martin)	465,000	1,780	687	261	677	Overseas department (France)	C	Basse-Terre	French, Creole
Guam	175,000	549	212	319	825	Unincorporated territory (U.S.)	C	Hagåtña	English, Chamorro, Japanese
Guangdong	94,205,000	177,800	68,649	530	1,372	Province (China)	D	Guangzhou (Canton)	Chinese dialects, Miao-Yao
Guangxi Zhuangzu	47,780,000	236,300	91,236	202	524	Autonomous region (China)	D	Nanning	Chinese dialects, Thai, Miao-Yao
† Guatemala	13,140,000	108,889	42,042	121	313	Republic	A	Guatemala	Spanish, Amerindian
Guernsey (incl. Dependencies)	66,000	78	30	846	2,200	Crown dependency (U.K. protection)	B	St. Peter Port	English, French
† Guinea	9,930,000	245,857	94,926	40	105	Republic	A	Conakry	French, indigenous
† Guinea-Bissau	1,520,000	36,125	13,948	42	109	Republic	A	Bissau	Portuguese, Crioulo, indigenous
Guizhou	38,040,000	170,000	65,637	224	580	Province (China)	D	Guiyang	Chinese (Mandarin), Thai, Miao-Yao
† Guyana	770,000	214,969	83,000	3.6	9.3	Republic	A	Georgetown	English, indigenous
Hainan	8,465,000	34,200	13,205	248	641	Province (China)	D	Haikou	Chinese, Min, Tai
† Haiti	8,955,000	27,750	10,714	323	836	Republic	A	Port-au-Prince	Creole, French
Hawaii	1,295,000	28,311	10,931	46	118	State (U.S.)	D	Honolulu	English, Hawaiian, Japanese
Hebei	69,845,000	190,000	73,359	368	952	Province (China)	D	Shijiazhuang	Chinese (Mandarin)
Heilongjiang	38,710,000	469,000	181,082	83	214	Province (China)	D	Harbin	Chinese dialects, Mongolian, Tungus
Henan	95,095,000	167,000	64,479	569	1,475	Province (China)	D	Zhengzhou	Chinese (Mandarin)
Holland *see Netherlands*									
† Honduras	7,715,000	112,088	43,277	69	178	Republic	A	Tegucigalpa	Spanish, indigenous
Hong Kong (Xianggang)	7,035,000	1,100	425	6,395	16,553	Special administrative region (China)	C	Hong Kong (Xianggang)	Chinese (Cantonese), English, Putonghua
Hubei	57,645,000	187,400	72,356	308	797	Province (China)	D	Wuhan	Chinese dialects
Hunan	64,215,000	210,000	81,082	306	792	Province (China)	D	Changsha	Chinese dialects, Miao-Yao
† Hungary	9,920,000	93,030	35,919	107	276	Republic	A	Budapest	Hungarian
† Iceland	305,000	103,000	39,769	3.0	7.7	Republic	A	Reykjavík	Icelandic
Idaho	1,530,000	216,446	83,570	7.1	18	State (U.S.)	D	Boise	English
Illinois	12,970,000	149,998	57,914	86	224	State (U.S.)	D	Springfield	English
† India (incl. part of Jammu and Kashmir)	1,157,055,000	3,166,285	1,222,510	365	946	Federal republic	A	New Delhi	English, Hindi, Telugu, Bengali, indigenous
Indiana	6,410,000	94,321	36,418	68	176	State (U.S.)	D	Indianapolis	English
† Indonesia	238,910,000	1,904,443	735,310	125	325	Republic	A	Jakarta	Bahasa Indonesia (Malay), English, Dutch, indigenous
Iowa	3,020,000	145,743	56,272	21	54	State (U.S.)	D	Des Moines	English

Region or political division	Est. Pop. 1/1/09	Area sq. km.	Area sq. mi.	Pop. per sq. km.	Pop. per sq. mi.	Form of Government and Ruling Power	Political Status	Capital	Predominant Languages
† **Iran**	66,135,000	1,648,195	636,372	40	104	Islamic republic	A	Tehrān	Farsi, Turkish dialects, Kurdish
† **Iraq**	28,585,000	438,317	169,235	65	169	Republic	A	Baghdād	Arabic, Kurdish, Assyrian, Armenian
† **Ireland**	4,180,000	70,273	27,133	59	154	Republic	A	Dublin (Baile Átha Cliath)	English, Irish Gaelic
Isle of Man	76,000	572	221	133	344	Crown dependency (U.K. protection)	B	Douglas	English, Manx Gaelic
† **Israel (excl. Occupied Areas)**	7,175,000	20,770	8,019	345	895	Republic	A	Jerusalem (Yerushalayim)	Hebrew, Arabic
† **Italy**	58,140,000	301,323	116,342	193	500	Republic	A	Rome (Roma)	Italian, German, French, Slovene
Ivory Coast *see Cote d'Ivoire*									
† **Jamaica**	2,815,000	10,991	4,244	256	663	Parliamentary state	A	Kingston	English, Creole
† **Japan**	127,200,000	377,750	145,850	337	872	Constitutional monarchy	A	Tōkyō	Japanese
Jersey	92,000	116	45	793	2,044	Crown dependency (U.K. protection)	B	St. Helier	English, French
Jiangsu	76,445,000	102,600	39,614	745	1,930	Province (China)	D	Nanjing	Chinese dialects
Jiangxi	43,935,000	166,600	64,325	264	683	Province (China)	D	Nanchang	Chinese dialects
Jilin	27,570,000	187,000	72,201	147	382	Province (China)	D	Changchun	Chinese (Mandarin), Mongolian, Korean
† **Jordan**	6,270,000	89,342	34,495	70	182	Constitutional monarchy	A	'Ammān	Arabic
Kansas	2,815,000	213,096	82,277	13	34	State (U.S.)	D	Topeka	English
† **Kazakhstan**	15,370,000	2,717,300	1,049,156	5.7	15	Republic	A	Astana (Aqmola)	Kazakh, Russian
Kentucky	4,290,000	104,659	40,409	41	106	State (U.S.)	D	Frankfort	English
† **Kenya**	38,475,000	582,646	224,961	66	171	Republic	A	Nairobi	English, Swahili, indigenous
† **Kiribati**	110,000	811	313	136	351	Republic	A	Bairiki	English, Gilbertese
† **Korea, North**	22,620,000	120,538	46,540	188	486	Socialist republic	A	P'yŏngyang	Korean
† **Korea, South**	48,445,000	99,268	38,328	488	1,264	Republic	A	Seoul (Sŏul)	Korean
Kosovo *(4)*	1,800,000	10,887	4,203	165	428	Republic	A	Priština	Albanian, Serbian
† **Kuwait**	2,645,000	17,818	6,880	148	384	Constitutional monarchy	A	Kuwait (Al-Kuwayt)	Arabic, English
† **Kyrgyzstan**	5,395,000	199,900	77,182	27	70	Republic	A	Bishkek	Kirghiz, Russian
† **Laos**	6,755,000	236,800	91,429	29	74	Socialist republic	A	Vientiane (Viangchan)	Lao, French, English
† **Latvia**	2,240,000	64,600	24,942	35	90	Republic	A	Rīga	Latvian, Russian, Lithuanian
† **Lebanon**	3,995,000	10,400	4,016	384	995	Republic	A	Beirut (Bayrūt)	Arabic, French, Armenian, English
† **Lesotho**	2,130,000	30,355	11,720	70	182	Constitutional monarchy	A	Maseru	English, Sesotho, Zulu, Xhosa
Liaoning	43,245,000	145,700	56,255	297	769	Province (China)	D	Shenyang	Chinese (Mandarin), Mongolian
† **Liberia**	3,395,000	111,369	43,000	30	79	Republic	A	Monrovia	English, indigenous
† **Libya**	6,240,000	1,759,540	679,362	3.5	9.2	Socialist republic	A	Tripoli (Ţarābulus)	Arabic
† **Liechtenstein**	35,000	160	62	219	565	Constitutional monarchy	A	Vaduz	German
† **Lithuania**	3,560,000	65,300	25,213	55	141	Republic	A	Vilnius	Lithuanian, Polish, Russian
Louisiana	4,435,000	134,264	51,840	33	86	State (U.S.)	D	Baton Rouge	English
† **Luxembourg**	490,000	2,586	999	189	490	Constitutional monarchy	A	Luxembourg	French, Luxembourgish, German
Macau (Aomen)	555,000	18	6.9	30,833	80,435	Special administrative region (China)	C	Macau (Aomen)	Chinese (Cantonese), Portuguese
† **Macedonia**	2,065,000	25,713	9,928	80	208	Republic	A	Skopje	Macedonian, Albanian
† **Madagascar**	20,345,000	587,041	226,658	35	90	Republic	A	Antananarivo	Malagasy, French
Maine	1,325,000	91,646	35,385	14	37	State (U.S.)	D	Augusta	English
† **Malawi**	14,100,000	118,484	45,747	119	308	Republic	A	Lilongwe	Chichewa, English
† **Malaysia**	25,495,000	329,758	127,320	77	200	Federal constitutional monarchy	A	Kuala Lumpur and Putrajaya	Malay, Chinese dialects, English, Tamil
† **Maldives**	395,000	298	115	1,326	3,435	Republic	A	Male'	Divehi
† **Mali**	12,490,000	1,240,192	478,841	10	26	Republic	A	Bamako	French, Bambara, indigenous
† **Malta**	405,000	316	122	1,282	3,320	Republic	A	Valletta	English, Maltese
Manitoba	1,210,000	647,797	250,116	1.9	4.8	Province (Canada)	D	Winnipeg	English
† **Marshall Islands**	64,000	181	70	354	914	Republic (U.S. protection)	A	Majuro (island)	English, indigenous, Japanese
Martinique	445,000	1,100	425	405	1,047	Overseas department (France)	C	Fort-de-France	French, Creole
Maryland	5,665,000	32,133	12,407	176	457	State (U.S.)	D	Annapolis	English
Massachusetts	6,535,000	27,336	10,555	239	619	State (U.S.)	D	Boston	English
† **Mauritania**	3,090,000	1,030,700	397,956	3.0	7.8	Republic	A	Nouakchott	Arabic, Pular, Soninke, Wolof
† **Mauritius (incl. Dependencies)**	1,280,000	2,040	788	627	1,624	Republic	A	Port Louis	English, Creole, Bhojpuri, French, Hindi, Tamil, others
Mayotte *(5)*	220,000	374	144	588	1,528	Territorial collectivity (France)	C	Mamoudzou	French, Swahili (Mahorian)
† **Mexico**	110,585,000	1,964,382	758,452	56	146	Federal republic	A	Mexico City (Ciudad de México)	Spanish, indigenous
Michigan	10,060,000	250,494	96,716	40	104	State (U.S.)	D	Lansing	English
† **Micronesia, Federated States of**	110,000	702	271	157	406	Republic (U.S. protection)	A	Palikir	English, indigenous
Midway Islands	*(1)*	5.2	2.0			Unincorporated territory (U.S.)	C		English
Minnesota	5,250,000	225,171	86,939	23	60	State (U.S.)	D	St. Paul	English
Mississippi	2,955,000	125,434	48,430	24	61	State (U.S.)	D	Jackson	English
Missouri	5,945,000	180,533	69,704	33	85	State (U.S.)	D	Jefferson City	English
† **Moldova**	4,320,000	33,851	13,070	128	331	Republic	A	Chişinău	Romanian (Moldovan), Russian
† **Monaco**	33,000	2.0	0.8	16,500	41,250	Constitutional monarchy	A	Monaco	French, English, Italian, Monegasque
† **Mongolia**	3,020,000	1,566,500	604,829	1.9	5.0	Republic	A	Ulaanbaatar	Khalkha Mongol, Turkish dialects, Russian, Chinese
Montana	970,000	380,838	147,042	2.5	6.6	State (U.S.)	D	Helena	English
† **Montenegro**	675,000	13,812	5,333	49	127	Republic	A	Podgorica	Serbian, Albanian
Montserrat	5,000	102	39	49	128	Overseas territory (U.K.)	C	Plymouth (abandoned)	English
† **Morocco (excl. Western Sahara)**	34,600,000	446,550	172,414	77	201	Constitutional monarchy	A	Rabat	Arabic, Berber dialects, French
† **Mozambique**	21,475,000	801,590	309,496	27	69	Republic	A	Maputo (Lourenço Marques)	Portuguese, indigenous
† **Myanmar (Burma)**	47,950,000	676,578	261,228	71	184	Provisional military government	A	Yangon (Rangoon) and Nay Pyi Taw	Burmese, indigenous
† **Namibia**	2,100,000	823,144	317,818	2.6	6.6	Republic	A	Windhoek	English, Afrikaans, German, indigenous
† **Nauru**	14,000	21	8.1	667	1,728	Republic	A	Yaren District	Nauruan, English
Nebraska	1,795,000	200,345	77,354	9.0	23	State (U.S.)	D	Lincoln	English
Nei Mongol (Inner Mongolia)	24,270,000	1,183,000	456,759	21	53	Autonomous region (China)	D	Hohhot	Mongolian
† **Nepal**	28,380,000	147,181	56,827	193	499	Federal republic	A	Kathmandu (Kāṭhmāṇḍāu)	Nepali, Maithali, Bhojpuri, other indigenous
† **Netherlands**	16,680,000	41,864	16,164	398	1,032	Constitutional monarchy	A	Amsterdam and The Hague ('s-Gravenhage)	Dutch
Netherlands Antilles	225,000	800	309	281	728	Self-governing territory (Netherlands protection)	B	Willemstad	Dutch, Papiamento, English
Nevada	2,615,000	286,351	110,561	9.1	24	State (U.S.)	D	Carson City	English
New Brunswick	750,000	72,908	28,150	10	27	Province (Canada)	D	Fredericton	English, French
New Caledonia	225,000	18,575	7,172	12	31	Overseas territory (France)	C	Nouméa	French, indigenous
New Hampshire	1,325,000	24,216	9,350	55	142	State (U.S.)	D	Concord	English
New Hebrides *see Vanuatu*									
New Jersey	8,730,000	22,588	8,721	386	1,001	State (U.S.)	D	Trenton	English
New Mexico	1,995,000	314,915	121,590	6.3	16	State (U.S.)	D	Santa Fe	English, Spanish

Region or political division	Est. Pop. 1/1/09	Area sq. km.	Area sq. mi.	Pop. per sq. km.	Pop. per sq. mi.	Form of Government and Ruling Power	Political Status	Capital	Predominant Languages
New South Wales	6,960,000	800,640	309,129	8.7	23	State (Australia)	D	Sydney	English
New York	19,595,000	141,299	54,556	139	359	State (U.S.)	D	Albany	English
† New Zealand	4,195,000	270,534	104,454	16	40	Parliamentary state	A	Wellington	English, Maori
Newfoundland and Labrador	510,000	405,212	156,453	1.3	3.3	Province (Canada)	D	St. John's	English
† Nicaragua	5,840,000	129,640	50,054	45	117	Republic	A	Managua	Spanish, English, indigenous
† Niger	15,025,000	1,267,000	489,192	12	31	Provisional military government	A	Niamey	French, Hausa, Djerma, indigenous
† Nigeria	147,735,000	923,768	356,669	160	414	Transitional military government	A	Abuja	English, Hausa, Fulani, Yorbua, Ibo, indigenous
Ningxia Huizu	6,115,000	66,400	25,637	92	239	Autonomous region (China)	D	Yinchuan	Chinese (Mandarin)
Niue	1,500	259	100	5.8	15	Self-governing territory (New Zealand protection)	B	Alofi	English, indigenous
Norfolk Island	2,000	36	14	56	143	External territory (Australia)	C	Kingston	English, Norfolk
North America	531,180,000	24,700,000	9,500,000	22	56	...		...	...
North Carolina	9,270,000	139,389	53,819	67	172	State (U.S.)	D	Raleigh	English
North Dakota	645,000	183,112	70,700	3.5	9.1	State (U.S.)	D	Bismarck	English
Northern Ireland	1,760,000	13,576	5,242	130	336	Administrative division (U.K.)	D	Belfast	English
Northern Mariana Islands	88,000	464	179	190	492	Commonwealth (U.S. protection)	B	Saipan (island)	English, Chamorro, Carolinian
Northern Territory	215,000	1,349,130	520,902	0.2	0.4	Territory (Australia)	D	Darwin	English, indigenous
Northwest Territories	43,000	1,346,106	519,735	0.03	0.08	Territory (Canada)	D	Yellowknife	English, indigenous
† Norway (incl. Jan Mayen and Svalbard)	4,655,000	323,877	125,050	14	37	Constitutional monarchy	A	Oslo	Norwegian, Lapp, Finnish
Nova Scotia	945,000	55,284	21,345	17	44	Province (Canada)	D	Halifax	English
Nunavut	30,000	2,093,190	808,185	0.01	0.04	Territory (Canada)	D	Iqaluit	English, indigenous
Oceania (incl. Australia)	34,605,000	8,500,000	3,300,000	4.1	10	...		...	...
Ohio	11,550,000	116,096	44,825	99	258	State (U.S.)	D	Columbus	English
Oklahoma	3,660,000	181,036	69,898	20	52	State (U.S.)	D	Oklahoma City	English
† Oman	3,365,000	309,500	119,499	11	28	Monarchy	A	Muscat (Masqaṭ)	Arabic, English, Baluchi, Urdu, Indian dialects
Ontario	12,950,000	1,076,395	415,599	12	31	Province (Canada)	D	Toronto	English
Oregon	3,810,000	254,805	98,381	15	39	State (U.S.)	D	Salem	English
† Pakistan (incl. part of Jammu and Kashmir)	174,525,000	879,902	339,732	198	514	Federal Islamic republic	A	Islāmābād	English, Urdu, Punjabi, Sindhi, Pashto
† Palau (Belau)	21,000	487	188	43	112	Republic	A	Melekeok	Angaur, English, Japanese, Palauan, Sonsorolese, Tobi
† Panama	3,335,000	75,517	29,157	44	114	Republic	A	Panamá	Spanish, English
† Papua New Guinea	5,995,000	462,840	178,704	13	34	Parliamentary state	A	Port Moresby	English, Motu, Pidgin, indigenous
† Paraguay	6,915,000	406,752	157,048	17	44	Republic	A	Asunción	Spanish, Guarani
Pennsylvania	12,515,000	119,282	46,055	105	272	State (U.S.)	D	Harrisburg	English
† Peru	29,365,000	1,285,216	496,225	23	59	Republic	A	Lima	Quechua, Spanish, Aymara
† Philippines	97,020,000	300,000	115,831	323	838	Republic	A	Manila	English, Pilipino, Tagalog
Pitcairn Islands (incl. Dependencies)	100	49	19	2.0	5.3	Overseas territory (U.K.)	C	Adamstown	English, Tahitian
† Poland	38,490,000	312,685	120,728	123	319	Republic	A	Warsaw (Warszawa)	Polish
† Portugal	10,695,000	91,985	35,516	116	301	Republic	A	Lisbon (Lisboa)	Portuguese
Prince Edward Island	140,000	5,660	2,185	25	64	Province (Canada)	D	Charlottetown	English
Puerto Rico	3,965,000	9,104	3,515	436	1,128	Commonwealth (U.S. protection)	B	San Juan	Spanish, English
† Qatar	830,000	11,427	4,412	73	188	Monarchy	A	Doha (Ad-Dawḥah)	Arabic, English
Qinghai	5,550,000	720,000	277,994	7.7	20	Province (China)	D	Xining	Tibetan dialects, Mongolian, Turkish dialects, Chinese (Mandarin)
Quebec	7,775,000	1,542,056	595,391	5.0	13	Province (Canada)	D	Québec	French, English
Queensland	4,180,000	1,730,650	668,208	2.4	6.3	State (Australia)	D	Brisbane	English
Reunion	815,000	2,510	969	325	841	Overseas department (France)	C	Saint-Denis	French, Creole
Rhode Island	1,055,000	4,002	1,545	264	683	State (U.S.)	D	Providence	English
Rhodesia *see Zimbabwe*	...	...	...	...	...	...		...	...
† Romania	22,230,000	237,500	91,699	94	242	Republic	A	Bucharest (București)	Romanian, Hungarian, German
† Russia	140,370,000	17,075,400	6,592,849	8.2	21	Federal republic	A	Moscow (Moskva)	Russian, Tatar, Ukrainian
† Rwanda	10,330,000	26,338	10,169	392	1,016	Republic	A	Kigali	French, Kinyarwanda, Kiswahili
St. Helena (incl. Dependencies)	7,500	314	121	24	62	Overseas territory (U.K.)	C	Jamestown	English
† St. Kitts and Nevis	40,000	261	101	153	396	Parliamentary state	A	Basseterre	English
† St. Lucia	160,000	616	238	260	672	Parliamentary state	A	Castries	English, French
St. Pierre and Miquelon	7,000	242	93	29	75	Territorial collectivity (France)	C	Saint-Pierre	French
† St. Vincent and the Grenadines	105,000	388	150	271	700	Parliamentary state	A	Kingstown	English, French
† Samoa	220,000	2,831	1,093	78	201	Constitutional monarchy	A	Apia	English, Samoan
† San Marino	30,000	61	24	492	1,250	Republic	A	San Marino	Italian
† Sao Tome and Principe	210,000	964	372	218	565	Republic	A	São Tomé	Portuguese, Fang
Saskatchewan	1,015,000	651,036	251,366	1.6	4.0	Province (Canada)	D	Regina	English
† Saudi Arabia	28,420,000	2,149,690	830,000	13	34	Monarchy	A	Riyadh (Ar-Riyāḍ)	Arabic
Scotland	5,150,000	78,133	30,167	66	171	Administrative division (U.K.)	D	Edinburgh	English, Scots Gaelic
† Senegal	13,525,000	196,712	75,951	69	178	Republic	A	Dakar	French, Wolof, Fulani, Serer, indigenous
† Serbia (excl. Kosovo)	7,395,000	77,474	29,913	95	247	Republic	A	Belgrade (Beograd)	Serbian
† Seychelles	87,000	455	176	191	494	Republic	A	Victoria	English, French, Creole
Shaanxi	37,815,000	205,000	79,151	184	478	Province (China)	D	Xi'an	Chinese (Mandarin)
Shandong	94,255,000	153,000	59,074	616	1,596	Province (China)	D	Jinan	Chinese (Mandarin)
Shanghai	18,375,000	6,200	2,394	2,964	7,675	Autonomous city (China)	D	Shanghai	Chinese (Wu)
Shanxi	34,170,000	156,000	60,232	219	567	Province (China)	D	Taiyuan	Chinese (Mandarin)
Sichuan	82,710,000	487,600	188,263	170	439	Province (China)	D	Chengdu	Chinese (Mandarin), Tibetan dialects, Miao-Yao
† Sierra Leone	6,365,000	71,740	27,699	89	230	Transitional military government	A	Freetown	English, Krio, Mende, Temne, indigenous
† Singapore	4,635,000	683	264	6,786	17,557	Republic	A	Singapore	Chinese (Mandarin), English, Malay, Tamil
† Slovakia	5,460,000	49,012	18,924	111	289	Republic	A	Bratislava	Slovak, Hungarian
† Slovenia	2,005,000	20,256	7,821	99	256	Republic	A	Ljubljana	Slovenian
† Solomon Islands	590,000	28,370	10,954	21	54	Parliamentary state	A	Honiara	English, indigenous
† Somalia	9,695,000	637,657	246,201	15	39	None	A	Mogadishu (Muqdisho)	Arabic, Somali, English, Italian
† South Africa	48,985,000	1,219,090	470,693	40	104	Republic	A	Pretoria (Tshwane), Cape Town and Bloemfontein	Afrikaans, English, Xhosa, Zulu, other indigenous
South America	391,890,000	17,800,000	6,900,000	22	57	...		...	...
South Australia	1,600,000	983,480	379,724	1.6	4.2	State (Australia)	D	Adelaide	English
South Carolina	4,505,000	82,932	32,020	54	141	State (U.S.)	D	Columbia	English
South Dakota	810,000	199,731	77,117	4.1	11	State (U.S.)	D	Pierre	English
South Georgia and the South Sandwich Islands *(3)*	*(1)*	3,755	1,450	...	...	Overseas territory (U.K.)	C	Grytviken Harbour	English

Region or political division	Est. Pop. 1/1/09	Area sq. km.	Area sq. mi.	Pop. per sq. km.	Pop. per sq. mi.	Form of Government and Ruling Power	Political Status	Capital	Predominant Languages
South West Africa *see Namibia*		...	...	...	...	...		...	...
† Spain	40,510,000	504,750	194,885	80	208	Constitutional monarchy	A	Madrid	Spanish (Castilian), Catalan, Galician, Basque
Spanish North Africa *(6)*	150,000	32	12	4,688	12,500	Five possessions (Spain)	C	...	Spanish, Arabic, Berber dialects
Spanish Sahara *see Western Sahara*		...	...	...	...	...		...	...
† Sri Lanka	21,230,000	65,610	25,332	324	838	Socialist republic	A	Colombo and Sri Jayewardenepura Kotte	English, Sinhala, Tamil
† Sudan	40,650,000	2,505,813	967,500	16	42	Provisional military government	A	Khartoum (Al-Kharṭūm)	Arabic, Nubian and other indigenous, English
† Suriname	480,000	163,265	63,037	2.9	7.6	Republic	A	Paramaribo	Dutch, Sranan Tongo, English, Hindustani, Javanese
† Swaziland	1,125,000	17,364	6,704	65	168	Monarchy	A	Mbabane and Lobamba	English, siSwati
† Sweden	9,050,000	449,964	173,732	20	52	Constitutional monarchy	A	Stockholm	Swedish, Lapp, Finnish
† Switzerland	7,595,000	41,293	15,943	184	476	Federal republic	A	Bern	German, French, Italian, Romansch
† Syria	19,965,000	185,180	71,498	108	279	Socialist republic	A	Damascus (Dimashq)	Arabic, Kurdish, Armenian, Aramaic, Circassian
Taiwan	22,950,000	36,002	13,901	637	1,651	Republic	A	T'aipei	Chinese (Mandarin), Taiwanese (Min), Hakka
† Tajikistan	7,280,000	143,100	55,251	51	132	Republic	A	Dushanbe	Tajik, Uzbek, Russian
† Tanzania	40,630,000	945,087	364,900	43	111	Republic	A	Dar es Salaam and Dodoma	English, Swahili, indigenous
Tasmania	500,000	68,400	26,409	7.3	19	State (Australia)	D	Hobart	English
Tennessee	6,250,000	109,151	42,143	57	148	State (U.S.)	D	Nashville	English
Texas	24,460,000	695,621	268,581	35	91	State (U.S.)	D	Austin	English, Spanish
† Thailand	65,705,000	513,115	198,115	128	332	Constitutional monarchy	A	Bangkok (Krung Thep)	Thai, indigenous
Tianjin (Tientsin)	10,885,000	11,300	4,363	963	2,495	Autonomous city (China)	D	Tianjin	Chinese (Mandarin)
Timor-Leste *see East Timor*		...	...	...	...	...		...	...
† Togo	5,940,000	56,785	21,925	105	271	Provisional military government	A	Lomé	French, Ewe, Mina, Kabye, Dagomba
Tokelau	1,500	12	4.6	125	326	Island territory (New Zealand)	C	...	English, Tokelauan
† Tonga	120,000	650	251	185	478	Constitutional monarchy	A	Nuku'alofa	Tongan, English
† Trinidad and Tobago	1,230,000	5,128	1,980	240	621	Republic	A	Port of Spain	English, Hindi, French, Spanish
Tristan da Cunha	300	104	40	2.9	7.5	Dependency (St. Helena)	C	Edinburgh	English
† Tunisia	10,435,000	163,610	63,170	64	165	Republic	A	Tunis	Arabic, French
† Turkey	76,300,000	783,577	302,541	97	252	Republic	A	Ankara	Turkish, Kurdish, Arabic
† Turkmenistan	4,855,000	488,100	188,457	9.9	26	Republic	A	Aşgabat	Turkmen, Russian, Uzbek
Turks and Caicos Islands	23,000	430	166	53	139	Overseas territory (U.K.)	C	Grand Turk	English
† Tuvalu	12,000	26	10	462	1,200	Parliamentary state	A	Funafuti	Tuvaluan, English
† Uganda	31,935,000	241,038	93,065	132	343	Republic	A	Kampala	English, Luganda, Swahili, indigenous
† Ukraine	45,845,000	603,700	233,090	76	197	Republic	A	Kiev (Kyïv)	Ukrainian, Russian, Romanian, Polish
† United Arab Emirates	4,710,000	83,600	32,278	56	146	Federation of monarchs	A	Abu Dhabi (Abū Ẓaby)	Arabic, Farsi, English, Hindi, Urdu
† United Kingdom	61,030,000	242,910	93,788	251	651	Parliamentary monarchy	A	London	English, Welsh, Scots Gaelic
† United States	305,710,000	9,826,630	3,794,083	31	81	Federal republic	A	Washington	English, Spanish
Upper Volta *see Burkina Faso*		...	...	...	...	...		...	...
† Uruguay	3,485,000	175,016	67,574	20	52	Republic	A	Montevideo	Spanish
Utah	2,750,000	219,887	84,899	13	32	State (U.S.)	D	Salt Lake City	English
† Uzbekistan	27,475,000	447,400	172,742	61	159	Republic	A	Tashkent (Toshkent)	Uzbek, Russian
† Vanuatu	215,000	12,190	4,707	18	46	Republic	A	Port Vila	Bislama, English, French
Vatican City	800	0.4	0.2	2,000	4,000	Monarchical-sacerdotal state	A	Vatican City	Italian, Latin, other
† Venezuela	26,615,000	912,050	352,145	29	76	Federal republic	A	Caracas	Spanish, Amerindian
Vermont	625,000	24,901	9,614	25	65	State (U.S.)	D	Montpelier	English
Victoria	5,235,000	227,420	87,807	23	60	State (Australia)	D	Melbourne	English
† Vietnam	86,545,000	331,689	128,066	261	676	Socialist republic	A	Ha Noi	Vietnamese, French, Chinese, English, Khmer, indigenous
Virginia	7,810,000	110,785	42,774	70	183	State (U.S.)	D	Richmond	English
Virgin Islands (U.S.)	110,000	347	134	317	821	Unincorporated territory (U.S.)	C	Charlotte Amalie	English, Spanish, Creole
Wake Island	*(1)*	7.8	3.0	...	...	Unincorporated territory (U.S.)	C	...	English
Wales	2,985,000	20,779	8,023	144	372	Administrative division (U.K.)	D	Cardiff	English, Welsh Gaelic
Wallis and Futuna	15,000	255	99	59	152	Overseas territory (France)	C	Matâ'utu	French, Wallisian
Washington	6,585,000	184,665	71,300	36	92	State (U.S.)	D	Olympia	English
West Bank (incl. East Jerusalem)	2,435,000	5,860	2,263	416	1,076	Israeli territory with limited self-government		...	Arabic, Hebrew
Western Australia	2,105,000	2,529,880	976,792	0.8	2.2	State (Australia)	D	Perth	English
Western Sahara	400,000	266,000	102,703	1.5	3.9	Occupied by Morocco		...	Arabic
West Virginia	1,825,000	62,755	24,230	29	75	State (U.S.)	D	Charleston	English
Wisconsin	5,660,000	169,639	65,498	33	86	State (U.S.)	D	Madison	English
Wyoming	535,000	253,336	97,814	2.1	5.5	State (U.S.)	D	Cheyenne	English
Xianggang *see Hong Kong*		...	...	...	...	...		...	...
Xinjiang Uygur (Sinkiang)	20,755,000	1,600,000	617,764	13	34	Autonomous region (China)	D	Ürümqi	Turkish dialects, Mongolian, Tungus, English
Xizang (Tibet)	2,845,000	1,220,000	471,045	2.3	6.0	Autonomous region (China)	D	Lhasa	Tibetan dialects
† Yemen	23,410,000	527,968	203,850	44	115	Republic	A	Sanaa (Şan'ā')	Arabic
Yugoslavia *see Serbia*		...	...	...	...	...		...	...
Yukon	32,000	482,443	186,272	0.07	0.2	Territory (Canada)	D	Whitehorse	English, Inuktitut, indigenous
Yunnan	45,390,000	394,000	152,124	115	298	Province (China)	D	Kunming	Chinese (Mandarin), Tibetan dialects, Khmer, Miao-Yao
Zaire *see Congo, Democratic Republic of the*		...	...	...	...	...		...	...
† Zambia	11,765,000	752,614	290,586	16	40	Republic	A	Lusaka	English, Tonga, Lozi, other indigenous
Zhejiang	50,425,000	101,800	39,305	495	1,283	Province (China)	D	Hangzhou	Chinese dialects
† Zimbabwe	11,305,000	390,759	150,873	29	75	Republic	A	Harare (Salisbury)	English, Shona, Sindebele
WORLD	6,750,375,000	150,100,000	57,900,000	45	117	...		...	...

(1) No Permanent Population.

(2) Population estimate includes 26,760,000 people not included in any province.

(3) Claimed by Argentina.

(4) Kosovo unilaterally declared its independence from Serbia in 2008.

(5) Claimed by Comoros.

(6) Comprises Ceuta, Melilla and several small islands.

† Member of the United Nations (2008)

... None, or not applicable.

General Information

Equatorial diameter of Earth . . . . 12,756 km (7,926 mi.)
Polar diameter of Earth . . . . 12,713 km (7,900 mi.)
Mean diameter of Earth . . . . 12,742 km (7,918 mi.)
Equatorial circumference of Earth . . . . 40,075 km (24,901 mi.)
Mean distance from Earth to Sun . . . . 149,598,000 km (92,955,900 mi.)
Mean distance from Earth to Moon . . . . . 384,403 km (238,857 mi.)
Total area of Earth . . . . 510,100,000 sq. km (197,000,000 sq. mi.)

Highest elevation on Earth's surface, Mt. Everest, Asia . . . . 8,848 m (29,028 ft.)
Lowest elevation on Earth's land surface, shores of the Dead Sea, Asia. . . . . 408 m (1,339 ft.) below sea level
Greatest known depth of the ocean, southwest of Guam, Pacific Ocean . . . . 10,924 m (35,840 ft.)
Total land area of Earth (incl. inland water and Antarctica) . . . . 150,100,000 sq. km (57,900,000 sq. mi.)

Area of Africa . . . . 30,300,000 sq. km (11,700,000 sq. mi.)
Area of Antarctica . . . . 14,000,000 sq. km (5,400,000 sq. mi.)
Area of Australia and Oceania . . 8,500,000 sq. km (3,300,000 sq. mi.)
Area of Asia . . . . 44,900,000 sq. km (17,300,000 sq. mi.)
Area of Europe . . . . 9,900,000 sq. km (3,800,000 sq. mi.)
Area of North America . . . . 24,700,000 sq. km (9,500,000 sq. mi.)
Area of South America . . . . 17,800,000 sq. km (6,900,000 sq. mi.)
World Population . . . . (est. 1/1/09), 6,750,375,000

Principal Islands Area in sq. km (sq. mi.)

Baffin I., Nu., Can. 507,451 (195,928)
Banks I., N.T., Can. 70,028 (27,038)
Borneo, Asia 748,168 (288,869)
Bougainville, Pap. N. Gui. 9,317 (3,597)
Cape Breton I., N.S., Can. 10,311 (3,981)
Celebes, Indon. 180,680 (69,761)
Ceram, Indon. 17,454 (6,739)
Corsica, Fr. 8,741 (3,375)
Crete, Grc. 8,349 (3,224)
Cuba, Cuba 105,805 (40,852)
Cyprus, Cyp. 9,234 (3,565)
Devon I., Nu., Can. 55,247 (21,331)
Ellesmere I., Nu., Can. 196,236 (75,767)
Flores, Indon. 14,154 (5,465)
Great Britain, U.K. 226,000 (87,259)
Greenland, Green. 2,166,086 (836,330)
Guadalcanal, Sol. Is. 5,352 (2,066)
Hainan Dao, China 33,209 (12,822)
Hawai' i, Hi., U.S. 10,500 (4,054)
Hispaniola, N.A. 73,929 (28,544)
Hokkaidō, Japan 78,719 (30,394)
Honshū, Japan 225,800 (87,182)
Iceland, Ice. 101,826 (39,315)
Ireland, Ire.-U.K. 81,638 (31,521)
Jamaica, Jam. 11,189 (4,320)
Java, Indon. 138,793 (53,588)
Kodiak I., Ak., U.S. 9,578 (3,698)
Kyūshū, Japan 37,437 (14,455)
Leyte, Phil. 7,367 (2,844)
Long I., N.Y., U.S. 3,502 (1,352)
Luzon, Phil. 109,964 (42,457)
Madagascar, Madag. 587,713 (226,917)
Melville I., Can. 42,149 (16,274)
Mindanao, Phil. 97,530 (37,657)
Mindoro, Phil. 10,571 (4,081)
Negros, Phil. 13,074 (5,048)
New Britain, Pap. N. Gui. 35,144 (13,569)
New Caledonia, N. Cal. . . 16,648 (6,428)
Newfoundland, Nf., Can. 108,860 (42,031)
New Guinea, Asia-Oc. . . 785,753 (303,381)
North East Land, Nor. 14,247 (5,501)
North I., N.Z. 111,582 (43,082)
Palawan, Phil. 12,188 (4,706)
Panay, Phil. 12,011 (4,637)
Prince of Wales I., Nu., Can. 33,339 (12,872)
Puerto Rico, P.R. 8,733 (3,372)
Sakhalin, Russia. 72,492 (27,989)
Samar, Phil. 12,849 (4,961)
Sardinia, Italy. 23,949 (9,247)
Shikoku, Japan 18,544 (7,160)
Sicily, Italy. 25,662 (9,908)
Southampton I., Nu., Can. 41,214 (15,913)
South I., N.Z. 145,836 (56,308)
Spitsbergen, Nor. 38,980 (15,050)
Sri Lanka, Sri L. 67,654 (26,121)
Sumatra, Indon. 443,065 (171,068)
Taiwan, Tai. 34,506 (13,323)
Tasmania, Austl. 65,519 (25,297)
Tierra del Fuego, S.A. . . . 47,401 (18,302)
Timor, Asia 28,418 (10,972)
Vancouver I., B.C., Can. 31,285 (12,079)
Victoria I., B.C., Can. 217,291 (83,896)
Vrangelya, Ostrov (Wrangel I.), Russia 7,865 (3,037)

Principal Lakes, Oceans and Seas Area in sq. km (sq. mi.)

Arabian Sea, Afr.-Asia 3,864,000 (1,492,000)
Aral Sea, Kaz.-Uzb. 17,158 (6,625)
Arctic Ocean . . . 14,056,000 (5,400,000)
Athabasca, L., Can. 7,935 (3,064)
Atlantic Ocean 76,762,000 (29,600,000)
Baikal, L., Russia 31,500 (12,162)
Balkhash, L., Kaz. 17,580 (6,788)
Baltic Sea, Eur. 422,000 (163,000)
Bering Sea, Asia-N.A. 2,291,900 (884,900)
Black Sea, Eur.-N.A. . . . 461,000 (178,000)
Caribbean Sea, N.A.-S.A. 2,753,000 (1,063,000)
Caspian Sea, Asia-Eur. 371,000 (143,244)
Chad, L., Afr. 1,540 (595)
Erie, L., Can.-U.S. 25,667 (9,910)
Eyre, L., Austl. 9,500 (3,668)
Great Bear Lake, Can. 31,328 (12,096)
Great Salt Lake, U.S. 5,483 (2,117)
Great Slave Lake, Can. . . . 28,568 (11,030)
Hudson Bay, Can. . . 1,230,000 (475,000)
Huron, L., Can.-U.S. 59,570 (23,000)
Indian Ocean 68,556,000 (26,500,000)
Japan, Sea of, Asia 1,007,800 (389,100)
Kok Nor (Qinghai Hu), China 4,460 (1,722)
Ladoga, L., Russia 16,400 (6,332)
Manitoba, L., Can. 4,624 (1,785)
Maracaibo, L., Ven. 13,010 (5,023)
Mediterranean Sea 2,505,000 (967,000)
Mexico, Gulf of, N.A. 1,500,000 (600,000)
Michigan, L., U.S. 57,757 (22,300)
Nicaragua, Lago de, Nicaragua 8,150 (3,147)
North Sea, Eur. 575,000 (222,000)
Nyasa, L., Afr. 29,500 (11,390)
Onega, L., Russia 9,890 (3,819)
Ontario, L., Can.-U.S. 19,011 (7,340)
Pacific Ocean . . 155,557,000 (60,100,000)
Red Sea, Afr.-Asia 438,000 (169,000)
Rudolf, L., Eth.-Kenya 6,750 (2,606)
Southern Ocean . . 20,327,000 (7,800,000)
Superior, L., Can.-U.S. . . . 82,103 (31,700)
Tanganyika. L., Afr. 32,600 (12,587)
Titicaca, Lago, Bol.-Peru . . . 8,372 (3,232)
Torrens, L., Austl. 5,745 (2,218)
Vänern (L.), Swe. 5,648 (2,181)
Van Gölü (L.), Tur. 3,740 (1,444)
Victoria, L., Afr. 68,870 (26,591)
Winnipeg, L., Can. 24,387 (9,416)
Winnipegosis, L., Can. 5,374 (2,075)
Yellow Sea, Asia . . . 1,243,000 (480,000)

Principal Mountains Elevation in m (ft.)

Aconcagua, Cerro, Arg. . . 6,959 (22,831)
Annapūrṇa, Nepal 8,091 (26,545)
Aoraki (Mt. Cook), N.Z. . . 3,754 (12,316)
Apo, Mt., Phil. 2,954 (9,692)
Ararat, Mt., Tur. 5,137 (16,854)
Barú, Volcán, Pan. 3,475 (11,401)
Belukha, Mt., Asia 4,506 (14,783)
Bia, Phou, Laos 2,820 (9,252)
Blanc, Mont, Eur. 4,807 (15,771)
Blanca Pk., Co., U.S. 4,372 (14,345)
Bolívar, Pico, Ven. 5,007 (16,427)
Bonete Grande, Cerro, Arg. 6,872 (22,546)
Borah Pk., Id., U.S. 3,859 (12,662)
Boundary Pk., Nv., U.S. 4,006 (13,143)
Cameroon Mtn., Camm. . . 4,100 (13,451)
Carrauntoohil, Ire. 1,038 (3,406)
Chaltel, Cerro, S.A. 3,340 (10,958)
Chimborazo, Ec. 6,310 (20,702)
Chirripó, Cerro, C.R. 3,819 (12,530)
Colima, Nevado de, Mex. 4,240 (13,911)
Cotopaxi, Ec. 5,897 (19,347)
Cristóbal Colón, Pico, Col. 5,775 (18,947)
Damāvand, Kūh-e, Iran . . 5,604 (18,386)
Dhawalāgiri, Nepal 8,167 (26,795)
Duarte, Pico, Dom. Rep. . . 3,175 (10,417)
Dufourspitze, Eur. 4,634 (15,203)
Elbert, Mt., Co., U.S. 4,399 (14,433)
El'brus, Gora, Russia 5,642 (18,510)
Elgon, Mt., Afr. 4,321 (14,177)
Erciyes Dağı, Tur. 3,917 (12,851)
Etna, Monte, Italy 3,323 (10,902)
Everest, Mt., Asia 8,848 (29,028)
Fairweather, Mt., N.A. . . . 4,663 (15,300)
Folādī, Koh-e, Afg. 5,135 (16,847)
Fuji-san, Japan 3,776 (12,388)
Galdhøpiggen, Nor. 2,469 (8,100)
Gannett Pk., Wy., U.S. 4,207 (13,804)
Gerlachovský štít, Slvk. . . . 2,655 (8,711)
Giluwe, Mt., Pap. N. Gui. . 4,368 (14,331)
Gongga Shan, China 7,590 (24,902)
Grand Teton, Wy., U.S. . . . 4,197 (13,770)
Großglockner, Aus. 3,797 (12,457)
Gunnbjørn Fjeld, Green. 3,700 (12,139)
Hekla, Ice. 1,491 (4,892)
Hkakabo Razi, Mya. 5,881 (19,295)
Hood, Mt., Or., U.S. 3,426 (11,239)
Huascarán, Nevado, Peru 6,746 (22,133)
Huila, Nevado del, Col. . . 5,750 (18,865)
Hvannadalshnúkur, Ice. . . 2,119 (6,952)
Illampu, Nevado, Bol. . . . 6,421 (21,066)
Illimani, Nevado de, Bol. 6,457 (21,184)
Imeni Ismail Samani, Pik (Communism Pk.), Taj. 7,495 (24,590)
Inthanon, Doi, Thai. 2,600 (8,530)
Jaya, Puncak, Indon. 5,030 (16,503)
Jungfrau, Switz. 4,158 (13,642)
K2 (Qogir Feng), Asia . . . 8,611 (28,250)
Kāmet, Asia 7,756 (25,446)
Kānchenjunga, Asia 8,598 (28,208)
Karisimbi, Volcan, Afr. . . . 4,507 (14,787)
Kebnekaise, Swe. 2,111 (6,926)
Kenya,Mt.,(Kirinyaga), Kenya 5,199 (17,058)
Kerinci, Gunung, Indon. . . 3,800 (12,467)
Kilimanjaro, Tan. 5,895 (19,340)
Kinabalu, Gunong, Malay. 4,101 (13,455)
Kinyeti, Sudan 3,187 (10,456)
Klyuchevskaya Sopka, Vulkan, Russia 4,750 (15,584)
Kosciuszko, Mt., Austl. 2,229 (7,313)
Koussi, Emi, Chad 3,415 (11,204)
Kula Kangri, Bhu. 7,554 (24,784)
La Selle, Morne, Haiti 2,674 (8,773)
Lassen Pk., Ca., U.S. 3,187 (10,457)
Llullaillaco, Volcán, S.A. 6,739 (22,110)
Logan, Mt., Yk., Can. 5,959 (19,551)
Longs Pk., Co., U.S. 4,345 (14,255)
Margherita Pk., Afr. 5,109 (16,763)
Maromokotro, Madag. 2,876 (9,436)
Massive, Mt., Co., U.S. . . . 4,396 (14,421)
Matterhorn, Eur. 4,478 (14,692)
Mauna Kea, Hi., U.S. 4,205 (13,796)
Mauna Loa, Hi., U.S. 4,169 (13,677)
Mayon Volcano, Phil. 2,462 (8,077)
McKinley, Mt. (Denali), Ak., U.S. 6,194 (20,320)
Meru, Mt., Tan. 4,565 (14,977)
Misti, Volcán, Peru 5,822 (19,101)
Mitchell, Mt., N.C., U.S. . . . 2,037 (6,684)
Môco, Morro de, Ang. . . . 2,620 (8,596)
Moldoveanu, Vârful, Rom. 2,544 (8,346)
Mulhacén, Spain 3,482 (11,424)
Musala, Blg. 2,925 (9,596)
Muztag, China 7,723 (25,338)
Namjagbarwa Feng, China 7,755 (25,443)
Nanda Devi, India 7,817 (25,645)
Nanga Parbat, Pak. 8,126 (26,660)
Nevis, Ben, Scot., U.K. 1,343 (4,406)
Ojos del Salado, Nevado, S.A. 6,893 (22,615)
Olympus, Mt., Grc. 2,917 (9,570)
Paektu-san, Asia 2,744 (9,003)
Paricutín, Mex. 2,800 (9,186)
Parnassós, Grc. 2,457 (8,061)
Pelée, Montagne, Mart. . . 1,397 (4,583)
Pico de Orizaba, Volcán, Mex. 5,610 (18,406)
Pidurutalagala, Sri L. 2,524 (8,281)
Pikes Pk., Co., U.S. 4,301 (14,110)
Pinatubo, Mt., Phil. 1,780 (5,840)
Pobeda, Gora Russia 3,147 (10,325)
Popocatépetl, Volcán, Mex. 5,465 (17,930)
Pulog, Mt., Phil. 2,934 (9,626)
Rainier, Mt., Wa., U.S. 4,392 (14,411)
Ramm, Jabal, Jord. 1,754 (5,755)
Ras Dejen, Eth. 4,620 (15,158)
Rinjani, Gunung, Indon. . . 3,726 (12,224)
Robson, Mt., B.C., Can. . . . 3,954 (12,972)
Roraima, Mt., S.A. 2,875 (9,432)
Ruapehu, Mt., N.Z. 2,797 (9,177)
Ruiz, Nevado del, Col. . . . 5,400 (17,717)
Saint Elias, Mt., N.A. 5,489 (18,009)
Saint Helens, Mt., Wa., U.S. 2,549 (8,364)
Sajama, Nevado, Bol. 6,542 (21,463)
Semeru, Gunung, Indon. 3,676 (12,060)
Shām, Jabal ash-, Oman . . 3,035 (9,957)
Shasta, Mt., Ca., U.S. 4,317 (14,162)
Snowdon, Wales, U.K. 1,085 (3,560)
Tahat, Alg. 2,908 (9,541)
Tajumulco, Volcán, Guat. 4,220 (13,845)
Tirich Mīr, Pak. 7,690 (25,230)
Toubkal, Jebel, Mor. 4,165 (13,665)
Triglav, Slvn. 2,864 (9,396)
Trikora, Puncak (Wilhelmina Pk.), Indon. 4,750 (15,584)
Tupungato, Cerro, S.A. . . 6,750 (22,146)
Turquino, Pico, Cuba 1,972 (6,470)
Uluru (Ayers Rock), Austl. . . 863 (2,831)
Uncompahgre Pk., Co., U.S. 4,361 (14,309)
Vesuvius, Italy 1,281 (4,203)
Vinson Massif, Ant. 4,897 (16,066)
Waddington, Mt., B.C., Can. 4,015 (13,173)
Washington, Mt., N.H., U.S. 1,917 (6,288)
Whitney, Mt., Ca., U.S. . . 4,418 (14,494)
Wilhelm, Mt., Pap. N. Gui. 4,509 (14,793)
Yü Shan, Tai. 3,997 (13,114)
Zugspitze, Eur. 2,962 (9,718)

Principal Rivers Length in km (mi.)

Albany, N.A. 982 (610)
Aldan, Asia 2,209 (1,373)
Amazonas-Ucayali, S.A. . . 6,437 (4,000)
Amu Darya, Asia 1,687 (1,048)
Amur, Asia 2,820 (1,752)
Araguaia, S.A. 1,969 (1,223)
Arkansas, N.A. 2,350 (1,460)
Atchafalaya-Red, N.A. 2,285 (1,420)
Athabasca, N.A. 1,231 (765)
Ayeyarwady, Asia 1,573 (977)
Brahmaputra, Asia 3,235 (2,010)
Brazos, N.A. 2,060 (1,280)
Canadian, N.A. 1,458 (906)
Churchill, N.A. 1,609 (1,000)
Colorado, N.A. (U.S.-Mex.) . . 2,334 (1,450)
Colorado, N.A. (TX) 1,387 (862)
Columbia, N.A. 2,000 (1,243)
Congo, Afr. 4,370 (2,715)
Danube, Eur. 2,860 (1,777)
Darling, Austl. 1,472 (915)
Dnieper, Eur. 2,285 (1,420)
Don, Eur. 1,970 (1,224)
Elbe, Eur. 1,091 (678)
Essequibo, S.A. 970 (603)
Euphrates, Asia 2,412 (1,499)
Fraser, N.A. 1,370 (851)
Ganges, Asia 3,000 (1,864)
Gila, N.A. 1,044 (649)
Godāvari, Asia 1,500 (932)
Huang (Yellow), Asia 4,670 (2,902)
Indigirka, Asia 1,726 (1,072)
Indus, Asia 3,180 (1,976)
Juruá, S.A. 2,758 (1,714)
Kama, Eur. 1,685 (1,047)
Kasai, Afr. 1,968 (1,223)
Kolyma, Asia 2,130 (1,324)
Lena, Asia 4,400 (2,734)
Limpopo, Afr. 1,212 (753)
Loire, Eur. 1,110 (690)
Mackenzie, N.A. 4,241 (2,635)
Madeira, S.A. 3,381 (2,101)
Magdalena, S.A. 1,530 (951)
Marañón, S.A. 1,546 (961)
Mekong, Asia 4,500 (2,796)
Mississippi, N.A. 3,766 (2,340)
Mississippi-Missouri, N.A. 6,420, (3,989)
Missouri, N.A. 4,088 (2,540)
Murray, Austl. 2,375 (1,476)
Negro, S.A. 1,341 (833)
Nelson, N.A. 2,575 (1,600)
Niger, Afr. 4,160 (2,585)
Nile, Afr. 6,650 (4,132)
Ob', Asia 3,650 (2,268)
Oder, Eur. 906 (563)
Ohio, N.A. 2,108 (1,310)
Oka, Eur. 1,304 (810)
Orange, Afr. 2,300 (1,429)
Orinoco, S.A. 2,740 (1,703)
Ottawa, N.A. 1,271 (790)
Paraguay, S.A. 2,297 (1,427)
Paranaíba, S.A. 1,450 (901)
Peace, N.A. 1,923 (1,195)
Pechora, Eur. 1,810 (1,125)
Pecos, N.A. 1,490 (926)
Plata-Paraná, S.A. 4,700 (2,920)
Platte, N.A. 1,593 (990)
Purús, S.A. 2,588 (1,608)
Red, N.A. 2,076 (1,290)
Rhine, Eur. 1,320 (820)
Rhône, Eur 810 (503)
Rio Grande, N.A. 3,058 (1,900)
St. Lawrence, N.A. 3,058 (1,900)
Salado, S.A. 1,156 (718)
São Francisco, S.A. 2,800 (1,740)
Saskatchewan-Bow, N.A. 1,979 (1,230)
Severnaya Dvina (N. Dvina), Eur. 711 (442)
Snake, N.A. 1,674 (1,040)
Songhua (Sungari), Asia 872 (542)
Syr Darya, Asia 1,590 (988)
Tagus, Eur. 1,100 (684)
Tarim, Asia 964 (599)
Tennessee, N.A. 1,426 (886)
Tigris, Asia 1,752 (1,089)
Tisa, Eur. 881 (547)
Tocantins, S.A. 2,124 (1,320)
Ucayali, S.A. 1,484 (922)
Ural, Asia 2,102 (1,306)
Uruguay, S.A. 1,616 (1,004)
Vilyuy, Asia 2,446 (1,520)
Volga, Eur. 3,660 (2,274)
Volta, Afr. 1,600 (994)
Xiang, Asia 934 (580)
Xingu, S.A. 1,883 (1,170)
Yangtze (Chang), Asia 6,301 (3,915)
Yellowstone, N.A. 1,114 (692)
Yenisey, Asia 3,490 (2,169)
Yukon, N.A. 3,187 (1,980)
Zambezi, Afr. 2,660 (1,653)

- **Abidjan,** Cote d'Ivoire 1,929,079
- **Abu Dhabi (Abū Z̧aby),** United Arab Emirates 552,000
- **Accra,** Ghana (1,390,000) 949,113
- **Ad-Dammām,** Saudi Arabia (1,250,000) 525,000
- **Addis Ababa (Ādīs Ābeba),** Ethiopia (2,200,000) 2,084,588
- **Ahmadābād,** India (4,519,278) 3,515,361
- **Aleppo (Ḩalab),** Syria (1,640,000) 1,591,400
- **Alexandria (El-Iskandarîya),** Egypt (3,350,000) 2,926,859
- **Algiers (Alger),** Algeria (2,547,983) 1,507,241
- **Almaty,** Kazakhstan (1,190,000) 1,156,200
- **'Ammān,** Jordan (1,500,000) 963,490
- **Amsterdam,** Netherlands (1,121,303) 727,053
- **Ankara,** Turkey (2,650,000) 2,559,471
- **Antananarivo,** Madagascar 1,103,304
- **Antwerp (Antwerpen),** Belgium (1,135,000) 453,030
- **Aşgabat,** Turkmenistan 557,600
- **Asunción,** Paraguay (700,000) 502,426
- **Athens (Athína),** Greece (3,150,000) 772,072
- **Atlanta,** United States (4,112,198) 416,474
- **Auckland,** New Zealand (1,129,800) 380,154
- **Baghdād,** Iraq 3,841,268
- **Baku (Bakı),** Azerbaijan (2,020,000) 1,080,500
- **Bamako,** Mali 658,275
- **Bandung,** Indonesia (2,300,000) 2,136,260
- **Banghāzī,** Libya (472,000) 446,250
- **Bangkok (Krung Thep),** Thailand (7,360,000) 6,355,144
- **Bangui,** Central African Republic 451,690
- **Barcelona,** Spain (4,000,000) 1,496,266
- **Barranquilla,** Colombia (1,260,000) 990,547
- **Beijing,** China (7,320,000) 6,690,000
- **Beirut (Bayrūt),** Lebanon (1,675,000) 509,000
- **Belfast,** United Kingdom (730,000) 296,700
- **Belgrade (Beograd),** Serbia (1,554,826) 1,136,786
- **Belo Horizonte,** Brazil (4,055,000) 1,366,301
- **Bengalūru (Bangalore),** India (5,686,844) 4,292,223
- **Berlin,** Germany (4,220,000) 3,425,759
- **Birmingham,** United Kingdom (2,705,000) 965,928
- **Bishkek,** Kyrgyzstan 631,300
- **Bogotá,** Colombia (5,290,000) 4,931,796
- **Bonn,** Germany (600,000) 304,841
- **Boston,** United States (5,819,100) 589,141
- **Brasília,** Brazil 1,947,133
- **Brazzaville,** Congo 1,050,000
- **Brisbane,** Australia (1,627,535) 888,449
- **Brussels (Bruxelles),** Belgium (2,390,000) 133,845
- **Bucharest (București),** Romania (2,300,000) 2,067,545
- **Budapest,** Hungary (2,450,000) 1,906,798
- **Buenos Aires,** Argentina (11,460,575) 2,776,138
- **Bulawayo,** Zimbabwe 621,742
- **Cairo (El-Qâhira),** Egypt (9,300,000) 6,068,695
- **Calgary,** Canada (1,079,310) 987,969
- **Cali,** Colombia (1,735,000) 1,641,498
- **Cape Town,** South Africa (1,900,000) 854,616
- **Caracas,** Venezuela (4,000,000) 1,822,465
- **Cardiff,** United Kingdom (645,000) 272,129
- **Casablanca,** Morocco (3,200,000) 2,761,975
- **Changchun,** China 2,470,000
- **Chelyabinsk,** Russia (1,310,000) 1,077,174
- **Chengdu,** China 2,760,000
- **Chennai (Madras),** India (6,424,624) 4,216,268
- **Chicago,** United States (9,157,540) 2,896,016
- **Chișinău,** Moldova 676,700
- **Chittagong,** Bangladesh (2,342,662) 1,566,070
- **Chongqing,** China 3,870,000
- **Cincinnati,** United States (1,979,202) 331,285
- **Cleveland,** United States (2,945,831) 478,403
- **Cologne (Köln),** Germany (1,830,000) 964,311
- **Colombo,** Sri Lanka (2,250,000) 642,163
- **Conakry,** Guinea 950,000
- **Copenhagen (København),** Denmark (2,030,000) 499,148
- **Córdoba,** Argentina (1,368,301) 1,267,521
- **Cotonou,** Benin (605,000) 536,827
- **Curitiba,** Brazil (2,595,000) 1,586,848
- **Dakar,** Senegal 1,490,450
- **Dalian,** China 2,400,000
- **Dallas,** United States (5,221,801) 1,188,580
- **Damascus (Dimashq),** Syria (2,230,000) 1,549,932
- **Dar es Salaam,** Tanzania 2,497,940
- **Delhi,** India (12,791,458) 9,817,439
- **Denver,** United States (2,581,506) 554,636
- **Detroit,** United States (5,456,428) 951,270
- **Dhaka,** Bangladesh (6,537,308) 3,637,892
- **Dnipropetrovs'k,** Ukraine (1,590,000) 1,147,000
- **Donets'k,** Ukraine (2,090,000) 1,088,000
- **Douala,** Cameroon 712,251
- **Dubai (Dubayy),** United Arab Emirates 1,171,000
- **Dublin (Baile Átha Cliath),** Ireland (1,175,000) 481,854
- **Durban,** South Africa (1,740,000) 715,669
- **Dushanbe,** Tajikistan (800,000) 562,000
- **Düsseldorf,** Germany (1,200,000) 529,062
- **Edinburgh,** United Kingdom (640,000) 401,910
- **Edmonton,** Canada (1,034,945) 730,372
- **El-Gîza,** Egypt 1,883,189
- **Eşfahān,** Iran (1,525,000) 1,266,072
- **Essen,** Germany (5,040,000) 608,732
- **Faisalābād,** Pakistan 2,008,861
- **Fortaleza,** Brazil (2,780,000) 788,956
- **Frankfurt am Main,** Germany (1,960,000) 643,469
- **Freetown,** Sierra Leone (525,000) 469,776
- **Fukuoka,** Japan (2,200,000) 1,302,454
- **Glasgow,** United Kingdom (1,870,000) 662,954
- **Goiânia,** Brazil 1,075,761
- **Guadalajara,** Mexico (4,095,853) 1,600,894
- **Guangzhou (Canton),** China 3,750,000
- **Guatemala,** Guatemala (1,500,000) 823,301
- **Guayaquil,** Ecuador 1,985,379
- **Hamburg,** Germany (2,460,000) 1,704,731
- **Hannover,** Germany (1,015,000) 520,670
- **Ha Noi,** Vietnam (1,275,000) 905,939
- **Harare (Salisbury),** Zimbabwe (1,470,000) 1,189,103
- **Harbin,** China 3,120,000
- **Havana (La Habana),** Cuba (2,285,000) 2,189,716
- **Helsinki,** Finland (1,075,000) 512,686
- **Hiroshima,** Japan (1,700,000) 1,113,786
- **Ho Chi Minh City (Saigon),** Vietnam (3,300,000) 2,796,229
- **Hong Kong (Xianggang),** China (4,770,000) 1,250,993
- **Honolulu,** United States (876,156) 371,657
- **Houston,** United States (4,669,571) 1,953,631
- **Hyderābād,** India (5,533,640) 3,449,878
- **Ibadan,** Nigeria 1,144,000
- **Islāmābād,** Pakistan 529,180
- **İstanbul,** Turkey (7,550,000) 6,620,241
- **İzmir,** Turkey (1,900,000) 1,757,414
- **Jaipur,** India 2,324,319
- **Jakarta,** Indonesia (11,500,000) 8,347,083
- **Jerusalem (Yerushalayim),** Israel (740,000) 680,500
- **Jiddah,** Saudi Arabia 2,200,000
- **Jinan,** China 2,150,000
- **Johannesburg,** South Africa (4,000,000) 712,507
- **Kabul (Kābol),** Afghanistan 1,424,400
- **Kampala,** Uganda 1,208,544
- **Kānpur,** India (2,690,486) 2,540,069
- **Kaohsiung,** Taiwan (2,400,000) 1,509,510
- **Karāchi,** Pakistan 9,339,023
- **Kathmandu (Kāṭhmāṇḍāu),** Nepal (1,150,000) 671,846
- **Katowice,** Poland (2,755,000) 327,032
- **Kharkiv,** Ukraine (1,950,000) 1,555,000
- **Khartoum (Al-Kharṭūm),** Sudan (2,950,000) 947,483
- **Kiev (Kyïv),** Ukraine (3,250,000) 2,630,000
- **Kigali,** Rwanda 603,049
- **Kingston,** Jamaica (830,000) 516,500
- **Kinshasa (Léopoldville),** Congo, Dem. Rep. of the 3,000,000
- **Kolkata (Calcutta),** India (13,216,546) 4,580,544
- **Kuala Lumpur,** Malaysia (2,500,000) 1,297,526
- **Kuwait (Al-Kuwayt),** Kuwait (1,126,000) 28,747
- **Lagos,** Nigeria (3,800,000) 1,213,000
- **Lahore,** Pakistan 5,143,495
- **La Paz,** Bolivia (1,487,854) 789,585
- **Leeds,** United Kingdom (1,530,000) 424,194
- **León,** Mexico (1,425,210) 1,137,465
- **Lilongwe,** Malawi 435,964
- **Lima,** Peru (6,321,173) 340,422
- **Lisbon (Lisboa),** Portugal (2,350,000) 663,394
- **Liverpool,** United Kingdom (1,515,000) 481,786
- **Lomé,** Togo 450,000
- **London,** United Kingdom (12,000,000) 7,650,944
- **Los Angeles,** United States (16,373,645) 3,694,820
- **Luanda,** Angola 1,459,900
- **Lucknow,** India (2,266,933) 2,207,340
- **Lusaka,** Zambia 1,084,703
- **Lyon,** France (1,648,216) 445,452
- **Madrid,** Spain (4,690,000) 2,882,860
- **Managua,** Nicaragua 864,201
- **Manaus,** Brazil 1,394,724
- **Manchester,** United Kingdom (2,760,000) 402,889
- **Manila,** Philippines (11,200,000) 1,654,761
- **Mannheim,** Germany (1,525,000) 310,475
- **Maputo (Lourenço Marques),** Mozambique 966,837
- **Maracaibo,** Venezuela 1,249,670
- **Marrakech,** Morocco (760,000) 672,506
- **Marseille,** France (1,516,340) 798,430
- **Mashhad,** Iran 1,887,405
- **Mecca (Makkah),** Saudi Arabia 1,025,000
- **Medan,** Indonesia 1,904,273
- **Medellín,** Colombia (2,290,000) 1,551,160
- **Melbourne,** Australia (3,366,542) 67,784
- **Mexico City (Ciudad de México),** Mexico (19,231,829) 8,720,916
- **Miami,** United States (3,876,380) 362,470
- **Milan (Milano),** Italy (3,790,000) 1,305,591
- **Milwaukee,** United States (1,689,572) 596,974
- **Minneapolis,** United States (2,968,806) 382,618
- **Minsk,** Belarus (1,722,000) 1,661,000
- **Mogadishu (Muqdisho),** Somalia 600,000
- **Mombasa,** Kenya 665,018
- **Monrovia,** Liberia 465,000
- **Monterrey,** Mexico (3,664,331) 1,133,070
- **Montevideo,** Uruguay (1,610,000) 1,269,552
- **Montréal,** Canada (3,635,571) 1,620,693
- **Moscow (Moskva),** Russia (13,500,000) 10,126,424
- **Mumbai (Bombay),** India (16,368,084) 11,914,398
- **Munich (München),** Germany (1,930,000) 1,205,923
- **Nagoya,** Japan (5,280,000) 2,109,681
- **Nāgpur,** India (2,122,965) 2,051,320
- **Nairobi,** Kenya 2,143,254
- **Nanjing,** China 2,490,000
- **Naples (Napoli),** Italy (3,150,000) 1,046,987
- **N'Djamena (Fort-Lamy),** Chad 546,572
- **Newcastle upon Tyne,** United Kingdom (1,350,000) 189,150
- **New Delhi,** India 294,783
- **New York,** United States (21,199,865) 8,008,278
- **Nizhniy Novgorod (Gorky),** Russia (1,900,000) 1,311,252
- **Nouakchott,** Mauritania 558,195
- **Novosibirsk,** Russia (1,530,000) 1,425,508
- **Nürnberg,** Germany (1,065,000) 489,758
- **Odesa,** Ukraine (1,150,000) 1,046,000
- **Omdurman (Umm Durmān),** Sudan 1,271,403
- **Omsk,** Russia (1,175,000) 1,134,016
- **Oran (Ouahran),** Algeria 628,558
- **Ōsaka,** Japan (16,500,000) 2,484,326
- **Oslo,** Norway (773,498) 504,040
- **Ottawa,** Canada (1,130,761) 648,480
- **Ouagadougou,** Burkina Faso 709,700
- **Palembang,** Indonesia 1,430,627
- **Panamá,** Panama (995,000) 415,964
- **Paris,** France (11,174,743) 2,125,246
- **Perm',** Russia (1,100,000) 1,001,653
- **Perth,** Australia (1,333,993) 13,463
- **Philadelphia,** United States (6,188,463) 1,517,550
- **Phnom Penh (Phnum Pénh),** Cambodia 570,155
- **Phoenix,** United States (3,251,876) 1,321,045
- **Port-au-Prince,** Haiti (1,425,594) 846,247
- **Portland,** United States (2,265,223) 529,121
- **Port Louis,** Mauritius (500,000) 144,303
- **Port Moresby,** Papua New Guinea 246,664
- **Porto,** Portugal (1,230,000) 302,472
- **Porto Alegre,** Brazil (3,375,000) 1,304,998
- **Prague (Praha),** Czech Republic (1,328,000) 1,214,174
- **Pretoria (Tshwane),** South Africa (1,100,000) 525,583
- **Puebla de Zaragoza,** Mexico (2,109,049) 1,399,519
- **Pune,** India (3,755,525) 2,540,069
- **Pusan,** Korea, South 3,797,566
- **P'yŏngyang,** Korea, North 2,741,260
- **Qingdao,** China 2,300,000
- **Québec,** Canada (715,515) 490,614
- **Quezon City,** Philippines 1,989,419
- **Quito,** Ecuador (1,650,000) 1,399,378
- **Rabat,** Morocco (1,210,000) 623,457
- **Recife,** Brazil (3,160,000) 1,421,993
- **Rīga,** Latvia (1,000,000) 874,200
- **Rio de Janeiro,** Brazil (10,465,000) 5,851,914
- **Riyadh (Ar-Riyāḍ),** Saudi Arabia 2,950,000
- **Rome (Roma),** Italy (3,235,000) 2,649,765
- **Rosario,** Argentina (1,161,188) 908,163
- **Rostov-na-Donu,** Russia (1,220,000) 1,068,267
- **Rotterdam,** Netherlands (1,089,979) 539,000
- **Sacramento,** United States (1,796,857) 407,018
- **St. Louis,** United States (2,603,607) 348,189
- **St. Petersburg (Leningrad),** Russia (5,950,000) 4,661,219
- **Salvador,** Brazil (2,855,000) 2,439,823
- **Samara,** Russia (1,440,000) 1,157,880
- **San Diego,** United States (2,813,833) 1,223,400
- **San Francisco,** United States (7,039,362) 776,733
- **San José,** Costa Rica (996,194) 309,672
- **San Juan,** Puerto Rico (2,450,292) 421,958
- **San Salvador,** El Salvador (1,250,000) 415,346
- **Santiago,** Chile (4,740,000) 4,295,593
- **Santo Domingo,** Dominican Republic (2,005,000) 913,540
- **São Paulo,** Brazil (17,380,000) 9,713,692
- **Sapporo,** Japan (2,200,000) 1,822,992
- **Saratov,** Russia (1,130,000) 873,055
- **Seattle,** United States (3,554,760) 563,374
- **Seoul (Sŏul),** Korea, South (15,850,000) 10,627,790
- **Shanghai,** China (11,010,000) 8,930,000
- **Shenyang,** China 4,050,000
- **Singapore,** Singapore (4,800,000) 4,185,200
- **Sofia (Sofiya),** Bulgaria (1,280,000) 1,190,126
- **Stockholm,** Sweden (1,491,726) 674,452
- **Stuttgart,** Germany (2,020,000) 585,274
- **Surabaya,** Indonesia 2,599,796
- **Sūrat,** India (2,811,466) 2,433,787
- **Sydney,** Australia (3,997,321) 47,204
- **Tabrīz,** Iran 1,191,043
- **T'aipei,** Taiwan (6,800,000) 2,641,856
- **Tallinn,** Estonia 403,981
- **Tashkent (Toshkent),** Uzbekistan (2,325,000) 2,113,300
- **Tbilisi,** Georgia (1,350,000) 1,081,678
- **Tegucigalpa,** Honduras 769,061
- **Tehrān,** Iran (8,800,000) 6,758,845
- **Tel Aviv-Yafo,** Israel (2,000,000) 360,500
- **Tianjin,** China 5,000,000
- **Tijuana,** Mexico (1,483,992) 1,286,187
- **Tōkyō,** Japan (32,000,000) 8,025,508
- **Toronto,** Canada (5,113,149) 2,503,281
- **Tripoli (Ţarābulus),** Libya (960,000) 591,062
- **Tunis,** Tunisia (1,350,000) 702,330
- **Turin (Torino),** Italy (1,550,000) 921,485
- **Ufa,** Russia (1,110,000) 1,042,437
- **Ulaanbaatar,** Mongolia 649,797
- **Ürümqi,** China 1,130,000
- **València,** Spain (1,340,000) 739,014
- **Vancouver,** Canada (2,116,581) 578,041
- **Vienna (Wien),** Austria (1,950,000) 1,609,631
- **Vientiane (Viangchan),** Laos 464,000
- **Vilnius,** Lithuania 578,639
- **Volgograd,** Russia (1,375,000) 1,011,417
- **Warsaw (Warszawa),** Poland (2,400,000) 1,707,147
- **Washington,** United States (7,608,070) 572,059
- **Winnipeg,** Canada (694,668) 631,774
- **Wuhan,** China 3,870,000
- **Xi'an,** China 2,410,000
- **Yangon (Rangoon),** Myanmar (2,800,000) 2,705,039
- **Yekaterinburg,** Russia (1,550,000) 1,293,537
- **Yerevan,** Armenia (1,320,000) 1,103,488
- **Yokohama,** Japan 3,433,612
- **Zagreb,** Croatia 867,865
- **Zürich,** Switzerland (870,000) 365,043

Values are latest available city populations or recent estimates.
Metropolitan area populations are shown in parentheses.

Annam	Annamese
Arab	Arabic
Bantu	Bantu
Bur	Burmese
Camb	Cambodian
Celt	Celtic
Chn	Chinese
Czech	Czech
Dan	Danish
Du	Dutch
Fin	Finnish
Fr	French
Ger	German
Gr	Greek
Hung	Hungarian
Ice	Icelandic
India	India
Indian	American Indian
In don	Indonesian
It	Italian
Jap	Japanese
Kor	Korean
Mal	Malayan
Mong	Mongolian
Nor	Norwegian
Per	Persian
Pol	Polish
Port	Portuguese
Rom	Romanian
Rus	Russian
Serb	Serbian
Siam	Siamese
So. Slav	Southern Slavonic
Sp	Spanish
Swe	Swedish
Tib	Tibetan
Tur	Turkish

å, Nor., Swe. ... brook, river
aa, Dan., Nor ... brook
āb, Per ... water, river
abad, India, Per ... town, city
ada, Tur ... island
adrar, Berber ... mountain
ákra, Gr ... cape
älf, Swe ... river
alp, Ger ... mountain
altipiano, It ... plateau
alto, Sp ... height
archipel, Fr ... archipelago
archipiélago, Sp ... archipelago
arquipélago, Port ... archipelago
arroyo, Sp ... brook, stream
as, Nor., Swe ... ridge
austral, Sp ... southern
baai, Du ... bay
bab, Arab ... gate, port
bach, Ger ... brook, stream
backe, Swe ... Hill
bad, Ger ... bath, spa
bahía, Sp ... bay, gulf
bahr, Arab ... river, sea, lake
baia, It ... bay, gulf
baía, Port ... bay
baie, Fr ... bay, gulf
bajo, Sp ... depression
bak, Indon ... stream
bakke, Dan., Nor ... hill
balkan, Tur ... mountain range
bana, Jap ... point, cape
banco, Sp ... bank
bandao, Chn ... peninsula
bandar, Mal., Per ... town, port, harbor
bang, Siam ... village
bassin, Fr ... basin
batang, Indon., Mal ... river
bei, Chn ... north
ben, Celtic ... mountain, summit
bender, Arab ... harbor, port
bereg, Rus ... coast, shore
berg, Du., Ger., Nor., Swe ... mountain, hill
bir, Arab ... well
birkat, Arab ... lake, pond, pool
bit, Arab ... house
bjaerg, Dan., Nor ... mountain
bocche, It ... mouth
boğazı, Tur ... strait
bois, Fr ... forest, wood
bolsón, Sp ... flat-floored desert valley
boreal, Sp ... northern
borg, Dan., Nor., Swe ... castle, town
borgo, It ... town, suburb
bosch, Du ... forest, wood
bouche, Fr ... river mouth
bourg, Fr ... town, borough
bro, Dan., Nor., Swe ... bridge
brücke, Ger ... bridge
bucht, Ger ... bay, bight
bugt, Dan., Nor., Swe ... bay, gulf
bulu, Indon ... mountain
burg, Du., Ger ... castle, town
buri, Siam ... town
burun, burnu, Tur ... cape
by, Dan., Nor., Swe ... village
caatinga, Port. (Brazil) ... open brushland
cabezo, Sp ... summit
cabo, Port., Sp. ... cape
campo, It., Port., Sp ... plain, field
campos, Port. (Brazil) ... plains
cañón, Sp ... canyon
cap, Fr ... cape
capo, It ... cape
casa, It., Port., Sp ... house
castello, It., Port ... castle, fort
castillo, Sp ... castle
càte, Fr ... hill
çay, Tur ... stream, river
cayo, Sp ... rock, shoal, islet
cerro, Sp ... mountain, hill
champ, Fr ... field
château, Fr ... castle
chott, Arab ... salt lake
chu, Tib ... water, stream
cidade, Port ... town, city
cima, Sp ... summit, peak
città, It ... town, city
ciudad, Sp ... town, city
cochilha, Port ... ridge
col, Fr ... pass
colina, Sp ... hill
cordillera, Sp ... mountain chain
costa, It., Port., Sp ... coast
côte, Fr ... coast
cuchilla, Sp ... mountain ridge
dağ, Tur ... mountain(s)
dake, Jap ... peak, summit
dal, Dan., Du., Nor., Swe ... valley
dan, Kor ... point, cape
danau, Indon ... lake
dao, Chn ... island
dar, Arab ... house, abode, country
darya, Per ... river, sea
dasht, Per ... plain, desert
deniz, Tur ... sea
désert, Fr ... desert
deserto, It ... desert
desierto, Sp ... desert
détroit, Fr ... strait
dijk, Du ... dam, dike
djebel, Arab ... mountain
do, Kor ... island
dong, Chn ... east
dorf, Ger ... village
dorp, Du ... village
duin, Du ... dune
dzong, Tib . fort, administrative capital
eau, Fr ... water
ecuador, Sp ... equator
eiland, Du ... island
elv, Dan., Nor ... river, stream
embalse, Sp ... reservoir
erg, Arab ... dune, sandy desert
est, Fr., It ... east
estado, Sp ... state
este, Port., Sp ... east
estrecho, Sp ... strait
étang, Fr ... pond, lake
état, Fr ... state
eyjar, Ice ... islands
feld, Ger ... field, plain
festung, Ger ... fortress
fiume, It ... river
fjäll, Swe ... mountain
fjärd, Swe ... bay, inlet
fjeld, Nor ... mountain, hill
fjord, Dan., Nor ... fiord, inlet
fjördur, Ice ... fiord, inlet
fleuve, Fr ... river
flod, Dan., Swe ... river
flói, Ice ... bay, marshland
fluss, Ger ... river
foce, It ... river mouth
fontein, Du ... a spring
forêt, Fr ... forest
fors, Swe ... waterfall
forst, Ger ... forest
fos, Dan., Nor ... waterfall
fu, Chn ... town, residence
fuente, Sp ... spring, fountain
fuerte, Sp ... fort
furt, Ger ... ford
gang, Kor ... stream, river
gangri, Tib ... mountain
gat, Dan., Nor ... channel
gàve, Fr ... stream
gawa, Jap ... river
gebergte, Du ... mountain range
gebiet, Ger ... district, territory
gebirge, Ger ... mountains
ghat, India ... pass, mountain range
gobi, Mong ... desert
gol, Mong ... river
göl, gölü, Tur ... lake
golfe, Fr ... gulf, bay
golfo, It., Port., Sp ... gulf, bay
gomba, gompa, Tib ... monastery
gora, Rus., So. Slav ... mountain
góra, Pol ... mountain
gorod, Rus ... town
grad, Rus., So. Slav ... town
guba, Rus ... bay, gulf
gundung, Indon ... mountain
guntō, Jap ... archipelago
gunung, Mal ... mountain
haf, Swe ... sea, ocean
hafen, Ger ... port, harbor
haff, Ger ... gulf, inland sea
hai, Chn ... sea, lake
hama, Jap ... beach, shore
hamada, Arab ... rocky plateau
hamn, Swe ... harbor
hāmūn, Per ... swampy lake, plain
hantō, Jap ... peninsula
hassi, Arab ... well, spring
haus, Ger ... house
haut, Fr ... summit, top
hav, Dan., Nor ... sea, ocean
havn, Dan., Nor ... harbor, port
havre, Fr ... harbor, port
háza, Hung ... house, dwelling of
heim, Ger ... hamlet, home
hem, Swe ... hamlet, home
higashi, Jap ... east
hisar, Tur ... fortress
hissar, Arab ... fort
ho, Chn ... river
hoek, Du ... cape
hof, Ger ... court, farmhouse
höfn, Ice ... harbor
hoku, Jap ... north
holm, Dan., Nor., Swe ... island
hora, Czech ... mountain
horn, Ger ... peak
hoved, Dan., Nor ... cape
hu, Chn ... lake
huang, Chn ... yellow
hügel, Ger ... hill
huk, Dan., Swe ... point
hus, Dan., Nor., Swe ... house
île, Fr ... island
ilha, Port ... island
indsö, Dan., Nor ... lake
insel, Ger ... island
insjö, Swe ... lake
irmak, irmagi, Tur ... river
isla, Sp ... island
isola, It ... island
istmo, It., Sp ... isthmus
jarvi, jaur, Fin ... lake
jebel, Arab ... mountain
jiang, Chn ... river
jima, Jap ... island
jökel, Nor ... glacier
joki, Fin ... river
jökuli, Ice ... glacier
kaap, Du ... cape
kai, Jap ... bay, gulf, sea
kaikyō, Jap ... channel, strait
kalat, Per ... castle, fortress
kale, Tur ... fort
kali, Mal ... creek, river
kand, Per ... village
kap, Dan., Ger ... cape
kapp, Nor, Swe ... cape
kasr, Arab ... fort, castle
kawa, Jap ... river
kefr, Arab ... village
kei, Jap ... creek, river
ken, Jap ... prefecture
khor, Arab ... bay, inlet
khrebet, Rus ... mountain range
kita, Jap ... north
ko, Jap ... lake
köbstad, Dan ... market-town
kol, Mong ... lake
kólpos, Gr ... gulf
kong, Chn ... river
kopf, Ger ... head, summit, peak
köpstad, Swe ... market town
körfezi, Tur ... gulf
kosa, Rus ... spit
kou, Chn ... river mouth
köy, Tur ... village
kraal, Du. (Africa) ... native village
ksar, Arab ... fortified village
kuala, Mal ... bay, river mouth
kuh, Per ... mountain
kum, Tur ... sand
kuppe, Ger ... summit
küste, Ger ... coast
kyo, Jap ... town, capital
la, Tib ... mountain pass
labuan, Mal ... anchorage, port
lac, Fr ... lake
lago, It., Port., Sp ... lake
lagoa, Port ... lake, bay
laguna, It., Port., Sp ... lagoon, lake
lahti, Fin ... bay, gulf
lan, Swe ... county
landsby, Dan., Nor ... village
liman, Tur ... bay, port
ling, Chn ... pass, ridge, mountain
llanos, Sp ... plains
loch, Celt. (Scotland) ... lake, bay
loma, Sp ... long, low hill
lough, Celt. (Ireland) ... lake, bay
machi, Jap ... town
man, Kor ... bay
mar, Port., Sp ... sea
mare, It., Rom ... sea
marisma, Sp ... marsh, swamp
mark, Ger ... boundary limit
massif, Fr ... block of mountains
mato, Port ... forest, thicket
me, Siam ... river
meer, Du., Ger ... lake, sea
mer, Fr ... sea
mesa, Sp ... flat-topped mountain
meseta, Sp ... plateau
mina, Port., Sp ... mine
minami, Jap ... south
minato, Japan ... harbor, haven
misaki, Jap ... cape, headland
mont, Fr ... mount, mountain
montagna, It ... mountain
montagne, Fr ... mountain
montaña, Sp ... mountain
monte, It., Port., Sp .. mount, mountain
more, Rus., So. Slav ... sea
morro, Port., Sp ... hill, bluff
mühle, Ger ... mill
mund, Ger ... mouth, opening
mündung, Ger ... river mouth
mura, Jap ... township
myit, Bur ... river
mys, Rus ... cape
nada, Jap ... sea
nadi, India ... river, creek
naes, Dan., Nor ... cape
nafud, Arab ... desert of sand dunes
nagar, India ... town, city
nahr, Arab ... river
nam, Siam ... river, water
nan, Chn., Jap ... south
näs, Nor., Swe ... cape
nez, Fr ... point, cape
nishi, nisi, Jap ... west
njarga, Fin ... peninsula
nong, Siam ... marsh
noord, Du ... north
nor, Mong ... lake
nord, Dan., Fr., Ger., It., Nor., Swe .. north
norte, Port., Sp ... north
nos, Rus ... cape
nyasa, Bantu ... lake
ö, Dan., Nor., Swe ... island
occidental, Sp ... western
ocna, Rom ... salt mine
odde, Dan., Nor ... point, cape
oeste, Port., Sp ... west
oka, Jap ... hill
oost, Du ... east
oriental, Sp ... eastern
óros, Gr ... mountain
ost, Ger., Swe ... east
öster, Dan., Nor., Swe ... eastern
ostrov, Rus ... island
oued, Arab ... river, stream
ouest, Fr ... west
ozero, Rus ... lake
pää, Fin ... mountain
padang, Mal ... plain, field
pampas, Sp. (Argentina) .. grassy plains
pará, Indian (Brazil) ... river
pas, Fr ... channel, passage
paso, Sp ... mountain pass, passage
passo, It., Port ... mountain pass, passage, strait
patam, India ... city, town
pélagos, Gr ... open sea
pegunungan, Indon ... mountains
peña, Sp ... rock
pendi, Chn ... basin
pertuis, Fr ... strait
pic, Fr ... mountain peak
pico, Port., Sp ... mountain peak
piedra, Sp ... stone, rock
ping, Chn ... plain, flat
planalto, Port ... plateau
planina, Serb ... mountains
playa, Sp ... shore, beach
ploskogor'ye, Rus ... mountains
pnom, Camb ... mountain
pointe, Fr ... point
polder, Du., Ger ... reclaimed marsh
polje, So. Slav ... plain, field
poluostrov, Rus ... peninsula
pont, Fr ... bridge
ponta, Port ... point, headland
ponte, It., Port ... bridge
pore, India ... city, town
porthmós, Gr ... strait
porto, It., Port ... port, harbor
potamós, Gr ... river
prado, Sp ... field, meadow
presqu'île, Fr ... peninsula
proliv, Rus ... strait
pueblo, Sp ... town, village
puerto, Sp ... port, harbor
pulau, Indon ... island
punkt, Ger ... point
punt, Du ... point
punta, It., Sp ... point
pur, India ... city, town
puy, Fr ... peak
qal'a, qal'at, Arab ... fort, village
qasr, Arab ... fort, castle
rann, India ... wasteland
ra's, Arab ... cape, head
reka, Rus., So. Slav ... river
represa, Port ... reservoir
rettō, Jap ... island chain
ría, Sp ... estuary
ribeira, Port ... stream
riberão, Port ... river
rio, It., Port ... stream, river
río, Sp ... river
rivière, Fr ... river
roca, Sp ... rock
rt, Serb ... cape
rūd, Per ... river
saari, Fin ... island
sable, Fr ... sand
sahara, Arab ... desert, plain
saki, Jap ... cape
sal, Sp ... salt
salar, Sp ... salt flat, salt lake
salto, Sp ... waterfall
san, Jap., Kor ... mountain, hill
sat, satul, Rom ... village
schloss, Ger ... castle
sebkha, Arab ... salt marsh
see, Ger ... lake, sea
şehir, Tur ... town, city
selat, Indon ... strait
selvas, Port., (Brazil) . tropical rain forests
seno, Sp ... bay
serra, Port ... mountain chain
serrania, Sp ... mountain ridge
seto, Jap ... strait
severnaya, Rus ... northern
shahr, Per ... town, city
shamo, Chn ... desert
shan, Chn ... mountain, hill, island
shatt, Arab ... river
shi, Jap, Chn ... city
shima, Jap ... island
shōtō, Jap ... archipelago
sierra, Sp ... mountain range
sjö, Nor., Swe ... lake, sea
sö, Dan., Nor ... lake, sea
söder, södra, Swe ... south
song, Annam ... river
sopka, Rus ... peak, volcano
source, Fr ... a spring
spitze, Ger ... summit, point
staat, Ger ... state
stad, Dan., Du., Nor., Swe ... city, town
stadt, Ger ... city, town
stato, It ... state
step', Rus ... treeless plain, steppe
straat, Du ... strait
strand, Dan., Du., Ger., Nor., Swe ... shore, beach
stretto, It ... strait
strom, Ger ... river, stream
ström, Dan., Nor., Swe .. stream, river
stroom, Du ... stream, river
su, suyu, Tur ... water, river
sud, Fr., Sp ... south
süd, Ger ... south
suidō, Jap ... channel
sul, Port ... south
sund, Dan., Nor., Swe ... sound
sungai, sungei, Indon., Mal ... river
sur, Sp ... south
syd, Dan., Nor., Swe ... south
tafelland, Ger ... plateau
take, Jap ... peak, summit
tal, Ger ... valley
tanjung, tanjong, Mal ... cape
târg, târgul, Rom ... market, town
tell, Arab ... hill
teluk, Indon ... bay, gulf
terra, It ... land
terre, Fr ... earth, land
thal, Ger ... valley
tierra, Sp ... earth, land
tō, Jap ... east; island
tonle, Camb ... river, lake
top, Du ... peak
torp, Swe ... hamlet, cottage
tsangpo, Tib ... river
tso, Tib ... lake
tsu, Jap ... harbor, port
tundra, Rus ... treeless arctic plains
tuz, Tur ... salt
udde, Swe ... cape
ufer, Ger ... shore, riverbank
ujung, Indon ... point, cape
umi, Jap ... sea, gulf
ura, Jap ... bay, coast, creek
ust'ye, Rus ... river mouth
valle, It., Port., Sp ... valley
vallée, Fr ... valley
valli, It ... lake
vár, Hung ... fortress
város, Hung ... town
varoš, So. Slav ... town
veld, Du ... open plain, field
verkh, Rus ... top, summit
ves, Czech ... village
vest, Dan., Nor., Swe ... west
vik, Swe ... cove, bay
vila, Port ... town
villa, Sp ... town
villar, Sp ... village, hamlet
ville, Fr ... town, city
vodokhranilishche, Rus ... reservoir
vostok, Rus ... east
wad, wādī, Arab . intermittent stream
wald, Ger ... forest, woodland
wan, Chn., Jap ... bay, gulf
weiler, Ger ... hamlet, village
westersch, Ger ... western
wüste, Ger ... desert
xi, Chn ... west, western
yama, Jap ... mountain
yarimada, Tur ... peninsula
yug, Rus ... south
zaki, Jap ... cape
zaliv, Rus ... bay, gulf
zapad, Rus ... west
zee, Du ... sea
zemlya, Rus ... land
zuid, Du ... south

Abbreviations of Geographic Names and Terms

Ab., Can.	Alberta, Can.
Afg.	Afghanistan
Afr.	Africa
Ak., U.S.	Alaska, U.S.
Al., U.S.	Alabama, U.S.
Alb.	Albania
Alg.	Algeria
Am. Sam.	American Samoa
And.	Andorra
Ang.	Angola
Ant.	Antarctica
Antig.	Antigua and Barbuda
Ar., U.S.	Arkansas, U.S.
Arg.	Argentina
Arm.	Armenia
Aus.	Austria
Austl.	Australia
Az., U.S.	Arizona, U.S.
Azer.	Azerbaijan
b.	Bay, Gulf, Inlet, Lagoon
Bah.	Bahamas
Bahr.	Bahrain
Barb.	Barbados
bas.	Basin
B.C.,	British Columbia, Can.
Bdi.	Burundi
Bel.	Belgium
Bela.	Belarus
Ber.	Bermuda
Bhu.	Bhutan
B.I.O.T.	British Indian Ocean Territory
Blg.	Bulgaria
Bngl.	Bangladesh
Bol.	Bolivia
Bos.	Bosnia and Herzegovina
Bots.	Botswana
Braz.	Brazil
Bru.	Brunei
Br. Vir. Is.	British Virgin Islands
Burkina	Burkina Faso
c.	Cape, Point
Ca., U.S.	California, U.S.
Camb.	Cambodia
Camrn.	Cameroon
can.	Canal
Can.	Canada
C.A.R.	Central African Republic
Cay. Is.	Cayman Islands
C. Iv.	Cote d'Ivoire
clf.	Cliff, Escarpment
co.	County, Parish
Co., U.S.	Colorado, U.S.
Col.	Colombia
Com.	Comoros
cont.	Continent
Cook Is.	Cook Islands
C.R.	Costa Rica
Cro.	Croatia
cst.	Coast, Beach
Ct., U.S.	Connecticut, U.S.
C.V.	Cape Verde
Cyp.	Cyprus
Czech Rep.	Czech Republic
d.	Dam
D.C., U.S.	District of Columbia, U.S.
De., U.S.	Delaware, U.S.
del.	Delta
Den.	Denmark
dep.	Dependency, Colony
depr.	Depression
des.	Desert
Dji.	Djibouti
Dom.	Dominica
Dom. Rep.	Dominican Republic
D.R.C.	Democratic Republic of the Congo
Ec.	Ecuador
El Sal.	El Salvador
Eng., U.K.	England, U.K.
Eq. Gui.	Equatorial Guinea
Erit.	Eritrea
Est.	Estonia
est.	Estuary
Eth.	Ethiopia
E. Timor	East Timor
Eur.	Europe
Falk. Is.	Falkland Islands
Far. Is.	Faroe Islands
Fin.	Finland
Fl., U.S.	Florida, U.S.
for.	Forest, Moor
Fr.	France
Fr. Gu.	French Guiana
Fr. Poly.	French Polynesia
Ga., U.S.	Georgia, U.S.
Gam.	The Gambia
Gaza	Gaza Strip
Geor.	Georgia
Ger.	Germany
Gib.	Gibraltar
Grc.	Greece
Green.	Greenland
Gren.	Grenada
Guad.	Guadeloupe
Guat.	Guatemala
Guern.	Guernsey
Gui.	Guinea
Gui.-B.	Guinea-Bissau
Guy.	Guyana
Hi., U.S.	Hawaii, U.S.
hist.	Historic Site, Ruins
hist. reg.	Historic Region
Hond.	Honduras
Hung.	Hungary
i.	Island
Ia., U.S.	Iowa, U.S.
ice	Ice Feature, Glacier
Ice.	Iceland
Id., U.S.	Idaho, U.S.
Il., U.S.	Illinois, U.S.
In., U.S.	Indiana, U.S.
Indon.	Indonesia
ind. res.	Indian Reservation
I. of Man	Isle of Man
Ire.	Ireland
is.	Islands
Isr.	Israel
isth.	Isthmus
Jam.	Jamaica
Jord.	Jordan
Kaz.	Kazakhstan
Kir.	Kiribati
Kor., N.	Korea, North
Kor., S.	Korea, South
Ks., U.S.	Kansas, U.S.
Kuw.	Kuwait
Ky., U.S.	Kentucky, U.S.
Kyrg.	Kyrgyzstan
La., U.S.	Louisiana, U.S.
Lat.	Latvia
Leb.	Lebanon
Leso.	Lesotho
Lib.	Liberia
Liech.	Liechtenstein
Lith.	Lithuania
lk.	Lake
Lux.	Luxembourg
Ma., U.S.	Massachusetts, U.S.
Mac.	Macedonia
Madag.	Madagascar
Malay.	Malaysia
Mald.	Maldives
Marsh. Is.	Marshall Islands
Mart.	Martinique
Maur.	Mauritania
May.	Mayotte
Mb., Can.	Manitoba, Can.
Md., U.S.	Maryland, U.S.
Me., U.S.	Maine, U.S.
Mex.	Mexico
Mi., U.S.	Michigan, U.S.
Micron.	Micronesia, Federated States of
Mn., U.S.	Minnesota, U.S.
Mo., U.S.	Missouri, U.S.
Mol.	Moldova
Mong.	Mongolia
Mont.	Montenegro
Mor.	Morocco
Moz.	Mozambique
Ms., U.S.	Mississippi, U.S.
Mt., U.S.	Montana, U.S.
mth.	River Mouth or Channel
mtn.	Mountain
mts.	Mountains
Mya.	Myanmar
N.A.	North America
nat. cap.	National Capital
N.B., Can.	New Brunswick, Can.
N.C., U.S.	North Carolina, U.S.
N. Cal.	New Caledonia
N.D., U.S.	North Dakota, U.S.
Ne., U.S.	Nebraska, U.S.
Neth.	Netherlands
Neth. Ant.	Netherlands Antilles
Nf., Can.	Newfoundland, Can.
N.H., U.S.	New Hampshire, U.S.
Nic.	Nicaragua
Nig.	Nigeria
N. Ire., U.K.	Northern Ireland, U.K.
N.J., U.S.	New Jersey, U.S.
N.M., U.S.	New Mexico, U.S.
N. Mar. Is.	Northern Mariana Islands
Nmb.	Namibia
Nor.	Norway
n.p.	National Park or Monument
N.S., Can.	Nova Scotia, Can.
N.T., Can.	Northwest Territories, Can.
Nu., Can.	Nunavut, Can.
Nv., U.S.	Nevada, U.S.
N.Y., U.S.	New York, U.S.
N.Z.	New Zealand
oc.	Ocean
Oc.	Australia and Oceania
Oh., U.S.	Ohio, U.S.
Ok., U.S.	Oklahoma, U.S.
On., Can.	Ontario, Can.
Or., U.S.	Oregon, U.S.
p.	Pass
Pa., U.S.	Pennsylvania, U.S.
Pak.	Pakistan
Pan.	Panama
Pap. N. Gui.	Papua New Guinea
P.E., Can.	Prince Edward I., Can.
Para.	Paraguay
pen.	Peninsula
Phil.	Philippines
Pit.	Pitcairn
pk.	Park, Reserve
pl.	Plain, Flat
plat.	Plateau, Highland
p.o.i.	Point of Interest
Pol.	Poland
Port.	Portugal
P.R.	Puerto Rico
Qc., Can.	Québec, Can.
r.	Rock
rec.	Recreational Site, Park
reg.	Physical Region
res.	Reservoir
Reu.	Reunion
rf.	Reef, Shoal
R.I., U.S.	Rhode Island, U.S.
Rom.	Romania
Rw.	Rwanda
s.	Sea
S.A.	South America
S. Afr.	South Africa
Sau. Ar.	Saudi Arabia
S.C., U.S.	South Carolina, U.S.
Scot., U.K.	Scotland, U.K.
S.D., U.S.	South Dakota, U.S.
Sen.	Senegal
Serb.	Serbia
Sey.	Seychelles
S. Geor.	South Georgia
Sing.	Singapore
Sk., Can.	Saskatchewan, Can.
S.L.	Sierra Leone
Slvk.	Slovakia
Slvn.	Slovenia
S. Mar.	San Marino
Sol. Is.	Solomon Islands
Som.	Somalia
Sp. N. Afr.	Spanish North Africa
Sri L.	Sri Lanka
state	State, Province, Department, Region, etc.
St. Hel.	St. Helena
St. K./N.	St. Kitts and Nevis
St. Luc.	St. Lucia
stm.	River, Creek, Stream
St. P./M.	St. Pierre and Miquelon
S. Tom./P.	Sao Tome and Principe
strt.	Strait, Channel, Sound
St. Vin.	St. Vincent and the Grenadines
Sur.	Suriname
sw.	Swamp, Marsh
Swaz.	Swaziland
Swe.	Sweden
Switz.	Switzerland
Tai.	Taiwan
Taj.	Tajikistan
Tan.	Tanzania
T./C. Is.	Turks and Caicos Islands
Thai.	Thailand
Tn., U.S.	Tennessee, U.S.
Tok.	Tokelau
Trin.	Trinidad and Tobago
Tun.	Tunisia
Tur.	Turkey
Turkmen.	Turkmenistan
Tx., U.S.	Texas, U.S.
U.A.E.	United Arab Emirates
Ug.	Uganda
U.K.	United Kingdom
Ukr.	Ukraine
Ur.	Uruguay
U.S.	United States
Ut., U.S.	Utah, U.S.
Uzb.	Uzbekistan
Va., U.S.	Virginia, U.S.
val.	Valley, Watercourse
Ven.	Venezuela
Viet.	Vietnam
V.I.U.S.	Virgin Islands (U.S.)
vol.	Volcano
Vt., U.S.	Vermont, U.S.
Wa., U.S.	Washington, U.S.
Wal./F.	Wallis and Futuna
W.B.	West Bank
Wi., U.S.	Wisconsin, U.S.
W. Sah.	Western Sahara
wtfl.	Waterfall
W.V., U.S.	West Virginia, U.S.
Wy., U.S.	Wyoming, U.S.
Yk., Can.	Yukon, Can.
Zam.	Zambia
Zimb.	Zimbabwe

Pronunciation of Geographic Names

Key to the sound values of letters and symbols used in the index to indicate pronunciation

ă	ăt; băttle
ă	finăl; appeăl
ā	rāte; elāte
å	senåte; inanimåte
ä	ärm; cälm
à	àsk; bàth
ȧ	sofȧ; mȧrine (short neutral or indeterminate sound)
â	fâre; prepâre
ch	choose; church
dh	as th in other; either
ē	bē; ēve
ė	ėvent; crėate
ĕ	bĕt; ĕnd
ĕ	recĕnt (short neutral or indeterminate sound)
ẽ	cratẽr; cindẽr
g	gō; gāme
gh	guttural g
ĭ	bĭt; wĭll
ĭ	(short neutral or indeterminate sound)
ī	rīde; bīte
ĸ	gutteral k as *ch* n German *ich*
ng	sing
ŋ	baŋk; liŋger
ɴ	indicates nasalized
ŏ	nŏd; ŏdd
ŏ	cŏmmit; cŏnnect
ō	ōld; bōld
ô	ôbey; hôtel
ô	ôrder; lông
oi	boil
o͞o	fo͞od; ro͞ot
ȯ	as oo in foot; wood
ou	out; thou
s	soft; so; sane
sh	dish; finish
th	thin; thick
ū	pūre; cūre
û	ûnite; ûsûrp
û	ûrn; fûr
ŭ	stŭd; ŭp
ŭ	circŭs; sŭbmit
ü	as in French tu
zh	as z in azure
'	indeterminate vowel sound

In many cases the spelling of foreign geographical names does not even remotely indicate the pronunciation to an American, e.g., Słupsk in Poland is pronounced swôpsk; Jujuy in Argentina is pronounced ho͞ohwē; La Spezia in Italy is lä-spĕ'zyä.

This condition is hardly surprising, however, when we consider that in our own language Worcester, Massachusetts, is pronounced wȯs'tẽr; Sioux City, Iowa, so͞o sĭ'tė; Schuylkill Haven, Pennsylvania, sko͞ol'kĭl hā-vĕn; Poughkeepsie, New York, pŏ-kĭp'sė.

The indication of pronunciation of geographic names presents several peculiar problems:

1. Many foreign languages use sounds that are not present in the English language and which an American cannot normally articulate. Thus, though the nearest English equivalent sound has been indicated, only approximate results are possible.

2. There are several dialects in each foreign language that cause variation in the local pronunciation of names. This also occurs in identical names in the various divisions of a great language group.

3. Within the United States there are marked differences in pronunciation, not only of local geographic names, but also of common words, indicating that the sound and tone values for letters as well as the placing of the emphasis vary considerably from one part of the country to another.

4. A number of different letters and diacritical combinations could be used to indicate essentially the same or approximate pronunciations.

Some variation in pronunciation other than that indicated in this index may be encountered, but such a difference does not necessarily indicate that either is in error, and in many cases it is a matter of individual choice as to which is preferred. In fact, an exact indication of pronunciation of many foreign names using English letters and diacritical marks is extremely diffiicult and sometimes impossible.

The following sources have been consulted during the process of creating and updating the thematic maps and statistics for the 22nd Edition.

Andreassen, L., M. Beedle, E. Berthier, F. Cawkwell, N. Dickmann, E. Dolgova, A. Fountain, N. Glasser, E. Hansson, U. Haritashya, G. Hartman, C. Helm, L. Iacovelli, H. Jiskoot, G. Kapustin, T. Khromova, J. Kincaid, S. Kutuzov, I. Lavrentiev, X. Li, L. Mabileau, J. Meyer, P. Mool, A. Muravyev, G. Nosenko, F. Paul, A. Racoviteanu, F. Rau, A. Rivera, M. Schnirch, Y. Seliverstov, O. Sigurdsson, S. Taschner, P. Zenteno, and N. Zheltyhina. (2001-2008). *GLIMS Glacier Database*. National Snow and Ice Data Center/ World Data Center for Glaciology.

American Wind Energy Association (AWEA). (http://www.awea.org/)

Brohan, P., J.J. Kennedy, I. Harris, S.F.B. Tett, and P.D. Jones. (2006). Uncertainty estimates in regional and global observed temperature changes: A new dataset from 1850. *Journal of Geophysical Research*, 111, D12106, doi:10.1029/2005JD006548. (http://www.cru.uea.ac.uk/cru/data/temperature)

Brown, J., O.J. Ferrians, Jr., J.A. Heginbottom, and E.S. Melnikov. (1998). *Circum-Arctic Map of Permafrost and Ground Ice Conditions*. National Snow and Ice Data Center/ World Data Center for Glaciology.

Census of Canada. *Population Counts, for Canada, Provinces and Territories, and Census Divisions by Urban and Rural, 2001 Census.*

Center for International Earth Science Information Network (CIESIN), Columbia University; International Food Policy Research Institute (IFPRI); The World Bank; and Centro Internacional de Agricultural Tropical (CIAT). (2005). *Global Rural-Urban Mapping Project (GRUMP), Alpha Version.* Socioeconomic Data and Applications Center (SEDAC), Columbia University. (http://sedac.ciesin.columbia.edu/gpw/)

Center for Systemic Peace. *Major Episodes of Political Violence 1946-2007.*

Central Intelligence Agency (CIA). *World Factbook.* (https://www.cia.gov/library/publications/the-world-factbook/)

Chorlton, L B. (2007). *Generalized Geology of the World: Bedrock Domains and Major Faults in GIS Format: A Small-Scale World Geology Map with an Extended Geological Attribute Database.* Geological Survey of Canada, Open File 5529.

Coastal Services Center, National Oceanic and Atmospheric Administration. (http://www.csc.noaa.gov/)

Energy Information Administration (EIA), United States Department of Energy. *Coal Production and Number of Mines by State and Mine Type, 2007-2008.*

Energy Information Administration (EIA), United States Department of Energy. *International Energy Annual.*

Energy Information Administration (EIA), United States Department of Energy. *Natural Gas Annual 2006.*

Energy Information Administration (EIA), United States Department of Energy. *Petroleum Supply Annual 2006.*

Energy Information Administration (EIA), United States Department of Energy. *World Anthracite Coal Production, Most Recent Annual Estimates, 1980-2006.*

Energy Information Administration (EIA), United States Department of Energy. *World Bituminous Coal Production, Most Recent Annual Estimates, 1980-2006.*

Energy Information Administration (EIA), United States Department of Energy. *World Coal Production, Most Recent Annual Estimates, 1980-2007.*

Energy Information Administration (EIA), United States Department of Energy. *World Crude Oil Reserves, January 1, 1980 - January 1, 2008 Estimates.*

Energy Information Administration (EIA), United States Department of Energy. *World Dry Natural Gas Production, Most Recent Annual Estimates, 1980-2007.*

Energy Information Administration (EIA), United States Department of Energy. *World Production of Crude Oil, NGPL, and Other Liquids, and Refinery Processing Gain, Most Recent Annual Estimates, 1980-2007.*

Energy Information Administration (EIA), United States Department of Energy. *World Proved Natural Gas Reserves, January 1, 1980 - January 1, 2008 Estimates.*

Farr, T.G., P.A. Rosen, E. Caro, R. Crippen, R. Duren, S. Hensley, M. Kobrick, M. Paller, E. Rodriguez, L. Roth, D. Seal, S. Shaffer, J. Shimada, J. Umland, M. Werner, M. Oskin, D. Burbank, and D. Alsdorf. (2007). The Shuttle Radar Topography Mission. *Reviews of Geophysics*, 45, RG2004, doi:10.1029/2005RG000183.

Federal Aviation Administration (FAA). *CY 2007 Passenger Boarding and All-Cargo Data.*

Federation of American Scientists. *Status of World Nuclear Forces 2009.*

Fetterer, F. and K. Knowles. (2002). *Sea Ice Index.* National Snow and Ice Data Center.

Food and Agriculture Organization of the United Nations (FAO). *FAOSTAT.*

Global Volcanism Program, Smithsonian Institution. *Volcanoes of the World.*

Halpern, B.S., S. Walbridge, K.A. Selkoe, C.V. Kappel, F. Micheli, C. D'Agrosa, J.F. Bruno, K.S. Casey, C. Ebert, H.E. Fox, R. Fujita, D. Heinemann, H.S. Lenihan, E.M. P. Madin, M.T. Perry, E.R. Selig, M. Spalding, R. Steneck, and R. Watson. (2008). A global map of human impact on marine ecosystems. *Science*, 319(5865), pp. 948-952. doi: 10.1126/science.1149345.

Hansen, M., R. DeFries, J.R.G. Townshend, and R. Sohlberg. (2000). Global land cover classification at 1km resolution using a decision tree classifier. *International Journal of Remote Sensing*, 21, pp. 1331-1365.

Hansen, M.C., S.V. Stehman, P.V. Potapov, T.R. Loveland, J.R.G. Townshend, R.S. DeFries, K.W. Pittman, F. Stolle, M.K. Steininger, M. Carroll, and C. Dimiceli. (2008). Humid tropical forest clearing from 2000 to 2005 quantified using multi-temporal and multi-resolution remotely sensed data. *PNAS*, 105(27), pp. 9439-9444.

Heidelberg Institute for International Conflict Research. *Conflict Barometer.*

Heinrich J. , C. Klinke, and C.B. Schmidt. (1994). The Hop Atlas: *The History and Geography of the Cultivated Plant.* Nuremberg, Germany: Jon. Barth & Sohn.

Hulme, M. (1998). *Global Land Precipitation Dataset, Version 1.0.* Climatic Research Unit, University of East Anglia.

International Center for Prison Studies. (2009). *World Prison Population List, 8th Edition.* King's College, London.

International Lake Environment Committee. *World Lakes Database.*

International Water Power & Dam Construction. *Yearbook 2008.*

Inter-Parliamentary Union. *Women in National Parliaments.*

Inter-Parliamentary Union. *Women's Suffrage: A World Chronology of the Recognition of Women's Rights to Vote and to Stand for Election.*

Keeling, C.D., S.C. Piper, R.B. Bacastow, M. Wahlen, T.P. Whorf, M. Heimann, and H.A. Meijer. (2001). Exchanges of atmospheric CO_2 and $13CO_2$ with the terrestrial biosphere and oceans from 1978 to 2000. *I. Global Aspects, SIO Reference Series, No. 01-06.* Scripps Institution of Oceanography.

Kelly, T.D. and M.D. Fenton. (2003). *Iron and Steel Statistics.* United States Geological Survey.

Küchler, A.W. (1949). A physiognomic classification of vegetation. *Annals of the Association of American Geographers*, 39(3), pp. 201-210.

Laboratory for Satellite Altimetry, Satellite Oceanography and Climatology Division, National Oceanic and Atmospheric Administration. *Altimetry Data.* (http://ibis.grdl.noaa.gov/)

LakeNet Global Lake Database.

Mackay, J., M. Eriksen, and O. Shafey. (2006). *The Tobacco Atlas, 2nd Edition.* American Cancer Society.

McDaniel, P. (2008). *The Twelve Soil Orders.* (http://soils.ag.uidaho.edu/soilorders/index.htm/)

Murphy, R.E. (1968). Annals map supplement number 9. Landforms of the world. *Annals of the Association of American Geographers*, 58(1), pp. 198-200.

National Aeronautics and Space Administration (NASA), Atmospheric Science Data Center. (http://eosweb.larc.nasa.gov/)

National Aeronautics and Space Administration (NASA). (http://www.nasa.gov/)

National Oceanic and Atmospheric Administration (NOAA). (http://www.noaa.gov/)

National Snow and Ice Data Center. (http://nsidc.org/)

National Weather Service, National Oceanic and Atmospheric Administration. (http://www.spc.noaa.gov/)

Natural Resources Canada. *The Atlas of Canada.*

New, M., D. Lister, M. Hulme, and I. Makin. (2000). A high-resolution data set of surface climate over global land areas. *Climate Research*, 21, pp. 1-25.

Olson, D.M., E. Dinerstein, E.D. Wikramanayake, N.D. Burgess, G.V.N. Powell, E.C. Underwood, J.A. D'Amico, I. Itoua, H.E. Strand, J.C. Morrison, C.J. Loucks, T.F. Allnutt, T.H. Ricketts, Y. Kura, J.F. Lamoreux, W.W. Wettengel, P. Hedao, and K.R. Kassem. (2001). Terrestrial ecoregions of the world: A new map of life on earth. *BioScience*, 51(11), pp. 933-938.

Rand McNally. *The Rand McNally Road Atlas 2009.*

Statistics Canada. *Air Carrier Traffic at Canadian Airports.*

Stockholm International Peace Research Institute. *The Financial Value of National Arms Exports, 1998-2006.*

Tapley, B., J. Ries, S. Bettadpur, D. Chambers, M. Cheng, F. Condi, B. Gunter, Z. Kang, P.Nagel, R. Pastor, T. Pekker, S.Poole, and F. Wang, (2005). GGM02 - An improved Earth gravity field model from GRACE. *Journal of Geodesy*, doi 10.1007/s00190-005-0480-z. (http://www.csr.utexas.edu/grace/gravity/)

TeleGeography Research. (http://www.telegeography.com/)

The Canadian Wind Energy Association (CanWEA). (http://www.canwea.ca/)

Thornthwaite, C.W. (1944). Report of committee on transpiration and evaporation. *American Geophysical Union Transactions*, 25(5), pp. 683-693.

Transport Canada. *Transportation in Canada 2007.*

Trewartha, G.T. (1968). *An Introduction to Climate, 4th Edition.* New York: McGraw-Hill Book Company.

United Nations (UN). *Comtrade Database, SITC Rev.3. 2006.*

United Nations (UN). *Human Development Reports 2007/2008.*

United Nations (UN). *World Contraceptive Use 2007.*

United Nations (UN). *World Population Prospects Database: The 2006 Revision.*

United Nations Children's Fund (UNICEF). *Report on the Global AIDS Epidemic.*

United Nations Educational, Scientific and Cultural Organization (UNESCO), Institute for Statistics Data Centre. *Public Reports.*

United Nations Educational, Scientific and Cultural Organization (UNESCO), International Hydrological Programme. *World Water Resources and Their Use.*

United Nations Environment Program. *Global Resource Information Database.*

United Nations Environment Program. *Islands Directory.*

United Nations High Commissioner for Refugees (UNHCR). *2007 Global Trends: Refugees, Asylum-Seekers, Returnees, Internally Displaced and Stateless Persons.*

United Nations Peacekeeping. *List of Operations 1948-2008.*

United Nations, Department of Economic and Social Affairs. *2005 UN Energy Statistics Yearbook.*

United Nations, Department of Economic and Social Affairs. *World Urbanization Prospects, 2007 Revision.*

United Nations, Office on Drugs and Crime. *2008 World Drug Report.*

United Nations, Organization for Economic Co-operation and Development (OECD). *Uranium 2007: Resources, Production and Demand.*

United States Bureau of Labor Statistics. *Quarterly Census of Employment and Wages.*

United States Census Bureau. American FactFinder. *2006 Annual Survey of Manufactures.*

United States Census Bureau. *Census 2000 Summary Files.*

United States Census Bureau. *International Database.*

United States Census Bureau. *Statistical Abstract of the United States, 2008 Edition.*

United States Census Bureau. *Statistical Abstract of the United States, 2009 Edition.*

United States Department of Agriculture (USDA), Economic Research Service. *Sugar and Sweeteners Outlook, 2008.*

United States Department of Agriculture (USDA), Natural Resource Conservation Service. *Soil Taxonomy: A Basic System of Soil Classification for Making and Interpreting Soil Surveys.*

United States Department of Agriculture (USDA). *Chickens and Eggs-2007 Summary.*

United States Department of Agriculture (USDA). *Feed Grains Database.*

United States Department of Agriculture (USDA). *Poultry-Production and Value-2007 Summary.*

United States Department of Energy. *Landscan 2001 High Resolution Global Population Data Set.* © 2003 UT-Battelle, LLC. All rights reserved. Notice: These data were produced by UT-Battelle, LLC under Contract No. DE-AC05-00OR22725 with the Department of Energy. The Government has certain rigths in this data. Neither UT-Battelle, LLC nor the United States Department of Energy, nor any of their employees, makes any warranty, express or implied, or assumes any legal liability or responsibility for the accuracy, completeness, or usefulness of any data, apparatus, product, or process disclosed, or represents that its use would not infringe privately owned rights.

United States Environmental Protection Agency (EPA). (http://www.epa.gov/)

United States Environmental Protection Agency (EPA). *Great Lakes Factsheet No. 1.*

United States Geological Survey (USGS). *Lengths of Major Rivers.*

United States Geological Survey (USGS). *Mineral Commodity Summaries.*

United States Geological Survey (USGS). *Mineral Resources Data System (MRDS).*

United States Geological Survey (USGS). *Minerals Yearbook 2006.*

United States Geological Survey (USGS). *National Atlas of the United States.* (http://www.nationalatlas.gov/)

United States Geological Survey (USGS). *Significant Earthquakes of the World.*

Woodworth, P.L. and R. Player. (2003). The Permanent Service for Mean Sea Level: An update to the 21st century. *Journal of Coastal Research*, 19, pp. 287-295. (http://www.pol.ac.uk/psmsl/)

World Energy Council. *2007 Survey of Energy Resources.*

World Health Organization (WHO). *Global Atlas of the Health Workforce.*

World Health Organization (WHO). *Global Information System on Alcohol and Health.*

World Health Organization (WHO). *Global Status Report on Alcohol 2004.*

World Health Organization (WHO). *Statistical Information System (WHOSIS).* (http://www.who.int/whosis/)

World Health Organization (WHO). *World Malaria Report 2008.*

World Wind Energy Association. *Worldwide Wind Energy Installation Figures per Continent.*

The editor wishes to thank the individual scientists, research units, and organizations that made their datasets and research available for this edition of Goode's World Atlas.

Listed below are page references for major topics covered by the thematic maps and graphs, the introductory text, and the tables.

Agriculture 94-95, 237, 242
Agricultural production 58
Agricultural yield 58
Air quality 106
Air travel 100
Alcohol consumption 65
Alliances 80
Aluminum 68
Amphetamines 78
Anthracite . . . . See coal
Apples 63
Aquifers 106
Arable land 58
Armed forces 79
Arms exports 79
Atmospheric heat gain 40-41
Atmospheric pressure 34, 35
Bananas 63
Barley 61
Bauxite . . . 68, 89, 162, 181, 215, 237, 242, 257, 275
Beef 66
Beer 65
Birth rate 49, 50
Bituminous coal . . . . See Coal
Cannabis 78
Canola oil 64
Capitals 289-93
Carbon dioxide, atmospheric concentration 39
Carbon dioxide, emissions 39, 106
Cartography and geospatial technology 9
Cassava 62
Cattle 66
Chickens 67
Chromite 70, 215, 237, 257
Cigarette consumption 63
Citrus fruit 63
Climate regions 30-31
Coal 75, 89, 101, 162, 181, 214, 257, 274
Cobalt 70, 257
Cocaine 78
Cocoa beans 61
Coconuts 64
Coffee 60
Coltan 69
Commuting time 104
Conflicts 81
Contraception use 50
Copper . . . 68, 89, 162, 181, 215, 237, 242, 257, 275
Corn 60, 64
Corn oil 64
Cotton 65
Cottonseed 64
Currents . . . . See Ocean currents
Dates 63
Death rate 49, 50
Deciduous fruit 63
Desertification 89, 162, 181, 215, 257, 274
Diamonds 257
Drug use 78
Ducks 67
Earnings . . . . See Income
Earth properties 8
Earthquakes 22-23, 89, 162, 181, 215, 257, 274
Earth-sun relationships 17
Economic alliances . . . . See Alliances
Education 52, 104
Eggs 67
Electricity . . . . See Energy
El Niño/ENSO 91
Energy . . . . 72, 73, 89, 162, 181, 214, 257, 274. Also see specific energy generation methods (geothermal, hydroelectric, nuclear, thermal, and wind).
Environments 127
Ethnic groups . . . . See Race and ethnicity
Evapotranspiration 90
Exports 77
Federal lands 98-99
Fisheries 63
Flax 65
Floods 89, 162, 181, 215
Food aid 57
Forest cover 76
Forest loss 76
Fruit 63
Fuels . . . . 74-75, 89, 101, 162, 181, 214, 257, 274. Also see specific fuel types (coal, natural gas, petroleum, and uranium).
GDP . . . . See Gross Domestic Product
Geology 96-97, 182-83
Geothermal energy 73, 89, 181, 214, 257, 274
Glaciation 24-25
Gold 71, 257, 275
Government 289-293
Grain sorghum 61
Grapes 65
Gross Domestic Product 39, 52
Ground water 106
Hazardous waste sites 106
Highways 98-99
HIV infection 54
Hops 65
Hours of daylight 84
Housing 104
Hurricanes 91
Hydroelectric energy 73, 89, 162, 181, 214, 257, 274
Imports 77
Income 105
Indian reservations 98-99
Infant mortality 50
Internet 82
Interstate highways 98-99
Iron ore 70, 89, 101, 162, 181, 215, 237, 242, 257, 275
Jute 65
Kyoto Protocol 39
Labor force 107
Land areas 18-19
Land cover 87, 160, 178-79, 210-11, 255, 274
Land elevations 20-21
Landforms 24-25, 88, 161, 180, 212, 256, 275
Languages 53, 105
Lead 69, 89, 162, 181, 215, 242, 257, 275
Life expectancy 56
Lignite . . . . See Coal
Literacy 52
Lithium 69
Maize . . . . See Corn
Malaria 55
Manganese 70, 215, 237, 242, 257, 275
Manufacturing 107
Map legend 16
Map projections 11-13
Map scale 10
Map symbols . . . . See Map legend
Marine productivity 40-41
Military alliances . . . . See Alliances
Military expenditures 79
Military installations 98-99
Millet 61
Minerals . . . . 68, 69, 70, 71, 89,162, 181 215, 237, 242, 257, 275. Also see specific minerals.
Molybdenum 70
National parks 98-99
Natural gas 74, 89, 101, 162, 181, 214, 257, 274
Natural hazards 22-23, 89, 162, 181, 215, 257, 274
Natural increase 51
Natural vegetation . . . . See vegetation
Nickel 70, 89, 215, 275
Nuclear energy 72, 89, 162, 181, 214, 257
Nuclear weapons 79
Oats 60
Ocean currents 36-37, 40-41
Ocean depths 20-21
Ocean floor 26-27, 28, 29
Oil . . . . See petroleum
Oil palm fruit 64
Olive oil 64
Olives 64
Opiates 78
Palm oil 64
Paper 76
Peacekeeping operations 81
Peanut oil 64
Peanuts 64
Petroleum 74, 89, 101, 162, 181, 214, 257, 274
Physical features 294
Physicians 55
Pigs 66
Pineapples 63
Plate tectonics 22-23, 24-25
Platinum 71
Political alliances . . . . See Alliances
Population 18-19, 48-49, 51, 52, 57, 102, 103, 104, 105
Population change 104
Population density . . . 48-49, 89, 102, 127, 162, 181, 214, 216, 217, 237, 242, 257, 275
Population growth . . . . See Natural increase
Population, age 104
Population, age-sex composition 49
Population, ethnic . . . . See Race and ethnicity
Population, urban . . . . See Urban population
Population, world 19, 51
Pork 66
Potatoes 62
Poultry 67
Poverty 105
Precipitation 34, 35, 36-37, 88, 90, 127, 161, 180, 213, 256, 275
Precipitation change 36
Precipitation variability 37
Prison population 78
Race and ethnicity 103, 216, 217
Radioactive waste 106
Railroads 100
Rapeseed 64
Refugees 81
Religions 53, 237, 242
Rice 61
Roundwood 76
Rubber 65
Rye 59
Satellite images 88, 161, 180, 212, 256, 275
Sea ice 40-41, 89, 162, 181, 215, 274
Sea level change 38, 91
Sheep 67
Shipping lanes 83
Silk 65
Silver 71, 275
Sisal 65
Soils 44-45
Solar insolation 91
Solar system 17
Soybean oil 64
Soybeans 64
Spices 62
Steel 71
Submarine cables 82-83
Sugar 62
Sunflower oil 64
Sunflower seeds 64
Superfund sites 106
Surface water 106
Tea 59
Telecommunications 82-83
Temperature 32, 33, 90
Temperature change 38
Temperature range 33
Temperature, polar 33
Terrestrial biomes 46-47
Territorial evolution 100
Thematic map types 14-15
Thermal energy 72
Time zones 84
Tin 68, 162, 215, 242, 257
Tobacco 63
Tornadoes 89, 91
Toxic chemicals 106
Tractors 58
Transportation 98-99, 100
Tropical storms 89, 91, 215, 257, 274
Tsunamis 89, 162, 181, 215, 257, 274
Tuberculosis 54
Tungsten 70, 162, 215, 242
Turkeys 67
Undernourishment 56
Unemployment 105
Uranium 75, 89, 162, 181, 214, 257, 274
Urban population 51, 295
Vanadium 70
Veal 66
Vegetable oils 64
Vegetation 42-43, 88, 92-93, 127, 161, 180, 213, 256, 275
Volcanoes 22-23, 89, 162, 181, 215, 257, 274
Water resources 106
Water, agricultural withdrawals 58
Water, drinking 57
Waterways 100
Wheat 59
Wind energy 73, 91
Winds 34, 35, 91
Wine 65
Women heads of household 105
Women's economic activity rate 79
Women's legislative participation rate 79
Women's voting rights 79
Wood production 76
Wool 67
Zinc 69, 89, 162, 181, 215, 242, 257, 275

This universal index includes in a single alphabetical list the names of selected features that appear on the reference maps. Each name is followed by a page number and geographical coordinates.

Abbreviation and Capitalization. Abbreviations of names on the maps have been standardized as much as possible. Names that are abbreviated on the maps are generally spelled out in full in the index.

Most initial letters of names are capitalized, except for a few Dutch names such as "s-Gravenhage". Capitalization of non-initial words in a name generally follows local practice.

Alphabetization. Names are alphabetized in the order of the letters of the English alphabet. Spanish *ll* and *ch*, for example, are not treated as separate letters. Furthermore, diacritical marks are disregarded in alphabetization – German or Scandinavian *ä* or *ö* are treated as *a* or *o*.

The names of physical features may appear inverted, since they are always alphabetized under the proper, not the generic, part of the name, thus: "Gibraltar, Strait of", not "Strait of Gibraltar". In this case "Gibraltar" is the proper part of the name and "Strait of" is the generic. Otherwise every entry, whether consisting of one word or more, is alphabetized as a single continuous entity on the basis of the proper part of the name. "Lakeland", for example, appears after "Lake Havasu City" and before "La Luz".

In the case of identical names, towns are listed first, then political divisions, then physical features.

Generic Terms. Except for cities, the names of all features are followed by terms that represent broad classes of features, for example, "Mississippi, stm." or "Alabama, state". A list of all abbreviations used in the index is on page 297.

Country names and the names of features that extend beyond the boundaries of one country are followed by the name of the continent in which each is located. Country designations follow the names of all other places in the index. The locations of places in the United States, Canada and the United Kingdom are further defined by abbreviations that include the state or political division in which each is located.

Pronunciations. Pronunciations are included for many of the names listed. An explanation of the pronunciation system used appears on page 297.

Page References and Geographical Coordinates. The page references and geographical coordinates are found in the last columns of each entry.

If a page contains several maps or insets, a lowercase letter identifies the specific map or inset.

Latitude and longitude coordinates for point features, such as cities and mountain peaks, indicate the location of the symbols. For extensive areal features, such as countries or mountain ranges, the locations are for the approximate center of the feature. For rivers, locations are given for the mouth.

A

Feature (Pronunciation)	Page	Lat.	Long.
Aachen, Ger. (ä′kĕn)	**194-95**	50°46′N	6°06′E
Aalborg, Den. (ôl′bŏr)	**192-93**	57°02′N	9°55′E
Aalen, Ger. (ä′lĕn)	**194-95**	48°50′N	10°06′E
Aali, Sadd el-, d., Egypt *see* Aswan High Dam	**268b**	23°59′N	32°53′E
Aarau, Switz. (ärou)	**194-95**	47°24′N	8°03′E
Aba, China	**238-39**	33°06′N	101°59′E
Aba, Nig.	**260a**	5°07′N	7°22′E
Abacaxis, stm., Braz.	**166-67**	3°54′S	58°46′W
Abaco, i., Bah.	**142-43**	26°28′N	77°05′W
Ābādān, Iran (ä-bŭ-dän′)	**228-29**	30°21′N	48°17′E
Abadla, Alg.	**188-89**	31°01′N	2°41′W
Abaetetuba, Braz. (ä′båĕ-tĕ-to͞o′bȧ)	**166-67**	1°44′S	48°53′W
Abagnar Qi, China *see* Xilinhot	**240-41**	43°56′N	116°03′E
Abag Qi, China	**240-41**	43°43′N	114°39′E
Abakan, Russia (ŭ-bȧ-kän′)	**218-19**	53°43′N	91°27′E
Abancay, Peru (ä-bän-kä′ė)	**170**	13°37′S	72°53′W
Abashiri, Japan (ä-bä-shē′rė)	**244**	44°01′N	144°16′E
Abasolo, Mex. (ä-bä-sō′lô)	**146-47**	24°04′N	98°22′W
Ābay, stm., Afr. *see* Blue Nile	**254**	15°38′N	32°30′E
Ābaya Hāyk', lk., Eth. (ȧ-bä′yȧ)	**262-63**	6°18′N	37°52′E
Abbé, Lac, lk., Afr. *see* Abe, Lake	**262-63**	11°10′N	41°48′E
Abbeville, Fr. (ȧb-vēl′)	**196-97**	50°07′N	1°50′E
Abbeville, Al., U.S. (ăb′ė-vĭl)	**124-25**	31°34′N	85°15′W
Abbeville, Ga., U.S. (ăb′ė-vĭl)	**124-25**	31°60′N	83°18′W
Abbeville, La., U.S. (ăb′ė-vĭl)	**122-23**	29°58′N	92°08′W
Abbeville, S.C., U.S. (ăb′ė-vĭl)	**124-25**	34°11′N	82°23′W
Abbotsford, B.C., Can. (ăb′ŭts-fĕrd)	**132-33**	49°03′N	122°17′W
Abbottābād, Pak.	**232-33**	34°09′N	73°13′E
'Abd al-Kūrī, i., Yemen (ăbd-ĕl-kȯ′rė)	**220-21**	12°12′N	52°13′E
Abdulino, Russia (äb-dȯ-lē′nō)	**186-87**	53°41′N	53°40′E
Abe, Lake, lk., Afr.	**262-63**	11°10′N	41°48′E
Abéché, Chad	**258-59**	13°50′N	20°50′E
Abemama, at., Kir.	**280-81**	0°26′N	173°54′E
Abengourou, C. Iv.	**260-61**	6°44′N	3°29′W
Abeokuta, Nig. (ä-bå-ô-ko͞o′tä)	**260a**	7°09′N	3°21′E
Aberdare, Wales, U.K. (ăb-ẽr-dâr′)	**190-91**	51°43′N	3°28′W
Aberdare National Park, n.p., Kenya	**267**	0°30′S	36°45′E
Aberdeen, Scot., U.K. (ăb-ẽr-dēn′)	**190-91**	57°09′N	2°06′W
Aberdeen, Id., U.S. (ăb-ẽr-dēn′)	**112-13**	42°57′N	112°51′W
Aberdeen, Md., U.S. (ăb-ẽr-dēn′)	**116-17**	39°31′N	76°10′W
Aberdeen, Ms., U.S. (ăb-ẽr-dēn′)	**124-25**	33°50′N	88°33′W
Aberdeen, S.D., U.S. (ăb-ẽr-dēn′)	**114-15**	45°28′N	98°29′W
Aberdeen, Wa., U.S. (ăb-ẽr-dēn′)	**112-13**	46°59′N	123°49′W
Aberdeen Lake, lk., Nu., Can.	**130-31**	64°27′N	99°00′W
Aberystwyth, Wales, U.K. (ă-bẽr-ĭst′wĭth)	**190-91**	52°25′N	4°05′W
Abez', Russia	**186-87**	66°32′N	61°44′E
Abhā, Sau. Ar.	**266**	18°13′N	42°30′E
Abhé Bad, lk., Afr. *see* Abe, Lake	**262-63**	11°10′N	41°48′E
Ābhē Bid Hāyk', lk., Afr. *see* Abe, Lake	**262-63**	11°10′N	41°48′E
Abidjan, nat. cap., C. Iv. (ä-bėd-zhäɴ′)	**260-61**	5°20′N	4°01′W
Abilene, Ks., U.S. (ăb′ĭ-lēn)	**120-21**	38°55′N	97°13′W
Abilene, Tx., U.S. (ăb′ĭ-lēn)	**120-21**	32°27′N	99°44′W
Abingdon, Il., U.S. (ăb′ĭng-dŭn)	**114-15**	40°48′N	90°23′W
Abingdon, Va., U.S. (ăb′ĭng-dŭn)	**124-25**	36°43′N	81°59′W
Abitibi, stm., On., Can.	**130-31**	51°03′N	80°55′W
Abitibi, Lac, lk., Can. (läk ăb-ĭ-tĭb′ĭ) *see* Abitibi, Lake	**136-37**	48°41′N	79°35′W
Abitibi, Lake, lk., Can. (lāk ăb-ĭ-tĭb′ĭ)	**136-37**	48°41′N	79°35′W
Ābīyata Hāyk', lk., Eth.	**269d**	7°36′N	38°36′E
Abkhazeti Autonomis Respublika, state, Geor. *see* Abkhazia	**227**	43°10′N	41°00′E
Abkhazia, state, Geor.	**227**	43°10′N	41°00′E
Åbo, Fin. *see* Turku	**192-93**	60°27′N	22°16′E
Abou-Deïa, Chad	**258-59**	11°27′N	19°17′E
Abou Simbel, hist., Egypt *see* Abu Simbel	**266**	22°22′N	31°38′E
Abovyan, Arm.	**227**	40°15′N	44°35′E
Abrantes, Port. (ȧ-brän′tĕs)	**198-99**	39°28′N	8°12′W
Abra Pampa, Arg.	**168-69**	22°43′S	65°42′W
Abruka saar, i., Est. (ȧ-brȯ′kȧ-så′år)	**192-93**	58°09′N	22°31′E
Abū 'Alī, i., Sau. Ar.	**230-31**	27°19′N	49°35′E
Abū 'Arīsh, Sau. Ar. (ä-bo͞o ȧ-rēsh′)	**266**	16°58′N	42°50′E
Abu Dhabi, nat. cap., U.A.E. (ä′bo͞o dä′bē)	**230-31**	24°28′N	54°22′E
Abuja, nat. cap., Nig. (ä-bū′jȧ)	**260-61**	9°12′N	7°11′E
Abū Kamāl, Syria	**228-29**	34°27′N	40°56′E
Abū Mūsā, i., Asia	**230-31**	25°52′N	55°02′E
Abū Mūsā, Jazīreh-ye, i., Asia *see* Abū Mūsā	**230-31**	25°52′N	55°02′E
Abunã, Braz.	**166-67**	9°41′S	65°22′W
Abuná, stm., S.A.	**166-67**	9°40′S	65°26′W
Abunã, stm., S.A. (ȧ-bo͞o-nä′)	**166-67**	9°40′S	65°26′W
Ābu Road, India (ȧ′bo͞o rōd)	**234-35**	24°30′N	72°49′E
Abū Shajarah, Ra's, c., Sudan	**266**	21°05′N	37°13′E
Abu Simbel, hist., Egypt	**266**	22°22′N	31°38′E
Abū Sunbul, hist., Egypt *see* Abu Simbel	**266**	22°22′N	31°38′E
Abū Ẕaby, nat. cap., U.A.E. *see* Abu Dhabi	**230-31**	24°28′N	54°22′E
Abyaḍ, Al-Baḥr al-, stm., Sudan *see* White Nile	**254**	15°38′N	32°31′E
Abyssinia, nation, Afr. *see* Ethiopia	**253**	9°0′N	39°00′E
Acacías, Col. (ȧ-kȧ-sē′äs)	**163c**	4°00′N	73°45′W
Acadia National Park, n.p., Me., U.S. (ă-kā′dĭ-ȧ năsh′ŭn-ăl pärk)	**117a**	44°20′N	68°14′W
Acajutla, El Sal. (ä-kä-ho͞ot′lä)	**148**	13°35′N	89°50′W
Acámbaro, Mex. (ä-käm′bä-rō)	**146-47**	20°03′N	100°43′W
Acaponeta, Mex. (ä-kä-pô-nā′tă)	**146-47**	22°29′N	105°22′W
Acaponeta, stm., Mex. (ä-kä-pô-nā′tă)	**146-47**	22°23′N	105°38′W
Acapulco de Juárez, Mex.	**146-47**	16°51′N	99°54′W
Acaraí, Serra, mts., S.A. *see* Acarai Mountains	**164-65**	1°30′N	58°15′W
Acarai Mountains, mts., S.A.	**164-65**	1°30′N	58°15′W
Acaraú, Braz.	**166-67**	2°53′S	40°07′W
Acarigua, Ven. (äkä-rē′gwä)	**164-65**	9°34′N	69°12′W
Acatlán de Osorio, Mex. (ä-kät-län′dā ô-sō′rē-ō)	**146-47**	18°13′N	98°03′W
Acayucan, Mex. (ä-kä-yo͞o′kän)	**146-47**	17°57′N	94°54′W
Accra, nat. cap., Ghana (ă′krȧ)	**260-61**	5°34′N	0°12′W
Acerra, Italy (ä-chĕ′r-rä)	**200-01**	40°57′N	14°22′E
Achacachi, Bol. (ä-chä-kä′chė)	**168-69**	16°02′S	68°41′W
Achalpur, India	**234-35**	21°18′N	77°31′E
Acheng, China	**240-41**	45°32′N	126°59′E
Achinsk, Russia (ȧ-chėnsk′)	**218-19**	56°16′N	90°30′E
Acireale, Italy (ä-chė-rā-ä′lā)	**200-01**	37°37′N	15°10′E
Acklins, i., Bah. (ăk′lĭns)	**142-43**	22°26′N	73°58′W
Acklins, Bight of, b., Bah. (bīt ŭv ăk′lĭns)	**142-43**	22°32′N	74°08′W
Aconcagua, Cerro, mtn., Arg. (sĕ′r-rô ä-kôn-kä′gwä)	**163e**	32°39′S	70°02′W

n-sing; ŋ-baŋk; ɴ-nasalized n; nŏd; cŏmmit; ōld; ôbey; ôrder; oi-boil; fo͞od; ȯ-as oo in foot; ou-out; s-soft; sh-dish; th-thin; pūre; ůnite; ûrn; stŭd; circŭs; ü-as in French tu; ′-indeterminate vowel.

Feature (Pronunciation)	Page	Lat.	Long.
Açores, is., Port. (ä-zō′rĕs) *see* Azores	**199c**	38°30′N	28°00′W
A Coruña, Spain	**198-99**	43°22′N	8°25′W
Acoyapa, Nic. (ä-kō-yä′pä)	**149**	11°58′N	85°11′W
Acre, Isr. *see* ′Akko	**228-29**	32°55′N	35°06′E
Acre, state, Braz. (ä′krä)	**166-67**	9°0′S	70°00′W
Acre, stm., S.A. (ä′krä)	**170**	8°45′S	67°24′W
Actopan, Mex. (äk-tô-pän′)	**146-47**	20°16′N	98°57′W
Ada, Mn., U.S. (ā′dŭ)	**114-15**	47°18′N	96°31′W
Ada, Oh., U.S. (ā′dŭ)	**116-17**	40°46′N	83°49′W
Ada, Ok., U.S. (ā′dŭ)	**120-21**	34°47′N	96°41′W
Adak Island, i., Ak., U.S. (ă-dăk′ ī′l*ă*nd)	**126a**	51°43′N	176°43′W
Adam, Oman	**220-21**	22°23′N	57°31′E
Adama, Eth. *see* Nazrēt	**269d**	8°32′N	39°16′E
Adamaoua, mts., Afr.	**260-61**	7°0′N	12°00′E
Adamawa, mts., Afr. *see* Adamaoua	**260-61**	7°0′N	12°00′E
Adams, Ma., U.S. (ăd′*ă*mz)	**116-17**	42°38′N	73°07′W
Adams, Wi., U.S. (ăd′*ă*mz)	**116-17**	43°57′N	89°49′W
Adams, stm., B.C., Can. (ăd′*ă*mz)	**132-33**	50°54′N	119°33′W
Adams, Mount, vol., Wa., U.S.	**112-13**	46°13′N	121°29′W
Adams Lake, lk., B.C., Can.	**132-33**	51°13′N	119°33′W
′Adan, Yemen *see* Aden	**266**	12°49′N	45°02′E
Adana, Tur. (ä′dä-nä)	**228-29**	37°00′N	35°20′E
Adapazarı, Tur. (ä-dä-pä-zä′rê) *see* Sakarya	**186-87**	40°47′N	30°24′E
Adare, Cape, c., Ant.	**287**	71°20′S	170°08′E
Adavale, Austl.	**276**	25°55′S	144°36′E
Ad-Dahnā′, des., Sau. Ar.	**220-21**	24°30′N	48°10′E
Ad-Dammām, Sau. Ar.	**230-31**	26°26′N	50°07′E
Ad-Dawādimī, Sau. Ar.	**266**	24°28′N	44°18′E
Ad-Dawḥah, nat. cap., Qatar *see* Doha	**230-31**	25°17′N	51°32′E
Ad-Dilam, Sau. Ar.	**230-31**	23°56′N	47°06′E
Addis Ababa, nat. cap., Eth. (ä′dĭs ä′bä-bä)	**269d**	9°02′N	38°45′E
Ad-Dīwānīyah, Iraq	**228-29**	31°59′N	44°55′E
Addo Elephant National Park, n.p., S. Afr.	**264-65**	33°29′S	25°46′E
Ad-Du′ayn, Sudan	**266**	11°26′N	26°10′E
Ad-Duwaym, Sudan (ad-dò-ām′)	**266**	13°59′N	32°18′E
Adel, Ga., U.S. (ā-dĕl′)	**124-25**	31°08′N	83°25′W
Adel, Ia., U.S. (ā-dĕl′)	**114-15**	41°37′N	94°01′W
Adelaide, Austl. (ăd′ē-lād)	**276**	34°55′S	138°35′E
Adelaide Peninsula, pen., Nu., Can.	**130-31**	68°09′N	97°45′W
Aden, Yemen (ä′dĕn)	**266**	12°49′N	45°02′E
Aden, Gulf of, b., (gŭlf ŭv ä′dĕn)	**220-21**	12°40′N	48°07′E
Ādīgrat, Eth.	**266**	14°17′N	39°27′E
Ādilābād, India (ŭ-dĭl-ä-bäd′)	**234-35**	19°41′N	78°33′E
Adīrī, Libya	**258-59**	27°32′N	13°13′E
Adirondack Mountains, mts., N.Y., U.S. (ăd-ĭ-rŏn′dăk moun′tĭnz)	**116-17**	44°0′N	74°00′W
Ādīs Ābeba, nat. cap., Eth. (ā-dēs′ ā′bā-bā) *see* Addis Ababa	**269d**	9°02′N	38°45′E
Adjud, Rom. (äd′zhòd)	**202-03**	46°06′N	27°11′E
Adjuntas, Presa de las, res., Mex. *see* Vicente Guerrero, Presa	**146-47**	23°57′N	98°46′W
Admiralty Island National Monument, n.p., U.S. (ăd′mĭ-r*ă*l-tê ī′l*ă*nd năsh′ŭn-*ă*l mŏn′*ŭ*-m*ĕ*nt)	**126**	57°40′N	134°16′W
Admiralty Islands, is., Pap. N. Gui. (ăd′mĭ-r*ă*l-tê ī′l*ă*ndz)	**277**	2°10′S	147°00′E
Adolfo Gonzales Chaves, Arg.	**173**	38°01′S	60°08′W
Adonara, Pulau, i., Indon.	**248-49**	8°20′S	123°10′E
Ādoni, India	**236**	15°38′N	77°16′E
Adra, Spain (ä′drä)	**198-99**	36°45′N	3°01′W
Adrano, Italy (ä-drä′nō)	**200-01**	37°40′N	14°50′E
Adrar, Alg.	**258-59**	27°52′N	0°18′W
Adrâr, reg., Maur.	**258-59**	20°26′N	12°46′W
Adria, Italy (ä′drē-ä)	**200-01**	45°03′N	12°04′E
Adrian, Mi., U.S. (ā′drĭ-*ă*n)	**116-17**	41°53′N	84°02′W
Adrian, Mn., U.S. (ā′drĭ-*ă*n)	**114-15**	43°38′N	95°56′W
Adrianople, Tur. *see* Edirne	**200-01**	41°41′N	26°34′E
Adriatico, Mare, s., Eur. *see* Adriatic Sea	**200-01**	42°30′N	16°00′E
Adriatic Sea, s., Eur. (ā-drē-ă′tĭc sē)	**200-01**	42°30′N	16°00′E
Adriatik, Deti, s., Eur. *see* Adriatic Sea	**200-01**	42°30′N	16°00′E
Ādwa, Eth.	**266**	14°11′N	38°53′E
Adycha, stm., Russia (ä′dĭ-chá)	**218-19**	68°13′N	134°48′E
Adygea, state, Russia *see* Adygheya	**186-87**	45°0′N	40°00′E
Adygheya, state, Russia	**186-87**	45°0′N	40°00′E
Adz′va, stm., Russia (ädz′vá)	**186-87**	66°36′N	59°24′E
Aegean Sea, s., (ê-jē′*ă*n sē)	**200-01**	38°30′N	25°00′E
Afars and Issas, nation, Afr. *see* Djibouti	**253**	11°30′N	43°00′E
Affon, stm., Benin *see* Ouémé	**260a**	6°27′N	2°33′E
Afghānestān, nation, Asia *see* Afghanistan	**206-07**	33°0′N	65°00′E
Afghanistan, nation, Asia (ăf-găn-ĭ-stăn′)	**206-07**	33°0′N	65°00′E
′Afīf, Sau. Ar.	**266**	23°55′N	42°56′E
Afikpo, Nig.	**260a**	5°55′N	7°55′E
Aflou, Alg. (ä-flōō′)	**188-89**	34°07′N	2°06′E
Afognak Island, i., Ak., U.S. (ä-fŏg-nák′ ī′l*ă*nd)	**126**	58°14′N	152°39′W
Africa, cont., (ăf′rĭ-k*ă*)	**254**	10°0′N	22°00′E
Afton, Ok., U.S. (ăf′t*ŭ*n)	**120-21**	36°42′N	94°58′W
Afton, Wy., U.S. (ăf′t*ŭ*n)	**112-13**	42°43′N	110°56′W
′Afula, Isr. (ä-fò′lá)	**228-29**	32°36′N	35°18′E
Afyon, Tur. (ä-fē-ōn)	**186-87**	38°46′N	30°33′E
Afyonkarahisar, Tur. *see* Afyon	**186-87**	38°46′N	30°33′E
Agadez, Niger (ä′g*á*-dĕs)	**258-59**	16°58′N	7°59′E
Agadir, Mor. (ä-g*á*-dēr)	**258-59**	30°28′N	9°39′W
Agadyr, Kaz.	**226**	48°16′N	72°53′E
Agana, nat. cap., Guam *see* Hagåtña	**279c**	13°28′N	144°45′E
Āgaro, Eth.	**262-63**	7°50′N	36°40′E
Agartala, India	**234-35**	23°50′N	91°16′E
Agate Fossil Beds National Monument, n.p., Ne., U.S.	**114-15**	42°25′N	103°43′W
Ağdam, Azer. (ăg′dăm)	**227**	39°59′N	46°56′E
Agde, Fr. (ägd)	**196-97**	43°19′N	3°28′E
Agen, Fr. (á-zhäɴ′)	**196-97**	44°12′N	0°38′E
Āghā Jārī, Iran	**230-31**	30°42′N	49°50′E
Agno, Phil. (äg′nō)	**250**	16°07′N	119°48′E
Agno, stm., Phil. (äg′nō)	**250**	16°02′N	120°09′E
Āgra, India (ä′grä)	**234-35**	27°11′N	78°00′E
Ağrı, Tur.	**227**	39°43′N	43°04′E
Ağrı Dağı, vol., Tur. *see* Ararat, Mount	**227**	39°42′N	44°18′E
Agrigento, Italy	**200-01**	37°18′N	13°35′E
Agrihan, i., N. Mar. Is.	**280-81**	18°46′N	145°40′E
Agryz, Russia	**186-87**	56°31′N	53°01′E
Aguadas, Col. (ä-gwä′dás)	**163c**	5°38′N	75°27′W
Aguadilla, P.R. (ä-gwä-dēl′yä)	**142a**	18°26′N	67°09′W
Aguadulce, Pan. (ä-gwä-dōōl′sä)	**150**	8°15′N	80°31′W
Aguán, stm., Hond. (ä-gwá′n)	**149**	15°58′N	85°44′W
Aguanaval, stm., Mex. (ä-guä-nä-väl′)	**144-45**	25°25′N	102°49′W
Aguanish, stm., Qc., Can.	**138-39**	50°15′N	62°07′W
Agua Prieta, Mex.	**144-45**	31°19′N	109°33′W
Aguarico, stm., S.A.	**170**	0°58′S	75°11′W
Aguascalientes, Mex.	**146-47**	21°53′N	102°18′W
Aguascalientes, state, Mex. (ä′gwäs-käl-yĕn′täs)	**146-47**	22°0′N	102°30′W
Água Vermelha, Represa de, res., Braz.	**168-69**	20°0′S	50°00′W
Águilas, Spain (ä′gē-läs)	**198-99**	37°25′N	1°35′W
Aguililla, Mex. (ä-gē-lēl-yä)	**146-47**	18°44′N	102°44′W
Agulhas, c., S. Afr. (ä-gōōl′yäs)	**264-65**	34°49′S	20°03′E
Agusan, stm., Phil. (ä-gōō′sän)	**250**	9°01′N	125°31′E
Ahaggar, mts., Alg. (á-há-gär′)	**258-59**	23°0′N	6°30′E
Ahaggar, Tassili oua-n-, plat., Alg.	**258-59**	21°0′N	6°00′E
Ahar, Iran	**227**	38°28′N	47°04′E
Ahlen, Ger. (ä′lĕn)	**194-95**	51°46′N	7°54′E
Ahmadābād, India (ŭ-mĕd-ā-bād′)	**234-35**	23°02′N	72°35′E
Ahmadnagar, India (ä′mûd-nû-gûr)	**236**	19°05′N	74°45′E
Aḥmar, Al-Baḥr al-, s., *see* Red Sea	**266**	20°0′N	38°00′E
Ahmar Mountains, mts., Eth.	**262-63**	9°14′N	41°25′E
Ahoskie, N.C., U.S. (ā-hŏs′kē)	**124-25**	36°17′N	76°59′W
Ahuacatlán, Mex. (ä-wä-kät-län′)	**146-47**	21°05′N	104°29′W
Ahumada, Mex.	**144-45**	30°37′N	106°31′W
Ahunui, at., Fr. Poly.	**280-81**	19°39′S	140°25′W
Åhus, Swe. (ô′hòs)	**192-93**	55°55′N	14°18′E
Ahvāz, Iran	**228-29**	31°19′N	48°42′E
Ahvenanmaa, is., Fin. (ä′vĕ-nán-mô) *see* Aland Islands	**192-93**	60°14′N	19°46′E
Aidar, stm., Eur.	**202-03**	48°44′N	39°16′E
Aigaíon Pélagos, s., *see* Aegean Sea	**200-01**	38°30′N	25°00′E
Aiken, S.C., U.S. (ā′kĕn)	**124-25**	33°33′N	81°43′W
Ailao Shan, mts., China	**238-39**	24°13′N	101°20′E
Aimorés, Braz.	**172**	19°30′S	41°05′W
Aïn Beni Mathar, Mor.	**188-89**	34°01′N	2°01′W
Aïn Sefra, Alg.	**188-89**	32°46′N	0°34′W
Ainsworth, Ne., U.S. (ānz′wûrth)	**114-15**	42°33′N	99°52′W
′Aïn Temouchent, Alg. (ä′ĕntĕ-mōō-shaɴ′)	**198-99**	35°18′N	1°09′W
Aipe, Col. (ī′pĕ)	**163c**	3°13′N	75°14′W
Aïr, Massif de l′, mts., Niger	**258-59**	18°0′N	8°30′E
Air Force Island, i., Nu., Can.	**130-31**	67°55′N	74°10′W
Aïssa, Djebel, mtn., Alg.	**258-59**	32°51′N	0°30′W
Aitape, Pap. N. Gui. (ä-ê-tä′pā)	**277**	3°09′S	142°20′E
Aitkin, Mn., U.S. (āt′kĭn)	**114-15**	46°32′N	93°43′W
Aitutaki, at., Cook Is. (ī-tōō-tä′kē)	**280-81**	18°52′S	159°45′W
Aiud, Rom. (ä′ê-ōd)	**194-95**	46°19′N	23°44′E
Aix-en-Provence, Fr. (ĕks-prô-văɴs)	**196-97**	43°32′N	5°27′E
Aix-la-Chapelle, Ger. *see* Aachen	**194-95**	50°46′N	6°06′E
Aix-les-Bains, Fr. (ĕks′-lā-bɑɴ′)	**196-97**	45°42′N	5°55′E
Āīzawl, India	**234-35**	23°44′N	92°43′E
Aizu-wakamatsu, Japan	**245**	37°30′N	139°56′E
Ajaccio, Fr. (ä-yät′chō)	**184-85**	41°56′N	8°44′E
Ajdābiyā, Libya	**188-89**	30°45′N	20°14′E
Ajjer, Tassili-n-, plat., Alg.	**258-59**	25°41′N	7°29′E
′Ajmān, U.A.E.	**230-31**	25°24′N	55°28′E
Ajmer, India (ŭj-mēr′)	**234-35**	26°27′N	74°38′E
Ajo, Az., U.S. (ä′hō)	**118-19**	32°23′N	112°52′W
Akagera, stm., Afr. *see* Kagera	**267**	0°56′S	31°47′E
Akan-kokuritsu-kōen, n.p., Japan	**244**	43°30′N	144°15′E
Akashi, Japan (ä′kä-shē)	**245**	34°39′N	134°59′E
Akdeniz, s., *see* Mediterranean Sea	**188-89**	35°0′N	20°00′E
Aketi, D.R.C. (ä-kå-tē)	**262-63**	2°45′N	23°46′E
Akhaltsikhe, Geor. (äkä′l-tsĭ-kĕ)	**227**	41°38′N	42°59′E
Akhisar, Tur. (äk-hĭs-sär′)	**200-01**	38°56′N	27°50′E
Akhtuba, stm., Russia	**186-87**	46°40′N	48°08′E
Akhtubinsk, Russia	**186-87**	48°16′N	46°10′E
Akimiski Island, i., Nu., Can. (ä-kĭ-mĭ′skĭ ī′l*ă*nd)	**130-31**	53°0′N	81°20′W
Akita, Japan (ä′kê-tä)	**244**	39°43′N	140°07′E
Akkerman, Ukr. *see* Bilhorod-Dnistrovs′kyi	**202-03**	46°12′N	30°18′E
′Akko, Isr.	**228-29**	32°55′N	35°06′E
Aklavik, N.T., Can. (äk′lä-vĭk)	**128-29**	68°15′N	135°06′W
Ākobo, stm., Afr.	**262-63**	7°47′N	33°03′E
Akola, India (á-kō′lä)	**234-35**	20°43′N	77°00′E
Akordat, Erit.	**266**	15°32′N	37°53′E
Akpatok Island, i., Nu., Can. (ák′pȧ-tŏk ī′l*ă*nd))	**130-31**	60°25′N	67°60′W
Akron, Co., U.S. (ăk′r*ŭ*n)	**120-21**	40°10′N	103°13′W
Akron, Oh., U.S. (ăk′r*ŭ*n)	**116-17**	41°05′N	81°31′W
Aksaray, Tur. (äk-sá-rī′)	**186-87**	38°23′N	34°03′E
Akşehir, Tur. (äk′shå-hĕr)	**186-87**	38°21′N	31°25′E
Akşehir Gölü, lk., Tur. (äk′shå-hĕr)	**186-87**	38°32′N	31°27′E
Aksu, China (ä-kŭ-sōō)	**226**	41°08′N	80°15′E
Āksum, Eth.	**266**	14°08′N	38°43′E
Aktyubinsk, Kaz. *see* Aqtöbe	**226**	50°18′N	57°10′E
Akūbū, stm., Afr.	**262-63**	7°47′N	33°03′E
Akune, Japan (ä′kò-nå)	**245**	32°01′N	130°12′E
Akure, Nig.	**260a**	7°15′N	5°11′E
Akureyri, Ice. (ä-kò-rå′rê)	**190a**	65°39′N	18°07′W
Akyab, Mya. *see* Sittwe	**246-47**	20°09′N	92°54′E
Al-′Amārah, Iraq	**228-29**	31°50′N	47°09′E
Al-′Aqabah, Jord.	**228-29**	29°32′N	35°01′E
Al-′Arabīyah as-Su′ūdīyah, nation, Asia *see* Saudi Arabia	**206-07**	25°0′N	45°00′E
Al-′Ayn, U.A.E.	**230-31**	24°13′N	55°45′E
Al-′Azīzīyah, Libya	**188-89**	32°32′N	13°01′E
Al-′Irāq, nation, Asia *see* Iraq	**206-07**	33°0′N	44°00′E
Al-′Uqaylah, Libya	**188-89**	30°15′N	19°12′E
Alabama, state, U.S. (ăl-*á*-băm′*á*)	**108-09**	32°50′N	87°00′W
Alabama, stm., Al., U.S. (ăl-*á*-băm′*á*)	**124-25**	31°08′N	87°57′W
Alabat Island, i., Phil. (ä-lä-bät′ ī′l*ă*nd)	**250**	14°07′N	122°03′E
Alacant, Spain	**198-99**	38°21′N	0°30′W
Alagoa Grande, Braz.	**163d**	7°03′S	35°38′W
Alagoas, state, Braz. (ä-lä-gō′äzh)	**163d**	9°0′S	36°00′W
Alagoinhas, Braz. (ä-lä-gō-ēn′yäzh)	**166-67**	12°08′S	38°25′W
Alagón, stm., Spain (ä-lä-gōn′)	**198-99**	39°45′N	6°52′W
Alajuela, C.R. (ä-lä-hwa′lä)	**149**	10°01′N	84°13′W
Alajuela, Lago, res., Pan. (lä′gô-ä-lä-hwa′lä)	**150**	9°15′N	79°35′W
Alaköl köli, lk., Kaz.	**226**	46°10′N	81°45′E
Alamagan, i., N. Mar. Is.	**280-81**	17°36′N	145°50′E
Alamein, Egypt *see* El-Alamein	**188-89**	30°49′N	28°58′E
Alamo, Nv., U.S. (ă′lá-mō)	**118-19**	37°22′N	115°10′W
Alamogordo, N.M., U.S. (ăl-*á*-m*á*-gôr′dō)	**120-21**	32°54′N	105°57′W
Alamosa, Co., U.S. (ăl-*á*-mō′s*á*)	**118-19**	37°28′N	105°52′W
Aland Islands, is., Fin. (ô′länd ī′l*ă*ndz)	**192-93**	60°14′N	19°46′E
Alanya, Tur.	**228-29**	36°33′N	32°01′E
Alaotra, Farihy, lk., Madag. (ä-lä-ō′tr*á*)	**264-65**	17°25′S	48°33′E
Alappuzha, India	**236**	9°29′N	76°20′E
Alashanyouqi, China	**240-41**	40°04′N	103°33′E
Alaska, state, U.S. (á-lăs′k*á*)	**108-09**	65°0′N	153°00′W
Alaska, Gulf of, b., Ak., U.S. (gŭlf ŭv á-lăs′k*á*)	**126**	58°0′N	146°00′W
Alaska Peninsula, pen., Ak., U.S. (á-lăs′k*á* pĕ-nĭn′sūl*ă*)	**126**	57°0′N	158°00′W
Alaska Range, mts., Ak., U.S. (á-lăs′k*á* rānj)	**126**	63°26′N	149°07′W
Alataw Shan, mts., Asia	**226**	45°0′N	81°00′E
Alatyr′, Russia (ä′lä-tür)	**186-87**	54°51′N	46°34′E
Alausí, Ec.	**170**	2°13′S	78°51′W
Alayskiy khrebet, mts., Kyrg.	**226**	39°51′N	72°07′E

ăt; fin*ă*l; rāte; senåte; ärm; åsk; sof*á*; fâre; ch-choose; dh-as th in other; bē; ēvent; bĕt; recĕnt; cratẽr; g-gō; gh-guttural g; bĭt; ĭ-short neutral; rīde; ĸ-guttural k as ch in German ich;

Feature (Pronunciation)	Page	Lat.	Long.
Alazeya, stm., Russia	218-19	70°51'N	153°39'E
Alba, Italy (äl'bä)	200-01	44°42'N	8°02'E
Albacete, Spain (äl-bä-thā'tā)	198-99	38°59'N	1°52'W
Al-Baḥrayn, nation, Asia *see* Bahrain	230-31	26°0'N	50°30'E
Albania, nation, Eur. (äl-bā'nĭ-*ȧ*)	174-75	41°0'N	20°00'E
Albano Laziale, Italy (äl-bä'nō lät-zē-ä'lä)	200-01	41°44'N	12°39'E
Albany, Austl. (ôl'b*ȧ*-nĭ)	270-71	35°01'S	117°53'E
Albany, Ga., U.S. (ôl'b*ȧ*-nĭ)	124-25	31°34'N	84°09'W
Albany, Ky., U.S. (ôl'b*ȧ*-nĭ)	124-25	36°41'N	85°08'W
Albany, Mo., U.S. (ôl'b*ȧ*-nĭ)	120-21	40°15'N	94°20'W
Albany, N.Y., U.S. (ôl'b*ȧ*-nĭ)	116-17	42°40'N	73°47'W
Albany, Or., U.S. (ôl'b*ȧ*-nĭ)	112-13	44°38'N	123°05'W
Albany, Tx., U.S. (ôl'b*ȧ*-nĭ)	120-21	32°44'N	99°17'W
Albany, stm., On., Can. (ôl'b*ȧ*-nĭ)	130-31	52°17'N	81°32'W
Al-Baṣrah, Iraq *see* Basra	228-29	30°30'N	47°48'E
Al-Batrūn, Leb. (äl-bä-trōōn')	228-29	34°15'N	35°40'E
Al-Bayḍā', Libya	188-89	32°45'N	21°37'E
Albemarle, N.C., U.S. (ăl'bē-märl)	124-25	35°14'N	80°12'W
Albemarle Island, i., Ec. *see* Isabela, Isla	170a	0°30'S	91°06'W
Albemarle Sound, strt., N.C., U.S. (ăl'bē-märl sound)	124-25	36°03'N	76°12'W
Albenga, Italy (äl-bĕŋ'gä)	200-01	44°03'N	8°13'E
Alberga Creek, stm., Austl. (ăl-bŭr'g*ȧ* krēk)	272-73	27°07'S	135°30'E
Albert, Fr. (ȧl-bâr')	196-97	49°60'N	2°39'E
Albert, Lac, lk., Afr. (läk ȧl-bâr') *see* Albert, Lake	267	1°40'N	31°00'E
Albert, Lake, lk., Afr. (lāk *ă*l'bērt)	267	1°40'N	31°00'E
Alberta, state, Can. (ăl-bûr'*tȧ*)	128-29	54°0'N	113°00'W
Alberta, Mount, mtn., Ab., Can. (mount ăl-bûr'*tȧ*)	132-33	52°18'N	117°28'W
Albert Edward, Mount, mtn., Pap. N. Gui. (mount ăl'bērt ĕd'wērd)	277	8°24'S	147°22'E
Albert Lea, Mn., U.S. (ăl'bērt lē')	114-15	43°39'N	93°22'W
Albert Nile, stm., Ug. (ȧl-bâr' nĭl)	267	3°36'N	32°02'E
Alberton, P.E., Can. (ăl'bēr-t*ŭ*n)	138-39	46°49'N	64°04'W
Albertville, D.R.C. *see* Kalemie	267	5°55'S	29°11'E
Albertville, Fr. (ȧl-bâr-vēl')	196-97	45°40'N	6°23'E
Albertville, Al., U.S. (ăl'bērt-vĭl)	124-25	34°16'N	86°13'W
Albi, Fr. (äl-bē')	196-97	43°55'N	2°08'E
Albia, Ia., U.S. (ăl-bĭ-ȧ)	114-15	41°02'N	92°48'W
Albion, Il., U.S. (ăl'bĭ-*ŭ*n)	116-17	38°22'N	88°04'W
Albion, In., U.S. (ăl'bĭ-*ŭ*n)	116-17	41°23'N	85°24'W
Albion, Mi., U.S. (ăl'bĭ-*ŭ*n)	116-17	42°15'N	84°45'W
Albion, Ne., U.S. (ăl'bĭ-*ŭ*n)	114-15	41°41'N	98°00'W
Al-Birkah, Libya	258-59	24°52'N	10°12'E
Alborán, Isla de, i., Spain (ē's-lä-dāl-äl-bō-rä'n)	198-99	35°57'N	3°02'W
Alborz, Reshteh-ye Kūhhā-ye, mts., Iran *see* Elburz Mountains	232-33	36°0'N	53°00'E
Albuquerque, N.M., U.S. (ăl-bů-kûr'kē)	118-19	35°05'N	106°38'W
Albury, Austl. (ôl'bēr-ē)	276	36°04'S	146°56'E
Alcalá de Henares, Spain (äl-kä-lä' dā ā-na'rās)	198-99	40°29'N	3°22'W
Alcalá la Real, Spain (äl-kä-lä'lä rā-äl')	198-99	37°28'N	3°56'W
Alcamo, Italy (äl'kȧ-mō)	200-01	37°59'N	12°58'E
Alcanar, Spain (äl-kä-när')	198-99	40°33'N	0°29'E
Alcañiz, Spain (äl-kän-yēth')	198-99	41°03'N	0°08'W
Alcântara, Braz. (äl-kän'tȧ-rȧ)	166-67	2°24'S	44°24'W
Alcázar de San Juan, Spain (äl-kä'thär dā sän hwän')	198-99	39°23'N	3°12'W
Alcazarquivir, Mor. *see* Er-Rachidia	188-89	31°57'N	4°26'W
Alcazarquivir, Mor. *see* Ksar-el-Kebir	269a	35°01'N	5°54'W
Alcira, Spain (ä-thē'rä) *see* Alzira	198-99	39°09'N	0°26'W
Alcobaça, Braz.	172	17°31'S	39°13'W
Alcobendas, Spain (äl-kō-bĕn'dȧs)	198-99	40°33'N	3°38'W
Alcoi, Spain	198-99	38°42'N	0°28'W
Alcoy, Spain *see* Alcoi	198-99	38°42'N	0°28'W
Aldabra, Groupe d', is., Sey. (grūp-dāl-dä'brä)	264-65	9°24'S	46°27'E
Aldama, Mex. (äl-dä'mä)	146-47	22°55'N	98°04'W
Aldama, Mex. (äl-dä'mä)	122-23	28°51'N	105°54'W
Aldan, Russia	218-19	58°36'N	125°24'E
Aldan, stm., Russia	218-19	63°26'N	129°26'E
Aldan Plateau, plat., Russia (ŭl-dän') *see* Aldanskoye Nagor'ye	218-19	57°0'N	127°00'E
Aldanskoye Nagor'ye, plat., Russia	218-19	57°0'N	127°00'E
Alderney, i., Guern. (ôl'dēr-nĭ)	196-97	49°43'N	2°13'W
Aldershot, Eng., U.K. (ôl'dēr-shŏt)	190-91	51°15'N	0°46'W
Aledo, Il., U.S. (*ȧ*-le'dō)	114-15	41°12'N	90°45'W
Alegranza, Isla, i., Spain	199d	29°24'N	13°30'W
Alegrete, Braz. (ä-lå-grā'tä)	173	29°47'S	55°47'W
Aleksandrov, Russia (ä-lyĕk-sän' drôf)	202-03	56°24'N	38°43'E
Aleksandrov-Gay, Russia	186-87	50°08'N	48°34'E
Aleksandrovsk-Sakhalinskiy, Russia	218-19	50°54'N	142°10'E
Aleksandrów Kujawski, Pol. (ä-lĕk-säh'drōōv kōō-yav'skē)	194-95	52°52'N	18°43'E
Alekseevka, Kaz.	226	52°00'N	70°57'E
Alekseyevka, Russia (ä-lyĕk-sā-yĕf'k*ȧ*)	202-03	50°38'N	38°41'E
Aleksin, Russia (ăb'ĭng-t*ŭ*n)	202-03	54°30'N	37°05'E
Além Paraíba, Braz. (ä-lĕ'm-pȧ-räē'bȧ)	172	21°52'S	42°40'W
Alençon, Fr. (ȧ-län-sôɴ')	196-97	48°26'N	0°05'E
Alenquer, Braz. (ä-lĕŋ-kĕr')	166-67	1°56'S	54°46'W
Alep, Syria *see* Aleppo	228-29	36°13'N	37°10'E
Aleppo, Syria (*ȧ*-lĕp-ō)	228-29	36°13'N	37°10'E
Alès, Fr. (ä-lĕs')	196-97	44°08'N	4°05'E
Alessandria, Italy (ä-lĕs-sän'drē-ä)	200-01	44°55'N	8°37'E
Ålesund, Nor. (ô'lĕ-sòn')	184-85	62°28'N	6°10'E
Aleutian Islands, is., Ak., U.S. (*ȧ*-lu'sh*ă*n ī'l*ȧ*ndz)	126a	52°0'N	176°00'W
Alexander City, Al., U.S. (äl-ĕg-zăn'dēr sĭ'tē)	124-25	32°57'N	85°57'W
Alexandra, N.Z.	278	45°15'S	169°23'E
Alexandretta, Tur. *see* İskenderun	228-29	36°35'N	36°11'E
Alexandretta, Gulf of, b., Tur. *see* İskenderun Körfezi	228-29	36°30'N	35°40'E
Alexandria, On., Can. (ăl-ĕg-zăn'drĭ-*ȧ*)	136-37	45°18'N	74°38'W
Alexandria, Egypt (ăl-ĕg-zăn'drĭ-*ȧ*)	268b	31°11'N	29°54'E
Alexandria, Rom. (ăl-ĕg-zăn'drĭ-*ȧ*)	200-01	43°59'N	25°21'E
Alexandria, In., U.S. (ăl-ĕg-zăn'drĭ-*ȧ*)	116-17	40°15'N	85°40'W
Alexandria, La., U.S. (ăl-ĕg-zăn'drĭ-*ȧ*)	122-23	31°18'N	92°27'W
Alexandria, Mn., U.S. (ăl-ĕg-zăn'drĭ-*ȧ*)	114-15	45°53'N	95°23'W
Alexandria, S.D., U.S. (ăl-ĕg-zăn'drĭ-*ȧ*)	114-15	43°39'N	97°47'W
Alexandria, Va., U.S. (ăl-ĕg-zăn'drĭ-*ȧ*)	116-17	38°48'N	77°03'W
Alexandria Bay, N.Y., U.S. (ăl-ĕg-zăn'drĭ-*ȧ* bā)	116-17	44°20'N	75°55'W
Alexandrina, Lake, lk., Austl.	276	35°26'S	139°10'E
Alexandroúpoli, Grc.	200-01	40°52'N	25°53'E
Aleysk, Russia	226	52°29'N	82°46'E
Alfaro, Spain (äl-färō)	198-99	42°11'N	1°45'W
Al-Fāshir, Sudan (äl-fä'shēr)	266	13°38'N	25°21'E
Alfenas, Braz. (äl-fĕ'nȧs)	172	21°27'S	45°57'W
Al-Furāt, stm., Asia *see* Euphrates	208-09	30°60'N	47°27'E
Algeciras, Spain (äl-hä-thē'räs)	198-99	36°08'N	5°27'W
Alger, nat. cap., Alg. *see* Algiers	269b	36°46'N	3°03'E
Algeria, nation, Afr. (ăl-gē'rĭ-*ȧ*)	253	28°0'N	3°00'E
Algérie, nation, Afr. *see* Algeria	253	28°0'N	3°00'E
Al-Ghayḍah, Yemen	220-21	16°13'N	52°12'E
Alghero, Italy (äl-gâ'rō)	200-01	40°34'N	8°19'E
Algiers, nat. cap., Alg. (ăl-jērs)	269b	36°46'N	3°03'E
Al-Ḥamād, pl., Sau. Ar.	228-29	32°0'N	39°30'E
Al-Ḥasakah, Syria	228-29	36°30'N	40°46'E
Al-Ḥawrah, Yemen	220-21	13°50'N	47°34'E
Al-Ḥayy, Iraq	228-29	32°10'N	46°03'E
Al-Ḥijāz, reg., Sau. Ar.	266	24°30'N	38°30'E
Al-Ḥillah, Iraq	228-29	32°29'N	44°26'E
Al-Hoceima, Mor.	198-99	35°15'N	3°56'W
Al-Ḥudaydah, Yemen	266	14°48'N	42°57'E
Al-Hufūf, Sau. Ar.	230-31	25°22'N	49°34'E
Alicante, Spain *see* Alacant	198-99	38°21'N	0°30'W
Alice, Tx., U.S. (ăl'ĭs)	122-23	27°45'N	98°05'W
Alice Springs, Austl. (ăl'ĭs sprĭngz)	270-71	23°42'S	133°52'E
Alīgarh, India (ä-lē-gŭr')	234-35	27°54'N	78°04'E
Alima, stm., Congo	262-63	1°31'S	16°40'E
Al-Imārāt al-'Arabīyah al-Muttaḥidah, nation, Asia *see* United Arab Emirates	206-07	24°0'N	54°00'E
Alingsås, Swe. (ȧ'lĭŋ-sôs)	192-93	57°56'N	12°32'E
'Ali Sabieh, Dji.	266	11°08'N	42°42'E
Aliwal North, S. Afr. (ä-lē-wäl' nôrth)	264-65	30°42'S	26°43'E
Al-Jawf, Libya	258-59	24°12'N	23°17'E
Al-Jawf, Sau. Ar.	228-29	29°48'N	39°52'E
Al-Jazaïr, nat. cap., Alg. *see* Algiers	269b	36°46'N	3°03'E
Al-Jazīrah, reg., Sudan	266	14°17'N	32°53'E
Aljezur, Port. (äl-zhä-zōōr')	198-99	37°18'N	8°48'W
Al-Jubayl, Sau. Ar.	230-31	27°01'N	49°40'E
Al-Jufrah, well, Libya	258-59	29°06'N	15°57'E
Al-Junaynah, Sudan	262-63	13°27'N	22°27'E
Al-Karak, Jord. (äl-kĕ-räk')	228-29	31°11'N	35°42'E
Al-Khābūrah, Oman	230-31	23°58'N	57°06'E
Al-Khalīl, W.B. *see* Hebron	228-29	31°32'N	35°06'E
Al-Kharṭūm, nat. cap., Sudan *see* Khartoum	266	15°35'N	32°32'E
Al-Khaṣab, Oman	230-31	26°12'N	56°15'E
Al-Khubar, Sau. Ar.	230-31	26°17'N	50°12'E
Al-Khums, Libya	188-89	32°39'N	14°16'E
Alkmaar, Neth. (älk-mär')	190-91	52°38'N	4°45'E
Al-Kūt, Iraq	228-29	32°30'N	45°49'E
Al-Kuwayt, nation, Asia *see* Kuwait	206-07	29°30'N	47°45'E
Al-Kuwayt, nat. cap., Kuw. (äl-kōō-wit) *see* Kuwait	228-29	29°19'N	47°60'E
Al-Lādhiqīyah, Syria *see* Latakia	228-29	35°31'N	35°48'E
Allahābād, India (ŭl-ŭ-hä-bäd')	234-35	25°26'N	81°51'E
Allakaket, Ak., U.S.	126	66°33'N	152°38'W
'Allāq, Bi'r, well, Libya	258-59	31°05'N	11°58'E
Allegan, Mi., U.S. (ăl'ē-găn)	116-17	42°32'N	85°51'W
Allegheny Plateau, plat., U.S. (ăl-ē-gā'nĭ plä-tō')	116-17	41°30'N	78°00'W
Allendale, S.C., U.S. (ăl'ĕn-dāl)	124-25	33°00'N	81°19'W
Allende, Mex. (äl-yĕn'dȧ)	122-23	28°20'N	100°50'W
Allende, Mex. (äl-yĕn'dȧ)	122-23	25°17'N	100°01'W
Allenstein, Pol. *see* Olsztyn	194-95	53°47'N	20°29'E
Allentown, Pa., U.S.	116-17	40°37'N	75°29'W
Alleppey, India (ȧ-lĕp'ē) *see* Alappuzha	236	9°29'N	76°20'E
Aller, stm., Ger. (äl'ēr)	194-95	52°57'N	9°11'E
Alliance, Ne., U.S. (*ȧ*-lī'*ă*ns)	114-15	42°06'N	102°52'W
Alliance, Oh., U.S. (*ȧ*-lī'*ă*ns)	116-17	40°55'N	81°06'W
Allier, stm., Fr. (ȧ-lyā')	196-97	46°58'N	3°04'E
Allinge, Den. (äl'ĭŋ-*ĕ*)	194-95	55°16'N	14°48'E
Al-Lubnān, nation, Asia *see* Lebanon	206-07	34°0'N	36°00'E
Alma, N.B., Can. (ăl'm*ȧ*)	138-39	45°36'N	64°57'W
Alma, Qc., Can. (ăl'm*ȧ*)	136-37	48°33'N	71°39'W
Alma, Ga., U.S. (ăl'm*ȧ*)	124-25	31°33'N	82°28'W
Alma, Mi., U.S. (ăl'm*ȧ*)	116-17	43°23'N	84°39'W
Alma, Ne., U.S. (ăl'm*ȧ*)	120-21	40°06'N	99°21'W
Alma, Wi., U.S. (ăl'm*ȧ*)	114-15	44°20'N	91°54'W
Almadén, Spain (äl-mȧ-dhān)	198-99	38°46'N	4°50'W
Al-Madīnah, Sau. Ar. *see* Medina	266	24°28'N	39°37'E
Al-Maghrib, nation, Afr. *see* Morocco	253	32°0'N	5°00'W
Almagro, Spain (äl-mä'grō)	198-99	38°53'N	3°43'W
Al-Majma'ah, Sau. Ar.	232-33	25°55'N	45°21'E
Al-Makhā', Yemen *see* Mocha	266	13°19'N	43°15'E
Al-Manāmah, nat. cap., Bahr. (äl-mä-nä'm*ȧ*)	230-31	26°13'N	50°35'E
Almansa, Spain (äl-män'sä)	198-99	38°52'N	1°06'W
Al-Marj, Libya	188-89	32°30'N	20°53'E
Almas, Pico das, mtn., Braz.	166-67	13°33'S	41°56'W
Almaty, Kaz.	226	43°17'N	76°56'E
Al-Mawṣil, Iraq *see* Mosul	228-29	36°20'N	43°08'E
Almazán, Spain (äl-mä-thän')	198-99	41°29'N	2°32'W
Almelo, Neth. (äl'mē-lō)	190-91	52°22'N	6°39'E
Almenara, Braz.	168-69	16°12'S	40°41'W
Almendralejo, Spain (äl-män-drä-lā'hō)	198-99	38°41'N	6°24'W
Almería, Spain (äl-mā-rē'ä)	198-99	36°51'N	2°27'W
Almería, Golfo de, b., Spain (gôl-fô-dĕ-äl-māī-rɛɴ')	198-99	36°46'N	2°30'W
Al'met'yevsk, Russia	186-87	54°54'N	52°19'E
Älmhult, Swe. (ĕlm'hōōlt)	192-93	56°33'N	14°09'E
Almirante, Pan. (äl-mē-rän'tä)	150	9°17'N	82°24'W
Almonte, On., Can. (ăl-mŏn'tē)	136-37	45°13'N	76°11'W
Almora, India	234-35	29°36'N	79°40'E
Al-Mubarraz, Sau. Ar.	230-31	25°25'N	49°35'E
Al-Muḥarraq, Bahr.	230-31	26°16'N	50°37'E
Al-Mukallā, Yemen	220-21	14°32'N	49°08'E
Almuñécar, Spain (äl-mōōn-yā'kär)	198-99	36°45'N	3°41'W
Alnön, i., Swe.	192-93	62°25'N	17°26'E
Alor, Pulau, i., Indon. (pōō-lou ä'lôr)	248-49	8°15'S	124°45'E
Alor Setar, Malay. (ä'lôr stär)	246-47	6°07'N	100°23'E
Alpen, mts., Eur. *see* Alps	184-85	46°25'N	10°00'E
Alpena, Mi., U.S. (ăl-pē'n*ȧ*)	116-17	45°03'N	83°26'W
Alpes, mts., Eur. *see* Alps	184-85	46°25'N	10°00'E
Alpi, mts., Eur. *see* Alps	184-85	46°25'N	10°00'E
Alpine, Az., U.S. (ăl'pīn)	118-19	33°50'N	109°08'W
Alpine, Tx., U.S. (ăl'pīn)	122-23	30°21'N	103°40'W
Alpine National Park, n.p., Austl.	276	36°57'S	147°12'E
Alps, mts., Eur. (ălps)	194-95	46°25'N	10°00'E
Al-Qaḍārif, Sudan	266	14°02'N	35°23'E
Al-Qaṭīf, Sau. Ar.	230-31	26°33'N	50°00'E
Al-Quds, nat. cap., Isr. *see* Jerusalem	228-29	31°47'N	35°14'E
Als, i., Den. (äls)	194-95	54°59'N	9°55'E
Alsace, hist. reg., Fr. (äl-sȧ's)	196-97	48°30'N	7°30'E
Alta Gracia, Arg. (äl'tä grä'sē-ä)	173	31°40'S	64°26'W
Altagracia, Ven.	164-65	10°43'N	71°30'W
Altamaha, stm., Ga., U.S. (ôl-tȧ-mȧ-hô')	124-25	31°19'N	81°18'W
Altamira, Braz. (äl-tä-mē'rä)	166-67	3°11'S	52°14'W
Altamira, Chile	168-69	25°48'S	69°51'W
Altamura, Italy (äl-tä-mōō'rä)	200-01	40°50'N	16°33'E
Altavista, Va., U.S. (äl-tä-vĭs'tȧ)	124-25	37°07'N	79°18'W

n-sing; ŋ-baŋk; ɴ-nasalized n; nŏd; c*ŏ*mmit; ōld; ôbey; ôrder; oi-boil; fōōd; ò-as oo in foot; ou-out; s-soft; sh-dish; th-thin; pūre; ůnite; ûrn; stŭd; circ*ŭ*s; ü-as in French tu; '-indeterminate vowel.

Feature (Pronunciation)	Page	Lat.	Long.
Altay, Mong.	240-41	46°24′N	96°15′E
Altay, Mong.	240-41	49°42′N	96°24′E
Altay, state, Russia	226	51°0′N	86°00′E
Altay, mts., Asia *see* Altay Mountains	222-23	48°0′N	90°00′E
Altay Mountains, mts., Asia (äl′tī′ moun′tĭnz)	222-23	48°0′N	90°00′E
Altay Shan, mts., Asia *see* Altay Mountains	222-23	48°0′N	90°00′E
Altenburg, Ger. (äl-tĕn-bo͝orgh)	194-95	50°59′N	12°26′E
Altiplano, plat., S.A. (äl-tē-plá′nō)	168-69	18°0′S	68°00′W
Alto Araguaia, Braz.	168-69	17°19′S	53°13′W
Alton, Il., U.S. (ôl′tŭn)	120-21	38°55′N	90°12′W
Altona, Mb., Can.	134-35	49°06′N	97°35′W
Altoona, Ia., U.S. (ăl-to͞o′nȧ)	114-15	41°38′N	93°28′W
Altoona, Pa., U.S. (ăl-to͞o′nȧ)	116-17	40°29′N	78°24′W
Altoona, Wi., U.S. (ăl-to͞o′nȧ)	114-15	44°48′N	91°26′W
Alto Parnaíba, Braz.	166-67	9°07′S	45°57′W
Alto Río Senguer, Arg.	171	45°03′S	70°51′W
Altun Shan, mts., China (äl-tòn shän)	222-23	38°0′N	88°00′E
Alturas, Ca., U.S. (ăl-to͞o′rȧs)	112-13	41°30′N	120°32′W
Altus, Ok., U.S. (ăl′tŭs)	120-21	34°38′N	99°20′W
Alu, i., Sol. Is. *see* Shortland Island	279e	7°04′S	155°43′E
Al-Ubayyiḍ, Sudan	266	13°11′N	30°13′E
Alūksne, Lat. (ä′lòks-nĕ)	192-93	57°25′N	27°04′E
Alula, Som. *see* Caluula	262-63	11°57′N	50°46′E
Al-Urdun, nation, Asia *see* Jordan	206-07	31°0′N	36°00′E
Al-Urdunn, stm., Asia *see* Jordan	228-29	31°46′N	35°34′E
Alushta, Ukr. (ä′lsho-tȧ)	202-03	44°41′N	34°24′E
Alva, Ok., U.S. (ăl′vȧ)	120-21	36°48′N	98°40′W
Alvarado, Mex. (äl-vä-rä′dhō)	146-47	18°46′N	95°46′W
Älvdalen, Swe. (ĕlv′dä-lĕn)	192-93	61°14′N	14°02′E
Alvear, Arg.	173	29°03′S	56°33′W
Alvesta, Swe. (äl-vĕs′tä)	192-93	56°54′N	14°33′E
Alwar, India (ŭl′wŭr)	234-35	27°34′N	76°37′E
Alxa Zuoqi, China	240-41	38°49′N	105°35′E
Al-Yaman, nation, Asia *see* Yemen	206-07	15°0′N	44°00′E
Alytus, Lith. (ä′lē-tòs)	194-95	54°24′N	24°04′E
Alzira, Spain	198-99	39°09′N	0°26′W
Amadjuak Lake, lk., Nu., Can. (ä-mädj′wäk lāk)	130-31	65°0′N	71°00′W
Amahai, Indon.	248-49	3°20′S	128°56′E
Amakuso-Shimo-shima, i., Japan (ämä-ko͞o′sä shē-mō shē′mä)	245	32°20′N	130°05′E
Åmål, Swe. (ô′môl)	192-93	59°03′N	12°42′E
Amalfi, Col. (ä-mä′l-fē)	164-65	6°55′N	75°05′W
Amambaí, Braz.	168-69	23°07′S	55°13′W
Amami-Ō-shima, i., Japan	244a	28°15′N	129°20′E
Amami-shotō, is., Japan	244a	27°58′N	129°02′E
Amapá, Braz.	166-67	2°02′N	50°46′W
Amapá, state, Braz.	166-67	1°0′N	52°00′W
Amarante, Braz. (ä-mä-rän′tä)	166-67	6°14′S	42°50′W
Amarillo, Tx., U.S. (ăm-ȧ-rĭl′ō)	120-21	35°13′N	101°50′W
Amarkantak, India	234-35	22°40′N	81°46′E
Amaro, Monte, mtn., Italy (môn-tĕ ä-mä′rō)	200-01	42°05′N	14°06′E
Amasya, Tur. (ä-mä′sē-à)	186-87	40°39′N	35°50′E
Amazon, stm., S.A. (ă′mȧ-zŏn)	164-65	0°04′S	49°15′W
Amazonas, state, Braz. (ä-mä-thō′näs)	166-67	5°0′S	63°00′W
Amazonas, stm., S.A. (ä-mä-thō′näs) *see* Amazon	164-65	0°04′S	49°15′W
Ambāla, India (ŭm-bä′lŭ)	234-35	30°21′N	76°49′E
Ambanja, Madag.	264-65	13°41′S	48°27′E
Ambargasta, Salinas de, pl., Arg.	168-69	29°15′S	64°30′W
Ambato, Ec. (äm-bä′tō)	170	1°15′S	78°37′W
Ambatondrazaka, Madag.	264-65	17°52′S	48°24′E
Ambelau, Pulau, i., Indon.	248-49	3°51′S	127°12′E
Amberg, Ger. (äm′bĕrgh)	194-95	49°27′N	11°52′E
Ambergris Cay, i., Belize (ăm′bẽr-grēs kā)	148	18°03′N	87°55′W
Ambert, Fr. (äN-bĕr′)	196-97	45°33′N	3°45′E
Ambikāpur, India	234-35	23°07′N	83°12′E
Amboina, Indon. *see* Ambon	248-49	3°44′S	128°11′E
Amboise, Fr. (äN-bwăz′)	196-97	47°24′N	0°60′E
Ambon, Indon.	248-49	3°44′S	128°11′E
Ambon, Pulau, i., Indon.	248-49	3°40′S	128°05′E
Amboseli National Park, n.p., Kenya	267	2°36′S	37°12′E
Ambositra, Madag. (äN-bō-sē′trä)	264-65	20°32′S	47°15′E
Ambovombe, Madag.	264-65	25°11′S	46°05′E
Amboy, Il., U.S. (ăm′boi)	116-17	41°43′N	89°20′W
Ambre, Cap d', c., Madag. *see* Bobaomby, Tanjona	264-65	11°58′S	49°15′E
Ambridge, Pa., U.S. (ăm′brĭdj)	116-17	40°36′N	80°13′W
Ambriz, Ang.	262-63	7°51′S	13°10′E
Ambrym, i., Vanuatu	279g	16°15′S	168°10′E
Amchitka Pass, strt., Ak., U.S. (ăm-chĭt′kȧ pās)	126a	51°30′N	179°30′W
Amderma, Russia	186-87	69°45′N	61°39′E
Amdo, China	234-35	32°17′N	91°44′E
Ameca, Mex. (ä-mĕ′kä)	146-47	20°33′N	104°02′W
Amecameca de Juárez, Mex.	146-47	19°07′N	98°46′W
Ameland, i., Neth.	190-91	53°27′N	5°45′E
Amelia Island, i., Fl., U.S.	124-25	30°37′N	81°27′W
American Falls, Id., U.S. (ȧ-mĕr′ĭ-kăn fôlz)	112-13	42°47′N	112°51′W
American Falls Reservoir, res., Id., U.S. (ȧ-mĕr′ĭ-kăn fôlz rĕ′sẽr-vwär)	112-13	42°57′N	112°44′W
American Fork, Ut., U.S. (ȧ-mĕr′ĭ-kăn fôrk)	118-19	40°24′N	111°48′W
American Highland, plat., Ant. (ȧ-mĕr′ĭ-kăn)	287	72°30′S	78°00′E
Americanos, Barra de los, i., Mex.	122-23	24°53′N	97°35′W
American Samoa, dep., Oc. (ȧ-mĕr′ĭ-kăn sȧ-mō′ȧ)	279b	14°20′S	170°00′W
Americus, Ga., U.S. (ȧ-mĕr′ĭ-kŭs)	124-25	32°05′N	84°14′W
Amerika Samoa, dep., Oc. *see* American Samoa	279b	14°20′S	170°00′W
Amersfoort, Neth. (ä′mĕrz-fōrt)	190-91	52°10′N	5°24′E
Ames, Ia., U.S. (āmz)	114-15	42°01′N	93°37′W
Amesbury, Ma., U.S. (āmz′bẽr-ē)	116-17	42°52′N	70°56′W
Ámfissa, Grc. (àm-fĭ′sȧ)	200-01	38°32′N	22°23′E
Amga, Russia (ŭm-gä′)	218-19	60°54′N	131°58′E
Amga, stm., Russia (ŭm-gä′)	218-19	62°35′N	135°04′E
Amgun, stm., Russia	218-19	52°56′N	139°41′E
Amherst, N.S., Can. (ăm′hẽrst)	138-39	45°50′N	64°12′W
Amherst, Ma., U.S. (ăm′hẽrst)	116-17	42°23′N	72°31′W
Amherst, N.Y., U.S. (ăm′hẽrst)	116-17	42°58′N	78°47′W
Amherst, Va., U.S. (ăm′hẽrst)	116-17	37°35′N	79°04′W
Amiens, Fr. (ä-myăN′)	196-97	49°54′N	2°18′E
Amistad, Presa de la, res., N.A. *see* Amistad Reservoir	122-23	29°28′N	101°07′W
Amistad Reservoir, res., N.A.	122-23	29°28′N	101°07′W
Amite, La., U.S. (ă-mēt′)	124-25	30°43′N	90°31′W
Amite, stm., La., U.S. (ă-mēt′)	124-25	30°13′N	90°36′W
Amlia Island, i., Ak., U.S. (ȧ′mlēȧ ī′lȧnd)	126a	52°07′N	173°34′W
'Ammān, nat. cap., Jord. (äm′mȧn)	228-29	31°57′N	35°56′E
Amnok-kang, stm., Asia *see* Yalu	243	39°57′N	124°22′E
Āmol, Iran	232-33	36°28′N	52°21′E
Amorgós, i., Grc. (ä-môr′gōs)	200-01	36°52′N	25°56′E
Amory, Ms., U.S. (āmō-rē)	124-25	33°59′N	88°29′W
Amos, Qc., Can. (ā′mŭs)	136-37	48°34′N	78°08′W
Amoy, China *see* Xiamen	225a	24°27′N	118°07′E
Amposta, Spain (äm-pōs′tä)	198-99	40°44′N	0°35′E
Amraoti, India *see* Amrāvati	234-35	20°56′N	77°46′E
Amrāvati, India	234-35	20°56′N	77°46′E
Amreli, India	234-35	21°36′N	71°13′E
Amritsar, India (ŭm-rĭt′sŭr)	234-35	31°38′N	74°52′E
Amroha, India	234-35	28°54′N	78°28′E
Amsterdam, N.Y., U.S. (ăm′stẽr-dăm)	116-17	42°57′N	74°11′W
Amsterdam, nat. cap., Neth. (äm-stẽr-däm′)	190-91	52°22′N	4°54′E
Amstetten, Aus. (äm′stĕt-ĕn)	194-95	48°07′N	14°52′E
Am Timan, Chad (äm′tē-män′)	258-59	11°02′N	20°17′E
Amu Darya, stm., Asia (ä-mò-dä′rēä)	226	44°14′N	59°41′E
Āmū Daryā, stm., Asia *see* Amu Darya	226	44°14′N	59°41′E
Amukta Pass, strt., Ak., U.S. (ä-mo͞ok′tȧ pās)	126a	52°26′N	171°51′W
Amundsen Gulf, b., Can. (ä′mŭn-sĕn-gŭlf′)	86	71°0′N	124°00′W
Amundsen Sea, s., Ant. (ä′mŭn-sĕn sē)	287	72°30′S	112°00′W
Amuntai, Indon.	248-49	2°25′S	115°15′E
Amur, stm., Asia (ä-mo͞or′)	218-19	52°57′N	141°10′E
Anaa, at., Fr. Poly.	280-81	17°26′S	145°31′W
Anabar, stm., Russia (ăn-à-bär′)	218-19	73°13′N	113°32′E
Anaco, Ven. (ä-nä′kô)	163b	9°25′N	64°28′W
Anaconda, Mt., U.S. (ăn-ȧ-kŏn′dȧ)	112-13	46°08′N	112°58′W
Anacortes, Wa., U.S. (ăn-ȧ-kôr′tĕz)	112-13	48°30′N	122°37′W
Anadarko, Ok., U.S. (ăn-à-där′kō)	120-21	35°04′N	98°15′W
Anadyr', Russia (ŭ-nȧ-dĭr′)	218-19	64°44′N	177°30′E
Anadyr', stm., Russia (ŭ-nȧ-dĭr′)	218-19	64°52′N	176°15′E
Anadyr, Gulf of, b., Russia (gŭlf ŭv ä-nȧ-dyĭr′)	218-19	64°0′N	179°00′W
Anadyr Mountains, plat., Russia (ä-nȧ-dyĭr′ moun′tĭnz) *see* Anadyrskoye Ploskogor'ye	218-19	67°0′N	174°00′E
Anadyrskiy Zaliv, b., Russia *see* Anadyr, Gulf of	218-19	64°0′N	179°00′W
Anadyrskoye Ploskogor'ye, plat., Russia	218-19	67°0′N	174°00′E
Anaheim, Ca., U.S. (ăn′ȧ-hīm)	118-19	33°50′N	117°55′W
Ānai Mudi, mtn., India	236	10°10′N	77°04′E
Anaktuvuk Pass, Ak., U.S.	126	68°09′N	151°43′W
Anambas, Kepulauan, is., Indon. (ä-näm-bäs)	246-47	3°0′N	106°00′E
Anamosa, Ia., U.S. (ăn-ȧ-mō′sȧ)	114-15	42°06′N	91°16′W
Anamur, Tur.	228-29	36°04′N	32°50′E
Anantapur, India	236	14°41′N	77°36′E
Anantnāg, India	234-35	33°45′N	75°08′E
Anapa, Russia (à-nä′pä)	202-03	44°54′N	37°20′E
Anápolis, Braz. (ä-nä′pō-lēs)	168-69	16°21′S	48°57′W
Anatahan, i., N. Mar. Is.	280-81	16°22′N	145°40′E
Anatom, i., Vanuatu	279g	20°12′S	169°48′E
Añatuya, Arg. (à-nyä-to͞o′yá)	173	28°27′S	62°49′W
Anauá, stm., Braz.	166-67	0°58′N	61°22′W
Anbanjing, China	238-39	23°57′N	100°54′E
Anchiang, China *see* Qianyang	238-39	27°11′N	110°02′E
Anchorage, Ak., U.S. (ăŋ′kẽr-āj)	126	61°12′N	149°53′W
Ancona, Italy (än-kō′nä)	200-01	43°37′N	13°31′E
Ancud, Chile (äŋ-ko͞odh′)	171	41°53′S	73°49′W
Ancud, Golfo de, b., Chile (gôl-fô-dĕ-äŋ-ko͞odh′)	171	42°05′S	73°00′W
Anda, China	240-41	46°24′N	125°19′E
Andalgalá, Arg.	168-69	27°35′S	66°19′W
Andalucía, hist. reg., Spain (än-dä-lo͞o-sē′ä)	198-99	37°15′N	4°30′W
Andalusia, Al., U.S. (ăn-dȧ-lo͞o′zhĭȧ)	124-25	31°19′N	86°29′W
Andaman and Nicobar Islands, state, India	246-47	11°0′N	93°00′E
Andaman Islands, is., India (ăn-dà-măn′ ī′lȧndz)	246-47	12°0′N	92°45′E
Andaman Sea, s., Asia (ăn-dá-măn′ sē)	246-47	10°0′N	95°00′E
Anderson, Ca., U.S. (ăn′dẽr-sŭn)	112-13	40°29′N	122°22′W
Anderson, In., U.S. (ăn′dẽr-sŭn)	116-17	40°05′N	85°41′W
Anderson, S.C., U.S. (ăn′dẽr-sŭn)	124-25	34°30′N	82°39′W
Anderson, stm., N.T., Can. (ăn′dẽr-sŭn)	130-31	69°42′N	128°54′W
Andes, mts., S.A. (ăn′dēz) (än′dās)	159	20°0′S	67°00′W
Andhra Pradesh, state, India	236	16°0′N	79°00′E
Andijon, Uzb.	232-33	40°47′N	72°21′E
Andkhvoy, Afg.	232-33	36°55′N	65°07′E
Andong, Kor., S. (än′dŭng′)	243	36°34′N	128°43′E
Andorra, nation, Eur. (än-dŏr′rä)	196-97	42°30′N	1°30′E
Andorra la Vella, nat. cap., And.	198-99	42°30′N	1°31′E
Andover, Ma., U.S. (ăn′dō-vẽr)	114-15	45°14′N	93°17′W
Andøya, i., Nor. (änd-ûê)	184-85	69°08′N	15°54′E
Andradina, Braz.	168-69	20°55′S	51°23′W
Andrews, N.C., U.S. (ăn′dro͞oz)	124-25	35°12′N	83°49′W
Andrews, S.C., U.S. (ăn′dro͞oz)	124-25	33°27′N	79°34′W
Andrews, Tx., U.S. (ăn′dro͞oz)	120-21	32°19′N	102°33′W
Andria, Italy (än′drē-ä)	200-01	41°13′N	16°17′E
Andros, i., Bah. (ăn′drŏs)	142-43	24°26′N	77°57′W
Ándros, i., Grc. (àn′drŏs)	200-01	37°50′N	24°53′E
Anegada, i., Br. Vir. Is.	143b	18°45′N	64°20′W
Aneto, mtn., Spain (ä-nĕ′tō)	198-99	42°38′N	0°40′E
Angamos, Punta, c., Chile	168-69	23°02′S	70°31′W
Ang'angxi, China (äŋ-äŋ-shyē)	240-41	47°09′N	123°48′E
Angara, stm., Russia	218-19	58°06′N	93°02′E
Angarsk, Russia	222-23	52°35′N	103°55′E
Ángel, Salto, wtfl., Ven. (säl′tō-á′n-hĕl) *see* Angel Falls	164-65	6°01′N	62°28′W
Ángel de la Guarda, Isla, i., Mex. (ĕ′s-lä-á′n-hĕl-dĕ-lä-gwä′r-dä)	144-45	29°22′N	113°28′W
Angeles, Phil. (än′hå-lās)	250	15°08′N	120°36′E
Angel Falls, wtfl., Ven. (ān′jĕl fôlz)	164-65	6°01′N	62°28′W
Ängelholm, Swe. (ĕng′ĕl-hôlm)	192-93	56°15′N	12°53′E
Angers, Fr.	196-97	47°28′N	0°33′W
Angicos, Braz.	163d	5°40′S	36°36′W
Angijak Island, i., Nu., Can.	130-31	65°40′N	62°15′W
Angkor Wat, hist., Camb. (ăng′kôr)	246-47	13°26′N	103°52′E
Anglesey, i., Wales, U.K. (ăŋ′g'l-sē)	190-91	53°17′N	4°22′W
Angleton, Tx., U.S. (aŋ′g'l-tŭn)	122-23	29°10′N	95°26′W
Angmagssalik, Green. (àŋ-mà′sà-lĭk)	284-85	65°35′N	37°50′W
Angoche, Moz.	264-65	16°14′S	39°55′E
Angoche, Ilha, i., Moz. (ē′lä-ăn-gō′chå)	264-65	16°21′S	39°51′E
Angol, Chile (aŋ-gōl′)	171	37°48′S	72°43′W
Angola, In., U.S. (ăŋ-gō′lȧ)	116-17	41°38′N	84°59′W
Angola, nation, Afr. (ăŋ-gō′lȧ)	253	12°30′S	18°30′E
Angontsy, Tanjona, c., Madag.	264-65	15°13′S	50°27′E
Angora, nat. cap., Tur. *see* Ankara	186-87	39°56′N	32°53′E
Angostura, Ven. *see* Ciudad Bolívar	164-65	8°07′N	63°33′W
Angostura, Presa de la, res., Mex.	144-45	16°02′N	92°22′W
Angoulême, Fr. (äŋ′go͞o-lâm′)	196-97	45°39′N	0°09′E
Angra dos Reis, Braz. (aŋ′grä dōs rā′ēs)	172	23°01′S	44°19′W
Angren, Uzb.	232-33	41°01′N	70°08′E
Anguilla, dep., N.A. (ăŋ-gwĭl′ȧ)	140-41	18°15′N	63°05′W

Feature (Pronunciation)	Page	Lat.	Long.
Anguilla Cays, is., Bah. (ăŋ-gwĭl´ȧ kēs)	142-43	23°31'N	79°33'W
Anguille, Cape, c., Nf., Can. (kăp´-äŋ-gē´yĕ)	138-39	47°55'N	59°24'W
Anholt, i., Den. (än´hōlt)	192-93	56°42'N	11°34'E
Anhui, state, China (än-hwā)	238-39	32°0'N	117°00'E
Anhwei, state, China *see* Anhui	238-39	32°0'N	117°00'E
Aniak, Ak., U.S. (ä-nyȧ´k)	126	61°35'N	159°33'W
Anina, Rom. (ä-nē´nä)	200-01	45°05'N	21°51'E
Anita, Pa., U.S. (ȧ-nē´ȧ)	116-17	41°0'N	78°58'W
Aniva, Zaliv, b., Russia (zä´lĭf ȧ-nē´vȧ)	244	46°16'N	142°48'E
Anjār, India	234-35	23°07'N	70°02'E
Anjouan, i., Com. *see* Nzwani	264-65	12°15'S	44°25'E
Anju-ŭp, Kor., N.	243	39°37'N	125°40'E
Ankaboa, Tanjona, c., Madag.	264-65	21°55'S	43°18'E
Ankang, China (än-käŋ)	238-39	32°41'N	109°01'E
Ankara, nat. cap., Tur. (än´kȧ-rȧ)	186-87	39°56'N	32°53'E
Ankazoabo, Madag.	264-65	22°18'S	44°31'E
Ānkober, Eth.	269d	9°35'N	39°44'E
Anlong, China (än-loŋ)	238-39	25°07'N	105°28'E
Anlu, China (än´lōō´)	238-39	31°16'N	113°41'E
Anmyŏn-do, i., Kor., S.	243	36°30'N	126°22'E
Anna, Il., U.S. (ăn´ȧ)	116-17	37°27'N	89°14'W
Annaba, Alg.	184-85	36°54'N	7°46'E
An-Nafūd, des., Sau. Ar.	228-29	28°30'N	41°00'E
An-Najaf, Iraq (än nä-jäf´)	228-29	32°00'N	44°20'E
Annamitique, Chaîne, mts., Asia	246-47	17°0'N	106°00'E
Annapolis, Md., U.S. (ă-năp´ô-lĭs)	116-17	38°58'N	76°31'W
Annapūrṇa, mtn., Nepal	234-35	28°34'N	83°50'E
Ann Arbor, Mi., U.S. (ăn är´bẽr)	116-17	42°16'N	83°43'W
An-Nāṣirīyah, Iraq	228-29	31°03'N	46°15'E
An-Nawfalīyah, Libya	188-89	30°46'N	17°50'E
Annecy, Fr. (ȧn sē´)	196-97	45°54'N	6°07'E
Annemasse, Fr. (än'mäs´)	196-97	46°12'N	6°14'E
An Nhon, Viet.	246-47	13°54'N	109°05'E
Anniston, Al., U.S. (ăn´ĭs-tŭn)	124-25	33°39'N	85°50'W
Annobón, i., Eq. Gui.	260-61	1°26'S	5°37'E
Annonay, Fr. (än´ĭs-tsiün)	196-97	45°15'N	4°40'E
An-Nuhūd, Sudan	266	12°42'N	28°26'E
Anori, Braz.	166-67	3°45'S	61°42'W
Anorontany, Tanjona, c., Madag.	264-65	12°26'S	48°45'E
Anpu, China (än-pōō)	238-39	21°27'N	110°01'E
Anqing, China	238-39	30°30'N	117°02'E
Ansbach, Ger. (äns´bäk)	194-95	49°18'N	10°35'E
Anse-d'Hainault, Haiti (äɴs´dĕnō)	142-43	18°30'N	74°26'W
Anserma, Col. (ȧ´n-sĕ´r-mȧ)	163c	5°14'N	75°48'W
Anshan, China	243	41°08'N	122°60'E
Anshun, China (än-shōōn´)	238-39	26°15'N	105°56'E
Anson, Tx., U.S. (ăn´sŭn)	120-21	32°44'N	99°53'W
Ansongo, Mali	258-59	15°40'N	0°30'E
Antakya, Tur. *see* Antioch	228-29	36°12'N	36°10'E
Antalaha, Madag.	264-65	14°55'S	50°17'E
Antalya, Tur. (än-tä´lē-ä) (ä-dä´lē-ä)	186-87	36°54'N	30°42'E
Antalya, Gulf of, b., Tur. *see* Antalya Körfezi	186-87	36°30'N	30°60'E
Antalya Körfezi, b., Tur.	186-87	36°30'N	30°60'E
Antananarivo, nat. cap., Madag. (än-tä´nä-nä-rēv)	264-65	18°55'S	47°32'E
An tAonach, Ire. *see* Nenagh	190-91	52°52'N	8°12'W
Antarctica, cont., (ănt-ärk´tĭ-kȧ)	287	87°0'S	60°00'E
Antarctic Peninsula, pen., Ant.	287	70°15'S	65°55'W
Antequera, Spain (än-tĕ-kĕ´rä)	198-99	37°01'N	4°33'W
Anthony, Tx., U.S. (ăn´thō-nē)	118-19	31°60'N	106°36'W
Anti-Atlas, mts., Mor.	258-59	30°0'N	8°30'W
Antibes, Fr. (äɴ-tēb´)	196-97	43°35'N	7°07'E
Anticosti, Île d', i., Qc., Can. (ăn-tĭ-kŏs´tē)	138-39	49°30'N	63°00'W
Antigo, Wi., U.S. (ăn´tĭ-gō)	116-17	45°08'N	89°08'W
Antigonish, N.S., Can. (ăn-tĭ-gō-nĕsh´)	138-39	45°37'N	61°60'W
Antigua, i., Antig.	143b	17°05'N	61°49'W
Antigua and Barbuda, nation, N.A. (än-tē´gwä ănd bär-bōō´dȧ)	140-41	17°03'N	61°48'W
Antigua Guatemala, Guat.	148	14°33'N	90°44'W
Antillas, Archipiélago de las, is., *see* West Indies	140-41	19°0'N	70°00'W
Antillas, Mar de las, s., *see* Caribbean Sea	140-41	15°0'N	73°00'W
Antillas Mayores, is., N.A. *see* Greater Antilles	142-43	20°0'N	74°00'W
Antillen, Nederlandse, dep., N.A. *see* Netherlands Antilles	140-41	12°15'N	68°45'W
Antilles, Grandes, is., N.A. *see* Greater Antilles	142-43	20°0'N	74°00'W
Antilles, Mer des, s., *see* Caribbean Sea	140-41	15°0'N	73°00'W
Antilles, Petites, is., *see* Lesser Antilles	143b	15°0'N	61°00'W
Antioch, Tur.	228-29	36°12'N	36°10'E
Antioch, Il., U.S. (ăn´tĭ-ŏk)	116-17	42°29'N	88°06'W
Antioquia, Col. (än-tē-ō´kēä)	163c	6°34'N	75°49'W
Antipodes Islands, is., N.Z.	287	49°40'S	178°47'E
Antlers, Ok., U.S. (ănt´lẽrz)	120-21	34°13'N	95°37'W
Antofagasta, Chile (än-tô-fä-gäs´tä)	168-69	23°39'S	70°23'W
Antón, Pan. (än-tōn´)	150	8°24'N	80°14'W
Antongila, Helodrano, b., Madag.	264-65	15°45'S	49°50'E
António Enes, Moz. (ăn-to´nyō ĕn´ēs) *see* Angoche	264-65	16°14'S	39°55'E
Antsirabe, Madag. (änt-sē-rä´bā)	264-65	19°52'S	47°02'E
Antsirañana, Madag.	264-65	12°17'S	49°17'E
Antsirane, Madag. *see* Antsirañana	264-65	12°17'S	49°17'E
Antsla, Est. (änt´slȧ)	192-93	57°50'N	26°32'E
Antsohihy, Madag.	264-65	14°49'S	48°03'E
Antung, China *see* Dandong	243	40°07'N	124°21'E
Antwerp, Bel. (ăn´twûrp)	190-91	51°13'N	4°25'E
Antwerpen, Bel. *see* Antwerp	190-91	51°13'N	4°25'E
Anugul, India	234-35	20°51'N	85°06'E
Anūpgarh, India (ŭ-nōp´gŭr)	234-35	29°11'N	73°13'E
Anuradhapura, Sri L. (ŭ-nōō´rä-dŭ-pōō´rŭ)	236	8°21'N	80°24'E
Anvers, Bel. *see* Antwerp	190-91	51°13'N	4°25'E
Anvers Island, i., Ant.	287	64°33'S	63°35'W
Anxi, China (än-shyē)	225a	25°04'N	118°11'E
Anxi, China (än-shyē)	240-41	40°29'N	95°47'E
Anyang, China (än´yäng)	240-41	36°06'N	114°20'E
Anykščiai, Lith. (anīksh-chä´ē)	192-93	55°32'N	25°07'E
Anyuanyi, China *see* Tianzhu	240-41	36°60'N	103°07'E
Anzhero-Sudzhensk, Russia (än´zhâ-rô-sōd´zhĕnsk)	218-19	56°05'N	86°01'E
Anzio, Italy (änt´zē-ō)	200-01	41°27'N	12°37'E
Aoba, i., Vanuatu	279g	15°25'S	167°50'E
Aoga-shima, i., Japan	244	32°28'N	139°46'E
Aomen, China	238-39	22°13'N	113°33'E
Aomori, Japan (äô-mō´rē)	244	40°49'N	140°45'E
Aoraki/Mount Cook National Park, n.p., N.Z.	278	43°35'S	170°15'E
Aoraki, mtn., N.Z.	278	43°36'S	170°10'E
Aôral, Phnum, mtn., Camb.	246-47	12°02'N	104°10'E
Aouk, Bahr, stm., Afr. (bär ä-ōōk´)	262-63	8°51'N	18°52'E
Aoukâr, reg., Maur.	258-59	18°0'N	9°30'W
Aoulef, Alg.	258-59	26°58'N	1°04'E
Apalachicola, Fl., U.S. (ăp-ȧ-lăch-ĭ-kō´lȧ)	124-25	29°44'N	84°60'W
Apalachicola, stm., Fl., U.S. (ăpȧ-lăch´ĭ-cōlȧ)	124-25	29°44'N	84°59'W
Apaporis, stm., S.A. (ä-pä-pô´rĭs)	170	1°21'S	69°25'W
Aparri, Phil. (ä-pär´rē)	250	18°20'N	121°40'E
Apatin, Serb. (ŏ´pō-tĭn)	200-01	45°40'N	18°59'E
Apatity, Russia	184-85	67°34'N	33°23'E
Apeldoorn, Neth. (ä´pĕl-dōōrn)	190-91	52°13'N	5°58'E
Apennines, mts., Italy (ă´-pȧ-nīnz)	200-01	43°0'N	13°00'E
Apía, Col. (ȧ-pē´ä)	163c	5°06'N	75°58'W
Apia, nat. cap., Samoa (ä´-pē-ȧ) (ä-pē´-ȧ)	279b	13°50'S	171°45'W
Apiacás, Serra dos, plat., Braz.	166-67	10°15'S	57°15'W
Apizaco, Mex. (ä-pē-zä´kō)	146-47	19°25'N	98°08'W
Apo, Mount, mtn., Phil. (mount ä´pō)	250	6°59'N	125°16'E
Apolo, Bol.	166-67	14°43'S	68°31'W
Aporé, stm., Braz.	168-69	19°28'S	50°56'W
Apostle Islands, is., Wi., U.S. (ȧ-pŏs´l ī´lȧndz)	114-15	46°50'N	90°30'W
Apóstoles, Arg.	173	27°55'S	55°48'W
Apostolove, Ukr.	202-03	47°39'N	33°43'E
Appalaches, Les, mts., N.A. *see* Appalachian Mountains	110-11	41°0'N	77°00'W
Appalachia, Va., U.S. (ăpȧ-lăch´ĭ-ȧ)	124-25	36°54'N	82°48'W
Appalachian Mountains, mts., N.A. (ăp-ȧ-lăch´ĭ-ȧn moun´tĭnz)	110-11	41°0'N	77°00'W
Appennino, mts., Italy (äp-pĕn-nē´nō) *see* Apennines	200-01	43°0'N	13°00'E
Appleton, Mn., U.S. (ăp´l-tŭn)	114-15	45°12'N	96°01'W
Appleton, Wi., U.S. (ăp´l-tŭn)	116-17	44°15'N	88°25'W
Appleton City, Mo., U.S. (ăp´l-tŭn sĭ´tē)	120-21	38°11'N	94°02'W
Apt, Fr. (äpt)	196-97	43°52'N	5°24'E
Apucarana, Braz.	168-69	23°33'S	51°27'W
Apure, stm., Ven. (ä-pōō´rā)	164-65	7°37'N	66°23'W
Apurímac, stm., Peru (ä-pōō-rē-mäk´)	170	11°52'S	73°57'W
Aqaba, Gulf of, b., (gŭlf ŭv ä´kȧ-bȧ)	228-29	29°05'N	34°44'E
Aqmola, nat. cap., Kaz. *see* Astana	226	51°12'N	71°27'E
Aqtaū, Kaz.	186-87	43°38'N	51°11'E
Aqtöbe, Kaz.	226	50°18'N	57°10'E
Aquidauana, Braz. (ä-kē-däwä´nä)	168-69	20°29'S	55°48'W
Aquila, Italy *see* L'Aquila	200-01	42°21'N	13°24'E
Aquin, Haiti (ä-kăn´)	142-43	18°17'N	73°24'W
Ar'ar, Sau. Ar.	228-29	30°56'N	41°04'E
Ara, India	234-35	25°34'N	84°40'E
Ara, stm., Japan (ä-rä)	245	35°40'N	139°51'E
Ara, stm., Japan (ä-rä)	245	38°09'N	139°25'E
'Arab, Baḥr al-, stm., Sudan	262-63	9°02'N	29°28'E
'Arab, Shaṭṭ al-, stm., Asia	228-29	29°57'N	48°33'E
Arabian Desert, des., Egypt (ȧ-rā´bĭ-ăn dĕs´ẽrt)	266	28°0'N	32°00'E
Arabian Gulf, b., Asia *see* Persian Gulf	230-31	27°0'N	51°00'E
Arabian Peninsula, pen., Asia (ȧ-rā´bĭ-ăn pĕ-nĭn´sūlȧ)	220-21	25°0'N	45°00'E
Arabian Sea, s., (ȧ-rā´bĭ-ăn sē)	220-21	15°0'N	65°00'E
Aracaju, Braz. (ä-rä´kä-zhōō´)	163d	10°54'S	37°04'W
Aracati, Braz. (ä-rä´kä-tē´)	166-67	4°34'S	37°46'W
Araçatuba, Braz. (ä-rä-sȧ-tōō´bä)	168-69	21°12'S	50°27'W
Aracruz, Braz. (ä-rä-krōō´s)	172	19°49'S	40°16'W
Araçuai, Braz.	168-69	16°53'S	42°04'W
Arad, Rom. (ŏ´rŏd)	194-95	46°11'N	21°19'E
Arafura, Laut, s., *see* Arafura Sea	224-25	9°0'S	133°00'E
Arafura Sea, s., (ä-rä-fōō´rä sē)	224-25	9°0'S	133°00'E
Aragarças, Braz.	168-69	15°55'S	52°15'W
Aragón, hist. reg., Spain (ä-rä-gōn´)	198-99	41°30'N	1°00'W
Araguacema, Braz.	166-67	8°50'S	49°34'W
Aragua de Barcelona, Ven.	163b	9°27'N	64°50'W
Araguaia, stm., Braz. (ä-rä-gwä´yä)	166-67	5°20'S	48°42'W
Araguari, Braz. (ä-rä-gwä´rē)	172	18°39'S	48°12'W
Araguari, stm., Braz.	166-67	1°13'N	50°02'W
Araguatins, Braz. (ä-rä-gwä-tēns)	166-67	5°39'S	48°06'W
Arāk, Iran	232-33	34°05'N	49°41'E
Aral, Kaz.	226	46°48'N	61°40'E
Aral Sea, lk., Asia (ă´-rŭl sē)	226	45°0'N	60°00'E
Aral Tengizi, lk., Asia *see* Aral Sea	226	45°0'N	60°00'E
Aramac, Austl.	277	22°58'S	145°15'E
Aramberri, Mex. (ä-räm-bĕr-rē´)	146-47	24°06'N	99°49'W
Aranda de Duero, Spain (ä-rän´dä dā dwā´rō)	198-99	41°41'N	3°41'W
Arandas, Mex. (ä-rän´däs)	146-47	20°43'N	102°20'W
Aranjuez, Spain (ä-rän-hwāth´)	198-99	40°02'N	3°37'W
Aransas Pass, Tx., U.S. (ȧ-răn´sȧs pȧs)	122-23	27°54'N	97°09'W
Aranyaprathet, Thai.	246-47	13°41'N	102°31'E
Arapiraca, Braz.	163d	9°45'S	36°39'W
Araranguá, Braz.	172	28°56'S	49°29'W
Araraquara, Braz. (ä-rä-rä-kwȧ´rä)	172	21°47'S	48°10'W
Ararat, Austl. (ăr´ărăt)	276	37°17'S	142°56'E
Ararat, Mount, vol., Tur. (mount ăr´ărăt)	227	39°42'N	44°18'E
Araripe, Chapada do, plat., Braz. (shä-pä´dä-dô-ä-rä-rē´pĕ)	166-67	7°23'S	39°49'W
Araruama, Lagoa de, b., Braz. (lä-gô´ä-ä-rä-rōō-ä´mä)	172	22°53'S	42°12'W
Aras, stm., Asia (ȧ-räs)	227	40°01'N	48°28'E
Arauca, Col. (ä-rou´kä)	164-65	7°04'N	70°45'W
Arauca, stm., S.A. (ä-rou´kä)	164-65	7°25'N	66°31'W
Arāvalli Range, mts., India (ä-rä´vŭ-lē rānj)	234-35	24°42'N	73°19'E
Araxá, Braz.	172	19°36'S	46°55'W
Araya, Punta de, c., Ven. (pŭn´tȧ-dĕ-ä-rä´yȧ)	163b	10°38'N	64°17'W
Araz, stm., Asia.	227	40°01'N	48°28'E
Ārba Minch', Eth.	262-63	6°01'N	37°34'E
Arbīl, Iraq	228-29	36°11'N	44°01'E
Arboga, Swe. (är-bō´gä)	192-93	59°24'N	15°51'E
Arboréa, Italy (är-bō-rĕ´ä)	200-01	39°46'N	8°35'E
Arbroath, Scot., U.K. (är-brōth´)	190-91	56°34'N	2°36'W
Arcachon, Fr. (är-kä-shôɴ´)	196-97	44°40'N	1°10'W
Arcadia, Fl., U.S. (är-kā´dĭ-ȧ)	125a	27°13'N	81°51'W
Arcadia, La., U.S. (är-kā´dĭ-ȧ)	120-21	32°34'N	92°56'W
Arcadia, Wi., U.S. (är-kā´dĭ-ȧ)	114-15	44°15'N	91°29'W
Arcata, Ca., U.S. (är-kȧ´tȧ)	112-13	40°53'N	124°05'W
Arc Dome, mtn., Nv., U.S. (ärk dōm)	118-19	38°50'N	117°21'W
Arcelia, Mex. (är-sā´lē-ä)	146-47	18°18'N	100°17'W
Archangel, Russia *see* Arkhangel'sk	186-87	64°32'N	40°25'E
Archbald, Pa., U.S. (ärch´bôld)	116-17	41°30'N	75°33'W
Archer Bend National Park, n.p., Austl. *see* Mungkan Kandju National Park	277	13°32'S	142°37'E
Arches National Park, n.p., Ut., U.S. (är´ches năsh´ŭn-ăl pärk)	118-19	38°43'N	109°36'W
Arco, Id., U.S. (är´kō)	112-13	43°39'N	113°18'W
Arcoverde, Braz.	163d	8°25'S	37°04'W
Arctic Ocean, oc., (ärk´tĭk ōshŭn)	288	85°0'N	170°00'E
Arctic Red, stm., N.T., Can.	130-31	67°27'N	133°45'W
Arctic Village, Ak., U.S.	126	68°05'N	145°31'W
Ardabīl, Iran	227	38°15'N	48°18'E
Ardahan, Tur. (är-dȧ-hän´)	227	41°06'N	42°43'E

n-sing; ŋ-baŋk; ɴ-nasalized n; nŏd; cŏmmit; ōld; ôbey; ôrder; oi-boil; fōōd; ŏ-as oo in foot; ou-out; s-soft; sh-dish; th-thin; pūre; ûnite; ûrn; stŭd; circŭs; ü-as in French tu; ´-indeterminate vowel.

Feature (Pronunciation)	Page	Lat.	Long.
Ardebil, Iran *see* Ardabīl	227	38°15'N	48°18'E
Ardennen, reg., Eur. *see* Ardennes	190-91	50°10'N	5°45'E
Ardennes, reg., Eur. (ȧr-dĕn′)	190-91	50°10'N	5°45'E
Ardila, stm., Eur. (är-dē′lä)	198-99	38°10'N	7°29'W
Ardmore, Ok., U.S. (ärd′mōr)	120-21	34°10'N	97°09'W
Arecibo, P.R. (ä-rå-sē′bō)	142a	18°28'N	66°43'W
Areia Branca, Braz. (ä-rĕ′yä-brä′n-kă)	163d	4°56'S	37°07'W
Arena, Point, c., Ca., U.S. (point ä-rā′n*ȧ*)	118-19	38°57'N	123°44'W
Arena, Punta, c., Mex.	144-45	23°34'N	109°28'W
Arendal, Nor. (ä′rĕn-däl)	192-93	58°27'N	8°48'E
Arequipa, Peru (ä-rå-kē′pä)	170	16°24'S	71°32'W
Arezzo, Italy (ä-rĕt′sō)	200-01	43°28'N	11°53'E
Argentan, Fr. (ȧr-zhäɴ-täɴ′)	196-97	48°45'N	0°01'W
Argenteuil, Fr. (är-zhäɴ-tû′y′)	196-97	48°57'N	2°14'E
Argentina, nation, S.A. (är-jĕn-tē′n*ȧ*)	158	34°0'S	64°00'W
Argentino, Lago, lk., Arg. (lä′gō är-ĸĕn-tē′nō)	171	50°14'S	72°26'W
Argenton-sur-Creuse, Fr. (är-zhäɴ′tôɴ-sür-krôs)	196-97	46°35'N	1°31'E
Arghandāb, stm., Afg.	232-33	31°27'N	64°23'E
Argonne, reg., Fr. (ä′r-gôn)	196-97	49°07'N	5°14'E
Árgos, Grc. (är′gŏs)	200-01	37°38'N	22°44'E
Arguello, Point, c., Ca., U.S. (point är-gwĕl′yō)	118-19	34°35'N	120°38'W
Argun', stm., Asia (ȧr-gōōn′)	218-19	53°19'N	121°27'E
Århus, Den. (ôr′hōōs)	192-93	56°09'N	10°13'E
Ariake-kai, b., Japan (ä′rê-ä′kä)	245	33°0'N	130°20'E
Arica, Chile (ä-rē′kä)	168-69	18°29'S	70°19'W
Arica, Col.	164-65	2°07'S	71°44'W
Arīḥā, W.B. *see* Jericho	228-29	31°52'N	35°27'E
Arima, Trin.	143b	10°37'N	61°17'W
Arinos, stm., Braz. (ä-rē′nôzh)	166-67	10°26'S	58°20'W
Aripuanã, stm., Braz. (ȧ-rê-pwän′y*ȧ*)	166-67	5°07'S	60°23'W
Ariquemes, Braz.	166-67	9°54'S	63°05'W
Aristazabal Island, i., B.C., Can.	132-33	52°38'N	129°07'W
Arizona, Arg.	171	35°43'S	65°19'W
Arizona, state, U.S. (ăr-ĭ-zō′n*ȧ*)	108-09	34°0'N	112°00'W
Arkadelphia, Ar., U.S. (är-k*ȧ*-dĕl′fĭ-*ȧ*)	120-21	34°07'N	93°04'W
Arkansas, state, U.S. (är′kăn-sô) (är-kăn′s*ȧ*s)	108-09	34°50'N	92°30'W
Arkansas, stm., U.S. (är′kăn-sô) (är-kăn′s*ȧ*s)	110-11	33°47'N	91°04'W
Arkansas City, Ks., U.S.	120-21	37°04'N	97°02'W
Arkhangel'sk, Russia (ȧr-ĸän′gĕlsk)	186-87	64°32'N	40°25'E
Arkhangel'skoye, Russia (är-kän-gĕl′skô-yĕ)	202-03	53°16'N	37°42'E
Arles, Fr. (ärl)	196-97	43°41'N	4°38'E
Arlington, S.D., U.S. (är′lĕng-t*ŭ*n)	114-15	44°22'N	97°08'W
Arlington, Tx., U.S. (ȧr′lĭng-t*ŭ*n)	120-21	32°45'N	97°07'W
Arlington, Va., U.S. (ȧr′lĭng-t*ŭ*n)	116-17	38°52'N	77°07'W
Arlington, Vt., U.S. (ȧr′lĭng-t*ŭ*n)	116-17	43°04'N	73°09'W
Arlington Heights, Il., U.S. (är′lĕng-t*ŭ*n hīts)	116-17	42°05'N	87°59'W
Arlit, Niger	258-59	18°45'N	7°21'E
Armant, Egypt (är-mänt′)	268b	25°37'N	32°32'E
Armavir, Russia (ȧr-mȧ-vĭr′)	186-87	44°59'N	41°07'E
Armenia, Col. (ȧr-mĕ′nê*ȧ*)	163c	4°31'N	75°42'W
Armenia, nation, Asia (är-mē′nē-*ȧ*)	227	40°0'N	45°00'E
Armeniya, nation, Asia *see* Armenia	227	40°0'N	45°00'E
Armentières, Fr. (ȧr-mäɴ-tyâr′)	196-97	50°41'N	2°53'E
Armidale, Austl. (är′mĭ-dāl)	276	30°31'S	151°40'E
Armour, S.D., U.S. (ä′r′mẽr)	114-15	43°19'N	98°21'W
Armstrong, On., Can.	134-35	50°19'N	89°04'W
Arnaud, stm., Qc., Can.	130-31	59°58'N	69°58'W
Arnedo, Spain (är-nā′dō)	198-99	42°14'N	2°07'W
Arnhem, Neth. (ärn′hĕm)	190-91	51°59'N	5°55'E
Arnhem, Cape, c., Austl. (kāp ärn′hĕm)	272-73	12°22'S	136°57'E
Arnhem Land, reg., Austl. (ärn′hĕm-lănd)	272-73	13°13'S	133°50'E
Arnold, Mn., U.S. (ȧr′nŭld)	114-15	46°53'N	92°05'W
Arnprior, On., Can. (ărn-prī′ẽr)	136-37	45°26'N	76°21'W
Arnsberg, Ger. (ärns′bĕrgh)	194-95	51°24'N	8°04'E
Arnstadt, Ger. (ärn′shtät)	194-95	50°50'N	10°57'E
Aroma, Sudan	266	15°48'N	36°08'E
Aroostook, stm., N.A.	138-39	46°49'N	67°43'W
Arop Island, i., Pap. N. Gui.	277	5°20'S	147°05'E
Arorae, i., Kir.	280-81	2°38'S	176°49'E
Arqalyq, Kaz.	226	50°15'N	66°53'E
Arraias, Braz.	166-67	12°58'S	46°55'W
Ar-Ramādī, Iraq	228-29	33°26'N	43°19'E
Arran, Island of, i., Scot., U.K. (ĭ′l*ȧ*nd ŏv ă′răn)	190-91	55°35'N	5°15'W
Arras, Fr. (ȧ-räs′)	196-97	50°17'N	2°47'E
Ar-Rass, Sau. Ar.	266	25°52'N	43°30'E
Arrecife, Spain	199d	28°57'N	13°33'W
Arrecifes, Arg. (är-rå-sē′fās)	173	34°04'S	60°06'W
Arriaga, Mex. (är-rêä′gä)	146-47	16°14'N	93°53'W
Ar-Riyāḍ, nat. cap., Sau. Ar. *see* Riyadh	230-31	24°38'N	46°43'E
Ar-Rub'al-Khālī, des., Asia *see* Rub'al-Khali	220-21	20°0'N	51°00'E
Ar-Ruṣayriṣ, Sudan	266	11°48'N	34°22'E
Ar-Ruṭbah, Iraq	228-29	33°02'N	40°17'E
Arsen'yev, Russia	244	44°09'N	133°17'E
Art, Île, i., N. Cal.	279g	19°43'S	163°39'E
Árta, Grc. (är′tä)	200-01	39°09'N	20°59'E
Arteaga, Mex. (är-tā-ä′gä)	146-47	18°20'N	102°18'W
Arteaga, Mex. (är-tā-ä′gä)	122-23	25°28'N	100°51'W
Artëm, Russia (*ȧ*r-tyôm′)	244	43°21'N	132°11'E
Artemisa, Cuba (är-tå-mē′sä)	142-43	22°49'N	82°46'W
Artesia, N.M., U.S. (är-tē′sĭ-*ȧ*)	120-21	32°51'N	104°24'W
Artibonite, stm., Haiti (är-tê-bô-nē′tä)	142-43	19°15'N	72°46'W
Artigas, Ur.	173	30°24'S	56°28'W
Artillery Lake, lk., N.T., Can.	130-31	63°09'N	107°52'W
Artvin, Tur.	227	41°10'N	41°50'E
Artyk, Russia	218-19	64°09'N	145°12'E
Aru, Kepulauan, is., Indon.	224-25	6°0'S	134°30'E
Aru, Tanjung, c., Indon.	248-49	2°11'S	116°35'E
Arua, Ug. (ä′rōō-ä)	267	3°01'N	30°55'E
Aruanã, Braz.	166-67	14°54'S	51°05'W
Aruba, dep., N.A. (ä-rōō′b*ȧ*)	140-41	12°30'N	69°58'W
Arunāchal Pradesh, state, India	234-35	28°30'N	95°00'E
Aruppukkottai, India	236	9°31'N	78°06'E
Arusha, Tan. (ȧ-rōō′shä)	267	3°22'S	36°41'E
Aruwimi, stm., D.R.C.	262-63	1°13'N	23°36'E
Arvayheer, Mong.	240-41	46°15'N	102°48'E
Arviat, Nu., Can.	128-29	61°08'N	94°07'W
Arvidsjaur, Swe.	184-85	65°36'N	19°07'E
Arvika, Swe. (är-vē′kȧ)	192-93	59°40'N	12°38'E
Arxan, China	240-41	47°11'N	119°57'E
Arys, Kaz.	226	42°26'N	68°48'E
Arzamas, Russia (är-zä-mäs′)	186-87	55°23'N	43°50'E
Aš, Czech Rep. (äsh′)	194-95	50°13'N	12°12'E
Asad, Buḥayrat al-, res., Syria	228-29	36°00'N	38°10'E
Asahi-dake, vol., Japan	244	43°40'N	142°51'E
Asahigawa, Japan *see* Asahikawa	244	43°46'N	142°22'E
Asahikawa, Japan	244	43°46'N	142°22'E
Āsānsol, India	234-35	23°41'N	86°59'E
Asbestos, Qc., Can. (ăs-bĕs′tŏs)	136-37	45°46'N	71°57'W
Asbury Park, N.J., U.S. (ăz′bĕr-ĭ pärk)	116-17	40°13'N	74°01'W
Ascensión, Mex. (äs-sĕn-sê-ōn′)	144-45	31°06'N	107°60'W
Ascension, i., St. Hel. (*ȧ*-sĕn′sh*ŭ*n)	254	7°57'S	14°22'W
Aschaffenburg, Ger. (ä-shäf′ĕn-bōrgh)	194-95	49°59'N	9°09'E
Aschersleben, Ger. (äsh′ẽrs-lā-bĕn)	194-95	51°46'N	11°28'E
Ascoli Piceno, Italy (äs′kô-lēpê-chä′nō)	200-01	42°52'N	13°35'E
Aseb, Erit.	266	12°58'N	42°42'E
Āsela, Eth.	269d	7°58'N	39°08'E
Åsele, Swe.	184-85	64°10'N	17°21'E
Aşgabat, nat. cap., Turkmen.	232-33	37°57'N	58°23'E
Asha, Russia (ä′shä)	186-87	55°00'N	57°16'E
Ashburn, Ga., U.S. (ăsh′b*ŭ*rn)	124-25	31°42'N	83°39'W
Ashburton, stm., Austl. (ăsh′bûr-t*ŭ*n)	272-73	21°42'S	114°55'E
Ashdown, Ar., U.S. (ăsh′doun)	120-21	33°41'N	94°08'W
Asheboro, N.C., U.S. (ăsh′bŭr-ô)	124-25	35°42'N	79°49'W
Asheville, N.C., U.S. (ăsh′vĭl)	124-25	35°36'N	82°34'W
Ashgabat, nat. cap., Turkmen. *see* Aşgabat	232-33	37°57'N	58°23'E
Ashikaga, Japan (ä′shê-kä′gä)	245	36°20'N	139°27'E
Ashkhabad, nat. cap., Turkmen. *see* Aşgabat	232-33	37°57'N	58°23'E
Ashland, Ky., U.S. (ăsh′l*ă*nd)	116-17	38°28'N	82°39'W
Ashland, Me., U.S. (ăsh′l*ă*nd)	117a	46°38'N	68°28'W
Ashland, Ne., U.S. (ăsh′l*ă*nd)	114-15	41°02'N	96°22'W
Ashland, Oh., U.S. (ăsh′l*ă*nd)	116-17	40°52'N	82°18'W
Ashland, Or., U.S. (ăsh′l*ă*nd)	112-13	42°12'N	122°42'W
Ashland, Va., U.S. (ăsh′l*ă*nd)	116-17	37°45'N	77°29'W
Ashland, Wi., U.S. (ăsh′l*ă*nd)	114-15	46°36'N	90°53'W
Ashley, N.D., U.S. (ăsh′lê)	114-15	46°02'N	99°22'W
Ashmore and Cartier Islands, dep., Oc.	224-25	12°25'S	123°20'E
Ashqelon, Isr. (äsh′kĕ-lōn)	228-29	31°40'N	34°35'E
Ash-Shāriqah, U.A.E. *see* Sharjah	230-31	25°22'N	55°24'E
Ash-Shiḥr, Yemen	220-21	14°46'N	49°37'E
Ashtabula, Oh., U.S. (ăsh-t*ȧ*-bū′l*ȧ*)	116-17	41°51'N	80°48'W
Ashton, Id., U.S. (ăsh′t*ŭ*n)	112-13	44°05'N	111°27'W
Ashur, hist., Iraq	228-29	35°30'N	43°16'E
Asia, cont., (ā′zh*ȧ*)	208-09	50°0'N	100°00'E
Asinara, Golfo dell', b., Italy (gôl′fō-dĕl-ä-sē-nä′rä)	200-01	41°0'N	8°32'E
Asinara, Isola, i., Italy	200-01	41°04'N	8°16'E
Asino, Russia	218-19	56°60'N	86°08'E
'Asīr, reg., Sau. Ar. (ä-sēr′)	220-21	19°0'N	42°00'E
Askersund, Swe. (äs′kĕr-sŏnd)	192-93	58°53'N	14°54'E
Asmara, nat. cap., Erit. (äz-mä′-r*ȧ*)	266	15°20'N	38°55'E
Asmera, nat. cap., Erit. (äs-mä′rä) *see* Asmara	266	15°20'N	38°55'E
Asotin, Wa., U.S. (*ȧ*-sō′tĭn)	112-13	46°20'N	117°03'W
Aspen, Co., U.S. (ăs′pĕn)	118-19	39°12'N	106°49'W
Aspiring, Mount, mtn., N.Z.	278	44°23'S	168°44'E
Assab, Erit. *see* Aseb	266	12°58'N	42°42'E
Assam, state, India (ăs-săm′)	234-35	26°0'N	93°00'E
As-Samāwah, Iraq	228-29	31°19'N	45°17'E
Assateague Island, i., U.S.	116-17	38°05'N	75°12'W
Assens, Den. (äs′sĕns)	192-93	55°16'N	9°55'E
Assiniboia, Sk., Can.	134-35	49°38'N	105°59'W
Assiniboine, stm., Can. (ă-sĭn′ĭ-boin)	134-35	49°53'N	97°08'W
Assiniboine, Mount, mtn., Can. (mount ă-sĭn′ĭ-boin)	132-33	50°52'N	115°39'W
Assis, Braz. (ä-sē′s)	168-69	22°40'S	50°26'W
Assomption, i., Sey.	264-65	9°44'S	46°30'E
As-Sūdān, nation, Afr. *see* Sudan	253	15°0'N	30°00'E
As-Sudd, reg., Sudan	262-63	8°0'N	31°00'E
As-Sulaymānīyah, Iraq	228-29	35°34'N	45°27'E
As-Sūrīyah, nation, Asia *see* Syria	206-07	35°0'N	38°00'E
As-Suwaydā', Syria	228-29	32°42'N	36°34'E
Astana, nat. cap., Kaz. (ä′stä-nä′)	226	51°12'N	71°27'E
Astara, Azer.	227	38°28'N	48°52'E
Asterābād, Iran *see* Gorgān	232-33	36°51'N	54°26'E
Asti, Italy (äs′tē)	200-01	44°55'N	8°13'E
Astorga, Spain (äs-tôr′gä)	198-99	42°28'N	6°03'W
Astoria, Or., U.S. (ăs-tō′rĭ-*ȧ*)	112-13	46°11'N	123°50'W
Astove, i., Sey.	264-65	10°06'S	47°45'E
Astrakhan', Russia (äs-trȧ-kän′)	186-87	46°21'N	48°02'E
Asunción, nat. cap., Para. (ä-sōōn-syōn′)	168-69	25°16'S	57°39'W
Asuncion Island, i., N. Mar. Is.	280-81	19°42'N	145°24'E
Aswân, Egypt	268b	24°05'N	32°55'E
Aswan High Dam, d., Egypt	268b	23°59'N	32°53'E
Asyût, Egypt	268b	27°11'N	31°11'E
Atacama, Desierto de, des., Chile (dĕ-syĕ′r-tô-dĕ-ä-tä-kȧ′mä)	168-69	20°08'S	69°53'W
Atacama, Puna de, plat., S.A. (pōō′nä-dĕ-ä-tä-kȧ′mä)	168-69	23°46'S	67°45'W
Atacama, Salar de, pl., Chile (sȧ-lȧr′dĕ-ȧtä-kȧ′mä)	168-69	23°33'S	68°14'W
Atacama Desert, des., Chile (ä-tä-kȧ′mä) *see* Atacama, Desierto de	168-69	20°08'S	69°53'W
Ataco, Col. (ä-tá′kō)	163c	3°35'N	75°23'W
Atafu, at., Tok.	280-81	8°33'S	172°30'W
Atakpamé, Togo	260-61	7°32'N	1°09'E
Atamyrat, Turkmen.	232-33	37°50'N	65°13'E
Aṭar, Maur. (ä-tär′)	258-59	20°32'N	13°02'W
Atascadero, Ca., U.S. (ăt-ăs-k*ȧ*-dâ′rō)	118-19	35°30'N	120°39'W
Atasū, Kaz.	226	48°41'N	71°39'E
Atbara, stm., Afr. *see* 'Aṭbarah	266	17°40'N	33°58'E
'Aṭbarah, Sudan (ät′bȧ-rä)	266	17°42'N	33°59'E
'Aṭbarah, stm., Afr.	266	17°40'N	33°58'E
Atbasar, Kaz. (ät′bä-sär′)	226	51°48'N	68°21'E
Atchafalaya, stm., La., U.S. (ăch-*ȧ*-f*ȧ*-lī′*ȧ*)	124-25	29°28'N	91°16'W
Atchafalaya Bay, b., La., U.S. (ăch-*ȧ*-f*ȧ*-lī′*ȧ* bā)	124-25	29°27'N	91°23'W
Atchara Autonomis Respublika, state, Geor.	227	41°40'N	42°00'E
Atchison, Ks., U.S. (ăch′ĭ-s*ŭ*n)	120-21	39°34'N	95°07'W
Athabasca, Ab., Can. (ăth-*ȧ*-băs′k*ȧ*)	132-33	54°42'N	113°17'W
Athabasca, stm., Ab., Can. (ăth-*ȧ*-băs′k*ȧ*)	130-31	58°40'N	110°55'W
Athabasca, Lake, lk., Can. (lāk ăth-*ȧ*-băs′k*ȧ*)	130-31	59°07'N	109°59'W
Athens, Al., U.S. (ăth′ĕnz)	124-25	34°48'N	86°58'W
Athens, Ga., U.S. (ăth′ĕnz)	124-25	33°57'N	83°22'W
Athens, Oh., U.S. (ăth′ĕnz)	116-17	39°20'N	82°06'W
Athens, Tn., U.S. (ăth′ĕnz)	124-25	35°27'N	84°36'W
Athens, Tx., U.S. (ăth′ĕnz)	122-23	32°13'N	95°51'W
Athens, nat. cap., Grc. (ăth′ĕnz)	200-01	37°59'N	23°44'E
Atherton, Austl.	277	17°16'S	145°30'E
Athi, stm., Kenya (ä′tê)	262-63	2°58'S	38°31'E
Athína, nat. cap., Grc. (ä-thē′nĕ) *see* Athens	200-01	37°59'N	23°44'E
Athos, Mount, mtn., Grc.	200-01	40°09'N	24°19'E
Athy, Ire. (*ȧ*-thī)	190-91	52°60'N	6°59'W
Ati, Chad	258-59	13°12'N	18°19'E

Feature (Pronunciation)	Page	Lat.	Long.
Atikokan, On., Can.	**136-37**	48°45′N	91°37′W
Atikonak Lake, lk., Nf., Can.	**130-31**	52°38′N	64°30′W
Atiu, i., Cook Is.	**280-81**	20°02′S	158°07′W
Atka, Russia	**218-19**	60°45′N	151°46′E
Atka Island, i., Ak., U.S. (ăt′kȧ ī′lȧnd)	**126a**	52°15′N	174°08′W
Atkarsk, Russia (ắt-kärsk′)	**186-87**	51°53′N	45°00′E
Atkinson, Ne., U.S. (ăt′kĭn-sŭn)	**114-15**	42°32′N	98°59′W
Atlanta, Ga., U.S. (ăt-lăn′tȧ)	**124-25**	33°46′N	84°25′W
Atlanta, Il., U.S. (ăt-lăn′tȧ)	**116-17**	40°16′N	89°14′W
Atlanta, Tx., U.S. (ăt-lăn′tȧ)	**120-21**	33°07′N	94°11′W
Atlantic, Ia., U.S. (ăt-lăn′tĭk)	**114-15**	41°24′N	95°01′W
Atlantic City, N.J., U.S. (ăt-lăn′tĭk sĭ′tê)	**116-17**	39°21′N	74°26′W
Atlantic Ocean, oc., (ăt-lăn′tĭk ōshŭn)	**20-21**	5°0′S	25°00′W
Atlantis, S. Afr.	**264-65**	33°32′S	18°29′E
Atlas Mountains, mts., Afr. (ăt′lăs moun′tĭnz)	**258-59**	33°0′N	2°00′W
Atlin, B.C., Can.	**128-29**	59°34′N	133°41′W
Atmore, Al., U.S. (ăt′mōr)	**124-25**	31°01′N	87°30′W
Atoka, Ok., U.S. (ȧ-tō′kȧ)	**120-21**	34°23′N	96°08′W
Atoui, Khaṭṭ, stm., Afr. (ȧ-tōō-ē′)	**258-59**	20°03′N	15°58′W
Atoyac, stm., Mex. (ä-tô-yäk′)	**146-47**	18°07′N	98°44′W
Atoyac de Álvarez, Mex. (ä-tô-yäk′dā äl′vä-rāz)	**146-47**	17°11′N	100°25′W
Atrak, stm., Asia	**232-33**	37°26′N	53°53′E
Atrato, stm., Col. (ä-trä′tō)	**164-65**	8°11′N	76°56′W
Atrek, stm., Asia see Atrak	**232-33**	37°26′N	53°53′E
Aṭ-Ṭā'if, Sau. Ar.	**266**	21°16′N	40°25′E
Attapu, Laos	**246-47**	14°48′N	106°51′E
Attawapiskat, On., Can.	**128-29**	52°56′N	82°25′W
Attawapiskat, stm., On., Can. (ăt′ȧ-wȧ-pĭs′kăt)	**130-31**	52°57′N	82°18′W
Attawapiskat Lake, lk., On., Can.	**134-35**	52°17′N	87°55′W
Attica, N.Y., U.S. (ăt′ĭ-kȧ)	**116-17**	42°52′N	78°17′W
Attleboro, Ma., U.S. (ăt′'l-bŭr-ô)	**116-17**	41°57′N	71°17′W
Attu Island, i., Ak., U.S. (ät-tōō′ ī′lȧnd)	**126a**	52°55′N	173°00′E
Atuel, stm., Arg.	**171**	36°16′S	66°51′W
Åtvidaberg, Swe. (ôt-vē′dȧ-bĕrgh)	**192-93**	58°12′N	16°00′E
Atyraū, Kaz.	**186-87**	47°07′N	51°55′E
Aubagne, Fr. (ō-bän′y')	**196-97**	43°17′N	5°34′E
Aubry Lake, lk., N.T., Can.	**130-31**	67°22′N	126°27′W
Auburn, Al., U.S. (ô′bûrn)	**124-25**	32°37′N	85°29′W
Auburn, Il., U.S. (ô′bûrn)	**120-21**	39°35′N	89°45′W
Auburn, In., U.S. (ô′bûrn)	**116-17**	41°22′N	85°03′W
Auburn, Ma., U.S. (ô′bûrn)	**116-17**	42°12′N	71°50′W
Auburn, Ne., U.S. (ô′bûrn)	**120-21**	40°23′N	95°51′W
Auburn, N.Y., U.S.	**116-17**	42°56′N	76°34′W
Auburn, Wa., U.S. (ô′bûrn)	**112-13**	47°19′N	122°12′W
Aubusson, Fr. (ō-bü-sôN′)	**196-97**	45°57′N	2°10′E
Auch, Fr. (ōsh)	**196-97**	43°39′N	0°35′E
Auckland, N.Z. (ôk′lănd)	**278**	36°51′S	174°45′E
Auckland Islands, is., N.Z. (ôk′lănd ī′lăndz)	**287**	50°46′S	166°12′E
Audubon, Ia., U.S. (ô′dŏŏ-bŏn)	**114-15**	41°43′N	94°56′W
Augathella, Austl. (ôr′gȧ′thĕ-lȧ)	**276**	25°48′S	146°34′E
Augrabies Falls National Park, n.p., S. Afr.	**264-65**	28°35′S	20°19′E
Augsburg, Ger. (ouks′bŏŏrgh)	**194-95**	48°23′N	10°53′E
Augusta, Austl.	**270-71**	34°19′S	115°10′E
Augusta, Ar., U.S. (ô-gŭs′tȧ)	**124-25**	35°17′N	91°22′W
Augusta, Ga., U.S. (ô-gŭs′tȧ)	**124-25**	33°28′N	81°59′W
Augusta, Ky., U.S. (ô-gŭs′tȧ)	**116-17**	38°46′N	84°00′W
Augusta, Me., U.S.	**117a**	44°19′N	69°47′W
Augusta, Wi., U.S. (ô-gŭs′tȧ)	**114-15**	44°41′N	91°07′W
Augustus Island, i., Austl.	**272-73**	15°21′S	124°31′E
Auob, stm., Afr. (ä′wŏb)	**264-65**	26°26′S	20°37′E
Aurangābād, India (ou-rŭŋ-gä-bäd′)	**234-35**	19°53′N	75°20′E
Aurillac, Fr. (ō-rē-yȧk′)	**196-97**	44°55′N	2°26′E
Aurora, On., Can. (ô-rō′rȧ)	**136-37**	43°60′N	79°28′W
Aurora, Co., U.S. (ô-rō′rȧ)	**120-21**	39°44′N	104°52′W
Aurora, Il., U.S. (ô-rō′rȧ)	**116-17**	41°45′N	88°20′W
Aurora, In., U.S. (ô-rō′rȧ)	**116-17**	39°03′N	84°55′W
Aurora, Mn., U.S. (ô-rō′rȧ)	**114-15**	47°32′N	92°14′W
Aurora, Mo., U.S. (ô-rō′rȧ)	**120-21**	36°58′N	93°43′W
Aurora, Ne., U.S. (ô-rō′rȧ)	**114-15**	40°52′N	98°01′W
Au Sable, stm., Mi., U.S. (ô-sā′b'l)	**116-17**	44°24′N	83°19′W
Aussig, Czech Rep. see Ústí nad Labem	**194-95**	50°40′N	14°02′E
Austin, Mn., U.S. (ôs′tĭn)	**114-15**	43°40′N	92°58′W
Austin, Nv., U.S. (ôs′tĭn)	**118-19**	39°31′N	117°07′W
Austin, Tx., U.S. (ôs′tĭn)	**122-23**	30°16′N	97°42′W
Austin, Lake, lk., Austl.	**272-73**	27°40′S	118°00′E
Australia, nation, Oc. (ôs-trā′lĭ-ȧ)	**270-71**	25°0′S	135°00′E
Australian Capital Territory, state, Austl. (ôs-trā′lĭ-ăn)	**276**	35°30′S	149°00′E
Austral Islands, is., Fr. Poly. see Tubuaï, Îles	**280-81**	23°0′S	150°00′W
Austria, nation, Eur. (ôs′trĭ-ȧ)	**174-75**	47°20′N	13°20′E
Austvågøya, i., Nor.	**184-85**	68°21′N	14°38′E
Autlán de Navarro, Mex.	**146-47**	19°47′N	104°22′W
Autun, Fr. (ō-tŭN′)	**196-97**	46°57′N	4°18′E
Auxerre, Fr. (ō-sâr′)	**196-97**	47°48′N	3°34′E
Auyán Tepuy, mtn., Ven.	**164-65**	5°51′N	62°25′W
Auzangate, Nevado, mtn., Peru	**170**	13°48′S	71°14′W
Ava, Mo., U.S. (ā′vȧ)	**120-21**	36°57′N	92°40′W
Avallon, Fr. (ȧ-vȧ-lôN′)	**196-97**	47°30′N	3°54′E
Avalon, Ca., U.S. (ăv′ȧ-lŏn)	**118-19**	33°20′N	118°19′W
Avaré, Braz.	**172**	23°07′S	48°55′W
Aveiro, Port. (ä-vā′rŏ)	**198-99**	40°38′N	8°39′W
Avellaneda, Arg. (ä-vĕl-yä-nā′dhä)	**173**	29°07′S	59°40′W
Avellaneda, Arg. (ä-vĕl-yä-nā′dhä)	**173**	34°40′S	58°23′W
Avellino, Italy (ä-vĕl-lē′nō)	**200-01**	40°55′N	14°47′E
Avesta, Swe. (ä-vĕs′tä)	**192-93**	60°09′N	16°11′E
Avezzano, Italy (ä-vät-sä′nō)	**200-01**	42°02′N	13°25′E
Avignon, Fr. (ȧ-vē-nyôN′)	**196-97**	43°57′N	4°49′E
Ávila, Spain (ä-vē-lä)	**198-99**	40°40′N	4°42′W
Avilés, Spain (ä-vē-lās′)	**198-99**	43°34′N	5°54′W
Avon, Ct., U.S. (ā′vŏn)	**116-17**	41°49′N	72°50′W
Avon, stm., Eng., U.K. (ā′vŭn)	**190-91**	50°44′N	1°47′W
Avon, stm., Eng., U.K. (ā′vŭn)	**190-91**	51°59′N	2°11′W
Avon Park, Fl., U.S. (ā′vŏn pärk′)	**125a**	27°36′N	81°30′W
Avranches, Fr. (ȧ-vräNsh′)	**196-97**	48°41′N	1°22′W
Awaji-shima, i., Japan	**245**	34°21′N	134°51′E
Āwasa, Eth.	**269d**	6°56′N	38°32′E
Āwash, stm., Eth.	**262-63**	11°09′N	41°41′E
Awjilah, Libya	**188-89**	29°08′N	21°18′E
Axiós, stm., Eur.	**200-01**	40°31′N	22°43′E
Ax-les-Thermes, Fr. (äks′lā tĕrm′)	**196-97**	42°43′N	1°51′E
Ayacucho, Arg. (ä-yä-kōō′chō)	**173**	37°09′S	58°29′W
Ayacucho, Peru	**170**	13°08′S	74°14′W
Ayaköz, Kaz.	**226**	47°57′N	80°26′E
Ayamonte, Spain (ä-yä-mô′n-tĕ)	**198-99**	37°13′N	7°24′W
Ayan, Russia (ȧ-yän′)	**218-19**	56°26′N	138°13′E
Ayan, stm., Russia	**218-19**	70°10′N	95°47′E
Ayapel, Col.	**164-65**	8°19′N	75°09′W
Ayaviri, Peru (ä-yä-vē′rē)	**170**	14°53′S	70°35′W
Aydar, stm., Eur. (ī-där′) see Aidar	**202-03**	48°44′N	39°16′E
Aydar Kŭl, lk., Uzb.	**232-33**	40°49′N	67°20′E
Ayden, N.C., U.S. (ā′dĕn)	**124-25**	35°28′N	77°25′W
Aydın, Tur. (āīy-dĕn)	**200-01**	37°51′N	27°50′E
Ayers Rock, mtn., Austl. see Uluru	**272-73**	25°20′S	130°60′E
Ayeyarwady, stm., Mya.	**220-21**	15°51′N	95°05′E
Aylesbury, Eng., U.K. (ālz′bēr-ĭ)	**190-91**	51°49′N	0°50′W
Aylmer Lake, lk., N.T., Can. (āl′mēr lāk)	**130-31**	64°05′N	108°30′W
Ayon, Ostrov, i., Russia (ôs-trôf′ ī-ôn′)	**218-19**	69°47′N	168°41′E
Ayr, Austl.	**277**	19°34′S	147°24′E
Ayr, Scot., U.K. (âr)	**190-91**	55°28′N	4°38′W
Ayvalık, Tur. (āīy-wă-lĭk)	**200-01**	39°20′N	26°42′E
Azaouagh, stm., Afr.	**258-59**	15°30′N	3°18′E
Azärbaycan, nation, Asia see Azerbaijan	**227**	40°30′N	47°30′E
Azare, Nig.	**260-61**	11°41′N	10°11′E
Azemmour, Mor. (ȧ-zē-mōōr′)	**269a**	33°18′N	8°21′W
Azerbaidzhan, nation, Asia see Azerbaijan	**227**	40°30′N	47°30′E
Azerbaijan, nation, Asia (ä′zēr-bä-ê-ján′)	**227**	40°30′N	47°30′E
Azogues, Ec. (ä-sō′gās)	**170**	2°44′S	78°50′W
Azores, is., Port. (ȧ′zōrz) (ȧ-zōrz′)	**199c**	38°30′N	28°00′W
Azov, Russia (ȧ-zôf′) (ä-zôf)	**202-03**	47°07′N	39°26′E
Azov, Sea of, s., Eur. (sē ŭv ä-zôf′)	**202-03**	46°0′N	36°00′E
Azovs'ke more, s., Eur. (ȧ-zôf′skô-yĕ mô′rĕ) see Azov, Sea of	**202-03**	46°0′N	36°00′E
Azovskoye More, s., Eur. see Azov, Sea of	**202-03**	46°0′N	36°00′E
Azraq, Al-Baḥr al-, stm., Afr. see Blue Nile	**254**	15°38′N	32°30′E
Azrou, Mor.	**269a**	33°26′N	5°12′W
Aztec, N.M., U.S. (ăz′tĕk)	**118-19**	36°49′N	108°00′W
Azua, Dom. Rep. (ä′swä)	**142-43**	18°27′N	70°44′W
Azuaga, Spain (ä-thwä′gä)	**198-99**	38°15′N	5°40′W
Azuero, Península de, pen., Pan.	**150**	7°40′N	80°35′W
Azul, Arg. (ä-sōōl′)	**173**	36°47′S	59°52′W
Aẓ-Ẓahrān, Sau. Ar. see Dhahran	**230-31**	26°18′N	50°08′E
Az-Zarqā', Jord.	**228-29**	32°03′N	36°05′E
Az-Zāwiyah, Libya	**188-89**	32°47′N	12°44′E
Az-Zilfī, Sau. Ar.	**266**	26°18′N	44°49′E

B

Feature (Pronunciation)	Page	Lat.	Long.
Ba'qūbah, Iraq	**228-29**	33°45′N	44°40′E
Baaba, Île, i., N. Cal.	**279g**	20°03′S	163°58′E
Babadaýhan, Turkmen.	**232-33**	37°42′N	60°24′E
Babaeski, Tur. (bä′bä-ĕs′kĭ)	**200-01**	41°26′N	27°06′E
Babahoyo, Ec. (bä-bä-ō′yō)	**170**	1°48′S	79°32′W
Babar, Pulau, i., Indon. (pōō-lou bä′bär)	**224-25**	7°55′S	129°45′E
Babeldaob, i., Palau	**280-81**	7°30′N	134°35′E
Bab el Mandeb, strt.,	**266**	12°44′N	43°21′E
Babelthuap, i., Palau see Babeldaob	**280-81**	7°30′N	134°35′E
Babi, Pulau, i., Indon.	**246-47**	2°05′N	96°39′E
Babine Lake, lk., B.C., Can. (băb′ēn lăk)	**132-33**	54°45′N	126°00′W
Bābol, Iran	**232-33**	36°33′N	52°41′E
Babrujsk, Bela.	**202-03**	53°08′N	29°14′E
Babuyan Island, i., Phil.	**250a**	19°32′N	121°57′E
Babuyan Islands, is., Phil. (bä-bōō-yän′ ī′lȧndz)	**250a**	19°15′N	121°40′E
Bacabal, Braz.	**166-67**	4°14′S	44°47′W
Bacan, Pulau, i., Indon.	**248-49**	0°35′S	127°30′E
Bacău, Rom.	**202-03**	46°34′N	26°55′E
Bac Bo, Vinh, b., Asia see Tonkin, Gulf of	**246-47**	20°0′N	108°00′E
Back, stm., Nu., Can. (băk)	**130-31**	67°09′N	95°21′W
Bačka Palanka, Serb. (băch′kä pälän-kä)	**200-01**	45°15′N	19°24′E
Bac Lieu, Viet.	**246-47**	9°17′N	105°43′E
Bac Ninh, Viet. (băk′nēn′')	**246-47**	21°12′N	106°05′E
Baco, Mount, mtn., Phil. (mount bä′kō)	**250**	12°49′N	121°10′E
Bacolod, Phil. (bä-kō′lôd)	**250**	10°40′N	122°57′E
Badajoz, Spain (bȧ-dhä-hōth′)	**198-99**	38°53′N	6°58′W
Badalona, Spain (bä-dhä-lō′nä)	**198-99**	41°28′N	2°16′E
Bad Axe, Mi., U.S. (băd′ ăks)	**116-17**	43°48′N	82°59′W
Baden-Baden, Ger. (bä′dĕn-bä′dĕn)	**194-95**	48°46′N	8°14′E
Bad Hersfeld, Ger. (bät hĕrsh′fĕlt)	**194-95**	50°52′N	9°42′E
Bad Kissingen, Ger. (bät kĭs′ĭng-ĕn)	**194-95**	50°12′N	10°05′E
Badlands, hills, U.S. (băd′ lăndz)	**114-15**	46°14′N	103°37′W
Badlands National Park, n.p., S.D., U.S. (băd′ lănds năsh′ŭn-ăl pärk)	**114-15**	43°50′N	102°21′W
Bad Reichenhall, Ger. (bät rī′kĕn-häl)	**194-95**	47°44′N	12°53′E
Bad River Indian Reservation, ind. res., Wi., U.S. (băd rĭv′ĕr ĭn′dĭ-ăn rĕ-sēr-vā′shĕn)	**114-15**	46°33′N	90°40′W
Bad Tölz, Ger. (bät tûltz)	**194-95**	47°45′N	11°35′E
Badu Island, i., Austl.	**277**	10°07′S	142°08′E
Baena, Spain (bä-ā′nä)	**198-99**	37°37′N	4°19′W
Bafatá, Gui.-B.	**260-61**	12°10′N	14°41′W
Baffin Bay, b., N.A. (băf′ĭn bā)	**86**	73°0′N	66°00′W
Baffin Bugt, b., N.A. see Baffin Bay	**86**	73°0′N	66°00′W
Baffin Island, i., Nu., Can. (băf′ĭn ī′lȧnd)	**86**	68°0′N	70°00′W
Bafia, Camrn.	**260-61**	4°45′N	11°16′E
Bafing, stm., Afr.	**260-61**	13°47′N	10°50′W
Bafoussam, Camrn.	**260-61**	5°29′N	10°25′E
Bāfq, Iran (bäfk)	**232-33**	31°35′N	55°24′E
Bafra, Tur. (bäf′rä)	**186-87**	41°34′N	35°53′E
Bafwasende, D.R.C.	**262-63**	1°06′N	27°16′E
Bagan, hist., Mya.	**246-47**	21°13′N	94°54′E
Bagansiapiapi, Indon.	**246-47**	2°09′N	100°48′E
Bagata, D.R.C.	**262-63**	3°44′S	17°59′E
Bagdad, nat. cap., Iraq see Baghdād	**228-29**	33°21′N	44°25′E
Bagé, Braz.	**173**	31°19′S	54°06′W
Baghdād, nat. cap., Iraq (bágh-dăd′) (băg′dăd)	**228-29**	33°21′N	44°25′E
Bagheria, Italy (bä-gä-rē′ä)	**200-01**	38°05′N	13°30′E
Baghlān, Afg.	**232-33**	36°08′N	68°42′E
Bagley, Mn., U.S. (băg′lê)	**114-15**	47°30′N	95°23′W
Bagnères-de-Bigorre, Fr. (băn-yâr′dē-bê-gor′)	**196-97**	43°04′N	0°09′E
Bago, Mya.	**246-47**	17°20′N	96°29′E
Bagoé, stm., Afr. (bȧ-gō′ā)	**260-61**	12°35′N	6°34′W
Baguio, Phil. (bä-gê-ō′)	**250**	16°25′N	120°36′E
Bahama, Canal Viejo de, strt., N.A. see Old Bahama Channel	**142-43**	22°40′N	78°41′W
Bahama Islands, is., Bah.	**20-21**	24°15′N	76°00′W
Bahamas, nation, N.A. (bȧ-hä′mȧs)	**140-41**	24°15′N	76°00′W
Baharampur, India	**234-35**	24°06′N	88°15′E
Bahāwalpur, Pak. (bŭ-hä′wŭl-pōōr)	**232-33**	29°23′N	71°40′E
Bäherden, Turkmen.	**232-33**	38°26′N	57°26′E
Bahia, Braz. (bä-ē′ȧ) see Salvador	**166-67**	12°59′S	38°30′W
Bahia, state, Braz.	**166-67**	12°0′S	42°00′W
Bahía, Islas de la, is., Hond. (ē′s-läs-dĕ-lä-bä-ē′ä)	**149**	16°20′N	86°30′W
Bahía Blanca, Arg. (bä-ē′ä bläŋ′kä)	**173**	38°43′S	62°17′W

n-sing; ŋ-baŋk; N-nasalized n; nŏd; cŏmmit; ōld; ôbey; ôrder; oi-boil; fōōd; ŏ-as oo in foot; ou-out; s-soft; sh-dish; th-thin; pūre; ûnite; ûrn; stŭd; circŭs; ü-as in French tu; ′-indeterminate vowel.

Feature (Pronunciation)	Page	Lat.	Long.
Bahía de Caráquez, Ec. (bä-e′ä dä kä-rä′kĕz)	170	0°37′S	80°26′W
Bahir Dar, Eth.	266	11°35′N	37°24′E
Bahraich, India	234-35	27°35′N	81°36′E
Bahrain, nation, Asia (bä-rān′)	230-31	26°0′N	50°30′E
Baḩrānī, Ḩālat al-, i., U.A.E.	230-31	24°28′N	54°21′E
Baia Mare, Rom. (bä′yä mä′rā)	194-95	47°39′N	23°35′E
Baicheng, China	240-41	45°37′N	122°51′E
Baidoa, Som.	262-63	3°07′N	43°39′E
Baie-Comeau, Qc., Can.	138-39	49°13′N	68°10′W
Baie-Saint-Paul, Qc., Can. (bā′săn′-pôl′)	138-39	47°27′N	70°30′W
Baikal, Lake, lk., Russia (lăk bī-käl′)	218-19	53°0′N	107°40′E
Baile Átha Cliath, nat. cap., Ire. *see* Dublin	190-91	53°21′N	6°15′W
Bailén, Spain (bä-ē-lān′)	198-99	38°06′N	3°46′W
Băileşti, Rom. (bă-ĭ-lĕsh′tē)	200-01	44°02′N	23°21′E
Bailong, stm., China	238-39	32°21′N	105°43′E
Bainbridge, Ga., U.S. (bān′brĭj)	124-25	30°54′N	84°34′W
Bainbridge, Oh., U.S. (bān′brĭj)	116-17	39°14′N	83°16′W
Baiquan, China (bī-chyuäṇ)	240-41	47°36′N	126°05′E
Baird, Tx., U.S. (bârd)	120-21	32°24′N	99°23′W
Bairin Zuoqi, China	240-41	43°59′N	119°23′E
Bairnsdale, Austl. (bârnz′dāl)	276	37°50′S	147°37′E
Baishuijiang, China	238-39	33°29′N	106°02′E
Baitou Shan, mtn., Asia *see* Paektu-san	243	41°59′N	128°07′E
Baiyin, China	240-41	36°33′N	104°12′E
Baja, Hung. (bŏ′yŏ)	194-95	46°11′N	18°57′E
Baja California, state, Mex. (bä-hä käl-ĭ-fôr′nĭ-*á*)	144-45	30°0′N	115°00′W
Baja California, pen., Mex. (bä-hä käl-ĭ-fôr′nĭ-*á*)	144-45	27°53′N	113°28′W
Baja California Norte, state, Mex. *see* Baja California	144-45	30°0′N	115°00′W
Baja California Sur, state, Mex. (bä-hä käl-ĭ-fôr′nĭ-*á* sōōr′)	144-45	26°0′N	112°00′W
Bajestān, Iran	232-33	34°31′N	58°11′E
Bājil, Yemen	266	15°04′N	43°17′E
Bajo Boquete, Pan.	150	8°47′N	82°26′W
Baker, Ca., U.S. (bā′kẽr)	118-19	35°16′N	116°04′W
Baker, La., U.S. (bā′kẽr)	124-25	30°35′N	91°10′W
Baker, Mt., U.S. (bā′kẽr)	112-13	46°22′N	104°17′W
Baker, Or., U.S. (bā′kẽr)	112-13	44°47′N	117°50′W
Baker, Mount, vol., Wa., U.S. (mount bā′kẽr)	112-13	48°47′N	121°49′W
Baker Island, dep., Oc. (bā′kẽr ī′l*ă*nd)	280-81	0°15′N	176°27′W
Baker Island, i., Oc.	280-81	0°12′N	176°29′W
Baker Lake, Nu., Can.	128-29	64°18′N	95°55′W
Baker Lake, lk., Nu., Can. (bā′kẽr lāk)	130-31	64°10′N	95°30′W
Bakersfield, Ca., U.S. (bā′kẽrz-fēld)	118-19	35°22′N	119°01′W
Bakhmach, Ukr. (bȧĸ-mäch′)	202-03	51°11′N	32°50′E
Bākhtarān, Iran *see* Kermānshāh	228-29	34°18′N	47°04′E
Bakhtegān, Daryācheh-ye, lk., Iran	230-31	29°20′N	54°05′E
Bakı, nat. cap., Azer. *see* Baku	227	40°23′N	49°51′E
Bakony, mts., Hung. (bȧ-kōn′y′)	194-95	47°01′N	17°45′E
Bakoy, stm., Afr. (bȧ-kô′ē)	258-59	13°48′N	10°49′W
Baku, nat. cap., Azer. (bȧ-kōō′)	227	40°23′N	49°51′E
Bakwanga, D.R.C. *see* Mbuji-Mayi	262-63	6°08′S	23°39′E
Balabac, Selat, strt., Asia *see* Balabac Strait	250	7°35′N	117°00′E
Balabac Island, i., Phil. (bä′lä-bäk ī′l*ă*nd)	250	7°57′N	117°01′E
Balabac Strait, strt., Asia (bä′lä-bäk strāt)	250	7°35′N	117°00′E
Balabanovo, Russia (bȧ-lȧ-bä′nô-vô)	202-03	55°11′N	36°40′E
Balabio, Île, i., N. Cal.	279g	20°07′S	164°11′E
Bālāghāt, India	234-35	21°49′N	80°11′E
Balaguer, Spain (bä-lä-gĕr′)	198-99	41°48′N	0°49′E
Balakovo, Russia (bȧ′lȧ-kô′vô)	186-87	52°01′N	47°47′E
Balambangan, Pulau, i., Malay.	248-49	7°16′N	116°55′E
Balāngīr, India	234-35	20°43′N	83°30′E
Balaözen, stm., Eur.	186-87	48°58′N	49°38′E
Balashov, Russia (bȧ′lȧ-shôf)	186-87	51°32′N	43°10′E
Balasore, India (bä-lȧ-sōr′) *see* Bāleshwar	234-35	21°29′N	86°57′E
Balassagyarmat, Hung. (bô′lôsh-shô-dyôr′môt)	194-95	48°04′N	19°19′E
Balaton, lk., Hung. (bô′lô-tôn)	194-95	46°50′N	17°45′E
Balayan, Phil. (bä-lä-yän′)	250	13°57′N	120°44′E
Balbina, Represa, res., Braz.	166-67	1°20′S	59°40′W
Balcarce, Arg. (bäl-kär′sā)	173	37°51′S	58°15′W
Baldock Lake, lk., Mb., Can.	134-35	56°33′N	97°57′W
Baldwinsville, N.Y., U.S. (bôld′wĭns-vĭl)	116-17	43°10′N	76°20′W
Baldy Peak, mtn., Az., U.S. (bôl′dē pēk)	118-19	33°55′N	109°35′W
Bâle, Switz. *see* Basel	194-95	47°33′N	7°36′E
Baleares, Islas, is., Spain *see* Balearic Islands	198-99	39°29′N	3°01′E
Balearic Islands, is., Spain (bă-lē-ă′-rĭk ī′l*ă*ndz)	198-99	39°29′N	3°01′E
Balears, Illes, is., Spain *see* Balearic Islands	198-99	39°29′N	3°01′E
Baler, Phil. (bä-lar′)	250	15°46′N	121°34′E
Bāleshwar, India	234-35	21°29′N	86°57′E
Baley, Russia (b*ȧ*l-yā′)	222-23	51°34′N	116°38′E
Bali, i., Indon. (bä′lê)	248-49	8°20′S	115°00′E
Bali, Laut, s., Indon. *see* Bali Sea	248-49	7°45′S	115°30′E
Balıkesir, Tur. (balĭk′ĭysĭr)	200-01	39°39′N	27°53′E
Balikpapan, Indon. (bä′lĕk-pä′pän)	248-49	1°16′S	116°50′E
Balimo, Pap. N. Gui.	277	8°03′S	142°56′E
Balin, China	240-41	48°19′N	122°19′E
Balintang Channel, strt., Phil. (bä-lĭn-täng′ chăn′ĕl)	250a	19°59′N	121°51′E
Bali Sea, s., Indon. (bä′lē sē)	248-49	7°45′S	115°30′E
Balkanabat, Turkmen.	232-33	39°31′N	54°23′E
Balkan Peninsula, pen., Eur. (bôl′k*ă*n p*ĕ*-nĭn′sūl*ă*)	200-01	44°0′N	23°00′E
Balkaria, state, Russia	227	43°30′N	43°30′E
Balkh, Afg. (bälk)	232-33	36°45′N	66°54′E
Balkh, stm., Afg.	232-33	36°38′N	66°56′E
Balkhash, Kaz. *see* Balqash	226	46°51′N	74°58′E
Balkhash, Lake, lk., Kaz. (lăk bȧl-kăsh′)	226	46°0′N	74°00′E
Ballarat, Austl. (băl′*ă*-răt)	276	37°34′S	143°51′E
Ballard, Lake, lk., Austl. (lăk băl′*ă*rd)	272-73	29°27′S	120°55′E
Ballia, India	234-35	25°45′N	84°09′E
Ballina, Austl. (băl-ĭ-nä′)	276	28°52′S	153°33′E
Ballinasloe, Ire. (băl′ĭ-n*á*-slō′)	190-91	53°20′N	8°14′W
Ballinger, Tx., U.S. (băl′ĭn-jẽr)	122-23	31°44′N	99°57′W
Ballston Spa, N.Y., U.S. (bôls′t*ŭ*n spä′)	116-17	43°00′N	73°51′W
Balonne, stm., Austl. (băl-ōn′)	276	28°37′S	148°10′E
Balqash, Kaz.	226	46°51′N	74°58′E
Balqash köli, lk., Kaz. *see* Balkhash, Lake	226	46°0′N	74°00′E
Balranald, Austl. (băl′-r*ă*n-ăld)	276	34°38′S	143°33′E
Balsas, Braz. (bäl′säs)	166-67	7°33′S	46°04′W
Balsas, stm., Braz.	166-67	7°14′S	44°34′W
Balsas, stm., Mex.	146-47	17°54′N	102°11′W
Balta, Ukr. (bäl′tȧ)	202-03	47°56′N	29°40′E
Baltasar Brum, Ur.	173	30°42′S	57°19′W
Bălţi, Mol.	202-03	47°46′N	27°55′E
Baltic Sea, s., Eur. (bôl′tĭk sē)	192-93	57°0′N	19°00′E
Baltijas jūra, s., Eur. *see* Baltic Sea	192-93	57°0′N	19°00′E
Baltijos jūra, s., Eur. *see* Baltic Sea	192-93	57°0′N	19°00′E
Baltim, Egypt (băl-tēm′)	268b	31°34′N	31°05′E
Baltimore, Md., U.S. (bôl′tĭ-môr)	116-17	39°17′N	76°37′W
Baltiysk, Russia (bäl-tēysk′)	194-95	54°39′N	19°55′E
Baltiyskoye More, s., Eur. *see* Baltic Sea	192-93	57°0′N	19°00′E
Bałtyckie, Morze, s., Eur. *see* Baltic Sea	192-93	57°0′N	19°00′E
Balūchestān, hist. reg., Asia *see* Baluchistan	232-33	28°0′N	63°00′E
Baluchistan, hist. reg., Asia	232-33	28°0′N	63°00′E
Baluchistān, hist. reg., Asia (bȧ-lò-chĭ-stän′) *see* Baluchistan	232-33	28°0′N	63°00′E
Balykchy, Kyrg.	226	42°28′N	76°12′E
Balyqshy, Kaz.	186-87	47°05′N	51°54′E
Bam, Iran	230-31	29°07′N	58°21′E
Bama, Nig.	260-61	11°32′N	13°41′E
Bamako, nat. cap., Mali (bä-mä-kō′)	258-59	12°39′N	7°60′W
Bambari, C.A.R. (bäm-bȧ-rē)	262-63	5°46′N	20°39′E
Bamberg, Ger. (bäm′bĕrgh)	194-95	49°54′N	10°54′E
Bamberg, S.C., U.S. (băm′bûrg)	124-25	33°18′N	81°02′W
Bamenda, Camrn.	260-61	5°58′N	10°09′E
Bāmīān, Afg.	232-33	34°50′N	67°49′E
Bamingui, stm., C.A.R.	262-63	8°34′N	19°04′E
Bamingui-Bangoran, Parc National du, n.p., C.A.R.	262-63	7°54′N	19°42′E
Bampūr, Iran (bŭm-pōōr′)	230-31	27°11′N	60°26′E
Banaba, i., Kir.	280-81	0°52′S	169°33′E
Banaras, India *see* Vārānasi	234-35	25°20′N	82°59′E
Banās, stm., India (băn-äs′)	234-35	25°55′N	76°44′E
Banâs, Râs, c., Egypt	266	23°54′N	35°47′E
Ban Bat, Viet.	246-47	13°13′N	108°40′E
Bancroft, On., Can. (băn′krŏft)	136-37	45°03′N	77°51′W
Bānda, India (bän′dä)	234-35	25°29′N	80°20′E
Banda, Laut, s., Indon. *see* Banda Sea	224-25	5°0′S	128°00′E
Banda Aceh, Indon.	246-47	5°33′N	95°19′E
Bandama, stm., C. Iv.	260-61	5°08′N	4°60′W
Bandar, India *see* Machilīpatnam	236	16°11′N	81°09′E
Bandar 'Abbās, Iran	230-31	27°11′N	56°16′E
Bandar Beheshtī, Iran	230-31	25°18′N	60°38′E
Bandar-e Anzalī, Iran	227	37°28′N	49°28′E
Bandar-e Khomeynī, Iran	228-29	30°26′N	49°06′E
Bandar-e Lengeh, Iran	230-31	26°34′N	54°53′E
Bandar-e Pahlavī, Iran *see* Bandar-e Anzalī	227	37°28′N	49°28′E
Bandar-e Shāhpūr, Iran *see* Bandar-e Khomeynī	228-29	30°26′N	49°06′E
Bandar-e Torkeman, Iran	232-33	36°54′N	54°04′E
Bandar Lampung, Indon.	246-47	5°26′S	105°16′E
Bandar Maharani, Malay. (bän-där′ mä-hä-rä′nê) *see* Muar	246-47	2°02′N	102°34′E
Bandar Penggaram, Malay. *see* Batu Pahat	246-47	1°51′N	102°56′E
Bandar Seri Begawan, nat. cap., Bru. (bän′där sĕr′ē bŭ′gä-wän)	248-49	4°56′N	114°56′E
Banda Sea, s., Indon. (bän′-dä sē)	224-25	5°0′S	128°00′E
Bandeira, Pico da, mtn., Braz. (pē′kò dä bän dā′rä)	172	20°26′S	41°47′W
Bandelier National Monument, n.p., N.M., U.S. (băn-dĕ-lēr′ năsh′ŭn-*ă*l mŏn′*ū*-m*ĕ*nt)	118-19	35°45′N	106°20′W
Bandera, Arg.	173	28°53′S	62°16′W
Banderas, Bahía de, b., Mex. (bä-ē′ä dĕ bän-dĕ′räs)	146-47	20°38′N	105°27′W
Bandiantaolehai, China	240-41	41°47′N	104°05′E
Bandırma, Tur. (bän-dĭr′mȧ)	200-01	40°22′N	27°59′E
Ban Don, Thai. *see* Surat Thani	246-47	9°06′N	99°18′E
Bandon, Or., U.S. (băn′d*ŭ*n)	112-13	43°07′N	124°23′W
Bandundu, D.R.C.	262-63	3°16′S	17°21′E
Bandung, Indon.	248-49	6°54′S	107°36′E
Banes, Cuba (bä′nās)	142-43	20°58′N	75°42′W
Banff, Ab., Can. (bănf)	132-33	51°10′N	115°36′W
Banff National Park, n.p., Ab., Can. (bănf năsh′ŭn-*ă*l pärk)	132-33	51°38′N	116°22′W
Banfora, Burkina	260-61	10°39′N	4°45′W
Bangalore, India (băŋ′gȧ′lôr) *see* Bengalūru	236	12°59′N	77°36′E
Bangassou, C.A.R. (bäɴ-gȧ-sōō′)	262-63	4°44′N	22°49′E
Banggai, Indon.	248-49	1°35′S	123°30′E
Banggai, Kepulauan, is., Indon. (bäng-gī′)	248-49	1°30′S	123°15′E
Banggai, Pulau, i., Indon.	248-49	1°37′S	123°33′E
Banggi, Pulau, i., Malay.	248-49	7°16′N	117°10′E
Banggong Co, lk., Asia (bäŋ-gōŋ tswo) *see* Pangong Tso	234-35	33°45′N	78°42′E
Banghāzī, Libya (bĕn-gä′zē)	258-59	32°07′N	20°04′E
Bangka, Pulau, i., Indon. (pōō-lou băŋ′k*á*)	246-47	2°15′S	106°00′E
Bangka, Selat, strt., Indon.	246-47	2°20′S	105°45′E
Bangkalan, Indon. (bäng-kȧ-län′)	248-49	7°02′S	112°45′E
Bangkok, nat. cap., Thai. (băɴ′kŏk)	246-47	13°45′N	100°31′E
Bangkulu, Pulau, i., Indon.	248-49	1°50′S	123°06′E
Bangladesh, nation, Asia (bän′-gl*ă*-dĕsh′)	206-07	24°0′N	90°00′E
Bangor, N. Ire., U.K. (băŋ′ŏr)	190-91	54°39′N	5°41′W
Bangor, Wales, U.K. (băŋ′ŏr)	190-91	53°14′N	4°09′W
Bangor, Me., U.S.	117a	44°48′N	68°47′W
Bangor, Mi., U.S. (băn′gẽr)	116-17	42°19′N	86°06′W
Bangor, Pa., U.S. (băn′gẽr)	116-17	40°51′N	75°13′W
Bangued, Phil. (bän-gād′)	250	17°36′N	120°37′E
Bangui, nat. cap., C.A.R. (bäɴ-gē′)	262-63	4°22′N	18°33′E
Bangweulu, Lake, lk., Zam. (lăk băng-wē-ōō′lōō)	264-65	11°04′S	29°53′E
Ban Hat Yai, Thai. *see* Hat Yai	246-47	7°01′N	100°28′E
Ban Houayxay, Laos	246-47	20°15′N	100°24′E
Baní, Dom. Rep. (bä′-nē)	142-43	18°17′N	70°19′W
Banifing, stm., Mali	260-61	14°29′N	4°13′W
Banja Luka, Bos. (bän-yä-lōō′kȧ)	200-01	44°46′N	17°12′E
Banjarbaru, Indon.	248-49	3°24′S	114°50′E
Banjarmasin, Indon. (bän-jẽr-mä′sên)	248-49	3°20′S	114°36′E
Banjul, nat. cap., Gam. (bôn-jōōl′)	260-61	13°27′N	16°36′W
Banks, Îles, is., Vanuatu *see* Banks Islands	279g	13°25′S	167°42′E
Banks Island, i., B.C., Can. (bănks ī′l*ă*nd)	132-33	53°25′N	130°10′W
Banks Island, i., N.T., Can. (bănks ī′l*ă*nd)	86	73°15′N	121°30′W
Banks Islands, is., Vanuatu	279g	13°25′S	167°42′E
Banks Peninsula, pen., N.Z. (bănks p*ĕ*-nĭn′sūl*ă*)	278	43°45′S	173°00′E
Banks Strait, strt., Austl. (bănks strāt)	276	40°40′S	148°07′E
Banningville, D.R.C. *see* Bandundu	262-63	3°16′S	17°21′E
Bannu, Pak.	232-33	32°59′N	70°37′E
Baños, Ec. (bä′-nyôs)	170	1°24′S	78°25′W

ăt; fināl; rāte; senāte; ärm; ȧsk; sofȧ; fâre; ch-choose; dh-as th in other; bē; ēvent; bĕt; recĕnt; cratẽr; g-gō; gh-guttural g; bĭt; ĭ-short neutral; rīde; ĸ-guttural k as ch in German ich;

Feature (Pronunciation)	Page	Lat.	Long.
Bānswāra, India	234-35	23°33′N	74°27′E
Bantaeng, Indon.	248-49	5°32′S	119°56′E
Bantayan Island, i., Phil.	250	11°13′N	123°44′E
Bantry, Ire. (băn′trĭ)	190-91	51°41′N	9°27′W
Banyak, Kepulauan, is., Indon.	246-47	2°10′N	97°15′E
Banyuwangi, Indon. (bän-jò-wäŋ′gē)	248-49	8°12′S	114°21′E
Baode, China	240-41	39°01′N	111°05′E
Baoding, China (bou-dĭŋ)	240-41	38°51′N	115°29′E
Baoji, China (bou-jyē)	238-39	34°23′N	107°09′E
Bao Lac, Viet.	246-47	11°33′N	107°47′E
Baoshan, China (bou-shän)	238-39	25°07′N	99°10′E
Baoting, China	238-39	18°38′N	109°47′E
Baotou, China (bou-tō)	240-41	40°35′N	109°58′E
Baoying, China (bou-yĭŋ)	238-39	33°14′N	119°19′E
Baquedano, Chile	168-69	23°20′S	69°50′W
Bar, Mont.	200-01	42°05′N	19°06′E
Baraboo, Wi., U.S. (bâr′ȧ-bōō)	116-17	43°28′N	89°44′W
Baracoa, Cuba (bä-rä-kō′ä)	142-43	20°21′N	74°30′W
Baradero, Arg. (bä-rä-dĕ′ō)	173	33°48′S	59°31′W
Baragaon, India *see* Nālanda	234-35	25°08′N	85°24′E
Barahona, Dom. Rep. (bä-rä-ô′nä)	142-43	18°12′N	71°06′W
Baranaviĉy, Bela.	194-95	53°08′N	26°01′E
Baranof Island, i., Ak., U.S. (bä-rä′nôf ī′lȧnd)	126	57°0′N	135°00′W
Barão de Melgaço, Braz. (bä-rouN-dĕ-mĕl-gä′sō)	168-69	16°13′S	55°58′W
Barat Daya, Kepulauan, is., Indon.	248-49	7°25′S	128°00′E
Baraya, Col. (bä-rà′yä)	163c	3°10′N	75°04′W
Barbacena, Braz. (bär-bä-sā′nȧ)	172	21°13′S	43°45′W
Barbacoas, Col. (bär-bä-kō′äs)	164-65	1°41′N	78°09′W
Barbados, nation, N.A. (bär-bā′dōz)	140-41	13°10′N	59°32′W
Barbas, Cap, c., W. Sah.	258-59	22°18′N	16°40′W
Barbastro, Spain (bär-bäs′trō)	198-99	42°02′N	0°08′E
Barberton, Oh., U.S. (bär′bĕr-tŭn)	116-17	41°02′N	81°36′W
Barbosa, Col. (bär-bô′-sȧ)	163c	6°26′N	75°20′W
Barboursville, W.V., U.S. (bär′bĕrs-vĭl)	116-17	38°25′N	82°18′W
Barbuda, i., Antig. (bär-bōō′dȧ)	143b	17°38′N	61°48′W
Barcaldine, Austl. (bär′kôl-dīn)	277	23°34′S	145°18′E
Barce, Libya *see* Al-Marj	188-89	32°30′N	20°53′E
Barcelona, Spain (bär-thå-lō′nä)	198-99	41°24′N	2°10′E
Barcelona, Ven. (bär-så-lō′nä)	163b	10°08′N	64°41′W
Barcelos, Braz. (bär-sĕ′lôs)	166-67	0°59′S	62°54′W
Barcelos, Port. (bär-thå′lōs)	198-99	41°32′N	8°37′W
Barcoo, stm., Austl.	277	25°12′S	142°50′E
Bardaï, Chad	258-59	21°22′N	16°59′E
Barddhamān, India	234-35	23°14′N	87°52′E
Bardsey Island, i., Wales, U.K. (bärd′sē ī′lȧnd)	190-91	52°46′N	4°48′W
Bardstown, Ky., U.S. (bärds′toun)	116-17	37°49′N	85°28′W
Bardwell, Ky., U.S. (bärd′wĕl)	124-25	36°52′N	89°01′W
Bareilly, India	234-35	28°21′N	79°25′E
Barentsevo More, s., Eur. *see* Barents Sea	218-19	74°0′N	36°00′E
Barentshavet, s., Eur. *see* Barents Sea	218-19	74°0′N	36°00′E
Barents Sea, s., Eur. (bä′rĕnts sē)	218-19	74°0′N	36°00′E
Barentu, Erit. (bä-rĕn′tōō)	266	15°07′N	37°35′E
Barfleur, Pointe de, c., Fr. (pwănt′ dĕ bȧr-flûr′)	196-97	49°42′N	1°16′W
Barguzin, stm., Russia	218-19	53°25′N	108°59′E
Bar Harbor, Me., U.S. (bär här′bĕr)	117a	44°23′N	68°13′W
Bari, Italy (bä′rē)	200-01	41°07′N	16°52′E
Bariloche, Arg. *see* San Carlos de Bariloche	171	41°09′S	71°18′W
Barinas, Ven. (bä-rē′näs)	164-65	8°38′N	70°13′W
Baring, Cape, c., N.T., Can. (kāp bâr′ĭng)	130-31	70°03′N	117°16′W
Bāripada, India	234-35	21°56′N	86°43′E
Barisāl, Bngl.	234-35	22°42′N	90°22′E
Barito, stm., Indon. (bä-rē′tō)	248-49	3°20′S	114°32′E
Barkley, Lake, res., U.S.	124-25	36°44′N	87°57′W
Barkley Sound, strt., B.C., Can.	132-33	48°53′N	125°20′W
Barkly Tableland, plat., Austl. (bär′klē tā′-bel-lănd)	272-73	18°0′S	136°00′E
Barkol, China (bär-kůl)	240-41	43°33′N	93°02′E
Bar-le-Duc, Fr. (băr-lē-dük′)	196-97	48°47′N	5°10′E
Barlee, Lake, lk., Austl. (lăk bär-lē′)	272-73	29°10′S	119°30′E
Barletta, Italy (bär-lĕt′tä)	200-01	41°19′N	16°17′E
Bārmer, India	234-35	25°44′N	71°24′E
Barnaul, Russia (bär-nä-ōl′)	226	53°22′N	83°45′E
Barnesville, Ga., U.S. (bärnz′vĭl)	124-25	33°03′N	84°10′W
Barnesville, Mn., U.S. (bärnz′vĭl)	114-15	46°39′N	96°25′W
Barnsley, Eng., U.K. (bärnz′lĭ)	190-91	53°34′N	1°29′W
Barnstaple, Eng., U.K. (bärn′stā-p'l)	190-91	51°05′N	4°03′W
Barnwell, S.C., U.S. (bärn′wĕl)	124-25	33°14′N	81°22′W
Baro, stm., Afr.	262-63	8°26′N	33°13′E
Baroda, India (bär-rō′dä) *see* Vadodara	234-35	22°18′N	73°11′E
Barpeta, India	234-35	26°19′N	91°00′E
Barqah, hist. reg., Libya *see* Cyrenaica	258-59	31°0′N	22°30′E
Barquisimeto, Ven. (bär-kē-sē-mā′tō)	164-65	10°05′N	69°19′W
Barra, Braz. (bär′rä)	166-67	11°05′S	43°09′W
Barra, Ponta da, c., Moz.	264-65	23°48′S	35°31′E
Barra do Corda, Braz. (bär′rä dò cōr-dä)	166-67	5°31′S	45°15′W
Barra Falsa, Ponta da, c., Moz.	264-65	22°54′S	35°34′E
Barra Mansa, Braz. (bär′rä män′sä)	172	22°33′S	44°10′W
Barranca, Peru	170	10°45′S	77°46′W
Barrancabermeja, Col. (bär-räŋ′kä-bēr-mā′hä)	164-65	7°04′N	73°51′W
Barrancas, Ven.	164-65	8°44′N	62°11′W
Barranquilla, Col. (bär-rän-kēl′yä)	164-65	10°59′N	74°48′W
Barras, Braz. (bá′r-räs)	166-67	4°15′S	42°18′W
Barre, Vt., U.S. (băr′ē)	116-17	44°12′N	72°30′W
Barreiras, Braz. (bär-rā′räs)	166-67	12°09′S	45°01′W
Barreiro, Port. (bär-rĕ′ē-rò)	198-99	38°39′N	9°04′W
Barreiros, Braz.	163d	8°49′S	35°12′W
Barren, Nosy, is., Madag.	264-65	18°30′S	43°53′E
Barretos, Braz. (bär-rā′tōs)	172	20°34′S	48°34′W
Barrhead, Ab., Can. (băr′ĭd)	132-33	54°07′N	114°24′W
Barrie, On., Can. (băr′ĭ)	136-37	44°23′N	79°41′W
Barrington Tops, mtn., Austl. (bă-rĕng-tŏn tŏps)	276	32°0′S	151°28′E
Barron, Wi., U.S. (băr′ŭn)	114-15	45°24′N	91°51′W
Barrow, Ak., U.S. (băr′ō)	126	71°18′N	156°38′W
Barrow, stm., Ire. (bȧ-rå)	190-91	52°17′N	7°00′W
Barrow, Point, c., Ak., U.S. (point băr′ō)	126	71°23′N	156°29′W
Barrow Creek, Austl.	270-71	21°31′S	133°55′E
Barrow Island, i., Austl.	272-73	20°48′S	115°23′E
Bārsi, India	236	18°14′N	75°42′E
Barstow, Ca., U.S. (bär′stō)	118-19	34°54′N	117°01′W
Bartica, Guy. (bär′tĭ-kȧ)	164-65	6°24′N	58°37′W
Bartın, Tur. (bär′tĭn)	186-87	41°38′N	32°21′E
Bartle Frere, mtn., Austl. (bärt′'l frēr′)	277	17°23′S	145°49′E
Bartlesville, Ok., U.S. (bär′tlz-vil)	120-21	36°45′N	95°59′W
Bartlett, Tn., U.S. (bärt′lĕt)	124-25	35°13′N	89°52′W
Bartlett, Tx., U.S. (bärt′lĕt)	122-23	30°48′N	97°26′W
Bartoszyce, Pol. (bär-tô-shĭ′tsȧ)	194-95	54°15′N	20°49′E
Bartow, Fl., U.S. (bär′tō)	125a	27°54′N	81°50′W
Bārū, stm., Afr. *see* Baro	262-63	8°26′N	33°13′E
Barú, Volcán, vol., Pan.	150	8°48′N	82°33′W
Baruun-Urt, Mong.	240-41	46°41′N	113°17′E
Barwon, stm., Austl. (bär′wŭn)	276	30°08′S	147°23′E
Barycz, stm., Pol. (bä′rĭch)	194-95	51°41′N	16°15′E
Barysau, Bela.	192-93	54°14′N	28°31′E
Barysh, Russia	186-87	53°39′N	47°07′E
Basankusu, D.R.C. (bä-sän-kōō′sōō)	262-63	1°13′N	19°48′E
Basarabia, hist. reg., Eur. *see* Bessarabia	202-03	46°53′N	28°44′E
Basco, Phil.	250a	20°27′N	121°58′E
Bascuñán, Cabo, c., Chile	168-69	28°52′S	71°29′W
Basel, Switz. (bä′z'l)	194-95	47°33′N	7°36′E
Basey, Phil.	250	11°18′N	125°04′E
Bashi Channel, strt., Asia (bäsh′ē chăn′ĕl)	222-23	22°0′N	121°00′E
Bashkortostan, state, Russia	186-87	54°0′N	56°00′E
Bashtanka, Ukr. (bȧsh-tän′kȧ)	202-03	47°24′N	32°27′E
Basilaki Island, i., Pap. N. Gui.	277	10°37′S	150°60′E
Basilan, Phil. *see* Isabela	250	6°41′N	121°58′E
Basilan Island, i., Phil.	250	6°34′N	122°03′E
Basin, Wy., U.S. (bā′s'n)	112-13	44°23′N	108°03′W
Basingstoke, Eng., U.K. (bā′zĭng-stōk)	190-91	51°16′N	1°07′W
Başkale, Tur. (bäsh-kä′lĕ)	227	38°03′N	44°01′E
Baskatong, Réservoir, res., Qc., Can.	136-37	46°46′N	75°50′W
Basoko, D.R.C. (bȧ-sō′kō)	262-63	1°14′N	23°36′E
Bas Qafqaz Silsilasi, mts., *see* Caucasus Mountains	227	42°38′N	45°00′E
Basra, Iraq (bŭs′-rä)	228-29	30°30′N	47°48′E
Bassano, Ab., Can. (bäs-sän′ō)	132-33	50°47′N	112°27′W
Bassein, Mya. *see* Pathein	246-47	16°46′N	94°44′E
Basse-Terre, i., Guad. (bȧs′ târ′)	143b	16°10′N	61°40′W
Basse-Terre, nat. cap., Guad. (bȧs′ târ′)	143b	16°00′N	61°43′W
Basseterre, nat. cap., St. K./N.	143b	17°18′N	62°44′W
Bassett, Ne., U.S. (băs′sĕt)	114-15	42°35′N	99°32′W
Bassett, Va., U.S. (băs′sĕt)	124-25	36°46′N	79°59′W
Bass Strait, strt., Austl. (băs strāt)	276	39°20′S	145°30′E
Båstad, Swe. (bô′stät)	192-93	56°25′N	12°52′E
Bastia, Fr. (bäs′tē-ä)	184-85	42°42′N	9°27′E
Bastrop, La., U.S. (băs′trŭp)	120-21	32°47′N	91°55′W
Bastrop, Tx., U.S. (băs′trŭp)	122-23	30°06′N	97°18′W
Basutoland, nation, Afr. *see* Lesotho	253	29°30′S	28°30′E
Bata, Eq. Gui. (bä′tä)	260-61	1°52′N	9°46′E
Batabanó, Golfo de, b., Cuba (gôl-fō-dĕ-bä-tä-bä′nô)	142-43	22°15′N	82°30′W
Batagay, Russia	218-19	67°40′N	134°40′E
Batagay-Alyta, Russia	218-19	67°48′N	130°25′E
Batala, India	234-35	31°49′N	75°13′E
Batang, China (bä-täŋ)	238-39	30°02′N	99°11′E
Batangafo, C.A.R.	262-63	7°19′N	18°18′E
Batangas, Phil. (bä-tän′gäs)	250	13°46′N	121°04′E
Batan Island, i., Phil.	250	13°15′N	124°00′E
Batan Island, i., Phil.	250a	20°27′N	121°59′E
Batan Islands, is., Phil. (bä-tän′ ī′lăndz)	250a	20°30′N	121°50′E
Batanta, Pulau, i., Indon.	224-25	0°52′S	130°39′E
Batavia, Il., U.S. (bȧ-tā′vĭ-ȧ)	116-17	41°51′N	88°19′W
Batavia, Oh., U.S. (bȧ-tā′vĭ-ȧ)	116-17	39°05′N	84°11′W
Batavia, nat. cap., Indon. *see* Jakarta	248-49	6°11′S	106°50′E
Bataysk, Russia (bȧ-tīsk′)	202-03	47°08′N	39°46′E
Bătdâmbâng, Camb. (bȧt-tȧm-bāng′)	246-47	13°06′N	103°12′E
Batesville, Ar., U.S. (bāts′vĭl)	124-25	35°47′N	91°39′W
Batesville, In., U.S. (bāts′vĭl)	116-17	39°18′N	85°13′W
Batesville, Ms., U.S. (bāts′vĭl)	124-25	34°19′N	89°57′W
Bath, N.B., Can. (bȧth)	138-39	46°31′N	67°35′W
Bath, Eng., U.K. (bȧth)	190-91	51°23′N	2°22′W
Bathurst, Austl. (băth′ŭrst)	276	33°25′S	149°35′E
Bathurst, N.B., Can.	138-39	47°36′N	65°39′W
Bathurst, nat. cap., Gam. *see* Banjul	260-61	13°27′N	16°36′W
Bathurst, Cape, c., N.T., Can. (kāp bath′-ûrst)	130-31	70°35′N	128°00′W
Bathurst Island, i., Austl. (bȧth′ŭrst ī′lȧnd)	272-73	11°37′S	130°17′E
Batna, Alg. (bät′nä)	269b	35°34′N	6°11′E
Baton Rouge, La., U.S. (băt′ŭn rōōzh′)	124-25	30°27′N	91°08′W
Battambang, Camb. *see* Bătdâmbâng	246-47	13°06′N	103°12′E
Batticaloa, Sri L.	236	7°43′N	81°42′E
Battle, stm., Can.	130-31	52°42′N	108°15′W
Battle Creek, Mi., U.S. (băt′'l krēk′)	116-17	42°19′N	85°11′W
Battle Creek, Ne., U.S. (băt′'l krēk′)	114-15	42°00′N	97°36′W
Battle Harbour, Nf., Can. (băt′'l här′bĕr)	128-29	52°16′N	55°35′W
Battle Mountain, Nv., U.S. (băt′'l moun′tĭn)	112-13	40°39′N	116°55′W
Batu, mtn., Eth.	269d	6°55′N	39°46′E
Batu, Kepulauan, is., Indon. (bä′tōō)	246-47	0°18′S	98°28′E
Batuata, Pulau, i., Indon.	248-49	6°12′S	122°42′E
Batumi, Geor. (bŭ-tōō′mē)	227	41°39′N	41°39′E
Batu Pahat, Malay.	246-47	1°51′N	102°56′E
Baturaja, Indon.	246-47	4°08′S	104°09′E
Baturité, Braz.	166-67	4°20′S	38°53′W
Baubau, Indon.	248-49	5°28′S	122°37′E
Bauchi, Nig. (bȧ-ōō′chē)	260-61	10°19′N	9°50′E
Bauld, Cape, c., Nf., Can.	138-39	51°38′N	55°26′W
Bauru, Braz. (bou-rōō′)	172	22°19′S	49°04′W
Bauska, Lat. (bou′skȧ)	192-93	56°24′N	24°14′E
Bautzen, Ger. (bout′sĕn)	194-95	51°11′N	14°26′E
Bavaria, hist. reg., Ger. (bȧ-vâ-rĭ-ȧ)	194-95	48°30′N	11°30′E
Bawdwin, Mya.	246-47	23°07′N	97°15′E
Bawean, Pulau, i., Indon. (pōō-lou bȧ′vē-än)	248-49	5°46′S	112°40′E
Bawiti, Egypt	188-89	28°21′N	28°52′E
Bawku, Ghana	260-61	11°04′N	0°15′W
Baxley, Ga., U.S. (băks′lĭ)	124-25	31°47′N	82°21′W
Bay, Laguna de, lk., Phil. (lä-gōō′nä dȧ bä′ē)	250	14°23′N	121°15′E
Bayamo, Cuba (bä-yä′mō)	142-43	20°23′N	76°38′W
Bayan Har Shan, mts., China	238-39	33°47′N	97°54′E
Bayanhongor, Mong.	240-41	46°10′N	100°42′E
Bayano, Lago, res., Pan.	150	9°12′N	78°44′W
Bayan Obo, China	240-41	41°59′N	110°08′E
Bayard, Ne., U.S. (bā′ĕrd)	114-15	41°46′N	103°20′W
Bayard, N.M., U.S. (bā′ĕrd)	118-19	32°46′N	108°08′W
Bayburt, Tur. (bä′ī-bòrt)	227	40°16′N	40°14′E
Bay City, Mi., U.S. (bā sĭ′tē)	116-17	43°36′N	83°53′W
Bay City, Tx., U.S. (bā sĭ′tĭ)	122-23	28°59′N	95°58′W
Baydhabo, Som. *see* Baidoa	262-63	3°07′N	43°39′E
Baydrag, stm., Mong.	240-41	45°37′N	99°15′E
Bayern, hist. reg., Ger. (bī′ĕrn) *see* Bavaria	194-95	48°30′N	11°30′E
Bayeux, Fr. (bȧ-yû′)	196-97	49°17′N	0°42′W
Baykal, Ozero, lk., Russia *see* Baikal, Lake	218-19	53°0′N	107°40′E
Baykit, Russia (bī-kēt′)	218-19	61°41′N	96°25′E
Baykonur, Kaz. *see* Bayqongyr	226	45°38′N	63°18′E
Bay Minette, Al., U.S. (bā′mĭn-ĕt′)	124-25	30°53′N	87°47′W
Bayombong, Phil. (bä-yôm-bŏng′)	250	16°29′N	121°09′E
Bayonne, Fr. (bȧ-yôn′)	196-97	43°30′N	1°29′W
Bayonne, N.J., U.S. (bā-yōn′)	116-17	40°40′N	74°07′W

n-sing; ŋ-baŋk; N-nasalized n; nŏd; cŏmmit; ōld; ôbey; ôrder; oi-boil; fōōd; ò-as oo in foot; ou-out; s-soft; sh-dish; th-thin; pūre; ůnite; ûrn; stŭd; circŭs; ü-as in French tu; ′-indeterminate vowel.

Feature (Pronunciation)	Page	Lat.	Long.
Bayou Bodcau Reservoir, res., La., U.S. (bī´yōō bŏd´kō rĕ´sẽr-vwär)	120-21	32°48'N	93°27'W
Bayqongyr, Kaz.	226	45°38'N	63°18'E
Bayram-Ali, Turkmen.	232-33	37°37'N	62°10'E
Bayreuth, Ger. (bī-roit´)	194-95	49°57'N	11°34'E
Bay Roberts, Nf., Can. (bā rŏb´ẽrts)	138-39	47°35'N	53°18'W
Bayrūt, nat. cap., Leb. *see* Beirut	228-29	33°53'N	35°30'E
Bays, Lake of, lk., On., Can. (lāk ŭv bās)	136-37	45°15'N	78°60'W
Bay Saint Louis, Ms., U.S. (bā´ sānt lōō´ĭs)	124-25	30°19'N	89°20'W
Bayt Laḥm, W.B. *see* Bethlehem	228-29	31°43'N	35°12'E
Baytown, Tx., U.S. (bā´town)	122-23	29°44'N	94°59'W
Baza, Spain (bä´thä)	198-99	37°29'N	2°46'W
Bazaruto, Ilha do, i., Moz. (ē´lä-dô-bä-zä-rōō´tō)	264-65	21°41'S	35°28'E
Be, Nosy, i., Madag.	264-65	13°20'S	48°15'E
Beach, N.D., U.S. (bēch)	114-15	46°55'N	104°00'W
Beachy Head, c., Eng., U.K. (bēchē hĕd)	190-91	50°45'N	0°15'E
Beacon, N.Y., U.S. (bē´kŭn)	116-17	41°31'N	73°58'W
Beagle Gulf, b., Austl.	272-73	12°0'S	130°20'E
Beardmore, On., Can.	134-35	49°36'N	87°58'W
Beardstown, Il., U.S. (bērds´toun)	120-21	40°01'N	90°26'W
Bear Island, i., Nor. (bâr ī´lănd) *see* Bjørnøya	218-19	74°27'N	19°02'E
Bear Lake, lk., Mb., Can. (bâr lāk)	134-35	55°08'N	96°00'W
Bear Lake, lk., U.S. (bâr lāk)	112-13	42°0'N	111°20'W
Bear River Range, mts., U.S. (bâr rĭv´ẽr rānj)	112-13	41°29'N	111°41'W
Beata, Cabo, c., Dom. Rep. (kä´bô-bĕ-ä´tä)	142-43	17°37'N	71°25'W
Beata, Isla, i., Dom. Rep.	142-43	17°35'N	71°31'W
Beatrice, Ne., U.S. (bē´ȧ-trĭs)	120-21	40°16'N	96°45'W
Beatton, stm., B.C., Can.	132-33	56°05'N	120°22'W
Beatty, Nv., U.S. (bēt´ē)	118-19	36°55'N	116°45'W
Beattyville, Ky., U.S. (bēt´ē-vĭl)	116-17	37°34'N	83°43'W
Beaucaire, Fr. (bō-kâr´)	196-97	43°48'N	4°39'E
Beaufort, Malay.	248-49	5°22'N	115°44'E
Beaufort, N.C., U.S. (bō´fōrt)	124-25	34°43'N	76°40'W
Beaufort, S.C., U.S. (bō´fōrt)	124-25	32°24'N	80°44'W
Beaufort Sea, s., N.A. (bō´fōrt sē)	86	73°0'N	140°00'W
Beaufort West, S. Afr.	264-65	32°21'S	22°35'E
Beaumont, Tx., U.S. (bō´mŏnt)	122-23	30°05'N	94°08'W
Beaune, Fr. (bōn)	196-97	47°01'N	4°50'E
Beauport, Qc., Can. (bō-pôr´)	136-37	46°52'N	71°10'W
Beaupré, Qc., Can.	138-39	47°02'N	70°54'W
Beausejour, Mb., Can.	134-35	50°04'N	96°31'W
Beauvais, Fr. (bō-vĕ´)	196-97	49°26'N	2°05'E
Beaver, Ok., U.S. (bē´vẽr)	120-21	36°49'N	100°31'W
Beaver, Ut., U.S. (bē´vẽr)	118-19	38°17'N	112°38'W
Beaver, stm., Can.	130-31	55°26'N	107°47'W
Beaver Dam, Wi., U.S. (bē´vẽr dăm)	116-17	43°27'N	88°50'W
Beaverhead Mountains, mts., U.S. (bē´vẽr-hĕd moun´tĭnz)	112-13	44°58'N	113°26'W
Beaver Island, i., Mi., U.S. (bē´vẽr ī´lănd)	116-17	45°40'N	85°32'W
Beaverton, Or., U.S. (bē´vẽr-tŭn)	112-13	45°29'N	122°49'W
Beāwar, India	234-35	26°06'N	74°19'E
Bečej, Serb. (bĕ´chā)	200-01	45°37'N	20°03'E
Béchar, Alg.	258-59	31°37'N	2°14'W
Bechuanaland, nation, Afr. *see* Botswana	253	22°0'S	24°00'E
Beckley, W.V., U.S. (bĕk´lĭ)	116-17	37°48'N	81°11'W
Bedford, Qc., Can. (bĕd´fẽrd)	136-37	45°07'N	72°59'W
Bedford, In., U.S. (bĕd´fẽrd)	116-17	38°51'N	86°29'W
Bedford, Va., U.S. (bĕd´fẽrd)	124-25	37°20'N	79°31'W
Beebe, Ar., U.S. (bē´bē)	124-25	35°04'N	91°53'W
Beecroft Head, c., Austl. (bē´krŭft hĕd)	276	35°00'S	150°51'E
Beersheba, Isr.	228-29	31°14'N	34°48'E
Be'er Sheva', Isr. (bēr-shē´bȧ) *see* Beersheba	228-29	31°14'N	34°48'E
Beeville, Tx., U.S. (bē´vĭl)	122-23	28°24'N	97°45'W
Bega, Austl. (bā´gȧ)	276	36°41'S	149°51'E
Beggs, Ok., U.S. (bĕgz)	120-21	35°45'N	96°05'W
Behbahān, Iran	230-31	30°35'N	50°14'E
Bei, stm., China (bā)	238-39	23°09'N	112°49'E
Bei'an, China (bā-än)	240-41	48°14'N	126°31'E
Beibu Wan, b., Asia *see* Tonkin, Gulf of	246-47	20°0'N	108°00'E
Beida, Libya *see* Al-Bayḍā'	188-89	32°45'N	21°37'E
Beihai, China (bā-hī)	238-39	21°27'N	109°05'E
Beijing, state, China	240-41	40°15'N	116°30'E
Beijing, nat. cap., China (bā-jyĭŋ)	240-41	39°55'N	116°22'E
Beipan, stm., China	238-39	25°01'N	106°04'E
Beipiao, China	240-41	41°48'N	120°46'E
Beira, Moz. (bā´rȧ)	264-65	19°50'S	34°50'E
Beirut, nat. cap., Leb. (bā-rōōt´)	228-29	33°53'N	35°30'E
Beitbridge, Zimb.	264-65	22°12'S	30°01'E
Beja, Port. (bā´zhä)	198-99	38°01'N	7°52'W
Béja, Tun.	184-85	36°44'N	9°11'E
Bejaïa, Alg.	269b	36°45'N	5°04'E
Bejuco, Pan. (bĕ-ĸōō´kō)	150	8°36'N	79°53'W
Bekdash, Turkmen. *see* Karabogaz	186-87	41°32'N	52°35'E
Békés, Hung. (bā´kāsh)	194-95	46°46'N	21°08'E
Békéscsaba, Hung. (bā´kāsh-chô´bô)	194-95	46°40'N	21°05'E
Bekobod, Uzb.	232-33	40°13'N	69°11'E
Bela, Pak.	232-33	26°13'N	66°18'E
Bela-Bela, S. Afr.	269c	24°53'S	28°19'E
Bela Crkva, Serb. (bĕ´lä tsĕrk´vä)	200-01	44°54'N	21°26'E
Belaga, Malay.	248-49	2°43'N	113°47'E
Belarus, nation, Eur. (byĭ-lă-rōōs´) (bē-lă-rōōs´)	174-75	53°50'N	28°00'E
Belau, nation, Oc. *see* Palau	280-81	5°0'N	137°00'E
Bela Vista, Braz.	168-69	22°06'S	56°32'W
Belaya, stm., Russia (byĕ´lä-yȧ)	227	45°06'N	39°29'E
Belaya, stm., Russia (byĕ´lä-yȧ)	186-87	55°47'N	54°04'E
Belaya Tserkov, Ukr. *see* Bila Tserkva	202-03	49°48'N	30°08'E
Belcher Islands, is., Nu., Can. (bĕl´chẽr ī´lăndz)	130-31	56°20'N	79°30'W
Belding, Mi., U.S. (bĕl´dĭng)	116-17	43°06'N	85°13'W
Belebey, Russia (byĕ´lĕ-bā´ĭ)	186-87	54°06'N	54°08'E
Belém, Braz. (bå-lĕɴ)	166-67	1°27'S	48°29'W
Belén, Para. (bā-lān´)	168-69	23°28'S	57°15'W
Belen, N.M., U.S. (bĕ-lån´)	118-19	34°40'N	106°46'W
Belëv, Russia (byĕl´yĕf)	202-03	53°48'N	36°09'E
Belfast, S. Afr.	269c	25°43'S	30°04'E
Belfast, N. Ire., U.K. (bĕl´fȧst)	190-91	54°36'N	5°56'W
Belfort, Fr. (bā-fôr´)	196-97	47°38'N	6°51'E
Belgaum, India	236	15°51'N	74°31'E
België, nation, Eur. *see* Belgium	174-75	50°50'N	4°00'E
Belgique, nation, Eur. *see* Belgium	174-75	50°50'N	4°00'E
Belgium, nation, Eur. (bĕl´jĭ-ŭm)	174-75	50°50'N	4°00'E
Belgorod, Russia (byĕl´gŭ-rut)	202-03	50°37'N	36°35'E
Belgrade, nat. cap., Serb. (bĕl´grād)	200-01	44°50'N	20°28'E
Belhaven, N.C., U.S. (bĕl´hā-vĕn)	124-25	35°32'N	76°37'W
Beliliou, i., Palau	280-81	7°00'N	134°15'E
Belitung, i., Indon.	246-47	2°50'S	107°55'E
Belize, nation, N.A. (bĕ-lēz´)	85	17°15'N	88°45'W
Belize, stm., Belize (bĕ-lēz´)	148	17°30'N	88°11'W
Belize City, Belize (bĕ-lēz´ sĭ´tē)	148	17°30'N	88°11'W
Bel'kovskiy, Ostrov, i., Russia (ôs-trôf´ byĕl-kôf´skĭ)	218-19	75°32'N	135°44'E
Bella Bella, B.C., Can.	132-33	52°09'N	128°07'W
Bella Coola, B.C., Can.	132-33	52°21'N	126°46'W
Bellary, India (bĕl-lä´rē)	236	15°09'N	76°55'E
Bella Unión, Ur. (bĕ´l-yä-ōō-nyô´n)	173	30°15'S	57°35'W
Bella Vista, Arg. (bā´lyȧ vēs´tä)	168-69	27°02'S	65°18'W
Bella Vista, Arg. (bā´lyȧ vēs´tȧ)	173	28°30'S	59°02'W
Bellavista, Peru	170	7°04'S	76°35'W
Belle Bay, b., Nf., Can. (bĕl bā)	138-39	47°36'N	55°18'W
Bellefontaine, Oh., U.S. (bel-fŏn´tån)	116-17	40°21'N	83°45'W
Belle Fourche, S.D., U.S. (bĕl´ fōōrsh´)	114-15	44°40'N	103°51'W
Belle Glade, Fl., U.S. (bĕl glād)	125a	26°42'N	80°40'W
Belle Isle, Strait of, strt., Nf., Can.	130-31	51°36'N	56°28'W
Belle Plaine, Ia., U.S. (bĕl plān´)	114-15	41°54'N	92°17'W
Belleville, On., Can. (bĕl´vĭl)	136-37	44°10'N	77°23'W
Belleville, Il., U.S. (bĕl´vĭl)	120-21	38°31'N	89°59'W
Belleville, Ks., U.S. (bĕl´vĭl)	120-21	39°49'N	97°38'W
Bellevue, Ia., U.S. (bĕl´vū)	114-15	42°15'N	90°25'W
Bellevue, Id., U.S. (bĕl´vū)	112-13	43°28'N	114°16'W
Bellevue, Ne., U.S. (bĕl´vū)	114-15	41°09'N	95°55'W
Bellevue, Oh., U.S. (bĕl´vū)	116-17	41°16'N	82°50'W
Bellevue, Wa., U.S. (bĕl´vū)	112-13	47°37'N	122°12'W
Belley, Fr. (bĕ-lē´)	196-97	45°46'N	5°41'E
Bellingham, Wa., U.S. (bĕl´ĭng-hăm)	112-13	48°46'N	122°29'W
Bellingshausen Sea, s., Ant. (bĕl´ĭngz houz'n sē)	287	71°0'S	85°00'W
Bellinzona, Switz. (bĕl-ĭn-tsō´nä)	194-95	46°11'N	9°01'E
Bell Island, i., Nf., Can. (bĕl ī´lănd)	138-39	50°44'N	55°35'W
Bello, Col. (bĕ´l-yô)	163c	6°20'N	75°34'W
Bell Peninsula, pen., Nu., Can. (bĕl pĕ-nĭn´sūlȧ)	130-31	63°50'N	81°60'W
Belluno, Italy (bĕl-lōō´nō)	200-01	46°09'N	12°13'E
Bell Ville, Arg. (bĕl vĕl´)	173	32°39'S	62°41'W
Belmond, Ia., U.S. (bĕl´mŏnd)	114-15	42°51'N	93°37'W
Belmonte, Braz. (bĕl-mōn´tå)	168-69	15°54'S	38°53'W
Belmopan, nat. cap., Belize (bĕl-mō-pän´)	148	17°14'N	88°47'W
Belogorsk, Russia	222-23	50°54'N	128°30'E
Belo Horizonte, Braz. (bĕ´lôre-sô´n-tĕ)	172	19°55'S	43°56'W
Beloit, Ks., U.S. (bē-loit´)	120-21	39°27'N	98°06'W
Beloit, Wi., U.S. (bē-loit´)	116-17	42°31'N	89°02'W
Belomorsk, Russia (byĕl-ô-môrsk´)	186-87	64°32'N	34°45'E
Belorechensk, Russia	186-87	44°46'N	39°52'E
Beloretsk, Russia (byĕ´lō-rĕtsk)	226	53°58'N	58°24'E
Belorussia, nation, Eur. *see* Belarus	174-75	53°50'N	28°00'E
Belorussiya, nation, Eur. *see* Belarus	174-75	53°50'N	28°00'E
Belo Tsiribihina, Madag.	264-65	19°42'S	44°33'E
Belovo, Russia (bvĕ´lŭ-vû)	218-19	54°25'N	86°19'E
Beloye, Ozero, lk., Russia	186-87	60°11'N	37°37'E
Beloye More, s., Russia *see* White Sea	186-87	65°37'N	37°52'E
Beloz'orsk, Russia	186-87	60°02'N	37°48'E
Belton, Mo., U.S. (bĕl´tŭn)	120-21	38°48'N	94°32'W
Belton, Tx., U.S. (bĕl´tŭn)	122-23	31°03'N	97°27'W
Belts, Mol. *see* Bălți	202-03	47°46'N	27°55'E
Beltsy, Mol. *see* Bălți	202-03	47°46'N	27°55'E
Belukha, Gora, mtn., Asia *see* Belukha, Mount	226	49°51'N	86°29'E
Belukha, Mount, mtn., Asia (mount byĭ-lōō´-khŭ)	226	49°51'N	86°29'E
Belvidere, Il., U.S. (bĕl-vĕ-dēr´)	116-17	42°15'N	88°49'W
Belzoni, Ms., U.S. (bĕl-zō´nē)	124-25	33°11'N	90°29'W
Bembézar, stm., Spain (bĕm-bā-thär´)	198-99	37°45'N	5°12'W
Bemidji, Mn., U.S. (bē-mĭj´ĭ)	114-15	47°29'N	94°54'W
Benalla, Austl. (bĕn-ăl´ȧ)	276	36°33'S	145°59'E
Benares, India *see* Vārānasi	234-35	25°20'N	82°59'E
Benavente, Spain (bā-nä-vĕn´tä)	198-99	42°00'N	5°40'W
Bend, Or., U.S. (bĕnd)	112-13	44°04'N	121°18'W
Bender, Mol. *see* Tighina	202-03	46°50'N	29°29'E
Bender Cassim, Som.	262-63	11°17'N	49°11'E
Bendery, Mol. *see* Tighina	202-03	46°50'N	29°29'E
Bendigo, Austl. (bĕn´dĭ-gō)	276	36°46'S	144°17'E
Benedito Leite, Braz.	166-67	7°13'S	44°34'W
Benešov, Czech Rep. (bĕn´ĕ-shôf)	194-95	49°47'N	14°43'E
Benevento, Italy (bā-nā-vĕn´tō)	200-01	41°08'N	14°46'E
Bengal, Bay of, b., Asia (bā ŭv bĕn-gôl´)	220-21	15°0'N	90°00'E
Bengalūru, India	236	12°59'N	77°36'E
Bengbu, China (bŭŋ-bōō)	238-39	32°57'N	117°21'E
Bengkulu, Indon.	246-47	3°48'S	102°16'E
Benguela, Ang. (bĕn-gĕl´ȧ)	264-65	12°35'S	13°25'E
Beni, D.R.C.	267	0°30'N	29°28'E
Beni, stm., Bol. (bā´nē)	166-67	10°59'S	66°07'W
Beni Abbes, Alg.	188-89	30°08'N	2°10'W
Beni Mazâr, Egypt	268b	28°29'N	30°48'E
Beni-Mellal, Mor.	269a	32°21'N	6°22'W
Benin, nation, Afr. (bĕn-ēn´)	253	9°30'N	2°15'E
Benin, stm., Nig. (bĕn-ēn´)	260a	5°45'N	5°04'E
Benin, Bight of, b., Afr. (bīt ŭv bĕn-ēn´)	260-61	5°30'N	3°00'E
Bénin, Golfe de, b., Afr. *see* Benin, Bight of	260-61	5°30'N	3°00'E
Benin City, Nig. (bĕn-ēn´ sĭ´tē)	260a	6°20'N	5°38'E
Beni Saf, Alg. (bā´nē säf´)	198-99	35°18'N	1°23'W
Beni Suef, Egypt	268b	29°04'N	31°06'E
Benito Juárez, Arg.	173	37°41'S	59°48'W
Benjamín, Isla, i., Chile	171	44°40'S	74°08'W
Benjamin Constant, Braz.	166-67	4°28'S	70°01'W
Benkelman, Ne., U.S. (bĕn-kĕl-măn)	120-21	40°03'N	101°32'W
Bennetta, Ostrov, i., Russia	218-19	76°41'N	149°06'E
Bennettsville, S.C., U.S. (bĕn´ĕts vĭl)	124-25	34°37'N	79°41'W
Benoni, S. Afr. (bĕ-nō´nī)	269c	26°11'S	28°19'E
Bénoué, stm., Afr.	260-61	7°48'N	6°46'E
Benson, Az., U.S. (bĕn-sŭn)	118-19	31°58'N	110°18'W
Benson, Mn., U.S. (bĕn-sŭn)	114-15	45°19'N	95°36'W
Bentinck Island, i., Austl.	277	17°04'S	139°30'E
Benton, Ar., U.S. (bĕn´tŭn)	120-21	34°34'N	92°36'W
Benton, Il., U.S. (bĕn´tŭn)	116-17	37°60'N	88°55'W
Benton, Ky., U.S. (bĕn´tŭn)	124-25	36°52'N	88°21'W
Benton, La., U.S. (bĕn´tŭn)	120-21	32°42'N	93°45'W
Benton Harbor, Mi., U.S. (bĕn´tŭn här´bẽr)	116-17	42°06'N	86°28'W
Bentonville, Ar., U.S. (bĕn´tŭn-vĭl)	120-21	36°22'N	94°12'W
Benue, stm., Afr. (bā´nōō-å)	260-61	7°48'N	6°46'E
Benxi, China (bŭn-shyē)	243	41°18'N	123°45'E
Beograd, nat. cap., Serb. (bĕ-ō´grȧd) *see* Belgrade	200-01	44°50'N	20°28'E
Beppu, Japan (bĕ´pōō)	245	33°17'N	131°30'E
Bequia, i., St. Vin. (bĕk-ē´ä)	143b	13°02'N	61°13'W
Berau, Teluk, b., Indon.	224-25	2°30'S	132°30'E
Berbera, Som. (bûr´bûr-ȧ)	262-63	10°26'N	45°01'E
Berbérati, C.A.R.	262-63	4°14'N	15°48'E
Berbice, stm., Guy.	164-65	6°15'N	57°32'W
Berck, Fr. (bĕrk)	196-97	50°25'N	1°35'E
Berdians'k, Ukr.	202-03	46°45'N	36°49'E
Berdigestyakh, Russia	218-19	62°06'N	126°41'E

ăt; fināl; rāte; senåte; ärm; åsk; sofȧ; fâre; ch-choose; dh-as th in other; bē; ēvent; bĕt; recĕnt; cratẽr; g-gō; gh-guttural g; bĭt; ĭ-short neutral; rīde; ĸ-guttural k as ch in German ich;

Feature (Pronunciation)	Page	Lat.	Long.
Berdychiv, Ukr.	202-03	49°54'N	28°37'E
Berea, Ky., U.S. (bẽ-rē´*a*)	116-17	37°34'N	84°18'W
Berens River, Mb., Can. (bĕrĕnz rĭv´ẽr)	134-35	52°22'N	97°02'W
Beresford, S.D., U.S. (bĕr´ĕs-fẽrd)	114-15	43°05'N	96°47'W
Berettyóújfalu, Hung. (bĕ´rĕt-tyō-ōō´y'fō-lōō)	194-95	47°13'N	21°33'E
Berezhany, Ukr. (bĕr-yĕ´zhȧ-nê)	194-95	49°27'N	24°57'E
Berezniki, Russia (bĕr-yôz´nyê-kê)	186-87	59°24'N	56°46'E
Berga, Spain (bĕr´gä)	198-99	42°06'N	1°51'E
Bergama, Tur. (bĕr´gä-mä)	200-01	39°08'N	27°11'E
Bergamo, Italy (bĕr´gä-mō)	200-01	45°42'N	9°41'E
Bergantín, Ven. (bĕr-gän-tē´n)	163b	10°01'N	64°21'W
Bergen, Nor. (bĕr´gĕn)	192-93	60°22'N	5°21'E
Bergerac, Fr. (bĕr-zhē-rȧk´)	196-97	44°51'N	0°29'E
Bergville, S. Afr. (bẽrg´vĭl)	269c	28°44'S	29°21'E
Berhala, Selat, strt., Indon.	246-47	0°48'S	104°25'E
Beringa, Ostrov, i., Russia	218-19	54°54'N	166°24'E
Beringovo More, s., *see* Bering Sea	126	60°0'N	175°00'W
Beringov Proliv, strt., *see* Bering Strait	126	65°30'N	169°00'W
Beringovskiy, Russia	218-19	63°04'N	179°22'E
Bering Sea, s., (bē´rĭng sē)	126	60°0'N	175°00'W
Bering Strait, strt., (bē´rĭng strāt)	126	65°30'N	169°00'W
Berja, Spain (bĕr´hä)	198-99	36°51'N	2°57'W
Berkakit, Russia	218-19	56°34'N	124°48'E
Berkane, Mor.	198-99	34°56'N	2°19'W
Berkeley, Ca., U.S. (bûrk´lĭ)	118-19	37°52'N	122°16'W
Berkeley Springs, W.V., U.S. (bûrk´lĭ sprĭngz)	116-17	39°37'N	78°15'W
Berlenga, i., Port. (bĕr-lĕn´gäzh)	198-99	39°25'N	9°30'W
Berlin, Md., U.S. (bûr-lĭn)	116-17	38°19'N	75°13'W
Berlin, N.H., U.S. (bûr-lĭn)	116-17	44°28'N	71°10'W
Berlin, Wi., U.S. (bûr-lĭn´)	116-17	43°58'N	88°57'W
Berlin, nat. cap., Ger. (bĕr-lēn´)	194-95	52°31'N	13°26'E
Bermejo, stm., Arg.	168-69	32°17'S	67°22'W
Bermejo, stm., S.A. (bĕr-mā´hō)	168-69	26°52'S	58°22'W
Bermejo, Paso del, p., S.A.	163e	32°49'S	70°05'W
Bermeo, Spain (bĕr-mā´yō)	198-99	43°25'N	2°44'W
Bermuda, dep., N.A. (bẽr-mū´-d*a*)	140-41	32°20'N	64°45'W
Bermuda Islands, is., Ber.	140-41	32°21'N	64°46'W
Bern, nat. cap., Switz. (bĕrn)	194-95	46°57'N	7°26'E
Bernasconi, Arg.	173	37°54'S	63°44'W
Berne, In., U.S. (bûrn)	116-17	40°39'N	84°57'W
Berne, nat. cap., Switz. *see* Bern	194-95	46°57'N	7°26'E
Bernier Island, i., Austl. (bĕr-nēr´ ī´l*a*nd)	272-73	24°52'S	113°08'E
Bernina, Piz, mtn., Eur.	194-95	46°22'N	9°51'E
Bernina, Pizzo, mtn., Eur. *see* Bernina, Piz	194-95	46°22'N	9°51'E
Beroun, Czech Rep. (bā´rōn)	194-95	49°58'N	14°04'E
Berryville, Ar., U.S. (bĕr´ê-vĭl)	120-21	36°22'N	93°35'W
Bershad', Ukr. (byĕr´shät)	202-03	48°23'N	29°34'E
Bertoua, Camrn.	260-61	4°35'N	13°41'E
Beru, i., Kir.	280-81	1°20'S	176°00'E
Berwick, Pa., U.S. (bûr´wĭk)	116-17	41°04'N	76°15'W
Berwick-upon-Tweed, Eng., U.K. (bûr´ĭk-ŭp´ŏn-twēd)	190-91	55°47'N	2°01'W
Besalampy, Madag. (bĕz-ȧ-lȧm-pē´)	264-65	16°45'S	44°30'E
Besançon, Fr. (bē-säɴ-sôn)	196-97	47°15'N	6°02'E
Bessarabia, hist. reg., Eur.	202-03	46°53'N	28°44'E
Bessemer, Al., U.S. (bĕs´ê-mẽr)	124-25	33°23'N	86°57'W
Bessemer, Mi., U.S. (bĕs´ê-mẽr)	114-15	46°29'N	90°03'W
Bessemer City, N.C., U.S. (bĕs´ê-mẽr sĭ´tê)	124-25	35°17'N	81°17'W
Betanzos, Spain (bĕ-tän´thōs)	198-99	43°17'N	8°12'W
Bethal, S. Afr. (bĕth´*a*l)	269c	26°27'S	29°28'E
Bethel, Ak., U.S. (bĕth´*e*l)	126	60°48'N	161°46'W
Bethlehem, S. Afr.	269c	28°14'S	28°19'E
Bethlehem, Pa., U.S. (bĕth´lê-hĕm)	116-17	40°38'N	75°23'W
Bethlehem, W.B. (bĕth´lê-hĕm)	228-29	31°43'N	35°12'E
Béthune, Fr. (bā-tün´)	196-97	50°32'N	2°39'E
Betioky, Madag.	264-65	23°43'S	44°23'E
Betpaqdala, des., Kaz.	226	46°0'N	70°00'E
Betsiamites, stm., Qc., Can.	138-39	48°56'N	68°37'W
Betsiboka, stm., Madag. (bĕt-sĭ-bō´kȧ)	264-65	16°03'S	46°35'E
Betūl, India	234-35	21°54'N	77°54'E
Beuthen, Pol. *see* Bytom	194-95	50°21'N	18°55'E
Beverley, Austl.	270-71	32°07'S	116°55'E
Bexhill, Eng., U.K. (bĕks´hĭl)	190-91	50°51'N	0°28'E
Beyneū, Kaz.	226	45°25'N	55°09'E
Beypazarı, Tur. (bā-pȧ-zä´rĭ)	186-87	40°10'N	31°56'E
Beyşehir Gölü, lk., Tur.	186-87	37°40'N	31°30'E
Bezhetsk, Russia (byĕ-zhĕtsk´)	202-03	57°48'N	36°42'E
Béziers, Fr. (bā-zyā´)	196-97	43°21'N	3°13'E
Bezwada, India *see* Vijayawāda	236	16°31'N	80°37'E
Bhadrak, India	234-35	21°03'N	86°30'E
Bhadrāvati, India	236	13°50'N	75°42'E
Bhāgalpur, India (bä´g*ŭ*l-pòr)	234-35	25°15'N	86°59'E
Bhaktapur, Nepal	234-35	27°41'N	85°26'E
Bhamo, Mya. (bŭ-mō´)	238-39	24°17'N	97°15'E
Bhandāra, India	234-35	21°09'N	79°39'E
Bharat, nation, Asia *see* India	206-07	20°0'N	77°00'E
Bharatpur, India (bẽrt´pòr)	234-35	27°13'N	77°29'E
Bharūch, India	234-35	21°43'N	72°60'E
Bhatinda, India (b*ŭ*-tĭn-dȧ)	234-35	30°12'N	74°57'E
Bhātpāra, India	234-35	22°52'N	88°24'E
Bhāvnagar, India	234-35	21°46'N	72°08'E
Bhawānipatna, India	234-35	19°55'N	83°10'E
Bhilai, India	234-35	21°13'N	81°26'E
Bhilainagar, India *see* Bhilai	234-35	21°13'N	81°26'E
Bhīlwāra, India	234-35	25°21'N	74°38'E
Bhīma, stm., India (bē´mȧ)	236	16°24'N	77°17'E
Bhind, India	234-35	26°34'N	78°47'E
Bhiwāni, India	234-35	28°47'N	76°08'E
Bhopāl, India (bô-päl)	234-35	23°15'N	77°24'E
Bhubaneshwar, India	234-35	20°14'N	85°50'E
Bhuj, India (bōōj)	234-35	23°15'N	69°40'E
Bhusāwal, India	234-35	21°03'N	75°47'E
Bhutan, nation, Asia (bōō-tän´)	206-07	27°30'N	90°30'E
Bia, Phou, mtn., Laos	246-47	18°59'N	103°09'E
Biafra, Bahía de, b., Afr. *see* Biafra, Bight of	260-61	4°0'N	8°00'E
Biafra, Bight of, b., Afr.	260-61	4°0'N	8°00'E
Biafra, Golfe de, b., Afr. *see* Biafra, Bight of	260-61	4°0'N	8°00'E
Biak, i., Indon. (bē´äk)	224-25	1°0'S	136°00'E
Biała Podlaska, Pol. (byä´wä pōd-läs´kä)	194-95	52°02'N	23°08'E
Białystok, Pol. (byä-wĭs´tōk)	194-95	53°08'N	23°09'E
Bianco, Monte, mtn., Eur. *see* Blanc, Mont	196-97	45°50'N	6°52'E
Biarritz, Fr. (byä-rēts´)	196-97	43°29'N	1°33'W
Bickerton, Cape, c., Ant.	287	66°20'S	136°56'E
Bicknell, In., U.S. (bĭk´n*e*l)	116-17	38°46'N	87°18'W
Bida, Nig. (bē´dä)	260-61	9°05'N	5°60'E
Bīdar, India	236	17°54'N	77°31'E
Biddeford, Me., U.S. (bĭd´ê-fẽrd)	116-17	43°30'N	70°27'W
Bié, Planalto do, plat., Ang.	264-65	13°30'S	17°02'E
Biebrza, stm., Pol. (byĕb´zhȧ)	194-95	53°13'N	22°26'E
Bielefeld, Ger. (bē´lē-fĕlt)	194-95	52°01'N	8°32'E
Biella, Italy (byĕl´lä)	200-01	45°34'N	8°04'E
Bielsko-Biała, Pol.	194-95	49°49'N	19°03'E
Bielsk Podlaski, Pol. (byĕlsk pŭd-lä´skĭ)	194-95	52°46'N	23°12'E
Bien Dong, s., Asia *see* South China Sea	224-25	10°0'N	113°00'E
Bien Hoa, Viet.	246-47	10°57'N	106°50'E
Bienville, Lac, lk., Qc., Can.	130-31	55°05'N	72°40'W
Biga, Tur. (bē´ghȧ)	200-01	40°14'N	27°15'E
Big Belt Mountains, mts., Mt., U.S. (bĭg bĕlt moun´tĭnz)	112-13	46°40'N	111°25'W
Big Bend National Park, n.p., Tx., U.S. (bĭg bĕnd năsh´ŭn-*a*l pärk)	122-23	29°12'N	103°12'W
Big Black, stm., Ms., U.S. (bĭg blăk)	124-25	32°03'N	91°04'W
Big Cypress Indian Reservation, ind. res., Fl., U.S. (bĭg sī´prĕs ĭn´dĭ-*a*n rĕ-sẽr-vā´shĕn)	125a	26°17'N	80°59'W
Big Cypress Swamp, sw., Fl., U.S. (bĭg sī´prĕs swŏmp)	125a	26°10'N	81°38'W
Big Delta, Ak., U.S. (bĭg dĕl´t*a*)	126	64°09'N	145°47'W
Biggar, Sk., Can.	134-35	52°03'N	107°58'W
Bigge Island, i., Austl.	272-73	14°35'S	125°10'E
Bighorn, stm., U.S. (bĭg hôrn)	110-11	46°09'N	107°29'W
Bighorn Mountains, mts., U.S. (bĭg hôrn moun´tĭnz)	112-13	43°59'N	107°04'W
Big Island, i., Nu., Can. (bĭg ī´l*a*nd)	130-31	62°43'N	70°43'W
Big Island, i., On., Can. (bĭg ī´l*a*nd)	134-35	49°09'N	94°37'W
Big Lake, Tx., U.S. (bĭg lāk)	122-23	31°11'N	101°28'W
Big Quill Lake, lk., Sk., Can.	134-35	51°55'N	104°22'W
Big Rapids, Mi., U.S. (bĭg răp´ĭdz)	116-17	43°42'N	85°29'W
Big River, Sk., Can. (bĭg rĭv´ẽr)	134-35	53°51'N	107°01'W
Big Sandy, stm., Wy., U.S. (bĭg sănd´ê)	112-13	41°52'N	109°47'W
Big Sioux, stm., U.S. (bĭg sōō)	114-15	42°29'N	96°28'W
Big Spring, Tx., U.S. (bĭg sprĭng)	120-21	32°15'N	101°29'W
Big Stone Gap, Va., U.S. (bĭg stōn)	124-25	36°52'N	82°47'W
Big Timber, Mt., U.S. (bĭg tĭm´-bẽr)	112-13	45°50'N	109°57'W
Big Trout Lake, lk., On., Can.	134-35	53°44'N	89°57'W
Bihār, state, India (bê-här´)	234-35	25°0'N	86°00'E
Bīhar Sharīf, India	234-35	25°12'N	85°32'E
Bijagós, Arquipélago dos, is., Gui.-B.	260-61	11°22'N	16°18'W
Bij-Chem, stm., Russia *see* Bol'shoy Yenisey	218-19	51°44'N	94°28'E
Bijie, China (bē-jyĕ)	238-39	27°18'N	105°17'E
Bīkāner, India (bĭ-kä´nûr)	234-35	28°01'N	73°19'E
Bikar, at., Marsh. Is.	280-81	12°14'N	170°08'E
Bikeqi, China	240-41	40°43'N	111°17'E
Bikin, Russia (bê-kēn´)	244	46°49'N	134°17'E
Bikin, stm., Russia (bê-kēn´)	244	46°51'N	134°02'E
Bikini, at., Marsh. Is.	280-81	11°35'N	165°23'E
Bilāspur, India (bê-läs´pōōr)	234-35	22°05'N	82°10'E
Bilāspur, India (bê-läs´pōōr)	234-35	31°18'N	76°46'E
Bila Tserkva, Ukr.	202-03	49°48'N	30°08'E
Bilauktaung Range, mts., Asia	246-47	13°0'N	99°00'E
Bilbao, Spain (bĭl-bä´ō)	198-99	43°15'N	2°56'W
Bilecik, Tur. (bē-lĕd-zhĕk´)	200-01	40°10'N	29°59'E
Biłgoraj, Pol. (bēw-gō´rī)	194-95	50°33'N	22°42'E
Bilhorod-Dnistrovs'kyi, Ukr.	202-03	46°12'N	30°18'E
Bili, stm., D.R.C.	262-63	4°08'N	22°29'E
Bilibino, Russia	218-19	68°04'N	166°21'E
Biliran Island, i., Phil.	250	11°36'N	124°29'E
Billabong Creek, stm., Austl. (bĭl´ȧ-bŏng krēk)	276	35°06'S	144°02'E
Billings, Mt., U.S. (bĭl´ĭngz)	112-13	45°47'N	108°31'W
Billiton, i., Indon. *see* Belitung	246-47	2°50'S	107°55'E
Bilma, Niger (bēl´mä)	258-59	18°41'N	12°56'E
Biloela, Austl.	277	24°24'S	150°30'E
Biloxi, Ms., U.S. (bĭ-lŏk´sĭ)	124-25	30°24'N	88°53'W
Biltine, Chad	258-59	14°32'N	20°55'E
Bimberi Peak, mtn., Austl. (bĭm´bĕrĭ pēk)	276	35°40'S	148°47'E
Bimini Islands, is., Bah.	142-43	25°42'N	79°15'W
Binga, D.R.C.	262-63	2°22'N	20°31'E
Binga, Monte, mtn., Afr.	264-65	19°47'S	33°03'E
Binga, Mount, mtn., Afr. *see* Binga, Monte	264-65	19°47'S	33°03'E
Binghamton, N.Y., U.S. (bĭng´*a*m-t*ŭ*n)	116-17	42°06'N	75°55'W
Binhai, China	238-39	34°00'N	119°50'E
Binjai, Indon.	246-47	3°36'N	98°30'E
Binongko, Pulau, i., Indon.	248-49	5°57'S	124°02'E
Bintimani, mtn., S.L.	260-61	9°13'N	11°07'W
Bintulu, Malay. (bēn´tōō-lōō)	248-49	3°10'N	113°02'E
Binxian, China (bĭn-shyän)	238-39	35°02'N	108°06'E
Binxian, China (bĭn-shyän)	240-41	37°28'N	117°58'E
Binzert, Tun. *see* Bizerte	184-85	37°16'N	9°52'E
Bioko, i., Eq. Gui. (bē-ō´-kō)	260-61	3°30'N	8°40'E
Bīr, India	236	18°60'N	75°46'E
Bira, Russia (bē´rȧ)	240-41	48°60'N	132°28'E
Bira, stm., Russia (bē´rȧ)	240-41	48°10'N	133°17'E
Birao, C.A.R.	262-63	10°17'N	22°48'E
Birch Mountains, hills, Ab., Can. (bûrch moun´tĭnz)	130-31	57°34'N	113°07'W
Birdsville, Austl. (bûrdz´vĭl)	276	25°54'S	139°21'E
Birecik, Tur. (bē-rĕd-zhĕk´)	228-29	37°03'N	38°03'E
Bīrjand, Iran (bēr´jănd)	232-33	32°53'N	59°13'E
Bîrlad, Rom.	202-03	46°14'N	27°40'E
Birmingham, Eng., U.K.	190-91	52°28'N	1°53'W
Birmingham, Al., U.S. (bûr´mĭng-häm)	124-25	33°31'N	86°49'W
Bîr Mogreïn, Maur.	258-59	25°14'N	11°35'W
Birnie, at., Kir.	280-81	3°35'S	171°31'W
Birnin-Kebbi, Nig.	260-61	12°28'N	4°12'E
Birnin Konni, Niger	258-59	13°48'N	5°15'E
Birobidzhan, Russia (bē´rō-bê-jän´)	240-41	48°47'N	132°55'E
Birsk, Russia (bĭrsk)	186-87	55°25'N	55°34'E
Biryusa, stm., Russia (bēr-yōō´sä)	218-19	57°43'N	95°27'E
Biržai, Lith. (bêr-zhä´ê)	192-93	56°12'N	24°46'E
Bisbee, Az., U.S. (bĭz´bē)	118-19	31°27'N	109°55'W
Biscay, Bay of, b., Eur. (bĭs´kā´ bā)	196-97	44°0'N	4°00'W
Biscayne Bay, b., Fl., U.S. (bĭs-kān´ bā)	125a	25°33'N	80°15'W
Biscayne National Park, n.p., Fl., U.S.	125a	25°25'N	80°12'W
Biscoe Islands, is., Ant.	287	65°60'S	66°30'W
Bishkek, nat. cap., Kyrg. (bĭsh-kĕk´)	226	42°52'N	74°35'E
Bishop, Ca., U.S. (bĭsh´*ŭ*p)	118-19	37°22'N	118°23'W
Bishop, Tx., U.S. (bĭsh´*ŭ*p)	122-23	27°35'N	97°47'W
Bishop's Falls, Nf., Can.	138-39	49°02'N	55°30'W
Bishopville, S.C., U.S. (bĭsh´*ŭ*p-vĭl)	124-25	34°13'N	80°15'W
Biskra, Alg.	269b	34°50'N	5°43'E
Bislig, Phil.	250	8°13'N	126°19'E
Bismarck, N.D., U.S. (bĭz´märk)	114-15	46°48'N	100°48'W
Bismarck Archipelago, is., Pap. N. Gui. (bĭz´märk är´kȧ-pĕ´-ȧ-gō)	277	5°0'S	150°00'E
Bismarck Range, mts., Pap. N. Gui. (bĭz´märk rānj)	277	5°30'S	144°45'E
Bismarck Sea, s., Pap. N. Gui. (bĭz´märk sē)	277	4°0'S	148°00'E
Bissagos, is., Gui.-B. *see* Bijagós, Arquipélago dos	260-61	11°22'N	16°18'W

n-sing; ŋ-baŋk; ɴ-nasalized n; nŏd; c*ŏ*mmit; ōld; ôbey; ôrder; oi-boil; fōōd; ŏŏ-as oo in foot; ou-out; s-soft; sh-dish; th-thin; pūre; ûnite; ûrn; stŭd; circ*ŭ*s; ü-as in French tu; ´-indeterminate vowel.

Feature (Pronunciation)	Page	Lat.	Long.
Bissau, nat. cap., Gui.-B. (bė-sa´ōō)	**260-61**	11°52´N	15°36´W
Bissett, Mb., Can.	**134-35**	51°02´N	95°40´W
Bistcho Lake, lk., Ab., Can.	**130-31**	59°44´N	118°46´W
Bistineau, Lake, res., La., U.S. (lăk bĭs-tĭ-nō´)	**120-21**	32°25´N	93°22´W
Bistrița, Rom. (bĭs-trĭt-sȧ)	**194-95**	47°08´N	24°30´E
Bistrița, stm., Rom. (bĭs-trĭt-sȧ)	**188-89**	46°28´N	26°57´E
Bitlis, Tur. (bĭt-lēs´)	**227**	38°22´N	42°06´E
Bitola, Mac. (bē´tô-lä) (mô´nä-stĕr)	**200-01**	41°02´N	21°20´E
Bitolj, Mac. *see* Bitola	**200-01**	41°02´N	21°20´E
Bitonto, Italy (bē-tôn´tō)	**200-01**	41°06´N	16°42´E
Bitra Island, i., India	**236**	11°36´N	72°11´E
Bitterfeld, Ger. (bĭt´ēr-fĕlt)	**194-95**	51°37´N	12°19´E
Bitterroot Range, mts., U.S. (bĭt´ēr-ōōt rānj)	**110-11**	47°06´N	115°10´W
Bitung, Indon.	**248-49**	1°26´N	125°08´E
Bityug, stm., Russia (bĭt´yōōg)	**186-87**	50°38´N	39°56´E
Biwa-ko, lk., Japan (bē-wä´kō)	**245**	35°15´N	136°05´E
Biya, stm., Russia (bĭ´yȧ)	**218-19**	52°26´N	85°00´E
Biysk, Russia (bēsk)	**226**	52°34´N	85°15´E
Bizerte, Tun. (bė-zĕrt´)	**184-85**	37°16´N	9°52´E
Bjarèzina, stm., Bela. (bĕr-yĕ´zē-nȧ)	**202-03**	52°33´N	30°14´E
Bjelovar, Cro. (byĕ-lō´vär)	**200-01**	45°54´N	16°50´E
Björneborg, Fin. *see* Pori	**192-93**	61°29´N	21°47´E
Bjørnøya, i., Nor.	**218-19**	74°27´N	19°02´E
Black, stm., Asia (blăk)	**246-47**	21°15´N	105°21´E
Black, stm., Ar., U.S. (blăk)	**120-21**	35°38´N	91°19´W
Blackall, Austl. (blăk´ŭl)	**277**	24°26´S	145°28´E
Black Bay, b., On., Can. (blăk bā)	**136-37**	48°34´N	88°32´W
Blackburn, Eng., U.K. (blăk´bûrn)	**190-91**	53°45´N	2°29´W
Black Canyon of the Gunnison National Park, n.p., Co., U.S.	**118-19**	38°34´N	107°44´W
Blackduck, Mn., U.S. (blăk´dŭk)	**114-15**	47°43´N	94°32´W
Blackfeet Indian Reservation, ind. res., Mt., U.S. (blăk´fēt ĭn´dĭ-ăn rĕ-sēr-vā´shĕn)	**112-13**	48°40´N	113°00´W
Blackfoot, Id., U.S. (blăk´fŏt)	**112-13**	43°12´N	112°21´W
Black Forest, mts., Ger. *see* Schwarzwald	**194-95**	48°21´N	8°11´E
Black Hills, mts., U.S. (blăk hĭlz)	**114-15**	44°0´N	104°00´W
Black Lake, Qc., Can. (blăk lāk)	**136-37**	46°03´N	71°22´W
Blackpool, Eng., U.K. (blăk´pōōl)	**190-91**	53°50´N	3°02´W
Black Range, mts., N.M., U.S. (blăk rānj)	**118-19**	33°20´N	107°50´W
Black River Falls, Wi., U.S. (blăk rĭv´ēr fôlz)	**114-15**	44°18´N	90°50´W
Black Rock Desert, des., Nv., U.S. (blăk rŏk dĕs´ērt)	**112-13**	41°06´N	118°51´W
Blacksburg, Va., U.S. (blăks´bûrg)	**124-25**	37°14´N	80°25´W
Black Sea, s., (blăk sē)	**186-87**	43°0´N	35°00´E
Blackshear, Ga., U.S. (blăk´shĭr)	**124-25**	31°18´N	82°15´W
Blackstone, Va., U.S. (blăk´stôn)	**124-25**	37°05´N	77°60´W
Blacktown, Austl. (blăk´toun)	**276**	33°46´S	150°54´E
Black Volta, stm., Afr. (blăk vôl´tȧ)	**260-61**	8°41´N	0°60´W
Blackwell, Ok., U.S. (blăk´wĕl)	**120-21**	36°48´N	97°18´W
Blagodarnoye, Russia (blä´gô-där-nō´yĕ)	**186-87**	45°06´N	43°25´E
Blagoveshchensk, Russia	**240-41**	50°17´N	127°33´E
Blaine, Mn., U.S. (blān)	**114-15**	45°11´N	93°15´W
Blaine, Wa., U.S. (blān)	**112-13**	48°60´N	122°45´W
Blair, Ne., U.S. (blâr)	**114-15**	41°33´N	96°08´W
Blairsville, Pa., U.S. (blârs´vĭl)	**116-17**	40°26´N	79°16´W
Blakely, Ga., U.S. (blāk´lė)	**124-25**	31°23´N	84°56´W
Blanc, Cap, c., Afr. *see* Nouâdhibou, Râs	**258-59**	20°47´N	17°03´W
Blanc, Mont, mtn., Eur. (môn blän)	**196-97**	45°50´N	6°52´E
Blanca, Bahía, b., Arg. (bä-ē´ä-blán´kä)	**173**	38°55´S	62°10´W
Blanca Peak, mtn., Co., U.S. (blăŋ´kȧ pēk)	**120-21**	37°35´N	105°29´W
Blanche, Lake, lk., Austl. (lăk blănch)	**276**	29°15´S	139°39´E
Blanco, Cabo, c., C.R. (ká´bō-blăŋ´kō)	**149**	9°34´N	85°07´W
Blanco, Cape, c., Or., U.S. (kāp blăn´kō)	**112-13**	42°50´N	124°33´W
Blanquilla, Isla, i., Ven.	**164-65**	11°51´N	64°37´W
Blantyre, Malawi (blăn-tīyr)	**264-65**	15°47´S	35°01´E
Blenheim, N.Z.	**278**	41°31´S	173°58´E
Bleus, Monts, mts., D.R.C.	**267**	1°37´N	30°28´E
Blida, Alg.	**269b**	36°29´N	2°49´E
Blind River, On., Can. (blīnd rĭv´ēr)	**136-37**	46°11´N	82°56´W
Blissfield, Mi., U.S. (blĭs-fēld)	**116-17**	41°50´N	83°52´W
Blitar, Indon.	**248-49**	8°06´S	112°10´E
Bloemfontein, S. Afr.	**269c**	29°07´S	26°12´E
Blois, Fr. (blwä)	**196-97**	47°35´N	1°20´E
Bloodvein, stm., Can.	**134-35**	51°48´N	96°53´W
Bloomer, Wi., U.S. (blōōm´ēr)	**114-15**	45°06´N	91°29´W
Bloomfield, Ia., U.S. (blōōm´fēld)	**114-15**	40°46´N	92°25´W
Bloomfield, In., U.S. (blōōm´fēld)	**116-17**	39°01´N	86°56´W
Bloomfield, Mo., U.S. (blōōm´fēld)	**124-25**	36°53´N	89°56´W
Bloomfield, Ne., U.S. (blōōm´fēld)	**114-15**	42°36´N	97°39´W
Blooming Prairie, Mn., U.S. (blōōm´ĭng prā´rĭ)	**114-15**	43°52´N	93°03´W
Bloomington, Il., U.S. (blōōm´ĭng-tŭn)	**116-17**	40°29´N	88°60´W
Bloomington, In., U.S. (blōōm´ĭng-tŭn)	**116-17**	39°09´N	86°32´W
Bloomington, Mn., U.S. (blōōm´ĭng-tŭn)	**114-15**	44°50´N	93°19´W
Bloomsburg, Pa., U.S. (blōōmz´bûrg)	**116-17**	40°60´N	76°27´W
Blossburg, Pa., U.S. (blŏs´bûrg)	**116-17**	41°41´N	77°05´W
Blountstown, Fl., U.S. (blŭnts´tun)	**124-25**	30°27´N	85°03´W
Bludenz, Aus. (blōō-dĕnts´)	**194-95**	47°09´N	9°50´E
Blue Earth, Mn., U.S. (blōō ûrth)	**114-15**	43°38´N	94°06´W
Bluefield, W.V., U.S. (blōō´fēld)	**124-25**	37°15´N	81°14´W
Bluefields, Nic. (blōō´fēldz)	**149**	12°01´N	83°46´W
Blue Mountain, mtn., Nf., Can. (blōō moun´tĭn)	**138-39**	50°24´N	57°10´W
Blue Mountain Peak, mtn., Jam.	**142-43**	18°03´N	76°35´W
Blue Mountains, mts., Austl. (blōō moun´tĭnz)	**276**	33°37´S	150°17´E
Blue Mountains, mts., U.S. (blōō moun´tĭnz)	**112-13**	45°16´N	118°42´W
Blue Mountains National Park, n.p., Austl.	**276**	33°47´S	150°23´E
Blue Nile, stm., Afr. (blōō nīl)	**254**	15°38´N	32°30´E
Bluenose Lake, lk., Nu., Can.	**130-31**	68°25´N	119°45´W
Blue Ridge, mts., U.S. (blōō rĭj)	**124-25**	37°0´N	82°00´W
Blue River, B.C., Can. (blōō rĭv´ēr)	**132-33**	52°06´N	119°20´W
Bluff, Ut., U.S. (blŭf)	**118-19**	37°17´N	109°33´W
Bluffton, In., U.S. (blŭf-tŭn)	**116-17**	40°44´N	85°10´W
Blumenau, Braz. (blōō´mĕn-ou)	**172**	26°56´S	49°05´W
Blyth, Eng., U.K. (blĭth)	**190-91**	55°08´N	1°31´W
Blytheville, Ar., U.S. (blĭth´vĭl)	**124-25**	35°56´N	89°55´W
Bo, S.L.	**260-61**	7°59´N	11°44´W
Boaco, Nic. (bô-ä´kō)	**149**	12°29´N	85°39´W
Bo'ai, China (bwo-ī)	**240-41**	35°10´N	113°04´E
Boano, Pulau, i., Indon.	**248-49**	2°58´S	127°56´E
Boa Vista, i., C.V. (bō-ä-vēsh´tȧ)	**260-61**	16°05´N	22°50´W
Bobaomby, Tanjona, c., Madag.	**264-65**	11°58´S	49°15´E
Bobbili, India	**236**	18°35´N	83°22´E
Bobo-Dioulasso, Burkina (bō´bō-dyōō-lás-sō´)	**260-61**	11°11´N	4°18´W
Bobruysk, Bela. *see* Babrujsk	**202-03**	53°08´N	29°14´E
Boby, mtn., Madag.	**264-65**	22°13´S	46°55´E
Boca do Acre, Braz.	**166-67**	8°45´S	67°23´W
Bocas del Toro, Pan. (bō´käs dĕl tō´rō)	**150**	9°20´N	82°15´W
Bochnia, Pol. (bōκ´nyä)	**194-95**	49°58´N	20°25´E
Bocholt, Ger. (bō´κōlt)	**194-95**	51°50´N	6°37´E
Bodaybo, Russia (bō-dī´bō)	**218-19**	57°51´N	114°11´E
Bodélé, reg., Chad (bō-dȧ-lā´)	**258-59**	16°30´N	16°30´E
Boden, Swe.	**184-85**	65°50´N	21°43´E
Bodh Gaya, India	**234-35**	24°42´N	84°58´E
Bodmin, Eng., U.K. (bŏd´mĭn)	**190-91**	50°29´N	4°43´W
Bodø, Nor.	**184-85**	67°17´N	14°24´E
Bodrum, Tur.	**200-01**	37°02´N	27°26´E
Boende, D.R.C. (bō-ĕn´dȧ)	**262-63**	0°14´S	20°52´E
Boerne, Tx., U.S. (bō´ĕrn)	**122-23**	29°47´N	98°43´W
Bōfu, Japan (bō´fōō) *see* Hōfu	**245**	34°03´N	131°35´E
Bogal, Lagh, stm., Kenya	**262-63**	0°46´N	40°50´E
Bogale, Mya.	**246-47**	16°17´N	95°24´E
Bogalusa, La., U.S. (bō-gȧ-lōō´sȧ)	**124-25**	30°47´N	89°51´W
Bogan, stm., Austl. (bŏ´gĕn)	**276**	29°58´S	146°20´E
Bogo, Phil.	**250**	11°02´N	124°01´E
Bogong, Mount, mtn., Austl.	**276**	36°44´S	147°18´E
Bogor, Indon.	**248-49**	6°35´S	106°47´E
Bogoroditsk, Russia (bō-gō´rō-dĭtsk)	**202-03**	53°47´N	38°08´E
Bogotá, nat. cap., Col. (bō-gō-tä´)	**163c**	4°37´N	74°06´W
Bogra, Bngl.	**234-35**	24°50´N	89°22´E
Boguchany, Russia	**218-19**	58°23´N	97°29´E
Bogué, Maur.	**258-59**	16°35´N	14°16´W
Bo Hai, b., China	**240-41**	38°30´N	120°02´E
Bohai Haixia, strt., China (bwo-hī hī-shyä)	**240-41**	38°15´N	121°00´E
Bohain-en-Vermandois, Fr. (bô-ăn-ŏn-vâr-män-dwä´)	**196-97**	49°59´N	3°27´E
Bohea Hills, mts., China *see* Wuyi Shan	**238-39**	27°42´N	117°09´E
Bohemia, hist. reg., Czech Rep.	**194-95**	49°50´N	14°00´E
Bohol, i., Phil. (bô-hôl´)	**250**	9°55´N	123°44´E
Bohol Sea, s., Phil.	**250**	9°10´N	124°25´E
Boipeba, Ilha de, i., Braz.	**166-67**	13°38´S	38°56´W
Bois, Lac des, lk., N.T., Can.	**130-31**	66°46´N	125°08´W
Boise, Id., U.S. (boi´zē)	**112-13**	43°37´N	116°13´W
Boise City, Ok., U.S. (boi´zē sĭ´tė)	**120-21**	36°44´N	102°30´W
Boissevain, Mb., Can. (bois´vān)	**134-35**	49°14´N	100°03´W
Bojeador, Cape, c., Phil.	**250**	18°30´N	120°35´E
Bojnūrd, Iran	**232-33**	37°29´N	57°20´E
Boksitogorsk, Russia	**186-87**	59°28´N	33°52´E
Bokurdak, Turkmen.	**232-33**	38°46´N	58°29´E
Bolbec, Fr. (bôl-bĕk´)	**196-97**	49°34´N	0°29´E
Bolgatanga, Ghana	**260-61**	10°48´N	0°51´W
Boli, China (bwo-lē)	**244**	45°45´N	130°34´E
Bolívar, Col.	**163c**	4°21´N	76°10´W
Bolivar, Mo., U.S. (bŏl´ĭ-vȧr)	**120-21**	37°37´N	93°25´W
Bolivar, Tn., U.S. (bŏl´ĭ-vȧr)	**124-25**	35°16´N	88°59´W
Bolívar, Cerro, mtn., Ven.	**164-65**	7°28´N	63°25´W
Bolívar, Pico, mtn., Ven.	**164-65**	8°33´N	71°01´W
Bolivar Peninsula, pen., Tx., U.S. (bŏl´ĭ-vȧr pĕ-nĭn´sūlȧ)	**122-23**	29°27´N	94°39´W
Bolivia, nation, S.A. (bō-lĭv´ĭ-ȧ)	**158**	17°0´S	65°00´W
Bolkhov, Russia (bŏl-κôf´)	**202-03**	53°27´N	36°00´E
Bollnäs, Swe. (bôl´nĕs)	**192-93**	61°21´N	16°24´E
Bolmen, lk., Swe. (bôl´mĕn)	**192-93**	56°55´N	13°40´E
Bolobo, D.R.C. (bō´lō-bō)	**262-63**	2°11´S	16°15´E
Bologna, Italy (bō-lōn´yä)	**200-01**	44°30´N	11°20´E
Bologoye, Russia (bŏ-lō-gô´yĕ)	**202-03**	57°54´N	34°03´E
Bol'shevik, Ostrov, i., Russia	**218-19**	78°40´N	102°30´E
Bol'shezemel'skaya Tundra, reg., Russia	**186-87**	67°30´N	55°60´E
Bol'shoy Begichëv, Ostrov, i., Russia	**218-19**	74°20´N	112°30´E
Bol'shoy Kavkaz, mts., *see* Caucasus Mountains	**227**	42°38´N	45°00´E
Bol'shoy Lyakhovskiy, Ostrov, i., Russia	**218-19**	73°35´N	142°00´E
Bol'shoy Uzen', stm., Eur. *see* Ülkenözen	**186-87**	48°60´N	49°59´E
Bol'shoy Yenisey, stm., Russia	**218-19**	51°44´N	94°28´E
Bolu, Tur. (bō´lò)	**186-87**	40°44´N	31°36´E
Bolzano, Italy (bôl-tsä´nō)	**200-01**	46°30´N	11°21´E
Boma, D.R.C. (bō´mä)	**262-63**	5°51´S	13°04´E
Bombala, Austl. (bŭm-bä´lă)	**276**	36°55´S	149°14´E
Bombay, India *see* Mumbai	**236**	18°57´N	72°50´E
Bom Jesus da Lapa, Braz.	**166-67**	13°15´S	43°25´W
Bømlo, i., Nor. (bûmlô)	**192-93**	59°47´N	5°12´E
Bomokandi, stm., D.R.C.	**262-63**	3°39´N	26°08´E
Bomu, stm., Afr.	**262-63**	4°09´N	22°29´E
Bon, Cap, c., Tun. (kȧp bôn)	**258-59**	37°05´N	11°03´E
Bonaire, i., Neth. Ant. (bō-nâr´)	**140a**	12°10´N	68°15´W
Bonaventure, Qc., Can.	**138-39**	48°02´N	65°30´W
Bonavista, Nf., Can. (bŏ-nȧ-vĭs´tȧ)	**138-39**	48°39´N	53°07´W
Bonavista Bay, b., Nf., Can. (bŏ-nȧ-vĭs´tȧ bā)	**138-39**	48°50´N	53°21´W
Bondo, D.R.C. (bôn´dȯ)	**262-63**	3°49´N	23°41´E
Bondoc Peninsula, pen., Phil. (bôn-dŏk´ pĕ-nĭn´sūlȧ)	**250**	13°30´N	122°30´E
Bondoukou, C. Iv. (bŏn-dōō´kōō)	**260-61**	8°02´N	2°48´W
Bône, Alg. *see* Annaba	**184-85**	36°54´N	7°46´E
Bone, Indon. *see* Watampone	**248-49**	4°32´S	120°19´E
Bone, Teluk, b., Indon.	**248-49**	4°0´S	120°40´E
Bonerate, Pulau, i., Indon.	**248-49**	7°21´S	121°07´E
Bonete Grande, Cerro, mtn., Arg. (sĕ´r-rô bô´nĕtĕh grän´dĕ)	**168-69**	27°57´S	68°45´W
Bongo, Massif des, mts., C.A.R.	**262-63**	8°36´N	22°50´E
Bonham, Tx., U.S. (bŏn´ăm)	**120-21**	33°35´N	96°11´W
Bonifacio, Bouches de, strt., Eur. *see* Bonifacio, Strait of	**200-01**	41°18´N	9°15´E
Bonifacio, Strait of, strt., Eur. (străt ŭv bō-nē-fä´chō)	**200-01**	41°18´N	9°15´E
Bonifay, Fl., U.S. (bŏn-ĭ-fā´)	**124-25**	30°47´N	85°41´W
Bonin Islands, is., Japan (bō´nĭn ī´lăndz)	**282-83**	26°58´N	142°14´E
Bonn, Ger. (bōn)	**194-95**	50°44´N	7°05´E
Bonners Ferry, Id., U.S. (bonērz fĕr´ē)	**112-13**	48°41´N	116°19´W
Bonne Terre, Mo., U.S. (bŏn târ´)	**120-21**	37°55´N	90°33´W
Bonny, Nig. (bŏn´ē)	**260a**	4°26´N	7°10´E
Bonnyville, Ab., Can. (bŏnē-vĭl)	**132-33**	54°16´N	110°44´W
Bontang, Indon.	**248-49**	0°08´N	117°30´E
Bontoc, Phil. (bŏn-tŏk´)	**250**	17°05´N	120°60´E
Boodjamulla National Park, n.p., Austl.	**277**	18°45´S	138°27´E
Booker T. Washington National Monument, n.p., Va., U.S. (bŏk´ēr tē wŏsh´ĭng-tŭn năsh´ŭn-ăl mŏn´ū-mĕnt)	**124-25**	37°01´N	79°45´W
Boonah, Austl.	**276**	28°00´S	152°41´E
Boone, Ia., U.S. (bōōn)	**114-15**	42°04´N	93°53´W
Boone, N.C., U.S. (bōōn)	**124-25**	36°13´N	81°41´W
Booneville, Ar., U.S. (bōōn´vĭl)	**120-21**	35°08´N	93°56´W
Booneville, Ms., U.S. (bōōn´vĭl)	**124-25**	34°39´N	88°34´W
Boonville, In., U.S. (bōōn´vĭl)	**116-17**	38°03´N	87°17´W
Boonville, Mo., U.S. (bōōn´vĭl)	**120-21**	38°58´N	92°45´W

Feature (Pronunciation)	Page	Lat.	Long.
Boorama, Som.	262-63	9°58'N	43°09'E
Boosaaso, Som. *see* Bender Cassim	262-63	11°17'N	49°11'E
Boothbay Harbor, Me., U.S. (bo͞oth'bā här'bẽr)	117a	43°51'N	69°38'W
Boothia, Gulf of, b., Nu., Can. (gŭlf ŭv bo͞o'thĭ-*ȧ*)	86	71°0'N	91°00'W
Boothia Peninsula, pen., Nu., Can.	130-31	70°30'N	95°00'W
Booué, Gabon	260-61	0°06'S	11°57'E
Borđoy, i., Far. Is.	190b	62°17'N	6°33'W
Bora-Bora, i., Fr. Poly.	280-81	16°30'S	151°45'W
Borah Peak, mtn., Id., U.S. (bō'rä pēk)	112-13	44°08'N	113°48'W
Borås, Swe. (bŏ'rōs)	192-93	57°44'N	12°57'E
Borāzjān, Iran (bō-räz-jän')	230-31	29°16'N	51°12'E
Borba, Braz. (bôr'bä)	166-67	4°23'S	59°35'W
Bordeaux, Fr. (bôr-dō')	196-97	44°50'N	0°34'W
Bordentown, N.J., U.S. (bôr'dĕn-toun)	116-17	40°08'N	74°44'W
Bordertown, Austl.	276	36°19'S	140°46'E
Bordj Bou Arreridj, Alg. (bôrj-bo͞o-ȧ-rā-rēj')	269b	36°04'N	4°46'E
Borgå, Fin. *see* Porvoo	192-93	60°24'N	25°40'E
Borgarnes, Ice.	190a	64°34'N	21°55'W
Borger, Tx., U.S. (bôr'gẽr)	120-21	35°40'N	101°24'W
Borgholm, Swe. (bôrg-hōlm')	192-93	56°52'N	16°40'E
Borgne, Lake, b., La., U.S. (lāk bôrn'y')	124-25	30°05'N	89°35'W
Borgomanero, Italy (bôr'gō-mä-nâ'rō)	200-01	45°42'N	8°28'E
Borgo Val di Taro, Italy (bô'r-zhō-väl-dē-tá'rō)	200-01	44°29'N	9°46'E
Borisoglebsk, Russia (bŏ-rē sô-glyĕpsk')	186-87	51°22'N	42°06'E
Borken, Ger. (bōr'kĕn)	194-95	51°51'N	6°52'E
Borkum, i., Ger. (bōr'ko͝om)	194-95	53°36'N	6°42'E
Borlänge, Swe. (bôr-lĕŋ'gĕ)	192-93	60°29'N	15°27'E
Borneo, i., Asia (bôr'-nē-ō)	248-49	0°30'N	114°00'E
Bornholm, i., Den. (bôrn-hôlm)	192-93	55°09'N	14°55'E
Boro, stm., Sudan	262-63	8°51'N	26°11'E
Borogontsy, Russia	218-19	62°41'N	131°09'E
Borovichi, Russia (bŏ-rō-vē'chē)	202-03	58°23'N	33°55'E
Borroloola, Austl. (bŏr-rô-lo͞o'l*ȧ*)	270-71	16°05'S	136°17'E
Borūjerd, Iran	228-29	33°53'N	48°45'E
Borzna, Ukr. (bôrz'ná)	202-03	51°15'N	32°26'E
Borzya, Russia (bôrz'yá)	240-41	50°22'N	116°31'E
Bosa, Italy (bō'sä)	200-01	40°18'N	8°30'E
Bosanska Gradiška, Bos. (bō'sän-skä grä-dĭsh'kä)	200-01	45°09'N	17°15'E
Bosanski Novi, Bos. (bō's sän-skĭ nō'vē)	200-01	45°04'N	16°23'E
Boscobel, Wi., U.S. (bŏs'kô-bĕl)	114-15	43°08'N	90°42'W
Bose, China (bwo-sŭ)	238-39	23°55'N	106°38'E
Boshan, China (bwo-shan)	240-41	36°29'N	117°51'E
Bosna, stm., Bos.	200-01	45°04'N	18°28'E
Bosna i Hercegovina, nation, Eur. *see* Bosnia and Herzegovina	174-75	44°15'N	17°50'E
Bosnia and Hercegovina, nation, Eur. *see* Bosnia and Herzegovina	174-75	44°15'N	17°50'E
Bosnia and Herzegovina, nation, Eur. (bŏs'nĭ-ȧ ănd hĕr-tsĕ-gō'vē-ná)	174-75	44°15'N	17°50'E
Bosporus, strt., Tur. (bŏs'p*ŏ*-rŭs)	200-01	41°06'N	29°04'E
Bossembélé, C.A.R.	262-63	5°16'N	17°39'E
Bossier City, La., U.S. (bŏsh'ẽr sĭ'tĭ)	120-21	32°31'N	93°44'W
Bosso, Dallol, stm., Niger	258-59	12°24'N	2°52'E
Bosten Hu, lk., China (bwo-stŭn ho͞o)	222-23	42°0'N	87°00'E
Boston, Ma., U.S. (bôs't*ŭ*n)	116-17	42°22'N	71°03'W
Boston Mountains, mts., Ar., U.S. (bôs't*ŭ*n moun'tĭnz)	120-21	35°50'N	93°20'W
Boteti, stm., Bots.	264-65	20°09'S	23°23'E
Bothaville, S. Afr. (bō'tä-vĭl)	269c	27°24'S	26°37'E
Bothnia, Gulf of, b., Eur. (gŭlf ŭv bŏth'nĭ-ȧ)	184-85	63°0'N	20°00'E
Botoșani, Rom. (bô-tô-shän'ĭ)	194-95	47°45'N	26°40'E
Botswana, nation, Afr. (bŏtswănă)	253	22°0'S	24°00'E
Bottineau, N.D., U.S. (bŏt-ĭ-nō')	114-15	48°49'N	100°27'W
Bottniska Viken, b., Eur. *see* Bothnia, Gulf of	184-85	63°0'N	20°00'E
Botucatu, Braz.	172	22°52'S	48°27'W
Botwood, Nf., Can. (bŏt'wŏd)	138-39	49°09'N	55°22'W
Bouaké, C. Iv.	260-61	7°42'N	5°02'W
Bouar, C.A.R. (bo͞o-är)	262-63	5°57'N	15°36'E
Boufarik, Alg. (bo͞o-fá-rēk')	269b	36°35'N	2°54'E
Bougainville, i., Pap. N. Gui. (bo͞o-găn-vēl')	279e	6°0'S	155°00'E
Bougie, Alg. *see* Bejaïa	269b	36°45'N	5°04'E
Bouira, Alg. (boo-ē'rá)	269b	36°23'N	3°54'E
Boujdour, Cap, c., W. Sah.	258-59	26°08'N	14°29'W
Boulder, Co., U.S. (bōl'dẽr)	118-19	40°02'N	105°15'W
Boulder, Mt., U.S. (bōl'dẽr)	112-13	46°14'N	112°08'W
Boulder, stm., Mt., U.S. (bōld'ẽr)	112-13	45°52'N	111°57'W
Boulder City, Nv., U.S. (bōld'ẽr sĭ'tē)	118-19	35°59'N	114°50'W
Boulia, Austl.	277	22°55'S	139°55'E
Boulogne-Billancourt, Fr. (bo͞o-lôn'y'-bē-yän-ko͞or')	196-97	48°51'N	2°15'E
Boulogne-sur-Mer, Fr. (bo͞o-lôn'y-sür-mâr')	196-97	50°43'N	1°36'E
Boundary Peak, mtn., Nv., U.S.	118-19	37°51'N	118°21'W
Bountiful, Ut., U.S. (boun'tĭ-fŏl)	112-13	40°53'N	111°53'W
Bounty Islands, is., N.Z.	287	47°42'S	179°04'E
Bourg-en-Bresse, Fr. (bo͞or-gĕn-brĕs')	196-97	46°13'N	5°13'E
Bourges, Fr. (bo͞orzh)	196-97	47°05'N	2°24'E
Bourke, Austl. (bŭrk)	276	30°06'S	145°56'E
Bournemouth, Eng., U.K. (bôrn'm*ŭ*th)	190-91	50°44'N	1°52'W
Bou Saâda, Alg. (bo͞o-sä'dä)	269b	35°13'N	4°11'E
Boutilimit, Maur.	258-59	17°33'N	14°42'W
Bouvetøya, i., Afr.	284-85	54°26'S	3°24'E
Bow, stm., Ab., Can. (bō)	132-33	49°56'N	111°42'W
Bowbells, N.D., U.S. (bō'bĕls)	114-15	48°48'N	102°15'W
Bowen, Austl. (bō'ĕn)	277	20°01'S	148°14'E
Bowie, Md., U.S. (bo͞o'ĭ) (bō'ē)	116-17	39°00'N	76°46'W
Bowie, Tx., U.S. (bo͞o'ĭ) (bō'ē)	120-21	33°34'N	97°51'W
Bowling Green, Ky., U.S. (bōlĭng grēn)	124-25	36°59'N	86°27'W
Bowling Green, Mo., U.S. (bōlĭng grēn)	120-21	39°21'N	91°12'W
Bowling Green, Oh., U.S. (bōlĭng grēn)	116-17	41°22'N	83°39'W
Bowling Green, Va., U.S. (bōlĭng grēn)	116-17	38°03'N	77°21'W
Bowman, N.D., U.S. (bō'm*ă*n)	114-15	46°11'N	103°24'W
Bowral, Austl.	276	34°28'S	150°26'E
Bowron, stm., B.C., Can. (bō'r*ŭ*n)	132-33	54°03'N	121°50'W
Boxing, China (bwo-shyĭŋ)	240-41	37°08'N	118°07'E
Boyang, China (bwo-yäŋ)	238-39	28°60'N	116°40'E
Boyle, Ire. (boil)	190-91	53°59'N	8°18'W
Boyoma Falls, wtfl., D.R.C. *see* Stanley Falls	262-63	0°29'N	25°13'E
Boysun, Uzb.	232-33	38°12'N	67°12'E
Bozeman, Mt., U.S. (bōz'm*ă*n)	112-13	45°41'N	111°03'W
Bozen, Italy *see* Bolzano	200-01	46°30'N	11°21'E
Bozhen, China (bwo-jŭn)	240-41	38°05'N	116°33'E
Bozhou, China	238-39	33°52'N	115°46'E
Bra, Italy (brä)	200-01	44°41'N	7°51'E
Bracebridge, On., Can. (brās'brĭj)	136-37	45°02'N	79°18'W
Brackettville, Tx., U.S. (brăk'*ĕ*t-vĭl)	122-23	29°18'N	100°25'W
Bradano, stm., Italy (brä-dä'nō)	200-01	40°23'N	16°51'E
Bradenton, Fl., U.S. (brā'd*ĕ*n-tŭn)	125a	27°30'N	82°33'W
Bradford, Eng., U.K. (brăd'fẽrd)	190-91	53°48'N	1°45'W
Bradley, Il., U.S. (brăd'lĭ)	116-17	41°08'N	87°51'W
Brady, Tx., U.S. (brā'dĭ)	122-23	31°08'N	99°20'W
Braga, Port. (brä'gä)	198-99	41°33'N	8°26'W
Bragado, Arg. (brä-gä'dō)	173	35°08'S	60°30'W
Bragança, Braz. (brä-gän'sä)	166-67	1°03'S	46°46'W
Bragança, Port.	198-99	41°49'N	6°45'W
Brāhmanbāria, Bngl.	234-35	23°59'N	91°07'E
Brāhmani, stm., India.	234-35	20°47'N	87°01'E
Brahmapur, India	236	19°18'N	84°49'E
Brahmaputra, stm., Asia (brä'ma-po͞o'trá)	234-35	24°02'N	91°00'E
Braidwood, Il., U.S. (brād'wŏd)	116-17	41°16'N	88°12'W
Brăila, Rom. (brē'ēlá)	202-03	45°16'N	27°58'E
Brainerd, Mn., U.S. (brān'ẽrd)	114-15	46°21'N	94°12'W
Brampton, On., Can. (brămp't*ŭ*n)	136-37	43°42'N	79°45'W
Branco, stm., Braz. (brăŋ'kō)	164-65	1°24'S	61°52'W
Brandberg, mtn., Nmb.	264-65	21°10'S	14°33'E
Brandenburg, Ger. (brän'dĕn-bŏrgh)	194-95	52°25'N	12°33'E
Brandfort, S. Afr. (brăn'd-fôrt)	269c	28°42'S	26°28'E
Brandon, Mb., Can. (brăn'd*ŭ*n)	134-35	49°50'N	99°58'W
Brandon, Ms., U.S. (brăn'd*ŭ*n)	124-25	32°16'N	89°59'W
Brandon, S.D., U.S. (brăn'd*ŭ*n)	114-15	43°36'N	96°34'W
Brandon, Vt., U.S. (brăn'd*ŭ*n)	116-17	43°48'N	73°05'W
Braniewo, Pol. (brä-nyĕ'vô)	194-95	54°23'N	19°50'E
Brantford, On., Can. (brănt'fẽrd)	136-37	43°09'N	80°15'W
Bras d'Or Lake, lk., N.S., Can. (brä-dôr' lāk)	138-39	45°52'N	60°50'W
Brasil, nation, S.A. *see* Brazil	158	10°0'S	55°00'W
Brasiléia, Braz.	166-67	10°60'S	68°45'W
Brasília, nat. cap., Braz. (brä-sē'lvä)	168-69	15°48'S	47°53'W
Brașov, Rom.	194-95	45°39'N	25°37'E
Brass, Nig. (brăs)	260a	4°19'N	6°15'E
Brassó, Rom. *see* Brașov	194-95	45°39'N	25°37'E
Bratislava, nat. cap., Slvk. (brä'tĭs-lä-vä)	194-95	48°09'N	17°07'E
Bratsk, Russia (brätsk)	218-19	56°08'N	101°39'E
Bratskoye Vodokhranilishche, res., Russia	218-19	55°57'N	101°52'E
Bratsk Reservoir, res., Russia (brätsk rĕ'sẽr-vwär) *see* Bratskoye Vodokhranilishche	218-19	55°57'N	101°52'E
Bratslav, Ukr. (brät'sláf)	202-03	48°49'N	28°57'E
Brattleboro, Vt., U.S. (brăt''l-bŭr-ô)	116-17	42°51'N	72°34'W
Braunschweig, Ger. (broun'shvīgh)	194-95	52°16'N	10°31'E
Brava, i., C.V.	260-61	14°52'N	24°43'W
Bravo, stm., N.A. *see* Rio Grande	110-11	25°57'N	97°09'W
Bravo del Norte, stm., N.A. *see* Rio Grande	110-11	25°57'N	97°09'W
Brawley, Ca., U.S. (brô'lĭ)	118-19	32°59'N	115°33'W
Brazeau, stm., Ab., Can.	132-33	52°55'N	115°14'W
Brazeau, Mount, mtn., Ab., Can. (mount bră-zō')	132-33	52°33'N	117°21'W
Brazil, In., U.S. (br*ȧ*-zĭl')	116-17	39°31'N	87°07'W
Brazil, nation, S.A. (br*ȧ*-zĭl')	158	10°0'S	55°00'W
Brazos, stm., Tx., U.S. (brä'zōs)	122-23	33°15'N	100°00'W
Brazos, Salt Fork, stm., U.S. (sôlt fôrk)	120-21	33°16'N	100°01'W
Brazzaville, nat. cap., Congo (brá-zá-vēl')	262-63	4°16'S	15°17'E
Brčko, Bos. (bĕrch'kô)	200-01	44°52'N	18°49'E
Breckenridge, Mn., U.S. (brĕk'*ĕ*n-rĭj)	114-15	46°15'N	96°34'W
Breckenridge, Tx., U.S. (brĕk'*ĕ*n-rĭj)	120-21	32°45'N	98°55'W
Břeclav, Czech Rep. (brzhĕl'láf)	194-95	48°46'N	16°54'E
Breda, Neth. (brā-dä')	190-91	51°35'N	4°46'E
Bregenz, Aus. (brā'gĕnts)	194-95	47°30'N	9°46'E
Bregovo, Blg. (brĕ'gô-vô)	200-01	44°10'N	22°39'E
Breiðafjörður, b., Ice.	190a	65°15'N	23°15'W
Brejo, Braz. (brå'zhò)	166-67	3°41'S	42°47'W
Bremen, Ger. (brā-m*ĕ*n)	194-95	53°04'N	8°51'E
Bremen, Ga., U.S. (brē'm*ĕ*n)	124-25	33°43'N	85°09'W
Bremen, In., U.S. (brē'm*ĕ*n)	116-17	41°27'N	86°08'W
Bremerhaven, Ger. (brām-ẽr-hä'f*ĕ*n)	194-95	53°32'N	8°36'E
Bremerton, Wa., U.S. (brĕm'ẽr-t*ŭ*n)	112-13	47°34'N	122°39'W
Brenham, Tx., U.S. (brĕn'*ă*m)	122-23	30°10'N	96°24'W
Brentwood, N.Y., U.S. (brĕnt'wŏd)	116-17	40°47'N	73°15'W
Brentwood, Tn., U.S. (brĕnt'wŏd)	124-25	36°02'N	86°47'W
Brescia, Italy (brā'shä)	200-01	45°33'N	10°13'E
Breslau, Pol. *see* Wrocław	194-95	51°07'N	17°02'E
Bressanone, Italy (brĕs-sä-nō'nä)	200-01	46°43'N	11°39'E
Bressuire, Fr. (grĕ-swēr')	196-97	46°50'N	0°29'W
Brèst, Bela.	194-95	52°07'N	23°42'E
Brest, Fr. (brĕst)	196-97	48°24'N	4°30'W
Bretagne, hist. reg., Fr. (brĕ-tän'yĕ) *see* Brittany	196-97	48°0'N	3°00'W
Breton Sound, strt., La., U.S. (brĕt'*ŭ*n sound)	124-25	29°34'N	89°16'W
Brevard, N.C., U.S. (brē-värd')	124-25	35°14'N	82°44'W
Breves, Braz. (brā'vĕzh)	166-67	1°40'S	50°29'W
Brewarrina, Austl. (bro͞o-ẽr-rē'n*ȧ*)	276	29°57'S	146°52'E
Brewster, Wa., U.S. (bro͞o'stẽr)	112-13	48°06'N	119°47'W
Brewster, Kap, c., Green.	86	70°09'N	22°06'W
Brewton, Al., U.S. (bro͞o't*ŭ*n)	124-25	31°07'N	87°05'W
Brezhnev, Russia *see* Naberezhnye Chelny	186-87	55°42'N	52°19'E
Bria, C.A.R.	262-63	6°33'N	21°58'E
Briançon, Fr. (brē-äN-sôN')	196-97	44°54'N	6°37'E
Bridgeport, Al., U.S. (brĭj'pôrt)	124-25	34°57'N	85°43'W
Bridgeport, Ca., U.S. (brĭj'pôrt)	118-19	38°16'N	119°14'W
Bridgeport, Ct., U.S. (brĭj'pôrt)	116-17	41°11'N	73°14'W
Bridgeport, Il., U.S. (brĭj'pôrt)	116-17	38°42'N	87°46'W
Bridgeport, Ne., U.S. (brĭj'pôrt)	114-15	41°40'N	103°05'W
Bridgeport, Tx., U.S. (brĭj'pôrt)	120-21	33°13'N	97°46'W
Bridgetown, Austl.	270-71	33°58'S	116°08'E
Bridgetown, N.S., Can. (brĭj' toun)	138-39	44°52'N	65°16'W
Bridgetown, nat. cap., Barb. (brĭj' toun)	143b	13°06'N	59°37'W
Bridgewater, N.S., Can.	138-39	44°22'N	64°31'W
Bridgton, Me., U.S. (brĭj't*ŭ*n)	116-17	44°04'N	70°42'W
Bridlington, Eng., U.K. (brĭd'lĭng-t*ŭ*n)	190-91	54°05'N	0°12'W
Brig, Switz. (brĕg)	194-95	46°19'N	8°00'E
Brigham City, Ut., U.S. (brĭg'*ă*m sĭ'tē)	112-13	41°31'N	112°01'W
Bright, Austl. (brĭt)	276	36°44'S	146°58'E
Brighton, Eng., U.K. (brī't*ŭ*n)	190-91	50°50'N	0°08'W
Brighton, Co., U.S. (brī't*ŭ*n)	120-21	39°59'N	104°49'W
Brighton, N.Y., U.S. (brī't*ŭ*n)	116-17	43°09'N	77°33'W
Brindisi, Italy (brēn'dē-zē)	200-01	40°38'N	17°56'E
Brinkley, Ar., U.S. (brĭŋk'lĭ)	124-25	34°54'N	91°11'W
Brioude, Fr. (brē-o͞od')	196-97	45°18'N	3°23'E
Brisbane, Austl. (brĭz'bân)	276	27°28'S	153°02'E
Bristol, Eng., U.K.	190-91	51°27'N	2°36'W
Bristol, Ct., U.S. (brĭs'tŭl)	116-17	41°41'N	72°57'W
Bristol, R.I., U.S. (brĭs'tŭl)	116-17	41°40'N	71°16'W

n-sing; ŋ-baŋk; N-nasalized n; nŏd; c*ŏ*mmit; ōld; ŏbey; ôrder; oi-boil; fo͞od; ŏ-as oo in foot; ou-out; s-soft; sh-dish; th-thin; pūre; ûnite; ûrn; stŭd; circ*ŭ*s; ü-as in French tu; '-indeterminate vowel.

Feature (Pronunciation)	Page	Lat.	Long.
Bristol, Tn., U.S. (brĭs′tŭl)	124-25	36°35′N	82°11′W
Bristol, Va., U.S. (brĭs′tŭl)	124-25	36°36′N	82°11′W
Bristol Bay, b., Ak., U.S. (brĭs′tŭl bā)	126	58°0′N	159°00′W
Bristol Channel, strt., U.K.	190-91	51°23′N	4°01′W
Bristow, Ok., U.S. (brĭs′tō)	120-21	35°50′N	96°24′W
British Columbia, state, Can. (brĭt′ĭsh kŏl′ŭm-bĭ-ȧ)	128-29	54°0′N	125°00′W
British Guiana, nation, S.A. *see* Guyana	158	5°0′N	59°00′W
British Honduras, nation, N.A. *see* Belize	85	17°15′N	88°45′W
British Indian Ocean Territory, dep., Afr.	206-07	7°0′S	72°00′E
British Solomon Islands, nation, Oc. *see* Solomon Islands	279e	8°0′S	159°00′E
British Virgin Islands, dep., N.A.	140-41	18°30′N	64°30′W
Britt, Ia., U.S. (brĭt)	114-15	43°06′N	93°49′W
Brittany, hist. reg., Fr.	196-97	48°0′N	3°00′W
Britton, S.D., U.S. (brĭt′ŭn)	114-15	45°48′N	97°45′W
Brive-la-Gaillarde, Fr. (brēv-lä-gī-yärd′ĕ)	196-97	45°09′N	1°32′E
Brixen, Italy *see* Bressanone	200-01	46°43′N	11°39′E
Brno, Czech Rep. (b′r′nô)	194-95	49°12′N	16°37′E
Brockport, N.Y., U.S. (brŏk′pōrt)	116-17	43°13′N	77°56′W
Brockton, Ma., U.S. (brŏk′tŭn)	116-17	42°05′N	71°01′W
Brockville, On., Can. (brŏk′vĭl)	136-37	44°36′N	75°41′W
Brodnica, Pol. (brôd′nĭt-sȧ)	194-95	53°15′N	19°24′E
Brody, Ukr. (brô′dĭ)	194-95	50°05′N	25°10′E
Broken Arrow, Ok., U.S. (brō′kĕn ăr′ō)	120-21	36°03′N	95°47′W
Broken Bow, Ne., U.S. (brō′kĕn bō)	114-15	41°24′N	99°39′W
Broken Bow, Ok., U.S. (brō′kĕn bō)	120-21	34°02′N	94°44′W
Broken Hill, Austl. (brōk′ĕn hĭl)	276	31°58′S	141°27′E
Broken Hill, Zam. *see* Kabwe	264-65	14°27′S	28°27′E
Brokopondo, Sur.	164-65	5°04′N	54°59′W
Brokopondo Stuwmeer, res., Sur. *see* W.J. van Blommestein Meer	164-65	4°49′N	55°04′W
Bromberg, Pol. *see* Bydgoszcz	194-95	53°07′N	18°01′E
Bronlund Peak, mtn., B.C., Can.	130-31	57°26′N	126°38′W
Brookfield, Mo., U.S. (brŏok′fēld)	120-21	39°48′N	93°05′W
Brookfield, Wi., U.S. (brŏok′fēld)	116-17	43°04′N	88°07′W
Brookhaven, Ms., U.S. (brŏok′hāv′n)	124-25	31°34′N	90°27′W
Brookings, Or., U.S. (brŏok′ĭngs)	112-13	42°04′N	124°17′W
Brookings, S.D., U.S. (brŏok′ĭngs)	114-15	44°19′N	96°48′W
Brooklyn Park, Mn., U.S. (brŏok′lĭn pärk)	114-15	45°07′N	93°20′W
Brooks, Ab., Can. (brŏoks)	132-33	50°34′N	111°54′W
Brooks Range, mts., Ak., U.S. (brŏoks rānj)	126	68°0′N	154°00′W
Brooksville, Fl., U.S. (brŏoks′vĭl)	125a	28°33′N	82°24′W
Brookton, Austl.	270-71	32°22′S	117°01′E
Broome, Austl. (brōōm)	270-71	17°58′S	122°14′E
Brownfield, Tx., U.S. (broun′fēld)	120-21	33°11′N	102°16′W
Browning, Mt., U.S. (broun′ĭng)	112-13	48°33′N	112°60′W
Brownstown, In., U.S. (brounz′toun)	116-17	38°53′N	86°03′W
Brownsville, Tn., U.S. (brounz′vĭl)	124-25	35°36′N	89°16′W
Brownsville, Tx., U.S. (brounz′vĭl)	122-23	25°56′N	97°29′W
Brownwood, Tx., U.S. (broun′wŏod)	122-23	31°43′N	98°59′W
Bruce, Mount, mtn., Austl. (mount brōōs)	272-73	22°35′S	118°08′E
Bruchsal, Ger. (brŏok′zäl)	194-95	49°08′N	8°36′E
Bruit, Pulau, i., Malay.	248-49	2°35′N	111°20′E
Bruneau, stm., U.S. (brōō-nō′)	112-13	42°57′N	115°57′W
Brunei, nation, Asia (brŏo-nī′)	206-07	4°30′N	114°40′E
Brunei, nat. cap., Bru. *see* Bandar Seri Begawan	248-49	4°56′N	114°56′E
Brünn, Czech Rep. *see* Brno	194-95	49°12′N	16°37′E
Brunswick, Ger. *see* Braunschweig	194-95	52°16′N	10°31′E
Brunswick, Ga., U.S. (brŭnz′wĭk)	124-25	31°11′N	81°30′W
Brunswick, Md., U.S. (brŭnz′wĭk)	116-17	39°19′N	77°38′W
Brunswick, Me., U.S. (brŭnz′wĭk)	117a	43°55′N	69°58′W
Brunswick, Península, pen., Chile	171	53°11′S	71°11′W
Brush, Co., U.S. (brŭsh)	120-21	40°15′N	103°38′W
Brusque, Braz. (brōō′s-kōōĕ)	172	27°07′S	48°56′W
Brussel, nat. cap., Bel. *see* Brussels	190-91	50°50′N	4°22′E
Brussels, nat. cap., Bel. (brŭs′ĕls)	190-91	50°50′N	4°22′E
Brüx, Czech Rep. *see* Most	194-95	50°31′N	13°39′E
Bruxelles, nat. cap., Bel. (brü-sĕl′) *see* Brussels	190-91	50°50′N	4°22′E
Bryan, Oh., U.S. (brī′ăn)	116-17	41°28′N	84°33′W
Bryan, Tx., U.S. (brī′ăn)	122-23	30°41′N	96°23′W
Bryansk, Russia	202-03	53°14′N	34°22′E
Bryce Canyon National Park, n.p., Ut., U.S. (brīs kăn′yŭn năsh′ŭn-ăl pärk)	118-19	37°29′N	112°15′W
Bryson City, N.C., U.S. (brīs′ŭn sĭ′tē)	124-25	35°26′N	83°27′W
Bryukhovetskaya, Russia (b′ryŭk′ô-vyĕt-skä′yä)	202-03	45°49′N	39°00′E
Bua Yai, Thai.	246-47	15°35′N	102°26′E
Būbiyān, i., Kuw.	230-31	29°45′N	48°15′E
Bucaramanga, Col. (bōō-kä′rä-mäŋ′gä)	164-65	7°03′N	73°05′W
Buchach, Ukr. (bŏo′chȧch)	194-95	49°04′N	25°25′E
Buchanan, Lib. (bů-kăn′ăn)	260-61	5°53′N	10°02′W
Buchanan, Mi., U.S. (bû-kăn′ăn)	116-17	41°49′N	86°22′W
Buchanan, Va., U.S. (bû-kăn′ăn)	116-17	37°31′N	79°41′W
Buchanan, Lake, lk., Tx., U.S. (lāk bû-kăn′ăn)	122-23	30°48′N	98°25′W
Buchans, Nf., Can.	138-39	48°49′N	56°52′W
Bucharest, nat. cap., Rom. (bōō-kȧ-rĕst′)	200-01	44°26′N	26°06′E
Buckhannon, W.V., U.S. (bŭk-hăn′ŭn)	116-17	38°59′N	80°14′W
Buckhaven, Scot., U.K. (bŭk-hā′v′n)	190-91	56°11′N	3°03′W
Buckie, Scot., U.K. (bŭk′ĭ)	190-91	57°40′N	2°59′W
Bucksport, Me., U.S. (bŭks′pôrt)	117a	44°34′N	68°48′W
București, nat. cap., Rom. (bōō-kô-rĕsh′tĭ) *see* Bucharest	200-01	44°26′N	26°06′E
Bucyrus, Oh., U.S. (bů-sī′rŭs)	116-17	40°48′N	82°58′W
Budapest, nat. cap., Hung. (bōō′dȧ-pĕsht′)	194-95	47°30′N	19°05′E
Budaun, India	234-35	28°02′N	79°08′E
Budennovsk, Russia	186-87	44°47′N	44°09′E
Budweis, Czech Rep. *see* České Budějovice	194-95	48°59′N	14°28′E
Buena Esperanza, Arg.	171	34°45′S	65°16′W
Buenaventura, Col. (bwā′nä-vĕn-tōō′rȧ)	163c	3°53′N	77°04′W
Buenaventura, Mex.	144-45	29°51′N	107°28′W
Buena Vista, Bol.	168-69	17°27′S	63°40′W
Buena Vista, Co., U.S. (bū′nȧ vĭs′tȧ)	118-19	38°50′N	106°09′W
Buena Vista, Va., U.S. (bū′nȧ vĭs′tȧ)	116-17	37°44′N	79°21′W
Buenos Aires, state, Arg. (bwā′nōs ī′rās)	173	36°0′S	60°00′W
Buenos Aires, nat. cap., Arg. (bwā′nōs ī′rās)	173	34°37′S	58°23′W
Buenos Aires, Lago, lk., S.A. (lä′gô-bwā′nōs ī′rās)	171	46°26′S	71°40′W
Buffalo, Mn., U.S. (bŭf′ȧ-lō)	114-15	45°11′N	93°53′W
Buffalo, Mo., U.S. (bŭf′ȧ-lō)	120-21	37°39′N	93°06′W
Buffalo, N.Y., U.S. (bŭf′ȧ-lō)	116-17	42°53′N	78°52′W
Buffalo, Ok., U.S. (bŭf′ȧ-lō)	120-21	36°50′N	99°38′W
Buffalo, Tx., U.S. (bŭf′ȧ-lō)	122-23	31°27′N	96°04′W
Buffalo, Wy., U.S. (bŭf′ȧ-lō)	112-13	44°22′N	106°42′W
Buffalo, stm., Tn., U.S. (bŭf′ȧ-lō)	124-25	35°60′N	87°50′W
Buffalo Lake, lk., N.T., Can.	130-31	60°10′N	115°30′W
Buford, Ga., U.S. (bū′fērd)	124-25	34°07′N	84°00′W
Buga, Col. (bōō′gä)	163c	3°54′N	76°18′W
Bugojno, Bos. (bŏo-gō′ĭ nô)	200-01	44°03′N	17°27′E
Bugsuk Island, i., Phil.	250	8°15′N	117°18′E
Bugt, China	240-41	48°46′N	121°54′E
Bugul'ma, Russia (bŏo-gŏol′mȧ)	186-87	54°31′N	52°47′E
Buguma, Nig.	260a	4°43′N	6°53′E
Buguruslan, Russia (bŏo-gŏo-rŏos-lȧn′)	186-87	53°39′N	52°27′E
Buhl, Id., U.S. (būl)	112-13	42°37′N	114°46′W
Buin, Chile (bŏo-ēn′)	163e	33°42′S	70°43′W
Buir Nur, lk., Asia (bōō-ēr nōōr)	240-41	47°48′N	117°42′E
Buitenzorg, Indon. *see* Bogor	248-49	6°35′S	106°47′E
Bujumbura, nat. cap., Bdi. (bōō-jŭm-bōō′rȧ)	267	3°23′S	29°22′E
Buka Island, i., Pap. N. Gui.	279e	5°15′S	154°35′E
Bukama, D.R.C. (bōō-kä′mä)	262-63	9°12′S	25°51′E
Bukavu, D.R.C.	267	2°30′S	28°51′E
Bukhara, Uzb. (bŏo-kä′rä) *see* Buxoro	232-33	39°46′N	64°26′E
Bukittinggi, Indon.	246-47	0°18′S	100°22′E
Bukoba, Tan.	267	1°19′S	31°48′E
Bulan, Phil.	250	12°40′N	123°53′E
Bulawayo, Zimb. (bōō-lȧ-wā′yō)	264-65	20°10′S	28°35′E
Bulgan, Mong.	240-41	48°49′N	103°33′E
Bulgaria, nation, Eur. (bŏol-gā′rĭ-ă)	174-75	43°0′N	25°00′E
Bŭlgariya, nation, Eur. *see* Bulgaria	174-75	43°0′N	25°00′E
Bulkley Ranges, mts., B.C., Can. (bŭlk′lē rānjĕz)	132-33	54°30′N	127°30′W
Bulloo, stm., Austl.	272-73	28°40′S	142°31′E
Bull Shoals Lake, res., U.S. (bŏol shōlz lāk)	120-21	36°29′N	92°47′W
Bultfontein, S. Afr. (bŏolt′fŏn-tān′)	269c	28°17′S	26°09′E
Bulungu, D.R.C. (bōō-lŏŋ′gōō)	262-63	6°03′S	21°53′E
Bumba, D.R.C. (bŏom′bȧ)	262-63	2°11′N	22°28′E
Bumbire Island, i., Tan.	267	1°39′S	31°53′E
Bunbury, Austl. (bŭn′bŭrĭ)	270-71	33°19′S	115°38′E
Bundaberg, Austl. (bŭn′dȧ-bûrg)	277	24°52′S	152°20′E
Būndi, India	234-35	25°27′N	75°38′E
Bungo-suidō, strt., Japan	245	33°0′N	132°13′E
Bunia, D.R.C.	267	1°32′N	30°15′E
Bunkie, La., U.S. (bŭŋ′kĭ)	124-25	30°57′N	92°11′W
Buntok, Indon.	248-49	1°44′S	114°50′E
Buon Ma Thuot, Viet.	246-47	12°40′N	108°03′E
Buqayq, Sau. Ar.	230-31	25°56′N	49°40′E
Burang, China	234-35	30°14′N	81°11′E
Buraydah, Sau. Ar.	266	26°19′N	43°59′E
Burayk, Libya	258-59	26°37′N	13°07′E
Burbank, Ca., U.S. (bûr′bănk)	118-19	34°11′N	118°19′W
Burco, Som.	262-63	9°32′N	45°33′E
Burdur, Tur. (bōōr-dŏor′)	186-87	37°43′N	30°17′E
Bureya, stm., Russia (bŏo-rā′yä)	240-41	49°25′N	129°32′E
Burgas, Blg. (bŏor-gäs′)	200-01	42°31′N	27°28′E
Burgaw, N.C., U.S. (bûr′gô)	124-25	34°33′N	77°56′W
Burgos, Spain (bōō′r-gōs)	198-99	42°21′N	3°42′W
Burgsvik, Swe. (bŏrgs′vĭk)	192-93	57°02′N	18°18′E
Burhānpur, India (bŏor′hän-pōōr)	234-35	21°18′N	76°14′E
Burias Island, i., Phil. (bōō′rē-äs ī′lănd)	250	12°57′N	123°08′E
Burica, Punta, c., N.A. (pōō′n-tä-bōō′rē-kä)	150	8°03′N	82°52′W
Burin, Nf., Can. (bûr′ĭn)	138-39	47°02′N	55°11′W
Burkburnett, Tx., U.S. (bûrk-bûr′nĕt)	120-21	34°06′N	98°34′W
Burke, stm., Austl.	277	23°12′S	139°34′E
Burketown, Austl. (bûrk′toun)	277	17°44′S	139°33′E
Burkina Faso, nation, Afr. (bōōr-kē′-nȧ fä′sō)	253	13°0′N	1°30′W
Burley, Id., U.S. (bûr′lĭ)	112-13	42°33′N	113°47′W
Burlington, On., Can. (bûr′lĭng-tŭn)	136-37	43°19′N	79°48′W
Burlington, Co., U.S. (bûr′lĭng-tŭn)	120-21	39°18′N	102°16′W
Burlington, Ia., U.S. (bûr′lĭng-tŭn)	114-15	40°48′N	91°06′W
Burlington, N.C., U.S. (bûr′lĭng-tŭn)	124-25	36°06′N	79°26′W
Burlington, N.D., U.S. (bûr′lĭng-tŭn)	114-15	48°16′N	101°25′W
Burlington, Vt., U.S.	116-17	44°29′N	73°12′W
Burlington, Wi., U.S. (bûr′lĭng-tŭn)	116-17	42°41′N	88°16′W
Burma, nation, Asia *see* Bhutan	206-07	22°0′N	98°00′E
Burnie, Austl. (bûr′nē)	276	41°04′S	145°54′E
Burnley, Eng., U.K. (bûrn′lĕ)	190-91	53°48′N	2°15′W
Burns, Or., U.S. (bûrnz)	112-13	43°36′N	119°03′W
Burnside, stm., Nu., Can.	130-31	66°51′N	108°12′W
Burns Lake, B.C., Can. (bûrnz lāk)	132-33	54°14′N	125°46′W
Burntwood, stm., Mb., Can.	134-35	56°08′N	96°20′W
Burqin, China	226	47°43′N	86°54′E
Burra, Austl.	276	33°40′S	138°55′E
Bursa, Tur. (bōōr′sȧ)	200-01	40°12′N	29°04′E
Bûr Sa'îd, Egypt *see* Port Said	268b	31°16′N	32°18′E
Būr Sūdān, Sudan *see* Port Sudan	266	19°37′N	37°13′E
Burton, Mi., U.S. (bûr′tŭn)	116-17	43°00′N	83°35′W
Burton upon Trent, Eng., U.K. (bŭr′tŭn-ŭp′-ŏn-trĕnt)	190-91	52°49′N	1°38′W
Buru, i., Indon.	248-49	3°24′S	126°40′E
Burundi, nation, Afr. (bū-rŭn′-dē)	253	3°15′S	30°00′E
Burun-Shibertuy, Gora, mtn., Russia	240-41	49°42′N	109°58′E
Burwell, Ne., U.S. (bûr′wĕl)	114-15	41°46′N	99°08′W
Buryatia, state, Russia	222-23	53°0′N	109°00′E
Buryatiya, state, Russia *see* Buryatia	222-23	53°0′N	109°00′E
Bury Saint Edmunds, Eng., U.K. (bĕr′ĭ-sānt ĕd′mŭndz)	190-91	52°15′N	0°42′E
Busan, Kor., S. *see* Pusan	243	35°05′N	129°03′E
Būshehr, Iran	230-31	28°58′N	50°51′E
Bushire, Iran *see* Būshehr	230-31	28°58′N	50°51′E
Bushnell, Il., U.S. (bŏosh′nĕl)	120-21	40°33′N	90°30′W
Businga, D.R.C. (bŏo-sĭŋ′gȧ)	262-63	3°20′N	20°53′E
Busselton, Austl. (bûs′l-tŭn)	270-71	33°39′S	115°21′E
Busto Arsizio, Italy (bōōs′tō är-sēd′zē-ō)	200-01	45°37′N	8°51′E
Busuanga Island, i., Phil. (bōō-swäŋ′gä ī′lănd)	250	12°05′N	120°05′E
Buta, D.R.C. (bōō′tȧ)	262-63	2°49′N	24°45′E
Butare, Rw.	267	2°36′S	29°44′E
Butaritari, at., Kir.	280-81	3°06′N	172°50′E
Bute Inlet, b., B.C., Can.	132-33	50°37′N	124°53′W
Butembo, D.R.C.	267	0°08′N	29°18′E
Butere, Kenya	267	0°13′N	34°30′E
Butha-Buthe, Leso. (bōō-thȧ-bōō′thȧ)	269c	28°45′S	28°16′E
Butha Qi, China *see* Zalantun	240-41	47°60′N	122°45′E
Butler, In., U.S. (bŭt′lēr)	116-17	41°25′N	84°52′W
Butler, Mo., U.S. (bŭt′lēr)	120-21	38°15′N	94°20′W
Butler, Pa., U.S.	116-17	40°51′N	79°54′W
Buton, Pulau, i., Indon.	248-49	5°02′S	122°53′E
Butte, Mt., U.S. (būt)	112-13	45°60′N	112°32′W
Butterworth, Malay.	246-47	5°24′N	100°23′E
Butuan, Phil. (bōō-tōō′än)	250	8°57′N	125°32′E
Buṭwal, Nepal	234-35	27°43′N	83°28′E
Buxoro, Uzb.	232-33	39°46′N	64°26′E
Buy, Russia (bwē)	186-87	58°29′N	41°33′E
Buyant-Uhaa, Mong.	240-41	44°55′N	110°09′E
Buynaksk, Russia	227	42°50′N	47°06′E
Buyr nuur, lk., Asia *see* Buir Nur	240-41	47°48′N	117°42′E

Feature (Pronunciation)	Page	Lat.	Long.
Büyük Ağrı Dağı, vol., Tur. see Ararat, Mount	227	39°42'N	44°18'E
Buzău, Rom. (bōō-zĕ'ò)	202-03	45°09'N	26°50'E
Búzi, stm., Moz.	264-65	19°53'S	34°45'E
Buzuluk, Russia (bò-zò-lòk')	186-87	52°47'N	52°15'E
Byala Slatina, Blg. (byä'la slä'tēnä)	200-01	43°28'N	23°58'E
Byblos, Leb. see Jubayl	228-29	34°08'N	35°40'E
Bydgoszcz, Pol. (bĭd'gôshch)	194-95	53°07'N	18°01'E
Byelorussia, nation, Eur. see Belarus	174-75	53°50'N	28°00'E
Bytantay, stm., Russia (byän'tāy)	218-19	68°45'N	134°27'E
Bytom, Pol. (bĭ'tŭm)	194-95	50°21'N	18°55'E
Byumba, Rw.	267	1°36'S	30°04'E
Byzantium, Tur. see İstanbul	200-01	41°02'N	28°59'E

C

Feature (Pronunciation)	Page	Lat.	Long.
Ca, stm., Asia	246-47	18°44'N	105°45'E
Caacupé, Para.	168-69	25°22'S	57°08'W
Caála, Ang.	264-65	12°51'S	15°33'E
Caazapá, Para.	168-69	26°11'S	56°22'W
Cabanatuan, Phil. (kä-bä-nä-twän')	250	15°29'N	120°59'E
Cabano, Qc., Can. (kä-bä-nō')	138-39	47°41'N	68°53'W
Cabedelo, Braz. (kä-bē-dā'lò)	163d	6°58'S	34°50'W
Cabeza del Buey, Spain (kȧ-bā'thä dĕl bwā')	198-99	38°43'N	5°13'W
Cabimas, Ven. (kä-bē'mäs)	164-65	10°24'N	71°26'W
Cabinda, Ang. (kä-bĭn'dȧ)	260-61	5°33'S	12°12'E
Cabinet Mountains, mts., U.S. (kăb'ĭ-nĕt moun'tĭnz)	112-13	48°19'N	116°12'W
Cabo, Braz.	163d	8°17'S	35°02'W
Cabo Frio, Braz. (kä'bō-frē'ô)	172	22°53'S	42°02'W
Cabonga, Réservoir, res., Qc., Can.	136-37	47°17'N	76°33'W
Caborca, Mex.	144-45	30°43'N	112°09'W
Cabot Strait, strt., Can. (kăb'ŭt strāt)	138-39	47°20'N	59°30'W
Cabo Verde, nation, Afr. see Cape Verde	253	16°0'N	24°00'W
Cabra, Spain (käb'rä)	198-99	37°29'N	4°27'W
Cabrera, Illa de, i., Spain	198-99	39°09'N	2°57'E
Cabrera, Isla de, i., Spain see Cabrera, Illa de	198-99	39°09'N	2°57'E
Cabriel, stm., Spain (kä-brē-ĕl')	198-99	39°14'N	1°03'W
Caçador, Braz.	168-69	26°47'S	51°01'W
Čačak, Serb. (chä'chäk)	200-01	43°54'N	20°21'E
Cáceres, Braz. (kä'sĕ-rĕs)	168-69	16°04'S	57°42'W
Cáceres, Spain (kä'thå-rās)	198-99	39°28'N	6°22'W
Cache, stm., Ar., U.S. (kăsh)	124-25	34°42'N	91°20'W
Cache Creek, B.C., Can. (kăsh krēk)	132-33	50°49'N	121°19'W
Cachimbo, Serra do, mts., Braz.	166-67	8°25'S	55°45'W
Cachoeira do Sul, Braz. (kä-shô-ā'rä-dô-sōō'l)	173	30°02'S	52°54'W
Cachoeiras de Macacu, Braz. (kä-shô-ā'räs-dĕ-mä-kȧ'kōō)	172	22°28'S	42°39'W
Cachoeiro de Itapemirim, Braz.	172	20°51'S	41°08'W
Cadereyta Jiménez, Mex. (kä-dä-rā'tä hē-mā'näz)	122-23	25°35'N	99°60'W
Cadillac, Mi., U.S. (kăd'ĭ-lăk)	116-17	44°15'N	85°24'W
Cádiz, Spain (kȧ'dēz)	198-99	36°31'N	6°17'W
Cadiz, Ky., U.S. (kā'dĭz)	124-25	36°52'N	87°50'W
Cadiz, Oh., U.S. (kā'dĭz)	116-17	40°16'N	80°60'W
Cádiz, Golfo de, b., Eur. (gôl-fô-dĕ-kȧ'dēz) see Cadiz, Gulf of	198-99	36°50'N	7°10'W
Cadiz, Gulf of, b., Eur. (gŭlf ŭv kā'dĭz)	198-99	36°50'N	7°10'W
Caen, Fr. (käN)	196-97	49°11'N	0°21'W
Caetité, Braz.	166-67	14°04'S	42°29'W
Cafayate, Arg.	168-69	26°04'S	65°59'W
Cagayan, stm., Phil.	250	18°22'N	121°37'E
Cagayan de Oro, Phil.	250	8°29'N	124°38'E
Cagayan Islands, is., Phil.	250	9°40'N	121°16'E
Cagayan Sulu Island, i., Phil.	250	7°01'N	118°30'E
Cagliari, Italy (käl'yä-rē)	200-01	39°14'N	9°07'E
Cagliari, Golfo di, b., Italy (gôl-fô-dē-käl'yä-rē)	200-01	39°08'N	9°11'E
Cagua, Ven. (kä'gwä)	163b	10°12'N	67°26'W
Caguas, P.R. (kä'gwäs)	142a	18°14'N	66°02'W
Cahaba, stm., Al., U.S. (kȧ hä-bä)	124-25	32°20'N	87°06'W
Cahors, Fr. (kȧ-ôr')	196-97	44°27'N	1°26'E
Cahul, Mol.	202-03	45°55'N	28°12'E
Caibarién, Cuba (kī-bä-rē-ĕn')	142-43	22°31'N	79°28'W
Caicedonia, Col. (kī-sĕ-dô-nēä)	163c	4°19'N	75°48'W
Caicó, Braz.	163d	6°27'S	37°06'W
Caicos Islands, is., T./C. Is. (kī'kōs ī'lăndz)	142-43	21°42'N	71°54'W
Caicos Passage, strt., N.A. (kī'kōs păs'ĭj)	142-43	22°00'N	72°30'W
Caimanera, Cuba (kī-mä-nā'rä)	142-43	19°59'N	75°10'W
Cairns, Austl. (kârnz)	277	16°56'S	145°45'E
Cairo, Ga., U.S. (kā'rō)	124-25	30°53'N	84°13'W
Cairo, Il., U.S. (kā'rō)	124-25	37°00'N	89°11'W
Cairo, nat. cap., Egypt (kī'rō)	268b	30°03'N	31°14'E
Cajamarca, Peru (kä-hä-mär'kä)	170	7°10'S	78°31'W
Cajazeiras, Braz.	166-67	6°54'S	38°34'W
Čakovec, Cro. (chȧ'kō-vĕts)	200-01	46°23'N	16°26'E
Calabar, Nig. (kăl-ȧ-bär')	260a	4°58'N	8°19'E
Calabozo, Ven. (kä-lä-bō'zō)	163b	8°55'N	67°26'W
Calafat, Rom. (kȧ-lȧ-fät')	200-01	43°59'N	22°57'E
Calagua Islands, is., Phil. (kä-gä-yän ī'lăndz)	250	14°27'N	122°55'E
Calahorra, Spain (kä-lä-ôr'rä)	198-99	42°18'N	1°58'W
Calais, Fr. (kȧ-lĕ')	196-97	50°58'N	1°51'E
Calais, Me., U.S.	117a	45°11'N	67°16'W
Calais, Pas de, strt., Eur. see Dover, Strait of	190-91	50°59'N	1°31'E
Calama, Chile (kä-lä'mä)	168-69	22°27'S	68°55'W
Calamar, Col. (kä-lä-mär')	164-65	1°58'N	72°42'W
Calamian Group, is., Phil. (kä-lä-myän' grōōp)	250	12°0'N	120°00'E
Calapan, Phil. (kä-lä-pän')	250	13°24'N	121°11'E
Călăraşi, Rom. (kŭ-lŭ-rash'ĭ)	202-03	44°13'N	27°19'E
Calatayud, Spain (kä-lä-tä-yōōdh')	198-99	41°21'N	1°38'W
Calayan Island, i., Phil.	250a	19°20'N	121°27'E
Calbayog, Phil.	250	12°04'N	124°34'E
Calcasieu, stm., La., U.S. (kăl'kȧ-shū)	122-23	30°03'N	93°19'W
Calcasieu Lake, lk., La., U.S. (kăl'kȧ-shū lāk)	122-23	29°56'N	93°16'W
Calçoene, Braz.	166-67	2°30'N	50°57'W
Calcutta, India (kăl-kŭt'ȧ) see Kolkata	234-35	22°32'N	88°22'E
Caldas, Col. (kȧ'l-däs)	163c	6°04'N	75°38'W
Caldas da Rainha, Port. (käl'däs dä rīn'yȧ)	198-99	39°24'N	9°08'W
Caldera, Chile (käl-dā'rä)	168-69	27°04'S	70°50'W
Caldwell, Id., U.S. (kôld'wĕl)	112-13	43°40'N	116°41'W
Caldwell, Oh., U.S. (kôld'wĕl)	116-17	39°44'N	81°31'W
Caldwell, Tx., U.S. (kôld'wĕl)	122-23	30°31'N	96°42'W
Caledonia, Mn., U.S. (kăl-ē-dō'nĭ-ȧ)	114-15	43°39'N	91°31'W
Calella, Spain (kä-lĕl'yä)	198-99	41°37'N	2°40'E
Calexico, Ca., U.S. (kȧ-lĕk'sĭ-kō)	118-19	32°41'N	115°30'W
Calgary, Ab., Can. (kăl'gȧ-rĭ)	132-33	51°03'N	114°05'W
Calhoun, Ga., U.S. (kăl-hōōn')	124-25	34°30'N	84°58'W
Calhoun, Ky., U.S. (kăl-hōōn')	116-17	37°32'N	87°15'W
Cali, Col. (kä'lē)	164-65	3°27'N	76°31'W
Calicut, India see Kozhikode	236	11°16'N	75°47'E
California, Mo., U.S. (kăl-ĭ-fôr'nĭ-ȧ)	120-21	38°38'N	92°34'W
California, state, U.S. (kăl-ĭ-fôr'nĭ-ȧ)	108-09	37°30'N	119°30'W
California, Golfo de, b., Mex. (gôl-fô-dĕ-kä-lē-fôr-nyä)	144-45	28°0'N	112°00'W
California, Gulf of, b., Mex. (gŭlf ŭv kăl-ĭ-fôr'nĭ-ȧ) see California, Golfo de	144-45	28°0'N	112°00'W
Calimere, Point, c., India	236	10°17'N	79°52'E
Calipatria, Ca., U.S. (kăl-ĭ-pàt'rĭ-ȧ)	118-19	33°08'N	115°31'W
Calkiní, Mex. (käl-kē-nē')	148	20°23'N	90°02'W
Callabonna, Lake, lk., Austl. (lāk căl ă'bŏnȧ)	276	29°41'S	140°03'E
Callao, Peru (käl-yä'ô)	163a	12°04'S	77°08'W
Calling Lake, lk., Ab., Can. (kôl'ĭng lāk)	132-33	55°13'N	113°15'W
Calmar, Swe. see Kalmar	192-93	56°40'N	16°22'E
Caloosahatchee, stm., Fl., U.S. (kȧ-loo-sȧ-hăch'ē)	125a	26°32'N	82°01'W
Caltagirone, Italy (käl-tä-jē-rō'nā)	200-01	37°14'N	14°31'E
Caltanissetta, Italy (käl-tä-nē-sĕt'tä)	200-01	37°29'N	14°04'E
Caluula, Som.	262-63	11°57'N	50°46'E
Calvert Island, i., B.C., Can.	132-33	51°33'N	128°02'W
Calvillo, Mex. (käl-vēl'yō)	146-47	21°51'N	102°43'W
Calvinia, S. Afr. (kăl-vĭn'ĭ-ȧ)	264-65	31°28'S	19°46'E
Camacupa, Ang.	264-65	12°01'S	17°28'E
Camagüey, Cuba (kä-mä-gwā')	142-43	21°22'N	77°55'W
Camagüey, state, Cuba (kä-mä-gwā')	142-43	21°30'N	78°00'W
Camaná, Peru	170	16°37'S	72°42'W
Camaquã, Braz.	168-69	30°51'S	51°49'W
Camará, Braz.	166-67	3°55'S	62°44'W
Camarón, Cabo, c., Hond. (kȧ'bô-kä-mä-rōn')	149	15°59'N	85°02'W
Camarones, Arg.	171	44°48'S	65°43'W
Camas, Wa., U.S. (kăm'ȧs)	112-13	45°35'N	122°24'W
Ca Mau, Viet.	246-47	9°11'N	105°09'E
Ca Mau, Mui, c., Viet.	246-47	8°37'N	104°43'E
Cambodia, nation, Asia (kăm-bō'dē-ȧ)	206-07	13°0'N	105°00'E
Camborne, Eng., U.K. (kăm'bôrn)	190-91	50°13'N	5°18'W
Cambrai, Fr. (käN-brĕ')	196-97	50°11'N	3°15'E
Cambrian Mountains, mts., Wales, U.K. (kăm'brĭ-ăn moun'tĭnz)	190-91	52°35'N	3°35'W
Cambridge, On., Can. (kām'brĭj)	136-37	43°21'N	80°18'W
Cambridge, Eng., U.K. (kām'brĭj)	190-91	52°13'N	0°08'E
Cambridge, Il., U.S. (kām'brĭj)	114-15	41°18'N	90°11'W
Cambridge, Ma., U.S.	116-17	42°22'N	71°06'W
Cambridge, Md., U.S.	116-17	38°33'N	76°04'W
Cambridge, Mn., U.S. (kām'brĭj)	114-15	45°34'N	93°13'W
Cambridge, Ne., U.S. (kām'brĭj)	120-21	40°17'N	100°10'W
Cambridge, Oh., U.S. (kām'brĭj)	116-17	40°02'N	81°35'W
Cambridge Bay, Nu., Can.	128-29	69°07'N	105°04'W
Cambridge City, In., U.S. (kām'brĭj sĭ'tê)	116-17	39°49'N	85°11'W
Camden, Al., U.S. (kăm'dĕn)	124-25	31°59'N	87°17'W
Camden, Ar., U.S. (kăm'dĕn)	120-21	33°36'N	92°50'W
Camden, Me., U.S. (kăm'dĕn)	117a	44°13'N	69°05'W
Camden, N.J., U.S.	116-17	39°56'N	75°07'W
Camden, S.C., U.S. (kăm'dĕn)	124-25	34°15'N	80°36'W
Cameron, Mo., U.S. (kăm'ĕr-ŭn)	120-21	39°44'N	94°14'W
Cameron, Tx., U.S. (kăm'ĕr-ŭn)	122-23	30°51'N	96°59'W
Cameron, Wi., U.S. (kăm'ĕr-ŭn)	114-15	45°25'N	91°45'W
Cameroon, nation, Afr. (kă'mȧ-rōōn')	253	6°0'N	12°00'E
Cameroon Mountain, vol., Camrn.	260-61	4°12'N	9°11'E
Cameroun, nation, Afr. see Cameroon	253	6°0'N	12°00'E
Cametá, Braz.	166-67	2°15'S	49°31'W
Camiguin Island, i., Phil.	250a	18°56'N	121°55'E
Camiling, Phil. (kä-mē-lĭng')	250	15°41'N	120°25'E
Camilla, Ga., U.S. (kȧ-mĭl'ȧ)	124-25	31°14'N	84°12'W
Caminha, Port. (kä-mĭn'yȧ)	198-99	41°52'N	8°49'W
Camiranga, Braz.	166-67	1°49'S	46°16'W
Camiri, Bol.	168-69	20°03'S	63°31'W
Camocim, Braz. (kä-mô-sēN')	166-67	2°54'S	40°51'W
Camooweal, Austl.	277	19°55'S	138°08'E
Campana, Arg. (käm-pä'nä)	173	34°10'S	58°57'W
Campana, Isla, i., Chile (ē's-lä-käm-pän'yä)	171	48°20'S	75°15'W
Campbell Island, i., N.Z.	287	52°33'S	169°08'E
Campbell River, B.C., Can.	132-33	50°01'N	125°15'W
Campbellsville, Ky., U.S. (kăm'bĕlz-vĭl)	124-25	37°21'N	85°21'W
Campbellton, N.B., Can. (kăm'bĕl-tŭn)	138-39	47°60'N	66°41'W
Campbelltown, Austl. (kăm'bĕl-toun)	276	34°04'S	150°49'E
Campeche, Mex. (käm-pā'chȧ)	148	19°50'N	90°31'W
Campeche, state, Mex. (käm-pā'chȧ)	148	19°0'N	90°30'W
Campechuela, Cuba (käm-på-chwā'lä)	142-43	20°14'N	77°17'W
Cam Pha, Viet.	246-47	21°02'N	107°21'E
Campina Grande, Braz. (käm-pē'nä grän'dĕ)	163d	7°13'S	35°53'W
Campinas, Braz. (käm-pē'näzh)	172	22°55'S	47°05'W
Campo Alegre de Goiás, Braz.	172	17°38'S	47°46'W
Campobasso, Italy (käm'pô-bäs'sō)	200-01	41°34'N	14°40'E
Campo Belo, Braz.	172	20°54'S	45°16'W
Campo de Criptana, Spain (käm'pō dā krēp-tä'nä)	198-99	39°24'N	3°07'W
Campo Gallo, Arg.	173	26°34'S	62°50'W
Campo Grande, Braz. (käm-pò grän'dĕ)	168-69	20°28'S	54°38'W
Campo Maior, Braz. (käm-pò mä-yôr')	166-67	4°49'S	42°10'W
Campo Mourão, Braz.	168-69	24°02'S	52°24'W
Campos, Braz. (kä'm-pôs)	172	21°45'S	41°21'W
Camrose, Ab., Can. (kăm-rōz)	132-33	53°01'N	112°50'W
Canada, nation, N.A. (kăn'ȧ-dȧ)	85	60°0'N	95°00'W
Cañada de Gómez, Arg. (kä-nyä'dä-dĕ-gô'mĕz)	173	32°49'S	61°24'W
Canadian, Tx., U.S. (kȧ-nā'dĭ-ăn)	120-21	35°55'N	100°23'W
Canadian, stm., U.S. (kȧ-nā'dĭ-ăn)	110-11	35°27'N	95°05'W
Canajoharie, N.Y., U.S. (kăn-ȧ-jō-hăr'ê)	116-17	42°54'N	74°35'W
Çanakkale, Tur. (chä-näk-kä'lĕ)	200-01	40°09'N	26°25'E
Çanakkale Boğazı, strt., Tur. see Dardanelles	200-01	40°17'N	26°33'E
Canandaigua, N.Y., U.S. (kăn-ăn-dā'gwȧ)	116-17	42°53'N	77°17'W
Cananea, Mex. (kä-nä-nĕ'ä)	144-45	30°59'N	110°18'W
Canarias, Islas, is., Spain (ē's-läs-kä-nä'ryäs) see Canary Islands	199d	28°01'N	15°35'W
Canary Islands, is., Spain (kȧ-nā'-rē ī'lăndz)	199d	28°01'N	15°35'W
Cañas, C.R. (kä'-nyäs)	149	10°25'N	85°06'W
Canastota, N.Y., U.S. (kăn-ȧs-tō'tȧ)	116-17	43°05'N	75°46'W
Canatlán, Mex. (kä-nät-län')	146-47	24°31'N	104°46'W
Canaveral, Cape, c., Fl., U.S.	125a	28°27'N	80°32'W

n-sing; ŋ-baŋk; N-nasalized n; nŏd; cŏmmit; ōld; ôbey; ôrder; oi-boil; fōōd; ò-as oo in foot; ou-out; s-soft; sh-dish; th-thin; pūre; ûnite; ûrn; stŭd; circŭs; ü-as in French tu; '-indeterminate vowel.

Feature (Pronunciation)	Page	Lat.	Long.
Canavieiras, Braz. (kä-ná-vē-ā′räs)	168-69	15°39′S	38°57′W
Canberra, nat. cap., Austl. (kăn′bĕr-*á*)	276	35°17′S	149°08′E
Canby, Mn., U.S. (kăn′bĭ)	114-15	44°43′N	96°17′W
Cancún, Mex.	148	21°08′N	86°51′W
Candala, Som. see Qandala	262-63	11°28′N	49°52′E
Candeias, Braz.	166-67	12°40′S	38°32′W
Candelaria, Cuba (kän-dě-lä′ryä)	142-43	22°45′N	82°58′W
Candelaria, stm., Mex. (kän-dě-lä-ryä)	148	18°38′N	91°17′W
Cando, N.D., U.S. (kăn′dō)	114-15	48°29′N	99°13′W
Candon, Phil. (kän-dōn′)	250	17°11′N	120°27′E
Canea, Grc. see Chaniá	200a	35°31′N	24°01′E
Canelones, Ur. (kä-nĕ-lô-nĕs)	173	34°32′S	56°17′W
Cangas, Spain (kän′gäs)	198-99	42°16′N	8°47′W
Cangas de Narcea, Spain (kä′n-gäs-dě-när-sě-ä)	198-99	43°11′N	6°33′W
Cangkuang, Tanjung, c., Indon.	248-49	6°50′S	105°15′E
Canguçu, Braz.	173	31°21′S	52°37′W
Cangzhou, China (tsäŋ-jō)	240-41	38°18′N	116°52′E
Caniapiscau, stm., Qc., Can.	130-31	57°41′N	69°29′W
Caniapiscau, Réservoir de, res., Qc., Can. see Caniapiscau, Lac	130-31	0°0′	0°00′
Caniapiscau, Lac, res., Qc., Can.	130-31	54°09′N	69°51′W
Canicattì, Italy (kä-nē-kät′tē)	200-01	37°21′N	13°51′E
Çankırı, Tur.	186-87	40°36′N	33°37′E
Cannanore, India	236	11°52′N	75°22′E
Cannelton, In., U.S. (kăn′*ĕ*l-t*ŭ*n)	116-17	37°55′N	86°45′W
Cannes, Fr. (kȧn)	196-97	43°33′N	7°01′E
Canning, N.S., Can. (kăn′ĭng)	138-39	45°09′N	64°25′W
Canoas, stm., Braz.	168-69	27°37′S	51°26′W
Canon City, Co., U.S. (kăn′y*ŭ*n sĭ′tē)	120-21	38°27′N	105°15′W
Canonsburg, Pa., U.S. (kăn′*ŭ*nz-bûrg)	116-17	40°16′N	80°11′W
Canora, Sk., Can. (k*á*-nō′r*á*)	134-35	51°37′N	102°26′W
Canouan, i., St. Vin.	143b	12°43′N	61°20′W
Canso, N.S., Can. (kăn′sō)	138-39	45°20′N	61°00′W
Cantabrian Mountains, mts., Spain (kăn-tă′brē-*ă*n moun′tĭnz) see Cantábrica, Cordillera	198-99	43°0′N	5°00′W
Cantábrica, Cordillera, mts., Spain	198-99	43°0′N	5°00′W
Cantandica, Moz.	264-65	18°02′S	33°08′E
Cantanhede, Port. (kän-tän-yā′dȧ)	198-99	40°21′N	8°36′W
Cantaura, Ven.	163b	9°18′N	64°21′W
Canterbury, Eng., U.K. (kăn′tĕr-bĕr-ē)	190-91	51°17′N	1°05′E
Canterbury Bight, b., N.Z.	278	44°15′S	171°38′E
Can Tho, Viet.	246-47	10°02′N	105°47′E
Canton, China see Guangzhou	238-39	23°08′N	113°16′E
Canton, Ms., U.S.	124-25	32°37′N	90°02′W
Canton, Oh., U.S.	116-17	40°48′N	81°23′W
Canton, i., Kir.	280-81	2°49′S	171°41′W
Cañuelas, Arg. (kä-nyòĕ′-läs)	173	35°03′S	58°45′W
Canutama, Braz.	166-67	6°31′S	64°21′W
Canyon, Tx., U.S. (kăn′y*ŭ*n)	120-21	34°59′N	101°55′W
Canyon de Chelly National Monument, n.p., Az., U.S.	118-19	36°07′N	109°27′W
Canyonlands National Park, n.p., Ut., U.S. (kăn′y*ŭ*n-lăndz năsh′ŭn-*ă*l pärk)	118-19	38°10′N	110°00′W
Cao Bang, Viet.	246-47	22°40′N	106°15′E
Capanaparo, stm., S.A.	164-65	7°03′N	67°04′W
Cap aux Meules, Île du, i., Qc., Can.	138-39	47°23′N	61°55′W
Cap-Chat, Qc., Can. (kȧp-shä′)	138-39	49°05′N	66°41′W
Cap-de-la-Madeleine, Qc., Can. (kȧp dē lȧ mȧ-d′lĕn′)	136-37	46°22′N	72°31′W
Cape Barren Island, i., Austl.	276	40°25′S	148°12′E
Cape Breton Highlands National Park, n.p., N.S., Can.	138-39	46°45′N	60°45′W
Cape Breton Island, i., N.S., Can. (kāp brĕt′*ŭ*n ī′l*ă*nd)	138-39	46°04′N	60°30′W
Cape Charles, Va., U.S. (kāp chärlz)	124-25	37°16′N	76°01′W
Cape Coast, Ghana	260-61	5°07′N	1°16′W
Cape Dorset, Nu., Can.	128-29	64°14′N	76°33′W
Cape Fear, stm., N.C., U.S. (kāp fēr)	124-25	33°53′N	78°01′W
Cape Girardeau, Mo., U.S. (kāp jē-rär-dō′)	120-21	37°18′N	89°32′W
Cape May, N.J., U.S. (kāp mā)	116-17	38°56′N	74°56′W
Cape Town, nat. cap., S. Afr. (kāp toun)	264-65	33°55′S	18°30′E
Cape Verde, nation, Afr. (kāp vērd′) (kāp vĕr′dē)	253	16°0′N	24°00′W
Cape York Peninsula, pen., Austl. (kāp yôrk pĕ-nĭn′sūl*ă*)	277	14°0′S	142°30′E
Cap-Haïtien, Haiti (kȧp ȧ-ē-syăn′)	142-43	19°45′N	72°12′W
Capim, stm., Braz.	166-67	1°41′S	47°47′W
Capitol Reef National Park, n.p., Ut., U.S. (kăp′ĭ-tŏl rēf năsh′ŭn-*ă*l pärk)	118-19	38°15′N	111°10′W
Capiz, Phil. see Roxas	250	11°35′N	122°45′E
Caprara, Punta, c., Italy (pōō′n-tä-kä-prä′rä)	200-01	41°07′N	8°19′E
Capreol, On., Can.	136-37	46°42′N	80°55′W
Capri, Isola di, i., Italy (ē′-sō-lä-dē-kä′prē)	200-01	40°33′N	14°13′E
Caprivi Strip, hist. reg., Nmb.	264-65	17°59′S	23°00′E
Cap Saint Jacques, Viet. see Vung Tau	246-47	10°21′N	107°05′E
Capulin Volcano National Monument, n.p., N.M., U.S. (kȧ-pū′lĭn vŏl-kā′nō năsh′ŭn-*ă*l mŏn′*ŭ*-m*ĕ*nt)	120-21	36°47′N	103°56′W
Caquetá, stm., S.A.	166-67	3°08′S	64°46′W
Caracal, Rom. (kȧ-rȧ-kȧl′)	200-01	44°07′N	24°22′E
Caracaraí, Braz.	166-67	1°50′N	61°08′W
Caracas, nat. cap., Ven. (kä-rä′käs)	164-65	10°30′N	66°56′W
Caraguatatuba, Braz. (kä-rä-gwä-tä-tōō′bä)	172	23°37′S	45°25′W
Caraïbes, Îles des, is., see West Indies	140-41	19°0′N	70°00′W
Caraïbes, Mer des, s., see Caribbean Sea	140-41	15°0′N	73°00′W
Carajás, Braz.	166-67	6°06′S	50°23′W
Carajás, Serra dos, hills, Braz. (sĕ′r-rä-dôs-kä-rä-zhá′s)	166-67	6°16′S	51°21′W
Carangola, Braz. (kä-rán′gō′lä)	172	20°43′S	42°02′W
Caraquet, N.B., Can. (kă-rȧ-kĕt′)	138-39	47°47′N	64°57′W
Caratasca, Laguna de, b., Hond. (lä-gò′nä-dĕ-kä-rä-täs′kä)	149	15°24′N	83°54′W
Caratinga, Braz.	172	19°47′S	42°09′W
Carauari, Braz.	166-67	4°52′S	66°52′W
Caravelas, Braz. (kȧ-rä-vĕl′äzh)	172	17°44′S	39°15′W
Carazinho, Braz. (kä-rä′zē-ny*ŏ*)	168-69	28°17′S	52°46′W
Carballo, Spain (kär-bäl′yō)	198-99	43°13′N	8°41′W
Carberry, Mb., Can.	134-35	49°52′N	99°21′W
Carbonara, Capo, c., Italy (kä′pō är-bō-nä′rä)	200-01	39°06′N	9°31′E
Carbondale, Il., U.S. (kär′bŏn-dāl)	116-17	37°43′N	89°13′W
Carbondale, Pa., U.S. (kär′bŏn-dāl)	116-17	41°35′N	75°30′W
Carbonear, Nf., Can. (kär-bô-nēr′)	138-39	47°45′N	53°14′W
Carbon Hill, Al., U.S. (kär′bŏn hĭl)	124-25	33°54′N	87°32′W
Carcassonne, Fr. (kȧr-kȧ-sôn′)	196-97	43°13′N	2°21′E
Carcross, Yk., Can. (kär′krôs)	128-29	60°11′N	134°42′W
Cárdenas, Cuba (kär′dȧ-näs)	142-43	23°02′N	81°12′W
Cárdenas, Mex. (ká′r-dĕ-näs)	146-47	18°00′N	93°22′W
Cárdenas, Mex. (ká′r-dĕ-näs)	146-47	21°60′N	99°39′W
Cardiel, Lago, lk., Arg.	171	48°55′S	71°15′W
Cardiff, Wales, U.K. (kär′dĭf)	190-91	51°29′N	3°11′W
Cardigan, Wales, U.K. (kär′dĭ-g*ă*n)	190-91	52°05′N	4°39′W
Cardston, Ab., Can. (kärds′t*ŭ*n)	132-33	49°12′N	113°18′W
Carei, Rom. (kä-rē′)	194-95	47°41′N	22°28′E
Careiro, Braz.	166-67	3°14′S	59°46′W
Careiro, Ilha do, i., Braz.	166-67	3°09′S	59°48′W
Carey, Oh., U.S. (kā′rē)	116-17	40°57′N	83°23′W
Carey, Lake, lk., Austl. (lāk kâr′ē)	272-73	29°04′S	122°19′E
Caribbean Sea, s., (kăr-ĭ-bē′*ă*n sē)	140-41	15°0′N	73°00′W
Caribe, Mar, s., see Caribbean Sea	140-41	15°0′N	73°00′W
Caribische Zee, s., see Caribbean Sea	140-41	15°0′N	73°00′W
Cariboo Mountains, mts., B.C., Can. (kă′rĭ-bōō moun′tĭnz)	132-33	53°0′N	121°00′W
Caribou, Me., U.S.	117a	46°51′N	68°00′W
Caribou Mountains, mts., Ab., Can.	130-31	59°06′N	115°10′W
Caricyn, Russia see Volgograd	186-87	48°44′N	44°25′E
Carinhanha, Braz. (kä-rĭ-nyän′yä)	166-67	14°19′S	43°48′W
Caripito, Ven.	164-65	10°06′N	63°06′W
Carleton, Mount, mtn., N.B., Can.	138-39	47°23′N	66°53′W
Carleton Place, On., Can. (kärl′t*ŭ*n plās)	136-37	45°09′N	76°09′W
Carletonville, S. Afr.	269c	26°21′S	27°24′E
Carlinville, Il., U.S. (kär′lĭn-vĭl)	120-21	39°16′N	89°53′W
Carlisle, Eng., U.K. (kär-līl′)	190-91	54°54′N	2°56′W
Carlos Casares, Arg. (kär-lôs-kä-sä′rĕs)	173	35°38′S	61°21′W
Carlow, Ire. (kär′lō)	190-91	52°50′N	6°55′W
Carlsbad, Czech Rep. see Karlovy Vary	194-95	50°14′N	12°53′E
Carlsbad, Ca., U.S. (kärlz′băd)	118-19	33°09′N	117°20′W
Carlsbad, N.M., U.S. (kärlz′băd)	120-21	32°25′N	104°14′W
Carlsbad Caverns National Park, n.p., N.M., U.S. (kärlz′băd kăv′ĕrnz năsh′ŭn-*ă*l pärk)	120-21	32°08′N	104°35′W
Carlyle, Il., U.S. (kärlīl′)	116-17	38°36′N	89°22′W
Carmacks, Yk., Can.	128-29	62°05′N	136°15′W
Carman, Mb., Can. (kär′mȧn)	134-35	49°31′N	97°59′W
Carmarthen, Wales, U.K. (kär-mär′th*ĕ*n)	190-91	51°52′N	4°19′W
Carmaux, Fr. (kȧr-mō′)	196-97	44°03′N	2°10′E
Carmel, In., U.S. (kär′m*ĕ*l)	116-17	39°58′N	86°07′W
Carmelo, Ur. (kär-mĕ′lo)	173	33°60′S	58°17′W
Carmen, Mex. see Ciudad del Carmen	148	18°39′N	91°49′W
Carmen, Isla, i., Mex.	144-45	26°00′N	111°08′W
Carmen, Isla del, i., Mex. (ē′s-lä-dĕl-kȧ′r-mĕn)	148	18°43′N	91°40′W
Carmen de Areco, Arg. (kär′mĕn′ dā ä-rā′kō)	173	34°23′S	59°50′W
Carmi, Il., U.S. (kär′mī)	116-17	38°05′N	88°10′W
Carnarvon, Austl. (kär-när′v*ŭ*n)	270-71	24°52′S	113°40′E
Carnarvon, S. Afr.	264-65	30°58′S	22°08′E
Carnarvon National Park, n.p., Austl.	277	24°42′S	147°55′E
Carnegie, Ok., U.S. (kär-nĕg′ĭ)	120-21	35°07′N	98°35′W
Carnegie, Lake, lk., Austl.	272-73	26°11′S	122°31′E
Carnot, C.A.R.	262-63	4°56′N	15°53′E
Carnsore Point, c., Ire. (kärn′sôr point)	190-91	52°11′N	6°22′W
Caro, Mi., U.S. (kâ′rō)	116-17	43°29′N	83°23′W
Carolina, Braz. (kä-rō-lē′nä)	166-67	7°21′S	47°25′W
Carolina, S. Afr. (kăr-ô-lī′n*á*)	269c	26°04′S	30°08′E
Caroline, at., Kir.	280-81	9°58′S	150°13′W
Caroline Islands, is., Oc. (kă′-rô-līn ī′l*ă*ndz)	280-81	8°0′N	147°00′E
Caroní, stm., Ven. (kä-rō′nē)	164-65	8°21′N	62°48′W
Carora, Ven. (kä-rō′rä)	164-65	10°10′N	70°05′W
Carpathian Mountains, mts., Eur. (kär-pā′thĭ-*ă*n moun′tĭnz)	186-87	48°0′N	24°00′E
Carpaţii, mts., Eur. see Carpathian Mountains	186-87	48°0′N	24°00′E
Carpaţii Meridionali, mts., Rom. see Transylvanian Alps	200-01	45°25′N	23°33′E
Carpentaria, Gulf of, b., Austl. (gŭlf ŭv kär-pĕn-târ′ĭ*á*)	272-73	14°0′S	139°00′E
Carpentras, Fr. (kär-pän-träs′)	196-97	44°04′N	5°03′E
Carranza, Cabo, c., Chile	171	35°36′S	72°38′W
Carrara, Italy (kä-rä′rä)	200-01	44°05′N	10°06′E
Carrauntoohil, mtn., Ire.	190-91	51°59′N	9°45′W
Carreta, Punta, c., Peru (pōō′n-tä-kär-rĕ′tĕ′rȧ)	170	14°11′S	76°17′W
Carriacou, i., Gren.	143b	12°30′N	61°26′W
Carrington, N.D., U.S. (kăr′ĭng-t*ŭ*n)	114-15	47°27′N	99°07′W
Carrizal Bajo, Chile	168-69	28°06′S	71°09′W
Carrizozo, N.M., U.S. (kär-rē-zō′zō)	120-21	33°39′N	105°53′W
Carroll, Ia., U.S. (kăr′*ŭ*l)	114-15	42°04′N	94°52′W
Carrollton, Ga., U.S. (kăr-*ŭ*l-t*ŭ*n)	124-25	33°35′N	85°05′W
Carrollton, Il., U.S. (kăr-*ŭ*l-t*ŭ*n)	120-21	39°18′N	90°24′W
Carrollton, Ky., U.S. (kăr-*ŭ*l-t*ŭ*n)	116-17	38°41′N	85°11′W
Carrollton, Mi., U.S. (kăr-*ŭ*l-t*ŭ*n)	116-17	43°27′N	83°57′W
Carrollton, Mo., U.S. (kăr-*ŭ*l-t*ŭ*n)	120-21	39°22′N	93°30′W
Carrollton, Tx., U.S. (kăr-*ŭ*l-t*ŭ*n)	120-21	32°58′N	96°53′W
Carrot, stm., Can.	134-35	53°50′N	101°19′W
Carson City, Nv., U.S.	118-19	39°10′N	119°46′W
Cartagena, Col. (kär-tä-hā′nä)	164-65	10°25′N	75°30′W
Cartagena, Spain (kär-tä-kĕ′nä)	198-99	37°37′N	0°59′W
Cartago, Col. (kär-tä′gō)	163c	4°45′N	75°55′W
Cartago, C.R.	149	9°51′N	83°55′W
Cartersville, Ga., U.S. (kär′tĕrs-vĭl)	124-25	34°10′N	84°48′W
Carthage, Il., U.S. (kär′thåj)	120-21	40°25′N	91°08′W
Carthage, Mo., U.S. (kär′thåj)	120-21	37°11′N	94°19′W
Carthage, Ms., U.S. (kär′thåj)	124-25	32°44′N	89°32′W
Carthage, N.Y., U.S. (kär′thåj)	116-17	43°59′N	75°37′W
Carthage, Tx., U.S. (kär′thåj)	122-23	32°09′N	94°22′W
Cartwright, Nf., Can. (kärt′rīt)	128-29	53°41′N	56°60′W
Caruaru, Braz. (kä-rò-å-rōō′)	163d	8°17′S	35°58′W
Carúpano, Ven. (kä-rōō′pä-nō)	164-65	10°40′N	63°15′W
Carutapera, Braz.	166-67	1°13′S	46°00′W
Caruthersville, Mo., U.S. (k*á*-rŭdh′ērz-vĭl)	124-25	36°11′N	89°40′W
Carvoeiro, Braz.	166-67	1°26′S	61°60′W
Carvoeiro, Cabo, c., Port. (kȧ′bō-kär-vô-ĕ′y-rō)	198-99	39°21′N	9°24′W
Cary, N.C., U.S. (kă′rē)	124-25	35°47′N	78°47′W
Casablanca, Mor. (kä-sä-blän′kä)	269a	33°36′N	7°36′W
Casa Branca, Braz. (ká′sä-brȧ′n-kä)	172	21°48′S	47°04′W
Casa Grande, Az., U.S. (kä′sä grän′dā)	118-19	32°53′N	111°45′W
Casa Grande Ruins National Monument, n.p., Az., U.S.	118-19	32°59′N	111°32′W
Casamance, stm., Sen. (kä-sä-mäns′)	260-61	12°33′N	16°45′W
Casanare, stm., Col.	164-65	6°02′N	69°51′W
Cascade Mountains, mts., N.A. (kăs-kād′ moun′tĭnz)	110-11	45°14′N	121°56′W
Cascade Point, c., N.Z. (kăs-kād′ point)	278	44°00′S	168°22′E
Cascade Range, mts., N.A. (kăs-kād′ rānj) see Cascade Range	110-11	45°14′N	121°56′W
Cascais, Port. (käs-k*á*-ēzh)	198-99	38°42′N	9°25′W
Cascavel, Braz.	168-69	24°58′S	53°27′W
Caserta, Italy (kä-zĕr′tä)	200-01	41°05′N	14°19′E
Casey, Il., U.S. (kā′sĭ)	116-17	39°18′N	87°60′W

Feature (Pronunciation)	Page	Lat.	Long.
Caseyr, c., Som. *see* Gwardafuy, Gees	**262-63**	11°50'N	51°17'E
Cashmere, Wa., U.S. (kăsh'mĭr)	**112-13**	47°31'N	120°27'W
Casilda, Arg. (kä-sē'l-dä)	**173**	33°03'S	61°11'W
Casino, Austl. (kȧ-sē'nō)	**276**	28°52'S	153°03'E
Casiquiare, stm., Ven. (kä-sē-kyä'rä)	**164-65**	1°60'N	67°08'W
Caspe, Spain (käs'pā)	**198-99**	41°14'N	0°03'W
Casper, Wy., U.S. (kăs'pēr)	**112-13**	42°51'N	106°20'W
Caspian Depression, pl., (kăs'pĭ-ăn dĭ-prĕ'shŭn)	**186-87**	48°0'N	52°00'E
Caspian Sea, lk., (kăs'pĭ-ăn sē)	**226**	41°18'N	50°59'E
Cass, W.V., U.S. (kăs)	**116-17**	38°24'N	79°55'W
Cassai, stm., Afr. (kä-sä'ē)	**262-63**	3°02'S	16°56'E
Cass City, Mi., U.S. (kăs sĭ'tē)	**116-17**	43°36'N	83°10'W
Casselman, On., Can. (kăs''l-măn)	**136-37**	45°19'N	75°06'W
Casselton, N.D., U.S. (kăs''l-tŭn)	**114-15**	46°54'N	97°13'W
Cássia, Braz. (kȧ'syä)	**172**	20°33'S	46°56'W
Cassiar, B.C., Can.	**128-29**	59°16'N	129°43'W
Cassiar Mountains, mts., Can.	**130-31**	59°0'N	129°00'W
Cassino, Italy (käs-sē'nō)	**200-01**	41°30'N	13°51'E
Cass Lake, Mn., U.S. (kăs lāk)	**114-15**	47°22'N	94°37'W
Cassopolis, Mi., U.S. (kăs-ŏ'pō-lĭs)	**116-17**	41°55'N	86°01'W
Cassville, Mo., U.S. (kăs'vĭl)	**120-21**	36°41'N	93°52'W
Cassville, Wi., U.S. (kăs'vĭl)	**114-15**	42°43'N	90°59'W
Castelli, Arg. (kȧs-tĕ'zhē)	**173**	36°06'S	57°49'W
Castelli, Arg. (kȧs-tĕ'zhē)	**168-69**	25°57'S	60°37'W
Castelló de la Plana, Spain	**198-99**	39°59'N	0°02'W
Castellón de la Plana, Spain *see* Castelló de la Plana	**198-99**	39°59'N	0°02'W
Castelnaudary, Fr. (kȧs'tĕl-nō-dȧ-rē')	**196-97**	43°19'N	1°57'E
Castelo, Braz. (käs-tĕ'lō)	**172**	20°35'S	41°13'W
Castelo Branco, Port.	**198-99**	39°49'N	7°29'W
Castelsarrasin, Fr. (kȧs'tĕl-sȧ-rȧ-zăN')	**196-97**	44°02'N	1°06'E
Castelvetrano, Italy (käs'tĕl-vē-trä'nō)	**200-01**	37°41'N	12°47'E
Castilla, Peru (kȧs-tē'l-yä)	**170**	5°12'S	80°37'W
Castillo de San Marcos National Monument, n.p., Fl., U.S. (käs-tē'lyā de-sän mär-kōs năsh'ŭn-ăl mŏn'ŭ-mĕnt)	**124-25**	29°55'N	81°19'W
Castillos, Ur.	**173**	34°13'S	53°50'W
Castle Dale, Ut., U.S. (kȧs'l dāl)	**118-19**	39°24'N	110°27'W
Castlegar, B.C., Can. (kȧs''l-gär)	**132-33**	49°19'N	117°40'W
Castlemaine, Austl. (kȧs''l-mān)	**276**	37°04'S	144°13'E
Castle Peak, mtn., Co., U.S. (kȧs''l pēk)	**118-19**	39°01'N	106°52'W
Castle Peak, mtn., Id., U.S. (kȧs''l pēk)	**112-13**	44°02'N	114°35'W
Castlereagh, stm., Austl.	**276**	30°12'S	147°31'E
Castle Rock, Co., U.S. (kȧs''l rŏk)	**120-21**	39°23'N	104°51'W
Castle Rock, Wa., U.S. (kȧs''l rŏk)	**112-13**	46°17'N	122°54'W
Castres, Fr. (kȧs'tr')	**196-97**	43°36'N	2°15'E
Castries, nat. cap., St. Luc. (kȧs-trē')	**143b**	14°01'N	60°59'W
Castro, Braz. (käs'trō)	**168-69**	24°47'S	49°60'W
Castro, Chile (käs'trō)	**171**	42°29'S	73°46'W
Castro Verde, Port. (käs-trō vĕr'dĕ)	**198-99**	37°41'N	8°05'W
Castrovillari, Italy (käs'trō-vēl-lyä'rē)	**200-01**	39°49'N	16°13'E
Catacamas, Hond. (kä-tä-kȧ'mäs)	**149**	14°51'N	85°54'W
Catahoula Lake, lk., La., U.S. (kăt-ȧ-hō'lȧ lāk)	**124-25**	31°30'N	92°08'W
Catalão, Braz. (kä-tä-louN')	**172**	18°11'S	47°56'W
Catalina, Chile	**168-69**	25°13'S	69°44'W
Catalina, i., Ca., U.S. *see* Santa Catalina Island	**118-19**	33°23'N	118°24'W
Catamarca, state, Arg. (kä-tä-mär'kä)	**168-69**	27°0'S	67°00'W
Catanduanes Island, i., Phil. (kä-tän-dwä'nĕs ī'lănd)	**250**	13°45'N	124°15'E
Catanduva, Braz. (kä-tȧn-dōō'vä)	**172**	21°08'S	48°58'W
Catania, Italy (kä-tä'nyä)	**200-01**	37°30'N	15°06'E
Catanzaro, Italy (kä-tän-dzä'rō)	**200-01**	38°54'N	16°36'E
Catarman, Phil.	**250**	12°30'N	124°38'E
Catbalogan, Phil. (kät-bä-lō'gän)	**250**	11°50'N	124°51'E
Cathedral Mountain, mtn., Tx., U.S. (kȧ-thē'drăl moun'tĭn)	**122-23**	30°10'N	103°40'W
Cat Island, i., Bah.	**142-43**	24°26'N	75°32'W
Catlettsburg, Ky., U.S. (kăt'lĕts-bûrg)	**116-17**	38°25'N	82°37'W
Catoche, Cabo, c., Mex. (kȧ'bô kä-tō'chĕ)	**148**	21°36'N	87°06'W
Catonsville, Md., U.S. (kā'tŭnz-vĭl)	**116-17**	39°17'N	76°44'W
Catorce, Mex. (kä-tôr'sȧ)	**146-47**	23°42'N	100°56'W
Catriló, Arg.	**173**	36°24'S	63°25'W
Catrimani, stm., Braz.	**166-67**	0°28'N	61°43'W
Catskill, N.Y., U.S. (kăts'kĭl)	**116-17**	42°14'N	73°52'W
Catskill Mountains, mts., N.Y., U.S. (kăts'kĭl moun'tĭnz)	**116-17**	42°10'N	74°30'W
Cattaraugus Indian Reservation, ind. res., N.Y., U.S. (kăt'tȧ-rȧ-gŭs ĭn'dĭ-ăn rĕ-sēr-vā'shĕn)	**116-17**	42°32'N	78°59'W
Catumbela, stm., Ang. (kä'tŏm-bĕl'ȧ)	**264-65**	12°27'S	13°30'E
Caubvick, Mount, mtn., Can.	**130-31**	58°53'N	63°43'W
Cauca, stm., Col. (kou'kä)	**164-65**	8°54'N	74°28'W
Caucasus Mountains, mts., (kô'kȧ-sŭs moun'tĭnz)	**227**	42°38'N	45°00'E
Caungula, Ang.	**262-63**	8°26'S	18°38'E
Cauquenes, Chile (kou-kā'näs)	**171**	35°58'S	72°19'W
Caura, stm., Ven. (kou'rä)	**164-65**	7°38'N	64°53'W
Caution, Cape, c., B.C., Can. (kăp kô'shŭn)	**132-33**	51°10'N	127°46'W
Cauto, stm., Cuba (kou'tō)	**142-43**	20°33'N	77°14'W
Cavalcante, Braz. (kä-väl-kän'tā)	**166-67**	13°47'S	47°30'W
Cavalier, N.D., U.S. (kăv-ȧ-lēr')	**114-15**	48°47'N	97°37'W
Cavalla, stm., Afr.	**260-61**	4°21'N	7°31'W
Cavally, stm., Afr.	**260-61**	4°21'N	7°31'W
Cavan, Ire. (kăv'ăn)	**190-91**	53°59'N	7°22'W
Caviana de Fora, Ilha, i., Braz.	**166-67**	0°10'N	50°10'W
Cavite, Phil. (kä-vē'tä)	**250**	14°29'N	120°54'E
Cawnpore, India *see* Kānpur	**234-35**	26°28'N	80°19'E
Caxias, Braz. (kä'shē-äzh)	**166-67**	4°50'S	43°21'W
Caxias do Sul, Braz. (kä'shē-äzh-dô-sōō'l)	**168-69**	29°11'S	51°11'W
Caxito, Ang. (kä-shē'tō)	**262-63**	8°33'S	13°36'E
Cayambe, Ec. (kȧ-īä'm-bĕ)	**170**	0°03'N	78°09'W
Cayambe, vol., Ec.	**170**	0°02'N	77°59'W
Cayenne, nat. cap., Fr. Gu. (kā-ĕn')	**164-65**	4°56'N	52°19'W
Cayman Brac, i., Cay. Is. (kā-män' brȧk)	**142-43**	19°43'N	79°49'W
Cayman Islands, dep., N.A. (kā'măn ī'lăndz) (kī-män' ī'lăndz)	**140-41**	19°30'N	80°40'W
Ceará, Braz. *see* Fortaleza	**166-67**	3°44'S	38°30'W
Ceará, state, Braz.	**166-67**	5°0'S	40°00'W
Ceará-Mirim, Braz. (sā-ä-rä'mē-rē'N)	**163d**	5°38'S	35°26'W
Ceatharlach, Ire. *see* Carlow	**190-91**	52°50'N	6°55'W
Cebaco, Isla de, i., Pan. (ē's-lä-dĕ-sȧ-bä'kō)	**150**	7°32'N	81°09'W
Cebu, Phil. (sā-bōō')	**250**	10°19'N	123°54'E
Cebu, i., Phil.	**250**	10°20'N	123°45'E
Čechy, hist. reg., Czech Rep. *see* Bohemia	**194-95**	49°50'N	14°00'E
Cedar, stm., U.S. (sē'dēr)	**114-15**	41°17'N	91°20'W
Cedar Breaks National Monument, n.p., Ut., U.S. (sē'dēr brāks năsh'ŭn-ăl mŏn'ŭ-mĕnt)	**118-19**	37°38'N	112°50'W
Cedarburg, Wi., U.S. (sē'dēr bûrg)	**116-17**	43°17'N	87°59'W
Cedar City, Ut., U.S. (sē'dēr sĭ'tē)	**118-19**	37°41'N	113°04'W
Cedar Falls, Ia., U.S. (sē'dēr fôlz)	**114-15**	42°31'N	92°27'W
Cedar Lake, res., Mb., Can. (sē'dēr lāk)	**134-35**	53°15'N	100°10'W
Cedar Rapids, Ia., U.S. (sē'dēr răp'ĭdz)	**114-15**	41°58'N	91°40'W
Cedar Springs, Mi., U.S. (sē'dēr sprĭngz)	**116-17**	43°13'N	85°33'W
Cedartown, Ga., U.S. (sē'dēr-toun)	**124-25**	34°02'N	85°14'W
Cedros, Isla, i., Mex.	**144-45**	28°11'N	115°13'W
Ceduna, Austl. (sē-dō'nȧ)	**270-71**	32°07'S	133°41'E
Ceerigaabo, Som.	**262-63**	10°37'N	47°22'E
Cegléd, Hung. (tsā'glād)	**194-95**	47°10'N	19°48'E
Celaya, Mex. (sā-lä'yä)	**146-47**	20°31'N	100°49'W
Celebes, i., Indon. (sĕ'-lĕ-bēz)	**248-49**	2°0'S	121°00'E
Celebes Sea, s., Asia (sĕ'-lĕ-bēz sē)	**248-49**	3°0'N	122°00'E
Çeleken, Turkmen.	**232-33**	39°26'N	53°07'E
Celestún, Mex. (sĕ-lĕs-tōō'n)	**148**	20°52'N	90°23'W
Celina, Oh., U.S. (sēlī'nȧ)	**116-17**	40°33'N	84°34'W
Celje, Slvn. (tsĕl'yĕ)	**200-01**	46°14'N	15°16'E
Celle, Ger. (tsĕl'ē)	**194-95**	52°37'N	10°05'E
Celtic Sea, s., Eur.	**190-91**	51°0'N	6°30'W
Cenderawasih, Teluk, b., Indon.	**224-25**	2°30'S	135°20'E
Center, Tx., U.S. (sĕn'tēr)	**122-23**	31°48'N	94°10'W
Centerville, Ia., U.S. (sĕn'tēr-vĭl)	**114-15**	40°44'N	92°52'W
Centerville, Pa., U.S. (sĕn'tēr-vĭl)	**116-17**	40°03'N	79°59'W
Centerville, S.D., U.S. (sĕn'tēr-vĭl)	**114-15**	43°07'N	96°60'W
Central, Cordillera, mts., Phil. (kôr-dēl-yĕ'rä-sĕn-träl)	**250**	17°02'N	120°53'E
Central, Massif, mts., Fr.	**196-97**	44°42'N	3°19'E
Central, Sistema, mts., Spain	**198-99**	40°34'N	4°29'W
Central African Republic, nation, Afr. (sĕn'trăl ăf'rĭ-kăn rē-pŭb'lĭk)	**253**	7°0'N	21°00'E
Central City, Ky., U.S. (sĕn'trăl sĭ'tĭ)	**124-25**	37°18'N	87°08'W
Central City, Ne., U.S. (sĕn'trăl sĭ'tē)	**114-15**	41°07'N	97°60'W
Centralia, Il., U.S. (sĕn-trā'lĭ-ȧ)	**116-17**	38°32'N	89°08'W
Centralia, Mo., U.S. (sĕn-trā'lĭ-ȧ)	**120-21**	39°13'N	92°08'W
Centralia, Wa., U.S. (sĕn-trā'lĭ-ȧ)	**112-13**	46°43'N	122°57'W
Central Kalahari Game Reserve, pk., Bots.	**264-65**	22°15'S	23°45'E
Central Russian Upland, plat., Russia (sĕn'trăl rŭsh'ăn ŭp'lănd) *see* Srednerusskaya Vozvyshennost'	**202-03**	52°0'N	38°00'E
Centreville, Al., U.S. (sĕn'tēr-vĭl)	**124-25**	32°57'N	87°08'W
Century, Fl., U.S. (sĕn'tů-rĭ)	**124-25**	30°58'N	87°16'W
Ceram, i., Indon. (sārăm') (sā'räm)	**224-25**	3°0'S	129°00'E
Ceram Sea, s., Indon. (sārăm' sē) (sā'räm sē)	**224-25**	2°30'S	128°00'E
Cerignola, Italy (chā-rē-nyô'lä)	**200-01**	41°17'N	15°54'E
Cernăuți, Ukr. *see* Chernivtsi	**194-95**	48°17'N	25°58'E
Cerralvo, Mex. (sĕr-räl'vō)	**122-23**	26°06'N	99°36'W
Cerralvo, Isla, i., Mex. (ē's-lä-sĕr-räl'vō)	**144-45**	24°14'N	109°52'W
Cerritos, Mex. (sĕr-rē'tôs)	**146-47**	22°26'N	100°17'W
Cerro de Pasco, Peru (sĕr'rō dā päs'kō)	**170**	10°41'S	76°16'W
Cervino, mtn., Eur. *see* Matterhorn	**194-95**	45°59'N	7°43'E
Cesena, Italy (chĕ'sĕ-nä)	**200-01**	44°09'N	12°15'E
Cēsis, Lat. (sā'sĭs)	**192-93**	57°19'N	25°17'E
Česká Lípa, Czech Rep. (chĕs'kä lē'pa)	**194-95**	50°41'N	14°32'E
Česká Republika, nation, Eur. *see* Czech Republic	**174-75**	49°40'N	15°10'E
České Budějovice, Czech Rep. (chĕs'kā bōō'dyĕ-yô-vēt-sĕ)	**194-95**	48°59'N	14°28'E
Çeşme, Tur. (chĕsh'mĕ)	**200-01**	38°18'N	26°19'E
Cessnock, Austl.	**276**	32°50'S	151°21'E
Cestos, stm., Lib.	**260-61**	5°29'N	9°33'W
Cetatea Albă, Ukr. *see* Bilhorod-Dnistrovs'kyi	**202-03**	46°12'N	30°18'E
Cetinje, Mont.	**200-01**	42°24'N	18°56'E
Ceuta, Sp. N. Afr. (thā-ōō'tä)	**269a**	35°54'N	5°19'W
Cévennes, reg., Fr. (sā-vĕn')	**196-97**	44°07'N	3°32'E
Ceylon, nation, Asia *see* Sri Lanka	**206-07**	7°0'N	81°00'E
Chacabuco, Arg. (chä-kä-bōō'kō)	**173**	34°38'S	60°29'W
Chachapoyas, Peru (chä-chä-poi'yäs)	**170**	6°13'S	77°52'W
Chaco, state, Arg. (chä'kō)	**173**	26°0'S	60°30'W
Chad, nation, Afr. (chăd)	**253**	15°0'N	19°00'E
Chad, Lake, lk., Afr. (lāk chăd)	**258-59**	13°03'N	14°33'E
Chadbourn, N.C., U.S. (chăd'bŭn)	**124-25**	34°19'N	78°50'W
Chadileuvu, stm., Arg. *see* Salado	**171**	38°49'S	64°59'W
Chadron, Ne., U.S. (chăd'rŭn)	**114-15**	42°50'N	103°00'W
Chaffee, Mo., U.S. (chăf'ē)	**124-25**	37°11'N	89°39'W
Chagos Archipelago, is., B.I.O.T. (chä'-gōs är'kȧ-pĕ'-ȧ-gō)	**208-09**	6°0'S	72°00'E
Chahanwusu, China *see* Dulan	**240-41**	36°10'N	98°22'E
Chaiyaphum, Thai.	**246-47**	15°48'N	102°02'E
Chalatenango, El Sal. (chäl-ä-tĕ-näŋ'gō)	**148**	14°02'N	88°56'W
Chalbi Desert, des., Kenya	**267**	3°00'N	37°20'E
Chalcis, Grc. *see* Chalkída	**200-01**	38°28'N	23°36'E
Chalkída, Grc.	**200-01**	38°28'N	23°36'E
Chalmette, La., U.S. (shăl-mĕt')	**124-25**	29°56'N	89°58'W
Chaltel, Cerro, mtn., S.A. (sĕ'r-rô-chäl'tĕl)	**171**	49°17'S	73°05'W
Chālūs, Iran	**232-33**	36°39'N	51°25'E
Chaman, Pak. (chŭm-än')	**232-33**	30°55'N	66°27'E
Chambal, stm., India (chŭm-bäl')	**234-35**	26°29'N	79°15'E
Chamberlain, S.D., U.S. (chām'bēr-lĭn)	**114-15**	43°49'N	99°19'W
Chambersburg, Pa., U.S.	**116-17**	39°56'N	77°40'W
Chambéry, Fr. (shäm-bā-rē')	**196-97**	45°35'N	5°55'E
Chambi, Jebel, mtn., Tun.	**258-59**	35°13'N	8°40'E
Chamical, Arg.	**168-69**	30°21'S	66°18'W
Ch'amo Hāyk', lk., Eth.	**262-63**	5°50'N	37°34'E
Chamonix-Mont-Blanc, Fr. (shȧ-mô-nē'-môN-bläN)	**196-97**	45°56'N	6°52'E
Champagne, hist. reg., Fr. (shăm-pän'yĕ)	**196-97**	49°0'N	4°30'E
Champaign, Il., U.S. (shăm-pān')	**116-17**	40°07'N	88°15'W
Champdoré, Lac, lk., Qc., Can.	**130-31**	55°55'N	65°48'W
Champerico, Guat. (chăm-pȧ-rē'kō)	**148**	14°17'N	91°55'W
Champlain, Lac, lk., N.A. *see* Champlain, Lake	**116-17**	44°45'N	73°15'W
Champlain, Lake, lk., N.A. (lāk shăm-plān')	**116-17**	44°45'N	73°15'W
Champotón, Mex. (chäm-pō-tōn')	**148**	19°21'N	90°43'W
Chañaral, Chile (chän-yä-räl')	**168-69**	26°21'S	70°37'W
Chan Chan, hist., Peru	**170**	8°03'S	79°08'W
Chanchan, Ruinas de, hist., Peru *see* Chan Chan	**170**	8°03'S	79°08'W
Chandalar, stm., Ak., U.S.	**126**	66°38'N	146°02'W
Chandeleur Islands, is., La., U.S. (shăn-dē-lōōr' ī'lăndz)	**124-25**	29°49'N	88°54'W
Chandīgarh, India	**234-35**	30°44'N	76°54'E
Chandīgarh, state, India	**234-35**	30°45'N	76°50'E
Chandler, Qc., Can. (chăn'dlēr)	**138-39**	48°21'N	64°41'W
Chandler, Az., U.S. (chăn'dlēr)	**118-19**	33°18'N	111°53'W
Chandler, Ok., U.S. (chăn'dlēr)	**120-21**	35°42'N	96°53'W
Chāndpur, Bngl.	**234-35**	23°13'N	90°40'E
Chandrapur, India	**234-35**	19°57'N	79°18'E
Chang, stm., China *see* Yangtze	**238-39**	31°24'N	121°54'E
Changan, China *see* Xi'an	**238-39**	34°15'N	108°52'E

Feature (Pronunciation)	Page	Lat.	Long.
Changane, stm., Moz.	264-65	24°44'S	33°32'E
Changbaek-sanjulgi, mts., Asia see Changbai Shan	243	41°53'N	128°02'E
Changbai Shan, mts., Asia	243	41°53'N	128°02'E
Chang Cheng, p.o.i., China see Great Wall	240-41	40°0'N	112°30'E
Changchun, China (chäŋ-chōn)	240-41	43°53'N	125°19'E
Changde, China (chäŋ-dŭ)	238-39	29°02'N	111°41'E
Changhua, Tai. (chäng'hwä')	225a	24°04'N	120°30'E
Changji, China	222-23	44°01'N	87°18'E
Changjiang, China	238-39	19°16'N	109°02'E
Changkiakow, China see Zhangjiakou	240-41	40°49'N	114°53'E
Changli, China (chäŋ-lē)	240-41	39°42'N	119°10'E
Changmar, China	234-35	34°16'N	79°57'E
Changning, China (chäŋ-nĭŋ)	238-39	24°58'N	99°43'E
Changning, China (chäŋ-nĭŋ)	238-39	26°19'N	112°21'E
Changning, China (chäŋ-nĭŋ)	238-39	28°21'N	104°53'E
Changqing, China (chäŋ-chyĭŋ)	240-41	36°33'N	116°44'E
Changsha, China	238-39	28°12'N	112°58'E
Changshu, China (chäŋ-shōō)	238-39	31°38'N	120°44'E
Changting, China	238-39	25°50'N	116°21'E
Changyi, China (chäŋ-yē)	240-41	36°52'N	119°24'E
Changzhi, China (chäŋ-jr)	240-41	36°11'N	113°07'E
Changzhou, China (chäŋ-jō)	238-39	31°47'N	119°57'E
Chaniá, Grc.	200a	35°31'N	24°01'E
Chankiang, China see Zhanjiang	238-39	21°12'N	110°23'E
Channel Islands, is., Eur (chăn'ĕl ī'lăndz)	196-197	49°20'N	2°20'W
Channel Islands, is., Ca., U.S. (chăn'ĕl ī'lăndz)	118-19	34°0'N	120°00'W
Channel Islands National Park, n.p., Ca., U.S.	118-19	33°28'N	119°02'W
Channel-Port aux Basques, Nf., Can.	138-39	47°35'N	59°10'W
Chanthaburi, Thai.	246-47	12°37'N	102°07'E
Chantilly, Fr. (shäN-tē-yē')	196-97	49°12'N	2°28'E
Chanute, Ks., U.S. (shȧ-nōōt')	120-21	37°41'N	95°27'W
Chany, Ozero, lk., Russia (ô'zĕ-rô chä'nê)	226	54°50'N	77°30'E
Chao'an, China (chou-än)	238-39	23°40'N	116°39'E
Chao Hu, lk., China	238-39	31°31'N	117°33'E
Chao Phraya, stm., Thai.	246-47	13°32'N	100°36'E
Chaor, stm., China (chou-r)	240-41	46°48'N	123°35'E
Chaoxian, China (chou shyĕn)	238-39	31°35'N	117°51'E
Chaoyang, China (chou-yäŋ)	238-39	23°16'N	116°35'E
Chaoyang, China (chou-yäŋ)	240-41	41°35'N	120°28'E
Chapala, Mex. (chä-pä'lä)	146-47	20°17'N	103°11'W
Chapala, Laguna de, lk., Mex. (lä-ö'nä-dĕ-chä-pä'lä)	146-47	20°15'N	103°00'W
Chaparral, Col. (chä-pär-rä'l)	163c	3°44'N	75°28'W
Chapayevsk, Russia (châ-pī'ĕfsk)	186-87	52°58'N	49°42'E
Chapecó, Braz.	168-69	27°06'S	52°38'W
Chapel Hill, N.C., U.S. (chăp''l hĭl)	124-25	35°55'N	79°04'W
Chapleau, On., Can. (chăp-lō')	136-37	47°51'N	83°25'W
Chapman, Mount, mtn., B.C., Can. (mount chăp'măn)	132-33	51°50'N	118°20'W
Chappell, Ne., U.S. (chă-pĕl')	114-15	41°06'N	102°28'W
Charadai, Arg.	173	27°39'S	59°52'W
Chār Borjak, Afg.	230-31	30°18'N	62°01'E
Charcas, Mex. (chär'käs)	146-47	23°08'N	101°08'W
Chärdjew, Turkmen. see Türkmenabat	232-33	39°05'N	63°35'E
Chardzhou, Turkmen. see Türkmenabat	232-33	39°05'N	63°35'E
Chari, stm., Afr. (shä-rē')	262-63	12°56'N	14°34'E
Chārīkār, Afg.	232-33	35°01'N	69°10'E
Chariton, Ia., U.S. (châr'ĭ-tŭn)	114-15	41°01'N	93°19'W
Charkhlik, China see Ruoqiang	220-21	39°01'N	88°11'E
Charleroi, Bel. (shȧr-lē-rwä')	190-91	50°25'N	4°26'E
Charleroi, Pa., U.S. (shär'lē-roi)	116-17	40°08'N	79°54'W
Charles, Cape, c., Va., U.S. (kāp chärlz)	124-25	37°08'N	75°58'W
Charles City, Ia., U.S. (chärlz sĭ'tē)	114-15	43°04'N	92°41'W
Charleston, Il., U.S. (chärlz'tŭn)	116-17	39°29'N	88°11'W
Charleston, Mo., U.S. (chärlz'tŭn)	124-25	36°55'N	89°21'W
Charleston, S.C., U.S. (chärlz'tŭn)	124-25	32°47'N	79°56'W
Charleston, W.V., U.S. (chärlz'tŭn)	116-17	38°21'N	81°38'W
Charlestown, In., U.S. (chärlz'toun)	116-17	38°27'N	85°40'W
Charleville, Austl. (chär'lē-vĭl)	276	26°24'S	146°14'E
Charlevoix, Mi., U.S. (shär'lē-voi)	116-17	45°18'N	85°15'W
Charlotte, Mi., U.S. (shär'lŏt)	116-17	42°33'N	84°50'W
Charlotte, N.C., U.S. (shär'lŏt)	124-25	35°14'N	80°51'W
Charlotte Amalie, nat. cap., V.I.U.S. (shär-lŏt'ĕ ä-mä'lĭ-ȧ)	143b	18°21'N	64°56'W
Charlotte Harbor, b., Fl., U.S. (shär'lŏt här'bēr)	125a	26°45'N	82°11'W
Charlottenberg, Swe. (shär-lŭt'ĕn-bĕrg)	192-93	59°54'N	12°18'E
Charlottesville, Va., U.S. (shär'lŏtz-vĭl)	116-17	38°02'N	78°29'W
Charlottetown, P.E., Can.	138-39	46°14'N	63°08'W
Charlton Island, i., Nu., Can.	130-31	52°0'N	79°30'W
Chārsadda, Pak. (chŭr-sä'dä)	232-33	34°09'N	71°44'E
Charters Towers, Austl.	277	20°04'S	146°16'E
Chartres, Fr. (shärt'r')	196-97	48°27'N	1°29'E
Chascomús, Arg. (chäs-kō-mōōs')	173	35°35'S	58°01'W
Chase City, Va., U.S. (chās sĭ'tĭ)	124-25	36°47'N	78°28'W
Châteaudun, Fr. (shä-tō-dăN')	196-97	48°04'N	1°20'E
Château-Gontier, Fr. (chȧ-tō'gôN'tyā')	196-97	47°50'N	0°42'W
Châteauguay, Qc., Can. (chȧ-tō-gā)	136-37	45°23'N	73°44'W
Châteauroux, Fr. (shȧ-tō-rōō')	196-97	46°48'N	1°42'E
Château-Thierry, Fr. (shȧ-tō'ty-ĕr-rē')	196-97	49°03'N	3°24'E
Châtellerault, Fr. (shä-tĕl-rō')	196-97	46°49'N	0°33'E
Chatham, On., Can. (chăt'ăm)	136-37	42°24'N	82°11'W
Chatham, Il., U.S. (chăt'ăm)	116-17	39°40'N	89°42'W
Chatham, i., Ec. see San Cristóbal, Isla	170a	0°50'S	89°26'W
Chatham, Isla, i., Chile	171	50°38'S	74°27'W
Chatham Sound, strt., B.C., Can. (chăt'ăm sound)	132-33	54°32'N	130°35'W
Chatham Strait, strt., Ak., U.S. (chăt'ăm strāt)	126	57°30'N	134°45'W
Chatrapur, India	236	19°21'N	84°60'E
Chattahoochee, Fl., U.S. (chăt-tȧ-hōō' cheē)	124-25	30°42'N	84°51'W
Chattahoochee, stm., U.S. (chăt-tȧ-hōō' cheē)	110-11	30°46'N	84°52'W
Chattanooga, Tn., U.S. (chăt-ȧ-nōō'gȧ)	124-25	35°03'N	85°18'W
Chaudière, stm., Qc., Can. (shō-dyĕr')	138-39	46°45'N	71°17'W
Chauk, Mya.	246-47	20°54'N	94°49'E
Chaumont, Fr. (shō-môN')	196-97	48°06'N	5°08'E
Chauny, Fr. (shō-nē')	196-97	49°37'N	3°13'E
Chaves, Port. (chä'vĕzh)	198-99	41°44'N	7°28'W
Chavin, hist., Peru	170	9°37'S	77°14'W
Chavin de Huantar, hist., Peru see Chavin	170	9°37'S	77°14'W
Chaykovskij, Russia	186-87	56°46'N	54°06'E
Cheb, Czech Rep. (KĕB)	194-95	50°05'N	12°22'E
Cheboksary, Russia (chyĕ-bôk-sä'rê)	186-87	56°08'N	47°15'E
Cheboygan, Mi., U.S. (shē-boi'găn)	116-17	45°38'N	84°29'W
Chech, 'Erg, des., Afr.	258-59	24°43'N	2°31'W
Chechen', Ostrov, i., Russia (ôs-trôf' chyĕch'ĕn)	227	43°59'N	47°41'E
Chechnya, state, Russia	227	43°20'N	45°45'E
Checotah, Ok., U.S. (chê-kō'tȧ)	120-21	35°29'N	95°31'W
Chedabucto Bay, b., N.S., Can. (chĕd-ȧ-bŭk-tō bā)	138-39	45°23'N	61°10'W
Cheduba Island, i., Mya.	246-47	18°48'N	93°38'E
Cheektowaga, N.Y., U.S. (chēk-tō-wä'gȧ)	116-17	42°54'N	78°45'W
Chefoo, China see Yantai	240-41	37°32'N	121°21'E
Chegdomyn, Russia	218-19	51°08'N	133°05'E
Chehalis, Wa., U.S. (chê-hā'lĭs)	112-13	46°40'N	122°58'W
Cheju, Kor., S. (chĕ'jōō')	240-41	33°30'N	126°32'E
Cheju-do, i., Kor., S. (chĕ'jōō')	240-41	33°22'N	126°30'E
Chekiang, state, China see Zhejiang	238-39	29°0'N	120°00'E
Chelan, Wa., U.S. (chê-lăn')	112-13	47°50'N	120°00'W
Chelif, Oued, stm., Alg. (wĕd shä-lĕf)	198-99	36°03'N	0°08'E
Chełm, Pol. (KĕIm)	194-95	51°08'N	23°30'E
Chełmno, Pol. (Kĕlm'nō)	194-95	53°21'N	18°27'E
Chelmsford, Eng., U.K. (chĕlm's-fērd)	190-91	51°44'N	0°28'E
Chelsea, Mi., U.S. (chĕl'sē)	116-17	42°19'N	84°01'W
Chelsea, Ok., U.S. (chĕl'sē)	120-21	36°32'N	95°26'W
Cheltenham, Eng., U.K. (chĕlt'năm)	190-91	51°54'N	2°04'W
Chelyabinsk, Russia (chĕl-yä-bĕnsk')	226	55°10'N	61°26'E
Chelyuskin, Mys, c., Russia (mĭs chĕl-yôs'-kĭn)	218-19	77°45'N	104°20'E
Chemnitz, Ger.	194-95	50°50'N	12°56'E
Chemulpo, Kor., S. see Inch'ŏn	243	37°28'N	126°38'E
Chenāb, stm., Asia (chê-näb)	232-33	29°21'N	71°02'E
Cheney, Wa., U.S. (chē'nā)	112-13	47°29'N	117°35'W
Chengchow, China see Zhengzhou	238-39	34°46'N	113°39'E
Chengde, China (chŭŋ-dŭ)	240-41	40°58'N	117°56'E
Chengdu, China (chŭŋ-dōō)	238-39	30°39'N	104°04'E
Chengshan Jiao, c., China (jyou chŭŋ-shän)	240-41	37°23'N	122°42'E
Chennai, India	236	13°06'N	80°15'E
Chenyang, China see Shenyang	243	41°48'N	123°24'E
Chenzhou, China	238-39	25°48'N	112°59'E
Chepén, Peru (chĕ-pĕ'n)	170	7°14'S	79°25'W
Chepo, Pan. (chā'pō)	150	9°10'N	79°06'W
Cheraw, S.C., U.S. (chē'rô)	124-25	34°42'N	79°54'W
Cherbourg, Fr. (shär-bōr')	196-97	49°39'N	1°38'W
Cheremkhovo, Russia (chĕr'yĕm-kô-vō)	218-19	53°09'N	103°04'E
Cherepanovo, Russia (chĕr'yĕ pä-nô'vō)	226	54°13'N	83°21'E
Cherepovets, Russia (chĕr-yĕ-pô'vyĕtz)	202-03	59°08'N	37°55'E
Chergui, Chott ech, lk., Alg. (chĕr-gē)	258-59	34°13'N	0°26'E
Cherkassy, Ukr. see Cherkasy	202-03	49°26'N	32°05'E
Cherkasy, Ukr.	202-03	49°26'N	32°05'E
Cherkessia, state, Russia	227	44°0'N	42°00'E
Cherkessk, Russia	227	44°13'N	42°04'E
Cherlak, Russia (chĭr-läk')	226	54°09'N	74°49'E
Chermoz, Russia (chĕr-môz')	186-87	58°47'N	56°09'E
Chernigov, Ukr. see Chernihiv	202-03	51°30'N	31°17'E
Chernihiv, Ukr.	202-03	51°30'N	31°17'E
Chernivtsi, Ukr.	194-95	48°17'N	25°58'E
Chernobyl, see Chornobyl', Ukr.	202-203	51°17'N	30°14'E
Cherno More, s., see Black Sea	186-87	43°0'N	35°00'E
Chernovtsy, Ukr. see Chernivtsi	194-95	48°17'N	25°58'E
Chernoye More, s., see Black Sea	186-87	43°0'N	35°00'E
Chernyakhovsk, Russia	194-95	54°38'N	21°49'E
Chernyanka, Russia (chĕrn-yăŋ'kä)	202-03	50°56'N	37°49'E
Cherokee, Ia., U.S. (chĕr-ô-kē')	114-15	42°45'N	95°33'W
Cherokee, Ok., U.S. (chĕr-ô-kē')	120-21	36°45'N	98°21'W
Cherrapunji, India	234-35	25°13'N	91°42'E
Cherryville, N.C., U.S. (chĕr'ĭ-vĭl)	124-25	35°23'N	81°23'W
Cherskiy, Russia	218-19	68°46'N	161°24'E
Cherskiy Mountains, mts., Russia (chĕr'skē moun'tĭnz) see Cherskogo, Khrebet	218-19	65°0'N	144°00'E
Cherskogo, Khrebet, mts., Russia	218-19	65°0'N	144°00'E
Cherson, Ukr. see Kherson	202-03	46°38'N	32°35'E
Chervonohrad, Ukr.	194-95	50°23'N	24°14'E
Chesaning, Mi., U.S. (chĕs'ȧ-nĭng)	116-17	43°11'N	84°07'W
Chesapeake, Va., U.S. (chĕs'ȧ-pēk)	124-25	36°48'N	76°16'W
Chesapeake Bay, b., U.S. (chĕs'ȧ-pēk bā)	116-17	38°38'N	76°27'W
Chester, Eng., U.K. (chĕs'tēr)	190-91	53°12'N	2°54'W
Chester, Il., U.S. (chĕs'tēr)	120-21	37°55'N	89°49'W
Chester, Mt., U.S. (chĕs'tēr)	112-13	48°32'N	110°57'W
Chester, Pa., U.S. (chĕs'tēr)	116-17	39°51'N	75°21'W
Chester, S.C., U.S. (chĕs'tēr)	124-25	34°42'N	81°13'W
Chester, Va., U.S. (chĕs'tēr)	116-17	37°21'N	77°26'W
Chester, W.V., U.S. (chĕs'tēr)	116-17	40°36'N	80°33'W
Chesterfield, Eng., U.K. (chĕs'tēr-fēld)	190-91	53°14'N	1°26'W
Chesterfield, Îles, is., N. Cal.	280-81	19°30'S	158°00'E
Chesterfield, Nosy, i., Madag.	264-65	16°20'S	43°58'E
Chesterfield Inlet, Nu., Can.	128-29	63°21'N	90°43'W
Chetek, Wi., U.S. (chĕ'tĕk)	114-15	45°19'N	91°39'W
Chettlatt Island, i., India	236	11°41'N	72°42'E
Chetumal, Mex.	148	18°30'N	88°18'W
Chetumal, Bahía, b., N.A. (bă-ē-ă-chĕt-ōō-mäl')	148	18°39'N	88°06'W
Chevak, Ak., U.S.	126	61°39'N	165°17'W
Cheviot, Oh., U.S. (shĕv'ĭ-ŭt)	116-17	39°09'N	84°37'W
Ch'ew Bahir, lk., Afr.	267	4°40'N	36°50'E
Chewelah, Wa., U.S. (chē-wē'lä)	112-13	48°17'N	117°43'W
Cheyenne, Wy., U.S. (shī-ĕn')	114-15	41°10'N	104°48'W
Cheyenne, stm., U.S. (shī-ĕn')	110-11	44°47'N	100°44'W
Cheyenne River Indian Reservation, ind. res., S.D., U.S. (shī-ĕn' rĭv'ēr ĭn'dĭ-ăn rĕ-sēr-vā'shĕn)	114-15	45°0'N	100°40'W
Cheyenne Wells, Co., U.S. (shī-ĕn' wĕls)	120-21	38°49'N	102°21'W
Chhapra, India	234-35	25°47'N	84°45'E
Chhatarpur, India	234-35	24°55'N	79°36'E
Chhattisgarh, state, India	234-35	21°30'N	82°00'E
Chhindwāra, India	234-35	22°03'N	78°57'E
Chi, stm., Thai.	246-47	15°11'N	104°43'E
Chiai, Tai. (chī'ī')	225a	23°29'N	120°27'E
Chiang Mai, Thai.	246-47	18°48'N	99°00'E
Chiang Rai, Thai.	246-47	19°55'N	99°50'E
Chiapa de Corzo, Mex. (chê-ä'pä dä kôr'zō)	146-47	16°41'N	92°60'W
Chiapas, state, Mex. (chê-ä'päs)	146-47	16°30'N	92°30'W
Chiavari, Italy (kyä-vä'rē)	200-01	44°20'N	9°20'E
Chiba, Japan (chē'bä)	245	35°36'N	140°08'E
Chiba, state, Japan (chē'bä)	245	35°30'N	140°20'E
Chibougamau, Qc., Can. (chē-bōō'gä-mou)	138-39	49°55'N	74°22'W
Chibougamau, Lac, lk., Qc., Can. (läk chē-bōō'gä-mou)	138-39	49°50'N	74°15'W
Chibuto, Moz.	264-65	24°42'S	33°34'E
Chicago, Il., U.S. (shĭ-kô-gō) (chĭ-kä'gō)	116-17	41°52'N	87°38'W

ăt; finăl; rāte; senåte; ärm; åsk; sofȧ; fâre; ch-choose; dh-as th in other; bē; ėvent; bĕt; recĕnt; cratēr; g-gō; gh-guttural g; bĭt; ĭ-short neutral; rīde; K-guttural k as ch in German ich;

Feature (Pronunciation)	Page	Lat.	Long.
Chicago Heights, Il., U.S. (shĭ-kô-gō hīts)	**116-17**	41°30′N	87°39′W
Chicapa, stm., Afr. (chē-kä′pä)	**262-63**	6°25′S	20°48′E
Chichagof Island, i., Ak., U.S. (chē-chä′gôf ī′lănd)	**126**	57°07′N	135°12′W
Chichén Itzá, hist., Mex.	**148**	20°40′N	88°35′W
Chichester, Eng., U.K. (chĭch′ĕs-tẽr)	**190-91**	50°51′N	0°47′W
Chichimilá, Mex. (chē-chē-mê′lä)	**148**	20°37′N	88°13′W
Chichiriviche, Ven. (chē-chē-rē-vē-chĕ)	**163b**	10°56′N	68°17′W
Chickamauga, Ga., U.S. (chĭk-*à*-mô′g*à*)	**124-25**	34°53′N	85°18′W
Chickasawhay, stm., Ms., U.S. (chĭk-*à*-sô′wā)	**124-25**	31°0′N	88°45′W
Chickasha, Ok., U.S. (chĭk′*à*-shā)	**120-21**	35°03′N	97°57′W
Chiclana de la Frontera, Spain (chē-klä′nä dĕ-lä-frôn-tĕ′rä)	**198-99**	36°25′N	6°08′W
Chiclayo, Peru (chê-klä′yō)	**170**	6°46′S	79°51′W
Chico, Ca., U.S. (chē′kō)	**118-19**	39°44′N	121°49′W
Chico, stm., Arg.	**171**	43°49′S	66°26′W
Chico, stm., Arg.	**171**	49°52′S	68°35′W
Chicopee, Ma., U.S. (chĭk′ô-pē)	**116-17**	42°09′N	72°37′W
Chicoutimi, Qc., Can. (shē-kōō′tē-mē′)	**136-37**	48°26′N	71°04′W
Chicxulub, Mex. (chēk-sōō-lōō′b)	**148**	21°08′N	89°31′W
Chidambaram, India	**236**	11°24′N	79°42′E
Chiefland, Fl., U.S. (chēf′lănd)	**124-25**	29°29′N	82°52′W
Chieri, Italy (kyâ′rē)	**200-01**	45°01′N	7°49′E
Chieti, Italy (kyĕ′tē)	**200-01**	42°21′N	14°10′E
Chifeng, China (chr-fŭŋ)	**240-41**	42°16′N	118°58′E
Chignecto Bay, b., Can. (shĭg-nĕk′tō bā)	**138-39**	45°35′N	64°45′W
Chihli, Gulf of, b., China *see* Bo Hai	**240-41**	38°30′N	120°02′E
Chihuahua, Mex. (chê-wä′wä)	**144-45**	28°38′N	106°05′W
Chihuahua, state, Mex.	**122-23**	28°30′N	106°00′W
Chihuahua, Desierto de, des., N.A. *see* Chihuahuan Desert	**144-45**	35°0′N	106°00′W
Chihuahuan Desert, des., N.A.	**144-45**	35°0′N	106°00′W
Chikmagalūr, India	**236**	13°19′N	75°47′E
Chikoy, stm., Asia	**240-41**	51°02′N	106°38′E
Chilcotin, stm., B.C., Can. (chĭl-kō′tĭn)	**132-33**	51°44′N	122°24′W
Childers, Austl.	**276**	25°14′S	152°17′E
Childress, Tx., U.S. (chīld′rĕs)	**120-21**	34°26′N	100°13′W
Chile, nation, S.A. (chē′lā)	**158**	30°0′S	71°00′W
Chile Chico, Chile	**171**	46°33′S	71°44′W
Chilecito, Arg. (chê-lâ-sē′tō)	**168-69**	29°10′S	67°30′W
Chilecito, Arg. (chê-lâ-sē′tō)	**163e**	33°53′S	69°04′W
Chilika Lake, lk., India	**234-35**	19°45′N	85°25′E
Chilko, stm., B.C., Can. (chĭl′kō)	**132-33**	52°06′N	123°28′W
Chilko Lake, lk., B.C., Can.	**132-33**	51°17′N	124°04′W
Chillán, Chile (chēl-yän′)	**171**	36°36′S	72°07′W
Chillicothe, Il., U.S. (chĭl-ĭ-kŏth′ê)	**116-17**	40°55′N	89°29′W
Chillicothe, Mo., U.S. (chĭl-ĭ-kŏth′ê)	**120-21**	39°48′N	93°33′W
Chillicothe, Oh., U.S. (chĭl-ĭ-kŏth′ê)	**116-17**	39°20′N	82°59′W
Chilliwack, B.C., Can. (chĭl′ĭ-wăk)	**132-33**	49°10′N	121°57′W
Chiloé, Isla Grande de, i., Chile	**171**	42°30′S	73°55′W
Chilpancingo de los Bravo, Mex.	**146-47**	17°33′N	99°30′W
Chilton, Wi., U.S. (chĭl′tŭn)	**116-17**	44°02′N	88°09′W
Chilung, Tai. (chĭ′lung)	**225a**	25°08′N	121°44′E
Chilwa, Lake, lk., Afr.	**264-65**	15°20′S	35°43′E
Chimaltenango, Guat. (chê-mäl-tå-näŋ′gō)	**148**	14°40′N	90°49′W
Chimborazo, vol., Ec. (chêm-bô-rä′zō)	**170**	1°28′S	78°48′W
Chimbote, Peru (chêm-bō′tå)	**170**	9°04′S	78°35′W
Chimboy, Uzb.	**226**	42°56′N	59°47′E
Chimkent, Kaz. *see* Shymkent	**226**	42°18′N	69°36′E
Chimoio, Moz.	**264-65**	19°09′S	33°30′E
China, Mex. (chē′nä)	**122-23**	25°42′N	99°14′W
China, nation, Asia (chī′n*à*)	**206-07**	35°0′N	105°00′E
Chinandega, Nic. (chē-nän-dā′gä)	**149**	12°38′N	87°08′W
China Selatan, Laut, s., Asia *see* South China Sea	**224-25**	10°0′N	113°00′E
Chincha Alta, Peru (chĭn′chä äl′tä)	**163a**	13°26′S	76°08′W
Chinchilla, Austl. (chĭn-chĭl′*à*)	**276**	26°45′S	150°38′E
Chin-do, i., Kor., S.	**243**	34°27′N	126°15′E
Chindwinn, stm., Mya. (chĭn-dwĭn)	**238-39**	21°24′N	95°16′E
Chingola, Zam. (chĭng-gōlä)	**264-65**	12°32′S	27°52′E
Chinhae, Kor., S.	**243**	35°08′N	128°40′E
Chin Hills, hills, Mya.	**234-35**	22°30′N	93°30′E
Chinhoyi, Zimb.	**264-65**	17°22′S	30°11′E
Chiniot, Pak.	**232-33**	31°43′N	72°59′E
Chinju, Kor., S. (chĭn′jōō)	**243**	35°10′N	128°05′E
Chinko, stm., C.A.R. (shĭn′kô)	**262-63**	4°51′N	23°53′E
Chinmen Tao, i., Tai.	**225a**	24°27′N	118°23′E
Chinnampo, Kor., N. *see* Namp′o	**243**	38°45′N	125°23′E
Chinon, Fr. (shē-nôɴ′)	**196-97**	47°11′N	0°15′E
Chinook, Mt., U.S. (shĭn-ŏk′)	**112-13**	48°36′N	109°14′W
Chinsali, Zam.	**264-65**	10°33′S	32°04′E
Chioggia, Italy (kyôd′jä)	**200-01**	45°13′N	12°17′E
Chíos, Grc. (kē′ôs)	**200-01**	38°23′N	26°09′E
Chíos, i., Grc.	**200-01**	38°23′N	26°04′E
Chipata, Zam.	**264-65**	13°37′S	32°38′E
Chipley, Fl., U.S. (chĭp′lĭ)	**124-25**	30°47′N	85°32′W
Chipman, N.B., Can. (chĭp′m*ă*n)	**138-39**	46°10′N	65°53′W
Chippewa, stm., Wi., U.S. (chĭp′ê-wä)	**114-15**	44°24′N	92°04′W
Chippewa Falls, Wi., U.S. (chĭp′ê-wä fôlz)	**114-15**	44°56′N	91°24′W
Chiquimula, Guat. (chē-kê-mōō′lä)	**148**	14°48′N	89°33′W
Chiquimulilla, Guat. (chē-kē-mōō-lê′l-yä)	**148**	14°05′N	90°23′W
Chiquinquirá, Col. (chê-kēŋ′kê-rä′)	**163c**	5°37′N	73°48′W
Chirchiq, Uzb.	**232-33**	41°28′N	69°35′E
Chire, stm., Afr. *see* Shire	**264-65**	17°42′S	35°19′E
Chiricahua National Monument, n.p., Az., U.S. (chĭ-rä-cä′hwä năsh′ŭn-*ă*l mŏn′*ŭ*-m*ĕ*nt)	**118-19**	31°59′N	109°22′W
Chiricahua Peak, mtn., Az., U.S.	**118-19**	31°52′N	109°20′W
Chiriquí Grande, Pan. (chē-rê-kē′ grän′dā)	**150**	8°57′N	82°08′W
Chiri-san, mtn., Kor., S. (chĭ′rĭ-sän′)	**243**	35°20′N	127°44′E
Chirripó, Cerro, mtn., C.R.	**149**	9°29′N	83°30′W
Chirua, Lago, lk., Afr. *see* Chilwa, Lake	**264-65**	15°20′S	35°43′E
Chisasibi, Qc., Can.	**128-29**	53°48′N	79°02′W
Chishima-rettō, is., Russia *see* Kuril Islands	**218-19**	47°14′N	152°18′E
Chisholm, Mn., U.S. (chĭz′*ŭ*m)	**114-15**	47°29′N	92°53′W
Chisimayu, Som.	**262-63**	0°22′S	42°32′E
Chişinău, nat. cap., Mol.	**202-03**	47°02′N	28°50′E
Chistopol, Russia (chĭs-tô′pôl-y′)	**186-87**	55°22′N	50°37′E
Chistyakovo, Ukr. *see* Torez	**202-03**	48°02′N	38°38′E
Chita, Russia (chê-tá′)	**222-23**	52°02′N	113°29′E
Chitato, Ang.	**262-63**	7°20′S	20°46′E
Chitradurga, India	**236**	14°14′N	76°24′E
Chitrāl, Pak. (chē-träl′)	**232-33**	35°52′N	71°49′E
Chitré, Pan.	**150**	7°58′N	80°26′W
Chittagong, Bngl. (chĭt-á-gông′)	**246-47**	22°20′N	91°50′E
Chittaurgarh, India	**234-35**	24°54′N	74°37′E
Chittoor, India	**236**	13°13′N	79°06′E
Chiumbe, stm., Afr. (chê-ŏm′bå)	**262-63**	6°59′S	21°11′E
Chivasso, Italy (kê-väs′sō)	**200-01**	45°12′N	7°54′E
Chivilcoy, Arg. (chē-vêl-koi′)	**173**	34°54′S	60°02′W
Chkalov, Russia	**186-87**	51°48′N	55°06′E
Chlef, Alg.	**198-99**	36°10′N	1°20′E
Chŏâm Khsant, Camb.	**246-47**	14°13′N	104°56′E
Chochis, Cerro, mtn., Bol.	**168-69**	18°08′S	59°54′W
Chodzież, Pol. (kōj′yĕsh)	**194-95**	52°60′N	16°55′E
Choele Choel, Arg. (chô-ĕ′lā-chôĕ′l)	**171**	39°17′S	65°39′W
Choiseul, i., Sol. Is. (shwä-zûl′)	**279e**	7°05′S	157°00′E
Chojnice, Pol. (kōĭ-nē-tsĕ)	**194-95**	53°42′N	17°34′E
Chokurdakh, Russia	**218-19**	70°38′N	147°53′E
Chókwe, Moz.	**264-65**	24°33′S	32°60′E
Cholet, Fr. (shô-lĕ′)	**196-97**	47°04′N	0°53′W
Choluteca, Hond. (chō-lōō-tā′kä)	**149**	13°18′N	87°12′W
Choma, Zam.	**264-65**	16°49′S	26°59′E
Chomutov, Czech Rep. (kō′mö-tôf)	**194-95**	50°28′N	13°25′E
Chona, stm., Russia (chō′nä)	**218-19**	62°54′N	111°06′E
Ch′ŏnan, Kor., S.	**243**	36°48′N	127°10′E
Chon Buri, Thai.	**246-47**	13°22′N	101°00′E
Chone, Ec. (chô′nĕ)	**170**	0°42′S	80°05′W
Ch′ŏngjin, Kor., N. (chŭng-jĭn′)	**243**	41°47′N	129°48′E
Ch′ŏngju, Kor., S. (chŭng-jōō′)	**243**	36°38′N	127°30′E
Chongqing, China	**238-39**	29°34′N	106°34′E
Chongqing, state, China	**238-39**	30°0′N	108°00′E
Ch′ŏngsan-do, i., Kor., S.	**243**	34°11′N	126°54′E
Chongzuo, China	**238-39**	22°24′N	107°22′E
Chŏnju, Kor., S. (chŭn-jōō′)	**243**	35°49′N	127°09′E
Chonos, Archipiélago de los, is., Chile	**171**	45°0′S	74°00′W
Chorne more, s., *see* Black Sea	**186-87**	43°0′N	35°00′E
Chornobyl′, Ukr.	**202-03**	51°17′N	30°14′E
Ch′ŏrwŏn, Kor., S.	**243**	38°17′N	127°14′E
Chōshi, Japan (chō′shē)	**245**	35°44′N	140°50′E
Chos Malal, Arg.	**171**	37°23′S	70°16′W
Chosŏn minjujuŭi-inmīn-konghwaguk, nation, Asia *see* North Korea	**206-07**	40°0′N	127°00′E
Choszczno, Pol. (chôsh′chnô)	**194-95**	53°10′N	15°25′E
Choteau, Mt., U.S. (shō′tō)	**112-13**	47°49′N	112°12′W
Chouk′ou, China *see* Shangshui	**238-39**	33°33′N	114°34′E
Choushan Islands, is., China *see* Zhoushan Qundao	**238-39**	30°0′N	122°00′E
Chown, Mount, mtn., Ab., Can. (mount choun)	**132-33**	53°24′N	119°22′W
Choybalsan, Mong.	**240-41**	48°04′N	114°32′E
Choyr, Mong.	**240-41**	46°22′N	108°22′E
Christchurch, N.Z. (krīst′chûrch)	**278**	43°32′S	172°39′E
Christiansburg, Va., U.S. (krĭs′chănz-bûrg)	**124-25**	37°08′N	80°25′W
Christina, stm., Ab., Can.	**134-35**	56°40′N	111°04′W
Christmas Island, dep., Oc. (krĭs′-m*ă*s ī′lănd)	**248-49**	10°30′S	105°40′E
Christmas Island, at., Kir. *see* Kiritimati	**280-81**	1°48′N	157°19′W
Chrudim, Czech Rep. (kroo′dyêm)	**194-95**	49°57′N	15°48′E
Chrzanów, Pol. (kzhä′nôf)	**194-95**	50°07′N	19°26′E
Chu, stm., Asia	**246-47**	19°53′N	105°45′E
Chubut, state, Arg. (chò-bōōt′)	**171**	44°0′S	69°00′W
Chubut, stm., Arg. (chò-bōōt′)	**171**	43°21′S	65°03′W
Chucunaque, stm., Pan. (chōō-kōō-nä′kå)	**150**	8°08′N	77°44′W
Chudovo, Russia (chò′dô-vô)	**192-93**	59°07′N	31°41′E
Chudskoye Ozero, lk., Eur. (chòt′skô-yĕ ôzê-rô) *see* Peipus, Lake	**192-93**	58°45′N	27°25′E
Chuguyevka, Russia (chò-gōō′yĕf-kå)	**244**	44°09′N	133°52′E
Chukchi Sea, s., (chōōk′chē sē)	**126**	69°0′N	171°00′W
Chukotskiy, Mys, c., Russia	**126**	64°16′N	173°07′W
Chukotskiy Poluostrov, pen., Russia *see* Chukotsk Peninsula	**218-19**	66°0′N	175°00′W
Chukotskoye More, s., *see* Chukchi Sea	**126**	69°0′N	171°00′W
Chukotsk Peninsula, pen., Russia	**218-19**	66°0′N	175°00′W
Chula Vista, Ca., U.S. (chōō′lá vĭs′tá)	**118-19**	32°38′N	117°05′W
Chul′man, Russia	**218-19**	56°51′N	124°53′E
Chulucanas, Peru	**170**	5°06′S	80°09′W
Chulym, stm., Russia	**218-19**	57°42′N	83°52′E
Chumbicha, Arg.	**168-69**	28°52′S	66°14′W
Chumphon, Thai.	**246-47**	10°30′N	99°08′E
Chumysh, stm., Russia	**226**	53°31′N	83°10′E
Chuna, stm., Russia	**218-19**	57°44′N	95°27′E
Ch′unch′ŏn, Kor., S. (chòn-chŭn′)	**243**	37°52′N	127°44′E
Ch′ungju, Kor., S. (chŭng′jōō′)	**243**	36°58′N	127°56′E
Chungking, China *see* Chongqing	**238-39**	29°34′N	106°34′E
Chungking, state, China *see* Chongqing	**238-39**	30°0′N	108°00′E
Chunya, stm., Russia (chòn′yä′)	**218-19**	61°37′N	96°32′E
Chuquicamata, Chile (chōō-kê-kä-mä′tä)	**168-69**	22°19′S	68°56′W
Chur, Switz. (kōōr)	**194-95**	46°51′N	9°31′E
Churchill, Mb., Can. (chûrch′ĭl)	**128-29**	58°47′N	94°10′W
Churchill, stm., Can. (chûrch′ĭl)	**130-31**	58°49′N	94°11′W
Churchill, Cape, c., Mb., Can. (kāp chûrch′ĭl)	**130-31**	58°46′N	93°14′W
Churchill Lake, lk., Sk., Can. (chûrch′ĭl lāk)	**134-35**	55°55′N	108°20′W
Chūru, India	**234-35**	28°18′N	74°58′E
Chusovaya, stm., Russia (chōō-sô-vä′yá)	**176-77**	58°09′N	57°03′E
Chusovoy, Russia (chōō-sô-vôy′)	**186-87**	58°18′N	57°49′E
Chuuk, is., Micron.	**280-81**	7°16′N	151°44′E
Chuvashia, state, Russia	**186-87**	55°30′N	47°00′E
Chuvashiya, state, Russia *see* Chuvashia	**186-87**	55°30′N	47°00′E
Chuxian, China (chōō shyĕn)	**238-39**	32°19′N	118°18′E
Chuxiong, China (chōō-shyôŋ)	**238-39**	25°03′N	101°33′E
Chüy, stm., Asia *see* Shū	**226**	45°00′N	67°45′E
Cianjur, Indon.	**248-49**	6°48′S	107°08′E
Cicero, Il., U.S. (sĭs′ẽr-ō)	**116-17**	41°51′N	87°45′W
Cicia, i., Fiji	**279f**	17°45′S	179°18′W
Ciechanów, Pol. (tsyĕ-kä′nôf)	**194-95**	52°53′N	20°37′E
Ciego de Ávila, Cuba (syā′gō-dĕ-ä′vê-lä)	**142-43**	21°51′N	78°46′W
Ciego de Ávila, state, Cuba (syā′gō-dĕ-ä′vê-lä)	**142-43**	22°0′N	78°40′W
Ciempozuelos, Spain (thyĕm-pô-thwā′lōs)	**198-99**	40°10′N	3°37′W
Ciénaga, Col. (syā′nä-gä)	**164-65**	11°00′N	74°15′W
Cienfuegos, Cuba (syĕn-fwā′gōs)	**142-43**	22°10′N	80°26′W
Cienfuegos, state, Cuba (syĕn-fwā′gōs)	**142-43**	22°10′N	80°30′W
Cieszyn, Pol. (tsyĕ′shĕn)	**194-95**	49°45′N	18°38′E
Cieza, Spain (thyā′thä)	**198-99**	38°14′N	1°25′W
Cihuatlán, Mex. (sē-wä-tlá′n)	**146-47**	19°14′N	104°34′W
Cikobia, i., Fiji	**279f**	15°43′S	179°58′W
Cilacap, Indon.	**248-49**	7°44′S	109°01′E
Cill Chainnigh, Ire. *see* Kilkenny	**190-91**	52°39′N	7°15′W
Cimarron, stm., U.S. (sĭm-*à*-rōn′)	**120-21**	36°10′N	96°17′W
Cina Selatan, Laut, s., Asia *see* South China Sea	**224-25**	10°0′N	113°00′E
Cincinnati, Oh., U.S. (sĭn-sĭ-năt′ĭ)	**116-17**	39°11′N	84°28′W
Cintalapa, Mex. (sĕn-tä-lä′pä)	**146-47**	16°41′N	93°43′W

Feature (Pronunciation)	Page	Lat.	Long.
Cinto, Monte, mtn., Fr. (môɴ chēn´tō)	184-85	42°23′N	8°56′E
Cipolletti, Arg.	171	38°56′S	67°59′W
Circleville, Oh., U.S. (sûr´k'lvĭl)	116-17	39°36′N	82°57′W
Circleville, Ut., U.S. (sûr´k'lvĭl)	118-19	38°10′N	112°16′W
Cirebon, Indon.	248-49	6°45′S	108°34′E
Cisco, Tx., U.S. (sĭs´kō)	120-21	32°24′N	98°59′W
Cisneros, Col. (sēs-nĕ´rôs)	163c	6°33′N	75°04′W
Cisterna di Latina, Italy (chēs-tĕ´r-nä-dē-lä-tē´nä)	200-01	41°36′N	12°49′E
Citlaltépetl, Volcán, vol., Mex. see Pico de Orizaba, Volcán	146-47	19°01′N	97°16′W
Citronelle, Al., U.S. (cĭt-rō´nĕl)	124-25	31°06′N	88°14′W
Città di Castello, Italy (chēt-tä´dē käs-tĕl´lō)	200-01	43°28′N	12°15′E
Ciudad Acuña, Mex.	122-23	29°19′N	100°56′W
Ciudad Altamirano, Mex. (syōō-dä´d-äl-tä-mē-rä´nō)	146-47	18°21′N	100°39′W
Ciudad Bolívar, Ven. (syōō-dhädh´ bô-lē´vär)	164-65	8°07′N	63°33′W
Ciudad Camargo, Mex.	122-23	27°42′N	105°10′W
Ciudad Cortés, C.R.	149	8°58′N	83°32′W
Ciudad Darío, Nic. (syōō-dhädh´dä´rē-ō)	149	12°43′N	86°07′W
Ciudad del Carmen, Mex. (syōō-dä´d-dĕl-ká´r-mĕn)	148	18°39′N	91°49′W
Ciudad del Maíz, Mex. (syōō-dhädh´del mä-ēz´)	146-47	22°24′N	99°36′W
Ciudad de México, nat. cap., Mex. see Mexico City	146-47	19°24′N	99°09′W
Ciudad de Nutrias, Ven.	164-65	8°05′N	69°17′W
Ciudad Guayana, Ven.	164-65	8°21′N	62°39′W
Ciudad Hidalgo, Mex. (syōō-dä´d-ê-dä´l-gô)	146-47	19°41′N	100°34′W
Ciudad Jiménez, Mex. see Jiménez	122-23	27°08′N	104°56′W
Ciudad Juárez, Mex. (syōō-dhädh hwä´răz)	122-23	31°45′N	106°28′W
Ciudad Lerdo, Mex. see Lerdo	122-23	25°32′N	103°31′W
Ciudad Madero, Mex. (syōō-dä´d-mä-dĕ´rô)	146-47	22°16′N	97°50′W
Ciudad Mante, Mex. (syōō-dä´d-mán´tĕ)	146-47	22°44′N	98°58′W
Ciudad Netzahualcóyotl, Mex.	146-47	19°27′N	99°03′W
Ciudad Obregón, Mex. (syōō-dhädh-ô-brĕ-gô´n)	144-45	27°29′N	109°57′W
Ciudad Ojeda, Ven.	164-65	10°13′N	71°19′W
Ciudad Real, Spain (thyōō-dhädh´rā-äl´)	198-99	38°59′N	3°55′W
Ciudad Rodrigo, Spain (thyōō-dhädh´rô-drē´gō)	198-99	40°36′N	6°32′W
Ciudad Valles, Mex.	146-47	21°59′N	99°00′W
Ciudad Victoria, Mex. (syōō-dhädh´vêk-tō´rē-ä)	146-47	23°44′N	99°08′W
Civitavecchia, Italy (chē´vê-tä-vĕk´kyä)	200-01	42°06′N	11°48′E
Clairton, Pa., U.S. (klârtŭn)	116-17	40°18′N	79°53′W
Clanton, Al., U.S. (klăn´tŭn)	124-25	32°51′N	86°38′W
Clanwilliam, S. Afr.	264-65	32°10′S	18°54′E
Clare, Mi., U.S. (klâr)	116-17	43°49′N	84°45′W
Claremont, N.H., U.S. (klâr´mŏnt)	116-17	43°22′N	72°20′W
Claremore, Ok., U.S. (klâr´mōr)	120-21	36°19′N	95°37′W
Claremorris, Ire. (klâr-mŏr´ĭs)	190-91	53°43′N	8°60′W
Clarence, Isla, i., Chile	171	54°11′S	71°49′W
Clarence Island, i., Ant.	287	61°11′S	54°03′W
Clarence Strait, strt., Austl. (klăr´ĕns strāt)	272-73	12°0′S	131°00′E
Clarendon, Tx., U.S. (klâr´ĕn-dŭn)	120-21	34°56′N	100°54′W
Claresholm, Ab., Can. (klâr´ĕs-hōlm)	132-33	50°01′N	113°35′W
Clarinda, Ia., U.S. (klȧ-rĭn´dȧ)	120-21	40°44′N	95°02′W
Clarines, Ven. (klä-rē´nĕs)	163b	9°58′N	65°09′W
Clarion, Ia., U.S. (klăr´ĭ-ŭn)	114-15	42°44′N	93°44′W
Clarion, Pa., U.S. (klăr´ĭ-ŭn)	116-17	41°13′N	79°23′W
Clarión, Isla, i., Mex.	144-45	18°22′N	114°44′W
Clark, S.D., U.S. (klärk)	114-15	44°53′N	97°44′W
Clarke Island, i., Austl.	276	40°33′S	148°10′E
Clarksburg, W.V., U.S. (klärkz´bûrg)	116-17	39°16′N	80°20′W
Clarksdale, Ms., U.S. (klärks-dāl)	124-25	34°12′N	90°34′W
Clark's Harbour, N.S., Can. (klärks här´bĕr)	138-39	43°28′N	65°38′W
Clarks Hill Lake, res., U.S. (klärks hĭl lăk) see J. Strom Thurmond Reservoir	124-25	33°45′N	82°16′W
Clarkston, Wa., U.S. (klärks´tŭn)	112-13	46°25′N	117°04′W
Clarksville, Ar., U.S. (klärks-vĭl)	120-21	35°28′N	93°28′W
Clarksville, Tn., U.S. (klärks-vĭl)	124-25	36°31′N	87°21′W
Clarksville, Tx., U.S. (klärks-vĭl)	120-21	33°37′N	95°04′W
Claxton, Ga., U.S. (klăks´tŭn)	124-25	32°10′N	81°54′W
Clay Center, Ks., U.S. (klā sĕn´tēr)	120-21	39°23′N	97°07′W
Clay City, Ky., U.S. (klā sĭ´tĭ)	116-17	37°52′N	83°57′W
Clayton, Ga., U.S. (klā´tŭn)	124-25	34°53′N	83°23′W
Clayton, N.C., U.S. (klā´tŭn)	124-25	35°39′N	78°28′W
Clayton, N.M., U.S. (klā´tŭn)	120-21	36°27′N	103°11′W
Clearfield, Ut., U.S. (klēr-fēld)	112-13	41°07′N	112°01′W
Clear Lake, Ia., U.S. (klēr lāk)	114-15	43°08′N	93°23′W
Clear Lake, S.D., U.S. (klēr lāk)	114-15	44°45′N	96°42′W
Clear Lake, lk., Ca., U.S. (klēr lāk)	118-19	39°02′N	122°50′W
Clearwater, Fl., U.S. (klēr-wô´tēr)	125a	27°58′N	82°47′W
Clearwater, stm., Can. (klēr-wô´tēr)	134-35	56°45′N	111°23′W
Clearwater, stm., Ab., Can. (klēr-wô´tēr)	132-33	52°22′N	114°57′W
Clearwater Mountains, mts., Id., U.S. (klēr-wô´tēr moun´tĭnz)	112-13	46°00′N	115°30′W
Cleburne, Tx., U.S. (klē´bŭrn)	120-21	32°22′N	97°24′W
Cle Elum, Wa., U.S. (klē ĕl´ŭm)	112-13	47°12′N	120°56′W
Clermont, Austl. (klēr´mŏnt)	277	22°49′S	147°40′E
Clermont-Ferrand, Fr. (klĕr-môɴ´fĕr-răɴ´)	196-97	45°47′N	3°06′E
Cleveland, Ms., U.S. (klēv´lȧnd)	124-25	33°45′N	90°43′W
Cleveland, Oh., U.S. (klēv´lȧnd)	116-17	41°29′N	81°42′W
Cleveland, Ok., U.S. (klēv´lȧnd)	120-21	36°19′N	96°28′W
Cleveland, Tn., U.S. (klēv´lȧnd)	124-25	35°10′N	84°52′W
Cleveland, Tx., U.S. (klēv´lȧnd)	122-23	30°21′N	95°05′W
Cleveland, Cape, c., Austl.	277	19°13′S	147°02′E
Cleveland Heights, Oh., U.S. (klēv´lȧnd hīts)	116-17	41°30′N	81°36′W
Cleves, Ger. see Kleve	194-95	51°47′N	6°09′E
Clewiston, Fl., U.S. (klē´wis-tŭn)	125a	26°45′N	80°54′W
Clifden, Ire. (klĭf´dĕn)	190-91	53°29′N	10°01′W
Clifton, Az., U.S. (klĭf´tŭn)	118-19	33°04′N	109°18′W
Clifton, Il., U.S. (klĭf´tŭn)	116-17	40°56′N	87°56′W
Clifton, N.J., U.S. (klĭf´tŭn)	116-17	40°53′N	74°10′W
Clifton, Tx., U.S. (klĭf´tŭn)	122-23	31°46′N	97°35′W
Clifton Forge, Va., U.S. (klĭf´tŭn fôrj)	116-17	37°49′N	79°50′W
Clinch, stm., U.S. (klĭnch)	124-25	35°53′N	84°30′W
Clingmans Dome, mtn., U.S. (klĭng´măns dōm)	124-25	35°35′N	83°30′W
Clinton, B.C., Can. (klĭn-´tŭn)	132-33	51°05′N	121°37′W
Clinton, On., Can. (klĭn-´tŭn)	136-37	43°37′N	81°32′W
Clinton, Ar., U.S. (klĭn-´tŭn)	120-21	35°36′N	92°28′W
Clinton, Ia., U.S. (klĭn-´tŭn)	114-15	41°51′N	90°12′W
Clinton, Il., U.S. (klĭn-´tŭn)	116-17	40°09′N	88°57′W
Clinton, In., U.S. (klĭn-´tŭn)	116-17	39°39′N	87°24′W
Clinton, Ky., U.S. (klĭn-´tŭn)	124-25	36°40′N	89°00′W
Clinton, Mo., U.S. (klĭn-´tŭn)	120-21	38°22′N	93°46′W
Clinton, Ms., U.S. (klĭn-´tŭn)	124-25	32°21′N	90°19′W
Clinton, N.C., U.S. (klĭn-´tŭn)	124-25	34°59′N	78°19′W
Clinton, Ok., U.S. (klĭn-´tŭn)	120-21	35°31′N	98°58′W
Clinton, S.C., U.S. (klĭn-´tŭn)	124-25	34°28′N	81°53′W
Clinton, Tn., U.S. (klĭn-´tŭn)	124-25	36°06′N	84°08′W
Clintonville, Wi., U.S. (klĭn´tŭn-vĭl)	116-17	44°36′N	88°46′W
Clio, Mi., U.S. (klē´ō)	116-17	43°10′N	83°44′W
Clipperton, Île, at., Oc.	140-41	10°18′N	109°13′W
Clipperton Island, at., Oc. see Clipperton, Île	140-41	10°18′N	109°13′W
Clodomira, Arg.	173	27°33′S	64°07′W
Cloncurry, Austl. (klŏn-kŭr´ē)	277	20°43′S	140°30′E
Cloquet, Mn., U.S. (klô-kā´)	114-15	46°43′N	92°28′W
Cloud Peak, mtn., Wy., U.S. (kloud pēk)	112-13	44°25′N	107°10′W
Clover, S.C., U.S. (klō´vēr)	124-25	35°07′N	81°14′W
Cloverdale, Ca., U.S. (klō´vēr-dāl)	118-19	38°48′N	123°01′W
Clovis, Ca., U.S. (klō´vĭs)	118-19	36°50′N	119°42′W
Clovis, N.M., U.S. (klō´vĭs)	120-21	34°24′N	103°13′W
Cluj-Napoca, Rom.	194-95	46°47′N	23°36′E
Cluny, Fr. (klü-nē´)	196-97	46°26′N	4°39′E
Clyde, Oh., U.S. (klīd)	116-17	41°18′N	82°58′W
Clyde, Tx., U.S. (klīd)	120-21	32°25′N	99°29′W
Cnossus, hist., Grc. see Knossos	200a	35°17′N	25°12′E
Côa, stm., Port. (kô´ä)	198-99	41°05′N	7°06′W
Coahuila, state, Mex. (kō-ä-wē´lä)	122-23	27°20′N	102°00′W
Coalcomán de Matamoros, Mex.	146-47	18°47′N	103°09′W
Coaldale, Ab., Can. (kōl´dāl)	132-33	49°44′N	112°37′W
Coalgate, Ok., U.S. (kōl´gāt)	120-21	34°32′N	96°14′W
Coalinga, Ca., U.S. (kō-ȧ-lĭŋ´gȧ)	118-19	36°09′N	120°22′W
Coari, Braz. (kō-är´ē)	166-67	4°06′S	63°07′W
Coari, stm., Braz.	166-67	4°27′S	63°29′W
Coast Mountains, mts., N.A. (kōst moun´tĭnz)	130-31	55°0′N	129°00′W
Coast Ranges, mts., U.S. (kōst rānjĕz)	110-11	40°46′N	123°38′W
Coatesville, Pa., U.S. (kōts´vĭl)	116-17	39°59′N	75°49′W
Coaticook, Qc., Can. (kō´tĭ-kŏk)	136-37	45°08′N	71°48′W
Coats Island, i., Nu., Can. (kōts ī´lȧnd)	130-31	62°30′N	82°60′W
Coats Land, reg., Ant. (kōts lănd)	287	77°0′S	28°00′W
Coatzacoalcos, Mex.	146-47	18°08′N	94°26′W
Cobá, hist., Mex. (kô´bä)	148	20°36′N	87°35′W
Cobalt, On., Can. (kō´bôlt)	136-37	47°24′N	79°41′W
Cobán, Guat. (kō-bän´)	148	15°28′N	90°22′W
Cobar, Austl.	276	31°30′S	145°50′E
Cobija, Bol. (kô-bē´hä)	166-67	11°02′S	68°44′W
Coblenz, Ger. see Koblenz	194-95	50°21′N	7°35′E
Cobourg, On., Can. (kō´bŏrgh)	136-37	43°58′N	78°09′W
Cobourg Peninsula, pen., Austl.	272-73	11°22′S	132°17′E
Coburg, Ger. (kō´bōōrg)	194-95	50°16′N	10°58′E
Cocanada, India see Kākināda	236	16°57′N	82°15′E
Cochabamba, Bol.	168-69	17°23′S	66°10′W
Cochin, India	236	9°56′N	76°15′E
Cochran, Ga., U.S. (kŏk´răn)	124-25	32°23′N	83°21′W
Cochrane, Ab., Can. (kŏk´răn)	132-33	51°11′N	114°29′W
Cochrane, On., Can. (kŏk´răn)	136-37	49°04′N	81°02′W
Cochrane, Lago, lk., S.A.	171	47°21′S	71°56′W
Cockburn Island, i., On., Can. (kŏk-bûrn ī´lȧnd)	136-37	45°55′N	83°22′W
Cockburn Town, nat. cap., T./C. Is. see Grand Turk	142-43	21°27′N	71°08′W
Coco, stm., N.A. (kô-kô)	149	14°59′N	83°11′W
Coco, Cayo, i., Cuba (kä´-yō-kô´kô)	142-43	22°29′N	78°28′W
Coco, Isla del, i., C.R. (ê´s-lä-dĕl-kô-kô)	86	5°32′N	87°04′W
Cocoa, Fl., U.S. (kō´kō)	125a	28°22′N	80°44′W
Cocoa Beach, Fl., U.S. (kō´kō bēch)	125a	28°19′N	80°37′W
Coco Channel, strt., Asia	246-47	13°45′N	93°01′E
Coco Islands, is., Mya.	246-47	14°09′N	93°25′E
Cocos Islands, dep., Oc. (kō´kôs ī´lȧndz)	224-25	12°10′S	96°55′E
Cocula, Mex. (kō-kōō´lä)	146-47	20°23′N	103°50′W
Cod, Cape, pen., Ma., U.S. (kăp kŏd)	116-17	41°42′N	70°15′W
Codajás, Braz. (kō-dä-häzh´)	166-67	3°49′S	62°05′W
Codera, Cabo, c., Ven. (ká´bô-kō-dĕ´rä)	163b	10°34′N	66°03′W
Codó, Braz.	166-67	4°29′S	43°53′W
Cody, Wy., U.S. (kō´dĭ)	112-13	44°32′N	109°03′W
Coeur d'Alene, Id., U.S. (kûr dȧ-lān´)	112-13	47°41′N	116°47′W
Coeur d'Alene Indian Reservation, ind. res., Id., U.S. (kûr dȧ-lān´ ĭn´dĭ-ăn rĕ-sĕr-vā´shĕn)	112-13	47°18′N	116°45′W
Coffeyville, Ks., U.S. (kŏf´ĭ-vĭl)	120-21	37°02′N	95°37′W
Coffs Harbour, Austl.	276	30°19′S	153°08′E
Cognac, Fr. (kôn-yak´)	196-97	45°42′N	0°20′W
Cohoes, N.Y., U.S. (kô-hōz´)	116-17	42°47′N	73°42′W
Coiba, Isla de, i., Pan.	150	7°27′N	81°45′W
Coig, stm., Arg. (kô´ĕk)	171	50°57′S	69°09′W
Coihaique, Chile	171	45°34′S	72°04′W
Coimbatore, India (kô-ēm-bá-tôr´)	236	10°60′N	76°58′E
Coimbra, Port. (kô-ēm´brä)	198-99	40°13′N	8°25′W
Coín, Spain (kô-ēn´)	198-99	36°40′N	4°46′W
Coire, Switz. see Chur	194-95	46°51′N	9°31′E
Cojutepeque, El Sal. (kô-hô-tĕ-pā´kå)	148	13°43′N	88°56′W
Cokato, Mn., U.S. (kô-kā´tō)	114-15	45°04′N	94°12′W
Colac, Austl. (kŏ´lȧc)	276	38°21′S	143°35′E
Colatina, Braz. (kô-lä-tē´nä)	172	19°32′S	40°39′W
Colbeck, Cape, c., Ant.	287	77°25′S	157°33′W
Colby, Ks., U.S. (kōl´bĭ)	120-21	39°24′N	101°03′W
Colchester, Eng., U.K. (kōl´chĕs-tēr)	190-91	51°53′N	0°54′E
Cold Lake, Ab., Can.	132-33	54°27′N	110°10′W
Cold Lake, lk., Can. (kōld lāk)	132-33	54°33′N	110°05′W
Coldwater, Mi., U.S. (kōld´wô-tēr)	116-17	41°57′N	85°00′W
Coleman, Tx., U.S. (kōl´mȧn)	122-23	31°50′N	99°25′W
Colenso, S. Afr. (kŏ-lĕnz´ō)	269c	28°45′S	29°50′E
Coleraine, Mn., U.S. (kōl-rān´)	114-15	47°17′N	93°26′W
Colesberg, S. Afr.	264-65	30°43′S	25°06′E
Colfax, Ia., U.S. (kōl´făks)	114-15	41°41′N	93°15′W
Colfax, La., U.S. (kōl´făks)	122-23	31°31′N	92°42′W
Colfax, Wa., U.S. (kōl´făks)	112-13	46°53′N	117°22′W
Colhué Huapi, Lago, lk., Arg. (lä´gô kôl-wā´ôá´pê)	171	45°32′S	68°46′W
Colima, Mex.	146-47	19°14′N	103°44′W
Colima, state, Mex. (kōlē´mä)	146-47	19°10′N	104°00′W
Colima, Nevado de, vol., Mex. (nĕ-vä´dô-dĕ-kô-lē´mä)	146-47	19°33′N	103°38′W
Colinas, Braz.	166-67	6°02′S	44°14′W
Coll, i., Scot., U.K. (kōl)	190-91	56°38′N	6°34′W
College, Ak., U.S. (kŏl´ĕj)	126	64°51′N	147°46′W
College Park, Ga., U.S. (kŏl´ĕj pärk)	124-25	33°39′N	84°27′W
Collie, Austl. (kŏl´ē)	270-71	33°21′S	116°09′E
Collier Bay, b., Austl. (kŏl-yēr bā)	272-73	16°10′S	124°15′E
Collingwood, On., Can. (kŏl´ĭng-wŏd)	136-37	44°29′N	80°12′W
Collins, Ms., U.S. (kŏl´ĭns)	124-25	31°38′N	89°34′W
Collinsville, Austl.	277	20°34′S	147°51′E
Collinsville, Il., U.S. (kŏl´ĭnz-vĭl)	120-21	38°41′N	89°58′W
Collinsville, Ok., U.S. (kŏl´ĭnz-vĭl)	120-21	36°22′N	95°50′W
Collipulli, Chile	171	37°57′S	72°26′W
Colmar, Fr. (kŏl´mär)	196-97	48°05′N	7°22′E

Feature (Pronunciation)	Page	Lat.	Long.
Colmenar Viejo, Spain (kôl-mä-när´vyä´hō)	198-99	40°40'N	3°46'W
Cologne, Ger. (kŭ-lōn´)	194-95	50°56'N	6°57'E
Colomb-Béchar, Alg. see Béchar	258-59	31°37'N	2°14'W
Colombia, Col. (kô-lôm´bē-ä)	163c	3°24'N	74°48'W
Colombia, nation, S.A. (kô-lôm´bē-ä)	158	4°0'N	72°00'W
Colombo, nat. cap., Sri L. (kô-lôm´bō)	236	6°55'N	79°52'E
Colón, Arg. (kō-lōn´)	173	32°16'S	58°08'W
Colón, Arg. (kō-lōn´)	173	33°55'S	61°04'W
Colón, Cuba (kô-lô´n)	142-43	22°43'N	80°54'W
Colón, Pan. (kō-lô´n)	150	9°22'N	79°54'W
Colón, Archipiélago de, is., Ec.	170a	0°43'N	91°30'W
Colonia Alvear Norte, Arg. see General Alvear	171	34°59'S	67°42'W
Colonia del Sacramento, Ur.	173	34°29'S	57°50'W
Colonia Dora, Arg.	173	28°37'S	62°57'W
Colonia Suiza, Ur. (kô-lô´nēä-sōē´zä)	173	34°18'S	57°14'W
Colorado, state, U.S. (kŏl-ô-rä´dō)	108-09	39°0'N	105°30'W
Colorado, stm., Arg.	171	39°52'S	62°09'W
Colorado, stm., N.A. (kŏl-ô-rä´dō)	110-11	31°55'N	114°58'W
Colorado, stm., Tx., U.S. (kŏl-ô-rä´dō)	122-23	28°36'N	95°59'W
Colorado City, Tx., U.S. (kŏl-ô-rä´dō sĭ´tĭ)	120-21	32°23'N	100°52'W
Colorado National Monument, n.p., Co., U.S. (kŏl-ô-rä´dō năsh´ŭn-ăl mŏn´ū-mĕnt)	118-19	39°03'N	108°41'W
Colorado Plateau, plat., U.S. (kŏl-ô-rä´dō plă-tō´)	118-19	38°0'N	109°00'W
Colorado Springs, Co., U.S. (kŏl-ô-rä´dō springz)	120-21	38°50'N	104°49'W
Colotepec, stm., Mex. (kô-lô´tĕ-pĕk)	146-47	15°48'N	97°01'W
Colotlán, Mex. (kô-lô-tlän´)	146-47	22°06'N	103°15'W
Colquechaca, Bol. (kôl-kä-chä´kä)	168-69	18°41'S	66°02'W
Colstrip, Mt., U.S. (kōl´strip)	112-13	45°53'N	106°39'W
Columbia, Il., U.S. (kô-lŭm´bĭ-á)	120-21	38°26'N	90°11'W
Columbia, Ky., U.S. (kô-lŭm´bĭ-á)	124-25	37°06'N	85°18'W
Columbia, Md., U.S. (kô-lŭm´bĭ-á)	116-17	39°14'N	76°50'W
Columbia, Mo., U.S. (kô-lŭm´bĭ-á)	120-21	38°57'N	92°21'W
Columbia, Ms., U.S. (kô-lŭm´bĭ-á)	124-25	31°15'N	89°50'W
Columbia, S.C., U.S. (kô-lŭm´bĭ-á)	124-25	34°00'N	81°02'W
Columbia, Tn., U.S. (kô-lŭm´bĭ-á)	124-25	35°37'N	87°02'W
Columbia, stm., N.A. (kô-lŭm´bĭ-á)	130-31	46°14'N	124°06'W
Columbia, Mount, mtn., Ab., Can. (mount kô-lŭm´bĭ-á)	132-33	52°09'N	117°25'W
Columbia City, In., U.S. (kô-lŭm´bĭ-á sĭ´tē)	116-17	41°09'N	85°28'W
Columbia Icefield, ice, Can. (kô-lŭm´bĭ-á ī´fēld)	132-33	52°08'N	117°27'W
Columbia Mountains, mts., N.A. (kô-lŭm´bĭ-á moun´tĭnz)	130-31	52°0'N	119°00'W
Columbiana, Al., U.S. (kô-ŭm-bĭ-ă´ná)	124-25	33°11'N	86°36'W
Columbus, Ga., U.S. (kô-lŭm´bŭs)	124-25	32°28'N	84°58'W
Columbus, In., U.S. (kô-lŭm´bŭs)	116-17	39°12'N	85°55'W
Columbus, Ms., U.S. (kô-lŭm´bŭs)	124-25	33°29'N	88°25'W
Columbus, Mt., U.S. (kô-lŭm´bŭs)	112-13	45°39'N	109°16'W
Columbus, Ne., U.S. (kô-lŭm´bŭs)	114-15	41°26'N	97°22'W
Columbus, N.M., U.S. (kô-lŭm´bŭs)	118-19	31°50'N	107°38'W
Columbus, Oh., U.S.	116-17	39°58'N	82°60'W
Columbus, Tx., U.S. (kô-lŭm´bŭs)	122-23	29°42'N	96°32'W
Columbus, Wi., U.S. (kô-lŭm´bŭs)	116-17	43°21'N	89°01'W
Colusa, Ca., U.S. (kô-lū´sá)	118-19	39°12'N	122°01'W
Colville, Wa., U.S. (kŏl´vĭl)	112-13	48°33'N	117°54'W
Colville, stm., Ak., U.S. (kŏl´vĭl)	126	70°27'N	150°18'W
Colville Indian Reservation, ind. res., Wa., U.S. (kŏl´vĭl ĭn´dĭ-ăn rĕ-sĕr-vā´shĕn)	112-13	48°15'N	119°00'W
Comacchio, Italy (kô-mäk´kyō)	200-01	44°43'N	12°11'E
Comala, Mex. (kō-mä-lä´)	146-47	19°19'N	103°45'W
Comalcalco, Mex. (kō-mäl-käl´kō)	146-47	18°16'N	93°12'W
Comanche, Ok., U.S. (kô-măn´chē)	120-21	34°21'N	97°58'W
Comanche, Tx., U.S. (kô-măn´chē)	122-23	31°54'N	98°36'W
Comandante Fontana, Arg.	168-69	25°19'S	59°40'W
Comayagua, Hond. (kō-mä-yä´gwä)	149	14°27'N	87°39'W
Combarbalá, Chile	168-69	31°11'S	71°03'W
Comilla, Bngl. (kô-mĭl´ä)	234-35	23°27'N	91°11'E
Comino, Capo, c., Italy (kä´pō kô-mē´nō)	200-01	40°32'N	9°49'E
Comitán de Domínguez, Mex.	148	16°15'N	92°07'W
Commentry, Fr. (kō-mäN-trē´)	196-97	46°17'N	2°44'E
Commerce, Ga., U.S. (kŏm´ērs)	124-25	34°12'N	83°27'W
Commerce, Ok., U.S. (kŏm´ērs)	120-21	36°56'N	94°52'W
Commerce, Tx., U.S. (kŏm´ērs)	120-21	33°16'N	95°54'W
Committee Bay, b., Nu., Can.	130-31	68°30'N	86°30'W
Communism Peak, mtn., Taj. see Imeni Ismail Samani, Pik	226	38°57'N	72°01'E
Como, Italy (kō´mō)	200-01	45°47'N	9°05'E
Comodoro Rivadavia, Arg.	171	45°52'S	67°30'W
Comoé, Parc National de la, n.p., C. Iv.	260-61	9°0'N	3°30'W
Comores, nation, Afr. see Comoros	264-65	12°10'S	44°15'E
Comores, Archipel des, is., Afr.	264-65	12°07'S	44°03'E
Comorin, Cape, c., India (kāp kŏ´mō-rĭn)	236	8°05'N	77°34'E
Comoros, nation, Afr. (kŏm´ō-rōz) (ká-mō´-rōz)	264-65	12°10'S	44°15'E
Comox, B.C., Can. (kō´mŏks)	132-33	49°41'N	124°56'W
Compiègne, Fr. (kôN-pyĕn´y')	196-97	49°25'N	2°49'E
Compostela, Mex. (kōm-pô-stā´lä)	146-47	21°14'N	104°53'W
Conakry, nat. cap., Gui. (kô-ná-krē´)	260-61	9°31'N	13°43'W
Concarneau, Fr. (kôN-kär-nō´)	196-97	47°52'N	3°55'W
Conceição do Araguaia, Braz.	166-67	8°15'S	49°19'W
Concepción, Bol. (kôn-sĕp´syōn´)	166-67	11°29'S	66°36'W
Concepción, Bol. (kôn-sĕp´syōn´)	168-69	16°15'S	62°04'W
Concepción, Chile	171	36°49'S	73°05'W
Concepción, Para.	168-69	23°24'S	57°25'W
Concepción del Oro, Mex. (kôn-sĕp-syōn´ dĕl ō´rō)	146-47	24°37'N	101°24'W
Concepción del Uruguay, Arg. (kôn-sĕp-syô´n-dĕl-ōō-rōō-gwī´)	173	32°29'S	58°14'W
Conception Bay, b., Nf., Can. (kŏn-sĕp´shŭn bā)	138-39	47°44'N	52°59'W
Conchos, stm., Mex. (kōn´chōs)	144-45	24°56'N	97°38'W
Conchos, stm., Mex. (kōn´chōs)	144-45	29°34'N	104°24'W
Concord, Ca., U.S. (kŏŋ´kŏrd)	118-19	37°59'N	122°02'W
Concord, N.C., U.S. (kŏŋ´kŏrd)	124-25	35°24'N	80°36'W
Concord, N.H., U.S.	116-17	43°12'N	71°32'W
Concordia, Arg. (kŏn-kôr´dĭ-á)	173	31°23'S	58°01'W
Concórdia, Braz.	168-69	27°14'S	52°02'W
Concordia, Mex. (kôn-kô´r-dyä)	146-47	23°16'N	106°04'W
Concordia, Ks., U.S. (kŏn-kô´r-dyä)	120-21	39°34'N	97°40'W
Condega, Nic. (kŏn-dĕ´gä)	149	13°22'N	86°24'W
Condobolin, Austl.	276	33°05'S	147°09'E
Condoto, Col.	163c	5°05'N	76°38'W
Conegliano, Italy (kō-nāl-yä´nō)	200-01	45°53'N	12°18'E
Conghua, China (tsôŋ-hwä)	238-39	23°33'N	113°35'E
Congo, nation, Afr. (kŏn´gō)	253	1°0'S	15°00'E
Congo, stm., Afr. (kŏn´gō)	262-63	5°58'S	12°44'E
Congo, Democratic Republic of the, nation, Afr. (dĕ-mō-krä´tĭc rĕ-pŭb´lĭk ŭv thá kŏn´gō)	253	4°0'S	25°00'E
Congo, République démocratique du, nation, Afr. see Congo, Democratic Republic of the	253	4°0'S	25°00'E
Conjeeveram, India see Kānchipuram	256	12°50'N	79°43'E
Conneaut, Oh., U.S. (kŏn-ē-ôt´)	116-17	41°57'N	80°34'W
Connecticut, state, U.S. (kŏ-nĕt´ĭ-kŭt)	108-09	41°45'N	72°45'W
Connecticut, stm., U.S.	110-11	41°16'N	72°20'W
Connellsville, Pa., U.S. (kŏn´nĕlz-vĭl)	116-17	40°01'N	79°36'W
Connersville, In., U.S. (kŏn´ērz-vĭl)	116-17	39°39'N	85°08'W
Connors Range, mts., Austl. (kŏn´nŏrs rānj)	277	21°40'S	149°10'E
Conrad, Mt., U.S. (kŏn´răd)	112-13	48°10'N	111°56'W
Conroe, Tx., U.S. (kŏn´rō)	122-23	30°19'N	95°28'W
Conselheiro Lafaiete, Braz.	172	20°40'S	43°47'W
Conshohocken, Pa., U.S. (kŏn-shô-hŏk´ĕn)	116-17	40°05'N	75°18'W
Constance, Ger. see Konstanz	194-95	47°40'N	9°10'E
Constance, Lake, lk., Eur. (lāk kŏn´stăns)	194-95	47°39'N	8°54'E
Constanța, Rom. (kōn-stän´tsá)	202-03	44°11'N	28°38'E
Constantine, Alg. (kôN-stäN´tēn´)	269b	36°22'N	6°37'E
Constantine, Mi., U.S. (kŏn´stăn-tēn)	116-17	41°50'N	85°39'W
Constantinople, Tur. see İstanbul	200-01	41°02'N	28°59'E
Constitución, Chile (kōn´stĭ-tōō-syōn´)	171	35°19'S	72°25'W
Contreras, Isla, i., Chile	171	51°52'S	74°57'W
Contwoyto Lake, lk., Can.	130-31	65°42'N	110°50'W
Converse, Tx., U.S. (kŏn´vērs)	122-23	29°31'N	98°18'W
Conway, Ar., U.S. (kŏn´wā)	120-21	35°05'N	92°27'W
Conway, N.H., U.S. (kŏn´wā)	116-17	43°59'N	71°07'W
Conway, S.C., U.S. (kŏn´wā)	124-25	33°50'N	79°03'W
Cook, Cape, c., B.C., Can. (kāp kŏok)	132-33	50°07'N	127°54'W
Cook, Mount, mtn., N.Z. (mount kŏok) see Aoraki	278	43°36'S	170°10'E
Cookeville, Tn., U.S. (kŏok´vĭl)	124-25	36°10'N	85°31'W
Cook Inlet, b., Ak., U.S. (kŏok ĭn´lĕt)	126	60°32'N	151°40'W
Cook Islands, dep., Oc. (kŏok ī´lándz)	280-81	20°0'S	158°00'W
Cook Strait, strt., N.Z. (kŏok strāt)	278	41°15'S	174°30'E
Cooktown, Austl. (kŏok´toun)	277	15°29'S	145°15'E
Coolgardie, Austl. (kōōl-gär´dē)	270-71	30°57'S	121°10'E
Cooma, Austl. (kōō´má)	276	36°14'S	149°08'E
Coonabarabran, Austl.	276	31°16'S	149°17'E
Coonamble, Austl. (kōō-năm´b'l)	276	30°58'S	148°23'E
Coonoor, India	236	11°20'N	76°48'E
Coon Rapids, Mn., U.S. (kōōn răp´ĭdz)	114-15	45°10'N	93°20'W
Cooper, Tx., U.S. (kōōp´ēr)	120-21	33°23'N	95°41'W
Cooper Creek, stm., Austl.	276	28°18'S	137°29'E
Cooperstown, N.D., U.S. (kōōp´ērs-toun)	114-15	47°27'N	98°07'W
Coosa, stm., U.S. (kōō´sá)	124-25	32°30'N	86°16'W
Coos Bay, Or., U.S. (kōōs bā)	112-13	43°22'N	124°13'W
Cootamundra, Austl. (kŏotá-mŭnd´rá)	276	34°39'S	148°01'E
Copan, hist., Hond.	149	14°50'N	89°09'W
Copenhagen, nat. cap., Den. (kō´pĕn-hä´gĕn)	192-93	55°41'N	12°34'E
Copiapó, Chile (kō-pyä-pō´)	168-69	27°22'S	70°20'W
Copley, Austl.	276	30°33'S	138°26'E
Copper, stm., Ak., U.S. (kŏp´ēr)	126	60°33'N	144°52'W
Copper Harbor, Mi., U.S. (kŏp´ēr här´bēr)	114-15	47°28'N	87°55'W
Coppermine, Nu., Can. see Kugluktuk	128-29	67°47'N	115°11'W
Coppermine, stm., Can. (kŏp´ēr mīn)	130-31	67°48'N	115°08'W
Coquilhatville, D.R.C. see Mbandaka	262-63	0°02'N	18°15'E
Coquimbo, Chile (kô-kēm´bō)	168-69	29°58'S	71°20'W
Corabia, Rom. (kô-rä´bĭ-á)	200-01	43°47'N	24°31'E
Corail, Mer de, s., Oc. see Coral Sea	272-73	20°0'S	158°00'E
Coral Harbour, Nu., Can.	128-29	64°08'N	83°12'W
Coral Sea, s., Oc. (kŏr´ăl sē)	272-73	20°0'S	158°00'E
Corangamite, Lake, lk., Austl. (lāk cōr-ăŋg´á-mīt)	276	38°10'S	143°25'E
Corato, Italy (kô´rä-tô)	200-01	41°09'N	16°25'E
Corbin, Ky., U.S. (kôr´bĭn)	124-25	36°57'N	84°06'W
Corby, Eng., U.K. (kôr´bĭ)	190-91	52°29'N	0°40'W
Corcaigh, Ire. see Cork	190-91	51°54'N	8°28'W
Corcovado, Golfo, b., Chile (gôl-fô-kôr-kô-vä´dhō)	171	43°30'S	73°30'W
Corcovado, Volcán, vol., Chile	171	43°12'S	72°48'W
Cordele, Ga., U.S. (kôr-dēl´)	124-25	31°58'N	83°47'W
Cordell, Ok., U.S. (kôr-dĕl´)	120-21	35°18'N	98°59'W
Córdoba, Arg. (kôr´dô-vä)	173	31°24'S	64°12'W
Córdoba, Mex. (kô´r-dô-bä)	146-47	18°53'N	96°56'W
Córdoba, Spain (kô´r-dô-bä)	198-99	37°54'N	4°47'W
Córdoba, state, Arg. (kôr´dô-vä)	173	32°0'S	64°00'W
Cordova, Spain see Córdoba	198-99	37°54'N	4°47'W
Cordova, Ak., U.S. (kôr´dô-vä)	126	60°33'N	145°45'W
Cordova, Al., U.S. (kôr´dô-á)	124-25	33°46'N	87°11'W
Corfu, i., Grc. see Kérkyra	200-01	39°40'N	19°45'E
Corinth, Grc.	200-01	37°56'N	22°58'E
Corinth, Ms., U.S. (kŏr´ĭnth)	124-25	34°56'N	88°31'W
Corinto, Braz. (kô-rē´n-tō)	172	18°23'S	44°27'W
Cork, Ire. (kôrk)	190-91	51°54'N	8°28'W
Corleone, Italy (kôr-lå-ō´nā)	200-01	37°49'N	13°18'E
Cornelia, Ga., U.S. (kôr-nē´lyá)	124-25	34°31'N	83°32'W
Cornell, Wi., U.S. (kôr-nĕl´)	114-15	45°10'N	91°09'W
Corner Brook, Nf., Can. (kôr´nēr brŏok)	138-39	48°57'N	57°58'W
Corning, Ar., U.S. (kôr´nĭng)	124-25	36°25'N	90°35'W
Corning, Ia., U.S. (kôr´nĭng)	114-15	40°60'N	94°45'W
Corning, N.Y., U.S. (kôr´nĭng)	116-17	42°09'N	77°03'W
Corno Grande, mtn., Italy (kôr´nō grän´dĕ)	200-01	42°28'N	13°34'E
Cornwall, On., Can. (kôrn´wôl)	136-37	45°02'N	74°44'W
Coro, Ven. (kō´rô)	164-65	11°27'N	69°40'W
Corocoro, Bol. (kō-rô-kō´rô)	168-69	17°11'S	68°27'W
Coromandel Coast, cst., India (kŏr-ô-man´dĕl kōst)	236	13°30'N	80°30'E
Coronado, Ca., U.S. (kŏr-ō-nä´dō)	118-19	32°41'N	117°11'W
Coronado, Bahía de, b., C.R. (bä-ē´ä-dĕ-kô-rô-nä´dō)	149	9°0'N	83°50'W
Coronation Gulf, b., Nu., Can. (kŏr-ô-nā´shŭn gŭlf)	130-31	68°24'N	109°56'W
Coronel, Chile (kō-rô-nĕl´)	171	37°01'S	73°08'W
Coronel Dorrego, Arg. (kŏ-rô-nĕl-dôr-rĕ´gô)	173	38°43'S	61°17'W
Coronel Fabriciano, Braz.	172	19°31'S	42°39'W
Coronel Oviedo, Para. (kŏ-rô-nĕl-ô-vē´dô)	168-69	25°27'S	56°26'W
Coronel Pringles, Arg. (kŏ-rô-nĕl-prĕn´glĕs)	173	37°59'S	61°22'W
Coronel Suárez, Arg. (kŏ-rô-nĕl-swä´räs)	173	37°27'S	61°56'W
Coropuna, Nevado, vol., Peru (nĕ-vä-dô-kō-rō-pōō´nä)	170	15°31'S	72°42'W
Corozal, Belize (cŏr-ŏth-ăl´)	148	18°24'N	88°24'W
Corpus Christi, Tx., U.S. (kôr´pŭs krĭstē)	122-23	27°48'N	97°24'W
Corpus Christi, Lake, res., Tx., U.S. (lāk kôr´pŭs krĭstē)	122-23	28°10'N	97°55'W
Corpus Christi Bay, b., Tx., U.S. (kôr´pŭs krĭstē bā)	122-23	27°48'N	97°20'W

n-sing; ŋ-baŋk; N-nasalized n; nŏd; cŏmmit; ōld; ôbey; ôrder; oi-boil; fōōd; ŏo-as oo in foot; ou-out; s-soft; sh-dish; th-thin; pūre; ûnite; ûrn; stŭd; circŭs; ü-as in French tu; ´-indeterminate vowel.

Feature (Pronunciation)	Page	Lat.	Long.
Corral, Chile (kô-räl′)	171	39°53′S	73°28′W
Corral de Almaguer, Spain (kô-räl′dā äl-mä-gâr′)	198-99	39°45′N	3°10′W
Corrente, stm., Braz.	166-67	13°08′S	43°28′W
Correntina, Braz. (kô-rĕn-tē-nȧ)	166-67	13°21′S	44°39′W
Corrientes, Arg. (kō-ryĕn′täs)	173	27°29′S	58°50′W
Corrientes, state, Arg. (kō-ryĕn′täs)	173	29°0′S	58°00′W
Corrientes, stm., S.A.	170	3°44′S	74°33′W
Corrientes, Cabo, c., Col. (kȧ′bô-kō-ryĕn′tās)	163c	5°30′N	77°32′W
Corrientes, Cabo, c., Cuba (kȧ′bô-kôr-rē-ĕn′tĕs)	142-43	21°46′N	84°31′W
Corrientes, Cabo, c., Mex.	146-47	20°24′N	105°41′W
Corriverton, Guy.	164-65	5°53′N	57°09′W
Corry, Pa., U.S. (kŏr′ĭ)	116-17	41°55′N	79°38′W
Corse, i., Fr. see Corsica	184-85	42°0′N	9°00′E
Corse, Cap, c., Fr. (kȧp kôrs)	184-85	43°01′N	9°25′E
Corsica, i., Fr. (kôr′sē-kä)	184-85	42°0′N	9°00′E
Corsicana, Tx., U.S. (kôr-sĭ-kăn′ȧ)	122-23	32°06′N	96°29′W
Cortazar, Mex. (kôr-tä-zär)	146-47	20°29′N	100°56′W
Cortés, Mar de, b., Mex. see California, Golfo de	144-45	28°0′N	112°00′W
Cortez, Co., U.S.	118-19	37°21′N	108°35′W
Cortez, Sea of, b., Mex. see California, Golfo de	144-45	28°0′N	112°00′W
Corubal, stm., Afr.	258-59	11°57′N	15°03′W
Coruche, Port. (kô-rōō′she)	198-99	38°58′N	8°31′W
Çorum, Tur. (chô-rōōm′)	186-87	40°33′N	34°57′E
Corumbá, Braz.	168-69	19°01′S	57°39′W
Corumbá, stm., Braz.	168-69	18°19′S	48°54′W
Corunna, Spain see A Coruña	198-99	43°22′N	8°25′W
Corunna, Mi., U.S. (kô-rŭn′ȧ)	116-17	42°59′N	84°07′W
Coruripe, Braz. (kō-rò-rē′pĭ)	163d	10°09′S	36°10′W
Corvallis, Or., U.S. (kôr-văl′ĭs)	112-13	44°35′N	123°16′W
Corvo, i., Port.	199c	39°42′N	31°06′W
Corydon, In., U.S. (kŏr′ĭ-dŭn)	116-17	38°13′N	86°07′W
Cos, i., Grc. see Kos	200-01	36°50′N	27°10′E
Cosenza, Italy (kô-zĕnt′sä)	200-01	39°17′N	16°15′E
Coshocton, Oh., U.S. (kô-shŏk′tŭn)	116-17	40°16′N	81°51′W
Cosmoledo, Atoll de, at., Sey. (kŏs-mô-lā′dō)	264-65	9°42′S	47°31′E
Cosne-sur-Loire, Fr. (kōn-sür-lwär′)	196-97	47°24′N	2°56′E
Costa Rica, nation, N.A. (kŏs′tȧ rē′kȧ)	85	10°0′N	84°00′W
Costermansville, D.R.C. see Bukavu	267	2°30′S	28°51′E
Cotabato, Phil. (kō-tä-bä′tō)	250	7°12′N	124°14′E
Cote d'Ivoire, nation, Afr. (kōt-dē-vwär)	253	8°0′N	5°00′W
Cotija de la Paz, Mex. (kô-tē′-кä-dĕ-lä-pȧ′z)	146-47	19°49′N	102°43′W
Cotonou, nat. cap., Benin (kô-tô-nōō′)	260a	6°22′N	2°26′E
Cotopaxi, vol., Ec. (kō-tô-păk′sē)	170	0°41′S	78°27′W
Cotswold Hills, hills, Eng., U.K. (kŭtz′wōld hĭlz)	190-91	51°49′N	1°57′W
Cottage Grove, Or., U.S. (kŏt′āj grōv)	112-13	43°48′N	123°03′W
Cottbus, Ger. (kŏtt′bōōs)	194-95	51°45′N	14°19′E
Cotulla, Tx., U.S. (kō-tŭl′lȧ)	122-23	28°26′N	99°14′W
Coudersport, Pa., U.S. (koŭ′dērz-port)	116-17	41°46′N	78°01′W
Coudres, Île aux, i., Qc., Can.	138-39	47°24′N	70°23′W
Coulommiers, Fr. (kōō-lô-myā′)	196-97	48°49′N	3°05′E
Council Bluffs, Ia., U.S. (koun′sĭl blŭfs)	114-15	41°16′N	95°51′W
Courtenay, B.C., Can. (cōōrt-nā′)	132-33	49°41′N	124°60′W
Coushatta, La., U.S. (kou-shăt′ȧ)	122-23	32°01′N	93°21′W
Couture, Lac, l., Qc., Can.	130-31	60°07′N	75°20′W
Coventry, Eng., U.K. (kŭv′ĕn-trĭ)	190-91	52°25′N	1°30′W
Covington, Ga., U.S. (kŭv′ĭng-tŭn)	124-25	33°36′N	83°51′W
Covington, In., U.S. (kŭv′ĭng-tŭn)	116-17	40°08′N	87°23′W
Covington, Ky., U.S. (kŭv′ĭng-tŭn)	116-17	39°05′N	84°31′W
Covington, La., U.S. (kŭv′ĭng-tŭn)	124-25	30°28′N	90°06′W
Covington, Tn., U.S. (kŭv′ĭng-tŭn)	124-25	35°34′N	89°39′W
Covington, Va., U.S. (kŭv′ĭng-tŭn)	116-17	37°48′N	79°60′W
Cowan, Lake, l., Austl. (lāk kou′ăn)	272-73	31°50′S	121°50′E
Cowes, Eng., U.K. (kouz)	190-91	50°46′N	1°18′W
Cowra, Austl. (kou′rȧ)	276	33°50′S	148°41′E
Coxim, Braz. (kō-shēɴ′)	168-69	18°30′S	54°45′W
Cox's Bāzār, Bngl.	246-47	21°26′N	91°58′E
Coyame, Mex. (kō-yä′mā)	122-23	29°28′N	105°07′W
Coyle, stm., Arg. see Coig	171	50°57′S	69°09′W
Coyuca de Benítez, Mex. (kō-yōō′kä dā bā-nē′tāz)	146-47	17°01′N	100°05′W
Coyuca de Catalán, Mex. (kō-yōō′kä dā kä-tä-län′)	146-47	18°19′N	100°42′W
Cozad, Ne., U.S. (kō′zăd)	114-15	40°52′N	99°59′W
Cozumel, Mex. (kŏ-zōō-mĕ′l)	148	20°31′N	86°55′W
Cozumel, Isla, i., Mex. (ē′s-lä-kō-zōō-mĕ′l)	148	20°25′N	86°55′W
Cracow, Pol. see Kraków	194-95	50°04′N	19°58′E
Cradock, S. Afr. (krā′dŭk)	264-65	32°10′S	25°37′E
Craig, Ak., U.S. (krāg)	126	55°28′N	133°06′W
Craig, Co., U.S. (krāg)	112-13	40°31′N	107°32′W
Craiova, Rom. (krȧ-yō′vȧ)	200-01	44°19′N	23°48′E
Cranbrook, B.C., Can. (krăn′brók)	132-33	49°31′N	115°46′W
Crandon, Wi., U.S. (krăn′dŭn)	116-17	45°34′N	88°54′W
Cranston, R.I., U.S. (krăns′tŭn)	116-17	41°48′N	71°26′W
Crater Lake, l., Or., U.S. (krā′tēr lāk)	112-13	42°56′N	122°06′W
Crater Lake National Park, n.p., Or., U.S. (krā′tēr lāk năsh′ŭn-ăl pärk)	112-13	42°52′N	122°10′W
Craters of the Moon National Monument and Preserve, n.p., Id., U.S. (krā′tērz ŭv thȧ mōōn năsh′ŭn-ăl mŏn′ū-mĕnt ănd prĭ-zûrv)	112-13	43°25′N	113°33′W
Crateús, Braz. (krä-tȧ-ōōzh′)	166-67	5°10′S	40°40′W
Crato, Braz. (krä′tò)	166-67	7°14′S	39°23′W
Crawford, Ne., U.S. (krô′fērd)	114-15	42°41′N	103°25′W
Crawfordsville, In., U.S. (krô′fērdz-vĭl)	116-17	40°02′N	86°54′W
Crazy Mountains, mts., Mt., U.S. (krā′zĭ moun′tĭnz)	112-13	46°08′N	110°20′W
Cree, stm., Sk., Can.	130-31	58°55′N	105°46′W
Cree Lake, l., Sk., Can. (krē lāk)	130-31	57°30′N	106°30′W
Creil, Fr. (krĕ′y′)	196-97	49°16′N	2°29′E
Crema, Italy (krā′mä)	200-01	45°22′N	9°42′E
Cremona, Italy (krā-mō′nä)	200-01	45°09′N	10°01′E
Crépy-en-Valois, Fr. (krā-pē′ĕɴ-vä-lwä′)	196-97	49°14′N	2°54′E
Crescent City, Ca., U.S.	112-13	41°46′N	124°12′W
Cresco, Ia., U.S. (krĕs′kō)	114-15	43°22′N	92°07′W
Crestline, Oh., U.S. (krĕst-līn)	116-17	40°47′N	82°44′W
Creston, B.C., Can. (krĕs′tŭn)	132-33	49°06′N	116°31′W
Creston, Ia., U.S. (krĕs′tŭn)	114-15	41°04′N	94°22′W
Crestview, Fl., U.S. (krĕst′vū)	124-25	30°46′N	86°34′W
Crestwood, Ky., U.S. (krĕst′wòd)	116-17	38°19′N	85°28′W
Crete, Ne., U.S. (krēt)	120-21	40°37′N	96°58′W
Crete, i., Grc. (krēt)	200a	35°13′N	25°00′E
Crete, Sea of, s., Grc. (sē ŭv krēt)	188-89	35°54′N	25°01′E
Crewe, Eng., U.K. (krōō)	190-91	53°06′N	2°27′W
Criciúma, Braz.	172	28°41′S	49°24′W
Crimea, pen., Ukr. see Crimean Peninsula	202-03	45°0′N	34°00′E
Crimean Peninsula, pen., Ukr.	202-03	45°0′N	34°00′E
Cripple Creek, Co., U.S. (krĭp′'l krēk)	120-21	38°45′N	105°11′W
Crisfield, Md., U.S. (krĭs-fēld)	116-17	37°59′N	75°51′W
Cristalândia, Braz.	166-67	10°36′S	49°12′W
Cristóbal Colón, Pico, mtn., Col. (pē′kô-krēs-tô′bäl-kō-lôn′)	164-65	10°50′N	73°41′W
Crna Gora, nation, Eur. see Montenegro	174-75	42°30′N	19°18′E
Croatia, nation, Eur. (krō-ā′-shȧ)	174-75	45°10′N	15°30′E
Crockett, Tx., U.S. (krŏk′ĕt)	122-23	31°18′N	95°28′W
Crocodile, stm., S. Afr.	269c	24°11′S	26°53′E
Crooked, stm., B.C., Can. (krōōk′ĕd)	132-33	54°50′N	122°53′W
Crooked Island, i., Bah.	142-43	22°45′N	74°13′W
Crooked Island Passage, strt., Bah. (krōōk′ĕd ī′lȧnd păs′ĭj)	142-43	22°43′N	74°35′W
Crookston, Mn., U.S. (kròks′tŭn)	114-15	47°47′N	96°37′W
Crosby, Mn., U.S. (krôz′bĭ)	114-15	46°29′N	93°58′W
Crosby, N.D., U.S. (krôz′bĭ)	114-15	48°55′N	103°18′W
Cross, stm., Afr. (krôs)	260a	4°49′N	8°15′E
Cross City, Fl., U.S. (krôs sĭ′tĭ)	124-25	29°38′N	83°07′W
Crossett, Ar., U.S. (krôs′ĕt)	120-21	33°08′N	91°58′W
Cross Lake, res., Mb., Can.	134-35	54°45′N	97°30′W
Cross Sound, strt., Ak., U.S. (krŏs sound)	126	58°10′N	136°30′W
Crotone, Italy (krō-tô′nĕ)	200-01	39°05′N	17°07′E
Crow Creek Indian Reservation, ind. res., S.D., U.S. (krō krēk ĭn′dĭ-ăn rĕ-sēr-vā′shĕn)	114-15	44°11′N	99°30′W
Crow Indian Reservation, ind. res., Mt., U.S. (krō ĭn′dĭ-ăn rĕ-sēr-vā′shĕn)	112-13	45°27′N	108°00′W
Crowley, La., U.S. (krou′lē)	122-23	30°13′N	92°23′W
Crown Point, N.Y., U.S. (kroun point′)	116-17	43°57′N	73°26′W
Crowsnest Pass, Ab., Can.	132-33	49°37′N	114°25′W
Crozet, Archipel, is., Afr. see Crozet, Îles	287	46°0′S	52°00′E
Crozet, Îles, is., Afr. (ēl-krô-zĕ′)	287	46°0′S	52°00′E
Cruz, Cabo, c., Cuba (kȧ′-bô-krōōz)	142-43	19°51′N	77°44′W
Cruz Alta, Braz. (krōōz äl′tä)	173	28°37′S	53°36′W
Cruz del Eje, Arg. (krōō′s-dĕl-ĕ-kĕ)	168-69	30°43′S	64°49′W
Cruzeiro, Braz. (krōō-zā′rò)	172	22°35′S	44°58′W
Cruzeiro do Sul, Braz. (krōō-zā′rò dò sōōl)	170	7°39′S	72°41′W
Crystal City, Tx., U.S. (krĭs′tăl sĭ′tĭ)	122-23	28°41′N	99°50′W
Crystal Falls, Mi., U.S. (krĭs′tăl fôlz)	114-15	46°06′N	88°19′W
Crystal Lake, Il., U.S. (krĭs′tăl lăk lāk)	116-17	42°14′N	88°17′W
Crystal Springs, Ms., U.S. (krĭs′tăl sprĭngz)	124-25	31°59′N	90°22′W
Csongrád, Hung. (chôn′grăd)	194-95	46°43′N	20°09′E
Cúa, Ven. (kōō′ä)	163b	10°09′N	66°53′W
Cua Lo, Viet.	246-47	18°49′N	105°43′E
Cuamba, Moz.	264-65	14°47′S	36°32′E
Cuando, stm., Afr.	264-65	18°30′S	23°36′E
Cuango, stm., Afr.	262-63	3°13′S	17°23′E
Cuanza, stm., Ang. (kwän′zä)	264-65	9°21′S	13°09′E
Cuatrociénegas, Mex. (kwä′trō syā′nȧ-gās)	122-23	26°60′N	102°04′W
Cuauhtémoc, Mex. (kwä-ōō-tē-mōk′)	146-47	22°33′N	98°09′W
Cuauhtémoc, Mex. (kwä-ōō-tē-mōk′)	144-45	28°25′N	106°52′W
Cuautitlán, Mex. (kwä-ōō-tēt-län′)	146-47	19°27′N	104°21′W
Cuautla, Mex. (kwä-ōō′tlä)	146-47	18°49′N	98°57′W
Cuba, nation, N.A. (kū′bȧ)	140-41	21°30′N	80°00′W
Cubagua, Isla, i., Ven. (ē′s-lä-kōō-bä′gwä)	163b	10°49′N	64°11′W
Cubal, Ang.	264-65	13°03′S	14°17′E
Cubango, stm., Afr. (kōō-băŋ′gō)	264-65	18°57′S	22°25′E
Cubia, Ang.	264-65	16°01′S	21°43′E
Cúcuta, Col. (kōō′kōō-tä)	164-65	7°53′N	72°29′W
Cuddalore, India (kŭd ȧ-lōr′)	236	11°45′N	79°46′E
Cuddapah, India (kŭd′ȧ-pä)	236	14°28′N	78°49′E
Cue, Austl. (kū)	270-71	27°26′S	117°54′E
Cuenca, Ec. (kwĕn′kä)	170	2°53′S	79°00′W
Cuenca, Spain	198-99	40°05′N	2°08′W
Cuencamé de Ceniceros, Mex.	122-23	24°52′N	103°42′W
Cuernavaca, Mex. (kwĕr-nä-vä′kä)	146-47	18°55′N	99°14′W
Cuero, Tx., U.S. (kwā′rō)	122-23	29°06′N	97°17′W
Cuiabá, Braz.	168-69	15°36′S	56°05′W
Cuiabá, stm., Braz.	166-67	17°54′S	57°28′W
Cuicatlán, Mex. (kwē-kä-tlän′)	146-47	17°46′N	96°58′W
Cuilapa, Guat. (kò-ē-lä′pä)	148	14°17′N	90°18′W
Cuilo, stm., Afr.	262-63	5°53′S	16°35′E
Cuilo, stm., Afr.	260-61	3°23′S	17°23′E
Cuíto, stm., Ang. (kōō-ē-′tō)	264-65	18°01′S	20°47′E
Cuitzeo, Lago de, l., Mex. (lä′gô-dĕ-kwēt′zā-ō)	146-47	19°55′N	101°05′W
Culebra, Isla de, i., P.R. (ē′s-lä-dĕ-kōō-lā′brä)	142a	18°19′N	65°17′W
Culfa, Azer.	227	38°58′N	45°38′E
Culgoa, stm., Austl. (kŭl-gō′ȧ)	276	29°59′S	146°07′E
Culiacán, Mex. (kōō-lyä-kȧ′n)	146-47	24°04′N	107°05′W
Culiacán, Mex.	144-45	24°49′N	107°24′W
Culion Island, i., Phil.	250	11°51′N	119°57′E
Cullera, Spain (kōō-lyĕ′rä)	198-99	39°10′N	0°14′W
Cullinan, S. Afr. (kò′lĭ-nȧn)	269c	25°40′S	28°33′E
Cullman, Al., U.S. (kŭl′măn)	124-25	34°10′N	86°51′W
Culpeper, Va., U.S. (kŭl′pĕp-ēr)	116-17	38°27′N	77°60′W
Culuene, stm., Braz.	166-67	12°56′S	52°50′W
Culver, In., U.S. (kŭl′vēr)	116-17	41°13′N	86°25′W
Cumaná, Ven.	163b	10°27′N	64°11′W
Cumbal, Nevado, vol., Col.	164-65	0°57′N	77°52′W
Cumberland, Md., U.S.	116-17	39°39′N	78°46′W
Cumberland, Wi., U.S. (kŭm′bēr-lănd)	114-15	45°32′N	92°01′W
Cumberland, stm., U.S. (kŭm′bēr-lănd)	110-11	37°08′N	88°25′W
Cumberland, Lake, res., Ky., U.S. (lāk kŭm′bēr-lănd)	124-25	36°57′N	84°55′W
Cumberland Peninsula, pen., Nu., Can. (kŭm′bēr-lănd pē-nĭn′sūlȧ)	130-31	66°32′N	64°13′W
Cumberland Plateau, plat., U.S. (kŭm′bēr-lănd plä-tō′)	124-25	36°0′N	85°00′W
Cumberland Sound, strt., Nu., Can. (kŭm′bēr-lănd sound)	130-31	65°10′N	65°30′W
Cumbres de Monterrey, Parque Nacional, n.p., Mex.	122-23	25°30′N	100°25′W
Cunani, Braz.	164-65	2°52′N	51°06′W
Cunco, Chile	171	38°56′S	72°02′W
Cunene, stm., Afr.	264-65	17°15′S	11°45′E
Cunnamulla, Austl. (kŭn-ȧ-mŭl-ȧ)	276	28°04′S	145°41′E
Curaçao, i., Neth. Ant. (kōō-rä-sä′ō)	140a	12°11′N	69°00′W
Curacautín, Chile (kä-rä-käōō-tē′n)	171	38°25′S	71°56′W
Curacó, stm., Arg. see Salado	171	38°49′S	64°59′W
Curanilahue, Chile	171	37°28′S	73°21′W
Curaray, stm., S.A.	170	2°26′S	74°04′W
Curepipe, Mauritius	265a	20°19′S	57°31′E
Curicó, Chile	171	34°59′S	71°14′W
Curitiba, Braz. (kōō-rē-tē′bȧ)	172	25°26′S	49°16′W
Currais Novos, Braz. (kōōr-rä′ēs nō-vōs)	163d	6°15′S	36°31′W
Curralinho, Braz.	166-67	1°48′S	49°47′W
Current, stm., U.S. (kûr′ĕnt)	120-21	36°15′N	90°55′W
Curtis Island, i., Austl.	277	23°38′S	151°09′E

Feature (Pronunciation)	Page	Lat.	Long.
Curuá, stm., Braz.	166-67	1°57'S	55°08'W
Curuá, stm., Braz.	166-67	5°21'S	54°28'W
Cururupu, Braz. (ko͞o-ro̤-ro̤-po͞o')	166-67	1°50'S	44°52'W
Curuzú Cuatiá, Arg.	173	29°48'S	58°03'W
Curvelo, Braz. (kȯr-vĕl'ȯ)	172	18°45'S	44°27'W
Cusco, Peru	170	13°31'S	71°59'W
Cushing, Ok., U.S. (kŭsh'ĭng)	120-21	35°59'N	96°46'W
Custer, S.D., U.S. (kŭs'tẽr)	114-15	43°45'N	103°36'W
Cut, Nuhu, i., Indon.	224-25	5°29'S	133°06'E
Cut Bank, Mt., U.S. (kŭt bănk)	112-13	48°38'N	112°20'W
Cuthbert, Ga., U.S. (kŭth'bẽrt)	124-25	31°46'N	84°47'W
Cuttack, India (kŭ-tăk')	234-35	20°30'N	85°50'E
Cutzamalá, stm., Mex. (ko͞o-tzä-mä-lä')	146-47	18°23'N	100°41'W
Cuvo, stm., Ang. (ko͞o'vō)	264-65	10°52'S	13°48'E
Cuyabá, Braz. *see* Cuiabá	168-69	15°36'S	56°05'W
Cuyahoga Falls, Oh., U.S. (kī-à-hō'gà fôlz)	116-17	41°08'N	81°29'W
Cuyo Islands, is., Phil. (ko͞o'yō ī'lăndz)	250	11°04'N	120°57'E
Cuyuni, stm., S.A. (ko͞o-yo͞o'nē)	164-65	6°23'N	58°40'W
Cyangugu, Rw.	267	2°29'S	28°54'E
Cyclades, is., Grc. (sĭ'clà-dēz)	200-01	37°30'N	25°00'E
Cynthiana, Ky., U.S. (sĭn-thĭ-ăn'à)	116-17	38°23'N	84°18'W
Cyprus, nation, Asia (sī'prŭs)	228-29	35°0'N	33°00'E
Cyrenaica, hist. reg., Libya (sĭr-à-nā'ĭ-kà)	258-59	31°0'N	22°30'E
Czech Republic, nation, Eur. (chĕk rĕ-pŭb'lĭk)	174-75	49°40'N	15°10'E
Czernowitz, Ukr. *see* Chernivtsi	194-95	48°17'N	25°58'E
Częstochowa, Pol. (chăN-stô kô'vá)	194-95	50°49'N	19°08'E

D

Feature (Pronunciation)	Page	Lat.	Long.
Da, Song, stm., Asia *see* Black	246-47	21°15'N	105°21'E
Da'an, China (dä-än)	240-41	45°30'N	124°18'E
Dabhoi, India	234-35	22°08'N	73°25'E
Dabie Shan, mts., China (dä-bĭĕ shän)	238-39	31°06'N	115°56'E
Dacca, nat. cap., Bngl. *see* Dhaka	234-35	23°43'N	90°25'E
Dac Glei, Viet.	246-47	15°11'N	107°48'E
Dachau, Ger. (dä'kou)	194-95	48°15'N	11°27'E
Dade City, Fl., U.S. (dād sĭ'tĭ)	125a	28°22'N	82°11'W
Dadeville, Al., U.S. (dād'vĭl)	124-25	32°50'N	85°46'W
Dadiangas, Phil. *see* General Santos	250	6°07'N	125°10'E
Dādra and Nagar Haveli, state, India	234-35	20°05'N	73°00'E
Dādu, Pak.	232-33	26°44'N	67°46'E
Dadu, stm., China (dä-do͞o)	238-39	29°33'N	103°46'E
Daegu, Kor., S. *see* Taegu	243	35°52'N	128°35'E
Daejeon, Kor., S. *see* Taejŏn	243	36°20'N	127°26'E
Daet, Phil.	250	14°07'N	122°57'E
Dagestan, state, Russia (dä-gĕs-tän')	227	43°0'N	47°00'E
Dagupan, Phil. (dä-go͞o'pän)	250	16°02'N	120°21'E
Da Hinggan Ling, mts., China (dä hĭŋ-gän lĭŋ) *see* Greater Khingan Range	222-23	49°0'N	122°00'E
Dahlak Archipelago, is., Erit. (dä-läk' är'kà-pĕ'-à-gō)	266	15°45'N	40°30'E
Dahlak Kebir, i., Erit. *see* Dehalak' Desēt	266	15°40'N	40°06'E
Dahomey, nation, Afr. *see* Benin	253	9°30'N	2°15'E
Dahy, Nafūd ad-, des., Sau. Ar.	266	22°20'N	45°35'E
Daireaux, Arg.	173	36°36'S	61°45'W
Dairen, China *see* Dalian	240-41	38°54'N	121°34'E
Dairût, Egypt	268b	27°33'N	30°49'E
Dai-sen, vol., Japan (dī'sĕn')	245	35°22'N	133°33'E
Dajabón, Dom. Rep. (dä-kä-bô'n)	142-43	19°33'N	71°42'W
Dajian Shan, mtn., China	238-39	26°42'N	103°34'E
Dakar, nat. cap., Sen. (dà-kär')	260-61	14°41'N	17°27'W
Dakhla, W. Sah.	258-59	23°43'N	15°56'W
Dalälven, stm., Swe.	192-93	60°38'N	17°27'E
Dalandzadgad, Mong.	240-41	43°35'N	104°25'E
Da Lat, Viet.	246-47	11°58'N	108°27'E
Dālbandin, Pak.	232-33	28°53'N	64°25'E
Dalby, Austl. (dôl'bē)	276	27°11'S	151°16'E
Dale, Nor. (dä'lĕ)	192-93	61°22'N	5°25'E
Dale Hollow Lake, res., U.S. (dāl hŏl'ō lāk)	124-25	36°37'N	85°20'W
Dalhart, Tx., U.S. (dăl härt)	120-21	36°04'N	102°31'W
Dalhousie, N.B., Can. (dăl-ho͞o'zē)	138-39	48°04'N	66°24'W
Dalhousie, India	234-35	32°34'N	75°58'E
Dalhousie, Cape, c., N.T., Can.	130-31	70°14'N	129°42'W
Dali, China (dä-lē)	238-39	34°47'N	109°57'E
Dali, China (dä-lē)	238-39	25°36'N	100°13'E
Dalian, China (dä-lĭĕn)	240-41	38°54'N	121°34'E
Daliang Shan, mts., China	238-39	28°0'N	103°00'E
Dallas, Or., U.S. (dăl'làs)	112-13	44°55'N	123°18'W
Dallas, Tx., U.S. (dăl'làs)	120-21	32°47'N	96°46'W
Dall Island, i., Ak., U.S. (dăl ī'ănd ī'lănd)	126	54°57'N	133°01'W
Dalmā, i., U.A.E.	230-31	24°31'N	52°19'E
Dal'negorsk, Russia	244	44°33'N	135°34'E
Dalnerechensk, Russia	244	45°56'N	133°44'E
Daloa, C. Iv.	260-61	6°53'N	6°27'W
Dalrymple, Mount, mtn., Austl. (mount dăl'rĭm-p'l)	277	21°02'S	148°38'E
Dāltenganj, India	234-35	24°03'N	84°04'E
Dalton, Ga., U.S. (dôl'tŭn)	124-25	34°46'N	84°58'W
Dalupiri Island, i., Phil.	250a	19°05'N	121°14'E
Daly, stm., Austl. (dā'lĭ)	272-73	13°19'S	130°16'E
Daly Waters, Austl.	270-71	16°15'S	133°22'E
Damān, India	234-35	20°25'N	72°51'E
Damān and Diu, state, India	234-35	20°25'N	72°50'E
Damanhûr, Egypt	268b	31°02'N	30°28'E
Damar, Pulau, i., Indon.	248-49	1°01'S	128°23'E
Damaraland, hist. reg., Nmb. (dă'ná-rà-lănd)	264-65	22°11'S	17°35'E
Damas, nat. cap., Syria *see* Damascus	228-29	33°31'N	36°18'E
Damascus, nat. cap., Syria (dà-măs'kŭs)	228-29	33°31'N	36°18'E
Damāvand, Kūh-e, vol., Iran	232-33	35°56'N	52°08'E
Dāmghān, Iran (dām-gän')	232-33	36°10'N	54°21'E
Damietta, Egypt	268b	31°25'N	31°49'E
Dāmodar, stm., India	234-35	22°17'N	88°06'E
Damoh, India	234-35	23°51'N	79°27'E
Dampier, Selat, strt., Indon. (så-lät' däm'pēr)	224-25	0°40'S	130°40'E
Danakil, reg., Afr.	266	13°0'N	41°00'E
Da Nang, Viet.	246-47	16°03'N	108°12'E
Danbury, Ct., U.S. (dăn'bēr-ĭ)	116-17	41°23'N	73°27'W
Dandenong, Austl. (dăn'dê-nông)	276	38°00'S	145°13'E
Dandong, China (dän-dôŋ)	243	40°07'N	124°21'E
Dang, stm., China	240-41	40°26'N	94°34'E
Dângrêk, Chuŏr Phnum, mts., Asia *see* Phanom Dongrak Range	246-47	14°25'N	103°30'E
Dangriga, Belize	148	16°58'N	88°13'W
Danilov, Russia (dä'nê-lôf)	202-03	58°11'N	40°11'E
Dankov, Russia (däŋ'kôf)	202-03	53°16'N	39°09'E
Danmark, nation, Eur. *see* Denmark	174-75	56°0'N	10°00'E
Danmarksstrædet, strt., *see* Denmark Strait	86	67°0'N	25°00'W
Dannemora, N.Y., U.S. (dăn-ê-mō'rà)	116-17	44°43'N	73°43'W
Dannhauser, S. Afr. (dän'hou-zēr)	269c	28°02'S	30°04'E
Dansville, N.Y., U.S. (dănz'vĭl)	116-17	42°33'N	77°41'W
Danube, stm., Eur. (dăn'ūb)	188-89	45°23'N	29°36'E
Danville, Ar., U.S. (dăn'vĭl)	120-21	35°04'N	93°23'W
Danville, Il., U.S. (dăn'vĭl)	116-17	40°08'N	87°38'W
Danville, In., U.S. (dăn'vĭl)	116-17	39°45'N	86°31'W
Danville, Ky., U.S. (dăn'vĭl)	116-17	37°39'N	84°47'W
Danville, Va., U.S. (dăn'vĭl)	124-25	36°35'N	79°24'W
Danxian, China (dän shyĕn)	238-39	19°31'N	109°33'E
Danzig, Pol. *see* Gdańsk	194-95	54°21'N	18°38'E
Daocheng, China	238-39	29°04'N	100°36'E
Daoxian, China (dou shyĕn)	238-39	25°35'N	111°27'E
Da Qaidam, China	240-41	37°53'N	95°19'E
Darabani, Rom. (dă-ră-băn'ĭ)	194-95	48°11'N	26°36'E
Darbhanga, India (dŭr-bŭŋ'gä)	234-35	26°09'N	85°54'E
Dardanelles, strt., Tur. (där-dà-nĕlz')	200-01	40°17'N	26°33'E
Dar-el-Beida, Mor. *see* Casablanca	269a	33°36'N	7°36'W
Dar es Salaam, nat. cap., Tan. (där ĕs sà-läm')	262-63	6°49'S	39°16'E
Dārfūr, hist. reg., Sudan (där-fo͞or')	262-63	13°30'N	23°30'E
Darganata, Turkmen.	232-33	40°29'N	62°10'E
Darhan, Mong.	240-41	49°29'N	105°56'E
Darién, Col. (dä-rĭ-ĕn')	163c	3°56'N	76°30'W
Darlag, China	238-39	33°46'N	99°40'E
Darling, stm., Austl. (där'lĭng)	276	34°07'S	141°55'E
Darling Downs, reg., Austl. (där'lĭng dounz)	276	27°30'S	150°30'E
Darling Range, mts., Austl. (där'lĭng rānj)	272-73	32°31'S	116°21'E
Darlington, Eng., U.K. (där'lĭng-tŭn)	190-91	54°32'N	1°34'W
Darlington, S.C., U.S. (där'lĭng-tŭn)	124-25	34°18'N	79°52'W
Darlington, Wi., U.S. (där'lĭng-tŭn)	116-17	42°41'N	90°07'W
Darmstadt, Ger. (därm'shtät)	194-95	49°52'N	8°39'E
Darnah, Libya	188-89	32°46'N	22°39'E
Darnley, Cape, c., Ant.	287	67°43'S	69°30'E
Daror, Eth.	262-63	8°14'N	44°41'E
Dart, Cape, c., Ant.	287	73°06'S	126°20'W
Dartmoor, for., Eng., U.K. (därt'mo͞or)	190-91	50°35'N	3°55'W
Dartmouth, N.S., Can. (därt'mŭth)	138-39	44°41'N	63°34'W
Daru, Pap. N. Gui. (dä'ro͞o)	277	9°05'S	143°14'E
Daruvar, Cro. (där'ro͞o-vär)	200-01	45°36'N	17°13'E
Darvaza, Turkmen.	232-33	40°10'N	58°29'E
Darwin, Austl. (där'wĭn)	224-25	12°27'S	130°50'E
Dasht, stm., Pak. (dŭsht)	230-31	25°13'N	61°42'E
Daşköpri, Turkmen.	232-33	36°17'N	62°38'E
Daşoguz, Turkmen.	226	41°50'N	59°58'E
Datia, India	234-35	25°40'N	78°27'E
Datong, China (dä-tôŋ)	240-41	37°03'N	101°36'E
Datong, China (dä-tôŋ)	240-41	40°05'N	113°17'E
Datong, stm., China	240-41	36°20'N	102°50'E
Datu Piang, Phil.	250	7°02'N	124°29'E
Daua, stm., Afr. *see* Dawa	262-63	4°10'N	42°06'E
Daugava, stm., Eur. *see* Western Dvina	192-93	57°04'N	24°03'E
Daugavpils, Lat. (dä'ò-gäv-pêls)	192-93	55°53'N	26°32'E
Daule, stm., Ec.	170	2°18'S	79°50'W
Daung Kyun, i., Mya.	246-47	12°14'N	98°05'E
Dauphin, Mb., Can. (dô'fĭn)	134-35	51°09'N	100°03'W
Dauphin Lake, lk., Mb., Can. (dô'fĭn lăk)	134-35	51°17'N	99°48'W
Dāvangere, India	236	14°28'N	75°55'E
Davao, Phil. (dä'vä-ō)	250	7°04'N	125°35'E
Davao Gulf, b., Phil.	250	6°40'N	125°55'E
Davenport, Ia., U.S. (dăv'ĕn-pōrt)	114-15	41°31'N	90°35'W
Davenport, Wa., U.S. (dăv'ĕn-pōrt)	112-13	47°39'N	118°08'W
David, Pan. (dà-vēdh')	150	8°26'N	82°26'W
David City, Ne., U.S. (dā'vĭd sĭ'tĕ)	114-15	41°15'N	97°08'W
Davis, Ca., U.S. (dā'vĭs)	118-19	38°33'N	121°44'W
Davis, Ok., U.S. (dā'vĭs)	120-21	34°30'N	97°07'W
Davis Mountains, mts., Tx., U.S. (dā'vĭs moun'tĭnz)	122-23	30°42'N	104°10'W
Davis Sea, s., Ant.	287	66°0'S	92°00'E
Davisstrædet, strt., N.A. *see* Davis Strait	86	67°0'N	57°00'W
Davis Strait, strt., N.A. (dā'vĭs strāt)	86	67°0'N	57°00'W
Davlekanovo, Russia	186-87	54°13'N	55°02'E
Davos, Switz. (dä'vōs)	194-95	46°48'N	9°49'E
Dawa, stm., Afr.	262-63	4°10'N	42°06'E
Dawāsir, Wādī ad-, stm., Sau. Ar.	220-21	20°26'N	45°35'E
Dawei, Mya.	246-47	14°05'N	98°13'E
Dawna Range, mts., Mya. (dô'ná rānj)	246-47	16°50'N	98°15'E
Dawson, Yk., Can. (dô'sŭn)	128-29	64°03'N	139°24'W
Dawson, Ga., U.S. (dô'sŭn)	124-25	31°47'N	84°26'W
Dawson, Mn., U.S. (dô'sŭn)	114-15	44°56'N	96°04'W
Dawson, stm., Austl. (dô'sŭn)	277	23°38'S	149°46'E
Dawson, Isla, i., Chile	171	53°55'S	70°43'W
Dawson Creek, B.C., Can. (dô'sŭn krēk)	132-33	55°47'N	120°20'W
Dawson Springs, Ky., U.S. (dô'sŭn sprĭngz)	124-25	37°10'N	87°41'W
Dawu, China (dä-wo͞o)	238-39	31°33'N	114°07'E
Dax, Fr. (dàks)	196-97	43°43'N	1°03'W
Daxian, China (dä-shyĕn)	238-39	31°13'N	107°30'E
Daxue Shan, mts., China	238-39	30°30'N	101°46'E
Dayr az-Zawr, Syria (dà-ērēz-zôr')	228-29	35°21'N	40°09'E
Dayton, Oh., U.S. (dā'tŭn)	116-17	39°45'N	84°12'W
Dayton, Tn., U.S. (dā'tŭn)	124-25	35°30'N	85°01'W
Dayton, Tx., U.S. (dā'tŭn)	122-23	30°03'N	94°54'W
Dayton, Wa., U.S. (dā'tŭn)	112-13	46°19'N	117°58'W
Daytona Beach, Fl., U.S. (dā-tō'nà bēch)	124-25	29°13'N	81°02'W
Dayu, China (dä-yo͞o)	238-39	25°24'N	114°22'E
Da Yunhe, can., China (dä yòn-hŭ) *see* Grand Canal	240-41	32°11'N	119°33'E
De Aar, S. Afr. (dē-är')	264-65	30°39'S	24°01'E
Dead Sea, lk., Asia (dĕd sē)	228-29	31°30'N	35°30'E
Deadwood, S.D., U.S. (dĕd'wòd)	114-15	44°23'N	103°44'W
Deal Island, i., Austl.	276	39°28'S	147°20'E
Dean, stm., B.C., Can. (dēn)	132-33	52°48'N	126°59'W
Dean Channel, strt., B.C., Can. (dēn chăn'ĕl)	132-33	52°33'N	127°13'W
Deán Funes, Arg. (dĕ-á'n-fo͞o-nĕs)	173	30°25'S	64°21'W
Dearborn, Mi., U.S. (dēr'bŭrn)	116-17	42°19'N	83°10'W
Dearg, Beinn, mtn., Scot., U.K. (bĕn dŭrg)	190-91	57°47'N	4°56'W
Dease, stm., B.C., Can.	130-31	59°55'N	128°29'W
Death Valley, Ca., U.S. (dĕth văl'ê)	118-19	36°19'N	116°25'W
Death Valley, val., Ca., U.S. (dĕth văl'ê)	118-19	36°30'N	117°00'W
Death Valley National Park, n.p., U.S. (dĕth văl'ê năsh'ŭn-ăl pärk)	118-19	36°25'N	116°56'W
Debao, China (dŭ-bou)	238-39	23°20'N	106°37'E
Dęblin, Pol. (dăN'blĭn)	194-95	51°35'N	21°53'E
Dębno, Pol. (dĕb-nō')	194-95	52°44'N	14°43'E
Debrecen, Hung. (dĕ'brĕ-tsĕn)	194-95	47°32'N	21°38'E
Debre Mark'os, Eth.	269d	10°20'N	37°44'E
Decatur, Al., U.S. (dê-kā'tŭr)	124-25	34°36'N	86°59'W
Decatur, Il., U.S. (dê-kā'tŭr)	116-17	39°50'N	88°57'W

Feature (Pronunciation)	Page	Lat.	Long.
Decatur, In., U.S. (dē-kā′tŭr)	116-17	40°50′N	84°55′W
Decatur, Mi., U.S. (dē-kā′tŭr)	116-17	42°06′N	85°58′W
Decatur, Tx., U.S. (dē-kā′tŭr)	120-21	33°14′N	97°35′W
Decazeville, Fr. (dē-käz′vēl′)	196-97	44°34′N	2°15′E
Deccan, plat., India (dĕk′ăn)	236	17°0′N	78°00′E
Deception Lake, lk., Sk., Can. (dē-sĕp′shŭn lāk)	134-35	56°33′N	104°10′W
Decorah, Ia., U.S. (dē-kō′rȧ)	114-15	43°18′N	91°47′W
Dee, stm., Scot., U.K. (dē)	190-91	57°08′N	2°05′W
Deep River, On., Can. (dēp rĭv′ẽr)	136-37	46°06′N	77°30′W
Deerfield, Ma., U.S. (dēr′fēld)	116-17	42°33′N	72°36′W
Deer Lake, Nf., Can. (dēr lāk)	138-39	49°11′N	57°26′W
Deer Lake, lk., On., Can. (dēr lāk)	134-35	52°39′N	94°29′W
Deer Lodge, Mt., U.S. (dēr lŏj)	112-13	46°24′N	112°44′W
Deer Park, Wa., U.S. (dēr pärk)	112-13	47°57′N	117°28′W
Deer River, Mn., U.S. (dēr rĭv′ẽr)	114-15	47°19′N	93°48′W
Defiance, Oh., U.S. (dē-fī′ăns)	116-17	41°16′N	84°22′W
Defuniak Springs, Fl., U.S. (dē fū′nĭ-ăk sprĭngz)	124-25	30°43′N	86°07′W
Dêgê, China	238-39	31°50′N	98°40′E
Degeh Bur, Eth.	262-63	8°13′N	43°33′E
Deggendorf, Ger. (dĕ′ghĕn-dôrf)	194-95	48°51′N	12°58′E
Dehalak' Desēt, i., Erit.	266	15°40′N	40°06′E
Dehiwala-Mount Lavinia, Sri L.	236	6°52′N	79°52′E
Dehra Dūn, India (dā′rŭ dōōn)	234-35	30°19′N	78°02′E
Dehri, India	234-35	24°54′N	84°11′E
Dehua, China (dŭ-hwä)	225a	25°30′N	118°14′E
Dehui, China	240-41	44°32′N	125°43′E
Dej, Rom. (dāzh)	194-95	47°09′N	23°53′E
De Kalb, Il., U.S. (dē kălb′)	116-17	41°56′N	88°45′W
DeLand, Fl., U.S. (dē lănd′)	124-25	29°02′N	81°18′W
Delano, Ca., U.S. (dĕl′ȧ-nō)	118-19	35°47′N	119°15′W
Delavan, Wi., U.S. (dĕl′ȧ-văn)	116-17	42°38′N	88°39′W
Delaware, Oh., U.S. (dĕl′ȧ-wâr)	116-17	40°18′N	83°04′W
Delaware, state, U.S. (dĕl′ȧ-wâr)	108-09	39°10′N	75°30′W
Delaware, stm., Ks., U.S. (dĕl′ȧ-wâr)	120-21	39°04′N	95°25′W
Delaware, stm., U.S.	116-117	39°20′N	75°25′E
Delaware Bay, b., U.S. (dĕl′ȧ-wâr bā)	116-17	39°05′N	75°15′W
De Leon, Tx., U.S. (dē lē-ŏn′)	122-23	32°07′N	98°33′W
Delft, Neth. (dĕlft)	190-91	52°01′N	4°22′E
Delgado, Cabo, c., Moz. (kȧ′bô-dĕl-gä′dō)	264-65	10°41′S	40°38′E
Delger, stm., Mong.	240-41	49°17′N	100°42′E
Delhi, India (dĕl′hī)	234-35	28°40′N	77°14′E
Delhi, La., U.S. (dĕl′hī)	124-25	32°28′N	91°30′W
Delhi, state, India (dĕl′hī)	234-35	28°37′N	77°10′E
Delicias, Mex.	122-23	28°12′N	105°29′W
Déline, N.T., Can.	128-29	65°11′N	123°25′W
Delingha, China	240-41	37°15′N	97°11′E
Dell Rapids, S.D., U.S. (dĕl răp′ĭdz)	114-15	43°49′N	96°43′W
Delmarva Peninsula, pen., U.S.	116-17	38°30′N	75°30′W
Delmenhorst, Ger. (dĕl′mĕn-hôrst)	194-95	53°04′N	8°38′E
De Long Mountains, mts., Ak., U.S. (dē′lŏng moun′tĭnz)	126	68°20′N	162°00′W
Delphi, In., U.S. (dĕl′fī)	116-17	40°35′N	86°40′W
Delphos, Oh., U.S. (dĕl′fŏs)	116-17	40°51′N	84°21′W
Del Rio, Tx., U.S. (dĕl rē′ō)	122-23	29°22′N	100°54′W
Delta Junction, Ak., U.S.	126	64°02′N	145°44′W
Dëma, stm., Russia (dyĕm′ä)	176-77	54°44′N	55°54′E
Dembī Dolo, Eth.	262-63	8°32′N	34°48′E
Demidov, Russia (dzyĕ′mē-dô′f)	192-93	55°16′N	31°32′E
Deming, N.M., U.S. (dĕm′ĭng)	118-19	32°16′N	107°45′W
Demini, stm., Braz.	166-67	0°46′S	62°56′W
Demopolis, Al., U.S. (dē-mŏp′ō-lĭs)	124-25	32°31′N	87°50′W
Dempo, Gunung, vol., Indon. (gōō-nŏng dĕm′pō)	246-47	4°00′S	103°09′E
Demyanka, stm., Russia (dyĕm-yän′kä)	218-19	59°31′N	69°05′E
Demyansk, Russia (dyĕm-yänsk′)	192-93	57°38′N	32°28′E
Denakil, reg., Afr. *see* Danakil	266	13°0′N	41°00′E
Denali, mtn., Ak., U.S. *see* McKinley, Mount	126	63°04′N	151°00′W
Denali National Park and Preserve, n.p., Ak., U.S.	126	63°15′N	150°30′W
Dêngqên, China	238-39	31°27′N	95°27′E
Denham, Austl.	270-71	25°55′S	113°33′E
Denham, Mount, mtn., Jam.	142-43	18°13′N	77°32′W
Denham Range, mts., Austl.	277	21°41′S	147°55′E
Deniliquin, Austl. (dĕ-nĭl′ĭ-kwĭn)	276	35°32′S	144°58′E
Denison, Ia., U.S. (dĕn′ĭ-sŭn)	114-15	42°01′N	95°22′W
Denison, Tx., U.S. (dĕn′ĭ-sŭn)	120-21	33°45′N	96°33′W
Denizli, Tur. (dĕn-ĭz-lē′)	200-01	37°47′N	29°05′E
Denmark, S.C., U.S. (dĕn′märk)	124-25	33°20′N	81°09′W
Denmark, nation, Eur. (dĕn′märk)	174-75	56°0′N	10°00′E
Denmark Strait, strt., (dĕn′märk strāt)	86	67°0′N	25°00′W
Denov, Uzb.	232-33	38°16′N	67°54′E
Denpasar, Indon.	248-49	8°39′S	115°13′E
Denton, Tx., U.S. (dĕn′tŭn)	120-21	33°14′N	97°08′W
D'Entrecasteaux Islands, is., Pap. N. Gui. (däɴ-tr′-kȧs-tō′ ī′lăndz)	277	9°27′S	150°32′E
Denver, Co., U.S. (dĕn′vẽr)	120-21	39°44′N	104°58′W
Deolāli, India	234-35	19°57′N	73°50′E
De Pere, Wi., U.S. (dē pēr′)	116-17	44°26′N	88°03′W
Dêqên, China	238-39	28°38′N	98°52′E
De Queen, Ar., U.S. (dē kwēn′)	120-21	34°03′N	94°21′W
De Quincy, La., U.S. (dē kwĭn′sĭ)	122-23	30°27′N	93°26′W
Dera, Lach, stm., Afr. (dā′rä)	262-63	0°13′N	42°18′E
Dera Ghāzi Khān, Pak. (dā′rŭ gä-zē′ ĸän′)	232-33	30°03′N	70°38′E
Dera Ismāīl Khān, Pak. (dā′rŭ ĭs-mä-ēl′ ĸän′)	232-33	31°49′N	70°55′E
Derbent, Russia (dĕr-bĕnt′)	227	42°03′N	48°17′E
Derby, Austl. (där′bē) (dûr′bē)	270-71	17°20′S	123°38′E
Derby, Eng., U.K. (där′bē)	190-91	52°55′N	1°29′W
Derby, Ks., U.S. (dûr′bē)	120-21	37°32′N	97°15′W
De Ridder, La., U.S. (dē rĭd′ẽr)	122-23	30°51′N	93°18′W
Dermott, Ar., U.S. (dûr′mŏt)	124-25	33°32′N	91°26′W
Derry, N. Ire., U.K. *see* Londonderry	190-91	54°59′N	7°20′W
Derry, N.H., U.S. (dâr′ĭ)	116-17	42°53′N	71°20′W
Derventa, Bos. (dĕr′ven-tȧ)	200-01	44°59′N	17°54′E
Derzhavinsk, Kaz.	226	51°06′N	66°19′E
Desaguadero, stm., S.A.	170	18°04′S	67°06′W
Désappointement, Îles du, is., Fr. Poly.	280-81	14°10′S	141°20′W
Deschambault Lake, lk., Sk., Can.	134-35	54°40′N	103°35′W
Deschutes, stm., Or., U.S. (dā-shōōt′)	112-13	45°39′N	120°55′W
Desē, Eth.	266	11°09′N	39°38′E
Deseado, stm., Arg. (dā-sā-ä′dhō)	171	47°46′S	65°53′W
Desengaño, Punta, c., Arg.	171	49°15′S	67°37′W
De Smet, S.D., U.S. (dē smĕt′)	114-15	44°23′N	97°33′W
Des Moines, Ia., U.S. (dē moin′)	114-15	41°34′N	93°36′W
Des Moines, stm., U.S. (dē moin′)	110-11	40°22′N	91°26′W
Desna, stm., Eur. (dyĕs-nȧ′)	202-03	50°33′N	30°34′E
Desolación, Isla, i., Chile (ē′s-lä-dĕ-sô-lä-syô′n)	171	53°00′S	74°09′W
De Soto, Mo., U.S. (dē sō′tō)	120-21	38°08′N	90°34′W
Dessau, Ger. (dĕsou)	194-95	51°50′N	12°14′E
Destruction Bay, Yk., Can.	128-29	61°14′N	138°50′W
Detmold, Ger. (dĕt′mōld)	194-95	51°56′N	8°53′E
Detroit, Mi., U.S. (dē-troit′)	116-17	42°21′N	83°04′W
Detroit Lakes, Mn., U.S. (dē-troit′ lākz)	114-15	46°49′N	95°51′W
Detva, Slvk. (dyĕt′vȧ)	194-95	48°32′N	19°29′E
Deutschland, nation, Eur. *see* Germany	174-75	51°0′N	10°00′E
Deva, Rom. (dā′vä)	194-95	45°53′N	22°55′E
Deventer, Neth. (dĕv′ĕn-tẽr)	190-91	52°16′N	6°10′E
Devils, stm., Tx., U.S. (dĕv′′lz)	122-23	29°32′N	100°59′W
Devils Lake, N.D., U.S. (dĕv′′lz lāk)	114-15	48°07′N	98°52′W
Devils Postpile National Monument, n.p., Ca., U.S. (dĕv′′lz pōst-pīl năsh′ŭn-ăl mŏn′ū-mĕnt)	118-19	37°37′N	119°05′W
Devils Tower National Monument, n.p., Wy., U.S. (dĕv′′lz tou′ẽr năsh′ŭn-ăl mŏn′ū-mĕnt)	112-13	44°36′N	104°43′W
Devon Island, i., Nu., Can.	86	75°0′N	87°00′W
Devonport, Austl. (dĕv′ŭn-pôrt)	276	41°11′S	146°21′E
Dewās, India	234-35	22°58′N	76°03′E
Dewey, Ok., U.S. (dū′ĭ)	120-21	36°48′N	95°56′W
De Witt, Ar., U.S. (dē wĭt′)	124-25	34°18′N	91°20′W
De Witt, Ia., U.S. (dē wĭt′)	114-15	41°49′N	90°32′W
Dexter, Me., U.S. (dĕks′tẽr)	117a	45°02′N	69°17′W
Dexter, Mo., U.S. (dĕks′tẽr)	124-25	36°48′N	89°58′W
Dexter, N.M., U.S. (dĕks′tẽr)	120-21	33°12′N	104°22′W
Dezfūl, Iran	228-29	32°23′N	48°24′E
Dezhnëva, Mys, c., Russia (mĭs dyĕzh′nyĭf)	126	66°08′N	169°41′W
Dezhou, China (dŭ-jō)	240-41	37°27′N	116°17′E
Dhahran, Sau. Ar.	230-31	26°18′N	50°08′E
Dhaka, nat. cap., Bngl. (dä′kä) (dăk′ȧ)	234-35	23°43′N	90°25′E
Dhamār, Yemen	266	14°33′N	44°24′E
Dhanbād, India	234-35	23°48′N	86°26′E
Dhār, India	234-35	22°36′N	75°18′E
Dhawalāgiri, mtn., Nepal	234-35	28°42′N	83°30′E
Dhuburi, India	234-35	26°01′N	89°59′E
Dhule, India	234-35	20°54′N	74°46′E
Diablo, Pico del, mtn., Mex.	144-45	30°59′N	115°45′W
Diablo Range, mts., Ca., U.S. (dyä′blō rānj)	118-19	37°0′N	121°20′W
Diamante, Arg.	173	32°04′S	60°39′W
Diamantina, Braz.	172	18°15′S	43°37′W
Diamantina, stm., Austl. (dī′man-tē′nȧ)	277	26°58′S	138°49′E
Diamantina National Park, n.p., Austl.	277	23°43′S	141°11′E
Diamantino, Braz. (dä-à-män-tē′no)	166-67	14°25′S	56°27′W
Dian Chi, lk., China (dĭĕn chē)	238-39	24°50′N	102°42′E
Dianópolis, Braz.	166-67	11°38′S	46°50′W
D' Iberville, Mont, mtn., Can. *see* Caubvick, Mount	130-31	58°53′N	63°43′W
Dibrugarh, India	234-35	27°29′N	94°55′E
Dickinson, N.D., U.S. (dĭk′ĭn-sŭn)	114-15	46°53′N	102°47′W
Dickson, Tn., U.S. (dĭk′sŭn)	124-25	36°04′N	87°22′W
Dickson City, Pa., U.S. (dĭk′sŭn sĭ′tē)	116-17	41°28′N	75°37′W
Dicle, stm., Asia *see* Tigris	208-09	30°60′N	47°27′E
Diego de Almagro, Isla, i., Chile	171	51°28′S	75°11′W
Diégo-Suarez, Madag. *see* Antsiranana	264-65	12°17′S	49°17′E
Dien Bien, Viet.	246-47	21°23′N	103°01′E
Dien Bien Phu, Viet. *see* Dien Bien	246-47	21°23′N	103°01′E
Dieppe, N.B., Can. (dē-ĕp′)	138-39	46°06′N	64°44′W
Dieppe, Fr.	196-97	49°56′N	1°05′E
Difuri, i., Mald.	236	5°24′N	73°38′E
Digboi, India	234-35	27°23′N	95°38′E
Digby, N.S., Can. (dĭg′bĭ)	138-39	44°37′N	65°46′W
Digul, stm., Indon.	277	7°10′S	138°41′E
Dijlah, stm., Asia *see* Tigris	208-09	30°60′N	47°27′E
Dijon, Fr. (dē-zhôɴ′)	196-97	47°19′N	5°02′E
Dikhil, Dji.	266	11°06′N	42°22′E
Dikson, Russia (dĭk′sŏn)	218-19	73°30′N	80°33′E
Dīla, Eth.	262-63	6°19′N	38°14′E
Dili, nat. cap., E. Timor (dĭl′ē)	248-49	8°35′S	125°35′E
Dilling, Sudan	266	12°02′N	29°40′E
Dillingham, Ak., U.S. (dĭl′ĕng-hăm)	126	59°03′N	158°28′W
Dillon, Mt., U.S. (dĭl′ŭn)	112-13	45°13′N	112°38′W
Dillon, S.C., U.S. (dĭl′ŭn)	124-25	34°25′N	79°22′W
Dilolo, D.R.C. (dē-lō′lō)	262-63	10°42′S	22°21′E
Dimāpur, India	234-35	25°55′N	93°44′E
Dimashq, nat. cap., Syria *see* Damascus	228-29	33°31′N	36°18′E
Dimitrovgrad, Blg.	200-01	42°03′N	25°37′E
Dimitrovgrad, Russia	186-87	54°13′N	49°36′E
Dimlang, mtn., Nig.	260-61	8°24′N	11°47′E
Dinagat Island, i., Phil.	250	10°12′N	125°35′E
Dinājpur, Bngl.	234-35	25°38′N	88°38′E
Dinan, Fr. (dē-näɴ′)	196-97	48°28′N	2°03′W
Dinant, Bel. (dē-näɴ′)	190-91	50°16′N	4°55′E
Dinara Planina, mts., Eur. (dē′nä-rä plä′nē-na) *see* Dinaric Alps	200-01	43°55′N	16°38′E
Dinaric Alps, mts., Eur.	200-01	43°55′N	16°38′E
Dinariche, Alpi, mts., Eur. *see* Dinaric Alps	200-01	43°55′N	16°38′E
Dindigul, India	236	10°22′N	77°59′E
Dingalan Bay, b., Phil. (dĭŋ-gä′län bā)	250	15°18′N	121°25′E
Dinggyê, China	234-35	28°35′N	86°37′E
Dinghai, China	238-39	30°01′N	122°06′E
Dingwall, Scot., U.K. (dĭng′wôl)	190-91	57°36′N	4°26′W
Dingxi, China	240-41	35°33′N	104°32′E
Dingxian, China (dĭŋ shyĕn)	240-41	38°31′N	114°60′E
Dingyuan, China (dĭŋ-yůän)	238-39	32°32′N	117°40′E
Dinosaur National Monument, n.p., U.S. (dī′nō-sôr năsh′ŭn-ăl mŏn′ū-mĕnt)	112-13	40°32′N	108°58′W
Dipolog, Phil.	250	8°35′N	123°21′E
Dirē Dawa, Eth.	262-63	9°35′N	41°52′E
Diriamba, Nic. (dēr-yäm′bä)	149	11°51′N	86°15′W
Dirj, Libya	188-89	30°10′N	10°28′E
Dirranbandi, Austl. (dĭ-rȧ-băn′dē)	276	28°35′S	148°14′E
Disappointment Islands, is., Fr. Poly. (dĭs′ȧ-point′ment ī′lăndz) *see* Désappointement, Îles du	280-81	14°10′S	141°20′W
Dispur, India	234-35	26°08′N	91°48′E
Disraëli, Qc., Can. (dĭs-rā′lĭ)	136-37	45°54′N	71°21′W
District of Columbia, state, U.S.	108-09	38°54′N	77°01′W
Distrito Federal, state, Braz. (dēs-trē′tô-fĕ-dĕ-rä′l)	168-69	15°45′S	47°45′W
Distrito Federal, state, Mex.	146-47	19°15′N	99°10′W
Dis ûq, Egypt	268b	31°08′N	30°39′E
Diu, India (dē′ōō)	234-35	20°43′N	70°59′E
Divinópolis, Braz. (dē-vē-nô′pō-lēs)	172	20°09′S	44°53′W
Divnoye, Russia	186-87	45°55′N	43°22′E
Divo, C. Iv.	260-61	5°50′N	5°22′W
Dixon, Il., U.S. (dĭks′ŭn)	116-17	41°50′N	89°29′W
Dixon Entrance, strt., N.A.	126	54°25′N	132°30′W
Diyarbakır, Tur. (dē-yär-bĕk′ĭr)	227	37°55′N	40°14′E
Dja, stm., Afr.	260-61	2°02′N	15°12′E
Djajapura, Indon. *see* Jayapura	277	2°32′S	140°43′E
Djakarta, nat. cap., Indon. *see* Jakarta	248-49	6°11′S	106°50′E
Djambala, Congo	262-63	2°33′S	14°46′E
Djedi, Oued, stm., Alg.	258-59	34°28′N	6°06′E
Djelfa, Alg.	269b	34°41′N	3°15′E

Feature (Pronunciation)	Page	Lat.	Long.
Djenné, Mali	258-59	13°54'N	4°33'W
Djérem, stm., Camrn.	260-61	5°19'N	13°24'E
Djibouti, nation, Afr. (jē-bo͞o-tē′)	253	11°30'N	43°00'E
Djibouti, nat. cap., Dji. (jē-bo͞o-tē′)	266	11°34'N	43°09'E
Djokjakarta, Indon. *see* Yogyakarta	248-49	7°48'S	110°22'E
Djougou, Benin	260-61	9°42'N	1°40'E
Djugu, D.R.C.	267	1°58'N	30°30'E
Dmitrov, Russia (d'mē′trôf)	202-03	56°21'N	37°31'E
Dnepr, stm., Eur. *see* Dnieper	202-03	46°31'N	32°22'E
Dneprodzerzhinsk, Ukr. *see* Dniprodzerzhyns'k	202-03	48°29'N	34°41'E
Dnestr, stm., Eur. *see* Dniester	188-89	46°19'N	30°17'E
Dnieper, stm., Eur. (nē′pûr)	202-03	46°31'N	32°22'E
Dniester, stm., Eur. (nēs′-tēr)	188-89	46°19'N	30°17'E
Dnipro, stm., Eur. *see* Dnieper	202-03	46°31'N	32°22'E
Dniprodzerzhyns'k, Ukr.	202-03	48°29'N	34°41'E
Dnipropetrovs'k, Ukr.	202-03	48°28'N	34°58'E
Dnister, stm., Eur. *see* Dniester	188-89	46°19'N	30°17'E
Dnjapro, stm., Eur. *see* Dnieper	202-03	46°31'N	32°22'E
Dno, Russia (d'nô′)	192-93	57°49'N	29°59'E
Doba, Chad	262-63	8°39'N	16°51'E
Doberai, Jazirah, pen., Indon.	224-25	1°30'S	132°30'E
Doboj, Bos. (dō′boi)	200-01	44°44'N	18°06'E
Dobrich, Blg.	200-01	43°35'N	27°50'E
Dobryanka, Russia (dôb-ryän′kȧ)	186-87	58°28'N	56°25'E
Doce, stm., Braz. (dō′sä)	172	19°39'S	39°49'W
Doctor Arroyo, Mex. (dōk-tōr′ är-rō′yō)	146-47	23°42'N	100°11'W
Dodecanese, is., Grc. (dō′dĕ-cȧ-nēs′)	200-01	36°30'N	27°00'E
Dodekanisoy, is., Grc. *see* Dodecanese	200-01	36°30'N	27°00'E
Dodge City, Ks., U.S. (dŏj sĭ′tē)	120-21	37°45'N	100°01'W
Dodgeville, Wi., U.S. (dŏj′vĭl)	116-17	42°58'N	90°08'W
Dodola, Eth.	269d	7°00'N	39°07'E
Dodoma, nat. cap., Tan. (dō′dō-mȧ)	267	6°11'S	35°45'E
Dogai Coring, lk., China	234-35	34°35'N	88°59'E
Dog Island, i., Anguilla	143b	18°17'N	63°15'W
Dog Lake, lk., On., Can. (dŏg lāk)	136-37	48°46'N	89°33'W
Dōgo, i., Japan	245	36°15'N	133°16'E
Doğu Karadeniz Dağları, mts., Tur.	227	40°30'N	40°30'E
Doha, nat. cap., Qatar (dō′hȧ)	230-31	25°17'N	51°32'E
Dolbeau-Mistassini, Qc., Can.	136-37	48°53'N	72°13'W
Dole, Fr. (dōl)	196-97	47°06'N	5°29'E
Dolgaya Kosa, spit, Russia (dôl-gä′yä kô′sȧ)	202-03	46°41'N	37°43'E
Dolinsk, Russia (dȧ-lēnsk′)	222-23	47°20'N	142°48'E
Dolisie, Congo *see* Loubomo	262-63	4°11'S	12°40'E
Dolores, Arg. (dô-lō′rĕs)	173	36°20'S	57°41'W
Dolores, Ur.	173	33°32'S	58°12'W
Dolores Hidalgo, Mex. (dô-lō′rĕs-ē-däl′gō)	146-47	21°09'N	100°56'W
Domažlice, Czech Rep. (dō′mäzh-lē-tsĕ)	194-95	49°27'N	12°56'E
Dombarovskiy, Russia	226	50°46'N	59°32'E
Dombås, Nor.	192-93	62°05'N	9°08'E
Dombóvár, Hung. (dŏm′bō-vär)	194-95	46°22'N	18°08'E
Domeyko, Chile	168-69	28°57'S	70°54'W
Domeyko, Cordillera, mts., Chile (kôr-dēl-yĕ′rä-dô-mā′kō)	168-69	24°45'S	69°09'W
Dominica, nation, N.A. (dŏ-mĭ-nē′kȧ)	140-41	15°30'N	61°20'W
Dominican Republic, nation, N.A. (dô-mĭn′ĭ-kăn rē-pŭb′lĭk)	140-41	19°0'N	70°40'W
Dominion, Cape, c., Nu., Can.	130-31	66°09'N	74°27'W
Dominique, Canal de la, strt., N.A. *see* Martinique Passage	143b	15°10'N	61°15'W
Domodedovo, Russia (dô-mô-dyĕ′do-vô)	202-03	55°26'N	37°47'E
Dom Pedrito, Braz.	173	30°59'S	54°40'W
Domuyo, Volcán, vol., Arg.	171	36°38'S	70°26'W
Don, stm., Russia (dŏn)	186-87	47°05'N	39°15'E
Don, stm., Scot., U.K. (dŏn)	190-91	57°11'N	2°06'W
Donaldsonville, La., U.S. (dŏn′ăld-sŭn-vĭl)	124-25	30°06'N	90°60'W
Donau, stm., Eur. *see* Danube	188-89	45°23'N	29°36'E
Don Benito, Spain (dōn′bȧ-nē′tō)	198-99	38°57'N	5°52'W
Doncaster, Eng., U.K. (doŋ′kăs-tēr)	190-91	53°32'N	1°08'W
Dondo, Moz.	264-65	19°35'S	34°44'E
Dondra Head, c., Sri L.	236	5°55'N	80°35'E
Donegal, Ire. (dŏn-ē-gôl′)	190-91	54°39'N	8°07'W
Donegal Bay, b., Ire. (dŏn-ĕ-gôl′ bā)	190-91	54°30'N	8°30'W
Donets'k, Ukr.	202-03	47°60'N	37°48'E
Dong, stm., China (dôŋ)	238-39	23°05'N	113°60'E
Dong, stm., China (dôŋ)	240-41	42°17'N	101°06'E
Dongara, Austl. (dôn-gä′rȧ)	270-71	29°15'S	114°58'E
Dongfang, China (dôŋ-fäŋ)	238-39	19°05'N	108°38'E
Donggala, Indon. (dôn-gä′lä)	248-49	0°41'S	119°44'E
Dongguan, China (dôŋ-gûän)	238-39	23°03'N	113°44'E
Dong Hai, s., Asia *see* East China Sea	222-23	30°0'N	126°00'E
Donghai Dao, i., China	238-39	21°02'N	110°25'E
Dong Hoi, Viet. (dông-hô-ĕ′)	246-47	17°29'N	106°37'E
Dong Nai, stm., Viet.	246-47	10°44'N	106°46'E
Dongola, Sudan *see* Dunqulah	266	19°11'N	30°28'E
Dong San Shen, hist. reg., China *see* Manchuria	240-41	47°0'N	125°00'E
Dongting Hu, lk., China (dôŋ-tĭŋ ho͞o)	238-39	29°20'N	112°54'E
Dongyang, China	238-39	29°16'N	120°14'E
Dongzhi, China	238-39	30°07'N	117°01'E
Doniphan, Mo., U.S. (dŏn′ĭ-făn)	124-25	36°37'N	90°50'W
Doniphan, Ne., U.S. (dŏn′ĭ-făn)	114-15	40°46'N	98°23'W
Donostia, Spain *see* Donostia-San Sebastián	198-99	43°19'N	1°60'W
Donostia-San Sebastián, Spain	198-99	43°19'N	1°60'W
Door Peninsula, pen., Wi., U.S. (dōr pĕ-nĭn′sŭlȧ)	116-17	44°55'N	87°20'W
Dorchester, Eng., U.K. (dôr′chĕs-tēr)	190-91	50°43'N	2°27'W
Dorchester, Cape, c., Nu., Can.	130-31	65°27'N	77°26'W
Dordogne, stm., Fr. (dôr-dôn′yĕ)	196-97	45°02'N	0°35'W
Dore Lake, lk., Sk., Can.	134-35	54°46'N	107°17'W
Dores do Indaiá, Braz.	172	19°28'S	45°36'W
Dori, Burkina	260-61	14°02'N	0°02'W
Dornbirn, Aus. (dōrn′bêrn)	194-95	47°25'N	9°45'E
Dorogobuzh, Russia (dôrôgô′-bo͞o′zh)	202-03	54°55'N	33°18'E
Dorohoi, Rom. (dō-rô-hoi′)	194-95	47°57'N	26°24'E
Dorpat, Est. *see* Tartu	192-93	58°23'N	26°43'E
Dorre Island, i., Austl. (dôr ī′lȧnd)	272-73	25°07'S	113°06'E
Dortmund, Ger. (dôrt′mȯnt)	194-95	51°31'N	7°28'E
Dörtyol, Tur. (dûrt′yōl)	228-29	36°52'N	36°13'E
Dosatuy, Russia	240-41	50°22'N	118°34'E
Dos Bahías, Cabo, c., Arg. (kȧ′bô-dôs-bä-ē′äs)	171	44°56'S	65°33'W
Dos Hermanas, Spain (dōsĕr-mä′näs)	198-99	37°17'N	5°55'W
Dosso, Niger (dôs-ō′)	258-59	13°03'N	3°12'E
Dossor, Kaz.	186-87	47°32'N	52°59'E
Dothan, Al., U.S. (dō′thăn)	124-25	31°13'N	85°24'W
Douai, Fr. (do͞o-ā′)	196-97	50°22'N	3°05'E
Douala, Camrn. (do͞o-ä′lä)	260-61	4°03'N	9°42'E
Douarnenez, Fr. (do͞o-ár nĕ-nĕs′)	196-97	48°06'N	4°20'W
Douglas, Az., U.S. (dŭg′lăs)	118-19	31°20'N	109°34'W
Douglas, Ga., U.S. (dŭg′lăs)	124-25	31°30'N	82°51'W
Douglas, Mi., U.S. (dŭg′lăs)	116-17	42°39'N	86°12'W
Douglas, Wy., U.S. (dŭg′lăs)	112-13	42°45'N	105°23'W
Douglas, nat. cap., I. of Man (dŭg′lăs)	190-91	54°10'N	4°29'W
Douglas Channel, strt., B.C., Can. (dŭg′lăs chăn′ĕl)	132-33	53°30'N	129°12'W
Douglasville, Ga., U.S. (dŭg′lăs-vĭl)	124-25	33°45'N	84°45'W
Dourada, Serra, plat., Braz. (sĕ′r-rä-dôo͞o-rá′dä)	166-67	13°10'S	48°34'W
Dourados, Braz.	168-69	22°13'S	54°49'W
Douro, stm., Eur. (dô′ò-rô)	198-99	41°09'N	8°41'W
Dover, Eng., U.K. (dō vēr)	190-91	51°08'N	1°18'E
Dover, De., U.S.	116-17	39°09'N	75°31'W
Dover, N.H., U.S.	116-17	43°12'N	70°53'W
Dover, N.J., U.S. (dō vēr)	116-17	40°54'N	74°32'W
Dover, Oh., U.S. (dō vēr)	116-17	40°31'N	81°28'W
Dover, Strait of, strt., Eur. (strāt ŭv dō vēr)	190-91	50°59'N	1°31'E
Dover-Foxcroft, Me., U.S. (dō′vēr fŏks′krôft)	117a	45°11'N	69°14'W
Dovrefjell, mts., Nor. (dŏv′rĕ fyĕl′)	192-93	62°06'N	9°25'E
Dowagiac, Mi., U.S. (dô-wô′jăk)	116-17	41°59'N	86°06'W
Drâa, Hamada du, des., Alg.	258-59	29°0'N	6°45'W
Drâa, Oued, stm., Afr. (wĕd drá)	258-59	28°41'N	11°07'W
Drăgășani, Rom. (drä-gȧ-shän′ĭ)	200-01	44°40'N	24°16'E
Draguignan, Fr. (drä-gēn-yäN′)	196-97	43°33'N	6°28'E
Drake, Pasaje de, strt., *see* Drake Passage	287	58°0'S	70°00'W
Drakensberg, mts., Afr. (drä′kĕnz-bĕrgh)	264-65	27°0'S	30°00'E
Drake Passage, strt., (drāk păs′ĭj)	287	58°0'S	70°00'W
Dráma, Grc. (drä′mä)	200-01	41°09'N	24°09'E
Drammen, Nor. (dräm′ĕn)	192-93	59°45'N	10°13'E
Drau, stm., Eur. (drou) *see* Drava	200-01	45°33'N	18°56'E
Drava, stm., Eur. (drä′vä)	200-01	45°33'N	18°56'E
Drayton Valley, Ab., Can. (drā′tăn văl′ē)	132-33	53°14'N	114°59'W
Dresden, Ger. (drās′dĕn)	194-95	51°03'N	13°44'E
Dreux, Fr. (drû)	196-97	48°44'N	1°22'E
Drin, stm., Alb. (drēn)	200-01	41°45'N	19°35'E
Drina, stm., Eur. (drē′nä)	200-01	44°54'N	19°21'E
Drøbak, Nor. (drû′bäk)	192-93	59°39'N	10°39'E
Drohobych, Ukr.	194-95	49°21'N	23°31'E
Druc', stm., Bela. (dro͞ot)	202-03	53°04'N	30°02'E
Druk-Yul, nation, Asia *see* Bhutan	206-07	27°30'N	90°30'E
Drumheller, Ab., Can. (drŭm-hĕl-ēr)	132-33	51°28'N	112°42'W
Drummond Island, i., Mi., U.S. (drŭm′ŭnd ī′lȧnd)	116-17	46°0'N	83°40'W
Drummondville, Qc., Can. (drŭm′ŭnd-vĭl)	136-37	45°53'N	72°30'W
Drumright, Ok., U.S. (drŭm′rīt)	120-21	35°60'N	96°36'W
Drwęca, stm., Pol. (d'r-vän′tsä)	194-95	52°60'N	18°41'E
Dryden, On., Can. (drī-dĕn)	134-35	49°47'N	92°51'W
Dry Tortugas, is., Fl., U.S. (drī tôr-to͞o′gäz)	125a	24°38'N	82°55'W
Dry Tortugas National Park, n.p., Fl., U.S. (drī tôr-to͞o′gäz năsh′ŭn-ăl pärk)	125a	24°37'N	82°54'W
Duala, Camrn. *see* Douala	260-61	4°03'N	9°42'E
Duarte, Pico, mtn., Dom. Rep. (pēcô dŭŭ′ärtĕh)	142-43	19°02'N	70°59'W
Dubai, U.A.E.	230-31	25°16'N	55°19'E
Dubawnt, stm., Can. (do͞o-bônt′)	130-31	64°31'N	100°05'W
Dubawnt Lake, lk., Can. (do͞o-bônt′ lāk)	130-31	63°08'N	101°30'W
Dubayy, U.A.E. *see* Dubai	230-31	25°16'N	55°19'E
Dubbo, Austl. (dŭb′ō)	276	32°15'S	148°36'E
Dublin, Ga., U.S. (dŭb′lĭn)	124-25	32°32'N	82°54'W
Dublin, Oh., U.S. (dŭb′lĭn)	116-17	40°06'N	83°07'W
Dublin, Tx., U.S. (dŭb′lĭn)	122-23	32°05'N	98°20'W
Dublin, nat. cap., Ire. (dŭb′lĭn)	190-91	53°21'N	6°15'W
Dubno, Ukr. (do͞o′b-nô)	194-95	50°24'N	25°45'E
Du Bois, Pa., U.S. (dȯ-bois′)	116-17	41°07'N	78°46'W
Dubovka, Russia (dȯ-bôf′kȧ)	186-87	49°04'N	44°49'E
Dubrovka, Russia (do͞o-brôf′kȧ)	202-03	53°42'N	33°31'E
Dubrovnik, Cro. (dȯ′brôv-nĕk)	200-01	42°39'N	18°06'E
Dubuque, Ia., U.S. (dȯ-būk′)	114-15	42°30'N	90°41'W
Duchesne, Ut., U.S. (dȯ-shān′)	118-19	40°10'N	110°24'W
Duck Lake, Sk., Can. (dŭk lāk)	134-35	52°49'N	106°14'W
Duck Valley Indian Reservation, ind. res., U.S. (dŭk văl′ē ĭn′dĭ-ăn rĕ-sēr-vā′shĕn)	112-13	42°00'N	116°10'W
Dudinka, Russia (do͞o-dĭn′kȧ)	218-19	69°24'N	86°11'E
Dudley, Eng., U.K. (dŭd′lĭ)	190-91	52°30'N	2°05'W
Duero, stm., Eur.	198-99	41°09'N	8°41'W
Dufourspitze, mtn., Eur.	194-95	45°55'N	7°52'E
Dugi Otok, i., Cro. (do͞o′gē o′tôk)	200-01	43°59'N	15°04'E
Duisburg, Ger. (do͞o′ĭs-bȯrgh)	194-95	51°26'N	6°47'E
Duitama, Col.	164-65	5°50'N	73°02'W
Dukhān, Qatar	230-31	25°25'N	50°48'E
Dukhovshchina, Russia (do͞o-κôfsh-′chēnȧ)	202-03	55°12'N	32°25'E
Dulan, China	240-41	36°10'N	98°22'E
Dulce, Golfo, b., C.R. (gōl′fô do͞ol′sā)	149	8°32'N	83°14'W
Duluth, Mn., U.S. (dȯ-lo͞oth′)	114-15	46°46'N	92°09'W
Dumaguete, Phil.	250	9°18'N	123°18'E
Dumai, Indon.	246-47	1°40'N	101°27'E
Dumali Point, c., Phil. (do͞o-mä′lē point)	250	13°07'N	121°33'E
Dumaran Island, i., Phil.	250	10°33'N	119°51'E
Dumaring, Indon.	248-49	1°32'N	118°13'E
Dumfries, Scot., U.K. (dŭm-frēs′)	190-91	55°04'N	3°37'W
Duna, stm., Eur. *see* Danube	188-89	45°23'N	29°36'E
Dünaburg, Lat. *see* Daugavpils	192-93	55°53'N	26°32'E
Dunai, stm., Eur. *see* Danube	188-89	45°23'N	29°36'E
Dunaj, stm., Eur. *see* Danube	188-89	45°23'N	29°36'E
Dunărea, stm., Eur. *see* Danube	188-89	45°23'N	29°36'E
Dunav, stm., Eur. *see* Danube	188-89	45°23'N	29°36'E
Duncan, B.C., Can. (dŭŋ′kăn)	132-33	48°47'N	123°42'W
Duncan, Ok., U.S. (dŭŋ′kăn)	120-21	34°30'N	97°57'W
Duncan, stm., B.C., Can. (dŭŋ′kăn)	132-33	50°11'N	116°57'W
Duncansby Head, c., Scot., U.K. (dŭn′kănz-bĭ hĕd)	190-91	58°39'N	3°02'W
Dundalk, Ire. (dŭn′kôk)	190-91	54°01'N	6°24'W
Dundas, Lake, lk., Austl. (lāk dŭn-dȧs)	272-73	32°35'S	121°50'E
Dundas Island, i., B.C., Can. (dŭn-dăs′ ī′lȧnd)	132-33	54°33'N	130°55'W
Dún Dealgan, Ire. *see* Dundalk	190-91	54°01'N	6°24'W
Dundee, S. Afr.	269c	28°09'S	30°15'E
Dundee, Scot., U.K.	190-91	56°29'N	2°59'W
Dund-Us, Mong.	240-41	47°60'N	91°38'E
Dunedin, N.Z.	278	45°52'S	170°29'E
Dunfermline, Scot., U.K. (dŭn-fĕrm′lĭn)	190-91	56°04'N	3°29'W
Dūngarpur, India	234-35	23°50'N	73°43'E
Dungarvan, Ire. (dŭn-gȧr′văn)	190-91	52°06'N	7°38'W
Dungu, D.R.C.	267	3°37'N	28°34'E
Dunhua, China (dȯn-hwä)	244	43°22'N	128°14'E
Dunhuang, China	240-41	40°08'N	94°40'E
Dunkerque, Fr. (dŭN-kĕrk′)	196-97	51°03'N	2°23'E
Dunkirk, Fr. *see* Dunkerque	196-97	51°03'N	2°23'E
Dunkirk, In., U.S. (dŭn′kûrk)	116-17	40°22'N	85°12'W

Feature (Pronunciation)	Page	Lat.	Long.
Dunkirk, N.Y., U.S. (dŭn'kûrk)	116-17	42°29'N	79°19'W
Dún Laoghaire, Ire. (dŭn-lā'rĕ)	190-91	53°17'N	6°08'W
Dunlap, Ia., U.S. (dŭn'lăp)	114-15	41°51'N	95°36'W
Dunlap, Tn., U.S. (dŭn'lăp)	124-25	35°22'N	85°23'W
Dunleary, Ire. see Dún Laoghaire	190-91	53°17'N	6°08'W
Dunmore, Pa., U.S. (dŭn'mōr)	116-17	41°27'N	75°38'W
Dunn, N.C., U.S. (dŭn)	124-25	35°18'N	78°37'W
Dunqulah, Sudan	266	19°11'N	30°28'E
Dunsmuir, Ca., U.S. (dŭnz'mūr)	112-13	41°13'N	122°17'W
Duolun, China (dwô-lo͞on)	240-41	42°12'N	116°28'E
Duomula, China	234-35	34°07'N	82°30'E
Duque de York, Isla, i., Chile	171	50°40'S	75°20'W
Duquesne, Pa., U.S. (dö-kān')	116-17	40°22'N	79°51'W
Du Quoin, Il., U.S. (dö-kwoin')	116-17	38°01'N	89°15'W
Durand, Mi., U.S. (dů-rănd')	116-17	42°55'N	83°59'W
Durand, Wi., U.S. (dů-rănd')	114-15	44°37'N	91°57'W
Durango, Mex. (do͞o-rä'n-gô)	146-47	24°02'N	104°41'W
Durango, Co., U.S. (dö-răŋ'gō)	118-19	37°17'N	107°52'W
Durango, state, Mex. (do͞o-rä'n-gô)	122-23	24°50'N	104°50'W
Durant, Ms., U.S. (dů-rănt')	124-25	33°05'N	89°52'W
Durant, Ok., U.S. (dů-rănt')	120-21	33°60'N	96°23'W
Durazno, Ur. (do͞o-räz'nō)	173	33°25'S	56°30'W
Durazzo, Alb. see Durrës	200-01	41°19'N	19°27'E
Durban, S. Afr. (dûr'băn)	264-65	29°55'S	30°56'E
Durbe, Lat. (do͞or'bĕ)	192-93	56°35'N	21°21'E
Düren, Ger. (dü'rĕn)	194-95	50°48'N	6°29'E
Durg, India	234-35	21°11'N	81°17'E
Durham, Eng., U.K. (dûr'ăm)	190-91	54°47'N	1°34'W
Durham, N.C., U.S.	124-25	35°60'N	78°54'W
Durmitor, mtn., Mont.	200-01	43°08'N	19°01'E
Durrës, Alb. (dör'ĕs)	200-01	41°19'N	19°27'E
Durrësi, Alb. see Durrës	200-01	41°19'N	19°27'E
D'Urville, Tanjung, c., Indon.	277	1°28'S	137°54'E
D'Urville Island, i., N.Z.	278	40°50'S	173°52'E
Dushan, China (do͞o-shän)	238-39	25°50'N	107°32'E
Dushanbe, nat. cap., Taj. (dū-shän'-bà) (dū-shän-bă')	232-33	38°34'N	68°47'E
Düsseldorf, Ger. (düs'ĕl-dôrf)	194-95	51°14'N	6°48'E
Duyfken Point, c., Austl.	277	12°34'S	141°38'E
Duyun, China (do͞o-yòn)	238-39	26°16'N	107°30'E
Dwārka, India	234-35	22°14'N	68°59'E
Dwight, Il., U.S. (dwīt)	116-17	41°05'N	88°25'W
Dyat'kovo, Russia (dyät'kō-vō)	202-03	53°35'N	34°21'E
Dyersburg, Tn., U.S. (dī'ērz-bûrg)	124-25	36°02'N	89°23'W
Dyersville, Ia., U.S. (dī'ērz-vĭl)	114-15	42°29'N	91°07'W
Dzavhan, stm., Mong.	240-41	48°53'N	93°26'E
Dzerzhinsk, Russia	186-87	56°15'N	43°24'E
Dzhambul, Kaz. see Taraz	226	42°54'N	71°21'E
Dzhankoi, Ukr.	202-03	45°43'N	34°23'E
Dzharylhach, ostriv, i., Ukr.	202-03	46°02'N	32°55'E
Dzhebariki-Khaya, Russia	218-19	62°11'N	135°46'E
Dzhezkazgan, Kaz. see Zhezqazghan	226	47°47'N	67°41'E
Dzhugdzhur, Khrebet, mts., Russia	218-19	58°0'N	136°00'E
Dzhugdzhur Range, mts., Russia (jo͝og-jo͝or' rānj) see Dzhugdzhur, Khrebet	218-19	58°0'N	136°00'E
Dzhungarian Alatau Mountains, mts., Asia see Alataw Shan	226	45°0'N	81°00'E
Dzibalchén, Mex. (zē-bäl-chĕ'n)	148	19°28'N	89°44'W
Dzilam González, Mex. (zē-lä'm-gôn-zä'lĕz)	148	21°17'N	88°56'W
Dzitás, Mex. (zē-tà's)	148	20°51'N	88°31'W
Dzuunmod, Mong.	240-41	47°43'N	106°57'E

E

Feature (Pronunciation)	Page	Lat.	Long.
Eagle, Id., U.S. (ē'gl)	112-13	43°41'N	116°19'W
Eagle Grove, Ia., U.S. (ē'gl grōv)	114-15	42°40'N	93°54'W
Eagle Lake, Me., U.S. (ē'gl lāk)	117a	47°02'N	68°36'W
Eagle Lake, Tx., U.S. (ē'gl lāk)	122-23	29°35'N	96°20'W
Eagle Pass, Tx., U.S. (ē'gl pås)	122-23	28°42'N	100°29'W
Earle, Ar., U.S. (ûrl)	124-25	35°16'N	90°28'W
Earlington, Ky., U.S. (ûr'lĭng-tŭn)	124-25	37°16'N	87°31'W
Easley, S.C., U.S. (ēz'lĭ)	124-25	34°50'N	82°35'W
East Angus, Qc., Can. (ēst ăŋ'gŭs)	136-37	45°29'N	71°40'W
Eastbourne, Eng., U.K. (ēst'bôrn)	190-91	50°46'N	0°17'E
East Caicos, i., T./C. Is. (ēst kī'kōs)	142-43	21°42'N	71°29'W
East Cape, c., N.Z.	278	37°41'S	178°33'E
East Chicago, In., U.S. (ēst shĭ-kô'gō)	116-17	41°39'N	87°26'W
East China Sea, s., Asia (ēst chī'nà sē)	222-23	30°0'N	126°00'E
East Detroit, Mi., U.S. (ēst dē-troit') see Eastpointe	116-17	42°28'N	82°57'W
Easter Island, i., Chile (ēs'tēr ī'lànd) see Pascua, Isla de	282-83	27°07'S	109°22'W
Eastern Desert, des., Egypt see Arabian Desert	266	28°0'N	32°00'E
Eastern Ghāts, mts., India (ē'stērn ghäts) (ē'stērn ghôts)	236	14°0'N	78°50'E
East Falkland, i., Falk. Is.	171	51°53'S	59°11'W
East Grand Forks, Mn., U.S. (ēst grănd fôrks)	114-15	47°55'N	97°00'W
East Helena, Mt., U.S. (ēst hĕ-hē'nà)	112-13	46°35'N	111°55'W
East Jordan, Mi., U.S. (ēst jôr'dăn)	116-17	45°09'N	85°07'W
Eastland, Tx., U.S. (ēst'lănd)	120-21	32°24'N	98°49'W
East Lansing, Mi., U.S. (ēst lăn'sĭng)	116-17	42°44'N	84°29'W
East Liverpool, Oh., U.S. (ēst lĭv'ēr-po͞ol)	116-17	40°38'N	80°35'W
East London, S. Afr. (ēst ŭn'dŭn)	264-65	32°60'S	27°54'E
Eastmain, Qc., Can.	128-29	52°13'N	78°33'W
Eastmain, stm., Qc., Can. (ēst'mān)	130-31	52°15'N	78°35'W
Eastmain-Opinaca, Réservoir, res., Qc., Can.	130-31	52°23'N	76°35'W
Eastman, Ga., U.S. (ēst'măn)	124-25	32°12'N	83°10'W
East Moline, Il., U.S. (ēst mô-lēn')	114-15	41°32'N	90°25'W
East Nishnabotna, stm., Ia., U.S. (ēst nĭsh-nà-bŏt'nà)	120-21	40°39'N	95°38'W
East Orange, N.J., U.S. (ēst ŏr'ĕnj)	116-17	40°46'N	74°12'W
East Pakistan, nation, Asia see Bangladesh	206-07	24°0'N	90°00'E
East Peoria, Il., U.S. (ēst pē-ō'rĭ-à)	116-17	40°40'N	89°35'W
Eastpointe, Mi., U.S.	116-17	42°28'N	82°57'W
Eastport, Me., U.S. (ēst'pōrt)	117a	44°54'N	66°60'W
East Providence, R.I., U.S. (ēst prŏv'ĭ-dĕns)	116-17	41°49'N	71°23'W
East Saint Louis, Il., U.S.	120-21	38°37'N	90°09'W
East Sea, s., Asia see Japan, Sea of	222-23	40°0'N	135°00'E
East Siberian Sea, s., Russia (ēst sī-bĭr'y'n sē)	218-19	74°0'N	166°00'E
East Stroudsburg, Pa., U.S. (ēst stroudz'bûrg)	116-17	41°00'N	75°11'W
East Tawas, Mi., U.S. (ēst tô'wăs)	116-17	44°17'N	83°28'W
East Timor, nation, Asia (ēst tē-mōr')	248-49	8°35'S	126°00'E
Eaton, Oh., U.S. (ē'tŭn)	116-17	39°45'N	84°38'W
Eaton Rapids, Mi., U.S. (ē'tŭn răp'ĭdz)	116-17	42°30'N	84°39'W
Eatonton, Ga., U.S. (ē'tŭn-tŭn)	124-25	33°20'N	83°23'W
Eau Claire, Wi., U.S. (ō klâr')	114-15	44°49'N	91°30'W
Eau Claire, Lac à l', lk., Qc., Can.	130-31	56°11'N	74°26'W
Eauripik, at., Micron.	280-81	6°41'N	143°03'E
Ebinur Hu, lk., China	226	44°56'N	82°52'E
Eboli, Italy (ĕb'ô-lē)	200-01	40°38'N	15°04'E
Ebolowa, Camrn.	260-61	2°55'N	11°09'E
Echoing, stm., Can. (ĕk'ō-ĭng)	134-35	55°51'N	92°04'W
Echuca, Austl. (ê-chó'kà)	276	36°08'S	144°45'E
Écija, Spain (ā'thē-hä)	198-99	37°32'N	5°05'W
Ecuador, nation, S.A. (ĕk'wà-dôr)	158	2°0'S	77°30'W
Eddyville, Ia., U.S. (ĕd'ĭ-vĭl)	114-15	41°09'N	92°38'W
Eddyville, Ky., U.S. (ĕd'ĭ-vĭl)	124-25	37°06'N	88°05'W
Ede, Nig.	260a	7°44'N	4°25'E
Edéa, Camrn. (ĕ-dä'ä)	260-61	3°48'N	10°08'E
Eden, Austl.	276	37°04'S	149°54'E
Eden, N.C., U.S. (ē'dĕn)	124-25	36°29'N	79°46'W
Eden, Tx., U.S. (ē'dĕn)	122-23	31°13'N	99°51'W
Eden Prairie, Mn., U.S. (ē'dĕn prâr'ĭ)	114-15	44°51'N	93°28'W
Edenton, N.C., U.S. (ē'dĕn-tŭn)	124-25	36°04'N	76°36'W
Edenville, S. Afr. (ē'd'n-vĭl)	269c	27°34'S	27°41'E
Edfu, Egypt	268b	24°58'N	32°52'E
Edgefield, S.C., U.S. (ĕj'fēld)	124-25	33°47'N	81°56'W
Edgeley, N.D., U.S. (ĕj'lĭ)	114-15	46°21'N	98°43'W
Edgemont, S.D., U.S. (ĕj'mŏnt)	114-15	43°18'N	103°48'W
Edgerton, Wi., U.S. (ĕj'ēr-tŭn)	116-17	42°50'N	89°04'W
Edinburg, Tx., U.S. (ĕd''n-bûrg)	122-23	26°18'N	98°10'W
Edinburgh, Scot., U.K. (ĕd''n-bŭr-ô)	190-91	55°57'N	3°13'W
Edirne, Tur.	200-01	41°41'N	26°34'E
Edmond, Ok., U.S. (ĕd'mŭnd)	120-21	35°39'N	97°29'W
Edmonton, Ab., Can. (ĕd'mŭn-tŭn)	132-33	53°33'N	113°30'W
Edmundston, N.B., Can. (ĕd'mŭn-stŭn)	138-39	47°22'N	68°19'W
Edna, Tx., U.S. (ĕd'nà)	122-23	28°59'N	96°39'W
Édouard, Lac, lk., Afr. see Edward, Lake	267	0°23'S	29°36'E
Edremit, Tur. (ĕd-rĕ-mēt')	200-01	39°36'N	27°01'E
Edson, Ab., Can. (ĕd'sŭn)	132-33	53°35'N	116°26'W
Eduardo Castex, Arg.	173	35°55'S	64°18'W
Edward, Lake, lk., Afr. (lāk ĕd'wērd)	267	0°23'S	29°36'E
Edwards Plateau, plat., Tx., U.S.	122-23	30°46'N	100°47'W
Edwardsville, Il., U.S. (ĕd'wērdz-vĭl)	120-21	38°48'N	89°57'W
Edward VII Peninsula, pen., Ant.	287	77°40'S	154°60'W
Eel, stm., Ca., U.S. (ēl)	110-11	40°38'N	124°20'W
Eesti, nation, Eur. see Estonia	174-75	59°0'N	26°00'E
Éfaté, i., Vanuatu (ā-fä'tā)	279g	17°40'S	168°25'E
Effigy Mounds National Monument, n.p., Ia., U.S. (ĕf'ĭ-jŭ mounds năsh'ŭn-ăl mŏn'ŭ-mĕnt)	114-15	43°06'N	91°13'W
Effingham, Il., U.S. (ĕf'ĭng-hăm)	116-17	39°07'N	88°33'W
Eg, stm., Mong.	240-41	49°23'N	103°38'E
Egadi, Isole, is., Italy (ĕ'sō-lĕ-ĕ'gä-dē)	200-01	37°58'N	12°16'E
Ege Denizi, s., see Aegean Sea	200-01	38°30'N	25°00'E
Eger, Czech Rep. see Cheb	194-95	50°05'N	12°22'E
Eger, Hung. (ĕ gĕr)	194-95	47°54'N	20°23'E
Egersund, Nor. (ĕ'ghĕr-sòn')	192-93	58°27'N	6°00'E
Egmont, Cape, c., N.Z. (kāp ĕg'mŏnt)	278	39°17'S	173°45'E
Eğridir Gölü, lk., Tur.	186-87	38°02'N	30°53'E
Egum Atoll, at., Pap. N. Gui.	277	9°25'S	151°55'E
Egvekinot, Russia	218-19	66°18'N	179°11'W
Egypt, nation, Afr. (ē'jĭpt)	253	27°0'N	30°00'E
Eichstätt, Ger. (īk'shtät)	194-95	48°54'N	11°11'E
Eidfjord, Nor. (ĕīd'fyôr)	192-93	60°28'N	7°05'E
Eidsvoll, Nor. (īdhs'vôl)	192-93	60°19'N	11°14'E
Eifel, mts., Ger. (ī'fĕl)	194-95	50°14'N	6°42'E
Eight Degree Channel, strt., Asia	236	8°0'N	73°00'E
Einbeck, Ger. (īn'bĕk)	194-95	51°49'N	9°52'E
Eindhoven, Neth. (īnd'hō-vĕn)	190-91	51°26'N	5°29'E
Éire, nation, Eur. see Ireland	174-75	53°0'N	8°00'W
Eirunepé, Braz.	166-67	6°39'S	69°52'W
Eisenach, Ger. (ī'zĕn-äκ)	194-95	50°59'N	10°19'E
Eivissa, Spain	198-99	38°55'N	1°25'E
Eivissa, i., Spain	198-99	39°0'N	1°25'E
Ejin Qi, China	240-41	41°52'N	100°56'E
Ejmiatsin, Arm.	227	40°10'N	44°18'E
Ejutla de Crespo, Mex. (å-hòt'lä dā krās'pō)	146-47	16°34'N	96°44'W
Ekenäs, Fin. (ĕ'kĕ-nâs)	192-93	59°58'N	23°26'E
Ekibastuz, Kaz.	226	51°43'N	75°20'E
Ekimchan, Russia	218-19	53°04'N	132°57'E
Eko, Nig. see Lagos	260a	6°27'N	3°24'E
Ekwan, stm., On., Can.	130-31	53°12'N	82°14'W
El Aaiún, nat. cap., W. Sah. see Laayoune	258-59	27°10'N	13°12'W
El Abiodh Sidi Cheikh, Alg.	188-89	32°53'N	0°32'E
El Affroun, Alg. (ĕl ȧf-froun')	269b	36°28'N	2°37'E
El-Agheila, Libya see Al-'Uqaylah	188-89	30°15'N	19°12'E
El-Alamein, Egypt	188-89	30°49'N	28°58'E
Elands, stm., S. Afr. (ĕlånds)	269c	25°17'S	27°32'E
El-Arish, Egypt	268b	31°08'N	33°50'E
El Asnam, Alg. see Chlef	198-99	36°10'N	1°20'E
Elat, Isr.	228-29	29°33'N	34°57'E
Elat, Gulf of, b., see Aqaba, Gulf of	228-29	29°05'N	34°44'E
Elath, Isr. see Elat	228-29	29°33'N	34°57'E
Elazığ, Tur. (ĕl-ä'zĕz)	186-87	38°41'N	39°15'E
Elba, Al., U.S. (ĕl'bà)	124-25	31°25'N	86°04'W
Elba, Isola d', i., Italy (ê-sō lä-d-ĕl'bà)	200-01	42°46'N	10°17'E
El Banco, Col. (ĕl băn'cô)	164-65	9°01'N	73°58'W
Elbe, stm., Eur. (ĕl'bĕ)	194-95	53°53'N	9°01'E
Elbert, Mount, mtn., Co., U.S. (mount ĕl'bērt)	118-19	39°07'N	106°27'W
Elberton, Ga., U.S. (ĕl'bēr-tŭn)	124-25	34°07'N	82°51'W
Elbeuf, Fr. (ĕl-bûf')	196-97	49°17'N	1°01'E
Elbing, Pol. see Elbląg	194-95	54°10'N	19°24'E
Elbistan, Tur. (ĕl-bē-stän')	186-87	38°13'N	37°12'E
Elbląg, Pol. (ĕl'blăng)	194-95	54°10'N	19°24'E
El Bonillo, Spain (ĕl bō-nēl'yō)	198-99	38°57'N	2°33'W
Elbow Lake, Mn., U.S. (ĕl'bō lāk)	114-15	45°60'N	95°59'W
El'brus, Gora, mtn., Russia (gà-rä'ĕl'brŏs')	227	43°21'N	42°26'E
Elburz Mountains, mts., Iran (ĕl'bŏrz' moun'tĭnz)	232-33	36°0'N	53°00'E
El Cajon, Ca., U.S.	118-19	32°48'N	116°57'W
El Calafate, Arg.	171	50°21'S	72°17'W
El Campo, Tx., U.S. (ĕl-kăm'pō)	122-23	29°11'N	96°16'W
El Carmen de Bolívar, Col.	164-65	9°43'N	75°07'W
El Centro, Ca., U.S. (ĕl-sĕn'trô)	118-19	32°48'N	115°33'W
Elche, Spain see Elx	198-99	38°15'N	0°42'W
Elda, Spain (ĕl'dä)	198-99	38°29'N	0°47'W
El'dikan, Russia	218-19	60°46'N	135°09'E
El Djazaïr, nation, Afr. see Algeria	253	28°0'N	3°00'E
El Djazaïr, nat. cap., Alg. see Algiers	269b	36°46'N	3°03'E
Eldon, Mo., U.S. (ĕl-dŭn)	120-21	38°21'N	92°35'W
Eldora, Ia., U.S. (ĕl-dō'rà)	114-15	42°22'N	93°06'W
Eldorado, Arg.	173	26°24'S	54°38'W
El Dorado, Ar., U.S. (ĕl dô-rä'dō)	120-21	33°12'N	92°41'W
El Dorado, Ks., U.S. (ĕl dô-rä'dō)	120-21	37°49'N	96°51'W
Eldoret, Kenya (ĕl-dô-rĕt')	267	0°31'N	35°17'E
Electra, Tx., U.S. (ê-lĕk'trà)	120-21	34°03'N	98°55'W
Elek, stm., Asia	186-87	51°30'N	53°20'E
Elektrostal', Russia (ĕl-yĕk'trō-stàl)	202-03	55°47'N	38°28'E

ăt; finăl; rāte; senåte; ärm; åsk; sofà; fâre; ch-choose; dh-as th in other; bē; ēvent; bĕt; recĕnt; cratēr; g-gō; gh-guttural g; bĭt; ĭ-short neutral; rīde; κ-guttural k as ch in German ich;

Feature (Pronunciation)	Page	Lat.	Long.
El Encanto, Col.	164-65	1°42′S	73°14′W
Elephant Butte Reservoir, res., N.M., U.S. (ĕl′ĕ-fănt būt rĕ′sĕr-vwär)	118-19	33°17′N	107°10′W
Elets, Russia	202-03	52°37′N	38°30′E
Eleuthera, i., Bah. (ē-lū′thēr-*à*)	142-43	25°11′N	76°13′W
El-Fayoum, Egypt	268b	29°19′N	30°50′E
El Ferrol del Caudillo, Spain *see* Ferrol	198-99	43°29′N	8°14′W
El Galpón, Arg.	168-69	25°24′S	64°38′W
Elgin, Scot., U.K. (ĕl′jĭn)	190-91	57°39′N	3°19′W
Elgin, Il., U.S. (ĕl′jĭn)	116-17	42°02′N	88°17′W
Elgin, Or., U.S. (ĕl′jĭn)	112-13	45°34′N	117°55′W
Elgin, Tx., U.S. (ĕl′jĭn)	122-23	30°21′N	97°22′W
El-Gîza, Egypt *see* Giza	268b	30°01′N	31°13′E
El Golea, Alg.	188-89	30°33′N	2°54′E
Elgon, Mount, mtn., Afr. (mount ĕl′gŏn)	267	1°08′N	34°33′E
El Guapo, Ven. (ĕl-gwá′pô)	163b	10°09′N	65°58′W
El Ḩank, clf., Afr.	258-59	24°23′N	6°36′W
El Hierro, i., Spain	199d	27°45′N	18°00′W
Elila, stm., D.R.C. (ĕ-lē′lȧ)	262-63	2°44′S	25°53′E
Élisabethville, D.R.C. *see* Lubumbashi	262-63	11°41′S	27°28′E
Elisenvaara, Russia (ā-lē′sĕn-vä′rȧ)	192-93	61°24′N	29°46′E
El-Iskandarîya, Egypt *see* Alexandria	268b	31°11′N	29°54′E
Elista, Russia	186-87	46°19′N	44°16′E
Elizabeth, Austl.	276	34°43′S	138°40′E
Elizabeth City, N.C., U.S. (ê-lĭz′*à*-bĕth sĭ′tĭ)	124-25	36°18′N	76°13′W
Elizabethton, Tn., U.S. (ê-lĭz-à-bĕth′tŭn)	124-25	36°21′N	82°14′W
Elizabethtown, Ky., U.S. (ê-lĭz′*ă*-bĕth-toun)	116-17	37°42′N	85°52′W
Elizabethtown, Pa., U.S. (ê-lĭz′*ă*-bĕth-toun)	116-17	40°09′N	76°37′W
El-Jadida, Mor.	269a	33°15′N	8°31′W
Elk, stm., B.C., Can. (ĕlk)	132-33	49°10′N	115°14′W
Elk City, Ok., U.S. (ĕlk sĭ′tê)	120-21	35°25′N	99°25′W
El Kef, Tun. (ĕl-kĕf′)	184-85	36°11′N	8°43′E
El-Khârga, Egypt	266	25°27′N	30°33′E
Elkhart, In., U.S. (ĕlk′härt)	116-17	41°42′N	85°57′W
Elkhorn, Wi., U.S. (ĕlk′hôrn)	116-17	42°40′N	88°33′W
Elkin, N.C., U.S. (ĕl′kĭn)	124-25	36°15′N	80°51′W
Elk Island National Park, n.p., Ab., Can. (ĕlk ī′l*ă*nd năsh′ŭn-*ă*l pärk)	132-33	53°36′N	112°54′W
Elko, Nv., U.S. (ĕl′kō)	112-13	40°50′N	115°46′W
Elk Point, S.D., U.S. (ĕlk point)	114-15	42°41′N	96°41′W
Elk Rapids, Mi., U.S. (ĕlk răp′ĭdz)	116-17	44°53′N	85°24′W
Elk River, Mn., U.S. (ĕlk rĭv′ĕr)	114-15	45°18′N	93°35′W
Elkton, Ky., U.S. (ĕlk′tŭn)	124-25	36°49′N	87°09′W
Elkton, S.D., U.S. (ĕlk′tŭn)	114-15	44°14′N	96°29′W
Ellás, nation, Eur. *see* Greece	174-75	39°0′N	22°00′E
Ellendale, N.D., U.S. (ĕl′ĕn-dāl)	114-15	46°00′N	98°32′W
Ellensburg, Wa., U.S. (ĕl′ĕnz-bûrg)	112-13	46°60′N	120°32′W
Ellesmere Island, i., Nu., Can. (ĕlz′mēr ī′l*ă*nd)	86	81°0′N	80°00′W
Ellice Islands, nation, Oc. *see* Tuvalu	280-81	8°0′S	178°00′E
Elliot Lake, On., Can.	136-37	46°23′N	82°39′W
Elliston, Austl.	270-71	33°39′S	134°54′E
Ellisville, Ms., U.S. (ĕl′ĭs-vĭl)	124-25	31°36′N	89°12′W
Ellore, India *see* Elūru	236	16°43′N	81°07′E
Ellsworth, Ks., U.S. (ĕlz′wûrth)	120-21	38°44′N	98°13′W
Elma, Wa., U.S. (ĕl′m*à*)	112-13	47°00′N	123°24′W
El-Mahalla el-Kubra, Egypt	268b	30°58′N	31°10′E
El-Mansûra, Egypt	268b	31°02′N	31°23′E
Elmhurst, Il., U.S. (ĕlm′hûrst)	116-17	41°53′N	87°57′W
El-Minya, Egypt	268b	28°06′N	30°45′E
Elmira, N.Y., U.S.	116-17	42°05′N	76°48′W
Elmira Heights, N.Y., U.S. (ĕl-mī′r*à* hīts)	116-17	42°09′N	76°50′W
El Nevado, Cerro, mtn., Arg. *see* Nevado, Cerro	171	35°34′S	68°28′W
El-Obeid, Sudan *see* Al-Ubayyiḍ	266	13°11′N	30°13′E
El Oued, Alg.	188-89	33°21′N	6°53′E
El Pao, Ven. (ĕl pá′ō)	163b	9°38′N	68°08′W
El Paso, Tx., U.S. (ĕl-pas′ō)	118-19	31°48′N	106°27′W
El Paso de Robles, Ca., U.S. *see* Paso Robles	118-19	35°38′N	120°41′W
El Pital, Cerro, mtn., N.A.	148	14°23′N	89°08′W
El Porvenir, Pan. (ĕl-pôr-vä-nēr′)	150	9°33′N	78°59′W
El-Qâhira, nat. cap., Egypt *see* Cairo	268b	30°03′N	31°14′E
El Reno, Ok., U.S. (ĕl-rē′nō)	120-21	35°32′N	97°57′W
El Salto, Mex. (ĕl-säl′tō)	146-47	23°46′N	105°21′W
El Salvador, nation, N.A. (ĕl säl′vä-dôr)	85	13°50′N	88°55′W
Elsaß, hist. reg., Fr. *see* Alsace	196-97	48°30′N	7°30′E
Elsberry, Mo., U.S. (ĕlz′bĕr-ĭ)	120-21	39°10′N	90°48′W
Elsinore, Den. *see* Helsingør	192-93	56°02′N	12°37′E
El-Suweis, Egypt *see* Suez	268b	29°58′N	32°33′E
El Tajín, hist., Mex.	146-47	20°27′N	97°23′W
El Tigre, Ven. (ĕl-tē′grĕ)	163b	8°54′N	64°15′W
El-Uqsor, Egypt *see* Luxor	268b	25°42′N	32°39′E
Elūru, India	236	16°43′N	81°07′E
Elvas, Port. (ĕl′väzh)	198-99	38°53′N	7°10′W
Elverum, Nor. (ĕl′vĕ-ròm)	192-93	60°53′N	11°34′E
El Viejo, Nic. (ĕl-vyĕ′κō)	149	12°40′N	87°10′W
Elwood, In., U.S. (ĕ′wòd)	116-17	40°16′N	85°50′W
Elx, Spain	198-99	38°15′N	0°42′W
Ely, Mn., U.S. (ē′lĭ)	114-15	47°54′N	91°52′W
Ely, Nv., U.S.	118-19	39°15′N	114°53′W
Elyria, Oh., U.S. (ê-lĭr′ĭ-*à*)	116-17	41°22′N	82°06′W
Émaé, i., Vanuatu	279g	17°04′S	168°22′E
Emāmshahr, Iran *see* Shāhrūd	232-33	36°25′N	54°58′E
Embi, Kaz.	226	48°50′N	58°09′E
Embira, stm., Braz. *see* Envira	166-67	6°42′S	69°48′W
Embu, Kenya	267	0°32′S	37°27′E
Emden, Ger. (ĕm′dĕn)	194-95	53°22′N	7°12′E
Emerald, Austl.	277	23°31′S	148°10′E
Emiliano Zapata, Mex. (ĕ-mē-lyá′nô-zä-pá′tȧ)	148	17°45′N	91°46′W
Emiliano Zapata, Mex. (ĕ-mē-lyá′nô-zä-pá′tȧ)	146-47	18°51′N	99°11′W
Eminence, Ky., U.S. (ĕm′*ĭ*-nĕns)	116-17	38°22′N	85°11′W
Emmen, Neth. (ĕm′*ĕ*n)	190-91	52°47′N	6°54′E
Emmetsburg, Ia., U.S. (ĕm′ĕts-bûrg)	114-15	43°07′N	94°41′W
Emmett, Id., U.S. (ĕm′ĕt)	112-13	43°53′N	116°30′W
Emory Peak, mtn., Tx., U.S. (ĕ′mō-rē pēk)	122-23	29°14′N	103°19′W
Empalme, Mex.	144-45	27°58′N	110°49′W
Empedrado, Arg.	173	27°57′S	58°48′W
Empoli, Italy (ām′pô-lē)	200-01	43°43′N	10°57′E
Emporia, Ks., U.S. (ĕm-pō′rĭ-*à*)	120-21	38°24′N	96°11′W
Emporia, Va., U.S. (ĕm-pō′rĭ-*à*)	124-25	36°42′N	77°33′W
Emporium, Pa., U.S. (ĕm-pō′rĭ-*ŭ*m)	116-17	41°30′N	78°15′W
Empty Quarter, des., Asia *see* Rub'al-Khali	220-21	20°0′N	51°00′E
En, stm., Eur. *see* Inn	194-95	48°34′N	13°28′E
Encarnación, Para. (ĕn-kär-nä-syōn′)	173	27°20′S	55°52′W
Encinal, Tx., U.S. (ĕn′sĭ-nôl)	122-23	28°02′N	99°21′W
Encontrados, Ven. (ĕn-kōn-trä′dōs)	164-65	9°03′N	72°14′W
Encounter Bay, b., Austl. (ĕn-koun′tēr bā)	276	35°35′S	138°44′E
Ende, Indon.	248-49	8°50′S	121°40′E
Enderbury, at., Kir. (ĕn′dĕr-bûrĭ)	280-81	3°08′S	171°05′W
Enderby Land, reg., Ant. (ĕn′dĕr-bĭ lănd)	287	68°05′S	52°53′E
Enderlin, N.D., U.S. (ĕn′dĕr-lĭn)	114-15	46°37′N	97°36′W
Endicott, N.Y., U.S. (ĕn′dĭ-k*ŏ*t)	116-17	42°07′N	76°04′W
Ene, stm., Peru	170	11°10′S	74°15′W
Enewetak, at., Marsh. Is.	280-81	11°39′N	162°17′E
Enfield, N.C., U.S. (ĕn′fēld)	124-25	36°11′N	77°40′W
Engaño, Cabo, c., Dom. Rep. (ká′-bô- ĕn-gä-nô)	142-43	18°37′N	68°20′W
Engel's, Russia (ĕn′gĕls)	186-87	51°29′N	46°08′E
Enggano, Pulau, i., Indon. (pōō-lou ĕng-gä′nō)	246-47	5°24′S	102°16′E
England, Ar., U.S. (ĭŋ′gl*ă*nd)	124-25	34°33′N	91°58′W
England, state, U.K. (ĭŋ′gl*ă*nd)	190-91	52°30′N	1°30′W
Englehart, On., Can.	136-37	47°49′N	79°52′W
Englewood, Co., U.S. (ĕn′g′l-wòd)	120-21	39°39′N	104°59′W
English, stm., On., Can. (ĭn′glĭsh)	134-35	50°11′N	95°03′W
English Bāzār, India *see* Ingrāj Bāzār	234-35	24°60′N	88°09′E
English Channel, strt., Eur. (ĭn′glĭsh chăn′ĕl)	190-91	50°13′N	2°20′W
Enguri, stm., Geor. (ĕn-gòr′)	227	42°24′N	41°33′E
Enid, Ok., U.S. (ē′nĭd)	120-21	36°24′N	97°52′W
Eniwetok, at., Marsh. Is. *see* Enewetak	280-81	11°39′N	162°17′E
eNjesuthi, mtn., Afr.	269c	29°09′S	29°23′E
Enköping, Swe. (ĕn′kû-pĭng)	192-93	59°38′N	17°05′E
Ennedi, plat., Chad (ĕn-nĕd′ê)	258-59	17°15′N	22°00′E
Enniskillen, N. Ire., U.K. (ĕn-ĭs-kĭl′ĕn)	190-91	54°21′N	7°38′W
Enriquillo, Dom. Rep. (ĕn-rē-kē′l-yô)	142-43	17°53′N	71°16′W
Enriquillo, Lago, lk., Dom. Rep. (lä′gô-ĕn-rē-kē′l-yô)	142-43	18°29′N	71°38′W
Enschede, Neth. (ĕns′κȧ-d*ĕ*)	190-91	52°13′N	6°54′E
Ensenada, Mex. (ĕn-sĕ-nä′dä)	144-45	31°52′N	116°37′W
Enshi, China (ŭn-shr)	238-39	30°14′N	109°27′E
Entebbe, Ug.	267	0°04′N	32°28′E
Enterprise, Or., U.S. (ĕn′tĕr-prīz)	112-13	45°26′N	117°17′W
Enugu, Nig. (ĕ-nōō′gōō)	260a	6°27′N	7°27′E
Enumclaw, Wa., U.S. (ĕn′ŭm-klô)	112-13	47°12′N	121°59′W
Enurmino, Russia	126	66°55′N	171°48′W
Envigado, Col. (ĕn-vē-gá′dō)	163c	6°10′N	75°35′W
Envira, stm., Braz.	166-67	6°42′S	69°48′W
Épernay, Fr. (ā-pĕr-nĕ′)	196-97	49°03′N	3°58′E
Ephraim, Ut., U.S. (ē′frā-ĭm)	118-19	39°22′N	111°35′W
Ephrata, Pa., U.S. (ĕfrā′t*à*)	116-17	40°11′N	76°11′W
Ephrata, Wa., U.S. (ĕfrā′t*à*)	112-13	47°19′N	119°33′W
Épi, i., Vanuatu	279g	16°43′S	168°16′E
Épinal, Fr. (ā-pē-nál′)	196-97	48°11′N	6°26′E
Equatorial Guinea, nation, Afr. (ĕ-kwä-tô′rĭ-*ă*l gĭn′ê)	253	2°0′N	9°00′E
Erciyes Daği, vol., Tur.	186-87	38°32′N	35°28′E
Erdenebulgan, Mong.	240-41	50°06′N	101°35′E
Erding, Ger. (ĕr′dēng)	194-95	48°18′N	11°55′E
Erechim, Braz. (ĕ-rĕ-shē′ɴ)	168-69	27°38′S	52°17′W
Ereğli, Tur. (ĕ-rā′ĭ-le)	186-87	37°31′N	34°04′E
Ereğli, Tur. (ĕ-rā′ĭ-le)	186-87	41°17′N	31°26′E
Erenhot, China	240-41	43°39′N	111°60′E
Erfoud, Mor.	188-89	31°26′N	4°14′W
Erfurt, Ger. (ĕr′fòrt)	194-95	50°58′N	11°01′E
Erguig, Bahr, stm., Chad	260-61	11°21′N	15°24′E
Ergun, stm., Asia	240-41	53°19′N	121°27′E
Er Hai, lk., China	238-39	25°48′N	100°11′E
Erick, Ok., U.S. (âr′ĭk)	120-21	35°14′N	99°52′W
Erie, Pa., U.S.	116-17	42°07′N	80°03′W
Erie, Lake, lk., N.A. (lāk ē′rĭ)	116-17	42°15′N	80°60′W
Erimo-misaki, c., Japan (ā′rē-mō mē′sä-kê)	244	41°55′N	143°15′E
Eritrea, nation, Afr. (ā-rē-trā′ȧ)	266	15°20′N	39°00′E
Erlangen, Ger. (ĕr′läng-*ĕ*n)	194-95	49°36′N	11°01′E
Ermelo, S. Afr.	269c	26°31′S	29°59′E
Erne, Lower Lough, lk., N. Ire., U.K. (lō′ĕr lŏk ûrn)	190-91	54°28′N	7°45′W
Erode, India	236	11°21′N	77°44′E
Eromanga, Austl.	276	26°40′S	143°16′E
Er-Rachidia, Mor.	188-89	31°57′N	4°26′W
Erromango, i., Vanuatu	279g	18°45′S	169°05′E
Ertis, Kaz.	226	53°20′N	75°28′E
Ertis, stm., Asia *see* Irtysh	218-19	61°05′N	68°47′E
Ertix, stm., Asia *see* Irtysh	218-19	61°05′N	68°47′E
Ērtra, nation, Afr. *see* Eritrea	266	15°20′N	39°00′E
Erwin, Tn., U.S. (ûr′wĭn)	124-25	36°08′N	82°25′W
Erzin, Russia	240-41	50°15′N	95°09′E
Erzincan, Tur. (ĕr-zĭn-jän′)	227	39°44′N	39°31′E
Erzurum, Tur. (ĕrz′rōōm′)	227	39°54′N	41°17′E
Esashi, Japan (ĕs′ä-shê)	244	41°52′N	140°10′E
Esbjerg, Den. (ĕs′byĕrgh)	192-93	55°28′N	8°27′E
Escalante, Ut., U.S. (ĕs-kȧ-lăn′tē)	118-19	37°46′N	111°36′W
Escambia, stm., Fl., U.S. (ĕs-kăm′bĭ-*à*)	124-25	30°32′N	87°11′W
Escanaba, Mi., U.S. (ĕs-k*à*-nô′b*à*)	116-17	45°45′N	87°04′W
Escarpada Point, c., Phil.	250	18°31′N	122°13′E
Escondido, Ca., U.S. (ĕs-k*ŏ*n-dē′dō)	118-19	33°07′N	117°05′W
Escuinapa de Hidalgo, Mex.	146-47	22°51′N	105°48′W
Escuintla, Guat. (ĕs-kwēn′tlä)	148	14°18′N	90°47′W
Esenguly, Turkmen.	232-33	37°27′N	54°01′E
Eşfahān, Iran	232-33	32°39′N	51°40′E
Esh-Sham, nat. cap., Syria *see* Damascus	228-29	33°31′N	36°18′E
Esil, Kaz.	226	51°57′N	66°24′E
Esil, stm., Asia *see* Ishim	218-19	57°43′N	71°12′E
Eskifjörđur, Ice. (ĕs′kē-fyûr′dōōr)	190a	65°04′N	13°57′W
Eskilstuna, Swe. (å′shĕl-stü-na)	192-93	59°22′N	16°31′E
Eskimo Point, Nu., Can. *see* Arviat	128-29	61°08′N	94°07′W
Eskişehir, Tur. (ĕs-kē-shĕ′h'r)	186-87	39°47′N	30°31′E
Esla, stm., Spain (ĕs-lä)	198-99	41°29′N	6°03′W
Eslāmshahr, Iran	232-33	35°33′N	51°14′E
Eslöv, Swe. (ĕs′lûv)	192-93	55°50′N	13°19′E
Esmeralda, Isla, i., Chile	171	48°57′S	75°25′W
Esmeraldas, Ec. (ĕs-mā-räl′däs)	170	0°57′N	79°39′W
España, nation, Eur. *see* Spain	174-75	40°0′N	4°00′W
Espanola, On., Can. (ĕs-p*à*-nō′l*à*)	136-37	46°16′N	81°47′W
Española, Isla, i., Ec.	170a	1°23′S	89°42′W
Esperance, Austl. (ĕs′pē-r*ă*ns)	270-71	33°51′S	121°53′E
Esperanza, Arg.	173	31°27′S	60°56′W
Espichel, Cabo, c., Port. (ká′bō-ĕs-pē-shĕl′)	198-99	38°25′N	9°13′W
Espinal, Col. (ĕs-pē-näl′)	163c	4°09′N	74°53′W
Espinhaço, Serra do, mts., Braz. (sĕ′r-rä-dô-ĕs-pē-nȧ-sô)	172	17°25′S	43°40′W
Espírito Santo, Braz. (ĕs-pē′rē-tô-sän′tô) *see* Vila Velha	172	20°20′S	40°17′W
Espírito Santo, state, Braz. (ĕs-pē′rē-tô-sän′tô)	172	19°30′S	40°30′W
Espiritu Santo, i., Vanuatu (ĕs-pē′rē-tōō sän′tō)	279g	15°15′S	166°50′E
Espíritu Santo, Isla del, i., Mex.	144-45	24°29′N	110°21′W
Espita, Mex. (ĕs-pē′tä)	148	21°02′N	88°18′W
Esposende, Port. (ĕs-pō-zĕn′dā)	198-99	41°32′N	8°47′W
Esquel, Arg. (ĕs-kĕ′l)	171	42°54′S	71°19′W

n-sing; ŋ-baŋk; ɴ-nasalized n; nŏd; c*ŏ*mmit; ōld; ôbey; ôrder; oi-boil; fōōd; ò-as oo in foot; ou-out; s-soft; sh-dish; th-thin; pūre; ůnite; ûrn; stŭd; circ*ŭ*s; ü-as in French tu; ′-indeterminate vowel.

Feature (Pronunciation)	Page	Lat.	Long.
Esquimalt, B.C., Can. (ěs-kwī´mŏlt)	132-33	48°26'N	123°24'W
Esquina, Arg.	173	30°01'S	59°32'W
Essaouira, Mor.	258-59	31°30'N	9°45'W
Essen, Ger. (ěs´sěn)	194-95	51°28'N	7°01'E
Essequibo, stm., Guy. (ěs-ā-kē´bō)	164-65	7°04'N	58°26'W
Essex, Md., U.S. (ěs´ěks)	116-17	39°19'N	76°29'W
Essexville, Mi., U.S. (ěs´ěks-vĭl)	116-17	43°37'N	83°50'W
Est, Pointe de l', c., Qc., Can.	138-39	49°08'N	61°41'W
Estación Colonia Alvear Norte, Arg. see General Alvear	171	34°59'S	67°42'W
Estación Foguista J. F. Juárez, Arg. see El Galpón	168-69	25°24'S	64°38'W
Estación Gobernador Vera, Arg. see Vera	173	29°28'S	60°13'W
Estación J. J. Castelli, Arg. see Castelli	168-69	25°57'S	60°37'W
Estados, Isla de los, i., Arg.	171	54°48'S	64°33'W
Estância, Braz. (ěs-tän´sĭ-ä)	166-67	11°16'S	37°26'W
Estarreja, Port. (ě-tär-rā´zhä)	198-99	40°45'N	8°34'W
Estcourt, S. Afr. (ěst-coort)	269c	29°00'S	29°53'E
Estelí, Nic.	149	13°05'N	86°22'W
Estepona, Spain (ěs-tå-pō´nä)	198-99	36°26'N	5°08'W
Esterhazy, Sk., Can. (ěs´těr-hā-zē)	134-35	50°39'N	102°05'W
Estevan, Sk., Can. (ě-stē´vǎn)	134-35	49°08'N	103°00'W
Estherville, Ia., U.S. (ěs´těr-vĭl)	114-15	43°25'N	94°50'W
Estill, S.C., U.S. (ěs´tĭl)	124-25	32°45'N	81°14'W
Eston, Sk., Can.	134-35	51°10'N	108°46'W
Estonia, nation, Eur. (ěs-tō´nĭ-å)	174-75	59°0'N	26°00'E
Estrela, mtn., Port. (ěs-trā´lȧ)	198-99	40°19'N	7°37'W
Estremoz, Port. (ěs-trā-mōzh´)	198-99	38°51'N	7°35'W
Estrondo, Serra do, plat., Braz. (sěr´-rȧ dò ěs-trōn´-dò)	166-67	9°0'S	48°45'W
Esumba, Île, i., D.R.C.	262-63	2°0'N	21°12'E
Eszék, Cro. see Osijek	200-01	45°33'N	18°42'E
Étampes, Fr. (ā-täNp´)	196-97	48°26'N	2°10'E
Étaples, Fr. (ā-täp´l')	196-97	50°31'N	1°38'E
Etāwah, India	234-35	26°46'N	79°01'E
Ethiopia, nation, Afr. (ē-thē-ō´pē-å)	253	9°0'N	39°00'E
Etna, Monte, vol., Italy (mōn-tå ět´nȧ)	200-01	37°45'N	15°00'E
Etolin Strait, strt., Ak., U.S. (ět ō lĭn strāt)	126	60°20'N	165°15'W
Etorofu-tō, i., Russia see Iturup, Ostrov	218-19	44°51'N	147°27'E
Etosha National Park, n.p., Nmb.	264-65	18°60'S	15°07'E
Etosha Pan, pl., Nmb. (ětō´shä)	264-65	18°45'S	16°15'E
Etowah, Tn., U.S. (ět´ô-wä)	124-25	35°19'N	84°32'W
Et Tîdra, i., Maur.	258-59	19°44'N	16°24'W
Eua, i., Tonga	280-81	21°22'S	174°56'W
Euboea, i., Grc. see Évvoia	200-01	38°34'N	23°50'E
Eucla, Austl. (ū´klȧ)	270-71	31°42'S	128°54'E
Euclid, Oh., U.S. (ū´klĭd)	116-17	41°35'N	81°31'W
Eufaula, Al., U.S. (ů-fô´lȧ)	124-25	31°54'N	85°09'W
Eufaula, Ok., U.S. (ů-fô´lȧ)	120-21	35°16'N	95°36'W
Eugene, Or., U.S. (ů-jēn´)	112-13	44°03'N	123°05'W
Eugenia, Punta, c., Mex.	144-45	27°51'N	115°05'W
Eunice, La., U.S. (ū´nĭs)	122-23	30°30'N	92°25'W
Eunice, N.M., U.S. (ū´nĭs)	120-21	32°26'N	103°10'W
Euphrates, stm., Asia (û-frā´tēz)	208-09	30°60'N	47°27'E
Eureka, Ca., U.S.	112-13	40°47'N	124°09'W
Eureka, Il., U.S. (û-rē´kȧ)	116-17	40°43'N	89°16'W
Eureka, Ks., U.S. (û-rē´kȧ)	120-21	37°50'N	96°18'W
Eureka, Mt., U.S. (û-rē´kȧ)	112-13	48°53'N	115°03'W
Eureka, S.D., U.S. (û-rē´kȧ)	114-15	45°46'N	99°38'W
Eureka Springs, Ar., U.S. (û-rē´kȧ sprĭngz)	120-21	36°24'N	93°45'W
Europa, Île, i., Reu.	264-65	22°20'S	40°21'E
Europa Island, i., Reu. see Europa, Île	264-65	22°20'S	40°21'E
Europe, cont., (ū´rŭp)	20-21	50°0'N	28°00'E
Eustis, Fl., U.S. (ūs´tĭs)	124-25	28°51'N	81°41'W
Eutaw, Al., U.S. (ū-tå)	124-25	32°50'N	87°53'W
Eutsuk Lake, lk., B.C., Can. (ōōt´sŭk lăk)	132-33	53°19'N	126°44'W
Evans, Lac, lk., Qc., Can.	130-31	50°55'N	77°00'W
Evanston, Il., U.S. (ěv´ǎn-stŭn)	116-17	42°01'N	87°41'W
Evanston, Wy., U.S. (ěv´ǎn-stŭn)	112-13	41°17'N	110°58'W
Evansville, In., U.S. (ěv´ǎnz-vĭl)	116-17	37°59'N	87°35'W
Evansville, Wi., U.S. (ěv´ǎnz-vĭl)	116-17	42°47'N	89°17'W
Evansville, Wy., U.S. (ěv´ǎnz-vĭl)	112-13	42°50'N	106°15'W
Eva Perón, Arg. see La Plata	173	34°55'S	57°57'W
Eveleth, Mn., U.S. (ěv´ê-lěth)	114-15	47°28'N	92°32'W
Evensk, Russia	218-19	61°57'N	159°15'E
Everard, Lake, lk., Austl. (lāk ěv´ēr-ård)	272-73	31°25'S	135°06'E
Everest, Mount, mtn., Asia (mount ěv´ēr-ěst)	234-35	27°59'N	86°56'E
Everett, Wa., U.S. (ěv´ēr-ět)	112-13	47°58'N	122°12'W
Everglades, The, sw., Fl., U.S. (thȧ ěv´ēr-glādz swŏmp)	125a	26°0'N	80°40'W
Everglades National Park, n.p., Fl., U.S. (ěv´ēr-glādz năsh´ŭn-ǎl pärk)	125a	25°27'N	80°53'W
Evergreen, Al., U.S. (ěv´ēr-grēn)	124-25	31°26'N	86°57'W
Evergreen, Co., U.S. (ěv´ēr-grēn)	118-19	39°38'N	105°19'W
Evergreen, Mt., U.S. (ěv´ēr-grēn)	112-13	48°13'N	114°17'W
Évora, Port. (ěv´ô-rä)	198-99	38°34'N	7°54'W
Evpatoria, Ukr. see Yevpatoriia	202-03	45°12'N	33°22'E
Évreux, Fr. (ā-vrû´)	196-97	49°01'N	1°09'E
Évvoia, i., Grc.	200-01	38°34'N	23°50'E
Exe, stm., Eng., U.K. (ěks)	190-91	50°42'N	3°29'W
Exeter, Eng., U.K. (ěk´sê-tēr)	190-91	50°43'N	3°32'W
Exmoor, plat., Eng., U.K. (ěks´mòr)	190-91	51°09'N	3°44'W
Exmouth, Austl.	270-71	21°56'S	114°07'E
Exmouth, Eng., U.K. (ěks´mŭth)	190-91	50°38'N	3°24'W
Exmouth Gulf, b., Austl.	272-73	22°0'S	114°20'E
Exploits, stm., Nf., Can. (ěks-ploits´)	138-39	49°05'N	55°18'W
Exuma Sound, strt., Bah. (ěk-sōō´mä sound)	142-43	24°12'N	76°01'W
Eyasi, Lake, lk., Tan. (lāk å-yä´sê)	267	3°40'S	35°05'E
Eyl, Som.	262-63	7°59'N	49°49'E
Eyre Creek, stm., Austl.	277	26°38'S	138°59'E
Eyre North, Lake, lk., Austl. (lāk âr nôrth)	272-73	28°33'S	137°15'E
Eyre Peninsula, pen., Austl.	272-73	33°15'S	135°48'E
Eyre South, Lake, lk., Austl. (lāk âr south)	276	29°18'S	137°26'E
Eysturoy, i., Far. Is.	190b	62°12'N	6°55'W
Ezequiel Ramos Mexía, Embalse, res., Arg.	171	39°27'S	69°01'W
Ezine, Tur. (å´zĭ-nå)	200-01	39°47'N	26°20'E

F

Feature (Pronunciation)	Page	Lat.	Long.
Faaborg, Den. (fô´bòrg)	194-95	55°06'N	10°15'E
Fabriano, Italy (fä-brē-ä´nò)	200-01	43°21'N	12°54'E
Fada, Chad (fä´dä)	258-59	17°12'N	21°35'E
Fada-Ngourma, Burkina (fä´dä´n gōōr´mä)	260-61	12°03'N	0°22'E
Faddeyevskiy, Ostrov, i., Russia (ôs-trôf´ fȧd-yä´skĭ)	218-19	75°27'N	144°20'E
Faenza, Italy (fä-ěnd´zä)	200-01	44°17'N	11°53'E
Færøerne, dep., Eur. see Faroe Islands	174-75	62°0'N	7°00'W
Fafen, stm., Eth.	262-63	5°39'N	44°08'E
Făgăraş, Rom. (fă-gä´räsh)	194-95	45°50'N	24°59'E
Fagernes, Nor.	192-93	60°59'N	9°15'E
Fagnano, Lago, lk., S.A. (lä´gô fäk-nä´nô)	171	54°34'S	67°58'W
Faguibine, Lac, lk., Mali	258-59	16°49'N	3°50'W
Faial, i., Port. (fä-yä´l)	199c	38°34'N	28°42'W
Fairbanks, Ak., U.S. (fâr´bănks)	126	64°51'N	147°42'W
Fairbury, Ne., U.S. (fâr´bĕr-ĭ)	120-21	40°09'N	97°11'W
Fairfax, Mn., U.S. (fâr´făks)	114-15	44°31'N	94°43'W
Fairfax, S.C., U.S. (fâr´făks)	124-25	32°57'N	81°14'W
Fairfield, Ia., U.S. (fâr´fēld)	114-15	41°01'N	91°58'W
Fairfield, Il., U.S. (fâr´fēld)	116-17	38°22'N	88°22'W
Fairfield, Oh., U.S. (fâr´fēld)	116-17	39°21'N	84°34'W
Fairfield, Tx., U.S. (fâr´fēld)	122-23	31°43'N	96°10'W
Fair Haven, Vt., U.S. (fâr ā´vĕn)	116-17	43°36'N	73°16'W
Fair Isle, i., Scot., U.K. (fâr ĭl)	190c	59°32'N	1°39'W
Fairmont, Mn., U.S. (fâr´mŏnt)	114-15	43°40'N	94°28'W
Fairmont, W.V., U.S. (fâr´mŏnt)	116-17	39°29'N	80°08'W
Fair Ness, c., Nu., Can.	130-31	63°25'N	72°02'W
Fairview, Ok., U.S. (fâr´vū)	120-21	36°16'N	98°29'W
Fairweather, Mount, mtn., N.A. (mount fâr-wědh´ēr)	126	58°54'N	137°32'W
Fairweather Mountain, mtn., N.A. see Fairweather, Mount	126	58°54'N	137°32'W
Faisalābād, Pak.	232-33	31°25'N	73°05'E
Faith, S.D., U.S. (fāth)	114-15	45°01'N	102°01'W
Faiyum, Egypt see El-Fayoum	268b	29°19'N	30°50'E
Faizābād, India	234-35	26°47'N	82°08'E
Fakaofo, at., Tok.	280-81	9°22'S	171°14'W
Fakfak, Indon.	224-25	2°56'S	132°18'E
Faku, China (fä-kōō)	243	42°30'N	123°25'E
Falam, Mya.	246-47	22°54'N	93°41'E
Falémé, stm., Afr.	258-59	14°46'N	12°15'W
Falfurrias, Tx., U.S. (făl´fōō-rē´ȧs)	122-23	27°14'N	98°09'W
Falher, Ab., Can. (făl´ēr)	132-33	55°44'N	117°13'W
Falkenberg, Swe. (fäl´kěn-běrgh)	192-93	56°54'N	12°29'E
Falkensee, Ger. (fäl´kěn-zā)	194-95	52°34'N	13°04'E
Falkirk, Scot., U.K. (fôl´kûrk)	190-91	56°00'N	3°47'W
Falkland Islands, dep., S.A. (fôk´lǎnd ī´lǎndz)	158	51°45'S	59°00'W
Falkland Islands, is., Falk. Is. (fôk´lǎnd ī´lǎndz)	171	51°41'S	59°08'W
Falkland Sound, strt., Falk. Is.	171	51°45'S	59°25'W
Falköping, Swe. (fäl´chûp-ĭng)	192-93	58°10'N	13°32'E
Fall River, Ma., U.S. (fôl rĭv´ēr)	116-17	41°42'N	71°10'W
Falls City, Ne., U.S. (fôlz sĭ´tê)	120-21	40°04'N	95°37'W
Falmouth, Jam. (făl´mŭth)	142-43	18°29'N	77°39'W
Falmouth, Eng., U.K. (făl´mŭth)	190-91	50°09'N	5°03'W
Falmouth, Ky., U.S. (făl´mŭth)	116-17	38°41'N	84°20'W
False Divi Point, c., India	236	15°43'N	80°50'E
Falster, i., Den. (fäls´tēr)	192-93	54°48'N	11°58'E
Fălticeni, Rom. (fŭl-tê-chān´y')	194-95	47°28'N	26°19'E
Falun, Swe. (fä-lōōn´)	192-93	60°36'N	15°38'E
Famagusta, Cyp. see Gazimağusa	228-29	35°07'N	33°57'E
Fanch'eng, China see Xiangfan	238-39	32°02'N	112°09'E
Fangxian, China (fäŋ-shyĕn)	238-39	32°03'N	110°44'E
Fanning Island, at., Kir. see Tabuaeran	280-81	3°51'N	159°18'W
Fano, Italy (fä´nō)	200-01	43°51'N	13°01'E
Fanø, i., Den. (făn´û)	192-93	55°25'N	8°25'E
Fan Si Pan, mtn., Viet.	246-47	22°15'N	103°46'E
Faradofay, Madag. see Tôlañaro	264-65	25°02'S	47°00'E
Farafangana, Madag. (fä-rä-fäŋ-gä´nä)	264-65	22°49'S	47°50'E
Farāh, Afg. (fä-rä´)	232-33	32°22'N	62°04'E
Farāh, stm., Afg.	232-33	31°27'N	61°27'E
Farallon de Medinilla, i., N. Mar. Is.	280-81	16°01'N	146°04'E
Farallon de Pajaros, i., N. Mar. Is.	280-81	20°32'N	144°54'E
Farasān, Jazā'ir, is., Sau. Ar.	266	16°48'N	41°54'E
Farewell, Cape, c., Green.	86	59°46'N	43°60'W
Farewell, Cape, c., N.Z. (kāp fâr-wĕl´)	278	40°30'S	172°41'E
Farg'ona, Uzb.	232-33	40°23'N	71°48'E
Fargo, N.D., U.S. (fär´gō)	114-15	46°52'N	96°49'W
Faribault, Mn., U.S. (fâ´rĭ-bō)	114-15	44°18'N	93°17'W
Farīdpur, Bngl.	234-35	23°36'N	89°51'E
Farmersburg, In., U.S. (fär´mērz-bûrg)	116-17	39°15'N	87°23'W
Farmersville, Tx., U.S. (fär´mērz-vĭl)	120-21	33°10'N	96°22'W
Farmington, Il., U.S. (färm-ĭng-tŭn)	114-15	40°42'N	90°00'W
Farmington, Mo., U.S. (färm-ĭng-tŭn)	120-21	37°47'N	90°25'W
Farmington, N.M., U.S. (färm-ĭng-tŭn)	118-19	36°44'N	108°12'W
Farmville, N.C., U.S. (färm-vĭl)	124-25	35°35'N	77°36'W
Farmville, Va., U.S. (färm-vĭl)	124-25	37°18'N	78°24'W
Farnham, Qc., Can. (fär´nǎm)	136-37	45°17'N	72°59'W
Faro, Braz. (fä´rò)	166-67	2°10'S	56°45'W
Faro, Port. (fä´rò)	198-99	37°01'N	7°56'W
Faro, stm., Afr.	260-61	9°20'N	12°54'E
Faroe Islands, dep., Eur. (fâr´ō ī´lǎndz)	174-75	62°0'N	7°00'W
Fårön, i., Swe.	192-93	57°56'N	19°08'E
Farquhar, Atoll de, at., Sey.	264-65	10°10'S	51°10'E
Farrukhābād, India (fŭ-ròk-hä-bäd´)	234-35	27°24'N	79°35'E
Farsund, Nor. (fär´sòn)	192-93	58°05'N	6°48'E
Fartak, Ra's, c., Yemen	220-21	15°39'N	52°12'E
Farukolu, i., Mald.	236	6°12'N	73°16'E
Farvel, Kap, c., Green. see Farewell, Cape	86	59°46'N	43°60'W
Farwell, Tx., U.S. (fär´wĕl)	120-21	34°24'N	103°01'W
Fasano, Italy (fä-zä´nō)	200-01	40°50'N	17°21'E
Fatehpur Sīkri, India	234-35	27°06'N	77°40'E
Fauro Island, i., Sol. Is.	279e	6°55'S	156°04'E
Fauske, Nor.	184-85	67°16'N	15°24'E
Fawn, stm., On., Can.	134-35	55°22'N	88°20'W
Faxaflói, b., Ice.	190a	64°25'N	23°00'W
Faya-Largeau, Chad	258-59	17°56'N	19°07'E
Fayette, Al., U.S. (fā-yĕt´)	124-25	33°41'N	87°50'W
Fayette, Mo., U.S. (fā-yĕt´)	120-21	39°09'N	92°41'W
Fayette, Ms., U.S. (fā-yĕt´)	124-25	31°43'N	91°04'W
Fayetteville, Ar., U.S. (fā-yĕt´vĭl)	120-21	36°05'N	94°10'W
Fayetteville, N.C., U.S. (fā-yĕt´vĭl)	124-25	35°03'N	78°53'W
Fayetteville, Tn., U.S. (fā-yĕt´vĭl)	124-25	35°09'N	86°34'W
Fayetteville, W.V., U.S. (fā-yĕt´vĭl)	116-17	38°03'N	81°06'W
Faylakah, i., Kuw.	230-31	29°26'N	48°20'E
Fayyum, Egypt see El-Fayoum	268b	29°19'N	30°50'E
Fazzān, hist. reg., Libya see Fezzan	258-59	26°0'N	14°00'E
Fear, Cape, c., N.C., U.S. (kāp fēr)	124-25	33°57'N	77°56'W
Fécamp, Fr. (fā-käN´)	196-97	49°46'N	0°23'E
Fedala, Mor. see Mohammedia	269a	33°42'N	7°23'W
Federal, Arg.	173	30°57'S	58°47'W
Federated States of Micronesia, nation, Oc. see Micronesia, Federated States of	280-81	5°0'N	152°00'E
Feia, Lagoa, b., Braz. (lä´gō-å fē´yä)	172	22°0'S	41°20'W
Feijó, Braz.	166-67	8°11'S	70°24'W

Feature (Pronunciation)	Page	Lat.	Long.
Feira de Santana, Braz. (fĕ´ȧ-rä dä sänt-än´ä)	166-67	12°15′S	38°58′W
Feixian, China (fā-shyĕn)	240-41	35°16′N	117°58′E
Fejaj, Chott, lk., Tun.	258-59	33°53′N	9°10′E
Felanitx, Spain (fā-lä-nēch´)	198-99	39°29′N	3°09′E
Feldkirch, Aus. (fĕlt´kĭrk)	194-95	47°14′N	9°36′E
Felipe Carrillo Puerto, Mex.	148	19°35′N	88°02′W
Felix, Cape, c., Nu., Can.	130-31	69°53′N	97°57′W
Feltre, Italy (fĕl´trā)	200-01	46°01′N	11°54′E
Fen, stm., China	240-41	35°28′N	110°34′E
Fengcheng, China (fŭŋ-chŭŋ)	238-39	28°10′N	115°46′E
Fengcheng, China (fŭŋ-chŭŋ)	243	40°27′N	124°04′E
Fengdu, China (fŭŋ-dōō)	238-39	29°58′N	107°46′E
Fengfeng, China	240-41	36°29′N	114°14′E
Fengtien, China see Shenyang	243	41°48′N	123°24′E
Fengxian, China (fŭŋ-shyĕn)	238-39	33°57′N	106°40′E
Fengyang, China (fŭng´yäng´)	238-39	32°52′N	117°33′E
Fengzhen, China (fŭŋ-jŭn)	240-41	40°26′N	113°09′E
Feni Islands, is., Pap. N. Gui.	277	4°04′S	153°38′E
Fenton, Mi., U.S. (fĕn-tŭn)	116-17	42°48′N	83°42′W
Fenyang, China	240-41	37°16′N	111°47′E
Feodosiia, Ukr.	202-03	45°02′N	35°22′E
Ferdows, Iran	232-33	34°01′N	58°10′E
Fergus Falls, Mn., U.S. (fûr´gŭs fôlz)	114-15	46°17′N	96°05′W
Fergusson Island, i., Pap. N. Gui.	277	9°31′S	150°39′E
Ferlo, Vallée du, stm., Sen.	260-61	15°50′N	15°43′W
Fermo, Italy (fĕr´mō)	200-01	43°10′N	13°43′E
Fermoy, Ire. (fûr-moi´)	190-91	52°08′N	8°17′W
Fernandina, Isla, i., Ec.	170a	0°26′S	91°30′W
Fernandina Beach, Fl., U.S. (fûr-năn-dē´nȧ bēch)	124-25	30°40′N	81°27′W
Fernando de Noronha, Ilha, i., Braz.	159	3°51′S	32°25′W
Fernandópolis, Braz.	168-69	20°16′S	50°15′W
Fernando Póo, i., Eq. Gui. see Bioko	260-61	3°30′N	8°40′E
Fernie, B.C., Can. (fûr´nĭ)	132-33	49°30′N	115°03′W
Ferrara, Italy (fĕr-rä´rä)	200-01	44°51′N	11°36′E
Ferrat, Cap, c., Alg. (kăp fĕr-rät)	198-99	35°55′N	0°23′W
Ferreñafe, Peru (fĕr-rĕn-yȧ´fĕ)	170	6°38′S	79°47′W
Ferriday, La., U.S. (fĕr´ĭ-dā)	124-25	31°38′N	91°33′W
Ferro, i., Spain see El Hierro	199d	27°45′N	18°00′W
Ferrol, Spain	198-99	43°29′N	8°14′W
Fès, Mor. (fĕs)	269a	34°03′N	5°00′W
Fessenden, N.D., U.S. (fĕs´ĕn-dĕn)	114-15	47°39′N	99°38′W
Festus, Mo., U.S. (fĕst´ŭs)	120-21	38°14′N	90°24′W
Fethiye, Tur. (fĕt-hē´yĕ)	200-01	36°37′N	29°08′E
Feuilles, stm., Qc., Can.	130-31	58°39′N	70°25′W
Feyẕābād, Afg.	232-33	37°08′N	70°34′E
Fez, Mor. see Fès	269a	34°03′N	5°00′W
Fezzan, hist. reg., Libya	258-59	26°0′N	14°00′E
Fianarantsoa, Madag. (fyȧ-nä´rȧn-tsō´ȧ)	264-65	21°25′S	47°07′E
Fianga, Chad	262-63	9°55′N	15°09′E
Ficksburg, S. Afr. (fĭks´bûrg)	269c	28°52′S	27°53′E
Fife Ness, c., Scot., U.K. (fīf´nes´)	190-91	56°17′N	2°36′W
Figeac, Fr. (fē-zhȧk´)	196-97	44°37′N	2°02′E
Figueira da Foz, Port. (fē-gwĕy-rä-dȧ-fô´z)	198-99	40°09′N	8°51′W
Figuig, Mor.	188-89	32°08′N	1°13′W
Fiji, nation, Oc. (fē´jē)	279f	18°0′S	178°00′E
Fiji, nation, Oc. (fē´jē)	279f	18°0′S	178°00′E
Filicudi, Isola, i., Italy (ē´-sō-lä fē´le-kōō´dē)	200-01	38°34′N	14°34′E
Fillmore, Ut., U.S. (fĭl´môr)	118-19	38°58′N	112°20′W
Fimi, stm., D.R.C.	262-63	3°02′S	16°56′E
Findlay, Oh., U.S. (fĭnd´lȧ)	116-17	41°02′N	83°38′W
Finisterre, Cabo de, c., Spain see Fisterra, Cabo de	198-99	42°53′N	9°16′W
Finland, nation, Eur. (fĭn´lănd)	174-75	64°0′N	26°00′E
Finland, Gulf of, b., Eur. (gŭlf ŭv fĭn´lănd)	192-93	60°0′N	27°00′E
Finskiy Zaliv, b., Eur. see Finland, Gulf of	192-93	60°0′N	27°00′E
Fiordland National Park, n.p., N.Z.	278	45°30′S	167°20′E
Fırat, stm., Asia see Euphrates	208-09	30°60′N	47°27′E
Firenze, Italy see Florence	200-01	43°47′N	11°14′E
Fīrozābād, India	234-35	27°09′N	78°24′E
Fīrozpur, India	234-35	30°55′N	74°37′E
Fisterra, Cabo de, c., Spain	198-99	42°53′N	9°16′W
Fitchburg, Ma., U.S. (fĭch´bûrg)	116-17	42°35′N	71°48′W
Fitzgerald, Ga., U.S. (fĭts-jĕr´ăld)	124-25	31°43′N	83°15′W
Fitz Roy, Arg.	171	47°03′S	67°14′W
Fitz Roy, Monte, mtn., S.A.	171	49°17′S	73°05′W
Fitzroy Crossing, Austl.	270-71	18°12′S	125°34′E
Fiume, Cro. see Rijeka	200-01	45°20′N	14°27′E
Fizi, D.R.C.	267	4°19′S	28°56′E
Flagstaff, Az., U.S. (flăg-stȧf)	118-19	35°11′N	111°39′W
Flaherty Island, i., Nu., Can.	130-31	56°14′N	79°17′W
Flåm, Nor. (flôm)	192-93	60°51′N	7°07′E
Flaming Gorge Reservoir, res., U.S. (flā´mĭng gôrj rĕ´sĕr-vwär)	112-13	41°14′N	109°35′W
Flandreau, S.D., U.S. (flăn´drō)	114-15	44°03′N	96°36′W
Flathead Indian Reservation, ind. res., Mt., U.S. (flăt´hĕd ĭn´dĭ-ăn rĕ-sĕr-vā´shĕn)	112-13	47°30′N	114°25′W
Flathead Lake, lk., Mt., U.S. (flăt´hĕd lāk)	112-13	47°52′N	114°08′W
Flat Rock, Mi., U.S. (flăt rŏk)	116-17	42°06′N	83°17′W
Flattery, Cape, c., Wa., U.S. (kāp flăt´ẽr-ĭ)	112-13	48°23′N	124°43′W
Flekkefjord, Nor. (fläk´kē-fyôr)	192-93	58°18′N	6°41′E
Flemingsburg, Ky., U.S. (flĕm´ĭngz-bûrg)	116-17	38°25′N	83°45′W
Flensburg, Ger. (flĕns´bōrgh)	194-95	54°47′N	9°26′E
Flers, Fr. (flĕr)	196-97	48°45′N	0°34′W
Flinders, stm., Austl. (flĭn´dẽrz)	277	17°36′S	140°36′E
Flinders Chase National Park, n.p., Austl.	276	35°58′S	136°44′E
Flinders Island, i., Austl. (flĭn´dẽrz ī´lȧnd)	276	40°0′S	148°00′E
Flin Flon, Mb., Can. (flĭn flŏn)	134-35	54°46′N	101°53′W
Flint, Mi., U.S. (flĭnt)	116-17	43°00′N	83°41′W
Flint, i., Kir.	280-81	11°25′S	151°48′W
Flint, stm., Ga., U.S. (flĭnt)	124-25	30°46′N	84°48′W
Flora, Il., U.S. (flō´rȧ)	116-17	38°40′N	88°29′W
Florala, Al., U.S. (flŏr-ăl´ȧ)	124-25	31°00′N	86°20′W
Florence, Italy (flŏr´ĕns)	200-01	43°47′N	11°14′E
Florence, Al., U.S. (flŏr´ĕns)	124-25	34°48′N	87°41′W
Florence, Az., U.S. (flŏr´ĕns)	118-19	33°02′N	111°23′W
Florence, Ky., U.S. (flŏr´ĕns)	116-17	38°60′N	84°38′W
Florence, Or., U.S. (flŏr´ĕns)	112-13	43°59′N	124°06′W
Florence, S.C., U.S. (flŏr´ĕns)	124-25	34°11′N	79°46′W
Florencia, Col. (flō-rĕn´sē-ȧ)	164-65	1°36′N	75°36′W
Flores, i., Indon. (flō´rĕs)	248-49	8°38′S	120°56′E
Flores, i., Port.	199c	39°26′N	31°13′W
Flores, Laut, s., Indon. see Flores Sea	248-49	8°0′S	120°00′E
Flores Island, i., B.C., Can.	132-33	49°20′N	126°10′W
Flores Sea, s., Indon. (flō´rĕs sē)	248-49	8°0′S	120°00′E
Floresville, Tx., U.S. (flō´rĕs-vĭl)	122-23	29°08′N	98°09′W
Floriano, Braz. (flō-rȧ-ä´nō)	166-67	6°47′S	43°01′W
Florianópolis, Braz. (flō-rē-ä-nō´pō-lĕs)	172	27°35′S	48°32′W
Florida, Col. (flō-rē´dä)	163c	3°21′N	76°15′W
Florida, Ur. (flō-rē-dhä)	173	34°06′S	56°12′W
Florida, state, U.S. (flŏr´ĭ-dȧ)	108-09	28°0′N	82°00′W
Florida, Estrecho de la, strt., N.A. see Florida, Straits of	142-43	24°59′N	79°45′W
Florida, Straits of, strt., N.A. (strāts ŭv flŏr´ĭ-dȧ)	142-43	24°59′N	79°45′W
Florida Bay, b., Fl., U.S. (flŏr´ĭ-dȧ bā)	125a	24°58′N	80°48′W
Florida Keys, is., Fl., U.S. (flŏr´ĭ-dȧ kēs)	125a	24°47′N	81°06′W
Florido, stm., Mex. (flō-rē´dō)	144-45	27°43′N	105°11′W
Flórina, Grc. (flō-rē´nä)	200-01	40°47′N	21°24′E
Florissant, Mo., U.S. (flŏr´ĭ-sănt)	120-21	38°48′N	90°20′W
Florø, Nor.	192-93	61°35′N	5°01′E
Floydada, Tx., U.S. (floi-dā´dȧ)	120-21	33°59′N	101°20′W
Flushing, Mi., U.S. (flûsh´ĭng)	116-17	43°04′N	83°50′W
Fly, stm., (flī)	277	8°14′S	142°09′E
Foča, Bos. (fō´chä)	200-01	43°30′N	18°47′E
Fochville, S. Afr. (fŏk´vĭl)	269c	26°29′S	27°31′E
Focșani, Rom. (fōk-shä´nē)	202-03	45°42′N	27°12′E
Fogang, China (fwo-gäŋ)	238-39	23°52′N	113°32′E
Foggia, Italy (fôd´jä)	200-01	41°28′N	15°32′E
Fogo, Nf., Can. (fō´gō)	138-39	49°43′N	54°18′W
Fogo, i., C.V.	260-61	14°54′N	24°23′W
Fogo Island, i., Nf., Can. (fō´gō ī´lănd)	138-39	49°39′N	54°11′W
Foguista J. F. Juárez, Arg. see El Galpón	168-69	25°24′S	64°38′W
Foix, Fr. (fwä)	196-97	42°58′N	1°37′E
Fokino, Russia	202-03	53°26′N	34°26′E
Folādī, Koh-e, mtn., Afg.	232-33	34°38′N	67°32′E
Foley Island, i., Nu., Can.	130-31	68°32′N	75°07′W
Foligno, Italy (fō-lēn´yō)	200-01	42°58′N	12°42′E
Fond-du-Lac, Sk., Can.	128-29	59°20′N	107°10′W
Fond du Lac, Wi., U.S. (fŏn dū lăk´)	116-17	43°46′N	88°27′W
Fond du Lac Indian Reservation, ind. res., Mn., U.S. (fŏn dū lăk´ ĭn´dĭ-ăn rĕ-sĕr-vā´shĕn)	114-15	46°45′N	92°37′W
Fondi, Italy (fōn´dē)	200-01	41°22′N	13°26′E
Fonseca, Golfo de, b., N.A. (gôl-fô-dĕ-fôn-sä´kä)	149	13°10′N	87°40′W
Fontainebleau, Fr. (fôn-tĕn-blō´)	196-97	48°24′N	2°42′E
Fontana, Ca., U.S. (fŏn-tă´nȧ)	118-19	34°06′N	117°26′W
Fonte Boa, Braz. (fōn´tȧ bō´ȧ)	166-67	2°32′S	66°01′W
Fontenay-le-Comte, Fr. (fônt-nĕ´lē-kônt´)	196-97	46°28′N	0°48′W
Fontur, c., Ice.	190a	66°22′N	14°35′W
Foochow, China see Fuzhou	225a	26°06′N	119°17′E
Forbach, Fr. (fôr´bäk)	196-97	49°12′N	6°54′E
Forbes, Austl. (fôrbz)	276	33°23′S	148°00′E
Forchheim, Ger. (fôrk´hīm)	194-95	49°43′N	11°04′E
Fordyce, Ar., U.S. (fôr´dīs)	120-21	33°49′N	92°25′W
Forest, Ms., U.S. (fŏr´ĕst)	124-25	32°22′N	89°28′W
Forest City, Ia., U.S. (fŏr´ĕst sĭ´tē)	114-15	43°16′N	93°39′W
Forest City, N.C., U.S. (fŏr´ĕst sĭ´tĭ)	124-25	35°20′N	81°52′W
Forest City, Pa., U.S. (fŏr´ĕst sĭ´tē)	116-17	41°39′N	75°28′W
Forestville, Qc., Can. (fŏr´ĕst-vĭl)	138-39	48°45′N	69°06′W
Forfar, Scot., U.K. (fôr´fȧr)	190-91	56°38′N	2°54′W
Forlì, Italy (fôr-lē´)	200-01	44°13′N	12°03′E
Formentera, i., Spain (fôr-mĕn-tā´rä)	198-99	38°42′N	1°28′E
Formiga, Braz. (fôr-mē´gȧ)	172	20°28′S	45°26′W
Formosa, Arg. (fôr-mō´sä)	168-69	26°10′S	58°12′W
Formosa, Braz.	168-69	15°32′S	47°20′W
Formosa, nation, Asia (fôr-mō´sȧ) see Taiwan	206-07	23°30′N	121°00′E
Formosa, state, Arg. (fôr-mō´sä)	168-69	25°0′S	60°00′W
Formosa, Serra, plat., Braz. (sĕ´r-rä fôr-mō´sä)	166-67	12°0′S	55°00′W
Formosa Strait, strt., Asia see Taiwan Strait	225a	24°0′N	119°00′E
Føroyar, dep., Eur. see Faroe Islands	174-75	62°0′N	7°00′W
Forrest City, Ar., U.S. (fŏr´ĕst sĭ´tĭ)	124-25	35°00′N	90°48′W
Forsyth, Ga., U.S. (fôr-sīth´)	124-25	33°02′N	83°57′W
Forsyth, Mt., U.S. (fôr-sīth´)	112-13	46°16′N	106°41′W
Fort Albany, On., Can. (fôrt ôl´bȧ nĭ)	128-29	52°13′N	81°40′W
Fortaleza, Braz. (fôr´tä-lā´zȧ)	166-67	3°44′S	38°30′W
Fort-Archambault, Chad see Sarh	262-63	9°09′N	18°23′E
Fort Atkinson, Wi., U.S. (fôrt ăt´kĭn-sŭn)	116-17	42°55′N	88°51′W
Fort Bayard, China see Zhanjiang	238-39	21°12′N	110°23′E
Fort Benton, Mt., U.S. (fôrt bĕn´tŭn)	112-13	47°49′N	110°41′W
Fort Berthold Indian Reservation, ind. res., N.D., U.S. (fôrt bĕrth´ôld ĭn´dĭ-ăn rĕ-sĕr-vā´shĕn)	114-15	47°40′N	102°25′W
Fort Bragg, Ca., U.S.	118-19	39°27′N	123°48′W
Fort Branch, In., U.S. (fôrt brănch)	116-17	38°15′N	87°35′W
Fort Chipewyan, Ab., Can.	128-29	58°43′N	111°10′W
Fort Collins, Co., U.S. (fôrt kŏl´ĭns)	120-21	40°35′N	105°05′W
Fort-Dauphin, Madag. see Tôlañaro	264-65	25°02′S	47°00′E
Fort-de-France, nat. cap., Mart. (dē fräns)	143b	14°36′N	61°04′W
Fort Dodge, Ia., U.S. (fôrt dŏj)	114-15	42°30′N	94°11′W
Fort Edward, N.Y., U.S. (fôrt wĕrd)	116-17	43°16′N	73°35′W
Fortescue, stm., Austl. (fôr´tĕs-kū)	272-73	21°00′S	116°06′E
Fort-Foureau, Camrn. see Kousséri	260-61	12°05′N	15°02′E
Fort Frances, On., Can. (fôrt frăn´sĕs)	134-35	48°37′N	93°24′W
Fort Franklin, N.T., Can. see Déline	128-29	65°11′N	123°25′W
Fort Frederica National Monument, n.p., Ga., U.S. (fôrt frĕd´ē-rĭ-kȧ năsh´ŭn-ăl mŏn´ŭ-mĕnt)	124-25	31°12′N	81°26′W
Fort-George, Qc., Can. see Chisasibi	128-29	53°48′N	79°02′W
Fort Gibson, Ok., U.S. (fôrt gĭb´sŭn)	120-21	35°49′N	95°15′W
Fort Good Hope, N.T., Can. (fôrt gŏŏd hōp)	128-29	66°15′N	128°37′W
Forth, Firth of, b., Scot., U.K. (fûrth ŏv fôrth)	190-91	56°07′N	2°58′W
Fort Johnston, Malawi see Mangochi	264-65	14°28′S	35°15′E
Fort Kent, Me., U.S. (fôrt kĕnt)	117a	47°15′N	68°35′W
Fort-Lamy, nat. cap., Chad see N'Djamena	258-59	12°07′N	15°03′E
Fort Lauderdale, Fl., U.S. (fôrt lô´dĕr-dāl)	125a	26°07′N	80°09′W
Fort Liard, N.T., Can.	128-29	60°14′N	123°27′W
Fort Lupton, Co., U.S. (fôrt lŭp´tŭn)	120-21	40°05′N	104°49′W
Fort Macleod, Ab., Can. (fôrt mȧ-kloud´)	132-33	49°43′N	113°25′W
Fort Madison, Ia., U.S. (fôrt măd´ĭ-sŭn)	120-21	40°38′N	91°19′W
Fort McMurray, Ab., Can. (fôrt mȧk-mŭr´ĭ)	134-35	56°44′N	111°25′W
Fort McPherson, N.T., Can. (fôrt mȧk-fŭr´s'n)	128-29	67°25′N	134°52′W
Fort Meade, Fl., U.S. (fôrt mēd)	125a	27°46′N	81°48′W
Fort Mill, S.C., U.S. (fôrt mĭl)	124-25	35°00′N	80°57′W
Fort Mojave Indian Reservation, ind. res., Az., U.S. (fôrt mô-hä´vâ ĭn´dĭ-ăn rĕ-sĕr-vā´shĕn)	118-19	34°55′N	114°35′W
Fort Morgan, Co., U.S. (fôrt môr´gȧn)	120-21	40°15′N	103°48′W
Fort Myers, Fl., U.S. (fôrt mī´ẽrz)	125a	26°38′N	81°52′W

n-sing; ŋ-baŋk; ɴ-nasalized n; nŏd; cŏmmit; ōld; ôbey; ôrder; oi-boil; fōōd; ŏŏ-as oo in foot; ou-out; s-soft; sh-dish; th-thin; pūre; ůnite; ûrn; stŭd; circŭs; ü-as in French tu; ´-indeterminate vowel.

Feature (Pronunciation)	Page	Lat.	Long.
Fort Nelson, B.C., Can. (fôrt nĕl´sŭn)	128-29	58°49′N	122°41′W
Fort Nelson, stm., B.C., Can. (fôrt nĕl´sŭn)	130-31	59°31′N	124°03′W
Fort Norman, N.T., Can. *see* Tulita	128-29	64°54′N	125°34′W
Fort Payne, Al., U.S. (fôrt pān)	124-25	34°27′N	85°43′W
Fort Peck Indian Reservation, ind. res., Mt., U.S. (fôrt pĕk ĭn´dĭ-ăn rĕ-sĕr-vā´shĕn)	112-13	48°22′N	105°40′W
Fort Peck Lake, res., Mt., U.S. (fôrt pĕk lāk)	112-13	47°45′N	106°45′W
Fort Pierce, Fl., U.S. (fôrt pērs)	125a	27°27′N	80°20′W
Fort Portal, Ug. (fôrt pôr´tál)	267	0°40′N	30°17′E
Fort Providence, N.T., Can. (fôrt prŏv´ĭ-dĕns)	128-29	61°21′N	117°35′W
Fort Pulaski National Monument, n.p., Ga., U.S. (fôrt pu-lăs´kĭ năsh´ŭn-ăl mŏn´ŭ-mĕnt)	124-25	32°01′N	80°55′W
Fort Qu'Appelle, Sk., Can.	134-35	50°46′N	103°48′W
Fort Resolution, N.T., Can. (fôrt rĕz´ô-lū´shŭn)	128-29	61°10′N	113°38′W
Fort Rosebery, Zam. *see* Mansa	264-65	11°12′S	28°53′E
Fort Saint James, B.C., Can. (fôrt sânt jāmz)	132-33	54°28′N	124°16′W
Fort Saint John, B.C., Can. (fôrt sânt jŏn)	132-33	56°17′N	120°54′W
Fort Saskatchewan, Ab., Can. (fôrt săs-kăt´choo-ân)	132-33	53°43′N	113°14′W
Fort Severn, On., Can. (fôrt sĕv´ĕrn)	134-35	55°60′N	87°38′W
Fort-Shevchenko, Kaz. (fôrt shĕv-chĕn´kô)	186-87	44°30′N	50°16′E
Fort Simpson, N.T., Can. (fôrt sĭmp´sŭn)	128-29	61°51′N	121°22′W
Fort Smith, N.T., Can. (fôrt smĭth)	128-29	60°01′N	111°54′W
Fort Smith, Ar., U.S. (fôrt smĭth)	120-21	35°23′N	94°25′W
Fort Stockton, Tx., U.S. (fôrt stŏk´tŭn)	122-23	30°54′N	102°53′W
Fort Sumner, N.M., U.S. (fôrt sŭm´nĕr)	120-21	34°29′N	104°14′W
Fort Sumter National Monument, n.p., S.C., U.S. (fôrt sŭm´tĕr năsh´ŭn-ăl mŏn´ŭ-mĕnt)	124-25	32°45′N	79°52′W
Fortuna, Ca., U.S. (fôr-tū´ná)	112-13	40°36′N	124°09′W
Fortune, Nf., Can. (fôr´tŭn)	138-39	47°04′N	55°50′W
Fortune Bay, b., Nf., Can. (fôr´tŭn bā)	138-39	47°25′N	55°25′W
Fort Union National Monument, n.p., N.M., U.S. (fôrt ūn´yŭn năsh´ŭn-ăl mŏn´ŭ-mĕnt)	120-21	35°56′N	105°03′W
Fort Valley, Ga., U.S. (fôrt văl´ē)	124-25	32°33′N	83°53′W
Fort Vermilion, Ab., Can. (fôrt vĕr-mĭl´yŭn)	128-29	58°23′N	116°02′W
Fort Walton Beach, Fl., U.S.	124-25	30°25′N	86°36′W
Fort Wayne, In., U.S. (fôrt wān)	116-17	41°04′N	85°07′W
Fort William, Scot., U.K. (fôrt wĭl´yăm)	190-91	56°49′N	5°06′W
Fort Worth, Tx., U.S. (fôrt wûrth)	120-21	32°45′N	97°21′W
Fort Yukon, Ak., U.S. (fôrt yōō´kŏn)	126	66°34′N	145°15′W
Forūr, Jazīreh-ye, i., Iran	230-31	26°17′N	54°31′E
Foshan, China	238-39	23°03′N	113°07′E
Fossano, Italy (fôs-sä´nō)	200-01	44°33′N	7°44′E
Fossil Butte National Monument, n.p., Wy., U.S.	112-13	41°50′N	110°40′W
Fosston, Mn., U.S. (fôs´tŭn)	114-15	47°34′N	95°45′W
Foster, Austl.	276	38°39′S	146°12′E
Foster, stm., Sk., Can.	134-35	55°47′N	105°48′W
Fostoria, Oh., U.S. (fŏs-tō´rĭ-á)	116-17	41°10′N	83°24′W
Fougères, Fr. (fōō-zhâr´)	196-97	48°21′N	1°12′W
Foulwind, Cape, c., N.Z. (kāp foul´wĭnd)	278	41°45′S	171°28′E
Fouriesburg, S. Afr. (fō´rēz-bûrg)	269c	28°37′S	28°13′E
Fourmies, Fr. (fōōr-mē´)	196-97	50°01′N	4°03′E
Foveaux Strait, strt., N.Z. (fō-vō´ strāt)	278	46°35′S	168°00′E
Fowler, In., U.S. (foul´ĕr)	116-17	40°37′N	87°19′W
Foxe Basin, b., Nu., Can. (fŏks bā´s'n)	130-31	68°25′N	76°60′W
Foxe Peninsula, pen., Nu., Can. (fŏks pĕ-nĭn´sūlă)	130-31	65°0′N	76°00′W
Foz do Iguaçu, Braz.	168-69	25°33′S	54°35′W
Fraga, Spain (frä´gä)	198-99	41°32′N	0°21′E
Franca, Braz. (frä´n-kä)	172	20°32′S	47°24′W
France, nation, Eur. (fräns)	174-75	46°0′N	2°00′E
Francés Viejo, Cabo, c., Dom. Rep. (ká´bô-frän´sás vyā´hô)	142-43	19°39′N	69°55′W
Franceville, Gabon (fräɴs-vēl´)	260-61	1°38′S	13°35′E
Francis Case, Lake, res., S.D., U.S. (lāk frán´sĭs kās)	114-15	43°15′N	98°57′W
Francistown, Bots. (frán´sĭs-toun)	264-65	21°10′S	27°30′E
Francois Lake, lk., B.C., Can.	132-33	54°02′N	125°43′W
Francs Peak, mtn., Wy., U.S.	112-13	43°58′N	109°20′W
Frankfort, S. Afr. (frănk´fŏrt)	269c	27°17′S	28°31′E
Frankfort, In., U.S. (frăŋk´fŭrt)	116-17	40°17′N	86°30′W
Frankfort, Ky., U.S.	116-17	38°12′N	84°50′W
Frankfort, Mi., U.S. (frăŋk´fŭrt)	116-17	44°38′N	86°14′W
Frankfurt, Ger.	194-95	52°21′N	14°32′E
Frankfurt am Main, Ger.	194-95	50°07′N	8°40′E
Franklin, In., U.S. (frănk´lĭn)	116-17	39°28′N	86°03′W
Franklin, Ky., U.S. (frănk´lĭn)	124-25	36°43′N	86°35′W
Franklin, La., U.S. (frănk´lĭn)	124-25	29°47′N	91°30′W
Franklin, N.C., U.S. (frănk´lĭn)	124-25	35°11′N	83°23′W
Franklin, N.H., U.S. (frănk´lĭn)	116-17	43°27′N	71°40′W
Franklin, Tx., U.S. (frănk´lĭn)	122-23	31°01′N	96°29′W
Franklin, Va., U.S. (frănk´lĭn)	124-25	36°41′N	76°56′W
Franklin, Wi., U.S. (frănk´lĭn)	116-17	42°52′N	87°60′W
Franklin, W.V., U.S. (frănk´lĭn)	116-17	38°39′N	79°21′W
Franklin Mountains, mts., N.T., Can. (frănk´lĭn moun´tĭnz)	130-31	62°59′N	123°43′W
Franklinton, La., U.S. (frăŋk´lĭn-tŭn)	124-25	30°51′N	90°09′W
Frantsa-Iosifa, Zemlya, is., Russia *see* Franz Josef Land	218-19	81°0′N	55°00′E
Franz Josef Land, is., Russia	218-19	81°0′N	55°00′E
Frascati, Italy (fräs-kä´tē)	200-01	41°48′N	12°41′E
Fraser, stm., B.C., Can.	132-33	49°06′N	123°11′W
Fraserburgh, Scot., U.K. (frā´zĕr-bûrg)	190-91	57°42′N	2°01′W
Fraser Island, i., Austl.	276	25°15′S	153°10′E
Fraser Plateau, plat., B.C., Can.	132-33	52°0′N	123°00′W
Fray Bentos, Ur.	173	33°08′S	58°18′W
Frazee, Mn., U.S. (frā-zē´)	114-15	46°44′N	95°42′W
Fredericia, Den. (frĕdh-ĕ-rē´tsē-à)	192-93	55°35′N	9°46′E
Frederick, Ok., U.S. (frĕd´ĕr-ĭk)	120-21	34°23′N	99°01′W
Fredericksburg, Tx., U.S. (frĕd´ĕr-ĭkz-bûrg)	122-23	30°16′N	98°52′W
Fredericksburg, Va., U.S. (frĕd´ĕr-ĭkz-bûrg)	116-17	38°18′N	77°28′W
Fredericktown, Mo., U.S. (frĕd´ĕr-ĭk-toun)	120-21	37°34′N	90°18′W
Fredericton, N.B., Can. (frĕd´-ĕr-ĭk-tŭn)	138-39	45°57′N	66°39′W
Frederikshavn, Den. (frĕdh´ĕ-rēks-houn)	192-93	57°26′N	10°32′E
Fredonia, Col. (frĕ-dō´nyà)	163c	5°56′N	75°40′W
Fredonia, N.Y., U.S. (frē-dō´nĭ-á)	116-17	42°26′N	79°20′W
Fredrikstad, Nor. (frĕdh´rēks-städ)	192-93	59°12′N	10°56′E
Freels, Cape, c., Nf., Can. (kāp frēlz)	138-39	49°15′N	53°28′W
Freeport, Bah. (frē´pōrt)	142-43	26°31′N	78°39′W
Freeport, Il., U.S. (frē´pōrt)	116-17	42°18′N	89°37′W
Freeport, N.Y., U.S. (frē´pōrt)	116-17	40°39′N	73°35′W
Freeport, Tx., U.S. (frē´pōrt)	122-23	28°57′N	95°22′W
Freetown, nat. cap., S.L. (frē´toun)	260-61	8°29′N	13°13′W
Freiberg, Ger. (frī´bĕrgh)	194-95	50°55′N	13°21′E
Freirina, Chile (frā-ĭ-rē´nä)	168-69	28°30′S	71°06′W
Freising, Ger. (frī´zĭng)	194-95	48°24′N	11°44′E
Fréjus, Fr. (frā-zhüs´)	196-97	43°26′N	6°45′E
Fremantle, Austl. (frē´măn-t'l)	270-71	32°03′S	115°45′E
Fremont, Ca., U.S. (frē-mŏnt´)	118-19	37°33′N	121°59′W
Fremont, Mi., U.S. (frē´-mŏnt)	116-17	43°28′N	85°57′W
Fremont, Ne., U.S. (frē´-mŏnt)	114-15	41°27′N	96°30′W
Fremont, Oh., U.S. (frē´-mŏnt)	116-17	41°21′N	83°07′W
French Guiana, dep., S.A. (frĕnch gē-ä´nä)	158	4°0′N	53°00′W
French Lick, In., U.S. (frĕnch lĭk)	116-17	38°33′N	86°37′W
French Polynesia, dep., Oc. (frĕnch pŏl-ĭ-nē´zhă)	280-81	15°0′S	140°00′W
French Somaliland, nation, Afr. *see* Djibouti	253	11°30′N	43°00′E
Fresco, stm., Braz.	166-67	6°40′S	52°00′W
Freshfield, Mount, mtn., Can. (mount frĕsh´fēld)	132-33	51°44′N	116°57′W
Fresnillo, Mex. (frås-nēl´yō)	146-47	23°10′N	102°52′W
Fresno, Col. (frĕs´nō)	163c	5°09′N	75°01′W
Fresno, Ca., U.S.	118-19	36°45′N	119°46′W
Fria, Cape, c., Nmb. (kāp frĭá)	264-65	18°29′S	12°02′E
Frías, Arg. (frē´äs)	168-69	28°38′S	65°07′W
Friedberg, Ger. (frēd´bĕrgh)	194-95	48°22′N	10°59′E
Friedrichshafen, Ger. (frē-drēks-häf´ĕn)	194-95	47°40′N	9°29′E
Friend, Ne., U.S. (frĕnd)	120-21	40°39′N	97°17′W
Friesische Inseln, is., Eur. *see* Frisian Islands	190-91	53°27′N	5°50′E
Frio, Cabo, c., Braz. (ká´bō-frē´ō)	172	22°53′S	42°00′W
Frisian Islands, is., Eur. (frē´zhăn ī´lăndz)	190-91	53°27′N	5°50′E
Frobisher Bay, Nu., Can. *see* Iqaluit	128-29	63°44′N	68°28′W
Frobisher Bay, b., Nu., Can. (frŏb´ĭsh´ĕr bā)	130-31	62°30′N	65°60′W
Frobisher Lake, lk., Sk., Can. (frŏb´ĭsh´ĕr lāk)	134-35	56°22′N	108°17′W
Frolovo, Russia	186-87	49°47′N	43°39′E
Frome, Lake, lk., Austl. (lāk frōōm)	276	30°42′S	139°48′E
Frontera, Mex. (frôn-tā´rä)	146-47	18°32′N	92°38′W
Frontera, Mex. (frôn-tā´rä)	122-23	26°56′N	101°27′W
Front Royal, Va., U.S. (frŭnt roi´ăl)	116-17	38°55′N	78°12′W
Frosinone, Italy (frō-zē-nō´nå)	200-01	41°39′N	13°21′E
Frostburg, Md., U.S. (frôst´bûrg)	116-17	39°39′N	78°55′W
Frøya, i., Nor.	184-85	63°43′N	8°42′E
Fruita, Co., U.S. (frōōt-á)	118-19	39°10′N	108°44′W
Fuchun, stm., China (fōō-chòn)	238-39	30°06′N	120°10′E
Fuego, Volcán de, vol., Guat. (vôl-ká´n-dĕ-fwā´gō)	148	14°29′N	90°53′W
Fuente de Cantos, Spain (fwĕn´tå dā kän´tōs)	198-99	38°15′N	6°18′W
Fuerte, stm., Mex. (fōō-ĕ´r-tĕ)	144-45	25°51′N	109°25′W
Fuerte Olimpo, Para. (fwĕr´tå ō-lēm-pō)	168-69	21°05′S	57°52′W
Fuerteventura, i., Spain (fwĕr´tå-vĕn-tōō´rä)	199d	28°20′N	14°00′W
Fuga Island, i., Phil.	250a	18°52′N	121°22′E
Fuji, Japan (fōō´jē)	245	35°09′N	138°40′E
Fuji, stm., Japan (fōō´jē)	245	35°07′N	138°39′E
Fujian, state, China (fōō-jyĕn)	225a	26°0′N	118°00′E
Fujin, China (fōō-jyĭn)	240-41	47°15′N	132°02′E
Fuji-san, vol., Japan (fōō´jē-sän)	245	35°22′N	138°44′E
Fujiyama, vol., Japan *see* Fuji-san	245	35°22′N	138°44′E
Fukien, state, China *see* Fujian	225a	26°0′N	118°00′E
Fukuchiyama, Japan (fō´kò-chē-yä´ma)	245	35°18′N	135°07′E
Fukue-jima, i., Japan (fò-kōō´ä jē´má)	244	32°40′N	128°45′E
Fukui, Japan (fōō´kōō-ē)	245	36°04′N	136°13′E
Fukuoka, Japan	245	33°35′N	130°25′E
Fukushima, Japan (fōō´kò-shē´mà)	245	37°45′N	140°28′E
Fulaga, i., Fiji	279f	19°09′S	178°36′W
Fulaga Passage, strt., Fiji	279f	18°56′S	178°36′W
Fulda, Ger. (fòl´dä)	194-95	50°33′N	9°41′E
Fuling, China (fōō-lĭŋ)	238-39	29°42′N	107°25′E
Fullerton, Ne., U.S. (fòl´ĕr-tŭn)	114-15	41°22′N	97°58′W
Fulton, Il., U.S. (fŭl´tŭn)	114-15	41°52′N	90°10′W
Fulton, Ky., U.S. (fŭl´tŭn)	124-25	36°31′N	88°53′W
Fulton, Mo., U.S. (fŭl´tŭn)	120-21	38°51′N	91°57′W
Fulton, Ms., U.S. (fŭl´tŭn)	124-25	34°14′N	88°24′W
Fulton, N.Y., U.S. (fŭl´tŭn)	116-17	43°19′N	76°25′W
Funafuti, at., Tuvalu	280-81	8°29′S	179°11′E
Funan, China *see* Fushun	243	41°52′N	123°54′E
Funchal, Port. (fòn-shäl´)	258-59	32°39′N	16°54′W
Fundación, Col. (fōōn-dä-syō´n)	164-65	10°31′N	74°11′W
Fundy, Bay of, b., Can. (bā ŭv fŭn´dĭ)	138-39	45°0′N	66°00′W
Fundy National Park, n.p., N.B., Can. (fŭn´dĭ năsh´ŭn-ăl pärk)	138-39	45°38′N	65°00′W
Fünfkirchen, Hung. *see* Pécs	194-95	46°04′N	18°13′E
Funing, China (fōō-nĭŋ)	238-39	23°34′N	105°37′E
Furnas, Represa de, res., Braz.	172	21°12′S	45°57′W
Furneaux Group, is., Austl. (fûr´nō grōōp)	276	40°10′S	148°05′E
Fürstenwalde, Ger. (für´stĕn-väl-dĕ)	194-95	52°21′N	14°04′E
Fürth, Ger. (fürt)	194-95	49°28′N	10°59′E
Fusan, Kor., S. *see* Pusan	243	35°05′N	129°03′E
Fushun, China (fōō´shōōn´)	243	41°52′N	123°54′E
Fusong, China (fōō-soŋ)	243	42°20′N	127°17′E
Fusui, China	238-39	22°38′N	107°55′E
Futuna, Île, i., Wal./F.	280-81	14°18′S	178°09′W
Fuxian, China (fōō shyĕn)	240-41	36°02′N	109°22′E
Fuxian, China (fōō shyĕn) *see* Wafangdian	240-41	39°37′N	122°01′E
Fuxian Hu, lk., China	238-39	24°29′N	102°53′E
Fuxin, China (fōō-shyĭn)	240-41	42°08′N	121°45′E
Fuyang, China (fōō-yäŋ)	238-39	32°54′N	115°49′E
Fuyang, stm., China (fōō-yäŋ)	240-41	38°11′N	116°04′E
Fuyu, China (fōō-yōō)	240-41	45°10′N	124°49′E
Fuyu, China (fōō-yōō)	240-41	47°49′N	124°28′E
Fuzhou, China	225a	26°06′N	119°17′E
Fuzhou, China (fōō-jō)	238-39	28°01′N	116°20′E
Fyn, i., Den. (fü´n)	192-93	55°20′N	10°30′E

G

Feature (Pronunciation)	Page	Lat.	Long.
Gaalkacyo, Som.	262-63	6°46′N	47°26′E
Gabela, Ang.	264-65	10°51′S	14°22′E
Gaberones, nat. cap., Bots. *see* Gaborone	264-65	24°40′S	25°56′E
Gabès, Tun. (gä´bĕs)	188-89	33°54′N	10°06′E
Gabès, Golfe de, b., Tun. (gôlf-dĕ-gä´bĕs)	258-59	34°14′N	10°30′E

Feature (Pronunciation)	Page	Lat.	Long.
Gabon, nation, Afr. (gȧ-bôɴ′)	253	1°0′S	11°45′E
Gaborone, nat. cap., Bots. (gä-bō-rō′-nä) (gä′bô-rōō-nä)	264-65	24°40′S	25°56′E
Gabrovo, Blg. (gäb′rô-vō)	200-01	42°51′N	25°19′E
Gachsārān, Iran	230-31	30°12′N	50°47′E
Gacko, Bos. (gäts′kô)	200-01	43°10′N	18°32′E
Gadag, India	236	15°25′N	75°37′E
Gadsden, Al., U.S. (gădz′dĕn)	124-25	34°00′N	86°01′W
Găeşti, Rom. (gä-yĕsh′tê)	200-01	44°43′N	25°20′E
Gaeta, Italy (gä-â′tä)	200-01	41°13′N	13°34′E
Gaferut, i., Micron.	280-81	9°12′N	145°23′E
Gaffney, S.C., U.S. (găf′nĭ)	124-25	35°04′N	81°39′W
Gafsa, Tun. (gäf′sä)	188-89	34°24′N	8°49′E
Gagnoa, C. Iv.	260-61	6°07′N	5°56′W
Gagra, Geor.	227	43°20′N	40°15′E
Gaillimh, Ire. *see* Galway	190-91	53°16′N	9°03′W
Gainesville, Fl., U.S. (gānz′vĭl)	124-25	29°39′N	82°18′W
Gainesville, Ga., U.S. (gānz′vĭl)	124-25	34°18′N	83°49′W
Gainesville, Tx., U.S. (gānz′vĭl)	120-21	33°38′N	97°09′W
Gainesville, Va., U.S. (gānz′vĭl)	116-17	38°47′N	77°38′W
Gairdner, Lake, lk., Austl. (lāk gârd′nẽr)	272-73	31°33′S	135°57′E
Gaithersburg, Md., U.S. (gā′thẽrs′bûrg)	116-17	39°08′N	77°12′W
Gaixian, China (gī-shyĕn)	240-41	40°24′N	122°22′E
Galán, Cerro, mtn., Arg.	168-69	25°57′S	66°54′W
Galana, stm., Kenya	262-63	3°09′S	40°08′E
Galapagos Islands, is., Ec. (gä-lä′-pä-gōs ī′lȧndz) *see* Colón, Archipiélago de	170a	0°43′N	91°30′W
Galashiels, Scot., U.K. (găl-ȧ-shēlz)	190-91	55°38′N	2°50′W
Galaţi, Rom.	202-03	45°26′N	28°03′E
Galatina, Italy (gä-lä-tē′nä)	200-01	40°10′N	18°10′E
Galatz, Rom. *see* Galaţi	202-03	45°26′N	28°03′E
Galdhøpiggen, mtn., Nor.	192-93	61°37′N	8°17′E
Galeana, Mex. (gä-lå-ä′nä)	122-23	24°50′N	100°04′W
Galela, Indon.	248-49	1°50′N	127°50′E
Galena, Ak., U.S. (gȧ-lē′nȧ)	126	64°44′N	156°57′W
Galena, Il., U.S. (gȧ-lē′nȧ)	114-15	42°25′N	90°25′W
Galera, Punta, c., Chile	171	39°58′S	73°40′W
Galera, Punta, c., Ec.	170	0°49′N	80°03′W
Galesburg, Il., U.S. (gālz′bûrg)	114-15	40°57′N	90°22′W
Galeton, Pa., U.S. (gāl′tŭn)	116-17	41°44′N	77°39′W
Galich, Russia (gål′ĭch)	186-87	58°23′N	42°22′E
Galicia, hist. reg., Eur. (gȧ-lĭsh′ĭ-ȧ)	194-95	49°0′N	22°00′E
Galicja, hist. reg., Eur. *see* Galicia	194-95	49°0′N	22°00′E
Galilee, Sea of, lk., Isr. (sē ŭv găl′ĭ-lē)	228-29	32°48′N	35°35′E
Galion, Oh., U.S. (găl′ĭ-ŭn)	116-17	40°44′N	82°47′W
Galkynyş, Turkmen.	232-33	39°16′N	63°11′E
Gallatin, Mo., U.S. (găl′ȧ-tĭn)	120-21	39°55′N	93°58′W
Gallatin, Tn., U.S. (găl′ȧ-tĭn)	124-25	36°24′N	86°27′W
Galle, Sri L. (gäl)	236	6°02′N	80°13′E
Gallinas, Punta, c., Col. (pōō′n-tä-gä-lyē′näs)	164-65	12°28′N	71°40′W
Gallipoli, Italy (gäl-lē′pô-lē)	200-01	40°03′N	17°59′E
Gallipoli, Tur.	200-01	40°26′N	26°41′E
Gallipoli Peninsula, pen., Tur. (gäl-lē′pô-lē pĕ-nĭn′sūlȧ)	200-01	40°20′N	26°30′E
Gallipolis, Oh., U.S. (găl-ĭ-pô-lēs)	116-17	38°49′N	82°12′W
Gällivare, Swe. (yĕl-ĭ-vär′ĕ)	184-85	67°08′N	20°41′E
Gallup, N.M., U.S. (găl′ŭp)	118-19	35°32′N	108°45′W
Galva, Il., U.S. (găl′vȧ)	114-15	41°10′N	90°02′W
Galveston, Tx., U.S. (găl′vĕs-tŭn)	122-23	29°18′N	94°48′W
Galveston Bay, b., Tx., U.S. (găl′vĕs-tŭn bā)	122-23	29°36′N	94°57′W
Galveston Island, i., Tx., U.S. (găl′vĕs-tŭn ī′lȧnd)	122-23	29°13′N	94°55′W
Gálvez, Arg.	173	32°02′S	61°13′W
Galway, Ire. (gôl′wā)	190-91	53°16′N	9°03′W
Gamba, China (gäm-bä)	234-35	28°17′N	88°31′E
Gambell, Ak., U.S.	126	63°47′N	171°44′W
Gambia, The, nation, Afr. (thȧ găm′bē-ȧ)	253	13°30′N	15°30′W
Gambier, Îles, is., Fr. Poly.	280-81	23°08′S	134°57′W
Gamboma, Congo (gäm-bō′mä)	262-63	1°53′S	15°51′E
Gamlakarleby, Fin. *see* Kokkola	184-85	63°50′N	23°09′E
Gamleby, Swe. (gäm′lĕ-bü)	192-93	57°55′N	16°23′E
Gan, stm., China (gän)	222-23	49°11′N	125°10′E
Gananoque, On., Can.	136-37	44°20′N	76°10′W
Gäncä, Azer.	227	40°41′N	46°21′E
Ganda, Ang.	264-65	13°02′S	14°39′E
Gandajika, D.R.C.	262-63	6°44′S	23°57′E
Gander, Nf., Can. (găn′dẽr)	138-39	48°57′N	54°35′W
Gander, stm., Nf., Can. (găn′dẽr)	138-39	49°29′N	54°24′W
Gander Lake, lk., Nf., Can. (găn′dẽr lāk)	138-39	48°57′N	54°39′W
Gāndhinagar, India	234-35	23°13′N	72°40′E
Gandía, Spain	198-99	38°58′N	0°11′W
Ganga, stm., Asia *see* Ganges	234-35	21°58′N	90°57′E
Gangānagar, India	234-35	29°55′N	73°52′E
Gangaw, Mya.	246-47	22°11′N	94°09′E
Gangdisê Shan, mts., China	234-35	31°0′N	82°00′E
Ganges, stm., Asia (găn′jēz)	234-35	21°58′N	90°57′E
Ganges, Mouths of the, mth., Asia (mouthz ŭv thȧ găn′jēz)	234-35	22°0′N	89°00′E
Gangneung, Kor., S. *see* Kangnŭng	243	37°46′N	128°54′E
Gangotri, India	234-35	30°60′N	78°59′E
Gangtok, India	234-35	27°19′N	88°38′E
Gangu, China	238-39	34°45′N	105°20′E
Gannan, China (gän-nän)	240-41	47°56′N	123°30′E
Gannett Peak, mtn., Wy., U.S. (găn′ĕt pēk)	112-13	43°11′N	109°39′W
Gansu, state, China (gän-sōō)	240-41	37°0′N	103°00′E
Ganzê, China	238-39	31°38′N	100°01′E
Ganzhou, China (gän-jō)	238-39	25°53′N	114°55′E
Gao, Mali (gä′ō)	258-59	16°16′N	0°02′W
Gao'an, China (gou-än)	238-39	28°26′N	115°23′E
Gaoyi, China (gou-yē)	240-41	37°37′N	114°36′E
Gaoyou, China (gou-yō)	238-39	32°47′N	119°26′E
Gaoyou Hu, lk., China (kä′ō-yōō′hōō)	238-39	32°50′N	119°20′E
Gap, Fr. (gȧp)	196-97	44°34′N	6°05′E
Garabogazköl Aylagy, b., Turkmen. *see* Kara-Bogaz-Gol Gulf	232-33	41°15′N	53°24′E
Garagum, des., Turkmen. *see* Kara Kum	226	39°0′N	60°00′E
Garagum Kanaly, can., Turkmen. *see* Kara-Kum Canal	232-33	37°34′N	65°41′E
Garanhuns, Braz. (gä-rän-yòɴsh′)	163d	8°54′S	36°29′W
Garber, Ok., U.S. (gär′bẽr)	120-21	36°26′N	97°35′W
Garden City, Ga., U.S. (gär′d′n sĭ′tê)	124-25	32°07′N	81°09′W
Garden City, Ks., U.S. (gär′d′n sĭ′tê)	120-21	37°58′N	100°52′W
Gardeyz, Afg.	232-33	33°36′N	69°13′E
Gardiner, Mt., U.S. (gärd′nẽr)	112-13	45°02′N	110°42′W
Gardner, Ks., U.S. (gärd′nẽr)	120-21	38°49′N	94°56′W
Gardner, Ma., U.S. (gärd′nẽr)	116-17	42°35′N	71°60′W
Gardner Canal, b., B.C., Can. (gärd′nẽr kȧ′năl)	132-33	53°28′N	128°18′W
Gardner Pinnacles, r., Hi., U.S. (gärd′nẽr pĭn′ȧ-k′lz)	127	25°0′N	167°55′W
Gargždai, Lith. (gärgzh′dī)	192-93	55°43′N	21°24′E
Garibaldi, Mount, vol., B.C., Can. (mount gâr-ĭ-băl′dê)	132-33	49°51′N	122°59′W
Garissa, Kenya	262-63	0°27′S	39°39′E
Garland, Tx., U.S. (gär′lănd)	120-21	32°56′N	96°38′W
Garmisch-Partenkirchen, Ger. (gär′mēsh pär′tĕn-kẽr′kĕn)	194-95	47°30′N	11°06′E
Garnett, Ks., U.S. (gär′nĕt)	120-21	38°17′N	95°14′W
Garoua, Camrn. (gär′wä)	260-61	9°19′N	13°23′E
Garqu Yan, China	238-39	33°54′N	92°19′E
Garrett, In., U.S. (găr′ĕt)	116-17	41°21′N	85°07′W
Garrison, N.D., U.S. (găr′ĭ-sŭn)	114-15	47°39′N	101°25′W
Garry Lake, lk., Nu., Can. (găr′ĭ lāk)	130-31	66°0′N	100°00′W
Garut, Indon.	248-49	7°12′S	107°54′E
Garwolin, Pol. (gär-vō′lĕn)	194-95	51°54′N	21°38′E
Gary, In., U.S. (gā′rĭ)	116-17	41°36′N	87°21′W
Garyarsa, China	234-35	31°43′N	80°20′E
Garzón, Col. (gär-thōn′)	164-65	2°12′N	75°38′W
Gas City, In., U.S. (găs sĭ′tê)	116-17	40°29′N	85°37′W
Gascogne, Golfe de, b., Eur. *see* Biscay, Bay of	196-97	44°0′N	4°00′W
Gash, stm., Afr.	266	16°45′N	35°54′E
Gaspé, Qc., Can.	138-39	48°49′N	64°29′W
Gasteiz, Spain	198-99	42°51′N	2°40′W
Gastonia, N.C., U.S. (găs-tô′nĭ-ȧ)	124-25	35°16′N	81°11′W
Gastre, Arg. (gäs-trĕ)	171	42°17′S	69°14′W
Gata, Cabo de, c., Spain (kȧ′bô-dĕ-gä′tä)	198-99	36°44′N	2°11′W
Gata, Sierra de, mts., Spain (syĕr′rȧ dā gä′tä)	198-99	40°16′N	6°44′W
Gatchina, Russia (gä-chē′nȧ)	192-93	59°33′N	30°08′E
Gates of the Arctic National Park and Preserve, n.p., Ak., U.S.	126	67°45′N	153°30′W
Gatesville, Tx., U.S. (gāts′vĭl)	122-23	31°25′N	97°44′W
Gatineau, Qc., Can. (gȧ′tê-nō)	136-37	45°29′N	75°38′W
Gatineau, stm., Qc., Can. (gȧ′tê-nō)	136-37	45°27′N	75°42′W
Gauer Lake, lk., Mb., Can.	134-35	57°0′N	97°50′W
Gauja, stm., Eur. (gå′ô-yä)	192-93	57°09′N	24°17′E
Gaustatoppen, mtn., Nor.	192-93	59°50′N	8°35′E
Gávdos, i., Grc. (gäv′dôs)	200a	34°50′N	24°06′E
Gävle, Swe. (yĕv′lĕ)	192-93	60°40′N	17°10′E
Gavrilov-Yam, Russia (gá′vrê-lôf yäm′)	202-03	57°18′N	39°52′E
Gaxun Nur, lk., China	240-41	42°22′N	100°34′E
Gaya, India (gŭ′yä)(gī′ȧ)	234-35	24°48′N	85°00′E
Gaylord, Mi., U.S. (gā′lôrd)	116-17	45°02′N	84°40′W
Gaylord, Mn., U.S. (gā′lôrd)	114-15	44°33′N	94°14′W
Gayndah, Austl. (gān′däh)	276	25°38′S	151°36′E
Gayny, Russia	186-87	60°18′N	54°19′E
Gaza, Gaza (gä′zȧ) (gā′zȧ)	228-29	31°30′N	34°28′E
Gazanjyk, Turkmen.	232-33	39°15′N	55°32′E
Gaziantep, Tur. (gä-zē-än′tĕp)	228-29	37°04′N	37°23′E
Gazimağusa, Cyp.	228-29	35°07′N	33°57′E
Gbadolite, D.R.C.	262-63	4°15′N	21°00′E
Gbanga, Lib.	260-61	7°00′N	9°29′W
Gboko, Nig.	260-61	7°20′N	8°60′E
Gdańsk, Pol. (g′dänsk)	194-95	54°21′N	18°38′E
Gdov, Russia (g′dôf′)	192-93	58°45′N	27°49′E
Gdynia, Pol. (g′dĕn′yä)	194-95	54°32′N	18°31′E
Geary, Ok., U.S. (gē′rĭ)	120-21	35°38′N	98°19′W
Gediz, stm., Tur.	200-01	38°36′N	26°48′E
Geelong, Austl. (jē-lông′)	276	38°08′S	144°21′E
Geeveston, Austl.	276	43°10′S	146°55′E
Gefle, Swe. *see* Gävle	192-93	60°40′N	17°10′E
Geita, Tan.	267	2°52′S	32°10′E
Gejiu, China (gŭ-jĭo)	238-39	23°22′N	103°09′E
Gela, Italy	200-01	37°04′N	14°15′E
Gelasa, Selat, strt., Indon.	246-47	2°55′S	107°13′E
Gelibolu, Tur. (gĕ-lĭb′ô-lò) *see* Gallipoli	200-01	40°26′N	26°41′E
Gelibolu Yarımadası, pen., Tur. *see* Gallipoli Peninsula	200-01	40°20′N	26°30′E
Gemena, D.R.C.	262-63	3°14′N	19°47′E
Gemlik, Tur. (gĕm′lĭk)	200-01	40°26′N	29°09′E
Gemsbok National Park, n.p., Bots.	264-65	25°15′S	21°10′E
Gen, stm., China	240-41	50°15′N	119°21′E
Genalē, stm., Afr.	262-63	0°15′S	42°39′E
General Acha, Arg.	173	37°23′S	64°36′W
General Alvear, Arg. (hĕ-nĕ-räl′ ăl-vĕ-ȧ′r)	171	34°59′S	67°42′W
General Alvear, Arg. (hĕ-nĕ-räl′ ăl-vĕ-ȧ′r)	173	36°01′S	60°01′W
General Belgrano, Arg. (hĕ-nĕ-räl′ bĕl-grä′nô)	173	35°46′S	58°29′W
General Carrera, Lago, lk., S.A.	171	46°26′S	71°40′W
General Cepeda, Mex. (hĕ-nĕ-räl′ sĕ-pĕ′dä)	122-23	25°23′N	101°28′W
General Conesa, Arg. (hĕ-nĕ-räl′ kô-nĕ′sä)	171	40°07′S	64°26′W
General Eugenio A. Garay, Para.	168-69	20°30′S	62°11′W
General Guido, Arg. (hĕ-nĕ-räl′ gē′dô)	173	36°40′S	57°48′W
General Juan Madariaga, Arg.	173	37°00′S	57°09′W
General La Madrid, Arg.	173	37°15′S	61°17′W
General Lavalle, Arg. (hĕ-nĕ-räl′ lȧ-vá′l-yĕ)	173	36°25′S	56°57′W
General Levalle, Arg.	173	34°01′S	63°55′W
General Manuel Belgrano, Cerro, mtn., Arg.	168-69	29°01′S	67°50′W
General Pico, Arg. (hĕ-nĕ-räl′ pē′kô)	173	35°40′S	63°46′W
General Pinedo, Arg.	173	27°19′S	61°17′W
General Roca, Arg. (hĕ-nĕ-räl′ rô-kä)	171	39°01′S	67°35′W
General San Martín, Arg. (hĕ-nĕ-räl′ sän-mär-tē′n)	173	34°34′S	58°33′W
General Santos, Phil.	250	6°07′N	125°10′E
General Viamonte, Arg. (hĕ-nĕ-räl′ vēȧ′môn-tĕ)	173	34°60′S	61°02′W
General Villegas, Arg.	173	35°02′S	63°01′W
Geneseo, Il., U.S. (jĕ-nĕsêō)	114-15	41°27′N	90°09′W
Geneva, Switz. (jĕ-nē′vȧ)	194-95	46°12′N	6°09′E
Geneva, Al., U.S. (jĕ-nē′vȧ)	124-25	31°02′N	85°52′W
Geneva, In., U.S. (jĕ-nē′vȧ)	116-17	40°35′N	84°57′W
Geneva, Ne., U.S. (jĕ-nē′vȧ)	120-21	40°32′N	97°36′W
Geneva, N.Y., U.S. (jĕ-nē′vȧ)	116-17	42°52′N	76°59′W
Geneva, Oh., U.S. (jĕ-nē′vȧ)	116-17	41°48′N	80°56′W
Geneva, Lake, lk., Eur. (lāk jĕ-nē′vȧ)	194-95	46°24′N	6°22′E
Genève, Switz. *see* Geneva	194-95	46°12′N	6°09′E
Genève, Lac de, lk., Eur. *see* Geneva, Lake	194-95	46°24′N	6°22′E
Genf, Switz. *see* Geneva	194-95	46°12′N	6°09′E
Genil, stm., Spain (hå-nēl′)	198-99	37°42′N	5°19′W
Genoa, Italy (jen′ô-ȧ)	200-01	44°25′N	8°57′E
Genova, Italy *see* Genoa	200-01	44°25′N	8°57′E
Genova, Golfo di, b., Italy (gôl-fô-dē-jĕn′ō-vä)	200-01	44°10′N	8°55′E
Genovesa, Isla, i., Ec. (ê′s-lä-gĕ-nō-vĕ-sä)	170a	0°20′N	89°57′W
Gensan, Kor., N. *see* Wŏnsan	243	39°09′N	127°26′E
Geographe Bay, b., Austl. (jē-ô-grăf′ bā)	272-73	33°35′S	115°15′E

n-sing; ŋ-baŋk; ɴ-nasalized n; nŏd; cŏmmit; ōld; ôbey; ôrder; oi-boil; fōōd; ò-as oo in foot; ou-out; s-soft; sh-dish; th-thin; pūre; ůnite; ûrn; stŭd; circŭs; ü-as in French tu; ′-indeterminate vowel.

Feature (Pronunciation)	Page	Lat.	Long.
George, S. Afr.	264-65	33°58'S	22°27'E
George, stm., Qc., Can.	130-31	58°46'N	66°08'W
George, Lake, lk., Ug. (lăk jôrg)	267	0°02'N	30°12'E
George, Lake, lk., Fl., U.S. (lăk jôr-ĭj)	124-25	29°17'N	81°36'W
Georgetown, On., Can. (jôr-ĭj-toun)	136-37	43°39'N	79°55'W
Georgetown, P.E., Can. (jôr-ĭj-toun)	138-39	46°11'N	62°32'W
Georgetown, Gam.	260-61	13°33'N	14°46'W
George Town, Malay.	246-47	5°25'N	100°20'E
Georgetown, De., U.S. (jôrg-toun)	116-17	38°41'N	75°23'W
Georgetown, Il., U.S. (jôrg-toun)	116-17	39°58'N	87°38'W
Georgetown, Ky., U.S. (jôrg-toun)	116-17	38°12'N	84°34'W
Georgetown, Oh., U.S. (jôrg-toun)	116-17	38°51'N	83°52'W
Georgetown, S.C., U.S. (jôr-ĭj-toun)	124-25	33°23'N	79°18'W
Georgetown, Tx., U.S. (jôr-toun)	122-23	30°38'N	97°41'W
George Town, nat. cap., Cay. Is. (jôr-ĭj-toun)	142-43	19°18'N	81°22'W
Georgetown, nat. cap., Guy. (jôrj´toun)	164-65	6°48'N	58°09'W
George Washington Birthplace National Monument, n.p., Va., U.S. (jôrj wŏsh´ĭng-tŭn bûrth´plās năsh´ŭn-ăl mŏn´ū-mĕnt)	116-17	38°11'N	76°56'W
George Washington Carver National Monument, n.p., Mo., U.S. (jôrg wăsh-ĭng-tŭn kär´vēr năsh´ŭn-ăl mŏn´ū-mĕnt)	120-21	37°00'N	94°21'W
George West, Tx., U.S. (jôrg wĕst)	122-23	28°20'N	98°07'W
Georgia, nation, Asia (jôr´ji-ă)	227	42°0'N	44°00'E
Georgia, state, U.S. (jôr´jĭ-ă)	108-09	32°50'N	83°15'W
Georgiana, Al., U.S. (jôr-jē-ăn´ȧ)	124-25	31°38'N	86°45'W
Georgian Bay, b., On., Can.	136-37	45°15'N	80°50'W
Georgiyevsk, Russia (gyôr-gyĕfsk´)	227	44°09'N	43°29'E
Gera, Ger. (gā´rä)	194-95	50°52'N	12°05'E
Geral, Serra, mts., Braz. (sĕr´rȧ zhȧ-räl´)	168-69	26°30'S	50°30'W
Geraldton, Austl. (jĕr´ăld-tŭn)	270-71	28°46'S	114°37'E
Geraldton, On., Can.	128-29	49°41'N	86°60'W
Gereshk, Afg.	232-33	31°49'N	64°34'E
Gering, Ne., U.S. (gē´rĭng)	114-15	41°48'N	103°40'W
Gerlachovský štít, mtn., Slvk.	194-95	49°11'N	20°09'E
Germantown, Tn., U.S. (jûr´mȧn-toun)	124-25	35°06'N	89°49'W
Germantown, Wi., U.S. (jûr´mȧn-toun)	116-17	43°14'N	88°07'W
Germany, nation, Eur. (jûr´mȧ-nĭ)	174-75	51°0'N	10°00'E
Germiston, S. Afr. (jûr´mĭs-tŭn)	269c	26°13'S	28°11'E
Gerona, Spain *see* Girona	198-99	41°59'N	2°49'E
Getafe, Spain (hā-tä´fā)	198-99	40°19'N	3°44'W
Gettysburg, S.D., U.S. (gĕt´ĭs-bûrg)	114-15	45°01'N	99°57'W
Ghaapplato, plat., S. Afr.	264-65	27°29'S	24°19'E
Ghadāmis, Libya	188-89	30°12'N	9°33'E
Ghāghara, stm., Asia	234-35	25°45'N	84°48'E
Ghāghra, stm., Asia *see* Ghāghara	234-35	25°45'N	84°48'E
Ghana, nation, Afr. (găn´ȧ)	253	8°0'N	1°00'W
Ghanzi, Bots. (găn´zē)	264-65	21°42'S	21°39'E
Ghardaïa, Alg. (gär-dä´ē-ä)	258-59	32°33'N	3°40'E
Gharm, Taj.	232-33	39°02'N	70°23'E
Gharyān, Libya	188-89	32°10'N	13°01'E
Ghāt, Libya	258-59	24°56'N	10°12'E
Ghawdex, i., Malta *see* Gozo	200b	36°03'N	14°15'E
Ghazal, Bahr el, stm., Chad (băr ĕl ghă-zäl´)	258-59	13°05'N	15°20'E
Ghāziābād, India	234-35	28°40'N	77°26'E
Ghaznī, Afg.	232-33	33°33'N	68°25'E
Ghazzah, Gaza (gä´ziä) *see* Gaza	228-29	31°30'N	34°28'E
Ghijduwon, Uzb.	232-33	40°06'N	64°41'E
Ghorīān, Afg.	232-33	34°21'N	61°29'E
Gibara, Cuba (hē-bä´rä)	142-43	21°07'N	76°08'W
Gibraleón, Spain (hē-brä-lå-ōn´)	198-99	37°23'N	6°58'W
Gibraltar, dep., Eur. (gĭ-bräl-tä´r)	174-75	36°08'N	5°21'W
Gibraltar, nat. cap., Gib. (gĭ-bräl-tä´r)	198-99	36°08'N	5°21'W
Gibraltar, Estrecho de, strt., *see* Gibraltar, Strait of	198-99	35°57'N	5°36'W
Gibraltar, Strait of, strt., (stāt ŭv gĭ-bräl-tä´r)	198-99	35°57'N	5°36'W
Gibson City, Il., U.S. (gĭb´sŭn sĭ´tē)	116-17	40°28'N	88°22'W
Gibson Desert, des., Austl. (gĭb´sŭn dĕs´ērt)	272-73	24°30'S	126°00'E
Giddings, Tx., U.S. (gĭd´ĭngz)	122-23	30°11'N	96°56'W
Gien, Fr. (zhē-ăN´)	196-97	47°41'N	2°38'E
Gießen, Ger. (gēs´sĕn)	194-95	50°35'N	8°40'E
Gifu, Japan (gē´fōō)	245	35°25'N	136°45'E
Gijón, Spain (hē-hōn´)	198-99	43°32'N	5°40'W
Gila, stm., U.S. (hē´lȧ)	110-11	32°43'N	114°33'W
Gila Bend, Az., U.S. (hē´lȧ bĕnd)	118-19	32°57'N	112°43'W
Gila Cliff Dwellings National Monument, n.p., N.M., U.S. (hē´lȧ klĭf dwĕl´ĭngz năsh´ŭn-ăl mŏn´ū-mĕnt)	118-19	33°02'N	108°16'W
Gilbert, Mn., U.S. (gĭl´bērt)	114-15	47°29'N	92°28'W
Gilbert, Mount, mtn., B.C., Can. (mount gĭl-bĕrt)	132-33	50°54'N	124°17'W
Gilbert Islands, nation, Oc. *see* Kiribati	280-81	5°0'S	170°00'W
Gilbert Islands, is., Kir. (gĭl-bĕrt ī´lăndz) *see* Kiribati	280-81	0°30'S	174°00'E
Gilbués, Braz.	166-67	9°50'S	45°21'W
Gilford Island, i., B.C., Can. (gĭl´fērd ī´lȧnd)	132-33	50°45'N	126°20'W
Gilgandra, Austl.	276	31°43'S	148°40'E
Gilgit, Pak. (gĭl´gĭt)	232-33	35°53'N	74°21'E
Gilgit, stm., Pak.	232-33	35°42'N	74°38'E
Gil Island, i., B.C., Can. (gĭl ī´lȧnd)	132-33	53°11'N	129°15'W
Gillam, Mb., Can.	134-35	56°21'N	94°43'W
Gillette, Wy., U.S. (jĭ-lĕt´)	112-13	44°18'N	105°30'W
Gillingham, Eng., U.K. (gĭl´ĭng ăm)	190-91	51°23'N	0°34'E
Gilman, Il., U.S. (gĭl´măn)	116-17	40°46'N	87°59'W
Gilmer, Tx., U.S. (gĭl´mēr)	120-21	32°44'N	94°57'W
Gilo, stm., Eth.	262-63	8°07'N	33°11'E
Gilroy, Ca., U.S. (gĭl-roi´)	118-19	37°01'N	121°34'W
Giluwe, Mount, mtn., Pap. N. Gui.	277	6°02'S	143°51'E
Gilyuy, stm., Russia	218-19	53°59'N	127°27'E
Gimcheon, Kor., S. *see* Kimch'ŏn	243	36°07'N	128°07'E
Gimli, Mb., Can. (gĭm´lē)	134-35	50°38'N	96°59'W
Gioia del Colle, Italy (jô´yä dĕl kôl´lā)	200-01	40°48'N	16°55'E
Girardot, Col. (hē-rär-dōt´)	163c	4°18'N	74°47'W
Giresun, Tur. (ghēr´ĕ-sōn´)	186-87	40°55'N	38°24'E
Girga, Egypt	268b	26°20'N	31°53'E
Girīdīh, India (jē-rē-dē)	234-35	24°11'N	86°18'E
Girona, Spain	198-99	41°59'N	2°49'E
Girvan, Scot., U.K. (gûr´văn)	190-91	55°15'N	4°52'W
Gisborne, N.Z. (gĭz´bŭrn)	278	38°40'S	178°01'E
Gisenyi, Rw.	267	1°42'S	29°16'E
Gisors, Fr. (zhē-zôr´)	196-97	49°17'N	1°47'E
Gitarama, Rw.	267	2°04'S	29°44'E
Gitega, Bdi.	267	3°21'S	29°54'E
Giurgiu, Rom. (jòr´jò)	200-01	43°54'N	25°58'E
Givet, Fr. (zhē-vĕ´)	196-97	50°08'N	4°50'E
Giyon, Eth.	269d	8°32'N	37°59'E
Giza, Egypt	268b	30°01'N	31°13'E
Gizo, Sol. Is.	279e	8°06'S	156°50'E
Giżycko, Pol. (gĭ´zhĭ-ko)	194-95	54°02'N	21°46'E
Gjoa Haven, Nu., Can.	128-29	68°39'N	95°55'W
Gjøvik, Nor. (gyû´vĕk)	192-93	60°48'N	10°41'E
Glace Bay, N.S., Can. (glās bā)	138-39	46°13'N	59°58'W
Glacier Bay National Park and Preserve, n.p., Ak., U.S.	126	59°04'N	136°36'W
Glacier National Park, n.p., B.C., Can. (glā´shēr năsh´ŭn-ăl pärk)	132-33	51°15'N	117°35'W
Glacier National Park, n.p., Mt., U.S.	112-13	48°35'N	113°40'W
Glacier Peak, vol., Wa., U.S. (glā´shēr pēk) (glā´shēr pēk)	112-13	48°07'N	121°07'W
Gladstone, Austl. (glăd´stōn)	277	23°51'S	151°15'E
Gladstone, Austl. (glăd´stōn)	276	33°17'S	138°21'E
Gladstone, Mi., U.S. (glăd´stōn)	116-17	45°51'N	87°01'W
Gladstone, Mo., U.S. (glăd´stōn)	120-21	39°14'N	94°35'W
Gladwin, Mi., U.S. (glăd´wĭn)	116-17	43°59'N	84°29'W
Glåma, stm., Nor. *see* Glomma	184-85	59°11'N	10°58'E
Glasgow, Scot., U.K. (glås´gō)	190-91	55°53'N	4°15'W
Glasgow, Ky., U.S.	124-25	37°00'N	85°55'W
Glasgow, Mt., U.S.	112-13	48°12'N	106°38'W
Glauchau, Ger. (glou´κou)	194-95	50°49'N	12°33'E
Glazov, Russia (glä´zôf)	186-87	58°08'N	52°39'E
Gleiwitz, Pol. *see* Gliwice	194-95	50°17'N	18°40'E
Glen Canyon, val., U.S. (glĕn kăn´yŭn)	118-19	37°10'N	110°50'W
Glencoe, S. Afr. (glĕn-cō)	269c	28°12'S	30°06'E
Glendale, Az., U.S. (glĕn´dāl)	118-19	33°32'N	112°12'W
Glendale, Ca., U.S. (glĕn´dāl)	118-19	34°08'N	118°14'W
Glendive, Mt., U.S. (glĕn´dīv)	112-13	47°06'N	104°43'W
Glen Innes, Austl. (glĕn ĭn´ĕs)	276	29°44'S	151°44'E
Glenns Ferry, Id., U.S. (glĕns fĕr´ē)	112-13	42°58'N	115°18'W
Glenrock, Wy., U.S. (glĕn´rŏk)	112-13	42°52'N	105°53'W
Glens Falls, N.Y., U.S. (glĕnz fôlz)	116-17	43°19'N	73°39'W
Glittertinden, mtn., Nor.	192-93	61°39'N	8°33'E
Gliwice, Pol. (gwĭ-wĭt´sĕ)	194-95	50°17'N	18°40'E
Globe, Az., U.S. (glōb)	118-19	33°24'N	110°47'W
Glomma, stm., Nor.	184-85	59°11'N	10°58'E
Glorieuses, Îles, is., Reu.	264-65	11°30'S	47°20'E
Glorioso Islands, is., Reu. *see* Glorieuses, Îles	264-65	11°30'S	47°20'E
Gloucester, Eng., U.K. (glŏs´tēr)	190-91	51°53'N	2°14'W
Gloversville, N.Y., U.S. (glŭv´ērz-vĭl)	116-17	43°03'N	74°21'W
Glovertown, Nf., Can. (glŭv´ĕr-toun)	138-39	48°40'N	54°03'W
Glückstadt, Ger. (glük-shtät)	194-95	53°47'N	9°26'E
Gmunden, Aus. (g'mòn´dĕn)	194-95	47°55'N	13°47'E
Gnesen, Pol. *see* Gniezno	194-95	52°32'N	17°37'E
Gniezno, Pol. (g'nyåz´nô)	194-95	52°32'N	17°37'E
Gnjilane, Serb. (gnyē´lå-nĕ)	200-01	42°28'N	21°29'E
Goa, state, India (gō´ȧ)	236	15°20'N	74°00'E
Goālpāra, India	234-35	26°10'N	90°37'E
Goba, Eth. (gō´bä)	269d	7°00'N	39°59'E
Gobabis, Nmb. (gō-bä´bĭs)	264-65	22°27'S	18°58'E
Gobernador Gregores, Arg.	171	48°46'S	70°15'W
Gobernador Vera, Arg. *see* Vera	173	29°28'S	60°13'W
Gobi Desert, des., Asia (gō´be dĕs´ērt)	240-41	43°0'N	105°00'E
Goch, Ger. (gôk)	194-95	51°41'N	6°09'E
Godāvari, stm., India (gô-dä´vŭ-rē)	236	16°59'N	81°47'E
Goderich, On., Can. (gŏd´rĭch)	136-37	43°45'N	81°42'W
Godfrey, Il., U.S. (gŏd´frē)	120-21	38°57'N	90°11'W
Godhavn, Green. (gōdh´håvn)	284-85	69°15'N	53°33'W
Godhra, India	234-35	22°46'N	73°37'E
Godoy Cruz, Arg.	163e	32°55'S	68°50'W
Gods, stm., Mb., Can. (gŏdz)	134-35	56°23'N	92°51'W
Gods Lake, lk., Mb., Can.	134-35	54°43'N	94°14'W
Godthåb, nat. cap., Green. (gŏt´hòb)	284-85	64°11'N	51°44'W
Godwin Austen, mtn., Asia *see* K2	232-33	35°53'N	76°30'E
Goeie Hoop, Kaap die, c., S. Afr. *see* Good Hope, Cape of	264-65	34°21'S	18°28'E
Goiana, Braz.	163d	7°33'S	34°59'W
Goiânia, Braz. (gô-vá´nyä)	168-69	16°40'S	49°16'W
Goiás, Braz. (gô-yá´s)	168-69	15°55'S	50°07'W
Goiás, state, Braz. (gô-yá´s)	166-67	16°0'S	50°00'W
Gökçeada, i., Tur.	200-01	40°10'N	25°50'E
Gökova Körfezi, b., Tur.	200-01	36°54'N	27°51'E
Göksu, stm., Tur. (gûk´sōō´)	228-29	36°19'N	34°03'E
Gol, Nor. (gûl)	192-93	60°42'N	8°57'E
Gold Coast, Austl. *see* Southport	276	27°58'S	153°25'E
Golden, B.C., Can. (gōl´dĕn)	132-33	51°18'N	116°58'W
Golden, Co., U.S. (gōl´dĕn)	118-19	39°45'N	105°13'W
Goldendale, Wa., U.S. (gōl´dĕn-dāl)	112-13	45°49'N	120°50'W
Golden Hinde, mtn., B.C., Can. (gōl´dĕn hīnd)	132-33	49°40'N	125°45'W
Goldsboro, N.C., U.S. (gōldz-bûr´ô)	124-25	35°23'N	77°60'W
Goldthwaite, Tx., U.S. (gōld´thwāt)	122-23	31°27'N	98°34'W
Golfito, C.R. (gôl-fē´tō)	149	8°38'N	83°10'W
Goliad, Tx., U.S. (gō-lĭ-ăd´)	122-23	28°40'N	97°23'W
Golmud, China	240-41	36°25'N	94°54'E
Goma, D.R.C.	267	1°41'S	29°13'E
Gombe, Nig.	260-61	10°17'N	11°10'E
Gomel', Bela. *see* Homel'	202-03	52°26'N	30°59'E
Gómez Palacio, Mex. (gō´mĕz pä-lä´syō)	122-23	25°35'N	103°30'W
Gonābād, Iran	232-33	34°21'N	58°41'E
Gonaïves, Haiti (gō-nà-ēv´)	142-43	19°27'N	72°41'W
Gonam, stm., Russia	218-19	57°19'N	131°15'E
Gonarezhou National Park, n.p., Zimb.	264-65	21°34'S	31°56'E
Gonâve, Île de la, i., Haiti (ēl-dē-lá-gō-náv´)	142-43	18°51'N	73°03'W
Gonbad-e Kāvūs, Iran	232-33	37°15'N	55°10'E
Gonda, India	234-35	27°08'N	81°58'E
Gondar, Eth. *see* Gonder	266	12°37'N	37°28'E
Gonder, Eth.	266	12°37'N	37°28'E
Gondia, India	234-35	21°28'N	80°12'E
Gongbo'gyamda, China	238-39	29°55'N	93°26'E
Gongga Shan, mtn., China (gôŋ-gä shän)	238-39	29°35'N	101°51'E
Gongola, stm., Nig.	260-61	9°29'N	12°03'E
Gongxian, China	240-41	34°48'N	113°03'E
Gongzhuling, China	240-41	43°30'N	124°49'E
Gonzales, La., U.S. (gŏn-zä´lĕz)	124-25	30°14'N	90°55'W
Gonzales, Tx., U.S. (gŏn-zä´lĕz)	122-23	29°30'N	97°27'W
Goodenough Island, i., Pap. N. Gui.	277	9°20'S	150°15'E
Good Hope, Cape of, c., S. Afr. (kāp ŏv gŏŏd hōp)	264-65	34°21'S	18°28'E
Good Hope Mountain, mtn., B.C., Can. (gŏŏd hōp moun´tĭn)	132-33	51°09'N	124°10'W
Gooding, Id., U.S. (gòd´ĭng)	112-13	42°57'N	114°43'W
Goodland, Ks., U.S. (gòd´lănd)	120-21	39°20'N	101°43'W
Goole, Eng., U.K. (gōōl)	190-91	53°42'N	0°53'W
Goondiwindi, Austl.	276	28°32'S	150°19'E
Goose Lake, lk., U.S. (gōōs lāk)	112-13	41°57'N	120°25'W
Goqên, China	238-39	29°09'N	97°14'E
Gorakhpur, India (gō´rŭk-pōōr)	234-35	26°46'N	83°22'E
Gorda, Punta, c., Cuba (pōō´n-tä-gôr-dä)	142-43	22°23'N	82°09'W
Gorgān, Iran	232-33	36°51'N	54°26'E
Gorgona, Isla, i., Col.	164-65	2°58'N	78°11'W

ăt; finăl; rāte; senåte; ärm; åsk; sofȧ; fâre; ch-choose; dh-as th in other; bē; ēvent; bĕt; recĕnt; cratẽr; g-gō; gh-guttural g; bĭt; ĭ-short neutral; rīde; κ-guttural k as ch in German ich;

Feature (Pronunciation)	Page	Lat.	Long.
Gorgona, Isola di, i., Italy (gôr-gō´nä)	200-01	43°26′N	9°54′E
Gori, Geor. (gō´rē)	227	41°59′N	44°06′E
Gorica, Italy *see* Gorizia	200-01	45°57′N	13°38′E
Gorinchem, Neth. (gō´rĭn-kĕm)	190-91	51°50′N	5°01′E
Gorizia, Italy (gô-rē´tsē-yä)	200-01	45°57′N	13°38′E
Gorkhā, Nepal	234-35	28°0′N	84°37′E
Gorky, Russia *see* Nizhniy Novgorod	186-87	56°19′N	44°01′E
Gorki Reservoir, res., Russia (gôr´kē rĕ´sĕr-vwär) *see* Gor'kovskoye Vodokhranilishche	186-87	57°02′N	43°10′E
Gor'kovskoye Vodokhranilishche, res., Russia	186-87	57°02′N	43°10′E
Gorlice, Pol. (gôr-lē´tsĕ)	194-95	49°39′N	21°10′E
Görlitz, Ger. (gŭr´lĭts)	194-95	51°09′N	14°59′E
Gorlovka, Ukr. *see* Horlivka	202-03	48°20′N	38°03′E
Gorna Oryakhovitsa, Blg. (gôr´nä-ôr-yĕk´ô-vē-tsà)	200-01	43°08′N	25°42′E
Gornji Milanovac, Serb. (gôrn´yē-mē´la-nô-väts)	200-01	44°01′N	20°27′E
Gorno-Altaysk, Russia (gôr´nŭ´ŭl-tīsk´)	226	51°58′N	85°51′E
Gornozavodsk, Russia	244	46°33′N	141°51′E
Gorodets, Russia	186-87	56°39′N	43°28′E
Goroka, Pap. N. Gui.	277	6°05′S	145°24′E
Gorontalo, Indon. (gō-rōn-tä´lo)	248-49	0°32′N	123°04′E
Görz, Italy *see* Gorizia	200-01	45°57′N	13°38′E
Gorzów Wielkopolski, Pol. (gō-zhōōv´vyĕl-ko-pōl´skē)	194-95	52°44′N	15°14′E
Gosford, Austl.	276	33°25′S	151°21′E
Goshen, In., U.S. (gō´shĕn)	116-17	41°35′N	85°49′W
Goslar, Ger. (gôs´lär)	194-95	51°55′N	10°26′E
Gostivar, Mac. (gos´tē-vär)	200-01	41°48′N	20°55′E
Gostynin, Pol. (gôs-tē´nĭn)	194-95	52°26′N	19°28′E
Göta, stm., Swe. (gōē´tä)	192-93	57°41′N	11°53′E
Göteborg, Swe.	192-93	57°43′N	11°58′E
Gotha, Ger. (gō´tȧ)	194-95	50°57′N	10°42′E
Gothenburg, Swe. *see* Göteborg	192-93	57°43′N	11°58′E
Gothenburg, Ne., U.S. (gŏth´ĕn-bûrg)	114-15	40°56′N	100°10′W
Gotland, i., Swe.	192-93	57°30′N	18°33′E
Gotō-rettō, is., Japan	244	32°50′N	129°00′E
Gotska Sandön, i., Swe.	192-93	58°22′N	19°16′E
Göttingen, Ger. (gŭt´ĭng-ĕn)	194-95	51°32′N	9°56′E
Gouda, Neth. (gou´dä)	190-91	52°01′N	4°42′E
Gouin, Réservoir, res., Qc., Can.	136-37	48°37′N	74°55′W
Goulburn, Austl. (gōl´bŭrn)	276	34°45′S	149°43′E
Goundam, Mali (gōōn-däN´)	258-59	16°25′N	3°40′W
Gouverneur, N.Y., U.S. (gŭv-ēr-nōōr´)	116-17	44°20′N	75°28′W
Governador Valadares, Braz. (gô-vĕr-nä-dō-´r vä-lä-dä´rĕs)	172	18°53′S	41°58′W
Goya, Arg. (gō´yä)	173	29°09′S	59°15′W
Goyania, Braz. *see* Goiânia	168-69	16°40′S	49°16′W
Göyçay, Azer. (gě-ôk´chī)	227	40°39′N	47°45′E
Gozo, i., Malta	200b	36°03′N	14°15′E
Graaff-Reinet, S. Afr. (gräf´rī´nĕt)	264-65	32°16′S	24°33′E
Gračac, Cro. (grä´chäts)	200-01	44°18′N	15°50′E
Graceville, Fl., U.S. (grās´vĭl)	124-25	30°57′N	85°31′W
Gracias a Dios, Cabo, c., N.A.	149	14°60′N	83°10′W
Graciosa, i., Port. (grä-syō´sä)	199c	39°04′N	28°00′W
Gradačac, Bos. (gra-dä´chats)	200-01	44°53′N	18°26′E
Gradaús, Braz.	166-67	7°43′S	51°10′W
Grænlandshav, s., *see* Greenland Sea	288	77°0′N	1°00′W
Grænlandssund, strt., *see* Denmark Strait	86	67°0′N	25°00′W
Grafton, Austl. (graf´tŭn)	276	29°42′S	152°56′E
Grafton, N.D., U.S. (graf´tŭn)	114-15	48°25′N	97°25′W
Grafton, W.V., U.S. (graf´tŭn)	116-17	39°20′N	80°01′W
Grafton, Cape, c., Austl.	277	16°53′S	145°56′E
Graham, N.C., U.S. (grā´ăm)	124-25	36°04′N	79°24′W
Graham, Tx., U.S. (grā´ăm)	120-21	33°07′N	98°35′W
Graham Island, i., B.C., Can. (grā´ăm ī´lănd)	132-33	53°47′N	132°34′W
Grahamstad, S. Afr. *see* Grahamstown	264-65	33°18′S	26°31′E
Grahamstown, S. Afr. (grā´ăms´toun)	264-65	33°18′S	26°31′E
Grajaú, Braz.	166-67	5°47′S	46°07′W
Grajaú, stm., Braz.	166-67	3°41′S	44°49′W
Grajewo, Pol. (grä-yä´vo)	194-95	53°39′N	22°28′E
Grampian Mountains, mts., Scot., U.K. (grăm´pĭ-ăn moun´tĭnz)	190-91	56°55′N	4°00′W
Grampians National Park, n.p., Austl.	276	37°15′S	142°25′E
Granada, Nic. (grä-nä´dhä)	149	11°56′N	85°58′W
Granada, Spain (grä-nä´dä)	198-99	37°11′N	3°36′W
Granbury, Tx., U.S. (grăn´bĕr-ĭ)	120-21	32°27′N	97°48′W
Granby, Qc., Can. (grăn´bĭ)	136-37	45°24′N	72°43′W
Granby, Co., U.S. (grăn´bĭ)	118-19	40°06′N	105°57′W
Granby, Mo., U.S. (grăn´bĭ)	120-21	36°55′N	94°15′W
Gran Canaria, i., Spain (grän´kä-nä´rē-ä)	199d	27°52′N	15°37′W
Gran Chaco, reg., S.A. (grän´chä´kō)	168-69	23°0′S	60°00′W
Grand, stm., On., Can. (grănd)	136-37	42°52′N	79°34′W
Grand, stm., Mi., U.S. (grănd)	116-17	43°04′N	86°14′W
Grand Bahama, i., Bah. (grănd bȧ-hā´mȧ)	142-43	26°38′N	78°25′W
Grand Bank, Nf., Can. (grănd băngk)	138-39	47°06′N	55°45′W
Grand-Bassam, C. Iv. (grän bá-sän´)	260-61	5°13′N	3°45′W
Grand-Bourg, Guad. (grän bōōr´)	143b	15°54′N	61°19′W
Grand Canal, can., China (grănd kȧ´năl)	240-41	32°11′N	119°33′E
Grand Canyon, Az., U.S. (grănd kăn´yŭn)	118-19	36°02′N	112°10′W
Grand Canyon, val., Az., U.S. (grănd kăn´yŭn)	118-19	36°22′N	112°30′W
Grand Canyon National Park, n.p., Az., U.S. (grănd kăn´yŭn năsh´ŭn-ăl pärk)	118-19	36°20′N	112°53′W
Grand Canyon-Parashant National Monument, n.p., Az., U.S.	118-19	36°20′N	113°44′W
Grand Cayman, i., Cay. Is. (grănd kā´măn) (grănd kī-män´)	142-43	19°20′N	81°15′W
Grand Coulee Dam, d., Wa., U.S. (grănd kōō´lē dăm)	112-13	47°57′N	119°01′W
Grande, stm., Bol. (grän´dĕ)	168-69	15°50′S	64°47′W
Grande, stm., Braz. (grän´dĕ)	166-67	11°05′S	43°09′W
Grande, stm., Braz. (grän´dĕ)	168-69	20°08′S	51°00′W
Grande, Bahía, b., Arg. (bä-ē´ä-grän´dĕ)	171	51°15′S	68°31′W
Grande, Cuchilla, mts., Ur. (kōō-chē´l-yä grän´dĕ)	173	33°25′S	55°06′W
Grande, Ilha, i., Braz. (ē´lä-grän´dĕ)	172	23°09′S	44°14′W
Grande, Rio, stm., N.A. (rē´ō grän´dā) *see* Rio Grande	110-11	25°57′N	97°09′W
Grande Cayemite, i., Haiti	142-43	18°37′N	73°45′W
Grande Comore, i., Com. *see* Njazidja	264-65	11°35′S	43°20′E
Grande de Santiago, stm., Mex. (grä´n-dĕ-dĕ-sän-tyä´gô)	146-47	21°37′N	105°28′W
Grande do Gurupá, Ilha, i., Braz.	166-67	1°0′S	51°30′W
Grande Prairie, Ab., Can. (grănd prâr´ĭ)	132-33	55°10′N	118°48′W
Grand Erg de Bilma, des., Niger	258-59	18°30′N	14°00′E
Grand Erg Occidental, des., Alg.	258-59	30°56′N	1°35′E
Grand Erg Oriental, des., Alg.	258-59	30°30′N	7°00′E
Grandes, Salinas, pl., Arg.	168-69	30°06′S	65°14′W
Grandes Antillas, Islas, is., N.A. *see* Greater Antilles	142-43	20°0′N	74°00′W
Grande-Terre, i., Guad.	143b	16°19′N	61°22′W
Grand Falls, N.B., Can. (grănd fôlz)	138-39	47°03′N	67°44′W
Grand Falls-Windsor, Nf., Can.	138-39	48°56′N	55°39′W
Grandfather Mountain, mtn., N.C., U.S. (grănd-fä-thẽr moun´tĭn)	124-25	36°07′N	81°48′W
Grandfield, Ok., U.S. (grănd´fēld)	120-21	34°14′N	98°41′W
Grand Forks, B.C., Can. (grănd fôrks)	132-33	49°02′N	118°27′W
Grand Forks, N.D., U.S. (grănd fôrks)	114-15	47°55′N	97°03′W
Grand Haven, Mi., U.S. (grănd hā´v′n)	116-17	43°04′N	86°13′W
Grand Island, Ne., U.S. (grănd ī´lănd)	114-15	40°55′N	98°21′W
Grand Island, i., Mi., U.S. (grănd ī´lănd)	114-15	46°30′N	86°40′W
Grand Junction, Co., U.S. (grănd jŭngk´shŭn)	118-19	39°04′N	108°34′W
Grand Lake, lk., N.B., Can. (grănd lāk)	138-39	45°53′N	66°03′W
Grand Lake, lk., Nf., Can. (grănd lāk)	138-39	48°59′N	57°22′W
Grand Lake, lk., La., U.S. (grănd lāk)	122-23	29°53′N	92°45′W
Grand Ledge, Mi., U.S. (grănd lĕj)	116-17	42°45′N	84°44′W
Grand Manan Island, i., N.B., Can. (grănd mȧ-năn ī´lănd)	138-39	44°43′N	66°49′W
Grand-Mère, Qc., Can. (grän mâr´)	136-37	46°37′N	72°42′W
Grândola, Port. (grän´dô-lȧ)	198-99	38°10′N	8°34′W
Grand Portage Indian Reservation, ind. res., Mn., U.S. (grănd pōr´tĭj ĭn´dĭ-ăn rĕ-sĕr-vā´shĕn)	114-15	47°58′N	89°47′W
Grand Rapids, Mb., Can. (grănd răp´ĭdz)	134-35	53°12′N	99°17′W
Grand Rapids, Mi., U.S. (grănd răp´ĭdz)	116-17	42°58′N	85°40′W
Grand Rapids, Mn., U.S. (grănd răp´ĭdz)	114-15	47°14′N	93°31′W
Grand-Sault, N.B., Can. *see* Grand Falls	138-39	47°03′N	67°44′W
Grand Staircase-Escalante National Monument, n.p., Ut., U.S.	118-19	37°30′N	111°30′W
Grand Teton, mtn., Wy., U.S. (grănd tē´tŏn)	112-13	43°44′N	110°48′W
Grand Teton National Park, n.p., Wy., U.S. (grănd tē´tŏn năsh´ŭn-ăl pärk)	112-13	43°56′N	110°46′W
Grand Traverse Bay, b., Mi., U.S. (grănd trăv´ẽrs bā)	116-17	45°02′N	85°30′W
Grand Turk, nat. cap., T./C. Is. (grănd tûrk)	142-43	21°27′N	71°08′W
Grandview, Mb., Can.	134-35	51°10′N	100°42′W
Grandview, Wa., U.S. (grănd´vyōō)	112-13	46°15′N	119°54′W
Grangeville, Id., U.S. (grānj´vĭl)	112-13	45°56′N	116°07′W
Granite City, Il., U.S. (grăn´ĭt sĭ´tē)	120-21	38°42′N	90°09′W
Granite Falls, Mn., U.S. (grăn´ĭt fôlz)	114-15	44°49′N	95°33′W
Granite Falls, N.C., U.S. (grăn´ĭt fôlz)	124-25	35°48′N	81°26′W
Granite Peak, mtn., Mt., U.S.	112-13	45°10′N	109°48′W
Gränna, Swe. (grĕn´à)	192-93	58°00′N	14°28′E
Granollers, Spain (grä-nôl-yĕrs´)	198-99	41°37′N	2°17′E
Grantham, Eng., U.K. (grăn´tȧm)	190-91	52°55′N	0°39′W
Grants, N.M., U.S.	118-19	35°10′N	107°51′W
Grants Pass, Or., U.S. (grȧnts pȧs)	112-13	42°26′N	123°19′W
Granville, Fr. (grän-vēl´)	196-97	48°51′N	1°35′W
Granville, N.Y., U.S. (grăn´vĭl)	116-17	43°24′N	73°16′W
Granville Lake, lk., Mb., Can.	134-35	56°17′N	100°30′W
Gräsö, i., Swe.	192-93	60°24′N	18°25′E
Grasse, Fr. (gräs)	196-97	43°40′N	6°55′E
Grasslands National Park, n.p., Sk., Can.	134-35	49°04′N	106°58′W
Grates Point, c., Nf., Can. (grāts point)	138-39	48°10′N	52°57′W
Graudenz, Pol. *see* Grudziądz	194-95	53°29′N	18°44′E
Gravatá, Braz.	163d	8°12′S	35°34′W
Gravelbourg, Sk., Can. (grăv´ĕl-bôrg)	134-35	49°52′N	106°34′W
Gravenhage, 's-, nat. cap., Neth. *see* Hague, The	190-91	52°06′N	4°18′E
Gray, Fr. (grå)	196-97	47°26′N	5°35′E
Grayling, Mi., U.S. (grā´lĭng)	116-17	44°39′N	84°42′W
Grays Peak, mtn., Co., U.S. (grāz pēk)	118-19	39°37′N	105°45′W
Graz, Aus. (gräts)	194-95	47°05′N	15°27′E
Great Artesian Basin, bas., Austl. (grāt är-tēzh-ăn bā´s′n)	272-73	25°0′S	143°00′E
Great Australian Bight, b., Austl. (grāt ôs-trā´lĭ-ăn bīt)	272-73	35°0′S	130°00′E
Great Barrier Island, i., N.Z. (grāt băr´ĭ-ēr ī´lănd)	278	36°10′S	175°25′E
Great Barrier Reef, rf., Austl.	277	18°0′S	146°50′E
Great Barrier Reef Marine Park, n.p., Austl.	277	18°0′S	146°50′E
Great Basin, bas., U.S. (grāt bā´s′n)	110-11	40°0′N	117°00′W
Great Basin National Park, n.p., Nv., U.S.	118-19	38°55′N	114°14′W
Great Bear Lake, lk., N.T., Can. (grāt bâr lāk)	130-31	66°0′N	120°00′W
Great Bend, Ks., U.S. (grāt bĕnd)	120-21	38°22′N	98°45′W
Great Britain, nation, Eur. *see* United Kingdom	174-75	54°0′N	2°00′W
Great Britain, i., U.K. (grāt brĭt´′n)	190-91	54°0′N	2°00′W
Great Channel, strt., Asia	246-47	6°25′N	94°20′E
Great Dismal Swamp, sw., U.S. (grāt dĭz´măl swŏmp)	124-25	36°28′N	76°28′W
Great Divide Basin, bas., Wy., U.S. (grāt dĭ-vīd´ bā´s′n)	112-13	42°0′N	108°10′W
Great Dividing Range, mts., Austl. (grāt dĭ-vī-dĭng rānj)	270-71	25°0′S	147°00′E
Greater Antilles, is., N.A. (grāt´ẽr ăn-tĭ´lēz)	142-43	20°0′N	74°00′W
Greater Khingan Range, mts., China (grāt´ẽr hĭŋ-gän rānj)	222-23	49°0′N	122°00′E
Greater Sunda Islands, is., Asia (grāt´ẽr sōōn´dȧ ī´lăndz)	248-49	2°0′S	110°00′E
Great Exuma, i., Bah. (grāt ĕk-sōō´mä)	142-43	23°32′N	75°50′W
Great Falls, Mt., U.S. (grāt fôlz)	112-13	47°30′N	111°18′W
Great Falls, S.C., U.S. (grāt fôlz)	124-25	34°34′N	80°54′W
Great Grimsby, Eng., U.K. *see* Grimsby	190-91	53°35′N	0°05′W
Grimsby, Eng., U.K.	190-91	53°35′N	0°05′W
Great Guana Cay, i., Bah. (grāt gwä´nä kē)	142-43	24°0′N	76°20′W
Great Inagua, i., Bah. (grāt ē-nä´gwä)	142-43	21°05′N	73°18′W
Great Indian Desert, des., Asia (grāt ĭn´dĭ-ăn dĕs´ẽrt)	232-33	27°0′N	71°00′E
Great Karoo, plat., S. Afr.	264-65	32°47′S	22°32′E
Great Limpopo Transfrontier Park, n.p., Afr.	264-65	23°0′S	31°30′E
Great Namaqualand, hist. reg., Nmb.	264-65	25°0′S	17°00′E
Great Nicobar, i., India (grāt nĭk-ô-bär´)	246-47	7°0′N	93°50′E
Great Palm Island, i., Austl.	277	18°43′S	146°37′E
Great Pee Dee, stm., S.C., U.S. (grāt pē-dē´)	124-25	33°18′N	79°17′W

n-sing; ŋ-baŋk; N-nasalized n; nŏd; cŏmmit; ōld; ôbey; ôrder; oi-boil; fōōd; ō-as oo in foot; ou-out; s-soft; sh-dish; th-thin; pūre; ûnite; ûrn; stŭd; circŭs; ü-as in French tu; ´-indeterminate vowel.

Feature (Pronunciation)	Page	Lat.	Long.
Great Plains, pl., U.S. (grāt plāns)	110-11	42°0'N	100°00'W
Great Ruaha, stm., Tan.	267	7°56'S	37°48'E
Great Salt Lake, lk., Ut., U.S. (grāt sôlt lāk)	112-13	41°10'N	112°30'W
Great Salt Lake Desert, des., Ut., U.S. (grāt sôlt lāk dĕs'ẽrt)	110-11	40°40'N	113°30'W
Great Sand Dunes National Park and Preserve, n.p., Co., U.S.	118-19	37°46'N	105°33'W
Great Sandy Desert, des., Austl. (grāt săn'dē dĕs'ẽrt)	272-73	21°30'S	125°00'E
Great Sandy National Park, n.p., Austl.	277	24°55'S	153°16'E
Great Slave Lake, lk., N.T., Can. (grāt slāv lāk)	130-31	61°30'N	114°00'W
Great Smoky Mountains National Park, n.p., U.S. (grāt smōk-ē moun'tĭnz năsh'ŭn-ăl pärk)	124-25	35°39'N	83°30'W
Great Victoria Desert, des., Austl. (grāt vĭk-tō'rĭ-ȧ dĕs'ẽrt)	272-73	28°30'S	127°45'E
Great Wall, p.o.i., China	240-41	40°0'N	112°30'E
Gréboun, mtn., Niger	258-59	20°0'N	8°35'E
Gredos, Sierra de, mts., Spain (syĕr'rä dā grā'dōs)	198-99	40°20'N	4°51'W
Greece, nation, Eur. (grēs)	174-75	39°0'N	22°00'E
Greeley, Co., U.S. (grē'lĭ)	114-15	40°24'N	104°41'W
Greeley, Ne., U.S. (grē'lĭ)	114-15	41°33'N	98°32'W
Green, stm., U.S. (grēn)	110-11	38°11'N	109°53'W
Green, stm., Ky., U.S. (grēn)	116-17	37°54'N	87°31'W
Green Bay, Wi., U.S. (grēn bā)	116-17	44°30'N	87°60'W
Green Bay, b., U.S. (grēn bā)	116-17	44°58'N	87°35'W
Greencastle, In., U.S. (grēn-kȧs''l)	116-17	39°38'N	86°51'W
Green Cove Springs, Fl., U.S. (grēn kōv sprĭngz)	124-25	29°60'N	81°42'W
Greenfield, Ca., U.S. (grēn'fēld)	118-19	36°19'N	121°15'W
Greenfield, Ia., U.S. (grēn'fēld)	114-15	41°18'N	94°28'W
Greenfield, In., U.S. (grēn'fēld)	116-17	39°47'N	85°46'W
Greenfield, Oh., U.S. (grēn'fēld)	116-17	39°20'N	83°23'W
Greenfield, Tn., U.S. (grēn'fēld)	124-25	36°09'N	88°48'W
Greenfield, Wi., U.S. (grēn'fēld)	116-17	42°57'N	88°01'W
Green Islands, is., Pap. N. Gui.	277	4°30'S	154°10'E
Greenland, dep., N.A. (grēn'lănd)	85	70°0'N	40°00'W
Greenland, i., Green. (grēn'lănd)	86	70°0'N	40°00'W
Greenland Sea, s., (grēn'lănd sē)	288	77°0'N	1°00'W
Green Mountains, mts., N.A. (grēn moun'tĭnz)	116-17	43°45'N	72°45'W
Greenock, Scot., U.K. (grēn'ŭk)	190-91	55°57'N	4°45'W
Green River, Ut., U.S. (grēn rĭv'ẽr)	118-19	38°60'N	110°09'W
Green River, Wy., U.S. (grēn rĭv'ẽr)	112-13	41°32'N	109°28'W
Greensboro, Al., U.S. (grēnz'bŭro)	124-25	32°42'N	87°36'W
Greensboro, Ga., U.S. (grēns-bûr'ô)	124-25	33°35'N	83°11'W
Greensboro, N.C., U.S. (grēns-bûr'ô)	124-25	36°04'N	79°48'W
Greensburg, In., U.S. (grēnz'bûrg)	116-17	39°20'N	85°29'W
Greensburg, Ky., U.S. (grēns-bûrg)	124-25	37°16'N	85°30'W
Greenville, Lib. (grēn'vĭl)	260-61	5°02'N	9°03'W
Greenville, Al., U.S. (grēn'vĭl)	124-25	31°50'N	86°37'W
Greenville, Il., U.S. (grēn'vĭl)	116-17	38°53'N	89°25'W
Greenville, Ky., U.S. (grēn'vĭl)	124-25	37°12'N	87°11'W
Greenville, Me., U.S. (grēn'vĭl)	117a	45°27'N	69°34'W
Greenville, Mi., U.S. (grēn'vĭl)	116-17	43°11'N	85°15'W
Greenville, Ms., U.S. (grēn'vĭl)	124-25	33°25'N	91°03'W
Greenville, N.C., U.S. (grēn'vĭl)	124-25	35°37'N	77°22'W
Greenville, Oh., U.S. (grēn'vĭl)	116-17	40°06'N	84°38'W
Greenville, Pa., U.S. (grēn'vĭl)	116-17	41°24'N	80°23'W
Greenville, S.C., U.S. (grēn'vĭl)	124-25	34°51'N	82°24'W
Greenville, Tx., U.S. (grēn'vĭl)	120-21	33°09'N	96°07'W
Greenwood, Ar., U.S. (grēn-wŏŏd)	120-21	35°12'N	94°16'W
Greenwood, In., U.S. (grēn-wŏŏd)	116-17	39°36'N	86°05'W
Greenwood, La., U.S. (grēn-wŏŏd)	120-21	32°27'N	93°58'W
Greenwood, Ms., U.S. (grēn-wŏŏd)	124-25	33°31'N	90°11'W
Greenwood, S.C., U.S. (grēn-wŏŏd)	124-25	34°12'N	82°09'W
Greer, S.C., U.S. (grēr)	124-25	34°56'N	82°14'W
Gregory, S.D., U.S. (grĕg'ō-rĭ)	114-15	43°14'N	99°26'W
Gregory, Lake, lk., Austl. (lāk grĕg'ô-rê)	276	28°55'S	139°00'E
Gregory Range, mts., Austl.	277	19°0'S	143°05'E
Greifswald, Ger. (grīfs'vält)	194-95	54°05'N	13°23'E
Greiz, Ger. (grīts)	194-95	50°39'N	12°13'E
Gremyachinsk, Russia (grâ'myȧ-chĭnsk)	186-87	58°35'N	57°51'E
Grenada, Ms., U.S. (grē-nā'da)	124-25	33°46'N	89°49'W
Grenada, nation, N.A. (grĕ-nā'-dȧ)	140-41	12°07'N	61°40'W
Grenadines, is., N.A. (grĕn'ȧ-dēnz)	143b	12°40'N	61°15'W
Grenoble, Fr. (grē-nô'bl')	196-97	45°11'N	5°42'E
Grenville, Cape, c., Austl.	277	11°58'S	143°14'E
Gresham, Or., U.S. (grĕsh'ăm)	112-13	45°29'N	122°26'W
Gresik, Indon.	248-49	7°10'S	112°39'E
Gretna, La., U.S. (grĕt'nȧ)	124-25	29°55'N	90°04'W
Gretna, Ne., U.S. (grĕt'nȧ)	114-15	41°08'N	96°15'W
Grey, stm., Nf., Can. (grā)	138-39	47°33'N	57°08'W
Greybull, Wy., U.S. (grā'bŏŏl)	112-13	44°30'N	108°03'W
Grey Islands, is., Nf., Can.	138-39	50°50'N	55°37'W
Greymouth, N.Z. (grā'mouth)	278	42°28'S	171°13'E
Grey Range, mts., Austl. (grā rānj)	276	27°0'S	143°35'E
Greytown, S. Afr. (grā'toun)	269c	29°04'S	30°35'E
Gribanovskiy, Russia	186-87	51°27'N	41°58'E
Gribbel Island, i., B.C., Can.	132-33	53°25'N	129°00'W
Griffin, Ga., U.S. (grĭf'ĭn)	124-25	33°15'N	84°16'W
Griffith, Austl. (grĭf-ĭth)	276	34°17'S	146°03'E
Grim, Cape, c., Austl. (kāp grĭm)	276	40°39'S	144°43'E
Grímsey, i., Ice. (grĭms'ā)	190a	66°33'N	18°01'W
Grimstad, Nor. (grĭm-städh)	192-93	58°20'N	8°36'E
Groesbeck, Tx., U.S. (grōs'bĕk)	122-23	31°31'N	96°32'W
Gronau, Ger. (grō'nou)	194-95	52°13'N	7°01'E
Groningen, Neth. (grō'nĭng-ĕn)	190-91	53°13'N	6°34'E
Grønland, dep., N.A. *see* Greenland	85	70°0'N	40°00'W
Grønland, i., Green. *see* Greenland	86	70°0'N	40°00'W
Grønlandshavet, s., *see* Greenland Sea	288	77°0'N	1°00'W
Groote Eylandt, i., Austl. (grō'tē ī'länt)	272-73	13°60'S	136°38'E
Grootfontein, Nmb. (grōt'fōn-tān')	264-65	19°34'S	18°06'E
Groot Karroo, plat., S. Afr. *see* Great Karoo	264-65	32°47'S	22°32'E
Groot Namaland, hist. reg., Nmb. *see* Great Namaqualand	264-65	25°0'S	17°00'E
Gros Morne, mtn., Nf., Can. (grō môrn')	138-39	49°36'N	57°48'W
Gros Morne National Park, n.p., Nf., Can. (grō môrn' năsh'ŭn-ăl pärk)	138-39	49°40'N	57°45'W
Grosseto, Italy (grôs-sā'tō)	200-01	42°46'N	11°07'E
Großglockner, mtn., Aus.	194-95	47°04'N	12°42'E
Grosswardein, Rom. *see* Oradea	194-95	47°04'N	21°56'E
Groton, Ct., U.S. (grŏt'ŭn)	116-17	41°21'N	72°05'W
Groton, S.D., U.S. (grŏt'ŭn)	114-15	45°27'N	98°06'W
Grouard Mission, Ab., Can.	132-33	55°32'N	116°09'W
Groveton, N.H., U.S. (grōv'tŭn)	116-17	44°36'N	71°31'W
Growa Point, c., Lib.	260-61	4°21'N	7°36'W
Groznyy, Russia (grôz'nĭ)	227	43°19'N	45°41'E
Grudziądz, Pol. (grō'jyŏnts)	194-95	53°29'N	18°44'E
Grünberg, Pol. *see* Zielona Góra	194-95	51°56'N	15°31'E
Grundy Center, Ia., U.S. (grŭn'dĭ sĕn'tẽr)	114-15	42°21'N	92°47'W
Gruziya, nation, Asia *see* Georgia	227	42°0'N	44°00'E
Gryazi, Russia (gryä'zĭ)	186-87	52°29'N	39°57'E
Gryazovets, Russia (gryä'zô-vĕts)	186-87	58°53'N	40°15'E
Gryfice, Pol. (grĭ'fĭ-tsĕ)	194-95	53°55'N	15°12'E
Guacanayabo, Golfo de, b., Cuba (gôl-fô-dĕ-gwä-kä-nä-yä'bō)	142-43	20°28'N	77°30'W
Guacara, Ven. (gwȧ'kä-rä)	163b	10°14'N	67°53'W
Guadalajara, Mex. (gwä-dhä-lä-hä'rä)	146-47	20°40'N	103°20'W
Guadalajara, Spain (gwä-dä-lä-kä'rä)	198-99	40°38'N	3°10'W
Guadalcanal, i., Sol. Is. (gwä-dhäl-kä-näl')	279e	9°32'S	160°12'E
Guadalquivir, stm., Spain (gwä-dhäl-kê-vēr')	198-99	36°47'N	6°24'W
Guadalupe, Mex.	122-23	25°41'N	100°15'W
Guadalupe, stm., Tx., U.S. (gwä-dhä-lōō'pä)	122-23	28°27'N	96°49'W
Guadalupe, Isla, i., Mex.	140-41	29°03'N	118°21'W
Guadalupe, Sierra de, mts., Spain (syĕr'rä dā gwä-dhä-lōō'pä)	198-99	39°29'N	5°28'W
Guadalupe Mountains, mts., U.S. (gwä-dhä-lōō'pȧ moun'tĭnz)	122-23	32°24'N	105°04'W
Guadalupe Mountains National Park, n.p., Tx., U.S.	122-23	31°55'N	104°55'W
Guadalupe Peak, mtn., Tx., U.S. (gwä-dhä-lōō'pȧ pēk)	122-23	31°50'N	104°52'W
Guadarrama, Sierra de, mts., Spain (syĕr'rä dā gwä-dhär-rä'mä)	198-99	40°51'N	4°01'W
Guadeloupe, dep., N.A. (gwä-dĕ-lōōp)	140-41	16°15'N	61°35'W
Guadiana, stm., Eur. (gwä-dvä'nä)	198-99	37°10'N	7°24'W
Guadix, Spain (gwä-dēsh')	198-99	37°18'N	3°08'W
Guafo, Isla, i., Chile	171	43°36'S	74°43'W
Guaíra, Braz. (gwä-ē-rä)	172	20°19'S	48°18'W
Guaíra, Braz. (gwä-ē-rä)	168-69	24°06'S	54°15'W
Guajaba, Cayo, i., Cuba (kä'yō-gwä-hä'bä)	142-43	21°50'N	77°30'W
Guajará-Mirim, Braz. (gwä-zhä-rä'mē-rēɴ')	166-67	10°48'S	65°22'W
Gualeguay, Arg. (gwä-lĕ-gwä'y)	173	33°09'S	59°20'W
Gualeguay, stm., Arg. (gwä-lĕ-gwä'y)	173	33°19'S	59°39'W
Gualeguaychú, Arg.	173	33°01'S	58°31'W
Guam, dep., Oc. (gwäm)	279c	13°28'N	144°47'E
Guam, i., Guam (gwäm)	279c	13°28'N	144°47'E
Guaminí, Arg.	173	37°01'S	62°25'W
Guamo, Col. (gwȧ'mō)	163c	4°02'N	74°58'W
Guanaja, Isla de, i., Hond.	149	16°29'N	85°53'W
Guanajuato, Mex. (gwä-nä-hwä'tō)	146-47	21°01'N	101°16'W
Guanajuato, state, Mex. (gwä-nä-hwä'tō)	146-47	21°0'N	101°00'W
Guanambi, Braz.	166-67	14°14'S	42°47'W
Guanare, Ven. (gwä-nä'rå)	164-65	9°03'N	69°46'W
Guandacol, Arg.	168-69	29°31'S	68°32'W
Guane, Cuba (gwä'nå)	142-43	22°12'N	84°05'W
Guang'an, China	238-39	30°28'N	106°38'E
Guangchang, China (gůäŋ-chäŋ)	238-39	26°51'N	116°19'E
Guangdong, state, China (gůäŋ-dôŋ)	287	23°0'N	113°00'E
Guanghua, China *see* Laohekou	238-39	32°25'N	111°36'E
Guangnan, China	238-39	24°10'N	105°06'E
Guangxi, state, China	238-39	24°0'N	109°00'E
Guangyuan, China	238-39	32°25'N	105°49'E
Guangzhou, China	238-39	23°08'N	113°16'E
Guanta, Ven. (gwän'tä)	163b	10°14'N	64°36'W
Guantánamo, Cuba (gwän-tä'nä-mô)	142-43	20°08'N	75°13'W
Guantánamo, state, Cuba (gwän-tä'nä-mô)	142-43	20°20'N	75°00'W
Guanxian, China (gůän-shyĕn)	238-39	31°00'N	103°36'E
Guanxian, China (gůän-shyĕn)	240-41	36°28'N	115°26'E
Guapí, Col.	164-65	2°36'N	77°54'W
Guápiles, C.R. (gwä-pē-lĕs)	149	10°12'N	83°47'W
Guaporé, stm., S.A. (gwä-pô-rā')	166-67	11°55'S	65°00'W
Guarabira, Braz. (gwä-rä-bē'rȧ)	163d	6°51'S	35°29'W
Guaranda, Ec. (gwä-rän'dä)	170	1°36'S	78°60'W
Guarapari, Braz. (gwä-rä-pä'rē)	172	20°40'S	40°30'W
Guarapuava, Braz. (gwä-rä-pwä'vȧ)	168-69	25°23'S	51°29'W
Guarda, Port. (gwär'dä)	198-99	40°33'N	7°15'W
Guarulhos, Braz. (gwȧ-rò'l-yôs)	172	23°29'S	46°32'W
Guasave, Mex.	144-45	25°34'N	108°28'W
Guasdualito, Ven.	164-65	7°15'N	70°45'W
Guasipati, Ven. (gwä-sē-pä'tē)	164-65	7°29'N	61°53'W
Guastalla, Italy (gwäs-täl'lä)	200-01	44°55'N	10°39'E
Guatemala, nation, N.A. (guä-tå-mä'lä)	85	15°30'N	90°15'W
Guatemala, nat. cap., Guat. (guä-tå-mä'lä)	148	14°38'N	90°32'W
Guaviare, stm., Col.	164-65	4°04'N	67°43'W
Guaxupé, Braz.	172	21°18'S	46°42'W
Guayama, P.R. (gwä-yä'mä)	142a	17°59'N	66°07'W
Guayana, Ven. *see* Ciudad Guayana	164-65	8°21'N	62°39'W
Guayana, nation, S.A. *see* Guyana	158	5°0'N	59°00'W
Guayaquil, Ec. (gwī-ä-kēl')	170	2°12'S	79°54'W
Guayaquil, Golfo de, b., S.A. (gôl-fô-dĕ gwī-ä-kēl')	170	2°57'S	80°36'W
Guaymas, Mex. (gwȧ'y-mäs)	144-45	27°55'N	110°55'W
Gûbâl, Madîq, strt., Egypt	268b	27°40'N	33°55'E
Gubakha, Russia (gōō-bä'kȧ)	186-87	58°52'N	57°33'E
Gubbio, Italy (gōōb'byô)	200-01	43°21'N	12°34'E
Gubkin, Russia	202-03	51°17'N	37°33'E
Gucheng, China (gōō-chŭŋ)	238-39	32°18'N	111°35'E
Gudermes, Russia	227	43°21'N	46°06'E
Guebwiller, Fr. (gĕb-vē-lâr')	196-97	47°55'N	7°12'E
Guelph, On., Can. (gwĕlf)	136-37	43°33'N	80°15'W
Guercif, Mor.	184-85	34°14'N	3°20'W
Guéret, Fr. (gā-rĕ')	196-97	46°10'N	1°52'E
Guernesey, dep., Eur. *see* Guernsey	196-97	49°28'N	2°35'W
Guernsey, dep., Eur. (gûrn'zĭ)	196-97	49°28'N	2°35'W
Guerrero, Mex. (gĕr-rā'rō)	144-45	28°33'N	107°30'W
Guerrero, state, Mex. (gĕr-rā'rō)	146-47	17°40'N	100°00'W
Gugē, mtn., Eth.	262-63	6°11'N	37°24'E
Guguan, i., N. Mar. Is.	280-81	17°19'N	145°51'E
Guide, China	240-41	36°01'N	101°27'E
Guilin, China (gwā-lĭn)	238-39	25°17'N	110°17'E
Guimarães, Port. (gē-mä-rȧɴsh')	198-99	41°27'N	8°18'W
Guimaras Island, i., Phil.	250	10°35'N	122°37'E
Guiné, Golfo da, b., Afr. *see* Guinea, Gulf of	260-61	2°0'N	2°30'E
Guinea, nation, Afr. (gĭn'ē)	253	11°0'N	10°00'W
Guinea, Golfo de, b., Afr. *see* Guinea, Gulf of	260-61	2°0'N	2°30'E
Guinea, Gulf of, b., Afr. (gŭlf ŭv gĭn'ē)	260-61	2°0'N	2°30'E
Guinea-Bissau, nation, Afr. (gĭn'ē bê-sa'ōō)	253	12°0'N	15°00'W
Guinea Ecuatorial, nation, Afr. *see* Equatorial Guinea	253	2°0'N	9°00'E
Guiné-Bissau, nation, Afr. *see* Guinea-Bissau	253	12°0'N	15°00'W
Guinée, nation, Afr. *see* Guinea	253	11°0'N	10°00'W
Guinée, Golfe de, b., Afr. *see* Guinea, Gulf of	260-61	2°0'N	2°30'E
Güines, Cuba	142-43	22°51'N	82°02'W

ăt; finăl; rāte; senāte; ärm; ȧsk; sofȧ; fâre; ch-choose; dh-as th in other; bē; ēvent; bĕt; recĕnt; cratẽr; g-gō; gh-guttural g; bĭt; ĭ-short neutral; rīde; ᴋ-guttural k as ch in German ich;

Feature (Pronunciation)	Page	Lat.	Long.
Guingamp, Fr. (găn-găn´)	196-97	48°34′N	3°09′W
Guiping, China	238-39	23°23′N	110°04′E
Güira de Melena, Cuba (gwē´rä dĕ må-lā´nä)	142-43	22°48′N	82°30′W
Guiratinga, Braz.	168-69	16°21′S	53°45′W
Güiria, Ven. (gwē-rē´ä)	143b	10°35′N	62°18′W
Guiuan, Phil.	250	11°02′N	125°44′E
Guixian, China	238-39	23°06′N	109°39′E
Guiyang, China	238-39	26°35′N	106°43′E
Guizhou, state, China (gwā-jō)	238-39	27°0′N	107°00′E
Gujarāt, state, India	234-35	22°0′N	72°00′E
Gujrānwāla, Pak.	232-33	32°09′N	74°11′E
Gujrāt, Pak.	232-33	32°34′N	74°05′E
Gulbarga, India (gŏl-bûr´gȧ)	236	17°20′N	76°50′E
Gulbene, Lat. (gŏl-bă´nĕ)	192-93	57°10′N	26°46′E
Gulfport, Ms., U.S. (gŭlf´pōrt)	124-25	30°22′N	89°06′W
Gulian, China	222-23	52°56′N	122°19′E
Guliston, Uzb.	232-33	40°30′N	68°46′E
Gulja, China *see* Yining	226	43°55′N	81°18′E
Gull Lake, Sk., Can. (gŭl lāk)	134-35	50°08′N	108°27′W
Gull Lake, lk., Ab., Can. (gŭl lāk)	132-33	52°33′N	114°02′W
Gulu, Ug.	267	2°47′N	32°18′E
Gumaca, Phil. (gōō-mä-kä´)	250	13°55′N	122°06′E
Gumdag, Turkmen.	232-33	39°12′N	54°36′E
Gummersbach, Ger. (gŏm´ĕrs-bäk)	194-95	51°02′N	7°34′E
Gumti, stm., India	234-35	25°31′N	83°10′E
Gümüşhane, Tur.	227	40°28′N	39°28′E
Guna, India	234-35	24°39′N	77°18′E
Gundagai, Austl.	276	35°04′S	148°06′E
Gunisao, stm., Mb., Can. (gŭn-*i*-sä´ō)	134-35	53°53′N	97°60′W
Gunisao Lake, lk., Mb., Can. (gŭn-*i*-sä´ō lăk)	134-35	53°33′N	96°15′W
Gunnbjørn Fjeld, mtn., Green.	86	68°53′N	30°00′W
Gunnedah, Austl. (gŭ´nē-dä)	276	30°59′S	150°15′E
Gunnison, Co., U.S. (gŭn´ĭ-s*ŭ*n)	118-19	38°33′N	106°55′W
Gunnison, Ut., U.S. (gŭn´ĭ-s*ŭ*n)	118-19	39°09′N	111°49′W
Gunsan, Kor., S. *see* Kunsan	243	35°59′N	126°43′E
Guntersville, Al., U.S. (gŭn´tērz-vĭl)	124-25	34°22′N	86°18′W
Guntūr, India (gŏn´tōōr)	236	16°18′N	80°27′E
Gunungsitoli, Indon.	246-47	1°15′N	97°37′E
Guoyang, China (gwō-yäŋ)	238-39	33°30′N	116°12′E
Gurara, stm., Nig.	260-61	8°11′N	6°42′E
Gurdon, Ar., U.S. (gûr´d*ŭ*n)	120-21	33°55′N	93°10′W
Guri, Embalse de, res., Ven.	164-65	7°30′N	62°50′W
Gurnee, Il., U.S. (gûr´nē)	116-17	42°22′N	87°54′W
Gurué, Moz.	264-65	15°27′S	36°59′E
Gurupá, Braz.	166-67	1°25′S	51°39′W
Gurupi, Braz.	166-67	11°43′S	49°02′W
Gurupi, stm., Braz.	166-67	1°16′S	46°09′W
Guryev, Kaz. *see* Atyraū	186-87	47°07′N	51°55′E
Gusau, Nig. (gōō-zä´ōō)	260-61	12°10′N	6°40′E
Guşgy, Turkmen.	232-33	35°16′N	62°21′E
Gushi, China (gōō-shr)	238-39	32°11′N	115°41′E
Gusinoozërsk, Russia	240-41	51°17′N	106°31′E
Gus'-Khrustal'nyy, Russia (gōōs-krōō-stäl'ny')	186-87	55°37′N	40°40′E
Gütersloh, Ger. (gü´tērs-lo)	194-95	51°54′N	8°23′E
Guthrie, Ok., U.S. (gŭth´rĭ)	120-21	35°53′N	97°26′W
Gutian, China	238-39	26°36′N	118°46′E
Gutiérrez Zamora, Mex. (gōō-tī-âr´räz zä-mō´rä)	146-47	20°27′N	97°05′W
Guttenberg, Ia., U.S. (gŭt´*ĕ*n-bûrg)	114-15	42°47′N	91°06′W
Guwāhāti, India	234-35	26°11′N	91°44′E
Guyana, nation, S.A. (gŭy´ănă)	158	5°0′N	59°00′W
Guyane, dep., S.A. *see* French Guiana	158	4°0′N	53°00′W
Guyang, China (gōō-yäŋ)	240-41	41°02′N	110°04′E
Guymon, Ok., U.S. (gī´mŏn)	120-21	36°41′N	101°29′W
Guyuan, China	240-41	35°59′N	106°18′E
Guzhen, China (gōō-jŭn)	238-39	33°19′N	117°19′E
Guzmán, Mex.	146-47	19°42′N	103°28′W
G'uzor, Uzb.	232-33	38°36′N	66°15′E
Gvardeysk, Russia (gvár-dĕysk´)	194-95	54°39′N	21°05′E
Gwādar, Pak. (gwä´dŭr)	230-31	25°08′N	62°20′E
Gwalior, India	234-35	26°13′N	78°09′E
Gwangju, Kor., S. *see* Kwangju	243	35°09′N	126°54′E
Gwardafuy, Gees, c., Som.	262-63	11°50′N	51°17′E
Gweru, Zimb.	264-65	19°27′S	29°49′E
Gwinn, Mi., U.S. (gwĭn)	114-15	46°17′N	87°25′W
Gyandzha, Azer. *see* Gäncä	227	40°41′N	46°21′E
Gyangzê, China	234-35	28°56′N	89°34′E
Gyaring Co, lk., China	234-35	31°05′N	88°24′E
Gyaring Hu, lk., China	240-41	34°54′N	97°15′E
Gyeongju, Kor., S. *see* Kyŏngju	243	35°51′N	129°13′E
Gympie, Austl. (gĭm´pē)	276	26°12′S	152°40′E
Gyöngyös, Hung. (dyûn´dyûsh)	194-95	47°47′N	19°56′E
Győr, Hung. (dyûr)	194-95	47°41′N	17°39′E
Gyula, Hung. (dyŏ´lä)	194-95	46°39′N	21°20′E
Gyumri, Arm.	227	40°47′N	43°51′E
Gyzylarbat, Turkmen.	232-33	38°59′N	56°16′E
Gyzyletrek, Turkmen.	232-33	37°36′N	54°47′E

H

Feature (Pronunciation)	Page	Lat.	Long.
Haapamäki, Fin. (häp´ä-mĕ-kē)	192-93	62°15′N	24°27′E
Haapsalu, Est. (häp´sä-lŏ)	192-93	58°56′N	23°33′E
Haar, Ger. (här)	194-95	48°06′N	11°44′E
Haarlem, Neth. (här´lĕm)	190-91	52°23′N	4°38′E
Hachijō-jima, i., Japan	244	33°05′N	139°48′E
Hachinohe, Japan (hä´chē-nō´hå)	244	40°30′N	141°29′E
Ḥadd, Ra's al-, c., Oman	230-31	22°32′N	59°48′E
Hadejia, stm., Nig.	260-61	12°50′N	10°51′E
Ḥadera, Isr. (ká-dĕ´rá)	228-29	32°27′N	34°55′E
Haderslev, Den. (hä´dhĕrs-lĕv)	192-93	55°15′N	9°30′E
Hadībū, Yemen	220-21	12°39′N	54°02′E
Ḥadīthah, Iraq	228-29	34°02′N	42°22′E
Ḥaḍramawt, reg., Yemen	220-21	15°0′N	50°00′E
Hadyai, Thai. *see* Hat Yai	246-47	7°01′N	100°28′E
Haeju, Kor., N. (hä´ĕ-jŭ)	243	38°03′N	125°43′E
Haft Gel, Iran	232-33	31°26′N	49°32′E
Hagåtña, nat. cap., Guam	279c	13°28′N	144°45′E
Hagen, Ger. (hä´gĕn)	194-95	51°22′N	7°28′E
Hagerstown, Md., U.S. (hā´gērz-toun)	116-17	39°39′N	77°43′W
Haggin, Mount, mtn., Mt., U.S.	112-13	46°05′N	113°05′W
Hagi, Japan (hä´gĭ)	245	34°24′N	131°24′E
Hague, Cap de la, c., Fr. (kȧp dē lȧ ȧg´)	196-97	49°43′N	1°56′W
Hague, The, nat. cap., Neth.	190-91	52°06′N	4°18′E
Haguenau, Fr. (ȧg´nō´)	196-97	48°49′N	7°47′E
Hahajima-rettō, is., Japan	280-81	26°37′N	142°10′E
Haicheng, China (hī-chŭŋ)	240-41	40°51′N	122°46′E
Haifa, Isr. (hä´ē-fȧ)	228-29	32°49′N	35°00′E
Haifeng, China (hä´ê-fĕng´)	238-39	22°58′N	115°20′E
Haikang, China	238-39	20°55′N	110°05′E
Haikou, China	238-39	20°03′N	110°22′E
Ḥā'il, Sau. Ar.	266	27°31′N	41°42′E
Hailar, China	240-41	49°11′N	119°44′E
Hailar, stm., China	240-41	49°30′N	117°51′E
Hailey, Id., U.S. (hā´lĭ)	112-13	43°32′N	114°19′W
Hailun, China (hä´ê-lōōn´)	240-41	47°27′N	126°58′E
Hailuoto, i., Fin.	184-85	65°02′N	24°42′E
Hainan, state, China (hī-nän´)	238-39	19°0′N	109°30′E
Hainan Dao, i., China (hī-nän dou)	238-39	19°0′N	109°30′E
Hainan Strait, strt., China (hī´-nän´ strāt) *see* Qiongzhou Haixia	238-39	20°10′N	110°15′E
Haines, Ak., U.S. (hānz)	126	59°14′N	135°27′W
Haines City, Fl., U.S. (hānz sĭ´tĭ)	125a	28°07′N	81°38′W
Haines Junction, Yk., Can.	128-29	60°45′N	137°28′W
Hai Ninh, Viet.	246-47	21°32′N	107°56′E
Hai Phong, Viet. (hī´fông´)(hä´ĕp-hŏng)	246-47	20°52′N	106°41′E
Haiti, nation, N.A. (hā´tĭ)	140-41	19°0′N	72°25′W
Haïti, i., N.A. *see* Hispaniola	142-43	19°0′N	71°00′W
Haizhou, China	238-39	34°35′N	119°08′E
Hajdúböszörmény, Hung. (hô´dŏ-bû´sûr-mān')	194-95	47°40′N	21°31′E
Hajdúnánás, Hung. (hô´ĭ-dŏ-nä´nāsh)	194-95	47°50′N	21°27′E
Ḥajjah, Yemen	266	15°42′N	43°36′E
Ḥakīm, Abyār al-, well, Libya	258-59	31°36′N	23°29′E
Hakodate, Japan (hä-kō-dä´t å)	244	41°45′N	140°43′E
Haku-san, vol., Japan (hä´kōō-sän´)	245	36°09′N	136°46′E
Ḥalab, Syria *see* Aleppo	228-29	36°13′N	37°10′E
Ḥalā'ib, Sudan	266	22°13′N	36°38′E
Halberstadt, Ger. (häl´bĕr-shtät)	194-95	51°53′N	11°03′E
Halcon, Mount, mtn., Phil. (mount häl-kōn´)	250	13°16′N	121°00′E
Halden, Nor. (häl´dĕn)	192-93	59°08′N	11°23′E
Haleakalā National Park, n.p., Hi., U.S. (hä´lā-ä´kä-lä năsh´ŭn-*ă*l pärk)	127a	20°44′N	156°13′W
Haleyville, Al., U.S. (hā´lĭ-vĭl)	124-25	34°14′N	87°38′W
Halfway, stm., B.C., Can.	132-33	56°13′N	121°26′W
Halifax, N.S., Can. (hăl´ĭ-făks)	138-39	44°39′N	63°36′W
Halifax Bay, b., Austl. (hăl´ĭ-făx bā)	277	18°50′S	146°30′E
Halla-san, mtn., Kor., S. (häl´lȧ-sän´)	240-41	33°22′N	126°32′E
Halle, Ger.	194-95	51°29′N	11°58′E
Hallettsville, Tx., U.S. (hăl´ĕts-vĭl)	122-23	29°27′N	96°56′W
Hall Islands, is., Micron.	280-81	8°35′N	151°59′E
Hallock, Mn., U.S. (hăl´*ŭ*k)	114-15	48°46′N	96°57′W
Hall Peninsula, pen., Nu., Can. (hôl pĕ-nĭn´sūl*ă*)	130-31	63°30′N	66°00′W
Hallsberg, Swe. (häls´bĕrgh)	192-93	59°05′N	15°08′E
Halls Creek, Austl. (hôlz krēk)	270-71	18°15′S	127°40′E
Halmahera, i., Indon. (häl-mä-hä´rä)	248-49	1°0′N	128°00′E
Halmahera, Laut, s., Indon. *see* Halmahera Sea	224-25	1°0′S	129°00′E
Halmahera Sea, s., Indon. (häl-mä-hä´rä sē)	224-25	1°0′S	129°00′E
Halmstad, Swe. (hälm´städ)	192-93	56°40′N	12°53′E
Hälsingborg, Swe. *see* Helsingborg	192-93	56°03′N	12°42′E
Haltern, Ger. (häl´tĕrn)	194-95	51°45′N	7°11′E
Haltiatunturi, mtn., Eur.	184-85	69°18′N	21°16′E
Halton Hills, On., Can. *see* Georgetown	136-37	43°39′N	79°55′W
Halys, stm., Tur. *see* Kızılırmak	186-87	41°44′N	35°58′E
Hamada, Japan	245	34°53′N	132°05′E
Hamadān, Iran (hŭ-mŭ-dăn´)	228-29	34°48′N	48°31′E
Ḥamāh, Syria (hä´mä)	228-29	35°08′N	36°45′E
Hamamatsu, Japan (hä´mä-mät´sŏ)	245	34°43′N	137°42′E
Hamar, Nor. (hä´mär)	192-93	60°48′N	11°05′E
Hamburg, Ger. (häm´bŏorgh)	194-95	53°33′N	9°59′E
Hamburg, Ar., U.S. (hăm´bûrg)	124-25	33°14′N	91°48′W
Hamburg, N.Y., U.S. (hăm´bûrg)	116-17	42°43′N	78°50′W
Hamden, Ct., U.S. (hăm´dĕn)	116-17	41°24′N	72°54′W
Hämeenlinna, Fin. (hĕ´măn-lĭn-nȧ)	192-93	60°58′N	24°31′E
HaMelaḥ, Yam, lk., Asia *see* Dead Sea	228-29	31°30′N	35°30′E
Hameln, Ger. (hä´m*ĕ*ln)	194-95	52°06′N	9°22′E
Hamersley Range, mts., Austl. (hăm´ērz-lē rānj)	272-73	22°24′S	117°34′E
Hamhŭng, Kor., N. (häm´hŏng´)	243	39°55′N	127°32′E
Hami, China (hä-mē)	240-41	42°50′N	93°31′E
Hamilton, Austl. (hăm´ĭl-t*ŭ*n)	276	37°45′S	142°01′E
Hamilton, On., Can. (hăm´ĭl-t*ŭ*n)	136-37	43°15′N	79°51′W
Hamilton, N.Z. (hăm´ĭl-t*ŭ*n)	278	37°47′S	175°17′E
Hamilton, Al., U.S. (hăm´ĭl-t*ŭ*n)	124-25	34°08′N	87°59′W
Hamilton, Mo., U.S. (hăm´ĭl-t*ŭ*n)	120-21	39°45′N	94°00′W
Hamilton, Mt., U.S. (hăm´ĭl-t*ŭ*n)	112-13	46°14′N	114°10′W
Hamilton, Oh., U.S. (hăm´ĭl-t*ŭ*n)	116-17	39°24′N	84°34′W
Hamilton, Tx., U.S. (hăm´ĭl-t*ŭ*n)	122-23	31°41′N	98°08′W
Hamilton, nat. cap., Ber. (hăm´ĭl-t*ŭ*n)	140-41	32°18′N	64°48′W
Hamina, Fin. (hȧ´mē-nȧ)	192-93	60°34′N	27°19′E
Hamlet, N.C., U.S. (hăm´lĕt)	124-25	34°54′N	79°42′W
Hamlin, Tx., U.S. (hăm´lĭn)	120-21	32°53′N	100°08′W
Hamm, Ger. (häm)	194-95	51°41′N	7°49′E
Hammamet, Tun.	184-85	36°24′N	10°37′E
Hammamet, Golfe de, b., Tun.	258-59	36°05′N	10°40′E
Hammerfest, Nor. (hä´mĕr-fĕst)	184-85	70°40′N	23°42′E
Hammond, In., U.S. (hăm´*ŭ*nd)	116-17	41°35′N	87°30′W
Hammond, La., U.S. (hăm´*ŭ*nd)	124-25	30°30′N	90°28′W
Hammonton, N.J., U.S. (hăm´ŭn-t*ŭ*n)	116-17	39°38′N	74°48′W
Hampton, N.B., Can. (hămp´t*ŭ*n)	138-39	45°32′N	65°51′W
Hampton, Ia., U.S. (hămp´t*ŭ*n)	114-15	42°45′N	93°12′W
Hampton, S.C., U.S. (hămp´t*ŭ*n)	124-25	32°52′N	81°07′W
Hampton, Va., U.S. (hămp´t*ŭ*n)	124-25	37°02′N	76°21′W
Ḥamrā', Al-Ḥamādah al-, des., Libya	188-89	30°0′N	12°00′E
Hāmūn, Daryācheh-ye, lk., Iran	230-31	30°43′N	61°07′E
Han, stm., China (hän)	238-39	23°41′N	116°38′E
Han, stm., China (hän)	238-39	30°34′N	114°17′E
Hāna, Hi., U.S. (hä´nä)	127a	20°45′N	155°59′W
Hancheng, China	240-41	35°29′N	110°25′E
Hancock, Mi., U.S. (hăn´kŏk)	114-15	47°08′N	88°36′W
Handan, China (hän-dän)	240-41	36°37′N	114°28′E
Hanford, Ca., U.S. (hăn´fērd)	118-19	36°20′N	119°38′W
Hangayn nuruu, mts., Mong.	240-41	47°32′N	98°42′E
Hangchow, China *see* Hangzhou	238-39	30°15′N	120°10′E
Hanggin Houqi, China	240-41	40°57′N	107°14′E
Hanggin Qi, China	240-41	39°55′N	108°52′E
Hangö, Fin. (hăŋ´gû) *see* Hanko	192-93	59°50′N	22°58′E
Hangzhou, China (hăng´chō´)	238-39	30°15′N	120°10′E
Hanjiang, China	225a	25°30′N	119°06′E
Hankinson, N.D., U.S. (hăŋ´kĭn-s*ŭ*n)	114-15	46°04′N	96°55′W
Hanko, Fin.	192-93	59°50′N	22°58′E
Hankow, China *see* Wuhan	238-39	30°34′N	114°17′E
Hanna, Ab., Can. (hăn´ȧ)	132-33	51°39′N	111°56′W
Hannibal, Mo., U.S. (hăn´ĭ b*ă*l)	120-21	39°42′N	91°22′W
Hannover, Ger. (hän-ō´vĕr)	194-95	52°24′N	9°44′E
Ha Noi, nat. cap., Viet. (hä-noi´)	246-47	21°02′N	105°50′E
Hanover, On., Can. (hăn´ô-vēr)	136-37	44°09′N	81°01′W
Hanover, Ger. *see* Hannover	194-95	52°24′N	9°44′E
Hanover, N.H., U.S. (hăn´ô-vēr)	116-17	43°42′N	72°17′W
Hanover, Pa., U.S. (hăn´ô-vēr)	116-17	39°48′N	76°59′W
Hanover, Va., U.S. (hăn´ô-vēr)	116-17	37°46′N	77°23′W
Hanover, Isla, i., Chile	171	50°58′S	74°45′W
Hantsport, N.S., Can. (hănts´pōrt)	138-39	45°04′N	64°12′W
Hanuy, stm., Mong.	240-41	49°21′N	102°22′E
Hanzhong, China (hän-jŏŋ)	238-39	33°04′N	107°02′E

Feature (Pronunciation)	Page	Lat.	Long.
Hāora, India	234-35	22°35′N	88°20′E
Haparanda, Swe. (hä-pa-rän′dä)	184-85	65°50′N	24°06′E
Happy Valley-Goose Bay, Nf., Can.	128-29	53°20′N	60°25′W
Hāpur, India	234-35	28°44′N	77°47′E
Harare, nat. cap., Zimb. (hä-rä′-rē)	264-65	17°50′S	31°03′E
Ḥarash, Bi'r al-, well, Libya	258-59	25°39′N	22°08′E
Harbin, China	240-41	45°45′N	126°38′E
Harbor Beach, Mi., U.S. (här′bẽr bēch)	116-17	43°51′N	82°39′W
Harbour Breton, Nf., Can. (här′bẽr brĕt′ŭn) (brĕ-tôn′)	138-39	47°30′N	55°49′W
Harbour Grace, Nf., Can. (här′bẽr grās)	138-39	47°44′N	53°15′W
Hardangerfjorden, b., Nor.	192-93	60°10′N	6°00′E
Hardin, Mt., U.S. (här′dĭn)	112-13	45°44′N	107°37′W
Hardoi, India	234-35	27°23′N	80°09′E
Hare Bay, b., Nf., Can. (hâr bā)	138-39	51°16′N	55°51′W
Hareidlandet, i., Nor.	184-85	62°21′N	5°57′E
Hārer, Eth.	262-63	9°18′N	42°08′E
Hargeysa, Som. (här-gā′ē-sá)	262-63	9°34′N	44°04′E
Har Horin, hist., Mong. see Karakorum	240-41	47°14′N	102°50′E
Har Hu, lk., China	240-41	38°15′N	97°40′E
Hari, stm., Indon.	246-47	1°04′S	104°12′E
Haridwār, India	234-35	29°56′N	78°07′E
Harīrūd, stm., Asia	232-33	37°24′N	60°31′E
Harlan, Ia., U.S. (här′lăn)	114-15	41°39′N	95°20′W
Harlan, Ky., U.S. (här′lăn)	124-25	36°51′N	83°19′W
Harlem, Mt., U.S. (hȧr′lĕm)	112-13	48°32′N	108°47′W
Harlingen, Neth. (här′lĭng-ĕn)	190-91	53°10′N	5°26′E
Harlingen, Tx., U.S. (här′lĭng-ĕn)	122-23	26°12′N	97°42′W
Harlow, Eng., U.K. (här′lō)	190-91	51°47′N	0°07′E
Harlowton, Mt., U.S. (hȧr′lō-tun)	112-13	46°26′N	109°50′W
Harney Basin, bas., Or., U.S. (här′nĭ bā′s'n)	112-13	43°15′N	119°00′W
Harney Peak, mtn., S.D., U.S. (här′nĭ pēk)	114-15	43°52′N	103°32′W
Härnösand, Swe. (hĕr-nû-sänd)	184-85	62°38′N	17°56′E
Haro, Spain (ä′rō)	198-99	42°36′N	2°52′W
Hārot, stm., Afg.	232-33	31°29′N	61°16′E
Harper, Lib. (här′pẽr)	260-61	4°23′N	7°43′W
Harpers Ferry, W.V., U.S. (här′pẽrz fĕr′ē)	116-17	39°19′N	77°45′W
Harricana, stm., Can.	130-31	51°10′N	79°47′W
Harriman, Tn., U.S. (hă′ĭ-măn)	124-25	35°56′N	84°33′W
Harrington, De., U.S. (hăr′ĭng-tŭn)	116-17	38°56′N	75°34′W
Harris, Lake, lk., Austl.	272-73	31°06′S	135°11′E
Harrisburg, Il., U.S. (hăr′ĭs-bûrg)	116-17	37°44′N	88°32′W
Harrisburg, Pa., U.S.	116-17	40°16′N	76°54′W
Harrismith, S. Afr. (hă-rĭs′mĭth)	269c	28°17′S	29°08′E
Harrison, Ar., U.S. (hăr′ĭ-sŭn)	120-21	36°14′N	93°07′W
Harrison, Mi., U.S. (hăr′ĭ-sŭn)	116-17	44°01′N	84°48′W
Harrison, Cape, c., Nf., Can.	130-31	54°55′N	57°56′W
Harrisonburg, Va., U.S. (hăr′ĭ-sŭn-bûrg)	116-17	38°27′N	78°52′W
Harrison Lake, lk., B.C., Can. (hăr′ĭ-sŭn lăk)	132-33	49°33′N	121°52′W
Harrisonville, Mo., U.S. (hăr-ĭ-sŭn-vĭl)	120-21	38°39′N	94°21′W
Harrisville, Mi., U.S. (hăr′ĭs-vĭl)	116-17	44°39′N	83°18′W
Harrisville, W.V., U.S. (hăr′ĭs-vĭl)	116-17	39°13′N	81°03′W
Harrodsburg, Ky., U.S. (hăr′ŭdz-bûrg)	116-17	37°46′N	84°51′W
Harstad, Nor. (här′städh)	184-85	68°47′N	16°34′E
Hart, Mi., U.S. (härt)	116-17	43°42′N	86°22′W
Hartford, Ct., U.S. (härt′fẽrd)	116-17	41°46′N	72°41′W
Hartford, Ky., U.S. (härt′fẽrd)	116-17	37°27′N	86°54′W
Hartford, Mi., U.S. (härt′fẽrd)	116-17	42°12′N	86°10′W
Hartford, S.D., U.S. (härt′fẽrd)	114-15	43°38′N	96°58′W
Hartford City, In., U.S. (härt′fẽrd sĭ′tē)	116-17	40°27′N	85°22′W
Hartlepool, Eng., U.K. (här′t'l-pōōl)	190-91	54°42′N	1°12′W
Hart Mountain, mtn., Mb., Can. (härt moun′tĭn)	134-35	52°29′N	101°25′W
Harts, stm., S. Afr.	264-65	28°24′S	24°17′E
Hartselle, Al., U.S. (härt′sĕl)	124-25	34°27′N	86°56′W
Hartshorne, Ok., U.S. (härts′hôrn)	120-21	34°51′N	95°34′W
Hartsville, S.C., U.S. (härts′vĭl)	124-25	34°22′N	80°05′W
Hartwell, Ga., U.S. (härt′wĕl)	124-25	34°21′N	82°55′W
Hartwell Lake, res., U.S. (härt′wĕl lăk)	124-25	34°28′N	82°51′W
Har Us nuur, lk., Mong.	240-41	48°0′N	92°10′E
Harvard, Il., U.S. (här′vȧrd)	116-17	42°25′N	88°37′W
Harvey, N.D., U.S.	114-15	47°46′N	99°55′W
Harwich, Eng., U.K. (här′wĭch)	190-91	51°57′N	1°17′E
Haryāna, state, India	234-35	29°20′N	76°20′E
Harz, mts., Ger. (härts)	194-95	51°45′N	10°30′E
Haskell, Tx., U.S. (hăs′kĕl)	120-21	33°09′N	99°44′W
Hassan, India	236	12°60′N	76°06′E
Hassi Messaoud, Alg.	188-89	31°41′N	6°04′E
Hässleholm, Swe. (hĕs′lĕ-hōlm)	192-93	56°09′N	13°46′E
Hastings, N.Z. (hās′tĭngz)	278	39°38′S	176°51′E
Hastings, Eng., U.K. (hās′tĭngz)	190-91	50°52′N	0°35′E
Hastings, Mi., U.S. (hās′tĭngz)	116-17	42°39′N	85°17′W
Hastings, Mn., U.S. (hās′tĭngz)	114-15	44°44′N	92°51′W
Hastings, Ne., U.S. (hās′tĭngz)	120-21	40°35′N	98°24′W
Hatay, Tur. see Antioch	228-29	36°12′N	36°10′E
Haţeg, Rom. (kät-säg′)	200-01	45°36′N	22°57′E
Hāthras, India	234-35	27°36′N	78°03′E
Ha Tinh, Viet.	246-47	18°20′N	105°54′E
Hatteras, Cape, c., N.C., U.S. (kăp hăt′ẽr-ȧs)	124-25	35°13′N	75°32′W
Hatteras Island, i., N.C., U.S.	124-25	35°25′N	75°29′W
Hattiesburg, Ms., U.S. (hăt′ĭz-bûrg)	124-25	31°20′N	89°17′W
Hatvan, Hung. (hôt′vôn)	194-95	47°40′N	19°41′E
Hat Yai, Thai.	246-47	7°01′N	100°28′E
Haugesund, Nor. (hou′gẽ-sŏŏn′)	192-93	59°25′N	5°18′E
Hauraki Gulf, b., N.Z. (hä-ōō-rä′kē gŭlf)	278	36°35′S	175°05′E
Haut Atlas, mts., Mor.	258-59	31°47′N	6°04′W
Haute-Volta, nation, Afr. see Burkina Faso	253	13°0′N	1°30′W
Havana, Il., U.S. (hȧ-vă′nȧ)	120-21	40°18′N	90°03′W
Havana, nat. cap., Cuba (hȧ-vă′nȧ)	142-43	23°06′N	82°27′W
Havel, stm., Ger. (hä′fĕl)	194-95	52°53′N	12°01′E
Haverhill, Ma., U.S. (hā′vẽr-hĭl)	116-17	42°47′N	71°05′W
Havre, Fr. see Le Havre	196-97	49°29′N	0°08′E
Havre, Mt., U.S. (hăv′ẽr)	112-13	48°33′N	109°41′W
Havre Aubert, Île du, i., Qc., Can.	138-39	47°14′N	61°57′W
Havre de Grace, Md., U.S. (hăv′ẽr dē grȧs′)	116-17	39°33′N	76°06′W
Havre-Saint-Pierre, Qc., Can.	138-39	50°15′N	63°36′W
Hawai'i, i., Hi., U.S. (hä-wī′ē)	127a	19°29′N	155°30′W
Hawai'ian Islands, is., Hi., U.S. (hä-wī′ȧn ī′lȧndz)	127	24°0′N	157°00′W
Hawai'i Volcanoes National Park, n.p., Hi., U.S.	127a	19°23′N	155°17′W
Hawaii, state, U.S. (häw wī′ē)	108-09	20°0′N	157°45′W
Hawi, Hi., U.S. (hä′wē)	127a	20°15′N	155°50′W
Hawick, Scot., U.K. (hô′ĭk)	190-91	55°25′N	2°47′W
Hawke Bay, b., N.Z. (hôk bā)	278	39°20′S	177°30′E
Hawker, Austl. (hô′kẽr)	276	31°53′S	138°25′E
Hawkesbury, On., Can. (hôks′bẽr-ĭ)	136-37	45°37′N	74°36′W
Hawkesbury Island, i., B.C., Can.	132-33	53°38′N	129°00′W
Hawkinsville, Ga., U.S. (hô′kĭnz-vĭl)	124-25	32°17′N	83°28′W
Hawley, Mn., U.S. (hô′lĭ)	114-15	46°53′N	96°19′W
Hawthorne, Nv., U.S.	118-19	38°32′N	118°37′W
Haxtun, Co., U.S. (hăks′tŭn)	114-15	40°38′N	102°38′W
Hay, Austl.	276	34°31′S	144°50′E
Hay, stm., Can. (hā)	130-31	60°52′N	115°44′W
HaYarden, stm., Asia see Jordan	228-29	31°46′N	35°34′E
Hayastan, nation, Asia see Armenia	227	40°0′N	45°00′E
Hayden, Id., U.S. (hā′dĕn)	112-13	47°46′N	116°47′W
Hayes, stm., Mb., Can. (hāz)	134-35	57°03′N	92°14′W
Hayes, Mount, mtn., Ak., U.S. (mount hāz)	126	63°37′N	146°43′W
Haynesville, La., U.S. (hānz′vĭl)	120-21	32°58′N	93°08′W
Hay River, N.T., Can. (hā rĭv′ẽr)	128-29	60°49′N	115°48′W
Hays, Ks., U.S. (hāz)	120-21	38°52′N	99°18′W
Hazard, Ky., U.S. (hăz′ȧrd)	124-25	37°15′N	83°12′W
Hazārībāg, India	234-35	23°60′N	85°22′E
Hazelton, B.C., Can. (hā′z'l-tŭn)	132-33	55°15′N	127°42′W
Hazelton Mountains, mts., B.C., Can. (hā′z'l-tŭn moun′tĭnz)	132-33	54°51′N	128°00′W
Hazleton, Pa., U.S. (hā′z'l-tŭn)	116-17	40°57′N	75°59′W
Headland, Al., U.S. (hĕd′lănd)	124-25	31°21′N	85°21′W
Healdsburg, Ca., U.S. (hēldz′bûrg)	118-19	38°37′N	122°52′W
Healdton, Ok., U.S. (hēld′tŭn)	120-21	34°14′N	97°29′W
Heard and McDonald Islands, dep., Oc.	282-83	53°05′S	73°00′E
Heard Island, i., Austl. (hûrd ī′lȧnd)	286	53°06′S	73°30′E
Hearne, Tx., U.S. (hûrn)	122-23	30°53′N	96°36′W
Hearst, On., Can. (hûrst)	128-29	49°41′N	83°42′W
Heavener, Ok., U.S. (hēv′nẽr)	120-21	34°54′N	94°36′W
Hebbronville, Tx., U.S. (hē′brŭn-vĭl)	122-23	27°18′N	98°41′W
Hebei, state, China (hŭ-bā)	240-41	38°0′N	116°00′E
Heber City, Ut., U.S. (hē′bẽr sĭ′tē)	118-19	40°31′N	111°25′W
Heber Springs, Ar., U.S. (hē′bẽr sprĭngz)	124-25	35°29′N	92°02′W
Hebi, China	240-41	35°58′N	114°09′E
Hebrides, is., Scot., U.K.	190-91	57°0′N	6°30′W
Hebron, Nf., Can. (hĕb′rŭn)	128-29	58°12′N	62°38′W
Hebron, N.D., U.S. (hĕb′rŭn)	114-15	46°54′N	102°03′W
Hebron, Ne., U.S. (hĕb′rŭn)	120-21	40°09′N	97°35′W
Hebron, W.B.	228-29	31°32′N	35°06′E
Hecate Strait, strt., B.C., Can. (hĕk′ȧ-tē strāt)	132-33	53°0′N	131°00′W
Hecelchakán, Mex. (ā-sĕl-chä-kän′)	148	20°10′N	90°08′W
Hechi, China (hŭ-chr)	238-39	24°42′N	108°02′E
Hechuan, China (hŭ-chyuän)	238-39	29°60′N	106°16′E
Hedemora, Swe. (hĭ-dĕ-mō′rä)	192-93	60°17′N	15°59′E
Ḥefa, Isr. see Haifa	228-29	32°49′N	35°00′E
Hefei, China (hŭ-fā)	238-39	31°51′N	117°17′E
Heflin, Al., U.S. (hĕf′lĭn)	124-25	33°39′N	85°35′W
Hegang, China	240-41	47°19′N	130°16′E
Heho, Mya.	246-47	20°43′N	96°49′E
Heidelberg, Ger. (hīdĕl-bĕrgh)	194-95	49°25′N	8°42′E
Heihe, China	240-41	50°14′N	127°30′E
Heijō, nat. cap., Kor., N. see P'yŏngyang	243	39°01′N	125°44′E
Heilbron, S. Afr. (hīl′brōn)	269c	27°17′S	27°59′E
Heilbronn, Ger. (hīl′brōn)	194-95	49°08′N	9°12′E
Heilong, stm., Asia see Amur	218-19	52°57′N	141°10′E
Heilongjiang, state, China (hā-lôŋ-jyäŋ)	240-41	48°0′N	128°00′E
Heilongjiang, stm., Asia see Amur	218-19	52°57′N	141°10′E
Heilungkiang, state, China see Heilongjiang	240-41	48°0′N	128°00′E
Heinola, Fin. (hå-nō′lȧ)	192-93	61°12′N	26°03′E
Hejaz, reg., Sau. Ar. (hĕ-jăz′) (hē-jăz′) see Al-Ḥijāz	266	24°30′N	38°30′E
Hejian, China (hŭ-jyĕn)	240-41	38°26′N	116°05′E
Hekla, vol., Ice.	190a	64°0′N	19°39′W
Hel, Pol. (hāl)	194-95	54°37′N	18°48′E
Helagsfjället, mtn., Swe.	184-85	62°55′N	12°27′E
Helena, Ar., U.S. (hē-lē′nȧ)	124-25	34°32′N	90°36′W
Helena, Mt., U.S.	112-13	46°36′N	112°02′W
Helen Island, i., Palau	280-81	2°58′N	131°49′E
Helgoland, i., Ger. (hĕl′gō-länd)	194-95	54°11′N	7°52′E
Hellín, Spain (ĕl-yĕn′)	198-99	38°30′N	1°42′W
Hells Canyon, val., U.S. (hĕls kăn′yŭn)	112-13	45°17′N	116°40′W
Helmand, stm., Asia (hĕl′mŭnd)	232-33	31°19′N	61°29′E
Helmond, Neth. (hĕl′mōnt) (ĕl′môn′)	190-91	51°29′N	5°40′E
Helmstedt, Ger. (hĕlm′shtĕt)	194-95	52°13′N	11°01′E
Helsingborg, Swe. (hĕl′sĭng-bôrgh)	192-93	56°03′N	12°42′E
Helsingfors, nat. cap., Fin. see Helsinki	192-93	60°10′N	24°57′E
Helsingør, Den. (hĕl-sĭng-ûr′)	192-93	56°02′N	12°37′E
Helsinki, nat. cap., Fin. (hĕl′sĕn-kē)	192-93	60°10′N	24°57′E
Helvetia, nation, Eur. see Switzerland	174-75	47°0′N	8°00′E
Hemingford, Ne., U.S. (hĕm′ĭng-fẽrd)	114-15	42°19′N	103°05′W
Hempstead, N.Y., U.S. (hĕmp′stĕd)	116-17	40°42′N	73°41′W
Hempstead, Tx., U.S. (hĕmp′stĕd)	122-23	30°05′N	96°05′W
Hemse, Swe. (hĕm′sĕ)	192-93	57°14′N	18°23′E
Henan, state, China (hŭ-nän)	238-39	34°0′N	114°00′E
Henderson, Ky., U.S. (hĕn′dẽr-sŭn)	116-17	37°50′N	87°35′W
Henderson, N.C., U.S. (hĕn′dẽr-sŭn)	124-25	36°20′N	78°24′W
Henderson, Nv., U.S. (hĕn′dẽr-sŭn)	118-19	36°02′N	114°59′W
Henderson, Tn., U.S. (hĕn′dẽr-sŭn)	124-25	35°26′N	88°38′W
Henderson, Tx., U.S. (hĕn′dẽr-sŭn)	122-23	32°09′N	94°48′W
Henderson Island, i., Pit.	280-81	24°22′S	128°19′W
Hendersonville, N.C., U.S. (hĕn′dẽr-sŭn-vĭl)	124-25	35°19′N	82°28′W
Hendersonville, Tn., U.S. (hĕn′dẽr-sŭn-vĭl)	124-25	36°18′N	86°37′W
Hendorābī, Jazīreh-ye, i., Iran	230-31	26°41′N	53°38′E
Hendrina, S. Afr. (hĕn-drē′nȧ)	269c	26°10′S	29°44′E
Hendū Kosh, mts., Asia see Hindu Kush	232-33	36°0′N	71°30′E
Hengām, Jazīreh-ye, i., Iran	230-31	26°39′N	55°53′E
Hengelo, Neth. (hĕng′ē-lō)	190-91	52°16′N	6°48′E
Hengshan, China (hĕng′shän′)	238-39	27°15′N	112°51′E
Hengshan, China (hĕng′shän′)	240-41	37°57′N	109°18′E
Hengshui, China (hĕng′shōō-ē′)	240-41	37°44′N	115°42′E
Hengxian, China (hŭŋ shyĕn)	238-39	22°41′N	109°12′E
Hengyang, China	238-39	26°54′N	112°36′E
Henlopen, Cape, c., De., U.S. (kăp hĕn-lō′pĕn)	116-17	38°47′N	75°06′W
Hennebont, Fr. (ĕn-bôn′)	196-97	47°48′N	3°16′W
Hennessey, Ok., U.S. (hĕn′ĕ-sĭ)	120-21	36°07′N	97°54′W
Henrietta, Tx., U.S. (hĕn-rĭ-ĕ′tȧ)	120-21	33°49′N	98°12′W
Henrietta Maria, Cape, c., On., Can. (kăp hĕn-rĭ-ĕt′ȧ mȧ-rē′ȧ)	130-31	55°08′N	82°20′W
Henzada, Mya.	246-47	17°38′N	95°28′E
Hepu, China (hŭ-pōō)	238-39	21°41′N	109°11′E
Herāt, Afg. (hĕ-rät′)	232-33	34°21′N	62°12′E
Heredia, C.R. (ā-rā′dhē-ä)	149	9°59′N	84°07′W
Hereford, Eng., U.K. (hĕrĕ′fẽrd)	190-91	52°04′N	2°43′W
Hereford, Tx., U.S. (hĕr′ĕ-fẽrd)	120-21	34°50′N	102°24′W
Herford, Ger. (hĕr′fôrt)	194-95	52°07′N	8°40′E
Herkimer, N.Y., U.S. (hûr′kĭ-mẽr)	116-17	43°02′N	74°58′W
Herlen, stm., Asia see Kerulen	240-41	48°44′N	117°03′E
Hermannstadt, Rom. see Sibiu	194-95	45°47′N	24°09′E
Hermansville, Mi., U.S. (hûr′mănz-vĭl)	116-17	45°42′N	87°36′W

Feature (Pronunciation)	Page	Lat.	Long.
Hermit Islands, is., Pap. N. Gui. (hûr′mĭt ī′lăndz)	277	1°30′S	145°05′E
Hermon, Mount, mtn., Asia	228-29	33°25′N	35°51′E
Hermosillo, Mex. (ĕr-mô-sē′l-yô)	144-45	29°05′N	110°58′W
Hermus, stm., Tur. *see* Gediz	200-01	38°36′N	26°48′E
Herning, Den. (hĕr′nĭng)	192-93	56°08′N	8°59′E
Herrin, Il., U.S. (hĕr′ĭn)	116-17	37°48′N	89°02′W
Herstal, Bel. (hĕr′stäl)	190-91	50°40′N	5°38′E
Hertford, N.C., U.S. (hûrt′fẽrd)	124-25	36°11′N	76°29′W
Hervey Bay, Austl.	276	25°17′S	152°50′E
Hervey Bay, b., Austl.	272-73	25°0′S	153°00′E
Hessen, hist. reg., Ger. (hĕs′ĕn)	194-95	50°30′N	9°15′E
Hettinger, N.D., U.S. (hĕt′ĭn-jẽr)	114-15	46°00′N	102°38′W
Hexian, China (hŭ shyĕn)	238-39	24°18′N	111°39′E
Heyuan, China (hŭ-yůän)	238-39	23°43′N	114°42′E
Heze, China (hŭ-dzŭ)	240-41	35°15′N	115°27′E
Hialeah, Fl., U.S. (hī-ȧ-lē′ăh)	125a	25°51′N	80°17′W
Hiawatha, Ia., U.S. (hī-ȧ-wô′thȧ)	114-15	42°02′N	91°41′W
Hibbing, Mn., U.S. (hĭb′ĭng)	114-15	47°25′N	92°56′W
Hickman, Ky., U.S. (hĭk′măn)	124-25	36°34′N	89°11′W
Hickory, N.C., U.S. (hĭk′ô-rĭ)	124-25	35°44′N	81°21′W
Hicks, Point, c., Austl.	276	37°48′S	149°16′E
Hidalgo, Mex. (ē-dhäl′gō)	146-47	24°15′N	99°26′W
Hidalgo, Mex. (ē-dhäl′gō)	122-23	25°58′N	100°27′W
Hidalgo, state, Mex. (ē-dhäl′gō)	146-47	20°30′N	99°00′W
Hidalgo del Parral, Mex. (ē-dä′l-gō-dĕl-pär-rä′l)	122-23	26°57′N	105°40′W
Higasi Sina Kai, s., Asia *see* East China Sea	222-23	30°0′N	126°00′E
Higginsville, Mo., U.S. (hĭg′ĭnz-vĭl)	120-21	39°05′N	93°43′W
Highland, Il., U.S. (hī′lănd)	120-21	38°44′N	89°41′W
Highland Park, Il., U.S. (hī′lănd pärk)	116-17	42°11′N	87°48′W
Highland Park, Mi., U.S. (hī′lănd pärk)	116-17	42°24′N	83°07′W
High Level, Ab., Can.	128-29	58°31′N	117°08′W
Highmore, S.D., U.S. (hī′mōr)	114-15	44°31′N	99°26′W
High Point, N.C., U.S. (hī point)	124-25	35°57′N	80°01′W
High Prairie, Ab., Can. (hī prā′rĭ)	132-33	55°26′N	116°29′W
High River, Ab., Can. (hī rĭv′ẽr)	132-33	50°35′N	113°52′W
Hightstown, N.J., U.S. (hīts-toun)	116-17	40°17′N	74°32′W
High Wycombe, Eng., U.K. (hī wĭ-kŭm)	190-91	51°38′N	0°45′W
Higüey, Dom. Rep. (ē-gwĕ′y)	142-43	18°37′N	68°43′W
Hiiumaa, i., Est. (hē′ŏm-ô)	192-93	58°55′N	22°38′E
Hikone, Japan (hē′kô-nĕ)	245	35°15′N	136°15′E
Hildesheim, Ger. (hĭl′dĕs-hīm)	194-95	52°09′N	9°57′E
Hilla, Iraq *see* Al-Ḩillah	228-29	32°29′N	44°26′E
Hillaby, Mount, mtn., Barb. (mount hĭl′ȧ-bĭ)	143b	13°12′N	59°35′W
Hillerød, Den. (hē′lĕ-rûdh hĭl)	192-93	55°56′N	12°19′E
Hillsboro, Il., U.S. (hĭlz′bŭr-ō)	116-17	39°09′N	89°29′W
Hillsboro, N.D., U.S. (hĭlz′bŭr-ō)	114-15	47°24′N	97°04′W
Hillsboro, Oh., U.S. (hĭlz′bŭr-ō)	116-17	39°12′N	83°37′W
Hillsboro, Or., U.S. (hĭlz′bŭr-ō)	112-13	45°31′N	122°59′W
Hillsboro, Tx., U.S. (hĭlz′bŭr-ō)	122-23	32°01′N	97°08′W
Hillsboro, W.V., U.S. (hĭlz′bŭr-ō)	116-17	38°08′N	80°13′W
Hillsdale, Mi., U.S. (hĭls-dāl hĭlz)	116-17	41°55′N	84°38′W
Hillston, Austl.	276	33°29′S	145°33′E
Hilo, Hi., U.S. (hē′lō)	127a	19°43′N	155°05′W
Himāchal Pradesh, state, India	234-35	32°0′N	77°00′E
Himalayas, mts., Asia (hĭ-mä′lȧ-yȧz) (hĭ-mȧ-lā′-yȧz)	234-35	28°0′N	84°00′E
Himalaya Shan, mts., Asia *see* Himalayas	234-35	28°0′N	84°00′E
Himatnagar, India	234-35	23°35′N	72°58′E
Himeji, Japan (hē′mȧ-jē)	245	34°50′N	134°42′E
Ḩimş, Syria	228-29	34°45′N	36°44′E
Hinche, Haiti (hēn′chȧ) (ănsh)	142-43	19°09′N	72°00′W
Hinchinbrook Island, i., Austl. (hĭn-chĭn-brook ī′lănd)	277	18°23′S	146°17′E
Hindenburg, Pol. *see* Zabrze	194-95	50°18′N	18°46′E
Hindu Kush, mts., Asia (hĭn′doo koosh′)	232-33	36°0′N	71°30′E
Hindupur, India (hĭn′doo-poor)	236	13°49′N	77°30′E
Hinnøya, i., Nor.	184-85	68°32′N	15°59′E
Hinton, Ab., Can. (hĭn′tŭn)	132-33	53°25′N	117°34′W
Hinton, W.V., U.S. (hĭn′tŭn)	116-17	37°40′N	80°53′W
Hirado-shima, i., Japan (hē′rä-dō shē′mä)	245	33°20′N	129°30′E
Hirara, Japan	279a	24°47′N	125°17′E
Hīrmand, stm., Asia *see* Helmand	232-33	31°19′N	61°29′E
Hirosaki, Japan (hē′rô-sä′kē)	244	40°36′N	140°29′E
Hiroshima, Japan (hē-rô-shē′mȧ)	245	34°24′N	132°28′E
Hirosima, Japan *see* Hiroshima	245	34°24′N	132°28′E
Hirschberg, Pol. *see* Jelenia Góra	194-95	50°54′N	15°44′E
Hirson, Fr. (ēr-sôN′)	196-97	49°56′N	4°05′E

Feature (Pronunciation)	Page	Lat.	Long.
Hisār, India	234-35	29°09′N	75°44′E
Hispaniola, i., N.A. (hĭ′spăn-ĭ-ō-lȧ)	142-43	19°0′N	71°00′W
Hitachi, Japan (hē-tä′chē)	245	36°36′N	140°39′E
Hitoyoshi, Japan (hē′tô-yô′shē)	245	32°12′N	130°46′E
Hitra, i., Nor. (hĭträ)	184-85	63°33′N	8°45′E
Hiu, i., Vanuatu	279g	13°08′S	166°33′E
Hiva Oa, i., Fr. Poly.	280-81	9°45′S	139°00′W
Hjo, Swe. (yō)	192-93	58°18′N	14°17′E
Hjørring, Den. (jûr′ĭng)	192-93	57°28′N	9°59′E
Hkakabo Razi, mtn., Mya	238-39	28°17′N	97°46′E
Hlohovec, Slvk. (hlô′ho-vĕts)	194-95	48°26′N	17°48′E
Ho, Ghana	260-61	6°36′N	0°28′E
Hoa Binh, Viet	246-47	20°50′N	105°20′E
Hobart, Austl. (hō′bȧrt)	276	42°52′S	147°18′E
Hobart, Ok., U.S. (hō′bȧrt)	120-21	35°02′N	99°06′W
Hobbs, N.M., U.S. (hŏbs)	120-21	32°42′N	103°08′W
Hobro, Den. (hô-brô′)	192-93	56°38′N	9°48′E
Ho Chi Minh City, Viet. (hō-chē-mĭn sĭ′tē)	246-47	10°45′N	106°40′E
Hodeida, Yemen *see* Al-Ḩudaydah	266	14°48′N	42°57′E
Hodgenville, Ky., U.S. (hŏj′ĕn-vĭl)	116-17	37°34′N	85°44′W
Hódmezővásárhely, Hung. (hōd′mĕ-zû-vô′shôr-hĕl-y′)	194-95	46°25′N	20°20′E
Hodna, Chott el, lk., Alg.	269b	35°25′N	4°45′E
Hodonín, Czech Rep. (hē′dô-nēn)	194-95	48°51′N	17°08′E
Hoei, Bel. *see* Huy	190-91	50°31′N	5°14′E
Hof, Ger. (hōf)	194-95	50°18′N	11°55′E
Hofsjökull, ice, Ice. (hôfs′yü′kool)	190a	64°50′N	18°54′W
Hōfu, Japan	245	34°03′N	131°35′E
Hofuf, Sau. Ar. *see* Al-Hufūf	230-31	25°22′N	49°34′E
Hogansville, Ga., U.S. (hō′gănz-vĭl)	124-25	33°11′N	84°55′W
Hoggar, mts., Alg. *see* Ahaggar	258-59	23°0′N	6°30′E
Hohensalza, Pol. *see* Inowrocław	194-95	52°48′N	18°15′E
Hohe Tauern, mts., Aus. (hō′ĕ tou′ẽrn)	194-95	47°06′N	12°56′E
Hohhot, China (hŭ-hoo-tŭ)	240-41	40°49′N	111°39′E
Hoihow, China *see* Haikou	238-39	20°03′N	110°22′E
Hoisington, Ks., U.S. (hoi′zĭng-tŭn)	120-21	38°31′N	98°46′W
Hokitika, N.Z. (hō-kĭ-tē′kä)	278	42°44′S	170°58′E
Hokkaidō, i., Japan (hŏk′kī-dō)	244	44°0′N	143°00′E
Hola, Kenya	262-63	1°30′S	40°01′E
Holbrook, Az., U.S. (hōl′brŏk)	118-19	34°54′N	110°10′W
Holden, Mo., U.S. (hōl′dĕn)	120-21	38°43′N	93°60′W
Holden, W.V., U.S. (hōl′dĕn)	116-17	37°49′N	82°04′W
Holdenville, Ok., U.S. (hōl′dĕn-vĭl)	120-21	35°05′N	96°24′W
Holdrege, Ne., U.S. (hōl′drĕj)	120-21	40°27′N	99°22′W
Holguín, Cuba (ôl-gēn′)	142-43	20°53′N	76°15′W
Holguín, state, Cuba (ôl-gēn′)	142-43	20°55′N	75°50′W
Holland, Mi., U.S. (hŏl′ănd)	116-17	42°47′N	86°06′W
Holland, nation, Eur. *see* Netherlands	174-75	52°15′N	5°30′E
Hollandia, Indon. *see* Jayapura	277	2°32′S	140°43′E
Hollandsbird Island, i., Nmb.	264-65	24°39′S	14°32′E
Hollick-Kenyon Plateau, plat., Ant.	287	79°0′S	97°00′W
Hollis, Ok., U.S. (hŏl′ĭs)	120-21	34°41′N	99°55′W
Hollister, Mo., U.S. (hŏl′ĭs-tẽr)	120-21	36°37′N	93°13′W
Holly Springs, Ms., U.S. (hŏl′ĭ sprĭngz)	124-25	34°46′N	89°27′W
Hollywood, Fl., U.S. (hŏl′ē-wŏŏd)	125a	26°01′N	80°09′W
Holmestrand, Nor. (hôl′mĕ-strän)	192-93	59°29′N	10°18′E
Holstebro, Den. (hôl′stĕ-brô)	192-93	56°22′N	8°38′E
Holyhead, Wales, U.K. (hŏl′ē-hĕd)	190-91	53°19′N	4°38′W
Holyoke, Co., U.S. (hōl′yōk)	114-15	40°35′N	102°18′W
Holyoke, Ma., U.S. (hōl′yōk)	116-17	42°12′N	72°37′W
Homa Bay, Kenya	267	0°31′S	34°27′E
Homalin, Mya	238-39	24°51′N	94°56′E
Homel', Bela	202-03	52°26′N	30°59′E
Homer, Ak., U.S. (hō′mẽr)	126	59°39′N	151°31′W
Homer, La., U.S. (hō′mẽr)	120-21	32°48′N	93°04′W
Homestead, Fl., U.S. (hōm′stĕd)	125a	25°29′N	80°28′W
Homestead National Monument of America, n.p., Ne., U.S. (hōm′stĕd)	120-21	40°16′N	96°48′W
Homewood, Al., U.S. (hōm′wŏŏd)	124-25	33°28′N	86°48′W
Hominy, Ok., U.S. (hŏm′ĭ-nĭ)	120-21	36°25′N	96°24′W
Homs, Libya *see* Al-Khums	188-89	32°39′N	14°16′E
Homs, Syria *see* Ḩimş	228-29	34°45′N	36°44′E
Honan, state, China *see* Henan	238-39	34°0′N	114°00′E
Honda, Col. (hōn′dȧ)	163c	5°13′N	74°45′W
Hondo, stm., N.A. (hon-dō′)	148	18°29′N	88°18′W
Honduras, nation, N.A. (hŏn-doo′rȧs)	85	15°0′N	86°30′W
Honduras, Golfo de, b., N.A. *see* Honduras, Gulf of	148	16°05′N	87°58′W
Honduras, Gulf of, b., N.A. (gŭlf ŭv hŏn-doo′rȧs)	148	16°05′N	87°58′W
Hønefoss, Nor. (hē′nē-fôs)	192-93	60°11′N	10°15′E
Honesdale, Pa., U.S. (hōnz′dāl)	116-17	41°34′N	75°16′W
Honfleur, Fr. (ôN-flûr′)	196-97	49°25′N	0°14′E

Feature (Pronunciation)	Page	Lat.	Long.
Hong, Song, stm., Asia *see* Red	238-39	20°18′N	106°32′E
Hon Gai, Viet	246-47	21°03′N	107°04′E
Hongjiang, China	238-39	27°04′N	109°58′E
Hong Kong, China (hông kông) (hŏng kŏng)	238-39	22°16′N	114°10′E
Hongliuyuan, China	240-41	41°02′N	95°25′E
Hongshui, stm., China (hôŋ-shwā)	238-39	23°48′N	109°32′E
Hongtong, China	240-41	36°17′N	111°40′E
Honguedo, Détroit d', strt., Qc., Can.	138-39	49°15′N	64°00′W
Hongze Hu, lk., China	238-39	33°16′N	118°34′E
Honiara, nat. cap., Sol. Is. (hō-nē-ä′-rä)	279e	9°26′S	159°57′E
Honiton, Eng., U.K. (hŏn′ĭ-tŏn)	190-91	50°48′N	3°12′W
Honolulu, Hi., U.S. (hŏn-ô-loo′loo)	127a	21°19′N	157°52′W
Honshū, i., Japan (hŏn′-shoo)	244	36°0′N	138°00′E
Hood, i., Ec. *see* Española, Isla	170a	1°23′S	89°42′W
Hood, Mount, vol., Or., U.S. (mount hŏŏd)	112-13	45°23′N	121°42′W
Hood River, Or., U.S. (hŏŏd rĭv′ẽr)	112-13	45°42′N	121°31′W
Hooker, Ok., U.S. (hŏŏk′ẽr)	120-21	36°52′N	101°13′W
Hook Island, i., Austl.	277	20°08′S	148°55′E
Hoonah, Ak., U.S. (hoo′nȧ)	126	58°07′N	135°26′W
Hooper Bay, Ak., U.S. (hŏŏp′ẽr bā)	126	61°31′N	166°06′W
Hoopeston, Il., U.S. (hoops′tŭn)	116-17	40°28′N	87°39′W
Hoosick Falls, N.Y., U.S. (hoo′sĭk fôlz)	116-17	42°54′N	73°21′W
Hoover Dam, d., U.S. (hoo′vẽr dăm)	118-19	36°02′N	114°43′W
Hope, B.C., Can.	132-33	49°23′N	121°26′W
Hope, Ar., U.S. (hōp)	120-21	33°40′N	93°35′W
Hope, Ben, mtn., Scot., U.K. (bĕn hōp)	190-91	58°24′N	4°37′W
Hopedale, Nf., Can. (hōp′dāl)	128-29	55°28′N	60°13′W
Hopeh, state, China *see* Hebei	240-41	38°0′N	116°00′E
Hopelchén, Mex. (o-pĕl-chĕ′n)	148	19°46′N	89°51′W
Hopes Advance, Cap, c., Qc., Can. (kȧp hōps ăd-vans′)	130-31	61°04′N	69°34′W
Hopetoun, Austl. (hōp′toun)	276	35°44′S	142°22′E
Hopetown, S. Afr. (hōp′toun)	264-65	29°35′S	24°04′E
Hopewell, Va., U.S. (hōp′wĕl)	124-25	37°18′N	77°17′W
Hopi Indian Reservation, ind. res., Az., U.S. (hō′pē ĭn′dĭ-ăn rĕ-sẽr-vā′shĕn)	118-19	35°45′N	110°35′W
Hopkinsville, Ky., U.S. (hŏp′kĭns-vĭl)	124-25	36°52′N	87°29′W
Hoquiam, Wa., U.S. (hō′kwĭ-ăm)	112-13	46°59′N	123°53′W
Horicon, Wi., U.S. (hŏr′ĭ-kŏn)	116-17	43°27′N	88°37′W
Horlivka, Ukr.	202-03	48°20′N	38°03′E
Hormoz, Jazīreh-ye, i., Iran	230-31	27°04′N	56°28′E
Hormuz, Strait of, strt., Asia (strāt ŭv hôr′mŭz′)	230-31	26°34′N	56°15′E
Horn, c., Ice.	190a	66°28′N	22°28′W
Horn, Cape, c., Chile (kāp hôrn)	171	55°59′S	67°16′W
Hornavan, lk., Swe.	184-85	66°10′N	17°46′E
Hornell, N.Y., U.S. (hôr-nĕl′)	116-17	42°20′N	77°39′W
Hornepayne, On., Can.	136-37	49°12′N	84°47′W
Horn Island, i., Ms., U.S.	124-25	30°15′N	88°43′W
Hornos, Cabo de, c., Chile *see* Horn, Cape	171	55°59′S	67°16′W
Horn Plateau, plat., N.T., Can.	130-31	62°08′N	120°16′W
Horqin Youyi Qianqi, China *see* Ulanhot	240-41	46°04′N	122°04′E
Horqueta, Para. (ŏr-kĕ′tä)	168-69	23°20′S	57°03′W
Horse Islands, is., Nf., Can. (hôrs ī′lăndz)	138-39	50°13′N	55°45′W
Horsens, Den. (hôrs′ĕns)	192-93	55°52′N	9°52′E
Horsham, Austl. (hôr′shăm) (hôrs′ăm)	276	36°43′S	142°12′E
Horten, Nor. (hôr′tĕn)	192-93	59°25′N	10°29′E
Horton, stm., N.T., Can.	130-31	69°55′N	127°02′W
Hosa'ina, Eth.	269d	7°37′N	37°56′E
Hoséré Vokré, mtn., Camrn.	260-61	8°20′N	13°15′E
Hoshangābād, India	234-35	22°45′N	77°43′E
Hoshiārpur, India	234-35	31°32′N	75°55′E
Hospet, India	236	15°16′N	76°23′E
Hoste, Isla, i., Chile (ĕ′s-lä-ŏs′tä)	171	55°05′S	69°15′W
Hotan, China (hwŏ-tän)	226	37°07′N	79°55′E
Hotan, stm., China (hwŏ-tän)	226	40°30′N	80°56′E
Hot Springs, Ar., U.S. (hŏt sprĭngz)	120-21	34°30′N	93°04′W
Hot Springs, N.M., U.S. (hŏt sprĭngz) *see* Truth or Consequences	118-19	33°08′N	107°15′W
Hot Springs, S.D., U.S. (hŏt sprĭngz)	114-15	43°26′N	103°29′W
Hot Springs, Va., U.S. (hŏt sprĭngz)	116-17	37°60′N	79°49′W
Hot Springs National Park, n.p., Ar., U.S. (hŏt sprĭngz năsh′ŭn-ăl pärk)	120-21	34°31′N	93°02′W
Hottah Lake, lk., N.T., Can.	130-31	65°04′N	118°29′W
Houghton, Mi., U.S. (hō′tŭn)	114-15	47°06′N	88°36′W
Houghton Lake, lk., Mi., U.S. (hō′tŭn lăk)	116-17	44°20′N	84°45′W
Houlton, Me., U.S. (hōl′tŭn)	117a	46°08′N	67°50′W
Houma, China	240-41	35°37′N	111°21′E
Houma, La., U.S. (hoo′mȧ)	124-25	29°35′N	90°43′W

Feature (Pronunciation)	Page	Lat.	Long.
Houston, Ms., U.S. (hūs′tŭn)	124-25	33°54′N	88°60′W
Houston, Tx., U.S. (hūs′tŭn)	122-23	29°45′N	95°22′W
Hovd, Mong. *see* Dund-Us	240-41	47°60′N	91°38′E
Hovd, stm., Mong.	218-19	48°05′N	92°13′E
Hövsgöl nuur, lk., Mong.	240-41	51°0′N	100°30′E
Howard, S.D., U.S. (hou′ărd)	114-15	44°00′N	97°32′W
Howe, Cape, c., Austl. (kāp hou)	276	37°30′S	149°58′E
Howell, Mi., U.S. (hou′ĕl)	116-17	42°36′N	83°56′W
Howe Sound, strt., B.C., Can. (hou sound)	132-33	49°22′N	123°18′W
Howland Island, dep., Oc. (hou′lănd ī′lănd)	280-81	0°51′N	176°38′W
Howland Island, i., Oc. (hou′lănd ī′lănd)	280-81	0°48′N	176°38′W
Hoxie, Ar., U.S. (kŏh′sĭ)	124-25	36°03′N	90°59′W
Hradec Králové, Czech Rep.	194-95	50°12′N	15°50′E
Hranice, Czech Rep. (hrän′yĕ-tsĕ)	194-95	49°33′N	17°45′E
Hrodna, Bela.	194-95	53°41′N	23°50′E
Hrubieszów, Pol. (hrōō-byă′shōōf)	194-95	50°49′N	23°56′E
Hrvatska, nation, Eur. (hr-väts′kä) *see* Croatia	174-75	45°10′N	15°30′E
Hsinchu, Tai. (hsĭn′chōō′)	225a	24°48′N	120°58′E
Hsinhailien, China *see* Lianyungang	238-39	34°37′N	119°11′E
Hsipaw, Mya.	246-47	22°37′N	97°18′E
Huacho, Peru	170	11°08′S	77°37′W
Huadian, China (hwä-dĭĕn)	240-41	42°58′N	126°45′E
Hua Hin, Thai.	246-47	12°35′N	99°57′E
Huai'an, China (hwī-än)	238-39	33°31′N	119°08′E
Huai'an, China (hwī-än)	240-41	40°40′N	114°25′E
Huaicheng, China *see* Huai'an	238-39	33°31′N	119°08′E
Huaide, China *see* Gongzhuling	240-41	43°30′N	124°49′E
Huailai, China	240-41	40°23′N	115°34′E
Huainan, China	238-39	32°40′N	117°01′E
Huaiyang, China (hōōaī′yang)	238-39	33°44′N	114°53′E
Huajuapan de León, Mex. (wäj-wä′päm dā lā-ōn′)	146-47	17°49′N	97°45′W
Hualfín, Arg.	168-69	27°14′S	66°50′W
Hualien, Tai. (hwä′lyĕn′)	225a	23°58′N	121°35′E
Huallaga, stm., Peru (wäl-yä′gä)	170	5°06′S	75°36′W
Huallanca, Peru	170	8°49′S	77°52′W
Huambo, Ang.	264-65	12°46′S	15°44′E
Huancavelica, Peru (wän′kä-vä-lē′kä)	163a	12°47′S	75°01′W
Huancayo, Peru (wän-kä′yō)	163a	12°05′S	75°13′W
Huang, stm., China (hůäŋ)	222-23	37°49′N	118°53′E
Huangchuan, China (hůäŋ-chůän)	238-39	32°08′N	115°03′E
Huang Hai, s., Asia *see* Yellow Sea	222-23	36°0′N	123°00′E
Huangho, stm., China *see* Huang	222-23	37°49′N	118°53′E
Huanghua, China (hůäŋ-hwä)	240-41	38°22′N	117°21′E
Huangshan, China	238-39	29°45′N	118°18′E
Huangshi, China	238-39	30°13′N	115°05′E
Huangyuan, China (hůäŋ-yůän)	240-41	36°41′N	101°16′E
Huanren, China (hůän-rŭn)	243	41°13′N	125°20′E
Huánuco, Peru (wä-nōō′kō)	170	9°56′S	76°15′W
Huanuni, Bol. (wä-nōō′nē)	168-69	18°17′S	66°50′W
Huaral, Peru (wä-rä′l)	163a	11°30′S	77°12′W
Huaraz, Peru	170	9°32′S	77°33′W
Huascarán, Nevado, mtn., Peru (nĕ-vä′dō wäs-kä-rän′)	170	9°07′S	77°37′W
Huasco, Chile (wäs′kō)	168-69	28°28′S	71°15′W
Huatabampo, Mex.	144-45	26°50′N	109°38′W
Huauchinango, Mex. (wä-ōō-chē-näŋ′gō)	146-47	20°11′N	98°03′W
Huautla, Mex. (wä-ōō′tlä)	146-47	18°08′N	96°50′W
Huaxian, China (hwä shyĕn)	238-39	23°23′N	113°12′E
Huaynamota, stm., Mex. (wäy-nä-mō′tä)	146-47	21°57′N	104°32′W
Hubbard, Tx., U.S. (hŭb′ẽrd)	122-23	31°50′N	96°48′W
Hubbard Creek Reservoir, lk., Tx., U.S. (hŭb′ẽrd krēk rĕ′sẽr-vwär)	120-21	32°48′N	99°01′W
Hubei, state, China (hōō-bā)	238-39	31°0′N	112°00′E
Hubli-Dhārwār, India	236	15°21′N	75°09′E
Huddersfield, Eng., U.K. (hŭd′ẽrz-fēld)	190-91	53°39′N	1°47′W
Hudiksvall, Swe. (hōō′dĭks-väl)	192-93	61°44′N	17°07′E
Hudson, Mi., U.S. (hŭd′sŭn)	116-17	41°51′N	84°21′W
Hudson, Wi., U.S. (hŭd′sŭn)	114-15	44°59′N	92°45′W
Hudson, stm., U.S. (hŭd′sŭn)	110-11	40°41′N	74°02′W
Hudson, Détroit d', strt., Can. *see* Hudson Strait	130-31	62°30′N	71°60′W
Hudson Bay, Sk., Can. (hŭd′sŭn bā)	134-35	52°52′N	102°23′W
Hudson Bay, b., Can. (hŭd′sŭn bā)	130-31	60°0′N	86°00′W
Hudson Falls, N.Y., U.S. (hŭd′sŭn fôlz)	116-17	43°19′N	73°35′W
Hudson Strait, strt., Can. (hŭd′sŭn strāt)	130-31	62°30′N	71°60′W
Hue, Viet. (ü-ā′)	246-47	16°28′N	107°35′E
Huehuetenango, Guat. (wā-wå-tå-näŋ′gō)	148	15°21′N	91°27′W
Huejuquilla El Alto, Mex. (wā-hōō-kēl′yä ĕl äl′tō)	146-47	22°38′N	103°54′W
Huelva, Spain (wĕl′vä)	198-99	37°16′N	6°57′W
Huesca, Spain (wĕs-kä)	198-99	42°08′N	0°25′W
Huéscar, Spain (wās′kär)	198-99	37°48′N	2°33′W
Huetamo de Núñez, Mex.	146-47	18°35′N	100°53′W
Hughenden, Austl. (hū′ĕn-dĕn)	277	20°51′S	144°13′E
Hugli, stm., India (hōōg′lĭ)	234-35	21°36′N	87°60′E
Hugo, Ok., U.S. (hū′gō)	120-21	34°01′N	95°31′W
Hugoton, Ks., U.S. (hū′gō-tŭn)	120-21	37°11′N	101°21′W
Huhehot, China *see* Hohhot	240-41	40°49′N	111°39′E
Huichapan, Mex. (wē-chä-pän′)	146-47	20°22′N	99°40′W
Hŭich'ŏn, Kor., N.	243	40°10′N	126°17′E
Huila, Nevado del, vol., Col. (nĕ-vä-dô-del-wē′lä)	163c	2°59′N	75°58′W
Huili, China	238-39	26°40′N	102°14′E
Huimin, China (hōōĭ mĭn)	240-41	37°29′N	117°32′E
Huinan, China	243	42°41′N	126°02′E
Huixtla, Mex.	148	15°08′N	92°27′W
Huize, China	238-39	26°25′N	103°18′E
Huizhou, China	238-39	23°05′N	114°24′E
Hujirt, Mong.	240-41	48°53′N	101°14′E
Hukuoka, Japan *see* Fukuoka	245	33°35′N	130°25′E
Hulan, China (hōō′län′)	240-41	45°59′N	126°36′E
Hulan Ergi, China	240-41	47°12′N	123°38′E
Hulin, China (hōō′lĭn′)	244	45°46′N	132°59′E
Hulun, China *see* Hailar	240-41	49°11′N	119°44′E
Hulun Nur, lk., China (hōō-lòn nòr)	240-41	49°01′N	117°32′E
Humacao, P.R. (ōō-mä-kä′ō)	142a	18°09′N	65°49′W
Humahuaca, Arg.	168-69	23°12′S	65°21′W
Humaitá, Braz.	166-67	7°30′S	63°02′W
Humaitá, Para.	173	27°05′S	58°32′W
Humble, Tx., U.S. (hŭm′b'l)	122-23	29°59′N	95°16′W
Humboldt, Sk., Can. (hŭm′bōlt)	134-35	52°11′N	105°07′W
Humboldt, Ia., U.S. (hŭm′bōlt)	114-15	42°43′N	94°13′W
Humboldt, Tn., U.S. (hŭm′bōlt)	124-25	35°50′N	88°55′W
Humboldt, stm., Nv., U.S. (hŭm′bōlt)	110-11	40°01′N	118°33′W
Hume, Lake, res., Austl.	276	36°08′S	147°02′E
Humphreys Peak, mtn., Az., U.S. (hŭm′frĭs pēk)	118-19	35°20′N	111°40′W
Húnaflói, b., Ice. (hōō′nä-flō′ĭ)	190a	65°50′N	20°50′W
Hunan, state, China (hōō′nän′)	238-39	28°0′N	111°00′E
Hunchun, China (hòn-chůn)	243	42°52′N	130°22′E
Hunedoara, Rom. (kōō′nĕd-wä′rä)	200-01	45°46′N	22°55′E
Hungary, nation, Eur. (hŭŋ′gá-rĭ)	174-75	47°0′N	20°00′E
Hŭngdŏki-dong, Kor., N.	243	39°50′N	127°38′E
Hungerford, Austl. (hŭn′gẽr-fẽrd)	276	28°59′S	144°24′E
Hŭngnam, Kor., N. *see* Hŭngdŏki-dong	243	39°50′N	127°38′E
Hunjiang, China	243	41°57′N	126°28′E
Hunsrück, mts., Ger. (hōōns′rûk)	194-95	49°43′N	7°08′E
Hunter, Île, i., N. Cal.	280-81	22°24′S	172°06′E
Hunter Island, i., Austl.	276	40°32′S	144°45′E
Hunter Island, i., B.C., Can.	132-33	51°54′N	128°03′W
Huntingburg, In., U.S. (hŭnt′ĭng-bûrg)	116-17	38°18′N	86°57′W
Huntingdon, Qc., Can. (hŭnt′ĭng-dŭn)	136-37	45°06′N	74°10′W
Huntingdon, Tn., U.S. (hŭnt′ĭng-dŭn)	124-25	36°00′N	88°26′W
Huntington, In., U.S. (hŭnt′ĭng-tŭn)	116-17	40°52′N	85°28′W
Huntington, Ut., U.S. (hŭnt′ĭng-tŭn)	118-19	39°20′N	110°58′W
Huntington, W.V., U.S. (hŭnt′ĭng-tŭn)	116-17	38°25′N	82°26′W
Huntington Beach, Ca., U.S. (hŭnt′ĭng-tŭn bēch)	118-19	33°39′N	117°59′W
Huntsville, On., Can. (hŭnts′vĭl)	136-37	45°19′N	79°12′W
Huntsville, Al., U.S. (hŭnts′vĭl)	124-25	34°43′N	86°36′W
Huntsville, Mo., U.S. (hŭnts′vĭl)	120-21	39°26′N	92°33′W
Huntsville, Tx., U.S. (hŭnts′vĭl)	122-23	30°43′N	95°33′W
Hunucmá, Mex.	148	21°01′N	89°52′W
Hunyuan, China	240-41	39°42′N	113°41′E
Huong Thuy, Viet.	246-47	16°25′N	107°40′E
Huon Gulf, b., Pap. N. Gui.	277	7°10′S	147°25′E
Huonville, Austl.	276	43°01′S	147°02′E
Huoqiu, China (hwô-chyô)	238-39	32°20′N	116°16′E
Huoshan, China (hwô-shän)	238-39	31°24′N	116°20′E
Hurd, Cape, c., On., Can. (kāp hûrd)	136-37	45°14′N	81°42′W
Hurghada, Egypt	268b	27°14′N	33°50′E
Hurley, Wi., U.S. (hûr′lĭ)	114-15	46°27′N	90°11′W
Huron, S.D., U.S. (hū′rŏn)	114-15	44°22′N	98°13′W
Huron, Lake, lk., N.A. (lāk hū′rŏn)	116-17	44°30′N	82°15′W
Hurricane, Ut., U.S. (hŭr′ĭ-kān)	118-19	37°11′N	113°17′W
Húsavík, Ice.	190a	66°03′N	17°19′W
Huși, Rom. (kòsh')	202-03	46°41′N	28°04′E
Husum, Ger. (hōō′zòm)	194-95	54°28′N	9°04′E
Hutchinson, Ks., U.S. (hŭch′ĭn-sŭn)	120-21	38°03′N	97°55′W
Hutchinson, Mn., U.S. (hŭch′ĭn-sŭn)	114-15	44°53′N	94°23′W
Huy, Bel. (û-ē′) (hû′ē)	190-91	50°31′N	5°14′E
Huzhou, China	238-39	30°52′N	120°06′E
Hvannadalshnúkur, mtn., Ice.	190a	64°01′N	16°41′W
Hvar, Otok, i., Cro. (ŏ′tŏk ĸhvär)	200-01	43°09′N	16°45′E
Hwaining, China *see* Anqing	238-39	30°30′N	117°02′E
Hwange, Zimb.	264-65	18°22′S	26°30′E
Hwange National Park, n.p., Zimb.	264-65	19°0′S	26°35′E
Hwang-hae, s., Asia *see* Yellow Sea	222-23	36°0′N	123°00′E
Hyargas nuur, lk., Mong.	240-41	49°12′N	93°24′E
Hyde Park, Guy.	164-65	6°30′N	58°16′W
Hyde Park, N.Y., U.S. (hīd pärk)	116-17	41°47′N	73°56′W
Hyderābād, India (hī-dẽr-å-băd′)	236	17°23′N	78°29′E
Hyderābād, Pak. (hī-dẽr-å-băd′)	232-33	25°23′N	68°21′E
Hyères, Fr. (ē-âr′)	196-97	43°08′N	6°08′E
Hyesan, Kor., N.	243	41°24′N	128°10′E
Hyndman Peak, mtn., Id., U.S. (hīnd′măn pēk)	112-13	43°45′N	114°08′W
Hyōgo, state, Japan (hĭyō′gō)	245	35°0′N	135°00′E

I

Feature (Pronunciation)	Page	Lat.	Long.
Iaco, stm., S.A.	170	9°02′S	68°35′W
Iași, Rom. (yä′shê)	202-03	47°10′N	27°36′E
Iba, Phil. (ē′bä)	250	15°20′N	119°58′E
Ibadan, Nig. (ê-bä′dän)	260a	7°23′N	3°54′E
Ibagué, Col.	163c	4°27′N	75°15′W
Iban, Pegunungan, mts., Asia *see* Iran Mountains	248-49	2°05′N	114°55′E
Ibarra, Ec. (ê-bär′rä)	170	0°22′N	78°08′W
Ibb, Yemen	266	13°58′N	44°11′E
Iberian Peninsula, pen., Eur. (ī-bĕr′ĭ-ăn pĕ-nĭn′sūlă)	176-77	40°0′N	5°00′W
Ibérica, Península, pen., Eur. *see* Iberian Peninsula	176-77	40°0′N	5°00′W
Iberville, Qc., Can. (ê-bår-vēl′) (ī′bẽr-vĭl)	136-37	45°19′N	73°14′W
Ibiá, Braz.	172	19°29′S	46°32′W
Ibicaraí, Braz.	166-67	14°51′S	39°37′W
Ibicuí, stm., Braz.	173	29°25′S	56°47′W
Ibiza, Spain *see* Eivissa	198-99	38°55′N	1°25′E
Ibiza, i., Spain (ê-bē′thä) *see* Eivissa	198-99	39°0′N	1°25′E
Ica, Peru (ē′kä)	170	14°04′S	75°45′W
Içá, stm., S.A.	170	3°07′S	67°56′W
Içana, Braz. (ê-sä′nä)	166-67	0°21′N	67°19′W
İçel, Tur.	228-29	36°49′N	34°38′E
Iceland, nation, Eur. (īs′lănd)	174-75	65°0′N	18°00′W
Ichalkaranji, India	236	16°41′N	74°28′E
Ichilo, stm., Bol.	168-69	15°50′S	64°47′W
Icó, Braz.	166-67	6°24′S	38°51′W
Idabel, Ok., U.S. (ī′dȧ-bĕl)	120-21	33°53′N	94°49′W
Ida Grove, Ia., U.S. (ī′dȧ-grōv)	114-15	42°21′N	95°28′W
Idah, Nig. (ē′dä)	260a	7°07′N	6°45′E
Idaho, state, U.S. (ī′dȧ-hō)	108-09	45°0′N	115°00′W
Idaho Falls, Id., U.S. (ī′dȧ-hō fôlz)	112-13	43°30′N	112°03′W
Idaho Springs, Co., U.S. (ī′dȧ-hō sprĭngz)	118-19	39°44′N	105°31′W
Ider, stm., Mong.	240-41	49°16′N	100°41′E
Idi, Indon. (ē′dê)	246-47	4°57′N	97°46′E
Idi Amin Dada, Lac, lk., Afr. *see* Edward, Lake	267	0°23′S	29°36′E
Idiofa, D.R.C.	262-63	5°01′S	19°35′E
Ídi Óros, mtn., Grc.	200a	35°18′N	24°43′E
Idlib, Syria	228-29	35°56′N	36°39′E
Idoûkâl-en-Taghès, mtn., Niger	258-59	17°43′N	8°45′E
Iesi, Italy (yä′sê) *see* Jesi	200-01	43°31′N	13°14′E
Ife, Nig.	260a	7°28′N	4°33′E
Ifôghas, Adrar des, mts., Afr.	258-59	20°0′N	2°00′E
Igarka, Russia (ē-gär′kȧ)	218-19	67°28′N	86°38′E
Iglesias, Italy (ē-lĕ′syôs)	200-01	39°19′N	8°32′E
Igloolik, Nu., Can.	128-29	69°23′N	81°48′W
Igluligaarjuk, Nu., Can. *see* Chesterfield Inlet	128-29	63°21′N	90°43′W
Iglulik, Nu., Can. *see* Igloolik	128-29	69°23′N	81°48′W
Igombe, stm., Tan.	267	4°43′S	31°23′E
Iguaçu, stm., S.A. (ê-gwä-sōō′)	168-69	25°36′S	54°36′W
Iguaçu, Cataratas do, wtfl., S.A. *see* Iguassu Falls	168-69	25°42′S	54°27′W
Iguaçu, Saltos do, wtfl., S.A. *see* Iguassu Falls	168-69	25°42′S	54°27′W
Iguala, Mex. (ê-gwä′lä)	146-47	18°21′N	99°32′W
Igualada, Spain (ê-gwä-lä′dä)	198-99	41°35′N	1°39′E
Iguape, Braz.	172	24°43′S	47°34′W
Iguassu Falls, wtfl., S.A. (ê-gwä-sōō′ fôlz)	168-69	25°42′S	54°27′W

Feature (Pronunciation)	Page	Lat.	Long.
Iguatu, Braz. (ê-gwä-tōō´)	166-67	6°22'S	39°18'W
Iguazú, stm., S.A. *see* Iguaçu	168-69	25°36'S	54°36'W
Iguazú, Cataratas del, wtfl., S.A. *see* Iguassu Falls	168-69	25°42'S	54°27'W
Ihosy, Madag.	264-65	22°24'S	46°07'E
Iida, Japan (ē´ē-dà)	245	35°31'N	137°50'E
Iisalmi, Fin.	184-85	63°32'N	27°17'E
Iizuka, Japan (ē´ē-zò-kà)	245	33°38'N	130°41'E
Ijâfene, des., Maur.	258-59	22°04'N	7°42'W
Ijebu-Ode, Nig. (ê-jĕ´bōō ōdā)	260a	6°49'N	3°56'E
IJsselmeer, lk., Neth. (ī´sĕl-mār)	190-91	52°45'N	5°25'E
Ijuí, Braz.	173	28°23'S	53°55'W
Ikaalinen, Fin. (ē´kä-lĭ-nĕn)	192-93	61°45'N	23°04'E
Ikaluktutiak, Nu., Can. *see* Cambridge Bay	128-29	69°07'N	105°04'W
Ikaría, i., Grc. (ē-kä´ryà)	200-01	37°36'N	26°09'E
Ikela, D.R.C.	262-63	1°11'S	23°17'E
Ikhtiman, Blg. (ēk´tē-män)	200-01	42°26'N	23°50'E
Iki, i., Japan (ē´kē)	245	33°47'N	129°43'E
Ikom, Nig.	260a	5°58'N	8°42'E
Ikoma, Tan. (ê-kō´mä)	267	2°05'S	34°38'E
Ikopa, stm., Madag.	264-65	17°00'S	46°45'E
Iksan, Kor., S.	243	35°56'N	126°57'E
Ilagan, Phil.	250	17°08'N	121°53'E
Ilam, nation, Asia *see* Sri Lanka	206-07	7°0'N	81°00'E
Ilan, Tai. (ē´län´)	225a	24°46'N	121°45'E
Ile, stm., Asia	226	45°20'N	74°05'E
Île-à-la-Crosse, Sk., Can.	134-35	55°26'N	107°55'W
Île-à-la-Crosse, Lac, lk., Sk., Can.	134-35	55°42'N	107°41'W
Ilebo, D.R.C.	262-63	4°20'S	20°36'E
Ilek, stm., Asia *see* Elek	186-87	51°30'N	53°20'E
Ilesha, Nig.	260a	7°38'N	4°45'E
Ilfracombe, Austl.	277	23°29'S	144°30'E
Ilfracombe, Eng., U.K. (ĭl-frȧ-kōōm´)	190-91	51°12'N	4°08'W
Ilha de Moçambique, Moz.	264-65	15°02'S	40°41'E
Ilha Grande, Baía da, b., Braz. (bäē´ä dē ēl´yà grän´dĕ)	172	23°09'S	44°30'W
Ílhavo, Port. (ēl´yà-vò)	198-99	40°36'N	8°40'W
Ilhéos, Braz. *see* Ilhéus	166-67	14°47'S	39°03'W
Ilhéus, Braz. (ē-lĕ´ōōs)	166-67	14°47'S	39°03'W
Ili, stm., Asia *see* Ile	226	45°20'N	74°05'E
Iliamna Lake, lk., Ak., U.S. (ē-lê-ăm´nà lăk)	126	59°37'N	154°49'W
Iligan, Phil.	250	8°15'N	124°16'E
Ilion, N.Y., U.S. (ĭl´ĭ-ŭn)	116-17	43°01'N	75°03'W
Ilizi, Alg.	258-59	26°29'N	8°29'E
Illampu, Nevado, mtn., Bol. (nĕ-vá´dô-ēl-yäm-pōō´)	168-69	15°50'S	68°34'W
Illapel, Chile (ē-zhä-pĕ´l)	168-69	31°37'S	71°09'W
Illimani, Nevado de, mtn., Bol. (nĕ-vá´dô-dĕ-ēl-yê-mä´nē)	168-69	16°50'S	67°54'W
Illinois, state, U.S. (ĭl-ĭ-noi´) (ĭl-ĭ-noiz´)	108-09	40°0'N	89°00'W
Illinois, stm., Il., U.S. (ĭl-ĭ-noi´) (ĭl-ĭ-noiz´)	110-11	38°58'N	90°25'W
Il'men', Ozero, lk., Russia (ô´zĕ-rô el´men´) (ĭl´mĕn)	192-93	58°17'N	31°20'E
Ilo, Peru	170	17°38'S	71°20'W
Iloilo, Phil. (ē-lô-ē´lō)	250	10°42'N	122°34'E
Ilorin, Nig. (ē-lô-rēn´)	260a	8°30'N	4°33'E
Ilwaki, Indon.	248-49	7°55'S	126°25'E
Imabari, Japan (ē´mä-bä´rē)	245	34°04'N	133°00'E
Imandra, Ozero, lk., Russia (ô´zĕ-rô ē-män´drà)	186-87	67°33'N	33°00'E
Imatra, Fin.	192-93	61°10'N	28°46'E
Imbituba, Braz.	172	28°14'S	48°41'W
Imeni Ismail Samani, Pik, mtn., Taj.	226	38°57'N	72°01'E
Imlay City, Mi., U.S. (ĭm´lā sĭ´tē)	116-17	43°01'N	83°05'W
Imola, Italy (ē´mô-lä)	200-01	44°21'N	11°43'E
Imotski, Cro. (ê-môts´kē)	200-01	43°27'N	17°13'E
Imperatriz, Braz.	166-67	5°31'S	47°28'W
Imperia, Italy (ēm-pā´rē-ä)	200-01	43°54'N	8°03'E
Imperial, Ne., U.S. (ĭm-pē´rĭ-ăl)	120-21	40°31'N	101°39'W
Impfondo, Congo (ĭmp-fōn´dô)	262-63	1°37'N	18°04'E
Imphāl, India (ĭmp´hŭl)	234-35	24°47'N	93°57'E
Inari, Fin.	184-85	68°54'N	26°60'E
Inari, lk., Fin. *see* Inarijärvi	186-87	69°0'N	28°00'E
Inarijärvi, lk., Fin.	186-87	69°0'N	28°00'E
Inca, Spain (ēŋ´kä)	198-99	39°43'N	2°55'E
Inca de Oro, Chile	168-69	26°45'S	69°54'W
İnce Burun, c., Tur. (ĭn´jå)	186-87	42°05'N	34°58'E
Inch'ŏn, Kor., S. (ĭn´chŭn)	243	37°28'N	126°38'E
Indefatigable, i., Ec. *see* Santa Cruz, Isla	170a	0°38'S	90°23'W
Independence, Ia., U.S. (ĭn-dê-pĕn´dĕns)	114-15	42°29'N	91°54'W

Feature (Pronunciation)	Page	Lat.	Long.
Independence, Ks., U.S. (ĭn-dê-pĕn´dĕns)	120-21	37°13'N	95°42'W
Independence, Ky., U.S. (ĭn-dê-pĕn´dĕns)	116-17	38°56'N	84°32'W
Independence, Mo., U.S. (ĭn-dê-pĕn´dĕns)	120-21	39°05'N	94°25'W
Independence Mountains, mts., Nv., U.S. (ĭn-dê-pĕn´dĕns moun´tĭnz)	112-13	41°18'N	116°00'W
Inderbor, Kaz.	186-87	48°32'N	51°42'E
India, nation, Asia (ĭn´dĭ-ȧ)	206-07	20°0'N	77°00'E
Indiana, Pa., U.S. (ĭn-dĭ-ăn´ȧ)	116-17	40°35'N	79°09'W
Indiana, state, U.S. (ĭn-dĭ-ăn´ȧ)	108-09	40°0'N	86°15'W
Indianapolis, In., U.S. (ĭn-dĭ-ăn-ăp´ô-lĭs)	116-17	39°46'N	86°08'W
Indian Head, Sk., Can. (ĭn´dĭ-ăn hĕd)	134-35	50°32'N	103°40'W
Indian Ocean, oc., (ĭn´dĭ-ăn ōshŭn)	20-21	10°0'S	70°00'E
Indianola, Ia., U.S. (ĭn-dĭ-ăn-ō´lȧ)	114-15	41°21'N	93°33'W
Indianola, Ms., U.S. (ĭn-dĭ-ăn-ō´lȧ)	124-25	33°27'N	90°39'W
Indian Springs, Nv., U.S. (ĭn´dĭ-ăn sprĭngz)	118-19	36°34'N	115°41'W
Indigirka, stm., Russia (ēn-dê-gēr´kà)	218-19	70°49'N	148°54'E
Indochina, reg., Asia (ĭn-dô-chĭ´nà)	246-47	16°0'N	107°00'E
Indonesia, nation, Asia (ĭn´dô-nē-zhȧ)	206-07	5°0'S	120°00'E
Indore, India (ĭn-dōr´)	234-35	22°43'N	75°52'E
Indragiri, stm., Indon. (ĭn-drà-jē´rē)	246-47	0°22'S	103°26'E
Indrāvati, stm., India (ĭn-drŭ-vä´tē)	236	18°44'N	80°17'E
Indus, stm., Asia (ĭn´dŭs)	208-09	24°60'N	68°16'E
Inferior, Laguna, b., Mex. (lä-gò´nä-ēn-fĕr-rôr)	146-47	16°17'N	94°40'W
Infiernillo, Presa del, res., Mex.	146-47	18°37'N	101°46'W
Ingende, D.R.C.	262-63	0°13'S	18°58'E
Ingersoll, On., Can. (ĭn´gēr-sŏl)	136-37	43°02'N	80°53'W
Ingham, Austl. (ĭng´ăm)	277	18°39'S	146°09'E
Ingoda, stm., Russia (ēn-gō´dȧ)	222-23	51°43'N	115°48'E
Ingolstadt, Ger. (ĭŋ´gôl-shtät)	194-95	48°46'N	11°26'E
Ingrāj Bāzār, India	234-35	24°60'N	88°09'E
I-n-Guezzâm, Alg.	258-59	19°27'N	5°48'E
Ingushetia, state, Russia	227	43°15'N	45°00'E
Ingushetiya, state, Russia *see* Ingushetia	227	43°15'N	45°00'E
Inhaca, Ilha da, i., Moz.	264-65	26°01'S	32°57'E
Inhambupe, Braz. (ēn-yäm-bōō´pä)	166-67	11°49'S	38°20'W
Inírida, stm., Col. (ē-nē-rē´dä)	164-65	3°55'N	67°51'W
Injune, Austl. (ĭn´jòn)	276	25°51'S	148°34'E
Inland Sea, s., Japan (ĭn´lănd sē) *see* Seto-naikai	245	34°22'N	133°37'E
Inn, stm., Eur. (ĭn)	194-95	48°34'N	13°28'E
Innamincka, Austl. (ĭnn-ȧ´mĭn-kȧ)	276	27°45'S	140°44'E
Inner Mongolia, state, China *see* Nei Mongol	240-41	43°0'N	115°00'E
Innisfail, Austl.	277	17°33'S	146°02'E
Innisfail, Ab., Can.	132-33	52°02'N	113°58'W
Innoko, stm., Ak., U.S.	126	62°11'N	159°44'W
Innsbruck, Aus. (ĭns´bròk)	194-95	47°16'N	11°24'E
Inongo, D.R.C. (ê-nôŋ´gō)	262-63	1°55'S	18°18'E
Inowrocław, Pol. (ē-nô-vrôts´läf)	194-95	52°48'N	18°15'E
I-n-Salah, Alg.	258-59	27°11'N	2°29'E
Inta, Russia	186-87	66°02'N	60°09'E
International Falls, Mn., U.S. (ĭn´tēr-năsh´ŭn-ăl fôlz)	114-15	48°36'N	93°25'W
Inthanon, Doi, mtn., Thai.	246-47	18°35'N	98°29'E
Intiyaco, Arg.	173	28°40'S	60°04'W
Inukjuak, Qc., Can.	128-29	58°28'N	78°06'W
Inuvik, N.T., Can.	128-29	68°20'N	133°39'W
Invercargill, N.Z.	278	46°25'S	168°22'E
Inverell, Austl.	276	29°47'S	151°07'E
Inverness, Scot., U.K. (ĭn-vēr-nĕs´)	190-91	57°28'N	4°15'W
Inverness, Fl., U.S. (ĭn-vēr-nĕs´)	124-25	28°50'N	82°20'W
Investigator Strait, strt., Austl. (ĭn-vĕst´ĭ´gā-tôr strāt)	276	35°25'S	137°10'E
Inyangani, mtn., Zimb. (ēn-yän-gä´nē)	264-65	18°17'S	32°50'E
Inza, Russia	186-87	53°51'N	46°22'E
Ioánnina, Grc. (yô-ä´nē-nà)	200-01	39°39'N	20°51'E
Iō-jima, i., Japan *see* Iwo Jima	280-81	24°47'N	141°20'E
Ionia, Mi., U.S. (ī-ō´nĭ-ȧ)	116-17	42°59'N	85°04'W
Ionian Islands, is., Grc. (ī-ō´nĭ-ăn ī´lăndz)	200-01	38°30'N	20°30'E
Ionian Sea, s., Eur. (ī-ō´nĭ-ăn sē)	200-01	39°0'N	19°00'E
Ionio, Mar, s., Eur. *see* Ionian Sea	200-01	39°0'N	19°00'E
Iónioi Nísoi, is., Grc. *see* Ionian Islands	200-01	38°30'N	20°30'E
Iónion Pélagos, s., Eur. *see* Ionian Sea	200-01	39°0'N	19°00'E
Íos, i., Grc. (ī´ŏs)	200-01	36°43'N	25°20'E
Iō-tō, i., Japan *see* Iwo Jima	280-81	24°47'N	141°20'E
Iowa, state, U.S. (ī´ô-wȧ)	108-09	42°15'N	93°15'W
Iowa, stm., Ia., U.S. (ī´ô-wȧ)	114-15	41°10'N	91°01'W

Feature (Pronunciation)	Page	Lat.	Long.
Iowa City, Ia., U.S. (ī´ô-wȧ sĭ´tē)	114-15	41°40'N	91°32'W
Iowa Falls, Ia., U.S. (ī´ô-wȧ fôlz)	114-15	42°32'N	93°16'W
Iowa Park, Tx., U.S. (ī´ô-wȧ pärk)	120-21	33°58'N	98°40'W
Ipameri, Braz.	172	17°43'S	48°09'W
Ipiales, Col. (ē-pê-ä´lås)	164-65	0°50'N	77°38'W
Ipiaú, Braz.	166-67	14°08'S	39°44'W
Ipoh, Malay.	246-47	4°36'N	101°04'E
Iporã, Braz.	168-69	16°27'S	51°07'W
Ippy, C.A.R.	262-63	6°16'N	21°12'E
Ipswich, Austl. (ĭps´wĭch)	276	27°37'S	152°47'E
Ipswich, Eng., U.K. (ĭps´wĭch)	190-91	52°04'N	1°09'E
Ipswich, S.D., U.S. (ĭps´wĭch)	114-15	45°27'N	99°02'W
Ipu, Braz. (ē-pōō)	166-67	4°20'S	40°42'W
Iqaluit, Nu., Can.	128-29	63°44'N	68°28'W
Iquique, Chile (ē-kē´kĕ)	168-69	20°13'S	70°09'W
Iquitos, Peru (ê-kē´tōs)	170	3°47'S	73°15'W
Irákleio, Grc.	200a	35°20'N	25°08'E
Iran, nation, Asia (ē-rän´)	206-07	32°0'N	53°00'E
Īrān, nation, Asia *see* Iran	206-07	32°0'N	53°00'E
Iran, Pegunungan, mts., Asia *see* Iran Mountains	248-49	2°05'N	114°55'E
Iran Mountains, mts., Asia (ē-rän´ moun´tĭnz)	248-49	2°05'N	114°55'E
Īrānshahr, Iran	230-31	27°13'N	60°42'E
Irapuato, Mex. (ē-rä-pwä´tō)	146-47	20°41'N	101°21'W
Iraq, nation, Asia (ē-räk´)	206-07	33°0'N	44°00'E
Irazú, Volcán, vol., C.R. (vôl-kà´n ē-rä-zōō´)	149	9°58'N	83°53'W
Irbid, Jord. (ēr-bēd´)	228-29	32°34'N	35°51'E
Irbīl, Iraq *see* Arbīl	228-29	36°11'N	44°01'E
Ireland, nation, Eur. (īr-lănd)	174-75	53°0'N	8°00'W
Irian, i., *see* New Guinea	277	5°0'S	140°00'E
Iringa, Tan. (ê-rĭŋ´gä)	267	7°47'S	35°42'E
Iriomote-jima, i., Japan (ērē´-ō-mō-tä jē´mȧ)	279a	24°20'N	123°50'E
Iriri, stm., Braz.	166-67	3°48'S	52°36'W
Irish Sea, s., Eur. (ī´rĭsh sē)	190-91	53°30'N	5°20'W
Irkutsk, Russia (ĭr-kòtsk´)	222-23	52°18'N	104°17'E
Iron Knob, Austl. (ī-ăn nŏb)	276	32°44'S	137°08'E
Iron Mountain, Mi., U.S. (ī´ērn moun´tĭn)	116-17	45°49'N	88°03'W
Iron River, Mi., U.S. (ī´ērn rĭv´ēr)	114-15	46°06'N	88°39'W
Ironton, Oh., U.S. (ī´ērn-tŭn)	116-17	38°32'N	82°41'W
Ironwood, Mi., U.S. (ī´ērn-wòd)	114-15	46°27'N	90°10'W
Ironwood Forest National Monument, n.p., Az., U.S.	118-19	32°27'N	111°30'W
Iroquois Falls, On., Can. (ĭr´ô-kwoi fôlz)	136-37	48°46'N	80°40'W
Irrawaddy, stm., Mya. (ĭr-ȧ-wäd´ē) *see* Ayeyarwady	220-21	15°51'N	95°05'E
Irtysh, stm., Asia (ĭr-tĭsh´)	218-19	61°05'N	68°47'E
Irumu, D.R.C. (ē-rò´mōō)	267	1°27'N	29°52'E
Irún, Spain (ê-rōōn´)	198-99	43°21'N	1°48'W
Iruña, Spain *see* Pamplona	198-99	42°49'N	1°39'W
Irvine, Ky., U.S. (ûr´vĭn)	116-17	37°42'N	83°58'W
Irving, Tx., U.S. (ûr´vĕng)	120-21	32°49'N	96°57'W
Isabela, Phil.	250	6°41'N	121°58'E
Isabela, Cabo, c., Dom. Rep. (kȧ´bô-ê-sä-bĕ´lä)	142-43	19°55'N	71°01'W
Isabela, Isla, i., Ec. (ê´s-lä-ē-sä-bā´lä)	170a	0°30'S	91°06'W
Isabelia, Cordillera, mts., Nic. (kôr-dēl-yĕ´rä-ē-sä-bēlyä)	149	13°30'N	85°32'W
Isabella Indian Reservation, ind. res., Mi., U.S. (ĭs-à-bĕl´-lȧ ĭn´dĭ-ăn rĕ-sēr-vā´shĕn)	116-17	43°41'N	84°48'W
Ísafjörður, Ice. (ēs´ä-fŷr-dòr)	190a	66°03'N	23°07'W
Ischia, Isola d', i., Italy (ē´-sō-lä-dē´sh-kyä)	200-01	40°43'N	13°54'E
Ise, Japan (ĭs´hĕ) (û´gē-yă´mä´dà)	245	34°30'N	136°42'E
Isernia, Italy (ê-zĕr´nyä)	200-01	41°36'N	14°14'E
Iset', stm., Russia	218-19	56°36'N	66°17'E
Ise-wan, b., Japan (ē´sĕ wän)	245	34°43'N	136°43'E
Iseyin, Nig.	260a	7°58'N	3°36'E
Isfahan, Iran *see* Eṣfahān	232-33	32°39'N	51°40'E
Ishigaki-shima, i., Japan	279a	24°24'N	124°12'E
Ishikari, stm., Japan	244	43°16'N	141°23'E
Ishikari-wan, b., Japan (ē´shē-kä-rē wän)	244	43°25'N	141°01'E
Ishim, Russia (ĭsh-ĕm´)	218-19	56°07'N	69°30'E
Ishim, stm., Asia (ĭsh-ĕm´)	218-19	57°43'N	71°12'E
Ishimskaya Ravnina, pl., Asia	226	55°0'N	70°00'E
Ishinomaki, Japan (ĭsh-nō-mä´kē)	244	38°25'N	141°18'E
Ishpeming, Mi., U.S. (ĭsh´pê-mĭng)	114-15	46°29'N	87°39'W
Isil'kul', Russia	226	54°54'N	71°16'E
Isiolo, Kenya	267	0°21'N	37°35'E
Isiro, D.R.C.	262-63	2°46'N	27°37'E
İskenderun, Tur. (ĭs-kĕn´dĕr-ōōn)	228-29	36°35'N	36°11'E

Feature (Pronunciation)	Page	Lat.	Long.
İskenderun Körfezi, b., Tur.	228-29	36°30'N	35°40'E
Iskitim, Russia	226	54°39'N	83°18'E
Iskŭr, stm., Blg. (ĭs'k'r)	200-01	43°45'N	24°26'E
Isla Cristina, Spain (ī'lä-krē-stē'nä)	198-99	37°12'N	7°19'W
Islāmābād, nat. cap., Pak. (ĭs'lä-mä-bäd') (ĭs-lä'-mä-bäd')	232-33	33°39'N	73°05'E
Isla Mujeres, Mex. (ē's-lä-mōō-kĕ'rĕs)	148	21°12'N	86°43'W
Ísland, nation, Eur. see Iceland	174-75	65°0'N	18°00'W
Island Lake, lk., Mb., Can. (ī'lănd lāk ī'lănd)	134-35	53°47'N	94°25'W
Islands, Bay of, b., Nf., Can. (bā ŭv ī'lăndz)	138-39	49°10'N	58°15'W
Íslandshaf, s., Eur. see Norwegian Sea	184-85	70°0'N	2°00'E
Islay, i., Scot., U.K. (ī'lā)	190-91	55°49'N	6°17'W
Isle of Man, dep., Eur. (īl ŭv măn)	190-91	54°15'N	4°30'W
Isle Royale National Park, n.p., Mi., U.S. (īl'roi-ăl' năsh'ŭn-ăl pärk)	114-15	47°58'N	88°55'W
Ismailia, Egypt (ēs-mä-ēl'ēá)	268b	30°36'N	32°16'E
Isparta, Tur. (ē-spär'tá)	186-87	37°46'N	30°33'E
Israel, nation, Asia (ĭz'rē-ĕl)	206-07	31°30'N	34°45'E
Isrā'īl, nation, Asia see Israel	206-07	31°30'N	34°45'E
Issoire, Fr. (ē-swár')	196-97	45°33'N	3°15'E
Issoudun, Fr. (ē-sōō-dăN')	196-97	46°57'N	2°00'E
Issyk-Kul, Lake, lk., Kyrg. (lāk ē'-sĭk-kōōl')	226	42°25'N	77°15'E
İstanbul, Tur. (ē-stän-bōōl')	200-01	41°02'N	28°59'E
İstanbul Boğazı, strt., Tur. see Bosporus	200-01	41°06'N	29°04'E
Istaravshan, Taj.	232-33	39°54'N	69°00'E
Istiaía, Grc. (ĭs-tyī'yä)	200-01	38°57'N	23°09'E
Istmina, Col. (ēst-mē'nä)	163c	5°09'N	76°41'W
Istra, pen., Eur. (ē-strä)	200-01	45°17'N	13°57'E
Istria, pen., Eur. see Istra	200-01	45°17'N	13°57'E
Itabaiana, Braz. (ē-tä-bä-yá-nä)	163d	7°20'S	35°20'W
Itabaiana, Braz. (ē-tä-bä-yá-nä)	166-67	10°41'S	37°26'W
Itabapoana, Braz. (ē-tä'-bä-pôä'nä)	172	21°18'S	40°59'W
Itaberaí, Braz.	168-69	16°01'S	49°48'W
Itabira, Braz.	172	19°38'S	43°14'W
Itabuna, Braz. (ē-tä-bōō'nä)	166-67	14°47'S	39°17'W
Itacoatiara, Braz. (ē-tá-kwá-tyä'rá)	166-67	3°08'S	58°26'W
Itagüí, Col. (ē-tä'gwē)	163c	6°10'N	75°38'W
Itaipu, Represa de, res., S.A.	168-69	24°56'S	54°26'W
Itaipu Reservoir, res., S.A. (ē-tī'pōō rĕ'sĕr-vwär) see Itaipu, Represa de	168-69	24°56'S	54°26'W
Itaituba, Braz. (ē-tä'ī-tōō'bà)	166-67	4°15'S	55°59'W
Itajaí, Braz. (ē-tä-zhī')	172	26°54'S	48°40'W
Itajubá, Braz.	172	22°26'S	45°27'W
Italia, nation, Eur. see Italy	174-75	43°0'N	13°00'E
Italy, nation, Eur. (ĭt'ȧ-lē)	174-75	43°0'N	13°00'E
Itämeri, s., Eur. see Baltic Sea	192-93	57°0'N	19°00'E
Itānagar, India	234-35	27°09'N	93°33'E
Itaparica, Ilha de, i., Braz.	166-67	13°0'S	38°42'W
Itapecuru-Mirim, Braz. (ē-tä-pĕ'kōō-rōō-mē-rēN')	166-67	3°24'S	44°20'W
Itapemirim, Braz.	172	21°01'S	40°49'W
Itaperuna, Braz. (ē-tá'pâ-rōō'nä)	172	21°12'S	41°54'W
Itapetinga, Braz.	168-69	15°15'S	40°16'W
Itapetininga, Braz. (ē-tä-pĕ-tē-nē'N-gä)	172	23°35'S	48°02'W
Itapicuru, stm., Braz.	166-67	2°51'S	44°12'W
Itapicuru, stm., Braz.	166-67	11°45'S	37°31'W
Itaquari, Braz.	172	20°20'S	40°23'W
Itaqui, Braz.	173	29°08'S	56°32'W
Itararé, Braz.	172	24°07'S	49°21'W
Itārsi, India	234-35	22°36'N	77°46'E
Itasca, Tx., U.S. (ī-tăs'kȧ)	122-23	32°10'N	97°09'W
Itaúna, Braz. (ē-tä-ōō'nä)	172	20°04'S	44°34'W
Itbayat Island, i., Phil.	250a	20°46'N	121°50'E
Itenes, stm., S.A. see Iténez	166-67	11°55'S	65°00'W
Iténez, stm., S.A.	166-67	11°55'S	65°00'W
Ithaca, Mi., U.S. (ĭth'ȧ-kȧ)	116-17	43°17'N	84°36'W
Ithaca, N.Y., U.S. (ĭth'ȧ-kȧ)	116-17	42°26'N	76°30'W
Itu, Braz. (ē-tōō')	172	23°16'S	47°18'W
Ituango, Col. (ē-twäŋ'gō)	164-65	7°06'N	75°44'W
Ituí, stm., Braz.	164-65	4°39'S	70°15'W
Ituiutaba, Braz. (ē-tōō-ēōō-tä'bä)	168-69	18°58'S	49°27'W
Itumbiara, Braz.	172	18°25'S	49°12'W
Iturbide, Mex. (ē'tōōr-bē'dhâ)	148	19°38'N	89°36'W
Ituri, stm., D.R.C.	262-63	1°40'N	27°02'E
Iturup, Ostrov, i., Russia (ôs-trôf' ē-tōō-rōōp')	218-19	44°51'N	147°27'E
Ituxi, stm., Braz.	166-67	7°18'S	64°51'W
Ituzaingó, Arg. (ē-tōō-zä-ē'n-gô)	173	27°36'S	56°40'W
Ītyop'iya, nation, Afr. see Ethiopia	253	9°0'N	39°00'E
Iuka, Ms., U.S. (ī-ū'kȧ)	124-25	34°49'N	88°11'W
Iul'tin, Russia	218-19	67°43'N	178°51'W
Ivaí, stm., Braz.	168-69	23°18'S	53°44'W
Ivalo, Fin.	184-85	68°40'N	27°32'E
Ivanhoe, Austl. (ĭv'ăn-hō)	276	32°55'S	144°19'E
Ivano-Frankivs'k, Ukr.	194-95	48°55'N	24°44'E
Ivanovo, Russia (ē-vä'nô-vō)	186-87	57°01'N	40°59'E
Ivanovo-Voznesensk, Russia see Ivanovo	186-87	57°01'N	40°59'E
Ivdel', Russia (ĭv'dyĕl)	186-87	60°41'N	60°27'E
Iviza, Spain see Eivissa	198-99	38°55'N	1°25'E
Ivory Coast, nation, Afr. see Cote d'Ivoire	253	8°0'N	5°00'W
Ivrea, Italy (ē-vrĕ'ä)	200-01	45°28'N	7°53'E
Ivujivik, Qc., Can.	128-29	62°23'N	77°55'W
Iwaki, Japan	245	37°03'N	140°55'E
Iwo, Nig.	260a	7°38'N	4°11'E
Iwo Jima, i., Japan (ē'wō jē'mȧ)	280-81	24°47'N	141°20'E
Ixmiquilpan, Mex. (ēs-mē-kēl'pän)	146-47	20°29'N	99°13'W
Ixtepec, Mex. (ēks-tĕ'pĕk)	146-47	16°32'N	95°05'W
Ixtlán de Juárez, Mex. (ēs-tlän' dā hwä'räz)	146-47	17°20'N	96°30'W
Ixtlán del Río, Mex. (ēs-tlän'dĕl rē'ō)	146-47	21°02'N	104°22'W
Iyo-nada, s., Japan (ē'yō nä-dä)	245	33°40'N	132°20'E
Izabal, Lago de, lk., Guat. (lä'gô-dĕ-ē'zä-bäl')	148	15°30'N	89°10'W
Izamal, Mex. (ē-zä-mä'l)	148	20°56'N	89°01'W
Izberbash, Russia	227	42°33'N	47°52'E
Izhevsk, Russia (ē-zhyĕfsk')	186-87	56°50'N	53°12'E
Izhma, stm., Russia	186-87	65°19'N	52°55'E
Izium, Ukr.	202-03	49°13'N	37°17'E
Izmaïl, Ukr.	200-01	45°21'N	28°50'E
İzmir, Tur. (ĭz-mēr')	200-01	38°26'N	27°09'E
İzmit, Tur. (ĭz-mēt')	200-01	40°47'N	29°57'E
Izuhara, Japan (ē'zōō-hä'rä)	243	34°12'N	129°17'E
Izumo, Japan (ē'zōō-mō)	245	35°22'N	132°46'E
Izu-shotō, is., Japan	244	32°0'N	140°00'E

J

Feature (Pronunciation)	Page	Lat.	Long.
Jabal, Baḩr al-, stm., Sudan see Mountain Nile	262-63	9°30'N	30°30'E
Jabalpur, India	234-35	23°10'N	79°56'E
Jaboatão, Braz. (zhä-bô-á-touN)	163d	8°07'S	35°01'W
Jaca, Spain (hä'kä)	198-99	42°35'N	0°34'W
Jacala, Mex. (hä-kä'lä)	146-47	21°01'N	99°11'W
Jacaltenango, Guat. (hä-kàl-tĕ-näŋ'gō)	148	15°40'N	91°44'W
Jacareí, Braz.	172	23°19'S	45°58'W
Jacarezinho, Braz. (zhä-kä-rĕ'zē-nyô)	168-69	23°10'S	49°59'W
Jacksboro, Tx., U.S. (jăks'bŭr-ô)	120-21	33°13'N	98°09'W
Jackson, Al., U.S. (jăk'sŭn)	124-25	31°31'N	87°54'W
Jackson, Ga., U.S. (jăk'sŭn)	124-25	33°18'N	83°58'W
Jackson, Ky., U.S. (jăk'sŭn)	116-17	37°33'N	83°24'W
Jackson, La., U.S. (jăk'sŭn)	124-25	30°50'N	91°13'W
Jackson, Mi., U.S. (jăk'sŭn)	116-17	42°15'N	84°24'W
Jackson, Mn., U.S. (jăk'sŭn)	114-15	43°37'N	94°59'W
Jackson, Mo., U.S. (jăk'sŭn)	120-21	37°23'N	89°40'W
Jackson, Ms., U.S. (jăk'sŭn)	124-25	32°19'N	90°11'W
Jackson, Oh., U.S. (jăk'sŭn)	116-17	39°03'N	82°39'W
Jackson, Tn., U.S. (jăk'sŭn)	124-25	35°37'N	88°49'W
Jackson, Wy., U.S. (jăk'sŭn)	112-13	43°29'N	110°45'W
Jackson Lake, lk., Wy., U.S. (jăk'sŭn lāk)	112-13	43°55'N	110°40'W
Jacksonville, Al., U.S. (jăk'sŭn-vĭl)	124-25	33°49'N	85°46'W
Jacksonville, Ar., U.S. (jăk'sŭn-vĭl)	120-21	34°52'N	92°07'W
Jacksonville, Fl., U.S. (jăk'sŭn-vĭl)	124-25	30°21'N	81°39'W
Jacksonville, Il., U.S. (jăk'sŭn-vĭl)	120-21	39°44'N	90°14'W
Jacksonville, N.C., U.S. (jăk'sŭn-vĭl)	124-25	34°45'N	77°25'W
Jacksonville, Tx., U.S. (jăk'sŭn-vĭl)	122-23	31°58'N	95°16'W
Jacksonville Beach, Fl., U.S. (jăk'sŭn-vĭl bēch)	124-25	30°17'N	81°24'W
Jacmel, Haiti (zhák-mĕl')	142-43	18°14'N	72°32'W
Jacobābād, Pak.	232-33	28°17'N	68°26'E
Jacobina, Braz. (zhä-kô-bē'nȧ)	166-67	11°11'S	40°31'W
Jacques-Cartier, Détroit de, strt., Qc., Can.	138-39	49°53'N	62°45'W
Jacques-Cartier, Mont, mtn., Qc., Can.	138-39	48°59'N	65°57'W
Jacquet River, N.B., Can. (zhá-kĕ' rĭv'ẽr) (jăk'ĕt rĭv'ẽr)	138-39	47°55'N	66°01'W
Jacuí, stm., Braz.	168-69	30°02'S	51°15'W
Jadotville, D.R.C. see Likasi	262-63	10°59'S	26°43'E
Jadransko more, s., Eur. see Adriatic Sea	200-01	42°30'N	16°00'E
Jadransko morje, s., Eur. see Adriatic Sea	200-01	42°30'N	16°00'E
Jaén, Peru (kä-ĕ'n)	170	5°43'S	78°47'W
Jaén, Spain	198-99	37°46'N	3°48'W
Jaffa, Cape, c., Austl. (kāp jăf'ȧ)	276	36°58'S	139°40'E
Jaffna, Sri L. (jäf'nȧ)	236	9°40'N	80°01'E
Jagādhri, India	234-35	30°10'N	77°18'E
Jagdalpur, India	236	19°05'N	82°02'E
Jägerndorf, Czech Rep. see Krnov	194-95	50°05'N	17°42'E
Jaguarão, Braz.	173	32°34'S	53°23'W
Jaguariaíva, Braz.	168-69	24°15'S	49°42'W
Jaguaribe, stm., Braz.	166-67	4°25'S	37°46'W
Jagüey Grande, Cuba (hä'gwā grän'dā)	142-43	22°32'N	81°08'W
Jahrom, Iran	230-31	28°29'N	53°33'E
Jaipur, India	234-35	26°55'N	75°48'E
Jaisalmer, India	234-35	26°55'N	70°55'E
Jajce, Bos. (yī'tsĕ)	200-01	44°20'N	17°17'E
Jājpur, India	234-35	20°51'N	86°20'E
Jakarta, nat. cap., Indon. (yä-kär'tä)	248-49	6°11'S	106°50'E
Jakobstad, Fin. (yá'kôb-stádh)	184-85	63°41'N	22°43'E
Jalālābād, Afg. (jŭ-lä-lá-bäd)	232-33	34°26'N	70°27'E
Jalal-Abad, Kyrg.	226	40°56'N	73°00'E
Jalandhar, India	234-35	31°19'N	75°35'E
Jalapa, Guat. (hä-lä'pá)	148	14°38'N	89°59'W
Jalapa, Mex. see Xalapa	146-47	19°32'N	96°55'W
Jālgaon, India	234-35	21°01'N	75°34'E
Jalisco, state, Mex. (hä-lēs'kō)	146-47	20°20'N	103°40'W
Jālna, India	234-35	19°51'N	75°54'E
Jalón, stm., Spain (hä-lōn')	198-99	41°47'N	1°03'W
Jālor, India	234-35	25°21'N	72°37'E
Jalostotitlán, Mex. (hä-lōs-tē-tlán')	146-47	21°11'N	102°27'W
Jalpa, Mex. (häl'pä)	146-47	21°38'N	102°60'W
Jalpāiguri, India	234-35	26°31'N	88°42'E
Jamaame, Som.	262-63	0°01'N	42°42'E
Jamaica, nation, N.A. (jȧ-mā'kȧ)	140-41	18°15'N	77°30'W
Jamanxim, stm., Braz.	166-67	4°45'S	56°27'W
Jambi, Indon. (mäm'bē)	246-47	1°37'S	103°36'E
Jambongan, Pulau, i., Malay.	248-49	6°40'N	117°27'E
James, stm., U.S. (jāmz)	110-11	42°52'N	97°19'W
James, stm., Va., U.S. (jāmz)	116-17	36°56'N	76°26'W
James Bay, b., Can. (jāmz bā)	130-31	53°30'N	80°30'W
Jamestown, Austl.	276	33°12'S	138°36'E
Jamestown, Ky., U.S. (jāmz'toun)	124-25	36°59'N	85°04'W
Jamestown, N.D., U.S. (jāmz'toun)	114-15	46°54'N	98°42'W
Jamestown, N.Y., U.S. (jāmz'toun)	116-17	42°06'N	79°14'W
Jammu, India (jŭmōō)	234-35	32°43'N	74°51'E
Jammu and Kashmir, state, India (jŭmōō ănd kash-mēr')	234-35	34°0'N	76°00'E
Jammu and Kashmir, hist. reg., Asia	234-35	34°0'N	76°00'E
Jamnagar, India (jäm-nû'gŭr)	234-35	22°28'N	70°04'E
Jamshedpur, India (jäm'shād-pōōr)	234-35	22°48'N	86°11'E
Jamuna, stm., Bngl.	234-35	23°43'N	89°49'E
Janaucu, Ilha, i., Braz.	166-67	0°30'N	50°10'W
Janesville, Ca., U.S. (jānz'vĭl)	118-19	40°18'N	120°32'W
Janesville, Wi., U.S. (jānz'vĭl)	116-17	42°41'N	89°02'W
Jangīpur, India	234-35	24°28'N	88°04'E
Jan Mayen, dep., Eur. (yän mī'ĕn)	288	71°02'N	8°19'W
Jan Mayen, i., Nor. (yän mī'ĕn)	288	71°03'N	8°19'W
Januária, Braz. (zhä-nwä'rē-ä)	168-69	15°29'S	44°22'W
Japan, nation, Asia (jȧ-păn')	206-07	36°0'N	138°00'E
Japan, Sea of, s., Asia (sē ŭv jȧ-păn')	222-23	40°0'N	135°00'E
Japurá, stm., S.A.	166-67	3°08'S	64°46'W
Jaraguá do Sul, Braz.	172	26°29'S	49°05'W
Jarama, stm., Spain (hä-rä'mä)	198-99	40°02'N	3°39'W
Jari, stm., Braz. (zhä-rē)	166-67	1°09'S	51°53'W
Jarkand, China see Shache	226	38°25'N	77°15'E
Jarocin, Pol. (yä-rō'tsyēn)	194-95	51°58'N	17°30'E
Jarosław, Pol. (yá-rôs-wáf)	194-95	50°01'N	22°41'E
Jarud Qi, China (jya-lōō-tŭ shyē)	240-41	44°34'N	120°54'E
Jarvis Island, dep., Oc.	280-81	0°19'S	160°01'W
Jarvis Island, i., Oc.	280-81	0°23'S	160°01'W
Jāsk, Iran (jäsk)	230-31	25°39'N	57°47'E
Jasło, Pol. (yäs'wō)	194-95	49°45'N	21°28'E
Jason Islands, is., Falk. Is.	171	51°09'S	60°54'W
Jasper, Ab., Can. (jăs'pẽr)	132-33	52°52'N	118°05'W
Jasper, Al., U.S. (jăs'pẽr)	124-25	33°50'N	87°17'W
Jasper, Fl., U.S. (jăs'pẽr)	124-25	30°31'N	82°57'W
Jasper, Ga., U.S. (jăs'pẽr)	124-25	34°28'N	84°25'W
Jasper, In., U.S. (jăs'pẽr)	116-17	38°23'N	86°56'W
Jasper, Tx., U.S. (jăs'pẽr)	122-23	30°54'N	94°00'W
Jasper National Park, n.p., Ab., Can. (jăs'pẽr năsh'ŭn-ăl pärk)	132-33	52°53'N	118°03'W
Jassy, Rom. see Iaşi	202-03	47°10'N	27°36'E
Jataí, Braz.	168-69	17°53'S	51°45'W
Jaú, Braz.	172	22°18'S	48°33'W

Feature (Pronunciation)	Page	Lat.	Long.
Jauja, Peru (kä-ō´κ)	163a	11°47'S	75°29'W
Jaumave, Mex. (hou-mä´vå)	146-47	23°25'N	99°23'W
Jaunpur, India	234-35	25°44'N	82°41'E
Java, i., Indon. (jä´vǎ)	248-49	7°30'S	109°59'E
Javari, stm., S.A. (kå-vä-rē)	170	4°21'S	70°02'W
Java Sea, s., Indon. (jä´vǎ sē)	248-49	5°0'S	110°00'E
Javhlant, Mong. *see* Uliastay	240-41	47°44'N	96°51'E
Jawa, i., Indon. *see* Java	248-49	7°30'S	109°59'E
Jawa, Laut, s., Indon. *see* Java Sea	248-49	5°0'S	110°00'E
Jawhar, Som.	262-63	2°47'N	45°31'E
Jaworzno, Pol. (yä-vôzh´nô)	194-95	50°12'N	19°15'E
Jaya, Puncak, mtn., Indon.	224-25	4°05'S	137°11'E
Jayapura, Indon.	277	2°32'S	140°43'E
Jaz Mūrīān, Hāmūn-e, lk., Iran	230-31	27°14'N	58°49'E
Jeanerette, La., U.S. (jěn-ēr-et´) (zhän-rět´)	124-25	29°55'N	91°40'W
Jeddah, Sau. Ar. *see* Jiddah	266	21°30'N	39°12'E
Jędrzejów, Pol. (yǎɴ-dzhā´yôf)	194-95	50°39'N	20°19'E
Jefferson, Ia., U.S. (jěf´ēr-sǔn)	114-15	42°01'N	94°23'W
Jefferson, Oh., U.S. (jěf´ēr-sǔn)	116-17	41°44'N	80°46'W
Jefferson, Tx., U.S. (jěf´ēr-sǔn)	120-21	32°45'N	94°21'W
Jefferson, Wi., U.S. (jěf´ēr-sǔn)	116-17	43°00'N	88°48'W
Jefferson, Mount, mtn., Nv., U.S. (mount jěf´ēr-sǔn)	118-19	38°46'N	116°55'W
Jefferson City, Mo., U.S.	120-21	38°33'N	92°10'W
Jefferson City, Tn., U.S. (jěf´ēr-sǔn sĭ´tē)	124-25	36°07'N	83°30'W
Jeffersontown, Ky., U.S. (jěf´ēr-sǔn-toun)	116-17	38°13'N	85°35'W
Jeffersonville, In., U.S. (jěf´ēr-sǔn-vĭl)	116-17	38°17'N	85°44'W
Jeju, Kor., S. *see* Cheju	240-41	33°30'N	126°32'E
Jēkabpils, Lat. (yěk´áb-pĭls)	192-93	56°30'N	25°52'E
Jelenia Góra, Pol. (yě-lěn´yà gô´rà)	194-95	50°54'N	15°44'E
Jelgava, Lat.	192-93	56°39'N	23°44'E
Jellico, Tn., U.S. (jěl´ĭ-kō)	124-25	36°35'N	84°08'W
Jemaja, Pulau, i., Indon.	246-47	2°55'N	105°45'E
Jember, Indon.	248-49	8°10'S	113°42'E
Jena, Ger. (yā´nä)	194-95	50°56'N	11°35'E
Jengish Chokusu, mtn., Asia	226	42°02'N	80°05'E
Jenkins, Ky., U.S. (jěŋ´kĭnz)	124-25	37°10'N	82°39'W
Jennings, La., U.S. (jěn´ĭngz)	122-23	30°13'N	92°39'W
Jeonju, Kor., S. *see* Chŏnju	243	35°49'N	127°09'E
Jequié, Braz.	166-67	13°52'S	40°05'W
Jequitinhonha, stm., Braz. (zhě-kē-těn-ō´n-yä)	168-69	15°51'S	38°53'W
Jerada, Mor.	184-85	34°19'N	2°10'W
Jerba, Île de, i., Tun.	258-59	33°48'N	10°54'E
Jérémie, Haiti (zhā-rā-mē´)	142-43	18°39'N	74°07'W
Jeremoabo, Braz. (zhě-rā-mō-á´bō)	166-67	10°06'S	38°19'W
Jerevan, nat. cap., Arm. *see* Yerevan	227	40°11'N	44°30'E
Jerez de la Frontera, Spain	198-99	36°42'N	6°08'W
Jericho, W.B.	228-29	31°52'N	35°27'E
Jerid, Chott, lk., Tun. (shôt jěr´ĭd)	258-59	33°42'N	8°26'E
Jerome, Id., U.S. (jě-rōm´)	112-13	42°44'N	114°31'W
Jersey, dep., Eur. (jûr´zĭ)	196-97	49°15'N	2°10'W
Jersey City, N.J., U.S. (jûr´zĭ sĭ´tē)	116-17	40°44'N	74°04'W
Jersey Shore, Pa., U.S. (jûr´zĭ shōr)	116-17	41°12'N	77°15'W
Jerseyville, Il., U.S. (jěr´zē-vĭl)	120-21	39°07'N	90°20'W
Jerusalem, nat. cap., Isr. (jě-rōō´så-lěm)	228-29	31°47'N	35°14'E
Jesi, Italy	200-01	43°31'N	13°14'E
Jesselton, Malay. *see* Kota Kinabalu	248-49	5°58'N	116°05'E
Jessore, Bngl.	234-35	23°10'N	89°13'E
Jesup, Ga., U.S. (jěs´ǔp)	124-25	31°36'N	81°53'W
Jesús Carranza, Mex. (hē-sōō´s-kär-rä´n-zä)	146-47	17°24'N	95°02'W
Jesús María, Arg.	173	30°59'S	64°05'W
Jewel Cave National Monument, n.p., S.D., U.S. (jū´ěl kāv)	114-15	43°45'N	103°51'W
Jhālāwār, India	234-35	24°36'N	76°10'E
Jhang Sadar, Pak.	232-33	31°16'N	72°19'E
Jhānsi, India (jän´sē)	234-35	25°27'N	78°35'E
Jharkhand, state, India	234-35	23°30'N	85°00'E
Jhelum, Pak.	232-33	32°56'N	73°43'E
Jhelum, stm., Asia (jā´lŭm)	232-33	31°12'N	72°08'E
Jhunjhunūn, India	234-35	28°08'N	75°24'E
Jiading, China (jyä-dĭŋ)	238-39	31°23'N	121°14'E
Jiali, China	238-39	30°45'N	93°20'E
Jialing, China *see* Guangyuan	238-39	32°25'N	105°49'E
Jialing, stm., China (jyä-lĭŋ)	238-39	29°34'N	106°35'E
Jiamusi, China	244	46°48'N	130°22'E
Ji'an, China (jyē-än)	238-39	27°07'N	114°59'E
Ji'an, China (jyē-än)	243	41°06'N	126°10'E
Jianchuan, China	238-39	26°34'N	99°53'E
Jiangjin, China	238-39	29°17'N	106°15'E
Jiangkou, China	238-39	23°35'N	110°11'E
Jiangling, China (jyäŋ-lĭŋ)	238-39	30°19'N	112°12'E
Jiangmen, China	238-39	22°34'N	113°05'E
Jiangsu, state, China (jyäŋ-sōō)	238-39	33°0'N	120°00'E
Jiangxi, state, China (jyäŋ-shyē)	238-39	28°0'N	116°00'E
Jiangyin, China (jyäŋ-yĭn)	238-39	31°54'N	120°15'E
Jianli, China (jyěn-lē)	238-39	29°49'N	112°54'E
Jianning, China (jyěn-nĭŋ)	238-39	26°50'N	116°49'E
Jian'ou, China (jyěn-ō)	238-39	27°02'N	118°19'E
Jianshi, China (jyěn-shr)	238-39	30°36'N	109°44'E
Jianshui, China	238-39	23°37'N	102°49'E
Jiaohe, China (jyou-hŭ)	240-41	43°43'N	127°20'E
Jiaoxian, China (jyou shyěn)	240-41	36°17'N	119°60'E
Jiaozuo, China (jyou-dzwô)	240-41	35°15'N	113°14'E
Jiashan, China (jyä-shän)	238-39	32°46'N	117°59'E
Jiashun Hu, lk., China	234-35	34°24'N	85°47'E
Jiaxing, China (jyä-shyĭŋ)	238-39	30°46'N	120°45'E
Jiayu, China (jyä-yōō)	238-39	29°58'N	113°55'E
Jibuti, nat. cap., Dji. *see* Djibouti	266	11°34'N	43°09'E
Jicarilla Apache Indian Reservation, ind. res., N.M., U.S. (κē-kà-rēl´yà ĭn´dĭ-ǎn rě-sēr-vā´shěn)	118-19	36°40'N	107°00'W
Jicarón, Isla, i., Pan. (ē´s-lä-κē-kä-rōn´)	150	7°16'N	81°49'W
Jiddah, Sau. Ar.	266	21°30'N	39°12'E
Jieyang, China (jyě-yäŋ)	238-39	23°33'N	116°21'E
Jiguaní, Cuba (κē-gwä-nē´)	142-43	20°22'N	76°25'W
Jijiga, Eth.	262-63	9°21'N	42°48'E
Jilin, China (jyē-lĭn)	240-41	43°51'N	126°33'E
Jilin, state, China	240-41	44°0'N	126°00'E
Jīma, Eth.	262-63	7°38'N	36°50'E
Jiménez, Mex. (κê-mā´nâz)	122-23	27°08'N	104°56'W
Jiménez, Mex. (κê-mā´nâz)	122-23	29°02'N	100°41'W
Jiménez del Téul, Mex. (κê-mā´nâz děl tě-ōō´l)	146-47	23°14'N	103°49'W
Jimeta, Nig.	260-61	9°16'N	12°26'E
Jim Thorpe, Pa., U.S. (jĭm´ thôrp´)	116-17	40°52'N	75°44'W
Jinan, China (jyē-nän)	240-41	36°40'N	116°59'E
Jincheng, China (jyĭn-chŭŋ)	240-41	35°30'N	112°50'E
Jindřichuv Hradec, Czech Rep. (yěn´d'r-zhĭ-kōōf hrä´děts)	194-95	49°09'N	15°01'E
Jing, stm., China (jyĭŋ)	238-39	34°28'N	109°05'E
Jingdezhen, China (jyĭn-dŭ-jŭn)	238-39	29°17'N	117°12'E
Jinggangshan, China	238-39	26°36'N	114°05'E
Jinghong, China	238-39	21°59'N	100°49'E
Jingning, China (jyĭŋ-nĭŋ)	240-41	35°32'N	105°44'E
Jingxian, China (jyĭŋ shyěn)	238-39	26°40'N	109°25'E
Jingxian, China (jyĭŋ shyěn)	238-39	30°41'N	118°24'E
Jingxian, China (jyĭŋ shyěn)	240-41	37°41'N	116°16'E
Jinhae, Kor., S. *see* Chinhae	243	35°08'N	128°40'E
Jinhua, China (jyĭn-hwä)	238-39	29°07'N	119°39'E
Jining, China (jyē-nĭŋ)	240-41	35°24'N	116°34'E
Jining, China (jyē-nĭŋ)	240-41	41°02'N	113°06'E
Jinja, Ug. (jĭn´jä)	267	0°26'N	33°13'E
Jinju, Kor., S. *see* Chinju	243	35°10'N	128°05'E
Jinmu Jiao, c., China	238-39	18°11'N	109°35'E
Jinning, China	238-39	24°40'N	102°35'E
Jinotega, Nic. (κē-nô-tā´gä)	149	13°05'N	85°60'W
Jinotepe, Nic. (κē-nô-tā´pâ)	149	11°51'N	86°12'W
Jinsen, Kor., S. *see* Inch'ŏn	243	37°28'N	126°38'E
Jinsha, stm., China *see* Yangtze	238-39	31°24'N	121°54'E
Jinshi, China	238-39	29°38'N	111°52'E
Jinta, China (jyĭn-tä)	240-41	40°00'N	98°53'E
Jinxi, China	240-41	40°45'N	120°50'E
Jinyun, China (jyĭn-yòn)	238-39	28°40'N	120°03'E
Jinzhai, China (jyĭn-jī)	238-39	31°45'N	115°55'E
Jinzhou, China (jyĭn-jō)	240-41	39°06'N	121°43'E
Jinzhou, China (jyĭn-jō)	240-41	41°07'N	121°08'E
Ji-Paraná, Braz.	166-67	10°52'S	61°57'W
Jiparaná, stm., Braz. *see* Machado	166-67	8°02'S	62°53'W
Jipijapa, Ec. (κē-pē-hä´pä)	170	1°21'S	80°35'W
Jiujiang, China (jyô-jyän)	238-39	29°43'N	115°59'E
Jiulian Shan, mts., China	238-39	24°17'N	114°36'E
Jiuling Shan, mts., China	238-39	28°46'N	114°45'E
Jiuquan, China (jyô-chyän)	240-41	39°45'N	98°30'E
Jiutai, China	240-41	44°09'N	125°50'E
Jixi, China	244	45°17'N	130°58'E
Jixian, China (jyē shyěn)	244	46°43'N	131°08'E
Jixian, China (jyē shyěn)	240-41	35°25'N	114°04'E
Jixian, China (jyē shyěn)	240-41	40°02'N	117°24'E
Jīzān, Sau. Ar.	266	16°54'N	42°36'E
Jizzax, Uzb.	232-33	40°08'N	67°51'E
J. J. Castelli, Arg. *see* Castelli	168-69	25°57'S	60°37'W
João Belo, Moz. *see* Xai-Xai	264-65	25°03'S	33°39'E
João Pessoa, Braz.	163d	7°07'S	34°52'W
Joaquín V. González, Arg.	168-69	25°05'S	64°09'W
Jódar, Spain (hō´där)	198-99	37°50'N	3°21'W
Jodhpur, India (hōd´pōōr)	234-35	26°17'N	73°01'E
Joensuu, Fin. (yô-ěn´sōō)	184-85	62°36'N	29°47'E
Joetsu, Japan	245	37°09'N	138°15'E
Joffre, Mount, mtn., Can. (mount jō´f'r)	132-33	50°32'N	115°13'W
Jõgeva, Est. (yû´gě-vå)	192-93	58°45'N	26°24'E
Jogjakarta, Indon. *see* Yogyakarta	248-49	7°48'S	110°22'E
Johannesburg, S. Afr. (yô-hän´ěs-bòrgh)	269c	26°12'S	28°05'E
John Day, stm., Or., U.S. (jŏn´ dā)	112-13	45°44'N	120°39'W
Johnsonburg, Pa., U.S. (jŏn´sŭn-bûrg)	116-17	41°29'N	78°41'W
Johnson City, N.Y., U.S. (jŏn´sǔn sĭ´tē)	116-17	42°07'N	75°58'W
Johnson City, Tn., U.S. (jŏn´sǔn sĭ´tē)	124-25	36°19'N	82°22'W
Johnson City, Tx., U.S. (jŏn´sǔn sĭ´tē)	122-23	30°16'N	98°24'W
Johnston, Lake, lk., Austl.	272-73	32°18'S	120°46'E
Johnston Atoll, dep., Oc.	280-81	16°45'N	169°32'W
Johnston Atoll, at., Oc. (jŏn´stǔn ǎ´tôl)	280-81	16°45'N	169°32'W
Johnstown, Pa., U.S. (jonz´toun)	116-17	40°19'N	78°55'W
Johor Bahru, Malay.	246-47	1°28'N	103°45'E
Joigny, Fr. (zhwán-yē´)	196-97	47°59'N	3°24'E
Joinville, Braz.	172	26°18'S	48°50'W
Jokkmokk, Swe.	184-85	66°37'N	19°50'E
Joliet, Il., U.S. (jō-lĭ-ět´)	116-17	41°31'N	88°04'W
Joliette, Qc., Can. (zhō-lyět´)	136-37	46°02'N	73°25'W
Jolo, Phil. (hō-lō)	250	6°02'N	120°60'E
Jolo Group, is., Phil.	250	6°01'N	121°18'E
Jolo Island, i., Phil. (hō-lō ī´lǎnd)	250	5°58'N	121°06'E
Jomda, China	238-39	31°27'N	98°15'E
Jon, Deti, s., Eur. *see* Ionian Sea	200-01	39°0'N	19°00'E
Jonava, Lith. (yō-nä´vå)	192-93	55°05'N	24°17'E
Jonesboro, Ar., U.S. (jōnz´bûro)	124-25	35°51'N	90°42'W
Jonesboro, La., U.S. (jōnz´bûro)	120-21	32°14'N	92°43'W
Jonesville, La., U.S. (jōnz´vĭl)	124-25	31°37'N	91°49'W
Joniškis, Lith. (yô´nĭsh-kĭs)	192-93	56°14'N	23°38'E
Jönköping, Swe. (yûn´chû-pĭng)	192-93	57°47'N	14°11'E
Jonquière, Qc., Can. (zhôɴ-kyâr´)	136-37	48°26'N	71°11'W
Jonuta, Mex. (hô-nōō´tä)	148	18°06'N	92°07'W
Joplin, Mo., U.S. (jŏp´lĭn)	120-21	37°05'N	94°31'W
Jordan, Mt., U.S.	112-13	47°20'N	106°57'W
Jordan, nation, Asia (jôr´dǎn)	206-07	31°0'N	36°00'E
Jordan, stm., Asia (jôr´dǎn)	228-29	31°46'N	35°34'E
Jorhāt, India (jôr-hät´)	234-35	26°46'N	94°13'E
Jos, Nig.	260-61	9°56'N	8°53'E
José Batlle y Ordóñez, Ur.	173	33°29'S	55°08'W
José de San Martín, Arg.	171	44°02'S	70°29'W
Joseph Bonaparte Gulf, b., Austl. (jō´sěf bō´nà-pärt gŭlf)	272-73	14°15'S	128°30'E
Joshua Tree National Park, n.p., Ca., U.S. (jô´shū-à trē năsh´ŭn-ǎl pärk)	118-19	33°55'N	116°00'W
Jostedalsbreen, ice, Nor.	192-93	61°40'N	7°00'E
Jovellanos, Cuba (hō-věl-yä´nōs)	142-43	22°48'N	81°11'W
J. Strom Thurmond Reservoir, res., U.S.	124-25	33°45'N	82°16'W
Juan Aldama, Mex. (kòá´n-äl-dá´mä)	146-47	24°19'N	103°19'W
Juan de Fuca, Strait of, strt., N.A. (strāt ŭv hwän´ dā fōō´kä)	112-13	48°18'N	124°00'W
Juan de Fuca Strait, strt., N.A. *see* Juan de Fuca, Strait of	112-13	48°18'N	124°00'W
Juan Fernández, Archipiélago, is., Chile	159	33°0'S	80°00'W
Juanjuí, Peru	170	7°10'S	76°45'W
Juárez, Arg. (hōōá´rěz) *see* Benito Juárez	173	37°41'S	59°48'W
Juazeiro, Braz. (zhōōá´zā´rô)	166-67	9°25'S	40°30'W
Juazeiro do Norte, Braz. (zhōōá´zā´rô-dô-nôr-tě)	166-67	7°12'S	39°20'W
Juba, Sudan	267	4°51'N	31°37'E
Jubal, Strait of, strt., Egypt *see* Gûbâl, Madīq	268b	27°40'N	33°55'E
Jubayl, Leb. (jōō-bīl´)	228-29	34°08'N	35°40'E
Jubba, stm., Afr.	262-63	0°15'S	42°39'E
Juby, Cap, c., Mor. (kăp yōō´bē)	258-59	27°57'N	12°55'W
Júcaro, Cuba (hōō´kä-rô)	142-43	21°38'N	78°51'W
Juchipila, Mex. (hōō-chē-pē´lä)	146-47	21°24'N	103°07'W
Juchitán de Zaragoza, Mex.	146-47	16°26'N	95°01'W
Juchitlán, Mex. (hōō-chē-tlän)	146-47	20°05'N	104°06'W
Juddah, Sau. Ar. *see* Jiddah	266	21°30'N	39°12'E
Juidongshan, China	238-39	23°44'N	117°30'E
Juigalpa, Nic. (hwē-gäl´pä)	149	12°06'N	85°22'W
Juiz de Fora, Braz. (zhò-ēzh´ dā fō´rä)	172	21°45'S	43°22'W
Jujuy, Arg. (hōō-hwē´) *see* San Salvador de Jujuy	168-69	24°12'S	65°18'W
Jujuy, state, Arg. (hōō-hwē´)	168-69	23°0'S	66°00'W
Julesburg, Co., U.S. (jōōlz´bûrg)	114-15	40°59'N	102°17'W

Feature (Pronunciation)	Page	Lat.	Long.
Juliaca, Peru (hōō-lē-ä´kä)	170	15°30'S	70°08'W
Juliana Top, mtn., Sur.	164-65	3°39'N	56°32'W
Julianehåb, Green.	284-85	60°44'N	46°02'W
Jumentos Cays, is., Bah. (hōō-mĕn´tōs kēs)	142-43	22°42'N	75°55'W
Jumilla, Spain (hōō-mēl´yä)	198-99	38°28'N	1°20'W
Jūnāgadh, India (jò-nä´gŭd)	234-35	21°31'N	70°27'E
Junction, Tx., U.S. (jŭŋk´shŭn)	122-23	30°29'N	99°47'W
Junction City, Ks., U.S. (jŭŋk´shŭn sĭ´tē)	120-21	39°02'N	96°50'W
Junction City, Or., U.S. (jŭŋk´shŭn sĭ´tē)	112-13	44°14'N	123°11'W
Jundiaí, Braz.	172	23°11'S	46°53'W
Juneau, Ak., U.S. (jōō´nō)	126	58°20'N	134°25'W
Junee, Austl.	276	34°52'S	147°35'E
Jungar Qi, China	240-41	39°49'N	111°10'E
Jungfrau, mtn., Switz. (yòng´frou)	194-95	46°32'N	7°58'E
Junín, Arg. (hōō-nē´n)	173	34°36'S	60°58'W
Junín de los Andes, Arg.	171	39°55'S	71°05'W
Jūniyah, Leb. (jōō-nē´ĕ)	228-29	33°60'N	35°39'E
Junxian, China	238-39	32°32'N	111°31'E
Juquiá, Braz.	172	24°19'S	47°38'W
Jur, stm., Sudan (jòr)	262-63	8°39'N	29°17'E
Jura, i., Scot., U.K. (jōō´rȧ)	190-91	56°01'N	5°56'W
Jura, mts., Eur. (zhü-rȧ´)	194-95	47°06'N	6°50'E
Jurbarkas, Lith. (yōōr-bär´käs)	192-93	55°05'N	22°47'E
Jūrmala, Lat.	192-93	56°58'N	23°42'E
Juruá, stm., S.A.	166-67	2°35'S	65°47'W
Juruena, stm., Braz. (zhōō-rōōĕ´nä)	166-67	7°21'S	58°09'W
Justo Daract, Arg.	171	33°52'S	65°11'W
Jutaí, stm., Braz.	166-67	2°44'S	66°48'W
Jutiapa, Guat. (hōō-tē-ä´pä)	148	14°18'N	89°54'W
Juticalpa, Hond. (hōō-tē-käl´pä)	149	14°40'N	86°13'W
Jutland, reg., Den. *see* Jylland	192-93	56°0'N	9°15'E
Juventud, Isla de la, i., Cuba	142-43	21°40'N	82°50'W
Jyekundo, China *see* Yushu	238-39	33°00'N	97°00'E
Jylland, reg., Den.	192-93	56°0'N	9°15'E
Jyväskylä, Fin.	192-93	62°15'N	25°45'E

K

Feature (Pronunciation)	Page	Lat.	Long.
K2, mtn., Asia (kä-tōō)	232-33	35°53'N	76°30'E
Ka'ena Point, c., Hi., U.S. (kä´ä-nä point)	127a	21°35'N	158°17'W
Kaapstad, nat. cap., S. Afr. *see* Cape Town	264-65	33°55'S	18°30'E
Kaarlela, Fin. *see* Kokkola	184-85	63°50'N	23°09'E
Kaba, stm., Afr. *see* Little Scarcies	260-61	8°51'N	13°07'W
Kabaena, Pulau, i., Indon. (pōō-lou kä-bȧ-ä´nȧ)	248-49	5°15'S	121°55'E
Kabala, S.L. (kȧ-bä´lȧ)	260-61	9°35'N	11°33'W
Kabale, Ug.	267	1°11'S	29°56'E
Kabalega Falls, wtfl., Ug.	267	2°17'N	31°42'E
Kabalo, D.R.C. (kä-bä´lō)	262-63	6°03'S	26°55'E
Kabara, i., Fiji	279f	18°57'S	178°57'W
Kabardin-Balkaria, state, Russia *see* Balkaria	227	43°30'N	43°30'E
Kabardino-Balkariya, state, Russia *see* Balkaria	227	43°30'N	43°30'E
Kabīr Kūh, mts., Iran	228-29	33°36'N	46°12'E
Kābol, nat. cap., Afg. (kä´bōōl) *see* Kabul	232-33	34°32'N	69°10'E
Kābol, stm., Asia	232-33	33°55'N	72°14'E
Kabompo, stm., Zam. (kȧ-bôm´pō)	264-65	14°12'S	23°11'E
Kabul, nat. cap., Afg. (kä´bōōl)	232-33	34°32'N	69°10'E
Kabul, stm., Asia (kä´bòl) *see* Kābol	232-33	33°55'N	72°14'E
Kaburuang, Pulau, i., Indon.	248-49	3°48'N	126°48'E
Kabwe, Zam.	264-65	14°27'S	28°27'E
Kachchh, Gulf of, b., India	234-35	22°37'N	69°30'E
Kachchh, Rann of, reg., Asia *see* Kutch, Rann of	234-35	24°15'N	70°46'E
Kachul, Mol. *see* Cahul	202-03	45°55'N	28°12'E
Kadamatt Island, i., India	236	11°13'N	72°47'E
Kadan Kyun, i., Mya.	246-47	12°30'N	98°22'E
Kadéï, stm., Afr.	260-61	3°31'N	16°03'E
Kadina, Austl.	276	33°58'S	137°43'E
Kadiyevka, Ukr. *see* Stakhanov	202-03	48°34'N	38°40'E
Kadoma, Zimb.	264-65	18°21'S	29°54'E
Kaduna, Nig. (kä-dōō´nä)	260-61	10°32'N	7°25'E
Kaduna, stm., Nig. (kä-dōō´nä)	260-61	8°45'N	5°48'E
Kāduqlī, Sudan	266	10°60'N	29°43'E
Kadzherom, Russia	186-87	64°41'N	55°55'E
Kaédi, Maur. (kä-ā-dē´)	258-59	16°10'N	13°29'W
Kaesŏng, Kor., N. (kä´ĕ-sŭng) (kī´jō)	243	37°59'N	126°34'E
Kafue, stm., Zam. (kä´fōō)	264-65	15°56'S	28°57'E
Kafue National Park, n.p., Zam. (kä´fōō năsh´ŭn-ăl pärk)	264-65	15°22'S	25°25'E
Kaga Bandoro, C.A.R.	262-63	6°59'N	19°12'E
Kagera, stm., Afr. (kä-gā´rȧ)	267	0°56'S	31°47'E
Kagoshima, Japan (kä´gō-shē´mȧ)	245	31°36'N	130°33'E
Kagoshima-wan, b., Japan (kä´gō-shē´mä wän)	245	31°24'N	130°38'E
Kahama, Tan.	267	3°49'S	32°36'E
Kahayan, stm., Indon.	248-49	3°16'S	114°06'E
Kahemba, D.R.C.	262-63	7°18'S	18°59'E
Kaho'olawe, i., Hi., U.S. (kä-hōō-lä´wē)	127a	20°33'N	156°37'W
Kahoka, Mo., U.S. (kȧ-hō´kȧ)	120-21	40°26'N	91°43'W
Kahramanmaraş, Tur.	186-87	37°35'N	36°57'E
Kahuku Point, c., Hi., U.S. (kä-hōō´kōō point)	127a	21°43'N	157°59'W
Kai, Kepulauan, is., Indon.	224-25	5°35'S	132°45'E
Kaibab Indian Reservation, ind. res., Az., U.S. (kä´ē-băb ĭn´dĭ-ăn rĕ-sēr-vā´shĕn)	118-19	36°55'N	112°40'W
Kaidu, stm., China (kī-dōō)	222-23	41°58'N	86°44'E
Kaifeng, China (kī-fŭŋ)	238-39	34°47'N	114°21'E
Kaijo, Kor., N. *see* Kaesŏng	243	37°59'N	126°34'E
Kai Kecil, i., Indon.	224-25	5°46'S	132°43'E
Kailas, mtn., China *see* Kangrinboqê Feng	234-35	31°04'N	81°18'E
Kailas Range, mts., China (kī-läs rānj) *see* Gangdisê Shan	234-35	31°0'N	82°00'E
Kailu, China	240-41	43°36'N	121°19'E
Kailua, Hi., U.S. (kä´ē-lōō´ä)	127a	21°24'N	157°45'W
Kailua, Hi., U.S. (kä´ē-lōō´ä)	127a	19°39'N	155°58'W
Kailua Kona, Hi., U.S. *see* Kailua	127a	19°39'N	155°58'W
Kaiping, China	238-39	22°22'N	112°37'E
Kairouan, Tun.	184-85	35°41'N	10°07'E
Kaiserslautern, Ger. (kī-zĕrs-lou´tĕrn)	194-95	49°26'N	7°45'E
Kaiyuan, China (kū-yuän)	243	42°32'N	124°02'E
Kaiyuan, China (kū-yuän)	238-39	23°42'N	103°14'E
Kajaani, Fin. (kä´yȧ-nē)	184-85	64°14'N	27°45'E
Kaka, Turkmen.	232-33	37°20'N	59°37'E
Kakabia, Pulau, i., Indon.	248-49	6°54'S	122°13'E
Kakamas, S. Afr.	264-65	28°46'S	20°36'E
Kakamega, Kenya	267	0°17'N	34°45'E
Kakhovka, Ukr. (kä-kôf´kȧ)	202-03	46°49'N	33°30'E
Kakhovka Reservoir, res., Ukr *see* Kakhovs'ke vodoskhovyshche	202-03	47°28'N	34°06'E
Kakhovs'ke vodoskhovyshche, res., Ukr.	202-03	47°28'N	34°06'E
Kakhul, Mol. *see* Cahul	202-03	45°55'N	28°12'E
Kākināda, India	236	16°57'N	82°15'E
Kakshaal-Too, mts., Asia	226	41°0'N	78°00'E
Kaktovik, Ak., U.S. (kăk-tō´vĭk)	126	70°08'N	143°38'W
Kakuma, Kenya	267	3°42'N	34°52'E
Kalaallit Nunaat, dep., N.A. *see* Greenland	85	70°0'N	40°00'W
Kalabahi, Indon.	248-49	8°15'S	124°32'E
Kalach, Russia (kȧ-lăch´)	186-87	50°25'N	41°00'E
Kalachinsk, Russia	226	55°02'N	74°35'E
Kalach-na-Donu, Russia	186-87	48°43'N	43°28'E
Kalae, c., Hi., U.S.	127a	18°55'N	155°41'W
Kalahari Desert, des., Afr. (kä-lä-hä´rē dĕs´ērt)	264-65	24°0'S	21°30'E
Kalahari Gemsbok National Park, n.p., S. Afr.	264-65	25°30'S	20°30'E
Kalama, Wa., U.S. (kȧ-lăm´ȧ)	112-13	46°01'N	122°50'W
Kalamáta, Grc.	200-01	37°03'N	22°07'E
Kalamazoo, Mi., U.S. (kăl-ȧ-mȧ-zōō´)	116-17	42°17'N	85°35'W
Kalamazoo, stm., Mi., U.S. (kăl-ȧ-mȧ-zōō´)	116-17	42°40'N	86°12'W
Kalanchak, Ukr. (kä-län-chăk´)	202-03	46°15'N	33°18'E
Kalao, Pulau, i., Indon.	248-49	7°18'S	120°58'E
Kalaotoa, Pulau, i., Indon.	248-49	7°22'S	121°47'E
Kalāt, Pak. (kŭ-lät´)	232-33	29°02'N	66°35'E
Kalaw, Mya.	246-47	20°38'N	96°34'E
Kalbarri, Austl.	270-71	27°42'S	114°10'E
Kaledupa, Pulau, i., Indon.	248-49	5°32'S	123°47'E
Kalemie, D.R.C.	267	5°55'S	29°11'E
Kalemyo, Mya.	246-47	23°13'N	94°07'E
Kalevala, Russia	186-87	65°11'N	31°11'E
Kalewa, Mya.	246-47	23°12'N	94°18'E
Kalgoorlie-Boulder, Austl. (kăl-gōōr´lē-bōld´ēr)	270-71	30°44'S	121°27'E
Kalibo, Phil.	250	11°43'N	122°23'E
Kalima, D.R.C.	262-63	2°36'S	26°37'E
Kalimantan, i., Asia *see* Borneo	248-49	0°30'N	114°00'E
Kālimpang, India	234-35	27°04'N	88°28'E
Kaliningrad, Russia (kä-lē-nēn´grät)	194-95	54°43'N	20°30'E
Kalisch, Pol. *see* Kalisz	194-95	51°46'N	18°06'E
Kalispell, Mt., U.S. (kăl´ĭ-spĕl)	112-13	48°12'N	114°19'W
Kalisz, Pol. (kä´lēsh)	194-95	51°46'N	18°06'E
Kalixälven, stm., Swe.	184-85	65°53'N	23°03'E
Kalmar, Swe. (käl´mär)	192-93	56°40'N	16°22'E
Kalmarsund, strt., Swe. (käl´mär)	192-93	56°40'N	16°25'E
Kal'mius, stm., Ukr. (käl´myōōs)	202-03	47°05'N	37°34'E
Kalmykia, state, Russia	186-87	46°30'N	45°30'E
Kalmykiya, state, Russia *see* Kalmykia	186-87	46°30'N	45°30'E
Kalpeni Island, i., India	236	10°05'N	73°38'E
Kalsūbai, mtn., India	234-35	19°36'N	73°43'E
Kaluga, Russia (kȧ-lò´gä)	202-03	54°32'N	36°17'E
Kalundborg, Den. (kȧ-lòn'´bôr´)	192-93	55°41'N	11°07'E
Kalush, Ukr. (kä´lòsh)	194-95	49°02'N	24°22'E
Kalyān, India	236	19°16'N	73°08'E
Kalyazin, Russia (käl-yȧ´zēn)	202-03	57°14'N	37°54'E
Kálymnos, i., Grc.	200-01	37°0'N	27°00'E
Kama, stm., Russia (kä´mä)	186-87	55°35'N	51°29'E
Kamaishi, Japan (kä´mä-ē´shē)	244	39°16'N	141°53'E
Kamakura, Japan (kä´mä-kōō´rä)	245	35°19'N	139°33'E
Kama Reservoir, res., Russia (kä´mä rĕ´sēr-vwär) *see* Kamskoye Vodokhranilishche	186-87	58°52'N	56°15'E
Kambarka, Russia	186-87	56°15'N	54°13'E
Kamchatka, Poluostrov, pen., Russia *see* Kamchatka Peninsula	218-19	56°0'N	160°00'E
Kamchatka Peninsula, pen., Russia (kăm-chăt´kȧ pĕ-nĭn´sūlȧ)	218-19	56°0'N	160°00'E
Kamenjak, Rt, c., Cro. (kä´mĕ-nyäk)	200-01	44°46'N	13°55'E
Kamenka, Russia	186-87	53°11'N	44°03'E
Kamen'-na-Obi, Russia (kä-mĭny´nŭ ô´bē)	226	53°48'N	81°20'E
Kāmet, mtn., Asia	234-35	30°54'N	79°37'E
Kam'ianets'-Podil's'kyi, Ukr.	194-95	48°40'N	26°36'E
Kamina, D.R.C.	262-63	8°44'S	25°00'E
Kaminak Lake, lk., Nu., Can.	130-31	62°09'N	95°07'W
Kamino-shima, i., Japan	243	34°35'N	129°25'E
Kaminuriak Lake, lk., Nu., Can.	130-31	62°59'N	95°35'W
Kamituga, D.R.C.	267	3°02'S	28°14'E
Kamloops, B.C., Can. (kăm´lōōps)	132-33	50°40'N	120°20'W
Kampala, nat. cap., Ug. (käm-pä´lä)	267	0°19'N	32°34'E
Kampar, stm., Indon. (käm´pär)	246-47	0°14'N	102°42'E
Kamphaeng Phet, Thai.	246-47	16°28'N	99°32'E
Kâmpóng Cham, Camb.	246-47	12°0'N	105°27'E
Kâmpóng Chhnăng, Camb.	246-47	12°15'N	104°40'E
Kâmpóng Saôm, Camb.	246-47	10°38'N	103°31'E
Kâmpóng Saôm, Chhâk, b., Camb.	246-47	10°50'N	103°32'E
Kâmpóng Thum, Camb. (kŏm´pŏng-tŏm)	246-47	12°42'N	104°54'E
Kâmpôt, Camb. (käm´pōt)	246-47	10°37'N	104°11'E
Kampuchea, nation, Asia *see* Cambodia	206-07	13°0'N	105°00'E
Kamsack, Sk., Can. (kăm´săk)	134-35	51°34'N	101°54'W
Kamskoye Vodokhranilishche, res., Russia	186-87	58°52'N	56°15'E
Kamuela, Hi., U.S. *see* Waimea	127a	20°02'N	155°40'W
Kámuk, Cerro, mtn., C.R. (sĕ´r-rō-kä-mōō´k)	149	9°17'N	83°01'W
Kamyshin, Russia (kä-mwēsh´ĭn)	186-87	50°07'N	45°24'E
Kanaaupscow, stm., Qc., Can.	130-31	53°40'N	76°44'W
Kanab, Ut., U.S. (kăn´ăb)	118-19	37°03'N	112°32'W
Kanab Plateau, plat., U.S. (kăn´ăb plă-tō´)	118-19	36°36'N	112°45'W
Kanagawa, state, Japan (kä´nä-gä´wä)	245	35°30'N	139°15'E
Kananga, D.R.C.	262-63	5°54'S	22°25'E
Kanash, Russia	186-87	55°31'N	47°29'E
Kanawha, stm., W.V., U.S. (kȧ-nô´wȧ)	116-17	38°50'N	82°09'W
Kanazawa, Japan (kä´nȧ-zä´wȧ)	245	36°34'N	136°39'E
Kanchanjanggā, mtn., Asia *see* Kānchenjunga	234-35	27°41'N	88°10'E
Kānchenjunga, mtn., Asia (kĭn-chĭn-jòn´gȧ)	234-35	27°41'N	88°10'E
Kānchipuram, India	236	12°50'N	79°43'E
Kandahār, Afg.	232-33	31°37'N	65°43'E
Kandalaksha, Russia (kȧn-dȧ-lák´shȧ)	184-85	67°09'N	32°24'E
Kandangan, Indon.	248-49	2°48'S	115°16'E
Kandavu, i., Fiji	279f	19°00'S	178°11'E
Kandy, Sri L. (kän´dē)	236	7°18'N	80°38'E
Kane, Pa., U.S. (kān)	116-17	41°40'N	78°48'W
Kāne'ohe, Hi., U.S. (kä-nā-ō´hā)	127a	21°25'N	157°48'W
Kanevskaya, Russia (kȧ-nyĕf´skȧ)	202-03	46°05'N	38°58'E
Kangar, Malay.	246-47	6°27'N	100°12'E
Kangaroo Island, i., Austl. (kăŋ-gȧ-rō´ ī´lănd)	276	35°50'S	137°05'E

Feature (Pronunciation)	Page	Lat.	Long.
Kangāvar, Iran (kŭn´gā-vär)	228-29	34°30'N	47°58'E
Kangding, China	238-39	30°04'N	102°01'E
Kangean, Kepulauan, is., Indon. (käŋ´gē-än)	248-49	6°55'S	115°30'E
Kangean, Pulau, i., Indon.	248-49	6°54'S	115°20'E
Kanggye, Kor., N. (käng´gyĕ)	243	40°58'N	126°36'E
Kangiqsliniq, Nu., Can. see Rankin Inlet	128-29	62°49'N	92°10'W
Kangiqsualujjuaq, Qc., Can.	128-29	58°42'N	65°59'W
Kangiqsujuaq, Qc., Can.	128-29	61°35'N	71°58'W
Kangirsuk, Qc., Can.	128-29	60°02'N	70°01'W
Kangnŭng, Kor., S. (käng´nŏ ng)	243	37°46'N	128°54'E
Kango, Gabon (käɴ-gō)	260-61	0°11'N	10°05'E
Kangrinboqê Feng, mtn., China	234-35	31°04'N	81°18'E
Kangto, mtn., Asia	234-35	27°52'N	92°30'E
Kanhsien, China see Ganzhou	238-39	25°53'N	114°55'E
Kaniama, D.R.C.	262-63	7°33'S	24°10'E
Kanin, Poluostrov, pen., Russia	186-87	68°0'N	45°00'E
Kanin Nos, Mys, c., Russia	186-87	68°39'N	43°17'E
Kankakee, Il., U.S. (kăŋ-kȧ-kē´)	116-17	41°07'N	87°51'W
Kankan, Gui. (käɴ-käɴ) (kän-kän´)	260-61	10°23'N	9°18'W
Kankō, Kor., N. see Hamhŭng	243	39°55'N	127°32'E
Kanmaw Kyun, i., Mya.	246-47	11°40'N	98°28'E
Kannapolis, N.C., U.S. (kăn-ăp´ō-lĭs)	124-25	35°29'N	80°37'W
Kannur, India see Cannanore	236	11°52'N	75°22'E
Kano, Nig. (kä´nō)	260-61	12°01'N	8°30'E
Kānpur, India (kän´pûr)	234-35	26°28'N	80°19'E
Kansas, state, U.S. (kăn´zȧs)	108-09	38°45'N	98°15'W
Kansas City, Ks., U.S. (kăn´zȧs sĭ´tē)	120-21	39°07'N	94°38'W
Kansas City, Mo., U.S. (kăn´zȧs sĭ´tē)	120-21	39°06'N	94°34'W
Kansk, Russia	218-19	56°12'N	95°43'E
Kansu, state, China see Gansu	240-41	37°0'N	103°00'E
Kantang, Thai. (kän´täng´)	246-47	7°24'N	99°32'E
Kanton, i., Kir. see Canton	280-81	2°49'S	171°41'W
Kantunilkin, Mex. (kän-tōō-nēl-kē´n)	148	21°06'N	87°29'W
Kanye, Bots.	264-65	24°59'S	25°19'E
Kaohsiung, Tai. (kä-ô-syông´)	225a	22°38'N	120°17'E
Kaoko Veld, plat., Nmb.	264-65	20°0'S	14°00'E
Kaolack, Sen.	260-61	14°09'N	16°04'W
Kaoma, Zam.	264-65	14°47'S	24°48'E
Kapenguria, Kenya	267	1°09'N	35°01'E
Kapfenberg, Aus. (kȧp´fĕn-bĕrgh)	194-95	47°27'N	15°17'E
Kapingamarangi, at., Micron.	280-81	1°04'N	154°46'E
Kapit, Malay.	248-49	2°00'N	112°56'E
Kapoeta, Sudan	267	4°47'N	33°35'E
Kaposvár, Hung. (kô´pŏsh-vär)	194-95	46°22'N	17°48'E
Kapuas, stm., Indon.	248-49	0°09'S	109°08'E
Kapuas Hulu, Pegunungan, mts., Asia see Upper Kapuas Mountains	248-49	1°15'N	113°30'E
Kapuas Hulu, Pergunungan, mts., Asia see Upper Kapuas Mountains	248-49	1°15'N	113°30'E
Kapuskasing, On., Can.	136-37	49°25'N	82°25'W
Kapuskasing, stm., On., Can.	136-37	49°38'N	82°16'W
Kara, Togo	260-61	9°33'N	1°12'E
Kara, stm., Russia (kärȧ)	186-87	69°07'N	64°45'E
Kara-Balta, Kyrg.	226	42°48'N	73°51'E
Karabogaz, Turkmen.	186-87	41°32'N	52°35'E
Kara-Bogaz-Gol Gulf, b., Turkmen. (kȧ-rä´ bŭ-gäs´ gôl gŭlf)	232-33	41°15'N	53°24'E
Karabük, Tur.	186-87	41°13'N	32°37'E
Karachay, state, Russia see Cherkessia	227	44°0'N	42°00'E
Karachay-Cherkessia, state, Russia see Cherkessia	227	44°0'N	42°00'E
Karachayevo-Cherkesiya, state, Russia see Cherkessia	227	44°0'N	42°00'E
Karachev, Russia (kȧ-rȧ-chôf´)	202-03	53°07'N	34°59'E
Karāchi, Pak. (kȧ-rä´chē)	232-33	24°54'N	67°01'E
Kara Deniz, s., see Black Sea	186-87	43°0'N	35°00'E
Karaginskiy, Ostrov, i., Russia	218-19	58°50'N	164°00'E
Karaginskiy Zaliv, b., Russia	218-19	58°50'N	164°00'E
Karaj, Iran	232-33	35°50'N	50°59'E
Karakax, stm., China	226	38°03'N	80°32'E
Karakelong, Pulau, i., Indon.	248-49	4°16'N	126°49'E
Karakol, Kyrg.	226	42°29'N	78°23'E
Karakoram Range, mts., Asia (kä´rä kō´rŏm rănj)	234-35	35°30'N	77°00'E
Karakorum, hist., Mong.	240-41	47°14'N	102°50'E
Karakorum Shan, mts., Asia see Karakoram Range	234-35	35°30'N	77°00'E
Kara Kum, des., Turkmen. (kärä-kōōm´)	226	39°0'N	60°00'E
Kara-Kum Canal, can., Turkmen. (kärä-kōōm´ kȧ´näl)	232-33	37°34'N	65°41'E
Karakumy, des., Turkmen. see Kara Kum	226	39°0'N	60°00'E
Karaman, Tur. (kä-rä-män´)	186-87	37°11'N	33°13'E
Karamay, China (kär-äm-ā)	226	45°36'N	84°51'E

Feature (Pronunciation)	Page	Lat.	Long.
Karamea Bight, b., N.Z. (kȧ-rȧ-mē´ȧ bīt)	278	41°30'S	171°40'E
Karasburg, Nmb.	264-65	28°01'S	18°45'E
Kara Sea, s., Russia (kärȧ sē)	218-19	76°0'N	80°00'E
Karasuk, Russia	226	53°43'N	78°03'E
Karatau Range, mts., Kaz.	226	43°36'N	68°52'E
Karatsu, Japan (kä´rȧ-tsōō)	245	33°26'N	129°59'E
Karawang, Indon.	248-49	6°18'S	107°18'E
Karbalā', Iraq	228-29	32°37'N	44°02'E
Karcag, Hung. (kär´tsäg)	194-95	47°19'N	20°56'E
Kardeljevo, Cro.	200-01	43°04'N	17°26'E
Kärdla, Est. (kĕrd´lȧ)	192-93	58°60'N	22°45'E
Kargasok, Russia	218-19	59°04'N	80°50'E
Kargopol', Russia (kär-gō-pōl´')	186-87	61°30'N	38°58'E
Kariba, Zimb.	264-65	16°31'S	28°48'E
Kariba, Lake, res., Afr.	264-65	17°0'S	28°00'E
Karimata, Kepulauan, is., Indon. (kä-rē-mä´tȧ)	246-47	1°25'S	109°05'E
Karimata, Pulau, i., Indon.	246-47	1°36'S	108°55'E
Karimata, Selat, strt., Indon.	248-49	2°05'S	108°40'E
Karīmnagar, India	236	18°26'N	79°09'E
Karimunjawa, Kepulauan, is., Indon. (kä´rē-mōōn-yä´vä)	248-49	5°50'S	110°25'E
Karisimbi, Volcan, vol., Afr.	267	1°30'S	29°27'E
Karkar Island, i., Pap. N. Gui. (kär´kär ī´lȧnd)	277	4°40'S	146°00'E
Karkūk, Iraq see Kirkuk	228-29	35°28'N	44°24'E
Karleby, Fin. see Kokkola	184-85	63°50'N	23°09'E
Karl-Marx-Stadt, Ger. see Chemnitz	194-95	50°50'N	12°56'E
Karlovac, Cro. (kär´lō-väts)	200-01	45°29'N	15°33'E
Karlovo, Blg. (kär´lō-vō)	200-01	42°39'N	24°48'E
Karlovy Vary, Czech Rep. (kär´lō-vē vä´rē)	194-95	50°14'N	12°53'E
Karlshamn, Swe. (kärls´häm)	192-93	56°09'N	14°51'E
Karlskrona, Swe. (kärls´krō-nä)	192-93	56°10'N	15°36'E
Karlsruhe, Ger. (kärls´rōō-ĕ)	194-95	49°01'N	8°23'E
Karlstad, Swe. (kärl´städ)	192-93	59°23'N	13°31'E
Karmøy, i., Nor. (kärm-ûe)	192-93	59°15'N	5°15'E
Karnāl, India	234-35	29°41'N	76°59'E
Karnātaka, state, India	236	14°0'N	76°00'E
Karonga, Malawi (kȧ-rōŋ´gȧ)	264-65	9°55'S	33°56'E
Kárpathos, i., Grc.	188-89	35°41'N	27°09'E
Karpaty, mts., Eur. see Carpathian Mountains	186-87	48°0'N	24°00'E
Karpinsk, Russia (kär´pĭnsk)	186-87	59°46'N	60°00'E
Karpogory, Russia	186-87	64°00'N	44°23'E
Karratha, Austl.	270-71	20°43'S	116°48'E
Kars, Tur. (kärs)	227	40°36'N	43°05'E
Karshi, Uzb. (kär´shē) see Qarshi	232-33	38°52'N	65°48'E
Karskoye More, s., Russia see Kara Sea	218-19	76°0'N	80°00'E
Kartaly, Russia (kȧr´tȧ lē)	226	53°03'N	60°39'E
Karumba, Austl.	277	17°28'S	140°51'E
Karūr, India	236	10°57'N	78°05'E
Kārwār, India	236	14°48'N	74°08'E
Kasai, stm., Afr.	262-63	3°02'S	16°56'E
Kasama, Zam. (kȧ-sä´mȧ)	264-65	10°12'S	31°11'E
Kasar, Ras, c., Afr. see Kasr, Ra's	266	18°01'N	38°34'E
Kasba Lake, l., Can.	130-31	60°18'N	102°07'W
Kasba-Tadla, Mor. (käs´bȧ-täd´lȧ)	269a	32°37'N	6°16'W
Kaschau, Slvk. see Košice	194-95	48°43'N	21°16'E
Kasenga, D.R.C. (kȧ-sĕŋ´gȧ)	262-63	10°22'S	28°37'E
Kasese, Ug.	267	0°10'N	30°05'E
Kāshān, Iran (kä-shän´)	232-33	33°59'N	51°26'E
Kashgar, China (käsh-gär) see Kashi	226	39°28'N	75°59'E
Kashi, China (kä-shr)	226	39°28'N	75°59'E
Kashihara, Japan (kä´shē-hä´rä)	245	34°30'N	135°48'E
Kashin, Russia (kä-shēn´)	202-03	57°22'N	37°37'E
Kashira, Russia (kä-shē´rȧ)	202-03	54°51'N	38°10'E
Kashiwazaki, Japan (kä´shē-wä-zä´kē)	245	37°22'N	138°33'E
Kāshmar, Iran	232-33	35°13'N	58°28'E
Kasia, India	234-35	26°45'N	83°55'E
Kasimov, Russia (kȧ-sē´môf)	186-87	54°56'N	41°23'E
Kaskaskia, stm., Il., U.S. (kăs-kăs´kĭ-ȧ)	120-21	37°58'N	89°57'W
Kaskattama, stm., Mb., Can. (kăs-kȧ-tä´mȧ)	134-35	57°03'N	90°05'W
Kasongo, D.R.C. (kä-sôŋ´gō)	262-63	4°27'S	26°40'E
Kasongo-Lunda, D.R.C.	262-63	6°29'S	16°50'E
Kaspīy Mangy oypaty, pl., see Caspian Depression	186-87	48°0'N	52°00'E
Kaspiysk, Russia	227	42°53'N	47°38'E
Kaspiyskiy, Russia	186-87	45°24'N	47°21'E
Kaspiyskoye More, lk., see Caspian Sea	226	41°18'N	50°59'E
Kasr, Ra's, c., Afr.	266	18°01'N	38°34'E
Kassa, Slvk. see Košice	194-95	48°43'N	21°16'E

Feature (Pronunciation)	Page	Lat.	Long.
Kassalā, Sudan	266	15°27'N	36°23'E
Kassel, Ger. (käs´ĕl)	194-95	51°19'N	9°29'E
Kasserine, Tun.	184-85	35°09'N	8°50'E
Kasson, Mn., U.S. (kăs´ŭn)	114-15	44°02'N	92°46'W
Kastamonu, Tur. (kä-stȧ-mō´nōō)	186-87	41°23'N	33°47'E
Kastellorizo, i., Grc. see Megísti	188-89	36°08'N	29°36'E
Kastoría, Grc. (kȧs-tō´rĭ-ȧ)	200-01	40°32'N	21°17'E
Kasulu, Tan.	267	4°34'S	30°06'E
Kasungu, Malawi	264-65	13°03'S	33°28'E
Kasūr, Pak.	232-33	31°07'N	74°27'E
Katahdin, Mount, mtn., Me., U.S. (mount kȧ-tä´dĭn)	117a	45°55'N	68°55'W
Katanda, D.R.C.	267	0°50'S	29°22'E
Katanga, hist. reg., D.R.C. (kȧ-täŋ´gȧ)	262-63	10°0'S	26°00'E
Katanga, stm., Russia	218-19	60°09'N	102°14'E
Katanning, Austl. (kȧ-tăn´ĭng)	270-71	33°42'S	117°33'E
Katchall Island, i., India	246-47	7°55'N	93°23'E
Katha, Mya.	238-39	24°10'N	96°20'E
Katherine, Austl. (kăth´ĕr-ĭn)	270-71	14°29'S	132°16'E
Kāthiāwār Peninsula, pen., India (kä´tyȧ-wär´ pē-nĭn´sūlȧ)	234-35	22°0'N	71°00'E
Kāṭhmāṇḍāu, nat. cap., Nepal see Kathmandu	234-35	27°42'N	85°19'E
Kathmandu, nat. cap., Nepal (kăt-män-dōō´)	234-35	27°42'N	85°19'E
Katihār, India	234-35	25°33'N	87°34'E
Katima Mulilo, Nmb.	264-65	17°30'S	24°16'E
Ka Tiriti o te Moana, mts., N.Z. see Southern Alps	278	43°30'S	170°30'E
Katmai National Park and Preserve, n.p., Ak., U.S. (kăt´mī năsh´ŭn-ăl pärk ănd prĭ-zûrv´)	126	58°30'N	155°05'W
Kātmāndu, nat. cap., Nepal see Kathmandu	234-35	27°42'N	85°19'E
Katni, India see Murwāra	234-35	23°50'N	80°24'E
Katoomba, Austl.	276	33°43'S	150°18'E
Katowice, Pol.	194-95	50°16'N	19°01'E
Katrineholm, Swe. (kȧ-trē´nĕ-hōlm)	192-93	58°59'N	16°12'E
Katsina, Nig. (kät´sē-nȧ)	260-61	12°60'N	7°36'E
Kattaqo'rg'on, Uzb.	232-33	39°55'N	66°16'E
Kattegat, strt., Eur. (kăt´ĕ-gät)	192-93	57°0'N	11°00'E
Kattegatt, strt., Eur. see Kattegat	192-93	57°0'N	11°00'E
Kattowitz, Pol. see Katowice	194-95	50°16'N	19°01'E
Katun', stm., Russia (kȧ-tōn´)	226	52°26'N	85°00'E
Kaua'i, i., Hi., U.S.	127a	22°0'N	159°30'W
Kaufbeuren, Ger. (kouf´boi-rĕn)	194-95	47°53'N	10°37'E
Kaufman, Tx., U.S. (kôf´măn)	120-21	32°35'N	96°20'W
Kaukauna, Wi., U.S. (kô-kô´nȧ)	116-17	44°16'N	88°16'W
Kaukau Veld, plat., Afr.	264-65	19°30'S	20°30'E
Kaunakakai, Hi., U.S. (kä´ōō-nȧ-kä´kī)	127a	21°06'N	157°01'W
Kaunas, Lith. (kou´nȧs) (kôv´nō)	192-93	54°54'N	23°54'E
Kauriālā, stm., Asia see Ghāghara	234-35	25°45'N	84°48'E
Kau-ye Kyun, i., Mya.	246-47	10°60'N	98°31'E
Kavála, Grc. (kä-vä´lä)	200-01	40°57'N	24°24'E
Kavalerovo, Russia	244	44°16'N	135°03'E
Kavaratti Island, i., India	236	10°34'N	72°38'E
Kavieng, Pap. N. Gui. (kä-vē-ĕng´)	277	2°34'S	150°48'E
Kavīr, Dasht-e, des., Iran (dŭsht-ē-ka-vēr´)	232-33	34°40'N	54°30'E
Kavkasioni, mts., see Caucasus Mountains	227	42°38'N	45°00'E
Kawaguchi, Japan (kä-wä-gōō-chē)	245	35°48'N	139°43'E
Kawambwa, Zam.	264-65	9°48'S	29°05'E
Kawasaki, Japan (kä-wä-sä´kē)	245	35°32'N	139°42'E
Kaxgar, stm., China	226	39°25'N	76°26'E
Kayak Island, i., Ak., U.S.	126	59°54'N	144°27'W
Kayan, stm., Indon.	248-49	2°55'N	117°35'E
Kaycee, Wy., U.S. (kā-sē´)	112-13	43°43'N	106°40'W
Kayes, Mali (kāz)	258-59	14°27'N	11°26'W
Kayoa, Pulau, i., Indon.	248-49	0°04'S	127°24'E
Kayseri, Tur. (kī´sĕ-rē)	186-87	38°44'N	35°29'E
Kayuagung, Indon.	246-47	3°23'S	104°50'E
Kazakh Hills, hills, Kaz. (kä-zäk´ hĭlz)	226	49°0'N	72°00'E
Kazakhstan, nation, Asia (kä-zäk-stän´)	206-07	47°0'N	76°00'E
Kazan', Russia (kȧ-zän´)	186-87	55°51'N	49°04'E
Kazan, stm., Can.	130-31	64°10'N	95°22'W
Kazanka, Ukr. (kȧ-zän´kȧ)	202-03	47°50'N	32°50'E
Kazanlŭk, Blg. (kȧ´zȧn-lĕk)	200-01	42°37'N	25°24'E
Kazan-rettō, is., Japan	280-81	25°0'N	141°00'E
Kazbek, Gora, vol., (gȧ-rä´ kȧz-bĕk´)	227	42°42'N	44°31'E
Kāzerūn, Iran	230-31	29°37'N	51°39'E
Kazincbarcika, Hung. (kô´zĭnts-bôr-tsĭ-ko)	194-95	48°15'N	20°39'E
Kazvin, Iran see Qazvīn	232-33	36°16'N	49°58'E
Kazym, stm., Russia (kä-zēm´)	218-19	63°53'N	65°53'E

n-sing; ŋ-baŋk; ɴ-nasalized n; nŏd; cŏmmit; ōld; ôbey; ôrder; oi-boil; fōōd; ȯ-as oo in foot; ou-out; s-soft; sh-dish; th-thin; pūre; ůnite; ûrn; stŭd; circŭs; ü-as in French tu; ´-indeterminate vowel.

Feature (Pronunciation)	Page	Lat.	Long.
Kearney, Ne., U.S. (kär′nĭ)	114-15	40°42′N	99°05′W
Keban Barajı, res., Tur.	186-87	38°56′N	38°55′E
Kebnekaise, mtn., Swe. (kĕp′nĕ-kà-ēs′ĕ)	184-85	67°55′N	18°35′E
K'ebrī Dehar, Eth.	262-63	6°45′N	44°17′E
Kech, stm., Pak.	232-33	25°59′N	62°44′E
Kecskemét, Hung. (kĕch′kĕ-māt)	194-95	46°54′N	19°42′E
Kėdainiai, Lith. (kē-dī′nĭ-ĭ)	192-93	55°18′N	23°59′E
Kedgwick, N.B., Can. (kĕdj′wĭk)	138-39	47°38′N	67°23′W
Kediri, Indon.	248-49	7°49′S	112°01′E
Kédougou, Sen.	260-61	12°33′N	12°11′W
Keele Peak, mtn., Yk., Can.	130-31	63°26′N	130°19′W
Keeling Islands, dep., Oc. (kē′lĭng ī′lǎndz) *see* Cocos Islands	224-25	12°10′S	96°55′E
Keelung, Tai. *see* Chilung	225a	25°08′N	121°44′E
Keene, N.H., U.S.	116-17	42°56′N	72°17′W
Keer-Weer, Cape, c., Austl.	277	13°51′S	141°29′E
Keetmanshoop, Nmb. (kāt′mȧns-hōp)	264-65	26°35′S	18°09′E
Keewatin, Mn., U.S. (kē-wä′tĭn)	114-15	47°24′N	93°04′W
Keflavík, Ice.	190a	64°01′N	22°35′W
Ke-hsi Mānsām, Mya.	246-47	21°56′N	97°51′E
Keijō, nat. cap., Kor., S. *see* Seoul	243	37°33′N	127°01′E
Keila, Est. (kā′lȧ)	192-93	59°18′N	24°26′E
Kelang, Pulau, i., Indon.	248-49	3°12′S	127°44′E
Kellogg, Id., U.S. (kĕl′ŏg)	112-13	47°32′N	116°08′W
Kelmė, Lith. (kĕl-mȧ)	192-93	55°38′N	22°55′E
Kélo, Chad	262-63	9°19′N	15°49′E
Kelowna, B.C., Can.	132-33	49°53′N	119°29′W
Keluang, Malay.	246-47	2°02′N	103°20′E
Kem', Russia (kĕm)	186-87	64°57′N	34°36′E
Kemano, B.C., Can.	132-33	53°32′N	127°57′W
Kemerovo, Russia	218-19	55°23′N	86°03′E
Kemi, Fin. (kā′mē)	184-85	65°46′N	24°34′E
Kemijärvi, Fin. (kā′mê-yĕr-vē)	184-85	66°43′N	27°24′E
Kemijoki, stm., Fin. (kā′mê-yô′kê)	184-85	65°47′N	24°30′E
Kemmerer, Wy., U.S. (kĕm′ẽr-ẽr)	112-13	41°48′N	110°33′W
Kemper, Fr. *see* Quimper	196-97	47°60′N	4°06′W
Kempsey, Austl. (kĕmp′sê)	276	31°06′S	152°50′E
Kempt, Lac, lk., Qc., Can. (läk kĕmpt)	136-37	47°25′N	74°15′W
Kemul, Kong, mtn., Indon.	248-49	1°52′N	116°13′E
Kenadsa, Alg.	188-89	31°33′N	2°25′W
Kenai, Ak., U.S. (kĕ-nī′)	126	60°34′N	151°13′W
Kenai Fjords National Park, n.p., Ak., U.S. (kĕ-nī′ fē-ôrdz′ nǎsh′ŭn-ǎl pärk)	126	59°50′N	150°09′W
Kenai Peninsula, pen., Ak., U.S. (kĕ-nī′ pĕ-nĭn′sūlȧ)	126	60°10′N	150°00′W
Kendal, Eng., U.K. (kĕn′dǎl)	190-91	54°20′N	2°44′W
Kendall, Cape, c., Nu., Can.	130-31	63°36′N	87°09′W
Kendallville, In., U.S. (kĕn′dǎl-vĭl)	116-17	41°26′N	85°15′W
Kendari, Indon.	248-49	3°57′S	122°36′E
Kenedy, Tx., U.S. (kĕn′ĕ-dĭ)	122-23	28°49′N	97°50′W
Kenema, S.L.	260-61	7°52′N	11°11′W
Kenge, D.R.C.	262-63	4°52′S	16°56′E
Kēng Tung, Mya.	246-47	21°17′N	99°36′E
Kenhardt, S. Afr.	264-65	29°20′S	21°09′E
Kénitra, Mor. (kĕ-nē′trȧ)	269a	34°16′N	6°35′W
Kenmare, N.D., U.S. (kĕn-mâr′)	114-15	48°42′N	102°05′W
Kennebec, stm., Me., U.S. (kĕn-ê-bĕk′)	117a	43°45′N	69°46′W
Kennebunk, Me., U.S. (kĕn-ê-bŭnk′)	116-17	43°23′N	70°33′W
Kennedy, Cape, c., Fl., U.S. *see* Canaveral, Cape	125a	28°27′N	80°32′W
Kenner, La., U.S. (kĕn′ẽr)	124-25	29°59′N	90°15′W
Kennett, Mo., U.S. (kĕn′ĕt)	124-25	36°14′N	90°02′W
Kennewick, Wa., U.S. (kĕn′ê-wĭk)	112-13	46°12′N	119°08′W
Kenney Dam, d., B.C., Can.	132-33	53°37′N	124°58′W
Kenora, On., Can. (kê-nō′rȧ)	134-35	49°46′N	94°29′W
Kenosha, Wi., U.S. (kĕ-nō′shȧ)	116-17	42°34′N	87°50′W
Kenova, W.V., U.S. (kê-nō′vȧ)	116-17	38°24′N	82°35′W
Kent, Oh., U.S. (kĕnt)	116-17	41°09′N	81°21′W
Kent, Wa., U.S. (kĕnt)	112-13	47°23′N	122°12′W
Kentaū, Kaz.	226	43°31′N	68°31′E
Kentland, In., U.S. (kĕnt′lǎnd)	116-17	40°46′N	87°27′W
Kenton, Oh., U.S. (kĕn′tŭn)	116-17	40°39′N	83°36′W
Kent Peninsula, pen., Nu., Can. (kĕnt pĕ-nĭn′sūlȧ)	130-31	68°30′N	107°00′W
Kentucky, state, U.S. (kĕn-tŭk′ĭ)	108-09	37°30′N	85°15′W
Kentucky, stm., Ky., U.S. (kĕn-tŭk′ĭ)	116-17	38°41′N	85°11′W
Kentucky Lake, res., U.S. (kĕn-tŭk′ĭ lāk)	124-25	36°41′N	88°04′W
Kentville, N.S., Can.	138-39	45°05′N	64°29′W
Kentwood, La., U.S. (kĕnt′wo͝od)	124-25	30°56′N	90°31′W
Kentwood, Mi., U.S. (kĕnt′wo͝od)	116-17	42°55′N	85°35′W
Kenya, nation, Afr. (kēn′yȧ)	253	1°0′N	38°00′E
Kenya, Mount, mtn., Kenya (mount kēn′yȧ)	267	0°09′S	37°19′E

Feature (Pronunciation)	Page	Lat.	Long.
Kenyon, Mn., U.S. (kĕn′yŭn)	114-15	44°17′N	93°00′W
Keokuk, Ia., U.S. (kē′ô-kŭk)	120-21	40°24′N	91°23′W
Kępno, Pol. (kǎn′pnō)	194-95	51°17′N	18°00′E
Kerala, state, India	236	10°0′N	76°30′E
Keramian, Pulau, i., Indon.	248-49	5°05′S	114°36′E
Kerang, Austl. (kê-rǎng′)	276	35°44′S	143°55′E
Kerbela, Iraq *see* Karbalā'	228-29	32°37′N	44°02′E
Kerch, Ukr.	202-03	45°21′N	36°28′E
Kerchens'ka protoka, strt., Eur. *see* Kerch Strait	202-03	45°23′N	36°41′E
Kerchenskiy Proliv, strt., Eur. (kĕr-chĕn′skĭ prô′lĭf) *see* Kerch Strait	202-03	45°23′N	36°41′E
Kerch Strait, strt., Eur.	202-03	45°23′N	36°41′E
Keren, Erit.	266	15°46′N	38°27′E
Kerguelen, Îles, i., Afr.	286	49°20′S	69°16′E
Kerguélen, Îles, is., Afr. (ĕl-kĕr′gå-lĕn)	286	49°15′S	69°10′E
Kericho, Kenya	267	0°22′S	35°16′E
Kerinci, Gunung, vol., Indon.	246-47	1°42′S	101°16′E
Keriya, stm., China (kĕ′rê-yä)	234-35	35°17′N	81°34′E
Kerkenna, Îles, is., Tun. (ēl-dĕ-kĕr′kĕn-nä)	258-59	34°44′N	11°12′E
Kerki, Turkmen. (kĕr′kê) *see* Atamyrat	232-33	37°50′N	65°13′E
Kérkyra, Grc.	200-01	39°37′N	19°55′E
Kérkyra, i., Grc.	200-01	39°40′N	19°45′E
Kermadec Islands, is., N.Z. (kẽr-mǎd′ĕk ī′lǎndz)	280-81	30°10′S	178°15′W
Kermān, Iran (kĕr-män′)	230-31	30°17′N	57°04′E
Kermānshāh, Iran	228-29	34°18′N	47°04′E
Kerme, Gulf of, b., Tur. *see* Gökova Körfezi	200-01	36°54′N	27°51′E
Kerrobert, Sk., Can.	134-35	51°55′N	109°08′W
Kerrville, Tx., U.S. (kûr′vĭl)	122-23	30°03′N	99°08′W
Kerulen, stm., Asia (kĕr′o͞o-lĕn)	240-41	48°44′N	117°03′E
Keşan, Tur. (kĕ′shȧn)	200-01	40°52′N	26°39′E
Kesennuma, Japan	244	38°54′N	141°35′E
Keshan, China (kŭ-shän)	240-41	48°01′N	125°52′E
Keshod, India	234-35	21°18′N	70°14′E
Kestell, S. Afr. (kĕs′tĕl)	269c	28°18′S	28°38′E
Keszthely, Hung. (kĕst′hĕl-lĭ)	194-95	46°46′N	17°15′E
Ket', stm., Russia (kyĕt)	218-19	58°55′N	81°32′E
Keta, Ozero, lk., Russia	218-19	68°44′N	90°00′E
Ketapang, Indon. (kĕ-tȧ-päng′)	248-49	1°52′S	109°58′E
Ketchikan, Ak., U.S. (kĕch-ĭ-kǎn′)	126	55°20′N	131°35′W
Ketchum, Id., U.S.	112-13	43°41′N	114°23′W
Ketoy, Ostrov, i., Russia	218-19	47°20′N	152°28′E
Kettering, Eng., U.K. (kĕt′ẽr-ĭng)	190-91	52°24′N	0°45′W
Kettering, Oh., U.S. (kĕt′ẽr-ĭng)	116-17	39°42′N	84°10′W
Kewanee, Il., U.S. (kê-wä′nê)	116-17	41°14′N	89°55′W
Kewaunee, Wi., U.S. (kê-wô′nê)	116-17	44°27′N	87°30′W
Keweenaw Peninsula, pen., Mi., U.S. (kē′wê-nô pĕ-nĭn′sūlȧ)	114-15	47°12′N	88°25′W
Keweenaw Point, c., Mi., U.S.	114-15	47°27′N	87°50′W
Key Largo, i., Fl., U.S.	125a	25°11′N	80°22′W
Keyser, W.V., U.S. (kī′sẽr)	116-17	39°26′N	78°59′W
Key West, Fl., U.S. (kē wĕst′)	125a	24°33′N	81°47′W
Kgalagadi Transfrontier Park, n.p., Afr.	264-65	25°23′S	20°44′E
Khabarovsk, Russia (kä-bä′rôfsk)	240-41	48°26′N	135°08′E
Khairpur, Pak.	232-33	27°32′N	68°45′E
Khajurāho, India	234-35	24°50′N	79°58′E
Khakhea, Bots.	264-65	24°45′S	23°31′E
Khal'mer-Yu, Russia (kŭl-myĕr′-yo͞o′)	186-87	67°57′N	64°45′E
Khambhāt, India	234-35	22°19′N	72°37′E
Khambhāt, Gulf of, b., India	234-35	20°57′N	72°26′E
Khamīs Mushayṭ, Sau. Ar.	266	18°18′N	42°44′E
Khammam, India	236	17°15′N	80°09′E
Khānābād, Afg.	232-33	36°41′N	69°07′E
Khandwa, India	234-35	21°49′N	76°21′E
Khandyga, Russia	218-19	62°40′N	135°32′E
Khānewāl, Pak.	232-33	30°18′N	71°56′E
Khanka, Lake, lk., Asia (läk кän′kȧ)	244	45°11′N	132°25′E
Khanka, Ozero, lk., Asia *see* Khanka, Lake	244	45°11′N	132°25′E
Khānpur, Pak.	232-33	28°38′N	70°40′E
Khantayskoye Vodokhranilishche, res., Russia	218-19	68°0′N	88°00′E
Khanty-Mansiysk, Russia (кŭn-te′mŭn-sēsk′)	218-19	60°59′N	69°01′E
Khao Laem Reservoir, res., Thai.	246-47	14°55′N	98°33′E
Khapcheranga, Russia	240-41	49°42′N	112°23′E
Kharagpur, India (kŭ-rŭg′po͝or)	234-35	22°20′N	87°20′E
Kharg Island, i., Iran *see* Khārk, Jazīreh-ye	230-31	29°15′N	50°19′E
Khargon, India	234-35	21°50′N	75°36′E

Feature (Pronunciation)	Page	Lat.	Long.
Khārk, Jazīreh-ye, i., Iran	230-31	29°15′N	50°19′E
Kharkiv, Ukr.	202-03	49°60′N	36°14′E
Kharkov, Ukr. *see* Kharkiv	202-03	49°60′N	36°14′E
Kharmanli, Blg. (kȧr-män′lê)	200-01	41°56′N	25°55′E
Khartoum, nat. cap., Sudan (kär-to͞om′)	266	15°35′N	32°32′E
Khasavyurt, Russia	227	43°15′N	46°35′E
Khāsh, stm., Afg.	232-33	30°48′N	61°46′E
Khashm al-Qirbah, Sudan	266	14°58′N	35°55′E
Khaskovo, Blg. (кȧs′kô-vô)	200-01	41°56′N	25°34′E
Khatanga, Russia (кȧ-tän′gȧ)	218-19	71°58′N	102°30′E
Khatangskiy Zaliv, b., Russia (kä-täŋ′g-skê zä′lĭf)	208-09	73°35′N	109°45′E
Khatt, Oued al, stm., W. Sah.	258-59	26°55′N	13°03′W
Khaybar, Kowtal-e, p., Asia *see* Khyber Pass	232-33	34°06′N	71°07′E
Khazar, Daryā-ye, lk., *see* Caspian Sea	226	41°18′N	50°59′E
Khemis Miliana, Alg.	269b	36°16′N	2°13′E
Khersān, stm., Iran	232-33	31°34′N	50°22′E
Kherson, Ukr. (кĕr-sôn′)	202-03	46°38′N	32°35′E
Kherson, co., Ukr. (кĕr-sôn′)	202-03	46°45′N	33°30′E
Kheta, stm., Russia	218-19	71°55′N	102°06′E
Khilok, Russia	240-41	51°21′N	110°27′E
Khilok, stm., Russia	240-41	51°19′N	106°59′E
Khimki, Russia (кēm′kĭ)	202-03	55°54′N	37°26′E
Khiwa, Uzb.	232-33	41°24′N	60°22′E
Khmel'nyts'kyi, Ukr.	194-95	49°26′N	27°01′E
Khodzheyli, Uzb. *see* Khŭjayli	226	42°48′N	59°25′E
Kholm, Afg.	232-33	36°41′N	67°42′E
Kholm, Russia (кôlm)	192-93	57°09′N	31°11′E
Kholmsk, Russia (кŭlmsk)	244	47°03′N	142°03′E
Khomeynīshahr, Iran	232-33	32°41′N	51°32′E
Khong, stm., Asia *see* Mekong	246-47	10°33′N	105°27′E
Khong, stm., Asia *see* Salween	208-09	16°33′N	97°40′E
Salween, stm., Asia	208-09	16°33′N	97°40′E
Khon Kaen, Thai.	246-47	16°27′N	102°50′E
Khoper, stm., Russia (кô′pẽr)	186-87	49°37′N	42°19′E
Khor, stm., Russia (кôr′)	240-41	47°49′N	134°41′E
Khorixas, Nmb.	264-65	20°21′S	14°59′E
Khorol, Ukr. (кô′rôl)	202-03	49°47′N	33°17′E
Khorol, stm., Ukr. (кô′rôl)	202-03	49°28′N	33°47′E
Khorramābād, Iran	228-29	33°29′N	48°21′E
Khorramshahr, Iran (kô-ram′shär)	228-29	30°25′N	48°11′E
Khorugh, Taj.	232-33	37°29′N	71°33′E
Khouribga, Mor.	269a	32°53′N	6°55′W
Khromtaū, Kaz.	226	50°15′N	58°27′E
Khudzhand, Taj. *see* Khujand	232-33	40°17′N	69°39′E
Khujand, Taj.	232-33	40°17′N	69°39′E
Khŭjayli, Uzb.	226	42°48′N	59°25′E
Khulna, Bngl.	234-35	22°49′N	89°34′E
Khunjerab Pass, p., Asia	232-33	36°52′N	75°28′E
Khust, Ukr. (kôst)	194-95	48°11′N	23°18′E
Khuzdār, Pak.	232-33	27°48′N	66°37′E
Khvalynsk, Russia (кvȧ-lĭnsk′)	186-87	52°29′N	48°05′E
Khvoy, Iran	227	38°33′N	44°58′E
Khyber Pass, p., Asia (kī′bẽr pȧs)	232-33	34°06′N	71°07′E
Kiamba, Phil.	250	5°60′N	124°37′E
Kiambi, D.R.C. (kyäm′bê)	267	7°19′S	28°01′E
Kiamichi, stm., Ok., U.S. (kyȧ-mē′chê)	120-21	33°57′N	95°14′W
Kiamusze, China *see* Jiamusi	244	46°48′N	130°22′E
Kiangarow, Mount, mtn., Austl.	276	26°50′S	151°32′E
Kiangsi, state, China *see* Jiangxi	238-39	28°0′N	116°00′E
Kiangsu, state, China *see* Jiangsu	238-39	33°0′N	120°00′E
Kiantajärvi, lk., Fin. (kyȧn′tȧ-yĕr-vē)	186-87	65°03′N	29°07′E
Kibombo, D.R.C.	262-63	3°54′S	25°55′E
Kibre Mengist, Eth.	262-63	5°52′N	39°00′E
Kıbrıs, nation, Asia *see* Cyprus	228-29	35°0′N	33°00′E
Kičevo, Mac. (kē′chĕ-vô)	200-01	41°32′N	20°57′E
Kıcık Qafqaz daqları, mts., Asia *see* Lesser Caucasus	227	40°60′N	44°35′E
Kicking Horse Pass, p., Can.	132-33	51°27′N	116°20′W
Kidal, Mali (kê-dȧl′)	258-59	18°26′N	1°24′E
Kiel, Ger. (kēl)	194-95	54°19′N	10°07′E
Kiel Canal, can., Ger. (kēl kä-näl′)	194-95	53°54′N	9°10′E
Kielce, Pol. (kyĕl′tsĕ)	194-95	50°53′N	20°38′E
Kiev, nat. cap., Ukr. (kē′ĕf) (kē′ĕv)	202-03	50°26′N	30°30′E
Kiev Reservoir, res., Ukr. (kē′ĕf rĕ′sẽr-vwär) (kē′ĕv rĕ′sẽr-vwär) *see* Kyïvs'ke vodoskhovyshche	202-03	50°51′N	30°32′E
Kiffa, Maur. (kēf′ȧ)	258-59	16°37′N	11°24′W
Kigali, nat. cap., Rw. (kê-gä′lê)	267	1°56′S	30°04′E
Kigoma, Tan. (kê-gō′mä)	267	4°53′S	29°37′E
Kiirun, Tai. *see* Chilung	225a	25°08′N	121°44′E
Kii-suidō, strt., Japan (kē so͞o-ê′dō)	245	33°55′N	134°55′E
Kikládes, is., Grc. *see* Cyclades	200-01	37°30′N	25°00′E
Kikori, stm., Pap. N. Gui.	277	7°22′S	144°14′E
Kikwit, D.R.C. (kē′kwēt)	262-63	5°02′S	18°49′E

Feature (Pronunciation)	Page	Lat.	Long.
Kil, Swe. (kēl)	192-93	59°31'N	13°19'E
Kīlauea, Hi., U.S. (kē-lä-ōō-ā'ä)	127a	22°13'N	159°25'W
Kili, i., Marsh. Is.	280-81	5°39'N	169°07'E
Kilimanjaro, mtn., Tan. (kyl-ē-măn-jä'rō)	267	3°04'S	37°22'E
Kilimatinde, Tan. (kĭl-ē-mä-tĭn'då)	267	5°52'S	34°58'E
Kilingi-Nõmme, Est. (kē'lĭn-gē-nôm'mĕ)	192-93	58°09'N	24°58'E
Kilis, Tur. (kē'lēs)	228-29	36°43'N	37°07'E
Kilkenny, Ire. (kĭl-kĕn-ĭ)	190-91	52°39'N	7°15'W
Kilkís, Grc. (kĭl'kĭs)	200-01	40°59'N	22°53'E
Killala, Ire. (kĭ-lă'lȧ)	190-91	54°13'N	9°13'W
Killarney, Mb., Can.	134-35	49°12'N	99°42'W
Killeen, Tx., U.S.	122-23	31°06'N	97°42'W
Kilmarnock, Scot., U.K. (kĭl-mär'nŭk)	190-91	55°36'N	4°30'W
Kilombero, stm., Tan.	262-63	8°31'S	37°22'E
Kilosa, Tan.	267	6°50'S	36°59'E
Kilrush, Ire. (kĭl'rŭsh)	190-91	52°39'N	9°29'W
Kilttān Island, i., India	236	11°29'N	73°00'E
Kimamba, Tan.	267	6°46'S	37°08'E
Kimba, Austl. (kĭm'bȧ)	276	33°07'S	136°26'E
Kimball, Ne., U.S. (kĭm-bȧl)	114-15	41°14'N	103°40'W
Kimball, S.D., U.S. (kĭm-bȧl)	114-15	43°45'N	98°57'W
Kimberley, B.C., Can. (kĭm'bēr-lĭ)	132-33	49°41'N	115°59'W
Kimberley, S. Afr. (kĭm'bēr-lĭ)	264-65	28°44'S	24°45'E
Kimberley, plat., Austl.	272-73	17°0'S	127°00'E
Kimberley, reg., Austl.	272-73	16°0'S	127°00'E
Kimch'aek, Kor., N.	243	40°41'N	129°12'E
Kimch'ŏn, Kor., S.	243	36°07'N	128°07'E
Kimmirut, Nu., Can.	128-29	62°51'N	69°53'W
Kimovsk, Russia	202-03	53°58'N	38°32'E
Kimry, Russia (kĭm'rē)	202-03	56°52'N	37°22'E
Kinabalu, Gunong, mtn., Malay.	248-49	6°05'N	116°33'E
Kincardine, On., Can. (kĭn-kär'dĭn)	136-37	44°10'N	81°38'W
Kincolith, B.C., Can.	132-33	55°00'N	129°57'W
Kinder, La., U.S. (kĭn'dēr)	122-23	30°29'N	92°51'W
Kindersley, Sk., Can. (kĭn'dērz-lē)	134-35	51°29'N	109°10'W
Kindia, Gui. (kĭn'dē-ȧ)	260-61	10°04'N	12°51'W
Kindu, D.R.C.	262-63	2°57'S	25°55'E
Kinel', Russia	186-87	53°14'N	50°38'E
Kineshma, Russia (kē-nĕsh'mȧ)	186-87	57°27'N	42°08'E
Kingaroy, Austl. (kĭŋ'gă-roi)	276	26°33'S	151°51'E
King City, Ca., U.S. (kĭng sĭ'tĭ)	118-19	36°12'N	121°08'W
Kingfisher, Ok., U.S. (kĭng'fĭsh-ēr)	120-21	35°52'N	97°56'W
Kingisepp, Russia (kĭŋ-gē-sep')	192-93	59°22'N	28°37'E
King Island, i., Austl. (kĭng ī'lȧnd)	276	39°50'S	144°00'E
King Island, i., B.C., Can.	132-33	52°12'N	127°42'W
Kingman, Az., U.S. (kĭng'mȧn)	118-19	35°12'N	114°02'W
Kingman, Ks., U.S. (kĭng'mȧn)	120-21	37°39'N	98°07'W
Kings Canyon National Park, n.p., Ca., U.S. (kĭngz kăn'yŭn năsh'ŭn-ăl pärk)	118-19	36°56'N	118°35'W
Kingscote, Austl. (kĭngz'kŭt)	276	35°39'S	137°37'E
King's Lynn, Eng., U.K. (kĭngz lĭn')	190-91	52°46'N	0°24'E
Kings Mountain, N.C., U.S. (kĭngz moun'tĭn)	124-25	35°15'N	81°20'W
Kings Peak, mtn., Ut., U.S. (kĭngz pēk)	112-13	40°46'N	110°22'W
Kingsport, Tn., U.S. (kĭngz'pōrt)	124-25	36°33'N	82°34'W
Kingston, On., Can. (kĭngz'tŭn)	136-37	44°14'N	76°30'W
Kingston, N.Y., U.S.	116-17	41°56'N	74°00'W
Kingston, Pa., U.S. (kĭngz'tŭn)	116-17	41°15'N	75°54'W
Kingston, nat. cap., Jam. (kĭngz'tŭn)	142-43	18°00'N	76°48'W
Kingston Southeast, Austl.	276	36°50'S	139°51'E
Kingston upon Hull, Eng., U.K.	190-91	53°45'N	0°20'W
Kingstown, Ire. *see* Dún Laoghaire	190-91	53°17'N	6°08'W
Kingstown, nat. cap., St. Vin. (kĭngz'toun)	143b	13°09'N	61°14'W
Kingstree, S.C., U.S. (kĭngz'trē)	124-25	33°40'N	79°50'W
Kingsville, Tx., U.S. (kĭngz'vĭl)	122-23	27°31'N	97°51'W
King William Island, i., Nu., Can. (kĭng wĭl'yăm ī'lȧnd)	130-31	69°0'N	97°30'W
King William's Town, S. Afr. (kĭng-wĭl'-yŭmz-toun)	264-65	32°51'S	27°22'E
Kinkala, Congo	262-63	4°22'S	14°46'E
Kinkony, Farihy, lk., Madag.	264-65	16°08'S	45°50'E
Kinnaird Head, c., Scot., U.K. (kĭn-ård'hĕd)	190-91	57°42'N	2°01'W
Kinneret, Yam, lk., Isr. *see* Galilee, Sea of	228-29	32°48'N	35°35'E
Kinngait, Nu., Can. *see* Cape Dorset	128-29	64°14'N	76°33'W
Kinsale, Old Head of, c., Ire. (ōld hĕd ŏv kĭn-săl)	190-91	51°37'N	8°32'W
Kinshasa, nat. cap., D.R.C. (kĭn-shä'sä)	262-63	4°21'S	15°18'E
Kinsley, Ks., U.S. (kĭnz'lĭ)	120-21	37°56'N	99°24'W
Kinston, N.C., U.S. (kĭnz'tŭn)	124-25	35°16'N	77°35'W
Kinyeti, mtn., Sudan	267	3°57'N	32°54'E
Kipawa, Lac, res., Qc., Can.	136-37	46°54'N	78°59'W
Kipengere Range, mts., Tan.	264-65	9°23'S	34°26'E
Kípros, nation, Asia *see* Cyprus	228-29	35°0'N	33°00'E
Kirby, Tx., U.S. (kûr'bĭ)	122-23	29°29'N	98°22'W
Kirbyville, Tx., U.S. (kûr'bĭ-vĭl)	122-23	30°40'N	93°54'W
Kirenga, stm., Russia (kē-rĕŋ'gȧ)	218-19	57°46'N	108°06'E
Kirensk, Russia (kē-rĕnsk')	218-19	57°48'N	108°10'E
Kirghizia, nation, Asia *see* Kyrgyzstan	226	41°30'N	75°00'E
Kirgiziya, nation, Asia *see* Kyrgyzstan	226	41°30'N	75°00'E
Kirgiz Range, mts., Asia	226	42°29'N	73°50'E
Kiribati, nation, Oc. (kē-rä-bäs)	280-81	5°0'S	170°00'W
Kiribati, is., Kir. (kē-rä-bäs)	280-81	0°30'S	174°00'E
Kırıkkale, Tur.	186-87	39°51'N	33°31'E
Kirin, state, China *see* Jilin	240-41	44°0'N	126°00'E
Kirinyaga, mtn., Kenya *see* Kenya, Mount	267	0°09'S	37°19'E
Kiritimati, at., Kir.	280-81	1°48'N	157°19'W
Kiriwina Islands, is., Pap. N. Gui.	277	8°35'S	151°05'E
Kirkcaldy, Scot., U.K. (kēr-kô'dĭ)	190-91	56°07'N	3°10'W
Kirkenes, Nor.	184-85	69°43'N	30°02'E
Kirkland, Wa., U.S. (kûrk'lănd)	112-13	47°40'N	122°12'W
Kirkland Lake, On., Can.	136-37	48°10'N	80°01'W
Kırklareli, Tur. (kêrk'lár-ĕ'lē)	200-01	41°45'N	27°14'E
Kirksville, Mo., U.S. (kûrks'vĭl)	120-21	40°12'N	92°34'W
Kirkuk, Iraq	228-29	35°28'N	44°24'E
Kirkwall, Scot., U.K. (kûrk'wôl)	190c	58°59'N	2°58'W
Kirov, Russia	202-03	54°04'N	34°19'E
Kirov, Russia	186-87	58°36'N	49°40'E
Kirov Bay, b., Azer. (kē'rŭf bā) *see* Qızılağac körfäzi	227	39°05'N	49°01'E
Kirovohrad, Ukr.	202-03	48°31'N	32°16'E
Kirovsk, Russia (kē-rôfsk')	184-85	67°37'N	33°40'E
Kirs, Russia	186-87	59°21'N	52°15'E
Kirsanov, Russia (kēr-sá'nôf)	186-87	52°39'N	42°45'E
Kırşehir, Tur. (kêr-shĕ'hēr)	186-87	39°09'N	34°10'E
Kīrthar Range, mts., Pak. (kĭr-tŭr rănj)	232-33	27°0'N	67°10'E
Kiruna, Swe. (kē-rōō'nä)	184-85	67°51'N	20°16'E
Kirzhach, Russia (kêr-zhák')	202-03	56°09'N	38°52'E
Kisaki, Tan. (kē-sä'kē)	267	7°27'S	37°37'E
Kisangani, D.R.C.	262-63	0°32'N	25°12'E
Kisar, Pulau, i., Indon.	248-49	8°05'S	127°10'E
Kīsh, Jazīreh-ye, i., Iran	230-31	26°32'N	53°56'E
Kishinev, nat. cap., Mol. *see* Chişinău	202-03	47°02'N	28°50'E
Kisii, Kenya	267	0°40'S	34°46'E
Kiska Island, i., Ak., U.S. (kĭs'kä ī'lȧnd)	126a	51°60'N	177°31'E
Kiskitto Lake, lk., Mb., Can. (kĭs-kĭ'tō lāk)	134-35	54°16'N	98°34'W
Kiskunfélegyháza, Hung. (kĭsh'kòn-fā'lĕd-y'hä'zô)	194-95	46°43'N	19°50'E
Kiskunhalas, Hung. (kĭsh'kòn-hô'lôsh)	194-95	46°25'N	19°30'E
Kislovodsk, Russia	227	43°55'N	42°44'E
Kismaayo, Som. *see* Chisimayu	262-63	0°22'S	42°32'E
Kiso, stm., Japan (kē'sō)	245	35°04'N	136°44'E
Kissidougou, Gui. (kē'sē-dōō'gōō)	260-61	9°11'N	10°06'W
Kissimmee, Fl., U.S. (kĭ-sĭm'ē)	125a	28°18'N	81°24'W
Kissimmee, stm., Fl., U.S. (kĭ-sĭm'ē)	125a	27°08'N	80°52'W
Kisumu, Kenya (kē'sōō-mōō)	267	0°05'S	34°46'E
Kita, Mali (kē'tá)	258-59	13°02'N	9°30'W
Kita-Daitō-jima, i., Japan	222-23	25°57'N	131°18'E
Kita-Iō-jima, i., Japan	280-81	25°26'N	141°17'E
Kitakyūshū, Japan	245	33°54'N	130°51'E
Kitale, Kenya	267	1°01'N	34°60'E
Kitami, Japan	244	43°48'N	143°54'E
Kitchener, On., Can. (kĭch'ĕ-nēr)	136-37	43°27'N	80°29'W
Kitega, Bdi. *see* Gitega	267	3°21'S	29°54'E
Kitgum, Ug. (kĭt'gòm)	267	3°17'N	32°52'E
Kitimat, B.C., Can. (kĭ'tĭ-măt)	132-33	54°01'N	128°42'W
Kitimat Ranges, mts., B.C., Can. (kĭ'tĭ-măt rānjĕz)	132-33	53°30'N	128°50'W
Kittanning, Pa., U.S. (kĭ-tăn'ĭng)	116-17	40°49'N	79°31'W
Kittery, Me., U.S. (kĭt'ēr-ĭ)	116-17	43°06'N	70°45'W
Kitty Hawk, N.C., U.S. (kĭt'tē hôk)	124-25	36°04'N	75°44'W
Kitui, Kenya	267	1°22'S	38°01'E
Kitwe, Zam.	264-65	12°49'S	28°13'E
Kivalina, Ak., U.S.	126	67°44'N	164°32'W
Kivu, Lac, lk., Afr.	267	2°03'S	28°54'E
Kiyev, nat. cap., Ukr. *see* Kiev	202-03	50°26'N	30°30'E
Kizel, Russia (kē'zĕl)	186-87	59°04'N	57°39'E
Kızılırmak, stm., Tur.	186-87	41°44'N	35°58'E
Kizlyar, Russia (kĭz-lyär')	227	43°51'N	46°42'E
Kladno, Czech Rep. (kläd'nō)	194-95	50°08'N	14°06'E
Klagenfurt, Aus. (klä'gĕn-fōrt)	194-95	46°38'N	14°19'E
Klaipėda, Lith. (klī'på-då)	192-93	55°43'N	21°08'E
Klamath, stm., U.S. (klăm'ăth)	110-11	41°33'N	124°06'W
Klamath Falls, Or., U.S. (klăm'ăth fôlz)	112-13	42°14'N	121°48'W
Klamath Mountains, mts., U.S. (klăm'ăth moun'tĭnz)	112-13	41°31'N	123°14'W
Klang, Malay.	246-47	3°03'N	101°27'E
Klatovy, Czech Rep. (klá'tô-vê)	194-95	49°24'N	13°18'E
Klausenburg, Rom. *see* Cluj-Napoca	194-95	46°47'N	23°36'E
Klein Karroo, plat., S. Afr. *see* Little Karoo	264-65	33°45'S	21°30'E
Klerksdorp, S. Afr. (klĕrks'dôrp)	269c	26°52'S	26°39'E
Kletnya, Russia (klyĕt'nyȧ)	202-03	53°23'N	33°13'E
Kleve, Ger. (klĕ'fĕ)	194-95	51°47'N	6°09'E
Klimovsk, Russia (klĭ'môfsk)	202-03	55°22'N	37°32'E
Klin, Russia (klĕn)	202-03	56°20'N	36°43'E
Klintehamn, Swe. (klēn'tē-häm)	192-93	57°24'N	18°12'E
Klintsy, Russia (klĭn'tsĭ)	202-03	52°45'N	32°15'E
Klip, stm., S. Afr. (klĭp)	269c	27°03'S	29°04'E
Klosterneuburg, Aus. (klōs-tēr-noi'bōōrgh)	194-95	48°18'N	16°21'E
Kluane National Park and Reserve, n.p., Yk., Can.	128-29	60°45'N	139°30'W
Kluczbork, Pol. (klōōch'bōrk)	194-95	50°58'N	18°14'E
Klyazma, stm., Russia (klyäz'mȧ)	186-87	56°10'N	42°58'E
Klyuchevskaya Sopka, Vulkan, vol., Russia (klyōō-chĕfskä'yä)	218-19	56°04'N	160°38'E
Klyuchi, Russia (klyōō'chĭ)	218-19	56°19'N	160°51'E
Knee Lake, lk., Mb., Can.	134-35	55°06'N	94°36'W
Knight Inlet, b., B.C., Can. (nīt ĭn'lĕt)	132-33	50°42'N	125°43'W
Knin, Cro. (knēn)	200-01	44°02'N	16°12'E
Knossos, hist., Grc.	200a	35°17'N	25°12'E
Knox, In., U.S. (nŏks)	116-17	41°17'N	86°37'W
Knox, Cape, c., B.C., Can.	132-33	54°11'N	133°04'W
Knoxville, Ia., U.S. (nŏks'vĭl)	114-15	41°19'N	93°06'W
Knoxville, Il., U.S. (nŏks'vĭl)	114-15	40°54'N	90°17'W
Knoxville, Tn., U.S. (nŏks'vĭl)	124-25	35°58'N	83°55'W
Kōbe, Japan (kō'bĕ)	245	34°41'N	135°10'E
København, nat. cap., Den. (kû-b'n-houn') *see* Copenhagen	192-93	55°41'N	12°34'E
Koblenz, Ger. (kō'blĕntz)	194-95	50°21'N	7°35'E
Kobroor, Pulau, i., Indon.	224-25	6°12'S	134°32'E
Kobryn, Bela. (kō'brĕn')	194-95	52°13'N	24°21'E
Kobuk, stm., Ak., U.S. (kō'bŭk)	126	66°34'N	161°33'W
Kobuk Valley National Park, n.p., Ak., U.S. (kō'bŭk văl'ē năsh'ŭn-ăl pärk)	126	67°20'N	159°00'W
Kobuleti, Geor. (kô-bò-lyā'tē)	227	41°49'N	41°48'E
Kocaeli, Tur. *see* İzmit	200-01	40°47'N	29°57'E
Kočevje, Slvn. (kô'chāv-ye)	200-01	45°38'N	14°52'E
Koch Bihār, India	234-35	26°19'N	89°27'E
Kochechum, stm., Russia	218-19	64°17'N	100°11'E
Kochi, India *see* Cochin	236	9°56'N	76°15'E
Kōchi, Japan	245	33°33'N	133°32'E
Kodiak, Ak., U.S. (kō'dyăk)	126	57°49'N	152°22'W
Kodiak Island, i., Ak., U.S. (kō'dyăk ī'lȧnd)	126	57°30'N	153°30'W
Koforidua, Ghana (kō fō-rĭ-dōō'ȧ)	260-61	6°05'N	0°17'W
Kōfu, Japan	245	35°39'N	138°34'E
Køge, Den. (kû'gĕ)	192-93	55°27'N	12°11'E
Kogon, Uzb.	232-33	39°43'N	64°33'E
Kōgŭm-do, i., Kor., S.	243	34°27'N	127°11'E
Kohāt, Pak.	232-33	33°35'N	71°27'E
Kohīma, India (kô-ē'mȧ)	234-35	25°39'N	94°06'E
Kohtla-Järve, Est.	192-93	59°24'N	27°15'E
Koidu-Sefagu, S.L.	260-61	8°38'N	10°59'W
Kōje-do, i., Kor., S. (kû'jĕ)	243	34°52'N	128°37'E
Kokiu, China *see* Gejiu	238-39	23°22'N	103°09'E
Kokkola, Fin. (kô'kô-lȧ)	184-85	63°50'N	23°09'E
Kokomo, In., U.S. (kō'kô-mō)	116-17	40°29'N	86°07'W
Koko Nor, lk., China (kō'kô nor) *see* Qinghai Hu	240-41	36°48'N	100°06'E
Kokopo, Pap. N. Gui. (kô-kô'pō)	277	4°20'S	152°14'E
Kökshetaū, Kaz.	226	53°17'N	69°24'E
Koksoak, stm., Qc., Can. (kôk'sô-ăk)	130-31	58°31'N	68°10'W
Kokstad, S. Afr. (kôk'shtät)	264-65	30°33'S	29°26'E
Kokubu, Japan (kō'kōō-bōō)	245	31°44'N	130°46'E
Kolaka, Indon.	248-49	4°05'S	121°37'E
Kola Peninsula, pen., Russia (kō'lȧ pĕ-nĭn'sūlȧ) *see* Kol'skiy Poluostrov	186-87	67°18'N	36°21'E
Kolār, India (kôl-är')	236	13°08'N	78°08'E
Kolberg, Pol. *see* Kołobrzeg	194-95	54°11'N	15°34'E
Kolchugino, Russia (kôl-chó'gē-nô)	202-03	56°18'N	39°23'E
Kolda, Sen.	260-61	12°53'N	14°57'W
Kolding, Den. (kŭl'dĭng)	192-93	55°30'N	9°28'E
Kolguyev, Ostrov, i., Russia (ŏs-trôf' kôl-gó'yĕf)	186-87	69°05'N	49°15'E

n-sing; ŋ-baŋk; ɴ-nasalized n; nŏd; cŏmmit; ōld; ōbey; ôrder; oi-boil; fōōd; ò-as oo in foot; ou-out; s-soft; sh-dish; th-thin; pūre; ûnite; ûrn; stŭd; circŭs; ü-as in French tu; '-indeterminate vowel.

Feature (Pronunciation)	Page	Lat.	Long.
Kolhāpur, India	236	16°42′N	74°13′E
Koliba, stm., Afr. *see* Corubal	258-59	11°57′N	15°03′W
Kolín, Czech Rep. (kō′lēn)	194-95	50°01′N	15°12′E
Kolkasrags, c., Lat. (kôl-käs′rágz)	192-93	57°45′N	22°36′E
Kolkata, India	234-35	22°32′N	88°22′E
Kollam, India	236	8°53′N	76°35′E
Köln, Ger. *see* Cologne	194-95	50°56′N	6°57′E
Koło, Pol. (kô′wô)	194-95	52°12′N	18°39′E
Kołobrzeg, Pol. (kô-lôb′zhĕk)	194-95	54°11′N	15°34′E
Kolombangara Island, i., Sol. Is.	279e	8°0′S	157°05′E
Kolomea, Ukr. *see* Kolomyia	194-95	48°32′N	25°03′E
Kolomna, Russia (kál-ôm′nȧ)	202-03	55°05′N	38°49′E
Kolomyia, Ukr.	194-95	48°32′N	25°03′E
Kolozsvár, Rom. *see* Cluj-Napoca	194-95	46°47′N	23°36′E
Kolpashevo, Russia (kŭl pȧ shô′vȧ)	218-19	58°20′N	82°56′E
Kolpino, Russia (kôl′pē-nô)	192-93	59°45′N	30°36′E
Kol'skiy Poluostrov, pen., Russia	186-87	67°18′N	36°21′E
Kolwezi, D.R.C. (kōl-wĕ′zē)	262-63	10°43′S	25°28′E
Kolyma, stm., Russia (kŭ-lĭ-mä′)	218-19	69°38′N	161°18′E
Kom, Iran *see* Qom	232-33	34°39′N	50°53′E
Komadugu Gana, stm., Nig.	260-61	13°04′N	12°24′E
Komandorski Islands, is., Russia (kŭ-mŭn-dôr′-skē ī′lȧndz)	218-19	55°0′N	167°00′E
Komandorskiye Ostrova, is., Russia *see* Komandorski Islands	218-19	55°0′N	167°00′E
Komárno, Slvk. (kô′mär-nô)	194-95	47°45′N	18°09′E
Komárom, Hung. (kô′mä-rôm)	194-95	47°45′N	18°07′E
Komatsu, Japan (kō-mät′sōō)	245	36°24′N	136°27′E
Kome Island, i., Ug.	267	0°05′S	32°45′E
Komi, state, Russia (kômē)	186-87	64°0′N	54°00′E
Komodo, Pulau, i., Indon.	248-49	8°33′S	119°29′E
Komoé, stm., Afr.	260-61	5°12′N	3°43′W
Kom Ombo, Egypt	268b	24°28′N	32°57′E
Komorn, Slvk. *see* Komárno	194-95	47°45′N	18°09′E
Kompasberg, mtn., S. Afr.	264-65	31°45′S	24°32′E
Komsomolets, Ostrov, i., Russia	218-19	80°30′N	95°00′E
Komsomol'skiy, Russia	186-87	67°32′N	63°59′E
Komsomol'sk-na-Amure, Russia	218-19	50°34′N	137°01′E
Konan, Kor., N. *see* Hŭngdŏki-dong	243	39°50′N	127°38′E
Konārak, India	234-35	19°54′N	86°07′E
Konda, stm., Russia (kôn′dȧ)	218-19	60°43′N	69°40′E
Kondoa, Tan. (kôn-dō′ȧ)	267	4°53′S	35°48′E
Kondopoga, Russia	186-87	62°12′N	34°17′E
Kondoz, Afg.	232-33	36°44′N	68°51′E
Kondoz, stm., Afg.	232-33	37°00′N	68°16′E
Kong, stm., Asia	246-47	13°32′N	105°57′E
Kongolo, D.R.C. (kôŋ′gō′lō)	262-63	5°23′S	27°01′E
Kongsberg, Nor. (kŭngs′bĕrg)	192-93	59°40′N	9°39′E
Kongsvinger, Nor. (kŭngs′vĭŋ-gĕr)	192-93	60°12′N	12°00′E
Kongur Shan, mtn., China	226	38°37′N	75°20′E
Königgrätz, Czech Rep. *see* Hradec Králové	194-95	50°12′N	15°50′E
Königsberg, Russia *see* Kaliningrad	194-95	54°43′N	20°30′E
Konin, Pol. (kô′nyĕn)	194-95	52°14′N	18°16′E
Kónitsa, Grc. (kô′nyĕ′tsȧ)	200-01	40°03′N	20°45′E
Konjic, Bos. (kôn′yēts)	200-01	43°39′N	17°57′E
Konkouré, stm., Gui.	260-61	9°57′N	13°41′W
Konosha, Russia	186-87	60°58′N	40°16′E
Konotop, Ukr. (kô-nô-tôp′)	202-03	51°14′N	33°12′E
Końskie, Pol. (koin′′skyĕ)	194-95	51°12′N	20°25′E
Konstanz, Ger. (kôn′shtänts)	194-95	47°40′N	9°10′E
Kontagora, Nig. (kōn-tȧ-gō′rä)	260-61	10°24′N	5°27′E
Kon Tum, Viet.	246-47	14°21′N	108°01′E
Konya, Tur. (kōn′yȧ)	186-87	37°52′N	32°31′E
Konzhakovskiy Kamen, mtn., Russia	186-87	59°38′N	59°08′E
Kootenay Lake, lk., B.C., Can. (kōō′tē-nȧ lāk)	132-33	49°35′N	116°50′W
Kootenay National Park, n.p., B.C., Can. (kōō′tē-nȧ năsh′ŭn-ăl pärk)	132-33	51°0′N	116°00′W
Kopervik, Nor. (kô′pĕr-vēk)	192-93	59°17′N	5°18′E
Kōpetdag, Gershi, mts., Asia *see* Koppeh Dāgh	232-33	37°50′N	58°00′E
Kopet Mountains, mts., Asia (kō-pĕt′ moun′tĭnz) *see* Koppeh Dāgh	232-33	37°50′N	58°00′E
Köping, Swe. (chû′pĭng)	192-93	59°31′N	15°60′E
Koppeh Dāgh, mts., Asia	232-33	37°50′N	58°00′E
Koprivnica, Cro. (kô′prēv-nē′tsȧ)	200-01	46°10′N	16°50′E
Korāput, India	236	18°49′N	82°43′E
Korat, Thai. *see* Nakhon Ratchasima	246-47	14°58′N	102°06′E
Korça, Alb. *see* Korçë	200-01	40°37′N	20°47′E
Korçë, Alb.	200-01	40°37′N	20°47′E
Korčula, Otok, i., Cro. (ō′tŏk kôr′chōō-lȧ)	200-01	42°57′N	16°50′E

Feature (Pronunciation)	Page	Lat.	Long.
Kordestān, hist. reg., Asia *see* Kurdistan	228-29	37°0′N	45°00′E
Korea, North, nation, Asia (nôrth kô-rē′ ȧ)	206-07	40°0′N	127°00′E
Korea, South, nation, Asia (south kô-rē′ ȧ)	206-07	36°30′N	128°00′E
Korea Bay, b., Asia (kô-rē′ ȧ bā)	240-41	39°0′N	124°00′E
Korea Strait, strt., Asia (kô-rē-ȧ strāt)	243	34°0′N	129°00′E
Korf, Russia	218-19	60°21′N	165°56′E
Korhogo, C. Iv. (kôr-hō′gō)	260-61	9°27′N	5°38′W
Kórinthos, Grc. (kô-rēn′thôs) (kŏr′ĭnth) *see* Corinth	200-01	37°56′N	22°58′E
Koritsa, Alb. *see* Korçë	200-01	40°37′N	20°47′E
Kōriyama, Japan (kō′rē-yä′mä)	245	37°24′N	140°23′E
Korla, China (kōr-lä)	222-23	41°44′N	86°09′E
Koro, i., Fiji	279f	17°16′S	179°24′E
Korocha, Russia (kô-rō′chȧ)	202-03	50°49′N	37°11′E
Koromere, c., N.Z. *see* East Cape	278	37°41′S	178°33′E
Korop, Ukr. (kō′rôp)	202-03	51°34′N	32°56′E
Koro Sea, s., Fiji	279f	18°00′S	179°50′E
Korosten', Ukr. (kô′rôs-tĕn)	202-03	50°57′N	28°39′E
Korsakov, Russia (kôr′sȧ-kôf′)	244	46°38′N	142°47′E
Korsør, Den. (kôrs′ûr′)	192-93	55°21′N	11°09′E
Koryak Mountains, mts., Russia *see* Koryakskoye Nagor'ye	218-19	62°30′N	172°00′E
Koryakskoye Nagor'ye, mts., Russia	218-19	62°30′N	172°00′E
Kos, i., Grc.	200-01	36°50′N	27°10′E
Kosa, Russia	186-87	59°57′N	54°59′E
Kościan, Pol. (kŭsh′tsyán)	194-95	52°05′N	16°39′E
Kosciusko, Ms., U.S. (kŏs-ĭ-ŭs′kō)	124-25	33°03′N	89°35′W
Kosciuszko, Mount, mtn., Austl. (mount kŏs-ĭ-ŭs′kō)	276	36°27′S	148°16′E
Kosciuszko National Park, n.p., Austl.	276	36°15′S	148°24′E
Kosh-Agach, Russia	222-23	50°01′N	88°45′E
Koshu, Kor., S. *see* Kwangju	243	35°09′N	126°54′E
Košice, Slvk. (kō′shē-tsĕ′)	194-95	48°43′N	21°16′E
Koslan, Russia	186-87	63°29′N	48°41′E
Köslin, Pol. *see* Koszalin	194-95	54°11′N	16°12′E
Kosovska Mitrovica, Serb. (kô′sôv-skä′ mē′trô-vē-tsä′)	200-01	42°53′N	20°52′E
Kosrae, i., Micron.	280-81	5°19′N	162°59′E
Kossou, Lac de, res., C. Iv.	260-61	7°15′N	5°42′W
Kostiantynivka, Ukr.	202-03	48°32′N	37°44′E
Kostroma, Russia (kôs-trô-má′)	186-87	57°46′N	40°57′E
Koszalin, Pol. (kô-shä′lĭn)	194-95	54°11′N	16°12′E
Kota, India	234-35	25°11′N	75°50′E
Kotabaru, Indon. *see* Jayapura	277	2°32′S	140°43′E
Kotabaru, Indon.	248-49	3°15′S	116°14′E
Kota Belud, Malay.	248-49	6°29′N	116°33′E
Kota Bharu, Malay.	246-47	6°08′N	102°15′E
Kotabumi, Indon.	246-47	4°49′S	104°53′E
Kotadabok, Indon.	246-47	0°33′S	104°31′E
Kota Kinabalu, Malay.	248-49	5°58′N	116°05′E
Kotamobagu, Indon.	248-49	0°43′N	124°18′E
Kotel, Blg. (kō-tĕl′)	200-01	42°54′N	26°28′E
Kotelnich, Russia (kô-tyĕl′nêch)	186-87	58°18′N	48°19′E
Kotel'nyy, Ostrov, i., Russia (ôs-trôf′ kô-tyĕl′nē)	218-19	75°45′N	138°44′E
Kotka, Fin. (kŏt′kȧ)	192-93	60°28′N	26°56′E
Kotlas, Russia (kôt′lȧs)	186-87	61°15′N	46°39′E
Kotovs'k, Ukr.	202-03	47°45′N	29°32′E
Kottagūdem, India	236	17°32′N	80°38′E
Kottayam, India	236	9°35′N	76°32′E
Kotte, nat. cap., Sri L. *see* Sri Jayewardenepura Kotte	236	6°54′N	79°54′E
Kotto, stm., C.A.R.	262-63	4°14′N	22°03′E
Kotuy, stm., Russia (kô-tōō′)	218-19	71°55′N	102°06′E
Kotzebue, Ak., U.S. (kŏt′sē-bōō)	126	66°53′N	162°36′W
Kotzebue Sound, strt., Ak., U.S. (kŏt′sē-bōō sound)	126	66°20′N	163°00′W
Koudougou, Burkina (kōō-dōō′gōō)	260-61	12°15′N	2°22′W
Kourou, Fr. Gu.	164-65	5°09′N	52°39′W
Kousséri, Camr.	260-61	12°05′N	15°02′E
Koussi, Emi, mtn., Chad (ā′mē ᴋōō-sē′)	258-59	19°50′N	18°30′E
Koutiala, Mali (kōō-tē-ä′lä)	258-59	12°23′N	5°28′W
Kouvola, Fin. (kō′ŏ-vô-lä)	192-93	60°52′N	26°41′E
Kovel', Ukr. (kō′vĕl)	194-95	51°14′N	24°42′E
Kovrov, Russia (kôv-rôf′)	186-87	56°22′N	41°20′E
Kowel, Ukr. *see* Kovel'	194-95	51°14′N	24°42′E
Kowie, S. Afr. *see* Port Alfred	264-65	33°36′S	26°54′E
Kowkcheh, stm., Afg.	232-33	37°10′N	69°24′E
Koygorodok, Russia	186-87	60°26′N	50°59′E
Koyukuk, stm., Ak., U.S. (kō-yōō′kŏk)	126	64°56′N	157°30′W
Kozáni, Grc.	200-01	40°18′N	21°49′E
Kozelets', Ukr. (kôzĕ-lyĕts)	202-03	50°55′N	31°07′E

Feature (Pronunciation)	Page	Lat.	Long.
Kozhikode, India	236	11°16′N	75°47′E
Kozienice, Pol. (kō-zyĕ-nē′tsĕ)	194-95	51°35′N	21°33′E
Kozyrëvsk, Russia	218-19	56°05′N	159°53′E
Kpalimé, Togo	260-61	6°54′N	0°38′E
Kra, Isthmus of, isth., Asia	246-47	10°02′N	98°52′E
Kraai, stm., S. Afr. (krä′ē)	264-65	30°40′S	26°45′E
Krâchéh, Camb.	246-47	12°30′N	106°02′E
Kragujevac, Serb. (krä′gōō′yĕ-váts)	200-01	44°01′N	20°55′E
Krakatoa, i., Indon.	248-49	6°10′S	105°26′E
Kraków, Pol. (krä′kôf)	194-95	50°04′N	19°58′E
Kralendijk, Neth. Ant.	140a	12°09′N	68°16′W
Kraljevo, Serb. (kräl′ye-vô)	200-01	43°44′N	20°41′E
Kramators'k, Ukr.	202-03	48°43′N	37°32′E
Kramfors, Swe. (kräm′fôrs)	184-85	62°56′N	17°47′E
Kraśnik, Pol. (kräsh′nĭk)	194-95	50°56′N	22°13′E
Krasnoarmeysk, Kaz.	226	53°51′N	69°45′E
Krasnoarmeysk, Russia (kräs′nô-ár-mäsk′)	186-87	51°02′N	45°42′E
Krasnoarmeysk, Russia (kräs′nô-ár-mäsk′)	202-03	56°08′N	38°08′E
Krasnodar, Russia (kräs′nô-dár)	202-03	45°02′N	38°59′E
Krasnokamsk, Russia (kräs-nô-kämsk′)	186-87	58°05′N	55°53′E
Krasnoslobodsk, Russia (kräs′nô-slôbôtsk′)	186-87	48°43′N	44°34′E
Krasnoufimsk, Russia (krŭs-nŭ-ōō-fĕmsk′)	186-87	56°37′N	57°46′E
Krasnovishersk, Russia (kräs-nô-vêshersk′)	186-87	60°24′N	57°05′E
Krasnovodsk, Turkmen. *see* Türkmenbaşy	232-33	40°01′N	52°58′E
Krasnoyarsk, Russia (kräs-nô-yársk′)	218-19	56°01′N	92°53′E
Krasnyi Luch, Ukr.	202-03	48°09′N	38°55′E
Krasnystaw, Pol. (kräs-nē-stáf′)	194-95	50°59′N	23°11′E
Krasnyy Kut, Russia (krás-nē kōōt′)	186-87	50°57′N	46°58′E
Kremenchug, Ukr. *see* Kremenchuk	202-03	49°05′N	33°25′E
Kremenchug Reservoir, res., Ukr. (krĕ-mĕn-chōōk′ rĕ′sĕr-vwär) *see* Kremenchuts'ke vodoskhovyshche	202-03	49°20′N	32°30′E
Kremenchuk, Ukr.	202-03	49°05′N	33°25′E
Kremenchuts'ke vodoskhovyshche, res., Ukr.	202-03	49°20′N	32°30′E
Kresttsy, Russia (krâst′sê)	192-93	58°15′N	32°31′E
Kresttsy, Russia (krâst′sê)	202-03	58°23′N	38°59′E
Kretinga, Lith. (krĕ-tĭŋ′gȧ)	192-93	55°53′N	21°15′E
Kribi, Camrn. (krē′bē)	260-61	2°55′N	9°54′E
Krishna, stm., India	236	15°51′N	80°52′E
Krishnanagar, India	234-35	23°24′N	88°30′E
Kristiania, nat. cap., Nor. *see* Oslo	192-93	59°55′N	10°45′E
Kristiansand, Nor. (krĭs-tyán-sän′′)	192-93	58°10′N	8°00′E
Kristianstad, Swe. (krĭs-tyán-städ′)	192-93	56°02′N	14°09′E
Kristiansund, Nor. (krĭs-tyán-sòn′′)	184-85	63°07′N	7°47′E
Kristiinankaupunki, Fin. *see* Kristinestad	192-93	62°16′N	21°22′E
Kristinehamn, Swe. (krês-tē′nĕ-häm′)	192-93	59°20′N	14°08′E
Kristinestad, Fin. (krĭs-tē′nĕ-städh)	192-93	62°16′N	21°22′E
Kríti, i., Grc. *see* Crete	200a	35°13′N	25°00′E
Kritikón Pélagos, s., Grc. *see* Crete, Sea of	188-89	35°54′N	25°01′E
Kriva Palanka, Mac. (krē-vȧ-pȧ-läŋ′kȧ)	200-01	42°12′N	22°20′E
Krivoy Rog, Ukr. *see* Kryvyi Rih	202-03	47°54′N	33°22′E
Križevci, Cro. (krē′zhĕv-tsĭ)	200-01	46°04′N	16°34′E
Krnov, Czech Rep. (k′r′nôf)	194-95	50°05′N	17°42′E
Krokodil, stm., S. Afr. (krŏ′kō-dĭl) *see* Crocodile	269c	24°11′S	26°53′E
Kromy, Russia (krô′mē)	202-03	52°41′N	35°46′E
Krŏng Kaôh Kŏng, Camb.	246-47	11°37′N	102°59′E
Kronshtadt, Russia (krōn′shtät)	192-93	59°59′N	29°47′E
Kronstadt, Rom. *see* Brașov	194-95	45°39′N	25°37′E
Kroonstad, S. Afr. (krōn′shtät)	269c	27°40′S	27°14′E
Kropotkin, Russia (krá-pôt′kĭn)	186-87	45°26′N	40°34′E
Krosno, Pol. (krôs′nô)	194-95	49°42′N	21°46′E
Krotoszyn, Pol. (krô-tō′shĭn)	194-95	51°42′N	17°26′E
Krško, Slvn. (k′rsh′kô)	200-01	45°58′N	15°29′E
Kruger National Park, n.p., S. Afr.	264-65	23°56′S	31°33′E
Krugersdorp, S. Afr. (krōō′gĕrz-dôrp)	269c	26°05′S	27°47′E
Krui, Indon.	246-47	5°14′S	103°56′E
Krung Thep, nat. cap., Thai. *see* Bangkok	246-47	13°45′N	100°31′E
Kruševac, Serb. (krŏ′shĕ-váts)	200-01	43°35′N	21°20′E
Kruzof Island, i., Ak., U.S.	126	57°10′N	135°40′W
Kryms'kyi pivostriv, pen., Ukr. *see* Crimean Peninsula	202-03	45°0′N	34°00′E
Kryvyi Rih, Ukr.	202-03	47°54′N	33°22′E
Ksar-el-Kebir, Mor.	269a	35°01′N	5°54′W

Feature (Pronunciation)	Page	Lat.	Long.
Kualakapuas, Indon.	248-49	3°02'S	114°25'E
Kuala Lumpur, nat. cap., Malay. (kwä'lä lòm-pōōr')	246-47	3°10'N	101°42'E
Kuala Terengganu, Malay.	246-47	5°19'N	103°09'E
Kuandian, China (kůän-dĭěn)	243	40°44'N	124°47'E
Kuantan, Malay.	246-47	3°49'N	103°20'E
Kuban', stm., Russia	186-87	45°21'N	37°25'E
Kuching, Malay. (kōō'chĭng)	248-49	1°34'N	110°20'E
Kudat, Malay. (kōō-dät')	248-49	6°53'N	116°46'E
Kudus, Indon.	248-49	6°48'S	110°50'E
Kudymkar, Russia (kōō-dĭm-kär')	186-87	59°01'N	54°39'E
Kufstein, Aus. (kōōf'shtīn)	194-95	47°35'N	12°10'E
Kugluktuk, Nu., Can.	128-29	67°47'N	115°11'W
Kuiseb, stm., S. Afr.	264-65	22°58'S	14°29'E
Kuito, Ang.	264-65	12°23'S	16°56'E
Kuiu Island, i., Ak., U.S.	126	56°45'N	134°10'W
Kujū-san, vol., Japan (kōō'jò-sän')	245	33°05'N	131°15'E
Kula Kangri, mtn., Bhu.	234-35	28°03'N	90°22'E
Kular, Russia	218-19	70°42'N	134°12'E
Kuldīga, Lat. (kòl'dē-gȧ)	192-93	56°58'N	21°60'E
Kuldja, China see Yining	226	43°55'N	81°18'E
Kulmbach, Ger. (klòlm'bäk)	194-95	50°06'N	11°27'E
Kŭlob, Taj.	232-33	37°55'N	69°47'E
Kuloy, Russia	186-87	61°02'N	42°29'E
Kulsary, Kaz.	186-87	46°59'N	53°59'E
Kulundinskaya Ravnina, pl., Asia see Qulyndy Zhazyghy	226	53°0'N	79°00'E
Kulundinskoye, Ozero, lk., Russia	226	53°0'N	79°36'E
Kuma, stm., Russia (kōō'mä)	186-87	44°57'N	46°27'E
Kumamoto, Japan (kōō'mä-mō'tò)	245	32°48'N	130°43'E
Kumanovo, Mac. (kò-mä'nò-vò)	200-01	42°08'N	21°43'E
Kumasi, Ghana (kōō-mä'sē)	260-61	6°41'N	1°38'W
Kumayri, Arm. see Gyumri	227	40°47'N	43°51'E
Kumba, Camrn. (kòm'bȧ)	260-61	4°38'N	9°26'E
Kumbakonam, India (kòm'bŭ-kō'nŭm)	236	10°57'N	79°23'E
Kumertau, Russia	186-87	52°46'N	55°48'E
Kŭm-gang, stm., Kor., S. (kòm gäng')	243	35°60'N	126°42'E
Kumo, Nig.	260-61	10°01'N	11°13'E
Kumul, China see Hami	240-41	42°50'N	93°31'E
Kunashir, Ostrov, i., Russia (òs-trôf' kōō-nū-shēr')	244	44°10'N	146°00'E
Kunashiri-tō, i., Russia see Kunashir, Ostrov	244	44°10'N	146°00'E
Kunene, stm., Afr.	264-65	17°15'S	11°45'E
Kungälv, Swe. (kŭng'ĕlf)	192-93	57°52'N	11°60'E
Kunghit Island, i., B.C., Can.	132-33	52°06'N	131°04'W
Kungsbacka, Swe. (kŭngs'bä-kȧ)	192-93	57°28'N	12°05'E
Kungur, Russia (kòn-gōōr')	186-87	57°26'N	56°57'E
Kunjirap Daban, p., Asia see Khunjerab Pass	232-33	36°52'N	75°28'E
Kunlun Mountains, mts., China (kōōn-lōōn moun'tĭnz) see Kunlun Shan	222-23	36°30'N	88°00'E
Kunlun Shan, mts., China (kōōn-lōōn shän)	222-23	36°30'N	88°00'E
Kunming, China (kōōn-mĭŋ)	238-39	25°03'N	102°43'E
Kunsan, Kor., S. (kòn'sän')	243	35°59'N	126°43'E
Kununurra, Austl.	270-71	15°46'S	128°44'E
K'uo-k'o-sha-lo Ling, mts., Asia see Kakshaal-Too	226	41°0'N	78°00'E
Kuopio, Fin. (kò-ô'pē-ŏ)	184-85	62°54'N	27°43'E
Kupang, Indon.	248-49	10°11'S	123°35'E
Kup'ians'k, Ukr.	202-03	49°43'N	37°38'E
Kupino, Russia (kōō-pĭ'nò)	226	54°21'N	77°17'E
Kupiškis, Lith. (kò-pĭsh'kĭs)	192-93	55°50'N	24°59'E
Kupreanof Island, i., Ak., U.S.	126	56°45'N	133°31'W
Kuqa, China (kōō-chyä)	226	41°11'N	83°28'E
Kür, stm., Asia	227	39°17'N	49°26'E
Kura, stm., Asia	227	39°17'N	49°26'E
Kurashiki, Japan (kōō'rä-shē'kē)	245	34°36'N	133°46'E
Kuraymah, Sudan	266	18°34'N	31°51'E
Kurayoshi, Japan (kōō'rȧ-yō'shē)	245	35°26'N	133°49'E
Kurchatov, Russia	202-03	51°39'N	35°36'E
Kurdistan, hist. reg., Asia (kûrd'ĭ-stăn)	228-29	37°0'N	45°00'E
Kure, Japan (kōō'rĕ)	245	34°15'N	132°34'E
Kure Atoll, at., Hi., U.S.	127	28°25'N	178°25'W
Kuressaare, Est. (kò'rĕ-sä'rĕ)	192-93	58°15'N	22°30'E
Kureyka, stm., Russia	218-19	66°29'N	87°15'E
Kurgan, Russia (kòr-gän')	226	55°27'N	65°20'E
Kurgan-Tyube, Taj. (kòr-gän' työ'bĕ) see Qŭrghonteppa	232-33	37°50'N	68°47'E
Kuria, i., Kir.	280-81	0°12'N	173°24'E
Kuril Islands, is., Russia (kōō'rĭl ī'lăndz)	218-19	47°14'N	152°18'E
Kuril'skiye Ostrova, is., Russia see Kuril Islands	218-19	47°14'N	152°18'E

Feature (Pronunciation)	Page	Lat.	Long.
Kurmuk, Sudan (kòr'mōōk)	262-63	10°33'N	34°17'E
Kurnool, India (kòr-nōōl')	236	15°50'N	78°02'E
Kuršėnai, Lith. (kòr'shå-nĭ)	192-93	55°59'N	22°56'E
Kursk, Russia (kòrsk)	202-03	51°44'N	36°11'E
Kuruman, S. Afr. (kōō-rōō-män')	264-65	27°28'S	23°26'E
Kurume, Japan (kōō'rò-mĕ)	245	33°19'N	130°31'E
Kurunegala, Sri L.	236	7°29'N	80°22'E
Kushiro, Japan (kōō'shē-rō)	244	42°59'N	144°24'E
Kushka, Turkmen. see Guşgy	232-33	35°16'N	62°21'E
Kushtia, Bngl.	234-35	23°55'N	89°08'E
Kushui, China	240-41	42°10'N	94°22'E
Kuskokwim, stm., Ak., U.S. (kŭs'kô-kwĭm)	126	60°16'N	162°29'W
Kuskokwim Bay, b., Ak., U.S. (kŭs'kô-kwĭm bā)	126	59°32'N	162°52'W
Kuskokwim Mountains, mts., Ak., U.S. (kŭs'kô-kwĭm moun'tĭnz)	126	62°24'N	157°07'W
Kustanay, Kaz. see Qostanay	226	53°12'N	63°37'E
Kūstī, Sudan	266	13°10'N	32°40'E
Kütahya, Tur. (kû-tä'hyȧ)	186-87	39°26'N	29°58'E
Kutaisi, Geor. (kōō-tŭ-ē'sē)	227	42°16'N	42°42'E
Kūt al-Imāra, Iraq see Al-Kūt	228-29	32°30'N	45°49'E
Kutaradja, Indon. see Banda Aceh	246-47	5°33'N	95°19'E
Kutch, Gulf of, b., India see Kachchh, Gulf of	234-35	22°37'N	69°30'E
Kutch, Rann of, reg., Asia	234-35	24°15'N	70°46'E
Kutina, Cro. (kōō'tē-nȧ)	200-01	45°29'N	16°46'E
Kutno, Pol. (kòt'nô)	194-95	52°14'N	19°22'E
Kuujjuaq, Qc., Can.	128-29	58°06'N	68°25'W
Kuusamo, Fin. (kōō'sȧ-mô)	184-85	65°59'N	29°10'E
Kuvshinovo, Russia (kòv-shē'nô-vô)	202-03	57°01'N	34°11'E
Kuwait, nation, Asia (kōō-wāt')	206-07	29°30'N	47°45'E
Kuwait, nat. cap., Kuw. (kōō-wāt')	228-29	29°19'N	47°60'E
Kuybyshev, Russia see Samara	186-87	53°11'N	50°07'E
Kuybyshev, Russia	218-19	55°27'N	78°18'E
Kuybyshev Reservoir, res., Russia (kōō'ē-bĭ-shĭf rĕ'sĕr-vwär) see Kuybyshevskoye Vodokhranilishche	186-87	54°30'N	48°30'E
Kuybyshevskoye Vodokhranilishche, res., Russia	186-87	54°30'N	48°30'E
Kuzneck, Russia see Novokuznetsk	226	53°45'N	87°07'E
Kuznetsk, Russia (kōōz-nyĕtsk')	186-87	53°07'N	46°36'E
Kuznetsovka, Russia (kòz-nyĕt'sôf-kȧ)	192-93	56°19'N	28°34'E
Kvaløya, i., Nor.	184-85	69°40'N	18°30'E
Kwajalein, at., Marsh. Is.	280-81	9°06'N	167°21'E
Kwando, stm., Afr.	264-65	18°30'S	23°36'E
Kwangchow, China see Guangzhou	238-39	23°08'N	113°16'E
Kwangju, Kor., S.	243	35°09'N	126°54'E
Kwango, stm., Afr. (kwäng'ō')	262-63	3°13'S	17°23'E
Kwangsi Chuang, state, China see Guangxi	238-39	24°0'N	109°00'E
Kwangtung, state, China see Guangdong	287	23°0'N	113°00'E
Kweichow, state, China see Guizhou	238-39	27°0'N	107°00'E
Kweihwa, China see Hohhot	240-41	40°49'N	111°39'E
Kweiyang, China see Guiyang	238-39	26°35'N	106°43'E
Kwekwe, Zimb.	264-65	18°56'S	29°49'E
Kwilu, stm., Afr. (kwē'lōō) see Cuilo	262-63	5°53'S	16°35'E
Kwilu, stm., Afr. (kwē'lōō)	260-61	3°23'S	17°23'E
Kyakhta, Russia (kyäk'ta)	240-41	50°21'N	106°27'E
Kyaukpyu, Mya. (chouk'pyoo')	246-47	19°25'N	93°33'E
Kyauktaw, Mya.	246-47	20°49'N	92°59'E
Kyïv, nat. cap., Ukr. (kē'yĕf) see Kiev	202-03	50°26'N	30°30'E
Kyïvs'ke vodoskhovyshche, res., Ukr.	202-03	50°51'N	30°32'E
Kyoga, Lake, lk., Ug.	267	1°30'N	33°00'E
Kyŏmip'o, Kor., N. see Songnim	243	38°44'N	125°38'E
Kyŏngju, Kor., S. (kyŭng'yōō)	243	35°51'N	129°13'E
Kyŏngsŏng, nat. cap., Kor., S. see Seoul	243	37°33'N	127°01'E
Kyōto, Japan (kyō'tō')	245	34°60'N	135°45'E
Kyrgyz Ala Too, mts., Asia see Kirgiz Range	226	42°29'N	73°50'E
Kyrgyzstan, nation, Asia (kûr'-gĭ-stän')	226	41°30'N	75°00'E
Kyūshū, i., Japan (kyōō-shōō)	245	33°0'N	131°00'E
Kyustendil, Blg. (kyòs-tĕn-dĭl')	200-01	42°18'N	22°41'E
Kyzyl, Russia (kĭ zĭl)	222-23	51°43'N	94°24'E
Kyzylkum, des., Asia see Qyzylqum	226	42°0'N	64°00'E
Kyzyl-Kyya, Kyrg.	232-33	40°15'N	72°07'E

L

Feature (Pronunciation)	Page	Lat.	Long.
Läänemeri, s., Eur. see Baltic Sea	192-93	57°0'N	19°00'E

Feature (Pronunciation)	Page	Lat.	Long.
Laascaanood, Som.	262-63	8°29'N	47°21'E
La Asunción, Ven. (lä ä-sōōn-syōn')	164-65	11°01'N	63°52'W
Laayoune, nat. cap., W. Sah. (lā-yōōn') (lā-yōōn')	258-59	27°10'N	13°12'W
La Baie, Qc., Can.	136-37	48°20'N	70°53'W
La Banda, Arg. (lä bän'dä)	168-69	27°44'S	64°15'W
La Barca, Mex. (lä bär'kȧ)	146-47	20°17'N	102°33'W
Labé, Gui.	260-61	11°19'N	12°17'W
Labe, stm., Eur. (lä'bĕ) see Elbe	194-95	53°53'N	9°01'E
Labian, Tanjong, c., Malay.	248-49	5°19'N	119°16'E
Labinsk, Russia	186-87	44°38'N	40°44'E
Labis, Malay. (läb'ĭs)	246-47	2°23'N	103°01'E
Labouheyre, Fr. (lȧ-bōō-âr')	196-97	44°13'N	0°55'W
Laboulaye, Arg. (lä-bô'ōō-lä-yĕ)	173	34°07'S	63°24'W
Labrador City, Nf., Can.	128-29	52°56'N	66°54'W
Labrador Sea, s., N.A. (lăb'rȧ-dôr sē)	86	57°0'N	53°00'W
Lábrea, Braz. (lä-brā'ä)	166-67	7°16'S	64°47'W
Labuan, Malay.	248-49	5°17'N	115°15'E
Labuha, Indon.	248-49	0°39'S	127°30'E
Labuk, stm., Malay.	248-49	5°53'N	117°30'E
La Calera, Chile (lä-kä-lĕ-rä)	163e	32°46'S	71°12'W
Lacantum, stm., Mex. (lä-kän-tōō'm)	148	16°33'N	90°41'W
La Carlota, Arg.	173	33°26'S	63°17'W
La Carolina, Spain (lä kä-rô-lē'nä)	198-99	38°17'N	3°37'W
Laccadive, Minicoy and Amīndīvi Islands, state, India see Lakshadweep	236	10°0'N	73°00'E
Laccadive Islands, is., India see Lakshadweep	236	10°0'N	73°00'E
La Ceiba, Hond. (lȧ sēbä)	149	15°46'N	86°48'W
Lacha, Ozero, lk., Russia (ô'zĕ-rô lȧ'chä)	186-87	61°20'N	38°48'E
La Chaux-de-Fonds, Switz. (lȧ shō dē-fôN')	194-95	47°06'N	6°50'E
Lachlan, stm., Austl. (läk'lăn)	276	34°21'S	143°53'E
La Chorrera, Col.	164-65	1°12'S	72°55'W
La Chorrera, Pan. (lȧchŏr-rā'rä)	150	8°53'N	79°47'W
Lachute, Qc., Can. (lȧ-shōōt')	136-37	45°39'N	74°21'W
La Ciotat, Fr. (lȧ syô-tȧ')	196-97	43°11'N	5°36'E
Lackawanna, N.Y., U.S. (lak-ȧ-wŏn'ȧ)	116-17	42°49'N	78°51'W
Lac La Biche, Ab., Can.	132-33	54°46'N	111°58'W
Lac-Mégantic, Qc., Can.	138-39	45°34'N	70°53'W
La Columna, mtn., Ven. see Bolívar, Pico	164-65	8°33'N	71°01'W
Lacombe, Ab., Can.	132-33	52°28'N	113°45'W
Laconia, N.H., U.S. (lȧ-kō'nĭ-ȧ)	116-17	43°32'N	71°28'W
La Coruña, Spain see A Coruña	198-99	43°22'N	8°25'W
La Crosse, Wi., U.S. (lȧ-krôs')	114-15	43°48'N	91°14'W
La Désirade, i., Guad.	143b	16°19'N	61°03'W
Ladispoli, Italy (lä-dē's-pô-lē)	200-01	41°57'N	12°05'E
Lādīz, Iran	230-31	28°55'N	61°18'E
Ladoga, Lake, lk., Russia (läk lä'-dȧ-gȧ)	192-93	61°0'N	31°30'E
La Dorada, Col.	163c	5°27'N	74°41'W
Ladozhskoye Ozero, lk., Russia see Ladoga, Lake	192-93	61°0'N	31°30'E
Ladybrand, S. Afr.	269c	29°11'S	27°27'E
Ladysmith, B.C., Can. (lā'dĭ-smĭth)	132-33	48°58'N	123°48'W
Ladysmith, S. Afr. (lā'dĭ-smĭth)	269c	28°34'S	29°46'E
Ladysmith, Wi., U.S. (lā'dĭ-smĭth)	114-15	45°28'N	91°06'W
Lae, Pap. N. Gui. (lä'ā)	277	6°43'S	146°59'E
Læsø, i., Den.	192-93	57°16'N	11°01'E
La Esperanza, Hond. (lä ĕs-pȧ-rän'zä)	149	14°19'N	88°11'W
Lafayette, Al., U.S. (lä-fā-yĕt')	124-25	32°54'N	85°24'W
Lafayette, Co., U.S. (lä-fā-yĕt')	120-21	39°60'N	105°05'W
Lafayette, In., U.S. (lä-fā-yĕt')	116-17	40°25'N	86°53'W
Lafayette, La., U.S. (lä-fā-yĕt')	124-25	30°13'N	92°02'W
Lafia, Nig.	260-61	8°30'N	8°31'E
La Flèche, Fr. (lȧ flāsh')	196-97	47°42'N	0°04'W
Lagan, stm., Swe.	192-93	56°33'N	12°56'E
Lågen, stm., Nor. (lô'ghĕn)	192-93	59°02'N	10°04'E
Lages, Braz.	168-69	27°49'S	50°18'W
Laghouat, Alg. (lä-gwät')	188-89	33°51'N	2°51'E
Lagoa da Prata, Braz. (lȧ-gô'ä-dȧ-prä'tȧ)	172	20°02'S	45°33'W
La Gomera, i., Spain	199d	28°07'N	17°11'W
Lagos, Nig. (lä'gōs)	260a	6°27'N	3°24'E
Lagos, Port. (lä'gōzh)	198-99	37°06'N	8°40'W
Lagos de Moreno, Mex. (lä'gōs dā mō-rā'nō)	146-47	21°22'N	101°54'W
La Grand'Combe, Fr. (lȧ grän kôNb')	196-97	44°13'N	4°01'E
La Grande, Or., U.S. (lȧ grănd')	112-13	45°20'N	118°05'W
La Grande Deux, Réservoir, res., Qc., Can.	130-31	53°40'N	76°55'W
La Grande Quatre, Réservoir, res., Qc., Can.	130-31	54°0'N	73°15'W
Lagrange, Ga., U.S.	124-25	33°02'N	85°02'W
La Grange, Ky., U.S. (lä grānj)	116-17	38°24'N	85°23'W
La Gran Sabana, pl., Ven.	164-65	5°21'N	62°04'W

n-sing; ŋ-baŋk; N-nasalized n; nŏd; cŏmmit; ōld; ôbey; ôrder; oi-boil; fōōd; ò-as oo in foot; ou-out; s-soft; sh-dish; th-thin; pūre; ûnite; ûrn; stŭd; circŭs; ü-as in French tu; '-indeterminate vowel.

Feature (Pronunciation)	Page	Lat.	Long.
La Guajira, Península de, pen., S.A.	**164-65**	12°0'N	71°40'W
Laguna, Braz. (lä-gōō'nä)	**172**	28°28'S	48°47'W
Lagunillas, Bol. (lä-gōō-nēl'yäs)	**168-69**	19°38'S	63°43'W
La Habana, nat. cap., Cuba (lä-ä-bä'nä) *see* Havana	**142-43**	23°06'N	82°27'W
Lahad Datu, Malay.	**250**	5°02'N	118°20'E
Lahaina, Hi., U.S. (lä-hä'ē-nä)	**127a**	20°53'N	156°40'W
Lahat, Indon.	**246-47**	3°48'S	103°32'E
Lāhījān, Iran	**232-33**	37°12'N	50°00'E
Laholm, Swe. (lä'hôlm)	**192-93**	56°31'N	13°03'E
Lahore, Pak. (lä-hōr')	**232-33**	31°35'N	74°20'E
Lahr, Ger. (lär)	**194-95**	48°21'N	7°52'E
Lahti, Fin. (lä'tē)	**192-93**	60°59'N	25°40'E
Laibach, nat. cap., Slvn. *see* Ljubljana	**200-01**	46°03'N	14°31'E
Laibin, China (lī-bĭn)	**238-39**	23°42'N	109°14'E
Laichow Bay, b., China *see* Laizhou Wan	**240-41**	37°20'N	119°19'E
L'Aigle, Fr. (lĕ'gl')	**196-97**	48°46'N	0°38'E
Laiwui, Indon.	**248-49**	1°21'S	127°39'E
Laiyang, China (lāī'yäng)	**240-41**	36°58'N	120°43'E
Laizhou Bay, b., China (lī-jō bā) *see* Laizhou Wan	**240-41**	37°20'N	119°19'E
Laizhou Wan, b., China (lī-jō wän)	**240-41**	37°20'N	119°19'E
Lajeado, Braz. (lä-zhĕä'dô)	**168-69**	29°24'S	51°57'W
Lajes, Braz. (lá'zhĕs)	**163d**	5°41'S	36°14'W
Lajinha, Braz. (lä-zhē'nyä)	**172**	20°09'S	41°37'W
La Junta, Co., U.S. (lä hōōn'tà)	**120-21**	37°59'N	103°33'W
Lake Arthur, La., U.S. (lāk är'thŭr)	**122-23**	30°05'N	92°41'W
Lakeba, i., Fiji	**279f**	18°13'S	178°47'W
Lakeba Passage, strt., Fiji	**279f**	17°55'S	178°45'W
Lake Cargelligo, Austl.	**276**	33°19'S	146°22'E
Lake Charles, La., U.S. (lāk chärlz')	**122-23**	30°14'N	93°13'W
Lake City, Fl., U.S. (lāk sĭ'tĭ)	**124-25**	30°12'N	82°38'W
Lake City, Mn., U.S. (lāk sĭ'tē)	**114-15**	44°27'N	92°17'W
Lake City, S.C., U.S. (lāk sĭ'tĭ)	**124-25**	33°52'N	79°45'W
Lake Cowichan, B.C., Can. (lāk kou'ĭ-chán)	**132-33**	48°50'N	124°03'W
Lake Crystal, Mn., U.S. (lāk krĭs'tál)	**114-15**	44°07'N	94°13'W
Lake Geneva, Wi., U.S. (lāk jĕ-nē'vá)	**116-17**	42°36'N	88°26'W
Lake Harbour, Nu., Can. *see* Kimmirut	**128-29**	62°51'N	69°53'W
Lake Havasu City, Az., U.S. (lāk hăv'â-sōō sĭ'tē)	**118-19**	34°29'N	114°21'W
Lakeland, Fl., U.S. (lāk wûrth')	**125a**	28°03'N	81°58'W
Lake Linden, Mi., U.S. (lāk lĭn'dĕn)	**114-15**	47°12'N	88°24'W
Lake Louise, Ab., Can. (lāk lōō-ēz')	**132-33**	51°27'N	116°13'W
Lake Mills, Ia., U.S. (lāk mĭlz')	**114-15**	43°25'N	93°32'W
Lake Oswego, Or., U.S. (lāk ŏs-wē'go)	**112-13**	45°25'N	122°43'W
Lake Placid, N.Y., U.S. (lāk plăs'ĭd)	**116-17**	44°17'N	73°59'W
Lake Preston, S.D., U.S. (lāk prĕs'tŭn)	**114-15**	44°22'N	97°23'W
Lake Providence, La., U.S. (lāk prŏv'ĭ-dĕns)	**124-25**	32°49'N	91°11'W
Lakeview, Or., U.S.	**112-13**	42°12'N	120°21'W
Lake Village, Ar., U.S. (lāk vĭl'āj)	**124-25**	33°19'N	91°17'W
Lake Wales, Fl., U.S. (lāk wālz')	**125a**	27°54'N	81°35'W
Lakewood, Co., U.S. (lāk'wŏŏd)	**120-21**	39°44'N	105°07'W
Lakewood, N.J., U.S. (lāk'wŏŏd)	**116-17**	40°07'N	74°14'W
Lakewood, Oh., U.S. (lāk'wŏŏd)	**116-17**	41°29'N	81°48'W
Lakewood, Wa., U.S. (lāk'wŏŏd)	**112-13**	47°11'N	122°31'W
Lake Worth, Fl., U.S. (lāk wûrth')	**125a**	26°37'N	80°03'W
Lakhdenpokh'ya, Russia (l'âk-dīe'npôkyâ)	**192-93**	61°31'N	30°12'E
Lakhīmpur, India	**234-35**	27°57'N	80°47'E
Lakota, N.D., U.S. (lá-kō'tá)	**114-15**	48°02'N	98°21'W
Lakshadweep, state, India	**236**	10°0'N	73°00'E
Lakshadweep, is., India	**236**	10°0'N	73°00'E
Lakshadweep Sea, s., Asia	**236**	7°0'N	76°00'E
La Libertad, Guat. (lä lē-bĕr-tädh')	**148**	16°47'N	90°07'W
La Ligua, Chile (lä lē'gwä)	**163e**	32°27'S	71°15'W
Lalitpur, India	**234-35**	24°41'N	78°25'E
Lalitpur, Nepal	**234-35**	27°40'N	85°19'E
La Loche, Sk., Can.	**134-35**	56°29'N	109°26'W
La Louvière, Bel. (là lōō-vyâr')	**190-91**	50°29'N	4°12'E
La Luz, Mex. (lä lōōz')	**146-47**	24°12'N	97°52'W
Lama, Ozero, lk., Russia	**218-19**	69°32'N	90°27'E
La Madrid, Arg.	**168-69**	27°39'S	65°15'W
La Malbaie, Qc., Can. (lá mäl-bā')	**138-39**	47°40'N	70°09'W
La Mancha, reg., Spain (lä män'chä)	**198-99**	39°21'N	2°28'W
La Manche, strt., Eur. *see* English Channel	**190-91**	50°13'N	2°20'W
Lamar, Co., U.S. (lá-mär')	**120-21**	38°05'N	102°37'W
Lamar, Mo., U.S. (lá-mär')	**120-21**	37°30'N	94°16'W
La Marmora, Punta, mtn., Italy (pò'n-tä-lä-mä'r-mô-rä)	**200-01**	39°59'N	9°20'E
Lamas, Peru (lä'más)	**170**	6°25'S	76°35'W

Feature (Pronunciation)	Page	Lat.	Long.
Lamballe, Fr. (läɴ-bäl')	**196-97**	48°28'N	2°32'W
Lambayeque, Peru (läm-bä-yā'kå)	**170**	6°41'S	79°54'W
Lambertsbaai, S. Afr. *see* Lambert's Bay	**264-65**	32°06'S	18°19'E
Lambert's Bay, S. Afr.	**264-65**	32°06'S	18°19'E
Lame Deer, Mt., U.S. (lām dēr')	**112-13**	45°39'N	106°41'W
La Méditerranée, s., *see* Mediterranean Sea	**188-89**	35°0'N	20°00'E
Lamego, Port. (lä-mā'gō)	**198-99**	41°06'N	7°49'W
Lamesa, Tx., U.S.	**120-21**	32°45'N	101°58'W
Lamía, Grc. (lá-mē'á)	**200-01**	38°54'N	22°26'E
Lamon Bay, b., Phil. (lä-mōn' bā)	**250**	14°28'N	122°01'E
Lamoni, Ia., U.S.	**120-21**	40°38'N	93°56'W
Lamotrek, at., Micron.	**280-81**	7°21'N	146°20'E
La Moure, N.D., U.S. (lá mōōr')	**114-15**	46°20'N	98°17'W
Lampang, Thai.	**246-47**	18°17'N	99°29'E
Lampasas, Tx., U.S. (lăm-păs'ás)	**122-23**	31°04'N	98°11'W
Lampazos de Naranjo, Mex.	**122-23**	27°02'N	100°31'W
Lamphun, Thai.	**246-47**	18°35'N	99°01'E
Lamu, Kenya (lä'mōō)	**262-63**	2°17'S	40°53'E
Lan', stm., Bela. (lán')	**194-95**	52°09'N	27°17'E
Lāna'i, i., Hi., U.S. (lä-nä'ē)	**127a**	20°50'N	156°55'W
Lanark, Scot., U.K. (lăn'árk)	**190-91**	55°41'N	3°47'W
Lancang, stm., Asia *see* Mekong	**246-47**	10°33'N	105°27'E
Lancaster, Eng., U.K.	**190-91**	54°03'N	2°50'W
Lancaster, Ca., U.S. (lăŋ'kăs-tē)	**118-19**	34°42'N	118°08'W
Lancaster, Ky., U.S. (lăŋ'kăs-tē)	**116-17**	37°37'N	84°35'W
Lancaster, Oh., U.S. (lăŋ'kăs-tē)	**116-17**	39°43'N	82°36'W
Lancaster, Pa., U.S.	**116-17**	40°03'N	76°19'W
Lancaster, S.C., U.S. (lăŋ'kăs-tē)	**124-25**	34°43'N	80°46'W
Lancaster, Wi., U.S. (lăŋ'kăs-tē)	**114-15**	42°51'N	90°43'W
Lanchow, China *see* Lanzhou	**240-41**	36°04'N	103°43'E
Lander, Wy., U.S. (lăn'dēr)	**112-13**	42°50'N	108°44'W
Landerneau, Fr. (läɴ-dĕr-nō')	**196-97**	48°27'N	4°16'W
Landes, reg., Fr. (länd)	**196-97**	44°10'N	0°52'W
Landsberg, Pol. *see* Gorzów Wielkopolski	**194-95**	52°44'N	15°14'E
Landsberg an der Warthe, Pol. *see* Gorzów Wielkopolski	**194-95**	52°44'N	15°14'E
Land's End, c., Eng., U.K. (lăndz ĕnd)	**190-91**	50°03'N	5°44'W
Landshut, Ger. (länts'hōōt)	**194-95**	48°33'N	12°09'E
Landskrona, Swe. (láns-krô'ná)	**192-93**	55°52'N	12°50'E
Lanett, Al., U.S. (lá-nĕt')	**124-25**	32°52'N	85°11'W
La'nga Co, lk., China (läŋ-lä tswo)	**234-35**	30°43'N	81°13'E
Langano Hāyk', lk., Eth.	**269d**	7°36'N	38°46'E
Langdon, N.D., U.S.	**114-15**	48°46'N	98°23'W
Langeland, i., Den.	**192-93**	55°00'N	10°51'E
Langeoog, i., Ger.	**194-95**	53°45'N	7°32'E
Langjökull, ice, Ice. (läng-yû'kōōl)	**190a**	64°42'N	20°12'W
Langkawi, Pulau, i., Malay.	**246-47**	6°24'N	99°50'E
Langley, B.C., Can. (lăng'lĭ)	**132-33**	49°06'N	122°39'W
Langon, Fr. (läɴ-gôɴ')	**196-97**	44°33'N	0°15'W
Langøya, i., Nor.	**184-85**	68°44'N	14°50'E
Langqên, stm., Asia *see* Sutlej	**234-35**	29°21'N	71°02'E
Langres, Fr. (läɴ'gr')	**196-97**	47°52'N	5°19'E
Langsa, Indon. (läng'sá)	**246-47**	4°28'N	97°58'E
Lang Son, Viet. (lăng'sŏn')	**246-47**	21°51'N	106°45'E
Langzhong, China (läŋ-jôŋ)	**238-39**	31°34'N	105°59'E
Lanigan, Sk., Can. (lăn'ĭ-gán)	**134-35**	51°52'N	105°02'W
Länkäran, Azer. (lĕn-kô-rän')	**227**	38°45'N	48°51'E
Lansdale, Pa., U.S. (lănz'dāl)	**116-17**	40°15'N	75°17'W
L'Anse, Mi., U.S. (láns)	**114-15**	46°45'N	88°26'W
Lansing, Mi., U.S.	**116-17**	42°45'N	84°33'W
Lanta Yai, Ko, i., Thai.	**246-47**	7°34'N	99°03'E
Lanxi, China	**238-39**	29°12'N	119°28'E
Lanzarote, i., Spain (län-zá-rō'tä)	**199d**	29°0'N	13°40'W
Lanzhou, China (län-jō)	**240-41**	36°04'N	103°43'E
Lao, nation, Asia *see* Laos	**206-07**	18°0'N	105°00'E
Laoag, Phil. (lä-wäg')	**250**	18°12'N	120°36'E
Laoang, Phil.	**250**	12°35'N	125°02'E
Lao Cai, Viet.	**246-47**	22°30'N	103°58'E
Laoha, stm., China	**240-41**	43°25'N	120°45'E
Laohekou, China	**238-39**	32°25'N	111°36'E
Laon, Fr. (läɴ)	**196-97**	49°34'N	3°39'E
La Orchila, Isla, i., Ven.	**164-65**	11°48'N	66°09'W
La Oroya, Peru (lä-ô-rō'yä)	**163a**	11°30'S	75°56'W
Laos, nation, Asia (lā-ōs) (lá-ōs')	**206-07**	18°0'N	105°00'E
La Palma, Pan. (lä-päl'mä)	**150**	7°42'N	80°11'W
La Palma, Pan.	**150**	8°24'N	78°09'W
La Palma, i., Spain (lä-päl'mä)	**199d**	28°40'N	17°52'W
La Paloma, Ur.	**173**	34°40'S	54°10'W
La Paragua, Ven.	**164-65**	6°51'N	63°19'W
La Paz, Arg. (lä päz')	**173**	30°44'S	59°38'W
La Paz, Arg. (lä päz')	**171**	33°27'S	67°34'W
La Paz, Hond. (lä-pá'z)	**149**	14°19'N	87°41'W
La Paz, Mex. (lä-pá'z)	**146-47**	23°41'N	100°43'W

Feature (Pronunciation)	Page	Lat.	Long.
La Paz, Mex.	**144-45**	24°10'N	110°18'W
La Paz, nat. cap., Bol. (lä-pá'z)	**168-69**	16°30'S	68°09'W
Lapeer, Mi., U.S. (lá-pēr')	**116-17**	43°03'N	83°18'W
Lapland, reg., Eur. (lăp'lánd)	**184-85**	68°0'N	25°00'E
La Plata, Arg. (lä plä'tä)	**173**	34°55'S	57°57'W
La Plata, Mo., U.S. (lá plā'tá)	**120-21**	40°02'N	92°29'W
La Pocatière, Qc., Can. (lá pô-ká-tyâr')	**138-39**	47°22'N	70°02'W
La Porte, In., U.S. (lá pōrt')	**116-17**	41°37'N	86°42'W
La Porte City, Ia., U.S. (lá pōrt' sĭ'tē)	**114-15**	42°19'N	92°11'W
Lappland, reg., Eur. *see* Lapland	**184-85**	68°0'N	25°00'E
Laptev Sea, s., Russia (läp'tyĭf sē)	**208-09**	76°0'N	126°00'E
Laptevykh, More, s., Russia *see* Laptev Sea	**208-09**	76°0'N	126°00'E
La Quiaca, Arg. (lä kē-ä'kä)	**168-69**	22°07'S	65°36'W
L'Aquila, Italy (lá'kē-lä)	**200-01**	42°21'N	13°24'E
Lār, Iran (lär)	**230-31**	27°40'N	54°20'E
Larache, Mor. (lä-räsh')	**269a**	35°12'N	6°09'W
Lārak, Jazīreh-ye, i., Iran	**230-31**	26°52'N	56°22'E
Laramie, Wy., U.S. (lăr'á-mĭ)	**112-13**	41°19'N	105°35'W
Larantuka, Indon.	**248-49**	8°19'S	122°58'E
Larat, Pulau, i., Indon.	**224-25**	7°08'S	131°50'E
Laredo, Spain (lä-rā'dhō)	**198-99**	43°25'N	3°25'W
Laredo, Tx., U.S. (lá-rā'dhō)	**122-23**	27°31'N	99°28'W
Largo, Cayo, i., Cuba (kä'yō-lär'gō)	**142-43**	21°38'N	81°28'W
Larimore, N.D., U.S. (lăr'ĭ-môr)	**114-15**	47°54'N	97°38'W
La Rioja, Arg. (lä rē-ōhä)	**168-69**	29°25'S	66°51'W
La Rioja, state, Arg. (lä-rē-ô'kä)	**168-69**	30°0'S	67°30'W
Lárisa, Grc. (lä'rē-sá)	**200-01**	39°38'N	22°25'E
Larissa, Grc. *see* Lárisa	**200-01**	39°38'N	22°25'E
Lārkāna, Pak.	**232-33**	27°33'N	68°13'E
Larnaca, Cyp. *see* Larnaka	**228-29**	34°55'N	33°38'E
Larnaka, Cyp.	**228-29**	34°55'N	33°38'E
Larned, Ks., U.S. (lär'nĕd)	**120-21**	38°11'N	99°05'W
La Rochelle, Fr. (lá rô-shĕl')	**196-97**	46°10'N	1°10'W
La Roche-sur-Yon, Fr. (lá rôsh'sûr-yôɴ')	**196-97**	46°40'N	1°26'W
La Roda, Spain (lä rō'dä)	**198-99**	39°12'N	2°09'W
La Ronge, Sk., Can.	**134-35**	55°06'N	105°17'W
La Rubia, Arg.	**173**	30°08'S	61°48'W
Larvik, Nor. (lär'vēk)	**192-93**	59°04'N	10°01'E
La Salle, Il., U.S. (lá sål')	**116-17**	41°20'N	89°06'W
La Sarre, Qc., Can.	**136-37**	48°48'N	79°12'W
Las Aves, Isla, i., Ven.	**143b**	15°41'N	63°37'W
Lascano, Ur.	**173**	33°40'S	54°13'W
Las Cruces, N.M., U.S. (läs-krōō'sĕs)	**118-19**	32°19'N	106°47'W
La Selle, Morne, mtn., Haiti (mŏrn lä sĕl')	**142-43**	18°22'N	71°59'W
La Serena, Chile (lä-sĕ-rĕ'nä)	**168-69**	29°54'S	71°15'W
Las Flores, Arg. (läs flo'rĕs)	**173**	36°01'S	59°06'W
Las Heras, Arg.	**171**	46°31'S	68°56'W
Lashio, Mya. (lăsh'ē-ō)	**246-47**	22°57'N	97°45'E
Lashkar, India *see* Gwalior	**234-35**	26°13'N	78°09'E
Las Lajas, Arg.	**171**	38°30'S	70°22'W
Las Lomitas, Arg.	**168-69**	24°43'S	60°36'W
Las Minas, Cerro, mtn., Hond.	**149**	14°33'N	88°39'W
La Solana, Spain (lä-sô-lä-nä)	**198-99**	38°57'N	3°14'W
Las Palmas de Gran Canaria, Spain (läs päl'mäs)	**199d**	28°07'N	15°26'W
La Spezia, Italy (lä-spĕ'zyä)	**200-01**	44°07'N	9°50'E
Las Piedras, Ur. (läs-pyĕ'drás)	**173**	34°44'S	56°13'W
Las Piedras, stm., Peru	**170**	12°31'S	69°14'W
Las Plumas, Arg.	**171**	43°43'S	67°14'W
Las Rosas, Mex. (läs rō'thäs)	**146-47**	16°22'N	92°22'W
Lassen Peak, vol., Ca., U.S. (lăs'ĕn pēk)	**112-13**	40°29'N	121°31'W
Lassen Volcanic National Park, n.p., Ca., U.S. (lăs'ĕn vŏl-kăn'ĭk năsh'ŭn-ăl pärk)	**112-13**	40°30'N	121°27'W
Las Tablas, Pan. (läs tä'bläs)	**150**	7°46'N	80°17'W
Last Mountain Lake, lk., Sk., Can. (làst moun'tĭn lāk)	**134-35**	51°06'N	105°15'W
Las Tórtolas, Cerro, mtn., S.A.	**168-69**	29°57'S	69°53'W
Lastoursville, Gabon (lás-tōōr-vēl')	**260-61**	0°48'S	12°42'E
Las Tunas, Cuba	**142-43**	20°58'N	76°57'W
Las Varillas, Arg.	**173**	31°52'S	62°42'W
Las Vegas, N.M., U.S. (läs vā'gäs)	**120-21**	35°36'N	105°13'W
Las Vegas, Nv., U.S. (läs vā'gäs)	**118-19**	36°11'N	115°08'W
Latacunga, Ec. (lä-tä-kōŋ'gä)	**170**	0°56'S	78°36'W
Latakia, Syria	**228-29**	35°31'N	35°48'E
La Teste-de-Buch, Fr. (lä-tĕst-dĕ-büsh)	**196-97**	44°38'N	1°09'W
Lathrop, Mo., U.S. (lā'thrŭp)	**120-21**	39°33'N	94°20'W
La Tortuga, Isla, i., Ven. (ē's-lä-lä-tôr-tōō'gä)	**163b**	10°56'N	65°20'W
Latouche Treville, Cape, c., Austl.	**272-73**	18°28'S	121°50'E
La Tremblade, Fr. (lä-trĕɴ-bläd')	**196-97**	45°46'N	1°08'W

Feature (Pronunciation)	Page	Lat.	Long.
Latrobe, Pa., U.S. (lȧ-trōb′)	116-17	40°18′N	79°22′W
La Tuque, Qc., Can. (lȧ′tük′)	136-37	47°26′N	72°47′W
Lātūr, India (lä-to͞or′)	236	18°24′N	76°35′E
Latvia, nation, Eur. (lăt′vē-ȧ)	174-75	57°0′N	25°00′E
Latvija, nation, Eur. *see* Latvia	174-75	57°0′N	25°00′E
Lauenburg, Pol. *see* Lębork	194-95	54°32′N	17°46′E
Lau Group, is., Fiji	279f	18°20′S	178°30′W
Lauis, Switz. *see* Lugano	194-95	46°01′N	8°57′E
Launceston, Austl. (lôn′sĕs-tŭn)	276	41°25′S	147°08′E
La Unión, Chile (lä-o͞o-nyō′n)	171	40°18′S	73°05′W
La Unión, El Sal.	148	13°20′N	87°51′W
La Unión, Mex. (lä o͞on-nyōn′)	146-47	17°58′N	101°49′W
Laura, Austl. (lôrȧ)	277	15°33′S	144°26′E
Laurel, De., U.S. (lô′rĕl)	116-17	38°33′N	75°34′W
Laurel, Md., U.S. (lô′rĕl)	116-17	39°06′N	76°51′W
Laurel, Ms., U.S. (lô′rĕl)	124-25	31°42′N	89°08′W
Laurel, Mt., U.S. (lô′rĕl)	112-13	45°40′N	108°46′W
Laurel, Ne., U.S. (lô′rĕl)	114-15	42°26′N	97°06′W
Laurens, S.C., U.S. (lô′rĕnz)	124-25	34°30′N	82°01′W
Laurentides, Les, plat., Qc., Can.	130-31	48°0′N	71°00′W
Laurinburg, N.C., U.S. (lô′rĭn-bûrg)	124-25	34°47′N	79°28′W
Laurium, Mi., U.S. (lô′rĭ-ŭm)	114-15	47°14′N	88°26′W
Lausanne, Switz. (lō-zán′)	194-95	46°31′N	6°38′E
Lausitzer Neiße, stm., Eur. *see* Neisse	194-95	52°04′N	14°46′E
Laut, Pulau, i., Indon.	248-49	3°40′S	116°10′E
Lautaro, Chile (lou-tä′rō)	171	38°31′S	72°26′W
Laut Kecil, Kepulauan, is., Indon.	248-49	4°49′S	115°44′E
Lava, Nosy, i., Madag.	264-65	14°33′S	47°36′E
Lava Beds National Monument, n.p., Ca., U.S. (lä′vȧ bĕds năsh′ŭn-ăl mŏn′ū-mĕnt)	112-13	41°45′N	121°32′W
Laval, Qc., Can.	136-37	45°33′N	73°44′W
Laval, Fr. (lä-väl′)	196-97	48°04′N	0°46′W
Lāvān, Jazīreh-ye, i., Iran	230-31	26°49′N	53°15′E
Lavapié, Punta, c., Chile	171	37°09′S	73°35′W
La Vega, Dom. Rep. (lä-vĕ′gä)	142-43	19°13′N	70°31′W
Laverton, Austl. (lā′vēr-tŭn)	270-71	28°37′S	122°24′E
La Victoria, Ven. (lä vēk-tō′rē-ä)	163b	10°13′N	67°20′W
Lavras, Braz. (lä′vräzh)	172	21°14′S	45°00′W
Lavrentiya, Russia	126	65°35′N	171°01′W
Lawas, Malay.	248-49	4°51′N	115°24′E
Lawn Hill National Park, n.p., Austl. *see* Boodjamulla National Park	277	18°45′S	138°27′E
Lawrence, In., U.S. (lô′rĕns)	116-17	39°50′N	86°01′W
Lawrence, Ks., U.S. (lô′rĕns)	120-21	38°57′N	95°15′W
Lawrence, Ma., U.S. (lô′rĕns)	116-17	42°42′N	71°10′W
Lawrenceburg, In., U.S. (lô′rĕnsbûrg)	116-17	39°05′N	84°52′W
Lawrenceburg, Ky., U.S. (lô′rĕnsbûrg)	116-17	38°03′N	84°54′W
Lawrenceburg, Tn., U.S. (lô′rĕnsbûrg)	124-25	35°15′N	87°20′W
Lawrenceville, Ga., U.S. (lô-rĕns-vĭl)	124-25	33°58′N	83°59′W
Lawrenceville, Il., U.S. (lô-rĕns-vĭl)	116-17	38°43′N	87°41′W
Lawrenceville, Va., U.S. (lô-rĕns-vĭl)	124-25	36°46′N	77°51′W
Lawton, Ok., U.S. (lô′tŭn)	120-21	34°36′N	98°24′W
Lawz, Jabal al-, mtn., Sau. Ar.	228-29	28°40′N	35°18′E
La'youn, nat. cap., W. Sah. *see* Laayoune	258-59	27°10′N	13°12′W
Laysan Island, i., Hi., U.S.	127	25°50′N	171°50′W
Layton, Ut., U.S. (lā′tŭn)	112-13	41°05′N	111°58′W
Lazarev, Russia	218-19	52°12′N	141°30′E
Lázaro Cárdenas, Mex.	146-47	17°57′N	102°12′W
Lazdijai, Lith. (läzh′dē-yī′)	194-95	54°14′N	23°32′E
Lead, S.D., U.S. (lēd)	114-15	44°21′N	103°46′W
Leader, Sk., Can.	134-35	50°53′N	109°31′W
Leadville, Co., U.S. (lĕd′vĭl)	118-19	39°15′N	106°18′W
Leaf, stm., Ms., U.S. (lēf)	124-25	31°0′N	88°45′W
League City, Tx., U.S. (lēg sĭ′tĭ)	122-23	29°30′N	95°06′W
Leamington, On., Can. (lĕm′ĭng-tŭn)	136-37	42°02′N	82°36′W
Leavenworth, Ks., U.S. (lĕv′ĕn-wûrth)	120-21	39°18′N	94°56′W
Leavenworth, Wa., U.S. (lĕv′ĕn-wûrth)	112-13	47°36′N	120°40′W
Łeba, Pol. (lā′bä)	194-95	54°45′N	17°33′E
Lebak, Phil.	250	6°31′N	124°02′E
Lebanon, In., U.S. (lĕb′ȧ-nŭn)	116-17	40°03′N	86°28′W
Lebanon, Ky., U.S. (lĕb′ȧ-nŭn)	116-17	37°34′N	85°15′W
Lebanon, Mo., U.S. (lĕb′ȧ-nŭn)	120-21	37°41′N	92°40′W
Lebanon, N.H., U.S. (lĕb′ȧ-nŭn)	116-17	43°39′N	72°15′W
Lebanon, Oh., U.S. (lĕb′ȧ-nŭn)	116-17	39°26′N	84°13′W
Lebanon, Or., U.S. (lĕb′ȧ-nŭn)	112-13	44°32′N	122°54′W
Lebanon, Tn., U.S. (lĕb′ȧ-nŭn)	124-25	36°13′N	86°17′W
Lebanon, Va., U.S. (lĕb′ȧ-nŭn)	124-25	36°54′N	82°05′W
Lebanon, nation, Asia (lĕb′ȧ-nŭn)	206-07	34°0′N	36°00′E
Lebedyan', Russia (lyĕ′bĕ-dyän′)	202-03	53°01′N	39°08′E
Lębork, Pol. (lăɴ-bòrk′)	194-95	54°32′N	17°46′E
Lebrija, Spain (lå-brē′hä)	198-99	36°56′N	6°04′W
Lebu, Chile	171	37°37′S	73°39′W
Lecce, Italy (lĕt′chā)	200-01	40°21′N	18°10′E

Feature (Pronunciation)	Page	Lat.	Long.
Lecco, Italy (lĕk′kō)	200-01	45°51′N	9°23′E
Le Creusot, Fr. (lĕkrû-zō)	196-97	46°48′N	4°26′E
Ledo, India	234-35	27°17′N	95°44′E
Ledu, China	240-41	36°28′N	102°24′E
Leduc, Ab., Can. (lĕ-do͞ok′)	132-33	53°15′N	113°32′W
Ledyanaya, Gora, mtn., Russia	218-19	61°53′N	171°09′E
Ledyard Bay, b., Ak., U.S.	126	69°14′N	164°31′W
Leech Lake, lk., Mn., U.S. (lēch lāk)	114-15	47°09′N	94°23′W
Leeds, Eng., U.K. (lēdz)	190-91	53°50′N	1°35′W
Leeds, N.D., U.S. (lēdz)	114-15	48°17′N	99°27′W
Leesburg, Fl., U.S. (lēz′bûrg)	124-25	28°49′N	81°53′W
Leesburg, Va., U.S. (lēz′bûrg)	116-17	39°06′N	77°34′W
Leesville, La., U.S. (lēz′vĭl)	122-23	31°08′N	93°16′W
Leeton, Austl.	276	34°33′S	146°24′E
Leeuwarden, Neth. (lā′wär-dĕn)	190-91	53°12′N	5°47′E
Leeuwin, Cape, c., Austl. (kāp o͞o′wĭn)	272-73	34°23′S	115°08′E
Leeward Islands, is., N.A. (lē′wērd ī′lăndz)	143b	17°0′N	63°00′W
Lefkáda, i., Grc.	200-01	38°42′N	20°39′E
Lefkoşa, nat. cap., Cyp. *see* Nicosia	228-29	35°10′N	33°22′E
Lefroy, Lake, lk., Austl. (lāk lē-froi′)	272-73	31°15′S	121°40′E
Leganés, Spain (lå-gä′nås)	198-99	40°20′N	3°46′W
Legaspi, Phil.	250	13°08′N	123°45′E
Leghorn, Italy *see* Livorno	200-01	43°34′N	10°19′E
Legnica, Pol. (lĕk-nĭt′sȧ)	194-95	51°13′N	16°10′E
Leh, India (lā)	234-35	34°10′N	77°35′E
Le Havre, Fr. (lē áv′r′)	196-97	49°29′N	0°08′E
Leicester, Eng., U.K. (lĕs′tēr)	190-91	52°39′N	1°08′W
Leikanger, Nor. (lī′kän′gēr)	192-93	61°12′N	6°50′E
Leine, stm., Ger. (lī′nĕ)	194-95	52°43′N	9°36′E
Leipzig, Ger. (līp′tsĭk)	194-95	51°20′N	12°23′E
Leiria, Port. (lā-rē′ä)	198-99	39°45′N	8°48′W
Leitchfield, Ky., U.S. (lēch′fēld)	116-17	37°29′N	86°18′W
Leizhou Bandao, pen., China (lā-jō bän-dou)	238-39	20°47′N	110°05′E
Leksand, Swe. (lĕk′sänd)	192-93	60°43′N	15°01′E
Leland, Mi., U.S. (lē′lănd)	116-17	45°01′N	85°46′W
Leland, Ms., U.S. (lē′lănd)	124-25	33°25′N	90°54′W
Leli Shan, mtn., China	234-35	33°26′N	81°42′E
Le Maire, Estrecho de, strt., Arg. (ĕs-trĕ′chô-dĕ-lĕ-mī′rĕ)	171	54°50′S	64°60′W
Léman, Lac, lk., Eur. *see* Geneva, Lake	194-95	46°24′N	6°22′E
Le Mans, Fr. (lē mäɴ′)	196-97	48°00′N	0°12′E
Le Mars, Ia., U.S. (lē märz′)	114-15	42°48′N	96°10′W
Lemesós, Cyp.	228-29	34°41′N	33°03′E
Lemhi Range, mts., Id., U.S. (lĕm′hī rānj)	112-13	44°33′N	113°36′W
Lemmon, S.D., U.S. (lĕm′ŭn)	114-15	45°56′N	102°10′W
Lemnos, i., Grc. *see* Límnos	200-01	39°55′N	25°18′E
Lempa, stm., N.A. (lĕm′pä)	148	13°15′N	88°49′W
Lena, stm., Russia (lē′nȧ) (lyĕ′nŭ)	218-19	72°25′N	126°40′E
Lençóis, Braz.	166-67	12°34′S	41°23′W
Lenexa, Ks., U.S. (lē′nĕx-ă)	120-21	38°58′N	94°44′W
Lenghu, China	220-21	38°50′N	93°26′E
Lenin, Qullai, mtn., Asia *see* Lenin Peak	232-33	39°20′N	72°55′E
Lenina, Pik, mtn., Asia *see* Lenin Peak	232-33	39°20′N	72°55′E
Lenin Atyndagy Choku, mtn., Asia *see* Lenin Peak	232-33	39°20′N	72°55′E
Leningrad, Russia *see* Saint Petersburg	192-93	59°57′N	30°15′E
Leningradskaya, Russia (lyĕ-nĭn-gräd′skȧ-yȧ)	202-03	46°19′N	39°23′E
Leninogor, Kaz. *see* Ridder	226	50°21′N	83°30′E
Leninogorsk, Russia	186-87	54°35′N	52°29′E
Lenin Peak, mtn., Asia	232-33	39°20′N	72°55′E
Leninsk, Kaz. *see* Bayqongyr	226	45°38′N	63°18′E
Leninsk-Kuznetskiy, Russia	218-19	54°41′N	86°12′E
Leninskoye, Russia	240-41	47°60′N	132°38′E
Lennox, S.D., U.S. (lĕn′ŭks)	114-15	43°21′N	96°55′W
Lennox, Isla, i., Chile	171	55°18′S	66°50′W
Lenoir, N.C., U.S. (lē-nōr′)	124-25	35°55′N	81°32′W
Lensk, Russia	218-19	60°44′N	114°56′E
Léo, Burkina	260-61	11°06′N	2°06′W
Leoben, Aus. (lå-ō′bĕn)	194-95	47°22′N	15°06′E
Léogâne, Haiti (lā-ô-gan′)	142-43	18°31′N	72°38′W
Leominster, Ma., U.S. (lĕm′ĭn-stēr)	116-17	42°32′N	71°45′W
León, Mex. (lå-ōn′)	146-47	21°07′N	101°42′W
León, Nic. (lĕ-ō′n)	149	12°26′N	86°52′W
León, Spain (lĕ-ō′n)	198-99	42°36′N	5°34′W
Leon, Ia., U.S. (lē′ŏn)	114-15	40°45′N	93°44′W
León, hist. reg., Spain (lĕ-ō′n)	198-99	42°0′N	6°00′W
Leon, stm., Tx., U.S. (lē′ŏn)	122-23	30°59′N	97°24′W
León de los Aldamas, Mex. *see* León	146-47	21°07′N	101°42′W
Leonforte, Italy (lā-ōn-fōr′tā)	200-01	37°38′N	14°23′E

Feature (Pronunciation)	Page	Lat.	Long.
Leonora, Austl.	270-71	28°53′S	121°20′E
Léopold II, Lac, lk., D.R.C. *see* Mai-Ndombe, Lac	260-61	2°25′S	18°18′E
Leopoldina, Braz. (lā-ô-pôl-dē′nä)	172	21°32′S	42°38′W
Léopoldville, nat. cap., D.R.C. *see* Kinshasa	262-63	4°21′S	15°18′E
Lepe, Spain (lā′pā)	198-99	37°15′N	7°12′W
Leping, China (lŭ-pĭŋ)	238-39	28°57′N	117°06′E
Le Port, Reu.	265a	20°55′S	55°18′E
Le Puy, Fr. (lē pwē′)	196-97	45°03′N	3°53′E
Lerdo, Mex. (lĕr′dō)	122-23	25°32′N	103°31′W
Lérida, Spain *see* Lleida	198-99	41°37′N	0°38′E
Lerma, stm., Mex. (lĕr′mä)	146-47	20°13′N	102°41′W
Le Roy, N.Y., U.S. (lē roi′)	116-17	42°58′N	77°59′W
Lerwick, Scot., U.K. (lĕr′ĭk) (lûr′wĭk)	190c	60°09′N	1°09′W
Lesbos, i., Grc. *see* Lésvos	200-01	39°10′N	26°20′E
Les Cayes, Haiti	142-43	18°12′N	73°45′W
Leshan, China (lŭ-shän)	238-39	29°34′N	103°45′E
Leshukonskoye, Russia	186-87	64°53′N	45°42′E
Leskovac, Serb. (lĕs′kô-vàts)	200-01	43°01′N	21°57′E
Leslie, Mi., U.S. (lĕz′lĭ)	116-17	42°27′N	84°25′W
Lesosibirsk, Russia	218-19	58°14′N	92°29′E
Lesotho, nation, Afr. (lĕsō′thō)	253	29°30′S	28°30′E
Lesozavodsk, Russia (lyĕ-sô-zá-vôdsk′)	244	45°28′N	133°24′E
Les Sables-d'Olonne, Fr. (lā sá′bl′dō-lŭn′)	196-97	46°30′N	1°47′W
Les Saintes, is., Guad. (lā-săɴt′)	143b	15°52′N	61°37′W
Lesser Antilles, is. (lĕs′ēr ăn-tĭ′lēz)	143b	15°0′N	61°00′W
Lesser Caucasus, mts., Asia	227	40°60′N	44°35′E
Lesser Khingan Range, mts., China	240-41	48°45′N	127°00′E
Lesser Slave, stm., Ab., Can. (lĕs′ēr slāv)	132-33	55°10′N	114°03′W
Lesser Slave Lake, lk., Ab., Can. (lĕs′ēr slāv lāk)	132-33	55°29′N	115°10′W
Lesser Sunda Islands, is., Asia (lĕs′ēr so͞on′dȧ ī′lăndz)	248-49	9°0′S	120°00′E
Le Sueur, Mn., U.S. (lē so͞or′)	114-15	44°28′N	93°55′W
Lésvos, i., Grc.	200-01	39°10′N	26°20′E
Leszno, Pol. (lĕsh′nô)	194-95	51°51′N	16°35′E
Lethbridge, Ab., Can. (lĕth′brĭj)	132-33	49°42′N	112°49′W
Lethem, Guy.	164-65	3°23′N	59°48′W
Leti, Kepulauan, is., Indon.	248-49	8°13′S	127°50′E
Leticia, Col. (lĕ-tē′syȧ)	164-65	4°10′S	69°56′W
Leucas, i., Grc. *see* Lefkáda	200-01	38°42′N	20°39′E
Levanger, Nor. (lĕ-väng′ĕr)	184-85	63°45′N	11°18′E
Leveque, Cape, c., Austl. (kāp lê-vĕk′)	272-73	16°26′S	122°56′E
Leverkusen, Ger. (lĕ′fĕr-ko͞o-zĕn)	194-95	51°03′N	6°59′E
Levice, Slvk. (lā′vĕt-sĕ)	194-95	48°13′N	18°37′E
Le Vigan, Fr. (lē vē-gäɴ′)	196-97	43°59′N	3°35′E
Lévis, Qc., Can. (lā-vē′) (lē′vĭs)	136-37	46°48′N	71°11′W
Levittown, Pa., U.S. (lĕ′vĭt-toun)	116-17	40°09′N	74°51′W
Levkosía, nat. cap., Cyp. *see* Nicosia	228-29	35°10′N	33°22′E
Lewes, De., U.S. (lo͞o′ĭs)	116-17	38°46′N	75°08′W
Lewis, Isle of, i., Scot., U.K. (īl ŏv lo͞o′ĭs)	190-91	58°08′N	6°45′W
Lewisburg, Tn., U.S. (lū′ĭs-bûrg)	124-25	35°27′N	86°48′W
Lewisburg, W.V., U.S. (lū′ĭs-bûrg)	116-17	37°48′N	80°27′W
Lewisporte, Nf., Can. (lū′ĭs-pōrt)	138-39	49°16′N	55°05′W
Lewiston, Id., U.S. (lū′ĭs-tŭn)	112-13	46°24′N	117°00′W
Lewiston, Me., U.S. (lū′ĭs-tŭn)	116-17	44°06′N	70°13′W
Lewistown, Il., U.S. (lū′ĭs-toun)	120-21	40°23′N	90°09′W
Lewistown, Mt., U.S. (lū′ĭs-toun)	112-13	47°04′N	109°26′W
Lexington, Il., U.S. (lĕk′sĭng-tŭn)	116-17	40°38′N	88°46′W
Lexington, Ky., U.S. (lĕk′sĭng-tŭn)	116-17	38°03′N	84°31′W
Lexington, Ma., U.S. (lĕk′sĭng-tŭn)	116-17	42°27′N	71°14′W
Lexington, Mo., U.S. (lĕk′sĭng-tŭn)	120-21	39°11′N	93°53′W
Lexington, N.C., U.S. (lĕk′sĭng-tŭn)	124-25	35°49′N	80°15′W
Lexington, Ne., U.S. (lĕk′sĭng-tŭn)	114-15	40°47′N	99°44′W
Lexington, Oh., U.S. (lĕk′sĭng-tŭn)	116-17	40°41′N	82°34′W
Lexington, Tn., U.S. (lĕk′sĭng-tŭn)	124-25	35°39′N	88°24′W
Lexington, Va., U.S. (lĕk′sĭng-tŭn)	116-17	37°46′N	79°27′W
Leyte, i., Phil. (lā′tā)	250	10°50′N	124°50′E
Leyte Gulf, b., Phil.	250	10°50′N	125°25′E
Leżajsk, Pol. (lĕ′zhä-ĭsk)	194-95	50°16′N	22°26′E
L'gov, Russia (lgôf)	202-03	51°39′N	35°16′E
Lhasa, China (läs′ä)	234-35	29°39′N	91°08′E
Lhasa, stm., China	234-35	29°20′N	90°46′E
Lhokseumawe, Indon.	246-47	5°11′N	97°08′E
Lhorong, China	238-39	30°47′N	95°51′E
Li, stm., China	238-39	29°12′N	112°11′E
Liangzhou, China *see* Wuwei	240-41	37°56′N	102°38′E
Lianjiang, China (lĭĕn-jyäŋ)	225a	26°12′N	119°31′E
Lianxian, China	238-39	24°47′N	112°21′E
Lianyungang, China (lĭĕn-yòn-gäŋ)	238-39	34°37′N	119°11′E
Lianzhou, China *see* Hepu	238-39	21°41′N	109°11′E
Liao, stm., China	240-41	40°41′N	122°09′E

n-sing; ŋ-baŋk; ɴ-nasalized n; nŏd; cŏmmit; ōld; ôbey; ôrder; oi-boil; fo͞od; ŏ-as oo in foot; ou-out; s-soft; sh-dish; th-thin; pūre; ûnite; ûrn; stŭd; circŭs; ü-as in French tu; ′-indeterminate vowel.

Feature (Pronunciation)	Page	Lat.	Long.
Liaocheng, China (lîou-chŭŋ)	240-41	36°27′N	115°59′E
Liaodong Bandao, pen., China (lîou-dôŋ bän-dou)	240-41	39°55′N	122°19′E
Liaodong Wan, b., China (lîou-dôŋ wäŋ)	240-41	40°30′N	121°30′E
Liaoning, state, China	243	41°0′N	123°00′E
Liaotung, Gulf of, b., China *see* Liaodong Wan	240-41	40°30′N	121°30′E
Liaotung Peninsula, pen., China *see* Liaodong Bandao	240-41	39°55′N	122°19′E
Liaoyang, China (lyä′ō-yäng′)	243	41°16′N	123°10′E
Liaoyuan, China (lîou-yůän)	240-41	42°55′N	125°08′E
Liard, stm., Can. (lē-är′)	130-31	61°51′N	121°19′W
Lib, i., Marsh. Is.	280-81	8°19′N	167°25′E
Libagon, Phil.	250	10°18′N	125°03′E
Líbano, Col. (lē′bá-nō)	163c	4°56′N	75°04′W
Libau, Lat. *see* Liepāja	192-93	56°31′N	21°01′E
Libby, Mt., U.S. (lĭb′ê)	112-13	48°23′N	115°33′W
Libenge, D.R.C. (lē-bĕŋ′gā)	262-63	3°39′N	18°38′E
Liberal, Ks., U.S. (lĭb′ẽr-ăl)	120-21	37°02′N	100°56′W
Liberec, Czech Rep. (lē′bĕr-ĕts)	194-95	50°46′N	15°04′E
Liberia, C.R.	149	10°37′N	85°26′W
Liberia, nation, Afr. (lī-bē′rĭ-ȧ)	253	6°30′N	9°30′W
Liberty, Ky., U.S. (lĭb′ẽr-tĭ)	124-25	37°19′N	84°56′W
Liberty, Mo., U.S. (lĭb′ẽr-tĭ)	120-21	39°15′N	94°25′W
Liberty, N.Y., U.S. (lĭb′ẽr-tĭ)	116-17	41°48′N	74°44′W
Liberty, S.C., U.S. (lĭb′ẽr-tĭ)	124-25	34°47′N	82°42′W
Liberty, Tx., U.S. (lĭb′ẽr-tĭ)	122-23	30°04′N	94°48′W
Lībiyā, nation, Afr. *see* Libya	253	27°0′N	17°00′E
Lībīyah, Aṣ-Ṣaḥrā' al-, des., Afr. *see* Libyan Desert	254	24°0′N	25°00′E
Libourne, Fr. (lē-bo͞orn′)	196-97	44°55′N	0°14′W
Libres, Mex. (lē′brās)	146-47	19°28′N	97°41′W
Libreville, nat. cap., Gabon (lē-br′vēl′)	260-61	0°24′N	9°28′E
Libya, nation, Afr. (lĭb′ê-ä)	253	27°0′N	17°00′E
Libyan Desert, des., Afr. (lĭb′ê-ăn dĕs′ẽrt)	254	24°0′N	25°00′E
Licancábur, Volcán, vol., S.A.	168-69	22°50′S	67°50′W
Licantén, Chile (lē-kän-tĕ′n)	171	34°59′S	72°06′W
Lichinga, Moz.	264-65	13°17′S	35°15′E
Lichtenburg, S. Afr. (lĭk′tĕn-bĕrgh)	269c	26°10′S	26°10′E
Licking, stm., Ky., U.S. (lĭk′ĭng)	116-17	39°05′N	84°31′W
Licungo, stm., Moz.	264-65	17°38′S	37°22′E
Lida, Bela. (lē′dȧ)	194-95	53°54′N	25°18′E
Lidköping, Swe. (lēt′chû-pĭng)	192-93	58°30′N	13°11′E
Lidzbark, Pol. (lĭts′bärk)	194-95	53°16′N	19°50′E
Liechtenstein, nation, Eur. (lēk′tĕn-shtīn)	194-95	47°09′N	9°35′E
Liège, Bel.	190-91	50°38′N	5°34′E
Liegnitz, Pol. *see* Legnica	194-95	51°13′N	16°10′E
Lienz, Aus. (lē-ĕnts′)	194-95	46°50′N	12°46′E
Liepāja, Lat. (le′pä-yä′)	192-93	56°31′N	21°01′E
Lietuva, nation, Eur. *see* Latvia	174-75	56°0′N	24°00′E
Lièvre, stm., Qc., Can.	136-37	45°31′N	75°26′W
Lifou, i., N. Cal.	279g	20°44′S	167°14′E
Ligao, Phil. (lē-gä′ō)	250	13°14′N	123°34′E
Ligonha, stm., Moz. (lē-gō′nyȧ)	264-65	16°53′S	39°08′E
Ligonier, In., U.S. (lĭg-ô-nēr)	116-17	41°28′N	85°34′W
Lihir Group, is., Pap. N. Gui.	277	2°55′S	152°36′E
Līhu'e, Hi., U.S. (lē-ho͞o′ā)	127a	21°59′N	159°22′W
Liivi laht, b., Eur. *see* Riga, Gulf of	192-93	57°30′N	23°35′E
Lijiang, China (lē-jyäŋ)	238-39	26°52′N	100°14′E
Likasi, D.R.C.	262-63	10°59′S	26°43′E
Likhoslavl, Russia (lyê-kôsläv′'l)	202-03	57°07′N	35°28′E
Likouala, stm., Congo	262-63	1°12′S	16°49′E
Lille, Fr. (lēl)	196-97	50°39′N	3°07′E
Lillehammer, Nor. (lēl′ĕ-häm′mẽr)	192-93	61°07′N	10°28′E
Lillesand, Nor. (lēl′ĕ-sän′)	192-93	58°15′N	8°24′E
Lillestrøm, Nor. (lēl′ĕ-strûm)	192-93	59°58′N	11°04′E
Lillooet, B.C., Can. (lĭ′lo͞o-ĕt)	132-33	50°42′N	121°56′W
Lillooet, stm., B.C., Can. (lĭ′lo͞o-ĕt)	132-33	49°45′N	122°08′W
Lilongwe, nat. cap., Malawi (lē-lô-än)	264-65	13°59′S	33°44′E
Liloy, Phil.	250	8°07′N	122°40′E
Lima, Oh., U.S. (lī′mȧ)	116-17	40°45′N	84°07′W
Lima, nat. cap., Peru (lē′mä)	170	12°04′S	77°03′W
Limassol, Cyp. *see* Lemesós	228-29	34°41′N	33°03′E
Limay, stm., Arg. (lē-mä′ê)	171	38°59′S	68°00′W
Limbaži, Lat. (lēm′bä-zi)	192-93	57°31′N	24°43′E
Limeira, Braz. (lē-mā′rä)	172	22°34′S	47°24′W
Limerick, Ire. (lĭm′nẽk)	190-91	52°40′N	8°38′W
Límnos, i., Grc.	200-01	39°55′N	25°18′E
Limoges, Fr.	196-97	45°50′N	1°15′E
Limón, Hond. (lē-mō′n)	149	15°51′N	85°31′W
Limon, Co., U.S. (lī′mŏn)	120-21	39°16′N	103°42′W
Limoux, Fr. (lē-mo͞o′)	196-97	43°04′N	2°12′E
Limpopo, stm., Afr. (lĭm-pō′pō)	264-65	25°12′S	33°31′E
Limpopo, Grande Parque Transfronteiriço do, n.p., Afr. *see* Great Limpopo Transfrontier Park	264-65	23°0′S	31°30′E
Limpopo, Parque Nacional do, n.p., Moz.	264-65	23°21′S	31°54′E
Linapacan Island, i., Phil.	250	11°27′N	119°49′E
Linares, Chile (lē-nä′rās)	171	35°50′S	71°36′W
Linares, Mex.	122-23	24°51′N	99°34′W
Linares, Spain (lē-nä′rĕs)	198-99	38°06′N	3°38′W
Lincoln, Arg. (lĭŋ′kŭn)	173	34°52′S	61°31′W
Lincoln, Eng., U.K. (lĭŋ′kŭn)	190-91	53°14′N	0°33′W
Lincoln, Il., U.S. (lĭŋ′kŭn)	116-17	40°09′N	89°22′W
Lincoln, Ks., U.S. (lĭŋ′kŭn)	120-21	39°02′N	98°09′W
Lincoln, Me., U.S. (lĭŋ′kŭn)	117a	45°22′N	68°31′W
Lincoln, Ne., U.S.	114-15	40°48′N	96°43′W
Lincoln, Mount, mtn., Co., U.S. (mount lĭŋ′kŭn)	118-19	39°21′N	106°07′W
Lincolnton, N.C., U.S. (lĭŋ′kŭn-tŭn)	124-25	35°28′N	81°15′W
Lindale, Ga., U.S. (lĭn′dāl)	124-25	34°12′N	85°11′W
Linden, Guy.	164-65	6°05′N	58°17′W
Linden, Al., U.S. (lĭn′dĕn)	124-25	32°19′N	87°48′W
Linden, Tx., U.S. (lĭn′dĕn)	120-21	33°01′N	94°22′W
Lindesberg, Swe. (lĭn′dĕs-bĕrgh)	192-93	59°36′N	15°13′E
Lindesnes, c., Nor. (lĭn′ĕs-nĕs)	192-93	58°0′N	7°02′E
Lindi, Tan. (lĭn′dê)	264-65	9°60′S	39°42′E
Lindi, stm., D.R.C.	262-63	0°33′N	25°05′E
Lindian, China (lĭn-dĭĕn)	240-41	47°11′N	124°52′E
Lindley, S. Afr. (lĭnd′lê)	269c	27°53′S	27°56′E
Lindsay, On., Can. (lĭn′zê)	136-37	44°21′N	78°44′W
Lindsay, Ok., U.S. (lĭn′zê)	120-21	34°50′N	97°37′W
Line Islands, is., Oc. (lĭn ī′lăndz)	280-81	0°05′N	157°00′W
Linfen, China	240-41	36°05′N	111°31′E
Lingao, China (lĭn-gou)	238-39	19°54′N	109°40′E
Lingayen, Phil. (lĭŋ′gä-yān′)	250	16°01′N	120°14′E
Lingayen Gulf, b., Phil.	250	16°15′N	120°14′E
Lingen, Ger. (lĭŋ′gĕn)	194-95	52°31′N	7°19′E
Lingga, Kepulauan, is., Indon.	246-47	0°05′S	104°35′E
Lingling, China (lĭŋ-lĭŋ) *see* Yongzhou	238-39	26°13′N	111°37′E
Lingyuan, China (lĭŋ-yůän)	240-41	41°15′N	119°16′E
Linh, Ngoc, mtn., Viet.	246-47	15°04′N	107°59′E
Linhai, China	238-39	28°51′N	121°07′E
Linhe, China (lĭn-hŭ)	240-41	40°49′N	107°30′E
Linjiang, China (lĭn-jyäŋ)	243	41°49′N	126°55′E
Linköping, Swe. (lĭn′chû-pĭng)	192-93	58°25′N	15°37′E
Linkou, China	244	45°19′N	130°16′E
Linqing, China (lĭn-chyĭŋ)	240-41	36°51′N	115°42′E
Linqu, China (lĭn-chyo͞o)	240-41	36°31′N	118°32′E
Linru, China	238-39	34°10′N	112°50′E
Lins, Braz. (lē′ns)	168-69	21°40′S	49°45′W
Lintao, China	240-41	35°22′N	103°51′E
Linton, In., U.S. (lĭn′tŭn)	116-17	39°02′N	87°10′W
Linton, N.D., U.S. (lĭn′tŭn)	114-15	46°16′N	100°14′W
Linxi, China (lĭn-shyē)	240-41	43°36′N	118°03′E
Linxia, China	240-41	35°36′N	103°13′E
Linyi, China (lĭn-yē)	240-41	35°04′N	118°22′E
Linyi, China (lĭn-yē)	240-41	37°11′N	116°52′E
Linz, Aus. (lĭnts)	194-95	48°18′N	14°18′E
Lion, Golfe du, b., Fr.	196-97	43°0′N	4°00′E
Lipa, Phil. (lē-pä′)	250	13°56′N	121°10′E
Lipari, Isola, i., Italy (ê′-sō-lä-lē′pä-rê)	200-01	38°29′N	14°56′E
Lipetsk, Russia (lyē′pĕtsk)	202-03	52°37′N	39°37′E
Lípez, Cerro, mtn., Bol.	168-69	21°55′S	66°53′W
Liping, China (lē-pĭŋ)	238-39	26°17′N	108°60′E
Lippe, stm., Ger. (lĭp′ĕ)	194-95	51°39′N	6°37′E
Lippstadt, Ger. (lĭp′shtät)	194-95	51°40′N	8°20′E
Lipu, China (lē-po͞o)	238-39	24°25′N	110°29′E
Lira, Ug.	267	2°15′N	32°54′E
Lisakovsk, Kaz.	226	52°32′N	62°33′E
Lisala, D.R.C. (lê-sä′lä)	262-63	2°09′N	21°31′E
Lisboa, nat. cap., Port. (lêzh-bō′ä) *see* Lisbon	198-99	38°43′N	9°08′W
Lisbon, N.D., U.S. (lĭz′bŭn)	114-15	46°26′N	97°41′W
Lisbon, Oh., U.S. (lĭz′bŭn)	116-17	40°46′N	80°46′W
Lisbon, nat. cap., Port. (lĭz′bŭn)	198-99	38°43′N	9°08′W
Lisbon Falls, Me., U.S. (lĭz′bŭn fôlz)	116-17	44°00′N	70°03′W
Lisburn, N. Ire., U.K. (lĭs′bŭrn)	190-91	54°31′N	6°03′W
Lisburne, Cape, c., Ak., U.S.	126	68°52′N	166°14′W
Lishui, China (lĭ′shwĭ′)	238-39	28°27′N	119°54′E
Lisianski Island, i., Hi., U.S.	127	26°02′N	174°00′W
Lisichansk, Ukr. *see* Lysychans'k	202-03	48°55′N	38°26′E
Lisieux, Fr. (lē-zyû′)	196-97	49°09′N	0°14′E
Liski, Russia (lyēs′kê)	202-03	50°59′N	39°31′E
Lismore, Austl. (lĭz′môr)	276	28°49′S	153°17′E
Litang, China	238-39	23°12′N	109°09′E
Litang, China	238-39	29°60′N	100°16′E
Litang, stm., China	238-39	28°03′N	101°32′E
Litchfield, Il., U.S. (lĭch′fēld)	116-17	39°10′N	89°39′W
Litchfield, Mn., U.S. (lĭch′fēld)	114-15	45°08′N	94°32′W
Lithgow, Austl. (lĭth′gō)	276	33°30′S	150°09′E
Lithuania, nation, Eur. (lĭth-û-ā-′nĭ-ȧ)	174-75	56°0′N	24°00′E
Litoměřice, Czech Rep. (lē′tô-myĕr′zhĭ-tsĕ)	194-95	50°33′N	14°08′E
Litovko, Russia	240-41	49°15′N	135°10′E
Little Abaco, i., Bah. (lĭt′'l ä′bä-kō)	142-43	26°54′N	77°43′W
Little Andaman, i., India (lĭt′'l ăn-dȧ-măn′)	246-47	10°45′N	92°30′E
Little Belt Mountains, mts., Mt., U.S. (lĭt′'l bĕlt moun′tĭnz)	112-13	46°45′N	110°35′W
Little Bighorn, stm., U.S. (lĭt′'l bĭg-hôrn)	112-13	45°44′N	107°34′W
Little Bighorn Battlefield National Monument, n.p., Mt., U.S. (lĭt′'l bĭg-hôrn băt′'l-fēld năsh′ŭn-ăl mŏn′ū-mĕnt)	112-13	45°32′N	107°20′W
Little Cayman, i., Cay. Is. (lĭt′'l kā′mȧn) (lĭt′'l kī-män′)	142-43	19°42′N	80°02′W
Little Current, On., Can.	136-37	45°58′N	81°55′W
Little Exuma, i., Bah. (lĭt′'l ĕk-so͞o′mä)	142-43	23°27′N	75°37′W
Little Falls, Mn., U.S. (lĭt′'l fôlz)	114-15	45°59′N	94°22′W
Little Falls, N.Y., U.S. (lĭt′'l fôlz)	116-17	43°03′N	74°52′W
Littlefield, Tx., U.S. (lĭt′'l-fēld)	120-21	33°55′N	102°20′W
Little Inagua, i., Bah. (lĭt′'l ê-nä′gwä)	142-43	21°30′N	72°60′W
Little Karoo, plat., S. Afr.	264-65	33°45′S	21°30′E
Little Karroo, plat., S. Afr. (lĭt′'l kä-ro͞o) *see* Little Karoo	264-65	33°45′S	21°30′E
Little Missouri, stm., U.S. (lĭt′'l mĭ-so͞o′rĭ)	110-11	47°36′N	102°17′W
Little Nicobar, i., India	246-47	7°20′N	93°40′E
Little Powder, stm., U.S. (lĭt′'l pou′dẽr)	112-13	45°28′N	105°21′W
Little Rock, Ar., U.S.	120-21	34°43′N	92°19′W
Little Scarcies, stm., Afr.	260-61	8°51′N	13°07′W
Little Sioux, stm., U.S. (lĭt′'l so͞o)	114-15	41°49′N	96°06′W
Little Smoky, stm., Ab., Can. (lĭt′'l smōk′ĭ)	132-33	55°40′N	117°38′W
Littleton, Co., U.S. (lĭt′'l-tŭn)	120-21	39°35′N	105°01′W
Littleton, N.H., U.S. (lĭt′'l-tŭn)	116-17	44°18′N	71°46′W
Litzmannstadt, Pol. *see* Łódź	194-95	51°47′N	19°31′E
Liuaniua, at., Sol. Is. *see* Ontong Java	279e	5°19′S	159°16′E
Liubliana, nat. cap., Slvn. *see* Ljubljana	200-01	46°03′N	14°31′E
Liubotyn, Ukr.	202-03	49°56′N	35°57′E
Liuchow, China *see* Liuzhou	238-39	24°19′N	109°23′E
Liuyang, China (lyo͞o′yäng′)	238-39	28°08′N	113°38′E
Liuzhou, China (liô-jō)	238-39	24°19′N	109°23′E
Live Oak, Fl., U.S. (līv ōk)	124-25	30°18′N	82°59′W
Livermore, Ca., U.S. (lĭv′ẽr-mōr)	118-19	37°41′N	121°46′W
Livermore, Ky., U.S. (lĭv′ẽr-mōr)	116-17	37°29′N	87°08′W
Liverpool, N.S., Can. (lĭv′ẽr-po͞ol)	138-39	44°03′N	64°43′W
Liverpool, Eng., U.K. (lĭv′ẽr-po͞ol)	190-91	53°25′N	2°57′W
Liverpool Range, mts., Austl. (lĭv′ẽr-po͞ol rānj)	276	31°51′S	150°18′E
Livingston, Guat.	148	15°50′N	88°46′W
Livingston, Al., U.S. (lĭv′ĭng-stŭn)	124-25	32°36′N	88°12′W
Livingston, Mt., U.S. (lĭv′ĭng-stŭn)	112-13	45°40′N	110°34′W
Livingston, Tn., U.S. (lĭv′ĭng-stŭn)	124-25	36°23′N	85°19′W
Livingston, Tx., U.S. (lĭv′ĭng-stŭn)	122-23	30°42′N	94°57′W
Livingston, Lake, res., Tx., U.S.	122-23	30°43′N	95°08′W
Livingstone, Zam. (lĭv-ĭng-stôn)	264-65	17°52′S	25°51′E
Livingstone, Chutes de, wtfl., Afr. *see* Livingstone Falls	260-61	4°51′S	14°29′E
Livingstone Falls, wtfl., Afr.	260-61	4°51′S	14°29′E
Livno, Bos. (lēv′nô)	200-01	43°50′N	17°00′E
Livny, Russia (lēv′nê)	202-03	52°25′N	37°37′E
Livorno, Italy (lê-vôr′nō) (lĕg′hôrn)	200-01	43°34′N	10°19′E
Livramento, Braz. (lē-vrá-mĕ′n-tô) *see* Santana do Livramento	173	30°53′S	55°31′W
Lixi, China	238-39	29°15′N	114°47′E
Lixian, China (lē shyĕn)	238-39	29°30′N	111°38′E
Lixian, China (lē shyĕn)	238-39	34°09′N	105°07′E
Lixian, stm., Asia *see* Black	246-47	21°15′N	105°21′E
Lizard Point, c., Eng., U.K. (lĭz′ȧrd point)	190-91	49°58′N	5°13′W
Ljubljana, nat. cap., Slvn. (lyo͞o′blyä′na)	200-01	46°03′N	14°31′E
Ljungby, Swe. (lyông′bü)	192-93	56°50′N	13°56′E
Ljusdal, Swe. (lyo͞os′däl)	192-93	61°50′N	16°06′E
Ljusnan, stm., Swe.	184-85	61°09′N	17°10′E
Llandudno, Wales, U.K. (lăn-düd′nō)	190-91	53°19′N	3°50′W
Llanelli, Wales, U.K. (lȧ-nĕl′ĭ)	190-91	51°41′N	4°09′W
Llanes, Spain (lyä′nås)	198-99	43°25′N	4°45′W

ăt; finăl; rāte; senâte; ärm; ȧsk; sofȧ; fâre; ch-choose; dh-as th in other; bē; ėvent; bĕt; recĕnt; cratẽr; g-gō; gh-guttural g; bĭt; ĭ-short neutral; rīde; ĸ-guttural k as ch in German ich;

Feature (Pronunciation)	Page	Lat.	Long.
Llano, Tx., U.S. (lä´nō) (lyä´nō)	122-23	30°46′N	98°40′W
Llano, stm., Tx., U.S. (lä´nō) (lyä´nō)	122-23	30°39′N	98°25′W
Llanos, pl., S.A. (lyá´nōs)	164-65	5°0′N	70°00′W
Lleida, Spain	198-99	41°37′N	0°38′E
Lloydminster, Sk., Can.	132-33	53°17′N	110°01′W
Llullaillaco, Cerro, vol., S.A. see Llullaillaco, Volcán	168-69	24°43′S	68°33′W
Llullaillaco, Volcán, vol., S.A. (vôl-ká´n lyōō-lyī-lyä´kō)	168-69	24°43′S	68°33′W
Loa, stm., Chile	168-69	21°26′S	70°03′W
Loanda, Braz.	168-69	23°00′S	53°11′W
Loanda, nat. cap., Ang. see Luanda	264-65	8°49′S	13°14′E
Loange, stm., Afr. (lô-äŋ´gä)	262-63	4°17′S	20°02′E
Lobamba, nat. cap., Swaz. (lōō´-bäm-bä) (lō-bäm´-bä)	264-65	26°27′S	31°12′E
Lobaye, stm., C.A.R.	262-63	3°41′N	18°35′E
Lobería, Arg. (lô-bĕ′rē´ä)	173	38°09′S	58°47′W
Lobito, Ang. (lô-bē´tō)	264-65	12°21′S	13°33′E
Lobos, Arg. (lō´bōs)	173	35°11′S	59°06′W
Loches, Fr. (lôsh)	196-97	47°08′N	0°60′E
Lockhart, Tx., U.S. (lŏk´härt)	122-23	29°53′N	97°40′W
Lock Haven, Pa., U.S. (lŏk´hā-vĕn)	116-17	41°07′N	77°27′W
Loc Ninh, Viet. (lŏk´nĭng´)	246-47	11°53′N	106°37′E
Lodève, Fr. (lô-dĕv´)	196-97	43°43′N	3°19′E
Lodeynoye Pole, Russia (lô-dĕy-nô´yĕ)	192-93	60°44′N	33°34′E
Lodi, Italy (lô´dē)	200-01	45°19′N	9°30′E
Lodi, Ca., U.S. (lō´dī)	118-19	38°08′N	121°16′W
Lodi, Wi., U.S. (lō´dī)	116-17	43°19′N	89°31′W
Lodja, D.R.C.	262-63	3°26′S	23°27′E
Lodsch, Pol. see Łódź	194-95	51°47′N	19°31′E
Lodwar, Kenya	267	3°08′N	35°38′E
Łódź, Pol.	194-95	51°47′N	19°31′E
Loei, Thai.	246-47	17°27′N	101°31′E
Lofa, stm., Afr.	260-61	6°39′N	11°04′W
Loffa, stm., Afr. see Lofa	260-61	6°39′N	11°04′W
Lofoten, is., Nor. (lô´fō-tĕn)	184-85	68°08′N	14°10′E
Logan, N.M., U.S. (lō´gán)	120-21	35°22′N	103°25′W
Logan, Oh., U.S. (lō´gán)	116-17	39°32′N	82°25′W
Logan, Ut., U.S. (lō´gán)	112-13	41°45′N	111°50′W
Logan, W.V., U.S. (lō´gán)	116-17	37°51′N	81°60′W
Logan, Mount, mtn., Yk., Can. (mount lō´gán)	130-31	60°34′N	140°24′W
Logansport, In., U.S. (lō´gănz-pōrt)	116-17	40°45′N	86°21′W
Logone, stm., Afr. (lō-gō´nä) (lō-gôn´)	260-61	12°05′N	15°02′E
Logroño, Spain (lô-grō´nyō)	198-99	42°28′N	2°27′W
Løgstør, Den. (lügh-stûr´)	192-93	56°58′N	9°16′E
Loi-kaw, Mya.	246-47	19°40′N	97°13′E
Loire, stm., Fr. (lwâr)	196-97	47°18′N	2°00′W
Loja, Ec. (lō´hä)	170	3°59′S	79°12′W
Loja, Spain (lô´-kä)	198-99	37°10′N	4°09′W
Lokoro, stm., D.R.C.	262-63	1°43′S	18°22′E
Lol, stm., Sudan (lōl)	262-63	9°13′N	28°59′E
Lolland, i., Den. (lôl´än')	192-93	54°47′N	11°16′E
Lom, Blg. (lōm)	200-01	43°50′N	23°15′E
Lom, stm., Afr.	260-61	5°19′N	13°24′E
Lomami, stm., D.R.C.	262-63	0°47′N	24°17′E
Lomas de Zamora, Arg. (lō´mäs dā zä-mō´rä)	173	34°46′S	58°24′W
Lomblen, Pulau, i., Indon. (pōō-lou lŏm-blĕn´)	248-49	8°25′S	123°30′E
Lombok, i., Indon. (lŏm-bŏk´)	248-49	8°45′S	116°30′E
Lomé, nat. cap., Togo (lō´mē)	260-61	6°08′N	1°13′E
Lomela, stm., D.R.C. (lô-mā´lá)	262-63	0°18′S	20°45′E
Lomond, Loch, lk., Scot., U.K. (lŏk lō´mŭnd)	190-91	56°06′N	4°37′W
Lomonosov, Russia (lô-mô´nô-sof)	192-93	59°55′N	29°48′E
Lompoc, Ca., U.S. (lŏm-pōk´)	118-19	34°39′N	120°27′W
Łomża, Pol. (lôm´zhá)	194-95	53°11′N	22°05′E
Lonaconing, Md., U.S. (lō-nă-kō´nĭng)	116-17	39°34′N	78°58′W
Loncoche, Chile	171	39°22′S	72°38′W
London, On., Can. (lŭn´dŭn)	136-37	42°59′N	81°14′W
London, Ky., U.S. (lŭn´dŭn)	124-25	37°07′N	84°05′W
London, Oh., U.S. (lŭn´dŭn)	116-17	39°53′N	83°27′W
London, nat. cap., Eng., U.K. (lŭn´dŭn)	190-91	51°30′N	0°10′W
Londonderry, N. Ire., U.K.	190-91	54°59′N	7°20′W
Londonderry, Cape, c., Austl.	272-73	13°45′S	126°56′E
Londonderry, Isla, i., Chile	171	55°03′S	70°35′W
Londrina, Braz. (lôn-drē´nä)	168-69	23°19′S	51°10′W
Long Beach, Ca., U.S. (lông bēch)	118-19	33°46′N	118°12′W
Long Beach, Ms., U.S. (lông bēch)	124-25	30°21′N	89°10′W
Long Beach, Wa., U.S. (lông bēch)	112-13	46°22′N	124°03′W
Long Branch, N.J., U.S. (lông branch)	116-17	40°18′N	73°60′W
Long Cay, i., Bah. (lông kē)	142-43	22°35′N	74°22′W
Longchang, China	238-39	29°21′N	105°17′E
Long Eaton, Eng., U.K. (lông ē´tŭn)	190-91	52°53′N	1°16′W
Longford, Ire. (lŏng´fērd)	190-91	53°44′N	7°48′W
Long Island, i., Bah.	142-43	23°15′N	75°07′W
Long Island, i., N.S., Can. (lông ī´lănd)	138-39	44°20′N	66°16′W
Long Island, i., Nu., Can.	130-31	54°52′N	79°21′W
Long Island, i., Pap. N. Gui. (lông ī´lănd) see Arop Island	277	5°20′S	147°05′E
Long Island, i., N.Y., U.S. (lông ī´lănd)	116-17	40°47′N	73°17′W
Long Island Sound, strt., U.S. (lông ī´lănd sound)	116-17	41°05′N	72°58′W
Longjiang, China	240-41	47°20′N	123°11′E
Longkou, China (lôŋ-kō)	240-41	37°39′N	120°21′E
Long Lake, lk., On., Can. (lông lāk)	136-37	49°30′N	86°50′W
Longli, China	238-39	26°28′N	106°58′E
Longmont, Co., U.S. (lông´mŏnt)	120-21	40°10′N	105°06′W
Longnawan, Indon.	248-49	1°48′N	114°53′E
Long Point, c., Nf., Can. (lông point)	138-39	48°47′N	58°47′W
Long Point Bay, b., On., Can. (lông point bā)	136-37	42°40′N	80°14′W
Longquan, China	238-39	28°03′N	119°06′E
Long Range Mountains, mts., Nf., Can. (lông rānj moun´tĭnz)	138-39	49°20′N	57°39′W
Longreach, Austl. (lông´rēch)	277	23°27′S	144°15′E
Longs Peak, mtn., Co., U.S. (lôngz pēk)	112-13	40°16′N	105°37′W
Longueuil, Qc., Can. (lôn-gû´y')	136-37	45°32′N	73°30′W
Longview, Tx., U.S. (lông-vū)	120-21	32°30′N	94°44′W
Longview, Wa., U.S. (lông-vū)	112-13	46°08′N	122°57′W
Longwy, Fr. (lôN-wē´)	196-97	49°31′N	5°47′E
Longxi, China (lôŋ-shyē)	238-39	34°57′N	104°42′E
Long Xuyen, Viet. (loung´ sōō´yĕn)	246-47	10°23′N	105°26′E
Longzhou, China (lôŋ-jō)	238-39	22°21′N	106°51′E
Lonoke, Ar., U.S. (lō´nōk)	124-25	34°48′N	91°54′W
Lons-le-Saunier, Fr. (lôN-lē-sō-nyā´)	196-97	46°41′N	5°32′E
Lookout, Cape, c., N.C., U.S. (kāp căp lŏk-ôut)	124-25	34°36′N	76°32′W
Loop Head, c., Ire. (lōōp hĕd)	190-91	52°34′N	9°56′W
Lopatka, Mys, c., Russia (mĭs lô-pät´kà)	218-19	50°53′N	156°40′E
Lop Buri, Thai.	246-47	14°48′N	100°37′E
Lopévi, i., Vanuatu	279g	16°31′S	168°20′E
Lopez, Cap, c., Gabon	260-61	0°38′S	8°42′E
Lop Nor, lk., China see Lop Nur	222-23	40°29′N	90°16′E
Lop Nur, lk., China	222-23	40°29′N	90°16′E
Lopori, stm., D.R.C. (lō-pō´rē)	262-63	1°14′N	19°49′E
Lora, Hāmūn-i-, lk., Asia	232-33	29°17′N	64°47′E
Lorain, Oh., U.S. (lô-rān´)	116-17	41°28′N	82°11′W
Loralai, Pak. (lō-rŭ-lī´)	232-33	30°22′N	68°36′E
Lorca, Spain (lôr´kä)	198-99	37°41′N	1°41′W
Lord Howe Island, i., Austl. (lôrd hou ī´lănd)	272-73	31°34′S	159°06′E
Lordsburg, N.M., U.S. (lôrdz´bûrg)	118-19	32°21′N	108°42′W
Loreto, Mex.	144-45	26°01′N	111°21′W
Lorica, Col. (lô-rē´kä)	164-65	9°14′N	75°49′W
Lorient, Fr. (lô-rē´äN´)	196-97	47°45′N	3°22′W
Lörrach, Ger. (lûr´äk)	194-95	47°37′N	7°40′E
Lorraine, hist. reg., Fr.	196-97	49°0′N	6°00′E
Los Alamos, N.M., U.S. (lōs ăl-ȧ-mōs´)	118-19	35°53′N	106°18′W
Los Andes, Chile (lôs än´dĕs)	163e	32°50′S	70°36′W
Los Angeles, Chile (lôs äŋ´hā-lās)	171	37°27′S	72°19′W
Los Angeles, Ca., U.S. (lōs ăŋ´gĕl-ĕs)	118-19	34°03′N	118°14′W
Losap Atoll, at., Micron.	280-81	6°58′N	152°39′E
Los Gatos, Ca., U.S. (lōs gä´tōs)	118-19	37°13′N	121°59′W
Los Lagos, Chile	171	39°52′S	72°48′W
Los Mochis, Mex.	144-45	25°47′N	108°60′W
Los Roques, Islas, is., Ven.	164-65	11°50′N	66°45′W
Los Teques, Ven. (lōs tĕ´kĕs)	163b	10°21′N	67°02′W
Lost River Range, mts., Id., U.S. (lŏst rĭv´ēr rānj)	112-13	44°10′N	113°35′W
Losuia, Pap. N. Gui.	277	8°33′S	151°03′E
Los Vilos, Chile (lôs vē´lôs)	168-69	31°53′S	71°29′W
Lota, Chile (lō´tä)	171	37°05′S	73°09′W
Lothringen, hist. reg., Fr. see Lorraine	196-97	49°0′N	6°00′E
Lotung, Tai.	225a	24°41′N	121°46′E
Louangphrabang, Laos (lōō-ang´prä-bäng´)	246-47	19°52′N	102°08′E
Loubomo, Congo	262-63	4°11′S	12°40′E
Loudon, Tn., U.S. (lou´dŭn)	124-25	35°44′N	84°21′W
Louga, Sen.	260-61	15°37′N	16°13′W
Louisa, Ky., U.S. (lōō´ēz-à)	116-17	38°06′N	82°37′W
Louisa, Va., U.S. (lōō´ēz-à)	116-17	38°01′N	78°00′W
Louise Island, i., B.C., Can.	132-33	52°58′N	131°50′W
Louisiade Archipelago, is., Pap. N. Gui.	277	11°0′S	153°00′E
Louisiana, Mo., U.S. (lōō-ē-zē-ăn´ȧ)	120-21	39°27′N	91°04′W
Louisiana, state, U.S. (lōō-ē-zē-ăn´ȧ)	108-09	31°15′N	92°15′W
Louis Trichardt, S. Afr. (lōō´ĭs trĭchărt) see Makhado	264-65	23°03′S	29°55′E
Louisville, Ga., U.S. (lōō´ē-vĭl)	124-25	32°60′N	82°24′W
Louisville, Il., U.S. (lōō´ē-vĭl)	116-17	38°46′N	88°31′W
Louisville, Ky., U.S. (lōō´ē-vĭl)	116-17	38°15′N	85°46′W
Louisville, Ms., U.S. (lōō´ē-vĭl)	124-25	33°07′N	89°03′W
Louis-XIV, Pointe, c., Qc., Can.	130-31	54°38′N	79°45′W
Loukhi, Russia	186-87	66°04′N	33°03′E
Louny, Czech Rep. (lō´nē)	194-95	50°21′N	13°48′E
Lourdes, Fr. (lōōrd)	196-97	43°06′N	0°03′W
Lourenço Marques, nat. cap., Moz. see Maputo	264-65	25°58′S	32°35′E
Louviers, Fr. (lōō-vyā´)	196-97	49°13′N	1°10′E
Lovat', stm., Russia	192-93	58°13′N	31°27′E
Lovech, Blg. (lō´vĕts)	200-01	43°08′N	24°43′E
Loveland, Co., U.S. (lŭv´lănd)	120-21	40°25′N	105°06′W
Lovell, Wy., U.S. (lŭv´ĕl)	112-13	44°50′N	108°24′W
Lovelock, Nv., U.S. (lŭv´lŏk)	118-19	40°11′N	118°28′W
Loviisa, Fin. (lô´vē-sà)	192-93	60°27′N	26°13′E
Lovisa, Fin. see Loviisa	192-93	60°27′N	26°13′E
Low, Cape, c., Nu., Can. (kāp lō)	130-31	63°07′N	85°18′W
Lowa, stm., D.R.C. (lō´wà)	262-63	1°25′S	25°52′E
Lowell, Ma., U.S.	116-17	42°38′N	71°19′W
Lower Brule Indian Reservation, ind. res., S.D., U.S. (lō´ēr brü´lā ĭn´dĭ-ăn rĕ-sĕr-vā´shĕn)	114-15	44°05′N	99°54′W
Lower California, pen., Mex. see Baja California	144-45	27°53′N	113°28′W
Lower Hutt, N.Z. (lō´ēr hŭt)	278	41°13′S	174°56′E
Lower Post, B.C., Can.	128-29	59°56′N	128°29′W
Lower Red Lake, lk., Mn., U.S. (lō´ēr rĕd lāk)	114-15	47°57′N	95°01′W
Lower Zambezi National Park, n.p., Zam.	264-65	15°32′S	29°56′E
Lowestoft, Eng., U.K. (lō´stŏf)	190-91	52°29′N	1°44′E
Łowicz, Pol. (lô´vĭch)	194-95	52°06′N	19°57′E
Loxton, Austl. (lôks´tŭn)	276	34°27′S	140°34′E
Loyalty Islands, is., N. Cal.	279g	21°0′S	167°00′E
Loyauté, Îles, is., N. Cal. see Loyalty Islands	279g	21°0′S	167°00′E
Loznica, Serb. (lōz´nē-tsá)	200-01	44°32′N	19°14′E
Lualaba, stm., D.R.C. (lōō-ȧ-lä´bȧ)	262-63	0°22′N	25°21′E
Luama, stm., D.R.C. (lōō´ä-má)	262-63	4°46′S	26°53′E
Lu'an, China (lōō-än)	238-39	31°44′N	116°29′E
Luan, stm., China	240-41	39°24′N	119°17′E
Luanda, nat. cap., Ang. (lōō-än´dä)	264-65	8°49′S	13°14′E
Luando, stm., Ang.	264-65	10°21′S	16°27′E
Luanginga, stm., Afr. see Luanguinga	264-65	15°12′S	22°55′E
Luang Prabang, Laos see Louangphrabang	246-47	19°52′N	102°08′E
Luangue, stm., Afr.	262-63	4°17′S	20°02′E
Luanguinga, stm., Afr. (lōō-ä-gĭŋ´gȧ)	264-65	15°12′S	22°55′E
Luangwa, Zam.	264-65	15°37′S	30°25′E
Luanshya, Zam.	264-65	13°08′S	28°24′E
Luapula, stm., Afr.	264-65	9°24′S	28°31′E
Lubaczów, Pol. (lōō-bä´chôf)	194-95	50°10′N	23°07′E
Lubań, Pol. (lōō´bän´)	194-95	51°07′N	15°18′E
Lubang, Phil. (lōō-bäng´)	250	13°51′N	120°07′E
Lubang Island, i., Phil.	250	13°46′N	120°11′E
Lubang Islands, is., Phil. (lōō-bäng´ ī´lăndz)	250	13°45′N	120°17′E
Lubango, Ang.	264-65	14°55′S	13°30′E
Lubāns, lk., Lat. (lōō-bä´nás)	192-93	56°46′N	26°53′E
Lubartów, Pol. (lōō-bär´tôf)	194-95	51°28′N	22°37′E
Lubbock, Tx., U.S. (lŭb´ŭk)	120-21	33°34′N	101°51′W
Lübeck, Ger. (lü´bĕk)	194-95	53°52′N	10°40′E
Lubiana, nat. cap., Slvn. see Ljubljana	200-01	46°03′N	14°31′E
Lubilash, stm., D.R.C. (lōō-bē-lâsh´)	262-63	6°03′S	23°45′E
Lublin, Pol. (lyō´blēn´)	194-95	51°14′N	22°35′E
Lubnān, nation, Asia see Lebanon	206-07	34°0′N	36°00′E
Lubny, Ukr. (lób´nē)	202-03	50°01′N	33°00′E
Lubuagan, Phil. (lò-bwä-gä´n)	250	17°22′N	121°10′E
Lubudi, D.R.C. (lò-bò´dē)	262-63	9°57′S	25°58′E
Lubudi, stm., D.R.C. (lò-bò´dē)	262-63	4°02′S	21°23′E
Lubudi, stm., D.R.C. (lò-bò´dē)	262-63	9°13′S	25°38′E
Lubumbashi, D.R.C. (lōō-bŭm-bä´shē)	262-63	11°41′S	27°28′E
Lucca, Italy (lōōk´kä)	200-01	43°51′N	10°30′E
Lucena, Phil. (lōō-sā´nä)	250	13°56′N	121°37′E
Lucena, Spain (lōō-thā´nä)	198-99	37°24′N	4°29′W
Lučenec, Slvk. (lōō´châ-nyĕts)	194-95	48°20′N	19°40′E
Lucera, Italy (lōō-châ´rä)	200-01	41°31′N	15°20′E
Lucerne, Switz. see Luzern	194-95	47°03′N	8°19′E
Lucipara, Kepulauan, is., Indon.	248-49	5°33′S	127°27′E
Lucknow, India (lŭk´nou)	234-35	26°52′N	80°55′E
Luçon, Fr. (lü-sôN´)	196-97	46°27′N	1°10′W
Lüda, China see Dalian	240-41	38°54′N	121°34′E
Lüderitz, Nmb. (lü´dĕr-ĭts) (lü´dĕ-rĭts)	264-65	26°39′S	15°09′E

n-sing; ŋ-baŋk; N-nasalized n; nŏd; cŏmmit; ōld; ôbey; ôrder; oi-boil; fōōd; ò-as oo in foot; ou-out; s-soft; sh-dish; th-thin; pūre; ůnite; ûrn; stŭd; circŭs; ü-as in French tu; ´-indeterminate vowel.

Feature (Pronunciation)	Page	Lat.	Long.
Ludhiāna, India	234-35	30°54'N	75°51'E
Ludington, Mi., U.S. (lŭd´ĭng-tŭn)	116-17	43°57'N	86°26'W
Ludlow, Eng., U.K. (lŭd´lō)	190-91	52°22'N	2°43'W
Ludvika, Swe. (loodh-vē´kȧ)	192-93	60°08'N	15°11'E
Ludza, Lat. (lōōd´zȧ)	192-93	56°32'N	27°44'E
Luena, Ang.	264-65	11°47'S	19°54'E
Luena, stm., Ang.	264-65	12°30'S	22°34'E
Lufeng, China	238-39	22°56'N	115°37'E
Lufira, stm., D.R.C. (lōō-fē´rȧ)	262-63	8°21'S	26°26'E
Lufkin, Tx., U.S. (lŭf´kĭn)	122-23	31°20'N	94°43'W
Luga, Russia (lōō´gȧ)	192-93	58°44'N	29°52'E
Luga, stm., Russia (lōō´gȧ)	192-93	59°40'N	28°18'E
Lugano, Switz. (lōō-gä´nō)	194-95	46°01'N	8°57'E
Lugenda, stm., Moz.	264-65	11°25'S	38°29'E
Lugo, Italy (lōō´gō)	200-01	44°25'N	11°55'E
Lugo, Spain (lōō´gō)	198-99	43°01'N	7°33'W
Luhans'k, Ukr.	202-03	48°34'N	39°20'E
Luik, Bel. *see* Liège	190-91	50°38'N	5°34'E
Luimneach, Ire. *see* Limerick	190-91	52°40'N	8°38'W
Lukanga Swamp, sw., Zam. (lōō-käŋ´gȧ swŏmp)	264-65	14°25'S	27°45'E
Lukenie, stm., D.R.C. (lōō-kā´ynȧ)	262-63	2°44'S	18°10'E
Lukolela, D.R.C.	262-63	1°04'S	17°11'E
Łuków, Pol. (wō´kōf)	194-95	51°56'N	22°23'E
Lukuga, stm., D.R.C. (lōō-kōō´gȧ)	262-63	5°40'S	26°55'E
Lukula, D.R.C.	262-63	5°22'S	12°57'E
Lulaka, stm., D.R.C.	262-63	0°53'S	20°11'E
Luleå, Swe.	184-85	65°36'N	22°10'E
Lüleburgaz, Tur. (lü´lĕ-bōr-gäs´)	200-01	41°25'N	27°22'E
Lüliang Shan, mts., China	240-41	37°25'N	111°20'E
Luling, Tx., U.S. (lū´lĭng)	122-23	29°41'N	97°39'W
Lulonga, stm., D.R.C.	262-63	0°38'N	18°21'E
Lulua, stm., D.R.C.	262-63	5°02'S	21°06'E
Lumberton, Ms., U.S. (lŭm´bẽr-tŭn)	124-25	31°00'N	89°30'W
Lumberton, N.C., U.S. (lŭm´bẽr-tŭn)	124-25	34°38'N	79°01'W
Lumberton, Tx., U.S. (lŭm´bẽr-tŭn)	122-23	30°14'N	94°12'W
Lund, Swe. (lŭnd)	192-93	55°42'N	13°11'E
Lüneburg, Ger. (lü´nē-bōrgh)	194-95	53°16'N	10°25'E
Lunel, Fr. (lü-nĕl´)	196-97	43°40'N	4°08'E
Lunenburg, N.S., Can. (lōō´nĕn-bûrg)	138-39	44°23'N	64°19'W
Lunéville, Fr. (lü-nā-vel´)	196-97	48°36'N	6°30'E
Lunga, stm., Zam.	264-65	14°34'S	26°26'E
Lūni, stm., India	234-35	24°37'N	71°17'E
Lunsar, S.L.	260-61	8°41'N	12°32'W
Luo, stm., China	238-39	34°41'N	110°08'E
Luoding, China (lwô-dĭŋ)	238-39	22°47'N	111°33'E
Luohe, China (lwô-hŭ)	238-39	33°34'N	114°02'E
Luoyang, China (lwô-yäŋ)	238-39	34°41'N	112°27'E
Luqu, China	238-39	34°38'N	102°14'E
Luray, Va., U.S. (lū-rā´)	116-17	38°39'N	78°28'W
Lurgan, N. Ire., U.K. (lûr´găn)	190-91	54°28'N	6°20'W
Lurín, Peru	163a	12°17'S	76°52'W
Lúrio, stm., Moz.	264-65	13°30'S	40°32'E
Lusaka, nat. cap., Zam. (lō-sä´kȧ)	264-65	15°24'S	28°17'E
Lusambo, D.R.C. (lōō-säm´bō)	262-63	4°57'S	23°30'E
Lushan, China	238-39	30°15'N	102°58'E
Lu Shan, mtn., China	238-39	29°31'N	115°58'E
Lüshun, China (lü-shŭn)	240-41	38°49'N	121°15'E
Lusk, Wy., U.S. (lŭsk)	114-15	42°46'N	104°27'W
Lūt, Dasht-e, des., Iran (dä´sht-ē-lōōt)	232-33	32°0'N	58°00'E
Lutherstadt Wittenberg, Ger.	194-95	51°52'N	12°39'E
Luton, Eng., U.K. (lū´tŭn)	190-91	51°53'N	0°25'W
Lutong, Malay.	248-49	4°28'N	113°60'E
Luts'k, Ukr.	194-95	50°45'N	25°20'E
Lutzow-Holm Bay, b., Ant.	287	69°10'S	37°30'E
Luverne, Al., U.S. (lū-vûn´)	124-25	31°43'N	86°16'W
Luverne, Mn., U.S. (lū-vûn´)	114-15	43°39'N	96°13'W
Luvua, stm., D.R.C.	262-63	6°45'S	26°57'E
Luwegu, stm., Tan.	262-63	8°31'S	37°23'E
Luwuk, Indon.	248-49	0°56'S	122°47'E
Luxembourg, nation, Eur. *see* Luxemburg	190-91	49°45'N	6°05'E
Luxembourg, nat. cap., Lux.	190-91	49°37'N	6°07'E
Luxemburg, nation, Eur. (lŭk´-sŭm-bûrg)	190-91	49°45'N	6°05'E
Luxi, China	238-39	24°21'N	98°23'E
Luxor, Egypt	268b	25°42'N	32°39'E
Luza, Russia	186-87	60°37'N	47°16'E
Luzern, Switz. (lō-tsĕrn)	194-95	47°03'N	8°19'E
Luzhou, China (lōō-jō)	238-39	28°53'N	105°27'E
Luziânia, Braz. (lōō-zyȧ´nēȧ)	168-69	16°15'S	47°55'W
Lužická Nisa, stm., Eur. *see* Neisse	194-95	52°04'N	14°46'E
Luzon, i., Phil. (lōō-zŏn´)	250	16°0'N	121°00'E
Luzon Strait, strt., Asia (lōō-zŏn´ strāt)	238-39	20°30'N	121°00'E
L'viv, Ukr.	194-95	49°51'N	24°02'E
Lwów, Ukr. *see* L'viv	194-95	49°51'N	24°02'E
Lyallpur, Pak. *see* Faisalābād	232-33	31°25'N	73°05'E
Lycksele, Swe.	184-85	64°36'N	18°41'E
Lydenburg, S. Afr. (lī´dĕn-bûrg)	269c	25°08'S	30°27'E
Lykens, Pa., U.S. (lī´kĕnz)	116-17	40°34'N	76°43'W
Lynchburg, Va., U.S. (lĭnch´bûrg)	124-25	37°25'N	79°09'W
Lyndonville, Vt., U.S. (lĭn´dŭn-vĭl)	116-17	44°32'N	72°00'W
Lynn, Ma., U.S. (lĭn)	116-17	42°28'N	70°57'W
Lynn Lake, Mb., Can. (lĭn lāk)	134-35	56°51'N	101°00'W
Lyon, Fr. (lē-ôN´)	196-97	45°45'N	4°49'E
Lyons, Ga., U.S. (lī´ŭnz)	124-25	32°12'N	82°19'W
Lyons, Ks., U.S. (lī´ŭnz)	120-21	38°21'N	98°12'W
Lyons, Ne., U.S. (lī´ŭnz)	114-15	41°56'N	96°29'W
Lysekil, Swe. (lü´sĕ-kĕl)	192-93	58°17'N	11°27'E
Lys'va, Russia (lĭs´vȧ)	186-87	58°06'N	57°48'E
Lysychans'k, Ukr.	202-03	48°55'N	38°26'E
Lyuban', Russia (lyōō´bȧn)	192-93	59°21'N	31°15'E
Lyubertsy, Russia (lyōō´bẽr-tsê)	202-03	55°41'N	37°53'E
Lyudinovo, Russia (lū-dē´novō)	202-03	53°52'N	34°28'E

M

Feature (Pronunciation)	Page	Lat.	Long.
Ma, stm., Asia	246-47	19°47'N	105°52'E
Ma'ān, Jord. (mä-än´)	228-29	30°12'N	35°44'E
Ma'anshan, China	238-39	31°42'N	118°30'E
Maastricht, Neth. (mäs´trĭkt)	190-91	50°52'N	5°42'E
Mabank, Tx., U.S. (mā´bănk)	120-21	32°23'N	96°06'W
Macaé, Braz.	172	22°24'S	41°47'W
MacAlpine Lake, lk., Nu., Can.	130-31	66°38'N	102°51'W
Macapá, Braz.	166-67	0°03'N	51°03'W
Macará, Ec.	170	4°22'S	79°56'W
Macau, Braz. (mä-kà´ō)	163d	5°07'S	36°38'W
Macclesfield, Eng., U.K. (măk´'lz-fēld)	190-91	53°16'N	2°08'W
MacDonnell Ranges, mts., Austl. (măk-dŏn´ĕl rānjĕz)	272-73	23°52'S	133°14'E
MacDowell Lake, lk., On., Can. (măk-dou ĕl lāk)	134-35	52°15'N	92°45'W
Macdui, Ben, mtn., Scot., U.K. (bĕn măk-dōō´ê)	190-91	57°05'N	3°39'W
Macedonia, nation, Eur. (măs-ê-dō´nĭ-ȧ)	174-75	41°50'N	22°00'E
Macedonia, hist. reg., Eur. (măs-ê-dō´nĭ-ȧ)	200-01	41°0'N	23°00'E
Maceió, Braz.	163d	9°40'S	35°43'W
Macerata, Italy (mä-chȧ-rä´tä)	200-01	43°18'N	13°27'E
Macfarlane, Lake, lk., Austl. (lăk măc´fär-lān)	276	31°58'S	136°43'E
Machado, stm., Braz.	166-67	8°02'S	62°53'W
Machagai, Arg.	173	26°56'S	60°02'W
Machakos, Kenya	267	1°31'S	37°16'E
Machala, Ec. (mä-chä´lä)	170	3°16'S	79°57'W
Machilīpatnam, India	236	16°11'N	81°09'E
Machiques, Ven.	164-65	10°04'N	72°32'W
Machu Picchu, hist., Peru	170	13°07'S	72°34'W
Măcin, Rom. (mȧ-chēn´)	202-03	45°15'N	28°08'E
Mackay, Austl. (mȧ-kī´)	277	21°10'S	149°12'E
MacKay Lake, lk., N.T., Can. (măk-kā´ lāk)	130-31	63°54'N	110°23'W
Mackenzie, stm., N.T., Can. (mȧ-kĕn´zĭ)	130-31	58°60'N	111°25'W
Mackenzie Bay, b., Can. (mȧ-kĕn´zĭ bā)	126	69°0'N	136°30'W
Mackenzie Mountains, mts., Can. (mȧ-kĕn´zĭ moun´tĭnz)	130-31	64°0'N	130°00'W
Mackinaw City, Mi., U.S. (măk´ĭ-nô sĭ´tê)	116-17	45°46'N	84°43'W
Maclean, Austl.	276	29°28'S	153°13'E
Macleod, Lake, lk., Austl.	272-73	24°04'S	113°42'E
Macomb, Il., U.S. (mȧ-kōōm´)	120-21	40°28'N	90°40'W
Mâcon, Fr. (mä-kôN)	196-97	46°19'N	4°50'E
Macon, Ga., U.S. (mā´kŏn)	124-25	32°50'N	83°38'W
Macon, Mo., U.S. (mā´kŏn)	120-21	39°44'N	92°28'W
Macon, Ms., U.S. (mā´kŏn)	124-25	33°06'N	88°34'W
Macquarie, stm., Austl. (mȧ-kwŏr´ê)	276	30°08'S	147°23'E
Mada, stm., Nig.	260-61	7°59'N	7°58'E
Madagascar, nation, Afr. (măd-ȧ-găs´kȧr)	253	19°0'S	46°00'E
Madagasikara, nation, Afr. *see* Madagascar	253	19°0'S	46°00'E
Madame, Isle, i., N.S., Can. (ĭl mȧ-dȧm´)	138-39	45°33'N	61°02'W
Madang, Pap. N. Gui. (mä-däng´)	277	5°17'S	145°45'E
Madawaska, stm., On., Can. (măd-ȧ-wôs´kȧ)	136-37	45°27'N	76°21'W
Madeira, i., Port. (mä-dā´rä)	258-59	32°44'N	17°00'W
Madeira, stm., S.A. (mä-dā´-rȧ)	166-67	3°22'S	58°45'W
Madeira, Arquipélago da, is., Port. (är-kē-pĕ´lä-gō-dä-mädĕy´-rä) *see* Madeira Islands	258-59	32°40'N	16°45'W
Madeira Islands, is., Port. (mä-dā´rä ī´lăndz)	258-59	32°40'N	16°45'W
Madeleine, Îles de la, is., Qc., Can.	138-39	47°30'N	61°45'W
Madelia, Mn., U.S. (mȧ-dē´lĭ-ȧ)	114-15	44°03'N	94°25'W
Madhya Pradesh, state, India (mŭd´vŭ prŭ-dāsh´)	234-35	23°0'N	79°00'E
Madidi, stm., Bol.	166-67	12°31'S	66°58'W
Madikeri, India	236	12°25'N	75°45'E
Madill, Ok., U.S. (mȧ-dĭl´)	120-21	34°05'N	96°47'W
Madison, Al., U.S. (măd´ĭ-sŭn)	124-25	34°41'N	86°45'W
Madison, Fl., U.S. (măd´ĭ-sŭn)	124-25	30°28'N	83°25'W
Madison, Ga., U.S. (măd´ĭ-sŭn)	124-25	33°36'N	83°28'W
Madison, In., U.S. (măd´ĭ-sŭn)	116-17	38°44'N	85°23'W
Madison, Me., U.S. (măd´ĭ-sŭn)	116-17	44°48'N	69°53'W
Madison, Mn., U.S. (măd´ĭ-sŭn)	114-15	45°01'N	96°12'W
Madison, Ms., U.S. (măd´ĭ-sŭn)	124-25	32°28'N	90°07'W
Madison, N.C., U.S. (măd´ĭ-sŭn)	124-25	36°23'N	79°58'W
Madison, Ne., U.S. (măd´ĭ-sŭn)	114-15	41°50'N	97°27'W
Madison, S.D., U.S. (măd´ĭ-sŭn)	114-15	44°00'N	97°07'W
Madison, Wi., U.S. (măd´ĭ-sŭn)	116-17	43°05'N	89°22'W
Madison, W.V., U.S. (măd´ĭ-sŭn)	116-17	38°04'N	81°49'W
Madisonville, Ky., U.S. (măd´ĭ-sŭn-vĭl)	124-25	37°20'N	87°30'W
Madisonville, Tx., U.S. (măd´ĭ-sŭn-vĭl)	122-23	30°56'N	95°55'W
Madiun, Indon.	248-49	7°37'S	111°31'E
Madoi, China	240-41	34°55'N	98°12'E
Madona, Lat. (mȧ´dō´nȧ)	192-93	56°51'N	26°14'E
Madras, India *see* Chennai	236	13°06'N	80°15'E
Madras, state, India *see* Tamil Nādu	236	11°0'N	78°15'E
Madre, Laguna, b., Mex. (lä-ô´nä mä´drä)	122-23	25°01'N	97°40'W
Madre, Laguna, b., Tx., U.S.	122-23	26°58'N	97°26'W
Madre, Sierra, mts., Phil. (sē-ĕ´r-rä-mȧ´drĕ)	250	16°20'N	122°00'E
Madre de Dios, stm., S.A. (mä´drä dä dê-ōs´)	166-67	10°24'S	65°24'W
Madre de Dios, Isla, i., Chile (ê´s-lä-mȧ´drä dä dê-ōs´)	171	50°15'S	75°05'W
Madre del Sur, Sierra, mts., Mex. (sē-ĕ´r-rä-mä´drä dĕl-sōōr´)	146-47	17°0'N	100°00'W
Madre Occidental, Sierra, mts., Mex. (sē-ĕ´r-rä-mä´drē-äk-sĭ-dĕn´-tl)	144-45	25°0'N	105°00'W
Madre Oriental, Sierra, mts., Mex. (sē-ĕ´r-rä-mȧ´drĕ ō-rē-ĕn-täl´)	144-45	21°26'N	99°50'W
Madrid, nat. cap., Spain (mä-drē´d)	198-99	40°24'N	3°41'W
Madridejos, Spain (mä-dhrê-dhā´hōs)	198-99	39°28'N	3°32'W
Madurai, India (mä-dōō´rä)	236	9°55'N	78°08'E
Maebashi, Japan (mä-ē-bä´shê)	245	36°23'N	139°05'E
Mae Hong Son, Thai.	246-47	19°16'N	97°57'E
Mae Klong, stm., Thai.	246-47	13°22'N	99°60'E
Mae Sot, Thai.	246-47	16°43'N	98°35'E
Maestra, Sierra, mts., Cuba (sē-ĕ´r-rä-mä-ās´trä)	142-43	20°06'N	76°24'W
Maéwo, i., Vanuatu	279g	15°10'S	168°10'E
Mafeking, S. Afr. (măf´ĕ´kĭng) *see* Mafikeng	264-65	25°53'S	25°39'E
Mafia Island, i., Tan.	262-63	7°50'S	39°50'E
Mafikeng, S. Afr.	264-65	25°53'S	25°39'E
Mafra, Braz. (mä´frä)	168-69	26°08'S	49°49'W
Mafra, Port. (mäf´rȧ)	198-99	38°57'N	9°19'W
Magadan, Russia (mȧ-gȧ-dän´)	218-19	59°35'N	150°50'E
Magallanes, Chile *see* Punta Arenas	171	53°09'S	70°55'W
Magallanes, Estrecho de, strt., S.A.	171	54°0'S	71°00'W
Magangué, Col.	164-65	9°18'N	74°48'W
Magat, stm., Phil. (mä-gät´)	250	17°02'N	121°50'E
Magdagachi, Russia	222-23	53°27'N	125°49'E
Magdalena, Bol. (mäg-dä-lā´nä)	166-67	13°20'S	64°08'W
Magdalena, Mex. (mäg-dä-lā´nä)	146-47	20°54'N	103°57'W
Magdalena, N.M., U.S. (mäg-dä-lā´nä)	118-19	34°07'N	107°15'W
Magdalena, stm., Col. (mäg-dä-lā´nä)	164-65	11°06'N	74°51'W
Magdalena, Bahía, b., Mex. (bä-ē´ä-mäg-dä-lā´nä)	144-45	24°35'N	112°00'W
Magdalena, Isla, i., Chile (ê´s-lä-mäg-dä-lā´nä)	171	44°40'S	73°10'W
Magdalena de Kino, Mex.	144-45	30°38'N	110°58'W
Magdeburg, Ger. (mäg´dĕ-bōrgh)	194-95	52°08'N	11°38'E
Magelang, Indon.	248-49	7°28'S	110°13'E
Magellan, Strait of, strt., S.A. (strāt ŭv mȧ-gĕl´-ăn) *see* Magallanes, Estrecho de	171	54°0'S	71°00'W
Magerøya, i., Nor.	184-85	71°02'N	25°42'E
Magnesia, Tur. *see* Manisa	200-01	38°37'N	27°26'E
Magnetic Island, i., Austl.	277	19°08'S	146°50'E

ăt; finăl; rāte; senāte; ärm; åsk; sofȧ; fâre; ch-choose; dh-as th in other; bē; ēvent; bĕt; recĕnt; cratẽr; g-gō; gh-guttural g; bĭt; ĭ-short neutral; rīde; K-guttural k as ch in German ich;

Feature (Pronunciation)	Page	Lat.	Long.
Magnitogorsk, Russia (mág-nyē´tô-gôrsk)	226	53°26'N	59°04'E
Magnolia, Ar., U.S. (măg-nō´lĭ-à)	120-21	33°16'N	93°15'W
Magnolia, Ms., U.S. (măg-nō´lĭ-à)	124-25	31°09'N	90°28'W
Mago, i., Fiji	279f	17°27'S	179°09'W
Magog, Qc., Can. (mà-gŏg´)	136-37	45°16'N	72°09'W
Magpie, stm., On., Can. (Măg´pī)	136-37	47°56'N	84°50'W
Magpie, stm., Qc., Can. (Măg´pī)	138-39	50°19'N	64°27'W
Magpie, Lac, lk., Qc., Can. (läk măg´pī)	138-39	51°0'N	64°41'W
Maguari, Cabo, c., Braz.	166-67	0°18'S	48°22'W
Magway, Mya.	246-47	20°30'N	94°30'E
Magyarország, nation, Eur. see Hungary	174-75	47°0'N	20°00'E
Mahābād, Iran	228-29	36°46'N	45°44'E
Mahagi, D.R.C.	267	2°19'N	31°01'E
Mahajanga, Madag.	264-65	15°43'S	46°19'E
Mahakam, stm., Indon.	248-49	0°35'S	117°17'E
Mahalapye, Bots.	264-65	23°06'S	26°50'E
Mahalla el-Kubra, Egypt see El-Mahalla el-Kubra	268b	30°58'N	31°10'E
Mahānadi, stm., India	234-35	20°19'N	86°47'E
Mahanoro, Madag. (má-há-nō´rō)	264-65	19°55'S	48°48'E
Mahārāshtra, state, India	236	19°0'N	76°00'E
Maha Sarakham, Thai.	246-47	16°11'N	103°18'E
Mahbūbnagar, India	236	16°44'N	77°59'E
Mahe, India (mä-ā´)	236	11°42'N	75°32'E
Mahébourg, Mauritius	265a	20°24'S	57°42'E
Mahendra Giri, mtn., India	236	18°58'N	84°21'E
Mahendranagar, Nepal	234-35	28°58'N	80°10'E
Mahenge, Tan. (mä-hĕŋ´gå)	267	7°38'S	36°16'E
Mahesāna, India	234-35	23°36'N	72°23'E
Mahilëu, Bela.	192-93	53°56'N	30°21'E
Mahnomen, Mn., U.S. (mô-nō´mĕn)	114-15	47°19'N	95°59'W
Mahón, Spain see Maó	198-99	39°53'N	4°16'E
Mahone Bay, b., N.S., Can. (mà-hōn´ bā)	138-39	44°30'N	64°15'W
Maicuru, stm., Braz.	166-67	2°12'S	54°18'W
Maiduguri, Nig. (mä´ē-dä-gōō´rē)	260-61	11°51'N	13°09'E
Maiko, stm., D.R.C.	262-63	0°11'N	25°32'E
Maikop, Russia see Maykop	186-87	44°36'N	40°06'E
Mai-Ndombe, Lac, lk., D.R.C.	260-61	2°25'S	18°18'E
Maine, state, U.S. (mān)	108-09	45°15'N	69°15'W
Maine, Gulf of, b., N.A.	138-39	43°0'N	68°00'W
Mainland, i., Scot., U.K. (măn-lănd)	190c	60°16'N	1°16'W
Maintenon, Fr. (mäN-tĕ-nôN´)	196-97	48°35'N	1°35'E
Maintirano, Madag. (mä´ēn-tē-rä´nō)	264-65	18°03'S	44°02'E
Mainz, Ger. (mīnts)	194-95	50°00'N	8°16'E
Maio, i., C.V. (mä´yo)	260-61	15°11'N	23°10'W
Maipo, stm., Chile (mī´pō)	163e	33°37'S	71°38'W
Maipo, Volcán, vol., S.A. (vôl-kä´n mī´pō)	163e	34°10'S	69°50'W
Maipú, Arg.	173	36°52'S	57°54'W
Maiquetía, Ven. (mī-kĕ-tē´ä)	163b	10°36'N	66°58'W
Maitland, Austl. (māt´lănd)	276	32°44'S	151°33'E
Maitland, Austl. (māt´lănd)	276	34°23'S	137°40'E
Maíz, Islas del, is., Nic.	149	12°15'N	83°00'W
Maizuru, Japan (mä-ĭ´zōō-rōō)	245	35°28'N	135°24'E
Majene, Indon.	248-49	3°32'S	118°57'E
Majī, Eth.	262-63	6°11'N	35°35'E
Majorca, i., Spain (mā-jôr´-kà) see Mallorca	198-99	39°30'N	3°00'E
Majuro, at., Marsh. Is.	280-81	7°05'N	171°09'E
Makanya, Tan. (mä-kän´yä)	267	4°21'S	37°50'E
Makarov, Russia	222-23	48°38'N	142°46'E
Makarska, Cro. (má´kär-ská)	200-01	43°17'N	17°01'E
Makasar, Selat, strt., Indon. see Makassar Strait	248-49	2°0'S	117°30'E
Makassar, Indon. (mà-kă´-sŭr)	248-49	5°08'S	119°25'E
Makassar Strait, strt., Indon. (mà-kă´-sŭr strāt)	248-49	2°0'S	117°30'E
Makatea, i., Fr. Poly.	280-81	15°50'S	148°16'W
Makedonija, nation, Eur. see Macedonia	174-75	41°50'N	22°00'E
Makedonija, hist. reg., Eur. see Macedonia	200-01	41°0'N	23°00'E
Makeni, S.L.	260-61	8°53'N	12°03'W
Makeyevka, Ukr. see Makiïvka	202-03	48°02'N	37°58'E
Makgadikgadi, pl., Bots.	264-65	20°17'S	25°43'E
Makhachkala, Russia (mäK´äch-kä´lä)	227	42°59'N	47°30'E
Makhado, S. Afr.	264-65	23°03'S	29°55'E
Makiïvka, Ukr.	202-03	48°02'N	37°58'E
Makindu, Kenya	267	2°16'S	37°50'E
Makinsk, Kaz.	226	52°39'N	70°25'E
Makkah, Sau. Ar. see Mecca	266	21°27'N	39°51'E
Makokou, Gabon (má-kô-kōō´)	260-61	0°35'N	12°51'E

Feature (Pronunciation)	Page	Lat.	Long.
Makona, stm., Afr. see Moa	260-61	6°60'N	11°34'W
Makoua, Congo	262-63	0°00'S	15°38'E
Makung, Tai.	225a	23°34'N	119°34'E
Makurdi, Nig.	260-61	7°44'N	8°31'E
Mala, Punta, c., Pan. (pó´n-tä-mä´lä)	150	7°28'N	80°01'W
Malabang, Phil.	250	7°38'N	124°04'E
Malabar Coast, cst., India (măl´á-bär kōst)	236	11°0'N	75°00'E
Malabo, nat. cap., Eq. Gui. (mä-lä´bō)	260-61	3°45'N	8°47'E
Malacca, Malay. see Kota Kinabalu	246-47	2°12'N	102°16'E
Malacca, Strait of, strt., Asia (strāt ŭv mà-lăk´à)	246-47	2°30'N	101°20'E
Malad City, Id., U.S. (mà-lăd´ sĭ´tē)	112-13	42°12'N	112°15'W
Maladzečna, Bela.	194-95	54°19'N	26°52'E
Málaga, Col. (má´lä-gà)	164-65	6°42'N	72°44'W
Málaga, Spain (má´lä-gà)	198-99	36°44'N	4°25'W
Malagasy Republic, nation, Afr. see Madagascar	253	19°0'S	46°00'E
Malaita, i., Sol. Is. (mà-lä´ē-tà)	279e	9°0'S	161°00'E
Malaka, Malay. see Kota Kinabalu	246-47	2°12'N	102°16'E
Malaka, Selat, strt., Asia see Malacca, Strait of	246-47	2°30'N	101°20'E
Malakāl, Sudan (má-lá-käl´)	262-63	9°31'N	31°39'E
Malakula, i., Vanuatu (mä-lä-kōō´lä)	279g	16°15'S	167°30'E
Malang, Indon.	248-49	7°59'S	112°38'E
Malanje, Ang. (mä-läŋ-gå)	264-65	9°32'S	16°20'E
Malanville, Benin	260-61	11°52'N	3°23'E
Mälaren, lk., Swe.	192-93	59°30'N	17°12'E
Malargüe, Arg.	171	35°28'S	69°35'W
Malartic, Qc., Can.	136-37	48°09'N	78°07'W
Malatya, Tur. (má-lä´tyä)	186-87	38°21'N	38°18'E
Malawi, nation, Afr. (mä-lä´-wē)	253	13°30'S	34°00'E
Malaŵi, Lake, lk., Afr. (läk mä-lä´-wē) see Nyasa, Lake	264-65	12°0'S	34°30'E
Malaya Vishera, Russia	192-93	58°51'N	32°14'E
Malaybalay, Phil.	250	8°09'N	125°08'E
Malay Peninsula, pen., Asia (mà-lā´ pĕ-nĭn´sūlă) (mā´lā)	246-47	6°0'N	101°00'E
Malaysia, nation, Asia (má-lā´zhá)	206-07	2°30'N	112°30'E
Malbork, Pol. (mäl´bôrk)	194-95	54°02'N	19°02'E
Malden, Mo., U.S. (môl´dĕn)	124-25	36°34'N	89°58'W
Malden, i., Kir. (môl´dĕn)	280-81	4°03'S	154°59'W
Maldive Islands, nation, Asia see Maldives	206-07	3°15'N	73°00'E
Maldives, nation, Asia (măl´dīvz) (môl´dēvz)	206-07	3°15'N	73°00'E
Maldonado, Ur. (mäl-dŏ-ná´dō)	173	34°55'S	54°57'W
Male', nat. cap., Mald. (mä-lā´)	236	4°10'N	73°30'E
Maléas, Ákra, c., Grc.	200-01	36°26'N	23°12'E
Male Atoll, at., Mald.	236	4°25'N	73°30'E
Mālegaon, India	234-35	20°33'N	74°32'E
Malheur Lake, lk., Or., U.S. (má-lōōr´ lāk)	112-13	43°20'N	118°45'W
Mali, nation, Afr. (mä´-lē)	253	17°0'N	4°00'W
Mali, stm., Mya.	238-39	25°43'N	97°31'E
Malik, Wādī al-, stm., Sudan	266	18°03'N	30°58'E
Mali Kyun, i., Mya.	246-47	13°06'N	98°16'E
Malinaltepec, Mex. (mä-lē-näl-tå-pĕk´)	146-47	17°05'N	98°39'W
Malindi, Kenya (mä-lēn´dē)	262-63	3°13'S	40°06'E
Malino, Bukit, mtn., Indon.	248-49	0°42'N	120°51'E
Malkara, Tur. (mäl´kà-rà)	200-01	40°52'N	26°55'E
Malko Tŭrnovo, Blg. (mäl´kō-t'r´nô-vä)	200-01	41°59'N	27°32'E
Mallawi, Egypt	268b	27°44'N	30°51'E
Mallery Lake, lk., Nu., Can.	130-31	63°55'N	98°25'W
Mallorca, i., Spain	198-99	39°30'N	3°00'E
Malmö, Swe.	192-93	55°36'N	13°01'E
Maloelap, at., Marsh. Is.	280-81	8°45'N	171°03'E
Malolos, Phil. (mä-lô´lôs)	250	14°51'N	120°49'E
Maloshuyka, Russia	186-87	63°44'N	37°25'E
Måløy, Nor.	184-85	61°56'N	5°08'E
Maloyaroslavets, Russia (mä´lô-yä-rô-slä-vyĕts)	202-03	55°01'N	36°28'E
Malpelo, Isla de, i., Col. (ē´s-lä-dĕ-mäl-pā´lō)	164-65	3°59'N	81°35'W
Malpeque Bay, b., P.E., Can. (môl-pĕk´ bā)	138-39	46°30'N	63°47'W
Malta, Mt., U.S. (môl´tà)	112-13	48°21'N	107°52'W
Malta, nation, Eur. (môl´tà)	174-75	35°50'N	14°35'E
Malta, i., Malta	200b	35°53'N	14°27'E
Maluku, is., Indon. see Moluccas	248-49	2°0'S	128°00'E
Maluku, Laut, s., Indon. see Molucca Sea	248-49	0°13'N	125°10'E
Malvern, Ar., U.S. (măl´vērn)	120-21	34°22'N	92°49'W
Malyy Anyuy, stm., Russia	218-19	68°31'N	160°55'E

Feature (Pronunciation)	Page	Lat.	Long.
Malyye Derbety, Russia	186-87	47°58'N	44°43'E
Malyy Kavkaz, mts., Asia see Lesser Caucasus	227	40°60'N	44°35'E
Malyy Shantar, Ostrov, i., Russia	218-19	54°30'N	137°36'E
Malyy Taymyr, Ostrov, i., Russia	218-19	78°08'N	107°12'E
Malyy Uzen', stm., Eur. see Balaözen	186-87	48°58'N	49°38'E
Mamberamo, stm., Indon.	277	1°35'S	137°52'E
Mambéré, stm., C.A.R.	262-63	3°32'N	16°03'E
Mamburao, Phil. (mäm-bōō´rä-ō)	250	13°15'N	120°35'E
Mammoth Cave National Park, n.p., Ky., U.S. (măm´ŏth kāv năsh´ŭn-ăl pärk)	124-25	37°11'N	86°08'W
Mamoré, stm., S.A.	166-67	10°24'S	65°23'W
Mamoudzou, nat. cap., May.	264-65	12°47'S	45°14'E
Mamry, Jezioro, lk., Pol. (mäm´rī)	194-95	54°07'N	21°44'E
Man, C. Iv.	260-61	7°24'N	7°33'W
Man, Isle of, dep., Eur. see Isle of Man	190-91	54°15'N	4°30'W
Manacapuru, Braz.	166-67	3°17'S	60°36'W
Manacor, Spain (mä-nä-kôr´)	198-99	39°34'N	3°12'E
Manado, Indon.	248-49	1°29'N	124°51'E
Managua, nat. cap., Nic. (mä-nä´gwä)	149	12°09'N	86°17'W
Managua, Lago de, lk., Nic. (lá´gô-dĕ-mä-nä´gwä)	149	12°20'N	86°20'W
Manakara, Madag. (mä-nä-kä´rŭ)	264-65	22°09'S	48°01'E
Manāli, India	234-35	32°16'N	77°09'E
Manama, nat. cap., Bahr. (mă-nă´mà) see Al-Manāmah	230-31	26°13'N	50°35'E
Manam Island, i., Pap. N. Gui.	277	4°05'S	145°02'E
Mananara, stm., Madag. (mä-nä-nä´rŭ)	264-65	23°21'S	47°42'E
Mananara Avaratra, Madag.	264-65	16°10'S	49°46'E
Mananjary, Madag. (mä-nän-zhä´rē)	264-65	21°14'S	48°21'E
Manáos, Braz. see Manaus	166-67	3°07'S	60°01'W
Mana Pools National Park, n.p., Zimb.	264-65	15°52'S	29°15'E
Manas Hu, lk., China	226	45°43'N	85°54'E
Manassas, Va., U.S. (mà-năs´às)	116-17	38°45'N	77°28'W
Manaus, Braz. (mä-nä´ōōzh)	166-67	3°07'S	60°01'W
Mancelona, Mi., U.S. (măn-sê-lō´ná)	116-17	44°54'N	85°04'W
Manchester, Eng., U.K. (măn´chĕs-tēr)	190-91	53°27'N	2°15'W
Manchester, Ct., U.S. (măn´chĕs-tēr)	116-17	41°47'N	72°31'W
Manchester, Ga., U.S. (măn´chĕs-tēr)	124-25	32°51'N	84°37'W
Manchester, Ia., U.S. (măn´chĕs-tēr)	114-15	42°29'N	91°28'W
Manchester, Ky., U.S. (măn´chĕs-tēr)	124-25	37°09'N	83°47'W
Manchester, N.H., U.S.	116-17	42°59'N	71°28'W
Manchester, Tn., U.S. (măn´chĕs-tēr)	124-25	35°29'N	86°05'W
Manchuria, hist. reg., China (măn-chōō´rē-ä)	240-41	47°0'N	125°00'E
Mand, stm., Iran	230-31	28°09'N	51°16'E
Manda Island, i., Kenya	262-63	2°15'S	40°57'E
Mandal, Nor. (män´däl)	192-93	58°02'N	7°27'E
Mandala, Puncak, mtn., Indon.	277	4°43'S	140°18'E
Mandalay, Mya. (măn´dà-lā)	246-47	21°58'N	96°05'E
Mandalgovĭ, Mong.	240-41	45°46'N	106°16'E
Mandalī, Iraq	228-29	33°45'N	45°32'E
Mandan, N.D., U.S. (măn´dăn)	114-15	46°49'N	100°55'W
Mandara, Monts, mts., Afr. see Mandara Mountains	260-61	10°45'N	13°40'E
Mandara Mountains, mts., Afr. (män-dä´rä moun´tĭnz)	260-61	10°45'N	13°40'E
Mandeb, Bab el, strt., (bäb´ĕl män-dĕb´) see Bab el Mandeb	266	12°44'N	43°21'E
Mandera, Kenya	262-63	3°56'N	41°52'E
Mandioli, Pulau, i., Indon.	248-49	0°44'S	127°14'E
Mandla, India	234-35	22°35'N	80°23'E
Mandsaur, India	234-35	24°03'N	75°05'E
Manduria, Italy (män-dōō´rē-ä)	200-01	40°24'N	17°38'E
Māndvi, India (mŭnd´vē)	234-35	22°51'N	69°22'E
Manfalût, Egypt	268b	27°19'N	30°58'E
Manfredonia, Italy (män-frå-dô´nyä)	200-01	41°38'N	15°55'E
Mangabeiras, Chapada das, hills, Braz.	166-67	9°55'S	46°32'W
Mangaia, i., Cook Is.	280-81	21°55'S	157°54'W
Mangalore, India (mŭŋ-gŭ-lōr´)	236	12°52'N	74°51'E
Mangchang, China	238-39	25°08'N	107°31'E
Mangkalihat, Tanjung, c., Indon.	248-49	1°02'N	118°59'E
Mangochi, Malawi	264-65	14°28'S	35°15'E
Mangoky, stm., Madag. (män-gō´kē)	264-65	21°20'S	43°32'E
Mangole, Pulau, i., Indon.	248-49	1°51'S	125°51'E
Mangshi, China see Luxi	238-39	24°21'N	98°23'E
Mangueira, Lagoa, b., Braz.	173	33°06'S	52°48'W
Mangum, Ok., U.S. (măŋ´gŭm)	120-21	34°53'N	99°30'W
Mangya, China	220-21	37°40'N	90°50'E
Manhattan, Ks., U.S. (măn-hăt´ăn)	120-21	39°11'N	96°34'W
Manhattan, Mt., U.S. (măn-hăt´ăn)	112-13	45°51'N	111°20'W
Manhuaçu, Braz. (män-óá´sōō)	172	20°15'S	42°02'W

n-sing; ŋ-baŋk; N-nasalized n; nŏd; cŏmmit; ōld; ŏbey; ôrder; oi-boil; fōōd; ó-as oo in foot; ou-out; s-soft; sh-dish; th-thin; pūre; ûnite; ûrn; stŭd; circŭs; ü-as in French tu; ´-indeterminate vowel.

Feature (Pronunciation)	Page	Lat.	Long.
Manicoré, Braz.	166-67	5°49'S	61°16'W
Manicouagan, stm., Qc., Can.	138-39	49°10'N	68°09'W
Manicouagan, Réservoir, res., Qc., Can.	130-31	51°22'N	68°44'W
Manihiki, at., Cook Is. (mä′nē-hē′kē)	280-81	10°24'S	161°01'W
Manila, nat. cap., Phil. (mȧ-nĭl′ȧ)	250	14°35'N	120°60'E
Manila Bay, b., Phil. (mȧ-nĭl′ȧ bā)	250	14°30'N	120°45'E
Manipa, Pulau, i., Indon.	248-49	3°18'S	127°33'E
Manipur, state, India	234-35	25°0'N	94°00'E
Manisa, Tur. (mä′nē-sȧ)	200-01	38°37'N	27°26'E
Manistee, Mi., U.S. (măn-ĭs-tē′)	116-17	44°15'N	86°19'W
Manistique, Mi., U.S. (măn-ĭs-tēk′)	116-17	45°58'N	86°14'W
Manitoba, state, Can. (măn-ĭ-tō′bȧ)	128-29	54°0'N	97°00'W
Manitoba, Lake, lk., Mb., Can. (lāk măn-ĭ-tō′bȧ)	134-35	50°47'N	98°43'W
Manitoulin Island, i., On., Can. (măn-ĭ-to͞o′lĭn ī′lȧnd)	136-37	45°47'N	82°20'W
Manitou Springs, Co., U.S. (măn′ĭ-to͞o sprĭngz)	120-21	38°52'N	104°54'W
Manitowoc, Wi., U.S. (măn-ĭ-tô-wŏk′)	116-17	44°06'N	87°39'W
Maniwaki, Qc., Can.	136-37	46°23'N	75°59'W
Manizales, Col. (mä-nē-zä′lås)	163c	5°04'N	75°31'W
Mānjra, stm., India	236	18°49'N	77°52'E
Mankanza, D.R.C.	262-63	1°33'N	19°04'E
Mankato, Ks., U.S. (măn-kā′tō)	120-21	39°47'N	98°13'W
Mankato, Mn., U.S. (măn-kā′tō)	114-15	44°10'N	93°59'W
Manlleu, Spain (män-lyā′o͞o)	198-99	42°0'N	2°17'E
Manna, Indon.	246-47	4°28'S	102°55'E
Mannar, Sri L. (mȧ-när′)	236	8°59'N	79°55'E
Mannar, Gulf of, b., Asia	236	8°30'N	79°00'E
Mannar Island, i., Sri L.	236	9°03'N	79°50'E
Mannheim, Ger. (män′hīm)	194-95	49°30'N	8°28'E
Manning, S.C., U.S. (măn′ĭng)	124-25	33°42'N	80°13'W
Mannington, W.V., U.S. (măn′ĭng-tŭn)	116-17	39°31'N	80°23'W
Manokwari, Indon. (mȧ-nŏk-wä′rē)	224-25	0°51'S	134°05'E
Manono, D.R.C.	262-63	7°18'S	27°25'E
Manosque, Fr. (mȧ-nôsk′)	196-97	43°50'N	5°47'E
Manouane, Lac, res., Qc., Can.	138-39	50°42'N	70°46'W
Manra, at., Kir.	280-81	4°27'S	171°15'W
Manresa, Spain (män-rā′sä)	198-99	41°44'N	1°49'E
Mansa, Zam.	264-65	11°12'S	28°53'E
Mansel Island, i., Nu., Can. (măn′sĕl ī′lȧnd)	130-31	61°60'N	79°50'W
Mansfield, Eng., U.K. (mănz′fēld)	190-91	53°09'N	1°12'W
Mansfield, La., U.S. (mănz′fēld)	122-23	32°02'N	93°43'W
Mansfield, Mo., U.S. (mănz′fēld)	120-21	37°06'N	92°35'W
Mansfield, Oh., U.S. (mănz′fēld)	116-17	40°45'N	82°31'W
Mansfield, Pa., U.S. (mănz′fēld)	116-17	41°48'N	77°04'W
Mansura, Egypt *see* El-Mansûra	268b	31°02'N	31°23'E
Manta, Ec. (män′tä)	170	0°57'S	80°43'W
Mantes-la-Jolie, Fr. (mänt-ĕ-lä-zhŏ-lē′)	196-97	48°59'N	1°43'E
Mantiqueira, Serra da, mts., Braz.	172	22°14'S	44°53'W
Manturovo, Russia	186-87	58°20'N	44°47'E
Manuae, at., Cook Is.	280-81	19°21'S	158°56'W
Manuae, at., Fr. Poly.	280-81	16°30'S	154°40'W
Manua Islands, is., Am. Sam.	279b	14°13'S	169°35'W
Manuel Rodríguez, Isla, i., Chile	171	52°34'S	73°51'W
Manui, Pulau, i., Indon. (po͞o-lou mä-no͞o′ē)	248-49	3°35'S	123°08'E
Manukau, N.Z.	278	37°02'S	174°54'E
Manus Island, i., Pap. N. Gui. (mä′no͞os ī′lȧnd)	277	2°05'S	147°00'E
Manyame, stm., Afr.	264-65	15°37'S	30°39'E
Manyara, Lake, lk., Tan.	267	3°35'S	35°50'E
Manych, stm., Russia	186-87	47°15'N	40°15'E
Manyoni, Tan.	267	5°45'S	34°50'E
Manzanillo, Cuba (män′zä-nēl′yō)	142-43	20°21'N	77°07'W
Manzanillo, Mex.	146-47	19°03'N	104°20'W
Manzhouli, China (män-jō-lē)	240-41	49°35'N	117°27'E
Mao, Chad (mä′ō)	258-59	14°07'N	15°19'E
Maó, Spain	198-99	39°53'N	4°16'E
Maoke, Pegunungan, mts., Indon.	277	4°0'S	138°00'E
Maoming, China	238-39	21°41'N	110°51'E
Mapastepec, Mex. (ma-päs-tå-pĕk′)	146-47	15°26'N	92°54'W
Mapi, Indon.	277	7°05'S	139°24'E
Mapimí, Mex. (mä-pē-mē′)	122-23	25°49'N	103°51'W
Mapimí, Bolsón de, des., Mex. (bôl-sŏ′n-dĕ-mä-pē′mē)	122-23	26°30'N	104°00'W
Maple Creek, Sk., Can. (mā′p'l crēk krēk)	134-35	49°55'N	109°29'W
Maplewood, Mn., U.S. (mā′p'l wo͝od)	114-15	45°01'N	93°04'W
Mapuera, stm., Braz.	166-67	1°05'S	57°03'W
Maputo, nat. cap., Moz. (mä-po͞o′-tō)	264-65	25°58'S	32°35'E
Maqat, Kaz.	186-87	47°39'N	53°22'E
Maquan, stm., China	234-35	29°33'N	84°07'E
Maquinchao, Arg.	171	41°15'S	68°41'W
Maquoketa, Ia., U.S. (mȧ-kō-kē-tȧ)	114-15	42°04'N	90°40'W
Mar, Serra do, mts., Braz. (sĕr′rȧ dō mär′)	172	23°30'S	45°30'W
Mara, stm., Afr.	267	1°32'S	33°59'E
Marabá, Braz.	166-67	5°21'S	49°06'W
Maracá, Ilha de, i., Braz.	166-67	2°05'N	50°25'W
Maracaibo, Ven. (mä-rä-kī′bō)	164-65	10°40'N	71°38'W
Maracaibo, Lago de, lk., Ven. (lä′gô-dĕ-mä-rä-kī′bō)	164-65	9°43'N	71°50'W
Maracaibo, Lake, lk., Ven. (lāk mä-rä-kī′bō) *see* Maracaibo, Lago de	164-65	9°43'N	71°50'W
Maracay, Ven. (mä-rä-käy′)	163b	10°16'N	67°37'W
Marādah, Libya	188-89	29°13'N	19°12'E
Maradi, Niger (mȧ-rȧ-dē′)	258-59	13°29'N	7°06'E
Marāgheh, Iran	228-29	37°23'N	46°15'E
Maragogipe, Braz.	166-67	12°46'S	38°55'W
Marahuaca, Cerro, mtn., Ven.	164-65	3°35'N	65°27'W
Marajó, Baía de, b., Braz.	166-67	1°0'S	48°30'W
Maralal, Kenya	267	1°05'N	36°41'E
Maramasike, i., Sol. Is.	279e	9°32'S	161°27'E
Marand, Iran	227	38°26'N	45°46'E
Maranhão, state, Braz. (mä-rän-youn)	166-67	5°0'S	45°00'W
Maranoa, stm., Austl. (mä-rä-nō′ä)	272-73	27°44'S	148°44'E
Marañón, stm., Peru (mä-rä-nyōn′)	170	4°29'S	73°30'W
Maraş, Tur. *see* Kahramanmaraş	186-87	37°35'N	36°57'E
Marathon, On., Can. (măr′ȧ-thŏn)	136-37	48°43'N	86°23'W
Marathon, Fl., U.S. (măr′ȧ-thŏn)	125a	24°42'N	81°06'W
Marathon, N.Y., U.S. (măr′ȧ-thŏn)	116-17	42°27'N	76°02'W
Marawwaḩ, i., U.A.E.	230-31	24°17'N	53°15'E
Marble Bar, Austl. (märb′'l bär)	270-71	21°10'S	119°45'E
Marble Hall, S. Afr.	269c	24°57'S	29°14'E
Marburg an der Drau, Slvn. *see* Maribor	200-01	46°33'N	15°39'E
Marca, Ponta da, c., Ang.	264-65	16°31'S	11°42'E
Marceline, Mo., U.S. (mär-sē-lēn′)	120-21	39°43'N	92°57'W
Marchena, Isla, i., Ec. (ē′s-lä-mär-chĕ′nä)	170a	0°21'N	90°29'W
Mar Chiquita, Laguna, lk., Arg. (lä-go͞o′nä-mär-chē-kē′tä)	173	30°42'S	62°36'W
Marcus Island, i., Japan (mär′kŭs ī′lȧnd)	280-81	24°18'N	153°58'E
Marcy, Mount, mtn., N.Y., U.S. (mount mär′sē)	116-17	44°07'N	73°56'W
Mardān, Pak.	232-33	34°12'N	72°03'E
Mar del Plata, Arg. (mär dĕl-plä′ta)	173	37°60'S	57°34'W
Mardin, Tur. (mär-dēn′)	228-29	37°18'N	40°45'E
Maré, i., N. Cal. (mȧ-rā′)	279g	21°30'S	167°59'E
Mareeba, Austl.	277	16°60'S	145°24'E
Marengo, Il., U.S. (mȧ-rĕŋ′gō)	116-17	42°15'N	88°37'W
Marfa, Tx., U.S. (mär′fȧ)	122-23	30°19'N	104°02'W
Margarita, Isla de, i., Ven. (ē′s-lä dĕ mä-gȧ-rē′tä)	163b	11°0'N	64°00'W
Margate, Eng., U.K. (mär′gāt)	190-91	51°23'N	1°25'E
Margelan, Uzb. *see* Marghilon	232-33	40°28'N	71°44'E
Margherita, Som. *see* Jamaame	262-63	0°01'N	42°42'E
Margherita Peak, mtn., Afr.	267	0°22'N	29°51'E
Marghilon, Uzb.	232-33	40°28'N	71°44'E
Mārgow, Dasht-e, des., Afg.	232-33	30°45'N	63°10'E
Marguerite, Pic, mtn., Afr. *see* Margherita Peak	267	0°22'N	29°51'E
Marhanets', Ukr.	202-03	47°39'N	34°38'E
Maria, Îles, is., Fr. Poly.	280-81	21°44'S	154°38'W
María Cleofas, Isla, i., Mex.	146-47	21°18'N	106°15'W
María Elena, Chile	168-69	22°20'S	69°40'W
Maria Island, i., Austl.	276	42°39'S	148°04'E
María Madre, Isla, i., Mex.	146-47	21°37'N	106°35'W
María Magdalena, Isla, i., Mex.	146-47	21°27'N	106°26'W
Mariana Islands, is., Oc. (mä-ryä′nä ī′lȧndz)	280-81	15°60'N	145°44'E
Marianna, Ar., U.S. (mā-rĭ-ăn′ȧ)	124-25	34°46'N	90°46'W
Marianna, Fl., U.S. (mā-rĭ-ăn′ȧ)	124-25	30°46'N	85°15'W
Mariánské Lázně, Czech Rep. (mär′yán-skĕ′läz′nyĕ)	194-95	49°58'N	12°42'E
Maria Theresiopel, Serb. *see* Subotica	200-01	46°06'N	19°41'E
Mariato, Punta, c., Pan.	150	7°13'N	80°53'W
Maribo, Den. (mä′rē-bō)	194-95	54°46'N	11°31'E
Maribor, Slvn. (mä′re-bôr)	200-01	46°33'N	15°39'E
Marīdī, Sudan	267	4°55'N	29°28'E
Marie Byrd Land, reg., Ant.	287	80°0'S	120°00'W
Marie-Galante, i., Guad. (mȧ-rē′ gȧ-länt′)	143b	15°56'N	61°16'W
Mari El, state, Russia	186-87	56°30'N	48°00'E
Marienbad, Czech Rep. *see* Mariánské Lázně	194-95	49°58'N	12°42'E
Marienburg, Pol. *see* Malbork	194-95	54°02'N	19°02'E
Mariental, Nmb.	264-65	24°37'S	17°58'E
Mariestad, Swe. (mä-rē′ĕ-städ′)	192-93	58°43'N	13°51'E
Marietta, Ga., U.S. (mā-rĭ-ĕt′ȧ)	124-25	33°57'N	84°33'W
Marietta, Oh., U.S. (mā-rĭ′-ĕt′ȧ)	116-17	39°25'N	81°28'W
Marietta, Ok., U.S. (mā-rĭ′-ĕt′ȧ)	120-21	33°56'N	97°07'W
Marília, Braz. (mä-rē′lyȧ)	168-69	22°13'S	49°57'W
Marimba, Ang.	262-63	8°22'S	16°59'E
Marinduque, i., Phil. (mä-rēn-do͞o′kä)	250	13°24'N	121°58'E
Marine City, Mi., U.S. (mȧ-rēn′ sĭ′tē)	116-17	42°43'N	82°29'W
Marinette, Wi., U.S. (măr-ĭ-nĕt′)	116-17	45°06'N	87°37'W
Maringá, Braz.	168-69	23°25'S	51°56'W
Maringa, stm., D.R.C. (mä-riŋ′gä)	262-63	1°13'N	19°50'E
Marinha Grande, Port. (mä-rēn′yȧ grän′dĕ)	198-99	39°45'N	8°56'W
Marion, Al., U.S. (măr′ĭ-ŭn)	124-25	32°38'N	87°19'W
Marion, Ar., U.S. (măr′ĭ-ŭn)	124-25	35°13'N	90°12'W
Marion, Ia., U.S. (măr′ĭ-ŭn)	114-15	42°02'N	91°36'W
Marion, Il., U.S. (măr′ĭ-ŭn)	116-17	37°44'N	88°56'W
Marion, In., U.S. (măr′ĭ-ŭn)	116-17	40°33'N	85°39'W
Marion, Ky., U.S. (măr′ĭ-ŭn)	124-25	37°20'N	88°05'W
Marion, N.C., U.S. (măr′ĭ-ŭn)	124-25	35°41'N	82°01'W
Marion, Oh., U.S. (măr′ĭ-ŭn)	116-17	40°35'N	83°07'W
Marion, S.C., U.S. (măr′ĭ-ŭn)	124-25	34°11'N	79°24'W
Marion, Va., U.S. (măr′ĭ-ŭn)	124-25	36°50'N	81°31'W
Marion, Lake, res., S.C., U.S. (lāk măr′ĭ-ŭn)	124-25	33°32'N	80°29'W
Mariquita, Col. (mä-rē-kē′tä)	163c	5°11'N	74°54'W
Mariscal Estigarribia, Para.	168-69	22°02'S	60°37'W
Maritzburg, S. Afr. *see* Pietermaritzburg	264-65	29°36'S	30°23'E
Mariupol', Ukr.	202-03	47°06'N	37°34'E
Mariy-El, state, Russia *see* Mari El	186-87	56°30'N	48°00'E
Marka, Som.	262-63	1°43'N	44°46'E
Markaryd, Swe. (mär′kä-rüd)	192-93	56°28'N	13°35'E
Marked Tree, Ar., U.S. (märkt trē)	124-25	35°32'N	90°25'W
Markha, Russia	218-19	60°36'N	123°19'E
Markha, stm., Russia	218-19	63°27'N	118°53'E
Markham, On., Can. (märk′ȧm)	136-37	43°52'N	79°16'W
Markovo, Russia (mär′kô-vô)	218-19	64°40'N	170°27'E
Marks, Russia	186-87	51°42'N	46°44'E
Marksville, La., U.S. (märks′vĭl)	124-25	31°07'N	92°05'W
Marlette, Mi., U.S. (mär-lĕt′)	116-17	43°20'N	83°04'W
Marlin, Tx., U.S. (mär′lĭn)	122-23	31°17'N	96°53'W
Marlinton, W.V., U.S. (mär′lĭn-tŭn)	116-17	38°14'N	80°06'W
Marlow, Ok., U.S. (mär′lō)	120-21	34°38'N	97°58'W
Marmande, Fr. (mär-mänd′)	196-97	44°30'N	0°10'E
Marmara, Sea of, s., Tur. (mär′mȧ-rȧ)	200-01	40°40'N	28°15'E
Marmara Denizi, s., Tur. *see* Marmara, Sea of	200-01	40°40'N	28°15'E
Marmarth, N.D., U.S. (mär′mȧrth)	114-15	46°18'N	103°55'W
Marmelos, stm., Braz.	166-67	6°05'S	61°46'W
Maroa, Ven. (mä-rō′ä)	164-65	2°44'N	67°33'W
Maromokotro, mtn., Madag.	264-65	14°01'S	48°58'E
Marondera, Zimb.	264-65	18°10'S	31°32'E
Marosvásárhely, Rom. *see* Târgu Mureș	194-95	46°33'N	24°34'E
Marotiri, Îles, is., Fr. Poly.	280-81	27°53'S	143°21'W
Maroua, Camrn. (mär′wä)	260-61	10°37'N	14°19'E
Marovoay, Madag.	264-65	16°07'S	46°39'E
Marquesas Islands, is., Fr. Poly. (mär-kĕ′säs ī′lȧndz)	280-81	8°59'S	139°31'W
Marquesas Keys, is., Fl., U.S. (mȧr-kē′zȧs kēs)	125a	24°34'N	82°08'W
Marquette, Mi., U.S.	114-15	46°32'N	87°23'W
Marquises, Îles, is., Fr. Poly. *see* Marquesas Islands	280-81	8°59'S	139°31'W
Marrah, Jabal, vol., Sudan (jĕb′ĕl mär′ä)	262-63	13°03'N	24°21'E
Marrakech, Mor. (mȧr-rä′kĕsh)	269a	31°38'N	8°01'W
Marrakesh, Mor. *see* Marrakech	269a	31°38'N	8°01'W
Marree, Austl. (mär′rē)	276	29°39'S	138°04'E
Marromeu, Moz.	264-65	18°16'S	35°52'E
Marsá al-Burayqah, Libya	188-89	30°23'N	19°36'E
Marsabit, Kenya	267	2°20'N	37°60'E
Marsala, Italy (mär-sä′lä)	200-01	37°48'N	12°26'E
Marseille, Fr. (mȧr-sâ′y')	196-97	43°18'N	5°24'E
Marseilles, Il., U.S. (mär-sĕlz′)	116-17	41°20'N	88°42'W
Marshall, Il., U.S. (mär′shȧl)	116-17	39°23'N	87°42'W
Marshall, Mi., U.S. (mär′shȧl)	116-17	42°16'N	84°57'W
Marshall, Mn., U.S. (mär′shȧl)	114-15	44°27'N	95°48'W
Marshall, Mo., U.S. (mär′shȧl)	120-21	39°07'N	93°12'W
Marshall, Tx., U.S. (mär′shȧl)	120-21	32°33'N	94°22'W

Feature (Pronunciation)	Page	Lat.	Long.
Marshall Islands, nation, Oc. (mär′shăl ī′lándz)	**280-81**	11°0′N	168°00′E
Marshalltown, Ia., U.S. (mär′shál-toun)	**114-15**	42°03′N	92°54′W
Marshfield, Mo., U.S. (märsh′fēld)	**120-21**	37°20′N	92°54′W
Marshfield, Wi., U.S. (märsh′fēld)	**116-17**	44°40′N	90°10′W
Marsh Harbour, Bah. (mär′sh här′bēr)	**142-43**	26°32′N	77°04′W
Marsh Island, i., La., U.S.	**124-25**	29°35′N	91°53′W
Mart, Tx., U.S. (märt)	**122-23**	31°32′N	96°50′W
Martaban, Gulf of, b., Mya. (gŭlf ŭv mär-tŭ-bän′)	**246-47**	16°46′N	97°01′E
Martha's Vineyard, i., Ma., U.S. (mär′tház vĭn′yárd)	**116-17**	41°24′N	70°38′W
Martigny, Switz. (már-tē-nyē′)	**194-95**	46°06′N	7°04′E
Martin, S.D., U.S. (mär′tĭn)	**114-15**	43°10′N	101°44′W
Martin, Tn., U.S. (mär′tĭn)	**124-25**	36°21′N	88°51′W
Martina Franca, Italy (mär-tē′nä fräŋ′kä)	**200-01**	40°42′N	17°20′E
Martinez, Ga., U.S. (mär-tē′nĕz)	**124-25**	33°31′N	82°05′W
Martinique, dep., N.A. (már-tē-nēk′)	**140-41**	14°40′N	61°00′W
Martinique Passage, strt., N.A.	**143b**	15°10′N	61°15′W
Martinsburg, W.V., U.S. (mär′tĭnz-bûrg)	**116-17**	39°27′N	77°57′W
Martinsville, In., U.S. (mär′tĭnz-vĭl)	**116-17**	39°25′N	86°25′W
Martinsville, Va., U.S. (mär′tĭnz-vĭl)	**124-25**	36°41′N	79°52′W
Martin Vaz, Ilhas, is., Braz.	**159**	20°30′S	28°51′W
Martos, Spain (mär′tōs)	**198-99**	37°43′N	3°58′W
Martre, Lac la, lk., N.T., Can. (läk lä märtr)	**130-31**	63°15′N	117°55′W
Marungu, mts., D.R.C.	**267**	7°42′S	30°01′E
Mary, Turkmen. (mä′rē)	**232-33**	37°35′N	61°49′E
Maryborough, Austl. (mā′rĭ-bŭr-ō)	**276**	25°32′S	152°42′E
Maryborough, Austl. (mā′rĭ-bŭr-ō)	**276**	37°03′S	143°44′E
Maryland, state, U.S. (mĕr′ĭ-lănd)	**108-09**	39°0′N	76°45′W
Marystown, Nf., Can. (mâr′ĭz-toun)	**138-39**	47°11′N	55°10′W
Marysville, Ca., U.S.	**118-19**	39°09′N	121°35′W
Marysville, Ks., U.S. (mā′rĭz-vĭl)	**120-21**	39°50′N	96°39′W
Marysville, Oh., U.S. (mā′rĭz-vĭl)	**116-17**	40°14′N	83°22′W
Marysville, Wa., U.S. (mā′rĭz-vĭl)	**112-13**	48°04′N	122°10′W
Maryville, Mo., U.S. (mā′rĭ-vĭl)	**120-21**	40°21′N	94°52′W
Maryville, Tn., U.S. (mā′rĭ-vĭl)	**124-25**	35°46′N	83°58′W
Masai Mara Game Reserve, pk., Kenya	**267**	1°15′S	35°15′E
Masai Steppe, plat., Tan.	**267**	4°45′S	37°00′E
Masaka, Ug.	**267**	0°20′S	31°44′E
Masalembu Besar, Pulau, i., Indon.	**248-49**	5°34′S	114°26′E
Masan, Kor., S. (mä-sän′)	**243**	35°12′N	128°34′E
Masatepe, Nic. (mä-sä-tĕ′pĕ)	**149**	11°54′N	86°09′W
Masaya, Nic. (mä-sä′yä)	**149**	11°58′N	86°06′W
Masbate, Phil. (mäs-bä′tā)	**250**	12°22′N	123°38′E
Masbate, i., Phil. (mäs-bä′tā)	**250**	12°15′N	123°30′E
Mascara, Alg.	**198-99**	35°23′N	0°08′E
Mascareignes, Îles, is., Afr.	**265a**	21°0′S	57°00′E
Mascarene Islands, is., Afr. *see* Mascareignes, Îles	**265a**	21°0′S	57°00′E
Mascota, Mex. (mäs-kō′tä)	**146-47**	20°31′N	104°47′W
Mascoutah, Il., U.S. (măs-kū′tā)	**120-21**	38°29′N	89°48′W
Maseru, nat. cap., Leso. (măz′ēr-ōō)	**269c**	29°19′S	27°29′E
Mashābih, i., Sau. Ar.	**266**	25°38′N	36°31′E
Mashhad, Iran	**232-33**	36°17′N	59°36′E
Māshkel, Hāmūn-i-, lk., Pak. (hā-mōōn′ē māsh-kĕl′)	**232-33**	28°15′N	63°00′E
Masi-Manimba, D.R.C.	**262-63**	4°46′S	17°57′E
Masindi, Ug. (mä-sēn′dē)	**267**	1°41′N	31°43′E
Masira, Gulf of, b., Oman *see* Maşīrah, Khalīj	**220-21**	20°10′N	58°15′E
Maşīrah, i., Oman	**220-21**	20°27′N	58°48′E
Maşīrah, Khalīj, b., Oman	**220-21**	20°10′N	58°15′E
Masjed-e Soleymān, Iran	**232-33**	31°58′N	49°18′E
Masoala, Saikanosy, pen., Madag.	**264-65**	15°26′S	50°04′E
Mason, Mi., U.S. (mā′sŭn)	**116-17**	42°35′N	84°26′W
Mason, Tx., U.S. (mā′sŭn)	**122-23**	30°45′N	99°15′W
Mason City, Ia., U.S. (mā′sŭn sĭ′tĭ)	**114-15**	43°09′N	93°12′W
Masqaţ, nat. cap., Oman *see* Muscat	**230-31**	23°36′N	58°32′E
Massa, Italy (mäs′sä)	**200-01**	44°03′N	10°09′E
Massachusetts, state, U.S. (măs-á-chōō′sĕts)	**108-09**	42°15′N	71°50′W
Massafra, Italy (mäs-sä′frä)	**200-01**	40°35′N	17°08′E
Massakory, Chad	**258-59**	12°60′N	15°44′E
Massawa, Erit.	**266**	15°37′N	39°26′E
Massena, N.Y., U.S. (mă-sē′ná)	**116-17**	44°56′N	74°53′W
Masset, B.C., Can. (măs′ĕt)	**132-33**	54°02′N	132°08′W
Massillon, Oh., U.S. (măs′ĭ-lŏn)	**116-17**	40°48′N	81°31′W
Massinga, Moz. (mä-sĭn′gä)	**264-65**	23°20′S	35°24′E
Massive, Mount, mtn., Co., U.S. (mount más′ĭv)	**118-19**	39°12′N	106°28′W
Maştağa, Azer.	**227**	40°32′N	49°59′E
Mastung, Pak.	**232-33**	29°48′N	66°52′E
Masuda, Japan (mä-sōō′dȧ)	**245**	34°41′N	131°51′E
Masulipatam, India *see* Machilīpatnam	**236**	16°11′N	81°09′E
Masvingo, Zimb.	**264-65**	20°04′S	30°49′E
Matadi, D.R.C. (má-tä′dē)	**262-63**	5°49′S	13°29′E
Matagalpa, Nic. (mä-tä-gäl′pä)	**149**	12°60′N	85°44′W
Matagami, Qc., Can.	**128-29**	49°45′N	77°39′W
Matagorda Island, i., Tx., U.S.	**122-23**	28°15′N	96°37′W
Mataiva, at., Fr. Poly.	**280-81**	14°53′S	148°40′W
Matamoros, Mex. (mä-tä-mō′rōs)	**122-23**	25°52′N	97°30′W
Matamoros, Mex. (mä-tä-mō′rōs)	**122-23**	25°32′N	103°14′W
Matandu, stm., Tan.	**262-63**	8°43′S	39°22′E
Matane, Qc., Can. (má-tån′)	**138-39**	48°50′N	67°31′W
Matanzas, Cuba (mä-tän′zäs)	**142-43**	23°03′N	81°34′W
Matanzas, state, Cuba (mä-tän′zäs)	**142-43**	22°40′N	81°20′W
Matapalo, Cabo, c., C.R. (ká′bō-mä-tä-pä′lō)	**149**	8°23′N	83°17′W
Matapan, Cape, c., Grc. *see* Taínaro, Ákra	**200-01**	36°23′N	22°29′E
Matapédia, Qc., Can. (mă-tá-pā′dē-á)	**138-39**	47°58′N	66°56′W
Matapédia, Lac, lk., Qc., Can. (läk mă-tá-pā′dē-á)	**138-39**	48°33′N	67°33′W
Matara, Sri L. (mä-tä′rä)	**236**	5°57′N	80°34′E
Mataram, Indon.	**248-49**	8°35′S	116°07′E
Mataró, Spain	**198-99**	41°32′N	2°26′E
Matasiri, Pulau, i., Indon.	**248-49**	4°48′S	115°49′E
Matâ'utu, nat. cap., Wal./F.	**280-81**	13°17′S	176°09′W
Matehuala, Mex. (mä-tå-wä′lä)	**146-47**	23°40′N	100°38′W
Matera, Italy (mä-tå′rä)	**200-01**	40°41′N	16°36′E
Mathura, India (mu-tŏ′rŭ)	**234-35**	27°30′N	77°41′E
Mathurai, India *see* Madurai	**236**	9°55′N	78°08′E
Matias Barbosa, Braz. (mä-tē′äs-bár-bô-sä)	**172**	21°53′S	43°19′W
Mato, Cerro, mtn., Ven.	**164-65**	7°16′N	65°15′W
Mato Grosso, state, Braz. (mät′ŏ grōs′ŏ)	**166-67**	12°0′S	57°00′W
Mato Grosso, Planalto do, plat., Braz. (plä-nál′tō-dō mät′ŏ grōs′ŏ)	**166-67**	14°59′S	53°37′W
Mato Grosso do Sul, state, Braz.	**168-69**	20°0′S	55°00′W
Matola, Moz.	**264-65**	25°49′S	32°27′E
Matosinhos, Port.	**198-99**	41°11′N	8°41′W
Maţraḩ, Oman (má-trä′)	**230-31**	23°37′N	58°31′E
Matsue, Japan (mät′sŏ-ĕ)	**245**	35°28′N	133°04′E
Matsumoto, Japan (mät′sŏ-mō′tō)	**245**	36°14′N	137°58′E
Matsu Tao, i., Tai.	**225a**	26°09′N	119°56′E
Matsuyama, Japan (mät′sŏ-yä′mä)	**245**	33°50′N	132°46′E
Mattawa, On., Can. (măt′á-wá)	**136-37**	46°18′N	78°41′W
Matterhorn, mtn., Eur. (mät′ēr-hôrn)	**194-95**	45°59′N	7°43′E
Matthew Town, Bah. (măth′ū toun)	**142-43**	21°01′N	73°42′W
Mattoon, Il., U.S. (mă-tōōn′)	**116-17**	39°29′N	88°23′W
Maturín, Ven. (mä-tōō-rēn′)	**164-65**	9°44′N	63°11′W
Maubeuge, Fr. (mō-bûzh′)	**196-97**	50°17′N	3°58′E
Maués, Braz. (má-wĕ′s)	**166-67**	3°22′S	57°43′W
Maui, i., Hi., U.S. (mä′ōō-ē)	**127a**	20°45′N	156°15′W
Maumee, Oh., U.S. (mô-mē′)	**116-17**	41°34′N	83°39′W
Maumee, stm., U.S. (mô-mē′)	**116-17**	41°42′N	83°27′W
Maun, Bots. (mä-ŏn′)	**264-65**	19°60′S	23°25′E
Mauna Kea, vol., Hi., U.S. (mä′ŏ-näkā′ä)	**127a**	19°50′N	155°28′W
Mauna Loa, vol., Hi., U.S. (mä′ŏ-nälō′ä)	**127a**	19°29′N	155°36′W
Maunoir, Lac, lk., N.T., Can.	**130-31**	67°30′N	125°00′W
Maurepas, Lake, lk., La., U.S. (lāk mō-rē-pä′)	**124-25**	30°15′N	90°30′W
Mauritania, nation, Afr. (mô-rē-tā′nĭ-á)	**253**	20°0′N	12°00′W
Mauritanie, nation, Afr. *see* Mauritania	**253**	20°0′N	12°00′W
Mauritius, nation, Afr. (mô-rĭsh′ĭ-ŭs)	**253**	20°17′S	57°33′E
Mauston, Wi., U.S. (môs′tŭn)	**116-17**	43°47′N	90°04′W
Mawlamyaing, Mya. *see* Mawlamyine	**246-47**	16°30′N	97°38′E
Mawlamyine, Mya.	**246-47**	16°30′N	97°38′E
Maxixe, Moz.	**264-65**	23°52′S	35°21′E
Maya, stm., Russia (mä′yä)	**218-19**	60°25′N	134°34′E
Mayaguana, i., Bah.	**142-43**	22°23′N	72°57′W
Mayagüez, P.R. (mä-yä-gwāz′)	**142a**	18°12′N	67°09′W
Mayfield, Ky., U.S. (mā′fēld)	**124-25**	36°45′N	88°38′W
Maykop, Russia	**186-87**	44°36′N	40°06′E
Maymyo, Mya. (mī′myō)	**246-47**	22°02′N	96°28′E
Mayo, Yk., Can. (mā-yō′)	**128-29**	63°36′N	135°51′W
Mayodan, N.C., U.S. (mā-yō′dăn)	**124-25**	36°24′N	79°59′W
Mayon Volcano, vol., Phil. (mä-yōn′ vŏl-kā′nō)	**250**	13°15′N	123°41′E
Mayotte, dep., Afr. (má-yŏt′)	**264-65**	12°50′S	45°10′E
Maysville, Ky., U.S. (māz′vĭl)	**116-17**	38°38′N	83°46′W
Mayumba, Gabon	**260-61**	3°22′S	10°40′E
Māyūram, India	**236**	11°06′N	79°39′E
Mayville, N.D., U.S. (mā′vĭl)	**114-15**	47°30′N	97°19′W
Mayville, Wi., U.S. (mā′vĭl)	**116-17**	43°30′N	88°32′W
Mayyit, Al-Baḩr al-, lk., Asia *see* Dead Sea	**228-29**	31°30′N	35°30′E
Maza, Arg.	**173**	36°48′S	63°20′W
Mazabuka, Zam. (mä-zä-bōō′kä)	**264-65**	15°51′S	27°46′E
Mazagan, Mor. *see* El-Jadida	**269a**	33°15′N	8°31′W
Mazagão, Braz. (mä-zá-gou′ɴ)	**166-67**	0°07′S	51°17′W
Mazara del Vallo, Italy (mät-sä′rä dĕl väl′lō)	**200-01**	37°39′N	12°36′E
Mazār-e Sharīf, Afg.	**232-33**	36°42′N	67°07′E
Mazarrón, Spain (mä-zär-rô′n)	**198-99**	37°36′N	1°19′W
Mazaruni, stm., Guy.	**164-65**	6°26′N	58°36′W
Mazatenango, Guat. (mä-zä-tå-näŋ′gō)	**148**	14°32′N	91°30′W
Mazatlán, Mex. (mä-zä-tlän′)	**146-47**	23°13′N	106°25′W
Mažeikiai, Lith. (má-zhā′kē-ī)	**192-93**	56°19′N	22°21′E
Mazoe, stm., Afr. *see* Mazowe	**264-65**	16°32′S	33°26′E
Mazowe, stm., Afr.	**264-65**	16°32′S	33°26′E
Mazyr, Bela.	**202-03**	52°03′N	29°16′E
Mbabane, nat. cap., Swaz. (m'bä-bä′nĕ)	**264-65**	26°20′S	31°09′E
Mbaïki, C.A.R. (m'bá-ē′kē)	**262-63**	3°52′N	17°60′E
Mbala, Zam.	**264-65**	8°51′S	31°22′E
Mbale, Ug.	**267**	1°05′N	34°10′E
Mbandaka, D.R.C.	**262-63**	0°02′N	18°15′E
M'banza Congo, Ang.	**262-63**	6°16′S	14°15′E
Mbanza-Ngungu, D.R.C.	**262-63**	5°14′S	14°53′E
Mbarara, Ug.	**267**	0°36′S	30°38′E
Mbari, stm., C.A.R.	**262-63**	4°36′N	22°44′E
Mbeya, Tan.	**264-65**	8°54′S	33°30′E
Mbinda, Congo	**262-63**	2°07′S	12°53′E
Mbini, stm., Afr.	**260-61**	1°35′N	9°38′E
Mbomou, stm., Afr. (m'bō′mōō)	**262-63**	4°09′N	22°29′E
Mbour, Sen.	**260-61**	14°25′N	16°58′W
Mbuji-Mayi, D.R.C.	**262-63**	6°08′S	23°39′E
Mbuji-Mayi, stm., D.R.C.	**262-63**	6°02′S	23°44′E
McAdam, N.B., Can. (măk-ăd′ăm)	**138-39**	45°35′N	67°20′W
McAlester, Ok., U.S. (măk ăl′ĕs-tēr)	**120-21**	34°56′N	95°46′W
McAllen, Tx., U.S. (măk-ăl′ĕn)	**122-23**	26°12′N	98°14′W
McBride, B.C., Can. (măk-brīd′)	**132-33**	53°18′N	120°10′W
McCamey, Tx., U.S. (mă-kā′mĭ)	**122-23**	31°08′N	102°13′W
McCauley Island, i., B.C., Can.	**132-33**	53°40′N	130°15′W
McColl, S.C., U.S.	**124-25**	34°40′N	79°33′W
McComb, Ms., U.S. (má-kōm′)	**124-25**	31°14′N	90°27′W
McCook, Ne., U.S. (má-kŏk′)	**120-21**	40°12′N	100°37′W
McGehee, Ar., U.S. (má-gē′)	**124-25**	33°38′N	91°24′W
McGill, Nv., U.S. (má-gĭl′)	**118-19**	39°25′N	114°49′W
McGrath, Ak., U.S. (măk′gráth)	**126**	62°58′N	155°38′W
McGregor, Tx., U.S. (măk-grĕg′ēr)	**122-23**	31°26′N	97°24′W
McGregor, stm., B.C., Can. (măk-grĕg′ēr)	**132-33**	54°10′N	122°01′W
McKeesport, Pa., U.S. (má-kez′pōrt)	**116-17**	40°21′N	79°52′W
McKenzie, Tn., U.S. (má-kĕn′zĭ)	**124-25**	36°08′N	88°31′W
McKinley, Mount, mtn., Ak., U.S. (mount má-kĭn′lĭ)	**126**	63°04′N	151°00′W
McKinney, Tx., U.S. (má-kĭn′ĭ)	**120-21**	33°12′N	96°37′W
McLaughlin, S.D., U.S. (mák-lŏf′lĭn)	**114-15**	45°49′N	100°48′W
McLennan, Ab., Can. (măk-lĭn′nán)	**132-33**	55°41′N	116°52′W
McLeod, stm., Ab., Can.	**132-33**	54°09′N	115°42′W
McLoughlin, Mount, mtn., Or., U.S. (mount mák-lŏk′lĭn)	**112-13**	42°27′N	122°19′W
McMinnville, Or., U.S. (măk-mĭn′vĭl)	**112-13**	45°13′N	123°11′W
McMinnville, Tn., U.S. (măk-mĭn′vĭl)	**124-25**	35°41′N	85°47′W
McPherson, Ks., U.S. (mák-fŭr′s'n)	**120-21**	38°22′N	97°40′W
McRae, Ga., U.S. (mák-rā′)	**124-25**	32°04′N	82°54′W
Mead, Lake, res., U.S. (lāk mēd)	**118-19**	36°08′N	114°26′W
Meade, stm., Ak., U.S.	**126**	70°55′N	156°00′W
Meadow Lake, Sk., Can. (mĕd′ō lāk)	**134-35**	54°08′N	108°26′W
Meadville, Pa., U.S.	**116-17**	41°39′N	80°09′W
Meaford, On., Can. (mē′fērd)	**136-37**	44°36′N	80°35′W
Meaux, Fr. (mō)	**196-97**	48°58′N	2°53′E
Mecca, Sau. Ar. (mĕk′á)	**266**	21°27′N	39°51′E
Mechanic Falls, Me., U.S. (mē-kăn′ĭk fôlz)	**116-17**	44°07′N	70°24′W
Mechanicsburg, Pa., U.S. (mē-kăn′ĭks-bŭrg)	**116-17**	40°12′N	77°01′W
Mechanicsville, Va., U.S. (mē-kăn′ĭks-vĭl)	**116-17**	37°36′N	77°22′W
Mecubúri, stm., Moz.	**264-65**	14°10′S	40°32′E
Medan, Indon. (må-dän′)	**246-47**	3°35′N	98°41′E
Medanosa, Punta, c., Arg. (pōō′n-tä-mĕ-dä-nô′sä)	**171**	48°06′S	65°55′W
Médéa, Alg.	**269b**	36°12′N	2°51′E
Medellín, Col. (må-dhĕl-yēn′)	**164-65**	6°15′N	75°35′W
Medenine, Tun. (mā-dē-nēn′)	**188-89**	33°20′N	10°30′E
Medford, Ok., U.S. (mĕd′fērd)	**120-21**	36°49′N	97°43′W
Medford, Or., U.S. (mĕd′fērd)	**112-13**	42°20′N	122°52′W

n-sing; ŋ-baŋk; ɴ-nasalized n; nŏd; cŏmmit; ōld; ōbey; ôrder; oi-boil; fōōd; ŏ-as oo in foot; ou-out; s-soft; sh-dish; th-thin; pūre; ůnite; ûrn; stŭd; circŭs; ü-as in French tu; ′-indeterminate vowel.

Feature (Pronunciation)	Page	Lat.	Long.
Medford, Wi., U.S. (mĕd′fẽrd)	116-17	45°08′N	90°20′W
Medgyes, Rom. *see* Mediaș	194-95	46°10′N	24°22′E
Mediaș, Rom. (mĕd-yäsh′)	194-95	46°10′N	24°22′E
Medical Lake, Wa., U.S. (mĕd′ĭ-kăl lāk)	112-13	47°37′N	117°43′W
Medicine Hat, Ab., Can. (mĕd′ĭ-sĭn hăt)	132-33	50°03′N	110°41′W
Medicine Lodge, Ks., U.S. (mĕd′ĭ-sĭn lŏj)	120-21	37°17′N	98°35′W
Medina, Sau. Ar. (må-dē′nȧ)	266	24°28′N	39°37′E
Medina, N.Y., U.S. (mē-dī′nȧ)	116-17	43°13′N	78°23′W
Medina, Oh., U.S. (mē-dī′nȧ)	116-17	41°08′N	81°51′W
Medina del Campo, Spain (må-dē′nä dĕl käm′pō)	198-99	41°19′N	4°55′W
Medina de Ríoseco, Spain (må-dē′nä dā rē-ō-sā′kō)	198-99	41°52′N	5°02′W
Medinīpur, India	234-35	22°26′N	87°20′E
Medio, Punta, c., Chile	168-69	27°07′S	70°56′W
Mediterranean Sea, s., (mĕd-ĭ-tẽr-ā′nē-ăn sē)	188-89	35°0′N	20°00′E
Méditerranée, Mer, s., *see* Mediterranean Sea	188-89	35°0′N	20°00′E
Mediterráneo, Mar, s., *see* Mediterranean Sea	188-89	35°0′N	20°00′E
Mediterraneo, Mar, s., *see* Mediterranean Sea	188-89	35°0′N	20°00′E
Mediterrània, Mar, s., *see* Mediterranean Sea	188-89	35°0′N	20°00′E
Mednogorsk, Russia	226	51°25′N	57°35′E
Médouneu, Gabon	260-61	0°59′N	10°55′E
Medveditsa, stm., Russia (mĕd-vyĕ′dē tså)	186-87	49°35′N	42°39′E
Medvezhyegorsk, Russia	186-87	62°55′N	34°28′E
Medyn', Russia (mĕ-dēn′)	202-03	54°57′N	35°53′E
Meekatharra, Austl. (mē-kȧ-thăr′ȧ)	270-71	26°35′S	118°30′E
Meeker, Co., U.S. (mēk′ẽr)	118-19	40°03′N	107°55′W
Meelpaeg Lake, res., Nf., Can. (mēl′pȧ-ĕg lāk)	138-39	48°16′N	56°35′W
Meerut, India (mē′rŭt)	234-35	28°59′N	77°42′E
Meghālaya, state, India	234-35	25°30′N	91°15′E
Meghna, stm., Bngl.	234-35	22°50′N	90°42′E
Megísti, i., Grc.	188-89	36°08′N	29°36′E
Mehun-sur-Yèvre, Fr. (mē-ŭɴ-sür-yĕvr′)	196-97	47°09′N	2°13′E
Meiganga, Camrn.	260-61	6°34′N	14°07′E
Meiktila, Mya.	246-47	20°52′N	95°52′E
Meixian, China *see* Meizhou	238-39	24°20′N	116°07′E
Meizhou, China	238-39	24°20′N	116°07′E
Mejillones, Chile (må-ĸē-lyō′nås)	168-69	23°06′S	70°27′W
Mek'elē, Eth.	266	13°30′N	39°28′E
Meknès, Mor. (mĕk′nĕs) (mĕk-nĕs′)	269a	33°54′N	5°33′W
Mekong, stm., Asia (mā-kông′)	246-47	10°33′N	105°27′E
Mékôngk, stm., Asia *see* Mekong	246-47	10°33′N	105°27′E
Mékrou, stm., Afr.	260-61	12°24′N	2°50′E
Melaka, Malay.	246-47	2°12′N	102°16′E
Melaka, Selat, strt., Asia *see* Malacca, Strait of	246-47	2°30′N	101°20′E
Melanesia, is., Oc. (mĕl-ȧ-nē′-zhȧ)	280-81	13°0′S	164°00′E
Mélanésie, is., Oc. *see* Melanesia	280-81	13°0′S	164°00′E
Melawi, stm., Indon.	248-49	0°05′N	111°29′E
Melbourne, Austl. (mĕl′bŭrn)	276	37°49′S	144°57′E
Melbourne, Fl., U.S. (mĕl′bŭrn)	125a	28°05′N	80°37′W
Melbourne Island, i., Nu., Can.	130-31	68°30′N	104°45′W
Melchor, Isla, i., Chile	171	45°08′S	73°57′W
Melekeok, nat. cap., Palau	280-81	7°29′N	134°37′E
Meleuz, Russia	186-87	52°58′N	55°56′E
Mélèzes, stm., Qc., Can.	130-31	57°41′N	69°29′W
Melfi, Chad	258-59	11°03′N	17°56′E
Melfort, Sk., Can. (mĕl′fôrt)	134-35	52°52′N	104°36′W
Melilla, Sp. N. Afr. (mā-lēl′yä)	198-99	35°18′N	2°57′W
Melipilla, Chile (må-lē-pē′lyä)	163e	33°41′S	71°13′W
Melita, Mb., Can.	134-35	49°16′N	100°59′W
Melitopol', Ukr. (mā-lē-tô′pôl-y′)	202-03	46°51′N	35°21′E
Mellen, Wi., U.S. (mĕl′ĕn)	114-15	46°20′N	90°40′W
Mellerud, Swe. (mâl′ĕ-rōōdh)	192-93	58°42′N	12°28′E
Melo, Ur. (mā′lō)	173	32°22′S	54°11′W
Melos, i., Grc. (mē′lŏs) *see* Mílos	200-01	36°41′N	24°28′E
Melrhir, Chott, lk., Alg.	258-59	34°18′N	6°17′E
Melrose, Mn., U.S. (mĕl′rōz)	114-15	45°40′N	94°49′W
Melton Mowbray, Eng., U.K. (mĕl′tŭn mō′brå)	190-91	52°46′N	0°53′W
Melun, Fr. (mē-lŭɴ′)	196-97	48°32′N	2°40′E
Melville, Sk., Can. (mĕl′vĭl)	134-35	50°55′N	102°48′W
Melville, Cape, c., Austl. (kāp mĕl′vĭl)	277	14°11′S	144°30′E
Melville, Lake, lk., Nf., Can. (lāk mĕl′vĭl)	130-31	53°40′N	59°44′W
Melville Island, i., Austl. (mĕl′vĭl ī′lȧnd)	272-73	11°40′S	131°00′E
Melville Island, i., Can.	86	75°15′N	109°59′W
Melville Peninsula, pen., Nu., Can. (mĕl′vĭl pĕ-nĭn′sūlȧ)	130-31	68°0′N	84°00′W
Memel, Lith. *see* Klaipėda	192-93	55°43′N	21°08′E
Memel, S. Afr. (mĕ′mĕl)	269c	27°41′S	29°34′E
Memmingen, Ger. (mĕm′ĭng-ĕn)	194-95	47°59′N	10°11′E
Mempawah, Indon.	246-47	0°20′N	108°58′E
Memphis, Mo., U.S. (mĕm′fĭs)	120-21	40°28′N	92°10′W
Memphis, Tn., U.S. (mĕm′fĭs)	124-25	35°09′N	90°03′W
Memphis, Tx., U.S. (mĕm′fĭs)	120-21	34°44′N	100°33′W
Mena, Ukr. (mē-ná′)	202-03	51°31′N	32°14′E
Mena, Ar., U.S. (mē′nȧ)	120-21	34°35′N	94°15′W
Menado, Indon. *see* Manado	248-49	1°29′N	124°51′E
Ménaka, Mali	258-59	15°55′N	2°24′E
Menard, Tx., U.S. (mē-närd′)	122-23	30°55′N	99°47′W
Menasha, Wi., U.S. (mē-năsh′ȧ)	116-17	44°12′N	88°26′W
Mendawai, stm., Indon.	248-49	3°14′S	113°19′E
Mende, Fr. (mäɴd)	196-97	44°30′N	3°30′E
Mendi, Pap. N. Gui.	277	6°10′S	143°40′E
Mendocino, Cape, c., Ca., U.S. (kāp mĕn′dō-sē′nō)	112-13	40°25′N	124°23′W
Mendota, Ca., U.S. (mĕn-dō′tȧ)	118-19	36°46′N	120°23′W
Mendota, Il., U.S. (mĕn-dō′tȧ)	116-17	41°33′N	89°07′W
Mendoza, Arg. (mĕn-dō′sä)	163e	32°53′S	68°49′W
Mendoza, state, Arg. (mĕn-dō′sä)	163e	34°30′S	68°30′W
Mengcheng, China (mŭŋ-chŭŋ)	238-39	33°16′N	116°33′E
Menggala, Indon.	246-47	4°29′S	105°15′E
Menghai, China	238-39	21°59′N	100°27′E
Menindee, Austl. (mĕ-nĭn-dē)	276	32°24′S	142°26′E
Menominee, Mi., U.S. (mē-nŏm′ĭ-nē)	116-17	45°08′N	87°37′W
Menominee, stm., U.S. (mē-nŏm′ĭ-nē)	114-15	45°06′N	87°36′W
Menongue, Ang.	264-65	14°39′S	17°41′E
Menorca, i., Spain (mĕ-nô′r-kä)	198-99	40°0′N	4°00′E
Mentawai, Selat, strt., Indon.	246-47	1°45′S	100°00′E
Menzel Bourguiba, Tun.	184-85	37°10′N	9°48′E
Meoqui, Mex.	122-23	28°16′N	105°29′W
Meppel, Neth. (mĕp′ĕl)	190-91	52°42′N	6°12′E
Meppen, Ger. (mĕp′ĕn)	194-95	52°42′N	7°18′E
Merauke, Indon. (må-rou′kä)	277	8°30′S	140°24′E
Merca, Som. *see* Marka	262-63	1°43′N	44°46′E
Merced, Ca., U.S. (mẽr-sĕd′)	118-19	37°18′N	120°29′W
Mercedario, Cerro, mtn., Arg. (sĕ′r-rō mĕr-så-dhä′rē-ō)	168-69	31°59′S	70°08′W
Mercedes, Arg. (mĕr-sā′dhās)	173	29°11′S	58°03′W
Mercedes, Arg. (mĕr-sā′dhās)	173	34°40′S	59°26′W
Mercedes, Ur.	173	33°15′S	58°02′W
Mercy, Cape, c., Nu., Can.	130-31	64°54′N	63°35′W
Merefa, Ukr. (må-rĕf′á)	202-03	49°51′N	36°05′E
Mergui, Mya. (mẽr-gē′)	246-47	12°26′N	98°37′E
Mergui Archipelago, is., Mya. (mẽr-gē′ är′kȧ-pĕ′-å-gō)	246-47	12°0′N	98°00′E
Mérida, Mex.	148	20°59′N	89°37′W
Mérida, Spain	198-99	38°55′N	6°20′W
Mérida, Ven. (mĕ′rē-dhä)	164-65	8°37′N	71°09′W
Meriden, Ct., U.S. (mĕr′ĭ-dĕn)	116-17	41°32′N	72°48′W
Meridian, Id., U.S. (mē-rĭd-ĭ-ăn)	112-13	43°36′N	116°21′W
Meridian, Ms., U.S. (mē-rĭd-ĭ-ăn)	124-25	32°22′N	88°42′W
Meridian, Tx., U.S. (mē-rĭd-ĭ-ăn)	122-23	31°55′N	97°40′W
Merikarvia, Fin. (mā′rē-kàr′vē-å)	192-93	61°51′N	21°30′E
Merín, Laguna, b., S.A. *see* Mirim, Lagoa	173	32°45′S	52°50′W
Merir, i., Palau	280-81	4°19′N	132°19′E
Merkel, Tx., U.S. (mûr′kĕl)	120-21	32°28′N	100°01′W
Merrill, Mi., U.S. (mĕr′ĭl)	116-17	43°25′N	84°20′W
Merrill, Wi., U.S. (mĕr′ĭl)	116-17	45°11′N	89°41′W
Merritt, B.C., Can. (mĕr′ĭt)	132-33	50°06′N	120°46′W
Merryville, La., U.S. (mĕr′ĭ-vĭl)	122-23	30°45′N	93°33′W
Mersa Matruh, Egypt	188-89	31°21′N	27°14′E
Merseburg, Ger. (mĕr′zē-bŏŏrgh)	194-95	51°21′N	11°60′E
Mersin, Tur. *see* İçel	228-29	36°49′N	34°38′E
Merthyr Tydfil, Wales, U.K. (mûr′thẽr tĭd′vĭl)	190-91	51°46′N	3°23′W
Méru, Fr. (mā-rü′)	196-97	49°14′N	2°08′E
Meru, Kenya (mā′rōō)	267	0°03′N	37°39′E
Meru, Mount, vol., Tan.	267	3°14′S	36°45′E
Merzifon, Tur. (mĕr′ze-fōn)	186-87	40°52′N	35°27′E
Mesa, Az., U.S. (mā′sȧ)	118-19	33°24′N	111°49′W
Mesabi Range, hills, Mn., U.S. (mā-sŏb′bē rānj)	114-15	47°30′N	92°50′W
Mesagne, Italy (mā-sän′yå)	200-01	40°34′N	17°49′E
Mesa Verde National Park, n.p., Co., U.S. (mā′sȧ vẽr′dē năsh′ŭn-ăl pärk)	118-19	37°15′N	108°26′W
Mescalero Apache Indian Reservation, ind. res., N.M., U.S. (mĕs-kä-lā′rō ȧ-păch′ē ĭn′dĭ-ăn rĕ-sẽr-vā′shĕn)	120-21	33°12′N	105°40′W
Mesewa, Erit. *see* Massawa	266	15°37′N	39°26′E
Meshchovsk, Russia (myĕsh′chĕfsk)	202-03	54°19′N	35°17′E
Meshed, Iran *see* Mashhad	232-33	36°17′N	59°36′E
Mesogéios Thálassa, s., *see* Mediterranean Sea	188-89	35°0′N	20°00′E
Mesopotamia, hist. reg., Asia	228-29	34°0′N	44°00′E
Mesoyéios Thálassa, s., *see* Mediterranean Sea	188-89	35°0′N	20°00′E
Messalo, stm., Moz.	264-65	11°41′S	40°26′E
Messina, Italy (mĕ-sē′nȧ)	200-01	38°11′N	15°33′E
Messina, Stretto di, strt., Italy (stĕ′t-tô dē mĕ-sē′nȧ)	200-01	38°09′N	15°35′E
Meta, stm., S.A.	164-65	6°11′N	67°28′W
Métabetchouane, stm., Qc., Can. (mĕ-tá-bĕt-chōō-än′)	136-37	48°26′N	71°58′W
Meta Incognita Peninsula, pen., Nu., Can.	130-31	62°45′N	68°30′W
Metán, Arg. (mĕ-tá′n)	168-69	25°30′S	64°57′W
Metapán, El Sal. (må-täpän′)	148	14°20′N	89°26′W
Metković, Cro. (mĕt′kô-vĭch)	200-01	43°03′N	17°39′E
Metlakatla, Ak., U.S. (mĕt-lȧ-kăt′lȧ)	126	55°07′N	131°35′W
Metropolis, Il., U.S. (mē-trŏp′ô-lĭs)	124-25	37°09′N	88°44′W
Metter, Ga., U.S. (mĕt′ẽr)	124-25	32°24′N	82°04′W
Metz, Fr. (mĕtz)	196-97	49°08′N	6°10′E
Meulaboh, Indon.	246-47	4°09′N	96°08′E
Mexia, Tx., U.S. (må-hē′ä)	122-23	31°40′N	96°29′W
Mexiana, Ilha, i., Braz.	166-67	0°02′S	49°35′W
Mexicali, Mex. (mâk-sē-kä′lē)	144-45	32°39′N	115°30′W
Mexicana, Altiplanicie, plat., Mex.	20-21	25°29′N	104°00′W
Mexican Hat, Ut., U.S. (mĕk′sĭ-kăn hăt)	118-19	37°12′N	109°52′W
Mexico, Me., U.S. (mĕk′sĭ-kō)	116-17	44°34′N	70°33′W
Mexico, Mo., U.S. (mĕk′sĭ-kō)	120-21	39°10′N	91°53′W
Mexico, nation, N.A. (mĕk′sĭ-kō)	85	23°0′N	102°00′W
México, state, Mex.	146-47	19°20′N	99°45′W
México, Golfo de, b., N.A. *see* Mexico, Gulf of	140-41	25°0′N	90°00′W
Mexico, Gulf of, b., N.A. (gŭlf ŭv mĕk′sĭ-kō)	140-41	25°0′N	90°00′W
Mexico City, nat. cap., Mex. (mĕk′sĭ-kō sĭ′tē)	146-47	19°24′N	99°09′W
Meyersdale, Pa., U.S. (mī′ẽrz-dāl)	116-17	39°49′N	79°02′W
Meymaneh, Afg.	232-33	35°56′N	64°48′E
Mezen', Russia	186-87	65°50′N	44°15′E
Mezen', stm., Russia	186-87	65°53′N	44°09′E
Mézenc, Mont, mtn., Fr. (mōɴ-mā-zĕɴ′)	196-97	44°55′N	4°11′E
Mezha, stm., Russia (myā′zhá)	202-03	55°43′N	31°31′E
Mezhdurechensk, Russia	226	53°41′N	88°07′E
Mezőkövesd, Hung. (mĕ′zû-kû′vĕsht)	194-95	47°48′N	20°35′E
Mezőtúr, Hung. (mĕ′zû-tōōr)	194-95	47°00′N	20°37′E
Mezquital, Mex. (mâz-kē-täl′)	146-47	23°29′N	104°22′W
Mfangano Island, i., Kenya	267	0°28′S	34°01′E
M'Goun, Irhil, mtn., Mor.	258-59	31°31′N	6°25′W
Miahuatlán de Porfirio Díaz, Mex.	146-47	16°19′N	96°36′W
Miajadas, Spain (mē-ä-hä′däs)	198-99	39°09′N	5°54′W
Miami, Fl., U.S. (mī-ă′-mē)	125a	25°47′N	80°13′W
Miami Beach, Fl., U.S.	125a	25°48′N	80°08′W
Miāneh, Iran	228-29	37°26′N	47°42′E
Mianyang, China	238-39	31°28′N	104°44′E
Miaoli, Tai. (mē-ou′lĭ)	225a	24°33′N	120°49′E
Miass, Russia (mī-äs′)	226	54°59′N	60°06′E
Michalovce, Slvk. (mē′kä-lôf′tsĕ)	194-95	48°46′N	21°56′E
Michelson, Mount, mtn., Ak., U.S. (mount mĭch′ĕl-sŭn)	126	69°19′N	144°17′W
Michigan, state, U.S. (mĭsh-′ĭ-găn)	108-09	44°0′N	85°00′W
Michigan, Lake, lk., U.S. (lāk mĭsh-′ĭ-găn)	116-17	44°0′N	87°00′W
Michigan City, In., U.S. (mĭsh-′ĭ-găn sĭ′tē)	116-17	41°43′N	86°53′W
Michipicoten Island, i., On., Can.	136-37	47°45′N	85°45′W
Michoacán, state, Mex.	146-47	19°10′N	101°50′W
Michurinsk, Russia (mĭ-chōō-rĭnsk′)	186-87	52°54′N	40°29′E
Micronesia, is., Oc. (mī-krō-nē′zhȧ)	280-81	11°0′N	159°00′E
Micronesia, Federated States of, nation, Oc. (fĕ′ẽr-ā′ĕd stāts ŭv mī-krō-nē′zhȧ)	280-81	5°0′N	152°00′E
Middelburg, S. Afr.	264-65	31°30′S	25°00′E
Middelfart, Den. (mēd′ĭl-fårt)	192-93	55°30′N	9°45′E
Middle, stm., B.C., Can. (mĭd′ĭl)	132-33	54°52′N	125°08′W
Middle Andaman, i., India (mĭd′ĭl ăn-dȧ-măn′)	246-47	12°30′N	92°50′E

Feature (Pronunciation)	Page	Lat.	Long.
Middle Caicos, i., T./C. Is.	142-43	21°48'N	71°47'W
Middlesboro, Ky., U.S. (mĭd''lz-bŭr-ô)	124-25	36°36'N	83°43'W
Middlesbrough, Eng., U.K. (mĭd''lz-brŭ)	190-91	54°34'N	1°14'W
Middleton, N.S., Can. (mĭd''l-tŭn)	138-39	44°57'N	65°04'W
Middleton Island, i., Ak., U.S.	126	59°26'N	146°19'W
Middletown, Oh., U.S.	116-17	39°31'N	84°23'W
Midland, On., Can. (mĭd'lănd)	136-37	44°45'N	79°52'W
Midland, Mi., U.S.	116-17	43°36'N	84°14'W
Midland, Tx., U.S.	122-23	32°00'N	102°05'W
Midway, Ky., U.S. (mĭd'wā)	116-17	38°08'N	84°42'W
Midway Islands, dep., Oc. (mĭd'wā ī'lăndz)	280-81	28°13'N	177°22'W
Międzyrzecz, Pol. (myăn-dzŭ'zhĕch)	194-95	52°27'N	15°35'E
Mier, Mex. (myâr)	122-23	26°26'N	99°09'W
Mieres, Spain (myā'rās)	198-99	43°16'N	5°46'W
Mier y Noriega, Mex. (myâr'ē nô-rē-ā'gä)	146-47	23°25'N	100°08'W
Miguel Alemán, Presa, res., Mex. (prā'sä-mē-găl'-ălĕ-mà'n)	146-47	18°13'N	96°32'W
Mikhaylov, Russia (mē-käy'lôf)	202-03	54°14'N	39°02'E
Mikhaylovka, Russia	186-87	50°04'N	43°15'E
Mikkeli, Fin. (měk'ĕ-lĭ)	192-93	61°42'N	27°16'E
Mikun', Russia	186-87	62°22'N	50°05'E
Mikura-jima, i., Japan (mē'kōō-rà jē'mä)	244	33°52'N	139°36'E
Milaca, Mn., U.S. (mē-lăk'à)	114-15	45°45'N	93°39'W
Milagro, Arg.	168-69	31°01'S	65°60'W
Milagro, Ec.	170	2°08'S	79°36'W
Milan, Italy (mē-län')	200-01	45°28'N	9°12'E
Milan, Mi., U.S. (mī'lăn)	116-17	42°05'N	83°41'W
Milan, Mo., U.S. (mī'lăn)	120-21	40°12'N	93°07'W
Milan, Oh., U.S. (mī'lăn)	116-17	41°18'N	82°37'W
Milan, Tn., U.S. (mī'lăn)	124-25	35°55'N	88°46'W
Milano, Italy (mē-lä'nō) see Milan	200-01	45°28'N	9°12'E
Milâs, Tur. (mē'läs)	200-01	37°19'N	27°47'E
Milbank, S.D., U.S. (mĭl'băŋk)	114-15	45°13'N	96°38'W
Mildura, Austl. (mĭl-dū'rà)	276	34°12'S	142°10'E
Mile, China	238-39	24°26'N	103°27'E
Miles, Austl.	276	26°40'S	150°11'E
Miles City, Mt., U.S. (mīlz sĭ'tē)	112-13	46°25'N	105°50'W
Milford, Ct., U.S. (mĭl'fērd)	116-17	41°13'N	73°04'W
Milford, De., U.S. (mĭl'fērd)	116-17	38°55'N	75°26'W
Milford, Ne., U.S. (mĭl'fērd)	114-15	40°46'N	97°03'W
Milford, Ut., U.S. (mĭl'fērd)	118-19	38°24'N	113°01'W
Milford Sound, b., N.Z.	278	44°31'S	167°48'E
Milk, stm., N.A.	112-13	48°03'N	106°19'W
Mil'kovo, Russia	218-19	54°42'N	158°38'E
Millau, Fr. (mē-yō')	196-97	44°06'N	3°05'E
Milledgeville, Ga., U.S. (mĭl'ĕj-vĭl)	124-25	33°05'N	83°14'W
Mille Lacs, Lac des, lk., On., Can. (läk dĕ mēl läks)	136-37	48°50'N	90°30'W
Mille Lacs Lake, lk., Mn., U.S.	114-15	46°15'N	93°40'W
Millen, Ga., U.S. (mĭl'ĕn)	124-25	32°48'N	81°56'W
Millenium, at., Kir. see Caroline	280-81	9°58'S	150°13'W
Miller, S.D., U.S. (mĭl'ĕr)	114-15	44°31'N	98°59'W
Millerovo, Russia (mĭl'ĕ-rô-vô)	186-87	48°56'N	40°24'E
Millersburg, Ky., U.S. (mĭl'ērz-bûrg)	116-17	38°18'N	84°09'W
Millersburg, Oh., U.S. (mĭl'ērz-bûrg)	116-17	40°33'N	81°54'W
Millicent, Austl. (mĭl-ĭ-sĕnt)	276	37°36'S	140°20'E
Millinocket, Me., U.S. (mĭl-ĭ-nŏk'ĕt)	117a	45°40'N	68°42'W
Mills Lake, lk., N.T., Can.	130-31	61°30'N	118°10'W
Mílos, i., Grc. (mē'lŏs)	200-01	36°41'N	24°28'E
Milton, On., Can. (mĭl'tŭn)	136-37	43°31'N	79°53'W
Milton, Fl., U.S. (mĭl'tŭn)	124-25	30°38'N	87°02'W
Milton, Pa., U.S. (mĭl'tŭn)	116-17	41°01'N	76°51'W
Milton, Wi., U.S. (mĭl'tŭn)	116-17	42°47'N	88°56'W
Milwaukee, Wi., U.S. (mĭl-wô'kē)	116-17	43°01'N	87°56'W
Min, stm., China (mēn)	222-23	26°04'N	119°33'E
Min, stm., China (mēn)	238-39	28°46'N	104°38'E
Minami-Daitō-jima, i., Japan	222-23	25°50'N	131°15'E
Minami-Iō-jima, i., Japan	280-81	24°14'N	141°28'E
Minami-Tori-shima, i., Japan see Marcus Island	280-81	24°18'N	153°58'E
Minas, Cuba (mē'näs)	142-43	21°29'N	77°36'W
Minas, Ur. (mē'näs)	173	34°23'S	55°14'W
Minas Basin, b., N.S., Can. (mī'năs bā's'n)	138-39	45°20'N	64°00'W
Minas Channel, strt., N.S., Can. (mī'năs chăn'ĕl)	138-39	45°15'N	64°45'W
Minas de Oro, Hond. (mē'näs-dĕ-dĕ-ŏ-rô)	149	14°46'N	87°20'W
Minas Gerais, state, Braz.	172	18°0'S	44°00'W
Minas Novas, Braz. (mē'näzh nō'väzh)	172	17°15'S	42°36'W
Minatitlán, Mex. (mē-nä-tē-tlän')	146-47	17°59'N	94°32'W
Mindanao, i., Phil. (mĭn-dä-nou')	250	8°0'N	125°00'E
Mindanao Sea, s., Phil. (mĭn-dä-nou' sē) see Bohol Sea	250	9°10'N	124°25'E
Mindelo, C.V.	260-61	16°52'N	24°60'W
Minden, Ger. (mĭn'dĕn)	194-95	52°18'N	8°55'E
Mindoro, i., Phil. (mĭn-dō'rō)	250	12°50'N	121°05'E
Mindoro Strait, strt., Phil. (mĭn-dō'rō strāt)	250	12°20'N	120°40'E
Mineiros, Braz.	168-69	17°34'S	52°34'W
Mineola, Tx., U.S. (mĭn-ē-ō'là)	120-21	32°40'N	95°29'W
Mineral'nyye Vody, Russia	227	44°12'N	43°08'E
Mineral Point, Wi., U.S. (mĭn'ēr-ăl point)	116-17	42°52'N	90°10'W
Mineral Wells, Tx., U.S. (mĭn'ēr-ăl wĕlz)	120-21	32°48'N	98°07'W
Minfeng, China	226	37°04'N	82°39'E
Mingäçevir, Azer.	227	40°46'N	47°02'E
Mingäçevir su anbarı, res., Azer.	227	40°55'N	46°48'E
Mingāora, Pak.	232-33	34°49'N	72°21'E
Mingechaur, Azer. see Mingäçevir	227	40°46'N	47°02'E
Mingechaur Reservoir, res., Azer. see Mingäçevir su anbarı	227	40°55'N	46°48'E
Minicoy Island, i., India	236	8°16'N	73°03'E
Minigwal, Lake, lk., Austl.	272-73	29°35'S	123°12'E
Minle, China	240-41	38°28'N	100°56'E
Minna, Nig. (mĭn'à)	260-61	9°37'N	6°33'E
Minneapolis, Mn., U.S. (mĭn-ē-ăp'ō-lĭs)	114-15	44°59'N	93°17'W
Minnedosa, Mb., Can. (mĭn-ē-dō'sà)	134-35	50°14'N	99°49'W
Minneota, Mn., U.S. (mĭn-ē-ō'tà)	114-15	44°34'N	96°00'W
Minnesota, state, U.S. (mĭn-ē-sō'tà)	108-09	46°0'N	94°15'W
Minnesota, stm., Mn., U.S.	114-15	44°54'N	93°11'W
Minnitaki Lake, lk., On., Can. (mĭ'nĭ-tä'kē lāk)	134-35	49°58'N	92°00'W
Minonk, Il., U.S. (mī'nŏnk)	116-17	40°54'N	89°02'W
Minorca, i., Spain see Menorca	198-99	40°0'N	4°00'E
Minot, N.D., U.S.	114-15	48°14'N	101°18'W
Minsk, state, Bela. (mēnsk)	194-95	53°45'N	27°45'E
Minsk, nat. cap., Bela. (mēnsk)	194-95	53°54'N	27°33'E
Mińsk Mazowiecki, Pol. (mēn'sk mä-zō-vyĕt'skĭ)	194-95	52°11'N	21°34'E
Minto, Lac, lk., Qc., Can.	130-31	57°12'N	74°58'W
Minturno, Italy (mēn-tōōr'nō)	200-01	41°16'N	13°45'E
Minxian, China	238-39	34°26'N	104°02'E
Minya, Egypt see El-Minya	268b	28°06'N	30°45'E
Minya Konka, mtn., China see Gongga Shan	238-39	29°35'N	101°51'E
Min'yar, Russia	186-87	55°03'N	57°33'E
Miracema do Tocantins, Braz.	166-67	9°33'S	48°24'W
Mirador, Braz. (mē-rà-dōr')	166-67	6°22'S	44°22'W
Miraflores, Col. (mē-rä-flō'răs)	164-65	1°25'N	72°17'W
Miramar, Arg.	173	38°16'S	57°51'W
Miramichi, N.B., Can.	138-39	47°02'N	65°28'W
Miramichi Bay, b., N.B., Can. (mĭr'à-mē'shē bā)	138-39	47°08'N	65°08'W
Miranda, stm., Braz.	168-69	19°25'S	57°20'W
Miranda de Ebro, Spain (mē-rá'n-dä-dĕ-ĕ'brō)	198-99	42°42'N	2°56'W
Miranda do Douro, Port. (mē-rän'dä dô-dwĕ'rô)	198-99	41°30'N	6°16'W
Mirandela, Port. (mē-rän-dā'là)	198-99	41°29'N	7°11'W
Mirecourt, Fr. (mēr-kōōr')	196-97	48°18'N	6°08'E
Miri, Malay. (mē'rē)	248-49	4°23'N	113°59'E
Mirim, Lagoa, b., S.A. (lä-gô'ä-mē-rêN')	173	32°45'S	52°50'W
Mirnyy, Russia	218-19	62°31'N	113°59'E
Mīrpur Khās, Pak. (mēr'pōōr käs)	232-33	25°31'N	69°01'E
Mirzāpur, India (mēr'zä-pōōr)	234-35	25°08'N	82°34'E
Misāha, Bîr, well, Egypt	266	22°12'N	27°57'E
Misantla, Mex. (mē-sän'tlä)	146-47	19°56'N	96°50'W
Miscou Island, i., N.B., Can. (mĭs'kō ī'lănd)	138-39	47°57'N	64°32'W
Mishan, China (mĭ'shäN)	244	45°32'N	131°52'E
Mishawaka, In., U.S. (mĭsh-à-wôk'à)	116-17	41°40'N	86°10'W
Mishmi Hills, hills, Asia	238-39	29°0'N	96°00'E
Misima Island, i., Pap. N. Gui.	277	10°42'S	152°45'E
Misiones, state, Arg. (mē-syō'nās)	173	27°0'S	55°00'W
Miskitos, Cayos, is., Nic.	149	14°23'N	82°46'W
Miskolc, Hung. (mĭsh'kôlts)	194-95	48°06'N	20°47'E
Misool, Pulau, i., Indon. (pōō-lou mē-sōl')	224-25	1°52'S	130°10'E
Mişr, nation, Afr. see Egypt	253	27°0'N	30°00'E
Mişrātah, Libya	188-89	32°22'N	15°06'E
Missinaibi, stm., On., Can. (mĭs'ĭn-ä'ē-bē)	130-31	50°45'N	81°31'W
Missinaibi Lake, lk., On., Can. (mĭs'ĭn-ä'ē-bē lāk)	136-37	48°21'N	83°43'W
Mission, S.D., U.S. (mĭsh'ŭn)	114-15	43°18'N	100°38'W
Mission, Tx., U.S. (mĭsh'ŭn)	122-23	26°13'N	98°19'W
Mississippi, state, U.S. (mĭs-ĭ-sĭp'ē)	108-09	32°50'N	89°30'W
Mississippi, stm., U.S. (mĭs-ĭ-sĭp'ē)	110-11	28°60'N	89°08'W
Mississippi River Delta, del., La., U.S.	110-11	29°10'N	89°15'W
Mississippi Sound, strt., U.S. (mĭs-ĭ-sĭp'ē sound)	124-25	30°15'N	88°40'W
Missoula, Mt., U.S. (mĭ-zōō'là)	112-13	46°52'N	114°00'W
Missouri, state, U.S. (mĭ-sōō'rē)	108-09	38°30'N	93°30'W
Missouri, stm., U.S. (mĭ-sōō'rē)	110-11	38°49'N	90°07'W
Missouri City, Tx., U.S. (mĭ-sōō'rē sĭ'tĭ)	122-23	29°37'N	95°31'W
Missouri Valley, Ia., U.S. (mĭ-sōō'rē văl'ē)	114-15	41°33'N	95°54'W
Mistassibi, stm., Qc., Can.	136-37	48°53'N	72°14'W
Mistassini, Lac, lk., Qc., Can. (läk mĭs-tá-sĭ'nē)	130-31	51°0'N	73°37'W
Misti, Volcán, vol., Peru	170	16°18'S	71°24'W
Mita, Punta de, c., Mex. (pōō'n-tä-dĕ-mē'tä)	146-47	20°47'N	105°32'W
Mitau, Lat. see Jelgava	192-93	56°39'N	23°44'E
Mitchell, Austl.	276	26°29'S	147°58'E
Mitchell, In., U.S. (mĭch'ĕl)	116-17	38°44'N	86°29'W
Mitchell, Ne., U.S. (mĭch'ĕl)	114-15	41°57'N	103°48'W
Mitchell, S.D., U.S. (mĭch'ĕl)	114-15	43°43'N	98°02'W
Mitchell, Mount, mtn., N.C., U.S. (mount mĭch'ĕl)	124-25	35°46'N	82°16'W
Mitiaro, i., Cook Is.	280-81	19°48'S	157°43'W
Mito, Japan	245	36°22'N	140°29'E
Mitsio, Nosy, i., Madag.	264-65	12°54'S	48°36'E
Mitsiwa, Erit. see Massawa	266	15°37'N	39°26'E
Mittellandkanal, can., Ger. (mĭt'ĕl-länd kä-näl')	194-95	52°14'N	11°43'E
Mitú, Col.	164-65	1°08'N	70°03'W
Mitumba, Monts, mts., D.R.C.	267	6°0'S	29°00'E
Mitzic, Gabon	260-61	0°47'N	11°34'E
Miyake-jima, i., Japan (mē'yä-kå jē'mä)	244	34°05'N	139°32'E
Miyako, Japan	244	39°38'N	141°57'E
Miyako-jima, i., Japan	279a	24°47'N	125°20'E
Miyakonojō, Japan	245	31°43'N	131°04'E
Miyazaki, Japan	245	31°54'N	131°26'E
Miyazu, Japan	245	35°32'N	135°11'E
Miyoshi, Japan (mē-yō'shē')	245	34°49'N	132°51'E
Miyun, China	240-41	40°22'N	116°50'E
Mizdah, Libya (mēz'dä)	188-89	31°26'N	12°59'E
Mizen Head, c., Ire.	190-91	51°27'N	9°49'W
Mizil, Rom. (mē'zēl)	200-01	44°59'N	26°27'E
Mizoram, state, India	246-47	23°30'N	93°00'E
Mizque, Bol.	168-69	17°57'S	65°20'W
Mjölby, Swe. (myûl'bü)	192-93	58°20'N	15°09'E
Mjøsa, lk., Nor. (myûsä)	192-93	60°40'N	11°00'E
Mkinvartsveri, Mt'a, vol., see Kazbek, Gora	227	42°42'N	44°31'E
Mladá Boleslav, Czech Rep. (mlä'dä bô'lĕ-slàf)	194-95	50°25'N	14°54'E
Mlanje Peak, mtn., Malawi see Sapitwa	264-65	15°57'S	35°36'E
Mława, Pol. (mwä'vá)	194-95	53°07'N	20°22'E
Moa, stm., Afr.	260-61	6°60'N	11°34'W
Moa, Pulau, i., Indon.	248-49	8°10'S	127°56'E
Moab, Ut., U.S. (mō'ăb)	118-19	38°35'N	109°33'W
Moa Island, i., Austl.	277	10°12'S	142°16'E
Moala, i., Fiji	279f	18°36'S	179°53'E
Moanda, Gabon	260-61	1°34'S	13°13'E
Moba, D.R.C.	267	7°04'S	29°44'E
Moberly, Mo., U.S. (mō'bēr-lĭ)	120-21	39°25'N	92°26'W
Mobile, Al., U.S. (mō-bēl')	124-25	30°41'N	88°03'W
Mobile Bay, b., Al., U.S. (mō-bēl' bā)	124-25	30°34'N	87°60'W
Mobridge, S.D., U.S. (mō'brĭj)	114-15	45°32'N	100°26'W
Mobutu Sese Seko, Lac, lk., Afr. see Albert, Lake	267	1°40'N	31°00'E
Moca, Dom. Rep. (mô'kä)	142-43	19°24'N	70°31'W
Moçambique, Moz. (mō-säN-bē'kĕ) see Ilha de Moçambique	264-65	15°02'S	40°41'E
Moçambique, nation, Afr. see Mozambique	253	18°15'S	35°00'E
Moçambique, Canal de, strt., Afr. see Mozambique Channel	264-65	19°0'S	41°00'E
Moçâmedes, Ang. (mô-zà-mĕ-dĕs) see Namibe	264-65	15°12'S	12°10'E
Mocha, Yemen	266	13°19'N	43°15'E
Mocha, Isla, i., Chile	171	38°22'S	73°55'W
Mochudi, Bots. (mō-chōō'dē)	269c	24°23'S	26°09'E
Mocímboa da Praia, Moz. (mô-sē'ĕm-bô-à prä'ēà)	264-65	11°20'S	40°22'E
Môco, Morro de, mtn., Ang.	264-65	12°28'S	15°10'E
Mococa, Braz. (mô-kō'kà)	172	21°28'S	46°60'W

n-sing; ŋ-baŋk; N-nasalized n; nŏd; cŏmmit; ōld; ôbey; ôrder; oi-boil; fōōd; ŏ-as oo in foot; ou-out; s-soft; sh-dish; th-thin; pūre; ûnite; ûrn; stŭd; circŭs; ü-as in French tu; '-indeterminate vowel.

Feature (Pronunciation)	Page	Lat.	Long.
Mocorito, Mex.	**144-45**	25°29′N	107°55′W
Moctezuma, Mex. (môk′tå-zōō′mä)	**144-45**	29°48′N	109°42′W
Mocuba, Moz.	**264-65**	16°51′S	36°60′E
Modder, stm., S. Afr.	**264-65**	29°03′S	24°38′E
Modena, Italy (mô′dě-nä)	**200-01**	44°39′N	10°55′E
Modesto, Ca., U.S.	**118-19**	37°39′N	120°60′W
Modimolle, S. Afr.	**269c**	24°42′S	28°25′E
Mödling, Aus. (mûd′lĭng)	**194-95**	48°05′N	16°18′E
Moe, Austl.	**276**	38°11′S	146°15′E
Moengo, Sur.	**164-65**	5°38′N	54°24′W
Moeris, Lake, lk., Egypt			
see Qârûn, Birket	**268b**	29°28′N	30°39′E
Moero, Lac, lk., Afr. *see* Mweru, Lake	**262-63**	9°0′S	28°45′E
Mogadishu, nat. cap., Som.			
(mŏg′*à*-dĭ′shōō)	**262-63**	2°03′N	45°20′E
Mogador, Mor. *see* Essaouira	**258-59**	31°30′N	9°45′W
Mogaung, Mya. (mô-gä′ông)	**238-39**	25°18′N	96°56′E
Mogilno, Pol. (mô-gēl′nô)	**194-95**	52°40′N	17°59′E
Mogocha, Russia	**218-19**	53°44′N	119°45′E
Mogok, Mya. (mô-gōk′)	**246-47**	22°56′N	96°31′E
Mogotón, mtn., N.A.	**149**	13°45′N	86°23′W
Moguer, Spain (mô-gĕr′)	**198-99**	37°16′N	6°50′W
Mohács, Hung. (mô′häch)	**194-95**	46°00′N	18°41′E
Mohall, N.D., U.S. (mō′hôl)	**114-15**	48°46′N	101°30′W
Mohammedia, Mor.	**269a**	33°42′N	7°23′W
Mohe, China (mwo-hŭ)	**222-23**	53°29′N	122°20′E
Mohéli, i., Com. *see* Mwali	**264-65**	12°18′S	43°42′E
Mohyliv-Podil's'kyi, Ukr.	**202-03**	48°28′N	27°47′E
Mo i Rana, Nor.	**184-85**	66°19′N	14°10′E
Moisie, stm., Qc., Can.	**138-39**	50°15′N	66°05′W
Moissac, Fr. (mwä-såk′)	**196-97**	44°06′N	1°05′E
Mojave, Ca., U.S. (mô-hä′vä)	**118-19**	35°04′N	118°10′W
Mojave Desert, des., Ca., U.S.			
(mô-hä′vä dĕs′ẽrt)	**118-19**	35°0′N	117°00′W
Mojiguaçu, stm., Braz. (mô-gê-gwá′sōō)	**172**	20°54′S	48°11′W
Moknine, Tun.	**184-85**	35°38′N	10°54′E
Mokp'o, Kor., S. (môk′pō′)	**243**	34°48′N	126°24′E
Moksha, stm., Russia	**186-87**	54°45′N	41°53′E
Moldavia, nation, Eur. *see* Moldova	**174-75**	47°0′N	29°00′E
Molde, Nor. (môl′dĕ)	**184-85**	62°45′N	7°11′E
Moldova, nation, Eur. (mäl-dō′v*à*)	**174-75**	47°0′N	29°00′E
Moldoveanu, Vârful, mtn., Rom.	**200-01**	45°36′N	24°44′E
Molepolole, Bots. (mō-lå-pô-lō′lå)	**264-65**	24°25′S	25°31′E
Molfetta, Italy (môl-fĕt′tä)	**200-01**	41°12′N	16°36′E
Molina, Chile (mô-lē′nä)	**171**	35°07′S	71°17′W
Molina de Aragón, Spain			
(mô-lē′nä dĕ ä-rä-gô′n)	**198-99**	40°51′N	1°53′W
Molina de Segura, Spain			
(mô-lē′nä dĕ sĕ-gōō′rä)	**198-99**	38°03′N	1°13′W
Moline, Il., U.S. (mô-lēn′)	**114-15**	41°30′N	90°29′W
Mollendo, Peru (mô-lyĕn′dō)	**170**	17°01′S	72°02′W
Mölndal, Swe. (mûln′däl)	**192-93**	57°41′N	11°56′E
Moloka'i, i., Hi., U.S. (mō-lô kä′ē)	**127a**	21°07′N	157°00′W
Molopo, stm., Afr. (mō-lō-pô)	**264-65**	28°31′S	20°13′E
Molson Lake, lk., Mb., Can.			
(mōl′s*ŭ*n lāk)	**134-35**	54°12′N	96°45′W
Moluccas, is., Indon. (mō-lŭk′ŭz)	**248-49**	2°0′S	128°00′E
Molucca Sea, s., Indon. (mō-lŭk′*à* sē)	**248-49**	0°13′N	125°10′E
Moma, Moz.	**264-65**	16°50′S	39°09′E
Mombasa, Kenya (mōm-bä′sä)	**262-63**	4°03′S	39°40′E
Mombetsu, Japan (môm′bĕt-sōō′)	**244**	44°21′N	143°21′E
Mompós, Col. (mōm-pōs′)	**164-65**	9°12′N	74°25′W
Møn, i., Den. (mûn)	**192-93**	55°00′N	12°20′E
Mona, Canal de la, strt., N.A.			
see Mona Passage	**143b**	18°30′N	67°45′W
Mona, Isla de, i., P.R.	**143b**	18°05′N	67°54′W
Monaco, nation, Eur. (mŏn′*à*-kō)	**196-97**	43°45′N	7°25′E
Mona Passage, strt., N.A.			
(mō′nä păs′ĭj)	**143b**	18°30′N	67°45′W
Monarch Mountain, mtn., B.C., Can.			
(mŏn′ẽrk moun′tĭn)	**132-33**	51°54′N	125°53′W
Monastir, Mac. *see* Bitola	**200-01**	41°02′N	21°20′E
Monastyrshchina, Russia			
(mô-nás-tērsh′chĭ-ná)	**202-03**	54°21′N	31°51′E
Monchegorsk, Russia			
(mōn′chĕ-gôrsk)	**184-85**	67°55′N	32°50′E
Monclova, Mex. (môn-klō′vä)	**122-23**	26°54′N	101°25′W
Moncton, N.B., Can. (mŭŋk′t*ŭ*n)	**138-39**	46°06′N	64°48′W
Mondego, stm., Port. (mōn-dĕ′gō)	**198-99**	40°08′N	8°45′W
Mondego, Cabo, c., Port.			
(ká′bō mōn-dā′gō)	**198-99**	40°11′N	8°54′W
Mondovi, Wi., U.S. (mŏn-dō′vĭ)	**114-15**	44°34′N	91°39′W
Monett, Mo., U.S. (mô-nĕt′)	**120-21**	36°55′N	93°56′W
Monforte de Lemos, Spain			
(mōn-fôr′tä dĕ lĕ′môs)	**198-99**	42°32′N	7°30′W
Mongala, stm., D.R.C. (mŏn-gäl′*à*)	**262-63**	1°53′N	19°50′E
Möng Hsat, Mya.	**246-47**	20°31′N	99°13′E
Mongibello, vol., Italy			
see Etna, Monte	**200-01**	37°45′N	15°00′E
Mongo, Chad	**258-59**	12°11′N	18°42′E
Mongol Altayn nuruu, mts., Asia	**222-23**	46°30′N	93°00′E
Mongol Ard Uls, nation, Asia			
see Mongolia	**206-07**	46°0′N	105°00′E
Mongolia, nation, Asia (mŏŋ-gō′lĭ-*à*)	**206-07**	46°0′N	105°00′E
Mongu, Zam. (mŏŋ-gōō′)	**264-65**	15°17′S	23°08′E
Monkoto, D.R.C. (môn-kō′tô)	**262-63**	1°37′S	20°40′E
Monmouth, Il., U.S.			
(mŏn′m*ŭ*th)(mŏn′mouth)	**114-15**	40°54′N	90°39′W
Monmouth, Or., U.S.			
(mŏn′m*ŭ*th)(mŏn′mouth)	**112-13**	44°51′N	123°14′W
Monmouth Mountain, mtn.,			
B.C., Can. (mŏn′m*ŭ*th moun′tĭn)	**132-33**	51°0′N	123°47′W
Mono, stm., Afr.	**260-61**	6°16′N	1°49′E
Mono Island, i., Sol. Is.	**279e**	7°22′S	155°33′E
Mono Lake, lk., Ca., U.S. (mō′nō lāk)	**118-19**	38°0′N	119°00′W
Monon, In., U.S. (mō′nŏn)	**116-17**	40°52′N	86°52′W
Monongahela, Pa., U.S.			
(mô-nŏn-gá-hē′lá)	**116-17**	40°11′N	79°55′W
Monopoli, Italy (mô-nô′pô-lē)	**200-01**	40°57′N	17°18′E
Monroe, Ga., U.S. (m*ŭ*n-rō′)	**124-25**	33°48′N	83°43′W
Monroe, La., U.S. (m*ŭ*n-rō′)	**120-21**	32°31′N	92°07′W
Monroe, Mi., U.S. (m*ŭ*n-rō′)	**116-17**	41°55′N	83°25′W
Monroe, N.C., U.S. (m*ŭ*n-rō′)	**124-25**	34°59′N	80°33′W
Monroe, Ut., U.S. (m*ŭ*n-rō′)	**118-19**	38°38′N	112°07′W
Monroe, Wi., U.S. (m*ŭ*n-rō′)	**116-17**	42°36′N	89°38′W
Monroe City, Mo., U.S. (m*ŭ*n-rō′ sĭ′tē)	**120-21**	39°39′N	91°44′W
Monroeville, Al., U.S. (m*ŭ*n-rō′vĭl)	**124-25**	31°31′N	87°20′W
Monrovia, nat. cap., Lib.			
(mŏn-rō′vĭ-*à*)	**260-61**	6°19′N	10°47′W
Mönsterås, Swe. (mŭn′stĕr-ôs)	**192-93**	57°02′N	16°27′E
Montague, P.E., Can. (mŏn′t*à*-gū)	**138-39**	46°10′N	62°39′W
Montague, Ca., U.S. (mŏn′t*à*-gū)	**112-13**	41°44′N	122°31′W
Montague, Mi., U.S. (mŏn′t*à*-gū)	**116-17**	43°25′N	86°21′W
Montague, Isla, i., Mex.	**144-45**	31°43′N	114°44′W
Montague Island, i., Ak., U.S.			
(mŏn′t*à*-gū ī′l*à*nd)	**126**	60°10′N	147°18′W
Montana, state, U.S. (mŏn-tăn′*à*)	**108-09**	47°0′N	110°00′W
Montargis, Fr. (môn-tár-zhē′)	**196-97**	48°00′N	2°44′E
Montauban, Fr.	**196-97**	44°01′N	1°21′E
Montauk, N.Y., U.S. (mŏn-tôk′)	**116-17**	41°03′N	71°57′W
Montauk Point, c., N.Y., U.S.			
(mŏn-tôk′ point)	**116-17**	41°04′N	71°52′W
Montbard, Fr. (môɴ-bár′)	**196-97**	47°38′N	4°20′E
Montbéliard, Fr. (môɴ-bā-lyár′)	**196-97**	47°31′N	6°46′E
Montbrison, Fr. (môɴ-brē-zôɴ′)	**196-97**	45°37′N	4°04′E
Mont-de-Marsan, Fr. (môɴ-dē-már-säɴ′)	**196-97**	43°53′N	0°30′W
Montdidier, Fr. (môɴ-dē-dyā′)	**196-97**	49°39′N	2°34′E
Monte Alegre, Braz.	**166-67**	2°00′S	54°05′W
Monte Azul, Braz.	**168-69**	15°09′S	42°52′W
Monte Caseros, Arg.			
(mō′n-tĕ-kä-sĕ′rôs)	**173**	30°15′S	57°39′W
Monte Comán, Arg.	**163e**	34°35′S	67°53′W
Monte Cristi, Dom. Rep.			
(mô′n-tĕ-krĕ′s-tē)	**142-43**	19°51′N	71°38′W
Monte Escobedo, Mex.			
(mōn′tå ĕs-kô-bā′dhō)	**146-47**	22°18′N	103°32′W
Montego Bay, Jam. (mŏn-tē′gō bā)	**142-43**	18°28′N	77°55′W
Montélimar, Fr. (môɴ-tā-lē-mär′)	**196-97**	44°34′N	4°45′E
Monte Lindo, stm., Para.	**168-69**	23°54′S	57°17′W
Montello, Wi., U.S. (mŏn-tĕl′ō)	**116-17**	43°48′N	89°20′W
Montemorelos, Mex.			
(mōn′tå-mō-rā′lōs)	**122-23**	25°11′N	99°50′W
Montemor-o-Novo, Port.			
(mōn-tĕ-môr′ô-nô′vô)	**198-99**	38°38′N	8°13′W
Montenegro, nation, Eur.			
(môn-tå-nā′grō) (mŏn-tĕ-nē′grō)	**174-75**	42°30′N	19°18′E
Montepuez, Moz.	**264-65**	13°07′S	38°60′E
Montepulciano, Italy			
(môn′tå-pōōl-chä′nō)	**200-01**	43°06′N	11°47′E
Monte Quemado, Arg.	**168-69**	25°48′S	62°49′W
Montereau-Faut-Yonne, Fr.			
(môn-t′rō′fō-yôn′)	**196-97**	48°23′N	2°57′E
Monterey, Ca., U.S. (mŏn-t*ĕ*-rā′)	**118-19**	36°36′N	121°54′W
Monterey, Tn., U.S. (mŏn-t*ĕ*-rā′)	**124-25**	36°09′N	85°16′W
Monterey, Va., U.S. (mŏn-t*ĕ*-rā′)	**116-17**	38°24′N	79°35′W
Monterey Bay, b., Ca., U.S.			
(mŏn-t*ĕ*-rā′ bā)	**118-19**	36°48′N	121°55′W
Montería, Col. (mōn-tā-rā′ä)	**164-65**	8°45′N	75°53′W
Monteros, Arg. (môn-tĕ′rôs)	**168-69**	27°10′S	65°30′W
Monterotondo, Italy			
(môn-tĕ-rô-tô′n-dō)	**200-01**	42°03′N	12°36′E
Monterrey, Mex. (mōn-tĕ-rā′)	**122-23**	25°41′N	100°19′W
Montesano, Wa., U.S. (mŏn-tē-sä′nō)	**112-13**	46°59′N	123°35′W
Monte Sant'Angelo, Italy			
(mô′n-tĕ sän ä′n-gzhĕ-lô)	**200-01**	41°43′N	15°57′E
Montes Claros, Braz.			
(môn-tĕs-klä′rôs)	**168-69**	16°44′S	43°51′W
Montevallo, Al., U.S. (môn-tê-văl′ō)	**124-25**	33°06′N	86°51′W
Montevarchi, Italy (mōn-tå-vär′kē)	**200-01**	43°32′N	11°35′E
Montevideo, Mn., U.S.			
(mŏn′tå-vê-dhā′ō)	**114-15**	44°57′N	95°43′W
Montevideo, nat. cap., Ur.			
(mōn′tå-vê-dhā′ō)	**173**	34°54′S	56°11′W
Monte Vista, Co., U.S. (mŏn′tē vĭs′t*à*)	**118-19**	37°35′N	106°09′W
Montezuma, Ga., U.S.			
(mŏn-tē-zōō′m*à*)	**124-25**	32°18′N	84°03′W
Montgomery, Pak. *see* Sāhīwāl	**232-33**	30°40′N	73°06′E
Montgomery, Al., U.S.			
(mŏnt-gŭm′ẽr-ĭ)	**124-25**	32°23′N	86°18′W
Monticello, Ar., U.S. (mŏn-tĭ-sĕl′ō)	**124-25**	33°38′N	91°47′W
Monticello, Fl., U.S. (mŏn-tĭ-sĕl′ō)	**124-25**	30°32′N	83°52′W
Monticello, Ga., U.S. (mŏn-tĭ-sĕl′ō)	**124-25**	33°18′N	83°41′W
Monticello, Ia., U.S. (mŏn-tĭ-sĕl′ō)	**114-15**	42°14′N	91°12′W
Monticello, Il., U.S. (mŏn-tĭ-sĕl′ō)	**116-17**	40°00′N	88°35′W
Monticello, In., U.S. (mŏn-tĭ-sĕl′ō)	**116-17**	40°45′N	86°45′W
Monticello, Ky., U.S. (mŏn-tĭ-sĕl′ō)	**124-25**	36°50′N	84°52′W
Monticello, Mn., U.S. (mŏn-tĭ-sĕl′ō)	**114-15**	45°18′N	93°48′W
Monticello, Ut., U.S. (mŏn-tĭ-sĕl′ō)	**118-19**	37°53′N	109°21′W
Montijo, Port. (môn-tē′zhō)	**198-99**	38°42′N	8°58′W
Montijo, Spain (môn-tē′hō)	**198-99**	38°55′N	6°37′W
Montijo, Golfo de, b., Pan.			
(gôl-fô-dĕ-môn-tē′hō)	**150**	7°40′N	81°07′W
Mont-Joli, Qc., Can. (môɴ zhô-lē′)	**138-39**	48°37′N	68°07′W
Mont-Laurier, Qc., Can.	**136-37**	46°32′N	75°30′W
Montluçon, Fr. (môɴ-lü-sôɴ′)	**196-97**	46°20′N	2°36′E
Montmagny, Qc., Can.			
(môɴ-mán-yē′)	**138-39**	46°59′N	70°33′W
Montmorillon, Fr. (môɴ′mô-rē-yôɴ′)	**196-97**	46°26′N	0°52′E
Montpelier, Id., U.S. (mŏnt-pēl′yẽr)	**112-13**	42°20′N	111°18′W
Montpelier, Oh., U.S. (mŏnt-pēl′yẽr)	**116-17**	41°34′N	84°36′W
Montpelier, Vt., U.S.	**116-17**	44°16′N	72°35′W
Montpellier, Fr. (môɴ-pĕ-lyā′)	**196-97**	43°37′N	3°52′E
Montréal, Qc., Can. (mŏn-trē-ôl′)	**136-37**	45°29′N	73°34′W
Montreal, stm., On., Can.			
(mŏn-trē-ôl′)	**136-37**	47°08′N	79°26′W
Montreal, stm., On., Can.			
(mŏn-trē-ôl′)	**136-37**	47°15′N	84°39′W
Montreal Lake, lk., Sk., Can.			
(mŏn-trē-ôl′ lāk)	**134-35**	54°20′N	105°40′W
Montreux, Switz. (môn-trû′)	**194-95**	46°26′N	6°55′E
Montrose, Scot., U.K. (mŏn-trōz′)	**190-91**	56°43′N	2°28′W
Montrose, Co., U.S. (mŏn-trōz′)	**118-19**	38°29′N	107°53′W
Monts, Pointe des, c., Qc., Can.			
(pwăɴt′ dā môɴ′)	**138-39**	49°20′N	67°23′W
Montserrat, dep., N.A. (mŏnt-sĕ-răt′)	**140-41**	16°45′N	62°12′W
Monywa, Mya. (mŏn′yōō-wå)	**246-47**	22°06′N	95°08′E
Monza, Italy (mōn′tsä)	**200-01**	45°35′N	9°17′E
Monze, Zam.	**264-65**	16°17′S	27°29′E
Monzón, Spain (mōn-thōn′)	**198-99**	41°55′N	0°12′E
Moody, Tx., U.S. (mōō′dĭ)	**122-23**	31°18′N	97°21′W
Mooi, stm., S. Afr. (mōō′ĭ)	**269c**	26°52′S	26°57′E
Mooi, stm., S. Afr. (mōō′ĭ)	**269c**	28°46′S	30°34′E
Moon, Mountains of the, mts., Afr.			
see Ruwenzori Range	**267**	0°20′N	29°53′E
Moonta, Austl. (mōōn′tä)	**276**	34°04′S	137°35′E
Moora, Austl. (môr′*à*)	**270-71**	30°38′S	116°00′E
Moore, Lake, lk., Austl. (lāk mōr)	**272-73**	29°44′S	117°32′E
Moorea, i., Fr. Poly.	**279d**	17°32′S	149°50′W
Mooresville, In., U.S. (mōrz′vĭl)	**116-17**	39°36′N	86°22′W
Mooresville, N.C., U.S. (mōrz′vĭl)	**124-25**	35°35′N	80°49′W
Moorhead, Mn., U.S. (mōr′hĕd)	**114-15**	46°53′N	96°45′W
Moorhead, Ms., U.S. (mōr′hĕd)	**124-25**	33°28′N	90°31′W
Moosehead Lake, lk., Me., U.S.	**117a**	45°38′N	69°39′W
Moose Jaw, Sk., Can. (mōōs jô)	**134-35**	50°23′N	105°32′W
Moose Jaw, stm., Sk., Can. (mōōs jô)	**134-35**	50°34′N	105°17′W
Moose Lake, Mb., Can. (mōōs lāk)	**134-35**	53°42′N	100°21′W
Moosomin, Sk., Can. (mōō′sô-mĭn)	**134-35**	50°08′N	101°41′W
Moosonee, On., Can. (mōō′sô-nē)	**128-29**	51°17′N	80°40′W
Moppo, Kor., S. *see* Mokp'o	**243**	34°48′N	126°24′E
Mopti, Mali (mŏp′tē)	**258-59**	14°29′N	4°12′W
Moquegua, Peru (mô-kā′gwä)	**170**	17°12′S	70°57′W
Mora, Spain (mō-rä)	**198-99**	39°41′N	3°46′W
Mora, Swe. (mō′rä)	**192-93**	61°00′N	14°35′E
Mora, Mn., U.S. (mō′r*à*)	**114-15**	45°53′N	93°18′W
Mora, N.M., U.S. (mō′r*à*)	**120-21**	35°58′N	105°22′W
Morādābād, India (mô-rä-dä-bäd′)	**234-35**	28°50′N	78°47′E
Moraleda, Canal, strt., Chile	**171**	44°30′S	73°30′W
Morant Cays, is., Jam.	**142-43**	17°24′N	75°59′W

ăt; finăl; rāte; senåte; ärm; åsk; sofȧ; fâre; ch-choose; dh-as th in other; bē; ēvent; bĕt; recĕnt; cratẽr; g-gō; gh-guttural g; bĭt; ĭ-short neutral; rīde; ᴋ-guttural k as ch in German ich;

Feature (Pronunciation)	Page	Lat.	Long.
Morant Point, c., Jam. (mô-rănt′ point)	142-43	17°55′N	76°11′W
Moratuwa, Sri L.	236	6°48′N	79°53′E
Morava, hist. reg., Czech Rep.	194-95	49°30′N	16°60′E
Moravská Ostrava, Czech Rep. see Ostrava	194-95	49°50′N	18°17′E
Morawhanna, Guy. (mô-rà-hwä′nà)	164-65	8°17′N	59°44′W
Moray Firth, b., Scot., U.K. (mŭr′å fûrth)	190-91	58°02′N	3°05′W
Morbi, India	234-35	22°49′N	70°50′E
Morden, Mb., Can. (môr′dĕn)	134-35	49°11′N	98°05′W
Moreau, stm., S.D., U.S. (mô-rō′)	114-15	45°19′N	100°20′W
Moree, Austl. (mō′rē)	276	29°28′S	149°51′E
Morehead, Ky., U.S. (mōr′hĕd)	116-17	38°11′N	83°27′W
Morehead City, N.C., U.S. (mōr′hĕd sĭ′tĭ)	124-25	34°43′N	76°45′W
Morelia, Mex. (mô-rā′lyä)	146-47	19°42′N	101°12′W
Morella, Spain (mô-rāl′yä)	198-99	40°37′N	0°07′W
Morelos, Mex. (mô-rā′lōs)	122-23	28°25′N	100°53′W
Morelos, state, Mex.	146-47	18°45′N	99°00′W
Morena, Sierra, mts., Spain (syĕr′rä mô-rā′nä)	198-99	38°0′N	5°00′W
Morenci, Mi., U.S. (mô-rĕn′sĭ)	116-17	41°43′N	84°13′W
Moresby Island, i., B.C., Can. (mōrz′bĭ ī′lănd)	132-33	52°50′N	131°55′W
Moreton Island, i., Austl. (môr′tŭn ī′lănd)	276	27°11′S	153°24′E
Morgan City, La., U.S. (môr′găn sĭ′tĭ)	124-25	29°42′N	91°12′W
Morganfield, Ky., U.S. (môr′găn-fēld)	116-17	37°41′N	87°55′W
Morganton, N.C., U.S. (môr′găn-tŭn)	124-25	35°45′N	81°41′W
Morgantown, Ky., U.S. (môr′găn-toun)	124-25	37°13′N	86°41′W
Morgantown, W.V., U.S. (môr′găn-toun)	116-17	39°38′N	79°57′W
Morgenzon, S. Afr. (môr′gănt-sŏn)	269c	26°44′S	29°37′E
Morghāb, stm., Asia	232-33	38°38′N	61°10′E
Morioka, Japan (mō′rē-ō′kà)	244	39°42′N	141°09′E
Morkoka, stm., Russia (môr-kô′kà)	218-19	65°11′N	115°51′E
Morlaix, Fr. (mŏr-lĕ′)	196-97	48°35′N	3°50′W
Mornington, Isla, i., Chile	171	49°45′S	75°23′W
Mornington Island, i., Austl.	277	16°33′S	139°24′E
Morocco, nation, Afr. (mô-rŏk′ō)	253	32°0′N	5°00′W
Morogoro, Tan. (mō-rô-gō′rō)	267	6°49′S	37°40′E
Moro Gulf, b., Phil.	250	6°51′N	123°00′E
Moroleón, Mex. (mô-rô-lā-ōn′)	146-47	20°08′N	101°12′W
Morombe, Madag. (mo͞o-ro͞om′bā)	264-65	21°45′S	43°22′E
Morón, Arg. (mo-rŏ′n)	173	34°39′S	58°37′W
Morón, Cuba (mô-rōn′)	142-43	22°07′N	78°38′W
Mörön, Mong.	240-41	49°38′N	100°10′E
Morón, Ven. (mô-rŏ′n)	163b	10°29′N	68°12′W
Morona, stm., Peru	170	4°45′S	77°04′W
Morondava, Madag. (mô-rōn-dá′và)	264-65	20°18′S	44°17′E
Morón de la Frontera, Spain (mô-rōn′dā läf rôn-tā′rä)	198-99	37°07′N	5°27′W
Moroni, nat. cap., Com.	264-65	11°42′S	43°15′E
Moron Us, stm., China	234-35	34°40′N	94°56′E
Morozovsk, Russia	186-87	48°21′N	41°50′E
Morrill, Ne., U.S. (mŏr′ĭl)	114-15	41°57′N	103°57′W
Morrilton, Ar., U.S. (mŏr′ĭl-tŭn)	120-21	35°09′N	92°45′W
Morrinhos, Braz. (mô-rēn′yōzh)	172	17°44′S	49°06′W
Morris, Mb., Can. (mŏr′ĭs)	134-35	49°21′N	97°23′W
Morris, Il., U.S. (mŏr′ĭs)	116-17	41°22′N	88°26′W
Morris, Mn., U.S. (mŏr′ĭs)	114-15	45°35′N	95°55′W
Morris, stm., Mb., Can. (mŏr′ĭs)	134-35	49°21′N	97°21′W
Morrison, Il., U.S. (mŏr′ĭ-sŭn)	116-17	41°49′N	89°57′W
Morristown, Tn., U.S. (mŏr′rĭs-toun)	124-25	36°13′N	83°17′W
Morro do Chapéu, Braz. (môr-ò dò-shä-pĕ′o͞o)	166-67	11°33′S	41°09′W
Morshansk, Russia (môr-shánsk′)	186-87	53°26′N	41°49′E
Morskoy araly, i., Kaz.	186-87	44°59′N	50°18′E
Morteros, Arg. (mòr-tĕ′tôs)	173	30°42′S	62°00′W
Mortes, stm., Braz.	166-67	11°43′S	50°43′W
Mortlock Islands, is., Micron.	280-81	5°28′N	153°41′E
Morwell, Austl.	276	38°14′S	146°24′E
Mosal′sk, Russia (mō-zálsk′)	202-03	54°30′N	34°58′E
Moscow, Id., U.S. (mŏs′kō)	112-13	46°44′N	117°00′W
Moscow, nat. cap., Russia (mŏs′kō)	202-03	55°45′N	37°38′E
Moscow, stm., Russia (mŏs′kō) see Moskva	202-03	55°04′N	38°51′E
Mosel, stm., Eur. (mō′sĕl) (mô-zĕl)	194-95	50°22′N	7°37′E
Moselle, stm., Eur.	194-95	50°22′N	7°37′E
Moshi, Tan. (mō′shē)	267	3°20′S	37°20′E
Mosjøen, Nor.	184-85	65°50′N	13°12′E
Moskenesøya, i., Nor.	184-85	67°60′N	13°06′E
Moskva, nat. cap., Russia (mŏs-kvä′) see Moscow	202-03	55°45′N	37°38′E
Moskva, stm., Russia (mŏs-kvä′)	202-03	55°04′N	38°51′E
Mosquera, Col.	164-65	2°30′N	78°26′W
Mosquito Coast, hist. reg., Nic. see Mosquitos, Costa de	149	13°0′N	83°45′W
Mosquitos, Costa de, hist. reg., Nic. (kôs-tä-dĕ-mŏs-kē′tō)	149	13°0′N	83°45′W
Mosquitos, Golfo de los, b., Pan.	150	9°0′N	81°15′W
Moss, Nor. (môs)	192-93	59°26′N	10°42′E
Mossaka, Congo	262-63	1°09′S	16°50′E
Mosselbaai, S. Afr. (mô′sŭl bā)	264-65	34°11′S	22°08′E
Mossel Bay, S. Afr. see Mosselbaai	264-65	34°11′S	22°08′E
Mossendjo, Congo	262-63	2°53′S	12°40′E
Mossoró, Braz.	163d	5°11′S	37°20′W
Moss Point, Ms., U.S. (môs point)	124-25	30°23′N	88°33′W
Most, Czech Rep. (môst)	194-95	50°31′N	13°39′E
Mostaganem, Alg.	198-99	35°56′N	0°05′E
Mostar, Bos. (môs′tär)	200-01	43°20′N	17°48′E
Mostardas, Braz.	168-69	31°07′S	50°57′W
Mosul, Iraq (mō′sŭl) (mōso͞ol′)	228-29	36°20′N	43°08′E
Mot′a, Eth.	266	11°04′N	37°53′E
Motagua, stm., N.A. (mô-tä′gwä)	148	15°43′N	88°13′W
Motala, Swe. (mô-tô′lä)	192-93	58°32′N	15°04′E
Motherwell, Scot., U.K. (mŭdh′ẽr-wĕl)	190-91	55°48′N	3°60′W
Motril, Spain (mô-trēl′)	198-99	36°45′N	3°31′W
Motygino, Russia	218-19	58°12′N	94°39′E
Mouhoun, stm., Afr. see Black Volta	260-61	8°41′N	0°60′W
Mouila, Gabon	260-61	1°52′S	11°00′E
Moulins, Fr. (mo͞o-lăn′)	196-97	46°34′N	3°20′E
Moulmein, Mya. see Mawlamyine	246-47	16°30′N	97°38′E
Moulouya, Oued, stm., Mor. (wĕd mo͞o-lo͞o′yà)	258-59	35°08′N	2°21′W
Moultrie, Ga., U.S. (mōl′trĭ)	124-25	31°11′N	83°47′W
Mound City, Il., U.S. (mound sĭ′tē)	124-25	37°05′N	89°10′W
Mound City, Mo., U.S. (mound sĭ′tē)	120-21	40°08′N	95°14′W
Moundou, Chad	262-63	8°34′N	16°05′E
Moundsville, W.V., U.S. (moundz′vĭl)	116-17	39°55′N	80°44′W
Mountain Brook, Al., U.S. (moun′tĭn brŏŏk)	124-25	33°29′N	86°42′W
Mountain Grove, Mo., U.S. (moun′tĭn grōv)	120-21	37°08′N	92°16′W
Mountain Home, Ar., U.S. (moun′tĭn hōm)	120-21	36°20′N	92°23′W
Mountain Home, Id., U.S. (moun′tĭn hōm)	112-13	43°08′N	115°41′W
Mountain Nile, stm., Sudan (moun′tĭn nīl)	262-63	9°30′N	30°30′E
Mountain View, Mo., U.S. (moun′tĭn vū)	124-25	36°60′N	91°42′W
Mountain Village, Ak., U.S.	126	62°05′N	163°44′W
Mount Airy, N.C., U.S. (mount âr′ĭ)	124-25	36°30′N	80°37′W
Mount Ayr, Ia., U.S. (mount âr)	120-21	40°43′N	94°14′W
Mount Barker, Austl.	270-71	34°38′S	117°40′E
Mount Carmel, Il., U.S. (mount kär′mĕl)	116-17	38°24′N	87°46′W
Mount Carmel, Pa., U.S. (mount kär′mĕl)	116-17	40°48′N	76°25′W
Mount Cook National Park, n.p., N.Z. see Aoraki/Mount Cook National Park	278	43°35′S	170°15′E
Mount Desert Island, i., Me., U.S. (mount dĕ-zûrt′ ī′lănd)	117a	44°20′N	68°20′W
Mount Dora, Fl., U.S. (mount dō′rà)	124-25	28°48′N	81°38′W
Mount Forest, On., Can. (mount fŏr′ĕst)	136-37	43°59′N	80°44′W
Mount Gambier, Austl. (mount găm′bēr)	276	37°50′S	140°47′E
Mount Gilead, Oh., U.S. (mount gĭl′ēăd)	116-17	40°33′N	82°49′W
Mount Hagen, Pap. N. Gui.	277	5°52′S	144°14′E
Mount Isa, Austl. (mount ī′zà)	277	20°44′S	139°29′E
Mount Kenya National Park, n.p., Kenya	267	0°09′S	37°19′E
Mount Magnet, Austl. (mount măg-nĕt)	270-71	28°04′S	117°51′E
Mount McKinley National Park, n.p., Ak., U.S. see Denali National Park and Preserve	126	63°15′N	150°30′W
Mount Morgan, Austl. (mount môr-găn)	277	23°39′S	150°23′E
Mount Morris, Mi., U.S. (mount mŏr′ĭs)	116-17	43°07′N	83°42′W
Mount Morris, N.Y., U.S. (mount mŏr′ĭs)	116-17	42°44′N	77°52′W
Mount Olive, N.C., U.S. (mount ŏl′ĭv)	124-25	35°12′N	78°04′W
Mount Pleasant, Ia., U.S. (mount plĕz′ănt)	114-15	40°58′N	91°32′W
Mount Pleasant, Mi., U.S. (mount plĕz′ănt)	116-17	43°36′N	84°46′W
Mount Pleasant, S.C., U.S. (mount plĕz′ănt)	124-25	32°47′N	79°52′W
Mount Pleasant, Tn., U.S. (mount plĕz′ănt)	124-25	35°32′N	87°12′W
Mount Pleasant, Tx., U.S. (mount plĕz′ănt)	120-21	33°09′N	94°58′W
Mount Pleasant, Ut., U.S. (mount plĕz′ănt)	118-19	39°33′N	111°27′W
Mount Rainier National Park, n.p., Wa., U.S. (mount rå-nēr′ năsh′ŭn-ăl pärk)	112-13	46°52′N	121°43′W
Mount Shasta, Ca., U.S. (mount shăs′tà)	112-13	41°19′N	122°18′W
Mount Sterling, Il., U.S. (mount stûr′lĭng)	120-21	39°59′N	90°46′W
Mount Sterling, Ky., U.S. (mount stûr′lĭng)	116-17	38°03′N	83°57′W
Mount Stewart, P.E., Can. (mount stū′ărt)	138-39	46°22′N	62°53′W
Mount Vernon, Il., U.S. (mount vûr′nŭn)	116-17	38°19′N	88°55′W
Mount Vernon, In., U.S. (mount vûr′nŭn)	116-17	37°56′N	87°54′W
Mount Vernon, Ky., U.S. (mount vûr′nŭn)	124-25	37°21′N	84°22′W
Mount Vernon, Mo., U.S. (mount vûr′nŭn)	120-21	37°06′N	93°49′W
Mount Vernon, N.Y., U.S. (mount vûr′nŭn)	116-17	40°55′N	73°50′W
Mount Vernon, Oh., U.S. (mount vûr′nŭn)	116-17	40°23′N	82°29′W
Mount Vernon, Wa., U.S. (mount vûr′nŭn)	112-13	48°25′N	122°20′W
Moura, Braz. (mō′rà)	166-67	1°29′S	61°37′W
Mourne Mountains, mts., N. Ire., U.K. (môrn moun′tĭnz)	190-91	54°10′N	6°04′W
Moussoro, Chad	258-59	13°38′N	16°30′E
Moûtiers, Fr. (mo͞o-tyâr′)	196-97	45°29′N	6°31′E
Moutong, Indon.	248-49	0°29′N	121°14′E
Moyahua, Mex. (mô-yä′wä)	146-47	21°16′N	103°10′W
Moyale, Kenya (mô-yä′lä)	262-63	3°32′N	39°03′E
Moyen Atlas, mts., Mor.	258-59	33°30′N	5°00′W
Moyero, stm., Russia	218-19	68°44′N	103°38′E
Moyo, Pulau, i., Indon.	248-49	8°15′S	117°34′E
Moyobamba, Peru (mō-yô-bäm′bä)	170	6°04′S	76°56′W
Mozambique, nation, Afr. (mō-zăm-bēk′)	253	18°15′S	35°00′E
Mozambique, Canal du, strt., Afr. see Mozambique Channel	264-65	19°0′S	41°00′E
Mozambique Channel, strt., Afr. (mō-zăm-bek′ chăn′ĕl)	264-65	19°0′S	41°00′E
Mozdok, Russia (môz-dôk′)	227	43°44′N	44°39′E
Mozhaysk, Russia (mô-zhäysk′)	202-03	55°30′N	36°01′E
Mozhga, Russia	186-87	56°27′N	52°12′E
Mozyr, Bela. see Mazyr	202-03	52°03′N	29°16′E
Mpika, Zam.	264-65	11°51′S	31°27′E
Mpwapwa, Tan. (′m-pwä′pwä)	267	6°19′S	36°26′E
M′Sila, Alg. (m′sē′lá)	269b	35°42′N	4°33′E
Msta, stm., Russia (m′stá′)	202-03	58°29′N	31°27′E
Mtkvari, stm., Asia see Kür	227	39°17′N	49°26′E
Mtsensk, Russia (m′tsĕnsk)	202-03	53°17′N	36°35′E
Mtwara, Tan.	264-65	10°21′S	40°15′E
Muanda, D.R.C.	262-63	5°57′S	12°22′E
Muang Khammouan, Laos	246-47	17°25′N	104°49′E
Muang Không, Laos	246-47	14°07′N	105°51′E
Muang Ngoy, Laos	246-47	20°42′N	102°40′E
Muang Pak-Lay, Laos	246-47	18°13′N	101°24′E
Muang Pakxan, Laos	246-47	18°25′N	103°39′E
Muang Sing, Laos	246-47	21°11′N	101°09′E
Muang Vangviang, Laos	246-47	18°55′N	102°26′E
Muang Xaignabouri, Laos	246-47	19°17′N	101°43′E
Muar, Malay.	246-47	2°02′N	102°34′E
Muaratewe, Indon.	248-49	0°56′S	114°52′E
Mubende, Ug.	267	0°35′N	31°24′E
Mucajaí, stm., Braz.	164-65	2°24′N	60°50′W
Muchinga Mountains, mts., Zam.	264-65	11°40′S	31°44′E
Mucuri, stm., Braz.	172	18°05′S	39°34′W
Mudan, stm., China (mo͞o-dän)	244	46°18′N	129°32′E
Mudanjiang, China (mo͞o-dän-jyäŋ)	244	44°35′N	129°36′E
Mudgee, Austl. (mŭ-jē)	276	32°36′S	149°35′E
Mueda, Moz.	264-65	11°40′S	39°34′E
Mufulira, Zam.	264-65	12°33′S	28°14′E
Muğla, Tur. (mo͞og′lä)	200-01	37°13′N	28°22′E
Mühlhausen, Ger. (mül′hou-zĕn)	194-95	51°13′N	10°28′E
Muhlig-Hofmann Mountains, mts., Ant.	287	72°10′S	4°53′E

n-sing; ŋ-baŋk; ɴ-nasalized n; nŏd; cŏmmit; ōld; ôbey; ôrder; oi-boil; fo͞od; ŏŏ-as oo in foot; ou-out; s-soft; sh-dish; th-thin; pūre; ůnite; ûrn; stŭd; circŭs; ü-as in French tu; ′-indeterminate vowel.

Feature (Pronunciation)	Page	Lat.	Long.
Muhu, i., Est. (mo͞o′ho͞o)	**192-93**	58°37′N	23°13′E
Mukacheve, Ukr.	**194-95**	48°26′N	22°45′E
Mukah, Malay.	**248-49**	2°54′N	112°06′E
Mukalla, Yemen *see* Al-Mukallā	**220-21**	14°32′N	49°08′E
Mukden, China *see* Shenyang	**243**	41°48′N	123°24′E
Mukry, Turkmen.	**232-33**	37°36′N	65°43′E
Mula, Spain (mo͞o′lä)	**198-99**	38°03′N	1°30′W
Muladu, i., Mald.	**236**	7°01′N	72°59′E
Mulchatna, stm., Ak., U.S.	**126**	59°39′N	157°08′W
Mulhacén, mtn., Spain	**198-99**	37°03′N	3°19′W
Mulhouse, Fr. (mü-lo͞oz′)	**196-97**	47°45′N	7°20′E
Muling, China (mo͞o-lĭŋ)	**244**	44°31′N	130°16′E
Muling, China (mo͞o-lĭŋ)	**244**	44°56′N	130°32′E
Muling, stm., China (mo͞o-lĭŋ)	**244**	45°52′N	133°30′E
Mull, Island of, i., Scot., U.K. (ī′lǎnd ŏv mŭl)	**190-91**	56°27′N	6°00′W
Mullan, Id., U.S. (mŭl′ǎn)	**112-13**	47°28′N	115°48′W
Muller, Pegunungan, mts., Indon. (mŭl′ẽr)	**248-49**	0°40′N	113°50′E
Mullewa, Austl.	**270-71**	28°33′S	115°31′E
Mullins, S.C., U.S. (mŭl′ĭnz)	**124-25**	34°12′N	79°15′W
Mulongo, D.R.C.	**262-63**	7°49′S	26°60′E
Multān, Pak. (mo̅o-tän′)	**232-33**	30°11′N	71°27′E
Mulvane, Ks., U.S. (mŭl-vān′)	**120-21**	37°28′N	97°15′W
Mumbai, India	**236**	18°57′N	72°50′E
Mumbwa, Zam. (mo̅om′bwä)	**264-65**	14°59′S	27°04′E
Mun, stm., Thai.	**246-47**	15°19′N	105°31′E
Muna, Mex. (mo͞o′nȧ)	**148**	20°29′N	89°43′W
Muna, stm., Russia	**218-19**	67°53′N	123°05′E
Muna, Pulau, i., Indon.	**248-49**	4°53′S	122°27′E
München, Ger. *see* Munich	**194-95**	48°08′N	11°35′E
Muncie, In., U.S. (mŭn′sĭ)	**116-17**	40°11′N	85°22′W
Munger, India	**234-35**	25°23′N	86°28′E
Mungindi, Austl. (mŭn-gĭn′dē)	**276**	28°59′S	148°59′E
Mungkan Kandju National Park, n.p., Austl.	**277**	13°32′S	142°37′E
Munich, Ger. (mū′nĭk)	**194-95**	48°08′N	11°35′E
Munising, Mi., U.S. (mū′nĭ-sĭng)	**114-15**	46°24′N	86°39′W
Munkács, Ukr. *see* Mukacheve	**194-95**	48°26′N	22°45′E
Münster, Ger.	**194-95**	51°57′N	7°37′E
Muntok, Indon. (mo̅on-tŏk′)	**246-47**	2°04′S	105°10′E
Muonio, Fin.	**184-85**	67°58′N	23°40′E
Muqayshiṭ, i., U.A.E.	**230-31**	24°10′N	53°45′E
Muqdisho, nat. cap., Som. *see* Mogadishu	**262-63**	2°03′N	45°20′E
Muradiye, Tur. (mo͞o-rä′dē-yě)	**227**	38°59′N	43°50′E
Murashi, Russia	**186-87**	59°24′N	48°58′E
Murat, stm., Tur. (mo͞o-rȧt′)	**227**	38°40′N	39°53′E
Murchison, stm., Austl. (mûr′chĭ-sŭn)	**272-73**	27°42′S	114°08′E
Murchison Falls, wtfl., Ug. *see* Kabalega Falls	**267**	2°17′N	31°42′E
Murchison Falls National Park, n.p., Ug.	**267**	2°15′N	31°50′E
Murcia, Spain (mo͞or′thyä)	**198-99**	37°59′N	1°08′W
Mur-de-Barrez, Fr.	**196-97**	44°51′N	2°39′E
Murdo, S.D., U.S. (mûr′dō)	**114-15**	43°53′N	100°41′W
Muret, Fr. (mü-rě′)	**196-97**	43°28′N	1°19′E
Murfreesboro, N.C., U.S. (mûr′frēz-bŭr-ō)	**124-25**	36°27′N	77°06′W
Murfreesboro, Tn., U.S. (mûr′frēz-bŭr-ō)	**124-25**	35°50′N	86°23′W
Murgap, stm., Asia (mo͞or-gäp′)	**232-33**	38°38′N	61°10′E
Murgon, Austl.	**276**	26°15′S	151°57′E
Mūrītāniyā, nation, Afr. *see* Mauritania	**253**	20°0′N	12°00′W
Murmansk, Russia (mo͞or-mänsk′)	**184-85**	68°58′N	33°05′E
Murom, Russia (mo͞o′rôm)	**186-87**	55°34′N	42°02′E
Muroran, Japan (mo͞o′rō-rän)	**244**	42°19′N	140°59′E
Muros, Spain (mo͞o′rōs)	**198-99**	42°47′N	9°04′W
Murphy, N.C., U.S. (mŭr′fĭ)	**124-25**	35°05′N	84°02′W
Murphysboro, Il., U.S. (mŭr′fĭz-bŭr-ō)	**116-17**	37°46′N	89°20′W
Murray, Ky., U.S. (mŭr′ĭ)	**124-25**	36°37′N	88°19′W
Murray, Ut., U.S. (mŭr′ĭ)	**112-13**	40°39′N	111°54′W
Murray, stm., Austl. (mŭr′ĭ)	**276**	35°22′S	139°21′E
Murray, stm., B.C., Can. (mŭr′ĭ)	**132-33**	55°43′N	121°13′W
Murray, Lake, lk., Pap. N. Gui.	**277**	7°0′S	141°30′E
Murray Bridge, Austl. (mŭr′ĭ brĭj)	**276**	35°08′S	139°16′E
Murray Harbour, P.E., Can. (mŭr′ĭ här′bẽr)	**138-39**	45°60′N	62°32′W
Murray-Sunset National Park, n.p., Austl.	**276**	34°45′S	141°30′E
Murrumbidgee, stm., Austl. (mŭr-ŭm-bĭd′jē)	**276**	34°42′S	143°08′E
Murska Sobota, Slvn. (mo͞or′skä sô′bô-tä)	**200-01**	46°40′N	16°10′E
Murua Island, i., Pap. N. Gui.	**277**	9°06′S	152°45′E
Murud, Gunong, mtn., Malay.	**248-49**	3°52′N	115°30′E
Mururoa, at., Fr. Poly.	**280-81**	21°52′S	138°55′W

Feature (Pronunciation)	Page	Lat.	Long.
Murwāra, India	**234-35**	23°50′N	80°24′E
Murwillumbah, Austl. (mŭr-wĭl′lŭm-bŭ)	**276**	28°21′S	153°24′E
Murzuq, Libya	**258-59**	25°56′N	13°55′E
Murzūq, Idhān, des., Libya	**258-59**	24°30′N	13°00′E
Mürzzuschlag, Aus. (mürts′tso͞o-shlägh)	**194-95**	47°36′N	15°41′E
Muş, Tur. (mo͞osh)	**227**	38°43′N	41°29′E
Musala, mtn., Blg.	**200-01**	42°11′N	23°34′E
Musay'īd, Qatar	**230-31**	24°59′N	51°33′E
Muscat, nat. cap., Oman (mŭs-kăt′)	**230-31**	23°36′N	58°32′E
Muscat and Oman, nation, Asia *see* Oman	**206-07**	22°0′N	58°00′E
Muscatine, Ia., U.S. (mŭs-kȧ-tēn)	**114-15**	41°25′N	91°02′W
Muscle Shoals, Al., U.S. (mŭs′'l shōlz)	**124-25**	34°44′N	87°40′W
Mushin, Nig.	**260a**	6°31′N	3°21′E
Musi, stm., Indon. (mo͞o′sē)	**246-47**	2°22′S	104°55′E
Muskegon, Mi., U.S. (mŭs-kē′gŭn)	**116-17**	43°14′N	86°15′W
Muskegon, stm., Mi., U.S. (mŭs-kē′gŭn)	**116-17**	43°13′N	86°19′W
Muskegon Heights, Mi., U.S. (mŭs-kē′gŭn hīts)	**116-17**	43°12′N	86°14′W
Muskingum, stm., Oh., U.S. (mŭs-kĭŋ′gŭm)	**116-17**	39°24′N	81°28′W
Muskogee, Ok., U.S. (mŭs-kō′gē)	**120-21**	35°44′N	95°22′W
Muskoka, Lake, lk., On., Can. (lāk mŭs-kō′kȧ)	**136-37**	45°02′N	79°25′W
Musoma, Tan.	**267**	1°30′S	33°48′E
Mussau Island, i., Pap. N. Gui. (mo͞o-sä′o͞o ī′lǎnd)	**277**	1°27′S	149°37′E
Musselshell, stm., Mt., U.S. (mŭs′'l-shěl)	**112-13**	47°27′N	107°55′W
Mustvee, Est. (mo͞ost′vě-ě)	**192-93**	58°51′N	26°56′E
Musu-dan, c., Kor., N. (mo͞o′so͞o dän)	**243**	40°51′N	129°43′E
Muswellbrook, Austl. (mŭs′wŭnl-brŏk)	**276**	32°16′S	150°54′E
Mutare, Zimb.	**264-65**	18°58′S	32°40′E
Mutsamudu, Com.	**264-65**	12°08′S	44°26′E
Mutsu, Japan	**244**	41°17′N	141°10′E
Mutsu-wan, b., Japan (mo͞ot′so͞o wän)	**244**	41°05′N	140°55′E
Mutton Bay, Qc., Can. (mŭt′'n bā)	**138-39**	50°47′N	59°02′W
Muttra, India *see* Mathura	**234-35**	27°30′N	77°41′E
Mutum, Braz. (mo͞o-to͞o′m)	**172**	19°48′S	41°27′W
Muynak, Uzb. *see* Mŭynoq	**226**	43°46′N	59°02′E
Mŭynoq, Uzb.	**226**	43°46′N	59°02′E
Muyua Island, i., Pap. N. Gui. *see* Murua Island	**277**	9°06′S	152°45′E
Muzaffarnagar, India	**234-35**	29°28′N	77°42′E
Muzaffarpur, India	**234-35**	26°07′N	85°23′E
Muztag, mtn., China	**234-35**	36°03′N	80°07′E
Muztag, mtn., China	**234-35**	36°25′N	87°25′E
Muztaū biīgi, mtn., Asia *see* Belukha, Mount	**226**	49°51′N	86°29′E
Mwali, i., Com.	**264-65**	12°18′S	43°42′E
Mwanza, Tan. (mwän′zä)	**267**	2°31′S	32°54′E
Mweka, D.R.C.	**262-63**	4°51′S	21°34′E
Mwene-Ditu, D.R.C.	**262-63**	7°03′S	23°27′E
Mweru, Lake, lk., Afr. (lāk mwē′ro͞o)	**262-63**	9°0′S	28°45′E
Mweru Wantipa, Lake, lk., Zam.	**262-63**	8°45′S	29°40′E
Mwokil, at., Micron.	**280-81**	6°53′N	159°47′E
Myanaung, Mya.	**246-47**	18°17′N	95°19′E
Myanmar, nation, Asia (myän-mär)	**206-07**	22°0′N	98°00′E
Myaundzha, Russia	**218-19**	63°03′N	147°11′E
Myaungmya, Mya.	**246-47**	16°35′N	94°55′E
Myingyan, Mya. (myĭng-yŭn′)	**246-47**	21°27′N	95°23′E
Myitkyinā, Mya. (myĭ′chē-nȧ)	**238-39**	25°23′N	97°24′E
Mykolaïv, Ukr.	**202-03**	46°58′N	31°59′E
Mymensingh, Bngl.	**234-35**	24°45′N	90°24′E
Myohyang-san, mtn., Kor., N. (myō′hyang-sän′)	**243**	40°01′N	126°21′E
Mýrdalsjökull, ice, Ice. (mür′däls-yû′kôl)	**190a**	63°40′N	19°05′W
Myrtle Beach, S.C., U.S. (mûr′t'l bēch)	**124-25**	33°42′N	78°54′W
Mysore, India (mī-sōr′)	**236**	12°18′N	76°39′E
Mysore, state, India *see* Karnātaka	**236**	14°0′N	76°00′E
Mys Shmidta, Russia	**126**	68°52′N	179°37′W
My Tho, Viet.	**246-47**	10°22′N	106°22′E
Mytilíni, Grc.	**200-01**	39°06′N	26°33′E
Mytishchi, Russia (mê-tēsh′chi)	**202-03**	55°55′N	37°46′E
Mzuzu, Malawi	**264-65**	11°24′S	33°57′E

N

Feature (Pronunciation)	Page	Lat.	Long.
Naantali, Fin. (nän′tȧ-lē)	**192-93**	60°30′N	22°04′E
Naberezhnye Chelny, Russia	**186-87**	55°42′N	52°19′E
Nabeul, Tun. (nā-bûl′)	**184-85**	36°27′N	10°46′E
Nabī Shu'ayb, Jabal an-, mtn., Yemen	**220-21**	15°17′N	43°59′E
Nābulus, W.B.	**228-29**	32°14′N	35°17′E
Nacala, Moz. (nă-kȧ′lȧ)	**264-65**	14°33′S	40°40′E
Náchod, Czech Rep. (näk′ôt)	**194-95**	50°25′N	16°11′E
Nacogdoches, Tx., U.S. (năk′ō-dō′chěz)	**122-23**	31°35′N	94°39′W
Nacozari de García, Mex.	**144-45**	30°24′N	109°39′W
Nadadores, Mex. (nä-dä-dō′räs)	**122-23**	27°02′N	101°35′W
Nadiād, India	**234-35**	22°41′N	72°52′E
Nador, Mor.	**198-99**	35°11′N	2°56′W
Nadym, Russia	**218-19**	65°35′N	72°39′E
Nadym, stm., Russia (ná′dĭm)	**218-19**	66°13′N	72°00′E
Næstved, Den. (něst′vĭdh)	**194-95**	55°14′N	11°46′E
Naga, Phil. (nä′gä)	**250**	13°38′N	123°11′E
Nāgāland, state, India	**234-35**	26°0′N	95°00′E
Nagano, Japan (nä′gä-nō)	**245**	36°39′N	138°12′E
Nagaoka, Japan (nä′gȧ-ō′kȧ)	**245**	37°27′N	138°51′E
Nagaon, India	**234-35**	26°21′N	92°41′E
Nāgappattinam, India	**236**	10°46′N	79°51′E
Nagarote, Nic. (nä-gä-rô′tě)	**149**	12°16′N	86°34′W
Nagasaki, Japan (nä′gȧ-sä′kē)	**245**	32°45′N	129°53′E
Nāgaur, India	**234-35**	27°12′N	73°44′E
Nāgercoil, India	**236**	8°10′N	77°26′E
Nagorno-Karabakh, hist. reg., Azer. (nu-gôr′nŭ-kŭ-rŭ-bäk′)	**227**	40°00′N	46°40′E
Nagoya, Japan	**245**	35°10′N	136°55′E
Nāgpur, India (näg′po͞or)	**234-35**	21°09′N	79°05′E
Nagqu, China	**234-35**	31°31′N	92°05′E
Nagua, Dom. Rep. (ná′gwä)	**142-43**	19°23′N	69°51′W
Naguna, Île, i., Vanuatu *see* Nguna, Île	**279g**	17°27′S	168°21′E
Nagybánya, Rom. *see* Baia Mare	**194-95**	47°39′N	23°35′E
Nagykanizsa, Hung. (nôd′y′kô′nē-shô)	**194-95**	46°27′N	16°60′E
Nagykőrös, Hung. (nôd′y′kŭ-rŭsh)	**194-95**	47°02′N	19°46′E
Nagyvarad, Rom. *see* Oradea	**194-95**	47°04′N	21°56′E
Naha, Japan (nä′hä)	**244a**	26°13′N	127°42′E
Nāhan, India	**234-35**	30°32′N	77°17′E
Nahanni National Park Reserve, n.p., N.T., Can.	**128-29**	61°35′N	125°45′W
Nahe, China	**240-41**	48°29′N	124°53′E
Nahr al-Urdunn, stm., Asia *see* Jordan	**228-29**	31°46′N	35°34′E
Nahuel Huapi, Lago, lk., Arg. (lä′gô nä′wěl wä′pē)	**171**	40°58′S	71°30′W
Naica, Mex. (nä-ē′kä)	**122-23**	27°51′N	105°30′W
Nain, Nf., Can. (nīn)	**128-29**	56°33′N	61°43′W
Nā'īn, Iran	**232-33**	32°52′N	53°05′E
Naini Tāl, India	**234-35**	29°24′N	79°26′E
Nairn, Scot., U.K. (nârn)	**190-91**	57°35′N	3°53′W
Nairobi, nat. cap., Kenya (nī-rō′bē)	**267**	1°16′S	36°49′E
Naitauba, i., Fiji	**279f**	17°01′S	179°16′W
Naivasha, Kenya (nī-vä′shȧ)	**267**	0°45′S	36°26′E
Najafābād, Iran	**232-33**	32°38′N	51°22′E
Najasa, stm., Cuba (nä-hä′sä)	**142-43**	20°43′N	77°59′W
Najd, hist. reg., Sau. Ar.	**266**	26°07′N	44°40′E
Najin, Kor., N. (nä′jĭn)	**243**	42°15′N	130°18′E
Naju, Kor., S. (nä′jo͞o′)	**243**	35°02′N	126°43′E
Nakambé, stm., Afr. *see* White Volta	**260-61**	8°57′N	1°10′W
Nakanbe, stm., Afr. *see* White Volta	**260-61**	8°57′N	1°10′W
Nakano-shima, i., Japan	**244**	29°50′N	129°52′E
Nakhichevan, Azer. *see* Naxçıvan	**227**	39°13′N	45°25′E
Nakhodka, Russia (nŭ-kôt′kŭ)	**244**	42°49′N	132°53′E
Nakhon Pathom, Thai.	**246-47**	13°49′N	100°04′E
Nakhon Phanom, Thai.	**246-47**	17°24′N	104°47′E
Nakhon Ratchasima, Thai.	**246-47**	14°58′N	102°06′E
Nakhon Sawan, Thai.	**246-47**	15°42′N	100°06′E
Nakhon Si Thammarat, Thai.	**246-47**	8°26′N	99°58′E
Nakskov, Den. (näk′skou)	**194-95**	54°50′N	11°08′E
Nakuru, Kenya	**267**	0°17′S	36°04′E
Nālanda, India	**234-35**	25°08′N	85°24′E
Nalchik, Russia (nål-chēk′)	**227**	43°29′N	43°37′E
Nalgonda, India	**236**	17°03′N	79°16′E
Nalubaale Dam, d., Ug.	**267**	0°27′N	33°11′E
Nālūt, Libya (nä-lo͞ot′)	**188-89**	31°53′N	10°60′E
Namak, Daryācheh-ye, lk., Iran	**232-33**	34°30′N	51°50′E
Namangan, Uzb. (nȧ-mȧn-gän′)	**232-33**	40°60′N	71°40′E
Namapa, Moz.	**264-65**	13°42′S	39°49′E
Nambour, Austl. (năm′bôr)	**276**	26°38′S	152°58′E
Namcha Barwa, mtn., China *see* Namjagbarwa Feng	**238-39**	29°38′N	95°04′E
Nam Co, lk., China (näm tswo)	**234-35**	30°41′N	90°32′E
Nam Dinh, Viet. (näm dĕnκ′)	**246-47**	20°26′N	106°10′E
Namhae-do, i., Kor., S. (näm′hī′)	**243**	34°48′N	127°57′E
Namhkam, Mya.	**238-39**	23°50′N	97°41′E

ăt; finăl; rāte; senâte; ärm; ȧsk; sofȧ; fâre; ch-choose; dh-as th in other; bē; ēvent; bět; recěnt; cratẽr; g-gō; gh-guttural g; bĭt; ĭ-short neutral; rīde; κ-guttural k as ch in German ich;

Feature (Pronunciation)	Page	Lat.	Long.
Namib Desert, des., Nmb. (nä-mēb′ dĕs′ĕrt)	264-65	23°0′S	15°00′E
Namibe, Ang.	264-65	15°12′S	12°10′E
Namibia, nation, Afr. (nä-mĭ′-bē-*ȧ*)	253	22°0′S	17°00′E
Namib Naukluft Park, pk., Nmb.	264-65	24°40′S	15°17′E
Namjagbarwa Feng, mtn., China	238-39	29°38′N	95°04′E
Namlea, Indon.	248-49	3°16′S	127°06′E
Nam Ngum Reservoir, res., Laos	246-47	18°33′N	102°37′E
Namoi, stm., Austl. (nămôi)	276	30°00′S	148°04′E
Namolok Atoll, at., Micron.	280-81	5°55′N	153°08′E
Nampa, Id., U.S. (năm′*pȧ*)	112-13	43°35′N	116°33′W
Namp'o, Kor., N.	243	38°45′N	125°23′E
Nampula, Moz.	264-65	15°07′S	39°16′E
Namsang, Mya.	246-47	20°53′N	97°43′E
Namsos, Nor. (näm′sôs)	184-85	64°28′N	11°32′E
Namu, B.C., Can.	132-33	51°50′N	127°52′W
Namuka-i-Lau, i., Fiji	279f	18°51′S	178°38′W
Namyit Island, i., Asia	224-25	10°24′N	114°27′E
Nan, Thai.	246-47	18°46′N	100°46′E
Nan, stm., Thai.	246-47	15°42′N	100°09′E
Nanaimo, B.C., Can. (nȧ-nī′mō)	132-33	49°10′N	123°57′W
Nanam, Kor., N. (nä′nän′)	243	41°43′N	129°42′E
Nanango, Austl.	276	26°41′S	152°00′E
Nanao, Japan (nä′nä-ō)	245	37°03′N	136°58′E
Nanchang, China	238-39	28°41′N	115°53′E
Nancheng, China (nän-chäŋ) *see* Hanzhong	238-39	33°04′N	107°02′E
Nancheng, China (nän-chäŋ)	238-39	27°34′N	116°39′E
Nanchong, China (nän-chôŋ)	238-39	30°47′N	106°05′E
Nancy, Fr. (näɴ-sē′)	196-97	48°41′N	6°10′E
Nanda Devi, mtn., India (nän′dä dā′vē)	234-35	30°23′N	79°59′E
Nānded, India	236	19°09′N	77°18′E
Nanga-Eboko, Camrn.	260-61	4°40′N	12°22′E
Nanga Parbat, mtn., Pak.	232-33	35°15′N	74°36′E
Nangis, Fr. (näɴ-zhē′)	196-97	48°34′N	3°01′E
Nanhai, China *see* Foshan	238-39	23°03′N	113°07′E
Nan Hai, s., Asia *see* South China Sea	224-25	10°0′N	113°00′E
Nanjing, China (nän-jyĭŋ)	238-39	24°31′N	117°23′E
Nanjing, China (nän-jyĭŋ)	238-39	32°03′N	118°47′E
Nanking, China *see* Nanjing	238-39	32°03′N	118°47′E
Nan Ling, mts., China	238-39	25°0′N	112°00′E
Nanliu, stm., China (nän-lĭō)	238-39	21°40′N	109°05′E
Nanning, China (nän′nĭng′)	238-39	22°48′N	108°20′E
Nanpan, stm., China (nän-pän)	238-39	24°57′N	106°08′E
Nanping, China (nän-pĭŋ)	238-39	26°38′N	118°10′E
Nansei-shotō, is., Japan *see* Ryukyu Islands	244a	25°44′N	126°58′E
Nanshan Island, i., Asia	224-25	10°44′N	115°49′E
Nansio, Tan.	267	2°06′S	33°03′E
Nantai-zan, vol., Japan (nän-täē-zän)	245	36°46′N	139°30′E
Nantes, Fr. (näɴt′)	196-97	47°14′N	1°33′W
Nanticoke, Pa., U.S. (năn′tĭ-kōk)	116-17	41°12′N	76°00′W
Nantong, China (nän-tôŋ)	238-39	32°01′N	120°51′E
Nantucket Island, i., Ma., U.S. (năn-tŭk′ĕt ī′l*ȧ*nd)	116-17	41°16′N	70°03′W
Nanumea, at., Tuvalu	280-81	5°42′S	176°09′E
Nanuque, Braz.	172	17°50′S	40°20′W
Nanxiong, China (nän-shôŋ)	238-39	25°07′N	114°20′E
Nanyang, China (nän-yäŋ)	238-39	33°00′N	112°32′E
Nanyuki, Kenya	267	0°01′N	37°05′E
Nao, Cabo de la, c., Spain *see* Nao, Cap de la	198-99	38°44′N	0°14′E
Nao, Cap de la, c., Spain	198-99	38°44′N	0°14′E
Náousa, Grc. (nä′o͞o-sȧ)	200-01	40°37′N	22°03′E
Napa, Ca., U.S. (năp′*ȧ*)	118-19	38°18′N	122°17′W
Napaktulik Lake, lk., Nu., Can.	130-31	66°15′N	113°05′W
Napanee, On., Can. (năp′*ȧ*-nē)	136-37	44°15′N	76°57′W
Naperville, Il., U.S. (nā′pẽr-vĭl)	116-17	41°46′N	88°09′W
Napier, N.Z. (nā′pĭ-ẽr)	278	39°29′S	176°54′E
Naples, Italy (nā′p′lz)	200-01	40°51′N	14°17′E
Naples, Fl., U.S. (nā′p′lz)	125a	26°09′N	81°48′W
Napo, stm., S.A. (nä′pō)	170	3°29′S	72°38′W
Napoleon, Oh., U.S. (n*ȧ*-pō′lē-*ŭ*n)	116-17	41°23′N	84°08′W
Napoli, Italy (nä′pô-lē) *see* Naples	200-01	40°51′N	14°17′E
Nappanee, In., U.S. (năp′*ȧ*-nē)	116-17	41°26′N	85°59′W
Nara, Japan (nä′rä)	245	34°41′N	135°50′E
Nara, state, Japan (nä′rä)	245	34°30′N	135°50′E
Naracoorte, Austl. (nȧ-rȧ-ko͞on′tē)	276	36°58′S	140°44′E
Naraụ, stm., Eur. *see* Narew	194-95	52°31′N	21°05′E
Nārāyanganj, Bngl.	234-35	23°37′N	90°30′E
Narbonne, Fr. (nȧr-bôn′)	196-97	43°11′N	2°60′E
Narborough, i., Ec. *see* Fernandina, Isla	170a	0°26′S	91°30′W
Nares Stræde, strt., N.A. *see* Nares Strait	86	80°30′N	68°00′W
Nares Strait, strt., N.A.	86	80°30′N	68°00′W
Narew, stm., Eur. (när′ĕf)	194-95	52°31′N	21°05′E
Narmada, stm., India	234-35	21°41′N	72°45′E
Nārnaul, India	234-35	28°03′N	76°06′E
Narodnaya, Gora, mtn., Russia (gȧ-rä′ nȧ-rôd′nȧ-yȧ)	186-87	65°04′N	60°09′E
Naro-Fominsk, Russia (nä′rô-mĕnsk′)	202-03	55°23′N	36°44′E
Narok, Kenya	267	1°05′S	35°52′E
Narrabri, Austl.	276	30°20′S	149°47′E
Narrandera, Austl. (nȧ-rȧn-dē′rȧ)	276	34°45′S	146°33′E
Narrogin, Austl. (năr′ô-gĭn)	270-71	32°56′S	117°11′E
Narromine, Austl.	276	32°14′S	148°14′E
Narsimhapur, India	234-35	22°57′N	79°12′E
Narva, Est. (nȧr′vȧ)	192-93	59°23′N	28°12′E
Narva laht, b., Eur.	192-93	59°30′N	27°40′E
Narvik, Nor. (när′vĕk)	184-85	68°26′N	17°25′E
Narvskiy Zaliv, b., Eur. (nȧr′vskĭ zä′lĭf) *see* Narva laht	192-93	59°30′N	27°40′E
Naryan-Mar, Russia (när′yȧn mär′)	186-87	67°37′N	52°60′E
Naryn, Kyrg.	226	41°26′N	75°59′E
Naryn, stm., Asia (nŭ-rĭn′)	226	41°46′N	73°13′E
Nasca, Peru	170	14°50′S	74°57′W
Nāshik, India	234-35	20°00′N	73°47′E
Nashua, Ia., U.S. (năsh′ŭ-*ȧ*)	114-15	42°57′N	92°33′W
Nashua, N.H., U.S.	116-17	42°46′N	71°28′W
Nashville, Ar., U.S. (năsh′vĭl)	120-21	33°57′N	93°51′W
Nashville, Ga., U.S. (năsh′vĭl)	124-25	31°12′N	83°15′W
Nashville, Il., U.S. (năsh′vĭl)	116-17	38°21′N	89°23′W
Nashville, Mi., U.S. (năsh′vĭl)	116-17	42°36′N	85°06′W
Nashville, Tn., U.S. (năsh′vĭl)	124-25	36°09′N	86°47′W
Našice, Cro. (nä′shē-tsĕ)	200-01	45°29′N	18°04′E
Nāṣir, Buḥayrat, res., Afr. *see* Nasser, Lake	266	22°40′N	32°00′E
Nâsir, Buheirat, res., Afr. *see* Nasser, Lake	266	22°40′N	32°00′E
Nasirābād, Bngl. *see* Mymensingh	234-35	24°45′N	90°24′E
Nass, stm., B.C., Can. (năs)	132-33	54°59′N	129°40′W
Nassau, nat. cap., Bah. (năs′ô)	142-43	25°04′N	77°20′W
Nassau Island, i., Cook Is.	280-81	11°34′S	165°24′W
Nasser, Lake, res., Afr. (lāk nä-sẽr)	266	22°40′N	32°00′E
Natagaima, Col. (nä-tä-gī′mä)	163c	3°38′N	75°06′W
Natal, Braz. (nä-täl′)	163d	5°47′S	35°13′W
Natal, Indon.	246-47	0°34′N	99°07′E
Natashquan, Qc., Can. (nä-tăsh′kwăn)	138-39	50°12′N	61°49′W
Natashquan, stm., Can.	138-39	50°11′N	61°35′W
Natchez, Ms., U.S. (năch′ĕz)	124-25	31°34′N	91°24′W
Natchitoches, La., U.S. (năk′ĭ-tŏsh)(nách-ĭ-tōsh′)	122-23	31°46′N	93°06′W
National City, Ca., U.S. (năsh′ŭn-*ă*l sĭ′tĭ)	118-19	32°41′N	117°06′W
Natividade, Braz. (nä-tē-vē-dä′dĕ)	166-67	11°42′S	47°47′W
Natron, Lake, lk., Afr. (lāk nā′trŏn)	267	2°25′S	35°60′E
Natuna Besar, Kepulauan, is., Indon.	246-47	4°40′N	108°00′E
Natuna Selatan, Kepulauan, is., Indon.	246-47	2°45′N	109°00′E
Natural Bridges National Monument, n.p., Ut., U.S. (năt′ů-răl brĭj′ĕs năsh′ŭn-*ă*l mŏn′*ŭ*-m*ĕ*nt)	118-19	37°37′N	109°59′W
Naturaliste, Cape, c., Austl. (kāp năt-ů-rȧ-lĭst′)	272-73	33°33′S	115°01′E
Naugatuck, Ct., U.S. (nô′g*ȧ*-tŭk)	116-17	41°29′N	73°03′W
Naujat, Nu., Can. *see* Repulse Bay	128-29	66°32′N	86°14′W
Naumburg, Ger. (noum′bôrgh)	194-95	51°10′N	11°48′E
Nauru, nation, Oc. (nä-o͞o′-ro͞o)	280-81	0°32′S	166°55′E
Nautla, Mex. (nä-o͞ot′lä)	146-47	20°13′N	96°47′W
Nava, Mex. (nä′vä)	122-23	28°25′N	100°46′W
Navadwīp, India	234-35	23°25′N	88°22′E
Navahermosa, Spain (nä-vä-ĕr-mō′sä)	198-99	39°38′N	4°28′W
Navajo Indian Reservation, ind. res., U.S. (năv′á-hō ĭn′dĭ-*ă*n rĕ-sẽr-vā′shĕn)	118-19	36°39′N	109°46′W
Navajo National Monument, n.p., Az., U.S. (năv′á-hō năsh′ŭn-*ă*l mŏn′*ŭ*-m*ĕ*nt)	118-19	36°44′N	110°29′W
Navanagar, India *see* Jamnagar	234-35	22°28′N	70°04′E
Navarin, Mys, c., Russia	218-19	62°17′N	179°06′E
Navarino, Isla, i., Chile (ĕ′s-lä-nȧ-vä-rē′nô)	171	55°05′S	67°49′W
Navasota, Tx., U.S. (năv-*ȧ*-sō′t*ȧ*)	122-23	30°23′N	96°06′W
Navasota, stm., Tx., U.S. (năv-*ȧ*-sō′t*ȧ*)	122-23	30°20′N	96°09′W
Navassa Island, dep., N.A.	140-41	18°24′N	75°01′W
Navassa Island, i., N.A. (n*ȧ*-văs′*ȧ* ī′l*ȧ*nd)	142-43	18°24′N	75°01′W
Navidad, Chile (nä-vē-dä′d)	163e	33°56′S	71°50′W
Naviti, i., Fiji	279f	17°07′S	177°15′E
Navoiy, Uzb.	232-33	40°07′N	65°23′E
Navojoa, Mex. (nä-vô-kô′ä)	144-45	27°05′N	109°27′W
Navsāri, India	234-35	20°57′N	72°56′E
Nawa, Japan *see* Naha	244a	26°13′N	127°42′E
Nawābshāh, Pak. (n*ȧ*-wäb′shä)	232-33	26°14′N	68°24′E
Naxçıvan, Azer.	227	39°13′N	45°25′E
Naxçıvan Muxtar Respublikası, state, Azer.	227	39°20′N	45°30′E
Náxos, i., Grc. (nák′sôs)	200-01	37°03′N	25°31′E
Nayarit, state, Mex. (nä-yä-rēt′)	146-47	22°0′N	105°00′W
Nayau, i., Fiji	279f	17°58′S	179°03′W
Nayoro, Japan	244	44°21′N	142°28′E
Nay Pyi Taw, nat. cap., Mya.	246-47	19°45′N	96°07′E
Nazaré, Port.	198-99	39°36′N	9°04′W
Nazaré da Mata, Braz. (nä-zä-rĕ′ dä-mä-tä)	163d	7°44′S	35°14′W
Nazas, Mex. (nä′zäs)	122-23	25°14′N	104°08′W
Nazas, stm., Mex. (nä′zäs)	122-23	25°35′N	105°03′W
Naze, Japan	244a	28°22′N	129°30′E
Naze, The, c., Nor. *see* Lindesnes	192-93	58°0′N	7°02′E
Nazilli, Tur. (nȧ-zĭ-lē′)	200-01	37°55′N	28°20′E
Nazrēt, Eth.	269d	8°32′N	39°16′E
N'dalatando, Ang.	264-65	9°18′S	14°54′E
Ndélé, C.A.R.	262-63	8°25′N	20°39′E
N'Djamena, nat. cap., Chad (ŭn-jä-mē-nä′)	258-59	12°07′N	15°03′E
Ndjolé, Gabon	260-61	0°08′S	10°45′E
Ndola, Zam. (n′dō′lä)	264-65	12°57′S	28°38′E
Neagh, Lough, lk., N. Ire., U.K. (lŏk nā)	190-91	54°37′N	6°23′W
Neagră, Marea, s., *see* Black Sea	186-87	43°0′N	35°00′E
Near Islands, is., Ak., U.S. (nēr ī′l*ȧ*ndz)	126a	52°37′N	173°03′E
Neath, Wales, U.K. (nēth)	190-91	51°40′N	3°48′W
Nebine Creek, stm., Austl. (nĕ-bēne′ krēk)	276	29°21′S	146°45′E
Neblina, Cerro de la, mtn., S.A. *see* Neblina, Pico da	164-65	0°50′N	65°59′W
Neblina, Pico da, mtn., S.A.	164-65	0°50′N	65°59′W
Nebraska, state, U.S. (nê-brăs′k*ȧ*)	108-09	41°30′N	100°00′W
Nebraska City, Ne., U.S. (nê-brăs′k*ȧ* sĭ′tē)	120-21	40°41′N	95°52′W
Nechako, stm., B.C., Can.	132-33	53°55′N	122°43′W
Nechako Plateau, plat., B.C., Can. (nĭ-chă′kō plă-tō′)	132-33	54°0′N	124°30′W
Nechako Range, mts., B.C., Can. (nĭ-chă′kō rānj)	132-33	53°21′N	124°37′W
Nechako Reservoir, res., B.C., Can. (nĭ-chă′kō rĕ′sẽr-vwär)	132-33	53°33′N	124°53′W
Neches, stm., Tx., U.S. (nĕch′ĕz)	122-23	29°59′N	93°52′W
Necker Island, i., Hi., U.S.	127	23°35′N	164°42′W
Necochea, Arg. (nā-kô-chā′ä)	173	38°34′S	58°44′W
Nederland, nation, Eur. *see* Netherlands	174-75	52°15′N	5°30′E
Nederlandse Antillen, dep., N.A. *see* Netherlands Antilles	140-41	12°15′N	68°45′W
Nêdong, China	234-35	29°13′N	91°47′E
Needles, Ca., U.S. (nē′d′lz)	118-19	34°50′N	114°36′W
Neenah, Wi., U.S. (nē′n*ȧ*)	116-17	44°11′N	88°28′W
Neepawa, Mb., Can.	134-35	50°14′N	99°28′W
Neftçala, Azer.	227	39°24′N	49°15′E
Negage, Ang.	262-63	7°45′S	15°17′E
Negapatam, India *see* Nāgappattinam	236	10°46′N	79°51′E
Negaunee, Mi., U.S. (nê-gô′nê)	114-15	46°30′N	87°36′W
Negēlē, Eth.	262-63	5°20′N	39°35′E
Negombo, Sri L.	236	7°13′N	79°51′E
Negotin, Serb. (nĕ′gô-tên)	200-01	44°13′N	22°33′E
Negra, Punta, c., Peru	170	6°05′S	81°06′W
Negritos, Peru	170	4°40′S	81°17′W
Negro, stm., Arg.	171	41°02′S	62°47′W
Negro, stm., S.A. (nā′grò)	164-65	3°08′S	59°55′W
Negro, stm., S.A. (nā′grò)	173	33°26′S	58°27′W
Negros, i., Phil. (nā′grōs)	250	10°0′N	123°00′E
Nehbandān, Iran	232-33	31°32′N	60°02′E
Neiba, Dom. Rep. (nå-ē′bä)	142-43	18°29′N	71°25′W
Neijiang, China (nā-jyäŋ)	238-39	29°35′N	105°03′E
Neillsville, Wi., U.S. (nēlz′vĭl)	114-15	44°34′N	90°35′W
Nei Mongol, state, China	240-41	43°0′N	115°00′E
Neiqiu, China (nā-chyō)	240-41	37°17′N	114°30′E
Neira, Col. (nā′rä)	163c	5°09′N	75°31′W
Neisse, stm., Eur. (nēs)	194-95	52°04′N	14°46′E
Neiva, Col. (nå-ē′vä)(nā′vä)	164-65	2°56′N	75°17′W
Nejd, hist. reg., Sau. Ar. *see* Najd	266	26°07′N	44°40′E
Nek'emtē, Eth.	262-63	9°02′N	36°29′E
Nekoosa, Wi., U.S. (nê-ko͞o′s*ȧ*)	116-17	44°18′N	89°54′W
Nelidovo, Russia	202-03	56°13′N	32°47′E
Neligh, Ne., U.S. (nē′-lē)	114-15	42°08′N	98°02′W
Nellore, India (nĕl-lōr′)	236	14°27′N	79°59′E

n-sing; ŋ-baŋk; ɴ-nasalized n; nŏd; cŏmmit; ōld; ôbey; ôrder; oi-boil; fo͞od; o̐-as oo in foot; ou-out; s-soft; sh-dish; th-thin; pūre; ůnite; ûrn; stŭd; circ*ŭ*s; ü-as in French tu; ′-indeterminate vowel.

Feature (Pronunciation)	Page	Lat.	Long.
Nelson, B.C., Can. (nĕl′sŭn)	132-33	49°29′N	117°18′W
Nelson, N.Z. (nĕl′sŭn)	278	41°18′S	173°15′E
Nelson, stm., Mb., Can. (nĕl′sŭn)	130-31	57°08′N	92°21′W
Nelson, Cape, c., Austl. (kāp nĕl′sŭn)	276	38°25′S	141°32′E
Nelspruit, S. Afr.	264-65	25°28′S	30°59′E
Neman, Russia (ŋĕ′-man)	192-93	55°02′N	22°02′E
Neman, stm., Eur. (ŋĕ′-man)	192-93	55°21′N	21°16′E
Nëman, stm., Eur.	192-93	55°21′N	21°16′E
Nemunas, stm., Eur.	192-93	55°21′N	21°16′E
Nemuro, Japan (nā′mo-rō)	244	43°20′N	145°35′E
Nen, stm., China (nŭn)	240-41	45°26′N	124°39′E
Nenagh, Ire. (nē′nȧ)	190-91	52°52′N	8°12′W
Nendo, i., Sol. Is.	272-73	10°45′S	165°54′E
Neosho, Mo., U.S. (nė-ō′shō)	120-21	36°52′N	94°23′W
Neosho, stm., U.S. (nė-ō′shō)	120-21	35°48′N	95°18′W
Nepal, nation, Asia (nĕ-pôl′)	206-07	28°0′N	84°00′E
Nepāl, nation, Asia *see* Nepal	206-07	28°0′N	84°00′E
Nepālganj, Nepal	234-35	28°04′N	81°37′E
Nephi, Ut., U.S. (nē′fī)	118-19	39°43′N	111°50′W
Nercha, stm., Russia	218-19	51°56′N	116°39′E
Nerchinsk, Russia (nyĕr′ chĕnsk)	222-23	51°59′N	116°35′E
Nerekhta, Russia (nyĕ-rĕk′tá)	202-03	57°28′N	40°34′E
Nerja, Spain (nĕr′hä)	198-99	36°45′N	3°52′W
Neskaupstađur, Ice.	190a	65°09′N	13°42′W
Ness, Loch, lk., Scot., U.K. (lŏk nĕs)	190-91	57°17′N	4°29′W
Ness City, Ks., U.S. (nĕs sĭ′tė)	120-21	38°27′N	99°54′W
Nesterov, Russia (nyĕs-tă′rôf)	192-93	54°38′N	22°35′E
Netanya, Isr.	228-29	32°21′N	34°52′E
Netherlands, nation, Eur. (nĕdh′ẽr-lăndz)	174-75	52°15′N	5°30′E
Netherlands Antilles, dep., N.A. (nĕdh′ẽr-lăndz ăn-tĭ′lēz)	140-41	12°15′N	68°45′W
Netherlands Guiana, nation, S.A. *see* Suriname	158	4°0′N	56°00′W
Nettuno, Italy (nĕt-tōō′nô)	200-01	41°28′N	12°39′E
Neubrandenburg, Ger. (noi-brän′dĕn-bòrgh)	194-95	53°33′N	13°15′E
Neudamm, Pol. *see* Dębno	194-95	52°44′N	14°43′E
Neufchâtel-en-Bray, Fr. (nû-shä-tĕl′ĕN-brā′)	196-97	49°44′N	1°27′E
Neumarkt, Rom. *see* Târgu Mureș	194-95	46°33′N	24°34′E
Neumünster, Ger. (noi′münstēr)	194-95	54°04′N	9°59′E
Neunkirchen, Aus. (noin′kĭrk-ĕn)	194-95	47°44′N	16°06′E
Neuquén, Arg. (nĕ-ò-kān′)	171	38°57′S	68°04′W
Neuquén, state, Arg. (nĕ-ò-kān′)	171	39°0′S	70°00′W
Neuquén, stm., Arg. (nĕ-ò-kān′)	171	38°59′S	68°00′W
Neuruppin, Ger. (noi′rōō-pēn)	194-95	52°55′N	12°49′E
Neusatz, Serb. *see* Novi Sad	200-01	45°15′N	19°50′E
Neuse, stm., N.C., U.S. (nūz)	124-25	35°09′N	76°31′W
Neustrelitz, Ger. (noi-strā′lĭts)	194-95	53°22′N	13°04′E
Neuwied, Ger. (noi′vēdt)	194-95	50°26′N	7°28′E
Nevada, Ia., U.S. (nė-vā′dȧ)	114-15	42°02′N	93°27′W
Nevada, Mo., U.S. (nė-vä′dȧ)	120-21	37°51′N	94°21′W
Nevada, state, U.S. (nė-vá′dá)	108-09	39°0′N	117°00′W
Nevada, Sierra, mts., Spain (syĕr′rä nā-vä′dhä)	198-99	37°05′N	3°10′W
Nevada, Sierra, mts., Ca., U.S. (sē-ĕ′r-rä nē-vä′dá)	118-19	38°0′N	119°15′W
Nevado, Cerro, mtn., Arg.	171	35°34′S	68°28′W
Nevado, Cerro, mtn., Col. (sĕ′r-rô-nĕ-vä′dô)	163c	3°59′N	74°04′W
Nevel', Russia (nyĕ′vĕl)	192-93	56°01′N	29°56′E
Nevel'sk, Russia	244	46°40′N	141°52′E
Nevers, Fr. (nē-vâr′)	196-97	46°60′N	3°09′E
Nevinnomyssk, Russia	186-87	44°38′N	41°56′E
Nevis, i., St. K./N. (nē′vĭs)	143b	17°10′N	62°34′W
Nevis, Ben, mtn., Scot., U.K. (bĕn nē′vĭs)	190-91	56°48′N	5°01′W
Nevşehir, Tur. (nĕv-shĕ′hĕr)	186-87	38°37′N	34°43′E
New, stm., U.S. (nū)	110-11	38°09′N	81°12′W
New Albany, In., U.S. (nū ôl′bȧ-nĭ)	116-17	38°17′N	85°50′W
New Albany, Ms., U.S. (nū ôl′bȧ-nĭ)	124-25	34°30′N	89°01′W
New Amsterdam, Guy. (nū ăm′stēr-dăm)	164-65	6°15′N	57°30′W
Newark, De., U.S. (nōō′ärk)	116-17	39°41′N	75°45′W
Newark, N.J., U.S.	116-17	40°43′N	74°10′W
Newark, N.Y., U.S. (nū′ĕrk)	116-17	43°03′N	77°05′W
Newark, Oh., U.S. (nōō′ûrk)	116-17	40°03′N	82°24′W
Newaygo, Mi., U.S. (nū′wā-go)	116-17	43°25′N	85°48′W
New Bedford, Ma., U.S. (nū bĕd′fērd)	116-17	41°38′N	70°56′W
Newberg, Or., U.S. (nū′bûrg)	112-13	45°18′N	122°58′W
New Bern, N.C., U.S. (nū bûrn)	124-25	35°06′N	77°04′W
Newberry, Mi., U.S. (nū′bĕr-ĭ)	114-15	46°22′N	85°28′W
Newberry, S.C., U.S. (nū′bĕr-ĭ)	124-25	34°17′N	81°37′W
New Boston, Oh., U.S. (nū bôs′tŭn)	116-17	38°45′N	82°56′W
New Boston, Tx., U.S. (nū bôs′tŭn)	120-21	33°28′N	94°25′W
New Braunfels, Tx., U.S. (nū broun′fĕls)	122-23	29°42′N	98°07′W
New Britain, Ct., U.S. (nū brĭt′′n)	116-17	41°40′N	72°46′W
New Britain, i., Pap. N. Gui. (nū brĭt′′n)	277	6°0′S	150°00′E
New Brunswick, state, Can. (nū brŭnz′wĭk)	128-29	46°30′N	66°15′W
Newburgh, N.Y., U.S.	116-17	41°30′N	74°02′W
Newbury, Eng., U.K. (nū′bĕr-ĭ)	190-91	51°24′N	1°19′W
Newburyport, Ma., U.S. (nū′bĕr-ĭ-pōrt)	116-17	42°48′N	70°52′W
New Caledonia, dep., Oc. (nū kăl-ē-dō′nĭ-ȧ)	279g	21°30′S	165°30′E
New Caledonia, i., N. Cal. (nū kăl-ē-dō′nĭ-ȧ) *see* Nouvelle-Calédonie	279g	21°33′S	165°42′E
New Carlisle, Qc., Can. (nū kär-līl′)	138-39	48°01′N	65°21′W
Newcastle, Austl. (nū-kás′'l)	276	32°56′S	151°45′E
Newcastle, S. Afr.	269c	27°46′S	29°55′E
New Castle, De., U.S. (nū kás′'l)	116-17	39°40′N	75°33′W
New Castle, In., U.S. (nū kás′'l)	116-17	39°55′N	85°22′W
Newcastle, Ok., U.S. (nū-kás′'l)	120-21	35°14′N	97°36′W
New Castle, Pa., U.S.	116-17	40°60′N	80°21′W
Newcastle, Wy., U.S. (nū-kás′'l)	114-15	43°51′N	104°13′W
Newcastle upon Tyne, Eng., U.K.	190-91	54°59′N	1°40′W
New Delhi, nat. cap., India (nū dĕl′hī)	234-35	28°36′N	77°13′E
Newell, S.D., U.S. (nū′ĕl)	114-15	44°43′N	103°25′W
New England Range, mts., Austl. (nů ĭŋ′glănd rānj)	276	29°52′S	151°44′E
Newenham, Cape, c., Ak., U.S. (kāp ū-ĕn-hăm)	126	58°39′N	162°10′W
Newfoundland, i., Nf., Can. (nū-fŭn′lănd′) (nū′fŭnd-lănd) (nū′found-lănd′)	138-39	48°30′N	56°00′W
Newfoundland and Labrador, state, Can.	128-29	52°0′N	56°00′W
New Georgia, i., Sol. Is. (nū ôr′jĭ-ȧ)	279e	8°09′S	157°26′E
New Glasgow, N.S., Can. (nū glás′gō)	138-39	45°36′N	62°38′W
New Guinea, i., (nū gĭne)	277	5°0′S	140°00′E
New Hampshire, state, U.S. (nū hămp′shĭr)	108-09	43°35′N	71°40′W
New Hampton, Ia., U.S. (nū hămp′tŭn)	114-15	43°03′N	92°19′W
New Hanover, S. Afr. (nū hăn′ôvēr)	269c	29°21′S	30°31′E
New Hanover, i., Pap. N. Gui. (nū hăn′ôvēr)	277	2°30′S	150°15′E
New Harmony, In., U.S. (nū här′mō-nĭ)	116-17	38°08′N	87°56′W
New Haven, Ct., U.S. (nū hā′vĕn)	116-17	41°19′N	72°56′W
New Haven, In., U.S. (nū hāv′'n)	116-17	41°04′N	85°02′W
New Hazelton, B.C., Can.	132-33	55°15′N	127°35′W
New Hebrides, nation, Oc. *see* Vanuatu	279g	16°0′S	167°00′E
New Hebrides, is., Vanuatu (nū hĕ′brĭ-dēz)	279g	16°0′S	167°00′E
New Iberia, La., U.S. (nū jī-bē′rĭ-ȧ)	124-25	30°00′N	91°49′W
New Jersey, state, U.S. (nū jûr′zĭ)	108-09	40°15′N	74°30′W
Newkirk, Ok., U.S. (nū′kûrk)	120-21	36°53′N	97°03′W
New Kowloon, China *see* Xinjiulong	238-39	22°21′N	114°10′E
New Lexington, Oh., U.S. (nū lĕk′sĭng-tŭn)	116-17	39°42′N	82°13′W
New Lisbon, Wi., U.S. (nū lĭz′bŭn)	116-17	43°52′N	90°10′W
New Liskeard, On., Can.	136-37	47°31′N	79°40′W
New London, Ct., U.S. (nū lŭn′dŭn)	116-17	41°21′N	72°07′W
New London, Wi., U.S. (nū lŭn′dŭn)	116-17	44°23′N	88°44′W
New Madrid, Mo., U.S. (nū măd′rĭd)	124-25	36°35′N	89°32′W
Newmarket, On., Can. (nū′mär-kĕt)	136-37	44°03′N	79°27′W
New Martinsville, W.V., U.S. (nū mär′tĭnz-vĭl)	116-17	39°38′N	80°52′W
New Mexico, state, U.S. (nū mĕk′sĭ-kō)	108-09	34°30′N	106°00′W
Newnan, Ga., U.S. (nū′năn)	124-25	33°23′N	84°48′W
New Norfolk, Austl. (nū nôr′fŏk)	276	42°47′S	147°03′E
New Orleans, La., U.S. (nū ôr′lănz)	124-25	29°59′N	90°05′W
New Philadelphia, Oh., U.S. (nū fĭl-ȧ-dĕl′fĭ-ȧ)	116-17	40°30′N	81°27′W
New Plymouth, N.Z. (nū plĭm′ŭth)	278	39°04′S	174°05′E
Newport, Eng., U.K. (nū-pôrt)	190-91	50°42′N	1°18′W
Newport, Wales, U.K. (nū-pôrt)	190-91	51°35′N	3°00′W
Newport, Ar., U.S. (nū′pōrt)	124-25	35°37′N	91°17′W
Newport, In., U.S. (nū′pōrt)	116-17	39°53′N	87°25′W
Newport, Or., U.S. (nū′pōrt)	112-13	44°39′N	124°03′W
Newport, R.I., U.S. (nū′pōrt)	116-17	41°29′N	71°19′W
Newport, Tn., U.S. (nū′pōrt)	124-25	35°58′N	83°11′W
Newport, Wa., U.S. (nū′pōrt)	112-13	48°11′N	117°07′W
Newport News, Va., U.S. (nū′pōrt nūz)	124-25	36°59′N	76°25′W
New Providence, i., Bah. (nū prŏv′ĭ-dĕns)	142-43	25°02′N	77°24′W
New Richmond, Wi., U.S. (nū rĭch′mŭnd)	114-15	45°07′N	92°32′W
New Roads, La., U.S. (nū rōds)	124-25	30°42′N	91°27′W
New Rochelle, N.Y., U.S. (nū rū-shĕl′)	116-17	40°54′N	73°49′W
New Rockford, N.D., U.S. (nū rŏk′fērd)	114-15	47°41′N	99°09′W
New Siberian Islands, is., Russia (nū sī-bĭr′y′n ī′lăndz)	218-19	75°0′N	142°00′E
New Smyrna Beach, Fl., U.S. (nū smûr′nȧ bēch)	124-25	29°02′N	80°56′W
New South Wales, state, Austl. (nū south wālz)	276	33°0′S	146°00′E
Newton, Ia., U.S. (nū′tŭn)	114-15	41°42′N	93°03′W
Newton, Il., U.S. (nū′tŭn)	116-17	38°59′N	88°10′W
Newton, Ks., U.S. (nū′tŭn)	120-21	38°03′N	97°21′W
Newton, Ma., U.S. (nū′tŭn)	116-17	42°20′N	71°13′W
Newton, Ms., U.S. (nū′tŭn)	124-25	32°20′N	89°10′W
Newton, N.C., U.S. (nū′tŭn)	124-25	35°40′N	81°13′W
Newton, Tx., U.S. (nū′tŭn)	122-23	30°50′N	93°46′W
New Ulm, Mn., U.S. (nū ŭlm)	114-15	44°19′N	94°28′W
New Waterford, N.S., Can. (nū wô′tēr-fērd)	138-39	46°15′N	60°06′W
New York, N.Y., U.S. (nū yôrk)	116-17	40°43′N	74°01′W
New York, state, U.S. (nū yôrk)	108-09	43°0′N	75°00′W
New Zealand, nation, Oc. (nū zē′lȧnd)	278	41°0′S	174°00′E
Neyshābūr, Iran	232-33	36°11′N	58°52′E
Nezahualcóyotl, Presa, res., Mex.	146-47	17°10′N	93°40′W
Nez Perce Indian Reservation, ind. res., Id., U.S. (nĕz′ pûrs′ ĭn′dĭ-ăn rĕ-sĕr-vā′shĕn)	112-13	46°20′N	116°30′W
Ngami, Lake, lk., Bots. (lāk n′gä′mē)	264-65	20°29′S	22°46′E
Ngangla Ringco, lk., China (ŋäŋ-lä rĭŋ-tswo)	234-35	31°34′N	83°01′E
Ngaoundéré, Camrn.	260-61	7°19′N	13°35′E
Ng'iro, Ewaso, stm., Kenya	262-63	0°28′N	39°55′E
Ngoring Hu, lk., China	240-41	34°53′N	97°41′E
Ngorongoro Conservation Area, pk., Tan.	267	3°0′S	35°30′E
Ngozi, Bdi.	267	2°54′S	29°53′E
Nguigmi, Niger ('n-gēg′mē)	258-59	14°15′N	13°07′E
Nguna, Île, i., Vanuatu	279g	17°27′S	168°21′E
Nguru, Nig. ('n-gōō′rōō)	260-61	12°52′N	10°27′E
Nhamundá, stm., Braz.	166-67	1°58′S	56°58′W
Nha Trang, Viet. (nyä-träng′)	246-47	12°16′N	109°12′E
Ni'ihau, i., Hi., U.S. (nē′ē-ha′ōō)	127a	21°54′N	160°09′W
Niagara, Wi., U.S. (nī-ăg′ȧ-rȧ)	116-17	45°46′N	88°01′W
Niagara Falls, On., Can. (nī-ăg′ȧ-rȧ fôlz)	136-37	43°05′N	79°02′W
Niah, Malay.	248-49	3°52′N	113°43′E
Niamey, nat. cap., Niger (nē-ä-mā′)	258-59	13°31′N	2°07′E
Niangara, D.R.C. (nē-äŋ-gá′rȧ)	267	3°42′N	27°54′E
Nias, Pulau, i., Indon. (pōō-lou nē′äs′)	246-47	1°05′N	97°35′E
Niassa, Lago, lk., Afr. *see* Nyasa, Lake	264-65	12°0′S	34°30′E
Nicaragua, nation, N.A. (nĭk-ȧ-rä′gwȧ)	85	13°0′N	85°00′W
Nicaragua, Lago de, lk., Nic. (lä′gô dĕ-nĭk-ȧ-rä′gwȧ)	149	11°39′N	85°26′W
Nicaragua, Lake, lk., Nic. (lăk nĭk-ȧ-rä′gwȧ) *see* Nicaragua, Lago de	149	11°39′N	85°26′W
Nice, Fr. (nēs)	196-97	43°43′N	7°16′E
Nichinan, Japan	245	31°36′N	131°23′E
Nicholas Channel, strt., N.A. (nĭk′ô-lȧs chăn′ĕl)	142-43	23°21′N	80°21′W
Nicholasville, Ky., U.S. (nĭk′ô-lȧs-vĭl)	116-17	37°53′N	84°35′W
Nicobar Islands, is., India (nĭk-ô-bär′ ī′lăndz)	246-47	8°0′N	93°30′E
Nicomedia, Tur. *see* İzmit	200-01	40°47′N	29°57′E
Nicosia, nat. cap., Cyp. (nē-kô-sē′á)	228-29	35°10′N	33°22′E
Nicosia, nat. cap., Cyp. (nē-kō-zē′ä)	228-29	35°10′N	33°22′E
Nicoya, C.R. (nē-kō′yä)	149	10°09′N	85°27′W
Nicoya, Golfo de, b., C.R. (gôl-fô dĕ nē-kō′yä)	149	9°47′N	84°48′W
Nicoya, Península de, pen., C.R.	149	10°01′N	85°25′W
Nictheroy, Braz. *see* Niterói	172	22°54′S	43°07′W
Nidzica, Pol. (nė-jĕt′sá)	194-95	53°22′N	20°26′E
Nienburg, Ger. (nē′ĕn-bòrgh)	194-95	52°38′N	9°13′E
Nieuw Nickerie, (nē′ū nē-nē′kĕ-rē′)	164-65	5°56′N	56°60′W
Niğde, Tur. (nĭg′dĕ)	186-87	37°60′N	34°44′E
Nigel, S. Afr. (nī′jĕl)	269c	26°26′S	28°28′E
Niger, nation, Afr. (nī′jẽr)	253	16°0′N	8°00′E
Niger, stm., Afr. (nī′jẽr) (nē-zhâr′)	260-61	4°17′N	6°04′E
Nigeria, nation, Afr. (nī-jē′rĭ-ȧ)	253	10°0′N	8°00′E
Nihoa, i., Hi., U.S.	127	23°03′N	161°56′W

Feature (Pronunciation)	Page	Lat.	Long.
Nihon, nation, Asia *see* Japan	206-07	36°0'N	138°00'E
Nihon-kai, s., Asia *see* Japan, Sea of	222-23	40°0'N	135°00'E
Niigata, Japan (nē'ė-gä'tȧ)	245	37°55'N	139°04'E
Nii-jima, i., Japan (nē jē'mȧ)	245	34°22'N	139°16'E
Nijmegen, Neth. (nī'mā-gĕn)	190-91	51°50'N	5°50'E
Nikel', Russia	184-85	69°25'N	30°15'E
Nikkō, Japan	245	36°45'N	139°37'E
Nikolayev, Ukr. *see* Mykolaïv	202-03	46°58'N	31°59'E
Nikolayevsk-na-Amure, Russia	218-19	53°09'N	140°44'E
Nikol'sk, Russia (nê-kôlsk')	186-87	53°42'N	46°05'E
Nikol'sk, Russia (nê-kôlsk')	186-87	59°32'N	45°27'E
Nikopol', Ukr.	202-03	47°34'N	34°24'E
Nikumaroro, at., Kir.	280-81	4°40'S	174°32'W
Nikunau, i., Kir.	280-81	1°22'S	176°27'E
Nîl, Bahr el-, stm., Afr. *see* Nile	266	30°10'N	31°07'E
Nîl, Nahr an-, stm., Afr. *see* Nile	266	30°10'N	31°07'E
Nile, stm., Afr. (nīl)	266	30°10'N	31°07'E
Niles, Mi., U.S. (nīlz)	116-17	41°50'N	86°15'W
Niles, Oh., U.S. (nīlz)	116-17	41°11'N	80°44'W
Nimba, Mont, mtn., Afr. (môn nĭm'bȧ)	260-61	7°37'N	8°25'W
Nimba, Mount, mtn., Afr. *see* Nimba, Mont	260-61	7°37'N	8°25'W
Nîmes, Fr. (nēm)	196-97	43°51'N	4°22'E
Nine Degree Channel, strt., India	236	9°0'N	73°00'E
Ninety Mile Beach, cst., Austl.	276	38°13'S	147°23'E
Ning'an, China (nĭŋ-än)	244	44°20'N	129°28'E
Ningbo, China (nĭŋ-bwo)	238-39	29°53'N	121°32'E
Ningcheng, China	240-41	41°33'N	119°20'E
Ningde, China (nĭŋ-dŭ)	238-39	26°43'N	119°33'E
Ningdu, China	238-39	26°31'N	115°58'E
Ningming, China	238-39	22°08'N	107°05'E
Ningshan, China	238-39	33°19'N	108°19'E
Ningsia Hui, state, China *see* Ningxia	240-41	37°0'N	106°00'E
Ningwu, China (nĭng'wōō')	240-41	39°03'N	112°12'E
Ningxia, state, China (nĭŋ-shyä)	240-41	37°0'N	106°00'E
Ninh Binh, Viet. (nēn bēnk')	246-47	20°15'N	105°59'E
Ninigo Group, is., Pap. N. Gui.	224-25	1°15'S	144°15'E
Ninnescah, stm., Ks., U.S. (nĭn'ĕs-kä)	120-21	37°20'N	97°10'W
Nioaque, Braz. (nēô-ä'-kĕ)	168-69	21°09'S	55°50'W
Niobrara, stm., U.S. (nī-ô-brâr'ȧ)	110-11	42°46'N	98°03'W
Nioki, D.R.C.	262-63	2°39'S	17°42'E
Nioro, Mali	258-59	15°14'N	9°35'W
Nipawin, Sk., Can.	134-35	53°22'N	104°00'W
Nipe, Bahía de, b., Cuba (bä-ē'ä-dĕ-nē'pā)	142-43	20°47'N	75°42'W
Nipigon, On., Can. (nĭp'ĭ-gŏn)	136-37	49°01'N	88°15'W
Nipigon, Lake, res., On., Can. (lăk nĭp'ĭ-gŏn)	134-35	49°40'N	88°34'W
Nipigon Bay, b., On., Can. (nĭp'ĭ-gŏn bā)	136-37	48°54'N	87°56'W
Nipissing, Lake, lk., On., Can. (lăk nĭp'ĭ-sĭng)	136-37	46°15'N	79°42'W
Niquero, Cuba (nē-kā'rō)	142-43	20°03'N	77°35'W
Niš, Serb.	200-01	43°19'N	21°54'E
Nisa, Port. (nē'sȧ)	198-99	39°30'N	7°39'W
Nish, Serb. *see* Niš	200-01	43°19'N	21°54'E
Nishapur, Iran *see* Neyshābūr	232-33	36°11'N	58°52'E
Nisser, lk., Nor. (nĭs'ĕr)	192-93	59°10'N	8°30'E
Nistru, stm., Eur. *see* Dniester	188-89	46°19'N	30°17'E
Niterói, Braz. (nē-tĕ-rô'ĭ)	172	22°54'S	43°07'W
Nitra, Slvk. (nē'trȧ)	194-95	48°19'N	18°06'E
Nitro, W.V., U.S. (nī'trô)	116-17	38°25'N	81°51'W
Niue, dep., Oc. (nĭ'ô)	280-81	19°02'S	169°52'W
Niulakita, i., Tuvalu	280-81	10°45'S	179°30'E
Niut, Gunung, mtn., Indon.	248-49	1°0'N	109°55'E
Niutao, i., Tuvalu	280-81	6°07'S	177°19'E
Nixon, Tx., U.S. (nĭk'sŭn)	122-23	29°16'N	97°46'W
Nizāmābād, India	236	18°40'N	78°06'E
Nizhnekamsk, Russia	186-87	55°33'N	51°58'E
Nizhnekamskoye Vodokhranilishche, res., Russia	186-87	55°50'N	53°00'E
Nizhneudinsk, Russia (nēzh'nyĭ-ōōdēnsk')	218-19	54°54'N	99°02'E
Nizhnevartovsk, Russia	218-19	60°56'N	76°34'E
Nizhniy Novgorod, Russia	186-87	56°19'N	44°01'E
Nizhniy Tagil, Russia (nyēzh'-nyē tŭgēl')	218-19	57°55'N	59°59'E
Nizhnyaya Tunguska, stm., Russia	218-19	65°46'N	87°54'E
Nizhyn, Ukr.	202-03	51°02'N	31°54'E
Njazidja, i., Com.	264-65	11°35'S	43°20'E
Njombe, stm., Tan.	267	6°56'S	35°06'E
Nkawkaw, Ghana	260-61	6°33'N	0°47'W
Nkongsamba, Camrn.	260-61	4°57'N	9°56'E
Nmai, stm., Mya.	238-39	25°43'N	97°31'E
Noākhāli, Bngl.	234-35	22°49'N	91°06'E
Noatak, stm., Ak., U.S. (nô-ȧ'tȧk)	126	67°00'N	162°30'W
Nobeoka, Japan (nō-bå-ō'kȧ)	245	32°35'N	131°41'E
Noblesville, In., U.S. (nō'bl'z-vĭl)	116-17	40°02'N	86°00'W
Nochistlán, Mex. (nô-chēs-tlän')	146-47	21°22'N	102°51'W
Nogales, Mex. (nō-gä'lĕs)	144-45	31°19'N	110°56'W
Nogales, Az., U.S. (nô-gä'lĕs)	118-19	31°21'N	110°56'W
Nogent-le-Rotrou, Fr. (nō-zhôn-lĕ-rō-trōō')	196-97	48°19'N	0°49'E
Noginsk, Russia (nô-gēnsk')	202-03	55°52'N	38°28'E
Nogoyá, Arg.	173	32°24'S	59°48'W
Noir, Isla, i., Chile	171	54°29'S	73°01'W
Noirmoutier, Île de, i., Fr. (ēl-dē-nwȧr-mōō-tyā')	196-97	47°0'N	2°15'W
Nokomis, Il., U.S. (nō-kō'mĭs)	116-17	39°18'N	89°17'W
Nolinsk, Russia (nô-lēnsk')	186-87	57°33'N	49°57'E
Nombre de Dios, Mex. (nôm-brĕ-dĕ-dyô's)	146-47	23°50'N	104°14'W
Nombre de Dios, Pan. (nô'm-brĕ dĕ-dyô's)	150	9°35'N	79°28'W
Nome, Ak., U.S. (nōm)	126	64°30'N	165°24'W
Nonacho Lake, lk., N.T., Can.	130-31	61°42'N	109°40'W
Nondalton, Ak., U.S.	126	60°01'N	154°49'W
Nong'an, China (nôŋ-än)	240-41	44°26'N	125°11'E
Nong Khai, Thai.	246-47	17°52'N	102°45'E
Nonouti, at., Kir.	280-81	0°38'S	174°26'E
Noord Zee, s., Eur. *see* North Sea	184-85	56°0'N	3°00'E
Noorvik, Ak., U.S.	126	66°53'N	160°59'W
Nootka Island, i., B.C., Can. (nōōt'kȧ ī'lȧnd)	132-33	49°44'N	126°46'W
Nordegg, Ab., Can. (nûr'dĕg)	132-33	52°28'N	116°05'W
Norderney, i., Ger. (nōr'dĕr-nĕy)	194-95	53°43'N	7°11'E
Nordhausen, Ger. (nôrt'hau-zĕn)	194-95	51°30'N	10°48'E
Nordhorn, Ger. (nôrt'hôrn)	194-95	52°26'N	7°04'E
Nordkapp, c., Nor.	184-85	71°10'N	25°47'E
Nord-Ostsee-Kanal, can., Ger. (nōrd-ōzt-zā kä-näl') *see* Kiel Canal	194-95	53°54'N	9°10'E
Nordsee, s., Eur. *see* North Sea	184-85	56°0'N	3°00'E
Nordsjøen, s., Eur. *see* North Sea	184-85	56°0'N	3°00'E
Norfolk, Ne., U.S. (nôr'fŏk)	114-15	42°02'N	97°25'W
Norfolk, Va., U.S. (nôr'fŏk)	124-25	36°51'N	76°16'W
Norfolk Island, dep., Oc. (nôr-fŏŏk ī'lȧnd)	280-81	29°02'S	167°57'E
Norge, nation, Eur. *see* Norway	174-75	62°0'N	10°00'E
Noril'sk, Russia (nô rēlsk')	218-19	69°19'N	88°14'E
Norin, stm., Asia *see* Naryn	226	41°46'N	73°13'E
Normal, Il., U.S. (nôr'mȧl)	116-17	40°30'N	88°59'W
Norman, Ok., U.S.	120-21	35°13'N	97°26'W
Norman, stm., Austl. (nôr'mȧn)	277	17°28'S	140°50'E
Normanby Island, i., Pap. N. Gui.	277	10°05'S	151°05'E
Normandie, hist. reg., Fr. (nŏr-män-dē') *see* Normandy	196-97	49°0'N	0°05'W
Normandy, hist. reg., Fr. (nôr-măn-dē')	196-97	49°0'N	0°05'W
Normanton, Austl. (nôr'mȧn-tŭn)	277	17°40'S	141°05'E
Norman Wells, N.T., Can.	128-29	65°17'N	126°42'W
Ñorquinco, Arg.	171	41°51'S	70°55'W
Norristown, Pa., U.S. (nŏr'ĭs-toun)	116-17	40°07'N	75°21'W
Norrköping, Swe. (nôr'chûp'ĭng)	192-93	58°36'N	16°11'E
Norrtälje, Swe. (nôr-tĕl'yĕ)	192-93	59°46'N	18°44'E
Norseman, Austl. (nôrs'mȧn)	270-71	32°12'S	121°48'E
Norskehavet, s., Eur. *see* Norwegian Sea	184-85	70°0'N	2°00'E
Norte, Serra do, plat., Braz. (sĕ'r-rä-dô-nôr'te)	166-67	11°20'S	59°00'W
North, Cape, c., N.S., Can.	138-39	47°02'N	60°24'W
North Adams, Ma., U.S. (nôrth ăd'ămz)	116-17	42°42'N	73°07'W
Northam, Austl. (nôr-dhăm)	270-71	31°39'S	116°40'E
Northam, S. Afr. (nôr'thăm)	269c	25°03'S	27°11'E
North America, cont., (nôrth ȧ-mĕr'ĭ-kȧ)	20-21	45°0'N	100°00'W
Northampton, Austl. (nôr-thămp'tŭn)	270-71	28°21'S	114°38'E
Northampton, Eng., U.K. (nôrth-ămp'tŭn)	190-91	52°15'N	0°54'W
North Andaman, i., India (nôrth ăn-dȧ-măn')	246-47	13°15'N	92°55'E
North Battleford, Sk., Can. (nôrth ăt''l-fērd)	134-35	52°46'N	108°16'W
North Bay, On., Can.	136-37	46°19'N	79°26'W
North Bend, Or., U.S. (nôrth bĕnd)	112-13	43°24'N	124°13'W
North Caicos, i., T./C. Is. (nôrth kī'kôs)	142-43	21°56'N	71°59'W
North Cape, c., N.Z. (nôrth kāp)	278	34°24'S	173°02'E
North Caribou Lake, lk., On., Can.	134-35	52°50'N	90°40'W
North Carolina, state, U.S. (nôrth kăr-ô-lī'nȧ)	108-09	35°30'N	80°00'W
North Cascades National Park, n.p., Wa., U.S.	112-13	48°30'N	121°00'W
North Channel, strt., On., Can.	136-37	46°02'N	82°50'W
North Channel, strt., U.K.	190-91	55°10'N	5°40'W
North Charleston, S.C., U.S. (nôrth chärlz'tŭn)	124-25	32°53'N	79°60'W
North Chicago, Il., U.S. (nôrth shĭ-kô'gō)	116-17	42°18'N	87°52'W
North Dakota, state, U.S. (nôrth dȧ-kō'tȧ)	108-09	47°30'N	100°15'W
Northeast Providence Channel, strt., Bah. (nôrth-ēst' prŏv'ĭ-dĕns chăn'ĕl)	142-43	25°40'N	77°09'W
Northeim, Ger. (nôrt'hīm)	194-95	51°42'N	10°00'E
Northern Cook Islands, is., Cook Is.	280-81	10°0'S	161°00'W
Northern Donets, stm., Eur. (nôrth'ērn dŏn-ĕts')	186-87	47°36'N	40°54'E
Northern Indian Lake, lk., Mb., Can.	130-31	57°21'N	97°19'W
Northern Ireland, state, U.K. (nôrth'ērn īr'lănd)	190-91	54°40'N	6°45'W
Northern Mariana Islands, dep., Oc. (nôrth'ērn mä-rē-ä'nȧ ī'lăndz)	280-81	16°0'N	149°00'E
Northern Sporades, is., Grc. *see* Vóreioi Sporades	200-01	39°15'N	23°55'E
Northern Territory, state, Austl.	270-71	20°0'S	134°00'E
Northfield, Mn., U.S. (nôrth'fēld)	114-15	44°28'N	93°10'W
North Island, i., N.Z. (nôrth ī'lȧnd)	278	39°0'S	176°00'E
North Judson, In., U.S. (nôrth jŭd'sŭn)	116-17	41°13'N	86°46'W
North Korea, nation, Asia (nôrth kô-rē'-ȧ)	206-07	40°0'N	127°00'E
North Lakhimpur, India	234-35	27°14'N	94°07'E
North Little Rock, Ar., U.S. (nôrth lĭt''l rŏk)	120-21	34°46'N	92°18'W
North Magnetic Pole, p.o.i.,	288	81°18'N	110°48'W
North Manchester, In., U.S. (nôrth măn'chĕs-tēr)	116-17	40°60'N	85°46'W
North Ogden, Ut., U.S. (nôrth ŏg'dĕn)	112-13	41°19'N	111°58'W
North Ossetia, state, Russia	227	43°0'N	44°15'E
North Platte, Ne., U.S. (nôrth plăt)	114-15	41°08'N	100°45'W
North Platte, stm., U.S. (nôrth plăt)	110-11	41°07'N	100°42'W
Northport, Al., U.S. (nôrth'pōrt)	124-25	33°14'N	87°34'W
North Saskatchewan, stm., Can. (nôrth săn-kăch'ê-wän)	130-31	53°14'N	105°05'W
North Sea, s., Eur. (nôrth sē)	184-85	56°0'N	3°00'E
North Shore City, N.Z.	278	36°48'S	174°47'E
North Siberian Lowland, pl., Russia (nôrth sī-bĭr'y'n lō'lȧnd) *see* Severo-Sibirskaya Nizmennost'	218-19	73°0'N	100°00'E
North Sydney, N.S., Can. (nôrth sĭd'nē)	138-39	46°13'N	60°16'W
North Thompson, stm., B.C., Can.	132-33	50°41'N	120°20'W
North Tonawanda, N.Y., U.S. (nôrth tŏn-ȧ-wŏn'dȧ)	116-17	43°02'N	78°52'W
Northumberland Strait, strt., Can. (nôr thŭm'bēr-lȧnd strāt)	138-39	46°0'N	63°30'W
North Vancouver, B.C., Can. (nôrth văn-kōō'vēr)	132-33	49°19'N	123°04'W
North Vernon, In., U.S. (nôrth vûr'nŭn)	116-17	39°00'N	85°38'W
North West Cape, c., Austl. (nôrth wĕst kāp)	272-73	21°48'S	114°10'E
Northwest Providence Channel, strt., Bah. (nôrth-wĕst' prŏv'ĭ-dĕns chăn'ĕl)	142-43	26°10'N	78°20'W
Northwest Territories, state, Can. (nôrth'wĕst tĕr'ĭ-tō'rĭs)	128-29	65°0'N	120°00'W
North Wilkesboro, N.C., U.S. (nôrth wĭlks'bûrô)	124-25	36°11'N	81°09'W
Northwood, Ia., U.S. (nôrth'wȯd)	114-15	43°27'N	93°13'W
Northwood, N.D., U.S. (nôrth'wȯd)	114-15	47°44'N	97°34'W
Norton, Ks., U.S. (nôr'tŭn)	120-21	39°50'N	99°53'W
Norton, Va., U.S. (nôr'tŭn)	124-25	36°56'N	82°38'W
Norton Sound, strt., Ak., U.S. (nôr'tŭn sound)	126	63°50'N	164°00'W
Norvegia, Cape, c., Ant.	287	71°25'S	12°18'W
Norwalk, Ct., U.S. (nôr'wôk)	116-17	41°07'N	73°25'W
Norwalk, Oh., U.S. (nôr'wôk)	116-17	41°15'N	82°36'W
Norway, Me., U.S. (nôr'wā)	116-17	44°13'N	70°32'W
Norway, nation, Eur. (nôr'wā)	174-75	62°0'N	10°00'E
Norway House, Mb., Can. (nôr'wā hous)	134-35	53°59'N	97°48'W
Norwegian Sea, s., Eur. (nôr-wē'jăn sē sē)	184-85	70°0'N	2°00'E
Norwich, Eng., U.K.	190-91	52°38'N	1°17'E
Norwich, Ct., U.S. (nôr'wĭch)	116-17	41°32'N	72°05'W
Norwood, Ma., U.S. (nôr'wŏŏd)	116-17	42°11'N	71°12'W
Norwood, Oh., U.S. (nôr'wŏŏd)	116-17	39°10'N	84°27'W
Noshiro, Japan (nō'shē-rô)	244	40°12'N	140°02'E

n-sing; ŋ-baŋk; N-nasalized n; nŏd; cŏmmit; ōld; ôbey; ôrder; oi-boil; fōōd; ŏŏ-as oo in foot; ou-out; s-soft; sh-dish; th-thin; pūre; ûnite; ûrn; stŭd; circŭs; ü-as in French tu; '-indeterminate vowel.

Feature (Pronunciation)	Page	Lat.	Long.
Nosivka, Ukr. (nô´sôf-kȧ)	202-03	50°56'N	31°36'E
Nosop, stm., Afr.	264-65	26°53'S	20°41'E
Nossob, stm., Afr. (nô´sŏb)	264-65	26°53'S	20°41'E
Nosy-Varika, Madag.	264-65	20°35'S	48°32'E
Noteć, stm., Pol. (nô´tĕcn)	194-95	52°44'N	15°25'E
Notodden, Nor. (nōt´ôd'n)	192-93	59°34'N	9°16'E
Noto-hantō, pen., Japan	245	37°20'N	137°00'E
Notozero, Ozero, lk., Russia	186-87	66°28'N	32°05'E
Notre-Dame, Monts, mts., Qc., Can.	138-39	48°10'N	68°00'W
Notre Dame Bay, b., Nf., Can. (nō´t'r dȧm´ bā)	138-39	49°46'N	55°15'W
Nottawasaga Bay, b., On., Can. (nŏt´ȧ-wȧ-sä´gȧ bā)	136-37	44°35'N	80°15'W
Nottingham, Eng., U.K. (nŏt´ĭng-ăm)	190-91	52°57'N	1°07'W
Nottingham Island, i., Nu., Can.	130-31	63°20'N	77°55'W
Nouâdhibou, Maur.	258-59	20°55'N	17°02'W
Nouâdhibou, Râs, c., Afr.	258-59	20°47'N	17°03'W
Nouakchott, nat. cap., Maur. (nü-äk´-shôt)	258-59	18°06'N	15°58'W
Nouméa, nat. cap., N. Cal. (nōō-mā´ä)	279g	22°17'S	166°27'E
Nouvelle, Qc., Can. (nōō-vĕl´)	138-39	48°08'N	66°19'W
Nouvelle-Calédonie, dep., Oc. see New Caledonia	279g	21°30'S	165°30'E
Nouvelle-Calédonie, i., N. Cal.	279g	21°33'S	165°42'E
Nouvelle-France, Cap de, c., Qc., Can.	130-31	62°27'N	73°42'W
Nouvelles-Hébrides, nation, Oc. see Vanuatu	279g	16°0'S	167°00'E
Nouvelles-Hébrides, is., Vanuatu see New Hebrides	279g	16°0'S	167°00'E
Nova Freixo, Moz. see Cuamba	264-65	14°47'S	36°32'E
Nova Friburgo, Braz. (nō´vȧ frē-bōōr´gô)	172	22°16'S	42°32'W
Nova Goa, India see Panaji	236	15°30'N	73°50'E
Nova Iguaçu, Braz. (nō´vä-ē-gwä-sōō´)	172	22°45'S	43°27'W
Nova Kakhovka, Ukr.	202-03	46°45'N	33°25'E
Nova Lima, Braz. (nō´vȧ lē´mä)	172	19°60'S	43°51'W
Nova Lisboa, Ang. see Huambo	264-65	12°46'S	15°44'E
Novara, Italy (nô-vä´rä)	200-01	45°27'N	8°37'E
Nova Scotia, state, Can. (nō´vȧ skō´shȧ)	128-29	45°0'N	63°00'W
Novaya Ladoga, Russia (nô´vȧ-ya lȧ-dô-gȧ)	192-93	60°05'N	32°15'E
Novaya Sibir', Ostrov, i., Russia (ôs-trôf´ nô´vȧ-ya sĕ-bēr´)	218-19	75°0'N	149°00'E
Novaya Zemlya, is., Russia (nô´vȧ-ya zĕm-lyȧ´)	218-19	74°0'N	57°00'E
Nova Zagora, Blg. (nō´vä zä´gô-rȧ)	200-01	42°30'N	26°01'E
Novelda, Spain (nō-vĕl´dȧ)	198-99	38°23'N	0°46'W
Nové Zámky, Slvk. (nô´vĕ zäm´kĕ)	194-95	47°60'N	18°11'E
Novgorod, Russia (nôv´gô-rŏt)	192-93	58°32'N	31°18'E
Novi Pazar, Blg. (nō´vĭ pä-zär´)	200-01	43°21'N	27°12'E
Novi Pazar, Serb. (nō´vĭ pá-zär´)	200-01	43°08'N	20°31'E
Novi Sad, Serb. (nō´vĭ säd´)	200-01	45°15'N	19°50'E
Novoanninskiy, Russia	186-87	50°32'N	42°41'E
Novo Aripuanã, Braz.	166-67	5°08'S	60°21'W
Novocherkassk, Russia (nô´vô-chĕr-kask´)	186-87	47°25'N	40°06'E
Novodvinsk, Russia	186-87	64°25'N	40°49'E
Novohrad-Volyns'kyi, Ukr.	194-95	50°36'N	27°38'E
Novokuybyshevsk, Russia	186-87	53°06'N	49°56'E
Novokuznetsk, Russia (nō´vô-kô´z-nyĕ´tsk)	226	53°45'N	87°07'E
Novo Mesto, Slvn. (nôvô mās´tô)	200-01	45°48'N	15°10'E
Novomoskovsk, Russia (nô´vô-môs-kôfsk´)	202-03	54°05'N	38°13'E
Novomoskovs'k, Ukr. (nō´vô-kô´z-nyĕ´tsk)	202-03	48°38'N	35°12'E
Novorossiysk, Russia (nô´vô-rô-sēsk´)	202-03	44°43'N	37°46'E
Novorzhev, Russia (nô´vô-rzhĕv´)	192-93	57°02'N	29°20'E
Novoshakhtinsk, Russia	202-03	47°48'N	39°54'E
Novosibirsk, Russia (nô´vô-sê-bērsk´)	218-19	55°01'N	82°53'E
Novosibirskiye Ostrova, is., Russia see New Siberian Islands	218-19	75°0'N	142°00'E
Novosibirskoye Vodokhranilishche, res., Russia	226	54°35'N	82°35'E
Novosil', Russia (nô´vô-sĭl)	202-03	52°58'N	37°03'E
Novosokol'niki, Russia (nô´vô-sô-kôl´nê-kê)	192-93	56°21'N	30°10'E
Novouzensk, Russia (nô-vô-ô-zĕnsk´)	186-87	50°29'N	48°10'E
Novovolyns'k, Ukr.	194-95	50°44'N	24°08'E
Novozybkov, Russia (nō´vô-zĕp´kôf)	202-03	52°32'N	31°56'E
Nový Jičín, Czech Rep. (nô´vē yē´chēn)	194-95	49°36'N	18°01'E
Novyy Oskol, Russia (nō´vē ôs-kôl´)	202-03	50°46'N	37°53'E
Novyy Uzen, Kaz. see Zhangaözen	186-87	43°19'N	52°47'E
Nowata, Ok., U.S. (nô-wä´tȧ)	120-21	36°42'N	95°38'W
Nowra, Austl. (nou´rȧ)	276	34°53'S	150°36'E
Nowshera, Pak.	232-33	34°01'N	71°59'E
Nowy Dwór Mazowiecki, Pol. (nō´vĭ dvōōr mä-zo-vyĕts´ke)	194-95	52°26'N	20°43'E
Nowy Targ, Pol. (nô´vê tärk´)	194-95	49°28'N	20°03'E
Noxubee, stm., U.S. (nŏks´ů-bē)	124-25	32°50'N	88°10'W
Nsanje, Malawi	264-65	16°58'S	35°12'E
Nsukka, Nig.	260a	6°51'N	7°24'E
Ntem, stm., Afr.	260-61	2°20'N	9°50'E
Ntomba, Lac, lk., D.R.C.	262-63	0°48'S	18°03'E
Nu, stm., Asia (nōō) see Salween	208-09	16°33'N	97°40'E
Nubian Desert, des., Sudan (nōō´bĭ-ăn dĕs´ērt)	266	20°30'N	33°00'E
Nueces, stm., Tx., U.S. (nů-ā´sås)	122-23	27°50'N	97°22'W
Nueltin Lake, lk., Can. (nwĕl´tin lăk)	130-31	60°19'N	99°40'W
Nueva, Isla, i., Chile	171	55°14'S	66°32'W
Nueva Gerona, Cuba (nwā´vä kĕ-rō´nä)	142-43	21°53'N	82°48'W
Nueva Imperial, Chile	171	38°44'S	72°57'W
Nueva Palmira, Ur. (nwā´vä päl-mē´rä)	173	33°52'S	58°23'W
Nueva Rosita, Mex. (nòĕ´vä rô-sê´tä)	122-23	27°57'N	101°13'W
Nueva Toltén, Chile	171	39°12'S	73°13'W
Nueve de Julio, Arg. (nwā´vå dā hōō´lyô)	173	35°27'S	60°53'W
Nuevitas, Cuba (nwā-vē´täs)	142-43	21°33'N	77°16'W
Nuevo, Cayo, i., Mex.	146-47	21°51'N	92°06'W
Nuevo, Golfo, b., Arg.	171	42°42'S	64°36'W
Nuevo Casas Grandes, Mex.	144-45	30°25'N	107°55'W
Nuevo Laredo, Mex. (nwā´vô lä-rā´dhō)	122-23	27°28'N	99°31'W
Nuevo León, state, Mex. (nwā´vô lå-ōn´)	122-23	25°40'N	100°00'W
Nuguria Islands, is., Pap. N. Gui.	277	3°21'S	154°41'E
Nui, at., Tuvalu	280-81	7°15'S	177°10'E
Nukha, Azer. see Şeki	227	41°10'N	47°10'E
Nuku'alofa, nat. cap., Tonga (nōō´-kōō-ä-lō´-fȧ)	280-81	21°08'S	175°13'W
Nukuoro, at., Micron.	280-81	3°51'N	154°58'E
Nukus, Uzb.	226	42°28'N	59°36'E
Nullarbor Plain, pl., Austl. (nŭ-lär´bôr plān)	272-73	31°0'S	129°00'E
Numara, i., Mald.	236	6°25'N	73°04'E
Numazu, Japan (nōō´mä-zōō)	245	35°06'N	138°52'E
Numedalslågen, stm., Nor. see Lågen	192-93	59°02'N	10°04'E
Numfoor, Pulau, i., Indon.	224-25	1°03'S	134°54'E
Nunavut, state, Can.	128-29	70°0'N	95°00'W
Nunivak Island, i., Ak., U.S. (nōō´nĭ-văk ī´lȧnd)	126	60°00'N	166°29'W
Nunjiang, China	240-41	49°10'N	125°14'E
Nuomin, stm., China	240-41	48°13'N	124°31'E
Nuoro, Italy (nwô´rō)	200-01	40°20'N	9°20'E
Nūra, stm., Kaz.	226	50°22'N	69°15'E
Nuremberg, Ger. see Nürnberg	194-95	49°27'N	11°04'E
Nürnberg, Ger. (nürn´bĕrgh)	194-95	49°27'N	11°04'E
Nushagak, stm., Ak., U.S. (nū-shă-găk´)	126	59°03'N	158°24'W
Nu Shan, mts., China	238-39	27°0'N	99°00'E
Nushki, Pak. (nŭsh´kē)	232-33	29°35'N	66°04'E
Nuuk, nat. cap., Green. see Godthåb	284-85	64°11'N	51°44'W
Nuweveldberge, mts., S. Afr.	264-65	32°14'S	21°48'E
Nyahururu Falls, Kenya	267	0°02'N	36°22'E
Nyainqêntanglha Shan, mts., China (nyä-ĭn-chyŭn-täŋ-lä shän)	234-35	30°0'N	90°00'E
Nyala, Sudan	266	12°03'N	24°54'E
Nyandoma, Russia	186-87	61°40'N	40°13'E
Nyanza, Rw.	267	2°21'S	29°45'E
Nyasa, Lake, lk., Afr. (lăk nyä´sä)	264-65	12°0'S	34°30'E
Nyasaland, nation, Afr. see Malawi	253	13°30'S	34°00'E
Nyborg, Den. (nü´bôr")	192-93	55°19'N	10°47'E
Nybro, Swe. (nü´brô)	192-93	56°45'N	15°55'E
Nyeri, Kenya	267	0°25'S	36°57'E
Nyíregyháza, Hung. (nyē´rĕd-y'hä´zä)	194-95	47°57'N	21°43'E
Nyköping, Swe. (nü´chû-pĭng)	192-93	58°45'N	16°60'E
Nylstroom, S. Afr. (nĭl´strōm) see Modimolle	269c	24°42'S	28°25'E
Nynäshamn, Swe. (nü-nĕs-hȧm'n)	192-93	58°55'N	17°57'E
Nyngan, Austl. (nĭŋ´gȧn)	276	31°33'S	147°10'E
Nyong, stm., Camrn. (nyông)	260-61	3°16'N	9°55'E
Nysa Łużycka, stm., Eur. see Neisse	194-95	52°04'N	14°46'E
Nyslott, Fin. see Savonlinna	192-93	61°52'N	28°54'E
Nytva, Russia	186-87	57°56'N	55°20'E
Nyunzu, D.R.C.	267	5°58'S	28°02'E
Nyurba, Russia	218-19	63°17'N	118°20'E
Nyuvchim, Russia	186-87	61°23'N	50°36'E
Nyuya, stm., Russia (nyōō´yȧ)	218-19	60°32'N	116°18'E
Nzérékoré, Gui.	260-61	7°45'N	8°49'W
Nzwani, i., Com. (äɴ-zhwäɴ)	264-65	12°15'S	44°25'E
O			
O'ahu, i., Hi., U.S. (ō-ä´hōō) (ō-ä´hü)	127a	21°30'N	158°00'W
Oahe, Lake, res., U.S.	114-15	45°29'N	100°20'W
Oak Bay, B.C., Can. (ōk bā)	132-33	48°27'N	123°18'W
Oak Creek, Wi., U.S. (ōk krēk´)	116-17	42°52'N	87°54'W
Oakdale, La., U.S. (ōk´dāl)	122-23	30°49'N	92°40'W
Oakes, N.D., U.S. (ōks)	114-15	46°08'N	98°06'W
Oak Grove, Ky., U.S. (ōk grōv)	124-25	36°40'N	87°26'W
Oak Harbor, Wa., U.S. (ōk här´bēr)	112-13	48°18'N	122°40'W
Oakland, Ca., U.S. (ōk´lănd)	118-19	37°48'N	122°17'W
Oakland, Md., U.S. (ōk´lănd)	116-17	39°24'N	79°24'W
Oakland, Ne., U.S. (ōk´lănd)	114-15	41°50'N	96°28'W
Oak Lawn, Il., U.S. (ōk lôn)	116-17	41°43'N	87°45'W
Oakley, Ks., U.S. (ōk´lĭ)	120-21	39°08'N	100°51'W
Oak Ridge, Tn., U.S. (ōk rĭj)	124-25	36°01'N	84°15'W
Oakville, On., Can. (ōk´vĭl)	136-37	43°27'N	79°40'W
Oakville, Mo., U.S. (ōk´vĭl)	120-21	38°28'N	90°19'W
Oaxaca, state, Mex. (wä-hä´kä)	146-47	17°0'N	96°30'W
Oaxaca de Juárez, Mex.	146-47	17°03'N	96°43'W
Ob', stm., Russia (ōb)	218-19	66°47'N	68°56'E
Oban, Scot., U.K. (ō´băn)	190-91	56°25'N	5°28'W
Oberlin, Ks., U.S. (o´bēr-lĭn)	120-21	39°49'N	100°32'W
Oberlin, Oh., U.S. (o´bēr-lĭn)	116-17	41°17'N	82°13'W
Obi, Kepulauan, is., Indon. (ō´bê)	248-49	1°27'S	127°38'E
Obi, Pulau, i., Indon.	248-49	1°30'S	127°45'E
Óbidos, Braz. (ō´bê-dòzh)	166-67	1°54'S	55°31'W
Obihiro, Japan (ō´bē-hē´rō)	244	42°55'N	143°12'E
Obluchye, Russia	240-41	49°01'N	131°04'E
Obninsk, Russia	202-03	55°06'N	36°37'E
Oboyan, Russia (ô-bô-yän´)	202-03	51°13'N	36°17'E
Observatoire, Caye de l', i., N. Cal.	272-73	21°25'S	158°50'E
Obsgchiy Syrt, mts., Eur. see Zhalpy Syrt	186-87	52°0'N	51°30'E
Obskaya Guba, b., Russia	218-19	69°0'N	73°00'E
Obuasi, Ghana	260-61	6°13'N	1°41'W
Ocala, Fl., U.S. (ô-kä´lȧ)	124-25	29°11'N	82°08'W
Ocampo, Mex. (ô-käm´pō)	144-45	28°11'N	108°23'W
Ocaña, Col. (ô-kän´yä)	164-65	8°14'N	73°21'W
Ocaña, Spain (ô-kä´n-yä)	198-99	39°57'N	3°30'W
Occidental, Cordillera, mts., Col.	164-65	5°0'N	76°00'W
Ocean City, Md., U.S. (ō´shăn sĭ´tê)	116-17	38°20'N	75°05'W
Ocean City, N.J., U.S. (ō´shăn sĭ´tê)	116-17	39°17'N	74°35'W
Ocean Falls, B.C., Can. (ō´shăn fôlz)	132-33	52°21'N	127°41'W
Ocean Grove, N.J., U.S. (ō´shăn grōv)	116-17	40°13'N	74°00'W
Ocean Island, i., Kir. see Banaba	280-81	0°52'S	169°33'E
Oceanside, Ca., U.S. (ō´shăn-sīd)	118-19	33°12'N	117°22'W
Ocean Springs, Ms., U.S. (ō´shăn springs sprĭngz)	124-25	30°26'N	88°50'W
Ochlockonee, stm., U.S. (ŏk-lô-kō´nē)	124-25	29°59'N	84°26'W
Ocilla, Ga., U.S. (ô-sĭl´ȧ)	124-25	31°36'N	83°15'W
Ockelbo, Swe. (ôk´ĕl-bô)	192-93	60°54'N	16°44'E
Ocmulgee National Monument, n.p., Ga., U.S. (ôk-mŭl´gē năsh´ŭn-ăl mŏn´ū-mĕnt)	124-25	32°43'N	83°38'W
Oconee, stm., Ga., U.S. (ô-kō´nē)	124-25	31°58'N	82°32'W
Oconomowoc, Wi., U.S. (ô-kŏn´ô-mô-wŏk´)	116-17	43°06'N	88°29'W
Oconto, Wi., U.S. (ô-kŏn´tō)	116-17	44°54'N	87°52'W
Oconto Falls, Wi., U.S. (ô-kŏn´tō fôlz)	116-17	44°52'N	88°08'W
Ocosingo, Mex.	148	16°55'N	92°06'W
Ocotal, Nic. (ō-kô-täl´)	149	13°38'N	86°28'W
Ocotlán, Mex. (ō-kô-tlän´)	146-47	20°21'N	102°47'W
Ocotlán de Morelos, Mex. (ô-kô-tlän´ dā mô-rā´lōs)	146-47	16°47'N	96°40'W
Ocracoke Island, i., N.C., U.S.	124-25	35°06'N	75°59'W
October Revolution Island, i., Russia see Oktyabr'skoy Revolyutsii, Ostrov	218-19	79°30'N	96°60'E
Ocumare del Tuy, Ven. (ô-kōō-mä´ra del twē´)	163b	10°07'N	66°46'W
Oda, Jabal, mtn., Sudan	266	20°21'N	36°39'E
Odda, Nor. (ôdh-ȧ)	192-93	60°04'N	6°32'E
Odemira, Port. (ô-dȧ-mē´rȧ)	198-99	37°35'N	8°38'W
Ödemiş, Tur. (û´dĕ-mēsh)	200-01	38°14'N	27°59'E
Odendaalsrus, S. Afr. (ō´dĕn-däls-rûs´)	269c	27°52'S	26°42'E
Odense, Den. (ō´dhĕn-sĕ)	192-93	55°24'N	10°23'E
Oder, stm., Eur. (ō´dēr)	194-95	53°55'N	14°17'E
Odesa, Ukr.	202-03	46°29'N	30°42'E
Odessa, De., U.S. (ô-dĕs´ȧ)	116-17	39°27'N	75°40'W
Odessa, Tx., U.S. (ô-dĕs´ȧ)	122-23	31°51'N	102°22'W
Odin, Mount, mtn., B.C., Can.	132-33	50°33'N	118°08'W
Odintsovo, Russia (ô-dēn´tsô-vô)	202-03	55°40'N	37°16'E
Odra, stm., Eur. (ô´drȧ) see Oder	194-95	53°55'N	14°17'E
Odrzywół, Pol.	194-95	51°32'N	20°33'E
Oeiras, Braz. (wå-ē-räzh´)	166-67	7°01'S	42°08'W
Oelwein, Ia., U.S. (ōl´wīn)	114-15	42°40'N	91°55'W

Feature (Pronunciation)	Page	Lat.	Long.
Oeno Atoll, at., Pit.	280-81	23°55'S	130°44'W
O'Fallon, Mo., U.S. (ō-făl'ŭn)	120-21	38°49'N	90°42'W
Offenburg, Ger. (ôf'ĕn-bòrgh)	194-95	48°28'N	7°57'E
Ofu, i., Am. Sam.	279b	14°10'S	169°40'W
Ogaadeen, reg., Afr. see Ogaden	262-63	8°0'N	44°00'E
Ogaden, reg., Afr. (ō-gä'dĕn)	262-63	8°0'N	44°00'E
Ogallala, Ne., U.S. (ō-gä-lä'lä)	114-15	41°08'N	101°43'W
Ogasawara-guntō, is., Japan see Bonin Islands	282-83	26°58'N	142°14'E
Ogasawara-shotō, is., Japan	280-81	26°0'N	142°00'E
Ogbomosho, Nig. (ŏg-bô-mō'shō)	260a	8°08'N	4°15'E
Ogden, Ia., U.S. (ŏg'dĕn)	114-15	42°03'N	94°02'W
Ogden, Ut., U.S. (ŏg'dĕn)	112-13	41°14'N	111°57'W
Ogdensburg, N.Y., U.S. (ŏg'dĕnz-bûrg)	116-17	44°42'N	75°30'W
Ogea Levu, i., Fiji	279f	19°08'S	178°24'W
Ogeechee, stm., Ga., U.S. (ô-gē'chē)	124-25	31°51'N	81°05'W
Ogilvie Mountains, mts., Yk., Can. (ō'g'l-vĭ moun'tĭnz)	130-31	65°03'N	139°29'W
Ogooué, stm., Afr.	260-61	0°49'S	9°00'E
Ogulin, Cro. (ō-gōō-lēn')	200-01	45°16'N	15°14'E
Ogurja Ada, i., Turkmen.	226	38°57'N	53°03'E
O'Higgins, Lago, lk., S.A.	171	48°53'S	72°39'W
Ohio, state, U.S. (ô'hī'ō)	108-09	40°15'N	82°45'W
Ohio, stm., U.S. (ô'hī'ō)	110-11	36°59'N	89°09'W
Ohlau, Pol. see Oława	194-95	50°56'N	17°19'E
Ohōtuku-kai, s., Asia see Okhotsk, Sea of	218-19	53°0'N	150°00'E
Ohrid, Mac. (ō'krēd)	200-01	41°07'N	20°49'E
Oiapoque, Braz.	164-65	3°51'N	51°49'W
Oiapoque, stm., S.A.	164-65	4°10'N	51°37'W
Oil City, Pa., U.S. (oil sĭ'tĭ sĭ'tē)	116-17	41°26'N	79°42'W
Ōita, Japan (ō'ē-tä)	245	33°14'N	131°37'E
Ojinaga, Mex. (ō-kē-nä'gä)	122-23	29°33'N	104°25'W
Ojocaliente, Mex. (ô-kō-kä-lyĕ'n-tĕ)	146-47	22°33'N	102°13'W
Ojos del Salado, Cerro, mtn., S.A. see Ojos del Salado, Nevado	168-69	27°06'S	68°32'W
Ojos del Salado, Nevado, mtn., S.A.	168-69	27°06'S	68°32'W
Oka, stm., Russia (ô-kä')	218-19	55°16'N	102°18'E
Oka, stm., Russia (ô-kä')	186-87	56°20'N	43°59'E
Okahandja, Nmb.	264-65	21°58'S	16°54'E
Okanagan, stm., N.A. (ō'kȧ-näg'ăn) see Okanogan	132-33	48°06'N	119°44'W
Okanagan Lake, lk., B.C., Can. (ō'kȧ-näg'ăn lāk)	132-33	49°55'N	119°31'W
Okanogan, stm., N.A.	132-33	48°06'N	119°44'W
Okāra, Pak.	232-33	30°48'N	73°27'E
Okavango, stm., Afr. (ō-kȧ-văn'gō)	264-65	18°57'S	22°25'E
Okavango Delta, del., Bots.	264-65	19°29'S	22°32'E
Okavango Swamp, del., Bots. see Okavango Delta	264-65	19°29'S	22°32'E
Okaya, Japan (ō'kȧ-yȧ)	245	36°04'N	138°03'E
Okayama, Japan (ō'kȧ-yä'mȧ)	245	34°40'N	133°55'E
Okazaki, Japan (ō'kȧ-zä'kē)	245	34°57'N	137°10'E
Okeechobee, Fl., U.S. (ō-kē-chō'bē)	125a	27°15'N	80°50'W
Okeechobee, Lake, lk., Fl., U.S. (lāk ō-kē-chō'bē)	125a	26°55'N	80°45'W
Okefenokee Swamp, sw., U.S. (ō'kē-fē-nō'kē swŏmp)	124-25	30°42'N	82°20'W
Okemah, Ok., U.S. (ō-kē'mä)	120-21	35°26'N	96°18'W
Okene, Nig.	260a	7°29'N	6°15'E
Okha, Russia (ŭ-kä')	218-19	53°35'N	142°57'E
Okhota, stm., Russia	218-19	59°20'N	143°04'E
Okhotsk, Russia (ô-kôtsk')	218-19	59°22'N	143°18'E
Okhotsk, Sea of, s., Asia (sē ŭv ô-kôtsk')	218-19	53°0'N	150°00'E
Okhotskoye More, s., Asia see Okhotsk, Sea of	218-19	53°0'N	150°00'E
Okhtyrka, Ukr.	202-03	50°18'N	34°54'E
Okinawa-jima, i., Japan	244a	26°32'N	127°60'E
Okino-Daitō-jima, i., Japan	222-23	24°28'N	131°11'E
Okino-Tori-shima, i., Japan	222-23	20°27'N	136°04'E
Oki-shotō, is., Japan	245	36°11'N	133°11'E
Oklahoma, state, U.S. (ō-klȧ-hō'mȧ)	108-09	35°30'N	98°00'W
Oklahoma City, Ok., U.S. (ō-klȧ-hō'mȧ sĭ'tĭ)	120-21	35°29'N	97°29'W
Okmulgee, Ok., U.S. (ōk-mŭl'gē)	120-21	35°36'N	95°58'W
Okolona, Ms., U.S. (ō-kô-lō'nȧ)	124-25	33°60'N	88°45'W
Okotoks, Ab., Can.	132-33	50°44'N	113°59'W
Oktyabr'sk, Kaz.	226	49°28'N	57°25'E
Oktyabrskiy, Russia	186-87	54°29'N	53°29'E
Oktyabr'skiy, Russia	218-19	52°41'N	156°13'E
Oktyabr'skoy Revolyutsii, Ostrov, i., Russia	218-19	79°30'N	96°60'E
Okushiri-tō, i., Japan (ō'koo-shē'rē tō)	244	42°10'N	139°27'E
Ola, Russia	218-19	59°37'N	151°20'E
Olanchito, Hond. (ō'län-chē'tô)	149	15°29'N	86°34'W
Öland, i., Swe. (û-länd')	192-93	56°45'N	16°38'E
Olary, Austl.	276	32°17'S	140°19'E
Olathe, Ks., U.S. (ô-lā'thē)	120-21	38°53'N	94°49'W
Olavarría, Arg. (ō-lä-vär-rē'ä)	173	36°54'S	60°19'W
Oława, Pol. (ô-lä'vä)	194-95	50°56'N	17°19'E
Olbia, Italy (ô'l-byä)	200-01	40°56'N	9°30'E
Old Bahama Channel, strt., N.A. (ōld bȧ-hā'mȧ chăn'ĕl)	142-43	22°40'N	78°41'W
Old Crow, Yk., Can. (ōld crō)	128-29	67°36'N	139°49'W
Oldenburg, Ger. (ôl'dĕn-bòrgh)	194-95	53°09'N	8°13'E
Old Forge, Pa., U.S. (ōld fôrj)	116-17	41°22'N	75°44'W
Olds, Ab., Can. (ōldz)	132-33	51°47'N	114°05'W
Old Wives Lake, lk., Sk., Can. (ōld wīvz lāk)	134-35	50°06'N	106°00'W
Olean, N.Y., U.S. (ō-lē-ăn')	116-17	42°05'N	78°26'W
Olekma, stm., Russia (ô-lyĕk-mȧ')	218-19	60°23'N	120°41'E
Olëkminsk, Russia (ô-lyĕk-mĕnsk')	218-19	60°22'N	120°26'E
Oleksandriia, Ukr.	202-03	48°40'N	33°07'E
Olenegorsk, Russia	184-85	68°09'N	33°14'E
Olenëk, stm., Russia (ô-lyĕ-nyôk')	208-09	73°00'N	119°45'E
Olga, Russia (ôl'gȧ)	244	43°44'N	135°17'E
Ölgiy, Mong.	222-23	48°58'N	89°58'E
Olhão, Port. (ôl-youɴ')	198-99	37°02'N	7°51'W
Ólimbos, mtn., Cyp.	228-29	34°56'N	32°51'E
Olímpia, Braz.	172	20°44'S	48°55'W
Olinda, Braz. (ô-lē'n-dä)	163d	8°01'S	34°51'W
Oliva, Arg.	173	32°03'S	63°34'W
Oliva, Spain (ô-lē'vä)	198-99	38°55'N	0°07'W
Olive Hill, Ky., U.S. (ŏl'ĭv hĭl)	116-17	38°18'N	83°11'W
Oliveira, Braz. (ô-lē-vā'rä)	172	20°42'S	44°49'W
Oliver, B.C., Can. (ŏ'lĭ-vẽr)	132-33	49°11'N	119°33'W
Olivia, Mn., U.S. (ō-lĭv'ē-ȧ)	114-15	44°47'N	94°59'W
Ollagüe, Chile (ô-lyä'gå)	168-69	21°13'S	68°16'W
Olmaliq, Uzb.	232-33	40°51'N	69°35'E
Olmos, Peru	170	5°56'S	79°46'W
Olmütz, Czech Rep. see Olomouc	194-95	49°36'N	17°16'E
Olney, Il., U.S. (ŏl'nĭ)	116-17	38°43'N	88°06'W
Olney, Tx., U.S. (ŏl'nē)	120-21	33°22'N	98°45'W
Olomane, stm., Qc., Can. (ō'lō mȧ'nē)	138-39	50°14'N	60°38'W
Olomouc, Czech Rep. (ô'lô-mōts)	194-95	49°36'N	17°16'E
Olonets, Russia (ô-lô'nĕts)	192-93	60°59'N	32°59'E
Olongapo, Phil.	250	14°52'N	120°17'E
Oloron-Sainte-Marie, Fr. (ō-lô-rôɴt'săɴt mȧ-rē')	196-97	43°11'N	0°36'W
Olot, Spain (ô-lōt')	198-99	42°11'N	2°29'E
Olovyannaya, Russia	240-41	50°57'N	115°34'E
Olsztyn, Pol. (ôl'shtĭn)	194-95	53°47'N	20°29'E
Olt, stm., Rom.	186-87	43°43'N	24°48'E
Olten, Switz. (ôl'tĕn)	194-95	47°21'N	7°54'E
Oltenița, Rom. (ôl-tā'nĭ-tsȧ)	200-01	44°05'N	26°38'E
Olutanga Island, i., Phil.	250	7°22'N	122°52'E
Olvera, Spain (ôl-vĕ'rä)	198-99	36°56'N	5°16'W
Olympia, Wa., U.S.	112-13	47°02'N	122°53'W
Olympic Mountains, mts., Wa., U.S. (ô-lĭm'pĭk moun'tĭnz)	112-13	47°50'N	123°45'W
Olympic National Park, n.p., Wa., U.S. (ô-lĭm'pĭk năsh'ŭn-ăl pärk)	112-13	47°51'N	123°44'W
Ólympos, mtn., Grc. see Olympus, Mount	200-01	40°05'N	22°21'E
Olympus, mtn., Cyp. see Ólimbos	228-29	34°56'N	32°51'E
Olympus, Mount, mtn., Grc. (mount ô-lĭm'pŭs)	200-01	40°05'N	22°21'E
Olympus, Mount, mtn., Wa., U.S. (mount ô-lĭm'pŭs)	112-13	47°48'N	123°43'W
Olyutorski, Cape, c., Russia (kāp ŭl-yōō'tôr-skē) see Olyutorskiy, Mys	218-19	59°57'N	170°22'E
Olyutorskiy, Mys, c., Russia (mĭs ŭl-yōō'tôr-skē)	218-19	59°57'N	170°22'E
Olyutorskiy Zaliv, b., Russia	218-19	60°15'N	168°30'E
Om', stm., Russia	218-19	54°59'N	73°22'E
Omagh, N. Ire., U.K. (ō'mä)	190-91	54°36'N	7°18'W
Omaha, Ne., U.S. (ō'mȧ-hä)	114-15	41°15'N	95°56'W
Omaha Indian Reservation, ind. res., Ne., U.S. (ō'mȧ-hä ĭn'dĭ-ăn rĕ-sẽr-vā'shĕn)	114-15	42°07'N	96°32'W
Omak, Wa., U.S.	112-13	48°25'N	119°31'W
Oman, nation, Asia (ō-män')	206-07	22°0'N	58°00'E
'Omān, Daryā-ye, b., Asia see Oman, Gulf of	230-31	24°30'N	58°30'E
Oman, Gulf of, b., Asia (gŭlf ŭv ō-män')	230-31	24°30'N	58°30'E
Omaruru, Nmb. (ō-mä-rōō'rōō)	264-65	21°26'S	15°57'E
Omatako, stm., Nmb.	264-65	17°57'S	20°28'E
Omboué, Gabon	260-61	1°37'S	9°16'E
Omdurman, Sudan (ŏm-dûr-măn')	266	15°39'N	32°29'E
Ometepe, Isla de, i., Nic. (ē's-lä-dĕ-ō-mĕ-tā'pȧ)	149	11°30'N	85°35'W
Ometepec, Mex. (ô-mā-tā-pĕk')	146-47	16°41'N	98°24'W
Ōminato, Japan see Mutsu	244	41°17'N	141°10'E
Omineca, stm., B.C., Can. (ô-mĭ-nĕk'ȧ)	132-33	56°07'N	124°28'W
Omineca Mountains, mts., B.C., Can.	130-31	56°0'N	125°00'W
Omo, stm., Afr. (ō'mō)	262-63	4°31'N	36°03'E
Omolon, stm., Russia (ō'mō)	218-19	68°42'N	158°43'E
Omro, Wi., U.S. (ŏm'rō)	116-17	44°02'N	88°45'W
Omsk, Russia (ômsk)	226	54°57'N	73°23'E
Omsukchan, Russia	218-19	62°30'N	155°46'E
Ōmura, Japan (ō'mōō-rȧ)	245	32°55'N	129°58'E
Ōmuta, Japan (ō-mò-tȧ)	245	33°01'N	130°27'E
Omutninsk, Russia (ō'mōō-tnēnsk)	186-87	58°39'N	52°11'E
Onawa, Ia., U.S. (ŏn-ȧ-wȧ)	114-15	42°02'N	96°06'W
Onda, Spain (ōn'dä)	198-99	39°58'N	0°15'W
Ondangwa, Nmb.	264-65	17°56'S	16°00'E
Ondo, Nig.	260a	7°06'N	4°50'E
Öndörhaan, Mong.	240-41	47°20'N	110°40'E
Onega, Russia (ô-nyĕ'gȧ)	186-87	63°55'N	38°06'E
Onega, stm., Russia (ô-nyĕ'gȧ)	186-87	63°57'N	37°57'E
Onega, Lake, lk., Russia (lāk ô-nyĕ'-gȧ)	186-87	61°30'N	35°45'E
Oneida, N.Y., U.S. (ō-nī'dȧ)	116-17	43°06'N	75°39'W
O'Neill, Ne., U.S. (ō-nēl')	114-15	42°28'N	98°39'W
Onekotan, Ostrov, i., Russia	218-19	49°21'N	154°42'E
Oneonta, Al., U.S. (ō-nē-ŏn'tȧ)	124-25	33°57'N	86°28'W
Oneonta, N.Y., U.S. (ō-nē-ŏn'tȧ)	116-17	42°28'N	75°04'W
Onezhskoye Ozero, lk., Russia see Onega, Lake	186-87	61°30'N	35°45'E
Ongi, stm., Mong.	240-41	44°31'N	103°40'E
Onitsha, Nig. (ô-nĭt'shȧ)	260a	6°09'N	6°47'E
Ono-i-Lau, i., Fiji	280-81	20°39'S	178°42'W
Onomichi, Japan (ō'nō-mē'chē)	245	34°25'N	133°12'E
Onon, stm., Asia (ō'nōn)	222-23	51°42'N	115°49'E
Onoto, Ven. (ô-nŏ'tô)	163b	9°36'N	65°11'W
Onotoa, at., Kir.	280-81	1°53'S	175°34'E
Onslow, Austl. (ŏnz'lō)	270-71	21°39'S	115°07'E
Onslow Bay, b., N.C., U.S. (ŏnz'lō bā)	124-25	34°20'N	77°20'W
Ontake-san, vol., Japan (ōn'tä-kå-sän)	245	35°53'N	137°29'E
Ontario, Or., U.S. (ŏn-tā'rĭ-ō)	112-13	44°02'N	116°57'W
Ontario, state, Can. (ŏn-tā'rĭ-ō)	128-29	51°0'N	85°00'W
Ontario, Lake, lk., N.A. (lāk ŏn-tā'rĭ-ō)	116-17	43°45'N	78°00'W
Ontonagon, Mi., U.S. (ŏn-tô-năg'ŏn)	114-15	46°52'N	89°19'W
Ontong Java, at., Sol. Is.	279e	5°19'S	159°16'E
Onverwacht, Sur.	164-65	5°36'N	55°12'W
Oodnadatta, Austl. (ōōd'nȧ-dȧ'tȧ)	270-71	27°33'S	135°27'E
Ooldea, Austl.	270-71	30°28'S	131°51'E
Oos-Londen, S. Afr. see East London	264-65	32°60'S	27°54'E
Oostende, Bel. (ōst-ĕn'dĕ)	190-91	51°14'N	2°55'E
Opalaca, Cordillera, mts., Hond. (kôr-dēl-yĕ'rä-ô-pä-lä'kä)	149	14°30'N	88°20'W
Oparino, Russia	186-87	59°51'N	48°17'E
Opasquia, On., Can. (ō-păs'kwē-ȧ)	134-35	53°16'N	93°35'W
Opelika, Al., U.S. (ŏp-ē-lī'kȧ)	124-25	32°38'N	85°23'W
Opelousas, La., U.S. (ŏp-ē-lōō'sȧs)	124-25	30°32'N	92°05'W
Opeongo Lake, lk., On., Can. (ŏp-ē-ôŋ'gō lāk)	136-37	45°42'N	78°24'W
Opobo, Nig.	260a	4°35'N	7°34'E
Opochka, Russia (ō-pôch'kȧ)	192-93	56°43'N	28°40'E
Opoczno, Pol. (ô-pôch'nô)	194-95	51°23'N	20°18'E
Opole, Pol. (ô-pôl'á)	194-95	50°40'N	17°57'E
Oporto, Port. see Porto	198-99	41°09'N	8°37'W
Opp, Al., U.S. (ŏp)	124-25	31°17'N	86°15'W
Oppeln, Pol. see Opole	194-95	50°40'N	17°57'E
Opportunity, Wa., U.S. (ŏp-ŏr tū'nĭ tĭ)	112-13	47°40'N	117°13'W
Opuwo, Nmb.	264-65	18°02'S	13°41'E
Oqsuqtooq, Nu., Can. see Gjoa Haven	128-29	68°39'N	95°55'W
Oradea, Rom. (ô-räd'yä)	194-95	47°04'N	21°56'E
Orai, India	234-35	25°59'N	79°28'E
Oral, Kaz.	186-87	51°13'N	51°22'E
Oral Dengizi, lk., Asia see Aral Sea	226	45°0'N	60°00'E
Oran, Alg. (ō-rän)(ō-rän')	198-99	35°41'N	0°39'W
Orange, Austl. (ŏr'ĕnj)	276	33°18'S	149°05'E
Orange, Fr. (ô-räɴzh')	196-97	44°08'N	4°49'E
Orange, Tx., U.S. (ŏr'ĕnj)	122-23	30°06'N	93°44'W
Orange, Va., U.S. (ŏr'ĕnj)	116-17	38°15'N	78°06'W
Orange, stm., Afr. (ŏr'ĕnj)	264-65	28°35'S	16°28'E
Orange, Cabo, c., Braz. (kå'bō-rȧ'n-zhĕ)	159	4°22'N	51°33'W
Orangeburg, S.C., U.S. (ŏr'ĕnj-bûrg)	124-25	33°30'N	80°51'W
Orange City, Ia., U.S. (ŏr'ĕnj sĭ'tē)	114-15	43°00'N	96°03'W
Orangeville, On., Can. (ŏr'ĕnj-vĭl)	136-37	43°55'N	80°05'W
Orange Walk, Belize (ŏr'ĕnj wôl'k)	148	18°06'N	88°33'W

n-sing; ŋ-baŋk; ɴ-nasalized n; nŏd; cŏmmit; ōld; ôbey; ôrder; oi-boil; fōōd; ò-as oo in foot; ou-out; s-soft; sh-dish; th-thin; pūre; ûnite; ûrn; stŭd; circŭs; ü-as in French tu; '-indeterminate vowel.

Feature (Pronunciation)	Page	Lat.	Long.
Orani, Phil. (ō-rä′nê)	250	14°49′N	120°31′E
Oranienburg, Ger. (ō-rä′nē-ĕn-boorgh)	194-95	52°45′N	13°15′E
Oranje, stm., Afr. see Orange	264-65	28°35′S	16°28′E
Oranjemund, Nmb.	264-65	28°34′S	16°28′E
Oranjestad, nat. cap., Aruba	140a	12°32′N	70°01′W
Orăştie, Rom. (ô-rûsh′tyå)	194-95	45°50′N	23°13′E
Oraşul Stalin, Rom. see Braşov	194-95	45°39′N	25°37′E
Orbetello, Italy (ôr-bå-tĕl′lō)	200-01	42°27′N	11°13′E
Orbost, Austl. (ôr′bŭst)	276	37°42′S	148°28′E
Ord, Ne., U.S. (ôrd)	114-15	41°36′N	98°56′W
Ordu, Tur. (ôr′dò)	186-87	40°59′N	37°52′E
Ordzhonikidze, Russia see Vladikavkaz	227	43°03′N	44°39′E
Örebro, Swe. (û′rē-brō)	192-93	59°17′N	15°12′E
Oregon, state, U.S.	108-09	44°0′N	121°00′W
Oregon City, Or., U.S.	112-13	45°21′N	122°36′W
Orekhovo-Zuyevo, Russia (ôr-yĕ′kô-vô zò′yĕ-vô)	202-03	55°48′N	38°58′E
Orël, Russia	202-03	52°59′N	36°04′E
Orem, Ut., U.S. (ō′rĕm)	118-19	40°16′N	111°41′W
Orense, Spain see Ourense	198-99	42°20′N	7°52′W
Organ Pipe Cactus National Monument, n.p., Az., U.S. (ôr′găn pīp kăk′tŭs năsh′ŭn-ăl mŏn′ū-mĕnt)	118-19	32°0′N	112°55′W
Orhon, stm., Mong.	240-41	50°14′N	106°08′E
Oriental, Cordillera, mts., Col. (kôr-dēl-yĕ′rä ō-rê-ĕn-täl′)	164-65	6°0′N	73°00′W
Oriental, Cordillera, mts., Peru	170	11°0′S	74°00′W
Orillia, On., Can. (ô-rĭl′ĭ-ȧ)	136-37	44°36′N	79°25′W
Orinoco, stm., S.A. (ō-rĭ-nō′kō)	164-65	8°47′N	60°40′W
Orissa, state, India (ō-rĭs′ȧ)	234-35	20°0′N	84°00′E
Oristano, Italy (ô-rês-tä′nō)	200-01	39°54′N	8°36′E
Oriximiná, Braz.	166-67	1°45′S	55°52′W
Orizaba, Mex. (ô-rê-zä′bä)	146-47	18°51′N	97°06′W
Orkla, stm., Nor. (ôr′klå)	184-85	63°19′N	9°51′E
Orkney, S. Afr. (ôrk′nĭ)	269c	26°59′S	26°41′E
Orkney Islands, is., Scot., U.K.	190c	59°0′N	3°00′W
Orlando, Fl., U.S. (ôr-lăn′dō)	125a	28°32′N	81°23′W
Orléans, Fr. (ôr-lā-äN′)	196-97	47°55′N	1°55′E
Orleans, In., U.S. (ôr-lēnz′)	116-17	38°40′N	86°27′W
Ormāra, Pak.	232-33	25°13′N	64°38′E
Ormoc, Phil.	250	11°01′N	124°37′E
Ormond Beach, Fl., U.S. (ôr′mŏnd bēch)	124-25	29°17′N	81°04′W
Örnsköldsvik, Swe. (ûrn′skôlts-vēk)	184-85	63°18′N	18°43′E
Orocué, Col.	164-65	4°49′N	71°20′W
Oroluk, at., Micron.	280-81	7°31′N	155°18′E
Oromocto, N.B., Can.	138-39	45°51′N	66°28′W
Orosháza, Hung. (ô-rôsh-hä′sô)	194-95	46°34′N	20°40′E
Orotukan, Russia	218-19	62°16′N	151°38′E
Oroville, Ca., U.S. (ōr′ô-vĭl)	118-19	39°31′N	121°33′W
Oroville, Wa., U.S. (ōr′ô-vĭl)	112-13	48°56′N	119°26′W
Oroville, Lake, res., Ca., U.S. (lāk ōr′ô-vĭl)	118-19	39°38′N	121°30′W
Orrville, Oh., U.S. (ôr′vĭl)	116-17	40°50′N	81°46′W
Orša, Bela.	192-93	54°31′N	30°25′E
Orsa, Swe. (ōr′sä)	192-93	61°08′N	14°37′E
Orsha, Bela. (ôr′shå) see Orša	192-93	54°31′N	30°25′E
Orsk, Russia (ôrsk)	226	51°12′N	58°34′E
Orşova, Rom. (ôr′shô-vä)	200-01	44°43′N	22°25′E
Ortega, Col. (ôr-tĕ′gä)	163c	3°56′N	75°13′W
Ortegal, Cabo, c., Spain (kå′bô-ôr-tå-gäl′)	198-99	43°46′N	7°54′W
Orthez, Fr. (ôr-tĕz′)	196-97	43°30′N	0°46′W
Orthon, stm., Bol.	166-67	10°49′S	66°04′W
Ortigueira, Spain (ôr-tê-gā′ê-rä)	198-99	43°41′N	7°50′W
Ortonville, Mn., U.S. (ôr-tŭn-vĭl)	114-15	45°19′N	96°27′W
Orūmīyeh, Iran	228-29	37°32′N	45°05′E
Orūmīyeh, Daryācheh-ye, lk., Iran	227	37°40′N	45°30′E
Oruro, Bol. (ô-rōō′rō)	168-69	17°58′S	67°07′W
Orust, i., Swe.	192-93	58°10′N	11°38′E
Orvieto, Italy (ôr-vyå′tō)	200-01	42°44′N	12°06′E
Orxon, stm., China	240-41	48°56′N	117°46′E
Osa, Península de, pen., C.R. (pĕ-nê′n-sōō-lä ō′sä)	149	8°34′N	83°31′W
Osage, Ia., U.S. (ō′sāj)	114-15	43°17′N	92°49′W
Osage, stm., Mo., U.S. (ō′sāj)	120-21	38°36′N	91°56′W
Ōsaka, Japan (ō′sä-kä)	245	34°41′N	135°31′E
Ōsaka-wan, b., Japan (ō′sä-kä wän)	245	34°30′N	135°18′E
Osakis, Mn., U.S. (ō-sā′kĭs)	114-15	45°52′N	95°09′W
Osceola, Ar., U.S. (ŏs-ê-ō′lå)	124-25	35°43′N	89°58′W
Osceola, Ia., U.S. (ŏs-ê-ō′lå)	114-15	41°02′N	93°46′W
Oscoda, Mi., U.S. (ŏs-kō′dȧ)	116-17	44°25′N	83°19′W
Osh, Kyrg. (ôsh)	232-33	40°32′N	72°48′E
Oshakati, Nmb.	264-65	17°47′S	15°41′E
Oshawa, On., Can. (ŏsh′ȧ-wȧ)	136-37	43°54′N	78°51′W
Ō-shima, i., Japan (ō′shē′mä)	245	34°44′N	139°25′E
Oshkosh, Ne., U.S. (ŏsh′kŏsh)	114-15	41°25′N	102°21′W
Oshkosh, Wi., U.S. (ŏsh′kŏsh)	116-17	44°00′N	88°33′W
Oshogbo, Nig.	260a	7°46′N	4°33′E
Oshwe, D.R.C.	262-63	3°22′S	19°30′E
Osijek, Cro. (ôs′ĭ-yĕk)	200-01	45°33′N	18°42′E
Osipenko, Ukr. see Berdians'k	202-03	46°45′N	36°49′E
Oskaloosa, Ia., U.S. (ŏs-kȧ-lōō′sȧ)	114-15	41°18′N	92°39′W
Oskarshamn, Swe.	192-93	57°16′N	16°29′E
Oskarström, Swe. (ôs′kärs-strûm)	192-93	56°48′N	12°58′E
Öskemen, Kaz.	226	49°57′N	82°38′E
Oslo, nat. cap., Nor. (ôs′lō)	192-93	59°55′N	10°45′E
Osmānābād, India	236	18°10′N	76°02′E
Osnabrück, Ger. (ôs-nä-brük′)	194-95	52°17′N	8°03′E
Osorno, Chile (ô-sō′r-nō)	171	40°35′S	73°07′W
Ossa, Mount, mtn., Austl. (mount ŏsȧ)	276	41°54′S	146°01′E
Osse, stm., Nig.	260a	5°55′N	5°16′E
Osseo, Wi., U.S. (ŏs′sē-ō)	114-15	44°35′N	91°14′W
Ossining, N.Y., U.S. (ŏs′ĭ-nĭng)	116-17	41°10′N	73°52′W
Ossipee, N.H., U.S. (ŏs′ĭ-pê)	116-17	43°42′N	71°07′W
Ossora, Russia	218-19	59°18′N	163°09′E
Ostashkov, Russia (ôs-täsh′kôf)	202-03	57°08′N	33°08′E
Ostende, Bel. see Oostende	190-91	51°14′N	2°55′E
Oster, Ukr. (ōs′tĕr)	202-03	50°57′N	30°53′E
Österreich, nation, Eur. see Austria	174-75	47°20′N	13°20′E
Östersjön, s., Eur. see Baltic Sea	192-93	57°0′N	19°00′E
Østersøen, s., Eur. see Baltic Sea	192-93	57°0′N	19°00′E
Östersund, Swe. (ûs′tĕr-sōōnd)	184-85	63°11′N	14°39′E
Östhammar, Swe. (ûst′häm′är)	192-93	60°15′N	18°22′E
Ostrau, Czech Rep. see Ostrava	194-95	49°50′N	18°17′E
Ostrava, Czech Rep.	194-95	49°50′N	18°17′E
Ostrogozhsk, Russia (ôs-tr-gôzhk′)	202-03	50°52′N	39°04′E
Ostrołęka, Pol. (ôs-trô-woN′kȧ)	194-95	53°05′N	21°35′E
Ostrov, Russia (ôs-trôf′)	192-93	57°21′N	28°20′E
Ostrowiec Świętokrzyski, Pol. (ōs-trō′vyĕts shvyĕN-tō-kzhĭ′ske)	194-95	50°56′N	21°24′E
Ostrów Mazowiecka, Pol. (ōs′trôf mä-zô-vyĕt′skä)	194-95	52°48′N	21°54′E
Ostrów Wielkopolski, Pol. (ōs′trōōf vyĕl-kō-pōl′skē)	194-95	51°39′N	17°48′E
Ostsee, s., Eur. see Baltic Sea	192-93	57°0′N	19°00′E
Ostuni, Italy (ôs-tōō′nē)	200-01	40°45′N	17°33′E
Ōsumi-shotō, is., Japan	244	30°29′N	130°39′E
Osuna, Spain (ô-sōō′nä)	198-99	37°14′N	5°06′W
Oswego, N.Y., U.S.	116-17	43°27′N	76°31′W
Otaru, Japan (ô′tȧ-rò)	244	43°12′N	140°60′E
Otavalo, Ec. (ōtä-vä′lō)	170	0°14′N	78°16′W
Othonoí, i., Grc.	200-01	39°51′N	19°24′E
Oti, stm., Afr.	260-61	8°30′N	0°06′E
Otjiwarongo, Nmb. (ŏt-jê-wȧ-rôn′gō)	264-65	20°27′S	16°38′E
Otočac, Cro. (ō′tō-chȧts)	200-01	44°52′N	15°13′E
Otoskwin, stm., On., Can.	134-35	52°11′N	87°45′W
Otra, stm., Nor.	192-93	58°09′N	8°01′E
Ótranto, Italy (ô′trän-tô) (ō-trän′tō)	200-01	40°09′N	18°28′E
Otranto, Canale d', strt., Eur. see Otranto, Strait of	200-01	40°0′N	19°00′E
Otranto, Strait of, strt., Eur. (strāt ŭv ō′trän-tô) (ô-trän′tō)	200-01	40°0′N	19°00′E
Otsego, Mi., U.S. (ŏt-sē′gō)	116-17	42°28′N	85°41′W
Ōtsu, Japan (ō′tsò)	245	35°00′N	135°52′E
Ottawa, Il., U.S. (ŏt′ȧ-wȧ)	116-17	41°21′N	88°51′W
Ottawa, Ks., U.S. (ŏt′ȧ-wȧ)	120-21	38°37′N	95°16′W
Ottawa, Oh., U.S. (ŏt′ȧ-wȧ)	116-17	41°01′N	84°02′W
Ottawa, nat. cap., On., Can. (ŏt′ȧ-wȧ)	136-37	45°25′N	75°41′W
Ottawa, stm., Can. (ŏt′ȧ-wȧ)	136-37	45°20′N	73°58′W
Ottawa Islands, is., Nu., Can.	130-31	59°30′N	80°10′W
Ottumwa, Ia., U.S. (ô-tŭm′wȧ)	114-15	41°01′N	92°24′W
Otway, Cape, c., Austl. (kāp ŏt′wā)	276	38°51′S	143°30′E
Otwock, Pol. (ôt′vôtsk)	194-95	52°07′N	21°16′E
Ou, stm., Laos	246-47	20°03′N	102°13′E
Ouachita Mountains, mts., U.S. (wŏsh′ĭ-tô moun′tĭnz)	120-21	34°40′N	94°25′W
Ouagadougou, nat. cap., Burkina (wä′gȧ-dōō′gōō)	260-61	12°23′N	1°32′W
Ouahigouya, Burkina (wä-ê-gōō′yä)	260-61	13°35′N	2°25′W
Ouahran, Alg. see Oran	198-99	35°41′N	0°39′W
Ouaka, stm., C.A.R.	262-63	4°59′N	19°56′E
Ouandja, stm., C.A.R.	262-63	9°34′N	21°39′E
Ouara, stm., C.A.R.	262-63	5°06′N	24°29′E
Ouarâne, reg., Maur.	258-59	21°0′N	10°30′W
Ouargla, Alg.	188-89	31°56′N	5°22′E
Ouarzazate, Mor.	258-59	30°56′N	6°54′W
Oubangui, stm., Afr. (ōō-bäN′gē)	262-63	0°25′S	17°47′E
Oudtshoorn, S. Afr. (outs′hōrn)	264-65	33°36′S	22°12′E
Oued-Zem, Mor. (wĕd-zĕm′)	269a	32°52′N	6°35′W
Ouémé, stm., Benin	260a	6°27′N	2°33′E
Ouesso, Congo	262-63	1°37′N	16°04′E
Ouezzane, Mor. (wĕ-zan′)	269a	34°48′N	5°34′W
Ouham, stm., Afr.	260-61	9°17′N	18°16′E
Oujda, Mor.	184-85	34°41′N	1°54′W
Oulu, Fin. (ō′lò)	184-85	65°01′N	25°28′E
Oulujärvi, lk., Fin.	184-85	64°18′N	27°08′E
Oulujoki, stm., Fin.	184-85	65°01′N	25°29′E
Oum er Rbia, Oued, stm., Mor.	269a	33°20′N	8°20′W
Ouray, Co., U.S. (ōō-rā′)	118-19	38°02′N	107°40′W
Ourense, Spain	198-99	42°20′N	7°52′W
Ourinhos, Braz. (ôò-rē′nyôs)	168-69	22°59′S	49°52′W
Ouro Fino, Braz. (ōū-rô-fē′nō)	172	22°17′S	46°22′W
Ouro Preto, Braz. (ō′rò prā′tò)	172	20°23′S	43°30′W
Outaouais, stm., Can. see Ottawa	136-37	45°20′N	73°58′W
Outardes, stm., Qc., Can.	138-39	49°03′N	68°30′W
Outlook, Sk., Can.	134-35	51°30′N	107°03′W
Ouvéa, i., N. Cal.	279g	20°33′S	166°34′E
Ouyen, Austl. (ōō-ĕn)	276	35°04′S	142°19′E
Ovalau, i., Fiji	279f	17°40′S	178°48′E
Ovalle, Chile (ō-väl′yä)	168-69	30°36′S	71°12′W
Ovar, Port. (ô-vär′)	198-99	40°52′N	8°37′W
Övertorneå, Swe.	184-85	66°23′N	23°39′E
Oviedo, Spain (ō-vê-ā′dhō)	198-99	43°22′N	5°51′W
Owando, Congo	262-63	0°29′S	15°55′E
Owase, Japan (ō′wä-shê)	245	34°04′N	136°12′E
Owego, N.Y., U.S. (ô-wē′gō)	116-17	42°07′N	76°16′W
Owen, Wi., U.S. (ō′ĕn)	114-15	44°57′N	90°33′W
Owen Falls Dam, d., Ug. see Nalubaale Dam	267	0°27′N	33°11′E
Owensboro, Ky., U.S. (ō′ĕnz-bŭr-ô)	116-17	37°46′N	87°06′W
Owen Sound, On., Can. (ō′ĕn sound)	136-37	44°34′N	80°56′W
Owen Stanley Range, mts., Pap. N. Gui. (ō′ĕn stăn′lē rānj)	277	9°20′S	147°55′E
Owensville, Mo., U.S. (ō′ĕnz-vĭl)	120-21	38°21′N	91°30′W
Owenton, Ky., U.S. (ō′ĕn-tŭn)	116-17	38°32′N	84°50′W
Owerri, Nig. (ô-wĕr′ê)	260a	5°29′N	7°01′E
Owo, Nig.	260a	7°12′N	5°35′E
Owosso, Mi., U.S. (ô-wŏs′ō)	116-17	42°60′N	84°10′W
Owyhee, stm., U.S. (ô-wī′hê)	112-13	43°48′N	117°02′W
Oxbow, Sk., Can.	134-35	49°14′N	102°11′W
Oxford, N.S., Can. (ŏks′fẽrd)	138-39	45°43′N	63°53′W
Oxford, Eng., U.K. (ŏks′fẽrd)	190-91	51°46′N	1°16′W
Oxford, Al., U.S. (ŏks′fẽrd)	124-25	33°36′N	85°50′W
Oxford, Ms., U.S. (ŏks′fẽrd)	124-25	34°21′N	89°33′W
Oxford, N.C., U.S. (ŏks′fẽrd)	124-25	36°19′N	78°35′W
Oxford, Oh., U.S. (ŏks′fẽrd)	116-17	39°30′N	84°45′W
Oxford Lake, lk., Mb., Can. (ŏks′fẽrd lāk)	134-35	54°49′N	95°29′W
Oxkutzcab, Mex. (ôx-kōō′tz-käb)	148	20°18′N	89°25′W
Oxnard, Ca., U.S. (ŏks′närd)	118-19	34°12′N	119°11′W
Oxus, stm., Asia see Amu Darya	226	44°14′N	59°41′E
Oyapok, stm., S.A. (ō-yȧ-pŏk′)	164-65	4°10′N	51°37′W
Oyem, Gabon	260-61	1°36′N	11°35′E
Oyo, Nig. (ō′yō)	260a	7°50′N	3°56′E
Oyonnax, Fr. (ô-yō-nȧks′)	196-97	46°16′N	5°39′E
Oyyl, stm., Kaz.	226	48°33′N	52°25′E
Ozamis, Phil.	250	8°09′N	123°49′E
Ozark, Al., U.S. (ō′zärk)	124-25	31°28′N	85°38′W
Ozark, Ar., U.S. (ō′zärk)	120-21	35°29′N	93°50′W
Ozark, Mo., U.S. (ō′zärk)	120-21	37°01′N	93°12′W
Ozark Plateau, plat., U.S. (ō′zärk plä-tō′)	120-21	37°0′N	93°00′W
Ozarks, Lake of the, res., Mo., U.S. (lāk ŭv thȧ ō′zärksz)	120-21	38°06′N	92°44′W
Ozernovskiy, Russia	218-19	51°30′N	156°31′E
Ozery, Russia (ô-zyô′rê)	202-03	54°51′N	38°33′E
Ozorków, Pol. (ô-zôr′kôf)	194-95	51°58′N	19°18′E

P

Feature (Pronunciation)	Page	Lat.	Long.
Paama, i., Vanuatu	279g	16°29′S	168°14′E
Paarl, S. Afr. (pärl)	264-65	33°44′S	18°58′E
Pabianice, Pol. (pä-byȧ-nē′tsĕ)	194-95	51°40′N	19°21′E
Pābna, Bngl.	234-35	24°00′N	89°14′E
Pacaraima, Serra, mts., S.A. (sĕr′rȧ pä-kä-rä-ē′mȧ) see Pakaraima Mountains	164-65	5°06′N	60°39′W
Pacaraima, Sierra de, mts., S.A. see Pakaraima Mountains	164-65	5°06′N	60°39′W
Pacasmayo, Peru (pä-käs-mä′yō)	170	7°24′S	79°33′W
Pachmarhi, India	234-35	22°28′N	78°26′E
Pachuca de Soto, Mex.	146-47	20°06′N	98°45′W
Pacific Ocean, oc., (pȧ-sĭf′ĭk ōshŭn)	20-21	10°0′S	150°00′W

ăt; finȧl; rāte; senåte; ärm; åsk; sofȧ; fâre; ch-choose; dh-as th in other; bē; ēvent; bĕt; recĕnt; cratẽr; g-gō; gh-guttural g; bĭt; ĭ-short neutral; rīde; ĸ-guttural k as ch in German ich;

Feature (Pronunciation)	Page	Lat.	Long.
Pacific Ranges, mts., B.C., Can. (pȧ-sĭf´ĭk rānj̆ĕz)	132-33	51°11'N	125°33'W
Pacific Rim National Park Reserve, n.p., B.C., Can. (pȧ-sĭf´ĭk rĭm năsh´ŭn-ăl pärk rĭ-zûrv´)	132-33	48°45'N	125°06'W
Padang, Indon.	246-47	0°57'S	100°22'E
Padang, Indon. (pä-däng´)	246-47	1°39'S	108°55'E
Padangsidempuan, Indon.	246-47	1°23'N	99°16'E
Paden City, W.V., U.S. (pā´dĕn sĭ´tĭ)	116-17	39°37'N	80°51'W
Paderborn, Ger. (pä-dĕr-bôrn´)	194-95	51°43'N	8°45'E
Padma, stm., Asia *see* Ganges	234-35	21°58'N	90°57'E
Padova, Italy (pä´dô-vä)	200-01	45°24'N	11°52'E
Padre Island, i., Tx., U.S. (pä´drā ī´lănd)	122-23	27°01'N	97°23'W
Padua, Italy (păd´û-ȧ) *see* Padova	200-01	45°24'N	11°52'E
Paducah, Ky., U.S.	124-25	37°05'N	88°37'W
Paektu-san, mtn., Asia (pâk´tōō-sän´)	243	41°59'N	128°07'E
Pagadian, Phil.	250	7°50'N	123°25'E
Pagalu, i., Eq. Gui. *see* Annobón	260-61	1°26'S	5°37'E
Pagan, i., N. Mar. Is.	280-81	18°07'N	145°46'E
Pago Pago, nat. cap., Am. Sam. (päŋ´-gō päŋ´-gō)	279b	14°16'S	170°42'W
Pagosa Springs, Co., U.S. (pȧ-gō´sȧ sprĭngz)	118-19	37°16'N	107°02'W
Pāhala, Hi., U.S. (pä-hä´lä)	127a	19°12'N	155°28'W
Pahang, stm., Malay.	246-47	3°30'N	103°24'E
Pahlevī, Iran *see* Bandar-e Anzalī	227	37°28'N	49°28'E
Paide, Est. (pī´dĕ)	192-93	58°54'N	25°35'E
Päijänne, lk., Fin. (pĕ´ē-yĕn-nĕ)	192-93	61°35'N	25°30'E
Painesville, Oh., U.S. (pānz´vĭl)	116-17	41°43'N	81°15'W
Painted Desert, des., Az., U.S. (pānt´ĕd dĕs´ērt)	118-19	35°45'N	111°07'W
Paintsville, Ky., U.S. (pānts´vĭl)	116-17	37°48'N	82°49'W
Paisley, Scot., U.K. (pāz´lĭ)	190-91	55°51'N	4°25'W
Paita, Peru (pä-ē´tä)	170	5°06'S	81°06'W
Pajala, Swe.	184-85	67°13'N	23°23'E
Pakaraima Mountains, mts., S.A.	164-65	5°06'N	60°39'W
Pakistan, nation, Asia (pä´-kĭ-stän)	206-07	30°0'N	70°00'E
Pakistan, East, nation, Asia *see* Bangladesh	206-07	24°0'N	90°00'E
Pakokku, Mya. (pȧ-kŏk´kò)	246-47	21°19'N	95°06'E
Paks, Hung. (pôksh)	194-95	46°39'N	18°53'E
Pak Sane, Laos *see* Muang Pakxan	246-47	18°25'N	103°39'E
Pakxé, Laos	246-47	15°08'N	105°48'E
Pala, Chad	262-63	9°21'N	14°54'E
Palacios, Tx., U.S. (pä-lä´syōs)	122-23	28°42'N	96°13'W
Palaiseau, Fr. (pȧ-lĕ-zō´)	196-97	48°43'N	2°16'E
Palana, Russia	218-19	59°05'N	159°59'E
Palangkaraya, Indon.	248-49	2°10'S	113°54'E
Palani, India	236	10°27'N	77°31'E
Pālanpur, India (pä´lŭn-pōōr)	234-35	24°10'N	72°27'E
Palapye, Bots. (pȧ-läp´yĕ)	264-65	22°34'S	27°07'E
Palatka, Russia	218-19	60°06'N	150°57'E
Palatka, Fl., U.S. (pȧ-lăt´kȧ)	124-25	29°39'N	81°39'W
Palau, nation, Oc. (pä-lä´ò)	280-81	5°0'N	137°00'E
Palauig, Phil. (pȧ-lou´ĕg)	250	15°26'N	119°56'E
Palawan, i., Phil. (pä-lä´wan)	250	9°30'N	118°30'E
Paldiski, Est. (päl´dĭ-skĭ)	192-93	59°20'N	24°06'E
Palembang, Indon. (pä-lĕm-bäng´)	246-47	2°58'S	104°46'E
Palencia, Spain (pä-lĕ´n-syä)	198-99	42°01'N	4°32'W
Palenque, Mex. (pä-lĕŋ´kå)	148	17°31'N	91°57'W
Palenque, hist., Mex.	148	17°30'N	91°60'W
Palenque, Punta, c., Dom. Rep. (pōō´n-tä pä-lĕŋ´kå)	142-43	18°15'N	70°09'W
Palermo, Italy (pä-lĕr´mô)	200-01	38°07'N	13°21'E
Palesse, reg., Eur. *see* Pripet Marshes	194-95	52°0'N	27°00'E
Palestine, Tx., U.S. (păl´ĕs-tīn)	122-23	31°45'N	95°38'W
Paletwa, Mya. (pŭ-lĕt´wä)	246-47	21°18'N	92°51'E
Pālghāt, India	236	10°46'N	76°39'E
Pāli, India	234-35	25°47'N	73°20'E
Palikir, nat. cap., Micron.	280-81	6°58'N	158°13'E
Pālitāna, India	234-35	21°31'N	71°49'E
Palizada, Mex. (pä-lē-zä´dä)	148	18°15'N	92°05'W
Palk Strait, strt., Asia (pôk strāt)	236	10°0'N	79°45'E
Palliser, Cape, c., N.Z.	278	41°37'S	175°17'E
Palma de Mallorca, Spain	198-99	39°34'N	2°39'E
Palmares, Braz. (päl-mä´rĕs)	163d	8°41'S	35°36'W
Palmas, Braz. (päl´mäs)	168-69	26°30'S	52°01'W
Palmas, Braz.	166-67	10°06'S	48°20'W
Palma Soriano, Cuba (päl´mä-sô-rē-ä´nō)	142-43	20°13'N	75°59'W
Palmeira dos Índios, Braz. (pä-mä´rä-dôs-ē´n-dyôs)	163d	9°25'S	36°37'W
Palmeirinhas, Ponta das, c., Ang.	264-65	9°05'S	12°60'E
Palmer, Ak., U.S. (päm´ēr)	126	61°32'N	149°05'W
Palmerston, at., Cook Is.	280-81	18°03'S	163°10'W
Palmerston, Cape, c., Austl.	277	21°33'S	149°28'E
Palmerston North, N.Z. (päm´ēr-stŭn nôrth)	278	40°21'S	175°37'E
Palmetto, Fl., U.S. (pȧl-mĕt´ō)	125a	27°31'N	82°35'W
Palmi, Italy (päl´mē)	200-01	38°21'N	15°51'E
Palmira, Col. (päl-mē´rä)	163c	3°33'N	76°18'W
Palm Springs, Ca., U.S.	118-19	33°50'N	116°32'W
Palmyra, Syria *see* Tudmur	228-29	34°33'N	38°17'E
Palmyra, Mo., U.S. (păl-mī´rȧ)	120-21	39°48'N	91°31'W
Palmyra, N.Y., U.S. (păl-mī´rȧ)	116-17	43°03'N	77°14'W
Palmyra Atoll, at., Oc.	280-81	5°51'N	162°05'W
Palo Alto, Ca., U.S. (pă´lō äl´tō)	118-19	37°26'N	122°08'W
Paloe, Pulau, i., Indon.	248-49	8°20'S	121°43'E
Palopo, Indon.	248-49	3°00'S	120°11'E
Palos, Cabo de, c., Spain (kä´bô-dĕ-pä´lôs)	198-99	37°38'N	0°41'W
Palu, Indon.	248-49	0°54'S	119°52'E
Palu, Tur. (pä-loo´)	227	38°41'N	39°60'E
Paluan, Phil. (pä-lōō´än)	250	13°26'N	120°27'E
Pāmban Island, i., India	236	9°16'N	79°19'E
Pamekasan, Indon.	248-49	7°10'S	113°29'E
Pamiers, Fr. (pȧ-myā´)	196-97	43°07'N	1°36'E
Pamir, mts., Asia *see* Pamirs	232-33	38°0'N	73°00'E
Pāmīr, Daryā-ye, mts., Asia *see* Pamirs	232-33	38°0'N	73°00'E
Pamirs, mts., Asia (pä-mērz)	232-33	38°0'N	73°00'E
Pamlico Sound, strt., N.C., U.S. (păm´lĭ-kō sound)	124-25	35°20'N	75°55'W
Pampa, Tx., U.S. (păm´pȧ)	120-21	35°32'N	100°58'W
Pampa, reg., Arg. (päm´pȧ) *see* Pampas	173	35°0'S	63°00'W
Pampanga, stm., Phil. (päm-päŋ´gä)	250	14°46'N	120°39'E
Pampas, reg., Arg. (päm´päs)	173	35°0'S	63°00'W
Pampas, stm., Peru	170	13°25'S	73°13'W
Pampeluna, Spain *see* Pamplona	198-99	42°49'N	1°39'W
Pamplona, Col. (päm-plō´nä)	164-65	7°22'N	72°38'W
Pamplona, Spain (päm-plô´nä)	198-99	42°49'N	1°39'W
Pana, Il., U.S. (pā´nȧ)	116-17	39°23'N	89°05'W
Panagyurishte, Blg. (pȧ-nȧ-gyōō´rêsh-tĕ)	200-01	42°30'N	24°12'E
Panaitan, Pulau, i., Indon.	248-49	6°36'S	105°12'E
Panaji, India	236	15°30'N	73°50'E
Panama, nation, N.A. (păn-ȧ-mä´ sĭ´tĭ)	85	9°0'N	80°00'W
Panamá, nat. cap., Pan. (păn-ȧ-mä´)	150	8°58'N	79°32'W
Panamá, Golfo de, b., Pan.	150	8°0'N	79°30'W
Panama, Gulf of, b., Pan. *see* Panamá, Golfo de	150	8°0'N	79°30'W
Panama, Isthmus of, isth., Pan. *see* Panamá, Istmo de	150	9°0'N	80°00'W
Panamá, Istmo de, isth., Pan.	150	9°0'N	80°00'W
Panama Canal, can., Pan.	150	9°23'N	79°56'W
Panama City, Fl., U.S. (păn-ȧ-mä´ sĭ´tĭ)	124-25	30°10'N	85°40'W
Panay, i., Phil. (pä-nī´)	250	11°15'N	122°30'E
Panay Gulf, b., Phil.	250	10°15'N	122°15'E
Pančevo, Serb. (pàn´chĕ-vô)	200-01	44°53'N	20°40'E
Panevėžys, Lith. (pä´nyĕ-vāzh´ēs)	192-93	55°44'N	24°23'E
Pangani, stm., Tan. (pän-gä´nē)	262-63	5°24'S	38°57'E
Pangkalanbuun, Indon.	248-49	2°42'S	111°38'E
Pangkalpinang, Indon. (päng-käl´pē-näng´)	246-47	2°08'S	106°06'E
Pangnirtung, Nu., Can.	128-29	66°08'N	65°43'W
Pangong Tso, lk., Asia	234-35	33°45'N	78°42'E
Panguitch, Ut., U.S. (pän´gwĭch)	118-19	37°50'N	112°26'W
Pangutaran Group, is., Phil.	250	6°14'N	120°39'E
Panhame, stm., Afr. *see* Manyame	264-65	15°37'S	30°39'E
Pānīpat, India	234-35	29°23'N	76°58'E
Panj, stm., Asia	232-33	37°00'N	68°16'E
Panjgūr, Pak.	232-33	26°58'N	64°05'E
Panjim, India *see* Panaji	236	15°30'N	73°50'E
Panna, India	234-35	24°43'N	80°11'E
Pannirtuuq, Nu., Can. *see* Pangnirtung	128-29	66°08'N	65°43'W
Pantar, Pulau, i., Indon. (pōō-lou pän´tär)	248-49	8°25'S	124°07'E
Pantelleria, Isola di, i., Italy (ē´sō-lä-dē-pän-tĕl-lå-rē´ä)	200-01	36°47'N	12°00'E
Pante Makasar, E. Timor	248-49	9°13'S	124°21'E
Pánuco, Mex. (pä´nōō-kô)	146-47	22°02'N	98°11'W
Pánuco, stm., Mex. (pä´nōō-kô)	146-47	22°16'N	97°47'W
Panxian, China	238-39	25°49'N	104°35'E
Panzós, Guat. (pän-zōs´)	148	15°24'N	89°39'W
Paoli, In., U.S. (pȧ-ō´lĭ)	116-17	38°33'N	86°28'W
Pápa, Hung. (pä´pô)	194-95	47°20'N	17°28'E
Papagayo, Golfo de, b., C.R. (gôl-fô-dĕ-pä-pä-gá´yō)	149	10°42'N	85°50'W
Papantla de Olarte, Mex. (pä-pän´tlä dā-ô-lä´r-tĕ)	146-47	20°27'N	97°19'W
Papeete, nat. cap., Fr. Poly. (pä-pē´-tē)	279d	17°32'S	149°34'W
Papenburg, Ger. (päp´ĕn-bòrgh)	194-95	53°06'N	7°24'E
Papua, Gulf of, b., Pap. N. Gui. (gŭlf ŭv păp-ōō-ȧ)	277	8°30'S	145°00'E
Papua New Guinea, nation, Oc. (păp-ōō-ȧ nū gĭne)	277	6°0'S	147°00'E
Papudo, Chile (pä-pōō´dô)	163e	32°31'S	71°28'W
Papun, Mya.	246-47	18°04'N	97°27'E
Pará, Braz. *see* Belém	166-67	1°27'S	48°29'W
Pará, state, Braz.	166-67	4°0'S	53°00'W
Pará, stm., Braz.	166-67	1°29'S	48°49'W
Paraburdoo, Austl.	270-71	23°12'S	117°44'E
Paracatu, Braz. (pä-rä-kä-tōō´)	172	17°14'S	46°52'W
Paracatu, stm., Braz.	168-69	16°35'S	45°06'W
Paracel Islands, is., China	224-25	15°46'N	112°17'E
Paraćin, Serb. (pá´rä-chēn)	200-01	43°52'N	21°25'E
Pāradwīp, India	234-35	20°17'N	86°41'E
Paragould, Ar., U.S. (păr´ȧ-gōōld)	124-25	36°04'N	90°30'W
Paraguá, stm., Bol.	166-67	13°32'S	61°49'W
Paragua, stm., Ven.	164-65	6°56'N	62°55'W
Paraguaçu, stm., Braz. (pä-rä-gwä-zōō´)	166-67	12°50'S	38°48'W
Paraguay, stm., S.A. (pä-rä-gwä´y)	173	27°19'S	58°36'W
Paraguai, stm., S.A. *see* Paraguay	173	27°19'S	58°36'W
Paraguaná, Península de, pen., Ven.	164-65	11°56'N	70°03'W
Paraguarí, Para.	168-69	25°37'S	57°09'W
Paraguay, nation, S.A. (păr´ȧ-gwā)	158	23°0'S	58°00'W
Parahyba, Braz. *see* João Pessoa	163d	7°07'S	34°52'W
Paraíba, Braz. *see* João Pessoa	163d	7°07'S	34°52'W
Paraíba, state, Braz. (pä-rä-ē´bä)	163d	7°15'S	36°30'W
Paraíba do Sul, stm., Braz.	172	21°37'S	41°02'W
Paraíso, Mex.	146-47	18°23'N	93°14'W
Paraiso, Pan. (pä-rä-ē´sō)	150	9°03'N	79°38'W
Parakou, Benin (pȧ-rȧ-kōō´)	260-61	9°20'N	2°37'E
Paramaribo, nat. cap., Sur. (pȧ-rȧ-má´rê-bō)	164-65	5°49'N	55°10'W
Paramirim, Braz.	166-67	13°27'S	42°14'W
Paramushir, Ostrov, i., Russia	218-19	50°25'N	155°50'E
Paraná, Arg. (pä-ä-nä´)	173	31°44'S	60°31'W
Paranã, Braz.	166-67	12°33'S	47°52'W
Paraná, state, Braz.	168-69	24°0'S	51°00'W
Paranã, stm., Braz.	166-67	12°30'S	48°14'W
Paraná, stm., S.A. (pä-ä-nä´)	168-69	33°48'S	59°14'W
Paranaguá, Braz.	172	25°31'S	48°31'W
Paranaguá, Baía de, b., Braz.	172	25°27'S	48°22'W
Paranaíba, Braz. (pä-rä-nȧ-ē´bȧ)	168-69	19°41'S	51°11'W
Paranaíba, stm., Braz. (pä-rä-nȧ-ē´bȧ)	168-69	20°08'S	51°00'W
Paranapanema, stm., Braz. (pä-rä´nȧ´pä-nĕ-mä)	168-69	22°42'S	53°10'W
Paranavaí, Braz.	168-69	23°04'S	52°29'W
Parapara, Ven. (pä-rä-pä-rä)	163b	9°44'N	67°17'W
Paray-le-Monial, Fr. (pȧ-rĕ´lē-mô-nyäl´)	196-97	46°27'N	4°07'E
Pārbat, stm., India	234-35	25°51'N	76°33'E
Parbhani, India	236	19°16'N	76°46'E
Pardo, stm., Braz. (pär´dō)	166-67	15°39'S	38°57'W
Pardo, stm., Braz. (pär´dō)	172	20°09'S	48°37'W
Pardubice, Czech Rep. (pär´dò-bĭt-sĕ)	194-95	50°02'N	15°46'E
Parece Vela, i., Japan *see* Okino-Tori-shima	222-23	20°27'N	136°04'E
Parent, Qc., Can.	136-37	47°56'N	74°37'W
Parepare, Indon.	248-49	4°01'S	119°38'E
Paria, Golfo de, b., (gôl-fô-dĕ-br-pä-rê-ä) *see* Paria, Gulf of	140-41	10°20'N	62°00'W
Paria, Gulf of, b.,	140-41	10°20'N	62°00'W
Paricutín, vol., Mex.	146-47	19°28'N	102°15'W
Parima, Serra, mts., S.A. (sĕr´rȧ pä-rē´mȧ) *see* Parima, Sierra	164-65	3°24'N	64°10'W
Parima, Sierra, mts., S.A.	164-65	3°24'N	64°10'W
Pariñas, Punta, c., Peru (pōō´n-tä-pä-rē´n-yäs)	170	4°40'S	81°20'W
Parintins, Braz. (pä-rĭn-tĭnzh´)	166-67	2°37'S	56°45'W
Paris, Ar., U.S. (păr´ĭs)	120-21	35°18'N	93°44'W
Paris, Il., U.S. (păr´ĭs)	116-17	39°37'N	87°42'W
Paris, Ky., U.S. (păr´ĭs)	116-17	38°12'N	84°16'W
Paris, Mo., U.S. (păr´ĭs)	120-21	39°29'N	92°00'W
Paris, Tn., U.S. (păr´ĭs)	124-25	36°18'N	88°20'W
Paris, Tx., U.S. (păr´ĭs)	120-21	33°40'N	95°33'W
Paris, nat. cap., Fr. (pȧ-rē´)	196-97	48°52'N	2°21'E
Parita, Bahía de, b., Pan. (bä-ē´ä-dĕ-pä-rē´tä)	150	8°08'N	80°24'W
Park City, Ks., U.S. (pärk sĭ´tê)	120-21	37°48'N	97°18'W
Parker, Co., U.S. (pär´kẽr pärk)	120-21	39°31'N	104°46'W
Parker, S.D., U.S. (pär´kẽr pärk)	114-15	43°24'N	97°08'W

n-sing; ŋ-baŋk; ɴ-nasalized n; nŏd; cŏmmit; ōld; ôbey; ôrder; oi-boil; fōōd; ò-as oo in foot; ou-out; s-soft; sh-dish; th-thin; pūre; ûnite; ûrn; stŭd; circŭs; ü-as in French tu; ´-indeterminate vowel.

Feature (Pronunciation)	Page	Lat.	Long.
Parkersburg, W.V., U.S. (pär′kẽrz-bûrg)	116-17	39°15′N	81°33′W
Parkes, Austl. (pärks)	276	33°09′S	148°10′E
Park Falls, Wi., U.S. (pärk fôlz)	114-15	45°56′N	90°26′W
Park Range, mts., Co., U.S. (pärk rānj)	112-13	40°40′N	106°40′W
Park Rapids, Mn., U.S. (pärk răp′ĭdz)	114-15	46°55′N	95°04′W
Park River, N.D., U.S. (pärk rĭv′ẽr)	114-15	48°24′N	97°45′W
Parkston, S.D., U.S. (pärks′tŭn)	114-15	43°24′N	97°59′W
Parla, Spain (pär′lä)	198-99	40°14′N	3°46′W
Parlākimidi, India	236	18°47′N	84°06′E
Parma, Italy (pär′mä)	200-01	44°49′N	10°20′E
Parnaguá, Braz.	166-67	10°13′S	44°38′W
Parnaíba, Braz. (pär-nä-ē′bä)	166-67	2°54′S	41°47′W
Parnaíba, stm., Braz. (pär-nä-ē′bä)	166-67	2°46′S	41°50′W
Parnassós, mtn., Grc.	200-01	38°32′N	22°35′E
Pärnu, Est. (pĕr′nōō)	192-93	58°22′N	24°33′E
Paroo, stm., Austl. (pä′rōō)	276	30°23′S	143°59′E
Parowan, Ut., U.S. (păr′ô-wăn)	118-19	37°51′N	112°50′W
Parral, Chile (pär-rä′l)	171	36°09′S	71°50′W
Parramatta, Austl.	276	33°49′S	151°00′E
Parras de la Fuente, Mex.	122-23	25°27′N	102°10′W
Parrsboro, N.S., Can. (pärz′bŭr-ô)	138-39	45°25′N	64°20′W
Parry, Cape, c., N.T., Can.	130-31	70°08′N	124°24′W
Parry, Mount, mtn., B.C., Can. (mount păr′ĭ)	132-33	52°53′N	128°45′W
Parry Sound, On., Can. (păr′ĭ sound)	136-37	45°20′N	80°02′W
Parsnip, stm., B.C., Can. (pärs′nĭp)	132-33	55°10′N	123°02′W
Parsons, Ks., U.S. (pär′s′nz)	120-21	37°20′N	95°16′W
Parsons, W.V., U.S. (pär′s′nz)	116-17	39°06′N	79°41′W
Parthenay, Fr. (pär-t′nĕ′)	196-97	46°39′N	0°15′W
Partinico, Italy (pär-tē′nē-kô)	200-01	38°03′N	13°07′E
Paru, stm., Braz.	166-67	1°35′S	52°31′W
Parys, S. Afr. (pa-rīs′)	269c	26°54′S	27°28′E
Pasadena, Ca., U.S. (păs-à-dē′nà)	118-19	34°09′N	118°09′W
Pasaje, Ec.	170	3°20′S	79°48′W
Pa Sak, stm., Thai.	246-47	14°21′N	100°35′E
Pascagoula, Ms., U.S. (păs-kȧ-gōō′lȧ)	124-25	30°22′N	88°33′W
Pascagoula, stm., Ms., U.S. (păs-kȧ-gōō′lȧ)	124-25	30°22′N	88°37′W
Paşcani, Rom. (päsh-kän′)	194-95	47°15′N	26°44′E
Pasco, Wa., U.S. (păs′kō)	112-13	46°14′N	119°05′W
Pascua, Isla de, i., Chile	282-83	27°07′S	109°22′W
Pasni, Pak.	232-33	25°16′N	63°27′E
Paso de Indios, Arg.	171	43°51′S	68°56′W
Paso de los Libres, Arg. (pä-sô-dĕ-lôs-lē′brĕs)	173	29°42′S	57°09′W
Paso de los Toros, Ur. (pä-sô-dĕ-lôs tō′rôs)	173	32°49′S	56°31′W
Paso Robles, Ca., U.S. (pä′sō rō′blĕs)	118-19	35°38′N	120°41′W
Passaic, N.J., U.S. (pă-sā′ĭk)	116-17	40°52′N	74°08′W
Passau, Ger. (päsou)	194-95	48°34′N	13°27′E
Passero, Capo, c., Italy (kä′pō päs-sĕ′rô)	200-01	36°40′N	15°09′E
Passo Fundo, Braz. (pä′sō fōn′dò)	168-69	28°15′S	52°25′W
Passos, Braz. (pä′s-sōs)	172	20°43′S	46°37′W
Pastaza, stm., S.A. (päs-tä′zä)	170	4°55′S	76°24′W
Pasto, Col. (päs′tô)	164-65	1°12′N	77°16′W
Pasuruan, Indon.	248-49	7°38′S	112°54′E
Pasvalys, Lith. (päs-vä-lēs′)	192-93	56°04′N	24°24′E
Patagonia, reg., Arg. (păt-à-gō′nĭ-à)	171	44°0′S	68°00′W
Pātan, India	234-35	23°51′N	72°07′E
Pate Island, i., Kenya	262-63	2°06′S	41°03′E
Paterson, N.J., U.S. (păt′ẽr-sŭn)	116-17	40°55′N	74°10′W
Pathānkot, India	234-35	32°16′N	75°39′E
Pathein, Mya.	246-47	16°46′N	94°44′E
Pathfinder Reservoir, res., Wy., U.S. (păth′fīn-dẽr rĕ′sẽr-vwär)	112-13	42°25′N	106°55′W
Patiāla, India (pŭt-ē-ä′lŭ)	234-35	30°19′N	76°23′E
Pātkai Range, mts., Asia	238-39	27°0′N	96°00′E
Patna, India	234-35	25°36′N	85°07′E
Patnanongan Island, i., Phil. (pät-nä-nôŋ′gän ī′lȧnd)	250	14°48′N	122°11′E
Pato Branco, Braz.	168-69	26°14′S	52°41′W
Patos, Braz. (pä′tōzh)	163d	7°01′S	37°16′W
Patos, Lagoa dos, b., Braz. (lä′gō-à dozh pä′tōzh)	168-69	31°06′S	51°15′W
Patos de Minas, Braz. (pä′tōzh dĕ-mē′nàzh)	172	18°35′S	46°31′W
Patquía, Arg.	168-69	30°02′S	66°52′W
Pátra, Grc.	200-01	38°14′N	21°44′E
Patricio Lynch, Isla, i., Chile	171	48°37′S	75°26′W
Patrocínio, Braz. (pä-trō-sē′nē-ò)	172	18°56′S	46°60′W
Pattani, Thai. (pät′à-nē)	246-47	6°52′N	101°15′E
Patten, Me., U.S. (păt′′n)	117a	45°59′N	68°27′W
Patterson, La., U.S. (păt′ẽr-sŭn)	124-25	29°42′N	91°18′W
Patuca, stm., Hond.	149	15°48′N	84°18′W
Patuca, Punta, c., Hond. (pōō′n-tä-pä-tōō′kä)	149	15°49′N	84°18′W
Pátzcuaro, Mex. (päts′kwä-rô)	146-47	19°31′N	101°37′W
Pau, Fr. (pō)	196-97	43°18′N	0°22′W
Pauini, stm., Braz.	166-67	7°47′S	67°05′W
Pauk, Mya.	246-47	21°27′N	94°28′E
Paulding, Oh., U.S. (pôl′dĭng)	116-17	41°08′N	84°35′W
Paulis, D.R.C. *see* Isiro	262-63	2°46′N	27°37′E
Paulistana, Braz.	166-67	8°09′S	41°09′W
Paulo Afonso, Braz.	166-67	9°21′S	38°14′W
Paul Roux, S. Afr. (pôrl rōō)	269c	28°18′S	27°58′E
Pauls Valley, Ok., U.S. (pôlz văl′ē)	120-21	34°44′N	97°13′W
Paungde, Mya.	246-47	18°29′N	95°30′E
Pavia, Italy (pä-vē′ä)	200-01	45°12′N	9°10′E
Pavlodar, Kaz. (pav-lô-dar′)	226	52°17′N	76°59′E
Pavlovo, Russia	186-87	55°57′N	43°04′E
Pavuvu Island, i., Sol. Is.	279e	9°03′S	159°06′E
Pawan, stm., Indon.	248-49	1°51′S	109°56′E
Pawhuska, Ok., U.S. (pô-hŭs′kȧ)	120-21	36°40′N	96°20′W
Pawnee, Ok., U.S. (pô-nē′)	120-21	36°20′N	96°48′W
Pawnee, stm., Ks., U.S. (pô-nē′)	120-21	38°10′N	99°06′W
Pawnee City, Ne., U.S. (pô-nē′ sĭ′tē)	120-21	40°07′N	96°09′W
Paw Paw, Mi., U.S. (pô pô)	116-17	42°13′N	85°53′W
Pawtucket, R.I., U.S. (pô-tŭk′ĕt)	116-17	41°53′N	71°23′W
Paxton, Il., U.S. (păks′tŭn)	116-17	40°27′N	88°05′W
Payakumbuh, Indon.	246-47	0°14′S	100°38′E
Payette, Id., U.S. (pà-ĕt′)	112-13	44°05′N	116°56′W
Pay-Khey, Khrebet, mts., Russia	186-87	69°0′N	63°00′E
Paynesville, Mn., U.S. (pānz′vĭl)	114-15	45°23′N	94°43′W
Paysandú, Ur. (pī-sän-dōō′)	173	32°20′S	58°05′W
Payson, Az., U.S. (pā′s′n)	118-19	34°10′N	111°19′W
Pazardzhik, Blg. (pä-zar-dzhek′)	200-01	42°12′N	24°20′E
Peabody, Ks., U.S. (pē′bŏd-ĭ)	120-21	38°10′N	97°06′W
Peace, stm., Can. (pēs)	130-31	58°60′N	111°25′W
Peace, stm., Fl., U.S. (pēs)	125a	26°58′N	82°01′W
Peace River, Ab., Can. (pēs rĭv′ẽr)	132-33	56°15′N	117°16′W
Pearl, stm., U.S. (pûrl)	124-25	30°11′N	89°32′W
Pearland, Tx., U.S. (pûrl′ănd)	122-23	29°33′N	95°17′W
Pearl and Hermes Atoll, at., Hi., U.S.	127	27°55′N	175°45′W
Pearl Harbor, b., Hi., U.S. (pûrl här′bẽr)	127a	21°22′N	157°59′W
Pearsall, Tx., U.S. (pēr′sôl)	122-23	28°54′N	99°06′W
Pebble Island, i., Falk. Is.	171	51°20′S	59°34′W
Peçanha, Braz. (pà-kän′yȧ)	172	18°32′S	42°34′W
Pechenga, Russia (pyĕ′chĕŋ-gà)	184-85	69°34′N	31°14′E
Pechora, Russia	186-87	65°08′N	57°09′E
Pechora, stm., Russia (pyĕ-chô′rà)	186-87	67°59′N	53°56′E
Pechorskoye More, s., Russia	186-87	70°0′N	54°00′E
Pecos, Tx., U.S. (pā′kôs)	122-23	31°25′N	103°30′W
Pecos, stm., U.S. (pā′kôs)	110-11	29°42′N	101°22′W
Pécs, Hung. (pāch)	194-95	46°04′N	18°13′E
Pedernales, Ven.	143b	9°57′N	62°15′W
Pedra Azul, Braz. (pā′drä-zōō′l)	168-69	15°60′S	41°17′W
Pedreiras, Braz. (pĕ-drā′räs)	166-67	4°34′S	44°39′W
Pedro Afonso, Braz.	166-67	8°60′S	48°10′W
Pedro II, Braz. (pā′drò sà-gòn′dò)	166-67	4°25′S	41°28′W
Pedro Juan Caballero, Para. (pĕ′drò hòà′n-kä-bäl-yĕ′rō)	168-69	22°33′S	55°45′W
Peebles, Scot., U.K. (pē′b′lz)	190-91	55°39′N	3°12′W
Peekskill, N.Y., U.S. (pēks′kĭl)	116-17	41°17′N	73°55′W
Peel, stm., Can.	130-31	67°42′N	134°31′W
Pegasus Bay, b., N.Z. (pĕg′à-sŭs bā)	278	43°20′S	173°00′E
Pegu, Mya. *see* Bago	246-47	17°20′N	96°29′E
Peiching, nat. cap., China *see* Beijing	240-41	39°55′N	116°22′E
Peipsi järv, lk., Eur. *see* Peipus, Lake	192-93	58°45′N	27°25′E
Peipus, Lake, lk., Eur. (lāk pī′pŭs)	192-93	58°45′N	27°25′E
Peiraiás, Grc.	200-01	37°57′N	23°39′E
Peixe, stm., Braz.	168-69	21°30′S	51°57′W
Pekalongan, Indon.	248-49	6°53′S	109°40′E
Pekanbaru, Indon.	246-47	0°31′N	101°27′E
Pekin, Il., U.S. (pē′kĭn)	116-17	40°34′N	89°39′W
Peking, nat. cap., China *see* Beijing	240-41	39°55′N	116°22′E
Pelagie, Isole, is., Italy	184-85	35°40′N	12°40′E
Pelat, Mont, mtn., Fr. (môɴ pē-là′)	196-97	44°16′N	6°42′E
Peleduy, Russia (pyĕl-yĭ-dōō′ē)	218-19	59°39′N	112°44′E
Pelée, Montagne, vol., Mart. (môɴ-pē-lā′ pȧ-lē′)	143b	14°48′N	61°10′W
Pelee Island, i., On., Can. (pē′lē ī′lȧnd)	136-37	41°46′N	82°39′W
Peleliu, i., Palau *see* Beliliou	280-81	7°00′N	134°15′E
Peleng, Pulau, i., Indon.	248-49	1°15′S	123°08′E
Pelham, Ga., U.S. (pĕl′hăm)	124-25	31°08′N	84°09′W
Pelican Rapids, Mn., U.S. (pĕl′ĭ-kăn răp′ĭdz)	114-15	46°35′N	96°04′W
Pella, Ia., U.S. (pĕl′ȧ)	114-15	41°25′N	92°55′W
Pellworm, i., Ger. (pĕl′vòrm)	194-95	54°31′N	8°38′E
Pelly, stm., Yk., Can. (pĕl′ĭ)	130-31	62°46′N	137°20′W
Pelly Crossing, Yk., Can.	128-29	62°50′N	136°35′W
Pelly Mountains, mts., Yk., Can. (pĕl′ĭ moun′tĭnz)	130-31	62°0′N	133°00′W
Peloponnesus, pen., Grc.	200-01	37°30′N	22°00′E
Pelopónnisos, pen., Grc. *see* Peloponnesus	200-01	37°30′N	22°00′E
Pelotas, Braz. (pà-lō′tăzh)	173	31°45′S	52°19′W
Pelotas, stm., Braz.	168-69	27°28′S	51°54′W
Pematangsiantar, Indon.	246-47	2°57′N	99°04′E
Pemba, Moz. (pĕm′bȧ)	264-65	13°01′S	40°32′E
Pemba, i., Tan. (pĕm′bȧ)	262-63	5°10′S	39°48′E
Pemberton, Austl.	270-71	34°27′S	116°01′E
Pembina, N.D., U.S. (pĕm′bĭ-nȧ)	114-15	48°58′N	97°15′W
Pembina, stm., Ab., Can. (pĕm′bĭ-nȧ)	132-33	54°45′N	114°17′W
Pembroke, On., Can. (pĕm′brōk)	136-37	45°49′N	77°07′W
Pembroke, Cape, c., Nu., Can.	130-31	62°56′N	81°56′W
Pembuang, stm., Indon.	248-49	3°21′S	112°33′E
Peñalara, Pico de, mtn., Spain (pē′kō-dĕ-pā-nyä-lä′rä)	198-99	40°51′N	3°57′W
Penang, Malay. *see* George Town	246-47	5°25′N	100°20′E
Peñarroya-Pueblonuevo, Spain (pĕn-yär-rō′yä-pwĕ′blô-nwĕ′vô)	198-99	38°18′N	5°16′W
Peñas, Cabo de, c., Spain (kȧ′bô-dĕ-pā′nyäs)	198-99	43°39′N	5°51′W
Penas, Golfo de, b., Chile (gôl-fô-dĕ-pĕ′n-äs)	171	47°22′S	74°50′W
Pender, Ne., U.S. (pĕn′dẽr)	114-15	42°07′N	96°43′W
Pendjari, stm., Afr.	260-61	10°55′N	0°50′E
Pendleton, Or., U.S. (pĕn′d′l-tŭn)	112-13	45°40′N	118°48′W
Pend Oreille, Lake, lk., Id., U.S. (lăk pŏn-dô-rā′) (lăk pĕn-dô-rĕl′)	112-13	48°10′N	116°17′W
Penedo, Braz. (pā-nā′dò)	163d	10°16′S	36°35′W
Penetanguishene, On., Can. (pĕn′ē-tăŋ-gĭ-shēn′)	136-37	44°46′N	79°56′W
Penganga, stm., India	236	19°54′N	79°10′E
P'enghu, Tai. *see* Makung	225a	23°34′N	119°34′E
P'enghu Ch'üntao, is., Tai.	225a	23°30′N	119°30′E
Penglai, China (pŭŋ-lī)	240-41	37°48′N	120°43′E
Pengshui, China	238-39	29°18′N	108°09′E
Pengxian, China	238-39	30°59′N	103°56′E
Peniche, Port. (pĕ-nē′chȧ)	198-99	39°21′N	9°22′W
Penida, Nusa, i., Indon.	248-49	8°44′S	115°32′E
Pennines, mts., Eng., U.K. (pĕn-īn′)	190-91	54°11′N	2°02′W
Pennsylvania, state, U.S. (pĕn-sĭl-vā′nĭ-ȧ)	108-09	40°45′N	77°30′W
Penn Yan, N.Y., U.S. (pĕn yăn′)	116-17	42°40′N	77°03′W
Penobscot, stm., Me., U.S.	117a	44°29′N	68°48′W
Penobscot Bay, b., Me., U.S. (pē-nŏb′skŏt bā)	117a	44°15′N	68°52′W
Penola, Austl.	276	37°23′S	140°50′E
Penonomé, Pan.	150	8°31′N	80°22′W
Penrhyn, at., Cook Is.	280-81	9°0′S	158°00′W
Pensacola, Fl., U.S. (pĕn-sȧ-kō′lȧ)	124-25	30°25′N	87°13′W
Pensacola Mountains, mts., Ant.	287	84°21′S	47°02′W
Pensilvania, Col. (pĕn-sēl-vá′nyä)	163c	5°32′N	75°03′W
Pentecost Island, i., Vanuatu (pĕn′tē-kŏst ī′lȧnd) *see* Pentecôte	279g	15°42′S	168°10′E
Pentecôte, i., Vanuatu	279g	15°42′S	168°10′E
Penticton, B.C., Can.	132-33	49°30′N	119°35′W
Pentland Firth, strt., Scot., U.K. (pĕnt′lănd fûrth)	190-91	58°44′N	3°07′W
Penyu, Kepulauan, is., Indon.	248-49	5°22′S	127°46′E
Penza, Russia (pĕn′zȧ)	186-87	53°12′N	45°00′E
Penzance, Eng., U.K. (pĕn-zăns′)	190-91	50°07′N	5°33′W
Penzhina, stm., Russia (pyĭn-zē-nŭ)	218-19	62°29′N	165°15′E
People's Democratic Republic of Korea, nation, Asia *see* Korea, North	206-07	40°0′N	127°00′E
Peoria, Il., U.S. (pē-ō′rĭ-ȧ)	116-17	40°41′N	89°36′W
Peotone, Il., U.S. (pē′ŏ-tôn)	116-17	41°20′N	87°47′W
Pequeñas Antillas, is., *see* Lesser Antilles	143b	15°0′N	61°00′W
Perabumulih, Indon.	246-47	3°27′S	104°15′E
Perak, stm., Malay.	246-47	3°58′N	100°53′E
Perdido, Monte, mtn., Spain (mōn-tȧ-pĕr-dē′dō)	198-99	42°40′N	0°05′E
Pereira, Col. (pȧ-rā′rä)	163c	4°50′N	75°42′W
Pereslavl'-Zalesskiy, Russia (pȧ-rȧ-slàv′′l zȧ-lyĕs′kĭ)	202-03	56°44′N	38°51′E
Pergamino, Arg. (pĕr-gä-mē′nō)	173	33°54′S	60°35′W
Perham, Mn., U.S. (pẽr′hăm)	114-15	46°36′N	95°35′W
Péribonka, stm., Qc., Can.	130-31	48°46′N	72°03′W
Périgueux, Fr. (pā-rē-gû′)	196-97	45°11′N	0°43′E
Perito Moreno, Arg.	171	46°36′S	70°55′W
Perlas, Laguna de, b., Nic. (lä-gò′nä-dĕ-läs-pĕr′läs)	149	12°30′N	83°40′W
Perleberg, Ger. (pĕr′lĕ-bĕrg)	194-95	53°05′N	11°52′E

ăt; finăl; rāte; senāte; ärm; ȧsk; sofȧ; fâre; ch-choose; dh-as th in other; bē; ēvent; bĕt; recĕnt; cratẽr; g-gō; gh-guttural g; bĭt; ĭ-short neutral; rīde; ĸ-guttural k as ch in German ich;

Feature (Pronunciation)	Page	Lat.	Long.
Perm', Russia (pĕrm)	**186-87**	58°00'N	56°16'E
Pernambuco, Braz. *see* Recife	**163d**	8°03'S	34°54'W
Pernambuco, state, Braz. (pĕr-näm-bōō'kō)	**166-67**	8°0'S	37°00'W
Pernik, Blg. (pĕr-nēk')	**200-01**	42°37'N	23°03'E
Péronne, Fr. (pā-rôn')	**196-97**	49°56'N	2°56'E
Perote, Mex. (pĕ-rō'tĕ)	**146-47**	19°34'N	97°15'W
Perpignan, Fr. (pĕr-pē-nyäN')	**196-97**	42°42'N	2°53'E
Perros, Bahía de, strt., Cuba (bä-ē'ä-dĕ-pā'rōs)	**142-43**	22°21'N	78°31'W
Perry, Fl., U.S. (pĕr'ĭ)	**124-25**	30°07'N	83°35'W
Perry, Ga., U.S. (pĕr'ĭ)	**124-25**	32°28'N	83°44'W
Perry, Ia., U.S. (pĕr'ĭ)	**114-15**	41°51'N	94°07'W
Perry, N.Y., U.S. (pĕr'ĭ)	**116-17**	42°43'N	78°01'W
Perry, Ok., U.S. (pĕr'ĭ)	**120-21**	36°17'N	97°17'W
Perrysburg, Oh., U.S. (pĕr ĭz-bûrg)	**116-17**	41°34'N	83°37'W
Perryton, Tx., U.S. (pĕr'ĭ-tŭn)	**120-21**	36°23'N	100°49'W
Perryville, Mo., U.S. (pĕr-ĭ-vĭl)	**120-21**	37°43'N	89°52'W
Persepolis, hist., Iran (pēr-sĕpô-lĭs)	**230-31**	29°57'N	52°52'E
Persia, nation, Asia *see* Iran	**206-07**	32°0'N	53°00'E
Persian Gulf, b., Asia (pûr'zhăn gŭlf)	**230-31**	27°0'N	51°00'E
Perth, Austl. (pûrth)	**270-71**	31°57'S	115°51'E
Perth, On., Can. (pûrth)	**136-37**	44°55'N	76°15'W
Perth, Scot., U.K. (pûrth)	**190-91**	56°24'N	3°27'W
Perth Amboy, N.J., U.S. (pûrth ăm'boi)	**116-17**	40°31'N	74°14'W
Pertuis, Fr. (pĕr-tüē')	**196-97**	43°42'N	5°30'E
Peru, Il., U.S. (pē-rōō')	**116-17**	41°20'N	89°07'W
Peru, In., U.S. (pē-rōō')	**116-17**	40°45'N	86°03'W
Peru, nation, S.A. (pē-rōō')	**158**	10°0'S	76°00'W
Perugia, Italy (pā-rōō'jä)	**200-01**	43°07'N	12°22'E
Pervomais'k, Ukr.	**202-03**	48°03'N	30°51'E
Pervouralsk, Russia (pĕr-vô-ô-rälsk')	**218-19**	56°55'N	59°57'E
Pesaro, Italy (pā'zä-rō)	**200-01**	43°55'N	12°55'E
Pescadores, is., Tai. *see* P'enghu Ch'üntao	**225a**	23°30'N	119°30'E
Pescara, Italy (pās-kä'rä)	**200-01**	42°29'N	14°12'E
Peshāwar, Pak. (pĕ-shä'wŭr)	**232-33**	33°60'N	71°33'E
Peshtigo, Wi., U.S. (pĕsh'tē-gō)	**116-17**	45°03'N	87°44'W
Pesqueira, Braz.	**163d**	8°22'S	36°42'W
Pesyakov, Ostrov, i., Russia	**186-87**	68°45'N	57°41'E
Petacalco, Bahía, b., Mex. (bä-ē'ä-dĕ-pē-tä-käl'kô)	**146-47**	17°56'N	101°57'W
Petah Tiqwa, Isr.	**228-29**	32°06'N	34°54'E
Petaluma, Ca., U.S. (pĕt-á-lō'má)	**118-19**	38°14'N	122°38'W
Petare, Ven. (pĕ-tä'rĕ)	**163b**	10°29'N	66°49'W
Petatlán, Mex. (pā-tä-tlän')	**146-47**	17°31'N	101°16'W
Peterborough, Austl.	**276**	32°58'S	138°50'E
Peterborough, On., Can. (pē'tēr-bŭr-ô)	**136-37**	44°18'N	78°20'W
Peterhead, Scot., U.K. (pē-tēr-hĕd')	**190-91**	57°31'N	1°47'W
Peter Pond Lake, lk., Sk., Can. (pē'tēr pŏnd lăk)	**134-35**	56°06'N	109°06'W
Petersburg, Ak., U.S. (pē'tērz-bûrg)	**126**	56°48'N	132°57'W
Petersburg, Il., U.S. (pē'tērz-bûrg)	**120-21**	40°00'N	89°50'W
Petersburg, In., U.S. (pē'tērz-bûrg)	**116-17**	38°29'N	87°17'W
Petersburg, Va., U.S. (pē'tērz-bûrg)	**124-25**	37°14'N	77°24'W
Petersburg, W.V., U.S. (pē'tērz-bûrg)	**116-17**	38°59'N	79°08'W
Peter the Great Bay, b., Russia *see* Petra Velikogo, Zaliv	**244**	42°40'N	132°00'E
Petitcodiac, N.B., Can. (pē-tē-kô-dyák')	**138-39**	45°56'N	65°11'W
Petit-Goâve, Haiti (pē-tē' gô-áv')	**142-43**	18°26'N	72°51'W
Petlalcingo, Mex. (pĕ-tläl-sĕŋ'gô)	**146-47**	18°04'N	97°56'W
Peto, Mex. (pĕ'tô)	**148**	20°08'N	88°54'W
Petorca, Chile (pā-tōr'kä)	**163e**	32°15'S	70°57'W
Petoskey, Mi., U.S. (pē-tŏs-kĭ)	**116-17**	45°22'N	84°57'W
Petra Velikogo, Zaliv, b., Russia	**244**	42°40'N	132°00'E
Petrich, Blg. (pā'trĭch)	**200-01**	41°24'N	23°13'E
Petrified Forest National Park, n.p., Az., U.S. (pĕt'rĭ-fīd fŏr'ĕst năsh'ŭn-ăl pärk)	**118-19**	34°54'N	109°47'W
Petrinja, Cro. (pā'trēn-yá)	**200-01**	45°27'N	16°16'E
Petrodvorets, Russia (pyĕ-trô-dvô-ryĕts')	**192-93**	59°53'N	29°52'E
Petrolia, On., Can. (pē-trō'lĭ-á)	**136-37**	42°53'N	82°09'W
Petrolina, Braz. (pĕ-trō-lē'ná)	**166-67**	9°24'S	40°30'W
Petropavlovsk, Kaz. (pyĕ-trô-päv'lôvsk)	**226**	54°52'N	69°09'E
Petropavlovsk-Kamchatskiy, Russia (pyĕ-trô-päv'lôvsk käm-chät'skĭ)	**218-19**	53°01'N	158°41'E
Petrópolis, Braz. (på-trô-pô-lēzh')	**172**	22°31'S	43°10'W
Petroșani, Rom.	**200-01**	45°25'N	23°23'E
Petrovgrad, Serb. *see* Zhovti Vody	**200-01**	45°23'N	20°24'E
Petrovsk, Russia (pyĕ-trôfsk')	**186-87**	52°19'N	45°23'E
Petrovsk-Zabaykal'skiy, Russia (pyĕ-trôfskzä-bĭ-käl'skĭ)	**240-41**	51°16'N	108°50'E
Petrozavodsk, Russia (pyä'trô-zá-vôtsk')	**186-87**	61°47'N	34°21'E
Petrozsény, Rom. *see* Petroșani	**200-01**	45°25'N	23°23'E
Petukhovo, Russia	**226**	55°04'N	67°54'E
Pevek, Russia	**218-19**	69°41'N	170°21'E
Peza, stm., Russia (pyå'zá)	**186-87**	65°36'N	44°37'E
Pézenas, Fr. (pā-zē-nä')	**196-97**	43°27'N	3°26'E
Pforzheim, Ger. (pfôrts'hīm)	**194-95**	48°54'N	8°42'E
Pha-an, Mya.	**246-47**	16°53'N	97°38'E
Phangan, Ko, i., Thai.	**246-47**	9°45'N	100°01'E
Phangnga, Thai.	**246-47**	8°27'N	98°32'E
Phanom Dong Rak, Thiu Khao, mts., Asia *see* Phanom Dongrak Range	**246-47**	14°25'N	103°30'E
Phanom Dongrak Range, mts., Asia	**246-47**	14°25'N	103°30'E
Phan Rang, Viet.	**246-47**	11°34'N	108°60'E
Phan Si Pan, mtn., Viet. *see* Fan Si Pan	**246-47**	22°15'N	103°46'E
Phan Thiet, Viet.	**246-47**	10°56'N	108°06'E
Phenix City, Al., U.S. (fē'nĭks sĭ'tĭ)	**124-25**	32°28'N	85°01'W
Phetchabun, Thiu Khao, mts., Thai.	**246-47**	16°32'N	100°55'E
Philadelphia, Ms., U.S. (fĭl-á-dĕl'phĭ-á)	**124-25**	32°46'N	89°07'W
Philadelphia, Pa., U.S. (fĭl-á-dĕl'phĭ-á)	**116-17**	39°57'N	75°10'W
Philip, S.D., U.S. (fĭl'ĭp)	**114-15**	44°02'N	101°39'W
Philippeville, Alg. *see* Skikda	**269b**	36°53'N	6°55'E
Philippines, nation, Asia (fĭl'ĭ-pēnz)	**206-07**	13°0'N	122°00'E
Philippine Sea, s., (fĭl'ĭ-pēn sē)	**222-23**	20°0'N	135°00'E
Philipsburg, Mt., U.S. (fĭl'lĭps-bērg)	**112-13**	46°20'N	113°18'W
Phillip Island, i., Austl. (fĭl'ĭp ī'lănd)	**276**	38°29'S	145°14'E
Phillips, Wi., U.S. (fĭl'ĭps)	**116-17**	45°42'N	90°24'W
Phillipsburg, Ks., U.S. (fĭl'lĭps-bērg)	**120-21**	39°45'N	99°19'W
Phillipsburg, N.J., U.S. (fĭl'lĭps-bērg)	**116-17**	40°42'N	75°11'W
Phitsanulok, Thai.	**246-47**	16°50'N	100°16'E
Phnom Penh, nat. cap., Camb. (nŏm'pĕn')	**246-47**	11°34'N	104°54'E
Phnum Pénh, nat. cap., Camb. (nŏm'pĕn') *see* Phnom Penh	**246-47**	11°34'N	104°54'E
Phoenix, Az., U.S. (fē'nĭks)	**118-19**	33°26'N	112°03'W
Phoenix Islands, is., Kir. (fē'nĭks ī'lăndz)	**280-81**	4°0'S	172°00'W
Phoenixville, Pa., U.S. (fē'nĭks-vĭl)	**116-17**	40°08'N	75°31'W
Phôngsali, Laos	**246-47**	21°43'N	102°07'E
Phra Chedi Sam Ong, p., Asia *see* Three Pagodas Pass	**246-47**	15°18'N	98°22'E
Phrae, Thai.	**246-47**	18°08'N	100°09'E
Phra Nakhon, nat. cap., Thai. *see* Bangkok	**246-47**	13°45'N	100°31'E
Phra Nakhon Si Ayutthaya, Thai.	**246-47**	14°21'N	100°34'E
Phuket, Thai.	**246-47**	7°52'N	98°23'E
Phuket, Ko, i., Thai.	**246-47**	8°0'N	98°22'E
Phu Ly, Viet.	**246-47**	20°31'N	105°56'E
Phu Quoc, Dao, i., Viet.	**246-47**	10°12'N	104°00'E
Piacenza, Italy (pyä-chĕnt'sä)	**200-01**	45°03'N	9°42'E
Piatra-Neamț, Rom.	**194-95**	46°57'N	26°24'E
Piauí, state, Braz.	**166-67**	7°0'S	43°00'W
Piazza Armerina, Italy (pyät'sä är-må-rē'nä)	**200-01**	37°23'N	14°22'E
Pic, stm., On., Can. (pĕk)	**136-37**	48°36'N	86°18'W
Picayune, Ms., U.S. (pĭk'à yōōn)	**124-25**	30°32'N	89°42'W
Pichanal, Arg.	**168-69**	23°18'S	64°14'W
Pichilemu, Chile (pē-chē-lĕ'mōō)	**163e**	34°23'S	72°00'W
Pichucalco, Mex. (pē-chōō-käl'kô)	**146-47**	17°30'N	93°10'W
Pickle Lake, On., Can.	**134-35**	51°30'N	90°04'W
Pico, i., Port. (pē'kô)	**199c**	38°28'N	28°20'W
Pico de Orizaba, Volcán, vol., Mex. (vôl-kä'n-pē'kô-dĕ-ô-rē-zä'bä)	**146-47**	19°01'N	97°16'W
Picos, Braz. (pē'kōzh)	**166-67**	7°05'S	41°28'W
Picton, On., Can. (pĭk'tŭn)	**136-37**	43°60'N	77°08'W
Picton, Isla, i., Chile	**171**	55°03'S	66°55'W
Pictou, N.S., Can. (pĭk-tōō')	**138-39**	45°41'N	62°42'W
Pidurutalagala, mtn., Sri L. (pē'dō-rô-tä'lá-gä'lá)	**236**	6°60'N	80°46'E
Piedmont, Al., U.S. (pēd'mŏnt)	**124-25**	33°55'N	85°37'W
Piedmont, Mo., U.S. (pēd'mŏnt)	**124-25**	37°09'N	90°42'W
Piedra del Águila, Arg.	**171**	40°03'S	70°03'W
Piedras, Punta, c., Arg. (pōō'n-tä-pyĕ'dräs)	**173**	35°26'S	57°07'W
Piedras Negras, Mex. (pyā'dräs nā'gräs)	**122-23**	28°42'N	100°31'W
Pierce, Ne., U.S. (pērs)	**114-15**	42°12'N	97°32'W
Pierre, S.D., U.S. (pēr)	**114-15**	44°22'N	100°21'W
Pietarsaari, Fin. *see* Jakobstad	**184-85**	63°41'N	22°43'E
Pietermaritzburg, S. Afr. (pē-tēr-mä-rĭts-bûrg)	**264-65**	29°36'S	30°23'E
Pietersburg, S. Afr. (pē'tērz-bûrg) *see* Polokwane	**269c**	23°53'S	29°26'E
Pigeon Lake, lk., Ab., Can. (pĭj'ŭn lăk)	**132-33**	53°0'N	114°00'W
Pigeon Lake, lk., On., Can. (pĭj'ŭn lăk)	**136-37**	44°30'N	78°30'W
Piggott, Ar., U.S. (pĭg-ŭt)	**124-25**	36°23'N	90°12'W
Pigüé, Arg.	**173**	37°37'S	62°25'W
Pihkva järv, lk., Eur. *see* Pskov, Lake	**192-93**	58°0'N	28°00'E
Pijijiapan, Mex. (pēĸē-kê-ä'pän)	**146-47**	15°42'N	93°13'W
Pikalëvo, Russia	**186-87**	59°31'N	34°11'E
Pikes Peak, mtn., Co., U.S. (pīks pēk)	**120-21**	38°51'N	105°03'W
Piketberg, S. Afr.	**264-65**	32°55'S	18°46'E
Pikeville, Ky., U.S. (pīk'vĭl)	**116-17**	37°30'N	82°33'W
Piła, Pol. (pē'lá)	**194-95**	53°09'N	16°44'E
Pilanesberg, hill, S. Afr. (pē'ăns'bûrg)	**269c**	25°12'S	27°05'E
Pilar, Arg. (pē'lär)	**173**	31°26'S	61°16'W
Pilar, Para.	**173**	26°54'S	58°19'W
Pilcomayo, stm., S.A. (pēl-cō-mī'ô)	**168-69**	25°17'S	57°40'W
Pili, Phil. (pē'lē)	**250**	13°32'N	123°17'E
Pīlībhīt, India	**234-35**	28°38'N	79°48'E
Pilica, stm., Pol. (pē-lēt'sä)	**194-95**	51°52'N	21°17'E
Pilipinas, nation, Asia *see* Philippines	**206-07**	13°0'N	122°00'E
Pilsen, Czech Rep. *see* Plzeň	**194-95**	49°45'N	13°23'E
Pinamalayan, Phil. (pē-nä-mä-lä'yän)	**250**	13°02'N	121°29'E
Pinang, Malay. *see* George Town	**246-47**	5°25'N	100°20'E
Pinar del Río, Cuba (pē-när' dĕl rē'ô)	**142-43**	22°25'N	83°41'W
Pinar del Río, state, Cuba (pē-när' dĕl rē'ô)	**142-43**	22°30'N	83°45'W
Pinatubo, Mount, vol., Phil. (mount pē-nä-tōō'bô)	**250**	15°08'N	120°21'E
Pincher Creek, Ab., Can. (pĭn'chēr krēk)	**132-33**	49°29'N	113°57'W
Pinckneyville, Il., U.S. (pĭnk'nĭ-vĭl)	**116-17**	38°05'N	89°23'W
Pindaré, stm., Braz.	**166-67**	3°18'S	44°47'W
Píndos Óros, mts., Grc.	**200-01**	39°49'N	21°14'E
Pindus Mountains, mts., Grc. (pĭn'dŭs moun'tĭnz) *see* Píndos Óros	**200-01**	39°49'N	21°14'E
Pine, stm., B.C., Can. (pīn)	**132-33**	56°09'N	120°44'W
Pine Bluff, Ar., U.S. (pīn blŭf)	**124-25**	34°14'N	92°02'W
Pine City, Mn., U.S. (pīn sĭ'tē)	**114-15**	45°50'N	92°58'W
Pine Creek, Austl. (pīn crēk krēk)	**270-71**	13°48'S	131°50'E
Pine Falls, Mb., Can. (pīn fôlz)	**134-35**	50°34'N	96°14'W
Pinega, stm., Russia (pê-nyĕ'gá)	**186-87**	64°08'N	41°54'E
Pinehouse Lake, lk., Sk., Can.	**134-35**	55°34'N	106°31'W
Pine Ridge, S.D., U.S. (pīn rĭj)	**114-15**	43°01'N	102°33'W
Pinerolo, Italy (pē-nå-rō'lō)	**200-01**	44°54'N	7°20'E
Pines, Isle of, i., Cuba (īl ŭv pīnz) *see* Juventud, Isla de la	**142-43**	21°40'N	82°50'W
Pineville, Ky., U.S. (pīn'vĭl)	**124-25**	36°45'N	83°42'W
Pineville, La., U.S. (pīn'vĭl)	**122-23**	31°20'N	92°26'W
Ping, stm., Thai.	**246-47**	15°42'N	100°09'E
Pingdingshan, China	**238-39**	33°45'N	113°18'E
Pingdu, China (pĭŋ-dōō)	**240-41**	36°47'N	119°56'E
Pingelap, at., Micron.	**280-81**	6°13'N	160°42'E
Pingjiang, China	**238-39**	28°42'N	113°35'E
Pingle, China (pĭŋ-lŭ)	**238-39**	24°38'N	110°40'E
Pingliang, China (pĭng'lyäng')	**240-41**	35°33'N	106°42'E
Pingquan, China (pĭŋ-chyůän)	**240-41**	40°59'N	118°39'E
Pingtan, China (pĭŋ-tän)	**225a**	25°31'N	119°47'E
Pingtan Dao, i., China (pĭŋ-tän dou)	**225a**	25°33'N	119°48'E
P'ingtung, Tai.	**225a**	22°40'N	120°29'E
Pingwu, China (pĭŋ-wōō)	**238-39**	32°25'N	104°33'E
Pingxiang, China (pĭŋ-shyäŋ)	**238-39**	22°08'N	106°44'E
Pingxiang, China (pĭŋ-shyäŋ)	**238-39**	27°38'N	113°50'E
Pingyao, China	**240-41**	37°16'N	112°14'E
Pingyi, China (pĭŋ-yē)	**240-41**	35°30'N	117°38'E
Pingyuan, China (pĭŋ-yůän)	**238-39**	24°36'N	115°55'E
Pinheiro, Braz.	**166-67**	2°31'S	45°05'W
Pinnacles National Monument, n.p., Ca., U.S. (pĭn'á-k'lz năsh'ŭn-ăl mŏn'ū-mĕnt)	**118-19**	36°30'N	121°11'W
Pinnaroo, Austl.	**276**	35°16'S	140°54'E
Pinos, Isla de, i., Cuba *see* Juventud, Isla de la	**142-43**	21°40'N	82°50'W
Pinrang, Indon.	**248-49**	3°48'S	119°39'E
Pins, Île des, i., N. Cal.	**279g**	22°37'S	167°28'E
Pinsk, Bela. (pĕn'sk)	**194-95**	52°07'N	26°07'E
Pinsk Marshes, reg., Eur. *see* Pripet Marshes	**194-95**	52°0'N	27°00'E
Pinta, Isla, i., Ec.	**170a**	0°35'N	90°44'W
Pinyug, Russia	**186-87**	60°15'N	47°47'E
Piombino, Italy (pyôm-bē'nō)	**200-01**	42°56'N	10°32'E
Pioneer Mountains, mts., Mt., U.S. (pī'ô-nēr' moun'tĭnz)	**112-13**	45°31'N	112°60'W
Pioner, Ostrov, i., Russia	**218-19**	79°50'N	92°30'E

n-sing; ŋ-baŋk; N-nasalized n; nŏd; cŏmmit; ōld; ôbey; ôrder; oi-boil; fōōd; ò-as oo in foot; ou-out; s-soft; sh-dish; th-thin; pūre; ûnite; ûrn; stŭd; circŭs; ü-as in French tu; '-indeterminate vowel.

Feature (Pronunciation)	Page	Lat.	Long.
Piorini, stm., Braz.	**166-67**	3°23'S	63°30'W
Piotrków Trybunalski, Pol. (pyōtr′ko͝ov trĭ-bo͞o-nal′skē)	**194-95**	51°24'N	19°42'E
Pipe Spring National Monument, n.p., Az., U.S. (pīp sprĭng năsh′ŭn-ăl mŏn′ū-mĕnt)	**118-19**	36°50'N	112°49'W
Pipestone, Mn., U.S. (pīp′stōn)	**114-15**	43°60'N	96°19'W
Pipestone, stm., On., Can.	**134-35**	52°54'N	89°15'W
Pipestone National Monument, n.p., Mn., U.S. (pīp′stōn năsh′ŭn-ăl mŏn′ū-mĕnt)	**114-15**	44°0'N	96°18'W
Pipinas, Arg.	**173**	35°32'S	57°19'W
Pipmuacan, Réservoir, res., Qc., Can. (pĭp-mä-kän′)	**138-39**	49°37'N	70°27'W
Piqua, Oh., U.S. (pĭk′wȧ)	**116-17**	40°09'N	84°15'W
Piracicaba, Braz. (pē-rä-sē-kä′bä)	**172**	22°43'S	47°36'W
Piraeus, Grc. see Peiraiás	**200-01**	37°57'N	23°39'E
Piran, Slvn. (pē-rȧ′n)	**200-01**	45°32'N	13°34'E
Pirané, Arg.	**168-69**	25°43'S	59°05'W
Pirapora, Braz. (pē-rȧ-pō′rȧ)	**172**	17°21'S	44°56'W
Pires do Rio, Braz.	**172**	17°18'S	48°17'W
Piriápolis, Ur.	**173**	34°52'S	55°16'W
Pirineos, mts., Eur. see Pyrenees	**196-97**	42°40'N	1°00'E
Pirna, Ger. (pĭr′nä)	**194-95**	50°58'N	13°57'E
Pirot, Serb. (pē′rōt)	**200-01**	43°10'N	22°35'E
Piru, Indon. (pē-ro͞o′)	**248-49**	3°03'S	128°11'E
Pisa, Italy (pē′sä)	**200-01**	43°44'N	10°24'E
Pisagua, Chile (pē-sä′gwä)	**168-69**	19°34'S	70°12'W
Pisco, Peru (pēs′kō)	**170**	13°42'S	76°12'W
Písek, Czech Rep. (pē′sĕk)	**194-95**	49°19'N	14°09'E
Pishan, China	**226**	37°37'N	78°16'E
Pisticci, Italy (pēs-tē′chē)	**200-01**	40°23'N	16°34'E
Pistoia, Italy (pēs-tô′yä)	**200-01**	43°56'N	10°55'E
Pisuerga, stm., Spain (pē-swĕr′gä)	**198-99**	41°33'N	4°52'W
Pitalito, Col. (pē-tä-lē′tō)	**164-65**	1°52'N	76°01'W
Pitanga, Braz.	**168-69**	24°44'S	51°45'W
Pitcairn Island, i., Pit.	**280-81**	25°04'S	130°06'W
Pitcairn Islands, dep., Oc. (pĭt′kârn ī′lȧndz)	**280-81**	25°04'S	130°05'W
Piteå, Swe.	**184-85**	65°19'N	21°29'E
Piteälven, stm., Swe.	**184-85**	65°23'N	21°19'E
Pitești, Rom. (pē-tĕsht′′)	**200-01**	44°51'N	24°52'E
Pithiviers, Fr. (pē-tē-vyā′)	**196-97**	48°10'N	2°15'E
Pitti Island, i., India	**236**	10°50'N	72°37'E
Pitt Island, i., B.C., Can. (pĭt ī′lȧnd)	**132-33**	53°35'N	129°45'W
Pittsburg, Ks., U.S. (pĭts′bûrg)	**120-21**	37°25'N	94°42'W
Pittsburg, Tx., U.S. (pĭts′bûrg)	**120-21**	33°00'N	94°58'W
Pittsburgh, Pa., U.S. (pĭts′bûrg)	**116-17**	40°27'N	80°01'W
Pittsfield, Il., U.S. (pĭts′fēld)	**120-21**	39°36'N	90°48'W
Pittston, Pa., U.S. (pĭts′tŭn)	**116-17**	41°20'N	75°47'W
Pium, Braz.	**166-67**	10°27'S	49°11'W
Piura, Peru (pē-o͞o′rä)	**170**	5°11'S	80°38'W
Pivdennyi Buh, stm., Ukr. see Southern Bug	**202-03**	46°39'N	31°56'E
Placentia Bay, b., Nf., Can.	**138-39**	47°15'N	54°30'W
Placerville, Ca., U.S. (plăs′ēr-vĭl)	**118-19**	38°44'N	120°48'W
Placetas, Cuba (plä-thā′täs)	**142-43**	22°19'N	79°39'W
Plainview, Mn., U.S. (plān′vū)	**114-15**	44°10'N	92°09'W
Plainview, Ne., U.S. (plān′vū)	**114-15**	42°21'N	97°47'W
Plainview, Tx., U.S. (plān′vū)	**120-21**	34°11'N	101°42'W
Plainwell, Mi., U.S. (plan′wĕl)	**116-17**	42°26'N	85°38'W
Plano, Il., U.S. (plā′nō)	**116-17**	41°40'N	88°32'W
Plano, Tx., U.S. (plā′nō)	**120-21**	33°03'N	96°41'W
Plant City, Fl., U.S. (plȧnt sĭ′tĭ)	**125a**	28°01'N	82°07'W
Plaquemine, La., U.S. (pläk′mēn′)	**124-25**	30°17'N	91°14'W
Plasencia, Spain (plä-sĕn′thē-ä)	**198-99**	40°02'N	6°05'W
Plaster Rock, N.B., Can. (plȧs′tēr rŏk)	**138-39**	46°55'N	67°24'W
Plastun, Russia (plȧs-to͞on′)	**244**	44°45'N	136°18'E
Plata, Río de la, est., S.A. (rē′ō dālä plä′tä)	**173**	35°0'S	57°00'W
Platinum, Ak., U.S. (plăt′ĭ-nŭm)	**126**	59°01'N	161°49'W
Plato, Col. (plä′tō)	**164-65**	9°48'N	74°47'W
Platte, S.D., U.S. (plăt)	**114-15**	43°23'N	98°51'W
Platte, stm., U.S. (plăt)	**120-21**	39°16'N	94°51'W
Platte, stm., Ne., U.S. (plăt)	**110-11**	41°04'N	95°53'W
Platteville, Wi., U.S. (plăt′vĭl)	**114-15**	42°44'N	90°29'W
Plattsburg, Mo., U.S. (plăts′bûrg)	**120-21**	39°34'N	94°27'W
Plattsburgh, N.Y., U.S.	**116-17**	44°42'N	73°28'W
Plattsmouth, Ne., U.S. (plăts′mŭth)	**114-15**	41°01'N	95°54'W
Plauen, Ger. (plou′ĕn)	**194-95**	50°30'N	12°08'E
Playa Vicente, Mex. (plä-yä vē-sĕn′tå)	**146-47**	17°48'N	95°49'W
Play Ku, Viet.	**246-47**	13°59'N	108°01'E
Pleasanton, Tx., U.S. (plĕz′ăn-tŭn)	**122-23**	28°58'N	98°29'W
Pleiku, Viet. see Play Ku	**246-47**	13°59'N	108°01'E
Plenty, Bay of, b., N.Z. (bā ŭv plĕn′tē)	**278**	37°40'S	177°00'E
Plentywood, Mt., U.S. (plĕn′tē-wo͝od)	**112-13**	48°46'N	104°32'W
Plesetsk, Russia	**186-87**	62°43'N	40°18'E
Plessisville, Qc., Can. (plĕ-sē′vēl′)	**136-37**	46°13'N	71°46'W
Pleszew, Pol. (plĕ′zhĕf)	**194-95**	51°54'N	17°47'E
Płock, Pol. (pwôtsk)	**194-95**	52°33'N	19°42'E
Ploërmel, Fr. (plô-ĕr-mĕl′)	**196-97**	47°56'N	2°24'W
Ploești, Rom. see Ploiești	**200-01**	44°57'N	26°02'E
Ploiești, Rom. (plô-yĕsht′′)	**200-01**	44°57'N	26°02'E
Plomb du Cantal, mtn., Fr. (plôn′dü-kän-täl′)	**196-97**	45°04'N	2°45'E
Plonge, Lac la, lk., Sk., Can. (läk lä plônzh)	**134-35**	55°08'N	107°17'W
Plovdiv, Blg. (plôv′dĭf) (fĭl-ĭp-ŏp′ō-lĭs)	**200-01**	42°09'N	24°45'E
Plungė, Lith. (plón′gā)	**192-93**	55°54'N	21°52'E
Plyeven, Blg.	**200-01**	43°25'N	24°37'E
Plymouth, Eng., U.K. (plĭm′ŭth)	**190-91**	50°23'N	4°10'W
Plymouth, In., U.S. (plĭm′ŭth)	**116-17**	41°20'N	86°18'W
Plymouth, N.C., U.S. (plĭm′ŭth)	**124-25**	35°52'N	76°45'W
Plymouth, N.H., U.S. (plĭm′ŭth)	**116-17**	43°46'N	71°41'W
Plymouth, Pa., U.S. (plĭm′ŭth)	**116-17**	41°14'N	75°58'W
Plymouth, Vt., U.S. (plĭm′ŭth)	**116-17**	43°33'N	72°44'W
Plymouth, Wi., U.S. (plĭm′ŭth)	**116-17**	43°46'N	87°59'W
Plzeň, Czech Rep.	**194-95**	49°45'N	13°23'E
Po, stm., Italy	**200-01**	44°59'N	12°03'E
Pobeda, Gora, mtn., Russia	**218-19**	65°12'N	146°12'E
Pobedy, Pik, mtn., Asia see Jengish Chokusu	**226**	42°02'N	80°05'E
Pocahontas, Ar., U.S. (pō-kȧ-hŏn′tȧs)	**124-25**	36°16'N	90°58'W
Pocahontas, Ia., U.S. (pō-kȧ-hŏn′tȧs)	**114-15**	42°44'N	94°40'W
Pocatello, Id., U.S. (pō-kȧ-tĕl′ō)	**112-13**	42°53'N	112°29'W
Pochëp, Russia (pô-chĕp′)	**202-03**	52°55'N	33°29'E
Pocomoke City, Md., U.S. (pō-kō-mōk′ sĭ′tē)	**116-17**	38°04'N	75°33'W
Poços de Caldas, Braz. (pō-sôs-dĕ-käl′dås)	**172**	21°47'S	46°34'W
Podgorica, nat. cap., Mont.	**200-01**	42°27'N	19°16'E
Podkamennaya Tunguska, stm., Russia	**218-19**	61°36'N	90°09'E
Podolsk, Russia (pô-dôl′′sk)	**202-03**	55°26'N	37°34'E
Podporozhye, Russia	**192-93**	60°55'N	34°10'E
Poggibonsi, Italy (pôd-jē-bôn′sē)	**200-01**	43°28'N	11°09'E
Pogranichnyy, Russia	**244**	44°24'N	131°23'E
P'ohang, Kor., S.	**243**	36°03'N	129°22'E
Pohjanlahti, b., Eur. see Bothnia, Gulf of	**184-85**	63°0'N	20°00'E
Pohnpei, i., Micron.	**280-81**	6°55'N	158°15'E
Poinsett, Cape, c., Ant.	**287**	65°48'S	113°10'E
Point Au Fer Island, i., La., U.S.	**124-25**	29°15'N	91°15'W
Pointe-à-Pitre, Guad. (pwănt′ ȧ pē-tr′)	**143b**	16°15'N	61°32'W
Pointe-des-Galets, Reu. see Le Port	**265a**	20°55'S	55°18'E
Pointe-Noire, Congo	**262-63**	4°48'S	11°52'E
Point Hope, Ak., U.S. (point hōp)	**126**	68°21'N	166°41'W
Point Pleasant, Oh., U.S. (point plĕz′ănt)	**116-17**	38°54'N	84°14'W
Point Pleasant, W.V., U.S. (point plĕz′ănt)	**116-17**	38°52'N	82°08'W
Poitiers, Fr. (pwȧ-tyā′)	**196-97**	46°35'N	0°20'E
Pokharā, Nepal	**234-35**	28°13'N	83°60'E
Pokhvistnevo, Russia	**186-87**	53°39'N	52°08'E
Pokrovsk, Russia	**218-19**	61°31'N	129°11'E
Pokrovskoye, Russia (pô-krôf′skô-yĕ)	**202-03**	52°37'N	36°51'E
Polack, Bela.	**192-93**	55°30'N	28°47'E
Poland, nation, Eur. (pō′lȧnd)	**174-75**	52°0'N	19°00'E
Polatlı, Tur.	**186-87**	39°36'N	32°10'E
Polessk, Russia (pô′lĕsk)	**192-93**	54°52'N	21°05'E
Polesye, reg., Eur. see Pripet Marshes	**194-95**	52°0'N	27°00'E
Polillo Island, i., Phil.	**250**	14°50'N	121°57'E
Polillo Islands, is., Phil. (pô-lēl′yō ī′lȧndz)	**250**	14°50'N	122°05'E
Polissya, reg., Eur. see Pripet Marshes	**194-95**	52°0'N	27°00'E
Pollāchi, India	**236**	10°39'N	77°01'E
Polokwane, S. Afr.	**269c**	23°53'S	29°26'E
Polonnaruwa, Sri L.	**236**	7°56'N	81°01'E
Polotsk, Bela. see Polack	**192-93**	55°30'N	28°47'E
Polska, nation, Eur. see Poland	**174-75**	52°0'N	19°00'E
Polson, Mt., U.S. (pōl′sŭn)	**112-13**	47°41'N	114°09'W
Poltava, Ukr. (pôl-tä′vä)	**202-03**	49°36'N	34°31'E
Põltsamaa, Est. (põlt′så-mä)	**192-93**	58°39'N	25°58'E
Polunochnoye, Russia (pô-lo͞o-nô′ch-nô′yĕ)	**186-87**	60°52'N	60°26'E
Polyarnyy, Russia (pŭl-yär′nē)	**218-19**	69°06'N	178°39'E
Polyarnyy, Russia (pŭl-yär′nē)	**184-85**	69°11'N	33°28'E
Polynesia, is., Oc. (pŏl-ĭ-nē′zhȧ)	**280-81**	4°0'S	156°00'W
Polynésie, is., Oc. see Polynesia	**280-81**	4°0'S	156°00'W
Polynésie française, dep., Oc. see French Polynesia	**280-81**	15°0'S	140°00'W
Pomerania, hist. reg., Eur. (pŏm-ê-rā′nĭ-ȧ)	**194-95**	54°0'N	16°00'E
Pomeroy, Oh., U.S. (pŏm′ēr-oi)	**116-17**	39°02'N	82°01'W
Pomeroy, Wa., U.S. (pŏm′ēr-oi)	**112-13**	46°28'N	117°35'W
Pommern, hist. reg., Eur. see Pomerania	**194-95**	54°0'N	16°00'E
Pomona, Ca., U.S. (pô-mō′nȧ)	**118-19**	34°03'N	117°45'W
Pomorze, hist. reg., Eur. see Pomerania	**194-95**	54°0'N	16°00'E
Pompano Beach, Fl., U.S. (pŏm′pȧ-nô bēch)	**125a**	26°14'N	80°08'W
Pompei, hist., Italy	**200-01**	40°45'N	14°30'E
Pompeii, hist., Italy see Pompei	**200-01**	40°45'N	14°30'E
Ponape, i., Micron. see Pohnpei	**280-81**	6°55'N	158°15'E
Ponca, Ne., U.S. (pŏn′kȧ)	**114-15**	42°33'N	96°43'W
Ponca City, Ok., U.S. (pŏn′kȧ sĭ′tĭ)	**120-21**	36°42'N	97°05'W
Ponce, P.R. (pōn′sā)	**142a**	18°01'N	66°37'W
Pondicherry, India	**236**	11°56'N	79°50'E
Pondicherry, state, India see Puducherry	**236**	11°56'N	79°50'E
Ponferrada, Spain (pôn-fĕr-rä′dhä)	**198-99**	42°33'N	6°35'W
Pongolo, stm., S. Afr.	**264-65**	26°52'S	32°21'E
Ponoka, Ab., Can. (pŏ-nō′kȧ)	**132-33**	52°40'N	113°35'W
Ponoy, stm., Russia	**186-87**	66°60'N	41°16'E
Ponta Delgada, Port. (pôn′tȧ dĕl-gä′dȧ)	**199c**	37°45'N	25°40'W
Ponta Grossa, Braz. (pôn′tä grō′sȧ)	**168-69**	25°05'S	50°10'W
Ponta Porã, Braz.	**168-69**	22°33'S	55°42'W
Pontarlier, Fr. (pôn′tȧr-lyā′)	**196-97**	46°54'N	6°22'E
Pont-Audemer, Fr. (pôn′tōd′mâr′)	**196-97**	49°21'N	0°31'E
Pontchartrain, Lake, lk., La., U.S. (lāk pôn-shȧr-trăn′)	**124-25**	30°10'N	90°10'W
Pontedera, Italy (pōn-tå-dä′rä)	**200-01**	43°40'N	10°38'E
Ponte Nova, Braz. (pô′n-tĕ-nô′vȧ)	**172**	20°25'S	42°54'W
Pontevedra, Spain (pôn-tĕ-vĕ-drä)	**198-99**	42°25'N	8°38'W
Pontiac, Il., U.S. (pŏn′tĭ-ăk)	**116-17**	40°53'N	88°37'W
Pontiac, Mi., U.S. (pŏn′tĭ-ăk)	**116-17**	42°38'N	83°18'W
Pontianak, Indon. (pôn-tê-ä′nák)	**246-47**	0°01'S	109°20'E
Pontine, Isole, is., Italy see Ponziane, Isole	**200-01**	40°55'N	12°57'E
Pontine Islands, is., Italy see Ponziane, Isole	**200-01**	40°55'N	12°57'E
Pontivy, Fr. (pôn-tê-vē′)	**196-97**	48°04'N	2°58'W
Pontoise, Fr. (pôn-twȧz′)	**196-97**	49°03'N	2°05'E
Pontotoc, Ms., U.S. (pŏn-tô-tŏk′)	**124-25**	34°15'N	88°60'W
Pontremoli, Italy (pôn-trĕm′ô-lē)	**200-01**	44°23'N	9°53'E
Pontus Mountains, mts., Tur. see Doğu Karadeniz Dağları	**227**	40°30'N	40°30'E
Ponziane, Isole, is., Italy	**200-01**	40°55'N	12°57'E
Poole, Eng., U.K. (po͞ol)	**190-91**	50°44'N	1°59'W
Poona, India see Pune	**236**	18°32'N	73°52'E
Poopó, Lago, lk., Bol.	**168-69**	18°47'S	67°05'W
Popayán, Col. (pō-pä-yän′)	**164-65**	2°27'N	76°36'W
Popigay, stm., Russia	**218-19**	72°57'N	106°10'E
Poplar, Mt., U.S. (pŏp′lēr)	**112-13**	48°06'N	105°13'W
Poplar, stm., Can.	**134-35**	53°02'N	97°27'W
Poplar Bluff, Mo., U.S. (pŏp′lēr blŭf)	**124-25**	36°45'N	90°24'W
Poplar Plains, Ky., U.S. (pŏp′lēr plānz)	**116-17**	38°22'N	83°40'W
Poplarville, Ms., U.S. (pŏp′lēr-vĭl)	**124-25**	30°51'N	89°34'W
Popocatépetl, Volcán, vol., Mex.	**146-47**	19°01'N	98°37'W
Popondetta, Pap. N. Gui.	**277**	8°47'S	148°13'E
Popovo, Blg. (pô′pô-vō)	**200-01**	43°21'N	26°14'E
Poprad, Slvk.	**194-95**	49°03'N	20°19'E
Porangatu, Braz.	**166-67**	13°26'S	49°12'W
Porbandar, India (pōr-bŭn′dŭr)	**234-35**	21°39'N	69°37'E
Porcher Island, i., B.C., Can. (pôr′kēr ī′lȧnd)	**132-33**	53°57'N	130°30'W
Porcupine, stm., N.A.	**126**	66°34'N	145°21'W
Pordenone, Italy (pōr-då-nō′nå)	**200-01**	45°58'N	12°39'E
Pori, Fin. (pô′rē)	**192-93**	61°29'N	21°47'E
Porkhov, Russia (pôr′kôf)	**192-93**	57°46'N	29°34'E
Porlamar, Ven. (pôr-lä-mär′)	**164-65**	10°57'N	63°51'W
Pornic, Fr. (pôr-nēk′)	**196-97**	47°07'N	2°06'W
Poronaysk, Russia (pô′rô-nīsk)	**218-19**	49°14'N	143°06'E
Porpoise Bay, b., Ant.	**287**	66°30'S	128°30'E
Porsgrunn, Nor. (pôrs′grön′)	**192-93**	59°09'N	9°40'E
Port, Reu. see Le Port	**265a**	20°55'S	55°18'E
Portachuelo, Bol. (pôrt-ä-chwā′lō)	**168-69**	17°21'S	63°24'W
Portage, In., U.S. (pōr′tāj)	**116-17**	41°35'N	87°11'W
Portage, Mi., U.S. (pôr′tåj)	**116-17**	42°12'N	85°36'W
Portage, Pa., U.S. (pôr′tåj)	**116-17**	40°23'N	78°40'W
Portage, Wi., U.S. (pôr′tåj)	**116-17**	43°33'N	89°28'W
Portage la Prairie, Mb., Can. (pōr′tĭj lä-prā′rĭ)	**134-35**	49°58'N	98°18'W
Port Alberni, B.C., Can. (pōr äl-bĕr-nē′)	**132-33**	49°15'N	124°48'W
Portalegre, Port. (pôr-tä-lā′grĕ)	**198-99**	39°17'N	7°25'W

ăt; finăl; rāte; senåte; ärm; ȧsk; sofȧ; fâre; ch-choose; dh-as th in other; bē; ēvent; bĕt; recĕnt; cratēr; g-gō; gh-guttural g; bĭt; ĭ-short neutral; rīde; ĸ-guttural k as ch in German ich;

Feature (Pronunciation)	Page	Lat.	Long.
Portales, N.M., U.S. (pôr-tä′lĕs)	**120-21**	34°11′N	103°20′W
Port Alfred, S. Afr.	**264-65**	33°36′S	26°54′E
Port Alice, B.C., Can. (pōrt ăl′ĭs)	**132-33**	50°23′N	127°26′W
Port Allegany, Pa., U.S. (pōrt ăl-ē-gā′nĭ)	**116-17**	41°49′N	78°17′W
Port Angeles, Wa., U.S. (pōrt ăn′jē-lĕs)	**112-13**	48°07′N	123°26′W
Port Antonio, Jam.	**142-43**	18°10′N	76°27′W
Port Arthur, Austl.	**276**	43°09′S	147°50′E
Port Arthur, China *see* Lüshun	**240-41**	38°49′N	121°15′E
Port Arthur, Tx., U.S.	**122-23**	29°54′N	93°56′W
Port Augusta, Austl. (pōrt ô-gŭs′tă)	**276**	32°30′S	137°46′E
Port au Port Bay, b., Nf., Can. (pôr′tō pōr′ bā)	**138-39**	48°40′N	58°45′W
Port-au-Prince, nat. cap., Haiti (pôr′tō präNs′)	**142-43**	18°32′N	72°20′W
Port Austin, Mi., U.S. (pōrt ôs′tĭn)	**116-17**	44°02′N	83°00′W
Port Blair, India (pōrt blâr)	**246-47**	11°39′N	92°45′E
Port Borden, P.E., Can. (pōrt bôr′dĕn)	**138-39**	46°15′N	63°42′W
Port-Cartier, Qc., Can.	**138-39**	50°02′N	66°52′W
Port Clinton, Oh., U.S. (pōrt klĭn′tŭn)	**116-17**	41°30′N	82°58′W
Port-de-Paix, Haiti (pôrt dē pĕ)	**142-43**	19°56′N	72°49′W
Port Dickson, Malay. (pōrt dĭk′sŭn)	**246-47**	2°31′N	101°49′E
Port Edward, China *see* Weihai	**240-41**	37°30′N	122°07′E
Portel, Braz.	**166-67**	1°58′S	50°48′W
Port Elgin, N.B., Can. (pōrt ĕl′jĭn)	**138-39**	46°03′N	64°06′W
Port Elgin, On., Can. (pōrt ĕl′jĭn)	**136-37**	44°26′N	81°23′W
Port Elizabeth, S. Afr. (pōrt ē-lĭz′ȧ-bĕth)	**264-65**	33°56′S	25°34′E
Porterville, Ca., U.S. (pōr′tĕr-vĭl)	**118-19**	36°05′N	119°02′W
Port Fairy, Austl.	**276**	38°23′S	142°14′E
Port-Francqui, D.R.C. *see* Ilebo	**262-63**	4°20′S	20°36′E
Port-Gentil, Gabon (pôr-zhäN-tē′)	**260-61**	0°43′S	8°47′E
Port-Harcourt, Nig. (pōrt här′kŭrt)	**260a**	4°47′N	7°01′E
Port Hardy, B.C., Can. (pōrt här′dĭ)	**132-33**	50°43′N	127°30′W
Port Hedland, Austl. (pōrt hĕd′lănd)	**270-71**	20°19′S	118°36′E
Port Hood, N.S., Can. (pōrt hŏŏd)	**138-39**	46°01′N	61°32′W
Port Hope, On., Can. (pōrt hōp)	**136-37**	43°57′N	78°17′W
Port Huron, Mi., U.S. (pōrt hū′rŏn)	**116-17**	42°58′N	82°26′W
Portimão, Port. (pôr-tē-mouN)	**198-99**	37°08′N	8°32′W
Port Jervis, N.Y., U.S. (pōrt jûr′vĭs)	**116-17**	41°23′N	74°42′W
Port Lairge, Ire. *see* Waterford	**190-91**	52°15′N	7°06′W
Portland, Austl. (pōrt′lănd)	**276**	38°21′S	141°36′E
Portland, In., U.S. (pōrt′lănd)	**116-17**	40°26′N	84°59′W
Portland, Me., U.S. (pōrt′lănd)	**116-17**	43°40′N	70°17′W
Portland, Mi., U.S. (pōrt′lănd)	**116-17**	42°52′N	84°54′W
Portland, Or., U.S. (pōrt′lănd)	**112-13**	45°32′N	122°40′W
Portland, Tn., U.S. (pōrt′lănd)	**124-25**	36°35′N	86°31′W
Portland, Tx., U.S. (pōrt′lănd)	**122-23**	27°53′N	97°19′W
Portland Bight, b., Jam. (pōrt′lănd bĭt)	**142-43**	17°50′N	77°06′W
Portland Inlet, b., B.C., Can. (pōrt′lănd ĭn′lĕt)	**132-33**	54°50′N	130°14′W
Portland Point, c., Jam. (pōrt′lănd point)	**142-43**	17°43′N	77°11′W
Port Lavaca, Tx., U.S. (pōrt lȧ-vä′kȧ)	**122-23**	28°37′N	96°38′W
Port Lincoln, Austl. (pōrt lĭŋ-kŭn)	**270-71**	34°44′S	135°52′E
Port Louis, nat. cap., Mauritius	**265a**	20°10′S	57°30′E
Port-Lyautey, Mor. *see* Kénitra	**269a**	34°16′N	6°35′W
Port Macquarie, Austl. (pōrt mȧ-kwŏ′rĭ)	**276**	31°27′S	152°55′E
Port Moresby, nat. cap., Pap. N. Gui. (pōrt mōrz′bē)	**277**	9°28′S	147°12′E
Port Neches, Tx., U.S. (pōrt nĕch′ĕz)	**122-23**	29°59′N	93°57′W
Porto, Port. (pōr′tò)	**198-99**	41°09′N	8°37′W
Porto Alegre, Braz. (pōr′tō ä-lā′grĕ)	**168-69**	30°03′S	51°12′W
Porto Amélia, Moz. *see* Pemba	**264-65**	13°01′S	40°32′E
Portobelo, Pan. (pôr′tō-bā′lō)	**150**	9°33′N	79°39′W
Porto de Moz, Braz.	**166-67**	1°44′S	52°14′W
Porto de Pedras, Braz. (pōr′tō pā′dräzh)	**163d**	9°09′S	35°17′W
Porto Esperança, Braz.	**168-69**	19°37′S	57°27′W
Porto Esperidião, Braz.	**168-69**	15°51′S	58°28′W
Portoferraio, Italy (pōr′tō-fĕr-rä′yō)	**200-01**	42°49′N	10°19′E
Port of Spain, nat. cap., Trin. (pōrt ŭv spān′)	**143b**	10°39′N	61°30′W
Portogruaro, Italy (pōr′tō-grò-ä′rō)	**200-01**	45°47′N	12°50′E
Porto Murtinho, Braz. (pōr′tō mòr-tēn′yò)	**168-69**	21°42′S	57°52′W
Porto Nacional, Braz. (pōr′tō nȧ-syō-näl′)	**166-67**	10°42′S	48°25′W
Porto-Novo, nat. cap., Benin (pōr′tō-nō′vō)	**260a**	6°29′N	2°37′E
Porto Santo, i., Port. (pōr′tò säN′tò)	**258-59**	33°04′N	16°20′W
Porto Seguro, Braz. (pōr′tō sā-gōō′rò)	**168-69**	16°26′S	39°04′W
Porto Torres, Italy (pōr′tò tôr′rĕs)	**200-01**	40°50′N	8°24′E
Porto União, Braz.	**168-69**	26°15′S	51°04′W

Feature (Pronunciation)	Page	Lat.	Long.
Porto-Vecchio, Fr. (pōr′tō-vĕk′ē-ô)	**184-85**	41°36′N	9°16′E
Porto Velho, Braz. (pōr′tō văl′yò)	**166-67**	8°46′S	63°54′W
Portoviejo, Ec. (pôr-tō-vyā′hō)	**170**	1°03′S	80°27′W
Port Phillip Bay, b., Austl. (pōrt fĭl′ĭp bā)	**276**	38°07′S	144°48′E
Port Pirie, Austl. (pōrt pĭ′rē)	**276**	33°12′S	138°00′E
Port Said, Egypt (pôrt sä-ēd′)	**268b**	31°16′N	32°18′E
Port Saint Lucie, Fl., U.S. (pôrt sānt lū′sē)	**125a**	27°20′N	80°20′W
Port Shepstone, S. Afr. (pōrt hĕps′tŭn)	**264-65**	30°45′S	30°25′E
Portsmouth, Eng., U.K. (pôrts′mŭth)	**190-91**	50°48′N	1°05′W
Portsmouth, N.H., U.S. (pôrts′mŭth)	**116-17**	43°04′N	70°46′W
Portsmouth, Oh., U.S. (pôrts′mŭth)	**116-17**	38°44′N	82°60′W
Portsmouth, Va., U.S. (pôrts′mŭth)	**124-25**	36°50′N	76°19′W
Port Stanley, nat. cap., Falk. Is. *see* Stanley	**171**	51°43′S	57°49′W
Port Sudan, Sudan (pôrt sōō-dän′)	**266**	19°37′N	37°13′E
Port Sulphur, La., U.S. (pōrt sŭl′fĕr)	**124-25**	29°29′N	89°42′W
Port Townsend, Wa., U.S. (pōrt tounz′ĕnd)	**112-13**	48°07′N	122°46′W
Portugal, nation, Eur. (pōr′tu-găl)	**174-75**	39°30′N	8°00′W
Portugalete, Spain (pōr-tōō-gä-lā′tā)	**198-99**	43°19′N	3°01′W
Portuguese Guinea, nation, Afr. *see* Guinea-Bissau	**253**	12°0′N	15°00′W
Port Vila, nat. cap., Vanuatu (port vē′lȧ)	**279g**	17°45′S	168°19′E
Port Wakefield, Austl. (pōrt wāk′fēld)	**276**	34°11′S	138°01′E
Port Washington, Wi., U.S. (pōrt wôsh′ĭng-tŭn)	**116-17**	43°23′N	87°53′W
Porvenir, Chile	**171**	53°18′S	70°22′W
Porvoo, Fin.	**192-93**	60°24′N	25°40′E
Posadas, Arg. (pō-sä′dhäs)	**173**	27°22′S	55°54′W
Posen, Pol. *see* Poznań	**194-95**	52°24′N	16°54′E
Poso, Indon.	**248-49**	1°23′S	120°46′E
Poso, Danau, lk., Indon. (pō′sō)	**248-49**	1°52′S	120°35′E
Posse, Braz.	**166-67**	14°06′S	46°22′W
Post, Tx., U.S. (pōst)	**120-21**	33°11′N	101°23′W
Postojna, Slvn. (pōs-tōyná)	**200-01**	45°47′N	14°13′E
Pos'yet, Russia (pos-yĕt′)	**243**	42°39′N	130°49′E
Potawatomi Indian Reservation, ind. res., Ks., U.S. (pŏt-ȧ-wă′tō-mē ĭn′dĭ-ăn rĕ-sĕr-vā′shĕn)	**120-21**	39°20′N	95°50′W
Potchefstroom, S. Afr. (pŏch′ĕf-strōm)	**269c**	26°43′S	27°07′E
Poteau, Ok., U.S. (pô-tō′)	**120-21**	35°04′N	94°38′W
Poteet, Tx., U.S. (pô-tēt)	**122-23**	29°02′N	98°34′W
Potenza, Italy (pô-tĕnt′sä)	**200-01**	40°39′N	15°48′E
Potgietersrus, S. Afr. (pôt-KĒ′tĕrs-rûs)	**269c**	24°11′S	29°01′E
Poti, Geor. (pô′tē)	**227**	42°09′N	41°40′E
Potomac, stm., U.S. (pô-tō′măk)	**110-11**	37°59′N	76°18′W
Potosí, Bol.	**168-69**	19°35′S	65°45′W
Potrerillos, Chile	**168-69**	26°26′S	69°29′W
Potsdam, Ger. (pôts′däm)	**194-95**	52°24′N	13°04′E
Potsdam, N.Y., U.S. (pŏts′dăm)	**116-17**	44°40′N	74°59′W
Pott, Île, i., N. Cal.	**279g**	19°35′S	163°35′E
Pottstown, Pa., U.S. (pŏts′toun)	**116-17**	40°16′N	75°39′W
Pottsville, Pa., U.S. (pŏts′vĭl)	**116-17**	40°41′N	76°12′W
Poughkeepsie, N.Y., U.S. (pŏ-kĭp′sē)	**116-17**	41°42′N	73°56′W
Pouso Alegre, Braz. (pō′zò ä-lā′grĕ)	**172**	22°14′S	45°56′W
Poŭthĭsăt, Camb.	**246-47**	12°32′N	103°56′E
Póvoa de Varzim, Port. (pō-vō′ȧ dā vär′zēN)	**198-99**	41°23′N	8°45′W
Povorino, Russia	**186-87**	51°12′N	42°16′E
Povungnituk, Qc., Can.	**128-29**	60°03′N	77°19′W
Povungnituk, stm., Qc., Can.	**130-31**	60°01′N	77°23′W
Powder, stm., U.S. (pou′dĕr)	**110-11**	46°44′N	105°27′W
Powell, Wy., U.S. (pou′ĕl)	**112-13**	44°45′N	108°45′W
Powell, Lake, res., U.S. (lāk pou′ĕl)	**118-19**	37°29′N	110°44′W
Powell Lake, lk., B.C., Can. (pou′ĕl lāk)	**132-33**	50°11′N	124°24′W
Powell River, B.C., Can. (pou′ĕl rĭv′ĕr)	**132-33**	49°53′N	124°33′W
Poxoréu, Braz.	**168-69**	15°50′S	54°23′W
Poyang Hu, lk., China	**238-39**	29°0′N	116°25′E
Poyarkovo, Russia	**240-41**	49°38′N	128°39′E
Požarevac, Serb. (pô′zhá′rĕ-váts)	**200-01**	44°38′N	21°11′E
Poznań, Pol.	**194-95**	52°24′N	16°54′E
Pozoblanco, Spain (pô-thō-bläŋ′kō)	**198-99**	38°23′N	4°51′W
Pozsony, nat. cap., Slvk. *see* Bratislava	**194-95**	48°09′N	17°07′E
Pozuelo de Alarcón, Spain (pô-thwā′lō dā ä-lär-kōn′)	**198-99**	40°27′N	3°46′W
Pra, stm., Ghana (prȧ)	**260-61**	5°01′N	1°38′W
Prachin Buri, Thai. (prä′chēn)	**246-47**	14°03′N	101°23′E
Prachuap Khiri Khan, Thai.	**246-47**	11°49′N	99°48′E
Pradera, Col. (prä-dĕ′rä)	**163c**	3°25′N	76°15′W
Prades, Fr. (pråd)	**196-97**	42°37′N	2°26′E
Prado, Braz.	**172**	17°18′S	39°15′W

Feature (Pronunciation)	Page	Lat.	Long.
Prague, nat. cap., Czech Rep. (präg)	**194-95**	50°05′N	14°26′E
Praha, nat. cap., Czech Rep. (prä′hȧ) *see* Prague	**194-95**	50°05′N	14°26′E
Praia, nat. cap., C.V. (prä′yȧ)	**260-61**	14°55′N	23°31′W
Prainha Nova, Braz.	**166-67**	7°29′S	60°38′W
Prairie du Chien, Wi., U.S. (prā′rĭ dò shēn′)	**114-15**	43°03′N	91°08′W
Pratas Island, i., Tai.	**222-23**	20°42′N	116°43′E
Prathet Thai, nation, Asia *see* Thailand	**206-07**	15°0′N	100°00′E
Prato, Italy (prä′tō)	**200-01**	43°53′N	11°06′E
Pratt, Ks., U.S. (prăt)	**120-21**	37°39′N	98°44′W
Prattville, Al., U.S. (prăt′vĭl)	**124-25**	32°28′N	86°28′W
Praya, Indon.	**248-49**	8°43′S	116°17′E
Pregolya, stm., Russia (prĕ-gô′lă)	**194-95**	54°41′N	20°23′E
Premont, Tx., U.S. (prē-mŏnt′)	**122-23**	27°22′N	98°07′W
Prenzlau, Ger. (prĕnts′lou)	**194-95**	53°19′N	13°52′E
Preparis Island, i., Mya.	**246-47**	14°53′N	93°41′E
Preparis North Channel, strt., Mya.	**246-47**	15°32′N	94°06′E
Preparis South Channel, strt., Mya.	**246-47**	14°37′N	93°53′E
Přerov, Czech Rep. (przhě′rôf)	**194-95**	49°27′N	17°28′E
Prescott, On., Can. (prĕs′kŭt)	**136-37**	44°43′N	75°31′W
Prescott, Ar., U.S. (prĕs′kŏt)	**120-21**	33°49′N	93°23′W
Prescott, Az., U.S. (prĕs′kŏt)	**118-19**	34°33′N	112°28′W
Presho, S.D., U.S. (prĕsh′ō)	**114-15**	43°55′N	100°04′W
Presidencia Roque Sáenz Peña, Arg.	**173**	26°47′S	60°26′W
Presidente Epitácio, Braz. (prā-sē-dĕn′tĕ å-pē-tä′syò)	**168-69**	21°46′S	52°07′W
Presidente Prudente, Braz.	**168-69**	22°07′S	51°24′W
Presidio, Tx., U.S. (prĕ-sĭ′dĭ-ō)	**122-23**	29°34′N	104°23′W
Presidio, stm., Mex. (prĕ-sē′dyô)	**146-47**	23°05′N	106°17′W
Prešov, Slvk. (prĕ′shôf)	**194-95**	48°59′N	21°15′E
Presque Isle, Me., U.S. (prĕsk-ēl′)	**117a**	46°41′N	68°01′W
Preßburg, nat. cap., Slvk. *see* Bratislava	**194-95**	48°09′N	17°07′E
Preston, Eng., U.K. (prĕs′tŭn)	**190-91**	53°46′N	2°42′W
Preston, Id., U.S. (prĕs′tŭn)	**112-13**	42°06′N	111°53′W
Preto, stm., Braz.	**166-67**	11°21′S	43°52′W
Pretoria, nat. cap., S. Afr. (prē-tō′rĭ-ȧ)	**269c**	25°45′S	28°11′E
Préveza, Grc. (prĕ′vá-zä)	**200-01**	38°57′N	20°45′E
Priboj, Serb. (prē′boi)	**200-01**	43°35′N	19°32′E
Price, Ut., U.S. (prīs)	**118-19**	39°36′N	110°48′W
Price Island, i., B.C., Can.	**132-33**	52°23′N	128°41′W
Prichard, Al., U.S. (prĭt′chârd)	**124-25**	30°44′N	88°05′W
Prienai, Lith. (prē-ĕn′ī)	**192-93**	54°38′N	23°57′E
Prieska, S. Afr. (prē-ĕs′kȧ)	**264-65**	29°40′S	22°44′E
Prijedor, Bos. (prē′yĕ-dôr)	**200-01**	44°59′N	16°42′E
Prijepolje, Serb. (prē′yĕ-pō′lyĕ)	**200-01**	43°24′N	19°39′E
Prikaspiyskaya Nizmennost', pl., *see* Caspian Depression	**186-87**	48°0′N	52°00′E
Prilep, Mac. (prē′lĕp)	**200-01**	41°21′N	21°34′E
Primorsk, Russia (prē-môrsk′)	**192-93**	60°22′N	28°38′E
Primrose Lake, lk., Can.	**132-33**	54°55′N	109°45′W
Prince Albert, Sk., Can. (prĭns ăl′bērt)	**134-35**	53°13′N	105°45′W
Prince Albert National Park, n.p., Sk., Can. (prĭns ăl′bērt năsh′ŭn-ăl pärk)	**134-35**	54°0′N	106°25′W
Prince Albert Sound, b., N.T., Can. (prĭns ăl′bērt sound)	**130-31**	70°27′N	114°50′W
Prince Charles Island, i., Nu., Can. (prĭns chärlz ī′lănd)	**130-31**	67°47′N	76°06′W
Prince Edward Island, state, Can. (prĭns ĕd′wērd ī′lănd)	**128-29**	46°20′N	63°20′W
Prince Edward Island, i., P.E., Can.	**130-31**	46°20′N	63°20′W
Prince Edward Island National Park, n.p., P.E., Can.	**138-39**	46°30′N	63°25′W
Prince Edward Islands, is., S. Afr. (prĭns ĕd′wērd ī′lăndz)	**287**	46°45′S	37°49′E
Prince George, B.C., Can. (prĭns jôrj)	**132-33**	53°54′N	122°46′W
Prince of Wales, Cape, c., Ak., U.S. (kāp prĭns ŭv wālz)	**126**	65°40′N	168°07′W
Prince of Wales Island, i., Austl. (prĭns ŭv wālz ī′lănd)	**277**	10°40′S	142°10′E
Prince of Wales Island, i., Nu., Can. (prĭns ŭv wālz ī′lănd)	**288**	72°40′N	99°00′W
Prince of Wales Island, i., Ak., U.S. (prĭns ŭv wālz ī′lănd)	**126**	55°47′N	132°50′W
Prince Rupert, B.C., Can. (prĭns rōō′pērt)	**132-33**	54°19′N	130°17′W
Princess Royal Island, i., B.C., Can.	**132-33**	52°57′N	128°49′W
Princeton, B.C., Can. (prĭns′tŭn)	**132-33**	49°27′N	120°31′W
Princeton, Il., U.S. (prĭns′tŭn)	**116-17**	41°22′N	89°28′W
Princeton, In., U.S. (prĭns′tŭn)	**116-17**	38°21′N	87°35′W
Princeton, Ky., U.S. (prĭns′tŭn)	**124-25**	37°07′N	87°53′W
Princeton, Mn., U.S. (prĭns′tŭn)	**114-15**	45°34′N	93°35′W
Princeton, Mo., U.S. (prĭns′tŭn)	**120-21**	40°24′N	93°35′W

Feature (Pronunciation)	Page	Lat.	Long.
Princeton, N.J., U.S. (prĭns′tŭn)	**116-17**	40°22′N	74°39′W
Princeton, W.V., U.S. (prĭns′tŭn)	**124-25**	37°22′N	81°06′W
Prince William Sound, strt., Ak., U.S. (prĭns wĭl′yăm sound)	**126**	60°42′N	147°07′W
Príncipe, i., S. Tom./P. (prēn′sē-pě)	**260-61**	1°37′N	7°25′E
Principe Channel, strt., B.C., Can. (prĭn′sĭ-pē chăn′ĕl)	**132-33**	53°28′N	130°00′W
Príncipe da Beira, Braz.	**166-67**	12°25′S	64°25′W
Prineville, Or., U.S. (prīn′vĭl)	**112-13**	44°18′N	120°51′W
Prinzapolka, stm., Nic. (prēn-zä-pōl′kä)	**149**	13°24′N	83°34′W
Priozërsk, Russia (prĭ-ô′zěrsk)	**192-93**	61°02′N	30°09′E
Pripet Marshes, reg., Eur.	**194-95**	52°0′N	27°00′E
Priština, Serb. (prēsh′tĭ-nä)	**200-01**	42°40′N	21°10′E
Pritzwalk, Ger. (prēts′välk)	**194-95**	53°09′N	12°10′E
Privas, Fr. (prē-väs′)	**196-97**	44°44′N	4°37′E
Privolzhskaya Vozvyshennost', plat., Russia	**186-87**	52°0′N	46°00′E
Privolzhskiy, Russia	**186-87**	51°24′N	46°02′E
Priyutovo, Russia	**186-87**	53°53′N	53°56′E
Prizren, Serb. (prē′zrěn)	**200-01**	42°13′N	20°45′E
Probolinggo, Indon.	**248-49**	7°45′S	113°13′E
Proctor, Mn., U.S. (prŏk′tēr)	**114-15**	46°44′N	92°14′W
Proddatūr, India	**236**	14°45′N	78°33′E
Progreso, Mex. (prô-grä′sō)	**148**	21°16′N	89°39′W
Prokopyevsk, Russia	**226**	53°54′N	86°44′E
Prokuplje, Serb. (prô′kôop′l-yě)	**200-01**	43°14′N	21°36′E
Prome, Mya.	**246-47**	18°49′N	95°13′E
Pronja, stm., Bela. (prô′nyä)	**192-93**	53°27′N	31°01′E
Propriá, Braz.	**163d**	10°13′S	36°50′W
Prosser, Wa., U.S. (prŏs′ēr)	**112-13**	46°12′N	119°46′W
Prostějov, Czech Rep. (prôs′tyě-yôf)	**194-95**	49°29′N	17°07′E
Protoka, stm., Russia (prôt′ô-kå)	**202-03**	45°44′N	37°47′E
Providence, R.I., U.S. (prŏv′ĭ-děns)	**116-17**	41°50′N	71°25′W
Providence, Atoll de, at., Sey.	**264-65**	9°14′S	51°03′E
Providencia, Isla de, i., Col.	**164-65**	13°21′N	81°22′W
Providenciales, i., T./C. Is.	**142-43**	21°47′N	72°17′W
Provideniya, Russia (prô-vĭ-dä′nĭ-yå)	**126**	64°23′N	173°18′W
Provo, Ut., U.S. (prō′vō)	**118-19**	40°13′N	111°38′W
Prudhoe Bay, b., Ak., U.S.	**126**	70°21′N	148°22′W
Prudnik, Pol. (prôd′nĭk)	**194-95**	50°19′N	17°35′E
Pruszków, Pol. (prôsh′kôf)	**194-95**	52°10′N	20°49′E
Prut, stm., Eur. (prōōt)	**186-87**	45°28′N	28°13′E
Prydz Bay, b., Ant.	**287**	69°0′S	76°00′E
Pryluky, Ukr.	**202-03**	50°36′N	32°23′E
Pryor, Ok., U.S. (prī′ēr)	**120-21**	36°19′N	95°19′W
Przemyśl, Pol. (pzhě′mĭsh′l)	**194-95**	49°47′N	22°47′E
Przhevalsk, Kyrg. (p′r-zhĭ-välsk′) *see* Karakol	**226**	42°29′N	78°23′E
Pskov, Russia (pskôf)	**192-93**	57°49′N	28°22′E
Pskov, Lake, lk., Eur. (lāk pskôf)	**192-93**	58°0′N	28°00′E
Pskovskoye Ozero, lk., Eur. (p′skôv′skô′yě ôzě-rô) *see* Pskov, Lake	**192-93**	58°0′N	28°00′E
Ptuj, Slvn. (ptōō′ē)	**200-01**	46°26′N	15°52′E
Pucallpa, Peru	**170**	8°23′S	74°32′W
Pucheng, China (pōō-chŭŋ)	**238-39**	27°55′N	118°32′E
Pucheng, China (pōō-chŭŋ)	**238-39**	34°58′N	109°35′E
Puck, Pol. (pôtsk)	**194-95**	54°43′N	18°24′E
Pudozh, Russia (pōō′dôzh)	**186-87**	61°48′N	36°34′E
Puducherry, India *see* Pondicherry	**236**	11°56′N	79°50′E
Puducherry, state, India	**236**	11°56′N	79°50′E
Pudukkottai, India	**236**	10°23′N	78°49′E
Puebla, state, Mex. (pwā′blä)	**146-47**	18°50′N	98°00′W
Puebla de Zaragoza, Mex.	**146-47**	19°03′N	98°12′W
Pueblo, Co., U.S. (pwā′blō)	**120-21**	38°16′N	104°38′W
Puente Genil, Spain (pwěn′tå-hå-nēl′)	**198-99**	37°24′N	4°47′W
Puerto Aisén, Chile (pwě′r-tō ä′y-sě′n)	**171**	45°15′S	72°15′W
Puerto Ángel, Mex. (pwě′r-tō äŋ′hål)	**146-47**	15°40′N	96°29′W
Puerto Armuelles, Pan. (pwe′r-tô är-mōō-ā′lyās)	**150**	8°17′N	82°52′W
Puerto Asís, Col.	**164-65**	0°31′N	76°31′W
Puerto Ayacucho, Ven.	**164-65**	5°40′N	67°38′W
Puerto Baquerizo Moreno, Ec.	**170a**	0°54′S	89°36′W
Puerto Barrios, Guat. (pwě′r-tô bär′rē-ôs)	**148**	15°43′N	88°35′W
Puerto Bermúdez, Peru (pwě′r-tō běr-mōō′dāz)	**170**	10°20′S	74°54′W
Puerto Berrío, Col. (pwě′r-tō běr-rē′ō)	**163c**	6°28′N	74°26′W
Puerto Cabello, Ven. (pwě′r-tō kä-běl′yō)	**163b**	10°28′N	68°01′W
Puerto Cabezas, Nic. (pwě′r-tô kä-bā′zäs)	**149**	14°01′N	83°23′W
Puerto Carreño, Col.	**164-65**	6°11′N	67°30′W
Puerto Chicama, Peru (pwě′r-tō chē-kä′mä)	**170**	7°42′S	79°25′W
Puerto Cortés, Hond. (pwě′r-tô kôr-tās′)	**149**	15°51′N	87°57′W

Feature (Pronunciation)	Page	Lat.	Long.
Puerto Cumarebo, Ven. (pwě′r-tō kōō-mä-rě′bô)	**164-65**	11°29′N	69°21′W
Puerto de la Cruz, Spain	**199d**	28°23′N	16°33′W
Puerto Deseado, Arg. (pwě′r-tō dā-så-ä′dhō)	**171**	47°44′S	65°54′W
Puerto Juárez, Mex.	**148**	21°10′N	86°49′W
Puerto la Cruz, Ven. (pwě′r-tō lä krōō′z)	**163b**	10°13′N	64°38′W
Puerto Leguízamo, Col.	**164-65**	0°11′S	74°46′W
Puerto Libertad, Mex.	**144-45**	29°55′N	112°41′W
Puerto Limón, C.R.	**149**	9°59′N	83°02′W
Puertollano, Spain (pwě-tôl-yä′nō)	**198-99**	38°41′N	4°06′W
Puerto Madryn, Arg. (pwě′r-tō mä-drěn′)	**171**	42°46′S	65°03′W
Puerto Maldonado, Peru (pwě′r-tō mäl-dō-nä′dô)	**170**	12°36′S	69°12′W
Puerto Montt, Chile (pwě′r-tō mô′nt)	**171**	41°28′S	72°57′W
Puerto Morazán, Nic.	**149**	12°50′N	87°11′W
Puerto Natales, Chile (pwě′r-tō nä-tá′lěs)	**171**	51°42′S	72°29′W
Puerto Padre, Cuba (pwě′r-tō pä′drā)	**142-43**	21°12′N	76°36′W
Puerto Peñasco, Mex. (pwě′r-tô pěn-yä′s-kô)	**144-45**	31°19′N	113°32′W
Puerto Pinasco, Para. (pwě′r-tō pē-ná′s-kô)	**168-69**	22°37′S	57°49′W
Puerto Pirámides, Arg.	**171**	42°34′S	64°15′W
Puerto Píritu, Ven. (pwě′r-tō pē′rē-tōō)	**163b**	10°02′N	65°02′W
Puerto Plata, Dom. Rep. (pwě′r-tō plä′tä)	**142-43**	19°45′N	70°39′W
Puerto Princesa, Phil.	**250**	9°44′N	118°45′E
Puerto Rico, Bol.	**166-67**	11°06′S	67°32′W
Puerto Rico, dep., N.A. (pwěr′tô rē′kō)	**140-41**	18°15′N	66°30′W
Puerto Rico, i., P.R. (pwěr′tô rē′kō)	**142a**	18°15′N	66°30′W
Puerto Salgar, Col. (pwě′r-tō säl-gär′)	**163c**	5°28′N	74°39′W
Puerto San José, Guat.	**148**	13°56′N	90°49′W
Puerto San Julián, Arg.	**171**	49°18′S	67°43′W
Puerto Santa Cruz, Arg. (pwě′r-tō sän′tä krōōz′)	**171**	50°01′S	68°34′W
Puerto Sastre, Para.	**168-69**	22°02′S	58°01′W
Puerto Suárez, Bol. (pwě′r-tō swä′råz)	**168-69**	18°57′S	57°51′W
Puerto Tejada, Col. (pwě′r-tō tě-κä′dä)	**163c**	3°14′N	76°25′W
Puerto Vallarta, Mex. (pwě′r-tō väl-yär′tä)	**146-47**	20°37′N	105°14′W
Puerto Varas, Chile (pwě′r-tō vä′räs)	**171**	41°20′S	72°58′W
Puerto Villamil, Ec.	**170a**	0°56′S	91°01′W
Puerto Wilches, Col. (pwě′r-tō vēl′c-hěs)	**164-65**	7°20′N	73°54′W
Pueyrredón, Lago, lk., S.A.	**171**	47°21′S	71°56′W
Puget Sound, b. Wa., U.S.	**112-113**	47°49′N	122°27′W
Pugachev, Russia (pōō′gá-chyôf)	**186-87**	52°02′N	48°49′E
Puhi-waero, c., N.Z. *see* South West Cape	**278**	47°17′S	167°28′E
Pukch'ŏng-ŭp, Kor., N.	**243**	40°14′N	128°19′E
Pukou, China	**238-39**	32°06′N	118°43′E
Pula, Cro. (pōō′lä)	**200-01**	44°52′N	13°51′E
Pulacayo, Bol. (pōō-lä-kä′yō)	**168-69**	20°23′S	66°42′W
Pulaski, N.Y., U.S. (pů-lăs′kĭ)	**116-17**	43°34′N	76°07′W
Pulaski, Tn., U.S. (pů-lăs′kĭ)	**124-25**	35°12′N	87°02′W
Pulaski, Va., U.S. (pů-lăs′kĭ)	**124-25**	37°03′N	80°47′W
Puławy, Pol. (pò-wä′vě)	**194-95**	51°25′N	21°59′E
Pullman, Wa., U.S. (pòl′măn)	**112-13**	46°44′N	117°10′W
Pulo Anna, i., Palau	**280-81**	4°40′N	131°58′E
Pulog, Mount, mtn., Phil. (mount pōō′lôg)	**250**	16°36′N	120°54′E
Puma Yumco, lk., China (pōō-mä yōōm-tswo)	**234-35**	28°33′N	90°24′E
Puná, Isla, i., Ec.	**170**	2°47′S	80°08′W
Punakha, Bhu. (pōō-nŭk′ŭ)	**234-35**	27°37′N	89°52′E
Punata, Bol. (pōō-nä′tä)	**168-69**	17°33′S	65°50′W
Pune, India	**236**	18°32′N	73°52′E
Punia, D.R.C.	**262-63**	1°28′S	26°27′E
Punjab, state, India (pŭn′jäb′)	**234-35**	31°0′N	75°30′E
Puno, Peru (pōō′nô)	**170**	15°51′S	70°02′W
Punta Alta, Arg.	**173**	38°53′S	62°04′W
Punta Arenas, Chile (pōō′n-tä-rě′näs)	**171**	53°09′S	70°55′W
Punta de Piedras, Ven. (pōō′n-tä dě pyě′dräs)	**163b**	10°54′N	64°06′W
Punta Gorda, Belize (pòn′tä gôr′dä)	**148**	16°06′N	88°48′W
Punta Gorda, Fl., U.S. (pŭn′tá gôr′dá)	**125a**	26°56′N	82°03′W
Puntarenas, C.R. (pònt-ä-rā′näs)	**149**	9°58′N	84°50′W
Punto Fijo, Ven. (pōō′n-tô fē′κô)	**164-65**	11°43′N	70°12′W
Punxsutawney, Pa., U.S. (pŭnk-sŭ-tô′nē)	**116-17**	40°56′N	78°58′W
Puqi, China	**238-39**	29°43′N	113°53′E
Puquio, Peru (pōō′kyô)	**170**	14°42′S	74°09′W

Feature (Pronunciation)	Page	Lat.	Long.
Pur, stm., Russia	**218-19**	67°21′N	77°55′E
Purcell, Ok., U.S. (pûr-sěl′)	**120-21**	35°02′N	97°22′W
Puri, India (pò′rē)	**234-35**	19°48′N	85°51′E
Purificación, Col. (pōō-rē-fē-kä-syōn′)	**163c**	3°51′N	74°55′W
Purificación, Mex. (pōō-rē-fē-kä-syô′n)	**146-47**	19°43′N	104°36′W
Pūrnia, India	**234-35**	25°47′N	87°29′E
Pursat, Camb. *see* Poŭthĭsăt	**246-47**	12°32′N	103°56′E
Purus, stm., S.A.	**166-67**	3°41′S	61°28′W
Purús, stm., S.A. (pōō-rōō′s)	**166-67**	3°41′S	61°28′W
Purwokerto, Indon.	**248-49**	7°25′S	109°14′E
Pusan, Kor., S. (pōō′sän′)	**243**	35°05′N	129°03′E
Pushkin, Russia (pòsh′kĭn)	**192-93**	59°43′N	30°26′E
Pustoshka, Russia (pûs-tôsh′kå)	**192-93**	56°20′N	29°22′E
Putaendo, Chile (pōō-tä-ěn-dô)	**163e**	32°37′S	70°44′W
Putao, Mya.	**238-39**	27°21′N	97°24′E
Putian, China (pōō-tǐěn)	**225a**	25°26′N	119°00′E
Puting, Tanjung, c., Indon.	**248-49**	3°32′S	111°49′E
Putla de Guerrero, Mex. (pōō′tlä-dě-gěr-rě′rō)	**146-47**	17°00′N	97°54′W
Putnam, Ct., U.S. (pŭt′năm)	**116-17**	41°55′N	71°54′W
Putorana, Gory, plat., Russia	**218-19**	69°0′N	95°00′E
Putrajaya, nat. cap., Malay.	**246-47**	2°56′N	101°43′E
Puttalam, Sri L.	**236**	8°01′N	79°51′E
Putumayo, stm., S.A. (pò-tōō-mä′yō)	**170**	3°07′S	67°56′W
Puyallup, Wa., U.S. (pū-ăl′ŭp)	**112-13**	47°11′N	122°17′W
Puyang, China (pōō-yäŋ)	**240-41**	35°42′N	115°00′E
Puyo, Ec.	**170**	1°29′S	77°59′W
Pweto, D.R.C. (pwä′tō)	**262-63**	8°28′S	28°54′E
Pyakupur, stm., Russia	**218-19**	64°56′N	77°44′E
Pyandzh, stm., Asia *see* Panj	**232-33**	37°00′N	68°16′E
Pyasina, stm., Russia (pyä-sē′nà)	**218-19**	73°52′N	87°09′E
Pyasino, Ozero, lk., Russia	**218-19**	69°45′N	87°45′E
Pyatigorsk, Russia (pyá-tē-gôrsk′)	**227**	44°04′N	43°04′E
Pyè, Mya. *see* Prome	**246-47**	18°49′N	95°13′E
Pyinmana, Mya. (pyěn-mä′nŭ)	**246-47**	19°44′N	96°13′E
P'yŏngyang, nat. cap., Kor., N. (pyŭng′gäng′)	**243**	39°01′N	125°44′E
Pyramid Lake, lk., Nv., U.S. (pĭ′rá-mĭd lāk)	**118-19**	40°01′N	119°35′W
Pyramid Lake Indian Reservation, ind. res., Nv., U.S. (pĭ′rá-mĭd lāk ĭn′dĭ-ăn rě-sěr-vā′shěn)	**118-19**	40°13′N	119°36′W
Pyrenees, mts., Eur. (pĭr-e-nēz′)	**196-97**	42°40′N	1°00′E
Pýrgos, Grc.	**200-01**	37°40′N	21°27′E
Pyritz, Pol. *see* Pyrzyce	**194-95**	53°09′N	14°53′E
Pyrzyce, Pol. (pězhĭ′tsě)	**194-95**	53°09′N	14°53′E

Q

Feature (Pronunciation)	Page	Lat.	Long.
Qā'en, Iran	**232-33**	33°44′N	59°10′E
Qaidam, stm., China	**240-41**	36°52′N	95°57′E
Qaidam Pendi, bas., China	**222-23**	37°0′N	95°00′E
Qal'at Bīshah, Sau. Ar.	**266**	19°60′N	42°36′E
Qalāt, Afg.	**232-33**	32°07′N	66°54′E
Qamani'tuaq, Nu., Can. *see* Baker Lake	**128-29**	64°18′N	95°55′W
Qamar, Ghubbat al-, b., Yemen	**220-21**	16°0′N	52°30′E
Qamdo, China (chyäm-dwō)	**238-39**	31°10′N	97°09′E
Qamea, i., Fiji	**279f**	16°46′S	179°46′W
Qandahār, Afg. *see* Kandahār	**232-33**	31°37′N	65°43′E
Qandala, Som.	**262-63**	11°28′N	49°52′E
Qapshaghay, Kaz.	**226**	43°52′N	77°04′E
Qapshaghay bögeni, res., Kaz.	**226**	43°49′N	77°42′E
Qaqortoq, Green. *see* Julianehåb	**284-85**	60°44′N	46°02′W
Qaraghandy, Kaz.	**226**	49°53′N	73°10′E
Qarataū, Kaz.	**226**	43°10′N	70°28′E
Qarataū zhotasy, mts., Kaz. *see* Karatau Range	**226**	43°36′N	68°52′E
Qaraton, Kaz.	**186-87**	46°26′N	53°31′E
Qarazhal, Kaz.	**226**	48°01′N	70°49′E
Qarqan, stm., China	**222-23**	39°26′N	88°22′E
Qarshi, Uzb.	**232-33**	38°52′N	65°48′E
Qârûn, Birket, lk., Egypt	**268b**	29°28′N	30°39′E
Qash, Nahr al-, stm., Afr. *see* Gash	**266**	16°45′N	35°54′E
Qatar, nation, Asia (kä′tår)	**206-07**	25°0′N	51°10′E
Qattâra, Munkhafad el-, depr., Egypt *see* Qattara Depression	**266**	30°0′N	27°30′E
Qattara Depression, depr., Egypt (kä-tä′rá dĭ-prě′shŭn)	**266**	30°0′N	27°30′E
Qazaqstan, nation, Asia *see* Kazakhstan	**206-07**	47°0′N	76°00′E
Qazaqtyng usaqshoqylyghy, hills, Kaz. *see* Kazakh Hills	**226**	49°0′N	72°00′E

ăt; finăl; rāte; senāte; ärm; åsk; sofá; fåre; ch-choose; dh-as th in other; bē; ēvent; bět; recěnt; cratēr; g-gō; gh-guttural g; bĭt; ĭ-short neutral; rīde; κ-guttural k as ch in German ich;

Feature (Pronunciation)	Page	Lat.	Long.
Qazımämmäd, Azer.	**227**	40°02'N	48°56'E
Qazvīn, Iran	**232-33**	36°16'N	49°58'E
Qena, Egypt	**268b**	26°11'N	32°43'E
Qeqertarsuaq, Green. *see* Godhavn	**284-85**	69°15'N	53°33'W
Qeshm, Jazīreh-ye, i., Iran	**230-31**	26°45'N	55°45'E
Qezel Owzan, stm., Iran	**228-29**	36°47'N	49°09'E
Qianyang, China	**238-39**	27°11'N	110°02'E
Qiemo, China	**220-21**	38°10'N	85°30'E
Qijiang, China (chyē-jyäŋ)	**238-39**	29°02'N	106°39'E
Qilian Shan, mts., China (chyē-lĭĕn shän)	**240-41**	39°06'N	98°40'E
Qing'an, China (chyĭŋ-än)	**240-41**	46°52'N	127°30'E
Qingdao, China (chyĭn-dou)	**240-41**	36°05'N	120°20'E
Qinghai, state, China (chyĭŋ-hī)	**240-41**	36°0'N	96°00'E
Qinghai Hu, lk., China (chyĭŋ-hī hōō)	**240-41**	36°48'N	100°06'E
Qingjiang, China (chyĭŋ-jyäŋ)	**238-39**	33°36'N	119°01'E
Qingshui, stm., China	**238-39**	27°08'N	109°37'E
Qingshui, stm., China	**240-41**	37°28'N	105°32'E
Qingtang, China	**238-39**	24°12'N	113°51'E
Qingyang, China (chyĭŋ-yäŋ)	**238-39**	30°38'N	117°51'E
Qingyang, China (chyĭŋ-yäŋ)	**240-41**	36°01'N	107°52'E
Qingyuan, China (chyĭŋ-yōän) *see* Baoding	**240-41**	38°51'N	115°29'E
Qingyuan, China (chyĭŋ-yōän)	**238-39**	23°42'N	113°02'E
Qingyuan, China (chyĭŋ-yōän)	**238-39**	27°37'N	119°06'E
Qingyuan, China (chyĭŋ-yōän)	**243**	42°07'N	124°58'E
Qingyuan, China (chyĭŋ-yōän)	**238-39**	24°35'N	108°45'E
Qing Zang Gaoyuan, plat., China *see* Tibet, Plateau of	**222-23**	33°0'N	92°00'E
Qinhuangdao, China (chyĭn-huaŋ-dou)	**240-41**	39°56'N	119°36'E
Qin Ling, mts., China (chyĭn lĭŋ)	**238-39**	34°0'N	108°00'E
Qinzhou, China (chyĭn-jō)	**238-39**	21°58'N	108°37'E
Qionghai, China (chyôŋ-hī)	**238-39**	19°16'N	110°28'E
Qionglai, China	**238-39**	30°25'N	103°28'E
Qiongzhong, China	**238-39**	19°02'N	109°48'E
Qiongzhou Haixia, strt., China	**238-39**	20°10'N	110°15'E
Qiqihar, China (chyē-chyē-här)	**240-41**	47°20'N	123°58'E
Qitai, China (chyē-tī)	**222-23**	44°01'N	89°35'E
Qixian, China (chyē-shyĕn)	**238-39**	34°33'N	114°47'E
Qiyang, China (chyē-yäŋ)	**238-39**	26°29'N	111°43'E
Qızılağaç körfäzi, b., Azer.	**227**	39°05'N	49°01'E
Qizilqum, des., Asia	**226**	42°0'N	64°00'E
Qo'qon, Uzb.	**232-33**	40°32'N	70°56'E
Qogir Feng, mtn., Asia *see* K2	**232-33**	35°53'N	76°30'E
Qom, Iran	**232-33**	34°39'N	50°53'E
Qomolangma Feng, mtn., Asia *see* Everest, Mount	**234-35**	27°59'N	86°56'E
Qomsheh, Iran	**232-33**	32°00'N	51°52'E
Qondūz, Afg. *see* Kondoz	**232-33**	36°44'N	68°51'E
Qorako'l, Uzb.	**232-33**	39°32'N	63°55'E
Qosshaghyl, Kaz.	**186-87**	46°51'N	53°48'E
Qostanay, Kaz.	**226**	53°12'N	63°37'E
Quadra Island, i., B.C., Can.	**132-33**	50°12'N	125°16'W
Quakertown, Pa., U.S. (kwā'kēr-toun)	**116-17**	40°27'N	75°21'W
Quanah, Tx., U.S. (kwä'nȧ)	**120-21**	34°18'N	99°45'W
Quang Ngai, Viet. (kwäng n'gä'ē)	**246-47**	15°07'N	108°47'E
Quan Long, Viet. *see* Ca Mau	**246-47**	9°11'N	105°09'E
Quanzhou, China (chyuän-jō)	**225a**	24°55'N	118°35'E
Qu'Appelle, stm., Can.	**134-35**	50°26'N	101°20'W
Quartu Sant'Elena, Italy (kwär-tōō' sänt a'lå-nä)	**200-01**	39°15'N	9°11'E
Quatsino Sound, strt., B.C., Can. (kwŏt-sē'nō sound)	**132-33**	50°26'N	127°59'W
Quba, Azer. (kōō'bä)	**227**	41°22'N	48°31'E
Qūchān, Iran	**232-33**	37°06'N	58°31'E
Queanbeyan, Austl.	**276**	35°21'S	149°14'E
Québec, Qc., Can. (kwē-bĕk') (kå-bĕk')	**136-37**	46°49'N	71°13'W
Québec, state, Can. (kwē-bĕk') (kē-bĕk')	**128-29**	52°0'N	72°00'W
Quedlinburg, Ger. (kvĕd'lĕn-bōōrgh)	**194-95**	51°47'N	11°09'E
Queen Charlotte, B.C., Can.	**132-33**	53°16'N	132°05'W
Queen Charlotte Islands, is., B.C., Can. (kwēn shär'lŏt ī'lǎndz)	**132-33**	53°0'N	132°00'W
Queen Charlotte Mountains, mts., B.C., Can. (kwēn shär'lŏt moun'tĭnz)	**132-33**	53°0'N	132°00'W
Queen Charlotte Sound, strt., B.C., Can. (kwēn shär'lŏt sound)	**132-33**	51°30'N	129°30'W
Queen Charlotte Strait, strt., B.C., Can. (kwēn shär'lŏt strāt)	**132-33**	50°50'N	127°25'W
Queen Fabiola Mountains, mts., Ant.	**287**	71°30'S	35°40'E
Queen Maud Gulf, b., Nu., Can. (kwēn mäd gŭlf)	**130-31**	68°25'N	102°30'W
Queen Maud Land, reg., Ant. (kwēn mäd lănd)	**287**	74°59'S	15°51'E
Queensland, state, Austl. (kwēnz'lǎnd)	**277**	22°0'S	145°00'E
Queenstown, Austl. (kwēnz'toun)	**276**	42°04'S	145°33'E
Queenstown, S. Afr.	**264-65**	31°54'S	26°53'E
Quelelevu, i., Fiji	**279f**	16°05'S	179°09'W
Quelimane, Moz. (kā-lē-mä'nĕ)	**264-65**	17°53'S	36°53'E
Quelpart Island, i., Kor., S. *see* Cheju-do	**240-41**	33°22'N	126°30'E
Quemado de Güines, Cuba (kā-mä'dhä-dĕ-gwē'nĕs)	**142-43**	22°48'N	80°15'W
Quemoy, i., Tai. *see* Chinmen Tao	**225a**	24°27'N	118°23'E
Querétaro, Mex. (kå-rā'tä-rō)	**146-47**	20°35'N	100°23'W
Querétaro, state, Mex. (kå-rā'tä-rō)	**146-47**	21°0'N	99°55'W
Quesnel, B.C., Can. (kā-nĕl')	**132-33**	52°58'N	122°29'W
Quesnel, stm., B.C., Can. (kā-nĕl')	**132-33**	52°58'N	122°30'W
Quesnel Lake, lk., B.C., Can. (kā-nĕl' lāk)	**132-33**	52°32'N	121°05'W
Quetta, Pak. (kwĕt'ä)	**232-33**	30°13'N	67°01'E
Quetzaltenango, Guat.	**148**	14°50'N	91°31'W
Quevedo, Ec.	**170**	1°01'S	79°27'W
Quezon City, Phil. (kā-zōn sē'tē)	**250**	14°38'N	121°03'E
Qufu, China (chyōō-fōō)	**240-41**	35°36'N	117°02'E
Quibdó, Col. (kēb'dō)	**163c**	5°42'N	76°39'W
Quila, Mex.	**146-47**	24°25'N	107°13'W
Quillacollo, Bol.	**168-69**	17°26'S	66°17'W
Quillota, Chile (kēl-yō'tä)	**163e**	32°51'S	71°14'W
Quilon, India (kwē-lōn') *see* Kollam	**236**	8°53'N	76°35'E
Quilpie, Austl. (kwĭl'pē)	**276**	26°37'S	144°16'E
Quimbaya, Col. (kēm-bä'yä)	**163c**	4°38'N	75°46'W
Quimilí, Arg.	**173**	27°38'S	62°25'W
Quimper, Fr. (kăN-pĕr')	**196-97**	47°60'N	4°06'W
Quince Mil, Peru	**170**	13°15'S	70°37'W
Quincy, Fl., U.S. (kwĭn'sē)	**124-25**	30°35'N	84°35'W
Quincy, Il., U.S. (kwĭn'sē)	**120-21**	39°56'N	91°24'W
Quincy, Ma., U.S. (kwĭn'sē)	**116-17**	42°15'N	71°02'W
Quincy, Mi., U.S. (kwĭn'sē)	**116-17**	41°57'N	84°53'W
Quincy, Wa., U.S. (kwĭn'sē)	**112-13**	47°14'N	119°51'W
Quines, Arg.	**168-69**	32°14'S	65°47'W
Quintana Roo, state, Mex. (kēn-tä-nä rō'ō)	**148**	19°40'N	88°30'W
Quintero, Chile (kēn-tē'rō)	**163e**	32°47'S	71°32'W
Quirihue, Chile	**171**	36°17'S	72°33'W
Quirimba, Ilha, i., Moz.	**264-65**	12°20'S	40°36'E
Quiroga, Mex. (kē-rō'gä)	**146-47**	19°40'N	101°32'W
Quitman, Ga., U.S. (kwĭt'mǎn)	**124-25**	30°47'N	83°34'W
Quitman, Ms., U.S. (kwĭt'mǎn)	**124-25**	32°02'N	88°44'W
Quito, nat. cap., Ec. (kē'tō)	**170**	0°12'S	78°30'W
Quixadá, Braz.	**166-67**	4°58'S	39°01'W
Qujing, China	**238-39**	25°35'N	103°50'E
Qulyndy Zhazyghy, pl., Asia	**226**	53°0'N	79°00'E
Qum, Iran *see* Qom	**232-33**	34°39'N	50°53'E
Qumarlêb, China	**238-39**	34°30'N	95°21'E
Qŭnghirot, Uzb.	**226**	43°03'N	58°51'E
Quorn, Austl. (kwôrn)	**276**	32°21'S	138°03'E
Qŭrghonteppa, Taj.	**232-33**	37°50'N	68°47'E
Quseir, Egypt	**266**	26°07'N	34°17'E
Quxian, China (chyōō-shyĕn)	**238-39**	30°51'N	106°58'E
Quy Nhon, Viet.	**246-47**	13°46'N	109°15'E
Quzhou, China (chyoō-jō)	**238-39**	28°57'N	118°52'E
Qyrghyz zhotasy, mts., Asia *see* Kirgiz Range	**226**	42°29'N	73°50'E
Qyzylorda, Kaz. (kzěl-ôr'dȧ)	**226**	44°51'N	65°30'E
Qyzylqum, des., Asia	**226**	42°0'N	64°00'E

R

Feature (Pronunciation)	Page	Lat.	Long.
Raab, Hung. *see* Győr	**194-95**	47°41'N	17°39'E
Raahe, Fin. (rä'ĕ)	**184-85**	64°41'N	24°31'E
Raba, Indon.	**248-49**	8°29'S	118°45'E
Rabat, nat. cap., Mor. (rȧ-bät')	**269a**	34°01'N	6°50'W
Rabaul, Pap. N. Gui. (rä'boul)	**277**	4°12'S	152°11'E
Rabi, i., Fiji	**279f**	16°30'S	179°58'W
Rābigh, Sau. Ar.	**266**	22°48'N	39°02'E
Race, Cape, c., Nf., Can. (kāp rās)	**138-39**	46°40'N	53°06'W
Rach Gia, Viet.	**246-47**	10°01'N	105°06'E
Racine, Wi., U.S. (rǎ-sēn')	**116-17**	42°43'N	87°47'W
Radford, Va., U.S. (răd'fērd)	**124-25**	37°07'N	80°35'W
Radom, Pol. (rä'dôm)	**194-95**	51°24'N	21°09'E
Radomsko, Pol. (rä-dôm'skô)	**194-95**	51°04'N	19°27'E
Radomyshl', Ukr. (rä-dô-mēsh''l)	**202-03**	50°30'N	29°15'E
Radviliškis, Lith. (rȧd'vē-lēsh'kēs)	**192-93**	55°49'N	23°33'E
Radzyń Podlaski, Pol. (räd'zēn-y' pŭd-lä'skĭ)	**194-95**	51°47'N	22°36'E
Rae, N.T., Can.	**128-29**	62°50'N	116°02'W
Rāe Bareli, India	**234-35**	26°13'N	81°14'E
Raeford, N.C., U.S. (rā'fērd)	**124-25**	34°58'N	79°16'W
Rafaela, Arg. (rä-fä-å'lä)	**173**	31°15'S	61°29'W
Rafḩā', Sau. Ar.	**228-29**	29°41'N	43°28'E
Rafsanjān, Iran	**230-31**	30°24'N	55°59'E
Raga, Sudan	**262-63**	8°28'N	25°41'E
Ragay Gulf, b., Phil.	**250**	13°30'N	122°45'E
Ragged Island, i., Bah.	**142-43**	22°14'N	75°44'W
Ragged Island Range, is., Bah.	**142-43**	22°34'N	75°52'W
Ragusa, Italy (rä-gōō'sä)	**200-01**	36°55'N	14°44'E
Rāichūr, India (rä'ē-chōōr')	**236**	16°12'N	77°21'E
Raigarh, India (rī'gŭr)	**234-35**	21°54'N	83°24'E
Rainbow Bridge National Monument, n.p., Ut., U.S. (rān'bō brĭj năsh'ŭn-ǎl mŏn'ū-měnt)	**118-19**	37°06'N	110°57'W
Rainier, Mount, vol., Wa., U.S. (mount rā-nēr')	**112-13**	46°51'N	121°45'W
Rainy Lake, lk., N.A. (rān'ē lāk)	**134-35**	48°39'N	93°17'W
Rainy River, On., Can. (rān'ē rĭv'ēr)	**134-35**	48°44'N	94°34'W
Raipur, India	**234-35**	21°15'N	81°39'E
Raivavae, i., Fr. Poly.	**280-81**	23°52'S	147°40'W
Rājahmundry, India (räj-ŭ-mŭn'drē)	**236**	17°01'N	81°47'E
Rajang, stm., Malay.	**248-49**	2°10'N	111°21'E
Rājapālaiyam, India	**236**	9°27'N	77°34'E
Rājasthān, state, India (rä'jŭs-tän)	**234-35**	27°0'N	74°00'E
Rājgarh, India	**234-35**	23°55'N	76°55'E
Rājkot, India (räj'kōt)	**234-35**	22°18'N	70°48'E
Rājshāhi, Bngl.	**234-35**	24°22'N	88°36'E
Rakaposhi, mtn., Pak.	**232-33**	36°10'N	74°30'E
Rakata, Pulau, i., Indon. *see* Krakatoa	**248-49**	6°10'S	105°26'E
Rakiura, i., N.Z. *see* Stewart Island	**278**	47°0'S	167°50'E
Rakvere, Est. (rȧk'vĕ-rĕ)	**192-93**	59°21'N	26°22'E
Raleigh, N.C., U.S.	**124-25**	35°47'N	78°39'W
Rambouillet, Fr. (räN-bōō-yĕ')	**196-97**	48°39'N	1°50'E
Rambutyo Island, i., Pap. N. Gui.	**277**	2°18'S	147°49'E
Rāmeswaram, India	**236**	9°17'N	79°19'E
Ramm, Jabal, mtn., Jord.	**228-29**	29°35'N	35°24'E
Ramos, Mex. (rä'mōs)	**146-47**	22°50'N	101°55'W
Ramos Arizpe, Mex. (rä'mōs ä-rēz'pā)	**122-23**	25°32'N	100°57'W
Rāmpur, India (räm'pōōr)	**234-35**	28°48'N	79°01'E
Rampur Boalia, Bngl. *see* Rājshāhi	**234-35**	24°22'N	88°36'E
Ramree Island, i., Mya. (räm'rē' ī'lǎnd)	**246-47**	19°06'N	93°48'E
Ramsey Lake, lk., On., Can. (răm'zē lāk)	**136-37**	47°15'N	82°16'W
Ramsgate, Eng., U.K. (rămz''gāt)	**190-91**	51°20'N	1°24'E
Ramu, stm., Pap. N. Gui. (rä'mōō)	**277**	4°03'S	144°40'E
Ranau, Malay.	**248-49**	5°57'N	116°41'E
Rancagua, Chile (rän-kä'gwä)	**163e**	34°10'S	70°44'W
Rānchi, India	**234-35**	23°21'N	85°20'E
Randers, Den. (rän'ĕrs)	**192-93**	56°28'N	10°03'E
Randleman, N.C., U.S. (răn'd'l-mǎn)	**124-25**	35°49'N	79°49'W
Randolph, Ne., U.S. (răn'dôlf)	**114-15**	42°23'N	97°21'W
Random Island, i., Nf., Can. (răn'dǔm ī'lǎnd)	**138-39**	48°08'N	53°45'W
Rāngāmāti, Bngl.	**246-47**	22°42'N	92°08'E
Rangeley, Me., U.S. (rānj'lē rānj)	**116-17**	44°58'N	70°39'W
Ranger, Tx., U.S. (rān'jēr rānj)	**120-21**	32°29'N	98°41'W
Rangoon, nat. cap., Mya. (răŋ-gōōn') *see* Yangon	**246-47**	16°47'N	96°12'E
Rangpur, Bngl. (rŭng'pōōr)	**234-35**	25°45'N	89°16'E
Rānīganj, India (rä-nē-gŭnj')	**234-35**	23°36'N	87°07'E
Rankin Inlet, Nu., Can.	**128-29**	62°49'N	92°10'W
Rann of Kutch, reg., Asia *see* Kutch, Rann of	**234-35**	24°15'N	70°46'E
Ranongga Island, i., Sol. Is.	**279e**	8°05'S	156°34'E
Rantauprapat, Indon.	**246-47**	2°06'N	99°49'E
Rantekombola, Bulu, mtn., Indon.	**248-49**	3°24'S	120°02'E
Rantoul, Il., U.S. (răn-tōōl')	**116-17**	40°18'N	88°09'W
Raoul Island, i., N.Z.	**280-81**	29°16'S	177°55'W
Rapa, i., Fr. Poly.	**280-81**	27°36'S	144°20'W
Rapallo, Italy (rä-päl'lō)	**200-01**	44°21'N	9°14'E
Rapel, stm., Chile (rä-pål')	**163e**	33°54'S	71°50'W
Rapid City, S.D., U.S. (răp'ĭd sĭ'tē)	**114-15**	44°04'N	103°13'W
Rapla, Est. (räp'lȧ)	**192-93**	59°00'N	24°47'E
Rappahannock, stm., Va., U.S. (răp'ȧ-hăn'ŭk)	**116-17**	37°35'N	76°18'W
Rapu Rapu Island, i., Phil.	**250**	13°12'N	124°09'E
Raraka, at., Fr. Poly.	**280-81**	16°10'S	144°54'W
Raroïa, i., Fr. Poly.	**280-81**	16°01'S	142°27'W
Rarotonga, i., Cook Is.	**280-81**	21°14'S	159°46'W
Ras Dashen Terara, mtn., Eth. (räs dä-shän') *see* Ras Dejen	**266**	13°16'N	38°24'E
Ras Dejen, mtn., Eth.	**266**	13°16'N	38°24'E
Raseiniai, Lith. (rä-syā'nyī)	**192-93**	55°23'N	23°08'E
Rashid, Egypt *see* Rosetta	**268b**	31°24'N	30°25'E
Rashin, Kor., N. *see* Najin	**243**	42°15'N	130°18'E

Feature (Pronunciation)	Page	Lat.	Long.
Rasht, Iran	228-29	37°17'N	49°35'E
Rasshua, Ostrov, i., Russia	218-19	47°45'N	153°01'E
Rasskazovo, Russia (räs-kä´sô-vô)	186-87	52°40'N	41°53'E
Rastatt, Ger. (rä-shtät)	194-95	48°51'N	8°12'E
Ratangarh, India (rŭ-tŭn´gŭr)	234-35	28°05'N	74°37'E
Rathenow, Ger. (rä´tĕ-nō)	194-95	52°36'N	12°20'E
Rat Islands, is., Ak., U.S. (răt ī´lăndz)	126a	52°0'N	178°00'E
Ratlām, India	234-35	23°20'N	75°02'E
Ratnāgiri, India	236	16°59'N	73°18'E
Raton, N.M., U.S. (rȧ-tōn´)	120-21	36°55'N	104°26'W
Rättvik, Swe. (rĕt´vēk)	192-93	60°53'N	15°07'E
Rauch, Arg. (rà´ōōch)	173	36°47'S	59°06'W
Rauma, Fin. (rä´ô-mȧ)	192-93	61°08'N	21°30'E
Raurkela, India	234-35	22°13'N	84°52'E
Ravenna, Italy (rä-vĕn´nä)	200-01	44°25'N	12°12'E
Ravenna, Ne., U.S. (rȧ-vĕn´ȧ)	114-15	41°02'N	98°55'W
Ravensburg, Ger. (rä´vĕns-bōōrgh)	194-95	47°47'N	9°37'E
Ravensthorpe, Austl. (rā´vĕns-thôrp)	270-71	33°35'S	120°03'E
Ravenswood, W.V., U.S. (rā´vĕnz-wo͝od)	116-17	38°57'N	81°46'W
Rāvi, stm., Asia	232-33	30°37'N	71°53'E
Rawaki, at., Kir.	280-81	3°43'S	170°43'W
Rāwalpindi, Pak. (rä-wŭl-pĕn´dē)	232-33	33°36'N	73°04'E
Rawicz, Pol. (rä´vēch)	194-95	51°37'N	16°52'E
Rawlinna, Austl.	270-71	31°02'S	125°18'E
Rawlins, Wy., U.S. (rô´lĭnz)	112-13	41°47'N	107°14'W
Rawson, Arg.	171	43°19'S	65°06'W
Raxaul, India	234-35	26°59'N	84°50'E
Ray, Cape, c., Nf., Can. (kāp rā)	138-39	47°38'N	59°18'W
Raya, Bukit, mtn., Indon.	248-49	0°40'S	112°41'E
Raychikhinsk, Russia	240-41	49°48'N	129°24'E
Raymond, N.H., U.S. (rā´mŭnd)	116-17	43°02'N	71°11'W
Raymond, Wa., U.S. (rā´mŭnd)	112-13	46°41'N	123°44'W
Raymondville, Tx., U.S. (rā´mŭnd-vĭl)	122-23	26°29'N	97°46'W
Rayne, La., U.S. (rān)	122-23	30°14'N	92°16'W
Raytown, Mo., U.S. (rā´toun)	120-21	38°59'N	94°28'W
Rayville, La., U.S. (rā-vĭl)	124-25	32°29'N	91°46'W
Raz, Pointe du, c., Fr. (pwănt dü rä)	196-97	48°03'N	4°44'W
Razdol'noye, Russia (räz-dôl´nô-yĕ)	244	43°30'N	131°49'E
Razlog, Blg. (räz´lôk)	200-01	41°53'N	23°29'E
Razorback Mountain, mtn., B.C., Can. (rā´zēr-băk moun´tĭn)	132-33	51°35'N	124°42'W
Ré, Île de, i., Fr.	196-97	46°12'N	1°24'W
Reading, Eng., U.K. (rĕd´ĭng)	190-91	51°28'N	0°59'W
Reading, Pa., U.S.	116-17	40°20'N	75°56'W
Real, Cordillera, mts., S.A.	168-69	16°50'S	66°34'W
Realicó, Arg.	173	35°02'S	64°14'W
Rebun-tō, i., Japan (rē´bōōn tō)	244	45°23'N	141°02'E
Recife, Braz. (rå-sē´fĕ)	166-67	8°03'S	34°54'W
Reconquista, Arg. (rā-kôn-kēs´tä)	173	29°09'S	59°38'W
Recreo, Arg.	168-69	29°17'S	65°04'W
Rector, Ar., U.S. (rĕk´tēr)	124-25	36°16'N	90°18'W
Rèçyča, Bela.	202-03	52°22'N	30°25'E
Red, stm., Asia (rĕd)	238-39	20°18'N	106°32'E
Red, stm., N.A. (rĕd)	110-11	50°25'N	96°47'W
Red, stm., U.S. (rĕd)	110-11	29°49'N	91°23'W
Red, stm., Ky., U.S. (rĕd)	116-17	37°50'N	84°06'W
Redang, Pulau, i., Malay.	246-47	5°46'N	103°01'E
Red Bank, Tn., U.S. (rĕd băngk)	124-25	35°07'N	85°17'W
Red Bluff, Ca., U.S.	118-19	40°11'N	122°14'W
Red Bluff Reservoir, res., U.S. (rĕd blŭf rĕ´sēr-vwär)	122-23	31°57'N	103°56'W
Redcliff, Ab., Can. (rĕd´clĭf)	132-33	50°05'N	110°47'W
Redcliffe, Austl. (rĕd´clĭf)	276	27°14'S	153°07'E
Red Cloud, Ne., U.S. (rĕd kloud)	120-21	40°05'N	98°31'W
Red Deer, Ab., Can. (rĕd dēr)	132-33	52°16'N	113°49'W
Red Deer, stm., Can. (rĕd dēr)	132-33	50°55'N	109°53'W
Red Deer, stm., Can. (rĕd dēr)	134-35	52°59'N	100°52'W
Red Deer Lake, lk., Mb., Can. (rĕd dēr lāk)	134-35	52°56'N	101°20'W
Redding, Ca., U.S.	112-13	40°35'N	122°23'W
Redfield, S.D., U.S. (rĕd´fēld)	114-15	44°52'N	98°31'W
Red Indian Lake, lk., Nf., Can. (rĕd ĭn´dĭ-ăn lāk)	138-39	48°39'N	56°50'W
Red Lake, On., Can. (rĕd lāk)	134-35	51°01'N	93°49'W
Red Lake, lk., On., Can.	134-35	51°01'N	94°05'W
Red Lake Falls, Mn., U.S. (rĕd lāk fôlz)	114-15	47°53'N	96°16'W
Red Lake Indian Reservation, ind. res., Mn., U.S. (rĕd lāk ĭn´dĭ-ăn rĕ-sēr-vā´shĕn)	114-15	48°03'N	94°59'W
Red Lion, Pa., U.S. (rĕd lī´ŭn)	116-17	39°54'N	76°36'W
Redmond, Or., U.S. (rĕd´mŭnd)	112-13	44°17'N	121°10'W
Redmond, Wa., U.S. (rĕd´mŭnd)	112-13	47°40'N	122°07'W
Red Oak, Ia., U.S. (rĕd ōk)	114-15	41°01'N	95°14'W
Redon, Fr. (rĕ-dôn´)	196-97	47°39'N	2°05'W
Redonda, i., Antig. (rĕ-dŏn´dȧ)	143b	16°56'N	62°21'W
Red Sea, s., (rĕd sē)	266	20°0'N	38°00'E
Red Sucker Lake, lk., Mb., Can. (rĕd sŭk´ēr lāk)	134-35	54°09'N	93°40'W
Red Wing, Mn., U.S.	114-15	44°34'N	92°32'W
Redwood Falls, Mn., U.S. (rĕd´wo͝od fôlz)	114-15	44°32'N	95°07'W
Redwood National Park, n.p., Ca., U.S. (rĕd´wo͝od năsh´ŭn-ăl pärk)	112-13	41°20'N	124°02'W
Reed City, Mi., U.S. (rēd sĭ´tē)	116-17	43°53'N	85°32'W
Reed Lake, lk., Mb., Can. (rēd lāk)	134-35	54°38'N	100°30'W
Reedley, Ca., U.S. (rēd´lē)	118-19	36°35'N	119°26'W
Reedsburg, Wi., U.S. (rēdz´bûrg)	116-17	43°32'N	89°60'W
Reedsport, Or., U.S. (rēdz´pôrt)	112-13	43°42'N	124°06'W
Reform, Al., U.S. (rē-fôrm´)	124-25	33°23'N	88°01'W
Refugio, Tx., U.S. (rå-fōō´hyô) (rē-fū´jō)	122-23	28°18'N	97°17'W
Rega, stm., Pol. (rĕ-gä)	194-95	54°09'N	15°17'E
Regensburg, Ger. (rā´ghĕns-bŏrgh)	194-95	49°01'N	12°06'E
Reggio di Calabria, Italy (rĕ´jô dē kä-lä´brē-ä)	200-01	38°07'N	15°39'E
Reghin, Rom. (rå-gēn´)	194-95	46°47'N	24°43'E
Regina, Sk., Can. (rē-jī´nȧ)	134-35	50°27'N	104°38'W
Registan, reg., Afg. see Rīgestān	232-33	31°0'N	65°00'E
Rehoboth, Nmb.	264-65	23°19'S	17°05'E
Reḥovot, Isr.	228-29	31°54'N	34°49'E
Reidsville, N.C., U.S. (rēdz´vĭl)	124-25	36°21'N	79°40'W
Reims, Fr. (răɴs)	196-97	49°15'N	4°02'E
Reindeer Lake, lk., Can. (rān´dēr lāk)	134-35	57°16'N	102°15'W
Reinosa, Spain (rå-ē-nō´sä)	198-99	42°60'N	4°08'W
Remada, Tun.	188-89	32°19'N	10°23'E
Remanso, Braz.	166-67	9°37'S	42°07'W
Remedios, Pan. (rĕ-mĕ´dyôs)	150	8°13'N	81°50'W
Remiremont, Fr. (rē-mēr-môɴ´)	196-97	48°01'N	6°36'E
Rendova Island, i., Sol. Is. (rĕn´dô-vȧ ī´lănd)	279e	8°32'S	157°20'E
Rendsburg, Ger. (rĕnts´bŏrgh)	194-95	54°18'N	9°40'E
Renfrew, On., Can. (rĕn´frōō)	136-37	45°29'N	76°42'W
Rengo, Chile (rĕn´gō)	163e	34°29'S	70°53'W
Reni, Ukr. (ran´)	200-01	45°28'N	28°17'E
Renmark, Austl. (rĕn´märk)	276	34°11'S	140°45'E
Rennell, i., Sol. Is. (rĕn-nĕl´)	272-73	11°33'S	160°05'E
Rennes, Fr. (rĕn)	196-97	48°07'N	1°41'W
Reno, Nv., U.S. (rē´nō)	118-19	39°32'N	119°49'W
Reno, Tx., U.S. (rē´nō)	120-21	32°56'N	97°35'W
Renovo, Pa., U.S. (rē-nō´vō)	116-17	41°20'N	77°45'W
Rensselaer, In., U.S. (rĕn´sē-lâr)	116-17	40°57'N	87°09'W
Rensselaer, N.Y., U.S. (rĕn´sē-lâr)	116-17	42°40'N	73°45'W
Renton, Wa., U.S. (rĕn´tŭn)	112-13	47°30'N	122°11'W
Reo, Indon.	248-49	8°19'S	120°29'E
Repetek, Turkmen.	232-33	38°34'N	63°11'E
Republic, Mo., U.S. (rē-pŭb´lĭk)	120-21	37°08'N	93°29'W
República Dominicana, nation, N.A. see Dominican Republic	140-41	19°0'N	70°40'W
Republican, stm., U.S. (rē-pŭb´lĭ-kăn)	110-11	39°03'N	96°48'W
Republican, South Fork, stm., U.S. (south fôrk rē-pŭb´lĭ-kăn)	120-21	40°04'N	101°31'W
Republic of Korea, nation, Asia	206-07	36°30'N	128°00'E
République centrafricaine, nation, Afr. see Central African Republic	253	7°0'N	21°00'E
Repulse Bay, Nu., Can.	128-29	66°32'N	86°14'W
Repulse Bay, b., Austl. (rē-pŭls´ bā)	277	20°36'S	148°43'E
Requena, Spain (rå-kā´nä)	198-99	39°29'N	1°06'W
Resht, Iran see Rasht	228-29	37°17'N	49°35'E
Resistencia, Arg. (rā-sēs-tĕn´syä)	173	27°27'S	59°00'W
Reşiţa, Rom. (rå´shē-tȧ)	200-01	45°18'N	21°53'E
Resolution Island, i., Nu., Can. (rĕz-ô-lū´shŭn ī´lănd)	130-31	61°30'N	65°00'W
Resolution Island, i., N.Z. (rĕz-ŏl-ûshŭn ī´lănd)	278	45°40'S	166°40'E
Restrepo, Col. (rĕs-trĕ´pô)	163c	3°48'N	76°31'W
Retalhuleu, Guat. (rā-täl-ōō-lān´)	148	14°32'N	91°41'W
Rethel, Fr. (r-tĕl')	196-97	49°31'N	4°22'E
Reunion, dep., Afr. (rā-ü-nyôn´)	265a	21°06'S	55°36'E
Réunion, dep., Afr. see Reunion	265a	21°06'S	55°36'E
Reus, Spain (rā´ōōs)	198-99	41°09'N	1°07'E
Reutlingen, Ger. (roit´lĭng-ĕn)	194-95	48°30'N	9°12'E
Reval, nat. cap., Est. see Tallinn	192-93	59°26'N	24°48'E
Revda, Russia (ryâv´dȧ)	186-87	67°58'N	34°34'E
Revelstoke, B.C., Can. (rĕv´ĕl-stōk)	132-33	50°59'N	118°11'W
Revillagigedo, Islas, is., Mex. (ē´s-läs-rĕ-vēl-yä-hĕ´gĕ-dô)	144-45	18°48'N	112°06'W
Revin, Fr. (rē-văɴ)	196-97	49°56'N	4°39'E
Rewa, India (rā´wä)	234-35	24°32'N	81°18'E
Rexburg, Id., U.S. (rĕks´bûrg)	112-13	43°50'N	111°47'W
Rey, Isla del, i., Pan. (ē´s-lä-dĕl-rā´ē)	150	8°22'N	78°55'W
Rey, Laguna del, lk., Mex. (lä-gŏ´nä-dĕl-rā)	122-23	27°01'N	103°24'W
Reyes, Bol. (rā´yĕs)	166-67	14°19'S	67°22'W
Reyes, Point, c., Ca., U.S. (point rā´yĕs)	118-19	38°0'N	123°01'W
Reykjanes, pen., Ice. (rā´kyȧ-nĕs)	190a	63°49'N	22°43'W
Reykjavík, nat. cap., Ice. (rā´kyȧ-vēk)	190a	64°08'N	21°56'W
Reynosa, Mex. (rā-ē-nō´sä)	122-23	26°05'N	98°17'W
Reẕā'īyeh, Iran see Orūmīyeh	228-29	37°32'N	45°05'E
Rēzekne, Lat. (rå´zĕk-nĕ)	192-93	56°30'N	27°20'E
Rheims, Fr. see Reims	196-97	49°15'N	4°02'E
Rhein, stm., Eur. see Rhine	194-95	51°53'N	6°02'E
Rheine, Ger. (rī´nĕ)	194-95	52°17'N	7°27'E
Rhin, stm., Eur. see Rhine	194-95	51°53'N	6°02'E
Rhine, stm., Eur. (rīn)	194-95	51°53'N	6°02'E
Rhinelander, Wi., U.S. (rīn´lăn-dēr)	116-17	45°38'N	89°24'W
Rhir, Cap, c., Mor.	258-59	30°38'N	9°53'W
Rhode Island, state, U.S. (rōd ī´lănd)	108-09	41°40'N	71°30'W
Rhodes, Grc. see Ródos	200-01	36°26'N	28°14'E
Rhodes, i., Grc. (rōdz) see Ródos	200-01	36°10'N	28°00'E
Rhodesia, nation, Afr. see Zimbabwe	253	20°0'S	30°00'E
Rhône, stm., Eur. (rōn)	196-97	43°53'N	4°39'E
Riachão, Braz. (rē-ä-chouɴ´)	166-67	7°22'S	46°39'W
Riau, Kepulauan, is., Indon.	246-47	1°0'N	104°30'E
Ribe, Den. (rē´bĕ)	192-93	55°20'N	8°46'E
Ribeirão Preto, Braz. (rē-bā-rouɴ-prĕ´tô)	172	21°10'S	47°48'W
Riberalta, Bol. (rē-bā-räl´tä)	166-67	11°00'S	66°05'W
Rib Lake, Wi., U.S. (rĭb lāk)	116-17	45°19'N	90°12'W
Rice Lake, Wi., U.S. (rīs lāk)	114-15	45°30'N	91°44'W
Rice Lake, lk., On., Can. (rīs lāk)	136-37	44°08'N	78°13'W
Richards Bay, S. Afr.	264-65	28°47'S	32°05'E
Richardson, Tx., U.S. (rĭch´ērd-sŭn)	120-21	32°58'N	96°44'W
Richardson Mountains, mts., Can. (rĭch´ērd-sŭn moun´tĭnz)	130-31	67°22'N	136°04'W
Richfield, Ut., U.S. (rĭch´fēld)	118-19	38°46'N	112°05'W
Rich Hill, Mo., U.S. (rĭch hĭl)	120-21	38°06'N	94°22'W
Richland, Ga., U.S. (rĭch´lănd)	124-25	32°05'N	84°40'W
Richland, Wa., U.S. (rĭch´lănd)	112-13	46°16'N	119°17'W
Richland Center, Wi., U.S. (rĭch´lănd sĕn´tēr)	114-15	43°20'N	90°23'W
Richmond, Austl. (rĭch´mŭnd)	277	20°44'S	143°08'E
Richmond, B.C., Can. (rĭch´mŭnd)	132-33	49°09'N	123°10'W
Richmond, Qc., Can. (rĭch´mŭnd)	136-37	45°40'N	72°09'W
Richmond, In., U.S. (rĭch´mŭnd)	116-17	39°49'N	84°54'W
Richmond, Ky., U.S. (rĭch´mŭnd)	116-17	37°45'N	84°18'W
Richmond, Mi., U.S. (rĭch´mŭnd)	116-17	42°49'N	82°45'W
Richmond, Mo., U.S. (rĭch´mŭnd)	120-21	39°17'N	93°59'W
Richmond, Va., U.S. (rĭch´mŭnd)	116-17	37°33'N	77°27'W
Richmond Hill, On., Can. (rĭch´mŭnd hĭl)	136-37	43°52'N	79°26'W
Richwood, La., U.S. (rĭch´wo͝od)	120-21	32°27'N	92°06'W
Richwood, W.V., U.S. (rĭch´wo͝od)	116-17	38°13'N	80°34'W
Ridā', Yemen	266	14°38'N	44°54'E
Ridder, Kaz.	226	50°21'N	83°30'E
Riding Mountain National Park, n.p., Mb., Can. (rīd´ĭng moun´tĭn năsh´ŭn-ăl pärk)	134-35	50°55'N	100°25'W
Riesa, Ger. (rē´zȧ)	194-95	51°18'N	13°18'E
Riesco, Isla, i., Chile	171	52°59'S	72°38'W
Rieti, Italy (rē-ā´tē)	200-01	42°24'N	12°52'E
Rif, mts., Mor.	258-59	35°0'N	4°00'W
Rift Valley, val., Afr. (rĭft văl´ē)	254	3°0'S	29°00'E
Rīga, nat. cap., Lat. (rē´gȧ)	192-93	56°57'N	24°06'E
Riga, Gulf of, b., Eur. (gŭlf ŭv rē´gȧ)	192-93	57°30'N	23°35'E
Rīgas jūras līcis, b., Eur. see Riga, Gulf of	192-93	57°30'N	23°35'E
Rigby, Id., U.S. (rĭg´bē)	112-13	43°40'N	111°56'W
Rīgestān, reg., Afg.	232-33	31°0'N	65°00'E
Rijeka, Cro. (rĭ-yĕ´kä)	200-01	45°20'N	14°27'E
Rijn, stm., Eur. see Rhine	194-95	51°53'N	6°02'E
Rima, stm., Nig.	260-61	13°04'N	5°07'E
Rimatara, i., Fr. Poly.	280-81	22°38'S	152°51'W
Rimavská Sobota, Slvk. (rē´måf-skȧ sô´bô-tȧ)	194-95	48°23'N	20°05'E
Rimbo, Swe. (rēm´bô)	192-93	59°45'N	18°22'E
Rimini, Italy (rē´mē-nē)	200-01	44°04'N	12°35'E
Rimouski, Qc., Can. (rē-mōōs´kē)	138-39	48°27'N	68°33'W
Rincón del Bonete, Lago Artificial de, res., Ur.	173	32°43'S	56°01'W
Rincón de Romos, Mex. (rĕn-kōn dā rô-mōs´)	146-47	22°14'N	102°18'W
Ringkøbing, Den. (rĭng´kûb-ĭng)	192-93	56°05'N	8°15'E
Ringsted, Den. (rĭng´stĕdh)	192-93	55°27'N	11°50'E
Ringvassøya, i., Nor. (rĭng´väs-ûê)	184-85	69°55'N	19°15'E
Rinjani, Gunung, vol., Indon.	248-49	8°24'S	116°28'E
Riobamba, Ec. (rē´ō-bäm-bä)	170	1°40'S	78°39'W

ăt; finăl; rāte; senâte; ärm; àsk; sofȧ; fâre; ch-choose; dh-as th in other; bē; ēvent; bĕt; recĕnt; cratēr; g-gō; gh-guttural g; bĭt; ĭ-short neutral; rīde; ĸ-guttural k as ch in German ich;

Feature (Pronunciation)	Page	Lat.	Long.
Rio Branco, Braz. (rē'ó brän'kó)	166-67	9°58's	67°48'w
Río Branco, Ur. (rīō brăncō)	173	32°36's	53°23'w
Rio Casca, Braz. (rē'ō-ká's-kä)	172	20°14's	42°39'w
Río Chico, Ven. (rē'ō chē'kō)	163b	10°18'n	65°59'w
Rio Claro, Braz. (rē'ó klä'rò)	172	22°26's	47°33'w
Río Colorado, Arg.	173	38°60's	64°07'w
Río Cuarto, Arg. (rē'ō kwär'tō)	173	33°08's	64°21'w
Rio de Janeiro, Braz. (rē'ó dā zhä-nå'ê-rò)	172	22°54's	43°14'w
Rio de Janeiro, state, Braz. (rē'ó dā zhä-nå'ê-rò)	172	22°0's	42°30'w
Rio do Sul, Braz.	168-69	27°13's	49°39'w
Río Gallegos, Arg. (rē'ō gä-lā'gōs)	171	51°38's	69°13'w
Río Grande, Arg.	171	53°49's	67°47'w
Rio Grande, Braz. (rē'ó grän'dě)	173	32°02's	52°06'w
Río Grande, Mex. (rē'ō grän'dā)	146-47	15°59'n	97°27'w
Río Grande, Mex. (rē'ō grän'dā)	146-47	23°50'n	103°03'w
Rio Grande, stm., N.A. (rē'ō grän'dā)	110-11	25°57'n	97°09'w
Rio Grande do Sul, Braz. *see* Rio Grande	173	32°02's	52°06'w
Rio Grande do Sul, state, Braz. (rē'ó grän'dě-dô-sōō'l)	173	30°0's	54°00'w
Ríohacha, Col. (rē'ō-ä'chä)	164-65	11°33'n	72°55'w
Río Hato, Pan. (rē'ō-ä'tô)	150	8°23'n	80°10'w
Rio Largo, Braz.	163d	9°29's	35°51'w
Riom, Fr. (rē-ôn')	196-97	45°54'n	3°07'e
Río Mayo, Arg.	171	45°41's	70°14'w
Rio Negro, Braz.	168-69	26°06's	49°47'w
Ríonegro, Col. (rē'ō-nĕ'grō)	163c	6°08'n	75°23'w
Río Negro, state, Arg. (rē'ō nā'grō)	171	40°0's	67°00'w
Rio Pardo, Braz.	168-69	29°59's	52°22'w
Rio Pardo de Minas, Braz. (rē'ō pär'dō-dě-mē'näs)	168-69	15°37's	42°33'w
Ríosucio, Col. (rē'ō-sōō'syō)	163c	5°25'n	75°42'w
Ríosucio, Col. (rē'ō-sōō'syō)	164-65	7°25'n	77°06'w
Río Tercero, Arg. (rē'ō děr-sě'rō)	173	32°11's	64°07'w
Rio Tinto, Braz.	163d	6°48's	35°05'w
Rio Verde, Braz. (rē'ō vĕr'dĕ)	168-69	17°47's	50°55'w
Ríoverde, Mex. (rē'ō-vĕr'då)	146-47	21°56'n	99°59'w
Ripley, Ms., U.S. (rĭp'lē)	124-25	34°45'n	88°57'w
Ripley, Tn., U.S. (rĭp'lē)	124-25	35°45'n	89°32'w
Ripley, W.V., U.S. (rĭp'lē)	116-17	38°48'n	81°44'w
Ripoll, Spain (rē-pōl'')	198-99	42°12'n	2°12'e
Ripon, Wi., U.S. (rĭp'ŏn)	116-17	43°51'n	88°50'w
Rishiri-tō, i., Japan (rē-shē'rē tō)	244	45°11'n	141°15'e
Rising Sun, In., U.S. (rīz'ĭng sŭn)	116-17	38°57'n	84°52'w
Risør, Nor. (rēs'ûr)	192-93	58°43'n	9°14'e
Rittman, Oh., U.S. (rĭt'năn)	116-17	40°58'n	81°47'w
Ritzville, Wa., U.S. (rĭts'vĭl)	112-13	47°07'n	118°22'w
Rivas, Nic. (rē'väs)	149	11°27'n	85°52'w
Rivera, Ur. (rē-vā'rä)	173	30°54's	55°33'w
River Falls, Wi., U.S. (rĭv'ēr fôlz)	114-15	44°52'n	92°37'w
Riverhead, N.Y., U.S. (rĭv'ēr hĕd)	116-17	40°55'n	72°40'w
Rivers, Mb., Can. (rĭv'ērz)	134-35	50°02'n	100°14'w
Riverside, Ca., U.S.	118-19	33°58'n	117°21'w
Rivers Inlet, B.C., Can.	132-33	51°42'n	127°15'w
Rivesaltes, Fr. (rēv'zält')	196-97	42°46'n	2°52'e
Riviera Beach, Fl., U.S. (rĭv-ĭ-ĕr'á bēch)	125a	26°46'n	80°04'w
Rivière-du-Loup, Qc., Can. (rē-vyâr' dü lōō')	138-39	47°50'n	69°32'w
Rivne, Ukr.	194-95	50°37'n	26°14'e
Riyadh, nat. cap., Sau. Ar. (rĭ-äd')	230-31	24°38'n	46°43'e
Rize, Tur. (rē'zĕ)	227	41°01'n	40°31'e
Rjukan, Nor. (ryōō'kän)	192-93	59°53'n	8°35'e
Road Town, nat. cap., Br. Vir. Is. (rōd toun)	143b	18°26'n	64°37'w
Roanne, Fr. (rō-än')	196-97	46°02'n	4°04'e
Roanoke, Al., U.S. (rō'á-nōk)	124-25	33°09'n	85°22'w
Roanoke, Va., U.S. (rō'á-nōk)	124-25	37°16'n	79°57'w
Roanoke, stm., U.S. (rō'á-nōk)	124-25	35°57'n	76°43'w
Roanoke Rapids, N.C., U.S. (rō'á-nōk răp'ĭdz)	124-25	36°28'n	77°39'w
Roan Plateau, plat., U.S. (rōn plä-tō')	118-19	39°30'n	109°40'w
Roatán, Hond. (rō-ä-tän')	149	16°20'n	86°32'w
Roatán, Isla de, i., Hond.	149	16°22'n	86°29'w
Roberval, Qc., Can. (rŏb'ēr-văl) (rô-bĕr-vál')	136-37	48°31'n	72°14'w
Robinson, Il., U.S. (rŏb'ĭn-sŭn)	116-17	39°01'n	87°45'w
Robinvale, Austl. (rŏb-ĭn'vāl)	276	34°36's	142°46'e
Roblin, Mb., Can.	134-35	51°15'n	101°23'w
Roboré, Bol.	168-69	18°20's	59°45'w
Robson, Mount, mtn., B.C., Can. (mount rŏb'sŭn)	132-33	53°07'n	119°09'w
Robstown, Tx., U.S. (rŏbz'toun)	122-23	27°47'n	97°40'w
Roca, Cabo da, c., Port. (ká'bō-dä-rô'kă)	198-99	38°47'n	9°29'w
Roca Partida, Isla, i., Mex.	144-45	19°00'n	112°04'w
Rocha, Ur. (rō'chås)	173	34°30's	54°19'w
Rochefort, Fr. (rôsh-fôr')	196-97	45°57'n	0°58'w
Rochelle, Il., U.S. (rô-shĕl')	116-17	41°55'n	89°04'w
Rochester, In., U.S. (rŏch'ĕs-tēr)	116-17	41°04'n	86°12'w
Rochester, Mn., U.S. (rŏch'ĕs-tēr)	114-15	44°00'n	92°29'w
Rochester, N.H., U.S. (rŏch'ĕs-tēr)	116-17	43°18'n	70°59'w
Rochester, N.Y., U.S.	116-17	43°09'n	77°36'w
Rock, stm., U.S. (rŏk)	114-15	41°29'n	90°38'w
Rockdale, Tx., U.S. (rŏk'dāl)	122-23	30°39'n	97°00'w
Rockefeller Plateau, plat., Ant.	287	80°0's	135°00'w
Rock Falls, Il., U.S. (rŏk fôlz)	116-17	41°47'n	89°41'w
Rockford, Il., U.S. (rŏk'fērd)	116-17	42°16'n	89°05'w
Rockford, Mi., U.S. (rŏk'fērd)	116-17	43°07'n	85°34'w
Rockhampton, Austl. (rŏk-hămp'tŭn)	277	23°23's	150°31'e
Rock Hill, S.C., U.S. (rŏk hĭl)	124-25	34°56'n	81°02'w
Rockingham, N.C., U.S. (rŏk'ĭng-hăm)	124-25	34°56'n	79°46'w
Rock Island, Il., U.S. (rŏk ī'lănd)	114-15	41°30'n	90°34'w
Rockland, On., Can. (rŏk'lănd)	136-37	45°33'n	75°17'w
Rockland, Me., U.S.	117a	44°07'n	69°07'w
Rockport, In., U.S. (rŏk'pōrt)	116-17	37°53'n	87°03'w
Rockport, Tx., U.S. (rŏk'pōrt)	122-23	28°01'n	97°03'w
Rock Rapids, Ia., U.S. (rŏk răp'ĭdz)	114-15	43°26'n	96°10'w
Rocksprings, Tx., U.S. (rŏk sprĭngs)	122-23	30°01'n	100°12'w
Rock Springs, Wy., U.S. (rŏk sprĭngz)	112-13	41°35'n	109°13'w
Rockstone, Guy. (rŏk'stôn)	164-65	5°59'n	58°32'w
Rock Valley, Ia., U.S. (rŏk văl'ĭ văl'ē)	114-15	43°12'n	96°18'w
Rockville, In., U.S. (rŏk'vĭl)	116-17	39°45'n	87°14'w
Rockwell City, Ia., U.S. (rŏk'wĕl sĭ'tē)	114-15	42°24'n	94°38'w
Rockwood, Me., U.S. (rŏk-wŏŏd)	116-17	45°40'n	69°45'w
Rockwood, Tn., U.S. (rŏk-wŏŏd)	124-25	35°52'n	84°41'w
Rocky Ford, Co., U.S. (rŏk'-ē fōrd)	120-21	38°03'n	103°43'w
Rocky Island Lake, res., On., Can. (rŏk'-ē ī'lănd lāk)	136-37	46°56'n	82°57'w
Rocky Mount, N.C., U.S. (rŏk'-ē mount)	124-25	35°57'n	77°48'w
Rocky Mount, Va., U.S. (rŏk'-ē mount)	124-25	37°00'n	79°54'w
Rocky Mountain House, Ab., Can. (rŏk'-ē moun'tĭn hous)	132-33	52°23'n	114°56'w
Rocky Mountain National Park, n.p., Co., U.S. (rŏk'-ē moun'tĭn năsh'ŭn-ăl pärk)	112-13	40°21'n	105°42'w
Rocky Mountains, mts., N.A. (rŏk'-ē moun'tĭnz)	86	48°0'n	116°00'w
Rodeo, Arg.	168-69	30°12's	69°06'w
Rodeo, Mex. (rô-dā'ō)	122-23	25°11'n	104°34'w
Rodez, Fr. (rô-dĕz')	196-97	44°21'n	2°34'e
Rodniki, Russia (rôd'nē-kē)	186-87	57°06'n	41°44'e
Ródos, Grc.	200-01	36°26'n	28°14'e
Ródos, i., Grc.	200-01	36°10'n	28°00'e
Roebourne, Austl. (rō'bŭrn)	270-71	20°46's	117°10'e
Rogagua, Laguna, lk., Bol.	166-67	13°42's	67°07'w
Rogaguado, Laguna, lk., Bol. (rō'gō-ä-gwä-dō)	166-67	12°52's	65°43'w
Rogers, Ar., U.S. (rŏj-ērz)	120-21	36°20'n	94°07'w
Rogers, Mount, mtn., Va., U.S.	124-25	36°39'n	81°33'w
Rogers City, Mi., U.S. (rŏj-ērz sĭ'tē)	116-17	45°25'n	83°49'w
Rohtak, India	234-35	28°53'n	76°36'e
Roi Georges, Îles du, is., Fr. Poly.	280-81	14°32's	145°08'w
Rojas, Arg. (rō'häs)	173	34°12's	60°44'w
Rojo, Cabo, c., Mex. (ká'bō rō'hō)	146-47	21°33'n	97°20'w
Rojo, Cabo, c., P.R. (ká'bō rō'hō)	142a	17°56'n	67°11'w
Rokan, stm., Indon.	246-47	1°50'n	100°55'e
Rokeby National Park, n.p., Austl. *see* Mungkan Kandju National Park	277	13°32's	142°37'e
Rokycany, Czech Rep. (rô'kĭ'tsä-nĭ)	194-95	49°45'n	13°36'e
Rolândia, Braz.	168-69	23°18's	51°23'w
Roldanillo, Col. (rôl-dä-nē'l-yō)	163c	4°24'n	76°09'w
Rolla, Mo., U.S.	120-21	37°57'n	91°46'w
Roma, Austl. (rō'má)	276	26°35's	148°47'e
Roma, nat. cap., Italy (rō'mä) *see* Rome	200-01	41°54'n	12°29'e
Romaine, stm., Can. (rô-mĕn')	138-39	50°18'n	63°48'w
Roman, Rom. (rō'män)	194-95	46°56'n	26°57'e
Romang, Pulau, i., Indon.	248-49	7°34's	127°26'e
Romania, nation, Eur. (rō-mā'nē-à)	174-75	46°0'n	25°30'e
Roman-Kosh, hora, mtn., Ukr.	202-03	44°37'n	34°15'e
Romano, Cape, c., Fl., U.S. (kāp rō-mä'nō)	125a	25°50'n	81°41'w
Romano, Cayo, i., Cuba (kä'yō-rô-mä'nō)	142-43	22°04'n	77°50'w
Romblon, Phil. (rōm-blōn')	250	12°34'n	122°16'e
Rome, Ga., U.S. (rōm)	124-25	34°16'n	85°10'w
Rome, N.Y., U.S. (rōm)	116-17	43°13'n	75°28'w
Rome, nat. cap., Italy (rōm)	200-01	41°54'n	12°29'e
Romeo, Mi., U.S. (rō'mē-ō)	116-17	42°48'n	83°00'w
Romilly-sur-Seine, Fr. (rô-mē-yē'sür-sān')	196-97	48°31'n	3°44'e
Romny, Ukr. (rôm'nĭ)	202-03	50°45'n	33°29'e
Rømø, i., Den. (rûm'û)	192-93	55°08'n	8°31'e
Romorantin-Lanthenay, Fr. (rô-mô-räN-tăN')	196-97	47°22'n	1°44'e
Rona, i., Scot., U.K.	184-85	59°07'n	5°49'w
Ronan, Mt., U.S. (rō'năn)	112-13	47°31'n	114°06'w
Roncador, Serra do, plat., Braz. (sĕr'rá dò rôn-kä-dôr')	166-67	12°0's	52°00'w
Ronda, Spain (rōn'dä)	198-99	36°44'n	5°10'w
Rondônia, state, Braz.	166-67	11°0's	63°00'w
Rondonópolis, Braz.	168-69	16°28's	54°38'w
Ronge, Lac la, lk., Sk., Can. (läk lä rônzh)	134-35	55°10'n	105°00'w
Rongelap, at., Marsh. Is.	280-81	11°20'n	166°50'e
Rongjiang, China (rôŋ-jyäŋ)	238-39	25°51'n	108°35'e
Rønne, Den. (rûn'ĕ)	194-95	55°06'n	14°42'e
Ronneby, Swe. (rōn'ĕ-bü)	192-93	56°12'n	15°18'e
Ronuro, stm., Braz.	166-67	11°56's	53°33'w
Roorkee, India	234-35	29°52'n	77°53'e
Roosendaal, Neth. (rō'zĕn-däl)	190-91	51°32'n	4°28'e
Roosevelt, Ut., U.S. (rōz''vĕlt)	118-19	40°19'n	109°59'w
Roosevelt, stm., Braz. (rô'sĕ-vĕlt)	166-67	7°34's	60°41'w
Roper, stm., Austl. (rōp'ēr)	272-73	14°44's	135°23'e
Roque Pérez, Arg. (rô'kĕ-pĕ'rĕz)	173	35°25's	59°20'w
Roraima, state, Braz. (rô'rīy-mă)	166-67	1°0'n	61°00'w
Roraima, Monte, mtn., S.A. *see* Roraima, Mount	164-65	5°13'n	60°44'w
Roraima, Mount, mtn., S.A. (mount rô-rä-ē'mä)	164-65	5°13'n	60°44'w
Røros, Nor. (rûr'ôs)	184-85	62°35'n	11°23'e
Ros', stm., Ukr. (rôs)	202-03	49°41'n	31°36'e
Rosales, Mex. (rō-zä'läs)	122-23	28°12'n	105°33'w
Rosamorada, Mex. (rō'zä-mō-rä'dhä)	146-47	22°08'n	105°12'w
Rosario, Arg. (rô-zä'rē-ō)	173	32°57's	60°40'w
Rosário, Braz. (rô-zä'rē-ò)	166-67	2°57's	44°14'w
Rosario, Mex. (rô-zä'rē-ō)	146-47	23°00'n	105°52'w
Rosario, Para.	168-69	24°25's	57°06'w
Rosario, Ur. (rô-zä'rē-ō)	173	34°19's	57°21'w
Rosario de la Frontera, Arg.	168-69	25°48's	64°58'w
Rosario de Lerma, Arg.	168-69	24°59's	65°35'w
Rosário do Sul, Braz. (rô-zä'rē-ò-dô-sōō'l)	173	30°15's	54°56'w
Rosário Oeste, Braz. (rô-zä'rē-ò ô'ĕst'ĕ)	166-67	14°50's	56°25'w
Roscoe, Tx., U.S. (rôs'kō)	120-21	32°27'n	100°33'w
Roseau, Mn., U.S. (rô-zō')	114-15	48°51'n	95°46'w
Roseau, nat. cap., Dom.	143b	15°18'n	61°23'w
Rosebud, stm., Ab., Can. (rōz'bŭd)	132-33	51°25'n	112°37'w
Rosebud Indian Reservation, ind. res., S.D., U.S. (rōz'bŭd ĭn'dĭ-ăn rĕ-sĕr-vā'shĕn)	114-15	43°08'n	100°33'w
Roseburg, Or., U.S.	112-13	43°14'n	123°20'w
Rosenheim, Ger. (rō'zĕn-hīm)	194-95	47°52'n	12°08'e
Rosetown, Sk., Can. (rōz'toun)	134-35	51°32'n	108°01'w
Rosetta, Egypt	268b	31°24'n	30°25'e
Roseville, Mn., U.S. (rōz'vĭl)	114-15	45°01'n	93°10'w
Roşiori de Vede, Rom. (rô-shôr'ē dĕ vĕ-dĕ)	200-01	44°07'n	24°60'e
Roskilde, Den. (rôs'kēl-dĕ)	192-93	55°39'n	12°06'e
Roslavl', Russia (rôs'läv'l)	202-03	53°57'n	32°52'e
Rossano, Italy (rô-sä'nō)	200-01	39°35'n	16°39'e
Rossiya, nation, Eur. *see* Russia	174-75	60°0'n	100°00'e
Rossland, B.C., Can. (rôs'lănd)	132-33	49°05'n	117°48'w
Rosso, Maur.	258-59	16°31'n	15°48'w
Rossosh', Russia (rôs'sŭsh)	186-87	50°12'n	39°35'e
Ross River, Yk., Can.	128-29	62°00'n	132°26'w
Ross Sea, s., Ant. (rôs sē)	287	76°0's	175°00'w
Rossville, Ga., U.S. (rôs'vĭl)	124-25	34°59'n	85°18'w
Rosthern, Sk., Can.	134-35	52°40'n	106°20'w
Rostock, Ger. (rôs'tŭk)	194-95	54°05'n	12°07'e
Rostov, Russia (rôstôv)	202-03	57°11'n	39°25'e
Rostov-na-Donu, Russia (rôstôv-nå-dô-nōō)	202-03	47°13'n	39°43'e
Roswell, Ga., U.S. (rôz'wĕl)	124-25	34°02'n	84°21'w
Roswell, N.M., U.S. (rôz'wĕl)	120-21	33°24'n	104°33'w
Rota, i., N. Mar. Is.	279c	14°10'n	145°12'e
Rotherham, Eng., U.K. (rŏdh'ēr-ăm)	190-91	53°25'n	1°23'w
Rothesay, Scot., U.K. (rŏth'så)	190-91	55°50'n	5°03'w
Roti, Pulau, i., Indon. (pōō-lou rō'tē)	248-49	10°45's	123°10'e
Rotorua, N.Z.	278	38°09's	176°14'e
Rotterdam, Neth. (rôt'ĕr-däm')	190-91	51°55'n	4°28'e
Rottweil, Ger. (rōt'vīl)	194-95	48°10'n	8°38'e
Rotuma, i., Fiji	280-81	12°30's	177°05'e

n-sing; ŋ-baŋk; N-nasalized n; nŏd; cŏmmit; ōld; ôbey; ôrder; oi-boil; fōōd; ŏ-as oo in foot; ou-out; s-soft; sh-dish; th-thin; pūre; ûnite; ûrn; stŭd; circŭs; ü-as in French tu; '-indeterminate vowel.

Feature (Pronunciation)	Page	Lat.	Long.
Roubaix, Fr. (rōō-bě´)	196-97	50°41´N	3°10´E
Rouen, Fr. (rōō-äN´)	196-97	49°27´N	1°07´E
Rouge, stm., Qc., Can. (rōōzh)	136-37	45°38´N	74°42´W
Round Mountain, mtn., Austl.	276	30°27´S	152°14´E
Round Rock, Tx., U.S. (round rŏk)	122-23	30°30´N	97°41´W
Roundup, Mt., U.S. (round´ŭp)	112-13	46°27´N	108°33´W
Rouyn-Noranda, Qc., Can.	136-37	48°14´N	79°01´W
Rovaniemi, Fin. (rô´vá-nyĕ´mĭ)	184-85	66°30´N	25°42´E
Rovereto, Italy (rō-vå-rā´tō)	200-01	45°54´N	11°02´E
Rovigo, Italy (rô-vē´gô)	200-01	45°05´N	11°47´E
Rovinj, Cro. (rô´ēn')	200-01	45°05´N	13°38´E
Rovira, Col. (rô-vē´rä)	163c	4°14´N	75°15´W
Rovno, Ukr. *see* Rivne	194-95	50°37´N	26°14´E
Rovuma, stm., Afr.	264-65	10°31´S	40°24´E
Rowley Island, i., Nu., Can.	130-31	69°05´N	78°52´W
Roxas, Phil.	250	11°35´N	122°45´E
Roy, Ut., U.S. (roi)	112-13	41°10´N	112°01´W
Royale, Isle, i., Mi., U.S.	114-15	48°0´N	89°00´W
Royal Oak, Mi., U.S. (roi´*ă*l ōk)	116-17	42°30´N	83°08´W
Royal Tunbridge Wells, Eng., U.K.	190-91	51°08´N	0°16´E
Royan, Fr. (rwä-yäN´)	196-97	45°38´N	1°01´W
Rožňava, Slvk. (rôzh´nyá-vá)	194-95	48°40´N	20°33´E
Rtishchevo, Russia ('r-tĭsh´chĕ-vô)	186-87	52°16´N	43°47´E
Ruaha National Park, n.p., Tan.	267	7°30´S	34°40´E
Ruapehu, Mount, vol., N.Z. (mount r´oo-á-pā´hōō)	278	39°17´S	175°34´E
Rub'al-Khali, des., Asia	220-21	20°0´N	51°00´E
Rubizhne, Ukr.	202-03	49°01´N	38°23´E
Rubondo Island, i., Tan.	267	2°20´S	31°52´E
Rubtsovsk, Russia	226	51°31´N	81°12´E
Ruby Mountains, mts., Nv., U.S. (rōō´bē moun´tĭnz)	118-19	40°25´N	115°31´W
Rudkøbing, Den. (rōōdh´kûb-ĭng)	194-95	54°56´N	10°44´E
Rūdnyy, Kaz.	226	52°59´N	63°07´E
Rudolf, Lake, lk., Afr. (lāk rōō´dŏlf)	267	3°30´N	36°00´E
Rudolf Häyk', lk., Afr. *see* Rudolf, Lake	267	3°30´N	36°00´E
Ruffec, Fr. (rü-fĕk´)	196-97	46°01´N	0°12´E
Rufiji, stm., Tan. (rò-fē´jê)	262-63	7°58´S	39°25´E
Rufino, Arg.	173	34°16´S	62°42´W
Rugao, China (rōō-gou)	238-39	32°24´N	120°33´E
Rugby, Eng., U.K. (rŭg´bē)	190-91	52°23´N	1°16´W
Rugby, N.D., U.S.	114-15	48°22´N	99°60´W
Rügen, i., Ger. (rü´ghĕn)	194-95	54°25´N	13°24´E
Rugufu, stm., Tan.	267	5°30´S	30°01´E
Ruhengeri, Rw.	267	1°30´S	29°38´E
Rui'an, China (rwā-än)	238-39	27°50´N	120°35´E
Ruijin, China	238-39	25°52´N	116°00´E
Ruiz, Mex. (ròē´z)	146-47	21°57´N	105°09´W
Ruiz, Nevado del, vol., Col. (nĕ-vá´dô-dĕl-rōōē´z)	163c	4°53´N	75°20´W
Rūjiena, Lat. (rò´yĭ-ă-nà)	192-93	57°54´N	25°20´E
Rukwa, Lake, lk., Tan. (lāk rōōk-wä´)	267	8°0´S	32°25´E
Ruma, Serb. (rōō´má)	200-01	45°00´N	19°49´E
Rumbek, Sudan (rŭm´bĕk)	262-63	6°48´N	29°41´E
Rum Cay, i., Bah. (rŭm kē)	142-43	23°41´N	74°53´W
Rumford, Me., U.S. (rŭm´fērd)	116-17	44°33´N	70°33´W
Rumoi, Japan	244	43°56´N	141°39´E
Runan, China (rōō-nän)	238-39	33°00´N	114°21´E
Runde, stm., Zimb.	264-65	21°18´S	32°24´E
Rundu, Nmb.	264-65	17°55´S	19°45´E
Rŭng, Kaôh, i., Camb.	246-47	10°44´N	103°14´E
Rungwa, stm., Tan.	267	7°37´S	31°49´E
Ruo, stm., China (rwô)	240-41	41°04´N	100°20´E
Ruoqiang, China	220-21	39°01´N	88°11´E
Rupat, Pulau, i., Indon. (pōō-lou rōō´pät)	246-47	1°50´N	101°35´E
Rupert, Id., U.S. (rōō´pērt)	112-13	42°38´N	113°41´W
Rurrenabaque, Bol.	166-67	14°28´S	67°30´W
Rurutu, i., Fr. Poly.	280-81	22°26´S	151°20´W
Rusape, Zimb.	264-65	18°32´S	32°08´E
Ruse, Blg. (rōō´sĕ) (rò´sĕ)	200-01	43°51´N	25°57´E
Rushville, Il., U.S. (rŭsh´vĭl)	120-21	40°07´N	90°33´W
Rushville, In., U.S. (rŭsh´vĭl)	116-17	39°36´N	85°27´W
Rushville, Ne., U.S. (rŭsh´vĭl)	114-15	42°43´N	102°28´W
Rusk, Tx., U.S. (rŭsk)	122-23	31°48´N	95°09´W
Russas, Braz. (rōō´s-säs)	166-67	4°56´S	37°58´W
Russell, Mb., Can. (rŭs´ĕl)	134-35	50°47´N	101°15´W
Russell, Ks., U.S. (rŭs´ĕl)	120-21	38°50´N	98°50´W
Russell, Ky., U.S. (rŭs´ĕl)	116-17	38°31´N	82°42´W
Russell Lake, lk., Mb., Can. (rŭs´ĕl lāk)	134-35	56°15´N	101°32´W
Russellville, Al., U.S. (rŭs´ĕl-vĭl)	124-25	34°30´N	87°44´W
Russellville, Ar., U.S. (rŭs´ĕl-vĭl)	120-21	35°17´N	93°09´W
Russellville, Ky., U.S. (rŭs´ĕl-vĭl)	124-25	36°51´N	86°53´W
Russia, nation, Eur. (rŭ´sh*ă*)	218-19	60°0´N	100°00´E
Rustavi, Geor.	227	41°32´N	45°02´E
Rustenburg, S. Afr. (rŭs´tĕn-bûrg)	269c	25°40´S	27°15´E
Ruston, La., U.S. (rŭs´t*ŭ*n)	120-21	32°32´N	92°38´W
Ruteng, Indon.	248-49	8°36´S	120°29´E
Rutherfordton, N.C., U.S. (rŭdh´ēr-fērd-tŭn)	124-25	35°22´N	81°58´W
Rutland, Vt., U.S.	116-17	43°37´N	72°59´W
Rutog, China	234-35	33°26´N	79°42´E
Rutshuru, D.R.C. (rōōt-shōō´rōō)	267	1°11´S	29°27´E
Ruvuma, stm., Afr.	264-65	10°31´S	40°24´E
Ruwenzori Range, mts., Afr.	267	0°20´N	29°53´E
Ruzayevka, Russia	186-87	54°04´N	44°57´E
Rwanda, nation, Afr. (rü-än´-d*ă*)	253	2°0´S	30°00´E
Ryazan', Russia (ryä-zän´')	202-03	54°38´N	39°44´E
Ryazhsk, Russia (ryäzh´sk')	186-87	53°42´N	40°05´E
Rybachiy, Poluostrov, pen., Russia	184-85	69°42´N	32°36´E
Rybachye, Kyrg. *see* Balykchy	226	42°28´N	76°12´E
Rybinsk, Russia	202-03	58°03´N	38°52´E
Rybnik, Pol. (rĭb´nêk)	194-95	50°06´N	18°33´E
Ryde, Eng., U.K. (rīd)	190-91	50°44´N	1°10´W
Ryeosu, Kor., S. *see* Yŏsu	243	34°44´N	127°44´E
Rylsk, Russia (rêl´'sk)	202-03	51°34´N	34°42´E
Ryojun, China *see* Lüshun	240-41	38°49´N	121°15´E
Ryōtsu, Japan (ryōt´sōō)	245	38°05´N	138°26´E
Ryukyu Islands, is., Japan (rū-kū ĭ´l*ă*ndz)	244a	25°44´N	126°58´E
Rzeszów, Pol. (zhå-shòf)	194-95	50°03´N	22°01´E
Rzhev, Russia ('r-zhĕf)	202-03	56°17´N	34°19´E

S

Feature (Pronunciation)	Page	Lat.	Long.
Saale, stm., Ger. (sä-lĕ)	194-95	51°57´N	11°55´E
Saalfeld, Ger. (säl´fĕlt)	194-95	50°39´N	11°22´E
Saarbrücken, Ger. (zähr´brü-kĕn)	194-95	49°14´N	6°60´E
Saaremaa, i., Est.	192-93	58°25´N	22°30´E
Saavedra, Arg. (sä-ä-vā´drä)	173	37°46´S	62°21´W
Saba, i., Neth. Ant. (sä´bä)	143b	17°38´N	63°14´W
Šabac, Serb. (shä´báts)	200-01	44°46´N	19°42´E
Sabadell, Spain (sä-bä-dhål´)	198-99	41°33´N	2°06´E
Sabah, hist. reg., Malay.	248-49	5°20´N	117°10´E
Sabanagrande, Hond. (sä-bä´nä-grä´n-dĕ)	149	13°49´N	87°17´W
Sabanalarga, Col. (sä-bá´nä-lär´gä)	164-65	10°38´N	74°55´W
Sabancuy, Mex. (sä-bäŋ-kwē´)	148	18°58´N	91°11´W
Sabang, Indon. (sä´bäng)	248-49	0°13´N	119°53´E
Sabang, Indon. (sä´bäng)	246-47	5°53´N	95°20´E
Şāberī, Hāmūn-e, lk., Asia	232-33	31°30´N	61°20´E
Sabhā, Libya	258-59	27°01´N	14°28´E
Sabi, stm., Afr. (sä´bē) *see* Save	264-65	20°58´S	35°04´E
Sabinal, Cayo, i., Cuba (kä´yō sä-bē-näl´)	142-43	21°40´N	77°18´W
Sabinas, Mex.	122-23	27°51´N	101°07´W
Sabinas, stm., Mex. (sä-bē´näs)	122-23	26°51´N	99°35´W
Sabinas, stm., Mex. (sä-bē´näs)	122-23	27°29´N	100°40´W
Sabinas Hidalgo, Mex. (sä-bē´näs ê-däl´gô)	122-23	26°30´N	100°10´W
Sabine, stm., U.S.	110-11	30°00´N	93°46´W
Sable, Cape, c., N.S., Can. (kāp sā´b'l)	138-39	43°25´N	65°37´W
Sable, Cape, pen., Fl., U.S. (kāp sā´b'l)	125a	25°12´N	81°05´W
Sable, Île de, i., N. Cal.	272-73	19°15´S	159°56´E
Sable Island, i., N.S., Can.	138-39	43°56´N	59°56´W
Sablé-sur-Sarthe, Fr. (säb-lā-sür-särt´)	196-97	47°50´N	0°20´W
Sabor, stm., Port. (sä-bōr´)	198-99	41°11´N	7°07´W
Sabzevār, Iran	232-33	36°13´N	57°40´E
Sac, stm., Mo., U.S. (sôk)	120-21	38°01´N	93°44´W
Sac City, Ia., U.S. (sŏk sĭ´tē)	114-15	42°25´N	94°60´W
Sachigo, stm., On., Can.	134-35	55°04´N	88°59´W
Sachigo Lake, lk., On., Can. (săch´ĭ-gō lāk)	134-35	53°49´N	92°08´W
Sachsen, hist. reg., Ger. (zäK´sĕn) *see* Saxony	194-95	52°45´N	9°30´E
Sackville, N.B., Can. (săk´vĭl)	138-39	45°54´N	64°22´W
Saco, Me., U.S. (sô´kô)	116-17	43°30´N	70°27´W
Sacramento, Ca., U.S. (săk-rá-mĕn´tō)	118-19	38°35´N	121°29´W
Sacramento, stm., Ca., U.S. (săk-rá-mĕn´tō)	110-11	38°03´N	121°53´W
Sacramento Mountains, mts., N.M., U.S.	120-21	32°42´N	105°37´W
Şa'dah, Yemen	266	16°49´N	43°48´E
Sadiya, India (sŭ-dē´yä)	234-35	27°50´N	95°40´E
Sado, i., Japan (sä´dō)	245	38°0´N	138°25´E
Saeki, Japan (sä´å-kĕ) *see* Saiki	245	32°58´N	131°55´E
Safâga, Egypt	268b	26°45´N	33°56´E
Safford, Az., U.S. (săf´fērd)	118-19	32°50´N	109°43´W
Safi, Mor. (sä´fē) (äs´fē)	258-59	32°18´N	9°13´W
Safīd Koh, Selseleh-ye, mts., Afg.	232-33	34°30´N	63°30´E
Safonovo, Russia	202-03	55°07´N	33°15´E
Saga, China	234-35	29°29´N	85°09´E
Saga, Japan	245	33°15´N	130°18´E
Sagaing, Mya.	246-47	21°53´N	95°59´E
Sagami-nada, b., Japan (sä´gä´mē nä-dä)	245	34°60´N	139°30´E
Saganaga Lake, lk., N.A. (sä-gá-nä´g*ă* lāk)	134-35	48°14´N	90°52´W
Sāgar, India	234-35	23°50´N	78°45´E
Sagarmāthā, mtn., Asia *see* Everest, Mount	234-35	27°59´N	86°56´E
Sagavanirktok, stm., Ak., U.S.	126	70°21´N	148°11´W
Saginaw, Mi., U.S. (săg´ĭ-nô)	116-17	43°26´N	83°58´W
Saginaw Bay, b., Mi., U.S. (săg´ĭ-nô bā)	116-17	43°50´N	83°40´W
Sagua de Tánamo, Cuba (sä-gwä dĕ tá´nä-mō)	142-43	20°35´N	75°14´W
Sagua la Grande, Cuba (sä-gwä lä grä´n-dĕ)	142-43	22°49´N	80°04´W
Saguaro National Park, n.p., Az., U.S. (säg-wä´rō năsh´ŭn-*ă*l pärk)	118-19	32°16´N	111°12´W
Saguenay, stm., Qc., Can. (săg-ĕ-nā´)	138-39	48°08´N	69°41´W
Sagunt, Spain	198-99	39°41´N	0°16´W
Sagunto, Spain (sä-gòn´tō) *see* Sagunt	198-99	39°41´N	0°16´W
Sa'gya, China	234-35	28°54´N	88°04´E
Sahara, des., Afr. (s*ă*-hä´r*ă*)	258-59	26°0´N	13°00´E
Sahāranpur, India (sŭ-hä´rŭn-pōōr´)	234-35	29°58´N	77°33´E
Sahel, reg., Afr.	258-59	12°0´N	17°00´E
Sāḩil, reg., Afr. *see* Sahel	258-59	12°0´N	17°00´E
Sāhīwāl, Pak.	232-33	30°40´N	73°06´E
Şaḩrā', des., Afr. *see* Sahara	258-59	26°0´N	13°00´E
Saïda, Alg.	184-85	34°50´N	0°09´E
Saidpur, Bngl.	234-35	25°47´N	88°54´E
Saigon, Viet. *see* Ho Chi Minh City	246-47	10°45´N	106°40´E
Saiki, Japan	245	32°58´N	131°55´E
Saimaa, lk., Fin. (sä´ĭ-mä)	192-93	61°15´N	28°15´E
Saín Alto, Mex. (sä-ēn´ äl´tō)	146-47	23°35´N	103°13´W
Saint Albans, Eng., U.K. (sånt ôl´b*ă*nz)	190-91	51°45´N	0°21´W
Saint Albans, Vt., U.S. (sānt ôl´b*ă*nz)	116-17	44°49´N	73°05´W
Saint Albans, W.V., U.S. (sānt ôl´b*ă*nz)	116-17	38°23´N	81°50´W
Saint Albert, Ab., Can. (sånt ăl´bērt)	132-33	53°38´N	113°38´W
Saint-Amand-Mont-Rond, Fr. (săN´t á-mäN´ môN-rôN´)	196-97	46°43´N	2°30´E
Saint-André, Cap, c., Madag. *see* Vilanandro, Tanjona	264-65	16°12´S	44°28´E
Saint Andrews, Scot., U.K.	190-91	56°20´N	2°48´W
Saint-Anselme, Qc., Can. (săN´ tän-sĕlm´)	138-39	46°37´N	70°57´W
Saint Anthony, Nf., Can. (săn ăn´thô-nē)	138-39	51°22´N	55°37´W
Saint Anthony, Id., U.S. (sānt ăn´thô-nē)	112-13	43°58´N	111°41´W
Saint-Augustin, Qc., Can.	138-39	51°14´N	58°38´W
Saint Augustine, Fl., U.S. (sånt ô´gŭs-tēn)	124-25	29°54´N	81°19´W
Saint-Barthélemy, i., Guad.	143b	17°54´N	62°50´W
Saint Bees Head, c., Eng., U.K. (sånt bēz´hĕd)	190-91	54°31´N	3°38´W
Saint Bride, Mount, mtn., Ab., Can. (mount sånt brīd)	132-33	51°31´N	115°57´W
Saint-Brieuc, Fr. (săN´ brēs´)	196-97	48°31´N	2°45´W
Saint Catharines, On., Can. (sånt kăth´á-rĭnz)	136-37	43°10´N	79°14´W
Saint-Chamond, Fr. (săN´ shá-môN´)	196-97	45°29´N	4°31´E
Saint Charles, Il., U.S. (sānt chärlz´)	116-17	41°55´N	88°19´W
Saint Charles, Md., U.S. (sānt chärlz´)	116-17	38°35´N	76°57´W
Saint Charles, Mi., U.S. (sānt chärlz´)	116-17	43°18´N	84°08´W
Saint Charles, Mn., U.S. (sānt chärlz´)	114-15	43°58´N	92°03´W
Saint Charles, Mo., U.S. (sānt chärlz´)	120-21	38°48´N	90°29´W
Saint Christopher, i., St. K./N.	143b	17°20´N	62°45´W
Saint Christopher and Nevis, nation, N.A. *see* Saint Kitts and Nevis	140-41	17°20´N	62°45´W
Saint Clair, Mi., U.S. (sānt klâr)	116-17	42°50´N	82°29´W
Saint Clair, Mo., U.S. (sānt klâr)	120-21	38°21´N	90°59´W
Saint-Claude, Fr. (săN´ klōd´)	196-97	46°23´N	5°51´E
Saint Cloud, Fl., U.S. (sānt kloud´)	125a	28°15´N	81°17´W
Saint Cloud, Mn., U.S. (sānt kloud)	114-15	45°33´N	94°10´W
Saint Croix, i., V.I.U.S. (sånt kroi´)	143b	17°45´N	64°45´W
Saint Croix, stm., N.A. (sānt kroi´)	138-39	45°10´N	67°09´W
Saint Croix, stm., U.S. (sānt kroi´)	114-15	44°45´N	92°48´W
Saint-Denis, Fr. (săN´dē-nē´)	196-97	48°57´N	2°21´E
Saint-Denis, nat. cap., Reu. (săN´dē-nē´)	265a	20°52´S	55°28´E

Feature (Pronunciation)	Page	Lat.	Long.
Saint-Dizier, Fr. (săN dē-zyā′)	196-97	48°39′N	4°57′E
Sainte-Agathe-des-Monts, Qc., Can.	136-37	46°03′N	74°17′W
Sainte-Foy, Qc., Can. (săN fwä)	136-37	46°47′N	71°17′W
Sainte Genevieve, Mo., U.S. (sānt jĕn′ĕ-vēv)	120-21	37°59′N	90°03′W
Saint Elias, Mount, mtn., N.A. (mount sānt ē-lī′ăs)	126	60°18′N	140°55′W
Saint-Élie, Fr. Gu.	164-65	4°50′N	53°17′W
Sainte-Lucie, Canal de, strt., N.A. *see* Saint Lucia Channel	143b	14°09′N	60°57′W
Sainte-Marguerite, stm., Qc., Can.	138-39	50°09′N	66°36′W
Sainte-Marie, Cap, c., Madag. *see* Vohimena, Tanjona	264-65	25°36′S	45°09′E
Sainte Marie, Nosy, i., Madag.	264-65	16°50′S	49°57′E
Saint-Étienne, Fr.	196-97	45°26′N	4°24′E
Saint-Eustache, Qc., Can. (săN′ tû-stásh′)	136-37	45°34′N	73°55′W
Saint-Félicien, Qc., Can. (săN fā-lē-syăN′)	136-37	48°39′N	72°27′W
Saint-Florent-sur-Cher, Fr. (săN′ flô-räN′sür-shâr′)	196-97	46°59′N	2°15′E
Saint-Flour, Fr. (săN flōōr′)	196-97	45°02′N	3°05′E
Saint Francis, Cape, c., S. Afr.	264-65	34°11′S	24°50′E
Saint-Gaudens, Fr. (săN gō-däNs′)	196-97	43°07′N	0°44′E
Saint George, Austl. (sānt jôrj′)	276	28°03′S	148°35′E
Saint George, N.B., Can. (săN jôrj′)	138-39	45°08′N	66°49′W
Saint George, S.C., U.S. (sānt jôrj′)	124-25	33°11′N	80°35′W
Saint George, Ut., U.S. (sānt jôrj′)	118-19	37°06′N	113°34′W
Saint George, Cape, c., Nf., Can. (kāp sānt jôr-jĕz)	138-39	48°29′N	59°15′W
Saint George, Cape, c., Fl., U.S. (kāp sānt jôr-jĕz)	124-25	29°35′N	85°04′W
Saint George Island, i., Fl., U.S.	124-25	29°39′N	84°53′W
Saint-Georges, Fr. Gu.	164-65	3°57′N	51°48′W
Saint George's, nat. cap., Gren. (sānt jôrj′ĕs)	143b	12°04′N	61°45′W
Saint George's Bay, b., Nf., Can. (sānt jôr-jĕz bā)	138-39	48°20′N	59°00′W
Saint Georges Bay, b., N.S., Can. (sānt jôr-jĕz bā)	138-39	45°50′N	61°45′W
Saint George's Channel, strt., Eur. (sānt jôr-jĕz chăn′ĕl)	190-91	52°0′N	6°00′W
Saint-Girons, Fr. (săN zhē-rôN′)	196-97	42°59′N	1°09′E
Saint Helena, dep., Afr. (sānt hē-lē′nȧ)	253	15°57′S	5°42′W
Saint Helena, i., St. Hel. (sānt hē-lē′nȧ)	254	15°57′S	5°43′W
Saint Helens, Or., U.S. (sānt hĕl′ĕnz)	112-13	45°52′N	122°48′W
Saint Helens, Mount, vol., Wa., U.S. (mount sānt hĕl′ĕnz)	112-13	46°12′N	122°11′W
Saint Helier, nat. cap., Jersey (sānt hyĕl′yēr)	196-97	49°12′N	2°07′W
Saint-Hyacinthe, Qc., Can.	136-37	45°38′N	72°57′W
Saint Ignace, Mi., U.S. (sānt ĭg′nās)	116-17	45°52′N	84°44′W
Saint Ignace Island, i., On., Can. (sānt ĭg′nās ī′lȧnd)	136-37	48°48′N	87°56′W
Saint James, Mn., U.S. (sānt jāmz′)	114-15	43°59′N	94°38′W
Saint James, Mo., U.S. (sānt jāmz′)	120-21	37°60′N	91°37′W
Saint James, Cape, c., B.C., Can. (kāp sānt jāmz′)	132-33	51°56′N	131°01′W
Saint-Jean, Lac, res., Qc., Can. (läk săN′ zhäN′)	136-37	48°35′N	72°05′W
Saint-Jean-d'Angély, Fr. (săN-zhäN′-däN-zhā-lē′)	196-97	45°57′N	0°31′W
Saint-Jean-de-Luz, Fr. (săN-zhäN′ dē lüz′)	196-97	43°24′N	1°39′W
Saint-Jean-sur-Richelieu, Qc., Can.	136-37	45°19′N	73°16′W
Saint-Jérôme, Qc., Can. (săN zhā-rōm′)	136-37	45°47′N	74°00′W
Saint John, N.B., Can. (sānt jŏn)	138-39	45°17′N	66°04′W
Saint John, i., V.I.U.S. (sānt jŏn)	143b	18°20′N	64°45′W
Saint John, stm., N.A. (sānt jŏn)	138-39	45°16′N	66°04′W
Saint John, Cape, c., Nf., Can. (kāp sānt jŏn)	138-39	49°59′N	55°32′W
Saint John's, Nf., Can. (sānt jŏns)	138-39	47°34′N	52°43′W
Saint Johns, Az., U.S. (sānt jŏnz)	118-19	34°30′N	109°22′W
Saint Johns, Mi., U.S. (sānt jŏnz)	116-17	42°60′N	84°33′W
Saint John's, nat. cap., Antig. (sānt jŏnz)	143b	17°07′N	61°51′W
Saint Johns, stm., Fl., U.S. (sānt jŏnz)	125a	30°24′N	81°23′W
Saint Johnsbury, Vt., U.S. (sānt jŏnz′bĕr-ē)	116-17	44°26′N	72°01′W
Saint Joseph, Mi., U.S. (sānt jō′sĕf)	116-17	42°05′N	86°29′W
Saint Joseph, Mo., U.S. (sānt jō′sĕf)	120-21	39°46′N	94°50′W
Saint Joseph, stm., U.S. (sānt jō′sĕf)	116-17	42°06′N	86°29′W
Saint Joseph, Lake, lk., On., Can.	134-35	51°03′N	90°52′W
Saint-Joseph-de-Beauce, Qc., Can. (sĕN zhō-zĕf′dē bōs)	138-39	46°18′N	70°52′W
Saint-Junien, Fr. (săN′zhü-nyăN′)	196-97	45°53′N	0°54′E
Saint Kilda, i., Scot., U.K. (sānt kĭl′dȧ)	190-91	57°49′N	8°36′W
Saint Kitts, i., St. K./N. (sānt kĭtts) *see* Saint Christopher	143b	17°20′N	62°45′W
Saint Kitts and Nevis, nation, N.A. (sānt kĭts ănd nē′vĭs)	140-41	17°20′N	62°45′W
Saint-Laurent, stm., N.A. *see* Saint Lawrence	86	49°14′N	67°01′W
Saint-Laurent, Golfe du, b., Can. *see* Saint Lawrence, Gulf of	138-39	48°0′N	62°00′W
Saint-Laurent du Maroni, Fr. Gu.	164-65	5°28′N	54°02′W
Saint Lawrence, Nf., Can. (sānt lô′rĕns)	138-39	46°56′N	55°24′W
Saint Lawrence, stm., N.A. (sānt lô′rĕns)	86	49°14′N	67°01′W
Saint Lawrence, Gulf of, b., Can. (gŭlf ŭv sānt lô′rĕns)	138-39	48°0′N	62°00′W
Saint Lawrence Island, i., Ak., U.S. (sānt lô′rĕns ī′lȧnd)	126	63°30′N	170°30′W
Saint-Louis, Sen.	260-61	16°01′N	16°29′W
Saint Louis, Mi., U.S. (sānt lōō′ĭs)	116-17	43°24′N	84°36′W
Saint Louis, Mo., U.S. (sānt lōō′ĭs) (lōō′ē)	120-21	38°39′N	90°13′W
Saint Lucia, nation, N.A. (sānt lōō′-shȧ)	140-41	13°53′N	60°58′W
Saint Lucia, Lake, lk., S. Afr.	264-65	28°04′S	32°28′E
Saint Lucia Channel, strt., N.A. (sānt lū′shĭ-ȧ chăn′ĕl)	143b	14°09′N	60°57′W
Saint-Malo, Fr. (săN′ mȧ-lō′)	196-97	48°39′N	2°01′W
Saint-Marc, Haiti (săN′ mȧrk′)	142-43	19°07′N	72°41′W
Saint Maries, Id., U.S. (sānt mā′rēs)	112-13	47°19′N	116°34′W
Saint-Martin, i., N.A. (săN-mär′tĭn)	143b	18°04′N	63°04′W
Saint Martinville, La., U.S. (sānt mär′tĭn-vĭl)	124-25	30°08′N	91°50′W
Saint Marys, Austl. (sānt mā′rēz)	276	41°35′S	148°11′E
Saint Marys, Ga., U.S. (sānt mā′rēz)	124-25	30°44′N	81°33′W
Saint Marys, Oh., U.S. (sānt mā′rēz)	116-17	40°33′N	84°24′W
Saint Marys, Pa., U.S. (sānt mā′rēz)	116-17	41°25′N	78°35′W
Saint Marys, W.V., U.S. (sānt mā′rēz)	116-17	39°23′N	81°12′W
Saint Mary's, Cape, c., Nf., Can.	138-39	46°50′N	54°12′W
Saint Mary's Bay, b., Nf., Can.	138-39	46°50′N	53°47′W
Saint Matthew Island, i., Ak., U.S. (sānt măth′ū ī′lȧnd)	218-19	60°29′N	172°53′W
Saint Matthews, S.C., U.S. (sānt măth′ūz)	124-25	33°40′N	80°47′W
Saint Matthias Group, is., Pap. N. Gui.	277	1°36′S	149°47′E
Saint-Maurice, stm., Qc., Can. (săN′ mō-rēs′) (sānt mô′rĭs)	136-37	46°21′N	72°31′W
Saint Michael, Ak., U.S. (sānt mī′kĕl)	126	63°29′N	162°02′W
Saint-Mihiel, Fr. (săN′ mē-yĕl′)	196-97	48°54′N	5°32′E
Saint-Nazaire, Fr. (săN′nȧ-zâr′)	196-97	47°17′N	2°13′W
Saint-Omer, Fr. (săN′tô-mâr′)	196-97	50°45′N	2°16′E
Saint Paul, Ab., Can. (sānt pôl′)	132-33	53°60′N	111°17′W
Saint-Paul, Reu.	265a	21°0′S	55°16′E
Saint Paul, Mn., U.S. (sānt pôl)	114-15	44°57′N	93°06′W
Saint Paul, Ne., U.S. (sānt pôl)	114-15	41°13′N	98°28′W
Saint Paul, stm., Lib.	260-61	6°25′N	10°44′W
Saint Pauls, N.C., U.S. (sānt pôls)	124-25	34°49′N	78°58′W
Saint Peter, Mn., U.S. (sānt pē′tēr)	114-15	44°20′N	93°58′W
Saint Peter Port, nat. cap., Guern. (sānt pē′tēr pôrt)	196-97	49°28′N	2°33′W
Saint Petersburg, Russia (sānt pē′tērz-bûrg)	192-93	59°57′N	30°15′E
Saint Petersburg, Fl., U.S. (sānt pē′tērz-bûrg)	125a	27°46′N	82°40′W
Saint-Pierre, Reu.	265a	21°19′S	55°29′E
Saint Pierre, i., Sey.	264-65	9°19′S	50°43′E
Saint-Pierre, nat. cap., St. P./M. (săN′pyâr′)	138-39	46°47′N	56°12′W
Saint Pierre and Miquelon, dep., N.A. (sānt pē-âr′ ănd mĭk-ē-lôn′)	138-39	46°55′N	56°20′W
Saint-Pierre-et-Miquelon, dep., N.A. *see* Saint Pierre and Miquelon	138-39	46°55′N	56°20′W
Saint-Pol-de-Léon, Fr. (săN-pô′dē-lā-ôN′)	196-97	48°41′N	3°59′W
Saint-Quentin, Fr. (săN′kän-tăN′)	196-97	49°51′N	3°18′E
Saint-Sébastien, Cap, c., Madag. *see* Anorontany, Tanjona	264-65	12°26′S	48°45′E
Saint Stephen, N.B., Can. (sānt stē′vĕn)	138-39	45°12′N	67°17′W
Saint Thomas, On., Can. (sānt tŏm′ȧs)	136-37	42°47′N	81°11′W
Saint Thomas, i., V.I.U.S.	143b	18°21′N	64°55′W
Saint-Tropez, Fr. (săN trô-pĕ′)	196-97	43°16′N	6°38′E
Saint Vincent, i., St. Vin.	143b	13°15′N	61°12′W
Saint-Vincent, Cap, c., Madag. *see* Ankaboa, Tanjona	264-65	21°55′S	43°18′E
Saint Vincent, Gulf, b., Austl. (gŭlf vĭn′sĕnt)	276	34°47′S	138°06′E
Saint Vincent and the Grenadines, nation, N.A. (sānt vĭn′sĕnt ănd thȧ grĕn′ȧ-dēnz)	140-41	13°15′N	61°12′W
Saipan, i., N. Mar. Is.	280-81	15°12′N	145°45′E
Saitama, state, Japan (sī′tä-mä)	245	36°0′N	139°30′E
Sajama, Nevado, mtn., Bol. (nĕ-vä′dô-sä-hȧ′mä)	168-69	18°06′S	68°54′W
Sak, stm., S. Afr.	264-65	30°06′S	20°42′E
Sakai, Japan (sä′kä-ē)	245	34°35′N	135°29′E
Sakākah, Sau. Ar.	228-29	29°58′N	40°13′E
Sakakawea, Lake, res., N.D., U.S.	114-15	47°44′N	102°18′W
Sakami, Lac, lk., Qc., Can.	130-31	53°15′N	76°45′W
Sakart'velo, nation, Asia *see* Georgia	227	42°0′N	44°00′E
Sakarya, Tur.	186-87	40°47′N	30°24′E
Sakarya, stm., Tur. (sȧ-kär′yȧ)	186-87	41°07′N	30°39′E
Sakata, Japan (sä′kä-tä)	244	38°55′N	139°51′E
Sakha, state, Russia *see* Yakutia	218-19	67°0′N	125°00′E
Sakhalin, i., Russia (sȧ-kȧ-lēn′)	218-19	51°0′N	143°00′E
Šakiai, Lith. (shä′kĭ-ī)	192-93	54°58′N	23°04′E
Sakishima-shotō, is., Japan (sä′kē-shē′ma gòn′tō′)	279a	24°33′N	124°26′E
Sal, i., C.V. (säal)	260-61	16°49′N	22°57′W
Sal, stm., Russia (sȧl)	186-87	47°31′N	40°44′E
Sal, Cay, i., Bah. (kē säl)	142-43	23°43′N	80°25′W
Sala, Swe. (sô′lä)	192-93	59°56′N	16°37′E
Salaberry-de-Valleyfield, Qc., Can.	136-37	45°15′N	74°08′W
Sala Consilina, Italy (sä′lä kôn-sē-lē′nä)	200-01	40°25′N	15°34′E
Salada, Laguna, lk., Mex. (lä-gò′nä-sä-lä′dä)	118-19	32°20′N	115°40′W
Saladas, Arg.	173	28°14′S	58°39′W
Saladillo, Arg. (sä-lä-dēl′yô)	173	35°38′S	59°47′W
Salado, stm., Arg. (sä-lä′dô)	168-69	31°41′S	60°44′W
Salado, stm., Arg. (sä-lä′dô)	173	35°45′S	57°23′W
Salado, stm., Arg. (sä-lä′dô)	171	38°49′S	64°59′W
Salado, stm., Mex. (sä-lä′dô)	122-23	26°52′N	99°19′W
Şalālah, Oman	220-21	17°01′N	54°06′E
Salamanca, Chile (sä-lä-mä′n-kä)	168-69	31°46′S	70°59′W
Salamanca, Mex. (sä-lä-mä′n-kä)	146-47	20°34′N	101°12′W
Salamanca, Spain (sä-lä-mä′n-kȧ)	198-99	40°58′N	5°39′W
Salamanca, N.Y., U.S. (săl-ȧ-măŋ′kȧ)	116-17	42°10′N	78°43′W
Salamat, Bahr, stm., Chad (bär sä-lä-mät′)	262-63	9°27′N	18°06′E
Salamina, Col. (sä-lä-mē′-nä)	163c	5°25′N	75°29′W
Salatiga, Indon.	248-49	7°20′S	110°31′E
Salavat, Russia	186-87	53°22′N	55°56′E
Salaverry, Peru (sä-lä-vā′rē)	170	8°14′S	78°58′W
Salawati, i., Indon. (sä-lä-wä′tē)	224-25	1°07′S	130°52′E
Sala y Gómez, Isla, i., Chile	282-83	26°26′S	105°26′W
Saldanha, S. Afr.	264-65	32°60′S	17°57′E
Saldus, Lat. (sȧl′dòs)	192-93	56°40′N	22°30′E
Sale, Austl. (sāl)	276	38°07′S	147°04′E
Salé, Mor.	269a	34°03′N	6°48′W
Salebabu, Pulau, i., Indon.	248-49	3°56′N	126°42′E
Salekhard, Russia (sŭ-lyĭ-kärt)	218-19	66°32′N	66°37′E
Salem, India (sä′lĕm)	236	11°39′N	78°10′E
Salem, Il., U.S. (sā′lĕm)	116-17	38°37′N	88°57′W
Salem, In., U.S. (sā′lĕm)	116-17	38°36′N	86°06′W
Salem, Mo., U.S. (sā′lĕm)	120-21	37°39′N	91°32′W
Salem, Oh., U.S. (sā′lĕm)	116-17	40°54′N	80°51′W
Salem, Or., U.S.	112-13	44°56′N	123°01′W
Salem, S.D., U.S. (sā′lĕm)	114-15	43°44′N	97°23′W
Salem, Va., U.S. (sā′lĕm)	124-25	37°18′N	80°03′W
Salem, W.V., U.S. (sā′lĕm)	116-17	39°17′N	80°34′W
Salerno, Italy (sä-lĕr′nô)	200-01	40°41′N	14°47′E
Salerno, Golfo di, b., Italy (gôl-fô-dē-sä-lĕr′nô)	200-01	40°32′N	14°42′E
Salgótarján, Hung. (shôl′gô-tôr-yän)	194-95	48°06′N	19°50′E
Salida, Co., U.S. (sȧ-lī′dȧ)	118-19	38°32′N	105°60′W
Salīmah, Wāḩat, well, Sudan	266	21°22′N	29°19′E
Salina, Ks., U.S. (sȧ-lī′nȧ)	120-21	38°50′N	97°36′W
Salina, Ut., U.S. (sȧ-lī′nȧ)	118-19	38°58′N	111°52′W
Salina, Isola, i., Italy (ē′-sō-lä-sä-lē′nä)	200-01	38°34′N	14°50′E
Salina Cruz, Mex. (sä-lē′nä krōōz′)	146-47	16°11′N	95°11′W
Salinas, Ec.	170	2°13′S	80°57′W
Salinas, Ca., U.S. (sȧ-lē′nȧs)	118-19	36°41′N	121°40′W
Salinas de Hidalgo, Mex.	146-47	22°38′N	101°44′W
Saline, stm., Ar., U.S. (sȧ-lēn′)	120-21	33°09′N	92°08′W
Salisbury, Md., U.S.	116-17	38°22′N	75°36′W
Salisbury, Mo., U.S. (sôlz′bē-rē)	120-21	39°26′N	92°48′W
Salisbury, N.C., U.S. (sôlz′bē-rē)	124-25	35°40′N	80°28′W
Salisbury, nat. cap., Zimb. (sôlz′bē-rē) *see* Harare	264-65	17°50′S	31°03′E
Salisbury Island, i., Nu., Can.	130-31	63°30′N	76°60′W
Salliq, Nu., Can. *see* Coral Harbour	128-29	64°08′N	83°12′W
Sallisaw, Ok., U.S. (săl′ĭ-sô)	120-21	35°28′N	94°48′W

Feature (Pronunciation)	Page	Lat.	Long.
Salluit, Qc., Can.	128-29	62°13′N	75°36′W
Salmon, Id., U.S. (săm′ŭn)	112-13	45°11′N	113°54′W
Salmon, stm., B.C., Can. (săm′ŭn)	132-33	54°04′N	122°33′W
Salmon, stm., N.B., Can. (săm′ŭn)	138-39	46°04′N	65°55′W
Salmon, stm., Id., U.S. (săm′ŭn)	112-13	45°51′N	116°47′W
Salmon Arm, B.C., Can. (săm′ŭn ärm)	132-33	50°42′N	119°19′W
Salmon River Mountains, mts., Id., U.S. (săm′ŭn rĭv′ẽr moun′tĭnz)	112-13	44°58′N	114°52′W
Salon-de-Provence, Fr. (så-lôn-dĕ-prô-väns′)	196-97	43°39′N	5°05′E
Salonika, Grc. *see* Thessaloníki	200-01	40°38′N	22°59′E
Salsk, Russia (sälsk)	186-87	46°28′N	41°33′E
Salt, stm., Az., U.S. (sôlt)	118-19	33°23′N	112°17′W
Salta, Arg. (säl′tä)	168-69	24°48′S	65°25′W
Salta, state, Arg. (säl′tä)	168-69	25°0′S	64°30′W
Saltillo, Mex. (säl-tēl′yô)	122-23	25°26′N	101°00′W
Salt Lake City, Ut., U.S. (sôlt lāk sĭ′tĭ sĭ′tė)	112-13	40°47′N	111°54′W
Salto, Arg. (säl′tō)	173	34°18′S	60°15′W
Salto, Ur.	173	31°23′S	57°58′W
Salto Grande, Embalse, res., S.A.	173	30°55′S	57°54′W
Salto Grande, Embalse de, res., S.A. *see* Salto Grande, Embalse	173	30°55′S	57°54′W
Salton Sea, lk., Ca., U.S. (sôlt′ŭn sē)	118-19	33°19′N	115°50′W
Saltville, Va., U.S. (sôlt′vĭl)	124-25	36°53′N	81°46′W
Saluda, S.C., U.S. (så-lo͞o′då)	124-25	34°00′N	81°47′W
Salûm, Egypt	188-89	31°34′N	25°09′E
Saluzzo, Italy (sä-lo͞ot′sō)	200-01	44°39′N	7°29′E
Salvador, Braz. (säl-vä-dōr′)	166-67	12°59′S	38°30′W
Salvador, El, nation, N.A. *see* El Salvador	85	13°50′N	88°55′W
Salvador, Lake, lk., La., U.S. (lāk săl′-vä-dōr lāk)	124-25	29°45′N	90°15′W
Salvatierra, Mex. (säl-vä-tyĕr′rä)	146-47	20°13′N	100°54′W
Salyan, Azer.	227	39°35′N	48°58′E
Salzburg, Aus. (sälts′bòrgh)	194-95	47°49′N	13°03′E
Salzwedel, Ger. (sälts-vā′dĕl)	194-95	52°51′N	11°09′E
Samâlût, Egypt (sä-mä-lo͞ot′)	268b	28°18′N	30°42′E
Samana Cay, i., Bah.	142-43	23°05′N	73°44′W
Samar, i., Phil. (sä′mär)	250	12°0′N	125°00′E
Samara, Russia (så-mä′rå)	186-87	53°11′N	50°07′E
Samara, stm., Russia (så-mä′rå)	186-87	53°10′N	50°04′E
Samara, stm., Ukr. (så-mä′rå)	202-03	48°28′N	35°06′E
Samarai, Pap. N. Gui. (sä-mä-rä′ė)	277	10°36′S	150°42′E
Samarinda, Indon.	248-49	0°30′S	117°09′E
Samarqand, Uzb.	232-33	39°40′N	66°56′E
Sāmarrā', Iraq	228-29	34°11′N	43°53′E
Samaúna, Braz.	166-67	7°56′S	60°01′W
Sambalpur, India (sŭm′bŭl-pòr)	234-35	21°28′N	83°59′E
Sambas, Indon.	246-47	1°19′N	109°16′E
Sambava, Madag.	264-65	14°16′S	50°09′E
Sambhal, India	234-35	28°35′N	78°34′E
Sāmbhar, India	234-35	26°54′N	75°13′E
Sambir, Ukr.	194-95	49°31′N	23°13′E
Samborombón, Bahía, b., Arg. (bä-ē′ä-säm-bô-rôm-bô′n)	173	36°0′S	57°12′W
Samch'ŏk, Kor., S.	243	37°27′N	129°10′E
Samch'ŏnp'o, Kor., S.	243	34°56′N	128°05′E
Same, Tan.	267	4°04′S	37°44′E
Samoa, nation, Oc. (sȧ-mō′ȧ)	279b	13°55′S	172°00′W
Samoa Islands, is., Oc. (sȧ-mō′ȧ ī′lȧndz)	279b	14°0′S	171°00′W
Samoded, Russia	186-87	63°37′N	40°30′E
Samokov, Blg. (sä′mô-kôf)	200-01	42°20′N	23°34′E
Sámos, i., Grc. (să′mŏs)	200-01	37°42′N	26°50′E
Samothrace, i., Grc. *see* Samothráki	200-01	40°29′N	25°36′E
Samothráki, i., Grc.	200-01	40°29′N	25°36′E
Sampit, Indon.	248-49	2°33′S	112°57′E
Sam Rayburn Reservoir, res., Tx., U.S.	122-23	31°13′N	94°17′W
Samsun, Tur. (såm′so͞on′)	186-87	41°17′N	36°20′E
Samtredia, Geor. (såm′trĕ-dė)	227	42°10′N	42°21′E
Samui, Ko, i., Thai.	246-47	9°32′N	100°01′E
San, Mali (sän)	258-59	13°18′N	4°54′W
Sanđoy, i., Far. Is.	190b	61°50′N	6°45′W
Ṣan'ā', nat. cap., Yemen (sän′ä) *see* Sanaa	266	15°21′N	44°12′E
Sanaa, nat. cap., Yemen (sän′ä)	266	15°21′N	44°12′E
Sanaga, stm., Camrn. (sä-nä′gä)	260-61	3°33′N	9°39′E
San Agustin, Cape, c., Phil.	250	6°18′N	126°12′E
Sanana, Pulau, i., Indon.	248-49	2°12′S	125°55′E
Sanandaj, Iran	228-29	35°19′N	47°00′E
San Andreas, Ca., U.S. (săn ăn′drē-ăs)	118-19	38°12′N	120°41′W
San Andrés, Col.	150	12°33′N	81°42′W
San Andrés, Isla de, i., Col. (ē′s-lä-dĕ-sän-än-drĕ′s)	164-65	12°33′N	81°43′W
San Andres Mountains, mts., N.M., U.S. (săn ăn′drē-ăs moun′tĭnz)	118-19	32°59′N	106°36′W
San Andrés Tuxtla, Mex. (sän-än-drā′s-to͞os′tlä)	146-47	18°26′N	95°13′W
San Angelo, Tx., U.S. (săn ăn-jĕ-lō)	122-23	31°29′N	100°26′W
San Antonio, Chile (sän-ån-tô′nyō)	163e	33°36′S	71°36′W
San Antonio, Col. (sän-ån-tô′nyō)	163c	3°55′N	75°29′W
San Antonio, Tx., U.S. (săn ăn-tō′nė-ō)	122-23	29°25′N	98°29′W
San Antonio, stm., Tx., U.S. (săn ăn-tō′nė-ō)	122-23	28°30′N	96°53′W
San Antonio, Cabo, c., Arg.	173	36°40′S	56°42′W
San Antonio, Cabo de, c., Cuba (kä′bô-dĕ-sän-än-tô′nyô)	142-43	21°52′N	84°57′W
San Antonio Bay, b., Tx., U.S. (săn ăn-tō′nė-ô bā)	122-23	28°20′N	96°45′W
San Antonio de los Cobres, Arg. (sän-án-tô′nyō då lōs kō′brås)	168-69	24°13′S	66°19′W
San Antonio Oeste, Arg. (sän-nä-tō′nyō ô-ĕs′tå)	171	40°45′S	64°58′W
San Augustine, Tx., U.S. (săn ô′gŭs-tēn)	122-23	31°31′N	94°07′W
San Benedetto del Tronto, Italy (sän bā′nå-dĕt′tô dĕl trōn′tô)	200-01	42°58′N	13°53′E
San Benedicto, Isla, i., Mex.	144-45	19°19′N	110°49′W
San Benito, Guat.	148	16°55′N	89°54′W
San Benito, Tx., U.S. (săn bĕ-nē′tō)	122-23	26°08′N	97°38′W
San Bernardino, Ca., U.S. (săn bûr-när-dē′nō)	118-19	34°06′N	117°17′W
San Bernardino Strait, strt., Phil.	250	12°32′N	124°10′E
San Bernardo, Chile (sän bĕr-när′dô)	163e	33°36′S	70°42′W
San Blas, Mex. (sän bläs′)	146-47	21°33′N	105°17′W
San Blas, Mex. (sän bläs′)	144-45	26°05′N	108°46′W
San Blas, Cape, c., Fl., U.S. (kāp săn bläs′)	124-25	29°40′N	85°22′W
San Borja, Bol.	166-67	14°49′S	66°51′W
San Buenaventura, Mex. (sän bwā′nå-vĕn-to͞o′rå)	122-23	27°04′N	101°33′W
San Buenaventura, Ca., U.S. *see* Ventura	118-19	34°17′N	119°17′W
San Carlos, Chile (sän-kå′r-lŏs)	171	36°26′S	71°57′W
San Carlos, Mex. (sän kär′lōs)	122-23	29°01′N	100°51′W
San Carlos, Nic. (sän-kä′r-lôs)	149	11°07′N	84°47′W
San Carlos, Phil.	250	10°30′N	123°25′E
San Carlos, Phil.	250	15°56′N	120°21′E
San Carlos, Az., U.S. (săn kär′lōs)	118-19	33°21′N	110°27′W
San Carlos, Ven.	164-65	9°40′N	68°35′W
San Carlos, stm., C.R. (sän kär′lōs)	149	10°47′N	84°12′W
San Carlos de Bariloche, Arg.	171	41°09′S	71°18′W
San Carlos de Bolívar, Arg.	173	36°13′S	61°07′W
San Carlos del Zulia, Ven.	164-65	9°02′N	71°56′W
San Carlos de Río Negro, Ven.	164-65	1°55′N	67°04′W
San Carlos Indian Reservation, ind. res., Az., U.S. (săn kär′lōs ĭn′dĭ-ăn rĕ-sẽr-vā′shĕn)	118-19	33°23′N	110°09′W
San Cataldo, Italy (sän kä-täl′dō)	200-01	37°29′N	13°59′E
Sánchez, Dom. Rep. (sän′chĕz)	142-43	19°14′N	69°37′W
San Clemente, Monte, mtn., Chile *see* San Valentín, Monte	171	46°36′S	73°20′W
San Clemente Island, i., Ca., U.S. (săn klå-mĕn′tå ī′lȧnd)	118-19	32°54′N	118°29′W
San Cristóbal, Arg.	173	30°19′S	61°13′W
San Cristóbal, Dom. Rep. (sän krēs-tô′bäl)	142-43	18°25′N	70°06′W
San Cristóbal, Ven. (sän krēs-tô′bäl)	164-65	7°45′N	72°13′W
San Cristobal, i., Sol. Is.	279e	10°36′S	161°45′E
San Cristóbal, Isla, i., Ec.	170a	0°50′S	89°26′W
San Cristóbal de las Casas, Mex.	146-47	16°45′N	92°38′W
Sancti Spíritus, Cuba (säŋk′tė spē′rė-to͞os)	142-43	21°56′N	79°27′W
Sancti Spíritus, state, Cuba (säŋk′tė spē′rė-to͞os)	142-43	22°0′N	79°20′W
Sancy, Puy de, mtn., Fr. (pwē-dĕ-sän-sē′)	196-97	45°32′N	2°49′E
Sandakan, Malay. (sån-dä′kån)	250	5°51′N	118°06′E
Sandefjord, Nor. (sän′dĕ-fyôr′)	192-93	59°08′N	10°14′E
Sanders, Az., U.S. (săn′dẽrz)	118-19	35°14′N	109°20′W
Sanderson, Tx., U.S. (săn′dẽr-sŭn)	122-23	30°09′N	102°24′W
Sandersville, Ga., U.S. (săn′dẽrz-vĭl)	124-25	32°59′N	82°49′W
Sand Hills, hills, Ne., U.S. (sănd hĭlz)	114-15	42°0′N	101°00′W
Sandia, Peru	170	14°16′S	69°27′W
San Diego, Ca., U.S. (săn dė-ā′gō)	118-19	32°43′N	117°08′W
San Diego, Tx., U.S. (săn dė-ā′gō)	122-23	27°46′N	98°14′W
San Diego, Cabo, c., Arg.	171	54°39′S	65°08′W
San Diego de la Unión, Mex. (sän dė-å-gô då lä o͞o-nyōn′)	146-47	21°28′N	100°52′W
Sandnes, Nor. (sänd′nĕs)	192-93	58°51′N	5°44′E
Sandomierz, Pol. (sån-dô′myĕzh)	194-95	50°41′N	21°46′E
San Donà di Piave, Italy (sän dô nà′ dė pyä′vĕ)	200-01	45°38′N	12°34′E
Sandoway, Mya. (sän-dô-wī′)	246-47	18°28′N	94°22′E
Sandpoint, Id., U.S. (sănd point)	112-13	48°17′N	116°33′W
Sand Springs, Ok., U.S. (sănd sprĭngz)	120-21	36°08′N	96°07′W
Sandstone, Mn., U.S. (sănd′stōn)	114-15	46°08′N	92°52′W
Sandusky, Mi., U.S. (săn-dŭs′kė)	116-17	43°25′N	82°49′W
Sandusky, Oh., U.S. (săn-dŭs′kė)	116-17	41°27′N	82°42′W
Sandwich, Il., U.S. (sănd′wĭch)	116-17	41°39′N	88°37′W
Sandy, Ut., U.S. (sănd′ė)	112-13	40°37′N	111°54′W
Sandy Cape, c., Austl.	277	24°42′S	153°16′E
Sandykgaçy, Turkmen.	232-33	36°33′N	62°33′E
Sandy Lake, lk., Nf., Can. (sănd′ė lāk)	138-39	49°16′N	57°00′W
Sandy Lake, lk., On., Can. (sănd′ė lāk)	134-35	53°02′N	93°00′W
Sandy Springs, Ga., U.S. (sănd′ė sprĭngz)	124-25	33°56′N	84°23′W
San Estanislao, Para. (sän ĕs-tä-nės-lå′ô)	168-69	24°39′S	56°29′W
San Felipe, Chile (sän fĕ-lē′på)	163e	32°45′S	70°43′W
San Felipe, Mex. (sän fĕ-lė′pĕ)	146-47	21°29′N	101°13′W
San Felipe, Mex. (sän fĕ-lė′pĕ)	144-45	31°02′N	114°51′W
San Felipe, Ven. (sän fĕ-lē′pĕ)	164-65	10°20′N	68°44′W
San Felipe, Cayos de, is., Cuba (kä′yōs-dĕ-sän-fĕ-lē′pĕ)	142-43	21°58′N	83°30′W
San Félix, Isla, i., Chile (ē′s-lä-dĕ-sän fā-lēks′)	159	26°17′S	80°06′W
San Fernando, Chile	163e	34°35′S	70°59′W
San Fernando, Mex. (sän fĕr-nän′dô)	122-23	24°51′N	98°10′W
San Fernando, Phil.	250	15°01′N	120°41′E
San Fernando, Phil.	250	16°37′N	120°19′E
San Fernando, Trin.	143b	10°17′N	61°27′W
San Fernando de Apure, Ven. (sän-fĕr-nä′n-dō-dĕ-ä-po͞o′rå)	164-65	7°53′N	67°27′W
San Fernando de Atabapo, Ven. (sän-fĕr-nä′n-dō-dĕ-ä-tä-bä′pô)	164-65	4°02′N	67°41′W
San Fernando del Valle de Catamarca, Arg.	168-69	28°28′S	65°47′W
Sanford, Fl., U.S. (săn′fōrd)	124-25	28°47′N	81°17′W
Sanford, Me., U.S. (săn′fẽrd)	116-17	43°27′N	70°47′W
Sanford, N.C., U.S. (săn′fẽrd)	124-25	35°29′N	79°11′W
San Francisco, Arg. (sän frän′sĭs′kô)	173	31°26′S	62°05′W
San Francisco, Ca., U.S. (săn frăn-sĭs′kō)	118-19	37°47′N	122°25′W
San Francisco del Oro, Mex. (sän frăn′sĭs′kô-dĕl ō′rō)	122-23	26°52′N	105°51′W
San Francisco del Rincón, Mex. (sän frăn′sĭs′kô-dĕl rēn-kōn′)	146-47	21°01′N	101°52′W
San Francisco de Macorís, Dom. Rep. (sän frăn′sĭs′kô-dā-mä-kō′rēs)	142-43	19°18′N	70°15′W
San Gabriel Chilac, Mex. (sän-gä-brē-ĕl-chē-läk′)	146-47	18°20′N	97°21′W
Sangar, Russia	218-19	63°55′N	127°29′E
Sangarius, stm., Tur. *see* Sakarya	186-87	41°07′N	30°39′E
Sangay, vol., Ec.	170	2°0′S	78°20′W
Sangeang, Pulau, i., Indon.	248-49	8°12′S	119°04′E
Sangerhausen, Ger. (säng′ẽr-hou-zĕn)	194-95	51°28′N	11°18′E
Sanggan, stm., China	240-41	40°21′N	115°25′E
Sanggau, Indon.	248-49	0°07′N	110°35′E
Sangha, stm., Afr.	260-61	1°12′S	16°50′E
Sangihe, Kepulauan, is., Indon.	248-49	3°0′N	125°30′E
Sangihe, Pulau, i., Indon.	248-49	3°35′N	125°32′E
San Gil, Col. (sän-кē′l)	164-65	6°33′N	73°08′W
San Giovanni in Fiore, Italy (sän jô-vän′nė ēn fyō′rå)	200-01	39°15′N	16°42′E
Sangju, Kor., S. (säng′jo͞o′)	243	36°25′N	128°10′E
Sāngli, India	236	16°52′N	74°34′E
San Gregorio, Ur.	173	32°38′S	55°50′W
Sangue, stm., Braz.	166-67	10°57′S	58°20′W
Sanibel Island, i., Fl., U.S. (săn′ĭ-bĕl ī′lȧnd)	125a	26°27′N	82°08′W
San Ignacio, Arg.	173	27°16′S	55°33′W
San Ignacio, Mex.	144-45	27°17′N	112°54′W
San Ignacio de Moxo, Bol.	166-67	14°56′S	65°37′W
San Ignacio de Velasco, Bol.	168-69	16°23′S	60°57′W
San Ildefonso, Cape, c., Phil. (kāp sän-ĕl-dĕ-fōn-sô)	250	16°02′N	122°00′E
San Ildefonso ó la Granja, Spain (sän-ĕl-dĕ-fōn-sô ō lä grän′khä)	198-99	40°54′N	4°00′W
San Isidro, Arg. (sän ē-sė′drô)	173	34°28′S	58°31′W
San Jacinto, Phil. (sän hä-sēn′tô)	250	12°34′N	123°44′E
San Javier, Arg.	173	30°34′S	59°56′W
San Javier, Bol.	168-69	16°20′S	62°38′W
San Joaquín, Bol.	166-67	13°04′S	64°49′W

Feature (Pronunciation)	Page	Lat.	Long.
San Joaquín, stm., Bol.	166-67	13°08′s	63°41′w
San Joaquin Valley, val., Ca., U.S.	118-19	36°55′n	120°29′w
San Jorge, Golfo, b., Arg. (gôl-fô-sän-κô′r-κě)	171	46°0′s	67°00′w
San Jorge Island, i., Sol. Is.	279e	8°27′s	159°35′e
San Jose, Phil.	250	12°21′n	121°04′e
San Jose, Ca., U.S.	118-19	37°21′n	121°54′w
San José, nat. cap., C.R. (sän hô-sā′)	149	9°56′n	84°05′w
San José, Isla, i., Mex. (ē′s-lä-sän κô-sě′)	144-45	25°00′n	110°38′w
San José, Isla, i., Pan. (ē′s-lä-sän hô-sā′)	150	8°15′n	79°07′w
San José de Chiquitos, Bol.	168-69	17°50′s	60°44′w
San José de Feliciano, Arg. (sän κô-sě′ dā lä ěs-kē′nä)	173	30°23′s	58°45′w
San José de Jáchal, Arg.	168-69	30°14′s	68°45′w
San José del Cabo, Mex.	144-45	23°03′n	109°41′w
San José del Guaviare, Col.	164-65	2°34′n	72°38′w
San José de Mayo, Ur.	173	34°21′s	56°42′w
San Jose Island, i., Tx., U.S.	122-23	28°02′n	96°55′w
San Juan, Arg. (sän hwän′)	168-69	31°32′s	68°32′w
San Juan, state, Arg. (sän hwän′)	168-69	31°0′s	69°00′w
San Juan, nat. cap., P.R. (sän hwän′)	142a	18°28′n	66°07′w
San Juan, stm., Arg.	168-69	32°17′s	67°22′w
San Juan, stm., Mex. (sän-hōō-än′)	122-23	26°22′n	98°51′w
San Juan, stm., N.A.	149	10°56′n	83°43′w
San Juan, stm., U.S. (săn hwän′)	110-11	37°11′n	110°43′w
San Juan, Pico, mtn., Cuba (pē′kô-sän-κôà′n)	142-43	21°59′n	80°09′w
San Juan Bautista, Para. (sän hwän′ bou-tēs′tä)	173	26°53′s	57°01′w
San Juan de la Maguana, Dom. Rep.	142-43	18°48′n	71°13′w
San Juan del Norte, Nic.	149	10°55′n	83°42′w
San Juan de los Morros, Ven. (sän-hōō-än′dě-lôs-mô′r-rôs)	163b	9°55′n	67°21′w
San Juan del Río, Mex. (sän hwän del rē′ô)	146-47	20°23′n	100°00′w
San Juan del Río, Mex. (sän hwän del rē′ô)	122-23	24°48′n	104°27′w
San Juan del Sur, Nic. (sän hwän děl sōōr)	149	11°15′n	85°52′w
San Juan Evangelista, Mex. (sän-hōō-ä′n-å-väŋ-kå-lěs′ta′)	146-47	17°54′n	95°07′w
San Juanito, Isla, i., Mex.	146-47	21°46′n	106°41′w
San Juan Mountains, mts., Co., U.S. (san hwän′ moun′tĭnz)	118-19	37°32′n	107°31′w
San Justo, Arg. (sän hōōs′tô)	173	30°47′s	60°35′w
Sankt Michel, Fin. *see* Mikkeli	192-93	61°42′n	27°16′e
Sankt Pölten, Aus. (zäŋkt-pûl′těn)	194-95	48°12′n	15°37′e
Sankuru, stm., D.R.C. (säŋ-kōō′rōō)	262-63	4°17′s	20°24′e
San Lázaro, Cabo, c., Mex. (kå′bô sän-lå′zä-rō)	144-45	24°48′n	112°18′w
Şanlıurfa, Tur.	228-29	37°10′n	38°48′e
San Lorenzo, Arg. (sän lô-rěn′zô)	173	32°44′s	60°45′w
San Lorenzo, Ec.	170	1°15′n	78°50′w
San Lorenzo, Cabo, c., Ec.	170	1°04′s	80°54′w
San Lorenzo, Isla, i., Peru	163a	12°05′s	77°14′w
Sanlúcar de Barrameda, Spain (sän-lōō′kär)	198-99	36°47′n	6°21′w
San Lucas, Bol. (sän lōō′kås)	168-69	20°06′s	65°08′w
San Lucas, Cabo, c., Mex.	144-45	22°52′n	109°54′w
San Luis, Arg. (sän lò-ēs′)	171	33°18′s	66°21′w
San Luis, Guat. (sän lò-ēs′)	148	16°13′n	89°27′w
San Luis, state, Arg. (sän lò-ēs′)	171	34°0′s	66°00′w
San Luís, Laguna, lk., Bol.	166-67	13°45′s	64°00′w
San Luis de la Paz, Mex. (sän lò-ēs′ dā lä päz′)	146-47	21°18′n	100°31′w
San Luis Obispo, Ca., U.S. (sän lò-ēes′ ô-bĭs′pō)	118-19	35°17′n	120°40′w
San Luis Potosí, Mex.	146-47	22°09′n	100°59′w
San Luis Potosí, state, Mex.	146-47	22°30′n	100°30′w
San Luis Río Colorado, Mex.	144-45	32°28′n	114°46′w
San Marcos, Mex. (sän mär′kôs)	146-47	16°49′n	99°23′w
San Marcos, Tx., U.S. (săn mär′kôs)	122-23	29°53′n	97°56′w
San Marcos de Colón, Hond. (sän-má′r-kōs-dě-kô-lô′n)	149	13°26′n	86°49′w
San Marino, nation, Eur. (sän mēr-ē′nô)	200-01	43°56′n	12°25′e
San Martín, Arg.	163e	33°05′s	68°29′w
San Martín, Col. (sän mär-tē′n)	164-65	3°42′n	73°42′w
San Martín, stm., Bol.	166-67	13°08′s	63°47′w
San Martín, Lago, lk., S.A. (lä′gô sän mär-tē′n)	171	48°53′s	72°39′w
San Martín de los Andes, Arg.	171	40°10′s	71°22′w
San Mateo, Ca., U.S. (săn mä-tā′ô)	118-19	37°34′n	122°19′w
San Mateo, Ven. (sän må-tě′ô)	163b	9°45′n	64°33′w
San Matías, Golfo, b., Arg. (gôl-fô-sän-mä-tē′äs)	171	41°30′s	64°15′w
Sanmenxia, China	238-39	34°47′n	111°12′e
San Miguel, El Sal. (sän mē-gål′)	148	13°28′n	88°11′w
San Miguel, Mex. (sän mē-gål′)	122-23	29°10′n	101°28′w
San Miguel, Pan. (sän mē-gål′)	150	8°27′n	78°56′w
San Miguel, stm., Bol. (sän-mē-gěl′)	166-67	13°53′s	63°54′w
San Miguel, Golfo de, b., Pan. (gôl-fô-dě-sän mē-gål′)	150	8°22′n	78°17′w
San Miguel del Monte, Arg.	173	35°27′s	58°49′w
San Miguel de Tucumán, Arg.	168-69	26°49′s	65°13′w
San Miguel El Alto, Mex. (sän mē-gål′ ěl äl′tô)	146-47	21°01′n	102°19′w
Sannār, Sudan	266	13°34′n	33°33′e
San Nicolas, Phil. (sän nē-kô-läs′)	250	18°10′n	120°36′e
San Nicolás, stm., Mex. (sän nē-kô-lá′s)	146-47	19°38′n	105°13′w
San Nicolás, Canal de, strt., N.A. *see* Nicholas Channel	142-43	23°21′n	80°21′w
San Nicolás de los Arroyos, Arg.	173	33°20′s	60°14′w
Sanok, Pol. (sä′nôk)	194-95	49°34′n	22°13′e
San Pablo, Phil. (sän-pä-blô)	250	14°04′n	121°19′e
San Pedro, Arg. (sän pā′drô)	168-69	24°15′s	64°52′w
San Pedro, Arg. (sän pā′drô)	173	33°41′s	59°41′w
San Pedro, Chile (sän pě′drô)	163e	33°54′s	71°26′w
San-Pédro, C. Iv.	260-61	4°45′n	6°38′w
San Pedro, Punta, c., Chile	168-69	25°31′s	70°38′w
San Pedro, Volcán, vol., Chile	168-69	21°53′s	68°25′w
San Pedro de Jujuy, Arg. *see* San Pedro	168-69	24°15′s	64°52′w
San Pedro de las Colonias, Mex. (sän pā′drô dě-läs-kô-lô′nyäs)	122-23	25°46′n	102°59′w
San Pedro de Macorís, Dom. Rep. (sän-pě′drō-dă mä-kô-rēs′)	142-43	18°28′n	69°18′w
San Pedro de Ycuamandiyú, Para.	168-69	24°05′s	57°08′w
San Pedro Sula, Hond. (sän pā′drô sōō′lä)	149	15°30′n	88°02′w
San Pietro, Isola di, i., Italy (ē′sō-lä-dē-sän pyā′trô)	200-01	39°08′n	8°16′e
San Quintín, Cabo, c., Mex.	144-45	30°22′n	115°60′w
San Rafael, Arg. (sän rä-fä-ĕl′)	163e	34°37′s	68°20′w
San Ramón de la Nueva Orán, Arg.	168-69	23°09′s	64°20′w
San Remo, Italy (sän rā′mô)	200-01	43°50′n	7°46′e
San Roque, Punta, c., Mex.	144-45	27°11′n	114°25′w
San Saba, Tx., U.S. (săn să′bà)	122-23	31°12′n	98°43′w
San Saba, stm., Tx., U.S. (săn să′bà)	122-23	31°15′n	98°36′w
San Salvador, i., Bah. (sän säl′vá-dôr)	142-43	24°02′n	74°27′w
San Salvador, nat. cap., El Sal. (sän säl-vä-dōr′)	148	13°40′n	89°13′w
San Salvador, Isla, i., Ec. (ē′s-lä-sän säl-vä-dōr′) *see* Santiago, Isla	170a	0°14′s	90°45′w
San Salvador de Jujuy, Arg.	168-69	24°12′s	65°18′w
San Sebastián, Spain (sän så-bås-tyän′) *see* Donostia-San Sebastián	198-99	43°19′n	1°60′w
San Severo, Italy (sän sě-vå′rō)	200-01	41°41′n	15°23′e
Sanshui, China (sän-shwā)	238-39	23°11′n	112°53′e
San Simon, stm., Az., U.S. (sån sī-mōn′)	118-19	32°52′n	109°33′w
Santa, stm., Peru	170	9°01′s	78°38′w
Santa Ana, Bol.	166-67	13°45′s	65°35′w
Santa Ana, El Sal. (sän′tä ä′nä)	148	13°59′n	89°34′w
Santa Ana, Mex. (sän′tä ä′nä)	146-47	24°04′n	100°30′w
Santa Ana, Mex. (sän′tä ä′nä)	144-45	30°32′n	111°07′w
Santa Ana, Ca., U.S. (săn′tá ăn′á)	118-19	33°45′n	117°53′w
Santa Anna, Tx., U.S. (săn′tá ăn′á)	122-23	31°44′n	99°19′w
Santa Bárbara, Hond. (sän-tä-bá′r-bä-rä)	149	14°55′n	88°14′w
Santa Bárbara, Mex. (sän-tä-bá′r-bä-rä)	122-23	26°49′n	105°48′w
Santa Barbara, Ca., U.S. (săn-tă-bä′r-bă-rä)	118-19	34°25′n	119°42′w
Santa Catalina Island, i., Ca., U.S. (săn′tá kă-tá-lē′ná ī′lánd)	118-19	33°23′n	118°24′w
Santa Catarina, Mex. (sän′tä kä-tä-rē′nä)	122-23	25°41′n	100°28′w
Santa Catarina, state, Braz. (sän-tä-kä-tä-rē′nà)	168-69	27°0′s	50°00′w
Santa Catarina, Ilha de, i., Braz.	172	27°36′s	48°30′w
Santa Clara, Cuba (sän′tá klä′rä)	142-43	22°25′n	79°58′w
Santa Cruz, Braz. (sän-tä-krōō′s)	163d	6°13′s	36°01′w
Santa Cruz, Ca., U.S. (săn′tá krōōz′)	118-19	36°59′n	122°02′w
Santa Cruz, stm., Arg. (sän′tá krōōz′)	171	50°08′s	68°21′w
Santa Cruz, Isla, i., Ec. (ē′s-lä-sän-tä-krōō′z)	170a	0°38′s	90°23′w
Santa Cruz de la Palma, Spain	199d	28°41′n	17°46′w
Santa Cruz de la Sierra, Bol.	168-69	17°48′s	63°10′w
Santa Cruz del Sur, Cuba (sän-tä-krōō′s-děl-sŏ′r)	142-43	20°43′n	77°59′w
Santa Cruz de Tenerife, Spain (sän′tä krōōz dā tā-nå-rē′fā)	199d	28°28′n	16°15′w
Santa Cruz do Sul, Braz.	168-69	29°43′s	52°26′w
Santa Cruz Islands, is., Sol. Is.	272-73	10°60′s	166°15′e
Santa Fe, Arg. (sän′tä fā′)	173	31°38′s	60°42′w
Santa Fe, Spain (sän′tä-fā′)	198-99	37°11′n	3°43′w
Santa Fe, N.M., U.S. (săn′tá fā′)	118-19	35°41′n	105°59′w
Santa Fe, state, Arg. (sän′tä fā′)	173	31°0′s	61°00′w
Santa Fe de Bogotá, nat. cap., Col. *see* Bogotá	164-65	4°37′n	74°06′w
Santa Fé do Sul, Braz.	168-69	20°13′s	50°56′w
Santai, China (san-tī)	238-39	31°09′n	105°01′e
Santa Inés, Isla, i., Chile (ē′s-lä-sän′tä ē-nās′)	171	53°46′s	72°44′w
Santa Isabel, Arg.	171	36°15′s	66°56′w
Santa Isabel, i., Sol. Is.	279e	8°0′s	159°00′e
Santa Isabel, nat. cap., Eq. Gui. *see* Malabo	260-61	3°45′n	8°47′e
Santa Magdalena, Isla, i., Mex.	144-45	24°54′n	112°13′w
Santa Margarita, Isla, i., Mex. (ē′s-lä-sän′tä mär-gä-rē′tä)	144-45	24°27′n	111°51′w
Santa Maria, Braz. (sän′tá mä-rē′á)	173	29°41′s	53°49′w
Santa Maria, Ca., U.S. (săn-tá má-rē′á)	118-19	34°57′n	120°26′w
Santa Maria, i., Port. (sän-tä-mä-rē′ä)	199c	36°58′n	25°06′w
Santa Maria, i., Vanuatu	279g	14°14′s	167°28′e
Santa María, stm., Mex. (sän′tä mä-rē′á)	146-47	21°48′n	99°10′w
Santa Maria, Cabo de, c., Ang.	264-65	13°25′s	12°32′e
Santa Maria, Cabo de, c., Port. (kå′bō-dě-sän-tä-mä-rē′ä)	198-99	36°58′n	7°54′w
Santa María, Isla, i., Ec.	170a	1°17′s	90°26′w
Santa María del Oro, Mex. (sän′tä-mä-rē′ä-děl-ô-rô)	122-23	25°56′n	105°23′w
Santa Marta, Col. (sän′tá mär′tä)	164-65	11°15′n	74°12′w
Santa Monica, Ca., U.S. (săn′tá mŏn′ĭ-ká)	118-19	34°01′n	118°29′w
Santana do Livramento, Braz.	173	30°53′s	55°31′w
Santander, Col. (sän-tän-děr′)	163c	3°03′n	76°29′w
Santander, Phil.	250	9°25′n	123°20′e
Santander, Spain (sän-tän-dâr′)	198-99	43°28′n	3°48′w
Sant'Antioco, Isola di, i., Italy (ē′sō-lä-dē-sän-än-tyŏ′kô)	200-01	39°02′n	8°25′e
Santarém, Braz. (sän-tä-rěɴ′)	166-67	2°26′s	54°43′w
Santarém, Port.	198-99	39°14′n	8°41′w
Santaren Channel, strt., Bah. (săn-tá-rěn′ chăn′ěl)	142-43	24°0′n	79°30′w
Santa Rita, Hond.	149	15°10′n	87°54′w
Santa Rosa, Arg.	173	36°37′s	64°17′w
Santa Rosa, Braz.	173	27°52′s	54°26′w
Santa Rosa, Ec.	170	3°27′s	79°57′w
Santa Rosa, Ca., U.S.	118-19	38°26′n	122°43′w
Santa Rosa, N.M., U.S. (săn′tá rō′sá)	120-21	34°56′n	104°41′w
Santa Rosa de Copán, Hond.	149	14°46′n	88°47′w
Santa Rosalía, Mex. (sän′tá rô-zä′lē-á)	144-45	27°20′n	112°17′w
Santa Rosa Range, mts., Nv., U.S. (săn′tá rō′zá rānj)	112-13	41°35′n	117°40′w
Santa Sylvina, Arg.	173	27°50′s	61°08′w
Santa Vitória do Palmar, Braz. (sän-tä-vē-tô′ryä-dô-päl-már)	173	33°31′s	53°22′w
Santee, Ca., U.S. (săn tē′)	118-19	32°50′n	116°57′w
Santee, stm., S.C., U.S. (săn tē′)	124-25	33°14′n	79°28′w
Santiago, Braz. (sän-tyá′gô)	173	29°11′s	54°52′w
Santiago, Pan. (sän-tyá′gô)	150	8°06′n	80°58′w
Santiago, i., C.V.	260-61	15°02′n	23°39′w
Santiago, nat. cap., Chile (sän-tē-á′gô)	171	33°27′s	70°40′w
Santiago, Isla, i., Ec.	170a	0°14′s	90°45′w
Santiago de Compostela, Spain	198-99	42°53′n	8°32′w
Santiago de Cuba, Cuba (sän-tyá′gô-dě kōō′bá)	142-43	20°02′n	75°49′w
Santiago de Cuba, state, Cuba (sän-tyá′gô-dě kōō′bá)	142-43	20°10′n	75°55′w
Santiago del Estero, Arg. (sän-tē-á′gō-děl ěs-tä-rô)	168-69	27°47′s	64°16′w
Santiago del Estero, state, Arg. (sän-tē-á′gō-děl ěs-tä-rô)	173	28°0′s	63°30′w
Santiago de los Caballeros, Dom. Rep.	142-43	19°27′n	70°42′w
Santiago Jamiltepec, Mex.	146-47	16°18′n	97°50′w
Santiago Papasquiaro, Mex.	122-23	25°03′n	105°25′w
Santiaguillo, Laguna, lk., Mex. (lä-oô′nä-sän-tē-ä-gēl′yô)	122-23	24°45′n	104°48′w
Santo Amaro, Braz. (sän′tō ä-mä′rò)	166-67	12°32′s	38°42′w
Santo André, Braz.	172	23°40′s	46°31′w
Santo Ângelo, Braz. (sän-tô-à′n-zhě-lô)	173	28°16′s	54°16′w

Feature (Pronunciation)	Page	Lat.	Long.
Santo Antão, i., C.V. (sän´tô ä´n-zhĕ-lô)	260-61	17°03'N	25°07'W
Santo Antônio de Jesus, Braz.	166-67	12°57'S	39°14'W
Santo Antônio do Içá, Braz.	166-67	3°04'S	67°56'W
Santo Domingo, Nic. (sän-tô-dô-mē´n-gō)	149	12°16'N	85°05'W
Santo Domingo, i., N.A. *see* Hispaniola	142-43	19°0'N	71°00'W
Santo Domingo, nat. cap., Dom. Rep. (sän´tô dô-mĭn´gô)	142-43	18°30'N	69°53'W
Santoña, Spain (sän-tō´nyä)	198-99	43°26'N	3°27'W
Santorini, i., Grc. *see* Thíra	200-01	36°26'N	25°27'E
Santos, Braz. (sän´tozh)	172	23°56'S	46°20'W
Santos Dumont, Braz. (sän´tôs-dô-mô´nt)	172	21°28'S	43°33'W
Santo Tomé, Arg.	173	28°33'S	56°02'W
Santo Tomé de Guayana, Ven. *see* Ciudad Guayana	164-65	8°21'N	62°39'W
San Valentín, Monte, mtn., Chile (mô´n-tĕ-sän-vä-lĕn-tē´n)	171	46°36'S	73°20'W
San Vicente, El Sal. (sän vē-sĕn´tå)	148	13°38'N	88°47'W
San Vicente de Cañete, Peru	163a	13°05'S	76°24'W
San Vicente del Caguán, Col.	164-65	2°07'N	74°47'W
San Xavier Indian Reservation, ind. res., Az., U.S. (sän x-ā´vĭēr ĭn´dĭ-*ă*n rĕ-sēr-vā´shĕn)	118-19	32°02'N	111°08'W
Sanya, China	238-39	18°14'N	109°30'E
Sanyuan, China	238-39	34°37'N	108°55'E
Sanza Pombo, Ang.	262-63	7°19'S	15°60'E
São Bento, Braz.	166-67	2°42'S	44°50'W
São Borja, Braz. (souɴ-bôr-zhä)	173	28°39'S	56°01'W
São Carlos, Braz. (souɴ kär´lōzh)	172	22°02'S	47°54'W
São Cristóvão, Braz. (souɴ-krēs-tō-vouɴ)	163d	11°01'S	37°12'W
São Domingos, Braz.	166-67	13°24'S	46°21'W
São Francisco, Braz. (souɴ frän-sēsh´kô)	168-69	15°57'S	44°52'W
São Francisco, stm., Braz. (souɴ frän-sēsh´kô)	159	10°30'S	36°24'W
São Francisco, Ilha de, i., Braz.	172	26°18'S	48°37'W
São Francisco do Sul, Braz. (souɴ frän-sēsh´kô-dô-sōō´l)	172	26°15'S	48°37'W
São Gabriel, Braz. (souɴ´gä-brē-ĕl´)	173	30°20'S	54°19'W
São João da Barra, Braz. (souɴ-zhôuɴ-dä-bä´rä)	172	21°38'S	41°02'W
São João da Boa Vista, Braz. (souɴ-zhôuɴ-dä-bôä-vē´s-tä)	172	21°59'S	46°48'W
São João Del Rei, Braz. (souɴ zhô-ouɴ´dĕl-rā)	172	21°08'S	44°15'W
São Jorge, i., Port. (souɴ zhôr´zhĕ)	199c	38°38'N	28°03'W
São José do Rio Preto, Braz. (souɴ zhô-zĕ´dô-re´ō-prĕ-tō)	168-69	20°49'S	49°23'W
São José dos Campos, Braz. (souɴ zhô-zā´dōzh käɴ pōzh´)	172	23°11'S	45°53'W
São Leopoldo, Braz. (souɴ-lĕ-ô-pôl´dô)	168-69	29°46'S	51°08'W
São Lourenço do Sul, Braz.	173	31°22'S	51°58'W
São Luís, Braz.	166-67	2°31'S	44°16'W
São Luís Gonzaga, Braz.	173	28°24'S	54°57'W
São Manuel, stm., Braz.	166-67	7°21'S	58°08'W
São Mateus, Braz. (souɴ mä-tā´ōzh)	172	18°44'S	39°52'W
São Miguel, i., Port.	199c	37°47'N	25°30'W
Saona, Isla, i., Dom. Rep. (ē´s-lä-sä-ô´nä)	142-43	18°09'N	68°40'W
Saône, stm., Fr. (sōn)	196-97	45°43'N	4°50'E
São Nicolau, i., C.V. (souɴ´ nē-kô-louɴ´)	260-61	16°36'N	24°11'W
São Paulo, Braz. (souɴ´ pou´lô)	172	23°33'S	46°38'W
São Paulo, state, Braz. (souɴ´ pou´lô)	172	22°0'S	49°00'W
São Paulo de Olivença, Braz. (souɴ´pou´lòdå ô-lē-vĕn´s*ä*)	166-67	3°28'S	68°57'W
São Raimundo Nonato, Braz. (souɴ´ rī-mō´n-do nô-nä´tò)	166-67	9°01'S	42°42'W
São Roque, Braz. (souɴ´ rō´kĕ)	172	23°32'S	47°08'W
São Roque, Cabo de, c., Braz. (kå´bō-dĕ-souɴ´ rō´kĕ)	163d	5°29'S	35°16'W
São Salvador, Braz. *see* Salvador	166-67	12°59'S	38°30'W
São Sebastião, Braz. (souɴ så-bäs-tē-ouɴ´)	172	23°49'S	45°25'W
São Sebastião, Ilha de, i., Braz.	172	23°51'S	45°20'W
São Sebastião, Ponta, c., Moz.	264-65	22°08'S	35°29'E
São Simão, Braz. (souɴ-sē-mouɴ)	172	21°29'S	47°33'W
São Simão, Represa de, res., Braz.	168-69	18°37'S	49°59'W
São Tiago, i., C.V. (souɴ tē-ä´gô) *see* Santiago	260-61	15°02'N	23°39'W
São Tomé, i., S. Tom./P.	260-61	0°12'N	6°36'E
São Tomé, nat. cap., S. Tom./P.	260-61	0°20'N	6°44'E
São Tomé, Cabo de, c., Braz.	172	21°59'S	40°59'W
Sao Tome and Principe, nation, Afr. (souɴ tòmā änd prēn´sē-pĕ)	260-61	1°0'N	7°00'E
São Tomé e Princípe, nation, Afr. *see* Sao Tome and Principe	260-61	1°0'N	7°00'E
Saoura, Oued, stm., Alg.	258-59	29°01'N	0°57'W
São Vicente, Braz. (souɴ ve-se´n-tĕ)	172	23°58'S	46°22'W
São Vicente, i., C.V. (souɴ vē-sĕn´tå)	260-61	16°49'N	24°55'W
São Vicente, Cabo de, c., Port. (kå´bō-dĕ-sän-vē-sĕ´n-tĕ)	198-99	37°01'N	8°59'W
Sap, Tonle, lk., Camb. (tôn´lā säp´)	246-47	13°0'N	104°00'E
Sapé, Braz.	163d	7°06'S	35°13'W
Sapele, Nig. (sä-pā´lā)	260a	5°54'N	5°40'E
Sapitwa, mtn., Malawi	264-65	15°57'S	35°36'E
Sapporo, Japan (säp-pô´rô)	244	43°04'N	141°21'E
Sapulpa, Ok., U.S. (s*ā*-pŭl´p*ā*)	120-21	36°00'N	96°06'W
Saqqez, Iran	228-29	36°14'N	46°18'E
Sarāb, Iran	227	37°56'N	47°31'E
Saragossa, Spain *see* Zaragoza	198-99	41°39'N	0°53'W
Sarajevo, nat. cap., Bos. (sä-rá-yĕv´ô) (sá-rä´ya-vô)	200-01	43°52'N	18°25'E
Sarakhs, Iran	232-33	36°32'N	61°10'E
Saranac Lake, N.Y., U.S. (săr´á-năk lāk)	116-17	44°20'N	74°08'W
Sarandí Grande, Ur. (sä-rän´dē-grän´dĕ)	173	33°45'S	56°20'W
Sarang, Kaz.	226	49°46'N	72°52'E
Sarangani Islands, is., Phil.	250	5°25'N	125°26'E
Saransk, Russia (sá-ränsk´)	186-87	54°11'N	45°09'E
Sarapul, Russia (sä-räpōl')	186-87	56°28'N	53°48'E
Sarasota, Fl., U.S. (săr-*á*-sōt*á*)	125a	27°20'N	82°32'W
Saratoga, Wy., U.S. (săr-*á*-tō´g*á*)	112-13	41°28'N	106°48'W
Saratoga Springs, N.Y., U.S. (săr-*á*-tō´g*á* springz)	116-17	43°05'N	73°47'W
Saratov, Russia (sá rä´tôf)	186-87	51°34'N	45°60'E
Saratovskoye Vodokhranilishche, res., Russia	186-87	52°47'N	48°26'E
Saravan, Laos	246-47	15°43'N	106°25'E
Sarawak, hist. reg., Malay. (sá-rä´wäk)	248-49	2°30'N	113°30'E
Sarayevo, nat. cap., Bos. *see* Sarajevo	200-01	43°52'N	18°25'E
Sardegna, i., Italy *see* Sardinia	200-01	40°0'N	9°00'E
Sardinia, i., Italy (sär-dĭn´*ĭá*)	200-01	40°0'N	9°00'E
Sardis Lake, res., Ms., U.S. (sär´dĭs lāk)	124-25	34°27'N	89°43'W
Sardis Lake, res., Ok., U.S. (sär´dĭs lāk)	120-21	34°42'N	95°21'W
Sargent, Ne., U.S. (sär´jĕnt)	114-15	41°38'N	99°22'W
Sargodha, Pak.	232-33	32°05'N	72°40'E
Sarh, Chad (är-chaɴ-bô´)	262-63	9°09'N	18°23'E
Sārī, Iran	232-33	36°34'N	53°04'E
Sarina, Austl.	277	21°26'S	149°14'E
Sariqamish Kuli, lk., Asia *see* Sarygamysh köli	226	41°56'N	57°25'E
Sariwŏn, Kor., N.	243	38°30'N	125°46'E
Şarköy, Tur. (shär´kû-ē)	200-01	40°38'N	27°07'E
Sarmi, Indon.	277	1°51'S	138°42'E
Sarmiento, Arg.	171	45°35'S	69°05'W
Sarmiento de Gambia, Cerro, mtn., Chile	171	54°27'S	70°50'W
Särna, Swe.	192-93	61°41'N	13°08'E
Sarnia, On., Can. (sär´nē-*ă*)	136-37	42°58'N	82°24'W
Sarny, Ukr. (sär´nē)	194-95	51°20'N	26°37'E
Sarpsborg, Nor. (särps´bôrg)	192-93	59°16'N	11°09'E
Sarrebourg, Fr. (sär-bōōr´)	196-97	48°44'N	7°03'E
Sarrebruck, Ger. *see* Saarbrücken	194-95	49°14'N	6°60'E
Sarreguemines, Fr. (sär-gē-mēn´)	196-97	49°07'N	7°04'E
Sartang, stm., Russia	218-19	67°27'N	133°15'E
Sárvár, Hung. (shär´vär)	194-95	47°15'N	16°56'E
Sarych, mys, c., Ukr. (mĭs sá-rēch´)	202-03	44°25'N	33°45'E
Sarygamysh köli, lk., Asia	226	41°56'N	57°25'E
Sarysū, stm., Kaz. (sä´rē-sōō)	226	45°11'N	66°39'E
Sary-Tash, Kyrg.	232-33	39°44'N	73°15'E
Sāsarām, India (sŭs-ŭ-räm´)	234-35	24°57'N	84°01'E
Sasebo, Japan (sä´sâ-bô)	245	33°10'N	129°43'E
Saskatchewan, state, Can. (săs-kăch´ē-wän)	128-29	54°0'N	105°00'W
Saskatchewan, stm., Can. (săs-kăch´ē-wän)	134-35	53°16'N	98°50'W
Saskatoon, Sk., Can. (săs-k*ă*-tōōn´)	134-35	52°08'N	106°39'W
Saskylakh, Russia	206-07	71°52'N	114°06'E
Sasovo, Russia (sás´ô-vô)	186-87	54°20'N	41°57'E
Sassandra, stm., C. Iv. (sás-sän´drá)	260-61	4°58'N	6°04'W
Sassari, Italy (säs´sä-rē)	200-01	40°43'N	8°33'E
Sata-misaki, c., Japan	245	30°60'N	130°40'E
Sātāra, India	236	17°41'N	74°00'E
Säter, Swe. (sĕ´tĕr)	192-93	60°21'N	15°45'E
Satilla, stm., Ga., U.S. (sá-tĭl´*á*)	124-25	30°59'N	81°29'W
Satīt, stm., Afr. *see* Tekezē	266	14°20'N	35°51'E
Satluj, stm., Asia *see* Sutlej	234-35	29°21'N	71°02'E
Satna, India	234-35	24°34'N	80°50'E
Sátoraljaújhely, Hung. (shä´tô-rô-lyô-ōō´yĕl´)	194-95	48°24'N	21°41'E
Sattahip, Thai.	246-47	12°40'N	100°54'E
Satu Mare, Rom. (sá´tōō-má´rĕ)	194-95	47°47'N	22°53'E
Sauđárkrókur, Ice.	190a	65°45'N	19°41'W
Sauce, Arg.	173	30°05'S	58°47'W
Saucillo, Mex.	122-23	28°01'N	105°16'W
Saudi Arabia, nation, Asia (sá-ó´dĭ á-rä´bĭ-á)	206-07	25°0'N	45°00'E
Saugatuck, Mi., U.S. (sô´g*á*-tŭk)	116-17	42°40'N	86°11'W
Saujbulagh, Iran *see* Mahābād	228-29	36°46'N	45°44'E
Sauk Centre, Mn., U.S. (sôk sĕn´tĕr)	114-15	45°44'N	94°57'W
Sauk City, Wi., U.S. (sôk sĭ´tē)	116-17	43°16'N	89°43'W
Sauk Rapids, Mn., U.S. (sôk răp´ĭdz)	114-15	45°36'N	94°10'W
Saül, Fr. Gu.	164-65	3°38'N	53°12'W
Sault Sainte Marie, On., Can. (sōō sânt má-rē´)	136-37	46°31'N	84°20'W
Sault Sainte Marie, Mi., U.S. (sōō sânt má-rē´)	114-15	46°29'N	84°21'W
Saunders Island, i., Falk. Is.	171	51°23'S	60°13'W
Saurimo, Ang.	264-65	9°40'S	20°23'E
Sava, stm., Eur. (sä´vä)	200-01	44°50'N	20°27'E
Savai'i, i., Samoa	279b	13°35'S	172°25'W
Savanna, Il., U.S. (s*á*-văn´*á*)	114-15	42°05'N	90°08'W
Savannah, Ga., U.S. (s*á*-văn´*á*)	124-25	32°03'N	81°06'W
Savannah, Mo., U.S. (s*á*-văn´*á*)	120-21	39°56'N	94°50'W
Savannah, Tn., U.S. (s*á*-văn´*á*)	124-25	35°14'N	88°14'W
Savannah, stm., U.S. (s*á*-văn´*á*)	110-11	32°01'N	80°53'W
Savannakhét, Laos	246-47	16°34'N	104°45'E
Savanna-la-Mar, Jam. (sá-văn´á lá mär´)	142-43	18°14'N	78°08'W
Savè, Benin	260a	8°01'N	2°25'E
Save, stm., Afr. (sä´vē)	264-65	20°58'S	35°04'E
Sāveh, Iran	232-33	35°01'N	50°21'E
Saverne, Fr. (sá-vĕrn´)	196-97	48°45'N	7°22'E
Savo Island, i., Sol. Is.	279e	9°08'S	159°49'E
Savona, Italy (sä-nō´nä)	200-01	44°19'N	8°28'E
Savonlinna, Fin. (sá´vôn-lēn´ná)	192-93	61°52'N	28°54'E
Savran', Ukr. (säv-rän´)	202-03	48°08'N	30°06'E
Savu Sea, s., Indon. (sä-vōō sē)	248-49	9°40'S	122°00'E
Sawahlunto, Indon.	246-47	0°40'S	100°46'E
Sawāi Mādhopur, India	234-35	25°59'N	76°22'E
Sawākin, Sudan	266	19°06'N	37°20'E
Sawdā', Jabal, mtn., Sau. Ar.	266	18°18'N	42°22'E
Sawdā', Qurnat as-, mtn., Leb.	228-29	34°18'N	36°07'E
Sawu, Laut, s., Indon. *see* Savu Sea	248-49	9°40'S	122°00'E
Sawu, Pulau, i., Indon.	248-49	10°30'S	121°54'E
Saxony, hist. reg., Ger.	194-95	52°45'N	9°30'E
Sayan Mountains, mts., Asia (sŭ-hän´ moun´tĭnz)	218-19	53°32'N	94°50'E
Sayanogorsk, Russia	218-19	53°06'N	91°24'E
Sayany, mts., Asia *see* Sayan Mountains	218-19	53°32'N	94°50'E
Şaydā, Leb. *see* Sidon	228-29	33°34'N	35°23'E
Saylac, Som.	266	11°20'N	43°28'E
Saynshand, Mong. *see* Buyant-Uhaa	240-41	44°55'N	110°09'E
Sayre, Ok., U.S. (sā´ēr)	120-21	35°18'N	99°38'W
Sayre, Pa., U.S. (sā´ēr)	116-17	41°59'N	76°31'W
Sayula, Mex. (sä-yōō´lä)	146-47	19°53'N	103°36'W
Saywūn, Yemen	220-21	15°56'N	48°45'E
Scandinavia, reg., Eur.	184-85	62°30'N	15°00'E
Scappoose, Or., U.S. (skă-pōōs´)	112-13	45°45'N	122°52'W
Scarborough, Trin.	143b	11°11'N	60°44'W
Scarborough, Eng., U.K. (skär´bŭr-ô)	190-91	54°17'N	0°25'W
Schässburg, Rom. *see* Sighişoara	194-95	46°14'N	24°48'E
Schefferville, Qc., Can.	128-29	54°48'N	66°50'W
Schenectady, N.Y., U.S. (sk*ĕ*-nĕk´tá-dē)	116-17	42°48'N	73°56'W
Schiermonnikoog, i., Neth.	190-91	53°29'N	6°11'E
Schio, Italy (skē´ô)	200-01	45°43'N	11°21'E
Schleswig, Ger. (shĕls´vĕgh)	194-95	54°32'N	9°33'E
Schneidemühl, Pol. *see* Piła	194-95	53°09'N	16°44'E
Schofield, Wi., U.S. (skō´fēld)	116-17	44°53'N	89°36'W
Schreiber, On., Can.	136-37	48°48'N	87°15'W
Schuyler, Ne., U.S. (slī´ler)	114-15	41°27'N	97°04'W
Schuylkill Haven, Pa., U.S. (skōōl´kĭl hā-vĕn)	116-17	40°38'N	76°10'W
Schwabach, Ger. (shvä´bäk)	194-95	49°20'N	11°02'E
Schwäbisch Hall, Ger. (shv*ă*´bēsh häl)	194-95	49°07'N	9°45'E
Schwaner, Pegunungan, mts., Indon. (skvän´ēr)	248-49	0°40'S	112°40'E

Feature (Pronunciation)	Page	Lat.	Long.
Schwarzwald, mts., Ger. (shvärts'väld)	194-95	48°21'N	8°11'E
Schwechat, Aus. (shvĕk'ät)	194-95	48°08'N	16°29'E
Schwedt, Ger. (shvĕt)	194-95	53°04'N	14°17'E
Schweinfurt, Ger. (shvīn'fōrt)	194-95	50°03'N	10°13'E
Schweiz, nation, Eur. *see* Switzerland	174-75	47°0'N	8°00'E
Schwerin, Ger. (shvĕ-rēn')	194-95	53°38'N	11°25'E
Sciacca, Italy (shē-äk'kä)	200-01	37°31'N	13°03'E
Scilly, Isles of, is., Eng., U.K. (īls ŏv sĭl'ē)	190-91	49°55'N	6°20'W
Scobey, Mt., U.S. (skō'bē)	112-13	48°48'N	105°25'W
Scotland, state, U.K. (skŏt'lănd)	190-91	57°0'N	4°00'W
Scotland Neck, N.C., U.S. (skŏt'lănd nĕk)	124-25	36°07'N	77°25'W
Scotstown, Qc., Can. (skŏts'toun)	138-39	45°31'N	71°16'W
Scott, Cape, c., B.C., Can. (kāp skŏt)	132-33	50°47'N	128°25'W
Scott City, Ks., U.S. (skŏt sĭ'tē)	120-21	38°29'N	100°54'W
Scottsbluff, Ne., U.S. (skŏts'blŭf)	114-15	41°52'N	103°40'W
Scottsboro, Al., U.S. (skŏts'bŭro)	124-25	34°41'N	86°01'W
Scottsburg, In., U.S. (skŏts'bûrg)	116-17	38°41'N	85°46'W
Scottsdale, Austl. (skŏts'dāl)	276	41°10'S	147°31'E
Scottsdale, Az., U.S.	118-19	33°35'N	111°52'W
Scottsville, Ky., U.S. (skŏts'vĭl)	124-25	36°45'N	86°11'W
Scottville, Mi., U.S. (skŏt'vĭl)	116-17	43°57'N	86°17'W
Scranton, Pa., U.S.	116-17	41°26'N	75°39'W
Scugog, Lake, lk., On., Can. (lāk skū'gôg)	136-37	44°10'N	78°50'W
Scunthorpe, Eng., U.K. (skŭn'thôrp)	190-91	53°36'N	0°40'W
Scutari, Alb. *see* Shkodër	200-01	42°04'N	19°31'E
Seaford, De., U.S. (sē'fērd)	116-17	38°39'N	75°37'W
Sea Islands, is., U.S. (sē ī'lăndz)	124-25	31°20'N	81°20'W
Seal Cays, is., T./C. Is. (sēl kēs)	142-43	21°10'N	71°38'W
Sealy, Tx., U.S. (sē'lē)	122-23	29°47'N	96°09'W
Searcy, Ar., U.S. (sûr'sē)	124-25	35°15'N	91°45'W
Seaside, Or., U.S. (sē'sīd)	112-13	45°59'N	123°55'W
Seattle, Wa., U.S. (sē-ăt''l)	112-13	47°36'N	122°20'W
Sébaco, Nic. (sĕ-bä'kô)	149	12°51'N	86°06'W
Sebastián Vizcaíno, Bahía, b., Mex.	144-45	28°0'N	114°30'W
Sebatik, Pulau, i., Asia	248-49	4°10'N	117°45'E
Sebewaing, Mi., U.S. (sē'bē-wăng)	116-17	43°44'N	83°26'W
Sebree, Ky., U.S. (sē-brē')	116-17	37°37'N	87°32'W
Sebring, Fl., U.S. (sē'brĭng)	125a	27°30'N	81°26'W
Sebuku, Pulau, i., Indon.	248-49	3°30'S	116°22'E
Sechura, Bahía de, b., Peru	170	5°39'S	81°01'W
Seda, China	238-39	32°20'N	100°41'E
Sedalia, Mo., U.S.	120-21	38°42'N	93°14'W
Sedan, Fr. (sē-däN)	196-97	49°42'N	4°56'E
Sedro-Woolley, Wa., U.S. (sē'drô-wòl'ē)	112-13	48°30'N	122°14'W
Segama, stm., Malay.	248-49	5°31'N	118°48'E
Segamat, Malay. (sā'gà-màt)	246-47	2°30'N	102°49'E
Segesvár, Rom. *see* Sighișoara	194-95	46°14'N	24°48'E
Segezha, Russia	186-87	63°44'N	34°18'E
Ségou, Mali (sā-gōō')	258-59	13°26'N	6°16'W
Segovia, Spain (så-gō'vē-ä)	198-99	40°57'N	4°07'W
Segre, stm., Eur. (så'grā)	198-99	41°22'N	0°18'E
Seguin, Tx., U.S. (sĕ-gēn')	122-23	29°34'N	97°58'W
Segura, stm., Spain (sĕ-gū'lä)	198-99	38°07'N	0°39'W
Segura, Sierra de, mts., Spain (sē-ĕ'r-rä-dĕ sē-gū'lä)	198-99	38°0'N	2°43'W
Seiland, i., Nor.	184-85	70°25'N	23°15'E
Seinäjoki, Fin. (så'ē-nĕ-yô'kē)	184-85	62°47'N	22°51'E
Seine, stm., On., Can. (sån)	134-35	48°38'N	92°58'W
Seine, stm., Fr.	196-97	49°29'N	0°29'E
Seishin, Kor., N. *see* Ch'ŏngjin	243	41°47'N	129°48'E
Seixas, Ponta do, c., Braz.	163d	7°09'S	34°47'W
Şeki, Azer.	227	41°10'N	47°10'E
Sekondi, Ghana	260-61	4°59'N	1°43'W
Selaru, Pulau, i., Indon.	224-25	8°10'S	130°59'E
Selatan, Tanjung, c., Indon. (tän'jông så-lä'tän)	248-49	4°10'S	114°38'E
Selawik, Ak., U.S. (sē-lá-wĭk)	126	66°40'N	160°07'W
Selawik Lake, lk., Ak., U.S.	126	66°30'N	160°40'W
Selayar, Pulau, i., Indon.	248-49	6°05'S	120°30'E
Selemdzha, stm., Russia (sâ-lĕmt-zhä')	218-19	51°44'N	128°53'E
Selenga, stm., Asia (sĕ-lĕŋ-gä') *see* Selenge	240-41	52°17'N	106°16'E
Selenge, stm., Asia	240-41	52°17'N	106°16'E
Selennyakh, stm., Russia (sĕl-yĭn-yäk)	218-19	67°51'N	144°53'E
Sélestat, Fr. (sē-lē-stä')	196-97	48°15'N	7°27'E
Seliger, Ozero, lk., Russia (ô'zĕ-rô sĕl'lē-gĕr)	202-03	57°13'N	33°03'E
Selizharovo, Russia (sâ'lē-zhä'rô-vô)	202-03	56°51'N	33°28'E
Selkirk, Mb., Can. (sĕl'kûrk)	134-35	50°09'N	96°52'W
Selma, Al., U.S. (sĕl'mȧ)	124-25	32°24'N	87°01'W
Selma, Ca., U.S. (sĕl'mȧ)	118-19	36°35'N	119°37'W
Selma, N.C., U.S. (sĕl'mȧ)	124-25	35°32'N	78°17'W
Selva, Arg.	173	29°46'S	62°03'W
Selvas, reg., Braz.	166-67	5°0'S	68°00'W
Selwyn Lake, lk., Can. (sĕl'wĭn lāk)	130-31	59°55'N	104°22'W
Selwyn Mountains, mts., Can. (sĕl'wĭn moun'tĭnz)	130-31	63°10'N	130°20'W
Selwyn Range, mts., Austl.	277	21°35'S	140°35'E
Semara, W. Sah.	258-59	26°44'N	11°41'W
Semarang, Indon. (sĕ-mä'räng)	248-49	6°58'S	110°25'E
Semënov, Russia	186-87	56°47'N	44°29'E
Semeru, Gunung, vol., Indon.	248-49	8°06'S	112°55'E
Semey, Kaz.	226	50°24'N	80°14'E
Seminole, Ok., U.S. (sĕm'ĭ-nōl)	120-21	35°13'N	96°40'W
Seminole, Tx., U.S. (sĕm'ĭ-nōl)	120-21	32°43'N	102°39'W
Seminole, Lake, res., U.S. (lāk sĕm'ĭ-nōl)	124-25	30°46'N	84°50'W
Semliki, stm., Afr. (sĕm'lē-kē)	267	1°12'N	30°31'E
Semnān, Iran	232-33	35°34'N	53°24'E
Semporna, Malay.	248-49	4°28'N	118°36'E
Senador Pompeu, Braz. (sĕ-nä-dōr-pôm-pĕ'ŏ)	166-67	5°35'S	39°22'W
Sena Madureira, Braz.	166-67	9°04'S	68°40'W
Senatobia, Ms., U.S. (sĕ-nȧ-tō'bē-ȧ)	124-25	34°38'N	89°58'W
Sendai, Japan (sĕn-dī')	245	31°49'N	130°19'E
Sendai, Japan	245	38°15'N	140°53'E
Seneca, Ks., U.S. (sĕn'ĕ-kȧ)	120-21	39°50'N	96°04'W
Seneca, S.C., U.S. (sĕn'ĕ-kȧ)	124-25	34°41'N	82°57'W
Seneca Falls, N.Y., U.S. (sĕn'ĕ-kȧ fôlz)	116-17	42°54'N	76°48'W
Senegal, nation, Afr. (sĕn-ĕ-gôl')	253	14°0'N	14°00'W
Sénégal, stm., Afr.	258-59	15°48'N	16°32'W
Senekal, S. Afr. (sĕn'ĕ-kȧl)	269c	28°19'S	27°38'E
Senftenberg, Ger. (zĕnf'tĕn-bĕrgh)	194-95	51°32'N	14°00'E
Senhor do Bonfim, Braz. (sĕn-yôr dô bôN-fē'N)	166-67	10°27'S	40°11'W
Senigallia, Italy (så-nē-gäl'lyä)	200-01	43°43'N	13°13'E
Senj, Cro. (sĕn')	200-01	44°59'N	14°55'E
Senja, i., Nor. (sĕnyä)	184-85	69°20'N	17°30'E
Senneterre, Qc., Can.	136-37	48°23'N	77°14'W
Senqu, stm., Afr. *see* Orange	264-65	28°35'S	16°28'E
Sens, Fr. (säNs)	196-97	48°12'N	3°17'E
Senta, Serb. (sĕn'tä)	200-01	45°56'N	20°06'E
Senyavin Islands, is., Micron.	280-81	6°54'N	158°04'E
Seoni, India	234-35	22°05'N	79°33'E
Seoul, nat. cap., Kor., S. (sōl)	243	37°33'N	127°01'E
Sepanjang, Pulau, i., Indon.	248-49	7°11'S	115°50'E
Sepetiba, Baía de, b., Braz. (bäē'ä dĕ så-pâ-tē'bȧ)	172	23°0'S	43°48'W
Sepik, stm., (sĕp-ēk')	277	3°53'S	144°28'E
Sept-Îles, Qc., Can. (sĕ-tēl')	138-39	50°12'N	66°22'W
Sequoia National Park, n.p., Ca., U.S. (sē-kwoi'ȧ năsh'ŭn-ăl pärk)	118-19	36°31'N	118°34'W
Serafimovich, Russia	186-87	49°35'N	42°45'E
Seram, i., Indon. *see* Ceram	224-25	3°0'S	129°00'E
Seram, Laut, s., Indon. *see* Ceram Sea	224-25	2°30'S	128°00'E
Serang, Indon. (så-räng')	248-49	6°07'S	106°09'E
Serayevo, nat. cap., Bos. *see* Sarajevo	200-01	43°52'N	18°25'E
Serbia, nation, Eur. (sĕr'bē-ăă)	174-75	44°0'N	21°00'E
Serdobsk, Russia (sĕr-dôpsk')	186-87	52°27'N	44°13'E
Seremban, Malay. (sĕr-ĕm-bän')	246-47	2°43'N	101°57'E
Serengeti National Park, n.p., Tan.	267	2°20'S	34°50'E
Serengeti Plain, pl., Tan.	267	2°50'S	35°00'E
Sergeyevka, Russia	244	43°20'N	133°21'E
Sergipe, state, Braz. (sĕr-zhĕ'pĕ)	166-67	10°30'S	37°30'W
Sergiyev Posad, Russia	202-03	56°19'N	38°09'E
Serian, Malay.	248-49	1°10'N	110°33'E
Sérifos, i., Grc.	200-01	37°10'N	24°29'E
Sermata, Pulau, i., Indon.	248-49	8°13'S	128°55'E
Serov, Russia (syĕ-rôf')	218-19	59°36'N	60°35'E
Serowe, Bots. (sĕ-rō'wĕ)	264-65	22°23'S	26°43'E
Serpukhov, Russia (syĕr'pò-kôf)	202-03	54°55'N	37°26'E
Serra do Navio, Braz.	166-67	0°55'N	52°01'W
Serra Talhada, Braz.	166-67	7°59'S	38°18'W
Sérres, Grc. (sĕr'rĕ) (sĕr'ĕs)	200-01	41°05'N	23°33'E
Serrinha, Braz. (sĕr-rēn'yȧ)	166-67	11°39'S	38°60'W
Sertã, Port. (sĕr'tȧ)	198-99	39°48'N	8°06'W
Sertânia, Braz. (sĕr-tȧ'nyä)	163d	8°05'S	37°16'W
Serutu, Pulau, i., Indon.	246-47	1°42'S	108°45'E
Sêrxü, China	238-39	33°08'N	97°55'E
Sesayap Lama, Indon.	248-49	3°35'N	116°60'E
Sese Islands, is., Ug.	267	0°20'S	32°20'E
Sesimbra, Port. (sĕ-sē'm-brä)	198-99	38°26'N	9°05'W
Sestri Levante, Italy (sĕs'trē lå-vän'tå)	200-01	44°16'N	9°24'E
Sestroretsk, Russia (sĕs-trô-rĕtsk)	192-93	60°06'N	29°58'E
Sète, Fr. (sĕt)	196-97	43°25'N	3°42'E
Sete Lagoas, Braz. (sĕ-tĕ lä-gô'äs)	172	19°28'S	44°15'W
Sétif, Alg.	269b	36°11'N	5°25'E
Seto-naikai, s., Japan (sĕ'tô nī'kī)	245	34°22'N	133°37'E
Settat, Mor. (sĕt-ät') (sē-tȧ')	269a	32°59'N	7°36'W
Settlers, S. Afr. (sĕt'lērs)	269c	25°01'S	28°28'E
Setúbal, Port. (så-tōō'bäl)	198-99	38°31'N	8°53'W
Setúbal, Baía de, b., Port. (bä-ē'ä-dĕ-så-tōō'bäl)	198-99	38°18'N	9°04'W
Seul, Lac, lk., On., Can. (läk sûl)	134-35	50°36'N	91°49'W
Sevana Lich, lk., Arm. (syĭ-vän')	227	40°18'N	45°19'E
Sevastopol', Ukr. (syĕ-vás-tô'pôl')	202-03	44°36'N	33°32'E
Severn, stm., On., Can. (sĕv'ērn)	134-35	55°59'N	87°36'W
Severna Park, Md., U.S. (sĕv'ērn-ȧ pärk)	116-17	39°04'N	76°33'W
Severnaya Dvina, stm., Russia	186-87	64°40'N	39°51'E
Severnaya Osetiya-Alaniya, state, Russia *see* North Ossetia	227	43°0'N	44°15'E
Severnaya Sos'va, stm., Russia	218-19	64°10'N	65°27'E
Severnaya Zemlya, is., Russia (sĕ-vyĭr-nīŭ zĭ-m'lyä')	218-19	79°30'N	98°00'E
Severnyye Uvaly, hills, Russia	186-87	59°28'N	48°13'E
Severodvinsk, Russia	186-87	64°34'N	39°50'E
Severo-Kuril'sk, Russia	218-19	50°42'N	156°07'E
Severomorsk, Russia	184-85	69°04'N	33°28'E
Severo-Sibirskaya Nizmennost', pl., Russia	218-19	73°0'N	100°00'E
Severouralsk, Russia (sĕ-vyĭ-rŭ-ōō-rälsk')	186-87	60°10'N	59°58'E
Severskiy Donets, stm., Eur. *see* Northern Donets	186-87	47°36'N	40°54'E
Sevier, stm., Ut., U.S. (sē-vēr')	118-19	39°02'N	113°06'W
Sevier Lake, lk., Ut., U.S. (sē-vēr' lāk)	118-19	38°56'N	113°09'W
Sevilla, Col. (sĕ-vē'l-yä)	163c	4°15'N	75°56'W
Sevilla, Spain (så-vēl'yä)	198-99	37°23'N	5°59'W
Seville, Spain *see* Sevilla	198-99	37°23'N	5°59'W
Sevsk, Russia (syĕfsk)	202-03	52°09'N	34°31'E
Seward, Ak., U.S. (sū'ărd)	126	60°07'N	149°27'W
Seward, Ne., U.S. (sū'ărd)	114-15	40°54'N	97°06'W
Seward Peninsula, pen., Ak., U.S. (sū'ărd pĕ-nĭn'sūlă)	126	65°0'N	164°00'W
Sewell, Chile (sĕ'ô-ĕl)	163e	34°05'S	70°23'W
Seyðisfjörður, Ice. (sā'dēs-fyûr-dòr)	190a	65°15'N	14°01'W
Seybaplaya, Mex. (sā-ē-bä-plä'yä)	148	19°40'N	90°40'W
Seychelles, nation, Afr. (sā-shĕl')	253	4°35'S	55°40'E
Seymchan, Russia	218-19	62°54'N	152°24'E
Seymour, In., U.S. (sē'mōr)	116-17	38°58'N	85°53'W
Seymour, Tx., U.S. (sē'mōr)	120-21	33°36'N	99°16'W
Sfax, Tun. (sfäks)	184-85	34°45'N	10°46'E
's-Gravenhage, nat. cap., Neth. ('s KRä'vĕn-hä'Kĕ) (häg) *see* Hague, The	190-91	52°06'N	4°18'E
Sha, stm., China (shä)	238-39	33°37'N	114°38'E
Shabeelle, stm., Afr. (shä'bā-lē)	262-63	0°10'N	42°46'E
Shabunda, D.R.C.	262-63	2°42'S	27°21'E
Shache, China (shä-chŭ)	226	38°25'N	77°15'E
Shagonar, Russia	222-23	51°32'N	92°49'E
Shahdol, India	234-35	23°18'N	81°22'E
Shāhjahānpur, India (shä-jŭ-hän'pōōr)	234-35	27°53'N	79°54'E
Shahrisabz, Uzb.	232-33	39°03'N	66°50'E
Shāhrūd, Iran	232-33	36°25'N	54°58'E
Shaker Heights, Oh., U.S. (shā'kēr hīts)	116-17	41°29'N	81°36'W
Shakhty, Russia (shäK'tē)	186-87	47°42'N	40°13'E
Shakhunya, Russia	186-87	57°40'N	46°38'E
Shaki, Nig.	260-61	8°40'N	3°24'E
Shaktoolik, Ak., U.S.	126	64°20'N	161°09'W
Shala Hāyk', lk., Eth. (shä'lȧ)	269d	7°28'N	38°31'E
Shalqar, Kaz.	226	47°50'N	59°37'E
Shām, Bādiyat ash-, des., Asia *see* Syrian Desert	228-29	32°0'N	40°00'E
Shām, Jabal ash-, mtn., Oman	230-31	23°13'N	57°16'E
Shamokin, Pa., U.S. (shȧ-mō'kĭn)	116-17	40°47'N	76°35'W
Shamrock, Tx., U.S. (shăm'rŏk)	120-21	35°13'N	100°15'W
Shandī, Sudan	266	16°41'N	33°26'E
Shandong, state, China (shän-dôŋ)	240-41	36°0'N	118°00'E
Shandong Bandao, pen., China (shän-dôŋ bän-dou)	240-41	37°0'N	121°00'E
Shand uul, mtn., Mong.	240-41	43°28'N	104°03'E
Shangani, stm., Zimb.	264-65	18°31'S	27°12'E
Shangcheng, China (shäŋ-chŭŋ)	238-39	31°48'N	115°24'E
Shangdu, China (shäŋ-dōō)	240-41	41°34'N	113°31'E
Shanghai, China (shăng'hī')	238-39	31°14'N	121°28'E
Shanghai, state, China	238-39	31°10'N	121°30'E
Shanglin, China (shäŋ-lĭn)	238-39	23°30'N	108°32'E
Shangqiu, China (shäŋ-chyô)	238-39	34°27'N	115°39'E
Shangrao, China (shäŋ-rou)	238-39	28°26'N	117°58'E
Shangshui, China	238-39	33°33'N	114°34'E

n-sing; ŋ-baŋk; N-nasalized n; nŏd; cŏmmit; ōld; ôbey; ôrder; oi-boil; fōōd; ò-as oo in foot; ou-out; s-soft; sh-dish; th-thin; pūre; ûnite; ûrn; stŭd; circŭs; ü-as in French tu; ′-indeterminate vowel.

Feature (Pronunciation)	Page	Lat.	Long.
Shangxian, China	238-39	33°52'N	109°56'E
Shangzhi, China (shäŋ-jr)	244	45°13'N	127°59'E
Shanhaiguan, China	240-41	40°01'N	119°45'E
Shannon, stm., Ire. (shăn´ŏn)	190-91	52°35'N	9°41'W
Shansi, state, China *see* Shanxi	240-41	37°0'N	112°00'E
Shantarskiye Ostrova, is., Russia (shän´tär-skyĕ ôs-trôf´)	218-19	54°54'N	137°33'E
Shantou, China (shän-tō)	238-39	23°21'N	116°40'E
Shantung, state, China *see* Shandong	240-41	36°0'N	118°00'E
Shantung Peninsula, pen., China *see* Shandong Bandao	240-41	37°0'N	121°00'E
Shanxi, state, China (shän-shyē)	240-41	37°0'N	112°00'E
Shanxian, China (shän shyĕn)	238-39	34°48'N	116°05'E
Shanyin, China	240-41	39°31'N	112°50'E
Shaoguan, China (shou-gûän)	238-39	24°49'N	113°36'E
Shaowu, China	238-39	27°19'N	117°30'E
Shaoxing, China (shou-shyĭŋ)	238-39	29°59'N	120°34'E
Shaoyang, China	238-39	27°15'N	111°28'E
Shar, Kaz.	226	49°35'N	81°03'E
Sharjah, U.A.E.	230-31	25°22'N	55°24'E
Shark Bay, b., Austl. (shärk bā)	272-73	25°30'S	113°30'E
Sharktooth Mountain, mtn., B.C., Can.	130-31	58°35'N	127°57'W
Sharon, Pa., U.S. (shăr´ŏn)	116-17	41°13'N	80°30'W
Sharon Springs, Ks., U.S. (shăr´ŏn sprĭngz)	120-21	38°54'N	101°45'W
Sharonville, Oh., U.S. (shăr´ŏn vĭl)	116-17	39°16'N	84°25'W
Sharpsburg, Md., U.S. (shärps´bûrg)	116-17	39°27'N	77°45'W
Sharqīyah, Aş-Şaḩrā' ash-, des., Egypt *see* Arabian Desert	266	28°0'N	32°00'E
Sharya, Russia	186-87	58°22'N	45°31'E
Shashe, stm., Afr.	264-65	22°11'S	29°21'E
Shashi, China (shä-shē)	238-39	30°19'N	112°14'E
Shasta, Mount, vol., Ca., U.S. (mount shăs´tȧ)	112-13	41°25'N	122°13'W
Shasta Lake, res., Ca., U.S. (shăs´tȧ lāk)	112-13	40°46'N	122°22'W
Shatt al-Arab, stm., Asia *see* 'Arab, Shaṭṭ al-	228-29	29°57'N	48°33'E
Shattuck, Ok., U.S. (shăt´ŭk)	120-21	36°17'N	99°53'W
Shatura, Russia	202-03	55°34'N	39°32'E
Shaunavon, Sk., Can.	134-35	49°39'N	108°24'W
Shaw, Ms., U.S. (shô)	124-25	33°37'N	90°46'W
Shawano, Wi., U.S. (shȧ-wô´nô)	116-17	44°46'N	88°36'W
Shawinigan, Qc., Can.	136-37	46°33'N	72°45'W
Shawnee, Ks., U.S. (shô-nē´)	120-21	39°01'N	94°44'W
Shawnee, Ok., U.S. (shô-nē´)	120-21	35°20'N	96°55'W
Shawneetown, Il., U.S. (shô´nē-toun)	116-17	37°42'N	88°11'W
Shaybārā, i., Sau. Ar.	266	25°26'N	36°50'E
Shay Gap, Austl.	270-71	20°30'S	120°05'E
Shaykh, Jabal ash-, mtn., Asia *see* Hermon, Mount	228-29	33°25'N	35°51'E
Shchekino, Russia	202-03	54°01'N	37°31'E
Shchelkovo, Russia (shchĕl´kô-vô)	202-03	55°54'N	38°01'E
Shchigry, Russia (shchē´grĕ)	202-03	51°52'N	36°55'E
Shchors, Ukr. (shchôrs)	202-03	51°49'N	31°57'E
Shchūchīnsk, Kaz.	226	52°56'N	70°11'E
Sheberghān, Afg.	232-33	36°40'N	65°45'E
Sheboygan, Wi., U.S. (shĕ-boi´găn)	116-17	43°45'N	87°43'W
Sheboygan Falls, Wi., U.S. (shĕ-boi´găn fôlz)	116-17	43°44'N	87°48'W
Shediac, N.B., Can. (shĕ´dē-ăk)	138-39	46°12'N	64°34'W
Shedin Peak, mtn., B.C., Can. (shĕd´ĭn pēk)	132-33	55°55'N	127°32'W
Sheenjek, stm., Ak., U.S.	126	66°45'N	144°34'W
Sheffield, Eng., U.K.	190-91	53°23'N	1°28'W
Sheffield, Al., U.S. (shĕf´fēld)	124-25	34°45'N	87°42'W
Shekhūpura, Pak.	232-33	31°42'N	73°59'E
Sheki, Azer. *see* Şeki	227	41°10'N	47°10'E
Shelagyote Peak, mtn., B.C., Can.	132-33	55°58'N	127°12'W
Shelbina, Mo., U.S. (shĕl-bī´nȧ)	120-21	39°41'N	92°03'W
Shelburn, In., U.S. (shĕl´bŭrn)	116-17	39°11'N	87°24'W
Shelburne, N.S., Can.	138-39	43°46'N	65°19'W
Shelby, Mi., U.S. (shĕl´bē)	116-17	43°37'N	86°22'W
Shelby, Ms., U.S. (shĕl´bē)	124-25	33°57'N	90°46'W
Shelby, Mt., U.S. (shĕl´bē)	112-13	48°30'N	111°51'W
Shelby, N.C., U.S. (shĕl´bē)	124-25	35°18'N	81°32'W
Shelby, Oh., U.S. (shĕl´bē)	116-17	40°53'N	82°39'W
Shelbyville, Il., U.S. (shĕl´bē-vĭl)	116-17	39°24'N	88°48'W
Shelbyville, In., U.S. (shĕl´bē-vĭl)	116-17	39°31'N	85°46'W
Shelbyville, Ky., U.S. (shĕl´bē-vĭl)	116-17	38°13'N	85°13'W
Shelbyville, Tn., U.S. (shĕl´bē-vĭl)	124-25	35°29'N	86°27'W
Shelbyville, Lake, res., Il., U.S. (lāk shĕl´bē-vĭl)	116-17	39°30'N	88°43'W
Sheldon, Ia., U.S. (shĕl´dŭn)	114-15	43°11'N	95°51'W
Shelikhova, Zaliv, b., Russia	218-19	60°0'N	158°00'E
Shelikof Strait, strt., Ak., U.S. (shĕ´lĕ-kôf strāt)	126	57°18'N	155°41'W
Shellbrook, Sk., Can.	134-35	53°14'N	106°23'W
Shelley, Id., U.S. (shĕl´lē)	112-13	43°23'N	112°07'W
Shelton, Ct., U.S. (shĕl´tŭn)	116-17	41°19'N	73°05'W
Shelton, Wa., U.S. (shĕl´tŭn)	112-13	47°12'N	123°06'W
Shemonaīkha, Kaz.	226	50°39'N	81°54'E
Shenandoah, Ia., U.S. (shĕn-ăn-dō´ȧ)	114-15	40°46'N	95°23'W
Shenandoah, Pa., U.S. (shĕn-ăn-dō´ȧ)	116-17	40°49'N	76°12'W
Shenandoah, Va., U.S. (shĕn-ăn-dō´ȧ)	116-17	38°29'N	78°37'W
Shenandoah National Park, n.p., Va., U.S. (shĕn-ăn-dō´ȧ năsh´ŭn-ăl pärk)	116-17	38°34'N	78°20'W
Sheng-li Feng, mtn., Asia *see* Jengish Chokusu	226	42°02'N	80°05'E
Shenkursk, Russia (shĕn-kōōrsk´)	186-87	62°07'N	42°53'E
Shenxian, China (shŭn shyĕn)	240-41	38°01'N	115°33'E
Shenyang, China (shŭn-yäŋ)	243	41°48'N	123°24'E
Shenzhen, China	238-39	22°34'N	114°07'E
Shepetivka, Ukr.	194-95	50°11'N	27°04'E
Shepparton, Austl. (shĕp´år-tŭn)	276	36°23'S	145°25'E
Sherbro Island, i., S.L.	260-61	7°34'N	12°43'W
Sherbrooke, Qc., Can.	136-37	45°24'N	71°54'W
Sheridan, Ar., U.S. (shĕr´ĭ-dăn)	120-21	34°18'N	92°24'W
Sheridan, Wy., U.S. (shĕr´ĭ-dăn)	112-13	44°48'N	106°58'W
Sherlovaya Gora, Russia	240-41	50°32'N	116°18'E
Sherman, Tx., U.S. (shĕr´măn)	120-21	33°35'N	96°36'W
Sherridon, Mb., Can.	134-35	55°07'N	101°05'W
Sherwood Park, Ab., Can.	132-33	53°31'N	113°18'W
Shetland Islands, is., Scot., U.K. (shĕt´lănd ī´lăndz)	190c	60°25'N	1°39'W
Shexian, China (shŭ shyĕn)	238-39	29°53'N	118°26'E
Sheyenne, stm., N.D., U.S. (shī-ĕn´)	114-15	47°01'N	96°50'W
Shiashkotan, Ostrov, i., Russia	218-19	48°52'N	154°10'E
Shibām, Yemen (shē´bäm)	220-21	15°54'N	48°40'E
Shidao, China	240-41	36°54'N	122°24'E
Shīeli, Kaz.	226	44°10'N	66°44'E
Shijiazhuang, China (shr-jyä-jûäŋ)	240-41	38°02'N	114°29'E
Shikārpur, Pak.	232-33	27°57'N	68°39'E
Shikoku, i., Japan (shē´kō´kōō)	245	33°45'N	133°30'E
Shikotan, Ostrov, i., Russia	218-19	43°47'N	146°45'E
Shikotan-tō, i., Russia *see* Shikotan, Ostrov	218-19	43°47'N	146°45'E
Shiliguri, India	234-35	26°43'N	88°26'E
Shilka, Russia	222-23	51°52'N	116°02'E
Shilka, stm., Russia (shĭl´kȧ)	222-23	53°21'N	121°27'E
Shillong, India (shĕl-lông´)	234-35	25°34'N	91°53'E
Shimanovsk, Russia	222-23	52°00'N	127°41'E
Shimber Berris, mtn., Som. *see* Shimbiris	262-63	10°44'N	47°15'E
Shimbiris, mtn., Som.	262-63	10°44'N	47°15'E
Shimian, China	238-39	29°16'N	102°17'E
Shimla, India	234-35	31°06'N	77°10'E
Shimoga, India	236	13°56'N	75°35'E
Shimonoseki, Japan	245	33°58'N	130°56'E
Shimono-shima, i., Japan	243	34°12'N	129°15'E
Shinano, stm., Japan (shē-nä´nô)	245	37°57'N	139°04'E
Shīndand, Afg.	232-33	33°18'N	62°08'E
Shingishū, Kor., N. *see* Sinŭiju	243	40°06'N	124°24'E
Shingū, Japan	245	33°43'N	136°00'E
Shinyanga, Tan. (shĭn-yäŋ´gä)	267	3°40'S	33°26'E
Shiono-misaki, c., Japan (shē-ô´nô mē´sä-kē)	245	33°26'N	135°46'E
Shiqizhen, China *see* Zhongshan	238-39	22°31'N	113°22'E
Shirati, Tan. (shē-rä´tē)	267	1°07'S	33°60'E
Shīrāz, Iran (shē-räz´)	230-31	29°36'N	52°32'E
Shire, stm., Afr. (shē´rå)	264-65	17°42'S	35°19'E
Shiretoko-misaki, c., Japan	244	44°21'N	145°20'E
Shishaldin Volcano, vol., Ak., U.S. (shī-shăl´dĭn vŏl-kā´nō)	126	54°45'N	163°57'W
Shively, Ky., U.S. (shĭv´lē)	116-17	38°12'N	85°49'W
Shivpuri, India	234-35	25°25'N	77°39'E
Shizuoka, Japan (shē´zōō´ōkȧ)	245	34°58'N	138°23'E
Shkodër, Alb. (shkô´dûr) (skōō´tärē)	200-01	42°04'N	19°31'E
Shkodra, Alb. *see* Shkodër	200-01	42°04'N	19°31'E
Shmidta, Ostrov, i., Russia	208-09	81°08'N	90°48'E
Shoal Lake, lk., Can. (shōl lāk)	134-35	49°32'N	95°00'W
Shoals, In., U.S. (shōlz)	116-17	38°40'N	86°47'W
Shōdo-shima, i., Japan (shō´dō shē´mä)	245	34°30'N	134°17'E
Shortland Island, i., Sol. Is.	279e	7°04'S	155°43'E
Shoshone, Id., U.S. (shô-shōn´tē)	112-13	42°56'N	114°25'W
Shostka, Ukr. (shôst´kȧ)	202-03	51°52'N	33°29'E
Shouguang, China (shō-gûäŋ)	240-41	36°53'N	118°44'E
Shouxian, China (shō shyĕn)	238-39	32°34'N	116°46'E
Shpola, Ukr. (shpô´lȧ)	202-03	49°00'N	31°24'E
Shqipëria, nation, Eur. *see* Albania	174-75	41°0'N	20°00'E
Shreveport, La., U.S. (shrēv´pôrt)	120-21	32°30'N	93°45'W
Shrewsbury, Eng., U.K. (shrōōz´bĕr-ĭ)	190-91	52°43'N	2°45'W
Shū, Kaz.	226	43°36'N	73°45'E
Shū, stm., Asia	226	45°00'N	67°45'E
Shuajingsi, China	238-39	32°00'N	103°17'E
Shuangcheng, China (shûäŋ-chŭŋ)	240-41	45°22'N	126°19'E
Shuangliao, China	240-41	43°30'N	123°30'E
Shuangyashan, China	244	46°35'N	131°19'E
Shubrâ el-Kheima, Egypt	268b	30°06'N	31°15'E
Shumagin Islands, is., Ak., U.S. (shōō´mȧ-gĕn ī´lăndz)	126	55°06'N	159°43'W
Shumen, Blg.	200-01	43°16'N	26°57'E
Shumerlya, Russia	186-87	55°29'N	46°25'E
Shunde, China (shòn-dŭ)	238-39	22°50'N	113°15'E
Shuqayyiqah, Nafūd, sand, Sau. Ar.	266	25°45'N	43°55'E
Shūshtar, Iran (shōōsh´tŭr)	228-29	32°03'N	48°51'E
Shuswap Lake, lk., B.C., Can. (shōōs´wŏp lāk)	132-33	50°57'N	119°15'W
Shuya, Russia (shōō´yȧ)	186-87	56°51'N	41°23'E
Shuyang, China (shōō yäng)	238-39	34°08'N	118°47'E
Shwangliao, China *see* Liaoyuan	240-41	42°55'N	125°08'E
Shwebo, Mya.	246-47	22°34'N	95°42'E
Shymkent, Kaz.	226	42°18'N	69°36'E
Shyok, stm., Asia	232-33	35°14'N	75°55'E
Siālkot, Pak. (sē-äl´kōt)	232-33	32°31'N	74°33'E
Siam, nation, Asia *see* Thailand	206-07	15°0'N	100°00'E
Siam, Gulf of, b., Asia *see* Thailand, Gulf of	246-47	10°0'N	101°00'E
Sian, China *see* Xi'an	238-39	34°15'N	108°52'E
Siargao Island, i., Phil.	250	9°53'N	126°02'E
Siasi Island, i., Phil.	250	5°33'N	120°51'E
Siau, Pulau, i., Indon.	248-49	2°46'N	125°23'E
Šiauliai, Lith. (shē-ou´lē-ī)	192-93	55°56'N	23°20'E
Sibay, Russia (sē´bāy)	226	52°42'N	58°40'E
Šibenik, Cro. (shē-bā´nēk)	200-01	43°44'N	15°54'E
Siberia, reg., Russia (sī-bĭr´ē-aȧ)	218-19	65°0'N	110°00'E
Sibi, Pak.	232-33	29°33'N	67°53'E
Sibir', reg., Russia *see* Siberia	218-19	65°0'N	110°00'E
Sibiryakova, Ostrov, i., Russia	218-19	72°50'N	79°00'E
Sibiti, Congo (sē-bē-tē´)	262-63	3°41'S	13°21'E
Sibiu, Rom. (sē-bĭ-ōō´)	194-95	45°47'N	24°09'E
Sibley, Ia., U.S. (sĭb´lē)	114-15	43°24'N	95°45'W
Sibolga, Indon. (sē-bō´gä)	246-47	1°45'N	98°47'E
Sibsāgar, India (sēb-sŭ´gŭr)	234-35	26°59'N	94°39'E
Sibu, Malay.	248-49	2°18'N	111°50'E
Sibut, C.A.R.	262-63	5°44'N	19°05'E
Sibutu Island, i., Phil.	250	4°46'N	119°29'E
Sibuyan Island, i., Phil. (sē-bōō-yän´ ī´lȧnd)	250	12°27'N	122°34'E
Sibuyan Sea, s., Phil. (sē-bōō-yän´ sē)	250	12°50'N	122°40'E
Sichuan, state, China (sz-chûän)	238-39	31°0'N	105°00'E
Sicilia, i., Italy *see* Sicily	200-01	37°30'N	14°00'E
Sicily, i., Italy (sĭs´ĭ-lē)	200-01	37°30'N	14°00'E
Sico Tinto, stm., Hond. (sē-kō tēn´tō)	149	15°50'N	85°03'W
Sicuani, Peru	170	14°16'S	71°13'W
Sidhi, India	234-35	24°24'N	81°53'E
Sîdi Barrâni, Egypt	188-89	31°37'N	25°56'E
Sidi Bel Abbès, Alg. (sē´dē-bĕl à-bĕs´)	198-99	35°12'N	0°11'W
Sidi-Bennour, Mor.	269a	32°39'N	8°25'W
Sidikalang, Indon.	246-47	2°44'N	98°20'E
Sidney, B.C., Can. (sĭd´nē)	132-33	48°39'N	123°24'W
Sidney, Mt., U.S. (sĭd´nē)	112-13	47°43'N	104°09'W
Sidney, Ne., U.S. (sĭd´nē)	114-15	41°09'N	102°59'W
Sidney, N.Y., U.S. (sĭd´nē)	116-17	42°19'N	75°23'W
Sidney, Oh., U.S. (sĭd´nē)	116-17	40°17'N	84°10'W
Sidney Lanier, Lake, res., Ga., U.S. (lăk sĭd´nē lăn´yĕr)	124-25	34°15'N	83°57'W
Sidon, Leb.	228-29	33°34'N	35°23'E
Sidra, Gulf of, b., Libya (gŭlf ŭv sĭ´drȧ) *see* Surt, Khalīj	258-59	31°30'N	18°00'E
Siedlce, Pol. (syĕd´'l-tsĕ)	194-95	52°10'N	22°17'E
Siegburg, Ger. (zēg´bōōrgh)	194-95	50°47'N	7°12'E
Siegen, Ger. (zē´ghĕn)	194-95	50°52'N	8°01'E
Siemiatycze, Pol. (syĕm´yä´tē-chĕ)	194-95	52°26'N	22°52'E
Siĕmréab, Camb.	246-47	13°22'N	103°51'E
Siena, Italy (sē-ĕn´ä)	200-01	43°19'N	11°20'E
Sienyang, China *see* Xianyang	238-39	34°20'N	108°42'E
Sieradz, Pol. (syĕ´rädz)	194-95	51°36'N	18°45'E
Sierpc, Pol. (syĕrpts)	194-95	52°51'N	19°40'E
Sierra Blanca, Tx., U.S. (sē-ĕ´rȧ blaŋ-kä)	122-23	31°11'N	105°21'W
Sierra Blanca Peak, mtn., N.M., U.S. (sē-ĕ´r-rä blăn´kȧ pēk)	120-21	33°23'N	105°48'W
Sierra Colorada, Arg.	171	40°36'S	67°45'W
Sierra Leone, nation, Afr. (sē-ĕr´rä lā-ō´nä)	253	8°30'N	11°30'W

Feature (Pronunciation)	Page	Lat.	Long.
Sierra Nevada, mts., Ca., U.S. see Nevada, Sierra	118-19	38°0'N	119°15'W
Sífnos, i., Grc.	200-01	36°58'N	24°43'E
Sighişoara, Rom. (sē-gē-shwä'rå)	194-95	46°14'N	24°48'E
Sigli, Indon.	246-47	5°23'N	95°57'E
Siglufjörđur, Ice.	190a	66°09'N	18°55'W
Sigsig, Ec. (sêg-sēg')	170	3°00'S	78°48'W
Siguatepeque, Hond. (sē-gwä'tě-pě-kě)	149	14°35'N	87°50'W
Sigüenza, Spain (sē-gwě'n-zä)	198-99	41°04'N	2°38'W
Siguiri, Gui. (sē-gē-rē')	260-61	11°26'N	9°10'W
Sihala, nation, Asia see Sri Lanka	206-07	7°0'N	81°00'E
Sihanoukville, Camb. see Kâmpóng Saôm	246-47	10°38'N	103°31'E
Sihong, China (sz-hôŋ)	238-39	33°28'N	118°12'E
Siirt, Tur. (sĭ-ērt')	227	37°56'N	41°57'E
Sīkar, India	234-35	27°36'N	75°08'E
Sikasso, Mali (sē-käs'sō)	258-59	11°19'N	5°40'W
Sikeston, Mo., U.S. (sīks'tŭn)	124-25	36°53'N	89°36'W
Sikhote-Alin', mts., Russia (se-kô'ta a-lēn')	218-19	48°0'N	138°00'E
Sikkim, state, India	234-35	27°35'N	88°35'E
Silao, Mex. (sē-lä'ō)	146-47	20°56'N	101°26'W
Silay, Phil.	250	10°48'N	122°59'E
Silchar, India (sĭl-chär')	234-35	24°49'N	92°48'E
Siler City, N.C., U.S. (sī'lēr sĭ'tĭ)	124-25	35°43'N	79°28'W
Silesia, hist. reg., Eur. (sĭ-lē'shå)	194-95	50°34'N	18°01'E
Siletitengiz köli, lk., Kaz.	226	53°15'N	73°15'E
Silifke, Tur.	228-29	36°22'N	33°56'E
Siling Co, lk., China	234-35	31°47'N	89°00'E
Silistra, Blg. (sē-lēs'trå)	200-01	44°07'N	27°16'E
Siljan, lk., Swe. (sēl'yän)	192-93	60°50'N	14°45'E
Silkeborg, Den. (sĭl'kĕ-bôr')	192-93	56°10'N	9°33'E
Siloam Springs, Ar., U.S. (sī-lōm sprĭngz)	120-21	36°11'N	94°33'W
Silsbee, Tx., U.S. (sĭlz' bē)	122-23	30°21'N	94°11'W
Silvânia, Braz. (sēl-vá'nyä)	168-69	16°42'S	48°37'W
Silver Bank Passage, strt., N.A. (sĭl'vēr bănk păs'ĭj)	142-43	20°53'N	70°18'W
Silver Bay, Mn., U.S. (sĭl'vēr bā)	114-15	47°18'N	91°15'W
Silver City, N.M., U.S. (sĭl'vēr sĭ'tē)	118-19	32°47'N	108°17'W
Silver Creek, N.Y., U.S. (sĭl'vēr krēk)	116-17	42°33'N	79°10'W
Silverthrone Mountain, vol., B.C., Can. (sĭl'vēr-thrōn moun'tĭn)	132-33	51°31'N	126°06'W
Silverton, Or., U.S. (sĭl'vēr-tŭn)	112-13	45°01'N	122°47'W
Silves, Port. (sēl'vězh)	198-99	37°11'N	8°26'W
Simbo Island, i., Sol. Is.	279e	8°17'S	156°31'E
Simcoe, On., Can. (sĭm'kō)	136-37	42°50'N	80°18'W
Simcoe, Lake, lk., On., Can. (lāk sĭm'kō)	136-37	44°28'N	79°19'W
Simeulue, Pulau, i., Indon.	246-47	2°37'N	96°04'E
Simferopol', Ukr.	202-03	44°57'N	34°05'E
Simití, Col.	164-65	7°56'N	73°57'W
Simojovel, Mex. (sē-mō-hō-věl')	146-47	17°07'N	92°39'W
Simonette, stm., Ab., Can. (sī-mŏn-ět')	132-33	55°09'N	118°16'W
Simonstad, S. Afr. see Simon's Town	264-65	34°15'S	18°27'E
Simon's Town, S. Afr.	264-65	34°15'S	18°27'E
Simoom Sound, B.C., Can.	132-33	50°46'N	126°26'W
Simplon Pass, p., Switz. (sĭm'plŏn păs) (săn-plôN' päs)	194-95	46°15'N	8°01'E
Simpson Desert, des., Austl. (sĭmp-sŭn děs'ērt)	272-73	25°0'S	137°00'E
Simpson Desert National Park, n.p., Austl.	276	25°40'S	138°15'E
Simpson Island, i., On., Can. (sĭmp-sŭn ī'lȧnd)	136-37	48°47'N	87°41'W
Simrishamn, Swe. (sēm'rēs-häm'n)	192-93	55°33'N	14°21'E
Simushir, Ostrov, i., Russia (ôs-trôf se-mōō'shēr)	218-19	46°58'N	152°02'E
Sinai, pen., Egypt see Sinai Peninsula	266	29°30'N	34°00'E
Sinaia, Rom. (sĭ-nä'yȧ)	200-01	45°21'N	25°33'E
Sinai Peninsula, pen., Egypt (sī'nī pě-nĭn'sūlȧ)	266	29°30'N	34°00'E
Sinaloa, state, Mex. (sē-nä-lô-ä)	144-45	25°0'N	107°30'W
Sinan, China (sz-nän)	238-39	27°56'N	108°14'E
Sīnāwin, Libya	188-89	31°02'N	10°36'E
Sincelejo, Col. (sēn-så-lā'hō)	164-65	9°18'N	75°24'W
Sindi, Est. (sēn'dē)	192-93	58°24'N	24°41'E
Sine, stm., Sen.	260-61	14°13'N	16°26'W
Sines, Port. (sē'nȧzh)	198-99	37°57'N	8°52'W
Singapore, nation, Asia (sĭn'gȧ-pōr')	206-07	1°22'N	103°48'E
Singapore, nat. cap., Sing. (sĭn'gȧ-pōr')	246-47	1°18'N	103°49'E
Singaraja, Indon.	248-49	8°07'S	115°06'E
Singida, Tan.	267	4°49'S	34°44'E
Singkawang, Indon.	246-47	0°55'N	108°59'E
Singkep, Pulau, i., Indon.	246-47	0°30'S	104°25'E
Singleton, Austl.	276	32°34'S	151°10'E
Singora, Thai. see Songkhla	246-47	7°12'N	100°36'E
Sinj, Cro. (sēn')	200-01	43°42'N	16°38'E
Sinjah, Sudan	266	13°08'N	33°55'E
Sinjai, Indon.	248-49	5°07'S	120°16'E
Sinkāt, Sudan	266	18°50'N	36°50'E
Sinkiang, state, China see Xinjiang	220-21	40°0'N	85°00'E
Sinmi-do, i., Kor., N.	243	39°33'N	124°53'E
Sinnamary, Fr. Gu.	164-65	5°23'N	52°57'W
Sinnûris, Egypt	268b	29°25'N	30°52'E
Sinop, Tur.	186-87	42°02'N	35°09'E
Sintang, Indon.	248-49	0°04'N	111°29'E
Sint Eustatius, i., Neth. Ant.	143b	17°30'N	62°59'W
Sint Maarten, i., N.A.	143b	18°04'N	63°04'W
Sinton, Tx., U.S. (sĭn'tŭn)	122-23	28°02'N	97°30'W
Sinú, stm., Col.	164-65	9°23'N	75°56'W
Sinŭiju, Kor., N. (sĭ'nōĭ-jōō)	243	40°06'N	124°24'E
Siocon, Phil.	250	7°42'N	122°08'E
Sion, Switz. (sē'ôN')	194-95	46°14'N	7°21'E
Sioux City, Ia., U.S. (sōō sĭ'tē)	114-15	42°30'N	96°23'W
Sioux Falls, S.D., U.S. (sōō fôlz)	114-15	43°33'N	96°44'W
Sioux Lookout, On., Can.	134-35	50°06'N	91°56'W
Siping, China (sz-pĭŋ)	240-41	43°10'N	124°23'E
Sipiwesk Lake, lk., Mb., Can.	134-35	55°05'N	97°35'W
Sipsey, stm., Al., U.S. (sĭp'sē)	124-25	33°0'N	88°10'W
Siquia, stm., Nic. (sē-kē'ä)	149	12°11'N	84°17'W
Siquijor Island, i., Phil.	250	9°11'N	123°34'E
Siracusa, Italy (sē-rä-koo'sä)	200-01	37°04'N	15°17'E
Sirājganj, Bngl. (sĭ-räj'gŭnj)	234-35	24°27'N	89°42'E
Şīr Banī Yās, i., U.A.E.	230-31	24°19'N	52°37'E
Sirdaryo, Uzb.	232-33	40°53'N	68°40'E
Sirdaryo, stm., Asia see Syr Darya	226	46°04'N	60°04'E
Sir Douglas, Mount, mtn., Can. (mount sûr dŭg'lăs)	132-33	50°44'N	115°20'W
Sirḥān, Wādī as-, val., Sau. Ar.	228-29	30°58'N	37°41'E
Sir James MacBrien, Mount, mtn., N.T., Can.	130-31	62°07'N	127°41'W
Sirohi, India	234-35	24°53'N	72°51'E
Sirrah, Nafūd as-, sand, Sau. Ar.	266	23°01'N	44°34'E
Sir Sandford, Mount, mtn., B.C., Can. (mount sûr sănd'fērd)	132-33	51°40'N	117°52'W
Sirte, Gulf of, b., Libya see Surt, Khalīj	258-59	31°30'N	18°00'E
Širvintos, Lith. (shēr'vĭn-tôs)	192-93	55°03'N	24°57'E
Sir Wilfrid Laurier, Mount, mtn., B.C., Can. (mount sûr wĭl'frĭd lôr'yēr)	132-33	52°47'N	119°45'W
Sisaba, mtn., Tan.	267	6°09'S	29°48'E
Sisak, Cro. (sē'sȧk)	200-01	45°29'N	16°23'E
Si Sa Ket, Thai.	246-47	15°08'N	104°20'E
Sishui, China (sz-shwā)	240-41	35°39'N	117°16'E
Sisseton, S.D., U.S. (sĭs'tŭn)	114-15	45°40'N	97°03'W
Sisteron, Fr. (sēst'rôN')	196-97	44°13'N	5°56'E
Sītāpur, India	234-35	27°34'N	80°41'E
Sitka, Ak., U.S. (sĭt'kȧ)	126	57°04'N	135°19'W
Sitten, Switz. see Sion	194-95	46°14'N	7°21'E
Sittoung, stm., Mya.	246-47	17°22'N	96°53'E
Sittwe, Mya.	246-47	20°09'N	92°54'E
Sivas, Tur. (sē'väs)	186-87	39°44'N	37°01'E
Siverek, Tur. (sē'vě-rěk)	186-87	37°45'N	39°19'E
Sivers'kyi Donets', stm., Eur. see Northern Donets	186-87	47°36'N	40°54'E
Siwa, Egypt	188-89	29°12'N	25°31'E
Sjælland, i., Den. (shěl'lån')	192-93	55°30'N	11°45'E
Skagen, Den. (skä'ghěn)	192-93	57°43'N	10°35'E
Skagerrak, strt., Eur. (skä-ghě-räk')	192-93	57°45'N	9°00'E
Skagway, Ak., U.S. (skăg-wā)	126	59°28'N	135°19'W
Skalistyy Golets, Gora, mtn., Russia	218-19	56°24'N	119°12'E
Skara, Swe. (skä'rȧ)	192-93	58°23'N	13°28'E
Skeena, stm., B.C., Can. (skē'nā)	130-31	54°08'N	130°07'W
Skeena Mountains, mts., B.C., Can. (skē'nā moun'tĭnz)	130-31	56°35'N	128°41'W
Skellefteå, Swe. (shěl'ěf-tě-å')	184-85	64°45'N	20°58'E
Skellefteälven, stm., Swe.	184-85	64°43'N	21°09'E
Skibbereen, Ire. (skĭb'ēr-ēn)	190-91	51°33'N	9°17'W
Skidegate Inlet, b., B.C., Can. (skĭ'-dě-gāt' ĭn'lět)	132-33	53°14'N	131°58'W
Skien, Nor. (skē'ěn)	192-93	59°12'N	9°36'E
Skierniewice, Pol. (skyěr-nyě-vēt'sě)	194-95	51°58'N	20°09'E
Skikda, Alg.	269b	36°53'N	6°55'E
Skive, Den. (skē'vě)	192-93	56°34'N	9°02'E
Skjálfandafljót, stm., Ice. (skyäl'fänd-ô)	190a	65°59'N	17°37'W
Škofja Loka, Slvn. (shkôf'yȧ lō'kȧ)	200-01	46°10'N	14°18'E
Skole, Ukr. (skô'lě)	194-95	49°02'N	23°29'E
Skopin, Russia (skô'pěn)	202-03	53°49'N	39°33'E
Skopje, nat. cap., Mac. (skôp'yě)	200-01	42°00'N	21°28'E
Skoplje, nat. cap., Mac. see Skopje	200-01	42°00'N	21°28'E
Skövde, Swe. (shûv'dě)	192-93	58°24'N	13°51'E
Skovorodino, Russia (skô'vô-rô'dĭ-nô)	222-23	53°59'N	123°56'E
Skowhegan, Me., U.S. (skou-hē'gǎn)	117a	44°47'N	69°43'W
Skunk, stm., Ia., U.S. (skŭnk)	114-15	40°42'N	91°07'W
Skuodas, Lith. (skwô'dȧs)	192-93	56°16'N	21°32'E
Skye, Island of, i., Scot., U.K. (ī'lȧnd ŏv skī)	190-91	57°25'N	6°28'W
Slamet, Gunung, vol., Indon. (gōō-nōng slä'mět)	248-49	7°14'S	109°12'E
Slantsy, Russia	192-93	59°07'N	28°05'E
Śląsk, hist. reg., Eur. see Silesia	194-95	50°34'N	18°01'E
Slater, Mo., U.S. (slāt'ēr)	120-21	39°13'N	93°04'W
Slatina, Rom. (slä'tē-nä)	200-01	44°26'N	24°22'E
Slaton, Tx., U.S. (slā'tŭn)	120-21	33°26'N	101°38'W
Slave, stm., Can. (slāv)	130-31	61°16'N	113°35'W
Slave Lake, Ab., Can.	132-33	55°17'N	114°48'W
Slavgorod, Russia (sláf'gô-rôt)	226	53°00'N	78°38'E
Slavonski Brod, Cro. (skä-vôn'skě brôd)	200-01	45°10'N	18°00'E
Slavuta, Ukr. (slá-vōō'tá)	194-95	50°17'N	26°52'E
Slavyansk-na-Kubani, Russia	202-03	45°15'N	38°08'E
Sławno, Pol. (swav'nō)	194-95	54°22'N	16°42'E
Slayton, Mn., U.S. (slā'tŭn)	114-15	43°59'N	95°45'W
Sleaford, Eng., U.K. (slē'fērd)	190-91	53°00'N	0°25'W
Sleepy Eye, Mn., U.S. (slēp'ĭ ī)	114-15	44°18'N	94°44'W
Slezsko, hist. reg., Eur. see Silesia	194-95	50°34'N	18°01'E
Slidell, La., U.S. (slī-děl')	124-25	30°16'N	89°47'W
Sligeach, Ire. see Sligo	190-91	54°17'N	8°28'W
Sligo, Ire. (slī'gō)	190-91	54°17'N	8°28'W
Sliven, Blg. (slē'věn)	200-01	42°41'N	26°20'E
Slobodskoy, Russia	186-87	58°44'N	50°11'E
Slonim, Bela. (swō'něm)	194-95	53°05'N	25°19'E
Slovakia, nation, Eur. (slô-vák'ĭ-ȧ)	174-75	48°30'N	20°00'E
Slovenia, nation, Eur. (slô-vē'nĭ-ȧ)	174-75	46°15'N	15°10'E
Slovenija, nation, Eur. (slô-vē'nê-yä) see Slovenia	174-75	46°15'N	15°10'E
Slovensko, nation, Eur. see Slovakia	174-75	48°30'N	20°00'E
Slov'ians'k, Ukr.	202-03	48°52'N	37°37'E
Sluck, Bela.	194-95	53°01'N	27°33'E
Slunj, Cro. (slòn')	200-01	45°08'N	15°33'E
Słupsk, Pol. (swòpsk)	194-95	54°28'N	17°02'E
Slutsk, Bela. (slòtsk) see Sluck	194-95	53°01'N	27°33'E
Smackover, Ar., U.S. (smăk'ô-vēr)	120-21	33°23'N	92°44'W
Smallwood Reservoir, res., Nf., Can.	130-31	54°09'N	64°24'W
Smederevska Palanka, Serb. (smě-dě-rěv'skä pä-län'kä)	200-01	44°22'N	20°58'E
Smethport, Pa., U.S. (směth'pōrt)	116-17	41°49'N	78°27'W
Smidovich, Russia	240-41	48°36'N	133°49'E
Smila, Ukr.	202-03	49°13'N	31°53'E
Smiltene, Lat. (směl'tě-ně)	192-93	57°26'N	25°54'E
Smithers, B.C., Can. (smĭth'ērs)	132-33	54°48'N	127°11'W
Smithfield, N.C., U.S. (smĭth'fēld)	124-25	35°30'N	78°21'W
Smithfield, Ut., U.S. (smĭth'fēld)	112-13	41°51'N	111°50'W
Smithfield, Va., U.S. (smĭth'fēld)	124-25	36°59'N	76°38'W
Smith Mountain Lake, res., Va., U.S. (smĭth moun'tĭn lāk)	124-25	37°07'N	79°39'W
Smiths Falls, On., Can. (smĭths fôlz)	136-37	44°54'N	76°01'W
Smithton, Austl. (smĭth'tŭn)	276	40°51'S	145°07'E
Smithville, Tx., U.S. (smĭth'vĭl)	122-23	30°00'N	97°09'W
Smoke Creek Desert, des., Nv., U.S. (smōk krēk děs'ērt)	112-13	40°31'N	119°47'W
Smoky, stm., Ab., Can. (smōk'ĭ)	132-33	56°11'N	117°20'W
Smøla, i., Nor. (smûlä)	184-85	63°24'N	8°00'E
Smolensk, Russia (smô-lyěnsk')	202-03	54°48'N	32°03'E
Smyrna, Tur. (smûr'nȧ) see İzmir	200-01	38°26'N	27°09'E
Smyrna, De., U.S. (smûr'nȧ)	116-17	39°18'N	75°36'W
Smyrna, Ga., U.S. (smûr'nȧ)	124-25	33°53'N	84°31'W
Smyrna, Tn., U.S. (smûr'nȧ)	124-25	35°59'N	86°31'W
Smythe, Mount, mtn., B.C., Can.	130-31	57°54'N	124°53'W
Snake, stm., U.S. (snāk)	110-11	46°11'N	119°01'W
Snares Islands, is., N.Z.	287	48°0'S	166°30'E
Śniardwy, Jezioro, lk., Pol. (snyärt'vĭ)	194-95	53°46'N	21°44'E
Snøhetta, mtn., Nor. (snû-hěttä)	184-85	62°20'N	9°17'E
Snøtinden, mtn., Nor.	184-85	66°38'N	14°00'E
Snov, stm., Eur. (snôf)	202-03	51°32'N	31°32'E
Snowdon, mtn., Wales, U.K.	190-91	53°04'N	4°05'W
Snowy Mountains, mts., Austl. (snō'ē moun'tĭnz)	276	36°15'S	148°18'E
Snyder, Ok., U.S. (snī'dēr)	120-21	34°40'N	98°57'W
Snyder, Tx., U.S. (snī'dēr)	120-21	32°43'N	100°55'W
Soasiu, Indon. see Tidore	248-49	0°38'N	127°24'E
Sobinka, Russia (sô-bĭŋ'kȧ)	202-03	55°58'N	40°02'E
Sobradinho, Represa de, res., Braz.	166-67	9°40'S	42°00'W
Sobral, Braz. (sô-brä'l)	166-67	3°42'S	40°21'W

n-sing; ŋ-baŋk; N-nasalized n; nŏd; cŏmmit; ōld; ôbey; ôrder; oi-boil; fōōd; ȯ-as oo in foot; ou-out; s-soft; sh-dish; th-thin; pūre; ûnite; ûrn; stŭd; circŭs; ü-as in French tu; ′-indeterminate vowel.

Feature (Pronunciation)	Page	Lat.	Long.
Sochaczew, Pol. (sô-κä´chĕf)	194-95	52°14'N	20°15'E
Sochi, Russia (sôch´ĭ)	227	43°35'N	39°44'E
Société, Archipel de la, is., Fr. Poly. *see* Society Islands	280-81	17°0'S	150°00'W
Society Islands, is., Fr. Poly. (sô-sī´ĕ-tē ī´lándz)	280-81	17°0'S	150°00'W
Socoltenango, Mex. (sô-kôl-tĕ-näŋ´gō)	148	16°12'N	92°14'W
Socorro, Col. (sô-kôr´rō)	164-65	6°29'N	73°16'W
Socorro, N.M., U.S. (sō-kô´r-rō)	118-19	34°04'N	106°54'W
Socorro, Isla, i., Mex.	144-45	18°45'N	110°58'W
Socotra, i., Yemen (sô-kō´trá) *see* Suquṭrā	220-21	12°31'N	53°54'E
Soc Trang, Viet.	246-47	9°36'N	105°58'E
Socuéllamos, Spain (sô-kōō-äl´yä-môs)	198-99	39°17'N	2°47'W
Sodankylä, Fin.	184-85	67°25'N	26°34'E
Soda Springs, Id., U.S. (sō´dá sprĭngz)	112-13	42°40'N	111°36'W
Söderhamn, Swe. (sû-dĕr-häm´'n)	192-93	61°19'N	17°05'E
Södertälje, Swe. (sû-dĕr-tĕl´yĕ)	192-93	59°12'N	17°37'E
Sodo, Eth.	269d	6°52'N	37°46'E
Soe, Indon.	248-49	9°52'S	124°17'E
Soerabaja, Indon. *see* Surabaya	248-49	7°15'S	112°45'E
Sofia, nat. cap., Blg. (sô´fē-á)	200-01	42°42'N	23°19'E
Sofia, stm., Madag.	264-65	15°25'S	47°14'E
Sofiya, nat. cap., Blg. (sô´fē-á) *see* Sofia	200-01	42°42'N	23°19'E
Sogamoso, Col. (sô-gä-mô´sō)	164-65	5°44'N	72°56'W
Sognefjorden, b., Nor.	192-93	61°06'N	5°10'E
Sogo Nur, lk., China	240-41	42°17'N	101°14'E
Sog Xian, China	238-39	31°50'N	93°47'E
Sōhu Gan, i., Japan	244	29°49'N	140°21'E
Soissons, Fr. (swä-sôɴ´)	196-97	49°23'N	3°20'E
Sŏjosŏn-man, b., Kor., N.	243	39°20'N	124°50'E
Sokal', Ukr. (sô´käl')	194-95	50°29'N	24°17'E
Sokch'o, Kor., S.	243	38°11'N	128°34'E
Söke, Tur. (sû´kĕ)	200-01	37°45'N	27°24'E
Sokhumi, Geor.	227	43°00'N	41°00'E
Sokodé, Togo	260-61	8°59'N	1°09'E
Sokol, Russia	186-87	59°28'N	40°07'E
Sokółka, Pol. (sô-kòl´ká)	194-95	53°24'N	23°31'E
Sokołów Podlaski, Pol. (sô-kô-wôf´ pŭd-lä´skĭ)	194-95	52°24'N	22°15'E
Sokoto, Nig. (sō´kô-tō)	260-61	13°04'N	5°15'E
Sokoto, stm., Nig.	260-61	11°24'N	4°08'E
Solano, Phil. (sō-lä´nō)	250	16°31'N	121°11'E
Solāpur, India	236	17°41'N	75°54'E
Soledad, Col. (sō-lĕ-dä´d)	164-65	10°56'N	74°46'W
Soledad Díez Gutiérrez, Mex.	146-47	22°12'N	100°56'W
Solikamsk, Russia (sô-lē-kámsk´)	186-87	59°40'N	56°46'E
Sol'-Iletsk, Russia	186-87	51°09'N	55°00'E
Solimões, stm., S.A. *see* Amazon	164-65	0°04'S	49°15'W
Solingen, Ger. (zō´lĭng-ĕn)	194-95	51°10'N	7°05'E
Sollefteå, Swe.	184-85	63°11'N	17°16'E
Solnechnogorsk, Russia	202-03	56°11'N	36°59'E
Solo, Indon. *see* Surakarta	248-49	7°34'S	110°50'E
Solomon Islands, nation, Oc. (sŏ´lō-mŭn ī´lándz)	279e	8°0'S	159°00'E
Solomon Sea, s., Oc. (sŏ´lō-mŭn sē)	277	8°0'S	155°00'E
Solon, China (swo-lōōn)	240-41	46°36'N	121°13'E
Solor, Pulau, i., Indon.	248-49	8°28'S	122°59'E
Solov'yëvsk, Russia	240-41	49°55'N	115°42'E
Soltau, Ger. (sôl´tou)	194-95	52°59'N	9°50'E
Sol'tsy, Russia (sôl´tsē)	192-93	58°07'N	30°20'E
Solvay, N.Y., U.S. (sŏl´vā)	116-17	43°04'N	76°14'W
Sölvesborg, Swe. (sûl´vĕs-bôrg)	192-93	56°03'N	14°35'E
Sol'vychegodsk, Russia (sôl´vĕ-chĕ-gôtsk´)	186-87	61°20'N	46°55'E
Solway Firth, b., U.K. (sŏl´wā fûrth´)	190-91	54°50'N	3°35'W
Solwezi, Zam.	264-65	12°11'S	26°25'E
Somalia, nation, Afr. (sō-ma´lē-ă)	253	6°0'N	48°00'E
Somaliland, nation, Afr. *see* Somalia	253	6°0'N	48°00'E
Somali Republic, nation, Afr. *see* Somalia	253	6°0'N	48°00'E
Sombor, Serb. (sôm´bôr)	200-01	45°47'N	19°07'E
Sombrerete, Mex. (sōm-brå-rā´tå)	146-47	23°41'N	103°39'W
Sombrero, i., St. K./N.	143b	18°36'N	63°26'W
Somerset, Ky., U.S. (sŭm´ẽr-sĕt)	124-25	37°05'N	84°37'W
Somerset, Oh., U.S. (sŭm´ẽr-sĕt)	116-17	39°48'N	82°18'W
Somerset East, S. Afr. (sŭm´ẽr-sĕt ēst)	264-65	32°44'S	25°35'E
Somersworth, N.H., U.S. (sŭm´ẽrz-wûrth)	116-17	43°16'N	70°52'W
Somerville, Tn., U.S. (sŭm´ẽr-vĭl)	124-25	35°15'N	89°21'W
Somerville, Tx., U.S. (sŭm´ẽr-vĭl)	122-23	30°20'N	96°32'W
Somoto, Nic. (sô-mō´tō)	149	13°28'N	86°35'W
Son, stm., India (sōn)	234-35	25°42'N	84°52'E
Sønderborg, Den. (sûn´'er-bôrgh)	194-95	54°55'N	9°48'E
Sonepur, India	234-35	20°49'N	83°54'E
Song Da, stm., Asia *see* Black	246-47	21°15'N	105°21'E
Songea, Tan. (sôn-gā´á)	264-65	10°41'S	35°39'E
Songhua, stm., China	240-41	47°43'N	132°31'E
Songhua Hu, res., China	240-41	43°25'N	127°10'E
Songjiang, China	238-39	31°01'N	121°14'E
Sŏngjin, Kor., N. (sŭng´jĭn´) *see* Kimch'aek	243	40°41'N	129°12'E
Songkhla, Thai. (sông´κlä´)	246-47	7°12'N	100°36'E
Songnim, Kor., N.	243	38°44'N	125°38'E
Songo, Moz.	264-65	15°39'S	32°43'E
Sonid Youqi, China	240-41	42°44'N	112°40'E
Sonmiāni Bay, b., Pak.	232-33	25°15'N	66°30'E
Sonneberg, Ger. (sôn´ĕ-bĕrgh)	194-95	50°21'N	11°11'E
Sonora, Ca., U.S. (sô-nō´rá)	118-19	37°59'N	120°22'W
Sonora, Tx., U.S. (sô-nō´rá)	122-23	30°34'N	100°39'W
Sonora, state, Mex. (sô-nō´rá)	144-45	29°20'N	110°40'W
Sonora, stm., Mex. (sô-nō´rá)	144-45	29°05'N	110°54'W
Sonora, Desierto de, des., N.A. *see* Sonoran Desert	144-45	30°0'N	113°00'W
Sonoran Desert, des., N.A.	144-45	30°0'N	113°00'W
Sonsón, Col. (sôn-sōn´)	163c	5°43'N	75°18'W
Sonsonate, El Sal. (sōn-sō-nä´tå)	148	13°43'N	89°43'W
Sonsorol Islands, is., Palau (sŏn-sô-rōl´ ī´lándz)	280-81	5°20'N	132°13'E
Son Tay, Viet.	246-47	21°08'N	105°30'E
Soomaaliya, nation, Afr. *see* Somalia	253	6°0'N	48°00'E
Soome laht, b., Eur. *see* Finland, Gulf of	192-93	60°0'N	27°00'E
Sora, Italy (sô´rä)	200-01	41°44'N	13°37'E
Sorell, Cape, c., Austl.	276	42°12'S	145°10'E
Sorel-Tracy, Qc., Can.	136-37	46°03'N	73°05'W
Soria, Spain (sō´rē-ä)	198-99	41°46'N	2°28'W
Sorocaba, Braz. (sô-rô-kä´bá)	172	23°30'S	47°28'W
Sorochinsk, Russia	186-87	52°26'N	53°10'E
Sorong, Indon. (sō-rông´)	224-25	0°53'S	131°15'E
Soroti, Ug. (sō-rō´tē)	267	1°43'N	33°36'E
Sørøya, i., Nor.	184-85	70°34'N	22°22'E
Sorrento, Italy (sôr-rĕn´tō)	200-01	40°38'N	14°22'E
Sor Rondane Mountains, mts., Ant.	287	72°0'S	25°00'E
Sorsogon, Phil. (sôr-sōgōn´)	250	12°59'N	124°01'E
Sortavala, Russia (sôr´tä-vä-lä)	192-93	61°42'N	30°40'E
Sosna, stm., Russia (sôs´ná)	202-03	52°42'N	38°55'E
Sosnogorsk, Russia	186-87	63°36'N	53°53'E
Sosnowiec, Pol. (sôs-nô´vyĕts)	194-95	50°18'N	19°08'E
Sos'va, stm., Russia (sôs´vá)	218-19	59°33'N	62°20'E
Soto la Marina, Barra, i., Mex.	146-47	24°10'N	97°44'W
Soufrière, vol., Guad. (sōō-frē-âr´)	143b	16°04'N	61°40'W
Sŏul, nat. cap., Kor., S. *see* Seoul	243	37°33'N	127°01'E
Sounding Creek, stm., Ab., Can. (soun´dĭng krēk)	132-33	52°06'N	110°28'W
Souris, Mb., Can. (sōō´rē´)	134-35	49°38'N	100°16'W
Souris, P.E., Can. (sōō´rē´)	138-39	46°21'N	62°15'W
Souris, stm., N.A. (sōō´rē´)	130-31	49°40'N	99°35'W
Sousa, Braz.	166-67	6°45'S	38°14'W
Sousse, Tun. (sōōs)	184-85	35°49'N	10°38'E
South, stm., N.C., U.S. (south)	124-25	34°35'N	78°16'W
South Africa, nation, Afr. (south ăf´rĭ-kă)	253	30°0'S	26°00'E
South America, cont., (south á-mĕr´ĭ-kă)	159	15°0'S	60°00'W
Southampton, Eng., U.K. (south-ămp´tŭn)	190-91	50°55'N	1°24'W
Southampton Island, i., Nu., Can.	130-31	64°20'N	84°40'W
South Andaman, i., India (south ăn-dá-măn´)	246-47	11°48'N	92°44'E
South Australia, state, Austl. (south ôs-trā´lĭ-á)	270-71	30°0'S	135°00'E
South Bend, In., U.S. (south bĕnd)	116-17	41°41'N	86°14'W
South Bend, Wa., U.S. (south bĕnd)	112-13	46°40'N	123°46'W
South Boston, Va., U.S. (south bôs´tŭn)	124-25	36°42'N	78°54'W
Southbridge, Ma., U.S. (south´brĭj)	116-17	42°05'N	72°03'W
South Carolina, state, U.S. (south kăr-ô-lī´ná)	108-09	34°0'N	81°00'W
South China Sea, s., Asia (south chī´ná sē)	224-25	10°0'N	113°00'E
South Dakota, state, U.S. (south dá-kō´tá)	108-09	44°15'N	100°00'W
South East Cape, c., Austl.	276	43°38'S	146°52'E
South East Point, c., Austl.	276	39°08'S	146°25'E
Southend-on-Sea, Eng., U.K. (south-ĕnd´-ŏn-sē)	190-91	51°33'N	0°45'E
Southern Alps, mts., N.Z. (sŭ-thûrn ălps)	278	43°30'S	170°30'E
Southern Bug, stm., Ukr. (sŭ-thûrn bōōg)	202-03	46°39'N	31°56'E
Southern Cook Islands, is., Cook Is.	280-81	20°0'S	159°00'W
Southern Cross, Austl.	270-71	31°14'S	119°19'E
Southern Indian Lake, lk., Mb., Can. (sŭth´ẽrn ĭn´dĭ-ăn lāk)	134-35	57°13'N	98°21'W
Southern Ocean, oc., (sŭ-thûrn ōshŭn)	20-21	50°0'S	135°00'E
Southern Pines, N.C., U.S. (sŭth´ẽrn pīnz)	124-25	35°10'N	79°24'W
Southern Ute Indian Reservation, ind. res., Co., U.S. (sŭth´ẽrn ūt ĭn´dĭ-ăn rĕ-sẽr-vā´shĕn)	118-19	37°05'N	107°45'W
South Georgia, i., S. Geor. (south jôr´já)	287	54°15'S	36°45'W
South Georgia and the South Sandwich Islands, dep., S.A.	287	54°0'S	38°00'W
South Haven, Mi., U.S. (south hāv´'n)	116-17	42°24'N	86°16'W
South Henik Lake, lk., Nu., Can.	130-31	61°30'N	97°27'W
South Indian Lake, Mb., Can.	134-35	56°48'N	98°57'W
Southington, Ct., U.S. (sŭdh´ĭng-tŭn)	116-17	41°36'N	72°53'W
South Island, i., India	236	10°03'N	72°17'E
South Island, i., N.Z. (south ī´lánd)	278	43°0'S	171°00'E
South Korea, nation, Asia (south kô-rē´-á)	206-07	36°30'N	128°00'E
South Luangwa National Park, n.p., Zam.	264-65	12°56'S	31°38'E
South Nahanni, stm., N.T., Can.	130-31	61°03'N	123°20'W
South Negril Point, c., Jam. (south nå-grēl´ point)	142-43	18°15'N	78°22'W
South Ogden, Ut., U.S. (south ŏg´dĕn)	112-13	41°12'N	111°59'W
South Orkney Islands, is., Ant.	287	60°35'S	44°07'W
South Paris, Me., U.S. (south păr´ĭs)	116-17	44°13'N	70°31'W
South Pittsburg, Tn., U.S. (south pĭs´bûrg)	124-25	35°01'N	85°43'W
South Platte, stm., U.S. (south plăt)	110-11	41°07'N	102°42'W
Southport, Austl. (south´pōrt)	276	27°58'S	153°25'E
Southport, Eng., U.K. (south´pôrt)	190-91	53°39'N	3°01'W
South River, On., Can.	136-37	45°50'N	79°22'W
South Sandwich Islands, is., S. Geor. (south sănd´wĭch ī´lándz)	287	57°31'S	26°37'W
South Saskatchewan, stm., Can. (south săs-kach´ĕ-wän)	130-31	53°14'N	105°04'W
South Shetland Islands, is., Ant.	287	62°0'S	58°00'W
South Shields, Eng., U.K. (south shēldz)	190-91	54°60'N	1°25'W
South Sioux City, Ne., U.S. (south sōō sĭt´ē sĭ´tē)	114-15	42°27'N	96°25'W
South Thompson, stm., B.C., Can. (south tŏmp´sŭn)	132-33	50°41'N	120°20'W
South West Africa, nation, Afr. *see* Namibia	253	22°0'S	17°00'E
South West Cape, c., Austl.	276	43°33'S	146°04'E
South West Cape, c., N.Z.	278	47°17'S	167°28'E
Southwest Miramichi, stm., N.B., Can. (south-wĕst´ mĭr á-mē´shē)	138-39	46°58'N	65°34'W
Southwest National Park, n.p., Austl.	276	43°05'S	146°09'E
Sovetsk, Russia (sô-vyĕtsk´)	192-93	55°05'N	21°53'E
Sovetsk, Russia (sô-vyĕtsk´)	186-87	57°35'N	48°58'E
Sovetskaya Gavan, Russia (sŭ-vyĕt´skī-u gä´vŭn')	222-23	48°58'N	140°18'E
Soweto, S. Afr.	269c	26°17'S	27°51'E
Sozopol, Blg. (sôz´ô-pôl')	200-01	42°25'N	27°42'E
Spa, Bel. (spä)	190-91	50°29'N	5°52'E
Spain, nation, Eur. (spān)	174-75	40°0'N	4°00'W
Spanish Fork, Ut., U.S. (spăn´ĭsh fôrk)	118-19	40°08'N	111°39'W
Spanish Sahara, dep., Afr. *see* Western Sahara	253	24°30'N	13°00'W
Spanish Town, Jam.	142-43	17°60'N	76°58'W
Sparks, Nv., U.S. (spärks)	118-19	39°32'N	119°44'W
Sparta, Grc. (spär´tá)	200-01	37°05'N	22°26'E
Sparta, Tn., U.S. (spär´tá)	124-25	35°56'N	85°28'W
Sparta, Wi., U.S. (spär´tá)	114-15	43°56'N	90°48'W
Spartanburg, S.C., U.S. (spär´tăn-bûrg)	124-25	34°57'N	81°56'W
Spartel, Cap, c., Mor. (kăp spär-tĕl´)	269a	35°48'N	5°55'W
Spárti, Grc. *see* Sparta	200-01	37°05'N	22°26'E
Spartivento, Capo, c., Italy (kä´pō spär-tē-vĕn´tô)	200-01	37°55'N	16°04'E
Spartivento, Capo, c., Italy (kä´pō spär-tē-vĕn´tô)	200-01	38°53'N	8°50'E
Spas-Demensk, Russia (spás dyĕ-mĕnsk´)	202-03	54°25'N	34°02'E
Spas-Klepiki, Russia (spás klĕp´ē-kē)	202-03	55°08'N	40°12'E
Spassk-Dal'niy, Russia (spŭsk´däl´nyē)	244	44°36'N	132°50'E
Spear, Cape, c., Nf., Can. (kăp spēr)	138-39	47°31'N	52°39'W
Spearfish, S.D., U.S. (spēr´fĭsh)	114-15	44°30'N	103°52'W
Speedway, In., U.S. (spēd´wā)	116-17	39°47'N	86°13'W
Spence Bay, Nu., Can. *see* Taloyoak	128-29	69°32'N	93°31'W
Spencer, Ia., U.S. (spĕn´sẽr)	114-15	43°09'N	95°09'W

ăt; finăl; rāte; senåte; ärm; åsk; sofá; fâre; ch-choose; dh-as th in other; bē; ēvent; bĕt; recĕnt; cratẽr; g-gō; gh-guttural g; bĭt; ĭ-short neutral; rīde; κ-guttural k as ch in German ich;

Feature (Pronunciation) | Page | Lat. | Long.

Spencer, In., U.S. (spĕn'sẽr) 116-17 39°17'N 86°46'W
Spencer, W.V., U.S. (spĕn'sẽr) 116-17 38°47'N 81°22'W
Spencer, Cape, c., Austl. 276 35°18'S 136°53'E
Spencer Gulf, b., Austl. (spĕn'sẽr gŭlf) 272-73 34°0'S 137°00'E
Speyer, Ger. (shpī'ẽr) 194-95 49°20'N 8°26'E
Spezia, Italy *see* La Spezia 200-01 44°07'N 9°50'E
Spinazzola, Italy (spē-nät'zō-lä) 200-01 40°58'N 16°05'E
Spires, Ger. *see* Speyer 194-95 49°20'N 8°26'E
Spirit Lake, Ia., U.S. (spĭr'ĭt lāk) 114-15 43°26'N 95°06'W
Spirit Lake, Id., U.S. (spĭr'ĭt lāk) 112-13 47°58'N 116°53'W
Spišská Nová Ves, Slvk. (spēsh'skä nō'vä vĕs) 194-95 48°57'N 20°34'E
Spitsbergen, i., Nor. (spĭts'bûr-gĕn) 218-19 78°45'N 16°00'E
Split, Cro. (splêt) 200-01 43°30'N 16°26'E
Split Lake, res., Mb., Can. 134-35 56°08'N 96°15'W
Spokane, Wa., U.S. (spōkăn') 112-13 47°39'N 117°24'W
Spokane Indian Reservation, ind. res., Wa., U.S. (spōkăn' ĭn'dĭ-*ă*n rĕ-sẽr-vā'shĕn) 112-13 47°55'N 118°00'W
Spoleto, Italy (spô-lā'tō) 200-01 42°44'N 12°44'E
Spooner, Wi., U.S. (spōōn'ẽr) 114-15 45°49'N 91°53'W
Spratly Islands, is., Asia 224-25 10°0'N 114°00'E
Springbok, S. Afr. (spring'bŏk) 264-65 29°43'S 17°55'E
Springdale, Nf., Can. (spring'dāl) 138-39 49°31'N 56°04'W
Springdale, Ar., U.S. (spring'dāl) 120-21 36°11'N 94°09'W
Springer, N.M., U.S. (spring'ẽr) 120-21 36°22'N 104°36'W
Springfield, Co., U.S. (spring'fēld) 120-21 37°24'N 102°37'W
Springfield, Fl., U.S. (spring'fēld) 124-25 30°12'N 85°37'W
Springfield, Il., U.S. 116-17 39°48'N 89°39'W
Springfield, Ky., U.S. (spring'fēld) 116-17 37°41'N 85°13'W
Springfield, Ma., U.S. 116-17 42°07'N 72°35'W
Springfield, Mn., U.S. (spring'fēld) 114-15 44°14'N 94°59'W
Springfield, Mo., U.S. (spring'fēld) 120-21 37°13'N 93°17'W
Springfield, Oh., U.S. (spring'fēld) 116-17 39°55'N 83°49'W
Springfield, Or., U.S. (spring'fēld) 112-13 44°03'N 123°01'W
Springfield, Tn., U.S. (spring'fēld) 124-25 36°30'N 86°53'W
Springfield, Vt., U.S. (spring'fēld) 116-17 43°18'N 72°29'W
Springhill, N.S., Can. (spring-hĭl') 138-39 45°39'N 64°03'W
Springs, S. Afr. (springs) 269c 26°15'S 28°26'E
Springsure, Austl. 277 24°07'S 148°05'E
Spring Valley, Il., U.S. (spring crēk văl'ē) 116-17 41°19'N 89°12'W
Spring Valley, Mn., U.S. (spring văl'ē) 114-15 43°41'N 92°25'W
Spruce Grove, Ab., Can. (sprōōs grōv) 132-33 53°32'N 113°55'W
Spruce Knob, mtn., W.V., U.S. 116-17 38°42'N 79°32'W
Squamish, B.C., Can. (skwŏ'mĭsh) 132-33 49°42'N 123°08'W
Squamish, stm., B.C., Can. (skwŏ'mĭsh) 132-33 49°39'N 123°14'W
Srbija, nation, Eur. (sr bē-yä) *see* Serbia 174-75 44°0'N 21°00'E
Sredinny Khrebet, mts., Russia 218-19 56°0'N 158°00'E
Srednekolymsk, Russia (s'rĕd'nyĕ kô-lĕmsk') 218-19 67°27'N 153°40'E
Srednerusskaya Vozvyshennost', plat., Russia 202-03 52°0'N 38°00'E
Śrem, Pol. (shrĕm) 194-95 52°05'N 17°02'E
Sremska Mitrovica, Serb. (srĕm'skä mē'trō-vē-tsä') 200-01 44°59'N 19°37'E
Sri Aman, Malay. 248-49 1°13'N 111°28'E
Sri Jayewardenepura Kotte, nat. cap., Sri L. 236 6°54'N 79°54'E
Srīkākulam, India 236 18°18'N 83°54'E
Sri Lanka, nation, Asia (shrē'-län-k*ȧ*) (srē'-län-k*ȧ*) 206-07 7°0'N 81°00'E
Srīnagar, India 234-35 34°05'N 74°48'E
Staaten River National Park, n.p., Austl. 277 16°40'S 143°00'E
Stafford, Eng., U.K. (stăf'fẽrd) 190-91 52°48'N 2°07'W
Stafford, Va., U.S. (stăf'fẽrd) 116-17 38°25'N 77°24'W
Stakhanov, Ukr. 202-03 48°34'N 38°40'E
Stalin, Rom. *see* Brașov 194-95 45°39'N 25°37'E
Stambaugh, Mi., U.S. (stăm'bô) 114-15 46°05'N 88°38'W
Stamford, Eng., U.K. (stăm'fẽrd) 190-91 52°39'N 0°29'W
Stamford, Ct., U.S. (stăm'fẽrd) 116-17 41°03'N 73°33'W
Stamford, Tx., U.S. (stăm'fẽrd) 120-21 32°55'N 99°49'W
Stamps, Ar., U.S. (stămps) 120-21 33°22'N 93°30'W
Standerton, S. Afr. (stăn'dẽr-t*ŭ*n) 269c 26°57'S 29°15'E
Standing Rock Indian Reservation, ind. res., U.S. (stănd'ĭng rŏk ĭn'dĭ-*ă*n rĕ-sẽr-vā'shĕn) 114-15 45°50'N 101°10'W
Stanford, Ky., U.S. (stăn'fẽrd) 116-17 37°31'N 84°41'W
Stanisławów, Ukr. *see* Ivano-Frankivs'k 194-95 48°55'N 24°44'E
Stanley, N.D., U.S. (stăn'lē) 114-15 48°19'N 102°24'W
Stanley, nat. cap., Falk. Is. (stăn'lē) 171 51°43'S 57°49'W
Stanley Falls, wtfl., D.R.C. 262-63 0°29'N 25°13'E
Stanovoye Nagor'ye, mts., Russia 218-19 56°0'N 114°00'E
Stanovoy Khrebet, mts., Russia (stŭn-ȧ-voi') 218-19 55°48'N 125°34'E
Stanovoy Mountains, mts., Russia (stŭn-ȧ-voi' moun'tĭnz) *see* Stanovoye Nagor'ye 218-19 56°0'N 114°00'E
Stanovoy Range, mts., Russia (stŭn-ȧ-voi' rānj) *see* Stanovoy Khrebet 218-19 55°48'N 125°34'E
Stanthorpe, Austl. 276 28°40'S 151°56'E
Stanton, Ky., U.S. (stăn't*ŭ*n) 116-17 37°50'N 83°55'W
Stanton, Tx., U.S. (stăn't*ŭ*n) 122-23 32°08'N 101°47'W
Staples, Mn., U.S. (stā'p'lz) 114-15 46°21'N 94°48'W
Staraya Russa, Russia (stä'rȧ-yȧ rōōsȧ) 192-93 57°60'N 31°21'E
Stara Zagora, Blg. (stä'rä zä'gô-rȧ) 200-01 42°26'N 25°39'E
Starbuck, Mb., Can. (stär'bŭk) 134-35 49°46'N 97°37'W
Starbuck, i., Kir. 280-81 5°37'S 155°53'W
Staritsa, Russia (stä'rē-tsȧ) 202-03 56°30'N 34°56'E
Starke, Fl., U.S. (stärk) 124-25 29°57'N 82°07'W
Starkville, Ms., U.S. (stärk'vĭl) 124-25 33°27'N 88°49'W
Starodub, Russia (stä-rô-drŏp') 202-03 52°35'N 32°46'E
Starominskaya, Russia (stä'rô mĭn'skȧ-yȧ) 202-03 46°32'N 39°02'E
Start Point, c., Eng., U.K. (stärt point) 190-91 50°14'N 3°39'W
Staryy Oskol, Russia (stä'rē ôs-kôl') 202-03 51°18'N 37°51'E
Staszów, Pol. (stä'shôf) 194-95 50°34'N 21°20'E
State College, Pa., U.S. (stāt kŏl'ĕj) 116-17 40°47'N 77°52'W
Staten Island, i., Arg. *see* Estados, Isla de los 171 54°48'S 64°33'W
Statesboro, Ga., U.S. (stāts'bŭr-ô) 124-25 32°27'N 81°47'W
Statesville, N.C., U.S. (stās'vĭl) 124-25 35°47'N 80°54'W
Staunton, Il., U.S. (stôn't*ŭ*n) 120-21 39°01'N 89°47'W
Staunton, Va., U.S. (stôn't*ŭ*n) 116-17 38°09'N 79°05'W
Staunton, stm., U.S. *see* Roanoke 124-25 35°57'N 76°43'W
Stavanger, Nor. (stä'väng'ẽr) 192-93 58°58'N 5°45'E
Stavropol', Russia 186-87 45°02'N 41°59'E
Stawell, Austl. 276 37°04'S 142°46'E
Steamboat Springs, Co., U.S. (stēm'bōt' springz) 112-13 40°29'N 106°50'W
Steel, stm., On., Can. (stēl) 136-37 48°47'N 86°54'W
Steens Mountain, mts., Or., U.S. (stēnz moun'tĭn) 112-13 42°35'N 118°40'W
Stefanie, Lake, lk., Afr. *see* Ch'ew Bahir 267 4°40'N 36°50'E
Steinamanger, Hung. *see* Szombathely 194-95 47°14'N 16°38'E
Steinbach, Mb., Can. 134-35 49°31'N 96°41'W
Steinkjer, Nor. (stēĭn-kyĕr) 184-85 64°01'N 11°29'E
Stellarton, N.S., Can. (stĕl'*ä*r-t*ŭ*n) 138-39 45°33'N 62°39'W
Stendal, Ger. (shtĕn'däl) 194-95 52°36'N 11°51'E
Stephens Island, i., B.C., Can. 132-33 54°10'N 130°45'W
Stephens Lake, res., Mb., Can. 134-35 56°26'N 95°07'W
Stephenville, Nf., Can. (stē'vĕn-vĭl) 138-39 48°33'N 58°37'W
Sterling, Ak., U.S. (stûr'lĭng) 126 60°32'N 150°48'W
Sterling, Co., U.S. (stûr'lĭng) 114-15 40°37'N 103°13'W
Sterling, Il., U.S. (stûr'lĭng) 116-17 41°48'N 89°42'W
Sterlitamak, Russia (styĕr'lē-ta-mȧk') 186-87 53°37'N 55°58'E
Šternberk, Czech Rep. (shtĕrn'bĕrk) 194-95 49°44'N 17°18'E
Stettin, Pol. *see* Szczecin 194-95 53°26'N 14°32'E
Steubenville, Oh., U.S. (stū'bĕn-vĭl) 116-17 40°22'N 80°38'W
Stevens Point, Wi., U.S. 116-17 44°31'N 89°34'W
Stevensville, Mt., U.S. (stē'vĕnz-vĭl) 112-13 46°30'N 114°05'W
Stewart, B.C., Can. 132-33 55°56'N 129°58'W
Stewart, Isla, i., Chile 171 54°52'S 71°12'W
Stewart Island, i., N.Z. 278 47°0'S 167°50'E
Steynsrus, S. Afr. (stīns'rōōs) 269c 27°57'S 27°34'E
Steyr, Aus. (shtīr) 194-95 48°03'N 14°25'E
Stikine, stm., N.A. (stĭ-kēn') 130-31 56°41'N 132°14'W
Stillwater, Ok., U.S. (stĭl'wô-tẽr) 120-21 36°08'N 97°05'W
Stillwater Range, mts., Nv., U.S. (stĭl'wô-tẽr rānj) 118-19 39°53'N 118°06'W
Štip, Mac. (shtĭp) 200-01 41°45'N 22°12'E
Stirling, Scot., U.K. (stûr'lĭng) 190-91 56°07'N 3°56'W
Stjørdalshalsen, Nor. (styûr-däls-hälsĕn) 184-85 63°29'N 10°56'E
Stockholm, nat. cap., Swe. (stôk'hôlm) 192-93 59°20'N 18°03'E
Stockport, Eng., U.K. (stŏk'pôrt) 190-91 53°25'N 2°10'W
Stockton, Ca., U.S. 118-19 37°57'N 121°17'W
Stockton, Ks., U.S. (stŏk't*ŭ*n) 120-21 39°26'N 99°16'W
Stockton Plateau, plat., Tx., U.S. (stŏk't*ŭ*n plä-tō') 122-23 30°30'N 102°30'W
Stœng Tréng, Camb. (stông'trĕng') 246-47 13°31'N 105°58'E
Stoke-on-Trent, Eng., U.K. (stōk-ŏn-trĕnt) 190-91 52°60'N 2°10'W
Stolbovoy, Ostrov, i., Russia 218-19 74°05'N 136°00'E
Stolin, Bela. (stô'lēn) 194-95 51°54'N 26°52'E
Stolp, Pol. *see* Słupsk 194-95 54°28'N 17°02'E
Stonehaven, Scot., U.K. (stōn'hā-v'n) 190-91 56°58'N 2°13'W
Stonewall, Mb., Can. (stōn'wôl) 134-35 50°08'N 97°19'W
Storm Bay, b., Austl. 276 43°10'S 147°32'E
Storm Lake, Ia., U.S. 114-15 42°39'N 95°13'W
Stornoway, Scot., U.K. (stôr'nō-wā) 190-91 58°13'N 6°24'W
Storsjøen, lk., Nor. (stôr-syûĕn) 192-93 60°21'N 11°41'E
Storsjön, lk., Swe. 184-85 63°12'N 14°18'E
Storuman, Swe. 184-85 65°05'N 17°05'E
Stosch, Isla, i., Chile 171 49°09'S 75°26'W
Strabane, N. Ire., U.K. (stră-băn') 190-91 54°50'N 7°27'W
Strahan, Austl. (strā'*ă*n) 276 42°09'S 145°19'E
Strakonice, Czech Rep. (strä'kô-nyĕ-tsĕ) 194-95 49°15'N 13°55'E
Stralsund, Ger. (shräl'sŏnt) 194-95 54°19'N 13°05'E
Stranraer, Scot., U.K. (străn-rär') 190-91 54°54'N 5°02'W
Strasbourg, Fr. (stràs-bōōr') 196-97 48°35'N 7°45'E
Stratford, On., Can. (străt'fẽrd) 136-37 43°22'N 80°58'W
Stratford, Ct., U.S. (străt'fẽrd) 116-17 41°13'N 73°08'W
Stratford, Tx., U.S. (străt'fẽrd) 120-21 36°20'N 102°04'W
Straubing, Ger. (strou'bĭng) 194-95 48°53'N 12°35'E
Strausberg, Ger. (strous'bĕrgh) 194-95 52°34'N 13°53'E
Streator, Il., U.S. (strē'tẽr) 116-17 41°08'N 88°49'W
Strehaia, Rom. (strĕ-кä'yȧ) 200-01 44°38'N 23°13'E
Streymoy, i., Far. Is. 190b 62°08'N 7°00'W
Strickland, stm., Pap. N. Gui. (strĭk'l*ă*nd) 277 7°35'S 141°23'E
Strongsville, Oh., U.S. (strôngz'vĭl) 116-17 41°18'N 81°50'W
Stronsay, i., Scot., U.K. (strŏn'sā) 190c 59°06'N 2°36'W
Stroudsburg, Pa., U.S. (stroudz'bûrg) 116-17 40°59'N 75°12'W
Strugi-Krasnyye, Russia (strōō'gĭ krä's-ny'yĕ) 192-93 58°16'N 29°07'E
Strumica, Mac. (strōō'mĭ-tsȧ) 200-01 41°26'N 22°37'E
Stryi, Ukr. 194-95 49°15'N 23°51'E
Strzelce Opolskie, Pol. (stzhĕl'tsĕ o-pŏl'skyĕ) 194-95 50°31'N 18°19'E
Strzelecki Creek, stm., Austl. 276 29°21'S 139°48'E
Stuart, Fl., U.S. (stū'ẽrt) 125a 27°12'N 80°15'W
Stuart, Ia., U.S. (stū'ẽrt) 114-15 41°30'N 94°20'W
Stuart, stm., B.C., Can. 132-33 53°59'N 123°33'W
Stuart Island, i., Ak., U.S. (stū'ẽrt ī'l*ă*nd) 126 63°35'N 162°31'W
Stuart Lake, lk., B.C., Can. (stū'ẽrt lāk) 132-33 54°32'N 124°35'W
Stuhlweissenburg, Hung. *see* Székesfehérvár 194-95 47°12'N 18°25'E
Stupino, Russia 202-03 54°54'N 38°05'E
Sturgeon, stm., On., Can. (stûr'jŭn) 136-37 46°16'N 79°56'W
Sturgeon Bay, Wi., U.S. (stûr'jŭn bā) 116-17 44°50'N 87°22'W
Sturgeon Bay, b., Mb., Can. (stûr'jŭn bā) 134-35 52°0'N 97°50'W
Sturgeon Falls, On., Can. (stûr'jŭn fôlz) 136-37 46°22'N 79°55'W
Sturt Stony Desert, des., Austl. 276 28°30'S 141°00'E
Stuttgart, Ger. (shtōōt'gärt) 194-95 48°48'N 9°11'E
Stuttgart, Ar., U.S. (stŭt'gärt) 124-25 34°30'N 91°33'W
Styr, stm., Eur. (stẽr) 194-95 52°07'N 26°36'E
Suđuroy, i., Far. Is. 190b 61°32'N 6°50'W
Suao, Tai. (sōōou) 225a 24°36'N 121°50'E
Subansiri, stm., Asia 238-39 26°46'N 93°45'E
Subarnarekha, stm., India 234-35 21°34'N 87°23'E
Sūbāṭ, stm., Sudan 262-63 9°22'N 31°33'E
Subotica, Serb. (sōō'bô'tē-tsȧ) 200-01 46°06'N 19°41'E
Suceava, Rom. (sōō-chȧ-ä'vȧ) 194-95 47°40'N 26°17'E
Sucre, nat. cap., Bol. (sōō'krā) 168-69 19°02'S 65°16'W
Sudan, nation, Afr. (sōō-dän') 253 15°0'N 30°00'E
Sudan, reg., Afr. (sōō-dän') *see* Sahel 258-59 12°0'N 17°00'E
Sudbury, On., Can. (sŭd'bĕr-ē) 136-37 46°29'N 80°59'W
Sudd, reg., Sudan *see* As-Sudd 262-63 8°0'N 31°00'E
Sudost', stm., Eur. (sò-dôst') 202-03 52°20'N 33°23'E
Sudzha, Russia (sòd'zhȧ) 202-03 51°11'N 35°18'E
Sue, stm., Sudan 262-63 7°40'N 28°02'E
Sueca, Spain (swā'kä) 198-99 39°12'N 0°19'W
Suez, Egypt (sōō-ĕz') 268b 29°58'N 32°33'E
Suez, Gulf of, b., Egypt (gŭlf ŭv sōō-ĕz') 268b 29°0'N 32°50'E
Suez Canal, can., Egypt (sōō-ĕz' kȧ'năl) 268b 29°57'N 32°35'E
Suffolk, Va., U.S. (sŭf'*ŭ*k) 124-25 36°44'N 76°35'W
Suhag, Egypt 268b 26°33'N 31°42'E
Ṣuḥār, Oman 230-31 24°20'N 56°44'E
Sühbaatar, Mong. 240-41 50°13'N 106°12'E
Suhl, Ger. (zōōl) 194-95 50°36'N 10°41'E
Suid-Afrika, nation, Afr. *see* South Africa 253 30°0'S 26°00'E
Suide, China (swā-dŭ) 240-41 37°31'N 110°15'E
Suifenhe, China (swā-fŭn-hŭ) 244 44°24'N 131°08'E
Suihua, China 240-41 46°39'N 126°59'E
Suining, China (sōō'ē-nĭng') 238-39 30°30'N 105°35'E

n-sing; ŋ-baŋk; ɴ-nasalized n; nŏd; c*ŏ*mmit; ōld; ôbey; ôrder; oi-boil; fōōd; ò-as oo in foot; ou-out; s-soft; sh-dish; th-thin; pūre; ûnite; ûrn; stŭd; circ*ŭ*s; ü-as in French tu; '-indeterminate vowel.

Feature (Pronunciation)	Page	Lat.	Long.
Suipacha, Arg. (swē-pä'chä)	173	34°47'S	59°42'W
Suisse, nation, Eur. *see* Switzerland	174-75	47°0'N	8°00'E
Suixian, China (swā shyěn)	238-39	34°26'N	115°04'E
Suizhong, China (swā-jôŋ)	240-41	40°20'N	120°20'E
Suizhou, China	238-39	31°42'N	113°22'E
Sukabumi, Indon.	248-49	6°55'S	106°55'E
Sukagawa, Japan (sōō'kä-gä'wä)	245	37°17'N	140°23'E
Sukarnapura, Indon. *see* Jayapura	277	2°32'S	140°43'E
Sukarno, Pegunungan, mtn., Indon. *see* Jaya, Puncak	224-25	4°05'S	137°11'E
Sukhinichi, Russia (sōō'kē'nê-chê)	202-03	54°07'N	35°22'E
Sukhona, stm., Russia (sò-kô'ná)	186-87	60°45'N	46°18'E
Sukhothai, Thai.	246-47	17°01'N	99°49'E
Sukhumi, Geor. (sò-kòm') *see* Sokhumi	227	43°00'N	41°00'E
Sukkozero, Russia	184-85	63°14'N	32°18'E
Sukkur, Pak. (sŭk'ŭr)	232-33	27°42'N	68°52'E
Sūknah, Libya	188-89	29°04'N	15°47'E
Sukumo, Japan (sōō'kò-mò)	245	32°56'N	132°44'E
Sula, i., Nor.	192-93	61°08'N	4°55'E
Sula, stm., Ukr. (sōō-lá')	202-03	49°38'N	32°43'E
Sula, Kepulauan, is., Indon.	248-49	1°52'S	125°22'E
Sulaimaniya, Iraq *see* As-Sulaymānīyah	228-29	35°34'N	45°27'E
Sulaimān Range, mts., Pak. (sò-lä-ê-män' rānj)	232-33	30°30'N	70°10'E
Sulawesi, i., Indon. *see* Celebes	248-49	2°0'S	121°00'E
Sulawesi, Laut, s., Asia *see* Celebes Sea	248-49	3°0'N	122°00'E
Sulina, Rom. (sōō-lē'ná)	202-03	45°09'N	29°40'E
Sulitelma, mtn., Eur. (sōō-lê-tyĕl'mà)	184-85	67°08'N	16°24'E
Sulitjelma, mtn., Eur. *see* Sulitelma	184-85	67°08'N	16°24'E
Sullana, Peru (sōō-lyä'nä)	170	4°54'S	80°41'W
Sulligent, Al., U.S. (sŭl'ĭ-jěnt)	124-25	33°54'N	88°08'W
Sullivan, Il., U.S. (sŭl'ĭ-văn)	116-17	39°36'N	88°37'W
Sullivan, In., U.S. (sŭl'ĭ-văn)	116-17	39°05'N	87°24'W
Sullivan, Mo., U.S. (sŭl'ĭ-văn)	120-21	38°12'N	91°10'W
Sulmona, Italy (sōōl-mō'nä)	200-01	42°04'N	13°55'E
Sulphur, La., U.S. (sŭl'fŭr)	122-23	30°14'N	93°22'W
Sulphur, Ok., U.S. (sŭl'fŭr)	120-21	34°30'N	96°58'W
Sulphur Springs, Tx., U.S. (sŭl'fŭr sprĭngz)	120-21	33°09'N	95°36'W
Sultanabad, Iran *see* Arāk	232-33	34°05'N	49°41'E
Sulu, Laut, s., Asia *see* Sulu Sea	250	8°0'N	120°00'E
Sulu Archipelago, is., Phil. (sōō'lōō är'kå-pě'-å-gō)	250	6°0'N	121°00'E
Sulūq, Libya	188-89	31°40'N	20°15'E
Sulu Sea, s., Asia (sōō'lōō sē)	250	8°0'N	120°00'E
Sumatera, i., Indon. *see* Sumatra	246-47	0°05'S	102°00'E
Sumatra, i., Indon. (sò-mä-trá)	246-47	0°05'S	102°00'E
Sumba, i., Indon. (sŭm'bá)	248-49	10°0'S	120°00'E
Sumba, Île, i., D.R.C.	262-63	1°44'N	19°32'E
Sumbawa, i., Indon. (sòm-bä'wä)	248-49	8°49'S	117°56'E
Sumbawa Besar, Indon.	248-49	8°30'S	117°24'E
Sumbawanga, Tan.	267	7°60'S	31°38'E
Sumbe, Ang.	264-65	11°14'S	13°51'E
Sumenep, Indon.	248-49	7°00'S	113°52'E
Summerland, B.C., Can. (sŭ'mēr-lănd)	132-33	49°36'N	119°41'W
Summerside, P.E., Can. (sŭm'ēr-sīd)	138-39	46°24'N	63°47'W
Summerville, Ga., U.S. (sŭm'ēr-vĭl)	124-25	34°29'N	85°21'W
Summerville, S.C., U.S. (sŭm'ēr-vĭl)	124-25	33°01'N	80°11'W
Summit Lake, B.C., Can.	132-33	54°17'N	122°37'W
Summit Peak, mtn., Co., U.S. (sŭm'mĭt pēk)	118-19	37°21'N	106°42'W
Šumperk, Czech Rep. (shòm'pěrk)	194-95	49°58'N	16°59'E
Sumqayıt, Azer.	227	40°35'N	49°38'E
Sumter, S.C., U.S. (sŭm'tēr)	124-25	33°55'N	80°21'W
Sumy, Ukr. (sōō'mĭ)	202-03	50°55'N	34°48'E
Sumzom, China	238-39	29°44'N	96°08'E
Sunchales, Arg.	173	30°56'S	61°34'W
Sunch'ŏn, Kor., S.	243	34°57'N	127°30'E
Sunda, Selat, strt., Indon. *see* Sunda Strait	248-49	6°00'S	105°46'E
Sundance, Wy., U.S. (sŭn'dăns)	114-15	44°24'N	104°23'W
Sunda Strait, strt., Indon. (sōōn'dá strāt)	248-49	6°00'S	105°46'E
Sunderland, Eng., U.K. (sŭn'dēr-lănd)	190-91	54°55'N	1°23'W
Sundsvall, Swe. (sònds'väl)	192-93	62°23'N	17°19'E
Sungaipenuh, Indon.	246-47	2°04'S	101°24'E
Sungari, stm., China *see* Songhua	240-41	47°43'N	132°31'E
Sungari Reservoir, res., China *see* Songhua Hu	240-41	43°25'N	127°10'E
Sunne, Swe. (sōōn'ě)	192-93	59°50'N	13°10'E
Sunnyvale, Ca., U.S. (sŭn-nê-văl)	118-19	37°22'N	122°01'W
Sunset Crater Volcano National Monument, n.p., Az., U.S. (sŭn-sět krā'tēr vŏl-kā'nō năsh'ŭn-ăl mŏn'ū-měnt)	118-19	35°22'N	111°31'W
Suntar, Russia (sòn-tár')	218-19	62°10'N	117°38'E
Sun Valley, Id., U.S.	112-13	43°43'N	114°23'W
Sunyani, Ghana	260-61	7°21'N	2°20'W
Suomenlahti, b., Eur. *see* Finland, Gulf of	192-93	60°0'N	27°00'E
Suomi, nation, Eur. *see* Finland	174-75	64°0'N	26°00'E
Suomussalmi, Fin.	184-85	64°53'N	29°02'E
Superior, Az., U.S. (su-pē'rĭ-ēr)	118-19	33°18'N	111°06'W
Superior, Ne., U.S. (su-pē'rĭ-ēr)	120-21	40°01'N	98°04'W
Superior, Wi., U.S. (su-pē'rĭ-ēr)	114-15	46°43'N	92°05'W
Superior, Laguna, b., Mex. (lä-gò'nä sōō-pā-rê-ōr')	146-47	16°21'N	94°55'W
Superior, Lake, lk., N.A. (lāk su-pē'rĭ-ēr)	114-15	48°0'N	88°00'W
Suphan Buri, Thai.	246-47	14°28'N	100°08'E
Suqian, China (sōō-chyěn)	238-39	33°57'N	118°18'E
Suquṭrā, i., Yemen	220-21	12°31'N	53°54'E
Şūr, Oman	230-31	22°34'N	59°30'E
Sura, stm., Russia	186-87	55°37'N	46°02'E
Surabaja, Indon. *see* Surabaya	248-49	7°15'S	112°45'E
Surabaya, Indon.	248-49	7°15'S	112°45'E
Surakarta, Indon.	248-49	7°34'S	110°50'E
Sūrat, India (sò'rŭt)	234-35	21°12'N	72°50'E
Surat Thani, Thai.	246-47	9°06'N	99°18'E
Surazh, Russia (sōō-rázh')	202-03	53°01'N	32°25'E
Surendranagar, India	234-35	22°43'N	71°38'E
Surgut, Russia (sòr-gòt')	218-19	61°16'N	73°12'E
Surigao, Phil.	250	9°46'N	125°29'E
Suriname, nation, S.A. (sōō-rê-näm')	158	4°0'N	56°00'W
Sūrīyah, nation, Asia *see* Syria	206-07	35°0'N	38°00'E
Sūrmaq, Iran	230-31	31°05'N	52°48'E
Surt, Libya	188-89	31°12'N	16°35'E
Surt, Khalīj, b., Libya	258-59	31°30'N	18°00'E
Suruga-wan, b., Japan (sōō'rōō-gä wän)	245	34°51'N	138°33'E
Susanville, Ca., U.S.	118-19	40°25'N	120°39'W
Susong, China (sōō-sôŋ)	238-39	30°09'N	116°07'E
Susquehanna, Pa., U.S. (sŭs'kwê-hăn'á)	116-17	41°56'N	75°36'W
Susquehanna, stm., U.S. (sŭs'kwê-hăn'á)	116-17	39°32'N	76°05'W
Susques, Arg.	168-69	23°25'S	66°30'W
Sussex, N.B., Can. (sŭs'ěks)	138-39	45°43'N	65°31'W
Susuman, Russia	218-19	62°46'N	148°10'E
Sutlej, stm., Asia (sŭt'lěj)	234-35	29°21'N	71°02'E
Sutton, W.V., U.S. (sut''n)	116-17	38°39'N	80°45'W
Sutton, Monts, mts., N.A. *see* Green Mountains	116-17	43°45'N	72°45'W
Suva, nat. cap., Fiji (sōō-vá)	279f	18°07'S	178°27'E
Suwałki, Pol. (sò-vou'kê)	194-95	54°06'N	22°56'E
Suwanose-jima, i., Japan	244	29°38'N	129°43'E
Suwarrow, at., Cook Is.	280-81	13°15'S	163°05'W
Suweis, Khalîg el-, b., Egypt *see* Suez, Gulf of	268b	29°0'N	32°50'E
Suweis, Qanâ el-, can., Egypt *see* Suez Canal	268b	29°57'N	32°35'E
Suwŏn, Kor., S.	243	37°16'N	127°01'E
Suzdal', Russia (sōōz'dál)	202-03	56°25'N	40°26'E
Suzhou, China (sōō-jō)	238-39	33°38'N	116°59'E
Suzhou, China (sōō-jō)	238-39	31°18'N	120°37'E
Svalbard, dep., Eur. (sväl'bärt)	208-09	78°0'N	17°00'E
Svay Riĕng, Camb.	246-47	11°04'N	105°49'E
Svelvik, Nor. (svěl'vêk)	192-93	59°37'N	10°24'E
Svendborg, Den. (svěn-bôrgh)	194-95	55°04'N	10°37'E
Sverdlovs'k, Ukr.	202-03	48°05'N	39°39'E
Sverige, nation, Eur. *see* Sweden	174-75	62°0'N	15°00'E
Svetlaya, Russia (svyět'lá-yá)	244	46°34'N	138°20'E
Svetlograd, Russia	186-87	45°20'N	42°50'E
Svilengrad, Blg. (svêl'ěn-grát)	200-01	41°46'N	26°13'E
Svir', stm., Russia	186-87	60°30'N	32°48'E
Svishtov, Blg. (svēsh'tôf)	200-01	43°37'N	25°21'E
Svizzera, nation, Eur. *see* Switzerland	174-75	47°0'N	8°00'E
Svobodnyy, Russia (svô-bôd'nĭ)	222-23	51°23'N	128°08'E
Svolvær, Nor. (svôl'věr)	184-85	68°15'N	14°33'E
Swainsboro, Ga., U.S. (swānz'bŭr-ô)	124-25	32°36'N	82°20'W
Swakop, stm., Nmb.	264-65	22°41'S	14°32'E
Swakopmund, Nmb. (svä'kôp-mònt) (swá'kôp-mònd)	264-65	22°40'S	14°32'E
Swan, stm., Can. (swŏn)	134-35	52°34'N	100°45'W
Swan Hill, Austl. (swŏn hĭl)	276	35°21'S	143°33'E
Swan Lake, lk., Mb., Can. (swŏn lāk)	134-35	52°31'N	100°45'W
Swan Range, mts., Mt., U.S. (swŏn rānj)	112-13	47°50'N	113°40'W
Swan River, Mb., Can. (swŏn rĭv'ēr)	134-35	52°05'N	101°16'W
Swansea, Wales, U.K. (swŏn'sē)	190-91	51°38'N	3°58'W
Swaziland, nation, Afr. (swä'zê-lănd)	253	26°30'S	31°30'E
Sweden, nation, Eur. (swē'děn)	174-75	62°0'N	15°00'E
Sweetwater, Tn., U.S. (swēt'wô-tēr)	124-25	35°36'N	84°28'W
Sweetwater, Tx., U.S. (swēt'wô-tēr)	120-21	32°28'N	100°24'W
Swellendam, S. Afr.	264-65	34°01'S	20°26'E
Świecie, Pol. (shvyän'tsyě)	194-95	53°24'N	18°27'E
Swift Current, Sk., Can. (swĭft kûr'ěnt)	134-35	50°17'N	107°47'W
Swindon, Eng., U.K. (swĭn'dŭn)	190-91	51°34'N	1°47'W
Swinemünde, Pol. *see* Świnoujście	194-95	53°54'N	14°15'E
Świnoujście, Pol. (shvĭ-nĭ-ô-wēsh'chyě)	194-95	53°54'N	14°15'E
Switzerland, nation, Eur. (swĭt'zēr-lănd)	174-75	47°0'N	8°00'E
Sycamore, Il., U.S. (sĭk'á-mōr)	116-17	41°59'N	88°41'W
Sychëvka, Russia (sē-chôf'ká)	202-03	55°49'N	34°17'E
Sydney, Austl. (sĭd'nê)	276	33°52'S	151°13'E
Sydney, N.S., Can. (sĭd'nê)	138-39	46°09'N	60°12'W
Sydney Mines, N.S., Can. (sĭd'nê mīns)	138-39	46°15'N	60°15'W
Syktyvkar, Russia (sük-tüf'kär)	186-87	61°39'N	50°49'E
Sylacauga, Al., U.S. (sĭl-á-kô'gá)	124-25	33°10'N	86°15'W
Sylhet, Bngl.	234-35	24°54'N	91°52'E
Sylvania, Ga., U.S. (sĭl-vā'nĭ-à)	124-25	32°45'N	81°38'W
Sylvester, Ga., U.S. (sĭl-věs'tēr)	124-25	31°32'N	83°50'W
Syracuse, Ks., U.S. (sĭr'á-kūs)	120-21	37°59'N	101°45'W
Syracuse, Ne., U.S. (sĭr'á-kūs)	120-21	40°39'N	96°11'W
Syracuse, N.Y., U.S.	116-17	43°03'N	76°09'W
Syrdarīya, stm., Asia *see* Syr Darya	226	46°04'N	60°04'E
Syr Darya, stm., Asia (sĭr-dä'rē-ä)	226	46°04'N	60°04'E
Syria, nation, Asia (sĭr'ĭ-á)	206-07	35°0'N	38°00'E
Syriam, Mya.	246-47	16°46'N	96°15'E
Syrian Desert, des., Asia	228-29	32°0'N	40°00'E
Sýros, i., Grc.	200-01	37°26'N	24°55'E
Syzran', Russia (sěz-rän')	186-87	53°09'N	48°26'E
Szabadka, Serb. *see* Subotica	200-01	46°06'N	19°41'E
Szamotuły, Pol. (shá-mô-tōō'wê)	194-95	52°37'N	16°35'E
Szatmárnémeti, Rom. *see* Satu Mare	194-95	47°47'N	22°53'E
Szczecin, Pol. (shchě'tsĭn)	194-95	53°26'N	14°32'E
Szechwan, state, China *see* Sichuan	238-39	31°0'N	105°00'E
Szeged, Hung. (sě'gěd)	194-95	46°16'N	20°10'E
Székesfehérvár, Hung. (sā'kěsh-fě'hār-vär)	194-95	47°12'N	18°25'E
Szekszárd, Hung. (sěk'särd)	194-95	46°21'N	18°43'E
Szentes, Hung. (sěn'těsh)	194-95	46°39'N	20°16'E
Szolnok, Hung.	194-95	47°11'N	20°12'E
Szombathely, Hung. (sôm'bôt-hěl')	194-95	47°14'N	16°38'E
Szydłowiec, Pol. (shid-wô'vyets)	194-95	51°14'N	20°52'E

T

Feature (Pronunciation)	Page	Lat.	Long.
Ta'izz, Yemen	266	13°35'N	44°01'E
Taal, Lake, lk., Phil. (tä-äl')	250	13°60'N	121°01'E
Tabaco, Phil. (tä-bä'kō)	250	13°23'N	123°43'E
Tabar Islands, is., Pap. N. Gui.	277	2°45'S	151°57'E
Ţabas, Iran	232-33	33°36'N	56°55'E
Tabasco, state, Mex. (tä-bäs'kô)	146-47	18°15'N	93°00'W
Taber, Ab., Can.	132-33	49°47'N	112°09'W
Tablas Island, i., Phil. (tä'bläs ī'lănd)	250	12°23'N	122°02'E
Tábor, Czech Rep. (tä'bôr)	194-95	49°25'N	14°41'E
Tabora, Tan. (tä-bō'rä)	267	5°01'S	32°50'E
Tabrīz, Iran (tá-brēz')	227	38°05'N	46°17'E
Tabuaeran, at., Kir.	280-81	3°51'N	159°18'W
Tabūk, Sau. Ar.	228-29	28°23'N	36°35'E
Tacheng, China (tä-chŭŋ)	226	46°45'N	82°58'E
Tacloban, Phil. (tä-klō'bän)	250	11°14'N	124°60'E
Tacna, Peru (täk'nä)	170	18°01'S	70°15'W
Tacoma, Wa., U.S. (tá-kō'má)	112-13	47°15'N	122°26'W
Taconic Range, mts., U.S. (tá-kŏn'ĭk rānj)	116-17	42°30'N	73°20'W
Tacotalpa, stm., Mex. (tä-kô-täl'pä)	146-47	17°48'N	92°51'W
Tacuarembó, Ur.	173	31°42'S	55°59'W
Tademaït, Plateau du, plat., Alg. (plä-tō' dü tä-dê-mä'ēt)	258-59	28°20'N	2°47'E
Tadoussac, Qc., Can. (tá-dōō-sák')	138-39	48°10'N	69°42'W
Tādpatri, India	236	14°54'N	78°00'E
Tadzhikistan, nation, Asia *see* Tajikistan	232-33	39°0'N	71°00'E
T'aebaek-sanmaek, mts., Asia (tī-bīk' sän-mīk')	243	37°30'N	128°31'E
Taedong-gang, stm., Kor., N. (tī-dŏng gäng')	243	38°43'N	125°07'E

Feature (Pronunciation)	Page	Lat.	Long.
Taegu, Kor., S. (tī´gōō´)	243	35°52′N	128°35′E
Taehan-min'guk, nation, Asia see South Korea	206-07	36°30′N	128°00′E
Taejŏn, Kor., S.	243	36°20′N	127°26′E
Tafalla, Spain (tä-fäl´yä)	198-99	42°31′N	1°40′W
Tafassasset, Oued, stm., Afr.	258-59	21°52′N	9°59′E
Taft, Ca., U.S. (tåft)	118-19	35°08′N	119°26′W
Taganrog, Russia (tá-gán-rôk´)	202-03	47°14′N	38°54′E
Taganrogskiy Zaliv, b., Eur. (tá-gán-rôk´skĭ zä´lĭf)	202-03	47°0′N	38°23′E
Tagbilaran, Phil.	250	9°40′N	123°52′E
Tagdempt, Alg. see Tiaret	198-99	35°28′N	1°21′E
Tagtabazar, Turkmen.	232-33	35°58′N	62°55′E
Taguatinga, Braz.	166-67	12°24′S	46°27′W
Taguke, China	234-35	32°06′N	84°43′E
Tagula Island, i., Pap. N. Gui. (tä´gōō-lå ī´lănd)	277	11°30′S	153°30′E
Tagus, stm., Eur. (tā´gŭs)	198-99	38°51′N	8°57′W
Tahan, Gunong, mtn., Malay.	246-47	4°38′N	102°14′E
Tahanroz's'ka zatoka, b., Eur. see Taganrogskiy Zaliv	202-03	47°0′N	38°23′E
Tahat, mtn., Alg. (tä-hät´)	258-59	23°17′N	5°32′E
Tahiti, i., Fr. Poly. (tä-hē´tē) (tä´ē-tē´)	279d	17°37′S	149°27′W
Tahlequah, Ok., U.S. (tă-lĕ-kwä´)	120-21	35°55′N	94°58′W
Tahoe, Lake, lk., U.S. (lāk tä´hō)	118-19	39°07′N	120°03′W
Tahoua, Niger (tä´ōō-ä)	258-59	14°53′N	5°16′E
Tahta, Egypt	268b	26°46′N	31°30′E
Tahulandang, Pulau, i., Indon.	248-49	2°20′N	125°25′E
Tahuna, Indon.	248-49	3°36′N	125°30′E
Tai'an, China (tī-än)	240-41	36°11′N	117°07′E
Taibai Shan, mtn., China (tī-bī shän)	238-39	33°54′N	107°46′E
Taibus Qi, China (tī-bōō-sz chyē)	240-41	41°53′N	115°17′E
T'aichung, Tai. (tī´chŏng)	225a	24°09′N	120°41′E
Taiden, Kor., S. see Taejŏn	243	36°20′N	127°26′E
Taif, Sau. Ar. see Aṭ-Ṭā'if	266	21°16′N	40°25′E
Taigu, China (tī-gōō)	240-41	37°25′N	112°33′E
Taihang Shan, mts., China (tī-häng shän)	240-41	38°0′N	114°00′E
Taihe, China (tī-hŭ)	238-39	26°49′N	114°54′E
Taihoku, nat. cap., Tai. see T'aipei	225a	25°03′N	121°30′E
Tai Hu, lk., China (tī hōō)	238-39	31°13′N	120°11′E
Taikyu, Kor., S. see Taegu	243	35°52′N	128°35′E
Tailai, China (tī-lī)	240-41	46°23′N	123°25′E
Tailem Bend, Austl. (tā-lĕm bĕnd)	276	35°16′S	139°27′E
T'ainan, Tai. (tī´nän´)	225a	23°0′N	120°12′E
Taínaro, Ákra, c., Grc.	200-01	36°23′N	22°29′E
Taining, China (tī´nĭng´)	238-39	26°54′N	117°09′E
T'aipei, nat. cap., Tai. (tī´pā´)	225a	25°03′N	121°30′E
Taiping, Malay.	246-47	4°51′N	100°44′E
Taira, Japan see Iwaki	245	37°03′N	140°55′E
Taishun, China	238-39	27°33′N	119°43′E
Taitao, Península de, pen., Chile	171	46°30′S	74°25′W
T'aitung, Tai. (tī´tōōng´)	225a	22°45′N	121°08′E
Taiwan, nation, Asia (tī-wän)	206-07	23°30′N	121°00′E
T'ai-wan Hai-hsia, strt., Asia see Taiwan Strait	225a	24°0′N	119°00′E
Taiwan Haixia, strt., Asia see Taiwan Strait	225a	24°0′N	119°00′E
Taiwan Strait, strt., Asia (tī-wän strāt strāt)	225a	24°0′N	119°00′E
Taiyuan, China (tī-yůän)	240-41	37°52′N	112°33′E
Taizhao, China	238-39	30°02′N	92°57′E
Taizhou, China (tī-jō)	238-39	32°30′N	119°55′E
Tajikistan, nation, Asia (tä-jĕk´-ĭ-stän´) (tä-jĭk´-ĭ-stän´)	232-33	39°0′N	71°00′E
Tajo, stm., Eur. see Tagus	198-99	38°51′N	8°57′W
Tajumulco, Volcán, vol., Guat. (vôl-kä´n tä-hōō-mōōl´kō)	148	15°02′N	91°55′W
Tajuña, stm., Spain (tä-kōō´n-yä)	198-99	40°07′N	3°35′W
Tak, Thai.	246-47	16°53′N	99°09′E
Takamatsu, Japan	245	34°20′N	134°03′E
Takao, Tai. see Kaohsiung	225a	22°38′N	120°17′E
Takaoka, Japan (ta´kä´ō-kä´)	245	36°45′N	137°01′E
Takasaki, Japan (tä´kät´sōō-kē´)	245	36°19′N	139°01′E
Takatsuki, Japan (tä´kät´sōō-kē´)	245	34°51′N	135°38′E
Takayama, Japan (tä´kä´yä´mä)	245	36°08′N	137°15′E
Takefu, Japan (tä´kĕ-fōō)	245	35°54′N	136°10′E
Takengon, Indon.	246-47	4°37′N	96°51′E
Takêv, Camb.	246-47	10°59′N	104°45′E
Takhli, Thai.	246-47	15°16′N	100°21′E
Takht-e Jamshīd, hist., Iran see Persepolis	230-31	29°57′N	52°52′E
Takla Lake, lk., B.C., Can.	132-33	55°25′N	125°54′W
Takla Makan Desert, des., China	222-23	39°0′N	83°00′E
Taklimakan Shamo, des., China see Takla Makan Desert	222-23	39°0′N	83°00′E
Tala, Mex. (tä´lä)	146-47	20°39′N	103°43′W
Talagante, Chile (tä-lä-gà´n-tĕ)	163e	33°39′S	70°55′W
Talara, Peru (tä-lä´rä)	170	4°35′S	81°16′W
Talas, Kyrg.	226	42°32′N	72°15′E
Talasea, Pap. N. Gui. (tä-lä-sā´ä)	277	5°18′S	150°02′E
Talaud, Kepulauan, is., Indon. (tä-lout´)	248-49	4°20′N	126°50′E
Talavera de la Reina, Spain	198-99	39°58′N	4°49′W
Talca, Chile (täl´kä)	171	35°25′S	71°39′W
Talcahuano, Chile (täl-kä-wä´nō)	171	36°42′S	73°07′W
Taldom, Russia (täl-dôm)	202-03	56°44′N	37°32′E
Taldyqorghan, Kaz.	226	45°01′N	78°23′E
Talghar, Kaz.	226	43°18′N	77°14′E
Talkeetna, Ak., U.S. (tål-kēt´nå)	126	62°20′N	150°07′W
Tall 'Afar, Iraq	228-29	36°22′N	42°27′E
Talladega, Al., U.S. (tăl-à-dē´gà)	124-25	33°26′N	86°06′W
Tallahassee, Fl., U.S. (tăl-à-hăs´ē)	124-25	30°26′N	84°17′W
Tallapoosa, stm., U.S. (tăl-à-pōō´sà)	124-25	32°30′N	86°16′W
Tallassee, Al., U.S. (tăl´à-sē)	124-25	32°33′N	85°55′W
Tallinn, nat. cap., Est. (tál´lĕn) (rä´väl)	192-93	59°26′N	24°48′E
Tallulah, La., U.S. (tă-lōō´lå)	124-25	32°25′N	91°12′W
Talo, mtn., Eth.	262-63	10°44′N	37°54′E
Talok, Indon.	248-49	1°03′N	118°49′E
Taloyoak, Nu., Can.	128-29	69°32′N	93°31′W
Talpa de Allende, Mex. (täl´pä dä äl-yĕn´då)	146-47	20°23′N	104°50′W
Talsi, Lat. (tal´sĭ)	192-93	57°15′N	22°37′E
Taltal, Chile (täl-täl´)	168-69	25°24′S	70°29′W
Talurqjuak, Nu., Can. see Taloyoak	128-29	69°32′N	93°31′W
Tama, Ia., U.S. (tā´mä)	114-15	41°58′N	92°35′W
Tamale, Ghana (tä-mä´lå)	260-61	9°24′N	0°50′W
Taman', Russia (tå-män'')	202-03	45°12′N	36°43′E
Tamanrasset, Alg.	258-59	22°48′N	5°31′E
Tamaqua, Pa., U.S. (tà-mô´kwă)	116-17	40°48′N	75°58′W
Tamaulipas, state, Mex. (tä-mä-ōō-lē´päs´)	146-47	24°0′N	98°45′W
Tamazunchale, Mex. (tä-mä-zōn-chä´lå)	146-47	21°16′N	98°47′W
Tambacounda, Sen. (täm-bä-kōōn´dä)	260-61	13°47′N	13°40′W
Tambelan, Kepulauan, is., Indon. (täm-bå-län´)	246-47	1°0′N	107°30′E
Tambo, Austl. (tăm´bō)	277	24°53′S	146°15′E
Tambo, stm., Peru	170	17°09′S	71°49′W
Tambora, Gunung, vol., Indon.	248-49	8°14′S	117°55′E
Tambov, Russia (tám-bôf´)	186-87	52°43′N	41°25′E
Tambura, Sudan (täm-bōō´rä)	262-63	5°36′N	27°28′E
Tame, Col.	164-65	6°28′N	71°44′W
Tamiahua, Mex. (tä-myä-wä)	146-47	21°16′N	97°27′W
Tamiahua, Laguna de, lk., Mex. (lä-gò´nä-dĕ-tä-myä-wä)	146-47	21°35′N	97°35′W
Tamiami Canal, can., Fl., U.S. (tă-mī-ăm´ĭ kå´näl)	125a	25°46′N	80°11′W
Tamil Nādu, state, India	236	11°0′N	78°15′E
Tam Ky, Viet.	246-47	15°33′N	108°30′E
Tammerfors, Fin. see Tampere	192-93	61°30′N	23°46′E
Tammisaari, Fin. see Ekenäs	192-93	59°58′N	23°26′E
Tampa, Fl., U.S. (tăm´på)	125a	27°58′N	82°27′W
Tampa Bay, b., Fl., U.S. (tăm´på bā)	125a	27°45′N	82°35′W
Tampere, Fin. (tàm´pĕ-rĕ)	192-93	61°30′N	23°46′E
Tampico, Mex. (täm-pē´kō)	146-47	22°13′N	97°51′W
Tamworth, Austl. (tăm´wûrth)	276	31°06′S	150°55′E
Tana, stm., Eur.	184-85	70°26′N	28°16′E
Tana, stm., Kenya (tä´nä)	262-63	2°31′S	40°31′E
Tanabe, Japan (tä-nä´bā)	245	33°44′N	135°24′E
T'ana Hāyk', lk., Eth.	262-63	11°57′N	37°17′E
Tanahjampea, Pulau, i., Indon.	248-49	7°05′S	120°42′E
Tanami Desert, des., Austl.	272-73	20°0′S	129°30′E
Tanana, Ak., U.S. (tă´nà-nô)	126	65°11′N	152°05′W
Tanana, stm., Ak., U.S. (tă´nà-nô)	126	65°09′N	152°04′W
Tananarive, nat. cap., Madag. see Antananarivo	264-65	18°55′S	47°32′E
Tanch'ŏn-ŭp, Kor., N.	243	40°27′N	128°54′E
Tancítaro, Pico de, mtn., Mex. (pē´kô-dĕ tän-sē´tä-rō)	146-47	19°23′N	102°13′W
Tandag, Phil.	250	9°04′N	126°12′E
Tandil, Arg. (tän-dēl´)	173	37°19′S	59°08′W
Tanega-shima, i., Japan (tä´nå-gä´ shē´mä)	244	30°40′N	131°00′E
Tang, stm., China (täŋ)	238-39	33°18′N	117°46′E
Tanga, Tan. (täŋ´gà)	262-63	5°04′S	39°06′E
Tanga Islands, is., Pap. N. Gui.	277	3°29′S	153°13′E
Tanganika, Lac, lk., Afr. see Tanganyika, Lake	262-63	6°0′S	29°30′E
Tanganyika, nation, Afr. see Tanzania	253	6°0′S	35°00′E
Tanganyika, Lac, lk., Afr. see Tanganyika, Lake	262-63	6°0′S	29°30′E
Tanganyika, Lake, lk., Afr. (lăk tăn´găn-yē´kà)	262-63	6°0′S	29°30′E
Tanger, Mor. (tăn-jēr´) see Tangier	269a	35°47′N	5°48′W
Tanggu, China (täŋ-gōō)	240-41	39°01′N	117°40′E
Tanggula Shan, mts., China (täŋ-gōō-lä shän)	238-39	33°0′N	92°00′E
Tangier, Mor. (tăn-jēr´)	269a	35°47′N	5°48′W
Tangipahoa, stm., U.S. (tăn´jē-pà-hō´à)	124-25	30°20′N	90°16′W
Tangra Yumco, lk., China (täŋ-rä yōōm-tswo)	234-35	31°01′N	86°34′E
Tangshan, China	240-41	39°37′N	118°12′E
Tanimbar, Kepulauan, is., Indon.	224-25	7°30′S	131°30′E
Tanjore, India see Thanjāvūr	236	10°47′N	79°09′E
Tanjungbalai, Indon. (tän´jông-bä´lå)	246-47	2°58′N	99°48′E
Tanjungkarang-Telukbetung, Indon. see Bandar Lampung	246-47	5°26′S	105°16′E
Tanjungpandan, Indon.	246-47	2°44′S	107°39′E
Tanjungpinang, Indon. (tän´jŏng-pē´näng)	246-47	0°55′N	104°28′E
Tanjungselor, Indon.	248-49	2°50′N	117°21′E
Tanna, i., Vanuatu	279g	19°30′S	169°20′E
Tannūrah, Ra's, c., Sau. Ar.	230-31	26°38′N	50°10′E
Tanout, Niger	258-59	14°58′N	8°53′E
Tanta, Egypt	268b	30°47′N	31°00′E
Tan-Tan, Mor.	258-59	28°26′N	11°06′W
Tantoyuca, Mex. (tän-tō-yōō´kä)	146-47	21°21′N	98°14′W
Tanzania, nation, Afr. (tăn-zä-nē´à)	253	6°0′S	35°00′E
Tao, stm., China (tou)	240-41	35°55′N	103°20′E
Tao'er, stm., China (tou-är)	240-41	45°41′N	123°49′E
Taongi, at., Marsh. Is.	280-81	14°37′N	168°58′E
Taormina, Italy (tä-ôr-mē´nä)	200-01	37°51′N	15°17′E
Taos, N.M., U.S. (tä´ōs)	120-21	36°25′N	105°35′W
Taoyüan, Tai.	225a	24°59′N	121°18′E
Tapa, Est. (tá´pà)	192-93	59°16′N	25°58′E
Tapachula, Mex.	148	14°54′N	92°17′W
Tapajós, stm., Braz. (tä-pä-zhô´s)	166-67	2°27′S	54°38′W
Tapalqué, Arg. (tä-päl-kĕ´)	173	36°21′S	60°02′W
Tapanahoni, stm., Sur.	164-65	4°21′N	54°26′W
Tapauá, stm., Braz.	166-67	5°47′S	64°24′W
Tāpi, stm., India	234-35	21°09′N	72°44′E
Tapuruquara, Braz.	166-67	0°23′S	65°05′W
Taquari Novo, stm., Braz.	168-69	19°15′S	57°14′W
Tara, Russia (tä´rà)	218-19	56°54′N	74°22′E
Tara, stm., Russia (tä´rà)	218-19	56°42′N	74°36′E
Taraba, stm., Nig.	260-61	8°33′N	10°14′E
Ṭarābulus, Leb. (tà-rä´bò-lōōs) see Tripoli	228-29	34°26′N	35°51′E
Ṭarābulus, hist. reg., Libya see Tripolitania	258-59	31°0′N	15°00′E
Ṭarābulus, nat. cap., Libya see Tripoli	258-59	32°52′N	13°10′E
Tarakan, Indon.	248-49	3°19′N	117°35′E
Tarancón, Spain (tä-rän-kōn´)	198-99	40°01′N	3°00′W
Taranto, Italy (tä´rän-tô)	200-01	40°28′N	17°15′E
Taranto, Golfo di, b., Italy (gôl-fō-dē tä´rän-tô)	200-01	40°10′N	17°20′E
Tarapoto, Peru (tä-rä-pô´tō)	170	6°30′S	76°24′W
Taraquá, Braz.	166-67	0°06′N	68°24′W
Tarare, Fr. (tá-rär´)	196-97	45°54′N	4°26′E
Tarashcha, Ukr. (tä´rásh-chá)	202-03	49°34′N	30°31′E
Tarauacá, stm., Braz.	166-67	7°29′S	70°04′W
Tarawa, at., Kir.	280-81	1°21′N	173°08′E
Taraz, Kaz.	226	42°54′N	71°21′E
Tarazona, Spain (tä-rä-thō´nä)	198-99	41°55′N	1°43′W
Tarbagatai Shan, mts., Asia see Tarbagatay, khrebet	226	47°12′N	83°00′E
Tarbagatay, khrebet, mts., Asia	226	47°12′N	83°00′E
Tarbes, Fr. (tárb)	196-97	43°14′N	0°05′E
Tarboro, N.C., U.S. (tär´bŭr-ô)	124-25	35°54′N	77°33′W
Taree, Austl. (tă-rē´)	276	31°54′S	152°28′E
Tarfaya, Mor.	258-59	27°57′N	12°55′W
Târgu Mureș, Rom.	194-95	46°33′N	24°34′E
Tari, Pap. N. Gui.	277	5°51′S	142°59′E
Tarija, Bol. (tär-rē´hä)	168-69	21°33′S	64°43′W
Tarim, stm., China (tá-rĭm´)	222-23	39°32′N	88°26′E
Tarim Pendi, bas., China	222-23	39°0′N	83°00′E
Taritatu, stm., Indon.	277	2°55′S	138°28′E
Tarkhankut, mys, c., Ukr. (mĭs tár-kän´kòt)	202-03	45°21′N	32°30′E
Tarkio, Mo., U.S. (tär´kĭ-ō)	120-21	40°26′N	95°23′W
Tarko-Sale, Russia	218-19	64°56′N	77°47′E
Tarlac, Phil. (tär´läk)	250	15°29′N	120°36′E
Tarma, Peru (tär´mä)	163a	11°24′S	75°44′W
Tarnopol, Ukr. see Ternopil'	194-95	49°33′N	25°37′E
Tarnów, Pol. (tär´nòf)	194-95	50°01′N	21°01′E
Taroom, Austl.	276	25°39′S	149°48′E

Feature (Pronunciation)	Page	Lat.	Long.
Tarpon Springs, Fl., U.S. (tär′pŏn sprĭngz)	**125a**	28°09′N	82°46′W
Tarquinia, Italy (tär-kwē′nē-ä)	**200-01**	42°15′N	11°45′E
Tarragona, Spain (tär-rä-gō′nä)	**198-99**	41°07′N	1°14′E
Tàrrega, Spain	**198-99**	41°39′N	1°09′E
Tárrega, Spain (tä rå-gä) *see* Tàrrega	**198-99**	41°39′N	1°09′E
Tarsus, Tur. (tår′sòs) (tär′sŭs)	**228-29**	36°54′N	34°55′E
Tartagal, Arg. (tär-tä-gå′l)	**168-69**	22°33′S	63°50′W
Tartu, Est. (tär′to͞o)	**192-93**	58°23′N	26°43′E
Ṭarṭūs, Syria	**228-29**	34°53′N	35°54′E
Tarutao, Ko, i., Thai.	**246-47**	6°35′N	99°40′E
Tarutung, Indon.	**246-47**	2°01′N	98°58′E
Taseyeva, stm., Russia	**218-19**	58°05′N	94°01′E
Tashauz, Turkmen. *see* Daşoguz	**226**	41°50′N	59°58′E
Ṭashk, Daryācheh-ye, lk., Iran	**230-31**	29°45′N	53°30′E
Tashkent, nat. cap., Uzb. (tåsh′kĕnt)	**232-33**	41°19′N	69°17′E
Tāshkurghān, Afg. *see* Kholm	**232-33**	36°41′N	67°42′E
Tashtagol, Russia	**226**	52°46′N	87°53′E
Tasiilaq, Green. *see* Angmagssalik	**284-85**	65°35′N	37°50′W
Tasikmalaya, Indon.	**248-49**	7°20′S	108°13′E
Tasman Bay, b., N.Z. (tăz′măn bā)	**278**	41°0′S	173°20′E
Tasmania, state, Austl.	**276**	43°0′S	147°00′E
Tasman Peninsula, pen., Austl. (tăz′măn pĕ-nĭn′sūlȧ)	**276**	43°05′S	147°50′E
Tasman Sea, s., Oc. (tăz′măn sē)	**282-83**	40°0′S	163°00′E
Tatabánya, Hung.	**194-95**	47°34′N	18°26′E
Tataouine, Tun.	**188-89**	32°55′N	10°28′E
Tatarskiy Proliv, strt., Russia	**218-19**	50°0′N	141°15′E
Tatar Strait, strt., Russia (tå-tär′ strāt) *see* Tatarskiy Proliv	**218-19**	50°0′N	141°15′E
Tateyama, Japan (tä′tē-yä′mä)	**245**	34°59′N	139°52′E
Tathlina Lake, lk., N.T., Can.	**130-31**	60°32′N	117°32′W
Tatnam, Cape, c., Mb., Can.	**134-35**	57°14′N	90°54′W
Tatta, Pak.	**232-33**	24°45′N	67°56′E
Tatvan, Tur.	**227**	38°31′N	42°18′E
Tau, i., Am. Sam.	**279b**	14°15′S	169°29′W
Taunggyi, Mya.	**246-47**	20°47′N	97°02′E
Taupo, N.Z.	**278**	38°41′S	176°06′E
Taupo, Lake, lk., N.Z. (läk tä′o͞o-pō)	**278**	38°49′S	175°55′E
Tauragė, Lith. (tou′rå-gā)	**192-93**	55°15′N	22°17′E
Tauranga, N.Z.	**278**	37°42′S	176°09′E
Tauroa Point, c., N.Z.	**278**	35°10′S	173°04′E
Taurus Mountains, mts., Tur. (tôr′ŭs moun′tĭnz)	**186-87**	37°0′N	33°00′E
Tavastehus, Fin. *see* Hämeenlinna	**192-93**	60°58′N	24°31′E
Tavda, stm., Russia (tåv-då′)	**218-19**	57°48′N	67°15′E
Tavira, Port. (tä-vē′rȧ)	**198-99**	37°07′N	7°39′W
Tavoy, Mya. *see* Dawei	**246-47**	14°05′N	98°13′E
Tavşanlı, Tur. (tåv′shän-lĭ)	**200-01**	39°32′N	29°29′E
Tawas City, Mi., U.S. (tô′wȧs sĭ′tĭ)	**116-17**	44°16′N	83°31′W
Tawau, Malay.	**248-49**	4°16′N	117°53′E
Tawitawi Island, i., Phil.	**250**	5°11′N	119°60′E
Ṭawkar, Sudan	**266**	18°25′N	37°44′E
Taxco de Alarcón, Mex. (täs′kō dē ä-lär-kō′n)	**146-47**	18°33′N	99°36′W
Taxkorgan Tajik Zizhixian, China	**234-35**	37°47′N	75°14′E
Tayabas Bay, b., Phil. (tä-yä′bäs bā)	**250**	13°45′N	121°45′E
Taylor, Tx., U.S. (tā′lẽr)	**122-23**	30°34′N	97°24′W
Taylorville, Il., U.S. (tā′lẽr-vĭl)	**116-17**	39°33′N	89°18′W
Taymura, stm., Russia	**218-19**	63°46′N	98°07′E
Taymyr, Ozero, lk., Russia (ô′zĕ-rô tĭ-mĭr′)	**208-09**	74°36′N	102°24′E
Taymyr, Poluostrov, pen., Russia (tĭ-mĭr′)	**218-19**	76°0′N	104°00′E
Taymyr Peninsula, pen., Russia (tĭ-mĭr′ pĕ-nĭn′sūlȧ) *see* Taymyr, Poluostrov	**218-19**	76°0′N	104°00′E
Tayshet, Russia (tī-shĕt′)	**218-19**	55°56′N	98°00′E
Taytay, Phil.	**250**	10°49′N	119°31′E
Taz, stm., Russia (táz)	**218-19**	67°30′N	78°44′E
Taza, Mor. (tä′zä)	**184-85**	34°14′N	4°01′W
Tazovskiy, Russia	**218-19**	67°29′N	78°42′E
Tbilisi, nat. cap., Geor. (′tbĭl-yē′sē)	**227**	41°44′N	44°47′E
Tchad, nation, Afr. *see* Chad	**253**	15°0′N	19°00′E
Tchad, Lac, lk., Afr. *see* Chad, Lake	**258-59**	13°03′N	14°33′E
Tchibanga, Gabon (chē-bäŋ′gä)	**260-61**	2°51′S	11°01′E
Teapa, Mex. (tā-ä′pä)	**146-47**	17°34′N	92°58′W
Tébessa, Alg.	**184-85**	35°24′N	8°07′E
Tebingtinggi, Indon.	**246-47**	3°19′N	99°10′E
Tecalitlán, Mex. (tā-kä-lē-tlän′)	**146-47**	19°28′N	103°17′W
Techiman, Ghana	**260-61**	7°36′N	1°56′W
Tecka, Arg.	**171**	43°28′S	70°50′W
Tecomán, Mex. (tā-kō-män′)	**146-47**	18°55′N	103°52′W
Tecpan de Galeana, Mex. (tĕk-pän′ dā gä-lā-ä′nä)	**146-47**	17°13′N	100°36′W
Tecuala, Mex. (tĕ-kwä-lä)	**146-47**	22°24′N	105°28′W
Tecuci, Rom. (ta-kòch′)	**202-03**	45°51′N	27°26′E
Tecumseh, Mi., U.S. (tē-kŭm′sĕ)	**116-17**	41°60′N	83°56′W
Tecumseh, Ne., U.S. (tē-kŭm′sĕ)	**120-21**	40°22′N	96°12′W
Tecumseh, Ok., U.S. (tē-kŭm′sĕ)	**120-21**	35°15′N	96°56′W
Tees, stm., Eng., U.K. (tēz)	**190-91**	54°36′N	1°15′W
Tefé, Braz.	**166-67**	3°23′S	64°43′W
Tefé, stm., Braz.	**166-67**	3°31′S	64°57′W
Tegal, Indon.	**248-49**	6°52′S	109°08′E
Tégua, i., Vanuatu	**279g**	13°15′S	166°37′E
Tegucigalpa, nat. cap., Hond. (tå-go͞o-sē-gäl′pä)	**149**	14°05′N	87°13′W
Tehek Lake, lk., Nu., Can.	**130-31**	64°56′N	95°37′W
Teheran, nat. cap., Iran *see* Tehrān	**232-33**	35°40′N	51°25′E
Tehrān, nat. cap., Iran (tĕ-hrän′)	**232-33**	35°40′N	51°25′E
Tehuacán, Mex. (tā-wä-kän′)	**146-47**	18°27′N	97°24′W
Tehuantepec, Golfo de, b., Mex. (gôl-fô dĕ tå-wän-tå-pĕk′)	**146-47**	15°60′N	94°50′W
Tehuantepec, Istmo de, isth., Mex. (ē′st-mô dĕ tå-wän-tå-pĕk′)	**146-47**	17°0′N	95°00′W
Teide, Pico del, mtn., Spain	**199d**	28°16′N	16°38′W
Tejen, Turkmen.	**232-33**	37°22′N	60°31′E
Tejen, stm., Asia	**232-33**	37°24′N	60°31′E
Tejo, stm., Eur. *see* Tagus	**198-99**	38°51′N	8°57′W
Tejupan, Punta, c., Mex. (po͞o′n-tä-tĕ-ĸo͞o-pä′n)	**146-47**	18°20′N	103°30′W
Tejupilco de Hidalgo, Mex. (tå-hō͞o-pēl′kō dā ē-dhäl′gō)	**146-47**	18°54′N	100°09′W
Tekamah, Ne., U.S. (tē-kā′mȧ)	**114-15**	41°47′N	96°13′W
Tekeli, Kaz.	**226**	44°48′N	78°51′E
Tekezē, stm., Afr.	**266**	14°20′N	35°51′E
Tekirdağ, Tur.	**200-01**	40°60′N	27°31′E
Tekit, Mex. (tĕ-kē′t)	**148**	20°32′N	89°20′W
Tela, Hond. (tā′lä)	**149**	15°46′N	87°28′W
Tel Aviv-Jaffa, Isr. *see* Tel Aviv-Yafo	**228-29**	32°03′N	34°47′E
Tel Aviv-Yafo, Isr. (tĕl-ä-vēv′já′já′fȧ)	**228-29**	32°03′N	34°47′E
Telegraph Creek, B.C., Can. (tĕl′ĕ-gráf krēk)	**128-29**	57°55′N	131°10′W
Telén, Arg.	**171**	36°16′S	65°31′W
Telen, stm., Indon.	**248-49**	0°09′S	116°41′E
Telescope Peak, mtn., Ca., U.S. (tĕl′ĕ-skōp pēk)	**118-19**	36°10′N	117°05′W
Teletskoye Ozero, lk., Russia	**226**	51°38′N	87°40′E
Tell City, In., U.S. (tĕl sĭ′tē)	**116-17**	37°57′N	86°46′W
Tello, Col. (tĕ′l-yô)	**163c**	3°04′N	75°08′W
Telluride, Co., U.S. (tĕl′ū-rīd)	**118-19**	37°57′N	107°48′W
Teloloapan, Mex. (tā′lō-lō-ä′pän)	**146-47**	18°21′N	99°52′W
Telos, i., Grc. *see* Tílos	**200-01**	36°26′N	27°23′E
Telsen, Arg.	**171**	42°27′S	66°58′W
Telšiai, Lith. (tĕl′sha′ē)	**192-93**	55°59′N	22°15′E
Teluk Intan, Malay.	**246-47**	4°01′N	101°02′E
Tema, Ghana	**260-61**	5°38′N	0°01′E
Temagami, Lake, lk., On., Can.	**136-37**	47°0′N	80°05′W
Temax, Mex. (tĕ′mäx)	**148**	21°09′N	88°56′W
Tembenchi, stm., Russia	**218-19**	64°37′N	99°56′E
Tembesi, stm., Indon.	**246-47**	1°42′S	103°06′E
Tembilahan, Indon.	**246-47**	0°16′S	103°13′E
Temesvár, Rom. *see* Timişoara	**200-01**	45°45′N	21°13′E
Temirtaū, Kaz.	**226**	50°03′N	72°57′E
Tempe, Az., U.S.	**118-19**	33°24′N	111°55′W
Tempio Pausania, Italy (tĕm′pē-ō pou-sä′nē-ä)	**200-01**	40°54′N	9°06′E
Temple, Tx., U.S. (tĕm′p′l)	**122-23**	31°06′N	97°21′W
Tempoal, stm., Mex. (tĕm-pô-ä′l)	**146-47**	21°46′N	98°27′W
Temryuk, Russia (tyĕm-ryòk′)	**202-03**	45°16′N	37°22′E
Temuco, Chile (tå-mo͞o′kō)	**171**	38°44′S	72°36′W
Tena, Ec.	**170**	0°59′S	77°49′W
Tenāli, India	**236**	16°15′N	80°35′E
Tenasserim, Mya. (tĕn-ăs′ẽr-ĭm)	**246-47**	12°05′N	99°01′E
Ten Degree Channel, strt., India	**246-47**	10°0′N	93°00′E
Tendrara, Mor.	**188-89**	33°04′N	2°00′W
Ténéré, des., Niger	**258-59**	18°43′N	10°51′E
Tenerife, i., Spain (tå-nå-rē′fā) (tĕn-ẽr-ĭf′)	**199d**	28°19′N	16°34′W
Ténès, Alg. (tā-nĕs′)	**198-99**	36°30′N	1°18′E
Tengchong, China	**238-39**	25°01′N	98°30′E
Tenggara, Nusa, is., Asia *see* Lesser Sunda Islands	**248-49**	9°0′S	120°00′E
Tengiz köli, lk., Kaz.	**226**	50°22′N	68°56′E
Tengxian, China (tŭŋ shyĕn)	**238-39**	23°20′N	110°53′E
Tengxian, China (tŭŋ shyĕn)	**240-41**	35°05′N	117°09′E
Tennant Creek, Austl. (tĕn′ănt krēk)	**270-71**	19°39′S	134°11′E
Tennessee, state, U.S. (tĕn-ĕ-sē′)	**108-09**	35°50′N	85°30′W
Tennessee, stm., U.S. (tĕn-ĕ-sē′)	**110-11**	37°05′N	88°34′W
Teno, stm., Eur.	**184-85**	70°26′N	28°16′E
Tenom, Malay.	**248-49**	5°07′N	115°56′E
Tenosique, Mex. (tā-nō-sē′kå)	**148**	17°29′N	91°26′W
Tenryū, stm., Japan (tĕn′ryo͞o′)	**245**	34°40′N	137°48′E
Tensas, stm., La., U.S. (tĕn′sô)	**124-25**	31°37′N	91°48′W
Tenterfield, Austl. (tĕn′tẽr-fēld)	**276**	29°04′S	152°01′E
Teocaltiche, Mex. (tā′ô-käl-tē′chå)	**146-47**	21°26′N	102°34′W
Teófilo Otoni, Braz. (tĕ-ô′fē-lō-tô′nē)	**172**	17°53′S	41°31′W
Teotihuacán, hist., Mex.	**146-47**	19°44′N	98°50′W
Tepalcatepec, Mex. (tā′päl-kä-tå′pĕk)	**146-47**	19°11′N	102°51′W
Tepatitlán de Morelos, Mex. (tā-pä-tē-tlän′ dā mō-rä′los)	**146-47**	20°48′N	102°44′W
Tepeaca, Mex. (tā-på-ä′kä)	**146-47**	18°58′N	97°54′W
Tepic, Mex. (tā-pēk′)	**146-47**	21°30′N	104°54′W
Tequila, Mex. (tå-kē′lä)	**146-47**	20°52′N	103°50′W
Tequisquiapan, Mex. (tå-kēs-kē-ä′pän)	**146-47**	20°32′N	99°54′W
Téra, Niger	**258-59**	14°00′N	0°46′E
Teraina, i., Kir.	**280-81**	4°42′N	160°45′W
Teramo, Italy (tĕ′rä-mô)	**200-01**	42°40′N	13°42′E
Terceira, i., Port. (tĕr-sā′rä)	**199c**	38°43′N	24°13′W
Tercero, stm., Arg.	**173**	32°55′S	62°20′W
Terek, stm., Russia	**227**	43°44′N	46°33′E
Teresina, Braz. (tĕr-å-sē′nȧ)	**166-67**	5°05′S	42°49′W
Teresópolis, Braz. (tĕr-ā-sô′pō-lēzh)	**172**	22°26′S	42°59′W
Tergüün Bogd uul, mtn., Mong.	**240-41**	44°57′N	100°15′E
Teriberka, Russia (tyĕr-ê-byôr′kå)	**186-87**	69°07′N	35°08′E
Términos, Laguna de, b., Mex. (lä-gò′nä dĕ ĕ′r-mē-nôs)	**148**	18°36′N	91°34′W
Termiz, Uzb.	**232-33**	37°14′N	67°16′E
Termoli, Italy (tĕr′mô-lē)	**200-01**	42°00′N	14°60′E
Ternate, Indon. (tĕr-nä′tā)	**248-49**	0°49′N	127°18′E
Terney, Russia	**244**	45°03′N	136°36′E
Terni, Italy (tĕr′nē)	**200-01**	42°34′N	12°39′E
Ternopil′, Ukr.	**194-95**	49°33′N	25°37′E
Terpeniya, Mys, c., Russia (mĭs tĕr-pă′nĭ-yȧ)	**218-19**	48°39′N	144°44′E
Terpeniya, Zaliv, b., Russia (zä′lĭf tĕr-pă′nĭ-yȧ)	**218-19**	49°0′N	143°30′E
Terrace, B.C., Can. (tĕr′ĭs)	**132-33**	54°32′N	128°35′W
Terracina, Italy (tĕr-rä-chē′nä)	**200-01**	41°17′N	13°15′E
Terranova di Sicilia, Italy *see* Gela	**200-01**	37°04′N	14°15′E
Terrebonne, Qc., Can. (tĕr-bŏn′)	**136-37**	45°42′N	73°37′W
Terre Haute, In., U.S. (tĕr-ē hōt′)	**116-17**	39°27′N	87°25′W
Teruel, Spain (tå-ro͞o-ĕl′)	**198-99**	40°21′N	1°06′W
Tes, stm., Asia	**222-23**	50°29′N	93°03′E
Teseney, Erit.	**266**	15°08′N	36°42′E
Tes-Khem, stm., Asia *see* Tes	**222-23**	50°29′N	93°03′E
Teslin, Yk., Can. (tĕs-lĭn)	**128-29**	60°11′N	132°43′W
Teslin, stm., Can. (tĕs-lĭn)	**130-31**	61°34′N	134°53′W
Teslin Lake, lk., Can. (tĕs-lĭn lāk)	**130-31**	60°15′N	132°57′W
Tessalit, Mali	**258-59**	20°12′N	1°00′E
Tessaoua, Niger (tĕs-sä′ô-ä)	**258-59**	13°45′N	7°59′E
Tete, Moz. (tā′tĕ)	**264-65**	16°09′S	33°36′E
Tetepare Island, i., Sol. Is.	**279e**	8°43′S	157°33′E
Teterow, Ger. (tā′tĕ-rō)	**194-95**	53°46′N	12°34′E
Tetiaroa, at., Fr. Poly.	**279d**	17°00′S	149°34′W
Tetouan, Mor.	**269a**	35°35′N	5°22′W
Tetovo, Mac. (tā′tô-vô)	**200-01**	42°00′N	20°59′E
Teuco, stm., Arg.	**168-69**	25°39′S	60°10′W
Tevere, stm., Italy *see* Tiber	**200-01**	41°45′N	12°14′E
Texarkana, Ar., U.S. (tĕk-sär-kăn′ȧ)	**120-21**	33°26′N	94°03′W
Texarkana, Tx., U.S. (tĕk-sär-kăn′ȧ)	**120-21**	33°26′N	94°04′W
Texas, state, U.S. (tĕk′sŭs)	**108-09**	31°30′N	99°00′W
Texas City, Tx., U.S. (tĕk′sŭs sĭ′tĭ)	**122-23**	29°23′N	94°54′W
Texoma, Lake, res., U.S. (lāk tĕk′ō-mă)	**120-21**	33°54′N	96°37′W
Teykovo, Russia (tĕy-kô-vô)	**202-03**	56°51′N	40°33′E
Teziutlán, Mex. (tå-zē-o͞o-tlän′)	**146-47**	19°49′N	97°21′W
Tezpur, India	**234-35**	26°37′N	92°48′E
Thabana-Ntlenyana, mtn., Leso.	**264-65**	29°28′S	29°16′E
Thabazimbi, S. Afr.	**269c**	24°37′S	27°24′E
Thai Lan, Vinh, b., Asia *see* Thailand, Gulf of	**246-47**	10°0′N	101°00′E
Thailand, nation, Asia (tī′lănd)	**206-07**	15°0′N	100°00′E
Thailand, Gulf of, b., Asia (gŭlf ŭv tī′lănd)	**246-47**	10°0′N	101°00′E
Thai Nguyen, Viet.	**246-47**	21°36′N	105°50′E
Thakhek, Laos *see* Muang Khammouan	**246-47**	17°25′N	104°49′E
Thal, Pak.	**232-33**	33°22′N	70°33′E
Thames, stm., On., Can. (tĕmz)	**136-37**	42°19′N	82°27′W
Thames, stm., Eng., U.K. (tĕmz)	**190-91**	51°27′N	0°21′E
Thāne, India	**236**	19°14′N	72°59′E
Thanh Hoa, Viet. (tän′hō′ȧ)	**246-47**	19°48′N	105°46′E
Thanh Pho Ho Chi Minh, Viet. *see* Ho Chi Minh City	**246-47**	10°45′N	106°40′E
Thanjāvūr, India	**236**	10°47′N	79°09′E
Thanlwin, stm., Asia *see* Salween	**208-09**	16°33′N	97°40′E

Feature (Pronunciation)	Page	Lat.	Long.
Thann, Fr. (tän)	196-97	47°49′N	7°05′E
Thar Desert, des., Asia (tär dĕs′ĕrt)	232-33	27°0′N	71°00′E
Thargomindah, Austl. (thar′gō-mĭn′dȧ)	276	28°0′S	143°49′E
Tharrawaddy, Mya.	246-47	17°39′N	95°47′E
Thásos, i., Grc. (thă′sôs)	200-01	40°39′N	24°40′E
Thayer, Mo., U.S. (thâ′ẽr)	124-25	36°31′N	91°33′W
Thayetmyo, Mya.	246-47	19°19′N	95°11′E
Thazi, Mya.	246-47	20°51′N	96°04′E
Thebes, Grc. (thēbz) *see* Thíva	200-01	38°20′N	23°19′E
Thebes, hist., Egypt (thēbz)	268b	25°42′N	32°39′E
The Coorong, b., Austl. (thȧ kō′rŏng)	276	35°46′S	139°15′E
The Dalles, Or., U.S. (thȧ dălz)	112-13	45°36′N	121°11′W
The Hague, nat. cap., Neth. (thȧ hāg) *see* Hague, The	190-91	52°06′N	4°18′E
The Minch, strt., Scot., U.K.	190-91	58°10′N	5°50′W
Theodore, Austl. (thēō′dôr)	277	24°57′S	150°05′E
Theodore Roosevelt National Park (North Unit), n.p., N.D., U.S. (thē-ô-dôr rōō-sȧ-vĕlt năsh′ŭn-ăl pärk)	114-15	47°34′N	103°24′W
Theodore Roosevelt National Park (South Unit), n.p., N.D., U.S. (thē-ô-dôr rōō-sȧ-vĕlt năsh′ŭn-ăl pärk)	114-15	46°58′N	103°25′W
The Pas, Mb., Can. (thȧ pä)	134-35	53°49′N	101°13′W
Thermopolis, Wy., U.S. (thẽr-mŏp′ô-lĭs)	112-13	43°39′N	108°13′W
The Snares, is., N.Z. *see* Snares Islands	287	48°0′S	166°30′E
Thessaloníki, Grc. (thĕs-sȧ-lô-nē′kē)	200-01	40°38′N	22°59′E
Thetford Mines, Qc., Can. (thĕt′fẽrd mīns)	136-37	46°05′N	71°18′W
The Valley, nat. cap., Anguilla	143b	18°13′N	63°04′W
Thibodaux, La., U.S. (tē-bô-dō′)	124-25	29°48′N	90°49′W
Thief River Falls, Mn., U.S. (thēf rĭv′ẽr fôlz)	114-15	48°07′N	96°11′W
Thiers, Fr. (tyâr)	196-97	45°51′N	3°32′E
Thiès, Sen. (tē-ĕs′)	260-61	14°48′N	16°56′W
Thika, Kenya	267	1°03′S	37°04′E
Thimphu, nat. cap., Bhu. (tĭm-pōō′)	234-35	27°28′N	89°39′E
Thingvellir, Ice.	190a	64°16′N	21°07′W
Thionville, Fr. (tyôN-vēl′)	196-97	49°22′N	6°10′E
Thíra, i., Grc.	200-01	36°26′N	25°27′E
Thiruvananthapuram, India	236	8°31′N	76°57′E
Thisted, Den. (tēs′tĕdh)	192-93	56°57′N	8°42′E
Thíva, Grc.	200-01	38°20′N	23°19′E
Thjórsá, stm., Ice. (tyûr′sä)	190a	63°55′N	20°40′W
Thomas, Ok., U.S. (tŏm′ăs)	120-21	35°45′N	98°45′W
Thomaston, Ga., U.S. (tŏm′ăs-tŭn)	124-25	32°53′N	84°19′W
Thomasville, Al., U.S. (tŏm′ăs-vĭl)	124-25	31°54′N	87°45′W
Thomasville, Ga., U.S. (tŏm′ăs-vĭl)	124-25	30°50′N	83°59′W
Thomasville, N.C., U.S. (tŏm′ăs-vĭl)	124-25	35°53′N	80°05′W
Thompson, Mb., Can. (tŏm-sŏn)	134-35	55°44′N	97°51′W
Thompson, stm., B.C., Can. (tŏm-sŏn)	132-33	50°14′N	121°35′W
Thompson, stm., U.S. (tŏm-sŏn)	120-21	39°45′N	93°37′W
Thompson Falls, Mt., U.S. (tŏm-sŏn fôlz)	112-13	47°35′N	115°21′W
Thomson, stm., Austl. (tŏm-sŏn)	277	25°11′S	142°50′E
Thomson's Falls, Kenya (tŏm-sŏns fôlz) *see* Nyahururu Falls	267	0°02′N	36°22′E
Thonon-les-Bains, Fr. (tô-nôN′lâ-băN′)	196-97	46°23′N	6°29′E
Thorn, Pol. *see* Toruń	194-95	53°01′N	18°37′E
Thorshavn, nat. cap., Far. Is. *see* Tórshavn	190b	62°01′N	6°46′W
Thouars, Fr. (tōō-är′)	196-97	46°59′N	0°13′W
Thrace, hist. reg., Eur. (thrās)	200-01	41°20′N	26°45′E
Thráki, hist. reg., Eur. *see* Thrace	200-01	41°20′N	26°45′E
Three Forks, Mt., U.S. (thrē fôrks)	112-13	45°53′N	111°34′W
Three Gorges Reservoir, res., China	238-39	31°0′N	110°30′E
Three Hummock Island, i., Austl.	276	40°26′S	144°55′E
Three Oaks, Mi., U.S. (thrē ōks)	116-17	41°48′N	86°36′W
Three Pagodas Pass, p., Asia	246-47	15°18′N	98°22′E
Three Points, Cape, c., Ghana	260-61	4°45′N	2°05′W
Three Rivers, Mi., U.S. (thrē rĭv′ẽrz)	116-17	41°56′N	85°37′W
Thrissur, India	236	10°31′N	76°13′E
Thu, Cu Lao, i., Viet.	246-47	10°32′N	108°57′E
Thule, Green.	288	76°41′N	68°51′W
Thunder Bay, On., Can. (thŭn′dẽr bā)	136-37	48°24′N	89°15′W
Thunder Bay, b., On., Can. (thŭn′dẽr bā)	136-37	48°24′N	89°00′W
Thursday Island, Austl.	277	10°36′S	142°15′E
Thurso, Scot., U.K.	190-91	58°35′N	3°32′W
Thysville, D.R.C. *see* Mbanza-Ngungu	262-63	5°14′S	14°53′E
Tiandong, China (tĭĕn-dôŋ)	238-39	23°36′N	107°08′E
Tianjin, China (tĭĕn-jyī)	240-41	39°08′N	117°11′E
Tianjin, state, China	240-41	39°30′N	117°15′E
Tianjun, China	240-41	37°20′N	98°57′E
Tianmen, China (tĭĕn-mŭn)	238-39	30°39′N	113°10′E
Tian Shan, mts., Asia (tĭĕn shän) *see* Tien Shan	226	42°0′N	80°00′E
Tianshui, China (tĭĕn-shwā)	238-39	34°32′N	105°54′E
Tiantai, China	238-39	29°08′N	121°00′E
Tianzhu, China	240-41	36°60′N	103°07′E
Tiaret, Alg.	198-99	35°28′N	1°21′E
Tibasti, Sarīr, des., Libya	258-59	24°0′N	17°00′E
Tibati, Camrn.	260-61	6°27′N	12°37′E
Tiber, stm., Italy (Itī′bẽr)	200-01	41°45′N	12°14′E
Tiberias, Lake, lk., Isr. *see* Galilee, Sea of	228-29	32°48′N	35°35′E
Tibesti, mts., Afr. (tī-bĕs′-tē)	258-59	21°30′N	17°30′E
Tibet, state, China (tĭ-bĕt′)	234-35	32°0′N	88°00′E
Tibet, Plateau of, plat., China (plä-tō′ ŭv tĭ-bĕt′)	222-23	33°0′N	92°00′E
Tibooburra, Austl.	276	29°26′S	142°01′E
Tiburón, Isla, i., Mex.	144-45	29°0′N	112°23′W
Tīchīt, Maur.	258-59	18°29′N	9°28′W
Ticonderoga, N.Y., U.S. (tī-kŏn-dẽr-ō′gȧ)	116-17	43°51′N	73°26′W
Ticul, Mex. (tē-kōō′l)	148	20°24′N	89°32′W
Tidaholm, Swe. (tē′dȧ-hōlm)	192-93	58°11′N	13°58′E
Tidjikja, Maur.	258-59	18°33′N	11°25′W
Tidore, Indon.	248-49	0°38′N	127°24′E
Tieli, China	240-41	46°59′N	128°04′E
Tieling, China (tĭĕ-liŋ)	243	42°18′N	123°51′E
Tien Giang, stm., Asia *see* Mekong	246-47	10°33′N	105°27′E
Tien Shan, mts., Asia (tĭĕn shän)	226	42°0′N	80°00′E
Tientsin, China *see* Tianjin	240-41	39°08′N	117°11′E
Tientsin, state, China *see* Tianjin	240-41	39°30′N	117°15′E
Tierp, Swe. (tyĕrp)	192-93	60°20′N	17°31′E
Tierra Blanca, Mex. (tyĕ′r-rä-blä′n-kä)	146-47	18°26′N	96°21′W
Tierra del Fuego, i., S.A. (tyĕr′rä dĕl fwā′gô)	171	54°0′S	69°00′W
Tietê, stm., Braz.	168-69	20°37′S	51°34′W
Tiffin, Oh., U.S. (tĭf′ĭn)	116-17	41°06′N	83°10′W
Tifton, Ga., U.S. (tĭf′tŭn)	124-25	31°27′N	83°30′W
Tiga, Île, i., N. Cal.	279g	21°08′S	167°48′E
Tighina, Mol.	202-03	46°50′N	29°29′E
Tigil', Russia	218-19	57°47′N	158°42′E
Tignish, P.E., Can. (tĭg′nĭsh)	138-39	46°58′N	64°02′W
Tigre, stm., Peru	170	4°29′S	74°05′W
Tigris, stm., Asia (tī-grĭs)	208-09	30°60′N	47°27′E
Tihuatlán, Mex. (tē-wä-tlän′)	146-47	20°44′N	97°34′W
Tijuana, Mex. (tē-hwä′nä)	144-45	32°32′N	117°01′W
Tikal, hist., Guat. (tē-käl′)	148	17°15′N	89°39′W
Tikei, Île, i., Fr. Poly.	280-81	14°58′S	144°33′W
Tikhoretsk, Russia (tē-kôr-yĕtsk′)	186-87	45°51′N	40°08′E
Tikhvin, Russia (tēk-vēn′)	192-93	59°39′N	33°32′E
Tikrīt, Iraq	228-29	34°36′N	43°42′E
Tiksi, Russia (tēk-sē′)	218-19	71°39′N	128°48′E
Tilburg, Neth. (tĭl′bŭrg)	190-91	51°34′N	5°05′E
Tilemsi, Vallée du, stm., Mali	258-59	16°18′N	0°01′E
Tillamook, Or., U.S. (tĭl′ȧ-mŏk)	112-13	45°27′N	123°50′W
Tillsonburg, On., Can. (tĭl′sŭn-bûrg)	136-37	42°52′N	80°43′W
Tilos, i., Grc.	200-01	36°26′N	27°23′E
Tilpa, Austl.	276	30°57′S	144°24′E
Tim, Russia (têm)	202-03	51°38′N	37°07′E
Timan Ridge, hills, Russia *see* Timanskiy Kryazh	186-87	65°0′N	51°00′E
Timanskiy Kryazh, hills, Russia	186-87	65°0′N	51°00′E
Timaru, N.Z. (tĭm′ȧ-rōō)	278	44°24′S	171°14′E
Timbalier Bay, b., La., U.S. (tĭm′bȧ-lẽr bā)	124-25	29°10′N	90°20′W
Timbuktu, Mali *see* Tombouctou	258-59	16°47′N	3°01′W
Timimoun, Alg. (tē-mē-mōōn′)	188-89	29°14′N	0°16′E
Timirist, Râs, c., Maur.	258-59	19°23′N	16°32′W
Timișoara, Rom.	200-01	45°45′N	21°13′E
Timmins, On., Can. (tĭm′ĭnz)	136-37	48°29′N	81°21′W
Timor, i., Asia (tē-mōr′)	248-49	9°0′S	125°00′E
Timor, Laut, s., *see* Timor Sea	272-73	11°0′S	128°00′E
Timor-Leste, nation, Asia *see* East Timor	248-49	8°35′S	126°00′E
Timor Sea, s., (tē-mōr′ sē)	272-73	11°0′S	128°00′E
Timor Timur, nation, Asia *see* East Timor	248-49	8°35′S	126°00′E
Timpanogos Cave National Monument, n.p., Ut., U.S. (tĭ-măn′ō-gōz kāv năsh′ŭn-ăl mŏn′ū-mĕnt)	118-19	40°26′N	111°44′W
Tinaca Point, c., Phil.	250	5°34′N	125°20′E
Tindouf, Alg. (tĕn-dōōf′)	258-59	27°49′N	8°08′W
Tinghert, Ḥamādat, plat., Afr.	258-59	29°0′N	9°00′E
Tingo María, Peru (tē′ngô-mä-rē′ä)	170	9°10′S	75°56′W
Tingri, China *see* Dinggyê	234-35	28°35′N	86°37′E
Tingsryd, Swe. (tĭngs′rüd)	192-93	56°32′N	14°59′E
Tinian, i., N. Mar. Is.	280-81	15°00′N	145°38′E
Tinkisso, stm., Gui.	260-61	11°21′N	9°11′W
Tinogasta, Arg. (tē-nô-gäs′tä)	168-69	28°03′S	67°34′W
Tínos, i., Grc.	200-01	37°36′N	25°10′E
Tinrhert, Hamada de, plat., Afr.	258-59	29°0′N	9°00′E
Tintina, Arg.	173	27°02′S	62°43′W
Tioman, Pulau, i., Malay.	246-47	2°48′N	104°10′E
Tipitapa, Nic. (tē-pē-tä′pä)	149	12°12′N	86°06′W
Tip Top Mountain, mtn., On., Can.	136-37	48°16′N	85°59′W
Tīrān, i., Sau. Ar.	228-29	27°57′N	34°33′E
Tīrân, Madîq, strt., *see* Tiran, Strait of	228-29	27°58′N	34°28′E
Tiran, Strait of, strt.	228-29	27°58′N	34°28′E
Tirana, nat. cap., Alb. *see* Tiranë	200-01	41°20′N	19°50′E
Tiranë, nat. cap., Alb. (tē-rä′nä)	200-01	41°20′N	19°50′E
Tirano, Italy (tē-rä′nō)	200-01	46°13′N	10°11′E
Tiraspol, Mol.	202-03	46°51′N	29°38′E
Tire, Tur. (tē′rĕ)	200-01	38°06′N	27°45′E
Tiree, i., Scot., U.K. (tī-rē′)	190-91	56°31′N	6°52′W
Tîrgu Mureș, Rom. *see* Târgu Mureș	194-95	46°33′N	24°34′E
Tirich Mīr, mtn., Pak.	232-33	36°15′N	71°50′E
Tirreno, Mar, s., Eur. *see* Tyrrhenian Sea	200-01	40°0′N	12°00′E
Tiruchchirāppalli, India (tĭr′ô-chĭ-rä′pȧ-lĭ)	236	10°49′N	78°42′E
Tirunelveli, India	236	8°44′N	77°41′E
Tiruppur, India	236	11°06′N	77°21′E
Tisa, stm., Eur.	200-01	45°08′N	20°17′E
Tisdale, Sk., Can. (tĭz′dāl)	134-35	52°51′N	104°03′W
Tisisat Falls, wtfl., Eth.	266	11°29′N	37°35′E
T'īs Isat Fwafwatē, wtfl., Eth. *see* Tisisat Falls	266	11°29′N	37°35′E
Tīsta, stm., Asia	234-35	25°31′N	89°42′E
Tisza, stm., Eur. (tē′sä)	200-01	45°08′N	20°17′E
Titicaca, Lago, lk., S.A. (lä′gō-tē-tē-kä′kä)	168-69	15°50′S	69°20′W
Titograd, nat. cap., Mont. *see* Podgorica	200-01	42°27′N	19°16′E
Titov Veles, Mac. (tē′tôv vĕ′lĕs)	200-01	41°43′N	21°47′E
Titusville, Fl., U.S. (tī′tŭs-vĭl)	125a	28°37′N	80°49′W
Titusville, Pa., U.S. (tī′tŭs-vĭl)	116-17	41°38′N	79°41′W
Tivoli, Italy (tē′vô-lê)	200-01	41°58′N	12°48′E
Tiwanaku, hist., Bol.	168-69	16°33′S	68°41′W
Tizimín, Mex. (tē-zē-mē′n)	148	21°10′N	88°10′W
Tizi Ouzou, Alg. (tē′zē-ōō-zōō′)	269b	36°48′N	4°02′E
Tiznit, Mor. (tēz-nēt)	258-59	29°42′N	9°43′W
Tjörn, i., Swe.	192-93	58°0′N	11°38′E
Tlacotalpan, Mex. (tlä-kô-täl′pän)	146-47	18°38′N	95°40′W
Tlacotepec, Mex. (tlä-kô-tȧ-pĕ′k)	146-47	17°47′N	99°59′W
Tlahualilo de Zaragoza, Mex.	122-23	26°06′N	103°26′W
Tlalnepantla, Mex. (tläl-nȧ-pȧn′tlä)	146-47	19°32′N	99°12′W
Tlaquepaque, Mex. (tlä-kĕ-pä′kĕ)	146-47	20°39′N	103°19′W
Tlaxcala, state, Mex.	146-47	19°25′N	98°10′W
Tlaxcala de Xicohténcatl, Mex.	146-47	19°19′N	98°14′W
Tlemcen, Alg.	198-99	34°53′N	1°18′W
Toamasina, Madag.	264-65	18°09′S	49°24′E
Tobago, i., Trin. (tô-bā′gō)	143b	11°15′N	60°40′W
Tobejuba, Isla, i., Ven.	143b	9°20′N	60°52′W
Tobelo, Indon.	248-49	1°44′N	128°00′E
Tobi, i., Palau	280-81	3°0′N	131°10′E
Tobol, stm., Asia (tô-bôl′)	218-19	58°09′N	68°13′E
Tobol'sk, Russia (tô-bôlsk′)	218-19	58°11′N	68°15′E
Tobruk, Libya	258-59	32°05′N	23°57′E
Tobyl, stm., Asia	218-19	58°09′N	68°13′E
Tocantinópolis, Braz. (tō-kän-tē-nō′pō-lēs)	166-67	6°19′S	47°25′W
Tocantins, state, Braz. (tō-kän-tēNs′)	166-67	10°0′S	48°00′W
Tocantins, stm., Braz. (tō-kän-tēNs′)	166-67	1°45′S	49°12′W
Toccoa, Ga., U.S. (tŏk′ō-ȧ)	124-25	34°35′N	83°20′W
Tocoa, Hond. (tō-kô′ä)	149	15°38′N	86°01′W
Tocopilla, Chile (tō-kô-pēl′yä)	168-69	22°06′S	70°11′W
Tocuyo de la Costa, Ven. (tô-kōō′yō-dĕ-lä-kôs′tä)	163b	11°02′N	68°22′W
Todos Santos, Bol.	168-69	16°48′S	65°08′W
Toemoek Hoemak Gebergte, mts., S.A. *see* Tumuc-Humac Mountains	164-65	2°19′N	54°35′W
Tofino, B.C., Can. (tō-fē′nō)	132-33	49°08′N	125°54′W
Toga, i., Vanuatu	279g	13°25′S	166°41′E
Togian, Kepulauan, is., Indon.	248-49	0°20′S	122°00′E
Togliatti, Russia	186-87	53°32′N	49°26′E
Togo, nation, Afr. (tō′gō)	253	8°0′N	1°10′E
Tok, Ak., U.S.	126	63°20′N	143°00′W
Tokachi, stm., Japan (tō-kä′chē)	244	42°44′N	143°43′E
Tokat, Tur. (tô-kät′)	186-87	40°19′N	36°34′E
Tokelau, dep., Oc. (tō-kē-lä′ô)	280-81	9°0′S	171°45′W
Tokmok, Kyrg.	226	42°50′N	75°18′E
Tokushima, Japan (tō′kô′shē-mä)	245	34°04′N	134°34′E
Tokuyama, Japan (tō′kô′yä-mä)	245	34°03′N	131°48′E

Feature (Pronunciation)	Page	Lat.	Long.
Tōkyō, nat. cap., Japan (tō´kē-ō)	245	35°42′N	139°47′E
Tôlañaro, Madag.	264-65	25°02′S	47°00′E
Toledo, Spain (tô-lĕ´dô)	198-99	39°53′N	4°03′W
Toledo, Ia., U.S. (tô-lē´dō)	114-15	41°60′N	92°35′W
Toledo, Oh., U.S. (tô-lē´dō)	116-17	41°39′N	83°33′W
Toledo, Or., U.S. (tô-lē´dō)	112-13	44°38′N	123°56′W
Toledo, Montes de, mts., Spain (mô´n-tĕs-dĕ-tô-lĕ´dô)	198-99	39°33′N	4°20′W
Toledo Bend Reservoir, res., U.S. (tô-lē´dō bĕnd rĕ´sĕr-vwär)	122-23	31°30′N	93°45′W
Toliara, Madag.	264-65	23°22′S	43°40′E
Tolima, Nevado del, vol., Col. (nĕ-vä-dô-dĕl-tô-lē´mä)	163c	4°40′N	75°19′W
Tolitoli, Indon.	248-49	1°02′N	120°49′E
Tolmezzo, Italy (tôl-mĕt´zô)	200-01	46°25′N	13°01′E
Tolo, Teluk, b., Indon. (tō´lō)	248-49	2°0′S	122°30′E
Tolosa, Spain (tô-lō´sä)	198-99	43°09′N	2°05′W
Tolsan-do, i., Kor., S.	243	34°38′N	127°45′E
Toluca, Il., U.S. (tô-lōō´kȧ)	116-17	40°60′N	89°08′W
Toluca, Nevado de, vol., Mex. (nĕ-vä-dô-dĕ-tô-lōō´kä)	146-47	19°05′N	99°44′W
Toluca de Lerdo, Mex.	146-47	19°17′N	99°39′W
Tolyatti, Russia see Togliatti	186-87	53°32′N	49°26′E
Tom', stm., Russia	218-19	56°53′N	84°27′E
Tomah, Wi., U.S. (tō´mȧ)	114-15	43°59′N	90°30′W
Tomahawk, Wi., U.S. (tŏm´ȧ-hôk)	116-17	45°28′N	89°44′W
Tomakomai, Japan	244	42°38′N	141°36′E
Tomar, Port. (tō-mär´)	198-99	39°36′N	8°25′W
Tomaszów Lubelski, Pol. (tô-mä´shōf lōō-bĕl´skĭ)	194-95	50°27′N	23°25′E
Tomaszów Mazowiecki, Pol. (tô-mä´shōf mä-zô´vyĕt-skĭ)	194-95	51°33′N	20°01′E
Tomatlán, Mex. (tô-mä-tlä´n)	146-47	19°55′N	105°14′W
Tombador, Serra do, plat., Braz. (sĕr´rȧ dô tōm-bä-dôr´)	166-67	12°0′S	57°40′W
Tombigbee, stm., U.S. (tŏm-bĭg´bē)	110-11	31°04′N	87°58′W
Tombouctou, Mali	258-59	16°47′N	3°01′W
Tombstone, Az., U.S. (tōōm´stōn)	118-19	31°43′N	110°04′W
Tombstone Mountain, mtn., Yk., Can.	130-31	64°25′N	138°30′W
Tombua, Ang. (ȧ-lĕ-zhän´drĕ)	264-65	15°48′S	11°49′E
Tomé, Chile	171	36°36′S	72°57′W
Tomea, Pulau, i., Indon.	248-49	5°45′S	123°56′E
Tomelilla, Swe. (tô´mĕ-lēl-lä)	192-93	55°33′N	13°57′E
Tomelloso, Spain (tō-mål-lyō´sō)	198-99	39°09′N	3°01′W
Tomini, Indon.	248-49	0°32′N	120°32′E
Tomini, Teluk, b., Indon.	248-49	0°20′S	121°00′E
Tommot, Russia (tŏm-mŏt´)	218-19	58°58′N	126°18′E
Tomo, stm., Col.	164-65	5°19′N	67°50′W
Tom Price, Austl.	270-71	22°41′S	117°48′E
Tomsk, Russia (tŏmsk)	218-19	56°30′N	84°58′E
Tonalá, Mex.	146-47	16°05′N	93°45′W
Tondano, Indon. (tôn-dä´nō)	248-49	1°18′N	124°55′E
Tønder, Den. (tûn´nĕr)	194-95	54°56′N	8°52′E
Tone, stm., Japan (tô´nĕ)	245	35°45′N	140°51′E
Tonga, nation, Oc. (tŏŋ´gȧ)	280-81	20°0′S	175°00′W
Tonga Islands, is., Tonga (tŏŋ´gȧ ī´lȧndz)	280-81	20°0′S	175°00′W
Tong'an, China (tôŋ-än)	225a	24°44′N	118°09′E
Tongatapu, i., Tonga	280-81	21°10′S	175°10′W
Tongbei, China (tôŋ-bā)	240-41	47°46′N	126°46′E
Tongcheng, China	238-39	31°03′N	116°57′E
Tongchuan, China	238-39	35°04′N	109°04′E
Tongguan, China (tôŋ-gūän)	238-39	34°36′N	110°17′E
Tonghai, China	238-39	24°07′N	102°47′E
Tonghe, China (tôŋ-hŭ)	244	45°58′N	128°45′E
Tonghua, China (tôŋ-hwä)	243	41°43′N	125°56′E
Tongjiang, China (tôŋ-jyäŋ)	238-39	31°56′N	107°14′E
Tongjiang, China (tôŋ-jyäŋ)	240-41	47°38′N	132°30′E
Tongjosŏn-man, b., Kor., N.	243	39°30′N	128°00′E
Tongliao, China (tôŋ-lĭou)	240-41	43°37′N	122°17′E
Tongoy, Chile (tōn-goi´)	168-69	30°16′S	71°29′W
Tongren, China (tôŋ-rŭn)	238-39	27°43′N	109°11′E
Tongsa Dzong, Bhu.	234-35	27°31′N	90°30′E
Tongxian, China (tôŋ shyĕn)	240-41	39°54′N	116°39′E
Tongyu, China	240-41	44°48′N	123°05′E
Tongzi, China	238-39	28°09′N	106°49′E
Tonk, India (Tŏŋk)	234-35	26°10′N	75°48′E
Tonkawa, Ok., U.S. (tŏŋ kȧ-wô)	120-21	36°41′N	97°19′W
Tonkin, Gulf of, b., Asia (gŭlf ŭv tôn-kăn´)	246-47	20°0′N	108°00′E
Tônlé Sab, Bœng, lk., Camb. see Sap, Tonle	246-47	13°0′N	104°00′E
Tonneins, Fr. (tô-năN´)	196-97	44°23′N	0°19′E
Tonopah, Nv., U.S. (tō-nô-pä´)	118-19	38°05′N	117°13′W
Tønsberg, Nor. (tûns´bĕrgh)	192-93	59°16′N	10°26′E

Feature (Pronunciation)	Page	Lat.	Long.
Tonto National Monument, n.p., Az., U.S. (tôn´tô)	118-19	33°34′N	111°02′W
Tooele, Ut., U.S. (tŏ-ĕl ē)	118-19	40°32′N	112°18′W
Toowoomba, Austl. (tŏ wōōm´bȧ)	276	27°34′S	151°57′E
Topeka, Ks., U.S.	120-21	39°02′N	95°41′W
Topol'čany, Slvk. (tô-pôl´chä-nü)	194-95	48°34′N	18°10′E
Topolobampo, Mex. (tō-pō-lô-bä´m-pô)	144-45	25°36′N	109°03′W
Topozero, Ozero, lk., Russia	186-87	65°40′N	32°00′E
Toppenish, Wa., U.S. (tôp´ĕn-ĭsh)	112-13	46°23′N	120°19′W
Torawitan, Tanjung, c., Indon.	248-49	1°45′N	124°60′E
Torbat-e Ḩeydarīyeh, Iran	232-33	35°17′N	59°13′E
Torbat-e Jām, Iran	232-33	35°15′N	60°38′E
Torbay, Nf., Can. (tôr-bā´)	138-39	47°40′N	52°45′W
Torbay, Eng., U.K. see Torquay	190-91	50°28′N	3°32′W
Torch, stm., Sk., Can.	134-35	53°52′N	103°06′W
Torch Lake, lk., Mi., U.S. (tôrch lāk)	116-17	45°03′N	85°20′W
Torda, Rom. see Turda	194-95	46°34′N	23°47′E
Torez, Ukr.	202-03	48°02′N	38°38′E
Torghay, stm., Kaz.	226	48°02′N	62°34′E
Torghay üstirti, plat., Kaz.	226	51°0′N	64°00′E
Torino, Italy see Turin	200-01	45°03′N	7°41′E
Tori Sima, i., Japan	244	30°29′N	140°19′E
Torit, Sudan	267	4°24′N	32°34′E
Tormes, stm., Spain (tōr´mās)	198-99	41°18′N	6°27′W
Torneälven, stm., Eur.	184-85	65°49′N	24°09′E
Torneträsk, lk., Swe. (tŏr´nĕ trĕsk)	184-85	68°20′N	19°23′E
Torngat, Monts, mts., Can. see Torngat Mountains	130-31	59°0′N	64°00′W
Torngat Mountains, mts., Can.	130-31	59°0′N	64°00′W
Tornio, Fin. (tôr´nĭ-ô)	184-85	65°51′N	24°10′E
Tornionjoki, stm., Eur.	184-85	65°49′N	24°09′E
Tornquist, Arg.	173	38°06′S	62°13′W
Toronto, On., Can. (tô-rŏn´tō)	136-37	43°38′N	79°24′W
Toropets, Russia (tô´rô-pyĕts)	192-93	56°30′N	31°40′E
Tororo, Ug.	267	0°42′N	34°11′E
Toros Dağları, mts., Tur. see Taurus Mountains	186-87	37°0′N	33°00′E
Torquay, Eng., U.K. (tôr-kē´)	190-91	50°28′N	3°32′W
Torrance, Ca., U.S. (tŏr´rănc)	118-19	33°51′N	118°20′W
Torrelavega, Spain (tôr-rā´lä-vā´gä)	198-99	43°21′N	4°03′W
Torremaggiore, Italy (tôr´rā mäd-jō´rā)	200-01	41°41′N	15°17′E
Torrens, Lake, lk., Austl. (lāk tŏr-ĕns)	276	31°03′S	137°51′E
Torreón, Mex. (tôr-rå-ōn´)	122-23	25°33′N	103°26′W
Torres, Îles, is., Vanuatu	279g	13°17′S	166°39′E
Torres Islands, is., Vanuatu (tôr´rĕs ī´lȧndz) (tŏr´ĕz ī´lȧndz) see Torres, Îles	279g	13°17′S	166°39′E
Torres Novas, Port. (tôr´rĕzh nō´väzh)	198-99	39°28′N	8°32′W
Torres Strait, strt., Oc. (tôr´rĕs strāt)	277	10°25′S	142°10′E
Torres Vedras, Port. (tôr´rĕsh vā´dräzh)	198-99	39°05′N	9°15′W
Torrevella, Spain	198-99	37°59′N	0°41′W
Torrevieja, Spain (tôr-rā-vyā´hä) see Torrevella	198-99	37°59′N	0°41′W
Torrington, Ct., U.S. (tŏr´ĭng-tŭn)	116-17	41°48′N	73°07′W
Torrington, Wy., U.S. (tŏr´ĭng-tŭn)	114-15	42°05′N	104°12′W
Torsby, Swe. (tôrs´bü)	192-93	60°08′N	13°01′E
Tórshavn, nat. cap., Far. Is. (tôrs-houn´)	190b	62°01′N	6°46′W
Tortola, i., Br. Vir. Is. (tôr-tō´lä)	143b	18°27′N	64°36′W
Tórtolas, Cerro de las, mtn., S.A. see Las Tórtolas, Cerro	168-69	29°57′S	69°53′W
Tortona, Italy (tôr-tō´nä)	200-01	44°54′N	8°52′E
Tortosa, Spain (tôr-tō´sä)	198-99	40°48′N	0°31′E
Tortue, Île de la, i., Haiti (ēl-dĕ-lȧ-tôr-tü´) see Tortuga Island	142-43	20°03′N	72°47′W
Tortuga Island, i., Haiti see Tortue, Île de la	142-43	20°03′N	72°47′W
Toruń, Pol.	194-95	53°01′N	18°37′E
Tõrva, Est. (t'r´vȧ)	192-93	58°01′N	25°56′E
Torzhok, Russia (tôr´zhôk)	202-03	57°03′N	34°58′E
Toscana, hist. reg., Italy (tôs-kä´nä) see Tuscany	200-01	43°25′N	11°00′E
Toshkent, nat. cap., Uzb. see Tashkent	232-33	41°19′N	69°17′E
Tosno, Russia (tôs´nô)	192-93	59°33′N	30°52′E
Tostado, Arg. (tôs-tá´dō)	173	29°14′S	61°46′W
Totana, Spain (tô-tä-nä)	198-99	37°46′N	1°30′W
Tot'ma, Russia (tôt´mȧ)	186-87	59°58′N	42°45′E
Totoras, Arg. (tô-tō´räs)	173	32°36′S	61°10′W
Totoya, i., Fiji	279f	18°56′S	179°51′W
Tottori, Japan (tô´tô-rē)	245	35°30′N	134°14′E
Toubkal, Jebel, mtn., Mor.	258-59	31°05′N	7°55′W
Toûil, Oued, stm., Alg. (wĕd tōō-ēl´)	269b	35°33′N	2°36′E
Toul, Fr. (tōōl)	196-97	48°41′N	5°53′E

Feature (Pronunciation)	Page	Lat.	Long.
Toulnustouc, stm., Qc., Can.	138-39	49°35′N	68°25′W
Toulon, Fr. (tōō-lôN´)	196-97	43°08′N	5°56′E
Toulouse, Fr. (tōō-lōōz´)	196-97	43°36′N	1°27′E
Toungoo, Mya. (tō-ŏŋ-gōō´)	246-47	18°56′N	96°26′E
Tourane, Viet. see Da Nang	246-47	16°03′N	108°12′E
Tourcoing, Fr. (tōr-kwăN´)	196-97	50°43′N	3°09′E
Tours, Fr. (tōōr)	196-97	47°24′N	0°43′E
Toussidé, Pic, vol., Chad (pĭk tōō-sē-dā´)	258-59	21°02′N	16°28′E
Towner, N.D., U.S. (tou´nĕr)	114-15	48°21′N	100°24′W
Townsend, Mt., U.S. (toun´zĕnd)	112-13	46°19′N	111°31′W
Townshend Island, i., Austl.	277	22°15′S	150°30′E
Townsville, Austl. (tounz´vĭl)	277	19°16′S	146°48′E
Towson, Md., U.S. (tou´sŭn)	116-17	39°24′N	76°36′W
Towuti, Danau, lk., Indon. (tô-wōō´tē)	248-49	2°45′S	121°32′E
Toxkan, stm., China	226	41°07′N	80°12′E
Toyama, Japan (tō´yä-mä)	245	36°41′N	137°13′E
Toyohashi, Japan (tō´yô-hä´shē)	245	34°46′N	137°23′E
Tozeur, Tun. (tô-zûr´)	188-89	33°55′N	8°08′E
Trabzon, Tur. (trȧb´zōn)	227	40°60′N	39°44′E
Tracy, Mn., U.S. (trā´sē)	114-15	44°14′N	95°37′W
Trafalgar, Cabo, c., Spain (kȧ´bô-trä-fäl-gä´r)	198-99	36°11′N	6°02′W
Trail, B.C., Can. (trāl)	132-33	49°06′N	117°42′W
Trakiya, hist. reg., Eur. see Thrace	200-01	41°20′N	26°45′E
Tranås, Swe. (trän´ôs)	192-93	58°03′N	14°59′E
Trancas, Arg.	168-69	26°13′S	65°17′W
Trang, Thai.	246-47	7°33′N	99°36′E
Trangan, Pulau, i., Indon. (pōō-lou träŋ´gän)	224-25	6°35′S	134°20′E
Trani, Italy (trä´nē)	200-01	41°16′N	16°25′E
Transylvania, hist. reg., Rom. (trăn-sĭl-vā´nĭ-ȧ)	194-95	46°44′N	23°37′E
Transylvanian Alps, mts., Rom. (trăn-sĭl-vā´nĭ-ȧn ălps)	200-01	45°25′N	23°33′E
Trapani, Italy	200-01	38°01′N	12°31′E
Traralgon, Austl. (trá´răl-gŏn)	276	38°12′S	146°32′E
Traverse City, Mi., U.S. (trăv´ērs sĭ´tē)	116-17	44°45′N	85°37′W
Travnik, Bos. (träv´nēk)	200-01	44°14′N	17°40′E
Trebinje, Bos. (trå´bēn-yĕ)	200-01	42°43′N	18°23′E
Trebišov, Slvk. (trĕ´bē-shôf)	194-95	48°38′N	21°44′E
Trebizond, Tur. see Trabzon	227	40°60′N	39°44′E
Treinta y Tres, Ur. (trå-ĕn´tä ē träs´)	173	33°14′S	54°23′W
Trelew, Arg. (trē´lū)	171	43°15′S	65°18′W
Trelleborg, Swe.	192-93	55°23′N	13°11′E
Tremblant, Mont, mtn., Qc., Can.	136-37	46°16′N	74°35′W
Trenčín, Slvk.	194-95	48°54′N	18°04′E
Trenque Lauquen, Arg. (trĕn´kĕ-lȧ´ô-kĕ´n)	173	35°58′S	62°45′W
Trent, Italy see Trento	200-01	46°04′N	11°08′E
Trent, stm., On., Can. (trĕnt)	136-37	44°06′N	77°34′W
Trento, Italy (trĕn´tô)	200-01	46°04′N	11°08′E
Trenton, N.S., Can. (trĕn´tŭn)	138-39	45°37′N	62°38′W
Trenton, On., Can. (trĕn´tŭn)	136-37	44°06′N	77°35′W
Trenton, Mo., U.S. (trĕn´tŭn)	120-21	40°05′N	93°37′W
Trenton, N.J., U.S.	116-17	40°13′N	74°45′W
Trenton, Tn., U.S. (trĕn´tŭn)	124-25	35°59′N	88°57′W
Tres Arroyos, Arg. (träs´är-rō´yōs)	173	38°22′S	60°16′W
Três Corações, Braz. (trĕ´s kō-rä-zô´ĕs)	172	21°42′S	45°15′W
Tres Esquinas, Col.	164-65	0°44′N	75°14′W
Três Lagoas, Braz. (trĕ´s lä-gô´ás)	168-69	20°47′S	51°43′W
Tres Marías, Islas, is., Mex.	146-47	21°32′N	106°32′W
Três Marias, Represa de, res., Braz.	172	18°14′S	45°16′W
Tres Picos, Cerro, mtn., Arg.	173	38°09′S	61°57′W
Tres Puntas, Cabo, c., Arg.	171	47°06′S	65°53′W
Três Rios, Braz. (trĕ´s rĕ´ōs)	172	22°07′S	43°12′W
Treviglio, Italy (trā-vē´lyô)	200-01	45°32′N	9°36′E
Treviso, Italy (trĕ-vē´sō)	200-01	45°40′N	12°14′E
Trichardt, S. Afr. (trĭ-kärt´)	269c	26°30′S	29°14′E
Trichinopoly, India see Tiruchchirāppalli	236	10°49′N	78°42′E
Trichūr, India see Thrissur	236	10°31′N	76°13′E
Trieste, Italy (trē-ĕs´tä)	200-01	45°40′N	13°46′E
Triglav, mtn., Slvn.	200-01	46°23′N	13°50′E
Trikora, Puncak, mtn., Indon.	277	4°18′S	138°40′E
Trincomalee, Sri L. (trĭŋ-kô-mȧ-lē´)	236	8°34′N	81°14′E
Trinidad, Bol. (trē-nē-dhädh´)	166-67	14°49′S	64°54′W
Trinidad, Col.	164-65	5°25′N	71°40′W
Trinidad, Cuba (trē-nē-dhädh´)	142-43	21°48′N	79°59′W
Trinidad, Co., U.S. (trĭn´ĭdăd)	120-21	37°10′N	104°30′W
Trinidad, Ur.	173	33°32′S	56°54′W
Trinidad, i., Trin. (trĭn´ĭ-dăd)	143b	10°30′N	61°15′W
Trinidad, Isla, i., Arg.	173	39°10′S	61°57′W
Trinidad and Tobago, nation, N.A. (trĭn´ĭ-dăd ănd tô-bā´gō)	140-41	11°0′N	61°00′W
Trinity, Tx., U.S. (trĭn´ĭ-tē)	122-23	30°56′N	95°23′W
Trinity, stm., Ca., U.S. (trĭn´ĭ-tē)	112-13	41°11′N	123°42′W

Feature (Pronunciation)	Page	Lat.	Long.
Trinity, stm., Tx., U.S. (trĭn´ĭ-tē)	122-23	29°46'N	94°41'W
Trinity Bay, b., Nf., Can. (trĭn´ĭ-tē bā)	138-39	47°59'N	53°31'W
Trinity Islands, is., Ak., U.S. (trĭn´ĭ-tē ī´lǎndz)	126	56°32'N	154°24'W
Tripoli, Leb.	228-29	34°26'N	35°51'E
Tripoli, nat. cap., Libya (trĭ´pō-lē)	258-59	32°52'N	13°10'E
Tripolitania, hist. reg., Libya	258-59	31°0'N	15°00'E
Tripura, state, India	234-35	24°0'N	92°00'E
Tristan da Cunha, i., St. Hel.	284-85	37°05'S	12°17'W
Trivandrum, India see Thiruvananthapuram	236	8°31'N	76°57'E
Trnava, Slvk. (t'r´nȧ-vȧ)	194-95	48°23'N	17°36'E
Trobriand Islands, is., Pap. N. Gui. (trō-brē-ǎnd´ ī´lǎndz) see Kiriwina Islands	277	8°35'S	151°05'E
Trogir, Cro. (trô´gēr)	200-01	43°31'N	16°15'E
Trois-Rivières, Qc., Can. (trwä´rē-vyâ´)	136-37	46°21'N	72°33'W
Troitsk, Russia (trô´ĕtsk)	226	54°06'N	61°32'E
Troitsko-Pechorsk, Russia (trô´ĭtsk-ô-pyĕ-chôrsk´)	186-87	62°42'N	56°12'E
Trollhättan, Swe. (trôl´hĕt-ĕn)	192-93	58°17'N	12°18'E
Trombetas, stm., Braz.	166-67	1°54'S	55°37'W
Tromsø, Nor.	184-85	69°39'N	18°57'E
Tronador, Cerro, mtn., S.A. (sĕ´r-rô trō-nä´dôr) see Tronador, Monte	171	41°10'S	71°54'W
Tronador, Monte, mtn., S.A.	171	41°10'S	71°54'W
Troncoso, Mex. (trôn-kô´sō)	146-47	22°44'N	102°22'W
Trondheim, Nor. (trôn´hâm)	184-85	63°26'N	10°24'E
Trosa, Swe. (trô´sä)	192-93	58°54'N	17°34'E
Trout Lake, lk., N.T., Can. (trout lāk)	130-31	60°35'N	121°10'W
Trout Lake, lk., On., Can. (trout lāk)	134-35	51°13'N	93°20'W
Troy, Al., U.S. (troi)	124-25	31°49'N	85°58'W
Troy, Mi., U.S. (troi)	116-17	42°37'N	83°09'W
Troy, Mo., U.S. (troi)	120-21	38°59'N	90°59'W
Troy, Mt., U.S. (troi)	112-13	48°29'N	115°55'W
Troy, N.C., U.S. (troi)	124-25	35°21'N	79°54'W
Troy, N.Y., U.S.	116-17	42°44'N	73°41'W
Troy, Oh., U.S. (troi)	116-17	40°02'N	84°13'W
Troyes, Fr. (trwä)	196-97	48°18'N	4°04'E
Trst, Italy see Trieste	200-01	45°40'N	13°46'E
Trstenik, Serb. (t'r´stĕ-nĕk)	200-01	43°37'N	20°60'E
Trubchevsk, Russia (tròp´chĕfsk)	202-03	52°35'N	33°48'E
Trucial States, nation, Asia see United Arab Emirates	206-07	24°0'N	54°00'E
Truckee, Ca., U.S. (trŭk´ē)	118-19	39°19'N	120°11'W
Trujillo, Col. (trò-ĸē´l-yō)	163c	4°10'N	76°18'W
Trujillo, Hond.	149	15°55'N	85°58'W
Trujillo, Peru (trò-ĸē´l-yō)	170	8°06'S	79°02'W
Trujillo, Spain (trōō-ĸē´l-yô)	198-99	39°27'N	5°53'W
Trujillo, Ven. (trò-ĸē´l-yō)	164-65	9°22'N	70°26'W
Truk Islands, is., Micron. see Chuuk	280-81	7°16'N	151°44'E
Trumann, Ar., U.S. (trōō´măn)	124-25	35°41'N	90°31'W
Truro, N.S., Can. (trōō´rō)	138-39	45°22'N	63°16'W
Truro, Eng., U.K. (trōō´rō)	190-91	50°16'N	5°03'W
Truth or Consequences, N.M., U.S. (trōōth ŏr kŏn´sē-kwĕn-sĭs)	118-19	33°08'N	107°15'W
Trutnov, Czech Rep. (tròt´nôf)	194-95	50°34'N	15°55'E
Trzcianka, Pol. (tchyän´kȧ)	194-95	53°02'N	16°28'E
Tsaidam Basin, bas., China (tsī-däm bā´s'n) see Qaidam Pendi	222-23	37°0'N	95°00'E
Tsaratanana, Madag.	264-65	16°46'S	47°37'E
Tsaritsyn, Russia see Volgograd	186-87	48°44'N	44°25'E
Tsavo East National Park, n.p., Kenya	262-63	2°31'S	38°44'E
Tsavo West National Park, n.p., Kenya	267	3°04'S	38°05'E
Tschenstochau, Pol. see Częstochowa	194-95	50°49'N	19°08'E
Tselinograd, nat. cap., Kaz. see Astana	226	51°12'N	71°27'E
Tsetserleg, Mong.	240-41	47°28'N	101°27'E
Tsévié, Togo	260-61	6°25'N	1°13'E
Tshabong, Bots.	264-65	26°02'S	22°25'E
Tshangalele, Lac, res., D.R.C.	262-63	10°55'S	27°03'E
Tshela, D.R.C. (tshā´lȧ)	262-63	4°59'S	12°57'E
Tshikapa, D.R.C. (tshē-kä´pä)	262-63	6°24'S	20°48'E
Tshikapa, stm., Afr. see Chicapa	262-63	6°25'S	20°48'E
Tshuapa, stm., D.R.C.	262-63	0°17'S	20°45'E
Tshumbe, stm., Afr. see Chiumbe	262-63	6°59'S	21°11'E
Tshwane, nat. cap., S. Afr. see Pretoria	269c	25°45'S	28°11'E
Tsimlyanskoye Vodokhranilishche, res., Russia	186-87	48°0'N	43°00'E
Tsinan, China see Jinan	240-41	36°40'N	116°59'E
Tsinghai, state, China see Qinghai	240-41	36°0'N	96°00'E
Tsingtao, China see Qingdao	240-41	36°05'N	120°20'E
Tsiombe, Madag.	264-65	25°19'S	45°29'E
Tsiribihina, stm., Madag. (tsē´rē-bē-hē-nä´)	264-65	19°50'S	44°27'E
Tsiroanomandidy, Madag.	264-65	18°47'S	46°03'E
Tskhinvali, Geor.	227	42°12'N	43°59'E
Tsöh Gol, stm., Asia see Chikoy	240-41	51°02'N	106°38'E
Tsu, Japan (tsōō)	245	34°43'N	136°31'E
Tsuchiura, Japan (tsōō´chē-ōō-rä)	245	36°05'N	140°12'E
Tsugaru-kaikyō, strt., Japan	244	41°35'N	141°00'E
Tsumeb, Nmb. (tsōō´mĕb)	264-65	19°15'S	17°43'E
Tsuruga, Japan (tsōō´rò-gä)	245	35°39'N	136°04'E
Tsurugi-san, mtn., Japan (tsōō´rò-gē- sän´)	245	33°51'N	134°06'E
Tsuruoka, Japan (tsōō´rò-ō´kä)	244	38°43'N	139°50'E
Tsushima, is., Japan (tsōō-shē´mä)	243	34°30'N	129°22'E
Tsuyama, Japan (tsōō´yä-mä´)	245	35°03'N	134°00'E
Tsymlyansk Reservoir, res., Russia (tsēm-lyänsk´ rĕ´sĕr-vwär) see Tsimlyanskoye Vodokhranilishche	186-87	48°0'N	43°00'E
Tual, Indon.	224-25	5°40'S	132°43'E
Tuamotu, Îles, is., Fr. Poly.	280-81	19°0'S	142°00'W
Tuapse, Russia (tò´áp-sĕ)	202-03	44°07'N	39°05'E
Tubarão, Braz. (tōō-bä-rouɴ´)	172	28°29'S	49°03'W
Tübingen, Ger. (tü´bĭng-ĕn)	194-95	48°31'N	9°04'E
Ţubruq, Libya see Tobruk	258-59	32°05'N	23°57'E
Tubuai, i., Fr. Poly.	280-81	23°18'S	149°30'W
Tubuaï, Îles, is., Fr. Poly.	280-81	23°0'S	150°00'W
Tucacas, Ven. (tōō-kä´käs)	163b	10°48'N	68°19'W
Tucano, Braz.	166-67	10°58'S	38°46'W
Tucson, Az., U.S. (tōō-sŏn´)	118-19	32°13'N	110°59'W
Tucumán, Arg. (tōō-kōō-män´) see San Miguel de Tucumán	168-69	26°49'S	65°13'W
Tucumán, state, Arg. (tōō-kōō-män´)	168-69	27°0'S	65°30'W
Tucumcari, N.M., U.S. (tò´kŭm-kâr-ē)	120-21	35°10'N	103°44'W
Tucupita, Ven. (tōō-kōō-pē´tä)	143b	9°03'N	62°02'W
Tucuruí, Braz.	166-67	3°43'S	49°43'W
Tucuruí, Represa de, res., Braz.	166-67	4°32'S	49°31'W
Tudela, Spain (tōō-dhā´lä)	198-99	42°04'N	1°36'W
Tudmur, Syria	228-29	34°33'N	38°17'E
Tugela, stm., S. Afr. (tōō-gel´ȧ)	264-65	29°13'S	31°30'E
Tuguegarao City, Phil.	250	17°37'N	121°43'E
Tuira, stm., Pan. (tōō-ē´rȧ)	150	8°21'N	78°04'W
Tukangbesi, Kepulauan, is., Indon.	248-49	5°40'S	123°50'E
Tukarak Island, i., Nu., Can.	130-31	56°18'N	78°45'W
Tūkrah, Libya	188-89	32°32'N	20°34'E
Tuktoyaktuk, N.T., Can.	128-29	69°26'N	133°01'W
Tuktut Nogait National Park, n.p., N.T., Can.	128-29	69°0'N	122°00'W
Tukums, Lat. (tò´kòms)	192-93	56°58'N	23°10'E
Tukuyu, Tan. (tōō-kōō´yȧ)	264-65	9°13'S	33°38'E
Tula, Mex. (tōō´lä)	146-47	23°00'N	99°42'W
Tula, Russia (tōō´lȧ)	202-03	54°12'N	37°37'E
Tulach Mhór, Ire. see Tullamore	190-91	53°16'N	7°30'W
Tulaghi, Sol. Is.	279e	9°06'S	160°09'E
Tulancingo, Mex. (tōō-län-sĭŋ´gō)	146-47	20°05'N	98°23'W
Tulangbawang, stm., Indon.	246-47	4°24'S	105°51'E
Tulare, Ca., U.S. (tōō-lä´rȧ) (tul-âr´)	118-19	36°12'N	119°20'W
Tulcán, Ec. (tōōl-kän´)	170	0°48'N	77°43'W
Tulcea, Rom. (tòl´chȧ)	202-03	45°11'N	28°48'E
Tulemalu Lake, lk., Nu., Can.	130-31	62°58'N	99°25'W
Tulia, Tx., U.S. (tōō´lĭ-ȧ)	120-21	34°33'N	101°46'W
Tulita, N.T., Can.	128-29	64°54'N	125°34'W
Ţūlkarm, W.B. (tōōl kärm)	228-29	32°19'N	35°02'E
Tullahoma, Tn., U.S. (tŭl-ȧ-hō´mȧ)	124-25	35°22'N	86°12'W
Tullamore, Ire. (tŭl-ȧ-mōr´)	190-91	53°16'N	7°30'W
Tulle, Fr. (tül)	196-97	45°16'N	1°46'E
Tuloma, stm., Russia	184-85	69°03'N	33°03'E
Tulsa, Ok., U.S. (tŭl´sȧ)	120-21	36°09'N	95°59'W
Tuluá, Col.	163c	4°05'N	76°12'W
Tuluksak, Ak., U.S.	126	61°06'N	160°58'W
Tulum, Mex. (tōō-lò´m)	148	20°14'N	87°27'W
Tulun, Russia (tò-lōōn´)	218-19	54°34'N	100°34'E
Tuma, stm., Nic. (tōō´mä)	149	13°03'N	84°44'W
Tumaco, Col.	164-65	1°47'N	78°48'W
Tuman-gang, stm., Asia	243	42°17'N	130°41'E
Tumannaya, stm., Asia see Tumen	243	42°17'N	130°41'E
Tumbes, Peru (tōō´m-bĕs)	170	3°34'S	80°27'W
Tumbler Ridge, B.C., Can.	132-33	55°08'N	120°57'W
Tumen, China (tōō-mŭn)	243	42°57'N	129°51'E
Tumen, stm., Asia (tōō-mŭn)	243	42°17'N	130°41'E
Tumeremo, Ven. (tōō-mȧ-rä´mō)	164-65	7°19'N	61°28'W
Tumkūr, India	236	13°20'N	77°06'E
Tumuc-Humac, Massif des, mts., S.A. see Tumuc-Humac Mountains	164-65	2°19'N	54°35'W
Tumuc-Humac Mountains, mts., S.A. (tōō-mòk´ōō-mäk´ moun´tĭnz)	164-65	2°19'N	54°35'W
Tumucumaque, Serra, mts., S.A. see Tumuc-Humac Mountains	164-65	2°19'N	54°35'W
Tumut, Austl.	276	35°18'S	148°13'E
Tunas de Zaza, Cuba (tōō´näs dĕ zä´zä)	142-43	21°38'N	79°33'W
Tunbridge Wells, Eng., U.K. (tŭn´brĭj welz´) see Royal Tunbridge Wells	190-91	51°08'N	0°16'E
Tungabhadra, stm., India	236	15°58'N	78°15'E
T'ungchou, China see Tongxian	240-41	39°54'N	116°39'E
Tung Hai, s., Asia see East China Sea	222-23	30°0'N	126°00'E
Tungsha Tao, i., Tai. see Pratas Island	222-23	20°42'N	116°43'E
Tunis, nation, Afr. see Tunisia	253	34°0'N	9°00'E
Tunis, nat. cap., Tun. (tū´nĭs)	184-85	36°48'N	10°11'E
Tunis, Golfe de, b., Tun. (gôlf-dĕ-tū´nĭs)	258-59	37°0'N	10°30'E
Tunis, Gulf of, b., Tun. (gŭlf ŭv tū´nĭs) see Tunis, Golfe de	258-59	37°0'N	10°30'E
Tunisia, nation, Afr. (tu-nĭzh´ē-ȧ)	253	34°0'N	9°00'E
Tunisie, nation, Afr. see Tunisia	253	34°0'N	9°00'E
Tunja, Col. (tōō´n-hȧ)	164-65	5°31'N	73°22'W
Tununak, Ak., U.S.	126	60°35'N	165°16'W
Tunuyán, Arg.	163e	33°33'S	69°00'W
Tunuyán, stm., Arg.	171	33°55'S	66°49'W
Tunxi, China see Huangshan	238-39	29°45'N	118°18'E
Tuo, stm., China	238-39	28°53'N	105°26'E
Tupã, Braz.	168-69	21°56'S	50°31'W
Tupaciguara, Braz.	172	18°36'S	48°41'W
Tupanciretã, Braz.	173	29°05'S	53°50'W
Tupelo, Ms., U.S. (tū´pê-lō)	124-25	34°15'N	88°42'W
Tupinambarana, Ilha, i., Braz.	166-67	3°0'S	58°00'W
Tupiza, Bol. (tōō-pē´zä)	168-69	21°25'S	65°44'W
Tupper Lake, N.Y., U.S. (tŭp´ēr lāk)	116-17	44°14'N	74°28'W
Tupungato, Cerro, mtn., S.A.	163e	33°25'S	69°45'W
Túquerres, Col. (tōō-kĕ´r-rĕs)	164-65	1°04'N	77°36'W
Tura, India	234-35	25°31'N	90°13'E
Tura, Russia (tòr´ȧ)	218-19	64°18'N	100°14'E
Tura, stm., Russia	218-19	57°13'N	66°57'E
Turbat, Pak.	232-33	25°60'N	63°04'E
Turbo, Col. (tōō´bō)	164-65	8°06'N	76°43'W
Turda, Rom. (tòr´dȧ)	194-95	46°34'N	23°47'E
Turfan Depression, depr., China	222-23	42°40'N	89°10'E
Turgay, stm., Kaz. see Torghay	226	48°02'N	62°34'E
Turia, stm., Spain (tōō´ryä)	198-99	39°27'N	0°19'W
Turin, Italy (tōō´rĭn)	200-01	45°03'N	7°41'E
Turka, Ukr. (tòr´kȧ)	194-95	49°09'N	23°03'E
Turkana, Lake, lk., Afr. (lāk tōōr-känȧ) see Rudolf, Lake	267	3°30'N	36°00'E
Turkey, nation, Asia (tûrk´ē)	206-07	39°0'N	35°00'E
Turkey, stm., Ia., U.S. (tûrk´ē)	114-15	42°41'N	90°56'W
Türkistan, Kaz.	226	43°18'N	68°16'E
Türkıye, nation, Asia see Turkey	206-07	39°0'N	35°00'E
Türkmenabat, Turkmen.	232-33	39°05'N	63°35'E
Türkmenbaşy, Turkmen.	232-33	40°01'N	52°58'E
Turkmenia, nation, Asia see Turkmenistan	206-07	40°0'N	60°00'E
Turkmenistan, nation, Asia (tûrk-mĕn´-ĭ-stän´)	206-07	40°0'N	60°00'E
Turkmeniya, nation, Asia see Turkmenistan	206-07	40°0'N	60°00'E
Turks and Caicos Islands, dep., N.A. (tûrks ănd kä´kōs ī´lǎndz) (tûrks ănd kī´kōs ī´lǎndz)	140-41	21°45'N	71°35'W
Turks Island Passage, strt., T./C. Is. (tûrks ī´lǎnd păs´ĭj)	142-43	21°25'N	71°18'W
Turks Islands, is., T./C. Is. (tûrks ī´lǎndz)	142-43	21°24'N	71°07'W
Turku, Fin. (tòrkò)	192-93	60°27'N	22°16'E
Turkwel, stm., Kenya	267	3°06'N	36°06'E
Turlock, Ca., U.S. (tûr´lŏk)	118-19	37°30'N	120°51'W
Turneffe Islands, is., Belize	148	17°22'N	87°51'W
Turnhout, Bel. (tŭrn-hout´)	190-91	51°19'N	4°57'E
Turnov, Czech Rep. (tòr´nôf)	194-95	50°35'N	15°09'E
Tŭrnovo, Blg. see Veliko Tŭrnovo	200-01	43°04'N	25°38'E
Turpan Pendi, depr., China see Turfan Depression	222-23	42°40'N	89°10'E
Turquino, Pico, mtn., Cuba (pē´kō tōōr-kē´nō)	142-43	19°59'N	76°50'W
Tursunzade, Taj.	232-33	38°31'N	68°14'E
Turukhansk, Russia (tōō-rōō-känsk´)	218-19	65°49'N	87°55'E
Tuscaloosa, Al., U.S. (tŭs-kȧ-lōō´sȧ)	124-25	33°12'N	87°34'W
Tuscany, hist. reg., Italy	200-01	43°25'N	11°00'E
Tuscola, Il., U.S. (tŭs-kō-lȧ)	116-17	39°48'N	88°17'W
Tuscumbia, Al., U.S. (tŭs-kŭm´bĭ-ȧ)	124-25	34°43'N	87°42'W
Tuskegee, Al., U.S. (tŭs-kē´gē)	124-25	32°26'N	85°42'W
Tutayev, Russia (tōō-tȧ-yĕf´)	202-03	57°53'N	39°34'E
Tuticorin, India (tōō-tē-kô-rĭn´)	236	8°48'N	78°09'E
Tutóia, Braz. (tōō-tō´yȧ)	166-67	2°48'S	42°17'W
Tutuila, i., Am. Sam.	279b	14°18'S	170°42'W
Tutupaca, Volcán, vol., Peru	170	17°01'S	70°22'W

Feature (Pronunciation)	Page	Lat.	Long.
Tuul, stm., Mong.	240-41	48°56'N	104°48'E
Tuvalu, nation, Oc. (to͞o-vä'-lo͞o)	280-81	8°0'S	178°00'E
Tuvuca, i., Fiji	279f	17°40'S	178°48'W
Ṭuwayq, Jabal, mts., Sau. Ar.	220-21	23°0'N	46°00'E
Tuxpan, Mex. (to͞os'pän)	146-47	21°56'N	105°17'W
Tuxpan de Rodríguez Cano, Mex.	146-47	20°58'N	97°24'W
Tuxtepec, Mex. (tòs-tå-pĕk')	146-47	18°05'N	96°07'W
Tuxtla Gutiérrez, Mex. (tòs'tlä go͞o-tyår'rĕs)	146-47	16°45'N	93°06'W
Tuyen Quang, Viet.	246-47	21°50'N	105°11'E
Tuy Hoa, Viet.	246-47	13°05'N	109°19'E
Tuymazy, Russia	186-87	54°36'N	53°43'E
Tuz Gölü, lk., Tur.	186-87	38°45'N	33°25'E
Tuzla, Bos. (tòz'lȧ)	200-01	44°33'N	18°40'E
Tver', Russia	202-03	56°52'N	35°55'E
Tvertsa, stm., Russia (tvĕr'tsȧ)	202-03	56°52'N	35°56'E
Tweed, stm., U.K. (twēd)	190-91	55°46'N	1°60'W
Tweeling, S. Afr. (twē'lĭng)	269c	27°39'S	28°30'E
Twin Falls, Id., U.S. (twĭn fôlz)	112-13	42°34'N	114°28'W
Two Rivers, Wi., U.S. (to͞o rĭv'ẽrz)	116-17	44°09'N	87°34'W
Tyan' Shan', mts., Asia *see* Tien Shan	226	42°0'N	80°00'E
Tyler, Mn., U.S. (tī'lẽr)	114-15	44°17'N	96°08'W
Tyler, Tx., U.S. (tī'lẽr)	120-21	32°21'N	95°19'W
Tylertown, Ms., U.S. (tī'lẽr-toun)	124-25	31°07'N	90°09'W
Tym, stm., Russia	218-19	59°26'N	80°01'E
Tymovskoye, Russia	218-19	50°51'N	142°38'E
Tynda, Russia	218-19	55°09'N	124°43'E
Tyndall, S.D., U.S. (tĭn'dȧl)	114-15	42°60'N	97°52'W
Tyrma, Russia	240-41	50°03'N	132°10'E
Tyrrhenian Sea, s., Eur. (tĭr-rē'nĭ-*ă*n sē)	200-01	40°0'N	12°00'E
Tyrrhénienne, Mer, s., Eur. *see* Tyrrhenian Sea	200-01	40°0'N	12°00'E
Tyul'gan, Russia	186-87	52°24'N	56°14'E
Tyumen', Russia (tyo͞o-mĕn')	218-19	57°10'N	65°33'E
Tyung, stm., Russia	218-19	63°46'N	121°32'E
Tyva, state, Russia	222-23	52°0'N	95°00'E
Tzaneen, S. Afr.	269c	23°49'S	30°10'E
Tzeliutsing, China *see* Zigong	238-39	29°22'N	104°45'E
Tzucacab, Mex. (tzo͞o-kä-kä'b)	148	20°04'N	89°02'W
Tzupo, China *see* Boshan	240-41	36°29'N	117°51'E

U

Feature (Pronunciation)	Page	Lat.	Long.
Uatumã, stm., Braz.	166-67	2°24'S	57°33'W
Uaupés, stm., S.A.	164-65	0°02'N	67°15'W
Ubá, Braz.	172	21°07'S	42°56'W
Ubangi, stm., Afr. (o͞o-bäŋ'gē)	262-63	0°25'S	17°47'E
Ubatuba, Braz. (o͞o-bȧ-to͞o'bȧ)	172	23°26'S	45°04'W
Ube, Japan	245	33°57'N	131°15'E
Úbeda, Spain (o͞o'bȧ-dä)	198-99	38°01'N	3°22'W
Uberaba, Braz. (o͞o-bå-rä'b*ȧ*)	172	19°46'S	47°56'W
Uberlândia, Braz. (o͞o-bĕr-lá'n-dyä)	172	18°54'S	48°15'W
Ubon Ratchathani, Thai. (o͞o'bŭn rä'chätä-nē)	246-47	15°14'N	104°52'E
Ubrique, Spain (o͞o-brē'kå)	198-99	36°41'N	5°27'W
Ubsu-Nur, Ozero, lk., Asia *see* Uvs Lake	222-23	50°20'N	92°45'E
Ubundu, D.R.C.	262-63	0°21'S	25°25'E
Ucayali, stm., Peru (o͞o'kä-yä'lē)	170	4°30'S	73°30'W
Uchaly, Russia (û-chä'lĭ)	226	54°18'N	59°27'E
Uchiura-wan, b., Japan (o͞o'chē-o͞o'rä wän)	244	42°20'N	140°40'E
Uchiza, Peru	170	8°25'S	76°25'W
Uchur, stm., Russia (o-chòr')	218-19	58°47'N	130°36'E
Uda, stm., Russia (o'dȧ)	222-23	51°49'N	107°34'E
Uda, stm., Russia (o'dȧ)	218-19	54°43'N	135°18'E
Uda, stm., Russia (o'dȧ)	218-19	56°02'N	99°38'E
Udachnyy, Russia	218-19	66°29'N	112°15'E
Udagamandalam, India	236	11°25'N	76°42'E
Udai, stm., Ukr. (o'dȧ)	202-03	50°04'N	33°07'E
Udaipur, India (o-dŭ'ē-po͞or)	234-35	24°35'N	73°42'E
Uddevalla, Swe. (o͞od'dĕ-väl-ȧ)	192-93	58°21'N	11°55'E
Udine, Italy (o͞o'dē-nå)	200-01	46°04'N	13°15'E
Udmurtia, state, Russia	186-87	57°0'N	53°00'E
Udmurtiya, state, Russia *see* Udmurtia	186-87	57°0'N	53°00'E
Udon Thani, Thai.	246-47	17°24'N	102°47'E
Ueda, Japan (wä'dä)	245	36°24'N	138°15'E
Uele, stm., D.R.C. (wä'lå)	262-63	4°07'N	22°26'E
Uelen, Russia	126	66°09'N	169°48'W
Uelzen, Ger. (ült'sĕn)	194-95	52°58'N	10°34'E
Uere, stm., D.R.C.	262-63	3°33'N	25°15'E
Ufa, Russia (o'fa)	186-87	54°42'N	55°58'E
Ufa, stm., Russia (o'fa)	186-87	54°41'N	56°02'E
Ugab, stm., Nmb. (o͞o'gäb)	264-65	21°11'S	13°38'E
Ugalla, stm., Tan. (o͞o-gä'lä)	267	5°17'S	30°58'E
Uganda, nation, Afr. (o͞o-gän'dä) (û-găn'd*ȧ*)	253	1°0'N	32°00'E
Uglegorsk, Russia (o͞o-glĕ-gôrsk)	218-19	49°04'N	142°03'E
Uglich, Russia (o͞og-lĕch')	202-03	57°32'N	38°20'E
Ugoma, mtn., D.R.C.	267	4°0'S	28°45'E
Uhrichsville, Oh., U.S. (ū'rĭks-vĭl)	116-17	40°24'N	81°21'W
Uíge, Ang.	262-63	7°38'S	15°04'E
Uina, stm., Camrn. *see* Vina	260-61	7°52'N	15°46'E
Uitenhage, S. Afr.	264-65	33°40'S	25°27'E
Uji, Japan (o͞o'jē)	245	34°54'N	135°49'E
Ujiji, Tan. (o͞o-jē'jē)	267	4°55'S	29°41'E
Uji-yamada, Japan *see* Ise	245	34°30'N	136°42'E
Ujjain, India (o͞o-jŭēn)	234-35	23°11'N	75°47'E
Ujungpandang, Indon. *see* Makassar	248-49	5°08'S	119°25'E
Újvidék, Serb. *see* Novi Sad	200-01	45°15'N	19°50'E
Ukara Island, i., Tan.	267	1°50'S	33°03'E
Ukerewe Island, i., Tan.	267	2°03'S	33°00'E
Ukhta, Russia (o͞ok'tȧ)	186-87	63°34'N	53°44'E
Ukiah, Ca., U.S.	118-19	39°09'N	123°12'W
Ukiah, Or., U.S. (ū-kī'ȧ)	112-13	45°09'N	118°56'W
Ukkusiksalik National Park, n.p., Nu., Can.	128-29	66°0'N	90°00'W
Ukmergė, Lith. (òk'mĕr-ghå)	192-93	55°15'N	24°46'E
Ukraïna, nation, Eur. *see* Ukraine	174-75	49°0'N	32°00'E
Ukraine, nation, Eur. (yo͞o-krān')	174-75	49°0'N	32°00'E
Ukyr, Russia	240-41	49°28'N	108°52'E
Ulaanbaatar, nat. cap., Mong. (o͞o'län-bä'tôr)	240-41	47°55'N	106°56'E
Ulaangom, Mong.	240-41	49°59'N	92°04'E
Ulaan-Uul, Mong.	240-41	44°23'N	111°12'E
Ulan Bator, nat. cap., Mong. *see* Ulaanbaatar	240-41	47°55'N	106°56'E
Ulanhad, China *see* Chifeng	240-41	42°16'N	118°58'E
Ulanhot, China	240-41	46°04'N	122°04'E
Ulan-Ude, Russia (o͞o'län o͞o'dȧ)	222-23	51°50'N	107°36'E
Ulawa Island, i., Sol. Is.	279e	9°47'S	161°57'E
Ulchin, Kor., S. (o͞ol'chĕn')	243	36°59'N	129°23'E
Uldz, stm., Asia	240-41	49°55'N	115°33'E
Uldza, stm., Asia *see* Uldz	240-41	49°55'N	115°33'E
Uleåborg, Fin. *see* Oulu	184-85	65°01'N	25°28'E
Ulhāsnagar, India	236	19°14'N	73°08'E
Uliast, Mong.	240-41	48°57'N	91°09'E
Uliastay, Mong.	240-41	47°44'N	96°51'E
Ulindi, stm., D.R.C. (o͞o-lĭn'dē)	262-63	1°40'S	25°52'E
Ulithi, at., Micron.	280-81	9°55'N	139°42'E
Ülkenözen, stm., Eur.	186-87	48°60'N	49°59'E
Ulm, Ger. (òlm)	194-95	48°24'N	9°59'E
Ulónguè, Moz.	264-65	14°37'S	34°19'E
Ulricehamn, Swe. (òl-rē'sĕ-häm)	192-93	57°48'N	13°25'E
Ulsan, Kor., S. (o͞ol'sän')	243	35°33'N	129°19'E
Ulúa, stm., Hond. (o͞o-lo͞o'ȧ)	149	15°52'N	87°44'W
Ulul, i., Micron.	280-81	8°36'N	149°40'E
Ulungur, stm., China (o͞o-lo͞on-gûr)	222-23	46°59'N	87°26'E
Ulungur Hu, lk., China	222-23	47°13'N	87°16'E
Uluru, mtn., Austl.	272-73	25°20'S	130°60'E
Ulverstone, Austl. (ŭl'vẽr-st*ŭ*n)	276	41°10'S	146°11'E
Ulyanovsk, Russia (o͞o-lyä'nôfsk)	186-87	54°19'N	48°22'E
Ulysses, Ks., U.S. (ū-lĭs'ēz)	120-21	37°35'N	101°21'W
Umán, Mex. (o͞o-män')	148	20°53'N	89°45'W
Uman', Ukr. (o-màn')	202-03	48°44'N	30°14'E
'Umān, nation, Asia *see* Oman	206-07	22°0'N	58°00'E
'Umān, Khalīj, b., Asia *see* Oman, Gulf of	230-31	24°30'N	58°30'E
Umarkot, Pak.	232-33	25°22'N	69°45'E
Umatilla Indian Reservation, ind. res., Or., U.S. (ū-mȧ-tĭl'ȧ ĭn'dĭ-*ă*n rĕ-sẽr-vā'shĕn)	112-13	45°41'N	118°31'W
Umba, Russia	186-87	66°41'N	34°18'E
Umboi Island, i., Pap. N. Gui.	277	5°36'S	147°53'E
Umeå, Swe.	184-85	63°50'N	20°16'E
Umeälven, stm., Swe.	184-85	63°47'N	20°19'E
Umm Durmān, Sudan *see* Omdurman	266	15°39'N	32°29'E
Umm Lajj, Sau. Ar.	266	25°07'N	37°16'E
Umm Ruwābah, Sudan	266	12°54'N	31°12'E
Umm Urūmah, i., Sau. Ar.	266	25°46'N	36°33'E
Umnak Island, i., Ak., U.S. (o͞om'nȧ ī'l*ă*nd)	126a	53°25'N	168°10'W
Umpqua, stm., Or., U.S. (ŭmp'kwȧ)	112-13	43°42'N	124°05'W
'Umrān, Yemen	266	15°58'N	43°58'E
Umtata, S. Afr. (òm-tä'tä)	264-65	31°35'S	28°47'E
Umuarama, Braz.	168-69	23°46'S	53°19'W
Unalakleet, Ak., U.S. (ū-nȧ-lăk'lēt)	126	63°53'N	160°47'W
Unalaska, Ak., U.S. (ū-nȧ-lăs'kȧ)	126a	53°52'N	166°32'W
Unalaska Island, i., Ak., U.S.	126a	53°45'N	166°45'W
Unauna, Pulau, i., Indon.	248-49	0°10'S	121°35'E
'Unayzah, Sau. Ar.	266	26°05'N	43°59'E
Uncia, Bol. (o͞on'sē-ä)	168-69	18°26'S	66°35'W
Uncompahgre Peak, mtn., Co., U.S. (ŭn-kŭm-pä'grē pēk)	118-19	38°04'N	107°28'W
Unecha, Russia (o-nĕ'chȧ)	202-03	52°51'N	32°42'E
Ungava, Baie d', b., Can. *see* Ungava Bay	130-31	59°30'N	67°30'W
Ungava, Péninsule d', pen., Qc., Can.	130-31	60°0'N	74°00'W
Ungava Bay, b., Can. (ŭŋ-gä'vȧ bā)	130-31	59°30'N	67°30'W
Ungava Peninsula, pen., Qc., Can. (ŭŋ-gä'vȧ pĕ-nĭn'sūl*ă*) *see* Ungava, Péninsule d'	130-31	60°0'N	74°00'W
Ungvár, Ukr. *see* Uzhhorod	194-95	48°37'N	22°19'E
União, Braz.	166-67	4°35'S	42°52'W
União dos Palmares, Braz.	163d	9°10'S	36°02'W
Unimak Island, i., Ak., U.S. (o͞o-nē-mák' ī'l*ă*nd)	126	54°43'N	164°27'W
Unini, stm., Braz.	166-67	1°41'S	61°31'W
Union, Mo., U.S. (ūn'yŭn)	120-21	38°27'N	91°01'W
Union, S.C., U.S. (ūn'yŭn)	124-25	34°43'N	81°37'W
Union City, In., U.S. (ūn'yŭn sĭ'tē)	116-17	40°11'N	85°00'W
Union City, Mi., U.S. (ūn'yŭn sĭ'tē)	116-17	42°04'N	85°09'W
Union City, Pa., U.S. (ūn'yŭn sĭ'tē)	116-17	41°54'N	79°50'W
Union City, Tn., U.S. (ūn'yŭn sĭ'tē)	124-25	36°25'N	89°03'W
Union Springs, Al., U.S. (ūn'yŭn sprĭngz)	124-25	32°09'N	85°43'W
Uniontown, Al., U.S. (ūn'y*ŭ*n-toun)	124-25	32°27'N	87°31'W
Uniontown, Pa., U.S.	116-17	39°54'N	79°44'W
Unionville, Mo., U.S. (ūn'y*ŭ*n-vĭl)	120-21	40°29'N	93°01'W
United Arab Emirates, nation, Asia (ū-nī'tĕd ăr'*ă*b ĕ'mĕr-*ĕ*ts)	206-07	24°0'N	54°00'E
United Arab Republic, nation, Afr. *see* Egypt	253	27°0'N	30°00'E
United Kingdom, nation, Eur. (ū-nī'tĕd kĭng'd*ŭ*m)	174-75	54°0'N	2°00'W
United States, nation, N.A. (ū-nī'tĕd stāts)	85	38°0'N	97°00'W
Unity, Sk., Can.	134-35	52°27'N	109°07'W
Upa, stm., Russia (o'pȧ)	202-03	54°02'N	36°21'E
Upata, Ven. (o͞o-pä'tä)	164-65	8°01'N	62°24'W
Upemba, Lac, lk., D.R.C.	262-63	8°36'S	26°26'E
Upington, S. Afr. (ŭp'ĭng-t*ŭ*n)	264-65	28°27'S	21°14'E
Upland, In., U.S. (ŭp'l*ă*nd)	116-17	40°28'N	85°29'W
Upolu, i., Samoa	279b	13°55'S	171°45'W
Upolu Point, c., Hi., U.S. (o͞o-pô'lo͞o point)	127a	20°16'N	155°51'W
Upper Arrow Lake, lk., B.C., Can. (ŭp'ẽr ăr'ō lāk)	132-33	50°31'N	117°56'W
Upper Kapuas Mountains, mts., Asia	248-49	1°15'N	113°30'E
Upper Klamath Lake, lk., Or., U.S. (ŭp'ẽr klăm'*ă*th lāk)	112-13	42°24'N	121°54'W
Upper Red Lake, lk., Mn., U.S. (ŭp'ẽr rĕd lāk)	114-15	48°10'N	94°40'W
Upper Sandusky, Oh., U.S. (ŭp'ẽr săn-dŭs'kē)	116-17	40°49'N	83°17'W
Upper Volta, nation, Afr. *see* Burkina Faso	253	13°0'N	1°30'W
Uppsala, Swe. (o͞op'sȧ-lä)	192-93	59°52'N	17°38'E
Upsala, Swe. *see* Uppsala	192-93	59°52'N	17°38'E
Ural, stm., (o-räl'') (ū-rôl)	176-77	46°50'N	51°33'E
Ural Mountains, mts., Russia (o-räl'' moun'tĭnz) (ū-rôl moun'tĭnz)	218-19	60°0'N	60°00'E
Ural'skiye Gory, mts., Russia *see* Ural Mountains	218-19	60°0'N	60°00'E
Ura-Tyube, Taj. *see* Istaravshan	232-33	39°54'N	69°00'E
Urbana, Il., U.S. (ûr-băn'ȧ)	116-17	40°07'N	88°13'W
Urbana, Oh., U.S. (ûr-băn'ȧ)	116-17	40°07'N	83°45'W
Urbino, Italy (o͞or-bē'nô)	200-01	43°44'N	12°39'E
Urdinarrain, Arg. (o͞or-dē-när-rȧē'n)	173	32°41'S	58°54'W
Urfa, Tur. *see* Şanlıurfa	228-29	37°10'N	38°48'E
Urganch, Uzb.	232-33	41°33'N	60°38'E
Urgench, Uzb. *see* Urganch	232-33	41°33'N	60°38'E
Urla, Tur. (òr'lȧ)	200-01	38°20'N	26°46'E
Urmi, stm., Russia (òr'mē)	240-41	48°36'N	135°01'E
Urmia, Iran *see* Orūmīyeh	228-29	37°32'N	45°05'E
Urmia, Lake, lk., Iran (lāk òr'mē*ă*) *see* Orūmīyeh, Daryācheh-ye	227	37°40'N	45°30'E
Urrao, Col. (o͞or-rá'ô)	163c	6°20'N	76°08'W
Uruapan del Progreso, Mex.	146-47	19°25'N	102°04'W
Urubamba, stm., Peru (o͞o-ro͞o-bäm'bä)	170	10°44'S	73°44'W
Uruguai, stm., S.A.	173	34°10'S	58°18'W
Uruguaiana, Braz.	173	29°46'S	57°04'W

ăt; fin*ă*l; rāte; senåte; ärm; ȧsk; sof*ȧ*; fâre; ch-choose; dh-as th in other; bē; ėvent; bĕt; recĕnt; cratẽr; g-gō; gh-guttural g; bĭt; ĭ-short neutral; rīde; ĸ-guttural k as ch in German ich;

Feature (Pronunciation)	Page	Lat.	Long.
Uruguay, nation, S.A. (o͞o-ro͞o-gwī´) (ū´ro͞o-gwā)	158	33°0'S	56°00'W
Uruguay, stm., S.A. (o͞o-ro͞o-gwī´)	173	34°10'S	58°18'W
Urumchi, China *see* Ürümqi	222-23	43°48'N	87°35'E
Ürümqi, China (ů-rům-chyē)	222-23	43°48'N	87°35'E
Urundi, nation, Afr. *see* Burundi	253	3°15'S	30°00'E
Urup, Ostrov, i., Russia (ôs-trôf´ o͝o´ro͝op´)	218-19	46°0'N	150°00'E
Uryupinsk, Russia (o͝or´yo͝o-pēn-sk´)	186-87	50°48'N	42°01'E
Urzhum, Russia	186-87	57°07'N	50°01'E
Urziceni, Rom. (o͝o-zē-chĕn´')	202-03	44°43'N	26°40'E
Usa, stm., Russia (o͞o´sȧ)	186-87	65°58'N	56°57'E
Uşak, Tur. (o͞o´shȧk)	200-01	38°41'N	29°24'E
Usborne, Mount, mtn., Falk. Is.	171	51°41'S	58°50'W
Üshtöbe, Kaz.	226	45°14'N	77°58'E
Ushuaia, Arg. (o͞o-sho͞o-ī´ä)	171	54°47'S	68°19'W
Usinsk, Russia	186-87	65°57'N	57°24'E
Üsküb, nat. cap., Mac. *see* Skopje	200-01	42°00'N	21°28'E
Usman', Russia (o͞os-mȧn´)	202-03	52°03'N	39°45'E
Uspanapa, stm., Mex. (o͞os-pä-nä´pä)	146-47	17°56'N	94°28'W
Ussel, Fr. (üs´ĕl)	196-97	45°33'N	2°18'E
Ussuri, stm., Asia (o͞o-so͞o´rē)	240-41	48°27'N	135°04'E
Ussuriysk, Russia	244	43°48'N	131°59'E
Ust'-Barguzin, Russia	218-19	53°25'N	109°02'E
Ust-Bolsheretsk, Russia	218-19	52°49'N	156°17'E
Ust'-Ilimsk, Russia	218-19	58°00'N	102°40'E
Ústí nad Labem, Czech Rep.	194-95	50°40'N	14°02'E
Ustinov, Russia *see* Izhevsk	186-87	56°50'N	53°12'E
Üstirt, plat., Asia *see* Ust-Urt Plateau	226	43°0'N	56°00'E
Ust-Kamchatsk, Russia	218-19	56°13'N	162°29'E
Ust'-Kut, Russia	218-19	56°47'N	105°43'E
Ust'-Kuyga, Russia	218-19	69°60'N	135°35'E
Ust-Maya, Russia (o͝ost mä´yȧ)	218-19	60°25'N	134°30'E
Ust'-Nera, Russia	218-19	64°34'N	143°18'E
Ust'-Omchug, Russia	218-19	61°08'N	149°37'E
Ust-Tsilma, Russia (o͝ost tsĭl´mȧ)	186-87	65°25'N	52°05'E
Ust-Urt Plateau, plat., Asia (o͞ost-o͞ort plä-tō´)	226	43°0'N	56°00'E
Ustyurt Platosi, plat., Asia *see* Ust-Urt Plateau	226	43°0'N	56°00'E
Ustyuzhna, Russia (yo͞ozh´nȧ)	202-03	58°51'N	36°28'E
Usu, China (ů-so͞o)	226	44°26'N	84°41'E
Usulután, El Sal. (o͞o-so͞o-lä-tän´)	148	13°20'N	88°26'W
Usumacinta, stm., N.A. (o͞o´so͞o-mä-sēn´tō)	148	18°23'N	92°39'W
Usumbura, nat. cap., Bdi. *see* Bujumbura	267	3°23'S	29°22'E
Utah, state, U.S. (ū´tô)	108-09	39°30'N	111°30'W
Utah Lake, lk., Ut., U.S. (ū´tô lāk)	118-19	40°13'N	111°49'W
Utembo, stm., Ang.	264-65	17°04'S	21°58'E
Utena, Lith. (o͞o´tā-nä)	192-93	55°30'N	25°37'E
Uthai Thani, Thai.	246-47	15°23'N	100°02'E
Utiariti, Braz.	166-67	13°02'S	58°17'W
Utica, N.Y., U.S.	116-17	43°06'N	75°15'W
Utiel, Spain (o͞o-tyål´)	198-99	39°34'N	1°13'W
Utila, Isla de, i., Hond. (ē´s-lä-dĕ-o͞o-tē´lä)	149	16°06'N	86°56'W
Uto, Japan (o͞o´tō´)	245	32°41'N	130°40'E
Utrecht, Neth. (ü´trĕkt) (ū´trĕkt)	190-91	52°05'N	5°08'E
Utrecht, S. Afr.	269c	27°40'S	30°19'E
Utrera, Spain (o͞o-trā´rä)	198-99	37°11'N	5°47'W
Utsunomiya, Japan (o͞ot´so͝o-nō-mē-yȧ´)	245	36°33'N	139°54'E
Uttaradit, Thai.	246-47	17°38'N	100°06'E
Uttarakhand, state, India	234-35	30°0'N	79°30'E
Uttaranchal, state, India *see* Uttarakhand	234-35	30°0'N	79°30'E
Uttar Pradesh, state, India (o͝ot-tär-prä-dĕsh)	234-35	27°0'N	80°00'E
Uummannarsuaq, c., Green. *see* Farewell, Cape	86	59°46'N	43°60'W
Uvá, stm., Col.	164-65	3°56'N	68°34'W
Uvalde, Tx., U.S. (ū-văl´dē)	122-23	29°13'N	99°47'W
Uvira, D.R.C. (o͞o-vē´rä)	267	3°23'S	29°09'E
Uvs Lake, lk., Asia	222-23	50°20'N	92°45'E
Uvs nuur, lk., Asia *see* Uvs Lake	222-23	50°20'N	92°45'E
Uwajima, Japan (o͞o-wä´jē-mä)	245	33°13'N	132°34'E
Uwayl, Sudan	262-63	8°46'N	27°24'E
'Uwaynāt, Jabal al-, mtn., Afr.	258-59	21°53'N	25°02'E
Uxmal, hist., Mex. (o͞o´x-mä´l)	148	20°22'N	89°46'W
Uyuni, Bol. (o͞o-yo͞o´nē)	168-69	20°28'S	66°50'W
Uyuni, Salar de, pl., Bol. (sä-lär-dĕ o͞o-yo͞o´nē)	168-69	20°17'S	68°07'W
Uzbekistan, nation, Asia (o͞oz-bĕk´-ē--stän´)	232-33	41°0'N	64°00'E
Ŭzbekiston, nation, Asia *see* Uzbekistan	232-33	41°0'N	64°00'E
Uzh, stm., Ukr. (o͝ozh)	202-03	51°15'N	30°15'E
Uzhhorod, Ukr.	194-95	48°37'N	22°19'E
Užice, Serb. (o͞o´zhĕ-tsĕ)	200-01	43°52'N	19°51'E
Uzlovaya, Russia	202-03	53°59'N	38°11'E

V

Feature (Pronunciation)	Page	Lat.	Long.
Vaal, stm., S. Afr. (väl)	264-65	29°04'S	23°38'E
Vaasa, Fin. (vä´sȧ)	184-85	63°06'N	21°37'E
Vache, Île à, i., Haiti	142-43	18°04'N	73°38'W
Vadodara, India	234-35	22°18'N	73°11'E
Vaduz, nat. cap., Liech. (vä´do͝ots)	194-95	47°09'N	9°32'E
Vaga, stm., Russia (va´gȧ)	186-87	62°49'N	42°53'E
Vágar, i., Far. Is.	190b	62°05'N	7°17'W
Vaghena Island, i., Sol. Is.	279e	7°26'S	157°46'E
Váh, stm., Slvk. (väκ)	194-95	47°45'N	18°09'E
Vaitupu, i., Tuvalu	280-81	7°28'S	178°41'E
Vakh, stm., Russia (vaκ)	218-19	60°49'N	76°48'E
Vākhān, hist. reg., Afg.	232-33	37°0'N	73°00'E
Vakhsh, stm., Taj.	232-33	37°07'N	68°19'E
Valcheta, Arg.	171	40°42'S	66°09'W
Valdai Hills, hills, Russia (vál-dī´ hĭlz) *see* Valdayskaya Vozvyshennost'	202-03	57°0'N	33°30'E
Valday, Russia (vål-dī´)	202-03	57°59'N	33°15'E
Valdayskaya Vozvyshennost', hills, Russia *see* Valdai Hills	202-03	57°0'N	33°30'E
Valdepeñas, Spain (väl-då-pān´yäs)	198-99	38°46'N	3°23'W
Valdés, Península, pen., Arg. (pĕ-nē´n-so͞o-lä vál-dĕ´s)	171	42°30'S	64°00'W
Valdez, Ak., U.S. (văl´dēz)	126	61°08'N	146°20'W
Valdivia, Chile (väl-dē´vä)	171	39°49'S	73°13'W
Val-d'Or, Qc., Can.	136-37	48°06'N	77°46'W
Valdosta, Ga., U.S. (văl-dŏs´tȧ)	124-25	30°50'N	83°16'W
Vale, Or., U.S. (vāl)	112-13	43°59'N	117°15'W
Valença, Braz. (vä-lĕn´sȧ)	166-67	13°22'S	39°05'W
Valença, Braz. (vä-lĕn´sȧ)	172	22°15'S	43°42'W
Valence, Fr.	196-97	44°56'N	4°54'E
València, Spain	198-99	39°28'N	0°22'W
Valencia, Spain *see* València	198-99	39°28'N	0°22'W
Valencia, Ven. (vä-lĕn´syä)	163b	10°11'N	68°00'W
Valenciennes, Fr. (vá-läN-syĕn´)	196-97	50°21'N	3°31'E
Valentine, Ne., U.S. (vá lăN-tē-nyē´)	114-15	42°53'N	100°33'W
Valera, Ven. (vä-lĕ´rä)	164-65	9°19'N	70°37'W
Valga, Est. (väl´gȧ)	192-93	57°47'N	26°03'E
Valjevo, Serb. (väl´yȧ-vô)	200-01	44°16'N	19°54'E
Valladolid, Mex. (väl-yä-dhō-lēdh´)	148	20°41'N	88°12'W
Valladolid, Spain (väl-yä-dhō-lēdh´)	198-99	41°39'N	4°43'W
Valle de Guanape, Ven. (vä´l-yĕ-dĕ-gwä-nä´pĕ)	163b	9°54'N	65°41'W
Valle de la Pascua, Ven. (väl´yā dā lä-pä´s-ko͞oä)	163b	9°13'N	66°00'W
Valle de Santiago, Mex. (väl´yā dā sän-tē-ä´gō)	146-47	20°24'N	101°12'W
Valledupar, Col. (vä´l-yĕ-do͞o-pär´)	164-65	10°28'N	73°15'W
Vallegrande, Bol. (väl´yā grän´dā)	168-69	18°29'S	64°06'W
Vallenar, Chile (väl-yå-när´)	168-69	28°34'S	70°46'W
Valletta, nat. cap., Malta (väl-lĕt´ä)	200b	35°54'N	14°31'E
Valley City, N.D., U.S. (văl´ē sĭ´tĭ)	114-15	46°56'N	98°00'W
Valleyview, Ab., Can.	132-33	55°05'N	117°17'W
Vallimanca, Arroyo, stm., Arg. (är-rō´yō väl-yē-mä´n-kä)	173	35°44'S	60°08'W
Valls, Spain (väls)	198-99	41°17'N	1°15'E
Valmiera, Lat. (väl´myĕ-rȧ)	192-93	57°32'N	25°26'E
Valognes, Fr. (vá-lôn´y')	196-97	49°31'N	1°28'W
Valona, Alb. *see* Vlorë	200-01	40°29'N	19°30'E
Vālpārai, India	236	10°19'N	76°54'E
Valparaíso, Chile (väl´pä-rä-ē´sō)	163e	33°03'S	71°37'W
Valparaíso, Mex.	146-47	22°46'N	103°35'W
Valparaiso, Fl., U.S. (văl-pȧ-rā´zō)	124-25	30°31'N	86°30'W
Valparaiso, In., U.S. (văl-pȧ-rā´zō)	116-17	41°28'N	87°03'W
Valréas, Fr. (vál-rā-ä´)	196-97	44°23'N	4°59'E
Vals, Tanjung, c., Indon.	277	8°24'S	137°38'E
Valuyki, Russia (vá-lo͞o-ē´kē)	202-03	50°12'N	38°08'E
Valverde del Camino, Spain (väl-vĕr-dĕ-dĕl-kä-mē´nō)	198-99	37°34'N	6°45'W
Van, Tur. (vŭn)	227	38°30'N	43°24'E
Van, Lake, lk., Tur. *see* Van Gölü	227	38°33'N	42°46'E
Vanadzor, Arm.	227	40°48'N	44°29'E
Vanavara, Russia	218-19	60°21'N	102°16'E
Van Buren, Ar., U.S. (văn bū´rĕn)	120-21	35°27'N	94°22'W
Van Buren, Me., U.S. (văn bū´rĕn)	117a	47°10'N	67°57'W
Vanceburg, Ky., U.S. (văns´bûrg)	116-17	38°36'N	83°19'W
Vancouver, B.C., Can. (văn-ko͞o´vẽr)	132-33	49°17'N	123°07'W
Vancouver, Wa., U.S. (văn-ko͞o´vẽr)	112-13	45°38'N	122°39'W
Vancouver Island, i., B.C., Can. (văn-ko͞o´vẽr ī´lȧnd)	130-31	49°45'N	126°00'W
Vancouver Island Ranges, mts., B.C., Can. (văn-ko͞o´vẽr ī´lȧnd rānjĕz)	132-33	49°25'N	125°25'W
Vandalia, Il., U.S. (văn-dā´lĭ-ȧ)	116-17	38°58'N	89°06'W
Vandalia, Mo., U.S. (văn-dā´lĭ-ȧ)	120-21	39°19'N	91°30'W
Vandalia, Oh., U.S. (văn-dā´lĭ-ȧ)	116-17	39°54'N	84°12'W
Vanderbijlpark, S. Afr.	269c	26°42'S	27°50'E
Vanderhoof, B.C., Can.	132-33	54°01'N	124°06'W
Vanderlin Island, i., Austl.	272-73	15°44'S	137°02'E
Van Diemen Gulf, b., Austl. (văn dē´mĕn gŭlf)	272-73	11°50'S	132°00'E
Vanegas, Mex. (vä-nĕ´gäs)	146-47	23°53'N	100°56'W
Vänern, lk., Swe.	192-93	58°55'N	13°30'E
Vänersborg, Swe. (vĕ´nĕrs-bôr´)	192-93	58°23'N	12°19'E
Vangaindrano, Madag.	264-65	23°21'S	47°36'E
Van Gölü, lk., Tur.	227	38°33'N	42°46'E
Vanikolo, i., Sol. Is.	272-73	11°37'S	166°52'E
Vanimo, Pap. N. Gui.	277	2°44'S	141°20'E
Vankarem, Russia	126	67°50'N	175°50'W
Van Lear, Ky., U.S. (văn lēr´)	116-17	37°47'N	82°48'W
Vannes, Fr. (vȧn)	196-97	47°39'N	2°46'W
Van Rees, Pegunungan, mts., Indon.	277	2°35'S	138°15'E
Vanua Balavu, i., Fiji	279f	17°14'S	178°57'W
Vanua Lava, i., Vanuatu	279g	13°45'S	167°28'E
Vanua Levu, i., Fiji	279f	16°33'S	179°15'E
Vanuatu, nation, Oc. (vä-no͞o-ä´-to͞o)	279g	16°0'S	167°00'E
Van Wert, Oh., U.S. (văn wûrt´)	116-17	40°52'N	84°35'W
Vārānasi, India	234-35	25°20'N	82°59'E
Varaždin, Cro. (vä´räzh´dĕn)	200-01	46°18'N	16°20'E
Varberg, Swe. (vär´bĕrg)	192-93	57°06'N	12°16'E
Vardar, stm., Eur. (vär´där) *see* Axiós	200-01	40°31'N	22°43'E
Vardø, Nor.	184-85	70°21'N	31°01'E
Varėna, Lith. (vä-rā´nȧ)	194-95	54°13'N	24°35'E
Vareš, Bos. (vä´rĕsh)	200-01	44°09'N	18°19'E
Varese, Italy (vä-rā´sȧ)	200-01	45°49'N	8°50'E
Varginha, Braz. (vär-zhē´n-yä)	172	21°34'S	45°26'W
Varkaus, Fin. (vär´kous)	184-85	62°19'N	27°54'E
Varna, Blg. (vär´nȧ)	200-01	43°13'N	27°54'E
Värnamo, Swe. (vĕr´nȧ-mô)	192-93	57°11'N	14°03'E
Vasa, Fin. *see* Vaasa	184-85	63°06'N	21°37'E
Vaslui, Rom. (vás-lo͞o´ē)	202-03	46°38'N	27°45'E
Vassar, Mi., U.S. (văs´ẽr)	116-17	43°22'N	83°35'W
Västerås, Swe. (vĕs´tĕr-ôs)	192-93	59°37'N	16°33'E
Västervik, Swe. (vĕs´tĕr-vēk)	192-93	57°45'N	16°39'E
Vasto, Italy (väs´tô)	200-01	42°07'N	14°43'E
Vasyugan, stm., Russia (vás-yo͞o-gán´)	218-19	59°07'N	80°46'E
Vasyugan'ye, sw., Russia	218-19	58°0'N	77°00'E
Vatican City, nation, Eur. (văt´ĭkăn sĭ´tē)	200-01	41°54'N	12°27'E
Vaticano, Città del, nation, Eur. *see* Vatican City	200-01	41°54'N	12°27'E
Vatnajökull, ice, Ice. (vät´nȧ-yû-kôl)	190a	64°25'N	16°50'W
Vatra Dornei, Rom. (vät´rȧ dôr´nå´)	194-95	47°21'N	25°22'E
Vättern, lk., Swe.	192-93	58°24'N	14°36'E
Vatu-i-ra Channel, strt., Fiji	279f	17°17'S	178°31'E
Vaughn, N.M., U.S.	120-21	34°36'N	105°15'W
Vaupés, stm., S.A. (vá´o͞o-pĕ´s)	164-65	0°02'N	67°15'W
Växjö, Swe. (vĕks´shû)	192-93	56°53'N	14°49'E
Vaygach, Ostrov, i., Russia (ôs-trôf´ vī-gách´)	186-87	70°0'N	59°30'E
Vedea, stm., Rom. (vå´dyȧ)	200-01	43°43'N	25°33'E
Vedia, Arg. (vĕ´dyä)	173	34°30'S	61°33'W
Vega, i., Nor.	184-85	65°39'N	11°50'E
Vegreville, Ab., Can.	132-33	53°30'N	112°03'W
Vejle, Den. (vī´lĕ)	192-93	55°42'N	9°32'E
Velebit, mts., Cro. (vā´lĕ-bĭt)	200-01	44°38'N	15°03'E
Vélez-Málaga, Spain (vā´lāth-mä´lä-gä)	198-99	36°47'N	4°06'W
Velhas, stm., Braz.	172	17°13'S	44°49'W
Velikaya, stm., Russia (vå-lē´kȧ-yȧ)	192-93	57°52'N	28°09'E
Velikaya, stm., Russia (vå-lē´kȧ-yȧ)	218-19	64°35'N	176°12'E
Veliki Bečkerek, Serb. *see* Zhovti Vody	200-01	45°23'N	20°24'E
Velikiye Luki, Russia (vyē-lē´-kyê lo͞o´ke)	192-93	56°20'N	30°33'E
Velikiy Ustyug, Russia (vå-lē´kĭ o͞os-tyôg´)	186-87	60°45'N	46°19'E
Veliko Tŭrnovo, Blg.	200-01	43°04'N	25°38'E
Velizh, Russia (vå´lēzh)	192-93	55°36'N	31°12'E
Vella Lavella, i., Sol. Is.	279e	7°45'S	156°40'E
Velletri, Italy (vĕl-lā´trē)	200-01	41°41'N	12°46'E
Vellore, India (vĕl-lōr´)	236	12°55'N	79°08'E
Vel'sk, Russia (vĕlsk)	186-87	61°04'N	42°06'E
Venadillo, Col. (vĕ-nä-dē´l-yō)	163c	4°43'N	74°55'W

Feature (Pronunciation)	Page	Lat.	Long.
Venado Tuerto, Arg. (vě-nä′dô-tōōě′r-tô)	173	33°45′S	61°58′W
Vendôme, Fr. (väɴ-dōm′)	196-97	47°47′N	1°04′E
Venëv, Russia (vĕn-ěf′)	202-03	54°21′N	38°16′E
Venezia, Italy *see* Venice	200-01	45°26′N	12°20′E
Venezuela, nation, S.A. (vĕn-ê-zwē′l*à*)	158	8°0′N	66°00′W
Venezuela, Golfo de, b., S.A. (gôl-fô-dě vĕn-ê-zwē′l*à*)	164-65	11°30′N	71°00′W
Venezuela, Gulf of, b., S.A. (gŭlf ŭv vĕn-ê-zwē′l*à*) *see* Venezuela, Golfo de	164-65	11°30′N	71°00′W
Venice, Italy (vĕn′ĭs)	200-01	45°26′N	12°20′E
Venice, Fl., U.S. (vĕn′ĭs)	125a	27°06′N	82°27′W
Venta, stm., Eur. (vĕn′tȧ)	192-93	57°24′N	21°34′E
Ventersburg, S. Afr. (vĕn-tĕrs′bûrg)	269c	28°06′S	27°09′E
Ventersdorp, S. Afr. (vĕn-tĕrs′dôrp)	269c	26°19′S	26°51′E
Ventimiglia, Italy (vĕn-tê-mēl′yä)	200-01	43°48′N	7°36′E
Ventspils, Lat. (vĕnt′spēls)	192-93	57°24′N	21°35′E
Ventuari, stm., Ven. (vĕn-tōōȧ′rē)	164-65	3°58′N	67°03′W
Ventura, Ca., U.S. (vĕn-tōō′r*à*)	118-19	34°17′N	119°17′W
Venustiano Carranza, Mex. (vě-nōōs-tyä′nō-kär-rä′n-zä)	146-47	16°22′N	92°34′W
Venustiano Carranza, Mex. (vě-nōōs-tyä′nō-kär-rä′n-zä)	146-47	19°45′N	103°46′W
Vera, Arg. (vě-rä)	173	29°28′S	60°13′W
Vera, Spain (vā′rä)	198-99	37°15′N	1°52′W
Veracruz, Mex.	146-47	19°12′N	96°08′W
Veracruz, state, Mex. (vā-rä-krōōz′)	146-47	19°20′N	96°40′W
Verāval, India (vĕr′vŭ-väl)	234-35	20°55′N	70°22′E
Vercelli, Italy (vĕr-chĕl′lê)	200-01	45°20′N	8°25′E
Verde, stm., Braz.	168-69	21°12′S	51°53′W
Verde, Cape, c., Sen. *see* Vert, Cap	260-61	14°44′N	17°30′W
Verden, Ger. (fĕr′dĕn)	194-95	52°55′N	9°14′E
Verdun-sur-Meuse, Fr.	196-97	49°10′N	5°23′E
Vereeniging, S. Afr. (vĕ-rā′nĭ-gĭng)	269c	26°40′S	27°57′E
Vereshchagino, Russia	186-87	58°05′N	54°39′E
Vereya, Russia (vĕ-râ′yȧ)	202-03	55°21′N	36°12′E
Verín, Spain (vå-rēn′)	198-99	41°57′N	7°26′W
Verkhnetulomskoye Vodokhranilishche, res., Russia	186-87	68°30′N	31°05′E
Verkhneudinsk, Russia *see* Ulan-Ude	222-23	51°50′N	107°36′E
Verkhniy Baskunchak, Russia	186-87	48°14′N	46°43′E
Verkhnyaya Inta, Russia	186-87	65°59′N	60°20′E
Verkhoyansk, Russia (vyĕr-ĸô-yänsk′)	218-19	67°32′N	133°25′E
Verkhoyanskiy Khrebet, mts., Russia	218-19	67°0′N	129°00′E
Verkhoyansk Mountains, mts., Russia (vyĕr-ĸô-yänsk′ moun′tĭnz) *see* Verkhoyanskiy Khrebet	218-19	67°0′N	129°00′E
Vermilion, Ab., Can. (vẽr-mĭl′y*ŭ*n)	132-33	53°21′N	110°51′W
Vermilion, stm., Ab., Can. (vẽr-mĭl′y*ŭ*n)	132-33	53°39′N	110°20′W
Vermilion Bay, b., La., U.S. (vẽr-mĭl′y*ŭ*n bā)	124-25	29°40′N	92°00′W
Vermilion Lake, lk., Mn., U.S. (vẽr-mĭl′y*ŭ*n lāk)	114-15	47°53′N	92°25′W
Vermillion, S.D., U.S. (vẽr-mĭl′y*ŭ*n)	114-15	42°47′N	96°56′W
Vermillion, stm., S.D., U.S. (vẽr-mĭl′y*ŭ*n)	114-15	42°44′N	96°53′W
Vermont, state, U.S. (vẽr-mŏnt′)	108-09	43°50′N	72°45′W
Vernal, Ut., U.S. (vûr′n*ă*l)	112-13	40°28′N	109°33′W
Vernon, B.C., Can. (vẽr-nôɴ′)	132-33	50°16′N	119°16′W
Vernon, Tx., U.S. (vûr′n*ŭ*n)	120-21	34°10′N	99°17′W
Vero Beach, Fl., U.S. (vē′rō bēch)	125a	27°38′N	80°24′W
Verona, Italy (vā-rō′nä)	200-01	45°27′N	10°60′E
Versailles, Fr. (vĕr-sī′y′)	196-97	48°48′N	2°08′E
Versailles, In., U.S. (vẽr-sālz′)	116-17	39°04′N	85°15′W
Versailles, Ky., U.S. (vẽr-sālz′)	116-17	38°03′N	84°44′W
Versec, Serb. *see* Vršac	200-01	45°07′N	21°18′E
Vershino-Shakhtaminskiy, Russia	240-41	51°18′N	117°52′E
Vert, Cap, c., Sen.	260-61	14°44′N	17°30′W
Verviers, Bel. (vĕr-vyā′)	190-91	50°35′N	5°52′E
Vesoul, Fr. (vē-sōōl′)	196-97	47°37′N	6°10′E
Vesterålen, is., Nor. (vĕs′tẽr ô′lĕn)	184-85	68°40′N	15°33′E
Vesterhavet, s., Eur. *see* North Sea	184-85	56°0′N	3°00′E
Vestfjorden, b., Nor.	184-85	68°08′N	15°00′E
Vestmannaeyjar, Ice. (vĕst′män-ä-ā′yär)	190a	63°26′N	20°17′W
Vestvågøya, i., Nor.	184-85	68°13′N	13°42′E
Vesuvio, vol., Italy (vĕ-sōō′vyä) *see* Vesuvius	200-01	40°49′N	14°26′E
Vesuvius, vol., Italy (vĕ-sōō′vy-*ŭ*s)	200-01	40°49′N	14°26′E
Ves'yegonsk, Russia (vĕ-syĕ-gônsk′)	202-03	58°40′N	37°16′E
Veszprém, Hung. (vĕs′prām)	194-95	47°06′N	17°54′E
Vet, stm., S. Afr. (vĕt)	264-65	27°41′S	25°39′E
Vetlanda, Swe. (vĕt-län′dä)	192-93	57°26′N	15°05′E
Vetluga, Russia (vyĕt-lōō′gȧ)	186-87	57°51′N	45°47′E
Vevay, In., U.S. (vē′vā)	116-17	38°45′N	85°04′W
Viacha, Bol. (vēȧ′chȧ)	168-69	16°39′S	68°18′W
Vian, Ok., U.S. (vī′*ă*n)	120-21	35°30′N	94°58′W
Viana, Braz. (vē-ä′nä)	166-67	3°13′S	45°01′W
Viana do Castelo, Port. (vē-ä′n*à* dò käs-tā′lò)	198-99	41°42′N	8°50′W
Viangchan, nat. cap., Laos *see* Vientiane	246-47	17°57′N	102°37′E
Viareggio, Italy (vê-ä-rĕd′jô)	200-01	43°53′N	10°15′E
Viborg, Den. (vē′bôr)	192-93	56°27′N	9°25′E
Viborg, Russia	192-93	60°43′N	28°46′E
Vibo Valentia, Italy (vē′bô-vä-lĕ′n-tyä)	200-01	38°40′N	16°06′E
Vicebsk, Bela.	192-93	55°12′N	30°12′E
Vicente Guerrero, Presa, res., Mex.	146-47	23°57′N	98°46′W
Vicenza, Italy (vê-chĕnt′sä)	200-01	45°33′N	11°33′E
Vichada, stm., Col.	164-65	4°56′N	67°50′W
Vichadero, Ur.	173	31°48′S	54°42′W
Vichuga, Russia (vē-chōō′gȧ)	186-87	57°13′N	41°55′E
Vichy, Fr. (vê-shē′)	196-97	46°08′N	3°26′E
Vicksburg, Ms., U.S. (vĭks′bûrg)	124-25	32°22′N	90°52′W
Viçosa, Braz. (vē-sô′sä)	163d	9°24′S	36°14′W
Viçosa, Braz. (vē-sô′sä)	172	20°46′S	42°52′W
Victor Harbor, Austl.	276	35°33′S	138°37′E
Victoria, Arg. (vēk-tô′rēä)	173	32°36′S	60°09′W
Victoria, Braz. *see* Vitória	172	20°19′S	40°21′W
Victoria, B.C., Can.	132-33	48°26′N	123°22′W
Victoria, Chile (vēk-tô-rēä)	171	38°14′S	72°20′W
Victoria, China *see* Hong Kong	238-39	22°16′N	114°10′E
Victoria, Malay. *see* Labuan	248-49	5°17′N	115°15′E
Victoria, Tx., U.S. (vĭk-tō′rĭ-*à*)	122-23	28°48′N	96°60′W
Victoria, Va., U.S. (vĭk-tō′rĭ-*à*)	124-25	36°59′N	78°14′W
Victoria, state, Austl.	276	38°0′S	145°00′E
Victoria, stm., Austl.	272-73	15°07′S	129°40′E
Victoria, Lake, lk., Afr. (lāk vĭk-tō′rĭ-*à*)	267	1°0′S	33°00′E
Victoria, Mount, mtn., Pap. N. Gui.	277	8°54′S	147°32′E
Victoria Falls, wtfl., Afr. (vĭk-tō′rĭ-*à* fôlz)	264-65	17°55′S	25°51′E
Victoria Falls National Park, n.p., Zimb.	264-65	17°55′S	25°40′E
Victoria Island, i., Can. (vĭk-tō′rĭ-*à* ī′l*à*nd)	86	71°0′N	110°00′W
Victoria Lake, res., Nf., Can. (vĭk-tō′rĭ-*à* lāk)	138-39	48°19′N	57°28′W
Victoria Land, reg., Ant. (vĭk-tō′rĭ-*à* lănd)	287	75°0′S	163°00′E
Victoria Nile, stm., Ug. (vĭk-tō′rĭ-*à* nīl)	267	2°14′N	31°26′E
Victoria Peak, mtn., Belize (vĕk-tōrĭ′*à* pēk)	148	16°48′N	88°37′W
Victoria Peak, mtn., B.C., Can. (vĭk-tō′rĭ-*à* pēk)	132-33	50°03′N	126°06′W
Victoriaville, Qc., Can. (vĭk-tō′rĭ-*à*-vĭl)	136-37	46°03′N	71°57′W
Victoria West, S. Afr.	264-65	31°24′S	23°07′E
Vicuña Mackenna, Arg.	173	33°54′S	64°24′W
Vidalia, Ga., U.S. (vĭ-dā′lĭ-*à*)	124-25	32°13′N	82°25′W
Vidalia, La., U.S. (vĭ-dā′lĭ-*à*)	124-25	31°33′N	91°26′W
Vidin, Blg. (vĭ′dĕn)	200-01	43°60′N	22°53′E
Vidisha, India	234-35	23°31′N	77°49′E
Vidzy, Bela. (vê′dz*ĭ*)	192-93	55°25′N	26°38′E
Viedma, Arg. (vyād′mä)	171	40°49′S	62°60′W
Viedma, Lago, lk., Arg.	171	49°35′S	72°35′W
Vienna, Ga., U.S. (vê-ĕn′*à*)	124-25	32°05′N	83°47′W
Vienna, Il., U.S. (vê-ĕn′*à*)	116-17	37°25′N	88°54′W
Vienna, W.V., U.S. (vê-ĕn′*à*)	116-17	39°19′N	81°33′W
Vienna, nat. cap., Aus. (vê-ĕn′*à*)	194-95	48°13′N	16°20′E
Vienne, Fr. (vyĕn′)	196-97	45°32′N	4°52′E
Vientiane, nat. cap., Laos (vyĕn′tän)	246-47	17°57′N	102°37′E
Vieques, Isla de, i., P.R. (ê′s-lä-dĕ-vyā′kås)	142a	18°08′N	65°25′W
Vierfontein, S. Afr. (vēr′fŏn-tān)	269c	27°05′S	26°45′E
Vierzon, Fr. (vyâr-zôɴ′)	196-97	47°13′N	2°05′E
Viesca, Mex. (vê-ās′kä)	122-23	25°21′N	102°48′W
Vieste, Italy (vyĕs′tâ)	200-01	41°53′N	16°11′E
Viet Nam, nation, Asia (vyĕt′näm′)	206-07	16°0′N	108°00′E
Vigan, Phil. (vēgän)	250	17°35′N	120°23′E
Vigevano, Italy (vê-jâ-vä′nô)	200-01	45°19′N	8°52′E
Vigo, Spain (vē′gō)	198-99	42°14′N	8°43′W
Vihti, Fin. (vē′tĭ)	192-93	60°25′N	24°19′E
Vijayawāda, India	236	16°31′N	80°37′E
Vikna, i., Nor.	184-85	64°57′N	10°58′E
Vila Cabral, Moz. *see* Lichinga	264-65	13°17′S	35°15′E
Vila Coutinho, Moz. *see* Ulóngue	264-65	14°37′S	34°19′E
Vila do Conde, Port. (vē′lȧ dò kôn′d*ĕ*)	198-99	41°21′N	8°45′W
Vila Franca de Xira, Port. (vē′l*à*-fräŋ′kȧ dä shē′rȧ)	198-99	38°56′N	8°60′W
Vila Gouveia, Moz. *see* Cantandica	264-65	18°02′S	33°08′E
Vilanandro, Tanjona, c., Madag.	264-65	16°12′S	44°28′E
Vilāni, Lat. (vē′lá-nĭ)	192-93	56°34′N	26°57′E
Vilankulo, Moz.	264-65	21°60′S	35°19′E
Vila Nova de Gaia, Port. (vē′l*à* nō′vȧ dā gä′yä)	198-99	41°08′N	8°37′W
Vila Pery, Moz. *see* Chimoio	264-65	19°09′S	33°30′E
Vila Real, Port. (vē′l*à* rā-äl′)	198-99	41°18′N	7°45′W
Vila Velha, Braz.	172	20°20′S	40°17′W
Vila Viçosa, Port. (vē′l*à*-vê-sō′zȧ)	198-99	38°47′N	8°13′W
Vilejka, Bela.	192-93	54°30′N	26°55′E
Vileyka, Bela. (vê-lā′ê-kȧ) *see* Vilejka	192-93	54°30′N	26°55′E
Vilhelmina, Swe.	184-85	64°38′N	16°39′E
Vilhena, Braz.	166-67	12°43′S	60°07′W
Viljandi, Est. (vēl′yän-dê)	192-93	58°22′N	25°36′E
Vilkaviškis, Lith. (vêl-kȧ-vēsh′kēs)	192-93	54°39′N	23°02′E
Villa Ángela, Arg. (vē′l-yä á′n-ĸĕ-lä)	173	27°35′S	60°43′W
Villa Bella, Bol. (vē′l-yä-bĕ′l-yä)	166-67	10°26′S	65°24′W
Villacañas, Spain (vêl-yä-kän′yäs)	198-99	39°37′N	3°20′W
Villach, Aus. (fē′läĸ)	194-95	46°36′N	13°50′E
Villacidro, Italy (vē-lä-chē′drô)	200-01	39°28′N	8°45′E
Villa Constitución, Arg. (vêl′yä-kōn-stê-tōō-syōn′)	173	33°14′S	60°20′W
Villa de Cura, Ven. (vêl′yä-dĕ-kōō′rä)	163b	10°02′N	67°29′W
Villa Dolores, Arg. (vēl′yä dô-lō′rās)	168-69	31°56′S	65°11′W
Villa Flores, Mex. (vēl′yä-flō′rās)	146-47	16°14′N	93°14′W
Villa Grove, Il., U.S. (vĭl′*à* grōv′)	116-17	39°51′N	88°10′W
Villaguay, Arg. (vē′l-yä-gwī)	173	31°52′S	59°02′W
Villa Hayes, Para. (vēl′yä äyås)(hāz)	168-69	25°05′S	57°34′W
Villahermosa, Mex. (vēl′yä-ĕr-mō′sä)	146-47	17°59′N	92°55′W
Villa Hidalgo, Mex. (vēl′yäē-däl′gō)	146-47	21°40′N	102°36′W
Villaldama, Mex. (vēl-yäl-dä′mä)	122-23	26°30′N	100°25′W
Villa María, Arg. (vē′l-yä-mä-rē′ä)	173	32°25′S	63°14′W
Villa Mercedes, Arg. (vēl′yä-mĕr-sā′dås)	171	33°40′S	65°28′W
Villa Montes, Bol. (vē′l-yä-mô′n-tĕs)	168-69	21°15′S	63°29′W
Villanueva, Mex. (vēl′yä-nòĕ′vä)	146-47	22°20′N	102°52′W
Villanueva de la Serena, Spain (vêl-yä-nwĕ′vä-dä lä sä-rä′nä)	198-99	38°58′N	5°48′W
Villa Ocampo, Arg.	173	28°29′S	59°21′W
Villa Ocampo, Mex. (vēl′yä-ô-käm′pō)	122-23	26°28′N	105°33′W
Villa Regina, Arg.	171	39°06′S	67°05′W
Villarrica, Para. (vēl-yä-rē′kä)	168-69	25°47′S	56°28′W
Villarrobledo, Spain (vēl-yär-rô-blā′dhō)	198-99	39°16′N	2°36′W
Villa Unión, Arg.	173	29°24′S	62°47′W
Villa Unión, Mex. (vēl′yä-ōō-nyōn′)	146-47	23°10′N	106°12′W
Villavicencio, Col. (vē′l-yä-vē-sĕ′n-syō)	164-65	4°10′N	73°38′W
Villazón, Bol. (vē′l-yä-zô′n)	168-69	22°05′S	65°35′W
Villena, Spain (vē-lyā′nȧ)	198-99	38°38′N	0°52′W
Villeneuve-sur-Lot, Fr. (vēl′nûv′sür-lō′)	196-97	44°25′N	0°42′E
Ville Platte, La., U.S. (vēl plăt′)	122-23	30°41′N	92°16′W
Villers-Cotterêts, Fr. (vē-ār′kô-trā′)	196-97	49°15′N	3°06′E
Villeta, Col. (vē′l-yĕ′tȧ)	163c	4°60′N	74°30′W
Villeurbanne, Fr. (vēl-ûr-bän′)	196-97	45°45′N	4°52′E
Villiers, S. Afr. (vĭl′ĭ-ĕrs)	269c	27°02′S	28°36′E
Vilnius, nat. cap., Lith. (vĭl′nê-òs)	192-93	54°40′N	25°17′E
Vilyuy, stm., Russia (vēl′yĭ)	218-19	64°23′N	126°26′E
Vilyuysk, Russia (vê-lyōō′ĭsk′)	218-19	63°45′N	121°37′E
Vilyuyskoye Vodokhranilishche, res., Russia	218-19	62°34′N	111°13′E
Vimmerby, Swe. (vĭm′ĕr-bü)	192-93	57°40′N	15°52′E
Vina, stm., Camrn.	260-61	7°52′N	15°46′E
Viña del Mar, Chile (vē′nyä dĕl mär′)	163e	33°01′S	71°33′W
Vincennes, In., U.S. (vĭn-zĕnz′)	116-17	38°41′N	87°32′W
Vincennes Bay, b., Ant.	287	66°18′S	108°49′E
Vindhya Range, mts., India (vĭnd′yä rănj)	234-35	23°0′N	77°00′E
Vineland, N.J., U.S. (vīn′l*ă*nd)	116-17	39°29′N	75°01′W
Vinh, Viet. (vĕn′y′)	246-47	18°40′N	105°41′E
Vinh Long, Viet.	246-47	10°14′N	105°59′E
Vinita, Ok., U.S. (vĭ-nē′t*à*)	120-21	36°38′N	95°09′W
Vinkovci, Cro. (vēn′kôv-tsê)	200-01	45°17′N	18°49′E
Vinnitsa, Ukr. *see* Vinnytsia	202-03	49°14′N	28°32′E
Vinnytsia, Ukr.	202-03	49°14′N	28°32′E
Vinson Massif, mtn., Ant. (vĭn′sŭn mă-sēf)	287	78°32′S	85°14′W
Vinton, Ia., U.S. (vĭn′tŭn)	114-15	42°10′N	92°01′W
Vinton, La., U.S. (vĭn′tŭn)	122-23	30°11′N	93°35′W
Virac, Phil. (vê-räk′)	250	13°36′N	124°14′E
Virden, Mb., Can. (vûr′d*ĕ*n)	134-35	49°50′N	100°55′W
Virgin Gorda, i., Br. Vir. Is.	143b	18°30′N	64°24′W
Virginia, S. Afr. (vẽr-jĭn′y*à*)	269c	28°05′S	26°53′E
Virginia, Mn., U.S. (vẽr-jĭn′y*à*)	114-15	47°31′N	92°32′W
Virginia, state, U.S. (vẽr-jĭn′y*à*)	108-09	37°30′N	78°45′W

Feature (Pronunciation)	Page	Lat.	Long.
Virginia Beach, Va., U.S. (vẽr-jĭn´yȧ bēch)	**124-25**	36°52′N	75°59′W
Virginia City, Nv., U.S.	**118-19**	39°19′N	119°39′W
Virgin Islands, dep., N.A. (vûr´jĭn ī´lȧndz)	**140-41**	18°20′N	64°50′W
Viroqua, Wi., U.S. (vĭ-rō´kwȧ)	**114-15**	43°34′N	90°53′W
Virovitica, Cro. (vē-rô-vē´tē-tsȧ)	**200-01**	45°50′N	17°24′E
Virrat, Fin. (vĭr´ät)	**192-93**	62°15′N	23°45′E
Virserum, Swe. (vĭr´sĕ-rŏom)	**192-93**	57°19′N	15°36′E
Virudunagar, India	**236**	9°35′N	77°57′E
Vis, Otok, i., Cro.	**200-01**	43°02′N	16°11′E
Visalia, Ca., U.S. (vĭ-sā´lĭ-ȧ)	**118-19**	36°20′N	119°18′W
Visayan Sea, s., Phil.	**250**	11°35′N	123°51′E
Visby, Swe. (vĭs´bü)	**192-93**	57°38′N	18°19′E
Viscount Melville Sound, strt., Can.	**86**	74°10′N	108°00′W
Višegrad, Bos. (vē´shĕ-gräd)	**200-01**	43°47′N	19°18′E
Vishākhapatnam, India	**236**	17°43′N	83°19′E
Vishera, stm., Russia (vĭ´shĕ-rȧ)	**186-87**	59°54′N	56°26′E
Visoko, Bos. (vē´sô-kô)	**200-01**	43°59′N	18°11′E
Vistula, stm., Pol. (vĭs´tů-lȧ)	**194-95**	54°21′N	18°56′E
Vitarte, Peru	**163a**	12°02′S	76°56′W
Viterbo, Italy (vē-tĕr´bō)	**200-01**	42°25′N	12°06′E
Viti, nation, Oc. see Fiji	**279f**	18°0′S	178°00′E
Viti, nation, Oc. see Fiji	**279f**	18°0′S	178°00′E
Viti Levu, i., Fiji	**279f**	18°0′S	178°00′E
Vitim, stm., Russia (vē´têm)	**218-19**	59°28′N	112°35′E
Vitória, Braz. (vē-tō´rē-ä)	**166-67**	2°53′S	52°00′W
Vitória, Braz.	**172**	20°19′S	40°21′W
Vitoria, Spain (vē-tô-ryä) see Gasteiz	**198-99**	42°51′N	2°40′W
Vitória da Conquista, Braz.	**166-67**	14°51′S	40°51′W
Vitry-le-François, Fr. (vē-trē´lĕ-frän-swȧ´)	**196-97**	48°44′N	4°36′E
Vivian, La., U.S. (vĭv´ĭ-ȧn)	**120-21**	32°53′N	93°59′W
Vizcaya, Golfo de, b., Eur. see Biscay, Bay of	**196-97**	44°0′N	4°00′W
Vize, Ostrov, i., Russia	**218-19**	79°33′N	76°50′E
Vizianagaram, India	**236**	18°07′N	83°25′E
Vladikavkaz, Russia	**227**	43°03′N	44°39′E
Vladimir, Russia (vlȧ-dyē´mēr)	**202-03**	56°08′N	40°24′E
Vladivostok, Russia (vlȧ-dē-vôs-tôk´)	**240-41**	43°08′N	131°56′E
Vlonë, Alb. see Vlorë	**200-01**	40°29′N	19°30′E
Vlora, Alb. see Vlorë	**200-01**	40°29′N	19°30′E
Vlorë, Alb.	**200-01**	40°29′N	19°30′E
Vogel Peak, mtn., Nig. see Dimlang	**260-61**	8°24′N	11°47′E
Voghera, Italy (vô-gā´rä)	**200-01**	44°60′N	9°01′E
Vohimena, Tanjona, c., Madag.	**264-65**	25°36′S	45°09′E
Voi, Kenya	**262-63**	3°23′S	38°34′E
Voinjama, Lib.	**260-61**	8°25′N	9°45′W
Voiron, Fr. (vwȧ-rôn´)	**196-97**	45°22′N	5°35′E
Volcano Islands, is., Japan (vŏl-kā´nō ī´lȧndz) see Kazan-rettō	**280-81**	25°0′N	141°00′E
Volga, stm., Russia (vôl´gä)	**186-87**	45°45′N	47°56′E
Volga Upland, plat., Russia (vôl´gä ŭp´lȧnd) see Privolzhskaya Vozvyshennost'	**186-87**	52°0′N	46°00′E
Volgodonsk, Russia	**186-87**	47°31′N	42°08′E
Volgograd, Russia (vŏl-gō-grä´t)	**186-87**	48°44′N	44°25′E
Volgograd Reservoir, res., Russia (vŏl-gō-grä´t rĕ´zĕr-vwär) see Volgogradskoye Vodokhranilishche	**186-87**	50°18′N	45°49′E
Volgogradskoye Vodokhranilishche, res., Russia	**186-87**	50°18′N	45°49′E
Volkhov, Russia (vôl´kôf)	**192-93**	59°55′N	32°19′E
Volksrust, S. Afr.	**269c**	27°22′S	29°54′E
Vologda, Russia (vô´lôg-dȧ)	**186-87**	59°14′N	39°55′E
Volokolamsk, Russia (vô-lô-kôlȧmsk)	**202-03**	56°02′N	35°58′E
Vólos, Grc.	**200-01**	39°22′N	22°57′E
Vol'sk, Russia (vôl´sk)	**186-87**	52°03′N	47°22′E
Volta, stm., Ghana (vôl´tȧ)	**260-61**	5°46′N	0°40′E
Volta Blanche, stm., Afr. (vôl´tȧ blänsh) see White Volta	**260-61**	8°57′N	1°10′W
Volta Lake, res., Ghana (vôl´tȧ lāk)	**260-61**	7°30′N	0°07′E
Volta Noire, stm., Afr. (vôl´tȧ nwär) see Black Volta	**260-61**	8°41′N	0°60′W
Volta Redonda, Braz. (vôl´tä-rä-dôn´dä)	**172**	22°32′S	44°07′W
Volzhsk, Russia	**186-87**	55°52′N	48°21′E
Volzhskiy, Russia	**186-87**	48°50′N	44°45′E
Vordingborg, Den. (vôr´dĭng-bôr)	**194-95**	55°01′N	11°55′E
Vóreioi Sporades, is., Grc.	**200-01**	39°15′N	23°55′E
Vorgashor, Russia	**186-87**	67°32′N	64°05′E
Vorkuta, Russia (vôr-kōō´tȧ)	**186-87**	67°29′N	64°03′E
Vormsi, i., Est. (vôrm´sĭ)	**192-93**	59°00′N	23°15′E
Vorona, stm., Russia (vô-rô´na)	**186-87**	51°21′N	42°02′E
Voronezh, Russia (vô-rô´nyĕzh)	**202-03**	51°40′N	39°10′E
Voronezh, stm., Russia (vô-rô´nyĕzh)	**186-87**	51°32′N	39°06′E
Voronya, stm., Russia (vô-rôny′ȧ)	**186-87**	69°00′N	35°42′E
Voroshilov, Russia see Ussuriysk	**244**	43°48′N	131°59′E
Voroshilovsk, Russia see Stavropol'	**186-87**	45°02′N	41°59′E
Võru, Est. (vô´rû)	**192-93**	57°50′N	27°01′E
Voskresensk, Russia (vôs-krĕ-sĕnsk´)	**202-03**	55°19′N	38°42′E
Voss, Nor. (vôs)	**192-93**	60°38′N	6°26′E
Vostochno-Sibirskoye More, s., Russia see East Siberian Sea	**218-19**	74°0′N	166°00′E
Vostok, i., Kir.	**280-81**	10°06′S	152°23′W
Votkinsk, Russia (vôt-kēnsk´)	**186-87**	57°03′N	53°59′E
Votuporanga, Braz.	**168-69**	20°26′S	49°58′W
Voyageurs National Park, n.p., Mn., U.S.	**114-15**	48°30′N	93°00′W
Voznesens'k, Ukr.	**202-03**	47°34′N	31°20′E
Vrangelya, Ostrov, i., Russia	**218-19**	71°14′N	179°21′W
Vranje, Serb. (vrän´yĕ)	**200-01**	42°34′N	21°55′E
Vratsa, Blg. (vrät´tsȧ)	**200-01**	43°12′N	23°34′E
Vrbas, Serb. (v'r´bäs)	**200-01**	45°34′N	19°39′E
Vrchlabí, Czech Rep. (v'r´chlä-bē)	**194-95**	50°37′N	15°37′E
Vrede, S. Afr. (vrĭ´dĕ)(vrēd)	**269c**	27°26′S	29°10′E
Vredefort, S. Afr. (vrĭ´dĕ-fôrt)(vrēd´fōrt)	**269c**	27°00′S	27°22′E
Vrindāvan, India	**234-35**	27°35′N	77°42′E
Vršac, Serb. (v'r´shàts)	**200-01**	45°07′N	21°18′E
Vryburg, S. Afr. (vrī´bûrg)	**264-65**	26°58′S	24°44′E
Vryheid, S. Afr. (vrī´hīt)	**264-65**	27°46′S	30°48′E
Vsetín, Czech Rep. (fsĕt´yēn)	**194-95**	49°20′N	18°00′E
Vukovar, Cro. (vŏ´kô-vär)	**200-01**	45°21′N	19°00′E
Vulcano, Isola, i., Italy (ê´-sō-lä-vōōl-kä´nô)	**200-01**	38°24′N	14°58′E
Vung Tau, Viet.	**246-47**	10°21′N	107°05′E
Vyatka, Russia see Kirov	**186-87**	58°36′N	49°40′E
Vyatka, stm., Russia (vyàt´kȧ)	**186-87**	55°35′N	51°29′E
Vyatskiye Polyany, Russia	**186-87**	56°14′N	51°05′E
Vyazemskiy, Russia (vyȧ-zĕm´skĭ)	**240-41**	47°33′N	134°46′E
Vyazma, Russia (vyàz´mȧ)	**202-03**	55°12′N	34°17′E
Vyazniki, Russia (vyàz´nē-kē)	**186-87**	56°15′N	42°08′E
Vychegda, stm., Russia (vē´chĕg-dȧ)	**186-87**	61°17′N	46°37′E
Vygozero, Ozero, lk., Russia	**186-87**	63°35′N	34°42′E
Vym', stm., Russia (vwēm)	**186-87**	62°13′N	50°24′E
Vyritsa, Russia (vē´rĭ-tsȧ)	**192-93**	59°24′N	30°20′E
Vyshniy Volochëk, Russia (vêsh´nyĭ vôl-ô-chĕk´)	**202-03**	57°35′N	34°34′E
Vyškov, Czech Rep. (vêsh´kôf)	**194-95**	49°17′N	16°60′E
Vysokogornyy, Russia	**218-19**	50°06′N	139°09′E
Vysokovsk, Russia (vĭ-sô´kôfsk)	**202-03**	56°19′N	36°34′E
Vytegra, Russia (vû´tĕg-rȧ)	**186-87**	61°00′N	36°28′E

W

Feature (Pronunciation)	Page	Lat.	Long.
Wa, Ghana	**260-61**	10°03′N	2°30′W
Wabana, Nf., Can.	**138-39**	47°39′N	52°57′W
Wabasca, stm., Ab., Can.	**130-31**	58°21′N	115°20′W
Wabasca-Desmarais, Ab., Can.	**132-33**	55°58′N	113°52′W
Wabash, In., U.S. (wô´băsh)	**116-17**	40°48′N	85°49′W
Wabash, stm., U.S. (wô´băsh)	**110-11**	37°48′N	88°01′W
Wabasha, Mn., U.S. (wä´bȧ-shô)	**114-15**	44°23′N	92°02′W
Wabē Gestro, stm., Eth.	**262-63**	4°17′N	42°03′E
Wabē Shebelē, stm., Afr.	**262-63**	0°10′N	42°46′E
Wabowden, Mb., Can. (wȧ-bō´d'n)	**134-35**	54°54′N	98°37′W
W.A.C. Bennett Dam, d., B.C., Can.	**132-33**	56°01′N	122°10′W
Waccasassa Bay, b., Fl., U.S. (wă-kȧ-să´sȧ bā)	**124-25**	29°06′N	82°52′W
Waco, Tx., U.S. (wā´kō)	**122-23**	31°33′N	97°09′W
Waddān, Libya	**188-89**	29°10′N	16°10′E
Waddeneilanden, is., Eur. see Frisian Islands	**190-91**	53°27′N	5°50′E
Waddington, Mount, mtn., B.C., Can. (mount wŏd´dĭng-tŭn)	**132-33**	51°22′N	125°16′W
Wadena, Sk., Can.	**134-35**	51°56′N	103°47′W
Wadena, Mn., U.S. (wŏ-dē´nȧ)	**114-15**	46°26′N	95°09′W
Wadesboro, N.C., U.S. (wādz´bûr-ô)	**124-25**	34°58′N	80°05′W
Wādī Ḥalfā', Sudan	**266**	21°48′N	31°20′E
Wadley, Ga., U.S. (wŭd´lē)	**124-25**	32°52′N	82°24′W
Wad Madanī, Sudan (wäd mĕ-dä´nē)	**266**	14°23′N	33°31′E
Wadsworth, Nv., U.S. (wŏdz´wûrth)	**118-19**	39°38′N	119°17′W
Wafangdian, China	**240-41**	39°37′N	122°01′E
Wagadugu, nat. cap., Burkina see Ouagadougou	**260-61**	12°23′N	1°32′W
Wager Bay, b., Nu., Can. (wā´jẽr bā)	**130-31**	65°26′N	88°40′W
Wagga Wagga, Austl. (wôg´ȧ wôg´ȧ)	**276**	35°07′S	147°21′E
Wagoner, Ok., U.S. (wăg´ŭn-ẽr)	**120-21**	35°58′N	95°23′W
Wągrowiec, Pol. (vôn-grô´vyĕts)	**194-95**	52°49′N	17°13′E
Wāh Cantonment, Pak.	**232-33**	33°48′N	72°41′E
Wahoo, Ne., U.S. (wä-hōō´)	**114-15**	41°13′N	96°37′W
Wahpeton, N.D., U.S. (wô´pē-tŭn)	**114-15**	46°16′N	96°36′W
Wahrān, Alg. see Oran	**198-99**	35°41′N	0°39′W
Wai'anae, Hi., U.S. (wä´ē-ȧ-nä´ä)	**127a**	21°27′N	158°11′W
Waigeo, Pulau, i., Indon. (pōō-lou wä-ē-gā´ô)	**224-25**	0°10′S	130°55′E
Waikabubak, Indon.	**248-49**	9°38′S	119°25′E
Waikato, stm., N.Z. (wä-ē-kä´to)	**278**	37°23′S	174°43′E
Waikerie, Austl. (wā´kẽr-ē)	**276**	34°11′S	139°59′E
Wailuku, Hi., U.S. (wä´ē-lōō´kōō)	**127a**	20°54′N	156°30′W
Waimea, Hi., U.S. (wä-ē-mā´ä)	**127a**	21°58′N	159°40′W
Waimea, Hi., U.S. (wä-ē-mā´ä)	**127a**	20°02′N	155°40′W
Waimea, Hi., U.S. (wä-ē-mā´ä)	**127a**	20°02′N	155°40′W
Wainganga, stm., India (wä-ēn-gŭŋ´gä)	**234-35**	19°37′N	79°48′E
Waingapu, Indon.	**248-49**	9°40′S	120°16′E
Wainwright, Ab., Can. (wān-rīt)	**132-33**	52°51′N	110°51′W
Wainwright, Ak., U.S. (wān-rīt)	**126**	70°39′N	159°59′W
Waitekere, N.Z.	**278**	36°55′S	174°40′E
Waitsburg, Wa., U.S. (wāts´bûrg)	**112-13**	46°16′N	118°09′W
Wajima, Japan (wä´jē-mȧ)	**245**	37°24′N	136°54′E
Wajir, Kenya	**262-63**	1°45′N	40°04′E
Waka, Eth.	**269d**	7°10′N	37°21′E
Wakamatsu, Japan see Aizu-wakamatsu	**245**	37°30′N	139°56′E
Wakasa-wan, b., Japan (wä´kä-sä wän)	**245**	35°45′N	135°40′E
Wakatipu, Lake, lk., N.Z. (lāk wä-kä-tē´pōō)	**278**	45°05′S	168°34′E
Wakayama, Japan (wä-kä´yä-mä)	**245**	34°13′N	135°10′E
WaKeeney, Ks., U.S. (wô-kē´nē)	**120-21**	39°01′N	99°53′W
Wakefield, Ne., U.S. (wāk-fēld)	**114-15**	42°16′N	96°52′W
Wake Forest, N.C., U.S. (wāk fŏr´ĕst)	**124-25**	35°58′N	78°31′W
Wake Island, dep., Oc. (wāk ī´lȧnd)	**280-81**	19°17′N	166°36′E
Wakhān, hist. reg., Afg. see Vākhān	**232-33**	37°0′N	73°00′E
Wakkanai, Japan (wä´kä-nä´ē)	**244**	45°24′N	141°41′E
Wakkerstroom, S. Afr. (vȧk´ẽr-strōm)(wăk´ẽr-strōōm)	**269c**	27°20′S	30°08′E
Wałbrzych, Pol. (väl´bzhûk)	**194-95**	50°46′N	16°17′E
Waldenburg, Pol. see Wałbrzych	**194-95**	50°46′N	16°17′E
Waldorf, Md., U.S. (wăl´dôrf)	**116-17**	38°37′N	76°54′W
Wales, Ak., U.S. (wālz)	**126**	65°36′N	168°04′W
Wales, state, U.K. (wālz)	**190-91**	52°30′N	3°30′W
Wales Island, i., Nu., Can.	**130-31**	68°0′N	86°43′W
Walgett, Austl. (wôl´gĕt)	**276**	30°02′S	148°07′E
Walhalla, N.D., U.S. (wŭl-hăl´ȧ)	**114-15**	48°56′N	97°55′W
Walhalla, S.C., U.S. (wŭl-hăl´ȧ)	**124-25**	34°46′N	83°04′W
Walikale, D.R.C.	**267**	1°25′S	28°03′E
Walker, Mi., U.S. (wôk´ẽr)	**116-17**	42°59′N	85°45′W
Walker, Mn., U.S. (wôk´ẽr)	**114-15**	47°06′N	94°35′W
Walker Lake, lk., Mb., Can. (wôk´ẽr lăk)	**134-35**	54°42′N	96°57′W
Walker Lake, lk., Nv., U.S.	**118-19**	38°44′N	118°43′W
Wallaceburg, On., Can.	**136-37**	42°36′N	82°23′W
Wallaroo, Austl. (wŏl-ȧ-rōō)	**276**	33°56′S	137°37′E
Walla Walla, Wa., U.S. (wŏl´ȧ wŏl´ȧ)	**112-13**	46°04′N	118°19′W
Wallis, Îles, is., Wal./F.	**280-81**	13°18′S	176°10′W
Wallis and Futuna, dep., Oc. (wŏl´ĭs ănd fōō-tōō´nȧ)	**280-81**	14°0′S	177°00′W
Wallis et Futuna, dep., Oc. see Wallis and Futuna	**280-81**	14°0′S	177°00′W
Wallowa Mountains, mts., Or., U.S. (wôl´ô-wȧ moun´tĭnz)	**112-13**	45°16′N	117°21′W
Walnut Creek, stm., Ks., U.S. (wôl´nŭt krēk)	**120-21**	38°21′N	98°41′W
Walnut Ridge, Ar., U.S. (wôl´nŭt rĭj)	**124-25**	36°04′N	90°58′W
Walsall, Eng., U.K. (wôl-sôl)	**190-91**	52°36′N	1°59′W
Walsenburg, Co., U.S. (wôl´sĕn-bûrg)	**120-21**	37°37′N	104°47′W
Walters, Ok., U.S. (wôl´tẽrz)	**120-21**	34°21′N	98°19′W
Walvisbaai, Nmb. see Walvis Bay	**264-65**	22°57′S	14°31′E
Walvis Bay, Nmb. (wôl´vĭs bā)	**264-65**	22°57′S	14°31′E
Walworth, Wi., U.S. (wôl´wŭrth)	**116-17**	42°32′N	88°37′W
Wamba, D.R.C.	**267**	2°09′N	28°00′E
Wamego, Ks., U.S. (wŏ-mē´gō)	**120-21**	39°12′N	96°18′W
Wami, stm., Tan. (wä´mē)	**262-63**	6°15′S	38°51′E
Wanfoxia, China	**240-41**	40°05′N	95°55′E
Wanganui, N.Z. (wôn´gȧ-nōō´ē)	**278**	39°56′S	175°02′E
Wangaratta, Austl. (wŏŋ´gȧ-răt´ȧ)	**276**	36°22′S	146°19′E
Wangiwangi, Pulau, i., Indon.	**248-49**	5°20′S	123°35′E
Wangpan Yang, b., China	**238-39**	30°30′N	121°46′E
Wanneroo, Austl.	**270-71**	31°45′S	115°48′E
Wanxian, China (wän-shyĕn)	**238-39**	30°49′N	108°22′E
Wanzai, China (wän-dzī)	**238-39**	28°05′N	114°27′E
Wapakoneta, Oh., U.S. (wä´pȧ-kō-nĕt´ȧ)	**116-17**	40°34′N	84°12′W
Wapawekka Lake, lk., Sk., Can. (wŏ´pä-wĕ´kȧ lăk)	**134-35**	54°55′N	104°40′W
Wapello, Ia., U.S. (wŏ-pĕl´ō)	**114-15**	41°10′N	91°12′W

n-sing; ŋ-baŋk; ɴ-nasalized n; nŏd; cŏmmit; ōld; ôbey; ôrder; oi-boil; fōōd; ŏ-as oo in foot; ou-out; s-soft; sh-dish; th-thin; pūre; ûnite; ûrn; stŭd; circŭs; ü-as in French tu; ´-indeterminate vowel.

Feature (Pronunciation)	Page	Lat.	Long.
Wapiti, stm., Can.	132-33	55°08'N	118°18'W
Wapusk National Park, n.p., Can.	128-29	58°0'N	93°30'W
Warangal, India (wŭ'răŋ-gâl)	236	18°00'N	79°35'E
Warburton Creek, stm., Austl.	276	27°59'S	137°25'E
Warden, S. Afr. (wôr'dĕn)	269c	27°51'S	28°58'E
Wardha, India (wŭr'dä)	234-35	20°45'N	78°37'E
Wardha, stm., India	234-35	19°36'N	79°47'E
Warialda, Austl.	276	29°33'S	150°35'E
Warmbad, S. Afr. *see* Bela-Bela	269c	24°53'S	28°19'E
Warm Baths, S. Afr. *see* Bela-Bela	269c	24°53'S	28°19'E
Warm Springs Indian Reservation, ind. res., Or., U.S. (wôrm sprĭngz ĭn'dĭ-ăn rĕ-sĕr-vā'shĕn)	112-13	44°53'N	121°23'W
Warrego, stm., Austl. (wôr'ē-gô)	270-71	30°25'S	145°21'E
Warren, Ar., U.S. (wŏr'ĕn)	120-21	33°37'N	92°04'W
Warren, Mi., U.S. (wŏr'ĕn)	116-17	42°30'N	83°02'W
Warren, Mn., U.S. (wŏr'ĕn)	114-15	48°12'N	96°47'W
Warren, Oh., U.S. (wŏr'ĕn)	116-17	41°14'N	80°49'W
Warrensburg, Mo., U.S. (wŏr'ĕnz-bûrg)	120-21	38°46'N	93°44'W
Warrenton, S. Afr.	264-65	28°07'S	24°51'E
Warrenton, Mo., U.S. (wŏr'ĕn-tŭn)	120-21	38°49'N	91°09'W
Warrenton, Or., U.S. (wŏr'ĕn-tŭn)	112-13	46°10'N	123°55'W
Warrenton, Va., U.S. (wŏr'ĕn-tŭn)	116-17	38°43'N	77°48'W
Warri, Nig. (wär'ē)	260a	5°31'N	5°46'E
Warrnambool, Austl. (wôr'năm-bōōl)	276	38°23'S	142°29'E
Warroad, Mn., U.S. (wôr'rōd)	114-15	48°54'N	95°19'W
Warsaw, In., U.S. (wôr'sô)	116-17	41°14'N	85°51'W
Warsaw, Ky., U.S. (wôr'sô)	116-17	38°47'N	84°54'W
Warsaw, Mo., U.S. (wôr'sô)	120-21	38°15'N	93°23'W
Warsaw, N.C., U.S. (wôr'sô)	124-25	34°59'N	78°05'W
Warsaw, Va., U.S. (wôr'sô)	116-17	37°57'N	76°46'W
Warsaw, nat. cap., Pol. (wôr'sô)	194-95	52°15'N	21°00'E
Warszawa, nat. cap., Pol. (vár-shä'vá) *see* Warsaw	194-95	52°15'N	21°00'E
Warwick, Austl. (wŏr'ĭk)	276	28°13'S	152°02'E
Warwick, R.I., U.S. (wŏr'ĭk)	116-17	41°43'N	71°23'W
Wasco, Ca., U.S. (wäs'kō)	118-19	35°36'N	119°20'W
Waseca, Mn., U.S. (wô-sē'kă)	114-15	44°05'N	93°30'W
Washburn, N.D., U.S. (wŏsh'bŭrn)	114-15	47°17'N	101°02'W
Washburn, Wi., U.S. (wŏsh'bŭrn)	114-15	46°40'N	90°54'W
Washington, Ga., U.S. (wŏsh'ĭng-tŭn)	124-25	33°44'N	82°44'W
Washington, Ia., U.S. (wŏsh'ĭng-tŭn)	114-15	41°18'N	91°41'W
Washington, Il., U.S. (wŏsh'ĭng-tŭn)	116-17	40°42'N	89°25'W
Washington, In., U.S. (wŏsh'ĭng-tŭn)	116-17	38°39'N	87°10'W
Washington, Mo., U.S. (wŏsh'ĭng-tŭn)	120-21	38°33'N	91°01'W
Washington, N.C., U.S. (wŏsh'ĭng-tŭn)	124-25	35°33'N	77°04'W
Washington, state, U.S. (wŏsh'ĭng-tŭn)	108-09	47°30'N	120°30'W
Washington, nat. cap., D.C., U.S. (wŏsh'ĭng-tŭn)	116-17	38°53'N	77°02'W
Washington, Mount, mtn., N.H., U.S. (mount wŏsh'ĭng-tŭn)	116-17	44°16'N	71°18'W
Washington Island, i., Kir. *see* Teraina	280-81	4°42'N	160°45'W
Washington Island, i., Wi., U.S. (wŏsh'ĭng-tŭn ī'lănd)	116-17	45°22'N	86°54'W
Waskaganish, Qc., Can.	128-29	51°29'N	78°45'W
Waskaiowaka Lake, lk., Mb., Can. (wŏ'skă-yō'wŏ-kă lāk)	134-35	56°31'N	96°18'W
Watampone, Indon.	248-49	4°32'S	120°19'E
Wataru, i., Mald.	236	5°43'N	73°23'E
Waterberge, mts., S. Afr. (wôrtēr'bûrg)	269c	24°28'S	27°58'E
Waterbury, Ct., U.S. (wô'tēr-bĕr-ē)	116-17	41°33'N	73°02'W
Waterbury, Vt., U.S. (wô'tēr-bĕr-ē)	116-17	44°20'N	72°45'W
Waterford, Ire. (wô'tēr-fērd)	190-91	52°15'N	7°06'W
Waterhen Lake, lk., Mb., Can.	134-35	52°06'N	99°34'W
Waterloo, Bel. (wô-tēr-lōō')	190-91	50°43'N	4°24'E
Waterloo, On., Can. (wô-tēr-lōō')	136-37	43°28'N	80°30'W
Waterloo, Qc., Can. (wô-tēr-lōō')	136-37	45°21'N	72°30'W
Waterloo, Ia., U.S. (wô-tēr-lōō')	114-15	42°30'N	92°21'W
Waterloo, Il., U.S. (wô-tēr-lōō')	120-21	38°20'N	90°09'W
Waterton Lakes National Park, n.p., Ab., Can.	132-33	49°06'N	114°01'W
Watertown, N.Y., U.S.	116-17	43°59'N	75°55'W
Watertown, S.D., U.S. (wô'tēr-toun)	114-15	44°54'N	97°07'W
Watertown, Wi., U.S. (wô'tēr-toun)	116-17	43°11'N	88°43'W
Water Valley, Ms., U.S. (văl'ē văl'ē)	124-25	34°10'N	89°38'W
Waterville, Me., U.S.	117a	44°33'N	69°38'W
Watervliet, N.Y., U.S. (wô'tēr-vlēt')	116-17	42°44'N	73°43'W
Watford, Eng., U.K. (wŏt'fŏrd)	190-91	51°40'N	0°25'W
Watford City, N.D., U.S.	114-15	47°48'N	103°18'W
Watling Island, i., Bah. *see* San Salvador	142-43	24°02'N	74°27'W
Watonga, Ok., U.S. (wŏ-tôŋ'gă)	120-21	35°51'N	98°25'W
Watrous, Sk., Can.	134-35	51°41'N	105°27'W
Watsa, D.R.C. (wät'sä)	267	3°02'N	29°32'E
Watseka, Il., U.S. (wŏt-sē'kă)	116-17	40°46'N	87°44'W
Watson Lake, Yk., Can. (wŏt'sŭn lāk)	128-29	60°04'N	128°44'W
Watsonville, Ca., U.S. (wŏt'sŭn-vĭl)	118-19	36°55'N	121°45'W
Wauchula, Fl., U.S. (wô-chōō'lă)	125a	27°33'N	81°49'W
Waukegan, Il., U.S. (wŏ-kē'găn)	116-17	42°21'N	87°51'W
Waukesha, Wi., U.S. (wô'kē-shô)	116-17	43°01'N	88°14'W
Waukon, Ia., U.S. (wô kŏn)	114-15	43°16'N	91°29'W
Waupaca, Wi., U.S. (wô-păk'ă)	116-17	44°21'N	89°05'W
Waupun, Wi., U.S. (wô-pŭn')	116-17	43°38'N	88°44'W
Waurika, Ok., U.S. (wô-rē'kȧ)	120-21	34°10'N	97°60'W
Wausau, Wi., U.S. (wô'sô)	116-17	44°57'N	89°37'W
Wausaukee, Wi., U.S. (wô-sô'kē)	116-17	45°23'N	87°59'W
Wauseon, Oh., U.S. (wô'sē-ŏn)	116-17	41°33'N	84°08'W
Wautoma, Wi., U.S. (wô-tō'mă)	116-17	44°04'N	89°18'W
Waverly, Ia., U.S. (wā'vēr-lē)	114-15	42°44'N	92°29'W
Waverly, Ne., U.S. (wā'vēr-lē)	114-15	40°55'N	96°32'W
Waverly, Oh., U.S. (wā'vēr-lē)	116-17	39°07'N	82°59'W
Waverly, Tn., U.S. (wā'vēr-lē)	124-25	36°05'N	87°48'W
Wāw, Sudan	262-63	7°41'N	27°59'E
Wawa, On., Can.	136-37	47°59'N	84°47'W
Waxahachie, Tx., U.S. (wăk-să-hăch'ē)	120-21	32°24'N	96°51'W
Waya, i., Fiji	279f	17°18'S	177°08'E
Wayabula, Indon.	248-49	2°18'N	128°12'E
Waycross, Ga., U.S.	124-25	31°13'N	82°22'W
Wayne, Ne., U.S. (wān)	114-15	42°14'N	97°01'W
Wayne, W.V., U.S. (wān)	116-17	38°13'N	82°27'W
Waynesboro, Ga., U.S. (wānz'bŭr-ô)	124-25	33°05'N	82°01'W
Waynesboro, Ms., U.S. (wānz'bŭr-ô)	124-25	31°40'N	88°39'W
Waynesboro, Pa., U.S. (wānz'bŭr-ô)	116-17	39°45'N	77°35'W
Waynesboro, Va., U.S. (wānz'bŭr-ô)	116-17	38°04'N	78°53'W
Waynesville, Mo., U.S. (wānz'vĭl)	120-21	37°50'N	92°12'W
Waynesville, N.C., U.S. (wānz'vĭl)	124-25	35°29'N	82°60'W
Waynoka, Ok., U.S. (wā-nō'kă)	120-21	36°35'N	98°53'W
Weagamow Lake, lk., On., Can. (wē'ăg-ă-mou lāk)	134-35	52°53'N	91°22'W
Weatherford, Ok., U.S. (wĕ-dhĕr-fērd)	120-21	35°32'N	98°43'W
Weatherford, Tx., U.S. (wĕ-dhĕr-fērd)	120-21	32°46'N	97°48'W
Weddell Island, i., Falk. Is.	171	51°53'S	61°05'W
Weddell Sea, s., Ant. (wĕd'ĕl sē)	287	72°0'S	45°00'W
Wedgeport, N.S., Can. (wĕj'pōrt)	138-39	43°45'N	65°60'W
Weed, Ca., U.S. (wēd)	112-13	41°25'N	122°23'W
Weenen, S. Afr. (vā'nĕn)	269c	28°51'S	30°05'E
Wei, stm., China (wā)	238-39	34°37'N	110°17'E
Wei, stm., China (wā)	240-41	36°49'N	115°41'E
Weichang, China (wā-chäŋ)	240-41	42°00'N	117°40'E
Weifang, China	240-41	36°42'N	119°06'E
Weihai, China (wa'hăī')	240-41	37°30'N	122°07'E
Weilheim, Ger. (vīl'hīm')	194-95	47°50'N	11°09'E
Weimar, Ger. (vī'mȧr)	194-95	50°59'N	11°19'E
Weinan, China	238-39	34°29'N	109°29'E
Weipa, Austl.	277	12°42'S	141°56'E
Weiser, Id., U.S. (wē'zēr)	112-13	44°15'N	116°58'W
Weißenfels, Ger. (vī'sĕn-fĕlz)	194-95	51°12'N	11°58'E
Weixi, China (wā-shyē)	238-39	27°11'N	99°17'E
Welch, W.V., U.S. (wĕlch)	124-25	37°26'N	81°35'W
Welkom, S. Afr. (wĕl'kŏm)	269c	27°58'S	26°44'E
Welland, On., Can. (wĕl'ănd)	136-37	42°59'N	79°15'W
Wellesley Islands, is., Austl.	277	16°42'S	139°30'E
Wellington, Austl. (wĕl'lĭng-tŭn)	276	32°34'S	148°57'E
Wellington, Co., U.S. (wĕl'lĭng-tŭn)	120-21	40°42'N	104°60'W
Wellington, Ks., U.S. (wĕl'lĭng-tŭn)	120-21	37°16'N	97°24'W
Wellington, Oh., U.S. (wĕl'lĭng-tŭn)	116-17	41°10'N	82°13'W
Wellington, Tx., U.S. (wĕl'lĭng-tŭn)	120-21	34°51'N	100°13'W
Wellington, nat. cap., N.Z. (wĕl'lĭng-tŭn)	278	41°18'S	174°46'E
Wellington, Isla, i., Chile (ĕ's-lä-ŏĕ'lĕng-tŏn)	171	49°20'S	74°40'W
Wells, Mn., U.S. (wĕlz)	114-15	43°45'N	93°44'W
Wells, Nv., U.S. (wĕlz)	112-13	41°07'N	114°58'W
Wells, Lake, lk., Austl. (lāk wĕlz)	272-73	26°41'S	123°11'E
Wellsboro, Pa., U.S. (wĕlz'bŭ-rô)	116-17	41°45'N	77°17'W
Wellsburg, W.V., U.S. (wĕlz'bûrg)	116-17	40°16'N	80°36'W
Wellston, Oh., U.S. (wĕlz'tŭn)	116-17	39°07'N	82°32'W
Wellsville, N.Y., U.S. (wĕlz'vĭl)	116-17	42°07'N	77°56'W
Wellsville, Oh., U.S. (wĕlz'vĭl)	116-17	40°36'N	80°39'W
Wellsville, Ut., U.S. (wĕlz'vĭl)	112-13	41°39'N	111°56'W
Wels, Aus. (vĕls)	194-95	48°10'N	14°01'E
Welshpool, Wales, U.K. (wĕlsh'pōōl)	190-91	52°40'N	3°09'W
Wembere, stm., Tan.	267	4°09'S	34°11'E
Wenatchee, Wa., U.S. (wē-năch'ē)	112-13	47°25'N	120°19'W
Wenatchee Mountains, mts., Wa., U.S. (wē-năch'ē moun'tĭnz)	112-13	47°20'N	120°45'W
Wenchang, China (wŭn-chäŋ)	238-39	19°33'N	110°45'E
Wenchow, China *see* Wenzhou	238-39	28°01'N	120°38'E
Wendover, Ut., U.S.	112-13	40°44'N	114°02'W
Wenlock, stm., Austl.	277	12°15'S	141°56'E
Wenshan, China	238-39	23°30'N	104°28'E
Wentworth, Austl. (wĕnt'wûrth)	276	34°06'S	141°55'E
Wenzhou, China (wŭn-jō)	238-39	28°01'N	120°38'E
Werdēr, Eth.	262-63	6°58'N	45°21'E
Wesel, Ger. (vā'zĕl)	194-95	51°40'N	6°38'E
Weser, stm., Ger. (vā'zĕr)	194-95	53°32'N	8°34'E
Wesermünde, Ger. *see* Bremerhaven	194-95	53°32'N	8°36'E
Weslaco, Tx., U.S. (wĕs-lā'kō)	122-23	26°10'N	97°59'W
Wessel, Cape, c., Austl.	272-73	11°02'S	136°45'E
Wessington Springs, S.D., U.S. (wĕs'ĭng-tŭn sprĭngz)	114-15	44°05'N	98°34'W
West Allis, Wi., U.S. (wĕst ăl'ĭs)	116-17	43°01'N	88°01'W
West Bend, Ia., U.S. (wĕst bĕnd)	114-15	42°57'N	94°26'W
West Bend, Wi., U.S. (wĕst bĕnd)	116-17	43°25'N	88°10'W
West Bengal, state, India (wĕst bĕn-gôl')	234-35	24°0'N	88°00'E
West Branch, Ia., U.S. (wĕst brȧnch)	114-15	41°40'N	91°21'W
West Branch, Mi., U.S. (wĕst brȧnch)	116-17	44°16'N	84°14'W
Westbrook, Me., U.S. (wĕst'brŏk)	116-17	43°41'N	70°21'W
West Caicos, i., T./C. Is. (wĕst kī'kōs) (wĕst käē'kō)	142-43	21°39'N	72°28'W
West Cape, c., N.Z. (wĕst kāp)	278	45°55'S	166°25'E
West Chester, Pa., U.S. (wĕst chĕs'tēr)	116-17	39°58'N	75°36'W
West Columbia, S.C., U.S. (wĕst cŏl'ŭm-bē-ă)	124-25	33°59'N	81°05'W
West Columbia, Tx., U.S. (wĕst cŏl'ŭm-bē-ă)	122-23	29°09'N	95°39'W
West Des Moines, Ia., U.S. (wĕst dē moin')	114-15	41°34'N	93°44'W
West End, Bah. (wĕst ĕnd)	142-43	26°42'N	78°59'W
Westerly, R.I., U.S. (wĕs'tēr-lē)	116-17	41°23'N	71°50'W
Western Australia, state, Austl. (wĕst'tērn ôs-trā'lĭ-ă)	270-71	25°0'S	122°00'E
Western Desert, des., Egypt (wĕst'tērn dĕs'ērt)	266	27°0'N	27°00'E
Western Dvina, stm., Eur.	192-93	57°04'N	24°03'E
Western Ghāts, mts., India (wĕst'tērn ghäts) (wĕst'tērn ghôts)	236	14°0'N	75°00'E
Westernport, Md., U.S. (wĕs'tērn pōrt)	116-17	39°29'N	79°03'W
Western Sahara, dep., Afr. (wĕst'tērn sȧ-hä'rȧ)	253	24°30'N	13°00'W
Western Samoa, nation, Oc. *see* Samoa	279b	13°55'S	172°00'W
Westerville, Oh., U.S. (wĕs'tēr-vĭl)	116-17	40°07'N	82°55'W
West Falkland, i., Falk. Is.	171	51°50'S	59°60'W
Westfield, Ma., U.S. (wĕst'fēld)	116-17	42°08'N	72°45'W
Westfield, N.Y., U.S. (wĕst'fēld)	116-17	42°19'N	79°35'W
Westfield, Wi., U.S. (wĕst'fēld)	116-17	43°53'N	89°30'W
West Frankfort, Il., U.S. (wĕst frăŋk'fŭrt)	116-17	37°54'N	88°56'W
West Helena, Ar., U.S. (wĕst hĕl'ĕn-ă)	124-25	34°33'N	90°39'W
West Indies, is., (wĕst ĭn'dēz)	140-41	19°0'N	70°00'W
West Lafayette, In., U.S. (wĕst lä-fā-yĕt')	116-17	40°25'N	86°54'W
West Liberty, Ia., U.S. (wĕst lĭb'ēr-tĭ)	114-15	41°34'N	91°16'W
West Liberty, Oh., U.S. (wĕst lĭb'ēr-tĭ)	116-17	40°15'N	83°47'W
Westlock, Ab., Can. (wĕst'lŏk)	132-33	54°09'N	113°52'W
Westminster, Co., U.S. (wĕst'mĭn-stēr)	120-21	39°51'N	105°04'W
West Nishnabotna, stm., Ia., U.S. (wĕst nĭsh-nă-bŏt'nă)	120-21	40°30'N	95°42'W
Weston, W.V., U.S. (wĕs'tŭn)	116-17	39°02'N	80°29'W
Weston-super-Mare, Eng., U.K. (wĕs'tŭn sū'pēr-mā'rē)	190-91	51°21'N	2°58'W
West Palm Beach, Fl., U.S. (wĕst päm bēch)	125a	26°44'N	80°08'W
West Pensacola, Fl., U.S. (wĕst pĕn-să-kō'lă)	124-25	30°25'N	87°16'W
West Plains, Mo., U.S. (wĕst-plānz')	124-25	36°44'N	91°52'W
West Point, Ga., U.S. (wĕst point)	124-25	32°54'N	85°09'W
West Point, Ms., U.S. (wĕst point)	124-25	33°36'N	88°39'W
West Point, Ne., U.S. (wĕst point)	114-15	41°51'N	96°43'W
West Point, N.Y., U.S. (wĕst point)	116-17	41°24'N	73°58'W
West Point, Va., U.S. (wĕst point)	116-17	37°32'N	76°48'W
West Point, c., Austl.	272-73	35°01'S	135°57'E
West Point Lake, res., U.S. (wĕst point lāk)	124-25	32°60'N	85°12'W

Feature (Pronunciation)	Page	Lat.	Long.
Westport, N.Z.	278	41°45′S	171°36′E
Westport, Wa., U.S. (wĕst′pôrt)	112-13	46°54′N	124°06′W
West Road, stm., B.C., Can. (wĕst rōd)	132-33	53°18′N	122°53′W
West Siberian Plain, pl., Russia (wĕst sī-bĭr′y′n plān)	218-19	60°0′N	75°00′E
West Union, Ia., U.S. (wĕst ūn′yŭn)	114-15	42°57′N	91°49′W
West Union, Oh., U.S. (wĕst ūn′yŭn)	116-17	38°47′N	83°33′W
Westville, N.S., Can. (wĕst′vĭl)	138-39	45°34′N	62°42′W
West Virginia, state, U.S. (wĕst vẽr-jĭn′ĭ-ȧ)	108-09	38°45′N	80°30′W
West Warwick, R.I., U.S. (wĕst wŏr′ĭk)	116-17	41°42′N	71°32′W
West Wyalong, Austl. (wĕst wīălông)	276	33°55′S	147°12′E
Wetar, Pulau, i., Indon. (pōō-lou wĕt′är)	248-49	7°48′S	126°18′E
Wetaskiwin, Ab., Can. (wĕ-tăs′kē-wŏn)	132-33	52°58′N	113°22′W
Wete, Tan.	262-63	5°02′S	39°44′E
Wetumpka, Al., U.S. (wē-tŭmp′kȧ)	124-25	32°33′N	86°12′W
Wetzlar, Ger. (vets′lär)	194-95	50°33′N	8°29′E
Wewak, Pap. N. Gui. (wā-wäk′)	277	3°35′S	143°39′E
Wewoka, Ok., U.S. (wē-wō′kȧ)	120-21	35°10′N	96°30′W
Weyburn, Sk., Can. (wā′bûrn)	134-35	49°40′N	103°51′W
Weymouth, Eng., U.K. (wā′mŭth)	190-91	50°37′N	2°28′W
Whangarei, N.Z.	278	35°43′S	174°19′E
Whapmagoostui, Qc., Can.	128-29	55°17′N	77°45′W
Wharton, Tx., U.S. (hwôr′tŭn)	122-23	29°19′N	96°06′W
Wharton Lake, lk., Nu., Can.	130-31	64°01′N	99°52′W
Wheatland, Wy., U.S. (hwēt′lănd)	112-13	42°03′N	104°57′W
Wheaton, Md., U.S. (hwē′tŭn)	116-17	39°02′N	77°03′W
Wheaton, Mn., U.S. (hwē′tŭn)	114-15	45°48′N	96°30′W
Wheeler Peak, mtn., N.M., U.S.	120-21	36°34′N	105°25′W
Wheeling, W.V., U.S. (hwēl′ĭng)	116-17	40°04′N	80°41′W
Whitby, On., Can. (hwĭt′bē)	136-37	43°53′N	78°56′W
White, stm., On., Can. (hwīt)	136-37	48°33′N	86°16′W
White, stm., N.A.	126	63°11′N	139°35′W
White, stm., U.S. (hwīt)	110-11	33°53′N	91°04′W
White, stm., U.S. (hwīt)	114-15	43°42′N	99°27′W
White, stm., In., U.S. (hwīt)	116-17	38°25′N	87°45′W
White, East Fork, stm., In., U.S. (ēst fôrk hwīt)	116-17	38°32′N	87°14′W
White Bay, b., Nf., Can. (hwīt bā)	138-39	50°07′N	56°27′W
White Cloud, Mi., U.S. (hwīt kloud)	116-17	43°33′N	85°46′W
Whitecourt, Ab., Can. (hwīt′côrt)	132-33	54°08′N	115°43′W
White Earth Indian Reservation, ind. res., Mn., U.S. (hwīt ûrth ĭn′dĭ-ăn rĕ-sẽr-vā′shĕn)	114-15	47°18′N	95°50′W
Whitefish, Mt., U.S.	112-13	48°25′N	114°20′W
Whitefish Lake, lk., N.T., Can.	130-31	62°41′N	106°48′W
White Hall, Ar., U.S. (hwīt hôl)	120-21	34°16′N	92°07′W
Whitehall, Mi., U.S. (hwīt′hôl)	116-17	43°24′N	86°20′W
Whitehall, Mt., U.S. (hwīt′hôl)	112-13	45°52′N	112°06′W
Whitehall, N.Y., U.S. (hwīt′hôl)	116-17	43°33′N	73°24′W
Whitehaven, Eng., U.K. (hwīt′hā-vĕn)	190-91	54°33′N	3°35′W
Whitehorse, Yk., Can. (hwīt′hôrs)	128-29	60°43′N	135°08′W
White Lake, lk., On., Can. (hwīt lāk)	136-37	45°17′N	76°32′W
White Lake, lk., La., U.S. (hwīt lāk)	122-23	29°43′N	92°28′W
White Mountain Peak, mtn., Ca., U.S. (hwīt moun′tĭn pēk)	118-19	37°38′N	118°16′W
White Mountains, mts., N.H., U.S. (hwīt moun′tĭnz)	116-17	44°10′N	71°35′W
White Nile, stm., Sudan (hwīt nīl)	254	15°38′N	32°31′E
White Pass, p., N.A. (hwīt pås)	126	59°38′N	135°10′W
White Russia, nation, Eur. *see* Belarus	174-75	53°50′N	28°00′E
Whitesail Lake, res., B.C., Can. (hwīt′sāl lāk)	132-33	53°30′N	127°00′W
White Sands National Monument, n.p., N.M., U.S.	118-19	32°46′N	106°20′W
White Sea, s., Russia (hwīt sē)	186-87	65°37′N	37°52′E
White Sulphur Springs, Mt., U.S. (hwīt sŭl′fŭr sprĭngz)	112-13	46°33′N	110°55′W
White Sulphur Springs, W.V., U.S. (hwīt sŭl′fŭr sprĭngz)	116-17	37°48′N	80°18′W
Whiteville, N.C., U.S. (hwīt′vĭl)	124-25	34°20′N	78°42′W
White Volta, stm., Afr. (hwīt vôl′tȧ)	260-61	8°57′N	1°10′W
Whitewater, Wi., U.S. (hwīt-wŏt′ẽr)	116-17	42°50′N	88°43′W
Whitewater Lake, lk., Mb., Can. (hwīt-wŏt′ẽr lāk)	134-35	49°15′N	100°20′W
Whitewright, Tx., U.S. (hwīt′rīt)	120-21	33°31′N	96°24′W
Whitney, Lake, res., Tx., U.S. (lāk hwĭt′nē)	122-23	31°56′N	97°26′W
Whitney, Mount, mtn., Ca., U.S. (mount hwĭt′nē)	118-19	36°35′N	118°18′W
Whitsunday Island, i., Austl. (hwĭt′s'n-dā ī′lȧnd)	277	20°15′S	148°59′E
Whittle, Cap, c., Qc., Can.	138-39	50°11′N	60°10′W
Wholdaia Lake, lk., N.T., Can.	130-31	60°41′N	104°18′W
Whyalla, Austl. (hwī-ăl′ȧ)	276	33°02′S	137°34′E
Wiarton, On., Can. (wī′ăr-tŭn)	136-37	44°44′N	81°08′W
Wichita, Ks., U.S. (wĭch′ĭ-tô)	120-21	37°41′N	97°19′W
Wichita Falls, Tx., U.S. (wĭch′ĭ-tô fôlz)	120-21	33°55′N	98°30′W
Wichita Mountains, mts., Ok., U.S. (wĭch′ĭ-tô moun′tĭnz)	120-21	34°45′N	98°40′W
Wick, Scot., U.K. (wĭk)	190-91	58°26′N	3°06′W
Wicklow Mountains, mts., Ire. (wĭk′lō moun′tĭnz)	190-91	53°02′N	6°24′W
Wieliczka, Pol. (vyě-lēch′kȧ)	194-95	49°59′N	20°04′E
Wien, nat. cap., Aus. (vēn) *see* Vienna	194-95	48°13′N	16°20′E
Wiener Neustadt, Aus. (vē′nĕr noi′shtät)	194-95	47°49′N	16°15′E
Wieprz, stm., Pol. (vyěpzh)	194-95	51°33′N	21°50′E
Wiesbaden, Ger. (vēs′bä-dĕn)	194-95	50°05′N	8°14′E
Wiggins, Ms., U.S. (wĭg′ĭnz)	124-25	30°51′N	89°09′W
Wight, Isle of, i., Eng., U.K. (īl ŏv wīt)	190-91	50°40′N	1°20′W
Wilber, Ne., U.S. (wĭl′bẽr)	120-21	40°29′N	96°58′W
Wilburton, Ok., U.S. (wĭl′bẽr-tŭn)	120-21	34°56′N	95°19′W
Wilcannia, Austl. (wĭl-căn-ĭȧ)	276	31°34′S	143°23′E
Wildcat Hill, mtn., Sk., Can. (wīld′kăt hĭl)	134-35	53°17′N	102°30′W
Wildhay, stm., Ab., Can. (wīld′hā)	132-33	53°59′N	117°17′W
Wilge, stm., S. Afr. (wĭl′jē)	269c	27°02′S	28°21′E
Wilhelm, Mount, mtn., Pap. N. Gui.	277	5°47′S	145°00′E
Wilhelmina Peak, mtn., Indon. *see* Trikora, Puncak	277	4°18′S	138°40′E
Wilhelmshaven, Ger. (vēl-hĕlms-hä′fĕn)	194-95	53°32′N	8°07′E
Wilkes-Barre, Pa., U.S. (wĭlks′băr-ē)	116-17	41°15′N	75°53′W
Wilkes Land, reg., Ant.	287	69°0′S	120°00′E
Wilkie, Sk., Can. (wĭlk′ē)	134-35	52°25′N	108°40′W
Willamette, stm., Or., U.S.	112-13	45°39′N	122°46′W
Willard, Oh., U.S. (wĭl′ȧrd)	116-17	41°03′N	82°43′W
Willcox, Az., U.S. (wĭl′kŏks)	118-19	32°15′N	109°50′W
Willemstad, nat. cap., Neth. Ant. (vĭ′lĕm-stät)	140a	12°06′N	68°56′W
Williams, Az., U.S. (wĭl′yămz)	118-19	35°15′N	112°12′W
Williamsburg, Ia., U.S. (wĭl′yămz-bûrg)	114-15	41°40′N	92°01′W
Williamsburg, Ky., U.S. (wĭl′yămz-bûrg)	124-25	36°44′N	84°10′W
Williamsburg, Va., U.S. (wĭl′yămz-bûrg)	124-25	37°16′N	76°43′W
Williams Lake, B.C., Can. (wĭl′yămz lāk)	132-33	52°07′N	122°09′W
Williamson, W.V., U.S. (wĭl′yăm-sŭn)	116-17	37°40′N	82°17′W
Williamson Head, c., Ant.	287	69°12′S	157°47′E
Williamsport, Pa., U.S.	116-17	41°14′N	77°01′W
Williamston, N.C., U.S. (wĭl′yămz-tŭn)	124-25	35°51′N	77°04′W
Williamston, S.C., U.S. (wĭl′yămz-tŭn)	124-25	34°37′N	82°29′W
Williamstown, Ky., U.S. (wĭl′yămz-toun)	116-17	38°38′N	84°34′W
Williamstown, W.V., U.S. (wĭl′yămz-toun)	116-17	39°24′N	81°27′W
Willimantic, Ct., U.S. (wĭl-ĭ-măn′tĭk)	116-17	41°43′N	72°12′W
Willis, Tx., U.S. (wĭl′ĭs)	122-23	30°25′N	95°29′W
Williston, Fl., U.S. (wĭl′ĭs-tŭn)	124-25	29°23′N	82°27′W
Williston, N.D., U.S. (wĭl′ĭs-tŭn)	114-15	48°09′N	103°37′W
Williston, S.C., U.S. (wĭl′ĭs-tŭn)	124-25	33°24′N	81°26′W
Williston Lake, res., B.C., Can. (wĭl′ĭs-tŭn lāk)	130-31	56°00′N	123°54′W
Willmar, Mn., U.S. (wĭl′mär)	114-15	45°07′N	95°03′W
Willow, Ak., U.S. (wĭl′ō)	126	61°45′N	150°03′W
Willowmore, S. Afr. (wĭl′ō-môr)	264-65	33°18′S	23°29′E
Willows, Ca., U.S. (wĭl′ōz)	118-19	39°31′N	122°12′W
Willow Springs, Mo., U.S. (wĭl′ō sprĭngz)	124-25	36°60′N	91°58′W
Wills Point, Tx., U.S. (wĭlz point)	120-21	32°43′N	96°01′W
Wilmington, De., U.S.	116-17	39°45′N	75°32′W
Wilmington, N.C., U.S. (wĭl′mĭng-tŭn)	124-25	34°13′N	77°57′W
Wilmington, Oh., U.S. (wĭl′mĭng-tŭn)	116-17	39°26′N	83°50′W
Wilmore, Ky., U.S. (wĭl′mōr)	116-17	37°46′N	84°37′W
Wilna, nat. cap., Lith. *see* Vilnius	192-93	54°40′N	25°17′E
Wilson, N.C., U.S. (wĭl′sŭn)	124-25	35°44′N	77°55′W
Wilson, Ok., U.S. (wĭl′sŭn)	120-21	34°10′N	97°25′W
Wilson, Cape, c., Nu., Can.	130-31	67°01′N	81°30′W
Wilson, Mount, mtn., Ca., U.S. (mount wĭl′sŭn)	118-19	34°14′N	118°04′W
Wilson, Mount, mtn., Co., U.S. (mount wĭl′sŭn)	118-19	37°51′N	107°59′W
Wilsons Promontory, pen., Austl.	276	38°55′S	146°20′E
Wilton, N.D., U.S. (wĭl′tŭn)	114-15	47°09′N	100°47′W
Winamac, In., U.S. (wĭn′ȧ măk)	116-17	41°03′N	86°36′W
Winburg, S. Afr. (wĭm-bûrg)	269c	28°31′S	27°01′E
Winchester, In., U.S. (wĭn′chĕs-tẽr)	116-17	40°10′N	84°59′W
Winchester, Ky., U.S. (wĭn′chĕs-tẽr)	116-17	37°60′N	84°11′W
Winchester, Tn., U.S. (wĭn′chĕs-tẽr)	124-25	35°11′N	86°06′W
Winchester, Va., U.S. (wĭn′chĕs-tẽr)	116-17	39°11′N	78°10′W
Windau, Lat. *see* Ventspils	192-93	57°24′N	21°35′E
Wind Cave National Park, n.p., S.D., U.S. (wĭnd kāv năsh′ŭn-ăl pärk)	114-15	43°34′N	103°29′W
Windhoek, nat. cap., Nmb. (vĭnt′hòk)	264-65	22°34′S	17°05′E
Windom, Mn., U.S. (wĭn′dŭm)	114-15	43°52′N	95°07′W
Windorah, Austl.	276	25°25′S	142°39′E
Wind River Indian Reservation, ind. res., Wy., U.S. (wĭnd rĭv′ẽr ĭn′dĭ-ăn rĕ-sẽr-vā′shĕn)	112-13	43°26′N	109°00′W
Wind River Range, mts., Wy., U.S. (wĭnd rĭv′ẽr rānj)	112-13	43°05′N	109°25′W
Windsor, Austl. (wĭn′zẽr)	276	33°37′S	150°49′E
Windsor, N.S., Can. (wĭn′zẽr)	138-39	44°59′N	64°08′W
Windsor, On., Can. (wĭn′zẽr)	136-37	42°18′N	83°02′W
Windsor, Eng., U.K. (wĭn′zẽr)	190-91	51°29′N	0°36′W
Windsor, Co., U.S. (wĭn′zẽr)	120-21	40°28′N	104°53′W
Windsor, Mo., U.S. (wĭn′zẽr)	120-21	38°32′N	93°32′W
Windsor, N.C., U.S. (wĭn′zẽr)	124-25	35°60′N	76°57′W
Windsor, Vt., U.S. (wĭn′zẽr)	116-17	43°29′N	72°23′W
Windward Islands, is., N.A. (wĭnd′wẽrd ī′lăndz)	143b	13°0′N	61°00′W
Windward Passage, strt., N.A. (wĭnd′wẽrd păs′ĭj)	142-43	19°56′N	73°52′W
Winfield, Ks., U.S.	120-21	37°14′N	96°60′W
Wingham, On., Can.	136-37	43°53′N	81°18′W
Winisk, stm., On., Can.	130-31	55°16′N	85°07′W
Winisk Lake, lk., On., Can.	134-35	52°55′N	87°22′W
Winkler, Mb., Can. (wĭnk′lẽr)	134-35	49°11′N	97°56′W
Winnebago, Mn., U.S. (wĭn′ē-bā′gō)	114-15	43°46′N	94°11′W
Winnebago, Lake, lk., Wi., U.S. (lāk wĭn′ē-bā′gō)	116-17	44°0′N	88°25′W
Winnebago Indian Reservation, ind. res., Ne., U.S. (wĭn′ē-bā′gō ĭn′dĭ-ăn rĕ-sẽr-vā′shĕn)	114-15	42°15′N	96°31′W
Winnemucca, Nv., U.S. (wĭn-ē-mŭk′ȧ)	112-13	40°58′N	117°44′W
Winner, S.D., U.S. (wĭn′ẽr)	114-15	43°23′N	99°52′W
Winnfield, La., U.S. (wĭn′fēld)	122-23	31°55′N	92°38′W
Winnipeg, Mb., Can. (wĭn′ĭ-pĕg)	134-35	49°52′N	97°10′W
Winnipeg, stm., Can. (wĭn′ĭ-pĕg)	134-35	50°41′N	96°24′W
Winnipeg, Lake, lk., Mb., Can. (lāk wĭn′ĭ-pĕg)	134-35	52°0′N	97°00′W
Winnipegosis, Lake, lk., Mb., Can.	134-35	52°30′N	99°59′W
Winnsboro, La., U.S. (wĭnz′bŭr′ô)	124-25	32°10′N	91°43′W
Winnsboro, S.C., U.S. (wĭnz′bŭr′ô)	124-25	34°23′N	81°05′W
Winnsboro, Tx., U.S. (wĭnz′bŭr′ô)	120-21	32°58′N	95°17′W
Winona, Mi., U.S. (wĭ-nō′nȧ)	114-15	46°52′N	88°55′W
Winona, Mn., U.S. (wĭ-nō′nȧ)	114-15	44°03′N	91°38′W
Winona, Ms., U.S. (wĭ-nō′nȧ)	124-25	33°29′N	89°44′W
Winslow, Az., U.S. (wĭnz′lō)	118-19	35°02′N	110°42′W
Winsted, Ct., U.S. (wĭn′stĕd)	116-17	41°56′N	73°04′W
Winston-Salem, N.C., U.S. (wĭn stŭn-sā′lĕm)	124-25	36°06′N	80°14′W
Winter Garden, Fl., U.S. (wĭn′tẽr gär′d'n)	125a	28°34′N	81°35′W
Winter Haven, Fl., U.S. (wĭn′tẽr hā′vĕn)	125a	28°02′N	81°44′W
Winters, Tx., U.S. (wĭn′tẽrz)	122-23	31°57′N	99°58′W
Winterset, Ia., U.S. (wĭn′tẽr-sĕt)	114-15	41°20′N	94°01′W
Winterthur, Switz. (vĭn′tẽr-tōōr)	194-95	47°30′N	8°43′E
Winthrop, Me., U.S. (wĭn′thrŭp)	116-17	44°18′N	69°59′W
Winton, Austl. (wĭn-tŭn)	277	22°23′S	143°02′E
Wirāṭnagar, Nepal	234-35	26°28′N	87°17′E
Wīrgañj, Nepal	234-35	27°01′N	84°52′E
Wisby, Swe. *see* Visby	192-93	57°38′N	18°19′E
Wisconsin, state, U.S. (wĭs-kŏn′sĭn)	108-09	44°45′N	89°30′W
Wisconsin, stm., Wi., U.S. (wĭs-kŏn′sĭn)	114-15	42°59′N	91°09′W
Wisconsin Dells, Wi., U.S. (wĭs-kŏn′sĭn dĕlz)	116-17	43°38′N	89°45′W
Wisconsin Rapids, Wi., U.S. (wĭs-kŏn′sĭn răp′ĭdz)	116-17	44°23′N	89°49′W
Wishek, N.D., U.S. (wĭsh′ĕk)	114-15	46°16′N	99°33′W

Feature (Pronunciation)	Page	Lat.	Long.
Wisła, stm., Pol. (vēs´wȧ) see Vistula	194-95	54°21'N	18°56'E
Wisłoka, stm., Pol. (vēs-wō´kȧ)	194-95	50°27'N	21°24'E
Wismar, Ger. (vĭs´mär)	194-95	53°53'N	11°28'E
Wisner, Ne., U.S. (wĭz´nẽr)	114-15	41°59'N	96°56'W
Wissembourg, Fr. (vē-säɴ-bo͞or´)	196-97	49°02'N	7°57'E
Witbank, S. Afr. (wĭt-băŋk)	269c	25°53'S	29°14'E
Withlacoochee, stm., U.S. (wĭth-lȧ-ko͞o´chē)	124-25	30°23'N	83°10'W
Witu Islands, is., Pap. N. Gui.	277	4°45'S	149°19'E
W.J. van Blommestein Meer, res., Sur.	164-65	4°49'N	55°04'W
Wkra, stm., Pol. (f´krȧ)	194-95	52°27'N	20°45'E
Włocławek, Pol. (vwô-tswä´vĕk)	194-95	52°39'N	19°04'E
Włodawa, Pol. (vwô-dä´vä)	194-95	51°33'N	23°34'E
Włoszczowa, Pol. (vwôsh-chô´vä)	194-95	50°51'N	19°58'E
Wodonga, Austl.	276	36°08'S	146°53'E
Wokam, Pulau, i., Indon.	224-25	5°37'S	134°30'E
Woleai, at., Micron.	280-81	7°21'N	143°53'E
Woleu, stm., Afr. see Mbini	260-61	1°35'N	9°38'E
Wolf, Volcán, vol., Ec.	170a	0°00'S	91°20'W
Wolf Point, Mt., U.S. (wo͝olf point)	112-13	48°05'N	105°37'W
Wolfsburg, Ger. (vōlfs´bo͝orgh)	194-95	52°26'N	10°47'E
Wolfville, N.S., Can. (wo͝olf´vĭl)	138-39	45°05'N	64°22'W
Wollaston, Islas, is., Chile	171	55°45'S	67°37'W
Wollaston Lake, lk., Sk., Can. (wo͝ol´ȧs-tŭn lāk)	130-31	58°15'N	103°20'W
Wollaston Peninsula, pen., Can. (wo͝ol´ȧs-tŭn pĕ-nĭn´sūlȧ)	130-31	70°0'N	115°00'W
Wollongong, Austl. (wo͝ol´ŭn-gông)	276	34°25'S	150°54'E
Wołomin, Pol. (vô-wō´mēn)	194-95	52°20'N	21°15'E
Wolseley, Sk., Can.	134-35	50°25'N	103°16'W
Wolverhampton, Eng., U.K. (wo͝ol´vẽr-hămp-tŭn)	190-91	52°35'N	2°08'W
Wondai, Austl.	276	26°19'S	151°53'E
Wŏnju, Kor., S.	243	37°21'N	127°57'E
Wŏnsan, Kor., N. (wŭn´sän´)	243	39°09'N	127°26'E
Wonthaggi, Austl. (wônt-hăg´ē)	276	38°37'S	145°35'E
Woodbine, Ia., U.S. (wo͝od´bīn)	114-15	41°44'N	95°43'W
Woodbridge, Va., U.S. (wo͝od´brĭj´)	116-17	38°39'N	77°15'W
Wood Buffalo National Park, n.p., Can. (wo͝od bŭf´ȧ-lō năsh´ŭn-ăl pärk)	128-29	59°06'N	112°58'W
Woodburn, Or., U.S. (wo͝od´bûrn)	112-13	45°08'N	122°51'W
Woodlark, i., Pap. N. Gui. (wo͝od´lärk) see Murua Island	277	9°06'S	152°45'E
Woodroffe, Mount, mtn., Austl. (mount wo͝od´rŭf)	272-73	26°20'S	131°45'E
Woodruff, S.C., U.S. (wo͝od´rŭf)	124-25	34°45'N	82°02'W
Woods, Lake of the, lk., N.A. (lāk ŭv thȧ wo͝odz)	134-35	49°15'N	94°45'W
Woodsfield, Oh., U.S. (wo͝odz-fēld)	116-17	39°46'N	81°07'W
Woodstock, N.B., Can. (wo͝od´stŏk)	138-39	46°09'N	67°34'W
Woodstock, On., Can. (wo͝od´stŏk)	136-37	43°08'N	80°45'W
Woodstock, Il., U.S. (wo͝od´stŏk)	116-17	42°19'N	88°26'W
Woodstock, N.Y., U.S. (wo͝od´stŏk)	116-17	42°03'N	74°07'W
Woodstock, Va., U.S. (wo͝od´stŏk)	116-17	38°53'N	78°30'W
Woodsville, N.H., U.S. (wo͝odz´vĭl)	116-17	44°09'N	72°02'W
Woodville, Ms., U.S. (wo͝od´vĭl)	124-25	31°07'N	91°18'W
Woodville, Tx., U.S. (wo͝od´vĭl)	122-23	30°46'N	94°25'W
Woodward, Ok., U.S. (wo͝od´wôrd)	120-21	36°27'N	99°23'W
Woomera, Austl. (wo͞om´ẽrȧ)	276	31°12'S	136°50'E
Woonsocket, R.I., U.S. (wo͞on-sŏk´ĕt)	116-17	42°01'N	71°31'W
Wooramel, stm., Austl.	272-73	25°51'S	114°16'E
Wooster, Oh., U.S. (wo͝os´tẽr)	116-17	40°48'N	81°56'W
Worcester, S. Afr. (wo͝os´tẽr)	264-65	33°39'S	19°27'E
Worcester, Eng., U.K. (wo͝o´stẽr)	190-91	52°11'N	2°14'W
Worcester, Ma., U.S.	116-17	42°16'N	71°48'W
Worden, Mt., U.S. (wôr´dĕn)	112-13	45°58'N	108°13'W
Workington, Eng., U.K. (wûr´kĭng-tŭn)	190-91	54°38'N	3°34'W
Worksop, Eng., U.K. (wûrk´sŏp) (wûr´sŭp)	190-91	53°19'N	1°07'W
Worland, Wy., U.S. (wûr´lănd)	112-13	44°01'N	107°57'W
Worms, Ger. (vōrms)	194-95	49°38'N	8°21'E
Worthing, Eng., U.K. (wûr´dhĭng)	190-91	50°49'N	0°23'W
Worthington, In., U.S. (wûr´dhĭng-tŭn)	116-17	39°07'N	86°59'W
Worthington, Mn., U.S. (wûr´dhĭng-tŭn)	114-15	43°38'N	95°36'W
Wotho, at., Marsh. Is.	280-81	10°06'N	166°01'E
Wotje, at., Marsh. Is.	280-81	9°27'N	170°02'E
Wowoni, Pulau, i., Indon. (po͞o-lou wō-wō´nē)	248-49	4°08'S	123°06'E
Wrangel Island, i., Russia (răn´gĕl ī´lănd) see Vrangelya, Ostrov	218-19	71°14'N	179°21'W
Wrangell, Ak., U.S. (răn´gĕl)	126	56°29'N	132°22'W
Wrangell, Cape, c., Ak., U.S. (kāp răn´gĕl)	126a	52°55'N	172°30'E
Wrangell, Mount, mtn., Ak., U.S. (mount răn´gĕl)	126	62°0'N	144°06'W
Wrangell Mountains, mts., Ak., U.S. (răn´gĕl moun´tĭnz)	126	62°0'N	143°00'W
Wrangell-Saint Elias National Park and Preserve, n.p., Ak., U.S.	126	61°37'N	142°57'W
Wrath, Cape, c., Scot., U.K. (kāp răth)	190-91	58°38'N	4°60'W
Wray, Co., U.S. (rā)	120-21	40°05'N	102°14'W
Wrens, Ga., U.S. (rĕnz)	124-25	33°12'N	82°20'W
Wrexham, Wales, U.K. (rĕk´săm)	190-91	53°03'N	2°60'W
Wrightsville, Ga., U.S. (rīts´vĭl)	124-25	32°44'N	82°43'W
Wrigley, N.T., Can.	128-29	63°16'N	123°38'W
Wrocław, Pol. (vrôtsläv) (brĕs´lou)	194-95	51°07'N	17°02'E
Września, Pol. (vzhāsh´nyȧ)	194-95	52°20'N	17°35'E
Wu, stm., China (wo͞o´)	238-39	24°49'N	113°35'E
Wu, stm., China (wo͞o´)	238-39	27°11'N	109°48'E
Wuchang, China (wo͞o-chäŋ)	240-41	44°55'N	127°10'E
Wuchang, China see Wuhan	238-39	30°34'N	114°17'E
Wudaoliang, China	234-35	35°12'N	93°05'E
Wudu, China	238-39	33°25'N	104°51'E
Wugang, China	238-39	26°44'N	110°38'E
Wugong Shan, mts., China	238-39	27°21'N	113°50'E
Wuhai, China	240-41	39°40'N	106°48'E
Wuhan, China (wo͞o-hän´)	238-39	30°34'N	114°17'E
Wuhu, China (wo͞o´ho͞o)	238-39	31°21'N	118°22'E
Wūjang, China	234-35	33°37'N	79°48'E
Wukari, Nig.	260-61	7°53'N	9°47'E
Wuliang Shan, mts., China	238-39	24°29'N	100°39'E
Wuliaru, Pulau, i., Indon.	224-25	7°27'S	131°04'E
Wunnummin Lake, lk., On., Can.	134-35	52°55'N	89°10'W
Wupatki National Monument, n.p., Az., U.S.	118-19	35°32'N	111°26'W
Wuppertal, Ger. (vo͝op´ẽr-täl)	194-95	51°17'N	7°11'E
Würzburg, Ger. (vürts´bo͝orgh)	194-95	49°48'N	9°56'E
Wushan, China	238-39	31°06'N	109°50'E
Wushenqi, China	240-41	38°58'N	109°01'E
Wusuli, stm., Asia see Ussuri	240-41	48°27'N	135°04'E
Wutai, China	240-41	38°44'N	113°21'E
Wutai Shan, mtn., China	240-41	39°04'N	113°35'E
Wutongqiao, China	238-39	29°24'N	103°49'E
Wutsin, China see Changzhou	238-39	31°47'N	119°57'E
Wuvulu Island, i., Pap. N. Gui.	277	1°45'S	142°50'E
Wuwei, China (wo͞o´wā´)	238-39	31°18'N	117°54'E
Wuwei, China (wo͞o´wā´)	240-41	37°56'N	102°38'E
Wuxi, China (wo͞o-shyē)	238-39	31°22'N	109°33'E
Wuxi, China (wo͞o-shyē)	238-39	31°35'N	120°18'E
Wuxing, China (wo͞o-shyĭŋ) see Huzhou	238-39	30°52'N	120°06'E
Wuyi Shan, mts., China (wo͞o-yē shän)	238-39	27°42'N	117°09'E
Wuyuan, China	240-41	41°03'N	108°22'E
Wuzhong, China	240-41	37°59'N	106°12'E
Wuzhou, China (wo͞o-jō)	238-39	23°30'N	111°21'E
Wyandotte, Mi., U.S. (wī´ăn-dŏt)	116-17	42°13'N	83°09'W
Wyandra, Austl.	276	27°16'S	145°59'E
Wymore, Ne., U.S. (wī´mōr)	120-21	40°07'N	96°40'W
Wyndham, Austl. (wĭnd´ăm)	270-71	15°29'S	128°07'E
Wynne, Ar., U.S. (wĭn)	124-25	35°13'N	90°48'W
Wynnewood, Ok., U.S. (wĭn´wo͝od)	120-21	34°39'N	97°10'W
Wynyard, Sk., Can. (wĭn´yẽrd)	134-35	51°47'N	104°10'W
Wyoming, Mi., U.S. (wī-ō´mĭng)	116-17	42°55'N	85°43'W
Wyoming, state, U.S. (wī-ō´mĭng)	108-09	43°0'N	107°30'W
Wyong, Austl.	276	33°17'S	151°25'E
Wyszków, Pol. (vēsh´ko͝of)	194-95	52°36'N	21°28'E
Wytheville, Va., U.S. (wĭth´vĭl)	124-25	36°57'N	81°06'W

X

Feature (Pronunciation)	Page	Lat.	Long.
Xaafuun, Raas, c., Som.	262-63	10°26'N	51°25'E
Xaidulla, China	234-35	36°26'N	77°58'E
Xainza, China	234-35	30°55'N	88°40'E
Xai-Xai, Moz.	264-65	25°03'S	33°39'E
Xalapa, Mex.	146-47	19°32'N	96°55'W
Xam, stm., Asia see Chu	246-47	19°53'N	105°45'E
Xam Nua, Laos	246-47	20°25'N	104°03'E
Xankändi, Azer.	227	39°49'N	46°45'E
Xapuri, Braz.	166-67	10°39'S	68°31'W
Xar Moron, stm., China	240-41	43°25'N	120°45'E
Xau, Lake, pl., Bots.	264-65	21°18'S	24°44'E
Xäzär, Dänizi, lk., see Caspian Sea	226	41°18'N	50°59'E
Xcalak, Mex. (sä-lä´k)	148	18°16'N	87°50'W
Xenia, Oh., U.S. (zē´nĭ-ȧ)	116-17	39°41'N	83°56'W
Xeres, Spain see Jerez de la Frontera	198-99	36°42'N	6°08'W
Xi, stm., China (shyē)	238-39	22°20'N	113°18'E
Xi, stm., China (shyē)	240-41	42°25'N	100°55'E
Xiaguan, China see Dali	238-39	25°36'N	100°13'E
Xiahe, China	240-41	35°24'N	102°32'E
Xiamen, China	225a	24°27'N	118°07'E
Xi'an, China (shyē-än)	238-39	34°15'N	108°52'E
Xiangfan, China	238-39	32°02'N	112°09'E
Xianggang, China see Hong Kong	238-39	22°16'N	114°10'E
Xiangkhoang, Laos	246-47	19°20'N	103°22'E
Xiangquan, stm., Asia see Sutlej	234-35	29°21'N	71°02'E
Xiangride, China	240-41	35°60'N	97°59'E
Xiangtan, China (shyäŋ-tän)	238-39	27°51'N	112°54'E
Xiantao, China	238-39	30°22'N	113°27'E
Xianyang, China (shyĕn-yäŋ)	238-39	34°20'N	108°42'E
Xianyou, China	225a	25°22'N	118°40'E
Xiaogan, China	238-39	30°55'N	113°54'E
Xiao Hinggan Ling, mts., China see Lesser Khingan Range	240-41	48°45'N	127°00'E
Xiapu, China (shyä-po͞o)	238-39	26°52'N	120°01'E
Xibaxa, stm., Asia see Subansiri	238-39	26°46'N	93°45'E
Xichang, China	238-39	27°54'N	102°16'E
Xicoténcatl, Mex. (sē-kō-tĕn-kät´l)	146-47	23°00'N	98°56'W
Xifeng, China (shyē-fŭŋ)	240-41	42°44'N	124°43'E
Xigazê, China	234-35	29°16'N	88°54'E
Xilinhot, China	240-41	43°56'N	116°03'E
Ximiao, China	240-41	41°07'N	100°17'E
Xinchang, China (shyĭn-chäŋ)	238-39	29°31'N	120°53'E
Xing'an, China (shyĭŋ-än)	238-39	25°37'N	110°31'E
Xinghua, China (shyĭŋ-hwä)	238-39	32°56'N	119°50'E
Xingkai Hu, lk., Asia see Khanka, Lake	244	45°11'N	132°25'E
Xingtai, China (shyĭŋ-tī)	240-41	37°04'N	114°30'E
Xingu, stm., Braz. (zhĕŋ-go͞o´)	166-67	1°30'S	51°50'W
Xingyi, China	238-39	25°05'N	104°54'E
Xinhua, China (shyĭn-hwä)	238-39	27°37'N	111°02'E
Xining, China (shyē-nĭŋ)	240-41	36°38'N	101°50'E
Xinjiang, China	240-41	35°37'N	111°13'E
Xinjiang, state, China (shyĭn-jyäŋ)	220-21	40°0'N	85°00'E
Xinjiulong, China	238-39	22°21'N	114°10'E
Xinmin, China (shyĭn-mĭn)	243	41°59'N	122°50'E
Xinpu, China see Lianyungang	238-39	34°37'N	119°11'E
Xintai, China (shyĭn-tī)	240-41	35°54'N	117°46'E
Xinxian, China (shyĭn shyĕn)	240-41	38°24'N	112°44'E
Xinxiang, China (shyĭn-shyäŋ)	240-41	35°18'N	113°52'E
Xinyang, China (shyĭn-yäŋ)	238-39	32°07'N	114°04'E
Xinye, China (shyĭn-yŭ)	238-39	32°33'N	112°21'E
Xiping, China (shyē-pĭŋ)	238-39	33°23'N	114°01'E
Xique-Xique, Braz.	166-67	10°50'S	42°43'W
Xırdalan, Azer.	227	40°28'N	49°46'E
Xisha Qundao, is., China see Paracel Islands	224-25	15°46'N	112°17'E
Xishui, China (shyē-shwā)	238-39	30°28'N	115°15'E
Xixian, China (shyē shyĕn)	238-39	32°21'N	114°44'E
Xizang, state, China (shyē-dzäŋ) see Tibet	234-35	32°0'N	88°00'E
Xongka, stm., Asia see Ca	246-47	18°44'N	105°45'E
Xuancheng, China (shyůän-chŭŋ)	238-39	30°57'N	118°45'E
Xuanhua, China (shyůän-hwä)	240-41	40°36'N	115°02'E
Xuchang, China (shyo͞o-chäŋ)	238-39	34°02'N	113°49'E
Xun, stm., China (shyo͝on)	238-39	23°26'N	111°30'E
Xuwen, China	238-39	20°20'N	110°11'E
Xuyong, China	238-39	28°10'N	105°25'E
Xuzhou, China	238-39	34°16'N	117°11'E

Y

Feature (Pronunciation)	Page	Lat.	Long.
Yaan, China (yä-än)	238-39	30°01'N	103°04'E
Yablonovy Range, mts., Russia (yȧ-blô-nô-vē´ fāŋj) see Yablonovyy Khrebet	218-19	53°30'N	115°00'E
Yablonovyy Khrebet, mts., Russia	218-19	53°30'N	115°00'E
Yaco, stm., S.A. see Iaco	170	9°02'S	68°35'W
Yacuiba, Bol. (yȧ-ko͞o-ē´bȧ)	168-69	22°02'S	63°42'W
Yacyretá, Isla, i., Para.	173	27°25'S	56°30'W
Yadong, China	234-35	27°29'N	88°54'E
Yafran, Libya	188-89	32°04'N	12°31'E
Yagodnoye, Russia	218-19	62°32'N	149°37'E
Yaguajay, Cuba (yä-guä-hä´ē)	142-43	22°20'N	79°14'W
Yahualica, Mex. (yä-wä-lē´kä)	146-47	21°09'N	102°51'W
Yaitopya, nation, Afr. see Ethiopia	253	9°0'N	39°00'E
Yakima, Wa., U.S. (yăk´ĭmȧ)	112-13	46°36'N	120°30'W

ăt; finăl; rāte; senâte; ärm; ȧsk; sofȧ; fâre; ch-choose; dh-as th in other; bē; ēvent; bĕt; recĕnt; cratẽr; g-gō; gh-guttural g; bĭt; ĭ-short neutral; rīde; ĸ-guttural k as ch in German ich;

Feature (Pronunciation)	Page	Lat.	Long.
Yakoma, D.R.C.	262-63	4°04'N	22°26'E
Yaku-shima, i., Japan (yä'kōō shē'mä)	244	30°20'N	130°30'E
Yakutat, Ak., U.S. (yăk'ô-tăt)	126	59°32'N	139°43'W
Yakutat Bay, b., Ak., U.S. (yōō-kū-tăt' bā)	126	59°40'N	140°00'W
Yakutia, state, Russia	218-19	67°0'N	125°00'E
Yakutsk, Russia (yà-kôtsk')	218-19	62°02'N	129°42'E
Yala, Thai.	246-47	6°33'N	101°17'E
Yalgoo, Austl.	270-71	28°21'S	116°41'E
Yalong, stm., China (yä-lôŋ)	238-39	26°36'N	101°48'E
Yalta, Ukr. (yäl'tà)	202-03	44°30'N	34°10'E
Yalu, stm., Asia	243	39°57'N	124°22'E
Yamagata, Japan	245	38°15'N	140°20'E
Yamaguchi, Japan	245	34°11'N	131°29'E
Yamal, Poluostrov, pen., Russia (yä-mäl')	218-19	70°0'N	70°00'E
Yamal Peninsula, pen., Russia (yŭ-mäl' pĕ-nĭn'sūlă) *see* Yamal, Poluostrov	218-19	70°0'N	70°00'E
Yamantau, Gora, mtn., Russia (gà-rä' yä'man-tàw)	226	54°15'N	58°06'E
Yamarovka, Russia	240-41	50°34'N	110°25'E
Yambio, Sudan	267	4°34'N	28°24'E
Yambol, Blg. (yàm'bôl)	200-01	42°29'N	26°31'E
Yamdena, Pulau, i., Indon.	224-25	7°36'S	131°25'E
Yamethin, Mya. (yŭ-mē'thĕn)	246-47	20°28'N	96°09'E
Yamma Yamma, Lake, lk., Austl. (lăk yäm'à yäm'à)	276	26°20'S	141°25'E
Yamoussoukro, nat. cap., C. Iv. (yä-mōō-sōō'-krō)	260-61	6°49'N	5°17'W
Yamuna, stm., India	234-35	25°26'N	81°54'E
Yamzho Yumco, lk., China (yäm-jwo yōōm-tswo)	234-35	28°58'N	90°45'E
Yana, stm., Russia (yä'nà)	218-19	71°32'N	136°38'E
Yanac, Austl. (yăn'ăk)	276	36°09'S	141°25'E
Yanam, India (yŭnŭm')	236	16°44'N	82°15'E
Yan'an, China (yän-än)	240-41	36°35'N	109°29'E
Yanbu'al-Baḥr, Sau. Ar.	266	24°05'N	38°05'E
Yanchang, China	240-41	36°35'N	110°01'E
Yancheng, China (yän-chŭŋ)	238-39	33°24'N	120°09'E
Yanchi, China	240-41	37°47'N	107°23'E
Yandé, Île, i., N. Cal.	279g	20°03'S	163°48'E
Yangambi, D.R.C.	262-63	0°46'N	24°27'E
Yangchun, China (yäŋ-chòn)	238-39	22°11'N	111°47'E
Yanggu, China (yäŋ-gōō)	240-41	36°07'N	115°46'E
Yangiyŭl, Uzb.	232-33	41°09'N	69°07'E
Yangjiang, China (yäŋ-jyäŋ)	238-39	21°53'N	111°58'E
Yangon, nat. cap., Mya. (yän'gŏn')	246-47	16°47'N	96°12'E
Yangquan, China (yäŋ-chyůän)	240-41	37°51'N	113°34'E
Yangtze, stm., China (yăng'tse)	238-39	31°24'N	121°54'E
Yangxin, China (yäŋ-shyĭn)	238-39	29°50'N	115°13'E
Yangzhou, China (yäŋ-jō)	238-39	32°24'N	119°25'E
Yanji, China (yän-jyē)	243	42°46'N	129°26'E
Yanji, China (yän-jyē)	243	42°54'N	129°30'E
Yankton, S.D., U.S. (yănk'tŭn)	114-15	42°52'N	97°24'W
Yanqi, China	222-23	42°03'N	86°34'E
Yanshan, China (yän-shän)	238-39	23°37'N	104°20'E
Yanshou, China (yän-shō)	244	45°26'N	128°21'E
Yantai, China	240-41	37°32'N	121°21'E
Yanzhou, China (yäŋ-jō)	240-41	35°33'N	116°49'E
Yaoundé, nat. cap., Camrn. (yà'-ōōn-dā')	260-61	3°52'N	11°31'E
Yap, i., Micron. (yăp)	280-81	9°31'N	138°06'E
Yapacaní, Bol.	168-69	16°45'S	64°18'W
Yapen, Pulau, i., Indon.	224-25	1°45'S	136°15'E
Yaponskoye More, s., Asia *see* Japan, Sea of	222-23	40°0'N	135°00'E
Yaque del Norte, stm., Dom. Rep. (yä'kå dĕl nôr'tå)	142-43	19°50'N	71°41'W
Yaqui, stm., Mex. (yä'kē)	144-45	27°40'N	110°38'W
Yaransk, Russia (yä-ränsk')	186-87	57°18'N	47°54'E
Yarensk, Russia	186-87	62°09'N	49°02'E
Yarí, stm., Col.	164-65	0°19'S	72°21'W
Yarkand, stm., China *see* Yarkant	226	40°28'N	80°51'E
Yarkant, China *see* Shache	226	38°25'N	77°15'E
Yarkant, stm., China	226	40°28'N	80°51'E
Yarlung, stm., Asia *see* Brahmaputra	234-35	24°02'N	91°00'E
Yarmouth, N.S., Can. (yär'mŭth)	138-39	43°50'N	66°06'W
Yaroslavl', Russia (yà-rô-släv''l)	202-03	57°37'N	39°52'E
Yar-Sale, Russia	218-19	66°51'N	70°53'E
Yartsevo, Russia (yär'tsyĕ-vô)	202-03	55°04'N	32°42'E
Yarumal, Col. (yä-rōō-mäl')	164-65	6°58'N	75°24'W
Yasawa, i., Fiji	279f	16°47'S	177°31'E
Yashiro-jima, i., Japan	245	33°55'N	132°15'E
Yass, Austl.	276	34°51'S	148°55'E
Yassy, Rom. *see* Iași	202-03	47°10'N	27°36'E
Yata, stm., Bol.	166-67	10°29'S	65°26'W
Yathkyed Lake, lk., Nu., Can. (yàth-kī-ĕd' lāk)	130-31	62°41'N	98°00'W
Yatsuga-take, mtn., Japan (yät'sōō-gä-dä'kä)	245	35°59'N	138°23'E
Yatsushiro, Japan (yät'sōō'shē-rô)	245	32°30'N	130°36'E
Yaundé, nat. cap., Camrn. *see* Yaoundé	260-61	3°52'N	11°31'E
Yautepec, Mex. (yä-ōō-tå-pĕk')	146-47	18°52'N	99°03'W
Yavari, stm., S.A.	170	4°21'S	70°02'W
Yaví, Cerro, mtn., Ven.	164-65	5°32'N	65°59'W
Yawata, Japan (yä'wä-tä) *see* Kitakyūshū	245	33°54'N	130°51'E
Yawatahama, Japan (yä'wä'tä'hä-mä)	245	33°27'N	132°26'E
Yaxian, China (yä shyĕn) *see* Sanya	238-39	18°14'N	109°30'E
Yazd, Iran	232-33	31°54'N	54°22'E
Yazoo, stm., Ms., U.S. (yă'zōō)	124-25	32°23'N	91°00'W
Yazoo City, Ms., U.S. (yă'zōō sĭ'tĭ)	124-25	32°51'N	90°25'W
Ye, Mya. (yā)	246-47	15°15'N	97°52'E
Yecla, Spain (yā'klä)	198-99	38°37'N	1°07'W
Yefremov, Russia (yĕ-frä'môf)	202-03	53°09'N	38°07'E
Yegor'yevsk, Russia (yĕ-gôr'yĕfsk)	202-03	55°23'N	39°02'E
Yei, Sudan	267	4°06'N	30°40'E
Yei, stm., Sudan	262-63	6°15'N	30°13'E
Yekaterinburg, Russia	218-19	56°51'N	60°36'E
Yekaterinoslav, Ukr. *see* Dnipropetrovs'k	202-03	48°28'N	34°58'E
Yelabuga, Russia (yĕ-lä'bŏ-gà)	186-87	55°46'N	52°05'E
Yelizavety, Mys, c., Russia (mĭs yĕ-lyĕ-sà-vyĕ'tĭ)	218-19	54°24'N	142°42'E
Yellow, stm., China *see* Huang	222-23	37°49'N	118°53'E
Yellowhead Pass, p., Can. (yĕl'ô-hĕd păs)	132-33	52°54'N	118°22'W
Yellowknife, N.T., Can. (yĕl'ô-nīf)	128-29	62°27'N	114°21'W
Yellow Sea, s., Asia (yĕl'ô sē)	222-23	36°0'N	123°00'E
Yellowstone, stm., U.S. (yĕl'ô-stōn)	110-11	47°59'N	103°60'W
Yellowstone Lake, lk., Wy., U.S.	112-13	44°28'N	110°23'W
Yellowstone National Park, n.p., U.S. (yĕl'ô-stōn năsh'ŭn-ăl pärk)	112-13	44°30'N	110°35'W
Yel'nya, Russia (yĕl'nyà)	202-03	54°34'N	33°11'E
Yemen, nation, Asia (yĕm'ĕn)	206-07	15°0'N	44°00'E
Yenangyaung, Mya. (yā'nän-d oung)	246-47	20°27'N	94°53'E
Yen Bai, Viet.	246-47	21°42'N	104°52'E
Yendi, Ghana (yĕn'dē)	260-61	9°26'N	0°00'W
Yenisey, stm., Russia (yĕ-nĕ-sĕ'ē)	218-19	71°54'N	82°20'E
Yeo Lake, lk., Austl. (yō lăk)	272-73	27°58'S	124°25'E
Yeppoon, Austl.	277	23°08'S	150°45'E
Yerevan, nat. cap., Arm. (yĕ-rĕ-vän')	227	40°11'N	44°30'E
Yergeni, hills, Russia	186-87	47°0'N	44°00'E
Yerington, Nv., U.S. (yĕ'rĭng-tŭn)	118-19	38°59'N	119°09'W
Yershov, Russia	186-87	51°21'N	48°16'E
Yerupaja, Nevado, mtn., Peru	170	10°16'S	76°54'W
Yerushalayim, nat. cap., Isr. *see* Jerusalem	228-29	31°47'N	35°14'E
Yessentuki, Russia	227	44°02'N	42°51'E
Ye-u, Mya.	246-47	22°46'N	95°26'E
Yeu, Île d', i., Fr. (ēl dyû)	196-97	46°42'N	2°20'W
Yevlax, Azer.	227	40°37'N	47°09'E
Yevpatoriia, Ukr.	202-03	45°12'N	33°22'E
Yeya, stm., Russia (yä'yà)	186-87	46°40'N	38°36'E
Yeysk, Russia (yĕysk)	202-03	46°42'N	38°16'E
Yezd, Iran *see* Yazd	232-33	31°54'N	54°22'E
Ygatimí, Para.	168-69	24°05'S	55°24'W
Yi'an, China	240-41	47°53'N	125°18'E
Yibin, China (yē-bĭn)	238-39	28°46'N	104°37'E
Yichang, China (yē-chäŋ)	238-39	30°42'N	111°17'E
Yichun, China	238-39	27°48'N	114°23'E
Yichun, China	240-41	47°43'N	128°55'E
Yidu, China (yē-dōō)	238-39	30°24'N	111°26'E
Yilan, China (yē-län)	244	46°18'N	129°32'E
Yiliang, China	238-39	24°57'N	103°08'E
Yinchuan, China (yĭn-chůän)	240-41	38°28'N	106°16'E
Ying, stm., China	238-39	32°30'N	116°31'E
Yingkou, China (yĭŋ-kō)	240-41	40°40'N	122°14'E
Yingtan, China	238-39	28°14'N	117°02'E
Yining, China (yē-nĭŋ)	226	43°55'N	81°18'E
Yirga 'Alem, Eth.	269d	6°52'N	38°24'E
Yirol, Sudan	262-63	6°33'N	30°30'E
Yishui, China (yē-shwā)	240-41	35°47'N	118°38'E
Yisra'el, nation, Asia *see* Israel	206-07	31°30'N	34°45'E
Yitulihe, China	240-41	50°37'N	121°33'E
Yiyang, China (yē-yäŋ)	238-39	28°24'N	117°26'E
Yiyang, China (yē-yäŋ)	238-39	28°35'N	112°20'E
Yoakum, Tx., U.S. (yō'kŭm)	122-23	29°17'N	97°09'W
Yockanookany, stm., Ms., U.S. (yŏk'á-nōō-kă-nĭ)	124-25	32°40'N	89°41'W
Yog Point, c., Phil. (yōg point)	250	14°06'N	124°12'E
Yogyakarta, Indon. (yŏg-yà-kär'tà)	248-49	7°48'S	110°22'E
Yokadouma, Camrn.	260-61	3°31'N	15°03'E
Yokkaichi, Japan (yô'kä'ē-chē)	245	34°58'N	136°39'E
Yokohama, Japan (yō'kô-hä'mä)	245	35°27'N	139°37'E
Yokosuka, Japan (yô-kō'sò-kä)	245	35°16'N	139°40'E
Yolöten, Turkmen.	232-33	37°17'N	62°22'E
Yom, stm., Thai.	246-47	15°52'N	100°16'E
Yonago, Japan (yō'nä-gō)	245	35°26'N	133°20'E
Yonezawa, Japan (yō'nĕ'zà-wä)	245	37°55'N	140°07'E
Yong'an, China (yôŋ-än)	238-39	25°58'N	117°22'E
Yongdeng, China	240-41	36°43'N	103°16'E
Yongding, China	238-39	24°44'N	116°44'E
Yongding, stm., China (yôŋ-dĭŋ)	240-41	39°16'N	117°04'E
Yŏngdŏk, Kor., S. (yŭng'dŭk')	243	36°26'N	129°23'E
Yongfeng, China	238-39	27°19'N	115°24'E
Yongnian, China (yôŋ-nĭĕn)	240-41	36°47'N	114°29'E
Yongren, China	238-39	26°06'N	101°48'E
Yongshan, China	238-39	28°09'N	103°32'E
Yongshun, China (yôŋ-shòn)	238-39	29°00'N	109°51'E
Yongzhou, China	238-39	26°13'N	111°37'E
Yonkers, N.Y., U.S. (yŏŋ'kẽrz)	116-17	40°56'N	73°54'W
York, Austl.	270-71	31°53'S	116°46'E
York, Eng., U.K.	190-91	53°58'N	1°05'W
York, Al., U.S. (yôrk)	124-25	32°30'N	88°18'W
York, Ne., U.S. (yôrk)	114-15	40°52'N	97°35'W
York, Pa., U.S.	116-17	39°58'N	76°44'W
York, S.C., U.S. (yôrk)	124-25	34°59'N	81°15'W
York, Cape, c., Austl. (kāp yôrk)	277	10°42'S	142°32'E
York Factory, Mb., Can.	134-35	57°03'N	92°15'W
Yorkton, Sk., Can. (yôrk'tŭn)	134-35	51°13'N	102°28'W
Yorktown, Tx., U.S. (yôrk'toun)	122-23	28°59'N	97°30'W
Yorktown, Va., U.S. (yôrk'toun)	124-25	37°14'N	76°31'W
Yoro, Hond. (yō'rô)	149	15°09'N	87°08'W
Yosemite National Park, n.p., Ca., U.S. (yô-sĕm'ĭ-tē năsh'ŭn-ăl pärk)	118-19	37°56'N	119°36'W
Yoshino, stm., Japan (yō'shē-nō)	245	34°04'N	134°37'E
Yoshkar-Ola, Russia (yôsh-kär'ô-lä')	186-87	56°39'N	47°52'E
Yösöbulag, Mong. *see* Altay	240-41	46°24'N	96°15'E
Yos Sudarso, Pulau, i., Indon.	277	7°50'S	138°30'E
Yŏsu, Kor., S. (yŭ'sōō')	243	34°44'N	127°44'E
You, stm., China (yō)	238-39	22°50'N	108°06'E
You, stm., China (yō)	225a	26°24'N	118°27'E
You, stm., China (yō)	238-39	28°27'N	110°23'E
Youghal, Ire. (yōō'ôl) (yôl)	190-91	51°57'N	7°51'W
Young, Austl. (yŭng)	276	34°19'S	148°18'E
Young, Ur. (yô-ōō'ng)	173	32°42'S	57°38'W
Youngstown, Oh., U.S.	116-17	41°06'N	80°39'W
Youssoufia, Mor.	269a	32°15'N	8°32'W
Youyang, China	238-39	28°58'N	108°41'E
Yozgat, Tur. (yôz'gàd)	186-87	39°49'N	34°48'E
Ypsilanti, Mi., U.S. (ĭp-sĭ-lăn'tĭ)	116-17	42°15'N	83°37'W
Yreka, Ca., U.S. (wī-rē'kå)	112-13	41°44'N	122°38'W
Ystad, Swe.	192-93	55°26'N	13°50'E
Ysyk-Köl, lk., Kyrg. *see* Issyk-Kul, Lake	226	42°25'N	77°15'E
Ytyk-Kyuyël', Russia	218-19	62°21'N	133°33'E
Yu, stm., China	238-39	23°24'N	110°06'E
Yuan, stm., Asia *see* Red	238-39	20°18'N	106°32'E
Yuan, stm., China (yůän)	238-39	28°45'N	111°20'E
Yuan'an, China (yůän-än)	238-39	31°04'N	111°25'E
Yuanling, China (yůän-lĭŋ)	238-39	28°20'N	110°16'E
Yuanmou, China	238-39	25°43'N	101°52'E
Yuba City, Ca., U.S. (yōō'bå sĭ'tĭ)	118-19	39°08'N	121°37'W
Yucatán, state, Mex. (yōō-kä-tän')	148	20°50'N	89°00'W
Yucatán, Canal de, strt., N.A. *see* Yucatan Channel	140-41	21°42'N	86°04'W
Yucatán, Península de, pen., N.A. *see* Yucatan Peninsula	148	19°30'N	89°00'W
Yucatan Channel, strt., N.A. (yōō-kä-tän' chăn'ĕl)	140-41	21°42'N	86°04'W
Yucatan Peninsula, pen., N.A. (yōō-kä-tän' pĕ-nĭn'sūlă)	148	19°30'N	89°00'W
Yuci, China (yōō-tsz)	240-41	37°41'N	112°44'E
Yudoma, stm., Russia (yōō-dō'mà)	218-19	59°10'N	135°14'E
Yueqing, China (yůĕ-chyĭn)	238-39	28°07'N	120°57'E
Yueyang, China (yůĕ-yäŋ)	238-39	29°22'N	113°06'E
Yug, stm., Russia (yòg)	186-87	60°43'N	46°19'E
Yukagirskoye Ploskogor'ye, plat., Russia	218-19	66°0'N	155°00'E
Yukhnov, Russia (yòk'nof)	202-03	54°44'N	35°15'E
Yukon, state, Can. (yōō'kŏn)	128-29	64°0'N	135°00'W
Yukon, stm., N.A. (yōō'kŏn)	126	62°36'N	164°49'W
Yulin, China (yōō-lĭn)	238-39	22°38'N	110°07'E

n-sing; ŋ-baŋk; N-nasalized n; nŏd; cŏmmit; ōld; ôbey; ôrder; oi-boil; fōōd; ò-as oo in foot; ou-out; s-soft; sh-dish; th-thin; pūre; ůnite; ûrn; stŭd; circŭs; ü-as in French tu; '-indeterminate vowel.

Feature (Pronunciation)	Page	Lat.	Long.
Yulin, China (yōō-lĭn)	240-41	38°17′N	109°45′E
Yuma, Az., U.S. (yōō´mä)	118-19	32°43′N	114°37′W
Yuma, Co., U.S. (yōō´mä)	120-21	40°07′N	102°44′W
Yumen, China (yōō-mŭn)	240-41	39°50′N	97°34′E
Yuncheng, China (yòn-chŭŋ)	238-39	35°01′N	110°59′E
Yuncheng, China (yòn-chŭŋ)	240-41	35°36′N	115°56′E
Yunnan, state, China (yun´nän´)	238-39	24°0′N	101°00′E
Yunxian, China (yòn shyĕn)	238-39	24°31′N	100°02′E
Yunxian, China (yòn shyĕn)	238-39	32°49′N	110°49′E
Yunxiao, China (yòn-shyou)	238-39	24°04′N	117°19′E
Yurga, Russia	218-19	55°43′N	84°55′E
Yurimaguas, Peru (yōō-rē-mä´gwäs)	170	5°54′S	76°06′W
Yuruá, stm., S.A. *see* Juruá	166-67	2°35′S	65°47′W
Yuryev, Est. *see* Tartu	192-93	58°23′N	26°43′E
Yuscarán, Hond. (yōōs-kä-rän´)	149	13°56′N	86°50′W
Yü Shan, mtn., Tai.	225a	23°28′N	120°57′E
Yushu, China (yōō-shōō)	238-39	33°00′N	97°00′E
Yushu, China (yōō-shōō)	240-41	44°50′N	126°34′E
Yutian, China (yōō-tĭĕn) (kŭ-r-yä)	234-35	36°51′N	81°40′E
Yuty, Para. (yōō-tē´)	173	26°38′S	56°11′W
Yuxian, China (yōō shyĕn)	238-39	34°10′N	113°28′E
Yuxian, China (yōō shyĕn)	240-41	38°06′N	113°24′E
Yuzha, Russia (yōō´zhà)	186-87	56°35′N	42°01′E
Yuzhno-Sakhalinsk, Russia (yōōzh´nô-sä-кä-lĭnsk´)	244	46°58′N	142°43′E
Yuzovka, Ukr. *see* Donets'k	202-03	47°60′N	37°48′E
Yvetot, Fr. (ēv-tō´)	196-97	49°37′N	0°46′E

Z

Feature (Pronunciation)	Page	Lat.	Long.
Zaandam, Neth. (zän´däm)	190-91	52°26′N	4°49′E
Zabaykal'sk, Russia	240-41	49°39′N	117°19′E
Zabīd, Yemen	266	14°11′N	43°19′E
Zābol, Iran	230-31	31°02′N	61°30′E
Zabrze, Pol. (zäb´zhě)	194-95	50°18′N	18°46′E
Zacapa, Guat. (sä-kä´pä)	148	14°58′N	89°32′W
Zacatecas, Mex. (sä-kä-tā´käs)	146-47	22°46′N	102°34′W
Zacatecas, state, Mex.	146-47	23°0′N	103°00′W
Zacatlán, Mex. (sä-kä-tlän´)	146-47	19°56′N	97°58′W
Zacoalco de Torres, Mex. (sä-kō-äl´kō dā tōr´rěs)	146-47	20°13′N	103°34′W
Zacualtipan, Mex. (sà-kò-äl-tē-pän´)	146-47	20°38′N	98°39′W
Zadar, Cro. (zä´där)	200-01	44°07′N	15°14′E
Zagazig, Egypt	268b	30°35′N	31°31′E
Zagreb, nat. cap., Cro. (zä´grěb)	200-01	45°49′N	15°58′E
Zāgros, Kūhhā-ye, mts., Iran *see* Zagros Mountains	220-21	33°03′N	48°33′E
Zagros Mountains, mts., Iran (zä´grōs moun´tĭnz)	220-21	33°03′N	48°33′E
Za'gya, stm., China	234-35	31°56′N	88°60′E
Zāhedān, Iran (zä´hå-dän)	230-31	29°30′N	60°52′E
Zaḩlah, Leb. (zä´lä´)	228-29	33°51′N	35°55′E
Zaḩodnjaja Dzvina, stm., Eur. *see* Western Dvina	192-93	57°04′N	24°03′E
Zaire, nation, Afr. *see* Congo, Democratic Republic of the	253	4°0′S	25°00′E
Zaire, stm., Afr. *see* Congo	262-63	5°58′S	12°44′E
Zaječar, Serb. (zä´yě-chär´)	200-01	43°55′N	22°17′E
Zakamensk, Russia	240-41	50°23′N	103°16′E
Zakopane, Pol. (zà-kô-pä´ně)	194-95	49°18′N	19°58′E
Zákynthos, i., Grc.	200-01	37°47′N	20°47′E
Zalaegerszeg, Hung. (zŏ´lô-ĕ´gĕr-sĕg)	194-95	46°51′N	16°51′E
Zalantun, China	240-41	47°60′N	122°45′E
Zalingei, Sudan	262-63	12°55′N	23°29′E
Zambeze, stm., Afr. *see* Zambezi	264-65	18°49′S	36°15′E
Zambezi, stm., Afr. (zàm-bā´zē)	264-65	18°49′S	36°15′E
Zambia, nation, Afr. (zăm´bē-å)	253	14°30′S	27°30′E
Zamboanga, Phil. (säm-bō-aŋ´gä)	250	6°54′N	122°04′E
Zamboanga Peninsula, pen., Phil.	250	7°32′N	122°16′E
Zambrów, Pol. (zäm´brōf)	194-95	52°58′N	22°15′E
Zamfara, stm., Nig.	260-61	12°02′N	4°03′E
Zamora, Spain (thä-mō´rä)	198-99	41°31′N	5°45′W
Zamora de Hidalgo, Mex.	146-47	19°59′N	102°17′W
Zanesville, Oh., U.S. (zānz´vĭl)	116-17	39°56′N	82°01′W
Zanjān, Iran	228-29	36°40′N	48°29′E
Zante, i., Grc. *see* Zákynthos	200-01	37°47′N	20°47′E
Zanzibar, Tan. (zăn´zĭ-bär)	262-63	6°10′S	39°12′E
Zanzibar, i., Tan. (zăn´zĭ-bär)	262-63	6°10′S	39°20′E
Zaozhuang, China (dzou-jůäŋ)	238-39	34°52′N	117°33′E
Zapadnaya Dvina, stm., Eur. *see* Western Dvina	192-93	57°04′N	24°03′E
Zapadno-Sibirskaya Ravnina, pl., Russia *see* West Siberian Plain	218-19	60°0′N	75°00′E
Zapala, Arg. (zä-pä´lä)	171	38°54′S	70°03′W
Zapata, Tx., U.S. (sä-pä´tä)	122-23	26°55′N	99°16′W
Zapata, Península de, pen., Cuba (pĕ-nē´n-sōō-lä-dĕ-zä-pá´tä)	142-43	22°20′N	81°35′W
Zapopan, Mex. (sä-pō´pän)	146-47	20°43′N	103°24′W
Zaporizhzhia, Ukr.	202-03	47°51′N	35°10′E
Zaporozhye, Ukr. *see* Zaporizhzhia	202-03	47°51′N	35°10′E
Zara, Cro. *see* Zadar	200-01	44°07′N	15°14′E
Zaragoza, Mex. (sä-rä-gō´sä)	146-47	23°56′N	99°46′W
Zaragoza, Mex. (sä-rä-gō´sä)	122-23	28°29′N	100°55′W
Zaragoza, Mex. (sä-rä-gō´sä)	122-23	31°43′N	106°23′W
Zaragoza, Spain (thä-rä-gō´thä)	198-99	41°39′N	0°53′W
Zaranj, Afg.	232-33	31°07′N	61°53′E
Zarasai, Lith. (zä-rä-sī´)	192-93	55°44′N	26°16′E
Zárate, Arg. (zä-rä´tā)	173	34°06′S	59°02′W
Zaraysk, Russia (zä-rä´ėsk)	202-03	54°45′N	38°53′E
Zard Kūh, mtn., Iran	232-33	32°22′N	50°04′E
Zarghon Shahr, Afg.	232-33	32°51′N	68°27′E
Zaria, Nig. (zä´rē-ä)	260-61	11°07′N	7°43′E
Zarzal, Col. (zär-zá´l)	163c	4°23′N	76°03′W
Žatec, Czech Rep. (zhä´těts)	194-95	50°20′N	13°33′E
Zavitinsk, Russia	240-41	50°07′N	129°27′E
Zawiercie, Pol. (zá-vyĕr´tsyĕ)	194-95	50°30′N	19°26′E
Zaysan, Kaz. (zī´sán)	226	47°28′N	84°52′E
Zaza, stm., Cuba (zá´zá)	142-43	21°39′N	79°33′W
Zbarazh, Ukr. (zbä-räzh´)	194-95	49°40′N	25°48′E
Zbruch, stm., Ukr. (zbròch)	194-95	48°32′N	26°27′E
Zduńska Wola, Pol. (zdōōn´skä vō´lä)	194-95	51°36′N	18°58′E
Zeehan, Austl.	276	41°53′S	145°20′E
Zeeland, Mi., U.S. (zē´lănd)	116-17	42°49′N	86°01′W
Zeerust, S. Afr.	269c	25°33′S	26°06′E
Zelënodol'sk, Russia	186-87	55°50′N	48°32′E
Zelenogorsk, Russia (zĕ-lă´nô-gôrsk)	192-93	60°12′N	29°43′E
Zémio, C.A.R. (za-myō´)	262-63	5°02′N	25°08′E
Zenica, Bos. (zĕ´nēt-sä)	200-01	44°12′N	17°54′E
Zernograd, Russia	186-87	46°50′N	40°18′E
Zeya, Russia (zâ´yá)	218-19	53°44′N	127°15′E
Zeya, stm., Russia	218-19	50°14′N	127°36′E
Zeyskoye Vodokhranilishche, res., Russia	218-19	54°28′N	127°45′E
Zêzere, stm., Port. (zĕ´zâ-rĕ)	198-99	39°28′N	8°20′W
Zgierz, Pol. (zgyĕzh)	194-95	51°52′N	19°25′E
Zhalpy Syrt, mts., Eur.	186-87	52°0′N	51°30′E
Zhambyl, Kaz. *see* Taraz	226	42°54′N	71°21′E
Zhangaözen, Kaz.	186-87	43°19′N	52°47′E
Zhangaqazaly, Kaz.	226	45°51′N	62°09′E
Zhangatas, Kaz.	226	43°34′N	69°44′E
Zhangdian, China *see* Zibo	240-41	36°47′N	118°03′E
Zhangjiakou, China	240-41	40°49′N	114°53′E
Zhangping, China	238-39	25°18′N	117°26′E
Zhangye, China (jäŋ-yu)	240-41	38°56′N	100°27′E
Zhangzhou, China (jäŋ-jō)	238-39	24°31′N	117°40′E
Zhänibek, Kaz.	186-87	49°26′N	46°50′E
Zhanjiang, China (jän-jyäŋ)	238-39	21°12′N	110°23′E
Zhao'an, China (jou-än)	238-39	23°44′N	117°09′E
Zhaodong, China (jou-dôŋ)	240-41	46°04′N	125°59′E
Zhaoqing, China	238-39	23°04′N	112°28′E
Zhaotong, China (jou-tôŋ)	238-39	27°21′N	103°43′E
Zhaoyuan, China (jou-yuän)	240-41	37°22′N	120°24′E
Zhaoyuan, China (jou-yuän)	240-41	45°31′N	125°09′E
Zharkent, Kaz.	226	44°10′N	79°60′E
Zhaxigang, China	234-35	32°31′N	79°41′E
Zhaysang köli, lk., Kaz.	226	48°0′N	84°00′E
Zhayyq, stm., *see* Ural	176-77	46°50′N	51°33′E
Zhdanov, Ukr. *see* Mariupol'	202-03	47°06′N	37°34′E
Zhejiang, state, China (jŭ-jyäŋ)	238-39	29°0′N	120°00′E
Zheleznogorsk-Ilimskiy, Russia	218-19	56°35′N	104°08′E
Zheltyye Vody, Ukr. *see* Zhovti Vody	202-03	48°21′N	33°31′E
Zhem, stm., Kaz.	226	46°43′N	53°08′E
Zhengding, China (jŭŋ-dĭŋ)	240-41	38°09′N	114°34′E
Zhengyang, China (jŭŋ-yäŋ)	238-39	32°36′N	114°23′E
Zhengzhou, China (jŭŋ-jō)	238-39	34°46′N	113°39′E
Zhenjiang, China (jŭn-jyäŋ)	238-39	32°12′N	119°25′E
Zhenping, China	238-39	33°08′N	112°19′E
Zhenyuan, China (jŭn-yůän)	238-39	26°52′N	108°19′E
Zhetiqara, Kaz.	226	52°11′N	61°12′E
Zhezqazghan, Kaz.	226	47°47′N	67°41′E
Zhidoi, China	238-39	33°08′N	94°50′E
Zhigansk, Russia (zhē-gånsk´)	218-19	66°48′N	123°21′E
Zhijiang, China (jr-jyäŋ)	238-39	27°26′N	109°40′E
Zhitomir, Ukr. *see* Zhytomyr	202-03	50°16′N	28°41′E
Zhob, Pak.	232-33	31°21′N	69°27′E
Zhob, stm., Pak.	232-33	32°03′N	69°49′E
Zhongba, China	234-35	29°54′N	83°40′E
Zhongdian, China	238-39	27°49′N	99°42′E
Zhongghar Alataū zhotasy, mts., Asia *see* Alataw Shan	226	45°0′N	81°00′E
Zhongguo, nation, Asia *see* China	206-07	35°0′N	105°00′E
Zhongning, China	240-41	37°28′N	105°39′E
Zhongshan, China	238-39	22°31′N	113°22′E
Zhongwei, China (jôŋ-wā)	240-41	37°31′N	105°11′E
Zhongxian, China (jôŋ shyĕn)	238-39	30°29′N	108°05′E
Zhongxiang, China	238-39	31°10′N	112°35′E
Zhosaly, Kaz.	226	45°29′N	64°06′E
Zhoucun, China (jō-tsōōn)	240-41	36°48′N	117°50′E
Zhoushan Dao, i., China	238-39	30°03′N	122°08′E
Zhoushan Qundao, is., China (jō-shän-chyòn-dou) *see* Choushan Islands	238-39	30°0′N	122°00′E
Zhovti Vody, Ukr.	202-03	48°21′N	33°31′E
Zhuanghe, China (jůäŋ-hŭ)	243	39°42′N	122°58′E
Zhucheng, China (jōō-chŭŋ)	240-41	36°0′N	119°24′E
Zhuji, China (jōō-jyē) *see* Shangqiu	238-39	34°27′N	115°39′E
Zhuji, China (jōō-jyē)	238-39	29°43′N	120°15′E
Zhujiang Kou, est., Asia (jōō-jyäŋ kō)	238-39	22°36′N	113°44′E
Zhukovskiy, Russia (zhò-kôf´skĭ)	202-03	55°36′N	38°09′E
Zhuozhou, China	240-41	39°29′N	115°58′E
Zhushan, China	238-39	32°10′N	110°19′E
Zhuzhou, China	238-39	27°50′N	113°09′E
Zhytomyr, Ukr.	202-03	50°16′N	28°41′E
Zi, stm., China (dzē)	238-39	28°44′N	112°32′E
Zi, stm., China (dzē)	238-39	31°58′N	96°57′E
Zibo, China (dzē-bwo)	240-41	36°47′N	118°03′E
Zielona Góra, Pol. (zhyĕ-lô´nä gōō´rä)	194-95	51°56′N	15°31′E
Zigong, China	238-39	29°22′N	104°45′E
Ziguinchor, Sen.	260-61	12°34′N	16°17′W
Zihuatanejo, Mex.	146-47	17°39′N	101°33′W
Zile, Tur. (zē-lĕ´)	186-87	40°18′N	35°54′E
Žilina, Slvk. (zhē´lĭ-nä)	194-95	49°13′N	18°44′E
Zillah, Libya	258-59	28°34′N	17°34′E
Zima, Russia (zē´má)	218-19	53°55′N	102°04′E
Zimapán, Mex. (sē-mä´pän)	146-47	20°44′N	99°23′W
Zimbabwe, nation, Afr. (zĭm-bäb´-wā) (zĭm-bäb´-wē)	253	20°0′S	30°00′E
Zimbabwe Ruins, hist., Zimb.	264-65	20°17′S	30°57′E
Zimnicea, Rom. (zėm-nē´chå)	200-01	43°40′N	25°23′E
Zinder, Niger (zĭn´dēr)	258-59	13°48′N	8°59′E
Zion, Il., U.S. (zī´ŭn)	116-17	42°27′N	87°50′W
Zion National Park, n.p., Ut., U.S.	118-19	37°18′N	113°02′W
Zipaquirá, Col.	163c	5°00′N	74°01′W
Zittau, Ger. (tsē´tou)	194-95	50°54′N	14°48′E
Ziway, Lake, lk., Eth.	269d	7°59′N	38°50′E
Ziway Hāyk', lk., Eth. *see* Ziway, Lake	269d	7°59′N	38°50′E
Ziya, stm., China (dzē-yä)	240-41	39°09′N	117°10′E
Ziyun, China	238-39	25°43′N	106°05′E
Zizhong, China	238-39	29°47′N	104°50′E
Zlatoust, Russia (zlá-tô-ôst´)	226	55°10′N	59°40′E
Zlynka, Russia (zlĕŋ´kå)	202-03	52°25′N	31°44′E
Znaim, Czech Rep. *see* Znojmo	194-95	48°51′N	16°03′E
Znamenka, Ukr. *see* Znam'ianka	202-03	48°43′N	32°40′E
Znam'ianka, Ukr.	202-03	48°43′N	32°40′E
Znojmo, Czech Rep. (znoi´mô)	194-95	48°51′N	16°03′E
Zolotonosha, Ukr. (zô´lô-tô-nô´shå)	202-03	49°40′N	32°02′E
Zomba, Malawi (zŏm´bå)	264-65	15°23′S	35°19′E
Zongo, D.R.C. (zŏŋ´gô)	262-63	4°21′N	18°37′E
Zonguldak, Tur. (zôn´gōōl´dák)	186-87	41°28′N	31°49′E
Zouérat, Maur.	258-59	22°40′N	12°27′W
Zrenjanin, Serb.	200-01	45°23′N	20°24′E
Zubtsov, Russia (zòp-tsôf´)	202-03	56°10′N	34°35′E
Zugdidi, Geor.	227	42°30′N	41°52′E
Zugspitze, mtn., Eur.	194-95	47°25′N	10°59′E
Zújar, stm., Spain (zōō´kär)	198-99	39°01′N	5°47′W
Zululand, hist. reg., S. Afr.	264-65	28°10′S	32°00′E
Zumbrota, Mn., U.S. (zŭm-brō´tå)	114-15	44°18′N	92°41′W
Zunhua, China (dzòn-hwä)	240-41	40°11′N	117°57′E
Zunyi, China	238-39	27°42′N	106°56′E
Zürich, Switz. (tsü´rĭk)	194-95	47°23′N	8°32′E
Zuyevka, Russia	186-87	58°24′N	51°09′E
Zvishavane, Zimb.	264-65	20°20′S	30°03′E
Zvolen, Slvk. (zvô´lĕn)	194-95	48°35′N	19°08′E
Zvornik, Bos. (zvôr´nĕk)	200-01	44°23′N	19°07′E
Zwedru, Lib.	260-61	6°04′N	8°07′W
Zweibrücken, Ger. (tsvī-brük´ĕn)	194-95	49°15′N	7°22′E
Zwickau, Ger. (tsvĭk´ou)	194-95	50°43′N	12°30′E
Zwolle, Neth. (zvôl´ĕ)	190-91	52°31′N	6°06′E
Zyryanka, Russia (zē-ryän´kå)	218-19	65°45′N	150°51′E
Zyryanovsk, Kaz.	226	49°44′N	84°18′E